Encyclopedia of the Solar System

SECOND EDITION

View From the Martian Rover *Spirit:* October 26, 2006

This was *Spirit's* view on its 1000th Martian-day (sol) of what was planned to be a 90-sol mission.
The robotic *Spirit* rover has stayed alive so long on Mars that it needed a place to wait out its
second cold and dim Martian winter. Earth scientists selected Low Ridge Hill, a place with sufficient
slant to give *Spirit's* solar panels enough sunlight to keep it powered and making scientific observations.
From its winter haven, *Spirit* photographed the above panorama, which has been digitally altered to exaggerate
colors. *Spirit's* track through the Martian hills can be seen at the center of the image. (Courtesy of NASA)

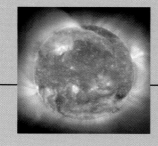

Encyclopedia of the Solar System

SECOND EDITION

EDITORS:

Lucy-Ann McFadden
Department of Astronomy
University of Maryland
College Park, Maryland

Paul R. Weissman
Jet Propulsion Laboratory
California Institute of Technology
Pasadena, California

Torrence V. Johnson
Jet Propulsion Laboratory
California Institute of Technology
Pasadena, California

ELSEVIER

Amsterdam • Boston • Heidelberg • London • New York • Oxford
Paris • San Diego • San Francisco • Singapore • Sydney • Tokyo
Academic Press is an Imprint of Elsevier

Cover: Images courtesy of NASA/JPL-Caltech and Planet Art

Academic Press is an imprint of Elsevier
525 B Street, Suite 1900, San Diego, CA 92101-4495, USA
84 Theobald's Road, London WC1X 8RR, UK
Radarweg 29, PO BOX 211, 1000 AE Amsterdam, The Netherlands
30 Corporate Drive, Suite 400, Burlington, MA 01803, USA

First edition 1999
Second edition 2007

British Library Cataloguing in Publication Data
A catalogue record for this book is available from the British Library

Library of Congress Catalog Number: 2006937972

ISBN-13: 978-0-12-088589-3
ISBN-10: 0-12-088589-1

For information on all Academic Press publications
visit our website at books.elsevier.com

Printed and bound in Canada

07 08 09 10 10 9 8 7 6 5 4 3 2 1

Contents

Contributors

Markus J. Aschwanden
The Sun
Lockheed-Martin ATC, Solar and Astrophysics Laboratory
Palo Alto, California

Fran Bagenal
Planetary Magnetospheres
University of Colorado
Boulder, Colorado

Anil Bhardwaj
X-Rays in the Solar System
Space Physics Laboratory
Vikram Sarabhai Space Centre
Trivandrum, India

Richard P. Binzel
Near-Earth Objects
Massachusetts Institute of Technology
Cambridge, Massachusetts

John Brandt
Physics and Chemistry of Comets
Institute for Astrophysics, Department of Physics
and Astronomy
University of New Mexico
Albuquerque, New Mexico

Daniel T. Britt
Main-Belt Asteroids
Department of Physics
University of Central Florida
Orlando, Florida

Bonnie J. Buratti
Planetary Satellites
Jet Propulsion Laboratory
California Institute of Technology
Pasadena, California

James D. Burke
Planetary Exploration Missions
Jet Propulsion Laboratory
California Institute of Technology
Pasadena, California

Michael H. Carr
Mars: Surface and Interior
U.S. Geological Survey
Menlo Park, California

David C. Catling
Mars Atmosphere: History and Surface Interaction
University of Washington
Seattle, Washington

John E. Chambers
The Origin of the Solar System
Carnegie Institution of Washington
Washington, D.C.

Mark J. Cintala
Planetary Impacts
NASA Johnson Space Center
Houston, Texas

William D. Cochran
Extra-Solar Planets
Department of Astronomy, McDonald Observatory
University of Texas
Austin, Austin, Texas

Geoffrey Collins
Ganymede and Callisto
Wheaton College
Norton, Massachusetts

Athena Coustenis
Titan
Observatoire de Paris-Meudon
Meudon, France

Guy Colsolmagno
Main-Belt Asteroids
Specola Vaticana
Castel Gandolfo, Italy

Wanda L. Davis
Astrobiology
NASA Ames Research Center, Space Science Division
Moffett Field, California

Imke de Pater
The Solar System at Radio Wavelengths
Department of Astronomy
University of California, Berkeley
Berkeley, California

Luke Dones
Comet Populations and Cometary Dynamics
Southwest Research Institute
Boulder, Colorado

Deborah L. Domingue
The Solar System at Ultraviolet Wavelengths
Johns Hopkins University Applied Physics Laboratory
Laurel, Maryland

Timothy E. Dowling
Earth as a Planet: Atmosphere and Oceans
Department of Mechanical Engineering
University of Louisville
Louisville, Kentucky

Adam M. Dziewonski
Earth as a Planet: Surface and Interior
Harvard University
Cambridge, Massachusetts

Michael Endl
Extra-Solar Planets
Department of Astronomy, McDonald Observatory
University of Texas, Austin
Austin, Texas

Jonathan Fortney
Interiors of the Giant Planets
NASA Ames Research Center
Moffett Field, California

Matthew P. Golombek
Mars: Landing Site Geology, Mineralogy, and Geochemistry
Jet Propulsion Laboratory
Mars Exploration Program
Pasadena, California

John T. Gosling
The Solar Wind
Laboratory for Atmospheric and Space Physics
University of Colorado
Boulder, Colorado

Richard A. F. Grieve
Planetary Impacts
Natural Resources Canada
Ottawa, Canada

Eberhard Grün
Solar System Dust
Max Planck Institute of Nuclear Physics
Heidelberg, Germany

Alex N. Halliday
The Origin of the Solar System
Department of Earth Sciences, University of Oxford
Oxford, United Kingdom

Douglas P. Hamilton
Planetary Rings
Department of Astronomy
University of Maryland
College Park, Maryland

Amanda R. Hendrix
The Solar System at Ultraviolet Wavelengths
Jet Propulsion Laboratory
California Institute of Technology
Pasadena, California

Donald M. Hunten
Venus: Atmosphere
Department of Planetary Sciences
Lunar and Planetary Laboratory, University of Arizona
Tucson, Arizona

Robert Jedicke
New Generation Optical/Infrared Telescopes
Institute for Astronomy
University of Hawaii
Honolulu, Hawaii

Torrence V. Johnson
Ganymede and Calllisto
Jet Propulsion Laboratory
California Institute of Technology
Pasadena, California

Randolph L. Kirk
Triton
U. S. Geological Survey
Flagstaff, Arizona

William S. Kurth
The Solar System at Radio Wavelengths
University of Iowa
Iowa City, Iowa

Margaret Galland Kivelson
Planetary Magnetospheres
Department of Earth and Space Sciences
University of California, Los Angeles
Los Angeles, California

Larry Lebofsky
Main-Belt Asteroids
Lunar and Planetary Laboratory, University of Arizona
Tucson, Arizona

Conway Leovy
Mars: Atmosphere: History and Surface Interaction
University of Washington
Seattle, Washington

David Leverington
A History of Solar System Studies
Author and Consultant
Stoke Lacy, Herefordshire, United Kingdom

Harold F. Levison
Comet Populations and Cometary Dynamics
Southwest Research Institute
Boulder, Colorado

Michael E. Lipschutz
Meteorites
Department of Chemistry, Purdue University
West Lafayette, Indiana

Jack J. Lissauer
Solar System Dynamics: Regular and Chaotic Motion
Space Science Division
NASA Ames Research Center
Moffett Field, California

Carey M. Lisse
X-Rays in the Solar System
Johns Hopkins University
Applied Physics Laboratory
Laurel, Maryland

Rosaly M. C. Lopes
Io: The Volcanic Moon
Jet Propulsion Laboratory
California Institute of Technology
Pasadena, California

Janet G. Luhmann
The Sun-Earth Connection
Space Sciences Laboratory
University of California, Berkeley
Berkeley, California

Mark S. Marley
Interiors of the Giant Planets
NASA Ames Research Center
Moffett Field, California

Lucy A. McFadden
Near-Earth Objects
Department of Astronomy
University of Maryland
College Park, Maryland

Christopher P. McKay
Astrobiology
NASA Ames Research Center, Space Science Division
Moffett Field, California

William B. McKinnon
Triton
Department of Earth and Planetary Sciences
Washington University
St. Louis, Missouri

Harry Y. McSween, Jr.
Mars: Landing Site Geology, Mineralogy, and Geochemistry
University of Tennessee, Knoxville
Knoxville, Tennessee

Alessandro Morbidelli
Kuiper Belt: Dynamics
Observatoirs de La Cöte d'Azur
Nice, France

Carl D. Murray
Solar System Dynamics: Regular and Chaotic Motion
Queen Mary, University of London
London, United Kingdom

Robert M. Nelson
The Solar System at Ultraviolet Wavelengths
Jet Propulsion Laboratory
California Institute of Technology
Pasadena, California

Steven J. Ostro
Planetary Radar
Jet Propulsion Laboratory
California Institute of Technology
Pasadena, California

Robert T. Pappalardo
Europa
University of Colorado
Boulder, Colorado

David C. Pieri
Earth as a Planet: Surface and Interior
Jet Propulsion Laboratory
California Institute of Technology
Pasadena, California

Carolyn C. Porco
Planetary Rings
Space Science Institute
Boulder, Colorado

Thomas H. Prettyman
Remote Elemental Sensing Using Nuclear Spectroscopy
Los Alamos National Laboratory, Space and
Atmospheric Sciences
Los Alamos, New Mexico

Louise M. Prockter
Europa
Planetary Exploration Group
Johns Hopkins University Applied Physics Laboratory
Laurel, Maryland

Ludolf Schultz
Meterorites
Max-Plank-Institut für Chemie
Mainz, Germany

Adam Showman
Earth as a Planet: Atmosphere and Oceans
University of Arizona
Tucson, Arizona

Suzanne E. Smrekar
Venus: Surface and Interior
Earth and Space Sciences Division
Jet Propulsion Laboratory
California Institute of Technology
Pasadena, California

Stanley C. Solomon
High Altitude Observatory
National Center for Atmospheric Research
Boulder, Colorado

S. Alan Stern
Pluto
Southwest Research Institute
Boulder, Colorado

Ellen R. Stofan
Venus: Surface and Interior
Proxemy Research
Rectortown, Maryland

Robert G. Strom
Mercury
Lunar and Planetary Laboratory
Department of Planetary Sciences
University of Arizona
Tucson, Arizona

Mark V. Sykes
Infrared Views of the Solar System from Space
Planetary Science Institute
Tucson, Arizona

Roald Tagle
Planetary Impacts
Humboldt University
Berlin, Germany

Stuart Ross Taylor
The Moon
Department of Earth and Marine Sciences, Emeritus
Australian National University
Canberra, Australia

Stephen C. Tegler
Kuiper Belt Objects: Physical Studies
Department of Physics and Astronomy
Northern Arizona University
Flagstaff, Arizona

Peter C. Thomas
Planetary Satellites
Cornell University
Ithaca, New York

Alan T. Tokunaga
New Generation Optical/Infrared Telescopes
Institute for Astronomy
University of Hawaii
Honolulu, Hawaii

Paul R. Weissman
The Solar System and its Place in the Galaxy
Jet Propulsion Laboratory
California Institute of Technology
Pasadena, California

Robert A. West
Atmospheres of the Giant Planets
Jet Propulsion Laboratory
California Institute of Technology
Pasadena, California

Lionel Wilson
Planetary Volcanism
Planetary Science Research Group
Institute of Environmental and Natural Sciences
Environmental Science Department
Lancaster, United Kingdom

About the Editors

Lucy-Ann McFadden is a planetary scientist at the University of Maryland. She was the founding director of the College Park Scholars Program, Science, Discovery, and the Universe. She has published over 75 articles in refereed journals and has been co-investigator on NASA's *NEAR*, *Deep Impact*, and *Dawn* missions, exploring asteroids and comets. McFadden has served on committees on solar system exploration for the National Academy of Sciences and on the editorial board of *Icarus*.

Paul R. Weissman is a Senior Research Scientist at the Jet Propulsion Laboratory, specializing in comets. He is the author of over 100 scientific papers and 30 popular articles. He is also co-author, along with Alan Harris, of a children's book on the *Voyager* mission. Dr. Weissman received his doctorate in planetary and space physics from the University of California, Los Angeles. His work includes both theoretical and observational studies of comets, investigating their orbital motion, their physical make-up, and the threat they pose due to possible impacts on the Earth. Dr. Weissman is an Interdisciplinary Scientist on ESA's *Rosetta* mission to comet Churyumov-Gerasimenko.

Torrence V. Johnson is a specialist on icy satellites in the solar system. He has written over 130 papers for scientific journals. He received a Ph.D. in planetary science from the California Institute of Technology and is now the Chief Scientist for Solar System Explorations at the Jet Propulsion Laboratory. Johnson was the Project Scientist for the *Galileo* mission and is currently an investigator on the *Cassini* mission. He is the recipient of two NASA Exceptional Scientific Achievement Medals and the NASA Outstanding Leadership Medal and has an honorary doctorate from the University of Padua, where Galileo made his first observations of the solar system.

Foreword

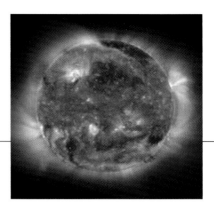

The solar system has become humankind's new backyard. It is the playground of robotic planetary spacecraft that have surveyed just about every corner of this vast expanse in space. Nowadays, every schoolchild knows what even the farthest planets look like. Fifty years ago, these places could only be imagined, and traveling to them was the realm of fiction. In just this short time in the long history of the human species we have leapt off the surface of our home planet and sent robotic extensions of our eyes, ears, noses, arms, and legs to the far reaches of the solar system and beyond.

In the early days of the 20th century, we were using airplanes to extend our reach to the last unexplored surface regions of our own planet. Now 100 years later, at the beginning of the 21st century, we are using spacecraft to extend our reach from the innermost planet Mercury to the outmost planet Neptune, and we have a spacecraft on the way to Pluto and the Kuiper Belt. Today, there are telescopes beyond imagination 100 or even 50 years ago that can image Pluto and detect planets around other stars! Now, Sol's planets can say "we are not alone"; there are objects just like us elsewhere in the universe. As humanity's space technology improves, perhaps in the next 100 years or so human beings also may be able to say "we are not alone."

When I was a kid 50 years ago, I was thrilled by the paintings of Chesley Bonestell and others who put their imagination on canvas to show us what it might be like "out there." Werner Von Braun's *Collier's* magazine articles of 1952–1954 superbly illustrated how we would go to the Moon and Mars using new rocket technologies. Reading those fabulous articles crystallized thoughts in my young mind about what to do with my life. I wanted to be part of the adventure to find out what these places were like. Not so long after the *Collier's* articles appeared, we did go to the Moon, and pretty much as illustrated, although perhaps not in such a grand manner. We have not sent humans to Mars—at least we haven't yet—but we have sent our robots to Mars and to just about every other place in the solar system as well.

This book is filled with the knowledge about our solar system that resulted from all this exploration, whether by spacecraft or by telescopes both in space and earth-bound. It could not have been written 50 years ago as almost everything in this Encyclopedia was unknown back then. All of this new knowledge is based on discoveries made in the interim by scientist-explorers who have followed their in-born human imperative to explore and to understand. Many old mysteries, misunderstandings, and fears that existed 50 years ago about what lay beyond the Earth have been eliminated.

We now know the major features of the landscape in our cosmic backyard and can look forward to the adventure, excitement, and new knowledge that will result from more in-depth exploration by today's spacecraft, such as those actually exploring the surface of these faraway places, including the *Huygens Titan* lander and the *Mars Exploration* rovers, doing things that were unimaginable before the Space Age began.

The *Encyclopedia of the Solar System* is filled with images, illustrations, and charts to aid in understanding. Every object in the solar system is covered by at least one chapter. Other chapters are devoted to the relationships among the objects in the solar system and with the galaxy beyond. The processes that operate on solar system objects, in their atmospheres, on their surfaces, in their interiors, and interactions with space itself are all described in detail. There are chapters on how we explore and learn about the solar system and about the investigations used to make new discoveries. And there are chapters on the history of solar system exploration and the missions that have carried out this enterprise. All written by an international set of world-class scientists using rigorous yet easy-to-understand prose.

Everything you want to know about the solar system is here. This is your highway to the solar system. It is as much fun exploring this Encyclopedia as all the exploration it took to get the information that it contains. Let your fingers be the spacecraft as you thumb through this book visiting all the planets, moons and other small objects in the solar system. Experience what it is like to look at our solar system with ultraviolet eyes, infrared eyes, radio eyes, and radar eyes.

It has been seven years since the first edition. The exploration of space has continued at a rapid pace since then, and many missions have flown in the interim. New discoveries are being made all the time. This second edition will catch you up on all that has happened since the first, including several new chapters based on information from our latest missions.

I invite you to enjoy a virtual exploration of the solar system by flipping through the pages in this volume. This book deserves a place in any academic setting and wherever there is a need to understand the cosmos beyond our home planet. It is the perfect solar system reference book, lavishly illustrated and well written. The editors and authors have done a magnificent job.

We live in a wonderful time of exploration and discovery. Here is your window to the adventure.

WESLEY T. HUNTRESS
Geophysical Laboratory
Carnegie Institution of Washington
Washington, D.C.

Preface to the Second Edition

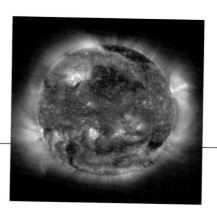

"Knowledge is not static. Science is a process, not a product. Some of what is presented in this volume will inevitably be out of date by the time you read it." From the Preface to the first edition, 1999.

Written on the eve of the new millennium, the statement above was our acknowledgment that one cannot simply "freeze" our knowledge of the solar system we inhabit, box it up, and display it like a collection of rare butterflies in a 19th century "cabinet of curiosities." Rather our goal was to provide our readers with an introduction to understanding the solar system as an interacting system, shaped by its place in the universe, its history, and the chemical and physical processes that operate from the extreme pressures and temperatures of the Sun's interior to the frigid realm of the Oort cloud. We aimed to provide a work that was useful to students, professionals, and serious amateurs at a variety of levels, containing both detailed technical material and clear expositions of general principles and findings. With the help of our extremely talented colleagues who agreed to author the chapters, we humbly believe we achieved at least some of these ambitious goals.

How to decide when to update a work whose subject matter is in a constant, exuberant state of flux? Difficult. Waiting for our knowledge of the solar system to be "complete" was deemed impractical, since our thesis is that this will never happen. Picking an anniversary date (30 years since this, or 50 years after that) seemed arbitrary. We compromised on taking an informal inventory of major events and advances in knowledge since that last edition whenever we got together at conferences and meetings. When we realized that virtually every chapter in the first edition needed major revisions and that new chapters would be called for to properly reflect new material, we decided to undertake the task of preparing a second edition with the encouragement and help from our friends and colleagues at Academic Press.

Consider how much has happened in the relatively short time since the first edition, published in 1999. An international fleet of spacecraft is now in place around Mars and two

rovers are roaming its surface, with more to follow. *Galileo* ended its mission of discovery at Jupiter with a spectacular fiery plunge into the giant planet's atmosphere. We have reached out and touched one comet with the *Deep Impact* mission and brought back precious fragments from another with *Stardust*. *Cassini* is sending back incredible data from the Saturn system and the *Huygens* probe descended to the surface of the giant, smog-shrouded moon Titan, revealing an eerily Earth-like landscape carved by methane rains. *NEAR* and *Hayabusa* each orbited and then touched down on the surface of near-Earth asteroids Eros and Itokawa, respectively. Scientists on the Earth are continually improving the capabilities of telescopes and instruments, while laboratory studies and advances in theory improve our ability to synthesize and understand the vast amounts of new data being returned.

What you have before you is far more than a minor tweak to add a few new items to a table here or a figure there. It is a complete revamping of the Encyclopedia to reflect the solar system as we understand it today. We have attempted to capture the excitement and breadth of all this new material in the layout of the new edition. The authors of existing chapters were eager to update them to reflect our current state of knowledge, and many new authors have been added to bring fresh perspectives to the work. To all of those authors who contributed to the second edition and to the army of reviewers who carefully checked each chapter, we offer our sincere thanks and gratitude.

The organization of the chapters remains based on the logic of combining individual surveys of objects and planets, reviews of common elements and processes, and discussions of the latest techniques used to observe the solar system. Within this context you will find old acquaintances and many new friends. The sections on our own home planet have been revised and a new chapter on the Sun-Earth connection added to reflect our growing understanding of the intimate relationship between our star and conditions here on Earth. The treatment of Mars has been updated and a new chapter included incorporating the knowledge gained from the rovers *Spirit* and *Opportunity* and new orbital exploration of the red planet. Galileo's remarkable discovery of evidence for subsurface oceans on the icy Galilean satellites is treated fully in new chapters devoted to Europa and to Ganymede and Callisto. New information from the *Deep Impact* mission and the *Stardust* sample return is included as well. We continue to find out more and more about the denizens of the most distant reaches of the solar system, and have expanded the discussion of the Kuiper belt with a new chapter on physical properties. The area of observational techniques and instrumentation has been expanded to include chapters covering the X-ray portion of the spectrum, new generation telescopes, and remote chemical analysis.

Finally, nothing exemplifies the dynamic character of our knowledge than the area of extra-solar planets, which completes the volume. In the first edition the chapter on extra-solar planets contained a section entitled, "What is a Planet?" which concluded with this: "The reader is cautioned that these definitions are not uniformly accepted." The chapter included a table of nineteen objects cautiously labeled "Discovered Substellar Companions." As this work goes to press, more than 200 extra-solar planets are known, many in multi-planet systems, with more being discovered every day. And at the 2006 General Assembly of the International Astronomical Union, the question of the definition of "planet" was still being hotly debated. The current IAU definition is discussed in the introductory chapter by one of us (PRW) and other views concerning the status of Pluto may be found in the chapter on that body.

In addition to the energy and hard work of all of our authors, this edition of the Encyclopedia is greatly enhanced by the vision and talents of our friends at Academic Press. Specifically, we wish to thank Jennifer Helé, our Publishing Editor, who oversaw the project and learned the hard truth that herding scientists and herding cats are one and the same thing. Jennifer was the task master who made us realize that we could not just keep adding exciting new results to the volume, but one day had to stop and actually publish it. Francine Ribeau was our very able Marketing Manager and Deena Burgess, our Publishing Services Manager in the U.K., handled all of the last minute loose ends and made certain that the book was published without a hitch yet on a very tight schedule. Frank Cynar was our Publishing Editor for the first edition and for the beginning of the second, assisted by Gail Rice who was the Developmental Editor early on for the second edition. At Techbooks, Frank Scott was the Project Manager who oversaw all the final chapter and figure submissions and proof checking. Finally, also at Techbooks, was Carol Field, our Developmental Editor, simply known as Fabulous Carol, who seemed to work 30-hour days for more than a year to see the volume through to fruition, while still finding time to get married in the midst of it all. This Encyclopedia would not exist without the tireless efforts of all of these extremely talented and dedicated individuals. To all of them we offer our eternal thanks.

Extensive use of color and new graphic designs have made the Encyclopedia even more beautiful and enhanced its readability while at the same time allowing the authors to display their information more effectively. The Encyclopedia you have before you is the result of all these efforts and we sincerely hope you will enjoy reading it as much as we enjoyed the process of compiling it.

Which brings us back to the quotation at the start of the Preface. We sincerely hope that this edition of the Encyclopedia will indeed also be out of date by the time you read it. The *New Horizons* spacecraft is on its way to the Pluto/Charon system, *MESSENGER* is on its way to Mercury, *Rosetta* is en route to a rendezvous with periodic comet Churyumov-Gerasimenko, new spacecraft are probing Venus and Mars, many nations are refocusing on exploration of the Moon, plans are being laid to study the deep interior of Jupiter and return to Europa, while the results from the Saturn system, Titan and Enceladus, have sparked a multitude of ideas for future exploration. We hope this Encyclopedia will help you, the reader, appreciate and enjoy this on-going process of discovery and change as much as we do.

Lucy-Ann McFadden
Paul R. Weissman
Torrence V. Johnson
November 1, 2006

Preface to the First Edition

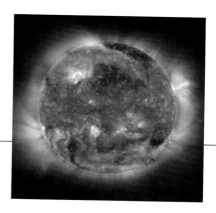

"This is what hydrogen atoms can accomplish
after four billion years of evolution."
—Carl Sagan, Cosmos, 1981

The quote above comes from the final episode of the public television series "Cosmos," which was created by Carl Sagan and several colleagues in 1981. Carl was describing the incredible accomplishments of the scientists and engineers who made the Voyager 1 and 2 missions to Jupiter and Saturn possible. But he just as easily could have been describing the chapters in this book.

This Encyclopedia is the product of the many scientists, engineers, technicians, and managers who produced the spacecraft missions which have explored our solar system over the past four decades. It is our attempt to provide to you, the reader, a comprehensive view of all we have learned in that 40 years of exploration and discovery. But we cannot take credit for this work. It is the product of the efforts of thousands of very talented and hard-working individuals in a score of countries who have contributed to that exploration. And it includes not only those involved directly in space missions, but also the many ground-based telescopic observers (both professional and amateur), laboratory scientists, theorists, and computer specialists who have contributed to creating that body of knowledge called solar system science. To all of these individuals, we say thank you.

Our goal in creating this Encyclopedia is to provide an integrated view of all we have learned about the solar system, at a level that is useful to the advanced amateur or student, to teachers, to non-solar system astronomers, and to professionals in other scientific and technical fields. What we present here is an introduction to the many different specialties that constitute solar system science, written by the world's leading experts in each field. A reader can start at the beginning and follow the course we have laid out, or delve into the volume at almost any point and pursue his or her own personal interests. If the reader wishes to go further, the lists of recommended reading at the end of each article

provide the next step in learning about any of the subjects covered.

Our approach is to have the reader understand the solar system not only as a collection of individual and distinct bodies, but also as an integrated, interacting system, shaped by its initial conditions and by a variety of physical and chemical processes. The Encyclopedia begins with an overview chapter which describes the general features of the solar system and its relationship to the Milky Way galaxy, followed by a chapter on the origin of the system. Next we proceed from the Sun outward. We present the terrestrial planets (Mercury, Venus, Earth, Mars) individually with separate chapters on their atmospheres and satellites (where they exist). For the giant planets (Jupiter, Saturn, Uranus, Neptune) our focus shifts to common areas of scientific knowledge: atmospheres, interiors, satellites, rings, and magnetospheres. In addition, we have singled out three amazing satellites for individual chapters: Io, Titan, and Triton. Next is a chapter on the planetary system's most distant outpost, Pluto, and its icy satellite, Charon. From there we move into discussing the small bodies of the solar system: comets, asteroids, meteorites, and dust. Having looked at the individual members of the solar system, we next describe the different view of those members at a variety of wavelengths outside the normal visual region. From there we consider the important processes that have played such an important role in the formation and evolution of the system: celestial dynamics, chaos, impacts, and volcanism. Last, we look at three topics which are as much in our future as in our past: life on other planets, space exploration missions, and the search for planets around other stars.

A volume like this one does not come into being without the efforts of a great number of very dedicated people. We express our appreciation to the more than 50 colleagues who wrote chapters, sharing their expertise with you, the reader. In addition to providing chapters that captured the excitement of their individual fields, the authors have endured revisions, rewrites, endless questions, and unforeseen delays.

For all of these we offer our humble apologies. To ensure the quality and accuracy of each contribution, at least two independent reviewers critiqued each chapter. The peer review process maintains its integrity through the anonymity of the reviewers. Although we cannot acknowledge them by name, we thank all the reviewers for their time and their conscientious efforts.

We are also deeply indebted to the team at Academic Press. Our executive editor, Frank Cynar, worked tirelessly with us to conceptualize and execute the encyclopedia, while allowing us to maintain the highest intellectual and scientific standards. We thank him for his patience and for his perseverance in seeing this volume through to completion. Frank's assistants, Daniela Dell'Orco, Della Grayson, Linda McAleer, Cathleen Ryan, and Suzanne Walters, kept the entire process moving and attended to the myriad of details and questions that arise with such a large and complex volume. Advice and valuable guidance came from Academic Press' director of major reference works, Chris Morris. Lori

Asbury masterfully oversaw the production and copy editing. To all of the people at Academic Press, we give our sincere thanks.

Knowledge is not static. Science is a process, not a product. Some of what is presented in this volume will inevitably be out of date by the time you read it. New discoveries seem to come every day from our colleagues using Earth-based and orbiting telescopes, and from the flotilla of new small spacecraft that are out there adding to our store of knowledge about the solar system. In this spirit we hope that you, the reader, will benefit from the knowledge and understanding compiled in the following pages. The new millennium will surely add to the legacy presented herein, and we will all be the better for it. Enjoy, wonder, and keep watching the sky.

Paul R. Weissman
Lucy-Ann McFadden
Torrence V. Johnson

The Solar System and Its Place in the Galaxy

Paul R. Weissman

Jet Propulsion Laboratory
California Institute of Technology
Pasadena, California

CHAPTER 1

1. Introduction

The origins of modern astronomy lie with the study of our solar system. When ancient humans first gazed at the skies, they recognized the same patterns of fixed stars rotating over their heads each night. They identified these fixed patterns, now called constellations, with familiar objects or animals, or stories from their mythologies and their culture. But along with the fixed stars, there were a few bright points of light that moved each night, slowly following similar paths through a belt of constellations around the sky (the Sun and Moon also appeared to move through the same belt of constellations). These wandering objects were the **planets** of our solar system. Indeed, the name "planet" derives from the Latin *planeta*, meaning wanderer.

The ancients recognized five planets that they could see with their naked eyes. We now know that the solar system consists of eight planets, at least three **dwarf planets**, plus a myriad of smaller objects: satellites, asteroids, comets, rings, and dust. Discoveries of new objects and new classes of objects are continuing even today. Thus, our view of the solar system is constantly changing and evolving as new data and new theories to explain (or anticipate) the data become available.

The solar system we see today is the result of the complex interaction of physical, chemical, and dynamical processes that have shaped the planets and other bodies. By studying each of the planets and other bodies individually as well as collectively, we seek to gain an understanding of those processes and the steps that led to the current solar system. Many of those processes operated most intensely early in the solar system's history, as the Sun and planets formed from an interstellar cloud of dust and gas, 4.56 billion years ago. The first billion years of the solar system's history was a violent period as the planets cleared their orbital zones of much of the leftover debris from the process of planet formation, flinging small bodies into planet-crossing, and often planet-impacting, **orbits** or out to interstellar space. In comparison, the present-day solar system is a much quieter place, though many of these processes continue today on a lesser scale.

Our knowledge of the solar system has exploded in the past four decades as interplanetary exploration spacecraft have provided close-up views of all of the planets, as well as of a diverse collection of satellites, asteroids, and comets. Earth-orbiting telescopes have provided an unprecedented view of the solar system, often at wavelengths not accessible from the Earth's surface. Ground-based observations have also continued to produce exciting new discoveries through the application of a variety of new technologies such as charge-coupled device (CCD) cameras, infrared detector arrays, adaptive optics, and powerful planetary radars. Theoretical studies have also contributed significantly to our understanding of the solar system, largely through the use of advanced computer codes and high-speed, dedicated computers. Serendipity has also played an important role in many new discoveries.

Along with this increased knowledge have come numerous additional questions as we attempt to explain the complexity and diversity that we observe on each newly encountered world. The increased spatial and spectral resolution of the observations, along with *in situ* measurements of atmospheres, surface materials, and **magnetospheres**, have revealed that each body is unique, the result of a different combination of physical, chemical, and dynamical processes that formed and shaped it, as well as its different initial composition. Yet, at the same time, there are broad systematic trends and similarities that are clues to the collective history that the solar system has undergone.

We have now begun an exciting new age of discovery with the detection of numerous planet-sized bodies around nearby stars. Although the properties and placement of these extra-solar planets appear to be very different from those in our solar system, they are likely the prelude to the discovery of other planetary systems that may more closely resemble our own.

We may also be on the brink of discovering evidence for life on other planets, in particular, Mars. There is an ongoing debate as to whether biogenic materials have been discovered in meteorites that were blasted off the surface of Mars and have found their way to Earth. Although still very controversial, this finding, if confirmed, would have profound implications for the existence of life elsewhere in the solar system and the galaxy.

The goal of this chapter is to provide the reader with an introduction to the solar system. It seeks to provide a broad overview of the solar system and its constituent parts, to note the location of the solar system in the galaxy, and to describe the local galactic environment. Detailed discussions of each of the bodies that make up the solar system, as well as the processes that have shaped those bodies and the techniques for observing the planetary system are provided in the following chapters of this Encyclopedia. The reader is referred to those chapters for more detailed discussions of each of the topics introduced.

Some brief notes about planetary nomenclature will likely be useful. The names of the planets are all taken from Greek and Roman mythology (with the exception of Earth), as are the names of their satellites, with the exception of the Moon and the Uranian satellites, the latter being named after Shakespearean characters. The Earth is occasionally referred to as Terra, and the Moon as Luna, each the Latin version of their names. The naming system for planetary rings is different at each planet and includes descriptive names of the structures (at Jupiter), letters of the Roman alphabet (at Saturn), Greek letters and Arabic numerals (at Uranus), and the names of scientists associated with the discovery of Neptune (at Neptune).

Asteroids were initially named after Greek and Roman goddesses. As their numbers have increased, asteroids have been named after the family members of the discoverers, after observatories, universities, cities, provinces, historical figures, scientists, writers, artists, literary figures, and, in at least one case, the astronomer's cat. Initial discoveries of asteroids are designated by the year of their discovery and a letter/number code. Once the orbits of the asteroids are firmly established, they are given official numbers in the asteroid catalog: over 136,500 asteroids have been numbered (as of September 2006). The discoverer(s) of an asteroid are given the privilege of suggesting its name, if done so within 10 years from when it was officially numbered.

Comets are generally named for their discoverers, though in a few well-known cases such as comets Halley and Encke, they are named for the individuals who first computed their orbits and linked several apparitions. Because some astronomers have discovered more than one short-period comet, a number is added at the end of the name in order to differentiate them, though this system is not applied to long-period comets. Comets are also designated by the year of their discovery and a letter code (a recently abandoned system used lowercase Roman letters and Roman numerals in place of the letter codes). The naming of newly discovered comets, asteroids, and satellites, as well as surface features on solar system bodies, is overseen by several working groups of the International Astronomical Union (IAU).

2. The Definition of A Planet

No formal definition of a planet existed until very recently. Originally, the ancients recognized five planets that could be seen with the naked eye, plus the Earth. Two more **jovian planets**, Uranus and Neptune, were discovered telescopically in 1781 and 1846, respectively.

The largest asteroid, Ceres, was discovered in 1801 in an orbit between Mars and Jupiter and was hailed as a new planet because it fit into Bode's law (see discussion later in this chapter). However, it was soon recognized that Ceres was much smaller than any of the known planets. As more and more asteroids were discovered in similar orbits between Mars and Jupiter, it became evident that Ceres was simply the largest body of a huge swarm of bodies between Mars and Jupiter that we now call the Asteroid Belt. A new term was coined, "**minor planet**," to describe these bodies.

Searches for planets beyond Neptune continued and culminated in the discovery of Pluto in 1930. As with Ceres, it was soon recognized that Pluto was much smaller than any of the neighboring jovian planets. Later, measurements of Pluto's diameter by stellar occultations showed that it was also smaller than any of the **terrestrial planets**, in fact, smaller even than the Earth's Moon. As a result, Pluto's status as a planet was called into question.

In the 1980s, dynamical calculations suggested the existence of a belt of many small objects in orbits beyond Neptune. In the early 1990s the first of these objects, 1992 QB$_1$ was discovered at a distance of 40.9 **astronomical units** (**AU**). More discoveries followed and over 1000 bodies have now been found in the trans-Neptunian zone. They are collectively known as the **Kuiper belt**. All of these bodies were estimated to be smaller than Pluto, though a few were found that were about half the diameter of Pluto.

The existence of the Kuiper belt suggested that Pluto, like Ceres, was simply the largest body among a huge swarm of bodies beyond Neptune, again calling Pluto's status into question. Then came the discovery of Eris (2003 UB$_{313}$), a Kuiper belt object in a distant orbit, which turned out to be slightly larger than Pluto.

In response, the IAU, the governing body for astronomers worldwide, formed a committee to create a formal definition of a planet. The definition was presented at the IAU's triennial gathering in Prague in 2006, where it was revised several times by the astronomers at the meeting. Eventually the IAU voted and passed a resolution that defined a planet.

That resolution states that a planet must have three qualities: (1) it must be round, indicating its interior is in hydrostatic equilibrium; (2) it must orbit the Sun; and (3) it must have gravitationally cleared its zone of other debris. The last requirement means that a planet must be massive enough to be gravitationally dominant in its zone in the solar system. Any round body orbiting the Sun that fails condition (3) is labeled a "dwarf planet" by the IAU.

The outcome left the solar system with the eight major planets discovered through 1846, and reclassified Ceres, Pluto, and Eris as dwarf planets. Other large objects in the asteroid and Kuiper belts may be added to the list of dwarf planets if observations show that they too are round.

Although most astronomers have accepted the new IAU definition, there are some who have not, and who are actively campaigning to change it. There are weaknesses in the definition, particularly in condition (3), which are likely to be modified by an IAU committee tasked with improving the definition. However, the likelihood of the definition being changed sufficiently to again classify Pluto as a planet is small.

In this chapter we will use the new IAU definition of a planet. For an alternative view of the new definition, the reader is directed to the chapter PLUTO.

3. The Architecture of the Solar System

The solar system consists of the Sun at its center, eight planets, three dwarf planets, 165 known natural **satellites** (or moons) of planets and dwarf planets (as of September 2006), four ring systems, approximately one million asteroids (greater than 1 km in diameter), trillions of comets (greater than 1 km in diameter), the **solar wind**, and a large cloud of interplanetary dust. The arrangement and nature of all these bodies are the result of physical and dynamical processes during their origin and subsequent evolution, and their complex interactions with one another.

At the center of the solar system is the Sun, a rather ordinary, **main sequence** star. The Sun is classified spectrally as a G2 dwarf, which means that it emits the bulk of its radiation in the visible region of the spectrum, peaking at yellow-green wavelengths. The Sun contains 99.86% of the mass in the solar system, but only about 0.5% of the angular momentum. The low angular momentum of the Sun results from the transfer of momentum to the accretion disk surrounding the Sun during the formation of the planetary system, and to a slow spin-down due to angular momentum being carried away by the solar wind.

The Sun is composed of hydrogen (70% by mass), helium (28%), and heavier elements (2%). The Sun produces energy through nuclear fusion at its center, hydrogen atoms combining to form helium and releasing energy that eventually makes its way to the Sun's surface as visible sunlight. The central temperature of the Sun where fusion takes place is 15.7 million kelvins, while the temperature at the visible surface, the photosphere, is ∼6400 K. The Sun has an outer atmosphere called the corona, which is only visible during solar eclipses, or through the use of specially designed telescopes called coronagraphs.

A star like the Sun is believed to have a typical lifetime of 9 billion to 10 billion years on the main sequence. The present age of the Sun (and the entire solar system) is estimated to be 4.56 billion years, so it is about halfway through its nominal lifetime. The age estimate comes from radioisotope dating of meteorites.

3.1 Dynamics

The planets all orbit the Sun in roughly the same plane, known as the **ecliptic** (the plane of the Earth's orbit), and in the same direction, counterclockwise as viewed from the north ecliptic pole. Because of gravitational torques from the other planets, the ecliptic is not inertially fixed in space, and so dynamicists often use the invariable plane, which is the plane defined by the summed angular momentum vectors of all of the planets.

To first order, the motion of any body about the Sun is governed by Kepler's laws of planetary motion. These laws state that (1) each planet moves about the Sun in an orbit that is an ellipse, with the Sun at one focus of the ellipse; (2) the straight line joining a planet and the Sun sweeps out equal areas in space in equal intervals of time; and (3) the squares of the sidereal periods of the planets are in direct proportion to the cubes of the semimajor axes of their orbits. The laws of planetary motion, first set down by J. Kepler in 1609 and 1619, are easily shown to be the result of the inverse-square law of gravity with the Sun as the

TABLE 1	Orbits of the Planets and Dwarf Planets[a]			
Name	Semimajor Axis (AU)	Eccentricity	Inclination (°)	Period (years)
Mercury	0.38710	0.205631	7.0049	0.2408
Venus	0.72333	0.006773	3.3947	0.6152
Earth	1.00000	0.016710	0.0000	1.0000
Mars	1.52366	0.093412	1.8506	1.8808
Ceres[b]	2.7665	0.078375	10.5834	4.601
Jupiter	5.20336	0.048393	1.3053	11.862
Saturn	9.53707	0.054151	2.4845	29.457
Uranus	19.1913	0.047168	0.7699	84.018
Neptune	30.0690	0.008586	1.7692	164.78
Pluto[b]	39.4817	0.248808	17.1417	248.4
Eris (2003 UB$_{313}$)[b]	68.1461	0.432439	43.7408	562.55

[a] J2000, Epoch: January 1, 2000.
[b] Dwarf planet.

central body, and the conservation of angular momentum and energy. Parameters for the orbits of the eight planets and three dwarf planets are listed in Table 1.

Because the planets themselves have finite masses, they exert small gravitational tugs on one another, which cause their orbits to depart from perfect ellipses. The major effects of these long-term or "secular" perturbations are to cause the **perihelion** point of each orbit to precess (rotate counterclockwise) in space, and the line of nodes (the intersection between the planet's orbital plane and the ecliptic plane) of each orbit to regress (rotate clockwise). Additional effects include slow oscillations in the **eccentricity** and **inclination** of each orbit, and the inclination of the planet's rotation pole to the planet's orbit plane (called the obliquity). For the Earth, these orbital oscillations have periods of 19,000 to 100,000 years. They have been identified with long-term variations in the Earth's climate, known as Milankovitch cycles, though the linking physical mechanism is not well understood.

Relativistic effects also play a small but detectable role. They are most evident in the precession of the perihelion of the orbit of Mercury, the planet deepest in the Sun's gravitational potential well. General relativity adds 43 arcsec/century to the precession rate of Mercury's orbit, which is 574 arcsec/century. Prior to Einstein's theory of general relativity in 1916, it was thought that the excess in the precession rate of Mercury was due to a planet orbiting interior to it. This hypothetical planet was given the name Vulcan, and extensive searches were conducted for it, primarily during solar eclipses. No planet was detected.

A more successful search for a new planet occurred in 1846. Two celestial mechanicians, U. J. J. Leverrier and J. C. Adams, independently used the observed deviations

of Uranus from its predicted orbit to successfully predict the existence and position of Neptune. Neptune was found by J. G. Galle on September 23, 1846, using Leverrier's prediction.

More complex dynamical interactions are also possible, in particular when the orbital period of one body is a small-integer ratio of another's orbital period. This is known as a mean-motion resonance and can have dramatic effects. For example, Pluto is locked in a 2:3 mean-motion resonance with Neptune, and although the orbits of the two bodies cross in space, the resonance prevents them from ever coming within 14 AU of each other. Also, when two bodies have identical perihelion precession rates or nodal regression rates, they are said to be in a secular resonance, and similarly interesting dynamical effects can result. In many cases, mean-motion and secular resonances can lead to chaotic motion, driving a body onto a planet-crossing orbit, which will then lead to its being dynamically scattered among the planets, and eventually either ejected from the solar system or impacted on the Sun or a planet. In other cases, such as Pluto and some asteroids, the mean-motion resonance is actually a stabilizing factor for the orbit.

Chaos has become a very exciting topic in solar system dynamics in the past 25 years and has been able to explain many features of the planetary system that were not previously understood. It should be noted that the dynamical definition of chaos is not always the same as the general dictionary definition. In celestial mechanics, the term "chaos" is applied to describe systems that are not perfectly predictable over time. That is, small variations in the initial conditions, or the inability to specify the initial conditions precisely, will lead to a growing error in predictions of the long-term behavior of the system. If the error grows exponentially, then the system is said to be chaotic. However, the chaotic zone, the allowed area in phase space over which an orbit may vary, may still be quite constrained. Thus, although studies have found that the orbits of the planets are chaotic, this does not mean that Jupiter may one day become Earth-crossing, or vice versa. It means that the precise position of the Earth or Jupiter in their orbits is not predictable over very long periods of time. Because this happens for all the planets, the long-term **secular perturbations** of the planets on one another are also not perfectly predictable and can vary.

On the other hand, chaos can result in some extreme changes in orbits, with sudden increases in eccentricity that can throw small bodies onto planet-crossing orbits. One well-recognized case occurs near mean-motion resonances in the asteroid belt, which causes small asteroids to be thrown onto Earth-crossing orbits, allowing for the delivery of **meteoroids** to the Earth.

The natural satellites of the planets and their ring systems (where they exist) are governed by the same dynamical laws of motion. Most major satellites and all ring systems are deep within their planets' gravitational potential wells and so they move, to first order, on Keplerian ellipses. The Sun, planets, and other satellites all act as perturbers on the satellite and ring particle orbits. Additionally, the equatorial bulges of the planets, caused by the planets' rotation, act as a perturber on the orbits. Finally, the satellites raise tides on the planets (and vice versa), and these result in yet another dynamical effect, causing the planets to transfer rotational angular momentum to the satellite orbits in the case of direct or prograde orbits (satellites in retrograde orbits lose angular momentum). As a result, satellites may slowly move away from their planets into larger orbits (or smaller orbits in the case of retrograde satellites).

The mutual gravitational interactions can be quite complex, particularly in multisatellite systems. For example, the three innermost Galilean satellites of Jupiter (so named because they were discovered by Galileo in 1610)—Io, Europa, and Ganymede—are locked in a 4:2:1 mean-motion resonance with one another. In other words, Ganymede's orbital period is twice that of Europa and four times that of Io. At the same time, the other jovian satellites (primarily Callisto), the Sun, and Jupiter's oblateness perturb the orbits, forcing them to be slightly eccentric and inclined to one another, while the tidal interaction with Jupiter forces the orbits to evolve outward. These competing dynamical processes result in considerable energy deposition in the satellites, which manifests itself as volcanic activity on Io, as a possible subsurface ocean on Europa, and as past tectonic activity on Ganymede.

This illustrates an important point in understanding the solar system. The bodies in the solar system do not exist as independent, isolated entities, with no physical interactions between them. Even these "action at a distance" gravitational interactions can lead to profound physical and chemical changes in the bodies involved. To understand the solar system as a whole, one must recognize and understand the processes that were involved in its formation and its subsequent evolution, and that continue to act today.

An interesting feature of the planetary orbits is their regular spacing. This is described by Bode's law, first discovered by J. B. Titius in 1766 and brought to prominence by J. E. Bode in 1772. The law states that the semimajor axes of the planets in **astronomical units** can be roughly approximated by taking the sequence 0, 3, 6, 12, 24, ... adding 4, and dividing by 10. The values for Bode's law and the actual semimajor axes of the planets and two dwarf planets are listed in Table 2. It can be seen that the law works very well for the planets as far as Uranus, but it then breaks down. It also predicts a planet between Mars and Jupiter, the current location of the asteroid belt. Yet Bode's law predates

TABLE 2	Bode's Law, $a_1 = 0.4$, $a_n = 0.3 \times 2^{n-2} + 0.4$		
Planet	**Semimajor Axis (AU)**	**n**	**Bode's Law**
Mercury	0.387	1	0.4
Venus	0.723	2	0.7
Earth	1.000	3	1.0
Mars	1.524	4	1.6
Ceres [a]	2.767	5	2.8
Jupiter	5.203	6	5.2
Saturn	9.537	7	10.0
Uranus	19.19	8	19.6
Neptune	30.07	9	38.8
Pluto[a]	39.48	10	77.2

[a] Dwarf planet.

the discovery of the first asteroid by 35 years, as well as the discovery of Uranus by 15 years.

The reason why Bode's law works so well is not understood. H. Levison has recently suggested that, at least for the giant planets, it is a result of their spacing themselves at distances where they are equally likely to scatter a smaller body inward or outward to the next planet in either direction.

However, it has also been argued that Bode's law may just be a case of numerology and not reflect any real physical principle at all. Computer-based dynamical simulations have shown that the spacing of the planets is such that a body placed in a circular orbit between any pair of neighboring planets will likely be dynamically unstable. It will not survive over the history of the solar system unless protected by some dynamical mechanism such as a mean-motion resonance with one of the planets. Over the history of the solar system, the planets have generally cleared their zones of smaller bodies through gravitational scattering. The larger planets, in particular Jupiter and Saturn, are capable of throwing small bodies onto hyperbolic orbits, allowing the objects to escape to interstellar space. In the course of doing this, the planets themselves "migrate" moving either closer or farther from the Sun as a result of the angular momentum exchange with many smaller bodies.

Thus, the comets and asteroids we now see in planet-crossing orbits must have been introduced into the planetary system relatively recently from storage locations either outside the planetary system, or in protected, dynamically stable reservoirs. Because of its position at one of the Bode's law locations, the asteroid belt is a relatively stable reservoir. However, the asteroid belt's proximity to Jupiter's substantial gravitational influence results in some highly complex dynamics. Mean-motion and secular resonances, as well as mutual collisions, act to remove objects from the asteroid belt and throw them into planet-crossing orbits. The failure of a major planet to grow in the asteroid belt is generally attributed to the gravitational effects of Jupiter disrupting the slow growth by accretion of a planetary-sized body in the neighboring asteroid belt region.

It is generally believed that comets originated as icy **planetesimals** in the outer regions of the **solar nebula**, at the orbit of Jupiter and beyond. Those proto-comets with orbits between the giant planets were gravitationally ejected, mostly to interstellar space. However, a fraction of the proto-comets were flung into distant but still bound orbits; the Sun's gravitational sphere of influence extends $\sim 2 \times 10^5$ AU, or about 1 **parsec** (**pc**). These orbits were sufficiently distant from the Sun that they were perturbed by random passing stars and by the tidal perturbation from the galactic disk. The stellar and galactic perturbations raised the perihelia of the comet orbits out of the planetary region. Additionally, the stellar perturbations randomized the inclinations of the comet orbits, forming a spherical cloud of comets around the planetary system and extending halfway

to the nearest stars. This region is now called the **Oort cloud**, after J. H. Oort who first suggested its existence in 1950.

The current population of the Oort cloud is estimated at several times 10^{12} comets, with a total mass of about 15 Earth masses of material. Between 50 and 80% of the Oort cloud population is in a dense core within $\sim 10^4$ AU of the Sun. Long-period comets (those with orbital periods greater than 200 years) observed passing through the planetary region come from the Oort cloud. Some of the short-period comets (those with orbital periods less than 200 years), such as comet Halley, may be long-period comets that have evolved to short-period orbits due to repeated planetary perturbations.

A second reservoir of comets is the Kuiper belt beyond the orbit of Neptune, named after G. P. Kuiper who in 1951 was one of the first to suggest its existence. Because no large planet grew beyond Neptune, there was no body to scatter away the icy planetesimals formed in that region. (The failure of a large planet to grow beyond Neptune is generally attributed to the increasing timescale for planetary accretion with increasing **heliocentric** distance.) This belt of remnant planetesimals may terminate at ~ 50 AU or may extend out several hundred AU from the Sun, analogous to the disks of dust that have been discovered around main sequence stars such as Vega and Beta Pictoris (Fig. 1).

The Kuiper belt actually consists of two different dynamical populations. The classical Kuiper belt is the population in low-inclination, low-eccentricity orbits beyond Neptune. Some of this population, including Pluto, is trapped in mean-motion resonances with Neptune at both the 3:2 and 2:1 resonances. The second population is objects in more eccentric and inclined orbits, typically with larger semimajor axes, called the **scattered disk**. These objects all have perihelia relatively close to Neptune's orbit, such that they continue to gravitationally interact with Neptune.

The Kuiper belt may contain many tens of Earth masses of comets, though the mass within 50 AU is currently estimated as ~ 0.1 Earth mass. A slow gravitational erosion of comets from the Kuiper belt, in particular from the scattered disk, due to the perturbing effect of Neptune, causes these comets to "leak" into the planetary region. Eventually, some fraction of the comets evolves due to gravitational scattering by the jovian planets into the terrestrial planets region where they are observed as short-period comets. Short-period comets from the Kuiper belt are often called Jupiter-family or ecliptic comets because most are in orbits that can have close encounters with Jupiter, and also are in orbits with inclinations close to the ecliptic plane. Based on the observed number of ecliptic comets, the number of comets in the Kuiper belt between 30 and 50 AU has been estimated at $\sim 10^9$ objects larger than 1 km diameter, with a roughly equal number in the scattered disk. Current studies suggest that the Kuiper belt has been collisionally

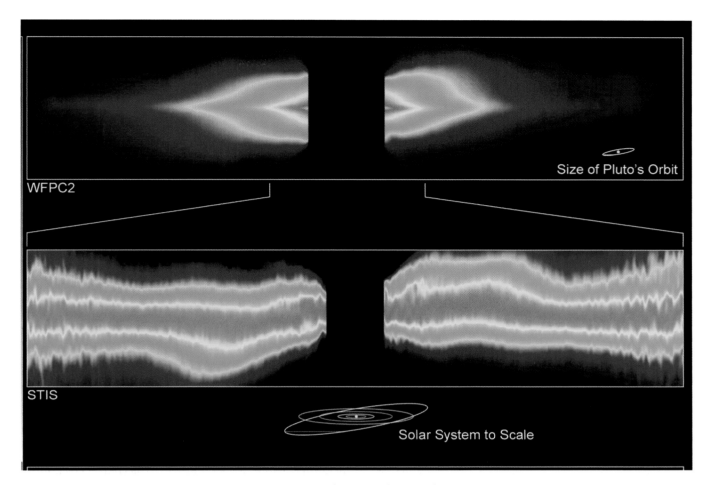

FIGURE 1 False color images of the dust disk around the star ß Pictoris, discovered by the *IRAS* satellite in 1983. The disk is viewed nearly edge on and is over 900 AU in diameter. The gaps in the center of each image are where the central star image has been removed. The top image shows the full disk as imaged with the Wide Field Planetary Camera 2 (WFPC2) camera onboard the *Hubble Space Telescope (HST)*. The lower image shows the inner disk as viewed by the Space Telescope Imaging Spectrograph (STIS) instrument on *HST*. The orbits of the outer planets of our solar system, including the dwarf planet Pluto, are shown to scale for comparison. There is evidence of a warping of the ß Pic disk, possibly caused by perturbations from a passing star. Infrared data show that the disk does not extend all the way in to the star, but that it has an inner edge at about 30 AU from ß Pic. The disk interior to that distance may have been swept up by the accretion of planets in the nebula around the star. This disk is a possible analog for the Kuiper belt around our own solar system.

eroded out to a distance of ~100 AU from the Sun, but that considerably more mass may still exist in orbits beyond that distance.

Although gravity is the dominant force in determining the motion of bodies in the solar system, other forces do come into play in special cases. Dust grains produced by asteroid collisions or liberated from the sublimating icy surfaces of comets are small enough to be affected by radiation pressure forces. For submicron grains, radiation pressure from sunlight is sufficient to blow the grains out of the solar system. For larger grains, radiation pressure causes the grains to depart from Keplerian orbits. Radiation effects can also cause larger grains to slowly spiral in toward the Sun through the Poynting–Robertson effect, and meter- to kilometer-sized bodies to spiral either inward or outward due to the Yarkovsky effect.

Electromagnetic forces play a role in planetary magnetospheres where ions are trapped and spiral back and forth along magnetic field lines, and in cometary Type I plasma tails where ions are accelerated away from the cometary coma by the solar wind. Dust grains trapped in planetary magnetospheres and in interplanetary space also respond to electromagnetic forces, though to a lesser extent than ions because of their much lower charge-to-mass ratios.

3.2 Nature and Composition

The solar nebula, the cloud of dust and gas out of which the planetary system formed, almost certainly exhibited a strong temperature gradient with heliocentric distance, hottest near the forming proto-Sun at its center, and cooling as one moved outward through the planetary region. This temperature gradient is reflected in the compositional arrangement of the planets and their satellites versus heliocentric distance. Parts of the gradient are also preserved in the asteroid belt between Mars and Jupiter and likely in the Kuiper belt beyond Neptune.

Physical parameters for the planets and dwarf planets are given in Table 3. The planets fall into two major compositional groups. The terrestrial or Earth-like planets are Mercury, Venus, Earth, and Mars and are shown in Fig. 2. The terrestrial planets are characterized by predominantly silicate compositions with iron cores. This results from the fact that they all formed close to the Sun where it was too warm for ices to condense. Also, the modest masses of the terrestrial planets and their closeness to the Sun did not allow them to capture and retain gas directly from the solar nebula. The terrestrial planets all have solid surfaces that are modified to varying degrees by both cratering and internal processes (tectonics, weather, etc.). Mercury is the most heavily cratered because it has no appreciable atmosphere to protect it from impacts or weather to erode the cratered terrain, and also because encounter velocities with Mercury are very high that close to the Sun. Additionally, tectonic processes on Mercury appear to have been modest at best. Mars is next in degree of cratering, in large part

because of its proximity to the asteroid belt. Also, Mars' thin atmosphere affords little protection against impactors. However, Mars also displays substantial volcanic and tectonic features, and evidence of erosion by wind and flowing water, the latter presumably having occurred early in the planet's history.

The surface of Venus is dominated by a wide variety of volcanic terrains. The degree of cratering on Venus is less than on Mercury or Mars for two reasons: (1) Venus' thick atmosphere (surface pressure = 94 bar) breaks up smaller asteroids and comets before they can reach the surface, and (2) vulcanism on the planet has covered over the older craters on the planet surface. The surface of Venus is estimated to be 500 million to 800 million years in age.

The Earth's surface is dominated by plate tectonics, in which large plates of the crust can move about the planet, and whose motions are reflected in such features as mountain ranges (where plates collide) and volcanic zones (where one plate dives under another). The Earth is the only planet with the right combination of atmospheric surface pressure and temperature to permit liquid water on its surface, and some 70% of the planet is covered by oceans. Craters on the Earth are rapidly erased by its active geology and weather, though the atmosphere only provides protection against very modest size impactors, on the order of 100 m diameter or less. Still, 172 impact craters or their remnants have been found on the Earth's surface or under its oceans.

The terrestrial planets each have substantially different atmospheres. Mercury has a tenuous atmosphere arising from its interaction with the solar wind. Hydrogen and helium ions are captured directly from the solar wind, whereas oxygen, sodium, and potassium are likely the product of sputtering. In contrast, Venus has a dense CO_2 atmosphere with a surface pressure 94 times the pressure at the Earth's

TABLE 3	Physical Parameters for the Sun, Planets, and Dwarf Planets					
Name	Mass (kg)	Equatorial Radius (km)	Density (g cm^{-3})	Rotation Period	Obliquity (o)	Escape Velocity (km sec^{-1})
Sun	1.989×10^{33}	695,500	1.41	25.38–35.	7.25	617.7
Mercury	3.302×10^{23}	2,440	5.43	56.646 d.	0.	4.25
Venus	4.869×10^{24}	6,052	5.24	243.018 d.	177.33	10.36
Earth	5.974×10^{24}	6,378	5.52	23.934 h.	23.45	11.18
Mars	6.419×10^{23}	3,397	3.94	24.623 h.	25.19	5.02
Ceres[a]	9.47×10^{20}	474	2.1	9.075 h.		0.52
Jupiter	1.899×10^{27}	71,492	1.33	9.925 h.	3.08	59.54
Saturn	5.685×10^{26}	60,268	0.70	10.656 h.	26.73	35.49
Uranus	8.662×10^{25}	25,559	1.30	17.24 h.	97.92	21.26
Neptune	1.028×10^{26}	24,764	1.76	16.11 h.	28.80	23.53
Pluto[a]	1.314×10^{22}	1,151	2.0	6.387 d.	119.6	1.23
Eris (2003 UB_{313})[a]	1.5×10^{22}	1,200	2.1			1.29

[a] Dwarf planet.

FIGURE 2 The terrestrial planets: the heavily cratered surface of Mercury as photographed by the *Mariner 10* spacecraft in 1974 (top left); false color image of clouds on the night side of Venus, backlit by the intense infrared radiation from the planet's hot surface, as seen by the *Galileo* Near-Infrared Mapping Spectrometer (NIMS) instrument in 1990 (top right); South America and Antarctica as imaged by the *Galileo* spacecraft during a gravity assist flyby of the Earth in 1990 (bottom left); Valles Marineris, a 3000 km long canyon on Mars as photographed by the *Viking 1* orbiter in 1980 (bottom right).

surface. Nitrogen is also present in the Venus atmosphere at a few percent relative to CO_2. The dense atmosphere results in a massive greenhouse on the planet, heating the surface to a mean temperature of 735 K. The middle and upper atmosphere contain thick clouds composed of H_2SO_4 and H_2O, which shroud the surface from view. However, thermal radiation from the surface does penetrate the clouds, making it possible to view surface features through infrared "windows."

The Earth's atmosphere is unique because of its large abundance of free oxygen, which is normally tied up in oxidized surface materials on other planets. The reason for this unusual state is the presence of life on the planet, which traps and buries CO_2 as carbonates and also converts the CO_2 to free oxygen. Still, the bulk of the Earth's atmosphere is nitrogen (78%), with oxygen making up 21% and argon about 1%. The water vapor content of the atmosphere varies from about 1 to 4%. Various lines of evidence suggest that the composition of the Earth's atmosphere has evolved considerably over the history of the solar system and that the original atmosphere was denser than the present-day atmosphere and dominated by CO_2.

Mars has a relatively modest CO_2 atmosphere with a mean surface pressure of only 6 mbar. The atmosphere also contains a few percent of N_2 and argon. Mineralogic and isotopic evidence and geologic features suggest that the past atmosphere of Mars may have been much denser and warmer, allowing liquid water to flow across the surface in massive floods.

The volatiles in the terrestrial planets' atmospheres (and the Earth's oceans) may have been contained in hydrated minerals in the planetesimals that originally formed the planets, and/or may have been added later due to asteroid and comet bombardment as the planets dynamically cleared their individual zones of leftover planetesimals. It appears most likely that all these reservoirs contributed some fraction of the volatiles on the terrestrial planets.

The jovian or Jupiter-like planets are Jupiter, Saturn, Uranus, and Neptune and are shown in Fig. 3. The jovian planets are also referred to as the gas giants. They are characterized by low mean densities and thick hydrogen–helium atmospheres, presumably captured directly from the solar nebula during the formation of these planets. The composition of the jovian planets is similar to that of the Sun, though more enriched in heavier elements. Because of their primarily gaseous composition and their high internal temperatures and pressures, the jovian planets do not have solid surfaces. However, they may each have silicate–iron cores of several to tens of Earth masses of material.

Because they formed at heliocentric distances where ices could condense, the giant planets may have initially had a much greater local density of solid material to grow from. This may, in fact, have allowed them to form before the terrestrial planets interior to them. Studies of the dissipation of nebula dust disks around nearby solar-type protostars suggest that the timescale for the formation of giant planets is on the order of 10 million years or less. This is very rapid as compared with the $\sim$100 million year timescale currently estimated for the formation of the terrestrial planets (though questions have now been raised as to the correctness of that accretionary timescale). Additionally, the higher uncompressed densities of Uranus and Neptune (0.5 g cm^{-3}) versus Jupiter and Saturn (0.3 g cm^{-3}), suggest that the outer two giant planets contain a significantly lower fraction of gas captured from the nebula. This may mean that the outer pair formed later than the inner two giant planets, consistent with the increasing timescale for planetary accretion at larger heliocentric distances.

Because of their heliocentric arrangement, the terrestrial and jovian planets are occasionally called the inner and outer planets, respectively, though sometimes the term "inner planets" is used only to denote Mercury and Venus, the planets interior to the Earth's orbit.

Among the dwarf planets, Ceres has a surface composition and density similar to carbonaceous chondrite meteorites. This is a primitive class of meteorites that shows only limited processing during and since formation. Water frost has also been detected on the surface of Ceres. Because of its large size, the interior of Ceres is likely differentiated.

Pluto and its largest satellite Charon are shown in Fig. 4. Pluto bears a strong resemblance to Triton, Neptune's large icy satellite (which is slightly larger than Pluto) and to other large icy planetesimals in the Kuiper belt beyond the orbit of Neptune. Pluto has a thin, extended atmosphere, probably methane and nitrogen, which is slowly escaping because of Pluto's low gravity. This puts it in a somewhat intermediate state between a freely outflowing cometary coma and a bound atmosphere. Spectroscopic evidence shows that methane frost covers much of the surface of Pluto, whereas its largest satellite Charon appears to be covered with water frost. Nitrogen frost has also been detected on Pluto. The density of Pluto is $\sim$2 g cm^{-3}, suggesting that the rocky component of the dwarf planet accounts for about 70% of its total mass.

The third dwarf planet, Eris, is a Kuiper belt object in a distant orbit that ranges from 37.8 to 97.5 AU from the Sun. It is slightly larger than Pluto, has a similar bulk density, and also displays evidence for methane frost on its surface.

There has been considerable speculation as to the existence of a major planet beyond Neptune, often dubbed "Planet X." The search program that found Pluto in 1930 was continued for many years afterward but failed to detect any other distant planet, even though the limiting magnitude was considerably fainter than Pluto's visual magnitude of $\sim$13.5. Other searches have been carried out, most notably by the *Infrared Astronomical Satellite (IRAS)* in 1983–1984. An automated algorithm was used to search for a distant planet in the *IRAS* data; it successfully "discovered" Neptune, but nothing else. Telescopic searches for Kuiper

FIGURE 3   The jovian planets: the complex, belted atmosphere of Jupiter with the Great Red Spot at the lower center, as photographed by *Cassini* during its gravity assist flyby in 2000 (top left); Saturn and its beautiful ring system, as photographed by *Cassini* in 2005 (top right); the featureless atmosphere of Uranus, obscured by a high-altitude methane haze, as imaged by *Voyager 2* in 1986 (bottom left); several large storm systems and a banded structure, similar to that of Jupiter, in Neptune's atmosphere, as photographed by *Voyager 2* in 1989 (bottom right).

FIGURE 4 Hubble Space Telescope image of the dwarf planet Pluto (center) with its large moon Charon (just above and to the right of Pluto), and two newly discovered small satellites (top). A NASA spacecraft mission, *New Horizons*, was launched in 2006 and will fly by Pluto and Charon in 2015. (Courtesy of NASA and the Space Telescope Science Institute.)

belt objects have found objects comparable to Pluto in size, but not significantly larger.

Gravitational analyses of the orbits of Uranus and Neptune show no evidence of an additional perturber at greater heliocentric distances. Studies of the trajectories of the *Pioneer 10* and *11* and *Voyager 1* and *2* spacecraft have also yielded negative results. Analyses of the spacecraft trajectories do provide an upper limit on the unaccounted mass within the orbit of Neptune of $< 3 \times 10^{-6}$ solar masses ($M_\odot$), equal to about one Earth mass.

The compositional gradient in the solar system is perhaps best visible in the asteroid belt, whose members range from stony bodies in the inner belt (inside of $\sim$2.6 AU), to volatile-rich carbonaceous bodies in the outer main belt (out to about 3.3 AU). (See Fig. 5.) There also exist thermally processed asteroids, such as Vesta, whose surface material resembles a basaltic lava flow, and iron–nickel objects, presumably the differentiated cores of larger asteroids that were subsequently disrupted by collisions. The thermal gradient that processed the asteroids appears to be very steep and likely cannot be explained simply by the individual distances of these bodies from the forming proto-Sun. Rather, various special mechanisms such as magnetic induction, short-lived radioisotopes, or massive solar flares have been invoked to explain the heating event that so strongly processed the inner half of the asteroid belt.

The largest asteroid is Ceres, now classified as a dwarf planet, at a mean distance of 2.77 AU from the Sun. Ceres was the first asteroid discovered, by G. Piazzi on January 1, 1801. Ceres is 948 km in diameter, rotates in 9.075 hours, and appears to have a surface composition similar to that of carbonaceous chondrite meteorites. The second largest as-

teroid is Pallas, also a carbonaceous type with a diameter of 532 km. Pallas is also at 2.77 AU, but its orbit has an unusually large inclination of 34.8°. Over 136,500 asteroids have had their orbits accurately determined and have been given official numbers in the asteroid catalog (as of September 2006). Another 204,700 asteroids have been observed well enough to obtain preliminary orbits, 137,300 of them at more than one opposition. Note that these numbers include all objects nominally classified as asteroids: main-belt, near-Earth, Trojans, Centaurs, and Kuiper belt objects (including Pluto and Eris). As a result of the large number of objects in the asteroid belt, impacts and collisions are frequent. Several "families" of asteroids have been identified by their closely grouped orbital elements and are likely fragments of larger asteroids that collided. Spectroscopic studies have shown that the members of these families often have very similar surface compositions, further evidence that they are related. The largest asteroids such as Ceres and Pallas are likely too large to be disrupted by impacts, but most of the smaller asteroids have probably been collisionally processed. Increasing evidence suggests that many asteroids may be "rubble piles," that is, asteroids that have been broken up but not dispersed by previous collisions, and that now form a single but poorly consolidated body.

Beyond the main asteroid belt there exist small groups of asteroids locked in dynamical resonances with Jupiter. These include the Hildas at the 3:2 mean-motion resonance, the Thule group at the 4:3 resonance, and the Trojans, which are in a 1:1 mean-motion resonance with Jupiter. The effect of the resonances is to prevent these asteroids from making close approaches to Jupiter, even though many of the asteroids are in Jupiter-crossing orbits.

The Trojans are particularly interesting. They are essentially in the same orbit as Jupiter, but they librate about points 60° ahead and 60° behind the planet in its orbit, known as the Lagrange L_4 and L_5 points. These are pseudo-stable points in the three-body problem (Sun–Jupiter–asteroid) where bodies can remain dynamically stable for extended periods of time. Some estimates have placed the total number of objects in the Jupiter L_4 and L_5 Trojan swarms as equivalent to the population of the main asteroid belt. Trojan-type 1:1 librators have also been found for the Earth and Mars (one each), and for Neptune (four). Searches at the L_4 and L_5 points of the other giant planets have been negative so far. Interestingly, the Saturnian satellites Dione and Tethys also have small satellites locked in Trojan-type librations in their respective orbits.

Much of what we know about the asteroid belt and about the early history of the solar system comes from meteorites recovered on the Earth. It appears that the asteroid belt is the source of almost all recovered meteorites. A modest number of meteorites that are from the Moon and from Mars, presumably blasted off of those bodies by asteroid and/or comet impacts, have been found. Cometary meteoroids are thought to be too fragile to survive atmospheric entry. In addition, cometary meteoroids typically encounter

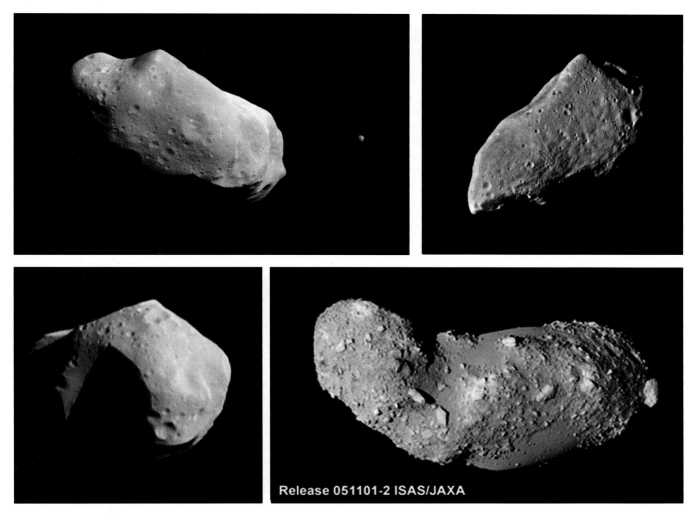

Release 051101-2 ISAS/JAXA

FIGURE 5   A sampling of main-belt and near-Earth asteroids: 243 Ida along with its small satellite Dactyl (top left), 951 Gaspra (top right), 253 Mathilde (bottom left), and 25143 Itokawa (bottom right). All these asteroids, with the exception of Mathilde, are stony types; Mathilde is a carbonaceous asteroid. Most of the asteroids exhibit heavily cratered surfaces, but Itokawa is an exception, appearing to be a complete rubble pile. Ida is 54 × 24 × 15 km in diameter, Dactyl is 1.5 km in diameter Gaspra is 18 × 10 × 9 km, Mathilde is roughly 53 km in diameter and Itokawa is only 550 × 300 × 260 m. The asteroids were photographed by the *Galileo* spacecraft while it was en route to Jupiter, in 1993 and 1991, the *NEAR* spacecraft while enroute to Eros in 1997, and the *Hayabusa* spacecraft while in orbit in 2005, respectively. Ida's tiny satellite, Dactyl, was an unexpected discovery of two of Galileo's remote sensing instruments, the Near Infrared Mapping Spectrometer and the Solid State Imaging system, during the flyby.

the Earth at higher velocities than asteroidal debris and thus are more likely to fragment and burn up during atmospheric entry. However, we may have cometary meteorites in our sample collections and simply not yet be knowledgeable enough to recognize them.

Recovered meteorites are roughly equally split between silicate and carbonaceous types, with a few percent being iron–nickel meteorites. The most primitive meteorites (i.e., the meteorites which appear to show the least processing in the solar nebula) are the volatile-rich carbonaceous

chondrites. However, even these meteorites show evidence of some thermal processing and aqueous alteration (i.e., processing in the presence of liquid water). Study of carbonaceous and ordinary (silicate) chondrites provides significant information on the composition of the original solar nebula, on the physical and chemical processes operating in the solar nebula, and on the chronology of the early solar system.

The other major group of primitive bodies in the solar system is the comets. Because comets formed farther from the Sun than the asteroids, in colder environments, they contain a significant fraction of volatile ices. Water ice is the dominant and most stable volatile. Typical comets also contain modest amounts of CO, CO_2, CH_4, NH_3, H_2CO, and CH_3OH, most likely in the form of ices, but possibly also contained within complex organic molecules and/or in clathrate hydrates. Organics make up a significant fraction of the cometary nucleus, as well as silicate grains. F. Whipple described this icy-conglomerate mix as "dirty snowballs," though the term "frozen mudball" may be more appropriate since the comets are more than 60% organics and silicates. It appears that the composition of comets is very similar to the condensed (solid) grains and ices observed in dense interstellar cloud cores, with little or no evidence of processing in the solar nebula. Thus, comets appear to be the most primitive bodies in the solar system. As a result, the study of comets is extremely valuable for learning about the origin of the planetary system and the conditions in the solar nebula 4.56 billion years ago.

Four cometary nuclei—periodic comets Halley, Borrelly, Wild 2, and Tempel 1—have been encountered by interplanetary spacecraft and imaged (Fig. 6). These irregular nuclei range from about 4 to 12 km in mean diameter and have low albedos, only 3–4%. The nuclei exhibit a variety of complex surface morphologies unlike any other bodies in the solar system. It has been suggested that cometary nuclei are weakly bound conglomerations of smaller dirty snowballs, assembled at low velocity and low temperature in the outer regions of the solar nebula. Thus, comets may be "primordial rubble piles," in some ways similar to the asteroids. Recent studies have suggested that cometary nuclei, like the asteroids, may have undergone intense collisional evolution, either while resident in the Kuiper belt, or in the giant planets region prior to their dynamical ejection to the Oort cloud.

Subtle and not-so-subtle differences in cometary compositions have been observed. However, it is not entirely clear if these differences are intrinsic or due to the physical evolution of cometary surfaces over many close approaches to the Sun. Because the comets that originated among the giant planets have all been ejected to the Oort cloud or to interstellar space, the compositional spectrum resulting from the heliocentric thermal profile is not spatially preserved as it has been in the asteroid belt. Although comets in the classical Kuiper belt are likely located close to their formation distances, physical studies of these distant objects are still in an early stage. There is an observed compositional trend, but it is associated with orbital eccentricity and inclination, rather than **semimajor axis**.

3.3 Satellites, Rings, and Things

The natural satellites of the planets, listed in the appendix to this volume, show as much diversity as the planets they orbit (see Fig. 7). Among the terrestrial planets, the only

known satellites are the Earth's Moon and the two small moons of Mars, Phobos and Deimos. The Earth's Moon is unusual in that it is so large relative to its primary. The Moon has a silicate composition similar to the Earth's mantle and a very small iron core.

It is now widely believed that the Moon formed as a result of a collision between the proto-Earth and another protoplanet about the size of Mars, late in the accretion of the terrestrial planets. Such "giant impacts" are now recognized as being capable of explaining many of the features of the solar system, such as the unusually high density of Mercury and the large obliquities of several of the planetary rotation axes. In the case of the Earth, the collision with another protoplanet resulted in the cores of the two planets merging, while a fraction of the mantles of both bodies was thrown into orbit around the Earth where some of the material reaccreted to form the Moon. The tidal interaction between the Earth and Moon then slowly evolved the orbit of the Moon outward to its present position, at the same time slowing the rotation of both the Earth and the Moon. The giant-impact hypothesis is capable of explaining many of the features of the Earth–Moon system, including the similarity in composition between the Moon and the Earth's mantle, the lack of a significant iron core within the Moon, and the high angular momentum of the Earth–Moon system.

Like most large natural satellites, the Moon has tidally evolved to where its rotation period matches its revolution period in its orbit. This is known as synchronous rotation. It results in the Moon showing the same face to the Earth at all times, though there are small departures from this because of the eccentricity of the Moon's orbit.

The Moon's surface displays a record of the intense bombardment all the planets have undergone over the history of the solar system. Returned lunar samples have been age-dated based on decay of long-lived radioisotopes. This has allowed the determination of a chronology of lunar bombardment by comparing the sample ages with the crater counts on the lunar plains where the samples were collected. The lunar plains, or maria, are the result of massive eruptions of lava during the first billion years of the Moon's history. The revealed chronology shows that the Moon experienced a massive bombardment between 4.2 billion and 3.8 billion years ago, known as the Late Heavy Bombardment. This time period is relatively late as compared with the 100 million to 200 million years required to form the terrestrial planets and to clear their orbital zones of most interplanetary debris. Similarities in crater size distributions on the Moon, Mercury, and Mars suggest that the Late Heavy Bombardment swept over all the terrestrial planets. Recent explanations for the Late Heavy Bombardment have focused on the possibility that it came from the clearing of the outer planets zones of their cometary debris. However, the detailed dynamical calculations of the timescales for that process are still being determined.

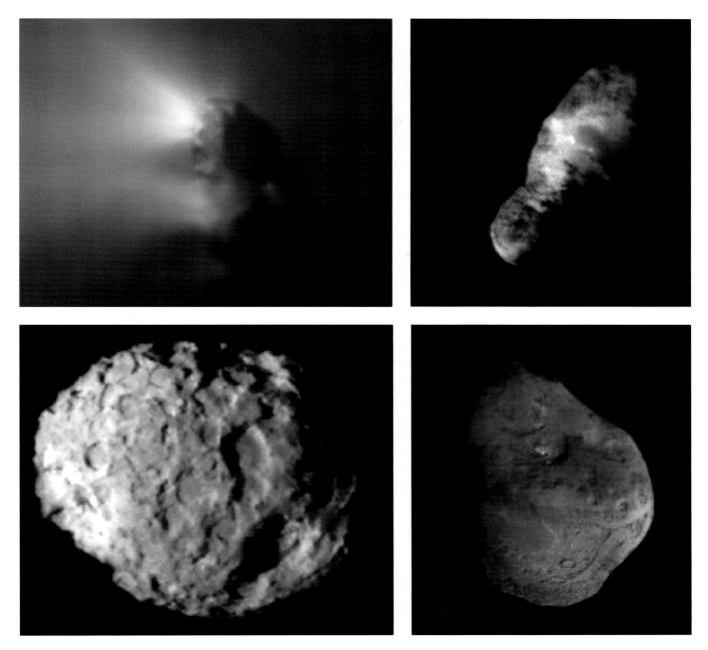

FIGURE 6 Four cometary nuclei photographed by flyby spacecraft: Halley's comet in 1986 (*Giotto*, top left), Borrelly in 2001 (*Deep Space 1*, top right), Wild 2 in 2004 (*Stardust*, bottom left), and Tempel 1 in 2005 (*Deep Impact*, bottom right). The nuclei show considerable diversity both in shape and in surface topography. The Halley nucleus is about 15 × 8 km in diameter, the Borrelly nucleus is 8 × 3.2 km, the Wild 2 nucleus is 5.2 × 4.0 km, and the Tempel 2 nucleus is 7.6 × 4.9 km. The Halley image shows bright dust jets emanating from active areas on the nucleus surface. The other three nuclei were also active during their respective flybys but the activity was too faint to show in these images.

Like almost all other satellites in the solar system, the Moon has no substantial atmosphere. There is a transient atmosphere due to helium atoms in the solar wind striking the lunar surface and being captured. Argon has been detected escaping from surface rocks and being temporarily cold-trapped during the lunar night. Also, sodium and potassium have been detected, likely the result of sputtering of surface materials due to solar wind particles, as on Mercury.

Unlike the Earth's Moon, the two natural satellites of Mars are both small, irregular bodies and in orbits relatively

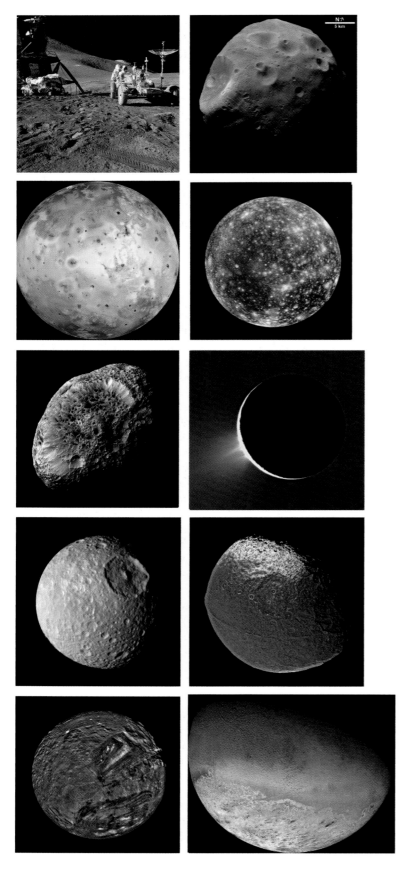

FIGURE 7   A sampling of satellites in the solar system: the dusty surface of the Earth's Moon, still the only other celestial body visited by humans (top row, left); Phobos, the larger of Mars' two moons showing the large crater Stickney at left (top row, right); the innermost Galilean satellite, Io, displays active vulcanism on its sulfur-rich surface (second row, left); the outermost Galilean satellite, Callisto, displays a heavily cratered surface, likely dating back to the origin of the solar system (second row, right); one of Saturn's smaller satellites, Hyperion, is irregularly shaped, in chaotic rotation, and displays a very unusual surface morphology (third row, left); Saturn's satellite Enceladus is one of several in the solar system that has active geysers on its surface (third row, right); another small Saturnian satellite, Mimas, displays an immense impact crater on one hemisphere (fourth row, left); Saturn's satellite Iapetus is black on one hemisphere and white on the other, and has a high ridge circling it at the equator (fourth row, right); Uranus' outermost major satellite, Miranda, has a complex surface morphology suggesting that the satellite was disrupted and reaccreted (bottom row, left); Neptune's one large satellite, Triton, displays a mix of icy terrains and ice vulcanism (bottom, right).

close to the planet. In fact, Phobos, the larger and closer satellite, orbits Mars faster than the planet rotates. Both of the martian satellites have surface compositions that appear to be similar to carbonaceous chondrites in composition. This has resulted in speculation that the satellites are captured asteroids. A problem with this hypothesis is that Mars is located close to the inner edge of the asteroid belt, where silicate asteroids dominate the population, and where carbonaceous asteroids are relatively rare. Also, both satellites are located very close to the planet and in near-circular orbits, which is unusual for captured objects.

In contrast to the satellites of the terrestrial planets, the satellites of the giant planets are numerous and are arranged in complex systems. Jupiter has four major satellites, easily visible in small telescopes from Earth, and 58 known lesser satellites. The discovery of the four major satellites by Galileo in 1610, now known as the Galilean satellites, was one of the early confirmations of the Copernican theory of a heliocentric solar system. The innermost Galilean satellite, Io, is about the same size as the Earth's Moon and has active vulcanism on its surface as a result of Jupiter's tidal perturbation and the gravitational interaction with Europa and Ganymede (see Section 2.1). The next satellite outward is Europa, somewhat smaller than Io, which appears to have a thin ice crust overlying a possible liquid water ocean, also the result of tidal heating by Jupiter and the satellite–satellite interactions. Estimates of the age of the surface of Europa, based on counting impact craters, are very young, suggesting that the thin ice crust may repeatedly break up and reform. The next satellite outward from Jupiter is Ganymede, the largest satellite in the solar system, even larger than the planet Mercury. Ganymede is another icy satellite and shows evidence of tectonic activity and of being partially resurfaced at some time(s) in its past. The final Galilean satellite is Callisto, another icy satellite that appears to preserve an impact record of comets and asteroids dating back to the origin of the solar system. As previously noted, the orbits of the inner three Galilean satellites are locked into a 4:2:1 mean-motion resonance.

The lesser satellites of Jupiter include 4 within the orbit of Io, and 54 at very large distance from the planet. The latter are mostly in retrograde orbits, which suggest that they are likely captured comets and asteroids. The orbital parameters of many of these satellites fall into several tightly associated groups. This suggests that each group consists of fragments of a larger object that was disrupted, most likely by a collision with another asteroid or comet. Possibly, the collision occurred within the gravitational sphere of Jupiter, which then could have led to the dynamical capture of the fragments.

All of the close-orbiting jovian satellites (out to the orbit of Callisto) appear to be in synchronous rotation with Jupiter. However, rotation periods have been determined for two of the outer satellites, Himalia and Elara, and these appear to be around 10 to 12 hours, much shorter than their ~250 day periods of revolution about the planet.

Saturn's satellite system is very different from Jupiter's in that it contains only one large satellite, Titan, comparable in size to the Galilean satellites, 8 intermediate-sized satellites, and 47 smaller satellites. Titan is the only satellite in the solar system with a substantial atmosphere. Clouds of organic residue in its atmosphere prevent easy viewing of the surface of that moon, though the *Cassini* spacecraft has had success in viewing the surface at infrared and radar wavelengths. The atmosphere is primarily nitrogen and also contains methane and possibly argon. The surface temperature on Titan has been measured at 94 K, and the surface pressure is 1.5 bar. *Cassini* has revealed on Titan a complex surface morphology that includes rivers, lakes, and possible cryo-vulcanism.

The intermediate and smaller satellites of Saturn all appear to have icy compositions and have undergone substantial processing, possibly as a result of tidal heating and also due to collisions. Orbital resonances exist between several pairs of satellites, and most are in synchronous rotation with Saturn. An interesting exception is Hyperion, which is a highly nonspherical body and which appears to be in chaotic rotation. Another moon, Enceladus, has a ring of material in its orbit that likely has come from the satellite, either as a result of a recent massive impact or as a result of active vulcanism on the icy satellite; *Cassini* has found ice geysers near Enceladus' south pole. Two other satellites, Dione and Tethys, have companion satellites in the same orbit, which oscillate about the Trojan-libration points for the Saturn–Dione (1 companion) and Saturn–Tethys (2 companions) systems, respectively. Yet another particularly interesting satellite of Saturn is Iapetus, which is dark on one hemisphere and bright on the other and has a narrow ridge circling the satellite at its equator. The reason(s) for the unusual dichotomy in surface albedos or the equatorial ridge are not known.

Saturn has one very distant, intermediate-sized satellite, Phoebe, which is in a retrograde orbit and which is suspected of being a captured early solar system planetesimal, albeit a very large one. Phoebe is not in synchronous rotation, but rather has a period of about 10 hours. The 47 known small satellites of Saturn include 10 embedded in or immediately adjacent to the planet's ring system, the three Trojan-type librators, and 34 in distant orbits. As with Jupiter, the majority of these distant objects are in retrograde orbits and some are in groups, which suggests that they are collisional fragments.

The Uranian system consists of five intermediate-sized satellites and 22 smaller ones. Again, these are all icy bodies. These satellites also exhibit evidence of past heating and possible tectonic activity. The satellite Miranda is particularly unusual in that it exhibits a wide variety of complex terrains. It has been suggested that Miranda, and possibly many other icy satellites, were collisionally disrupted at some time in their history, and the debris then reaccreted in orbit to form the currently observed satellites, but preserved some of the older morphology. Such disruption/reaccretion phases may have even reoccurred on several occasions for a particular satellite over the history of the solar system. Of the smaller Uranian satellites, 13 are embedded in the ring system and 9 are in distant, mostly retrograde orbits. Again, these are likely captured objects.

Neptune's satellite system consists of one large icy satellite, Triton, and 12 smaller ones. Triton is somewhat larger than Pluto and is unusual in that it is in a retrograde orbit. As a result, the tidal interaction with Neptune is causing the satellite's orbit to decay, and eventually Triton will be torn apart by the planet's gravity when it passes within the **Roche limit**. The retrograde orbit is often cited as evidence that Triton must have been captured from interplanetary space and did not actually form in orbit around the planet. Despite its tremendous distance from the Sun, Triton's icy surface displays a number of unusual terrain types that strongly suggest thermal processing and possibly even current activity. The *Voyager 2* spacecraft photographed what appeared to be plumes from "ice volcanoes" on Triton.

The lesser satellites of Neptune include 6 that are either in or adjacent to the ring system and 6 in distant orbits, evenly split between direct and retrograde.

Among the dwarf planets, Ceres has no known satellites. Pluto has one very large satellite, Charon, which is slightly more than half the size of Pluto, and two smaller satellites, Nix and Hydra, each estimated to be ~40–60 km in diameter. The Pluto–Charon system is fully tidally evolved. This means that Pluto and Charon each rotate with the same period, 6.38723 days, which is also the revolution period of the satellite in its orbit. As a result, Pluto and Charon always show the same faces to each other. It is suspected that the Pluto–Charon system was formed by a giant impact between two large Kuiper belt objects. The third dwarf planet, Eris, also has an intermediate-sized satellite, Dysnomia, about 300–400 km in diameter.

In addition to their satellite systems, all of the jovian planets have ring systems (Fig. 8). As with the satellite systems, each ring system is distinctly different from its neighbors. Jupiter has a single ring at 1.72–1.81 planetary radii, discovered by the Voyager 1 spacecraft. The ring has several components, related to the four small satellites in or close to the ring. The micron-sized ring particles appear to be sputtered material off the embedded satellites.

Saturn has an immense, broad ring system extending between 1.11 and 2.27 planetary radii, easily seen in a small telescope from Earth. The ring system consists of three major rings, known as A, B, and C ordered from the outside in toward the planet, a diffuse ring labeled D inside the C ring and extending down almost to the top of the Saturnian atmosphere, and several other narrow, individual rings.

Closer examination by the *Voyager* spacecraft revealed that the A, B, and C rings were each composed of thousands of individual ringlets. This complex structure is the result of mean-motion resonances with the many Saturnian satellites, as well as with small satellites embedded within the rings themselves. Some of the small satellites act as gravitational "shepherds," focusing the ring particles into narrow ringlets. Additional narrow and diffuse rings are located outside the main ring system.

The Uranian ring system was discovered accidentally in 1977 during observation of a stellar occultation by Uranus. A symmetric pattern of five narrow dips in the stellar signal was seen on both sides of the planet. Later observations of other stellar occultations found an additional five narrow rings. *Voyager 2* detected several more, fainter, diffuse rings and provided detailed imaging of the entire ring system.

The success with finding Uranus' rings led to similar searches for a ring system around Neptune using stellar occultations. Rings were detected but were not always symmetric about the planet, suggesting gaps in the rings. Subsequent *Voyager 2* imaging revealed large azimuthal concentrations of material in one of the six detected rings.

All of the ring systems are within the Roche limits of their respective planets, at distances where tidal forces from the planet will disrupt any solid body, unless it is small enough and strong enough to be held together by its own material strength. This has led to the general belief that the rings are disrupted satellites, or possibly material that could never successfully form into satellites. Ring particles have typical sizes ranging from micron-sized dust to meter-sized objects and appear to be made primarily of icy materials, though in some cases contaminated with carbonaceous materials. Jupiter's ring is an exception because it appears to be composed of carbonaceous and silicate materials, with no ice.

Another component of the solar system is the zodiacal dust cloud, a huge, continuous cloud of fine dust extending throughout the planetary region and generally concentrated toward the ecliptic plane. The cloud consists of dust grains liberated from comets as the nucleus ices sublimate and from collisions between asteroids. Comets are estimated to account for about two thirds of the total material in the **zodiacal cloud**, with asteroid collisions providing the rest. Dynamical processes tend to spread the dust uniformly around the Sun, though some structure is visible as a result of the most recent asteroid collisions. These structures, or bands as they are also known, are each associated with specific asteroid collisional families.

Dust particles will typically burn up due to friction with the atmosphere when they encounter the Earth, appearing as visible meteors. However, particles less than about 50 μm in radius have sufficiently large area-to-mass ratios that they can be decelerated high in the atmosphere at an altitude of about 100 km and can radiate away the energy generated by friction without vaporizing the particles. These particles then settle slowly through the atmosphere and

FIGURE 8   The ring systems of the jovian planets: Jupiter's single ring photographed in forward scattered light while the *Galileo* spacecraft was in eclipse behind the giant planet: the lit circle is sunlight filtering through the atmosphere of Jupiter (top); Saturn's rings break up into hundreds of ringlets when viewed at high resolution, as in this *Cassini* mosaic (middle); Uranus' system of narrow rings as viewed in forward scattered light by *Voyager 2* as it passed behind the planet (bottom left); two of Neptune's rings showing the unusual azimuthal concentrations, as photographed by *Voyager 2* as it passed behind the planet; the greatly overexposed crescent of Neptune is visible at lower right in the image (bottom right).

are eventually incorporated into terrestrial sediments. In the 1970s, NASA began experimenting with collecting interplanetary dust particles (IDPs, also known as Brownlee particles because of the pioneering work of D. Brownlee) using high altitude U2 reconnaissance aircraft. Terrestrial sources of particulates in the stratosphere are rare and consist largely of volcanic aerosols and aluminum oxide particles from solid rocket fuel exhausts, each of which are readily distinguishable from extraterrestrial materials.

The composition of the IDPs reflects the range of source bodies that produce them and include ordinary and carbonaceous chondritic material and suspected cometary par-ticles. Because the degree of heating during atmospheric deceleration is a function of the encounter velocity, recovered IDPs are strongly biased toward asteroidal particles from the main belt, which approach the Earth in lower eccentricity orbits. Nevertheless, suspected cometary particles are included in the IDPs. The cometary IDPs show a random, "botryoidal" (cluster-of-grapes) arrangement of submicron silicate grains similar in size to interstellar dust grains, intimately mixed in a carbonaceous matrix. Voids in IDPs may have once been filled by cometary ices. In 2006, the *Stardust* spacecraft returned samples of cometary dust collected during a flyby of comet Wild 2; these will provide

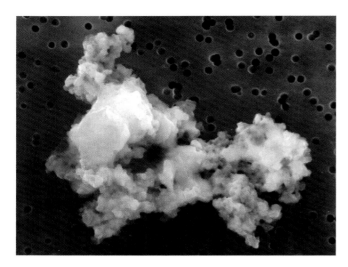

FIGURE 9   A suspected cometary interplanetary dust particle. The IDP is a highly porous, apparently random collection of submicron silicate grains embedded in a carbonaceous matrix. The particle is ~10 μm across. The voids in the IDP may have once been filled with cometary ices.

an important comparison with the IDPs collected by high-flying aircraft. An example of a suspected cometary IDP is shown in Fig. 9.

Extraterrestrial particulates are also collected on the Earth in Antarctic ice cores, in melt-ponds in Greenland, and as millimeter-sized silicate and nickel–iron melt products in sediments. The IDP component in terrestrial sediments can be determined by measuring the abundance of ^{3}He. ^{3}He has normal abundances in terrestrial materials of 10^{-6} or less. The ^{3}He is implanted in the IDP grains during their exposure to the solar wind. Using this technique, one can look for variations in the infall rate of extraterrestrial particulates over time, and such variations are seen, sometimes correlated with impact events on the Earth.

A largely unseen part of the solar system is the solar wind, an ionized gas that streams continuously into space from the Sun. The solar wind is composed primarily of protons (hydrogen nuclei) and electrons with some alpha particles (helium nuclei) and trace amounts of heavier ions. It is accelerated to supersonic speed in the solar corona and streams outward at a typical velocity of 400 km sec^{-1}. The solar wind is highly variable, changing with both the solar rotation period of ~25 days and with the 22 year solar cycle, as well as on much more rapid time scales. As the solar wind expands outward, it carries the solar magnetic field with it in a spiral pattern caused by the rotation of the Sun. The solar wind was first inferred in the late 1940s by L. Biermann based on observations of cometary plasma tails. The theory of the supersonic solar wind was first described by E. N. Parker in 1958, and the solar wind itself was detected in 1962 by the *Explorer 10* spacecraft in Earth orbit, and the *Mariner 2* spacecraft while en route to a flyby of Venus.

The solar wind interaction with the planets and the other bodies in the solar system is also highly variable, depending primarily on whether or not the body has its own intrinsic magnetic field. For bodies without a magnetic field, such as Venus and the Moon, the solar wind impinges directly on the top of the atmosphere or on the solid surface, respectively. For bodies like the Earth or Jupiter, which do have magnetic fields, the field acts as a barrier and deflects the solar wind around it. Because the solar wind is expanding at supersonic speeds, a shock wave, or bow shock, develops at the interface between the interplanetary solar wind and the planetary magnetosphere or ionosphere. The planetary magnetospheres can be quite large, extending out ~12 planetary radii upstream (sunward) of the Earth, and 50–100 radii from Jupiter. Solar wind ions can leak into the planetary magnetospheres near the poles, and these can result in visible aurora, which have been observed on the Earth, Jupiter (Fig. 10), and Saturn. As it flows past the planet, the interaction of the solar wind with the planetary magnetospheres results in huge magnetotail structures that often extend over interplanetary distances.

All the jovian planets, as well as the Earth, have substantial magnetic fields and thus planetary magnetospheres. Mercury has a weak magnetic field, but Venus has no detectable field. Mars has a patchy field, indicative of a past magnetic field at some point in the planet's history, but it has no organized magnetic field at this time. The *Galileo* spacecraft detected a magnetic field associated with Ganymede, the largest of the Galilean satellites. However, no magnetic field was detected for Europa or Callisto. The Earth's Moon has no magnetic field.

The most visible manifestation of the solar wind is cometary plasma tails, which result when the evolving gases in the cometary comae are ionized by sunlight and by charge exchange with the solar wind and then accelerated by the

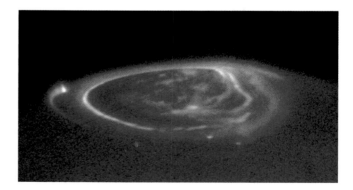

FIGURE 10   The auroral ring over the north polar region of Jupiter, as imaged by the *Hubble Space Telescope*. Several of the bright spots correspond to "footprints" of the Galilean satellites and their interaction with Jupiter's magnetosphere.

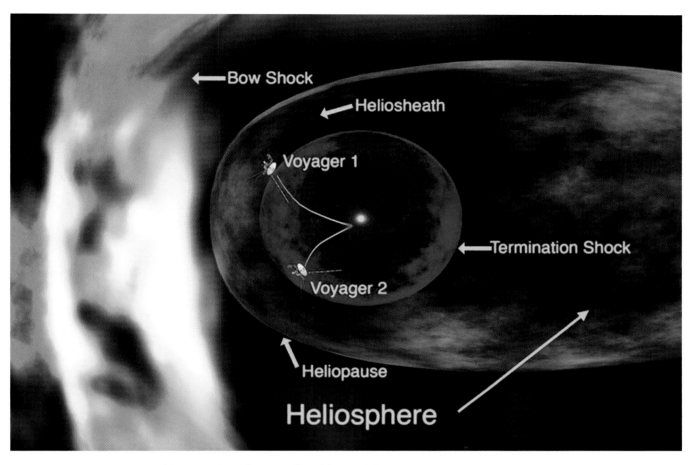

FIGURE 11   Artist's concept of the major boundaries predicted for the heliosphere and the trajectories of the two *Voyager* spacecraft. *Voyager 1* crossed the termination shock in 2004.

solar magnetic field. The ions stream away from the cometary comae at high velocity in the antisunward direction. Structures in the tail are visible as a result of fluorescence by CO^+ and other ions.

At some distance from the Sun, far beyond the orbits of the planets, the solar wind reaches a point where the ram pressure from the wind is equal to the external pressure from the local interstellar medium flowing past the solar system. A termination shock will develop upstream of that point, and the solar wind will be decelerated from supersonic to subsonic. Recently *Voyager 1* detected the termination shock at 94 AU. Beyond this distance is a region still dominated by the subsonic solar plasma, extending out another 30–50 AU or more. The outer boundary of this region is known as the heliopause and defines the limit between solar system–dominated plasma and the interstellar medium. It is not currently known if the flow of interstellar medium past the solar system is supersonic or subsonic. If it is supersonic, then there must additionally be a bow shock beyond the heliopause, where the interstellar medium encounters the obstacle presented by the **heliosphere**. A di-

agram of the major features of the heliosphere is shown in Fig. 11.

The *Voyager 1* and *2* spacecraft, which are currently leaving the planetary region on hyperbolic trajectories, continue to study the outermost regions of the heliosphere. *Voyager 1* is currently at 100.4 AU (as of September 2006) and *Voyager 2* is at 80.7 AU. The *Voyager* spacecraft are expected to continue to send measurements until the year 2015, when they are expected to be at about 130 and 106 AU from the Sun, respectively.

To many planetary scientists, the heliopause defines the boundary of the solar system because it marks the changeover from a solar wind to an interstellar medium dominated space. However, as already noted, the Sun's gravitational sphere of influence extends out much farther, to $\sim 2 \times 10^5$ AU, and there are bodies in orbit around the Sun at those distances. These include the Kuiper belt and scattered disk, which may each extend out to $\sim 10^3$ AU (possibly even farther for the scattered disk), and the Oort cloud which is populated to the limits of the Sun's gravitational field.

4. The Origin of the Solar System

Our knowledge of the origin of the Sun and the planetary system comes from two sources: study of the solar system itself and study of star formation in nearby giant molecular clouds. The two sources are radically different. In the case of the solar system, we have an abundance of detailed information on the planets, their satellites, and numerous small bodies. But the solar system we see today is highly evolved and has undergone massive changes since it first condensed from the natal interstellar cloud. We must learn to recognize which qualities reflect that often violent evolution and which truly record conditions at the time of solar system formation.

In contrast, when studying even the closest star-forming regions (which are about 140 pc from the Sun), we are handicapped by a lack of adequate resolution and detail. In addition, we are forced to take a "snapshot" view of many young stars at different stages in their formation, and from that attempt to generate a time-ordered sequence of those different stages and processes involved. When we observe the formation of other stars, we also need to recognize that some of the observed processes or events may not be applicable to the formation of our own Sun and planetary system.

Still, a coherent picture has emerged of the major events and processes in the formation of the solar system. That picture assumes that the Sun is a typical star and that it formed in a similar way to many of the low-mass protostars we see today.

The birthplace of stars is giant molecular clouds in the galaxy. These huge clouds of molecular hydrogen have masses of 10^5–10^6 $M_\odot$. Within these clouds are denser regions or cores where star formation actually takes place. Some process, perhaps the shockwave from a nearby supernova, triggers the gravitational collapse of a cloud core. Material falls toward the center of the core under its own self-gravity and a massive object begins to grow at the center of the cloud. Heated by the gravitational potential energy of the infalling matter, the object becomes self-luminous and is then described as a protostar. Although central pressures and temperatures are not yet high enough to ignite nuclear fusion, the protostar begins to heat the growing nebula around it. The timescale of the infall of the cloud material for a solar-mass cloud is about 10^6 years.

The infalling cloud material consists of both gas and dust. The gas is mostly hydrogen (75% by mass) and helium (22%). The dust (2%) is a mix of interstellar grains, including silicates, organics, and condensed ices. A popular model suggests that the silicate grains are coated with icy-organic mantles. As the dust grains fall inward, they experience a pressure from the increasing density of gas toward the center of the nebula. This slows and even halts the inward radial component of their motion. However, the dust grains can still move vertically with respect to the central plane of the nebula, as defined by the rotational angular

momentum vector of the original cloud core. As a result, the grains settle toward the central plane.

As the grains settle, they begin to collide with one another. The grains stick and quickly grow from microscopic to macroscopic objects, perhaps meters in size (initial agglomerations of grains may look very much like the suspected cometary IDP in Fig. 9). This process continues and even increases as the grains reach the denser environment at the central plane of the nebula. The meter-sized bodies grow to kilometer-sized bodies and the kilometer-sized bodies grow to 100 km-sized bodies. These bodies are known as planetesimals. As a planetesimal begins to acquire significant mass, its cross section for accretion grows beyond its physical cross section because it is now capable of gravitationally deflecting smaller planetesimals toward it. These larger planetesimals then "run away" from the others, growing at an ever increasing rate.

The actual process is far more complex than described here, and there are many details of this scenario that still need to be worked out. For example, the role of turbulence in the nebula is not well quantified. Turbulence would tend to slow or even prevent the accretion of grains into larger objects. Also, the role of electrostatic and magnetic effects in the nebula is not understood.

Nevertheless, it appears that accretion in the central plane of the solar nebula can account for the growth of planets from interstellar grains. An artist's concept of the accretion disk in the solar nebula is shown in Fig. 12. In the inner region of the solar nebula, close to the forming Sun, the higher temperatures would vaporize icy and organic grains, leaving only silicate grains to form the planetesimals, which eventually merged to form the terrestrial planets. At larger distances where the nebula was cooler, organic and icy grains would condense, and these would combine with the silicates to form the cores of the giant planets. Because the total mass of ice and organics may have been several times the mass of silicates, the cores of the giant planets may actually have grown faster than the terrestrial planets interior to them.

At some point, the growing cores of the giant planets became sufficiently massive to begin capturing hydrogen and helium directly from the nebula gas. Because of the lower temperatures in the outer planets zone, the giant planets were able to retain the gas and continue to grow even larger. The terrestrial planets close to the Sun may have acquired some nebula gas, but likely they could not hold on to it at their higher temperatures.

Observations of protostars in nearby molecular clouds have found substantial evidence for accretionary disks and gas nebulae surrounding these stars. The relative ages of these protostars can be estimated by comparing their luminosity and color with theoretical predictions of their location in the Hertzsprung–Russell diagram. One of the more interesting observations is that the nebula dust and gas around solar-mass protostars seem to dissipate after about

FIGURE 12 Artist's concept of the accretion disk in the solar nebula, showing dust, orbiting planetesimals and the proto-Sun at the center. (Painting by William Hartmann.)

10^7 years. It appears that the nebula and dust may be swept away by mass outflows, essentially super-powerful solar winds, from the protostars. If the Sun formed similarly to the protostars we see today, then these observations set strong limits on the likely formation times of Jupiter and Saturn.

An interesting process that must have occurred during the late stages of planetary accretion is "giant impacts" (i.e., collisions between very large protoplanetary objects). As noted in Section 2.3, a giant impact between a Mars-size protoplanet and the proto-Earth is now the accepted explanation for the origin of the Earth's Moon. Although it was previously thought that such giant impacts were low probability events, they are now recognized to be a natural consequence of the final stages of planetary accretion.

Another interesting process late in the accretion of the planets is the clearing of debris from the planetary zones. At some point in the growth of the planets, their gravitational spheres of influence grew sufficiently large that an encounter with a planetesimal would more likely lead to the planetesimal being scattered into a different orbit, rather than an actual collision. This would be particularly true for the massive jovian planets, both because of their stronger gravitational fields and because of their larger distances from the Sun.

Because it is just as likely that a planet will scatter objects inward as outward, the clearing of the planetary zones resulted in planetesimals being flung throughout the solar system and in a massive bombardment of all planets and satellites. Many planetesimals were also flung out of the planetary system to interstellar space or to distant orbits in the Oort cloud. Although the terrestrial planets are generally too small to eject objects out of the solar system, they can scatter objects to Jupiter-crossing orbits where Jupiter will quickly dispose of them.

The clearing of the planetary zones has several interesting consequences. The dynamical interaction between the planets and the remaining planetesimals results in an exchange of angular momentum. Computer-based dynamical simulations have shown that this causes the semimajor axes of the planets to migrate. In general, Saturn, Uranus, and Neptune are expected to first move inward and then later outward as the ejection of material progresses. Jupiter, which ejects the most material because of its huge mass, migrates inward but by only a few tenths of an astronomical unit.

This migration of the giant planets has significant consequences for the populations of small bodies in the planetary region. As the planets move, the locations of their mean-motion and secular resonances will move with them. This will result in some small bodies being captured into resonances while others will be thrown into chaotic orbits, leading to their eventual ejection from the system or possibly to impacts on the planets and the Sun. The radial migration of the giant planets has been invoked both in the clearing of the outer regions of the main asteroid belt, and the inner regions of the Kuiper belt.

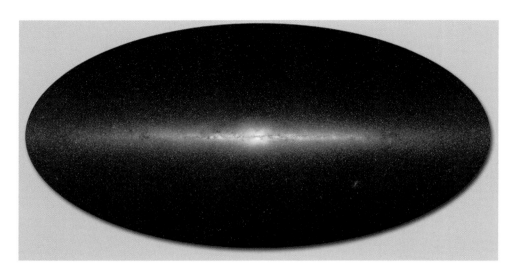

FIGURE 13   An image of the entire sky at infrared wavelengths, constructed from ground-based data by the 2MASS survey. The Milky Way galaxy is visible as the bright horizontal band through the image, with the galactic bulge at the center of the image. Lanes of interstellar dust obscure the view of the galactic center. The Magellanic clouds, two small, irregular companion galaxies to the Milky Way are visible below and to the right of the galactic center.

Another consequence of the clearing of the planetary zones is that rocky planetesimals formed in the terrestrial planets zone will be scattered throughout the jovian planets region, and vice versa for icy planetesimals formed in the outer planets zone. The bombardment of the terrestrial planets by icy planetesimals is of particular interest, both as an explanation for the Late Heavy Bombardment and as a means of delivering the volatile reservoirs of the terrestrial planets. Isotopic studies suggest that some fraction of the water in the Earth's oceans may have come from comets and/or volatile-rich asteroids, though not all of it. Also, the discovery of an asteroidal-appearing object, 1996 PW, on a long-period comet orbit has provided evidence that asteroids may indeed have been ejected to the Oort cloud, where they may make up 1–3% of the population there.

5. The Solar System's Place in the Galaxy

The Milky Way is a large, spiral galaxy, about 30 kpc in diameter. Some parts of the galactic disk can be traced out to 25 kpc from the galactic center, and the halo can be traced to 50 kpc. The galaxy contains approximately 10^{11}, stars and the total mass of the galaxy is estimated to be about 4×10^{11} solar masses ($M_\odot$). Approximately 25% of the mass of the galaxy is estimated to be in visible stars, about 15% in stellar remnants (white dwarfs, neutron stars, and black holes), 25% in interstellar clouds and interstellar material, and 35% in "dark matter." Dark matter is a general term used to describe unseen mass in the galaxy, which is needed to explain the observed dynamics of the galaxy (i.e., stellar motions, galactic rotation) but which has not been detected through any available means. There is considerable speculation about the nature of the dark matter, which includes everything from exotic nuclear particles to brown dwarfs (substellar objects, not capable of nuclear burning) and dark stars (the burned out remnants of old stars) to

massive black holes. The age of the galaxy is estimated to be 13 billion years, equal to the age of the universe.

The Milky Way galaxy consists of four major structures: the galactic disk, the central bar, the halo, and the corona (Fig. 13). As the name implies, the disk is a highly flattened, rotating structure about 15–25 kpc in radius and about 0.5–1.3 kpc thick, depending on which population of stars is used to trace the disk. The disk contains relatively young stars and interstellar clouds, arranged in a multiarm spiral structure (Figs. 14 and 15). At the center of the disk is the bar, a prolate spheroid about 3 kpc in radius in the plane of

FIGURE 14   Messier 33, a large spiral galaxy in the constellation Triangulum, as photographed by the *Galex* spacecraft. M33 is part of the local group of nearby galaxies. The Milky Way galaxy may appear similar to this.

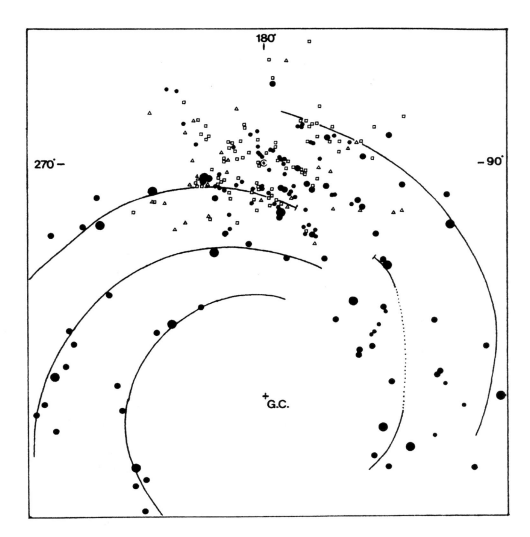

FIGURE 15   The spiral structure of the Milky Way galaxy as inferred from the positions of HII regions (clouds of ionized hydrogen) in the galaxy. The Sun and solar system are located at the upper center, as indicated by the ⊙ symbol. (Reprinted with kind permission from Kluwer Academic Publishers, Forbes and Shuter, in "Kinematics, Dynamics, and Structure of the Milky Way," p. 221, Fig. 3, copyright © 1983.)

the disk, and with a radius of about 1.5 kpc perpendicular to the disk. The bar rotates more slowly than the disk and consists largely of densely packed older stars and interstellar clouds. It does not display spiral structure. At the center of the bar is the nucleus, a complex region only 4–5 pc across, which appears to have a massive black hole at its center. The mass of the central black hole has been estimated at 2.6 million $M_\odot$.

The halo surrounds both of these structures and extends ~20–30 kpc from the galactic center. The halo has an oblate spheroid shape and contains older stars and globular clusters of stars. The corona appears to be a yet more distant halo extending 60–100 kpc and consists of dark matter, unobservable except for the effect it has on the dynamics of observable bodies in the galaxy. The corona may be several times more massive than the other three galactic components combined.

The galactic disk is visible in the night sky as the Milky Way, a bright band of light extending around the celestial sphere. When examined with a small telescope, the Milky Way is resolved into thousands or even millions of individual stars and numerous nebulae and star clusters. The direction to the center of the galaxy is in the constellation Sagittarius (best seen from the southern hemisphere in June), and the disk appears visibly wider in that direction, which is the view of the central bulge and bar.

The disk is not perfectly flat; there is evidence for warping in the outer reaches of the disk, between 15 and 25 kpc. The warp may be the result of gravitational perturbations due to encounters with other galaxies and/or with the Magellanic clouds, two nearby, irregular dwarf galaxies that appear to be in orbit around the Milky Way. In addition, the Milky Way's central bar appears to be tilted relative to the plane of the galactic disk. The nonspherical shape of the bar and the tilt have important implications for understanding stellar dynamics and the long-term evolution of the galaxy.

Stars in the galactic disk have different characteristic velocities as a function of their stellar classification, and hence age. Low mass, older stars, like the Sun, have relatively high random velocities and, as a result, can move farther out of the galactic plane. Younger, more massive stars have lower mean velocities and thus smaller scale heights above and below the plane. Giant molecular clouds, the birthplace of stars, also have low mean velocities and thus are confined to regions relatively close to the galactic plane. The galactic disk rotates clockwise as viewed from "galactic north," at a relatively constant velocity of 160–220 km sec^{-1}. This motion is distinctly non-Keplerian, the result of the very nonspherical mass distribution. The rotation velocity for a circular galactic orbit in the galactic plane defines the Local Standard of Rest (LSR). The LSR is then used as the reference frame for describing local stellar dynamics.

The Sun and the solar system are located approximately 8.5 kpc from the galactic center (though some estimates put it closer at ∼7 kpc), and 10–20 pc above the central plane of the galactic disk. The circular orbit velocity at the Sun's distance from the galactic center is 190–220 km sec^{-1}, and the Sun and the solar system are moving at approximately 17 to 22 km sec^{-1} relative to the LSR. The Sun's velocity vector is currently directed toward a point in the constellation of Hercules, approximately at right ascension 18^h 0^m, and declination +30°, known as the solar apex. Because of this motion relative to the LSR, the solar system's galactic orbit is not circular. The Sun and planets move in a quasi-elliptical orbit between about 8.4 and 9.7 kpc from the galactic center, with a period of revolution of about 240 million years. The solar system is currently close to and moving inward toward "perigalacticon," the point in the orbit closest to the galactic center. In addition, the solar system moves perpendicular to the galactic plane in a harmonic fashion, with an estimated period of 52 million to 74 million years, and an amplitude of ±49–93 pc out of the galactic plane. (The uncertainties in the estimates of the period and amplitude of the motion are caused by the uncertainty in the amount of dark matter in the galactic disk.) The Sun and planets passed through the galactic plane about 2 million to 3 million years ago, moving "northward."

The Sun and solar system are located at the inner edge of one of the spiral arms of the galaxy, known as the Orion or local arm. Nearby spiral structures can be traced by constructing a 3-dimensional map of stars, star clusters, and interstellar clouds in the solar neighborhood. Two well-defined neighboring structures are the Perseus arm, farther from the galactic center than the local arm, and the Sagittarius arm, toward the galactic center. The arms are about 0.5 kpc wide, and the spacing between the spiral arms is ∼1.2–1.6 kpc. The local galactic spiral arm structure is illustrated in Fig. 15.

The Sun's velocity relative to the LSR is low as compared with other G-type stars, which have typical velocities of 40–45 km sec^{-1} relative to the LSR. Stars are accelerated by encounters with giant molecular clouds in the galactic disk. Thus, older stars can be accelerated to higher mean velocities, as noted earlier. The reason(s) for the Sun's low velocity is not known. Velocity-altering encounters with gi-

ant molecular clouds occur with a typical frequency of once every 300 million to 500 million years.

The local density of stars in the solar neighborhood is about 0.11 pc^{-3}, though many of the stars are in binary or multiple star systems. The local density of binary and multiple star systems is 0.086 pc^{-3}. Most of these are low-mass stars, less massive and less luminous than the Sun. The nearest star to the solar system is Proxima Centauri, which is a low-mass ($M \simeq 0.1\ M_\odot$), distant companion to Alpha Centauri, which itself is a double star system of two close-orbiting solar-type stars. Proxima Centauri is currently about 1.3 pc from the Sun and about 0.06 pc (1.35×10^4 AU) from the Alpha Centauri pair it is orbiting. The second nearest star is Barnard's star, a fast-moving red dwarf at a distance of 1.83 pc. The brightest star within 5 pc of the Sun is Sirius, an A1 star ($M \simeq 2\ M_\odot$) about 2.6 pc away. Sirius also is a double star, with a faint, white dwarf companion. The stars in the solar neighborhood are shown in Fig. 16.

The Sun's motion relative to the LSR, as well as the random velocities of the stars in the solar neighborhood, will occasionally result in close encounters between the Sun and other stars. Using the value above for the density of stars in the solar neighborhood, one can predict that ∼12 star systems (single or multiple stars) will pass within 1 pc of the Sun per million years. The total number of stellar encounters scales as the square of the encounter distance. This rate has been confirmed in part by data from the *Hipparcos* astrometry satellite, which measured the distances and proper motions of ∼118,000 stars, and which was used to reconstruct the trajectories of stars in the solar neighborhood.

Based on this rate, the closest stellar approach over the lifetime of the solar system would be expected to be at ∼900 AU. Such an encounter would result in a major perturbation of the Oort cloud and would eject many comets to interstellar space. It would also send a shower of comets into the planetary region, raising the impact rate on the planets for a period of about 2 million to 3 million years, and having other effects that may be detectable in the stratigraphic record on the Earth or on other planets. A stellar encounter at 900 AU could also have a substantial perturbative effect on the orbits of comets in the Kuiper belt and scattered disk and would likely disrupt the outer regions of those populations. Obviously, the effect that any such stellar passage will have is a strong function of the mass and velocity of the passing star.

Because the Sun likely formed in a star cluster, and because the Sun will move through denser regions of the galactic disk (in particular, the spiral arms), the encounter rate mentioned above is likely a lower limit and was higher in the past. That also means that the closest stellar encounters may have been even closer to the planetary system.

The advent of space-based astronomy, primarily through Earth-orbiting ultraviolet and X-ray telescopes, has made it possible to study the local interstellar medium surrounding the solar system. The structure of the local interstellar

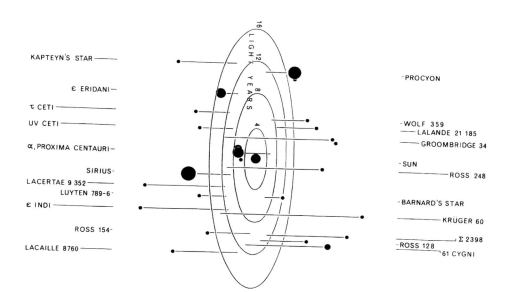

FIGURE 16 A 3-dimensional representation of the stars in the solar neighborhood. Horizontal lines indicate the relative distance of the stars north (to the right) or south (to the left) of the celestial equator. The size of the dot representing each star denotes its relative brightness. (From G. F. Gilmore, in "Astronomy and Astrophysics Encyclopedia," S. P. Maran, Ed. Copyright © 1992 John Wiley & Sons, New York. Reprinted by permission of John Wiley & Sons, Inc.)

medium has turned out to be quite complex. The solar system appears to be on the edge of an expanding bubble of hot plasma about 120 pc in radius, which appears to have originated from multiple supernovae explosions in the Scorpius-Centaurus OB association. The Sco-Cen association is a nearby star-forming region that contains many young, high-mass O- and B-type stars. Such stars have relatively short lifetimes and end their lives in massive supernova explosions, before collapsing into black holes. The expanding shells of hot gas blown off the stars in the supernova explosions are able to "sweep" material before them, leaving a low density "bubble" of hot plasma.

Within this bubble, known as the Local Bubble, the solar system is at this time within a small interstellar cloud, perhaps 2–5 pc across, known as the Local Interstellar Cloud. That cloud is apparently a fragment of the expanding shells of gas from the supernova explosions, and there appear to be a number of such clouds within the local solar neighborhood.

6. The Fate of the Solar System

Stars like the Sun are expected to have lifetimes on the main sequence of about 10^{10} years. The main sequence lifetime refers to the time period during which the star produces energy through hydrogen fusion in its core. As the hydrogen fuel in the core is slowly depleted over time, the core contracts to maintain the internal pressure. This raises the central temperature and as a result, the rate of nuclear fusion also increases and the star slowly brightens. Thus, temperatures throughout the solar system will slowly increase over time. Presumably, this slow brightening has already been going on since the formation of the Sun and solar system.

A 1 $M_\odot$ star like the Sun is expected to run out of hydrogen at its core in about 10^{10} years. As the production of energy declines, the core again contracts. The rising internal temperature and pressure are then able to ignite hydrogen burning in a shell surrounding the depleted core. The hydrogen burning in the shell heats the surrounding mass of the star and causes it to expand. The radius of the star increases and the surface temperature drops. The luminosity of the star increases dramatically, and it becomes a red giant. Eventually the star reaches a brightness about 10^3 times more luminous than the present-day Sun, a surface temperature of 3000 K, and a radius of 100–200 solar radii. One hundred solar radii is equal to 0.46 AU, larger than the orbit of Mercury. Two hundred radii is just within the orbit of the Earth. Thus, Mercury and likely Venus will be incorporated into the outer shell of the red giant Sun and will be vaporized.

The increased solar luminosity during the red giant phase will result in a fivefold rise in temperatures throughout the solar system. At the Earth's orbit this temperature increase will vaporize the oceans and roast the planet at a temperature on the order of ~1400 K or more. At Jupiter's orbit it will melt the icy Galilean satellites and cook them at a more modest temperature of about 600 K, about the same as current noon-time temperatures on the surface of Mercury. Typical temperatures at the orbit of Neptune will be about the same as they are today at the orbit of the Earth. Comets in the inner portion of the Kuiper belt will be warmed sufficiently to produce visible comae.

The lowered gravity at the surface of the greatly expanded Sun will result in a substantially increased solar wind, and the Sun will slowly lose mass from its outer envelope. Meanwhile, the core of the Sun will continue to contract until the central temperature and pressure are great enough to ignite helium burning in the core. During this time, hydrogen burning continues in a shell around the core. Helium burning continues during the red giant phase until the helium in the core is also exhausted. The star again contracts, and this permits helium burning to ignite in a shell around the core. This is an unstable situation, and the star can undergo successive contractions and reignition pulses, during which it will blow off part or all of its outer envelope into space. These huge mass ejections produce an expanding nebula around the star, known as a planetary nebula (because it looks somewhat like the disk of a jovian planet through a telescope). For a star with the mass of the Sun, the entire red giant phase lasts about 7×10^8 years.

As the Sun loses mass in this fashion, the orbits of the surviving planets will slowly spiral outward. This will also be true for comets in the Kuiper belt and Oort cloud. The gravitational sphere of influence of the Sun will shrink as a result of the Sun's decreasing mass, so comets will be lost to interstellar space at the outer limits of the Oort cloud.

As a red giant star loses mass, its core continues to contract. However, for an initially 1 $M_\odot$ star like the Sun, the central pressure and temperature cannot rise sufficiently to ignite carbon burning in the core, the next phase in nuclear fusion. With no way of producing additional energy other than gravitational contraction, the luminosity of the star plunges. The star continues to contract and cool, until the contraction is halted by degenerate electron pressure in the super-dense core. At this point, the mass of the star has been reduced to about 70% of its original mass and the diameter is about the same as the present-day Earth. Such a star is known as a white dwarf. The remnants of the previously roasted planets will be plunged into a deep freeze as the luminosity of the white dwarf slowly declines.

The white dwarf star will continue to cool over a period of about 10^9 years, to the point where its luminosity drops below detectable levels. Such a star is referred to as a black dwarf. A nonluminous star is obviously very difficult to detect. There is some suggestion that they may have been found through an observing technique known as micro-lensing events. Dark stars provide one of the possible explanations for the dark matter in the galaxy.

7. Concluding Remarks

This chapter has provided an introduction to the solar system and its varied members, viewing them as components of a large and complex system. Each of them (the Sun, the planets, their satellites, the comets and asteroids, etc.) is also a fascinating world in its own right. The ensuing chapters provide more detailed descriptions of each of these members of the solar system, as well as descriptions of important physical and dynamical processes, discussions of some of the more advanced ways we study the solar system, the search for life elsewhere in the solar system, and finally, the search for planetary systems around other stars.

Bibliography

Lewis, J. S. (2004). "Physics and Chemistry of the Solar System," 2nd Ed. Elsevier Academic Press, San Diego.

von Steiger, R., Lallement, R., and Lee, M. A., eds. (1996). "The Heliosphere in the Local Interstellar Medium." Kluwer, Dordrecht, The Netherlands.

Sparke, L. S., and Gallagher, J. S. (2000) "Galaxies in the Universe: An Introduction." Cambridge University Press, Cambridge, UK.

The Origin of the Solar System

John E. Chambers
Carnegie Institution of Washington
Washington, D.C.

Alex N. Halliday
University of Oxford
Oxford, United Kingdom

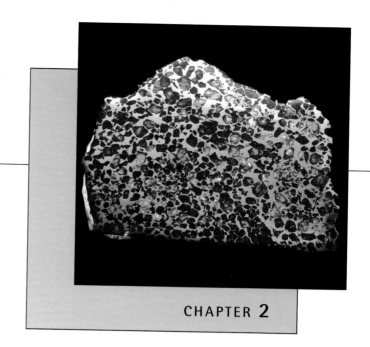

CHAPTER **2**

1. Introduction

The origin of the solar system has long been a fascinating subject posing difficult questions of deep significance. It takes one to the heart of the question of our origins, of how we came to be here and why our surroundings look the way they do. Unfortunately, we currently lack a self-consistent model for the origin of the solar system and other planetary systems. The early stages of planet formation are obscure, and we have only a modest understanding of how much the orbits of planets change during and after their formation. At present, we cannot say whether terrestrial planets similar to the Earth are commonplace or highly unusual. Nor do we know the source of the water that makes our planet habitable.

In the face of such uncertainty, one might ask whether we will ever understand how planetary systems form. In fact, the last 10 years have seen rapid progress in almost every area of planetary science, and our understanding of the origin of the solar system and other planetary systems has improved greatly as a result. Planetary science today is as exciting as it has been at any time since the *Apollo* landings on the Moon, and the coming decade looks set to continue this trend.

Some key recent developments follow:

1. A decade ago, the first planet orbiting another Sun-like star was discovered. Since then, new planets have been found at an astounding rate, and roughly 200 objects are known today. Most of these planets appear to be gas giants similar to Jupiter and Saturn. Recently, several smaller planets have been found, and these may be akin to Uranus and Neptune, or possibly large analogs of terrestrial planets like Earth.

2. In the last 10 years, there have been a number of highly successful space missions to other bodies in the solar system, including Mars, Saturn, Titan, and several asteroids and comets. Information and images returned from these missions have transformed our view of these objects and greatly enhanced our understanding of their origin and evolution.

3. The discovery that one can physically separate and analyze star dust–presolar grains that can be extracted from meteorites and that formed in the envelopes of other stars—has meant that scientists can for the first time test decades of theory on how stars work. The parallel development of methods for extracting isotopic information at the submicron scale has opened

up a new window to the information stored in such grains.

4. The development of multiple collector inductively coupled plasma mass spectrometry has made it possible to use new isotopic systems for determining the mechanisms and timescales for the growth of bodies early in the solar system.

5. Our theoretical understanding of planet formation has advanced substantially in several areas, including new models for the rapid growth of giant planets, a better understanding of the physical and chemical evolution of protoplanetary disks, and the growing realization that planets can migrate substantially during and after their formation.

6. The recent development of powerful new computer codes and equations of state has facilitated realistic, high-resolution simulations of collisions between planet-sized bodies. Scientists are discovering that the resolution of their models significantly changes the outcome, and the race is on to find reliable solutions.

Today, the formation of the solar system is being studied using three complementary approaches.

- Astronomical observations of protoplanetary disks around young stars are providing valuable information about probable conditions during the early history of the solar system and the timescales involved in planet formation. The discovery of new planets orbiting other stars is adding to the astonishing diversity of possible planetary systems and providing additional tests for theories of how planetary systems form.

- Physical, chemical, and isotopic analyses of meteorites and samples returned by space missions are generating important information about the formation and evolution of objects in the solar system and their constituent materials. This field of cosmochemistry has taken off in several important new directions in recent years, including the determination of timescales involved in the formation of the terrestrial planets and asteroids, and constraints on the origin of the materials that make up the solar system.

- Theoretical calculations and numerical simulations are being used to examine every stage in the formation of the solar system. These provide valuable insights into the complex interplay of physical and chemical processes involved, and help to fill in some of the gaps when astronomical and cosmochemical data are unavailable.

In this chapter, we will describe what we currently know about how the solar system formed and highlight some of the main areas of uncertainty that await future discoveries.

2. Star Formation and Protoplanetary Disks

The solar system formed 4.5–4.6 billion years (Ga) ago by collapse of a portion of a **molecular cloud** of gas and dust rather like the Eagle or Orion Nebulae. Some of the star dust from that ancient Solar Nebula has now been isolated from **primitive meteorites**. Their isotopic compositions are vastly different from those of our own solar system and provide fingerprints of nearby stars that preceded our Sun. These include red giants, asymptotic giant branch stars, supernovae, and novae. It has also become clear from studying modern molecular clouds that stars like our Sun can form in significant numbers in close proximity to each other. Such observation also provide clues as to how own solar system formed because they have provided us with images of circumstellar disks—the environments in which planetary objects are born.

Observations from space-based infrared telescopes such as the *Infrared Astronomical Satellite* (*IRAS*) have shown that many young stars give off more infrared radiation than would be expected for blackbodies of the same size. This infrared excess comes from micron-sized grains of dust orbiting the star in an optically thick (opaque) disk. Dark, dusty disks can be seen with the *Hubble Space Telescope* surrounding some young stars in the Orion Nebula (Fig. 1). These disks have been dubbed proplyds, short for protoplanetary disks. It is thought that protoplanetary disks are mostly composed of gas, and in a few cases this gas has been detected, although gas is generally much harder to see than dust. The fraction of stars having a massive disk declines with stellar age, and large infrared excesses are rarely seen in stars older than 10^7 years. In some cases, such as the disk surrounding the star HR 4796A, there are signs that the inner portion of a disk has been cleared of dust (Fig. 2), perhaps due to the presence of one or more planets.

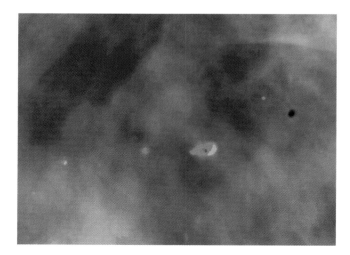

FIGURE 1 Proplyds are young stellar objects embedded in an optically dense envelope of gas and dust. The objects shown here are from the Orion Nebula.

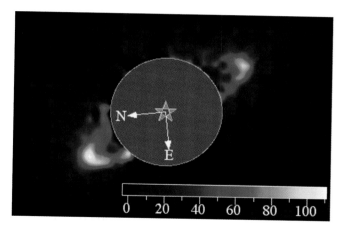

FIGURE 2   The circumstellar disk surrounding HR 4796A as revealed by interferometry measurements of the infrared excess. Note the area close into the star swept clear of dust, which has presumably been incorporated into planetary objects.

Roughly half of stars up to a few hundred million years old have low-mass, optically thin (nearly transparent) disks containing some dust but apparently little or no gas. In a few cases, such as the star Beta Pictoris, a disk can be seen at visible wavelengths when the light from the star itself is blocked. Dust grains in these disks will be quickly accelerated outward by radiation pressure or spiral inward due to Poynting–Robertson drag caused by collisions with photons from the central star. This dust should be either removed from the disk or destroyed in high-speed collisions with other dust grains on a timescale that is short compared to the age of the star. For this reason, the dust in these disks is thought to be second-generation material formed by collisions between asteroids or sublimation from comets orbiting these stars in more massive analogs of the Kuiper Belt in our own solar system. These are often referred to as debris disks as a result.

In the solar system, the planets all orbit the Sun in the same direction, and their orbits are very roughly coplanar. This suggests the solar system originated from a disk-shaped region of material referred to as the solar nebula, an idea going back more than 2 centuries to Kant and later Laplace. The discovery of disks of gas and dust around many young stars provides strong support for this idea and implies that planet formation is associated with the formation of stars themselves. Stars typically form in clusters of a few hundred to a few thousand objects in dense regions of the interstellar medium called molecular clouds (see Fig. 3). The gas in molecular clouds is cold (roughly 10 K) and dense compared to that in other regions of space (roughly 10^4 atoms/cm^3) but still much more tenuous than the gas in a typical laboratory "vacuum." Stars in these clusters are typically separated by about 0.1 pc (0.3 lightyears), much less than the distance between stars in the Sun's neighborhood.

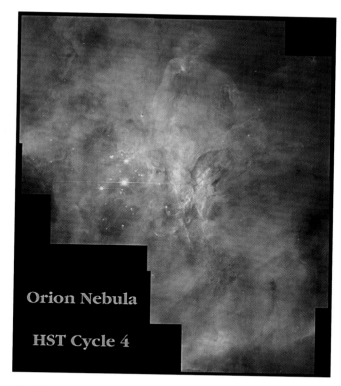

FIGURE 3   This *Hubble Space Telescope* image of the Orion Nebula shows molecular clouds of gas and dust illuminated by radiation from young stars. Some early stars appear shrouded in dusty disks (see Fig. 1). Scientists think that our solar system formed by collapse of a portion of a similar kind of molecular cloud leading to formation of a new star embedded in a dusty disk. How that collapse occurred is unclear. It may have been triggered by a shock wave carrying material being shed from another star such as an AGB star or supernova.

It is unclear precisely what causes the densest portions of a molecular cloud (called molecular cloud cores) to collapse to form stars. It may be that contraction of a cloud core is inevitable sooner or later due to the core's own gravity, or an external event may cause the triggered collapse of a core. The original triggered collapse theory was based on the sequencing found in the ages of stars in close proximity to one another in molecular clouds. This suggests that the formation and evolution of some stars triggered the formation of additional stars in neighboring regions of the cloud. However, several other triggering mechanisms are possible, such as energetic radiation and mass loss from other newly formed stars, the effects of a nearby, pulsating asymptotic giant branch (AGB) star, or a shock wave from the supernova explosion of a massive star.

Gas in molecular cloud cores is typically moving. When a core collapses, the gas has too much angular momentum for all the material to form a single, isolated star. In many cases, a binary star system forms. In others cases, a single protostar forms (called a T Tauri star or pre-main sequence star), while a significant fraction of the gas goes into orbit about the star forming a disk that is typically 100 astronomical units (AU) in diameter. Temperatures in T Tauri stars are initially too low for nuclear reactions to take place. However T Tauri stars are much brighter then older stars like the Sun due to the release of gravitational energy as the star contracts. The initial collapse of a molecular cloud core takes roughly 10^5 years, and material continues to fall onto both the star and its disk until the core is depleted.

The spectra of T Tauri stars contain strong ultraviolet and visible emission lines caused by hot gas falling onto the star. This provides evidence that disks lose mass over time as material moves inward through the disk and onto the star, a process called viscous accretion. This process provides one reason why older stars do not have disks, another reason being planet formation itself. Estimated disk accretion rates range from 10^{-6} to 10^{-9} solar masses per year. The mechanism responsible for viscous accretion is unclear. A promising candidate is magneto-rotational instability (MRI), in which partially ionized gas in the disk becomes coupled to the local magnetic field. Because stars rotate, the magnetic field sweeps around rapidly, increasing the orbital velocity of material that couples strongly to it and moving it outward. Friction causes the remaining material to move inward. As a result, a disk loses mass to its star and spreads outwards over time. This kind of disk evolution explains why the planets currently contain only 0.1% of the mass in the solar system but have retained more than 99% of its angular momentum. MRI requires a certain fraction of the gas to be ionized, and it may not be effective in all portions of a disk. Disks are also eroded over time by photo-evaporation. In this process, gas is accelerated when atoms absorb ultraviolet photons from the central star or nearby, energetic stars, until the gas is moving fast enough to escape into interstellar space.

T Tauri stars often have jets of material moving rapidly away from the star perpendicular to the plane of the disk. These jets are powered by the inward accretion of material through the disk coupled with the rotating magnetic field. Outward flowing winds also arise from the inner portions of a disk. It is possible that a wind arising from the very inner edge of the disk (called the x-wind) can entrain small solid particles with it. These objects will be heated strongly as they emerge from the disk's shadow. Many of these particles will return to the disk several AU from the star, and may drift inward again to repeat the process. Some of these particles may be preserved today in meteorites.

T Tauri stars are strong emitters of X-rays, generating fluxes up to 10^4 times greater than that of the Sun during the strongest solar flares. Careful sampling of large populations of young solar mass stars in the Orion Nebula shows that this is normal behavior in young stars. This energetic flare activity is strongest in the first million years and declines at later times, persisting for up to 10^8 years. From this it has been concluded that the young Sun generated

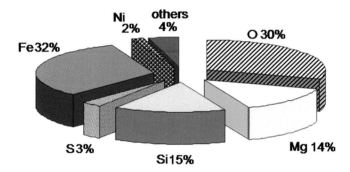

FIGURE 4   Pie chart showing the bulk composition of the Earth. Most of the iron (Fe), nickel (Ni), and sulfur (S) are in Earth's core, while the silicate Earth mostly contains magnesium (Mg), silicon (Si), and oxygen (O) together with some iron.

10^5 times as many energetic protons as today. It is thought that reactions between these protons and material in the disk may have provided some of the **short-lived isotopes** whose daughter products are seen today in meteorites although the formation of nearly all of these predate that of the solar system. (See Section 4.)

The minimum mass of material that passed through the solar nebula can be estimated from the total mass of the planets, asteroids, and comets in the solar system. However, all of these objects are depleted in hydrogen and helium relative to the Sun. Ninety percent of the mass of the terrestrial planets is made up of oxygen, magnesium, silicon, and iron (Fig. 4), and although Jupiter and Saturn are mostly composed of hydrogen and helium, they are enriched in the heavier elements compared to the Sun. When the missing hydrogen and helium is added, the minimum-mass solar nebula (MMSN) turns out to be 1–2% of the Sun's mass. The major uncertainties in this number come from the fact that the interior compositions of the giant planets and the initial mass of the Kuiper Belt are poorly known. Not all of this mass necessarily existed in the nebula at the same time, but it must have been present at some point. Current theoretical models predict that planet formation is an inefficient process, with some mass falling into the Sun or being ejected into interstellar space, so the solar nebula was probably more massive than the MMSN.

Gas in the solar nebula was heated as it viscously accreted toward the Sun, releasing gravitational energy. The presence of large amounts of dust meant the inner portions of the nebula were optically thick to infrared radiation so these regions became hot. Numerical disk models show that temperatures probably exceeded 1500 K in the terrestrial-planet forming region early in the disk's history. Viscous heating mainly took place at the disk midplane where most of the mass was concentrated. The surfaces of the disk would have been much cooler. The amount of energy generated by viscous accretion declined rapidly with distance from the Sun. In the outer nebula, solar irradiation was the more important effect. Protoplanetary disks are thought to be *flared*, so that their vertical thickness grows more rapidly than their radius. As a result, the surface layers are always irradiated by the central star. For this reason, the surface layers of the outer solar nebula may have been warmer than the midplane.

The nebula cooled over time as the viscous accretion rate declined and dust was swept up by larger bodies, reducing the optical depth. In the inner nebula, cooling was probably rapid. Models show that at the midplane at 1 AU, the temperature probably fell to about 300 K after 10^5 years. Because the energy generated by viscous accretion and solar irradiation declined with distance from the Sun, disk temperatures also declined with heliocentric distance. At some distance from the Sun, a location referred to as the ice line, temperatures became low enough for water ice to form. Initially, the ice line may have been 5–6 AU from the Sun, but it moved inward over time as the nebula cooled. Some asteroids contain hydrated minerals formed by reactions between water ice and dry rock. This suggests water ice was present when these asteroids formed, in which case the ice line would have been no more than 2–3 AU from the Sun at the time.

Meter-sized icy bodies drifted rapidly inward through the solar nebula due to **gas drag** (see Section 5). When these objects crossed the ice line, they would have evaporated, depositing water vapor in the nebular gas. As a result, the inner nebula probably became more oxidizing over time as the level of oxygen from water increased. When the flux of drifting particles dwindled, the inner nebula may have become chemically reducing again, as water vapor diffused outward across the ice line, froze to form ice, and became incorporated into growing planets.

3. Meteorites and the Origin of the Solar System

Much of the above is based on theory and observations of other stars. To find out how our own solar system formed, it is necessary to study meteorites and interplanetary dust particles (IDPs). These are fragments of rock and metal from other bodies in the solar system that have fallen to Earth and survived passage through its atmosphere. Meteorites and IDPs tend to have broadly similar compositions, and the difference is mainly one of size. IDPs are much the smaller of the two, typically 10–100 μm in diameter, while meteorites can range up to several meters in size. Most such objects are quite unlike any objects formed on Earth. Therefore, we cannot readily link them to natural present-day processes as earth scientists do when unraveling past geological history. Yet the approaches that are used are in some respects very similar. The research conducted on meteorites and IDPs is dominated by two fields: petrography and geochemistry.

Petrography is the detailed examination of mineralogical and textural features. Geochemistry uses the isotopic and chemical compositions. This combined approach to these fascinating archives has provided a vast amount of information on our Sun and solar system and how they formed. We know about the stars and events that predated formation of the Sun, the nature of the material from which the planets were built, the solar nebula the timescales for planetary accretion, and the interior workings and geological histories of other planets. Not only these, meteorites provide an essential frame of reference for understanding how our own planet Earth formed and differentiated.

The geochemistry of meteorites and IDPs provides evidence that the Sun's protoplanetary disk as well as the planets it seeded had a composition that was similar in some respects to that of the Sun itself (Fig. 5). In other respects however, it is clear the disk was a highly modified residuum that generated a vast range of planetary compositions. The composition of the Sun can be estimated from the depths of lines associated with each element in the Sun's spectra (although this is problematic for the lightest elements and the noble gases). The Sun contains almost 99.9% of the total mass of the solar system. A sizable fraction of this material passed through the solar nebula at some point, which tells us that the composition of the original nebula would have

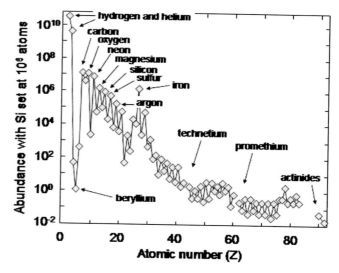

FIGURE 5   The abundances of elements in our Sun and solar system are estimated from the spectroscopic determination of the composition of the Sun and the laboratory analysis of primitive meteorites called carbonaceous chondrites—thought to represent unprocessed dust and other solid debris from the circumsolar disk. To compare the abundances of different elements, it is customary to scale the elements relative to one million atoms of silicon. The pattern provides powerful clues to how the various elements were created. See text for details. (Based on a figure in W. S. Broecker "How to Build a Habitable Planet," with kind permission.)

been similar to that of the Sun today. The challenge is therefore to explain how it is possible that a disk that formed gas giant objects like Jupiter and Saturn, also generated rocky terrestrial planets like the Earth (Fig. 4).

Most meteorites are thought to come from parent bodies in the Main Asteroid Belt that formed during the first few million years of the solar system. As a result, these objects carry a record of processes that occurred in the solar nebula during the formation of the planets. In a few cases, the trajectories of falling meteorites have been used to establish that they arrived on orbits coming from the Asteroid Belt. Most other meteorites are deduced to come from asteroids based on their age and composition. IDPs are thought to come from both asteroids and comets. A few meteorites did not originate in the Asteroid Belt. The young ages and noble-gas abundances of the Shergottite–Nakhlite–Chassignite (SNC) meteorites suggest they come from Mars. A few dozen SNC meteorites have been found to date, and a comparable number of lunar meteorites from the Moon are also known.

The Earth is currently accumulating meteoritic material at the rate of about 5×10^7 kg/year. At this rate, it would take more than 10^{17} years to obtain the Earth's current mass of 5.97×10^{24} kg, which is much longer than the age of the universe. Even though it is thought that the Earth did form as the result of the accumulation of smaller bodies, it is clear that the rate of impacts was much higher while the planets were forming than it is today.

Broadly speaking, meteorites can be divided into three types: chondrites, achondrites, and irons, which can be distinguished as follows:

1. **Chondrites** are mixtures of grains from submicron-sized dust to millimeter- to centimeter-sized particles of rock and metal, apparently assembled in the solar nebula. Most elements in chondrites are present in broadly similar ratios to those in the Sun, with the exception of carbon, nitrogen, hydrogen, and the noble gases, which are all highly depleted. For this reason, chondrites have long been viewed as representative of the dust and debris in the circumstellar disk from which the planets formed. So, for example, refractory elements that would have resided in solid phases in the solar nebula have chondritic (and therefore solar) relative proportions in the Earth, even though the volatile elements are vastly depleted. The nonmetallic components of chondrites are mostly silicates such as olivine and pyroxene. **Chondrules** are a major component of most chondrites (see Fig. 6). These are roughly millimeter-sized rounded beads of rock that formed by melting, either partially or completely. Their mineral-grain textures suggest they cooled over a period of a few hours, presumably in the nebula, with the heating possibly caused by passage

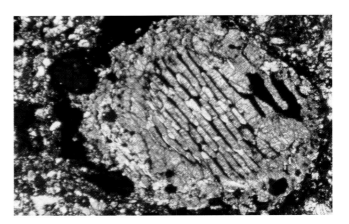

FIGURE 6 Chondrules are spherical objects, sometimes partly flattened and composed of mafic silicate minerals, metal, and oxides. They are thought to form by sudden (flash) heating in the solar nebula. Some formed as much as 3 Ma after the start of the solar system. (Photograph courtesy of Drs. M. Grady and S. Russell and the Natural History Museum, London.)

through shock waves in the nebular gas. Some chondrules are thought to have formed later in collisions between planetary objects. Most chondrites also contain **calcium-aluminum-rich inclusions** (**CAIs**, see Fig. 7), which have chemical compositions similar to those predicted for objects that condensed from a gas of roughly solar composition at very high temperatures. It is possible that CAIs formed in the very

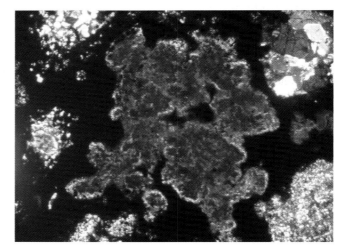

FIGURE 7 Calcium–aluminum refractory inclusions are found in chondrite meteorites and are thought to be the earliest objects that formed within our solar system. They have a chemical composition consistent with condensation from a hot gas of solar composition. How they formed exactly is unclear, but some have suggested they were produced close in to the Sun and then scattered across the disk. (Photograph courtesy of Drs. M. Grady and S. Russell and the Natural History Museum, London.)

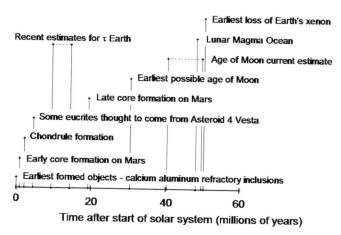

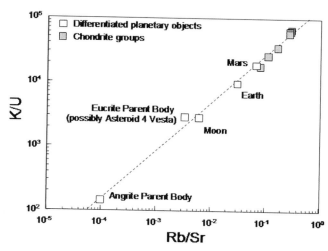

FIGURE 8   The current best estimates for the timescales over which very early inner solar system objects and the terrestrial planets formed. The approximated mean life of accretion is the time taken to achieve 63% growth at exponentially decreasing rates of growth. The dashed lines indicate the mean lives for accretion deduced for the Earth based on W isotopes. (Based on a figure that first appeared in A. N. Halliday and T. Kleine, 2006, Meteorites and the timing, mechanisms and conditions of terrestrial planet accretion and early differentiation, in "Meteorites and the Early Solar System II" (D. Lauretta, L. Leshin, and H. MacSween., eds.), pp. 775–801, Univ. Arizona Press, Tucson.)

FIGURE 9   Comparison between the K/U and Rb/Sr ratios of the Earth and other differentiated objects compared with chondrites. The alkali elements K and Rb are both relatively volatile compared with U and Sr, which are refractory. Therefore, these trace element ratios provide an indication of the degree of volatile element depletion in inner solar system differentiated planets relative to chondrites, which are relatively primitive. It can be seen that the differentiated objects are more depleted in moderately volatile elements than are chondrites. (Based on a figure that first appeared in A. N. Halliday and D. Porcelli, 2001, In search of lost planets—The paleocosmochemistry of the inner solar system, *Earth Planet. Sci. Lett.* **192**, 545–559.)

innermost regions of the solar nebula close to the Sun. Dating based on radioactive isotopes suggest that CAIs are the oldest surviving materials to have formed in the solar system. CAIs in the Efremovka chondrite are 4.5672 ± 0.0006 Ga old based on the $^{235/238}U–^{207/206}Pb$ system, and this date is often used to define the canonical start to the solar system. The oldest chondrules appear to have formed at about the same time, but most chondrules are 1–3 million years (Ma) younger than this (Fig. 8). The space between the chondrules and CAIs in chondrites is filled with fine-grained dust called matrix. Most chondrites are variably depleted in moderately volatile elements like potassium (K) and rubidium (Rb) (Fig. 9). This depletion is more a feature of the chondrules and CAIs rather than the matrix. Chondrites are subdivided into groups of like objects thought to come originally from the same parent body. Currently, about 15 groups are firmly established, 8 of which are collectively referred to as carbonaceous chondrites. These tend to be richer in highly volatile elements such as carbon and nitrogen compared to other chondrites, although as with all meteorites these elements are less abundant than they are in the Sun. Ordinary chondrites are more depleted in volatile elements than carbonaceous chondrites, and are largely made of silicates and metal grains. Enstatite chondrites are similar but

highly reduced. Chondrules are absent from the most primitive, volatile-rich group of carbonaceous chondrites (the CI group), either because their parent body formed entirely from matrix-like material or because chondrule structures have been erased by subsequent reactions with water in the parent body. Chondrites also contain presolar grains, which are submicron grains that are highly anomalous isotopically and have compositions that match those predicted to form by condensation in the outer envelopes of various stars. These represent a remarkable source of information on stellar nucleosyntheis and can be used to test theoretical models.

2. **Achondrites** are silicate-rich **mafic** and ultramafic igneous rocks not too dissimilar from those forming on Earth but with slightly different chemistry and isotopic compositions. They clearly represent the near-surface rocks of planets and asteroids that have melted and differentiated. A few achondrites come from asteroids that appear to have undergone only partial differentiation. In principle, it is possible to group achondrites and distinguish which planet or asteroid they came from. The oxygen isotopic composition of a meteorite is particularly useful in this respect. Isotopically, oxygen is extremely heterogeneous in the solar system, and planets that formed

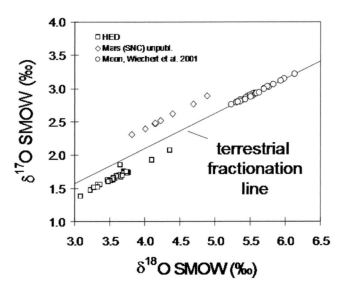

FIGURE 10 Oxygen isotopic composition of various bodies in the solar system. The x and y axes show increasing 17O and 18O abundances, respectively. The oxygen isotopic composition of the components in chondrites, in particular CAIs, is highly heterogeneous for reasons that are unclear. The net result of this variability is that different planets possess distinct oxygen isotopic compositions that define as individual mass fractionation lines as shown here for eucrites, howardites, and diogenites, which come from Vesta and SNC meteorites, thought to come from Mars. The Moon is thought to have formed from the debris produced in a giant impact between the proto-Earth when 90% formed and an impacting Mars-sized planet sometimes named "Theia." The fact that the data for lunar samples are collinear with the terrestrial fractionation line could mean that the Moon formed from the Earth, or the planet from which it was created was formed at the same heliocentric distance, or it could mean that the silicate reservoirs of the two planets homogenized during the impact process, for example by mixing in a vapor cloud from which lunar material condensed. (From A. N. Halliday, 2003, The origin and earliest history of the Earth, in "Meteorites, Comets and Planets" (A. M. Davis, ed.), Vol. 1, "Treatise of Geochemistry" (H. D. Holland and K. K. Turekian, eds.), pp. 509–557, Elsevier-Pergamon, Oxford.

FIGURE 11 Iron meteorites are the most abundant kind of meteorite found because they are distinctive and because they survive long after other kinds of meteorites are destroyed by weathering. In contrast, chondrites are the most abundant class of meteorite observed to fall. Some iron meteorites are thought to represent disrupted fragments of planetesimal cores. Others appear to have formed at low pressures, probably as metal-rich pools formed from impacts on asteroids. The Henbury meteorite shown here is a type IIIAB magmatic iron that fell near Alice Springs, Australia, about 5000 years ago. The texture shown on the sawn face are Windmanstatten patterns formed by slow cooling, consistent with an origin from a core located deep within a meteorite parent body. (Photograph courtesy of Drs. M. Grady and S. Russell and the Natural History Museum, London.)

in different parts of the nebula seem to have specific oxygen isotope compositions. This makes it possible to link all of martian meteorites together for example (Fig. 10). These meteorites are specifically linked to Mars because nearly all of them are too young to have formed on any asteroid; they had to come from an object that was large enough to be geologically active in the recent past. This was confirmed by a very close match between the composition of the atmosphere measured with the *Viking* lander and that measured in fluids trapped in alteration products in martian meteorites. In fact, martian meteorites provide an astonishing archive of information into how Mars formed and evolved as discussed in Section 6. To date, only one asteroidal source has been positively identified: Vesta, whose spectrum and orbital location strongly suggest it is the source of the howardite, eucrite, and diogenite (HED) meteorites.

3. **Irons** (see Fig. 11) are largely composed of iron, nickel (about 10% by mass), and sulfides, together with other elements that have a chemical affinity for iron, called **siderophile elements**. Like chondrites, irons can be grouped according to their likely parent body, and several dozen groups or unique irons have been found. The textures of mineral grains in iron meteorites have been used to estimate how quickly their parent bodies cooled, and thus the depth at which they formed. It appears that most irons are samples of metallic cores of small asteroidal parent bodies, 10–100 km in radius. These appear to have formed very early, probably within a million years of CAIs, when there was considerable heat available from decay of short-lived radioactive isotopes (see Section 6). Other irons appear to have formed by impact melting at the surface of asteroids, and these formed later. A rare class of stony-iron meteorite (amounting to about 5% of all nonchondritic meteorites) called pallasites contains an intricate mixture of metal and silicate (Fig. 12). It is thought these come from the core–mantle boundary regions of differentiated asteroids that broke up during collisions.

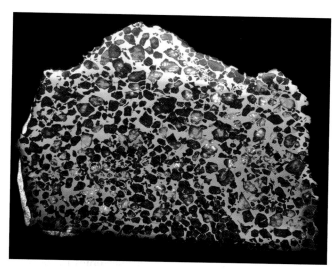

FIGURE 12   The pallasite Esquel is a mixture of silicate (olivine) and iron metal that may have formed at a planetary core-mantle boundary. (Photograph courtesy of Drs. M. Grady and S. Russell and the Natural History Museum, London.)

Note that there are no clear examples of mantle material within meteorite collections. The isotopic compositions of some elements in irons reveal that they have been exposed to cosmic rays for long periods—up to hundreds of millions of years. This means their parent bodies broke up a long time ago. Because they are extremely hard, they survived the collisions that destroyed their parent body as well as any subsequent impacts. In contrast, fragments of mantle material (as with samples excavated by volcanoes on the Earth) are extremely friable and would not survive collisions.

Survivability is also an issue for meteorites entering Earth's atmosphere and being recovered in recognizable form. Chondrites and achondrites are mainly composed of silicates that undergo physical and chemical alteration on the surface of Earth more rapidly than the material in iron meteorites. Furthermore, iron meteorites are highly distinctive, so they are easier to recognize than stony meteorites. For this reason, most meteorites found on the ground are irons, whereas most meteorites that are seen to fall from the sky (referred to as falls) are actually chondrites. Most falls are ordinary chondrites, which probably reflects the fact that they survive passage through the atmosphere better than the weaker carbonaceous chondrites. The parent bodies of ordinary chondrites may also have orbits in the Asteroid Belt that favor their delivery to Earth. IDPs are less prone to destruction during passage through the atmosphere than meteorites so they probably provide a less biased sample of the true population of interplanetary material. Most IDPs are compositionally similar to carbonaceous rather than ordinary chondrites and this suggests that the Asteroid Belt is dominated by carbonaceous-chondrite-like material.

Mass spectrometric measurements on meteorites and lunar samples provide evidence that the isotopes of most elements are present in similar proportions in the Earth, Moon, Mars, and the asteroids. The isotopes of elements heavier than hydrogen and helium were made by **nucleosynthesis** in stars that generate extremely varied isotopic compositions. Since the solar nebula probably formed from material from a variety of sources, the observed isotopic homogeneity was originally interpreted as indicative that the inner solar nebula was very hot and planetary material condensed from a ~2000 K gas of solar composition. However, a variety of observations including the preservation of presolar grains in chondrites suggest that the starting point of planet formation was cold dust and gas. This homogeneity is therefore nowadays interpreted as indicating that the inner nebula was initially turbulent, allowing dust to become thoroughly mixed. CAIs sometimes contain nucleosynthetic isotopic anomalies. This suggests that CAIs sampled varied proportions of the isotopes of the elements before they became homogenized in the turbulent disk. With improved mass spectrometric measurements evidence has been accumulating for small differences in isotopic composition in some elements between certain meteorites and those of the Earth and Moon. This area of study that searches for nucleosythetic isotopic heterogeneity in the solar system is ongoing and is now providing a method for tracking the provenance of different portions of the disk.

However, oxygen and the noble gases are very different in this respect. Extreme isotopic variations have been found for these elements. The different oxygen and noble gas isotope ratios provide evidence of mixing between compositions of dust and those of volatile (gaseous) components. Some of this mixing may have arisen later when the nebula cooled, possibly because large amounts of isotopically distinct material are thought to have arrived from the outer nebula in the form of water ice. There are also possibilities for generating some of the heterogeneity in oxygen by irradiation within the solar nebula itself.

The terrestrial planets and asteroids are not just depleted in nebular gas relative to the Sun. They are also depleted in moderately volatile elements (elements such as lead, potassium, and rubidium that condense at temperatures in the range 700–1350 K) (Figs. 9 and 13). In chondritic meteorites, the degree of depletion becomes larger as an element's condensation temperature decreases. It was long assumed that this is the result of the loss of gas from a hot nebula before it cooled. For example, by the time temperatures became cool enough for lead to condense, much of the lead had already accreted onto the Sun as a gas. However, it is clear that moderately volatile elements are depleted in chondrites at least in part because they contain CAIs and chondrules that lost volatiles by evaporation during heating events. The least depleted chondrites (CI carbonaceous chondrites) contain no CAIs or chondrules. Another mechanism for losing moderately volatile elements is planetary

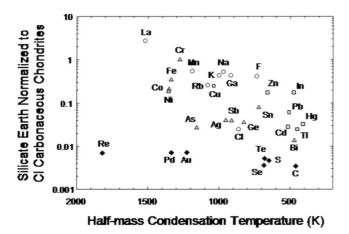

FIGURE 13 The estimated composition of the silicate portion of the Earth as a function of the calculated temperature at which half the mass of the element would have condensed. The concentrations of the various elements are normalized to the average composition of the solid matter in the disk as represented by CI carbonaceous chondrites. Open circles: lithophile elements; shaded squares: chalcophile elements; shaded triangles: moderately siderophile elements; solid diamonds: highly siderophile elements. It can be seen that refractory lithophile elements are enriched relative to CI concentrations. This is because of core formation and volatile losses compared with CI chondrites. The moderately volatile lithophile elements like K are depleted because of loss of volatiles. Siderophile elements are depleted by core formation. However, the pattern of depletion is not as strong as expected given the ease with which these elements should enter the core. The explanation is that there was addition of a late veneer of chondritic material to the silicate Earth after core formation. (From A. N. Halliday, 2003, The origin and earliest history of the Earth, in "Meteorites, Comets and Planets" (A. M. Davis, ed.), Vol. 1, "Treatise of Geochemistry" (H. D. Holland and K. K. Turekian, eds.), pp 509–557, Elsevier-Pergamon, Oxford.)

collisions. Energetic collisions between large bodies would have generated high temperatures and could have caused further loss of moderately volatile elements. For this reason, the terrestrial planets have compositions that differ from one another and also from chondritic meteorites. The Moon is highly depleted in moderately volatile elements (Fig. 9) and is thought to be the product of such an energetic planetary collision.

4. Nucleosynthesis and Short-lived Isotopes

With the exception of hydrogen and helium the elements were mainly made by stellar nucleosynthesis. If one examines Fig. 5, seven rather striking features stand out.

- The estimated abundances of the elements in the Sun and the solar nebula span a huge range of 13 orders of magnitude. For this reason, they are most easily compared by plotting on a log scale such that the number of atoms of Si is 10^6.

- Hydrogen and helium are by far the most abundant elements in the Sun as they are elsewhere in the universe. These two elements were made in the Big Bang.

- The abundances of the heavier elements generally decrease with increasing atomic number. This is because most of the elements are themselves formed from lighter elements by stellar nucleosynthesis.

- Iron is about 1000 times more abundant than its neighbors in the periodic table because of a peak in the binding energy providing enhanced stability during nucleosynthesis.

- Lithium, beryllium, and boron are all relatively underabundant compared to other light elements because they are unstable in stellar interiors.

- A saw-toothed variability is superimposed on the overall trend reflecting the relatively high stability of even-numbered isotopes compared to odd-numbered ones.

- All the elements in the periodic table are present in the solar system except those with no long-lived or stable isotopes, namely technetium (Tc), promethium (Pm), and the trans-uranic elements.

Those elements lighter than Fe can be made by fusion because the process of combining two nuclei to make a heavier nuclide releases energy. This produces the energy in stars and is activated when the pressure exceeds a critical threshold (i.e., when a star reaches a certain mass). Larger stars exert more pressure on their cores such that fusion reactions proceed more quickly. When a star has converted all of the hydrogen in its center to helium, it will either die out if it is small or proceed to the next fusion cycle such as the conversion of helium to carbon if it is sufficiently massive to drive this reaction. Lithium, beryllium, and boron are unstable at the temperatures and pressures of stellar interiors, hence the drop in abundance in Fig. 5. They are made by spallation reactions from heavier elements by irradiation in the outer portions of stars.

Nearly all nuclides heavier than Fe must be made by neutron irradiation because their synthesis via fusion would consume energy. Neutron addition continues until an unstable isotope is made; it will decay to an isotope of another element, which then receives more neutrons until another unstable nuclide is made and so forth. These are s-process isotopes (produced by a slow burst of neutrons). However, some of these isotopes cannot be made simply by adding a neutron to a stable nuclide because there is no stable isotope with a suitable mass. Such nuclides are instead created with a very high flux of neutrons such that unstable nuclides produced by neutron irradiation receive additional neutrons before they have time to decay, jumping the gap to very heavy nuclides. These are r-process isotopes (produced by

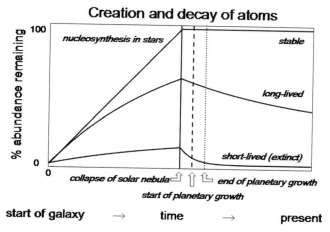

FIGURE 14 Most solar system nuclides heavier than hydrogen and helium were produced in stars over the history of our galaxy. This schematic figure shows the difference between nuclides that are stable, those that have very long half-lives (such as ^{238}U used for determining the ages of geological events and the solar system itself), and those that have short half-lives of $<10^8$ years, assuming all were produced at a constant rate through the history of the galaxy. The short-lived nuclides decay very fast and provide crucial insights into the timescales of events, including planet formation, immediately following their incorporation into the solar nebula.

a rapid burst of neutrons). Such extremely high fluxes of neutrons are generated in supernova explosions.

The composition of the Sun and solar system represents the cumulative ~8 Ga history of such stellar processes in this portion of the galaxy prior to collapse of the solar nebula (Fig. 14). It is unknown how constant these processes were. However, the isotopes of some elements in meteorites provide evidence that stellar nucleosynthesis was still going on just prior to the collapse of the solar nebula. In fact, the formation of the solar system may have been triggered by material being ejected from a massive star as it was exploding, seeding the solar nebula with freshly synthesized nuclides.

Chondrites show evidence that they once contained short-lived radioactive isotopes probably produced in massive stars shortly before the solar system formed. As already pointed out most stable isotopes are present in the same ratios in the Earth, the Moon, Mars, and different groups of meteorites, which argues that material in the solar nebula was thoroughly mixed at an early stage. However, a few isotopes such as ^{26}Mg are heterogeneously distributed in chondrites. In most cases, these isotopes are the daughter products of short-lived isotopes. In other words the excess ^{26}Mg comes from the radioactive decay of ^{26}Al. Every atom of ^{26}Al decays to a daughter atom of ^{26}Mg; therefore,

$$\left(^{26}\mathrm{Mg}\right)_{\mathrm{today}} = \left(^{26}\mathrm{Mg}\right)_{\mathrm{original}} + \left(^{26}\mathrm{Al}\right)_{\mathrm{original}} \quad (1)$$

Because it is easier to measure these effects using isotopic ratios rather than absolute numbers of atoms, we divide by another isotope of Mg:

$$\left(\frac{^{26}\mathrm{Mg}}{^{24}\mathrm{Mg}}\right)_{\mathrm{today}} = \left(\frac{^{26}\mathrm{Mg}}{^{24}\mathrm{Mg}}\right)_{\mathrm{original}} + \left(\frac{^{26}\mathrm{Al}}{^{24}\mathrm{Mg}}\right)_{\mathrm{original}} \quad (2)$$

However, the ^{26}Al is no longer extant and so cannot be measured. For this reason, we convert Eq. (2) to a form that includes a monitor of the amount of ^{26}Al that would be determined from the amount of Al today. Aluminum has only one stable nuclide ^{27}Al. Hence, Eq. (2) becomes

$$\left(\frac{^{26}\mathrm{Mg}}{^{24}\mathrm{Mg}}\right)_{\mathrm{today}} = \left(\frac{^{26}\mathrm{Mg}}{^{24}\mathrm{Mg}}\right)_{\mathrm{original}}$$
$$+ \left\{ \left(\frac{^{26}\mathrm{Al}}{^{27}\mathrm{Al}}\right)_{\mathrm{original}} \times \left(\frac{^{27}\mathrm{Al}}{^{24}\mathrm{Mg}}\right)_{\mathrm{today}} \right\} \quad (3)$$

which represents the equation for a straight line (Fig. 15). A plot of ^{26}Mg/^{24}Mg against ^{27}Al/^{24}Mg for a suite of co-genetic samples or minerals will define a straight line the slope of which gives the ^{26}Al/^{27}Al at the time the object formed. This can be related in time to the start of the solar system with Soddy and Rutherford's equation for radioactive decay:

$$\left(\frac{^{26}\mathrm{Al}}{^{27}\mathrm{Al}}\right)_{\mathrm{original}} = \left(\frac{^{26}\mathrm{Al}}{^{27}\mathrm{Al}}\right)_{\mathrm{BSSI}} \times e^{-\lambda t} \quad (4)$$

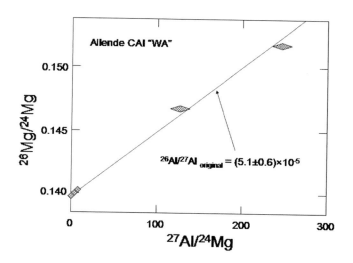

FIGURE 15 The decay of a short-lived nuclide such as ^{26}Al generates excess ^{26}Mg in proportion to the elemental ratio Al/Mg. The data here were produced for a CAI from the Allende meteorite. The slope of the line corresponds to the ^{26}Al/^{27}Al at the time of formation of the object. See text for discussion. (Based on a figure in T. Lee, D.A. Papanastassiou, and G. J. Wasserburg, 1976, *Astrophys. J.* **211**, L107.)

TABLE 1	Extinct Radionuclides			
Radionuclide	**Half-life (million years)**	**Ratio**	**Initial Ratio**	**Stable Daughter**
^{10}Be	1.5	^{10}Be/^{9}Be	1×10^{-3}	^{10}B
^{26}Al	0.71	^{26}Al/^{27}Al	6×10^{-5}	^{26}Mg
^{41}Ca	0.10	^{41}Ca/^{40}Ca	1×10^{-8}	^{41}K
^{53}Mn	3.7	^{53}Mn/^{55}Mn	6×10^{-6}	^{53}Cr
^{60}Fe	1.5	^{60}Fe/^{56}Fe	1×10^{-6}	^{60}Ni
^{92}Nb	36	^{92}Nb/^{93}Nb	3×10^{-5}	^{92}Zr
^{107}Pd	6.5	^{107}Pd/^{110}Pd	9×10^{-5}	^{107}Ag
^{129}I	15.7	^{129}I/^{127}I	1×10^{-4}	^{129}Xe
^{146}Sm	103	^{146}Sm/^{144}Sm	0.008	^{142}Nd
^{182}Hf	9	^{182}Hf/^{180}Hf	1×10^{-4}	^{182}W
^{244}Pu	80	^{244}Pu/^{238}U	0.007	^{131,132,134,136}Xe

in which BSSI is the bulk solar system initial ratio, $\lambda = \ln 2/$half life is the decay constant (or probability of decay in unit time), and t is the time that elapsed since the start of the solar system. Using this method and the $(^{26}$Al/^{27}Al$)_{BSSI}$ of $\sim 6 \times 10^{-5}$ (Table 1), it has been possible to demonstrate that many chondrules formed 1–3 Ma after CAIs.

Over the past 40 years, scientists have found evidence that about a dozen short-lived isotopes existed early in the solar system. These isotopes are listed in Table 1. Other isotopes such as ^{36}Cl and ^{205}Pb were probably present as well, but their initial abundances are currently uncertain.

These short-lived isotopes can be broken down into three types on the basis of their origin in the solar nebula:

1. The Sun and the other stars in its cluster inherited a mixture of isotopes from their parent molecular cloud that built up over time from a range of stellar sources.
2. Some short-lived isotopes were probably injected into the Sun's molecular cloud core or the solar nebula itself from at least one nearby star, possibly a supernova.
3. It is likely that some short-lived isotopes were also generated in the innermost regions of the solar nebula when material was bombarded with energetic particles from the Sun.

Determining the origin of a particular isotope and the timing of its production is often difficult. Isotopes with half-lives of less than 10^6 years must have come from a source close to the solar nebula in order to have survived, while isotopes with longer half-lives may have come from further away. Irradiation in the solar nebula could have produced a variety of light isotopes but the relative importance of local production versus external sources is still unclear. Formation in the nebula appears to be the most promising source for ^{10}Be. However, if all of the ^{26}Al had formed this way, it seems likely that some of the other isotopes, especially ^{41}Ca, would have been more abundant than they actually

were. In fact, there is mounting evidence that many of the short-lived isotopes were quite uniformly distributed in the solar system, which is hard to explain if they formed in a localized region close to the Sun.

Some of the heavier short-lived isotopes that existed in the early solar system (e.g., ^{107}Pd, ^{129}I) can only be produced in large amounts in a massive star. For example, a large flux of neutrons is required to produce ^{129}I, and this is achievable during the enormously energetic death throws of a massive star undergoing a type II supernova explosion. Many of the isotopic ratios in Table 1 are similar, lying in the range 10^{-6}–10^{-4} for isotopes with half-lives of 0.7×10^6 to 30×10^6 years. This is as expected if all of these isotopes were synthesized in roughly similar proportions just prior to the start of the solar system. Many of these isotopes have initial abundances similar to those that would be formed by an AGB star. However, models for AGB stars do not predict the amounts of ^{53}Mn and ^{182}Hf that once existed. In fact, ^{182}Hf (half-life = 9×10^6 years) requires a large flux of neutrons of the kind produced in the supernova explosion of a much larger star. It is possible that more than one kind of nucleosynthetic process gave rise to the short-lived isotopes in the early solar system. At present, it seems likely that a nearby supernova was involved because the abundance of ^{60}Fe, which has a fairly short half-life, is too high to be explained by alternative sources. Some isotopes that may have been present have yet to be found, including ^{126}Sn and ^{247}Cm with half-lives of 0.3 and 16 Ma. These are both r-process isotopes that should have been present in the early solar system if a supernova occurred nearby. The fact that ^{247}Cm has not been detected places strong constraints on a supernova source. Modeling these processes is complex, but it appears that the supernova explosion of a 25 solar-mass star may explain the correct relative abundances of many of the short-lived isotopes, including ^{182}Hf, provided that roughly 5 solar masses of material was left behind in the form of a supernova remnant or a black hole.

Supernovas are sufficiently energetic that they could tear apart a molecular cloud core rather than cause it to collapse. Shocks waves with a velocity of at least 20–45 km s^{-1} are capable of triggering collapse, but if the velocity exceeds ~100 km s^{-1}, a molecular cloud core will be shredded instead. If the supernova was sufficiently far away, the shock wave would have slowed by the time it reached the molecular cloud core. However, the supernova cannot have been more than a few tens of parsecs away; otherwise, ^{41}Ca (with a half life of only 0.104×10^6 years) would have decayed before it reached the solar nebula. The former presence of ^{41}Ca in CAIs may provide the best constraint on the time between nucleosynthesis of the short-lived isotopes and their incorporation into the solar system. To do this, it will be necessary to ascertain the particular stellar source(s) that gave rise to these isotopes, so that the initial amount of ^{41}Ca can be calculated.

5. Early Stages of Planetary Growth

Dust grains are a relatively minor constituent of protoplanetary disks, but they represent the starting point for the formation of rocky planets like Earth, and possibly also gas-rich planets like Jupiter. These grains are small, typically 1 μm in diameter or less. In a microgravity environment, electrostatic forces dominate interactions between such grains. Charge transfer during grain collisions can lead to the formation of grain dipoles that align with one another forming aggregates up to several centimeters in size. Freshly deposited frost surfaces make grains stickier and increase the ability of grain aggregates to hold together during subsequent collisions.

Laboratory experiments show that low-velocity collisions between grains tend to result in sticking, while faster collisions often cause grains to rebound. Irregularly shaped micron-sized grains often stick to one another at collision speeds of up to tens of meters per second. Fluffy aggregates may stick more readily than compact solids as some of the energy of impact goes into compaction. However, the primary components of chondritic meteorites are compact chondrules, so further compaction cannot have played a big role in the formation of their parent bodies. In general, sticking forces scale with the surface area of an object, while collisional energy scales with mass and hence volume. As a result, growth becomes more difficult, and breakup becomes more likely, as aggregates become larger. It is possible that early growth in the solar nebula took place mainly as the result of large objects sweeping up smaller ones. This idea is supported by recent experiments that found that small dust aggregates tend to embed themselves in larger ones if they collide at speeds above about 10 m/s.

Dust grains, grain aggregates, and chondrules would have been closely coupled to the motion of gas in the solar nebula. The smallest particles were mainly affected by Brownian motion—collisions with individual gas molecules, which caused the particles to move with respect to one another, leading to collisions. Particles also settled slowly toward the disk's midplane due to the vertical component of the Sun's gravitational field. Settling was opposed by gas drag so that each particle fell at its terminal velocity:

$$v_z = -\left(\frac{\rho}{\rho_{gas}}\right)\left(\frac{v_{kep}}{c_s}\right)\left(\frac{rz}{a^2}\right)v_{kep} \qquad (5)$$

where r and ρ are the radius and density of the particle, ρ_{gas} is the gas density, a is the orbital distance from the Sun, z is the height above the disk midplane, and c_s is the sound speed in the gas. Here v_{kep} is the speed of a solid body moving on a circular orbit, called the **Keplerian velocity:**

$$v_{kep} = \sqrt{\frac{GM_{sun}}{a}} \qquad (6)$$

where M_{sun} is the mass of the Sun. Large particles fell faster than small ones, sweeping up material as they went, increasing their vertical speed further. Calculations show that micron-sized particles would grow and reach the midplane in about 10^3–10^4 orbital periods if these were the only processes operating.

If the gas was turbulent, particles would have become coupled to turbulent eddies due to gas drag. Particles of a given size were coupled most strongly to eddies whose turnover (rotation) time was similar to the particle's stopping time, given by

$$t_s = \frac{\rho r}{\rho_{gas} c_s} \qquad (7)$$

Meter-sized particles would have coupled to the largest eddies, with turnover times comparable to the orbital period P. In a strongly turbulent nebula, meter-sized particles would have collided with one another and with smaller particles at high speeds, typically tens of meters per second.

Gas pressure in the nebula generally decreased with distance from the Sun. This means gas orbited the Sun more slowly than solid bodies, which moved at the Keplerian velocity. Large solid bodies thus experienced a headwind of up to 100 m/s. The resulting gas drag removed angular momentum from solid bodies, causing them to undergo radial drift toward the Sun. Small particles with $t_s \ll P$ drifted slowly at terminal velocity. Very large objects with $t_s \gg P$ were only weakly affected by gas drag and also drifted slowly. Drift rates were highest for meter-sized bodies with $t_s \sim P$ (see Fig. 16), and these drifted inward at rates of 1 AU every few hundred years. Rapid inward drift meant that these bodies collided with smaller particles at high speeds. Rapid drift also meant that meter-sized objects had very short lifetimes, and many were probably lost when they reached the hot innermost regions of the nebula and vaporized.

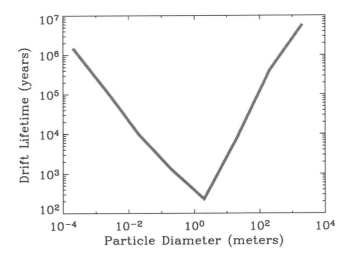

FIGURE 16 The lifetime of solid particles orbiting at 1 AU from the Sun in the minimum-mass solar nebula when the particles drift inward due to gas drag. Drift rates are fastest for meter-sized particles, which are lost in a few hundred years unless they rapidly grow larger.

The short drift lifetimes and high collision speeds experienced by meter-sized particles have led some researchers to conclude that particle growth stalled at this size because particles were destroyed as fast as they formed. This is often referred to as the meter-sized barrier. This remains an open question however, due to a shortage of experimental data regarding the physics of collisions in a microgravity environment and uncertainty about the level of turbulence in the solar nebula.

The presence of nebular gas was not entirely detrimental to growth. Experiments have shown that gas drag can reduce the effect of destructive impacts onto boulder-sized bodies, as collision fragments become entrained in the gas and blown back onto the surface of the larger body. Numerical simulations also show that chondrule-sized particles would be strongly concentrated in stagnant regions in a turbulent nebula, a process called **turbulent concentration**, thus increasing the chance of further growth.

Bodies larger than 1 km generally took a long time to drift inward due to gas drag. These objects were also large enough to have appreciable gravitational fields, making them better able to hold on to fragments generated in collisions. For these reasons, growth became easier once bodies became this large. Much effort has been devoted to seeing whether kilometer-sized bodies could have formed directly, avoiding the difficulties associated with the meter-sized barrier. **Gravitational instability** (GI) offers a possible way to do this. If the level of turbulence in the nebula was very low, solid particles would have settled close to the nebula midplane, increasing their local concentration. Radial drift of particles may also have concentrated particles at a particular location. If enough particles became concentrated in one place, their combined gravitational attraction

would render the configuration unstable, allowing the region to become gravitationally bound and collapse. If the particles were then able to contract enough to form a single solid body, the resulting object would be roughly 1–10 km in radius. Such an object is called a **planetesimal**.

Gravitational instability faces severe obstacles however. As solid particles accumulated near the nebula midplane, they would have begun to drag gas around the Sun at Keplerian speeds, while gas above and below the midplane continued to travel at sub-Keplerian speeds. The velocity difference between the layers generated turbulence, puffing up the particle layer until a balance between vertical sedimentation and turbulence was reached. This balance may have prevented particle concentrations from becoming high enough for GI to occur. Calculations suggest that the solid-to-gas ratio in a vertical column of nebula material had to become roughly unity before GI would take place. This means that the concentration of solid material had to become enhanced by 1–2 orders of magnitude compared to the nebula as a whole. If a region of the disk did start to undergo GI, it would only contract to form a planetesimal if the relative velocities of the particles in that region became low enough. Turbulence and radial drift both lead to large relative velocities between particles and may have rendered GI ineffective.

The difficulties associated with both the meter-sized barrier and gravitational instability mean that the question of how planetesimals formed remains open for now. However, the fact that roughly half of young stars have debris disks of dust thought to come from asteroids and comets implies that growth of large solid bodies occurs in many protoplanetary disks, even if the mechanism remains obscure.

6. Formation of Terrestrial Planets

The growth of bodies beyond 1 km in size is reasonably well understood. Gravitational interactions and collisions between pairs of planetesimals dominate the evolution from this point onward. A key factor in determining the rate of growth is **gravitational focusing**. The probability that two planetesimals will collide during a close approach depends on their cross-sectional area multiplied by a gravitational focusing factor F_g:

$$F_g = 1 + \frac{v_{esc}^2}{v_{rel}^2} \qquad (8)$$

where v_{rel} is the planetesimals' relative velocity, and v_{esc} is the escape velocity from a planetesimal, given by

$$v_{esc} = \sqrt{\frac{2GM}{r}} \qquad (9)$$

where M and r are the planetesimal's mass and radius, respectively. When planetesimals pass each other slowly, there is time for their mutual gravitational attraction to focus their trajectories toward each other, so F_g is large, and the chance of a collision is high. Fast moving bodies typically do not collide unless they are traveling directly toward each other because $F_g \cong 1$ in this case. The relative velocities of planetesimals depend on their orbits about the Sun. Objects with similar orbits are the most likely to collide with each other. In particular, planetesimals moving on nearly circular, coplanar orbits have high collision probabilities while ones with highly inclined, eccentric (elliptical) orbits do not.

Most close encounters between planetesimals did not lead to a collision, but bodies often passed close enough for their mutual gravitational tug to change their orbits. Statistical studies show that after many such close encounters, high-mass bodies tend to acquire circular, coplanar orbits, while low-mass bodies are perturbed onto eccentric, inclined orbits. This is called dynamical friction and is analogous to the equipartition of kinetic energy between molecules in a gas. Dynamical friction means that, on average, the largest bodies in a particular region experience the strongest gravitational focusing; therefore, they grow the fastest (Fig. 17). This state of affairs is called **runaway growth** for obvious reasons. Most planetesimals remained small, while a few objects, called **planetary embryos**, grew much larger.

Runaway growth continued as long as interactions between planetesimals determined their orbital distribution. However, once embryos became more than about a thousand times more massive than a typical planetesimal, gravi-

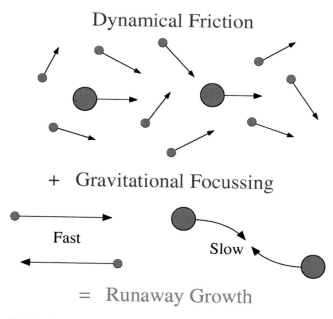

Dynamical Friction

+ Gravitational Focussing

Fast

Slow

= Runaway Growth

FIGURE 17 Runaway growth of a few large planetesimals takes place due to a combination of dynamical friction (which gives large planetesimals circular and coplanar orbits), and gravitational focusing (which increases the chance of a collision between bodies moving on similar orbits).

tational perturbations from the embryos became more important. The evolution now entered a new phase called **oligarchic growth**. The relative velocities of planetesimals were determined by a balance between perturbations from nearby embryos and damping due to gas drag. Embryos continued to grow faster than planetesimals, but growth was no longer unrestrained. Large embryos stirred up nearby planetesimals more than small embryos did, weakening gravitational focusing and slowing growth. As a result, neighboring embryos tended to grow at similar rates. Embryos spaced themselves apart at regular radial intervals, with each one staking out an annular region of influence in the nebula called a feeding zone.

As embryos became larger, they perturbed planetesimals onto highly inclined and eccentric orbits. The planetesimals began to collide with one another at high speeds, causing fragmentation and breakup. A huge number of sub-kilometer-sized collision fragments were generated, together with a second generation of fine dust particles. Gas drag operates efficiently on small fragments, so their orbits rapidly became almost circular and coplanar. As a result, many fragments were quickly swept up by embryos, increasing the embryos' growth rates still further.

Numerical calculations show that embryo feeding zones were typically about 10 Hill radii in width, where the Hill radius of an embryo with mass M and orbital radius a is given by

$$r_h = a \left(\frac{M}{3M_{sun}} \right)^{1/3} \tag{10}$$

If an embryo were to accrete all of the solid material in its feeding zone it would stop growing when its mass reached a value called the **isolation mass**, given by

$$M_{iso} \cong \left(\frac{8b^3 \pi^3 \Sigma_{solid}^3 a^6}{3M_{sun}} \right)^{1/2} \tag{11}$$

where Σ is the surface (column) density of solid material in that region of the disk, and $b \approx 10$ is the width of a feeding zone in Hill radii. The surface density in the Sun's protoplanetary nebula is not known precisely, but for plausible values, the isolation masses would have been about 0.1 Earth masses at 1 AU, and around 10 Earth masses in the outer solar system. Calculations suggest that bodies approached their isolation mass in the inner solar system roughly 10^5 years after planetesimals first appeared in large numbers. Growth was slower in the outer solar system, but bodies were probably nearing their isolation mass at 5 AU after 10^6 years.

Large embryos significantly perturbed nearby gas in the nebula forming spiral waves. Gas passing through these waves had a higher density than that in the surrounding region. Gravitational interactions between an embryo and its spiral waves transferred angular momentum between them. For conditions likely to exist in the solar nebula, the net result was that each embryo lost angular momentum and migrated inward towards the Sun. This is called **type-I migration.** Migration rates are proportional to an embryo's mass M and the local surface density of gas Σ_{gas}:

$$\frac{da}{dt} \approx -4 \left(\frac{M}{M_{sun}} \right) \left(\frac{\Sigma_{gas}a^2}{M_{sun}} \right) \left(\frac{v_{kep}}{c_s} \right)^2 v_{kep} \qquad (12)$$

where c_s is the sound speed in the gas and v_{kep} is the orbital velocity of a body moving on a circular, Keplerian orbit. Type-I migration became important after embryos grew to about 0.1 Earth masses. Migration rates can be uncomfortably fast, with a 10-Earth mass body at 5 AU migrating into the Sun in 10^5 years in a minimum-mass nebula. It is possible that many objects migrated all the way into the Sun and were lost in this way, and the question of how other bodies survived migration is one of the great unresolved questions of planet formation at present.

Oligarchic growth in the inner solar system ended when embryos had swept up roughly half of the solid material. However, these embryos were still an order of magnitude less massive than Earth. Further collisions were necessary to form planets the size of Earth and Venus. With the removal of most of the planetesimals, dynamical friction weakened. As a result, interactions between embryos caused their orbits to become more inclined and eccentric. The embryos' gravitational focusing factors became small, and this greatly reduced the collision rate. As a result, the last stage of planet formation was prolonged, and the Earth may have taken 100 Ma to finish growing.

Embryos underwent numerous close encounters with one another before colliding. Each encounter changed an embryo's orbit, with the result that embryos moved considerable distances radially in the nebula. Numerical calculations show that the orbital evolution must have been highly chaotic (Fig. 18). As a result, it is impossible to predict the precise characteristics of a planetary system based on observations of typical protoplanetary disks. Other stars with nebulas similar to the Sun may have formed terrestrial planets that are very different from those in the solar system.

The radial motions of embryos partially erased any chemical gradients that existed in the nebula during the early stages of planet formation. Mixing cannot have been complete however because Mars and Earth have distinct compositions. Mars is richer in the more volatile rock-forming elements, and the two planets have distinct oxygen isotope mixtures. Unfortunately, we have no confirmed samples of Mercury and Venus, so we know little about their composition. Mercury is known to have an unexpectedly high density, suggesting it has a large iron-rich core and a small mantle. This probably does not reflect compositional differences in the solar nebula because there is no known reason why iron-rich materials would preferentially form closer to

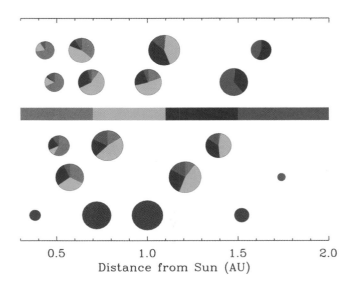

FIGURE 18 Four artificial planetary systems generated by numerical simulations of planetary accretion. Each horizontal row of symbols represents one planetary system, with symbol radius proportional to planetary radius, with the largest objects similar in size to the Earth. The shaded segments show the composition of each planet in terms of material that originated in four different portions of the nebula. Planets in these simulations typically contain material from many regions of the nebula. The row of gray symbols shows the terrestrial planets of the solar system for comparison.

the Sun than silicate materials. A more likely explanation is that Mercury suffered a near-catastrophic impact after it had differentiated, and this stripped away much of the silicate mantle. Mercury's location close to the Sun made it especially vulnerable in this respect because orbital velocities and hence impact speeds are highest close to the Sun.

Earth and Venus are probably composites of ten or more embryos so their chemical and isotopic compositions represent averages over a fairly large region of the inner solar system. Mars and Mercury are sufficiently small that they may be individual embryos that did not grow much beyond the oligarchic growth stage. It is currently a mystery why Earth and Venus continued to grow while Mars did not. It may be that Mars formed in a low-density region of the nebula or that all other embryos were removed from that region without colliding with Mars.

As embryos grew larger, their temperatures increased due to kinetic energy released during impacts and the decay of radioactive isotopes in their interiors. Short-lived isotopes such as ^{26}Al and ^{60}Fe, with half-lives of 0.7×10^6 and 1.5×10^6 years, respectively (Table 1), were particularly powerful heat sources early in the solar system. Bodies more than a few kilometers in radius would have melted if they had formed within the first 2 Ma when the short-lived isotopes were still abundant. Embryos that melted also differentiated, with iron and siderophile elements sinking to the center to form a core, while lighter silicates formed a mantle closer to the surface.

The abundances of the highly siderophile elements in Earth's mantle are higher than one would expect to find after the planet differentiated because most siderophile material should have been extracted into the core. The most likely explanation for these high abundances is that Earth continued to acquire some material after its core and mantle had finished separating. This late veneer amounted to about 1% of the total mass of the planet.

The origin of Earth's water is the subject of much debate at present. Earth's oceans contain about 0.03% of the planet's total mass. A roughly comparable amount of water exists in the mantle (with an uncertainty of a factor of 3 in either direction). Earth may have also lost an unknown fraction of its water early in its history due to reactions with iron. Temperatures at 1 AU are currently too high for water ice to condense, and this was probably also true for most of the history of the solar nebula (pressures were always too low for liquid water to condense). As a result, Earth probably received most of its water as the result of collisions with other embryos or planetesimals that contained water ice or hydrated minerals in their interiors.

Planetesimals similar to modern comets almost certainly delivered some water to Earth. However, a typical comet has a probability of only about one in a million of colliding with Earth, so it is unlikely that comets provided the bulk of the planet's water. The deuterium–hydrogen ratio (D/H) seen in comets is twice that of Earth's oceans, which suggests comets supplied at most about 10% of Earth's water. However, D/H has been measured in only 3 comets to date, so this conclusion is tentative. Planetesimals from the Asteroid Belt are another possible source of water. Carbonaceous chondrites are especially promising because they contain up to 10% water by mass in the form of hydrated silicates, and this water would be released upon impact with the Earth. Calculations suggest that if the early Asteroid Belt was several orders of magnitude more massive than today, it could have supplied the bulk of Earth's water. This water must have arrived before core formation was complete however, because carbonaceous chondrites and Earth's mantle have different osmium isotope ratios. As a result, the delivery of water to Earth and its acquisition of a late veneer were separate processes that occurred at different times in its history.

The origin of Earth's atmospheric constituents is also somewhat uncertain. When the solar nebula was still present, planetary embryos probably had thick atmospheres mostly composed of hydrogen and helium captured from the nebula. Most of this atmosphere was lost subsequently by hydrodynamic escape as hydrogen atoms were accelerated to escape velocity by ultraviolet radiation from the Sun, dragging other gases along with them. Much of Earth's current atmosphere was probably outgassed from the mantle at a later stage. Some noble gases currently escaping from Earth's interior are similar to those found in the Sun, which suggests they may have been captured into Earth's mantle from the nebula or were trapped in bodies that later

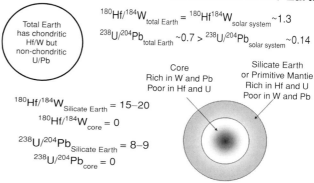

^{180}Hf/^{184}W and ^{238}U/^{204}Pb fractionation in the Earth

Total Earth has chondritic Hf/W but non-chondritic U/Pb

^{180}Hf/^{184}W$_{\text{total Earth}}$ = ^{180}Hf^{184}W$_{\text{solar system}}$ ~1.3

^{238}U/^{204}Pb$_{\text{total Earth}}$ ~0.7 > ^{238}U/^{204}Pb$_{\text{solar system}}$ ~0.14

Core
Rich in W and Pb
Poor in Hf and U

Silicate Earth or Primitive Mantle Rich in Hf and U Poor in W and Pb

^{180}Hf/^{184}W$_{\text{Silicate Earth}}$ = 15–20

^{180}Hf/^{184}W$_{\text{core}}$ = 0

^{238}U/^{204}Pb$_{\text{Silicate Earth}}$ = 8–9

^{238}U/^{204}Pb$_{\text{core}}$ = 0

FIGURE 19 Hafnium–tungsten chronometry provides insights into the rates and mechanisms of formation of the solar system whereas U–Pb chronometry provides us with an absolute age of the solar system. In both cases the radioactive parent/radiogenic daughter element ratio is fractionated by core formation, an early planetary process. It is this fractionation that is being dated. The Hf/W ratio of the total Earth is chondritic (average solar system) because Hf and W are both refractory elements. The U/Pb ratio of the Earth is enhanced relative to average solar system because approximately >80% of the Pb was lost by volatilization or incomplete condensation mainly at an early stage of the development of the circumstellar disk. The fractionation within the Earth for Hf/W and U/Pb is similar. In both cases, the parent (Hf or U) prefers to reside in the silicate portion of the Earth. In both cases the daughter (W or Pb) prefers to reside in the core.

collided with Earth. Most of the xenon produced by radioactive decay of plutonium (half-life 83 Ma) and ^{129}I has been lost, which implies that Earth's atmosphere was still being eroded 100 Ma after the start of the solar system, possibly by impacts.

Radioactive isotopes can be used to place constraints on the timing of planet formation. The hafnium–tungsten system is particularly useful in this respect because the parent nuclide ^{182}Hf is lithophile (tending to reside in silicate mantles) while the daughter nuclide ^{182}W is siderophile (tending to combine with iron during core formation) (Fig. 19). Isotopic data can be used in a variety of ways to define a timescale for planetary accretion. The simplest method uses a model age calculation, which corresponds to the calculated time when an object or sample would have needed to form from a simple average solar system reservoir, as represented by chondrites, in order generate its isotopic composition. For the ^{182}Hf–^{182}W system, this time is given as

$$t_{\text{CHUR}} = \frac{1}{\lambda} \ln \left[\left(\frac{^{182}\text{Hf}}{^{180}\text{Hf}} \right)_{\text{BSSI}} \right.$$

$$\left. \times \left(\frac{\left(\frac{^{182}W}{^{184}W} \right)_{\text{SAMPLE}} - \left(\frac{^{182}W}{^{184}W} \right)_{\text{CHONDRITES}}}{\left(\frac{^{180}\text{Hf}}{^{184}W} \right)_{\text{SAMPLE}} - \left(\frac{^{180}\text{Hf}}{^{184}W} \right)_{\text{CHONDRITES}}} \right) \right]$$

(13)

where t_{CHUR} is the time of separation from a CHondritic Uniform Reservoir, $\lambda = (\ln 2/\text{half-life})$ is the decay constant for ^{182}Hf (0.078 per million years) and $(^{182}\text{Hf}/^{180}\text{Hf})_{\text{BSSI}}$ is the bulk solar system initial ratio of ^{182}Hf to ^{180}Hf. Tungsten-182 excesses have been found in Earth, Mars, and the HED meteorites, which are thought to come from asteroid Vesta, indicating that all these bodies differentiated while some ^{182}Hf was still present. Iron meteorites, which come from the cores of differentiated planetesimals, have low Hf/W ratios and are deficient in ^{182}W. This means these planetesimals must have formed at a very early stage before most of the ^{182}Hf had decayed. New, very precise $^{182}\text{Hf}–^{182}\text{W}$ chronometry has shown that some of these objects formed within the first few hundred thousand years of the solar system (Fig. 8).

New modeling of the latest $^{182}\text{Hf}–^{182}\text{W}$ data for martian meteorites also provides evidence that Mars grew and started differentiating within about 1 Ma of the start of the solar system. This short timescale is consistent with runaway growth described earlier. So far, isotopic data for other silicate objects has not been so readily explicable in terms of very rapid growth. However, asteroid Vesta certainly formed within about 3 Ma of the start of the solar system (Fig. 8).

The existence of meteorites from differentiated asteroids suggests that core formation began early, and this is confirmed by $^{182}\text{Hf}–^{182}\text{W}$ chronometry. Therefore, most planetary embryos would have been differentiated when they collided with one another. Although Mars grew extremely rapidly, Earth does not appear to have reached its current size until the giant impact that was associated with the formation of the Moon (see Section 8). $^{182}\text{Hf}–^{182}\text{W}$ chronometry for lunar samples shows that this took place 35–50 Ma after the start of the solar system. Geochemical evidence has been used to argue that the formation of the Moon probably happened near the end of Earth's accretion, and this is consistent with the results of Moon-forming impact simulations. This is also consistent with the W isotopic composition of the silicate Earth itself (Fig. 20). This shows that the Earth accreted at least half of its mass within the first 3×10^7 years of the solar system. However, the data are fully consistent with the final stage of accretion being around the time of the Moon-forming impact. Because the Earth accreted over a protracted period rather than in a single event, it is simplest to model the W isotope data in terms of an exponentially decreasing rate of growth (Fig. 20).

$$F = 1 - e^{-(1/\tau) \times t} \qquad (14)$$

where F is the mass fraction of the Earth that has accumulated, τ is the mean life for accretion in millions of years (Fig. 20) and t is time in millions of years. This is consistent with the kinds of curves produced by the late George Wetherill who modeled the growth of the terrestrial planets using Monte Carlo simulations. The W isotope data are consistent with a mean life of between 10 and

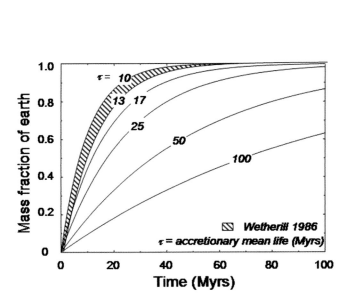

FIGURE 20 The mean life of accretion of the Earth (τ) is the inverse of the time constant for exponentially decreasing oligarchic growth from stochastic collisions between planetary embryos and planets. The growth curves corresponding to several such mean lives are shown including the one that most closely matches the calculation made by the late George Wetherill based on Monte Carlo simulations. The mean life determined from tungsten isotopes (Fig. 8) is in excellent agreement with Wetherill's predictions.

15 Ma, depending on the exact parameters used. This is fully consistent with the timescales proposed by Wetherill. From these protracted timescales, it is clear that Earth took much longer to approach its current size than Mars or Vesta, which probably formed from different mechanisms (Fig. 8).

7. The Asteroid Belt

The Asteroid Belt currently contains only enough material to make a planet 2000 times less massive than Earth, even though the spatial extent of the belt is huge. It seems likely that this region once contained much more mass than it does today. A smooth interpolation of the amount of solid material needed to form the inner planets and the gas giants would place about 2 Earth-masses in the Asteroid Belt. Even if most of this mass was lost at an early stage, the surface density of solid material must have been at least 100 times higher than it is today in order to grow bodies the size of Ceres and Vesta (roughly 900 and 500 km in diameter, respectively) in only a few million years.

Several regions of the Asteroid Belt contain clusters of asteroids with similar orbits and similar spectral features, suggesting they are made of the same material. These clusters are fragments from the collisional breakup of larger asteroids. There are relatively few of these asteroid families, which implies that catastrophic collisions are quite rare. This suggests the Asteroid Belt has contained relatively little mass for most of its history. The spectrum of asteroid Vesta, located 2.4 AU from the Sun, shows that it has a basaltic crust. The HED meteorites, which probably come from Vesta, show this crust formed only a few million years after the solar system, according to several isotopic systems. The survival of Vesta's crust suggests that the crust formed the impact rate in the belt has never been much higher than it is today. For these reasons, it is thought that most of the Asteroid Belt's original mass was removed at a very early stage by a dynamical process rather than by collisional erosion.

The Asteroid Belt currently contains a number of **orbital resonances** associated with the giant planets. Resonances occur when either the orbital period or precession period of an asteroid has a simple ratio with the corresponding period for one of the planets. Many resonances induce large changes in orbital eccentricity, causing asteroids to fall into the Sun, or to come close to Jupiter, leading to close encounters and ejection from the solar system. For this reason, there are very few asteroids that orbit the Sun twice every time Jupiter orbits the Sun once, for example. When the nebular gas was still present, small asteroids moving on eccentric orbits would have drifted inward rapidly due to gas drag. After the giant planets had formed, a combination of resonances and gas drag may have transferred most objects smaller than a few hundred kilometers from the Asteroid Belt into the terrestrial-planet region. Larger planetary embryos would not have drifted very far. However, once oligarchic growth ceased, embryos began to gravitationally scatter one another across the belt. Numerical simulations show that most or all of these bodies would eventually enter a resonance and be removed, leaving an Asteroid Belt greatly depleted in mass and containing no objects bigger than Ceres. The timescale for the depletion of the belt depends sensitively on the orbital eccentricities of the giant planets at the time, which are poorly known. The belt may have been cleared in only a few million years, but it may have required as much as several hundred million years if the giant planets had nearly circular orbits.

The albedos and spectral features of asteroids vary widely from one body to another, but clear trends are apparent as one moves across the Asteroid Belt. S-type asteroids, which generally lie in the inner Asteroid Belt, appear to be more thermally processed than the C-type asteroids that dominate the middle belt. These may include the parent bodies of ordinary and carbonaceous chondrites respectively. C-types in turn seem more processed than the P-type asteroids that mostly lie in the outer belt. These differences may reflect differences in the composition of solid materials in different parts of the nebula, or differences in the time at which asteroids formed. Ordinary and enstatite chondrites, which probably come from the inner Asteroid Belt, tend to be dry, while carbonaceous chondrites from the middle and outer belt contain up to 10% water by mass in the form of hydrated minerals. This suggests that temperatures were cold enough in the outer Asteroid Belt for water ice to form

and become incorporated into asteroids where it reacted with dry rock. Temperatures were apparently too high for water ice to condense in the inner Asteroid Belt. It is possible that some of the objects currently in the Asteroid Belt formed elsewhere. For example, it has been proposed that many of the parent bodies of the iron meteorites, and possibly Vesta, formed in the terrestrial-planet region and were later gravitationally scattered outward to their current orbits.

Iron meteorites from the cores of melted asteroids are common, whereas meteorites from the mantles of these asteroids are rarely seen. This suggests that a substantial amount of collisional erosion took place at an early stage, with only the strong, iron-rich cores of many bodies surviving. A number of other meteorites also show signs that their parent asteroids experienced violent collisions early in their history. Chondrites presumably formed somewhat later than the differentiated asteroids, when the main radioactive heat sources had mostly decayed. Chondrites are mostly composed of chondrules, which typically formed 1–3 Ma after CAIs. Chondrite parent bodies cannot be older than the youngest chondrules they contain, so they must have formed several million years after the start of the solar system. For this reason, it appears that the early stages of planet formation were prolonged in the Asteroid Belt. Chondrites have experienced some degree of thermal processing, but their late formation meant that their parent bodies never grew hot enough to melt, which has allowed chondrules, CAIs, and matrix grains to survive.

8. Growth of Gas and Ice Giant Planets

Jupiter and Saturn are mostly composed of hydrogen and helium. These elements do not condense at temperatures and pressures found in protoplanetary disks, so they must have been gravitationally captured from the gaseous component of the solar nebula. Observations of young stars indicate that protoplanetary disks survive for only a few million years, and this sets an upper limit for the amount of time required to form giant planets. Uranus and Neptune also contain significant amounts of hydrogen and helium (somewhere in the range 3–25%), and so they probably also formed quickly, before the solar nebula dispersed.

Jupiter and Saturn also contain elements heavier than helium and they are enriched in these elements compared to the Sun. The gravitational field of Saturn strongly suggests it has a core of dense material at its center, containing roughly one fifth of the planet's total mass. Jupiter may also have a dense core containing a few Earth masses of material. The interior structure of Jupiter remains quite uncertain because we lack adequate equations of state for the behavior of hydrogen at the very high pressures found in the planet's interior. The upper atmospheres of both planets are enriched in elements such as carbon, nitrogen, sulfur, and argon, compared to the Sun. It is thought likely that these enrichments extend deep into the planets' interiors, but this remains uncertain.

Giant planets may form directly by the contraction and collapse of gravitationally unstable regions of a protoplanetary disk. This disk instability is analogous to the gravitational instabilities that may have formed planetesimals, but instead the instability takes place in nebula gas rather than the solid component of the disk. Instabilities will occur if the Toomre stability criterion Q becomes close to or lower than 1, where

$$Q = \frac{M_{sun}c_s}{\Sigma \pi a^2 v_{kep}} \qquad (15)$$

where v_{kep} is the Keplerian velocity, c_s is the sound speed, and Σ is the local surface density of gas in the disk. Gas in an unstable region quickly becomes much denser than the surrounding material. Disk instability requires high surface densities and low sound speeds (cold gas), so it is most likely to occur in the outer regions of a massive protoplanetary disk. Numerical calculations suggest instabilities will occur beyond about 5 AU in a nebula a few times more massive than the minimum-mass solar nebula. What happens to an unstable region depends on how quickly the gas cools as it contracts, and this is the subject of much debate. If the gas remains hot, the dense regions will quickly become sheared out and destroyed by the differential rotation of the disk. If cooling is efficient, simulations show that gravitationally bound clumps will form in a few hundred years, and these may ultimately contract to form giant planets. Initially, such planets would be homogeneous and have the same composition as the nebula. Their structure and composition may change subsequently due to gravitational settling of heavier elements to the center and capture of rocky or icy bodies such as comets.

The evidence for dense cores at the centers of Jupiter and Saturn suggests to many scientists that giant planets form by core accretion rather than disk instability. In this model, the early stages of giant-planet formation mirror the growth of rocky planets, beginning with the formation of planetesimals, followed by runaway and oligarchic growth. However, planetary embryos would have grown larger in the outer solar system for two reasons. First, feeding zones here are larger because the Sun's gravity is weaker, so each embryo gravitationally holds sway over a larger region of the nebula. Second, temperatures here were cold enough for volatile materials such as tars, water ice, and other ices to condense, so more solid material was available to build large embryos.

In the outer solar system, bodies roughly ten times more massive than Earth would have formed via oligarchic growth in a million years, provided the disk was a few times more massive than the minimum-mass solar nebula. Bodies that grew larger than Mars would have captured substantial atmospheres of gas from the nebula. Such atmospheres

remain in equilibrium due to a balance between an embryo's gravity and an outward pressure gradient. However, there is a critical core mass above which an embryo can no longer support a static atmosphere. Above this limit, the atmosphere begins to collapse onto the planet forming a massive gas envelope that increases in mass over time as more gas is captured from the nebula. As gas falls toward the planet, it heats up as gravitational potential energy is released. The rate at which a planet grows depends on how fast this heat can be radiated away. The critical core mass depends on the opacity of the envelope and the rate at which planetesimals collide with the core, but calculations suggest it is in the range 3–20 Earth masses. The growth of the envelope is slow at first, but speeds up rapidly once an embryo reaches 20–30 Earth masses. Numerical simulations show that Jupiter-mass planets can form this way in 1–5 Ma. Such planets are mostly composed of hydrogen-rich nebular gas, but are also enriched in heavier elements due to the presence of a solid core. As with the disk instability, the planet's envelope may be further enriched in heavy elements by collisions with comets.

Measurements by the *Galileo* spacecraft showed that Jupiter's upper atmosphere is enriched in carbon, nitrogen, sulfur, and the noble gases argon, krypton, and xenon by factors of 2–3 compared to the Sun. If these enrichments are typical of Jupiter's envelope as a whole, it suggests the planet captured a huge number of comets. Argon can be trapped in cometary ices but only if these ices form at temperatures below about 30 K. Temperatures at Jupiter's current distance from the Sun were probably quite a lot higher than this. This suggests either that the comets came from colder regions of the nebula or that Jupiter itself migrated inward over a large distance. However, the fact that relatively refractory elements such as sulfur are present in the same enrichment as the noble gases suggests these elements may all have been captured as gases from the nebula along with hydrogen and helium. If so, Jupiter's envelope must be non-homogeneous, with the lower layers depleted in heavy elements, perhaps due to exclusion from high pressure phases of hydrogen, while the upper layers are enriched.

It is unclear why Jupiter and Saturn stopped growing when they reached their current masses. These planets are sufficiently massive that they would continue to grow very rapidly if a supply of gas was available nearby. It is possible, but unlikely, that they stopped growing because the nebula happened to disperse at this point. A more likely explanation is that the growth of these planets slowed because they each became massive enough to clear an annular gap in the nebula around their orbit. Gap clearing happens when a planet's Hill radius becomes comparable to the vertical thickness of the gas disk, which would have been the case for Jupiter and Saturn. Gas orbiting a little further from the Sun than Jupiter would have been sped up by the planet's gravitational pull, moving the gas away from the Sun. Gas orbiting closer to the Sun than Jupiter was slowed down, causing it to move inward. These forces open up a gap in the disk around Jupiter's orbit, balancing viscous forces that would cause gas to flow back into the gap. Numerical simulations show that generally gaps are not cleared completely, and some gas continues to cross a gap and accrete onto a planet. However, the accretion rate declines as a planet becomes more massive.

Uranus and Neptune are referred to as ice giant planets because they contain large amounts of materials such as water and methane that form ices at low temperatures. They contain some hydrogen and helium, but they did not acquire the huge gaseous envelopes that Jupiter and Saturn possess. This suggests the nebula gas had largely dispersed in the region where Uranus and Neptune were forming before they became massive enough to undergo rapid gas accretion. This may be because they formed in the outer regions of the protoplanetary disk, where embryo growth rates were slowest. It is also possible that the nebula dispersed more quickly in some regions than others. In particular, the outer regions of the nebula may have disappeared at an early stage as the gas escaped the solar system due to photoevaporation by ultraviolet radiation.

The presence of a gap modifies planetary migration. Planets massive enough to open a gap still generate spiral density waves in the gas beyond the gap, but these waves are located further away from the planet as a result, so migration is slower. As a planet with a gap migrates inward, gas tends to pile up at the inner edge of the gap and become rarified at the outer edge, slowing migration as a result. The migration of the planet now becomes tied to the inward viscous accretion of the gas toward the star. The planet, its gap, and the nebular gas all move inward at the same rate, given by

$$\frac{da}{dt} = -1.5\alpha \left(\frac{c_s}{v_{kep}} \right)^2 v_{kep} \qquad (16)$$

where $\alpha = \nu v_{kep}/(a c_s^2)$ and ν is the viscosity of the nebular gas. This is called **type-II migration**. Type-II migration slows when a planet's mass becomes comparable to that of the nebula, and migration ceases as the nebular gas disperses.

Giant planets in the solar system experienced another kind of migration as they interacted gravitationally with planetesimals moving on orbits between the giant planets and in the primordial Kuiper Belt. One consequence of this process was the formation of the Oort cloud of comets. Once Jupiter approached its current mass, many planetesimals that came close to the planet would have been flung far beyond the outer edge of the protoplanetary disk. Some were ejected from the solar system altogether, but others remained weakly bound to the Sun. Over time, gravitational interactions with molecular clouds, other nearby stars, and the galactic disk circularized the orbits of these objects so they no longer passed through the planetary system. Many of these objects are still present orbiting far from the Sun in the Oort cloud. The ultimate source of angular momentum for these objects came at the expense of Jupiter's orbit,

which shrank accordingly. Saturn, Uranus, and Neptune ejected some planetesimals, but they also perturbed inward many objects, which were then ejected by Jupiter. As a result, Saturn, Uranus, and Neptune probably moved outward rather than inward.

As Neptune migrated outward, it interacted dynamically with the primordial Kuiper Belt of comets orbiting in the very outer region of the nebula. Some of these comets were ejected from the solar system or perturbed inward toward Jupiter. Others were perturbed onto highly eccentric orbits with periods of hundreds or thousands of years, and now form the scattered disk, a region that extends out beyond the Kuiper Belt but whose objects are gradually being removed by close encounters with Neptune. A sizable fraction of the objects in the region beyond Neptune were trapped in external mean-motion resonances and migrated outward with the planet. Pluto, currently located in the 3:2 mean-motion resonance with Neptune, probably represents one of these objects.

As the giant planets migrated, it is possible that they passed through orbital resonances with one another. In particular, if Jupiter and Saturn passed through the 2:1 mean-motion resonance, their orbital eccentricities would have increased significantly, with important consequences throughout the solar system. The eccentricities of Uranus and Neptune would have briefly become large until they were damped by dynamical friction with the primordial Kuiper Belt. Many comets would have been perturbed into the inner solar system as a result. In addition, the changing orbits of the giant planets would have perturbed many main-belt asteroids into unstable resonances, also leading to a flux of asteroids into orbits crossing the inner planets. Currently, it is unclear whether Jupiter and Saturn passed through the 2:1 resonance, or when this may have happened. It has been proposed that passage through this resonance was responsible for the late heavy bombardment of the inner planets, which occurred 600–700 Ma after the start of the solar system and left a clear record of impacts on the Moon, Mars, and Mercury.

9. Planetary Satellites

Earth's moon possesses a number of unusual features. It has a low density compared to the inner planets, and it has only a very small core. The Moon is highly depleted in volatile materials such as water. In addition, the Earth–Moon system has a large amount of angular momentum per unit mass. If they were combined into a single body, the object would rotate once every 4 hours! All these features can be understood if the Moon formed as the result of an oblique impact between Earth and another large, differentiated body, sometimes referred to as Theia, late in Earth's formation.

Numerical simulations of this giant impact show that much of Theia's core would have sunk through Earth's mantle to coalesce with Earth's core. Molten and vaporized mantle material from both bodies was ejected outward. Gravitational torques from the highly nonspherical distribution of matter during the collision gave some of this mantle material enough angular momentum to go into orbit about Earth. This material quickly formed into a disk, from which the Moon accreted. Certain features of the Moon's composition are very similar to those of the Earth, which means that either ((1)) Theia was formed from similar material, (2) the resulting vapor and debris that condensed to form the Moon totally equilibrated with the outer portions of the Earth, or (3) the Moon is mostly composed of material from Earth rather than Theia, although numerical simulations tend to find that the opposite is true in this case.

The impact released huge amounts of energy, heating the disk sufficiently that many volatile materials escaped. As a result, the Moon formed mostly from volatile-depleted mantle materials, explaining its current composition. The simulations suggest Theia probably had a mass similar to Mars, which has roughly one tenth the mass of Earth. We know little about Theia's composition except that, like Mars, it seems to have been rich in geochemical volatile elements such as rubidium compared to Earth (Fig. 9). The Earth and the Moon have identical oxygen isotope characteristics (Fig. 10). It was once thought that this meant Earth and Theia had a similar isotopic composition, but this similarity now appears to be the result of exchange of material between the Earth and the protolunar disk while the Moon was forming.

The satellites of the giant planets are much smaller relative to their parent planet than the Moon is compared to the Earth. Whereas the Moon is roughly 1/80 of the mass of the Earth, the satellite systems of Jupiter, Saturn, and Uranus each contain about 1/10,000 of the mass of their respective planet. The satellites of the giant planets can be divided into two classes with different properties. Those close to their parent planet tend to have nearly circular orbits in the same plane as the planet's equator and orbiting in the same direction as the planet spins. These are referred to as regular satellites. Satellites orbiting further from the planet tend to have highly inclined and eccentric orbits, and these are called irregular satellites as a result. The regular satellites tend to be larger and include the Galilean satellites of Jupiter and Saturn's largest satellite Titan.

The orbits of the regular satellites suggest they formed from gas-rich circumplanetary disks orbiting each planet, while the irregular satellites are thought to have been captured later. Large satellites would have moved rapidly inward through a circumplanetary disk due to type-I migration, on a timescale that was short compared to the lifetime of the solar nebula. For this reason, it is likely that multiple generations of satellites formed, with the satellites we see today being the last to form. The satellites probably formed from planetesimals originating in the solar nebula that were slowed and captured when they passed through the relatively dense gas in the circumplanetary disk.

Orbital resonances involving two or more satellites are common. For example, the inner three Galilean satellites—Io, Europa, and Ganymede—have orbital periods in the ratio 1:2:4. This contrasts with the absence of resonances between the planets except for Neptune and Pluto. The ubiquity of satellite resonances suggests many of the satellites migrated considerable distances during or after their formation, becoming captured in a resonance en route. Some resonances may have arisen as the growing satellites migrated inward through their planet's accretion disk. Others could have arisen later as tidal interactions between a planet and its satellite caused the satellites to move outward at different rates.

The Neptunian satellite system is different from those of the other giant planets, having relatively few moons with most mass contained in a single large satellite Triton, which is larger than Pluto. Triton is unusual in that its orbit is retrograde, unlike all the other large satellites in the solar system. This suggests it was captured rather than forming in situ. Several capture mechanisms have been proposed, but most are low-probability events, which makes them unlikely to explain the origin of Triton. A more plausible idea is that Triton was once part of a binary planet like the Pluto–Charon system, orbiting around the Sun. During a close encounter with Neptune, the binary components were parted. Triton's companion remained in orbit about the Sun, taking with it enough kinetic energy to leave Triton in a bound orbit about Neptune. Triton's orbit would have been highly eccentric initially, but tidal interactions with Neptune caused its orbit to shrink and become more circular over time. As Triton's orbit shrank, it would have disturbed the orbits of smaller satellites orbiting Neptune, leading to their destruction by mutual collisions. This is presumably the reason for the paucity of regular satellites orbiting Neptune today.

10. Extrasolar Planets

At the time of writing, about 200 planets are known orbiting stars other than the Sun. These are referred to as extrasolar planets or exoplanets. Most of these objects have been found using the Doppler radial velocity technique. This makes use of the fact that the gravitational pull of a planet causes its star to move in an ellipse with the same period as the orbital period of the planet. As the star moves toward and away from the observer, lines in its spectra are alternately blue- and red-shifted by the Doppler effect, indicating the planet's presence. Current levels of precision allow the detection of gas giant planets and also ice giants in some cases, but not Earth-mass planets. The planet's orbital period P can be readily identified from the radial velocity variation. The mean radius of the planet's orbit a can then be found using Kepler's third law if the star's mass M_* is known:

$$a^3 = \frac{P^2 G M_*}{4\pi^2} \qquad (17)$$

Unfortunately, the Doppler method determines only one component of the star's velocity, so the orientation of the orbital plane is not known in general. This means one can obtain only a lower limit on the planet's mass. For randomly oriented orbits however, the true mass of the planet is most likely to lie within 30% of its minimum value.

Some extrasolar planets have been detected when they transit across the face of their star, typically causing the star to dim by 1–2% for a few hours. Only a small fraction of extrasolar planets generate a transit since their orbital plane must be almost edge on as seen from the Earth. When a planet is observed using both the Doppler and transit methods, its true mass can be obtained since the orientation of the orbital plane is known. If the stellar radius is also known, the degree of dimming yields the planet's radius and hence its density. The densities of extrasolar planets observed this way are generally comparable to that of Jupiter and substantially lower than that of Earth. This suggests these planets are composed mainly of gas rather than rock or ice. In one case, hydrogen has been detected escaping from an extrasolar planet. A few objects have been found whose minimum masses are below 15 Earth masses, and it is plausible that these are more akin to ice giants or even terrestrial planets than gas giants.

Stars with known extrasolar planets tend to have high metallicities; that is, they are enriched in elements heavier than helium compared to most stars in the Sun's neighborhood (Fig. 21). (The Sun also has a high metallicity.) The meaning of this correlation is hotly debated, but it is consistent with the formation of giant planets via core accretion (see Section 8). When a star has a high metallicity, its disk will contain large amounts of the elements needed to form a solid core, promoting rapid growth and increasing the likelihood that a gas giant can form before the gas disk disperses.

Both the Doppler velocity and transit techniques are biased toward finding massive planets since these generate a stronger signal. Both are also biased toward detecting planets lying close to their star. In the case of transits, the probability of suitable orbital alignment declines with increasing orbital distance, while for the Doppler velocity method, one generally needs to observe a planet for at least a full orbital period to obtain a firm detection. Despite these biases, it is clear that at least 10% of Sun-like stars have planets, and this fraction may be much higher. The fraction of planets with a given mass increases as the planetary mass grows smaller, despite the strong observational bias working in the opposite direction. Roughly 10% of known extrasolar planets have orbital periods of only a few days, which implies their orbits are several times smaller than Mercury's orbit about the Sun. These planets are often referred to as hot Jupiters due to their likely high temperatures. Theoretical models of planet formation suggest it is unlikely that planets will form this close to a star. Instead, it is thought that these planets formed at larger distances and moved inward due to type-I and/or type-II migration. Alternatively, they may have been scattered onto highly eccentric orbits following close encounters with other planets in the same system. In this case, subsequent tidal interactions with the star will circularize a planet's orbit and cause the orbit to shrink.

Roughly 20 stars are known to have two or more planets. In a sizable fraction of these cases, the planets are involved in orbital resonances where either the ratio of the orbital periods or precession periods of two planets is close to the ratio of two integers, such as 2:1. This state of affairs has a low probability of occurring by chance, which suggests these planets have been captured into a resonance when the orbits of one or both planets migrated inwards.

11. Summary and Future Prospects

Thanks to improvements in isotopic chronology, we now know the timescales over which the Earth, Moon, Mars, and some asteroids formed. Terrestrial-planet accretion started soon after the solar system formed, leading to the growth of some Mars-sized and smaller objects within the first few million, and in some cases only a few hundred thousand, years. This early accretionary phase was accompanied by widespread melting due to heat generated by short-lived isotopes and the formation of planetary cores. The Moon formed relatively late, 30–55 Ma after the start of the solar system, with the most likely date being 40–50 Ma. This was the last major event in Earth's formation. These isotopic timescales are consistent with theoretical models that predict rapid runaway and oligarchic growth at early times, to form asteroid-to-Mars-sized bodies within a million years, while predicting that Earth took tens of millions of years to grow to its final size.

The presence in Earth's mantle of nonnegligible amounts of siderophile elements such as platinum and osmium argues that roughly 1% of Earth's mass arrived after its core

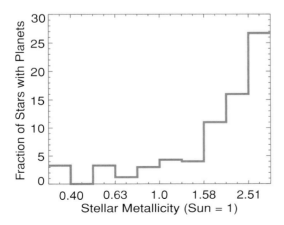

FIGURE 21 The fraction of stars that have planets as a function of the stellar metallicity (the abundance of elements heavier than helium compared to the Sun). Here the iron-to-hydrogen ratio relative to the Sun is used a proxy for metallicity.

had finished forming. For some time it has been postulated that Earth formed in a very dry environment and that its water was delivered along with these siderophile elements in a late veneer. This now appears unlikely given the composition of Earth's mantle. Instead, Earth probably acquired its water earlier, perhaps from carbonaceous–chondrite-like asteroids, before core formation was complete. This implies that the planet somehow held onto much of its water during the giant impact that led to the formation of the Moon.

It now seems that chondrites, the most primitive meteorites in our collection both physically and chemically, actually formed at a rather late stage, long after the parent bodies of the iron meteorites had formed. Chondrites escaped melting because the potent heat sources ^{26}Al and ^{60}Fe had largely decayed by that point. For a long time, it has been thought that chondrites, or something similar, provided the basic building blocks of Earth and the other terrestrial planets, but it now seems that the parent bodies of the iron meteorites provide a better analog in this respect. Currently, we do not have good dynamical or cosmochemical models for how chondrites and their constituents formed. Chondrules, CAIs, matrix grains, and presolar grains all survived in the nebula for several million years, undergoing different degrees of thermal processing, and then were collected together into large bodies. The refractory CAIs may have formed close to the Sun prior to being scattered across the disk, perhaps by an x-wind. Supporting evidence for this hypothesis comes from the recent discovery of high-temperature condensates in samples from comet Wild 2 returned by the *Stardust* mission. Where chondrules formed remains unclear, but these objects would have been highly mobile as long as nebular gas was present, and they may have drifted radially over large distances.

The origin of giant planets remains a subject of debate, but the observed correlation between stellar metallicity and the presence of giant planets, and the recent discovery of a Saturn-mass extrasolar planet that appears to have a very massive core, lend weight to the core accretion model. Recent simulations using plausible envelope opacities have found that giant planets can form within the typical lifetime of a protoplanetary disk, overcoming a longstanding obstacle for core accretion. It is becoming apparent that planetary migration is an important feature in the formation and early evolution of planetary systems. This presumably explains the fact that extrasolar planets are seen to orbit their stars at a wide range of distances. Planets also migrate when they clear away residual planetesimals. This may have led to a dramatic episode early in the history of the solar system associated with the late heavy bombardment of comets and asteroids onto the Moon and inner planets.

It is impressive to look back on the past 10 years of discovery in planetary science partly because the breakthroughs have involved so many diverse areas of research. Technology has been a key driver, be it in the form of more powerful computers, mass spectrometers, instrumentation for planetary missions, or new telescopes and detectors. The near future looks equally exciting. The Atacama Large Millimeter Array (ALMA) promises to transform our knowledge of protoplanetary disks with very high spatial resolution able to observe features as small as 1 AU in size and sufficient sensitivity to detect many new molecules including organic materials. Space missions will continue to expand our survey of the solar system, with the *Messenger* and *New Horizons* probes en route to Mercury and Pluto, and the *Rosetta* spacecraft heading for comet Churyumov–Gerasimenko. In addition to NASA and ESA, space agencies in Japan, China, and India are also becoming active players in space exploration. The Doppler radial velocity and transit techniques continue to be refined and are set to expand the catalogue of known extrasolar planets. The relatively new micro-lensing technique is opening up the possibility of finding Earth-mass planets. Within the next few years, the *Kepler* and *COROT* space missions should finally answer the question of whether Earth-sized planets are common or relatively rare. Here on Earth, continuing analysis of dust samples from comet Wild 2 returned by the *Stardust* mission, and solar wind samples from the *Genesis* mission, will enhance our understanding of the cosmochemical evolution of the solar system. New isotopic measurement techniques and a new generation of nanosims ion probes are sure to generate exciting discoveries at a rapid pace. All in all, we have much to look forward to.

Bibliography

de Pater, I., and Lissauer, J. J. (2001). "Planetary Sciences." Cambridge Univ. Press, New York, NY.

Lewis, J. S. (2004). "Physics and Chemistry of the Solar System." Academic Press, San Diego, CA.

Reipurth, B., Jewitt, D., and Keil, K. (2006). "Protostars and Planets V." Univ. Arizona Press, Tucson.

A History of Solar System Studies

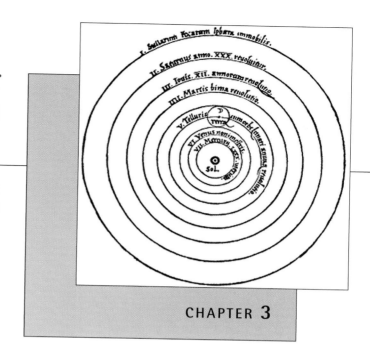

David Leverington

BAE Systems, England (retired)

CHAPTER 3

This chapter gives a brief overview of the history of solar system research from the earliest times up to the start of the space age.

1. Babylonians and Greeks

Many early civilizations studied the heavens, but it was the Babylonians of the first millennium B.C. who first used mathematics to try to predict the positions of the Sun, Moon, and visible planets (Mercury, Venus, Mars, Jupiter, and Saturn) in the sky. In this they differed from the Greeks, as the Babylonians were priests trying to predict the movement of the heavenly bodies for religious purposes, whereas the Greeks were philosophers trying to understand why they moved in the way they did. The Babylonians were fascinated by numbers, whereas the Greeks were more interested in geometrical figures.

The accuracy of the Babylonian predictions in the 2nd century B.C. is remarkable. For example, their estimate of the length of the **sidereal** year was within 6 minutes of its true value, and that of the average **anomalistic month** was within 3 seconds. In addition, Jupiter's sidereal and **synodic periods** were within 0.01% of their correct values.

Pythagoras (c. 580–500 B.C.) was a highly influential early Greek philosopher who set up a school of philosophers, now known as the Pythagoreans. None of Pythagoras' original writings survive, but later evidence suggests that the Pythagoreans were probably the first to believe that the Earth is spherical, and that the planets all move in separate orbits inclined to the celestial equator. But the Pythagorean spherical Earth did not spin and was surrounded by a series of concentric, crystalline spheres supporting the Sun, Moon, and individual planets. Each had its own sphere, which revolved around the Earth at different speeds, producing a musical sound, the "music of the spheres," as they went past each other.

Hicetus of Syracuse (fl. 5th century B.C.) was the first person to specifically suggest that the Earth spun on its axis, at the center of the universe. This model was further developed by Heracleides who proposed that Mercury and Venus orbited the Sun as it orbited the Earth. Then Aristarchus (c. 310–230 B.C.), who was one of the last of the Pythagoreans, went one step further and proposed a heliocentric (i.e., Sun-centered) universe in which the planets orbit the Sun in the (correct) order of Mercury, Venus, Earth, Mars, Jupiter and Saturn, with the Moon orbiting a spinning Earth. This was 1700 years before Copernicus came up with the same idea. Aristarchus was also the first to produce a realistic estimate for the Earth–Moon distance, although his estimate of the Earth–Sun distance was an order of magnitude too low.

While the Pythagoreans were developing their ideas, Plato (c. 427–347 B.C.) was developing a completely different school of thought. Plato, who was a highly respected philosopher, was not too successful with his geocentric (i.e., Earth-centered) model of the universe. His main legacy to astronomy was his teaching that all heavenly bodies must be spherical, as that is the perfect shape, and that they must move in uniform circular orbits, for the same reason. Aristotle (384–322 B.C.), a follower of Plato, was one of the greatest of Greek philosophers. His ideas were to hold sway in Europe until well into the Middle Ages. However, his geocentric model of the universe was highly complex, requiring a total of 56 spheres to explain the motions of the Sun, Moon, and planets. Unfortunately, many of its predictions were wrong, and it soon fell into disuse.

Hipparchus (c. 185–120 B.C.), who was the first person to quantify the **precession of the equinoxes,** was aware that the Sun's velocity along the ecliptic was not linear. This was known to the Babylonians and to Callippus of Cyzicus, but they did not seek an explanation. Hipparchus, on the other hand, in adopting Plato's philosophy of uniform circular motion in a geocentric universe, realized that this phenomenon could only be explained if the Sun was orbiting an off-center Earth. However, his estimate of the off-center amount was far too large, although his **apogee** position was in error by only 35′.

The mathematician Apollonius of Perga (c. 265–190 B.C.) appears to have been the first to examine the properties of epicycles. These were later adopted by Ptolemy (c. A.D. 100–170) in his geocentric model of the universe. In Ptolemy's scheme (Fig. 1), the Moon, Sun, and planets each describe a circular orbit called an epicycle, the center of which goes in a circle, called a deferent, around a nonspinning Earth. Because the inferior planets, Mercury and Venus, each appear almost symmetrically on both sides of the Sun at maximum **elongation**, he assumed that the centers of their epicycles were always on a line joining the Earth and Sun. For the superior planets he assumed that the lines linking these with the center of their epicycles were always parallel to the Earth–Sun line. Unfortunately, this simple system did not provide accurate enough position estimates, and so Ptolemy introduced a number of modifications. In the case of the Moon, he made the center of the Moon's deferent describe a circle whose center was the Earth. For the planets he introduced the concept of an equant, which was a point in space equidistant with the Earth from the center of the deferent (Fig. 2). The equant was the point about which the planet's angular velocity appeared to be uniform. Other modifications were also required, but by the time he had finished, he was able to make accurate position estimates for all but the Moon and Mercury. In addition, assuming that there were no gaps between the furthest part of one epicycle and the nearest part of the next, he was able to produce an estimate for the size of the solar system of about 20,000 times the radius of the Earth (or about 120 million km). Although this was a gross underestimate, it gave, for the first time, an idea of how large the solar system really was.

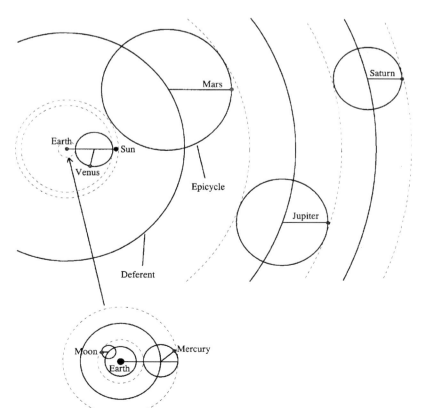

FIGURE 1 Ptolemy's model of the universe in which all bodies, except the Sun (and stars), describe epicycles, the centers of which orbit the Earth in deferents. He assumed that there were no gaps between the circle enclosing the furthest distance of one planet, and that just touching the epicycle of the next planet out from the Earth.

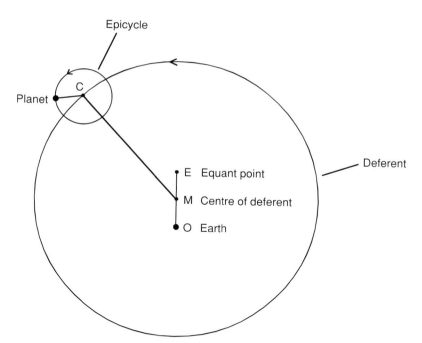

Epicycle

Planet

C

E Equant point

M Centre of deferent

O Earth

Deferent

FIGURE 2 Ptolemy modified his epicycle theory for the superior planets by moving the Earth *O* from the center *M* of the deferent, and by defining an equant point *E* such that the distance $EM = MO$. He then assumed that the angular velocity of *C*, the center of the epicycle, is uniform about the equant point *E*, rather than about the center *M* of the deferent.

2. Copernicus and Tycho

There was virtually no progress in astronomy over the next one thousand years, and during this time many of the Greek texts had been lost in Europe. But in the 12th century Arab translations found their way to Europe, mainly via Islamic Spain. Then in the 14th century Ibn al-Shātir (1304–1375), working in Damascus, improved Ptolemy's model by modifying his epicycles and deleting his equant. Interestingly, al-Shātir's system was very much like Copernicus' later system, but with the Earth, not the Sun, at the center.

Copernicus' heliocentric theory of the universe (Fig. 3) was published in his *De Revolutionibus Orbium Caelestium* in 1543, the year of his death. Interestingly, in the light of Galileo's later problems with the Church, the book was well received. This is probably because of the Foreword, which had been written by the theologian Andreas Osiander and explained that the book described a mathematical model of the universe, rather than the universe itself.

Copernicus (1473–1543) acknowledged that his idea of a spinning Earth in a heliocentric universe was not new, having been proposed by Aristarchus. In addition, Copernicus' theory was based on circular motion and still depended on epicycles, although he deleted the equant. But he had resurrected the heliocentric theory, which had not been seriously considered for almost two thousand years, at the height of the Renaissance, which was eager for new ideas.

In the Middle Ages, Aristotle's ideas were taught at all the European universities. But now Copernicus had broken with the Aristotelian concept of a nonspinning Earth at the center of the universe. Then in 1577 Tycho Brahe (1546–

1601) disproved another of Aristotle's ideas. Aristotle had believed that comets are in the Earth's atmosphere, but Tycho was unable to measure any clear parallax for the comet of that year. Finally, Tycho, in his book of 1588, rejected another of Aristotle's ideas, that the heavenly bodies are carried in their orbits on crystalline spheres. This is because,

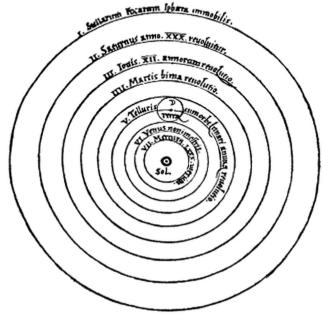

FIGURE 3 Copernicus' heliocentric universe, as described in his *De Revolutionibus*, in which the planets orbit the Sun (Sol) and the Moon orbits the Earth (Terra).

in Tycho's new model of the universe, all the planets, except the Earth, orbit the Sun as the Sun orbits the Earth. This meant that the sphere that carried Mars around the Sun would intercept that which carried the Sun around the Earth, which was clearly impossible if they were crystalline.

3. Kepler and Galileo

Johannes Kepler (1571–1630) looked at the universe in an entirely different way than his predecessors. The Babylonians had examined it arithmetically, and the Greeks and later astronomers had considered it in geometrical terms. Kepler, on the other hand, tried to understand the structure of the solar system by considering physical forces.

Kepler conceived of a force emanating from the Sun that pushed the planets around their orbit of the Sun such that planetary movement would stop if the force stopped. The magnitude of his force, and hence the linear velocity of the planets, decreased linearly with distance. This should have resulted in the period of the planets varying as their distance squared, but Kepler made a mathematical error and came up with another relationship. Fortuitously, however, his analysis produced remarkably accurate results.

Although Kepler was having some success with this and other theories, he thought he could improve them if he had access to Tycho Brahe's accurate observational data. So Kepler went to see Tycho; a visit that ended with him joining Tycho and eventually succeeding him after his death.

Tycho had initially asked Kepler to analyze Mars' orbit, a task that he continued well after Tycho's death. Kepler published his results in 1609 in his book *Astronomia Nova*, in which he reintroduced the equant, previously deleted by Copernicus. In Kepler's model, all the planets orbited the Sun in a circle, with the Sun off-center, but he could not find a suitable circle to match Mars' observations, even with an equant. So he decided to reexamine the Earth's orbit, as the Earth was the platform from which the observations had been made.

Copernicus had proposed that the Earth moved around the Sun in a circle at a uniform speed, with the Sun off-center. So there had been no need for an equant. But Kepler found that an equant was required to explain the Earth's orbit. However, even adding this, he could not fit a circle, or even a flattened circle to Mars' orbit. And so in desperation he tried an ellipse, with the Sun at one focus, and, much to his surprise, it worked.

Kepler now considered what type of force was driving the planets in their orbits, and concluded that the basic circular motion was produced by vortices generated by a rotating Sun. Magnetic forces then made the orbits elliptical. So Kepler thought that the Sun rotated on its axis, and that the planets and Sun were magnetic.

Initially, Kepler had only shown that Mars moved in an ellipse, but in his *Epitome* of 1618–1621 he showed that this was the case for all the planets, as well as the Moon

and the satellites of Jupiter. He also stated what we now know as his third law, that the square of the periods of the planets are proportional to the cubes of their mean distances from the Sun. Finally, in his *Rudolphine Tables*, he listed detailed predictions for planetary positions and predicted the transits of Mercury and Venus across the Sun's disc.

Galileo Galilei (1564–1642) made his first telescopes in 1609 and started his first telescopic observations of the Moon in November of that year. He noticed that the **terminator** had a very irregular shape and concluded that this was because the Moon had mountains and valleys. It was quite unlike the pure spherical body of Aristotle's cosmology.

Galileo undertook a series of observations of Jupiter in January 1610 and found that it had four moons that changed their positions from night to night (Fig. 4). Galileo presented his early Moon and Jupiter observations in his *Sidereus Nuncius* published in March 1610. By 1612, he had determined the periods of Jupiter's moons to within a few minutes.

Galileo's *Sidereus Nuncius* created quite a stir, with many people suggesting that Galileo's images of Jupiter's moons were an illusion. Kepler, who was in communication with Galileo, first saw the moons himself in August 1610 and supported Galileo against his doubters. The month before, Galileo had also seen what he took to be two moons on either side of Saturn, but for some reason they did not move. Finally in late 1610 he observed the phases of Venus, finally proving that Ptolemy's structure of the solar system was incorrect. As a result, Galileo settled on the Copernican heliocentric system.

Sunspots had been seen from time to time in antiquity, but most people took them to be something between the

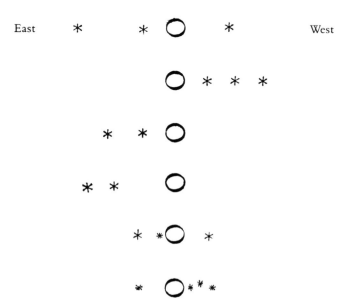

FIGURE 4 Galileo's observations of the moons of Jupiter on consecutive nights from 7 to 13 January (excluding 9 January) 1610, as shown in his book *Sidereus Nuncius*.

Earth and Sun. Although Thomas Harriot and Galileo had both seen sunspots telescopically in 1610, it was Johann Fabricius who first published his results in June 1611. He concluded that they were on the surface of the Sun, and that their movement indicated that the Sun was rotating. This was completely against Aristotle's teachings that the Sun was a perfect body.

In the meantime, Galileo had visited the Jesuits of the Roman College to get their support for his work and, in particular, their support for Copernicus' heliocentric cosmology. His reception was very warm, and he was even received in audience by the pope. But, although the Roman Catholic Church did not argue with his observations, outlined above, there was considerable unease at his interpretation. Initially, the Church was prepared to tolerate Galileo's support of the Copernican cosmology, provided he presented this cosmology as a working hypothesis, rather than as a universal truth. But Galileo was stubborn and tried to take on the Church in its interpretation of theology. In this he could not win, of course, and the Church put him on trial, where he was treated very well. Nevertheless, he was forced in 1633 to recant his views and was then placed under house arrest for the remaining nine years of his life.

4. Second Half of the 17th Century

4.1 The Moon

Thomas Harriot (1560–1621) was the first astronomer to record what we now know as the libration in latitude of the Moon, which has a period of one month. This occurs because the Moon's spin axis is not perpendicular to its orbit. A little later Galileo detected a libration in longitude, which he thought had a period of one day. In fact, it has a period of one month and is caused by the eccentricity of the Moon's orbit.

Although Galileo thought that the Moon has an atmosphere, he concluded that there was very little water on the surface as there were no clouds. His early telescopes were not sufficiently powerful, however, to show much surface detail. But over the next few decades, maps of the Moon were produced by a number of astronomers. The most definitive of which were published in 1647 by Johannes Hevelius (1611–1687). They were the first to show the effect of libration.

By midcentury, it was clear that there were numerous craters on the Moon, and in 1665 Robert Hooke (1635–1703) speculated on their cause in his *Micrographia*. He undertook laboratory-like experiments and noted that if round objects were dropped into a mixture of clay and water, features that resemble lunar craters were produced. But he could not think of the source of large objects hitting the Moon. However, he also found that he could produce crater-like features if he boiled dry alabaster powder in a container. As a result, he concluded that lunar craters are produced by the collapsed blisters of warm viscous lava.

4.2 Saturn

Christiaan Huygens (1629–1695) and his brother Constantyn finished building a state-of-the-art telescope in early 1655. Shortly afterwards Christiaan discovered Saturn's first Moon, Titan, which he announced in his *De Saturni* of 1656. The next four moons of Saturn were discovered by Gian Domenico Cassini (1625–1712); Iapetus in 1671, Rhea in 1672, and both Tethys and Dione in 1684.

Huygens had also mentioned in *De Saturni* that he had solved the problem of Saturn's two "moons" observed by Galileo. In fact, the behavior of these moons had been very odd, as they had both completely disappeared in November 1612, reappearing again in mid 1613. Since then, their shape had gradually changed. In 1650, Francesco Grimaldi discovered Saturn's polar flattening, but still the behavior of the moons, then called ansae, was unexplained. Finally, Huygens announced, in his *Systema Saturnium* of 1659, that the ansae were actually a thin, flat, solid ring, which was inclined to the ecliptic, and so changed its appearance with time. Then in 1675 Cassini noticed that Saturn's ring was divided in two by a dark line, now called the Cassini Division, going all the way around the planet. Cassini speculated that the two rings were not solid but composed of swarms of small satellites.

Other major observational discoveries of this period are listed in Table 1.

4.3 Newton

Kepler had thought that the planets were being pushed around their orbits by a vortex emanating from the Sun but attributed the tides on Earth to the combined attraction of the Sun and Moon by a gravitational force. It seems strange to us that he did not think of this attractive force as having some effect on the orbits of the planets.

René Descartes (1596–1650) also developed a vortex theory to explain the motion of the planets. In his theory, the vortices are in the ether, which is a frictionless fluid filling the universe. In his *Principia* of 1644, Descartes stated that each planet had two "tendencies": one tangential to its orbit and one away from the orbit's center. It is the pressure in the vortex that counterbalances the latter and keeps the planet in its orbit.

In 1664, Isaac Newton (1642–1727) started to consider the motion of a body in a circle. In the following year, he proved that the force acting radially on such a body is proportional to its mass multiplied by its velocity squared, and divided by the radius of the circle (i.e., mv^2/r). From this, he was able to prove that the force on a planet moving in a circular orbit is inversely proportional to the square of its distance from the center. Newton realized that this outward centrifugal force on a planet must be counterbalanced by an equal and opposite centripetal force, but it was not obvious at that time that this force was gravity.

TABLE 1	Key Solar System Discoveries and Observations, 1630–1700

Sun-Earth distance
| 1672 | Richer, Cassini, and Picard deduce a solar parallax of 9.5 minutes of arc from observations of the parallax of Mars. John Flamsteed independently deduces a similar value. This implied a Sun-Earth distance of about 22,000 earth radii, or 140 million km. |

Moon
| | See main text |

Mercury
| 1631 | First observation of a transit of Mercury by Gassendi, Remus, and Cysat—all independently. It occurred on the date predicted by Kepler. |
| 1639 | Phases of Mercury first observed by Zupus. |

Venus
1639	First observation of a transit of Venus by Horrocks and Crabtree.
1646	Fontana observes that Venus' terminator is uneven, attributing the cause to high mountains. (This is now known to be incorrect; Venus is covered in dense clouds.)
1667	Cassini deduces a rotation period of about 24 hours. (This is now known to be incorrect).

Mars
| 1659 | Huygens observes Syrtis Major and deduces a planetary rotation period of about 24 hours. |
| 1672 | Huygens first unambiguously records the south polar cap. |

Jupiter
c. 1630	Fontana, Torricelli, and Zucchi independently observe the main belts.
1643	Riccioli observes the shadows of the Galilean satellites on Jupiter's disc.
1663	Cassini deduces a Jupiter rotation period of 9 h 56 min.
1665	Cassini observes a prominent spot that may be an early appearance of the Great Red Spot.
1690	Cassini observes the differential rotation of Jupiter.
1691	Cassini observes Jupiter's polar flattening, which he estimates to be about 7%.

Saturn
| | See main text |

At this time, it was known that gravity acted on objects on the Earth's surface, but it was not known how far from Earth gravity extended. To get a better understanding of this, Newton devised his so-called Moon test. In this test, he compared the force acting on the Moon, because of its motion in a circle, with the force of the Earth's gravity at the Moon's orbit and found that they were not the same. The difference was not large, but it was sufficient to cause Newton to stop work on gravity. In fact, at that time, Newton appears to have thought that the centripetal force was a mixture of the gravitational force and the force created by vortices in the ether, so he may not have been too surprised by his result.

Newton was finally prompted to return to the subject of gravity by an exchange of letters with Robert Hooke in 1679. In the following year, Newton proved that, assuming an inverse square law of attraction, planets and moons will orbit a central body in an ellipse, with the central body at one focus. Then in 1684 he finally rejected the idea of etherial vortices and started to develop his theory of universal gravitation.

It was during this period that the comet of 1680 appeared. At that time, most astronomers, including New-ton, believed that comets described rectilinear orbits. John Flamsteed (1646–1719), on the other hand, believed that comets described closed orbits, and he suggested, in a letter to Edmond Halley (1656–1742), that the 1680 comet had passed in front of the Sun. Newton, who had been sent a copy of this letter, thought, like a number of astronomers, that there had been two comets, one approaching the Sun and one retreating. Further communications between Flamsteed and Newton in 1681 did not resolve their disagreements, causing Newton to drop the subject of cometary orbits. Eventually, Newton returned to the subject, and by 1686 he had changed his position entirely, as he proved that cometary orbits are highly elliptical or parabolic, to a first approximation. So the 1680 comet had been one comet after all. Newton now felt, having solved the problem of cometary orbits, that he could complete his *Principia*, which was published in 1687.

Newton developed his universal theory of gravitation in his *Principia*, which ran to three editions. For example, he used Venus to "weigh" the Sun, and planetary moons to weight their parent planets, and by the third edition he had deduced the masses and densities for the Earth, Jupiter, and Saturn relative to the Sun (Table 2).

TABLE 2	A Comparison of Newton's Results (Relative to the Sun) with Modern Values			
	Mass		Density	
	Principia	Modern Value	Principia	Modern Value
Sun	1	1	100	100
Earth	1/169,282	1/332,980	400	392
Jupiter	1/1,067	1/1,047	94.5	94.2
Saturn	1/3,021	1/3,498	67	49

Newton realized that if gravity was really universal, then not only would the Sun's gravity affect the orbit of a planet, and the planet's gravity affect the orbit of its moons, but the Sun would also affect the orbits of the moons, and one planet would affect the orbits of other planets. In particular, Newton calculated that Jupiter, at its closest approach to Saturn, would have about 1/217 times the gravitational attraction of the Sun. So he was delighted when Flamsteed told him that Saturn's orbit did not seem to fit exactly the orbit that it should if it was only influenced by the Sun. Gravity really did appear to be universal.

Richer, Cassini, and Picard had found evidence in 1672 that the Earth had an equatorial bulge. Newton was able to use his new gravitational theory to calculate a theoretical value for this **oblateness** of 1/230 (modern value 1/298). He then considered the gravitational attraction of the Moon and Sun on the oblate Earth and calculated that the Earth's spin axis should precess at a rate of about 50″.0 per annum (modern value 50″.3). This explained the precession of the equinoxes.

5. The 18th Century

5.1 Halley's Comet

Halley used Newton's methodology to determine the orbits of 24 comets that had been observed between 1337 and 1698. None of them appeared to be hyperbolic, and so the comets were all clearly permanent members of the solar system. Halley also concluded that the comets of 1531, 1607, and 1682 were successive appearances of the same comet as their orbital elements were very similar. But the time intervals between successive perihelia were not the same; a fact he attributed to the perturbing effect of Jupiter. Taking this into account, he predicted in 1717 that the comet would return in late 1758 or early 1759.

Shortly before the expected return of this comet, which we now called Halley's comet, Alexis Clairaut (1713–1765) attempted to produce a more accurate prediction of its **perihelion** date. He used a new approximate solution to the three-body problem that allowed him to take account of planetary perturbations. This showed that the return would be delayed by 518 days due to Jupiter and 100 days due to

Saturn. As a result, he predicted that Halley's comet would reach perihelion on about 15 April 1759 ± 1 month. It did so on 13 March 1759, so Clairaut was just 33 days out with his estimate.

5.2 The 1761 and 1769 Transits of Venus

James Gregory (1638–1675) had suggested in 1663 that observations of a transit of Mercury could be used to determine the **solar parallax**, and hence the distance of the Sun from Earth. Such a determination required observations from at least two different places on Earth, separated by as large a distance as possible. In 1677, Edmond Halley observed such a transit when he was on St. Helena observing the southern sky. But, when he returned, he found that Jean Gallet in Avignon seemed to have been the only other person who had recorded the transit. Unfortunately, there were too many problems in comparing their results, which resulted in a highly inaccurate **solar parallax.**

In 1678, Halley reviewed possible methods of measuring the solar parallax and suggested that transits of Venus would produce the most accurate results. The problem was, however, that these occur in pairs, 8 years apart, only every 120 years. The next pair were due almost one hundred years later, in 1761 and 1769.

Joseph Delisle (1688–1768) took up Halley's suggestion and tried to motivate the astronomical community to undertake coordinated observations of the 1761 transit. After much discussion, the French Academy of Sciences sent observers to Vienna, Siberia, India, and an island in the Indian Ocean, while other countries sent observers to St. Helena, Indonesia, Newfoundland, and Norway. Unfortunately, precise timing of the planetary contacts proved much more difficult than expected, resulting in solar parallaxes ranging from 8″.3 to 10″.6. Interestingly, several observers noticed that Venus appeared to be surrounded by a luminous ring when the planet was partially on the Sun. Mikhail Lomonsov (1711–1765) correctly concluded that this showed that Venus was surrounded by an extensive atmosphere.

The lessons learned from the 1761 transit were invaluable in observing the next transit in 1769. This was undertaken from over 70 different sites, and analysis of all the results eventually yielded a best estimate of 8″.6 (modern value 8″.79) for the solar parallax.

5.3 The Discovery of Uranus

On 13 March 1781, William Herschel (1738–1822), whilst looking for double stars, noticed what he thought was a comet. Four days later, when he next saw the object, it had clearly moved, confirming Herschel's suspicion that it was a comet. He then wrote to Nevil Maskelyne (1732–1811), the Astronomer Royal, notifying him of his discovery. As a result, Maskelyne observed the object on a number of occasions, but he was unsure as to whether it was a comet or a new planet.

Over the next few weeks a number of astronomers observed the object and calculated its orbit, which was found to be essentially circular. So it was a planet, now called Uranus. It was the first planet to be discovered since ancient times, and its discovery had a profound effect on the astronomical community, indicating that there may yet be more undiscovered planets in the solar system.

A few years later Herschel discovered the first two of Uranus' satellites, now called Titania and Oberon, with orbits at a considerable angle to orbit.

5.4 Origin of the Solar System

Immanuel Kant (1724–1804) outlined his theory of the origin of the solar system in his *Universal Natural History* of 1755. In this he suggested that the solar system had condensed out of a nebulous mass of gas, which had developed into a flat rotating disc as it contracted. As it continued to contract, it spun faster and faster, throwing off masses of gas that cooled to form the planets. However, Kant had difficulty in explaining how a nebula with random internal motions could start rotating when it started to contract.

Forty years later, Laplace (1749–1827) independently produced a similar but more detailed theory. In his theory, the mass of gas was rotating before it started contracting. As it contracted, it spun faster, progressively throwing from its outer edge rings of material that condensed to form the planets. Laplace suggested that the planetary satellites formed in a similar way from condensing rings of material around each of the protoplanets. Saturn's rings did not condense to form a satellite because they were too close to the planet. At face value, the theory seemed plausible, but it became clear in the 19th century that the original solar nebula did not have enough angular momentum to spin off the required material.

5.5 The First Asteroids

A number of astronomers had wondered why there was such a large gap in the solar system between the orbits of Mars and Jupiter. Then in 1766 Johann Titius (1729–1796) produced a numerical series that indicated that there should be an object orbiting the Sun with an orbital radius of 2.8 **astronomical units** (AUs). Johann Elert Bode (1747–1826) was convinced that this was correct and mentioned it in his book

of 1772. However, what is now known as the Titius–Bode series was not considered of any particular significance, until Uranus was found with an orbital radius of 18.9 AU. This was very close to the 19.6 AU required by the series.

In 1800, a group of astronomers, who came to be known as the Celestial Police, agreed to undertake a search for the missing planet. But before they could start Giuseppe Piazzi (1746–1826) found a likely candidate by accident in January 1801. Unfortunately, although he observed the object for about 6 weeks, he was unable to fit an orbit, and wondered if it was a comet. But Karl Gauss (1777–1855) had derived a new method of determining orbits from a limited amount of information, and in November of that year he was able to fit an orbit. It was clearly a planet, now called Ceres, at almost exactly the expected distance from the Sun. But it was much smaller than any other planet. Then in March 1802 Heinrich Olbers (1758–1840) found another, similar object, now called Pallas, at a similar distance from the Sun. At first Olbers thought that these two objects may be the remnants of an exploded planet. But he dropped the idea after the discovery of the fourth such asteroid, as they are now called, in 1807, because its orbit was inconsistent with his theory.

6. The 19th Century

6.1 The Sun

Sunspots were still an enigma in the 19th century. Many astronomers thought that they were holes in the photosphere, but because the Sun was presumably hotter beneath the photosphere, the Sunspots should appear bright rather than dark. Then in 1872 Angelo Secchi suggested that matter was ejected from the surface of the Sun at the edges of a sunspot. This matter then cooled and fell back into the center of the spot, so producing its dark central region.

In 1843, Heinrich Schwabe found that the number of sunspots varied with a period of about 10 years. A little later Rudolf Wolf analyzed historical records that showed periods ranging from 7 to 17 years, with an average of 11.1 years. Then in 1852, Sabine, Wolf, and Gautier independently concluded that there was a correlation between sunspots and disturbances in the Earth's magnetic field. There were also various unsuccessful attempts to link the sunspot cycle to the Earth's weather. But toward the end of the century, Walter Maunder pointed out that there had been a lack of sunspots between about 1645 and 1715. He suggested that this period, now called the Maunder Minimum, could have had a more profound effect on the Earth's weather than the 11-year solar cycle.

In 1858, Richard Carrington discovered that the latitude of sunspots changed over the solar cycle. In the following year, he found that sunspots near the solar equator moved faster than those at higher latitudes, showing that the Sun did not rotate as a rigid body. This so-called differential rotation of the Sun was interpreted by Secchi as indicating that the Sun was gaseous. In the same year, Carrington and Hodgson independently observed two white light solar flares moving over the surface of a large sunspot. About 36 hours later, this was followed by a major geomagnetic storm.

Astronomy was revolutionized in the 19th century by Kirchoff's and Bunsen's development of spectroscopy in the early 1860s, which, for the first time, enabled astronomers to determine the chemical composition of celestial objects. Kirchoff measured thousands of dark Fraunhofer lines in the solar spectrum and recognized the lines of sodium and iron. By the end of the century, about 40 different elements had been discovered on the Sun.

Solar prominences had been observed during a total solar eclipse in 1733, but it was not until 1860 that they were proved to be connected with the Sun rather than the Moon. Spectroscopic observations during and after the 1868 total eclipse showed that prominences were composed of hydrogen and an element that produced a bright yellow line. This was initially attributed to sodium, but Norman Lockyer suggested that it was caused by a new element that he called helium. This was confirmed when helium was found on Earth in 1895.

6.2 Vulcan

Newton's gravitational theory had been remarkably accurate in explaining the movement of the planets, but by the 19th century there appeared to be something wrong with the orbit of Mercury. In 1858, Le Verrier analyzed data from a number of transits and concluded that the perihelion of Mercury's orbit was precessing at about 565″/century, which was 38″/century more than could be accounted for using Newton's theory. As a result, Le Verrier suggested that there was an unknown planet called Vulcan, inside the orbit of Mercury, causing the extra precession. A number of astronomers reported seeing such a planet, but none of the observations stood up to detailed scrutiny, and the idea was eventually dropped.

Einstein finally solved the problem of Mercury's perihelion precession in 1915 with his general theory of relativity. No extra planets were required.

6.3 Mercury

There was considerable disagreement among astronomers in the 19th century on what could be seen on Mercury. Some thought that they could see an atmosphere around the planet, but others could not. Hermann Vogel detected water vapor lines in its spectrum, and Angelo Secchi saw clouds in its atmosphere. However, Friedrich Zöllner concluded, from his photometer measurements, that Mercury was more like the Moon with, at most, a very thin atmosphere.

A number of astronomers detected markings on Mercury's disc in the middle of the 19th century and concluded that the planet's period is about 24 hours. On the other hand, Daniel Kirkwood maintained that it should have a synchronous rotation period because of tidal effects of the Sun on its crust. In the 1880s, Giovanni Schiaparelli confirmed this synchronous rotation observationally, and in 1897 Percival Lowell came to the same conclusion. So at the end of the century, synchronous rotation was thought to be the most likely.

6.4 Venus

In the 18th century, Venus was thought to have an axial rotation rate of about 24 hours. In fact, a 24-hour period was generally accepted until in 1890 Schiaparelli and others concluded that it, like Mercury, has a synchronous rotation period.

Spectroscopic observations of Venus yielded conflicting results in the 19th century. A number of astronomers detected oxygen and water vapor lines in its atmosphere; however, W. W. Campbell, who used the powerful Lick telescopes, could find no such lines.

6.5 The Moon

The impact theory for the formation of lunar craters was resurrected at the start of the 19th century, after the discovery of the first asteroids and a number of meteorites. There now seemed to be a ready source of impacting bodies, which Hooke had been unaware of when he had abandoned his impact hypothesis. But both the impact and volcanic theories still had problems. Most meteorites would not hit the lunar surface vertically, and so the craters should be elliptical, but they were mostly circular. Also, as Grove K. Gilbert pointed out, the floors of lunar craters are generally below the height of their surrounding area, whereas on Earth the floors of volcanic craters are generally higher than their surroundings.

Edmond Halley had discovered in 1693 that the Moon's position in the sky was in advance of where it should be based on ancient eclipse records. This so-called secular acceleration of the Moon could be because the Moon was accelerating in its orbit, and/or because the Earth's spin rate was slowing down. In 1787, Laplace had shown that the observed effect, which was about 10″/century^2, could be completely explained by planetary perturbations. But in 1853, John Couch Adams included some of Laplace's second-order terms, which Laplace had omitted, so reducing the calculated figure from 10″/century^2 to just 6″/century^2. Charles Delaunay suggested that the missing amount was probably due to tidal friction, but it was impossible at that time to produce a reasonably accurate estimate of the effect. In the early 20th century, Taylor and Jeffreys produced the necessary calculations, showing that Delaunay was correct.

In 1879, George Darwin developed a theory of the origin of the Moon. In this the proto-Earth had gradually contracted and increased its spin rate as it cooled. Then, when the spin rate had reached about 3 hours per revolution, it had broken into two unequal parts: the Earth and the Moon.

After breakup, tidal forces had caused the Earth's spin rate to slow down and the Moon's orbit to gradually increase in size.

A major problem with this theory was that the Earth would have had a tendency to break up the Moon shortly after separation. It was not clear whether the Moon could have passed through the danger zone before this could have happened.

6.6 The Earth

Karl Friedrich Küstner undertook precise position measurements of a number of stars in 1884 and 1885 from the Berlin Observatory. When he analyzed his results, however, he found that the latitude of the observatory had apparently decreased by about 0.20″ in a year. Intrigued, the International Commission for Geodesy (ICG) decided to organize a series of observations around the world to define the effect more precisely. These results indicated that the Earth's spin axis was moving, relative to its surface, with a period of about 12 or 13 months.

Seth Chandler had also noticed slight variations in the latitude of the Harvard College Observatory, at about the same time as Küstner was making his measurements, but Chandler had not taken the matter further. Galvanized by Küstner's and the ICG's results, however, he undertook a thorough review of all available data. As a result, he concluded that the observed effect had two components. One had a period of 14 months, and was due to the nonrigid Earth not spinning around its shortest diameter. The other, which had a period of a year, was due to the seasonal movement of water and air from one hemisphere to the other and back.

6.7 Mars

The first systematic investigation of Mars' polar caps had been undertaken in the 18th century by Giacomo Maraldi, who found that the south polar cap had completely disappeared in late 1719, only to reappear later. William Herschel suggested that this was because it consisted of ice and snow that melted in the southern summer.

At the end of the 18th century, most astronomers thought that the reddish color of Mars was due to its atmosphere. But in 1830, John Herschel suggested that it was the true color of its surface. Camille Flammarion, on the other hand, hypothesized that it was the color of its vegetation.

It was generally believed by astronomers in the mid-19th century that there must be some form of life on Mars, even if it was only plant life, because the planet clearly had an atmosphere and a surface that exhibited seasonal effects. The polar caps were apparently made of ice or snow, and there were dark areas on the surface that may be seas.

Schiaparelli produced a map of Mars, following its 1877 **opposition**, that showed a network of linear features that he called *canali*. This was translated incorrectly into English as canals, which implied that they had been built by intelligent beings. Schiaparelli and others saw more *canali* in subsequent years (Fig. 5), but other, equally competent observers could not see them at all. Percival Lowell then went further than Schiaparelli in not only observing many canali, but interpreting them to be a network of artificial irrigation channels. At the end of the century, the debate as to whether these *canali* really existed was still in full swing.

Spectroscopic observations of Mars in the late 19th century yielded conflicting results. Some astronomers detected oxygen and water vapor lines, whereas Campbell at the Lick Observatory could find none. There was also a problem with the polar caps: Calculations showed that the average temperature of Mars should be about $-34°C$, yet both polar caps clearly melted substantially in summer, which they should not have done if they had been made of water ice or snow. In 1898, Ranyard and Stoney suggested that the caps could be made of frozen carbon dioxide. But there appeared to be a melt band at the edge of the caps in spring, yet carbon dioxide should sublimate directly into gas on Mars.

Two satellites of Mars, now called Phobos and Deimos, were discovered by Asaph Hall in 1877. Their orbits were extremely close to the planet, and the satellites were both very small. As a result, they were thought to be captured asteroids.

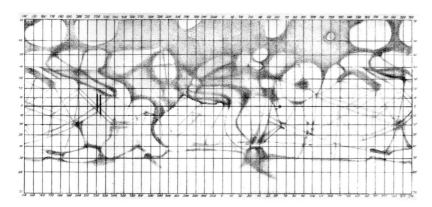

FIGURE 5 Schiaparelli's map of Mars produced following the 1881 opposition. A large number of *canali* are seen, many of them double. (From Robert Ball, 1897, "The Story of the Heavens," Plate XVIII.)

6.8 Jupiter

The Great Red Spot (GRS) was first clearly observed in the 1870s. Then in 1880 an unusually bright, white equatorial spot appeared; it rotated around Jupiter over 5 minutes faster than the GRS. This gave a differential velocity of about 400 km/h. But the rotation rates of both the white spot and the GRS were not constant, indicating that neither could be surface features as some astronomers had supposed.

White and dark spots were continuously appearing and disappearing on Jupiter, suggesting that they were probably clouds. But the GRS was completely different because, although it changed its appearance and size over time, it was still there at the end of the century. This longevity led astronomers to wonder if it could really be a cloud system.

In 1778, Leclerc, Compte de Buffon, had suggested that rapid changes in Jupiter's appearance showed that it had not completely cooled down since its formation. In the 19th century, Jupiter's differential rotation and low density, which were both similar in nature to those of the Sun, caused some astronomers to go even further and wonder if Jupiter was self-luminous. Although this was considered unlikely, the idea had not been completely ruled out by the end of the century.

William Herschel had concluded in 1797 that the axial rotation rates of the four Galilean satellites were synchronous. However, it was not until the 1870s that Engelmann and Burton independently confirmed this for Callisto and the 1890s that Pickering and Douglass confirmed it for Ganymede. The rotation rates of Io and Europa were still unclear.

In 1892, Edward Barnard discovered Jupiter's fifth satellite, now called Amalthea, very close to the planet, when he was observing Jupiter visually through the 36-in. Lick refractor. Amalthea was very small compared to the four Galilean satellites. It was the last satellite of any planet to be discovered visually.

6.9 Saturn

In 1837, Johann Encke found that the A ring was divided into two by a clear gap, now called the Encke Division. Then in 1850 W. C. and G. P. Bond discovered a third ring, now called the C ring, inside the B ring. The new ring was very dark (Fig. 6) and partly transparent. In 1867, Kirkwood pointed out that any particles in the Cassini Division would have periods of about one-half that of Mimas, one-third that of Enceladus, one-quarter that of Tethys, and one-sixth that of Dione. He concluded that these resonances had created the Cassini Division, which would be clear of particles.

The true nature of Saturn's rings had been a complete mystery in the 18th century. Cassini had thought that they may be composed of many small satellites, and Laplace had suggested that they were made of a number of thin

FIGURE 6 Trouvelot's 1874 drawing of Saturn. It clearly shows the dark C ring extending from the inner edge of the B ring to about half-way to the planet. (From Edmund Ledger, 1882, "The Sun: Its Planets and Their Satellites," Plate IX.)

solid rings. Others thought that they may be liquid. But in 1857, James Clerk Maxwell proved mathematically that they could not be solid or liquid. Instead, he concluded that they were composed of an indefinite number of small particles.

Two new satellites were found in the 19th century: Hyperion by G. P. Bond in 1848 and Phoebe by William Pickering 50 years later. Phoebe was the first satellite in the solar system to be discovered photographically. It was some 13 million kilometers from Saturn, in a highly eccentric, **retrograde** orbit. So it appeared to be a captured object.

6.10 Uranus

Little was know about Uranus in the 19th century. William Herschel had noticed that Uranus had a polar flattening, its orientation indicating that its axis of rotation was perpendicular to the plane of its satellites. But observations of apparent surface features produced very different orientations. Uranus' spectrum appeared to be clearly different from those of Jupiter and Saturn, but it was very difficult to interpret. There was even confusion about the discovery of new satellites. It was not until 1851 that William Lassell could be sure that he had discovered two new satellites, now called Ariel and Umbriel within the orbit of Titania. He had, in fact, seen them both some years before, but his earlier observations had been too infrequent to produce clear orbits.

6.11 The Discovery of Neptune

In 1821, Alexis Bouvard tried to produce an orbit for Uranus using both prediscovery and postdiscovery observations. But he could not find a single orbit to fit them. The best he could manage was an orbit based on only the postdiscovery observations; he published the result but admitted that it was less than ideal. However, it did not take long for Uranus to deviate more and more from even this orbit. One possible explanation was that Uranus was being disturbed by yet another planet, and if the Titius–Bode series was correct it would be about 38.8 AU from the Sun.

In 1843, the Englishman John Couch Adams set out to try to calculate the orbit of the planet that seemed to be disturbing the orbit of Uranus. By September 1845, he had calculated its orbital elements and its expected position in the sky, and over the next year, he progressively updated this prediction. Unfortunately, these predictions varied wildly, making it impossible to use them for a telescopic search of the real planet. In parallel, and unknown to both men, Urbain Le Verrier, a French astronomer, undertook the same task. He published his final results in August 1846 and asked Johann Galle of the Berlin Observatory if he would undertake a telescope search for it. Galle and his assistant d'Arrest found the planet within an hour of starting the search on 23 September 1846. There then followed a monumental argument between the English and French astronomical establishments on the priority for the orbital predictions. But much of the evidence on the English side was never published, and an "official line" was agreed. That evidence has recently come to light, however, and it is currently being analyzed to establish the exact sequence of events. What is clear, however, is that when Neptune's real orbit was calculated, it turned out to be quite different from either of the orbits predicted by Le Verrier or Adams. So its discovery had been somewhat fortuitous.

Less than a month after Neptune's discovery, William Lassell observed an object close to Neptune, which he thought may be a satellite. It was not until the following July that he was able to confirm his discovery of Neptune's first satellite, now called Triton. Triton was later found to have a retrograde orbit inclined at approximately 30° to the ecliptic.

6.12 Asteroids

The fourth asteroid, Vesta, had been discovered in 1807, but it was not until 1845 that the fifth asteroid was found. Then the discovery rate increased rapidly so that nearly 500 asteroids were known by the end of 1900. As the number of asteroids increased, Kirkwood noticed that there were none with certain fractional periods of Jupiter's orbital period. This he attributed to resonance interactions with Jupiter.

All the early asteroids had orbits between those of Mars and Jupiter, and even as late as 1898 astronomers had discovered only one that had part of its orbit inside that of Mars. But in 1898, Eros was found with an orbit that came very close to that of the Earth, with the next closest approach expected in 1931. This could be used to provide an accurate estimate of solar parallax.

In 1906, two asteroids were found at the Lagrangian points, 60° in front of and behind Jupiter in its orbit. They were the first of the so-called Trojan asteroids to be discovered.

6.13 Comets

Charles Messier discovered a comet that passed very close to the Earth in 1770. Anders Lexell was the first to fit an orbit to it, showing that it had a period of just 5.6 years. With such a short period it should have been seen a number of times before, but it had not. As Lexell explained, this comet had not been seen because it had passed very close to Jupiter in 1767, which had radically changed its orbit. In the late 19th century, Hubert Newton examined the effect of such planetary perturbations on the orbits of comets and found that, for a random selection of comets, they were remarkably inefficient. Lexell's comet appeared to be an exception.

Jean Louis Pons in 1818 discovered a comet that, on further investigation, proved to have been seen near previous perihelia. In the following year, Johann Encke showed that the comet, which now bears his name, has an orbit that takes it inside the orbit of Mercury. When the comet returned in 1822, Encke noticed that it was a few hours early and suggested that it was being affected by some sort of resistive medium close to the Sun. In 1882, however, a comet passed even closer to the Sun and showed no effect of Encke's medium. Then in 1933, Wolf's comet was late, rather than early. The problem of these cometary orbits was finally solved in 1950 when Fred Whipple showed that the change in period was caused by jetlike, vaporization emissions from the rotating cometary nucleus.

The first successful observation of a cometary spectrum was made by Giovanni Donati in 1864. When the comet was near the Sun, it had three faint luminous bands, indicating that it was self-luminous. Then four years later, William Huggins found that the bands were similar to those emitted by hydrocarbon compounds in the laboratory.

Quite a number of cometary spectra were recorded over the next 20 years. When they were first found, they generally exhibited a broad continuous spectrum like that of the Sun indicating that they were scattering sunlight. As they got closer to the Sun, however, the hydrocarbon bands appeared. Then in 1882 Wells' comet approached very close to the Sun. Near perihelion its bandlike structure disappeared to be replaced by a bright, double sodium line. In the second comet of 1882, this double sodium line was also accompanied by several iron lines when the comet was very near the Sun. As the comet receded, these lines faded and the hydrocarbon bands returned.

6.14 Meteor Showers

A spectacular display of shooting stars was seen in November 1799, and again in November 1833. They seemed to originate in the constellation Leo. In the following year, Denison Olmsted pointed out the similarities between these two meteor showers and a less intense one in 1832. These so-called Leonid meteors seemed to be an annual event occurring on or about 12 November. Olmsted explained that the radiant in Leo was due to a perspective

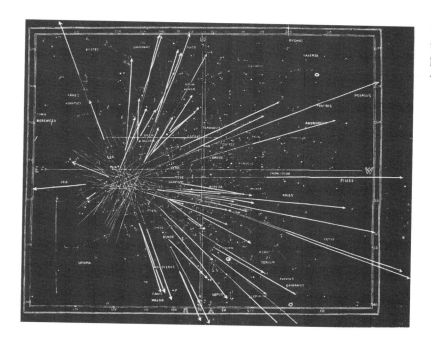

FIGURE 7 Paths of the Leonid meteors showing their apparent origin from a common radiant due to parallax. (From Simon Newcomb, 1898, "Popular Astronomy," p. 403.)

effect (Fig. 7). A similar effect was then observed for a meteor shower on 8 August 1834, which appeared to have a radiant in Perseus. Shortly afterward, Lambert Quetelet showed that these were also an annual event.

In 1839, Adolf Erman suggested that both the Leonid and Perseid meteor showers were produced by the Earth passing through swarms of small particles that were orbiting the Sun and spread out along Earth's orbit. But it was still unclear as to the size of the orbit. In 1864, Hubert Newton found that the node of the Leonids' orbit was precessing at about 52″/year. John Couch Adams then showed that only a particle in a 33.25-year orbit would have this nodal precession. So the Leonids were orbiting the Sun in a diffuse cloud every 33.25 years, which explained why the most intense showers occurred with this frequency. The stragglers all around the orbit explained why we saw the Leonids on an annual basis. In 1867, Carl Peters recognized that the source of the Leonid meteor stream was a periodic comet called Tempel–Tuttle. This was just after Schiaparelli had linked the Perseids to another periodic comet, Swift–Tuttle.

7. The 20th Century Prior to the Space Age

7.1 The Sun

In the 19th century, most physicists had thought that heat was transported from the interior to the exterior of the Sun by convection. But in 1894, R. A. Sampson suggested that the primary mechanism was radiation. Then, 30 years later, Arthur Eddington used the concept of radiative equilibrium to calculate the temperature at the center of the Sun and found it to be about 39 million K. At about the same time,

Cecilia Payne showed that hydrogen and helium were the most abundant elements in the stars. Although this idea was initially rejected, it was soon accepted for both the Sun and stars. As a result, in 1935 Eddington reduced his temperature estimate for the center of the Sun to 19 million K.

However, Eddington's calculations made no assumption on how the Sun's heat was produced, which was still unknown at the time. Earlier, in 1920, Eddington himself had proposed two alternative mechanisms. The heat could be produced either by the mutual annihilation of protons and electrons or by the fusion of hydrogen atoms into helium atoms in some unknown manner. There were other mechanisms suggested by other physicists, but the issue could not be resolved at the time because nuclear physics was still in its infancy. The breakthrough came in 1938 when Charles Critchfield explained how energy could be produced at high temperatures by a chain reaction starting with proton–proton collisions and ending with the synthesis of helium nuclei. Hans Bethe then collaborated with Critchfield to develop this idea. But Bethe also examined an alternative mechanism that relied on carbon as a catalyst to produce helium from hydrogen, in the so-called carbon cycle. Carl von Weizsäcker independently developed this same scheme. Which mechanism was predominant in the Sun depended crucially on temperature, and it was not until the 1950s that it became clear that the proton–proton chain is dominant in the Sun.

In the 19th century, the **corona** had been found to have a faint continuous spectrum crossed by Fraunhofer absorption lines, but the conditions in the corona were unclear. Of particular interest was a bright green emission line in the coronal spectrum; Young and Harkness found it in 1869 and originally attributed it to iron. In 1898, however, it was found to have a slightly different wavelength than the iron line. Because no known element generated the required line, it was attributed to a new element called coronium.

At that time, it was assumed that the temperature of the Sun and its corona gradually reduced from the center moving outwards. But in the early part of the 20th century, competing theories were put forward, one for a low-temperature corona and another for a high-temperature one. In 1934, Walter Grotrian analyzed the coronal spectrum and concluded that the temperature was an astonishing 350,000 K. A few years later Bengt Edlén, in a seminal paper, showed that coronal lines are produced by highly ionized iron, calcium, and nickel at a temperature of at least 2 million K. The "coronium" line, in particular, was the product of highly ionized iron. How the temperature of the corona could be so high, when the photosphere temperature is only of the order of 6,000 K, was a mystery, which has not been completely resolved even today.

Charles Young discovered in 1894 that, at very high dispersions, many absorption lines in sunspot spectra appeared to have a sharp bright line in their centers. In 1908, George Ellery Hale and Walter Adams found that photographs of the Sun taken in the light of the 656.3-nm hydrogen line showed patterns that looked like iron filings in a magnetic field. This caused Hale to examine sunspot spectra in detail. He found that the Young effect was actually caused by Zeeman splitting of spectral lines in a magnetic field, which was of the order of 3,000 gauss. So sunspots were the home of very high magnetic fields.

Hale then started to examine the polarities of sunspots, and found that spots generally occur in pairs, with the polarity of the lead spot, as they crossed the disc, being different in the two hemispheres. This pattern was well established by 1912 when the polarities were found to be reversed at the solar minimum. They reversed yet again at the next solar minimum in 1923. So the solar cycle was really 22 years, not 11.

Walter Maunder found in 1913 that large magnetic storms on Earth start about 30 hours after a large sunspot crosses the center of the solar disc. Later work showed that the most intense storms were often associated with solar flares. In 1927, Chree and Stagg found that smaller storms, which did not seem to be associated with sunspots, tended to recur at the Sun's synodic period of 27 days. Julius Bartels called the invisible source on the Sun of these smaller storms, M regions. Both the so-called flare storms and the M storms were assumed to be caused by particles ejected from the Sun. In 1951, Ludwig Biermann suggested that, to explain the behavior of cometary ion tails, there must be a continuous stream of charged particles emitted by the Sun. Then in 1957, Eugene Parker proposed his theory of the solar wind, which was later confirmed by early spacecraft.

Marconi noticed in 1927 that interference with radio signals in September and October of that year coincided with the appearance of large sunspots and intense aurorae. In the late 1930s, Howard Dellinger carried out a detailed examination of the timing of shortwave radio fadeouts, at numerous receiving stations, and solar flares. He found a reasonable but by no means perfect correlation. The fade-outs seemed to start almost instantaneously after the flare was seen, and they only occurred when the receiving station was in daylight. So Dellinger concluded that they were caused by some form of electromagnetic radiation from the Sun, rather than particles.

7.2 Mercury

The synchronous rotation period of Mercury was gradually accepted as a fact in the 20th century. But in 1962, W. E. Howard found that Mercury's dark side seemed to be warmer than it should be if it were permanently in shadow. Then 3 years later, Dyce and Pettengill found, using radar, that Mercury's rotation period was not synchronous, but represented two-thirds of its orbital rotation period.

7.3 Venus

There was considerable confusion in the first half of the 20th century about Venus' rotation period. All sorts of periods were proposed between about 24 hours and synchronous (225 days). Then in 1957 Charles Boyer found a distinctive V-shaped pattern of Venus' clouds that had a 4-day period. In 1962, however, Carpenter and Goldstein deduced a period of about 250 days retrograde using radar, which was modified to 243 days in 1965 for the rotation period of Venus' surface. So Venus has a 243-day period, whilst its clouds have a period of about 4 days, both periods being retrograde.

In 1932, Adams and Dunham concluded that there was no oxygen or water vapor on Venus, but carbon dioxide was clearly present. A few years later, Rupert Wildt calculated that the greenhouse heating of the latter could produce a surface temperature as high as 400 K. Then in 1956, Mayer, McCullough, and Sloanaker deduced a surface temperature of about 600 K by analyzing Venus' thermal radio emissions. The suggestion that Venus' surface temperature could be so high was naturally treated with caution. Shortly afterward, Carl Sagan estimated that the surface atmospheric pressure was an equally incredible 100 bar.

7.4 The Moon

The idea that there may be life on the Moon had fascinated people for centuries. Even respected astronomers like William Herschel had thought that there would be "lunarians" as he called them. But by the start of the 20th century, it was thought that the most complex lifeforms would be some sort of plant life. However, by the 1960s, when the Americans were planning their lunar landings, even this concept had been rejected. Nevertheless, it was thought that there may be some sort of very elemental life, like bacteria, on the Moon.

Bernard Lyot had concluded in 1929, from polarization measurements, that the Moon was probably covered by volcanic ash. Then in the 1950s, Thomas Gold suggested that the Moon may be covered with dust up to a few meters deep. If this was so, it would have provided a major problem for the manned *Apollo* missions.

At the end of the 19th century, the key objection to the impact theory for the formation of lunar craters had been that the craters were generally circular, when they should have been elliptical, because most of the impacts would not be vertical. However, after the First World War it was realized that the shape of the lunar craters resembled shell craters. The shell craters were formed by the shock wave of the impact or explosion, so a nonvertical impact could still produce a circular crater. Nevertheless, not all lunar craters have the same general appearance. So, by the start of the space age it was still unclear if they had been produced by volcanic action, meteorite impact, or both.

7.5 The Earth

It was known in the 19th century that temperatures in deep mines on Earth increased with depth. That, together with the existence of volcanoes, clearly indicated that the Earth has a molten interior. Calculations indicated that the rocks would be molten at a depth of only about 40 km.

In 1897, Emil Wiechert suggested that the Earth has a dense metallic core, mostly of iron, surrounded by a lighter rocky layer, now called the mantle. A little later, Richard Oldham found clear evidence for the existence of the core from earthquake data. Then in 1914, Beno Gutenberg showed that the interface between the mantle and the core, now called the Wiechert–Gutenberg discontinuity, is at about $0.545r$ from the center of the Earth (where r is its radius).

A little earlier, Andrija Mohorovičić had discovered the boundary between the crust and mantle, now called the Mohorovičić discontinuity, by analyzing records of the Croatian earthquake of 1909. The depth of this discontinuity was later found to vary from about 70 km under some mountains to only about 5 km under the deep oceans.

A number of theories were proposed to try to define and explain the internal structure of the Earth. In particular, Harold Jeffreys produced a theory that assumed that all the **terrestrial planets** and the Moon have a core of liquid metals, mostly iron, and a silicate mantle. But it could not explain how those planets with the smallest cores could have retained a higher percentage of lighter material in their mantles. In 1948, William Ramsey solved this problem when he proposed that the whole of the interior of the terrestrial planets consists of silicates, with the internal pressure in the largest planets causing the silicates near the center to become metallic. Unfortunately this idea became unviable when Eugene Rabe found in 1950 that Mercury's density was much higher than originally thought. It was even higher than that of Venus and Mars, which were much larger planets.

In the mid-20th century, most astronomers believed that the planets had been hot when first formed from the solar nebula, but in 1949 Harold Urey suggested that the nebula had been cold. According to Urey, the Earth had been heating up since it was formed because of radioactive decay. Internal convection had then started as iron had gradually settled into the core. Urey believed that the Moon was homogenous because it was relatively small.

At the turn of the 19th century, it was thought that radio waves generally traveled in a straight line. So it was a great surprise when Marconi showed in 1901 that radio waves could be successfully transmitted across the Atlantic. Refraction could have caused them to bend to a limited degree, but not enough to cross the ocean. In the following year, Heaviside and Kennelly independently suggested that the waves were being reflected off an electrically conducting layer in the upper atmosphere.

The structure of what we now call the E or Heaviside layer, and of other layers in the ionosphere, was gradually clarified over the next 20 years or so. The 80 km high D layer was found to largely disappear at night, and the higher E layer was found to maintain its reflectivity for only 4 or 5 hours after sunset. In addition, it was found that solar flares can cause a major disruption to the ionosphere (see Section 7.1). However, it was not until after the Second World War that the cause of these effects could be examined in detail by first sounding rockets and then by spacecraft. The first major discovery was made by Herbert Friedman in 1949 when he showed that the Sun emits X-rays, which have a major effect on the Earth's ionosphere.

7.6 Mars

There was a great deal of uncertainty about the surface of Mars in the first half of the 20th century. It was thought unlikely that the linear markings called *canali* really existed, but they were still recorded from time to time by respected observers. In addition, some astronomers thought that the bluish green areas on Mars were vegetation, while others thought that they were volcanic lava.

There was also considerable uncertainty about the spectroscopic observations of Mars. Some observers recognized water vapor and oxygen lines, whereas others found none. But in 1947 Gerard Kuiper clearly found evidence for a small amount of carbon dioxide, and in 1963 Andouin Dollfus found a trace amount of water vapor. Estimates of the surface atmospheric pressure varied from about 25 to 120 millibars. Then in 1963, shortly before the first spacecraft reached Mars, a figure of 25 ± 15 millibars was estimated by Kaplan, Münch, and Spinrad.

It seemed clear that the yellow clouds seen on Mars were dust. In 1909, Fournier and Antoniadi found that they appeared to cover the whole planet for a while. Later Antoniadi found that they tended to occur around perihelion when the solar heating is greatest, and so appeared to be produced by thermally generated winds. Thirty years later, De Vaucouleurs measured the wind velocities as being typically in the range of 60 to 90 km/h when the clouds first formed.

7.7 Internal Structures of the Giant Planets

It was known in the 19th century that the densities of Jupiter, Saturn, Uranus, and Neptune were similar to that of the Sun, and were much less than that of the terrestrial planets. At that time, it was thought that Jupiter, and probably Saturn, had not yet fully cooled down since their formation. As a result, they were probably emitting more energy than they received from the Sun.

In 1923, Donald Menzel found that the cloud top temperatures of Jupiter and Saturn were about 160 K. This compares with temperatures of 120 and 90 K for Jupiter and Saturn, respectively, that would be maintained solely by incident solar radiation. Three years later, Menzel produced modified observed temperatures of 140, 120, and 100 K, for Jupiter, Saturn, and Uranus. So any internally generated heat would be rather low.

In 1923, Harold Jeffreys pointed out that the ratio of the densities of Io and Europa, the innermost of Jupiter's large satellites, to that of Jupiter, was about the same as the ratio of the density of Titan, Saturn's largest satellite, to that of Saturn. He then assumed that the density of the cores of Jupiter and Saturn were the same as these their large satellites. In that case, the thickness of the planetary atmospheres would be about 20% of their radii.

In the following year, Jeffreys included consideration of the moments of inertia of Jupiter and Saturn in his analysis and concluded that their atmospheres would have depths of $0.09 R_J$ and $0.23 R_S$, respectively (where R_J and R_S are the radii of Jupiter and Saturn, respectively). He assumed that beneath their atmospheres there was a layer of ice and solid carbon dioxide, which in turn was surrounded a rocky core.

Various schemes were then produced by a number of physicists, of which those of Rupert Wildt in 1938 and William Ramsey in 1951 were probably the most significant. Wildt, who was particularly interested in internal pressures, wanted to find out if matter at the core of the large planets was **degenerate**. His calculations indicated that it was not. Ramsey, on the other hand, developed his theory assuming that the giant planets were made of hydrogen. He then added helium and other ingredients until their densities and moments of inertia were correct. On this basis, he concluded that Jupiter and Saturn were composed of 76% and 62% hydrogen, by mass, respectively, with central pressures of 32 and 6×10^6 bar. At these pressures, most of the hydrogen would be metallic.

The structures of Uranus and Neptune were a problem in Ramsey's analysis because the heavier planet, Neptune, was the smaller. So their constituents could not be the same. Then in 1961 William Porter produced a model that seemed to fit; in this model, Neptune had 74% ammonia and 26% heavier elements, whereas Uranus had less heavy elements and a small amount of hydrogen.

7.8 Atmospheres of the Giant Planets

Vesto Slipher undertook a detailed investigation of the spectra of Jupiter, Saturn, Uranus, and Neptune in the early decades of the 20th century. He recorded numerous bands for all the planets but had trouble interpreting them. In 1932, Rupert Wildt deduced that a number of the bands in all four planets were due to ammonia and methane. However, subsequent work by Mecke, Dunham, Adel, and Slipher showed that some of the lines had been misattributed, so there was no ammonia in the atmospheres of Uranus and Neptune. This was, presumably, because it had been frozen out at their lower temperatures. Adel and Slipher also concluded that the methane concentration reduced in going from Neptune to Uranus to Saturn to Jupiter.

7.9 Jupiter

In 1955, Burke and Franklin made the unexpected discovery that Jupiter was emitting radio waves at 22.2 MHz. Subsequently, it was found that Jupiter emitted energy at many radio frequencies. Some of it was thermal energy, with an effective temperature of 145 K, but some was clearly nonthermal. The latter was taken to indicate that Jupiter had an intense magnetic field, with radiation belts similar to those that had, by then, been found around the Earth.

Our knowledge of Jupiter's Galilean satellites changed little in the 20th century before the space age. In 1900, Bernard had observed that the poles of Io appeared to be reddish in color. Then in 1914 Paul Guthnik showed that all four Galilean satellites exhibited synchronous rotation. In the 19th century, it was thought that all four satellites probably had atmospheres, but this was considered more and more unlikely as the 20th century progressed.

7.10 Saturn

A prominent white equatorial spot had been observed on Saturn in 1876. Then in 1903 Edward Barnard discovered another temporary prominent white spot at about $36°N$, but its rotation period around Saturn was some 25 minutes slower. Another equatorial spot that had a similar period to the 1876 equatorial spot appeared in 1933, and another spot that had a similar period to the 1903 spot was observed at about $60°N$ in 1960. The velocities of these spots showed that there was an equatorial current on Saturn, similar to that on Jupiter. But the one on Saturn had a velocity of about 1400 km/h, compared with just 400 km/h for Jupiter. It was unclear why Saturn, which is farther from the Sun, and so receives less heat than Jupiter, should have a much faster equatorial current.

Markings on Saturn's rings were seen by a number of observers in the late 19th and early 20th centuries, including the respected observers Etienne Trouvelot and Eugène Antoniadi. In 1955, Guido Ruggieri noticed clear radial streaks at both ansae of the A ring, but after further investigation he concluded that they were an optical illusion. It is unclear whether any of these observations were early observations of spokes, of the sort discovered by the *Voyager* spacecraft on the B ring, or not.

In the winter of 1943–1944, Gerard Kuiper photographed the spectrum of the ten largest satellites of the solar system and found evidence for an atmosphere on Titan and possibly Triton. He could find no such evidence for the Galilean satellites of Jupiter, however.

7.11 Uranus and Neptune

In the 19th century, Triton had been found to orbit Neptune in a retrograde sense, and it was unclear at the time whether Neptune's spin was also retrograde. But in 1928 Moore and Menzel found, by observing the **Doppler shift** of its spectral lines, that Neptune's spin was **direct** or **prograde**. So Neptune's largest satellite was orbiting the planet in the opposite sense to the planet's spin. This phenomenon had not been observed before in the solar system for a major satellite.

Kuiper discovered Uranus' fifth satellite, now called Miranda, in 1948. It was orbiting the planet in an approximately circular orbit inside that of the other four satellites. Then in the following year he discovered Neptune's second satellite, now called Nereid, orbiting Neptune in the opposite sense to Triton. Nereid was in a highly elliptical orbit well outside the orbit of Triton. So Nereid was the "normal" satellite in orbiting Neptune direct or prograde, whereas the larger Triton, which was nearer to Neptune in an almost circular orbit, appeared to be the abnormal one.

7.12 The Discovery of Pluto

The discoveries of Uranus and Neptune made astronomers realize that there may well be planets even farther out from the Sun. As Neptune had only been discovered in 1846, and as it was moving very slowly, its orbit was not very well known in the second half of the 19th century. However astronomers had much better information on Uranus' orbit, and so they reexamined it to see if there were any unexplained deviations that might indicate the whereabouts of a new planet. Such deviations were soon found, and a number of possible locations for the new planet proposed by various astronomers, including Percival Lowell. A photographic search for the new planet was started at Lowell's observatory, but this was abandoned when Lowell died in 1916.

In 1929, Vesto Slipher, the new director of Lowell's observatory, recruited Clyde Tombaugh to undertake a search for the new planet using a photographic refractor that had been specifically purchased for the task. Tombaugh photographed the whole of the zodiac, and used a blink comparator to find objects that had moved over time. The task was very tedious, but he discovered Pluto in February 1930 after working for 10 months. However, although the planet's orbit was very similar to that predicted by Lowell (Fig. 8), it

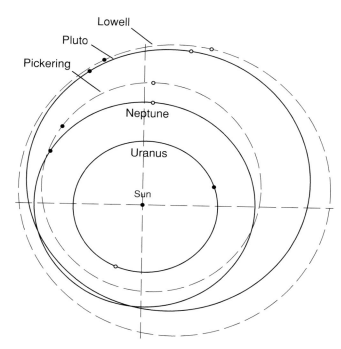

FIGURE 8 A comparison between the true orbit of Pluto and that predicted by Lowell and Pickering. Although Lowell's orbit was reasonably close to that of Pluto, the agreement was fortuitous. (The open circles show the positions of the planets in 1900, and the closed circles represent those in 1930.)

was far too small to have perturbed Uranus in the way that Lowell had estimated.

Over the years, the estimated mass of Pluto has gradually reduced from 6.6 M_E (M_E is the mass of the Earth) predicted by Lowell, to 0.7 M_E (maximum) at the time of its discovery, to 0.002 M_E now. Its orbit is highly eccentric, and it has the largest inclination of the traditional planets.

In 1955, Walker and Hardie deduced a rotation period of 6d 9h 17min from regular fluctuations in Pluto's intensity. Little more was known about the planet when the space age started.

7.13 Asteroids

In 1918, Kiyotsugu Hirayama identified families of asteroids based on their orbital radius, eccentricity, and inclination. Initially, he identified three families, Themis (22 members), Eos (21 members), and Koronis (13 members). Hirayama suggested that the three families were each the remnants of a larger asteroid that had fractured. This resurrected, in modified form, the theories of Thomas Wright and Wilhelm Olbers, in the 18th and 19th centuries. They both believed that there had been a planet between the orbits of Mars and Jupiter that had broken up.

In the 19th century, Eros had been discovered with a perihelion of 1.13 AU. In 1932, another asteroid, now called Amor, was found that had an orbit that came even closer to that of the Earth than Eros. Then, just 6 weeks later, the first asteroid, now called Apollo, whose orbit crossed that of the Earth, was discovered. The names of Amor and Apollo have now been given to families of asteroids with similar orbital characteristics.

7.14 Comets

Huggins had shown in the 19th century that there were hydrocarbon compounds in the heads of comets, but he was not able to specify exactly which hydrocarbons were involved. Molecular carbon, C_2, was first identified in the head of a comet just after the turn of the century, and by the mid-1950s C_3, CH, CN, OH, NH, and NH_2, had been found in the heads of comets.

Molecular bands were observed in the tail of Daniel's comet by Deslandres, Bernard, and Evershed in 1907 and in the tail of Morehouse's comet by Deslandres and Bernard the following year. These bands were later identified by Alfred Fowler as those of ionized carbon monoxide, (CO^+) and N_2^+. Later CO_2^+ as also found in the tail of a comet.

In the 1930s, Karl Wurm observed that many of the molecules found in comets were chemically very active, and so they cannot have been present there for very long. He suggested, instead, that they had come from the more stable so-called parent molecules $(CN)_2$, H_2O, and CH_4 (methane). In 1948, Pol Swings, in his study of Encke's comet, concluded that the parent molecules were water, methane, ammonia (NH_3), nitrogen, carbon monoxide and carbon dioxide, all of which had been in the form of ice before being heated by the Sun.

In 1950 and 1951, Fred Whipple proposed his icy-conglomerate model (better known as his dirty snowball theory) in which the nucleus is composed of ices, such as methane, with meteoric material embedded within it. Unfortunately, some of the parent molecules were highly volatile. But in 1952 Delsemme and Swings suggested that these highly volatile elements would be able to resist solar heating better if they were trapped within the crystalline structure of water ice, in what are known as clathrate hydrates.

It was difficult to determine the orbits of long-period comets because they were only observed for the fraction of their orbit when they were close to the Sun. However, a survey of about 400 cometary orbits observed up to 1910 showed that only a tiny minority appeared to be hyperbolic. Strömgren and Fayet then showed that none of these comets had hyperbolic orbits before they passed Saturn or Jupiter on their approach to the Sun. So the long-period comets appeared to be members of the solar system.

In 1932, Ernst Öpik concluded, from an analysis of stellar perturbations, that comets could remain bound to the Sun at distances of up to 10^6 AU. Some years later, Adrianus Van Woerkom showed that there must be a continuous source of new, near-parabolic comets to explain the relative numbers observed. Then in 1950 Jan Oort showed that the orbits of 10 comets, with near parabolic orbits, had an average **aphelion** distance of about 100,000 AU. As a result, he suggested that all long-period comets originate in what is called the Oort cloud about 50,000 to 150,000 AU from the Sun.

7.15 The Origin of the Solar System

In the early decades of the 20th century, theories of the origin of the solar system generally focused on the effect of collisions, and close encounters of another star to the Sun. But all the theories were found to have significant problems, so Laplace's theory of a condensing nebula was reconsidered.

Laplace's theory had been rejected in the 19th century because the original solar nebula did not appear to have had enough angular momentum. However, in the 1930s, McCrea showed that this would not be a problem if the original nebula had been turbulent.

In 1943, Carl von Weizsäcker produced a theory where cells of circulating convection currents, or vortices, formed in the solar nebula after the Sun had condensed. These vortices produced planetesimals that grew to form planets by accretion. Unfortunately, as Chandrasekhar and Kuiper showed, the vortices would not be stable enough to allow condensation to take place. Kuiper then produced his own theory, as did Safronov and others, with the common theme of planetesimals merging to form planets, but none was fully satisfactory.

Bibliography

Hoskin, M., ed. (1997). "Cambridge Illustrated History of Astronomy." Cambridge Univ. Press, Cambridge, England.

Hufbauer, K. (1993). "Exploring the Sun; Solar Science Since Galileo." Johns Hopkins Univ. Press, Baltimore.

Koestler, A. (1990). "The Sleepwalkers: A History of Man's Changing Vision of the Universe." Penguin Books.

Leverington, D. (2003). "Babylon to Voyager and Beyond: A History of Planetary Astronomy." Cambridge Univ. Press, Cambridge, England.

North, J. (1995). "The Norton History of Astronomy and Cosmology." Norton, New York.

Pannekoek, A. (1961). "A History of Astronomy." Interscience, New York (Dover reprint 1989).

Taton, R., and Wilson, C., eds. (1989). "The General History of Astronomy, Vol. 2, Planetary Astronomy from the Renaissance to the Rise of Astrophysics: Part A, Tycho Brahe to Newton." Cambridge Univ. Press, Cambridge, England.

Taton, R. and Wilson, C., eds. (1995). "The General History of Astronomy, Vol. 2, Planetary Astronomy from the Renaissance to the Rise of Astrophysics: Part B, The Eighteenth and Nineteenth Centuries." Cambridge Univ. Press, Cambridge, England.

The Sun

Markus J. Aschwanden

Lockheed Martin ATC
Solar and Astrophysics Laboratory
Palo Alto, Callifornia

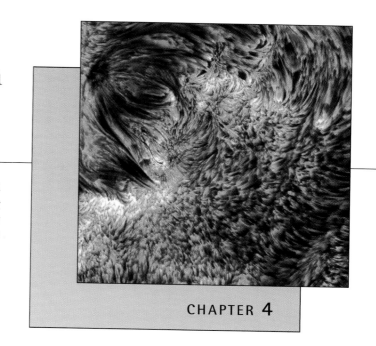

CHAPTER 4

1. Introduction

The Sun is the central body and energy source of our solar system. The Sun is our nearest star, but otherwise it represents a fairly typical star in our galaxy, classified as $G2$-V spectral type, with a radius of $r_\circ \approx 700{,}000$ km, a mass of $m_\circ \approx 2 \times 10^{33}$ g, a luminosity of $L_\circ \approx 3.8 \times 10^{26}$ W, and an age of $t_\circ \approx 4.6 \times 10^9$ years (Table 1). The distance from the Sun to our Earth is called an astronomical unit (AU) and amounts to $\sim\!150 \times 10^6$ km. The Sun lies in a spiral arm of our galaxy, the Milky Way, at a distance of 8.5 kiloparsecs from the galactic center. Our galaxy contains $\sim\!10^{12}$ individual stars, many of which are likely to be populated with similar solar systems, according to the rapidly increasing detection of extrasolar planets over the last years; the binary star systems are very unlikely to harbor planets because of their unstable, gravitationally disturbed orbits. The Sun is for us humans of particular significance, first because it provides us with the source of all life, and second because it furnishes us with the closest laboratory for astrophysical plasma physics, magneto-hydrodynamics (MHD), atomic physics, and particle physics. The Sun still represents the only star from which we can obtain spatial images, in many wavelengths.

The basic structure of the Sun is sketched in Fig. 1. The Sun and the solar system were formed together from an interstellar cloud of molecular hydrogen some 5 billion years ago. After gravitational contraction and subsequent collapse, the central object became the Sun, with a central temperature hot enough to ignite thermonuclear reactions, the ultimate source of energy for the entire solar system. The chemical composition of the Sun consists of 92.1% hydrogen and 7.8% helium by number (or 27.4% He by mass), and 0.1% of heavier elements (or 1.9% by mass, mostly C, N, O, Ne, Mg, Si, S, Fe). The central core, where hydrogen burns into helium, has a temperature of $\sim\!15$ million K (Fig. 1). The solar interior further consists of a radiative zone, where energy is transported mainly by radiative diffusion, a process where photons with hard X-ray (keV) energies get scattered, absorbed, and reemitted. The outer third of the solar interior is called the convective zone, where energy is transported mostly by convection. At the solar surface, photons leave the Sun in optical wavelengths, with an energy that is about a factor of 10^5 lower than the original hard X-ray photons generated in the nuclear core, after a random walk of $\sim\!10^5$–10^6 years.

The irradiance spectrum of the Sun is shown in Fig. 2, covering all wavelengths from gamma rays, hard X-rays, soft X-rays, extreme ultraviolet (EUV), ultraviolet, white light, infrared, to radio wavelengths. The quiet Sun irradiates most of the energy in visible (white-light) wavelengths, to which our human eyes have developed the prime sensitivity during the evolution. Emission in extreme ultraviolet is dominant in the solar **corona** because it is produced

TABLE 1	Basic Physical Properties of the Sun
Physical Parameter	**Numerical Value**
Solar radius, $R_\odot$	695,500 km
Solar mass, $m_\odot$	1.989×10^{33} g
Mean density, $\rho_\odot$	1.409 g cm^{-3}
Gravity at solar surface, $g_\odot$	274.0 m s^{-2}
Escape velocity at solar surface, v	617.7 km s^{-1}
Synodic rotation period, P	$P = 27.3$ days (equator)
Sidereal rotation period, P	$P = 25.4$ days (equator)
	$P = 35.0$ days (at latitude $\pm 70°$)
Mean distance from Earth	1 AU = 149,597,870 km
Solar luminosity, $L_\odot$	3.844×10^{26} W (or 10^{33} ergs s^{-1})
Solar age, $t_\odot$	4.57×10^{9} years
Temperature at Sun center, T_c	15.7×10^{6} K
Temperature at solar surface, T_{ph}	6400 K

Source: Cox, 2000.

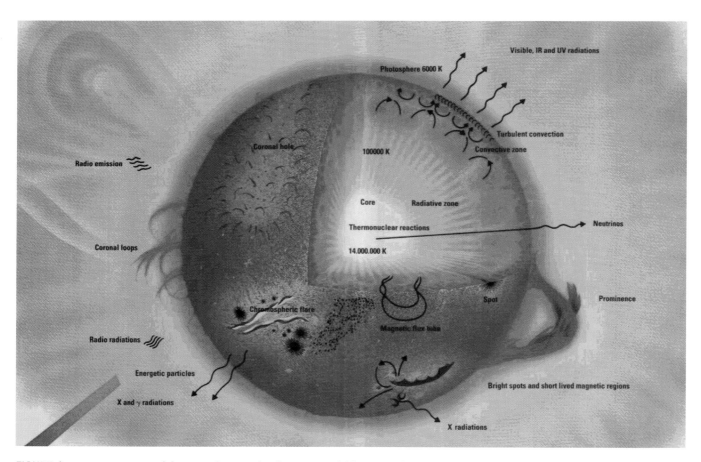

FIGURE 1   A cutaway view of the Sun, showing the three internal (thermonuclear, radiative, and convective) zones, the solar surface (photosphere), the lower (chromosphere) and upper atmosphere (corona), and a number of phenomena associated with the solar activity cycle (filaments, prominences, flares). (Courtesy of Calvin J. Hamilton and NASA/ESA.)

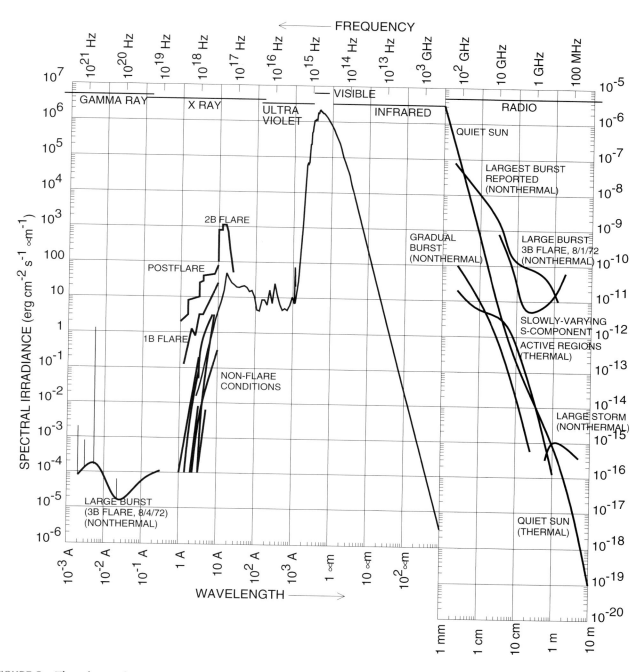

FIGURE 2   The solar irradiance spectrum from gamma rays to radio waves. The spectrum is shifted by 12 orders of magnitude in the vertical axis at λ = 1 mm to accommodate for the large dynamic range in spectral irradiance. (Courtesy of H. Malitson and NASA/NSSDC.)

by ionized plasma in the coronal temperature range of ~1–2 million K. Emissions in shorter wavelengths require higher plasma temperatures and thus occur during **flares** only. Flares also accelerate particles to nonthermal energies, which cause emission in hard X-rays, gamma rays, and radio wavelengths, but to a highly variable degree.

2. The Solar Interior

The physical structure of the solar interior is mostly based on theoretical models that are constrained (1) by global quantities (age, radius, luminosity, total energy output; see Table 1); (2) by the measurement of global oscillations (helioseismology); and (3) by the neutrino flux, which now constrains for the first time elemental abundances in the solar interior, since the neutrino problem has been solved in the year 2001.

2.1 Standard Models

There are two types of models of the solar interior: (1) hydrostatic equilibrium models and (2) time-dependent numerical simulations of the evolution of the Sun, starting from an initial gas cloud to its present state today, after ~8% of the hydrogen has been burned into helium.

The standard hydrostatic model essentially calculates the radial run of temperature, pressure, and density that fulfill the conservation of mass, momentum, and energy in all internal spherical layers of the Sun, constrained by the boundary conditions of radius, temperature, and radiation output (luminosity) at the solar surface, the total mass, and the chemical composition. Furthermore, the ideal gas law and thermal equilibrium is assumed, and thus the radiation is close to that of an ideal black body. The solar radius has been measured by triangulation inside the solar system (e.g., during a Venus transit) and by radar echo measurements. The mass of the Sun has been deduced from the orbital motions of the planets (Kepler's laws) and from precise laboratory measurements of the gravitational constant. The solar luminosity is measured by the heat flux received at Earth. From these standard models, a central temperature of ~15 million K, a central density of ~150 g cm^{-3}, and a central pressure of 2.3×10^{17} dyne cm^{-2} have been inferred. Fine-tuning of the standard model is obtained by including convective transport and by varying the (inaccurately known) helium abundance.

2.2 Thermonuclear Energy Source

The source of solar energy was understood in the 1920s, when Hans Bethe, George Gamow, and Carl Von Weizsäcker identified the relevant nuclear chain reactions that generate solar energy. The main nuclear reaction is the transformation of hydrogen into helium, where 0.7% of the mass is converted into radiation (according to Einstein's energy equivalence, $E = mc^2$), the so-called p-p chain, which starts with the fusion of two protons into a nucleus of deuterium (^{2}He), and, after chain reactions involving ^{3}He, ^{7}Be, and ^{7}Li, produces helium (^{4}He),

$$p + p \rightarrow {}^{2}\text{He} + e^+ + \nu_e$$
$$^{2}\text{He} + p \rightarrow {}^{3}\text{He} + \gamma$$
$$^{3}\text{He} + {}^{3}\text{He} \rightarrow {}^{4}\text{He} + p + p$$

or

$$^{3}\text{He} + {}^{4}\text{He} \rightarrow {}^{7}\text{Be} + \gamma$$
$$^{7}\text{Be} + e^- \rightarrow {}^{7}\text{Li} + \nu_e$$
$$^{7}\text{Li} + p \rightarrow {}^{8}\text{Be} + \gamma \rightarrow {}^{4}\text{He} + {}^{4}\text{He}$$

One can estimate the Sun's lifetime by dividing the available mass energy by the luminosity,

$$t_\circ \approx 0.1 \times 0.007 \, m_\circ c^2 / L_\circ \approx 10^{10} \text{ years}$$

where we assumed that only about a fraction of 0.1 of the total solar mass is transformed because only the innermost core of the Sun is sufficiently hot to sustain nuclear reactions.

An alternative nuclear chain reaction occurring in the Sun and stars is the carbon–nitrogen–oxygen (CNO) cycle,

$$^{12}\text{C} + p \rightarrow {}^{13}\text{N} + \gamma$$
$$^{13}\text{N} \rightarrow {}^{13}\text{C} + e^+ + \nu_e$$
$$^{13}\text{C} + p \rightarrow {}^{14}\text{N} + \gamma$$
$$^{14}\text{N} + p \rightarrow {}^{15}\text{O} + \gamma$$
$$^{15}\text{O} \rightarrow {}^{15}\text{N} + e^+ + \nu_e$$
$$^{15}\text{N} + p \rightarrow {}^{12}\text{C} + {}^{4}\text{He}$$

The p-p chain produces 98.5% of the solar energy, and the CNO cycle produces the remainder, but the CNO cycle is faster in stars that are more massive than the Sun.

2.3 Neutrinos

Neutrinos interact very little with matter, unlike photons, and thus most of the electronic neutrinos (ν_e), emitted by the fusion of hydrogen to helium in the central core, escape the Sun without interactions and a very small amount is detected at Earth. Solar neutrinos have been detected since 1967, pioneered by Raymond Davis, Jr., using a chlorine tank in the Homestake Gold Mine in South Dakota, but the observed count rate was about a third of the theoretically expected value, causing the puzzling neutrino problem that persisted for the next 35 years. However, Pontecorvo and Gribov predicted already in 1969 that low-energy solar neutrinos undergo a "personality disorder" on their travel to Earth and oscillate into other atomic flavors of muonic neutrinos (ν_μ) (from a process involving a muon particle) and tauonic neutrinos (ν_t) (from a process involving a tauon particle), which turned out to be the solution of the missing neutrino problem for detectors that are only sensitive to the highest-energy (electronic) neutrinos, such as the Homestake chlorine tank and the gallium detectors GALLEX in Italy and SAGE in Russia. Only the Kamiokande and Super-Kamiokande-I pure-water experiments and the Sudbury Neutrino Observatory (SNO; Ontario, Canada) heavy-water experiments are somewhat sensitive to the muonic and tauonic neutrinos. It was the SNO that measured in 2001 for the first time all three lepton flavors and, in this way, brilliantly confirmed the theory of neutrino (flavor) oscillations. Today, after the successful solution of the neutrino problem, the measured neutrino fluxes are sufficiently accurate to constrain the helium abundance and heavy element abundances in the solar interior.

2.4 Helioseismology

In the decade of 1960–1970, global oscillations were discovered on the solar surface in visible light, which became the field of helioseismology. Velocity oscillations were first measured by R. Leighton, and then interpreted in 1970 as standing sound waves in the solar convection zone by R. Ulrich, C. Wolfe, and J. Leibacher. These acoustic oscillations, also called p-modes (pressure-driven waves), are detectable from fundamental up to harmonic numbers of ~1000 and are most conspicuous in dispersion diagrams, $\omega(k)$, where each harmonic shows up as a separate ridge, when the oscillation frequency (ω) is plotted as function of the wavelength λ (i.e., essentially the solar circumference divided by the harmonic number). Frequencies of the p-mode correspond to periods of ~5 minutes. An example of a p-mode standing wave is shown in Fig. 3 (left), which appears like a standing wave on a drum skin. Each mode is characterized by the number of radial, longitudinal, and latitudinal nodes, corresponding to the radial quantum number n, the azimuthal number m, and the degree l of spherical harmonic functions. Since the density and temperature increase monotonically with depth inside the Sun, the sound speed varies as a function of radial distance from the Sun center. P-mode waves excited at the solar surface propagate downward and are refracted toward the surface, where the low harmonics penetrate very deep, whereas high harmonics are confined to the outermost layers of the solar interior. By measuring the frequencies at each harmonic, the sound speed can be inverted as a function of the depth; in this way, the density and temperature profile of the solar interior can be inferred and unknown parameters of theoretical standard models can be constrained, such as the abundance of helium and heavier elements. By exploiting the Doppler effect, frequency shifts of the p-mode oscillations can be used to measure the internal velocity rates as a function of depth and latitude, as shown in Fig. 3 (right). A layer of rapid change in the internal rotation rate was discovered this way at the bottom of the convection zone, the so-called tachocline (at 0.693 ± 0.002 solar radius, with a thickness of 0.039 ± 0.013 solar radius).

Besides the p-mode waves, gravity waves (g-modes), where buoyancy rather than pressure supplies the restoring force, are suspected in the solar core. These gravity waves are predicted to have long periods (hours) and very small velocity amplitudes, but they have not yet been convincingly detected.

Global helioseismology detects p-modes as a pattern of standing waves that encompass the entire solar surface;

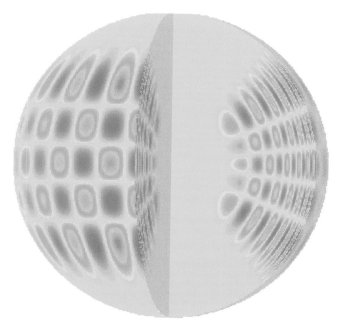 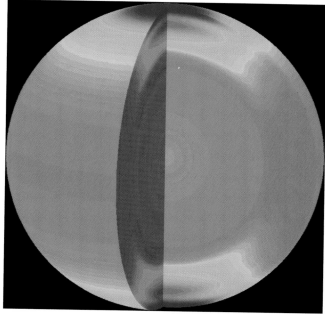

FIGURE 3 Left: A global acoustic p-mode wave is visualized: The radial order is $n = 14$, the angular degree is $l = 20$, the angular order is $m = 16$, and the frequency is $\nu = 2935.88 \pm 0.1\,\mu Hz$ with *SoHO*/MDI (Michelson Doppler Image). The red and blue zones show displacement amplitudes of opposite sign. Right: The internal rotation rate is shown with a color code, measured with SoHO/MDI during May 1996–April 1997. The red zone shows the fastest rotation rates ($P \approx 25$ days), dark blue the slowest ($P \approx 35$ days). Note that the rotation rate varies in latitude differently in the radiative and convective zones. (Courtesy of *SoHO*/MDI and NASA.)

however, local deviations of the sound speed can also be detected beneath sunspots and active regions, a diagnostic that is called local helioseismology. Near sunspots, p-modes are found to have oscillation periods in the order of 3 minutes, compared to 5 minutes in active region plages and quiet-Sun regions.

2.5 Solar Dynamo

The Sun is governed by a strong magnetic field (much stronger than those on planets), which is generated with a magnetic field strength of $B \approx 10^5$ G in the tachocline, the thin shear layer sandwiched between the radiative and the convective zone. Buoyant magnetic flux tubes rise through the convection zone (due to the convective instability obeying the Schwartzschild criterion) and emerge at the solar surface in active regions, where they form sunspots with magnetic field strengths of $B \approx 10^3$ G and coronal loops with field strengths of $B \approx 10^2$ G at the photospheric footpoints, and $B \approx 10$ G in larger coronal heights. The differential rotation on the solar surface is thought to wind up the surface magnetic field, which then fragments under the magnetic stress, circulates meridionally to the poles, and reorients from the toroidally stressed state (with field lines oriented in east–west direction) at solar maximum into a poloidal dipole field (connecting the North with the South Pole) in the solar minimum. This process is called the solar dynamo, which flips the magnetic polarity of the Sun every ~11 years (the solar cycle), or returns to the same magnetic configuration every ~22 years (the Hale cycle). The solar cycle controls the occurrence rate of all solar activity phenomena—from sunspot numbers, active regions, to flares, and **coronal mass ejections (CMEs)**.

3. The Photosphere

The **photosphere** is a thin layer at the solar surface that is observed in white light. The irradiance spectrum in Fig. 2 shows the maximum at visible wavelengths, which can be fitted with a black-body spectrum with a temperature of $T \approx 6400$ K at wavelengths of $\lambda \geq 2000$ Å, which is the solar surface temperature. The photosphere is defined as the range of heights from which photons directly escape, which encompasses an optical depth range of $0.1 \leq \tau \leq 3$ and translates into a height range of $h \approx 300$ km for the visible wavelength range.

3.1 Granulation and Convection

The photospheric plasma is only partially ionized, there are fewer than 0.001 electrons per hydrogen atom at the photospheric temperature of $T = 6400$ K at $\lambda = 5000$ Å. These few ionized electrons come mostly from less abundant elements with a low ionization potential, such as magnesium, while hydrogen and helium are almost completely atomic. The magnetic field is *frozen in* to the gas under these conditions. However, the temperature is rapidly increasing below the photospheric surface, exceeding the hydrogen ionization temperature of $T = 11,000$ K at a depth of 50 km, where the number of ionized electrons increases to 0.1 electrons per hydrogen atom, and the opacity increases by a similar factor. The high opacity of the partially ionized plasma impedes the heat flow. Moreover, a stratification with a temperature gradient steeper than an adiabatic gradient is unstable to convection (Schwartzschild criterion). Thus the partially ionized photosphere of the Sun, as well as of other low-mass stars (with masses $m < 2m_\odot$ are therefore convective.

The observational manifestation of subphotospheric convection is the granulation pattern (Fig. 4, right), which contains granules with typical sizes of ~1000 km and lifetimes of $\tau \approx 7$ min. The subphotospheric gas flows up in the bright centers of granulation cells, cools then by radiating away some heat at the optically thin photospheric surface, and, while cooling, becomes denser and flows down in the intergranular lanes. This convection process can now be reproduced with numerical simulations that include hydrodynamics, radiative transfer, and atomic physics of ionization and radiative processes (Fig. 4, left). The convection process is also organized on larger scales, exhibiting cellular patterns on scales of ~5,000–10,000 km (mesogranulation) and on scales of ~20,000 km (supergranulation).

3.2 Photospheric Magnetic Field

Most of what we know about the solar magnetic field is inferred from observations of the photospheric field, from the Zeeman effect of spectral lines in visible wavelengths (e.g., Fe 5250 Å). From two-dimensional (2D) maps of the photospheric magnetic field strength, we extrapolate the coronal three-dimensional (3D) magnetic field, or try to trace the subphotospheric origin from emerging magnetic flux elements. The creation of magnetic flux is thought to happen in the tachocline at the bottom of the convection zone, from where it rises upward in form of buoyant magnetic flux tubes and emerges at the photospheric surface. The strongest fields emerge in sunspots, amounting to several kilogauss field strengths, and fields with strengths of several 100 G emerge also all over in active regions, often in the form of a leading sunspot trailed by following groups of opposite magnetic polarity. Due to the convective motion, small magnetic flux elements that emerge in the center of granulation cells are then swept to the intergranular lanes, where often unresolved small concentrations are found, with sizes of less than a few 100 km. The flow velocities due to photospheric convection are on the order of

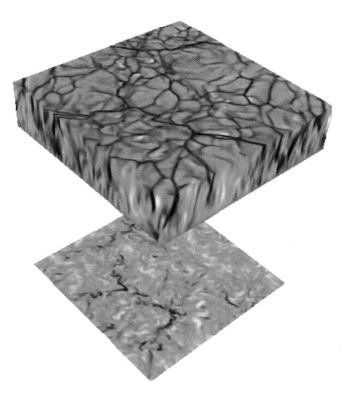

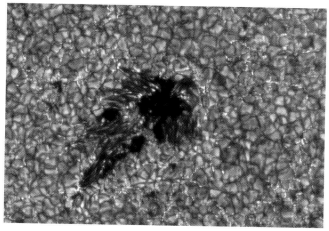

FIGURE 4 (Left) Numerical simulation of cellular convection at the solar surface, performed by Fausto Cattaneo and Andrea Malagoli; (Right) High-resolution observation of the granulation pattern in the solar photosphere. A granule has a typical size of 1000 km, representing the surface of an elementary convection cell. The large black area represents a sunspot, where the temperature is cooler than the surroundings. (This image was taken by Tom Berger with the Swedish Solar Observatory.)

~ 1 km s^{-1}. In the quiet Sun, away from active regions, the mean photospheric magnetic field amounts to a few Gauss.

3.3 Sunspots

Sunspots are the areas with the strongest magnetic fields, and therefore a good indicator of the solar activity (Fig. 5, bottom). The *butterfly diagram* shows that sunspots (or active regions) appear first at higher latitudes early in the solar cycle and then drift equatorward toward the end of the solar cycle (Fig. 5, top). Since all solar activity phenomena are controlled by the magnetic field, they have a similar solar cycle dependence as sunspots, such as the flare rate, active region area, global soft X-ray brightness, and radio emission. The appearance of dark sunspots lowers the total luminosity of the Sun only by about 0.15% at sunspot maximum, and thus the variation of the sunlight has a negligible effect on the Earth's climate. The variation of the EUV emission, which affects the ionization in the Earth's ionosphere, however, has a more decisive impact on the Earth's climate.

An individual sunspot consists of a very dark central umbra, surrounded by a brighter, radially striated penumbra. The darkness of sunspots is attributed to the inhibition of convective transport of heat, emitting only about 20% of the average solar heat flux in the umbra and being significantly cooler (~ 4500 K) than the surroundings (~ 6000 K). Their diameters range from 3600 to 50,000 km, and their lifetime ranges from a week to several months. The magnetic field in the umbra is mostly vertically oriented, but it is strongly inclined over the penumbra, nearly horizontally. Current theoretical models explain the interlocking comb structure of the filamentary penumbra with outward submerged field lines that are pumped down by turbulent, compressible convection of strong descending plumes.

Sunspots are used to trace the surface rotation, since Galileo in 1611. The average sidereal differential rotation rate is

$$\omega = 14.522 - 2.84 \sin^2 \Phi \ \mathrm{deg/day}$$

where Φ is the heliographic latitude. The rotation rate of an individual feature, however, can deviate from this average

DAILY SUNSPOT AREA AVERAGED OVER INDIVIDUAL SOLAR ROTATIONS

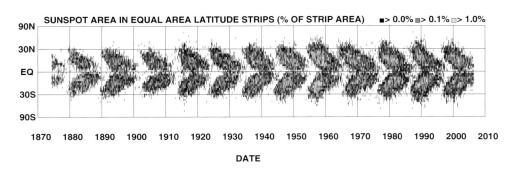

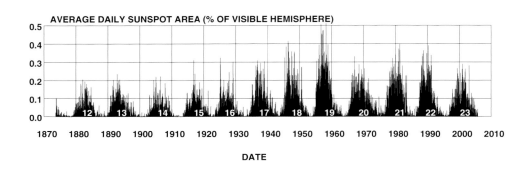

FIGURE 5 Top: Butterfly diagram of sunspot appearance, which marks the heliographic latitude of sunspot locations as a function of time, during the solar cycles 12–23 (covering the years 1880–2000). Bottom: Sunspot area as a function of time, which is a similar measure of the solar cycle activity as the sunspot number. (Courtesy of D. Hathaway and NASA/MSFC.)

by a few percent because it depends on the anchor depth to which the feature is rooted, since the solar internal differential rate varies radially (Fig. 3, right).

4. The Chromosphere and Transition Region

4.1 Basic Physical Properties

The **chromosphere** (from the Greek word $\chi\rho\omega\mu o\sigma$, color) is the lowest part of the solar atmosphere, extending to an average height of ~2000 km above the photosphere. The first theoretical concepts conceived the chromosphere as a spherical layer around the solar surface (in the 1950s; Fig. 6, left), while later refinements included the diverging magnetic fields (canopies) with height (in the 1980s; Fig. 6, middle), and finally ended up with a very inhomogeneous mixture of cool gas and hot plasma, as a result of the extremely dynamic nature of chromospheric phenomena (in the 2000s; Fig. 6, right). According to hydrostatic standard models assuming local thermodynamic equilibrium (LTE), the temperature reaches first a temperature minimum of $T = 4300$ K at a height of $h \approx 500$ km above the photosphere, and rises then suddenly to ~10,000 K in the upper chromosphere at $h \approx 2000$ km, but the hydrogen density drops by about a factor of 10^6 over the same chromospheric height range. These hydrostatic models have been criticized because they neglect the magnetic field, horizontal inhomogeneities, dynamic processes, waves, and non-LTE conditions.

Beyond the solar limb (without having the photosphere in the background), the chromospheric spectrum is characterized by emission lines; these lines appear dark on the disk as a result of photospheric absorption. The principal lines of the photospheric spectrum are called the Fraunhofer lines, including, for example, hydrogen lines (H I; with the Balmer series Hα (6563 Å), Hβ 4861 Å, Hγ 4341 Å, Hδ 4102 Å), calcium lines (Ca II; K 3934 Å, H 3968 Å), and helium lines (He I; D_3 5975 Å).

4.2 Chromospheric Dynamic Phenomena

The appearance and fine structure of the chromosphere varies enormously depending on which spectral line, wavelength, and line position (core, red wing, blue wing) is used because of their sensitivity to different temperatures (and thus altitudes) and Doppler shifts (and thus velocity ranges). In the H and K lines of Ca II, the chromospheric images show a bright network surrounding supergranulation cells, which coincide with the large-scale subphotospheric convection cells. In the Ca II K2 or in ultraviolet continuum lines (1600 Å), the network and internetwork appear grainier. The so-called bright grains have a high contrast in

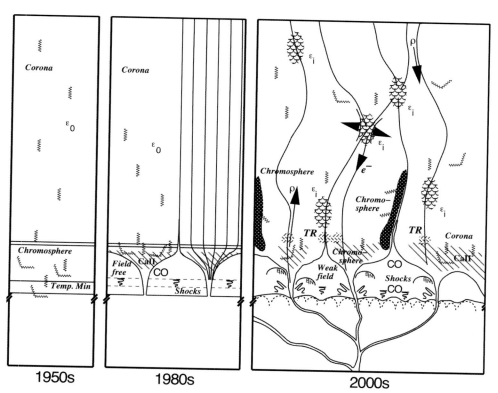

FIGURE 6 Cartoon of geometric concepts of the solar chromosphere, transition region, and corona: gravitationally stratified layers in the 1950s (left), vertical fluxtubes with chromospheric canopies in the 1980s (middle), and a fully inhomogeneous mixing of photospheric, chromospheric, and coronal zones by dynamic processes such as heated upflows, cooling downflows, intermittent heating (ε), nonthermal electron beams (e), field line motions and reconnections, emission from hot plasma, absorption and scattering in cool plasma, acoustic waves, and shocks (right). (Courtesy of Carolus J. Schrijver.)

wavelengths that are sensitive to the temperature minimum (4300 K), with an excessive temperature of 30–360 K, and with spatial sizes of $\sim$1000 km. The bright points in the network are generally associated with magnetic elements that collide, which then heat the local plasma after **magnetic reconnection.** In the intranetwork, bright grains result from chromospheric oscillations that produce shock waves. There are also very thin spaghetti-like elongated fine structures visible in Hα spectroheliograms (Fig. 7, left), which are called fibrils around sunspots. More vertically oriented fine structures are called mottles on the disk, or spicules above the limb. Mottles appear as irregular threads, localized in groups around and above supergranules, at altitudes of 700–3000 km above the photosphere, with lifetimes of 12–20 min, and are apparently signatures of upward and downward motions of plasmas with temperatures of $T = 8000$–15,000 K and velocities of $v \approx 5$–10 km s^{-1}. Spicules (Fig. 7, right) are jet-like structures of plasma with temperatures of $T \approx 10,000$ K that rise to a maximum height of $h \approx 10,000$ km into the lower corona, with velocities of $v \approx 20$ km s^{-1}. They carry a maximum flux of 100 times the solar wind into the low corona. Recent numerical simulations by DePontieu and Erdélyi show that global (helioseismic) p-mode oscillations leak sufficient energy from the global resonant cavity into the chromosphere to power shocks that drive upward flows and form spicules. There is also the notion that mottles, fibrils, and spicules could be unified, being different manifestations of the same

physical phenomenon at different locations (quiet Sun, active region, above the limb), in analogy to the unification of **filaments** (on the disk) and **prominences** (above the limb).

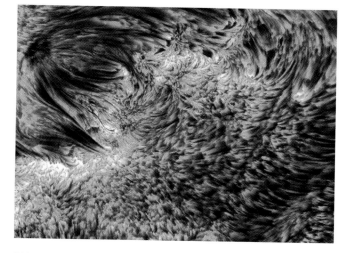

FIGURE 7 High-resolution image of Active Region 10380 on June 16, 2003, located near the limb, showing chromospheric spicules in the right half of the image. The image was taken with the Swedish 1-m Solar Telescope (SST) on La Palma, Spain, using a tunable filter, tuned to the blue-shifted line wing of the Hα 6563 Å line. The spicules are jets of moving gas, flowing upward in the chromosphere with a speed of $\sim$15 km s^{-1}. The scale of the image is 65,000 × 45,000 km. (Courtesy of Bart DePontieu.)

5. The Corona

It is customary to subdivide the solar corona into three zones, which all vary their size during the solar cycle: (1) active regions, (2) quiet-Sun regions, and (3) coronal holes.

5.1 Active Regions

Active regions are located in areas of strong magnetic field concentrations, visible as sunspot groups in optical wavelengths or magnetograms. Sunspot groups typically exhibit a strongly concentrated leading magnetic polarity, followed by a more fragmented trailing group of opposite polarity. Because of this bipolar nature, active regions are mainly made up of closed magnetic field lines. Due to the permanent magnetic activity in terms of magnetic flux emergence, flux cancellation, magnetic reconfigurations, and magnetic reconnection processes, a number of dynamic processes such as plasma heating, flares, and CMEs occur in active regions. A consequence of plasma heating in the chromosphere are upflows into coronal loops, which give active regions the familiar appearance of numerous filled loops, which are hotter and denser than the background corona, producing bright emission in soft X-rays and EUV wavelengths. In the EUV image shown in Fig. 8, active regions appear in white.

5.2 Quiet–Sun Regions

Historically, the remaining areas outside of active regions were dubbed quiet-Sun regions. Today, however, many dynamic processes have been discovered all over the solar surface, so that the term *quiet Sun* is considered to be a misnomer, only justified in relative terms. Dynamic processes in the quiet Sun range from small-scale phenomena such as network heating events, nanoflares, explosive events, bright points, and soft X-ray jets, to large-scale structures, such as transequatorial loops or coronal arches. The distinction between active regions and quiet-Sun regions becomes more and more blurred because most of the large-scale structures that overarch quiet-Sun regions are rooted in active regions. A good working definition is that quiet-Sun regions encompass all closed magnetic field regions (excluding active regions), which demarcates the quiet-Sun territory from coronal holes (that encompass open magnetic field regions).

5.3 Coronal Holes

The northern and southern polar zones of the solar globe have generally been found to be darker than the equatorial zones during solar eclipses. Max Waldmeier thus dubbed those zones as coronal holes (i.e., *Koronale Löcher* in German). Today it is fairly clear that these zones are dom-

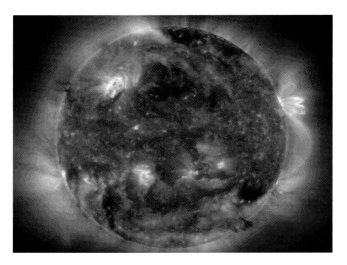

FIGURE 8   The multitemperature corona, recorded with the EIT (Extreme-ultraviolet Imaging Telescope) instrument on board the *SoHO* spacecraft. The representation shown here is a false-color composite of three images all taken in extreme ultraviolet light. Each individual image highlights a different temperature regime in the upper solar atmosphere and was assigned a specific color; red at 2 million, green at 1.5 million, and blue at 1 million degrees K. The combined image shows active regions in white color (according to Newton's law of color addition), because they contain many loops with different temperatures. Also, nested regions above the limb appear in white, because they contain a multitude of loops with different temperatures along a line-of-sight, while isolated loops on the disk show a specific color according to their intrinsic temperature. (Courtesy of EIT/*SoHO* and NASA.)

inated by open magnetic field lines, which act as efficient conduits for flushing heated plasma from the corona into the solar wind, whenever they are fed by chromospheric upflows at their footpoints. Because of this efficient transport mechanism, coronal holes are empty of plasma most of the time, and thus appear much darker than the quiet Sun, where heated plasma flowing upward from the chromosphere remains trapped, until it cools down and precipitates back to the chromosphere. A coronal hole is visible in Fig. 8 at the North Pole, where the field structures point radially away from the Sun and show a cooler temperature ($T \leq 1.0$ MK; dark blue in Fig. 8) than the surrounding quiet-Sun regions.

5.4 Hydrostatics of Coronal Loops

Coronal loops are curvilinear structures aligned with the magnetic field. The cross section of a loop is essentially defined by the spatial extent of the heating source because the heated plasma distributes along the coronal magnetic field lines without cross-field diffusion, since the thermal pressure is much less than the magnetic pressure in the solar corona. The solar corona consists of many thermally isolated loops, where each one has its own gravitational stratification, depending on its plasma temperature. A useful quantity is the hydrostatic pressure scale height λ_p, which depends only on the electron temperature T_e,

$$\lambda_p(T_e) = \frac{2k_B T_e}{\mu m_H g_\odot} \approx 47{,}000 \, \frac{T_e}{1 \, \text{MK}} \quad \text{(km)}.$$

Observing the solar corona in soft X-rays or EUV, which are both optically thin emissions, the line-of-sight integrated brightness intercepts many different scale heights, leading to a hydrostatic weighting bias toward systematically hotter temperatures in larger altitudes above the limb. The observed height dependence of the density needs to be modeled with a statistical ensemble of multihydrostatic loops. Measuring a density scale height of a loop requires careful consideration of projection effects, loop plane inclination angles, cross-sectional variations, line-of-sight integration, and the instrumental response functions. Hydrostatic solutions have been computed from the energy balance between the heating rate, the radiative energy loss, and the conductive loss. The major unknown quantity is the spatial heating function, but analysis of loops in high-resolution images indicate that the heating function is concentrated near the footpoints, say at altitudes of $h \leq 20{,}000$ km. Of course, a large number of coronal loops are found to be not in hydrostatic equilibrium, while nearly hydrostatic loops have been found preferentially in the quiet corona and in older dipolar active regions. An example of an active region [recorded with the Transition Region and Coronal Explorer (TRACE) about 10 hours after a flare] is shown in Fig. 9, which clearly shows superhydrostatic loops where the coronal plasma is distributed over up to four times larger heights than expected in hydrostatic equilibrium (Fig. 9, bottom).

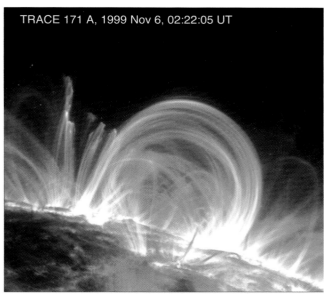

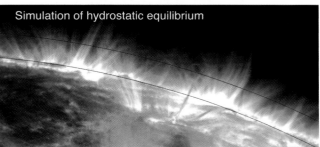

FIGURE 9   An active region with many loops that have an extended scale height of $\lambda_p / \lambda_T \leq 3$–4 (top) has been scaled to the hydrostatic thermal scale height of $T = 1$ MK (bottom). The pressure scale height of the 1 MK plasma is $\lambda_T = 47{,}000$ km, but the observed flux is proportional to the emission measure ($F \rightarrow EM \rightarrow n_e^2$), which has the half pressure scale height $\lambda_T / 2 = 23{,}000$ km.

5.5 Dynamics of the Solar Corona

Although the Sun appears lifeless and unchanging to our eyes, except for the monotonic rotation that we can trace from the sunspot motions, there are actually numerous vibrant dynamic plasma processes continuously happening in the solar corona, which can be detected mainly in EUV and soft X-rays. There is currently a paradigm shift stating that most of the apparently static structures seen in the corona are probably controlled by plasma flows and intermittent heating. It is, however, not easy to measure and track these flows with our remote sensing methods, like the apparently motionless rivers seen from an airplane. For slow flow speeds, the so-called laminar flows, there is no feature to track, while the turbulent flows may be easier to detect because they produce whirls and vortices that can be tracked. A similar situation happens in the solar corona. Occasionally, a moving plasma blob is detected in a coronal loop; it can be used as a tracer. Most of the flows in coronal loops seem to be subsonic (like laminar flows) and thus featureless. Occasionally, we observe turbulent flows, which clearly reveal motion, especially when cool and hot plasma becomes mixed by turbulence and thus yields contrast by emission and absorption in a particular temperature filter. Motion can also be detected with Doppler shift measurements, but this yields only the flow component along the line-of-sight. There is increasing evidence that flows are ubiquitous in the solar corona.

There are a number of theoretically expected dynamic processes. For instance, loops at coronal temperatures are thermally unstable when the radiative cooling time is shorter than the conductive cooling time, or when the heating scale height falls below one third of a loop half length. Recent observations show ample evidence for the presence of flows in coronal loops, as well as evidence for impulsive heating with subsequent cooling, rather than a stationary hydrostatic equilibrium. High-resolution observations of coronal loops reveal that many loops have a superhydrostatic density scale height, far in excess of hydrostatic equilibrium solutions (Fig. 9, top). Time-dependent hydrodynamic simulations are still in a very exploratory phase, and hydrodynamic modeling of the **transition region,** coronal holes, and the solar wind remains challenging due to the number of effects that cannot easily be quantified by observations, such as unresolved geometries, inhomogeneities, time-dependent dynamics, and MHD effects.

The coronal plasma is studied with regard to hydrostatic equilibria in terms of fluid mechanics (hydrostatics), with regard to flows in terms of fluid dynamics (hydrodynamics), and including the coronal magnetic field in terms of magneto-hydrodynamics (MHD). The coronal magnetic field has many effects on the hydrodynamics of the plasma. It can play a passive role in the sense that the magnetic geometry does not change (e.g., by channeling particles, plasma flows, heat flows, and waves along its field lines or by maintaining a thermal insulation between the plasmas of neighboring loops or fluxtubes). On the other hand, the magnetic field can play an active role (where the magnetic geometry changes), such as exerting a Lorentz force on the plasma, building up and storing nonpotential energy, triggering an instability, changing the topology (by various types of magnetic reconnection), and accelerating plasma structures (filaments, prominences, CMEs).

5.6 The Coronal Magnetic Field

The solar magnetic field controls the dynamics and topology of all coronal phenomena. Heated plasma flows along magnetic field lines and energetic particles can only propagate along magnetic field lines. Coronal loops are nothing other than conduits filled with heated plasma, shaped by the geometry of the coronal magnetic field, where cross-field diffusion is strongly inhibited. Magnetic field lines take on the same role for coronal phenomena as do highways for street traffic. There are two different magnetic zones in the solar corona that have fundamentally different properties: open-field and closed-field regions. Open-field regions (white zones above the limb in Fig. 10), which always exist in the polar regions, and sometimes extend toward the equator, connect the solar surface with the interplanetary field and are the source of the fast solar wind ($\sim$800 km s^{-1}). A consequence of the open-field configuration is efficient plasma transport out into the **heliosphere,** whenever chromospheric plasma is heated at the footpoints. Closed-field regions (gray zones in Fig. 10), in contrast, contain mostly closed-field lines in the corona up to heights of about one solar radius, which open up at higher altitudes and connect eventually to the heliosphere, but produce a slow solar wind component of $\sim$400 km s^{-1}. It is the closed-field regions that contain all the bright and overdense coronal loops, produced by filling with chromospheric plasma that stays trapped in these closed-field lines. For loops reaching altitudes higher than about one solar radius, plasma confinement starts to become leaky, because the thermal plasma pressure exceeds the weak magnetic field pressure that decreases with height (plasma-ß parameter < 1).

The magnetic field on the solar surface is very inhomogeneous. The strongest magnetic field regions are in sunspots, reaching field strengths of $B = 2000$–3000 G. Sunspot groups are dipolar, oriented in an east–west direction (with the leading spot slightly closer to the equator)

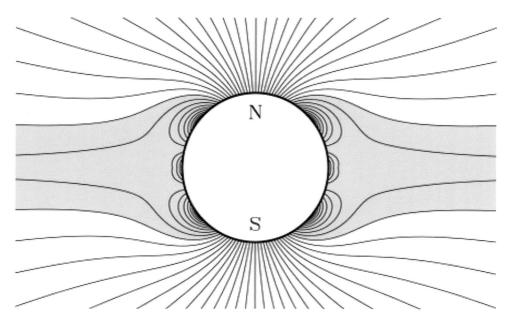

FIGURE 10 The global coronal magnetic field can be subdivided into open-field regions (mostly near the polar regions) and into closed-field regions (mostly in latitudes of $\Phi \leq 70°$). The analytical magnetic field model shown here, a multipole-current sheet coronal model of Banaszkiewicz, approximately outlines the general trends. The high-speed solar wind originates and leaves the Sun in the unshaded volume.

and with opposite leading polarity in both hemispheres, reversing every 11-year cycle (Hale's laws). Active regions and their plages comprise a larger area around sunspots, with average photospheric fields of $B \approx 100–300$ G, containing small-scale pores with typical fields of $B \approx 1000$ G. The background magnetic field in the quiet Sun and in coronal holes has a net field of $B \approx 0.1 – 0.5$ G, while the absolute field strengths in resolved elements amount to $B = 10–50$ G. Our knowledge of the solar magnetic field is mainly based on measurements of Zeeman splitting in spectral lines, whereas the coronal magnetic field is reconstructed by extrapolation from magnetograms at the lower boundary, using a potential or force-free field model. The extrapolation through the chromosphere and transition region is, however, uncertain due to unknown currents and non-force-free conditions. The fact that coronal loops exhibit generally much less expansion with height than potential-field models underscores the inadequacy of potential-field extrapolations. Direct measurements of the magnetic field in coronal heights are still in their infancy.

5.7 MHD Oscillations of Coronal Loops

Much like the discovery of helioseismology four decades ago, it was recently discovered that also the solar corona contains an impressively large ensemble of plasma structures that are capable of producing sound waves and harmonic oscillations. Thanks to the high spatial resolution, image contrast, and time cadence capabilities of the *Solar and Heliospheric Observatory (SoHO)* and *TRACE* spacecraft, oscillating loops, prominences, or sunspots, and propagating waves have been identified and localized in the corona and transition region, and studied in detail since 1999. These new discoveries established a new discipline that became known as coronal seismology. Even though the theory of MHD oscillations was developed several decades earlier,

only the new imaging observations provide diagnostics on length scales, periods, damping times, and densities that allow a quantitative application of the theoretical dispersion relations of MHD waves. The theory of MHD oscillations has been developed for homogeneous media, single interfaces, slender slabs, and cylindrical fluxtubes. There are four basic speeds in fluxtubes: (1) the Alfvén speed $v_A = B_0/\sqrt{4\pi\rho_0}$, (2) the sound speed $c_s = \sqrt{\gamma P_0/\rho_0}$, (3) the cusp or tube speed $c_T = (1/c_s^2 + 1/v_A^2)^{-1/2}$, and (4) the kink or mean Alfvén speed $c_k = [(\rho_0 v_A^2 + \rho_e v_{A_e}^2)/(\rho_0 + \rho_e)]^{1/2}$. For coronal conditions, the dispersion relation reveals a slow-mode branch (with acoustic phase speeds) and a fast-mode branch of solutions (with Alfvén speeds). For the fast-mode branch, a symmetric (sausage) mode and an asymmetric (kink) mode can be distinguished. The fast kink mode produces transverse amplitude oscillations of coronal loops, which have been detected with TRACE (Fig. 11), having periods in the range of $P = 2–10$ min, and can be used to infer the coronal magnetic field strength, thanks to its nondispersive nature. The fast sausage mode is highly dispersive and is subject to a long-wavelength cutoff, so that standing wave oscillations are only possible for thick and high-density (flare and postflare) loops, with periods in the range of $P \approx 1$ s to 1 min. Fast sausage-mode oscillations with periods of $P \approx 10$ s have recently been imaged for the first time with the Nobeyama radioheliograph, and there are numerous earlier reports on nonimaging detections with periods of $P \approx 0.5–5$ s. Finally, slow-mode acoustic oscillations have been detected in flare-like loops with Solar Ultraviolet Measurements of Emitted Radiation (SUMER) having periods in the range of $P \approx 5–30$ min. All loop oscillations observed in the solar corona have been found to be subject to strong damping, typically with decay times of only one or two periods. The relevant damping mechanisms are resonant absorption for fast-mode oscillations (or alternatively phase mixing, although requiring an extremely low Reynolds number), and thermal conduction for

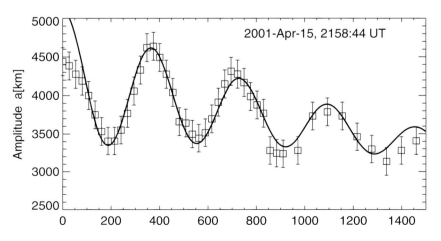

FIGURE 11 The transverse amplitude of a kink-mode oscillation measured in one loop of a postflare loop arcade observed with *TRACE* on April 15, 2001, 21:58:44 UT. The amplitudes are fitted by a damped sine plus a linear function, $a(t) = a_0 + a_1 \sin(2\pi^* (t - t_0)/P) \exp(-t/\tau_{\mathrm{D}}) + a_2^* t$, with a period of $P = 365$ s and a damping time of $t_{\mathrm{D}} = 1000$ s. (Courtesy of Ed DeLuca and Joseph Shoer.)

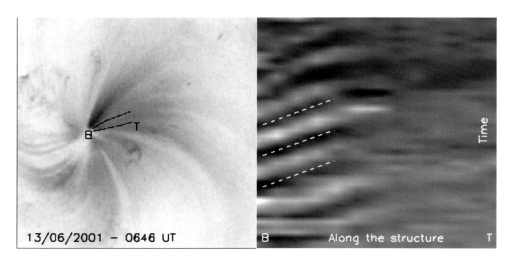

FIGURE 12 TRACE 171 Å observation of a slow-mode (acoustic) wave recorded on June 13, 2001, 06:46 UT. (Left) The diverging fan-like loop structures emerge near a sunspot, where the acoustic waves are launched and propagate upward. (Right) A running difference plot is shown for the loop segment marked in the left frame, with time running upward in the plot. Note the diagonal pattern, which indicates propagating disturbances. (Courtesy of Ineke De Moortel.)

slow-mode acoustic oscillations. Quantitative modeling of coronal oscillations offers exciting new diagnostics on physical parameters.

5.8 MHD Waves in Solar Corona

In contrast to standing modes (with fixed nodes), also propagating MHD waves (with moving nodes) have been discovered in the solar corona recently. Propagating MHD waves result mainly when disturbances are generated impulsively, on time scales faster than the Alfvénic or acoustic travel time across a structure.

Propagating slow-mode MHD waves (with acoustic speed) have been recently detected in coronal loops with *TRACE* and *SoHO*/EIT (Fig. 12); they are usually being launched with 3-minute periods near sunspots, or with 5-minute periods in plage regions. These acoustic waves propagate upward from a loop footpoint and are quickly damped; they have never been detected in downward direction at the opposite loop side. Propagating fast-mode MHD waves (with Alfvénic speeds) have recently been discovered in a loop in optical [Solar Eclipse Coronal Imaging System (SECIS) eclipse] data, as well as in (Nobeyama) radio images.

Besides from coronal loops, slow-mode MHD waves have also been detected in plumes in open-field regions in coronal holes, while fast-mode MHD waves have not yet been detected in open-field structures. However, spectroscopic observations of line broadening in coronal holes provide strong support for the detection of **Alfvén waves,** based on the agreement with the theoretically predicted height-dependent scaling between line broadening and density, $\Delta v(h) \propto n_e(h)^{-1/4}$.

The largest manifestation of propagating MHD waves in the solar corona are global waves that spherically propagate after a flare and/or CME over the entire solar surface. These global waves were discovered earlier in Hα, called Moreton waves, and recently in EUV, called EIT waves (Fig. 13), usually accompanied with a coronal dimming behind the wave front, suggesting evacuation of coronal plasma by the CME. The speed of Moreton waves is about three times faster than that of EIT waves, which still challenges dynamic MHD models of CMEs.

5.9 Coronal Heating

When Bengt Edlén and Walter Grotrian identified Fe IX (nine-times ionized iron) and Ca XIV (14-times ionized calcium) lines in the solar spectrum in 1943, a coronal temperature of $T \approx 1$ MK was first inferred from the formation temperature of these highly ionized atoms. A profound consequence of this measurement is the implication that the corona then consists of a fully ionized hydrogen plasma. Comparing this coronal temperature with the photospheric temperature of 6400 K, we are confronted with the puzzle of how the 200 times hotter coronal temperature can be maintained, the so-called coronal heating problem. Of course, there is also a chromospheric heating problem and a solar wind heating problem. If only thermal conduction were at work, the temperature in the corona should steadily drop down from the chromospheric value with increasing distance, according to the second law of thermodynamics. Moreover, since we have radiative losses by EUV emission, the corona would just cool off in a matter of hours to days, if the plasma temperature could not be maintained continuously by some heating source.

The coronal heating problem has been narrowed down by substantial progress in theoretical modeling with MHD codes, new high-resolution imaging with the SXT (*Yohkoh* Soft X-ray Telescope), EIT, *TRACE*, and Hinode telescopes, and with more sophisticated data analysis using automated pattern recognition codes. The total energy losses

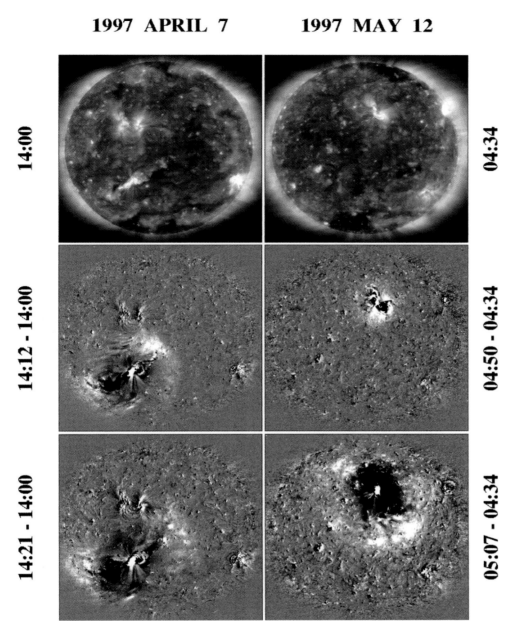

1997 APRIL 7   1997 MAY 12

14:00

14:12 – 14:00

14:21 – 14:00

04:34

04:50 – 04:34

05:07 – 04:34

FIGURE 13   Two global wave events observed with *SoHO*/EIT 195 Å, on April 7, 1997 (top row) and May 12, 1997 (bottom row). The intensity images (right) were recorded before the eruption, while the difference images (left and middle) show differences between the subsequent images, enhancing emission measure increases (white areas) and dimming (black areas). (Courtesy of Yi-Ming Wang.)

in the solar corona range from $F = 3 \times 10^5 \, \mathrm{erg\,cm^{-2}\,s^{-1}}$ in quiet-Sun regions to $F \approx 10^7 \, \mathrm{erg\,cm^{-2}\,s^{-1}}$ in active regions. Two main groups of DC (direct current) and AC (alternating current) models involve as a primary energy source chromospheric footpoint motion or upward leaking Alfvén waves, which are dissipated in the corona by magnetic reconnection, current cascades, MHD turbulence, Alfvén resonance, resonant absorption, or phase mixing. There is also strong observational evidence for solar wind heating by cyclotron resonance, while velocity filtration seems not to be consistent with EUV data. Progress in theoretical models has mainly been made by abandoning homogeneous fluxtubes, but instead including gravitational scale heights and more

realistic models of the transition region, and taking advantage of numerical simulations with 3D MHD codes (by Boris Gudiksen and Aake Nordlund). From the observational side we can now unify many coronal small-scale phenomena with flare-like characteristics, subdivided into microflares (in soft X-rays) and nanoflares (in EUV) solely by their energy content. Scaling laws of the physical parameters corroborate their unification. They provide a physical basis to understand the frequency distributions of their parameters and allow estimation of their energy budget for coronal heating. Synthesized data sets of microflares and nanoflares in EUV and soft X-rays have established that these impulsive small-scale phenomena match the radiative

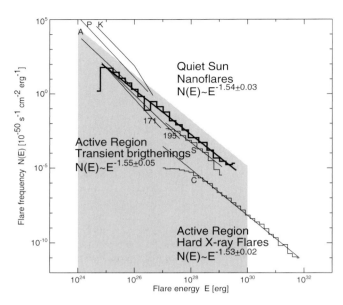

FIGURE 14 Compilation of frequency distributions of thermal energies from nanoflare statistics in the quiet Sun, active region transient brightenings, and hard X-ray flares. The overall slope of the synthesized nanoflare distribution, $N(E) \propto E^{-1.54 \pm 0.03}$, is similar to that of transient brightenings and hard X-ray flares. The grey area indicates the coronal heating requirement of $F = 3 \times 10^5$ erg cm^{-2} s^{-1} for quiet-Sun regions. Note that the observed distribution of nanoflare energies, which only includes the radiative losses, accounts for about one third of the heating rate requirement of the quiet Sun.

loss of the average quiet-Sun corona (Fig. 14), which points to small-scale magnetic reconnection processes in the transition region and lower corona as primary heating sources.

6. Solar Flares and Coronal Mass Ejections

Rapidly varying processes in the solar corona, which result from a loss of magnetic equilibrium, are called eruptive phenomena, such as flares, CMEs, or eruptive filaments and prominences. The fundamental process that drives all these phenomena is magnetic reconnection.

6.1 Magnetic Reconnection

The solar corona has dynamic boundary conditions: (1) The solar dynamo in the interior of the Sun constantly generates new magnetic flux from the bottom of the convection zone (i.e., the tachocline) which rises by buoyancy and emerges through the photosphere into the corona; (2) the differential rotation as well as convective motion at the solar surface continuously wrap up the coronal field; and (3) the connectivity to the interplanetary field has constantly to

break up to avoid excessive magnetic stress. These three dynamic boundary conditions are the essential reasons why the coronal magnetic field is constantly stressed and has to adjust by restructuring the large-scale magnetic field by topological changes, called magnetic reconnection processes. Of course, such magnetic restructuring processes occur wherever magnetic stresses build up (e.g., in filaments, in twisted sigmoid-shaped loops, and along sheared neutral lines). Topological changes in the form of magnetic reconnection always liberate free nonpotential energy, which is converted into heating of plasma, acceleration of particles, and kinematic motion of coronal plasma. Magnetic reconnection processes can occur in a slowly changing quasi-steady way, which may contribute to coronal heating (Section 5.9), but more often happen as sudden violent processes that are manifested as flares and CMEs.

Theory and numerical simulations of magnetic reconnection processes in the solar corona have been developed for steady 2D reconnection (Fig. 15, left), bursty 2D reconnection, and 3D reconnection. Only steady 2D reconnection models can be formulated analytically; they provide basic relations for inflow speed, outflow speed, and reconnection rate, but represent oversimplifications for most (if not all) observed flares. A more realistic approach seems to be bursty 2D reconnection models (Fig. 15, right), which involve the tearing-mode and coalescence instability and can reproduce the sufficiently fast temporal and small spatial scales required by solar flare observations. The sheared magnetic field configurations and the existence or coronal and chromospheric nullpoints, which are now inferred more commonly in solar flares, require ultimately 3D reconnection models, possibly involving nullpoint coalescence, spine reconnection, fan reconnection, and separator reconnection. Magnetic reconnection operates in two quite distinct physical parameter domains: in the chromosphere during magnetic flux emergence, magnetic flux cancellation, and so-called explosive events and under coronal conditions during microflares, flares, and CMEs.

6.2 Filaments and Prominences

Key elements in triggering flares and/or CMEs are erupting filaments. A filament is a current system above a magnetic neutral line that builds up gradually over days and erupts during a flare or CME process. The horizontal magnetic field lines overlying a neutral line (i.e., the magnetic polarity inversion line) of an active region are filled with cool gas (of chromospheric temperature), embedded in the much hotter tenuous coronal plasma. On the solar disk, these cool dense features appear dark in Hα or EUV images, in absorption against the bright background, and are called *filaments*, while the same structures appear bright above the limb, in emission against the dark sky background, where they are called prominences. Thus, filaments and prominences

Sweet-Parker model

Petschek model

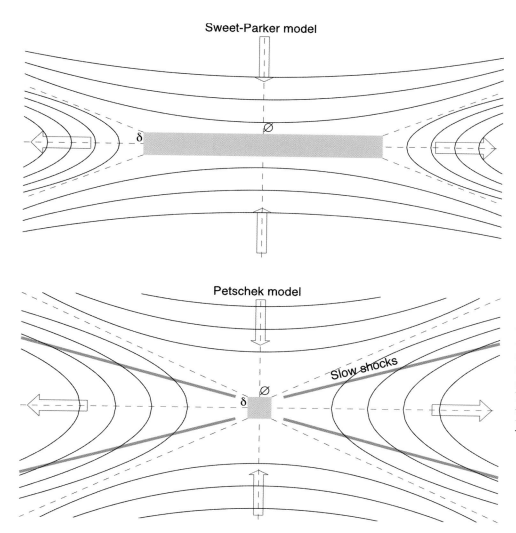

FIGURE 15 Left: Geometry of the Sweet–Parker (top) and Petschek reconnection model (bottom). The geometry of the diffusion region (gray box) is a long thin sheet ($\Delta \gg d$) in the Sweet–Parker model, but much more compact ($\Delta \approx d$) in the Petschek model. The Petschek model also considers slow-mode MHD shocks in the outflow region. Right: Numeric MHD simulation of a magnetic reconnection process in a sheared arcade. The grayscale represents the mass density difference ratio, and the dashed lines show the projected magnetic field lines in the vicinity of the reconnection region, at two particular times of the reconnection process. The location **a** corresponds to a thin compressed region along the slowly rising inner separatrix, **b** to a narrow downflow stream outside of the left outer separatrix, and **c** indicates a broader upflow that follows along the same field lines. (Courtesy of Judith Karpen.)

are identical structures physically, while their dual name just reflects a different observed location (inside or outside the disk). A further distinction is made regarding their dynamic nature: Quiescent filaments/prominences are long-lived stable structures that can last for several months, while eruptive filaments/prominences are usually associated with flares and CMEs (see example in Fig. 16).

6.3 Solar Flare Models

A flare process is associated with a rapid energy release in the solar corona, believed to be driven by stored nonpotential magnetic energy and triggered by an instability in the magnetic configuration. Such an energy release process results in acceleration of nonthermal particles and in heating of coronal/chromospheric plasma. These processes emit radiation in almost all wavelengths: radio,

white light, EUV, soft X-rays, hard X-rays, and even gamma rays during large flares. The energy range of flares extends over many orders of magnitude. Small flares that have an energy content of 10^{-6} to 10^{-9} of the largest flares fall into the categories of microflares and nanoflares (Fig. 14), which are observed not only in active regions but also in quiet-Sun regions. Some of the microflares and nanoflares have been localized above the photospheric network and are thus also dubbed network flares or network heating events. There are also a number of small-scale phenomena with rapid time variability for which it is not clear whether they represent miniature flare processes (e.g., active region transients, explosive events, blinkers). It is conceivable that some are related to photospheric or chromospheric magnetic reconnection processes, in contrast to flares that always involve coronal magnetic reconnection processes.

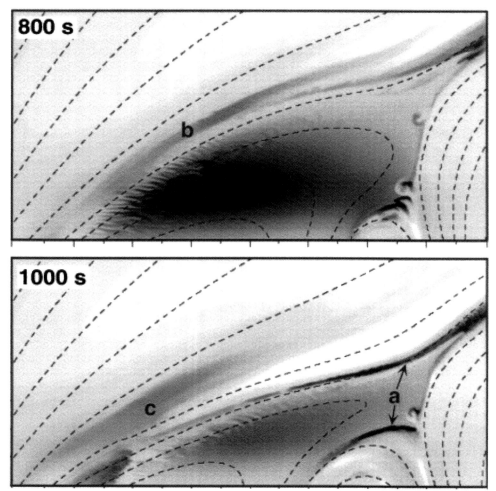

FIGURE 15   (*Continued*)

FIGURE 16   Erupting filament observed with *TRACE* at 171 Å on July 19, 2000, 23:30 UT, in Active Region 9077. The dark filament mass has temperatures around 20,000 K, while the hot kernels and threads contain plasma with temperatures of 1.0 MK or hotter. The erupting structure extends over a height of 75,000 km here. (Courtesy of *TRACE* and NASA.)

The best known flare/CME models entail magnetic reconnection processes that are driven by a rising filament/prominence, flux emergence, converging flows, or shear motion along the neutral line. Flare scenarios with a driver perpendicular to the neutral line (rising prominence, flux emergence, convergence flows) are formulated as 2D reconnection models, while scenarios that involve shear along the neutral line (tearing-mode instability, quadrupolar flux transfer, the magnetic breakout model, sheared arcade interactions) require 3D descriptions. A 2D reconnection model involving a magnetic X-point is shown in Fig. 17 (left); a generalized 3D version involving a highly sheared neutral line is sketched in Fig. 17 (right). There are more complex versions like the magnetic breakout model, where a second arcade triggers reconnection above a primary arcade. Observational evidence for magnetic reconnection in flares includes the 3D geometry, reconnection inflows, outflows, detection of shocks, jets, ejected plasmoids, and secondary effects like particle acceleration, conduction fronts, and chromospheric evaporation processes. Flare images in soft X-rays often show the cusp-shaped geometry of

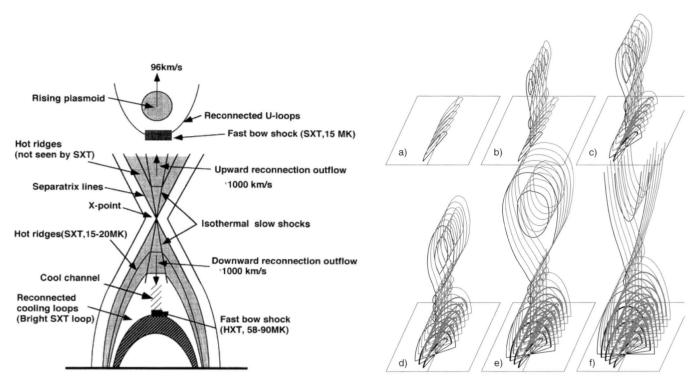

FIGURE 17   Left: A version of the standard 2D X-type reconnection model for two-ribbon flares, pioneered by Carmichael, Sturrock, Hirayama, and Kopp-Pneumann (CSHKP), which also includes the slow and fast shocks in the outflow region, the upward-ejected plasmoid, and the locations of the soft X-ray bright flare loops. (Courtesy of Saku Tsuneta.) Right: 3D version of the two-ribbon flare model, based on the observed evolution during the Bastille Day (July 14, 2000) flare: (*a*) low-lying, highly sheared loops above the neutral line first become unstable; (*b*) after loss of magnetic equilibrium the filament jumps upward and forms a current sheet according to the model by Forbes and Priest. When the current sheet becomes stretched, magnetic islands form and coalescence of islands occurs at locations of enhanced resistivity, initiating particle acceleration and plasma heating; (*c*) the lowest lying loops relax after reconnection and become filled due to chromospheric evaporation (loops with thick linestyle); (*d*) reconnection proceeds upward and involves higher lying, less sheared loops; (*e*) the arcade gradually fills up with filled loops; (*f*) the last reconnecting loops have no shear and are oriented perpendicular to the neutral line. At some point, the filament disconnects completely from the flare arcade and escapes into interplanetary space.

reconnecting field lines (Fig. 18, top), while EUV images invariably display the relaxed postreconnection field lines after the flare loops cooled down to EUV temperatures in the postflare phase (Fig. 18, middle and bottom).

6.4 Flare Plasma Dynamics

The flare plasma dynamics and associated thermal evolution during a flare consists of a number of sequential processes: plasma heating in coronal reconnection sites, chromospheric flare plasma heating (either by precipitating nonthermal particles and/or downward propagating heat conduction fronts), chromospheric evaporation in the form of upflowing heated plasma, and cooling of postflare loops. The initial heating of the coronal plasma requires anomalous resistivity because Joule heating with classical resistivity is unable to explain the observed densities, temperatures, and rapid timescales in flare plasmas. Other forms of coronal flare plasma heating, such as slow shocks, electron beams, proton beams, or inductive currents, are difficult to constrain with currently available observables. The second stage of chromospheric heating is more thoroughly explored, based on the theory of the thick-target model, with numeric hydrodynamic simulations, and with particle-in-cell simulations. Important diagnostics on chromospheric heating are also available from Hα, white light, and UV emission, but quantitative modeling is still quite difficult

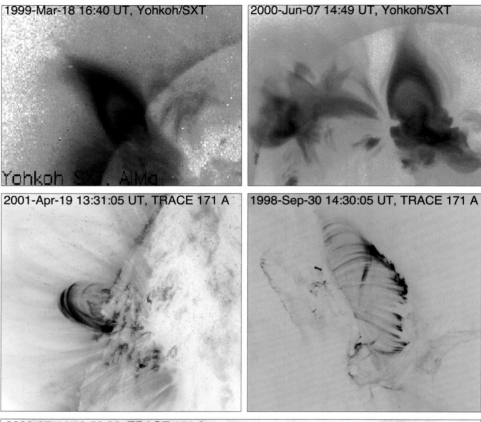

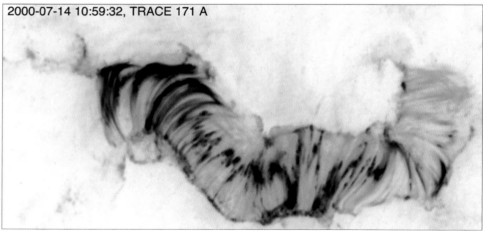

FIGURE 18   Soft X-ray and EUV images of flare loops and flare arcades with bipolar structure. Yohkoh/SXT observed flares (March 18, 1999, 16:40 UT, and June 7, 2000, 14:49 UT) with "candle-flame"-like cusp geometry during ongoing reconnection, while *TRACE* sees postflare loops once they cooled down to 1–2 MK, when they already relaxed into a near-dipolar state. Examples are shown for a small flare (the April 19, 2001, 13:31 UT, GOES class M2 flare), and for two large flares with long arcades, seen at the limb (September 30, 1998, 14:30 UT) and on the disk (the July 14, 2000, 10:59 UT, X5.7 flare). (Courtesy of *Yohkoh*/ISAS and *TRACE*/NASA.)

because of the chromospheric opacities and partial ionization. The third stage of chromospheric evaporation has been extensively explored with hydrodynamic simulations, in particular to explain the observed Doppler shifts in soft X-ray lines, while application of spatial models to imaging data is quite sparse. Also certain types of slow-drifting radio bursts seem to contain information on the motion of chromospheric evaporation fronts. The fourth stage of postflare loop cooling is now understood to be dominated by thermal conduction initially and by radiative cooling later on. How-

ever, spatiotemporal temperature modeling of flare plasmas (Fig. 19) has not yet been fitted to observations in detail.

6.5 Particle Acceleration and Kinematics

Particle acceleration in solar flares is mostly explored by theoretical models because neither macroscopic nor microscopic electric fields are directly measurable by remote-sensing methods. The motion of particles can be described in terms of acceleration by parallel electric fields, drift

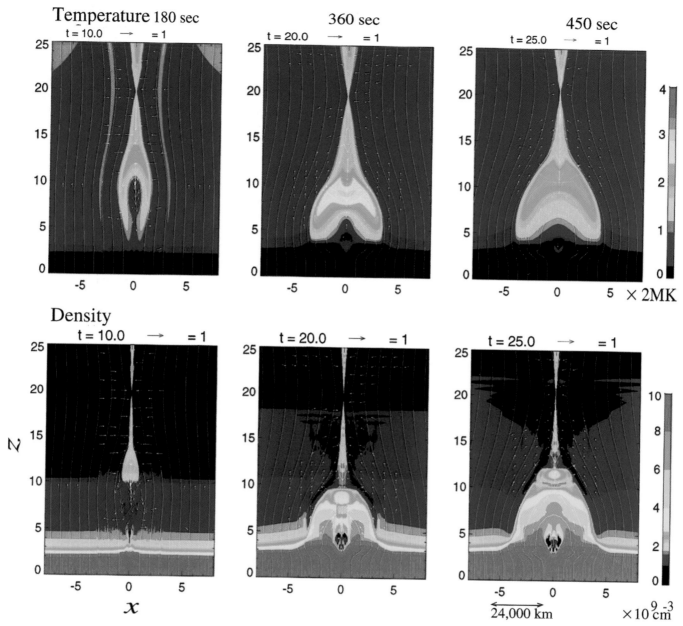

FIGURE 19   2D numerical MHD simulation of a solar flare with chromospheric evaporation and anisotropic heat conduction in the framework of a 2D magnetic reconnecting geometry. The temporal evolution of the plasma temperature (top row) and density (bottom row) is shown. The temperature and density scale is shown in the bars on the right side. The simulation illustrates the propagation of thermal conduction fronts and the upflows of chromospheric plasma in response. (Courtesy of Takaaki Yokoyama and Kazunari Shibata.)

velocities caused by perpendicular forces (i.e., $E \times B$-drifts), and gyromotion caused by the Lorentz force of the magnetic field. Theoretical models of particle acceleration in solar flares can be broken down into three groups: (1) DC electric field acceleration, (2) stochastic or second-order Fermi acceleration, and (3) shock acceleration. In the models of the first group, there is a paradigm shift from large-scale DC electric fields (of the size of flare loops) to small-scale electric fields (of the size of magnetic islands produced by the tearing mode instability). The acceleration and trajectories of particles is studied more realistically in the inhomogeneous and time-varying electromagnetic fields around magnetic X-points and O-points of magnetic reconnection sites, rather than in static, homogeneous, large-scale

Parker-type current sheets. The second group of models entails stochastic acceleration by gyroresonant wave–particle interactions, which can be driven by a variety of electrostatic and electromagnetic waves, supposed that wave turbulence is present at a sufficiently enhanced level and that the MHD turbulence cascading process is at work. The third group of acceleration models includes a rich variety of shock acceleration models, which is extensively explored in magnetospheric physics and could cross-fertilize solar flare models. Two major groups of models are studied in the context of solar flares (i.e., first-order Fermi acceleration or shock-drift acceleration, and diffusive shock acceleration). New aspects are that shock acceleration is now applied to the outflow regions of coronal magnetic reconnection sites, where first-order Fermi acceleration at the standing fast shock is a leading candidate. Traditionally, evidence for shock acceleration in solar flares came mainly from radio type II bursts. New trends in this area are the distinction of different acceleration sites that produce type II emission: flare blast waves, the leading edge of CMEs (bowshock), and shocks in internal and lateral parts of CMEs. In summary, we can say that (1) all three basic acceleration mechanisms seem to play a role to a variable degree in some parts of solar flares and CMEs, (2) the distinctions among the three basic models become more blurred in more realistic (stochastic) models, and (3) the relative importance and efficiency of various acceleration models can only be assessed by including a realistic description of the electromagnetic fields, kinetic particle distributions, and MHD evolution of magnetic reconnection regions pertinent to solar flares.

Particle kinematics, the quantitative analysis of particle trajectories, has been systematically explored in solar flares by performing high-precision energy-dependent time delay measurements with the large-area detectors of the *Compton Gamma-Ray Observatory (CGRO)*. There are essentially five different kinematic processes that play a role in the timing of nonthermal particles energized during flares: (1) acceleration, (2) injection, (3) free-streaming propagation, (4) magnetic trapping, and (5) precipitation and energy loss. The time structures of hard X-ray and radio emission from nonthermal particles indicate that the observed energy-dependent timing is dominated either by free-streaming propagation (obeying the expected electron time-of-flight dispersion) or by magnetic trapping in the weak-diffusion limit (where the trapping times are controlled by collisional pitch angle scattering). The measurements of the velocity dispersion from energy-dependent hard X-ray delays allows then to localize the acceleration region, which was invariably found in the cusp of postflare loops (Fig. 20).

6.6 Hard X-Ray Emission

Hard X-ray emission is produced by energized electrons via collisional **bremsstrahlung,** most prominently in the form of thick-target bremsstrahlung when precipitating electrons hit the chromosphere. Thin-target bremsstrahlung may be observable in the corona for footpoint-occulted flares. Thermal bremsstrahlung dominates only at energies of ≤ 15 keV. Hard X-ray spectra can generally be fitted with a thermal spectrum at low energies and with a single- or double-powerlaw nonthermal spectrum at higher energies. Virtually all flares exhibit fast (subsecond) pulses in hard X-rays, which scale proportionally with flare loop size and are most likely spatiotemporal signatures of bursty magnetic reconnection events. The energy-dependent timing of these fast subsecond pulses exhibit electron time-of-flight delays from the propagation between the coronal acceleration site and the chromospheric thick-target site. The inferred acceleration site is located about 50% higher than the soft X-ray flare loop height, most likely near X-points of magnetic reconnection sites (Fig. 20). The more gradually varying hard X-ray emission exhibits an energy-dependent time delay with opposite sign, which corresponds to the timing of the collisional deflection of trapped electrons. In many flares, the time evolution of soft X-rays roughly follows the integral of the hard X-ray flux profile, which is called the Neupert effect. Spatial structures of hard X-ray sources include: (1) footpoint sources produced by thick-target bremsstrahlung, (2) thermal hard X-rays from flare looptops, (3) above-the-looptop (Masuda-type) sources that result from nonthermal bremsstrahlung from electrons that are either trapped in the acceleration region or interact with reconnection shocks, (4) hard X-ray sources associated with upward soft X-ray ejecta, and (5) hard X-ray halo or albedo sources due to backscattering at the photosphere. In spatially extended flares, the footpoint sources assume ribbon-like morphology if mapped with sufficient sensitivity. The monthly hard X-ray flare rate varies about a factor of 20 during the solar cycle, similar to magnetic flux variations implied by the monthly sunspot number, as expected from the magnetic origin of flare energies.

6.7 Gamma–Ray Emission

The energy spectrum of flares (Fig. 21) in gamma-ray wavelengths (0.5 MeV–1 GeV) is more structured than in hard X-ray wavelengths (20–500 keV) because it exhibits both continuum emission as well as line emission. There are at least six different physical processes that contribute to gamma-ray emission: (1) electron bremsstrahlung continuum emission, (2) nuclear deexcitation line emission, (3) neutron capture line emission at 2.223 MeV, (4) positron annihilation line emission at 511 keV, (5) pion-decay radiation at ≥ 50 MeV, and (6) neutron production. The ratio of continuum to line emission varies from flare to flare, and gamma-ray lines can completely be overwhelmed in electron-rich flares or flare phases. When gamma-ray lines are present, they provide a diagnostic of the elemental abundances, densities, and temperatures of the ambient plasma

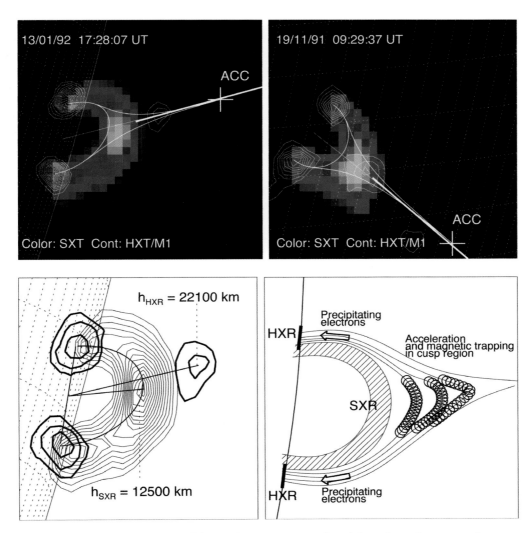

FIGURE 20 Top: The geometry of the acceleration region inferred from direct detections of above-the-looptop hard X-ray sources with *Yohkoh*/HXT (Hard X-Ray Telescope) (contours) and simultaneous modeling of electron time-of-flight distances based on energy-dependent time delays of 20–200 keV hard X-ray emission measured with Burst and Transient Source Experiment, BATSE/*CGRO* (crosses marked with ACC). Soft X-rays detected with *Yohkoh*/SXT or thermal hard X-ray emission from the low-energy channel of *Yohkoh*/HXT/Lo are shown in colors, outlining the flare loops. Bottom: The observations in the left panel show a *Yohkoh*/HXT 23–33 keV image (thick contours) and Be119 *SXT* image (thin contours) of the Masuda flare, January 13, 1992, 17:28 UT. The interpretation of the above-the-looptop source is that temporary trapping occurs in the acceleration region in the cusp region below the reconnection point (bottom right).

in the chromosphere, as well as of the directivity and pitch angle distribution of the precipitating protons and ions that have been accelerated in coronal flare sites, presumably in magnetic reconnection regions. Critical issues that have been addressed in studies of gamma-ray data are the maximum energies of coronal acceleration mechanisms, the ion/electron ratios (because selective acceleration of ions indicate gyroresonant interactions), the ion/electron timing (to distinguish between simultaneous or second-step acceleration), differences in ion/electron transport (e.g.,

neutron sources were recently found to be displaced from electron sources), and the first ionization potential (FIP) effect of chromospheric abundances (indicating enhanced abundances of certain ions that could be preferentially accelerated by gyroresonant interactions). Although detailed modeling of gamma-ray line profiles provides significant constraints on elemental abundances and physical properties of the ambient chromospheric plasma, as well as on the energy and pitch angle distribution of accelerated particles, little information or constraints could be retrieved about the

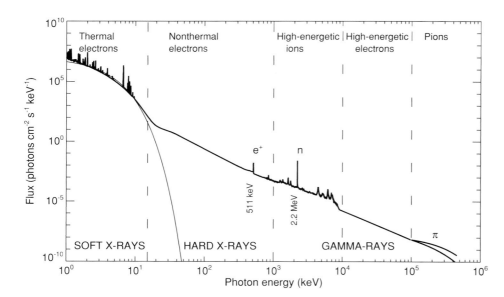

FIGURE 21 Composite photon spectrum of a large flare, extending from soft X-rays (1–10 keV), hard X-rays (10 keV–1 MeV), to gamma-rays (1 MeV–100 GeV). The energy spectrum is dominated by different processes: by thermal electrons (in soft X-rays), bremsstrahlung from nonthermal electrons (in hard X-rays), nuclear deexcitation lines (in ∼0.5–8 MeV gamma-rays), bremsstrahlung from high-energetic electrons (in ∼10–100 MeV gamma-rays), and pion-decay (in ≥100 MeV gamma rays). Note also the prominent electron-positron annihilation line (at 511 keV) and the neutron capture line (at 2.2 MeV).

timescales and geometry of the acceleration mechanisms, using gamma-ray data. Nevertheless, the high spectral and imaging resolution of the recently launched Ramaty High-Energy Spectroscopic Solar Imager (RHESSI) spacecraft facilitates promising new data for a deeper understanding of ion acceleration in solar flares.

6.8 Radio Emission

Radio emission in the solar corona is produced by thermal, nonthermal, up to high-relativistic electrons, and thus provides useful diagnostics complementary to EUV, soft X-rays, hard X-rays, and gamma rays. Thermal or Maxwellian distribution functions produce in radio wavelengths either free-free emission (bremsstrahlung) for low magnetic field strengths or gyroresonant emission in locations of high magnetic field strengths, such as above sunspots, which are both called incoherent emission mechanisms. Since EUV and soft X-ray emission occurs in the optically thin regime, the emissivity adds up linearly along the line-of-sight. Free-free radio emission is somewhat more complicated because the optical thickness depends on the frequency, which allows direct measurement of the electron temperature in optically thick coronal layers in metric and decimetric frequencies up to $\nu \leq 1$ GHz. Above ∼2 GHz, free-free emission becomes optically thin in the corona, but gyroresonance emission at harmonics of $s \approx 2, 3, 4$ dominates in strong-field regions. In flares, high-relativistic electrons are produced that emit gyrosynchrotron emission, which allows for detailed modeling of precipitating and trapped electron populations in time profiles recorded at different microwave frequencies.

Unstable non-Maxwellian particle velocity distributions, which have a positive gradient in parallel (beams) or perpendicular (losscones) direction to the magnetic field, drive gyroresonant wave-particle interactions that produce coherent wave growth, detectable in the form of coherent radio emission. Two natural processes that provide these conditions are dispersive electron propagation (producing beams) and magnetic trapping (producing losscones). The wave-particle interactions produce growth of Langmuir waves, upper-hybrid waves, and electron-cyclotron maser emission, leading to a variety of radio burst types (type I, II, III, IV, V, DCIM; Fig. 22), which have been mainly explored from (nonimaging) dynamic spectra, while imaging observations have been rarely obtained. Although there is much theoretical understanding of the underlying wave-particle interactions, spatiotemporal modeling of imaging observations is still in its infancy. A solar-dedicated, frequency-agile imager with many frequencies (FASR) is in planning stage and might provide more comprehensive observations.

6.9 Coronal Mass Ejections

As a result of phenomena in the atmosphere, every star is losing mass, caused by dynamic phenomena in its atmosphere, which accelerate plasma or particles beyond the escape speed. Inspecting the Sun, our nearest star, we observe two forms of mass loss: the steady solar wind outflow and the sporadic ejection of large plasma structures, or CMEs. The solar wind outflow amounts to ∼2×10^{-10} (g cm^{-2} s^{-1}) in coronal holes, and to ≤4×10^{-11} (g cm^{-2} s^{-1}) in active regions. The phenomenon of a CME occurs with a frequency of about one event per day, carrying a mass in the range of $m_{CME} \approx 10^{14} - 10^{16}$ g, which corresponds to an average mass loss rate of $m_{CME}/(\Delta t \cdot 4\pi R_\odot^2 \approx 2) \times 10^{-14} - 2 \times 10^{-12}$ (g cm^{-2} s^{-1}), which is ≤1% of the solar wind mass loss in

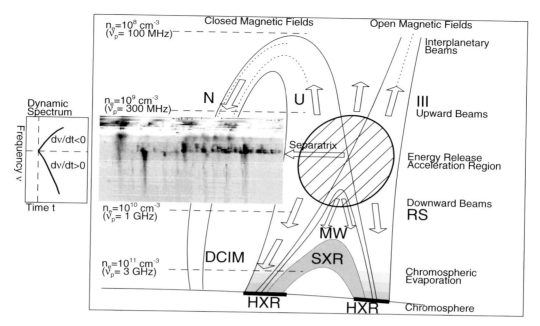

FIGURE 22 Radio burst types in the framework of the standard flare scenario: The acceleration region is located in the reconnection region above the soft X-ray–bright flare loop, accelerating electron beams in the upward direction (type III, U, N bursts) and in the downward direction (type RS, DCIM bursts). Downward moving electron beams precipitate to the chromosphere (producing hard X-ray emission and driving chromospheric evaporation), or remain transiently trapped, producing microwave (MW) emission. Soft X-ray loops become subsequently filled up, with increasing footpoint separation as the X-point rises. The insert shows a dynamic radio spectrum (*ETH* Zurich) of the September 6, 1992, 1154 UT, flare, showing a separatrix between type III and type RS bursts at ∼600 MHz, probably associated with the acceleration region.

coronal holes, or ≤10% of the solar wind mass in active regions. The transverse size of CMEs can cover a fraction up to more than a solar radius, and the ejection speed is in the range of $v_{CME} \approx 10^2$–10^3 (km s^{-1}). A CME structure can have the geometric shape of a fluxrope, a semishell, or a bubble (like a light bulb, see Fig. 24), which is the subject of much debate, because of ambiguities from line-of-sight projection effects and the optical thinness. There is a general consensus that a CME is associated with a release of magnetic energy in the solar corona, but its relation to the flare phenomenon is controversial. Even big flares [at least Geostationary Orbiting Earth Satellite (GOES) M-class] have no associated CMEs in 40% of the cases. A long-standing debate focused on the question of whether a CME is a byproduct of the flare process or vice versa. This question has been settled in the view that flares and CMEs are two aspects of a large-scale magnetic energy release, but the two terms evolved historically from two different observational manifestations (i.e., flares, which mainly denote the emission in hard X-rays, soft X-rays, and radio waves, and CMEs, which refer to the white-light emission of the erupting mass in the outer corona and heliosphere). Recent studies, however, clearly established the coevolution of both processes

triggered by a common magnetic instability. A CME is a dynamically evolving plasma structure, propagating outward from the Sun into interplanetary space, carrying a frozen-in magnetic flux and expanding in size. If a CME structure travels toward the Earth, which is mostly the case when launched in the western solar hemisphere, due to the curvature of the Parker spiral interplanetary magnetic field, such an Earth-directed event can engulf the Earth's magnetosphere and generate significant geomagnetic storms. Obviously such geomagnetic storms can cause disruptions of global communication and navigation networks, can cause failures of satellites and commercial power systems, and thus are the subject of high interest.

Theoretical models include five categories: (1) thermal blast models, (2) dynamo models, (3) mass loading models, (4) tether release models, and (5) tether straining models. Numerical MHD simulations of CMEs are currently produced by combinations of a fine-scale grid that entails the corona and a connected large-scale grid that encompasses propagation into interplanetary space, which can reproduce CME speeds, densities, and the coarse geometry. The trigger that initiates the origin of a CME seems to be related to previous photospheric shear motion and subsequent kink

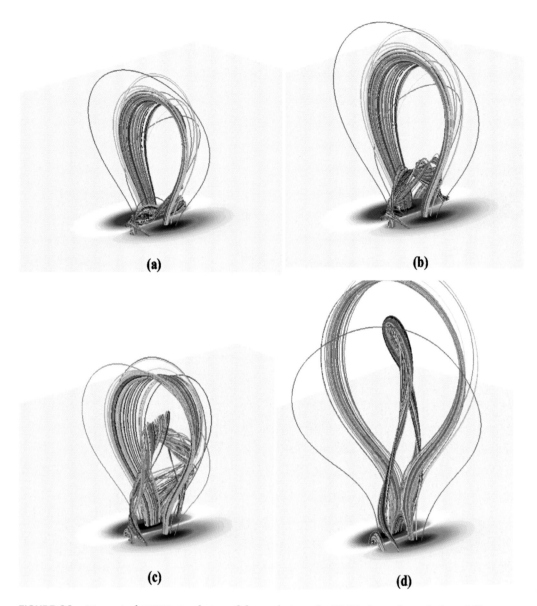

FIGURE 23 Numerical MHD simulation of the evolution of a CME, driven by turbulent diffusion. The four panels correspond to the times (a) $t = 850$, (b) $t = 950$, (c) $t = 1050$, and (d) $t = 1150$, where viscous relaxation is started at $t = 850$, triggering a global disruption involving opening, reconnection through the overlying arcade and below, and the formation of a current sheet, associated with a high dissipation of magnetic energy and a strong increase of kinetic energy. (Courtesy of T. Amari.)

instability of twisted structures (Fig. 23). The geometry of CMEs is quite complex, exhibiting a variety of topological shapes from spherical semishells to helical fluxropes (Fig. 24), and the density and temperature structure of CMEs is currently investigated with multiwavelength imagers. The height-time, velocity, and acceleration profiles of CMEs seem to establish two different CME classes: gradual CMEs associated with propagating interplanetary shocks and impulsive CMEs caused by coronal flares. The total energy of CMEs (i.e., the sum of magnetic, kinetic, and gravi-

tational energy) seems to be conserved in some events, and the total energy of CMEs is comparable to the energy range estimated from flare signatures. A phenomenon closely associated with CMEs is coronal dimming (Fig. 13), which is interpreted in terms of an evacuation of coronal mass during the launch of a CME. The propagation of CMEs in interplanetary space provides diagnostic information on the heliospheric magnetic field, the solar wind, interplanetary shocks, solar energetic particle (SEP) events, and interplanetary radio bursts.

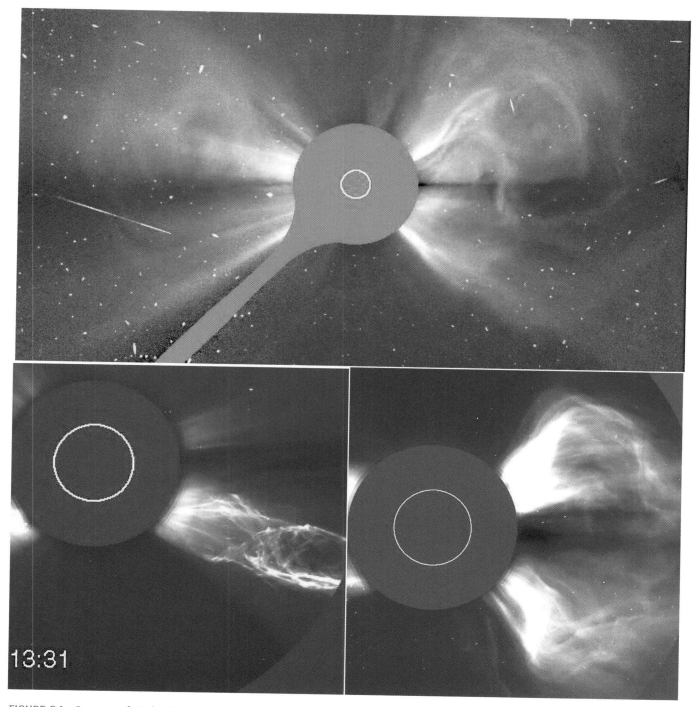

FIGURE 24   Large Angle Solar COronagraph (LASCO) C3 image of a halo CME of May 6, 1998 (top); an erupting prominence of June 2, 1998, 13:31 UT (bottom left); and a large CME of November 6, 1997, 12:36 UT (bottom right). (Courtesy of *SoHO*/LASCO and NASA.)

7. Final Comments

The study of the Sun, our nearest Star, is systematically moving from morphological observations (sunspots, active regions, filaments, flares, CMEs) to a more physics-based modeling and theoretical understanding, in terms of nuclear physics, magneto-convection, magneto-hydrodynamics, magnetic reconnection, and particle physics processes. The major impact of physics-based modeling came from the multiwavelength observations from solar-dedicated space-based (*Hinode*, SMM, *Yohkoh*, *CGRO*, *SoHO*, *TRACE*, RHESSI) and ground-based instruments (in radio, Hα, and white-light wavelengths). Major achievements over the last decades are the advancement of new disciplines such as helioseismology and coronal seismology, and the solution of the neutrino problem; however, there are still unsolved outstanding problems such as the coronal heating problem and particle acceleration mechanisms. We can optimistically expect substantial progress from future solar-dedicated space missions [Solar TErrestrial RElationships Observatory (*STEREO*), *Solar Dynamics Observatory* (*SDO*), *Solar Orbiter*, *Solar Probe*] and ground-based instruments [Synoptic Optical Long-Term Investigations of the Sun (SOLIS), The Advanced Technology Solar Telescope (ATST) FASR].

Bibliography

Aschwanden, M. J. (2004). "Physics of the Solar Corona—An Introduction." Praxis Publishing Ltd., Chichester, England, and Springer: New York.

Benz, A. O. (2003). "Plasma Astrophysics, Kinetic Processes in Solar and Stellar Coronae," 2nd edition. Kluwer Acad. Publ., Dordrecht, Netherlands.

Cox, A. N., ed. (2000). "Allen's Astrophysical Quantities," 4th edition. American Institute of Physics Press/Springer, New York.

Dwivedi, B. N. (2003). "The Dynamic Sun." Cambridge University Press, Cambridge, England.

Foukal, P. V. (2003). "Solar Astrophysics," 2nd edition. John Wiley and Sons, New York.

Golub, L., and Pasachoff, J. M. (2001). "Nearest Star: The Surprising Science of Our Sun." Harvard Univ. Press, Cambridge, Massachusetts.

Golub, L., and Pasachoff, J.M. (1997). "The Solar Corona." Cambridge Univ. Press, Cambridge, Massachusetts.

Lang, K. R. (2001). "The Cambridge Encyclopedia of the Sun." Cambridge Univ. Press, Cambridge, England.

Murdin, P. (ed.) 2000, *"Encyclopedia of Astronomy and Astrophysics,"* Institute of Physics Publishing/Grove's Dictionaries, New York.

Schrijver, C. J, and Zwaan, C. (2000). "Solar and Stellar Magnetic Activity." Cambridge Univ. Press, Cambridge, England.

Stix, M. (1989, 2002). "The Sun," 2nd edition. Springer, New York.

Zirker, J. B. (2002). "Journey from the Center of the Sun." Princeton Univ. Press, Princeton, New Jersey.

The Solar Wind

John T. Gosling

University of Colorado
Boulder, Colorado

The Solar Wind is a **plasma**, that is, an ionized gas, that permeates interplanetary space. It exists as a consequence of the supersonic expansion of the Sun's hot outer atmosphere, the **solar corona**. The solar wind consists primarily of electrons and protons, but **alpha particles** and many other ionic species are also present at low abundance levels. At the orbit of Earth, 1 astronomical unit (AU) from the Sun, typical solar wind densities, flow speeds, and temperatures are on the order of 8 protons cm^{-3}, 440 km/s, and 1.2×10^5 K, respectively; however, the solar wind is highly variable in both space and time. A weak magnetic field embedded within the solar wind plasma is effective both in excluding some low-energy cosmic rays from the solar system and in channeling energetic particles from the Sun into the **heliosphere.** The solar wind plays an essential role in shaping and stimulating planetary magnetospheres and the ionic tails of comets. [*See* PLANETARY MAGNETOSPHERES.]

1. Discovery

1.1 Early Indirect Observations

In 1859, R. Carrington made one of the first white light observations of a **solar flare**. He noted that a major geomagnetic storm began approximately 17 hours after the flare and tentatively suggested that a causal relationship might exist between the solar and geomagnetic events. Subsequent observations revealed numerous examples of associations between solar flares and large geomagnetic storms. In the early 1900s, F. Lindemann suggested that this could be explained if large geomagnetic storms result from an interaction between the geomagnetic field and plasma clouds ejected into interplanetary space by solar activity. Early studies of geomagnetic activity also noted that some geomagnetic storms tend to recur at the ~27 day rotation period of the Sun as observed from Earth, particularly during declining years of solar activity. This observation led to the suggestion that certain regions on the Sun, commonly called M (for magnetic)-regions, occasionally produce long-lived charged particle streams in interplanetary space. Further, because some form of auroral and geomagnetic activity is almost always present at high geomagnetic latitudes, it was inferred that charged particles from the Sun almost continuously impact and perturb the geomagnetic field.

Observations of modulations in galactic cosmic rays (highly energetic charged particles that originate outside the solar system) in the 1930s also suggested that plasma and magnetic fields are ejected from the Sun during intervals of high solar activity. For example, S. Forbush noted that cosmic ray intensity often decreases suddenly during large geomagnetic storms and then recovers slowly over a period of several days. Moreover, cosmic ray intensity varies in a cycle of ~11 years, but roughly 180° out of phase with the **solar activity cycle**. One possible explanation of these observations was that magnetic fields embedded in plasma clouds from the Sun sweep cosmic rays away from the vicinity of Earth.

In the early 1950s, L. Biermann concluded that there must be a continuous outflow of charged particles from the Sun to explain the fact that ionic tails of comets always point away from the Sun. He estimated that a continuous particle flux on the order of 10^{10} protons cm^{-2} s^{-1} was needed at 1 AU to explain the comet tail observations. He later revised his estimate downward to a value of ~10^9 protons cm^{-2} s^{-1}, closer to the average observed solar wind proton flux of ~3.8 $\times 10^8$ protons cm^{-2} s^{-1} at 1 AU.

1.2 Parker's Solar Wind Model

Apparently inspired by these diverse observations and interpretations, E. Parker, in 1958, formulated a radically new model of the solar corona in which the solar atmosphere is continually expanding outward. Prior to Parker's work most theories of the solar atmosphere treated the corona as static and gravitationally bound to the Sun except for sporadic outbursts of material into space at times of high solar activity. S. Chapman had constructed a model of a static solar corona in which heat transport was dominated by electron thermal conduction. For a 10^6 K corona, Chapman found that even a static solar corona must extend far out into space. Parker realized, however, that a static model leads to pressures at large distances from the Sun that are seven to eight orders of magnitude larger than estimated pressures in the interstellar plasma. Because of this mismatch in pressure at large heliocentric distances, he reasoned that the solar corona could not be in hydrostatic equilibrium and must therefore be expanding. His consideration of the hydrodynamic (i.e., fluid) equations for mass, momentum, and energy conservation for a hot solar corona led him to unique solutions for the coronal expansion that depended on the coronal temperature close to the surface of the Sun. Parker's model produced low flow speeds close to the Sun, supersonic flow speeds far from the Sun, and vanishingly small pressures at large heliocentric distances. In view of the fluid character of the solutions, Parker called this continuous, supersonic, coronal expansion the solar wind. The region of space filled by the solar wind is now known as the heliosphere.

1.3 First Direct Observations of the Solar Wind

Several Russian and American space probes in the 1959–1961 era penetrated interplanetary space and found tentative evidence for a solar wind. Firm proof of the wind's existence was provided by C. Snyder and M. Neugebauer, who flew a plasma experiment on *Mariner 2* during its epic 3-month journey to Venus in late 1962. Their experiment detected a continual outflow of plasma from the Sun that was highly variable, being structured into alternating streams of high- and low-speed flows that lasted for several days each. Several of the high-speed streams recurred at roughly the rotation period of the Sun. Average solar wind proton densities (normalized for a 1 AU heliocentric distance), flow speeds, and temperatures during this 3-month interval were 5.4 cm^{-3}, 504 km/s, and 1.7 $\times 10^5$ K, respectively, in essential agreement with Parker's predictions. The *Mariner 2* observations also showed that helium, in the form of alpha particles, is present in the solar wind in variable amounts; the average alpha particle abundance relative to protons of 4.6% is about a factor of 2 lower than estimates of the helium abundance within the Sun. Finally, measurements made by *Mariner 2* confirmed that the solar wind carried a magnetic field whose strength and orientation in the ecliptic plane were much as predicted by Parker (see Section 3).

Despite the good agreement of observations with Parker's model, we still do not fully understand the processes that heat the solar corona and accelerate the solar wind. Parker simply assumed that the corona is heated to a very high temperature, but he did not explain how the heating was accomplished. Moreover, it is now known that electron heat conduction is insufficient to power the coronal expansion. Present models for heating the corona and accelerating the solar wind generally fall into two classes: (1) heating and acceleration by waves generated by convective motions below the photosphere and (2) bulk acceleration and heating associated with transient events in the solar atmosphere such as **magnetic reconnection**. Present observations are incapable of distinguishing among these and other alternatives.

2. Statistical Properties in the Ecliptic Plane at 1 AU

Table 1 summarizes a number of statistical solar wind properties derived from spacecraft measurements in the

TABLE 1	Statistical Properties of the Solar Wind at 1 AU				
Parameter	Mean	STD	Most Probable	Median	5–95% Range
n (cm^{-3})	8.7	6.6	5.0	6.9	3.0–20.0
V_{sw} (km/s)	468	116	375	442	320–710
B (nT)	6.2	2.9	5.1	5.6	2.2–9.9
A(He)	0.047	0.019	0.048	0.047	0.017–0.078
T_p ($\times 10^5$ K)	1.2	0.9	0.5	0.95	0.1–3.0
T_e ($\times 10^5$ K)	1.4	0.4	1.2	1.33	0.9–2.0
T_α ($\times 10^5$ K)	5.8	5.0	1.2	4.5	0.6–15.5
T_e/T_p	1.9	1.6	0.7	1.5	0.37–5.0
T_α/T_p	4.9	1.8	4.8	4.7	2.3–7.5
nV_{sw} ($\times 10^8$/cm^{2}s)	3.8	2.4	2.6	3.1	1.5–7.8
C_s (km/s)	63	15	59	61	41–91
C_A (km/s)	50	24	50	46	30–100

ecliptic plane at 1 AU. The table includes mean values, standard deviations about the mean values, most probable values, median values, and the 5–95% range limits for the proton number density (n), the flow speed (V_{sw}), the magnetic field strength (B), the alpha particle abundance relative to protons [A(He)], the proton temperature (T_p), the electron temperature (T_e), the alpha particle temperature (T_a), the ratio of the electron and proton temperatures (T_e/T_p), the ratio of alpha particle and proton temperatures (T_a/T_p), the number flux (nV_{sw}), the sound speed (C_s), and the **Alfvén speed** (C_A) (the speed at which small amplitude perturbations in the magnetic field propagate through the plasma). All solar wind parameters exhibit considerable variability; moreover, variations in solar wind parameters are often coupled to one another. Proton temperatures are considerably more variable than electron temperatures, and alpha particle temperatures are almost always higher than electron and proton temperatures. Alpha particles and the protons tend to have nearly equal thermal speeds and therefore temperatures that differ by a factor of ~4. The solar wind flow is usually both supersonic and super-Alfvénic. Finally, we note that the Sun yearly loses ~6.8×10^{19} g to the solar wind, a very small fraction of the total solar mass of ~2×10^{33} g.

3. Nature of the Heliospheric Magnetic Field

In addition to being a very good thermal conductor, the solar wind plasma is an excellent electrical conductor. The electrical conductivity of the plasma is so high that the solar magnetic field is "frozen" into the solar wind flow as it expands away from the Sun. Because the Sun rotates, magnetic field lines in the equatorial plane of the Sun are bent into spirals (Fig. 1) whose inclinations relative to the radial direction depend on heliocentric distance and the speed of

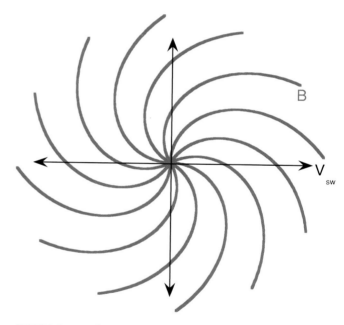

FIGURE 1 Configuration of the heliospheric magnetic field in the ecliptic plane for a uniform, radial solar wind flow.

the wind. At 1 AU, the average field line in the equatorial plane is inclined ~45° to the radial direction.

In Parker's simple model, the magnetic field lines out of the equatorial plane take the form of helices wrapped about the rotation axis of the Sun. These helices are ever more elongated at higher solar latitudes and eventually approach radial lines over the solar poles. The equations describing Parker's model of the magnetic field far from the Sun are

$$B_r(r, \phi, \theta) = B(r_0, \phi_0, \theta)(r_0/r)^2$$
$$B_\phi(r, \phi, \theta) = -B(r_0, \phi_0, \theta)(\omega r_0^2/V_{sw}r)\sin\theta$$
$$B_\theta = 0$$

Here r, ϕ, and θ are radial distance, longitude, and latitude in a Sun-centered spherical coordinate system, B_r, B_ϕ, and B_θ are the magnetic field components, ω is the Sun's angular velocity (2.9×10^{-6} radians sec^{-1}), V_{sw} is the flow speed (assumed constant with distance from the Sun), and ϕ_0 is an initial longitude at a reference distance r_0 from Sun center. This model is in reasonably good agreement with suitable averages of the **heliospheric magnetic field** measured over a wide range of heliocentric distances and latitudes. However, the instantaneous orientation of the field usually deviates substantially from that of the model field at all distances and latitudes. Moreover, there is evidence that the magnetic field lines wander in latitude as they extend out into the heliosphere. This appears to be a result of field line foot point motions associated with differential solar rotation (the surface of the Sun rotates at different rates at different latitudes) and convective motions in the solar atmosphere.

4. Coronal and Solar Wind Stream Structure

The solar corona is highly nonuniform, being structured by the complex solar magnetic field into arcades, rays, holes (regions relatively devoid of material), and streamers. [*See* The Sun.] The strength of the Sun's magnetic field falls off sufficiently rapidly with height above the solar surface that it is incapable of containing the coronal expansion at altitudes above ~0.5–1.0 solar radii. The resulting solar wind outflow produces the "combed-out" appearance of coronal structures above those heights in eclipse photographs.

The solar wind is also highly nonuniform. In the ecliptic plane, it tends to be organized into alternating streams of high- and low-speed flows. Figure 2, which shows solar wind flow speed, flow azimuth, the radial component of the heliospheric magnetic field, and the field strength at 1 AU for a 50-day interval in 2004, illustrates certain characteristic aspects of this **stream structure**. Four high-speed streams with flows exceeding 700 km/s are clearly evident in the figure. The third and fourth streams were actually reencounters with the first and second streams, respectively, on the following solar rotation. Each high-speed stream was asymmetric with the speed rising more rapidly than it fell, and each stream was essentially unipolar in the sense that B_r was either positive or negative throughout the stream. Reversals in field polarity occurred in the low-speed flows between the streams. Those polarity reversals correspond to crossings of the **heliospheric current sheet** (discussed in more detail in the following section) that separates solar wind regions of opposite magnetic polarity. The magnetic field and plasma density (not shown) peaked on the leading edges of the streams, and the solar wind flow there was deflected first westward (positive flow azimuth) and then eastward. This pattern of variability is highly repeatable from

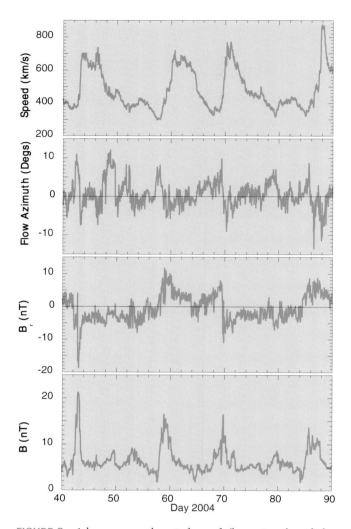

FIGURE 2   1-hr average solar wind speed, flow azimuth, radial component of the heliospheric magnetic field, and the field magnitude at 1 AU for a 50-day interval in 2004.

one stream to the next and is the inevitable consequence of the evolution of the streams as they progress outward from the Sun (see Section 6).

Recurrent high-speed streams originate primarily in coronal holes, which are large, nearly unipolar regions in the solar atmosphere having relatively low density. Low-speed flows, on the other hand, tend to originate in the coronal streamers that straddle regions of magnetic field polarity reversals in the solar atmosphere. Both coronal and solar wind stream structure evolve considerably from one solar rotation to the next as the solar magnetic field, which controls that structure, continuously evolves. It is now clear that the mysterious M-regions, hypothesized long before the era of satellite X-ray observations of the Sun, are to be identified with coronal holes, and the long-lived particle streams responsible for recurrent geomagnetic activity are

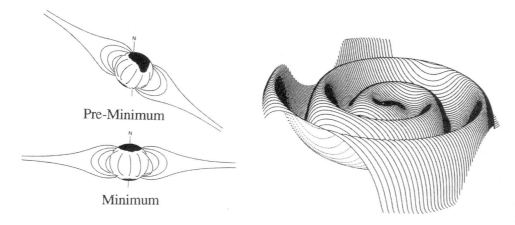

Pre-Minimum

Minimum

FIGURE 3 Right, schematic illustrating the configuration of the heliospheric current sheet when the solar magnetic dipole is tilted substantially relative to the rotation axis of the Sun. The heliospheric current sheet separates magnetic fields of opposite magnetic polarity and is the heliospheric extension of the solar magnetic equator. Left, schematic illustrating the changing tilt of coronal structure and the solar magnetic dipole relative to the rotation axis of the Sun as a function of the phase of the solar activity cycle. [Adapted from J. R. Jokipii and B. Thomas, 1981, *Astrophys. J.* **243**, 1115, and from A. J. Hundhausen, 1977, in "Coronal Holes and High Speed Wind Streams" (J. Zirker, ed.), Colorado Associated University Press, Boulder, Colorado.]

to be identified with high-speed solar wind streams. [*See* SUN–EARTH CONNECTION.]

5. The Heliospheric Current Sheet and Solar Latitude Effects

5.1 The Sun's Large–Scale Magnetic Field and the Ballerina Skirt Model

On the declining phase of the solar activity cycle and near solar activity minimum, the Sun's large-scale magnetic field well above the photosphere appears to be approximately that of a dipole. The solar magnetic dipole is tilted with respect to the Sun's rotation axis; this tilt changes with the advance of the solar cycle. As illustrated in the left-hand side of Fig. 3, near the solar activity minimum the solar magnetic dipole tends to be aligned nearly with the rotation axis, whereas on the declining phase of the activity cycle it is generally inclined at a considerable angle relative to the rotation axis. Near the solar maximum, the Sun's large-scale field is probably not well approximated by a dipole.

When the solar magnetic dipole and the solar rotation axis are closely aligned, the heliospheric current sheet, which is effectively the extension of the solar magnetic equator into the solar wind, coincides roughly with the solar equatorial plane. On the other hand, at times when the dipole is tilted substantially, the heliospheric current sheet is warped and resembles a ballerina's twirling skirt, as illustrated in the right-hand side of Fig. 3. Successive outward

ridges in the current sheet (folds in the skirt) correspond to successive solar rotations and are separated radially by about 4.7 AU when the flow speed at the current sheet is 300 km/s. The maximum solar latitude of the current sheet in this simple picture is equal to the tilt angle of the magnetic dipole axis relative to the rotation axis.

5.2 Solar Latitude Effects

On the declining phase of the solar activity cycle and near the solar activity minimum, stream structure and solar wind variability are largely confined to a relatively narrow latitude band centered on the solar equator. This is illustrated in the upper left portion of Fig. 4, which shows solar wind speed as a function of solar latitude measured by *Ulysses* on the declining phase of the most recent solar cycle. (*Ulysses* is in a solar orbit that takes it to solar latitudes of ±80° in its ~5.5-year journey about the Sun.) At this phase of the solar cycle, the solar wind is dominated by stream structure at low latitudes, but it flows at a nearly constant speed of ~850 km/s at high latitudes. This latitude effect is a consequence of the following: (1) Solar wind properties change rapidly with distance from the heliospheric current sheet, with flow speed being a minimum in the vicinity of the current sheet; and (2) the heliospheric current sheet is commonly tilted relative to the solar equator but is usually found within about ±30° of it during this phase of the solar cycle. The width of the band of solar wind variability changes as the solar magnetic dipole tilt changes. The upper right portion of Fig. 4 demonstrates that, in contrast, in the years surrounding the

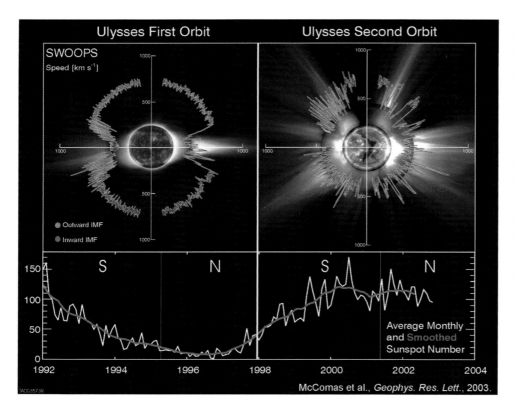

FIGURE 4 Solar wind speed as a function of heliographic latitude as measured by the *Ulysses* space probe during the declining phase of the solar activity cycle (left) and near solar activity maximum (right). The speed data are color-coded according to the polarity of the magnetic field and are superimposed on representative images of the solar corona at those phases of the solar cycle. Smoothed sunspot numbers are shown at the bottom. The S and N labels on the latter indicate the solar hemisphere that *Ulysses* was in at those times. (From D. J. McComas et al., 2003, *Geophys. Res. Lett.* **30**, 10 10.1029.2003GL017136.)

solar activity maximum, the band of solar wind variability extends up to the highest latitudes sampled by *Ulysses*.

6. Evolution of Stream Structure with Heliocentric Distance

6.1 Kinematic Stream Steepening and the Dynamic Response

Because the coronal expansion is spatially variable, alternately slow and fast plasma is directed outward along any radial line from the Sun as the Sun rotates (with a period of 27 days as seen from Earth). Faster-moving plasma overtakes slower-moving plasma ahead while outrunning slower-moving plasma behind. The result is that the leading edges of high-speed streams steepen with increasing distance from the Sun, producing the asymmetric stream profiles obvious in Fig. 2. Material within the streams is rearranged as the streams steepen; plasma and field on the leading edge of a stream are compressed, causing an increase in plasma density, temperature, field strength, and pressure there, while plasma and field on the trailing edge become increasingly rarefied. The buildup of pressure on the leading edge of a stream produces forces that accelerate the low-speed wind ahead and decelerate the high-speed wind within the stream itself. The net result is a transfer of momentum and energy from the fast-moving wind to the slow-moving wind.

6.2 Shock Formation

As long as the amplitude of a high-speed solar wind stream is sufficiently small, it gradually dampens with increasing heliocentric distance in the manner just described. However, when the difference in flow speed between the crest of a stream and the trough ahead is greater than about twice the local fast mode speed, C_f [the fast mode speed is the characteristic speed with which small amplitude pressure signals propagate in a plasma: $C_f = (C_s^2 + C_A^2)^{0.5}$], ordinary pressure signals do not propagate sufficiently fast to move the slow wind out of the path of the oncoming high-speed stream. In that case, the pressure eventually increases nonlinearly, and **shock** waves form on either side of the high-pressure region (see Fig. 5). The leading shock, known as a forward shock, propagates into the low-speed wind ahead, and the trailing shock, known as a reverse shock, propagates back through the stream. Both shocks are, however, convected away from the Sun by the high bulk flow of the wind. The major accelerations and decelerations associated with

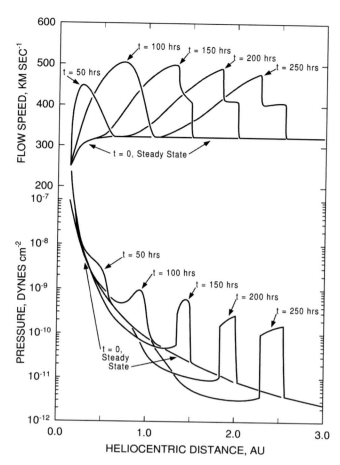

FIGURE 5 Snapshots of solar wind flow speed (above) and pressure (below) as functions of heliocentric distance at different times during the evolution of a large-amplitude, high-speed solar wind stream as calculated from a simple one-dimensional numerical model. (Adapted from A. J. Hundhausen, 1973, *J. Geophys. Res.* **78**, 1528.)

stream evolution occur discontinuously at the shocks, giving a stream speed profile the appearance of a double saw-tooth wave. The stream amplitude decreases and the compression region expands with increasing heliocentric distance as the shocks propagate. Observations indicate that the shocks typically do not form until the streams are well beyond 1 AU. Nevertheless, because C_f generally decreases with increasing heliocentric distance, virtually all large-amplitude solar wind streams steepen into shock wave structures at heliocentric distances beyond ∼3 AU. At heliocentric distances beyond the orbit of Jupiter (∼5.4 AU) a large fraction of the mass in the solar wind is found within compression regions bounded by shock waves on the rising portions of damped high-speed streams. The basic structure of the solar wind in the solar equatorial plane in the distant heliosphere thus differs considerably from that observed at 1 AU. Stream amplitudes are severely reduced, and short wavelength struc-

ture is damped out. The dominant structure in the solar equatorial plane in the outer heliosphere is the expanding compression region where most of the plasma and magnetic field are concentrated.

6.3 Stream Evolution in Two and Three Dimensions

When the coronal expansion is spatially variable but time-stationary, a steady flow pattern such as that sketched in Fig. 6 develops in the equatorial plane. This entire pattern corotates with the Sun, and the compression regions are known as corotating interaction regions (CIRs); however, only the pattern rotates—each parcel of solar wind plasma moves outward nearly radially as indicated by the black arrows. The region of high pressure associated with a CIR is nearly aligned with the magnetic field line spirals in the equatorial plane, and the pressure gradients are thus nearly perpendicular to those spirals. Consequently, at 1 AU, the pressure gradients that form on the rising speed portions of high-speed streams have transverse as well as radial components. In particular, not only is the low-speed plasma ahead of a high-speed stream accelerated to a higher speed, but it is also deflected in the direction of solar rotation. In contrast, the high-speed plasma near the crest of the stream is both decelerated and deflected in the direction opposite

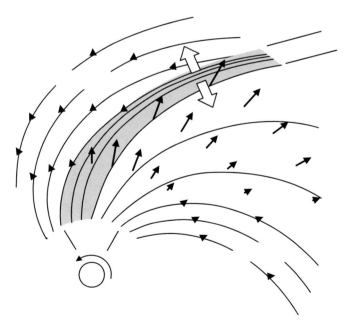

FIGURE 6 Schematic illustrating two-dimensional, quasi-stationary stream structure in the ecliptic plane in the inner heliosphere. The compression region on the leading edge of a stream is nearly aligned with the spiral magnetic field, and the forces associated with the pressure gradients have transverse as well as radial components. (From V. J. Pizzo, 1978, *J. Geophys. Res.* **83**, 5563.)

to solar rotation. These transverse deflections produce the systematic west–east flow direction changes observed near the leading edges of quasi-stationary, high-speed streams (see Fig. 2).

There is an interesting three-dimensional aspect to stream evolution, ultimately associated with the fact that the solar magnetic dipole is tilted relative to the solar rotation axis. That tilt causes CIRs in the northern and southern solar hemispheres to have opposed meridional tilts that, particularly in the outer heliosphere, can be discerned in plasma data as systematic north–south deflections of the flow at CIRs. The meridional tilts are such that the forward waves in both hemispheres initially propagate equatorward, whereas the reverse waves in both hemispheres propagate poleward. As a result, forward shocks in the outer heliosphere near the solar minimum are generally confined to the low-latitude band of solar wind variability, whereas the reverse shocks are commonly observed both within the band of variability and poleward of it. However, the reverse waves seldom reach latitudes more than ~15° above the low-latitude band of variability.

7. Coronal Mass Ejections and Transient Solar Wind Disturbances

7.1 Coronal Mass Ejections

The solar corona evolves on a variety of time scales closely connected with the evolution of the coronal magnetic field. [*See* THE SUN.] The most rapid and dramatic evolution in the corona occurs in events known as **coronal mass ejections,** or **CMEs** (Fig. 7a). CMEs originate in closed field regions in the corona where the magnetic field normally is sufficiently strong to constrain the coronal plasma from expanding outward. Typically these closed field regions are found in the coronal streamer belt that encircles the Sun and that underlies the heliospheric current sheet. The outer edges of CMEs often have the optical appearance of closed loops such as the event shown in Fig. 7a. Few, if any, CMEs ever appear to sever completely their magnetic connection with the Sun. During a typical CME, somewhere between 10^{15} and 10^{16} g of solar material is ejected into the heliosphere. Ejection speeds near the Sun range from less than

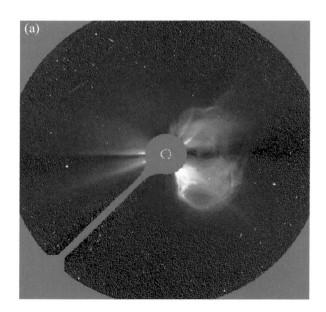
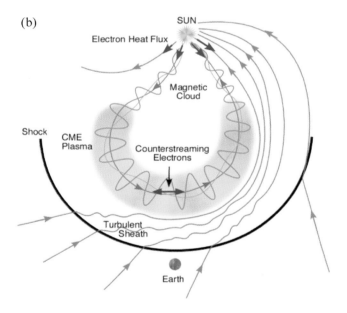

FIGURE 7   (a) A coronal mass ejection as imaged by the LASCO/C3 coronagraph on *SOHO* on April 20, 1998. The Sun, indicated by the white circle, has been occulted within the instrument. The field of view of the image is 30 solar diameters. [The SOHO/LASCO data are produced by a consortium of the Naval Research Laboratory (USA), Max-Planck-Institut fur Sonnensystemforschung (Germany), Laboratoire d'Astronomie (France), and the University of Birmingham (UK). SOHO is a project of international cooperation between ESA and NASA.] (b) A sketch of a solar wind shock disturbance produced by a fast ICME directed toward Earth. Red and magenta arrows indicate the ambient magnetic field and that threading the ICME, respectively. Blue arrows indicate the suprathermal electron strahl flowing away from the Sun along the magnetic field. The ambient magnetic field is compressed by its interaction with the ICME and is forced to drape around the ICME. [To appear in T. H. Zurbuchen and I. G. Richardson, 2006, in "Coronal Mass Ejections" (H. Kunow et al., eds.), Kluwer academic Publishers, Dordrecht.]

50 km/s in some of the slowest events to greater than 2500 km/s in the fastest ones. The average CME speed at ~5 solar radii is close to the median ecliptic solar wind speed of ~440 km/s. Since observed solar wind speeds near 1 AU are never less than ~280 km/s, the slowest CMEs are further accelerated enroute to 1 AU.

7.2 Origins, Associations with Other Forms of Solar Activity, and Frequency of Occurrence

The processes that trigger CMEs and that determine their sizes and outward speeds are only poorly understood; there is presently no consensus on the physical processes responsible for initiating or accelerating these events, although it is clear that stressed magnetic fields are the underlying cause of these events and that CMEs play a fundamental role in the long-term evolution of the structure of the solar corona. They appear to be an essential part of the way the corona responds to the evolution of the solar magnetic field associated with the advance of the solar activity cycle. Indeed, the release of a CME is one way that the solar atmosphere reconfigures itself in response to changes in the solar magnetic field. CMEs are commonly, but not always, observed in association with other forms of solar activity such as eruptive prominences and solar flares. From a historical perspective, one might be led to expect that large solar flares are the prime cause of CMEs; however, it is now clear that flares and CMEs are separate, but closely related, phenomena associated with magnetic disturbances on the Sun. Like other forms of solar activity, CMEs occur with a frequency that varies in a cycle of ~11 years. On average, the Sun emits about 3.5 CMEs/day near the peak of the solar activity cycle, but only about 0.1 CMEs/day near solar activity minimum.

7.3 Heliospheric Disturbances Driven by Fast Coronal Mass Ejections

As illustrated in Fig. 7b, fast CMEs produce transient solar wind disturbances that, in turn, often are the cause of large, geomagnetic storms. [See SUN–EARTH CONNECTIONS.] Figure 8 shows calculated radial speed and pressure profiles of a simulated solar wind disturbance driven by a fast CME at the time the disturbance first reaches 1 AU. As indicated by the insert in the top portion of the figure, the disturbance was initiated at the inner boundary of the one-dimensional fluid calculation by abruptly raising the flow speed from 275 to 980 km/s, sustaining it at this level for 6 hours, and then returning it to its original value of 275 km/s. The initial disturbance thus mimics a uniformly fast, spatially limited CME with an internal pressure equal to that of the surrounding solar wind plasma. A region of high pressure develops on the leading edge of the disturbance as the CME overtakes the slower wind ahead. This region of

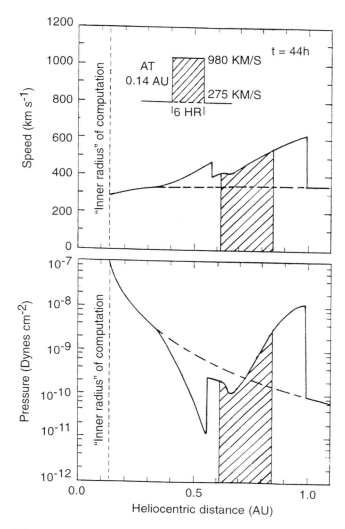

FIGURE 8   Solar wind speed and pressure as functions of heliocentric distance for a simple, one-dimensional gas-dynamic simulation of a CME-driven disturbance. The dashed line indicates the steady state prior to introduction of the temporal variation in flow speed imposed at the inner boundary of 0.14 AU and shown at the top of the figure. The hatching identifies material that was introduced with a speed of 980 km/s at the inner boundary, and therefore identifies the CME in the simulation. [Adapted originally from A. J. Hundhausen, 1985, in "Collisionless Shocks in the Heliosphere: A Tutorial Review" (R. G. Stone and B. T. Tsurutani, eds.), American Geophysical Union, Washington, D.C.].

higher pressure is bounded by a forward shock on its leading edge that propagates into the ambient solar wind ahead and by a reverse shock on its trailing edge that propagates backward into and eventually through the CME. Both shocks are, however, carried away from the Sun by the highly supersonic flow of the solar wind. Observations and more detailed calculations indicate, however, that reverse shocks in CME-driven disturbances are ordinarily present only near the central portions of the disturbances.

Except for the reverse shock, the simple calculation shown in Fig. 8 is consistent with observations of many solar wind disturbances obtained near 1 AU in the ecliptic plane and illustrates to first order the radial and temporal evolution of an interplanetary disturbance driven by a fast CME (now commonly called an interplanetary coronal mass ejection, ICME, when observed in the solar wind). The leading edge of the disturbance is a shock that stands off ahead of the ICME (see also Fig. 7b). The ambient solar wind ahead of the ICME is compressed, heated, and accelerated as the shock passes by, and the leading portion of the ICME is compressed, heated, and slowed as a result of the interaction. In the example illustrated, the ICME slows from an initial speed of 980 km/s to less than 600 km/s by the time the leading edge of the disturbance reaches 1 AU. This slowing is a result of momentum transfer to the ambient solar wind ahead and proceeds at an ever-slower rate as the disturbance propagates outward. Figure 9 displays selected plasma and magnetic field data from a solar wind disturbance driven by an ICME observed near 1 AU. The shock is distinguished in the data by discontinuous increases in flow speed, density, temperature, and field strength. The plasma identified as the ICME had a higher flow speed than the ambient solar wind ahead of the shock. In this case, it was also distinguished by counterstreaming suprathermal electrons (indicative of a closed magnetic field topology, see Section 10.3), anomalously low proton temperatures, somewhat elevated helium abundance, and a strong, smoothly rotating magnetic field that indicates the magnetic field topology was that of a nested helical structure (i.e., a flux rope; see Fig. 7b).

7.4 Characteristics of Interplanetary Coronal Mass Ejections

The identification of ICMEs in solar wind plasma and field data is still something of an art; however, shocks serve as useful fiducials for identifying fast ICMEs. Table 2 provides a summary of plasma and field signatures that qualify as being unusual compared to the normal solar wind, but that are commonly observed a number of hours after shock passage. Most of these anomalous signatures are also observed elsewhere in the solar wind where, presumably, they serve to identify those numerous, relatively low-speed ICMEs that do not drive shock disturbances. Few ICMEs at 1 AU exhibit all of these characteristics, and some of these signatures are more commonly observed than are others.

Most ICMEs expand as they propagate outward through the heliosphere. ICME radial thicknesses are variable; at 1 AU the typical ICME has a radial width of ~0.2 AU, whereas at Jupiter's orbit ICMEs can have radial widths as large as 2.5 AU. Approximately one third of all ICMEs

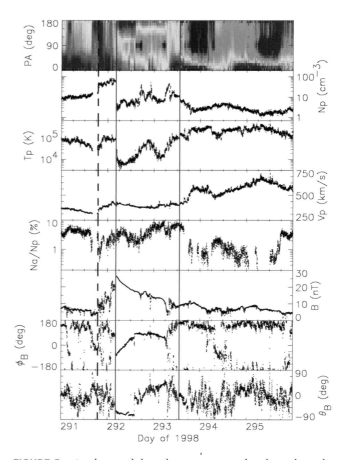

FIGURE 9 A solar wind disturbance associated with moderately fast ICME observed by the *Advanced Composition Explorer* in October 1998. From top to bottom the quantities plotted are color-coded pitch angle distributions of 256–288 eV electrons, proton density, proton temperature, bulk flow speed, alpha-proton density ratio, and magnetic field strength, azimuth, and polar angle in solar ecliptic coordinates. The color scale for $f(v)$ extends from 5×10^{-32} (dark purple) to 2×10^{-29} s^3 cm^{-6} (dark red). Dashed and solid vertical lines respectively mark the shock and the edges of the ICME. (From J. T. Gosling et al., 2002, *Geophys. Res. Lett.* **29**, 12 10.1029/2001GL013949.)

in the ecliptic plane have sufficiently high speeds relative to the ambient solar wind to drive shock disturbances at 1 AU; the remainder do not and simply coast along with the rest of the solar wind. Typically, ICMEs cannot be distinguished from the normal solar wind at 1 AU on the basis of either their speed or density (the event in Fig. 9 is an example). Near the solar activity maximum, ICMEs account for 15–20% of the solar wind in the ecliptic plane at 1 AU, while they account for less than 1% near the solar activity minimum. The Earth intercepts about 72 ICMEs/year near the solar activity maximum and ~8 ICMEs/year near the solar activity minimum. ICMEs are much less common at high heliographic latitudes, particularly near activity minimum

TABLE 2	Characteristics of Interplanetary Coronal Mass Ejections at 1 AU

Common signatures:
    Counterstreaming (along the field) suprathermal (energy >70 eV) electrons
    Counterstreaming (along the field) energetic (energy >20 keV) protons
    Helium abundance enhancement
    Anomalously low proton and electron temperatures
    Strong magnetic field
    Low **plasma beta**
    Low magnetic field strength variance
    Anomalous field rotation (flux rope)
    Anomalous ionic composition (e.g., Fe^{16+}, He^{+})
    Cosmic ray depression
Average radial thickness: 0.2 AU
Range of speeds: 300–2000 km/s
Single point occurrence frequency:
    ~72 events/year at solar activity maximum
    ~8 events/year at solar activity minimum
Magnetic field topology: Predominantly closed magnetic loops rooted in Sun
Fraction of events driving shocks: ~1/3
Fraction of earthward-directed events producing large geomagnetic storms: ~1/6

when ICMEs are confined largely to the low-latitude band of solar wind variability.

7.5 The Magnetic Field Topology of ICMEs and the Problem of Magnetic Flux Balance

The coronal expansion carries the solar magnetic field outward to form the heliospheric magnetic field. In the quiescent wind, these field lines are usually "open" in the sense that they connect to field lines of the opposite polarity only in the very distant heliosphere. CMEs, on the other hand, originate in the corona in closed field regions that have not previously participated directly in the solar wind expansion and carry new magnetic flux into the heliosphere. The magnetic flux that each CME apparently adds to the heliosphere must be balanced by removal of magnetic flux elsewhere since the overall heliospheric magnetic field strength is roughly (within a factor of 2) constant in time. Magnetic reconnection within the magnetic "legs" of a CME close to the Sun appears to be the prime way that this balance is achieved. Figure 10 illustrates that such reconnection is inherently three-dimensional in nature and initially produces helical magnetic field lines that are partially disconnected from the Sun (see, also, Fig. 7b). Sustained three-dimensional magnetic reconnection eventually produces open and disconnected field lines threading an ICME, both of which are sometimes observed. All of the types of reconnection illustrated in Fig. 10 reduce the amount of magnetic flux permanently added to the

heliosphere by an ICME. However, it is not presently clear what mix of reconnections within the magnetic legs of ICMEs and elsewhere in the solar atmosphere (e.g., at the heliospheric current sheet) is actually responsible for maintaining a rough long-term balance of magnetic flux in the heliosphere.

7.6 Field Line Draping About Fast Interplanetary Coronal Mass Ejections

Because the closed field nature of ICMEs effectively prevents any substantial interpenetration between the plasma within an ICME and that in the surrounding wind, the ambient plasma and magnetic field ahead must be deflected away from the path of a fast ICME. Figure 7b illustrates that such deflections cause the ambient magnetic field to drape about the ICME. The degree of draping and the resulting orientation of the field ahead of an ICME depend upon the relative speed between the ICME and the ambient plasma, the shape of the ICME, and the original orientation of the magnetic field in the ambient plasma. Draping plays an important role in reorienting the magnetic field ahead of a fast ICME. On the other hand, conditions and processes back at the Sun largely determine the field orientation within ICMEs. As a final point of interest, Figure 7b also illustrates that, just as the bow wave in front of a boat moving through water is considerably broader in extent than is the boat that produces it, so too is the shock in front of a fast ICME somewhat broader in extent than is the

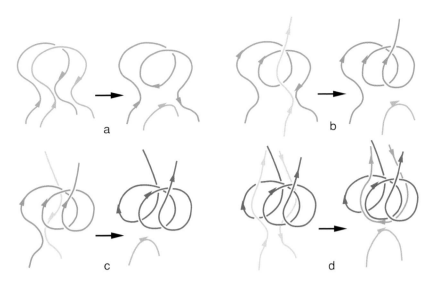

FIGURE 10 Sketches of successive steps in three-dimensional reconnection in the corona beneath a departing CME. The sketches are not to scale and are intended only to illustrate successive changes in CME magnetic topologies resulting from reconnection. (From J. T. Gosling et al., 1995, *Geophys. Res. Lett.* **22**, 869.)

ICME that drives it. As a result, spacecraft often encounter ICME-driven shocks without also encountering the ICMEs that drive them.

8. Variation with Distance from the Sun

For a structureless solar wind, the speed remains nearly constant beyond the orbit of Earth, the density falls off with heliocentric distance (r, as r^{-2}), and the magnetic field decreases with distance as described by the equations in Section 2. The temperature also decreases with increasing heliocentric distance due to the spherical expansion of the plasma; however, the precise nature of the decrease depends upon particle species and the relative importance of such things as collisions and heat conduction (e.g., protons and electrons have different temperatures and evolve differently with increasing heliocentric distance). For an adiabatic expansion of an isotropic plasma, the temperature falls off as $r^{-4/3}$; for a plasma dominated by heat conduction, the temperature falls as $r^{-2/7}$.

Of course, the solar wind is not structureless. The continual interaction of high- and low-speed flows with increasing heliocentric distance produces a radial variation of speed that differs considerably from that predicted for a structureless wind. High-speed flows decelerate and low-speed flows accelerate with increasing heliocentric distance as a result of momentum transfer (see Sections 6 and 7). Consequently, near the solar equatorial plane far from the Sun (beyond ~15 AU) the solar wind flows at 400 to 500 km/s most of the time (Fig. 11). Only rarely are substantial speed perturbations observed at these distances; these rare events usually are associated with disturbances driven by very large and fast ICMEs that require a greater-than-usual distance to share their momentum with the low-speed wind.

With increasing heliocentric distance an ever-greater fraction of the plasma and magnetic field in the wind becomes concentrated within the compression regions on the rising speed portions of high-speed flows; extended rarefaction regions relatively devoid of plasma and field follow these compressions. Thus, at low heliographic latitudes, solar wind density and magnetic field strength tend to vary over a wider range in the outer heliosphere than near the orbit of the Earth, although the average density falls roughly as r^{-2}, and the average magnetic field falls off roughly as predicted by the equations in Section 2. On the other hand, plasma heating associated with the compression regions causes the solar wind temperature to fall off with increasing heliocentric distance more slowly than it otherwise would. Observations reveal that both proton and electron temperatures decrease with distance somewhere between the adiabatic and conduction-dominated extremes.

9. Termination of the Solar Wind

Interstellar space is filled with a dilute gas of neutral and ionized particles and is threaded by a weak magnetic field. In the absence of the solar wind, the interstellar plasma would penetrate deep into the solar system. However, the interstellar and solar wind plasmas cannot interpenetrate one another because of the magnetic fields embedded in both. The result is that the solar wind creates a cavity in the interstellar plasma.

The details of the solar wind's interaction with the interstellar plasma are still somewhat speculative largely because, until recently, we lacked direct observations of this interaction. Figure 12 shows what are believed to be the major elements of the interaction. The Sun and heliosphere

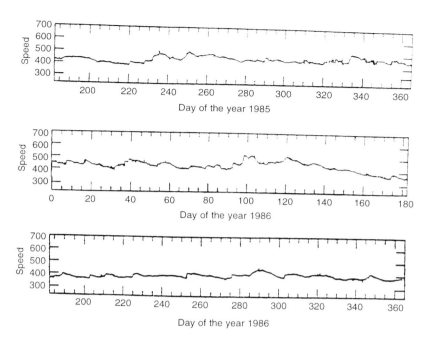

FIGURE 11   Solar wind speed as a function of time as measured by *Voyager 2* during a 1.5-year interval when the spacecraft was beyond 18 AU from the Sun. Because stream amplitudes are severely damped at large distances from the Sun, the solar wind speed there generally varies within a very narrow range of values. Compare with the speed variations evident in Fig. 2 that were obtained at 1 AU during a comparable period of the solar cycle. [Adapted from A. J. Lazarus and J. Belcher, 1988, in "Proceedings of the Sixth International Solar Wind Conference" (V. J. Pizzo, T. E. Holzer, and D. G. Sime, eds.), National Center for Atmospheric Research, Boulder, Colorado.]

move at a speed of ~23 km/s relative to the interstellar medium. If this relative motion exceeds the fast mode speed, C_f, in the interstellar plasma, then a bow shock must stand in the interstellar plasma upstream of the heliosphere to initiate the slowing and deflection of the interstellar plasma around the heliosphere. The **heliopause**, which is the outermost boundary of the heliosphere, separates the interstellar and solar wind plasmas. Sunward of the heliopause is a **termination shock** where the solar wind flow becomes subsonic so that it can be turned to flow roughly

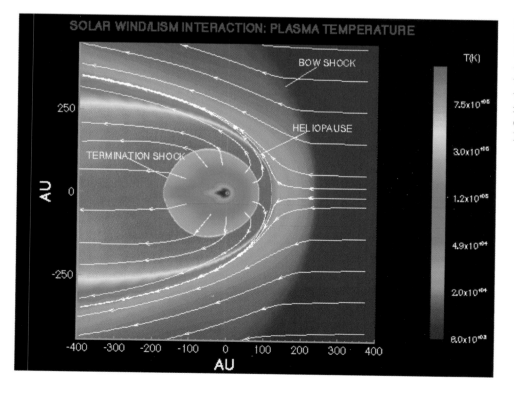

FIGURE 12   Simulated structure of the solar wind's interaction with the interstellar plasma. Color-coding represents the proton temperature and arrows indicate the direction of the solar wind and interplanetary plasma flows. (Courtesy of G. P. Zank and H. R. Mueller.)

parallel to the heliopause. The shape of the heliosphere is asymmetric because of its motion relative to the interstellar gas; it is compressed in the direction of that motion and is greatly elongated in the opposite direction. Observations in the outer heliosphere suggest that the termination shock is constantly in motion relative to the Sun owing to an ever-changing solar wind momentum flux; it may never truly achieve an equilibrium position. The size and shape of the heliosphere depend on the momentum flux carried by the solar wind, the pressure of the interstellar plasma, and the motion of the heliosphere relative to the interstellar medium. *Voyager 1* recently verified the existence of the termination shock, having crossed it in December 2004 at a heliocentric distance of 94 AU roughly in the direction of the heliosphere's motion relative to the interstellar medium. It is currently believed that the heliopause lies at a heliocentric distance of 115–150 AU and should be encountered by Voyager within the next decade.

10. Kinetic Properties of the Plasma

10.1 The Solar Wind as a Marginally Collisional Plasma

On a large scale, the solar wind behaves like a compressible fluid and is capable of supporting relatively thin structures such as shocks. It is perhaps not obvious why the solar wind should exhibit this fluid-like behavior since the wind is a dilute plasma in which collisions are relatively rare. For example, using values given in Table 1, we find that the time between collisions for a typical solar wind proton at 1 AU is several days. (These collisions do not result from direct particle impacts such as colliding billiard balls, but rather from the long distance **Coulomb interactions** characteristic of charged particles.) The time between collisions is thus comparable to the time for the solar wind to expand from the vicinity of the Sun to 1 AU; this is the basis for statements that the solar wind is a marginally collisional plasma.

There are several reasons why the solar wind behaves like a fluid even in the absence of particle collisions to effect fluid-like behavior. First, when the temperature is low and the density is high, collisions are more frequent than noted previously. Second, the presence of the heliospheric magnetic field causes charged particles to gyrate about the field, and they thus do not travel in straight lines between collisions. For typical conditions at 1 AU, solar wind electrons and protons have **gyro radii** of ~1.4 and ~60 km, respectively, which are small compared to the scale size of most structures in the solar wind. Third, the solar wind is subject to a variety of instabilities that are triggered whenever particle distribution functions depart significantly from thermal distributions (see Section 10.2). These instabilities produce collective interactions that mimic the effects of particle collisions. Finally, because the magnetic field is frozen

into the solar wind flow, parcels of plasma originating from different positions on the Sun cannot interpenetrate one another.

10.2 Kinetic Aspects of Solar Wind Ions

Collisional gases can usually be described by a single isotropic (i.e., the same in all directions) temperature (T) with the distribution of particle speeds (v) obeying $f(v) \sim \exp[-m(v - V_0)^2 2kT]$, where f is the number of particles per unit volume of velocity space, k is Boltzman's constant (1.38×10^{-16} erg/deg), m is the particle mass, and V_0 is the bulk speed of the gas. In contrast, proton distribution functions in the solar wind are usually anisotropic because of the paucity of collisions and because the magnetic field provides a preferred direction in space. At 1 AU, the proton temperature parallel to the field is generally greater than the temperature perpendicular to the field, on average by a factor of ~1.4. Moreover, solar wind proton and alpha particle distributions often exhibit significant non-Maxwellian features such as the double-peaked distributions illustrated in Fig. 13a. The secondary proton and alpha particle peaks are associated with beams streaming relative to the main solar wind component along the heliospheric magnetic field. The relative streaming speed of such beams is usually comparable to or less than the local Alfvén speed, suggesting that the streaming is limited by a kinetic instability. Closer to the Sun, where the Alfvén speed is higher, relative streaming speeds between the beams and the main components can be as large as several hundred km/s. Secondary proton beams are common in the solar wind in both low- and high-speed flows and may play a fundamental role in the overall acceleration and heating of the wind; however, their origin in solar and/or heliospheric processes is presently uncertain. Figure 13b illustrates that solar wind ion distributions in the low-speed wind also commonly have extended nonthermal tails of uncertain origin. Particles in these extended tails are easily accelerated to much higher energy when they encounter shocks (see Section 7).

10.3 Kinetic Aspects of Solar Wind Electrons

Electron distributions in the solar wind consist of a relatively cold and dense thermal "core" population that is electrically bound to the solar wind ion population and a much hotter and freer-running suprathermal population that becomes collisionless close to the Sun. At 1 AU, the breakpoint between these populations typically occurs at ~70 eV (Fig. 14a). This breakpoint moves steadily to lower energies with increasing heliocentric distance as the core population cools. Typically the core contains about 95% of the electrons, and at 1 AU has a temperature of ~1.3×10^5. The core electrons typically are mildly anisotropic, with the temperature parallel to the field exceeding the temperature

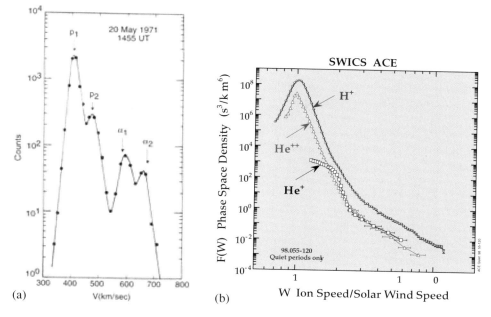

FIGURE 13 (a) A cut through a solar wind ion count spectrum parallel to the magnetic field. The first two peaks are protons, and the second two peaks are alpha particles. (The velocity scale for the alpha particles has been increased by a factor of 1.4.) Both the proton and alpha particle spectra show clear evidence for a secondary beam of particles streaming along the field relative to the main solar wind beam at about the Alfvén speed. Such secondary beams, not always well resolved, are common in both the low and the high-speed wind. (From J. R. Asbridge et al., 1974, *Solar Phys.* **37**, 451.) (b) Solar wind speed distributions of H+, He2+, and He+ observed in the low-speed solar wind at 1 AU, averaged over a 65-day period in 1998 and excluding intervals of shocks and other disturbances. Such extended suprathermal tails appear to be ubiquitous in the low-speed solar wind. The He+ ions are primarily of interstellar origin. (From G. Gloeckler et al., Acceleration and transport of energetic particles observed in the heliosphere, in Mewaldt et al., 2000.)

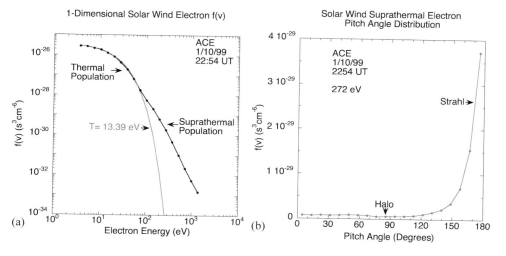

FIGURE 14 (a) One-dimensional cut through a solar wind electron distribution showing the thermal and suprathermal populations. (b) Suprathermal electron pitch angle distribution (relative to the magnetic field) showing the field-aligned strahl and the nearly isotropic halo components.

perpendicular to the field by a factor of ~1.1 on average at 1 AU. However, the temperature anisotropy for core electrons varies systematically with density such that at very low densities (<2 cm^{-3}) the temperature ratio often exceeds 2.0, while at very high densities (>10 cm^{-3}) the temperature ratio is often slightly less than 1.0. Such systematic variations of core electron temperature anisotropy with plasma density reflect the marginally collisional nature of the thermal electrons and their nearly adiabatic expansion in the spiral magnetic field.

The suprathermal electrons consist of a beam of variable width and intensity, known as the strahl, directed outward from the Sun along the heliospheric magnetic field and a more tenuous and roughly isotropic "halo" (Fig. 14b). The angular width of the strahl results from a competition between focusing associated with conservation of magnetic moment in the diverging heliospheric magnetic field and defocusing associated with particle scattering. The strahl carries the solar wind electron heat flux; variations in strahl intensity largely reflect spatial variations in the corona from which it arises. In addition, brief (hours) strahl intensifications often occur during solar electron bursts associated with solar activity (see Section 7). The strahl serves as an effective tracer of magnetic field topology in the interplanetary medium since its usual unidirectional nature arises because field lines in the normal solar wind are "open" (see Section 7.5) and are thus effectively connected to the solar corona at only one end. In contrast, field lines threading ICMEs are often attached to the Sun at both ends (see Section 7.4 and 7.5), and counterstreaming strahls are commonly observed there. Indeed, counterstreaming strahls are one of the more reliable signatures of ICMEs (see Figs. 7b and 9 and Table 2). Finally, the nearly isotropic electron halo results primarily from the scattering out of the strahl at distances beyond 1 AU and the subsequent reflection of those backscattered electrons inside 1 AU by the stronger magnetic fields that reside there.

11. Heavy Ion Content

Although the solar wind consists primarily of protons (hydrogen), electrons, and alpha particles (doubly ionized helium), it also contains traces of ions of a number of heavier elements. Table 3 provides estimates of the relative abundances of some of the more common solar wind elements summed over all ionization states. After hydrogen and helium, the most abundant elements are carbon and oxygen. The ionization states of all solar wind ions are "frozen in" close to the Sun because the characteristic times for ionization and recombination are long compared to the solar wind expansion time. Commonly observed ionization states include He^{2+}, C^{5+}, C^{6+}, O^{6+} to O^{8+}, Si^{7+} to Si^{10+}, and Fe^{8+} to Fe^{14+}. Ionization state temperatures in the low-speed wind are typically in the range 1.4 to 1.6×10^6 K, whereas ionization state temperatures in the high-speed wind are

TABLE 3   Average Elemental Abundances in the Solar Wind

Element	Abundance Relative to Oxygen
H	1900 ± 400
He	75 ± 20
C	0.67 ± 0.10
N	0.15 ± 0.06
O	1.00
Ne	0.17 ± 0.02
Mg	0.15 ± 0.02
Si	0.19 ± 0.04
Ar	0.0040 ± 0.0010
Fe	$019 + 0.10, - 0.07$

typically in the range 1.0 to 1.2×10^6 K. Unusual ionization states such as Fe^{+16} and He^{+1}, which are not common in the normal solar wind, are often abundant within ICMEs, reflecting the unusual coronal origins of those events.

The relative abundance values in Table 3 are long-term averages; however, abundances vary considerably with time. Such variations have been extensively studied for the He^{2+}/H$^+$ ratio, A(He), but are less well established for heavier elements. The most probable A(He) value is ~0.045, but the A(He) ranges from less than 0.01 to values of 0.35 on occasion. The average A(He) is about half that commonly attributed to the solar interior, for reasons presently unknown. Much of the variation in A(He) and in the abundance of heavier elements is related to the large-scale structure of the wind. For example, Fe/O and Mg/O ratios are systematically lower in high-speed streams than in low-speed flows. A(He) tends to be relatively constant at ~0.045 within quasi-stationary, high-speed streams but tends to be highly variable within low-speed flows. Particularly low (<0.02) abundance values are commonly observed at the heliospheric current sheet. A(He) values greater than about 0.10 are relatively rare and account for less than 1% of all the measurements. At 1 AU, enhancements in A(He) above 0.10 occur almost exclusively within ICME plasma. The physical causes of these variations are uncertain for the most part, although thermal diffusion, gravitational settling, and Coulomb friction in the chromosphere and corona all probably play roles.

12. Energetic Particles

A proton moving with a speed of 440 km/s has an energy of ~1 keV. Thus, by most measures, solar wind ions are low-energy particles. The heliosphere is, nevertheless, filled with a number of energetic ion populations of varying intensities with energies ranging upwards from ~1 to $\sim10^8$ keV/nucleon. These populations include galactic cosmic rays, anomalous cosmic rays (see discussion that follows), and energetic particles associated with CIRs, CMEs, solar flares, and the planetary bow shocks. All but the galactic cosmic rays are energized within the heliosphere.

Shocks are particularly effective particle accelerators, and all but one of the preceding populations have shock origins. The physical process by which a collisionless shock accelerates a small fraction of the ions it intercepts to high energy is reasonably well understood, although complex in detail. The effectiveness of the acceleration process depends upon factors such as shock speed and strength, the angle between the magnetic field and the shock, normal time of field line connection to the shock, and the local reservoir of particles available for acceleration. Recent work indicates that shocks in the solar wind most easily accelerate ions that already exceed solar wind thermal energies when they encounter the shocks. These so-called seed particles include the suprathermal ion tails always present in the low-speed wind (Fig. 13b), but also "pickup ions" (see discussion that follows), and energetic particles remaining in the heliosphere from previous solar flares and CME-driven disturbances.

Anomalous cosmic rays have energies that are lower than the galactic cosmic rays; are predominantly singly ionized H, He, N, O, and Ne; and, like galactic cosmic rays, have an intensity that varies slowly with time. They are associated with a particularly interesting seed population—neutral particles from the local interstellar cloud that penetrate deep into the heliosphere. As the neutrals approach the Sun, they are ionized by solar extreme ultraviolet radiation, electron impact, or charge exchange with solar wind protons; are then picked up by the solar wind magnetic field (the pickup process accelerates them to ~4 keV/nucleon); and are swept into the outer reaches of the heliosphere by the solar wind flow. As they encounter the heliosphere's termination shock, the pickup ions are accelerated to high energies. After acceleration, the energized particles diffuse back into the interior of the heliosphere as anomalous cosmic rays.

Of the energetic ion populations in the heliosphere, that associated directly with solar flares appears to be the only population that is not obviously shock-associated, although even in this case shock acceleration cannot be ruled out conclusively. Flare events are usually impulsive and short-lived (hours), are overabundant in ^{3}He, appear to originate low in the solar atmosphere, occur at a rate of ~1000 events/year near solar activity maximum, and generally occur in association with impulsive energetic solar electron bursts. The latter have energies ranging up to several hundred keV. Recent work suggests that solar electron bursts originate at a variety of altitudes in the solar atmosphere and can be triggered by more than one process.

13. Waves and Turbulence

The solar wind is filled with waves and turbulence having various spatial and temporal scales. Figure 15 illustrates that

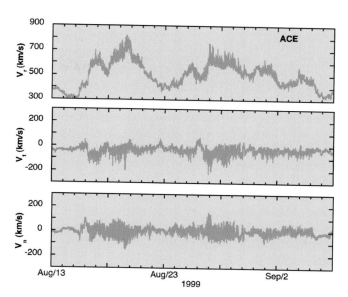

FIGURE 15   Solar wind velocity data sampled at a cadence of 64s in $r-$, $t-$, and n-coordinates, where the $+r$-direction is radial outward from the Sun, the $+t$-direction is in the direction of the Earth's motion about the Sun, and the $+n$-direction completes a right-handed system. The high-frequency fluctuations in this representative 25-day interval are caused by waves and turbulence in the wind.

fluctuations in solar wind velocity associated with waves and turbulence are observed throughout much of the solar wind, but fluctuation amplitudes tend to be greatest in high-speed streams. Many of these fluctuations are Alfvénic in nature (coupled changes in flow velocity and magnetic field vectors) and the waves and turbulence usually are propagating away from the Sun through the solar wind; together with the fact that fluctuation amplitudes generally decrease with distance from the Sun, this indicates that many of these fluctuations originate close to the Sun. Indeed, it is commonly thought that the fluctuations are largely remnants of waves and turbulence in the solar corona that heat and accelerate the solar wind. In addition to their probable, but poorly understood, role in heating and accelerating the solar wind plasma, waves and turbulence strongly affect energetic particle transport in the heliosphere and are essential elements of most current models of particle energization at shocks in the heliosphere.

14. Conclusion

The solar wind is a magnificent natural laboratory for studying and obtaining understanding of processes and phenomena that also occur in a variety of other astrophysical contexts. These include kinetic and fluid aspects of plasmas, plasma heating and acceleration, collisionless shock formation, particle acceleration and transport, magnetic reconnection, and turbulence and waves. Proof of the existence of the solar wind was one of the first great triumphs of the space age, and much has been learned about the physical nature of the wind and related processes in intervening years. Nevertheless, our understanding of the solar wind is far from complete. For example, we still do not know what physical processes heat and accelerate the solar wind or what determines its flow speed. We do not yet know if the low-speed wind arises primarily from quasi-stationary processes or from a series of small transient solar events. Likewise, the physical origins of coronal mass ejections are still being debated; we do not yet fully understand why they occur or how they relate to the long-term evolution of the solar magnetic field and the structure of the solar corona. We do not yet understand how a rough balance of magnetic flux is maintained in the solar wind in the presence of ICMEs or how the magnetic topologies of ICMEs evolve with time. In general, our ideas about the structure of the heliospheric magnetic field are still developing and need testing with observations. Ideas about the termination of the solar wind in the outer heliosphere and the role of the termination shock in accelerating anomalous cosmic rays are just now being tested by in situ observations for the first time. The physical origin of variations in elemental abundances in the solar wind is just beginning to be understood, as are temporal changes in the charge states of the heavier elements. Origins of double ion beams and suprathermal ion tails in the solar wind also remain unknown, and we do not yet fully understand why different ionic species have different speeds and temperatures in the solar wind. Further analysis of existing data, new types of measurements, and fresh theoretical insights should lead to understanding in these and other areas of solar wind research in the years ahead.

Bibliography

Balogh, A., Gosling, J. T., Jokipii, J. R., Kallenbach, R., and Kunow, H., eds. (1999). "Corotating Interaction Regions." Kluwer Acad. Pub., Dordrecht, Netherlands.

Balogh, A., Marsden, R. G., and Smith, E. J., eds. (2001). "The Heliosphere Near Solar Minimum: The Ulysses Perspective." Springer-Praxis, Chichester, England.

Crooker, N., Joselyn, J. A., and Feynman, J., eds. (1997). "Coronal Mass Ejections," Geophysical Monograph 99, American Geophysical Union, Washington, D.C.

Fleck, B., and Zurbuchen, T. H, eds. (2005). "Proceedings of Solar Wind 11//Soho 16, Connecting Sun and Heliosphere," ESA SP-592. Noordwijk, Netherlands.

Florinski, V., Pogorelov, N. V., and Zank, G. P., eds. (2004). "Physics of the Outer Heliosphere," AIP Conference Proceedings 719. American Institute of Physics, Melville, New York.

Marsden, R. G., ed. (1995). "The High Latitude Heliosphere." Kluwer, Boston.

Mewaldt, R. A., Jokipii, J. R., Lee, M. A., Mobius, E., and Zurbuchen, T. H., eds. (2000). "Acceleration and Transport of Energetic Particles Observed in the Heliosphere," AIP Conference Proceedings 528. American Institute of Physics, Melville, New York.

Schwenn, R., and Marsch, E., eds. (1991). "Physics of the Inner Heliosphere. 2. Particles, Waves and Turbulence." Springer-Verlag, Berlin/Heidelberg, Germany.

Velli, M., Bruno, R. and Malara, F., eds. (2003). "Solar Wind 10," AIP Conference Proceedings 679. American Institute of Physics, Melville, New York.

Mercury

Robert G. Strom

Department of Planetary Sciences
University of Arizona
Tucson, Arizona

CHAPTER 6

M ercury is the innermost and smallest planet in the solar system.[1] It has no known satellites. The exploration of Mercury has posed questions concerning fundamental issues of its origin and, therefore, the origin and evolution of all the terrestrial planets. The data obtained by *Mariner 10* on its three flybys of Mercury on March 29 and September 21, 1974, and on March 16, 1975, remain our best source of detailed information on this planet. However, recent ground-based observations have provided important new information on the topography, radar, and microwave characteristics of its surface; discovered new constituents in its atmosphere; and helped constrain its surface composition. The *MESSENGER* spacecraft is currently on its way to Mercury and will begin orbiting the planet in March 2011. The name *MESSENGER* is an acronym for Mercury, Surface, Space ENvironment, GEochemistry, and Ranging. To the ancients, Mercury was the messenger of the gods. Mercury is often compared with the Moon because it superficially resembles that satellite. However, major differences set Mercury apart from the Moon and, for that matter, all other planets and satellites in the solar system.

[1] Assuming Pluto, a Kuiper Belt object smaller than Mercury, is not a planet, which is controversial.

Mariner 10 imaged only about 45% of the surface at an average resolution of about 1 km, and less than 1% at resolutions between about 100 to 500 m (Fig. 1). This coverage and resolution is comparable to telescopic Earth-based coverage and resolution of the Moon before the advent of space flight. However, unlike the Moon in the early 1960s, only about 25% of the surface was imaged at Sun angles low enough to allow adequate terrain analyses. As a consequence, there are still many uncertainties and questions concerning the history and evolution of Mercury. *Mariner 10* also discovered a magnetic field, measured the temperature, and derived the physical properties of its surface.

On Mercury, the prime meridian (0°) was chosen to coincide with the subsolar point during the first perihelion passage after January 1, 1950. Longitudes are measured from 0° to 360°, increasing to the west. Craters are mostly named after famous authors, artists, and musicians such as Dickens, Michelangelo, and Beethoven, whereas valleys are named for prominent radio observatories such as Arecibo and Goldstone. Scarps are named for ships associated with exploration and scientific research such as Discovery and Victoria. Plains are named for the planet Mercury in various languages such as Odin (Scandinavian) and Tir (Germanic). Borealis Planitia (Northern Plains) and Caloris Planitia (Plains of Heat) are exceptions. The most

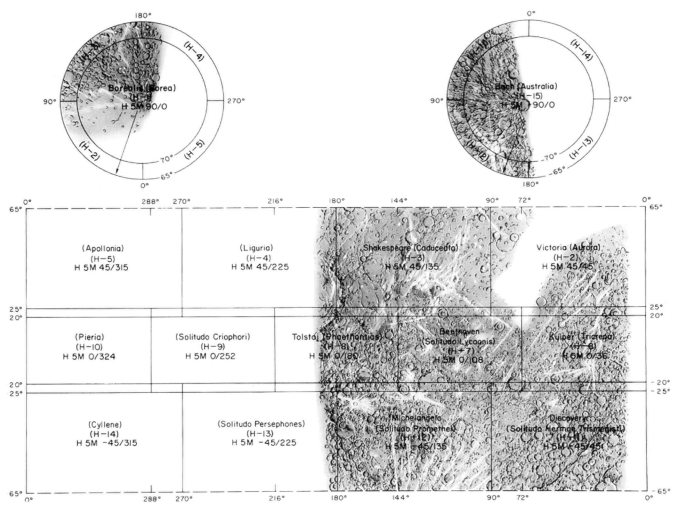

FIGURE 1 Shaded relief map of Mercury showing the quadrangle names and major features. About 55% of the planet is unknown.

prominent feature viewed by *Mariner 10* is named the **Caloris Basin** (Basin of Heat) because it nearly coincides with one of the "hot poles" of Mercury.

1. General Characteristics

Mercury's diameter is only 4878 km, but it has a relatively large mass of 3.301×10^{23} kg. Because of its large mass in relation to its volume, Mercury has an exceptionally high mean density of 5440 kg/m^3, second only to the density of the Earth (5520 kg/m^3). The manner in which it reflects light (its photometric properties) is very similar to the way light is reflected by the Moon. The brightness (albedo) of certain terrains is greater than comparable terrains on the Moon. Mercury is covered with a **regolith** consisting of fragmental material derived from the impact of meteoroids over billions of years. Mercury's surface is heavily cratered with

smooth plains that fill and surround large impact basins. Long **lobate scarps** traverse the surface for hundreds of kilometers, and large expanses of **intercrater plains** (the most extensive terrain type) fill regions between clusters of craters in the highlands. Also a peculiar terrain consisting of a jumble of large blocks and linear troughs occurs antipodal to the Caloris Basin.

2. Motion and Temperature

Mercury has the most eccentric (0.205) and inclined (7°) orbit of any planet. However, over periods of a few million years, its eccentricity may vary from about 0.1 to 0.28 and its inclination from about 0° to 11°. Its average distance from the Sun is 0.3871 AU (5.79×10^7 km). Because of its large eccentricity, however, the distance varies from 0.3075 AU (4.6×10^7 km) at perihelion to 0.4667 AU (6.98×10^7 km) at aphelion. As a consequence, Mercury's orbital velocity averages 47.6 km/s but varies from 56.6 km/s at perihelion to 38.7 km/s at aphelion. At perihelion the Sun's apparent diameter is over three times larger than its apparent diameter as seen from Earth.

Mercury's rotation period is 58.646 Earth days, and its orbital period is 87.969 Earth days. Therefore, it has a unique 3:2 resonance between its rotational and orbital periods: It makes exactly three rotations on its axis for every two orbits around the Sun. This resonance was apparently acquired over time as the natural consequence of the dissipative processes of tidal friction and the relative motion between a solid mantle and a liquid core. As a consequence of this resonance, a solar day (sunrise to sunrise) lasts two Mercurian years or 176 Earth days. The **obliquity** of Mercury is close to 0°; therefore, it does not experience seasons as do Earth and Mars. Consequently, the polar regions never receive the direct rays of sunlight and are always frigid compared to torrid sunlit equatorial regions.

Another effect of the 3:2 resonance between the rotational and orbital periods is that the same hemisphere always faces the Sun at alternate perihelion passages. This happens because the hemisphere facing the Sun at one perihelion will rotate one and a half times by the next perihelion, so that it faces away from the Sun; after another orbit, it rotates another one-and-a half times so that it directly faces the Sun again. Because the subsolar points of the 0° and 180° longitudes occur at perihelion they are called **hot poles**. The subsolar points at the 90° and 270° longitudes are called **warm poles** because they occur at aphelion. Yet another consequence of the 3:2 resonance and the large eccentricity is that an observer on Mercury (depending on location) would witness a double sunrise, or a double sunset, or the Sun would backtrack in the sky at noon during perihelion passage. Near perihelion Mercury's orbital velocity is so great compared to its rotation rate that it overcomes the Sun's apparent motion in the sky as viewed from Mercury.

Although Mercury is closest to the Sun, it is not the hottest planet. The surface of Venus is hotter because of its atmospheric greenhouse effect. However, Mercury experiences the greatest range (day to night) in surface temperatures (650° C = 1170° F) of any planet or satellite in the solar system because of its close proximity to the Sun, its peculiar 3:2 spin orbit coupling, its long solar day, and its lack of an insulating atmosphere. Its maximum surface temperature is about 467° C (873° F) at perihelion on the equator; hot enough to melt zinc. At night just before dawn, the surface temperature plunges to about −183° C (−297° F).

3. Exosphere

Although Mercury has an atmosphere, it is extremely tenuous with a surface pressure a trillion times less than Earth's.

TABLE 1	Mercury's Main Exospheric Constituents[a]	
Constituent	**Vertical Column Abundance (atoms/cm^2)**	
Hydrogen (H)	~5 ×	10^{10}
Helium (He)	~2 ×	10^{13}
Oxygen (O)	~7 ×	10^{12}
Sodium (Na)	~2 ×	10^{12}
Potassium (K)	~1 ×	10^{10}
Calcium (Ca)	~1 ×	10^{7}

[a] The Earth's atmosphere has ~2 × 10^{18} molecules/cm^2

The number density of atoms at the surface is only 10^5 atoms cm^{-3} for the known constituents (Table 1). It is, therefore, an exosphere where atoms rarely collide; their interaction is primarily with the surface. *Mariner 10*'s ultraviolet spectrometer identified hydrogen, helium, and oxygen and set upper limits on the abundance of argon, neon, and carbon in the exosphere. The hydrogen and helium are probably derived largely from the solar wind, although a portion of the helium may be of radiogenic origin, and some hydrogen could result from the photodissociation of H_2O. The interaction of high-energy particles with surface materials may liberate enough oxygen to be its principal source, but breakdown of water vapor molecules by sunlight could also be a possible source.

In 1985–1986, Earth-based telescopic observations detected sodium and potassium in the exosphere, and subsequent observations have detected calcium (Table 1). Sodium and potassium are also found in the Moon's exosphere. Both sodium and potassium have highly variable abundances 10^4–10^5 Na atoms/cm^3 and 10^2–10^4 K atoms/cm^3 near the surface on timescales of hours to years. Their abundances also vary between day and night by a factor of about 5, the dayside being greater. Often bright spots of emission are seen at high northern latitudes or over the Caloris Basin. The temperature of the gas is about 500 K, but a hotter more extended Na coma sometimes exists. Observed variations in the abundances of these elements are consistent with the photoionization timescale of 120 minutes for sodium and ~90 minutes for potassium. Photoionization of the gas will result in the exospheric ions being accelerated by the electric field in the planetary **magnetosphere.** Ions created on one hemisphere will be accelerated toward the planetary surface and recycled, but ions on the opposite hemisphere will be ejected away and lost. The total loss rate of sodium atoms is about 1.3×10^{22} atoms per second, so the atoms must be continuously supplied by the surface. The total fraction of ions lost to space from the planet is at least 30%. The atmosphere, therefore, is transient and exists in a steady state between its surface sources and sinks.

Although both sodium and potassium are probably derived from the surface of Mercury, the mechanism by which they are supplied is not well understood. The sodium and potassium in the Mercurian exosphere could be released from sodium- and potassium-bearing minerals by their interaction with solar radiation, or impact vaporization of micrometeoroid material. Both sodium and potassium show day-to-day changes in their global distribution.

If surface minerals are important sources for the exosphere, then a possible explanation is that their sodium/potassium ratio varies with location on Mercury. A possible explanation for some of the K and Na variations is that Na and K ion implantation into regolith grains during the long Mercurian night (88 Earth days), and subsequent diffusion to the exosphere when the enriched surface rotates into the intense sunlight. At least one area of enhanced exospheric potassium emission apparently coincides with the Caloris Basin whose floor is highly fractured. This exospheric enhancement has been attributed to increased diffusion and degassing in the surface and subsurface through fractures on the basin floor, although other explanations may be possible.

4. Polar Deposits

High-resolution, full-disk radar images of Mercury obtained from both the Arecibo and the linked Goldstone—Very Large Array radar facilities discovered unusual features at Mercury's poles. The radar signals show very high reflectivities centered on the poles. The reflectivity and ratio values are similar to outer planet icy satellites and the residual polar water ice cap of Mars. Therefore, Mercury's polar radar features have been interpreted to be water ice. The radar characteristics are consistent with the ice being covered by a few centimeters of regolith. It has also been proposed that the radar characteristics are the result of volume scattering by inhomogeneities in elemental sulfur deposits. In this case, it is proposed that sulfur volatilized from sulfides in the regolith was **cold-trapped** at the poles.

Mariner 10 images of Mercury's polar regions show cratered surfaces where ice or sulfur could be concentrated in permanently shadowed portions of the craters. Radar studies have shown that the anomalies are indeed concentrated in the permanently shadowed portions of these polar craters (Fig. 2). The south polar radar feature is centered at about 88°S and 150°W and is largely confined within a crater (Chao Meng-Fu) 150 km in diameter, but a few smaller features occur outside this crater. In the north polar region, the deposits reside in about 25 craters down to a latitude as low as 72° (Fig. 2). Because the obliquity of Mercury is near 0°, it does not experience seasons, and, therefore, temperatures in the polar regions should be <135 K. Water ice can be stable in the interiors of craters even down to 72° latitude if covered with only a few centimeters of regolith, or if

it is relatively new. This means that water ice would still be present in its perpetually shadowed craters or even in illuminated craters at high latitude, if covered with a veneer of regolith. In permanently shaded polar areas (i.e., the floors and sides of large craters), the temperatures should be less than 112 K, and water ice should be stable to evaporation on timescales of billions of years when covered with a thin veneer of regolith. The problem with sulfur being the deposits on Mercury is that sulfur is stable at much higher temperatures than water, and there are no highly radar-reflective deposits in the polar regions where temperatures are within the stability range of sulfur. A 1-m- thick layer of water ice is stable for one billion years at a temperature of −161° C while sulfur is stable at a considerably higher temperature of −55° C. Much of the region surrounding permanently shadowed craters is less than −55° C, but there are no radar-reflective deposits there.

The deposits are concentrated only in the freshest craters, and even in some craters less than 10 km in diameter. Degraded craters do not show the highly radar-reflective deposits, probably because there are no permanently shadowed regions in these low-rimmed and shallow craters. In fact, the permanently shadowed cold traps are essentially full. Furthermore, the strong radar signal indicates that the material is relatively pure. The thickness of the deposits has been estimated to be between ∼2 and maybe 20 m. The higher value is, in fact, arbitrary because the radar data cannot place upper limits on the thickness. The area covered by these deposits (both north and south) is estimated to be ∼30,000 km^2. This would be equivalent to 4×10^{16} to 8×10^{17} grams of ice, or 40–800 km^3 for a 2–20 meter thick deposit. Each meter thickness of ice would be equivalent to about 10^{13} kilograms of ice.

If the deposits are water ice, then they could originate from comet or water-rich asteroid impacts that released the water to be cold-trapped in the permanently shadowed craters. Because comets and asteroids also impact the Moon, similar deposits would be expected to occur in the permanently shadowed regions of lunar craters. The neutron and gamma ray spectrometers on the *Lunar Prospector* spacecraft discovered enhanced hydrogen signals in permanently shadowed craters in the polar regions of the Moon. This has been interpreted as water ice with a concentration of only $1.5 \pm 0.8\%$ weight fraction.

5. Interior and Magnetic Field

Mercury's internal structure is unique in the solar system. It also imposes severe constraints on any proposed origin of the planet. Mercury's mean density of 5440 kg/m^3 is only slightly less than Earth's (5520 kg/m^3) and larger than Venus' (5250 kg/m^3). Because of Earth's large internal pressures, however, its uncompressed density is only 4400 kg/m^3 compared to Mercury's uncompressed density

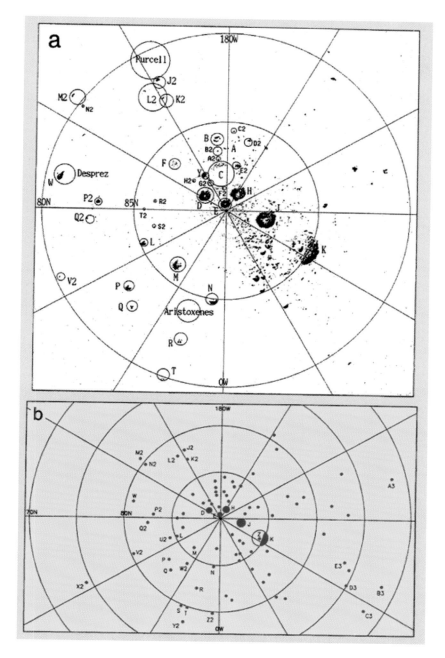

FIGURE 2   (a) A high-resolution radar image of the north polar deposits down to 80°N is shown. The deposits are within the permanently shaded areas of fresh craters. Degraded craters with low rims do not have the deposits. (b) A map of the deposits (shown in blue) down to 70°N. (Courtesy of John Harmon, Arecibo Observatory, Puerto Rico.)

of 5300 kg/m^3. This means that Mercury contains a much larger fraction of iron than any other planet or satellite in the solar system (Figs. 3 and 4). If this iron is concentrated in a core, then the core must be about 75% of the planet diameter, or some 42% of its volume. Thus, its silicate mantle and crust is only about 600 km thick. For comparison, Earth's iron core is only 54% of its diameter, or just 16% of its volume.

Aside from Earth, Mercury is the only other terrestrial planet with a significant magnetic field. *Mariner 10* first encountered Mercury's magnetosphere at a distance of 1.9 radii from its surface. It took measurements of the field for only 30 minutes; ~17 minutes during Mercury's first equatorial pass, and ~13 minutes during the third high latitude pass. These short observations are all we know about Mercury's magnetic field, magnetosphere, and particle environment. However, investigators constructed a picture of the magnetic field environment at Mercury based on analogy with that of Earth's magnetic field and particle environment. Because Mercury probably lacks the ionosphere and trapped radiation zones of Earth's magnetosphere, many comparisons are inappropriate.

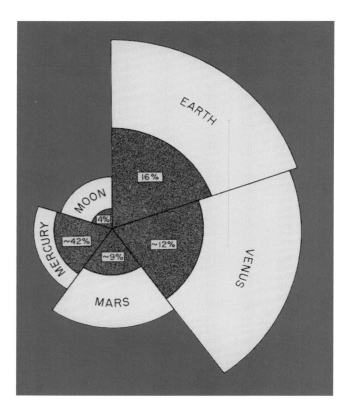

FIGURE 3   Comparison of terrestrial planet sizes and core radii. The percent of the total planetary volume of the cores is also shown in yellow. The size of the Moon's core is not known, but the maximum possible size is shown.

FIGURE 4   A *Mariner 10* photomosaic together with an accurate artist's rendition of the size of Mercury's core compared to the silicate portion. The outer part of the core is still in a liquid state. (From Strom, 1987.)

The measured magnetic field is strong enough to hold off the solar wind and form a bow shock. As the spacecraft approached the planet, it measured a sudden increase in the field strength that represented the bow shock. Also the instruments measured signals indicating the entrance to and exit from a **magnetopause** surrounding a magnetospheric cavity about 20 times smaller than the Earth's (Fig. 5). Also because of the small size of Mercury's magnetosphere, magnetic events happen more rapidly and repeat more often than in Earth's magnetosphere. The nominal magnetopause subsolar distance is estimated to be about 1.35 ± 0.2 Mercury radii, and the bow shock distance is about 1.9 ± 0.2 Mercury radii. The polarity of the field is the same as Earth's. The magnetic strength increased as the spacecraft approached the planet. The interplanetary field is about 25 nT (nano-Tesla) in the vicinity of Mercury, but it increased to 100 nT at closest approach to Mercury. If that rate of increase continued to the surface, the surface strength would be about 200–500 nT. This is about 1% of the Earth's strength.

Although other models may be possible, the maintenance of terrestrial planet magnetic fields is thought to require an electrically conducting fluid outer core surrounding a solid inner core. Therefore, Mercury's dipole magnetic field is taken as evidence that Mercury currently has a fluid outer core of unknown thickness. Recent high-resolution radar measurements of the magnitude of Mercury's **librations** indicate that the mantle is detached from the core confirming the outer core is fluid. Although the thickness of the outer fluid core is unknown at present, theoretical studies indicate that a dipole magnetic field can be generated and maintained even in a thin outer fluid core. Thermal history models strongly suggest that Mercury's core would have solidified long ago unless there was some way of maintaining high core temperatures throughout geologic history. Most theoretical studies consider the addition of a light, alloying element to be the most likely cause of a currently molten outer core. Although oxygen is such an element, it is not sufficiently soluble in iron at Mercury's low internal pressures. Metallic silicon has also been suggested, but sulfur is considered to be the most likely candidate. Some models require only a small amount of sulfur, whereas others support greater amounts. Currently we do not know how much sulfur is in the core, but it is possible that the *MESSENGER* mission will provide the answer. If sulfur is the cause of Mercury's outer fluid core, then estimates of its abundance can be used to estimate the thickness of the outer fluid core. For a sulfur abundance in the core of less than 0.2%, the entire core should be solidified at the present time, and for an abundance of 7% the core should be entirely fluid at present. Therefore, if sulfur is the alloying element, then Mercury could contain between 0.2 and 7% sulfur in its core. As discussed later, possible sulfur abundances can be estimated from the planetary radius decrease derived from the **tectonic framework**.

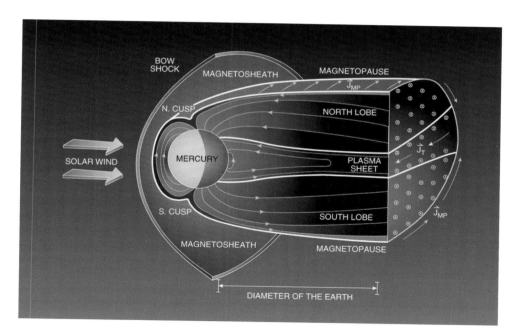

FIGURE 5   Artist's rendition of Mercury's dipole magnetic field showing the bow shock, **magnetosheath**, magnetopause, and a possible plasma sheet. The diameter of the Earth is also shown for comparison. (Courtesy of Jim Slavin, Goddard Spaceflight Center, Greenbelt, Maryland.)

6. Geology and Planet Evolution

Mercury has heavily cratered upland regions and large areas of younger smooth plains that surround and fill impact basins (Fig. 6). Thermal infrared measurements from *Mariner 10* indicate that the surface is a good insulator and, therefore, consists of a porous cover of fine-grained regolith. Earth-based microwave measurements indicate that this layer is a few centimeters thick and is underlain by a highly compact region extending to a depth of several meters. Mercury's heavily cratered terrain contains large areas of gently rolling intercrater plains, the major terrain type on the planet. Mercury's surface is also traversed by a unique system of contractional **thrust faults** called lobate scarps. The largest well-preserved structure viewed by *Mariner 10* is the Caloris impact basin some 1300 km in diameter. Antipodal to this basin is a large region of broken-up terrain called the **hilly and lineated terrain,** probably caused by focused seismic waves from the Caloris impact.

6.1 Geologic Surface Units

The origin of some of the major terrains and their inferred geologic history are somewhat uncertain because of the limited photographic coverage and resolution and the poor quality or lack of other remotely sensed data. In general, the surface of Mercury can be divided into four major terrains: (1) heavily cratered regions, (2) intercrater plains, (3) smooth plains, and (4) hilly and lineated terrain. Other relatively minor units have been identified, such as ejecta deposits exterior to the Caloris and other basins.

6.1.1 IMPACT CRATERS AND BASINS

The heavily cratered uplands probably record the period of late heavy meteoroid bombardment that ended about 3.8 billion years ago on the Moon, and presumably at about the same time on Mercury. This **period of late heavy bombardment** occurred throughout the inner solar system and is also recorded by the heavily cratered regions on the Moon and Mars. Based on chemical evidence from Apollo samples from the Moon, the bombardment may have been catastrophic lasting only about 100 million years or less. It appears to have peaked about 3.9 billion years ago. In the heavily cratered terrain on Mercury, there is an increasing paucity of craters with decreasing crater diameter relative to heavily cratered terrain on the Moon. This paucity of craters is probably due to obliteration of the smaller craters by emplacement of intercrater plains during the period of late heavy bombardment. Below a diameter of about 20 km, the abundance of craters increases sharply. These craters may represent secondary impact craters from large craters or basins. The crater population superimposed on the smooth plains within and surrounding the Caloris Basin has a size distribution intermediate between the heavily cratered and lightly cratered plains. This suggests that unlike the lunar maria the Caloris smooth plains formed near the end of late heavy bombardment.

Fresh impact craters on Mercury exhibit similar morphologies as those on the other terrestrial planets. Small craters are bowl-shaped, but with increasing size they develop central peaks, flat floors, and terraces on their inner walls. The transition from simple (bowl-shaped) to complex (central peak and terraces) craters is about 10 km. At

(a)

(b)

FIGURE 6   (a) Mercury as viewed by *Mariner 10* on its first approach in March 1974. (b) Mercury's opposite hemisphere viewed by *Mariner 10* as it left the planet on the first encounter, and (c) the southern hemisphere viewed on the second encounter in September 1974. (Courtesy of NASA.)

diameters between ~130 and 310 km Mercurian craters have an interior concentric ring, and at diameters larger than about 300 km they may have multiple inner rings. The freshest craters have extensive ray systems, some of which extend for distances over 1000 km. For a given rim diameter, the radial extent of Mercurian continuous ejecta is uniformly smaller by a factor of about 0.65 than that for the Moon. Furthermore, the maximum density of secondary impact craters occurs closer to the crater rim than for similarly sized lunar craters: The maximum density occurs at about 1.5 crater radii from the rim of Mercurian primaries, whereas the maximum density occurs at about 2–2.5 crater

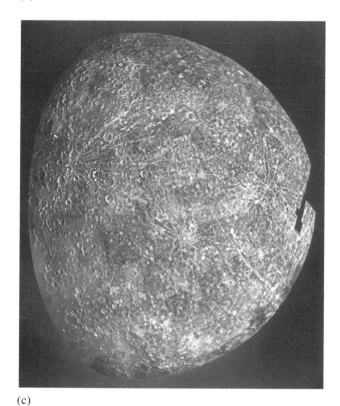

(c)

FIGURE 6   (*Continued*)

radii on the Moon. All of these differences are probably due to the larger surface gravity of Mercury (3.70 m/s^2) compared to the Moon (1.62 m/s^2).

Twenty-two multiring basins have been recognized on the part of Mercury viewed by *Mariner 10.* However, high-resolution radar images of the side not viewed by *Mariner 10* show several large circular features about 1000 km in diameter that may be impact basins. Based on the pattern and extent of **grabens** on the floor of Caloris, it is estimated that Mercury's **lithosphere** under Caloris was thicker (>100 km) than the Moon's (between 25 and >75 km depending on location) at the end of late heavy bombardment. The 1300-km-diameter Caloris impact basin is the largest well-preserved impact structure (Fig. 7), although the much more degraded Borealis Basin is larger (1530 km). The floor structure of the Caloris Basin is like no other basin floor structure in the solar system. It consists of closely spaced ridges and troughs arranged in both a concentric and radial pattern (Fig. 8a and 8b). The ridges are probably due to contraction, while the troughs are probably extensional grabens that postdate the ridges. The fractures get progressively deeper and wider toward the center of the basin. Near the edge of the basin there are very few fractures. This pattern may have been caused by subsidence and subsequent uplift of the basin floor.

6.1.2 HILLY AND LINEATED TERRAIN

Directly opposite the Caloris Basin on the other side of Mercury (the antipodal point of Caloris) is the unusual hilly and lineated terrain that disrupts preexisting landforms, particularly crater rims (Fig. 9a and 9b). The hills are 5–10 km wide and about 0.1–1.8 km high. Linear depressions that are probably extensional fault troughs form a roughly orthogonal pattern. Geologic relationships suggest that the age of this terrain is the same as that of the Caloris Basin. Similar, but smaller, terrains occur at the **antipodes** of the Imbrium and Orientale impact basins on the Moon. The hilly and lineated terrain is thought to be the result of shock waves generated by the Caloris impact and focused at the antipodal region (Fig. 10). Computer simulations of shock wave propagation indicate that focused shock waves from an impact of this size can cause vertical ground motions of about 1 km or more and tensile failure to depths of tens of kilometers below the antipode. Although the lunar Imbrium Basin (1400 km diameter) is larger than the Caloris Basin, the disrupted terrain at its antipode is much smaller than that at the Caloris antipode. The larger disrupted terrain on Mercury may be the result of enhanced shock wave focusing due to the large iron core.

6.1.3 INTERCRATER PLAINS

Mercury's two plains units have been interpreted as either impact basin ejecta or as lava plains. The older intercrater plains are the most extensive terrain on Mercury (Figs. 11

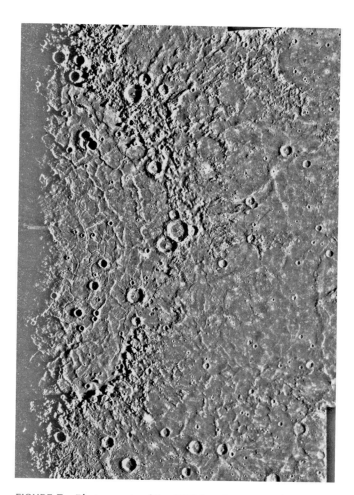

FIGURE 7   Photomosaic of the 1300-km-diameter Caloris impact basin showing the highly ridged and fractured nature of its floor. (Courtesy of NASA.)

and 12). They both partially fill and are superimposed by craters in the heavily cratered uplands. Furthermore, they have probably been responsible for obliterating a significant number of craters as evidenced by the paucity of craters less than about 40 km diameter compared to the highlands of the Moon. Therefore, intercrater plains were emplaced over a range of ages contemporaneous with the period of late heavy bombardment. There are no definitive features diagnostic of their origin. Because intercrater plains were emplaced during the period of late heavy bombardment, they are probably extensively fragmented and do not retain any signature of their original surface morphology. Although no landforms diagnostic of volcanic activity have been discovered, there are also no obvious source basins to provide ballistically emplaced ejecta. The global distribution of intercrater plains and the lack of source basins for ejecta deposits are indirect evidence for a volcanic origin. Additional evidence for a volcanic origin is recent *Mariner 10* enhanced color images showing color boundaries that coincide with geologic unit boundaries of some intercrater plains (Fig. 13). If intercrater plains are volcanic, then they are

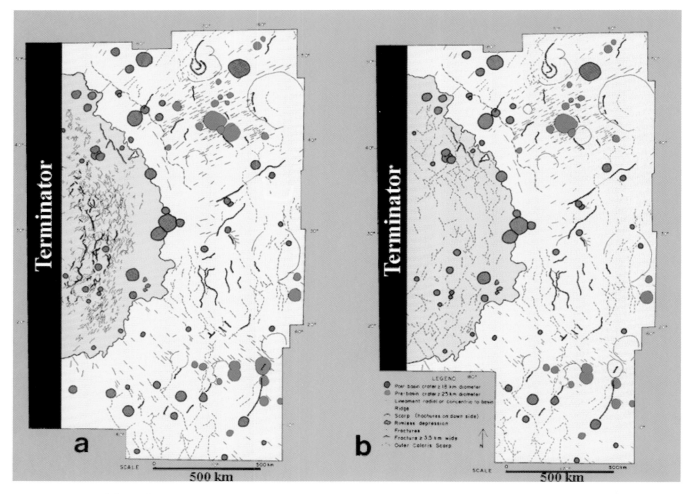

FIGURE 8 Map of the (a) fractures and (b) ridges on the floor of the Caloris Basin. The basin interior is shown in brown, and the dash-dot line to the northeast of the main ring is a faint outer ring. The floor fractures and ridges have both radial and concentric components. The red spots are post-Caloris craters, and the blue ones are pre-Caloris craters partly covered with basin ejecta. The lines radial to the basin are lineations due to the basin ejecta, and the thick black lines are lobate scarps. The small green spots at the eastern edge of the basin are rimless volcanic collapse pits. (From Strom et al., 1975, *J. Geophys. Res.* **80,** 2478–2507.)

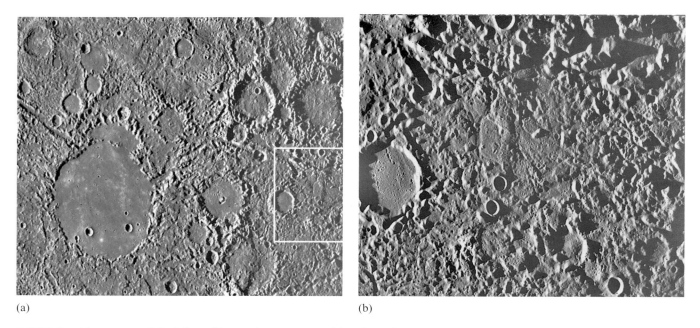

(a)   (b)

FIGURE 9 (a) A portion of the hilly and lineated terrain antipodal to the Caloris impact basin. The image is 543 km across. (b) Detail of the hilly and lineated terrain. The largest crater in (b) is 31 km in diameter. (Courtesy of NASA.)

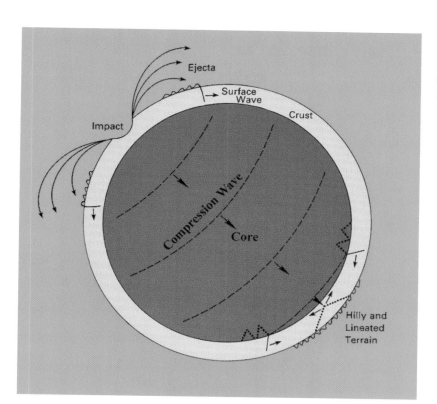

FIGURE 10   Diagrammatic representation of the formation of the hilly and lineated terrain by focused seismic waves from the Caloris impact. (From P. Schultz and D. Gault, 1975, "The Moon," **12,** pp. 159–177.)

FIGURE 11   View of the intercrater plains surrounding clusters of craters in the Mercurian highlands. Several lobate scarps (thrust faults) can also be seen. (Courtesy of NASA.)

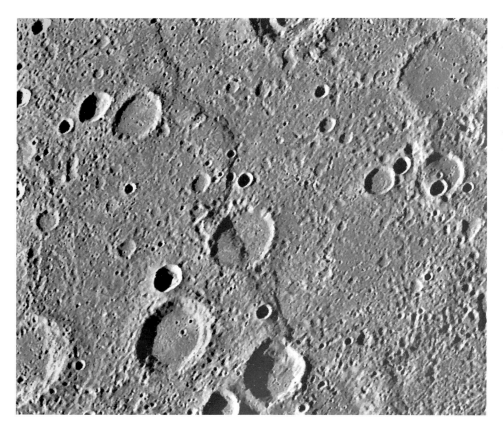

FIGURE 12 High-resolution view of the intercrater plains. The chains and clusters of small craters are secondaries from younger craters. The 90-km-diameter crater in the upper right-hand corner has been embayed by intercrater plains. The lobate scarp that diagonally crosses the image is a thrust fault. (Courtesy of NASA.)

probably lava flows erupted from fissures early in Mercurian history. Intercrater plains are probably ≥3.9 billions years old.

6.1.4 SMOOTH PLAINS

The younger smooth plains cover almost 40% of the total area imaged by *Mariner 10*. About 90% of the regional exposures of smooth plains are associated with large impact basins. They also fill smaller basins and large craters. The largest occurrence of smooth plains fill and surround the Caloris Basin (Fig. 7), and occupy a large circular area in the north polar region that is probably an old impact basin (Borealis Basin). They are similar in morphology and mode of occurrence to the lunar maria. Craters within the Borealis, Goethe, Tolstoy, and other basins have been flooded by smooth plains (Fig. 14). This indicates the plains are younger than the basins they occupy. This is supported by the fact that the density of craters superimposed on the smooth plains that surround the Caloris Basin is substantially less than the density of craters superimposed on the floors of all major basins including Caloris. Furthermore, several irregular rimless depressions that are probably of volcanic origin occur in smooth plains on the floors of the Caloris and the Tolstoy basins. The smooth plains' youth relative to the basins they occupy, their great areal extent, and other stratigraphic relationships suggest they are vol-

canic deposits erupted relatively late in Mercurian history. *Mariner 10* enhanced color images show the boundary of smooth plains within the Tolstoy Basin is also a color boundary, further strengthening the volcanic interpretation for the smooth plains. Based on the shape and density of the size/frequency distribution of superimposed craters, the smooth plains probably formed near the end of late heavy bombardment. They may have an average age of about 3.8 billion years as indicated by crater densities. If so, they are, in general, older than the lava deposits that constitute the lunar maria.

Three large radar-bright anomalies have been identified on the unimaged side of Mercury. They are designated as A (347°W longitude, −34° latitude), B (343°W longitude, 58° longitude), and C (246°W longitude, 11° N latitude). All features are relatively fresh impact craters with radar-bright ejecta blankets and/or rays similar to Kuiper crater (60 km diameter) on the imaged portion of Mercury. Feature A is 85 km in diameter with an extensive ray system and a rough radar-bright floor, consistent with a fresh impact crater (Figure 15). Feature B is 95 km diameter with radar-bright rays and a radar-dark floor (Figure 15). Unlike feature A, the radar-dark floor indicates it is smooth at the 12.6 cm wavelength of the image. Feature C is a fresh crater about 125 kilometers in diameter. Water-rich comets or asteroids responsible for one or more

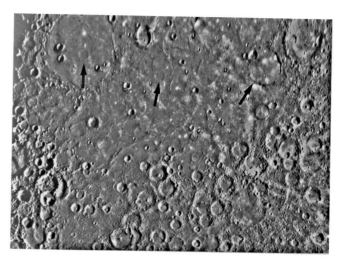

FIGURE 14   Photomosaic of the Borealis Basin showing numerous craters (arrows) that have been flooded by smooth plains. The largest crater is the Goethe Basin 340 km in diameter. (Courtesy NASA.)

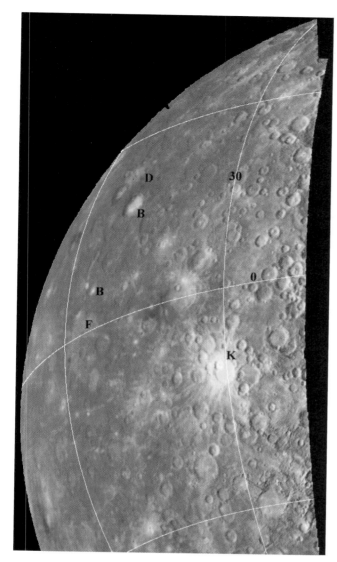

FIGURE 13   Enhanced color mosaic of a portion of the incoming side of Mercury as viewed by *Mariner 10*. The area at F has a sharp boundary that coincides with an intercrater plains boundary and may have a different composition. The relatively dark and blue unit at D is consistent with enhanced titanium content. The bright orange unit at B may represent primitive crustal material, and Kuiper crater at K shows a yellowish color representing fresh material excavated from a subsurface unit that may have an unusual composition. (Courtesy of Mark Robinson, Northwestern Univ., Evanston, Illinois.)

of these craters could be the source of the polar water-ice deposits.

6.2 Surface Composition

Little is known about the surface composition of Mercury. If the plains units (intercrater and smooth) are lava flows, then they must have been very fluid with viscosities similar to fluid flood **basalts** on the Moon, Mars, Venus, and Earth.

The way in which light is reflected from the surface is very similar to that of the Moon. However, at comparable **phase angles** and wavelengths in the visible part of the spectrum, Mercury appears to have systematically higher albedos than the Moon. Mercurian albedos range from 0.09 to 0.36 at 5° phase angle. The higher albedos are usually associated with rayed craters. However, the highest albedo (0.36) on *Mariner 10* images is not associated with a bright-rayed crater: It is a floor deposit in Tyagaraja Crater at 3° N latitude and 149° longitude. The lunar highlands/mare albedo ratio is almost a factor of 2 on the Moon, but it is only a factor of 1.4 on Mercury. Furthermore, at ultraviolet wavelengths (58–166 nm) Mercury's albedo is about 65% lower than the Moon's at comparable wavelengths. These differences in albedo suggest that there are systematic differences in the surface composition between the two bodies.

A recalibration and color ratioing of *Mariner 10* images have been used to derive the FeO abundance, the opaque mineral content, and the soil maturity over the region viewed by *Mariner 10*. The probably volcanic smooth plains have a FeO content of <6 weight percent that is similar to the rest of the planet imaged by *Mariner 10*. The surface of Mercury, therefore, may have a more homogeneous distribution of elements affecting color (e.g., more **alkali plagioclase**) than does the Moon. At least the smooth plains may be low iron or alkali basalts. Since the iron content of lavas is thought to be representative of their mantle source regions, it is estimated that Mercury's mantle has about the same FeO content (<6 weight percent) as the crust, indicating Mercury is highly reduced with most of the iron in the core. In contrast, the estimated FeO contents of the mantle of the bulk Moon is 11.4 %, of Venus and the Earth 8%, and of Mars ~18%. There are some low-albedo regions with spectral properties

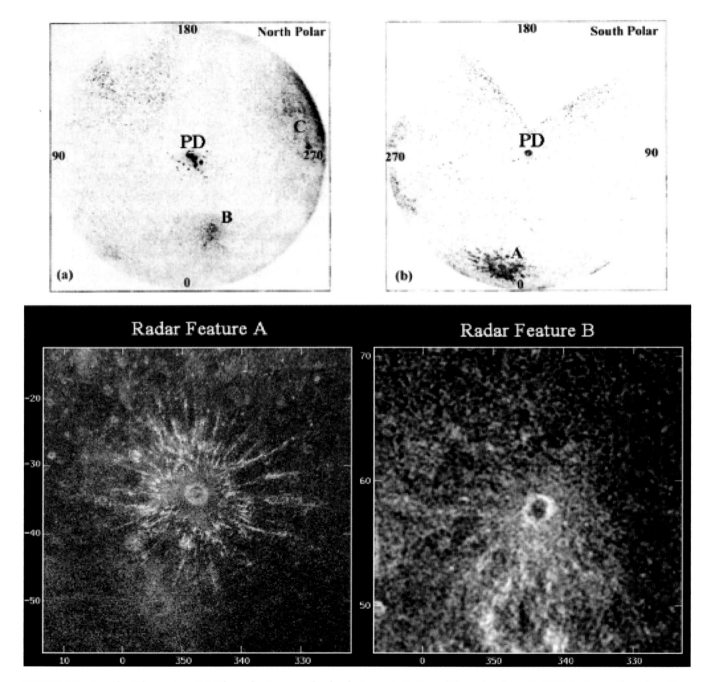

FIGURE 15   Arecibo Observatory 2.4-Ghz radar images of radar features A, B, C, and the polar deposits (PD) in the north and south hemispheres of Mercury (upper left and right). The resolution is 15 km (0.53°). The lower left and right images are high-resolution radar images of two impact craters seen in the top hemispheric rader images. Features A (85 km diameter) and B (95 km diameter) are two of the brightest (freshest?) radar features on the planet. (Courtesy of John Harmon, Arecibo Observatory, Puerto Rico.)

suggesting high opaque mineral areas. These have diffuse boundaries that may be associated with fractures (Fig. 13). These areas could be more mafic volcanic pyroclastic deposits. The bright-rayed craters on Mercury have a very low opaque mineral index that may indicate the craters have excavated into an anothositic crust. Color ratios of lunar

and Mercurian crater rays also suggest that the surface of Mercury is low in Ti^{4+}, Fe^{2+}, and metallic iron compared to the surface of the Moon. From spectroscopic measurements, the FeO content of Mercury's surface is less than 3%. This is consistent with Mercury's lower ultraviolet reflectivity and smaller albedo contrast. The FeO content is significantly less than many of the surfaces of the Moon and other terrestrial planets. Earth-based microwave and mid-infrared observations also indicate that Mercury's surface has less FeO plus TiO_2, and at least as much **feldspar** as the lunar highlands. This has been interpreted as indicating that Mercury's surface is largely devoid of basalt, but it could also mean that the basalts only have a low iron content or are fluid sodium-rich basalts. It has been suggested that eruption of highly differentiated basaltic magma may have produced alkaline lavas. On Earth there are low viscosity alkali basalts that could produce the type of volcanic morphology represented by Mercury's plains. Mercury could be the only body in the inner solar system that has not experienced substantial high-iron basaltic volcanism and, therefore, may have undergone a crustal petrologic evolution different from other terrestrial planets.

In summary, both Earth-based spectroscopic observations and calibrated *Mariner 10* images indicate that the surface composition of Mercury has a varied composition with a wide range of SiO_2 content. The FeO content appears to be between 1 and 3%. This is abnormally low compared to other terrestrial planets and the Moon. There is spectrographic evidence for the Mg-rich mineral pyroxene. The spectroscopic data are consistent with compositions ranging from low-iron basalts to **anorthosites.** We will have to await the *MESSENGER* mission data to discover the detailed composition of Mercury and its variation across the surface.

6.3 Tectonic Framework

No other planet or satellite in the solar system has a tectonic framework like Mercury's. It consists of a system of contractional thrust faults called lobate scarps (Figs. 16 and 17). Individual scarps vary in length from ~20 to >500 km and have heights from a few 100 m to about 3 km. They have a random spatial and azimuthal distribution over the imaged half of the planet and presumably occur on a global scale. Thus, at least in its latest history, Mercury was subjected to global contractional stresses. The only occurrences of features indicative of extensional stresses are localized fractures associated with the floor of the Caloris Basin and at its antipode, both of which are the direct or indirect result of the Caloris impact. No lobate scarps have been embayed by intercrater plains on the region viewed by *Mariner 10*, and they transect fresh as well as degraded craters. Few craters are superimposed on the scarps. Therefore, the system of thrust faults appears to postdate the formation of intercrater plains and to have been formed relatively late in Mercurian history. This tectonic framework was probably caused by crustal shortening resulting from a decrease in the planet radius due to cooling of the planet. The amount of radius decrease is estimated to have been anywhere between 0.5 and 2 km.

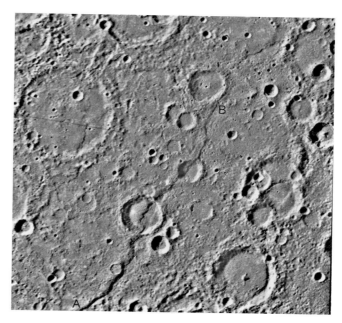

FIGURE 16 Photomosaic of Discovery scarp. This lobate scarp is a thrust fault about 1 km high and 500 km long. It cuts across two craters 55 and 35 km in diameter. (Courtesy NASA.)

Also there is apparently a system of structural lineaments consisting of ridges, troughs, and linear crater rims that have at least three preferred orientations trending in northeast, northwest, and north–south directions. The Moon also shows a similar lineament system. The Mercurian system has been attributed to modifications of ancient linear crustal joints formed in response to stresses induced by tidal spin down.

6.4 Thermal History

All thermal history models of planets depend on compositional assumptions, such as the abundance of uranium, thorium, and potassium in the planet. Since our knowledge of the composition of Mercury is so poor, these models can only provide a general idea of the thermal history for certain starting assumptions. Nevertheless, they are useful in providing insights into possible modes and consequences of thermal evolution. Starting from initially molten conditions for Mercury, thermal history models with from 0.2 to 5% sulfur in the core indicate that the total amount of planetary radius decrease due to cooling is from ~6 to 10 km depending on the amount of sulfur (Fig. 18). About 6 km of this contraction is solely due to mantle cooling during about the first 700 million years before the start of inner core formation. The amount of radius decrease due to inner core formation alone is about 1 km for 5% sulfur and about 4 km for 0.2% sulfur.

Thermal models suggest that inner core formation may have begun about 3 billion years ago, and, therefore, after

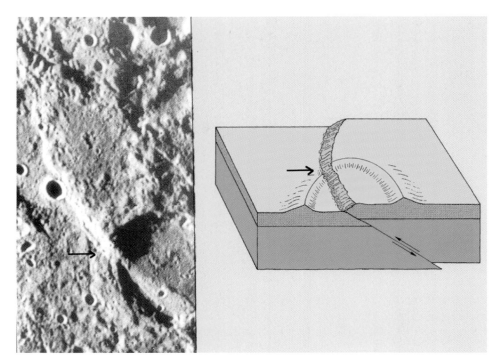

FIGURE 17   The 130-km-long Vostok scarp transects two craters 80 and 65 km in diameter. The northwest rim of the lower crater (Guido d'Arezzo) has been offset about 10 km by thrusting of the eastern part of the crater over the western part. The diagram to the right shows the geologic relationship of the thrust fault and the offset crater rim. (Modified from Strom and Sprague, 2003.)

the period of cataclysmic bombardment (Fig. 18). This would imply that the observed tectonic framework began at about the same time, and that smooth and intercrater plains were emplaced before inner core formation. Indeed, the geologic evidence indicates that at least the observed tectonic framework began to form relatively late in Mercury's history; certainly after intercrater plains formation and possibly after smooth plains formation. However, under initially molten conditions, the thermal history models indicate that the lithosphere has always been in contraction. The surface record of the period of intense contraction caused by mantle cooling has probably been erased by the period of cataclysmic bombardment and intercrater plains formation that occurred from about 3.9 to 3.8 billion years ago. That would explain why there is no evidence for old compressive structures.

If the upper value of a 2-km radius decrease, inferred from the thrust faults, was due solely to cooling and solidification of the inner core, then the core sulfur abundance is probably 2–3%, and the present fluid outer core is about 500 or 600 km thick. If the lower value of a 0.5-km radius decrease is correct, then there must be more than 5% sulfur in the core, and the present fluid outer core would be over 1000 km thick.

If the smooth and intercrater plains are volcanic flows, then they must have had some way to easily reach the surface to form such extensive deposits. Early lithospheric compressive stresses would make it difficult for lavas to reach the surface, but the lithosphere may have been rela-

tively thin at this time (<50 km). Large impacts would be expected to strongly fracture it, possibly providing egress for lavas to reach the surface and bury compressive structures.

6.5 Geologic History

Mercury's earliest history is very uncertain. If a portion of the mantle was stripped away, as invoked by most scenarios to explain its high mean density, then Mercury's earliest recorded surface history began after core formation, and a possible mantle-stripping event (see Section 7). The earliest events are the formation of intercrater plains (≥3.9 billion years ago) during the period of late heavy bombardment. These plains may have been erupted through fractures caused by large impacts in a thin lithosphere. Near the end of late heavy bombardment, the Caloris Basin was formed by a large impact that caused the hilly and lineated terrain from seismic waves focused at the antipodal region. Further eruption of lava within and surrounding the Caloris and other large basins formed the smooth plains about 3.8 billion years ago. The system of thrust faults formed after the intercrater plains, but how soon after is not known. If the observed thrust faults resulted only from core cooling, then they may have begun after smooth plains formation and resulted in a decrease in Mercury's radius. As the core continued to cool and the lithosphere thickened, compressive stresses closed off the magma sources, and volcanism ceased near the end of late heavy bombardment. All of Mercury's volcanic events probably took place very

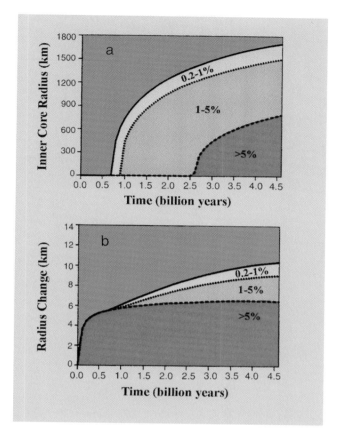

FIGURE 18   (a) A thermal history model for inner core radius as a function of time for three values of initial core sulfur content. The colors show the ranges in sulfur content from 0.2 to >5%, and the solid, dotted, and dashed lines are for sulfur contents of 0.2, 1, and 5%, respectively. (b) Decrease in Mercury's radius due to mantle cooling and inner core growth for three values of initial core sulfur content as in (a). (Modified from Vilas et al., 1988.)

early in its history, perhaps during the first 700 to 800 million years. Today the planet may still be contracting as the present fluid outer core continues to cool.

7. Origin

The origin of Mercury and how it acquired such a large fraction of iron compared to the other terrestrial planets is not well determined. Chemical condensation models for Mercury's present position in the innermost part of the solar nebula, from which the solar system formed, cannot account for the large fraction of iron that must be present to explain its high density. Although these early models are probably inaccurate, revised models that take into account material supplied from feeding zones in more distant regions of the inner solar system only result in a mean uncompressed density of about 4200 kg/m^3, rather than the observed 5300 kg/m^3. Furthermore, at Mercury's present distance, the models predict the almost complete absence of sulfur (100 parts per trillion FeS), which is apparently required to account for the presently molten outer core. Other volatile elements and compounds, such as water, should also be severely depleted (<1 part per billion of hydrogen).

Three hypotheses have been put forward to explain the discrepancy between the predicted and observed iron abundance. One (selective accretion) involves an enrichment of iron due to mechanical and dynamical accretion processes in the innermost part of the solar system; the other two (postaccretion vaporization and giant impact) invoke removal of a large fraction of the silicate mantle from a once larger proto-Mercury. In the selective accretion model, the differential response of iron and silicates to impact fragmentation and aerodynamic sorting leads to iron enrichment owing to the higher gas density and shorter dynamical timescales in the innermost part of the solar nebula. In this model, the removal process for silicates from Mercury's present position is more effective than for iron, leading to iron enrichment. The postaccretion vaporization hypothesis proposes that intense bombardment by solar electromagnetic and corpuscular radiation in the earliest phases of the Sun's evolution vaporized and drove off much of the silicate fraction of Mercury leaving the core intact. In the giant impact hypothesis, a planet-sized object impacts Mercury and essentially blasts away much of the planet's silicate mantle leaving the core largely intact.

Discriminating among these models is difficult, but may be possible from the chemical composition of the silicate mantle (Fig. 19). For the selective accretion model, Mercury's silicate portion should contain about 3.6–4.5% alumina, about 1% alkali oxides (Na and K), and between 0.5 and 6% FeO. Postaccretion vaporization should lead to very severe depletion of alkali oxides (~0%) and FeO (<0.1%), and extreme enrichment of refractory oxides (~40%). If a giant impact stripped away the crust and upper mantle late in accretion, then alkali oxides may be depleted (0.01–0.1%), with refractory oxides between ~0.1 and 1% and FeO between 0.5 and 6%. Unfortunately our current knowledge of Mercury's silicate composition is extremely poor, but near and mid-infrared spectroscopic measurements favor low FeO- and alkali-bearing feldspars. If the tenuous atmosphere of sodium and potassium is being outgassed from the interior, as suggested by some, then the postaccretion vaporization model may be unlikely. Deciding between the other two models is not possible with our current state of ignorance about the silicate composition. Since the selective accretion hypothesis requires Mercury to have formed near its present position, then sulfur should be nearly absent, unless the solar nebula temperatures in this region were considerably lower than predicted by the chemical equilibrium condensation model.

Support for the giant impact hypothesis comes from three-dimensional computer simulations of terrestrial planet formation for several starting conditions. Since these

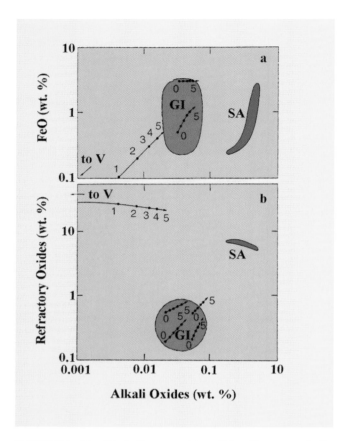

FIGURE 19   Possible bulk composition of the silicate mantle for the three models of Mercury's origin; selective accretion (SA), postaccretion vaporization (V), and giant impact (GI). The composition is parameterized for the FeO content, the alkali content (soda plus potash) in (a), and the refractory oxide content (calcium plus aluminum plus titanium oxides) in (b). The modifying effects of late infall of 0–5% of average chondritic meteorite material on several regolith compositions are indicated by arrows labeled 0–5. (Modified from Vilas et al., 1988.)

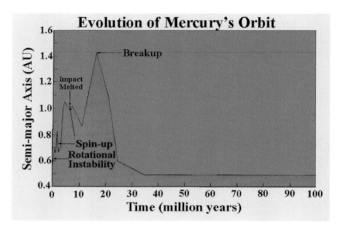

FIGURE 20   Results of a computer simulation of terrestrial planet evolution showing the change of "Mercury's" semimajor axis during its accretion. In this case "Mercury's" semimajor axis spans the entire terrestrial planet region (0.5–1.4 AU) during the planet's growth. (Modified from Vilas et al., 1988.)

ing result in homogenizing the entire terrestrial planet region, contrary to the observed large systematic density differences.

The simulations also indicate that byproducts of terrestrial planet formation are planet-sized objects up to three times the mass of Mars that become perturbed into eccentric orbits (mean e ~0.15 or larger) and eventually col-

simulations are by nature stochastic, a range of outcomes is possible. They suggest, however, that significant fractions of the terrestrial planets may have accreted from material formed in widely separated parts of the inner solar system. The simulations indicate that during its accretion Mercury may have experienced large excursions in its semimajor axis. These semimajor axis excursions may have ranged from as much as 0.4–1.4 AU due to energetic impacts during accretion (Fig. 20). Consequently, Mercury could have accumulated material originally formed over the entire terrestrial planet range of heliocentric distances. About half of Mercury's mass could have accumulated at distances between about 0.8 and 1.2 AU (Fig. 21). If so, then Mercury may have acquired its sulfur from material that formed in regions of the solar nebula where sulfur was stable. Plausible models estimate FeS contents of 0.1–3%. However, the most extreme models of accretional mix-

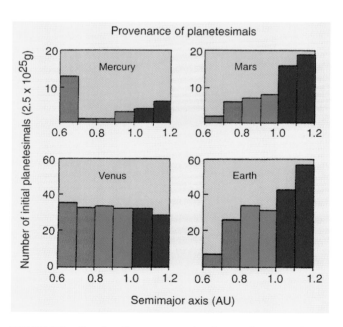

FIGURE 21   Results of a computer simulation of terrestrial planet evolution showing the region (semimajor axis) from which the terrestrial planets acquired their mass. In this simulation, "Mercury" acquires about half its mass from regions between 0.8 (green) and 1.2 AU (blue). (Modified from Vilas et al., 1988.)

lide with the terrestrial planets during their final stages of growth. The final growth and giant impacts occur within the first 50 million years of solar system history. Such large impacts may have resulted in certain unusual characteristics of the terrestrial planets, such as the slow retrograde rotation of Venus, the origin of the Moon, the martian crustal dichotomy, and Mercury's large iron core.

In computer simulations where proto-Mercury was 2.25 times the present mass of Mercury with an uncompressed density of about 4000 kg/m^3, nearly central collisions of large projectiles with iron cores impacting at 20 km/s, or noncentral collisions at 35 km/s resulted in a large silicate loss and little iron loss (Fig. 22). In the former case, although a large portion of Mercury's iron core is lost, an equally large part of the impactor's iron core is retained resulting in about the original core size. At Mercury's *present* distance from the Sun, the ejected material reaccretes back onto Mercury if the fragment sizes of the ejected material are greater than a few centimeters. However, if the ejected material is in the vapor phase or fine-grained ($\leq$1 cm), then it will be drawn into the Sun by the **Poynting–Robertson effect** in a time shorter than the expected collision time with Mercury (about 10^6 years). The proportion of fine-grained to large-grained material ejected from such an impact is uncertain. Therefore, it is not known if a large impact at Mercury's present distance could exclude enough mantle material to account for its large iron core. However, the disruption event need not have occurred at Mercury's present distance from the Sun. It could have occurred at a much greater distance (e.g., >0.8 AU; Fig. 20). In this case the ejected mantle material would be mostly swept up by the larger terrestrial planets, particularly Earth and Venus.

8. The *Messenger* Mission

Mercury is the least known of all the terrestrial planets, but it is probably the only planet that holds the key to understanding details of the origin and evolution of all these bodies. Because only half of the planet has been imaged at relatively low resolution, and because of the poor characterization of its magnetic field and almost complete ignorance of its silicate composition and variation across the surface, there is little hope of deciding between competing hypotheses of its origin and evolution until more detailed information is obtained. Fortunately, help in on the way.

The spacecraft *MESSENGER* is now on its way to orbit Mercury. This mission is one of NASA's Discovery series of planetary exploration missions. *MESSENGER* is managed by the Applied Physics Laboratory of Johns Hopkins University in Maryland, and the Carnegie Institution of Washington, D.C.

On August 3, 2004, the *MESSENGER* spacecraft was launched from Cape Canaveral, Florida, to explore Mercury

for the first time in over 30 years (Fig. 23). After the Earth flyby that took place in August 2005, it will make two flybys of Venus (October 2006 and June 2007) and three flybys of Mercury (January and October 2008, and September 2009) before it is inserted into Mercury orbit in March 2011. It will take 7 years to put the spacecraft in orbit around Mercury because the spacecraft must make six planetary encounters to slow it enough to put it in orbit with a conventional retrorocket. A direct flight to Mercury would get the spacecraft there in about 4 months, just like *Mariner 10*. However, it would be traveling at such a high speed at Mercury encounter that it would take the equivalent of a launch rocket to put it in orbit. That is the reason *Mariner 10* could not be captured into orbit around Mercury.

There are seven main objectives of the mission, all of which are important to understanding the origin and evolution of Mercury and the inner planets. One is to determine the nature of the polar deposits including their composition. Another objective is to determine the properties of Mercury's core including its diameter and the thickness of its outer fluid core. This is accomplished by accurately measuring Mercury's libration amplitude from the laser altimeter and radio science experiments. A third objective is to determine variations in the structure of the lithosphere and whether or not convection is currently taking place. A fourth objective is to determine the nature of the magnetic field and to confirm whether it is a dipole. There are several instruments to study the chemical and mineralogical composition of the crust that should place constraints on Mercury's origin and, we hope, help us decide among the three competing hypotheses. Also these data will be extremely useful to help us decipher Mercury's geology. The geologic evolution of Mercury will be addressed by the dual camera system that will image the entire surface at high resolution and at a variety of wavelengths. Finally, the exosphere will be studied to determine its composition and how it interacts with the magnetosphere and surface.

There are eight science experiments on board the spacecraft (Fig. 24). They are (1) a dual imaging system, (2) a gamma-ray and neutron spectrometer, (3) a magnetometer, (4) a laser altimeter, (5) atmospheric (0.155–0.6 μm) and surface (0.3–1.45 μm) spectrometers, (6) an energetic particle and plasma spectrometer, (7) an X-ray spectrometer, and (8) a radio science experiment that uses the telecommunication system. These instruments will be used to accomplish the objectives discussed previously. They are listed in Table 2 together with the measurements they will make.

MESSENGER will be placed in an elliptical orbit with a 200-km periapse altitude located at about 60°N latitude (Fig. 25). The orbit has a 12-hour period when data will be collected and read out. The spacecraft will also collect valuable data on its three flybys of Mercury prior to orbit insertion. *MESSENGER* should provide the data

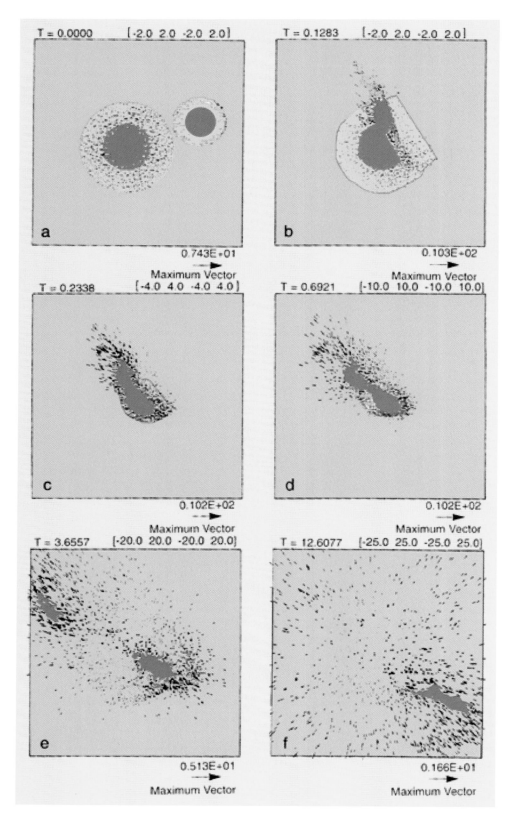

FIGURE 22   Computer simulation of a large, off-axis, 35 km/sec impact with Mercury. In this simulation, the mantle separates from the core. A portion of the mantle must reaccrete to form the present-day Mercury. (From Benz et al., 1988.)

Mariner 10, Nov. 3, 1973;12:45 am EST
Atlas/Centaur Launch Vehicle

MESSENGER, Aug. 3, 2004;2:16 am EDT
Delta Heavy Launch Vehicle

FIGURE 23 These images show the launches to Mercury of *Mariner 10* (left) and *MESSENGER* (right) almost 31 years apart.

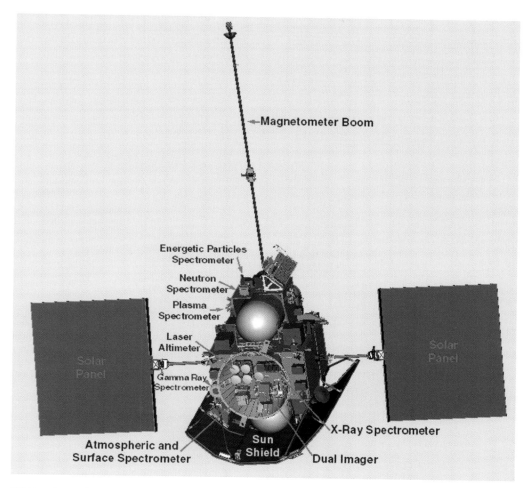

FIGURE 24 This drawing of the *MESSENGER* spacecraft shows the placement on the spacecraft of the science instruments listed in Table 2.

TABLE 2	*MESSENGER* Instruments and Measurements	
Instrument	**Observation**	
Duel Imaging System (1.5° and 10.5° field of view)	Surface mapping in stereo (10 color filters)	
Gamma-Ray and Neutron Spectrometer	Surface composition (O, Si, Fe, H, K)	
X-Ray Spectrometer (1– to 10 kKeV)	Surface composition (Mg, Al, Fe, Si, S, Ca, Ti)	
Atmospheric and Surface Spectrometer	Surface and Exosphere composition	
Magnetometer	Magnetic field	
Laser Altimeter	Topography of northern hemisphere	
Energetic Particles and Plasma Spectrometer	Energetic particles and plasma	
Radio Science (X-band transponder)	Gravity field and physical liberation	

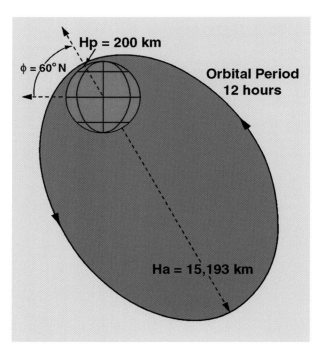

FIGURE 25 This drawing shows the orbit of *MESSENGER* in its 12-hour period. The elliptical orbit is required to maintain an acceptable operating temperature, and yet still obtain high-resolution data at and near closest approach. Hp is the closest approach, Ha is the farthest point in the orbit, and the orbit is inclined at an angle of 60° to the equator.

necessary to answer most of the questions raised in this chapter. The Europeans are also planning a Mercury mission called BepiColombo in the years immediately following the *MESSENGER* investigations.

Bibliography

Benz, W., Slattery, W., and Cameron, A. G. W. (1988). *Icarus* **74**, 516–528.

Harmon, J. K., Perillat, P. J., and Slade, M. A. (2001). *Icarus* **149**, 1–15.

Potter, A. E., and Morgan, T. H. (1988). *Science* **247**, 675.

Robinson, M. S., and Lucey, P. G. (1997). *Science* **275**, 197–200.

Sprague, A., Kozlowski, R. W. H., and Hunten, D. M. (1990). *Science* **249**, 1140–1142.

Strom, R. G. (1984). "The Geology of the Terrestrial Planets" (M. H. Carr, ed.). NASA SP-469, pp. 13–55. NASA, Washington, D.C.

Strom, R. G. (1987). "Mercury: The Elusive Planet." Smithsonian Institution Press, Washington, D.C.

Strom, R. G., and Sprague, A. L. (2003). "Exploring Mercury: the Iron Planet." Springer-Praxis, Chichester, England.

Various authors in a special issue on the *Mariner 10* encounter with Mercury. (1975). *J. Geophys. Res.* **80**.

F. Vilas, F. Chapman, C. R., and Matthews, M. S., eds. (1988). "Mercury," Univ. Arizona Press, Tucson.

Venus: Atmosphere

Donald M. Hunten

University of Arizona
Tucson, Arizona

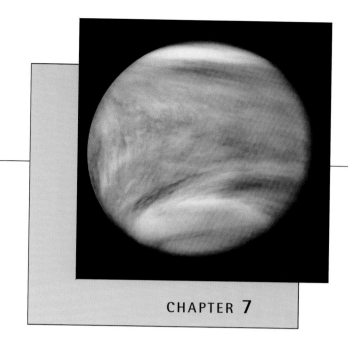

CHAPTER 7

Venus possesses a dense, hot atmosphere, primarily of carbon dioxide, with a pressure of 93 bars and a globally uniform temperature of 740 K at the surface. The surface is totally hidden at visible wavelengths by a cloud deck (really a deep haze) of concentrated sulfuric acid droplets that extends from 50 km altitude to a poorly defined top at 65 km (Fig. 1 and also Fig. 8). The clouds are thus located in the top part of the **troposphere**, which extends from 0 to 65 km. The middle atmosphere (**stratosphere and mesosphere**) extends from 65 to about 95 km, and the upper atmosphere (**thermosphere and exosphere**) from 95 km up. Although the rotation period of the solid planet is 243 Earth days (**sidereal**), the atmosphere in the cloud region rotates in about 4 days, and the upper atmosphere in about 6 days, all in the same **retrograde** direction.

1. Introduction

1.1 History

The study of Venus by Earth-based telescopes has been frustrated by the complete cloud cover. The presence of CO_2 was established in 1932, as soon as infrared-sensitive photographic plates could be applied to the problem. But establishment of the abundance was impossible because there was no way to determine the path length of the light as it scattered among the cloud particles. Moreover, it was assumed that nitrogen would also be abundant, as it is on Earth, and this gas cannot be detected in the spectral range available from the ground. Careful observation of the feeble patterns detectable in blue and near-ultraviolet images was able to establish the presence of the 4-day rotation at the cloud tops. These patterns are shown in the much more recent spacecraft images of Fig. 1. Radio astronomers, observing Venus's emission at the microwave wavelength of 3.15 cm, discovered in 1958 that it appears to be much hotter than expected, and this was confirmed by later results at other wavelengths. The most likely suggested explanation was that the radiation came from a hot surface, warmed by an extreme version of the **greenhouse effect**; but the required warming is so extreme that other hypotheses were debated. Spacecraft measurements, as will be described, finally settled the issue in favor of the greenhouse effect and showed that the pressure at the mean surface is 93 bars.

A large number of spacecraft experiments on 22 missions have been devoted to study of the atmosphere; along with that of Mars, it is better explored than that of any planet other than the Earth. United States missions, starting in 1962, were the flybys *Mariner 2, 5,* and *10* (which went on to Mercury); *Pioneer Venus Multiprobe* and *Orbiter* in 1978; the radar mapper *Magellan*; and the Jupiter-bound *Galileo*. Successful Soviet ones were *Venera 4–14*, which included entry and descent probes as well as flybys or orbiters,

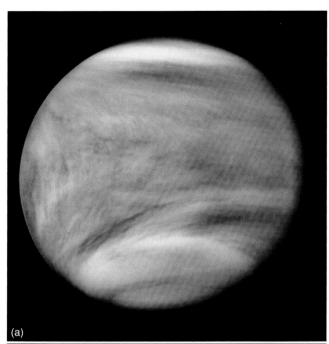

FIGURE 1   Ultraviolet images of Venus' clouds as seen from *Pioneer Venus Orbiter* on February 5, 1979 (a) and on February 26, 1979 (b). (Courtesy of NASA.)

Venera 15 and *16*, which were radar mappers, and *Vega 1* and *2*, which dropped both probes and balloon-borne payloads on their way to Halley's Comet. Early missions were devoted to reconnaissance, in particular to confirmation of the high surface pressure and temperature inferred

from the microwave radio measurements. [*See* PLANETARY EXPLORATION MISSIONS.]

The composition of the clouds was another important question, but it was actually answered first from analysis of ground-based observations of the polarization of light reflected from the planet. Although such measurements were first made in the 1930s, the computers and programs to carry out the analysis did not exist until the middle 1970s. This analysis pinned down the refractive index and showed that the particles are spherical; these two properties eventually led to the identification of supercooled droplets of concentrated sulfuric acid (H_2SO_4). Measurements from the *Pioneer Venus* probes confirmed this composition and gave much greater detail on the sizes and layering of the haze.

1.2 Measuring Techniques

Three principal techniques can be applied from Earth: spectroscopy, radiometry, and imaging. They can be used over a wide variety of wavelengths, from the ultraviolet to the shortest part of the radio spectrum. Spectroscopy, as mentioned earlier, was first applied in 1932 and led to the discovery of CO_2. Little more was done until the middle 1960s, when traces of water vapor were found and a tight upper limit was set on the amount of O_2. The development of Fourier spectroscopy permitted an extension further into the infrared, where CO, HCl, and HF were observed. Radiometry, and especially polarimetry, eventually led to the identification of the substance of the cloud particles. After the near-infrared "windows" were identified (see Section 1.4), starting in 1983, spectroscopy of deeper parts of the atmosphere provided important further information. Visual studies, followed more recently by photography and infrared imaging, disclosed the 4-day rotation of the cloud tops and the 6-day period of a deeper region. Similar remarks apply to radio astronomical studies. Radiometry gave the data that finally led to the establishment of the high surface temperature, and millimeter-wave spectroscopy has led to the interesting results on CO discussed in Section 3.4. Until the early 1990s, all ground-based radio work used radiation from the whole disk, but modest spatial resolution is beginning to be available by interferometry (the technique of combining the signals from several antennas).

Many of the same techniques have been applied from flyby and orbiting spacecraft, but an important addition is the radio occultation experiment, which tracks the effect of the atmosphere on the telemetry carrier as the spacecraft disappears behind the atmosphere or reappears from behind it. On Venus, the regions observed in this way are the ionosphere and the neutral atmosphere from about 34 to 90 km. At greater depths, the **refraction** of the waves by the atmosphere is so great that the beam strikes the surface and never reappears. In addition to carrying several instruments for remote sensing, *Pioneer Venus Orbiter* (1978–1992) actually penetrated the upper atmosphere once per orbit, and took advantage of this by carrying a suite of instruments to make measurements in situ. Two mass spectrometers measured individual gases and positive ions; a **Langmuir probe** and a retarding potential analyzer measured electron and ion densities, temperatures, and velocities; and a fifth instrument measured plasma waves. Higher-energy ions and electrons, both near the planet and in the solar wind, were measured by a plasma analyzer, and important auxiliary information was provided by a magnetometer. In addition, the atmospheric drag on the spacecraft gave an excellent measure of the density as a function of height.

A large number of probes have descended part or all the way through the atmosphere, and the Vega balloons carried out measurements in the middle of the cloud region. All of them have carried an "atmospheric structure" package measuring pressure, temperature, and acceleration; height was obtained on the early *Venera* probes by radar and on all probes by integration of the hydrostatic equation. Gas analyzers have increased in sophistication from the simple chemical cells on *Venera 4* to mass spectrometers and gas chromatographs on later Soviet and U.S. missions. In some cases, however, there are suspicions that the composition was significantly altered in passage through the sampling inlets, especially below 40 km, where the temperature is high. A variety of instruments have measured the clouds and their optical properties. Radiometers observed the loss of solar energy through the atmosphere, and others have observed the thermal infrared fluxes. Winds were obtained by tracking the horizontal drifts of the probes as they descended, and the balloons as they floated. *Venera 11–14* carried radio receivers to seek evidence of lightning activity.

1.3 Composition

The fact that carbon dioxide is indeed the major gas was established by a simple chemical analyzer on the *Venera 4* entry probe. The mole fraction was found to be about 97%, in reasonable agreement with the currently accepted value shown in Table 1. The next most abundant gas is nitrogen; though it is only 3.5% of the total, the absolute quantity is about three times that in the Earth's atmosphere. The temperature profile is illustrated in Fig. 2, along with a sketch of the cloud layers.

Many of the strange properties of the atmosphere can be traced to an extreme scarcity of water and its vapor and the total absence of liquid water. On Earth, carbon dioxide and sulfuric, hydrochloric, and hydrofluoric acids are all carried down by precipitation, a process that is absent in the hot, dry lower atmosphere of Venus. They all then react and are incorporated in geological deposits; the best estimates of the total amount of carbonate rocks in the Earth give a quantity of CO_2 almost equal to that seen in the atmosphere of Venus. Free oxygen is undetectable at the Venus cloud tops; one molecule in ten million could have been seen.

TABLE 1		Composition of the Venus Atmosphere	
	Species	Mole fraction at 70 km	Mole fraction at 40 km
%	CO_2	96.5	96.5
	N_2	3.5	3.5
ppm[a]	He	~12	~12
	Ne	7	7
	Ar	70	70
	Kr	~0.2	~0.2
	CO	5170	45
	H_2O	≤1	45
	SO_2	0.05	~100
	H_2S	?	1
	COS		0.25
	HCl	0.4	0.5
	HF	0.005	0.005
	O_2	<0.1	0–20
%	D/H	1.6	1.6

[a] Parts per million.

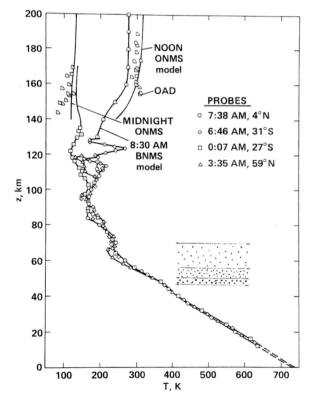

FIGURE 2 Temperature profile from the surface to 200 km altitude, obtained from different experiments in the *Pioneer Orbiter* and *Probes*. ONMS, orbiter neutral mass spectrometer; OAD, orbiter atmospheric drag; BNMS, bus neutral mass spectrometer. The cloud region with its three layers has been sketched in. (From Hunten et al., 1984.)

There is, of course, plenty of oxygen in carbon dioxide, and dissociation by sunlight liberates it in copious quantities. It is readily detected (as is CO) by spacecraft instruments orbiting through the upper atmosphere but is removed before it can reach the cloud level. Small quantities of O_2 are also found below the clouds, probably liberated by the thermal decomposition of the cloud particles. All these lines of evidence point to the action of a strong mechanism in the middle atmosphere that converts O_2 and CO back into CO_2. The observed HCl molecules are the key; they too are broken apart by solar radiation, and the free chlorine atoms enter a **catalytic cycle** that does the job. This chemistry is closely coupled to the sulfur chemistry (see Section 4) that maintains the clouds.

Carbon dioxide, aided by the other molecules listed in Table 1, makes the lower atmosphere opaque to thermal (infrared) radiation; it is this opacity that makes the extreme greenhouse effect possible. Only a few percent of the incident solar energy reaches the surface, but this is enough. Venus is a remarkable and extreme example of the large climatic effects that can be produced by seemingly small causes. One chlorine atom in two and a half million can completely eliminate free oxygen from the middle atmosphere, and ozone has no hope of surviving in significant quantities. The temperature increase caused by the greenhouse effect is almost 500° C. The idea that the 30° seen on Earth could become 32° or 33° if its atmospheric content of CO_2 should double seems entirely probable to experts on Venus's atmosphere, and so does significant loss of ozone from release of chlorinated refrigerants. It thus seems that the obvious differences between Earth and Venus are all traceable to the differences in their endowments of water (vapor or liquid). Although origin and evolution are discussed in Section 6, a short preview is given here. It is plausible that both planets started out with similar quantities, but that the greater solar flux at Venus caused all its water to evaporate (a "runaway greenhouse"). Solar ultraviolet photons could then dissociate it into hydrogen (which escaped) and oxygen (which reacted with surface materials). Strong evidence in favor of this scenario is the extreme enhancement of heavy hydrogen (deuterium, or D), almost exactly 100 times more abundant relative to H than it is on Earth. Such a fractionation is expected because the escape of H is much easier than that of D. [See Earth as a Planet: Atmospheres and Oceans.]

1.4 Near-Infrared Sounding

Study of the atmosphere below the clouds was revitalized in 1988 by the discovery of several narrow spectral windows in the near infrared, where the radiation from deep layers can be detected from above (Fig. 3). The two most prominent ones are at 1.74 and 2.3 μm (Fig. 4), and others are at 1.10, 1.18, 1.27, and 1.31 μm. As we have seen, at microwave radio wavelengths, radiation from the actual surface can escape to space. At other infrared wavelengths, the emission

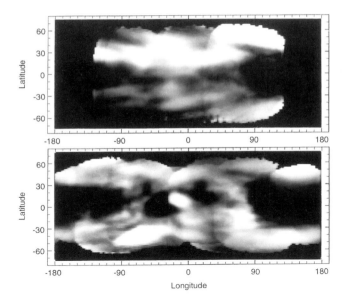

FIGURE 3   Near-infrared images (2.36-μm) of the night side combined into maps for (above) December 31 to January 7, 1991, and (below) February 7 to 15, 1991. Bright areas are thinner parts of the cloud through which thermal radiation from deeper layers can shine. (From Crisp et al., 1991.)

from the night side is characteristic of the temperature of the cloud tops, about 240 K. In the windows, the brightness, and therefore the temperature of the emitting region, is considerably higher. Images taken in a window reveal horizontally banded structures that appear to be silhouettes of the lowest part of the cloud (around 50 km) against the hotter atmosphere below (see Fig. 3 and Section 4). [See Infrared Views of the Solar System from Space.]

Numerous absorption lines and bands allow inferences about the composition to levels all the way to the surface. One such spectrum is shown in Fig. 4. Each "window" allows the composition to be obtained at a different level; this is particularly important for water vapor, discussed in the next section. The measurement of carbonyl sulfide (COS) shown in Table 1 was obtained by this analysis. This gas has resisted all attempts to measure it from entry probes, even though it has long been expected to be present. Other gases include CO, HF, HCl, and light and heavy water vapor, all in good agreement with prior results. These results are also included in Table 1.

2. Lower Atmosphere

2.1 Temperatures

It is convenient to regard the lower atmosphere as extending from the surface to about 65 km, the level of the visible cloud tops and also of the tropopause. This region has been measured in detail by many descent probes, with results in close agreement, and also by radio occultation. The temperature profile (see Fig. 2) is close to the **adiabat,** becoming

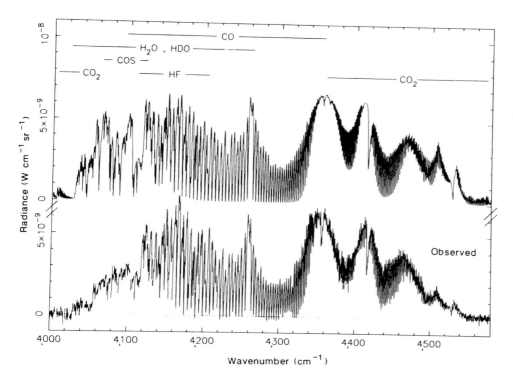

FIGURE 4 Near-infrared spectrum in the 2.3-μm window (bottom); the spectrum above it was calculated by making use of laboratory data for the sic different molecules shown. (From Bézard et al., 1990.)

noticeably less steep above the tropopause. As on Earth, the tropopause is a few kilometers lower at high latitudes than near the equator. The high surface temperature is maintained by the greenhouse effect, driven by the few percent of solar energy that reaches the surface. Converted to thermal infrared, this energy leaks out very slowly because of the opacity of the atmospheric gases at such long wavelengths. The molecules principally responsible are CO_2, SO_2, H_2O, and perhaps others. Quantitative calculations have shown that the greenhouse mechanism is adequate, and that the observed solar and infrared net fluxes can be reproduced. These models treat the temperatures as globally uniform, so that they can be restricted to considering vertical heat transport only.

The surface temperature is remarkably uniform with both latitude and longitude, largely because of a very long radiative time constant. A very slow atmospheric circulation is therefore adequate, but the details of how the nonuniform solar heating is converted to a uniform surface temperature are not understood. The 'runaway greenhouse' that may have operated early in the history of the planet is discussed in Section 6.

2.2 Water Vapor

Table 1 shows rather uncertain quantities of H_2O, but there is no doubt that there is a major difference in the mole fractions below and above the clouds. This is almost certainly because the concentrated sulfuric acid of the cloud particles is a powerful drying agent (see Section 4.3). A summary of the many attempts to measure the abundance

below the cloud is given in Fig. 5. Direct measurements have been made by several mass spectrometers and gas chromatographs, but the amounts are so small and the results so divergent that there remain many questions. Indirect measurements come from radiation fluxes, which are

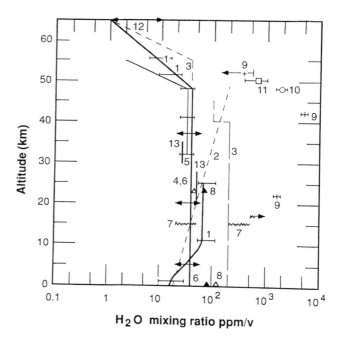

FIGURE 5 Water vapor mixing ratios or mole fractions from various experiments, mostly on *Pioneer Venus* and *Venera* probes. (From Donahue and Hodges, 1992.)

strongly affected by the opacity of water vapor. The four *Pioneer Venus* probes carried infrared net flux radiometers (points labeled "7" in Fig. 5), and *Venera 11* and *12* carried an instrument working with weaker absorptions in the near infrared (dashed line labeled "2"). These measurements relate to the atmosphere far from the probe and are not affected by the difficulty, encountered by the gas chromatograph and mass spectrometers, of obtaining an undistorted sample of the gas. It is likely that many of the divergences are due to the extreme difficulty of measuring such small quantities of a reactive molecule at the high temperatures of the lower atmosphere, but some of the variations may reflect real effects of latitude or height. Particularly puzzling has been the indication from the mass spectrometer on the *Pioneer Venus Large Probe* that the mole fraction falls off by nearly a factor of 10 between 10 km altitude and the surface (Fig. 5, line "1"). It is likely that this result is incorrect; it is not supported by remote sensing of this region in the near-infrared windows.

The ratio of heavy to light hydrogen (D/H) (150 times the value on Earth) was first measured on ions in the ionosphere and has been confirmed by the mass spectrometer just mentioned and by analysis of spectra taken from Earth in the near-infrared windows. In turn, the deuterium provides a valuable signature for distinguishing Venus water vapor in the mass spectrometer from any contaminants carried along from Earth. The likely enrichment process is discussed in Section 6.

3. Middle and Upper Atmosphere

3.1 Temperatures

The middle atmosphere (stratosphere and mesosphere) extends from the tropopause at 65 km to the temperature minimum or mesopause at about 95 km (see Fig. 2). The upper atmosphere lies above this level. Here, temperatures can no longer be measured directly, but are inferred from the **scale heights** of various gases with use of the **hydrostatic equation**. On Earth and most other bodies, this region is called the thermosphere because temperatures in the outer layer, or exosphere, are as high as 1000 K. The temperature is much more modest on Venus; the exospheric temperature is no more than 350 K on the day side. The corresponding region on the night side is sometimes called a cryosphere (cold sphere) because its temperature is not far above 100 K. Measurements of these temperatures by *Pioneer Venus Orbiter* are shown in Fig. 6. The large temperature difference translates into a pressure difference that drives strong winds from the day side to the night side, at all levels above 100 km.

On Earth, the exospheric temperature changes markedly with solar activity, being perhaps 700 K at sunspot minimum and 1400 K at maximum. The corresponding change at Venus is much more modest, perhaps 50 K. Many of these

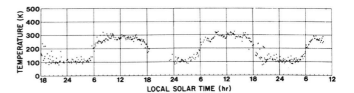

FIGURE 6 Diurnal variation of temperature in the upper thermosphere. Scale heights H were measured by the mass spectrometer on *Pioneer Venus Orbiter* and converted to temperatures by the formula $T = mgH/k$. The measurements sweep out the entire range of local solar time as Venus moves around the Sun about $2^1/_2$ times. (From Hunten et al., 1984.)

differences are traceable to the fact that CO_2, the principal radiator of heat, is just a trace constituent of Earth's atmosphere but is the major constituent for Venus (and also Mars). Venus's slow rotation is responsible for the very cold temperatures on the night side, although the atmosphere does rotate substantially faster than the solid planet. [*See* MARS: LANDING SITE GEOLOGY, MINERALOGY, AND GEOCHEMISTRY.]

3.2 Ionosphere

The principal heat source for the thermosphere is the production of ions and electrons by far-ultraviolet solar radiation. The most abundant positive ions are O_2^+, O^+, and CO_2^+. As part of these processes, CO_2 is dissociated into CO and O, and N_2 into N atoms. All of these ions, molecules, and atoms have been observed or directly inferred (Fig. 7). Some of the O^+ ions (with an equal number of electrons) flow around

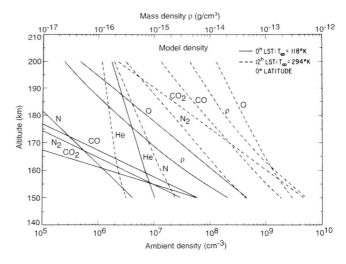

FIGURE 7 Daytime (dashed) and nighttime (solid) number densities of the major gases in the thermosphere obtained by fitting a large number of measurements by the mass spectrometer on *Pioneer Venus Orbiter.* (From Hunten et al., 1984.)

to the night side and help to maintain a weak ionosphere there. Venus lacks any detectable magnetic field, and the day side ionosphere is therefore impacted by the solar wind, a tenuous medium of ions (mostly H^+) and electrons flowing from the Sun at about 400 km/s. Electrical currents are induced in the ionosphere, and they divert the solar wind flow around the planet. The boundary between the two media, called the ionopause, is typically at an altitude of a few hundred kilometers near the subsolar point, flaring out to perhaps 1000 km above the terminators and forming a long, tail-like cavity behind the planet. [See THE SOLAR WIND.]

3.3 Winds

The thermospheric winds carry the photochemical products O, CO, and N from the day side to the night side, where they are almost as abundant as they are on the day side. However, as Fig. 7 illustrates, all gases fall off much more rapidly on the night side because of the low temperature. They descend into the middle atmosphere in a region perhaps 2000 km in diameter and generally centered near the equator at 2 A.M. local time. This region can be observed by the emission of airglow emitted during the recombination of N and O atoms into NO molecules, which then radiate in the ultraviolet, and O_2 molecules, which radiate in the near infrared. The light gases hydrogen and helium are also carried along and accumulate over the convergent point of the flow; for these gases, the peak density is observed at about 4 A.M. These offsets are the principal evidence that this part of the atmosphere rotates with a 6-day period, a rotation that is superposed on the rapid day-to-night flow.

3.4 Chemical Recombination

Oxidation of the CO back to CO_2 is much slower than the recombination of O and N atoms, but a very efficient process is required. This conclusion follows from Earth-based observations of a microwave (2.6-mm wavelength) absorption line of CO, from which a height distribution can be obtained from 80 to 110 km. It is found that the downward-flowing CO is substantially depleted on the night side below 95 km (as well as on the day side). The proposed solution involves reactions of chlorine atoms, as well as residual O atoms descending from the thermosphere. The chlorine acts as a catalyst, promoting reactions but not being consumed itself, and the reaction cycle works without the direct intervention of any solar photons other than the ones that produced the O atoms and CO molecules half a world away.

The availability of Cl atoms is assured by the observed presence of HCl at the cloud tops (Table 1). On Earth, any HCl emitted into the atmosphere is rapidly dissolved in water drops and rained out. Chlorine atoms reach the stratosphere only as components of molecules, such as the artificial ones CCl_4, CF_2Cl_2, and $CFCl_3$ and the natural

one CH_3Cl, none of which dissolve in water. Once they have been mixed to regions above the ozone layer, they are dissociated by solar ultraviolet photons. Because liquid water is absent on Venus, the abundance of HCl is large to start with, and it is not kept away from the stratosphere. Here again the atoms are released by solar ultraviolet. The chlorine abundance is nearly a thousand times greater than that on Earth, and Venus is an example and a warning of what chlorine can do to an atmosphere. The middle atmosphere is also the seat of important chemistry involving sulfur, which is discussed in the next section.

4. Clouds and Hazes

4.1 Appearance and Motions

The clouds are perhaps the most distinctive feature of Venus. They do show subtle structure in the blue and near ultraviolet, illustrated in Fig. 1, which has been processed to bring out the detail and flattened to remove the limb darkening. Although the level shown in the figure is conventionally called the "cloud top," it is not a discrete boundary at all. Similar cloud particles extend as a haze to much higher altitudes, at least 80 km; the "cloud top" is simply the level at which the **optical depth** reaches unity, and the range of visibility (the horizontal distance within which objects are still visible) is still several kilometers.

Study of daily images, first from Earth and later from spacecraft, reveals that the cloud top region is rotating with a period of about 4 days, corresponding to an equatorial east–west wind speed of about 100 m/sec. The speed varies somewhat with latitude; in some years, but not all, the rotation is almost like that of a solid body. Although there are not nearly as many near-infrared images like Fig. 3, they show a longer period consistent with the idea that the silhouettes are of the lower cloud, where entry probes have measured wind speeds of 70 to 80 m/s.

4.2 Cloud Layers

Several entry probes have made measurements of cloud scattering as they descended, but the most detailed results were obtained from *Pioneer Venus* and are shown in Fig. 8. Three regions (upper, middle, and lower) can be distinguished in the main cloud, and there is also a thin haze extending down to 30 km. Size distributions are shown in Fig. 9; it is these, more than the gross properties of Fig. 8, that distinguish the regions. In the upper cloud, the one that can be studied from Earth or from orbit, most particles ("Mode 1") are about 1 μm in diameter and should really be considered a haze rather than a cloud; there are also larger ("Mode 2") particles with diameters around 2 μm. The same particles extend throughout the clouds, but the Mode 2 ones become somewhat larger in the middle and lower clouds,

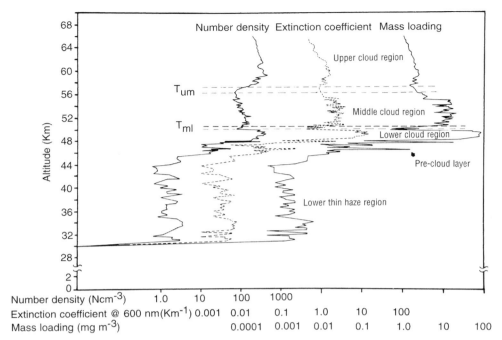

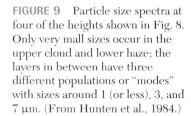

FIGURE 8   Cloud profiles obtained by the particle size spectrometer on *Pioneer Venus Large Probe.* The three curves indicate different properties: number of particles per cubic centimeter, extinction coefficient or optical depth per kilometer of height, and mass per cubic centimeter. (From Hunten et al., 1984.)

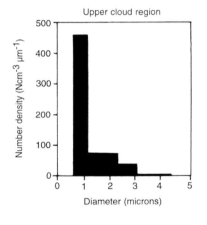

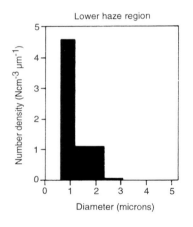

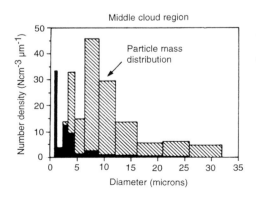

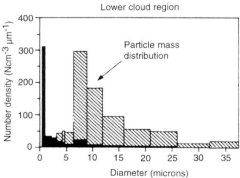

FIGURE 9   Particle size spectra at four of the heights shown in Fig. 8. Only very small sizes occur in the upper cloud and lower haze; the layers in between have three different populations or "modes" with sizes around 1 (or less), 3, and 7 μm. (From Hunten et al., 1984.)

and a third population ("Mode 3"), greater than 6 μm in diameter, is also found. The existence of distinct modes is still not understood; the optical properties of all three are generally consistent with sulfuric acid, although there is some suspicion that the rare Mode 3 particles might be solid crystals.

4.3 Cloud Chemistry

A cloud particle of diameter 1 μm has a sedimentation velocity of 7.5 m/day at 60 km; this velocity varies as the square of the size. Though small, these velocities eventually carry the particles out of the cloud to lower altitudes and higher temperatures, where they will evaporate. At still lower heights the hydrated H_2SO_4 must decompose into H_2O, SO_2, and oxygen, all of which are (at least probably) much more abundant beneath the clouds than above them (Table 1). Atmospheric mixing carries these gases back upward. Nearly all the water vapor is absorbed by the cloud particles. Above the clouds, solar ultraviolet photons attack the SO_2, starting the process that converts it back to H_2SO_4. An important intermediate is the reactive free radical SO, and probably some elemental sulfur is produced. Ultraviolet spectra (pertaining to the region above the clouds) reveal the presence of the small amounts of SO_2 shown in Table 1, but much less than has been measured below the clouds.

Sulfuric acid is perfectly colorless in the blue and near ultraviolet, and the yellow coloration that provides the contrasts of Fig. 1 must be caused by something else. Certainly the most likely thing is elemental sulfur, but yellow compounds are abundant in nature, and the identification remains tentative. The photochemical models do predict production of some sulfur, but it is a minor by-product, and the amount produced is uncertain. Probably the most likely alternative is ferric chloride, particularly for the Mode 3 particles in the lower cloud.

4.4 Lightning

Electromagnetic pulses have been observed by the entry probes *Venera 11, 12, 13,* and *14,* by *Pioneer Venus Orbiter,* and by *Galileo.* For many years it seemed that the most likely source was lightning, and many workers are convinced of its reality. However, some searches for the corresponding optical flashes have been negative, except for one ambiguous interval from *Venera 9.* A recent study from the Earth does seem to have turned up a few optical events. A close flyby by the *Cassini* spacecraft saw no evidence whatever of any impulses with a sensitive instrument that, in a later Earth encounter, found them in abundance (Gurnett et al., 2001). This is strong evidence against the presence of lightning on Venus, at least at the time and in the region that was observed. The negative results may simply be because the flashes are too faint, but another concern is that conditions on Venus do not seem propitious for large-scale charge separation. On Earth, lightning is seen during intense precipitation and in volcanic explosions. In thunderstorms, large drops are efficient at carrying charge of one sign away from the region where it is produced, and the gravitational force is large enough to resist the strong electric fields. This is not the case for small particles. There does not seem to be enough cloud mass on Venus to generate large, precipitating particles, although they are difficult to detect and may have been missed. As for volcanic explosions, most of them are driven by steam; on Venus, water is very scarce, and the 93-bar surface pressure means that, other things being equal, any explosion is damped by a factor of 93 compared with Earth. In spite of these concerns, lightning remains one of the more plausible explanations for the radio bursts, but it is important to seek others.

5. General Circulation

Careful tracking of entry probes, notably the four of Pioneer Venus, has shown that the entire atmosphere is superrotating, with a speed decreasing smoothly from the 100 m/s at the cloud top to near zero at 5–10 km. Winds in the meridional direction are much slower. Because the density increases by a large factor over this height range, the angular momentum is a maximum at 20 km. Small amounts of superrotation are observed in many atmospheres, especially thermospheres, but they are superposed on a rapid planetary rotation. (A familiar example is the midlatitude "prevailing westerlies" on the Earth.) In spite of a great deal of theoretical effort and a number of specific suggestions, there is still no accepted mechanism for the basic motion of the Venus atmosphere, nor is it given convincingly in any numerical general circulation model. What is needed is to convert the slow apparent motion of the Sun (relative to a fixed point on Venus) into a much more rapid motion of the atmosphere. There must also be a slow meridional (north–south) component, sometimes called a Hadley circulation, to transport heat from the equatorial to the polar regions.

There are no direct measurements above the cloud tops, but deductions from temperature measurements suggest a slowing of the 100 m/s flow up to perhaps the 100-km level. At still greater heights the dominant flow is a rapid day–night one, first suggested on theoretical grounds and confirmed by the large observed temperature difference. But the flow is not quite symmetrical; maxima in the hydrogen and helium concentrations, and in several airglow phenomena, are systematically displaced from the expected midnight location toward morning. Possible explanations are a wind of around 65 m/s or a wave-induced drag force that is stronger at the morning side than the evening.

6. Origin and Evolution

It is generally believed that the Sun, the planets, and their atmospheres condensed, about 4.6 billion years ago, from a "primitive solar nebula." The presumed composition of the nebula was that of the Sun, mostly hydrogen and helium with a small sprinkling of heavier elements. It is these impurities that must have condensed into dust and ice particles and accreted to form the planets. Evidently, the Jovian planets were also able to retain a substantial amount of the gas as well, but the terrestrial planets and many satellites must have been made from the solids. [See THE ORIGIN OF THE SOLAR SYSTEM.]

An intermediate stage in the accretion was the formation of "planetesimals," Moon-sized objects that merged to form the final planets. For the terrestrial planets (Mercury, Venus, Earth, Moon, and Mars), the number was probably about 500. These objects would not remain in near-circular orbits, and the ones in the inner solar system might end up as part of any of the terrestrial planets. One would therefore expect them to begin with similar atmospheric compositions, and indeed those of Venus and Earth have many interesting resemblances, as mentioned in Section 1. The smaller bodies appear to have lost all or most of their original gas (or never possessed much in the first place).

Many of the differences between the atmospheres of Earth and Venus can be traced to the near-total lack of water on Venus. These dry conditions have been attributed to the effects of a runaway greenhouse followed by massive escape of hydrogen. A runaway greenhouse might have occurred on Venus because it receives about twice as much solar energy as the Earth. If Venus started with a water inventory similar to that of the Earth, the enhanced heating would have evaporated additional water into the atmosphere. Because water vapor is an effective greenhouse agent, it would trap some of the thermal radiation emitted by the surface and deeper atmosphere, producing an enhanced greenhouse warming and raising the humidity still higher. This feedback may have continued until the oceans were gone and the atmosphere contained several hundred bars of steam. (This pressure would depend on the actual amount of water on primitive Venus.) Water vapor would probably be the major atmospheric constituent, extending to high altitudes where it would be efficiently dissociated into hydrogen and oxygen by ultraviolet sunlight. Rapid escape of hydrogen would ensue, accompanied by a much smaller escape of the heavier deuterium and oxygen. The oxygen would react with iron in the crust, and also with any hydrocarbons that might have been present. Although such a scenario is reasonable, it cannot be proved to have occurred. The enhanced D/H ratio certainly points in this general direction, but it could have been produced from a much smaller endowment of water (as little as 1%) than is in the Earth's oceans.

It used to be thought that Venus was a near twin of the Earth, perhaps a little warmer but perhaps able to sustain Earth-like life. It is still possible that the large divergences we now see could have arisen from different evolutionary paths; alternatively, the two planets may always have been very different.

Two important minor gases in the atmosphere are likely to be variable in time: water vapor H_2O and sulfur dioxide SO_2. Each one is an infrared absorber that contributes to the greenhouse effect, and together they make up the material of the clouds, which also are involved with the greenhouse and which reflect some of the solar energy that would otherwise help heat the planet. Both are likely to be released from large-scale volcanic flows and eruptions, and water may also be brought in by the impact of a large comet. Water is dissociated by solar ultraviolet radiation, and the light H atoms escape from the top of the atmosphere while the oxygen, as well as the sulfur dioxide, react chemically with materials of the surface.

These processes have been studied by Bullock and Grinspoon (2001) who find that the present situation is unstable and that after a billion years the clouds may disappear altogether. The predicted surface temperature may fall by about 50°C; although the planet will absorb more of the incoming solar energy, the effectiveness of the greenhouse will also be reduced. Rapid supply of gases from a volcanic event could raise the surface temperature by as much as 100°C for half a billion years, followed eventually by a return to conditions similar to present ones. A large number of other scenarios can be imagined, depending on the rate and timing of the events that might supply extra gases and the ratio of water to sulfur dioxide in each event. For example, the impact of a large comet would supply mostly water vapor, with relatively little sulfur dioxide.

Bibliography

Bézard, B., de Bergh, C., Crisp, D., and Maillard, J.-P. (1990). *Nature* **345,** 508–511.

Bougher, S. W., Phillips, R. J., and Hunten, D. M., eds. (1997). "Venus II." Univ. Arizona Press, Tucson.

Bullock, M. A., and Grinspoon, D. H. (2001). *Icarus* **150,** 19–37.

Crisp, D., McMuldrough, S., Stephens, S. K., Sinton, W. M., Ragent, B., Hodapp, K.-W., Probst, R. G., Doyle, L. R., Allen, D. A., and Elias, J. (1991). *Science* **253,** 1538–1541.

Donahue, T. M., and Hodges, R. R., Jr. (1992). *J. Geophys. Res.* **97,** 6083–6091.

Fox, J. L., and Bougher, S. W. (1991). *Space Sci. Rev.* **55,** 357–489.

Gurnett, D. A., et al. (2001). *Nature* **409,** 313–315.

Hunten, D. M., Colin, L., Donahue, T. M., and Moroz, V. I., eds. (1984). "Venus." Univ. Arizona Press, Tucson.

Krasnopolsky, V. I. (1986). "Photochemistry of the Atmospheres of Mars and Venus." Springer-Verlag, New York.

Russell, C. T., ed. (1991). "Venus Aeronomy." *Space Sci. Rev.* **55,** 1–489.

Yung, Y. L., and DeMore, W. B. (1982). *Icarus* **51,** 199–247.

Venus: Surface and Interior

Suzanne E. Smrekar

Jet Propulsion Laboratory
California Institute of Technology
Pasadena, California

Ellen R. Stofan

Proxemy Research
Rectortown, Virginia

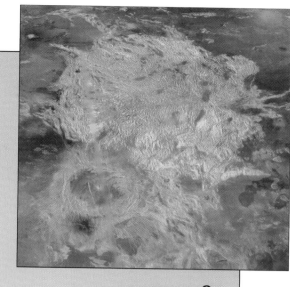

CHAPTER 8

Venus plays a pivotal role in understanding the evolution of the terrestrial planets, the four rocky bodies closest to the Sun. Venus is the planet most similar to the Earth in terms of radius and density, implying a very similar bulk composition. Because terrestrial planets have all formed via the same process, condensing out of the solar nebula, the primary factor that distinguishes them is their size, and to a lesser extent, distance from the Sun. The energy available to drive geologic evolution comes from the heat of accretion and from decay of radiogenic isotopes. Over time, radiogenic decay becomes more dominant. Thus, larger planets have a greater abundance of radiogenic elements and can be expected to be geologically active longer. Earth has abundant geologic activity today. We are uncertain about the present-day level of activity on Venus, but it has clearly been extremely active within the last billion years. The majority of geologic activity on Mars occurred over 3 Ga (billion years) ago. Mercury has not been active since the earliest part of solar system evolution, the heavy bombardment era. [See Planetary Impacts.] Most importantly, Venus has evolved without the system of plate tectonics that governs the pattern of geologic activity on the Earth. Clearly, size is important in determining the duration of geologic activity, but other factors must affect the overall style of geologic evolution. The atmospheric conditions on Venus are also wildly different from those on Earth. The greenhouse effect, in which abundant carbon dioxide causes the atmosphere to heat up, was discovered on Venus. Its thick, dense atmosphere gives Venus a surface temperature of about 468°C (874°F), and a pressure 90 times greater than Earth's. For this reason, Venus has been called Earth's "evil twin."

Volatiles on a planet are essentially the link between the atmosphere, the surface, and the interior, as well as an essential element in the habitability. A planet's atmosphere forms primarily through the outgassing of volatiles from the interior. Outgassing results from the eruption and degassing of lava onto the surface. The rate of resurfacing is a function of the broad-scale geologic processes operating on a planet. These processes are driven by heat loss from the interior, which is primarily fueled by decay of radioactive elements. The interiors of the larger terrestrial planets are hot enough to convect, allowing hot material to rise and cold material to sink on timescales of millions of years. On Earth, convection is linked to surface processes via the process of **plate tectonics**. The presence of water in the interior of Earth acts to reduce the strength of the rock, which in turn allows the exterior shell of the Earth to be broken up into plates. As plates are pushed back into the interior, water is recycled back into the interior. Volatiles on Earth are also strongly affected by both the hydrosphere and the biosphere, both lacking on Venus.

Although plate tectonics has controlled the evolution of Earth for at least 3 Ga, Venus has no trace of such a process.

Plate tectonics creates a bimodal topography distribution on Earth, with high continents and low ocean basins, as well as an interconnected system of ridges and mountain belts at plate boundaries. No such features are seen on Venus. Most explanations of why Venus never developed plate tectonics point to the very low amounts of water currently present on Venus. The water in the atmosphere is equivalent to a surface layer less than 10 cm thick. The abundance of heavy hydrogen, or deuterium, in the atmosphere relative to the normal hydrogen population indicates that a huge amount of water was lost from Venus atmosphere early in it history. The dry atmosphere implies a dry interior for Venus, which is believed to make the outer shell on Venus too stiff to break into the plates observed on Earth.

Although plate tectonics does not operate on Venus, it is clearly an active planet with a relatively young surface and a wealth of volcanic and tectonic features. The majority of the planet is covered with volcanic features such as shield volcanoes and lava plains directly analogous to Earth's volcanic features. Many of the highland areas appear to form over mantle plumes, where hot material from the interior rises to the surface creating "hot spots" on the surface similar to Hawaii. In contrast, many of the tectonic features are unique to Venus. Examples include **coronae**, which are believed to result from small-scale **plumes** deforming the surface, and **tessera**, which are intensely deformed regions with multiple intersecting fracture sets.

1. History of Venus Exploration

Venus has long been observed as one of the brightest objects in the evening or morning sky. Transits of Venus had been used to determine its orbital period and diameter, and Lomonsov discovered that Venus had an atmosphere during the transit of 1761. But it was not until the 1960s that the modern exploration of Venus began, with observation by Earth-based radio telescopes. Radio telescopes at Arecibo in Puerto Rico and at Goldstone in California were used to accurately measure the rotation period and diameter of Venus. They also produced images of the surface that showed large, continent-sized regions. However, Earth-based radio telescopes were hindered by only being able to image the same side of Venus that faced Earth at inferior conjunction. [See THE SOLAR SYSTEM AT RADIO WAVELENGTHS.]

Spacecraft observation of Venus began in 1962 with a flyby by the *Mariner 2* spacecraft. It observed Venus from 34,833 km, determining a 468°C (874°F) surface temperature and observing that Venus lacked a magnetic field. In 1967, *Mariner 5* flew by Venus at an altitude of 4023 km, returning data on atmospheric composition and surface temperature. Also in 1967, the first probe entered the Venus atmosphere, when the Soviet Union's *Venera 4* returned

data for 93 minutes. The *Venera 5* and *6* probes followed in 1969, sending back more atmospheric measurements. Two more *Venera* probes followed in 1970 and 1972 making soft landings on the surface, with *Venera 8* in 1972 transmitting data on surface temperature, pressure, and composition. The *Venera 8* measurements were initially thought to be consistent with a granitic composition (see Section 5 for more discussion).

The next U.S. mission to observe Venus was *Mariner 10* in 1973, which was on its way to Mercury. *Mariner 10* provided observations of the atmospheric circulation of Venus with both visible and ultraviolet wavelengths. In 1975, the Soviet Union landed two more probes on the surface of Venus, *Venera 9* and *10*, sending back panoramas of the surface for the first time (see Section 2) and making detailed geochemical measurements. These landers measured surface compositions similar to terrestrial **basalts**.

The U.S. *Pioneer Venus* mission in 1978 consisted of an orbiter plus four atmospheric probes. The probes returned data on atmospheric circulation, composition, pressure, and temperature. The orbiter provided radar images of the surface, as well as a detailed global topographic map with a resolution of about 150 km. Major topographic regions such as Aphrodite Terra and Bell Regio were mapped, as were the 11 km high Maxwell Montes. The spacecraft was also used to map the gravity field of Venus.

The Soviets followed with 4 more soft landers between 1978 and 1981, with three of the landers returning surface panoramas and surface compositional information. The last two soft landers (*Venera 13* and *14*) returned color panoramas (see Section 2) and drilled into the surface for samples. The next two Soviet missions were orbiters, *Venera 15* and *16*, and returned synthetic-aperture radar (SAR) images of the northern hemisphere of Venus in 1983, with resolutions of about 5–10 km. This rich data set revealed new types of features on the surface of Venus, including tessera terrain and coronae (discussed later). *Vega 1* and *2* in 1984 carried balloon probes into the atmosphere and were the Soviet Union's last missions to Venus.

NASA's *Magellan* mission to Venus was launched in 1989 from the space shuttle and arrived at Venus in August of 1990. It obtained SAR images and altimetry of the surface between 1990 and 1994, mapping over 98% of the surface. The spacecraft also obtained high-resolution gravity field measurements, especially after the orbit was lowered and circularized in 1993. The 120 m resolution SAR images and 1–10 km resolution altimetry data completely unveiled the surface of Venus and provided a global data set that could be used to test models of the interior and surface evolution of the planet.

In 2005, the European Space Agency will launch its first mission to Venus: *Venus Express*. The mission focused on studying the composition and the circulation of the atmosphere of Venus.

2. General Characteristics

2.1 Orbital Rotations and Motions

Venus orbits the Sun in a nearly circular path once every 224.7 Earth days. It is the second planet from the Sun, located between Mercury and Earth. The plane of Venus' orbit is inclined to that of the Earth by 3.4°. Analysis of the obliquity of Venus reveals that it has a liquid core, similar to that of Earth. One day on Venus lasts 116.7 Earth days. The rotation of Venus on its axis is not only extremely slow but occurs in the opposite direction from all the other planets (retrograde rotation), so that the Sun rises in the west.

When visible, Venus is the brightest planet in the night sky due to its size and proximity to both the Sun and to Earth. It's easy visibility and the unusual pattern it makes in the night sky have given Venus a special place in **astrology** and made it an easy target for stargazers. Its proximity to the Sun means that it never rises very high in the sky, but it can often be seen as either the "evening star" in the west or the "morning star" in the east.

2.2 Radius, Topography, and Physiography

The radius of Venus is 6052 km, only 5% less than the equatorial radius of the Earth, 6378 km. The average density of Venus is 5230 kg/m³, somewhat higher than Earth's density. Thus, the acceleration of gravity at the surface is 8.87 m/s², 90% of Earth's. The radius of the Earth measured at the poles is approximately 21 km less than the radius at the equator. This difference is called the rotational bulge. The Earth's spin accelerates the equator more than the pole, causing the pole to be flattened and the equator to bulge out. The very slow rotation of Venus means that no such flattening occurs, making it, on average, nearly spherical.

The topography on Venus is dominated by plains, which cover at least 80% of the planet. There are also major highlands, including plateaus and topographic rises, as well as rifts and ridge belts that stand out from the background plains (see Fig. 1). Based on available data, the topography of Venus is unique in our solar system. Most of the smaller solid bodies, such as Mercury, Mars, and our Moon, as well as many satellites, bear the mark of numerous small craters and large impact basins, left from an earlier period in the history of our solar system when large impactors were common. As we will discuss in more detail later, the absence of impact basins and the small number of craters indicate that the surface of Venus is relatively young. On average, it is comparable to the age of the surface of the Earth.

Venus and Earth both have a large topographic range, which results from the intense geologic activity that the two planets have experienced. However, the distribution of elevations on the two planets is very different (see Fig. 2). Earth's topography is bimodal, while Venus' topography is

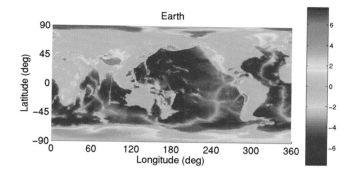

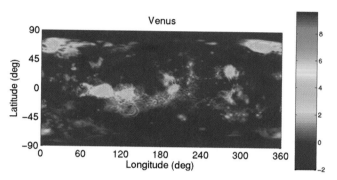

FIGURE 1 Topography of the Earth and Venus in a sinusoidal projection at a resolution of 1 pixel/degree. Note the long ridges that dissect many of Earth's ocean basins and the long mountain belts that are the signature of plate tectonics. Venus has numerous large highland regions, but the only long, quasi-linear mountain belts occur in the northernmost highland region, Isthar Terra.

unimodal. The two peaks on Earth reflect the division between oceans and continents. Venus has no ocean and, as we will discuss later, arguably no continents. The topography on Venus differs from that on Earth in other significant ways. Most importantly, Venus lacks the interconnected system of narrow midocean ridges and long linear mountain belts that are the hallmark of plate tectonics on Earth (see Fig. 1). The absence of these features on Venus reflects fundamental differences in evolution between the two planets, and will be discussed in greater detail later.

2.3 Surface Conditions

The surface conditions on Venus can best be described as hellish. The surface temperature at the mean planetary elevation is 437°C (867°F). The surface temperature at the highest elevations is approximately 10°C less. The surface pressure is 95 bars, equivalent to the pressure under almost 1 km of water. The surface temperature varies by only about 1°C over the course of a year due to the dense, insolating

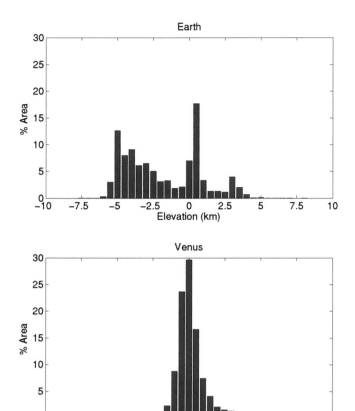

FIGURE 2 Histogram of the elevation in 0.5 km bins for Earth and Venus, normalized by area.

atmosphere. The atmosphere is 96.5% carbon dioxide, with lesser amounts of nitrogen, sulfur dioxide, argon, carbon monoxide, and water. The clouds are composed of 75% sulfuric acid and 25% water. [*See* VENUS: ATMOSPHERE.]

2.4 Views of the Surface

Four Soviet landers have returned views of the surface of Venus, *Venera 9, 10, 13,* and *14*. These panoramas showed relatively similar sites: rocky surfaces with varying amounts of sediment (Fig. 3). Rocks at each site tend to be relatively angular, suggesting minimal erosion and possible ejection from an impact crater. All the sites are consistent with a volcanic origin, showing platy lava flows that have been covered to varying extents by sediments. The sediments may be of impact origin, produced by aeolian erosion or by chemical weathering.

3. Impact Craters and Resurfacing History

There are approximately 940 identified impact craters on the surface of Venus. They range in diameter from approx-

imately 1.5 km to 268.7 km. The dense atmosphere on Venus causes impactors 1 km in diameter to break up before impacting the ground, reducing the number of craters 30 km in diameter. The shock waves that travel through these small objects can cause them to explode in a manner analogous to the Tunguska event on Earth. [*See* PLANETARY IMPACTS.] Atmospheric breakup and explosion, or other dynamic effects in the atmosphere, are believed to produce both radar-bright and radar-dark splotches on the surface (Fig. 4). The brightness of a radar image is primarily a function of how rough it is at the scale of the radar wavelength (for the *Magellan* radar, 12.6 cm). The darker the image, the smoother the surface. Very rough areas appear very bright. Rough areas reflect the signal back to the spacecraft, while smooth areas allow the radar waves to bounce off in a direction away from the spacecraft. Approximately 400 of these "splotch" regions have been identified. These regions are believed to be either areas where fine-grained material has been scoured away (radar-bright areas) or regions where relatively fine-grain material has settled out of the atmosphere (smooth, radar-dark areas). Additionally, most impact craters have associated with them dark parabolas, which are also part of fine-grained ejecta that are deposited out of the atmosphere.

In the absence of samples returned from planetary bodies, the only means of dating the surface is the analysis of the impact crater population. A great deal of work has been done on assessing the population of comets and asteroids available to impact the larger planetary bodies. [*See* COMET POPULATIONS AND COMETARY DYNAMICS; MAIN BELT ASTEROIDS; and PLANETARY IMPACTS.] Dating of samples returned from the Moon has been used to tie the record of lunar craters to an absolute age. The estimated flux of impactors on the Moon must be extrapolated to other bodies in the solar system, which have different dynamical environments and thus different expected rates for impacts. This introduces a major uncertainty into the estimated age of a surface based on impact crater counts. Another major factor is the history of the surface itself. Modification of a surface by erosion, deposition, or tectonism can decrease the number of identifiable craters. Erosion can also remove deposits that had covered a surface. Additionally, secondary craters can form when large blocks of material are ejected during an impact. Impactors can break up during entry to the atmosphere, producing multiple smaller impacts rather than a single large impact. Despite these issues and the resulting uncertainties, estimated surface age is a very important clue in deciphering the geologic history of a planet.

The estimated age of resurfacing on Venus is ∼750 million years (Ma). Given all the possible uncertainties in this age, estimates between 300 Ma and 1 Ga are permissible. This age is in contrast to ages of 3–4 Ga on average for Mars and the Moon. On Earth, new crust is continually forming along spreading centers in the oceanic crust. Continental

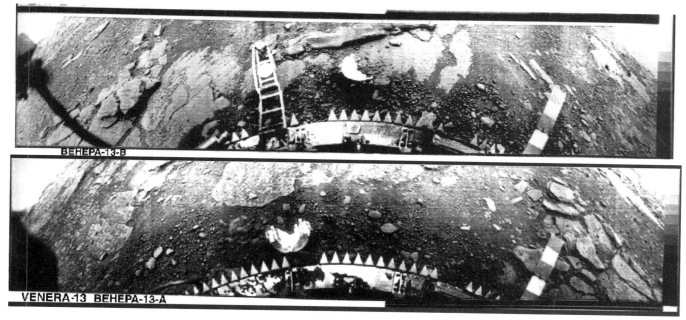

FIGURE 3   These photographs of the surface of Venus were obtained by the Soviet *Venera 13* spacecraft. *Venera 13* was the first of the *Venera* lander missions to include a color camera. The *Venera 13* lander touched down on March 3, 1982, near 305°E, 5°S, in the plains east of Phoebe Regio. The arm on the surface in the top image is a soil mechanics experiment. A color bar for calibration is visible in each image, as well as other spacecraft parts.

crust can be old as 4 Ga, but craters are erased by water and wind erosion much more rapidly than on the other terrestrial planets.

There are two highly intriguing characteristics of the Venus crater population. The first is that the distribution of craters cannot be distinguished from a random population. The second is that very few of the craters are modified by either volcanism or tectonism. Only ~17% of the total population is either volcanically embayed and/or tectonized. An example of a crater that is both embayed and tectonized is Baranamtarra (Fig. 5). The low number of modified craters on the surface of Venus means that there is little record of the process or processes that reset the surface age to be less than 1 Ga. If volcanic flows had covered the surface of Venus at a uniform rate, there would be more partially buried craters. This observation of the crater population initially led to the hypothesis of global, catastrophic resurfacing. Subsequent detailed modeling of resurfacing showed that the population is consistent with a wide range of resurfacing models, allowing for different size areas to resurface at different rates. The small number of modified craters is most consistent with resurfacing occurring as small, ~400 km diameter, regions. However, yet another variable is whether or not some craters with dark floors may in fact have been volcanically flooded. However, even if a larger number of craters have been modified by volcanism than initially estimated, the region covered by volcanism in these areas is small and still most consistent with resurfacing proceeding in small, local patches.

We can estimate the rate of volcanic resurfacing if we assume that craters have been removed by burial under volcanic flows. Crudely, if one takes the characteristic resurfacing age to be 750 Ma and the average crater rim height to be 0.5 km, then the rate of volcanic production is ~0.3 km³/year. Alternatively, if we consider the hypothesis that Venus resurfaced more quickly, in perhaps 50 Ma, the production rate is ~4.6 km³/year. The relative volume of lava extruded on the surface is believed to be small compared to the volume intruded into the subsurface, perhaps 10% of the total. Thus, the total volume of melt produced might be a factor of 10 larger. For comparison, the estimated rate of volcanism for intrusive and extrusive volcanism on Earth is 20 km³/year.

On planets with large numbers of craters, such as Mars, the surface age of local regions can be estimated from the crater populations. On Venus, some attempts have been made to determine the relative ages of either populations of specific types of geologic features or large areas on Venus. However, statistical analysis of this approach indicates that the very small number of craters on Venus makes attempts at dating particular landforms or even large areas not reliable. Although traditional crater-counting methods are not very useful, both the distribution of modified craters and the distribution of dark crater parabolas suggest some variation in surface age. In particular, the region with the highest

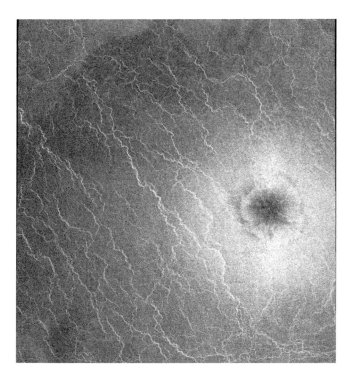

FIGURE 4   This radar image (approximately 125 × 140 km in size) shows an impact splotch with a dark center and a bright halo. The splotch is superimposed on a set of predominantly northwest-trending wrinkle ridges. The spacing between major ridges is roughly 10–20 km. These wrinkle ridges are part of the set of ridges that wraps around Western Eistla Regio.

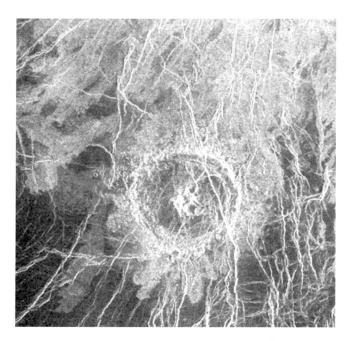

FIGURE 5   Crater Baranamtarra is both heavily embayed by volcanic flows and fractured. It is 25.5 km in diameter and centered at 17.94°N, 267.80°E.

density of volcanoes, coronae, and rifts appears to have a lower density of haloes and more modified craters, suggesting a younger age.

Overall, the crater population on Venus indicates it is a comparatively active planet, completely resurfaced within the last 1 Ga, possibly with resurfacing on-going today. Volcanic resurfacing rates are likely on the same order of magnitude as those on Earth, but are a function of the poorly constrained rate of resurfacing, which could be either constant or variable. The distribution and modification of the craters implies that there are limited differences in the ages of large regions on Venus, unlike the dichotomy between the age of oceanic and continental crust on the Earth. The small number of modified impact craters leaves few clues as to the process(es) that obliterated the earliest surface of Venus. Below we discuss the implications of resurfacing for the overall geologic evolution of Venus.

4. Interior Processes

One of the greatest curiosities about Venus is that its global-scale geologic processes are totally unlike that of Earth. On Earth, the system that shapes the Earth's large-scale physiography and the majority of geologic features is plate tectonics. The surface of the Earth is broken into dozens of plates that move over the surface of the Earth at rates of up to a few cm per year. The plates are tens to hundreds of kilometers thick. Mountain belts form where plates meet, such as where they collide, slide at an angle past each other, or where one plate is pushed into the **mantle** beneath another at subduction zones. Hot material wells up from the mantle below along narrow ridges in the ocean crust, creating new oceanic crust. These characteristics features are easily seen in the topography for Earth, even at the relatively low resolution available for Venus (Fig. 1). Venus clearly does not have plate tectonics. There is no evidence for this type of geologic process in the topography or in the radar images. [See EARTH AS A PLANET: SURFACE AND INTERIOR.]

The energy that drives plate tectonics and other geologic processes is predominantly generated by the decay of radioactive elements. For the terrestrial planets, the primary contributors to radioactive decay are uranium (U), thorium (Th), and potassium (K). Based on estimates of the abundance of these elements on Earth and in chondrites [see METEORITES], radioactive decay cannot account for the total amount of energy. In addition, a significant amount, perhaps 25%, of the heat lost from the interior results from cooling of the planet over time, with some additional contribution from the heat of initial planetary accretion. The heat in the interior of the planet is predominantly transmitted to the surface via convection in the interior. Convection in the mantle brings hot, low-density material from the interior to the surface, or near the surface, allowing it to cool.

Generally speaking, the larger a planet, the longer it will continue to lose energy and be geologically active. However, the details of the thermal evolution of a given body can be quite variable. Venus and Earth provide perhaps the quintessential example of variations in evolution. Most explanations of how Venus and Earth ended up on different geologic paths have to do with the history of volatiles. Volatiles, mainly in the form of water, play a key role in enabling plate tectonics on Earth. The presence of even a small amount of water in rock has a major effect on its strength and on the temperature at which it will melt. The water in the **lithosphere** is believed to be essential to making it weak enough to break into plates in response to the motions of convection in the interior. The **asthenosphere** is the upper part of the mantle, directly below the lithosphere, which has a lower viscosity than the rest of the mantle and acts to lubricate the motion of the plates at the surface of the Earth. The low viscosity of the asthenosphere may be a result of small amounts of melt. Melt would not be expected in the asthenosphere unless at least a small percentage of water is present. Thus, water appears to be an essential ingredient in the development of plate tectonics.

Measurements made to date indicate that the surface and atmosphere of Venus have very little water. In terms of the strength of the crust, the extremely high surface temperatures might be expected to offset the lack of water, making the crust extremely weak. However, laboratory studies of rock strength at Venus temperatures have shown that dry basalt (see Section 5) is stronger than wet basalt at Earth temperatures. This extreme strength of the crust on Venus likely contributes to the lack of lithosphere scale breaks that are required to form plates. As we discuss later, there is also evidence suggesting that Venus has no asthenosphere.

Recent studies have proposed that Venus exists in a "stagnant lid" convection mode rather than the "active lid" mode predicted for Earth. When convective stresses exceed the lithospheric strength, an active lid such as the terrestrial system of plate tectonics is predicted. On Earth, conditions such as weak, narrow fault zones, or the presence of a low-viscosity asthenosphere, allow the convective stresses to exceed the lithospheric strength. On Venus, the present-day lithospheric strength is apparently too high to allow plates to develop. This model is consistent with the loss of volatiles as key to differences on Venus and Earth.

Given the similarity in heat-producing elements and size between Earth and Venus and the absence of plate tectonics on Venus, how does Venus lose its heat? Venus must be convecting in its interior. As we will describe, although there is no evidence for plate tectonics, there is evidence that mantle plumes contribute to heat loss. On Earth, hot blobs of material form within the overall convecting pattern in the interior. These plumes form hot spots, such as the Hawaiian Island chain. The hot mantle material pushes up on the lithosphere, creating a broad topographic swell. The heat causes the lithosphere and crust to melt locally, thick-

ening the crust and forming surface volcanoes. On Earth, the majority of the heat is lost where the upwelling mantle creates new crust at midocean rises, and the cold lithosphere is pushed back into the mantle at subduction zones. Hot spots account for <10% of Earth's heat loss.

There are approximately 10 such hot spot features on Venus. These rises are Atla, Bell, Beta, Dione, W. Eistla, C. Eistla, E. Eistla, Imdr, Themis, and Laufey Regiones (Fig. 6). Those features believed to be active today, such as Atla, Beta, and Bell Regiones, have broad topographic swells, abundant volcanism, and strong, positive gravity signatures. Several rises also have rifts, such as Guor Linea at W. Eistla (Fig. 7). These features are characteristic of hot spots above a mantle plume. However, there are too few hot spot features on Venus (∼10 on Venus versus 10–30 on Earth) to account for a major portion of Venus' heat budget. In addition to the large-scale (1000–2000 km diameter) hot spots on Venus, there are also smaller scale (mean diameter of ∼250 km) features called coronae (see Section 7). There are ∼515 of these features, which are unique to Venus. There is considerable evidence that many, perhaps all, of these features form above small-scale plumes. However, even if all coronae represented small-scale plumes, they would not be able to account for more than about one quarter of the interior heat loss on Venus.

The relationship between the gravity and topography provides evidence that Venus does not have a low-viscosity asthenosphere. On Earth, a mantle plume must pass through the asthenosphere before reaching the lithosphere. (Note that there is some debate about the existence of an asthenospheric layer beneath the very thick continental lithosphere on Earth, but its existence below the oceanic lithosphere, where the majority of plumes are observed, is well accepted.) The plume tends to spread out in the relatively weak asthenospheric layer, resulting in a reduced amount of topographic uplift for a given plume size. Comparing the observed amount of uplift to the estimated size and depth of the low-density plume provides evidence for this behavior on Earth but not on Venus. On Venus, plumes strike the lithosphere directly, thus causing more uplift for a given plume size.

The relationship between the gravity and the topography provide some insight into interior structure and convection. The magnitude of variations in the gravity field as compared to a given topographic feature is an indication of the interior structure that supports a given topographic feature. The strength of the lithosphere can support topography. Variations in density in the interior can also support topography. A mountain can be supported by a thick 'root' of low-density crust, analogous to an iceberg floating in denser water. Variations in the mantle temperature associated with convection can also support topography. The gravity field of Venus has been carefully studied to estimate the thickness of the strong, or elastic, part of the lithosphere, the thickness of the crust, and the location of low–density,

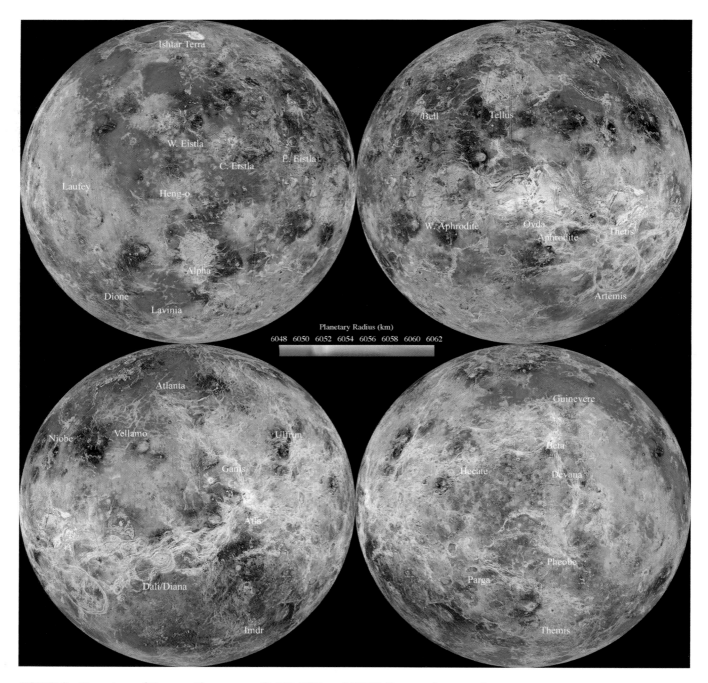

FIGURE 6   Four views of Venus, with centers at 0°, 90°, 180°, and 270° E. Topography is in color, with *Magellan* radar images overlain on top.

relatively hot regions in the mantle. Clearly some highlands, such as tessera plateaus (see section VII), are compensated by crustal roots. Many other highlands appear to be compensated by mantle plumes.

In addition to plumes, conduction through the lithosphere must contribute to the heat loss on Venus. The thinner the lithosphere, the more rapidly the planet loses heat. Estimates of the thickness of the lithosphere on Venus,

derived from gravity and topography, are typically 100–200+ km. This is comparable to the lithospheric thickness on Earth, and is too large to account for the majority of Venus' heat loss. There is growing evidence that the recycling of the lower lithosphere back into the mantle may help cool Venus, just as subduction helps cool the Earth. New models for corona formation show that at least some coronae may form above sites where the thickening, cold

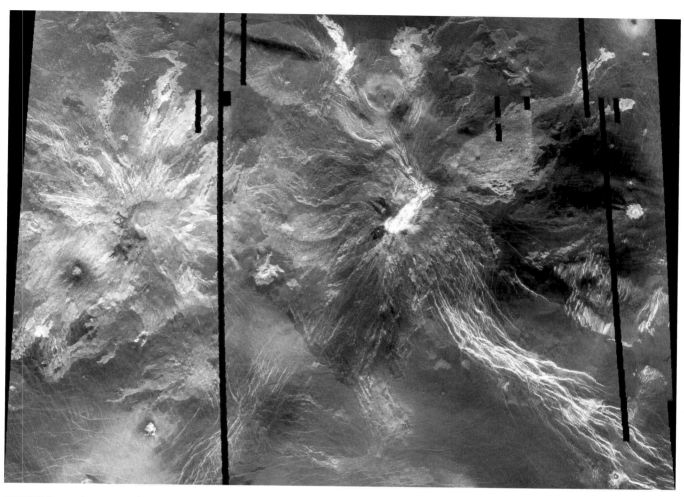

FIGURE 7 Radar image of W. Eistla Regio centered at 22°N, 354.5°E with dimensions of approximately 1725 × 1260 km. The western volcano is Sif Mons, 350 km in diameter, and the eastern volcano is Gula Mons, 450 km in diameter. Radial radar-bright and dark flows surround both volcanoes; radar-bright linear fractures of Guor Linea are seen in the southeastern corner.

lithosphere becomes too dense and breaks off into the mantle. A new estimate of lithospheric thickness variations also suggests that the lower lithosphere may thicken and become unstable locally. Although possibly important, such a process is not going to be nearly as efficient a cooling mechanism as subduction.

Volcanism, resulting from melting of the mantle and/or lithosphere and the rise of hot magma, can contribute to heat loss. As discussed earlier, Venus was completely resurfaced, most likely by volcanism within the last billion years. The cratering record shows that there has been less volcanism, on average, since resurfacing than the terrestrial average. Thus, volcanism cannot be a dominant mechanism of heat loss currently.

Another constraint on interior processes is the absence (or extremely low level) of a magnetic field. The *Mariner* flyby missions measured no magnetic field, indicating that, if present, the field must be <0.005 Gauss at the surface.

Most models of interior dynamos indicate that a planet must be losing large amounts of heat from the planet's metal core to provide enough energy for a dynamo. Some models have suggested that relatively rapid heat loss through plate tectonics is a good method of driving a dynamo. Thus, one possible scenario is that Venus had early plate tectonics and an active dynamo but eventually lost much of its water from the crust through volcanism to the atmosphere, where it was subsequently lost to space. This decrease in water increased the strength of the lithosphere to the point that tectonics ceased and the dynamo shut down. Heat is then lost primarily by conduction through the lithosphere, causing the mantle to heat up and increase the rate of volcanism, causing the planet to resurface. This idea is speculative, as there is no direct evidence for an early plate tectonic period.

The unusual cratering record on Venus indicates that the first 3.5 Ga of geologic history has been somehow erased, with a lower rate of resurfacing occurring subsequently.

In contrast, Mars, Mercury, and the Moon have surfaces that preserve the large impact basins from early bombardment and reflect a gradual loss of heat and decline in geologic activity. Some models have proposed that resurfacing on Venus occurs episodically. In one scenario, the lithosphere thickens and becomes denser due to both cooling and chemical phase transitions. The lithosphere is predicted to founder, or get mixed into the mantle, when it becomes gravitationally unstable. However, how the lithosphere actually breaks and initiates this process is unclear. In another scenario, the stagnant lid heat insulates the mantle, causing it to heat up to the point that widespread melting occurs, eventually erupting on the surface. Other models show that volcanism that is globally distributed and resurfaces small regions in each event can produce the observed distribution. High mantle temperatures could facilitate this kind of widespread volcanism.

5. Composition

5.1 Global Implications

The similarity between Venus and Earth in terms of size and location in the solar system indicates that their bulk compositions should be comparable. The exact composition of the crust is related to the composition and temperature in the interior of the planet when the rock melts, as well how much of the original rock is melted. The typical rock type that forms on Earth when the interior melts and erupts is basalt. Thus, it is not surprising that geochemical measurements on the surface of Venus have a gross composition similar to terrestrial basalts, with some variation. On Earth, basalts make up the majority of the oceanic crust and are found in volcanic regions of continents. When processes such as subduction remelt basalts, the resulting rocks are enriched in silica (SiO_2). Continental rocks are a result of billions of years of remelting of a basaltic crust driven by convective and plate tectonic processes. They are lower density than basalt due to the enrichment of silica relative to iron and magnesium. The presence of at least small amounts of water may be essential to the formation of such silica-rich rocks. Continents stand higher than the oceanic crust due to both their lower density and the greater thickness of continental crust. As we will discuss, there is limited evidence for silica-rich rock on Venus.

The abundances of primary mineral-forming and radiogenic elements were measured by spectrometers on *Venera* landers. *Venera 8, 9,* and *10* and *Vega 1* and *2* landers measured the amounts of uranium (U), thorium (Th), and potassium (K) using a gamma ray spectrometer (Table 1). The *Venera 13* and *14* landers measured these elements as well the major–element forming minerals (see Table 2). Due to the orbital dynamics of delivering probes to the surface of Venus in any given time period, *Venera 8–14* landers are

TABLE 1	Abundances of Primary Mineral-Forming and Radiogenic Elements[a]		
Lander	U (ppm)	Th (ppm)	K (weight percent)
Venera 8	2.2 ± 0.7	6.5 ± 0.2	4.0 ± 1.2
Venera 9	0.6 ± 0.2	3.6 ± 0.4	0.5 ± 0.1
Venera 10	0.5 ± 0.3	0.7 ± 0.3	0.3 ± 0.2
Vega 1	0.68 ± 0.47	1.5 ± 1.2	0.5 ± 0.3
Vega 2	0.68 ± 0.38	2.0 ± 1.0	0.4 ± 0.2

[a] The abundances for each element represent an interpretation of the most likely minerals on the surface of Venus, as is standard practice in geochemical analysis.

all located in a relatively small region on Venus within 270–330°E and 15°S–30°N. This area includes the eastern flank of Beta Regio, a major hot spot, and the plains to the east of Beta and Pheobe Regiones. The *Vega 1* and *2* landers, sent at an earlier time, are located near 170°E, 10°N and 180°E, 10°S to the west of Atla Regio.

The silica content and the relative abundances of iron and magnesium for rocks at the *Venera* lander sites (Table 1) are characteristic of basalt. Although some variations in composition do exist, when the overall abundance of elements is considered in the context of minerals that occur stably together, all the rock compositions are consistent with a basaltic composition. Early analysis of the relatively high value of U, Th, and K at *Venera 8* suggested that this location was composed of a more silica-rich rock, possibly even **granite**. However, subsequent analyses have discounted this idea and concluded that the *Venera 8* site is mostly basaltic, although it has more silica than other lander sites.

TABLE 2	Elements and Major–Element Forming Minerals Measured by the *Venera 13* and *14* Landers[a]		
Constituent	*Venera 13*	*Venera 14*	*Vega 2*
SiO_2	45.1 ± 3.0	48.7 ± 3.6	45.6 ± 3.2
TiO_2	1.59 ± 0.45	1.25 ± 0.41	0.2 ± 10.1
Al_2O_3	15.8 ± 3.0	17.9 ± 2.6	16.0 ± 1.8
FeO	9.3 ± 2.2	8.8 ± 1.8	7.74 ± 1.1
MnO	0.2 ± 0.1	0.16 ± 0.08	0.14 ± 0.12
MgO	11.4 ± 6.2	8.1 ± 3.3	11.5 ± 3.7
CaO	7.1 ± 0.96	10.3 ± 1.2	7/5 ± 0.7
K_2O	4.0 ± 0.63	0.2 ± 0.07	0.1 ± 0.08
S	0.65 ± 0.4	0.35 ± 0.31	1.9 ± 0.6
Cl	<0.3	<0.4	<0.3

[a] Values are in weight percent. The raw data is converted from measurements of elemental abundance into likely chemical combinations.

Variations in elemental abundance do suggest that some real differences exist. The bulk composition of Venus can be extrapolated from these measures. Within the uncertainties, the composition is similar to that of Earth. Similarly available data on Fe/Mg and Fe/Mn suggest that the core composition is similar to Earth's. Some variation may occur after the rock forms. For example, the amount of Al, Ti, Ca, or Si may change through chemical weathering or metamorphism when the rock experiences changes in pressure and/or temperature.

The initial chemical measurements of the surface have provided invaluable constraints on the surface composition. However, the overall number and geographic diversity of sites remains limited. The precision of the measurements that were possible with instrumentation built in the 1970s is very low compared with measurements possible today. The uncertainties in the measurements mean that numerous questions such as the size of the core (which is constrained by the ratios of Fe/Mn/Mg) and the amount of crustal recycling cannot be addressed. In fact, the uncertainties in the Venusian measurements are so large that they encompass the entire range of composition for basalts on Earth, Mars, the Moon, and meteorites. In contrast, basalts from the Moon and Mars (as represented in meteorites) have a distinct chemical signature from Earth. [See METEORITES.] These variations represent key differences in the formation and evolution of these bodies, such as the formation of a magma ocean on the Moon.

In addition to direct measurements of the composition, morphology can be used as a very crude indication of composition. For example, lavas with a basaltic composition tend to be very fluid, forming long, narrow flows and broad, low volcanoes. As the silica content increases, the viscosity of the lava increases. The thicknesses of flows increase, their lengths decrease, and the slopes of volcanoes formed increases. Terrestrial examples are Mauna Loa in Hawaii (basaltic) and Mt. St. Helens in Washington (more silica-rich). On Venus, the morphology of flows is generally consistent with low-viscosity, basaltic compositions. There are some features that appear to represent much thicker, shorter flows (see description of "pancakes" in Section 6). However, these morphologies cannot be considered diagnostic of composition as factors such as the volume and rate of material erupting, the atmospheric pressure during eruption, and the amount of gas in the lava also shape the morphology of the flow.

5.2 Surface Weathering

Although weathering of the surface by wind is relatively mild on Venus as compared to Earth, the environment for chemical weathering is extremely harsh. In addition to the searing temperature and high pressure, the atmosphere contains highly corrosive and chemically active gases such as SO_2 (sulfuric acid), CO, OCS, HCl (hydrochloric acid)

and CO_2. A variety of minerals form in laboratory experiments that simulate Venus conditions, such as wollastonite, anhydrite, and hematite, but no landers have measured actual minerals. Measurement of the specific minerals present and their abundances is highly desirable as they provide insight into the nature of the chemical interaction between the surface and the atmosphere. This information is a critical piece of understanding the larger problem of how Venus arrived at the hellish climate that now exists.

One of the key questions is how much CO_2 is trapped as carbonates on the surface of Venus. Most of the CO_2 found on Earth is trapped as carbonates via biological processes, specifically the formation and accumulation of seashells. This process is an important element of the overall balance that makes Earth habitable. Available information from surface composition and laboratory experiments suggests that significant amounts of carbonates could be present on the surface of Venus, perhaps up to 10%. If so, this would mean that CO_2 in surface rocks is an important part of determining the atmospheric pressure and composition. Another key question is how atmospheric SO_2 interacts with the surface. On Earth, most of the SO_2 is dissolved in the oceans. Rates of chemical reactions involving SO_2 are known for the conditions in the atmosphere of Venus and predict that the SO_2 in the present-day sulfuric acid clouds on Venus should react with other chemicals and disappear over time. This analysis indicates that SO_2 should disappear from the atmosphere within 2 Ma. The fact that sulfuric acid clouds are present today implies that new sulfur gases have been added to the atmosphere with this time by volcanic eruptions. Other important measurements for understanding the surface–atmosphere interactions are the oxidation state of iron minerals and minerals that react with hydrogen chloride (HCl) or hydrogen fluoride.

6. Volcanism

With the exception of Jupiter's moon Io, Venus is the most volcanic world in the solar system. Volcanic features of a broad range in morphology cover the surface, from sheetlike expanses of lava flows to volcanoes shaped like pancakes and ticks, as illustrated later. The high surface temperature and pressure on Venus make explosive volcanism less likely, though some possible deposits produced by explosive volcanism have been mapped. The extreme conditions on Venus also result in volcanoes that tend to be taller and broader than those on Earth or Mars. *Magellan* data illustrated that volcanic features do not occur in chains or specific patterns, indicating the lack of plate tectonics on Venus.

The plains or low-lying regions on Venus are covered by sheet and digitate deposits that are interpreted to be volcanic in origin (Fig. 8). These extensive deposits are likely to be flood basalts, formed in similar ways to the Columbia

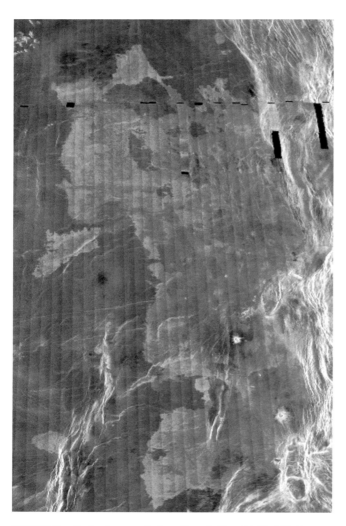

FIGURE 8   Radar image of a lava flow field at 60°N, 183°E in the plains of Venus. The flow field is approximately 540 × 900 km. The name of the flow field is Mamapacha Fluctus, and it is made up of lava flows of moderate radar brightness or moderate roughness.

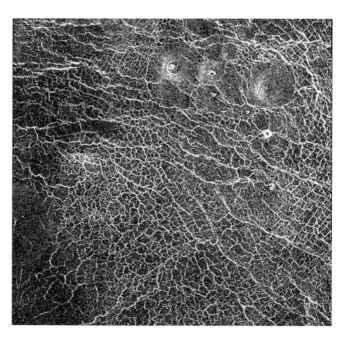

FIGURE 9   Radar image of small shield volcanoes and polygonal terrain, ~30 by 30 km, centered at 28.8°N, 142.2°E. Polygons range in size from the limit of resolution to several kilometers in diameter. The volcanoes at the north overlay the polygons. Polygons are superimposed on the volcanoes in center right of the image, where calderas indicate the top of the volcanoes. On the western side of the image, various volcanic flows bury polygons. Thus, the formation of the polygons appears synchronous with the volcanism in this region.

River basalts or the Deccan Traps on Earth. In some plains regions, the surface is clearly built up of multiple, superposed lava flow deposits, while other regions are more featureless. Lava flows have varying brightness in the *Magellan* SAR images. Most lava flows are of intermediate brightness. Comparisons to radar images of lava flows in Hawaii indicate that the venusian flows have similar roughness, though some flows on Venus are unusually smooth.

The plains are also covered with abundant small (<5 km across) shield and cone-shaped volcanoes (e.g. Fig. 9). Thousands of these volcanoes have been mapped, and they may contribute as much as 15% of the plains volcanic deposits. Other flows in the plains may have originated at fissures, which were then obscured by later eruptions. Timing of the plains flows is a subject of debate, with some advocating that the plains formed relatively synchronously across Venus in a single resurfacing event. Others argue that the data support a slower, nonsynchronous formation for the plains. Unfortunately, the impact crater population can be interpreted to support either hypothesis, and it will take future mission data to constrain plains formation on Venus.

Large volcanoes on Venus (those with diameters >100 km) are found at topographic rises, along rift zones, and concentrated in the region bounded by Beta Regio, Atla Regio, and Themis Regiones. Over 100 large volcanoes have been identified. Large volcanoes have average heights of about 1.5 km and aprons of lava flows that extend hundreds of kilometers from their summits. Maat Mons, the largest volcano on Venus is about 8.5 km high and 400 km across (Fig. 10). In comparison, Mauna Loa, the largest volcano on Earth, is about 9 km high and 100 km across. Detailed studies of individual large volcanoes have revealed their complex histories. Many volcanoes show evidence of multiple eruptions from their summits as well as sites on their flanks. Some large volcanoes have calderas at their summits similar to volcanoes on Earth and Mars, formed by collapse of the underlying magma chamber. Others have radially fractured summits, with the radial fractures interpreted

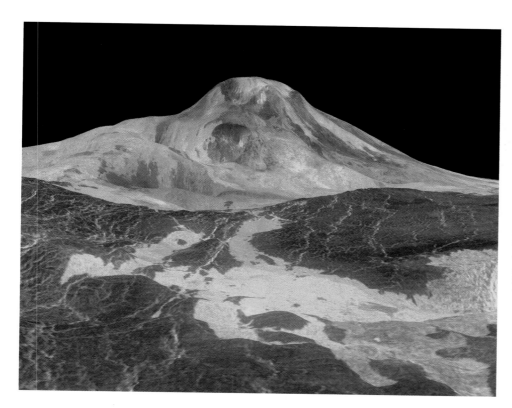

FIGURE 10   Lava flows extend for hundreds of kilometers across the fractured plains shown in the foreground to the base of Maat Mons, which is located at about 0.9°N latitude, 194.5°E longitude. Magellan data was combined with radar altimetry to develop a 3-dimensional map of the surface. The vertical scale in this perspective has been exaggerated 22.5 times. The simulated red color is based on images recorded by the Soviet *Venera 13* and *14* spacecraft that indicate the atmosphere on Venus would make the surface appear red to our eyes. (The image was produced at the JPL Multimission Image Processing Laboratory.)

as the surface expression of subsurface dikes. These dike sets provide evidence that many large volcanoes have undergone multiple episodes of intrusion and extrusion.

At the smaller end of the scale, volcanoes 5–50 km across are also abundant on the surface of Venus (Fig. 11). Many of the volcanoes resemble their terrestrial counterparts, with summit calderas and radiating digitate flows. Venus also has several types of volcanic features that differ from those on Earth and other planets. The steep-sided or pancake domes are flat-topped, with steep sides (Fig. 12), similar to the flat-topped Inyo domes in California that are formed by silicic lavas. The Venus domes may have a different composition though, as they are much larger and have smooth rather than blocky surfaces in comparison to the terrestrial domes. Other unusual volcanoes on Venus resemble ticks, or bottle caps. These small domes have scalloped margins and are interpreted to be steep-sided domes whose margins have collapsed.

The *Magellan* radar also imaged channels, a few kilometers wide and hundreds of kilometers long. The channels are found many places within the plains, tend to be very sinuous, and in places show evidence of levees and flow breakouts. The channels have formed by lava of some unusual composition, so fluid that it behaved like water and able to flow long distances without cooling. A number of compositions have been proposed, including carbonate or sulfur-rich lavas and ultramafic silicate melts. Others have suggested that the channels were formed by erosion of the surface by lava, similar to lunar rilles on the Moon. [*See* THE MOON.] Some of the channels extend for long distances, allowing them to be used as a time marker, as it can be assumed that the channel formed over a relatively short period of time. For example, the channel may superpose one feature, but be overlain or cut by another. Also, a few channels now trend uphill, indicating that the surface deformed after they formed.

Are volcanoes on Venus still active? There are over 1500 active volcanoes on Earth, but Venus probably has fewer active volcanoes. Gravity studies indicate that a number of volcanoes may be dynamically supported, and thus still active. A decline in SO_2 over time observed by the *Pioneer Venus* spacecraft has been interpreted to possibly indicate a relatively recent eruption. In addition, volcanism within the last 10–50 Ma is supported by climate models. Future missions to Venus monitoring the atmosphere may be able to detect a future venusian eruption.

7. Tectonics

For the larger terrestrial planets, Venus, Earth, and Mars, mantle convection is the primary driving force for tectonic processes. On Mars, most tectonic structures are associated with either the gigantic Tharsis rise or the global dichotomy. [*See* MARS SURFACE AND INTERIOR.] The global dichotomy divides the smoother northern lowlands from

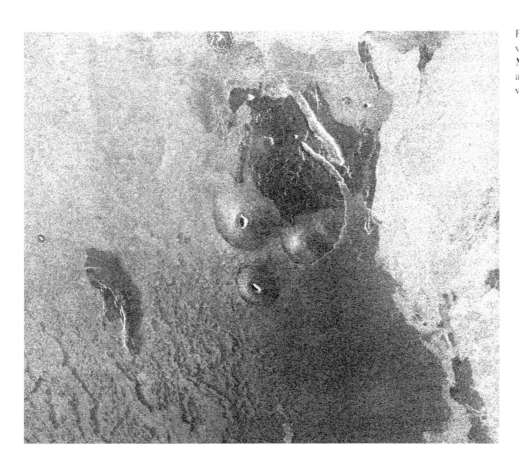

FIGURE 11   Radar image of small volcanoes on the flank of Maat Mons. The image is centered at about 3.2° N, 194.9° E, and is 90 km wide and 80 km long.

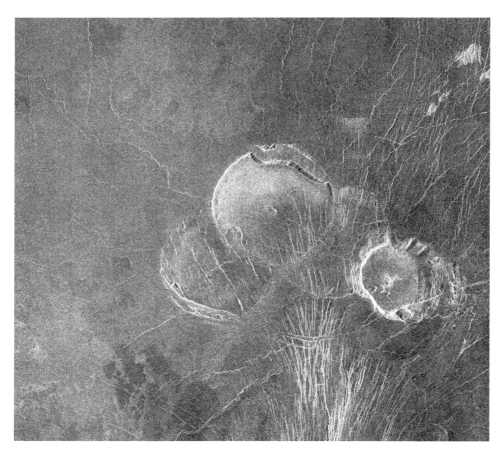

FIGURE 12   This image shows two steep-sided and one scalloped-margin domes in the plains of Venus. At the center of the image is a 50 km dome that overlaps another feature to the southwest that is about 45 km in diameter. This volcano is cut by many fractures. The southeastern volcano (25 km diameter) has scalloped edges that give this feature a bottle cap- or tick-like appearance. The scalloped edges are interpreted to form when material slides off the volcano margin.

the heavily cratered southern highlands. On Earth, plate tectonics is clearly dominant. Tectonic features on Venus are highly variable and enigmatic. Tessera terrains are unique to Venus and are defined as having multiple intersecting deformation directions. One possible factor in creating these highly deformed regions is that Venus experiences very little surface erosion. In contrast, most continental regions have experienced multiple episodes of deformation, but surface structures are often eroded between events, leaving evidence of only the most recent occurrence. Many of the tectonic features on Venus are continuous for thousands of kilometers and likely reflect underlying mantle processes including upwelling, downwelling, and horizontal flow. We describe next the characteristics and likely origins of the key types of tectonic features on Venus.

7.1 Tessera and Crustal Plateaus

Tessera terrains are highly deformed and thus stand out as very bright in radar images (Fig. 13). They are made up of both extensional and compressional deformational features. Each set of lineations may represent a separate deformation event, or two sets may form simultaneously if shear deformation is involved. In some cases, the sequence of events can be determined, but more often it is ambiguous. Tesserae occur both as isolated fragments embayed by later plains material and in major plateaus. There are 6 major crustal plateaus: Alpha, Ovda, Pheobe, Thetis, and Tellus Regiones plus Ishtar Terra. Figure 14 shows Alpha Regio, one of the smaller highland plateaus. Western Ovda Regio may be a relaxed crustal plateau. These plateaus are 1000–3000 km in diameter and 0.5–4 km higher than the surrounding plains. Their gravity signature indicates that they are supported by crustal roots rather than active mantle processes.

Ishtar Terra is unique among the highland plateaus. It is the largest of the crustal plateaus, and is surrounded by significant mountain belts on 3 sides, with large areas of tesserae occurring on their exterior flanks. They are Venus' only real mountain belts. Lakshmi Planum makes up the interior of Ishtar Terra. This smooth plateau is elevated 3–4 km above the surrounding plains and is covered by volcanic flows. The Maxwell Montes to the east of Lakshmi Planum contain the highest point on Venus, at approximately 11 km above the mean planetary radius (see Fig. 6). Although other crustal plateaus tend to have relatively flat interiors and rims of higher topography, no other crustal plateau is as extreme as Ishtar Terra in terms of its diameter, elevation, and circumferential deformation features.

Crustal plateaus have been proposed to form over mantle upwellings and over mantle downwellings. In the mantle upwelling scenario, a plume creates a crustal plateau though decompression melting above the plume head, analogously to plateaus formed on the terrestrial seafloor. Deformation occurs as the topography viscously relaxes. The

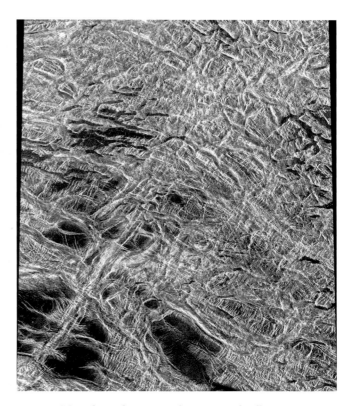

FIGURE 13 This radar image of a portion of Tellus Regio is centered at 36°N, 79.4°E and is approximately 340 × 420 km. The area is deformed by a northeast and a northwest set of lineations. Locally, each set contains both narrow, linear fractures resembling extensional graben and areas where the fractures coalesce into ridges and appear to be compressional ridges. In the northern section of the image, a third set of very narrow north northwest-trending fractures cross cuts the other sets. The dark regions are volcanically flooded valleys, with two small vents visible in the southwest corner of the image.

alternative model forms the plateaus above a cold, sinking mantle downwelling. On Earth, both subduction zones and local sites of downwelling form below cold mountain roots. Venusian crustal plateaus are proposed to form as a downwelling causes sinking of the lower lithosphere and accumulation and compression of the crust at the surface. The mechanism for forming small, local regions of tessera is not clear. In many cases, these regions are embayed and thus appear to be old and possibly inactive. There are few clues as to original processes that cause deformation. One possibility is that these areas represent sections of tessera plateaus that were once elevated but have topographically relaxed. If plateaus formed in an earlier, hotter time period, relaxation may have proceeded more rapidly, allowing for complete relaxation of plateaus. The semicircular rim of Western Ovda Terra could be the remnant of a relaxed plateau. Alternatively, small tessera terrains may be a result of an entirely different type of tectonic event, such as ridge belt formation.

FIGURE 14   This false color, perspective radar image of Alpha Regio is approximately 2000 km across and is centered at 25°S, 5°E. The blank strip is a data gap. The texture of the deformed regions is similar to that of Tellus Regio (Fig. 13). A corona is located at the southwest edge of Alpha. To the west are several small pancake domes. An impact crater is seen on the western margin of Alpha.

7.2 Chasmata and Fracture Belts

Chasmata (*chasma* means canyon) are regions of extensional deformation, as indicated by their locally low topography and **graben** or graben-like morphology. There are 5 major chasmata on Venus that extend for thousands of kilometers and are several kilometers deep: Parga, Hecate (see Fig. 15), Dali/Diana, Devana, and Ganis Chasmata. The fracture zones in these regions are typically ~200 km wide, with topographic troughs that are generally narrower, with widths of ~50–80 km. There are 7 smaller chasmata, with lengths of hundreds of kilometers and proportionately narrower fracture belts and troughs. Several of the chasmata occur on the flanks of hot spot rises and may be a result of topographic uplift above a plume. The majority of other chasmata form synchronously with coronae, as discussed earlier. Although chasmata are not required for coronae to form, nor vise versa, it is clear that the presence of one increases the likelihood of the other. Both extension and upwelling plumes can thin the lithosphere, which may focus additional extension and upwelling in an area.

Fracture belts appear similar to minor chasmata but are less intensely fractured, implying lesser amounts of extension. A curious feature of fracture belts is that they are topographically broad swells rather than topographic lows. The positive relief suggests that they went through a compressional stage, and that the fractures may be due to topographic uplift rather than regional extension.

7.3 Coronae

Coronae are large (>100 km across) circular features surrounded by concentric ridges and fractures (Fig. 16). Over 500 coronae have been identified on Venus; the largest one is Artemis Corona at 2500 km across. Coronae often have volcanoes in their interiors, and many are surrounded by extensive lava flows. Coronae tend to be raised at least 1 km above the surrounding plains, but others are depressions, rimmed depressions or rimmed plateaus. Most coronae are located along rift or chasmata systems, although some are at topographic rises, and others occur in the plains away from other features. Coronae are thought to form over thermal plumes or rising hot blobs, smaller in scale and probably rising from shallower depths than the plumes that form topographic rises. The wide range in corona topographic shapes indicates that coronae evolution also involves delamination or sinking of lithospheric material in its later stages. Studies of the gravity signatures of some large coronae indicate that many coronae are likely to be isostatically compensated, and thus probably inactive. The fact that we do not see coronae on Earth may be due to the lack of an asthenosphere on Venus.

7.4 Ridge Belts and Wrinkle Ridges

Ridge belts occur in a variety of morphologies and are distributed around the planet. Based on the morphology of

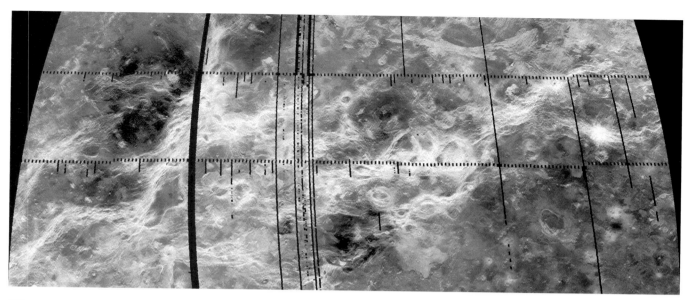

FIGURE 15   Radar image of Hecate Chasma, approximately 7000 × 3000 km, centered at 16°N, 240°E. Hecate Chasma is a huge tectonic feature, stretching from Atla Regio to Beta Regio. The rift is very bright in radar and has a wispy appearance. The rift comprises numerous branches at a range of orientations, with coronae present throughout the region, both on and off the rift.

FIGURE 16   Radar image of Heng-o and Beltis Coronae (B). The topographic rim of Heng-o, which corresponds approximately to the fracture annulus deforms the local regional plains. To the west lie extensive flow fields interpreted to originate from Beltis Corona (to the northwest of Heng-o) and the western annulus of Heng-o. Three volcanic centers of different ages lie within the annulus of Heng-o. Black lines and boxes are gaps in the data. Curved black edges result from the sinusoidal projection.

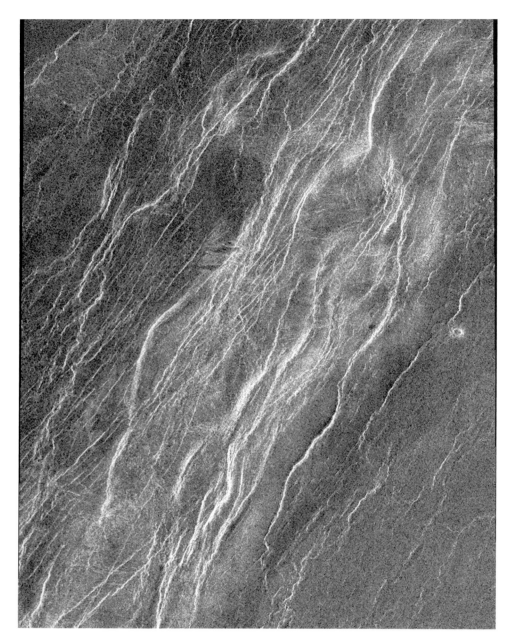

FIGURE 17   Radar image of a ridge belt in Atalanta/Vinmara Planitia, approximately 85 × 110 km in dimension and centered at 41°N, 196°E. The belt comprises a series of northeast-trending ridges. There are both very narrow ridges, down to the resolution of the data, with ridges several kilometers wide.

individual fractures and the long, narrow topographic highs that comprise individual ridges, ridge belts are interpreted to be a result of compressional stresses (see Fig. 17). Individual ridges are typically less than 0.5 km high, 10–20 km wide, and 100–200 km long with a spacing of ~25 km. The two largest concentrations of ridge belts occur in Atalanta/Vinmara Planitiae and Lavinia Plantia. The belts in Atalanta/Vinmara Planitiae are roughly an order of magnitude larger than those elsewhere. Belts in Livinia are unusual in that they have extensional fractures roughly parallel to compressional features within the same belt, possibly due to topographic uplift along the ridge. Larger belts are believed to result from mantle downwelling, similarly to the proposed downwelling origin for crustal plateaus, but with lower strain. Smaller belts may be associated with more local scale tectonics.

Wrinkle ridges are extremely common features on Venus and are also interpreted as simple compressional folds and or faults but are much narrower (~1 km or less in width) than ridges. They have positive relief, based on the fact that lava flows can be seen to pond against some wrinkle ridges, but that relief is too small to be seen in *Magellan* altimetry. Most ridges occur in evenly spaced set, 20–40 km apart. These sets of wrinkle ridges can be local in nature,

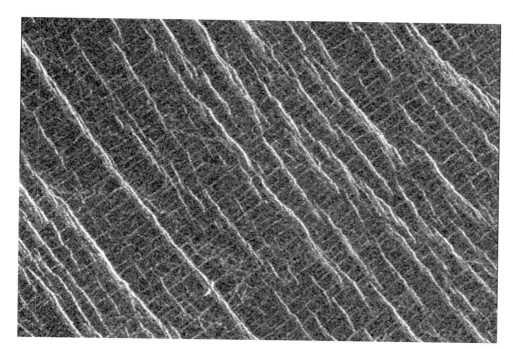

FIGURE 18   Radar image of a section of lineated plains approximately 35 km across, centered at 30°N, 333.3°E. These gridded plains are located in Guinevere Planitia and are incredible uniform in orientation, size, and space over nearly 1000 km.

associated with a corona for example, but more commonly they cover thousands of kilometers. These larger sets are likely to be gravitational spreading of high topography into lower regions and can be seen to form rings around some large topographic features (see Fig. 4). Other sets cannot be clearly associated with topographic highs. One hypothesis is that these features result from thermal contraction due to climate-change driven atmospheric temperature changes. In some regions, there are two sets of wrinkle ridges, although one set is usually better developed.

7.5 Plains Fractures, Grids, and Polygons

A wide range of long, narrow, approximately straight fractures occur in the plains. Some fractures are wide enough to be resolved as graben, but most are too narrow (less than 0.5 km) to be resolved as more than fractures. Most are interpreted as extensional fractures because they parallel resolvable graben and because of their shape. Some are clearly associated with local features such as volcanoes or corona and are probably due to extension above dikes. In some locations, there are either single sets or intersecting grids of fractures that cover hundreds of kilometers (Fig. 18). They are very regularly spaced, with separations of 1–2.5 km. The narrow spacing suggests that a thin layer is involved in the deformation. It is not obvious how a uniform stress can be transmitted to such a thin layer over such a broad regions. Shear deformation be required to produce grids of intersecting lineations.

Another type of extensional feature observed on Venus is polygons, which are found in over 200 locations on Venus.

These features are analogous to mud cracks in that they form in a uniform, extensional stress field. However, they form not as water is lost but instead when rock cools and contracts. The typical diameter is ~2 km, but some are up to 25 km across. Some areas have multiple scales of deformation. Again, some of these features can be associated with local events such as volcanoes, but others cover very broad regions and do not have an obvious origin. Polygons are most commonly associated with small volcanic edifices, and frequently appear to form synchronously (Fig. 9). Some may form by actual cooling of lava flows. Such basaltic columns are common on Earth, but the scale of the features found on Venus is orders of magnitude larger, implying that the flow thickness on Venus would probably to too large to be plausible. Another mechanism, as proposed for wrinkle ridges, is the possible heating and cooling of the upper crust due to climate change.

8. Summary

Venus provides a unique window in to the evolution of terrestrial planets. It is essentially identical to Earth in size and bulk composition, yet its geologic history is entirely different. Venus' level of geologic activity over the last billion years is comparable to that of Earth and exhibits many of the same geologic processes. The convecting interior drives geologic activity at the surface, creating a dozen major highlands. These highlands include hot spots, which form above mantle plumes, and the more enigmatic and intensely deformed highland plateaus. The majority of the surface is composed of vast volcanic plains along with nearly ubiquitous tectonic features. There are tens of thousands of volcanic features from small-scale (hundreds of meters) flows, vents, and shields, to hundreds of large-scale (>100 km) shield volcanoes that blanket the surface. The pervasive volcanism may have buried the earliest, heavily cratered surfaces, or they may have been destroyed through tectonic processes. Tectonic features range in scale from pervasive linear fractures and polygons at the limit of resolution to highland plateaus composed of tessera terrain 1000–2000 km in diameter.

Despite the similarities between Venus and Earth, Earth is the only body in our solar system that developed the system of plate tectonics that has so shaped the geologic and environmental evolution our planet. The atmosphere of Venus lost nearly all of its water early in its evolution. The loss appears to have affected the interior as well, causing the lithosphere to be too strong to break into the plates observed on Earth, and the asthenosphere to be too strong to facilitate rapid horizontal plate motion. This same loss of water has contributed to the dominance of CO_2 in the atmosphere and the resulting greenhouse effect that created the scorching surface conditions. Why Venus lost its water is not understood, but as with Mars, the absence of a magnetic field exposes the atmosphere to erosion by solar wind. In turn, a planet must be losing heat rapidly enough to drive the formation of a magnetic dynamo. The interior volatile content affects the processes through which planets lose heat and appears to be the key to whether or not plate tectonics develops. Was Venus originally on the same evolutionary path as Earth? What was the pivotal event or process that sent Venus down an alternate path to the hellish, uninhabitable planet we observe today? We can begin to address these questions, thus better understanding the evolution of our own planet, through future missions to understand the coupled evolution of the atmosphere, surface, and interior.

Bibliography

Brougher, S. W., Hunten, D. M., and Phillips. R. J., eds. (1997). "Venus II." Univ. Arizona Press, Tucson.

Fegley, Jr., B., and Treiman, A. H. (1992). Chemistry of atmosphere–surface interactions on Venus and Mars. In *"Venus and Mars: Atmospheres, Ionospheres, and Solar Wind Interactions"* (J. G. Luhmann, M. Tatrallyay, and R. G. Pepin, eds.), pp. 7–71. Geophysical Monograph No. 66. American Geophysical Union, Washington D.C.

Venus data are available through the Planetary Data System at pds.nasa.gov.

Earth as a Planet: Atmosphere and Oceans

Timothy E. Dowling
University of Louisville
Louisville, Kentucky

Adam P. Showman
University of Arizona
Tucson, Arizona

CHAPTER 9

Earth is the only planet that orbits the Sun in the distance range within which water occurs in all three of its phases at the surface (as solid ice caps, liquid oceans, and atmospheric water vapor), which results in several unusual characteristics. Earth is unique in the solar system in exhibiting a global ocean at the surface, which covers almost three quarters of the planet's area (such that the total amount of dry land is about equal to the surface area of Mars). The ocean exerts a strong control over the planet's climate by transporting heat from equator to pole, interacting with the atmosphere chemically and mechanically, and, on geological timescales, influencing the exchange of **volatiles** between the planet's atmosphere and interior. The Earth's atmosphere follows the general pattern of a troposphere at the bottom, a stratosphere in the middle, and a thermosphere at the top. There is the usual east–west organization of winds, but with north–south and temporal fluctuations that are larger than found in any other atmosphere. Many of the atmospheric weather patterns (jet streams, Hadley cells, vortices, thunderstorms) occur on other planets too, but their manifestation on Earth is distinct and unique. The Earth's climate has varied wildly over time, with atmospheric CO_2 and surface temperature fluctuating in response to ocean chemistry, planetary orbital variations, feedbacks between the atmosphere and interior, and a 30% increase in solar luminosity over the past 4.6 billion years (Ga). Despite these variations, the Earth's climate has

remained temperate, with at least partially liquid oceans, over the entire recorded ∼3.8 Ga geological record of the planet. Life has had a major influence on the ocean–atmosphere system, and as a result it is possible to discern the presence of life from remote spacecraft data. Global biological activity is indicated by the presence of atmospheric gases such as oxygen and methane that are in extreme thermodynamic disequilibrium, and by the widespread presence of a red-absorbing pigment (chlorophyll) that does not match the spectral signatures of any known rocks or minerals. The presence of intelligent life on Earth can be discerned from stable radio-wavelength signals emanating from the planet that do not match naturally occurring signals but do contain regular pulsed modulations that are the signature of information exchange.

1. Overview of Planetary Characteristics

Atmospheres are found on the Sun, 8 planets, and 7 of the 60-odd satellites, for a total count of 16—in addition to the atmospheres that exist around the ∼200 known gas giant planets orbiting other stars. Each has its own brand of weather and its own unique chemistry. They can be divided into two major classes: the terrestrial-planet atmospheres, which have solid surfaces or oceans as their lower boundary condition, and the gas giant atmospheres, which are essentially bottomless. Venus and Titan form one terrestrial subgroup that is characterized by a slowly rotating planet, and interestingly, both exhibit a rapidly rotating atmosphere. Mars, Io, Triton, and Pluto form a second terrestrial subgroup that is characterized by a thin atmosphere, which in large measure is driven by vapor-pressure equilibrium with the atmosphere's solid phase on the surface. Both Io and Triton have active volcanic plumes. Earth's weather turns out to be the most unpredictable in the solar system. Part of the reason is that its mountain ranges frustrate the natural tendency for winds to settle into steady east–west patterns, and a second reason is that its atmospheric eddies, the fluctuating waves and storm systems that deviate from the average, are nearly as big as the planet itself and as a result strongly interfere with each other. [See VENUS: ATMOSPHERE; IO: THE VOLCANIC MOON; TRITON; AND PLUTO.]

Earth has many planetary attributes that are important to the study of its atmosphere and oceans, and conversely there are several ways in which its physically and chemically active fluid envelope directly affects the solid planet. Earth orbits the Sun at a distance of only 108 times the diameter of the Sun. The warmth from the Sun that Earth receives at this distance, together with a 30 K increase in surface temperature resulting from the atmospheric greenhouse effect, is exactly what is needed for H_2O to appear in all three of its phases. This property of the semimajor axis of Earth's orbit is the most important physical characteristic of the planet that supports life. (One interesting consequence is that Earth is the only planet in the solar system where one can ski.)

Orbiting the Sun at just over 100 Sun diameters is not as close as it may sound; a good analogy is to view a basketball placed just past first base while standing at home plate on a baseball diamond. For sunlight, the Sun-to-Earth trip takes 499 s or 8.32 min. Earth's semimajor axis, $a_3 = 1.4960 \times 10^{11}$ m = 1 AU (astronomical unit), and orbital period, $\tau_3 = 365.26$ days = 1 year, where the subscript 3 denotes the third planet out from the Sun, are used as convenient measures of distance and time. When the orbital period of a body encircling the Sun, τ, is expressed in years, and its semimajor axis, a, is expressed in AU, then Kepler's third law is simply $\tau = a^{3/2}$, with a proportionality constant of unity. [See SOLAR SYSTEM DYNAMICS: REGULAR AND CHAOTIC MOTION.]

1.1 Length of Day

The Earth's rotation has an enormous effect on the motions of its fluid envelope that accounts for the circular patterns of large storms like hurricanes, the formation of **western boundary currents** like the Gulf Stream, the intensity of jet streams, the extent of the Hadley cell, and the nature of fluid instabilities. All of these processes are thoroughly discussed in Sections 2–5. Interestingly, the reverse is also true: The Earth's atmosphere and oceans have a measurable

effect on the planet's rotation rate. For all applications but the most demanding, the time Earth takes to turn once on its axis, the length of its day, is adequately represented by a constant value equal to 24 hours or 1440 minutes or 86,400 seconds. The standard second is the Système International (SI) second, which is precisely 9,192,631,770 periods of the radiation corresponding to the transition between two hyperfine levels of the ground state of the ^{133}Cs atom. When the length of day is measured with high precision, it is found that Earth's rotation is not constant. The same is likely to hold for any dynamically active planet. Information can be obtained about the interior of a planet, and how its atmosphere couples with its surface, from precise length-of-day measurements. Earth is the only planet to date for which we have achieved such accuracy, although we also have high-precision measurements of the rotation rate of pulsars, the spinning neutron stars often seen at the center of supernova explosions.

The most stable pulsars lose only a few seconds every million years and are the best-known timekeepers, even better than atomic clocks. In contrast, the rotating Earth is not an accurate clock. Seen from the ground, the positions as a function of time of all objects in the sky are affected by Earth's variable rotation. Because the Moon moves across the sky relatively rapidly and its position can be determined with precision, the fact that Earth's rotation is variable was first realized when a series of theories that should have predicted the motion of the Moon failed to achieve their expected accuracy. In the 1920s and 1930s, it was established that errors in the position of the Moon were similar to errors in the positions of the inner planets, and by 1939, clocks were accurate enough to reveal that Earth's rotation rate has both irregular and seasonal variations.

The quantity of interest is the planet's three-dimensional angular velocity vector as a function of time, $\Omega(t)$. Since the 1970s, time series of all three components of $\Omega(t)$ have been generated by using very long baseline interferometry (VLBI) to accurately determine the positions of quasars and laser ranging to accurately determine the positions of man-made satellites and the Moon, the latter with corner reflectors placed on the Moon by the *Apollo* astronauts. [See PLANETARY EXPLORATION MISSIONS.]

The theory of Earth's variable rotation combines ideas from geophysics, meteorology, oceanography, and astronomy. The physical causes fall into two categories: those that change the planet's moment of inertia (like a spinning skater pulling in her arms) and those that torque the planet by applying stresses (like dragging a finger on a spinning globe). Earth's moment of inertia is changed periodically by tides raised by the Moon and the Sun, which distort the solid planet's shape. Nonperiodic changes in the solid planet's shape occur because of fluctuating loads from the fluid components of the planet, namely, the atmosphere, the oceans, and, deep inside the planet, the liquid iron–nickel core. In addition, shifts of mass from earthquakes and melting ice cause nonperiodic changes. Over long timescales, plate tectonics and mantle convection significantly alter the moment of inertia and hence the length of day.

An important and persistent torque that acts on Earth is the gravitational pull of the Moon and the Sun on the solid planet's tidal bulge, which, because of friction, does not line up exactly with the combined instantaneous tidal stresses. This torque results in a steady lengthening of the day at the rate of about 1.4 ms per century and a steady outward drift of the Moon at the rate of 3.7 ± 0.2 cm/year, as confirmed by lunar laser ranging. On the top of this steady torque, it has been suggested that observed 5 ms variations that have timescales of decades are caused by stronger, irregular torques from motions in Earth's liquid core. Calculations suggest that viscous coupling between the liquid core and the solid mantle is weak, but that electromagnetic and topographic coupling can explain the observations. Mountains on the core–mantle boundary with heights around 0.5 km are sufficient to produce the coupling and are consistent with seismic tomography studies, but not much is known about the detailed topography of the core–mantle boundary. Detailed model calculations take into account the time variation of Earth's external magnetic field, which is extrapolated downward to the core–mantle boundary. New improvements to the determination of the magnetic field at the surface are enhancing the accuracy of the downward extrapolations.

Earth's atmosphere causes the strongest torques of all. The global atmosphere rotates faster than the solid planet by about 10 ms^{-1} on average. Changes in the global circulation cause changes in the pressure forces that act on mountain ranges and changes in the frictional forces between the wind and the surface. Fluctuations on the order of 1 ms in the length of day, and movements of the pole by several meters, are caused by these meteorological effects, which occur over seasonal and interannual timescales. General circulation models (GCMs) of the atmosphere routinely calculate the global atmospheric angular momentum, which allows the meteorological and nonmeteorological components of the length of day to be separated. All the variations in the length of day over weekly and daily timescales can be attributed to exchanges of angular momentum between Earth's atmosphere and the solid planet, and this is likely to hold for timescales of several months as well. Episodic reconfigurations of the coupled atmosphere–ocean system, such as the **El Niño-Southern Oscillation** (**ENSO**), cause detectable variations in the length of day, as do changes in the stratospheric jet streams.

2. Vertical Structure of the Atmosphere

Earth may differ in many ways from the other planets, but not in the basic structure of its atmosphere (Fig. 1). Planetary exploration has revealed that essentially every atmo-

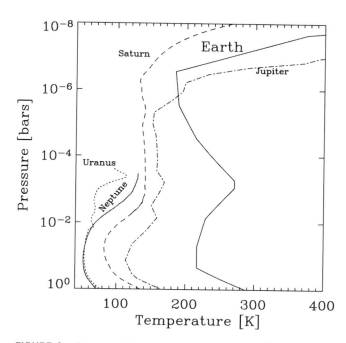

FIGURE 1 Representative temperature structure for the Earth (thick solid line) as compared with those of several other planets, including Jupiter (dash-dot), Saturn (dashed), Uranus (dotted), and Neptune (solid). For Earth, the altitude scale runs from the surface to about 130 km altitude. Atmospheres have high pressure at the bottom and low pressure at the top, so pressure is a proxy for altitude. Starting at the bottom of Earth's atmosphere and moving up, the troposphere, stratosphere, mesosphere, and thermosphere correspond to layers where temperature decreases, increases, decreases, and then increases with height, respectively. The top of Earth's troposphere, stratosphere, and mesosphere are at altitudes of about 10–15 km, 50 km, and 100 km, respectively. Note that other planets also generally have tropospheres and thermospheres, although the details of the intermediate layers (the stratosphere and mesosphere) differ from planet to planet.

sphere starts at the bottom with a **troposphere**, where temperature decreases with height at a nearly constant rate up to a level called the tropopause, and then has a **stratosphere**, where temperature usually increases with height or, in the case of Venus and Mars, decreases much less quickly than in the troposphere. It is interesting to note that atmospheres are warm both at their bottoms and their tops, but do not get arbitrarily cold in their interiors. For example, on Jupiter and Saturn there is significant methane gas throughout their atmospheres, but nowhere does it get cold enough for methane clouds to form, whereas in the much colder atmospheres of Uranus and Neptune, methane clouds do form. Details vary in the middle-atmosphere regions from one planet to another, where photochemistry is important, but each atmosphere is topped off by a high-temperature, low-density thermosphere that is sensitive to solar activity and an exobase, the official top of an atmosphere, where molecules float off into space when they achieve escape velocity. [*See* Atmospheres of the Giant Planets.]

Interestingly, the top of the troposphere occurs at about the same pressure, about 0.1–0.3 bar, on most planets (Fig. 1). This similarity is not coincidental but instead results from the pressure dependence of the atmospheric opacity to solar and especially infrared radiation. In the high-pressure regime of tropospheres, the gas is relatively opaque at infrared wavelengths, which inhibits heat loss by radiation from the deep levels and hence promotes a profile where temperature decreases strongly with altitude. In the low-pressure regime of stratospheres, the gas becomes relatively transparent at infrared wavelengths, which allows the temperature to become more constant—or in some cases even increase—with altitude. This transition from opaque to transparent tends to occur at pressures of 0.1–0.3 bar for the compositions of most planetary atmospheres in our solar system.

In the first 0.1 km of a terrestrial atmosphere, the effects of daily surface heating and cooling, surface friction, and topography produce a turbulent region called the planetary boundary layer, or PBL. Right at the surface, molecular viscosity forces the "no slip" boundary condition and the wind reduces to zero, such that even a weak breeze results in a strong vertical wind shear that can become turbulent near the surface. However, only a few millimeters above the surface, molecular viscosity ceases to play a direct role in the dynamics, except as a sink for the smallest eddies. The mixing caused by turbulent eddies is often represented as a viscosity with a strength that is a million times or more greater than the molecular viscosity.

Up to altitudes of about 80 km, Earth's atmosphere is composed of 78% N_2, 21% O_2, 0.9% Ar, and 0.002% Ne by volume, with trace amounts of CO_2, CH_4, and numerous other compounds. Diffusion, chemistry, and other effects substantially alter the composition at altitudes above ~90 km.

2.1 Troposphere

The troposphere is the lowest layer of the atmosphere, characterized by a temperature that decreases with altitude (Fig. 1). The top of the troposphere is called the tropopause, which occurs at an altitude of 18 km at the equator but only 8 km at the poles (the cruising altitude of commercial airliners is typically 10 km). Gravity, combined with the compressibility of air, causes the density of an atmosphere to fall off exponentially with height, such that Earth's troposphere contains 80% of the mass and most of the water vapor in the atmosphere, and consequently most of the clouds and stormy weather. Vertical mixing is an important process in the troposphere. Temperature falls off with height at a predictable rate because the air near the surface is heated and becomes light, and the air higher up cools to space and

becomes heavy, leading to an unstable configuration and convection. The process of convection relaxes the temperature profile toward the neutrally stable configuration, called the adiabatic temperature lapse rate, for which the decrease of temperature with decreasing pressure (and hence increasing height) matches the drop-off of temperature that would occur inside a balloon that conserves its heat as it moves, that is, moves adiabatically.

In the troposphere, water vapor, which accounts for up to ~1% of air, varies spatially and decreases rapidly with altitude. The water vapor mixing ratio in the stratosphere and above is almost 4 orders of magnitude smaller than that in the tropical lower troposphere.

2.2 Stratosphere

The nearly adiabatic falloff of temperature with height in Earth's troposphere gives way above the tropopause to an increase of temperature with height. This results in a rarified, stable layer called the stratosphere. Observations of persistent, thin layers of aerosol and of long residence times for radioactive trace elements from nuclear explosions are direct evidence of the lack of mixing in the stratosphere. The temperature continues to rise with altitude in Earth's stratosphere until one reaches the stratopause at about 50 km. The source of heating in Earth's stratosphere is the photochemistry of ozone, which peaks at about 25 km. Ozone absorbs ultraviolet (UV) light, and below about 75 km nearly all this radiation gets converted into thermal energy. The Sun's UV radiation causes stratospheres to form in other atmospheres, but instead of the absorber being ozone, which is plentiful on Earth because of the high concentrations of O_2 maintained by the biosphere, other gases absorb the UV radiation. On the giant planets, methane, hazes, and aerosols do the job.

The chemistry of Earth's stratosphere is complicated. Ozone is produced mostly over the equator, but its largest concentrations are found over the poles, meaning that dynamics is as important as chemistry to the ozone budget. Mars also tends to have ozone concentrated over its poles, particularly over the winter pole. The dry martian atmosphere has relatively few hydroxyl radicals to destroy the ozone. Some of the most important chemical reactions in Earth's stratosphere are those that involve only oxygen. Photodissociation by solar UV radiation involves the reactions $O_2 + h\nu \rightarrow O + O$ and $O_3 + h\nu \rightarrow O + O_2$, where $h\nu$ indicates the UV radiation. Three-body collisions, where a third molecule, M, is required to satisfy conservation of momentum and energy, include $O + O + M \rightarrow O_2 + M$ and $O + O_2 + M \rightarrow O_3 + M$, but the former reaction proceeds slowly and may be neglected in the stratosphere. Reactions that either destroy or create "odd" oxygen, O or O_3, proceed at much slower rates than reactions that convert between odd oxygen. The equilibrium between O and O_3 is controlled by fast reactions that have rates and concentrations that are altitude-dependent. Other reactions that are important to the creation and destruction of ozone involve minor constituents such as NO, NO_2, H, OH, HO_2, and Cl. An important destruction mechanism is the catalytic cycle $X + O_3 \rightarrow XO + O_2$ followed by $XO + O \rightarrow X + O_2$, which results in the net effect $O + O_3 \rightarrow 2O_2$. On Earth, human activity has led to sharp increases in the catalysts $X = Cl$ and NO and subsequent sharp decreases in stratospheric ozone, particularly over the polar regions. The Montreal Protocol is an international treaty signed in 1987 that is designed to stop and eventually reverse the damage to the stratospheric ozone layer; regular meetings of the parties, involving some 175 countries, continually update the protocol.

2.3 Mesosphere

Above Earth's stratopause, temperature again falls off with height, although at a slower rate than in the troposphere. This region is called the mesosphere. Earth's stratosphere and mesosphere are often referred to collectively as the middle atmosphere. Temperatures fall off in the mesosphere because there is less heating by ozone and emission to space by carbon dioxide is an efficient cooling mechanism. The mesopause occurs at an altitude of about 80 km, marking the location of a temperature minimum of about 130 K.

2.4 Thermosphere

As is the case for ozone in Earth's stratosphere, above the mesopause, atomic and molecular oxygen strongly absorb solar UV radiation and heat the atmosphere. This region is called the thermosphere, and temperatures rise with altitude to a peak that varies between about 500 and 2000 K depending on solar activity. Just as in the stratosphere, the thermosphere is stable to vertical mixing. At about 120 km, molecular diffusion becomes more important than turbulent mixing, and this altitude is called the homopause (or turbopause). Rocket trails clearly mark the homopause—they are rapidly mixed below this altitude but linger relatively undisturbed above it. Molecular diffusion is mass-dependent and each species falls off exponentially with its own scale height, leading to elemental fractionation that enriches the abundance of the lighter species at the top of the atmosphere.

For comparison with Earth, the structure of the thermospheres of the giant planets has been determined from *Voyager* spacecraft observations, and the principal absorbers of UV light are H_2, CH_4, C_2H_2, and C_2H_6. The thermospheric temperatures of Jupiter, Saturn, and Uranus are about 1000, 420, and 800 K, respectively. The high temperature and low gravity on Uranus allow its upper atmosphere to extend out appreciably to its rings. [See Atmospheres of the Giant Planets.]

2.5 Exosphere and Ionosphere

At an altitude of about 500 km on Earth, the mean free path between molecules grows to be comparable to the density scale height (the distance over which density falls off by a factor of $e \approx 2.7128$). This defines the exobase and the start of the exosphere. At these high altitudes, sunlight can remove electrons from atmospheric constituents and form a supply of ions. These ions interact with a planet's magnetic field and with the solar wind to form an ionosphere. On Earth, most of the ions come from molecular oxygen and nitrogen, whereas on Mars and Venus most of the ions come from carbon dioxide. Because of the chemistry, however, ionized oxygen atoms and molecules are the most abundant ion for all three atmospheres.

Mechanisms of atmospheric escape fall into two categories, thermal and nonthermal. Both processes provide the kinetic energy necessary for molecules to attain escape velocity. When escape velocity is achieved at or above the exobase, such that further collisions are unlikely, molecules escape the planet. In the thermal escape process, some fraction of the high-velocity wing of the Maxwellian distribution of velocities for a given temperature always has escape velocity; the number increases with increasing temperature. An important nonthermal escape process is dissociation, both chemical and photochemical. The energy for chemical dissociation is the excess energy of reaction, and for photochemical dissociation, it is the excess energy of the bombarding photon or electron, either of which is converted into kinetic energy in the dissociated atoms. A common effect of electrical discharges of a kilovolt or more is "sputtering," where several atoms can be ejected from the spark region at high velocities. If an ion is formed very high in the atmosphere, it can be swept out of a planet's atmosphere by the solar wind. Similarly at Io, ions are swept away by Jupiter's magnetic field. Other nonthermal escape mechanisms involve charged particles. Charged particles get trapped by magnetic fields and therefore do not readily escape. However, a fast proton can collide with a slow hydrogen atom and take the electron from the hydrogen atom. This charge–exchange process changes the fast proton into a fast, hydrogen atom that is electrically neutral and hence can escape.

Nonthermal processes account for most of the present-day escape flux from Earth, and the same is likely to be true for Venus. They are also invoked to explain the $62 \pm 16\%$ enrichment of the $^{15}N/^{14}N$ ratio in the martian atmosphere. If the current total escape flux from thermal and nonthermal processes is applied over the age of the solar system, the loss of hydrogen from Earth is equivalent to only a few meters of liquid water, which means that Earth's sea level has not been affected much by this process. However, the flux could have been much higher in the past, since it is sensitive to the structure of the atmosphere. [See Mars Atmosphere: History and Surface Interaction.]

3. Atmospheric Circulation

3.1 Processes Driving the Circulation

The atmospheric circulation on Earth, as on any planet, involves a wealth of phenomena ranging from global weather patterns to turbulent eddies only centimeters across and varies over periods of seconds to millions of years. All this activity is driven by absorbed sunlight and loss of infrared (heat) energy to space. Of the sunlight absorbed by Earth, most (∼70%) penetrates through the atmosphere and is absorbed at the surface; in contrast, the radiative cooling to space occurs not primarily from the surface but from the upper troposphere at an altitude of 5–10 km. This mismatch in the altitudes of heating and cooling means that, in the absence of air motions, the surface temperature would increase while the upper tropospheric temperature would decrease. However, such a trend produces an unstable density stratification, forcing the troposphere to overturn. The hot air rises, the cold air sinks, and thermal energy is thus transferred from the surface to the upper troposphere. This energy transfer by air motions closes the "energy loop," allowing the development of a quasi-steady state where surface and atmospheric temperatures remain roughly steady in time. This vertical mixing process is fundamentally responsible for near-surface convection, turbulence, cumulus clouds, thunderstorms, hurricanes, dust devils, and a range of other small-scale weather phenomena.

At global scales, much of Earth's weather results not simply from vertical mixing but from the atmosphere's response to horizontal temperature differences. Earth absorbs most of the sunlight at the equator, yet it loses heat to space everywhere over the surface. This mismatch makes the near-surface equatorial air hot and the polar air cold. This configuration is gravitationally unstable—the hot equatorial air has low density and the cold polar air has high density. Just as the cold air from an open refrigerator slides across your feet, the cold polar air slides under the hot equatorial air, lifting the hotter air upward and poleward while pushing the colder air downward and equatorward. This overturning process transfers energy between the equator and the poles and leads to a much milder equator-to-pole temperature difference (about 30 K at the surface) than would exist in the absence of such motions. On average, the equatorial regions gain more energy from sunlight than they lose as radiated heat, while the reverse holds for the poles; the difference is transported between equator and pole by the air and ocean. The resulting atmospheric overturning causes many of Earth's global-scale weather patterns, such as the 1000 km long fronts that cause much midlatitude weather and the organization of thunderstorms into clusters and bands. Horizontal temperature and density contrasts can drive weather at regional scales too; examples include air-sea breezes and monsoons.

3.2 Influence of Rotation

The horizontal pressure differences associated with horizontal temperature differences cause a force (the "pressure-gradient force") that drives most air motion at large scales. However, how an atmosphere responds to this force depends strongly on whether the planet is rotating. On a nonrotating planet, the air tends to directly flow from high to low pressure, following the "nature abhors a vacuum" dictum. If the primary temperature difference occurs between equator and pole, this would lead to a simple overturning circulation between the equator and pole. On the other hand, planetary rotation (when described in a noninertial reference frame rotating with the solid planet) introduces new forces into the equations of motion: the centrifugal force and the Coriolis force. The centrifugal force naturally combines with the gravitational force and the resultant force is usually referred to as simply the gravity. For rapidly rotating planets, the Coriolis force is the dominant term that balances the horizontal pressure-gradient force in large-scale circulations (a balance called geostrophy). Because the Coriolis force acts perpendicular to the air motion, this leads to a fascinating effect—the horizontal airflow is perpendicular to the horizontal pressure gradient. A north–south pressure gradient (resulting from a hot equator and a cold pole, for example) leads primarily not to north–south air motions but to east–west air motions! This is one reason why east–west winds dominate the circulation on most planets, including Earth. For an Earth-sized planet with Earth-like wind speeds, rotation dominates the large-scale dynamics as long as the planet rotates at least once every 10 days.

Physically, the Coriolis force acts in the following way. Air moving eastward (i.e., in the same direction as the planet's rotation, but faster) experiences a force that moves it away from the rotation axis—namely, equatorward—just as a child experiences an outward force on a spinning merry-go-round. Conversely, air moving westward (in the same direction as the planet's rotation, but slower) would experience a poleward force. And, just as an ice skater spins faster as she pulls in her arms, air that moves toward the planetary rotation axis—namely, poleward—spins faster, which is equivalent to saying that it deflects eastward. Conversely, air that moves away from the planetary rotation axis (equatorward) deflects westward. If one pays attention to the directions of the force in each of these cases, one sees that, in the northern hemisphere, this rotationally induced force is always to the right of the air motion, while in the southern hemisphere, it is always to the left of the air motion.

Two other important effects of rapid rotation are the suppression of motions in the direction parallel to the rotation axis, called the Taylor–Proudman effect, and the coupling of horizontal temperature gradients with vertical wind shear, a 3-dimensional relationship described by the thermal-wind equation.

3.3 Observed Global-Scale Circulation

As described earlier, the atmospheric circulation organizes primarily into a pattern of east–west winds, and perhaps the most notable feature is the eastward-blowing jet streams in the midlatitudes of each hemisphere (Fig. 2). In a longitudinal and seasonal average, the winter hemisphere wind maximum reaches 40 m s^{-1} at 30° latitude, and the summer hemisphere wind maximum reaches 20–30 m s^{-1} at 40–50° latitude. In between these eastward wind maxima, from latitude 20°N to 20°S the tropospheric winds blow weakly westward. The jet streams are broadly distributed in height, with peak speeds at about 12-km altitude. Although the longitudinally and seasonally averaged winds exhibit only a single tropospheric eastward-wind maximum in each hemisphere, instantaneous 3-dimensional snapshots of the atmosphere illustrate that there often exist two distinct jet streams, the subtropical jet at ~30° latitude and the polar jet at ~50° latitude. These jets are relatively narrow—a few hundred km in latitudinal extent—and can reach speeds up to 100 m s^{-1}. However, the intense jet cores are usually less than a few thousand kilometers in longitudinal extent (often residing over continental areas such as eastern Asia and eastern North America), and the jets typically exhibit wide, time-variable wavelike fluctuations in position. When averaged over longitude and time, these variations in the individual jet streams smear into the single eastward maximum evident in each hemisphere in Fig. 2.

Although the east–west winds dominate the time-averaged circulation, weaker vertical and latitudinal motion are required to transport energy from the equator to the poles. Broadly speaking, this transport occurs in two distinct modes. In the tropics exists a direct thermal overturning circulation called the *Hadley cell*, where, on average, air rises near the equator, moves poleward, and descends. This is an extremely efficient means of transporting heat and contributes to the horizontally homogenized temperatures that exist in the tropics. However, planetary rotation prevents the Hadley cell from extending all the way to the poles (to conserve angular momentum about the rotation axis, equatorial air would accelerate eastward to extreme speeds as it approached the pole, a phenomenon that is dynamically inhibited). On Earth, the Hadley cell extends to latitudes of ~30°. Poleward of ~30°, the surface temperatures decrease rapidly toward the pole; this is the location of the subtropical jet. Although planetary rotation inhibits the Hadley cell in this region, north–south motions still occur via a complex 3-dimensional process called **baroclinic instability**. Meanders on the jet stream grow, pushing cold high-latitude air under warm low-latitude air in confined regions ~1000–5000 km across. These instabilities grow, mature, and decay over ~5 day periods; new ones form as old ones disappear. These structures evolve to form regions of sharp thermal gradient called *fronts*, as well as 1000–5000 km long arc-shaped clouds and precipitation that dominate much of the winter weather in the United States, Europe, and other midlatitude regions.

Water vapor in Earth's troposphere greatly accentuates convective activity because latent heat is liberated when moist air is raised above its lifting–condensation level, and this further increases the buoyancy of the rising air, leading to moist convection. Towering thunderstorms get their energy from this process, and hurricanes are the most dramatic and best-organized examples of moist convection. Hurricanes occur only on Earth because only Earth provides the necessary combination of high humidity and surface friction. Surface friction is required to cause air to spiral into the center of the hurricane, where it is then forced upward past its lifting–condensation level.

The Hadley cell exerts a strong control over weather in the tropics. The upward transport in the ascending branch of the Hadley circulation occurs almost entirely in localized thunderstorms whose convective towers cover only a small fraction (perhaps ~1%) of the total horizontal area of the tropics. Because this ascending branch resides near the equator, equatorial regions receive abundant rainfall, allowing the development of tropical rainforests in Southeast Asia/Indonesia, Brazil, and central Africa.

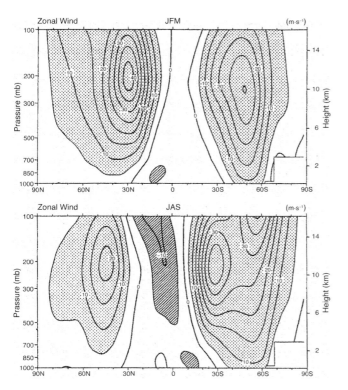

FIGURE 2 Longitudinally averaged zonal (i.e., east–west) winds in Earth's troposphere, showing the midlatitude maxima associated with the jet streams. (From Hurrell et al., 1998.)

On the other hand, this condensation and rainout of water dehydrates the air, so the descending branch of the Hadley cell, which occurs in the subtropics at ~20–30° latitude, is relatively dry. Because of the descending motion and dry conditions, little precipitation falls in these regions, which explains the abundance of arid biomes at 20–30° latitude, including the deserts of the African Sahara, southern Africa, Australia, central Asia, and the southwestern United States. However, the simple Hadley cell is to some degree a theoretical idealization, and many regional 3-dimensional time-variable phenomena—including monsoons, equatorial waves, El Niño, and longitudinal overturning circulations associated with continent–ocean and sea-surface-temperature contrasts—affect the locations of tropical thunderstorm formation and hence the climatic rainfall patterns.

Satellite images (Fig. 3) dramatically illustrate the signature of the Hadley cell and midlatitude baroclinic instabilities as manifested in clouds. In Fig. 3, the east–west band of clouds stretching across the disk of Earth just north of the equator corresponds to the rising branch of the Hadley cell (this cloud band is often called the intertropical convergence zone). These clouds are primarily the tops of thunderstorm anvils. In the midlatitude regions of both hemispheres (30–70° latitude), several arc-shaped clouds up to 3000–5000 km long can be seen. These are associated with baroclinic instabilities. These clouds, which

FIGURE 3   Visible-wavelength image of Earth from the *GOES* geostationary weather satellite, illustrating the clouds associated with the Hadley cell, baroclinic instabilities, and other weather systems. North America can be seen at the upper right and South America (mostly obscured by clouds) is at the lower right.

can often dominate midlatitude winter precipitation, form when large regions of warm air are forced upward over colder air masses during growing baroclinic instabilities. In many cases, the forced ascent associated with these instabilities produces predominantly sheet-like stratus clouds and steady rainfall lasting for several days, although sometimes the forced ascent can trigger local convection events (e.g., thunderstorms).

What causes the jet stream? This is a subtle question. At the crudest level, poleward-moving equatorial air deflects eastward due to the **Coriolis acceleration** (or, equivalently, due to the air's desire to conserve angular momentum about the planetary rotation axis), so the formation of eastward winds in the midlatitudes is a natural response to the Hadley circulation. These strong eastward winds in midlatitudes are also consistent with the large latitudinal thermal gradients in midlatitudes via the thermal-wind equation mentioned in Section 3.2. However, these processes alone would tend to produce a relatively broad zone of eastward flow rather than a narrow jet. Nonlinear turbulent motions, in part associated with baroclinic instabilities, pump momentum up-gradient into this eastward-flowing zone and help to produce the narrow jet streams.

Although the Earth's equator is hotter than the poles at the surface, it is noteworthy that, in the upper troposphere and lower stratosphere (~18 km altitude), the reverse is true. This seems odd because sunlight heats the equator much more strongly than the poles. In reality, the cold equatorial upper troposphere results from a dynamical effect: Large-scale ascent in the tropics causes air to expand and cool (a result of decreased pressure as the air rises), leading to the low temperatures despite the abundant sunlight. Descent at higher latitudes causes compression and heating, leading to warmer temperatures. Interestingly, this means that, in the lower stratosphere, the ascending air is actually denser than the descending air. Such a circulation, called a thermally indirect circulation, is driven by the absorption of atmospheric waves that are generated in the troposphere and propagate upward into the stratosphere. There is a strong planetary connection because all four giant planets—Jupiter, Saturn, Uranus, and Neptune—are also thought to have thermally indirect circulations in their stratospheres driven by analogous processes.

3.4 Insights from other Atmospheres

Planetary exploration has revealed that atmospheric circulations come in many varieties. Perhaps ironically, Earth is observed to have the most unpredictable weather of all. The goal of planetary meteorology is to understand what shapes and maintains these diverse circulations. The *Voyager* spacecraft provided close-up images of the atmospheres of Jupiter, Saturn, Uranus, and Neptune and detailed information on the three satellites that have atmospheres thick enough to sport weather—Io, Titan, and

Triton. The atmospheres of Venus and Mars have been sampled by entry probes, landers, orbiting spacecraft, and telescopic studies. Basic questions like why Venus' atmosphere rotates up to 60 times faster than does the planet, or why Jupiter and Saturn have superrotating equatorial jets, do not have completely satisfactory explanations. However, by comparing and contrasting each planet's weather, a general picture has begun to emerge.

Theoretical studies and comparative planetology show that planetary rotation rate and size exert a major control over the type of global atmospheric circulations that occur. When the rotation rate is small, Hadley cells are unconfined and stretch from the equator to pole. Venus, with a rotation period of 243 days, seems to reside in such a state. Titan rotates in 16 days and, according to circulation models, its Hadley cell extends to at least ~60° latitude, a transitional regime between Venus and Earth. On the other hand, fast rotation confines the Hadley cell to a narrow range of latitudes (0–30° on Earth) and forces baroclinic instabilities to take over much of the heat transport between low latitudes and the poles. Increasing the rotation rate still further—or making the planet larger—causes the midlatitudes to break into series of narrow latitudinal bands, each with their own east–west jet streams and baroclinic instabilities. The faster the rotation rate, the straighter and narrower are the bands and jets. This process helps explain the fact that Jupiter and Saturn, which are large and rapidly rotating, have ~30 and 20 jet streams, respectively (as compared to only a few jet streams for Earth). Fast rotation also contributes to smaller structures because it inhibits free movement of air toward or away from pressure lows and highs, instead causing the organization of vortices around such structures. Thus, a planet identical to Earth but with a faster or slower rotation rate would exhibit different circulations, equatorial and polar temperatures, rainfall patterns, and cloud patterns, and hence would exhibit a different distribution of deserts, rainforests, and other biomes.

The giant planets Jupiter and Saturn exhibit numerous oval-shaped windstorms that superficially resemble terrestrial hurricanes. However, hurricanes can generate abundant rainfall because friction allows near-surface air to spiral inward toward the low-pressure center, providing a source of moist air that then ascends inside thunderstorms; in turn, these thunderstorms release energy that maintains the hurricane's strength against the frictional energy losses. In contrast, windstorms like Jupiter's Great Red Spot and the hundreds of smaller ovals seen on Jupiter, as well as the dozens seen on Saturn and the couple seen on Neptune, do not directly require moist convection to drive them and hence are not hurricanes. Instead, they are simpler systems that are closely related to three types of long-lasting, high-pressure "storms," or coherent vortices, seen on Earth: blocking highs in the atmosphere and Gulf Stream rings and Mediterranean salt lenses ("meddies") in the ocean. Blocking highs are high-pressure centers that stubbornly

settle over continents, particularly in the United States and Russia, thereby diverting rain from its usual path for months at a time. For example, the serious 1988 drought in the U.S. Midwest was exacerbated by a blocking high. Gulf Stream rings are compact circulations in the Atlantic that break off from the meandering Gulf Stream, which is a river inside the Atlantic Ocean that runs northward along the eastern coast of the United States and separates from the coast at North Carolina, where it then jets into the Atlantic in an unsteady manner. Seen in three dimensions, the Gulf Stream has the appearance of a writhing snake. Similar western boundary currents occur in other ocean basins, for example, the Kuroshio Current off the coast of Japan and the Agulhas Current off the coast of South Africa. Jet streams in the atmosphere are a related phenomenon. When Gulf Stream rings form, they trap phytoplankton and zooplankton inside them, which are carried large distances. Over the course of a few months, the rings dissipate at sea, are reabsorbed into the Gulf Stream, or run into the coast, depending on which side of the Gulf Stream they formed. The ocean plays host to another class of long-lived vortices, Mediterranean salt lenses, which are organized high-pressure circulations that float under the surface of the Atlantic. They form when the extra-salty water that slips into the Atlantic from the shallow Mediterranean Sea breaks off into vortices. After a few years, these meddies eventually wear down as they slowly mix with the surrounding water. The mathematical description of these long-lasting vortices on Earth is the same as that used to describe the ovals seen on Jupiter, Saturn, and Neptune. [See ATMOSPHERES OF THE GIANT PLANETS.]

Given that we know that atmospheric motions are fundamentally driven by sunlight, and we know that the problem is governed by Newton's laws of motion, why then are atmospheric circulations difficult to understand? Several factors contribute to the complexity of observed weather patterns. In the first place, fluids move in an intrinsically nonlinear fashion that makes paper-and-pencil analysis formidable and often intractable. Laboratory experiments and numerical experiments performed on high-speed computers are often the only means for making progress on problems in geophysical fluid dynamics. In the second place, meteorology involves the intricacies of moist thermodynamics and precipitation, and we are only beginning to understand and accurately model the microphysics of these processes. And for the terrestrial planets, a third complexity arises from the complicated boundary conditions that the solid surface presents to the problem, especially when mountain ranges block the natural tendency for winds to organize into steady east–west jet streams. For oceanographers, even more restrictive boundary conditions apply, namely, the ocean basins, which strongly affect how currents behave. The giant planets are free of this boundary problem because they are completely fluid down to their small rocky cores. However, the scarcity of data for the giant planets, especially with respect to their vertical structure beneath the cloud tops, provides its own set of difficulties.

3.5 Ironic Unpredictability—an Anecdote

The fickleness of Earth's weather compared to that of the other planets provides many fascinating scientific problems for meteorologists. Trying to live on such a planet presents Earth's inhabitants with practical problems as well. On the lighthearted side, there are common bromides such as "If you don't like the weather, wait 15 minutes," and "Everybody complains about the weather, but nobody does anything about it." On the serious side, lightning storms and tornados wreak havoc every year, and before the advent of weather satellites, hurricanes once struck populated coastlines without warning, causing terrible loss of life.

Even now, the tracks of hurricanes are notoriously difficult to predict. The point is best made with an example, and the following is a lighthearted anecdote from the first author's personal experience: Perhaps he should have known better than to leave the windows of his apartment open on such a warm, breezy morning in the summer of 1991, but the apartment needed airing out, and the author was preoccupied with a desire to come up with a good way to illustrate to a group of distinguished terrestrial meteorologists that the weather on Jupiter is more predictable than the weather on Earth. And so, he left the windows open, locked the door, and headed out to Boston's Logan Airport to begin a 10 day trip to a symposium on "Vortex Dynamics in the Atmosphere and Ocean," which was being held in Vienna. His preoccupation was not helped by the use of the singular "atmosphere" in the symposium's title, which, one could argue, should have been written with the plural "atmospheres." To be sure, Earth has its great vortices, like Gulf Stream rings, Mediterranean salt lenses, and atmospheric blocking highs, and even more powerful storms, like hurricanes, which are driven by moist thermodynamics (in fact, Hurricane Bob was at that moment slowly heading toward the Carolina coast). Yet Jupiter's Great Red Spot and the hundreds of other long-lived vortices found on the gas giant planets are in many ways simpler systems to study, and we have excellent observations of them from spacecraft like *Voyager*, *Galileo*, and *Cassini*.

After arriving at the conference, the author decided to make his case by pointing out that a *Voyager*-style mission to track hurricanes on Earth would most likely end in failure. This is because the *Voyager* cameras had to be choreographed 30 days in advance of each encounter to give the flight engineers time to sort through the conflicting requests of the various scientists and time to program the onboard computer. For the atmospheric working group, this constraint meant that success or failure depended on the accuracy of 30-day weather forecasts for the precise locations of the drifting Great Red Spot and other targeted features. On Earth, storms rarely last 30 days, and much

less do they end up where they are predicted to be going a month in advance. The fact that the *Voyager* missions to Jupiter were a complete success, as were the subsequent Saturn, Uranus, and Neptune missions, illustrates in a practical way the remarkable predictability of the weather on the gas giant planets relative to on Earth. [*See* Planetary Exploration Missions.]

Having made his point, on the road back to the Vienna Airport after the conference the author was getting accustomed to the fact that taxis in German-speaking countries are Mercedes, when the driver explained that the announcer on the radio was saying that there had been a coup in Moscow. This left him worried about his Russian colleagues, several of whom he had just met in the preceding week. On the flight back across the Atlantic, he was thinking about this when the Lufthansa pilot announced, with resignation in his voice, that because of thunderstorms the plane could not land in Boston and was being redirected to Montreal. After about 2 hours in Montreal, where the plane was nestled between several other waylaid international planes that were littered across the tarmac, the go-ahead was given to finish the trip to Boston. The landing was bumpy, and the skyline was disturbingly dark, but there was a beautiful sunset that was framed with orange, red, and black clouds. It was only after getting off the plane that the author first learned that Hurricane Bob had just hit Boston. Boston? Wasn't Bob supposed to hit the Carolina coast? It was difficult not to take this egregious forecasting error personally. On returning home, jet-lagged, the author discovered that his apartment was dark, the electricity was out, the windows were of course still open, the curtains, carpet, and furniture were soaked, and wall hangings and broken glass were strewn about the floor. The irony of the situation is not hard to grasp. *Voyager* would have returned beautiful, fair weather images of North Carolina and South Carolina, and would have completely missed the hurricane, which ended up passing through this author's apartment 1000 km north of the previous week's prediction.

4. Oceans

Earth is the only planet in the solar system with a global ocean at the surface. The oceans have an average thickness of 3.7 km and cover 71% of Earth's area; the greatest thickness is 10.9 km, which occurs at the Marianas Trench. The total oceanic mass—1.4×10^{21} kg—exceeds the atmosphere mass of 5×10^{18} kg by nearly a factor of 300, implying that the oceans dominate Earth's surface inventory of volatiles. (One way of visualizing this fact is to realize that, if Earth's entire atmosphere condensed as ices on the surface, it would form a layer only ~10 m thick.) The Earth therefore sports a greater abundance of fluid volatiles at its surface than any other solid body in the solar system. Even Venus' 90 bar CO_2 atmosphere contains only one third the mass of Earth's oceans. On the other hand, Earth's oceans constitute only 0.02% of Earth's total mass; the mean oceanic thickness of 3.7 km pales in comparison to Earth's 6400 km radius, implying that the oceans span only 0.06% of Earth's width. The Earth is thus a relatively dry planet, and the oceans truly are only skin deep.

It is possible that Earth's solid mantle contains a mass of dissolved water (stored as individual water molecules inside and between the rock grains) equivalent to several oceans' worth of water. Taken together, however, the total water in Earth probably constitutes less than 1% of Earth's mass. In comparison, most icy satellites and comets in the outer solar system contain ~40–60% H_2O by mass, mostly in solid form. This lack of water on Earth in comparison to outer solar-system bodies reflects the relatively dry conditions in the inner solar system when the terrestrial planets formed; indeed, the plethora of water on Earth compared to Venus and Mars has raised the question of whether even the paltry amount of water on Earth must have been delivered from an outer-solar-system source such as impact of comets onto the forming Earth.

The modern oceans can be subdivided into the Pacific, Atlantic, Indian, and Arctic Oceans, but these four oceans are all connected, and this contiguous body of water is often simply referred to as the global ocean.

4.1 Oceanic Structure

The top meter of ocean water absorbs more than half of the sunlight entering the oceans; even in the sediment-free open ocean, only 20% of the sunlight reaches a depth of 10 m and only ~1% penetrates to a depth of 100 m (depending on the angle of the Sun from vertical). Photosynthetic single-celled organisms, which are extremely abundant near the surface, can thus only survive above depths of ~100 m; this layer is called the photic zone. The much thicker aphotic zone, which has too little light for photosynthetic production to exceed respiration, extends from ~100 m to the bottom of the ocean. Despite the impossibility of photosynthesis at these depths, the deep oceans nevertheless exhibit a wide variety of life fueled in part by dead organic matter that slowly sediments down from the photic zone.

From a dynamical point of view, the ocean can be subdivided into several layers. Turbulence caused by wind and waves homogenizes the top 20–200 m of the ocean (depending on weather conditions), leading to profiles of density, temperature, salinity, and composition that vary little across this layer, which is therefore called the mixed layer. Below the mixed layer lies the thermocline, where the temperature generally decreases with depth down to ~0.5–1 km. The salinity also often varies with depth between ~100–1000 m, a layer called the halocline. For example, regions of abundant precipitation but lesser evaporation, such as the North Pacific, have relatively fresh surface waters, so the salinity

increases with depth below the mixed layer in those regions. The variation of temperature and salinity between ~100–1000 m implies that density varies with depth across this layer too; this is referred to as the pycnocline. Below the thermocline, halocline, and pycnocline lies the deep ocean, where temperatures are usually relatively constant with depth at a chilly 0–4°C.

The temperature at the ocean surface varies strongly with latitude, with only secondary variations in longitude. Surface temperatures reach 25–30°C near the equator, where abundant sunlight falls, but plummet to 0°C near the poles. In contrast, the deep oceans (>1 km) are generally more homogeneous and have temperatures between 0–4°C all over the world. (When enjoying the bathtub-temperature water and coral reefs during a summer vacation to a tropical island, it is sobering to think that if one could only scuba dive deep enough, the temperature would approach freezing.) This latitude-dependent upper-ocean structure implies that the thermocline and pycnocline depths decrease with latitude: They are about ~1 km near the equator and reach zero near the poles.

Because warmer water is less dense than colder water, the existence of a thermocline over most of the ocean implies that the top ~1 km of the ocean is less dense than the underlying deep ocean. The implication is that, except for localized regions near the poles, the ocean is stable to vertical convective overturning.

4.2 Ocean Circulation

Ocean circulation differs in important ways from atmospheric circulation, despite the fact that the two are governed by the same dynamical laws. First, the confinement of oceans to discrete basins separated by continents prevents the oceanic circulation from assuming the common east–west flow patterns adopted by most atmospheres. (Topography can cause substantial north–south deflections in an atmospheric flow, which may help explain why Earth's atmospheric circulation involves more latitudinal excursions than that of the topography-free giant planets; nevertheless, air's ability to flow over topography means that atmospheres, unlike oceans, are still fundamentally unbounded in the east–west direction.) The only oceanic region unhindered in the east–west direction is the Southern Ocean surrounding Antarctica, and, as might be expected, a strong east–west current, which encircles Antarctica, has formed in this region.

Second, the atmosphere is heated from below, but the ocean is heated from above. Because air is relatively transparent to sunlight, sunlight penetrates through the atmosphere and is absorbed primarily at the surface, where it heats the near-surface air at the bottom of the atmosphere. In contrast, liquid water absorbs sunlight extremely well, so that 99% of the sunlight is absorbed in the top 3% of the ocean. This means, for example, that atmospheric convection—thunderstorms—predominate at low latitudes (where abundant sunlight falls) but are rare near the poles; in contrast, convection in the oceans is totally inhibited at low latitudes and instead can occur only near the poles.

Third, much of the large-scale ocean circulation is driven not by horizontal density contrasts, as in the atmosphere (although these do play a role in the ocean), but by the frictional force of wind blowing over the ocean surface. In fact, the first simple models of ocean circulation developed by Sverdrup, Stommel, and Munk in the 1940s and 1950s, which were based solely on forcing caused by wind stress, did a reasonably good job of capturing the large-scale horizontal circulations in the ocean basins.

As in the atmosphere, the Earth's rotation dominates the large-scale dynamics of the ocean. Horizontal Coriolis forces nearly balance pressure-gradient forces, leading to geostrophy. As in the atmosphere, this means that ocean currents flow perpendicular to horizontal pressure gradients. Rotation also means that wind stress induces currents in a rather unintuitive fashion. Because of the existence of the Coriolis force, currents do not simply form in the direction of the wind stress; instead, the three-way balance between Coriolis, pressure-gradient, and friction forces can induce currents that flow in directions distinct from the wind direction.

Averaged over time, the surface waters in most mid-latitude ocean basins exhibit a circulation consisting of a basin-filling gyre that rotates clockwise in the northern hemisphere and counterclockwise in the southern hemisphere. This circulation direction implies that the water in the western portion of the basin flows from the equator toward the pole, while the water in the eastern portion of the basin flows from the pole toward the equator. However, the flow is extremely asymmetric: The equatorward flow comprises a broad, slow motion that fills the eastern 90% of the ocean basin; in contrast, the poleward flow becomes concentrated into a narrow current (called a western boundary current) along the western edge of the ocean basin. The northward-flowing Gulf Stream off the U.S. eastern seaboard and the Kuroshio Current off Japan are two examples; these currents reach speeds up to ~ 1 m s^{-1} in a narrow zone 50–100 km wide. This extraordinary asymmetry in the ocean circulation results from the increasing strength of the Coriolis force with latitude; theoretical models show that in a hypothetical ocean where Coriolis forces are independent of latitude, the gyre circulations do not exhibit western intensification. These gyres play an important role in Earth's climate by transporting heat from the equator toward the poles. Their clockwise (counterclockwise) rotation in the northern (southern) hemisphere helps explain why the water temperatures tend to be colder along continental west coasts than continental east coasts.

In addition to the gyres, which transport water primarily horizontally, the ocean also experiences vertical overturning. Only near the poles does the water temperature become cold enough for the surface density to exceed the deeper density. Formation of sea ice helps this process, because sea ice contains relatively little salt, so when it forms, the remaining surface water is saltier (hence denser) than average. Thus, vertical convection between the surface and deep ocean occurs only in polar regions, in particular in the Labrador Sea and near parts of Antarctica. On average, very gradual ascending motion must occur elsewhere in the ocean for mass balance to be achieved. This overturning circulation, which transports water from the surface to the deep ocean and back over ~ 1000 year timescales, is called the thermohaline circulation.

The thermohaline circulation helps explain why deep-ocean waters have near-freezing temperatures worldwide: All deep-ocean water, even that in the equatorial oceans, originated at the poles and thus retains the signature of polar temperatures. Given the solar warming of low-latitude surface waters, the existence of a thermocline is thus naturally explained. However, the detailed dynamics that control the horizontal structure and depth of the thermocline are subtle and have led to major research efforts in physical oceanography over the past 4 decades.

Despite the importance of the basin-filling gyre and thermohaline circulations, much of the ocean's kinetic energy resides in small eddy structures only 10–100 km across. The predominance of this kinetic energy at small scales results largely from the natural interaction of buoyancy forces and rotation. Fluid flows away from pressure highs toward pressure lows, but Coriolis forces short-circuit this process by deflecting the motion so that fluid flows perpendicular to the horizontal pressure gradient. The stronger the influence of rotation relative to buoyancy, the better this process is short-circuited, and hence the smaller are the resulting eddy structures. In the atmosphere, this natural length scale (called the deformation radius) is 1000–2000 km, but in the oceans it is only 10–100 km. The rings and meddies described earlier provide striking examples of oceanic eddies in this size range.

4.3 Salinity

When one swims in the ocean, the leading impression is of saltiness. The ocean's global-mean salinity is 3.5% by mass but varies between 3.3 and 3.8% in the open oceans and can reach 4% in the Red Sea and Persian Gulf; values lower than 3.3% can occur on continental shelves near river deltas. The ocean's salt would form a global layer 150 m thick if precipitated into solid form. Sea salt is composed of 55% chlorine, 30% sodium, 8% sulfate, 4% magnesium, and 1% calcium by mass. The $\sim 15\%$ variability in the salinity of open-ocean waters occurs because evaporation and precipitation add or remove freshwater, which dilutes or concentrates the local salt abundance. However, this process cannot influence the relative proportions of elements in sea salt, which therefore remain almost constant everywhere in the oceans.

In contrast to seawater, most river and lake water is relatively fresh; for example, the salinity of Lake Michigan is ~200 times less than that of seawater. However, freshwater lakes always have both inlets and outlets. In contrast, lakes that lack outlets—the Great Salt Lake, the Dead Sea, the Caspian Sea—are always salty. This provides a clue about processes determining saltiness.

Why is the ocean salty? When rain falls on continents, enters rivers, and flows into the oceans, many elements leach into the water from the continental rock. These elements have an extremely low abundance in the continental water, but because the ocean has no outlet (unlike a freshwater lake), these dissolved trace components can build up over time in the ocean. Ocean-seafloor chemical interactions (especially after volcanic eruptions) can also introduce dissolved ions into the oceans. However, the composition of typical river water differs drastically from that of sea salt—typical river salt contains ~9% chlorine, 7% sodium, 12% sulfate, 5% magnesium, and 17% calcium by mass. Although sodium and chlorine comprise ~85% of sea salt, they make up only ~16% of typical river salt. The ratio of chlorine to calcium is 0.5 in river salt but 46 in sea salt. Furthermore, the abundance of sulfate and silica is much greater in river salt than in sea salt. These differences result largely from the fact that processes act to remove salt ions from ocean water, but the efficiency of these processes depends on the ion. For example, many forms of sea life construct shells of calcium carbonate or silica, so these biological processes remove calcium and silica from ocean water. Much magnesium and sulfate seems to be removed in ocean water–seafloor interactions. The relative inefficiency of such removal processes for sodium and chlorine apparently leads to the dominance of these ions in sea salt despite their lower proportion in river salt.

Circumstantial evidence suggests that ocean salinity has not changed substantially over the past billion years. This implies that the ocean is near a quasi-steady state where salt removal balances salt addition via rivers and seafloor–ocean interactions. These removal processes include biological sequestration in shell material, abiological seafloor–ocean water chemical interactions, and physical processes such as formation of evaporate deposits when shallow seas dry up, which has the net effect of returning the water to the world ocean while leaving salt behind on land.

4.4 Atmosphere–Ocean Interactions

Many weather and climate phenomena result from a coupled interaction between the atmosphere and ocean and would not occur if either component were removed. Two major examples are hurricanes and El Niño.

Hurricanes are strong vortices, 100–1000 km across, with warm cores and winds often up to ~70 m s^{-1}; the temperature difference between the vortex and the surrounding air produces the pressure differences that allow strong vortex winds to form. In turn, the strong winds lead to increased evaporation off the ocean surface, which provides an enhanced supply of water vapor to fuel the thunderstorms that maintain the warm core. This enhanced evaporation from the ocean must continue throughout the hurricane's lifetime because the thermal effects of condensation in thunderstorms inside the hurricane provides the energy that maintains the vortex against frictional losses. Thus, both the ocean and atmosphere play crucial roles. When the ocean component is removed—say, when the hurricane moves over land—the hurricane rapidly decays.

El Niño corresponds to an enhancement of ocean temperatures in the eastern equatorial Pacific at the expense of those in the western equatorial Pacific; increased rainfall in western North and South America result, and drought conditions often overtake Southeast Asia. El Niño events occur every few years and have global effects. At the crudest level, "normal" (non-El Niño) conditions correspond to westward-blowing equatorial winds that cause a thickening of the thermocline (hence producing warmer sea-surface temperatures) in the western equatorial Pacific; these warm temperatures promote evaporation, thunderstorms, and upwelling there, drawing near-surface air in from the east and thus helping to maintain the circulation. On the other hand, during El Niño, the westward-blowing trade winds break down, allowing the thicker thermocline to relax eastward toward South America, hence helping to move the warmer water eastward. Thunderstorm activity thus becomes enhanced in the eastern Pacific and reduced in the western Pacific compared to non-El Niño conditions, again helping to maintain the winds that allow those sea-surface temperatures. Although El Niño differs from a hurricane in being a hemispheric-scale long-period fluctuation rather than a local vortex, El Niño shares with hurricanes the fact that it could not exist were either the atmosphere or the ocean component prevented from interacting with the other. To successfully capture these phenomena, climate models need accurate representations of the ocean and the atmosphere and their interaction, which continues to be a challenge.

4.5 Oceans on other Worlds

The *Galileo* spacecraft provided evidence that subsurface liquid-water oceans exist inside the icy moons Callisto, Europa, and possibly Ganymede. The recent detection of a jet of water molecules and ice grains from the south pole of Enceladus raises the question of whether that moon has a subsurface reservoir of liquid water. Theoretical models suggest that internal oceans could exist on a wide range of other bodies, including Titan, the smaller moons of Saturn and Uranus, Pluto, and possibly even some larger Kuiper Belt objects. These oceans of course differ from Earth's ocean in that they are ice-covered; another difference is that they must transport the geothermal heat flux of those bodies and hence are probably convective throughout. Barring exotic chemical or fluid dynamical effects, then, one expects that such oceans lack thermoclines. In many cases, these oceans may be substantially thicker than Earth's oceans; estimates suggest that Europa's ocean thickness lies between 50 and 150 km.

The abundant life that occurs near deep-sea vents ("black smokers") in Earth's oceans has led to suggestions that similar volcanic vents may help power life in Europa's ocean. (In contrast to Europa, any oceans in Callisto and Ganymede would be underlain by high-pressure polymorphs of ice rather than silicate rock, so such silicate–water interactions would be weaker.) However, much of the biological richness of terrestrial deep-sea vents results from the fact that Earth's oceans are relatively oxygenated; when this oxidant-rich water meets the reducing water discharged from black smokers, sharp chemical gradients result, and the resulting disequilibrium provides a rich energy source for life. Thus, despite the lack of sunlight at Earth's ocean floor, the biological productivity of deep-sea vents results in large part from the fact that the oceans are communicating with an oxygen-rich atmosphere. If Europa's ocean is more reducing than Earth's ocean, then the energy source available from chemical disequilibrium may be smaller. Nevertheless, a range of possible disequilibrium reactions exist that could provide energy to drive a modest microbial biosphere on Europa.

5. Climate

Earth's climate results from a wealth of interacting physical, chemical, and biological effects, and an understanding of current and ancient climates has required a multidecadal research effort by atmospheric physicists, atmospheric chemists, oceanographers, glaciologists, astronomers, geologists, and biologists. The complexity of the climate system and the interdisciplinary nature of the problem have made progress difficult, and even today many aspects remain poorly understood. "Climate" can be defined as the mean conditions of the atmosphere/ocean system—temperature, pressure, winds/currents, cloudiness, atmospheric humidity, oceanic salinity, and atmosphere/ocean chemistry in three dimensions—when time-averaged over intervals longer than that of typical weather patterns. It also refers to the distribution of sea ice, glaciers, continental lakes and streams, coastlines, and the spatial distribution of ecosystems that result.

5.1 Basic Processes—Greenhouse Effect

Earth as a whole radiates with an effective temperature of 255 K, and therefore its flux peaks in the thermal infrared part of the spectrum. This effective temperature is 30 K colder than the average temperature on the surface, and quite chilly by human standards.

What ensures a warm surface is the wavelength-dependent optical properties of the troposphere. In particular, infrared light does not pass through the troposphere as readily as visible light. The Sun radiates with an effective temperature of 5800 K and therefore its peak flux is in the visible part of the spectrum (or stated more correctly in reverse, we have evolved such that the part of the spectrum that is visible to us is centered on the peak flux from the Sun). The atmosphere reflects about 31% of this sunlight directly back to space, and the rest is absorbed or transmitted to the ground. The sunlight that reaches the ground is absorbed and then reradiated at infrared wavelengths. Water vapor (H_2O) and carbon dioxide (CO_2), the two primary greenhouse gases, absorb some of this upward infrared radiation and then emit it in both the upward and downward directions, leading to an increase in the surface temperature to achieve balance. This is the greenhouse effect. Contrary to popular claims, the elevation of surface temperature by the greenhouse effect is not a situation where "the heat cannot get out." Instead, the heat must get out, and to do so in the presence of the blanketing effect of greenhouse gases requires an elevation of surface temperatures.

The greenhouse effect plays an enormous role in the climate system. A planet without a greenhouse effect, but otherwise identical to Earth, would have a global-mean surface temperature 17°C below freezing. The oceans would be mostly or completely frozen, and it is doubtful whether life would exist on Earth. We owe thanks to the greenhouse effect for Earth's temperate climate, liquid oceans, and abundant life.

Water vapor accounts for between one third and two thirds of the greenhouse effect on Earth (depending on how the accounting is performed), with the balance resulting from CO_2, methane, and other trace gases. Steady increases in carbon dioxide due to human activity seem to be causing the well-documented increase in global surface temperature over the past $\sim$100 years. On Mars, the primary atmospheric constituent is CO_2, which together with atmospheric dust causes a modest 5 K greenhouse effect. Venus has a much denser CO_2 atmosphere, which, along with atmospheric sulfuric acid and sulfur dioxide, absorbs essentially all the infrared radiation emitted by the surface, causing an impressive 500 K rise in the surface temperature. Interestingly, if all the carbon held in Earth's carbonate rocks were liberated into the atmosphere, Earth's greenhouse effect would approach that on Venus. [See MARS ATMOSPHERE: HISTORY AND SURFACE INTERACTION; VENUS: ATMOSPHERE.]

5.2 Basic Processes—Feedbacks

The Earth's climate evolves in response to volcanic eruptions, solar variability, oscillations in Earth's orbit, and changes in internal conditions such as the concentration of greenhouse gases. The Earth's response to these perturbations is highly nonlinear and is determined by feedbacks in the climate system. Positive feedbacks amplify a perturbation and, under some circumstances, can induce a runaway process where the climate shifts abruptly to a completely different state. In contrast, negative feedbacks reduce the effect of a perturbation and thereby help maintain the climate in its current state. Some of the more important feedbacks are as follows.

Thermal feedback: Increases in the upper tropospheric temperature lead to enhanced radiation to space, tending to cool the Earth. Decreases in the upper tropospheric temperature cause decreased radiation to space, causing warming. This is a negative feedback.

Ice-albedo feedback: Ice caps and glaciers reflect visible light easily, so the Earth's brightness (**albedo**) increases with an increasing distribution of ice and snow. Thus, a more ice-rich Earth absorbs less sunlight, promoting colder conditions and growth of even more ice. Conversely, melting of glaciers causes Earth to absorb more sunlight, promoting warmer conditions and even less ice. This is a positive feedback.

Water-vapor feedback: Warmer surface temperatures allow increased evaporation of water vapor from the ocean surface, increasing the atmosphere's absolute humidity. Because water vapor is a greenhouse gas, it promotes an increase in the strength of the greenhouse effect and hence even warmer conditions. Cooler conditions inhibit evaporation, lessen the greenhouse effect, and cause additional cooling. This is a positive feedback.

Cloud feedback: Changes in climate can cause changes in the spatial distribution, heights, and properties of clouds. Greater cloud coverage means a brighter Earth (higher albedo), leading to less sunlight absorption. Higher altitude clouds have colder tops that radiate heat to space less well, promoting a warmer Earth. For a given mass of condensed water in a cloud, clouds with smaller particles reflect light better, promoting a cooler Earth. Unfortunately, for a specified climate perturbation (e.g., increasing the CO_2 concentration), the extent to which the coverage, heights, and properties of clouds will change remains unclear. Thus, not only the magnitude but even the sign (positive or negative) of this feedback remains unknown.

The sum of these and other feedbacks determine how Earth's climate evolved during past epochs and how Earth will respond to current human activities such as emissions of CO_2. Much of the uncertainty in current climate projections results from uncertainty in these feedbacks. A related concept is that of thresholds, where the climate undergoes an abrupt shift in response to a gradual change. For example, Europe enjoys temperate conditions despite its high latitude in part because of heat transported poleward by the Gulf Stream. Some climate models have suggested that increases in CO_2 due to human activities could suddenly shift the ocean circulation in the North Atlantic into a regime that transports heat less efficiently, which could cause widespread cooling in Europe (although this might be overwhelmed by the expected global warming that will occur over the next century). The rapidity with which ice ages ended also suggests that major reorganizations of the ocean/atmosphere circulation occurred during those times. Although thresholds play a crucial role in past and possibly future climate change, they are notoriously difficult to predict because they involve subtle nonlinear interactions.

5.3 Recent Times

A wide range of evidence demonstrates that Earth's global-mean surface temperature rose by about 0.6°C between 1900 and 2000. Since the mid-1970s, the global-mean rate of temperature increase has been ∼0.17°C per decade (with a greater rate of warming over land than ocean). As of 2006, 20 of the hottest years measured since good instrumental records started in ∼1860 have occurred within the past 25 years, and the past 25 years has been the warmest 25 year period of the past 1000 years. There is widespread consensus among climate experts that the observed warming since ∼1950 has been caused primarily by the release of CO_2 due to human activities, primarily the burning of oil, coal, natural gas, and forests: The greater CO_2 concentration has increased the strength of the greenhouse effect, modified by the feedbacks discussed in Section 5.2. Before the Industrial Revolution, the CO_2 concentration was ∼280 ppm (i.e., a mole fraction of 2.8×10^{-4}), and in 2006 the CO_2 concentration was 380 ppm—a 36% increase. Interestingly, only half of the CO_2 released by human activities each year remains in the atmosphere; the remainder is currently absorbed by the biosphere and especially the oceans. The increase in mean surface temperature has been accompanied by numerous other climate changes, including retreat of glaciers worldwide, thawing of polar permafrost, early arrivals of spring, late arrivals of autumn, changes in the Arctic sea-ice thickness, approximately 0.1–0.2 m of sea-level increase since 1900, and various effects on natural ecosystems. These changes are expected to accelerate in the 21st century.

5.4 Ice Ages

The repeated occurrence of ice ages, separated by warmer interglacial periods, dominates Earth's climatic record of the past 2 million years. During an ice age, multi-kilometer-thick ice sheets grow to cover much of the high-latitude land area, particularly in North America and Europe; most or all of these ice sheets melt during the interglacial periods (however, ice sheets on Antarctica and Greenland have resisted melting during most interglacials, and these two ice sheets still exist today). The sea level varies by up to 120 m

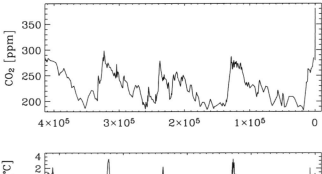

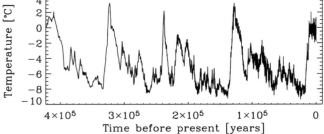

FIGURE 4 CO_2 concentrations (top) and temperature variations (bottom) over the last 420,000 years as obtained from ice cores at Vostok, Antarctica (data from Petit et al., 1999). The approximate 100,000-year period of the ice ages is evident, although many shorter period fluctuations are superimposed within the record. Prominent ice-age terminations occurred at ~410, 320, 240–220, 130, and 15 ka in the past. Also note the correlation between temperature and CO_2 concentration during these cycles, which shows the influence of changes in the greenhouse effect on ice ages. The vertical line at the right side of the top plot shows the increase in CO_2 caused by humans between ~1800 and 2006.

between glacial and interglacial periods, causing migration of coastlines by hundreds of kilometers in some regions. The time history of temperature, ice volume, and other variables can be studied using stable isotopes of carbon, hydrogen, and oxygen as recorded in glacial ice, deep-sea sediments, and land-based records such as cave calcite and organic material. This record shows that glacial/interglacial cycles over the past 800,000 years have a predominant period of ~100,000 years (Fig. 4). During this cycle, glaciers gradually increase in volume (and air temperature gradually decreases) over most of the 100,000 year period; the glaciers then melt, and the temperature increases over a relatively short ~5000 year interval. The cycle is thus extremely asymmetric and resembles a saw-tooth curve rather than a sinusoid. The last ice age peaked 18,000 years ago and ended by 10,000 years ago; the modern climate corresponds to an interglacial period. Analysis of ancient air trapped in air bubbles inside the Antarctic and Greenland ice sheets shows that the atmospheric CO_2 concentration is low during ice ages—typically about 200 ppm—and rises to ~280 ppm during the intervening interglacial periods (Fig. 4).

Ice ages seem to result from changes in the strength of sunlight caused by periodic variations in Earth's orbit, magnified by several of the feedbacks discussed in Section 5.2. A power spectrum of the time series in Fig. 4 shows that temperature, ice volume, and CO_2 vary predominantly on periods of 100, 41, 23, and 19 thousand years (ka; the summation of sinusoids at each of these periods leads to the saw-tooth patterns in Fig. 4). Interestingly, these periods match the periods over which northern hemisphere sunlight varies due to orbital oscillations. The Earth's orbital eccentricity oscillates on periods of 100 ka, the orbital obliquity (the tilt of Earth's rotation axis) oscillates on a period of 41 ka, and the Earth's rotation axis precesses on periods of 19 and 23 ka. These variables affect the difference in sunlight received at Earth between winter and summer and between the equator and pole. In turn, these sunlight variations determine the extent to which snowpack accumulates in high northern latitudes during winter, and the extent to which this snowpack resists melting during summer; glaciers build up when snow that falls during winter cannot melt the following summer. The idea that these orbital variations cause ice ages has become known as the Milankovitch theory of ice ages.

By themselves, however, orbital variations are only part of the story. Sunlight variations due to the 100 ka eccentricity variations are much weaker than sunlight variations due to the 41, 23, and 19 ka obliquity and precession variations. Thus, if the orbit-induced sunlight variations translated directly into temperature and ice variations, ice ages would be dominated by the 41, 23, and 19 ka periods, but instead, the 100 ka period dominates (as can be seen in Fig. 4). This means that some nonlinearity in the climate system amplifies the climatic response at 100 ka much better than at the shorter periods. Furthermore, the observed oscillations in CO_2 between glacial and interglacial periods (Fig. 4) indicates that ice ages are able to occur partly because the greenhouse effect is weak during ice ages but strong during interglacial periods. Most likely, atmospheric CO_2 becomes dissolved in ocean water during ice ages, allowing the atmospheric CO_2 levels to decrease; the ocean then rejects this CO_2 at the end of the ice age, increasing its atmospheric concentration. Recent analyses of Antarctic ice cores show that, at the end of an ice age, temperature rise precedes CO_2 rise in Antarctica by about 800 years, indicating that CO_2 variation is an amplifier rather than a trigger of ice-age termination. Interestingly, however, both of these events precede the initiation of deglaciation in the northern hemisphere. These observations suggest that the end of an ice age is first triggered by a warming event in the Antarctic region; this initiates the process of CO_2 rejection from the oceans to the atmosphere, and the resulting increase in the greenhouse effect, which is global, then allows deglaciation to commence across the rest of the planet. The ice-albedo and water-vapor feedbacks (Section 5.2) help amplify the transition. However, many details, including the exact mechanism that allows CO_2 to oscillate between the ocean and atmosphere, remain to be worked out.

Figure 4 shows how the increase in CO_2 caused by human activities compares to the natural variability in the past. The saw-toothed variations in CO_2 between 200 and 280 ppm over 100,000 year periods indicate the ice-age/interglacial cycles, and the vertical spike in CO_2 at the far right of Fig. 4 (from 280 to 380 ppm) shows the human-induced increase. The current CO_2 concentration far exceeds that at any previous time over the past 420,000 years, and is probably the greatest CO_2 level the Earth has seen since 20 million years ago. The fact that CO_2 rises by 30–40% at the end of an ice age indicates that very large magnitude climate changes can accompany modest CO_2 variations; it is noteworthy that human activities have so far increased CO_2 by an additional 36% beyond preindustrial values. The relationship between CO_2 and global temperature during ice ages may differ from the relationship these quantities will take over the next century of global warming; however, it is virtually certain that additional CO_2 will cause global temperature increases and widespread climate changes. Current economic and climate projections indicate that, because of continued fossil fuel burning, the atmospheric CO_2 will reach 500–1000 ppm by the year 2100 unless drastic measures are adopted to reduce fossil fuel use.

5.5 Volatile Inventories of Terrestrial Planets

Venus, Earth, and Mars have present-day atmospheres that are intriguingly different. The atmospheres of Venus and Mars are both primarily CO_2, but they represent two extreme fates in atmospheric evolution: Venus has a dense and hot atmosphere, whereas Mars has a thin and cold atmosphere. It is reasonable to ask whether Earth is ultimately headed toward one or the other of these fates, and whether these three atmospheres have always been so different.

The history of volatiles on the terrestrial planets includes their origin, their interactions with refractory (nonvolatile) material, and their rates of escape into space. During the initial accretion and formation of the terrestrial planets, it is thought that most or all of the original water reacted strongly with the iron to form iron oxides and hydrogen gas, with the hydrogen gas subsequently escaping to space. Until the iron cores in the planets were completely formed and this mechanism was shut down, the outflow of hydrogen probably took much of the other solar-abundance volatile material with it. Thus, one likely possibility is that the present-day atmospheres of Venus, Earth, and Mars are not primordial, but have been formed by outgassing and by cometary impacts that have taken place since the end of core formation.

The initial inventory of water that each terrestrial planet had at its formation is a debated question. One school of thought is that Venus formed in an unusually dry state compared with Earth and Mars; another is that each terrestrial planet must have started out with about the same amount of water per unit mass. The argument for an initially dry Venus is that water-bearing minerals would not condense in the high-temperature regions of the protoplanetary nebula inside of about 1 AU. Proponents of the second school of thought argue that gravitational scattering caused the terrestrial planets to form out of materials that originated over the whole range of terrestrial-planet orbits, and therefore that the original water inventories for Venus, Earth, and Mars should be similar.

An important observable that bears on the question of original water is the enrichment of deuterium (D) relative to hydrogen. A measurement of the D/H ratio yields a constraint on the amount of hydrogen that has escaped from a planet. For the D/H ratio to be useful, one needs to estimate the relative importance of the different hydrogen escape mechanisms and the original D/H ratio for the planet. In addition, one needs an idea of the hydrogen sources available to a planet after its formation, such as cometary impacts. The initial value of D/H for a planet is not an easy quantity to determine. A value of 0.2×10^{-4} has been put forward for the protoplanetary nebula, which is within a factor of 2 or so for the present-day values of D/H inferred for Jupiter and Saturn. However, the D/H ratio in Standard Mean Ocean Water (SMOW, a standard reference for isotopic analysis) on Earth is 1.6×10^{-4}, which is also about the D/H ratio in hydrated minerals in meteorites, and is larger by a factor of 8 over the previously mentioned value. At the extreme end, some organic molecules in carbonaceous chondrites have shown D/H ratios as high as 20×10^{-4}. The enrichment found in terrestrial planets and most meteorites over the protoplanetary nebula value could be the result of exotic high-D/H material deposited on the terrestrial planets, or it could be the result of massive hydrogen escape from the planets early in their lifetimes through the hydrodynamic blowoff mechanism (which is the same mechanism that currently drives the solar wind off the Sun).

6. Life in the Atmosphere–Ocean System

6.1 Interplanetary Spacecraft Evidence for Life

An ambitious but ever-present goal in astronomy is to detect or rule out life in other solar systems, and in planetary science that goal is to detect or rule out life in our own solar system apart from Earth. Water in its liquid phase is one of the few requirements shared by all life on Earth, and so the hunt for life is focused on the search for liquid water. We know that Mars had running water on its surface at some point in its history because we can see fluvial channels in high-resolution images, and because the Mars rovers *Spirit* and *Opportunity* have discovered aqueous geochemistry on the ground; there is even some evidence suggesting present-day seepage in recent orbiter images. Farther out in the solar system, we know that Europa, a satellite of Jupiter, has a smooth icy surface with cracks and flow features that resemble Earth's polar ice fields and suggest a liquid-water interior, while its larger sibling, Ganymede, exhibits a conductive reaction to Jupiter's magnetic field that is most easily explained by a salty liquid-water interior [See MARS: SURFACE AND INTERIOR; METEORITES; PLANETARY SATELLITES.]

However, to date we have no direct evidence for extraterrestrial life. This includes data from landers on Venus, Mars, and the Moon, and flyby encounters with 8 planets, a handful of asteroids, a comet (Halley in 1986), and over 60 moons. Are the interplanetary spacecraft we have sent out capable of fulfilling the goal of detecting life? This question has been tested by analyzing data from the *Galileo* spacecraft's two flyby encounters with Earth, which, along with a flyby encounter with Venus, were used by the spacecraft's navigation team to provide gravity assists to send *Galileo* to Jupiter. The idea was to compare ground-truth information to what we can learn solely from Galileo. [See ATMOSPHERES OF THE GIANT PLANETS; IO: THE VOLCANIC MOON; PLANETARY SATELLITES.]

Galileo's first Earth encounter occurred on December 8, 1990, with closest approach 960 km above the Caribbean Sea; its second Earth encounter occurred on December 8, 1992, with closest approach 302 km above the South Atlantic. A total of almost 6000 images were taken of Earth by *Galileo*'s camera system. Figure 5 shows the Earth–Moon system as seen by *Galileo*. Notice that the Moon is significantly darker than Earth. The spacecraft's instru-

ments were designed and optimized for Jupiter; nevertheless, they made several important observations that point to life on Earth. These strengthen the null results encountered elsewhere in the solar system. The evidence for life on Earth includes complex radio emissions, nonmineral surface pigmentation, disequilibrium atmospheric chemistry, and large oceans.

6.1.1 RADIO EMISSIONS

The only clear evidence obtained by *Galileo* for intelligent life on Earth was unusual radio emissions. Several natural radio emissions were detected, none of which were unusual, including solar radio bursts, auroral kilometric radiation, and narrowband electrostatic oscillations excited by thermal fluctuations in Earth's ionospheric plasma. The first unusual radio emissions were detected at 1800 UT and extended through 2025 UT, just before closest approach. These were detected by the plasma wave spectrometer (PWS) on the nightside, in-bound pass, but not on the day side, out-bound pass. The signal strength increased rapidly as Earth was approached, implying that Earth itself was the source of the emissions. The fact that the signals died off on the day side suggests that they were cut off by the day side ionosphere, which means we can place the source below the ionosphere.

The unusual signals were narrowband emissions that occurred in only a few distinct channels and had average frequencies that remained stable for hours. Naturally occurring radio emissions nearly always drift in frequency, but

FIGURE 5 The Earth–Moon system as observed by the *Galileo* spacecraft.

these emissions were steady. The individual components had complicated modulations in their amplitude that have never been detected in naturally occurring emissions. The simplest explanation is that these signals were transmitting information, which implies that there is advanced technological life on Earth. In fact, the radio, radar, and television transmissions that have been emanating from Earth over the last century result in a nonthermal radio emission spectrum that broadcasts our presence out to interstellar distances. [*See* THE SOLAR SYSTEM AT RADIO WAVELENGTHS.]

6.1.2 SURFACE FEATURES

During its first encounter with Earth, the highest-resolution mapping of the surface by *Galileo*'s Solid-State Imaging System (SSI) covered Australia and Antarctica with 1–2 km resolution. No usable images were obtained from Earth's night side on the first encounter. The second encounter netted the highest resolution images overall of Earth by *Galileo*, 0.3–0.5 km per pixel, covering parts of Chile, Peru, and Bolivia. The map of Australia from the first encounter includes 2.3% of Earth's total surface area, but shows no geometric patterns that might indicate an advanced civilization. In the second encounter, both the cities of Melbourne and Adelaide were photographed, and yet no geometric evidence is visible because the image resolution is only 2 km. The map of Antarctica, 4% of Earth's surface, reveals nearly complete ice cover and no signs of life. Only one image, taken of southeastern Australia during the second encounter, shows east–west and north–south markings that would raise suspicions of intelligent activity. The markings in fact were caused by boundaries between wilderness areas, grazing lands, and the border between South Australia and Victoria. Studies have shown that it takes nearly complete mapping of the surface at 0.1-km resolution to obtain convincing photographic evidence of an advanced civilization on Earth, such as roads, buildings, and evidence of agriculture.

On the other hand, many features are visible in the *Galileo* images that have not been seen on any other body in the solar system. The SSI camera took images in six different wavelength channels. A natural-color view of Earth was constructed using the red, green, and violet filters, which correspond to wavelengths of 0.670, 0.558, and 0.407 μm, respectively. The images reveal that Earth's surface is covered by enormous blue expanses that specularly reflect sunlight, and end in distinct coastlines, which are both easiest to explain if the surface is liquid. This implies that much of the planet is covered with oceans. The land surfaces show strong color contrasts that range from light brown to dark green.

The SSI camera has particular narrowband infrared filters that have never been used to photograph Earth before, and so they yielded new information for geological, biological, and meteorological investigations. The infrared filters allow the discrimination of H_2O in its solid, liquid, and gaseous forms; for example, clouds and surface snow can be distinguished spectroscopically with the 1 μm filter. False-color images made by combining the 1 μm channel with the red and green channels reveal that Antarctica strongly absorbs 1 μm light, establishing that it is covered by water ice. In contrast, large regions of land strongly reflect 1 μm without strongly reflecting visible colors, which conflicts with our experience from other planetary surfaces and is not typical of igneous or sedimentary rocks or soil. Spectra made with the 0.73 and 0.76 μm channels reveal several land areas that strongly absorb red light, which again is not consistent with rocks or soil. The simplest explanation is that some nonmineral pigment that efficiently absorbs red light has proliferated over the planet's surface. It is hard to say for certain if an interstellar explorer would realize that this is a biological mechanism for gathering energy from sunlight, probably so, but certainly we would recognize it on another planet as the signature of plant life. We know from ground truth that these unusual observations are caused by the green pigments chlorophyll a ($C_{55}H_{72}MgN_4O_5$) and chlorophyll b ($C_{55}H_{70}MgN_4O_6$), which are used by plants for photosynthesis. No other body in the solar system has the green and blue colorations seen on Earth. [*See* THE SOLAR SYSTEM AT ULTRAVIOLET WAVELENGTHS; INFRARED VIEWS OF THE SOLAR SYSTEM FROM SPACE.]

6.1.3 OXYGEN AND METHANE

Galileo's Near-Infrared Mapping Spectrometer (NIMS) detected the presence of molecular oxygen (O_2) in Earth's atmosphere with a volume mixing ratio of 0.19 ± 0.05. Therefore, we know that the atmosphere is strongly oxidizing. (It is interesting to note that Earth is the only planet in the solar system where one can light a fire.) In light of this, it is significant that NIMS also detected methane (CH_4) with a volume mixing ratio of $3 \pm 1.5 \times 10^{-6}$. Because CH_4 oxidizes rapidly into H_2O and CO_2, if thermodynamical equilibrium holds, then there should be no detectable CH_4 in Earth's atmosphere. The discrepancy between observations and the thermodynamic equilibrium hypothesis, which works well on other planets (e.g., Venus), is an extreme 140 orders of magnitude. This fact provides evidence that Earth has biological activity and that it is based on organic chemistry. We know from ground truth that Earth's atmospheric methane is biological in origin, with about half of it coming from nonhuman activity like methane bacteria and the other half coming from human activity like growing rice, burning fossil fuels, and keeping livestock. NIMS also detected a large excess of nitrous oxide (N_2O) that is most easily explained by biological activity, which we know from ground truth comes from nitrogen-fixing bacteria and algae.

The conclusion is that the interplanetary spacecraft we have sent out to explore our solar system are capable of detecting life on planets or satellites, both the intelligent and primitive varieties, if it exists in abundance on the surface. On the other hand, if there is life on a planet or satellite that does not have a strong signature on the surface, as would probably be the case if Europa or Ganymede harbor life, then a flyby mission may not be adequate to decide the question. With regard to abundant surface life, we have a positive result for Earth and a negative result for every other body in the solar system.

7. Conclusions

Viewing Earth as a planet is the most important change of consciousness that has emerged from the space age. Detailed exploration of the solar system has revealed its beauty, but it has also shown that the home planet has no special immunity to the powerful forces that continue to shape the solar system. The ability to remotely sense Earth's dynamic atmosphere, oceans, biosphere, and geology has grown up alongside our ever-expanding ability to explore distant planetary bodies. Everything we have learned about other planets influences how we view Earth. Comparative planetology has proven in practice to be a powerful tool for studying Earth's atmosphere and oceans. The lion's share of understanding still awaits us, and in its quest we continue to be pulled outward.

Bibliography

Dowling, T. E. (2001). Oceans. In "Encyclopedia of Astronomy and Astrophysics," pp. 1919–1928. IOP Publishing LTD and Nature Publishing Group, Bristol.

Geissler, P., Thompson, W. R., Greenberg, R., Moersch, J., McEwen, A., and Sagan, C. (1995). Galileo multispectral imaging of Earth. *J. Geophys. Res.* **100**, 16, 895–16, 906.

Hide, R., and Dickey, J. O. (1991). Earth's variable rotation. *Nature* **253**, 629–637.

Holton, J. R., Pyle, J., and Curry, J. A., eds. (2002). "Encyclopedia of Atmospheric Sciences." Academic Press.

Hurrell, J. W., van Loon, H., and Shea, D. J. (1998). The mean state of the troposphere. In "Meteorology of the Southern Hemisphere" (D. J. Karoly and D. G. Vincent, eds.), Met. Monograph 27 (49), pp. 1–46. American Meteorological Society, Boston.

Petit, J. R., et al. (1999). Climate and atmospheric history of the past 420,000 years from the Vostok ice core, Antarctica. *Nature* **399**, 429–436.

Showman, A. P., and Malhotra, R. (1999). The Galilean satellites. *Science* **286**, 77–84.

Earth as a Planet: Surface and Interior

David C. Pieri

Jet Propulsion Laboratory
California Institute of Technology
Pasadena, California

Adam M. Dziewonski

Harvard University
Cambridge, Massachusetts

CHAPTER 10

1. Introduction: the Earth as a Guide to Other Planets

The surface of the Earth is perhaps the most geochemically diverse and dynamic among the planetary surfaces of our solar system. Uniquely, it is the only one with liquid water oceans under a stable atmosphere, and—as far as we now know—it is the only surface in our solar system that has given rise to life. The Earth's surface is a dynamic union of its solid crust, its atmosphere, its hydrosphere, and its biosphere, all having acted in concert to produce a constantly renewing and changing symphony of form (Figure 1).

The unifying theme of the Earth's surficial system is water—in liquid, vapor, and solid phases—which transfers and dissipates solar, mechanical, chemical, and biological energy throughout global land and submarine landscapes. The surface is a window to the interior processes of the Earth, as well as the putty that atmospheric processes continually shape. It is also the Earth's interface with extraterrestrial processes and, as such, has regularly borne the scars of impacts by meteors, comets, and asteroids, and will continue to do so.

Our solar system has a variety of terrestrial planets and satellites in various hydrologic states with radically differing hydrologic histories. Some appear totally desiccated, such as the Moon, Mercury, and Venus. In some places where

water is very abundant now at the surface, such as on the Earth, and the Jovian Galilean satellite Europa (solid at the surface and possibly liquid underneath), the Saturnian satellite Enceladus (possibly erupting water vapor into space through an icy surface), and Titan, Saturn's largest moon (where a 94°K surface temperature makes water ice at least as hard as granite). In other places, such as Mars and Ganymede, it appears that water may have been very abundant in liquid form on the surface in the distant past. Also, in the case of Mars, water may yet be abundant in solid and/or liquid form in the subsurface today. Thus, for understanding geological (and, where applicable, biological) processes and environmental histories of terrestrial planets and satellites within our solar system, it is crucial to explore the geomorphology of surface and submarine landforms and the nature and history of the land–water interface where it existed. Such an approach and "lessons learned" from this solar system will also be key in future reconnaissance of extrasolar planets. [See MARS: SURFACE AND INTERIOR.]

2. Physiographic Provinces of Earth

2.1 Basic Divisions

From a geographic and geomorphologic point of view, especially when seen from space, the surface of the Earth

FIGURE 1 Blue Marble view of the Earth from Apollo 17. Earth as seen from the outbound Apollo 17, showing Mediterranean Sea to the north and Antarctica to the south. The Arabian Peninsula and the northeastern edge of Africa can also be seen. Asia is on the northeast (upper right) horizon. Most striking is the prevalence of liquid water (thus evidence of an average surface temperature >273°K), not now present in the arid landscapes of the other solid bodies within our solar system.(Courtesy of NASA)

is dominated by its oceans of liquid water: approximately 75% of the Earth's surface is covered by liquid or solid water. The remaining 25% of nonmarine subaerial land, the subject of nearly all historical geological and geomorphological study, lies mainly in its Northern Hemisphere, where most of the world's population lives. The Southern Hemisphere is dominated by oceans, some subaerial continental and archipelago land masses (mainly parts of Africa, South America, southeast Asia, and Australia), and the large, mainly subglacial, island continent of Antarctica (Fig. 2a).

Remarkably, despite the fact that geological and geographical sciences have been practiced on the Earth for about 200 years, it has only been during the last 40 or so that scientists have begun detailed mapping and geophysical explorations of the submarine land surface. Subsea remote-sensing technology has provided one of the most profound discoveries in the history of geological science: the paradigm of "plate tectonics." The extent, morphology, and dynamics of the Earth's massive tectonic plates were only realized after careful topographic and geomagnetic mapping of the intensely volcanic midoceanic ridges and their associated parallel-paired geomagnetic domains.

Similar topographic mapping of the corresponding submarine trenches along continental or island-arc margins was equally revealing. The midoceanic ridges were found to be sites of accretion of new volcanically generated plate mate-

rial, and the trenches the sites of deep subduction, where oceanic crust is consumed beneath other over-riding crustal plates. Tectonic plates represent the most fundamental and largest geomorphic provinces on Earth.

The Earth's crustal plates come in two varieties: oceanic and continental (Fig. 3).

Oceanic plates comprise nearly all of the Earth's ocean floors, and thus most of the Earth's crustal area. They are composed almost exclusively of iron and magnesium-rich rocks derived from volcanic processes (called "basalts"). Oceanic plates are created by volcanic eruptions along the apices of the Earth's midoceanic ridges: 1000-km-long sinuous ridges that rise from the flat ocean floor (called "abyssal plains") in the middle of oceans. Oceanic plates are typically less than 10 km thick. Here, nearly continuous volcanic activity from countless submarine volcanic centers (far more than the 1000 or so active subaerial volcanoes) provides a steady supply of new basalt, which is accreted and incorporated into the interior part of the plate.

At plate edges, roughly the reverse occurs, where the outer, oldest plate margins are forced below over-riding adjacent plate edges. Usually, when two oceanic plates collide, the resulting subduction zone forms an island arc along the trace of the collision. The islands, in this case, are the result of the eruption of lighter, more silica-rich magmas generated as part of the subduction process. The subducted plate margin is consumed along the axis of the resulting trench.

FIGURE 2   (a) Physiographic map of the Earth. This image was generated from digital data bases of land and sea-floor elevations on a 2-minute latitude/longitude grid. Assumed illumination is from the west, and the projection is Mercator. Spatial resolution of the gridded data varies from true 2-minute for the Atlantic, Pacific, and Indian Ocean floors and all land masses to 5 minutes for the Arctic Ocean floor. (Courtesy of NOAA). (b) Volcanoes and the Crustal Plates. Global map of the major tectonic plate boundaries and locations of the world's volcanoes (Courtesy of the U.S. Geological Survey).

Because the more silicic island arcs tend to be less dense and thus more resistant to subduction, they can be accreted onto plate margins and can thus increase the areal extent at the edges of oceanic plates or can enlarge the margins of existing continental plates.

Continental plates tend to consist of much more silicic material, and are thus lighter, as compared with oceanic plates. Because of their lower density and the fact that they are isostatically compensated, they are much thicker than oceanic plates (30–40 km thick) and tend to "float" over the denser, more mafic (ferromagnesian—of the metals iron and magnesium) subjacent material in the Earth's upper mantle. When continental plates collide with oceanic plates, deep subduction trenches, such as the Peru–Chile trench along the west coast of South America form, as the oceanic plate is forced under the much thicker and less dense continental plate. Usually, the landward side of the affair is marked by so-called Cordilleran belts of mountains, including andesitic-type volcanoes, which parallel the coast-line. The Andes Mountains are an example of this type of tectonic arrangement.

When continental plates collide, a very different tectonic and geomorphic regime ensues. Here, equally buoyant and thick continental plates crush against each other, resulting in the formation of massive fold belts and towering mountains, as long as the tectonic zone is active (e.g., the Himalayan Range in Asia). When aggregate stresses are tensional rather than compressive, extensional mountain ranges can form, as tectonic blocks founder and rotate. The western U.S. Basin and Range Province is a good example of that type of mountain terrane. Another large subaerial extensional tectonic landform is the axial rift valley and associated inward-facing fault scarps, which form when aggregate tensional stresses tend to pull a continental plate apart (e.g., the East African Rift Valley).

The geomorphic provinces just discussed generally tend to be very dynamic, with lifetimes that are intrinsically short (100–200 Myr) relative to the age of the Earth (4.56 Byr).

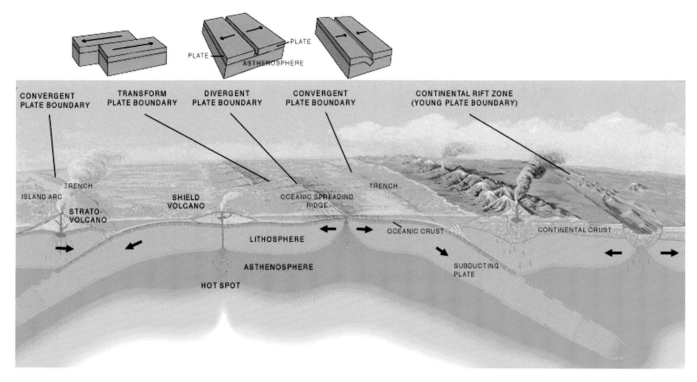

FIGURE 3 Tectonic plate interactions. Tectonic plate interactions and the three fundamental kinds of plate boundaries. (Left) A convergent boundary caused by the subduction of oceanic material as it is overridden by another oceanic plate. (Center left) A subplate hot spot capped by a shield volcano (e.g., Hawaiian Islands). (Center right) A divergent plate boundary, in particular, a midoceanic spreading ridge. (Right) Another kind of convergent plate boundary, where the oceanic crust is being subducted by overriding continental crust, producing a chain of volcanic mountains (e.g., Andes Mountains). (Far right) A continental rift zone, another kind of divergent plate boundary (e.g., East African Rift). Finally, a transform plate boundary is shown at the upper middle of the scene, where two plates are sliding past each other without subduction. The three relationships are shown as block diagrams at the top of the figure. (Courtesy of the U.S. Geological Survey.)

Some of the stable interior areas of continental plates, or cratons, however, do possess landforms and associated lithologic regions with ages comparable within a factor of two or three to the age of the Earth (2–3 Byr). The interior of the Canadian Shield and the Australian Continent are two such special areas. Despite having been scoured repeatedly by continental ice sheets, the granitic craton of the Canadian Shield possesses a record of giant asteroidal and cometary impacts that are about 2 Byr old. [*See* PLANETARY IMPACTS]. These interior cratonic areas, in contrast to most of the rest of the Earth, which is mobile and active, provide a chance to view a part of the long sweep of the Earth's surface history. They are thus important, particularly in trying to understand how the environmental history of the Earth compares to that of the other terrestrial planets.

The distribution of the earth's landscape altitudes, relative to the mean geoid, is bimodal—continental and sea floor (Fig. 4a). Although limited in percentage of surface area coverage, the interface between the two modes is a relatively high-energy place called the littoral or tidal zone. Ocean tides in this zone generate frequent (twice daily) environmental stresses on its residents that profoundly encourage evolution and natural selection, and may have been a key influence on the origin and early evolution of life here. It is interesting that Mars is another planet with a global bimodal highland/lowland dichotomy and may have had early oceans, although the absence of large lunar tides may be significant in this context. [*See* PLANETS AND THE ORIGIN OF LIFE.]

2.2 Landform Types

2.2.1 SUBMARINE LANDFORMS

Geomorphically, submarine oceanic basins comprise the areally dominant landform of the Earth, but, ironically,

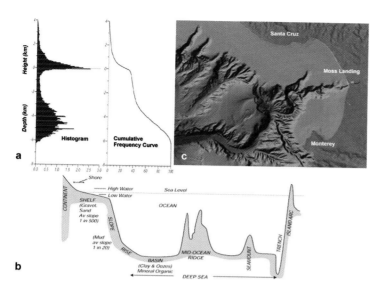

FIGURE 4 (a) Global altitude diagrams. At left are histograms of land altitudes and seafloor depth as a percentage of the Earth's surface area (50-m intervals), illustrating the classic continent–sea floor dichotomy. The interface between the two, subject to tidal and climatic fluctuation stress, is thought to have provided, in part, stimuli for biological evolutionary adaptations. At right is the global hypsometric curve, showing cumulative frequency of global topographic heights. (b) Ocean basin schematic Principal features of the ocean floor shown in schematic form—height is greatly exaggerated. (c) Topography of the submarine Monterey Canyon, California, USA. The continental shelf offshore of Monterey California showing the Monterey and other canyons. Such canyons are common on shelves on both Atlantic and Pacific margins, often cutting through the shelf and down the continental slope to deep water. Figures used with permission of the Monterey Bay Aquarium Research Institute (MBARI).

they are probably less well-explored than the well-imaged surfaces of Mars, Venus, and the satellites of the Outer Planets. Dominant features of oceanic basins are the oceanic ridge and rise systems, which have a total length of about 60,000 km (~1.5 times the equatorial circumference of the Earth), rise to 1–3 km above the average depth of the ocean, and can be locally rugged. In the Atlantic Ocean, oceanic rises exhibit a central rift valley that is at the center of the rise, whereas in the Pacific Ocean this is not always present (see Fig. 2a).

Older crust within oceanic basins can have gently rolling abyssal hills, which are generally smoother than the ridge and rise systems. These may have been much more rugged originally, but are now buried beneath accumulated sediment cover. Perhaps the most areally dominant feature of ocean basins (with the largest ones occurring in the Atlantic Ocean) is the predominantly flat abyssal plains that stretch for thousands of kilometers, usually also covered with accumulated marine sediments. Generally characterized by little topographic relief, in places they are punctuated by seamounts (Fig. 4b), which are conical topographic rises sometimes topped by coral lagoons, or which sometimes do not reach the oceans' surface. These features are subsea volcanoes associated with island arcs or with mid-plate hot spots, such as the famous Emperor Seamount chain, the southeastern end of which terminates in the Hawaiian Islands. Such large hotspots are probably the result of persistent hot upwelling plumes from the upper mantle. Smaller "petite spot" subsea volcanoes may form above flexure cracks in oceanic plates.

Oceanic margins represent another important, although more areally restricted, submarine landform province (Figs. 4b and c). "Atlantic style" continental margins tend to exhibit substantial ancient sediment accumulations and a shelf-slope-rise overall morphology, which probably represents submerged subaerial landscapes remnant from the last

Ice Age, when the sea level was lower (about 135 m below current sea level, worldwide). Continental shelves are usually less than about 100 km in width and have very shallow (~0.1°) topographic slopes. They typically end in a slope break that merges into the continental slope (~4° slope, about 50 km wide), which in turn merges into a gentle continental rise (~0.2° slope, about 50 km wide), which then typically transitions into an abyssal plain. Submarine canyons (also probably remnant from the last Ice Age, e.g., Hudson Canyon off the coast of New York) can deeply cut the continental shelf and slope and terminate in broad submarine sediment fan deposits at the seaward canyon outlet. "Pacific style" oceanic margins can be even more narrow. Along the margins of continents of the Pacific Rim, a short shelf and slope can terminate into deep submarine trenches, manifested by subduction zones (e.g., South America, Kamchatka), up to 10 km deep. Similar fore-arc submarine morphology is observed along the margins of Pacific island arcs (e.g., Aleutians and Kurile Islands). Much shallower "back-arc" basins occur behind the arcs, on the over-riding plate (e.g., Sea of Okhotsk).

2.2.2 SUBAERIAL LANDFORMS

The subject of classic geomorphological investigations, and historically far more well studied because they are where people on Earth live, are the "subaerial" landscapes—the quarter of the Earth's surface that is not submerged. These terranes exist almost exclusively on continents; however, some important subaerial landscapes (particularly volcanic ones, e.g., Hawaii, Galapagos Islands) exist on oceanic islands. Most continental landscapes are predominately Cenozoic to late Cenozoic in age, because over that time scale (65 Myr or so), the combined action of plate tectonics, constructive landscape processes (e.g., volcanism and sedimentary deposition), and destructive landscape processes (e.g., erosion and weathering) have tended to rearrange, bury, or destroy pre-existing continental landscapes at all spatial scales. Thus, while often retaining the imprint of pre-existing forms, subaerial landscapes on the Earth are constantly being reinvented.

Because the Earth's crust is so dynamic, one must realize from the planetary perspective that any geomorphic survey of the Earth's surface may be representative only of the current continental plate arrangement, and currently associated climatic and atmospheric circulation regimes. Plate tectonics is a powerful force in setting scenarios for continental geomorphology. For instance, during early Cenozoic times the global continental geography was characterized by the warm circumglobal Tethys Sea and higher sea levels than now (possibly linked to higher rates of midoceanic spreading), which strongly biased the overall terrestrial climate toward the tropical range (Fig. 5).

The rearrangement of continental landmasses in the later Cenozoic closed the Tethys Sea, produced a circum-Antarctic ocean, and set up predominantly north–south circulation regimes within the Atlantic and Pacific Oceans.

This global plate geography, combined with greater ocean basin volume (linked to lower ridge spreading rates) and the onset of continental glaciation, lowered sea levels, exposing large marine continental self-environments to subaerial erosion. Our current global surface environment reflects a kind of "oceanic recovery" after the last Ice Age, with somewhat higher sea levels. Thus, our current perception of the Earth's subaerial geomorphic landform inventory is strongly biased by our temporal observational niche in its environmental history. Hypothetical interstellar visitors who arrived here 50 Myr ago or may arrive 50 Myr in the future would likely have a much different perception because of this distinctive dynamic character.

Terrestrial subaerial landform suites are the classic landscapes studied in geomorphology. These are listed in Table 1. Currently, on the Earth, globally dominant subaerial geomorphic regimes are related to the surface transport of liquid water and sediment due to the action of rainfall. Thus drainage basins dominate terrestrial landscapes at nearly all scales, from the continental scale to sub-100-m scales. These include currently active drainage basins in humid and semiarid climatic zones, to only occasionally active or relict drainages in arid zones. Drainage basin topographies and network topologies, however, are strongly influenced by the interplay of the orogenic aspects of plate tectonics (i.e., mountain building) and prevailing climatic regimes, including the biogenic aspects of climate (e.g., vegetative ground cover). Clearly, areas of rapid uplift (e.g., San Gabriel Mountains, California), have characteristically steep bedrock drainages, where gravitational energies are high enough to scour stream valleys, generally have parallel or digitate (hand-like) drainage patterns, have high local flood potentials, and respond strongly to local weather (e.g., spatial scales 10–100 km in characteristic dimension). At the other spatial extreme, major continental drainages (e.g., Amazon River, Mississippi River, Ob River in Siberia—Table 1), with highly dendritic (tree-like) overall pattern organization, are low average gradient systems that integrate the effects of a variety of climatic regimes at different spatial scales and tend to respond to mesoscale and larger climatic and weather events (e.g., 100-to 1000-km scale).

Subaerial volcanic processes produce characteristic landforms in all terrestrial climate zones (see Fig. 2b). They tend to occur in belts, mainly at plate boundaries, with a few notable oceanic (e.g., Hawaiian Islands) and continental (e.g., the San Francisco volcanic field in Northern Arizona; the Columbia and Snake River volcanic plains in the U.S. Pacific Northwest; the Deccan Traps in India), exceptions that occur within plate interiors. Although not as massive or as topographically high as their planetary counterparts (e.g., Martian volcanoes such as Olympus Mons), they provide some of the most spectacular and graceful landforms on the Earth's surface (e.g., Mount Fujiyama, Japan; Mt. Kilimanjaro, Kenya). Our planet's central vent volcanic landforms range from the majestic strato-cone volcanic

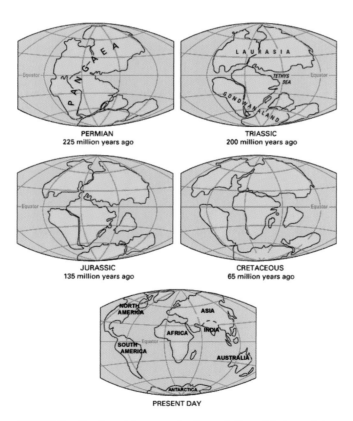

FIGURE 5   Continental geography through time. Modern plate tectonic theory is consistent provides the scientific framework for observations of continental drift. Geologic evidence records the breakup of the supercontinent Pangaea about 225–200 million years ago, eventually fragmenting over time to create our familiar continental geography. The Tethys Sea referred to in the text is labeled. (Courtesy of the United States Geological Survey).

TABLE 1		Classification of Terrestrial Geomorphological Features by Scale	
Order	Approximate Spatial Scale (km²)	Characteristic Units (with examples)	Approximate Time Scale of Persistence (years)
1	10^7	Continents, ocean basins	10^8–10^9
2	10^5–10^6	Physiographic provinces, shields, depositional plains, continental-scale river drainage basins (e.g., Amazon, Mississippi Rivers, Danube, Rio Grande)	10^8
3	10^4	Medium-scale tectonic units (sedimentary basins, mountain massifs, domal uplifts)	10^7–10^8
4	10^2	Smaller tectonic units (fault blocks, volcanoes, troughs, sedimentary sub-basins, individual mountain zones)	10^7
5	10–10^2	Large-scale erosional/depositional units (deltas, major valleys, piedmonts)	10^6
6	10^{-1}–10	Medium-scale erosional/depositional units or landforms (floodplains, alluvial fans, moraines, smaller valleys and canyons)	10^5–10^6
7	10^{-2}	Small-scale erosional/depositional units or landforms (ridges, terraces, and dunes)	10^4–10^5
8	10^{-4}	Larger geomorphic process units (hillslopes, sections of stream channels)	10^3
9	10^{-6}	Medium-scale geomorphic process units (pools and riffles, river bars, solution pits)	10^2
10	10^{-8}	Microscale geomorphic process units (fluvial and eolian ripples, glacial striations)	10^{-1}–10^4

Modified from Baker, 1986, and Bloom, 1998.

structures just mentioned to large collapse and resurgent caldrons or caldera features (e.g., Valles Caldera, New Mexico; Yellowstone Caldera, Wyoming; Campi Flegrei, Italy; Krakatau, Indonesia). More areally extensive and lower subaerial shield volcanoes, formed by more fluid lavas (and thus with topographic slopes generally less than 5°) exist in the Hawaiian Islands, at Piton de la Fournaise (Reunion Island), in Sicily at Mount Etna (compound shield with somewhat higher average slopes, up to ~20°), and the Galapagos Islands (Ecuador), for example. Often their areal extent corresponds strongly to the rate of their effusion. Subaerial and submarine volcanoes occur on the Earth at nearly all latitudes. Indeed some of the world's most active volcanoes occur along the Kurile-Kamchatka-Aleutian arc, in subarctic to arctic environments, often with significant volcano–ice interaction. High-altitude volcanoes that occur at more humid, lower latitudes (e.g., Andean volcanoes like Nevado del Ruiz in Columbia) can also have significant magma or lava–ice interactions. Volcanoes also occur in Antarctica, Mt. Erebus being the most active, with a perennial lava pond). [See PLANETARY VOLCANISM.]

2.3 Summary: Terrestrial vs. Planetary Landscapes

Overall, the Earth's geomorphic or physiographic provinces, as compared to those of the other planets in our solar system, are distinguished by their variety, their relative youth, and their extreme dynamism. Many of the other terrestrial-style bodies, such as the Moon, Mars, and

Mercury, are relatively static, with landscapes more or less unchanging for billions of years. Although this may not have been the case early in their histories, as far as we can tell from spacecraft exploration, this is the case now. Other landscapes, such as those on Venus and Europa and a few of the other outer planets' satellites, appear younger and appear to be the result of very dynamic planet-wide processes, and possibly for Venus, a planet-wide volcanic "event." Currently most of these bodies appear relatively static, although this point may be credibly debated. For instance, the Jovian satellite Io has vigorous ongoing volcanic activity as was first discovered in Voyager spacecraft imaging, and the Saturnian satellite Enceladus appears to be erupting water from relatively warm spots in its southern hemisphere, as seen in recent Cassini spacecraft data. Nevertheless, it seems that the crusts of all of these bodies are currently somewhat less variegated than that of the Earth. Be aware, however, that this last statement may turn out to be just another example of "Earth chauvinism," and will be proved wrong once we eventually know the lithologies and detailed environmental histories of these bodies as well as we know the Earth's. [See VENUS: SURFACE AND INTERIOR.]

3. Earth Surface Processes

The expenditure of energy in the landscape is what sculpts a planetary surface. Such energy is either "interior" (endogenic) or "exterior" (exogenic) in origin. The combined gravitational and radiogenic thermal energy of the Earth (endogenic processes) powers the construction of terrestrial landscapes. Thus, the Earth's main constructional landscape processes, plate tectonics, and resulting volcanism, are endogenic processes.

Destructional processes, such as rainfall-driven runoff and stream flow, are essentially exogenic processes. That is, the energy that drives the evaporation of water that eventually results in precipitation, and the winds that transport water vapor, comes from an exterior source—the Sun (with the possible exception of very local, but often hazardous, weather effects near explosive volcanic eruptions, and endogenic energy source). In familiar ways, such destructional geomorphic processes work to reduce the "gravitational disequilibria" that constructive landscapes represent. For instance, the relatively low and ancient Appalachian Mountains, pushed up during one of the collisions between the North American and European continental landmasses, were probably once as tall as the current Himalayan chain. Their formerly steep slopes and high altitudes represented a great deal of gravitational disequilibria, and thus a great deal of potential energy that was subsequently expended as kinetic energy by erosive downhill transport processes (e.g., rainfall runoff and stream flow). Once the processes of continental collision ebbed and tectonic uplift ceased, continuing erosion and surface transport processes (such as rainfall, associated runoff, snowfall, and glaciation) over only a few tens of millions of years reduced the proto-Appalachian Mountains to their present gently sloping and relatively low-relief state.

Volcanic landforms provide myriad illustrations of the competition between destructive and constructive processes in the landscape. For example, Mt. Fuji, the most sacred of Japanese mountains, is actually an active volcano that erupts on the order of every 100–150 years. Its perfectly symmetrical conical shape is the result of volcanic eruptions that deposit material faster than it can be transported away, on average. If Fuji stopped erupting, it would become deeply incised by stream erosion and it would lose its classic profile over a geologically short time interval (Fig. 6).

3.1 Constructive Processes in the Landscape

Over the geologic history of the Earth, volcanism has been one of the most ubiquitous processes shaping its surface. Molten rock (lava) erupts at the Earth's surface as a result of the upward movement of slightly less dense magma. Its melting and upward migration are triggered by convective instabilities within the upper mantle. Volcanic processes very likely dominated the earliest terrestrial landscapes and competed with meteorite impacts as the dominant surface process during the first billion years of Earth's history. With the advent of plate tectonics, multiphase melting of ultramafic rocks tended to distill more silicic lavas. Because silicate-rich rocks tend to be less dense than more mafic

FIGURE 6 Mt. Fuji, Japan, at sunrise from Lake Kawaguchi. Perhaps the world's quintessential volcano, the perfect conical shape of Mt. Fuji has inspired Japanese landscape artists for centuries. It is considered a sacred mountain in Japanese tradition and thousands of people hike to its summit every year. Volcanologically, Mt. Fuji is termed a "strato-volcano" and rises to an altitude of 3776 m above sea level. It erupts approximately every 150 years, on average.

varieties, they tend to "float" and resist subduction, thus continental cores (cratons) were generally created and enlarged by island-arc accretion.

Most volcanism tends to occur on plate boundaries. Subaerial plate boundary volcanism tends to produce island arcs (e.g., Aleutian Islands; Indonesian archipelago) when oceanic plates override one another or subaerial volcanic mountain chains (e.g., Andes) under-ride more buoyant continental plates. Such volcanism tends to be relatively silica-rich (e.g., andesites), producing lavas with higher viscosities, thus tending to produce steeper slopes. Rough lava flows on these volcanoes tend to be classified as aa or blocky lavas. High interior gas pressures contained by higher viscosity magmas can produce very explosive eruptions, some of which can send substantial amounts of dust, volcanic gas, and water vapor into the stratosphere.

Another kind of volcanic activity tends to occur within continental plates. As is thought to have been widespread on the Moon, Mars, and Venus and to a lesser degree within impact basins on Mercury, continental flood eruptions have erupted thousands of cubic kilometers of layered basalts (e.g., Deccan and Siberian Traps in India and Russia; Columbia River Basalt Group in the USA). These are among the largest single subcontinental landforms on the Earth. Such lavas were mafic, of relatively low viscosity, and are thought to have erupted from extended fissure vents at very high eruption rates over relatively short periods

(1–10 years). Recent work on the 100-km-long Carrizozo flow field in New Mexico, however, suggests that such massive deposits may have formed at much lower volume effusion rates over much longer periods than previously thought (10–100 years or more). The same may be true for lava flows of similar appearance on other planets. [*See* PLANETARY VOLCANISM].

Perhaps the most familiar kind of subaerial volcanism is the well-behaved, generally nonexplosive, Hawaiian-style low viscosity eruptions of tholeiitic basalts that form shield volcanoes, erupting in long sinuous flows. Typically such flows are either very rough ("aa") (Fig. 7a) with well-defined central channels and levees or very smooth, almost glassy ("pahoehoe") (Fig. 7b).

FIGURE 7   Aa flow from Mauna Loa Volcano, Hawaii, USA; Advancing flow of incandescent aa lava. Generally, aa flows are very rough and meters to tens of meters thick. They form broad toes and lobes and can advance kilometers per day, as often happens during eruptions of large a flows on Mauna Loa volcano in Hawaii (e.g., Mauna Loa 1984 eruption). (Courtesy of the U.S. Geological Survey). (b) Pahoehoe from Kilauea Volcano, Hawaii, USA cascading over scarp. Incandescent (∼1400K) fluid pahoehoe flows near the coast south of Kilauea Volcano, showing a lava breakout from an upstream lava tube cascading into two main branches. The cliff is approximately 15m high. Fields of Pahoehoe lava tend to form in a very complex intertwined fashion, and old cooled flows are often smooth enough to walk on in bare feet. (Courtesy of the U.S. Geological Survey).

These lavas are thought to be comparable to lavas observed in remote-sensing images of Martian central vent volcanoes (e.g., Alba Patera, Olympus Mons). Shield volcanoes on both planets tend to exhibit very low slopes (i.e., ∼5°). Active submarine basaltic volcanoes tend to occur along midoceanic ridges. Often the hot sulfide-rich waters circulating at erupting submarine venting sites provide habitats for a wide variety of exotic chalcophile (sulfur-loving) biota found nowhere else on Earth and proposed as a model for submarine life on Europa.

The transport of water across the land surface also has a hand in forming constructional landforms. Sediment erosion, transportation, and deposition can set the stage for a variety of landscapes, especially in concert with continental scale tectonic ("epirogenic") uplift. The Colorado Plateau in the southwestern United States is perhaps the best example of this type of landscape. The Grand Canyon of the Colorado River slices through the heart of the Colorado Plateau and exposes over 5000 vertical feet of sedimentary layers, the oldest of which date to the beginning of the Cambrian era (Fig. 8a).

Water itself can form constructive landforms on the Earth. In its solid form, water can be thought of as another solid component of the Earth's crust, essentially as just another rock. Under the present climatic regime, the Earth's great ice sheets—Antarctica and Greenland—along with numerous valley glaciers scattered in mountain ranges across the world in all climatic zones, compose a distinct suite of landforms. Massive (up to kilometers thick) deposits of perennial ice form smooth, crevassed, plastically deforming layers of glacial ice. Continental ice sheets depress the upper crust upon which they reside and can scour the subjacent rocky terrains to bedrock, as during the Wisconsin Era glaciation in Canada (i.e., last Ice Age in North America). Valley glaciers, mainly by mechanical and chemical erosion in concert, tend to carve out large hollows (cirques) in their source areas and have large outflows of meltwater at their termini (Fig. 8b).

3.2 Destructive Geomorphic Processes

Friction probably represents the largest expenditure of energy as geologic materials move through the landscape: friction of water (liquid or solid) on rock, friction of the wind, friction of rock on rock, or rock on soil. All of these processes are driven by the relentless force of gravity and generally express themselves as transport of material from a higher place to a lower one. Erosion (removal and transport of geologic materials) is the cumulative result, over time reducing the average altitude of the landscape and often resculpting or eliminating preexisting landforms of positive relief (e.g., mountains) and incising landforms of negative relief (e.g., river valleys or canyons). Overall, the source of potential energy for these processes (e.g., the height of mountain

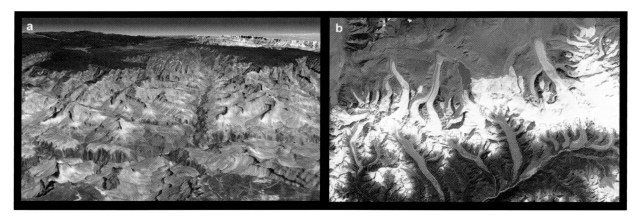

FIGURE 8 (a) Classic view of the Grand Canyon of the Colorado River in Arizona USA. The massive layering records the local geologic history for at least the last 500 Myr. Comparable layering has also been observed recently in canyons on Mars. This simulated true color perspective view over the Grand Canyon was created from Advanced Spaceborne Thermal Emission and Reflection Radiometer (ASTER) data acquired on May 12, 2000. The Grand Canyon Village is in the lower foreground; the Bright Angel Trail crosses the Tonto Platform, before dropping down to the Colorado Village and then to the Phantom Ranch (green area across the river). Bright Angel Canyon and the North Rim dominate the view. At the top center of the image the dark blue area with light blue haze is an active forest fire. (Courtesy NASA/GSFC/METI/ERSDAC/JAROS, and U.S./Japan ASTER Science Team). (b) Bhutan Glaciers, Himalayan Mountains, Asia. Classic Himalayan valley glaciers in Bhutan, showing theater-like "cirque" source areas, long debris-covered ice streams, and terminal meltwater lakes. ASTER data have revealed significant spatial variability in glacier flow, with velocities from 10–200m/yr. Meltwater volumes have been increasing in recent years and threaten to breach terminal moraine deposits with consequent dangerous downstream flooding. This ASTER scene was acquired 20 November 2001, is centered near 28.3 degrees north latitude, 90.1 degrees east longitude, and covers an area of 32.3 × 46.7 km. (Courtesy NASA/GSFC/METI/ERSDAC/JAROS, and U.S./Japan ASTER Science Team).

ranges) is provided by the tectonic activity of plates as they collide or subduct.

Subaerial landscapes on the Earth are most generally dominated by erosive processes, and subaqueous landscapes are generally dominated by depositional processes. Thus, from a planetary perspective, it is the ubiquitous availability and easy transport of water, mostly in liquid form, that makes it the predominant agent of sculpting terrestrial landscapes on Earth. Based on the geologic record of ancient landscapes, it appears that this has been the case for eons on the Earth. Such widespread and constant erosion does not appear to have happened for such a long time on any other planet in the solar system, although it appears that Mars may have had a period of time when aqueous erosion was important and even prevalent.

Fluvial erosion and transport systems (river and stream networks) dominate the subaerial landscapes of the Earth, including most desert areas. Even in deserts where aeolian (wind-driven, e.g., sand dunes) deposits dominate the current landscape, the bedrock signature of ancient river systems, relict from more humid past climatic epochs, can be detected in optical and radar images taken from orbiting satellites. Surface runoff, usually due to the direct action of rainfall occurs in nearly all climatic zones, except the very coldest.

On the Earth, such network forms resulting from this process tend to be scale-independent and take on a nearly fractal character. That is, network patterns tend to be replicated at nearly all scales, with regular geometric relationships that tend to be similar, no matter what the physical size of the network. In contrast to the situation on the Earth, the most visible and well-expressed Martian valley networks tend to be highly irregular in their network geometries, probably reflecting very restricted source areas of seepage or melt-driven runoff, rather than rainfall, and strong directional control by fractures and faults that was not overcome easily by river erosional processes. In addition, they are distributed very sparsely and are primitive in their branching, very much like the canyon networks arid areas of the world like Northern Africa (Fig. 9) and the desert Southwest of the United States (Fig. 8a). Thus, in contrast to Mars, for most of its discernable history, the Earth's landscapes have been distinguished, overall, by well-integrated and complexly branched fluvial drainage networks driven primarily by rainfall.

Uniquely on the Earth (within this solar system at least) it is the competition between constant fluvial erosion and constant tectonic uplift (and in some land areas, frequent volcanic eruptions) that is the predominant determinant of the landscape's appearance. For instance, the present terrestrial

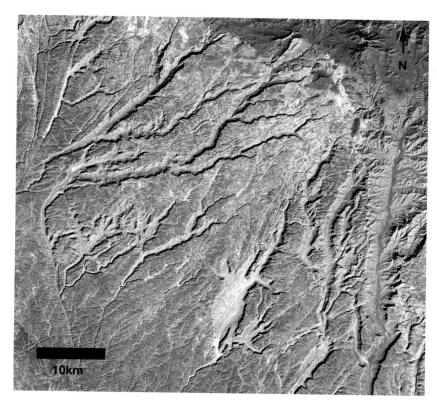

FIGURE 9 Desert drainage networks in Chad, North Africa. Shown here are deeply incised canyons on the southwest slope of the Tarso Voon Volcano located in the west-central part of the Tibesti Volcanic Range, in northern Chad (20.5°N latitude, 17°W longitude, approximately 3400 feet above sea level). Characteristic steep-walled theater-headed canyons form as overlying relatively soft Tarso Voon ignimbrites are stripped back over more resistant basement schists, through the action of groundwater seepage and surface runoff during infrequent storms, or during previous eras of wetter climate. Such differential erodability very likely also played a part in the formation of ancient complex ramified canyon networks on Mars of similar scale and appearance, and may reflect the former presence of more abundant supplies of near-surface water during warmer periods on Mars in its distant past. The ASTER image was acquired on 12 January 2003. Spatial resolution is 15m/pixel and the image as shown is a RGB composite of three visible bands (1N, 2N, 3N–0.52 to 0.82 μm). (Courtesy NASA/GSFC/METI/ERSDAC/JAROS, and U.S./Japan ASTER Science Team).

landscape is not dominated by impact scars. Plate tectonic processes are, in part, responsible; however, fluvial erosion is probably the dominant factor for subaerial landscapes in this regard. Also, without constant tectonic reinforcement, rainfall would probably reduce a Himalayan-style, or Alpine range to Appalachian-style mountains within 10 Myr or so. On the Earth, when tectonic forces subside, constant fluvial erosion wins out and hilly landscapes are flattened.

Other erosive processes, independently or in concert with fluvial activity, also clearly play a role on the Earth, including seepage-induced collapse (called "groundwater sapping") which can result in networks of steep-walled gulleys and canyons. In addition, the chemical action of groundwater can form landscapes of caves and sinkholes in limestone areas (called "karsts"). Whereas groundwater sapping and karst formation on the Earth may be relatively less important than fluvial erosion, the opposite case may be true for Mars. Another process regime that dominates arid and polar deserts on the Earth, and apparently is highly active, even today, on Mars, is that of wind-driven erosion and transport of fine dust and sand (called "aeolian" the Roman god of the winds). On the earth, aeolian processes are dominant only in certain restricted areas, such as the desert sand seas of Africa and Asia (Fig. 10a). On Mars, however, fine dust and sand dune and drift morphologies appear everywhere and can reveal important information on current wind regimes and on the constitution of the fine

material based on observations and models of terrestrial dune morphologies.

Another important terrestrial geomorphic process is weathering—the breakdown of consolidated material into constituent grains. Rock can be broken down in several ways. Chemical weathering can occur when natural acids act on carbonates in susceptible rocks, such as limestone or sandstones, releasing the residual silicate grains. Mechanical weathering of rock can occur when the hydrostatic pressures of ice in freeze-thaw cycles overcome rock brittle strength thresholds at microscopic and macroscopic scales. The formation of salt crystals also exerts mechanical energy to break up rocks and can chemically weather rocks. Oxidation of minerals, particularly iron-containing minerals, is another form of chemical weathering. Biological weathering occurs through chemical weathering caused by biogenic acids, particularly in tropical areas. It can also occur mechanically, by bioturbation of soils and sediments, as well as by the physical pressure of root and stem turgor in cracks and fissures within solid rock. It is of significance that on the Earth, all three major forms of weathering are enhanced or enabled by the ubiquitous presence of water.

Perhaps some of the most dramatic forms of nonvolcanic landscape alteration that we see on the Earth today fall into the category that geomorphologists call mass wasting. Generally, the term mass wasting is applied to processes such as landslides, creep, snow and debris avalanches, submarine

FIGURE 10   (a) Sand dunes in Namibia. Namib-Naukluft National Park is an ecological preserve in Namibia's vast Namib Desert, and is the largest game park in Africa. Coastal winds create the tallest sand dunes in the world here, with some dunes reaching 300 meters in height. This ASTER perspective view was created by draping an ASTER color image over an ASTER-derived digital elevation model. The image was acquired October 14, 2002. In the great deserts of the world, sand sheets are the dominant morphology and are wind driven. In open desert areas (e.g., Sahara or Arabian Peninsula), dune trains may stretch for tens or hundreds of miles.(Courtesy NASA/GSFC/METI/ERSDAC/JAROS, and U.S./Japan ASTER Science Team). (b) Deadly landslide in La Conchita, California. Large 1995 landslide and more recent 2005 debris flow that initiated from the slide above the town of La Conchita, California. It destroyed or seriously damaged 36 houses and killed 10 people. Loss of coherence in water-saturated ancient marine sediments was triggered by heavy rain. Landslides observed on Mars are typically 1 to as many as 3 orders of magnitude larger and may indicate the past presence of water. Alternatively substantial atmospheric lubrication is possible as is thought to have occurred during the ancient gigantic Blackhawk Slide on the slopes of the San Bernardino Mountains in California. (Courtesy of the U.S. Geological Survey).

slides and slumps, volcano-tectonic sector collapses, and scour related to the action of glaciers. Mass-wasting processes tend to affect a relatively minor proportion of the Earth's surface at any given time, however, such as volcanic eruptions (with which they are often associated), when they occur near population areas, their effects can be devastating (Fig. 10b). On Mars, massive landslides, similar in morphology and scale to the largest terrestrial submarine landslides, are commonly seen within Vallis Marineris and its tributary canyons.

4. Tools for Studying Earth's Deep Interior

In comparison with other planets, the interior of the Earth can be studied in unprecedented detail. This is because of the existence of sources of energy, such as earthquakes or magnetic and electric disturbances. Seismic waves, for example, can penetrate deep inside the Earth, and the time they travel between the source (earthquake or an explosion) and the receiver (seismographic station) depends on the physical properties of the Earth. The same is true

with respect to electromagnetic induction, although observations are different in this case.

Observation and interpretation of seismic waves provide the principal source of information on the structure of the deep interior of the Earth. Both compressional (P waves) and shear (S waves) can propagate in a solid, only P waves in a liquid. Compressional waves propagate faster than shear waves by, roughly, a ratio of $\sqrt{3}$. Velocities, generally, increase with depth because of the increasing pressure; hence the curved ray paths (Fig. 11).

At the discontinuities (which include the Earth's surface), waves may be converted from one type to another. Figure 11a shows P waves emanating from the source ("Focus"). The P waves can propagate downward (right part of the figure) and are observed as PP, PS, PPP, PPS, for example. They can also propagate upward, be reflected from the surface, and then observed as so-called "depth phases": PP, PPS. Depth phases are very helpful for a precise determination of the depth of focus.

Figure 11a shows rays in the mantle; there are also the outer core and inner core. The outer core is liquid and has distinctly different composition; the P-wave speed is some

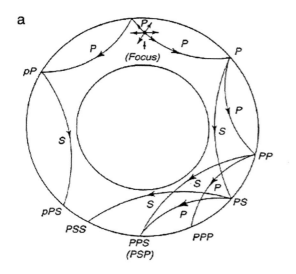

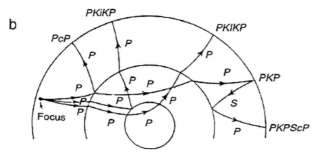

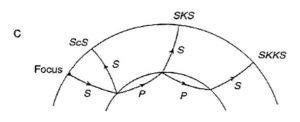

FIGURE 11 (a) Ray paths of the compressional waves (P) in the mantle, including their conversion to shear waves (S). (b) Ray paths of the P waves interacting with the outer and inner core. (c) Ray paths of the S waves interacting with the core; the S waves are converted into P waves in the outer core.

such that the "radial" component shows horizontal motion along the great circle from the earthquake to the station; "transverse" component is also horizontal motion but in the direction perpendicular to the ray path, and "vertical" component shows up-and-down motion.

Figure 13 compares observed travel times, reported by the International Seismological Centre with those predicted by an Earth model. The scatter around the predicted values is caused by the effects of lateral heterogeneity and measurement errors.

Measurements of the travel times of the waves such as shown in Figures 11 and 13 have led to the derivation of models of the seismic wave speed as a function of depth. These, in turn, were used to improve the location of earthquakes and further refine the models. The first models were constructed early in the 20th century; the models published by Beno Gutenberg and Sir Harold Jeffreys in the 1930s are very similar in most depth ranges to current ones. The model of Jeffreys is compared with a recent model (iasp91) in Figure 14. The upper mantle (the topmost 700 km) with its discontinuities and the inner core are exceptions.

In addition to the body waves, which propagate through the volume of the Earth, there are also surface waves, whose amplitude is the largest at the surface and decreases exponentially with depth. Surface waves are important in studying the crust and upper mantle and, in particular, their lateral variations, as the Earth is most inhomogeneous near the surface. There are Rayleigh waves with the particle motion in the vertical plane (perpendicular to the surface; second and third trace in Figure 12) and Love waves whose particle motion is in the horizontal plane (parallel to the surface). Surface waves are dispersed in the Earth because of the variation of the physical parameters with depth; notice that the longer period surface waves in Figure 12 arrive before shorter period waves.

Very long period surface waves (>100 sec) are sometimes called "mantle waves," have horizontal wavelengths in excess of 1000 km, and maintain substantial amplitudes (and, therefore, sensitivity to the physical properties) down to depths as large as 600–700 km. Because of their long periods, mantle waves are attenuated relatively slowly and can be observed at the same station as they travel around the world several times along the same great circle (both in the minor and major arc direction). Figure 15 shows a three-component recording of mantle waves (note the time scale); the observed seismograms are shown at the top of each pair of traces; the bottom trace is a synthetic seismogram computed for a three-dimensional Earth model.

Superposition of free oscillations of the Earth (known also as the normal modes) in the time domain will yield mantle waves. First spectra of the vibrations of the Earth were obtained following the Chilean earthquake of 1960; the largest seismic event ever recorded on seismographs. The measurements of the frequencies of free oscillations lead to the renewed interest in the Earth's structure. In particular, they, unlike body waves, are sensitive to the density

40% lower than at the bottom of the mantle; also, there are no S waves. The inner core is solid, with a composition similar to that of the outer core. Figure 11b shows the rays (mostly P waves) that are reflected from the core–mantle boundary (CMB; a letter c is inserted, e.g., PcP) or that are transmitted through the outer core (letter K: PKP) or also through the inner core (letter I: PKIKP). Figure 11c shows S-wave rays interacting with the CMB: reflected (ScS) or converted at the CMB into a P wave and then again reconverted into a S wave: SKS and SKKS. The latter indicates one internal reflection from the underside of the CMB.

Figure 12 shows an example of an earthquake recorded on a three-component seismograph system and then rotated

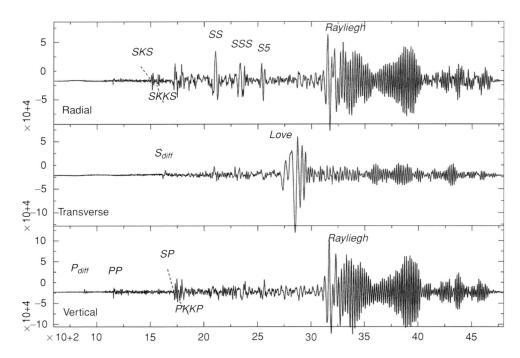

FIGURE 12   Three component recording at a GSN digital, high dynamic range station. Note identification of various phases. The dispersed Rayleigh waves are seen on the radial and vertical component and Love waves are seen on the transverse component.

distribution and thus provide additional constraints on the mass distribution other than the average density and moment of inertia. Figure 16 shows an example of a spectrum of a vertical component recording of a very large deep earthquake under Bolivia; the lowest frequency mode shown has a period of about 40 minutes.

Sometime in the 1970s it became clear that further refinements in one-dimensional Earth models cannot be achieved, and perhaps do not make much sense, without considering the three dimensionality of the Earth's structure. All three types of data described earlier are sensitive to the lateral heterogeneity. Travel times will be perturbed by slight variations of the structure along a particular ray path, compared to the prediction by a one-dimensional model. All

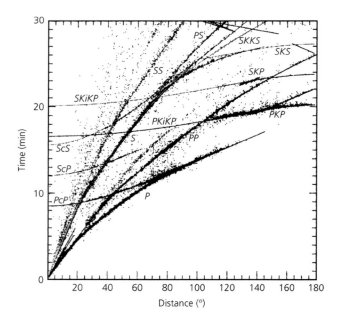

FIGURE 13   Observed travel times from a Bulletin of International Seismological Centre are compared with predictions for model IASP91. There are additional observed branches, such as PPP and SSS, for which travel times have not been computed.

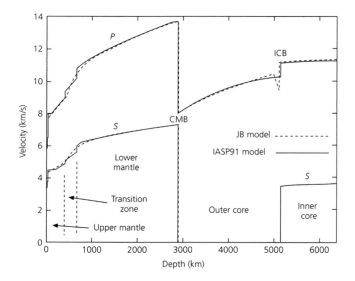

FIGURE 14   Comparison of a velocity model by Jeffreys (ca, 1937) with model IASP91. Notice that for the most part changes have been minor, except for the discontinuities in the transition zone, solidity of the inner core and structure just above inner core boundary.

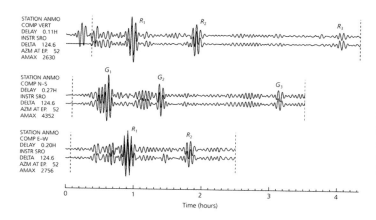

FIGURE 15 Mantle waves observed on multiple orbits around the Earth. The symbol "R" designates Rayleigh waves and "G" Love waves. Odd numbered (1,3) arrivals correspond to minor arc arrivals plus an integer number of complete paths around the Earth. Even numbered wavegroups correspond to initial propagation in the major arc direction. The signal between arrivals of the fundamental mode wavegroups represents contribution of overtones. Top traces are observed seismograms, bottom traces are synthetic seismogram computed for 3-D model of upper mantle M84C; if 1-D model (PREM) was used, there would be significant differences between observed and computed traces.

we need is many observations of travel times along criss-crossing paths. Many millions of such data are available from the routine process of earthquake location; they are assembled from some 6000 stations around the world by the National Earthquake Information Center in Golden, Colorado, and by the International Seismological Centre in England (see Figure 13). Surface waves, mantle waves, and periods of free oscillations in a three-dimensional Earth also depend on the location of the source and the receiver. Progress during the last decade in global seismographic instrumentation, in terms of the quality and distribution of the observatories and exchange and accessibility of the data, makes the required observations much more readily available.

5. Seismic Sources

Even though the field of seismology can be divided into studies of seismic sources (earthquakes, explosions) and of the Earth's structure, they are not fully separable. To ob-

tain information on an earthquake, we must know what happened to the waves along the path between the source and receiver, and this requires the knowledge of the elastic and anelastic Earth structure. The reverse is also true: in studying the Earth structure, we need information about the earthquake; at least its location in space and time, but sometimes also the model of forces acting at the epicenter.

Most of the earthquakes can be described as a process of release of shear stress on a fault plane. Sometimes the stress release can take place on a curved surface or involve multiple fault planes: the radiation of seismic waves is more complex in these cases. Also, explosions, such as those associated with nuclear tests, have a distinctly different mechanism and generate P and S waves in different proportions, which is the basis for distinguishing them from earthquakes.

Figure 17 shows three principal types of stress release, sometimes also called the earthquake mechanism. The top part of Figure 17a is a view in the horizontal plane of two blocks sliding with respect to each other in the direction shown by the arrows. Such a mechanism is called strike slip, and the sense of motion is left-lateral; there is also an

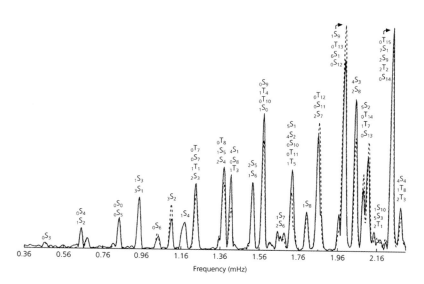

FIGURE 16 Amplitude spectrum of a vertical component seismogram of the great deep Bolivia earthquake of 1994. The peaks in the spectrum correspond to periods of free oscillations (vibrations) of the Earth. The symbols designate the specific normal modes. Some of them appear in groups which indicate a possibility of coupling between modes close in frequency. Usually the fundamental modes (pre-subscript "0") are excited most strongly.

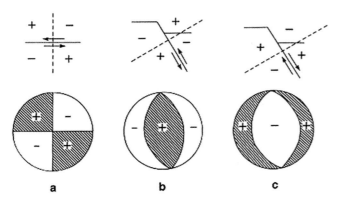

FIGURE 17 Three classical types of earthquakes (top) and the distribution of the signs of the P-wave arrivals: (a) strike slip, (b) thrust or reverse fault and (c) normal fault. The beachballs represent the equal area projection of the signs of first motion of the P-waves. The motion would be positive within the shaded areas. The lines separating shaded areas with the unfilled ones are called "nodal planes."

auxiliary plane, indicated by a dashed line; a ground motion generated by a slip on the auxiliary plane (right lateral) cannot be distinguished from that on the principal plane. The bottom part of Figure 17a is a stereographic projection of the sign of P-wave motion observed on the lower hemisphere of the focal sphere (a mathematical abstraction in which we encapsulate the point source in a small uniform sphere). The plus sign corresponds to compressive arrivals and minus sign to dilatational zones; quadrants with compressive arrivals are shaded.

The top part of Figure 17b is a section in the vertical plane. In this case, the block on the right moves upward on a plane that dips at a 45° angle with respect to the block on the left; this mechanism is called thrust and is associated with compression in the horizontal plane and tension in the vertical plane and corresponds to the convergence of the material on both sides of the fault. Such processes are responsible for mountain building. The shaded central region in the bottom part of Figure 17b, with the dilatational arrivals on the sides, is characteristic of the thrust—reverse faulting—events. Figure 17c illustrates the opposite mechanism, in which tension is horizontal and compression vertical; this is called normal faulting and is associated with extension, which can lead to the development of troughs or basins.

The "beach-ball" diagrams are commonly used as a graphic code to represent the tectonic forces. Some earthquakes are a combination of two different types of motion, e.g., thrust and strike slip; in this case the point at which the two planes intersect would be moved away from either the rim or the center of the beach-ball diagram.

The size of the earthquake is measured by magnitude. There are several different magnitude scales depending on the type of a wave whose amplitude is being measured. In general, magnitude is a linear function of the logarithm of the amplitude; thus a unit magnitude increase corresponds to a 10-fold increase in amplitude. Most commonly used magnitudes are the body-wave magnitude, m_b, and surface wave magnitude, M_S. The frequency of occurrence of earthquakes, i.e., a number of earthquakes per unit time (year) above a certain magnitude M, satisfies the Gutenberg–Richter law: $\log_{10} N = a \cdot M + b$. The value of a is close to -1, which means that there are, on average, 10 times more earthquakes above magnitude 5 than above magnitude 6. A new magnitude, M_W, based on the estimates of the released seismic moment [shear modulus × fault area × offset (slip) on the fault] is becoming increasingly popular; it is more informative for very large earthquakes, for which M_S may become saturated.

Figure 18 is a map of the principal tectonic plates, defined in plate tectonic theory. The direction of the arrows shows the relative motion of the plates; their length corresponds to the rate of motion. At a plate boundary where the blue arrows converge, we expect compression and, therefore, thrust faulting; one of the plates is subducted:

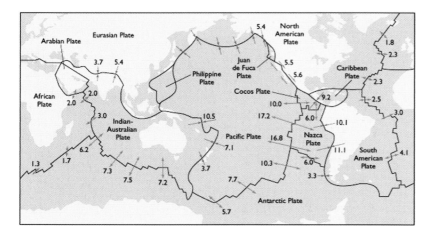

FIGURE 18 Principal tectonic plates and relative plate motion rates. Red arrows signify spreading, blue arrows–convergence, and green arrows–strike-slip motion.

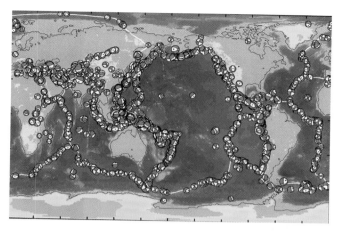

FIGURE 19   Source mechanisms of approximately 4,000 earthquakes from 1993 to 1997 obtained through the CMT analysis. The center of a beach ball is plotted at the epicenter. Only a small fraction of earthquakes are visible. Note the preponderance of earthquakes occurring on plate boundaries (Figure 18) and their mechanism corresponding closely to the type of the boundary (convergent, thrust faulting; divergent, normal faulting; transform, strike-slip faulting). Some earthquakes occur away from plate boundaries. They are particularly numerous in Asia and Africa along the east African rift system, but there are some in eastern North America and the center of the Pacific.

hence the term "subduction zones." At a plate boundary where the red arrows diverge, there is normal faulting and creation of a new crust: midocean ridges. For boundaries that slip past each other in the horizontal plane (green arrows), also called the transform faults, there is strike-slip faulting.

Figure 19 shows the source mechanism of approximately 4000 shallow earthquakes from 1993 through 1997 determined at Harvard University using the centroid-moment tensor (CMT) method; the center of each beach ball is at the epicenter—many earthquakes have been plotted on top of each other. It is easy to see that thrust faulting is dominant at the converging boundaries (subduction zones), there are exceptions related to bending of the plates, plate motion oblique to the boundary and other causes.

At midocean ridges, we see predominantly normal faulting, the faults where a midocean ridge is offset, show strike-slip faulting, in accordance with the plate tectonic theory. The exception is where the fault is complex. Along the San Andreas Fault, the most famous transform fault, we see many complexities that led to earthquakes other than the pure strike slip. For example, the Northridge earthquake of January 1994 was a thrust, and the Loma Prieta earthquake of October 1989 was half-thrust, half-strike slip. There are also earthquakes away from the plate boundaries. These are called intraplate earthquakes and their existence demonstrates the limits of the validity of the plate tectonic theory,

as there should be no deformation within the plates. A very wide zone of deformaton is observed in Asia; the rare large earthquakes in eastern North America are sometimes associated with isostatic adjustment following the last glaciation. If we compare the distribution of earthquakes along a midocean ridge, including its transform faults, with that of the Alpide belt, we notice that for the oceanic plates the region in which earthquakes occur is very narrow, while in Eurasia it may be 3000 km wide. A part of the reason that the theory of plate tectonics has been put forward is because of observations (bathymetry, magnetic stripes, and seismicity) in the oceans.

There are also deep earthquakes, with the deepest ones just above 700 km depth; earthquakes with a focal depth from 50 to 300 km are said to be of an intermediate depth and are called "deep" when the focal depth is greater than 300 km. Intermediate and deep earthquakes are explained as occurring in the subducted lithosphere and are used to map the position of the subducted slab at depth. Not all subduction zones have very deep earthquakes; for example, in Aleutians, Alaska, and Middle America the deepest earthquakes are above 300 km depth. The variability of the maximum depth and the mechanism of deep earthquakes have been attributed in the late 1960s to the variation in the resistance that the subducted plate encounters; more recent studies indicate more complex causes, often invoking the phase transformations (change in the crystal structure) that the slab material subjected to the relatively rapidly changing temperature and pressure may undergo.

6. Earth's Radial Structure

A spherically symmetric Earth model (SSEM) approximates the real Earth quite well; the relative size of the three-dimensional part with respect to SSEM varies from several percent in the upper mantle to a fraction of a percent in the middle mantle and increases again above the CMB.

A concept of an SSEM, often referred to as an "average" Earth model, is a necessary tool in seismology. Such models are used to compute functionals of the Earth's structure (such as travel times), and their differential kernels are needed to locate earthquakes and to determine their mechanism. Knowledge of the internal properties of the Earth is needed in geodesy and astronomy. Important inferences with respect to the chemical composition and physical conditions within the deep interior of the Earth are made using information on radial variations of the elastic and anelastic parameters and density.

An SSEM is a useful mathematical representation that is not necessarily completely representative of the real Earth. This is most obvious at the Earth's surface, where one must face the dilemma of how to reconcile the occurrence at the same depth, or elevation, of water and rocks; the systems

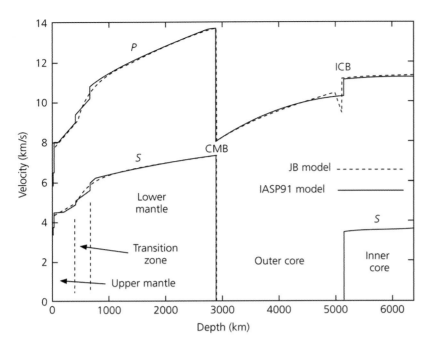

FIGURE 20   The preliminary Earth model (PREM) of Dziewonski and Anderson (1981) describing the compressional velocity(v_s), shear velocity (v_S), and density (ρ). (a) Model for the entire Earth and (b) an expansion of the uppermost 1000 km. From Moho to 220 km depth the model is characterized by transverse anisotropy, in which the waves propagating in the vertical (solid line) and horizontal (dashed lines) planes have different velocities. Parameter η, characterizing the propagation of P waves at intermediate angles, is unity in an isotropic medium and is about 0.95, just under the Moho. Below 220 km depth the model is isotropic.

of equations governing the wave propagation in liquid and in solid are different. The commonly adopted solution is to introduce a layer of water whose thickness is such that the total volume of water in all the oceans and that calculated for the SSEM are equal. It is a reasonable decision, but it will be necessary to introduce corrective measures even when constructing the model, as practically all seismographs that record ground motion are located on land.

This chapter uses the preliminary reference Earth model (PREM) published in 1981 by Dziewonski and Anderson as an example. It has been derived using a large assembly of body-wave travel time data, surface wave dispersion and periods of free oscillations, collected through the end of 1970s. An effort to revise it is now under way: a large body of very accurate data has been assembled in the nearly 20 years since the publication of PREM. However, with the exception of the upper mantle, no substantial differences are expected. A reference model designed to fit the travel times of body waves (ak135) has been developed by Kennett and Engdahl in 1995.

Figure 20a shows the density, compressional velocity, and shear velocity in the model PREM. To illustrate the complexities in the uppermost 800 km of the model, its expansion is shown in Figure 20b. In what follows, we shall give a brief summary of our knowledge and significance of the individual shells in the Earth's structure.

6.1 Crust

This is the most variable part of the Earth's structure, both in terms of its physical properties as well as history. Large

areas of the Earth's surface are covered by soils, water, and the sediments. These provide support for life and economic activity. However, the vast proportion of what is called "the crust of the Earth" consists of crystalline rocks, mostly of igneous origin.

The primary division is between the continental and the oceanic crust. The former can be very old, with a significant fraction being older than 1.5 Ga. It is light, with an abundance of calcium, potassium, sodium, and aluminum. Its average thickness is 40 km, but varies substantially, from about 25 km in the areas of continental thinning due to extension (the Basin and Range province in the Western United States, for example) to 70 km under Tibet, in the area of continent—continent collision.

The oceanic crust is thin (7 km, on average, covered by some 4.5 km of the ocean), young (from 0 to 200 Ma), and somewhat more dense, with a greater abundance of elements such as magnesium and iron. It is created at the midocean ridges and is consumed in subduction zones, with trenches being their surficial manifestation. The difference between oceanic and continental crusts is called by some the most important fact in Earth sciences, as it is related intimately to plate tectonics. The thinner, denser oceanic crust provides conditions more favorable for initiation of the subduction process.

Overall, crustal thickness follows the Airy's hypothesis of isostasy closely: thick roots under mountains and a thin crust under "depressed" areas—oceans. The seismic velocities in the crust increase with depth. It is a subject of a debate whether this increase is gradual or the crust is layered; recently, the latter view has begun to prevail.

6.2 Upper Mantle: Lithosphere and Asthenosphere (25–400 km Depth)

The boundary between the crust and the upper mantle was discovered in 1909 by a Yugoslavian geophysicist Andreiji Mohorovicic. It represents a 30% increase in seismic velocities and some 15% increase in density. It is a chemical boundary with the mantle material primarily composed of minerals olivine and pyroxene, being much richer in heavier elements, such as magnesium and iron.

The terms lithosphere and asthenosphere refer to the rheological properties of the material. The lithosphere, strong and brittle, is characterized by very high viscosity. It is often modeled as an elastic layer. It includes the crust and some 30 to 100 km of the upper mantle. The asthenosphere is hotter (>1573 K by convention), its viscosity much lower, and in modeling is represented by yielding. Under loads, such as glacial caps, the lithosphere bends elastically, whereas the asthenosphere flows. The difference of rheological properties is explained by differences in temperature: the viscosity is an exponential function of temperature. The lithosphere is relatively cool; the transport of heat is mostly through conduction. The asthenosphere is hotter, and the convective processes are believed to become important. Low viscosity of the asthenosphere is used to explain the mechanical decoupling between the plates (in the plate tectonic theory) and the underlying mantle. The depth of this decoupling varies with position: it is shallow near midocean ridges and increases as the plate cools with time and its lithosphere grows in thickness.

The continents, with its very old and cold shield regions, may be significantly different. If the hypothesis of the "tectosphere" is correct, it may have roots that are 400 km deep and move as coherent units over long periods of the Earth's history. The depth of roots is still subject to a debate (most recent results would indicate their depth extent as 200–250 km). As the seismic velocities decrease with increasing temperature, the vertical gradient of seismic velocities in the transition between the lithosphere and the asthenosphere may become negative. This is called the "low velocity zone"; its presence creates a shadow zone in seismic wave propagation, making interpretation of data complex and nonunique.

Measurements of attenuation of seismic waves led to the determination of models of Q (quality factor) for the shear and compressional energy. Anelastic dissipation of shear energy, due to grain boundary friction, is most important. Attenuation in the range of depths corresponding to the low velocity zone is several times stronger than in the lithosphere.

6.3 Transition Zone (400–660 km Depth)

Knowledge of the composition of the transition zone is essential to the understanding of the composition, evolution, and dynamics of the Earth. In seismic models, this depth range has been known for a long time to have a strong velocity gradient; much too steep for an increase under pressure of the elastic moduli and density of a homogeneous material. It was first postulated in the 1930s that this steep gradient may be due to phase transformations: changes in the crystal lattice that for a given material take place at certain temperatures and pressures.

In the 1960s, when major improvement in seismic instrumentation took place, two discontinuities were discovered: one at 400 km and the other at 670 km (the current best estimate of the global average of their depth is 410 and 660 km, respectively). Their existence has been well documented by nearly routine observations of reflected and converted waves. There is still some uncertainty of how abrupt the velocity changes are: the 410-km discontinuity is believed to be spread over some 5–10 km, whereas the 660-km discontinuity appears to be abrupt. The estimates of the velocity and density contrasts are still being studied by measuring the amplitudes of the reflected and converted waves; the values of these contrasts are important for understanding the mineralogical composition of the transition zone.

In general terms, the seismological models are consistent with the hypothesis that olivine is the main (up to 60%) constituent of the upper mantle. Laboratory experiments under pressures corresponding to depths up to 750 km show that olivine undergoes phase transformations to denser phases with higher seismic wave speeds. At pressures roughly corresponding to 400 km depth, the α-olivine transforms into β-spinel. The latter will transform to γ-spinel at about 500 km depth, with only a minor change in seismic velocities. Indeed, a seismic discontinuity at 520 km has been reported, although some studies indicate that in some parts of the world it may not be substantial enough to be detected. At 660 km γ-spinel transforms into perovskite and magnesio-wüstite.

Although olivine may be the dominant constituent, it is not the only one. The presence of other minerals complicates the issue. Also, there are other hypotheses of the bulk composition of the upper mantle: "piclogyte model," for example.

6.4 Lower Mantle (660–2890 km)

The uncertainties in the mineralogy of the upper mantle and the bulk composition of the Earth have created one of the most stubborn controversies in the Earth sciences: are the upper mantle and lower mantle chemically distinct? A "yes" answer means that there has not been an effective mixing between these two regions throughout the Earth's history, implying that the convection in the Earth is layered. The abrupt cessation of seismic activity at about 660 km depth, coinciding with the phase transformation described earlier, and geochemical arguments—mostly with respect to differences in isotopic composition of the midocean ridge basalts and ocean island basalts—are used as strong arguments in favor of the layered convection. New evidence, gathered within the tomographic studies to be discussed later, gives support to a significant impedance to the flow between the upper mantle and lower mantle.

The whole mantle convection is favored by geodynamicists who develop kinematic and dynamic models of the mantle flow. For example, the geometry and motions of the known motions of the plates are much easier to explain assuming whole mantle circulation. Evidence has been presented for penetration of slabs into the lower mantle, based on the presence of fast velocity anomalies in the regions of the past and current subduction. At the same time, there is evidence for stagnation and "ponding" in the transition zone of some of the subducted slabs. The recent results from seismic tomography seem to support the concept of at least partial separation of the upper and lower mantle flow.

In the early 1990s a model of mantle avalanches was developed: the subducted material is temporarily accumulated in the transition zone as the result of an endothermic phase transformation at the 660-km discontinuity. Once enough material with the negative buoyancy collects, however, a penetration can occur in a "flushing event," where most of the accumulated material sinks into the lower mantle. The calculations, originally performed in two-dimensional geometry, indicated the possibility of such events causing major upheavals in the Earth's history. However, when calculations were extended to three-dimensional spherical geometry, their distribution in space and time turned out to be rather uniform.

The computer models of the mantle convection are still tentative. There are many parameters that control the process. Some, such as the generation of the plates and plate boundaries at the surface, are difficult to model. Others, such as the variation of the thermal expansion coefficient with pressure-or temperature-dependent viscosity, are poorly known; even one-dimensional viscosity variation with depth is subject to major controversies.

The lower mantle appears mineralogically uniform, with the possible exception of the uppermost and lowermost 100–150 km. There is a region of a steeper velocity gradient in the depth range of 660–800 km, which may be an expression of the residual phase transformations. Also, at the bottom of the mantle, there is a region of a nearly flat, possibly slightly negative gradient. This region, just above the CMB, known as "D," is the subject of intense research. Its strongly varying properties, both radially and horizontally, are being invoked in modeling mantle convection, chemical interaction with the core, possible chemical heterogeneity (enrichment in iron), and as evidence for partial melting. In 2004, the existence of a new phase: "post-perovskite" has been proposed; its existence may affect the complexities in the "D" region. The seismic velocities and density throughout the bulk of the lower mantle appear to satisfy the Adams–Williamson law, describing the properties of the homogeneous material under an adiabatic increase in pressure.

6.5 Outer Core (2981–5151 km)

The outer core is liquid: it does not transmit shear waves. Consideration of the average density and the moment of inertia pointed to a structure with a core that would be considerably heavier, possibly made of iron, judging from cosmic abundances. We now know that the core is mostly made of iron, with some 10% admixture of lighter elements, needed to lower its density. It has formed relatively early in the Earth's history (first 50 Ma) in a melting event in which droplets of iron gravitationally moved toward the center. Though difficult to estimate, some current models place its temperature in the range of 3000–5000K.

The presence of a liquid with a very high electrical conductivity creates conditions favorable to self-excitation of a magnetic dynamo. It is important to know that the magnetic field we observe at the surface is only a small fraction of the fields present in the core. Actually we see only one class of the field: the poloidal, whereas the toroidal field, possibly much stronger, is confined to the core.

Numerical models of the dynamo predicted several key phenomena observed at the surface: the primary dipolar structure with the alignment of the dipole axis close to the axis or rotation of the Earth, the westward drift of secular variations, and reversals of the polarity of the magnetic field. The later phenomenon is the cause of the magnetic anomalies on the ocean floor, which allowed estimating the rate of ocean spreading. The two most widely known models of dynamo are quite different in detail, with one by Glatzmaier and Roberts having the strongest field deep in the core, and one by Kwang and Bloxham being the strongest near the surface of the outer core.

Seismological data are consistent with the model of the core as that of a homogeneous fluid under adiabatic temperature conditions. As often near major discontinuities, there is some difficulty with pinning down the values near the end of the interval: just below the CMB and just above the inner core boundary (ICB).

6.6 Inner Core (5251–6371 km)

An additional seismic discontinuity deep inside the core, which came to be called the inner core, was discovered by Inge Lehmann in 1936. The fact that it is solid was postulated soon afterwards, but satisfactory proof awaited another 35 years, when observations and analysis of the free oscillations of the Earth showed that it indeed must have a finite rigidity.

It is believed that the inner core formed during the history of the Earth, perhaps some 2 Ga ago. As the Earth was cooling, the temperatures at the Earth's center dropped below the melting point of iron (at the pressure of 330 GPa) and the inner core began to grow. The release of the gravitational energy associated with the precipitation of solid iron is believed to be an important source of the energy driving the dynamo. Again, estimates are difficult, but models yield a current temperature range of 5000–7500K.

Seismologically, the inner core has been considered quite uninteresting, with a very small variation of the physical parameters across the region. This all changed in the mid-1980s when it was discovered that this region is anisotropic, with the symmetry axis roughly parallel to the rotation axis. A deviation from that symmetry and an observation of temporal variation of travel times through the inner core brought forward an interpretation that the inner core rotates at a slightly (1°/year) higher rate than the mantle. This is being explained by the electromagnetic coupling with the dynamo field of the other core. However, this observation has soon become very controversial. Several studies now indicate that this differential rotation must be much less. In 2002, it was proposed that there exists an "inner-most inner core," the central region with some 300-km radius in which the anisotropy is distinctly different than in the bulk of the inner core. Since then, the anomalous properties of this region have been confirmed by other studies.

7. Earth in Three Dimensions

Figure 21 is an example of results obtained using global seismic tomography (GST). It shows a triangular cut into an Earth model of the shear velocity anomalies in the Earth's mantle and shows only deviations from the average: if the Earth were radially symmetric, this picture would be entirely featureless. The surface is the top of the mantle (Mohorovicic discontinuity, or Moho) and the bottom is the core–mantle boundary. Seismic wave speeds higher than average are shown with blue colors, whereas slower than average are shown as yellow and red colors. Seismic velocities decrease with increasing temperature: the inference is that the light areas are hotter than average and dark are colder. Seismic wave speeds also vary with chemical composition, but there are strong indications that the thermal effect is dominant.

Density is also a function of temperature. Material hotter than average is lighter and, in a viscous Earth, will tend to float to the surface, whereas colder material is denser and will tend to sink. Thus our picture can be thought to represent a snapshot of the temperature pattern in the convecting Earth's mantle. In particular, the picture implies a downwelling under the Indian Ocean and an upwelling originating at the core–mantle boundary under Africa; sections passing through this anomaly indicate that this upwelling may continue to the surface. This "window into the Earth" shows the outer core (blue), inner core (pink) and the

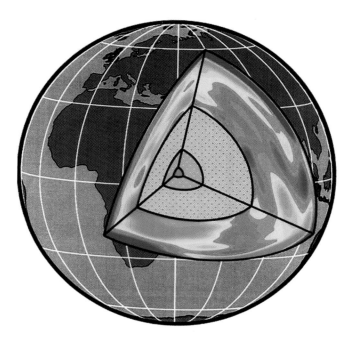

FIGURE 21 A three-dimensional model S362D1 of Gu et al. representing the lateral deviations of the shear velocities with respect to PREM. The sides represent a vertical cross section along three different profiles. Faster than average velocities (caused by colder than normal temperature, presumably) are shown in green/blue and slower (hotter) in yellow/red colors. The scale is ±1.5%; significant saturation of the scale occurs in the upper mantle. Note the lateral and vertical consistency of the sign of the anomalies over large distances and depths. The mantle underneath Asia and the Indian Ocean is fast at nearly all depths, whereas the mantle under central Africa is slow. The liquid outer core is shown in blue, inner core in red and the innermost inner core in red.

innermost inner core (red); the latter represents only 0.01% of the Earth's volume.

The GST is limited by the distribution of globally detected earthquakes and by the locations of seismographic stations. There is not much that we can do about the distribution of seismicity, except that now and then an earthquake occurs in an unexpected place, so the coverage is expected to improve with time. Generally, the earthquake distribution is more even in the Northern Hemisphere. Much has been done in the last decade to improve the distribution and the quality of the seismographic stations, and recent results show considerably better resolution of the details in the top 200 km, for example. However, even using the available oceanic islands (which are very noisy, because of the wave action), there are oceanic areas with dimensions of several thousand kilometers where no land exists. A series of experiments by Japanese, French, and American seismologists have demonstrated that the establishment of a permanent or semipermanent network of ocean bottom high-quality

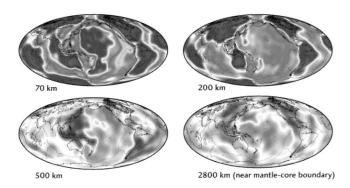

70 km   200 km

500 km   2800 km (near mantle-core boundary)

FIGURE 22 Maps of lateral variations of S velocities at four depths in a shear velocity model of Ekström and Dziewonski. The yellow/red colors indicate slower than average velocities and blue – the faster. The range of variations is about 7% at 70 km and 3% at near the core–mantle boundary.

seismographic stations is now a real, even though expensive, possibility.

Figure 22 is a collection of maps of the shear velocity anomalies from a recent model of the mantle by Ekström and Dziewonski published in 1998, built using a wide range of types of data (travel times, surface wave dispersion and waveforms). The nominal resolution of this model is about 100 km in depth and 1500 km horizontally near the surface.

At 70 km depth, the model agrees with the predictions of the plate tectonics and the thermal history of the continents. The stable continental areas (old and cool) are very fast (up to +7%), whereas material under the midocean ridges is much slower than normal (up to −7%). This negative anomaly decreases with the increasing age of the oceanic plate to become faster than average for ages greater than 100 Ma. The depth to which the anomalies associated with the midocean ridges persists in the tomographic maps (>200 km) puzzles geodynamicists who think that midocean ridges are passive features and that conditions below about 100 km depth are isothermal.

Figure 23 shows a cross-section through the upper mantle of the Pacific from the model of Ekström and

Dziewonski; the direction of the cross-section follows the direction of motion of the Pacific plate. Going from East to West, we see higher seismic velocities associated with subduction under South Africa; very slow lithosphere at the East Pacific rise; increase in velocities with the distance from the ridge, and subduction under the Mariana trench; the red dots are earthquakes. It is clear that the velocities change with age to depths below 200 km. The map in Figure 22 at 200 km depth shows diminished variability of velocities under ocean but still very strong anomalies under the continents; the old cratons, in particular.

The map in Figure 22 at 500 km depth represents average shear velocity anomalies in the transition zone. The most characteristic features are the fast anomalies in the western Pacific and Eastern Asia, under South America, the Atlantic, reaching to western Africa. In the western Pacific they can be associated with subduction zones, although they are much wider than an anomaly associated with a 100-km-thick slab. Studies of the topography of the 660-km discontinuity show that the areas of high seismic velocity are correlated with a depressed boundary, yielding credence to an interpretation that these anomalies are indicative of an accumulation (temporary, perhaps; see earlier discussion on the models of flow in the mantle) of the subducted material in the transition zone. Figure 24 shows comparison of lateral variation in velocities obtained in a model named S362D1. The two maps one just above and the other just below the 660-km discontinuity are very different; the map representing the transition zone shows features similar to that at 500-km map in Figure 22; the lower mantle map is quite different and has distinctly different spectral content: it is dominated by relatively short wavelength features. This result, and similar results obtained by other modeling groups, supports the concept of a separation–perhaps not absolute–between the upper mantle and lower mantle.

In the middle mantle the anomalies are not well organized. This observation contrasts with the results of "high-resolution" tomography which in this depth range shows two narrow high velocity features: one stretching from the Hudson Bay to Bolivia and the other from Indonesia to the

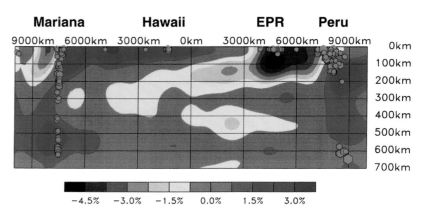

Mariana   Hawaii   EPR   Peru

9000km 6000km 3000km 0km 3000km 6000km 9000km

0km
100km
200km
300km
400km
500km
600km
700km

−4.5%   −3.0%   −1.5%   0.0%   1.5%   3.0%

FIGURE 23 Cross-section through the upper mantle of model of Ekström and Dziewonski. Note that velocities change as a function of distance from the East Pacific Rise (proportional to age of the plate) to depths greater than 200 km Red dots indicate earthquakes. There is vertical exaggeration by a factor of about 20.

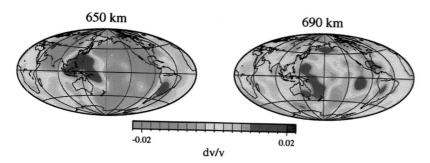

650 km   690 km

-0.02   0.02

dv/v

FIGURE 24   Shear velocity anomalies just above and below the boundary between the upper mantle and lower mantle in model S262D1 of Gu et al., The differences indicate a drastic change in the pattern of the anomalies, most likely associated with a serious impedance to flow.

Mediterranean. Even though elements of these two structures are present in our model, they are not equally well defined. Also, there are many other features of comparable amplitude. This is also true with respect to models published by scientists at Berkeley and at Scripps, who used parameterization similar to that in Figure 22. Intensive efforts are made to understand the differences between the results of two different approaches to tomography.

The map at 2800 km depth shows the velocity anomalies as the CMB is approached. The ring of high velocities circumscribing the Pacific basin is already visible at 2000 km; it strengthens considerably over the next 500 km and increases even further toward the CMB. In the wavenumber domain of spherical harmonics, the spectrum of lateral heterogeneities is very red, being dominated by degree, 2 and 3. This is the dominant signal in the lower mantle, very clear in properly displayed data. The location of the ring of fast velocities corresponds to the location of subduction zones during the past 200 Ma. The large red (slow) regions are sometimes called the African and the Pacific "superplumes." Their origin is unknown; they, most likely, represent both thermal and chemical heterogeneity. There is a good correlation between the location of the two superplumes and distribution of hotspots at the Earth's surface, indicating a degree of connection between the tectonics at the surface and conditions near the core–mantle boundary.

Figure 25 gives two views of low-pass filtered anomalies in the lower mantle in a model of by Ritsema et al. in 1999, plotted in Cartesian coordinates: the red is a 0.6% isosurface and blue is +0.6%. We see the circum-Pacific ring of fast anomalies and the two low velocity anomalies: one very concentrated under the Pacific and a more diffuse one under the Atlantic and Africa. Their radial continuity throughout the lower mantle indicates that they cannot be explained by processes at the core–mantle boundary alone. The origin of this large-amplitude, very large wavelength signal has not yet been explained by geodynamic modeling, although an assumption that the velocity and gravity anomalies are correlated leads to a good prediction of the geoid at the gravest harmonics.

It was believed since 1977, the time of publication of the first large-scale GST study, that three-dimensional images of lateral heterogeneity in the mantle will be an essential tool in addressing some of the fundamental problems in earth sciences. The results accumulated since then confirm that statement even though much progress is still to be made. Cooperation among the different fields of Earth sciences (geodynamics, mineral physics, geochemistry, seismology, geomagnetism) it the requisite condition to fulfill this goal.

top view

bottom view

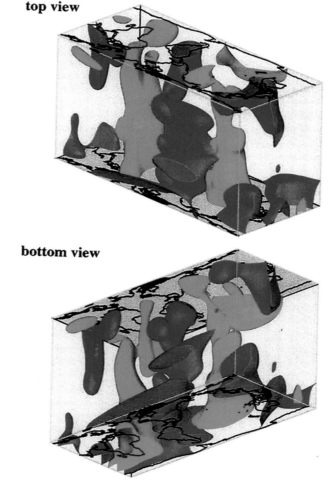

FIGURE 25   Low-pass filtered S-velocity model of Ritsema et al. in a three-dimensional projection; the top 800 km of the structure is removed.

8. Earth as a Rosetta Stone

The Earth is unique among its partners in our solar system in that it has had liquid water oceans for most of its history, has a highly mobile crust, and a dynamically convecting interior. This combination means that the surface is and has been constantly driven by the movement of the interior, such that the oldest terrestrial subaerial landscapes are at most ~10% of the age of the planet, and the oldest submarine landscapes are only a little more than 10% of that. Thus, the Earth not only has one of the most globally dynamic surfaces in the solar system, but its interior is also one of the most dynamic. Only the tidally wracked and volcanically incessant surface of Io, Jupiter's innermost satellite, may be younger and more active. Driven by internal forces, the periodic conglomeration and separation of continental landmasses, causing opening and closing of oceans, and construction and destruction of mountain ranges profoundly impact the global climate. The environmental stresses caused by such reshuffling of the surface may themselves have influenced the progress of evolution on the planet—evolution that was possibly reset every 100 Myr or so by devastating asteroidal impacts. In the final analysis, the Earth is the only planetary body with which the human species has had intimate experience—for millenia. Thus, beyond being our home, the Earth is for us a crucial yardstick—a Rosetta stone—by which we will measure and interpret the processes, internal structure, and overall histories of other planets in this solar system and, someday, of other planets around other stars. [*Also See* EARTH AS A PLANET: ATMOSPHERE AND OCEANS.]

Bibliography

Baker, V. R. (1986). Introduction: Regional landforms analysis. "Geomorphology from Space: A Global Overview of Regional Landforms" (N. M. Short and R. W. Blair, Jr., eds.), pp. 1–26. NASA Scientific and Technical Information Branch, Washington, DC.

Bloom, A. L. (1998). "Geomorphology: A Systematic Analysis of Late Cenozoic Landforms," 3rd Ed. Prentice Hall, Upper Saddle River, NJ.

Dziewonski, A. M., and Anderson, D. L. (1984). Seismic tomography of the Earth's interior. Am. Sci. 72 (5), 483–494.

Dziewonski, A. M., and J. H. Woodhouse (1987). Global images of the Earth's interior, Science, 236, 37–48.

Ekström, G., and A. M. Dziewonski. The unique anisotropy of the Pacific upper mantle, Nature, 394, 168–172, 1998.

Francis, P.W. and C. Oppenheimer 2004. "Volcanoes." Oxford University Press, Oxford, UK.

Gu, J. Y., A. M. Dziewonski, W.-J. Su and G. Ekström (2001). Shear velocity model of the mantle and discontinuities in the pattern of lateral heterogeneities, J. Geophys. Res., 106, 11169–11199.

Heezen, B., and Tharp, M. (1997). "Panoramic Maps of the Ocean Floor."

Ishii, M., and A.M. Dziewonski (2002). The innermost inner core of the earth: Evidence for a change in anisotropic behavior at the radius of about 300 km., Proc. Natl. Acad. Sci. USA, 99, 14026–14030

King, L. C. (1967). "Morphology of the Earth," 2nd Ed. Oliver and Boyd Ltd., Edinburgh.

Pieri, D. and M. Abrams (2004). ASTER Watches the World's Volcanoes: A New Paradigm for Volcanological Observations from Orbit, *Journal of Volcanology and Geothermal Research*, 135 (1–2), 13–28.

Ritter, Dale F., R. Craig Kochel, and Jerry R Miller (2002). "Process Geomorphology." Fourth Edition, McGraw-Hill, New York.

Schumm, S.A., (2005). "River Variability and Complexity." Cambridge University Press, New York, 234pp.

Short, N. M., and Blair, R. W., Jr. (eds.) (1968). "Geomorphology from Space: A Global Overview of Regional Landforms." NASA Scientific and Technical Information Branch, Washington, DC.

Snead, R. E. (1980). "World Atlas of Geomorphic Features." Robert E. Krieger Co., Huntington, NY, and Van Nostrand Reinhold, NY.

Stein, S. and M. Wysession (2003). An Introduction to Seismology, Earthquakes, and Earth Structure, Blackwell Publishing: Oxford, UK, 498pp.

Toomer, G. J. (translator) (1998). Ptolemy's "Almagest." Princeton University Press, Princeton, NJ.

Ward, P. and D. Brownlee (2002). "The Life and Death of Planet Earth: How the new science of astrobiology charts the ultimate fate of our world," Henry Holt and Company, NY.

The Sun–Earth Connection

J. G. Luhmann

University of California, Berkeley

S. C. Solomon

National Center for Atmospheric Research, Boulder

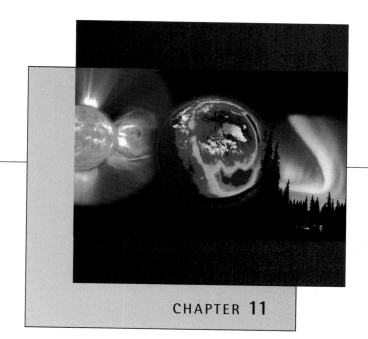

CHAPTER **11**

The Sun has profound effects on the Earth through its primarily visible and infrared photon emissions. This radiated energy, generated as a by-product of the nuclear reactions in the Sun's core [see THE SUN] is absorbed or reflected at different wavelengths by the sea and land surfaces and the atmosphere. The result is the atmospheric circulation system that generates tropospheric weather through the diurnal and seasonal cycles caused by Earth's rotation and axis tilt. [See EARTH AS A PLANET: ATMOSPHERE AND OCEANS.] The climate of the Earth is the result of the long-term interaction of solar radiation, weather, surface, oceans, and human activity.

These influences are not the only ways the Sun affects the Earth. Ultraviolet (UV) and X-ray light from the Sun are much less intense, but more energetic and variable than the visible emissions. The UV radiation is absorbed in the stratosphere where it affects the production of the ozone layer and other atmospheric chemistry, while the extreme ultraviolet (EUV) photons and X-rays are absorbed in the thermosphere (above ~90 km), creating the ionized component of the upper atmosphere known as the ionosphere. Even more variable is the emission of charged particles and magnetic fields by the Sun. One form of this output is the magnetized solar wind **plasma** and its gusty counterpart, the **coronal mass ejection** or **CME**. CMEs interact with the Earth to create major **geomagnetic storms**. These and other forms of matter, energy, and momentum transfer couple the physical domains of the connected Sun–Earth system, which is illustrated in Figure 1. A brief summary of the subject of this chapter, whose focus is this system, follows here.

Sun–Earth connection physics begins in the solar interior where dynamo activity [see THE SUN] generates the solar magnetic field. The solar magnetic field, coupled with the mechanical and radiative energy outputs from core fusion reactions, ultimately determines both the variability of the Sun's energetic (EUV, X-ray) photon outputs and the interplanetary conditions at the orbit of Earth. The latter include the solar wind plasma properties, the interplanetary magnetic field magnitude and orientation, and the energetic particle radiation environment. Both the energetic photon outputs and interplanetary conditions vary with the ~11-year solar cycle, which is characterized by changing frequencies of solar flares and CMEs, the two primary forms of solar activity. These in turn determine **space weather** conditions in near-Earth space or **geospace**, the region comprising of the magnetosphere, the upper atmosphere, and the ionosphere. Only in the 1960s was it appreciated that the interplanetary magnetic field orientation relative to Earth's own dipolar field plays a major role in solar

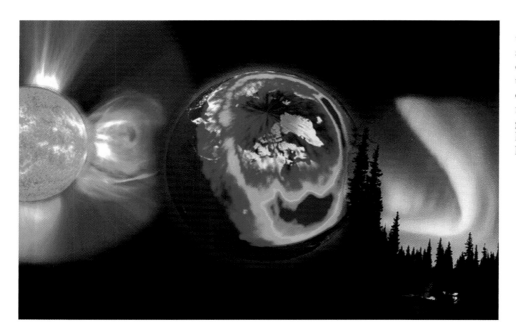

FIGURE 1   Triptych illustrating the coupled Sun–Earth system, showing from left to right an image of the erupting solar corona from the *SOHO* spacecraft, and images of the Earth's auroral emissions from space (center) and from the ground (right). (See http://sohowww.nascom.nasa.gov/hotshots/2003 03 14/.)

wind–magnetosphere couplings as described in more detail in the main text below.

The magnetosphere, the region of near-Earth space dominated by the magnetic field of the Earth and shaped by its interaction with the solar wind (see Fig. 1), organizes geospace. Various particle populations in the magnetosphere, including the plasmas originating in the solar wind and Earth's ionosphere, and the more energetic particles trapped in the **radiation belts,** are constantly modified by changing interplanetary conditions. The ionosphere acts as a conducting inner boundary affecting the magnetosphere's response to those conditions, but it is also a source of ions and electrons for the magnetosphere. Under the disturbed local interplanetary conditions that occur after an Earth-directed CME, a collection of major magnetospheric modifications called a geomagnetic storm occurs. The population of trapped energetic particles in the radiation, or Van Allen, belts surrounding the Earth undergoes enhancements, losses, and redistribution. Current systems and particle exchanges couple the magnetosphere and ionosphere to a greater than normal degree. The result is enhanced solar wind energy transfer into geospace, causing auroral emissions and related changes in the high-latitude dynamics of the ionosphere, as well as in the density and composition in the thermosphere. Evidence of atmospheric influences of geomagnetic storms and other Sun–Earth connection effects down to the stratosphere has been reported, although it remains controversial. On the other hand, induced currents in conductors on the ground from storm-associated magnetic field changes are unarguable proof of the depth of influence of extreme space weather.

Studies of the Sun–Earth connection investigate the physics that makes the solar wind, magnetosphere, and upper atmosphere/ionosphere a highly coupled system. Figure 2 shows an attempt to diagram its various components and their relationships. There are also practical aspects to understanding the connections shown. Specifications of radiation tolerances for spacecraft electronics components, designs of protective astronaut suits and on-orbit shielding, and definitions of the surge limits for power grids on the ground can be made with a better understanding of space weather effects. Forecast models can help predict the changes in the magnetosphere that alter the radiation belts and the changes in the ionosphere that disrupt radio communications and GPS navigation. Sun–Earth connection knowledge also increases our understanding of other areas of astronomy and astrophysics such as planet–solar wind interactions, extra-solar planetary systems, stellar activity, and the acceleration of particles in the universe.

1. The Solar and Heliospheric Roles in the Sun–Earth Connection

Solar radiation in the ultraviolet, EUV, and X-ray wavelengths are the primary sources of ionization in the Earth's atmosphere. Of these, solar EUV fluxes are the most important source of the ionosphere. Figure 3 illustrates the relatively large variability of this part of the solar spectrum, compared to the visible and infrared wavelengths that dominate the "solar constant." As mentioned before, this

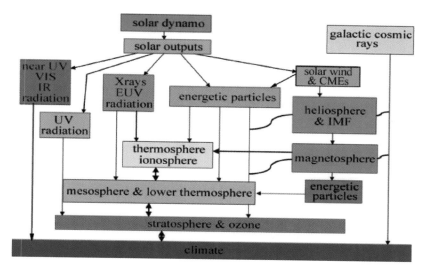

FIGURE 2 Flow diagram illustrating the connections in and complexity of the coupled Sun–Earth system. The solar dynamo (top) generates the solar magnetic field, which modulates the solar outputs of extreme ultraviolet and X-ray emissions, as well as the solar wind plasma. The solar wind and its gusty counterpart, CMEs (green box, upper right), directly determine the state of the local heliosphere, which controls the state of the magnetosphere (including its energetic particle or radiation belt populations). In the meantime, both solar photons (upper right boxes) and solar energetic particles directly affect the state of the upper atmosphere. The possible connection to climate, suggested at the bottom, is currently a matter of speculation. (Adapted from http://lws-trt.gsfc.nasa.gov/lika radtg.ppt.)

variability is a result of the control of these emissions by the solar magnetic field, which undergoes significant evolution during the course of the ~11-year solar activity cycle. [*See* THE SUN.] The extreme ultraviolet emissions come largely from bright plage areas seen on the photosphere and from the chromospheric network, while the X-rays come mainly from hot plasma-containing coronal loops structured by the coronal magnetic field. Both of these features can be seen in the composite solar image in Figure 1. The plages and X-ray bright loops are related to active regions, areas with the strongest photospheric magnetic fields, that are nonuniformly distributed over the solar surface. The changing numbers of active regions, and their areas, determine

the solar activity cycle. Thus, the solar EUV flux experienced at Earth undergoes variations on both the 27-day time scale of solar rotation and the near-decadal time scale of solar activity. [*See* THE SUN.] The transient brightenings in active regions called solar flares occasionally produce solar EUV and X-ray emission enhancements of up to several orders of magnitude at photon energies extending into the gamma ray range. These outbursts affect Earth's atmosphere and ionosphere at depths depending on their wavelengths as indicated in Fig. 4. The magnetosphere responds to changes in the ionosphere and upper atmosphere, but its primary solar controller is the magnetized solar wind plasma.

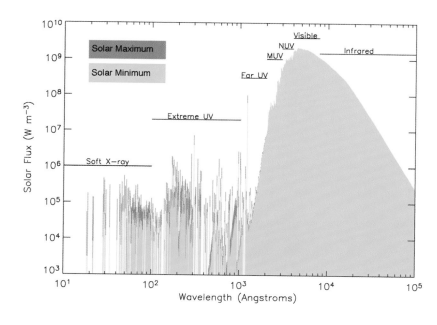

FIGURE 3 Illustration of the solar spectrum, showing the intensities of various wavelength emissions and their variation from active (red) to quiet times. Notice that order of magnitude variations from solar minimum to maximum occur at the short (¡ 1000 Å is equivalent to 0.1 Å) wavelengths. (1 nanometers.)

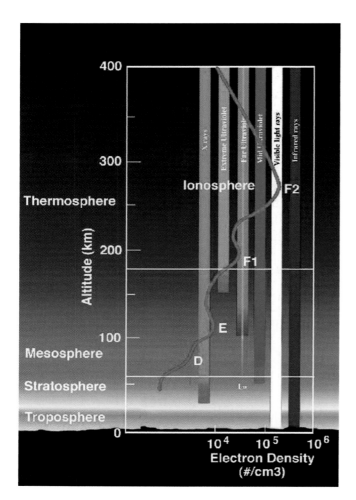

FIGURE 4   Atmospheric depths affected by various wavelengths in the solar spectrum. (Courtesy of Windows to the Universe, http://www.windows.ucar.edu.) The longer wavelength radiation affects mainly heating, while the shorter wavelengths can produce ionization and associated chemistry changes. The letters F1, F2, E, and D are used to designate different ionospheric layers.

The solar wind is the outflowing, ionized gas or plasma of the solar upper atmosphere. [*See* THE SOLAR WIND.] This outermost extension of the corona fills a space up to at least ~80 AU in radial extent, defining the **heliosphere,** the region surrounding the Sun. The mainly hydrogen solar wind flows primarily from places in the corona that are magnetically "open" to interplanetary space. These open field regions are often called coronal holes because of their dark appearance in soft X-ray and EUV images. [*See* THE SUN.] The solar wind also carries with it the stretched out coronal magnetic field that takes on an average outward or inward orientation depending on the magnetic field direction or polarity at its photospheric base. There is also a component of the quiet solar wind that comes from the edges of coronal closed magnetic field regions, producing the equivalent of a boundary layer between outflows from different open field

regions. On the average, the solar wind speed is slowest in these boundary layers and fastest where it flows from the center of large open field regions. Typical solar wind speeds range from ~300 km/s to ~800 km/s and are roughly constant with radial distance. Undisturbed solar wind magnetic field strengths at 1 AU range from ~5 to 10 nT, and densities range from ~5 to 15 particles/cm^3. Because the open and closed field regions change with the distribution of active regions on the Sun, the solar wind stream structure and field polarity pattern evolve with the solar activity cycle. They are simplest at the quietest times of the cycle, during which the corona usually exhibits two main solar wind sources near the Sun's polar regions, one with positive (outward) and one with negative (inward) magnetic polarity.

A critical aspect of the solar wind stream structure for the Sun–Earth connection is Earth's location near the solar rotational equator. This region is often dominated by the presence of the slow wind and the related heliospheric current sheet that separates the solar wind from open coronal field sources with outward and inward magnetic field polarities. This circumstance, together with the rotation of the Sun, produces local interplanetary conditions that at low solar activity exhibit repeating or corotating 27-day variations in solar wind speed and density, and interplanetary magnetic field polarity. E. Parker, who first proposed the existence of the solar wind, also recognized that the Sun's rotation would wind up the interplanetary magnetic field into a spiral shape. Figure 5 illustrates this "Parker Spiral"

SOLAR WIND

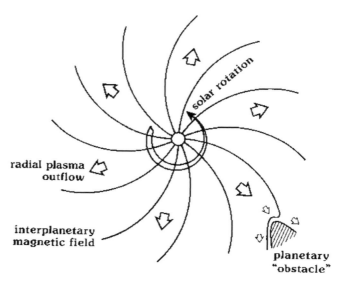

FIGURE 5   Illustration of the Parker Spiral interplanetary magnetic field carried in the outflowing solar wind, and wound up by solar rotation. At 1 AU the typical angle the field makes with respect to the Sun–Earth line is ~ 45°. (*See* SOLAR WIND for further information.)

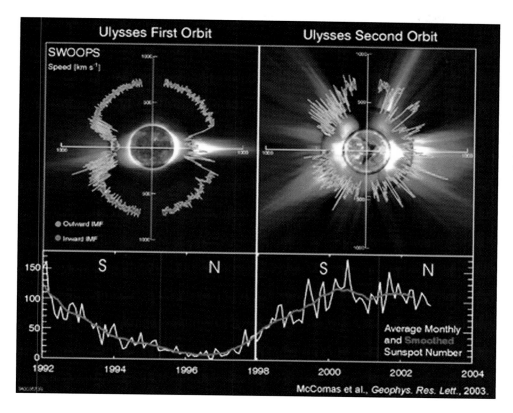

FIGURE 6 Solar wind velocities at solar minimum and maximum as measured on the *Ulysses* spacecraft which passes over the Sun's poles. The speed is shown in a polar coordinate system with zero speed at the center of the Sun. The blue and red indicate interplanetary magnetic field polarity. (McComas et al., 2003, *Geophys. Res. Lett.* v.30, doi 10.1029/2003 GLO 17136, 2003.)

orientation of the near-ecliptic field in the heliospheric, and its typical 45° (from radial) orientation at 1 AU. [*See* THE SOLAR WIND.] At solar minimum, adjacent streams of different speeds from different coronal source regions may interact, producing spiral density and field ridges at their boundaries. When these ridges, which are called stream interaction regions or corotating interaction regions (CIRs), rotate past the Earth, they can cause modest geomagnetic activity. At maximum solar activity, the solar wind conditions are more variable and structured, and less organized by solar latitude. They also exhibit many transient disturbances caused by rapid changes in coronal structure and CMEs, whose effects are described in more detail later. Figure 6 shows solar wind characteristics from periods around solar minimum and solar maximum. These interplanetary conditions shape the Earth's magnetosphere, control its responses to the solar wind, and regulate states of internal particles, energy, and stresses.

As the Sun becomes more active, as indicated by increasing numbers of sunspots, it produces greater numbers of both flares and CMEs. CMEs have the greatest effects on geospace, and so we focus on them here. The details of the CME initiation process, as well as CME structure, are subjects of intensive current research. As seen in white light images from coronagraphs like the *SOHO* LASCO instrument (an example of which is shown in Fig. 1), CMEs appear to be eruptions of a magnetic bubble or twisted "flux rope" of coronal magnetic fields. These structures, which are referred to as drivers or ejecta, travel outward at speeds ranging from tens to several thousand km/s. As they travel, they interact with the surrounding solar wind, compressing it ahead if they are moving faster. If they move fast enough relative to their surroundings, they create an interplanetary shock wave. Figure 7 illustrates the effect of a CME on the solar wind and interplanetary magnetic field at 1 AU. These propagating disturbances are experienced by the magnetosphere as sudden increases of solar wind density, velocity, and magnetic field at the shock passage, followed by several hours of enhanced solar wind parameters, and then the ejecta passage characterized by a period with normal densities but high magnetic field strengths and, often, inclinations. The entire structure may take hours to days to pass Earth depending on its speed. Enhanced solar wind pressure associated with a CME is usually from the sheath portion between the shock front and the ejecta. The ejecta fields can occasionally be modeled as a passing magnetic flux rope configuration as suggested in Fig. 7. Around the minimum of solar activity the local interplanetary medium is disturbed by one of these interplanetary CMEs or ICMEs once every few months, but at solar maximum they can occur about once a week. The most extreme (largest, fastest) ICMEs usually follow CMEs associated with large, complex active regions on the Sun that also produce solar flares.

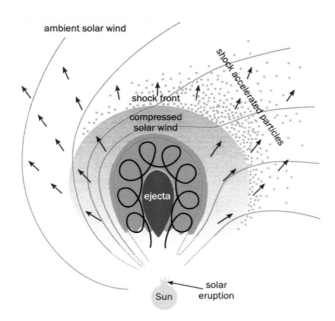

FIGURE 7   Illustration of the interplanetary effects of a CME. The CME produces an ejection of coronal material (ejecta) that may include a helical magnetic field structure or flux rope (illustrated by the black line). This structure plows into the ambient solar wind and may produce a shock in the solar wind plasma ahead of the ejecta. The region of compressed solar wind between the shock and ejecta is referred to as the sheath. Some solar wind particles are accelerated at the shock and speed out ahead of it along interplanetary field lines (red, green). (Luhmann, 2000, *Physics World*, p. 31–36, July 2000.)

In addition to the magnetized solar wind plasma, the heliosphere also contains a population of energetic (tens of kiloelectron volts (keV) to hundreds of megaelectron volts (MeV)) charged particles that varies with time. Ions and electrons are accelerated at both flare sites on the Sun and by the shock waves formed in the corona and interplanetary space by the fast-moving CME ejecta or by interacting high- and low-speed solar wind streams. CME shocks produce the most intense and long-lived (several day) episodes. The particles race ahead of their shock source along the spiral interplanetary field lines, surrounding the magnetosphere with a sea of potentially hazardous radiation within tens of minutes of the events at the Sun. Sometimes the fluxes of these particles increase by several orders of magnitude when the CME shock itself arrives several days after the event in the corona. The contributions of **solar energetic particles** to local interplanetary conditions are related to the level of flare and CME activity on the Sun. In an interesting opposite effect of solar activity, other more permanently present energetic charged particles called cosmic rays, which arrive at Earth from the heliospheric boundary and beyond, show locally decreased fluxes when solar activity is high. This is likely due to the sweeping action of the highly structured solar wind around the time of solar activity maximum, when interplanetary field disturbances carried outward present effective barriers to incoming charged particles. Under certain conditions solar energetic particles can enter the magnetosphere where they contribute to the radiation belts and produce layers of enhanced ionization deep in the Earth's polar atmosphere.

2. The Geospace Role in the Sun–Earth Connection

The Earth's space environment is determined by its nearly dipolar internal magnetic field that forms an obstacle to the solar wind, creating the magnetosphere [*see* PLANETARY MAGNETOSPHERES]. Spreiter and coworkers (1966) and Axford and Hines (1961) were among the pioneering researchers to recognize the fluid-like aspects of the solar wind interaction with Earth's compressible field, describing it in terms of a blunt body in a hypersonic flow. The size and shape of this blunt body, the magnetopause, can be calculated from the assumption of pressure balance between the Earth's internal magnetic field pressure and the incident dynamic pressure of the solar wind (Dynamic pressure = Mass density × Velocity squared). It typically occurs at ∼10 Earth radii along the line connecting the centers of the Earth and the Sun and at ∼15 Earth radii in the terminator plane. In contrast to the compressed, solar wind pressure-confined day side, the night side magnetosphere stretches out into an elongated structure called the magnetotail. These features, confirmed by decades of observations, are illustrated in Fig. 8. The magnetopause separates geospace and the solar wind plasma-dominated regions outside. As seen in Fig. 8, the outermost features associated with the Earth's magnetic obstacle are actually the bow shock that forms in the solar wind ∼5 Earth radii upstream of the day side magnetopause and the magnetosheath. The magnetosheath is the slowed, deflected solar wind between the bow shock and the magnetopause. Thus, when the Earth's field interacts with the solar wind, it does so through the altered solar wind in the magnetosheath.

Dungey (1961) first recognized that the magnetopause is not a complete barrier to the solar wind, and that magnetospheric field topology is also controlled by its interconnection, or **reconnection,** with the interplanetary field. This leap of understanding revolutionized the study of solar wind magnetosphere coupling and geomagnetic activity. Figure 9 reproduces Dungey's original cartoon suggesting the different appearances of the magnetospheric field topology for the extreme cases of steady northward and southward interplanetary fields. Similar pictures can be obtained by adding background uniform fields of both directions to a

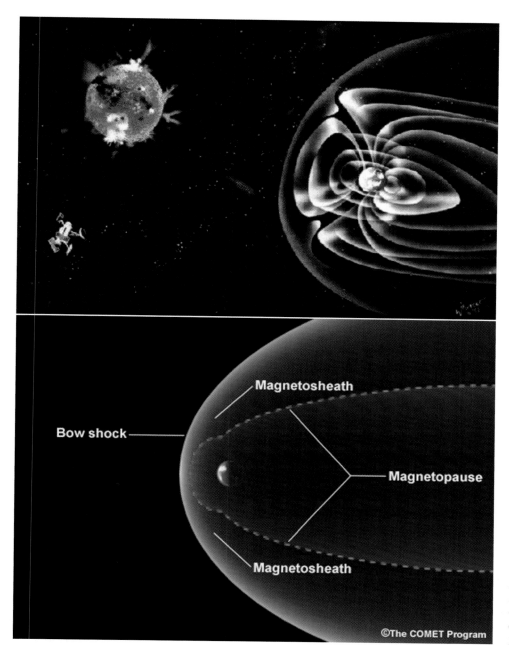

FIGURE 8   (a) Illustration of the blunt body shape of the magnetosphere, showing some gossamer "shells" of magnetic field surfaces along which charged particles drift, and the magnetopause. (Rice University) (b) Magnetospheric boundaries described in the text. The magnetopause nominally separates solar wind and magnetospheric domains. (The source of this material is the Cooperative Program for Operational Meteorology, Education, and Training (COMET®) Web site at http://meted.ucar.edu/ of the University Corporation for Atmospheric Research (UCAR) pursuant to a Cooperative Agreement with National Oceanic and Atmospheric Administration. Copyright 1997–2004 University Corporation for Atmospheric Research. All Rights Reserved.)

dipole field. The northward interplanetary field, which is parallel to the Earth's dipole field at the equator, produces a magnetically closed magnetosphere. The southward field, which is antiparallel to the Earth's dipole field at the equator, produces a magnetically open configuration with the polar region fields of the Earth connected to the interplanetary field. These differences greatly affect the transfer of both energy and particles from the solar wind into geospace. For the northward case, the solar wind interaction resembles a viscous boundary interaction at the magnetopause, and there is minimal exchange of energy and particles. For the southward case, the charged particles in interplanetary space have access to the polar regions along interplanetary field lines. An electric field associated with solar wind convection ($E = -V \times B$, where V is the solar wind velocity and B its magnetic field), maps along open field lines into polar regions where it drives vigorous magnetosphere and ionosphere circulation as in Fig. 10. The two-celled vortical convection pattern has been observed in the ionosphere by high-latitude radars and can be inferred from

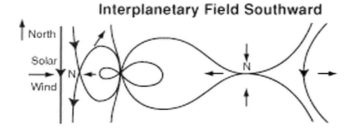

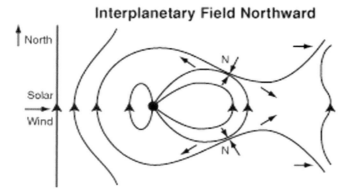

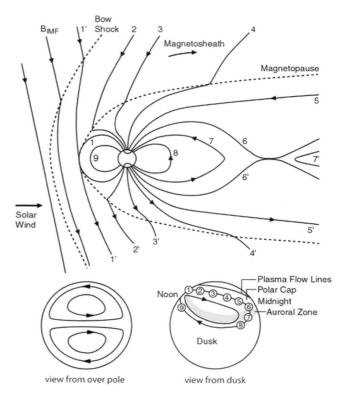

FIGURE 9 Dungey's original 1961 cartoon suggesting the reason for the association of greater geomagnetic activity with southward interplanetary magnetic fields. Southward interplanetary fields can reconnect or merge with the Earth's dipole field at the day side magnetopause. Another reconnection in the magnetotail returns the Earth's opened fields to their original "closed" state so that all the Earth's field is not permanently opened by day side reconnection. The process of reconnection drives magnetospheric circulation (see Fig. 10) and is thus a means by which solar wind energy is transferred to the magnetosphere. (Reprinted figure with permission from J. W. Dungey, 1961, *Phys. Rev. Lett.*, **6**, 47. Copyright 2005 by the American Physical Society.)

FIGURE 10 Illustration of magnetospheric circulation during periods of southward interplanetary magnetic field. Various key features of the solar wind interaction are shown, including magnetospheric field line connections to the interplanetary field and their mapping to the high-altitude atmosphere. The numbers indicate a time sequence. The driven circulation occurs all the way down to the polar ionosphere as shown by the inset, which shows the dusk half of the double-celled ionosphere convection pattern. The aurora occurs mainly in the regions of convection reversals. (Kivelson and Russell, 1995, "Introduction to Space Physics.")

magnetometer measurements. Solar wind–magnetosphere coupling is thus greatly enhanced at times when the interplanetary magnetic field is southward.

The physics of the reconnection or magnetic field merging process that results in this configuration change for southward interplanetary fields is still a subject of intensive research. Because space is not a vacuum, simple superposition of the external (interplanetary) and internal (Earth dipole) fields is not a physically correct explanation. Somehow the solar wind plasma that carries the interplanetary field "frozen" into its flow must allow the field to merge with the magnetospheric field at the magnetopause when the two have antiparallel components. The interested reader is referred to the review by Drake for further details on current theories of magnetic field reconnection in space plasmas. When CME effects reach the Earth, the solar wind dynamic pressure incident on the magnetosphere can in-

crease by an order of magnitude or more, primarily due to the compression of the ambient solar wind plasma by the driver or ejecta from the CME. The onset of this increase may be sudden if a leading shock is present. The solar wind magnetic field is also compressed with the plasma and can become significantly inclined with respect to the Earth's equatorial plane. The ejecta fields are also highly inclined, and often strong and steady, or slowly rotating over intervals of about a day. Thus, larger than normal northward and/or southward interplanetary field components result from both passing segments of the disturbance. The magnetosphere's response to these disturbed heliospheric conditions includes increased compression of the dayside magnetosphere, sometimes to within a few Earth radii of the surface, and increased reconnection between the Earth's magnetic field and interplanetary field during the passage of the southward-oriented portions. The time-dependent

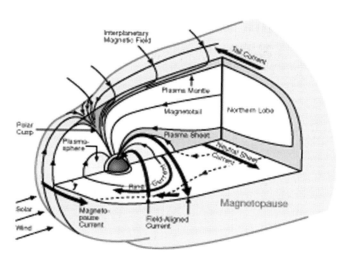

FIGURE 11   A more detailed illustration of the magnetosphere, showing features mentioned in the text, such as the ring current and magnetotail current sheet. The plasmasphere is a region of denser, corotating magnetospheric plasma of ionospheric origin. The plasma sheet is a denser region of magnetotail plasma that participates in the physics of magnetotail reconnection and ring-current formation. (Rice University.)

nature of these boundary conditions can introduce additional complexity into the solar wind–magnetosphere coupling.

Resulting geospace consequences of CMEs are numerous and varied, as illustrated by Fig. 11. The associated magnetospheric compression is accompanied by enhancements of the energetic radiation belt particles trapped in the Earth's dipole field, due to a combination of inward diffusion and energization of the existing particle populations. [See PLANETARY MAGNETOSPHERES.] Solar energetic particles accelerated at the CME-driven shock or in associated solar flares can also leak into the magnetosphere along newly reconnected field lines at the magnetopause or along open field lines into the polar regions, as these particles tend to stream along field lines. Magnetic reconnection between the stretched out, antiparallel fields in the magnetotail causes currents to flow through the high-latitude ionosphere. As magnetospheric charged particles move toward Earth with the field lines, they are accelerated, in some cases by electric fields parallel to the magnetic field. These energized particles include electrons, protons and other heavier ions. When they reach the upper atmosphere, they collide with neutral gases at altitudes of ~100 to ~200 km, causing ionization and excitation of atoms and molecules. The ionization enhances the flow of magnetosphere–ionosphere currents, and when the excited atoms and molecules decay back to their ground states, they emit the light known as the

aurora. (Further information about the aurora can be found in Section 3.)

Magnetotail reconnection also triggers injection of particles toward Earth at low latitudes that form a ring current at ~4 to ~7 Earth radii. In the polar regions, protons and ions, including ionospheric oxygen ions, O^+, are driven upward from the base of open field lines and flow into the magnetosphere, changing the composition of the magnetosphere and ring current ion populations. The ring current noticeably changes the magnetic field in the magnetosphere and at Earth's surface. Altogether these phenomena characterize a geomagnetic storm, whose magnitude is characterized by the ring current–related reduction of the field at the ground, defined by the index Dst. (Another index, Kp, is also used, but it is more a measure of the auroral current systems.) Eventually, the magnetosphere and ionosphere return to their prestorm states. Most effects are gone after a few days, but some trapped particle populations may last much longer. This complex geospace response to a CME has recently been simulated by several research groups using numerical models of geospace, with solar wind measurements defining the time-dependent boundary conditions on the magnetosphere. Some results from one of these models are illustrated in Fig. 12.

There are also weaker, more frequent geomagnetic disturbances known as substorms. Substorms may occur during storms, as periodic enhancements of the storm-time geospace responses, or as standalone disturbances when the quiet interplanetary magnetic field has a southward component. In some cases, they appear to follow a sudden change in the interplanetary north–south field component or a dynamic pressure pulse in the solar wind. Current ideas on the reasons for substorms, which have been debated for decades, include internal instabilities of the magnetosphere that occur in response to a variety of triggers. However, geomagnetic storms involve the largest episodic energy transfers from the solar wind and are thus responsible for the strongest Sun–Earth connection effects, collectively referred to as space weather.

3. Atmospheric Effects of the Sun–Earth Connection

The atmospheric responses to solar activity and its magnetospheric consequences are the closest counterparts of space weather to traditional weather. They are therefore of special interest in Sun–Earth connection research. Direct effects are largely confined to the thermosphere and ionosphere, above the mesopause at ~90 km. They fall into two main categories: the effects of particles entering or "precipitating" into the atmosphere and the effects of high-latitude ionospheric convection from magnetosphere–ionosphere coupling.

(a)

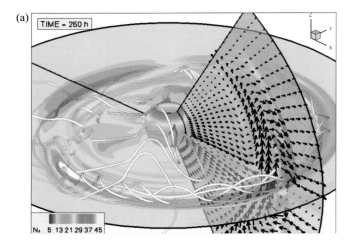

(b)

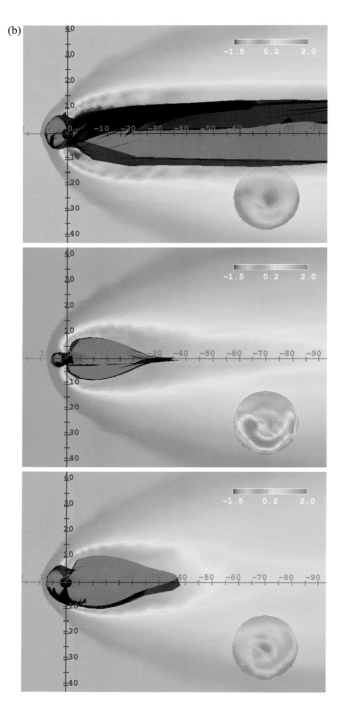

FIGURE 12 (a) Results from a numerical simulation of a CME, showing the distortion of the interplanetary plasma density (contours) and magnetic field (white lines) as it travels toward the Earth. The vectors indicate directions of the velocity in a selected meridional slice. Note the flux rope ejecta that drives the leading interplanetary shock (sharp red contour outer boundary). (Courtesy of D. Odstrcil, University of Colorado.) (b) Geospace response to the CME in (a), showing density contours (log scale), in the local solar wind, the surface of outermost closed magnetospheric field lines, and a view of the resulting energy input into the earth's high latitude atmosphere at three different times. (Luhmann et al., 2004, *J. Atmosph. Solar. Terr. Phys.* v. 66, p. 1243–1256, 2004.)

There are several types of precipitating particles: ∼1–20 keV auroral electrons, ∼10–100 keV ring current ions (protons and some oxygen ions), ∼1–10 MeV radiation belt electrons, and ∼1–100 MeV solar energetic particles (primarily protons). The more energetic the particles, the deeper they penetrate; the altitude ranges to which these various particles penetrate to deposit their energy are illustrated in Fig. 13. As mentioned earlier, when these particles encounter atmospheric atoms and molecules, they cause impact ionization and dissociate molecules into their atomic elements. They also excite bound electrons to unstable states, which then radiatively decay to produce photons with specific energies and thus wavelengths that give the aurora its colors. Chemical reactions caused by the interactions of ions with the dissociated and excited atomic products also excite particular emission features. The characteristic green and red auroral emissions at 557.7 and 630.0 nm are produced by the excitation of the upper atmosphere oxygen atoms; other auroral emission features in the blue and near-ultraviolet spectral regions are formed from excitation of molecular nitrogen and its ion.

Most auroral emissions occur in an oval-shaped band just equatorward of the open field lines at high latitude, giving the auroral oval (shown in Figs. 1 and 10) its name. Ring current ion precipitation can produce high-altitude red aurora at lower latitudes. In contrast, radiation-belt electrons and solar energetic protons leave mainly chemical signatures. Along with other chemical by-products, all particle precipitation produces nitric oxide (NO) from molecular nitrogen dissociation. Very energetic particles can produce NO in the mesosphere (50–90 km) and even the stratosphere

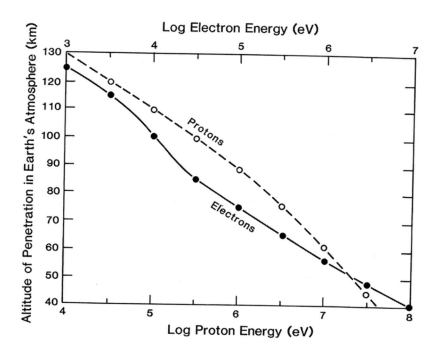

FIGURE 13   Plot showing the approximate depth of penetration into the atmosphere of energetic electrons and protons. (Kivelson and Russell, 1995, "Introduction to Space Physics.")

(15–50 km). Increases in nitric oxide affect ozone levels because chemical reactions involving nitric oxide, ozone, nitrogen dioxide, and atomic oxygen form a catalytic cycle that reduces ozone. For some major solar proton events ~30% depletions of ozone in the mesosphere and upper stratosphere have been detected. Even though the reduction of mesospheric ozone does not have the biological impact of reductions in the denser lower stratospheric ozone, it can modify the temperature and thus the dynamics of the mesosphere. These alterations can in turn modify the transmission of energy from the stratosphere and troposphere to the upper atmosphere. Whether these effects have significant consequences for the lower atmosphere, especially over the long term, is unknown.

As noted above, the solar wind electric field within the open magnetospheric field regions at high latitudes typically stirs the polar ionosphere in a twin vortex pattern, as illustrated in Fig. 10. Ionospheric ions and electrons are dragged antisunward over the polar caps and then forced into a return flow at lower latitudes. The differential motion of the ions and electrons caused by the competition between atmospheric and ionospheric interparticle collisions and the electric and magnetic fields for control of their motion leads to an ionospheric current. This auroral electrojet current has a strength dependent on the combination of the solar wind electric field and the level of ionization in the auroral ionosphere. Collisional dissipation or friction within the volume occupied by the electrojet heats the auroral zone atmosphere. This resistive "joule" heating results in large density perturbations called auroral gravity waves in the upper atmosphere. These travel equatorward, in some cases depositing significant energy and modifying upper atmosphere circulation globally. Traveling ionospheric disturbances are one manifestation of the passage of these waves. Magnetic field perturbations associated with the time-varying electrojet current and the ring current mentioned earlier are detected on the ground, giving the geomagnetic storm its name. A particular geomagnetic index called AE is a widely used measure of the level of ground magnetic field modifications by the auroral electrojet currents.

4. Practical Aspects of the Sun–Earth Connection

The Sun–Earth connection is a complex and fascinating physical system that also has many practical consequences. Society is increasingly dependent on space-based telecommunications and satellite systems that monitor tropospheric weather, global resources, and the results of human activity. The satellite environment is part of the design of these spacecraft, which can suffer radiation damage to electronics if the extremes they may encounter are not taken into account. Satellite orbits are affected by drag as they pass through the upper thermosphere, where changes in density caused by EUV radiation and geomagnetic storms affect their tracking and lifetime. This is particularly true of large, relatively low-altitude vehicles such as the *Hubble Space Telescope* and the *International Space Station*. In addition, satellite orientation controls often rely on Earth's magnetic field, which can be highly variable during geomagnetic storms. Changes in the ionosphere disrupt radio communications by changing ionospheric transmission or reflection characteristics. Global positioning system (GPS) navigational signals, which pass through and are altered by the ionosphere as they are transmitted from very high-altitude satellites to ground receivers, can degrade during disturbed conditions, giving inaccurate locations. On the ground, currents induced in power system transformers and in oil pipelines by storm-related magnetic field perturbations lead to overload and corrosion.

In the era of human space flight, there is also great concern over space radiation hazards from energetic particles. Human-occupied vehicle orbits at low latitude and low altitude are largely protected from the radiation belts and the less common but potentially dangerous major solar energetic particle events. However, the orbit of the *International Space Station* is sufficiently inclined with respect to Earth's equator that it is occasionally exposed to solar energetic particles at high latitudes, when the magnetosphere is disturbed as it often is during solar particle events. Astronaut radiation exposure is carefully monitored and is limited by NASA. To minimize it, plans for extravehicular activities take into account conditions on the Sun and the likelihood of a major solar event that might affect the Earth. Even commercial and military aircraft on polar routes monitor major solar events as a precaution. For future space travel outside of the effective but imperfect magnetospheric shield, protection from solar particle radiation is a major problem to be solved.

In response to both international civilian interests and military needs, the National Oceanic and Atmospheric Administration (NOAA) runs a Space Environment Center (SEC) that collects, analyzes, and distributes information on the Earth's space environment and solar activity. Space weather reports are regularly issued via the internet (http://sec.noaa.gov), where one can also find access to the archives of solar, heliospheric, magnetospheric, and upper atmosphere/ionosphere data that are used. Customers of these services seek information on subjects ranging from interference to radio transmissions by solar radio bursts or ionospheric scintillations, to satellite orbit decay rates based on solar EUV emission intensities. Alerts are posted when a forecaster interprets behavior in the relevant data to mean a solar energetic particle event, geomagnetic storm, or ionospheric disturbance will occur within the next minutes to days. One of the most useful geomagnetic storm–forecasting methods takes advantage of the *SOHO* spacecraft, which allows the forecaster to identify CMEs and the location of active regions on the solar disk. When a CME is headed toward Earth, it sometimes appears in the *SOHO* coronagraph images as a ring around the Sun called a halo CME. These events are known to have an increased probability of causing a geomagnetic storm. However storm forecasts are still extremely difficult, with false alarms, including

halo CMEs actually heading away from the Earth, having originated on the far side of the Sun, a major problem.

Geomagnetic indices, calculated from ground-based measurements of magnetic field perturbations, are routinely used as a measure of the level of space weather disturbance. Different indices emphasize particular Earth responses depending on how and from what stations they are calculated. Several of these were mentioned above. The auroral electrojet index AE is primarily a measure of auroral zone ionospheric currents obtained from high- and mid-latitude monitors, while the ring current index Dst is mainly a measure of the ring current obtained at lower latitudes. The planetary index Kp uses ground stations in both regions. These indices were developed and have been recorded since before the space age. They are used both to parameterize empirical models (e.g., of the auroral zone ionosphere), and to maintain a continuous long-term historical record of the Sun–Earth connection in concise form.

A major goal of Sun–Earth connection research today is a physics-based model of the coupled heliosphere–magnetosphere–upper atmosphere/ionosphere system, including CMEs. Such a model could provide both a forecast tool for space weather events based on solar observations, as well as a numerical experiment framework to gain greater insight into the physics of the coupled system and its extremes. For example, severe space weather events are occasionally observed, but even larger events have been inferred from records of the Earth's historical cosmic ray exposure present in ice cores. What is the worst that could happen to our planet and space assets after one or more of these greatest of solar activity episodes? It is an intriguing question. As a practical matter, the plan to again send humans into deep space, to the Moon and to Mars, also renews the concerns of space radiation hazard issues faced to a lesser degree on the *International Space Station.* The unpredictability of CMEs, and the fact that historically the strongest space weather events have not been at solar maximum, helps motivate applied space weather research.

For many decades, there has also been research on and discussion about the connection of solar activity and Earth's climate. It is possible that the very small changes in total solar irradiance, the so-called solar constant, or changes in UV radiation can have measurable climatic effects. It is also possible that some Sun–Earth connection–related phenomena can reach the troposphere. One highly cited example is the coincidence of the Maunder Minimum in solar activity with the Little Ice Age in Europe during the 15th and 16th centuries. Ideas on how nonradiative effects might play a role include mechanisms such as cloud cover alteration by low-altitude ionization effects of energetic cosmic rays reaching Earth (Tinsley, 2000). The Maunder Minimum climate change has also inspired speculation about the climatic and other consequences of geomagnetic field reversals. Analyses of ice age records and geomagnetic field reversals have as yet produced no definitive results, although it may be significant that the Earth's field still maintains an important higher order harmonic component (e.g., a quadrupole moment) during reversals.

5. Implications for Planetary Astronomy and Astrophysics

The Sun–Earth connection scenario, and the physics it encompasses, is often invoked in the planetary sciences in connection with solar wind interaction issues. Our knowledge of the solar wind coupling to the Earth's magnetosphere and upper atmosphere is far greater than our comparable knowledge for any other planet due to both the wealth of available observations and the efforts that have been put into their interpretation. Planetary spacecraft found that there are magnetospheres around Mercury, Jupiter, Saturn, Uranus, and Neptune. Mars does not have a strong dipolar internal field, but it has patchy crustal magnetism that makes a rather unique obstacle to the solar wind. Venus has essentially no planetary field and thus represents another extreme contrast in solar wind interaction styles. One of the main goals of the *Messenger* mission to Mercury is to better observe the magnetosphere there in terms of its response to solar wind conditions, and its particle content and dynamics. Mercury has no substantial atmosphere or ionosphere so it represents an interesting contrast to Earth that may tell us more about the atmosphere's role in the Sun–Earth connection. Jupiter's giant magnetosphere was found by the *Galileo* spacecraft to be dominated by the internal mass content contributed by the volcanic satellite Io, while Saturn's magnetosphere is currently under scrutiny by the *Cassini Orbiter*. Saturn, like Jupiter, has an aurora that was observed by the *Hubble Space Telescope* prior to the recent missions, but these auroras, and Earth's, may have different reasons for occurring where and when they do. The comparative analysis of planet–solar wind interactions and the related atmospheric effects is extremely valuable for achieving maximum understanding from necessarily limited planetary data.

In the world of astrophysics, extrasolar planetary system research strives to infer from poorly resolved observations the details of individual planets. One possibility for remote sensing is provided by the stellar wind interaction with the planets, which may produce detectable emissions from the planetary atmospheres. To be useful, these emissions must be interpretable in terms of familiar examples in our own solar system. Of particular interest is the detection of Earth-like planets. The Sun–Earth connection suggests a range of possible remote signatures for applications to these "origins" investigations. Similarly, the identification of the effects of stellar winds around other stars is enabled in part by our own heliospheric experience. The interaction of the stellar wind and the surrounding interstellar medium produces a feature like the magnetosheath that is remotely detectable in Lyman-alpha emission. Some stellar outbursts suggest the occurrence of CMEs, and the associated space weather around remote worlds.

Finally, fundamental astrophysical processes are involved in energetic particle acceleration as well as in magnetic reconnection in the Sun–Earth Connection system. Much of what has been learned about particle acceleration at shocks in plasmas has come from the analysis of the observations from the region around the Earth's bow shock. Similarly, reconnection processes at the magnetopause and in Earth's magnetotail have been examined using spacecraft data from both single spacecraft and small constellations. These difficult observations of a dynamical and nonuniform space plasma system with many scales are slowly yielding information about the process that suggests it occurs when and where electrons are no longer controlled by the magnetic field. In addition, numerical simulations have been carried out using both fluid, kinetic (particle), and hybrid (mixed particle and fluid) codes to shed light on the microphysics of how oppositely directed magnetic fields in space plasmas undergo major topological changes. Laboratory work has also contributed to these investigations, all under the umbrella of Sun–Earth connection research.

6. Epilogue

The term "Sun–Earth connection" is used to describe the physically rich and dynamic system by which processes at the Sun affect near-Earth space via other than the solar constant radiative emissions. The subject has most recently been given a new label at NASA, which perhaps better communicates its impact. The "Living with a Star" program seeks to investigate, through sponsored research and space missions, the ways in which the Sun controls the Earth's past, present, and future. To do this, it is necessary to use a combination of theory, measurements, and modeling to study the system components—the heliosphere, the magnetosphere, and the upper atmosphere and ionosphere—to learn how they are coupled. As described above, the couplings are numerous and diverse. They are sometimes subtle like cosmic ray effects on clouds and sometimes overt like CMEs and the related topological changes reconnection imposes on the magnetosphere in response to their associated large southward-oriented interplanetary magnetic fields. The consequences of these couplings are only partly understood. Practical applications of space environment knowledge are in the meantime growing in popularity and demand. Other fields are beneficiaries of the Sun–Earth connection planetary and astrophysical "laboratory." And in an era of new human exploration initiatives, space weather may one day become part of the weather report on your local news.

Bibliography

Axford, W. I., and C. O. Hines (1961). A unifying theory of high-latitude geophysical phenomena and geomagnetic storms. *Can. J. Phys.* **39,** 1433–1464.

Drake, J. F., and M. A. Shay (2005). The fundamentals of collisionless reconnection. In "Reconnection of Magnetic Fields: Magnetohydrodynamics and Collisionless Theory and Observations" (J. Birn and E. R. Priest, eds.). Cambridge Univ. Press, Cambridge, England.

Dungey, J. W. (1961). Interplanetary magnetic field and the auroral zones. *Phys. Rev. Lett.* **6,** 47–48.

Kivelson, M. G., and C. T. Russell (1995). "Introduction to Space Physics." Cambridge Univ. Press, New York.

Parker, E. N. (1963). "Interplanetary Dynamical Processes." Interscience, New York.

Spreiter, J. R., A. L. Summers, and A. Y. Alksne (1966). Hydromagnetic flow around the magnetosphere. *Planet. Space Sci.* **14,** 223–253.

Tinsley, B. A. (2000). Influence of solar wind on the global electric circuit and inferred effects on cloud microphysics, temperature and dynamics in the troposphere. *Space Sci. Rev.* 94, 231–258.

The Moon

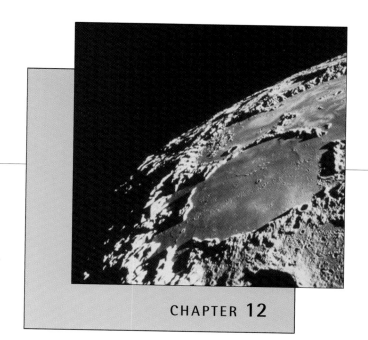

Stuart Ross Taylor

Australian National University
Canberra, Australia
Lunar and Planetary Institute
Houston, Texas

CHAPTER 12

The Moon is a unique satellite in the solar system, the largest relative to its planet. It has a radius of 1738 km, a density of 3.344 g/cm³(Earth density = 5.52 g/cm³), and a mass that is 1/81 that of the Earth. Its orbit is inclined at 5.09° to the plane of the ecliptic. It rotates on its axis once every 27 days. The **moment of inertia** value is 0.3931, consistent with a small increase of density toward the center. The current consensus is that the Moon formed as a consequence of the collision with the Earth of a Mars-sized body about 4.5 billion years ago. The rocky mantle of the impactor spun out to form the Moon, while the core of the impactor fell into the growing Earth. This model explains the high spin of the Earth–Moon system, the strange lunar orbit, the low density of the Moon relative to the Earth, and the bone-dry and refractory composition of our satellite. The model also provides a source of energy to melt the early Moon. The geochemical and petrological evidence is clear that the Moon was molten and floated a crust of feldspar about 4.45 billion years ago. This forms the present white highland crust. The interior crystallized into a sequence of mineral zones by about 4.4 billion years ago. Possibly a small metallic core formed, although the composition of the core is still under debate. Major impacts produced many craters and multiring basins, probably during a spike or "cataclysm" around 3.9–4.0 billion years ago. The oldest basin observed is the South Pole–Aitken Basin and the youngest is the Orientale Basin, which formed 3.85 billion years ago.

Beginning about 4.3 billion years ago, and peaking between 3.8 and 3.2 billion years ago, partial melting occurred in the lunar interior, and basaltic lavas flooded the low-lying basins on the surface. This occurred mostly on the nearside, where the crust is thinner. Major activity ceased around 3.0 billion years ago, although minor activity may have continued until 1.0–1.3 billion years ago and the Moon has suffered only a few major impacts (forming, for example, the young rayed craters such as Copernicus and Tycho) since that time.

1. Introduction

The Earth's Moon (Fig. 1) is a unique satellite in the solar system. None of the other terrestrial planets possesses comparable moons: Phobos and Deimos, the tiny satellites of Mars, are probably captured asteroids. Most of the 130 or so satellites of the outer planets are composed of low-density rock-ice mixtures and either formed in accretion disks around their parent planets or were captured. None of them resembles the Moon so that the origin of our unique satellite has been an outstanding problem. It is in plain sight, accessible even to naked-eye observation, yet it has remained until recently one of the most enigmatic objects in the solar system.

The Moon has played a pivotal role in human development. Both the axial tilt and the 24-hour rotation period

FIGURE 1   A composite full-Moon photograph that shows the contrast between the heavily cratered highlands and the smooth, dark basaltic plains of the maria. Mare Imbrium is prominent in the northwest quadrant. The dark, irregular, basalt-flooded area on the west is Oceanus Procellarum. Mare Crisium is the dark circular basalt patch on the eastern edge. (Courtesy of UCO/Lick Observatory, photograph L9.)

of the Earth may be a direct consequence of lunar formation. Indeed, the lunar tidal effects may have been crucial in providing an environment for life to develop. It is also possible that the Moon has stabilized the obliquity of the Earth, preventing large-scale excursions that might have had catastrophic effects on evolution. The other planets are so remote as to be only points of light, or enigmatic images in telescopes.

Without the presence of the Moon, with its distinctive surface features and its regular waxing and waning phases to stimulate the human imagination, it might have taken much longer for us to appreciate the true nature of the solar system. In many other ways, such as the development of calendars and the constant reminder that there are other rocky bodies in the universe, the Moon has had a profound effect on the human race. One of the outstanding human achievements of the latter half of the 20th century has been the exploration of the Moon, including the landing of astronauts on the lunar surface and our understanding both of lunar evolution and origin.

2. Physical Properties

2.1 Orbit and Rotation

The Moon revolves about the Earth in a counterclockwise sense, viewed from a north polar orientation. This is the same sense in which the Earth and the other planets rotate around the Sun. The orbit of the Moon around the Earth is elliptical with a very small eccentricity ($e = 0.0549$) so that it is nearly circular. The orbital speed of the Moon is 1.03 km/s. The Moon rotates on its axis once every 27.32166 days. This is the sidereal month and corresponds to the time taken for the average period of revolution of the Moon about the Earth.

The lunar synodical month or lunation (the time between successive new moons) is 29.5306 days, longer than the sidereal month, as the Earth has also moved in its orbit around the Sun during the interval. The lunar orbit is neither in the equatorial plane of the Earth nor in the plane of the ecliptic (the plane of the Earth's orbit around the Sun), but is closer to, but inclined at 5.09°, to the latter. The axis of rotation of the Moon, however, is nearly vertical to the plane of the ecliptic, being tilted only at 1°32′ from the ecliptic pole. The inclination of the lunar orbit to the equatorial plane of the Earth varies from 18.4° to 28.6°.

The mean Earth–Moon distance is 384,400 km or 60 Earth radii, but the distance varies from 363,000 to 406,000 km. The moon is closest to the Earth at **perigee** and farthest at **apogee**. The Moon is receding from the Earth, due to tidal interaction, at a rate of 3.74 cm/year.

Tidal calculations have often been used to assess the history of the lunar orbit, but attempts to determine whether the Moon was once very much closer to the Earth, for example, near the **Roche limit** (about 18,000 km), which would place significant constraints on lunar origins, produce nonunique solutions. The problem is that the past distribution of land and sea is not known precisely. The continents approached their present dimensions only about 2 billion years ago in the Proterozoic era; oceans with small scattered land masses dominated the first half of Earth history so that the extent of shallow seas, which strongly affect tidal dissipation, is uncertain. Work on tidal sequences in South Australia has shown that, in the late Precambrian (650 million years ago), the year had 13.1 ± 0.5 months and 400 ± 20 days. At that time, the mean lunar distance was 58.4 ± 1.0 Earth radii so that, during the Upper Proterozoic, the Moon was only marginally closer to the Earth.

Over 57% of the surface of the Moon is visible from the Earth, with variations of 6.8° in latitude and 8° in longitude. These variations in the lunar orbit are referred to as librations and are due to the combined effects of wobbles in the rotations of Earth and Moon.

The phases of the Moon as seen from the Earth are conventionally referred to as new moon, first quarter, full moon, and last quarter.

2.2 Eclipses

The presence of the Moon in orbit about the Earth close to the ecliptic plane produces two types of eclipses, so-called lunar and solar, that are visible from the Earth. Lunar eclipses occur at full moon, when the Earth lies between

Moon and Sun and so intercepts the light from the Sun. The Moon usually appears red or copper-colored during such events, as a portion of the red part of the visible solar spectrum is refracted by the Earth's atmosphere and faintly illuminates the Moon. When the Moon is partly shadowed, the border forms an arc of a circle, thus proving that the Earth has a spherical form. Typically there are two lunar eclipses a year, and they can be seen from all parts of the Earth where the Moon is visible.

In contrast, solar eclipses, in which the new moon comes between the Earth and the Sun, are visible only from small regions of the Earth. Between two and five occur each year, but they reoccur at a particular location only once in every 300 or 400 years. The basic cause of the variability in eclipses is that the lunar orbit is inclined at $5.09°$ to the plane of the orbit of the Earth about the Sun (the plane of the ecliptic). For this reason, a solar eclipse does not occur at every new moon. It is an extraordinary coincidence that, as seen from the Earth, the Moon and the Sun are very close to the same angular size of about $0.5°$ despite the factor of 389 in their respective distances, so that the two disks overlap nearly perfectly during solar eclipses. The Moon and Sun return to nearly the same positions every 6585.32 days (about 18 years), a period known to Babylonian astronomers as the saros, while other cycles occur up to periods of 23,000 years.

In past ages, eclipses were regarded mostly as ominous portents, and the ability to predict them gave priests, who understood their cyclical nature, considerable political power. There are many examples of the influence of eclipses on history, one notable example being the lunar eclipse of August 27, 413 B.C. This eclipse delayed the departure of the Athenians from Syracuse, resulting in the total destruction of their army and fleet by the Syracusans. Thus, there is a certain irony that the word eclipse is derived from the Greek term for "abandonment."

2.3 Albedo

Albedo is the fraction of incoming sunlight that is reflected from the surface. Values range from 5 to 10% for the maria to nearly 12–18% for the highlands. At full moon, the lunar surface is bright from limb to limb, with only marginal darkening toward the edges. This observation is not consistent with reflection from a smooth sphere, which should darken toward the edge. This led early workers to conclude that the surface was porous on a centimeter scale and had the properties of dust. The pulverized nature of the top surface of the **regolith** provides multiple reflecting surfaces, accounting for the brightness of the lunar disk.

2.4 Lunar Atmosphere

The Moon has an extremely tenuous atmosphere of about 2×10^5 molecules/cm^3 at night and only 10^4 molecules/cm^3 during the day. It has a mass of about 10^4 kg, about 14 orders of magnitude less than that of the terrestrial atmosphere.

The main components are hydrogen, helium, neon, and argon. Hydrogen and neon are derived from the solar wind, as is 90% of the helium. The remaining He and ^{40}Ar come from radioactive decay. About 10% of the argon is ^{36}Ar, derived from the solar wind.

2.5 Mass, Density, and Moment of Inertia

The mass of the Moon is 7.35×10^{25} g, which is 1/81 of the mass of the Earth. Although the Galilean satellites of Jupiter and Titan are comparable in mass, the Moon/Earth ratio is the largest satellite-to-parent ratio in the solar system. (The Charon/Pluto ratio is larger, but Pluto, an icy planetesimal, is less than 20% of the mass of the Moon and is the king of the Kuiper Belt, rather than a major planet.) The lunar radius is 1738 ± 0.1 km, which is intermediate between that of the two Galilean satellites of Jupiter, Europa ($r = 1561$ km) and Io ($r = 1818$ km). The Moon is much smaller than Ganymede ($r = 2634$ km), which is the largest satellite in the solar system.

The lunar density is 3.344 ± 0.003 g/cm^3, a fact that has always excited interest on account of the Moon's proximity to the Earth, which has a much higher density of 5.52 g/cm^3. The lunar density is also intermediate between that of Europa ($d = 3.014$ g/cm^3) and Io, the innermost of the Galilean satellites of Jupiter, with a density of 3.529 g/cm^3. The other 130-odd satellites in the solar system are ice-rock mixtures and so are much less dense.

The lunar moment of inertia is 0.3931 ± 0.0002. This requires a slight density increase toward the center, in addition to the presence of a low-density crust (a homogeneous sphere has a moment of inertia of 0.400; the value for the Earth, with its dense metallic core that constitutes 32.5% of the mass of the Earth, is 0.3315).

2.6 Angular Momentum

The spin **angular momentum** of the Earth–Moon system is anomalously high compared with that of Mars, Venus, or the Earth alone. Some event or process spun up the system relative to the other terrestrial planets. However, the angular momentum of the Earth–Moon system (3.41×10^{41} g·cm^2/s) is not sufficiently high for classic fission to occur. If all the mass of the Earth–Moon system were concentrated in the Earth, the Earth would rotate with a period of 4 hours. Yet this rapid rotation is not sufficient to induce fission, even in a fully molten Earth.

2.7 Center of Mass/Center of Figure Offset

The mass of the Moon is distributed in a nonsymmetrical manner, with the center of mass (CM) lying 1.8 km closer to the Earth than the geometrical center of figure (CF) (Fig. 2). This is a major factor in locking the Moon into synchronous orbit with the Earth so that the Moon always presents the same face to the Earth, although librations

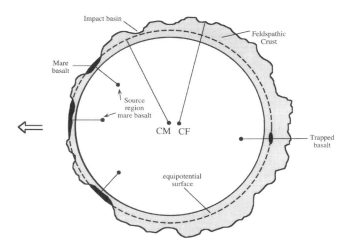

FIGURE 2   A cross section through the Moon in the equatorial plane that shows the displacement toward the Earth of the center of mass relative to the center of the figure, due to the presence of a thicker farside, low-density feldspathic crust. It also illustrates that an equipotential surface is closer to the surface on the nearside. Magmas that originate at *equal depths* below the surface will have greater difficulty in reaching the surface on the farside, a problem exacerbated by the greater farside crustal thickness. However, not all flooding of lunar basins is at the same level. Some magmas originate at different depths, whereas others come from different locations at different times. Others may extrude smaller or greater amounts of lava, leading to differences in the amount of basalt filling a particular basin. These factors contribute to filling of mare basins to differing depths not necessarily related to the equipotential surface.

allow a total of 57% of the surface to be visible at various times.

Various explanations have been advanced to account for the offset of the center of mass from the center of figure. Dense **mare** basalts are more common on the nearside, but their volume is insufficient by about an order of magnitude to account for the effect. It has also been suggested that this offset could arise if the lunar core is displaced from the center of mass. However, such a displacement would generate shear stresses that could not be supported by the hot interior. Another suggestion is that a density asymmetry developed in the mantle during crystallization of the magma ocean, with a greater thickness of low-density Mg-rich **cumulates** being concentrated within the farside mantle. It is unlikely that such density irregularities would survive stress relaxation in the hot interior, unless actively maintained by convection. The conventional explanation for the CM/CF offset is that the farside highland low-density crust is thicker, probably a consequence of asymmetry developed during crystallization of the magma ocean. The crust is massive enough and sufficiently irregular in thickness to account for the CM/CF offset. The scarcity of mare basalts on the farside (Fig. 3) is consistent with a thicker farside crust. Lavas rise owing to the relative low density of the melt and do not

FIGURE 3   The heavily cratered farside highlands. Note the scarcity of mare basalts. Mare Crisium is the dark circular patch of basalt on the northwest horizon. (Courtesy NASA, *Apollo 16* metric frame 3023.)

possess sufficient hydrostatic head to reach the surface on the farside, except in craters in some very deep basins (e.g., Ingenii).

2.8 Remote Spectral Observations

Spectral observations of the Moon from the Earth are limited to the visible and infrared portion of the electromagnetic spectrum between about 3000 and 25,000 Å. These studies identify plagioclase by a weak absorption band at 13,000 Å (1.3 μm) and pyroxene by two strong bands at about 9700–10,000 Å (0.97–1.0 μm), as well as olivine. This technique has enabled mapping of several distinctive mare basalt types on the lunar surface. In addition, mapping of pyroclastic glass deposits has been possible because of their characteristic absorption bands due to Fe^{2+} and Ti^{4+}. These features have also enabled the mapping of the FeO and TiO_2 contents of mare basalts, the amount of anorthosite in the lunar highland crust and the identification of olivine in a few central peaks of craters (e.g., Copernicus).

3. Geophysics

3.1 Gravity

The young ray craters have negative **Bouguer anomalies** because of the mass defect associated with excavation of the crater and the low density of the fallback rubble.

Craters less than 200 km in diameter have negative Bouguer anomalies for the same reason (e.g., Sinus Iridum has a negative Bouguer anomaly of −90 mgal). Volcanic domes such as the Marius hills have positive Bouguer anomalies (+65 mgal), indicating support by a rigid lithosphere. The younger, basalt-filled circular maria on the nearside have large positive Bouguer anomalies, referred to as **mascons** (e.g., Mare Imbrium, +220 mgal). These are due to the uplift of a central plug of denser mantle material during impact followed by the much later addition of dense mare basalt. The gravity signature of young, large, ringed basins, such as Mare Orientale, shows a "bull's-eye" pattern with a central positive Bouguer anomaly (+200 mgal) surrounded by a ring of negative Bouguer anomalies (−100 mgal) with an outer positive Bouguer anomaly collar (+30 to +50 mgal).

The lunar highland crust is strong. High mountains such as the Apennines (7 km high), formed during the Imbrium collision 3.85 billion years ago, are uncompensated and are supported by a strong cool interior. The gravity data are consistent with an initially molten Moon that cooled quickly and became rigid enough to support loads such as the circular mountainous rings around the large, younger, ringed basins as well as the mascons. Even if some farside lunar basins do not show mascons, this may merely be a consequence of the greater thickness of the farside crust. The South Pole–Aitken Basin is particularly significant in this respect. As the oldest (at least 4.1 billion years) and largest impact basin, the fact that it is uncompensated, with major mantle uplift preserved beneath it, this places considerable restrictions on lunar thermal models. It also indicates that melting in the deep interior to produce the mare basalts had no effect on the strength of the crust. The volume of mare basalts is only about 0.1% of the whole Moon so that the amount of melting required to produce them involved only a trivial volume of the Moon.

3.2 Seismology

The lunar seismic signals have a large degree of wave scattering and a very low attenuation so that during moonquakes the Moon "rings like a bell" owing to the absence of water and the very fractured nature of the upper few hundred meters. Observed moonquakes have been mostly less than 3 on the Richter scale; the largest recorded ones have a magnitude between 5 and 5.7. Many are repetitive and reoccur at fixed phases of the lunar tidal cycle. The *Apollo* seismometers recorded the impacts of 11 meteorites with masses of more than one ton. The Moon is seismically inert compared to the Earth, and tidal energy is the main driving force for the weak lunar seismic events.

3.3 Heat Flow and Lunar Temperature Profile

Two measurements of lunar heat flow are available: 2.1 μW/cm² at the *Apollo 15* site and 1.6 μW/cm² at the

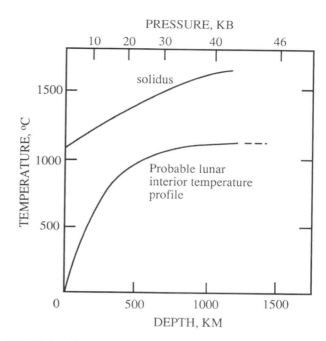

FIGURE 4 The present-day variation of lunar temperature with depth, showing that the temperature is well below that required for partial melting (**solidus** curve.)

Apollo 17 site. It is interesting that these observed heat flows are close to Earth-based estimates from microwave observations. However these values provide only mild constraints on the bulk lunar abundances of the heat-producing elements K, U, and Th as these are not distributed symmetrically. The lunar interior must have been stiff enough for the past 4.0 billion years to account for the support of the mountain rings and the mascons. The most probable lunar temperature profile is shown in Fig. 4, which indicates temperatures of 800°C at a depth of 300 km. Unlike the Earth, which dissipates most of its heat by volcanism at the midocean ridges, the Moon loses its heat by conduction. Most of its original internal heat has been lost, and **differentiation** has concentrated most of the K, U, and Th near the surface, albeit in a nonuniform manner. The present heat flow could indicate lunar U values as high as 45 ppb or over twice the terrestrial abundances. A more conservative value of 30 ppb U is adopted here, based on petrological and geochemical constraints, as well as accommodating the high heat flow values. This uranium abundance is still 50% higher than the well-established terrestrial value of 20 ppb U. Although there has been considerable controversy over the reality of a higher than terrestrial lunar uranium abundance, it appears to be confirmed by the requirement from the *Clementine* mission data for a higher than terrestrial lunar aluminum abundance. Both Al and U are refractory elements, not easily separated by nebular processes, and their abundances are generally considered to be correlated in the terrestrial planets.

3.4 Magnetic Field

The lunar rocks contain a stable natural remnant magnetism. Apparently between about 3.6 and 3.9 billion years ago, there was a planetary-wide magnetic field that has now vanished. The field appears to have been much weaker both before and after this period. The paleointensity of the field is uncertain, but perhaps was several tenths of an oersted. The most reasonable interpretation is that the Moon possessed a lunar dipole field of internal origin, all other suggested origins appearing less likely. The favored mechanism is that the field was produced by dynamo action in a liquid Fe core. A core 400 km in diameter could produce a field of about 0.1 oersted at the lunar surface.

Localized strong magnetic anomalies are associated with patterns of swirls, as at Reiner Gamma. These swirls have been suggested to have formed by some focusing effect of the seismic waves that resulted from the large impacts that formed the basins. More work is clearly needed to substantiate this hypothesis and to understand the association of swirls and magnetic fields. Other remnant fields, with field strengths of only about 1/100th of the terrestrial field, were measured at the *Apollo* landing sites.

4. Lunar Surface

The absence of plate tectonics, water, and life, and the essential absence of atmosphere, indicates that the present lunar surface is unaffected by the main agents that affect the surface of the Earth. Ninety-nine percent of the lunar surface is older than 3 billion years and more than 80% is older than 4 billion years. In contrast, 80% of the surface of the Earth is less than 200 million years old. The major agent responsible for modifying the lunar surface is the impact of objects ranging from micrometer-sized grains to bodies tens to hundreds of kilometers in diameter.

Because of the effective absence of a lunar atmosphere, the lunar surface is exposed to ultraviolet radiation with a flux of about 1300 W/m^2. The absence of a magnetic field allows the solar wind (1–100 eV) and solar (0.1–1 MeV) and galactic (0.1–10 GeV) cosmic rays to impinge directly on the surface. The relative fluxes are 3×10^8, 10^6, and 2–4 protons/cm^2/s, respectively. The penetration depths of these particles extend to micrometers, centimeters, and meters, respectively.

The maximum and minimum lunar surface temperatures are about 390 K and 104 K. At the *Apollo 17* site, the maximum temperature was 374 K (111°C), and the minimum was 102 K (−171°C). The temperatures at the *Apollo 15* site were about 10 K lower. The conductivity of the upper 1–2 cm of the surface is very low (1.5×10^{-5} W/cm^2). This increases about fivefold below 2 cm. A cover of about 30 cm of regolith is sufficient to damp out the surface temperature

FIGURE 5   *Apollo 16* astronaut John Young and the lunar rover at Station 4 on the slopes of Stone Mountain, illustrating the nature of the lunar surface and the absence of familiar landmarks. Smoky Mountain in the left background, with Ravine crater (1 km in diameter) on its flank, is 9 km distant. (Courtesy of NASA, AS16-110-17960.)

fluctuation of about 280 K to about ±3 K, so that structures on the Moon could be well insulated by a modest depth of burial. This in turn might produce difficulties in losing heat generated in buried structures. Impacts of micrometeoroids of about 1 mg mass could be expected about once a year on a lunar structure.

The combination of strong sunlight, low gravity, awkward space suits, and absence of familiar landmarks makes orientation difficult on the lunar surface. All astronauts have commented on the difficulty of judging distance (Fig. 5).

4.1 Regolith

The surface of the Moon is covered with a debris blanket, called the regolith, produced by the impacts of meteorites. It ranges from fine dust to blocks several meters across. The fine-grained fraction is usually referred to as the lunar soil (Fig. 6). This is an unfortunate use of the term "soil," which has organic connotations, but the term is as thoroughly entrenched as the astronomers' use of "metals" for all elements heavier than helium. Although there is much local variation, the average regolith thickness on the maria is 4–5 m, whereas the highland regolith is about 10 m thick.

Seismic velocities were only about 100 m/s at the surface, but increased to 4.7 km/s at a depth of 1.4 km at the

FIGURE 6 The nature of the lunar upper surface is illustrated in this view of small pebbles being collected by a rake near the *Apollo 16* landing site in the Descartes highlands. Lunar sample 60018 was taken from the top of the boulder. (Courtesy of NASA, AS16-116-18690.)

Apollo 17 site. The density is about 1.5 g/ cm^3 at the surface, increasing with compaction to about 1.7 g/cm^3 at a depth of 60 cm. The porosity at the surface is about 50% but is strongly compacted at depth. The regolith is continuously being turned over or **gardened** by meteorite impact. The near-surface structure, revealed by core samples (the deepest was nearly 3 m at the *Apollo 17* site), shows that the regolith is a complex array of overlapping ejecta blankets typically ranging in thickness from a few millimeters up to about 10 cm, derived from the multitude of meteorite impacts at all scales. These have little lateral continuity even on scales of a few meters. Most of the regolith is of local origin: Lateral mixing occurs only on a local scale so that the mare–highland contacts are relatively sharp over a kilometer or so. The rate of growth of the regolith is very slow, averaging about 1.5 mm/million years or 15 Å/year, but it was more rapid between 3.5 and 4 billion years ago.

Five components make up the lunar soil: mineral fragments, crystalline rock fragments, breccia fragments, impact glasses, and **agglutinates.** The latter are aggregates of smaller soil particles welded together by glasses. They may compose 25–30% of a typical soil and tend to an equilibrium size of about 60 μm. Their abundance in a soil is a measure of its maturity, or length of exposure to meteoritic bombardment. Most lunar soils have reached a steady state in particle size and thickness. Agglutinates contain metallic Fe droplets (typically 30–100 Å) referred to as "nanophase" iron, produced by reduction with implanted solar wind hydrogen, which acts as the reducing agent, during melting of soil by meteorite impact.

A *megaregolith* of uncertain thickness covers the heavily cratered lunar highlands. This term refers to the debris sheets from the craters and particularly those from the large impact basins that have saturated the highland crust. The aggregate volume of ejecta from the presently observable lunar craters amounts to a layer about 2.5 km thick. The postulated earlier bombardment may well have produced megaregolith thicknesses in excess of 10 km. Related to this question is the degree of fracturing and brecciation of the deeper crust due to the large basin collisions. Some estimates equate this fracturing with the leveling off in seismic velocities (V_p) to a constant 7 km/s at 20–25 km. In contrast to the highlands, bedrock is present at relatively shallow depths (tens of meters) in the lightly cratered maria.

4.2 Tectonics

The dominant features of the lunar surface are the old heavily cratered highlands and the younger basaltic maria, mostly filling the large impact basins (see Figs. 1 and 3). There is a general scarcity of tectonic features on the Moon, in great contrast to the dynamically active Earth. There are no large-scale tectonic features, and the lunar surface acts as a single thick plate that has been subjected to only small internal stresses. Attention has often been drawn to a supposed "lunar grid" developed by internal tectonic stresses. However, the lineaments that constitute the "grid" are formed by the overlap of ejecta blankets from the many multiringed basins and have no tectonic significance. Most of the lunar tectonic features are related to stresses associated with subsidence of the mare basins, following flooding with lava.

Wrinkle ridges (or mare ridges) are low-relief, linear to arcuate, broad ridges that commonly form near the edges of the circular maria. They are the result of compressional bending stresses, related to subsidence of the basaltic maria from cooling.

Rilles, which are extensional features similar to terrestrial grabens, are often hundreds of kilometers long and up to 5 km wide. Unlike the wrinkle ridges, they cut only the older maria as well as the highlands and indicate that some extensional stress existed in the outer regions of the Moon prior to about 3.6 billion years ago. They should probably be termed grabens so as to avoid confusion with the sinuous rilles, such as Hadley Rille, that are formed by flowing lava, presumably through thermal erosion. The set of three rilles, each about 2 km wide, that are concentric to Mare Humorum at about 250 km from the basin center are particularly instructive examples, showing a clear extensional relation to subsidence of the impact basin (Fig. 7).

4.3 Lunar Stratigraphy

The succession of events on the lunar surface has been determined by establishing a stratigraphic sequence based on

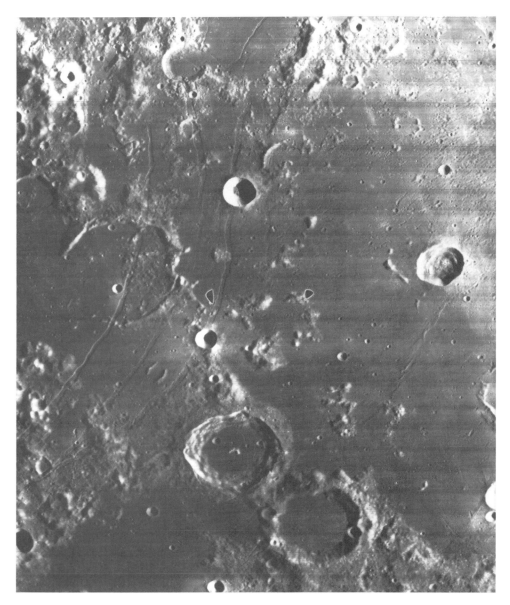

FIGURE 7 Three sets of curved rills or grabens, each about 2 km wide, concentric to Mare Humorum, the center of which is about 250 km distant. The ruined crater intersected by the rilles is Hippalus, 58 km in diameter. The crater at the bottom right, flooded with mare basalt, is Campanus 48 km in diameter. (Courtesy of NASA, Orbiter IV-132-H.)

the normal geological principle of superposition, a fundamental contribution due to Gene Shoemaker. Geological maps based on this concept have been made for the entire Moon, notably by Don Wilhelms. Relative ages have been established by crater counting, and isotopic dating of returned samples has enabled absolute ages to be assigned to the various units. The formal stratigraphic sequence is given in Table 1.

5. Lunar Structure

5.1 Lunar Crust

Reevaluation of the *Apollo* seismic data indicate that the lunar highland crust is 45 km thick (rather than 60 km) at the *Apollo* landing sites and the average crustal thickness lies between 54 and 62 km in thickness. The farside crust averages about 15 km thicker than that of the nearside. The crust thus constitutes about 9% of lunar volume. The maximum relief on the lunar surface is over 16 km. The deepest basin (South Pole–Aitken) has 12-km relief.

The mare basalts cover 17% of the lunar surface, mostly on the nearside (see Fig. 1). Although prominent visually, they are usually less than 1 or 2 km thick, except near the centers of the basins. These basalts constitute only about 1% of the volume of the crust and make up less than 0.1% of the volume of the Moon.

Seismic velocities increase steadily down to 20 km. At that depth, there is a change in velocities within the crust that probably represents the depth to which extensive

TABLE 1	Lunar Stratigraphy	
System	**Age (billion years)**	**Remarks**
Copernican	1.0 to present	The youngest system, which includes fresh ray craters (e.g., Tycho), begins with the formation of Copernicus.
Eratosthenian	1.0–3.1	Youngest mare lavas and craters without visible rays (e.g., Eratosthenes).
Imbrium	3.1–3.85	Extends from the formation of the Imbrium Basin to the youngest dated mare lavas. Includes Imbrium Basin deposits, Orientale and Schrödinger multiring basins, most visible basaltic maria, and many large impact craters, including those filled with mare lavas (e.g., Plato, Archimedes).
Nectarian	3.85–3.92	Extends from the formation of the Nectaris Basin to that of the Imbrium Basin. Contains 12 large, multiring basins and some buried maria.
Pre-Nectarian	Pre-3.92	Basins and craters formed before the Nectaris Basin. Includes 30 identified multiring basins.

fracturing, due to massive impacts, has occurred. At an earlier stage, this velocity change was thought to represent the base of the mare basalts, but these are now known to be much thinner. The main section of the crust from 20 to 60 km has rather uniform velocities of 6.8 km/s, corresponding to the velocities expected from the average anorthositic composition of the lunar samples.

5.2 Lunar Mantle

The structure of the mantle (Fig. 8) has been difficult to evaluate on account of the complexity of interpreting the lunar seismograms. The average **P-wave velocity** is 7.7 km/s and the average S-wave velocity is 4.45 km/s down to about 1100 km. Most models postulate a pyroxene-rich

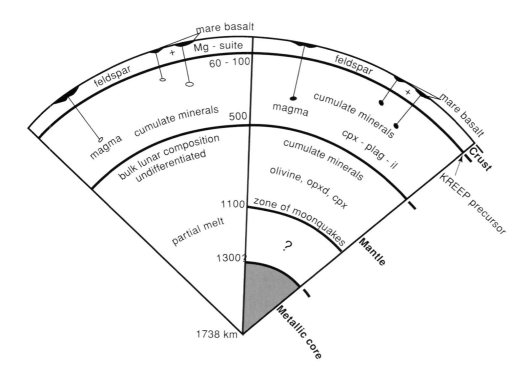

FIGURE 8   Two alternatives for the internal structure of the Moon. On the left, only half of the Moon melted and differentiated and the deep interior has primitive lunar composition. Some partial melting has occurred due to the presence of K, U, and Th. This model is consistent with the lunar free oscillation periods (Amir Khan, Univ. Copenhagen, personal communication). On the right, the Moon was totally melted and differentiated, forming a small metallic core. (Adapted from Taylor, 2001.)

upper mantle that is distinct from an olivine-rich lower mantle beneath about a depth of 500–600 km. Seismic data are ambiguous regarding the nature of the lunar mantle below 500 km. They may be interpreted as representing Mg-rich olivines or indicate the presence of garnet. If the latter is present, this has profound implications for the bulk Moon Al content. However this distinction cannot be made on the basis of the Apollo seismic data.

The main foci for moonquakes lie deep within the lower mantle at about 800–1000 km. The outer 800 km has a very low seismic attenuation, indicative of a volatile-free rigid lithosphere. Solid-state convection is thus extremely unlikely in the outermost 800 km.

Below about 800 km, P- and S-waves become attenuated ($V_S = 2.5$ km/s). P-waves are transmitted through the center of the Moon, but S-waves are missing, possibly suggesting the presence of a melt phase. It is unclear, however, whether the S-waves were not transmitted or were so highly attenuated that they were not recorded.

5.3 Lunar Core

The evidence for a metallic core is suggestive but inconclusive. Electromagnetic sounding data place an upper limit of a 400- to 500-km radius for a highly conducting core. The moment of inertia value of 0.3931 ± 0.0002 is low enough to require a small density increase in the deep interior, in addition to the low-density crust. Although a metallic core with radius about 400 km (4% of lunar volume) is consistent with the available data, denser silicate phases might be present. The resolution of these problems requires improved seismic data.

6. Impact Processes

6.1 Craters and Multiring Basins

One of the most diagnostic features of the lunar surface, that is in great contrast to the surface of the Earth, is the ubiquitous presence of impact craters at all scales, from micrometer-sized "zap pits" to multiring basins. The largest confirmed example is the South Pole–Aitken Basin (180°E, 56°S), 2500 km in diameter and 12 km deep. The presence of the larger Procellarum Basin (3200 km diameter, centered at 23°N, 15°W) covering much of the nearside is questionable. Although the correct explanation for the origin of the lunar craters had already been reached by G. K. Gilbert in 1893 and R. B. Baldwin in 1949, this topic was the subject of ongoing controversy until about 1960, and the question still surfaces occasionally in popular articles. Since meteorites and other impacting bodies could be expected to strike the Moon at all angles, the circularity of the lunar

FIGURE 9   An oblique view of crater Linné in northern Mare Serenitatis. The rim crest diameter is 2450 m. Note the ejecta blocks on the rim, dunelike features on the flanks, and secondary craters at 1–3 crater radii from the rim crest. Linné was famous in the 19th century as a "disappearing" lunar crater because it was not seen by several observers. This was a consequence of observations at the limits of Earth-based telescopic resolution. (Courtesy of NASA, *Apollo 15* pan photo 9353.)

craters was long used as an argument against impact and in favor of a volcanic origin. It was eventually realized that bodies impacting the Moon at velocities of several km/sec explode on impact and the explosion mostly forms a circular crater regardless of the angle of impact, except for very oblique impacts. The morphology of the craters resembles that of terrestrial explosion craters and is quite distinct from the landforms of terrestrial volcanic centers.

The smallest craters are simple bowl-shaped depressions, surrounded by an overturned rim and an ejecta blanket (e.g., Linné, 2450-m diameter, Fig. 9). With increasing size, more complex forms develop. At diameters greater than about 15–20 km, slump terraces appear on the crater walls. Central peaks formed by rebound appear at crater diameters greater than about 25–30 km (e.g., Copernicus, 93-km diameter, Fig. 10). Central-peak basins, in which a fragmentary ring of peaks surrounds a central peak (e.g., Compton, 162 km diameter), develop in the size range 140–180 km. Larger craters develop internal concentric peak rings in place of the central peak (e.g., Schrödinger, 320 km diameter, Figs. 11 and 12). Such central peaks and peak rings may develop from fluidized waves during impact.

The ultimate form resulting from impact is the multiring basin, which may have six or more rings. A classic lunar example is Orientale (Fig. 13). This structure is 920 km in diameter (about the size of France), with several concentric

FIGURE 10 Oblique view of crater Copernicus, 93 km in diameter, showing a central-peak complex and well-developed slump terraces on the inner walls. (Courtesy of NASA, AS17-151-23260.)

mountain rings having a typical relief of about 3 km with steep inward-facing scarps. These formed in a few minutes after the impact of a body perhaps 50 km in diameter. The central portion has been flooded with mare basalt. Thirty

FIGURE 11 The transition between central-peak craters and peak-ring craters. The large central basin is Schrödinger (320 km in diameter), which has a well-developed peak ring. Antoniadi (135 km in diameter), southeast from Schrödinger, has both a central peak and a peak ring. The small crater immediately southwest of Antoniadi has a central peak only. (Courtesy of NASA, Orbiter IV-8M.)

such basins have been recognized on the Moon (Fig. 14), with another 14 probable. There is much controversy over the origin of multiring basins. One possibility proposes that the crust is fluidized by the impact and the rings form like ripples on a pond into which a stone has been dropped. The most likely explanation is that the mountain rings are fault scarps, formed by collapse into a deep transient crater formed by the initial impact.

The depth of excavation of the lunar basins decreases with increasing basin diameter. A transient cavity forms during the initial stage of the impact, but most excavated material comes from shallower depths. Thus, no unequivocal lunar mantle material has been recognized in the returned samples from the lunar highland crust, and the transient depth of excavation of the largest basins do not appear to have exceeded 50 km. Ejecta blankets incorporate much local material as they travel across the surface in a manner analogous to a base surge. Apart from the ejecta blankets, numerous blocks from large impacts travel with sufficient velocity to produce secondary craters. These must be carefully distinguished from primary craters to avoid confusion in the dating of lunar surfaces by crater counting.

Shock pressures up to 100 GPa (1 GPa = 10 kbar) cause a variety of effects from the development of planar features in minerals (>10 GPa) to whole-rock melting (50–100 GPa). Above about 150 GPa, the rocks are vaporized. Vapor masses of a few times projectile mass and melt masses about 100 times the projectile mass may be formed. Impact melts compose 30–50% of all samples returned from the lunar highlands.

6.2 Lunar Cratering History and the Lunar Cataclysm

The intense cratering of the lunar highlands and the absence of a similar heavily cratered surface on the Earth were long recognized as due to an early "pregeological" bombardment. In contrast, the lightly cratered basaltic mare surfaces, on which the cratering rate is about 200 times less, had escaped this catastrophe and were clearly much younger. The ages of the mare surfaces, dated from the sample return to be between 3.3 and 3.8 billion years old, showed that the cratering flux was similar, within a factor of 2, to that observed terrestrially. It also established that the intense cratering of the highlands occurred more than 3.8 billion years ago. Most highland samples have ages in the range 3.8–4.3 billion years. The radiometric ages of the ejecta blankets from the large collisions tend to cluster around 3.9 billion years, with the dates for the Imbrium collision being 3.85 billion years and that for Nectaris, 3.90 or 3.92 billion years. This is a surprisingly narrow range and indicates a rapid increase in the cratering flux just before 3.8 billion years. This clustering has led to the concept of a "lunar cataclysm" or a spike in the collisional history at that time.

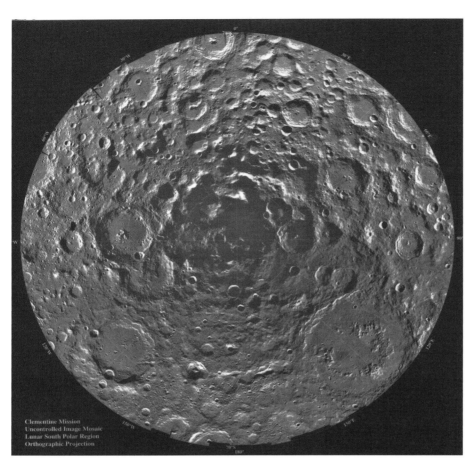

FIGURE 12 A mosaic of 1500 *Clementine* UV VIS images centered on the South Pole, showing the heavily cratered south polar region of the Moon. The Schrödinger Basin (320 km in diameter), which is the freshest peak basin on the Moon, is at four o'clock. Schrödinger is slightly older than Orientale. Note the small volcanic cone in the bottom left-hand sector. It is possible that some ice (from cometary impacts?) is trapped in the permanently shadowed craters at the South Pole. (LP1 *Clementine* press release.)

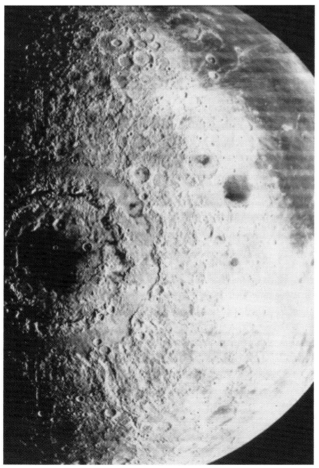

FIGURE 13 Orientale is a classic example of a multiring basin. The diameter of the outer mountain ring (Montes Cordillera) is 930 km, about the size of France. Note the radial structures resulting from the impact. It is the youngest major impact basin on the Moon. This structure was formed about 3800 million years ago in a few minutes following the impact of a planetsimal or asteroid about 50–100 km in diameter. Basalt has flooded the center of the Orientale Basin. The small circular patch of mare basalt northeast of Orientale is Grimaldi. The western edge of Oceanus Procellarum fills the northeastern horizon. (Courtesy NASA, Orbiter IV-187M.)

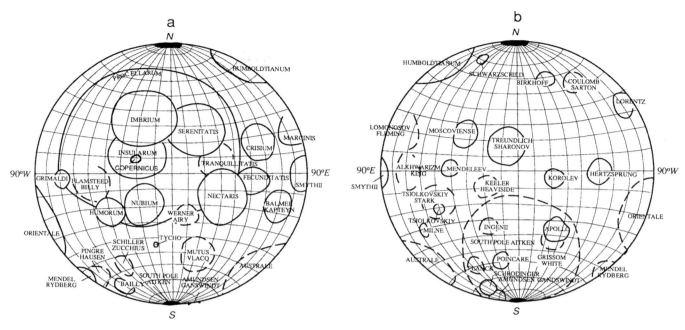

FIGURE 14 The distribution of major impact basins on (a) the nearside and (b) the farside of the Moon. (Courtesy of D. E. Williams.)

The noncataclysmic explanation is that the Imbrium and Orientale Basins formed during the tail end of the accretion of the planets and so represent the final sweep-up of large objects. The problem with this scenario is that extrapolation from the rate at 3.8 billion years back to 4.5 billion years results in the accretion of a Moon several orders of magnitude larger than observed. It seems probable that accretion of the Moon was essentially complete and that the Moon was at its present size by about 4450 million years ago, at the time of the crystallization of the feldspathic highland crust. Other arguments in favor of the cataclysm include the scarcity of impact melts older than 4 billion years and the lead isotope data, which indicate a major resetting of the lead ages at 3.86 billion years. Although it is often argued that the sampling from the *Apollo* missions is dominated by Imbrium ejecta, lunar meteorites have provided fresh insights. These provide a random sampling of the surface but display no impact melts older than 3.92 billion years, supporting the notion of a "cataclysm" although the storage for several hundred million years and supply of the massive impactors poses some interesting problems. Figure 15 shows a reconstruction of the lunar crater production rate with time.

7. The Maria

The maria make up the prominent dark areas that form the features of "the man in the Moon" (Figs. 1 and 16).

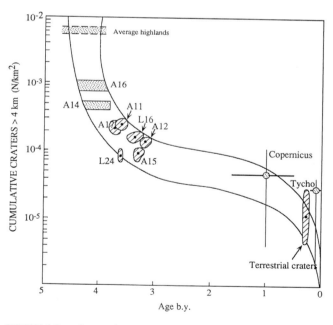

FIGURE 15 The production rate over geological time for lunar craters greater than 4 km in diameter. This illustrates the very high cratering prior to 3.8 billion years ago, but whether this represents the tail end of accretion or spike (cataclysm), as preferred here, in the bombardment history is unclear. The terrestrial rate is for the past 200 million years. (Updated from the Basaltic Volcanism Study Project, 1981.)

FIGURE 16 The distinction between the maria and highlands is clearly shown in this view of the lunar farside. The large circular crater, filled with dark mare basalt, is Thomson (112 km in diameter), within the partly visible Mare Ingenii (370 km in diameter, 34°S, 164°E). The large crater in the highland terrain in the right foreground is Zelinskiy (54 km in diameter). The stratigraphic sequence, from oldest to youngest, is (a) formation of the white highland crust, (b) excavation of the Ingenii Basin, (c) excavation of the Thomson Crater within the Ingenii Basin, (d) excavation of Zelinskiy Crater in the highland crust, (e) flooding of Ingenii Basin and Thomson Crater with mare basalt, and (f) excavation of small craters, including a probable chain of secondary craters, on the mare basalt surface. (Courtesy of NASA, AS15-87-11724.)

After centuries of speculation during which the maria were thought to be composed of sediments, dried lake beds, asphalt, or other unlikely materials, they were conclusively identified following the *Apollo 11* sample return in 1969 as being formed of basaltic lavas. This conclusion had already been reached by earlier workers such as R. B. Baldwin and G. Kuiper and was strongly suggested by the data from the *Surveyor* landers. These vast plains cover 17% (6.4×10^6 km²) of the surface of the Moon, and they are exceedingly smooth, with slopes of 1:500 to 1:200 and elevation differences of only 150 m over distances of 500 km. This smoothness and the lack of volcanic constructional forms, which litter many terrestrial volcanic fields, remind one of plateau basalts on Earth and are probably due to several factors. These include a combination of high eruption rates and the low viscosity of the lunar lavas, which is about an order of magnitude lower than that of their terrestrial counterparts and is close to that of engine oil at room temperature. The lava flows (Fig. 17) are thin (10–40 m) and up to 1200 km long, a consequence of the low viscosity and probable long duration of the eruption. Flow fronts are generally less than about 15 m in height. Occasional small volcanic domes and cones occur on the mare surface. The classic example is the region of the Marius Hills.

The maria are not all at the same level, and this is indicative of independent eruptions from diverse sources at differing depths in the interior. They are mostly subcircular in form owing to their filling of the multiring basins, originally excavated by impact. The dark basaltic lavas that fill

FIGURE 17 Mare basalt flows in southwestern Mare Imbrium. Flow thicknesses are in the range 10–30 m. The source of the flow is about 200 km southwest of crater La Hire (5 km in diameter), seen at right center. This crater is superimposed on Mount La Hire (about 30 m long at its base), a highland remnant that is partially submerged by lavas. Note the prominent concentric wrinkle ridges on Mare Imbrium. (Courtesy of NASA, AS15-1556.)

the basins form the maria, as in Mare Imbrium (see Figs. 1 and 14) or Mare Ingenii (see Fig. 16). The basins, as in the Imbrium Basin, were formed much earlier by impact and have nothing to do with the generation of the mare basalts. Thus, the mare basalt, which fills many basins, is unrelated to the formation of the basins, a common misconception; instead, it is derived from the deep lunar interior and merely floods into the low-lying depressions much later. Some impact melt, distinct in composition from the lavas, may be formed at the time of the impact, but it should not be confused with the basaltic mare lavas, which differ both in composition and age. Oceanus Procellarum (see Fig. 1) is the type example of an irregular mare, where the lavas have flooded widely over the highland crust. However, this mare may be filling parts of an old, large, and very degraded Procellarum Basin (3200 km in diameter), although the existence of this basin is questionable.

The mare lavas reach the surface because of the density difference between the melt and that of the overlying column of rock. The scarcity of maria on the farside of the Moon (see Fig. 3) is due to the greater crustal thickness. An exception is part of the area of the deep depression of the South Pole–Aitken multiring basin (2500 km in diameter), on which is superimposed the Ingenii impact basin (650 km in diameter), now occupied in part by the lavas of Mare Ingenii (see Fig. 16). However, most of the South Pole–Aitken Basin, which is deeper than the nearside maria, is not flooded with lava. This argues for mantle heterogeneity and localized sources for the mare basalts, rather than some moonwide melting of the interior, with consequent flooding of lava to a uniform level.

Dark mantle deposits, which represent pyroclastic deposits formed probably by "fire fountaining" during lunar eruptions, occur, for example, around the southern borders of Mare Serenitatis. These pyroclastic deposits are composed mainly of glass droplets and fragments and can be distinguished from the ubiquitous glasses of impact origin by their uniformity, homogeneous composition, and absence of meteoritic contamination. Over 25 distinct compositions have been recognized. They commonly have a superficial coating of volatile elements such as Pb, Zn, Cl, and F, derived from volcanic vapors during the eruption. The dominant gas, however, was probably CO. The source of the volatile elements is uncertain. They are rare in the lunar samples, and the Moon is generally thought to be strongly depleted in them. Possibly they come from local cumulate sources, and so do not imply an enrichment of the deep lunar interior in volatile elements. However, they may have originated at a greater depth than the crystalline mare basalts.

Some areas of mare basalts (so-called cryptomaria) are covered by ejecta blankets of highland material from multiring basins; their presence is revealed by the haloes of dark basalt ejected from impact craters that have punched through the light-colored highland plains

units of anorthositic composition into the underlying basalts.

Although they are prominent visually on the Moon, the maria typically form a thin veneer, mostly less than 1–2 km thick, except in the centers of the circular maria where they may reach maximum of 5 km as in the middle of Mare Imbrium. The basalt thickness in Orientale is estimated to be only 0.6 km. The total volume of mare basalt is usually estimated at between 6 and 7×10^6 km^3 or about 0.1% of lunar volume. Cooling rates for mare basalts range from 0.1°C to 30°C per hour, indicative of fast cooling in thin lava flows.

Sinuous rilles occur widely near the edges of the maria and are either lava channels or collapsed lava tubes. They have eroded into the surrounding lavas by a combination of thermal and mechanical erosion. The classic example is Hadley Rille (Fig. 18), visited by the *Apollo 15* mission. The rille is 135 km long and averages 1.2 km in width and 370 m in depth. Massive lava bedrock is exposed in the rille wall at the *Apollo 15* site. The sinuous rilles should not be confused with the straight or arcuate rilles, which are grabens of tectonic origin.

7.1 Mare Basalt Ages

The oldest ages for returned lunar mare basalts are from *Apollo 14* breccias; aluminous low-Ti basaltic clasts in these

FIGURE 18 Hadley Rille, a typically sinuous rille, about 1 km wide, at the *Apollo 15* landing site, close to the base of the Apennine Mountains. (Courtesy of NASA, Lunar Orbiter IV-102H3.)

breccias range in age from 3.9 to 4.3 billion years. The oldest basalt from a visible maria is *Apollo* sample number 10003, a low-K basalt from Mare Tranquilitatis with an age of 3.86 ± 0.03 billion years. This gives a younger limit for the age of the Imbrium collision because the lavas of Mare Tranquilitatis overlie the Imbrium ejecta blanket.

The youngest dated sample is number 12022, an ilmenite basalt with an age of 3.08 ± 0.05 billion years, although some doubtful younger ages are in the literature. Low-Ti basalts are generally younger than high-Ti basalts. Stratigraphically younger flows, some of which appear to embay young ray craters, may be as young as 1 billion years but are of very limited extent. The most voluminous period of eruption of lavas appears to have been between about 3.8 and 3.1 billion years ago. Isotopic measurements show that the mare basalt source regions formed at about 4.4 billion years, and this age must represent the solidification of much of the magma ocean.

7.2 Composition of the Mare Basalts

The basic classification is chemical, with finer subdivisions based on mineral composition. The basalts are divided into low-Ti, high-Ti, and high-Al basalts. The low-Ti basalts include VLT (very-low-Ti), olivine, pigeonite, and ilmenite basalts. The high-Ti basalts include high-K, low-K, and VHT (very-high-Ti) basalts. The *Clementine* data suggest that there is a continuous variation in Ti contents. The major minerals are pyroxene, olivine (Mg-rich), plagioclase (Ca-rich), and opaques, mainly ilmenite. The basalts are highly reduced, with oxygen fugacities of 10^{-14} at $1100°C$ or about a factor of 10^6 lower than those of terrestrial basalts at any given temperature. Ferric iron is effectively absent, and 90% of Cr and 70% of Eu in the Moon is divalent. An alloy of FeNi metal is a common late-stage crystallizing phase.

In comparison with terrestrial basalts, the silica contents of mare basalts are low (37–45%), and the lavas are iron-rich (18–22% FeO). The lunar basalts are notably high in Ti, Cr, and Fe/Mg ratios and low in Ni, Al, Ca, Na, and K compared with terrestrial counterparts (Table 2). They are depleted in volatile (e.g., K, Rb, Pb, Bi) and siderophile (e.g., Ni, Co, Ir, Au) elements. The ratio of volatile (e.g., K) to refractory elements (e.g., U) is low. Thus, lunar K/U ratios average about 2500 compared to terrestrial values of about 12,000. The rare earth elements (REEs) display a characteristic depletion in divalent Eu or europium anomaly (Fig. 19). This is one of the several pieces of evidence that the mare basalts come from a previously differentiated interior, rather than being melted from a primitive undifferentiated lunar composition. Even the lunar glasses that may come from deeper show the tell-tale evidence of depletion in Eu, indicating that they too come from a differentiated interior.

The differences in composition of the mare basalts are mostly due to source region heterogeneity, with only minor evidence for near-surface fractionation. Variations in the amount of partial melting from a uniform source, subsequent fractional crystallization, or assimilation cannot account for the observed diversity. Some mare basalts are vesicular, evidence for a now-vanished gas phase, usually thought to be CO.

7.3 Origin of the Mare Basalts

Mare basalts originate by partial melting, at temperatures of about $1200°C$, deep in the lunar interior (see Fig. 8), probably at depths between 200 and 400 km. The lunar volcanic glasses appear to come from greater depths, but still from a differentiated source. The basalts are derived from the zones and piles of cumulate minerals developed, at various depths, during crystallization of the magma ocean. The isotopic systematics of the mare basalts indicate that the source region had crystallized by 4.4 billion years. Partial melting occurred in these diverse mineral zones some hundreds of millions of years later due to the slow buildup of heat from the presence of the radioactive elements K, U, and Th. The melting was not extensive. Over 25 distinct types of mare basalt were erupted over an interval of more than 1 billion years, but the total amount of melt so generated amounted to only about 0.1% of the volume of the Moon. This forms a stark contrast to the state of the Moon at accretion, when it may have been entirely molten.

8. Lunar Highland Crust

Most of the rocks returned from the highlands are polymict breccias, pulverized by the massive bombardment. However, some monomict breccias have low siderophile element contents. These are considered to be "pristine" rocks that represent the original igneous components making up the highland crust. Three pristine constituents make up the lunar highland crust, namely, ferroan anorthosites that are the dominant component, the Mg suite, and KREEP.

8.1 Ferroan Anorthosite

Ferroan anorthosite is the single most common pristine highland rock type, making up probably 80% of the highland crust. The pristine clasts in lunar meteorites are mostly ferroan anorthosites. The major component (95%) is highly calcium-rich plagioclase, typically An_{95-97} with a pronounced enrichment in Eu (Eu/Eu* = 50). Low-Mg pyroxene is the next most abundant mineral, but the mafic minerals are only minor constituents in this nearly monomineralic feldspathic rock. The anorthosites are typically coarsely crystalline with cumulate textures. Reliable

TABLE 2	Elemental Abundances[a]					
Oxide (weight percent)	Cl	Earth Mantle + Crust	Bulk Moon	Highlands	Low-Ti Basalt	High-Ti Basalt
SiO_2	34.2	49.9	47	45.0	43.6	37.8
TiO_2	0.11	0.16	0.3	0.56	2.60	13.0
Al_2O_3	2.44	3.64	6.0	24.6	7.87	8.85
FeO	35.8	8.0	13.0	6.6	21.7	19.7
MgO	23.7	35.1	29	6.8	14.9	8.44
CaO	1.89	2.89	4.5	15.8	8.26	10.7
Na_2O	0.98	0.34	0.09	0.45	0.23	0.36
K_2O	0.10	0.02	0.01	0.03	0.05	0.05
Σ	99.2	100.1	99.9	100	100.4	99.5
Volatile elements						
K (ppm)	854	180	83	200	420	500
Rb (ppm)	3.45	0.55	0.28	0.7	1.0	1.2
Cs (ppb)	279	18	12	20	40	30
Moderately volatile element						
Mn (ppm)	2940	1000	1200	570	2150	2080
Moderately refractory element						
Cr (ppm)	3975	3000	4200	800	5260	3030
Refractory elements						
Sr (ppm)	11.9	17.8	30	130	101	121
U (ppb)	12.2	18	30	80	220	130
La (ppm)	0.367	0.55	0.90	2.0	6.0	5.22
Eu (ppm)	0.087	0.13	0.21	1.0	0.84	1.37
V (ppm)	85	128	150	30	175	50
Siderophile elements						
Ni (ppm)	16500	2000	400	100	64	2
Ir (ppb)	710	3.2	0.01	—	0.02	0.04
Mo (ppb)	1380	59	1.4	5	50	50
Ge (ppm)	48.3	1.2	0.0035	0.02	0.003	0.003

[a] Elemental abundances in Cl chondrites (volatile = primitive solar nebula). Earth mantle + crust = primitive Earth mantle; bulk Moon; average lunar highland crust; low-Ti mare basalt (12002) and high-Ti mare basalt (70215). Both of these later samples are probably primary basaltic magmas. Data sources from Taylor (1982) and Hartmann et al. (1986).

ages of 4440 ± 20 million years and 4460 ± 40 million years have been obtained, and their average of 4450 million years is taken to represent the crystallization age of ferroan anorthosites from the lunar magma ocean and the flotation of the feldspathic highland crust as "rockbergs." Alternatively, this date may represent the "isotopic closure age" during cooling of the crust.

8.2 Mg Suite

The Mg suite comprises norites, troctolites, dunites, spinel troctolites, and gabbroic anorthosites. They are characterized by higher, and so more primitive, $Mg/(Mg + Fe^{2+})$ ratios compared to the ferroan anorthosites. They range in

age from about 4.44 billion years down to about 4.2 billion years, but typical ages are 100–200 million years younger than those of the ferroan anorthosites. The Mg suite is petrographically distinct from the older ferroan anorthosites and does not appear to be related directly to the crystallization from the magma ocean. It makes up a minor but significant proportion (perhaps 10%) of the highland crust and has two distinct and contradictory components in terms of conventional petrology. It is Mg-rich, and so primitive in terms of igneous differentiation, but also contains high concentrations of incompatible elements, typical or highly evolved or differentiated igneous systems. These characteristics point to an origin by mixing of these two distinct components.

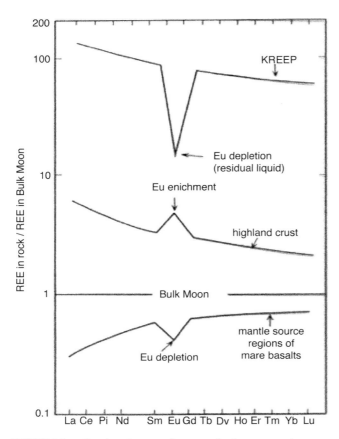

FIGURE 19 The abundances of rare earth elements in the source regions of the mare basalts, the highland crust, and KREEP, relative to bulk moon concentrations. These patterns result from the preferential entry of divalent europium (similar radius to strontium) into plagioclase feldspar. This mineral floats to form the highland crust, and so depletes the interior of Eu. Mare basalts that subsequently erupted from this region deep within the Moon bear the signature of this early depletion. KREEP is the final residue of the crystallization of the magma ocean. It is strongly depleted in Eu owing to prior crystallization of plagioclase and is enriched in the other rare earth elements (e.g., K, U, Th, Ba, Rb, Cs, Zr, P) that are excluded from olivine, pyroxene, and ilmenite during the crystallization of the major mineral phases of the magma ocean.

The source of the highly evolved component is clearly KREEP. The source of the "primitive" Mg-rich component is less obvious. If the primitive component came from deep cumulates, the concentrations of Ni in olivine of the Mg suite are low, not high as predicted. Conventional theories propose that the Mg suite arose as separate plutons that intruded the crust as separate igneous intrusions. However, all Mg suite rocks have parallel REE patterns, a characteristic compatible with mixing, but not expected to occur in separate igneous intrusions. This is a major constraint on the concept that the lunar highland crust formed through "serial magmatism." Furthermore, it is of interest that the Mg suite

contains Mg-rich orthopyroxene, a mineral that is lacking in mare basalts. Clearly the Mg suite originates in a location distinct from the source region of the mare basalts. During crystallization of the magma ocean, Mg-rich minerals (e.g., olivine and orthopyroxene) are among the first to crystallize and accumulate on the bottom of the magma chamber, in this case at depths exceeding 400 km. It is sometimes suggested that massive overturning has occurred to bring these within reach of the surface. However, the magma ocean had completed crystallization by 4400 million years with only the KREEP component remaining liquid until about 4360 million years; it was solid at the time of the formation of the Mg suite. There is no obvious source of energy for remelting early refractory Mg-rich cumulates. Such material may have been derived from a late infall of **planetesimals** that might provide both the primitive component and the energy for melting. Subsequent melting to produce mare basalts took place in more differentiated cumulates and produced lavas with a different mineralogy (e.g., lacking orthopyroxene), without the primitive characteristics of the Mg suite.

8.3 Alkali Suite

A rare component of the highlands crust is the Alkali suite. The largest sample is 1.6 gm and they seem to have undergone severe thermal metamorphism but their origin is not well understood. They are commonly 85% plagioclase feldspar, the remainder being mostly pyroxene. Their significant feature is an enrichment in the alkali elements so that they contain Na-rich rather than Ca-rich feldspar. They are probably related to KREEP as the trace element patterns are similar.

8.4 Kreep

KREEP is enriched in potassium, rare earth elements, and phosphorus, hence the name. It is commonly applied as an adjective to refer to highland rocks with an enhanced and characteristic trace element signature. KREEP originated as the final 2% or so melt phase from the crystallization of the magma ocean and is strongly enriched in those "incompatible" trace elements excluded from the major mineral phases (olivine, orthopyroxene, clinopyroxene, plagioclase, ilmenite) during crystallization of the bulk of the magma ocean. This residual phase was the last to crystallize, at about 4.36 billion years, and apparently pervaded the crust, with which it was intimately mixed by cratering. Its presence tends to dominate the trace element chemistry of the highland crust. Extreme REE enrichment up to 1000 times the chondritic (or solar nebula abundances) are known (see Fig. 19). This extreme concentration of trace elements amounts to a significant part of the total lunar budget and so provides strong evidence for the magma ocean hypothesis and for large-scale lunar melting.

8.5 Kreep Basalt

KREEP basalt, an enigmatic rock type with only a few undisputed examples, is highly enriched in incompatible elements (KREEP) but has a more primitive $Mg/(Mg + Fe^{2+})$ ratio. This combination of primitive and evolved components suggests that they are derived, like the members of the Mg suite, from different sources and may be impact melts. Probably the Apennine Bench formation is composed of KREEP basalt. This formation appears to have formed close in time to the excavation of the Imbrium Basin.

8.6 Breccias

A consequence of the massive bombardment that pulverized the lunar highlands is that the rocks returned from the lunar highlands are breccias, usually consisting of rock fragments or clasts set in a fine-grained matrix. Lunar breccias are usually divided into monomict, dimict, and polymict breccias, consisting, respectively, of a single rock type, two distinct components, and a variety of rock types and impact melts. Polymict breccias, usually involving several generations of breccias, are the most common rock type returned from the lunar highlands. They are further subdivided into fragmental breccias, glassy melt breccias, crystalline melt breccias (or impact melt breccias), clast-poor impact melts, granulitic breccias, and regolith breccias.

8.7 The Magma Ocean

The geochemical evidence is clear that at least half and possibly the whole Moon was molten at accretion. This stupendous mass of molten rock is referred to as the "magma ocean," and a very energetic mode of origin of the Moon, such as provided by the giant impact hypothesis, is required to account for it. The crystallization of such a body is difficult to constrain, or even imagine, from our limited terrestrial experience. A possible scenario is that initial crystallization of olivine and orthopyroxene formed deep cumulates. As the Al and Ca content of the magma increased, plagioclase crystallized and floated in the bone-dry melt, forming rockbergs that eventually coalesced to form the lunar highland crust, around 4450 ± 20 million years ago. The first-order variation in thickness from nearside to farside is probably a relic of primordial convection currents in the magma ocean. Excavation by large basin impacts has subsequently imposed additional substantial variations in crustal thickness.

Plagioclase was a very early phase to crystallize, as all lavas derived from the interior bear the signature of prior removal of Eu (and Sr) (see Fig. 19). Accordingly, the magma ocean was probably enriched in Ca and Al over typical terrestrial values, a conclusion reinforced by the more recent *Galileo*, *Clementine*, and *Lunar Prospector* data. The

implication is that the Moon was enriched in these and other refractory elements compared to our estimates of the terrestrial mantle. Continued crystallization of the magma ocean eventually produced KREEP, which appears to have pervaded and has been intimately mixed into the highland crust on the nearside. The crystallization of the magma ocean was probably asymmetric, as shown by the variations in crustal thickness and the apparent concentration of the residual KREEP melt under the nearside. Crystallization of the main phases was complete by 4400 million years ago, and the final KREEP residue was solid by about 4360 million years ago.

The crystallization sequence portrayed here was far from peaceful. During all this time, the outer portions of the Moon were subjected to a continuing bombardment, which broke up and mixed the various components of the highland crust. Perhaps coeval with these events was the intrusion into the crust of the Mg suite. Probably some local overturning of the deeper cumulate pile may have occurred, but such events did not homogenize the interior that later produced a wide variety of mare basalt compositions.

8.8 Lunar Crustal Terranes

Geochemical mapping carried out by the *Clementine* and *Lunar Prospector* missions has resulted in a significant advance in our understanding of the detailed structure of the lunar highland crust. Based on the FeO and Th abundances measured by the *Clementine* and *Lunar Prospector* missions, the crust can be divided into three major terranes: (1) the Feldspathic Highland Terrane (FHT), (2) the Procellarum KREEP Terrane (PKT), and (3) the South Pole–Aitken Terrane (SPAT) (Fig. 20). The Feldspathic Highland Terrane constitutes the feldspathic lunar crust formed by flotation from the magma ocean. The Procellarum KREEP Terrane results from the intrusion (or mixing in) of the residual KREEP liquid from the last stages of crystallization of the magma ocean. The South Pole–Aitken Terrane is the result of the subsequent excavation of the 2500 km diameter South Pole–Aitken Basin, that stripped off most of the upper crust over that region and whose ejecta contributed significantly to the thickness of the farside anorthositic crust, north of the basin. The interior of the South Pole–Aitken Basin, the deepest basin on the Moon, has a more mafic (Fe- and Mg-rich) composition relative to the more feldspathic lunar highlands, but it is not clear that the impact has uncovered the lunar mantle. It would be of great interest to study this area in detail, as no excavated mantle samples have ever been identified in the returned *Apollo* samples.

The South Pole–Aitken Basin, where most of the upper crust is missing, has been preserved for over 4.1 billion years without significant isostatic compensation occurring. As this is the oldest and largest recognized lunar basin, the

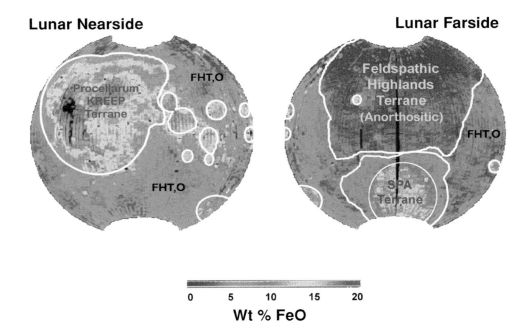

Lunar Nearside

Lunar Farside

FHT,O

Procellarum
KREEP
Terrane

FHT,O

Feldspathic
Highlands
Terrane
(Anorthositic)

FHT,O

SPA
Terrane

0   5   10   15   20

Wt % FeO

FIGURE 20   Major lunar crustal terranes. The Procellarum KREEP Terrane (PKT) is on the nearside, with Th greater than 3.5 ppm. The South Pole–Aitken Terrane (SPAT) has an outer region corresponding to basin ejecta. The Feldspathic Highland Terrane (FHT) corresponds to the thickest part of the crust, concentrated on the lunar farside. FHT,O consists of those regions where basin ejecta or cryptomare obscure the feldspathic surface. (Courtesy of Brad Joliff, Washington University.)

lack of compensation indicates that the crust and interior have been strong enough to support this structure for most of lunar history. There is little sign of the residual KREEP component in this location, despite the depth of excavation. This reinforces the notion that the residual KREEP melt was not uniformly distributed. Figure 21 shows the mantle uplift beneath the South Pole–Aitken Basin as well as that partially superimposed later uplift resulting from the excavation of the *Apollo* basin.

9. Lunar Composition

The Moon is bone-dry and highly reduced, no indigenous H_2O having been detected at ppb levels, and lacks ferric iron. It is strongly depleted to volatile elements (e.g., K, Pb, Bi) by a factor of about 50 compared to the Earth, or 200 relative to primordial solar nebula abundances. Compared to the Earth, the most striking difference is in the abundance of iron that is reflected in the low lunar density. The Earth

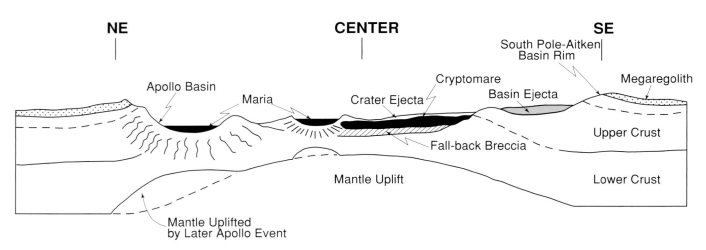

FIGURE 21   The South Pole–Aitken Basin (2500 km diameter and 12 km deep) on which are superimposed two later impact basins.

contains about 25% metallic Fe; the Moon, less than about 2–3%. However, the bulk Moon contains 12–13% FeO, 50% more than in current estimates of 8% FeO in the terrestrial mantle. Along with its depletion in iron, the Moon also has a low abundance of siderophile elements that are depleted in order of their metal-silicate distribution coefficients. This observation indicates that these elements have been segregated into a metallic core. However, this pattern may have been established in precursor planetesimals or in the impactor from which most of the Moon appears to have been derived, rather than, or as well as, into a lunar core.

The other major element abundances are mostly model-dependent. Si/Mg ratios are commonly assumed to be chondritic (CI), although the Earth and many meteorite classes differ from this value. The lunar Mg value is generally estimated to be about 0.80, lower than that of the terrestrial mantle value of 0.89.

The Moon is probably enriched in refractory elements such as Ti, U, Al, and Ca, a conclusion consistent with geophysical studies of the lunar interior. This conclusion is reinforced by the data from the *Galileo*, *Clementine*, and *Lunar Prospector* missions, which indicate that the highland crust is dominated by anorthositic rocks. This requires that the bulk lunar composition contains about 5–6% Al_2O_3, compared with a value of about 3.6% for the terrestrial mantle and so is probably enriched in refractory elements (e.g., Ca, Al, Ti, U) by a factor of about 1.5 compared to the Earth.

In the light of the caveats already given, the bulk composition of the Moon is only known to a first approximation. Data for the bulk composition of the Moon are given in Table 2 compared to CI, the terrestrial mantle abundances and to the bulk Earth.

Both the Cr and O isotopic compositions are identical in the Earth and Moon, probably indicating an origin in the same part of the nebula, consistent with the single impact hypothesis that derives most of the Moon from the silicate mantle of the impactor, Theia.

Clearly the Moon has a composition that cannot be made by any single-stage process from the material of the primordial solar nebula. The compositional differences from that of the primitive solar nebula, from the Earth, from Phobos and Deimos (almost certainly of carbonaceous chondritic composition), and from the satellites of the outer planets (rock-ice mixtures with the exception of Io) thus call for a distinctive mode of origin.

9.1 Lunar Minerals

Only about 100 minerals have been identified in lunar samples, in contrast to the several thousand species that have been identified on Earth. This lunar paucity is due to the dry nature of the Moon and the depletion in volatile and siderophile elements. Extensive summaries of lunar mineralogy can be found in Frondel (1975), Heiken et al. (1991), and Papike et al. (1999).

9.2 Lunar Meteorites

Our understanding of the lunar crust has been aided by the discovery of lunar meteorites of which about 20 are known. From their feldspar-rich and KREEP-poor composition, many appear to be from the lunar farside; they are distinct from the nearside highland samples returned by *Apollo 14, 15, 16,* and *17* and *Luna 20*. However, their major element composition is close to that of estimates of the average highland crust. They confirm, as do the *Galileo*, *Clementine*, and *Lunar Prospector* missions, the essentially anorthositic nature of the lunar highland crust.

9.3 Tektites

The notion that tektites were derived from the Moon enjoyed considerable support before the *Apollo* missions. However, the controversy that had raged, particularly in the 1960s, over a lunar versus a terrestrial origin was settled in favor of the latter source by the first sample return from the Moon in 1969. It has been decisively established from isotopic and chemical evidence that tektites are derived from the surface of the Earth by meteoritic or asteroidal impact. Because the debate still surfaces occasionally, readers interested in these glassy objects will find a useful review of the evidence for a terrestrial origin in Koeberl (1994).

10. The Origin of the Moon

10.1 The Nature of the Problem

Hypotheses for the origin of the Moon must explain the high value for the angular momentum of the Earth–Moon system, the strange lunar orbit inclined at 5.09° to the plane of the ecliptic, the high mass relative to that of its primary planet and the low bulk density of the Moon, much less than that of the Earth or the other inner planets. The chemical age and isotopic data revealed by the returned lunar samples added additional complexities to these classic problems because the lunar composition is unusual by either cosmic or terrestrial standards. It is perhaps not surprising that previous theories for the origin of the Moon failed to account for this diverse set of properties and that only recently has something approaching a consensus been reached.

Hypotheses for lunar origin can be separated into five categories:

1. Capture from an independent orbit
2. Formation as a double planet
3. Fission from a rapidly rotating Earth
4. Disintegration of incoming planetesimals
5. Earth impact by a Mars-sized planetesimal and capture of the resulting debris into Earth orbit

These are not all mutually exclusive, and elements of some hypotheses occur in others. For example:

1. Capture of an already formed Moon from an independent orbit has been shown to be highly unlikely on dynamic grounds. The hypothesis provides no explanation for the peculiar composition of our satellite. In addition, it could be expected that the Moon might be an example of a common and primitive early solar system object, similar to the captured rock-ice satellites of the outer planets. This indeed had been the expectation of Harold Urey, based on the similarity of the lunar density to that of primitive carbonaceous chondrites. It would be an extraordinary coincidence if the Earth had captured an object with a unique composition, in contrast to the many examples of icy satellites captured by the giant planets.

2. Formation of the Earth and the Moon in association as a double-planet system immediately encounters the problems of differing density and composition of the two bodies. Various attempts to overcome the density problem led to coaccretion scenarios in which disruption of incoming differentiated planetesimals formed from a ring of low-density silicate debris. Popular models to provide this ring involved the breakup of differentiated planetesimals as they come within a Roche limit (about 3 Earth radii). The denser and tougher metallic cores of the planetesimals survived and accreted to the Earth, while their mantles formed a circumterrestrial ring of broken-up silicate debris from which the Moon could accumulate. This attractive scenario has been shown to be flawed because the proposed breakup of planetesimals close to the Earth is unlikely to occur. It is also difficult to achieve the required high value for the angular momentum in this model. Such a process might be expected to have been common during the formation of the terrestrial planets, and Venus, in particular, could be expected to have a satellite.

3. In 1879, George Darwin proposed that the Moon was derived from the terrestrial mantle by rotational fission. Such fission hypotheses have been popular since they produced a low-density, metal-poor Moon. However, the angular momentum of the Earth–Moon system, although large, is insufficient by a factor of about 4 to allow for rotational fission. If the Earth had been spinning fast enough for fission to occur, there is no available mechanism for removing the excess angular momentum following lunar formation. The lunar sample return provided an opportunity to test these hypotheses because they predicted that the bulk composition of the Moon should provide some identifiable signature of the terrestrial mantle. The O and Cr isotopic compositions are similar, and this is sometimes used to argue for a lunar origin from the Earth's mantle. However, the enstatite chondrites also have identical O isotopic compositions in both bodies; however, both bodies differ significantly in major and trace element contents. Similarity does not constitute identity. Fission hypotheses failed to account for significant chemical differences between the compositions of the Moon and that of the terrestrial mantle or to provide a unique terrestrial signature in the lunar samples. The Moon contains, for example, 50% more FeO and has distinctly different trace siderophile element signatures. It also contains higher concentrations of refractory elements (e.g., Al, U) and lower amounts of volatile elements (e.g., Bi, Pb). The Moon and the Earth have distinctly different siderophile element patterns. The similarity in V, Cr, and Mn abundances in the Moon and the Earth is nonunique since CM, CO, and CV chondrites show the same pattern. These differences between the chemical compositions of the Earth's mantle and the Moon are fatal to theories that wish to derive the Moon from the Earth.

4. One proposed modification of the fission hypothesis uses multiple small impacts to place terrestrial mantle material into orbit. It is exceedingly difficult to obtain the required high angular momentum by such processes because multiple impacts should average out.

Most of these Moon-forming hypotheses should be general features of planetary and satellite formation and should produce Moon-like satellites around the other terrestrial planets. They either fail to account for the unique nature of the Earth–Moon system and the peculiar bone-dry composition of the Moon, or they do not account for the differences between the lunar composition and that of the terrestrial mantle. These earlier theories accounted neither for the lunar orbit nor for the high angular momentum, relative to the other terrestrial planets, of the Earth–Moon system, a rock on which all older hypotheses foundered.

10.2 The Single–Impact Hypothesis

The single-impact hypothesis was developed by A. G. W. Cameron basically to solve the angular momentum problem, but, in the manner of successful hypotheses, it has accounted for other parameters as well and has become virtually a consensus. The theory proposes that, during the final stages of accretion of the terrestrial planets, a body about the size of Mars collided with the Earth and spun out a disk of material from which the Moon formed. This giant impact theory resolves many of the problems associated with the origin of the Moon and its orbit. The following scenario is one of several possible, although restricted, variations on the theme.

In the closing stages of the accretion of the terrestrial planets 50–100 million years after T_0 (4567 million years ago), the Earth suffered a grazing impact with an object (named Theia) of about 0.10 Earth mass. This body is assumed to have differentiated into a silicate mantle and a metallic core. It came from the same general region of the nebula as the Earth (the oxygen and chromium signatures of Earth and Moon are identical and the impact velocities are required to be low in the models).

Theia was disrupted by the collision and mostly went into orbit about the Earth. Gravitational torques, due to the asymmetrical shape of the Earth following the impact, assisted in accelerating material into orbit. Expanding gases from the vaporized part of the impactor also promoted material into orbit. Following the impact, the mantle material from Theia was accelerated, but its metallic core remained as a coherent mass and was decelerated relative to the Earth, so that it fell into the Earth within about 4 hours. A metal-poor mass of silicate, mostly from the mantle of Theia, remained in orbit.

In some variants of the hypothesis, this material immediately coalesced to form a totally molten Moon. In others, it broke up into several moonlets that subsequently accreted to form a partly molten Moon. This highly energetic event accounts for the geochemical evidence that indicates that at least half the Moon was molten shortly after accretion. Figure 22 illustrates several stages of a computer simulation of the formation of the Moon according to one version of the single giant impact hypothesis.

Although the giant impact event vaporized much of the material, the material now in the Moon does not seem to have condensed from vapor. The extreme depletion of very volatile elements and the bone-dry nature of the Moon may be inherited from Theia and so have been a general feature of the early inner solar nebula (all primary meteorite minerals are anhydrous) with volatiles and water added later to the Earth from near Jupiter.

Unique events are notoriously difficult to accommodate in most scientific disciplines. An obvious requirement in this model is that a suitable population of impactors existed in the early solar system. Evidence in support of the previous existence of large objects in the early solar system comes from the ubiquitous presence of heavily cratered ancient planetary surfaces, from the large number of impact basins with diameters up to 2000 km or so, and from the obliquities or tilts of the planets, all of which demand collisions with large objects in the final stages of accretion. The extreme example is that an encounter between Uranus and an Earth-sized body is required to tip that planet on its side. Thus, the possibility of many large collisions in the early solar system is well established, one of which had the right parameters to form the Moon. The single impact scenario is thus consistent with the planetesimal hypothesis for the formation of the planets from a hierarchical sequence of smaller bodies.

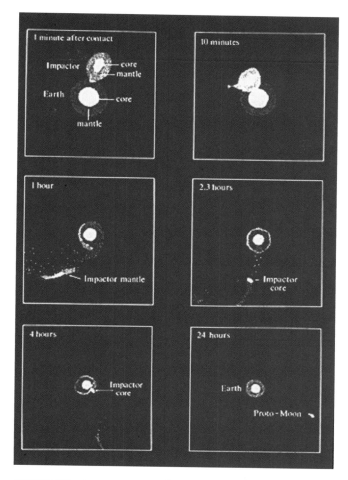

FIGURE 22   A computer simulation of the origin of the Moon by a glancing impact of a body larger than Mars with the early Earth. This event occurred about 4500 million years ago during the final stages of accretion of the terrestrial planets. Both the impactor and the Earth have differentiated into a metallic core and rocky silicate mantle. Following the collision, the mantle of the impactor is ejected into orbit. The metallic core of the impactor clumps together and falls into the Earth within about 4 hours in this simulation. Most terrestrial mantle material ejected by the impact follows a ballistic trajectory and is reaccreted by the Earth. The metal-poor, low-density Moon is thus derived mainly from the silicate mantle of the impactor. (Courtesy A. G. W. Cameron.)

This research was conducted in part at the Lunar and Planetary Institute, which is operated by the USRA under contract CAN-NCC5-679 with NASA. This is LPI Contribution 1260.

Bibliography

Basaltic Volcanism Study Project (1981). "Basaltic Volcanism on the Terrestrial Planets." Pergamon, New York.

Canup, R. M. (2004). Dynamics of lunar formation. *Annu. Rev. Astron. Astrophys.* **42,** 441–475.

Canup, R. M., and Righter, K. (2000). "Origin of the Earth and Moon." Arizona Univ. Press, Tucson.

Frondel, J. W. (1975). "Lunar Mineralogy." John Wiley & Sons, New York.

Fuller, M. J., and Cisowski, S. M. (1987). Lunar paleomagnetism. In "Geomagnetism 2" (J. A. Jacobs, ed.), pp. 307–455. Academic Press, San Diego.

Hartman, W. R., Phillips, R. J., and Taylor, G. L., eds. (1986). "Origin of the Moon." Lunar and Planetary Institute, Houston.

Heiken, G., Vaniman, D., and French, B. M. (1991). "The Lunar Sourcebook." Cambridge Univ. Press, Cambridge, England.

Jolliff, B. L., et al. (2000). Major lunar crustal terranes. *J. Geophys. Res.* **105,** 4197–4216.

Khan, A., and Mosegaard, K. (2001). New information on the deep lunar interior from an inversion of lunar free oscillation periods. *Geophys. Res. Lett.* **28,** 1791–1794.

Koeberl, C. (1994). "Tektite Origin by Hypervelocity Asteroidal or Cometary Impact: Target Rocks, Source Craters and Mechanisms," Geol. Soc. Am. Spec. Paper **293,** 133–151.

Papike, J. J., ed. (1999). "Planetary Materials," Miner. Soc. Amer. Rev. Miner. 36.

Taylor, S. R. (1982). "Planetary Science: A Lunar Perspective." Lunar and Planetary Institute, Houston.

Taylor, S. R. (2001). "Solar System Evolution: A New Perspective," 2nd Ed. Cambridge Univ. Press, Cambridge, England.

Wilhelms, D. E. (1987). "The Geologic History of the Moon," U.S. Geol. Surv. Prof. Paper No. 1348. U.S. Geological Survey, Washington, D.C.

Meteorites

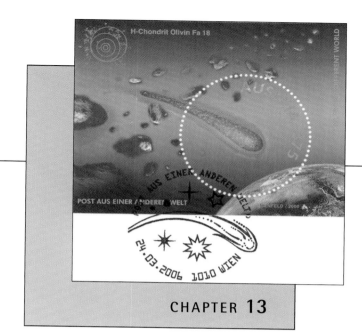

Michael E. Lipschutz

Purdue University
West Lafayette, Indiana

Ludolf Schultz

Max-Planck-Institut für Chemie,
Mainz, Germany

CHAPTER **13**

Meteorites, the "Poor Man's Space Probe," are important because they contain the oldest solar system materials for research and sample a wide range of parent body—exteriors and interiors—some primitive, some highly evolved. Meteorites record certain solar and galactic effects and yield otherwise unobtainable data relevant to the genesis, evolution, and composition of the Earth, other major planets, satellites, asteroids, and the Sun. Some contain inclusions created before solar system formation; others contain organic matter produced on grain boundaries in the early nebula and/or in giant interstellar clouds. Meteorites also constitute important "ground truth" in a chemical and physical sense, critical to interpreting planetary data obtained by remote sensing. Most importantly, meteorites are on Earth, available for laboratory study by the simplest to the most sophisticated analytical techniques. If one picture is worth 10,000 words, then one sample is worth 10,000 pictures. Even though meteorites are only tiny source-fragments, proper integration of data from them can better describe their sources, just as a more complete mosaic can be deduced from a few tesserae.

1. Introduction

1.1 General

In the Western world, 1492 marked the discovery of the New World by the Old, the Spanish Expulsion, and, the oldest documented, preserved, and scientifically studied meteorite *fall*—a 127 kg (LL6) stone that fell at Ensisheim in Alsace. [A meteorite is named for the nearest post office or geographic feature. The chemical-petrologic classification is the scheme by which Ensisheim, for example, is classified as an LL6 chondrite (see Section 1.2).] The oldest preserved meteorite fall might be Nogata (Japan), an L6, which allegedly fell in 861 (but all associated documentation is more recent) and is in a Shinto shrine there. Recovered meteorites, whose fall was unobserved, are *finds*, some having been discovered (occasionally artificially reworked) in archaeological excavations in such Old World locations as Ur, Egypt, and Poland, and in New World burial sites. Obviously, prehistoric and early historic man recognized meteorites as unusual, even venerable, objects.

---

* Actual meteriorite (chondrite) dust is embedded in the stamp reproduced above. The stamp was issued by the Austrian postal service in 2006.

TABLE 1	Numbers of Classified Non-Antarctic Meteorite Falls and Finds, Including Those from Hot and Cold Deserts.						
Meteorite	Falls[a]	Finds[a,b]	ANSMET[c]	Meteorite	Falls[a]	Finds[a,b]	ANSMET[c]
Chondrites	**797**	**>814**	**2925 (11557)**	**Achondrites**	**81**	**>73**	**184**
CI1	5	0(2)	0(0)	Acapulcoites	1	1 }(14)	5(12)
CM/C2	17	6(85)	51(200)	Lodranites	1	0	4(4)
C other	18	14(225)	58(182)	Winonaites	0	3(10)	1(1)
E	17	8(307)	43(103)	Angrites	1	1(3)	2(2)
H	316	405	*1048*(4194)	Aubrites	9	3(2)	7(38)
L	350	350	*1140*(4562)	Howardites	20	4(58)	26(44)
LL	72	30	*574*(2299)	Eucrites	29	12(137)	66(124)
Other	2	1(34)	11(17)	Diogenites	11	0(81)	22(24)
				Ureilites	5	3(90)	34(47)
Irons	**40**	**>690(34)**	**47(97)**	Lunar	0	1(22)	8(15)
				Martian	4	3(17)	8(8)
Stony-Irons	**12**	**>61**	**13**	Other	0	3(9)	1(4)
Mesosiderites	7	21(10)	11(29)				
Pallasites	5	40(4)	2(11)				

[a] Data from Grady (2000) updated to Nov. 2004 (J. N. Grossman, USGS, personal communication). These do not include 41 unclassified stony or 8 unclassified iron meteorites.

[b] Except for Lunar and Martian meteorites, numbers in parentheses indicate fragments (uncorrected for pairing) recovered as meteorite clusters from hot and cold deserts (ANSMET data not included): These are not combined with corresponding non-desert-cluster finds (Grady, 2000; Grossman, personal communication). The ~16,500 JARE samples are incompletely classified and, except for lunar and martian meteorites, are not included in this table: Ordinary chondrites from hot and other cold deserts (other than ANSMET) are also omitted Because of their special importance, numbers of lunar and martian meteorites (cf. http://epsc.wustl.edu/admin/resources/ meteorites/moon_meteorites.html and http://curator.jsc.nasa.gov/curator/antmet/marsmets/contents.htm, respectively) in parentheses are the meteorite falls corrected for pairing.

[c] Antarctic Search for Meteorites (ANSMET) recoveries from West Antarctica. Numbers in parentheses are fragments recovered: Associated numbers are corrected for known pairings or by estimating (italics) four fragments per fall. (Data from K. Righter, NASA—JSC.)

Despite this history, and direct evidence for meteorite falls, scientists generally began to accept them as genuine samples of other planetary bodies only at the beginning of the 19th century. Earlier, acceptance of meteorites as being extraterrestrial and, thus, of great scientific interest, was spotty. One might laboriously assemble a meteorite collection only to have someone later dispose of this invaluable material. This occurred, for example, when the noted mineralogist, Ignaz Edler von Born, discarded the imperial collection in Vienna as "useless rubbish" in the latter part of the 18th century. With the recognition that meteorites sample extraterrestrial planetary bodies, collections of them proved particularly important. In 1943, with the imminent invasion of Germany, the Russian government planned for "trophy brigades" to accompany their armies and collect artistic, scientific, and production materials as restitution for Russian property seized or destroyed by Nazi armies during their occupation of parts of Russia. Meteorites that fell in Russia, fragments of which were acquired by and housed in German collections, were explicitly identified as material to be seized. In late 2004, the price for a meteorite from Mars was at least $4000/g (the current price of gold is $14/g).

Apart from its recovery and preservation, Ensisheim is a typical fall. For finds, some peculiarity must promote recognition—hence, the high proportion of high-density, iron meteorites outside of Antarctica (Table 1). Observed falls are taken to best approximate the contemporary population of near-Earth meteoroids. Of course, bias may affect the fall population. Some data suggest that highly friable meteoroids are largely or totally disaggregated during atmospheric passage.

The initial entry velocities of meteorites range from 11 to 70 km/s, average 15 km/s, and cause surface material to melt and ablate by frictional heating during atmospheric passage. Heat generation and ablation rates are rapid and nearly equivalent, so detectable heat effects only affect a few millimeters below the surface: The meteorite's interior is preserved in its cool, preterrestrial state. Ablation and fragmentation—causing substantial (~90%) mass loss and deceleration, often to terminal velocity—leave a dark brown-to-black, sculpted fusion crust as the surface, diagnostic of a meteorite on Earth (Fig. 1a). If it is appropriately shaped perhaps by ablation, a meteoroid may assume a quasi-stable orientation late in its atmospheric traversal. In this case, material ablated from the front can redeposit as delicate droplets or streamlets on its sides and rear (Fig. 1b). The delicate droplets on Lafayette's fusion crust would have been erased in a few days' weathering: It must have been recovered almost immediately after it fell. Yet, when Lafayette was recognized as meteoritic during a 1931 visit to Purdue

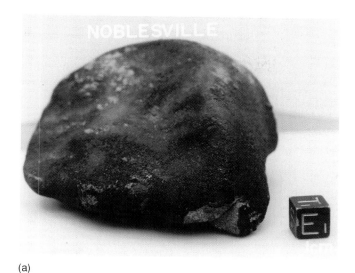

NOBLESVILLE

(a)

(b)

FIGURE 1   Fusion crusts: (a) Noblesville H chondrite; (b) Lafayette Martian meteorite. Noblesville, which fell on 31 August 1991, has nearly complete fusion crust but exposed surface at lower right next to the 1-cm cube shows a genomict (H6 in H4) breccia. (Photo courtesy of NASA Johnson Space Center.) Lafayette exhibits very delicate, redeposited droplets on its sides, indicating an orientation with its top pointing Earthward late in atmospheric traversal. (Photo courtesy of the Smithsonian Institution.)

University by O. C. Farrington (a prominent meteoriticist), the chemistry professor on whose desk it was found thought it a terrestrial glacial artifact. Who actually recovered this martian meteorite is a mystery.

Meteorites derive from asteroids and, less commonly, from larger parent bodies: 18 individual samples representing 31 separate falls (all but 5 from Antarctica) come from Earth's Moon; and 32 others (6 from Antarctica) almost certainly are from Mars [*see* MARS: SURFACE AND INTERIOR; THE MOON]. Some interplanetary dust particles may also come from these sources, and/or comets. Meteorites are rocks and therefore polymineralic (Table 2), with each of

TABLE 2   Common Meteoritic or Cited Minerals

Mineral	Formula	Mineral	Formula	Mineral	Formula
Anorthite	$CaAl_2Si_2O_8$	Hibonite	$CaAl_{12}O_{19}$	Pyroxene solid solution	
Clinopyroxene	$(Ca,Mg,Fe)SiO_3$	Ilmenite	$FeTiO_3$	enstatite (En)	$MgSiO_3$
Chromite	$FeCr_2O_4$	Kamacite	$\alpha\text{-}(Fe,Ni)$	ferrosilite (Fs)	$FeSiO_3$
Cohenite	$(Fe,Ni)_3C$	Lonsdaleite	C	wollastonite (Wo)	$CaSiO_3$
Cristobalite	SiO_2	Magnetite	Fe_3O_4	Schreibersite	$(Fe,Ni)_3P$
Diamond	C	Melilite solid solution		Serpentine (chlorite)	$(Mg,Fe)_6Si_4O_{10}(OH)_8$
Diopside	$CaMgSi_2O_6$	åkermanite (Åk)	$Ca_2MgSi_2O_7$	Spinel	$MgAl_2O_4$
Enstatite	$MgSiO_3$	gehlenite (Ge)	$Ca_2Al_2SiO_7$	Spinel solid solution	
Epsomite	$MgSO_4 \cdot 7H_2O$	Oldhamite	CaS	spinel	$MgAl_2O_4$
Fayalite	Fa_2SiO_4	Olivine	$(Mg,Fe)_2SiO_4$	hercynite	$FeAl_2O_4$
Feldspar solid solution		Olivine solid solution		chromite	$FeCr_2O_4$
albite (Ab)	$NaAlSi_3O_8$	fayalite (Fa)	Fe_2SiO_4	Taenite	$\gamma\text{-}(Fe,Ni)$
anorthite (An)	$CaAl_2Si_2O_8$	forsterite (Fo)	Mg_2SiO_4	Tridymite	SiO_2
orthoclase (Or)	$KAlSi_3O_8$	Orthopyroxene	$(Mg,Fe)SiO_3$	Troilite	FeS
Ferrosilite	$FeSiO_3$	Pentlandite	$(Fe,Ni)_9S_8$	Whitlockite	$Ca_3(PO_4)_2$
Forsterite	Mg_2SiO_4	Plagioclase			
Gehlenite	$Ca_2Al_2SiO_7$	albite (Ab)	$NaAl_2Si_2O_8$		
Graphite	C	anorthite (An)	$CaAl_2Si_2O_8$		

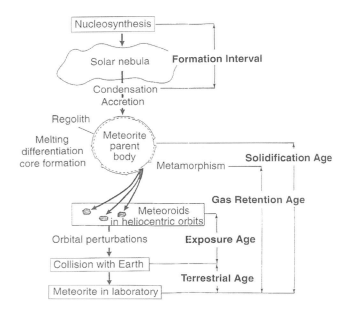

FIGURE 2 From nebula to meteorite: genetic processes and the corresponding age determinable for each process. Nuclides of nearly all elements were formed by nuclear reactions in interiors of large stars, which then ejected them in very energetic supernova events. Ejected nebular gas and dust subsequently nucleated, condensed, and accreted into primitive bodies. Source bodies for most meteorites were heated, causing solid-state metamorphism or, at higher temperatures, differentiation involving separation of solids, liquids, and gases. As a body evolved, it suffered numerous impacts, and, if atmosphere-free, its surface was irradiated by solar and galactic particles that embedded in the skins of small grains and/or caused nuclear reactions. Larger impacts ejected fragments that orbited the Sun. Subsequently, orbital changes caused by large-body gravitational attraction placed meteoroids into Earth-crossing orbits allowing their landing and immediate recovery (as a fall) or later (as a find). Each process can alter elemental and/or isotopic contents. Which of these processes affected a given meteorite and the time elapsed since it occurred are definable.

the hundred or so known meteoritic minerals generally having some chemical compositional range, reflecting its formation and/or subsequent alteration processes. Important episodes during meteorite genesis are in Fig. 2.

1.2 From Parent Body to Earth

To arrive on Earth, a meteoroid (meteorite-to-be) must be excavated and removed from the gravitational field of its parent body by an impact. This impact can generate short-lived but intense shocks, which provides the impulse necessary for the meteoroid to exceed the parent body's escape velocity. In general, the higher the shock pressure acting upon matter, the higher its ejection velocity and temperature, both the shock temperature derived from passage of the pressure wave and the postshock residual temperature

(from compressional, nonadiabatic heat) after decompression. Residual temperatures as high as 1250°C, have been recorded in stony meteorites and correspond to pressures >57 GPa or 570,000 atm (570,000 times the Earth's sea level pressure). Significantly higher temperatures (pressures) would vaporize matter, so there is a limit to the shock-induced ejection velocity of survivable meteoroids (i.e., Mars' escape velocity, 5.4 km/s).

In very special scenarios, ejecta can be accelerated by impact-jetting—especially during oblique impacts—thus acquiring a velocity higher than expected from the degree of shock-loading. At least some martian meteorites, the 7 nakhlites, are not heavily shocked and may signal this unusual case. In general, however, a parent body much larger than Mars is unlikely to provide meteorites to Earth.

The overwhelming majority of meteorites, those of asteroidal origin, seemingly sample a few hundred dominant asteroids, not the thousands known. These may include the near-Earth asteroids (NEA) already in Earth-crossing or-approaching orbits, ejected from Kirkwood Gap regions by chaotic motion and gravitational effects of Jupiter [*see* MAIN-BELT ASTEROIDS and NEAR-EARTH OBJECTS]. As discussed later, some types of meteorites and asteroids can be linked. The nine meteorite falls whose orbits were determined photographically seem NEA-like (Fig. 3). Some evidence suggests that co-orbital streams of meteorites and/or asteroids exist—perhaps arising from meteoroids' gentle

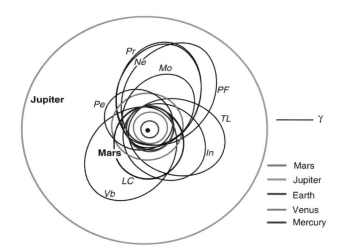

FIGURE 3 Orbits determined from overlapping camera coverage for nine recovered chondrite falls: Pr-Pribram (H5, 7 Apr. 1959); LC—Lost City (H5, 3 Jan. 1970); In—Innisfree (L5, 5 Feb. 1977); Pe—Peekskill (H6, 9 Oct. 1992); TL—Tagish Lake (C, 18 Jan. 2000); Mo—Moravka (H5-6, 6 May 2000); Ne—Neuschwanstein (EL6, 6 Apr. 2002); PF—Park Forest (L5, 26 Mar. 2003); Vb—Villalbeto de la Peña (L6, 4 Jan. 2004). The orbits shown are projections onto the ecliptic plane (orbits of the terrestrial planets and Jupiter in color are included with γ, the vernal equinox). Pribram and Neuschwanstein had identical orbits, but are of different chondritic types.

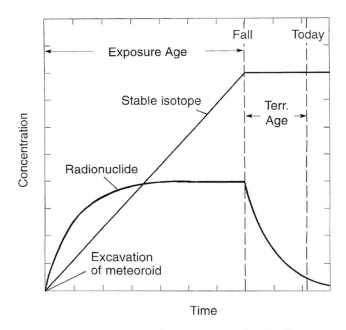

FIGURE 4 Concentrations of cosmic ray–produced radioactive and stable nuclides during cosmic ray exposure and after the meteorite's fall to Earth.

disruption in space—but this is very controversial. Evidence from temperature-sensitive components indicates that, in their orbits about the Sun, some meteorites have perihelia within 0.5 AU resulting in detectable solar heating.

Some meteorites contain regolithic material bombarded by very energetic particles. Once material is ejected from its parent body by an impact until it falls on Earth, meter-sized meteoroids are irradiated by cosmic rays (mainly protons) of solar or galactic origin. Solar cosmic rays have a power–law energy distribution with the particle flux increasing rapidly with decreasing energy: most solar particles have energies <1 MeV. Galactic and some solar particles have energies of hundreds of MeV to GeV and can induce nuclear reactions producing cosmogenic radioactive or stable nuclides. In larger meteoroids, cosmogenic nuclear reactions occur only in the meter-thick shell that cosmic ray primaries and secondaries penetrate. As discussed in Sections 6.1 and 6.2, levels of nuclides produced during cosmic ray exposure (CRE) establish the duration of energetic particle bombardment (the CRE age) and the time spent by a meteoritic find on Earth, the terrestrial age (Fig. 4).

1.3 Impact on Earth

If a meteoroid is small enough to be decelerated significantly during atmospheric passage, it may land as an individual or as a shower. A recovered individual can have a mass of ≤1 g [as in the 1965 fall of the Revelstoke stone (CI1) in British Columbia], or up to 60 metric tons (e.g., the Hoba IVB iron meteorite found in 1920 on a Namibian farm

where it remains). A meteorite shower results from a meteoroid fragmenting high in the atmosphere, usually leaving a particle trail down to dust size. Shower fragments striking the Earth define an ellipse whose long axis—perhaps extending for tens of kilometers—is a projection of the original trajectory. Typically, the most massive fragments travel farthest and fall at the farthest end of the ellipse.

Some falls are signaled by both light and sound displays; others, like the Peekskill meteorite (Fig. 5), exhibit a spectacular fireball trail observed over many states. Small falls, like Noblesville (Fig. 1a), fall silently and unspectacularly and, when recovered immediately, have cold to slightly warm surfaces. Meteorites can fall anywhere at any time. The 500-g Borodino stone (H5) fell on 5 September 1812—two days before the famous battle there—and was recovered by a Russian sentry.

U.S. Department of Defense data demonstrate that, at least since 1975, reconnaissance satellites have detected large explosions at random locations in the Earth's atmosphere. On average, about 9 of these mysterious explosions [with energies up to 1 megaton (Mt) equivalent of TNT] occur annually: no meteorite falls or fireballs are associated with any of these events.

Large meteoroids—tens of meters or larger—are not decelerated much by atmospheric transit and, with an appropriate trajectory, may ricochet off the Earth's atmosphere (Fig. 5a) or strike it at full geocentric velocity, >11 km/s. (Obviously, distinguishing a large meteoroid from a small asteroid is arbitrary.) Such explosive, crater-forming impacts can do considerable damage. The 1-km-diameter Meteor Crater (Fig. 5b) in northern Arizona, which formed 50,000 years (i.e., 50 ka) ago by the impact of a 25- to 86-m meteoroid, yielded fragments now surviving as Canyon Diablo iron meteorites. At least 40 terrestrial craters exhibit features believed to be produced only by intensive explosive impact of a large meteoroid (e.g., as in the 1908 event at Tunguska in Siberia) or perhaps even a comet nucleus. Another 269 features on Earth may be of impact origin. One expert classed 130 of them as definite impact craters. The 180-km-diameter Chicxulub feature in Yucatan, Mexico, is suspected as the impact site of a 10-km meteoroid. By consensus, this impact generated the climatic consequences responsible for the extinction of ~60% of then-known species of biota—including dinosaurs—ending the Cretaceous period and beginning the Tertiary, 65 Ma ago (the K-T event). Other, less well-established events are suggested as having caused extinctions at other times.

Some meteorites have struck man-made objects. The Peekskill stone meteorite (H6), with a recovered mass of 12.4 kg, ended its journey on the trunk of a car (Fig. 5d). Its descent in 1992 was videotaped over a five-state area of the eastern United States by many at Friday evening high school football games (Fig. 5c), yielding a well-determined orbit. Two authenticated reports of humans hit by meteorite falls exist. The first involved a 3.9 kg (H4) stone (the larger

(a)

(b)

(c)

(d)

FIGURE 5   Large meteoroids: (a) fireball of 80-m object (estimated mass, 1 Mt) on 10 August 1972 moving left to right (see arrow) over Grand Teton National Park that apparently skipped out of the atmosphere. (Photo by Dennis Milon.) (b) The 1-km-diameter Meteor Crater in Arizona formed by the explosive impact of the Canyon Diablo IA octahedrite meteoroid about 50 ka ago. (Photo by Allan E. Morton.) (c) From the videotape record of the Peekskill meteoroid during its atmospheric traverse on 9 October 1992. During fragmentation episodes such as this one (over Washington, D.C.), large amounts of material fell, but nothing was recovered. (d) Landing site of Peekskill chondrite in the right rear of an automobile. (Photo by Peter Brown, University of Western Ontario.)

of two fragments), which, after passing through her roof in Sylacauga, Alabama in 1954 struck a recumbent woman's thigh, badly bruising her. The second involved a 3.6-g piece of the Mbale (Uganda) L6 meteorite shower of 1992, which bounced off a banana tree's leaves and hit a boy on the head. Chinese records from 616 to 1915 claim numerous human and animal casualties, including many killed, by meteorite falls. Unauthenticated reports of human injuries or human deaths exist: One undocumented report tells of a dog being killed by a piece of the 40-kg Nakhla meteorite shower of 1911 near Alexandria, Egypt. This, incidentally, is one of the 32 martian meteorites. Despite the small number of casualties to date, the probability of dying in a meteoroid impact exceeds that of being killed in an airplane crash.

This arises because the impact of a large meteoroid, small asteroid, or comet nucleus is capable of causing devastating loss, indeed the total extinction of life. Such impacts seem rare.

Meteorites may impact anywhere on Earth, and, as of November 2004, the numbers of known falls and isolated, non-desert-cluster finds are 1046 and 1840, respectively (cf. Table 1). For these, it can readily be established whether meteorite fragments found nearby are from the same meteoroid; however, such linkages are difficult for the numerous meteorite pieces found clustered in hot or cold (Antarctic) deserts since 1969. So far, starting in 1969, but mainly since 1976, Antarctica has yielded over 31,000 fragments [16,500 collected by JARE (Japanese Antarctic Research

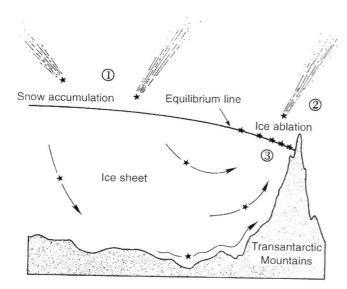

FIGURE 6 Cross section of Antarctic ice sheet and subice topography: meteorites fall (1), are collected by the ice sheet and buried (i.e., preserved), transported, and concentrated near a barrier to the ice sheet (2), and are exposed by strong South Polar winds that ablate the stagnant ice (3). [Reprinted from "Workshop on Antarctic Glaciology and Meteorites," C. Bull and M. E. Lipschutz (eds.), LPI Tech. Rept. 82-03. Copyright 1982 with kind permission from the Lunar and Planetary Institute, 3600 Bay Area Boulevard, Houston, TX 77058-1113.)

Expedition) in Queen Maud Land; 13,907 by ANSMET (Antarctic Search for Meteorites), the US-led team upstream of the Trans-Antarctic Mountain Range; and >677 by a European consortium, which is now an Italian-led effort]. Hot desert-clusters in Australia, North Africa (mainly Algeria and Libya), China, and the United States have yielded >5000 more to date. (These discoveries are possible in these areas because dark meteorites can be readily distinguished from the local, light-colored terrestrial rocks, "meteorwrongs.") The 14-million-km^2 ancient Antarctic ice sheet is a meteorite trove because of the continent's unique topography and its effect on ice motion, which promotes the meteorites' collection, preservation, transportation and concentration (Fig. 6). Assuming four fragments per meteoroid, Antarctic meteorites recovered thus far correspond to about 7500 different impact events; no one has estimated the number of fragments produced in a hot-desert meteorite fall. Desert meteorites are named for the nearest topographic feature, usually abbreviated by a one- or three-letter code, and number: the first two digits of Antarctic meteorites denote the expedition year.

To complicate matters, expeditions have taken two paths in characterizing their meteorite recoveries. ANSMET chooses to characterize each fragment by type. Other expeditions scan their collection to identify meteorites of rare type, which are of intrinsic interest for more complete study (see Table 1). The "pairing" of even these samples, let alone

the more common meteorites in these other collections, has not yet been addressed.

2. Meteorite Classification

2.1 General

Meteorites, like all solar system matter, ultimately derive from primitive materials that condensed and accreted from the gas- and dust-containing presolar disk. Most primitive materials were altered by postaccretionary processes—as in lunar, terrestrial, and martian samples—but some survived essentially intact, as specific chondrites or inclusions in them. Some primitive materials are recognizable unambiguously (albeit with considerable effort), usually from isotopic abundance peculiarities; others are conjectured as unaltered primary materials. Postaccretionary processes produced obvious characteristics that permit classification of the thousands of known meteorites into a much smaller number of types. Many classification criteria contain genetic implications, which we now summarize.

At the coarsest level, we class meteorites as irons, stones, or stony-irons from their predominant constituent (Figs. 7a and 8): each can then be classified by a scheme with genetic implications (Fig. 7b). Stones include the numerous, more-or-less primitive chondrites (Table 1; Figs. 8a and 8b) and the achondrites (Fig. 8d) of igneous origin. Irons (Fig. 8e), stony-irons (Fig. 8c), and achondrites are differentiated meteorites, presumably formed from melted chondritic precursors by secondary processes in parent bodies (Fig. 2). During melting, physical (and chemical) separation occurred, with high-density iron sinking to form pools or a core below the lower density achondritic parent magma. Ultimately, these liquids crystallized as parents of the differentiated meteorites, the irons forming parent body cores or, perhaps, dispersed "raisins" within their parent. Stony-iron meteorites are taken to represent metal-silicate interface regions. Pallasites (Fig. 8c), which have large (centimeter-sized) rounded olivines embedded in well-crystallized metal, resemble an "equilibrium" assemblage that may have solidified within a few years but that cooled slowly at iron meteorite formation-rates, a few degrees per million years (Ma). Mesosiderite structures suggest more rapid and violent metal and silicate mixing, possibly by impacts.

During differentiation, siderophilic elements are easily reduced to metal; they follow metallic iron geochemically and are extracted into metallic melts. Such elements (e.g., Ga, Ge, Ni, or Ir) are thus depleted in silicates and enriched in metal to concentrations well above those in precursor chondrites. Conversely, magmas become enriched in lithophilic elements—like rare earth elements (REE), Ca, Cr, Al, or Mg—above chondritic levels: concentrations of such elements approach zero in metallic iron.

CLASSIFICATION OF METEORITES

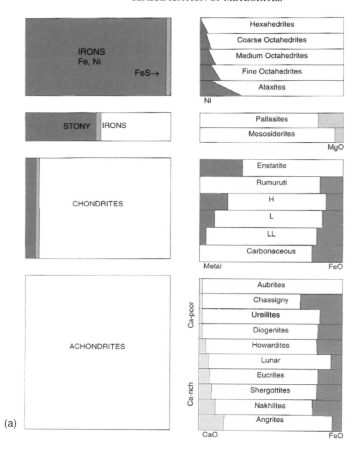

(a)

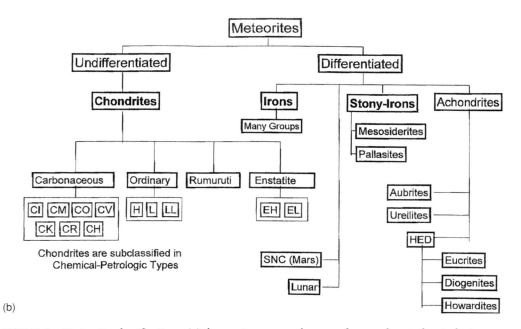

(b)

FIGURE 7   Meteorite classifications: (a) the most common classes and some chemical-petrologic classification criteria (in left-hand boxes, red denotes iron-nickle metal, orange indicates FeS and white signifies silicates); (b) genetic associations involving meteorites.

FIGURE 8   Common meteorite types (approximate longest dimension in cm): (a) Whitman, H5 (6 cm); (b) Allende; C3V (8 cm)—note 1-cm chondrule in center; (c) Springwater pallasite (18 cm); (d) Sioux County eucrite (8 cm); (e) Sanderson IIIB medium octahedrite (13 cm)—note large FeS inclusions.

During substantial heating, noble gases and other atmophile elements—like carbon and nitrogen—are vaporized and lost from metallic or siliceous regions. Chalcophilic elements that form sulfides like troilite (Table 2) include Se, Te, Tl, or Bi. Chalcophiles and a few siderophiles and lithophiles are also often quite easily mobilized (i.e., vaporized from condensed states of matter) so that they may be enriched in sulfides in the parent body or lost from it. Concentrations of these elements in specific meteorites then depend in part on the fractionation histories of their parents and are markers of heating.

2.2 Characteristics of Specific Classes

It is obvious, even to the naked eye, that most iron meteorites consist of large metallic iron crystals, which are usually single-crystal, bcc α-Fe (kamacite) lamellae 0.2–50 mm thick with decimeter to meter lengths (Fig. 8e). These relatively wide Ni-poor lamellae are bounded by thin, Ni-rich fcc γ-Fe (taenite). The solid-state nucleation and diffusive growth process by which kamacite grew at slow cooling rates from taenite previously nucleated from melt is quite well understood. The 1-atm Fe–Ni phase diagram and measurement of Ni-partitioning between kamacite and taenite permits cooling rate estimation between ~900 and 400°C. These typically are a few degrees or so per Ma, depending on the iron meteorite group, consistent with formation in objects of asteroidal size. The Ni concentration in the melt determines the temperature of incipient crystallization, and this, in turn, establishes kamacite orientation in the final meteorite. These orientations are revealed in iron meteorites by brief etching (with nitric acid in alcohol) of highly polished cut surfaces: Baron Alois von Widmanstätten discovered this in the 18th century, and the etched structure is called the "Widmanstätten pattern." (An Englishman, G. Thomsen, independently discovered this, but his contribution was unrecognized.)

Meteorites containing <6% Ni are called hexahedrites because they yield a hexahedral etch pattern of large, single-crystal (centimeter-thick) kamacite (Fig. 7a). Iron meteorites containing 6–16% Ni crystallize in an octahedral pattern and are octahedrites. Lower-Ni meteorites have the thickest kamacite lamellae (>3.3 mm) and yield the very coarsest Widmanstätten pattern, while those highest in Ni are composed of very thin (<0.2 mm) kamacite lamellae and are very fine octahedrites. Iron meteorites containing >16% Ni nucleate kamacite at such low temperatures that large single crystals could not form over the 4.57 billion years (Ga) of solar system history: they lack a Widmanstätten pattern and are called Ni-rich ataxites (i.e., without structure). The Ni-poor ataxites are hexahedrites or octahedrites that were reheated either in massive impacts or artificially after they fell on Earth.

As noted earlier, when primitive parent bodies differentiated, siderophilic elements were extracted into molten

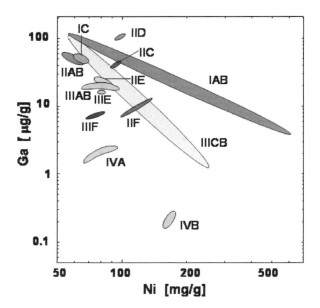

FIGURE 9   Contents of Ni and Ga in iron meteorites. (Some larger chemical groups are indicated by Roman numerals and letters.)

metal. During melt crystallization, fractionation or separation of siderophiles could occur. About 50 years ago, Ga and Ge contents of iron meteorites were found to be quantized, not continuous: They could then be used to classify irons into groups denoted as I to IV. Originally, these Ga–Ge groups, which correlate well with Ni content and the Widmanstätten pattern, were thought to sample core materials from a very few parent bodies. Subsequent studies of many additional meteorites and some additional elements, especially Ni and Ir, modified this view. At present, the chemical groups (Fig. 9) suggest that iron meteorites sample perhaps 100 parent bodies, although many, if not most, irons derive from but 5 parents (Fig. 9) represented by the IAB, IIAB, IIIABCD, IVA, and IVB irons. (The earlier Roman numeral notation for Ga–Ge groups was retained to semiquantitatively indicate the meteorite's Ga or Ge content. However, a letter suffix was added to indicate whether siderophiles fractionated from each other.) In addition to the major minerals (kamacite, taenite, and mixtures of them), minor amounts of other minerals like troilite, and graphite may be present. Also, silicates or other oxygen-containing inclusions exist in some iron meteorites.

In most cases, chondrites contain spherical millimeter-to centimeter-sized chondrules or their fragments. These chondrules were silicates that melted rapidly at temperatures near 1600°C and cooled rapidly at some ~1000°C/h early in the solar system's history; others cooled more slowly at 10–100°C/h. Rapid heating and cooling are relatively easy to do in the laboratory but are difficult on a larger, solar system–sized scale. Yet, large volumes of chondrules must have existed in the solar system because chondrites are numerous (Table 1). Chondrites (and many achondrites) date back to the solar system's formation—indeed, they provide chronometers for it (see Sections 6.4 and 6.5)—and represent accumulated primary nebular condensate and accretionary products. A portion of this condensate formed from the hot nebula as millimeter-sized Ca- and Al-rich inclusions (CAI), mineral aggregates predicted as vapor-deposition products by thermodynamic calculations. These CAI, found mainly in chondrites rich in carbonaceous (organic) material, exhibit many isotopic anomalies and contain atoms with distinct nucleosynthetic histories. Other inclusions (like SiC and extremely fine diamond) represent relict presolar material. Other condensates formed at much lower temperatures. Some—perhaps even many—CAI may be refractory residues, not condensates.

Although most chondrites contain the same minerals, the proportions of these and their compositions differ in the 6 or so principal chondritic chemical groups. The primary bases for chondrite classification involve proportions of iron as metal and silicate (in which oxidized iron—expressed as FeO—may be present), and total iron (from Fe, FeO, and FeS) content (Fig. 7a). The last (Fig. 10) defines meteorites with high and low total iron (H and L, respectively) or low total iron and low metal (LL). Numbers of H, L, and LL chondrites are so large (Table 1) that these are called

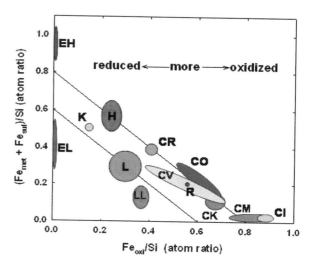

FIGURE 10 Silicon-normalized contents of Fe as metal and in FeS (ordinate) vs. Fe in ferromagnesian silicates (abscissa) in various chondritic groups. (Each diagonal defines constant total iron content.)

the ordinary chondrites. Obviously, chondrite compositions (typically, as in Table 3, with elements apportioned by chemical form) are not continuous but, rather, quantized. Table 3 lists major element ratios diagnostic of specific chondritic

TABLE 3	Average Chemical Compositions and Elemental Ratios of Carbonaceous and Ordinary Chondrites and Eucrites

Species[a]	C1	C2M	C3V	H	L	LL	EUC	Species[a]	C1	C2M	C3V	H	L	LL	EUC
SiO_2	22.69	28.97	34.00	36.60	39.72	40.60	48.56	NiO	1.33	1.71					
TiO_2	0.07	0.13	0.16	0.12	0.12	0.13	0.74	CoO	0.08	0.08					
Al_2O_3	1.70	2.17	3.22	2.14	2.25	2.24	12.45	NiS			1.72				
Cr_2O_3	0.32	0.43	0.50	0.52	0.53	0.54	0.36								
Fe_2O_3	13.55							CoS			0.08				
FeO	4.63	22.14	26.83	10.30	14.46	17.39	19.07	SO_3	5.63	1.59					
MnO	0.21	0.25	0.19	0.31	0.34	0.35	0.45	CO_2	1.50	0.78					
MgO	15.87	19.88	24.58	23.26	24.73	25.22	7.12								
CaO	1.36	1.89	2.62	1.74	1.85	1.92	10.33	Total	98.86	99.82	99.84	99.99	99.99	99.92	100.07
Na_2O	0.76	0.43	0.49	0.86	0.95	0.95	0.29	ΣFe	18.85	21.64	23.60	27.45	21.93	19.63	15.04
K_2O	0.06	0.06	0.05	0.09	0.11	0.10	0.03								
P_2O_5	0.22	0.24	0.25	0.27	0.22	0.22	0.05	Ca/Al	1.08	1.18	1.10	1.11	1.12	1.16	1.12
H_2O^+	10.80	8.73	0.15	0.32	0.37	0.51	0.30	Mg/Si	0.90	0.89	0.93	0.82	0.80	0.80	0.19
H_2O^-	6.10	1.67	0.10	0.12	0.09	0.20	0.08	Al/Si	0.085	0.085	0.107	0.066	0.064	0.062	0.29
Fe^0		0.14	0.16	15.98	7.03	2.44	0.13	Ca/Si	0.092	0.100	0.118	0.073	0.071	0.072	0.325
Ni			0.29	1.74	1.24	1.07	0.01	CaTi/Si	0.004	0.006	0.006	0.004	0.004	0.004	0.0019
Co			0.01	0.08	0.06	0.05	0.00	$\Sigma Fe/Si$	1.78	1.60	1.48	1.60	1.18	1.03	0.66
FeS	9.08	5.76	4.05	5.43	5.76	5.79	0.14	$\Sigma Fe/Ni$	18.12	16.15	16.85	15.84	17.73	18.64	
C	2.80	1.82	0.43	0.11	0.12	0.22	0.00	Fe^0/Ni			9.21	5.67	2.29		
S (elem)	0.10							$Fe^0/\Sigma Fe$			0.58	0.32	0.12		

[a] ΣFe includes all iron in the meteorite whether existing in metal (Fe^0), FeS, or in silicates as Fe^{2+} (FeO) or Fe^{3+} (Fe_2O_3). The symbol H_2O^- indicates loosely bound (adsorbed?) water removable by heating to 110°C: H_2O^+ indicates chemically bound .water that can be lost only above 110°C. (Data courtesy of Dr. E. Jarosewich, Smithsonian Institution.)

groups. The total iron in some enstatite (E) chondrites exceeds that in the H group of ordinary chondrites, denoting them as EH chondrites; the EL chondrite designation is self-evident.

Achondrites, formed at high-temperatures, contain essentially no metal or sulfide and are enriched in refractory lithophiles (cf. Table 3), which, with their constituent minerals, allow classification into specific groups (Fig. 7a). Most groups are named for a specific prototypical meteorite; others—howardites, eucrites, and diogenites (HED meteorites)—were named nonsystematically. At least 10 achondrite groups can be distinguished from their oxidized iron and calcium contents (FeO and CaO). Some apparently were associated in the same parent body but derive from different regions: the HED and the SNC (Shergottites–Nakhlites–Chassigny) associations. The HED meteorites are thought to come from 4 Vesta, and/or other V class asteroids produced from it. The consensus that the 32 SNC meteorites come from Mars is so strong, that these are often called martian meteorites, not SNCs.

2.3 Oxygen Isotopics and Interpretation

Meteorites "map" the solar system by isotopic composition of oxygen (Fig. 11), a major element in all but the irons. Because its high chemical reactivity causes oxygen to form numerous compounds, it exists in many meteoritic minerals, even in silicate inclusions in iron meteorites. In standard references, such as the Chart of the Nuclides, the terrestrial composition of its three stable (i.e., nonradioactive) isotopes is given as 99.756% ^{16}O, 0.039% ^{17}O, and 0.205% ^{18}O. In fact, any physical or chemical reaction alters its isotopic composition slightly by mass-fractionation. Since the mass difference between ^{16}O and ^{18}O is twice that existing between ^{16}O and ^{17}O, a mass-dependent reaction (e.g., physical changes and most chemical reactions) increases or decreases the $^{18}O/^{16}O$ ratio by some amount and will alter the $^{17}O/^{16}O$ ratio in the same direction, but by half as much. Accordingly, in a plot of $^{17}O/^{16}O$ vs. $^{18}O/^{16}O$ or units derived from these ratios (i.e., $\delta^{17}O$ and $\delta^{18}O$; cf. Fig. 11 caption), all mass-fractionated samples derived by chemical or physical processes from an oxygen reservoir with a fixed initial isotopic composition will lie along a line of slope ~1/2.

Data from terrestrial samples define the Terrestrial Fractionation Line (TFL) in Fig. 11, whose axes are like those described earlier, but normalized to a terrestrial reference material, Standard Mean Ocean Water (SMOW). Not only do all terrestrial data lie along the TFL line, but so too do the oxygen isotopic compositions of lunar samples, which occupy a small part of it. The single Earth–Moon line (defined by data covering the solid line's full length) suggests that both bodies sampled a common oxygen isotopic reservoir, thus supporting the idea that the Moon's matter spun

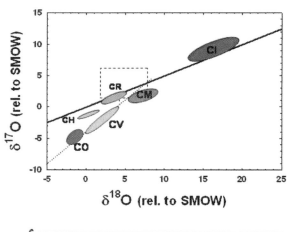

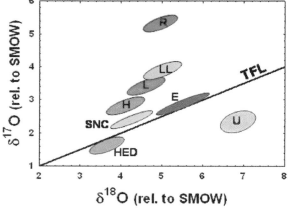

FIGURE 11   Relation between oxygen isotopic compositions in whole-rock and separated mineral samples from the Earth, Moon, and various meteorite classes. Units, $\delta^{17}O$ (‰) and $\delta^{18}O$ (‰), are those used by mass spectrometrists and are, in effect, $^{17}O/^{16}O$ and $^{18}O/^{16}O$ ratios, respectively. Both $\delta^{17}O$ (‰) and $\delta^{18}O$ (‰) are referenced to SMOW. Oxygen isotopic compositions for carbonaceous chondrites are much more variable than for other meteorite classes (dashed box in the upper part expanded in the lower one).

off during the massive impact of a Mars-sized projectile with a proto-Earth (see relevant chapters).

One important feature of Fig. 11 is that many chondrite and achondrite groups defined by major element composition and mineralogy (e.g., Figs. 7a and 7b) occupy their own regions in oxygen isotope space. These data suggest that at least eight major chondritic groups (H, L, LL, CH, CI, CM, CR, and E) and a minor one (R), acapulcoites and brachinites, the two achondrite associations (SNC and HED), ureilites (U) and the silicate inclusions in group IAB iron meteorites derive from different "batches" of nebular material. The HED region also includes data for most pallasites and many mesosiderites suggesting derivation from a common parent body. Extension of the HED region by a line with slope 1/2 passes through the isotopic region of the oxygen-containing silicate inclusions from IIIAB irons, suggesting that they, too, may be related to the HED association. Perhaps these irons come from deeper in the HED parent body, but this would imply more complete disruption than V-class asteroids (e.g., 4 Vesta) exhibit. Even though oxygen isotopic compositions of the rare angrites and brachinites resemble those of the HED association, differences in other properties weaken the connection. Other possible links indicating common nebular reservoirs (based upon limited oxygen isotopic data) are silicate inclusions in IIE irons with H chondrites, silicates in IVA irons with L or LL chondrites, aubrites with E chondrites, winonaites (primitive meteorites modified at high-temperatures) with silicates from IAB and IIICD irons, and the very rare, highly-metamorphosed—even melted—primitive acapulcoites and lodranites.

One interpretation of Fig. 11 is that the solar system was isotopically inhomogeneous because each batch of nebular matter seems to have its characteristic oxygen isotopic composition. Isotopic homogenization of gases is more facile than is chemical homogenization so that the isotopic inhomogeneity demonstrated by Fig. 11 implies that the solar system condensed and accreted from a chemically inhomogeneous presolar nebula (Fig. 2).

The other important feature to be noted from Fig. 11 is the "carbonaceous chondrite anhydrous minerals line," with slope near 1. A feature distinguishing C1 and C2 chondrites (Section 2.4.4.1) from all others (cf. Fig. 7b) is evidence for preterrestrial aqueous alteration or hydrolysis of some phases in them. (Evidence for hydrous alteration of minerals is also observed in some unequilibrated ordinary chondrites.) Anhydrous minerals (including CAI) in carbonaceous chondrites were seemingly never exposed to water so that these chondrites are regarded as a mixture of materials with different histories. As seen from Fig. 11, oxygen isotopic compositions of anhydrous minerals in CM, CV, and CO chondrites are consistent with a line defined by CAI whose slope cannot reflect the mass-fractionation process indicated by a slope 1/2 line like TFL. Instead, the anhydrous minerals line seems to represent a mixture of two end members (batches of nebular material), which, at the ^{16}O-rich (i.e., low ^{17}O, ^{18}O) end lie at or beyond the CO region. Ureilite oxygen isotopic compositions lie on an anhydrous minerals line near CM, suggesting a link. These achondrites contain carbon (as graphite-diamond mixtures) in amounts intermediate to those of CV or CO chondrites and CM. Ureilite data do not indicate formation by differentiation of material with uniform oxygen isotopic composition. Rather, ureilite formation may reflect carbonaceous chondrite-like components mixed in various proportions.

As originally interpreted, the anhydrous minerals line represented a mixture of nebular material containing pure ^{16}O with others higher in ^{17}O and ^{18}O. If so, the former reflected a unique nucleosynthetic history, perhaps material condensed from an expanding, He- and C-burning supernova shell. Subsequently, photochemical reactions of molecular oxygen with a given isotopic composition were shown to yield oxygen molecules with isotopic composition defining a slope 1 line as in Fig. 11.

Which process—nebular or photochemical—produced the trends in Fig. 11 is unknown. Even so, Fig. 11 still serves to link meteorites or groups of them produced from one batch of solar system matter. Moreover, the position of any sample(s) could reflect some combination of the mass-fractionated and mixing (slope 1) lines. For example, primary matter that ultimately yielded L chondrites (or any ordinary chondrite group) and HED meteorites could have had a single initial composition, subsequently mass-fractionated and/or mixed or reacted photochemically to produce meteorite groups with very different oxygen isotopic compositions. However, suitable meteorites with intermediate oxygen isotopic compositions are unknown.

2.4 Chondrites

The available data suggest that heat sources for melting primitive bodies (presumably compositionally chondritic) that formed differentiated meteorites were within rather than external to parent bodies. Important sources no doubt include radioactive heating from radionuclides—both extant (^{40}K, ^{232}Th, ^{235}U, and ^{238}U) and extinct (e.g., ^{26}Al)—which were more abundant in the early solar system, and impact heating. Calculations show that ^{26}Al was important in heating small (a few kilometers) primitive parents; other heat sources were effective in differentiating larger ones. Electrical inductive heating driven by dense plasma outflow along strong magnetic lines of force associated with the very early, pre-main-sequence (T-Tauri stage) Sun is possible but not proven.

2.4.1 PETROGRAPHIC PROPERTIES

Major element and/or oxygen isotope data demonstrate that differences between parent materials of chondrites of the various chemical groups (e.g., H, CM or EH) are of primary nebular—preaccretionary—origin. Parent body differentiation, on the other hand is secondary (postaccretionary). Such heating does not necessarily melt the entire parent body, and it is thus reasonable to expect an intermediate region between the primitive surface and the molten differentiated interior. Properties of many chondrites support this expectation and suggest that solid-state alteration of primary chondritic parent material (similar to type 3 chondrites) occurred during secondary heating. Eight characteristics observed during petrographic study of optically thin sections (Fig. 12) serve to estimate the degree of thermal metamorphism experienced by a chondrite and to categorize it into the major 3–6 types (Table 4). The absence of chondrules and the presence of abnormally large

(a)

(b)

(c)

FIGURE 12   Petrographic (2.5-mm-wide) thin sections in polarized transmitted light. Partial large chondrules are obvious in the H3 chondrite Sharps (a) but barely recognizable in the H6 chondrite Kernouve (b); Nakhla (c) is of martian origin. (Photos courtesy of Dr. Robert Hutchison, Natural History Museum, London.)

(>100 μm) feldspar characterize very rare type 7. These pigeonholes approximate a chondritic thermal metamorphic continuum. Petrographic properties (with bulk carbon and water contents) suggest increasing aqueous alteration of type 3 material into types 2 and 1.

Two of these characteristics are illustrated in Fig. 12: the opaque matrix and distinct chondrules of the type 3 chondrite Sharps (Fig. 12a) should be contrasted with the recrystallized matrix and poorly defined chondrules of extensively metamorphosed (type 6) Kernouve (Fig. 12b). Chemically, Fe^{2+} contents of the ferromagnesian silicates—olivine and pyroxene (Table 2)—are almost completely random in a chondrite like Sharps and quite uniform in

one like Kernouve. Chondrites of higher numerical types could acquire their petrographic characteristics (Table 4) by extended thermal metamorphism of a more primitive (i.e., lower type) chondrite of the same chemical group. Temperature ranges estimated for formation of types 3–7 are 300–600, 600–700, 700–750, 750–950, and >950°C, respectively.

The petrography of achondrites, like the martian meteorite Nakhla, clearly indicates igneous processes in parent bodies at temperatures ≫1000°C. The resultant melting and differentiation erased all textural characteristics of the presumed chondritic precursor (Fig. 12c) so its nature can only be inferred.

TABLE 4 Definitions of Chondrite Petrographic Types[a]

Uniform	1	2	3	4	5	6
(i) Homogeneity of olivine and pyroxene compositions	—	>5% mean deviations		>5% mean deviations to uniform	Uniform	
(ii) Structural state of low-Ca pyroxene	—	Predominantly monoclinic		Monoclinic >20%	Monoclinic <20%	Orthorhombic
(iii) Degree of development of secondary feldspar	—	Absent		<2 μm grains	<50 μm grains	50- to 100-μm grains
(iv) Igneous glass	—	Clear and isotropic primary glass; variable abundance		Turbid if present	Absent	
(v) Metallic minerals (maximum Ni content)	—	(<20%) Taenite absent or very minor	Kamacite and taenite present (>20%)			
(vi) Sulfide minerals (average Ni content)	—	>0.5%	<0.5%			
(vii) Overall texture	No chondrules	Very sharply defined chondrules		Well-defined chondrules	Chondrules readily delineated	Poorly defined chondrules
(viii) Texture of matrix	All fine-grained, opaque	Much opaque matrix	Opaque matrix	Transparent microcrystalline matrix	Recrystallized matrix	
(ix) Bulk carbon content	~3.5%	1.5–2.8%	0.1–1.1%	<0.2%		
(x) Bulk water content	~6%	3–11%	<2%			

[a]The strength of the vertical line is intended to reflect the sharpness of the type boundaries. A few ordinary chondrites of petrographic type 7 have abnormally large (≥100 μm) feldspar and no chondrules: These have been interpreted as reflecting higher metamorphic temperatures than those associated with type 6. Water contents do not include loosely bound, that is, terrestrial water (modified from Table 1.1.4 in "Meteorites and the Early Solar System").

Chemical changes involving loss of a constituent, like carbon or water in chondrites, require an open system; other changes in Table 4 could occur in open or closed systems. We emphasize that thermal metamorphism can only affect secondary (parent body) characteristics—those listed horizontally in Table 4—not primary ones. Postaccretionary processes by which H chondrite-like material can form from L or vice versa are unknown.

2.4.2 CHEMICAL–PETROLOGIC CLASSIFICATION

Because properties of a given chondrite reflect both its primary and its subsequent histories, a chondritic classification scheme reflecting both is used. Chondrites already mentioned are Ensisheim, LL6; Nogata, L6; Sharps, H3 (Fig. 12a); Sylacauga, H4; and Kernouve and Peekskill, H6 (Fig. 12b). No ordinary (or enstatite) type 1 or 2 chondrite is known. Type 3 ordinary chondrites, the unequilibrated ordinary chondrites (UOC), vary the most among themselves and from chondrites of other petrographic types. Within UOC, a variety of properties—for example, the chemical heterogeneity of ferromagnesian silicates, the contents of highly elements (mainly noble gases), and thermoluminescence (TL) sensitivity—subdivide UOC into subtypes 3.0 to 3.9. Sharps (Fig. 12a) is the most primitive H chondrite known, being an H3.0 or H3.4, depending on the classification criteria used. (A similar subclassification of CO chondrites also exists.)

Many properties of ordinary chondrites demonstrate that each group has its special history, even in something as simple as the numbers of each chemical-petrographic type (Table 1). For example, proportions of H3 or L3 are low, 1–2% (5 of 316 H and 7 of 350 L chondrite falls), whereas 13% of LL falls (9 of 72) are LL3. Proportions of more evolved chondrites also differ (Table 1). The plurality of H falls are H5 (138 of 316 or 44%) while type 6 dominates L and LL chondrites (239 of 350 or 68% and 35 of 72 or 49%, respectively). Non-desert-cluster chondrite finds generally exhibit similar trends. Stony-iron and, especially, iron finds are very numerous because they are obviously "strange," hence more likely to be brought to someone knowledgeable enough to identify them as meteoritic. Achondrites grossly resemble terrestrial igneous rocks and are less likely to be picked up: Only their fusion crust permits ready recognition of their exotic origin (Table 1).

2.4.3 BRECCIAS

Even though most chondrites are readily pigeonholed, a few consist of two or more meteorite types, each readily identifiable in the lithified breccia. Noblesville, for example, consists of light H6 clasts embedded in dark H4 matrix (Fig. 1a). Such an assemblage—two petrographic types of the same chondritic chemical group—is a genomict breccia. A polymict breccia contains two or more chemically distinct meteorite types, implying the mixing of materials from 2 (or more) parent bodies, each with its own history. The most striking such case is Cumberland Falls where black forsterite chondrite inclusions as large as 3 cm × 5 cm are embedded in an 8 cm × 11 cm white enstatite achondrite.

Of the other sorts of breccias, perhaps the most important is the regolith breccia. Noblesville (Fig. 1a) is such a meteorite, and its typically dark and fine-grained matrix contains large quantities of light noble gases—He and Ne—of solar origin (cf. Section 5.1). In addition to these gases, radiation damage in present as solar-flare tracks (linear solid-state dislocations) in a 10-nm-thick rim on the myriad matrix crystals. However, solar gases and flare tracks are absent in the larger, lighter-colored clasts of regolith breccias. Clearly, dark matrix is lithified fine dust originally dispersed on the very surface of the meter-thick regolith or fragmental rocky debris layer produced by repeated impacts on bodies with no protective atmosphere. (The lunar regolith is both thicker, ~1 km, and more mature and gardened, or better mixed by impacts than are asteroidal regoliths.) This dust acquired its gas- and track-component from particles with keV/nucleon energies streaming outward as solar wind or solar flares with MeV energies [see THE SOLAR WIND] so that the dust sampled the solar photospheric composition. The irradiated dust, often quite rich in volatile trace elements from another source, was mixed with coarser, unirradiated pebble-like material and formed into a breccia by mild impacts that did not heat or degas the breccia to any great extent. Regolith breccias occur in many meteoritic types but are especially encountered as R (and H) chondrites, aubrites, and howardites.

2.4.4 CARBONACEOUS CHONDRITES

2.4.4.1 Composition

The only type 1 or 2 chondrites are carbonaceous chondrites, of which nearly all non-Antarctic ones are observed falls. A dominant process recorded in them involves hydrolysis, the action of liquid water (in the nebula or on parent bodies) that altered preexisting grains, producing various hydrated, clay-like minerals. The chondrites' petrography and the decidedly nonterrestrial $^2H/^1H$ ratios in water from them show that this hydrolysis was preterrestrial. As noted earlier, oxygen isotopic compositions of hydrated minerals demonstrate that the two groups derive from different batches of nebular matter; thus, C1 (or CI) could not form C2 (or CM) by thermal metamorphism nor could C1 have formed by hydrolysis of C2 parent material. For this reason, some specialists prefer the CM designation: others prefer a hybrid classification like C2M or CM2 because other C2-like chondrites exist. Tagish Lake, although very primitive, is unique.

C1 chondrites contain *no* chondrules (Table 4), but their obvious compositional and mineralogic similarities to chondrule-containing meteorites prompt this classification. Compositionally, C1 (or CI) chondrites closely resemble the solar photosphere, the correlation between abundances in the solar photosphere and C1 chondrites exist over 10 orders of magnitude (10 billion). A few differences exist: Elements depleted in C1 chondrites relative to the Sun's surface (e.g., hydrogen, helium, or carbon) are gaseous or easily form volatile compounds that largely remained as vapor in the nebular region where C1 chondrite parent material condensed and accreted. Other elements (e.g., lithium, beryllium, and boron) are easily destroyed by low-temperature nuclear reactions during pre-main-sequence stellar evolution; consequently, they are depleted in the solar photosphere relative to C1 chondrites.

Because chemical analysis of C1 chondrites (or any planetary material) on Earth is more precise and accurate than is spectral analysis of the solar photosphere, "cosmic abundance" tables of chemical and isotopic data for most elements mainly derive from C1 chondrite analyses. Generally these data are used to estimate our solar system's composition. Only where such processes, as incomplete nebular condensation, are suspected do such compilations adopt solar photospheric values. Recall, however, that earlier we inferred chemical heterogeneity of the pre-solar nebula existed (Fig. 2), so cosmic abundances may not have been the same in all nebular regions.

2.4.4.2 Organic Constituents

Although chondrites are depleted in carbon, hydrogen, and nitrogen relative to the solar photosphere, C1 and, to a lesser extent, C2 chondrites contain large amounts of organic matter (Table 3). They are visible in situ (as globules) only in the unique carbonaceous chondrite Tagish Lake (Fig. 3). Over 400 different organic (C-based) molecules of very different types are known mainly in C1 and C2 chondrites, but their concentrations are very low. Molecular characteristics demonstrate that many are preterrestrial, but the problem of terrestrial contamination is ever-present.

Polycyclic aromatic hydrocarbons (PAH) were found inside 2 martian meteorites, but not near their surfaces, suggesting that the PAH are not terrestrial contaminants but, rather, originated on Mars. Particles identified as microfossils were reported in at least one martian meteorite and, decades earlier, in CI and CM chondrites. Some advocate biogenic formation of these, but their arguments fail to alter the consensus view that meteoritic organics formed abiogenically.

Since many organic compounds in meteorites can be altered or destroyed by even brief exposure to temperatures of 200–300°C, their presence in meteorites constitute a thermometer for postaccretionary heating during metamorphism, shock, or atmospheric transit.

2.4.5 SHOCK

A meteorite parent body cannot be disrupted by internal processes, but only by collision with another similarly sized object. Accordingly, many meteorites evidence exposure to significant shock. A few decades ago, chondrites were qualitatively classed "shocked" if the hand-specimen interior exhibited blackening, veining, or brecciation. Now, petrographic and mineralogic characteristics provide a semiquantitative estimate of the shock-exposure level. Such characteristics reflect changes induced directly, by the peak pressure wave, or indirectly, by the shock-associated, high residual temperature. Specific shock-pressure indicators ("shock barometers") have been calibrated against characteristics produced by laboratory shock-loading experiments. Using these criteria, the degree of shock-loading is known for almost 4300 ordinary chondrites (Table 5).

The current scheme to estimate shock histories of equilibrated ordinary chondrites involves the addition of S1, S2, ..., S6 to its chemical-petrographic classification. The peak shock pressures at the transitions are <5 GPa, S1/S2; 5–10 GPa, S2/S3; 15–20 GPa, S3/S4; 30–35 GPa, S4/S5; 45–55 GPa, S5/S6. Whole-rock melting and formation of impact melt rocks or melt breccias occur at 75–90 GPa. Thus, the Noblesville H4 regolith breccia (Fig. 1a) is S1 as a whole with some H6 clasts being S2. Other chondritic compositional data (radiogenic gases and thermally mobile trace elements, i.e., easily volatilized and lost in an open system) also give information on shock histories (cf. Sections 5.3 and 6.3). Equilibrated L chondrites exhibit the highest proportion of heavily shocked chondrites, almost half having been shocked above 20 GPa. Lesser, but significant, proportions of H and LL chondrites show substantial degrees of shock loading (Table 5). The only C

TABLE 5	Degrees of Shock Loading in Ordinary Chondrites as a Function of Specific Chemical-Petrologic Type						
Type	Total	S1	S2	S3	S4	S5	S6
H3	229	53	119	41	14	2	—
H4	511	91	257	125	33	4	1
H5	1021	94	444	403	75	3	2
H6	549	77	159	234	65	12	2
L3	53	9	23	12	9	—	—
L4	206	18	71	69	37	3	8
L5	412	23	108	151	99	19	12
L6	972	28	121	319	333	104	67
L7	5	—	—	1	2	2	—
LL3	50	6	27	10	6	1	—
LL4	49	4	27	16	1	1	—
LL5	83	7	42	21	9	4	—
LL6	143	2	44	55	33	6	3

TABLE 6	Extinct Radionuclides Whose Decay Products Are Detected in Meteorites		
Radio-nuclide	Half-life (Ma)	Daughter	Initial Ratio
^{41}Ca	0.13	^{41}K	^{41}Ca/^{40}Ca $= 1.4 \times 10^{-8}$
^{26}Al	0.7	^{26}Mg	^{26}Al/^{27}Al $= 5 \times 10^{-5}$
^{10}Be	1.5	^{10}Be	^{10}Be/^{9}Be $= 9.5 \times 10^{-4}$
^{60}Fe	1.5	^{60}Ni	^{60}Fe/^{56}Fe $= \sim6 \times 10^{-8}$
^{53}Mn	3.7	^{53}Cr	^{53}Mn/^{55}Mn $= 4.4 \times 10^{-5}$
^{107}Pd	6.5	^{107}Ag	^{107}Pd/^{108}Pd $= 4 \times 10^{-4}$
^{182}Hf	9	^{182}W	^{182}Hf/^{180}Hf $= 2 \times 10^{-4}$
^{129}I	16	^{129}Xe	^{129}I/^{127}I $= 1.4 \times 10^{-4}$
^{244}Pu	82	^{131}Xe–^{136}Xe	^{244}Pu/^{238}U $= \sim5 \times 10^{-3}$

and E chondrites exhibiting evidence for unusually strong shock (i.e., to S5) are a CK5 and an EL3.

The mineralogy (really, metallography) of iron meteorites is relatively simple and their shock classification is easy at <13, 13–75, and >75 GPa. Some iron meteorites that were shock-loaded at 13–75 GPa were subsequently annealed at 400–500°C for days or weeks, presumably by contact with massive chunks of collisional debris at these or higher temperatures; they are readily identified. About half of all iron meteorites were shocked at >13 GPa, nearly all during collisions that disrupted their parents. Only large meteoroids that formed explosion craters can generate pressures as high as 13 GPa when they hit Earth.

The best preserved, perhaps only, case of strong shock-loading during terrestrial impact involves the Canyon Diablo meteoroid that produced Meteor Crater, Arizona (Fig. 5b). Some Canyon Diablo fragments contain millimeter- to centimeter-sized graphite–diamond aggregates, indicating partial transformation of graphite to diamond. [Highly unequilibrated chondrites contain very tiny (~0.002 μm) vapor-deposited diamond grains that do not have a high-pressure origin; see Section 5.2.1.] These aggregates contain lonsdaleite, a hexagonal diamond polymorph produced, so far as is known, only by shock-transformation of graphite, which also is hexagonal. Diamond-containing Canyon Diablo specimens always show metallographic evidence for exposure to shock >13 GPa; are mainly on the crater rim, not in the surrounding plain; and contain low levels of cosmogenic stable nuclides and radionuclides indicating derivation from the interior near the front of the impacting meteoroid where the greater explosive shock existed. The mutual correlations between degree of shock-loading, depth in the impacting meteoroid, and geographic locations around Meteor Crater, argue that strongly shocked Canyon Diablo specimens experienced this during terrestrial impact.

The percentages of strongly shocked (i.e., >13 GPa) members of iron meteorite chemical groups differ widely. The IIIAB irons constitute the plurality of all known iron meteorites, and have virtually all been shocked preterrestrially in the 13–75 GPa range. Nearly 60% of IVA irons, the next largest group, show such shock. A similar proportion of IIB iron meteorites have been shocked at >13 GPa, but this group is small. No other chemical group of iron meteorites shows an especially high proportion of shocked members. Shock-loading experiments show that pressures of 13–75 GPa acting on metallic Fe impart a free-surface velocity of 1–3 km/s. This shock-impulse was important, maybe essential, in bringing large numbers of strongly shocked meteorites to Earth. The parent bodies of the IIIAB and IVA irons may have been in the Asteroid Belt: the shock-impulse could produce ejecta with more elliptical, perhaps Mars-crossing orbits. Mars could, with time, gravitationally perturb these fragments into Earth-crossing orbits.

Semiquantitative petrographic shock indicators in basaltic achondrites (i.e., mainly the HED association and shergottites) suggest a 6-stage shock scale corresponding to <5, 5–20, 20–45, 45–60, 60–80, and >80 GPa pressures. The full range is seen in HED samples—primarily in clasts in howardites, a solar gas-rich group (Section 5.1), that are mainly polymict breccias containing eucrite and diogenite fragments. These and other data, mainly compositional, suggest that howardites are a shock-produced, near-surface mixture of two deeper eucritic and diogenitic igneous layers in the HED parent body.

Nearly all ureilites show petrographic evidence for very substantial shock. Most also contain large graphite-diamond aggregates generally believed to have formed during preterrestrial impacts.

Lunar meteoroids, which were ejected by impacts somewhere on the 95% of the Moon's surface not sampled by the Apollo or Luna programs, are breccias in most cases (Section 4.1). Otherwise, they show no unusual evidence for shock greater than that evident in rocks returned by these programs.

Shergottites have been heavily shocked, in keeping with their accepted derivation from a massive object, like Mars with its 5-km/s escape velocity. Nakhlites, linked to shergottites by oxygen isotopic compositions (Fig. 11) and other properties, are less shocked (Section 4.2).

Metallic portions of few stony-irons indicate strong shock: No pallasites and only 3 of 18 mesosiderites were shocked >13 GPa. Somehow, parent bodies of the stony-irons and half of the iron meteorites were disrupted, and the meteoroids were excavated from appreciable depth without subjecting them to major shock-loading. More puzzling is the fact that silicate portions of mesosiderites contain much shocked material. Apparently, these stony-irons formed by intrusion of shock-loaded silicate into or onto preexisting, generally unshocked metal, possibly after excavation from parent body interiors.

3. Meteorites of Asteroidal Origin and their Parent Bodies

3.1 The Meteorite–Asteroid Connection

Two links have already been noted that suggest or imply an asteroidal origin for most meteorites. These are:

1. Photographically determined orbits for seven ordinary chondrites, one unique carbonaceous one and an EL6 (Fig. 3).

2. Mineralogic evidence indicating origin of most meteorites in asteroidal-sized objects. (Some chondrites could come from much smaller primary objects.) This evidence includes iron meteorite cooling rates (implying formation depths of asteroidal dimensions), the presence of minerals (e.g., tridymite), and phase relations (e.g., the Widmanstätten pattern) indicative of low-pressure ($\ll 1$ GPa) origin, and the absence of any mineral indicating high *lithostatic* (generated by the rocky overburden)—rather than *shock*-pressures.

Another property linking meteorites and certain asteroidal types, spectral reflectance, is a research area of strong current interest. The reflectivity (albedo)-wavelength variation for an asteroid, involving white (solar) incident light, can characterize its mineralogy and mineral chemistry somewhat [*see* MAIN-BELT ASTEROIDS]. To uncover possible links, asteroidal spectral reflectance can be compared with possible meteoritic candidates, both as-recovered or treated in the laboratory to simulate effects of extraterrestrial processes (Fig. 13).

The best matches exist between the HED association and rare V-class asteroids (4 Vesta and its smaller progeny), iron meteorites and numerous M-class asteroids; CI and CM chondrites thermally metamorphosed at temperatures up to 900°C with the very numerous C-class and apparently related B-, F-, and G-class asteroids; aubrites with the somewhat unusual E-class asteroids; pallasites with a few of the very abundant, and diverse S-class—which constitute a plurality of all classified asteroids—and/or rare A-type asteroids; and ordinary chondrites with the very rare Q-type asteroids, which are near-Earth asteroids, or 6 Hebe, an inner Belt object belonging to the S(IV) subclass of S asteroids. A typical good news/bad news situation results. The good news is that specific meteorite types are similar to (derive from) surface regions of identifiable asteroid types. The bad news is that relative frequencies with which meteorites of a given type and asteroids of a supposedly similar type are encountered do not agree. Specifically, there is the ordinary chondrite-S asteroid paradox (cf. Table 1): Why are there so few asteroidal candidates for the very numerous ordinary chondrites and so few olivine-dominated stony-irons from the very numerous S asteroids? One obvious answer is that "space weathering" (energetic dust impingement on a meteoroid surface causing metal reduction and dispersion)

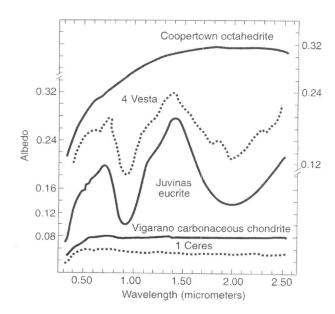

FIGURE 13 Spectral reflectances of the Coopertown IIIE coarse octahedrite, Juvinas eucrite and V-class asteroid 4 Vesta, and Vigarano C3V chondrite and G-class asteroid 1 Ceres. The albedo scale for all but Coopertown is on the left: Coopertown's is at right. Solid and dashed lines delineate meteorite and asteroid spectra, respectively. (Courtesy of Dr. Lucy-Ann McFadden, University of Maryland.)

could mask ordinary chondrite-like interiors. Another, is that Earth collects a biased meteorite sampling compared with the asteroid population, in either near-Earth space or in the Asteroid Belt. This may also account for the near absence of meteorites from the numerous D or P asteroids, located at >3 AU from the Sun. Alternatively, ejecta from such asteroids might not survive atmospheric passage because D and P surface materials are inferred to be very organic-rich and, presumably, very friable. Tagish Lake, the only meteorite that spectrally resembles a D-class asteroid, contains organic globules and is extremely friable.

3.2 Sampling Bias

The contemporary flux of meteorites is biased and unrepresentative of the meteoroid population in near-Earth space, let alone in the Asteroid Belt, so generalizations about parent body formation and evolution from studies of meteorites falling today may be incomplete. The contemporary flux of meteorites includes not only observed falls during mainly the past 200 years, but also by non-desert-cluster finds (i.e., omitting the many from hot and cold deserts). Most finds contain metallic iron, which should be readily oxidized on Earth, but they are surprisingly resistant to destructive oxidation, even in temperate climates. The smaller iron–nickel grains of chondrite finds are more readily oxidized.

Meteoritic terrestrial ages are generally based upon decay of cosmogenic radionuclides (Fig. 4). Non-desert-cluster finds have been on Earth for up to ~20 ka, but the oldest one actually dated is the Tamarugal IIIA octahedrite that has a terrestrial age of 3.6 Ma. As is discussed in Section 6.1, terrestrial ages for meteorite cluster finds from hot deserts usually range up to 50 ka: many Antarctic meteorites are much older. A few dozen fossil meteorites found in Ordovician seabed layers in several Swedish quarries have ~480 Ma terrestrial ages (Section 6.4).

The oldest Antarctic meteorite is Lazarev, an Antarctic octahedrite that is not part of any established iron–meteorite chemical group. Its terrestrial age is 5 Ma. Although an Antarctic chondrite has a terrestrial age of ~2 Ma, the more typical terrestrial ages for these are in the 0.1- to 1-Ma range (averaging 0.3 Ma for the population from the Allan Hills, Victoria Land; Section 6.1.1). Conceivably, the meteorite population landing on Earth during that time window could have differed from the contemporary one. The number of iron and stony-iron observed falls is comparable with those from Victoria Land (Table 1); however, Antarctic achondrites and chondrites are more numerous. Additional differences exist in the details. For example, samples from Victoria Land have, on average, smaller masses than do those from more contemporary falls; small samples are readily detected in Antarctica. Meteorites of rare types—like achondrites—are easily recognized, even in hand-specimen, and pieces can be readily paired with others of the same fall. Hence, in the Victoria Land (ANSMET) population, the numbers of different Antarctic achondrites are reliable (Table 1). When small populations are compared, the results are always suspect. We note that the number of aubrites and howardites are comparable but the number of ureilites and lunar meteorites are larger in the Victoria Land population. A difference may exist for C1 chondrites, but they are typically friable and might not survive pulverization in the Antarctic ice sheet. At face value, Antarctic ordinary chondrites seem very numerous, but their pairing uncertainties are particularly serious. Ordinary chondrites differ only subtly from each other—even as falls—so the apparent excess of Antarctic LL chondrites is clouded. Numerous studies of Antarctic meteorites reveal many preterrestrial genetic differences between them and falls, but detailed interpretations of these differences remain controversial.

The 16,500 fragments collected from Queen Maud Land, Antarctica, by Japanese meteorite recovery teams include quite a few fragments of rare or unique meteorite types. These include 6 different lunar meteorites (9 fragments), 4 martian meteorites (6 fragments), 6 thermally metamorphosed (open-system) C1–C3 chondrites, and a unique C1M or C2I chondrite. In general, Queen Maud Land samples have terrestrial ages of up to 0.3 Ma, averaging 0.1 Ma (i.e., intermediate between those of contemporary falls and Antarctic samples from Victoria Land) and are

of smaller mass, on average, than even those from Victoria Land. The Queen Maud Land population is less well characterized than the Victoria Land population, so, except for lunar and martian meteorites, we do not list any of them in Table 1.

4. Meteorites from Larger Bodies

During the early Apollo program, NASA decided to quarantine lunar samples and the astronauts that brought them to prevent contamination of Earth by some hypothetical "Andromeda Strain." This quarantine cost much and proved ineffective. Years before *Apollo 11* (in 1969), E. Anders argued that because lunar escape velocity was so low (2.38 km/s) and shock-induced ejecta velocities were so high, lunar samples must already be on Earth to contaminate us (if they were going to). To eliminate this unnecessary expenditure, he offered to eat the first gram of lunar sample brought by *Apollo 11*. His offer was not accepted: quarantine ended with *Apollo 12*, and the first meteorite recognized as lunar (ALH A81005) was found in Antarctica in 1982. Yamato 791197 was recovered in 1979 but its lunar origin was not recognized then.

Today, 31 lunar meteorites are known as are 32 martian meteorites (Table 1) and NASA plans an expensive quarantine to protect humankind from another hypothetical "Andromeda Strain" if and when they bring martian samples to Earth. One author of this chapter (MEL) offers to eat the first gram of that sample to demonstrate that an expenditure of between $10 million and $1 billion ($1,000,000,000) for quarantine is unwarranted.

These two are the only likely large-body sources for meteorites. Other solar system bodies have escape velocities comparable to that of Mars' (e.g., Mercury, Pluto, and the satellites of giant planets like Jupiter or Saturn), but their distances from Earth and their proximities to much larger objects with greater gravitational attraction makes Earth-capture of ejecta from them virtually impossible. Thus, we need only consider the Moon or Mars as meteorite sources. (For additional information, see the websites listed in the footnote to Table 1).

4.1 Lunar Meteorites

The minerals, textures, chemical compositions, and isotope ratios of these 31 individuals (each between 2 g and 1.8 kg) are similar to those of samples brought to Earth by the Apollo and Luna missions [see THE MOON] and unlike those of terrestrial rocks or martian and other meteorites (Fig. 14). Only their fusion crust differentiates them from Apollo and Luna samples. Most are regolith, fragmental, or melt breccias from the lunar highlands: 7 are Mare basalts, 3 of which include regolith breccias or cumulate clasts. Their cosmic ray exposure ages range up to 10 Ma

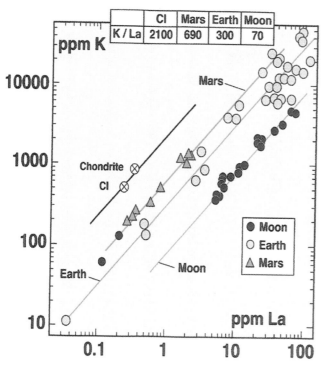

	CI	Mars	Earth	Moon
K / La	2100	690	300	70

FIGURE 14 Lithophile element concentrations (K vs. La) in ordinary and CI chondrites and samples of evolved bodies: lunar samples from various *Apollo* missions and lunar meteorites; terrestrial rocks; and martian meteorites. Data for HED achondrites parallel and lie between the Earth and Moon lines.

but, in principle, could be much less; they probably originated from ~20 impacts forming lunar explosion craters that are a few kilometers in diameter, and possibly as small as 0.5 km.

Lunar meteorites total ~11.2 kg, much less than the 382 kg of Apollo and Luna material, but they provide very important lunar information. Because Apollo and Luna landing sites (all Nearside) were chosen for safety reasons or as geologically interesting but unrepresentative, their regional sampling of the Moon is biased. Lunar meteorites represent random (but unknown) impact sites. Indeed, when compared with lunar spectral reflectance data from the *Clementine* spacecraft, the distribution of FeO contents, KREEP-associated U and Th contents, and, indeed, the highlands nature of lunar meteorites themselves parallel the overall lunar character. One meteorite, NWA 773, samples a Mare basalt region unlike any provided by the Apollo or Luna missions. Much will doubtless be learned about the Moon and its history from these lunar meteorites and others, yet discovered.

4.2 Martian Meteorites

The 32 martian meteorites are unusual igneous meteorites (of five different types), and all but ALH 84001 crystal-lized from parent melts ≤ 1.3 Ga ago (the youngest, 170 Ma ago). This alone suggests a large parent because only a planetary body could retain interior temperatures sufficient to maintain igneous melts that recently. Asteroid-sized objects could have been differentiated early but would have cooled rapidly, crystallizing igneous rocks 4.5 Ga ago. That is the age of ALH 84001, which must be a rare survivor of early martian differentiation. It is, of course, linked by oxygen isotopic composition to the other 31 rocks in the SNC portion of Fig. 11. They are linked to Mars specifically by gases (e.g., ^{20}Ne, ^{36}Ar, ^{40}Ar, ^{84}Kr, ^{131}Xe, N_2, CO_2) in shock-formed glass in EET A79001, the only meteorite showing a contact between two igneous regions. Contents of these gases in EET A79001 match those in the martian atmosphere measured in 1976 by the *Viking* landers. The Martian atmosphere apparently lost more light gases than did the Earth.

Because martian escape velocity is higher than that of the Moon, impacts intense enough to propel Martian meteorites Earthward must be greater, requiring larger explosion craters, 10–100 km in diameter. From cosmic ray histories, the 32 martian meteorites apparently derive from 6–8 events. All solidified near, but below, the martian surface; none were surface samples irradiated by cosmic rays or heavily weathered so that Fe^{2+} is present in them rather than the red Fe^{3+} of martian surface samples. Martian sedimentary rocks and soil may be too friable to survive impact-ejection.

Considering evidence for water flow on Mars' surface, Martian meteorites are surprisingly dry, and evidence for desiccated salts is slight. Curiously, low initial (radiogenic) Sr and Nd isotopic data indicate that parent magmas in the martian mantle were depleted in heat-producing radionuclides (Fig. 14) relative to chondrites (Section 6.4). Shergottites are also depleted in light REE. Not surprisingly, we cannot specify sites from which martian meteorites derive. From crystallization ages, ALH 84001 apparently originated in the old (heavily cratered) southern highlands, while the other 31 come from the young (less-cratered) volcanic northern plains areas.

5. Chemical and Isotopic Constituents of Meteorites

Earlier, we summarized meteorite compositions and genetic processes as necessary to understand general meteoritic properties. Here, we focus upon these topics in greater detail.

5.1 Noble Gases

The chemical inertness of noble gases allows their ready separation from all other chemical elements. Thus, gas mass spectrometers can determine very small noble gas concentrations in a meteorite and, in addition, measure the isotopic composition. Most analyses are carried out on meteorite samples of <100 mg, but, with effort, samples as small as 6 µg provide essential data. By 2004, about 7400 analyses of the light noble gases—He, Ne, and Ar—were reported for all meteorite types.

Noble (rare) gases in meteorites have different origins and each component has a specific isotopic or elemental composition. Some components like the radiogenic gases were produced in situ in meteorites. Radiogenic ^{40}Ar is produced by spontaneous, radioactive decay of long-lived, naturally occurring ^{40}K (half-life, $t_{1/2} = 1.28$ Ga), while ^{4}He is produced similarly from ^{232}Th, ^{235}U, and ^{238}U decay ($t_{1/2} = 14.1$, 0.704, and 4.47 Ga, respectively). Fission Kr and Xe components derive from spontaneous or induced fission of heavy nuclei (e.g., long-lived U isotopes), each with a characteristic fission-fragment distribution. In addition, decay products of extinct radionuclides (e.g., ^{129}I and ^{244}Pu: $t_{1/2} = 5.7$ and 81 Ma, respectively) exist in meteorites (Section 6.5).

Other in situ produced gases are cosmogenic nuclides formed by nuclear reactions of high-energy galactic or solar particles with meteoroids. The specific nuclear reaction depends on the particle energy and the chemical composition of the target material. Nuclear reactions of primary (GeV energies) particles involve initiation of a cascade of secondary particles with smaller energies so that the isotopic or elemental ratios of cosmogenic noble gas isotopes depend on the meteorite's position within the meteoroid and on its size. Cosmogenic nuclides are limited to the surface (<1 m depth) of larger bodies and to meter-sized objects in space. Inert gases found in iron meteorites are mainly cosmogenic, but stony meteorites contain a mix of many components.

Trapped gases include a whole family of noble gas components that were not produced in situ but incorporated in the meteoroid when it formed. Trapped gases are of three main varieties, solar, planetary, and "exotic." Elemental solar gas ratios are similar to those observed in the Sun. Solar gases are introduced into meteoritic mineral grains by direct implantation of solar wind ions or more energetic solar flare particles in the regolith of atmosphere-less surfaces of parent bodies (e.g., the Moon). The planetary noble gas pattern shows a systematic fractionation in which the lightest noble gases—He and Ne—are depleted relative to Ar, Kr, and Xe. Different meteorite types or individual mineral separates have characteristic isotopic and elemental signatures that differ, for example, from those of terrestrial atmospheric noble gases.

Each event depicted in Fig. 2 can, in principle, alter planetary matter mineralogically and/or chemically. To illustrate this qualitatively, consider an element like Ne, whose concentration in meteorites reflects any or all of the events in Fig. 2. As a light noble gas, it forms physical bonds in

meteorites rather than chemical bonds. The three stable Ne isotopes (20, 21, and 22) were created by several stellar nucleosynthetic processes, and a mixture of them was introduced into the presolar nebula with other nucleosynthesized nuclides. Some proportion of this Ne (with its characteristic ^{20}Ne/^{22}Ne and ^{21}Ne/^{22}Ne ratios) was trapped by condensing and accreting nebular material. Presolar grains incorporated into the material, and not subsequently destroyed, contain another component (Ne-E)—pure ^{22}Ne—produced by decay of the now-extinct, radionuclide ^{22}Na ($t_{1/2} = 2.60$ years).

Partial or total Ne degassing accompanied heating of the primitive parent body interior and transformation into a more evolved form, with the Ne components escaping into space or being redeposited into cooler parent body material nearer the surface. Fine-grained matter on the parent body's surface could acquire solar wind and solar flare Ne, which has distinct isotopic compositions that are implanted in regolith (Section 2.4.3). Impacts repeatedly churned ("gardened") the regolith so that a multisource Ne mixture (cosmogenic, solar, trapped) could be present in any sample. Finally, an impact occurred that removed a meter-sized meteoroid, thus starting the CRE "clock" that accumulated a new batch of cosmogenic Ne and other nuclides, including radionuclides.

Because a meteorite can sample Ne from any or all of these sources, its isotopic composition represents a weighted average of the isotopic compositions of its component sources. These can be recognized on a three-isotope plot (Fig. 15). A sample consisting of essentially one component is represented by one point in such a diagram, while a neon mix of two components will lie on a line connecting the isotopic compositions of these components. Included in Fig. 15, as an example, is the Ne isotopic composition of samples of the meteoritic breccia ALH 85151, which contains both solar and cosmogenic gas. Lunar soils also contain solar Ne, but this is a mixture of Ne from the low-energy solar wind and from more energetic solar particles, each differing in isotopic composition. The solar Ne isotopic composition extrapolated from the ALH 85151 data lies almost midway between the Ne isotopic components from these two solar sources. Addition of Ne from other sources, like Ne-E, can complicate this picture. A mixture of three Ne components will fall within a triangle whose apexes each have the Ne isotopic composition of a pure component. In addition, many chondrites contain one or more trapped Ne components, examples of which are in the Fig. 15 inset.

A similar picture can be drawn for Kr and Xe with many isotopes and several possibilities for three-isotope plots, but He and Ar have additional individual complications. For He, only two stable isotopes exist—^{3}He and ^{4}He—so no three-isotope plot is possible. Furthermore, an additional monoisotopic component, radiogenic ^{4}He, can exist in meteorites (see Section 6.3). Argon differs from He in having three stable isotopes. In most stony

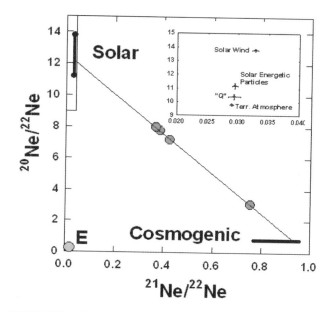

FIGURE 15   Three-isotope plot for stable Ne isotopes in ALH 85151 chondrite. Data for separated mineral grains lie on lines connecting average Solar Ne (composed of keV/nucleon solar wind and MeV solar energetic flare particles) with cosmogenic Ne produced by GeV galactic cosmic rays. Data points in the upper left of the line represent fine Ne in dust grains exposed on a regolith surface with constituent Ne being almost entirely solar. The point at the lower right is from grain interiors (with low surface-to-volume ratios) whose Ne is nearly all cosmogenic. The box in the upper left is expanded in the inset to show isotopic compositions of individual Ne components in meteorites with low $^{21}Ne/^{22}Ne$ ratios. Pure ^{22}Ne (so-called Ne-E) is formed by radioactive decay of very short-lived (2.60 years) ^{22}Na in the protoplanetary nebula or presolar grains. If ALH 85151 contained substantial Ne-E, data points would lie in the triangular region defined by E, solar and cosmogenic Ne. The inset depicts isotopic compositions of Ne from solar wind, solar energetic particles, the terrestrial atmosphere, and an absorbed presolar, planetary component (Q) that is released when mineral grains are etched with nitric acid.

meteorites, ^{40}Ar is mainly radiogenic, deriving from decay of ^{40}K ($t_{1/2} = 1.28$ Ga). This monoisotopic ^{40}Ar component limits the use of three-isotope plots for interpreting the trapped Ar component.

Krypton and Xe systematics are complicated for several reasons. The Kr and Xe isotopes derive from several nucleosynthetic sources, two of which are especially important. One is now-extinct ^{129}I, which decayed to produce ^{129}Xe and gives chronometric information (cf. Section 6.5). The second involved fission of now-extinct ^{244}Pu which produced a Xe component with a characteristic fission–yield curve. In addition to induced and spontaneous U fission products, different trapped components exist. Kr and Xe in presolar grains provide almost pure gas from individual nucleosynthetic events.

Each solar system body has its particular formation history and, thus, its own noble gas isotopic "fingerprint." Gases on the Earth, Moon, Venus, and Mars can be distinguished from each other and from those in chondrites. As discussed in Section 4.2, glass in the EET A79001 shergottite contains martian atmospheric gas indicating that it (and the other 31 martian meteorites) formed there.

5.2 Noble Gas Components and Mineral Sites

Our brief Ne discussion outlined, in principle, how to disentangle several Ne components from an average meteoritic datum. Actually, the situation is more complicated because each "component" may, in fact, be resolvable into constituents from specific sources, each with reproducible isotopic patterns involving more than one noble gas. Ingenious laboratory treatments can yield a phase enriched in one true gaseous constituent from others. These include investigation of individual grains, selective acid dissolution of specific minerals, enrichment by mineral density using heavy liquids, stepwise heating and mass-analysis of gases evolved in some temperature interval, or some combination of these steps (and others).

5.2.1 INTERSTELLAR GRAINS IN METEORITES

Until about 1970, the solar system was considered "isotopically homogeneous," objects in it having formed from a well-mixed and chemically and isotopically homogenized primordial nebula. (The later discovery of oxygen isotopic variations, e.g. Fig. 11, disproved this.) However, even then, rare samples extracted from meteorites exhibited anomalous contents of, for example, Ne or Xe isotopes. These anomalies cannot be explained by well-established processes like decay of naturally occurring radionuclides, cosmic ray interaction with matter, or mass-dependent physical or chemical fractionation.

These isotopic anomalies, usually orders of magnitude larger than in other solar system materials, are associated with very minor mineral phases of primitive chondrites distributed irregularly in unequilibrated meteorites. These minerals include diamond, graphite, silicon carbide, and aluminum oxide, with typical grain-sizes being 1–10 μm, with diamond being much smaller (~0.002 μm). Presolar SiC grains, at least, follow a power-law mass distribution dominated by submicron particles, with rare large ones. These minerals are rare in meteorites (e.g., SiC in the CM chondrite, Murchison, is about 5 ppm by mass). Figure 16 depicts such an anomaly, the Ba isotopic composition in presolar SiC separated from Murchison. The data are normalized to terrestrial values of ^{130}Ba and ^{132}Ba, the anomalous s- and r-process isotopes (see below) lying far above the horizontal line.

Since the isotopic composition of these grains differs wildly from those of ordinary solar system matter, they must

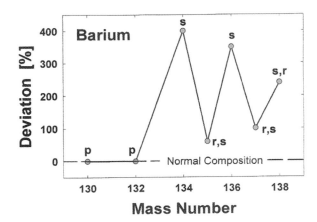

FIGURE 16   Stable Ba isotopes in a SiC separate from the Murchison CM chondrite normalized to those in normal terrestrial Ba. Letters indicate nucleosynthetic processes by which individual isotopes are produced. The presolar neutron-capture isotopes (on slow, s, and rapid, r, timescales that formed in presupernova and supernova stages, respectively) are anomalously high, by up to 4×.

derive from outside our solar system. These grains were incorporated into the solar nebula with intact memories of their individual nucleosynthetic sources, accreted into meteoritic matter and obviously survived all later episodes in their parent bodies' histories. The isotopic anomalies identified thus far point at specific genetic processes. Most SiC grains probably formed in stars on the asymptotic giant branch (i.e., AGB stars) in the Hertzsprung-Russell diagram. This is the source of isotopes produced by neutron capture on a slow timescale (or so-called s-process) nuclides, with rapid neutron capture (r-process) nuclides forming immediately prior to the supernova stage. Supernovae also seem required to explain the isotopic anomalies in tiny diamonds.

The isotopic anomalies of many trace elements in these presolar grains provide a wealth of unique information regarding the evolution of stars and nucleosynthesis. This information is only obtainable by exhaustive, detailed, highly sensitive, and highly accurate analyses of rare interstellar grains from primitive samples in terrestrial laboratories and requires both inspiration and perspiration. Undoubtedly, isotopic anomalies in these rare meteoritic constituents will tell us more about stellar formation and evolution, as well as the formation and early history of the Solar System.

5.2.2 CAI

In addition to low-temperature materials, like the matrix of C1 chondrites and presolar grains, refractory grains like CAI also record early solar history. The CAI are millimeter- to centimeter-sized refractory inclusions, especially recognizable in C2 and C3 chondrites but also identifiable in some UOC and in R and E3 chondrites. Typically, CAI

consist of refractory silicate and oxide mineral assemblages rimmed by thin multilayered bands of minerals. Major-element compositions of CAI agree with calculations by equilibrium vapor-deposition evaporation models to represent the first 5% of condensable nebular matter solidifying at ≥ 1400 K from a gas of cosmic (solar photospheric) composition at a pressure of 10^{-3} atm or at 0.3 atm, if the dust/gas ratio is 40-fold enriched. Most individual CAI contain tiny particles (usually <50 μm) very rich in refractory siderophiles (Re, W, Mo, Pt, Pd, Os, Ir, and Rh) and, occasionally, refractory lithophiles like Zr and Sc. Sometimes, even smaller (micrometer-sized) refractory metal nuggets are found consisting of single-phase pure noble metals or their alloys.

The textural and mineralogic complexities of CAI indicate a variety of formation and alteration processes in their history. Undoubtedly, CAI formed at high temperatures; properties of some suggest vapor condensation as crystalline solids, whereas others seemingly reflect liquid or amorphous intermediates. Volatilization, melting, solid-state metamorphism, and/or alteration in the nebula or after accretion may also have affected some to many CAI. Clearly, CAI had complicated histories that obscured their primary textural properties but left their chemical and isotopic properties relatively unaltered.

Volumetrically, fine-grained CAI are encountered more often than coarse-grained ones, but the latter are more easily studied. Coarse-grained CAI are grouped into four types, defined mainly by mineralogy, formed at progressively lower temperatures: Type A, dominated by melilite, compositionally Åkermanite (Åk) 0–70; Type B, a mixture of melilite, fassaitic pyroxene, spinel, and minor anorthite; anorthite-dominated Type C; and forsterite-bearing inclusions. Type A CAI seem the most diverse, having apparently condensed as solid from vapor with many heavily altered; thus, reconstruction of their original composition is difficult. The other three types formed from partly molten mixtures to melt droplets, respectively. Type B CAI are mineralogically the most complex and host a much wider array of isotopic anomalies. Compositionally, CAI reflect a high-temperature origin and are refractory-rich: refractory lithophiles like REE are generally enriched 20× or more relative to C1 compositions, although considerable variability occurs in individual CAI due to thermal history and oxygen fugacity variations. The oxygen isotopic compositions of CAIs help define the anhydrous minerals line (with slope 1) in Fig. 11.

The centimeter-sized Type B CAI in C3V chondrites attract the most interest, and their individual minerals have been probed by an array of very sophisticated instruments that put to shame conventional chemical microanalytical techniques. Many to all these CAI exhibit isotopic anomalies (both in positive and/or negative directions) for O, Ca, Ti, and Cr. A few CAI, mineralogically and texturally indistinguishable from others, are called FUN inclusions because they exhibit *F*ractionated and *U*nidentified *N*uclear isotopic effects involving not only Kr and Xe but also elements like Mg, Si, Sr, Ba, Nd, and Sm. Six FUN inclusions contain mass-fractionated oxygen (i.e., follow slope 1/2 lines in Fig. 11), and the two Type B inclusions of these six exhibit isotopic anomalies for every element thus far studied.

Although CAI, in general, and FUN inclusions, in particular, yield much information, we do not yet know why isotopic anomalies appear in some CAI but not others, and why some elements in a sample exhibit anomalies but others do not. The CAI apparently formed from unhomogenized matter early in the solar system's history just as or before chondrules did, by analogous processes.

5.3 Elements other than Noble Gases

Having briefly touched upon some important but complicated meteoritic constituents, let us consider information conveyed by trace elements in whole-rock samples. Most elements in the Periodic Table are present in a meteorite at very low levels—microgram/gram (ppm), nanogram/gram (ppb), or picogram/gram (ppt) concentrations. Such low concentrations exist because nucleosynthesis produced stable isotopes of trace elements in only small amounts, and because their geochemical and/or physical properties prevent enrichment—indeed, may cause significant depletion—during genetic episodes. Their geochemistry may cause some trace elements to be sited in specific hosts of particular meteorites [e.g., siderophiles like Ir, Ga, or Ge are enriched (relative to Cl levels) in iron meteorites (cf. Fig. 9)], whereas others are dispersed among a variety of minerals. The same element may be dispersed in one meteorite class but be sited in a particular host in another. For example, REE are found in phosphates in achondrites, but some are dispersed elements in chondrites. They concentrate in whitlockite in eucrites, and are even more enriched in CAI. Trace elements convey important information because a small absolute concentration change induced by a genetic process will result in a large relative effect.

This improvement in "signal-to-noise" is illustrated in explosive meteorite impact. Whatever the initial composition of proto-Earth material, much of its initial complement of refractory siderophiles was extracted into the core, thus depleting them in the crust. Fall of a massive chondrite or, even better, an iron meteorite enriched in siderophiles, followed by an explosion, will spread mixed projectile and target ejecta widely, redepositing the ejecta in a thin layer. Subsequent chemical analysis of a vertical slice that includes the deposition layer will reveal siderophile enrichment in that layer. Siderophile enrichments—especially of refractory Ir—in the K-T boundary layer around the Earth suggested that dinosaurs (and many other biota) died off from sudden environmental changes created by a meteoroid/asteroid impact 65 Ma ago. Initially controversial, this idea is now generally accepted. In many instances, enrich-

ments of several siderophiles in impact breccias at an explosion crater on Earth or the Moon provide a fingerprint of the meteoritic type that created the crater.

As discussed in Section 2.4.4.1, volatile elements condensable at very low temperatures may not have similar contents in C1 chondrites and the solar photosphere. Meteorite compositions are referenced to readily condensable material by normalization to a refractory lithophile—most commonly Si, sometimes Mg or Al—rather than hydrogen as in the solar photosphere. For meteorites, normalized ratios can be on a weight or atom basis: in the latter, trace element contents are usually referred to as atomic abundances and are often normalized to C1 contents. In the most primitive chondrites—EH or UOC—C1-normalized abundances approach or exceed C1 levels. On this basis, we say that moderately to highly refractory siderophiles are enriched in iron meteorites or that refractory lithophiles are enriched in achondrites. Contents of the more refractory trace elements are characteristic of, hence can define, achondrite associations (Fig. 14).

A priori identification of a trace element as refractory or volatile is impossible because its chemical form in a meteorite is usually unknown. For example, indium metal or gaseous oxygen are each quite volatile as elements, but refractory when chemically bonded in InO. Because In exists only at ppb levels in even the most volatile-rich meteorites, neither InO nor any other In compound is identifiable. Several approaches have been used to obtain at least a qualitative elemental volatility order: The orders obtained generally agree, with some minor differences. Criteria used include calculation of theoretical condensation temperatures in a nebular gas of solar composition at pressures of 10^{-3}–10^{-6} atm, determination of C1-normalized atomic abundances in equilibrated (petrographic types 5 and 6) ordinary chondrites, and laboratory studies of elemental mobility (ease of vaporization and loss) during week-long heating of primitive chondrites under conditions simulating parent body metamorphism (400–1000°C, 10^{-4} atm H_2). By these criteria, elements considered as moderately volatile include (in increasing order) Ni, Co, Au, Mn, As, P, Rb, Cu, K, Na, Ga, and Sb, whereas strongly volatile ones include Ag, Se, Cs, Te, Zn, Cd, Bi, Tl and In.

Small but real ($<2\times$) differences exist in contents of the more refractory trace elements in the various chondritic groups. Siderophile contents are higher in EH than in EL chondrites and decrease in ordinary chondrites as H > L > LL, in keeping with total iron contents. Naturally, achondrites are enriched in refractory lithophiles and depleted by orders of magnitude in siderophiles (whether refractory or volatile) and volatile elements of any geochemical character. In some achondrites (mainly HED meteorites and at least one lunar meteorite), high levels of volatile contents are evident, reflecting deposition of late volcanic emanations on their parents. As expected from our picture of how iron meteorites formed, these meteorites are rich only in the more refractory siderophiles: They contain essentially no volatiles, or lithophiles except in silicate inclusions.

More volatile elements exhibit much greater variability in stony meteorites. Concentrations of the three or four most volatile elements are several orders of magnitude higher in UOC than in their equilibrated analogues and decrease by one or two orders of magnitude with increasing UOC homogenization of ferromagnesian silicates. Contents of most strongly volatile elements in H and L chondrites of petrographic types 4–6 are highly variable and do not correlate with the petrographic types. However, in H chondrites, concentrations of many moderately volatile elements vary as H4 > H5 > H6, consistent with loss at progressively higher metamorphic temperatures in stratified parent(s). As discussed in Section 6.4, chronometric data also are consistent with this theory for H chondrite parent(s). Such a model cannot be established for the L chondrites because late shock (evident in the petrographic properties of many of them) affected other thermometric characteristics, thus obscuring earlier histories. In addition to the petrographic evidence, strongly shocked L4–L6 chondrites exhibit loss of some noble gases, highly mobile elements and siderophiles, and lithophile enrichments.

Mean contents of Ag, Te, Zn, Cd, Bi, Tl, and In decrease in L4–6 chondrites with increasing shock-loading (and, therefore, residual temperature) estimated from petrographic shock indicators. Trace element contents of H chondrites do not vary with shock. In unshocked chondrites, volatile contents are significantly lower in H than in L chondrites, suggesting that L chondrite parent material formed from the nebula at lower temperatures than did H. Apparently, nebular temperatures during H chondrite parent material formation were so high ($\sim$700 K) that only a very small complement of volatile trace elements could condense. Hence, essentially none was present to be lost later at high, shock-induced residual temperatures.

The H chondrite regolith breccias, like Noblesville (Fig. 1a), differ from "normal" H chondrites in that the dark, gas-rich portions of the breccias are quite rich in volatile trace elements, sometimes exceeding C1 levels. These volatiles, distributed very heterogeneously in the dark matrix, were apparently not implanted by the solar wind but rather occur in black clasts. These black clasts represent either volatile-rich nebular condensate or a sink for material degassed from the parent body interior. During exposure on the asteroidal surface, these dark clasts and light ones (containing "normal" levels of volatiles) were apparently gardened by repeated impacts, ultimately forming the regolith breccia matrix. Less is known about equilibrated LL chondrites: They may have a unique thermal history or one like that of H or L chondrites.

In contrast to ordinary chondrites, volatile trace elements in carbonaceous chondrites are very homogeneously distributed. These elements are unfractionated from each other in almost all carbonaceous chondrites, implying that their parent material incorporated greater or lesser amounts of C1-like matter during accretion.

The proportions define a continuum from 100% C1 down to about 20% in C5 or C6. As in enstatite chondrites, volatile-rich samples have higher proportions of more siderophile trace elements. These trends accord with oxygen isotope data, implying a continuum of formation conditions for parent materials of carbonaceous chondrites.

Contents of mobile trace elements and noble gases, and the petrography of 15 C1–C3 chondrites (14 from Antarctica and 1 from a hot desert) provide unambiguous evidence for open-system thermal metamorphism in their parent bodies. These properties permit a semiquantitative metamorphic temperature in the 400–900°C range to be estimated for each of the 15. Each was dehydrated during metamorphism and none (including the 14 Antarctic chondrites) was rehydrated during terrestrial residence. As noted in Section 3.1, spectral reflection properties of these 15 thermally metamorphosed carbonaceous chondrites (and none of the more numerous "normal" ones) link them to C, G, B, and F asteroids. Petrographic properties of C1–C6 chondrites were established during nebular condensation and accretion. If C4–C6 or CK chondrites experienced thermal metamorphism, it occurred under closed-system conditions.

Enstatite chondrites present a special problem because nonvolatile siderophiles in them define high (EH) and low (EL) groups established during primary nebular and accretion. Prior to discoveries of desert meteorites, volatile element contents in E3,4 chondrites were known to be orders of magnitude higher than in E6. Whether this difference reflected primary or secondary processes was unclear since E3–E5 chondrites were EH and E6 were EL.

Fortunately, Antarctic collections include previously unknown EL3 chondrites among others, and new data show that EL3 and EH3–EH4 chondrites contain comparable levels of the most volatile elements. These data suggest source regions of E3 and E4 chondrites, whether EL or EH, essentially reflect primary nebular condensation and/or accretion. Volatiles in E5 and especially E6 (whether EH or EL) are greatly depleted from E3 and E4 levels in a manner suggesting open-system loss during thermal metamorphism of their primitive parent(s). Data for these elements suggest further that enstatite achondrites derived from E6 chondrite-like material that previously experienced FeS–Fe eutectic loss (formation temperature, 980°C).

Oxygen isotopic data for all enstatite meteorites (i.e., chondrites and achondrites) are similar (Fig. 11) with $\delta^{18}O$ increasing systematically in E3–E6, independent of their being EH or EL. The oxygen isotopic compositions in EL chondrites lie along the terrestrial fractionation line (Fig. 11), but the distribution in EH chondrites falls along a line of slope 0.66, neither purely mass-dependent nor mass-independent. Two alternative explanations exist. Enstatite meteorites may derive from a single parent body, partitioned in refractories but not volatiles during primary accretion, which lost these volatiles by postaccretionary thermal metamorphism. Alternatively, Mother Nature may have been particularly perverse in providing samples of two parent bodies (EH and EL) with similar oxygen isotopic compositions and volatile element distributions, with primitive material coming mainly from the former and evolved portions mainly from the latter.

For meteorites of less common types, meteorites from hot deserts and, particularly Antarctica, doubtless provide a broader sampling of extraterrestrial materials than do contemporary falls (Section 3.2). Systematic and reproducible differences involving moderately to highly volatile elements suggest this may extend to ordinary chondrites. This suggestion, which remains highly controversial, receives some support by observed asteroid streams, comet stream formation by differential tidal disruption of Comet Shoemaker–Levy 9, failures of alternatives to explain Antarctic meteorite/fall compositional differences, identification of population differences of unambiguous preterrestrial origin, and so on. However, members of putative streams differ in cosmic ray exposure history. If differences between falls and desert meteorites exist, they reflect variations in the near-Earth meteoroid flux with time.

6. Meteorite Chronometry

How old are meteorites? An "age" is a time interval between two events marked by specific chronometers. An accurate chronometer must involve a mechanism operating on a predictable, but not necessarily constant rate. The "clock" starts by an event beginning the time interval and its end must be clearly and sharply recorded. Chronometers used in modern geo- and cosmochronology usually involve long-lived, naturally occurring radioactive isotopes such as the U-isotopes, ^{87}Rb, or ^{40}K. Radioactive decay allows calculation of an age if the concentrations of both parent and daughter nuclide are known, the time interval beginning is defined, and the system is not disturbed (i.e., it is a "closed system") during the time interval. Some meteorite ages involve production of particular stable or radioactive nuclides, or decay of the latter. Typically, the chronometer half-life should be comparable with the time interval being measured.

Meteorites yield a variety of ages, each reflecting a specific episode in its history. Some of these are shown in Fig. 2: the end of nucleosynthesis in a star, the first formation of solids in the solar system, melt crystallization in parent bodies, excavation of meteoroids from these bodies, and the meteorite's fall to Earth. Other events, like volcanism or metamorphism on parent objects can be established as can formation intervals (based on extinct radionuclides) mea-

suring the time between the last production of new nucleosynthetic material and mineral formation in early solar system materials. CRE ages date the exposure of a meteoroid as a small body (<1 m) in interplanetary space, where the meteorite's terrestrial age is the time elapsed since it landed on the Earth's surface. In the following sections we discuss some of these.

6.1 Terrestrial Ages

Terrestrial ages are determined from amounts of cosmogenic radionuclides found in meteorite falls and finds. The principles of the method are depicted in Fig. 4 with ^{14}C ($t_{1/2} = 5.73$ ka), ^{81}Kr ($t_{1/2} = 200$ ka), ^{36}Cl ($t_{1/2} = 301$ ka), and ^{26}Al ($t_{1/2} = 730$ ka) being the nuclides most frequently employed. In Section 3.2, we summarized the most important conclusions associated with meteorites' terrestrial ages. A meteorite's survival time during terrestrial residence is determined by the weathering conditions where the meteorite resides. Survival times (hence, terrestrial ages) for meteorites are much lower for warm and/or wet areas than for cold, arid ones.

Stony meteorites in Antarctica have terrestrial ages up to 2 Ma (Fig. 17), and age distributions depend on their locations, presumably reflecting ice sheet dynamic differences. Meteorites from the Allan Hills average ~300 ka, whereas those from Queen Maud Land have much younger ages (<300 ka), averaging 100 ka. Meteorites from other parts of western Antarctica have ages up to 400 ka, also averaging 100 ka.

Meteorites from hot deserts generally have terrestrial ages up to 50 ka but a few have maximum ages up to 150 ka. Age distributions tend to vary in the four sites, and although most maximum ages are 40–50 ka, those for lunar and

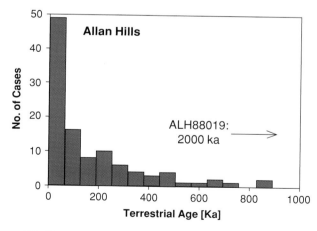

FIGURE 17   Terrestrial age distributions of meteorites from the Allan Hills region of Victoria Land, (west Antarctica), which, on average, are older than meteorites from any other part of Antarctica.

martian meteorites and one H4 are ~150 ka. Their terrestrial age distribution may depend on the hot desert recovery site.

6.2 CRE Ages

In principle, we need to determine both a cosmogenic radionuclide (or its decay product) and a stable nuclide to establish a CRE age. In practice, however, production rates of stable cosmogenic noble gas nuclides in stony meteorites are well-known, and it usually suffices to measure just their concentrations. Absent contrary evidence, we generally assume that irradiation by cosmic rays of solar and galactic origin is simple, that is, the meteoroid was completely shielded (buried in a parent body) until an impact ejected it as a meter-sized object that remained essentially undisturbed until collision with Earth. Some stones (e.g., the H4 chondrite Jilin) and irons (e.g., Canyon Diablo) exhibit complex irradiation histories involving preirradiation on the parent body surface, or secondary collisions in space that fractured the meteoroid and exposed new surfaces to CRE. In such cases (complex irradiation history), different samples of a meteorite exhibit different CRE ages. As noted in Section

1.2, meteoroids approaching the Sun to within 0.5 AU are warmed, causing diffusive loss of gases, especially ^{3}H (a contributor to ^{3}He production) from iron. Such cases are recognized by low isotopic ratios, particularly of ^{3}He/^{4}He or ^{3}He/^{21}Ne, or low natural TL.

The data for ordinary chondrites in Fig. 18 show that all groups have CRE ages ranging up to 90 Ma, but the distributions differ markedly. For H chondrites, there is at least one major peak, ~7 Ma and a smaller one at 33 Ma. For L chondrites, major peaks are not obvious, but clusters occur at 20–30 and 40 Ma, with a smaller one at 5 Ma. For LL chondrites, the major peak at 15 Ma includes ~30% of all measured samples, and another is at 30 Ma. Major peaks correspond to major collisional breakups on and of chondrite parent bodies. Contrary to the H-chondrite situation, nearly two thirds of the L chondrites have CRE ages >10 Ma.

Current data suggest that major CRE peaks vary with petrographic grade for each ordinary chondrite group but poor statistics may cloud the situation in some cases. For example, among L chondrites, the 40-Ma event mainly produced L5 and L6 types. The 7-Ma CRE age peak is particularly evident among H4 and H5 chondrites, and a cluster at 4 Ma is evident for H5 and H6. A possible 10 Ma peak for LL

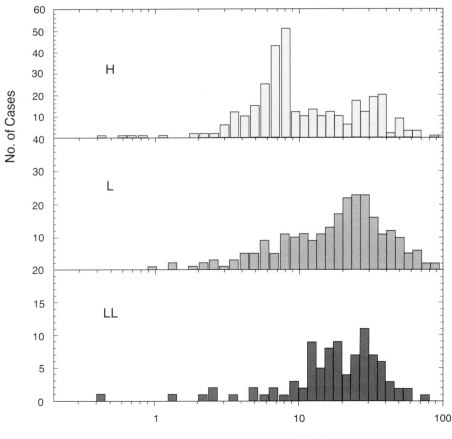

FIGURE 18 Cosmic ray exposure ages for ordinary H, L, and LL chondrites. Peaks in the histograms indicate major collisional events on parent bodies that generated many meter-sized fragments (see Section 6.2).

Cosmic-Ray Exposure Age [Ma]

chondrites includes mainly LL6. The fall frequency of most ordinary chondrites (except H5) has long been known to be twice as great between noon and midnight (i.e., P.M. falls) as between midnight and noon (A.M. falls). This cannot reflect some social cause so the difference must reflect meteoroids' orbits. Meteoroids with perihelia ~1 AU will be predominantly P.M. falls, whereas those having aphelia of ~1 AU will be A.M. falls. These A.M. falls result from the Earth's overtaking meteoroids or involve meteoroids that narrowly miss Earth, and subsequently are gravitationally perturbed, causing their landing on the Earth's forward hemisphere. Fall frequencies for H5 chondrites differ significantly, with A.M. and P.M. falls being about equal. Clearly, a fundamental difference exists between the orbital elements of H5 and other ordinary chondrites.

There are too few CRE ages for carbonaceous and enstatite chondrites to exhibit significant peaks. Carbonaceous chondrites tend to have short CRE ages (<20 Ma). For martian meteorites, exposure ages range from 0.5 to 16 Ma, with some clustering being apparent.

Clustering of exposure ages is also observed for HED meteorites with two diogenite clusters (at about 22 and 39 Ma) coinciding with those in the eucrite and howardite CRE distributions. As discussed in Section 2.2.4, Vesta or its daughters provided these three different achondritic classes.

Attempts to develop a reliable CRE age method for iron meteorites yielded unsatisfactory results except in one case, a difficult, tedious, and no longer practiced technique involving long-lived ^{40}K and stable ^{39}K and ^{41}K. About 70 iron meteorites were dated by the $^{40}K/^{41}K$ method and the resulting ages range from 100 Ma to 1.2 Ga. That CRE ages for iron meteorites greatly exceed those of stones is attributed to the greater resistance of iron meteoroids to preterrestrial destructive collisions (so-called space erosion). CRE exposure age peaks are evident for a few chemical groups. For group IIIAB, 13 of 14 meteorites have a CRE age of 650 ± 60 Ma; this age is also exhibited by 3 of 4 measured IIICD meteorites, suggesting a major collisional event involving the parent of the chemical group III iron meteorites. Otherwise, only the IVA irons exhibit a CRE age peak: 7 of the 9 dated samples have an age of 400 ± 60 Ma. From Section 2.4.5, recall that these iron meteorite groups are the two most numerous ones and contain the highest proportions of strongly shocked members. Either the parent asteroids of these groups were unusually large (requiring unusually large and violent breakup events) and/or the Earth preferentially sampled collisional fragments that had been strongly shocked (thus acquiring a significant shock-induced impulse).

6.3 Gas Retention age

As discussed in Section 5.1, the decay series initiated by the long-lived ^{232}Th, ^{235}U, and ^{238}U yield six, seven, and eight

α-particles, respectively, whereas long-lived ^{40}K produces ^{40}Ar. Thus, from measurements of U, Th, and radiogenic ^4He or of ^{40}K and radiogenic ^{40}Ar, one can calculate a gas retention age or the time elapsed since a meteorite sample cooled sufficiently low to retain these noble gases, if the system was closed during this period. This radiogenic age could record primary formation of the meteorite's parent material, but, in most cases, subsequent episodes (metamorphic and/or shock) were accompanied by substantial heating that partially or completely degassed the primary material. A variant of the K/Ar age, the $^{40}Ar–^{39}Ar$ method involves conversion of some stable ^{39}K to ^{39}Ar by fast-neutron bombardment, that is, $^{39}K(n, p)^{39}Ar$, followed by stepwise heating and mass-spectrometric analysis. From the $^{39}Ar/^{40}Ar$ ratio in each temperature step, it is possible to correct for later gas loss. This variant even permits analysis of small, inhomogeneous samples with a pulsed laser heat source.

Gas retention ages of many chondrites, achondrites, and even silicate inclusions in iron meteorites range up to about 4.6 Ga. Many meteorites, particularly L chondrites, have young gas retention ages, ~500 Ma, while H chondrites cluster at higher ages (Fig. 19). Meteorites with young gas retention ages generally exhibit petrographic evidence for strong shock-loading, implying diffusive gas loss from material having quite high residual temperatures generated in major destructive collisions. Almost always, meteorites having young K/Ar or $^{40}Ar–^{39}Ar$ ages have lower U, Th–He ages. This occurs because He is more easily lost from most minerals than is Ar. Diffusive loss of ^{40}Ar, incidentally, is much more facile than is loss of trapped ^{36}Ar or ^{38}Ar, enhancing its value as a chronometer. Preferential ^{40}Ar loss occurs because most of it is sited in feldspars and in minerals where K is and is associated with radiation damage that provides a ready diffusive escape path. Highly mobile trace elements are lost more readily than is even ^{40}Ar so that L chondrites with young gas retention ages have lower contents of such elements than do those with old ages. The similarity in the CRE age of group III iron meteorites and the gas retention age of L chondrites may be coincidental or, perhaps, may reflect a particularly massive collision of their parent(s).

The number of fossil meteorites discovered in ordinary limestone beds in Swedish quarries implies that the meteorite flux on Earth ~480 Ma ago was 100× higher than the contemporary flux. A recent study found that chromite grains (highly resistant to weathering) from fossil L chondrites have CRE ages of 0.1–1.2 Ma. This implies that these L chondrites arrived on Earth within 100–200 ka after the major collisional breakup that produced contemporary L chondrite falls.

As discussed in Section 4.2, solidification ages for most martian meteorites are ~1.3 Ga (that of ALH 84001 being ~3.7 Ga), implying the existence of parent magmas as recently as 1.3 Ga ago. The $^{40}Ar–^{39}Ar$ ages of essentially unshocked nakhlites accord with the 1.3 Ga age, but

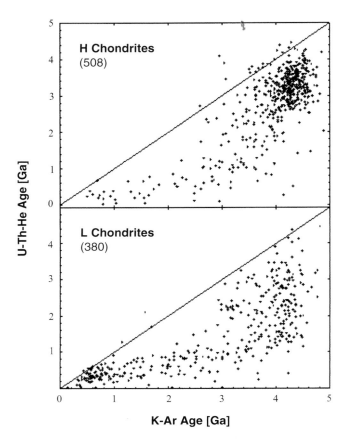

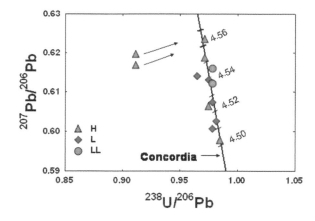

FIGURE 19   Gas retention ages of 508 H and 380 L chondrites. Data from the U, Th–He and K–Ar methods are plotted against each other. These data assume cosmogenic (^{4}He/^{3}He) = 5, K concentrations of 800 and 900 ppm for H and L, respectively, U concentrations of 13 and 15 ppb, respectively and (Th/U) = 3.6. The 45° line represents concordant ages. The two major chondrite types exhibit strong thermal history differences. The dominant concordant long ages of H chondrites suggest that their parent(s) generally remained thermally unaltered since formation 4–4.5 Ga ago. The concentration of data defining concordant short ages of L chondrites suggests strong shock-heating in a major collision(s) 0.1–1.0 Ga ago. Nearly all discordant meteorites lie below the 45° lines because radiogenic ^{4}He is lost far more easily than is radiogenic ^{40}Ar.

shergottites, which are heavily shocked, have gas retention ages probably indicating partial degassing of their parent material ≤250 Ma ago, consistent with Rb–Sr internal isochrons for shergottites at 180 Ma as discussed in the next section.

6.4 Solidification Age

Solidification ages establish the time elapsed since the last homogenization of parent and daughter nuclides, normally by crystallization of a rock or mineral. Nuclides used to

establish solidification ages are isotopes of nongaseous elements insensitive to events that might have affected gas retention. Some techniques, such as the Pb/Pb method, which involves the ultimate decay products of ^{235}U, ^{238}U, and ^{232}Th (^{207}Pb, ^{206}Pb and ^{208}Pb, respectively) involve relatively mobile Pb that should be more easily redistributed than would be the ^{147}Sm–^{143}Nd dating pair. Hence, in principle, a sample dated by several techniques might yield somewhat different ages depending upon its postformation thermal history.

Common techniques found to yield useful solidification ages include: the Pb–Pb method mentioned previously; ^{147}Sm ($t_{1/2}$ = 106 Ga)–^{143}Nd; ^{87}Rb ($t_{1/2}$ = 48 Ga)–^{87}Sr; and ^{187}Re ($t_{1/2}$ = 41 Ga)–^{187}Os. Generally, methods used to determine solidification ages depend upon data depicted in isochron diagrams, in which, for example, enrichment of radiogenic ^{87}Sr is proportional to the amount of ^{87}Rb, and ^{86}Sr is taken for normalization. The slope of such a line yields an "internal isochron" for a meteorite or a single inclusion of a meteorite, if minerals having various ^{87}Rb/^{86}Sr ratios are measured. The y-intercept provides the initial ^{87}Sr/^{86}Sr ratio—a relative measure of the time that nucleosynthetic products were present in the system prior to solidification (i.e., how "primitive" the system is). Clearly, the lower the ^{87}Sr/^{86}Sr ratio is, the less radiogenic (or evolved) was the source material. For some time, basaltic achondrites (e.g., HED meteorites) and the angrite, Angra dos Reis, competed as the source containing the most primitive (least radiogenic) Sr, but, more recently, Rb-poor CAI inclusions in the C3V chondrite, Allende, have become "champions" in this category, with ^{87}Sr/^{86}Sr = 0.69877 ± 2.

Solidification ages for most meteoritic samples are "old" (i.e., close to 4.56 to 4.57 Ga; Fig. 20). The results obtained by different methods agree quite well, although some "finestructure" can be detected. A large number of chondrites

FIGURE 20   U–Pb ages of phosphates from ordinary chondrites. Numbers on the Concordia line are in Ga. The oldest solidification age (for H chondrites) is ~4.563 Ga ago and thermal metamorphism occupied the next 60–70 Ma.

have been studied by the Pb–Pb, Rb–Sr, and Nd–Sm techniques, and results for them are consistent with an age of about 4.56 Ga. The U/Pb method used to date phosphates from ordinary chondrites (Fig. 20) produce ages for H6 chondrites exhibiting small, but significant, differences in Pb–Pb ages from H4 and H5 chondrites. These data suggest 4.563 Ga as the oldest ordinary chondrite solidification age with metamorphism requiring 60–70 Ma. The results are consistent with a stratified ("onion-shell") model for the H chondrite parent body and suggestive of a simple, progressive metamorphic alteration with increasing depth in it.

Most meteorites have solidification ages around 4.56 Ga; however, there is clear evidence of more recent disturbances of chronometric systems—particularly Pb–Pb and Rb–Sr—in many meteorites. For example, Rb–Sr internal isochrons for E chondrites (believed by some to have experienced open-system thermal metamorphism as discussed in Section 5.3) were disturbed 4.3–4.45 Ga ago. Of course, chronometers in heavily shocked L chondrites show clear evidence for late disturbance.

Four techniques (^{40}Ar–^{39}Ar, Rb–Sr, Pb–Pb, and Sm–Nd) yield an age for nakhlites of 1.3 Ga, implying their derivation from a large planet, Mars (Section 4.2). The heavily shocked shergottites seem to have derived from several magma reservoirs and Rb–Sb internal isochrons suggest a major shock-induced disturbance 180 Ma ago, before the martian meteoroids were ejected from their parent planet.

6.5 Extinct Radioactivities

Measurements of decay products of an extinct radionuclide do not provide absolute dates in the sense discussed in earlier sections, but they do permit relative chronologies on timescales comparable with the half-life of the radionuclide (Table 6). Thus far, clear positive evidence has been found in meteorites or their constituent minerals for the presence in the early solar system of the following nuclides: ^{41}Ca ($t_{1/2} = 110$ ka), ^{26}Al ($t_{1/2} = 730$ ka), ^{60}Fe ($t_{1/2} = 1.5$ Ma), ^{53}Mn ($t_{1/2} = 3.7$ Ma), ^{107}Pd ($t_{1/2} = 6.5$ Ma), ^{129}I ($t_{1/2} = 15.7$ Ma), ^{244}Pu ($t_{1/2} = 82$ Ma), and ^{146}Sm ($t_{1/2} = 103$ Ma). In most cases, relative ages are calculated from three-isotope plots involving decay products of the extinct radionuclide. However, in some cases, the relative chronologic information can be combined with data for absolute ages, allowing small time differences in the early solar system to be established. For example, combining the ^{53}Mn/^{55}Mn ratio measured in the Omolon pallasite with the absolute Pb–Pb age of the LEW 86010 angrite yields an absolute age of 4557.8 ± 0.4 Ga for Omolon.

In recent years, much effort has gone into this area so that we should focus upon only one set of results in concluding this chapter. The oldest technique used is that of I–Xe dating, which depends upon the decay of ^{129}I into ^{129}Xe. In this technique, a meteorite on Earth is bombarded with neutrons in a nuclear reactor as in ^{40}Ar–^{39}Ar dating (see Section 5.2) to convert some stable ^{127}I into short-lived ^{128}I ($t_{1/2} = 25$ m), which decays into stable ^{128}Xe. Stepwise heating releases Xe: a linear array with slope >0 on a three-isotope plot of ^{129}Xe/^{132}Xe $vs.$ ^{128}Xe/^{132}Xe indicates an iodine-correlated ^{129}Xe release, whose slope is proportional to ^{129}I/^{127}I at the last time ^{129}I and ^{129}Xe were in equilibrium. This ratio is a measure of the formation interval. Absolute age values, however, can only be obtained if the ratio ^{129}I/^{127}I at the time of the closure of the solar nebula is known. Because this number is not available, only relative ages can be given.

The I–Xe clock proves to be remarkably resistant to resetting by heating: the principal effect is to degrade the linearity, but not to destroy it completely. Shock seems quite effective in resetting this clock, and hydrolysis, which affected C1 and C2 chondrites, is even more effective.

Data for 79 chondrites, aubrites, and silicate inclusions in iron meteorites, relative to the Bjurböle L4 chondrite, give highly reproducible I–Xe intervals; therefore, the Bjurböle L4 chondrite is arbitrarily assumed to have an age of zero. Each meteorite class spans an I–Xe interval >10 Ma, whereas all meteoritic materials possessing isochrons span ~55 Ma. Apparently, the only systematic variation of the I–Xe formation interval with chondritic petrographic type involves E chondrites: EH chondrite parent material formed earlier than did EL. Clearly, while the nuclide ^{129}I was still alive (i.e., during or shortly after nucleosynthesis), primitive nebular matter condensed and evolved into essentially the materials that we now receive as meteorites. The conclusion is supported by other isotopic and charged-particle track evidence (see Sections 4.1, 4.2, 5.3, and 5.4).

As we have seen from the foregoing summary, the meteoritic record can be read best in an interdisciplinary light. Results of one type of study—say, trace element chemical analysis—provide insight to another—orbital dynamics, for example. Early experience gained from meteorite studies, provided guidance for proper handling, preservation, and analysis of Apollo lunar samples. Studies of these samples, in turn, led to the development of extremely sensitive techniques now being used to analyze meteorites and microgram-sized interplanetary dust particles of probable cometary origin collected in Antarctica (Fig. 21) and just successfully brought to Earth by the *Genesis* spacecraft, despite its hard landing. Undoubtedly, this experience will prove invaluable as samples from other planets, their satellites, and small solar system bodies are brought to Earth for study.

Previous studies of meteorites have provided an enormous amount of knowledge about the solar system, and there is no indication that the scientific growth curve in this area is beginning to level off. Indeed, work on the present version began late in 2004, and we were amazed to see how much had been learned about meteorites since 1998 when

FIGURE 21   A French Southern and Antarctic Territories stamp illustrating a micrometeorite or cosmic dust particle (left) collected by melting Antarctic ice cores, the coring drill being at the right. Representations of meteor trails (of cometary origin) and a fireball are at the top.

the first edition of this Encyclopedia was published. Predictions about future developments are very hazardous, but we can expect future surprises, probably from desert meteorites, which seem to include so many peculiar objects. As has been said in another connection, those who work with meteorites don't pray for miracles, they absolutely rely on them.

Further Reading

Buchwald, V. F. (1975). "Handbook of Iron Meteorites." Univ. California Press, Berkeley.

Dick, S. J. (1998). "Life on Other Worlds: The 20th-Century Extraterrestrial Life Debate." Cambridge Univ. Press, Cambridge, England.

Grady, M. M. (2000). "Catalogue of Meteorites," 5th edition. Cambridge Univ. Press, Cambridge, England.

Hewins, R. H., Jones, R. H., and Scott, E. R. D. (eds.) (1996). "Chondrules and the Protoplanetary Disk." Cambridge Univ. Press, Cambridge, England.

Hutchison, R. (2004). "Meteorites, A Petrologic, Chemical and Isotopic Synthesis." Cambridge Univ. Press, Cambridge, England.

Kerridge, J. F., and Matthews, M. S. (eds.) (1988). "Meteorites and the Early Solar System." Univ. Arizona Press, Tucson.

Papike, J. J. (ed.) (1998). "Planetary Materials." Mineralogical Society of America, Washington, D.C.

Porcelli, D. P., Ballentine, C. J., and Wieler, R. (eds.) (2002). "Noble Gases in Geochemistry and Cosmochemistry." Mineralogical Society of America and Geochemical Society, Washington, D.C.

Taylor, J., and Martel, L. (1996–present). Planetary Science Research Discoveries (PSRD). http://www.psrd.hawaii.edu

Wooten, H. A. (2004). The 125 reported interstellar and circumstellar molecules. National Radio Astronomy Observatory. http://www.cv.nrao.edu/~awootten/allmols.html

Near-Earth Objects

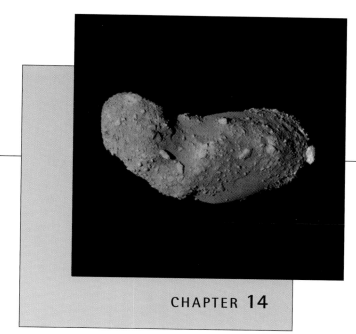

Lucy A. McFadden
University of Maryland

Richard P. Binzel
Massachusetts Institute of Technology

CHAPTER **14**

1. Introduction

Near-earth objects (NEOs) reside in the vicinity of Earth near 1.0 AU (the mean distance between Earth and the Sun). Any object, such as an asteroid or comet, orbiting the Sun with a perihelion, $q < 1.3$ AU, well inside the orbit of Mars, is defined as an NEO. Aphelia, Q, of NEOs generally lie within a sphere of radius 5.2 AU, defined by Jupiter's orbit. Among this broad group are four subgroups: Amors, Apollos, Atens, and interior Earth objects (IEOs). Comets, releasing gas and dust with $q < 1.3$ would be referred to as near-Earth comets (NECs) if they posed an impact threat to Earth. Amors approach but don't cross the orbit of Earth. They have a semimajor axis, $a > 1.0$ AU, and perihelion $1.017 \leq q = 1.3$ AU, between the aphelion of Earth's orbit and inside the perihelion of Mars (Fig. 1a). Those that actually cross Earth's orbit, Apollos, have $a > 1.0$ AU and $q = 1.017$ AU, Earth's aphelion distance (Fig. 1b). Atens have $a \geq 1.0$ AU and $q > 0.983$ AU, Earth's perihelion distance. An object with both a and $q < 0.983$ AU, is an IEO.

The Amor asteroid, 433 Eros, was the first NEO discovered in 1898, by D. Witt of Berlin, Germany, using a photographic plate to record its position. It is also one of the largest NEOs, being 33 km in its longest dimension, with two other axes of 10.2×10.2 km diameter. 1862 Apollo, the first Earth-crossing asteroid, and 1221 Amor, the namesake of that group, were both discovered in 1932. It wasn't until 44 years later that 2062 Aten, the first of the group orbiting within Earth's orbit, was discovered by Eleanor Helin, still using photographic plates for the search. 1998 DK36 was the first IEO discovered in 1998.

As the dynamical evolution of asteroids and their role in probably causing biological extinction events on the Earth was recognized in the 1980s, dedicated searches for NEOs resulted in increased discovery rates. Due to both increased sky coverage and availability of sensitive digital detectors, the known NEOs number >4100 at this writing, compared to 85 known in early 1980 (Fig. 2). About 25 of the NEOs found since the 1990s are binary objects orbiting around a common center of mass; 15% of all NEOs are estimated to be binaries. 1862 Apollo, an asteroid between 1.2 and 1.5 km in diameter, was reported to be a binary in 2005.

Most of the near-Earth objects originated in the Main Asteroid Belt, located between Mars and Jupiter, although some of them probably evolved into their current orbits from the reservoir of short-period comets extending beyond Jupiter and into the outer solar system. The range of composition and physical characteristics of asteroid-like near-Earth objects spans those found among the Main Belt, though 15% of them probably are derived from cometary reservoirs.

The *Near-Earth Asteroid Rendezvous (NEAR)* mission was the first designed to orbit an asteroid. 433 Eros was

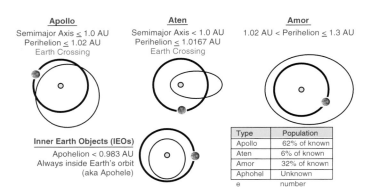

FIGURE 1 (a) Amors approach Earth but do not cross its orbit. (b) Apollo orbits cross that of Earth.

its target, and the spacecraft remained in orbit from 2000 to 2001, ending its mission with a controlled descent and successfully becoming the first spacecraft to land on an asteroid. *NEAR* accomplished the first detailed in situ measurements of an asteroid's surface morphology, mineralogy, chemistry, internal state, and magnetic properties. The Japanese-led *Hayabusa* mission was launched on May 9, 2003, on a 4 year mission to investigate asteroid 25143 Itokawa and to demonstrate the technology necessary to return samples to Earth. The spacecraft went into orbit around Itokawa in September 2005, performing remote sensing measurements for 3 months. The shape and surface morphology of this small near-Earth object is unlike any seen before. In November 2005, there were two scheduled touchdowns in which some surface material may have been collected. The return capsule is scheduled for a June 2010 return to Earth and will hopefully contain some surface material collected from Itokawa.

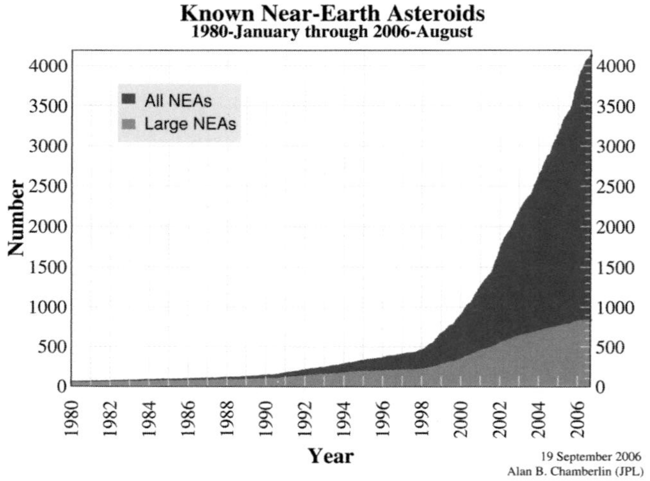

2. Significance

2.1 Remnants of the Early Solar System

From a scientific point of view, near-Earth objects are studied for the same reason as comets and main-belt asteroids: They are remnants of the early solar system (Fig. 3). As such, they contain information that has been lost in the planets through large-scale, planetary processes such as accretion, tectonism, volcanism, and metamorphism. Knowledge of the asteroids and comets as less processed material from the early solar nebula, studied together with direct samples in the form of meteorites, is critical to piecing together a scenario for the formation of the solar system. [*See* THE ORIGIN OF THE SOLAR SYSTEM.]

Most near-Earth objects are asteroid-like in their nature, being derived from the Main Belt. This region is a dividing point in the solar system, where the planets that formed closer to the Sun, the terrestrial planets, are dominated by rocky, **lithophile** material. Beyond the Asteroid Belt, the planets are composed predominantly of nebula gases. Perhaps 10–20% of all near-Earth objects originated elsewhere in the solar system, such as the cometary reservoirs lying at great distances from the Sun, beyond the gaseous planets. Knowing about material from these reservoirs reveals information about both chemical and physical processes that were active in the outer regions of the solar system, both in the near and distant past. The objective of scientific study of the near-Earth objects is to determine which of them might be derived from which regions of asteroidal and cometary reservoirs.

FIGURE 2 Cumulative total of discovered near-Earth objects versus time. Large NEOs are defined as those with an absolute magnitude (*H*) of 18 or brighter. (Data compiled by Alan Chamberlin, NASA/JPL.)

2.2 Hazard Assessment

Although chips, hand-sized rocks, and large boulders, all called meteorites, are continually landing on Earth [*See*

FIGURE 3   Clearing Out the Solar Nebula: The First Planetesimels. Painting by William K. Hartmann, reprinted with permission.

METEORITES], and astronomers find house-sized objects occasionally passing between Earth and the Moon, knowledge of the near-Earth objects, their locations, and physical and chemical characteristics is needed to inventory and assess their hazard potential to the Earth.

Disastrous impacts by asteroids and comets have been the popular subject of Hollywood movies, books, newspaper articles, and television shows. The recognition that a giant asteroid or comet perhaps 10 km across most likely caused the extinction of the dinosaurs in a geological episode known as the Cretaceous–Tertiary Event has highlighted the potential for destruction should an energetic collision occur again (Fig. 4). Furthermore, as scientists analyze the energy involved in collisions, they realize that the impacts are tremendous and larger than anything created by human activities (e.g., nuclear weapons) or naturally occurring phenomena on Earth (e.g., volcanoes, earthquakes, or tsunamis).

Scientists ponder the results of computer simulations that consider the interactions of colliding objects with various Earth systems both natural and civilized. Coupled with these computer simulations is the very real phenomenon of the collision of comet Shoemaker–Levy 9 with Jupiter, which was observed worldwide through telescopes in 1994. The possibly devastating hazard posed to Earth if hit by a high-energy asteroid or comet is now well recognized by scientists and policy makers.

One of the objectives of NASA's Deep Impact mission, which sent an impactor spacecraft to collide with comet 9P/Tempel 1 in July 2005, was to study a comet nucleus and its interior and to assess the hazard to Earth of impact by a comet. When that analysis is complete, additional basic knowledge of comets will be available to assess what would happen should a new comet be found on a collision course with Earth. The most hazardous cometary impact would be one with a large orbital velocity

FIGURE 4 "Dinosaur's Demise." (Painting by Don Davis. Reprinted with permission.)

relative to Earth's. [See PLANETARY IMPACTS; COMETARY DYNAMICS.]

2.3 Exploration Destinations and Resource Potential

NEOs come closer to Earth than any other planetary bodies. With low orbital inclinations and small semimajor axes, they are accessible targets for spacecraft. As humans extend their activities beyond low Earth orbit, relatively nearby destinations are attractive as training venues for missions to Mars. Considering the very long-term future in space one realizes that launching materials from Earth is expensive. As civilization moves beyond Earth, knowledge of materials in space is critical to their efficient use in situ. It is probably more economical to use space resources than transporting material from Earth (Fig. 5).

3. Origins

In the widely accepted scenario of the formation of the solar system, gas and dust collapse into a disk-shaped nebula from which planetesimals and eventually planets form. Planets grow after seeding conditions begin and molecules and dust grains form aggregates, which then form clumps that continue growing into objects large enough to be called planetesimals. This process starts with dust and ice grains about 1 mm in diameter. They behave at first as discrete particles sweeping up smaller grains as they grow. Both electromagnetic and gravitational forces come into play to overcome the destructive forces of erosion from particle collisions. Planet growth is gravitationally controlled and is

called accretion. Asteroids are planetesimals that were prevented from growing to the size of the major planets by pervasive eroding forces that counteract accretion, the net effect being to keep the asteroids relatively small.

The formation of Jupiter was a major force in interfering with the growth of a larger planet between Mars and Jupiter at ~2.8 AU. The details of the main-belt formation are not well known because it formed early in the history of the solar system, ~4.5 billion years ago. Since the earliest formation times, gravitational interactions between planets and small objects (asteroids and comets) have resulted in perturbations of their orbits. These perturbations result in

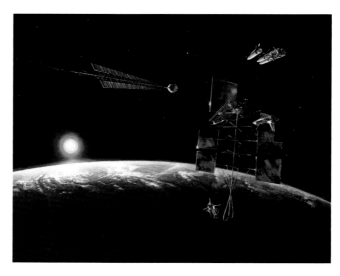

FIGURE 5 Painting showing the beginning of a mission to an Earth-approaching asteroid (Denise Watt, NASA).

the orbit, over time, evolving into one crossing a planet's orbit, the subject of this chapter.

3.1 Relationship to Main Belt Asteroids

Early asteroid studies in the 1940s revealed a range of colors (see Section 5). Techniques to study both reflected and emitted electromagnetic radiation from the asteroids were developed and used to derive information about their mineral and chemical composition. In the late 1970s, two scientists, Jonathan Gradie and Edward Tedesco, recognized that there is a relationship between the apparent composition of the asteroids and their distance from the Sun. This finding represented observational support for a model predicted by another astronomer, John Lewis, in which the solar nebula was in a state of chemical equilibrium when it formed. Asteroid composition changes as a function of temperature, and hence distance from the Sun. Therefore, one does not expect all asteroids to have the same composition. Furthermore, the exact nature of asteroidal material holds clues to the temperature and location where the material formed.

This information is valuable as scientists piece together the scenario leading to the formation of our solar system and look for evidence of the existence of other solar systems. Studies of the composition of near-Earth objects led to the conclusion that NEO composition spans the range found among the Main Asteroid Belt, thus establishing that many or most of the NEOs are derived from the main belt. Follow-on research has confirmed these findings and identified the proportion that is derived from comets as ~15%. Furthermore, physical information derived from NEOs can be reasonably considered to apply to Main Belt Asteroids.

Statistical analysis of the evolution of many asteroid orbits over the age of the solar system indicates that the lifetime of an Earth-crossing body against gravitational perturbations is relatively short, on the order of 10 million years or less. Within this time frame, the bodies will either collide with a planet or be dynamically ejected from the solar system. This time interval applies to the average of the entire population and does not refer to the exact lifetime of any particular asteroid. It turns out that the orbital evolution of a specific asteroid or comet cannot actually be determined very far into the future or the past owing to the difficulty of knowing the exact starting conditions and accurately predicting frequent close approaches between the NEO and the planets. [*See* SOLAR SYSTEM DYNAMICS: REGULAR AND CHAOTIC MOTION.]

3.2 Relationship to Meteorites

Exploring the relationship between NEOs and meteorites is motivated by the possibility of making a very rich connection between the geochemical, isotopic, and structural informa-

tion on meteorites available from laboratory studies and the near-Earth objects. Meteorites fall to Earth frequently, but most often land unnoticed in the oceans or in remote areas. In January 2000, an exceptionally bright **bolide** was seen by eyewitnesses in the Yukon, Northern British Columbia, parts of Alaska, and the Canadian Northwest Territories. Nearly 10 kg of precious samples were recovered from the surface of frozen Tagish Lake. Using eyewitness reports and the bolide's detection by military satellites, the orbit of the impacting body was traced back to the Asteroid Belt (Fig. 6). Prior to striking the Earth, the body is estimated to have been about 5 m across with a mass of 150 metric tons. [*See* METEORITES.]

The determination of meteorite orbits serves as a constraint on the mechanisms that result in meteoroid delivery to Earth. Numerical computer simulations reveal regions of the Asteroid Belt that act as "escape hatches" for delivering material to the terrestrial planets zone. One such region corresponds to a Kirkwood gap, located where an asteroid's orbital period is shorter than Jupiter's by the ratio of two small integers, such as 3:1, 5:2, or 2:1. Any asteroid or debris that migrates into this gap finds Jupiter to be especially effective in increasing its orbital eccentricity. As the orbit becomes increasingly elongated, it can intersect the orbit of the Earth. In the 1980s, work by Jim Williams, Jack Wisdom, and others illuminated the importance and efficiency of resonances in the Asteroid Belt and their role in supplying meteorites.

3.3 Relationship to Comets

Comets are predominantly icy and dusty objects that come from the outer reaches of the solar system. Their orbital periods are long, their orbital eccentricities are high, and they may have large or small orbital inclinations. What is their relationship to near-Earth objects? In the 1950s, Ernst Öpik concluded that comets must be a partial source of near-Earth objects because he could not produce the number of observed meteorites from the Asteroid Belt alone via his calculations. Building on Öpik's work, George Wetherill predicted that 20% of the near-Earth object population consists of extinct cometary nuclei. Some now find evidence that the fraction of comets is smaller, closer to 15%. The hypothesis that NEOs derive from comets continues to merit consideration as knowledge of comets and asteroids increases and simulations of the dynamical evolution of interacting small bodies under the gravitational influences of the planets continues to develop. [*See* COMETARY DYNAMICS; PHYSICS AND CHEMISTRY OF COMETS.]

Are there hints that any particular near-Earth object that looks like an asteroid was once a comet? If an object sometimes has a tail like a comet and sometimes looks just like an asteroid (no coma or tail), which is it: asteroid or comet? There is both dynamical and physical evidence that addresses this question.

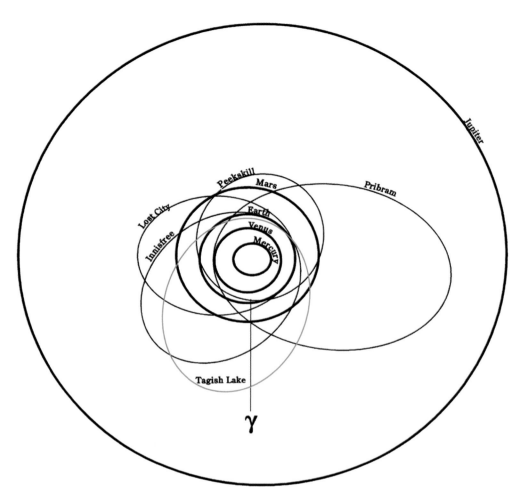

FIGURE 6   Orbit of Tagish Lake meteorite with other recovered meteorite orbits. (Credit: *AAAS Science* 13 October 2000, Vol. 290.)

3.3.1 TISSERAND PARAMETER

The first clue that an asteroid-like object may be a comet in disguise comes from its orbit. Examining orbital elements, asteroids and comets separate out readily when plotting orbital eccentricity versus semimajor axis (Fig. 7). Another way to characterize an orbit is to calculate its Tisserand parameter from the equation:

$$T = a_J/a + 2[(a/a_J)(1 - e^2)]^{1/2} \cos i$$

In this equation, a and a_J refer to the semimajor axis values for the object and Jupiter. The parameters i and e are the inclination and eccentricity of the object's orbit. The Tisserand parameter is useful because it is a constant even if the comet's orbit is perturbed by Jupiter. Also it helps describe whether an object is in an orbit that is strongly controlled by Jupiter or not. Most objects that display the characteristics of comets have a value $T < 3$, while most objects that are asteroid-like have $T > 3$. The value of $T = 3$ is represented by the solid line in Fig. 7. Objects with $T < 3$ are excellent candidates for being comets in disguise – they do

not currently display any telltale coma or tail because they are at present dormant or inactive.

3.3.2 DYNAMICAL AND PHYSICAL EVIDENCE FOR EXTINCT COMETS

A powerful way to investigate the mystery of how many extinct comets reside in the near-Earth object population is to explore both dynamical factors and physical measurements to identify possible candidates. For example, numerical simulations of the orbits of short-period comets can reveal how likely it is that gravitational interactions with Jupiter and the other planets can send them into the near-Earth object population. In these simulations, many thousands of hypothetical comets, each with slightly different initial orbits can be tracked for millions of years to see how they are tossed around chaotically by the gravitational tugs and pulls of the planets. In the same way, thousands of different starting places for main-belt asteroid orbits can be modeled to reveal the effectiveness of resonances for sending asteroids into near-Earth space. Alessandro Morbidelli, William Bottke,

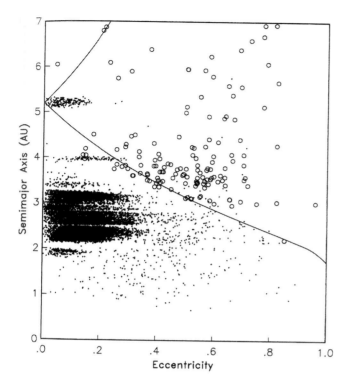

FIGURE 7 Tisserand parameter. The solid line represents the Tisserand parameter with a value of 3. (Graph provided by Jeff Bytof, NASA/JPL.)

and co-workers have done extensive computer calculations to assess the relative effectiveness of these dynamical processes. Their calculations suggest that, when considering NEOs of all sizes, about 15–20% of all NEOs have their origins as comets. Nearly all of these are currently inactive, showing no evidence of a coma or a tail. They are comets disguised as asteroids.

Spacecraft and telescopic measurements of known comets reveal what characteristics to look for when trying to determine if a given asteroid-like NEO is a comet in disguise. For example, the inactive surface regions of comets Halley, Borrelly, Wild 2, and Tempel 1 are very dark (low albedo) and have gray to reddish colors. Some other comets go through periods of very low activity, allowing astronomers to clearly see and measure the albedos and colors of the nucleus. All of these measurements consistently show low albedos (reflecting only about 4% or less of the incoming light) and gray or reddish colors. When observed in reflected sunlight, these objects exhibit featureless spectra with no absorption bands due to olivine or pyroxene (mineral types) on their surfaces.

Knowing the dynamical signature (Tisserand parameter, $T < 3$), low albedo and gray/red color, allows asteroid-like NEOs to be identified as extinct comet candidates. A survey of nearly 50 NEOs residing in orbits having $T < 3$ conducted by one author (RPB) shows about one half of them

exhibit the low albedo and color characteristics seen for comet nuclei. From the surveys searching for NEOs, correcting for the fact that for any given size, dark objects will be more difficult to detect than bright ones, about 30% of the all NEOs reside in $T < 3$ orbits. If one half of these are comet-like in their physical characteristics, this suggests up to 15% of all NEOs are extinct comet candidates. Other researchers find a smaller percentage of 5–15% derived from simulations of orbital dynamics.

Until 2001 only upper limits on cometary activity were derived for the extinct cometary candidates. Object 2001 OG_{108} has an orbital period of 50 years and inclination almost perpendicular to the ecliptic plane, similar to that of Comet Halley. Upon its discovery, there was no detectable coma. At a distance of 1.4 AU, the object became active as it passed through the inner solar system. Its bare nucleus has the characteristics of cometary nuclei, and when close to the Sun, it outgases like a comet.

3.4 Meteor Shower Associations

The near-Earth objects 2101 Adonis and 2201 Oljato have orbits similar to those of meteor showers. Adonis is very difficult to observe and not much is known about it. Oljato, also a difficult target for telescopes, has intrigued scientists since it was first observed in 1979. The jury is still out on whether or not this asteroid is an extinct comet, but the evidence now seems to suggest that it is asteroidal in its origin. One thing is certain: The object is not normal even when considered as an asteroid.

In 1983, Fred Whipple recognized the orbital elements of an asteroid found by an Earth-orbiting infrared telescope to be essentially the same as the Geminid meteor shower, which occurs in mid-December. [*See* INFRARED VIEWS OF THE SOLAR SYSTEM FROM SPACE]. There is little doubt that this asteroid, now named 3200 Phaethon, is the parent body of the Geminid meteors. But is Phaethon an extinct cometary nucleus? The supposition is yes, according to one line of thought based on similarities of orbital inclinations and the location of perihelion (longitude of perihelion relative to the ecliptic) of asteroids and comets compared to meteor showers. Its reflectance spectrum (see Section 5) is unlike other comet nuclei, however. There are currently nine NEOs that have orbital elements that, over the past 5000 years, may be associated with the path of existing meteor showers.

3.5 Dynamical History

Dynamicists have simulated the pathways that objects might take from unstable regions of the Asteroid Belt using computations of dynamical forces acting in the solar system. In some cases, fragments from asteroid collisions may be violently cast into these regions of instability. However, a softer touch may play an even bigger role. Constant warming by the Sun causes asteroids of all sizes to reradiate their heat back into space. Because the asteroids are rotating, the reradiation does not occur in the same direction as the incoming sunlight, resulting in a small force acting on the asteroid. This force acts as a very gentle push on the asteroid, which over many millions of years can cause the asteroid to slowly drift inward or outward from its original main-belt location. This is called Yarkovsky drift and is especially effective on small objects; it may be particularly important for supplying meteoroids to Earth. Cast away fragments or drifting bodies that enter regions where resonances with Jupiter's orbit are particularly strong, such as the 3:1 Kirkwood gap, find that small changes in the semimajor axis can result in large, exponential changes in other orbital elements, in particular eccentricity, changing the orbit significantly on a short timescale. Thus, the effects of chaotic regions are more than the sum of small changes in motion over long periods of time. These regions of **chaotic motion** are associated with resonances with both Jupiter and Saturn (Fig. 8). The two gas giant planets are believed to play a significant role in directing meteoroids to Earth, and presumably also many of the near-Earth objects.

Other objects evolve from Jupiter-family comets or Halley-type short-period comets. Life in the Jupiter family is not long-lived, as Jupiter imparts changes to the orbits on timescales of 10^4–10^6 years. Leaving Jupiter's gravitational sphere of influence, the soon-to-be near-Earth objects may

sometimes be perturbed by Mars and other terrestrial planets and also affected by the influences of nongravitational forces, such as volatile outgassing or splitting of the cometary nucleus. These phenomena also contribute to orbital changes that result in planet-crossing orbits.

4. Population

4.1 Search Programs and Techniques

Organized, telescopic search programs for near-Earth objects operate worldwide. The search programs supported by the National Aeronautics and Space Administration (NASA) include the Lincoln Near-Earth Asteroid Research (LINEAR) program, the Near-Earth Asteroid Tracking (NEAT) system, Lowell Observatory's Near-Earth Object Search (LONEOS), the Catalina Sky Survey, and Spacewatch, the last two operated by independent teams at the University of Arizona. International efforts and interests are also strong at Japan's National Space Development Agency (NASDA) and a joint venture among the Department of Astronomy of the University of Asiago, the Astronomical Observatory of Padua in Italy, and the DLR Institute of Space Sensor Technology and Planetary Exploration in Berlin-Adlershof, Germany. Though the objectives of these programs are all similar, to inventory the objects in the vicinity of Earth, each has its own design and approach. In the past, when astronomers imaged the sky with photographic plates, it was an eye-straining process to compare them and determine if something moved. Search programs today employ digital imaging devices known as **charge-coupled devices** or **CCDs** that cover large areas of the sky in a single exposure. Typically a given area of sky is imaged and reimaged 3–5 times at intervals of 10 minutes to an hour. With digital images, fast computers can compare the images, identify and subtract all of the "uninteresting" objects that remain fixed, leaving behind the tracks of a moving asteroid or comet. By rapidly repeating this process for many patches of sky throughout a night, nearly the whole sky can be scanned in the course of about 2 weeks. Increasingly rapid and increasingly sensitive search systems are expected to come on line by the end of the decade.

When a near-Earth object is first discovered, astronomers initially trace only a short piece of its orbit as measured over a few hours or even over a few weeks. With each new NEO discovery, astronomers wish to assess whether the object poses any immediate or future impact threat. Orbit calculations for most objects can be made reliably for many decades into the future, but of course if only a tiny part of the orbit has been observed, the extrapolation into the future becomes increasingly uncertain. Sometimes that extrapolation shows that the Earth itself resides within the overall uncertainty region for an NEO's future position. If the cross section of the Earth occupies 1/10,000th of this

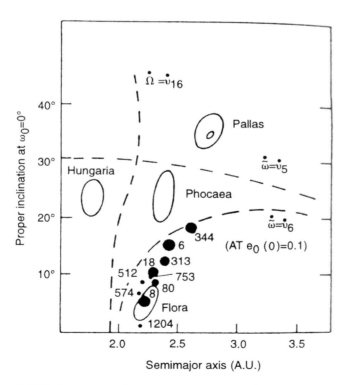

FIGURE 8 Dynamical resonances are regions where gravitational interactions either deplete or protect asteroids from changes in their orbit. (From Jim Williams, NASA/JPL.)

THE TORINO SCALE
Assessing Asteroid/Comet Impact Predictions

No Hazard	0	The likelihood of collision is zero, or is so low as to be effectively zero. Also applies to small objects such as meteors and bolides that burn up in the atmosphere as well as infrequent meteorite falls that rarely cause damage.
Normal	1	A routine discovery in which a pass near the Earth is predicted that poses no unusual level of danger. Current calculations show the chance of collision is extremely unlikely with no cause for public attention or public concern. New telescopic observations very likely will lead to re-assignment to Level 0.
Meriting Attention by Astronomers	2	A discovery, which may become routine with expanded searches, of an object making a somewhat close but not highly unusual pass near the Earth. While meriting attention by astronomers, there is no cause for public attention or public concern as an actual collision is very unlikely. New telescopic observations very likely will lead to re-assignment to Level 0.
Meriting Attention by Astronomers	3	A close encounter, meriting attention by astronomers. Current calculations give a 1% or greater chance of collision capable of localized destruction. Most likely, new telescopic observations will lead to re-assignment to Level 0. Attention by the public and by public officials is merited if the encounter is less than a decade away.
Meriting Attention by Astronomers	4	A close encounter, meriting attention by astronomers. Current calculations give a 1% or greater chance of collision capable of regional devastation. Most likely, new telescopic observations will lead to re-assignment to Level 0. Attention by the public and by public officials is merited if the encounter is less than a decade away.
Threatening	5	A close encounter posing a serious, but still uncertain threat of regional devastation. Critical attention by astronomers is needed to determine conclusively whether or not a collision will occur. If the encounter is less than a decade away, governmental contingency planning may be warranted.
Threatening	6	A close encounter by a large object posing a serious, but still uncertain threat of a global catastrophe. Critical attention by astronomers is needed to determine conclusively whether or not a collision will occur. If the encounter is less than three decades away, governmental contingency planning may be warranted.
Threatening	7	A very close encounter by a large object, which if occurring this century, poses an unprecedented but still uncertain threat of a global catastrophe. For such a threat in this century, international contingency planning is warranted, especially to determine urgently and conclusively whether or not a collision will occur.
Certain Collisions	8	A collision is certain, capable of causing localized destruction for an impact over land or possibly a tsunami if close offshore. Such events occur on average between once per 50 years and once per several 1000 years.
Certain Collisions	9	A collision is certain, capable of causing unprecedented regional devastation for a land impact or the threat of a major tsunami for an ocean impact. Such events occur on average between once per 10,000 years and once per 100,000 years.
Certain Collisions	10	A collision is certain, capable of causing a global climatic catastrophe that may threaten the future of civilization as we know it, whether impacting land or ocean. Such events occur on average once per 100,000 years, or less often.

space, then there is a 1 in 10,000 chance of an impact with the Earth. Even though headlines may proclaim the end of the world, statistically speaking, the odds are actually 10,000 to 1 in our favor that continued observations refining the orbit will show a collision is ultimately ruled out. Thus, daily activities should continue unchanged. Working with many colleagues, one of us (RPB) has developed the 10 point Torino scale (Fig. 9) to help the media and the public assess whether any NEO discovery merits public concern or response. Indeed, continued observations have ruled out any substantial threat from all previous headline makers. There are currently two objects with a rating of 1 meriting careful monitoring, according to the Near Earth Object Program posting at http://neo.jpl.nasa.gov. The value of the searches is to change our knowledge from probably being safe to being highly certain about any threat from impacts for many generations.

4.2 How Many?

It is difficult to quote the definitive size of the near-Earth object population. Search programs are constantly adding to the inventory, but there are inherent limitations in search techniques. Consider setting out to count the number of near-Earth objects. First, one can only look for them at night. At any one time, one can only search half the sky. Then there are limitations in how much sky one can cover in one night, controlled by the telescope field of view and the recording instrumentation. The realities of weather and equipment performance further hinder the search. The combination of these factors represents an estimate of what fraction of an expected population has been found for a range of size and brightness.

To date, search programs have found more than 4100 near-Earth objects of all sizes. The biggest objects appear brightest and are most easily found. Searchers know of 30 NEOs as large as 5 km across and believe all objects this size and larger have been found. Around 300 objects have been cataloged that are larger than 2 km, and NEO catalogs are nearly complete at this size. Catalogs are known to be incomplete for objects smaller than 2 km, but by knowing how much area of the sky has been searched and how sensitive these searches have been, it is possible to estimate how many objects are left to find. A recent Ph.D. thesis by J. Scott Stuart carefully analyzed the search statistics from the LINEAR program, taking into account the different colors and reflectivities (albedos) that are typical for NEOs. Based on Stuart's work, the best estimate is that there are about 1100 total NEOs larger than 1 km in diameter and up to 85,000 NEOs larger than 100 m (Fig. 10).

When considering impact hazards on Earth, most scientists consider 1 km as the size large enough for an impact to present a global threat to human survival. Thus, current search efforts have as their most immediate goal to find all objects larger than 1 km. The good news is that more than 870 of all cataloged NEOs are estimated to be 1 km or larger and thus astronomers are 80% toward completing the most immediate goal, and that may be reached in just a few more years. In the process, many smaller objects are found, and these begin to help bring completeness to all sizes. Searchers have a long way to go to complete the survey of all 85,000 objects that may be larger than 100 m; these may be capable of Tunguska-like (or somewhat greater) amounts of damage. Completing the surveys down to these sizes will require new, large, specialized telescopes with huge CCD arrays to scan the skies more frequently and with greater sensitivity. Another possibility would be to conduct the search using small telescopes in space.

5. Physical Properties

The first physical measurement after the position of a near-Earth object is established is its brightness measured on the astronomical **magnitude** scale. The changing cross section of an object as viewed from Earth affects its brightness and with time reflects the shape and rotation rate of the object. Analysis of this changing brightness, accounting for the observational geometry, results in constraints on its shape and the determination of its rotation rate and orientation in space. From analyses of reflected sunlight off main-belt asteroid surfaces at different wavelengths, NEO colors are classified into different taxonomic types. [*See* MAIN-BELT ASTEROIDS.] Further analysis can determine surface mineralogy, and, from that, constraints on the temperatures at which these objects formed can be made.

The *Near-Earth Asteroid Rendezvous* mission studied the physical and chemical properties of asteroid 433 Eros from orbit and at the spacecraft's landing site. From its shape and surface morphology, astronomers deduced information about its global structure. An X-ray and gamma ray spectrometer provided information about its surface chemistry. See Section 6 for details.

5.1 Brightness

The standard asteroid photometric magnitude system compensates for the distance and **phase angle** at which the object is observed. The magnitude scales by the inverse square law. As the distance from both the Sun and the observer increases, the brightness decreases by a factor equal to the inverse square of those distances. Scattering properties of the surface are expressed in the phase function, which is compensated for by extrapolating the magnitude to $0°$ phase. For comparison purposes, a magnitude measurement is converted to an absolute scale, H, which is defined as the brightness of an object at a distance of 1.0 AU from both the Earth and Sun, and viewed at $0°$ phase angle. The measured slope of brightness changes with phase, G, has

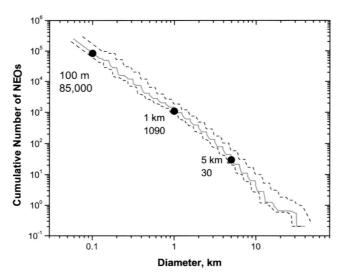

FIGURE 10   Estimated number of NEOs as a function of diameter.

been measured for some of the brighter near-Earth objects. Large phase coefficients indicate a very rough surface with significant effects due to shadowing, such that the magnitude changes significantly with changing phase angle. Low values of G indicate either a very dark surface, where the impact of shadows is not significant against a dark surface, or that few scattering centers exist and hence there is minimal shadowing. When observations are made over a range of phase angles, fits to theoretical models with multiple variables can be made. Combined with other observational techniques (e.g., radar, polarimetry, lidar), constraints on the physical characteristics of the surface regolith can be made.

5.2 Configuration

Lightcurves are measurements of brightness as a function of time (Fig. 11). If the object is perfectly spherical such that its cross section does not change with time, there will be no variation, and the lightcurve would be flat. There are no such objects known, although there are lightcurves with very small amplitudes (not commonly found among near-Earth asteroids). Lightcurves of NEOs often show two or more maxima and minima, often with inflections embedded within. The **triaxial ellipsoid** shape of each NEO can be modeled using observations. Inflections in the lightcurves represent changes in the object's cross section that reflect either the large-scale shape or albedo variations across the surface or both.

Radar measurements are also analyzed to produce images that reveal the shape of asteroids. Coded wave packets transmitted from Earth to an asteroid reflect back and are received as a radar echo. The bandwidth of the echo power spectrum is proportional to the cross section of the asteroid presented to Earth and normal to the line of sight at the time of interaction with the surface, convolved with Doppler shifts in the returned signals caused by the object's rotation. The signal can be built up as the asteroid rotates, producing an image that represents its shape. For those objects that have approached Earth at close enough range to employ this technique, such as 4769 Castalia, 4179 Toutatis, 1627 Ivar, 1620 Geographos, and 433 Eros, the results show shapes varying from slightly noncircular to very irregular. [*See* PLANETARY RADAR.]

Knowledge of the objects' shapes provides clues to the collisional history of this population. If all objects were spherical, astronomers would believe them to have formed from a viscous and rotating material that was not disturbed since formation. The fact that many near-Earth objects are irregularly shaped implies that they are products of collisions that have knocked off significant chunks of material from a larger body. Images of 433 Eros (Fig. 12) show it described as an ellipse measuring $33 \times 10.2 \times 10.2$ km. Its shape is irregular and controlled by large impact craters.

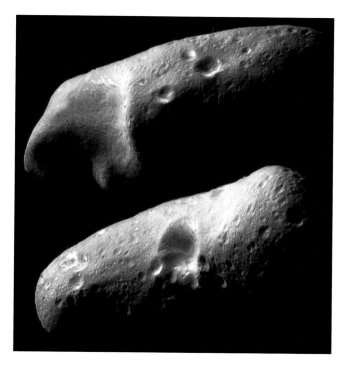

FIGURE 12 Asteroid 433 Eros's eastern and western hemispheres. Two mosaics created from 6 images when the *NEAR* spacecraft was orbiting 355 km (220 mi) above the surface. Smallest detail is 35 m (120 ft) across. The large depression on the top image is Himeros (10 km across). In the bottom image, the 5.3 km crater Psyche is prominent. Bright exposures can be seen on interior walls of craters. (Credit: NASA/JHU/APL.)

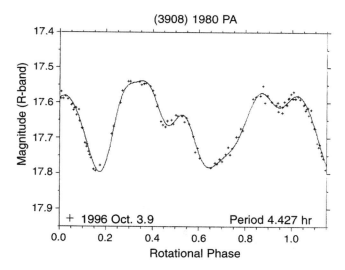

FIGURE 11 Lightcurve for Amor asteroid 3908 Nyx indicating its irregular shape. (Courtesy of Petr Pravec, Astronomical Institute, Academy of Sciences of the Czech Republic.)

Some near-Earth objects' shapes have been interpreted as being two bodies stuck together and are referred to as a contact binary. This interpretation is intriguing because it leads to speculation that the two components were brought together in a low-velocity collision and just stuck together instead of one or both being destroyed. An alternative interpretation is that the asteroid is so irregularly shaped that it appears to be two pieces, but really is continuous. Such a situation would imply a history of collisional fragmentation that kept the main body of the asteroid intact, albeit severely altering its shape, but not disrupting it totally. Measurements at different aspect angles are required to truly confirm the interpretation that some objects are contact binaries. About 16% of near-Earth objects larger than 200 m in diameter may be contact binary systems according to estimates.

5.3 Rotation Rates

Of 32 measured near-Earth objects with an average diameter of 3 km, the mean rotation rate is 4.94 ± 0.54 rev/day, whereas a sample of the same number of comparably sized, main-belt asteroids has a mean rotation rate of 4.30 ± 0.46 rev/day. Because the standard deviation of these means overlaps, no statistical significance is placed on these differences. The mean rotation rate of comets is larger than the mean of the NEOs. Comets rotate on average more slowly than NEOs. The implications of different rotation rates for the history of the object are discussed elsewhere. [See ASTEROIDS.] Because of their proximity to Earth, NEOs are the smallest objects in space for which we can measure their rotational properties. In some cases, the rotation rates for NEOs smaller than about 150 m are 100 rev/day or faster (i.e., they have rotation periods of just a few minutes). These objects are likely relatively strong and intact rock fragments. Larger objects that spin substantially slower, may be less strong "rubble piles" composed of individual fragments or fractured rock held together only by gravity. A rubble pile must spin at a rate slower than once every 2.2 hours, or else it will fly apart. Thus, near-Earth objects give us insights into the likely range of internal structures occurring within small bodies in our solar system.

5.4 Size

For an object illuminated by the Sun alone, the sum of the reflected and emitted (thermal) radiation from the object (assuming no internal energy sources or sinks) is equal to the total incident solar radiation upon it. Knowing where the object is, in terms of its distance from the Sun and the output of the Sun, the amount of incident energy on the object's surface can be calculated. By measuring the reflected and reemitted (thermal) components of radiation, and with some rudimentary knowledge of the nature of the body's surface materials determined from spectral measurements,

one can estimate its albedo and determine its diameter. The two parameters, diameter and albedo, are derived in tandem, with the requirement that the sum of reflected and emitted components is equal to the incident solar flux. This can be expressed mathematically as

$$\pi R^2 (F/r^2)(1 - A) = 4\pi R^2 \varepsilon \sigma T^4$$

In this equation, R is the object's mean radius and F is the solar flux, a constant. The distance from the Sun is r, and A is a term called the bolometric Bond albedo. The emissivity of the asteroid, ε, is assumed to be 1, and the parameter σ is the Stefan–Boltzmann constant. The temperature, T, is derived from the radiated flux from the asteroid measured in the thermal infrared spectral region. One can then solve for the bolometric Bond albedo, A, which is the integrated reflected light at all wavelengths. Albedo and diameter are calculated based on measurements of visible and infrared flux.

Another method of estimating the size of small asteroids is from their measured brightness and an assumed albedo. This method is referred to as a photometric diameter. It is used when no thermal measurements and only visual magnitudes are available. The diameter is given by the equation

$$\log d = 3.1295 - 0.5 \log p_{\mathrm{H}} - 0.2 \, H_{\mathrm{v}}$$

where p_{H}, the geometric albedo, is assumed, and H_{v} is the magnitude defined by the International Astronomical Union magnitude system for asteroids in the V, or visual bandpass. Unfortunately, the range of asteroid albedos is large, from only a few percent up to 50% or more, producing considerable uncertainty in the photometric diameters. However, the taxonomic type of the asteroid (see below), determined from brightness measurements at several different wavelengths, can be used to narrow the range of probable albedos. Notice that an object with a lower albedo, reflecting the same amount of light, will be significantly larger than a high-albedo object. For example, a 15th magnitude object (on the bright end of any NEO) with an albedo of 0.15, an average, "bright" asteroid, would have a diameter of 3.4 km, whereas an asteroid with a 0.06 albedo, at the high end of the range of dark asteroids, would be 1.6 [5.4/3.4 = 1.588] times as large at 5.4 km. Keep in mind that the plot showing the frequency of near-Earth objects as a function of brightness and size (Fig. 10) provides only an estimate of the size and frequency of objects and, except at the large end of the magnitude scale, is an extrapolation and estimate of the size of the complete population.

5.5 Mass

The mass of binary asteroids can be determined from Kepler's third law, $P^2/a^3 = 4\pi^2/G(M_{\mathrm{p}} + m_{\mathrm{s}})$, where P is the period of revolution, a is the semimajor axis, both observed quantities. G is the universal gravitational constant, and M_p and m_s are the mass of the primary and secondary and are solved for as a sum. Using this expression, the mass of the binary near-Earth object 2000 DP107, for example, is calculated to be $4.6 \pm 0.5 \times 10^{11}$ kg, a little more than $1/1000^{th}$ the mass of all living matter on Earth, estimated at 3.6×10^{14} kg. At least 25 other NEOs are known to be binaries. From Doppler and range measurements of the *NEAR–Shoemaker* spacecraft, the mass of Eros was measured to be $6.687 \pm 0.003 \times 10^{15}$ kg. For comparison, the Moon's mass is more than 10 million times greater at 7.348×10^{22} kg. While the range of measured NEO masses spans four orders of magnitude, their total mass is small compared to the solar system's total planetary mass.

5.6 Color and Taxonomy

Since the early part of the 20th century, astronomers have recognized that small bodies come in different colors. As observational techniques evolved and the ability to investigate them improved, the number of observable characteristics increased. Sorting objects into meaningful groups is the process of classification or taxonomy. Asteroid taxonomy developed in response to advances in observing techniques and new technology in the field of stellar photometric astronomy. Current taxonomy is based on the application of statistical clustering techniques to the parameters of color and albedo. The intention of the classification scheme is to reflect the compositional variations and thus their origin and evolution. Astronomers are constantly attempting to test and refine the asteroid taxonomy by employing new statistical methods and extending the number of meaningful parameters that are included in the classification process, while eliminating meaningless or redundant parameters. Today, the alphabet soup of asteroid taxonomy extends to about 12 letters with subtypes numbering up to 26. The taxonomy too has evolved, and one has to be aware of which system is being used and what the exact definitions are. Bobby Bus presented a taxonomy in 1999 that has 26 classes. [*See* ASTEROIDS.]

Near-Earth objects have representatives from all taxonomic types except one, indicating that many locations in the Asteroid Belt feed the near-Earth population. Ninety percent of NEOs fall in the S-, Q-, C-, and X-complexes (a complex is a grouping of taxa from different instrument types and different taxonomies combined into a general category that can encompass all available observations). Two thirds of NEOs are bright and members of the S- (40%) or Q- (25%) complexes. When considering the observed ratio of dark objects to bright, there are almost four times as many bright objects observed compared to dark ones in the NEO population. However, darker objects are more difficult to discover and measure. Accounting for this discovery bias against darker objects is especially important when estimating how many extinct comets may be present in the near-Earth object population (Section 3.2).

5.7 Mineralogy

By measuring the percentage of reflected sunlight from the surface of an object, it is possible to constrain its surface mineralogy. This technique was pioneered by Tom McCord and his students and colleagues in the 1970s. In 2006, spectral reflectance measurements of over 200 NEOs were available. The inventory is still growing.

Astronomers find that 65% of near-Earth objects contain two strong absorption bands, one in the ultraviolet with a band centered below 0.35 μm and the other in the near infrared near 1 μm. Sometimes a second near-infrared band is observed at a wavelength of 2 μm. Other objects do not have prominent absorption bands: They are found to be featureless and either flat or sloped. Most often these featureless objects also have a low albedo. Figure 13 shows spectral reflectance measurements of some near-Earth objects. Three spectra have prominent ultraviolet and near-infrared

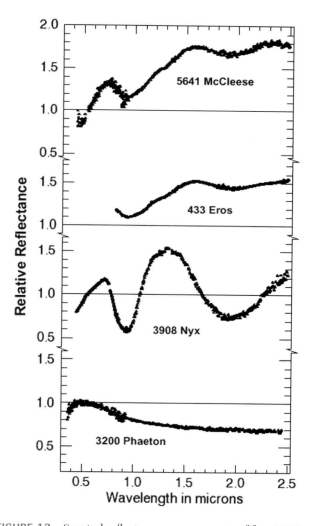

FIGURE 13 Spectral reflectance measurements of four NEOs. The range of spectra reflects the range of surface characteristics including mineralogy and particle sizes of the surface material.

absorption bands that are common in silicate minerals. The broad band at 1 μm of asteroid 5641 McCleese is diagnostic of a mineral called olivine, which consists of silicon oxide tetrahedra bound in eightfold symmetry by magnesium, calcium, and iron cations. Subtle differences in the position of the center of the band constrain the chemistry of the olivine, which can accommodate a range of magnesium and iron in its mineralogical structure. The presence of a second absorption near 2 μm indicates that a second silicate, pyroxene, is present.

The spectrum of 433 Eros contains both olivine and two types of pyroxene. Detailed spectral analysis and modeling suggest the presence of an additional component that may be a glassy material, or possibly vapor-deposited coatings of nanometer size iron grains. They are inferred because the brightness of the spectrum is lower than mixtures of only crystalline silicates. These mineral constituents are present in ordinary chondrite meteorites; the deviation from ordinary chondritic composition and the processes controlling that have been studied and ascribed to **space weathering** and/or partial melting.

The spectrum of asteroid 3908 Nyx (Fig. 13) is dominated by pyroxene and has the same spectral characteristics as the basaltic achondrite meteorites. This asteroid may have traveled to the near-Earth region of space over the age of the solar system and may be a fragment of the large main-belt asteroid, 4 Vesta. [See MAIN-BELT ASTEROIDS.]

The lower spectrum in Fig. 13 is characteristic of a subgroup of C-types, labeled B. There is no UV absorption and not much of an infrared absorption. Interpretation of this spectrum is uncertain. This asteroid, 3200 Phaethon, is a candidate for an extinct comet, though its albedo (9–11%) is higher than most comets observed to date (~4%).

Mineralogical studies of near-Earth objects show that they are not all alike. Nor are they alike in the Main Asteroid Belt. The range of variations in mineral composition reflects that seen in the Main Asteroid Belt, indicating that NEOs are mostly derived from the Main Belt. None of the NEOs are compositionally similar to any of the major planets because they do not share any of the spectral reflectance characteristics of the major planets or the Moon. NEOs with low albedo, featureless spectra with higher IR reflectance relative to the UV, might be extinct cometary nuclei.

6. In Situ Studies

6.1 NEAR

The *Near-Earth Asteroid Rendezvous* spacecraft was launched from Cape Canaveral, Florida, on February 16, 1996, on a 3 year journey to asteroid 433 Eros. *NEAR* orbited Eros for 1 year in 2000–2001, training its 6 scientific instruments on the asteroid's surface. It provided the first detailed characterization of a NEO's chemical and physical properties. The objective was to study Eros' relationship to meteorites, the nature of its surface and collisional history as well as aspects of its interior state and structure.

The spacecraft carried a complement of instruments covering the electromagnetic spectrum. The magnetometer measured no magnetic field down to its detection limit of 1–2 nano-Teslas (Earth's magnetic field measures 50,000 nano-Teslas). A possible explanation for this unexpected result is that magnetic material within Eros is randomly oriented to the point of canceling all fields. If this is the case, then there has been no heating of the asteroid to the point of producing any preferred orientation of any magnetic material.

Orbital imaging of Eros revealed an irregularly shaped body dominated at the global scale by both convex and concave forms, including a 10 km diameter depression named Himeros, and a 5.3 km bowl-shaped crater named Psyche (Fig. 12). At scales of 1 km to 100 m, (Fig. 14) there are grooves and ridge patterns superimposed on a heavily cratered surface, mostly covered by overlapping craters. At the <100 m scale, the surface is dominated by boulders, evidence of down-slope movement, and ponding of material in crater bottoms, all of which indicate regolith accumulation and transport. There are not many craters <100 m diameter. Evidence of structural strength on the asteroid includes chains of craters, sinuous and linear depressions, ridges and scarps, and rectilinear craters. The multiple orientations of these features indicate that they were formed in multiple events. The prominent ridge system, named Rahe Dorsum, spans the northern hemisphere and defines a plane through the asteroid. It cuts across Himeros and possibly Psyche, indicating that it predates the formation of these large craters. Segments form cliffs with slopes above the angle of repose indicating an interior structure with considerable cohesive strength.

Eros is not a gravitationally bound rubble pile. Rather, it is a fragment of a once larger body, possessing cohesive strength throughout. Both gravitational forces and mechanical strength play a role in the formation and evolution of Eros. Its density of 2670 ± 30 kg/m^3 is low compared to ordinary chondrite meteorites of 3400 kg/m^3 that have approximately the same composition as Eros. By considering the porosity of each meteorite and its relation to a much larger asteroid, the macroporosity of Eros is determined to be 20% and most likely due to collisional fragmentation throughout Eros' interior.

The X-ray spectrometer (XRS) onboard *NEAR* provided relative abundance ratios of six elements: Mg, Al, Si, S, Ca, and Fe, from <100 μm depth. For all these elements, except sulfur, the ratios are within the range of ordinary chondrite and some R-type (partially melted) meteorites. [See METEORITES.] The sulfur depletion is most likely due to a surface phenomenon, micrometeorite-induced, impact

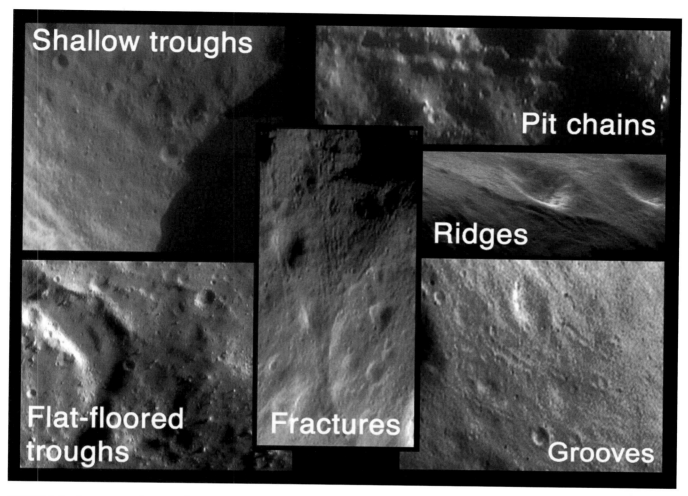

FIGURE 14 A montage of images showing structural features on 433 Eros. Shallow troughs are partially filled-in linear structures; pit chains are crater chains. A ridge winding almost around the entire asteroid is called Rahe Dorsum; fractures are at the end of Rahe Dorsum, and grooves are evenly spaced and may have raised rims. (Photo credit: NASA/JHU/APL.)

volatilization, and subsequent sputtering, though it could also represent a different starting composition for Eros as compared to ordinary chondrites, or partial melting.

The gamma ray spectrometer (GRS) produced observable and interpretable signal from the landing site of the *NEAR* spacecraft (Fig. 15). Five elements were detected: K, Mg, Si, O, and Fe. The abundance of potassium and the Mg/Si and Si/O ratios are chondritic. These results agree with the XRS findings and are consistent with the imaging and near-infrared spectrometer's findings. With the depleted S/Si ratio found with XRS, the abundance of potassium, a relatively volatile element, was expected to be low as well, but it isn't at the Eros landing site. Eros' Fe/Si and Fe/O ratios are low compared to chondritic values, and low compared to the XRS results, too. They fall within the region of differentiated and partially differentiated meteorites and above the values observed for meteorites. The most likely

explanation for low elemental abundances is for a regolith process where the iron migrates within the regolith. The GRS samples to greater depths than the XRS, so the different results between the two instruments is explained by real differences in the top 100 μm of the regolith compared to the 10s of centimeters depth sampled by GRS. Spatially resolved spectra from the *NEAR* mission indicate that the surface composition of Eros is uniform, showing very little compositional variation, except in some interior regions of craters where fresh material is exposed.

6.2 Sample Return Mission

The *Hayabusa* mission to asteroid Itokawa included in situ observations over a 3 month tour. This S-type asteroid, with a 12.1 hour rotation period, was observed with 4 instruments: an imaging camera with 8 filters, a near-infrared

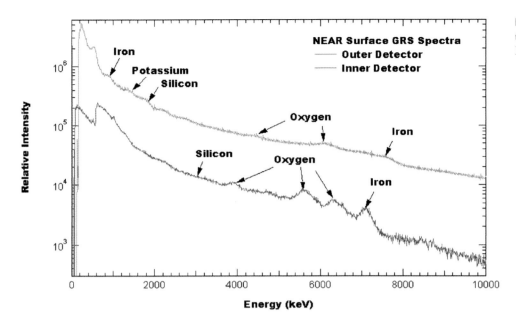

FIGURE 15 Gamma ray spectra of the surface of 433 Eros. (Credit: NASA/JHU/APL.)

spectrometer, a laser ranging instrument, and an X-ray fluorescence spectrometer. The asteroid has major axes of $x = 535$, $y = 294$, $z = 209 \pm 1$ m, which is the same as $0.33 \times 0.18 \times 0.13$ miles. The science team considered this object to be shaped like a sea otter with a head and a body (Fig. 16). Eighty percent of the surface is considered to be rough terrain; the smooth areas are called seas. The surface shows many boulders ranging in size from a few to 50 m, the largest, named Yoshinodai, is 1/10th the size of the whole asteroid. The boulders are likely relics from large impact events that produced Itokawa's current shape. Instead of counting craters on Itokawa, scientists count boulders to understand its impact history. Itokawa has experienced less processing in its rough terrain–including breaking, sorting and transporting of material in its lifetime–than other

small bodies such as Eros and Phobos, the Martian moon. Smooth terrain, covering 20% of the body, has few if any boulders and is featureless. At closest range, the spacecraft resolved centimeter- to millimeter-sized grains, the size of pebbles, in the smooth areas. Missing from the asteroid is a range of crater sizes, though their remnants are barely visible, almost erased by debris from both impacts and global shaking, which filled in craters with fine material over time.

With a mass of $3.510 \times 10^{10} \pm 0.105$ kg and a volume of $1.840 \times 10^{7} \pm 0.092$ m^3, Itokawa's density is 1900 ± 130 kg/m^3. This is lower than other S-type asteroids. Using the compositional knowledge from the near-infrared spectra and the X-ray spectrometer, indicating that the bulk composition is close to LL ordinary chondrites [see METEORITES], and the measured bulk density of those meteorites, the macroporosity of Itokawa is estimated to be 41%, twice as high as Eros and, in fact, soils on Earth. The absence of linear features extending the length of the body, the presence of local facets (flat areas) 10s of meters long and the high porosity indicate that Itokawa is a rubble pile with a composition closest to that of LL ordinary chondrite but also consistent with some primitive achondrites.

If all goes well with the spacecraft's return to Earth in 2010, scientists will eagerly open the return capsule and hopefully find precious samples of the asteroid's surface.

7. Impact Hazards

It does not take much imagination to envision an asteroid or comet hurtling through space that just happens to be on

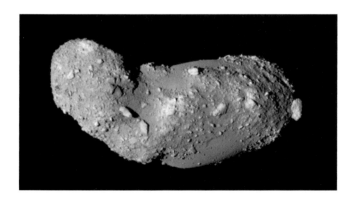

FIGURE 16 Global image of the western hemisphere of 25143 Itokawa imaged by the *Hayabusa* imaging camera during science mode imaging. (Courtesy: JAXA.)

a collision course with Earth. Arthur C. Clarke used this scenario in his 1994 book *The Hammer of God* and Hollywood movies followed. Consider the physics. Meteorites fall to Earth, and there are impact craters on Earth and the Moon. The largest lunar impact basins formed more than 3.3 billion years ago, and the largest impactors were swept up or ejected soon after the solar system formed. Most objects colliding today with Earth and the Moon are small and harmless. Fragments that are a meter to a centimeter in size appear as bright bolides in the sky and can deliver meteorites to the ground, though these are essentially harmless. There are two aspects of the collision hazard to be considered: the magnitude of the collision and their frequency in time.

7.1 Magnitude

The primary physical parameter of concern is the energy of the collision and particularly the energy transferred to Earth. The controlling parameters are mass and velocity according to the relation

$$E = \frac{1}{2}\, mv^2$$

If a massive body were to collide with Earth, the energy of impact would be proportional to its mass. Objects that are 10s of kilometers to a kilometer in size can cause significant damage to Earth as a whole by triggering changes in global climate that will affect human systems such as agriculture. Objects less than a kilometer in size still pose a significant regional threat, having impact energies that eclipse the world's arsenal of nuclear weapons.

Energy is also proportional to the square of the velocity, so a high-velocity object will have considerably higher impact energy than relatively slow-moving objects. Most near-Earth objects travel at similar orbital velocities to Earth when nearby, about 30 km/s. But because their orbits are often inclined or more eccentric than the Earth's orbit, there is still a measurable relative velocity. Objects in highly eccentric and/or inclined orbits, such as comets, can have tremendous impact energy.

Any object approaching the Earth is accelerated by the Earth's gravity. The minimum velocity of any object entering the Earth's atmosphere is equal to the Earth's escape velocity, 11.8 km/s. So even relatively slow-moving NEOs can have quite significant energy when they hit.

Assessment of the damage that a particular impact will impart to Earth is based on how much energy any particular location can absorb and whether or not that location can recover from an impact. Meteoroids enter Earth's atmosphere with energies estimated in the 10^{11}–10^{15} J in the 1–50 m size range breaking up and burning up in Earth's atmosphere, leaving perhaps only scattered dust. On the other hand, damage from meteorites has been documented

on various scales, from killing a dog in Egypt in 1911 to punching holes in roofs, bruising a human thigh, and going through the trunk of a car.

Craters are produced by impacts with energies on the order of 4.2×10^{16} J, or 10 megatons (MT) of TNT. [*See* PLANETARY IMPACTS; METEORITES.] Impacts of greater energies, by an order of magnitude or so, can impart regional damage. Studies have shown that an impact of 4.2×10^{17} J, or 100 MT, can destroy areas within a 25 km radius. Of further concern is that an impact into the ocean might induce tsunamis that would destroy coastal areas. The Cretaceous–Tertiary Event 65 million years ago has been estimated at $>4.2 \times 10^{23}$ J, or 100,000,000 MT! Such large impacts occur very infrequently. But they do occur.

7.2 Frequency

A complete assessment of the situation requires knowledge of the frequency of collisions by objects of different sizes. Objects in the range of 100s of kilometers in diameter were swept up and incorporated into the planets or dynamically ejected as the solar system formed during the period called the Late Heavy Bombardment. The lunar basins formed during this time, which ended ~3.8 billion years ago. No terrestrial collisions are expected from such large objects today. [*See* THE MOON.]

An impact by an object less than 50 m in diameter with energies $<4.2 \times 10^{16}$ J (<10 MT) occurs about once every 1000 years. An impact in Tunguska, Siberia in 1908 may have been an NEO about this size. Interestingly, that object did not make a crater because it probably was a weak (heavily cracked) rocky body that broke apart in the atmosphere. Only the shock wave of air from the ~12 MT explosion reached the ground, felling thousands of square kilometers of remote forest. The frequency of impacts increases exponentially with decreasing size and energy. Conversely, for larger objects the frequency decreases.

To assess the potential for any near-Earth object to collide with the Earth, it is imperative to have an accurate assessment of the numbers and locations of this population. It is then important to know the nature of the orbit for each object because that bears directly on the object's velocity relative to Earth in its motion around the Sun. The active asteroid search programs are designed to inventory the population of objects that may impact Earth. Upon discovery of a new NEO, its orbit is determined and its future orbital evolution is projected by computer simulations. If there is a potential threat of its impacting Earth, a call for follow-up observations is made, and the threat is evaluated carefully. The existence of potentially hazardous asteroids (PHAs) is monitored closely, worldwide. Even though impacts are not a likely occurrence, they remain a possibility.

Bibliography

Beatty, J. Kelly, Petersen, Carolyn Collins, and Chaikin, Andrew. (1999). "The New Solar System," p. 421. Sky Publishing, Cambridge, Massachusetts, and Cambridge Univ. Press, Cambridge, United Kingdom.

Bell, J. F., and Minton, Jacqueline, eds. (2002). "Asteroid Rendezvous: NEAR Shoemaker's Adventures at Eros." Cambridge Univ. Press, New York.

Bottke, William F., Cellino, Alberto, Paolicchi, Paolo, and Binzel, Richard P., eds. (2002). "Asteroids III, p. 785. Univ. Arizona Press, Tucson.

Gehrels, T., ed. (1994). "Hazards Due to Comets and Asteroids." Univ. Arizona Press, Tucson.

Lewis, J. S. (1996). "Mining the Sky, Untold Riches from the Asteroids, Comets, and Planets." Helix Books/Addison-Wesley, Reading, Massachusetts.

Lewis, J. S., Matthews, M. S., and Guerrieri, M. L., eds. (1993). "Resources of Near-Earth Space." Univ. Arizona Press, Tucson.

Mars Atmosphere: History and Surface Interactions

David C. Catling
University of Bristol
Bristol, United Kingdom
University of Washington
Seattle, Washington

Conway Leovy
University of Washington
Seattle, Washington

CHAPTER **15**

A fundamental question about the surface of Mars is whether it was ever conducive to life in the past, which is related to the broader questions of how the planet's atmosphere evolved over time and whether past climates supported widespread liquid water. Taken together, geochemical data and models support the view that much of the original atmospheric inventory was lost to space before about 3.5 billion years ago. It is widely believed that before this time the climate would have needed to be warmer in order to produce certain geological features, particularly valley networks, but exactly how the early atmosphere produced warmer conditions is still an open question. For the last 3.5 billion years, it is likely that Mars has been cold and dry so that geologically recent outflow channels and gullies were probably formed by fluid release mechanisms that have not depended upon a warm climate.

1. Introduction

The most interesting and controversial questions about Mars revolve around the history of water. Because temperatures are low, the very thin Martian atmosphere can contain only trace amounts of water as vapor or ice clouds, but water is present as ice and hydrated minerals near the surface. Some geological structures resemble dust-covered glaciers or rock glaciers. Others strongly suggest surface water flows relatively recently as well as in the distant past.

But the present climate does not favor liquid water near the surface. Surface temperatures range from about 140 to 310° K. Above freezing temperatures occur only under highly desiccating conditions in a thin layer at the interface between soil and atmosphere, and surface pressure over much of the planet is below the triple point of water [611 Pascals or 6.11 millibars (mbar)]. If liquid water is present near the surface of Mars today, it must be confined to thin adsorbed layers on soil particles or highly saline solutions. No standing or flowing liquid water, saline or otherwise, has been found.

Conditions on Mars may have been different in the past. Widespread geomorphic evidence for liquid flowing across the surface may indicate warmer and wetter past climates and massive releases of liquid water from subsurface aquifers. Hydrated minerals and sedimentary features interpreted to indicate liquid flow found by one of NASA's twin Mars Exploration Rovers (MERs), named *Opportunity*, in Terra Meridiani support the hypothesis that water once flowed at or near the surface, but the timing and circumstances of flow remain unknown. On the opposite side of Mars from *Opportunity*, instruments on the *Spirit Rover* have identified hydrated minerals in rocks in an apparent ancient volcanic setting in the Columbia Hills region of Gusev Crater. NASA's *Mars Odyssey* orbiter has also detected subsurface ice, mainly in high latitudes, while the *Mars Express* orbiter of the European Space Agency (ESA) has detected hydrated minerals in locations ranging from the northern circumpolar dunes to layered deposits in the equatorial regions. The extent and timing of the presence of liquid water are central to the question of whether microbial life ever arose and evolved on Mars.

Atmospheric volatiles are substances that tend to form gases or vapors at the temperature of a planet's surface. Consequently, such volatiles can influence climate. Here we review the current understanding of volatile reservoirs, the sources and sinks of volatiles, the current climate, and the evidence for different climates in the past. We focus on the hypothesis that there have been one or more extended warm and wet climate regimes in the past, the problems with that hypothesis, and the alternative possibility that Mars has had a cold, dry climate similar to the present climate over nearly all of its history, while still allowing for some fluid flow features to occur on the surface. Mars undergoes very large orbital variations (Milankovitch cycles), and the possible relevance of these differences to climate history will be discussed. Whether or not extended periods of warm, wet climates have occurred in the past, wind is certainly an active agent of surface modification at present and has probably been even more important in the past. We discuss the evidence for modification of the surface by wind erosion, burial, and exhumation and the resulting complications for interpreting Mars' surface history. We conclude with a brief overview of open questions.

2. Volatile Inventories and their History

2.1 Volatile Abundances

Mars' thin atmosphere is dominated by carbon dioxide (Table 1). In addition to the major gaseous components listed, the atmosphere contains a variable amount of water vapor (H_2O) up to 0.1%, minor concentrations of photodissociation products of carbon dioxide (CO_2) and water vapor (e.g., CO, O_2, H_2O_2, and O_3), and trace amounts of noble gases neon (Ne), argon (Ar), krypton (Kr), and xenon (Xe). Recently, trace amounts of methane (CH_4) have also been identified, averaging ∼10 parts per billion

by volume, although currently a wide range of methane values have been reported, and these differences have yet to be reconciled. The differences may represent measurement errors or variability in the source of methane and its transport.

Volatiles that can play important roles in climate are stored in the regolith and near-surface sediments. Crude estimates of some of these are given in Table 2. Water is stored in the permanent north polar cap, north polar cap layered terrains, and layered terrains surrounding the South Pole, and as ice, hydrated salts, or adsorbed water in the regolith. The regolith is a geologic unit that includes fine dust, sand, and rocky fragments made up of the Martian soil together with loose rocks, but excluding bedrock. Although the surface of the residual northern polar cap is water ice, the ∼5 km deep cap itself consists of a mixture of ice and fine soil with an unknown proportion of each. Layered south polar terrains may also contain an amount of water ice equivalent to a global ocean 20 m deep. Measurements of the energy of neutrons emanating from Mars into space have provided evidence for abundant water ice, adsorbed water, and/or hydrated minerals in the upper 1–2 m of regolith at high latitudes and in some low-latitude regions (Fig. 1). Cosmic rays enter the surface of Mars and cause neutrons to be ejected with a variety of energies depending on the elements in the subsurface and their distribution. Abundant hydrogen serves as a proxy for water and/or hydrated minerals. If water ice extends deep into the regolith, it could correspond to tens of meters of equivalent global ocean. It is also possible that Mars has liquid water aquifers beyond the depth where the temperature exceeds the freezing point (the so-called melting isotherm), but direct evidence is currently lacking.

Carbonate weathering of dust has occurred over billions of years in the prevailing cold dry climate, and as a consequence some CO_2 appears to have been irreversibly transferred from the atmosphere to carbonate weathered dust particles. The total amount depends on the global average depth of dust. Some CO_2 is likely to be adsorbed in the soil also, but the amount is limited by competition for **adsorption** sites with water. Despite an extensive search from

TABLE 1	Basic Properties of the Present Atmosphere
Average surface pressure	∼6.1 millibars (mbar), varying seasonally by ∼30%
Surface temperature	Average 215 K, range: 140–310 K
Major gases	CO_2 95.3%, $^{14}N_2$ 2.7%, ^{40}Ar 1.6%
Significant atmospheric isotopic ratios relative to the terrestrial values	D/H = 5
	$^{15}N/^{14}N$ = 1.7
	$^{38}Ar/^{36}Ar$ = 1.3
	$^{13}C/^{12}C$ = 1.07

TABLE 2	Volatile Reservoirs	

Water (H_2O) Reservoir	*Equivalent Global Ocean Depth*
Atmosphere	10^{-5} m
Polar caps and layered terrains	5–30 m
Ice, adsorbed water, and/or hydrated salts stored in the regolith	0.1–100 m
Deep aquifers	Unknown
Carbon Dioxide (CO_2) Reservoir	*Equivalent Surface Pressure*
Atmosphere	~6 mbar
Carbonate in weathered dust	~200 mbar per 100 m global average layer of weathered dust
Adsorbed in regolith	<200 mbar
Carbonate sedimentary rock	~0 (at surface)
Sulfur Dioxide (SO_2) Reservoir	*Equivalent Global Layer Depth*
Atmosphere	0
Sulfate in weathered dust	~8 m per 100 m global average layer of weathered dust
Sulfate sedimentary rock reservoirs	Extensive, but not yet quantifiable

orbit, no carbonate sedimentary rock outcrops have been identified down to a spatial resolution of about 100 m.

Table 2 also lists sulfates. Although there are no detectable sulfur-containing gases in the atmosphere at present, sulfur is an important volatile for climate because it may have briefly resided in the atmosphere in the past. Measurements by NASA's *Mars Pathfinder* and *Viking* landers showed that sulfur is a substantial component of soil dust (~7–8% by mass) and surface rocks. Hydrated sulfate salt deposits have also been recently identified in numerous deposits in the Martian tropics from near-infrared spectral data on the European Space Agency's *Mars Ex-*

press spacecraft. Observed sulfate minerals include gypsum ($CaSO_4 \cdot 2H_2O$) and kieserite ($MgSO_4 \cdot H_2O$), while jarosite has been found by the *Opportunity* rover. [Jarosite is $XFe_3(SO_4)_2(OH)_6$, where X is a singly charged species such as Na^+, K^+, or hydronium (H_3O^+).] Anhydrous sulfates, such as anhydrite ($CaSO_4$), are also likely to be present but would give no signature in the spectral region studied by *Mars Express*.

Evidence of volatile abundances also comes from analysis of a certain class of meteorites, the Shergotite, Nahkla, and Chassigny or SNC meteorites. [*See* METEORITES.] These meteorites are known to be of Martian origin from

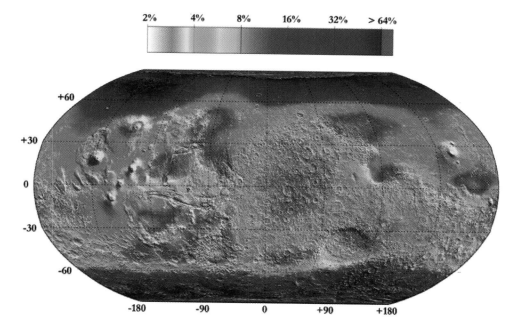

FIGURE 1   Water-equivalent hydrogen content of subsurface water-bearing soils derived from the *Mars Odyssey* neutron spectrometer. (From Feldman et al., 2004, *J. Geophys. Res.* **109**, E09006, doi:10.1029/2003JE002160.)

their relatively young ages, igneous composition, unique oxygen isotope ratios, and gaseous inclusions whose elemental and isotopic compositions closely match the present Martian atmosphere. Ages of crystallization of these basaltic rocks (i.e., the times when the rocks solidified from melts) range from ~1.35 billion years to ~0.16 million years. Many of the SNC meteorites contain salt minerals, up to 1% by volume, which include halite (NaCl), gypsum, anhydrite, and carbonates of calcium, magnesium, and iron. The bulk meteorite compositions are generally dry, 0.04–0.4 weight percent water. This is consistent with a relatively dry Martian mantle (<1.8 weight percent water for preeruptive magmas). On the other hand, the Martian mantle is inferred to be sulfur-rich compared with Earth (estimated as ~0.025 weight percent sulfur). Another type of Martian meteorite, identified as ALH84001, is a unique sample of very early crust, ~4.5 billion years old, which contains about 1% by volume of distributed, 3.9-billion-year-old carbonate. ALH84001 has been heavily studied because of a controversial investigation in which four features of the meteorite were argued to be of possible biological origin: the carbonates, traces of organic compounds, 0.1-micrometer-scale structures identified as microfossils, and crystals of the mineral magnetite (Fe_3O_4) (McKay et al; see Bibliography). However, the biological nature of all of these features has been strongly disputed, and many scientists have suggested that they were formed by abiological processes.

2.2 Sources and Losses of Volatiles

Volatile delivery began during formation of the planet. Planetary evolution models indicate that impacting bodies that condensed from the evolving solar nebula near Mars' orbit were highly depleted relative to solar composition in the atmospheric volatiles—carbon, nitrogen, hydrogen, and noble gases. Nonetheless, formation of Jupiter and the outer planets would have gravitationally deflected volatile-rich asteroids from the outer solar system and **Kuiper Belt** comets to the inner solar system. Analyses of the compositions of the SNC meteorites indicate that Mars acquired a rich supply of the relatively volatile elements during its formation. However, carbon, nitrogen, and noble gases are severely depleted compared with Earth and Venus, apparently because loss processes efficiently removed these elements from Mars, as they did for hydrogen.

Two processes, **hydrodynamic escape** and impact escape, must have removed much of any early Martian atmosphere. Hydrodynamic escape blowoff occurs when hydrogen flowing outward in a planetary wind (analogous to the solar wind) entrains and removes other gases. Since all atmospheric species can be entrained in this process, it is not very sensitive to atomic mass. Intense solar ultraviolet radiation and solar wind particle fluxes provide the energy needed to drive hydrodynamic escape. These fluxes would have been several orders of magnitude larger than at present during the first ~10^7 years after planet formation

as the evolving Sun moved toward the **main sequence**. Although the early Sun was 25–30% less luminous overall, studies of early stars suggest that the early Sun was rotating more than ten times faster than at present, which would have caused more magnetic activity, associated with over a hundred times more emission in the extreme ultraviolet portion of the spectrum than today. Consequently, hydrodynamic escape would have been a very efficient atmospheric removal mechanism if hydrogen had been a major atmospheric constituent during this period.

The amount of hydrogen in the early atmosphere of a terrestrial planet depends on the interactions between iron and water during accretion and separation of the core and mantle. If water brought in by impacting bolides could mix with free iron in this period, it would oxidize free iron, releasing large amounts of hydrogen to the atmosphere and fostering hydrodynamic escape. Interior modeling constrained by Mars' gravitational field and surface composition together with analyses of the composition of the SNC meteorites indicates that the mantle is rich in iron oxides relative to Earth, consistent with the hypothesis that a thick hydrogen-rich atmosphere formed at this early stage. It has been suggested that hydrodynamic escape removed the equivalent of an ocean at least 1 km deep together with most other atmospheric volatiles from Mars, although this estimate is based on extrapolation from the current value of the deuterium–hydrogen ratio (D/H), which is uncertain because D/H may reflect geologically recent volatile exchange rather than preferential loss of hydrogen compared to deuterium over the full history of Mars. Comets arriving after the completion of hydrodynamic escape may have brought in most of the atmospheric volatiles in the current inventory.

Mars is also vulnerable to impact-induced escape. Large impacting bodies release enough energy to accelerate all atmospheric molecules surrounding the impact site to speeds above the escape velocity. A large fraction of these fast molecules would escape. Since this mechanism is very sensitive to the gravitational acceleration, impact-induced escape would have been far more efficient on Mars than on Earth. The early history of the inner solar system is characterized by a massive flux of large asteroids and comets, many of which would have been capable of causing impact-induced escape at Mars. Based on dating of lunar rocks and impact features, this "massive early bombardment" is known to have declined rapidly after planet formation, and it terminated in the interval 4.0–3.5 Ga. The period on Mars prior to about 3.5 Ga is known as the Noachian epoch, so that massive bombardment effectively ceased around the end of the Noachian.

The late stage of massive early bombardment has left an obvious imprint in the form of impact basins (e.g., Hellas) and large impact craters that are still obvious features of roughly half of the surface (Fig. 2). More subtle "ghost" craters and basins that have been largely erased by erosion and/or filling in the relatively smooth northern plains provide further evidence of Noachian impact bombardment.

FIGURE 2 Elevation map of Mars derived from the Mars Orbiter Laser Altimeter (MOLA) on NASA's *Mars Global Surveyor*, with some major features labeled. (NASA/MOLA Science Team.)

Calculations suggest that impact escape should have removed all but ~1% of an early CO_2 rich atmosphere (Carr, 1996, p. 141; see Bibliography). Water in an ocean or in ice would have been relatively protected, however, and the efficiency of its removal by massive early bombardment is unknown.

What was the size of Mars' volatile reservoirs at the end of the massive impact bombardment period ~3.5 billion years ago? The isotopic ratios $^{13}C/^{12}C$, $^{18}O/^{16}O$, $^{38}Ar/^{36}Ar$, and $^{15}N/^{14}N$ are heavy compared with the terrestrial ratios (see Table 1). This has been interpreted to indicate that 50–90% of the initial reservoirs of CO_2, N_2, and cosmogenic argon may have been lost over the past 3.5 billion years by mass-selective **nonthermal escape** from the upper atmosphere (mainly **sputtering** produced by the impact of the solar wind on the upper atmosphere). Considering the possible current reservoirs of CO_2 in Table 2, the resulting CO_2 available 3.5 billion years ago could have been as much as ~1 bar and as little as a few tens of millibars.

Another approach to estimating the CO_2 abundance at the end of massive impact bombardment is based on the abundance of ^{85}Kr in the present atmosphere. Since this gas is chemically inert and too heavy to escape after the end of the period of massive impact bombardment, its current abundance probably corresponds closely to the abundance at the end of massive impact bombardment. Since impact escape would have effectively removed all gases independent of atomic mass, the ratio of ^{85}Kr abundance to C in plausible impacting bodies (Kuiper Belt comets or outer solar system asteroids) can then yield estimates of the total

available CO_2 reservoir at the end of the Noachian. The corresponding atmospheric pressure, if all CO_2 were in the atmosphere, would be only ~0.1 bar, in the lower range of estimates from the isotopic and escape flux analysis. This low estimate is consistent with the low modern nitrogen abundance after allowing for mass selective escape as indicated by the high ratio $^{15}N/^{14}N$ (Table 1). But early nitrogen abundance estimates are sensitive to uncertainties in modeling escape.

As mentioned previously, slow carbonate weathering of atmospheric dust has also removed CO_2 from the atmosphere. This irreversible mechanism may account for the fate of a large fraction of the CO_2 that was available in the late Noachian. Some CO_2 may also reside as adsorbed CO_2 in the porous regolith (Table 2). It has long been speculated that much of the CO_2 that was in the early atmosphere got tied up as carbonate sedimentary deposits beneath ancient water bodies. However, the failure to find carbonate sediments, in contrast to discovery of widespread sulfate sedimentary deposits, makes the existence of a large sedimentary carbonate reservoir doubtful (see further discussion later).

Escape of water in the form of its dissociation products H and O takes place now and must have removed significant amounts of water over the past 3.5 billion years. Isotopic ratios of D/H and $^{18}O/^{16}O$ in the atmosphere and in SNC meteorites and escape flux calculations provide rather weak constraints on the amount that has escaped over that period. Upper bounds on the estimates of water loss range up to 30–50 m of equivalent global ocean. These amounts are roughly comparable to estimates of the amounts currently stored in the polar caps and regolith.

Sulfur is not stable in the Martian atmosphere in either oxidized or reduced form, but significant amounts must have been introduced into the atmosphere by volcanism. Formation of the Tharsis ridge volcanic structure, believed to have been in the late Noachian period, must have corresponded with outgassing of large amounts of sulfur as well as water from the mantle and crust. Martian soils contain up to 7–8% by weight of sulfur in the form of sulfates, and Martian rocks are also rich in sulfates. SNC meteorites are $\sim$5 times as rich in sulfur as in water. It is likely that the regolith contains more sulfur than water. The volatile elements chlorine and bromine are also abundant in rocks and soils, but more than an order of magnitude less so than sulfur.

An important observation in SNC meteorites is that sulfur and oxygen isotopes in sulfates are found in relative concentrations that are mass-independently fractionated. Most kinetic processes fractionate isotopes in a mass-dependent way. For example, the mass difference between ^{34}S and ^{32}S means that twice as much fractionation between these isotopes is produced as between ^{33}S and ^{32}S in a mass-dependent isotopic discrimination process such as diffusive separation. Mass-independent fractionation (MIF) is a deviation from such proportionality. MIF is found to arise due to the interaction of ultraviolet radiation with atmospheric gases in certain photochemical processes. On Earth, the MIF of oxygen in sulfates in the extraordinarily dry Atacama Desert is taken to prove that these sulfates were deposited by photochemical conversion of atmospheric SO_2 to submicron particles and subsequent dry deposition. The MIF signature in sulfates in SNC meteorites suggests that a similar process may have produced these sulfates on Mars.

Recent discovery of methane in the atmosphere is a major surprise. Methane is removed from the atmosphere by photochemical processes that ultimately convert it to carbon dioxide and water, with a lifetime in the atmosphere of only a few hundred years. The maintenance of significant amounts of methane in the atmosphere therefore requires significant sources to replenish it. At present, sources of methane remain a matter of speculation. On Earth, methane production is almost entirely dominated by biological sources. Biogenic methane production cannot be ruled out for Mars, but abiotic production from geothermal processes (known as thermogenic methane) must be considered less speculative at this stage.

3. Present and Past Climates

3.1 Present Climate

The thin, predominantly carbon dioxide atmosphere produces a small greenhouse effect, raising the average surface temperature of Mars only about 5°C above the temperature that would occur in the absence of an atmosphere. Carbon dioxide condenses out during winter in the polar caps, causing a seasonal range in the surface pressure of about 30%. There is a small seasonal residual CO_2 polar cap at the South Pole but this cap is quite thin, and it probably represents a potential increase in carbon dioxide pressure of <2 mbar if it were entirely sublimated into the atmosphere. The atmospheric concentration of water vapor is controlled by saturation and condensation and so varies seasonally and probably daily as well. Water vapor exchanges with the polar caps over the course of the Martian year, especially with the north polar cap. During summer, the central portion of the cap surface is water ice, a residual left after sublimation of the winter CO_2 polar cap. Water vapor sublimates from this surface in northern spring to early summer, and is transported southward, but most of it is precipitated or adsorbed at the surface before it reaches southern high latitudes.

In addition to gases, the atmosphere contains a variable amount of icy particles that form clouds and dust. Dust loading can become quite substantial, especially during northern winter. Transport of dust from regions where the surface is being eroded by wind to regions of dust deposition occurs in the present climate. Acting over billions of years, wind erosion, dust transport, and dust deposition strongly modify the surface (see Section 3.5). Visible optical depths can reach $\sim$5 in global average and even more in local dust storms. A visible optical depth of 5 means that direct visible sunlight is attenuated by a factor of $1/e^5$, which is roughly 1/150. Much of the sunlight that is directly attenuated by dust reaches the surface as scattered diffuse sunlight. Median dust particle diameters are $\sim$1 micrometer, so this optical depth corresponds to a column dust mass $\sim$3 mg/m^2. Water ice clouds occur in a "polar hood" around the winter polar caps and over low latitudes during northern summer, especially over uplands. Convective carbon dioxide clouds occur at times over the polar caps, and they occur rarely as high-altitude carbon dioxide cirrus clouds.

Orbital parameters cause the cold, dry climate of Mars to vary seasonally in somewhat the same way as intensely continental climates on Earth. The present tilt of Mars' axis (25.2°) is similar to that of Earth (23.5°), and the annual cycle is 687 Earth days long or about 1.9 Earth years. Consequently, seasonality bears some similarity to that of the Earth, but Martian seasons last about twice as long on average. However, the **eccentricity** of Mars' orbit is much larger than Earth's (0.09 compared with 0.015), and perihelion (the closest approach to the Sun) currently occurs near northern winter solstice. As a consequence, asymmetries between northern and southern seasons are much more pronounced than on Earth. Mars' rotation rate is similar to Earth's, and like Earth, the atmosphere is largely transparent to sunlight so that heat is transferred upward from the solid surface into the atmosphere. These are the major factors that control the forces and motions in the atmosphere (i.e., atmospheric dynamics). Consequently, atmospheric dynamics of Mars and Earth are similar. Both are dominated by a single meandering midlatitude jet stream, strongest during winter, and a thermally driven **Hadley circulation** in lower latitudes. The Hadley circulation is strongest near the solstices, especially northern winter solstice, which is near perihelion, when strong rising motion takes place in the summer (southern) hemisphere and strong sinking motion occurs in the winter (northern) hemisphere.

Mars lacks an ozone layer, and the thin, dry atmosphere allows very short wavelength ultraviolet radiation to penetrate to the surface. In particular, solar ultraviolet radiation in the range 190–300 nm, which is largely shielded on Earth by the ozone layer, can reach the lower atmosphere and surface on Mars. This allows water vapor dissociation close to the Martian surface (H_2O + ultraviolet photon $\rightarrow$ H + OH). As a consequence of photochemical reactions, oxidizing free radicals (highly reactive species with at least one unpaired electron, such as OH or HO_2) are produced in near-surface air. In turn, any organic material near the surface rapidly decomposes, and the soil near the surface oxidizes. These conditions as well as the lack of liquid water probably preclude life at the surface on present-day Mars.

Although liquid water may not be completely absent from the surface, even in the present climate it is certainly very rare. This is primarily because of the low temperatures. Even though temperatures of the immediate surface rise above freezing at low latitudes near midday, above freezing temperatures occur only within a few centimeters or millimeters on either side of the surface in locales where the relatively high temperatures would be desiccating. A second factor is the relatively low pressure. Over large regions of Mars, the pressure is below the triple point at which exposed liquid water would rapidly boil away.

Because the present atmosphere and climate of Mars appear unsuitable for the development and survival of life, at least near the surface, there is great interest in the possibility that Mars had a thicker, warmer, and wetter atmosphere in the past. These possibilities are constrained by the volatile abundances, estimates of which are provided in Table 2.

3.2 Past Climates

Three types of features strongly suggest that fluids have shaped the surface during all epochs—Noachian (prior to about 3.5 billion years ago), Hesperian (roughly 3.5 to 2.5–2.0 billion years ago), and Amazonian (from roughly 2.5 to 2.0 billion years ago to the present). In terrains whose ages are estimated on the basis of crater distributions and morphology to be Noachian to early Hesperian, "valley network" features are abundant (Fig. 3). The morphology of valley networks is very diverse, but most consist of dendritic networks of small valleys, often with V-shaped profiles that have been attributed to surface water flows or groundwater sapping. Although generally much less well developed than valley network systems produced by fluvial erosion

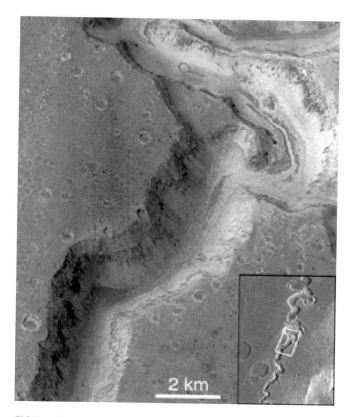

FIGURE 3a   An image of Nanedi Vallis (5.5°N, 48.4°W) from the Mars Orbiter Camera (MOC) on NASA's *Mars Global Surveyor* spacecraft. The sinuous path of this valley at the top of the image is suggestive of meanders. In the upper third of the image, a central channel is observed and large benches indicate earlier floor levels. These features suggest that the valley was incised by fluid flow. (The inset shows a lower-resolution *Viking Orbiter* image for context.) (From image MOC-8704, NASA/Malin Space Science Systems.)

on Earth, they are suggestive of widespread precipitation and/or subsurface water release (groundwater sapping) that would have required a much warmer climate, mainly but not entirely, contemporaneous with termination of massive impact events at the end of the Noachian (~3.5 billion years ago). In Fig. 3, we show two very different examples of valley network features. Fig. 3a is a high-resolution image that shows a valley without tributaries in this portion of its reach (although some tributary channels are found farther upstream), but its morphology strongly suggests repeated flow events. Figure 3b shows a fairly typical valley network at comparatively low resolution. Such images, from the *Viking* spacecraft, suggested a resemblance to drainage systems on Earth. However, at high resolution, morphology of the individual valleys in this system does not strongly suggest liquid flow, possibly due to subsequent modification of the surface.

A second class of features suggesting liquid flow is a system of immense channels apparently produced by fluid activity during the Hesperian to early Amazonian epochs

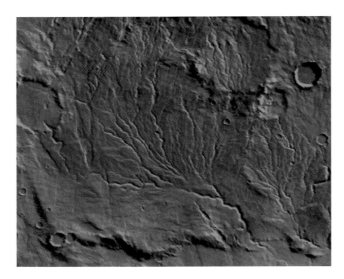

FIGURE 3b A valley network, centered near 42°S, 92°W. The image is about 200 km across. This false color mosaic was constructed from the *Viking* Mars Digital Image Map. (From NASA/Lunar and Planetary Institute Contribution No. 1130.)

(Figs. 4, 5). These features, referred to as outflow channels (or catastrophic outflow channels), are sometimes more than 100 km in width, up to ~1000 km in length, and as much as several kilometers deep. They are found mainly in low latitudes (between 20° north and south) around the periphery of major volcanic provinces such as Tharsis and Elysium, where they debauch northward toward the low-lying northern plains. The geomorphology of these channels has been compared with the scablands produced by outwash floods in eastern Washington State from ice age Lake Missoula, but if formed by flowing water, flow volumes must have been larger by an order of magnitude or more. It has been estimated that the amount of water required to produce them is equivalent to a global ocean at least 50 m deep. Many of these channels originate in large canyons or jumbled chaotic terrain that was evidently produced by collapse of portions of the plateau surrounding Tharsis. The origin of these features is unknown, but the dominant hypothesis is that the outflow channels were generated by catastrophic release of water from subsurface aquifers or rapidly melting subsurface ice. If water was released by these flows, its fate is unknown, although a number of researchers have proposed that water pooled in the northern plains and may still exist as ice beneath a dust-covered surface.

Gullies are a third piece of evidence and suggest that water has flowed in the very recent geologic past across the surface. Such features are commonly found on poleward-facing sloping walls of craters, plateaus, and canyons, mainly at southern midlatitudes (~35–55°S) (Fig. 6). These gullies typically have well-defined alcoves above straight or meandering channels that terminate in debris aprons. Their setting on steep slopes and their morphology suggest that they were produced in the same way as debris flows in terrestrial alpine regions. These flows are typically produced by rapid release of water from snow or ice barriers and consist typically of ~75% rock and silt carried by ~25% water. Several possible mechanisms have been suggested to generate local release of water or brines in debris flows from ice-rich layers on Mars (including slow heating variations due to Milankovitch cycles—see discussion later). Evidence for the active influence of Milankovitch-type cycles includes a thin, patchy mantle of material, apparently consisting of cemented dust, that has been observed within a 30–60° latitude band in each hemisphere, corresponding to places where near-surface ice has been stable in the

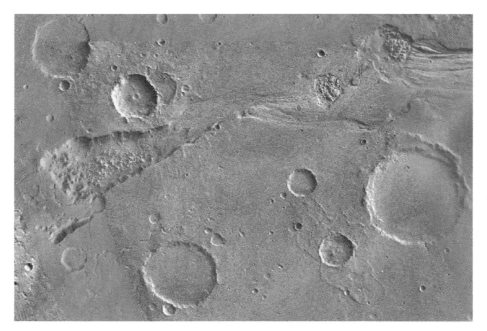

FIGURE 4 The head of the channel Ravi Vallis, about 300 km long. An area of chaotic terrain on the left of the image is the apparent source region for Ravi Vallis, which feeds into a system of channels that flow into Chryse Basin in the northern lowlands of Mars. Two further such regions of chaotic collapsed material are seen in this image, connected by a channel. The flow in this channel was from west to east (left to right). This false color mosaic was constructed from the *Viking* Mars Digital Image Map. (From NASA/Lunar and Planetary Institute Contribution No. 1130.)

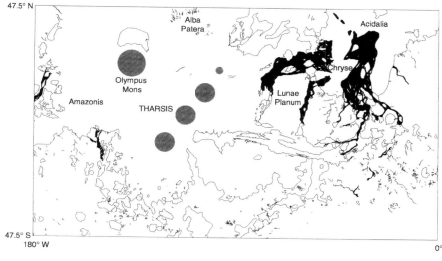

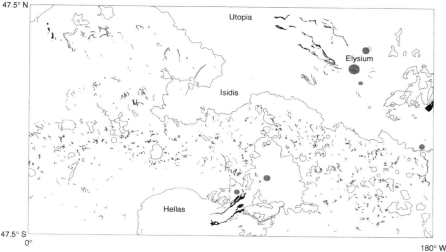

FIGURE 5 The distribution of outflow channels and valleys over ±47.5° latitude. The upper panel shows the western hemisphere and the lower panel the eastern hemisphere. Outflow channels are marked in black and drain into four regions: Amazonis and Arcadia, Chryse and Acidalia, Hellas, and Utopia; valley networks are marked as finer features. Volcanoes are shaded gray except for Alba Patera so that valleys on its flanks are not obscured. A thin line marks the boundary between Noachian and Hesperian units. (From Carr, 1996.)

last few million years due to orbital changes. The material is interpreted to be an atmospherically deposited ice–dust mixture from which the ice has sublimated. Gullies, which are probably associated with ice from past climate regimes, are found within these same latitude bands. Consequently, gullies do not require an early warm climate or enormous low-latitude reservoirs of subsurface water or ice, so we will not discuss gullies further.

The three geomorphic features listed previously (valleys, channels, and gullies) provide for a reasonably direct attribution for the cause of erosion. For completeness, we mention that relatively high erosion rates are evident in the Noachian from craters with heavily degraded rims and infilling or erosion. Some models of the degradation of craters suggest that the erosion and deposition was caused by fluvial activity, at least in part. However, the interpretation is necessarily complex because the image data suggests that craters were also degraded or obscured by impacts, eolian transport, mass wasting, and, in some places, airfall deposits such as volcanic ash or impact ejecta.

3.3 Mechanisms for Producing Warm Climates

Despite extensive investigation, the causes of early warm climates, if indeed they have existed since the late Noachian, remain to be identified. Here we review several possibilities.

1. Carbon dioxide greenhouse. An appealing suggestion put forward after the *Mariner* 9 orbiter mission in 1972 is that the early atmosphere contained much more CO_2 than it does now. The idea is that substantial CO_2 caused an enhanced greenhouse effect through its direct infrared radiative effect and the additional greenhouse effect of increased water vapor, which the atmosphere would have held at higher temperatures. Applied to the late Noachian period of valley network formation, this theory runs into difficulty because of the lower solar output at ~3.5 billion years ago (~75% of present output), and consequent large amount of CO_2 required to produce an adequate CO_2–H_2O greenhouse effect. At least several bars of CO_2 would have been required to produce widespread surface temperatures above freezing. However, such thick atmospheres are not

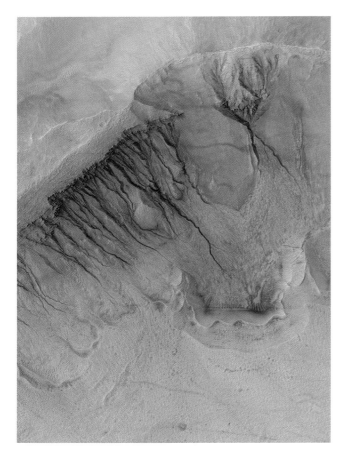

FIGURE 6   Gullies in the northern wall of an impact crater in Terra Sirenum at 39.1°S, 166.1°W. The image is approximately 3 km across. (Synthetic color portion of Mars Orbiter Camera image E11-04033; NASA/Malin Space Science Systems.)

physically possible because CO_2 condenses into clouds at ~1 bar. It has been suggested that such CO_2 ice clouds could have contributed to the greenhouse effect to the degree that made up for the loss of CO_2 total pressure. However, recent studies indicate that CO_2 ice clouds could not warm the surface above freezing because CO_2 particles would grow rapidly and precipitate, leading to rapid cloud dissipation. Warming may also be self-limiting: by heating the air, the clouds could cause themselves to dissipate.

If a massive CO_2 atmosphere ever existed, it could have persisted for tens of millions of years, but it would have eventually collapsed due to removal of the CO_2 by solution in liquid water and subsequent formation of carbonate sediments. However, despite extensive efforts, not a single outcrop of carbonate sediments has been found. The absence occurs even in areas in which water is interpreted to have flowed (the *Opportunity* rover site) and in which extensive erosion would be expected to have exposed carbonate sediments buried beneath regolith. In contrast, sulfate sedimentary deposits are widespread in the tropics (Fig. 7), some in terrains that have been exhumed by wind erosion. In retrospect, it is not surprising that carbonate reservoirs

have not been found. In the presence of abundant sulfuric acid, carbonate would be quickly converted to sulfate with release of CO_2 to the atmosphere, where it would be subject to various loss processes discussed earlier.

Although a future discovery of a large carbonate sediment reservoir cannot be ruled out, it now seems doubtful, and the amount of CO_2 available seems inadequate to have produced a warm enough climate to account by itself for the valley networks by surface runoff and/or groundwater sapping in the late Noachian.

2. Impact heating. The largest asteroid or comet impacts would vaporize large quantities of rock. Vaporized rock would immediately spread around the planet, condense, and, upon reentry into the atmosphere, would flash heat the surface to very high temperature. This would quickly release water from surface ice into the atmosphere. Upon precipitation, this water could produce flooding and rapid runoff over large areas. Water would be recycled into the atmosphere as long as the surface remained hot, anywhere from a few weeks to thousands of years depending on impact size. It has been proposed that this is an adequate mechanism for producing most of the observed valley networks. Although a very extended period of warm climate would not be produced this way, repeated short-term warm climate events could have occurred during the late Noachian to early Hesperian. Detailed questions of timing of large impact events and formation of the valley network features needed to test this hypothesis remain to be resolved, but impact heating must have released ice to the atmosphere and caused subsequent precipitation at some times during the Noachian.

3. Sulfur dioxide greenhouse. The high abundance of sulfur in surface rocks and dust as well as in the Martian meteorites suggests that Martian volcanism may have been very sulfur-rich. In contrast to Earth, Martian volcanoes may have released sulfur in amounts equal to or exceeding water vapor releases. In the atmosphere in the presence of water vapor, reduced sulfur would rapidly oxidize to SO_2 and perhaps some carbonyl sulfide, COS. Sulfur dioxide is a powerful greenhouse gas, but it would dissolve in liquid water and be removed from the atmosphere by precipitation very rapidly. SO_2 could only have been a significant greenhouse gas if it raised the average temperature to near freezing, making it easier for perturbations such as impacts to warm the climate. In this way, with a sufficient SO_2 volcanic flux, the amount of SO_2 would perhaps have been self-limiting. Detailed constraints on possible early SO_2 greenhouse conditions, including persistence and timing have yet to be worked out.

4. Methane-aided greenhouse. Methane is also a powerful greenhouse gas, but because of its instability in the atmosphere, it has not seemed an attractive option for contributing to an early warm climate until very recently. With the apparent detection of methane in the current atmosphere and the lack of definitive identification of its sources,

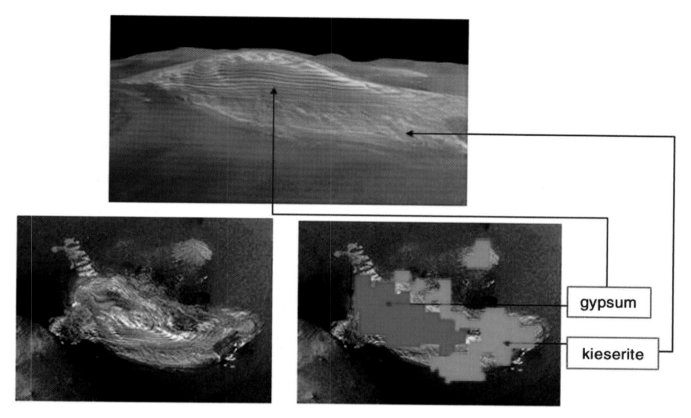

FIGURE 7   The upper three-dimensional view shows a 2.8-km-tall and 40-km-long sulfate-rich layered deposit that lies within Juventae Chasma, a deep chasm some 500 km north of Valles Marineris. Below are maps of sulfates on the deposit obtained by a near-infrared spectrometer, OMEGA (Observatoire pour la Minéralogie, l'Eau, les Glaces, et l'Activité), on the Mars Express spacecraft. Gypsum (blue) dominates in the layered bench-cliff topography, while kieserite (red) lies around and below. (Reprinted with permission from Bibring et al., 2005, *Science* **307**, 1576–1581. Copyright AAAS.)

the possibility of an early methane-aided greenhouse warrants further investigation. However, the required amount of methane to warm early Mars would require a global methane flux from the surface of Mars similar to that produced by the present-day biosphere on Earth.

5. Mechanisms for producing large flow features in cold climates. Although some precipitation must have occurred due to impacts and short-lived greenhouse warming is plausible, other factors may have produced valley network and outflow channel features. Hydrated sulfates are widespread at the surface today and must have been widespread on early Mars as well. Volcanic or impact heating could have caused rapid dehydration of sulfates and flow of the resulting brines across the surface. Under some circumstances, catastrophic dehydration of massive hydrated sulfate deposits could have occurred, and resulting high volume flows could have produced outflow channel features. It is also possible that fluids other than water or brine produced the outflow channels. For example, the abundance of sulfur indicated in mantle and crustal rocks suggests that Martian volcanism may have produced very fluid sulfur-rich magmas. Indeed, extensive fluid lava flows have been identified in high-resolution images of the Martian surface. Extensive outflow channels,

some of which strongly resemble Martian outflow channel features, are found on Venus. These unexpected features were apparently formed by highly fluid magma flows. The spatial relationship between the Martian outflow channels and the major volcanic constructs is consistent with the hypothesis that very fluid magmas may have played some role in the formation of outflow channels.

3.4 Milankovitch Cycles

As on Earth, Mars' orbital elements (**obliquity,** eccentricity, argument of perihelion) exhibit oscillations known as Milankovitch cycles at periods varying from 50,000 to several million years. The obliquity and eccentricity oscillations are much larger in amplitude on Mars than on Earth (Fig. 8). Milankovitch cycles cause climate variations in two ways. First, they control the distribution of incoming solar radiation (insolation) on both an annual average and seasonal basis as functions of latitude. Second, because Milankovitch cycle variations of insolation force variations of annual average surface temperature, they can drive exchanges of volatiles between various surface reservoirs and between surface reservoirs and the atmosphere. Water vapor can move between polar cap ice deposits, and ice and adsorbed water in the regolith. Carbon dioxide can move between the atmosphere, seasonal residual polar caps, and the surface adsorption reservoir. Milankovitch variations are believed to be responsible for the complex layered structures in both the north polar water ice cap and terrains surrounding the south polar residual carbon dioxide ice cap.

In general, annual average polar cap temperatures increase relative to equatorial temperatures as obliquity increases. At very low obliquity (<10–20° depending on the precise values of polar cap **albedo** and thermal **emissivity**), the carbon dioxide atmosphere collapses onto permanent carbon dioxide ice polar caps. Orbital calculations indicate that this collapse could occur ∼1–2% of the time. At high obliquity, atmospheric pressure may increase due to warming and release of adsorbed carbon dioxide from high-latitude regolith. Calculations indicate, however, that the maximum possible pressure increase is likely to be small, only a few millibars, so Milankovitch cycles are unlikely to have been responsible for significant climate warming.

3.5 Wind Modification of the Surface

Orbital and landed images of the surface show ubiquitous evidence of active wind modification of the surface, which complicates the interpretation of climate and volatile history. The action of wind erosion, dust transport, and dust deposition is modulated by Milankovitch cycles and must have strongly changed the surface over the last few billions of years and during the Noachian.

Today, dunes, ripples, and other aeolian bedforms are widespread. Wind-modified objects, known as ventifacts, are very evident in the grooves, facets, and hollows produced by the wind in rocks at the surface. Yardangs are also common, which are positive relief features in coherent materials sculpted by wind on scales from tens of meters to kilometers. Strong winds that exert stress on the surface can initiate saltation (hopping motion) of fine sand grains (diameter ∼100–1000 micrometers) and creep of larger particles. Saltating grains can dislodge and suspend finer dust particles (diameters ∼1–10 micrometers) in the atmosphere, thereby initiating dust storms. Minimum wind speeds required to initiate saltation are typically ∼30 m s^{-1} at the level 2 m above the surface, but this saltation threshold wind speed decreases with increasing surface pressure.

Such strong winds are rare on Mars. In the *Viking* lander, both wind observations and computer simulation models of the atmospheric circulation suggest that they occur at most sites <0.01% of the time. Nevertheless, over the planet as a whole, dust storms initiated by saltation are common; they tend to occur with greater frequency in the lower elevation regions rather than in the uplands because relatively high surface pressure in the lowlands lowers the saltation threshold wind speed. They are favored by topographic variations, including large- and small-scale slopes and are common over ice-free surfaces near the edges of the season-

ally varying polar caps and in "storm track" regions where the equator-to-pole gradient of atmospheric temperature is strong. Dust storms generated by strong winds and saltation are common in some tropical lowland regions, especially close to the season of perihelion passage when the Hadley circulation is strong (near the southern summer solstice at the current phase of the Milankovitch cycle). During some years, these perihelion season storms expand and combine to such an extent that high dust opacity spreads across almost the entire planet. These planet-wide dust events are fostered by positive feedbacks between dust-induced heating of the atmosphere, which contributes to driving wind systems, and the action of the wind in picking up dust.

Dust can also be raised at much lower wind speeds in small-scale quasi-vertical convective vortices called dust devils. Because the atmosphere is so thin, convective heating per unit mass of atmosphere is much greater on Mars than anywhere on Earth, and Martian dust devils correspondingly tend to be much larger sizes (diameters up to several hundred meters and depths up to several kilometers). Since the winds required to raise dust in the vortical dust devils are lower than saltation threshold winds, dust devils are common in some regions of Mars during the early afternoon and summer when convective heating is strongest. They are often associated with irregular dark tracks produced by the removal of a fine dust layer from an underlying darker stratum. The relative importance of large saltation-induced dust storms and dust devils to the overall dust balance is unclear, but modeling studies suggest that the former are substantially more important.

Over the four billion–year history of the observable surface of Mars, there must have been substantial systematic wind transport of fine soil particles from regions in which erosion is consistently favored to regions of net deposition. Models of Martian atmospheric circulation and the saltation process suggest that net erosion must have taken place in lowland regions, particularly in the northern lowlands, the Hellas basin, and some tropical lowlands (e.g., Isidis Planitia and Chryse Planitia), with net deposition in upland regions and in some moderate elevation regions where the regional slope is small and westward facing, such as portions of Arabia Terra and southern portions of Amazonis Planitia. The distribution of surface **thermal inertia** inferred from the measured surface diurnal temperature variation supports these distributions. Regions of high thermal inertia, corresponding to consolidated or coarse-grained soils, exposed surface rocks, and bedrock patches are found where the circulation–saltation models predict net erosion over Milankovitch cycles, and regions of very low thermal inertia corresponding to fine dust are found where net deposition is predicted by the models.

There are no terrestrial analogs of surfaces modified by wind erosion and deposition over four billion years, so it is difficult to comprehend fully the modifying effect of Martian winds extending over such a long time. However,

(A)

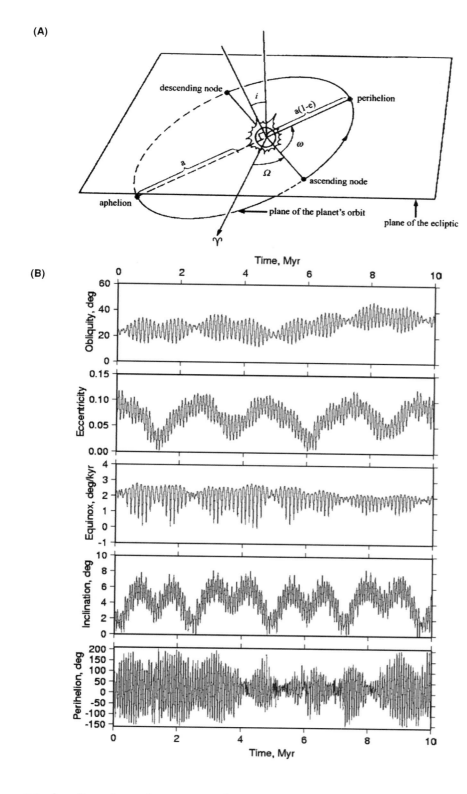

(B)

FIGURE 8 (a) Orbital elements. Mars, like other planets, moves in an elliptical orbit with a semimajor axis a. The eccentricity e defines how much the ellipse is elongated. The plane of the orbit is inclined by angle i to the ecliptic, which is the geometrical plane that contains the Earth's orbit. The ascending node is the point where the planet moves up across the ecliptic plane and the descending node is where the planet moves below it. The vernal equinox, marked γ, represents a reference direction that defines the longitude of the ascending node, Ω. Angle ω is the argument of perihelion. (b) Calculated variations in Martian orbital parameters over the last 10 million years. (Reprinted from Armstrong et al., 2004, *Icarus* **171**, 255–271, with permission from Elsevier.)

it is clear from the surface imagery that wind has played a large role in modifying the surface. In some areas, repeated burial and exhumation events must have taken place. Based on the heights of erosionally resistant mesas, the Meridiani Planum site of the *Opportunity* rover activities appears to have been exhumed from beneath at least ~50 meters and perhaps as much as several kilometers of soil. Many of the sulfate layer deposits described earlier appear to be undergoing exhumation. Since surface features can be repeatedly buried, exposed, and reburied over time, inferences of event sequences and surface ages from crater size distributions are rendered complex.

Because the saltation process operates on the extreme high-velocity tail of the wind speed distribution, it is very sensitive to surface density or pressure changes. Some model results have indicated that an increase in surface pressure up to only 40 mbar would increase potential surface erosion rates by up to two orders of magnitude. If, as is likely, Mars had a surface pressure $\sim$100 mbar or higher during the late Noachian, rates of surface modification by wind should have been orders of magnitude greater than today. Indeed, it has long been observed that late Noachian surfaces were undergoing much more rapid modification than during later periods. This has generally been attributed to precipitation and runoff under a warmer climate regime, as discussed earlier. But surface modification by winds under a denser atmosphere should also have contributed to the observed rapid modification of late Noachian age surfaces.

4. Concluding Remarks

Although ice is now known to be widespread near the surface and there is considerable evidence that liquid water once flowed across the surface in dendritic valley networks and immense outflow channels, we still do not know the exact conditions responsible for releasing water (or other fluids) at the surface. New observations point to the importance of sulfur compounds, particularly sulfates, in Martian surface and atmosphere evolution, and the high ratio of sulfur to water in Martian meteorites suggests that sulfates may have exerted an important control on the availability of water rather than conversely as on Earth. Recent spectroscopic identification of methane is a surprise because of its relatively short lifetime in the atmosphere, which requires a continuous source. Future measurements should aim to confirm this result and define the distribution of methane. If significant amounts of methane are indeed found to be present in the atmosphere, then the methane source and potential past climatic impact need to be understood.

It has always been difficult to understand how Mars could have had a sufficiently dense carbon dioxide atmosphere to produce a warm wet climate at any time from the late Noachian onward. The severity of the problem is that the early Martian atmosphere has to provide $\sim$80°C of greenhouse warming to raise the mean global temperature above freezing, which is more than double the greenhouse warming of 33°C of the modern Earth. So, despite new spectral data from orbit, the failure to find sedimentary carbonate rocks showing that exhumed sulfate deposits are widespread is noteworthy, though in retrospect it should not be surprising. If a large sedimentary carbonate reservoir is indeed absent, it is far less likely than previously thought that Mars has had extended episodes of warm wet climate due to a carbon dioxide greenhouse at any time from the late Noachian onward. In view of these new results, other candidate mechanisms for the release of fluids at the surface to form valley networks and outflow channels should be considered. During the Noachian, large impacts would have provided sufficient heat to vaporize subsurface volatiles, such as water and CO_2 ice. Consequently, impacts may have generated many temporary warm, wet climates, which would be accompanied by erosion from rainfall or the recharge of aquifers sufficient to allow groundwater flow and sapping. Such a scenario would explain why the end of massive impact bombardment is accompanied by an apparently large drop in erosion rates, as well as why valley networks are found predominantly on Noachian terrain.

Geochemical data and models suggest that most of Mars' original volatile inventory was lost early by hydrodynamic escape and impact erosion. However, we do not know the degree to which volatiles were sequestered into the subsurface as minerals or ices and protected. Future landed and orbital missions can refine our understanding of the distribution and properties of subsurface ices and hydrated minerals. Radar measurements could show the depth of water ice deposits and possibly the presence of any subsurface liquid water or brine aquifers, if subsurface ice extends deep enough to allow these. But determining the amount of sulfate and carbonate that has been sequestered into the subsurface will require drilling into the deep subsurface and extensive further exploration of Mars.

Bibliography

Carr, M. H. (1996). "Water on Mars." Oxford Univ. Press, Oxford.

Carr, M. H. (1981). "The Surface of Mars." Yale Univ. Press, New Haven, Connecticut.

Cattermole P. (2001). "Mars, The Mystery Unfolds." Oxford Univ. Press, Oxford.

Hartmann, W. K. (2003). "A Traveler's Guide to Mars." Univ. Arizona Press, Tucson.

Jakosky, B. M., and Phillips, R. J. (2001). Mars volatile and climate history. *Nature* **412**, 237–244.

Kallenbach, R., Geiss, J., and Hartmann, W. K., eds. (2001). "Chronology and Evolution of Mars," Kluwer Acad. Publ., Dordrecht, Netherlands.

Kieffer, H. H., Jakosky, B. M., Snyder, C. W., and Matthews, M. S., eds. (1992). "Mars." Univ. Arizona Press, Tucson.

Leovy, C. B. (2001). Weather and climate on Mars. *Nature* **412**, 245–249.

McKay, D.S., et al. (1996). Search for past life on Mars: Possible relic biogenic activity in Martian meteorite ALH84001. *Science* **273**, 924–930.

Mars: Surface and Interior

Michael H. Carr

U. S. Geological Survey
Menlo Park, California

CHAPTER 16

Mars, the outermost of the four terrestrial planets—Mercury, Venus, Earth, and Mars—is intermediate in size between Earth and the Moon. The terrestrial planets all have solid surfaces, and on these surfaces is preserved a partial record of how each planet has evolved. Successive events, such as volcanic eruptions or meteorite impacts, both create a new record and partly destroy the old. The task of the geologist is to reconstruct the history of the planet from what is preserved at the surface. Both Mercury and the Earth's moon appear to have become geologically inactive early in their history so most of the preserved record dates from very early in the history of the solar system prior to 3.5 billion years ago. The geologic record on Venus is relatively young, most of the surface apparently having formed in the last half billion years. The record on Earth is also mostly young although ancient records are preserved on some continents. On Mars we have a record that spans almost the entire history of the solar system. Although much of the martian surface dates back to the first billion years, volcanism, tectonism, fluvial activity, glaciation, and so forth appear to have continued at a low rate until the recent geologic past so that we can follow the evolution of the planet for almost its entire history.

Our knowledge of the geologic evolution of the Earth has been largely derived from the study of the lithology, chemistry, mineralogy, and distribution of rocks at the surface. Geomorphology has played a relatively minor role.

On Mars, however, much of what we know about its geology is derived from the morphology of the surface. Even though the geomorphologic data are being increasingly supplemented by information from martian meteorites and landers on the surface, our understanding is still largely based on the appearance of the surface from orbit, and this is the main subject of this chapter.

1. Mars Exploration

The modern era of Mars exploration began on July 14, 1965, when the *Mariner 4* spacecraft flew by the planet and transmitted to Earth 22 close-up pictures of the planet, with resolutions of several kilometers. Prior to that time, we were dependent on telescopic observations, whose resolution at best is 100–200 kilometers, and which reveal no topography, only surface markings. We knew from the telescopic observations that Mars has a thin CO_2 atmosphere, polar caps that advance and recede with the seasons, and surface markings that undergo annual and secular change, but geologic studies of the planet could realistically begin only when we acquired spacecraft data.

The *Mariner 4* pictures revealed an ancient surface that resembled the lunar highlands. These results were disappointing because it had been speculated that Mars, which has an atmosphere and is larger than the Moon, might be

TABLE 1	Mars Missions		
Mission	**Nation**	**Launch Date**	**Fate**
Mariner 4	US	11/18/1964	Flew by 7/15/1965; first close-up images
Mariner 6	US	2/24/1969	Flew by 7/31/1969; imaging and other data
Mariner 7	US	3/27/1969	Flew by 8/5/1969; imaging and other data
Mars 2	USSR	5/19/1971	Crash landed; no surface data
Mars 3	USSR	5/28/1971	Crash landed; no surface data
Mariner 8	US	5/8/1971	Fell into Atlantic Ocean
Mariner 9	US	5/30/1971	Into orbit 11/3/1971; mapped planet
Mars 4	USSR	7/21/1973	Failed to achieve Mars orbit
Mars 5	USSR	7/25/1973	Into orbit 2/12/1974; imaged surface
Mars 6	USSR	8/5/1973	Crash landed
Mars 7	USSR	8/9/1973	Passed by Mars
Viking 1	US	8/20/1975	Landed on surface 7/20/1976; orbiter mapping
Viking 2	US	9/9/1975	Landed on surface 9/3/1976; orbiter mapping
Phobos 1	USSR	7/7/1988	Lost 9/2/1988
Phobos 2	USSR	7/12/1988	Mars and Phobos remote sensing
Mars Observer	US	9/22/1992	Lost during Mars orbit insertion
Pathfinder	US	12/4/1996	Landed 7/4/1997; lander and rover data
Global Surveyor	US	11/7/1996	Into orbit 9/11/1997; imaging and other data
Mars Odyssey	US	4/7/2001	In orbit 10/24/2001; imaging, remote sensing
Spirit Rover	US	6/10/2003	Landed in Gusev 1/3/2004
Opportunity Rover	US	7/7/2003	Landed in Meridiani 1/24/2004
Mars Express	Europe	6/2/2003	In orbit 12/25/2003; imaging, remote sensing
Mars Reconnaissance Orbiter	US	8/12/2005	In orbit 3/10/2006; imaging, remote sensing

more Earth-like than Moon-like. *Mariner 4* was followed by two more *Mariner* spacecraft in 1969 (Table 1), which seemed to confirm Mars' lunar-like characteristics. However, our perception of Mars changed dramatically in 1972 when systematic mapping by the *Mariner 9* orbiter spacecraft revealed the planet that we know today. As mapping progressed, huge volcanoes, deep canyons, enormous dry riverbeds, and extensive dune fields came into view, and a complex, variegated geologic history became apparent. Exploration of Mars continued in the 1970s as both the USSR and the United States sent landers to the surface and other vehicles to the planet. Exploration in the 1970s culminated with the *Viking* mission, which successfully placed two landers on the surface and two other spacecraft into orbit. By the end of the *Viking* mission, almost all the surface had been photographed from orbit at a resolution of about 250 m/pixel and small fractions with resolutions as high as 10 m/pixels. In addition, the *Viking* landers had carried out a variety of experiments directed mostly toward detecting life and understanding the chemistry of the soil.

In the early 1980s, our understanding of Mars was further enhanced when it became clear that we had samples of Mars in our meteorite collections here on Earth. A group of meteorites called SNCs (which stands for Shergotty–Nakhla–Chassigny and is pronounced snicks) were initially suspected of being of martian origin because they were basaltic and were 1.3 billion years old. These meteorites could not have come from the Earth because their oxygen isotope ratios are distinctively different from terrestrial ratios. The only plausible body that could have been volcanically active at that time and supplied the meteorites was Mars. A martian origin was later confirmed by finding, trapped within the meteorites, gasses that are identical in composition to those in the martian atmosphere as measured by the *Viking* landers. The meteorites are believed to have been ejected from Mars by large impacts and subsequently captured by Earth after spending several million years in space. We have since added to the collection, and there are now about 30 known martian meteorites.

All these meteorites are basaltic, and all but one have ages significantly less than the age of the planet. The exception, ALH84001, is 4.5 billion years old. In 1996, it was tentatively suggested that carbonate globules within the meteorite, together with some disequilibrium mineral assemblages, polycyclic aromatic hydrocarbons (PAHs), and a number of different types of very small segmented rods that resemble some terrestrial nanofossils, might be the result of biologic activity. This suggestion has, however, received little support from subsequent investigations by the general science community.

The most recent stage of Mars exploration started in 1997 with the landing of *Mars Pathfinder* on Chryse Planitia. This has been followed by a series of long-lived missions that have precisely determined the topography and gravity field and returned a vast amount of imaging and spectral data. In addition, as of this writing, the two rovers, *Spirit* and *Opportunity*, were relaying from the surface data that showed definitive evidence of water-lain sediments and aqueous alteration.

2. General Characteristics

2.1 Orbital and Rotational Constants

The martian day is almost the same as a day on Earth, but the year is almost twice as long (Table 2). Because its rotational axis is inclined to the orbit plane, Mars, like the Earth, has seasons. But the Mars orbit has significant eccentricity. This causes one pole that tilts toward the Sun at perihelion to have warmer summers than the other pole. At present, the south has the warmer summers, but, because of a slow change in the direction of tilt of the rotational axis and a slow change in the orientation of perihelion, the hot and cold poles change on a 51,000-year cycle. The eccentricity also causes the seasons to have significantly different lengths (see Table 2). At present the Mars **obliquity** is similar to the Earth's. Yet the Earth experiences only minor changes in obliquity, while the obliquity of Mars changes chaotically, ranging from a low of 0° to a high in excess of 60°.

At low obliquities, the atmosphere thins as most of the CO_2 in the atmosphere condenses on the poles. At high obliquities, the water ice polar caps dissipate, and ice condenses at lower latitudes.

2.2 Surface Conditions

Mars has a thin atmosphere that provides almost no thermal blanketing. As a result, temperatures at the surface have a wide diurnal range, controlled largely by latitude, the reflectivity of the surface, and the thermal properties of the surface materials. Typically, surface temperatures in summer at latitudes ±60° range from 180 K at night to 290 K at midday but can range more widely if the surface consists of unusually low-density, fine-grained material. However, these temperatures are somewhat deceiving because, at depths of a few centimeters below the surface, temperatures are at the diurnal mean of 210–220 K. At the poles in winter, temperatures drop to 150 K at which point CO_2 condenses out of the atmosphere to form the seasonal cap. The atmospheric pressure at the surface ranges from about 14 millibars in the bottom of the Hellas basin to about 3 millibars at the top of the tallest volcanoes, and it changes annually as a result of formation of the polar caps. Winds are typically a few meters per second but there may be gusts up to 50 m/s. Dust devils and local dust storms are common, and almost every year regional or global-scale dust storms occur.

TABLE 2	Earth and Mars: General Characteristics Compared	
	Earth	**Mars**
Mean equatorial radius (km)	6378	3396
Mass ($\times 10^{24}$ kg)	5.98	0.624
Mean distance from Sun (10^6 kg)	150	228
Orbit eccentricity	0.017	0.093
Obliquity	23.5°	25.2°
Length of day	24 h	24 h 39 m 35 s
Length of year (Earth days)	365.3	686.9
Seasons (Earth days)		
Northern spring	92.9	199
Northern summer	93.6	183
Northern fall	89.7	147
Northern winter	89.1	158
Atmosphere	79% N_2, 21% O_2	95% CO_2, 3% N_2, 2% Ar
Surface pressure (mbar)	1000	7
Mean surface temperature (K)	288	215
Surface gravitational acceleration (cm/s^{-2})	981	371
Moons	1	2

The stability of water is of profound importance for understanding martian geology. Under the conditions just described, the planet has a thick **permafrost zone** that extends a few kilometers deep at the equator and several kilometers deep at the poles. Any unbound water present will exist as ice in this zone. There may be liquid water beneath the permafrost. Water ice caps are present at both poles, although that at the South Pole is largely masked by a remnant summer CO_2 cap. At latitudes between about 40° and the edge of the water ice cap, abundant ice has been detected just below a dehydrated zone a few tens of centimeters thick. At latitudes less than about 40°, ice is unstable at all depths. A block of ice placed in the ground at these latitudes will slowly sublimate into the atmosphere. The small amounts of water that have been detected at low latitudes may be water bound in minerals or water inherited from an earlier era when water ice was stable at these latitudes.

2.3 Planet Formation and Global Structure

Like the other planets, Mars formed from materials that condensed out of the early solar nebula, a disc of gas and dust that surrounded the early Sun. Carbonaceous chondrites, a class of meteorites that is almost identical in composition to the photosphere of the Sun, are believed to resemble closely the composition of the early nebula. Radioisotopes date the formation of the nebula at 4.567 billion years ago. The planets formed as the dust and gas accumulated into discrete bodies, and gravitational attraction favored growth of larger bodies over smaller bodies. Mars formed remarkably quickly. The evidence is from short-lived radioisotopes. The high rate of accretion resulted in global melting. Melting enabled settling of heavy iron-rich melts to the center of the planet to form a core separated from the silicate-rich mantle. During this process siderophile elements, which dissolve preferentially in iron-rich melts over coexisting silicate-rich melts, became depleted in the mantle and enriched in the core. As a result, formation of the core can be dated because the daughter products of some short-lived, strongly siderophile elements are present in the mantle, as indicated by the composition of martian meteorites. For example, ^{182}Hf decays to ^{182}W with a half-life of 9 million years. W is highly siderophilic so should mostly enter the core, yet there is an excess of ^{182}W in the mantle. Not all the Hf had decayed before the core formed. This and other isotopic evidence indicate that the core formed within 20 million years of the formation of the elements that comprise the solar system. Global melting may also have enabled some crust to form very early. This is supported both by isotopic evidence and by the finding of a martian meteorite (ALH84001) that has a 4.5-billion-year age. New crust, of course, has continued to form as indicated by volcanoes and extensive volcanic plains.

The Earth's core is inferred to be iron-rich from (1) the core's density as deduced from the core's size and the planet's moment of inertia, (2) modeling the bulk composition of the Earth and comparing it with the chondritic meteorites from which the Earth formed, and (3) depletion of siderophile elements in mantle-derived rocks as compared with chondritic meteorites. We can do similar reasoning for Mars except that we know the size of the Earth's core from seismic data but must infer the size of Mars' core. The best estimate is that the core radius is between 1300 and 1500 km. In addition, the martian core may be more sulfur-rich than the Earth's core because the Mars mantle is more depleted in chalcophile elements (those that preferentially dissolve in sulfur-rich melts) than is the Earth's.

One of the more surprising results of the *Mars Global Surveyor* mission was discovery of large magnetic anomalies in the crust despite the absence of a magnetic field today. Their presence indicates that Mars had a magnetic field in the past, but that it switched off at some time. The size of the anomalies suggests that they must result from sources in the outer few tens to several tens of kilometers of the crust and that their magnetizations are higher by an order of magnitude than magnetizations typically encountered in terrestrial rocks. The anomalies probably formed when rocks, containing iron-bearing minerals, crystallized in the presence of a magnetic field. Most of the anomalies, and all the largest are in the southern uplands. They are particularly prominent on either side of the 180° longitude where there are several broad, east-west stripes. One interpretation of the linear anomalies is that they result from injection of dikes or dike swarms several tens of kilometers wide and hundred of kilometers long in the presence of a strong magnetic field. Anomalies are mostly absent around the youngest large impact basins, Utopia, Hellas, Isidis, and Argyre. The simplest explanation is that there was no longer a magnetic field when these basins formed, formation of the basins destroyed any preexisting anomalies, and no new ones formed when the affected materials cooled after the basin-forming events. The ages of the basins are not known, but, by analogy with the Moon, they are likely to have formed toward the end of heavy bombardment around 3.8–4 billion years ago. Thus, the magnetic field may have turned off by around 4 billion years ago.

Earth's magnetic field is generated by convection within its core. Mars' early dynamo probably had a similar cause. Possible causes for cessation of the dynamo are loss of core heat, solidification of most of the core, and/or changes in the mantle convection regime. Magnetization of minerals within 3.9- to 4.1-billion-years-old carbonates in the martian meteorites ALH84001 suggests that there was still a magnetic field at this time. If true, it implies that Mars had a magnetic field for the first 500 million years of its history and that the field turned off around 4 billion years ago, just before formation of the youngest impact basins.

Like the Earth's, Mars' mantle is chondritic in composition except for the depletion of siderophile and chalcophile elements as noted earlier and the depletion of volatile elements, which would have been largely lost from the interior during the early global melting phase. It consists mainly of iron-magnesium silicates. One difference between the mantles on the two planets is that the Fe/Mg ratio is higher in the martian mantle.

The crust is essentially a melt extract from the mantle. It is probably mostly basaltic in composition. The thickness of the crust varies considerably, ranging from 5 to 100 km, as estimated from the relations between the global gravity field and the global topography. The thickest crust is under the high-standing cratered terrain in the southern hemisphere; the thinnest is under the large impact basins of Isidis and Hellas.

2.4 Global Topography and Physiography

The topography and physiography have a marked north-south asymmetry, which is referred to as the global dichotomy. (See Fig. 1.) The dichotomy is expressed three ways: as a change in elevation, a change in crustal thickness, and a change in crater density. The southern uplands have an average elevation 5.5 km higher than that in the northern plains, the crust is roughly 25 km thicker in the uplands, and most of the upland terrain is heavily cratered, dating back to the period of heavy bombardment. (All the terrestrial planets were heavily bombarded by meteoritic debris early in their history. The period ended around 3.8 billion years ago.) The plains are mostly younger surfaces, but there must be an older surface at some depth beneath the younger plains. The low-lying plains constitute roughly one third of the planet and are mostly in the north. The cause of the dichotomy is not known. Suggestions include a very large impact, soon after the planet formed, or internal convection sweeping most of the light, crustal material into one half of the planet.

Superimposed on the global dichotomy is the Tharsis bulge, which is more than 5000 km across and 10 km high and is centered on the equator at 100° W. Most of the planet's volcanic activity has been centered on the bulge, which has the planet's five largest volcanoes (Alba Patera, Montes Olympus, Arsia, Ascreus, and Pavonis) on its northwest flank. Tharsis is also at the center of a vast array of radial faults and circumferential ridges that affect over half the planet's surface. To the east of the center of the bulge are a series of vast canyons thousands of kilometers long and up to 10 km deep. They are roughly radial to the bulge and appear to have formed largely by faulting, although they also have been extensively modified by fluvial and mass-wasting processes. At the east end of the canyons, extensive areas of terrain have seemingly collapsed to form **chaotic terrain** out of which emerge large dry riverbeds that extend for thousands of kilometers downslope into the northern plains. The bulge appears to be a massive accumulation of volcanic rocks. Their accumulation started very early in the

FIGURE 1   Topographic map of Mars between latitudes 65° S and 65° N. The highest elevations (whites and grays) are in Tharsis. The lowest elevations (blues) are in Hellas and the northern plains. The dominant feature of the planet is the global dichotomy between the low-lying northern plains and the cratered southern uplands. The main positive features are the volcanic provinces of Tharsis and Elysium. The main negative features are the large impact basins Hellas, Argyre, Isidis, and the buried basin Utopia. The canyons extending eastward from Tharsis, and large outflow channels such as Ares Vallis, are visible even at this global scale. (Mars Orbiter Laser Altimeter.)

planet's history so that the bulge had largely formed at the end of heavy bombardment. A much smaller bulge, centered in Elysium at 25° N, 3213° W, has also been a center of volcanic, tectonic, and fluvial activity. Other prominent topographic features are large impact basins; the largest are Hellas (2600 km diameter), Isidis (1600 km), and Argyre (1500 km).

The physiography of the poles is distinctively different from that of the rest of the planet. At each pole, extending out to the 80° latitude circle, is a stack of finely layered deposits a few kilometers thick. In the north, they rest on plains; in the south, they rest on cratered uplands. The small number of superimposed impact craters suggests that they are only a few tens of millions of years old.

3. Impact Cratering

3.1 Cratering Rates

All solid bodies in the solar system are subject to impact by asteroidal and cometary debris. (See Fig. 2.) The cratering rates are low. On Earth, in an area the size of the United States, a crater larger than 10 km across is expected to form every 10–20 million years and one larger than 100 km across, every billion years. The rates on the other terrestrial planets are likely to be within a factor of 2 or 3 of these rates. As a consequence, any surface that has a large number of craters several tens of kilometers across or larger must date back to a time when cratering rates were higher. On the Moon, surfaces are either densely covered by large craters (lunar highlands) or sparsely affected by large craters (maria) with no surfaces of intermediate crater densities. This contrast arises because of the Moon's cratering history. Very early on, cratering rates were high. Around 3.8 billion years ago they declined rapidly to roughly the present rate. Accordingly, surfaces that formed prior to 3.8 billion years ago are heavily cratered, and those that formed afterward are much less cratered. Mars has had a similar cratering history, hence the contrast between the heavily cratered uplands and the sparsely cratered plains.

Craters provide a means of estimating the ages of surfaces. As we just saw, the most densely cratered surfaces formed prior to 3.8 billion years ago, and the cratering rate has been roughly constant since that time. Consequently, a 3-billion-year-old surface will have three times more craters on it than a 1-billion-year-old surface. There is considerable uncertainty in estimating absolute ages this way because we do not know exactly what the cratering rate on Mars has been for the past few billion years. Nevertheless, by counting craters, we can put surfaces in a time-ordered sequence and make rough estimates of their absolute ages.

3.2 Crater Morphology

Impact craters have similar morphologies on different planets. Small craters are simply bowl-shaped depressions with constant depth-to-diameter ratios. With increasing size, the craters become more complex as central peaks appear, terraces form on the walls, and the depth-to-diameter ratio decreases. At very large diameters, the craters become multiringed, and it is not clear which ring is the equivalent of the crater rim of smaller craters. On Mars the transition from simple to complex takes place at 8–10 km, and the transition from complex craters to multiringed basins takes place at 130–150 km diameter.

Although impact craters on Mars resemble those on the Moon, the patterns of ejecta are quite different. Lunar craters generally have continuous hummocky ejecta near the rim crest, outside of which is a zone of radial or concentric ridges, which merge outward into strings or loops of secondary craters, formed by material thrown out of the main crater. In contrast, the ejecta around most fresh-appearing martian craters, especially those in the 5–100 km size range, are disposed in discrete, clearly outlined lobes. Various patterns are observed. The ejecta around craters smaller than 15 km in diameter are enclosed in a single, continuous lobate ridge, or rampart, situated about one crater diameter from the rim. Around larger craters, there may be many

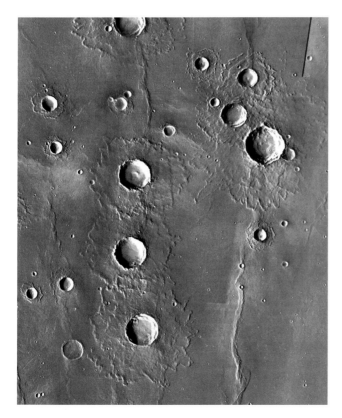

FIGURE 2   Impact craters in Lunae Planum. The ejecta are distributed around the craters in lobes, each surrounded by a low ridge or rampart. The largest crater is 35 km across. Thermal Emission Imaging System (THEMIS.)

lobes, some superimposed on others, but all surrounded by a rampart. Other craters have distinct mounds of ejecta around the rim, with more typical lobate ejecta outside the mounds. The distinctive martian ejecta patterns have been attributed to two possible causes. The first suggestion, based on experimental craters formed under low atmospheric pressures, is that the patterns are formed by interaction of the ejecta with the atmosphere. The second, and generally preferred, explanation is that the ejecta contained water and had a mudlike consistency and so continued to flow along the ground after ejection from the crater and ballistic deposition. This view is supported by the resemblance of martian craters to those produced by impacts into mud.

The previous discussion refers to fresh-appearing craters. Erosion rates at low latitudes for most of martian history are very low—typically 0.01–0.05 µm/year, although rates may be higher locally. However, early in the planet's history, erosion rates were much higher. As a consequence, in the cratered uplands, craters range in morphology from fresh-appearing craters to barely discernible, rimless depressions. In contrast, on volcanic plains in equatorial regions, almost all the craters are fresh-appearing even though they may be billions of years old. Obliteration rates have been higher at high latitudes. This has been attributed to ice-abetted creep of the near-surface materials, but other factors such as repeated burial and removal of material by the wind, may have contributed to modification of the craters. Such a process has been invoked to explain the so-called pedestal craters that are particularly common at high latitudes. These craters are inset into a platform or pedestal that has about the same areal extent as the ejecta. The simplest explanation is that the region in which these craters are found was formerly covered with a layer of loose material that has since been removed by the wind except around craters where the surface was armored by the ejecta.

4. Volcanism

Mars has had a long and varied volcanic history. Crystallization ages of martian meteorites are as young as 150 million years, and the scarcity of impact craters on some volcanic surfaces suggests that the planet is still volcanically active, although the rates must be very low compared with those found on the Earth. The tectonic framework within which martian volcanism occurs is very different from that in which most terrestrial volcanism occurs. Most terrestrial volcanism takes place at plate boundaries, which have no martian equivalents, there being no plate tectonics on Mars. Perhaps the closest terrestrial analogs to martian volcanoes are those, such as the Hawaiian volcanoes, that occur within plates rather than on the boundaries. Most martian volcanism is basaltic, but basaltic volcanism expresses itself somewhat differently on Mars because of the lower heat flow, the lower gravity, and the lower atmospheric pressure. Eruptions are expected to be larger and less frequent, more likely to produce ash, and ash clouds are more likely to collapse and produce ash-rich surface flows.

The large **shield volcanoes** of Tharsis and Elysium present the most spectacular evidence of volcanism. Shield volcanoes, such as those in Hawaii, are broad domes with shallow sloping flanks that form mainly by eruption of fluid basaltic lava. Each has a summit depression formed by collapse following eruptions on the volcano flanks or at the summit. In contrast, stratovolcanoes such as Mt. Fujiyama tend to be much smaller and have steeper flanks and a summit depression that is a true volcanic vent. Explosive, ash-rich eruptions tend to be more common in the building of a stratovolcano, and the lava tends to be more volatile rich, more siliceous, and more viscous than that which forms shields. In Tharsis, three large shield volcanoes form a northeast-southwest trending line, and 1500 km to the northwest of the line stands the largest shield of all, Olympus Mons, which is 550 km across and reaches a height of 21 km above the mars datum. (See Fig. 3.) The three aligned Tharsis Montes shields are only slightly smaller. Olympus Mons has a summit caldera 80 km across, and the flanks have a fine striated pattern caused by long linear flows, some with central leveed channels. The main edifice is surrounded by a cliff in places 8 km high. Outside the main edifice is the aureole that consists of several huge lobes with a distinctively ridged texture. It is thought to have formed as a result of the collapse of the periphery of the volcano in huge landslides that formed the lobes and left a cliff around the main edifice. The largest lobe has roughly the same area as France. The edifice is thought to have been built slowly over billions of years by large eruptions, widely spaced in time and fed from a large magma chamber within the edifice that was itself fed by a magma source deep within the mantle. Although huge, Olympus Mons is not the largest volcano in areal extent. Alba Patera, at the north end of Tharsis is 2000 × 3000 km across, almost the size of the United States. The large size of the martian shields results partly from the lack of plate tectonics. The largest shield volcanoes on Earth, those in Hawaii, are relatively short-lived. They sit on the Pacific plate, and the source of the lava is below the rigid plate. As a Hawaiian volcano grows, movement of the Pacific plate carries it away from the lava source so it becomes extinct within a few hundred thousand years. A trail of extinct volcanoes across the Pacific attests to the long-term supply of magma from the mantle source presently below Hawaii. On Mars, a volcano remains stationary and will continue to grow as long as magma continues to be supplied, so the volcanoes are correspondingly larger.

The Elysium province is much smaller than Tharsis, having only three sizeable volcanoes. One unique attribute of the Elysium province is the array of large channels that

FIGURE 3　View looking southeast across Tharsis. Olympus Mons, in the foreground, is 550 km across and 21.2 km high and is surrounded by a cliff 8 km high. Lobes of the aureole can be seen extending from the base of the cliff. 10× vertical exaggeration. (Mars Orbiter Laser Altimeter.)

start in graben around the volcanoes and extend thousands of kilometers to the northwest. They may have been formed by dikes injected into ice-rich frozen ground. Other volcanoes occur near Hellas and in the cratered uplands. Not all the volcanoes formed by fluid lava. Some appear to be surrounded by extensive ash deposits, and some have densely dissected flanks as though they were made of easily erodible materials such as ash.

Lava plains constitute the bulk of the planet's volcanic products. There are several kinds of volcanic plains. On some plains, found mostly between the volcanoes in Tharsis and Elysium, volcanic flows are clearly visible. On others, mostly found around the periphery of Tharsis and in isolated patches in the cratered uplands, ridges are common, but flows are rare. Others with numerous low cones may have formed when lava flowed over water-rich sediments. Finally, some young, level plains appear to consist of thin plates that have been pulled apart for they can be reconstructed like a jig-saw puzzle. The plates may indicate rafting of pieces of crust on a lava lake.

5. Tectonics

Most of the deformation of the Earth's surface results from the movement of the large lithospheric plates with respect to one another. Linear mountain chains, transcurrent fault zones, rift systems, and oceanic trenches all result directly from plate tectonics. There are no plate tectonics on Mars, so most of the deformational features familiar to us here on Earth are absent. The tectonics of Mars is dominated by the Tharsis bulge. The enormous pile of volcanics that constitute the Tharsis bulge has stressed the **lithosphere** and caused it to flex under the load. Modeling suggests that,

around the bulge, tensional stresses should be circumferential, and compressional stresses should be radial. This is entirely consistent with what is observed. The bulge is surrounded by arrays of radial tensional fractures and circumferential compressional ridges. Some of the tensional fractures, particularly those to the southwest of the bulge, extend for several thousand kilometers. Development of some fractures may have been accompanied by emplacement of dikes. The fractures clearly started to form very early in the planet's history, since many of the young lava plains are only sparsely fractured, whereas the underlying plains, visible in windows through the younger plains, are heavily fractured.

Not all the deformational features result from the Tharsis load. Ridges, suggestive of compression, are common on intercrater plains, such as Hesperia Planum and Syrtis Major, that are far removed from Tharsis. Some arcuate faults around Isidis and Hellas, clearly result from the presence of the large basins. Circular fractures around large volcanoes, such as Elysium Mons, and Ascreus Mons have formed as a result of bending of the lithosphere under the volcano's load. Finally, large areas of the northern plains are cut by fractures that form polygonal patterns at a variety of scales. Polygonal fracture patterns are common in the terrestrial arctic where they form as a result of seasonal contraction and expansion of ice-rich permafrost. Some of the polygonal patterns on Mars, those with polygons up to a few tens of meters across, may have also formed in this way. However, some polygons that are several kilometers across could not have formed in this way and may be the result of regional warping of the surface. Despite these examples, the variety of deformation features is rather sparse compared with those of Earth because of the lack of plate tectonics. In particular, folded rocks of any type are rare.

6. Canyons

On the eastern flanks of the Tharsis bulge is a vast system of interconnected canyons. They extend just south of the equator from Noctis Labyrinthus at the crest of the Tharsis bulge eastward for about 4000 km until they merge with some large channels and chaotic terrain. The characteristics of the canyons change from west to east. Noctis Labyrinthus at the western end consists of numerous intersecting closed, linear depressions. The depressions are generally aligned with faults in the surrounding plateau. Further east, the depressions become deeper, wider, and more continuous to form roughly east–west trending canyons. Still further east, the canyons become shallower; fluvial features become more common, both on the canyon floor and on the surrounding plateau; and finally the canyons end as the canyon walls merge into walls containing chaotic terrain. The canyons almost certainly formed largely by faulting and not by fluvial erosion, as is the case with the Grand Canyon, Arizona. Faulting is indicated by the partial merger of numerous closed depressions in the western end of the canyons and by straight walls in the east. While faulting created the initial relief, the canyons have been subsequently enlarged by failure of the walls in huge landslides and by fluvial action. The faulting was on such an enormous scale that it probably involved the entire lithosphere. (See Figs. 4 and 5.)

Thick sequences of layered (and unlayered) deposits are present in many places throughout the canyons, including some closed canyons completely isolated from main depression Fig. 5. The consensus is that the canyons formerly contained lakes and that the layered sediments were deposited in these lakes. The lakes drained to the east, hence the continuity eastward from the canyons into several large flood channels. Orbital detection of sulfates within the canyons

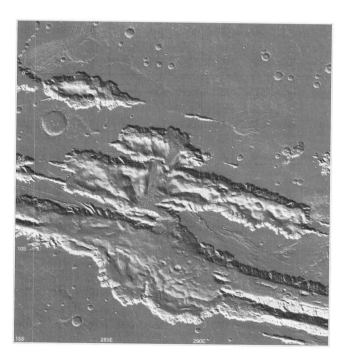

FIGURE 4 The middle section of the canyons. In the upper left is the completely enclosed Hebes Chasma, within which a mound of layered sediments is clearly visible. The main part of the canyon consists of three parallel canyons each 200 km across, also partly filled with mounds of sediments. The sediments are believed to have been deposited in lakes which drained catastrophically to the east. Candor Chasma (see Fig. 5) is the middle canyon. (Mars Orbiter Laser Altimeter.)

FIGURE 5 Detail of the layered sediments in Candor Chasma as seen from orbit. Such sediments are common throughout the canyon and in some of the adjacent depressions. (Mars Orbiter Camera.)

and some of the outlying depressions supports the lake hypothesis. If climatic conditions were similar to present-day conditions, such lakes would have frozen over, although the lake beneath the ice could have been sustained for extended periods if fed by groundwater. Even though the lake hypothesis is plausible, there are many unanswered questions, such as where the sediment in the layered deposits came from and what caused the layering.

7. Erosion and Deposition

7.1 Water

Water-worn features present some of the most puzzling problems of martian geology. Valley networks likely formed when the climate was significantly warmer than at present, yet how the climate might have changed is unclear. Huge floods have episodically moved across the surface, yet there is little trace left of the vast amounts of water that must have been involved, and gullies are forming on steep slopes during the present epoch despite the cold conditions. Perhaps most puzzling of all is whether there were ever oceans present, and, if so, how big they were, when they formed, and where all the water went.

7.1.1 BRANCHING VALLEY NETWORKS

Much of the ancient cratered uplands is dissected by branching valley networks that superficially resemble terrestrial river valleys. (See Fig. 6.) They are mostly 1–4 km across and tens to hundreds of kilometers long, although a

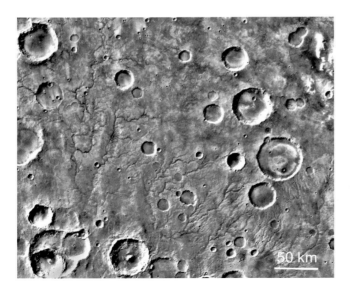

FIGURE 6   Valley networks in the ancient cratered terrain northeast of Hellas. The regional slope is to the southwest down into Hellas. The complex branching patterns indicate that the valleys formed by surface runoff following precipitation rather than seepage of groundwater. (Mars Orbiter Camera WA.)

few extend for thousands of kilometers. Large parts of the cratered uplands are heavily dissected, but other parts are sparsely dissected. Most of the younger plains are not dissected, although there are a few exceptions. A few volcanoes are also very heavily dissected. The distribution suggests that the rate of valley formation was high prior to about 3.8 billion years ago, and that it declined rapidly about that time. Seepage of groundwater has clearly contributed to formation of some of the valleys; however, most appear to have formed as a result of precipitation followed by surface runoff, which requires significantly warmer conditions than prevail at present.

No satisfactory explanation has been proposed for how early Mars could have been warmed to allow precipitation and stream flow. The output of the Sun is thought to have been less than at present during this early era. Greenhouse models suggest that even a very thick CO_2—H_2O atmosphere could not warm the surface enough. The lack of detection of carbonates from orbit also appears to rule out massive amounts of CO_2 as a cause. The possible role of other greenhouse gases is being explored. One possibility is that large impacts injected massive amounts of water into the atmosphere, which precipitated out as hot acid rain. The idea is attractive in that it might explain why the valley networks formed mainly in the old terrains when impact rates were high, but it may also explain the rarer localized occurrence of younger valleys.

7.1.2 OUTFLOW CHANNELS

Outflow channels are very different from valley networks. (See Figs. 7 and 8.) They are tens of kilometers wide and thousands of kilometers long, have streamlined walls and scoured floors, and contain teardrop-shaped islands. Most start full-size and have few if any tributaries. They closely resemble large terrestrial flood features and have almost universally been accepted to be the result of massive floods. Most start around the Chryse basin, emerging either from the canyons or from closed rubble-filled depressions and extending northward for thousands of kilometers until all traces are lost in the northern plains. The largest flood features are in the Chryse region, but others occur in Elysium, Hellas, and elsewhere, commonly starting at faults. As already indicated, the channels that merge with the canyons may have formed by catastrophic drainage of lakes within the canyons. Other outflow channels appear to have formed by massive eruptions of groundwater. Groundwater stored under pressure beneath a kilometers-thick permafrost may have been released when the permafrost seal was broken by impact, volcanic activity, or faulting. Most of the outflow channels formed in the middle of Mars' history, well after the time that the valley networks formed; some networks have formed much more recently. Cold surface conditions and a thick permafrost were probably required for their formation.

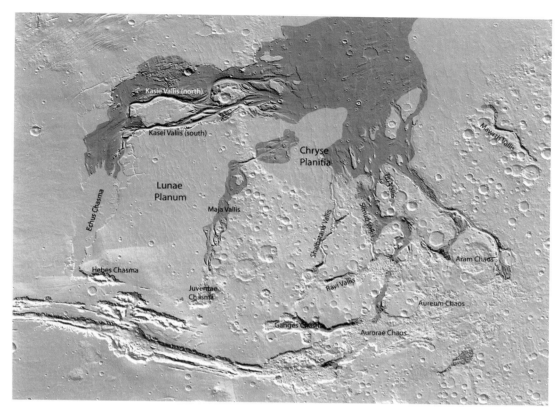

FIGURE 7   Outflow channels around the Chryse basin, to the north and east of the canyons. Dark areas are the scoured channel floors. Where the channels enter the Chryse basin they scour broad tracts several hundred kilometers across. (Mars Orbiter Laser Altimeter.)

Major issues are concerned with how much water was involved and where it all went. The size of the channels suggests that the discharges were enormous, 1000 to 10,000 times the discharge of the Mississippi River. But we do not know how long the floods lasted, so we do not know the total volume of each flood. Nevertheless, large bodies of water, or seas, must have been left in low-lying areas when the floods were over. Efforts to find evidence for these seas has had mixed results. Some researchers claim that Mars must have had oceans as extensive as those on Earth; others claim that seas larger than the Mediterranean were unlikely. Under present conditions, such seas would have frozen, and the ice would have slowly sublimed, thereby adding to the ice at the poles. However, estimates of the amount of water currently in the polar ice caps falls far short of even the lowest estimates of the amounts of water involved in the floods, so a mystery remains as to where the water went.

7.1.3 GULLIES

On many steep, poleward-facing slopes in the southern hemisphere are gullies several meters across and hundreds of meters long. They are forming during the present epoch. They are most likely formed by melting snow during periods of high obliquity. During these periods, water is driven off the poles and accumulates as ice at lower latitudes. That which accumulates on poleward-facing slopes may melt in midsummer when the slopes are directly facing at the Sun and almost permanently illuminated. Observations of gullies emerging from under smooth-surfaced deposits (ice?) on some crater walls support this suggestion.

7.2 Ice

As indicated in Section 2.1, ice has been detected on the surface at the poles and just below the surface at latitudes higher than 40°. Geomorphic evidence of ground ice is pervasive at latitudes higher than about 30°. At these latitudes, when viewed at resolutions better than 50 m/pixel, many surface features, such as ridges and crater rims appear rounded and subdued, as compared with the same features at lower latitudes. The rounding or softening has been attributed to slow, downhill movement of the near-surface materials as a result of the presence of ground ice. The rounding does not occur at lower latitudes because ground ice is unstable, and likely absent in significant quantities. Also at the higher latitudes, debris flows extend 20–30 km away from almost all cliffs and steep mountains. The

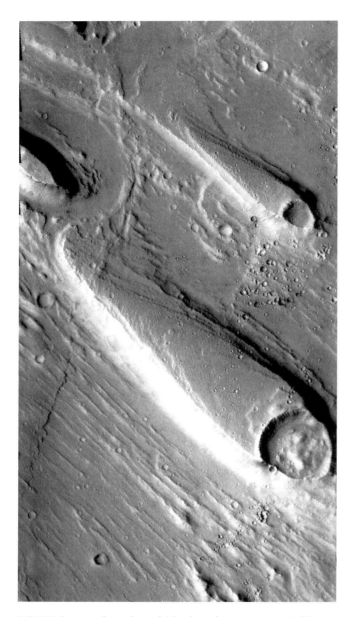

FIGURE 8   Teardrop-shaped islands and scour in Ares Vallis. The islands formed where flow was diverted around preexisting craters. Flow is from lower right to upper left. The image is 19 km across. (Mars Orbiter Camera.)

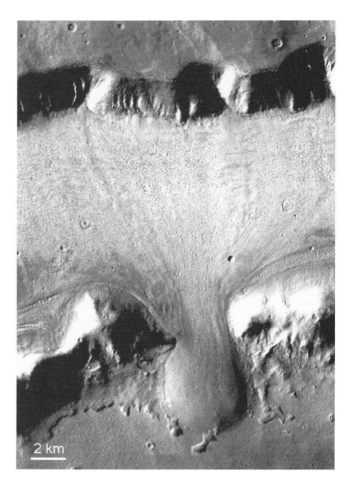

FIGURE 9   Ice-rich flow in the fretted terrain. Ice-rich material shed from the cliff at the top of the image has flowed away from the cliff and converged on a gap in hills to the south. At the latitude of this image (40° N), similar flows occur at the base of almost all cliffs and hills, which suggests that there is abundant ice in the ground. (THEMIS.)

simplest explanation is that debris shed from high ground contains ice that enables it to flow. At low latitudes, because of the absence of ice, debris flows do not form. Material eroded from the cliffs remains adjacent to the cliff and protects it from further erosion. The process is particularly evident in what has been termed fretted terrain. These are high-latitude sections of the plains/upland boundary in which wide, flat-floored valleys reach deep into the upland. Material has flowed away from the walls to widen the valleys and form the flat floors. Again, the simplest explanation is that the upland materials at these high latitudes contain ice that facilitates flow of eroded debris. (See Fig. 9.)

A wide range of other observations, particularly in the low-lying northern plains, have been interpreted to be the result of ground ice or glaciers. These include polygonally fractured ground (analogous to arctic-patterned ground?), closely spaced, curvilinear, parallel ridges (moraines?), local hollows (left by removal of ice?), branching ridges (sites of former subglacial streams?), and striated ground (glacial scour?). In addition, several features adjacent to volcanoes in Tharsis have been interpreted as glacial in origin. They are thought to have formed during periods of high obliquity when ice, driven off the poles, accumulated on the volcanoes.

7.3 Wind

We know that the wind redistributes material across the martian surface. We have observed dust storms from orbit and the changing patterns of surface markings that they cause. The 2004 rovers have made movies of dust devils, and tracks made by dust devils are visible on many high-resolution images taken from orbit. Dust can be seen draped over rocks in most lander images. Crater tails caused by eolian deposition or erosion in the lee of craters are common. Dunes are visible in almost all orbiter images with resolutions of a few m/pixel or better, and in some areas, such those as around the North Pole, dunes cover vast areas. Given all this evidence, it is somewhat surprising that wind erosion is not more widespread. The wind appears to mostly move loose material around the surface. Additions to the inventory of loose material by erosion of primary rocks must be proceeding extremely slowly.

Though the effects of wind erosion in most places are trivial, locally the effects may be substantial. This is particularly true where friable deposits are at the surface. In southern Amazonis and south of Elysium Planitia, thick, easily erodible deposits cover the plains/upland boundary. Eroded into these deposits are arrays of curvilinear, parallel grooves that resemble terrestrial wind-cut grooves called yardangs. Wherever such wind erosion is observed, other evidence indicates that what is being eroded is a deposit that blankets the bedrock. Erosion of bedrock units such as lava flows is minute. Wind may be ineffective as an erosion agent because of the lack of abrasive debris for the wind to move. On Earth, quartz sand is an effective erosion agent, but quartz sand is rare or absent on Mars. Most of the loose material blown around by the wind appears to be ground up basalt and its weathering products, and these materials have little abrasive capacity.

8. Poles

During fall and winter, CO_2 condenses onto the polar regions to form a seasonal cap that can extend as far equatorward as 40° latitude. In summer, the CO_2 cap sublimates. That in the north sublimates completely to expose a water ice cap, the temperature at the pole rises from the frost point of CO_2 (150 K) to the frost point of water (200 K), and the amount of water vapor over the pole rises dramatically. In the south, the CO_2 cap does not dissipate completely, but water ice has been detected under the seasonal cap.

At both poles, layered deposits several kilometers thick extend out to roughly the 80° latitude. Individual layers are best seen in the walls of valleys cut into the sediments, where layering is observed at a range of scales down to the resolution limit of our best pictures. The frequency of impact craters on the upper surface of the deposits suggests that the sediments are young, of the order of 10^8 years or less.

The poles act as a cold trap for water. Any water entering the atmosphere as a result of geologic processes such as volcanic eruptions or floods will ultimately be frozen out at the poles. The poles may also be a trap for dust, in that dust can be scavenged out of the atmosphere as CO_2 freezes

onto the poles each fall and winter. The layered deposits are, therefore, probably mixtures of dust and ice. The layering is thought to be caused in some way by periodic changes in the thermal regimes at the poles, induced by variations in the planet's orbital and rotational motions (see Section 2.1). These cyclical motions affect temperatures at the poles, the stability of CO_2 and H_2O, the pressure and circulation of the atmosphere, the incidence of dust storms and so forth; hence, the belief that they are responsible in some way for the observed layering.

9. The View from the Surface

At the time of writing, we had successfully landed at five locations on the martian surface: two *Viking* spacecraft in 1976, *Mars Pathfinder* in 1997, and two rovers in 2004. *Viking 1* landed on a rolling, rock-strewn plain partly covered with dunes in the Chryse basin. *Viking 2* landed on a level, rocky plain in Utopia. The main goal of the *Viking* landers was life detection. They carried a complex array of experiments designed to detect metabolism in different ways and to determine what organics might be in the soil. Neither metabolism nor organics were detected. The lack of organics was somewhat surprising since organics should have been there from meteorite infall. The soil, however, turned out to be oxidizing, which probably caused decomposition of any organics that might at one time have been present. *Pathfinder* also landed on a rock-strewn plain in Chryse. The site is at the mouth of one of the large outflow channels. It was hoped that evidence of floods might be observed there. However, the only sign of floods were some rocks stacked on edge and terraces on nearby hills that could have been shorelines.

The two rovers *Spirit* and *Opportunity*, launched in 2003, have been far more fruitful and have provided the first solid evidence from the surface for pooling of water and aqueous alteration. *Spirit* landed on the flat floor of the 160-km-diameter crater Gusev. The site was chosen because the southern wall of Gusev is breached by a large channel called Ma'adim Vallis. Water from the channel must at one time have pooled in Gusev, and it was hoped that the rover would be able to sample sediments from the postulated Gusev lake. The floor of Gusev turned out to be another rock-strewn plain. The rocks are basalts, but they have alteration rinds with varying amounts of water-soluble components such as S, Cl, and Br. The alteration is minor and has been attributed to the action of acid fogs. Erosion rates estimated from craters superimposed on the plains indicate that the rates have been several orders of magnitude less than typical terrestrial rates. These somewhat disappointing results spurred a move to some nearby hills, where it was hoped different materials would be found, and indeed they were. Most of the rocks on the Columbia Hills are very different from those on the plains. As of this writing, six different classes of rocks had been identified

FIGURE 10 View from the *Spirit Rover* in the Columbia Hills. The level plains of Gusev are in the background. The hills in the distance are part of a delta-like deposit at the mouth of a large channel that enters the crater from the south. The rocks in the foreground have been aqueously altered to varying degrees. The origin of the hills is unknown, but they may have been uplifted by an impact event that postdated the formation of Gusev itself. (Mars Exploration Rover.)

ranging from almost unaltered olivine basalts like those on the plains to almost completely altered, soft rocks enriched throughout with mobile elements such as S, Cl, and Br. Primary basalt minerals are almost absent having been replaced by secondary minerals such iron oxides and oxyhydroxides that have high Fe^{3+}/Fe^{2+} ratios compared with the unaltered rocks. A sulfate cement in some rocks suggests evaporation of sulfate-rich waters. On some of the rocks, there is a surface rind that is harder than the interior, so the rocks have been hollowed out by the wind. Layered rock is common, and a coarse stratification appears to follow the contours of the hills. The origin of the Columbia Hills rocks is still being debated. Some may have formed by aqueous alteration of newly deposited impact or volcanic debris. Some may have been hydrothermally altered long after deposition. For others, waters from the postulated Gusev lake may have been implicated. Whatever the cause was, aqueous processes were involved. (See Fig. 10.)

Opportunity landed in Meridiani Planum on a thick stack of layered rocks that had been observed from orbit. The site was chosen because a particular form of the iron mineral hematite that forms in aqueous environments had been detected there. The number of impact craters superimposed on the layered rocks suggests that they formed at the end of the heavy bombardment period around 3.8 billion years ago. The rovers demonstrated unequivocally that the local rocks are reworked evaporitic sandstones with roughly equal proportions of basaltic debris and evaporitic minerals such as Mg, Ca, Fe, and Na sulfates and chlorides. Although most of the rocks were deposited by the wind, there had to be a nearby source for the evaporites, which form by evaporation of bodies of water. The source had to be substantial because the layered sequence on which the rover landed extends for several hundred kilometers. A small fraction of the rocks have depositional textures that indicate that they were deposited in standing water. The environment in which the Meridiani sequence accumulated is thus thought to be one in which there were wind-blown dunes with interdune ponds. *Opportunity* spent much of its time examining the local rock section in a crater called Endurance. (See Figs. 11 and 12.) The relations in the crater indicated that there had been almost no aqueous activity since the crater formed. Thus, the two rovers, although landing on very different geologic materials, are telling a somewhat similar story. The oldest rocks, those that formed during heavy bombardment, have abundant evidence for aqueous processes, but the evidence for such processes after the end of heavy bombardment is sparse or absent.

10. Summary

Mars is a geologically variegated planet on which have operated many of the geologic processes familiar to us here on Earth. It has been volcanically active throughout its history; the crust has experienced extensive deformation, largely as a result of massive surface loads; and the surface has been eroded by wind, water, and ice. Despite these similarities, the evolutions of Mars and Earth have been very different. The lack of plate tectonics on Mars has prevented the formation of linear mountain chains and cycling of crustal

FIGURE 11 View of Endurance crater from the Opportunity Rover in Meridiani. The impact crater formed in a sequence of horizontally layered rocks, which are exposed in the foreground and in the walls of the crater. The horizon in the background gives and indication of how level the rock sequence is. The rover entered the crater and made measurements down section, almost to the center of the crater. Burns Cliff, seen in Fig. 12, is on the far wall. (Mars Exploration Rover.)

FIGURE 12 View of Burns Cliff from *Opportunity Rover* in Meridiani. The rocks consist of a mixture of evaporites, such as sulfates and chlorides, and basaltic debris. The bedding patterns indicate that they were mostly deposited by the wind. However, the evaporites must originally have been derived by evaporation of a nearby lake or sea.

material though the mantle, and climatic conditions that hindered the flow of water across the surface have limited erosion and deposition to almost negligible levels for most of the planet's history. Since the end of heavy bombardment around 3.8 billion years ago, the rate of all geologic processes has been orders of magnitude lower than on Earth, so that even ancient geologic features are well preserved. The result is a geologic record, preserved on the surface, that spans almost the entire history of the planet. For the heavy bombardment period, we have compelling chemical and mineralogic evidence for aqueous alteration and bodies of water at the surface. Similar evidence for later periods is sparse, although geomorphic evidence indicates that there were episodic large floods. The climatic implications of the geologic observations remain uncertain. Even though early Mars must have had at least warm climatic episodes, any warm episodes after the end of heavy bombardment must have been very short because the cumulative amounts of erosion and weathering are so small. What caused the early warm conditions remains a mystery.

Bibliography

Baker, V. R. (1982). "The Channels of Mars." Univ. Texas Press, Austin.

Beatty, J. K., and Chaikin, A. (1990). "The New Solar System." Sky Publishing, Cambridge, Massachusetts.

Carr, M. H. (1981). "The Surface of Mars." Yale Univ. Press, New Haven, Connecticut.

Carr, M. H. (1996). "Water on Mars." Oxford Univ. Press, New York.

Hamblin, W. K., and Christiansen, E. H. (1990). "Exploring the Planets." Macmillan, New York. *Nature* **412**, 207–253 (2001). A group of review articles on various aspects of the evolution of Mars.

Wilford, J. N. (1990). "Mars Beckons." Knopf, New York.

Mars: Landing Site Geology, Mineralogy and Geochemistry

Matthew P. Golombek

Jet Propulsion Laboratory
Pasadena, California

Harry Y. McSween, Jr.

University of Tennessee
Knoxville, Tennessee

CHAPTER 17

1. Introduction

Most of our detailed information about the materials that make up the martian surface comes from the in situ investigations accomplished by the five successful landers (Table 1). The focus of these landers and the era in which they explored Mars have varied. The first successful landings were the *Viking* landers in 1976, part of two orbiter/lander pairs that launched in 1975. The overriding impetus for the *Viking* landers was to determine if life existed on Mars. Both immobile, legged landers carried sophisticated life detection experiments as well as imagers, seismometers, atmospheric science packages, and magnetic and physical properties experiments. The *Viking* mission was done in the post-*Apollo* era (after 1972) and involved a massive mobilization of engineering and scientific talent (as well as a budget befitting a major mission). The life detection experiments found no unequivocal evidence for life in the soil (although gases released from the soil suggested a significant oxidizing component) but did image the landing sites and determine the chemistry of soils. The successful landings and operations of the orbiters (that lasted years) set the stage for the systematic study of Mars and left a legacy for landing using an aeroshell and a supersonic parachute that have been used by all subsequent landers. The *Viking* orbiters returned imaging data of valley networks and eroded ancient craters and terrain that suggested

an earlier wetter and possibly warmer environment, contrary to the present climate whose atmosphere is generally too cold and thin (and dry) to support liquid water (current atmospheric pressure and temperature are so low that water is typically stable in solid and vapor states).

The *Mars Pathfinder* mission, launched 20 years later in 1996, was an engineering demonstration of a low-cost lander and small mobile rover. The spacecraft was a small free flyer that used a *Viking*-derived aeroshell and parachute, but developed and used robust airbags surrounding a tetrahedral lander. The lander carried a stereoscopic color imager (IMP), which included a magnetic properties experiment and wind sock, and an atmospheric structure and meteorology experiment. The 10 kg rover (*Sojourner*) carried engineering cameras, 10 technology experiments, and an Alpha Proton X-ray Spectrometer (APXS) for measuring the chemical composition of surface materials (Table 2). The *Mars Pathfinder* lander and rover operated on the surface for about 3 months (well beyond their design lifetime), and the rover traversed about 100 m around the lander, exploring the landing site and characterizing surface materials in a couple of hundred square meter area. Rocks measured by the APXS appeared high in silica, similar to andesites; tracking of the lander fixed the spin pole and polar moment of inertia that requires a central metallic core and a differentiated planet; and the atmosphere was observed to be quite dynamic with water ice clouds, abruptly changing near

TABLE 1	Landing Sites on Mars			
Site	**Latitude (deg. +N)**	**Longitude (deg. +E)**	**Elevation (km, MOLA)**	**Region**
Viking 1	22.27	311.81	−3.6	Chryse Planitia
Viking 2	47.67	134.04	−4.5	Utopia Planitia
Mars Pathfinder	19.09	326.51	−3.7	Ares Vallis
MER *Spirit*	−14.57	175.47	−1.9	Gusev crater
MER *Opportunity*	−1.95	354.47	−1.4	Meridiani Planum

surface morning temperatures, and the first measurement of small wind vortices or dust devils. The mission captured the imagination of the public, garnered front-page headlines during the first week of operations, and became one of NASA's most popular missions by becoming the largest Internet event in history at the time. Much of the flight system, lander, and rover design were used for the next two successful landings.

The Mars Exploration Rover (MER) landed twin moderately sized rovers in early 2004, and they have explored over 6 km of the surface at two locations. Each rover carries a moderately sophisticated payload that includes multiple imaging systems including the color, stereo Panoramic Camera (Pancam) and the Miniature Thermal Emission Spectrometer (Mini-TES). The rovers also carry an arm that can brush and grind away the outer layer of rocks (the Rock Abrasion Tool or RAT) and can place an APXS, Mössbauer Spectrometer (MB), and Microscopic Imager (MI) against rock and soil targets (Table 2). The rover and payload partially mimic a field geologist (eyes, legs, rock hammer, and hand lens) in that they are able to identify interesting targets using the remote sensing instruments (a field geologist's eyes), can rove to those targets (legs), and can remove the outer weathering rind of a rock (equivalent to a rock hammer) and identify the rock type (equivalent, or better than a geologist's hand lens) using the chemical composition (APXS), iron mineralogy (MB), and rock texture (MI). These rovers have lasted years (well beyond their 3 month design lifetime) and returned a treasure trove of basic field observations along their traverses as well as sophisticated measurements of the chemistry, mineralogy, and physical properties of rocks and soils encountered. They have returned compelling information that indicates an early wet and likely warm environment on Mars.

2. Landing Sites on Mars

The five landing sites (Table 1) that constitute the "ground truth" for orbital remote sensing data on Mars were all selected primarily on the basis of safety considerations, in which surface characteristics appeared to match the engineering constraints based on the designed entry, descent, and landing system, with scientific desires being subsidiary. The most important factor controlling the selection of the five landing sites is elevation, as all landers used an aeroshell

TABLE 2	Instruments Used to Examine Rocks at Spacecraft Landing Sites

Alpha Particle X-ray Spectrometer (APXS) on Mars Exploration Rovers — measures rock chemistry, using interactions of alpha particles with the target

Alpha Proton X-ray Spectrometer (APXS) on Mars Pathfinder — measured rock chemistry, using interactions of alpha particles and protons with the target

Imager for Mars Pathfinder (IMP) — lander-mounted digital imaging system for stereo, color images and visible near-infrared reflectance spectra of minerals

Microscopic Imager (MI) on Mars Exploration Rovers — equivalent to a geologist's hand lens, a high-resolution (∼100 microns) camera used to image textures and fabrics

Miniature Thermal Emission Spectrometer (Mini-TES) on Mars Exploration Rovers — identifies minerals by thermal infrared spectra caused by crystal lattice vibrations

Mössbauer Spectrometer (MB) on Mars Exploration Rovers — identifies iron-bearing minerals and distribution of iron oxidation states by measuring scattered gamma rays

Panorama Camera (Pancam) on Mars Exploration Rovers — digital imaging system for stereo, color images and visible near-infrared reflectance spectra of minerals

Rock Abrasion Tool (RAT) on Mars Exploration Rovers — brushes or grinds rock surfaces to reveal fresh interiors

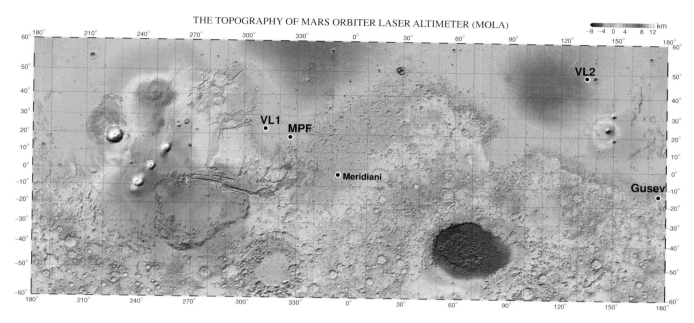

FIGURE 1   MOLA (Mars Orbiter Laser Altimeter on *Mars Global Surveyor*) topographic map of Mars showing the 5 successful landing sites. Elevations are reported with respect to the geoid (or geopotential surface) derived from the average equatorial radius extrapolated to the rest of the planet via a high order and degree gravity field. The resulting topography faithfully records downhill as the direction that liquid water would flow. Longitudes are measured positive to the east according to the most recent convention. The locations of the landers and their elevations are reported in Table 1. Prior to MOLA, which provided definitive topography and an accurate cartographic grid, elevations and locations were poorly known for landing spacecraft on Mars. The map shows 3 fundamental aspects of Mars: the southern highlands, northern lowlands, and Tharsis, an enormous elevated region of the planet (located southwest of VL1 on the map). Tharsis is surrounded by a system of generally radial extensional tectonic features (including the huge Valles Marineris canyon that extends to the east of Tharsis) and generally concentric compressional tectonic features that both imprint the entire western hemisphere of the planet. Located at the edges of Tharsis and the highland–lowland boundary are the catastrophic outflow channels that funneled huge volumes of water into the northern plains (including Chryse Planitia where the VL1 and MPF landing sites are located) intermediate in Mars history.

and parachute to slow them down and sufficient atmosphere and time were required to carry out entry and descent. This favored landing at low elevations as shown in Fig. 1, which shows the locations of the landing sites on a topographic map of Mars. The map shows the southern hemisphere is dominated by ancient heavily cratered terrain estimated to be more than 3.7 billion years old. The northern hemisphere is dominated by younger, smoother, less cratered terrain that is on average 5 km lower in elevation. Astride the hemispheric dichotomy is the enormous Tharsis volcanic province, which rises to an elevation of 10 km above the datum, covers one quarter of the planet, is surrounded by tectonic features that cover the entire western hemisphere, and is topped by five giant volcanoes and extensive volcanic plains. The elevated Tharsis province and the cratered highlands have been too high to land existing spacecraft. The *Viking* landers landed in the northern lowlands, as did *Mars Pathfinder*, and the Mars Exploration Rovers landed at relatively low elevations in the transition between the highlands and lowlands.

Landing site selection for the five landers included intense periods of data analysis of preexisting and incoming information. The *Viking* lander/orbiter pairs were captured into Mars orbit and the orbiter cameras started a concentrated campaign to image prospective landing sites (at tens to hundreds of meters per pixel) selected on the basis of previous *Mariner 9* images. A large site selection science group assembled mosaics (using paper cut outs pasted together by hand) in real time and after waiving off several landing sites on the basis of rough terrain and radar scattering results (and missing the intended July 4 landing), *Viking 1* landed on ridged plains in Chryse Planitia. The site is downstream from Maja and Kasei Valles, giant catastrophic outflow channels that originate north of Valles Marineris, the huge extensional rift or canyon that radiates from Tharsis (Fig. 1). Its low elevation and proximity to the channels

FIGURE 2 Regional color mosaic of Chryse Planitia, Ares Vallis, and the *Mars Pathfinder* landing ellipse. Mosaic shows catastrophic outflow channels cutting the heavily cratered (ancient) terrain to the south and flowing to the lower northern plains. Ares Vallis is about 100 km wide and 2 km deep and by analogy with similar features on Earth formed in about a 2 week period when roughly the volume of water in the Great Lakes carved the valley. Note streamlined islands produced during the flooding. The *Mars Pathfinder* landing ellipse shown is 200 × 100 km and lies about 100 km north of the mouth of the channel where it exits the highlands and thus was interpreted to be a depositional plain composed of materials deposited by the flood. Characterization of the surface after landing supports this interpretation.

suggested that water and near-surface ice might have accumulated there, possibly leading to organic molecules and life. *Viking 2* was sent to the middle northern latitudes where larger amounts of atmospheric water vapor existed, thereby ostensibly improving the chance for life. Landing was deferred for *Viking 2*, and the site selection team analyzed images and thermal observations, before landing in the mid-northern plains, just west of the crater Mie (Fig. 1). Although predictions of the surfaces and materials present at the *Viking* landing sites were incorrect (likely due to the newness of the data and the coarse resolution of the orbital images), the atmosphere was within specifications and both landed successfully.

The *Mars Pathfinder* site selection effort involved little new data since the *Viking* mission 20 years earlier, but there was a much better understanding of how the two *Viking* landing sites related to the remote sensing data. The site selection effort took place over a 2½ year period prior to launch and included extensive analysis of all existing data as well as the acquisition of Earth-based radar data. An Earth analog in the Ephrata fan near the mouth of a catastrophic outflow channel in the Channeled Scabland of western and central Washington State was identified and studied as an aid to understanding the surface characteristics of the selected site on Mars. Important engineering constraints, in addition to the required low elevation, were the narrow latitudinal band 15°N ± 5° for solar power and the large landing ellipse 300 × 100 km, which required a relatively smooth flat surface over a large area. This and the requirement to have the landing area covered by high-resolution *Viking Orbiter* images (<50 m/pixel) severely limited the number of possible sites to consider (~7). The landing site selected for *Mars*

Pathfinder was near the mouth of the catastrophic outflow channel, Ares Vallis, that drained into the Chryse Planitia lowlands from the highlands to the southeast (Fig. 2). Ares Vallis formed at an intermediate time in Mars history (after the early warm and wet period) and involved outpourings of huge volumes of water (roughly comparable to the water in the Great Lakes) in a relatively short period of time (few weeks). The surface appeared acceptably safe, and the site offered the prospect of analyzing a variety of rock types from the ancient cratered terrain and intermediate-aged ridged plains. Surface and atmospheric predictions were correct, and *Pathfinder* landed safely.

Landing site selection for the Mars Exploration Rovers took place over a 2½ year period involving an unprecedented profusion of new information from the *Mars Global Surveyor* (launched in 1996) and *Mars Odyssey* (launched in 2001) orbiters. These orbiters supplied targeted data of the prospective sites that made them the best-imaged, best-studied locations in the history of Mars exploration. For comparison, most of the ellipses were covered by ~3 m/pixel Mars Orbiter Camera (MOC) images, whereas the *Mars Pathfinder* ellipse was covered by ~40 m/pixel *Viking* images. All the major engineering constraints were addressed by data and scientific analyses that indicated the selected sites were safe. Important engineering requirements for landing sites for these rovers included relatively low elevation, a latitude band of 10°N to 15°S for solar power, and ellipse sizes that were ultimately less than 100 km long and 15 km wide. Because of the smaller ellipse size compared to *Pathfinder*, ~150 sites were initially possible from which high science priority sites were selected for further investigation. Both sites selected showed strong evidence for surface processes involving water to determine the aqueous, climatic, and geologic history of sites where conditions may have been favorable to the preservation of prebiotic or biotic processes. The site selected for the *Spirit* rover was within Gusev crater, an ancient 160 km diameter impact crater at the edge of the cratered highlands in the eastern hemisphere. The southern rim of Gusev is breached by Ma'adim Vallis, an 800 km long branching valley network that drains the ancient cratered highlands to the south (Fig. 3). The smooth flat floor of Gusev was interpreted as sediments deposited in a crater lake, so that the rover could analyze fluvial sediments deposited in a lacustrine environment (Fig. 4). The site selected for the

FIGURE 4 Mosaic of Gusev crater showing the landing ellipse, landing location for the *Spirit* lander, and the extensive data sets that were obtained to evaluate the Mars Exploration Rover landing sites. Ma'adim Valles breaches the southern rim, and hills immediately downstream have been interpreted as delta deposits. Blue ellipse is the final targeted ellipse and the red X is the landing location. Background of mosaic is *Viking* 230 m/pixel mosaic, overlain by MOLA elevations in color. Thin image strips mostly oriented to the north-northwest are MOC high-resolution images typically at 3 m/pixel. Wider image strips mostly oriented to the north-northeast are *Mars Odyssey* visible images at 18 m/pixel. Mosaic includes 13°S–16°S latitude and 174°E–177°E longitude; solid black lines are 0.5° (~30 km), and dashed black grid is 0.1° (~6 km).

FIGURE 3 Regional color mosaic of Ma'adim Valles and Gusev crater. The 800 km long Ma'adim Valles, one of the largest branching valley networks on Mars, drains the heavily cratered terrain to the south and breaches the southern rim of Gusev crater. Gusev crater, which formed much earlier, is 150 km in diameter, and the smooth flat floor strongly suggests it was a crater lake that filled with water and sediments. *Spirit* has yet to identify sediments associated with Ma'adim Valles. The cratered plains are underlain by basalt flows and so represent a late volcanic cover. Rocks in the Columbia Hills have been altered by water, but cannot be related to deposition in a lake associated with Ma'adim discharge.

Opportunity rover is in Meridiani Planum in which thermal spectra from orbit indicated an abundance (somewhat unique) of a dark gray, coarse-grained mineral (hematite) that typically forms in the presence of liquid water. Layers associated with the hematite deposit in Meridiani Planum suggested a sequence of sedimentary rocks that could be interrogated by the rover. Meridiani Planum is a unique portion of the ancient heavily cratered terrain in western Arabia Terra that was downwarped and heavily eroded early in Mars history and thus stands at a lower elevation than the adjacent southern highlands (Fig. 5). The atmospheric and surface characteristics inferred from the extensive remote sensing data were correct for both, and *Spirit* and *Opportunity* landed safely. Geologic interpretations of materials available for study were less successful (no fluvial or lake sediments were found in Gusev crater) underscoring the inherent ambiguity of understanding the geology and materials available for study from remotely sensed data.

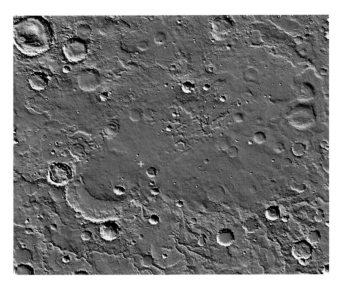

FIGURE 5 Regional setting of Meridiani Planum in MOLA shaded relief map (~850 km wide). Note smooth, lightly cratered plains on which *Opportunity* landed (cross), which bury underlying heavily cratered (ancient) terrain with valley networks to the south. The large degraded craters in the smooth plains indicate that the sulfate rocks below the basaltic sand surface are very old (>3.7 billion years). In contrast, the lightly cratered basaltic sand surface that *Opportunity* has traverse is young.

3. Mars Landing Sites in Remotely Sensed Data

3.1 Surface Characteristics

Understanding the relationship between orbital remote sensing data and the surface is essential for safely landing spacecraft and for correctly interpreting the surfaces and kinds of materials globally present on Mars. Safely landing spacecraft on the surface of Mars is obviously critically important for future landed missions. Understanding the surfaces and kinds of materials globally present on Mars is also fundamentally important to deciphering the erosional, weathering, and depositional processes that create and affect the Martian surface layer. This surface layer or regolith, composed of rocks and soils, although likely relatively thin, represents the key record of geologic processes that have shaped it, including the interaction of the surface and atmosphere through time via various alteration (weathering) and eolian (wind-driven) processes.

Remote sensing data available for selecting landing sites has varied for each of the landers, but most used visible images of the surface as well as thermal inertia and albedo. Thermal inertia is a measure of the resistance of surface materials to a change in temperature and can be related to particle size, bulk density, and cohesion. A surface composed of mostly rocks will change temperature more slowly, remaining warmer in the evening and night, than a surface composed of fine-grained loose material that will change temperature rapidly, thereby achieving higher and lower surface temperatures during the warmest part of the day and the coldest part of the night, respectively. As a result, surfaces with high thermal inertia will be composed of more rocks or cohesive material than surfaces with low thermal inertia. Thermal inertia can be determined by measuring the surface temperature using a spectrometer that measures the thermal emission (temperature) at multiple times of the day or by fitting a thermal model to a single temperature measurement. Thermal observations of Mars have been made by many orbiters, including the *Mariners*, *Viking*, *Mars Global Surveyor*, and *Mars Odyssey*, with increasingly high spatial resolution by the last three. In addition, measurement of different thermal wavelength emissions from the surface has been used to separate the area of the surface covered by high inertia materials or rocks from the area covered by lower inertia materials or soil. The albedo is a measure of the brightness of the material in which the viewing geometry has been taken into account.

Global thermal inertia and albedo data show that the surface of Mars exhibits particular combinations that cover most of the surface. One has high albedo and low thermal inertia and is likely dominated by substantial thicknesses (a meter or more) of bright red dust that is likely neither load-bearing nor trafficable. These areas have very few rocks and have been eliminated for landing solar-powered spacecraft, and they will likely be eliminated for rover missions interested in studying rocks or outcrop. Moderate to high thermal inertia and low albedo regions are likely relatively dust free and composed of dark eolian sand and/or rock. Moderate to high thermal inertia and intermediate to high albedo regions are likely dominated by cemented crusty, cloddy, and blocky soil units that have been referred to as duricrust with some dust and various abundances of rocks. Coarse resolution global abundance of rocks on Mars, derived by thermal differencing techniques that remove the high inertia (rocky) component, shows that the first type of surface has almost no rocks and the latter two types of surfaces have rock abundances that vary from about 8% (the global mode of rock abundance of Mars) to a maximum of about 35% of the surface covered by rocks.

The five landing sites sample the latter two types of surfaces in the thermal inertia and albedo combinations that cover most of Mars. Along with variations in their rock abundance, they sample the majority of likely safe surfaces that exist and are available for landing spacecraft on Mars. The *Viking* landing sites both have relatively high albedo and high rock abundance (~17%), in addition to intermediate thermal inertia. On the surface, these sites are consistent with these characteristics, with both being rocky and somewhat dusty plains with a variety of soils, some of which are cohesive and cemented (Figs. 6 and 7). Prior to landing, the *Mars Pathfinder* site was expected to be a rocky plain composed of materials deposited by the Ares Vallis catastrophic flood that was safe for landing and roving and was less dusty than the *Viking* landing sites based on the intermediate to

FIGURE 6   The *Viking 1* landing site. (a) Mosaic of the *Viking 1* landing site showing bright drifts and dark rocks. Large rock to left is Big Joe and is subrounded. Smaller angular dark rocks are sitting on soil and have been interpreted as impact ejecta blocks. Bright drift in the center of the image shows layers, and some particles may be large enough to require deposition by running water rather than the wind. (b) Color mosaic of the *Viking 1* landing site showing dusty reddish surface, darker pitted rocks nearby, and a crater rim on the left horizon. Jointed slightly lighter toned low rock mass in the middle distance appears to be outcrop. The location of the site on ridged plains suggests that the outcrop is basalt, with angular rocks as ejecta and drift materials deposited by either the wind or floodwaters from Maja or Kasei Valles.

high thermal inertia, high rock abundance (18%), slightly lower albedo, and relation to an analogous catastrophic outflow depositional plain in the Channel Scabland. All of these predictions were confirmed by data gathered by the *Mars Pathfinder* lander and rover (Fig. 8). The *Spirit* landing site in Gusev crater has comparable thermal inertia and fine component thermal inertia and albedo to the two *Viking* sites and so was expected to be similar to these locations, but with fewer rocks (8%). Dark dust devil tracks in orbital images suggested some of the surfaces would be lower albedo, where the dust has been preferentially removed (Fig. 9). *Spirit* has landed and traversed across both dusty (Fig. 10) and dust devil track surfaces and found that the average rock abundance is similar to expectations, that in darker dust devil tracks the albedo is low and the surface is relatively dust free (at the landing site), and that in areas outside of dust devil tracks the albedo is higher and the surface is more heavily coated with bright atmospheric

dust that has fallen from the sky (Fig. 10). The Meridiani Planum site has moderate thermal inertia, very low albedo and few rocks. This site was expected to look very different from the three landing sites with a dark surface, little bright red dust, and few rocks. *Opportunity* has traversed across a dark, basaltic sand surface with very few rocks and almost no dust (Fig. 11).

The slopes and relief at various length scales that were important to landing safely were also estimated at the five landing sites using a variety of altimetric, stereo, shape from shading, and radar backscatter remote sensing methods. Results estimated from these data are in accord with what was found at the surface. Of the five landing sites, Meridiani Planum was judged to be the smoothest, flattest location ever investigated at 1 km, 100 m, and several meter length scales, which is in agreement with the incredibly smooth flat plain traversed by *Opportunity* (Fig. 11). On the other extreme, the *Mars Pathfinder* landing site (Fig. 8)

FIGURE 7 Color mosaic of the *Viking 2* landing site showing flat rocky and dusty plain. Pitted rocks in foreground suggest they are volcanic basalts, and the angular homogeneous rock field suggests they are distal ejecta from the fresh crater Mie to the east of the landing site. Lighter toned trough in the middle of the image, in front of the large rocks, has been interpreted to result from the freezing and thawing of subsurface ground ice.

FIGURE 9 Color mosaic of the *Spirit* landing site on the cratered plains of Gusev. Note the soil-filled hollows that are impact craters filled in by sediment. Dark angular blocks are consistent with ejecta, and the pebble-rich surface is similar to a desert pavement in which the sand-sized particles have been moved by the wind leaving a lag deposit. The landing site is in a dust devil track explaining its lower albedo and less dusty surface. The plain is relatively flat with Grissom Hill in the background. Note dark wind tails behind rocks in lower middle foreground.

was expected to be the roughest at all three of these length scales, which agrees with the undulating ridge and rough terrain and the more distant streamlined islands visible from the lander. The other three landing sites are in between these extremes at the three length scales, with *Viking 2* (Fig. 7) and portions of Gusev (Fig. 9) fairly smooth at the 100 m and 1 km scale, *Viking 1* slightly rougher at all three length scales, and *Viking 2* and portions of Gusev (like the

Columbia Hills) in between in roughness at the several meter length scale. All these observations are consistent with the relief observed at the surface.

The close correspondence between surface characteristics inferred from orbital remote sensing data and that found at the landing sites argues that future efforts to select safe landing sites will be successful. Linking the five landing sites to their remote sensing signatures suggests that they span many of the important surfaces available for landing on Mars. Such surfaces that have moderate to high thermal inertia with low to high albedo (but not low albedo and low thermal inertia) constitute almost 80% of the planet, suggesting that to first order most of Mars is likely safe for suitably engineered landers. These results show that basic

FIGURE 8 Color mosaic of the *Mars Pathfinder* landing site showing undulating, ridge–trough moderately dusty and rocky plain. Large rocks in the middle left of the image appear stacked or imbricated on a ridge with a trough behind it that trends toward the northeast. Streamlined hills on the horizon, the ridge–trough topography, and angular to subrounded boulders are consistent with depositional plains deposited by catastrophic floods as expected from the setting of the site downstream from the mouth of Ares Valles outflow channel. Note dust coating the tops of rocks.

FIGURE 10 Color mosaic of the eastern part of Bonneville crater showing dusty and rocky surface of this part of the cratered plains. Note that wall of the crater is composed of dark rubble, suggesting that it formed in a regolith of basalt ejecta. This location is not in a dust devil track and so is much dustier with much higher albedo, consistent with inferences made from orbital images. Hills in the background are the Columbia Hills, which are 90 m high and composed of older rocks. *Spirit* traversed the cratered plains and climbed to the top of the Columbia Hills (highest peak shown is Husband Hill).

FIGURE 11 False color mosaic of the *Opportunity* landing site showing dark, basaltic sand plain and the rim of Eagle crater in the foreground (brighter). Note light toned pavement outcrop near the rim, which is slightly brighter and dustier than the plains. Parachute and 1 m high backshell that *Opportunity* used to land are 450 m away and demonstrate the exceptionally smooth, flat surface as expected from orbital data. The dust-free surface of the plains is in agreement with their very low albedo from orbital data. The ridge on the horizon to the left is the rim of Endurance crater about 800 m away that *Opportunity* traversed to and drove into to study the stratigraphic section. Even though dust has rapidly fallen on the solar panels, the basalt surface is relatively dust-free, indicating that the dust is being swept off the surface at a rate that roughly equals its deposition rate.

engineering parameters important for safely landing spacecraft such as elevation, atmospheric profile, bulk density, rock distribution and slope can be adequately constrained using available and targeted remote sensing data.

3.2 Global Geochemical Units

The compositions of surface materials on Mars can be inferred from measurements of heat emitted from the planet's surface. Thermal emission spectrometers on the *Mars Global Surveyor* and *Mars Odyssey* orbiting spacecraft reveal two distinct kinds of spectra (thermal energy emitted as a function of wavelength). Based on spectral similarity to rocks measured in the laboratory on Earth, Surface Type 1 material is interpreted as basaltic rock and/or sand derived from basalt (Fig. 12). Basalt consists mostly of silicate minerals—pyroxene, feldspar (plagioclase), and olivine—and forms by partial melting of the upper mantle producing a mafic (magnesium and iron rich) magma that erupts on the surface as a dark lava flow (or shallow intrusion). Basalt is the most abundant type of lava on Earth, comprising the floors of the oceans and significant flooded areas of the continents, and it is no surprise that it is common on Mars as well. The giant shield volcanoes of Olympus Mons and the Tharsis Montes are likely composed of basalts based on their similar morphology to shield volcanoes as well as the many plains that resemble basalt plains on Earth. Surface Type 2 material is variously interpreted as either andesite or partly weathered basalt; the spectrum is consistent with either possibility (Fig. 12). Andesite is another

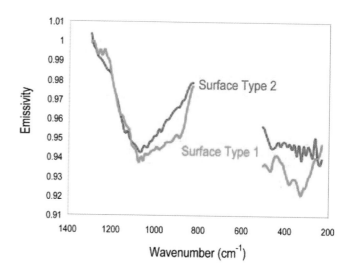

FIGURE 12 Examples of Surface Type 1 and Surface Type 2 thermal emission spectra, from the *Mars Global Surveyor* spacecraft. Surface Type 1 spectra match laboratory spectra of basalt. Surface Type 2 spectra could be either andesite, a more silica rich volcanic rock, or slightly weathered basalt.

common lava type on Earth, occurring primarily at subduction zones. Andesite contains pyroxene (or amphibole) and feldspar. Andesite can form when mafic crystals form in cooling basaltic magma and are extracted from the liquid, leaving an andesitic liquid behind. The spectra of Surface Type 2 can also be explained as a mixture of basaltic minerals plus clays, which commonly form when basalt is weathered by interaction with water.

The thermal emission spectrometers have fairly large footprints (one is about 5 km/pixel and the other is 100 m/pixel), so they cover big regions. Mars surface spectra (Fig. 12) represent mixtures of spectra for the individual minerals that comprise the rocks and soil. The spectrum can be unmixed ("deconvolved") into the spectra for constituent minerals, allowing not only their identification but also an estimate of their proportions. Because we know the chemical compositions of the minerals in the spectral library and the proportions needed to produce the measured spectra, it is possible to calculate the chemical composition of the mixture. That is important because volcanic rocks are usually classified based on their chemistry rather than their mineralogy (minerals in volcanic rocks are small and hard to identify, and quickly solidified magmas often form glass rather than crystalline minerals). The commonly used chemical classification for volcanic rocks, based on the measured abundances of the alkali elements (sodium and potassium, expressed as oxides) versus silica (silicon dioxide), is shown in Fig. 13. The estimated chemical compositions of Surface Type 1 and Surface Type 2 are illustrated in this figure.

In addition to these major units, a few areas on Mars show the distinctive thermal spectra of hematite, iron oxide

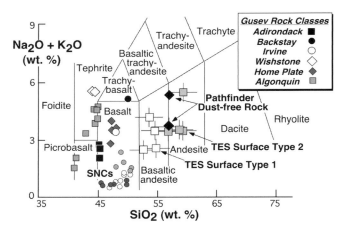

FIGURE 13 Alkalis ($Na_2O + K_2O$) versus silica (SiO_2) diagram, commonly used to classify volcanic rocks. Several estimates of the compositions of Surface Type 1 and Surface Type 2 materials are shown, along with the measured compositions of martian meteorites and APXS analyses of rocks from the *Mars Pathfinder* and *Spirit* landing sites (Gusev). See text for discussion of rock types and classes.

usually formed by interaction with water. The Meridiani Planum region has the highest concentration of hematite measured from orbit, which as discussed earlier led to its selection as a landing site for the *Opportunity* rover.

The global distribution of these spectrally identified units on Mars is distinctive (Fig. 14). The southern hemisphere of Mars is heavily pocked with impact craters, indicating that it is very ancient. This material is mapped mostly as Surface Type 1. In contrast, much of the northern hemisphere is topographically lower than the terrain to the south, and it is extremely smooth and relatively uncratered. The surface of the northern lowlands is inferred to be much younger than

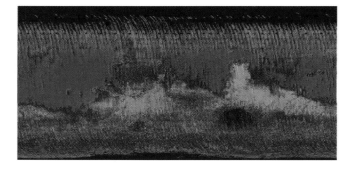

FIGURE 14 Global map showing the distributions of Surface Type 1 (green) and Surface Type 2 (red) materials, based on thermal emission spectroscopy from *Mars Global Surveyor*. Dust-covered areas where this technique cannot distinguish rock units are shown in blue. The preponderance of red spectra of Surface Type 2 in the northern lowlands is consistent with these materials being slightly weathered basalts.

the southern highlands, although the basement beneath this surface layer is also old. Within this northern basin are located most of the Surface Type 2 materials. The distribution of global geochemical units is illustrated in Fig. 14. About half of the surface of Mars is covered with a layer of dust, which precludes the thermal emission spectrometers from mapping the compositions of the rocks that underlie the dust. Unfortunately, most of the spacecraft landing sites on Mars are located near the equator (this constraint maximizes the solar energy received by landers or rovers), which is also where most dust is concentrated. Consequently, it is difficult to compare interpretations of orbital spectra with rocks actually on the ground. The two MER landing sites are exceptions—*Spirit* landed in a region mapped as Surface Type 1, and the *Mars Odyssey* site in Meridiani was selected because of its hematite spectral signature.

4. Landing Site Geology

4.1 Introduction

The geology of the five landing sites has been investigated from color, stereo, panoramic imaging that provides information on the morphology of the landing sites, on the lithology, texture, distribution, and shape of rocks and eolian and soil deposits and on local geologic features that are present. All landing sites that have been investigated on Mars are composed of rocks, outcrops, eolian bedforms and soils, many of which are cemented. Craters and eroded crater forms are also observed at almost all of the landing sites, and other hills have been observed at some of the landing sites. Our knowledge of how the surfaces at the different landing sites developed and the important geological processes that have acted on them is directly related to the mobility of the lander (arm) or rover and the ability of the lander or rover to make basic field geologic observations. The lack of mobility of the two *Viking* landers and the inability to analyze rocks at these sites hampered our ability to constrain their geologic evolution. In contrast, even the limited mobility of the *Sojourner* rover and its ability to make basic field observations over a couple of hundred square meter area resulted in a much better understanding of the geology and the events that shaped the Ares Vallis surface. The two Mars Exploration Rovers that traversed over 6 km each have collected a robust collection of geologic observations over a wide area that have resulted in a much better knowledge of the geologic evolution of the rocks and surfaces investigated. This section will review the basic geological materials found at the five landing sites and discuss the landforms present.

4.2 Rocks

Rocks are common at all of the landing sites (except Meridiani). At most sites, they are distinct dark, angular to subrounded clasts that range in size from several meters diameter down to small pebbles that are a centimeter or less in diameter. Most appear as float, or individual rocks not associated with a continuous outcrop or a body of rock. Many appear dust covered and there is evidence at Gusev for some surface chemical alteration as is common on Earth (see next section), where rocks exposed to the atmosphere develop an outer rind of weathered material. Although the composition of rocks could not be measured at the *Viking 1* and *2* landing sites (Figs. 5, 6 and 7), their dark angular and occasionally pitted appearance is consistent with a common igneous rock known as basalt. Rocks making up the cratered plains on which *Spirit* landed and traversed (Figs. 9 and 10) for the first few kilometers are clearly made up of basalts (see next section). The distribution and shape of many of the rocks at the *Viking 1* and *2* landing sites and the Gusev cratered plains are all consistent with a surface that has experienced impact cratering with the rocks constituting the ejected fragments. Many subrounded rocks at the *Mars Pathfinder* (Fig. 8) and *Viking 1* landing site have been attributed to deposition in the catastrophic floods in which motion in the water partially rounded the clasts. Some rocks at the Pathfinder site had textures that looked like layers (perhaps sedimentary or volcanic), one resembled a pillow basalt in which hot lava cools rapidly in the presence of water, and several rocks resembled conglomerates, in which rounded pebbles and cobbles were embedded in a rock, in which the cobbles were rounded by running water and later cemented in a finer grained matrix. At most of the landing sites, some rocks known as ventifacts, appear polished, fluted, and grooved as a result of sand-sized grains, entrained by the wind, that impact and erode the rocks.

4.3 Outcrop

Continuous expanses of rocks typically referred to as outcrop (or bedrock) have been observed at three of the landing sites. An area of continuous jointed rocks has been observed at the *Viking 1* landing site, but little else is known about it (Fig. 6). Outcrop has been discovered in the Columbia Hills by *Spirit* where there may be coherent stratigraphic layers in and nearby the Cumberland Ridge on the flank of Husband Hill (Fig. 15). These rocks, described in the next section, appear to be layers of ejecta or explosive volcanics deposited early in Mars history. In places, the rocks are finely layered, and in other places they appear massive. At Meridiani Planum light-toned outcrops are exposed in crater walls and areas where the covering dark, basaltic sand sheet is thin (Fig. 11). These outcrops appear to be thinly laminated evaporites that formed via evaporation of subareal salt water (see next section) early in Mars history. The layers are composed of sand-sized grains of fairly uniform composition that appear to have been reworked by the wind in sand dunes before being diagenetically altered by acid groundwater of differing compositions (see next section).

FIGURE 15   Color mosaic of the northeast flank of Husband Hill showing layered strata called Methuselah dipping to the northwest. These rocks are clastic rocks consistent with impact ejecta that have been highly altered by liquid water. Hills in the background are the rim of the 20 km diameter Thira crater near the eastern end of the landing ellipse shown in Fig. 4.

4.4 Soils

All the landing sites have soils composed of generally small fragments of granules, sand, and finer materials. Except where they have been sorted into bedforms by the wind, they have a variety of grain sizes and cohesion, even though their composition appears remarkably similar at all the landing sites (dominantly basaltic). Crusty to cloddy and blocky soils are also present at most of the landing sites and are distinguished as more cohesive and cemented materials. These materials appear to be the duricrust inferred to be present over much of Mars based on higher thermal inertia, but generally low rock abundance. Strong cemented light-toned duricrust was uncovered at the *Mars Pathfinder* site by *Sojourner* and may contribute to the higher thermal inertia at this site than at the others. Some bright soil deposits outside the reach of the arm at the *Viking 1* landing site (Fig. 6a) show layers and hints of coarse particles that could be fluvial materials deposited by the Maja or Kasei Valles floods.

4.5 Eolian Deposits

Most of the landing sites have examples of eolian bedforms, or materials that have been transported and typically sorted by the wind. Sand-sized particles that are several hundred micrometers in diameter can be moved by saltation in which they are picked up by the wind and hop in parabolic arcs across the surface. Because these particles can be preferentially moved by the wind, they are effectively sorted by the wind into bedforms. Sand dunes form when sand-sized particles are sorted into a large enough pile to move across the surface. Sand dunes take a variety of forms such as barchan or crescent-shaped (horns pointing downwind), star-shaped from reversing winds, transverse to the wind, and longitudinal or parallel to the wind, but they are generally diagnostic

FIGURE 16 False color mosaic of star sand dunes in the bottom of Endurance crater. Dark bluish surface is basalt with a surface lag of hematite spherules. Lighter sides of dunes are likely dust that has settled from the atmosphere. Note the light-toned outcrop in the foreground.

enough to be identified from orbit. Sand dunes have been identified at the *Mars Pathfinder* landing site where a small barchan dune was discovered in a trough by the rover and at Meridiani Planum where star dunes were found at the bottom of Endurance crater (Fig. 16).

Ripples are eolian bedforms formed by saltation-induced creep of granules, which are millimeter-sized particles. They typically have a coarse fraction of granules at the crest and poorly sorted interiors indicating a lag of coarser grains after the sand-sized particles have been removed (Fig. 17). Ripples have been found at the *Mars Pathfinder*, Gusev, and Meridiani sites. Drifts of eolian material have also been identified at many of the landing sites behind rocks as wind tails and other configurations. Finally the reddish dust on Mars is only several micrometers in diameter and is carried in suspension in the atmosphere giving rise to the omnipresent reddish color. Although it takes high winds to entrain dust-sized particles in the atmosphere, once it is in the atmosphere it takes a long time to settle out. Dust has been identified on the surface at all the landing sites (in addition to being in the atmosphere), giving everything a reddish color, and has fallen on the solar panels decreasing solar power. Dust devils, or wind vortices, have been observed at the *Mars Pathfinder* and Gusev sites and appear to be an important mechanism for lifting dust into the atmosphere.

4.6 Craters

Impact craters are ubiquitous on Mars, so it is no surprise that craters have been imaged at most of the landing sites. At *Viking 1* (Fig. 6) and the *Mars Pathfinder* landing sites, the uplifted rims of craters have been imaged from the side. At Gusev (Fig. 10) and Meridiani (Fig. 11), the rovers have investigated a number of craters of various sizes during their traverses, including the interiors of some. Because impact craters resemble nuclear explosion craters and because many fresh craters have been characterized on the Moon, much is known about the physics of impact cratering and

FIGURE 17 Large ripple called Serpent that was studied by *Spirit* on the cratered plains. (A) A Hazard Camera image showing the rover front wheels and the tracks produced by a wheel wiggle maneuver to section the drift. (B) Color image of the dusty (reddish) surface and darker more poorly sorted interior. (C) MI image of the brighter (dust cover) granule-rich surface (millimeter-sized particles) and poorly sorted, but generally finer grained, basaltic sand interior. The dusty, granule-rich surface indicates the eolian feature is an inactive (dust cover) ripple formed by the saltation induced creep of granules, which are left as a lag.

the resulting shape and characteristics of fresh craters. [*See* PLANETARY IMPACTS.] Primary impact craters less than 1 km in diameter have well-understood bowl-shaped interiors whose depth is about 0.2 times their diameters; they also have uplifted rims and ejecta deposits (Fig. 10) that get less rocky and thin with distance from the crater. As a result, imaging impact craters provides clues to the geomorphologic changes that have occurred at the site such as the amount of erosion and/or deposition.

5. Landing Site Mineralogy and Geochemistry

5.1 Rocks

Based on their appearance, rocks at the *Viking* landing sites (Figs. 6 and 7) were inferred to be basalts, but the *Viking* lander arms could not reach and collect small enough rocks to analyze, so little is known about their composition. Rocks at the *Mars Pathfinder*, *Spirit*, and *Odyssey* landing sites have been analyzed by a variety of rover-mounted instruments, as described in Table 2.

Pathfinder rock chemical compositions were analyzed by the APXS (Fig. 18), and partial mineralogy was inferred from IMP spectra on the lander. The APXS analyzes only the outer surface (generally just a few tens of micrometers) of rocks. IMP images showed that the rocks were variably coated with dust. Plots of different elements versus sulfur yield straight lines, with soils plotting at the sulfur-rich end best interpreted as mixing lines between the compositions of rocks and soil. The composition of the dust-free rock interior was inferred by extrapolating the rock composition trends to zero sulfur. The dust-free rocks have concentrations of alkalis and silica that would classify them as andesite (two different calibrations of the APXS instrument data are shown in Fig. 13), and it was inferred from the rocks'

appearance that these were volcanic rocks. However, because the APXS analyzes only the rock surface, it is also possible that this andesitic composition represents a silica-rich weathering rind beneath the dust rather than the composition of the rock interiors. The IMP spectra indicate the presence of iron oxides, but a more comprehensive spectral interpretation is hampered by the dust coatings.

Rocks at the *Spirit* landing site in Gusev crater were analyzed using a greater variety of analytical instruments (Table 2), aided by the RAT that can brush or grind the outer rock surface. Rocks on the plains in the vicinity of the *Spirit* lander are clearly basalts, in agreement with the location of Gusev crater within an area mapped by TES as Surface Type 1. Some of these rocks are vesicular—pocked with holes that were once gas bubbles exsolved from magma—and most rocks are coated with dust (Fig. 19). Spectra from Pancam, Mini-TES, and MB of relatively dust-free or RAT-abraded rocks provide a consistent picture of the minerals that comprise these basalts—olivine, pyroxene, and iron oxides. All the spectra from these instruments are dominated by minerals containing iron and magnesium. Chemical compositions of plains basalts measured by APXS support the presence of olivine, pyroxene, and oxides, but they also suggest abundant feldspar (plagioclase) and phosphate, which cannot be seen by other spectra. The APXS analyses confirm that the rocks on the plains of Gusev (Fig. 13) are basalts (Adirondack class) especially rich in olivine (and hence lower in silica). Abundant dark crystals interpreted to be olivine can be seen in MI images of RAT holes in the rocks (Fig. 19). Surface alteration rinds and veins cutting through the interiors of these rocks can also be clearly seen in some MI images, suggesting limited interactions of the rocks with water.

After analyzing rocks near the landing site, the *Spirit* rover traversed about 3 km across the plains and climbed

FIGURE 18   Color mosaic of *Sojourner* with APXS instrument measuring the chemical composition of the rock Yogi. Note dusty surface darkened by the rover wheels. Brighter toned soil in the wheel tracks is cemented soil or duricrust. Note tabular rock on the left horizon, called Couch, and other tabular and partially rounded boulders as expected if deposited by catastrophic floods.

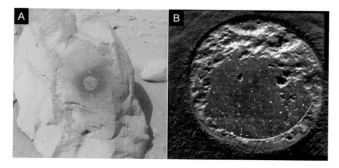

FIGURE 19   The Gusev cratered plains rock Humphrey studied by the *Spirit* rover. (A) Pancam color image of rock after RAT grinding showing darker interior and thus the presence of a dusty and slightly weathered surface. (B) Microscopic image of Humphrey RAT hole, illustrating dark grains thought to be olivine crystals and holes likely to be vesicles, consistent with the basaltic chemistry and mineralogy determined by the APXS and MB.

Columbia Hills Rocks

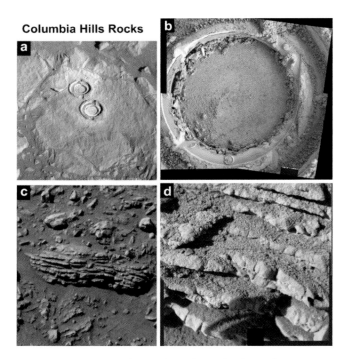

FIGURE 20   Images of rocks from the Columbia Hills in Gusev crater. (a) Pancam image of Wooly Patch, after several RAT grinds showing darker interior and natural dusty surface. (b) MI of Wooly Patch RAT hole, showing clastic texture. (c) Pancam image of Tetl, which exhibits fine layering. (d) MI of Tetl, illustrating coherent layers separated by finer grained material.

Husband Hill, a promontory within Gusev crater (one of the Columbia Hills in Fig. 10). The Hills outcrops are distinct from the plains basalts. Some are massive, others are laminated, and most are altered and deeply weathered (Fig. 20). Pancam, Mini-TES, and MB spectra suggest highly varying mineralogy. Some rocks contain combinations of olivine, pyroxene, feldspar, and iron oxides (as on the plains), whereas others contain large amounts of glass, sulfate, ilmenite, and phosphate. APXS analyses have been used to divide the rocks into several different classes according to their chemistry, but the mineralogy can vary considerably even within a class. Some rocks appear to be relatively unaltered, but most show very high contents of sulfur, phosphorus, and chlorine, suggesting a high degree of alteration. The chemical compositions of these rocks are not illustrated in Fig. 13 because that classification is only applicable to unaltered igneous rocks. The textures of Hills rocks, as revealed by the MI, are also highly variable but commonly indicate alteration of rocks composed of angular particles and clasts (Fig. 20). RAT grinding indicates that these rocks are much softer than the plains basalts. They have been interpreted as mixtures of materials formed by impacts or explosive volcanic eruptions, and subsequently altered by fluids. Two classes of rocks on the northwest flank of Husband Hill have what appear to be roughly concordant dips to the northwest suggesting a stratigraphy (Fig. 15). The lower rock has layered

materials and angular to rounded clasts in a matrix that compares favorably to impact ejecta that has been altered by water to various extents. The upper rock is a finely layered sedimentary rock that has been cemented by sulfate, except the aqueous alteration did not affect the basaltic character of the sediment. A few distinctive rock types found as loose stones (geologists call these "float") in the Hills include Backstay, Irvine, and Wishstone, which are dark, fine-grained basaltic rocks with compositions distinct from the plains basalts (Fig. 13), and only limited signs of alteration by water. These rocks appear to have formed by removal of crystals from magmas similar in composition to plains basalts.

Once *Spirit* gained the crest of Husband Hill, it traveled down the south face, encountering olivine-rich rocks of the Algonquin class (Fig. 13). Upon reaching the bottom, *Spirit* traversed an area containing highly vesicular rocks (scoria) to Home Plate, tentatively interpreted as a small volcanic edifice formed of ash. The compositions of all the relatively unaltered igneous rocks in Gusev crater are rich in alkalis and low in silica (Fig. 13), allowing their classification as alkaline rocks. These are the first alkaline rocks recognized on Mars.

Rocks at the *Opportunity* landing site are mostly exposed in the walls of impact craters and where the sand is thin. Outcrops in Eagle crater were studied extensively after landing (Fig. 11), and a thicker stratigraphic section in Endurance crater was analyzed later in the mission. Pancam and MI images (Fig. 21) show that the rocks are finely laminated, sometimes exhibiting cross-bedding (Fig. 22), and RAT grinds indicate that they are very soft. At the microscopic scale, they consist of sand grains bound together by fine-grained cement. Small gray spherules, called blueberries (Figs. 21 and 22), are embedded within the rock (the spherules are actually gray, but appear bluish in many false color images). Some parts of the outcrop also exhibit tabular voids (Fig. 21). APXS analyses of these rocks indicate very high concentrations of sulfur, chlorine, and bromine (highly water soluble elements), demonstrating that the cement and sand (partially) consists of sulfate and halide salts. MB spectra reveal the presence of iron sulfate, and Mini-TES spectra suggest magnesium and calcium sulfates also occur. The spherules are at least half hematite, the mineral seen from orbital TES spectra of the Meridiani region. The rocks are interpreted as sandstones composed of dirty evaporites of basaltic and sulfate composition formed by the evaporation of brines. Their textures suggest repeated cycles of flooding, exposure, and desiccation. Exposure and desiccation allowed some of the sediments to be mobilized into sand dunes (Fig. 23). After deposition, the rocks underwent a number of different phases of diagenesis by groundwater of varying composition that circulated through the rocks, mobilizing and reprecipitating iron in the form of hematite spherules (concretions) and dissolving highly soluble minerals to leave the voids.

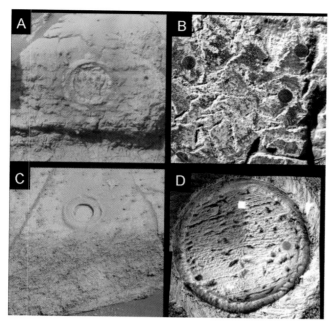

FIGURE 21   Images of Meridiani outcrops acquired by the *Opportunity* rover. (A) Pancam image of Guadalupe in Eagle crater, after RAT grinding. Notice slightly redder, dustier surface around the circular RAT hole and small hematite spherules protruding from the outcrop. (B) Microscopic image of Guadalupe RAT hole, showing blueberries (dark circles) and tabular voids produced by dissolution of soluble minerals. (C) Pancam image of Ontario in Endurance crater, after RAT grinding (circular smooth area). (D) Mosaic of microscopic images of Ontario, showing fine laminations, tabular voids, and a few blueberries (dark circles).

Two oddball rocks at the *Opportunity* site deserve special mention. Bounce Rock, so named because the lander bounced on it as it rolled to a stop, was discovered on the Meridiani plains as the rover exited Eagle crater. Its chemical composition, as measured by APXS, is remarkably like the compositions of a group of Martian basaltic meteorites

FIGURE 22   MI image mosaic of the Upper Dells in Eagle crater showing fine sand-sized particles making up the laminations, blueberries, and cuspate or curved cross laminations that indicate the sand-sized particles were deposited by running water.

FIGURE 23   Color image mosaic of evaporite outcrop of Burns Cliff at the rim of Endurance crater. The lower unit exposed in the lower left shows steeply dipping layers that are truncated by a middle-layered unit with shallow dipping beds. Uppermost unit is lighter toned. The lowermost unit has been interpreted as eolian cross beds that are truncated by the flatter beds of a sand sheet. The uppermost layer is interpreted as the unit deposited in running water of an ephemeral playa or salt water lake.

called shergottites (Fig. 13). Its mineralogy is dominated by pyroxenes and plagioclase, as are shergottites. This rock is obviously not in place and may have been lofted in as ejecta from a large impact crater to the south. Heat Shield Rock, named for its proximity to the heat shield discarded during descent of the *Opportunity* lander, is likewise an interloper in this terrain. The *Opportunity* instruments revealed that it is an iron meteorite, composed of iron–nickel alloys, similar to some iron meteorites that fall to Earth. [See METEORITES.]

5.2 Soils

In addition to numerous soil analyses by the *Mars Pathfinder*, *Spirit*, and *Opportunity* rovers, we have soils collected by scoops and analyzed at the two *Viking* landing sites. As defined by soil scientists on Earth, "soil" usually contains a component of organic matter formed by decayed organisms. Soils on Mars do not contain measurable organic materials, but the term "soil" is nonetheless commonly used in planetary science ("regolith" may be a more accurate term, meaning the surface layer formed by the destruction of rocks).

It is not easy to distinguish between soil and dust on Mars. Soil, normally dark, represents deposited materials, commonly of sand or silt grains (Fig. 17). Bright reddish dust is much finer grained (several micrometers in size) and can either be suspended in the atmosphere or deposited on the ground. The top surface of soil is usually a thin layer of reddish dust, as seen by the color change when it is disturbed in rover tracks (Fig. 17) or airbag bounces. Most measurements of soil mineralogy or chemistry represent a mixture of soil and dust, sometimes with an admixture of small particles of the local rocks.

At all these sites, the soils have broadly similar compositions, consisting of basaltic sands mixed with fine-grained dust and salts. Pancam, Mini-TES, and MB spectra of bright dust are dominated by nanophase ferric oxides, especially hematite. MB spectra of dark soils indicate abundant olivine, pyroxene, and magnetite at the MER landing sites. The degree of alteration appears to be limited, but fractionation of chlorine and bromine in some soils suggest some mobilization by water. APXS chemical analyses show that plagioclase is also an important component of soils, and that their compositions resemble basalts with extra sulfur, chlorine, and bromine. At the *Pathfinder* site, local andesitic rock fragments are present in varying amounts, and at the *Opportunity* site hematite spherules occur abundantly at the surface as a lag of granules. Trenches dug by the *Spirit* and *Opportunity* rovers reveal clods, suggesting greater proportions of salts that precipitated in the subsurface have bound sand into weakly cohesive clumps, and APXS analyses of some subsurface soils show high concentrations of magnesium sulfate salt. Soils also contain significant amounts of nickel, which may reflect admixture of meteorite material into the regolith. Dust appears similar in composition to the soil (basaltic). Analysis of dust adhered to magnets on the rovers indicates it contains olivine, magnetite, and a nanophase iron oxide (likely hematite) that suggests the dust is an oxidation or alteration product of fine-grained basalt. The presence of olivine in the dust suggests liquid water was not heavily involved in its formation because olivine would readily alter to other minerals (especially serpentine) in the presence of water.

6. Implications for the Evolution of Mars

6.1 Origin of Igneous Rocks

Igneous rocks form by partial melting of the planet's deep interior. The significance of the olivine-rich basaltic compositions found by *Spirit* on the Gusev plains is that they appear to represent "primitive" magmas formed by melting in the mantle. Most magmas partly crystallize as they ascend toward the surface, losing the crystals in the process, so that the liquid progressively changes composition. Primitive magmas retain their original compositions and thus reveal the nature of their mantle source regions.

It is unlikely that rocks with andesitic composition at the *Mars Pathfinder* landing site formed by partial melting of the mantle, unless the mantle contains large quantities of water-bearing minerals. More likely, andesite lavas would form by partial melting of previously formed basaltic crustal rocks (the crust forms an outer layer above the mantle). An

alternative, previously mentioned, is that these rocks are not really andesites at all, but instead are basalts with silica-rich weathering rinds. The latter idea seems especially plausible considering that Surface Type 2 (andesitic) rocks are found primarily in places (like the northern lowlands) where surface waters would collect and the sediments they carried would be deposited. If this is correct, the orbital data and the samples of rocks at the five landing sites strongly argue that Mars is a basalt-covered world. Basalts, sediments derived from basalts, and dust derived from mildly weathered basalts are confirmed or suspected at all the landing sites. Adding the thermal emission spectra of Type 1 and Type 2 materials as basalt and weathered basalt would suggest that most of Mars is made of this primitive volcanic rock.

The gamma ray spectrometer (GRS) on the *Mars Odyssey* orbiter has provided direct chemical measurements of large areas of the martian surface. These measurements are of the top meter or so of material, rather than the topmost few hundred micrometers of the surface analyzed by TES spectra. The measured silica contents of Surface Types 1 and 2 terrains are not significantly different, but the potassium content of Surface Type 2 is higher. These conflicting results do not clearly support either proposed origin.

6.2 Chemical Evolution and Surface Water

The minerals that form outcrops of evaporites at the *Opportunity* landing site could only have precipitated from highly acidic water ("acidic" means low pH, or hydrogen ion concentration). Any sea at Meridiani was more like battery acid than drinking water. Given the abundance of basaltic lavas on the martian surface, it is surprising that these waters would be so acidic. Reactions between water and basalt on Earth tend to produce neutral to basic solutions. On Mars, huge volumes of sulfur and chlorine emitted from volcanoes must have combined with water to make sulfuric and hydrochloric acids. Only a few locations on Earth—mostly areas devastated by acidic waters released by weathering of sulfides that drained from mines—mimic this kind of fluid. Acidic water dissolves and precipitates different minerals than the waters we are more familiar with on Earth. Carbonates are not precipitated, and iron sulfates are more common in acidic solutions.

If carbonates could not precipitate in an acid water environment, interesting additional constraints can be placed on the evolution of the atmosphere on Mars. If the early environment of Mars was wetter and warmer and liquid water was in equilibrium, then the atmospheric pressure and temperature both had to be higher. Higher atmospheric pressure requires much more carbon in the atmosphere (mostly composed of carbon dioxide). As the atmosphere thinned, substantial deposits of carbonate would normally be deposited in the crust (as occurs on Earth). In an acid environment, carbonate could not be deposited in the crust, which would then require that the carbon in the atmosphere be lost to space by either solar wind or impact erosion processes.

The presence of significant amounts of sulfate and chloride in soils from all the landing sites further suggests that acidic waters may have been common at one time in all parts of Mars. Either evaporites were abundant and have been redistributed as small particles throughout the planet's regolith, or they occur as cements formed by groundwater leaching all over Mars. Results from the OMEGA spectrometer on the *Mars Express* orbiter support the finding of abundant sulfates elsewhere on Mars.

6.3 Weathering on Mars

There is considerable controversy about the degree to which Mars rocks are weathered. Weathering by acidic water preferentially attacks olivine, and the surface layers of rocks at the MER sites appear to be depleted in olivine. However, remote sensing indicates that olivine is a common mineral in many places on Mars, and olivine appears to be a ubiquitous constituent of martian soils and dust. Perhaps weathering was more common in the distant past, when acidic waters were abundant and produced outcrops like those found by *Opportunity*. Then the acid waters disappeared, and since that time the lavas that were erupted and the soils that formed have only experienced limited weathering.

OMEGA data indicate that clay minerals occur in some localities in the ancient terrains of Mars, although clays have not been found definitively at any of the landing sites. Nevertheless, clays have been suggested to be present in some rocks on Husband Hill (Gusev crater), based on aspects of their chemistry. Clay minerals form by weathering, and they clearly demonstrate that weathering occurred on Mars in the distant past.

6.4 Eolian Processes

The remarkable uniformity of soil compositions at all the landing sites, some separated by thousands of kilometers, suggests an efficient homogenization process. Transport of rock particles by the wind has apparently mixed these materials very efficiently, so that the soil everywhere on Mars represents a globally distributed stratigraphic layer. A similar process must have occurred for the dust particles as well. A dust cycle can be inferred from the omnipresent dusty atmosphere being supplied by dust devils and other processes that occasionally lead to globe-encircling storms. Dust has been observed to be deposited on most of the spacecraft at a rate that is so high that it must be picked up at a similar rate (or the surfaces would be quickly buried by thick accumulations of dust). It may be that dust is deposited at a higher rate overall in broad areas of the planet that have very low thermal inertia and very high albedo. Sand-sized particles created by impact and other processes have been harnessed by the wind to form sand dunes and other eolian

bedforms observed. The consistent basaltic composition of the soil and dust all over Mars further argues that Mars is dominated by basaltic rocks and that the soil and dust forms by physical weathering and minor oxidation without large quantities of water. This further argues that these weathering products have formed and been mobilized by the wind in the current dry and desiccating environment.

6.5 Geologic Evolution of the Landing Sites and Climate

Study of the geology, geomorphology, and geochemistry of the five landing sites in context with their regional geologic setting allows constraints to be placed on the environmental and climatic conditions on Mars through time. The *Viking 1* landing site shows sedimentary drift and soil deposits over angular, dark, presumably volcanic rocks with local outcrops (Fig. 6). The location of this site on the ridged plains terrain downstream from the mouth of Maja and Kasei Valles, suggest that the site is on layered basalts (the preferred interpretation of the ridged plains) with rocks, soils, and drifts derived from impact ejecta, flood, and eolian processes. The rocks at the *Viking 2* landing site (Fig. 7), in the mid-northern plains, are angular and pitted consistent with their being volcanic rocks as part of the distal ejecta from Mie crater. High-resolution orbiter images show the surface has a small-scale hummocky character, and lander images show small polygonal sediment-filled troughs, both suggesting that the surface has been partially shaped by the presence of ground ice. The density of craters observed from orbit at both sites places them intermediate in Mars' history (roughly 3.7–3.0 billion years ago), and constraints on the geomorphologic development of the sites suggest very little erosion or change of the surfaces.

Many characteristics of the *Mars Pathfinder* landing site (Figs. 8 and 18) are consistent with its being a plain composed of materials deposited by catastrophic floods as suggested by its location near the mouth of the Ares Vallis catastrophic outflow channel. Some of the rocks potentially identified (conglomerate, pillow basalt) are suggestive of a wetter past. However, given that the surface still appears similar to that expected for a fresh depositional fan, any erosional and/or depositional processes appear to have been minimal since it formed around 3 billion years ago.

The cratered plains of Gusev that *Spirit* has traversed (exclusive of the Columbia Hills) are generally low-relief, moderately rocky plains dominated by hollows, which appear to be craters filled with soil (Fig. 9). Rocks are generally angular basalt fragments in an unconsolidated regolith greater than 10 m thick of likely impact origin (Fig. 10). The observed gradation and deflation of ejected fines and deposition in craters to form hollows thus provides a measure of the rate of erosion or redistribution since the plains formed about 3.5 billion years ago. These rates of erosion are so slow that they provide a broad indicator of a climate that has been cold and dry. Taking together the slow rates of change inferred from the *Viking*, *Pathfinder*, and Gusev cratered plains landing sites, argues for a dry and desiccating climate similar to today's for the past ~3.7 billion years.

Rocks in the Columbia Hills (Fig. 15) sampled by the *Spirit* rover reveals an earlier period in which liquid water was present. The Husband Hills appear to be older materials that were either uplifted and/or eroded before deposition of the basalts responsible for the cratered plains. The basalts of the cratered plains are intermediate in Mars history and so the Columbia Hills rocks are likely older than roughly 3.7 billion years. These rocks record impact and explosive volcanic processes in their deposition, but many have been heavily altered and/or deeply weathered by water. In contrast, soils in the Columbia Hills are similar to basaltic soils elsewhere, suggesting these formed and were deposited later in the cold and dry martian climate.

The geology and geomorphology of the Meridiani Planum landing site explored by the *Opportunity* rover shows clear evidence for an earlier wet and warm environment followed by a drier period dominated by eolian activity. The layered rocks examined by *Opportunity* are older that 3.7 billion years based on the density of highly eroded large craters observed in orbital images (Fig. 5). These rocks are dirty evaporites composed of materials that have precipitated from salty water and mobilized and moved by the wind (Fig. 23) before being deposited and altered by groundwater. On Earth, this sequence of events and resulting rocks is common in hot and dry salt water playa or sabkha environments such as the Persian Gulf, the Gulf of California, and some inland enclosed basins. By analogy, the environment on Mars was warm and wet when these rocks were deposited prior to 3.7 billion years ago. Because the evaporites are part of a sedimentary sequence that outcrops throughout the broad Meridiani region, these climatic conditions were operative over an area that was at least 1000 km wide, arguing that the environment was both warm and wet and the atmosphere was thicker. Latter on in Mars history, the environment changed, and Meridiani Planum was dominated by eolian activity that eroded and filled in impact craters and concentrated the hematite spherules as a lag on the top of the layer of basaltic sand. The presence of olivine in the basaltic sand suggests these materials were not weathered by liquid water, and the saltation of the sand appears to have efficiently eroded the weak sulfates.

6.6 Implications for a Habitable World

The Meridiani Planum evaporites and Columbia Hills rocks in Gusev crater indicate a warm and wet environment before about 3.7 billion years ago. This is consistent with a variety of coeval geomorphic indicators such as valley networks, degraded and filled ancient craters, highly eroded terrain, and layered sedimentary rocks that point to an early warm and wet climate. The warm and wet environment would also imply a thicker atmosphere capable of supporting liquid water. In contrast, the surficial geology of the landing sites younger than about 3.7 billion years old all indicate a dry and desiccating environment in which liquid water was not stable and eolian and impact processes dominate. This further indicates a major climatic change occurred around 3.7 billion years ago.

A warm and wet environment before 3.7 billion years suggests that Mars was habitable at a time when life started on the Earth. However, the highly acidic nature of water at some Mars landing sites may not have been conducive to the appearance of early organisms. In any case, the earliest chemical evidence for life on Earth is about 3.9 billion years old, and the most important ingredient for life on Earth is liquid water. If liquid water was stable on Mars when life began on Earth, could a second genesis on Mars have occurred? Is it possible that life actually started on Mars earlier when it was more clement than Earth that was subject to early giant possibly sterilizing impacts and was later transported to the Earth via meteorites ejected off the martian surface? Will life form anywhere that liquid water is stable or is it a rare occurrence? Are we alone in the universe? These are the compelling questions that can be addressed by upcoming landers and rovers in a Mars exploration program.

Bibliography

Additional Reading

Golombek, M. P., et al. (1997), Overview of the Mars Pathfinder mission and assessment of landing site predictions. *Science* **278**, 1743–1748. And the next five papers in *Science* (pp. 1734–1774) in which the scientific results of the *Mars Pathfinder* mission were first reported.

Golombek, M. P., et al. (1999). Overview of the *Mars Pathfinder* mission: Launch through landing, surface operations, data sets, and science results. *J. Geophys. Res.* **104**, 8523–8553. Special issues of the *Journal of Geophysical Research, Planets* (volume 104, pages 8521–9096, April 25, 1999; volume 105, pages 1719–1865, January 25, 2000) also featured the scientific results of the mission.

Kieffer, H. H., Jakosky, B. M., Snyder, C. W., and Matthews, M. S. (1992). "MARS." Univ. Arizona Press, Tucson.

Squyres, S. W., et al. (2004). The *Spirit* rover's Athena science investigation at Gusev crater, Mars. *Science* **305** (5685), 794–799, DOI: 10.1126/science.1100194. And the next ten papers (pp. 793–845) in which the first results of the *Spirit* rover were reported.

Squyres, S. W., et al. (2004). The *Opportunity* rover's Athena science investigation at Meridiani Planum, Mars. *Science* **306** (5702), 1698–1703, DOI: 10.1126/science.1106171. And the next ten papers (pp. 1697–1756) in which the first results of the *Opportunity* rover were reported.

Golombek, M., et al. (2005). Assessment of Mars Exploration Rover landing site predictions. *Nature* **436**, 44–48, DOI: 10.1038/nature03600. And the next five papers (pp. 42–70) in which futher results from the Mars Exploration Rovers were reported.

Main-Belt Asteroids

Daniel T. Britt

University of Central Florida
Orlando, Florida

Br. Guy Colsolmagno

Specola Vaticana
Castel Gandolfo, Italy

Larry Lebofsky

Lunar and Planetary Laboratory
University of Arizona
Tucson, Arizona

CHAPTER 18

1. Introduction to Asteroids

1.1 What Are Asteroids?

Asteroids are small, naturally formed solid bodies that orbit the Sun, are airless, and show no detectable outflow of gas or dust. Shown in Fig. 1 are four asteroids that have been imaged in detail by spacecraft: 243 Ida, 951 Gaspra, 253 Mathilde, and 433 Eros. The difference between asteroids and the other naturally formed Sun-orbiting bodies, planets and comets, is largely historical and to some extent arbitrary. To the ancient Greeks and other peoples, there were three kinds of bright objects populating the heavens. The first and most important group was the stars, or *astron* in Greek, which are fixed relative to each other. The English word "star" is an Old English and Germanic derivation of the Indo-European base word *stêr*, which provided the source of the Greek *astron* and the Latin *astralis*. The terms for the study of stars were based on the Greek root (i.e. astronomy or astrophysics). The second group of objects is planets, or Greek *planetos*, meaning wanderer, because the planets were not fixed but moved relative to the background of the stars. For the ancients, *planetos* included the Sun, Moon, Mercury, Venus, Mars, Jupiter, and Saturn. The final group is comets or *kometes*, meaning long-haired, because of their long tails or comas and their unpredictable paths and appearances.

Asteroids were not known to the ancients, and the first asteroid, 1 Ceres, was discovered in 1801 by the Sicilian astronomer Giuseppe Piazzi. He was searching in the gap between Mars and Jupiter for what theorists at the time speculated would be the location of a "missing planet." 1 Ceres was thought initially to be this new planet. However, other astronomers disputed this designation because of Ceres' apparently small size. Soon after William Olbers discovered the second such object, Pallas, in 1802, Sir William Herschel (who had discovered Uranus 20 years earlier) proposed that, because these new objects were planet-like in their sun-centered orbits, but star-like in that they were unresolvable points of light in a telescope, the disused Greek root for a single star *aster* should be used to describe this new addition to the celestial population. However, this term was not universally adopted at that time. By the mid 1800s, after several dozen of these bodies had been discovered, the French and Germans referred to them as "small" (*petit* or *kleine*) planets, while the British Royal Astronomical Society officially called them "minor planets." Until modern times, the term "asteroid" was only used by astronomers in America.

In 2006, the International Astronomical Union (IAU) added additional terms to the mix by defining a group of "dwarf planets." The IAU was attempting to precisely define a planet given the increasing evidence that Pluto was just one of the larger members of the **Kuiper Belt** and

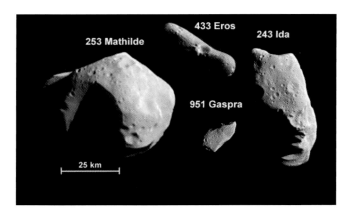

FIGURE 1   The four asteroids that have been imaged by spacecraft flyby: 243 Ida, 951 Gaspra, and 253 Mathilde. 433 Eros was imaged by a spacecraft that orbited the asteroid. (Photograph courtesy of Johns Hopkins University/Applied Physics Laboratory.)

substantially different from the terrestrial or gas giant planets. A dwarf planet orbits the Sun, is not a satellite of another body, has sufficient mass to assume a hydrostatic equilibrium (nearly spherical) shape, and does not have sufficient mass to have "cleared its neighborhood" of small bodies. Under this definition, 1 Ceres joins Pluto as a dwarf planet. However, for the purposes of this chapter, Ceres can also be considered a large asteroid.

Although asteroids share many of the characteristics of planets (Sun-centered orbits, seemingly solid bodies), the primary distinction is that they are simply much smaller than the known planets or dwarf planets. Similarly, the distinction between asteroids and comets is also based on their observational qualities rather than any inherent difference in physical properties or composition. Comets are characterized by their coma, or cloud of sublimating gas and expelled dust. This gives them their characteristic diffuse "fuzzy" halo and long streaming tail. [*See* PHYSICS AND CHEMISTRY OF COMETS.] Compared to the fuzzy look of comets, an asteroid is a "star-like" sharp point of light. But comets only become "cometary" when they enter the inner solar system and are heated sufficiently by the Sun to evaporate their volatile materials. The point at which frozen volatiles begin to sublimate can vary depending on composition, but for most comets this is approximately at 4 AU. A number of outer solar system objects that could be called asteroids may be composed of the same collection of volatile ices, dust, metal, and carbonaceous organics as comets. Because their orbits are less elliptical than currently active comets, they never travel close enough to the Sun to warm their surfaces, cause their ices for flash to gas and appear cometary. These objects are "solid" bodies only because their surfaces stay cold enough to keep their gases frozen.

In the final analysis, asteroids are defined by what they are not: They move against the celestial background so they are not stars. They are not large enough to be planets, dwarf or otherwise. They are not actively shedding gas and dust so are not comets.

1.2 Discoveries, Numbers, and Names

Because asteroids appear as relatively small and dim points of light moving slowly against the stellar background, finding and identifying an object as an asteroid is fundamentally a question of observation coupled with precise "bookkeeping." The field of view seen through a telescope at any one moment is filled with literally hundreds of points of light and rarely will one be an asteroid. The asteroid may move a small amount relative to the stars during the course of a night's observations, but the trick is to know the relative positions of all the viewed stars precisely enough to know when one of the points of light is out of place. Today the viewing through the telescope is done by extremely sensitive charge-coupled devices (CCD) that feed their digital data directly to computers to do the bookkeeping of the stars and known asteroids. In the days when Giuseppe Piazzi discovered 1 Ceres, all the observations were done with an eye to the telescope, and the bookkeeping was done by hand drawings of the star fields. Discoveries were made by visually comparing each point of light in the telescope field with a chart that was drawn on a previous observation. With these methods it is not surprising that only 4 more asteroids were found in the 45 years after Piazzi found 1 Ceres.

The application of photography to astronomy revolutionized the search for asteroids in the last half of the 19th century and the early part of the 20th century. A photographic plate is essentially an instant and precise local star chart that is far more light-sensitive than the human eye, and able to take advantage of long exposures that compensate for the Earth's rotation. As a result, stars appear as fixed as bright dots, while asteroids become streaks because they move relative to the stars. Modern searches have replaced photographic plates with highly sensitive electronic imaging and computers. As of this writing, there are about 120,437 numbered asteroids.

A newly discovered asteroid is given a temporary "name" based on the date of discovery. The first four characters are the year of discovery, followed by a letter indicating which half-month of the year the discovery took place. The final character is a letter assigned sequentially to the asteroids discovered in the half-month in question. Thus, asteroid 2006 CE would be the fifth asteroid discovered in the first two weeks of February in 2006. If a half month has more than 24 discoveries, then the letter sequence starts over with additional numerical characters added as a subscript. The 25th object discovered in the first half of February would be 2006 CA_1.

However, discovery is just the first step. Unless an asteroid is tracked and its orbit reliability determined, it will be "lost." This tracking process takes weeks and sometimes months of additional observations. Once an object has an accurate orbit, it is given a permanent number. The numbers are not assigned in order of discovery, but sequentially by order of orbit determination. With the assignment of a number, the asteroid's discoverer has the right to suggest a name for the object. Asteroids are unique in that they can be named after persons living or dead, real or imaginary, mythological characters or creatures, and in several cases, pets (though this is now discouraged); however, political and military leaders must have been dead 100 years before an asteroid can bear their names, and asteroids cannot be named to advertise commercial products.

1.3 Sizes and Shapes

Shown in Table 1 is a listing of the diameters of the 20 largest main-belt asteroids. Asteroids sizes drop rapidly, with the largest asteroid 1 Ceres being almost twice as large as the next largest. There are only 5 asteroids with diameters greater than 400 km and only 3 with diameters between 400 and 300 km. The asteroid population becomes relatively abundant only below 300 km diameter.

The number of asteroids increases exponentially as the size decreases in a "power-law" size distribution. This is consistent with an initial population of strong, solid bodies that have been ground down by repeated impacts over the age of the solar system. Today most asteroids are fragments of larger parent bodies that have been collisionally shattered into much smaller pieces. This power law is seen not only in the sizes of asteroids but also in the sizes of the craters on the Moon, Mars, and the moons of Jupiter and Saturn, reflecting the population of the asteroids whose impacts made those craters.

Given the conditions in the Asteroid Belt today, only the largest asteroids are large enough to have survived from the beginning of the solar system. The power law predicts, and observations confirm, that by far the most common asteroids are the smallest. Asteroid search programs using powerful telescopes, extremely sensitive CCD sensors, and state-of-the-art software regularly find asteroids in near-Earth space with diameters as small as only 5–10 m. The primary limitation on our ability to find asteroids is their size. Smaller objects reflect less light and, after a point, a small object is not observable because the light it reflects drops below the limiting sensitivity of the telescopic system trying to detect it. The good news is that we have probably discovered and tracked all asteroids in the Main Asteroid Belt larger than 20 km and all those in near-Earth space

TABLE 1	Diameters of the 20 Largest Asteroids		
Asteroid Name and Number	Asteroid Class	Semimajor Axis	Diameter (km)
1 Ceres	C	2.767	940
4 Vesta	V	2.362	576
2 Pallas	B	2.771	538
10 Hygeia	C	3.144	430
704 Interamnia	D	3.062	338
511 Davida	C	3.178	324
65 Cybele	C	3.429	308
52 Europa	C	3.097	292
87 Sylvia	P	3.486	282
451 Patientia	C	3.063	280
31 Euphrosyne	C	3.156	270
15 Eunomia	S	2.644	260
324 Bamberga	C	2.683	252
3 Juno	S	2.670	248
16 Psyche	M	2.922	246
48 Doris	C	3.112	246
13 Eugenia	C	2.576	244
624 Hector	D	5.201	232
24 Themis	C	3.133	228
95 Arethusa	C	3.068	228

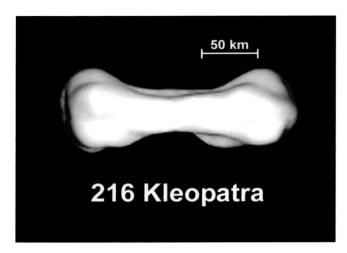

FIGURE 2 Radar image of the asteroid 216 Kleopatra. This irregularly shaped object resembles a 200 m long dog bone. Although this object is an extreme example, all the asteroids "seen" so far by either spacecraft or radar are very irregular in shape. (Photograph courtesy of the Jet Propulsion Laboratory.)

larger than 4 km. The bad news is that there are thousands of small asteroids in Earth-crossing orbits, a few as large as several kilometers in diameter, that remain undiscovered and potential threats to Earth. [*See* NEAR-EARTH OBJECTS.]

Because most asteroids are probably collisionally produced fragments of larger asteroids it should not be a surprise that they are not perfect spheres. Many asteroids that have been directly imaged optically or by radar tend to show very irregular shapes (Fig. 1). The exception is the largest asteroid (or dwarf planet), 1 Ceres, which is large enough for hydrodynamic forces to maintain a spherical shape. Other large asteroids are far from spherical. For example, shown in Fig. 2 is a radar image of the asteroid 216 Kleopatra, which has a strong resemblance to a 200 km long bone! Most asteroid shapes can be approximated as triaxial ellipsoids, which are objects that have different dimensions on each of their principle axes. In the case of Kleopatra, the long dimension in Fig. 2 is over four times greater than the short dimension.

Star/asteroid occultations provide a direct measurement of an asteroid's shape and an opportunity for amateur astronomers to become involved in significant scientific research. The principle is simple: When an asteroid passes through (or "occults") the light from a star, the asteroid creates a "shadow" in the starlight projected on the Earth. Observers in different locations time the disappearance of the occulted star and trace out the shape of this shadow by reconstructing their "chords" or time-tagged observations of the star disappearing behind the asteroid and reappearing on the other side. When done skillfully with modern equipment such as CCD detectors, computer-driven

imaging systems, precise time, and the Global Positioning System, these measurements can be taken with very high accuracy and provide an excellent "snapshot" of the two-dimensional shape of the asteroid at the moment of occultation.

1.4 Asteroid Density, Porosity, and Rotation Rates

A fundamental physical property of an asteroid is its density. To first order, asteroid density is related to its composition and should be similar to the densities of meteorites thought to be derived from those asteroids. [*See* METEORITES.] However, as is often the case, such expectations are often frustrated by unexpected results from direct measurements. Asteroids in general appear to be significantly under-dense relative to their meteorite analogs.

The primary complication is porosity. Asteroids appear to have significant porosity; some may be as much as 50% empty space, whereas their meteorite analogs have only small to moderate porosities. The observed power law of asteroid sizes and studies of the collisional dynamics of the asteroid belt have suggested a history of intense collisional evolution and that only the largest asteroids retain their primordial masses and surfaces. Asteroids below 300 km in diameter will have been shattered by energetic collisions. Some objects reaccrete to form gravitationally bound rubble piles, while the rest are broken into smaller fragments to be further shattered or fragmented. Thus, most asteroids may be shattered heaps of loosely bound rubble with significant porosity in the form of large fractures, vast internal voids, and loose-fitting joints between major fragments. Thus, it is not surprising that the average asteroid would have a very large porosity.

Another line of evidence supporting the rubble pile model for asteroids are the images of 253 Mathilde. This object, whose density is only half the density of typical meteorite material, has 6 identified impact craters that are larger than the size necessary to shatter the asteroid. The only way that Mathilde could have survived these repeated huge impacts is if it were already a shattered rubble pile that dissipates much of the energy of large impacts in the friction of the pieces of rubble grinding against each other.

In addition to the images of this one asteroid, Mathilde, and evidence from the densities available only for a few dozen asteroids to date, data to support this rubble-pile model of asteroids in general comes from the rotation rates of asteroids. For objects that have not been catastrophically disrupted by collisions, rotation rates are probably set by the accretion conditions of the solar nebula and would tend to be relatively slow (1 or 2 revolutions per day). For small asteroids that are fragments of catastrophic impacts, rotation rates are set by the conditions of angular momentum partitioning during the collision and should be much more rapid (>5 revolutions per day). However, the faster a rubble-pile asteroid spins, the more likely that the centrifugal acceleration of material at the surface will be greater than the acceleration of the object's gravity. The result is that above a critical rotation rate, material would be "spun off" the weak gravity of the surface. Observations of asteroid rotation rates show a "rotation rate barrier" where almost every asteroid rotates below this critical rate. The rotation rate barrier would not strongly affect monolithic asteroids, so this is evidence that most asteroids are limited by their coherent strength to rotate slower or shed material from centrifugal acceleration.

2. Locations and Orbits

2.1 Zones, Orbits, and Distributions

Minor planets can be found in almost any region of the solar system, but as shown in Fig. 3a, one of the largest concentrations of asteroids is located in the "belt" between 1.8 and 4.0 AU. A more detailed analysis of the *average* distances of asteroids from the Sun (the asteroids' semimajor axes) as shown in Fig. 4 reveals a subtle structure to the Asteroid Belt. First, there appears to be a sharp inner boundary to the Asteroid Belt at about 2.2 AU. But note that this boundary curves to higher AU for asteroids with higher orbital **inclinations**. Second, there is a sharp gap in the number of

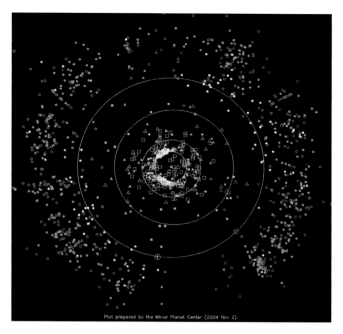

FIGURE 3b   The location of asteroids in the outer solar system. The outer circle is the orbit of Neptune with the location of the planet shown as a tick mark on the orbital path. The "swarms" before and after Jupiter are the Trojans and the thick asteroid belt outside of Neptune is the Kuiper Belt.

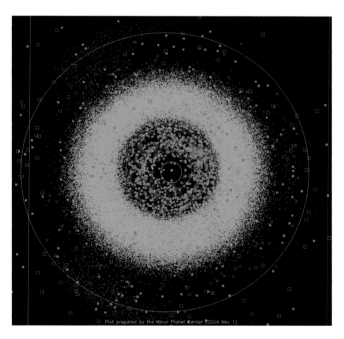

FIGURE 3a   The location of asteroids in the inner solar system. The outer circle is the orbit of Jupiter with the location of the planet shown as a tick mark on the orbital path. The "swarms" before and after Jupiter are the Trojans and the thick Main Asteroid Belt is readily visible just outside the orbit of Mars.

asteroids whose average distance from the sun (**semimajor axis**) is 3.28 AU. Asteroids orbiting here would have exactly half the orbital period of Jupiter and are said to be in a 1:2 *mean-motion resonance* with Jupiter. Similar gaps can be seen elsewhere in the asteroid population as well, most notably at the locations of the 1:3 and 2:5 mean-motion resonances. The gaps in the distribution of asteroid semimajor axes are called **Kirkwood gaps** for Daniel Kirkwood who first pointed them out in 1886. Unlike the gaps in Saturn's rings, however, these gaps are not directly visible within the Asteroid Belt in Fig. 3a because asteroid orbits have a wide range of eccentricities and are constantly crossing through the region of these gaps. Third, there is a dearth of asteroids in orbits with semimajor axes beyond 3.5 AU, with two exceptions: There are clusters of asteroids at 3.97 AU, corresponding to the 2:3 mean-motion resonance with Jupiter, and at 5.2 AU, where asteroids share the same orbit as Jupiter.

These boundaries and gaps are formed by the steady influence of the gravitational attraction of the planets on the orbits of the asteroids. In general, these interactions occur at random time intervals and at random locations of the asteroid's orbit, and on average they cancel out without causing a significant change in the asteroid's orbit. However, an asteroid whose orbital period is a simple fraction of Jupiter's 11.86 year period will be in *resonance* with Jupiter and have

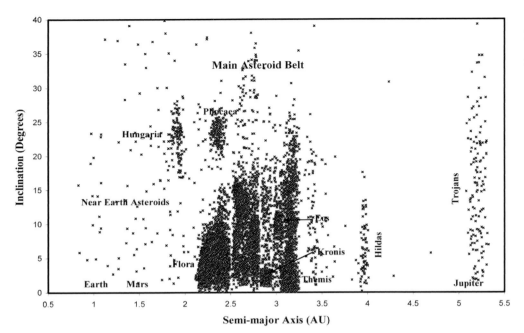

FIGURE 4   Plot of orbital parameters of numbered asteroids in semimajor axis vs. inclination space.

a close approach in the same place in its orbit over and over again. Jupiter's strongest pull will occur when it is closest; for an asteroid with a 6 year period (in a 1:2 resonance), this closest approach will occur at the same place every other asteroid orbit. (Similarly, asteroids in the 1:3 resonance encounter Jupiter at the same place in their orbits, every third orbit.) Jupiter's pull at this point, imparting some energy to the asteroid's orbit, will then compound itself, rather than cancel out. The largest effect of this sort of perturbation is to increase the **eccentricity** of the asteroid's orbit. This does not change its "average" distance from the Sun, but it makes the perihelion move closer to the Sun, and the aphelion move farther out. Once its eccentricity reaches a value of about 0.3, a Main Belt Asteroid's orbit begins to approach or even cross the orbit of Mars. Close encounters with Mars can further alter its orbit, leading to interactions with the other inner planets or with Jupiter, which eventually results in a collision with either a planet or the Sun, or ejection from the solar system. For asteroids, orbital life in the Kirkwood gaps is (relatively) short, but exciting.

This kind of resonance explains the Kirkwood gaps. But it does not explain the inner boundary and its dependence on the inclination of the asteroid orbit, the lack of asteroids with semimajor axes outside 3.5 AU, or the concentration of asteroids at the outer resonances. More indirect effects give rise to these patterns. The shape of the inner boundary is the result of a subtle but surprisingly powerful effect. Every asteroid has an orbit that is at least slightly eccentric, and the orientation of its perihelion slowly drifts with time. This *precession* of the perihelion is caused by the perturbations of the other planets. Likewise, the orientations of the major

planets' orbits, which are not perfectly circular, also drift with time. A subtle interaction arises when the precession of Saturn's orbit is in resonance with the precession of an asteroid's orbit. This *secular resonance* (so-called because it builds up over time, regardless of where the asteroid and Saturn are in their orbits) is called the ν_6 resonance; ν is the Greek letter that represents the precession rate, and the 6 represents Saturn, the sixth planet from the Sun. Its effect is to increase an asteroid orbit's eccentricity, as with the Jupiter mean-motion resonances. The position of this resonance depends on both the location and the inclination of the asteroid orbit. For asteroids orbiting in the plane of the planets, it occurs at around 2.2 AU; as the inclination of the asteroid orbit increases, the location of this resonance moves further from the Sun. This resonance sculpts the inner edge of the Asteroid Belt.

A possible inward migration of Jupiter's orbit early in the history of the solar system may have been responsible for clearing out the outer regions of the Asteroid Belt. If the solar nebula from which the planets were formed was a smooth cloud of gas and dust, there should have been nearly as much material in the region just inside where Jupiter was formed—the location of the Asteroid Belt today—as there was in Jupiter itself. But Jupiter's gravity has its strongest effect on material closest to it. If Jupiter formed first, its gravity would have stirred up the material nearby and stopped it from forming another planet. That material would have been ejected from the inner solar system by Jupiter's gravity. Some of that material may today be residing in the far-distant **Oort cloud.** [*See* PHYSICS AND CHEMISTRY OF COMETS.]

But if Jupiter were responsible for ejecting a large amount of material originally lying in the asteroid region, by conservation of energy it must have moved inward as this material moved outward. As Jupiter moved, the location of its resonances within the asteroid region also moved. Numerical models have shown that moving these resonances through the outer Asteroid Belt would effectively deplete it of material. However, modeling also shows that asteroids in the 2:3 resonance are stabilized against ejection, and thus carried along with Jupiter as it moves.

2.2 Special Orbital Classes

Even though most asteroids are found in the Main Asteroid Belt between Jupiter and Mars, there are a number of other asteroid groups. The "asteroids" beyond the orbit of Jupiter are probably volatile-rich and would become cometary if they were moved to the inner solar system, but for the purposes of this discussion we will list these groups of small bodies as asteroids here. The asteroids that circle the Sun at the same orbital distance as Jupiter are called *Trojan* asteroids. They reside in dynamically stable zones 60° ahead and behind Jupiter. These positions are the last two of the five Lagrangian points, "named by the 19th-century mathematician J. L. Lagrange. He first described the orbital behavior of small bodies affected by the gravitation pull of two large objects such as the Sun and Jupiter. He found that along with three unstable equilibrium points (L_1 through L_3), a small body like an asteroid could share Jupiter's orbit so long as both formed an equilateral triangle with the Sun. There are two such points; the L_4 point lies ahead of Jupiter, while L_5 trails behind it.

The Trojans derive their name from the first such asteroid discovered, named Achilles after the hero of the Trojan War. The L_4 region asteroids are named for Greek heroes of the Iliad, while Trojan heroes populate the L_5 region. (The exceptions, named before this rule was adopted, include two of the largest Trojans: 617 Patroclus, named for the Greek hero, orbits among the Trojans at L_5, while 624 Hektor, the largest Trojan and a hero of Troy, orbits at L_4 with the Greeks.) Nearly 2000 Jupiter Trojans have been discovered to date; oddly, the L_4 region is nearly twice as populated as the L_5 region.

Another major group of minor planets is the *Centaurs*. Named as a class after the discovery of Chiron, a small body orbiting between Saturn and Uranus, the term has eventually grown to include any noncometary body beyond Saturn whose orbit crosses the orbit of a major planet; even the noncomet part must be relaxed, as Chiron itself has been seen on occasion to have a comet-like coma. These "asteroids" are most likely large, volatile-rich objects (i.e., comets) perturbed inward from the Kuiper Belt (Fig. 3b). But because the Centaurs orbit deep in the outer solar system, they cannot warm sufficiently to allow volatiles to sublimate off

and show cometary activity, so they are considered asteroids until proven otherwise. In terms of their orbits, this group includes the classical Centaurs (some two dozen objects known to orbit like Chiron between Saturn and Uranus), roughly 50 objects whose orbits cross Uranus' or Neptune's orbit, and the 75 objects (discovered to date) that lie in highly eccentric orbits ranging out beyond the Kuiper Belt. All are considered scattered disk objects, which have been dynamically scattered by Neptune's gravity out of the disk of the Kuiper Belt. [*See* KUIPER BELT: DYNAMICS.]

The Kuiper Belt itself is the outermost set of minor bodies. It is made up of objects populating space beyond the orbit of Neptune but inside about 1000 AU. The first object was discovered in 1992 (1992 QB1) with a semimajor axis of 44 AU and an estimated diameter of several hundred kilometers. Besides the scattered disk objects noted earlier, other dynamical classes of Kuiper Belt objects include others like 1992 QB1 in low-inclination, low-eccentricity orbits (sometimes called "cubewanos" after their first example) and others orbiting like Pluto (and so called "plutinos") in a 2:3 resonance with Neptune. Again, all these objects are probably cometary. In fact, the existence of the Kuiper Belt was first suggested in 1949 as a source area for short-period comets. Given the nearly 1000 Kuiper Belt "asteroids" discovered so far, there are probably hundreds of thousands of objects larger than a kilometer populating this belt. [*See* KUIPER BELT: DYNAMICS.]

Inward from the main asteroid belt are the asteroids that cross the orbits of the inner planets: the *Amor*, *Apollo*, and *Aten* asteroids. Amor asteroids are asteroids whose eccentric orbits dip in from the Asteroid Belt to cross the orbit of Mars, but without reaching the orbit of the Earth. Apollos are those that do cross Earth's orbit, but whose semimajor axis is always ≥ 1 AU. This differentiates them from Atens, which also cross the Earth's orbit but that have semimajor axes inside of Earth's orbit. The Apollo and Amor objects are collectively called near-Earth objects or NEOs. They are relatively small objects; the largest known NEO is the Amor object 1036 Ganymed, with a diameter of 38.5 km. NEOs are also subject to a power law distribution; as the population increases, their sizes drop rapidly. As of September 2006, there are 830 NEOs with diameters >1 km out of a population of approximately 3800 known NEOs. It is estimated that there are approximately 1200 total NEOs that are larger than 1 km. These are the objects that can and (in the course of geologic time) do frequently collide with Earth. Indeed, computer calculations indicate that most NEOs could only survive in their present orbits for roughly 10 million years before falling into the Sun, colliding with a planet, or being ejected. Thus, the NEO population must be continually replenished from the Asteroid Belt. Compositional data indicates that NEOs are drawn from every zone of the Asteroid Belt and have been perturbed into the inner solar system by a variety of mechanisms including the

Yarkovsky effect described in the next section. [See NEAR-EARTH OBJECTS.]

2.3 The Evolution of Orbits: Yarkovsky and YORP

The gravitational perturbations of the planets are not the only forces acting on the asteroids. Although the Kirkwood gaps show that resonances are the most effective way to clear material from the Asteroid Belt, the generally low population of asteroids throughout the belt (even in its most heavily populated regions), the replenishment of asteroids into short-lived NEO orbits, and the constant delivery of meteorites from the Asteroid Belt to the Earth (see discussion that follows) all indicate that some other forces must be moving material from the main belt to the resonance regions.

One early hypothesis was that collisions between asteroids could impart enough momentum to scatter the collision products into a wide variety of new orbits, some of which would lie in resonance with Jupiter and thus be delivered out of the Asteroid Belt. However, detailed computer modeling of both the collisions and the ensuing orbits of the collisional products conclusively shows that this process alone fails by many orders of magnitude to move nearly enough material from the Asteroid Belt to match the observed population of NEOs or the meteorite flux. Some other force or forces must be involved.

One early suggestion apparently first proposed by the Russian theorist I. O. Yarkovsky in the late 19th century is that sunlight itself could provide a surprisingly effective way of changing the orbits of asteroids. The general idea is simple enough. Light carries momentum, so as sunlight is absorbed or reflected by an asteroid, there is a small momentum transfer from the light to the asteroid. However, because sunlight comes from the same direction as the force of the Sun's gravity (and, like gravity, varies as $1/r^2$) this effect by itself will merely change the effective pull of the Sun, without changing the energy (or semimajor axis) of an asteroid's orbit. (There is a small relativistic effect called Poynting–Robertson drag, but it is ineffectual for anything larger than small grains of dust.) However, when an asteroid absorbs sunlight, the energy of that light heats the asteroid, and that heat must eventually be reradiated to space as infrared photons. When each infrared photon is emitted, it exerts a tiny amount of recoil momentum to the asteroid itself. And, unlike the direct reflection of sunlight, this recoil is not necessarily in the same direction as the pull of the Sun's gravity because there is always a small time lag between the absorption and the reradiation of the energy.

For example, the afternoon side of a spinning body will always be slightly warmer than the morning side. This means that more infrared energy is radiated from the afternoon side; that side of the asteroid experiences a greater recoil from those photons' emissions than the morning side

does. The way the spin axis is tilted, or the differences in heating between perihelion and aphelion, is another example of situations that will lead to the asymmetric radiation of infrared photons. This difference can serve to constantly add or subtract (depending on how the asteroid spins) energy from the asteroid's orbit and thus change its semimajor axis. It can also change the way the asteroid itself spins. An elaborate theory based on the work of Yarkovsky, as further elaborated by O'Keefe, Radzievskii, and Paddack, dubbed the YORP effect, suggests a number of ways in which the momentum of emitted radiation can alter both the speed and the direction of an asteroid's spin. More than just a mathematical curiosity, the predictions of this work have been confirmed in a number of cases, including asteroids whose spin rates have been observed to change or be aligned in a way predicted by this theory.

2.4 Asteroid Families

As discoveries of asteroids accumulated in the early part of the 20th century, astronomers noted that it was common for several asteroids to have very similar orbital elements and that asteroids tended to cluster together in semimajor axis, eccentricity, and inclination space. In 1918, K. Hirayama suggested that these clusters were "families" of asteroids. Hirayama suggested 5 families, and this number has been greatly increased by the work of generations of orbital dynamacists.

These families are probably the result of the collisional breakup of a large parent asteroid into a cloud of smaller fragments sometime in the distant past. Time and the gravitational influence of other solar system objects has gradually dispersed the orbits of these fragments, but not enough to erase the characteristic clustering of families.

It has been suggested that families could provide a glimpse at geologic units that are usually deeply hidden in the interiors of planets. If a differentiated asteroid were broken into family members, for example, that family should have members that represent the metallic core; others coming from the metal–rock transition zone called the core–mantle boundary; yet others made of the dense, iron-rich units in the mantle; and others originating from the crust of the former planetesimal. In fact, however, no such elaborate collection of different asteroid types has been seen in a family. However, families may be relatively short-lived. The Yarkovsky effect has proved to be very effective in moving family members out of their original orbits. Understanding and defining the dynamics of asteroid families remains an active and rapidly changing field of study.

2.5 Asteroids and Meteorites

There are a number of lines of evidence that show the ultimate source region for meteorites is the Asteroid Belt. [See METEORITES.] The strongest evidence is the direct observations of a half-dozen recovered chondrites that were photographed falling by camera networks, or whose fireballs were recorded by many well-separated video images. These data show that each of the meteorites had its orbital origins in the Main Asteroid Belt. Other evidence includes the similarity of meteorite reflectance spectra to several classes of asteroids; the existence of xenoliths (pieces of other meteorite types included in meteorite breccias) in meteorites, which requires that the source region have the mineralogical diversity found in the Asteroid Belt; and the solar-wind implanted gases found in **regolith** meteorites indicating that implantation took place in regions consistent with the location of the Asteroid Belt.

Meteorites do not automatically provide the location and taxonomic class of their particular parent bodies. The very fact that a meteorite is "in our hands" suggests the occurrence of some violent event that may have fragmented and perhaps destroyed the parent body. The best that can be done is to link individual asteroid spectral classes with meteorite compositional groups. This task is somewhat speculative because most meteorites were originally buried beneath the surface of an asteroid, asteroid surface conditions are unknown, and the effects of space weathering on asteroids are poorly understood. All spectral matches between asteroids and meteorites, including the ones detailed here, should be viewed with healthy skepticism.

There are several factors that bias the population of meteorites arriving on Earth and therefore limit our sample of the Asteroid Belt. First, the dynamical processes that deliver meteorites from the Asteroid Belt to Earth are probably strongly biased toward sampling relatively narrow zones in the Asteroid Belt. Calculations demonstrate that the vast majority of meteorites and planet crossing asteroids originate from just two resonances in the belt, the 1:3 Kirkwood gap and ν_6 resonance. Both of these zones are in the inner Asteroid Belt where the asteroid population is dominated by S-type asteroids. However, the Yarkovsky effect significantly increases the chances of fragments from anywhere in the Asteroid Belt working their way into Earth-crossing orbits. A second factor is the relative strength of the meteorites. To survive the stress of impact, acceleration, and then deceleration when hitting the Earth's atmosphere, without being crushed into dust, the meteorite must have substantial cohesive strength. Large iron meteorites are more likely to survive until they hit the surface of the Earth; they may form a crater (like Meteor Crater in Arizona) when they hit, but in that process most of the iron is vaporized and lost. The Earth's atmosphere is probably the most potent filter for meteorites. The relatively weak, volatile-rich meteorites from the outer Asteroid Belt stand little chance of surviving the stress and heating of atmospheric entry. It is very likely that the meteorites available to us represent only a small fraction of the asteroids, and it is possible that most asteroids either cannot or only rarely contribute to the meteorite collections.

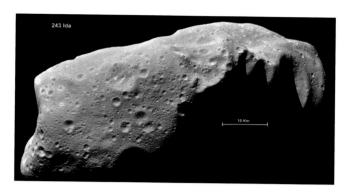

FIGURE 5 The surface of asteroid 243 Ida. (Photograph courtesy of the Jet Propulsion Laboratory.)

3. Physical Characteristics and Composition

3.1 The Surfaces of Asteroids

As shown in Figs. 1, 5, and 6, the surfaces of asteroids appear cratered, lined with fractures, and covered with regolith. These surfaces are dominated by impact processes. As discussed in earlier sections, asteroids are strongly affected by collisional disruption and have a complex history of impact fracturing and fragmentation. Objects in the size range shown in the figures are probably formed as disrupted fragments from larger objects, and some are likely rubble piles themselves. Because asteroids are far too small to retain an atmosphere that could offer some protection from the exposure to space, the surfaces of asteroids are exposed to an extremely harsh environment. There are a range of processes associated with exposure to the space environment; high levels of hard radiation, high-energy cosmic rays, ions and charged particles from the solar wind, impacts by micrometeorites, impacts by crater-forming objects, and finally impacts by other asteroids large enough to destroy the parent asteroid. The overall result of these processes is threefold: First, large impacts shatter the parent asteroid creating

FIGURE 6 Asteroid 25143 Itokawa. The asteroid is approximately 700 m in its longest dimension. The smooth areas in the center and on the lower left center are examples of ponding of fine regolith.

substantial internal fracturing, porosity, and an extremely rough and irregular surface. Second, small impacts and micrometeorites create a regolith that blankets the asteroid in a fine soil of debris form the bedrock. Finally, micrometeorites, radiation, and the solar wind produce chemical and spectral alteration in the regolith soil and exposed bedrock that "weathers" the surface of the asteroids.

All the small asteroids viewed by spacecraft show significant regoliths, and the power of radar waves reflected by asteroids large and small (especially those passing near the Earth, and so more easily observed by radar) also shows that their surfaces are comparable to dry soil or sand. [*See* PLANETARY RADAR.] On several of these asteroids the regoliths appear to have been altered by space-weathering processes, although just how this alteration affects asteroidal material is still not completely understood. In the asteroid population, there are general spectral trends that appear to be associated with the age of an asteroid's surface. The red continuum slope of S-class asteroids declines in magnitude with asteroid size. This effect appears to be related to the age of the asteroidal surface, with younger less altered surfaces tending to be less red. This effect is seen in the meteorite population. Meteorites that have evidence of residing on the surfaces of asteroids have strong spectral differences from meteorites that were not exposed on asteroid surfaces.

Another major surface effect is the development of small "ponds" of fine regolith material as shown in Fig. 6. These ponds have been seen on Eros and Itokawa and consist of very fine dust that has been somehow mobilized on the surface and accumulated in local "depressions" or gravitational lows. The actual magnitude and direction of gravity on a body as small and irregularly shaped as an asteroid is not at all intuitive, but the effect is still strong enough to drive surface processes. Ponds appear to develop over time and appear to bury the boulders and cobbles within them.

Another process that affects the surfaces of asteroids is the reaccretion of ejecta debris. Impacts of other small asteroids produce the abundant craters seen on all these objects. Although much of the impact debris escapes the low gravity of an asteroid, a large amount is reaccreted by the asteroid. The abundance and location of boulders on objects such as Eros, shown in Fig. 7, and Itokawa (Fig. 6) has been explained by the low-velocity ejecta debris slowly "falling" back onto the rotating asteroid.

3.2 Asteroid Satellites

It had been long suspected that some asteroids had satellites; this was spectacularly confirmed when the *Galileo* spacecraft flew by asteroid 243 Ida and discovered its moon Dactyl. As of this writing, 107 asteroid satellites have been announced in 103 systems including 2 triple systems and one quadruple. Shown in Fig. 8 is an image of asteroid 22 Kalliope and its satellite Linus.

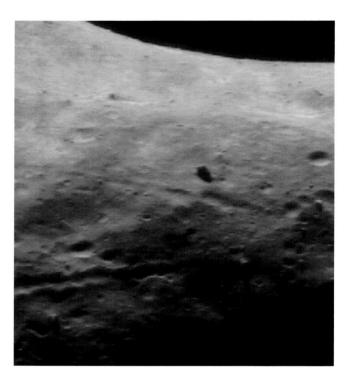

FIGURE 7   Fractures and boulders on the surface of asteroid 433 Eros. Even with the weak gravity of asteroids, low-velocity ejecta such as these boulders do reaccrete to the surface. (Photograph courtesy of APL/JHU.)

NEOs tend to have small separation distances from their satellites, which are probably the result of formation by "fission." Almost all NEOs are rubble piles and with a high enough rotation rate that centrifugal acceleration can throw boulders from the surface into orbit. Many NEOs have rotation rates close to the fission limit, and additional collisions or the YORP effect can enhance asteroid spin enough to cause fission. After fission occurs, the new satellite carries

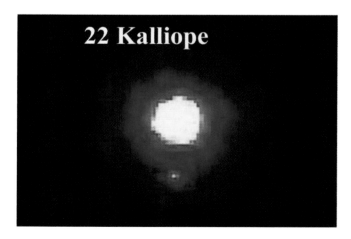

FIGURE 8   Asteroid 22 Kalliope and its satellite Linus.

away some of the primary's angular momentum, thus dropping the rotation back below the fission limit. The lack of distant NEO companions may be the result of gravitational encounters with planets. Distant satellites would be much more likely to be stripped from their primaries during close planetary encounters common with NEOs.

Although only about 100 have been discovered, asteroid satellites are thought to be fairly common with a few percent of all asteroids having satellites. With over 137,000 numbered asteroids, a large number of satellites remain to be discovered. This is another area where amateur astronomers can make a significant contribution to science. Some satellites have been discovered by direct imaging, either from spacecraft or **adaptive optics** (i.e. Fig., 8), but most satellites are discovered by analysis of asteroid **lightcurves**. The principle is that the satellite in its orbit will periodically add or subtract its illumination from the brightness of the asteroid. By precisely tracking the change in brightness, it is possible to identify the satellite and determine its orbit and period. With CCD imagers available commercially and modest-sized telescopes, a skilled amateur can successfully compete in discovering asteroid satellites.

3.3 Telescopic Observations of Composition

Our understanding of the composition of asteroids rests on two pillars: the detailed study of meteorite mineralogy and geochemistry and the use of remote sensing techniques to analyze asteroids. The meteorites provide, as discussed in a previous section, an invaluable but limited sample of asteroidal mineralogy. To extend this sample to what are effectively unreachable objects, remote sensing uses a variety of techniques to determine asteroid composition, size, shape, rotation, and surface properties. The best available technique for the remote study of asteroid composition is visible and near-infrared **reflectance spectroscopy** using ground-based and Earth-orbiting telescopes. Reflectance spectroscopy is fundamentally the analysis of the "color" of asteroids over the wavelength range 0.2–3.6 μm. An experienced rockhound limited to the three colors of the human eye can identify a surprisingly wide variety of rock-forming minerals. For example the silicate *olivine* is green, and important copper minerals such as *azurite* (blue) and *malachite* (green) are vividly colored. These colors are a fundamental diagnostic property of the mineralogy because the atoms of a mineral's crystal lattice interact with light and absorb specific wavelengths depending on its structural, ionic, and molecular makeup, producing a unique reflectance spectrum. The reflectance spectrum is essentially a set of colors, but instead of three colors, our remote sensing instruments "see" very precisely in 8, 52, or even several thousand colors. What can be seen are very precise details of the mineralogy of the major rock-forming minerals olivine, pyroxene, spinel, the presence of phyllosilicates,

organic compounds, hydrated minerals, and the abundance of free iron and opaque minerals.

In addition to a spectroscopic inventory of minerals, telescopic measurements yield several other critical pieces of information. The **albedo** or fundamental reflectivity of the asteroid can be determined by measurements of the visible reflected light and the thermal emission radiated at longer wavelengths. A dark asteroid will absorb much more sunlight than it reflects, but it will heat up and radiate that extra absorbed energy at thermal wavelengths. Ratioing the reflected and emitted flux at critical wavelengths provides an estimate of an asteroid's albedo. Reflectance measured at a series of phase angles can be used to model the photometric properties of the surface material and estimate physical properties like the surface roughness, surface soil compaction, and the light-scattering properties of the asteroidal material. Measurements of polarization as a function of solar phase angle can be used to infer albedo and also provide insight into the texture and mineralogy of the surface.

3.4 Composition, Taxonomy, and the Distribution of Classes

The basic knowledge of asteroids is primarily limited to ground-based telescopic data, usually broadband colors in the visible and near-infrared wavelengths and albedo that is indicative of composition; this forms the basis of asteroid taxonomy. Asteroids that have similar color and albedo characteristics are grouped together in a class denoted by a letter or group of letters. Asteroids in particularly large classes tend to be broken into subgroups with the first letter denoting the dominant group and the succeeding letters denoting less prominent spectral affinities or subgroups.

Asteroid taxonomy has developed in tandem with the increase in the range and detail of asteroid observational data sets. Early observations were often limited in scope to the larger and brighter asteroids and in wavelength range to filter sets used for stellar astronomy. As observations widened in scope and more specialized filter sets and observational techniques were applied to asteroids, our appreciation of the variety and complexity of asteroid spectra has also increased. The asteroid classification system has evolved to reflect this complexity, and the number of spectral classes has steadily increased. Shown in Table 2 is a listing of the expanded "Tholen" asteroid classes and the current mineralogical interpretation of their reflectance spectra. The Tholen classification is still widely used, but it is not by any means the only asteroid classification system. Other widely accepted classifications include the SMASSII system, the Barucci system, and the Howell system.

To explain the compositional meaning of asteroid reflectance spectra and color data, we can treat the Asteroid Belt as a series of zoned geologic units, starting at the outer

TABLE 2	Meteorite Parent Bodies	
Asteroid Class	**Inferred Major Surface Minerals**	**Meteorite Analogs**
D	Organics + anhydrous silicates? (+ ice??)	None (cosmic dust?)
P	Anhydrous silicates + organics? (+ ice??)	None (cosmic dust?)
C (dry)	Olivine, pyroxene, carbon (+ ice??)	"CM3" chondrites, gas-rich/blk chondrites?
K	Olivine, orthopyroxene, opaques	CV3, CO3 chondrites
Q	Olivine, pyroxene, metal	H, L, LL chondrites
C (wet)	Clays, carbon, organics	CI1, CM2 chondrites
B	Clays, carbon, organics	None (highly altered CI1, CM2??)
G	Clays, carbon, organics	None (highly altered CI1, CM2??)
F	Clays, opaques, organics	None (altered CI1, CM2??)
W	Clays, salts????	None (opaque-poor CI1, CM2??)
V	Pyroxene, fledspar	Basaltic achondrites
R	Olivine, pyroxene	None (olivine-rich achondrites?)
A	Olivine	Brachinites, pallasites
M	Metal, enstatite	Irons (+ EH, EL chondrites?)
T	Troilite?	Troilite-rich irons (Mundrabilla)?
E	Mg-pyroxene	Enstatite achondrites
S	Olivine, pyroxene, metal	Stony irons, IAB irons, lodranites, winonites, siderophyres, ureilites, H, L, LL chondrites

zones of the main belt and working inward toward the Sun. The outer asteroid belt is dominated by the low-albedo P and D classes. The analogs most commonly cited are cosmic dust or CI carbonaceous chondrites that are enriched in organics like the Tagish Lake meteorite. However, the spectral characteristics of these asteroids are difficult to duplicate with material that is delivered to the inner solar system. Probably P and D asteroids are composed of primitive materials that have experienced different geochemical evolution than cosmic dust or CI chondrites. Their spectra indicate increasing amounts of complex organic molecules with increasing distance from the Sun. These objects are also probably very rich in volatiles including water ice.

Dark inner asteroid belt asteroids include the B, C, F, and G classes whose meteorite analogs are the dark CI and CM carbonaceous chondrite meteorites. The spectral differences between these classes are thought to represent varying histories of aqueous alteration or thermal metamorphism. The CI carbonaceous chondrites, rich in water, clay minerals, volatiles, and carbon, represent primitive material that has been mildly heated and altered by the action of water. [See METEORITES.]

Sunward of 3 AU, differentiated bright asteroids become much more common. This zone was strongly affected by the early solar system heating event and contains those classes most likely to represent differentiated and metamorphosed meteorites. Perhaps the best asteroid/meteorite spectral matches are the V-class asteroids with the basaltic achondrite meteorites. V-types are interpreted to be a differentiated assemblage of primarily orthopyroxene with varying amounts of plagioclase, which makes them very close analogs to the basaltic howardite–eucrite–diogenite (HED) association of meteorites. These meteorites are basaltic partial melts, essentially surface lava flows and near-surface intrusions originating on asteroids that underwent extensive heating, melting, and differentiation.

While the V-class asteroids represent the surface and near-surface lava flows of a differentiated asteroid, the A-class asteroids are thought to represent the next zone deeper. These asteroids are interpreted to be nearly pure olivine and may be derived from the mantle of extensively differentiated parent bodies. The Earth's mantle is dominated by olivine and theoretical studies show that differentiation of asteroids with a bulk composition similar to ordinary chondrite meteorites should produce olivine-rich mantles. Another possible mantle-derived asteroid is the R class, which is a single-member class made up of the asteroid 349 Dembowska. Analysis of its reflectance spectra suggests a mineralogy that contains both olivine and pyroxene and may be a partial melt residue of incomplete differentiation.

A more common asteroid class is the M class, which has the spectral characteristics of almost pure iron-nickel metal and several show high radar reflections consistent with metal. These objects are thought to be direct analogs to the metallic meteorites and may represent the cores of differentiated asteroids. Isotopic and chemical studies indicate that iron meteorites could come from as many as

60 different parent bodies indicating a wide variety of differentiated bodies in the Asteroid Belt. However, some M-class asteroids have been shown to have hydrated minerals on their surfaces. The spectral characteristics of M asteroids can also be characteristics of some clay-rich silicates, and this raises the possibility that the "wet" M asteroids are assemblages of clays, like the CI carbonaceous chondrites, but without the carbon-rich opaques that darken the CIs. The W (or "wet") class was coined to classify these unusual objects.

The E-class asteroids are another example of the perils of extrapolation from limited information to a convenient meteorite analog. Looking at the spectrum of the "type" asteroid for the E-class, 44 Nysa, it was easy to assume that these asteroids were excellent analogs for the enstatite achondrites. The only problem was that enstatite meteorites are entirely anhydrous, and 44 Nysa was observed to be strongly hydrated. Although some E-class asteroids are probably composed of the same differentiated enstatite assemblages as the enstatite achondrites, about half of the observed Es are hydrated and cannot be composed of anhydrous enstatite. The "wet" E asteroids like Nysa may be related to the W asteroids and have surfaces rich in hydrated silicate clays.

Perhaps the most complex class of asteroids is the very large S class. S-class spectra, on average, indicate varying amounts of olivine and pyroxene with a substantial metallic component, but the mineralogy of these asteroids varies from almost pure olivine to almost pure pyroxene, to a variety of mixtures of these two end-members. With this wide range of mineralogies comes a wide range of meteorite analogs and possible formation scenarios. The S class probably represents a range of asteroid material from core–mantle boundary, the mantle, and the lower crust of differentiated asteroids and includes undifferentiated but metamorphosed asteroids that are the parent bodies of ordinary chondrite meteorites. Ordinary chondrites are by far the largest meteorite type, accounting for approximately 80% of observed meteorite falls, but so far only a few small asteroids have been identified as Q class, direct analogs for ordinary chondrites. A number of S-class asteroids have spectral absorption bands roughly similar to those of ordinary chondrites, but S asteroids typically have a moderate spectral red slope that is not seen in ordinary chondrites. However, it has been shown in laboratory experiments that ordinary chondrite material can redden in response to "space weathering" by micrometeorite bombardment. The small ordinary chondrite parent bodies are probably relatively young fragments that have not had enough time to redden their surfaces. The larger ordinary chondrite parents have older, reddened surfaces and are members of the S class.

In general, the differentiated asteroids of the V, A, R, S, and M classes may represent examples of a geologic transect from the crust to the core of differentiated asteroids

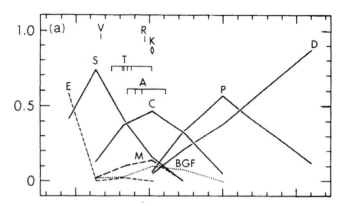

FIGURE 9 The distribution of taxonomic classes from Bell et al. (1989). Reproduced courtesy of Bell, Davis, Hartmann, and Gaffey, 1989, in "Asteroids II" (R. P. Binzel, T. Gehrels, and M. S. Matthews, eds.), Univ. Arizona Press, Tucson, p. 925.

and can tell us a great deal about the geochemical evolution of a differentiated body. In this scenario, the V-class asteroids would be the surface and crustal material. The A asteroids would be from a completely differentiated mantle, while the R asteroids would represent a mantle that experienced only partial differentiation. Some S asteroids, particularly the olivine-rich members, would be either material from some region in the mantle or the core–mantle boundary. And finally, M-class materials represent samples of the metallic cores of these asteroids. From the preceding discussion, it is clear that the asteroid classes were not uniformly distributed throughout the Asteroid Belt. The S class dominates the inner asteroid belt, while the C class is far more abundant in the outer Asteroid Belt. The most populous taxonomic classes (the E, S, C, P, and D classes) peak in abundance at different heliocentric distances. Shown in Fig. 9 is the distribution of taxonomic classes. If we assume that the spectral and albedo differences between the asteroid classes reflect real differences in mineralogy, then we are seeing rough compositional zones in the Asteroid Belt. According to models of solar system condensation, the high-to-moderate-temperature silicate minerals would tend to dominate the inner solar system, while lower-temperature carbonaceous minerals would be common in the cooler, outer regions of the solar system. The transition between moderate- and low-temperature nebular condensates is apparently what we are seeing in the taxonomic zonation of the Asteroid Belt. The innermost major group of asteroids, peaking at 2 AU, is the E class, which is rich in iron-free silicate enstatite, indicating formation under high-temperature, relatively reducing conditions. The next group out is the S class, thought to be rich in the moderate-temperature silicates olivine and pyroxene and to have large amounts of free iron–nickel, which indicate more oxidizing conditions. The C class peaks in abundance at 3 AU and shows a major transition in asteroid mineralogy to less free metal, more oxidized silicates, important low-temperature carbon minerals, and significant amounts of volatiles such as water. The P asteroids peak at about 4 AU, and the D asteroids, which peak at 5.2 AU, are probably richer in low-temperature materials such as carbon compounds, complex organics, clays, water, and volatiles and represent the transition between the rocky asteroids of the main belt and the volatile-rich comets in the Kuiper Belt and the Oort cloud.

Several processes have blurred the taxonomic imprint from the original condensation. Apparently, a thermal event heated much of the Asteroid Belt soon after accretion. Evidence from meteorites shows that some parent asteroids were completely melted (basaltic achondrites, irons, stony irons), some asteroids were strongly metamorphosed (ordinary chondrites), and some were heated only enough to boil off volatiles and produce aqueous alteration (CI and CM carbonaceous chondrites). This event seems to have been much more intense in the inner Asteroid Belt and strongly affected the E-, S-, A-, R-, V-, and M-class asteroids. The dynamical interaction of asteroids with each other and the planets, particularly Jupiter, has altered and blurred the original orbital distribution of the asteroids and cleared whole sections of the belt. The net result probably has been to expand the original compositional zones and produce orbital overlaps of zones that once may have been distinct from each other.

4. Puzzles and Promise

4.1 Telescopic Searches and Exploration

It is a rare but exciting event in science when a single idea by a small group of scientists ignites an entirely new field of study and redefines the scientific debate. That is exactly what happened to such diverse fields as impact physics, asteroid observations, and paleontology after Alvarez and colleagues hypothesized that the iridium anomaly found in Cretaceous–Tertiary (K/T) boundary sediments was the mark of an impact event that destroyed the dinosaurs. [See PLANETARY IMPACTS.]

Asteroid impacts are a consistent and steady-state fact in the solar system. One just has to look at the extensively cratered surface of any solid body to realize that impacts happen. To some extent, the fact that the Earth has active geological processes that erase the scars of impact craters rapidly and a thick atmosphere that filters out the smaller impactors has lulled us into a false sense of security.

The real question is not whether asteroids hit the Earth, but rather how often does it happen. Before they hit, these impactors are comets and asteroids with the same power law distribution of sizes that we see in the Asteroid Belt,

so small impacts will be more frequent and large "species-killing" impacts will be much rarer. However, as those who live near dormant volcanoes should realize, rare events on human timescales can be common and frequent events on geologic timescales.

There is plenty of evidence in the geologic and fossil record for repeated major impacts, some of which are associated with mass extinctions. For instance, there were 5 mass extinctions during the last 600 million years, about what would be predicted by a purely impact-driven extinction model. The bottom line is that asteroid impacts should be treated as one of the steady-state processes that results from a dynamic solar system. Although the chances of a cratering event like the one that dug the almost 1 mile diameter Meteor Crater in Arizona happening on any random day are small, the probability is 100% that it will happen sometime. The only question is when? When faced with predictable dangers, it is sensible to take precautions. In the same way that people who live on the Gulf coast of North America track hurricanes and people who live in tornado-prone Oklahoma build houses with cellars, it seems a reasonable precaution to identify, track, and study the asteroids in near-Earth space. [See NEAR-EARTH OBJECTS.]

4.2 Origins of Asteroids

As pieces of a planet that was never formed, the asteroids represent important chemical and physical clues about the origin of the planets. But these clues can only be interpreted by having a reliable theory for how the asteroids themselves were formed. The key questions to be addressed for the asteroids include: How much material was originally in the region of the solar system where the asteroids were formed? What interrupted the formation of a planet here? Where did all the missing material go? What processes shaped both the structure of the individual asteroids and the characteristics of the Asteroid Belt as a whole?

We do have a reasonably complete census of asteroids in the main belt, down to a size of a few kilometers, and from that we can infer how the perturbing gravity of Jupiter and Saturn has shaped the distribution of asteroids today. We know some asteroids come in distinct spectral classes, and that there is a tendency for S-type asteroids to be found in the inner belt and C-types to be found in the outer belt. But while we can infer compositions for those types, based on the meteorite sample, we recognize that those inferences are very uncertain, and that there could well be material in the Asteroid Belt that is not sampled in our meteorite collections. Still, with the data in hand, we can sketch out a testable scenario for the formation and evolution of the Asteroid Belt, knowing that this is not a final answer but rather a best-guess, which we will continue to test and refine as we learn more about the asteroids.

Our first guess is to assume that the asteroids, like the planets, formed in a solar nebula of gas and dust that smoothly varied in density and temperature from the hot, dense center where the Sun was forming to the thin, cold outer edges where the nebula bordered interstellar space. [See THE ORIGIN OF THE SOLAR SYSTEM.] It is possible to calculate the rate at which dust in this cloud would encounter and stick to other bits of dust. These calculations indicate that it is possible in the early solar nebula for very loose balls of dust (more than 90% empty space) as large as a kilometer across to be formed. Relatively low-speed collisions between such dust balls would lead to further compression and accretion into objects big enough to not be carried away with the gas when the last of the solar nebula was pulled into the Sun or ejected in a massive early solar wind.

But it seems probable that these proto-asteroids looked very different from the asteroids we see today. When Jupiter and the other major planets formed, their concentrated gravity would have begun to stir up the asteroidal material. This stirring, and the absence of a nebula gas to damp down their motions, would have added enough energy to the orbits in the Asteroid Belt that further collisions between the asteroids would lead to asteroids breaking apart instead of sticking together. If one takes the present-day masses of the planets, adding a solar proportion of hydrogen and helium to the rocky planets' compositions, and then imagines spreading this material in a disk around the Sun to simulate the smallest possible nebula capable of making planets, one can see that the amount of material in such a nebula varies smoothly from the center to the outer reaches of the solar system, with three notable exceptions. Inside Mercury and outside Neptune the nebula had distinct boundaries. And in the region of Mars and the Asteroid Belt, there appears to be a significant amount of mass missing today.

We saw in Section 2 how Jupiter and Saturn perturb asteroids out of the Asteroid Belt. But modeling the early solar nebula allows us to estimate just how much material was so perturbed. It suggests that Mars is made up of less than 10% of the material originally available in its region of the solar nebula, while the mass of the Asteroid Belt is less than 0.1% of the inferred original material present. The perturbations of asteroidal material by Jupiter and Saturn must have been extremely efficient, at least in the earliest stages of the solar system's history.

One inevitable result of having 99.9% of the mass of the Asteroid Belt excited into such orbits is that there must have been a very high collision rate among asteroids in the early solar system. These collisions would break larger asteroids into smaller pieces and destroy the smaller pieces entirely. But for the largest asteroids—many tens of kilometers in radius—impacts energetic enough to shatter them may not have enough energy to disperse the pieces completely.

Instead, the fragments were likely to reaccrete into piles of rubble, consistent with the structure that asteroids are inferred to have today.

As the Asteroid Belt is dissipated, the rate of collision likewise would have dropped. Given the present-day population of the Asteroid Belt, collisions that are capable of breaking pieces of an asteroid into earth-crossing orbits or creating families of asteroids where one asteroid once orbited still do occur. We do see young families of asteroids today. Likewise, by measuring short-lived radioactive isotopes formed in meteorites by cosmic rays, we can see peaks in the ages of certain meteorite classes that imply they were broken off a parent body at a specific moment some tens to hundreds of millions of years ago. But these events must be many, many times less frequent today than when the Asteroid Belt was much more heavily populated.

One result of this scattering of asteroids by Jupiter and Saturn may have been that a few rare bodies originally from the Asteroid Belt may have been captured into orbits around other planets. Among the moons suspected of being captured asteroids are the Martian moons Phobos and Deimos, and the irregular moons of the gas giant planets.

4.3 Spacecraft Missions to Asteroids

Although telescopic studies are by far the most prolific source of data on asteroids, critical science questions on asteroid composition, structure, and surface processes can only be addressed by spacecraft missions getting close to these objects. The range of spacecraft encounters includes flybys, rendezvous, and sample return missions, which provide information of ever-increasing detail and reliability. We have now seen the results of a number of flybys, starting with two by the *Galileo* spacecraft (243 Ida and 951 Gaspra) on its way to Jupiter.

The *NEAR* (*Near Earth Asteroid Rendezvous*) spacecraft, the first dedicated asteroid mission, flew past asteroid 253 Mathilde and arrived in orbit around 433 Eros in 2001. After orbiting Eros for one year and mapping its morphology, elemental abundances, and mineralogy with an X-ray/gamma ray spectrometer (XGRS), imaging camera, near-infrared reflectance spectrometer, laser rangefinder, and magnetometer, the spacecraft ended its mission by landing on the surface of Eros. [See NEAR-EARTH OBJECTS.]

The next mission to fly past an asteroid was *Deep Space 1* (*DS1*). Primarily a technology demonstration to test the new solar-electric propulsion ion drive system, it flew past asteroid 9969 Braille on its way to comet Borrelly, but unfortunately a camera-pointing error during the Braille encounter limited the amount of useful data from that mission.

In late 2005, the Japanese space agency's ambitious *Hayabusa* asteroid sample return mission rendezvoused with asteroid 25143 Itokawa. This NEA turned out to have an extremely rough surface, as shown in Fig. 7. After several months of mapping and analysis, the spacecraft collected samples by shooting a small projectile into the surface and collecting some of the fragments splashed off. As of this writing, the spacecraft is on its way back to Earth and will parachute the sample pod onto the Australian desert in 2010.

Even though we have made great strides in exploring asteroids, they are still largely unexplored; indeed, in the case of the smaller NEOs, they are still largely undiscovered. They have great potential for science, for destruction, as resources in space, and for exploration. We are only just starting to understand these numerous objects that share our solar system.

Bibliography

Bottke, W. F., Cellino, A., Paolicchi, P., and Binzel, R. P., eds. (2002). "Asteroids III." Univ. Arizona Press, Tucson.

Gehrels, T., ed. (1994). "Hazards due to Comets and Asteroids." Univ. Arizona Press, Tucson.

Planetary Satellites

Bonnie J. Buratti

Jet Propulsion Laboratory
California Institute of Technology
Pasadena, California

Peter C. Thomas

Cornell University
Ithaca, New York

CHAPTER 19

A planetary satellite (or **moon**) is any one of the celestial bodies in orbit around a planet, which is known as the primary body. They range from large, planet-like, geologically active worlds with significant atmospheres such as Neptune's satellite Triton and Saturn's satellite Titan to tiny irregularly shaped objects as small as a kilometer in diameter. Two satellites are larger than the planet Mercury: Titan and Jupiter's Ganymede, the largest moon with a radius of 2634 km. Six planetary satellites are larger than Pluto. The large and medium-sized satellites are thought to have been formed in place around their primaries at the same time the solar system condensed 4.6 billion years ago, while many of the smaller satellites are captured objects or remnants of collisions. Small satellites that have been found in Saturn's rings help form gaps in ring particles, while other "shepherd" satellites act to gravitationally define the edges of the rings. The satellites in the inner solar system—the two moons of Mars and the Earth's Moon—are composed primarily of rocky material. The satellites of the outer solar system—Jupiter and beyond—have as major components some type of frozen volatile, primarily water ice, but also methane, ammonia, nitrogen, carbon monoxide, carbon dioxide, or sulfur dioxide existing alone or in combination with other volatiles. As of July 2006, the planets have among them a total of 156 known satellites. There undoubtedly exist many more undiscovered small satellites in the outer solar system. The relative sizes of the main satellites are illustrated in Fig. 1. Table 1 is a summary of the characteristics of the main planetary satellites; a current list of all satellites and their physical and dynamical properties is maintained by the National Aeronautics and Space Administration at http://horizons.jpl.nasa.gov. This chapter covers the satellites of Mars, Jupiter, Saturn, Uranus, and Neptune, but not the Galilean satellites (the four largest moons of Jupiter), Triton, and Titan. Mercury and Venus are not expected to have any large satellites because of solar tides. [*See* Io: The Volcanic Moon; Europa; Ganymede and Callisto; Titan; Triton; Pluto).

1. Summary of Characteristics

1.1 Discovery

None of the satellites of the outer planets was known before the invention of the telescope. When Galileo turned his telescope to Jupiter in 1610, he discovered the four large satellites in the jovian system. His observations of their orbital motion around Jupiter in a manner analogous to the motion of the planets around the Sun provided important evidence for the acceptance of the heliocentric (Sun-centered) model of the solar system. These four moons—Io, Europa, Ganymede, and Callisto—are sometimes called the Galilean satellites.

In 1655, Christian Huygens discovered Titan, the giant satellite of Saturn. Later in the 17th century, Giovanni Cassini discovered the four next largest satellites of Saturn.

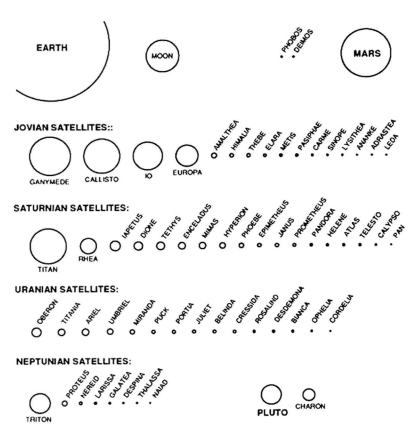

More than 100 years would pass before the next satellite discoveries were made: the Uranian satellites Titania and Oberon and 2 smaller moons of Saturn. As telescopes acquired more resolving power in the 19th century, the family of satellites grew (see Table 1). Observations obtained by the camera on the *Voyager* spacecraft led to the discovery of 3 small satellites of Jupiter, four of Saturn (S27 Pan was discovered 9 years after the data were obtained), 10 satellites of Uranus, and 6 of Neptune (see Table 1). The *Cassini* spacecraft revealed the existence of 4 small moons of Saturn. Many small satellites have been recently discovered by sensitive charge-coupled device (CCD) cameras attached to large ground-based telescopes and the Hubble Space Telescope, including the 2 small satellites of Pluto, Hydra and Nix, which brings the total of known satellites of Pluto to three. [*See* PLUTO.]

The natural planetary satellites are generally named after figures in world mythologies who were associated with the namesakes of their primaries. They are also designated by the first letter of their primary and an Arabic numeral assigned in order of discovery: Io is J1, Europa J2, and so on. When satellites are first discovered but not yet confirmed or officially named, they are known by the year in which they were discovered, the initial of the primary, and a number assigned consecutively for all solar system discoveries, for example, 2003 J23. Names for all satellites are assigned by the International Astronomical Union. See http://planetarynames.wr.usgs.gov/append7.html for a list of names and discovery circumstances maintained by the International Astronomical Union.

After planetary scientists were able to map geologic formations of the satellites from spacecraft images, they named many of the features after characters or locations from Western and Eastern mythologies. These names are also approved by the IAU.

1.2 Physical and Dynamical Properties

The motion of a satellite around the center of mass of itself and its primary defines an ellipse with the primary at one of the foci. The orbit is defined by three primary orbital elements: (1) the semimajor axis, which is the maximum distance between the ellipse and its center; (2) the eccentricity, and (3) the angle made by the intersection of the plane of the orbit and the plane of the primary's spin equator (the angle of inclination; for outer satellites the inclination is defined by the plane of the satellite's orbit to the orbital plane of the planet). The orbits are said to be regular if they are in the same sense of direction (the prograde sense) as that determined by the rotation of the primary, and if their eccentricities and inclinations are low. The orbit of a satellite is irregular if its motion is in the opposite (or retrograde) sense

TABLE 1 — Properties of the Main Planetary Satellites

Satellite	Distance from Primary (10³ km)	Revolution Period (days) R = retrograde	Orbital Eccentricity	Orbital Inclination (degrees)	Radius (km)	Density (g/cm³)	Visual Geometric Albedo	Discoverer	Year Discovered
Mars									
M1 Phobos	9.38	0.32	0.018	1.0	14 × 10	1.9	0.05	Hall	1877
M2 Deimos	23.50	1.26	0.002	2.8	8 × 6	2.1	0.05	Hall	1877
Jupiter									
J15 Adrastea	128	0.30	0.0	0.0	10		<0.1	Voyager	1979
J16 Metis	129	0.30	0.0	0.0	20		<0.1	Voyager	1979
J5 Amalthea	181	0.49	0.003	0.4	131 × 73 × 67	1.0	0.05	Barnard	1892
J14 Thebe	222	0.67	0.015	0.8	50		<0.1	Voyager	1979
J1 Io	422	1.77	0.004	.04	1,818	3.53	0.6	Galileo	1610
J2 Europa	671	3.55	0.010	0.5	1,560	2.99	0.6	Galileo	1610
J3 Ganymede	1,070	7.15	0.002	0.2	2,634	1.94	0.4	Galileo	1610
J4 Callisto	1,883	16.69	0.007	0.5	2,409	1.85	0.2	Galileo	1610
J13 Leda	11,094	239	0.148	26.7	5			Kowal	1974
J6 Himalia	11,480	251	0.163	27.6	85		0.03	Perrine	1904
J10 Lysithea	11,720	259	0.107	29.0	12			Nicholson	1938
J7 Elara	11,737	260	0.207	24.8	40		0.03	Perrine	1904
J12 Ananke	21,200	631R	0.17	147	10			Nicholson	1951
J11 Carme	22,600	692R	0.21	163	15			Nicholson	1938
J8 Pasiphae	23,500	735R	0.38	145	18			Melotte	1908
J9 Sinope	23,700	758R	0.28	153	14			Nicholson	1914
Saturn									
S18 Pan	134	0.57	0.0	0.0	10	—	—	Showalter	1990
S15 Atlas	138	0.60	0.000	0.0	19 × 17 × 14		0.4	Voyager	1980
S16 Prometheus	139	0.61	0.002	0.0	74 × 50 × 34		0.6	Voyager	1980
S17 Pandora	142	0.63	0.004	0.05	55 × 44 × 31		0.6	Voyager	1980
S10 Janus	151	0.69	0.007	0.14	97 × 95 × 77	0.65	0.6	Dollfus	1966
S11 Epimetheus	151	0.69	0.009	0.34	69 × 55 × 55	0.65	0.5	Fountain and Larson	1978
S1 Mimas	186	0.94	0.020	1.5	199	1.15	0.8	Herschel	1789
S2 Enceladus	238	1.37	0.004	0.0	252	1.61	1.4	Herschel	1789
S3 Tethys	295	1.89	0.000	1.1	536	0.96	0.8	Cassini	1684
S14 Calypso	295	1.89	0.0	1.1	15 × 8 × 8	1.0	0.6	Pascu et al.	1980
S13 Telesto	295	1.89	0.0	1.0	15 × 12 × 8	1.0	0.9	Smith et al.	1980
S4 Dione	377	2.74	0.002	0.02	563	1.47	0.55	Cassini	1684
S12 Helene	377	2.74	0.005	0.15	16	1.5	0.5	Laques and Lecacheux	1980
S5 Rhea	527	4.52	0.001	0.35	734	1.23	0.65	Cassini	1672
S6 Titan	1,222	15.94	0.029	0.33	2,575	1.88	0.2	Huygens	1655

(Continued)

TABLE 1   Properties of the Main Planetary Satellites (*Continued*)

Satellite	Distance from Primary (10³ km)	Revolution Period (days) R = retrograde	Orbital Eccentricity	Orbital Inclination (degrees)	Radius (km)	Density (g/cm³)	Visual Geometric Albedo	Discoverer	Year Discovered
S7 Hyperion	1,481	21.28	0.104	0.4	180 × 140 × 112	0.60	0.3	Bond and Lassell	1848
S8 Iapetus	3,561	79.33	0.028	14.7	718	1.09	0.4–0.08	Cassini	1671
S9 Phoebe	12,952	550.4R	0.163	150	107	1.6	0.06	Pickering	1898
Uranus									
U6 Cordelia	49.7	0.33	0.0005	0.14	13			*Voyager 2*	1986
U7 Ophelia	53.8	0.38	0.010	0.09	15			*Voyager 2*	1986
U8 Bianca	59.2	0.43	0.001	0.16	21			*Voyager 2*	1986
U9 Cressida	61.8	0.46	0.0002	0.04	31		~0.04	*Voyager 2*	1986
U10 Desdemona	62.7	0.47	0.0002	0.16	27		~0.04	*Voyager 2*	1986
U11 Juliet	64.4	0.49	0.0006	0.06	42		~0.06	*Voyager 2*	1986
U12 Portia	66.1	0.51	0.0002	0.09	54		~0.09	*Voyager 2*	1986
U13 Rosalind	69.9	0.56	0.00009	0.28	27		~0.04	*Voyager 2*	1986
U14 Belinda	75.3	0.62	0.0001	0.03	33			*Voyager 2*	1986
U15 Puck	86.0	0.76	0.00005	0.31	77		0.07	*Voyager 2*	1985
U5 Miranda	130	1.41	0.003	3.4	236	1.2	0.35	Kuiper	1948
U1 Ariel	191	2.52	0.003	0.0	579	1.6	0.36	Lassell	1851
U2 Umbriel	266	4.14	0.005	0.0	585	1.5	0.20	Lassell	1851
U3 Titania	436	8.71	0.002	0.0	789	1.7	0.30	Herschel	1787
U4 Oberon	583	13.46	0.001	0.0	761	1.6	0.22	Herschel	1787
U16 Caliban	7775	654	0.2	146	20?			Gladman et al.	1997
U17 Sycorax	8846	795	0.34	154	40?			Gladman et al.	1997
Neptune									
N8 Naiad	48.2	0.29	0.000	0.0	29			*Voyager 2*	1989
N7 Thalassa	50.1	0.31	0.0002	4.5	40			*Voyager 2*	1989
N5 Despina	52.5	0.33	0.0001	0.0	74		0.05	*Voyager 2*	1989
N6 Galatea	62.0	0.43	0.0001	0.0	79			*Voyager 2*	1989
N4 Larissa	73.6	0.55	0.000	0.0	104 × 89		0.06	*Voyager 2*	1989
N3 Proteus	117.6	1.12	0.0004	0.0	208		0.06	*Voyager 2*	1989
N1 Triton	354.8	5.87R	0.000015	157	1,353	2.08	0.73	Lassell	1846
N2 Nereid	5,513	360.1	0.751	29	170		0.16	Kuiper	1949

See web-based sources in the bibliography for additional information.

of motion, if it is highly eccentric, or if it has a high angle of inclination. Satellites with irregular (nonspherical) shapes are often called irregular satellites. Most of the outer planets' major satellites move in regular, prograde orbits, while most of the small satellites have irregular orbits. Satellites that move in irregular orbits are thought to be likely captured objects. Most of the major, regular planetary satellites present the same hemisphere toward their primaries, a state that is the result of tidal evolution.

When two celestial bodies orbit each other, the gravitational force exerted on the nearside is greater than that exerted on the farside. The result is an elongation of each body to form tidal bulges, which can consist of solid, liquid, or gaseous (atmospheric) material. The primary tugs on the satellite's tidal bulge to lock its longest axis onto the primary-satellite line. The satellite, which is said to be in a state of **synchronous rotation**, keeps the same face toward the primary. Since this despun state occurs rapidly (usually within a few million years), most large natural satellites with known rotational periods are in synchronous rotation. Tidal evolution is dependent on the size of the satellite and its distance from the primary. Satellites that are far away from the primary often maintain their original rotational period. One satellite of Saturn, Hyperion, rotates chaotically. This unusual state of rotation is due to gravitational forces acting on Hyperion by other close massive satellites.

The satellites of the outer solar system are unique worlds, each representing a vast panorama of physical processes. The two satellites of Mars and the small outer satellites of Jupiter, Saturn, Uranus, and Neptune are irregular chunks of rock, ice, or mixtures of the two. They are perhaps captured asteroids or even objects from the Kuiper Belt that have been subjected to intensive meteoritic bombardment. Several of the satellites, including the Saturnian satellite Phoebe and areas of the Uranian satellites, are covered with **C-type material**, the dark, unprocessed, carbon-rich material found on the C-class of asteroids. Iapetus presents a particular enigma: one hemisphere is 10 times more reflective than the other. The surfaces of other satellites such as Hyperion and the dark side of Iapetus contain primitive matter that is spectrally red and is thought to be rich in organic compounds. Because these materials, which are common in the outer solar system, represent the material from which the solar system formed, understanding their occurrence and origin will yield clues to the state and early evolution of the solar system. In addition, the transport of organic matter from the outer solar system to the inner solar system, perhaps by comets, is sometimes hypothesized to be an essential step in the formation of life. [*See* MAIN-BELT ASTEROIDS; KUIPER BELT.]

Before the advent of spacecraft exploration, planetary scientists expected satellites to be geologically dead worlds. They assumed that heat sources were not sufficient to have melted their mantles to provide a source of liquid or semiliquid ice or ice–silicate slurries. Reconnaissance of the icy

satellite systems of the four outer giant planets by the two *Voyager* spacecraft uncovered a wide range of geologic processes, including currently active volcanism on Io and Triton. *Cassini* discovered active tectonic processes on Enceladus, a small satellite of Saturn. At least one additional satellite (Europa) may have current activity. The medium-sized satellites of Saturn and Uranus are large enough to have undergone internal melting with subsequent **differentiation** and resurfacing. Among the Galilean satellites, only Callisto lacks evidence for periods of such activity after formation.

Recent work on the importance of tidal interactions and subsequent heating has provided the theoretical foundation to explain the existence of widespread activity in the outer solar system. Another factor is the presence of non-ice components, such as ammonia hydrate or methanol, which lower the melting point of near-surface materials. Partial melts of water ice and various contaminants—each with its own melting point and viscosity—provide material for a wide range of geologic activity. The realization that such partial melts are important to understanding the geologic history of the satellites has spawned an interest in the rheology (viscous properties and resulting flow behavior) of various ice mixtures and exotic phases of ices that exist at extreme temperatures or pressures. Conversely, the types of features observed on the surfaces provide clues to the likely composition of the satellites' interiors.

Because the surfaces of so many outer planet satellites exhibit evidence of geologic activity, planetary scientists have begun to think in terms of unified geologic processes that function throughout the solar system. For example, partial melts of water ice with various contaminants could provide flows of liquid or partially molten slurries that in many ways mimic terrestrial or lunar lava flows formed by the partial melting of mixtures of silicate rocks. The ridged and grooved terrains on satellites such as Ganymede, Enceladus, Tethys, and Miranda may all have resulted from similar tectonic activities. Finally, explosive volcanic eruptions occurring on Io, Triton, Earth, and Enceladus may all result from the escape of volatiles released as the pressure in upward-moving liquids decreases. [*See* PLANETARY VOLCANISM.]

2. Formation and Evolution of Satellites

2.1 Theoretical Models of Formation

Because the planets and their associated moons condensed from the same cloud of gas and dust at about the same time, the formation of the natural planetary satellites must be addressed within the context of the formation of the planets. The solar system formed 4.6 ± 0.1 billion years ago. This age is derived primarily from radiometric dating of meteorites, which are thought to consist of primordial, unaltered matter. In the radiometric dating technique, the fraction of a radioactive isotope (usually rubidium, argon, or uranium), which has decayed into its daughter isotope, is measured. Since the rate at which these isotopes decay has been measured in the laboratory, it impossible to infer the time elapsed since formation of the meteorites, and thus of the solar system. [*See* THE ORIGIN OF THE SOLAR SYSTEM.]

The Sun and planets formed from a disk-shaped rotating cloud of gas and dust known as the protosolar nebula. When the temperature in the nebula cooled sufficiently, small grains began to condense. The difference in solidification temperatures of the constituents of the protosolar nebular accounts for the major compositional differences of the satellites. Since there was a temperature gradient as a function of distance from the center of the nebula, only those materials with high melting temperatures (e.g., silicates, iron, aluminum, titanium, and calcium) solidified in the central (hotter) portion of the nebula. Earth's Moon consists primarily of these materials. Beyond the orbit of Mars, carbon, in combination with silicates and organic molecules, condensed to form the carbonaceous material found on C-type asteroids. Similar carbonaceous material is found on the surfaces of the martian moon Phobos, several of the jovian and Saturnian satellites, regions of the Uranian satellites, and possibly Triton and Charon. In the outer regions of the asteroid belt, formation temperatures were sufficiently cold to allow water ice to condense and remain stable. Thus, the jovian satellites are primarily ice–silicate admixtures (except for Io, which has apparently outgassed all its water). For the satellites of Saturn and Uranus, these materials are predicted to be joined by methane and ammonia, and their hydrated forms. For the satellites of Neptune and Pluto, formation temperatures were low enough for other volatiles, such as nitrogen, carbon monoxide, and carbon dioxide, to exist in solid form. In general, the satellites that formed in the inner regions of the solar system are denser than the outer planets' satellites because they retained a lower fraction of volatile materials.

After small grains of material condensed from the protosolar nebula, electrostatic forces caused them to stick together. Collisions between these larger aggregates caused meter-sized particles, or planetesimals, to be accreted. Finally, gravitational attraction between ever larger aggregates occurred to form kilometer-sized planetesimals. The largest of these bodies swept up much of the remaining material to create the protoplanets and their companion satellite systems. One important concept of planetary satellite formation is that a satellite cannot accrete within the planet's **Roche limit**, the distance at which the tidal forces of the primary become greater than gravitational forces that bind loose particles into a satellite.

The formation of the regular satellite systems of the outer giant planets is sometimes thought to be a smaller-scaled version of the formation of the solar system. A density gradient as a function of distance from the primary exists for the Galilean satellites (see Table 1); this pattern implies that more volatiles (primarily ice) are included in the bulk composition as the distance increases. However, the formation scenario must be more complicated for Saturn or Uranus because their regular satellites do not follow this pattern.

The retrograde satellites are probably captured asteroids, comets, Kuiper Belt Objects, or large planetesimals left over from the major episode of planetary formation. None of the satellites discussed in this chapter have appreciable atmospheres, although the large Saturnian satellite Titan has an atmosphere with a surface pressure higher than that of the Earth's. At least one satellite (Ganymede) has an internal magnetic field.

2.2 Evolution

Soon after the satellites accreted, they began to heat up from the release of gravitational potential energy. An additional heat source was provided by the release of mechanical energy during the heavy bombardment of their surfaces by remaining debris. The decay of radioactive elements found in silicate materials provided another major source of heat. The heat produced in the larger satellites was sufficient to cause melting and chemical fractionation; the dense material, such as silicates and iron, went to the center of the satellite to form a core, while ice and other volatiles remained in the crust. A fourth source of heat is provided by tidal interactions. When a satellite is being tidally despun, the resulting frictional energy is dissipated as heat. Because this process happens very quickly for most satellites (~10 million years), another mechanism involving orbital resonances among satellites is thought to cause the heat production required for more recent resurfacing events. Gravitational interactions tend to turn the orbital periods of the satellites within a system into multiples of each other. In the Galilean system, for example, Io and Europa complete four and two orbits, respectively, for each orbit completed by Ganymede. The result is that the satellites meet each other at the same point in their orbits. The resulting flexing of the tidal bulge induced on the bodies by their mutual gravitational attraction causes significant heat production in some cases. [*See* PLANETARY IMPACTS; SOLAR SYSTEM DYNAMICS: REGULAR AND CHAOTIC MOTION.]

Some satellites, such as the Earth's Moon, Ganymede, and several of the Saturnian and Uranian satellites, underwent periods of melting and active geology within a billion years of their formation and then became quiescent. The evolution of these objects was truncated because of limited amounts of radioactive material, efficient dissipation of internal heat, or the lack of an ongoing heat source. Others, such as Io, Triton, Enceladus, and possibly Europa, are currently geologically active.

For nearly a billion years after their formation, the satellites all underwent intense bombardment and cratering.

The satellites Phobos, Mimas, and Tethys all have impact craters caused by bodies that were nearly large enough to break them apart; probably such catastrophes did occur.

The bombardment tapered off to a slower rate and presently continues. By counting the number of craters on a satellite's surface and making certain assumptions about the flux of impacting material, geologists are able to estimate when a specific portion of a satellite's surface was formed. Continual bombardment of satellites causes the pulverization of both rocky and icy surfaces to form a covering of fine material known as a **regolith**.

Many planetary scientists expected that most of the craters formed on the outer planets' satellites would have disappeared owing to viscous relaxation. The two *Voyager* spacecraft revealed surfaces covered with craters that in many cases had morphological similarities to those found in the inner solar system, including central peaks, large ejecta blankets, and well-formed outer walls. Recent research has shown that the elastic properties of ice provide enough strength to offset viscous relaxation. Silicate mineral contaminants or other impurities in the ice may also provide extra strength to sustain impact structures.

Planetary scientists classify the erosional processes affecting satellites into two major categories: endogenic, which includes all internally produced geologic activity, and exogenic, which encompasses the changes brought by outside agents. The latter category includes the following processes: (1) meteoritic bombardment and resulting gardening and impact volatilization; (2) magnetospheric interactions, including sputtering and implantation of energetic particles; (3) alteration by high-energy ultraviolet photons; and (4) accretion of particles of dust and ice from sources such as planetary rings.

Meteoritic bombardment acts in two major ways to alter the optical characteristics of the surface. First, the impacts excavate and expose fresh material (cf. the bright ray craters on the Moon, Ganymede, and the Uranian satellites). Second, impact volatilization and subsequent escape of volatiles result in a lag deposit enriched in opaque, dark materials. The relative importance of the two processes depends on the flux, size distribution, and composition of the impacting particles, and on the composition, surface temperature, and mass of the satellite. For the Galilean satellites, older geologic regions tend to be darker and redder, but both the Galilean and Saturnian satellites tend to be brighter on the hemispheres that lead in the direction of orbital motion (the so-called "leading" side, as opposed to the "trailing" side); this effect is thought to be due to preferential micrometeoritic gardening on the leading side. The accretion of dust particles external to the satellites may be occurring on the leading side of Iapetus and Callisto and possibly on the Uranian satellites to cause their leading sides to be darker. Finally, bright icy particles from the E-ring of Saturn seem to be coating the surfaces of the inner Saturnian satellites.

For satellites that are embedded in planetary **magnetospheres**, their surfaces are affected by magnetospheric interactions in three ways: (1) chemical alterations; (2) selective erosion, or sputtering; and (3) deposition of magnetospheric ions. In general, volatile components are more susceptible to sputter erosion than refractory ones. The overall effect of magnetospheric erosion is thus to enrich surfaces in darker, redder opaque materials. A similar effect may be caused by the bombardment of UV photons, although much fundamental laboratory work remains to be done to determine the quantitative effects of this process. [*See* PLANETARY MAGNETOSPHERES.]

3. Observations of Satellites

3.1 Telescopic Observations

3.1.1 SPECTROSCOPY

Before the development of interplanetary spacecraft, all observations from Earth of objects in the solar system were obtained by ground-based telescopes. One particularly useful tool of planetary astronomy is spectroscopy, or the acquisition of spectra from a celestial body. Spectra consist of electromagnetic radiation that has been split by an optical device such as a prism into its component wavelength. The surface or atmosphere of a satellite has a characteristic pattern of absorption and emission bands. Comparison of the astronomical spectrum with laboratory spectra of materials that are possible components of the surface yields information on the composition of the satellite. For example, water ice has a series of absorption features between 1 and 4 μm. The detection of these bands on three of the Galilean satellites and several satellites of Saturn and Uranus demonstrated that water ice is a major constituent of their surfaces. Other examples are the detections of SO_2 frost on the surface of Io, methane in the atmosphere of Titan, nitrogen and carbon dioxide on Triton, and water ice on Charon.

3.1.2 PHOTOMETRY

Photometry of planetary satellites is the accurate measurement of radiation reflected to an observer from their surfaces or atmospheres. These measurements can be compared to light-scattering models that are dependent on physical parameters, such as the porosity of the optically active upper surface layer, the albedo of the material, and the degree of topographic roughness. These models predict brightness variations as a function of solar **phase angle** (the angle between the observer, the Sun, and the satellite). Like the Earth's Moon, the planetary satellites present changing phases to an observer on Earth. As the face of the satellite becomes fully illuminated to the observer, the integrated brightness exhibits a nonlinear surge in brightness that is thought to result from the disappearance of mutual shadowing among surface particles. The magnitude of this surge, known as the **opposition effect**, is greater for a more porous surface. Many planetary and satellite surfaces exhibit a large opposition surge at very small solar phase angles that has been attributed to constructive interference of sunlight.

One measure of how much radiation a satellite reflects is the **geometric albedo**, p, which is the disk-integrated brightness at "full moon" (or a phase angle of $0°$) compared to a perfectly reflecting, diffuse disk of the same size. The **phase integral**, q, defines the angular distribution of radiation over the sky:

$$q = 2 \int_0^\pi \Phi(\alpha)\sin \alpha \, d\alpha$$

where $\Phi(\alpha)$ is the disk-integrated brightness and α is the phase angle. The **Bond albedo**, which is given by $A = p \times q$, is the ratio of the integrated flux reflected by the satellite to the integrated flux received. The geometric albedo and phase integral are wavelength dependent, whereas a true (or bolometric) Bond albedo is integrated over all wavelengths.

Another ground-based photometric measurement that has yielded important information on the satellites' surfaces is the integrated brightness of a satellite as a function of orbital angle. For a satellite in synchronous rotation with its primary, the subobserver geographical longitude of the satellite is equal to the longitude of the satellite in its orbit. Observations showing significant albedo and color variegations for Io, Europa, Rhea, Dione, and especially Iapetus suggest that diverse geologic terrains coexist on these satellites. This view was confirmed by images obtained by the *Voyager* spacecraft.

Another important photometric technique is the measurement of reflected light as one celestial body occults, or blocks, another body. Time-resolved observations of occultations yield the light reflected from successive regions of the eclipsed body. This technique has been used to map albedo variations on Pluto and its satellite Charon and to map the distribution of infrared emission—and thus volcanic activity—on Io. Stellar occultations have been used to probe the diameters and atmospheres of many satellites, including Iapetus, Titan, and Triton.

3.1.3 RADIOMETRY

Satellite radiometry is the measurement of radiation that is absorbed and reemitted at thermal wavelengths. The distance of each satellite from the Sun determines the mean temperature for the equilibrium condition that the absorbed radiation is equal to the emitted radiation:

$$\pi R^2 (F/r^2)(1 - A) = 4\pi R^2 \varepsilon \sigma T^4$$
$$T = \left(\frac{(1 - A)F}{4\sigma \varepsilon r^2}\right)^{1/4}$$

where R is the radius of the satellite, r is the Sun-satellite distance, ε is the emissivity, σ is Stefan–Boltzmann's constant, A is the Bond albedo, and F is the incident solar flux (a slowly rotating body would radiate over $2\pi R^2$). Typical mean temperatures in Kelvins for the satellites are: the Earth's Moon, 280 K; Europa, 103 K; Iapetus, 89 K; the Uranian satellites, 60 K; and the Neptunian satellites, 45 K. For thermal equilibrium, measurements as a function of wavelength yield a blackbody curve characteristic of T: in general, the temperatures of the satellites closely follow the blackbody emission values. Some discrepancies are caused by a weak **greenhouse effect** (in the case of Titan), or the existence of volcanic activity (in the case of Io).

Another possible use of radiometric techniques, when combined with photometric measurements of the reflected portion of the radiation, is the estimate of the diameter of a satellite. A more accurate method of measuring the diameter of a satellite from Earth involves measuring the light from a star as it is occulted by the satellite. The time the starlight is dimmed is proportional to the satellite's diameter.

A third radiometric technique is the measurement of the thermal response of a satellite's surface as it is being eclipsed by its primary. The rapid loss of heat from a satellite's surface indicates a thermal conductivity consistent with a porous upper surface. Eclipse radiometry of Phobos, Callisto, and Ganymede suggests that these objects all lose heat rapidly and thus have porous regoliths created from eons of meteoritic bombardment.

3.1.4 POLARIMETRY

Polarimetry is the measurement of the degree of polarization of radiation reflected from a satellite's surface. The polarization characteristics depend on the shape, size, and optical properties of the surface particles. Generally, the radiation is linearly polarized and is said to be negatively polarized if it lies in the scattering plane and positively polarized if it is perpendicular to the scattering plane. Polarization measurements as a function of solar phase angle for atmosphereless bodies are negative at small phase angles; comparisons with laboratory measurements indicate that this is characteristic of complex, porous surfaces consisting of multisized particles. In 1970, ground-based polarimetry of Titan that showed it lacked a region of negative polarization led to the correct conclusion that it has a thick atmosphere.

3.1.5 RADAR

Planetary radar is a set of techniques that involve the transmittance of radio waves to a remote surface and the analysis of the echoed signal. Among the outer planets' satellites, the Galilean satellites, Titan, and several other Saturnian satellites have been observed with radar. [*See* PLANETARY RADAR.]

TABLE 2	Major Flyby Missions to Planetary Satellites	
Mission	**Objects**	**Encounter Dates**
Mariner 9	Martian satellites	1971
Viking 1 and *2*	Martian satellites	1976
Pioneer 10	Jovian satellites	1979
Pioneer 11	Jovian satellites	1979
	Saturnian satellites	1979
Voyager 1	Jovian satellites	1979
	Saturnian satellites	1980
Voyager 2	Jovian satellites	1979
	Saturnian satellites	1981
	Uranian satellites	1986
	Neptunian satellites	1989
Phobos 2	Martian satellites	1989
Galileo	Jovian satellites	1996–1998
Mars Global Surveyor	Martian satellites	1998–present
Cassini-Huygens	Saturnian satellites	2004–2008

3.2 Spacecraft Exploration

Interplanetary missions to the planets and their moons have enabled scientists to increase their understanding of the solar system more in the past 35 years than in all of previous scientific history. Analysis of data returned from spacecraft has led to the development of whole new fields of scientific endeavor, such as planetary geology. From the earliest successes of planetary imaging, which included the flight of a Soviet *Luna* spacecraft in 1959 to the far side of the Earth's Moon to reveal a surface devoid of smooth lunar plains, unlike that of the visible side, and the crash landing of three United States *Ranger* spacecraft, which sent back pictures in 1964 and 1965 which showed that the Earth's Moon was cratered down to meter scales, it was evident that interplanetary imaging experiments had immense capabilities. Table 2 summarizes the successful spacecraft missions to the planetary satellites.

The return of images from space is very similar to the transmission of television images. A camera records the level of intensity of radiation incident on its focal plane, which holds a 2-dimensional array of detectors. In the most modern cameras, this array consists of Charge-coupled devices (CCDs). A computer onboard the spacecraft records these numbers and sends them by means of a radio transmitter to Earth, where another computer reconstructs the image.

Although images are the most spectacular data returned by spacecraft, a whole array of equally valuable experiments are included in each scientific mission. For example, a gamma-ray spectrometer aboard the lunar orbiters was able to map the abundance of iron and titanium on the Moon's surface. The *Voyager* spacecraft included an infrared spectrometer capable of mapping temperatures; an ultravio-

let spectrometer; a photopolarimeter, which simultaneously measured the color, intensity, and polarization of light; and a radio science experiment that was able to measure the pressure of Titan's atmosphere by observing how radio waves passing through it were attenuated.

The *Pioneer* spacecraft, which were launched in 1972 and 1973 toward an encounter with Jupiter and Saturn, returned the first disk-resolved images of the Galilean satellites. But even greater scientific advancements were made by the *Voyager* spacecraft, which returned thousands of images of the satellite systems of all four outer planets, some of which are shown in Section 4. Color information for the objects was obtained by means of six broadband filters attached to the camera. The return of large numbers of images with resolution down to a kilometer has enabled geologists to construct geologic maps, to make detailed crater counts, and to develop realistic scenarios for the structure and evolution of the satellites.

Further advances were made by the *Galileo* spacecraft, which was launched in 1990 and began obtaining data at Jupiter in 1996. The mission consisted of a probe that explored the jovian atmosphere and an orbiter designed to make several close flybys of the Galilean satellites. The orbiter contained both visual and infrared imaging devices, an ultraviolet spectrometer, and a photopolarimeter. The visual camera was capable of obtaining images with better than 20-m resolution. The spacecraft was intentionally crashed into Jupiter in September 2003 to avoid possible contamination of Europa in the future. (Europa has a subsurface ocean that may be an appropriate habitat for primitive life.) The *Cassini–Huygens* mission to Saturn was launched in 1997 and entered into orbit around Saturn in 2004 for at least a 4-year in-depth study of the planet, its rings, satellites and magnetosphere. Its instruments include a camera, an imaging spectrometer, infrared and ultraviolet spectrometers, a radar system, and a suite of fields and particles experiments. In January 2005, the spacecraft jettisoned its *Huygens* probe onto the surface of Titan; valuable data on the **ionosphere**, atmosphere, and surface of this unique world were obtained. [*See* PLANETARY EXPLORATION MISSIONS AND TITAN.]

The moons of Mars, Phobos and Deimos, have been explored by spacecraft that have flown by or entered into orbit around Mars: *Mariner 9*, *Viking 1* and *2*, *Phobos 2*, and *Mars Global Surveyor* (see Table 2).

4. Individual Satellites

4.1 The Satellites of Mars: Phobos and Deimos

Mars has two small satellites, Phobos and Deimos (fear and terror), which were discovered by the American astronomer Asaph Hall in 1877. They were named after the attendants of Mars in Greek mythology. In Jonathan Swift's moral satire *Gullliver's Travels* (published in 1726), a fanciful but coincidentally accurate prediction of the existence and orbital characteristics of two small Martian satellites was made. The two bodies are barely visible in the scattered light from Mars in Earth-based telescopes. Most of what is known about Phobos and Deimos was obtained from the *Mariner 9* and the *Viking 1* and 2 missions to Mars (see Table 2). Their physical and orbital properties are listed in Table 1. Both satellites are shaped approximately like ellipsoids, and they are in synchronous rotation. Phobos, and possibly Deimos, has a regolith of dark material similar to that found on C-type (carbonaceous) asteroids common in the outer asteroid belt. Thus the satellites may have been asteroids or asteroidal fragments, which were perturbed into a Mars-crossing orbit and captured. The orbital period of Phobos around Mars is only 7.7 hours: An observer on Mars would see the moon rise and set twice in a single day.

Both satellites are heavily cratered, which indicates that their surfaces are at least 3 billion years old (Fig. 2). But only Deimos appears to be covered with a fine dust, which gives its surface a smoother appearance. This dust may exist because the surface is more easily pulverized by impacts, or it may be the result of a large impact the shaped the moon's southern hemisphere. The surface of Phobos is extensively scored by linear grooves that are deepest near the huge impact crater Stickney (named after the Asaph Hall's wife, Angeline Stickney Hall, who collaborated with him in many of his astronomical observations) but that define planes cutting through the satellite and parallel to its intermediate axis that points along its direction of orbital motion. The grooves are probably fractures, enhanced by the collision that produced Stickney. There is some evidence that tidal action is bringing Phobos, which is already inside Roche's limit, closer to Mars. The satellite will either disintegrate (perhaps to form a ring) or crash into Mars in about 100 million years.

4.2 The Small Satellites of Jupiter

Ten years ago, Jupiter had 12 known small satellites, including three discovered by the *Voyager* mission. As of July 2006, the planet had 63 known satellites (59 of them small); most of the new small satellites were discovered by sensitive CCD cameras on large telescopes. The small satellites are irregular in shape, and many of the outer satellites may be captured objects.

Within the orbit of Io are at least four satellites: Adrastea, Metis, Amalthea, and Thebe (see Fig. 3). Adrastea and Metis, both discovered by *Voyager*, are the closest known satellites to Jupiter and move in nearly identical orbits just outside the outer edge of the thin Jovian ring, for which they may be a source of particles. Between Amalthea and Io lies the orbit of Thebe, also discovered by *Voyager*. Little is known about the composition of these satellites, but they are most likely primarily rock–ice mixtures. The three inner

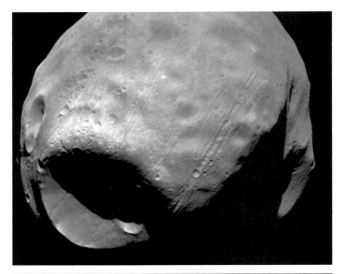
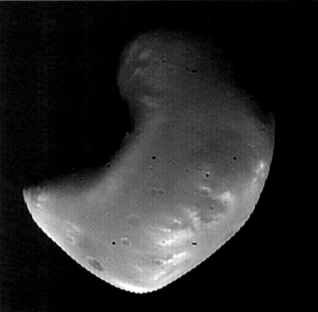

FIGURE 2 The two moons of Mars: (a) Phobos and (b) Deimos. Both pictures were obtained by the *Viking* spacecraft.

satellites sweep out particles in the jovian magnetosphere to form voids at their orbital positions.

Amalthea is a dark, reddish, heavily cratered object reflecting less than 5% of the visible radiation it receives; the red color is probably due to contamination by sulfur particles from Io. Little else is known about its composition except that the dark material may be carbonaceous. In addition to two large craters, Pan (100 km wide) and Gaea (80 km wide), Amalthea has mountains that are about 20 km high. In 2002, the *Galileo* spacecraft swooped to within 150 km of the moon's surface to find that its density is anomalously low, about that of water ice ($\sim$1 gm/cm^3).

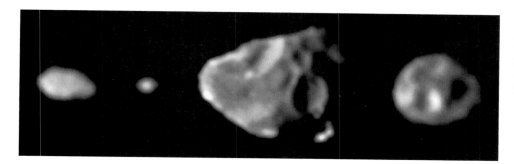

FIGURE 3   Four small satellites of Jupiter: Metis, Adrastea, Amalthea, and Thebe, shown to their correct relative sizes. All four satellites orbit between Jupiter's ring and the Galilean satellites.

This result implies Amalthea is probably a "rubble pile" composed of an agglomeration of debris reaccreted from a collision long ago. Thebe is a low albedo satellite with a reflectivity of 4–5%. It is also reddish in color, probably due to contamination by sulfur from Io.

Exterior to the Galilean satellites, there is a class of four satellites moving in inclined, prograde orbits (Lysithea, Elara, Himalia, and Leda). They are dark objects, reflecting only 2 or 3% of incident radiation, and may be similar to C- and D-type asteroids. Themisto is a prograde satellite moving in an inclined orbit between the Galilean satellites and this group of four. Beyond this family lies the prograde S2000 J11 and the retrograde Euporie. Another family of objects is represented by the outermost satellites, which have highly inclined retrograde orbits. They include Sinope, Pasiphae, Carme, and Ananke, which were all discovered in the first half of the last century, and 21 additional objects discovered in 1999–2001. These satellites orbit at distances of 21 million to 24 million km from Jupiter, and they may be captured asteroids.

The 23 additional small satellites discovered in 2003 are all small bodies orbiting at distances between 17 million and over 28 million km: They comprise an "extended family" of the outer retrograde satellites of Jupiter (one satellite, Karpo, is in an inclined prograde orbit). This large group of retrograde satellites appears to fall into smaller families that occupy three dynamical groups separated by their distances from Jupiter. The small outer satellites of Jupiter are probably captured asteroids, and they most likely represent a large reservoir of additional undiscovered satellites. [*See* APPENDIX TABLE SUMMARY.]

4.3 The Saturnian System

4.3.1 THE MEDIUM-SIZED ICY SATELLITES OF SATURN: RHEA, DIONE, TETHYS, MIMAS, ENCELADUS, AND IAPETUS

The Saturnian system contains 47 known satellites. Excluding the giant Titan, the six largest satellites of Saturn are smaller than the Galilean satellites but still sizable, with radii greater than 200 km—as such they represent a unique class of icy satellite. Earth-based telescopic measurements showed the spectral signature of ice for all six satellites. The satellites' low densities and high albedos (see Table 1) im-

ply that their bulk composition is largely water ice, possibly combined with ammonia or other volatiles. They have smaller amounts of rocky silicates than the Galilean satellites. Resurfacing has occurred on several of the satellites. Most of what is presently known of the Saturnian system was obtained from the *Voyager* flybys in 1980 and 1981 and the *Cassini–Huygens* exploration of Saturn's satellites. The six medium-sized icy satellites are shown to relative size in Fig. 4.

The innermost medium-sized satellite Mimas is covered with craters, including one (named Herschel) that is as large as a third of the satellite's diameter (see upper left of Fig. 4). The impacting body was probably nearly large enough to break Mimas apart; such disruptions may have occurred on other objects. There is a suggestion of surficial grooves

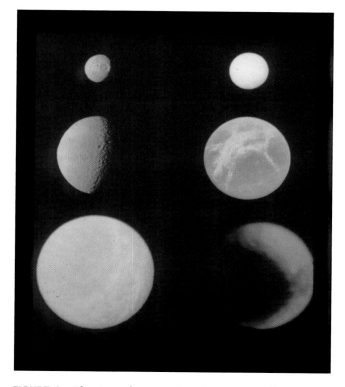

FIGURE 4   The six medium-sized icy Saturnian satellites. They are, left to right from the upper left, Mimas, Enceladus, Tethys, Dione, Rhea, and Iapetus.

that may be features caused by the impact. No components other than water ice have been detected on Mimas.

The next satellite outward from Saturn is Enceladus, one of three satellites in the solar system that currently exhibit volcanic or geyser-like activity. Enceladus was known from telescopic measurements to reflect nearly 100% of the visible radiation incident on it (for comparison, the Moon reflects only about 11%). Recent Hubble Space Telescope observations indicate the geometric albedo is 1.4, far higher than anything else in the solar system. The only likely composition consistent with this observation is almost pure water ice, or some other highly reflective volatile substance. *Voyager 2* obtained data that showed an object that had been subjected, in the recent geologic past, to extensive resurfacing; grooved formations similar to those on Ganymede were evident. The lack of impact craters on the grooved terrain is consistent with an age less than a billion years. It was thus thought likely that some form of ice volcanism was active in the recent past—or even currently—on Enceladus. About half of the surface observed by *Voyager* is extensively cratered and dates from nearly 4 billion years ago.

The *Cassini–Huygens* spacecraft made 3 close passes to Enceladus in February, March, and July 2005. The first 2 flybys showed regions of recent geologic activity, including both extensional and compressional faults, and very low crater counts. However, there was no evidence for current activity. During the July flyby, the spacecraft approached the south pole of the satellite to within 170 km, and found multiple pieces of evidence for the active transport of material from the surface of the body. The spacecraft's mass spectrometer, dust collector, and ultraviolet spectrometer all detected particles escaping from the south pole of Enceladus. The infrared detector mapped a large circular region extending from the pole to 60°S latitude that is at least 30 K higher than expected. The magnetometer detected evidence for a magnetic field, which would imply subsurface liquid. Finally, a series of well-defined linear features, which are ~50 K hotter than the surrounding regions, appear to be rich in fresh, recently produced ice. *Cassini* captured visible and infrared images of plumes of water ice extending from the linear features. Figure 5 shows a close-up image of the south pole of Enceladus, and Fig. 6 shows the ice-laden plumes. The heat source for this activity is not yet fully understood, but tidal forces may play a role, as the satellite is in a 2:1 orbital resonance with Dione.

A final element to the enigma of Enceladus is the possibility that it is responsible for the formation of the E-ring of Saturn, a tenuous collection of icy particles that extends from inside the orbit of Enceladus to past the orbit of Dione. The position of maximum thickness of the ring coincides with the orbital position of Enceladus. If some form of volcanism is presently active on the surface, it could provide a source of particles for the ring. An alternative source mechanism is an impact and subsequent escape of particles

FIGURE 5 *Cassini* photomosaic of Enceladus, which is geologically active and thought to be the source of particles in the E-ring. The south polar region off to the lower right contains groove-like features ("tiger stripes") that are more than 50 K hotter that the surrounding regions. More heavily cratered terrain is visible in the northern hemisphere of the satellite. The spacecraft was 112,000 km from the satellite when this image was obtained.

from the surface although the recent *cassini* results make this scenario unlikely. [*See* PLANETARY RINGS.]

Tethys is covered with impact craters, including Odysseus, the largest known impact structure in the solar system. The craters tend to be flatter than those on Mimas or the Moon, probably because of viscous relaxation and flow over the eons under the stronger gravitational field of Tethys. Evidence for resurfacing episodes is seen in regions that have fewer craters and higher albedos. In addition, there is a huge trench formation, the Ithaca Chasma, which may be a degraded form of the grooves found on Enceladus.

Dione, which is about the same size as Tethys, exhibits a wide diversity of surface morphology. Most of the surface is heavily cratered (Fig. 7), but gradations in crater density indicate that several periods of resurfacing occurred during the first billion years of its existence. The leading side of the satellite is about 25% brighter than the trailing side, possibly due to more intensive micrometeoritic bombardment on this hemisphere. Bright wispy streaks seen by the *Voyager* spacecraft were revealed by *Cassini* to be deep and extensive tectonic faults. Dione modulates the radio emission from Saturn, but the mechanism for this phenomenon is unknown.

Rhea appears to be superficially very similar to Dione (see Fig. 4). Bright wispy streaks—*Cassini* also showed

FIGURE 6   A view of the south pole of Enceladus, obtained by the *Cassini* camera. The backlit geyser-like fountains are likely to be particles of water ice erupting from high-pressure reservoirs of liquid water below the surface. The geysers appear to originate from the "tiger stripes."

they are tectonic features—cover one hemisphere. However, there is no evidence for any resurfacing events early in its history. There does seem to be a dichotomy between crater sizes—some regions lack large craters whereas other regions have a preponderance of such impacts. The larger craters may be due to a population of larger debris more prevalent during an earlier episode of collisions.

When Cassini discovered Iapetus in 1672, he noticed that at one point in its orbit around Saturn it was very bright, but that on the opposite side of the orbit it nearly disappeared from view. He correctly deduced that one hemisphere is composed of highly reflective material and the other side is much darker. *Voyager* images show that the bright side, which reflects nearly 50% of the incident radiation, is fairly typical of a heavily cratered icy satellite. The other side, which is centered on the direction of motion, is coated with a material with a reflectivity of about 3–4% (Fig. 8). Other

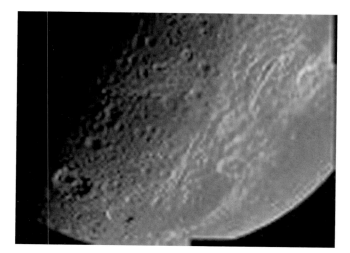

FIGURE 7   The heavily cratered face of Dione is shown in this *Cassini* image. Features that appeared as bright wispy streaks in *Voyager* images (see Fig. 4) are shown to be bright ice cliffs formed by tectonic fractures.

FIGURE 8   *Cassini* image of Iapetus, showing both bright and dark terrains. The image was obtained at a distance of 173,000 km, with a resolution of about 2 km. The equatorial band is clearly visible in the low-albedo terrain.

aspects of Iapetus are unusual. It is the only large Saturnian satellite in a highly inclined orbit, and it is less dense than objects of similar albedo. This latter fact implies a higher fraction of ice or possibly methane or ammonia in its interior.

Two models exist for the origin of the dark material: It derived from an exogenic (external) source, or it was endogenically (internally) created. One scenario for the exogenic deposit of material entails dark particles being ejected from Phoebe and drifting inward to coat Iapetus. The major problem with this model is that the dark material on Iapetus is redder than Phoebe, although the material could have undergone chemical changes after its expulsion from Phoebe to make it redder. Recent observations also show that many of the outer retrograde satellites are similar in color to the dark hemisphere of Iapetus, so the dust may have come from those satellites as well. One observation lending credence to an internal origin is the concentration of material on crater floors, which implies an infilling mechanism. The *Cassini* images obtained on December 31, 2004, reveal a gradual transition between the bright and dark regions, which is more consistent with an exogenic model. The *Cassini* visual and infrared spectrometer also detected carbon dioxide and organics in the dark material. The *Cassini* cameras also captured a unique geologic feature on Iapetus: an equatorial ridge 20 km wide and 13 km high extending over at least one hemisphere of the satellite.

4.3.2 HYPERION AND PHOEBE

Telescopic observations showed that the surface of Hyperion, which lies between the orbits of Iapetus and Titan, is covered with ice. Because Hyperion has a visual geometric albedo of 0.30, this ice must be mixed with a significant amount of darker, rocky material. Its composition may be similar to D-type asteroids. Although Hyperion is only slightly smaller than Mimas, it has a highly irregular shape (see Table 1). This suggests, along with the satellite's battered appearance, that it has been subjected to intense bombardment and fragmentation. *Cassini* images of the satellite (Fig. 9) show craters that appear to have been deeply eroded, perhaps by sublimation of ice by darker, hotter deposits in the crater floors. Hyperion is the only satellite known to be in chaotic rotation—perhaps a collision within the last few million years knocked it out of a tidally locked orbit. [*See* CHAOTIC MOTION IN THE SOLAR SYSTEM.]

Saturn's outer satellite Phoebe, a dark object (see Table 1) with a surface composition similar to that of C-type asteroids (but apparently with more organic material), moves in a highly inclined, retrograde orbit, suggesting that it is a captured object. The spectral signature of water ice was detected by ground-based telescopes. Although it is smaller than Hyperion, Phoebe has a more nearly spherical shape. Figure 10 shows an image of the satellite obtained by *Cassini–Huygens* in July 2004. The heavily battered sur-

FIGURE 9 *Cassini* image of Hyperion, showing deep craters and a mottled, porous surface. The image was obtained at a distance of about 62,000 km, with a resolution of about 700 m. The largest crater on its surface is approximately 120 km in diameter and 10 km deep.

face has a number of unusual conical craters, and the largest crater reveals higher albedo ice cliffs on its rims. Carbon dioxide and organic material was detected by *Cassini*, suggesting that the satellite formed in the outer solar system, perhaps as far out as the Kuiper Belt.

FIGURE 10 *Cassini* photomosaic of Phoebe, obtained at distances ranging from 16,000 to 12,000 km, with a corresponding resolution of about 150 m. The heavily cratered surface shows no hint of geological resurfacing. Icy cliffs are evident in the large crater at the top of the image, and the crater to the right of center shows evidence for layering near its rim.

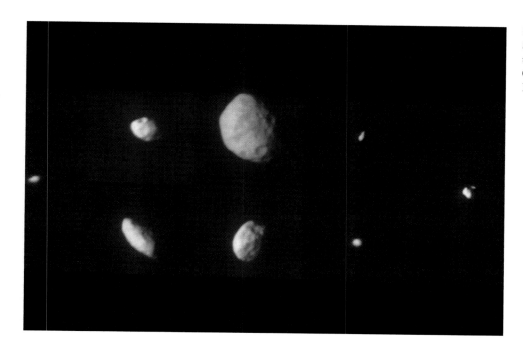

FIGURE 11 Six small satellites of Saturn. They are, clockwise from the top, Atlas, Pandora, Janus, Calypso, Helene, Telesto, Epimetheus, and Prometheus.

4.3.3 THE INNER SMALL SATELLITES

Four types of unusual inner small satellites have been found in the Saturnian system: the shepherding satellites, the co-orbitals, the Lagrangians, and the satellites that orbit in ring gaps. All of these objects are irregularly shaped (see Fig. 11) and probably consist primarily of ice. The three shepherds—Atlas, Pandora, and Prometheus—are modeled as playing a key role in defining the edges of Saturn's A- and F-rings. The orbit of Atlas, Saturn's innermost satellite that is not within the ring system, lies several hundred kilometers from the outer edge of the A-ring. The other 2 shepherds, which orbit on either side of the F-ring, not only constrain the width of this narrow ring but may cause its kinky appearance. The *Cassini* spacecraft also discovered Pallene and Methone, two small satellites that orbit between Mimas and Enceladus.

The co-orbital satellites Janus and Epimetheus, which were discovered in 1966 and 1978, exist in an unusual dynamical situation. They move in almost identical orbits at about 2.5 Saturn radii. Every four years the inner satellite (which orbits slightly faster than the outer one) overtakes its companion. Instead of colliding, the satellites exchange orbits. The 4-year cycle then begins over again. Perhaps these two satellites were once part of a larger body that disintegrated after a major collision.

Four other small satellites of Saturn orbit in the **La-grangian points** of larger satellites: Two are associated with Dione (Helene and Polydeuces, which was discovered by *Cassini* in October 2004) and two with Tethys (Telesto and Calypso). The Lagrangian points are locations within an object's orbit in which a less massive body can move in an identical, stable orbit. They lie about 60° in front of and in back of the larger body. Although no other known satellites

in the solar system are Lagrangians, the Trojan asteroids orbit in two of the Lagrangian points of Jupiter, Neptune, and Mars.

The final class of unusual Saturnian satellite is those that dwell in ring gaps and sweep and clear particles from the gaps. Pan, which was discovered in 1990 from *Voyager* images, sits in the Encke gap. The *Cassini* spacecraft discovered Daphnis, a small satellite in the Keeler gap, in May 2005.

4.3.4 SMALL OUTER SATELLITES

Like the other giant outer planets, Saturn has a large family of outer irregular satellites, most of which have been recently discovered with large telescopes. The 25 known outer small satellites move in eccentric inclined orbits, and most of their orbits are retrograde, implying that they are captured objects. The farthest satellites orbit more than 20 million km from Saturn.

4.4 The Satellites of Uranus

4.4.1 THE MEDIUM-SIZED SATELLITES OF URANUS: MIRANDA, ARIEL, UMBRIEL, TITANIA, AND OBERON

Uranus has a total of 27 known satellites. The main satellites are medium-sized bodies that orbit the planet from 130,000 to 583,000 km. The orbits of Ariel, Umbriel, Titania, and Oberon are regular, whereas Miranda's orbit is slightly inclined. Figure 12 is a telescopic image of the satellites typical of the quality attainable before the advent of spacecraft missions.

Theoretical models suggest that the satellites are composed of water ice, possibly in the form of methane

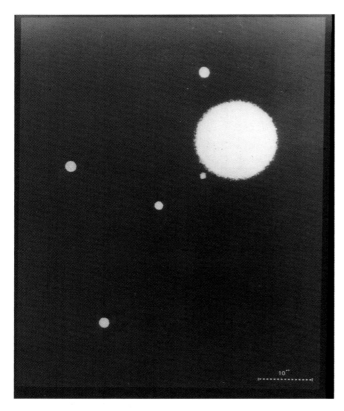

FIGURE 12   Telescopic view of Uranus and its five satellites obtained by Ch. Veillet on the 154-cm Danish-ESO telescope. Outward from Uranus they are Miranda, Ariel, Ubriel, Titania, and Oberon. (Photograph courtesy of Ch. Veillet.)

clathrates or ammonia hydrates, and silicate rock. Water ice has been detected spectroscopically on all five satellites. Carbon dioxide has been detected on Ariel. The relatively dark visual albedos of the satellites, ranging from 0.13 for Umbriel to 0.33 for Ariel (see Table 1), and gray spectra, indicate that their surfaces are contaminated by a dark component such as graphite or carbonaceous material. Another darkening mechanism that may be important is bombardment of the surface by ultraviolet radiation. The higher density of Umbriel implies that its bulk composition includes a larger fraction of rocky material than the other four satellites. Heating and differentiation have occurred on Miranda and Ariel, and possibly on some of the other satellites. Models indicate that tidal interactions may provide an important heat source in the case of Ariel.

Miranda, Ariel, Oberon, and Titania all exhibit large opposition surges, indicating that the regoliths of these bodies are composed of very porous material, perhaps resulting from eons of micrometeoritic "gardening." Umbriel lacks a significant surge, which suggests that its surface properties are in some way unusual. Perhaps its regolith is very compacted, or it is covered by a fine dust that scatters optical radiation in the forward direction.

The *Voyager 2* spacecraft encountered Uranus in January 1986 to reveal satellites that have undergone melting and resurfacing (Fig. 13). Three features on Miranda, known as coronae, consist of a series of ridges and valleys ranging from 0.5 to 5 km in height (Fig. 14). The origin of these features is uncertain: Some geologists favor a compressional folding interpretation, whereas others invoke a volcanic origin or a faulting origin. Both Ariel, which is the satellite that has had the most recent geologic activity, and Titania are covered with cratered terrain transected by grabens, which are fault-bounded valleys. Umbriel is heavily cratered, and it is the darkest of the satellites, both of which suggest that its surface is very old, although the moderate-resolution images obtained by *Voyager* cannot rule out heating or geologic activity. Some scientists have in fact interpreted small albedo variations on its surface as evidence for melting events early in its history. Oberon is similarly covered with craters, some of which have very dark deposits on their floors. On its surface are situated faults or rifts, suggesting resurfacing events (*Voyager* provided ambiguous, medium-resolution views of the satellite). In general, the Uranian satellites appear to have exhibited more geologic activity than the Saturnian satellites and Callisto, possibly because of the presence of methane, ammonia, nitrogen, or additional volatiles.

There is some evidence that Umbriel and Oberon, as well as certain regions of the other satellites, contain **D-type material**, the organic-rich primordial constituent that seems to be ubiquitous in the outer solar system. D-type material is seen in the dark, red D-type asteroids and may contain some of the same molecules that are seen on the dark side of Iapetus, on Hyperion, and on specific areas of the larger satellites.

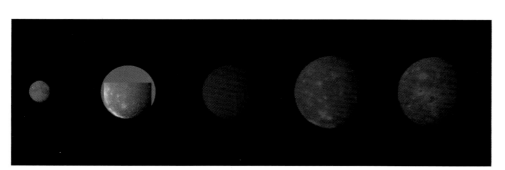

FIGURE 13   The five major satellites of Uranus, shown to relative size based on *Voyager 2* images. They are, from the left, Miranda, Ariel, Umbriel, Titania, and Oberon.

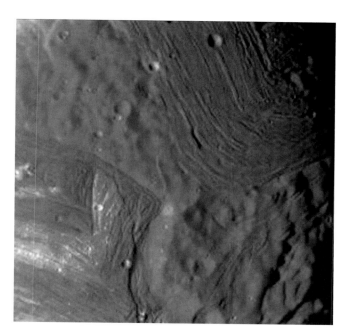

FIGURE 14   Image of Miranda obtained by the *Voyager 2* spacecraft at 30,000–40,000 km from the Moon. Resolution is 560–740 m. Older, cratered terrain is transected by ridges and valleys, indicating more recent geologic activity.

4.4.2 THE SMALL SATELLITES OF URANUS

Voyager 2 discovered 10 new small satellites of Uranus, including two that act as shepherding satellites for the outer (epsilon) ring of Uranus (see Table 1). All these satellites lie inside the orbit of Miranda. Images of two satellites, Puck and Cordelia, provided sufficient resolution to directly determine their radii (see Table 1). The sizes of the other bodies were derived by making the assumption that their surface brightnesses are equal to those of the other inner satellites and estimating the projected area required to yield their observed integral brightnesses. Puck appears to be only slightly nonspherical in shape. It is likely that the other small satellites are irregularly shaped. The satellites' visual geometric albedos range from 0.04 to 0.09, which is slightly higher than that of Uranus's dark ring system. No reliable color information was obtained by *Voyager 2* for any of the small satellites, although their low albedo suggests that they are C-type objects. Ground-based observers have discovered another 12 outer irregular satellites to bring the total of known Uranian satellites to 27. Most of the small outer satellites are moving in retrograde orbits, implying they are probably captured bodies.

4.5 The Satellites of Neptune

4.5.1 INTRODUCTION

Neptune has 13 known satellites: one is the large moon Triton and the remaining 12 are small, irregularly shaped bodies (see Table 1). The small satellites can be divided into two categories: the 6 inner bodies, which move in highly regular, circular orbits close to Neptune (<5 planetary radii), and the irregular outer satellites. At the time of the *Voyager 2* encounter, Nereid, which moves in an eccentric prograde orbit bringing it from 57 to 385 planetary radii from Neptune, was the only known satellite in the latter category. Five more moons were discovered in 2003, including one with a period of 26.3 years, which corresponds to a distance of 47 million km from Neptune. Triton has an appreciable atmosphere, seasons, and currently active geologic processing. [*See* TRITON]

Only Triton and the outer satellite Nereid were known before the reconnaissance of Neptune by the *Voyager 2* spacecraft in 1989. Nereid was discovered in 1949 by Gerard P. Kuiper at McDonald Observatory in Texas. In keeping with the theme of water and oceans for the Neptunian system, the satellite was named after the sea nymphs known in Greek mythology as Nereids. Reliable ground-based observations of Nereid were limited to estimates of its visual magnitude.

4.5.2 ORBITAL AND BULK PROPERTIES

The six inner satellites were all discovered within a few days during the *Voyager* encounter with Neptune in August 1989. They were given names of mythical nautical figures by the International Astronomical Union. For four of these satellites (Proteus, Larissa, Galatea, and Despina), as well as Nereid, *Voyager* images provided sufficient resolution to determine their dimensions (see Table 1). All five bodies are irregularly shaped. The sizes of Thalassa and Naiad were derived by making the assumption that their albedos are equal to those of the other inner satellites. The size of the satellites increases with the distance from Neptune. Proteus is the largest known irregularly shaped satellite in the solar system (see Table 1). The satellite has probably not been subjected to viscous relaxation; rather its mechanical properties have been determined by the physics of water ice, with an internal temperature below 110 K.

Spacecraft tracking of the six inner satellites, and ground-based observations of Nereid, provided accurate orbit determinations, which are listed in Table 1. All the small inner satellites except Proteus orbit inside the so-called synchronous distance, which is the distance from Neptune at which the planet's rotational spin period equals the satellite's orbital period. The rotational periods of these satellites are unknown, but they are most likely in synchronous rotation.

Voyager observations and more recent telescopic measurements suggest that Nereid is in nonsynchronous rotation, with a period of 11.5 hours.

The masses of the satellites were not measured directly by *Voyager*. Limits may be obtained by assuming reasonable values for their bulk densities. These values range from 0.7 g/cm^3, corresponding to water ice with a bulk porosity of about 30%, to 2 g/cm^3, corresponding to water ice with a significant fraction of rocky material. If the satellites were

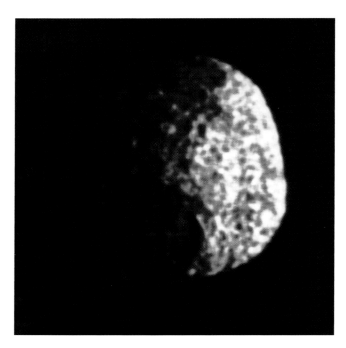

FIGURE 15   The best *Voyager* image of Proteus, with a resolution of 1.3 km/pixel.

formed from captured material, the higher density is more reasonable. In any case, the small satellites have less than 1% of the mass of Triton. The ring system of Neptune contains only a very small amount of mass, possibly one-millionth of the small satellites' combined masses.

4.5.3 APPEARANCE AND COMPOSITION

Figure 15 depicts the best *Voyager* images obtained for Proteus, with a resolution of 1.3 km/pixel. The large feature—possibly an impact basin—has a diameter of about 250 km. Close scrutiny of this image reveals a concentric structure within the impact basin. Possible ridgelike features appear to divide the surface. The regions of Proteus outside of the impact basin show signs of being heavily cratered.

The best image of Larissa was obtained at a resolution of 4.2 km/pixel and that of Nereid at a resolution of 43 km/pixel. Neither image has sufficient resolution to depict surface features. Analysis of calibrated, integral *Voyager* measurements of the four inner satellites reveals that their geometric albedos are about 0.06, in the *Voyager* clear filter with an effective wavelength of about 480 nm. The integral brightness of Nereid is almost 3 times that of Proteus, which is slightly smaller; its geometric albedo is therefore nearly 3 times as high.

The limited spectral data obtained by *Voyager* suggest that Proteus, Nereid, and Larissa are gray objects. The dark albedos and spectrally neutral character of the inner satellites suggest that they are carbonaceous objects, similar to the primitive C-type asteroids, possibly the Uranian satel-lite Puck, the satellites of Mars, and several other small satellites. Nereid, however, with its markedly higher albedo, probably has a surface of water frost contaminated by a dark, spectrally neutral material. It is more similar to the differentiated satellites of Uranus than to the dark C-type objects.

4.5.4 ORIGINS AND EVOLUTION

Three of the five outer satellites discovered in 2003 have retrograde orbits and may thus be captured objects. Jupiter, Saturn, Uranus, and Neptune all appear to have families of outer, captured satellites.

The evolution of the inner satellites was likely punctuated by the capture of Triton. Initially, the inclinations and eccentricities of the satellites would have been increased by the capture, and subsequent collisions would have occurred. The resulting debris would then have reaccreted to form the present satellites. Models of the collisional history of the satellites suggest that with the exception of Proteus they are much younger than the age of the solar system. The heavily cratered surface that appears in the one resolvable *Voyager* image of these bodies (see Fig. 14) does suggest that they have undergone vigorous bombardment.

The only satellite that has been shown to have a dynamical relationship with the rings of Neptune is Galetea, which confines the ring arcs. The orbits of the satellites have probably evolved under the influence of tidal evolution and resonances. For example, the inclination of Naiad is possibly due to its escape from an inclination resonance state with Despina.

Bibliography

Beatty, J. K., Petersen, C. C., and Chaikin, A., eds. (1999). "The New Solar System," 4th Ed. Sky Publishing, Cambridge, Massachusetts.

Belton, M. J. S., and the Galileo Science Teams (1996). *Science* **274**, 377–413.

Bergstralh, J., and Miner, E., eds. (1991). "Uranus." Univ. Arizona Press, Tucson.

Burns, J., and Matthews, M., eds. (1986). "Satellites." Univ. Arizona Press, Tucson.

de Pater, I., and Lissauer, J. (2001). "Planetary Sciences." Cambridge Univ. Press, Cambridge, England.

Gehrels, T., ed. (1984). "Saturn." Univ. Arizona Press, Tucson.

http://nssdc.gsfc.nasa.gov/planetary/planetfact.html

http://ssd.jpl.nasa.gov

http://planetarynames.wr.usgs.gov/append7.html

Up-to-date listings of satellite discoveries, names, and physical and dynamical properties.

Hartmann, W. K. (2004). "Moons and Planets," 4th Ed. Wadsworth, Belmont, California.

Peale, S. (1999). Origin and evolution of the natural satellites. *Annu. Rev. Astron. Astrophys.* **37**, 533.

Stone, E., and the Voyager Science Teams (1989). *Science* **246**, 1417–1501.

Atmospheres of the Giant Planets

Robert A. West

Jet Propulsion Laboratory
California Institute of Technology
Pasadena, California

CHAPTER 20

The atmospheres of the giant planets—Jupiter, Saturn, Uranus, and Neptune—are very unlike those of the Earth, Mars, and Venus. They are composed mainly of hydrogen and helium, with some trace species, the most abundant of which are water, methane, and ammonia. They are cold enough to form clouds of ammonia and hydrocarbon ices, which extend deep into the interior of the planet, and indeed a significant fraction of the planet's mass may be responsible for the near-surface winds. The winds are primarily east–west (zonal) jets that alternate with latitude. Superimposed on the jets are spots of all sizes up to about three Earth diameters. Some of them, like Jupiter's Great Red Spot, are remarkably long-lived. At the highest altitudes, powerful **auroras**, as well as some still mysterious processes, heat the atmospheres to temperatures higher than current models can explain.

1. Introduction

To be an astronaut explorer in Jupiter's atmosphere would be strange and disorienting. There is no solid ground to stand on. The temperature would be comfortable at an altitude where the pressure is eight times that of Earth's surface, but it would be perpetually hazy overhead, with variable conditions (dry or wet, cloudy or not) to the east, west, north, and south. One would need to carry oxygen as there is no free oxygen, and to wear special clothing

to protect the skin against exposure to ammonia, hydrogen sulfide, and ammonium hydrosulfide gases, which form clouds and haze layers higher in the atmosphere. A trip to high latitudes would offer an opportunity to watch the most powerful, vibrant, and continuous auroral displays in the solar system. On the way, one might pass through individual storm systems the size of Earth or larger and be buffeted by strong winds alternately from the east and west. One might be sucked into a dry downwelling sinkhole like the environment explored by the *Galileo* probe. The probe fell to depths where the temperature is hot enough to vaporize metal and rock. It is now a part of Jupiter's atmosphere.

Although the atmospheres of the giant planets share many common attributes, they are at the same time very diverse. The roots of this diversity can be traced to a set of basic properties, and ultimately to the origins of the planets. The most important properties that influence atmospheric behavior are listed in Table 1. The distance from the Sun determines how much sunlight is available to heat the upper atmosphere. The minimum temperature for all of these atmospheres occurs near the 100 mbar level and ranges from 110 K at Jupiter to 50 K at Neptune. The distance from the Sun and the total mass of the planet are the primary influences on the bulk composition. All the giant planets are enriched in heavy elements, relative to their solar abundances, by factors ranging from about 3 for Jupiter to 1000 for Uranus and Neptune. The latter two planets are sometimes called the ice giants because they have a large

TABLE 1	Physical Properties of the Giant Planets			
Property	Jupiter	Saturn	Uranus	Neptune
Distance from the Sun (Earth distance = 1^a)	5.2	9.6	19.2	30.1
Equatorial radius (Earth radius = 1^b)	11.3	9.4	4.1	3.9
Planet total mass (Earth mass = 1^c)	318.1	95.1	14.6	17.2
Mass of gas component (Earth mass = 1)	254–292	72–79	1.3–3.6	0.7–3.2
Orbital period (years)	11.9	29.6	84.0	164.8
Length of day (hours, for a point rotating with the interior	9.9	10.7	17.4	16.2
Axial inclination (degrees from normal to orbit plane)	3.1	26.7	97.9	28.8
Surface gravity (equator–pole, m s^{-2})	(22.5–26.3)	(8.4–11.6)	(8.2–8.8)	(10.8–11.0)
Ratio of emitted thermal energy to absorbed solar energy	1.7	1.8	~1	2.6
Temperature at the 100-mbar level (K)	110	82	54	50

^{a} Earth distance = 1.5×10^8 km.
^{b} Earth radius = 6378 km.
^{c} Earth mass = 6×10^{24} kg.

fraction of elements (O, C, N, and S) that were the primary constituents of ices in the early solar nebula.

The orbital period, axial tilt, and distance from the Sun determine the magnitude of seasonal temperature variations in the high atmosphere. Jupiter has weak seasonal variations; those of Saturn are much stronger. Uranus is tipped such that its poles are nearly in the orbital plane, leading to more solar heating at the poles than at the equator when averaged over an orbit. The ratio of radiated thermal energy to absorbed solar energy is diagnostic of how rapidly convection is bringing internal heat to the surface, which in turn influences the abundance of trace constituents and the morphology of eddies in the upper atmosphere. Vigorous convection from the deeper interior is responsible for unexpectedly high abundances of several trace species on Jupiter, Saturn, and Neptune, but convection on Uranus is sluggish. All these subjects are treated in more detail in the sections that follow.

2. Chemical Composition

This section is concerned with chemical abundances in the observable part of the atmosphere, a relatively thin layer of gas near the top (where pressures are between about 5 bar and a fraction of a microbar). To place the subject in context, some mention will be made of the composition of the interior. [See INTERIORS OF THE GIANT PLANETS.]

The bulk composition of a planet cannot be directly observed, but must be inferred from information on its mean density, its gravity field, and the abundances of constituents that are observed in the outer layers. The more massive planets were better able to retain the light ele-

ments during their formation, and so the bulk composition of Jupiter resembles that of the Sun. When the giant planets formed, they incorporated relatively more rock and ice fractions than a pure solar composition would allow, and the fractional amounts of rocky and icy materials increase from Jupiter through Neptune. [See THE ORIGIN OF THE SOLAR SYSTEM.] Most of the mass of the heavy elements is sequestered in the deep interior. The principal effects of this layered structure on the observable outer layers can be summarized as follows.

On Jupiter the gas layer (a fluid molecular envelope) extends down to about 40% of the planet's radius, where a phase transition to liquid metallic hydrogen occurs. Fluid motions that produce the alternating jets and vertically mix gas parcels may fill the molecular envelope but probably do not extend into the metallic region. Thus, the radius of the phase transition provides a natural boundary that may be manifest in the latitudinal extent of the zonal jets (see Section 4), whereas vertical mixing may extend to levels where the temperature is quite high. These same characteristics are found on Saturn, with the additional possibility that a separation of helium from hydrogen is occurring in the metallic hydrogen region, leading to enrichment of helium in the deep interior and depletion of helium in the upper atmosphere.

Uranus and Neptune contain much larger fractions of ice- and rock-forming constituents than do Jupiter and Saturn. A large water ocean may be present in the interiors of these planets. Aqueous chemistry in the ocean can have a profound influence on the abundances of trace species observed in the high atmosphere.

In the observable upper layers, the main constituents are molecular hydrogen and atomic helium, which are well mixed, up to the **homopause** level, where the mean free path for collisions becomes large enough that the lighter constituents are able to diffuse upward more readily than heavier ones. Other constituents are significantly less abundant than hydrogen and helium, and many of them condense in the coldest regions of the atmosphere. Figure 1 shows how temperature varies with altitude and pressure, and the locations of the methane, ammonia, and water cloud layers.

The giant planets have retained much of the heat generated by their initial collapse from the solar nebula. They cool by emitting thermal infrared radiation to space. Thermal radiation is emitted near the top of the atmosphere, where the opacity is low enough to allow infrared photons to escape to space. In the deeper atmosphere, heat is transported by convective fluid motions from the deep, hot interior to the colder outer layers. In this region, upwelling gas parcels expand and subsiding parcels contract adiabatically (e.g., with negligible transport of heat through their boundaries by radiation or conduction). Therefore temperature depends on altitude according to the adiabatic law $T = T_0 + C(z - z_0)$, where T_0 is the temperature at some reference altitude z_0, C is a constant (the adiabatic lapse rate) that depends on the gas composition, and z is altitude. The adiabatic lapse rate for dry hydrogen and helium on Jupiter is -2.2 K/km. On Uranus it is -0.8 K/km. The adiabatic lapse rate is different in regions where a gas is condensing or where heat is released as *ortho*-hydrogen and is converted to *para*-hydrogen. Both of these processes are important in the giant planet atmospheres at pressures between about 30 and 0.1 bar.

Hydrogen is the main constituent in the observable part of the giant planet atmospheres, but not until recently was it recognized as especially important for thermodynamics. The hydrogen molecule has two ground-state configurations for its two electrons. The electrons can have their spins either parallel or antiparallel, depending on whether the spins of the nuclei are parallel or antiparallel. These states are called the ortho and para states. Transitions between ortho and para states are slow because, unlike most molecules, the nuclear spin must change when the electron spin changes. At high temperature (about 270 K or higher), the ortho:para relative abundance is 3:1. At lower temperature, a larger fraction is converted to the para state. Heat release from conversion of *ortho*- to *para*-hydrogen can act in the same way as **latent heat** release from condensation. The relative fractions of *ortho*- and *para*-hydrogen are observed to be close to thermal equilibrium values in the giant planet atmospheres, leading to the question of how equilibrium is achieved. Catalytic reactions on the surfaces of **aerosol** particles are thought to be important in equilibrating the ortho and para states.

Temperature follows the adiabatic law at pressures deeper than about 2 bar. The atmospheric temperature would drop at the adiabatic rate to near absolute zero at the top of the atmosphere were it not for sunlight, which heats the upper atmosphere. Sunlight penetrates to pressure levels near 20 bar, depending on how much overlying cloud and haze opacity is present. The competition between convective cooling and solar heating produces a temperature minimum near the 100 mbar level (the tropopause). At pressures between about 100 and 0.1 mbar, the temperature is determined primarily by equilibrium between thermal radiative cooling and solar heating. At even lower pressures, other processes, including auroral heating, dump energy into the atmosphere and produce higher temperatures. More will be said about this in Section 5.

The current inventory of observed gaseous species is listed in Table 2. Molecular hydrogen and helium are the most abundant. Helium is in its ground state in the **troposphere** and **stratosphere** and therefore does not

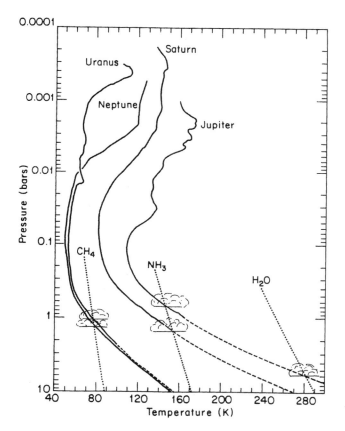

FIGURE 1   Profiles of temperature as a function of pressure in the outer planet atmospheres derived from measurements by the Voyager Radio Sciences experiment (solid curves). The dashed parts of the temperature profiles are extrapolations using the adiabatic lapse rate. At high altitudes (not shown), temperatures rise to about 1200 K for Jupiter, 800 K for Saturn and Uranus, and 300 K for Neptune. The dotted lines show vapor pressure curves divided by observed mixing ratios for water, ammonia, and methane. Condensate clouds are located where the solid and dotted curves cross. (From Gierasch and B. Conrath, 1993, *J. Geophys. Res.* **98**, 5459–5469. Copyright American Geophysical Union.)

TABLE 2	Abundances of Observed Species in the Atmospheres of the Giant Planets

	Peak mixing ratio (by number) or upper limit			
Constituent	Jupiter	Saturn	Uranus	Neptune
Species with constant mixing ratio below the homopause				
H_2	0.86	0.90	0.82	0.79
HD	4×10^{-5}	4×10^{-5}		
He	0.14	0.10	0.15	0.18
CH_4	2×10^{-3}	2×10^{-3}		
CH_3D	3.5×10^{-7}	2×10^{-7}		
^{20}Ne	2×10^{-5}			
^{36}Ar	1×10^{-5}			
Condensable species (estimated or measured below the condensation region)				
NH_3	2.5×10^{-4}	2×10^{-4}		
H_2S	7×10^{-5}			
H_2O	6×10^{-4}			
CH_4			0.025	0.02–0.03
CH_3D			2×10^{-5}	2×10^{-5}
Disequilibrium species in the troposphere				
PH_3	5×10^{-7}	2×10^{-6}		
GeH_4	7×10^{-10}	4×10^{-10}		
AsH_3	2.4×10^{-9}	3×10^{-9}		
CO	2×10^{-9}	$1–25 \times 10^{-9}$	$<1 \times 10^{-8}$	1×10^{-6}
HCN			$<1 \times 10^{-10}$	1×10^{-9}
Photochemical species (peak values)				
C_2H_2	1×10^{-7}	3×10^{-7}	1×10^{-8}	6×10^{-8}
C_2H_4	7×10^{-9}			
C_2H_6	7×10^{-6}	7×10^{-6}	$<1 \times 10^{-8}$	2×10^{-6}
C_3H_4	2.5×10^{-9}			
C_6H_6	2×10^{-9}			

produce spectral lines from which its abundance can be determined. The mixing ratio for Saturn, Uranus, and Neptune is inferred from its influence on the broad collision-induced hydrogen lines near the 45 μm wavelength, and from a combined analysis of the infrared spectrum and refractivity profiles retrieved from spacecraft radio occultation measurements. Helium on Jupiter is accurately known from measurements made by the *Galileo* probe, which descended through the atmosphere. It is a little smaller than the mixing ratio inferred for the primitive solar nebula from which the planets formed. Helium is substantially depleted in Saturn's upper atmosphere, consistent with the idea that helium is precipitating out in the metallic hydrogen region. For Uranus and Neptune, the helium mixing ratio is close to the mixing ratio (0.16) in the primitive solar nebula. There is still some uncertainty in the helium mixing ratio for Uranus, Neptune, and Saturn because additional factors, such as aerosol opacity and molecular nitrogen abundance, affect the shapes of the collision-induced spectral features, and we do not have a completely consistent set of values for all these parameters.

Mixing ratios of **deuterated** hydrogen and methane (HD and CH_3D) also provide information on the formation of the planets. **Deuterium**, which once existed in the Sun, has been destroyed in the solar atmosphere, and the best information on its abundance in the primitive solar nebula comes from measurements of the giant planet atmospheres. On Jupiter, the deuterium mixing ratio is thought to be close to that of the primitive solar nebula. On Uranus and Neptune, it is enhanced because those planets incorporated relatively more condensed material on which deuterium preferentially accumulated through isotopic fractionation. Isotopic fractionation (the enhancement of the heavier isotope over the lighter isotope during condensation) occurs because the heavier isotope has a lower energy than the lighter isotope in the condensed phase.

The elements oxygen, carbon, nitrogen, and sulfur are the most abundant molecule-forming elements in the Sun (after hydrogen), and all are observed in the atmospheres of the giant planets, mostly as H_2O, CH_4, NH_3, and (for Jupiter) H_2S. Water condenses even in Jupiter's atmosphere, at levels that are difficult to probe with infrared radiation (6 bars or deeper). A straightforward interpretation of Jupiter's spectrum indicated its abundance to be about a hundred times less than what is expected from solar composition. The *Galileo* probe measurements indicated that water was depleted relative to solar abundance by roughly a factor of two at the deepest level measured (near 20 bars of pressure) and even more depleted at higher altitude. However, the probe descended in a relatively dry region of the atmosphere, analogous to a desert on Earth, and the bulk water abundance on Jupiter may well be close to the solar abundance. Water is not observed on any of the other giant planets because of the optically thick overlying clouds and haze layers. It is thought to form a massive global ocean on Uranus and Neptune based on the densities and gravity fields of those planets, coupled with theories of their formation.

Methane is well mixed, up to the homopause level, in the atmospheres of Jupiter and Saturn, but it condenses as ice in the atmospheres of Uranus and Neptune. Its mixing ratio below the condensation level is enhanced over that expected for a solar-composition atmosphere by factors of 2.6, 5.1, 35, and 40 for Jupiter, Saturn, Uranus, and Neptune, respectively. These enhancements are consistent with ideas about the amounts of icy materials that were incorporated into the planets as they formed. The stratospheres of Uranus and Neptune form a cold trap, where methane ice condenses into ice crystals that fall out, making it difficult for methane to mix to higher levels. Nevertheless, the methane abundance in Neptune's stratosphere appears to be significantly higher than its vapor pressure at the temperature than the tropopause would allow (and also higher than the abundance in the stratosphere of Uranus), suggesting some mechanism such as convective penetration of the cold trap by rapidly rising parcels of gas. This mechanism does not appear to be operating on Uranus, and this difference between Uranus and Neptune is symptomatic of the underlying difference in internal heat that is available to drive convection on Neptune but not on Uranus.

Ammonia is observed on Jupiter and Saturn, but not on Uranus or Neptune. Ammonia condenses as an ammonia ice cloud near 0.6 bar on Jupiter and at higher pressures on the colder outer planets. Ammonia and H_2S in solar abundance would combine to form a cloud of NH_4SH (ammonium hydrosulfide) near the 2 bar level in Jupiter's atmosphere and at deeper levels in the colder atmospheres of the other giant planets. Hydrogen sulfide was observed in Jupiter's atmosphere by the mass spectrometer instrument on the *Galileo* probe. Another instrument (the nephelometer) on the probe detected cloud particles in the vicinity of the 1.6 bar pressure level, which would be consistent with the predicted ammonium hydrosulfide cloud. Evidence from thermal emission at radio wavelengths has been used to infer that H_2S is abundant on Uranus and Neptune. Ammonia condenses at relatively deep levels in the atmospheres of Uranus and Neptune and has not been spectroscopically detected. A dense cloud is evident at the level expected for ammonia condensation (2–3 bar) in near-infrared spectroscopic observations, but the microwave spectra of those planets are more consistent with a strong depletion of ammonia at those levels. An enhancement of H_2S relative to NH_3 could act to deplete ammonia by the formation of ammonium hydrosulfide in the deeper atmosphere. In that case, H_2S ice is the most likely candidate for the cloud near 3 bars.

Water, methane, and ammonia are in thermochemical equilibrium in the upper troposphere. Their abundances at altitudes higher than (and temperatures colder than) their condensation level are determined by temperature (according to the vapor–pressure law) and by meteorology, as is water in Earth's atmosphere. Some species (PH_3, GeH_4, and CO) are not in thermochemical equilibrium in the upper troposphere. At temperatures less than 1000 K, PH_3 would react with H_2O to form P_4O_6 if allowed to proceed to thermochemical equilibrium. Apparently the time scale for this reaction (about 10^7 s) is longer than the time to convect material from the 1000 K level to the tropopause. A similar process explains the detections of GeH_4. Yet another phenomenon (impact of a comet within the past 200 years) probably accounts for the detection of CO in the stratosphere.

Ammonia and phosphine are present in the stratospheres of Jupiter and Saturn, and methane is present in the stratospheres of all the giant planets. These species are destroyed at high altitudes by ultraviolet sunlight and by charged particles in auroras, producing N, P, and C, which can react to form other compounds. Ammonia photochemistry leads to formation of hydrazine (N_2H_4), and phosphine photochemistry leads to diphosphine (P_2H_4). These constituents condense in the cold troposphere of Jupiter and Saturn and may be responsible for much of the ultraviolet-absorbing haze seen at low latitudes. Nitrogen gas and solid P are other by-products of ammonia and phosphine chemistry. Solid phosphorus is sometimes red and has been proposed as the constituent responsible for the red color of Jupiter's Great Red Spot. That suggestion (one of several) has not been confirmed, and neither N_2H_4 nor P_2H_4 has been observed spectroscopically.

Organic compounds derived from dissociation of methane are present in the stratospheres of all the giant planets. The photochemical cycle leading to stable C_2H_2 (acetylene), C_2H_4 (ethylene), C_2H_6 (ethane), and C_4H_2 (diacetylene) is shown schematically in Fig. 2. The chain may progress further to produce polyacetylenes ($C_{2n}H_2$). These species form condensate haze layers in the cold stratospheres of Uranus and Neptune. More complex hydrocarbon species (C_3H_8, C_3H_4) are observed in Jupiter's atmosphere primarily in close proximity to high-latitude regions, where auroral heating is significant. The abundant polar aerosols in the atmospheres of Jupiter and Saturn

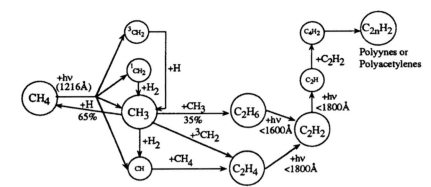

FIGURE 2   Summary of CH$_4$ (methane) photochemical processes in the stratospheres of the giant planets. Photodissociation by ultraviolet light is indicated by $+h\nu$ at the indicated wavelength. Methane photodissociation is the starting point in the production of a host of other hydrocarbons. (Revised by S. K. Atreya from Fig. 5-3 from J. B. Pollack and S. K. Atreya, 1992, in "Exobiology in Solar System Exploration" (G. Carle et al., eds.), NASA-SP 512, pp. 82–101.)

may owe their existence to the ions created by auroras in the upper atmosphere.

As instruments become more sensitive, new species are detected. These include C_2H_4, C_3H_4, and C_6H_6 in the atmospheres of Jupiter and Saturn, and C_3H_8 for Saturn. The methyl radical CH_3 (an unstable transition molecule in the reaction chain) has been detected on Jupiter, Saturn, and Neptune.

Hydrogen cyanide (HCN) is present in the stratospheres of Jupiter and Neptune, but for two very different reasons. On Jupiter, HCN was emplaced high in the stratosphere as a result of the 1994 impacts of comet Shoemaker–Levy 9. During the 3 years after the impacts, it was observed to spread north of the impact latitude (near 45°S), eventually to be globally distributed. It is expected to dissipate over the span of a decade or so. Cometary impact may also be responsible for HCN in Neptune's stratosphere.

Quantitative thermochemical and photochemical models are available for many of the observed constituents and provide predictions for many others that are not yet observed. These models solve a set of coupled equations that describe the balance between the abundances of species that interact and include important physical processes such as ultraviolet **photolysis**, condensation/sublimation, and vertical transport. Current models heuristically lump all the transport processes into an effective eddy mixing coefficient, and the value of that coefficient is derived as part of the solution of the set of equations. As we gain more detailed observations and more comprehensive laboratory measurements of reaction rates, we will be able to develop more sophisticated models. Some models are beginning to incorporate transport by vertical and horizontal winds. Figures 3 and 4 show vertical profiles calculated from models for a number of photochemically produced species.

3. Clouds and Aerosols

The appearance of the giant planets is determined by the distribution and optical properties of cloud and aerosol haze particles in the upper troposphere and stratosphere.

Cameras on the *Voyager* spacecraft provided detailed views of all the giant planets, whose general appearances can be compared in Fig. 5. Their atmospheres show a banded structure (which is difficult to see on Uranus) of color and shading parallel to latitude lines. These were historically named belts and zones on Jupiter and Saturn, with belts being relatively dark and zones relatively bright. Specific belts and zones were named in accordance with their approximate latitudinal location (Equatorial Belt, North and South Tropical Zones near latitudes ±20°, North and South Temperate Zones and Belts near ±35°, and polar regions).

The nomenclature should not be construed to mean that low latitudes are relatively warmer than high latitudes, as they are on Earth and Mars. Nor is it true that the reflectivities of these features remain constant with time. Some features on Jupiter, such as the North and South

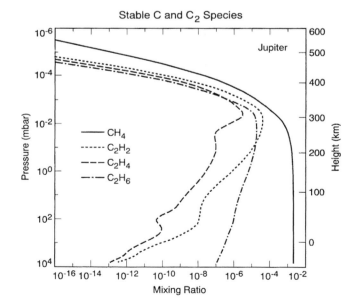

FIGURE 3   Vertical profiles of some photochemical species in Jupiter's stratosphere. The mixing ratios (horizontal axis) are plotted as a function of pressure. (From G. R. Gladstone et al., 1996, *Icarus* **119**, 1–52. Copyright Academic Press.)

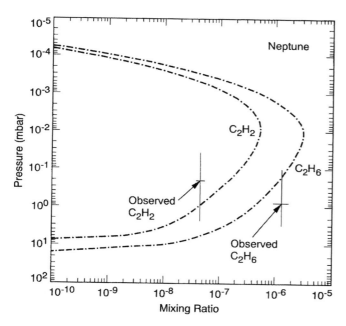

FIGURE 4 Vertical profiles of photochemical species in the Neptunian stratosphere. (From P. Romani et al., 1993, *Icarus* **106**, 442–462. Copyright by Academic Press.)

Tropical Zones, are persistently bright, whereas others, like the South Equatorial Belt, are sometimes bright and sometimes dark. On Jupiter, there is a correlation between visible albedo and temperature, such that bright zones are

usually cool regions and dark belts are usually warm near the tropopause. Cool temperatures are associated with adiabatic cooling of upwelling gas, and the correlation of cool temperatures with bright clouds points to enhanced condensation of ice particles as condensable gases flow upward and cool. This correlation does not hold completely on Jupiter and almost not at all on the other giant planets. The mechanisms responsible for producing reflectivity contrasts and color remain largely mysterious, although a number of proposals have been advanced. These will be discussed in more detail.

Our understanding of aerosols and clouds is rooted in thermochemical equilibrium models that predict the temperature (and hence pressure and altitude) of the bases of condensate clouds. The cloud base occurs where the vapor pressure of a condensable gas equals its partial pressure. Model predictions for the four giant planets are shown in Fig. 6. The deepest cloud to form is a solution of water and ammonia on Jupiter and Saturn, with dissolved H_2S as well on Uranus and Neptune. At higher altitudes, an ammonium hydrosulfide cloud forms, and its mass depends on both the amounts of H_2S and NH_3 available and the ratio of S to N. At still higher altitudes, an ammonia or hydrogen sulfide cloud can form if the S/N ratio is less than or greater than 1, respectively. If the ratio is greater than 1, all the N will be taken up as NH_4SH, with the remaining sulfur available to condense at higher altitudes. This seems to be the situation on Uranus and Neptune, but the reverse is true for Jupiter and Saturn. Only the atmospheres of Uranus

FIGURE 5 *Voyager* images of Jupiter, Saturn, Uranus, and Neptune, scaled to their relative sizes. Earth and Venus are also sown scaled to their relative sizes.

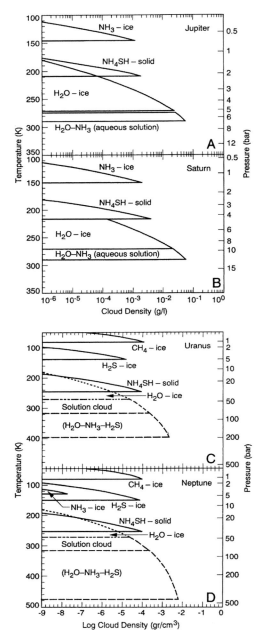

FIGURE 6 The diagrams in the four panels show the locations of condensate cloud layers on Jupiter, Saturn, Uranus, and Neptune. These figures indicate how much cloud material would condense at various temperatures (corresponding to altitude) if there were no advective motions in the atmosphere to move vapor and clouds. They are based on simple thermochemical equilibrium calculations, which assume, for Jupiter and Saturn, that the condensable species have mixing ratios equal to those for a solar composition atmosphere. (Figures for Jupiter and Saturn were constructed from models by S. K. Atreya and M. Wong, based on S. K. Atreya and P. N. Romani, 1985, in "Planetary Meteorology" (G. E. Hunt, ed.), pp. 17–68, Cambridge Univ. Press, Cambridge, United Kingdom. Those for Uranus and Neptune were first published by I. de Pater et al., 1991, *Icarus* **91**, 220–233. Copyright by Academic Press.)

and Neptune are cold enough to condense methane, which occurs at 1.3 bar in Uranus and about 2 bar in Neptune. It is predominantly the uppermost clouds that we see at visible wavelengths.

Observational evidence to support the cloud stratigraphy shown in Fig. 6 is mixed. The *Galileo* probe detected cloud particles near 1.6-bar pressure and sensed cloud opacity at higher altitudes corresponding to the ammonia cloud. With data only from remote-sensing experiments, it is difficult to probe to levels below the top cloud, and the evidence we have for deeper clouds comes from careful analyses of radio occultations and of gaseous absorption lines in the visible and near infrared, and from thermal emission at 5, 8.5, and 45 μm. Contrary to expectation, spectra of the planets show features due to ice in only a small fraction of the cloudy area. The *Voyager* radio occultation data showed strong refractivity gradients at locations predicted for methane ice clouds on Uranus and Neptune, essentially confirming their existence and providing accurate information on the altitude of the cloud base. Ammonia gas is observed spectroscopically in Jupiter's upper troposphere, and its abundance decreases with altitude above its cloud base in accordance with expectation. There is no doubt that ammonia ice is the major component of the visible clouds on Jupiter and Saturn, but it cannot be the only component and is not responsible for the colors (pure ammonia ice is white). In fact, all the ices shown in Fig. 6 are white at visible wavelengths. The colored material must be produced by some disequilibrium process like photochemistry or bombardment by energetic particles from the magnetosphere.

Colors on Jupiter are close to white in the brightest zones, gray yellow to light brown in the belts, and orange or red in some of the spots. The colors in Fig. 5 are slightly and unintentionally exaggerated owing to the difficulty of achieving accurate color reproduction on the printed page. Colors on Saturn are more subdued. Uranus and Neptune are gray-green. Neptune has a number of dark spots and white patchy clouds. Part of the green tint on Uranus and Neptune is caused by strong methane gas absorption at red wavelengths, and part is due to aerosols that also absorb preferentially at wavelengths longer than 0.6 μm.

Candidate materials for the **chromophore** material in outer planet atmospheres are summarized in Table 3. All candidate materials are thought to form by some nonequilibrium process such as photolysis or decomposition by protons or ions in auroras, which acts on methane, ammonia, or ammonium hydrosulfide. Methane is present in the stratospheres of all the giant planets. Ammonia is present in the stratosphere of Jupiter. Ammonium hydrosulfide is thought to reside near the 2-bar level and deeper in Jupiter's atmosphere, which is too deep for ultraviolet photons to penetrate.

There are two major problems in understanding which, if any, of the proposed candidate chromophores are

TABLE 3	Candidate Chromophore Materials in the Atmospheres of the Giant Planets

Material	Formation mechanism
Sulfur	Photochemical products of H_2S and NH_4SH. Red allotropes are unstable.
H_2S_x, $(NH_4)_2S_x$, $N_2H_4S_x$	Photochemical products of H_2S and NH_4SH.
N_2H_4	Hydrazine, a photochemical product of ammonia, a candidate for Jupiter's stratospheric haze.
Phosphorus (P_4)	Photochemical product of PH_3.
P_2H_4	Diphosphine, a photochemical product of phosphine, a candidate for Saturn's stratospheric haze.
Products of photo- or charged-particle decomposition of CH_4	Includes acetylene photopolymers (C_xH_2), proton-irradiated methane, and organics with some nitrogen and/or sulfur. Confined to stratospheric levels where ultraviolet photons and auroral protons or ions penetrate.

responsible for the observed colors. First, no features have been identified in spectra of the planets that uniquely identify a single candidate material. Spectra show broad slopes, with more absorption at blue wavelengths on Jupiter and Saturn and at red wavelengths on Uranus and Neptune. All the candidates listed in Table 3 produce broad blue absorption. None of them can account for the red and near-infrared absorption in the spectra of Uranus and Neptune. Second, our understanding of the detailed processes that lead to the formation of chromophores is inadequate. Gas-phase photochemical theory cannot account for the abundance of chromophore material. It is likely that ultraviolet photons or charged-particle bombardment of solid, initially colorless particles like acetylene and ethane ice in the stratospheres of Uranus and Neptune or ammonium hydrosulfide in Jupiter's atmosphere breaks chemical bonds in the solid state, paving a path to formation of more complex hydrocarbons or inorganic materials that seem to be required. Additional laboratory studies are needed to address these questions. [See THE SOLAR SYSTEM AT ULTRAVIOLET WAVELENGTHS.]

Haze particles are present in the stratospheres of all the giant planets, but their chemical and physical properties and spatial distributions are quite different. Jupiter and Saturn have ultraviolet (UV)-absorbing aerosols abundant at high latitudes and high altitudes (corresponding to pressures ranging from a fraction of a millibar to a few tens of millibars). The stratospheric aerosols on Uranus and Neptune do not absorb much in the UV and are not concentrated at high latitude. The polar concentration of UV-absorbing aerosols on Jupiter and Saturn suggests that their formation may be due to chemistry in auroral regions, where protons and/or ions from the magnetosphere penetrate the upper atmosphere and deposit energy. Association with auroral processes may help explain why UV absorbers

are abundant poleward of about 70° latitude on Saturn, extend to somewhat lower latitudes on Jupiter, and show a hemispheric asymmetry in Jupiter's atmosphere. Saturn's magnetic dipole is nearly centered and parallel to Saturn's spin axis, but Jupiter's magnetic dipole is both significantly offset and tilted with respect to its spin axis, producing asymmetric auroras at lower latitudes than on Saturn. Other processes, such as the **meridional circulation**, also influence the latitudinal distribution of aerosols, so more work needs to be done to establish the role of auroras in aerosol formation.

Photochemistry is responsible for the formation of diacetylene, acetylene, and ethane hazes in the stratospheres of Uranus and Neptune. The main steps in the life cycle of stratospheric aerosols are shown in Fig. 7. Methane gas mixes upward to the high stratosphere, where it is photolyzed by ultraviolet light. Diacetylene, acetylene, and ethane form from gas-phase photochemistry and diffuse downward. Temperature decreases downward in the stratosphere, so ice particles form when the vapor pressure equals the partial pressure of the gas. On Uranus, diacetylene ice forms at 0.1 mbar, acetylene at 2.5 mbar, and ethane at 14 mbar. The ice particles sediment to deeper levels on a time scale of years and evaporate in the upper troposphere at 600 mbar and deeper. Polymers that form from solid-state photochemistry in the ice particles are probably responsible for the little ultraviolet absorption that does occur. They are less volatile than the pure ices and probably mix down to the methane cloud and below.

Photochemical models predict formation of hydrazine in Jupiter's stratosphere and diphosphine in Saturn's atmosphere. If these are the only stratospheric haze constituents, it is not apparent why the ultraviolet absorbers are concentrated at high latitude. As discussed earlier, auroral bombardment of methane provides an attractive candidate

URANUS' STRATOSPHERIC AEROSOL CYCLE

PRESSURE	PROCESS	TRANSPORT
mb		
0.05	CH$_4$ PHOTOLYSIS	
⇩		EDDY DIFFUSION, IN SITU CONDENSATION (C$_4$H$_2$)
0.10	C$_4$H$_2$ ⎤	UV PHOTOLYSIS
2.50	C$_2$H$_2$ ⎥ ICES	TO VISIBLE
14.0	C$_2$H$_6$ ⎦	ABSORBING POLYMERS
⇩		SEDIMENTATION
600	C$_2$H$_6$ EVAPORATES	
900	C$_2$H$_2$ EVAPORATES	
900-1300	CH$_4$ CLOUD	
~3000	C$_4$H$_2$ EVAPORATES	
?	POLYMERS EVAPORATE	

FIGURE 7   Life cycle for stratospheric aerosols on Uranus. (From J. Pollack et al., 1987, *J. Geophys. Res.* **92**, 15,037–15,066. Copyright American Geophysical Union.)

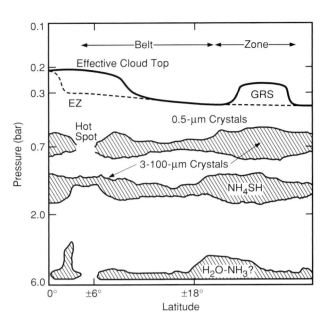

FIGURE 8   Observations of Jupiter at wavelengths that sense clouds lead to a picture of the jovian cloud stratigraphy shown here. There has been no direct evidence for a water–ammonia cloud near the 6 bar pressure level, but it is likely that such a cloud exists from indirect evidence. The hot spots are named from their visual appearance at a wavelength of 5 μm. They are not physically much warmer than their surroundings, but they are deficient in cloudy material (see Fig. 9). (From R. West et al., 2004, in "Jupiter: The Planet, Satellites and Magnetosphere" (F. Bagenal, T. Dowling, and W. McKinnon, eds.), pp. 79–104, Cambridge Univ. Press, Cambridge, United Kingdom.)

process for the abundant high-latitude aerosols on Jupiter and Saturn. However, we do not know enough to formulate a detailed chemical model of this process.

Thermochemical equilibrium theory serves as a guide to the location of the bases of tropospheric clouds, but meteorology and cloud microphysical processes determine the vertical and horizontal distribution of cloud material. These processes are too complex to let us predict to what altitudes clouds should extend, and so we must rely on observations. Several diagnostics are available to measure cloud and haze vertical locations. At short wavelengths, gas molecules limit the depth to which we can see. In the visible and near infrared are methane and hydrogen absorption bands, which can be used to probe a variety of depths depending on the absorption coefficient of the gas. There are a few window regions in the thermal infrared where cloud opacity determines the outgoing radiance. The deepest probing wavelength is 5 μm. At that wavelength, thermal emission from the water-cloud region near the 5 bar pressure level provides sounding for all the main clouds in Jupiter's atmosphere. [*See* INFRARED VIEWS OF THE SOLAR SYSTEM FROM SPACE.]

The results of cloud stratigraphy studies for Jupiter's atmosphere are summarized in Fig. 8. There is spectroscopic evidence for the two highest tropospheric layers in Jupiter's atmosphere. There is also considerable controversy surrounding the existence of the water-ammonia cloud on Jupiter. The *Galileo* probe descended into a dry region of

the atmosphere and did not find a water cloud, but water clouds may be present in moister regions of the atmosphere that are obscured by overlying clouds. There is evidence for a large range of particle sizes. Small particles (less than about 1 μm radius) provide most of the cloud opacity in the visible. They cover belts and zones, although their optical thickness in belts is sometimes less than in zones. Most of the contrast between belts and zones in the visible comes from enhanced abundance or greater visibility of chromophore material, which seems to be vertically, but not horizontally, well mixed in the ammonia cloud. The top of this small-particle layer extends up to about 200 mbar, depending on latitude. Jupiter's Great Red Spot is a location of relatively high-altitude aerosols, consistent with the idea that it is a region of upwelling gas.

Larger particles (mean radius near 6 μm) are also present, mostly in zones. This large-particle component appears to respond to rapid changes in the meteorology. It is highly variable in space and time and is responsible, together with the deeper clouds, for the richly textured appearance of the planet at 5 μm wavelength (Fig. 9). Some of the brightest regions seen in Fig. 9 are called 5 μm hot

FIGURE 9 At a wavelength of 5 μm, most of the light from Jupiter is thermal radiation emitted near the 6 bar pressure level below the visible cloud. Places where the clouds are thin permit the deep radiation to escape from space, making these regions appear bright. Thicker clouds block the radiation and these appear dark. Jupiter's Great Red Spot is the dark oval just below the center. This image was taken with the NASA Infrared Telescope Facility. (Courtesy of J. Spencer.)

spots, not because they are warmer than their surroundings but because thermal radiation from the 5 bar region emerges with little attenuation from higher clouds. The *Galileo* probe sampled one of these regions. The dark regions in the image are caused by optically thick clouds in the NH_4SH and NH_3 cloud regions. The thickest clouds are generally associated with upwelling, bright (at visible wavelengths) zones, but many exceptions to this rule are observed. Until we understand the chemistry and physics of chromophores, we should not expect to understand why or how well albedo is correlated with other meteorological parameters.

Most of Jupiter's spots are at nearly the same altitude. Some notable exceptions are the Great Red Spot (GRS), the three white ovals just south of the GRS, and some smaller ovals at other latitudes. These anticyclonic features extend to higher altitudes, probably up to the 200 mbar level, compared to a pressure level of about 300 mbar for the surrounding clouds. Some of the anticyclonic spots have remarkably long lifetimes compared to the terrestrial norm. The GRS was recorded in drawings in 1879, and reports of red spots extend back to the 17th century. The three white ovals in a latitude band south of the GRS formed from a bright cloud band that split into three segments in 1939. The segments shrunk in longitude over the course of a year, until the region (the South Temperate Belt) was mostly dark except for three high-albedo spots that remain to the

present. Whereas anticyclonic ovals tend to be stable and long-lived, cyclonic regions constantly change.

Similar features are observed in Saturn's atmosphere, although the color is much subdued compared to Jupiter, and Saturn has nothing that is as large or as long-lived as the GRS. The reduced contrast may be related to Saturn's colder tropopause temperature. The distance between the base of the ammonia cloud and the top of the troposphere (where the atmosphere becomes stable against convection) is greater on Saturn than on Jupiter. The ammonia-ice cloud on Saturn is both physically and optically thicker than it is on Jupiter. Occasionally (about two or three times each century), a large, bright cloud forms near Saturn's equator. One well-observed event occurred in 1990, but its cause is unknown. It appears to be a parcel of gas that erupts from deeper levels, bringing fresh condensate material to near the top of the troposphere. It becomes sheared out in the wind shear and dissipates over the course of a year.

Uranus as seen by *Voyager* was even more bland than Saturn, but recent images from the *Hubble Space Telescope* and from the ground show a much richer population of small clouds (see Fig. 10). Midlatitude regions on Uranus and Neptune are cool near the tropopause, indicating upwelling. But cloud optical thickness may be lower there than at other latitudes. The relation between cloud optical thickness and vertical motion is more complicated than the simple condensation model would predict.

FIGURE 10   Images of Uranus (left) and Neptune (right) taken in 2004 and 2000, respectively. Both were obtained at the Keck telescope with filters in the near-infrared. Many cloud features that were not seen during the *Voyager* flyby can be seen. The Uranus ring can also be seen (a red ellipse in this false-color representation). The Uranus image appeared on the cover of *Icarus* (December 15, 2005, issue) and was provided by L. Sromovsky. The Neptune image is from I. de Pater et al. (2005, *Icarus* **174**, 263–373. Copyright Academic Press).

Neptune's clouds are unique among the outer planet atmospheres. *Voyager* observed four large cloud features that persisted for the duration of the *Voyager* observations (months). The largest of these is the Great Dark Spot (GDS) and its white companion. Because of its size and shape, the GDS might be similar to Jupiter's Great Red Spot, but the GDS had a short life compared to the GRS.

There is no explanation yet of what makes the dark spot dark. The deepest cloud (near the 3 bar level) is probably H_2S ice, since ammonia is apparently depleted and NH_4SH would be sequestered at a deeper level. At higher altitudes there is an optically thin methane haze (near 2 bar) and stratospheric hazes of ethane, acetylene, and diacetylene. At high spatial resolution, the wispy white clouds associated with the companion to the GDS and found elsewhere on the planet form and dissipate in a matter of hours. It was difficult to estimate winds from these features because of their transitory nature. Individual wisps moved at a different speed than the GDS and its companion, suggesting that these features form and then evaporate high above the GDS as they pass through a local pressure anomaly, perhaps a standing wave caused by flow around the GDS. Cloud shadows were seen in some places, a surprise after none was seen on the other giant planets. The clouds casting the shadows are about 100 km higher than the lower cloud deck, suggesting that the lower cloud is near 3 bar and the shadowing clouds near 1 bar, in the methane condensation region. More recent Hubble and ground-based images show clouds not seen in *Voyager* images (Fig. 10).

4. Dynamical Meteorology of the Troposphere and Stratosphere

Our understanding of giant planet meteorology comes mostly from *Voyager* observations, with observations from *Galileo*, *Cassini*, the *Hubble Space Telescope*, and ground-based data adding to the picture. Although we have theories and models for many of the dynamical features, the fundamental nature of the dynamical meteorology on the giant planets remains puzzling chiefly because of our inability to probe to depths greater than a few bars in atmospheres that go to kilobar pressures and because of limitations in spatial and time sampling, which may improve with future missions to the planets.

Thermodynamic properties of atmospheres are at the heart of a variety of meteorological phenomena. In the terrestrial atmosphere, condensation, evaporation, and transport of water redistribute energy in the form of latent heat. The same is true for the outer planet atmospheres, where condensation of water, ammonia, ammonium hydrosulfide, hydrogen sulfide, and methane takes place. Condensables also influence the dynamics through their effects on density gradients. In the terrestrial atmosphere, moist air is less dense than dry air at the same temperature because the molecular weight of water vapor is smaller than that of the dry air. Because of this fact, and also because moist air condenses and releases latent heat as it rises, there can be a growing instability leading to the formation of convective plumes, thunderstorms, and anvil clouds at high altitudes. On the giant planets, water vapor is significantly heavier than the dry atmosphere and so the same type of instability will not occur unless a strongly upwelling parcel is already present. Some researchers proposed that the Equatorial Plumes on Jupiter and the elongated clouds on Uranus are the outer planet analogs to terrestrial anvil clouds.

Terrestrial lightning occurs most frequently over tropical oceans and over a fraction of the land surface. Its distribution in latitude, longitude, and season is indicative of certain properties of the atmosphere, especially the availability of liquid water. Lightning has been observed on the giant planets as well, either from imaging on the night side (Jupiter) or from signals recorded by plasma wave instruments. A somewhat mysterious radio emission from Saturn (the so-called Saturn Electrostatic Discharge events) has been interpreted as a lightning signature. Combined imaging and plasma wave observations from *Cassini* in 2004 revealed a large cloud complex associated with this source. The intensity and size of the lightning spots in the images imply that they are much more energetic than the average lightning bolt in the terrestrial atmosphere, and they occur in the water–ammonia cloud region as expected. The *Galileo* probe did not detect lightning in Jupiter's atmosphere within a range of about 10,000 km from its location at latitude 6.5°N. [*See* THE SOLAR SYSTEM AT RADIO WAVELENGTHS.]

The heat capacity of hydrogen, and therefore the dry adiabatic lapse rate of the convective part of the atmosphere, depends on the degree to which the ortho/para states equilibrate. The lapse rate is steepest when equilibration is operative. The observed lapse rate for Uranus, as measured by the *Voyager* radio occultation experiment, is close to the "frozen" lapse rate—the rate when the relative fractions of ortho and para hydrogen are fixed. How can the observed relative fractions be near equilibrium when the lapse rate points to nonequilibrium? One suggestion is that the atmosphere is layered. Each layer is separated from the next by an interface that is stable and that is thin compared to the layer thickness. The air within each layer mixes rapidly compared to the time for equilibration, but the exchange rate between layers is slow or comparable to the timescale for conversion of ortho to para and back.

How can layers be maintained in a convective atmosphere? In the terrestrial ocean, two factors influence buoyancy: temperature and salinity. If the water is warmer at depth, or if the convective amplitude is large, the different timescales for diffusion of heat and salinity lead to layering. In the atmospheres of the outer planets, the higher molecular weight of condensables acts much as salinity in ocean water. Layering can be established even without molecular weight gradients. Layering in the terrestrial stratosphere and mesosphere has been observed. Layers of rapidly convecting gas occur where gravity waves break or where other types of wave instabilities dump energy. Between layers of rapid stirring are stably stratified layers with transport by diffusion rather than convection.

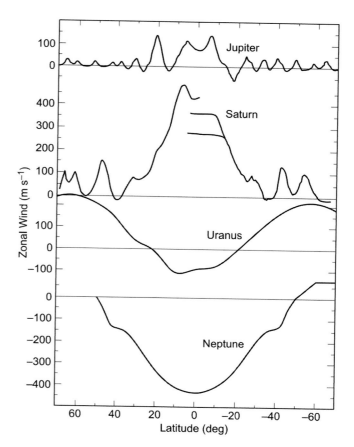

FIGURE 11 Zonal (east–west) wind velocity for the giant planets as a function of latitude. For Jupiter, the data are from Porco et al. (2003, *Science* **299**, 1541–1547. Copyright American Association for the Advancement of Science). For Saturn's northern hemisphere, the data are from P. Gierasch and B. Conrath (1993, *J. Geophys. Res.* **98**, 5459–5469. Copyright American Geophysical Union). For Saturn's southern hemisphere, data are from Porco et al. (2005, *Science* **307**, 1243–1247. Copyright American Association for the Advancement of Science). Two branches are shown for the southern low latitudes. Both are from *Cassini* observations, with similar values from *Hubble Space Telescope* images. The higher wind speeds were observed for deepest clouds, while the lower winds were observed for higher clouds. Both branches are moving more slowly than clouds at similar latitudes in the north observed by *Voyager*. This apparent change in the wind speed must have involved a large energy exchange. Data for Uranus and Neptune are mostly from analyses of Hubble and Keck data (L. Sromovsky and P. Fry, 2005, *Icarus* **179**, 459–484. Copyright Academic Press. L. Sromovsky et al., 2001, *Icarus* **150**, 244–260. Copyright Academic Press.)

Some of the variety of the giant planet meteorology, as well as our difficulty to understand it, is nicely illustrated by observations of the wind field at the cloud tops. Wind vectors of all the giant planet atmospheres are predominantly in the east–west (zonal) direction (Fig. 11). These are determined by tracking visible cloud features over hours, days, and months. Jupiter has an abundance of small features and the zonal winds are well mapped. Saturn has fewer features, and they are of less contrast than those on Jupiter, but there is still a large enough number to provide detail in the wind field. Only a few features were seen in *Voyager* images of the Uranus atmosphere, and all but one of these were between latitudes 20°S and 40°S. More recent images from the *Hubble Space Telescope* show new features at many other latitudes. The *Voyager 2* radio occultation provided an additional estimate for wind speed at the equator. Neptune has more visible features than Uranus, but most of them are transitory and difficult to follow long enough to gauge wind speed.

Figure 11 reveals a great diversity in the zonal flow among the giant planet atmospheres. Wind speed is relative to the rotation rate of the deep interior as revealed by the magnetic field and radio emissions. Jupiter has a series of jets that oscillate with latitude and are greatest in the prograde direction at latitude 23°N, and near ±10°. The pattern of east–west winds is approximately symmetric about the equator except at high latitude. Saturn has a very strong prograde jet at low latitudes (within the region ±15°). It also has alternating but mostly prograde jets at higher latitudes, with the scale of latitudinal variation being about 10°. Uranus appears to have a single prograde maximum near 60°S, and the equatorial region is retrograde. Neptune has an enormous differential rotation, mostly retrograde except at high latitude. Various theories have been advanced to explain the pattern of zonal jets. None of them can account for the great variety among the four planets.

The zonal jets are stable over long time periods (observations span many decades for Jupiter and Saturn), despite the many small-scale features that evolve with much shorter life times. An interesting exception to this rule occurred at equatorial latitudes on Saturn between the time of the *Voyager* observations (around 1981) and observations in the 1990s and later by the *Hubble Space Telescope* and beginning in 2004 by the *Cassini* cameras. Current equatorial jet speeds are significantly less than those measured on Saturn by *Voyager*. It is difficult to understand how such a large change of momentum could occur, and another explanation has been sought. Possibly the equatorial atmosphere was clearer (less haze) during the *Voyager* epoch, permitting observations to deeper levels where the wind speed is higher. Detailed analyses of haze altitudes show that the haze is thicker and higher in more recent times than it was in 1981, but probably not enough to account for the difference in wind speed.

Some of the key observations that any dynamical theory must address include: (1) the magnitude, direction, and latitudinal scale of the jets; (2) the stability of the jets, at least for Jupiter and Saturn, where observations over long periods show little or no change except for Saturn's equatorial jet, which was mentioned earlier; (3) the magnitude and latitu-

dinal gradients of heat flux; and (4) the interactions of the mean zonal flow with small spots and eddies. One of the controversies during the past two decades concerns how deep the flow extends into the atmosphere. It is possible to construct shallow-atmosphere models that have approximately correct jet scales and magnitudes. A shallow-atmosphere model is one in which the jets extend to relatively shallow levels (100 bar or less), and the deeper interior rotates as a solid body, or at least as one whose latitudinal wind shear is not correlated with the wind shear of the jets. The facts that the jets and some spots on Jupiter are very stable, that there is approximate hemispheric symmetry in the zonal wind pattern between latitudes ±60°, and that the Jovian interior has no density discontinuities down to kilobar levels suggested to some investigators that the jets extend deep into the atmosphere. A natural architecture for the flow in a rotating sphere with no density discontinuities is one in which the flow is organized on rotating cylinders (Fig. 12).

Apart from the stability and symmetry noted here, there is little evidence to suggest that the zonal wind pattern really does extend to the deep interior. The conductivity of Jupiter's atmosphere at depth is probably too high to allow the type of structure depicted in Fig. 12 to exist. The strength of the zonal jet at the location where the *Galileo* probe entered (6.5°N) increased with depth, consistent with the idea of a deeply rooted zonal wind field on Jupiter. One way to test that hypothesis is to make highly precise measurements of the gravity field close to the planet. There are density gradients associated with the winds, and these produce features in the gravity field close to the planet. The largest signature is produced by Neptune's remarkable differential rotation. The *Voyager 2* spacecraft flew just above Neptune's atmosphere and provided the first evidence that the differential rotation cannot extend deep into the atmosphere. Gravity-field tests of the deep-wind hypothesis for the other giant planets are more difficult because the differential rotation is much weaker. No spacecraft have come close enough to make the measurements but one is planned for Jupiter

What process maintains the zonal wind pattern? *Voyager* measurements shed some light on this question, but provided some puzzles as well. The ultimate energy source for maintaining atmospheric motions is the combination of internal thermal and solar energy absorbed by the atmosphere. Jupiter, Saturn, and Neptune all have significant internal energy sources, whereas Uranus has little or none. A measure of the amount of energy available for driving winds is the escaping radiative energy per square meter of surface area. Twenty times as much energy per unit area is radiated from Jupiter's atmosphere as from Neptune's, yet the wind speeds (measured relative to the interior as determined from the magnetic field rotation rate) on Neptune are about three times higher than those on Jupiter. Rather than driving zonal winds, the excess internal energy may go

(a)
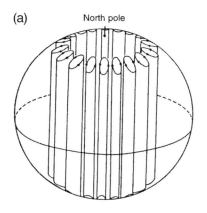

(b)
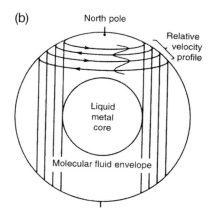

FIGURE 12   One model for the zonal wind fields of the giant planets has differential rotation organized on cylinders (a), exploiting the natural symmetry of a rotating deep fluid (b). (From F. Busse, 1976, *Icarus* **29**, 255–260. Copyright by Academic Press.)

into driving smaller scale eddies, which are most abundant to Jupiter.

What influence does the absorbed solar radiation have? Most planets receive more solar radiation at their equator than at their poles. For Uranus, the reverse is true. Yet the upper tropospheric and stratospheric temperatures on Uranus and Neptune are nearly identical, and the winds for both planets (as for Earth) are retrograde at the equator. According to one theory, deposition of solar energy may account for the fact that Uranus possesses very little internal energy today. Otherwise, it is hard to see how solar energy can be important for the tropospheric circulation of the giant planets.

What role do eddies have in maintaining the flow? Measurements of the small spots on Jupiter and Saturn have allowed an estimate of the energy flow between the mean zonal wind and the eddy motions. For Jupiter, the eddies at the cloud top appear to be pumping energy into the mean zonal flow, although that conclusion has been challenged on the grounds that the sampling may be biased. If further observation and analysis confirm the initial result, we need to explain why the jets are so stable when there is apparently enough energy in eddy motions to significantly modify the jovian wind field. At the same time, other observations imply dissipation and decay of zonal winds at altitudes just above the cloud tops.

The relationship known to atmospheric physicists as the thermal wind equation provides a means of estimating the rate of change of zonal wind with height (which is usually impossible to measure remotely) from observations of the latitudinal gradient of temperature (which is usually easy to measure). One of the common features of all the outer planet atmospheres is a decay of zonal wind with height in the stratosphere, tending toward solid-body rotation at high altitudes. The decay of wind velocity with height could be driven by eddy motions or by gravity wave breaking, which effectively acts as friction on the zonal flow.

Thermal contrasts on Jupiter are correlated with the horizontal shear and with cloud opacity as indicated by 5 μm

images (see Fig. 9). Cool temperatures at the tropopause level (near 100 mbar) are associated with upwelling and anticyclonic motion, and warmer temperatures are associated with subsidence. Jupiter's Great Red Spot is an anticyclonic oval with cool tropospheric temperatures, upwelling flow, and aerosols extending to relatively high altitudes. Enhanced cloud opacity and ammonia abundance in cooler anticyclonic latitudes (mostly the high-albedo zones on Jupiter) are predicted in upwelling regions. The correlation is best with cloud opacity in the 5 μm region. At shorter wavelengths (in the visible and near infrared), there is a weaker correlation between cloud opacity and **vorticity**. Perhaps the small aerosols near the top of the troposphere, sensed by the shorter wavelengths but not at 5 μm, are transported horizontally from zone to belt on a time scale that is short compared to their rainout time (several months).

The transport of heat may well be more complicated than the previous paragraph implies. There may be at least two regions, an upper troposphere where heat transport is determined by slow, large-scale motions as previously depicted, and a lower troposphere at pressures between 2 and 10 bar, where heat is transported upward mostly in the belts, by small convective storms which are seen in the belts. There is evidence from the *Galileo* and *Cassini* observations that this is the case.

The upwelling/subsidence pattern at the jet scale in the upper troposphere penetrates into the lower stratosphere. We have relatively little information on the stratospheric circulation for the giant planets. Most of it is based on the observed thermal contrasts and the idea that friction is a dominant driver for stratospheric dynamics. We are beginning to appreciate the role of forcing by gravity or other dissipative waves. A model for the Uranus stratospheric circulation is based on the frictional damping and the observed thermal contrast as a function of latitude. The coldest temperatures in the lower stratosphere are at midlatitudes, indicating upwelling there and subsidence at the equator and poles. A different pattern is expected if the deposition of solar energy controlled the circulation. Momentum forcing by vertically propagating waves from the deeper atmosphere is apparently more important than solar energy deposition.

The mean meridional circulation in Jupiter's stratosphere differs from that predicted by the frictional damping model at pressure levels less than about 80 mbar. The zonal pattern of upwelling/sinking extends to about 100 mbar, giving way at higher altitude to a two-cell structure with cross-equatorial flow. There is also a hemispheric asymmetry. The high latitudes (poleward of 60°S and 40°N) are regions of sinking motion at the tropopause. Recent analysis of images from the *Hubble Space Telescope* indicate that the optical depth of the ammonia cloud decreases rapidly with latitude poleward of 60°S and 40°N and is well correlated with the estimated downward velocity. The descending dry air inhibits cloud formation. To produce that circulation, there must be momentum forcing in the latitude range 40°S to 80°S and 30°N to 80°N at pressures between 2 and 8 mbar. Dissipation of gravity waves propagating from the deep interior is the most likely source of momentum forcing.

Superimposed on the long-term mean are much faster processes such as horizontal eddy mixing, which can transport material in the north–south direction in days or weeks. The impacts of comet Shoemaker–Levy 9 on Jupiter in 1994 provided a rare opportunity to see the effects of eddy transport on small dust particles and trace chemical constituents deposited in the stratosphere immediately after impact. Particles spread rapidly from the impact latitude (45°S) to latitude 20°S, but there has been almost no transport farther toward the equator. Trace constituents at higher altitude such as HCN were observed to move across the equator into the northern hemisphere. [*See* Physics and Chemistry of Comets.]

Long-term monitoring of the jovian stratosphere has yielded some interesting observations of an oscillating temperature cycle at low latitudes. At pressures between 10 and 20 mbar, the equator and latitudes ±20° cool and warm alternately on timescales of 2–4 years. The equator was relatively (1–2 K above the average 147 K) warm and latitudes ±20° were relatively (1–2 K below average) cool in 1984 and 1990. The reverse was true in 1986 and 1987. Changes in temperature must be accompanied by changes in the wind field, and these must be generated by stresses induced by wave forcing or convection. The similarities of the jovian temperature oscillations to low-latitude temperature oscillations in the terrestrial atmosphere led some researchers to propose that the responsible mechanism is similar to that driving the quasi-biennial oscillation (QBO) on Earth: forcing by vertically propagating waves. The period of the oscillation is about 4 (Earth) years and so the phenomenon has been called the quasi-quadrennial oscillation or QQO.

The *Voyager* cameras and more recently *Hubble* and ground-based images provided much information about the shapes, motions, colors, and lifetimes of small features in the atmospheres of the giant planets. In terms of the number of features and their contrast, a progression is evident from Jupiter, with thousands of visible spots, to Uranus, with only a few. Neptune has a few large spots that were seen for weeks and an abundance of small ephemeral white patches at a few latitudes. We do not have a good explanation for the contrasts and color because the thermochemical equilibrium ices that form these clouds (NH_3, NH_4SH, H_2O, CH_4, and H_2S) are colorless. We need to know more about the composition, origin, and location of the colored material before we can understand how the contrasts are produced.

Fortunately, it is not necessary to understand how the contrasts are produced to study the meteorology of these features. One of the striking attributes of some of the clouds is their longevity. Jupiter's Great Red Spot has been observed since 1879 and may have existed much earlier. A little to the south of the GRS are three white ovals, each about one third the diameter of the GRS. These formed in 1939–1940, beginning as three very elongated clouds (extending 90° in longitude) and rapidly shrinking in longitude. They survived as three distinct ovals until 1998 when two of them merged. In the year 2000, the remaining two merged, leaving one. There are many smaller, stable ovals at some other jovian latitudes. All these ovals are anticyclonic and reside in anticyclonic shear zones. Because they are anticyclonic features, there is upwelling and associated high and thick clouds, and cool temperatures at the tropopause. Sinking motion takes place in a thin boundary region at the periphery of the clouds. The boundary regions are bright at 5 μm wavelength, consistent with relatively cloud-free regions of sinking. The Great Red Spot as revealed by *Galileo* instruments is actually much more complex, with cyclonic flow and small regions of enhanced 5 μm emission (indicating reduced cloudiness) in its interior.

Another attribute of many of the ovals is the oscillatory nature of their positions and sometimes shape. The most striking example is Neptune's Great Dark Spot, whose aspect ratio (ratio of shortest to longest dimension) varied by more than 20% with a period of about 200 hours, with a corresponding oscillation in orientation angle. Neptune's Dark Spot 2 drifted in latitude and longitude, following a sinusoidal law with amplitude 5° in latitude (between 50°S and 55°S) and 90° (peak to peak) in longitude. Other spots on Neptune and Jupiter, including the GRS, show sinusoidal oscillations in position. The jovian spots largely remain at a fixed mean latitude, but the mean latitude of the GDS on Neptune drifted from 26°S to 17°S during the 5000 hours of observations by the *Voyager 2* camera. Ground-based observations in 1993 did not show a bright region at methane absorption wavelengths in the southern hemisphere, unlike the period during the *Voyager* encounter when the high-altitude white companion clouds were visible from Earth. The GDS may have drifted to the northern hemisphere and/or may have disappeared. *Hubble Space Telescope* images and ground-based images since the *Voyager* encounter show new spots at new latitudes.

Jupiter's Great Red Spot is often and incorrectly said to be the jovian analog of a terrestrial hurricane. Hurricanes are cyclonic vortices. The GRS and other stable ovals are anticyclones. Hurricanes owe their (relatively brief) stability to energy generated from latent heating (condensation) over a warm ocean surface, where water vapor is abundant. Upwelling occurs in a broad circular region, and subsidence is confined to a narrow core (the eye). The opposite is true for anticyclonic spots in the giant planet atmospheres, where subsidence takes pace in a narrow ring on the perimeter of the oval. The key to their stability is the long-lived, deep-seated background latitudinal shear of the jets. The stable shear in the jets provides an environment that is able to support the local vortices. Latent heat, so important for a terrestrial hurricane, seems to play no role. However, the ephemeral bright small clouds seen in some locations may be places where strong upwelling is reinforced by release of latent heat analogous to a terrestrian thunderstorm.

5. Energetic Processes in the High Atmosphere

At low pressure (less than about 50 μbar), the mean free path for collisions becomes sufficiently large that lighter molecules diffusively separate from heavier ones. The level where this occurs is called the homopause. The outer planet atmospheres are predominantly composed of H_2 and He, with molecular hydrogen dissociating to atomic hydrogen, which becomes the dominant constituent at the exobase (the level where the hottest atoms can escape to space). This is also the region where solar EUV (extreme ultraviolet) radiation can dissociate molecules and ionize molecules and atoms. Ion chemistry becomes increasingly important at high altitudes. Some reactions can proceed at a rapid rate compared to neutral chemistry. Ion chemistry may be responsible for the abundant UV-absorbing haze particles (probably hydrocarbons) in the polar stratospheres of Jupiter and Saturn.

The high atmospheres of the giant planets are hot (400–800 K for Jupiter to 300 K for Uranus and Neptune), much hotter than predicted on the assumption that EUV radiation is the primary energy source. Estimates prior to the *Voyager* observations predicted high-altitude temperatures closer to 250 K or less. One of the challenges of the post–*Voyager* era is to account for the energy balance of the high atmosphere. Possible sources of energy in addition to EUV radiation include (1) Joule heating, (2) currents induced by a planetary dynamo mechanism, (3) electron precipitation from the magnetosphere (and also proton and S and O ion precipitation in the jovian auroral region), and (4) breaking inertia-gravity waves.

Joule heating requires electric currents in the ionosphere that accelerate electrons and protons. It is a major source of heating in the terrestrial thermosphere. We do not have enough information on the magnetosphere to know how important this process or the others mentioned are for the giant planet atmospheres. The planetary dynamo current theory postulates that currents are established when electrons and ions embedded in the neutral atmosphere move through the magnetic field, forced by the neutral wind tied to the deeper atmosphere. Electric fields aligned with the magnetic field are generated by this motion and accelerate high-energy photoelectrons that collide with neutrals or induce plasma instabilities and dissipate energy. Similar mechanisms are believed to be important in the terrestrial atmosphere.

Electron precipitation in the high atmosphere was one of the first mechanisms proposed to account for bright molecular hydrogen UV emissions. There is recent evidence for supersonic pole-to-equator winds in the very high atmosphere on Jupiter driven by auroral energy. These winds collide at low latitudes, producing supersonic turbulence and heating. Electron and ion precipitation outside of the auroral regions undoubtedly contributes to the heating, but the details remain unclear. The possible contribution from breaking planetary waves is difficult to estimate, but *Galileo* probe measurements, details of the radio occultation profiles, and less direct lines of evidence point to a significant energy density in the form of inertia-gravity waves in the stratosphere and higher. How much of that is dissipated at pressures less than 50 μbar is unknown but could be significant to the energy budget of the high atmosphere.

The giant planets have extensive ionospheres. Like the neutral high atmospheres, they are hotter than predicted prior to the *Voyager* encounters. As for Earth, the ionospheres are highly structured, having a number of high-density layers. Layering in the terrestrial ionosphere is partially due to the deposition of metals from meteor ablation. The same mechanism is thought to be operative in the giant planet ionospheres. The Jupiter and Saturn ionospheres are dominated by the H_3^+ ion, whereas those of Uranus and Neptune are dominated by H^+.

Auroras are present on all the giant planets. Auroras on Earth (the only other planet in the solar system known to have auroras) are caused by energetic charged particles streaming down the high-latitude magnetic field lines. The most intense auroras on Earth occur when a solar flare disturbs the solar wind, producing a transient in the flow that acts on Earth's magnetosphere through ram pressure. As the magnetosphere responds to the solar wind forcing, plasma instabilities in the tail region accelerate particles along the high-latitude field lines.

The configuration of the magnetic field is one of the key parameters that determines the location of auroras. Jupiter's magnetosphere is enormous compared to Earth's. If its magnetosphere could be seen by the naked eye from Earth, it would appear to be the size of the Moon (about 30 arc minutes), whereas Jupiter's diameter is less than 1 arc minute.

TABLE 4	Magnetic Field Parameters (Offset Tilted Dipole Approximation)				
	Earth	**Jupiter**	**Saturn**	**Uranus**	**Neptune**
Tilt (degrees)	11.2	9.4	0.0	58.6	46.9
Offset (planetary radius)	0.076	0.119	0.038	0.352	0.485

To a first approximation, the magnetic fields of Earth and the giant planets can be described as tilted dipoles, offset from the planet center. Table 4 lists the strength, tilt, and radial offsets for each of these planets. Earth and Jupiter have relatively modest tilts and offsets, Saturn has virtually no tilt and almost no offset, whereas Uranus and Neptune have very large tilts and offsets. Such diversity presents a challenge to planetary dynamo modelers. [See PLANETARY MAGNETOSPHERES.]

The mapping of the magnetic fields onto the upper atmosphere determines where auroral particles intercept the atmosphere. Maps for Jupiter, Uranus, and Neptune are shown in Fig. 13, along with locations of field lines connected to the orbits of some satellites that may be important for auroral formation. The configuration for Saturn is not shown because contours of constant magnetic field magnitude are concentric with latitude circles owing to the field symmetry. Because of the large tilts and offsets for Uranus and Neptune, auroras on those planets occur far from the poles.

The jovian aurora is the most intense and has received the most scrutiny. The remainder of this section will focus on what is known about it. It has been observed over a remarkable range of wavelengths, from X-rays to the infrared, and possibly in the radio spectrum as well. Energetic electrons from the magnetosphere dominate the energy input, but protons and S and O ions contribute as well. Sulfur and oxygen k-shell emission seems to be the most plausible explanation for the X-rays. Models of energetic electrons impacting on molecular hydrogen provide a good fit to the observed molecular hydrogen emission spectra. Secondary electrons as well as UV photons are emitted when the primary impacting electrons dissociate the molecules, and these secondaries also contribute to the UV emissions. Some of the UV-emitted radiation is reabsorbed by other hydrogen molecules, and some is absorbed by methane molecules near the top of the homopause. From the detailed shape of the spectrum, it is possible to infer the depth of penetration of electrons into the upper atmosphere. In the near infrared (2–4 μm), emissions from the H_3^+ ion are prominent. Attempts to account for all the observations call for more than one type of precipitating particle and more than one type of aurora.

Ultraviolet auroras from atomic and molecular hydrogen emissions are brightest within an oval that is approximately bounded by the closed field lines connected to the middle magnetosphere (corresponding to a region some 10–30 Jupiter radii from the planet) rather than the orbit of Io or open field lines connected to the tail. Weaker diffuse and highly variable UV emissions appear closer to the pole. They are produced by precipitation of energetic particles originating from more distant regions in the magnetosphere. There is also an auroral hot spot at the location where magnetic field lines passing through Io enter the atmosphere (the Io flux tube footprint). All these features are evident in Fig. 14.

Io is a significant source of sulfur and oxygen, which come off its surface. The satellite and magnetosphere produce hot and cold plasma regions near the Io orbit, which may stimulate plasma instabilities. High spatial resolution, near-infrared H_3^+ images show emission from a region that maps to the last closed field lines far out in the magnetosphere (Fig. 15). This and evidence for auroral response to fluctuations in solar wind ram pressure indicate that at least some of the emission is caused by processes that are familiar to modelers of the terrestrial aurora. [See IO: THE VOLCANIC MOON.]

Auroral emission is strongest over a small range of longitudes. In the north, longitudes near 180°, System III coordinates (which rotate with the magnetic field) show enhanced emission in the UV and also in the thermal infrared. The spectrum of the aurora in the UV resembles electron impact on molecular hydrogen, except the shortest wavelengths are deficient. This deficit can be accounted for if the emission is occurring at some depth in the atmosphere (near 10 μbar) below the region where methane and acetylene absorb UV photons. By contrast, the Uranian high atmosphere is depleted in hydrocarbons and does not produce an emission deficit.

Energy deposition at depth is also required to explain the warm stratospheric temperatures seen in the 7.8 μm methane band. At 10 μbar of pressure, the hot spot region near longitude 180° appears to be 60–140 K warmer than the surrounding region, which is near 160 K. Undoubtedly such temperature contrasts drive the circulation of the high atmosphere. Auroral energy also contributes to anomalous chemistry. An enhancement is seen in acetylene emission in the hot spot region, whereas ethane emission decreases there. A significant part of the acetylene enhancement could be due simply to the higher emission from a warmer

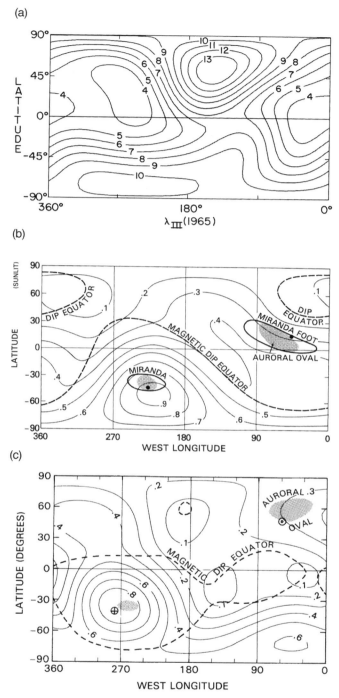

(a)

(b)

(c)

FIGURE 13 (a) Contours of magnetic field magnitude (gauss) on the surface of Jupiter (using the GSFC Model D_4). (b) Contours of constant magnetic field on the upper atmosphere of Uranus, along with the location of the auroral oval and the lines connected to the orbit of the satellite Miranda (Model Q_3). The magnetic dip equator is the location where the field lines are tangent to the surface. (c) Contours of constant magnetic field magnitude and pole locations (circled cross and dot) for Neptune (Model O_8). (From J. Connerney, 1993, *J. Geophys. Res.* **98**, 18,659–18,679. Copyright American Geophysical Union.)

FIGURE 14 (Top) Image of Jupiter at ultraviolet wavelengths taken with the Wide Field and Planetary Camera 2 on the *Hubble Space Telescope*. Bright auroral ovals can be seen against the dark UV-absorbing haze and in the polar regions. Jupiter's north magnetic pole is tilted toward Earth, making it easier to see the northern auroral oval as well as some diffuse emission inside the oval. Small bright spots just outside the oval in both hemispheres are at the location of the magnetic field lines connecting to Io, depicted by a blue curve. Io is dark at UV wavelengths. (Bottom) Image taken a few minutes after the one above in a filter that samples the violet part of the spectrum just within the range that the human eye can detect. The Great Red Spot appears dark at this wavelength and can just be seen in the top image as well. Io's small disk appears here along the blue curve, which traces the magnetic field lines in which it is embedded. (Courtesy of J. Trauger and J. Clarke.)

FIGURE 15 Auroral regions are bright in this image at wavelength 3.4 μm, where the H^{3+} ion emits light. Magnetic field lines connecting to Io and to the 30-Jupiter-radius equator crossing are shown. The brightest emissions are poleward of the 30R_J field line, which means the precipitating particles responsible for this emission come from more distant regions on the magnetosphere. (Reprinted with permission from J. Connerney et al., 1993, *Science* **262**, 1035–1038. Copyright 1993 American Association for the Advancement of Science.)

stratosphere, but a decrease in ethane requires a smaller ethane mole fraction.

Future work on the auroras of Jupiter and the other giant planets will focus on which types of particles are responsible for the emissions, the regions of the magnetosphere or torus from which they originate, the acceleration mechanisms, and how the deposited energy drives circulation and chemistry in the high atmosphere.

Bibliography

Atreya, S. K., Pollack, J. B., and Matthews, M. S., eds. (1989). "Origin and Evolution of Planetary and Satellite Atmospheres." Univ. Arizona Press, Tucson.

Bagenal, F., Dowling, T., and McKinnon, W., eds. (2004). "Jupiter: The Planet, Satellites and Magnetosphere" Cambridge Univ. Press, Cambridge, United Kingdom.

Beatty, J. K., and Chaikin, A., eds. (1990). "The New Solar System," 3rd Ed. Sky Publishing, Cambridge, Massachusetts.

Beebe, R. (1994). "Jupiter: The Giant Planet." Smithsonian Institution Press, Washington, D.C.

Bergstralh, J. T., Miner, E. D., and Matthews, M. S., eds. (1991). "Uranus." Univ. Arizona Press, Tucson.

Chamberlain, J. W., and Hunten, D. M. (1987). "Theory of Planetary Atmospheres: An Introduction to Their Physics and Chemistry," 2nd Ed. Academic Press, Orlando, Florida/San Diego.

Cruikshank, D. P., ed. (1995). "Neptune." Univ. Arizona Press, Tucson.

Gehrels, T., and Matthews, M. S., eds. (1984). "Saturn." Univ. Arizona Press, Tucson.

Rogers, J. H. (1995). "The Planet Jupiter." Cambridge Univ. Press, Cambridge, United Kingdom.

Interiors of the Giant Planets

Mark S. Marley

Jonathan J. Fortney

NASA Ames Research Center
Moffett Field, California

CHAPTER 21

T he giant or jovian planets—Jupiter, Saturn, Uranus, and Neptune—account for 99.5% of all the planetary mass in the solar system. An understanding of the formation and evolution of the solar system thus requires knowledge of the composition and physical state of the material in their interiors. But such information does not come easily. The familiar faces of these planets, such as the cloud-streaked disk of Jupiter, tell relatively little about what lies beneath. Knowledge of these planetary interiors must instead be gained from analysis of the mass, radius, shape, and gravitational fields of the planets. For giant planets around other stars, at best only the mass and radius can be determined. The study of the behavior of planetary materials at high densities and pressures further provides the experimental and theoretical framework upon which planetary interior models are subsequently based. Interior models provide a window into the internal structure of these planets and shed light on processes that led to planet formation in our solar system and others.

1. General Overview

Several lines of observational evidence provide information on the composition and structure of the giant planets. The first and most easily obtained quantities are the mass (known from the orbits of natural satellites), radius (polar and equatorial radii), and rotation period (obtained originally from telescopic observations, now derived from remote and in situ observations of planetary magnetic fields). By the 1940s, these fundamental observations, coupled with the advances in understanding the high-pressure behavior of matter in the 1920s and 1930s, constrained the composition of Jupiter and Saturn to be predominantly hydrogen. Direct measurement of the planets' high-order gravity fields, interior rotation states, and heat flow, along with spacecraft and ground-based spectroscopic detection of atmospheric elemental composition, has since allowed the construction of more detailed interior models.

These models divide the giant planets into two broad categories. Jupiter and Saturn are predominantly hydrogen–helium gas giants with a somewhat enhanced abundance of heavier elements and dense cores. Uranus and Neptune are ice giants with hydrogen–helium envelopes and dense cores. The following description of Jupiter's interior, as illustrated schematically in Fig. 1, is qualitatively valid for Saturn and serves as a point of departure for understanding the interiors of Uranus and Neptune. Individual planetary interior structures are discussed in Section 5.

The interior begins at the base of the outermost atmospheric envelope that we can see directly. The jovian atmospheres consist of a gaseous mixture of molecular hydrogen,

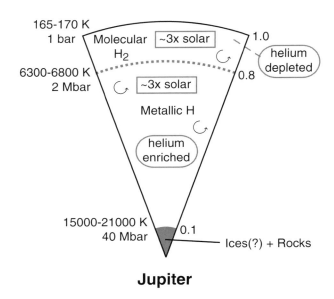

165-170 K
1 bar

Molecular H₂   ~3x solar   1.0

helium depleted

6300-6800 K
2 Mbar   ~3x solar   0.8

Metallic H

helium enriched

15000-21000 K
40 Mbar   0.1

Ices(?) + Rocks

Jupiter

FIGURE 1   Highly schematic, idealized cross section of the interior structure of Jupiter. The numbers to the right refer to the relative radius (r/R) of the core and the molecular-to-metallic hydrogen phase transition. On the left are listed the approximate temperatures and pressures at which these interfaces occur. Arrows indicate convection. Boxes denote approximate enhancement of elements other than H and He over the abundance found in the Sun. The core mass is uncertain, but likely has a mass between 0 and 10 Earth masses. The real Jupiter is undoubtedly more complex. It is likely that interfaces are gradual and the composition of the various regions is inhomogeneous.

helium, methane, ammonia, and water. At 1 **bar** pressure (the pressure at sea level on Earth), the temperature in Jupiter's atmosphere is 165 K. Near this level, the ammonia condenses into clouds; the water condensation level is even deeper. In the colder atmospheres of Uranus and Neptune, methane also condenses into clouds. Deeper into the planet the pressure of the overlying atmosphere compresses the gas, increasing its temperature and density. This process, **adiabatic compression**, is the same one responsible for the increase in temperature with decreasing altitude on Earth. One hundred kilometers beneath the cloud tops the temperature has reached 350 K.

As pressures and temperatures increase, the gas begins to take on the characteristics of a liquid. Since the **critical point** of the dominant constituent, molecular hydrogen, lies at 13 bars and 33 K, there is not a distinct gas–liquid phase boundary. By several hundred thousand bars the envelope closely resembles a hot liquid. This characteristic of the giant planets—they exist in the supercritical regime of their primary constituent—leads to their most fundamental property: these planets have essentially bottomless atmospheres.

Deeper into the planet, the temperature and pressure continue to increase steadily. By 20,000 km beneath the cloud tops, the temperature reaches 7000 K, and the pressure is 2 Mbar. Recent experiments suggest that by this point hydrogen, previously present as molecules of H₂, has undergone a phase transition to a liquid, metallic state. Most of the mass of Jupiter consists of this **metallic hydrogen**: protons embedded in a sea of electrons. Helium and other constituents exist as impurities in the hydrogen soup. For the remaining 50,000 km to Jupiter's core, the pressure and temperature continue to rise, reaching 40 Mbar and 20,000 K in the deep interior. Near the center of the planet, the composition changes, perhaps gradually, from a predominantly hydrogen–helium mixture to a combination of rock and ice. The density of this rock and ice core is 10,000–20,000 kg m⁻³, higher than the metallic hydrogen density of about 1000 kg m⁻³ (uncompressed water, like that which comes out of a tap, also has a density of 1000 kg m⁻³).

Throughout most of the interior, the transport of energy by radiation is severely hampered by the high opacity of compressed hydrogen. Other constituents such as methane and water effectively block energy transport by radiation in those regions of the spectrum where the hydrogen is a less powerful absorber. Because conduction of heat by the thermal motion of molecules is also inefficient, **convection** is the prevailing energy transport mechanism throughout the interior. It had been suggested in the 1990s that in a thin zone in Jupiter's interior at temperatures of 1000 to 3000 K energy transport by radiation was in fact dominant. However, more recent studies suggest this is not the case. The rising and sinking convective cells in the interior move slowly, at velocities of just centimeters per second or less. Because of the continuous nature of the atmosphere, the wind patterns seen in the belts and zones of Jupiter and Saturn may have roots that reach into the deep, convective interior of the planet. Indeed the winds measured by the *Galileo* spacecraft's atmosphere probe continued to blow steadily at the deepest levels reached by the probe, about 20 bars.

The interior of Saturn is much like that of Jupiter. Saturn's lower mass and consequently lower pressures produce a smaller metallic hydrogen region. Uranus and Neptune lack a metallic hydrogen region; instead, at about 80% of their radius, the abundance of methane, ammonia, and water increases markedly. In this region, temperatures of over 5000 to 10,000 K produce an ocean of electrically charged water, ammonia, and methane molecules, along with more complex compounds. Most of the mass of Uranus and Neptune exists in such a state. Deep in their interiors, all the planets likely have cores of primarily rocky material.

This picture of the interiors of the jovian planets has been painstakingly pieced together since the 1930s. This chapter discusses the components of observation, experiment, and theory that are combined to reach these conclusions.

TABLE 1	Observed Properties of Jovian Planets			
Quantity	Jupiter	Saturn	Uranus	Neptune
M (kg)	1.8986×10^{27}	5.6846×10^{26}	8.683×10^{25}	1.024×10^{26}
a (km)	$71,492 \pm 4$	$60,268 \pm 4$	$25,559 \pm 4$	$24,766 \pm 15$
P_s (hours)	9.92492	10.78	17.24	16.11
$J_2 \times 10^6$	$14,697 \pm 1$	$16,332 \pm 10$	$3,516 \pm 3$	$3,539 \pm 10$
$J_4 \times 10^6$	-584 ± 5	-919 ± 40	-35 ± 4	-28 ± 22
$J_6 \times 10^6$	31 ± 20	104 ± 50	—	—
q	0.0892	0.151	0.0295	0.026
Λ_2	0.1647	0.108	0.1191	0.136
$\bar{\rho}$(g cm^{-3})	1.328	0.688	1.27	1.64
Y	0.238 ± 0.007	$0.18 - 0.25$	0.26 ± 0.05	0.26 ± 0.05
T_1 (K)	165	135	76	74

2. Constraints on Planetary Interiors

2.1 Gravitational Field

A variety of observations yield information about the makeup and interior structure of the jovian planets. The mass of each of the four jovian planets (Table 1) has been known with some precision since the discovery of their natural satellites. The masses range from 318 times the Earth's mass ($M_\oplus$) for Jupiter to 14.5 $M_\oplus$ for Uranus. A second fundamental observable property is the radius of each planet measured at a specified pressure, typically the 1-bar pressure level. Radii are most accurately measured by the occultation technique, in which the attenuation of the radio signal from a spacecraft is measured as the spacecraft passes behind the planet. Jovian planet radii range from 11 times the Earth's radius ($R_\oplus$) for Jupiter to 3.9 $R_\oplus$ for Neptune. The combination of mass and radius allows calculation of mean planetary density, $\bar{\rho}$. Although a surprising amount can be learned about the bulk composition of a planet from just $\bar{\rho}$ (as we will later see for Extra-solar Giant Planets), more subtle observations are required to probe the detailed variation of composition and density with radius.

If the jovian planets did not rotate, they would assume a spherical shape, and their external gravitational field would be the same as that of a point of the same mass. No information about the variation in density with radius could be extracted. Fortunately, the planets do rotate, and their response to their own rotation provides a great deal more information. This response is observed in their external gravitational field.

For a uniformly rotating body in hydrostatic equilibrium, the external gravitational potential, Φ, is

$$\Phi = -\frac{GM}{r}\left(1 - \sum_{n=1}^{\infty}\left(\frac{a}{r}\right)^{2n} J_{2n} P_{2n}(\cos\theta)\right)$$

where G is the gravitational constant, M is the planetary mass, a is the equatorial radius, θ is the colatitude (the angle between the rotation axis and the radial vector $\mathbf{r}$), P_{2n} are the Legendre polynomials, and the dimensionless numbers J_{2n} are known as the gravitational moments. The assumption of hydrostatic equilibrium means that the planet is in a fluid state, responding only to its rotation, and there are no permanent, nonaxisymmetric lumps in the interior. This assumption is believed to be quite good for the jovian planets.

The gravitational harmonics are found from observations of the orbits of natural satellites, precession rates of elliptical rings, and perturbations to the trajectories of spacecraft. As a spacecraft flies by a planet, it samples the gravitational field at a variety of radii. Careful tracking of the spacecraft's radio signal reveals the Doppler shift due to its acceleration in the gravitational field of the planet. Inversion of these data yields an accurate determination of the planet's mass and gravitational harmonics (see Table 1). In practice, it is difficult to measure terms of order higher than J_4, and the value of J_6 is generally quite uncertain. Progressively higher order gravitational harmonics reflect the distribution of mass in layers progressively closer to the surface of the planet. Thus, even if they could be measured accurately, terms such as J_8 would not contribute greatly to an understanding of the deep interior.

A planet's response to its own rotation is characterized by how much a surface of constant total potential (including the effects of both gravity and rotation) is distorted. The amount of distortion on such a surface of constant potential, known as a level surface, depends on the distribution of mass inside the planet, the mean radius of the level surface, and the rotation rate. The distortion, or oblateness, of the outermost level surface is measured from direct observations of the planet and is given by $\varepsilon = (a - b)/a$, where b is the polar radius. The equatorial and polar radii can be found from direct telescope measurement or, more accurately,

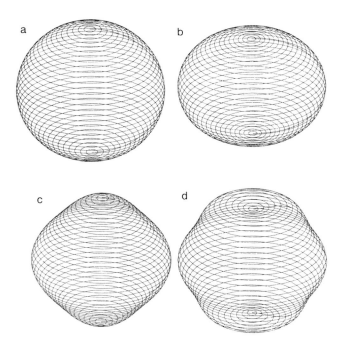

FIGURE 2 Illustration of the ways that a planet changes shape owing to its own rotation. A nonrotating planet (a) is purely spherical. Saturn's distortion due to its gravitational harmonic J_2 is shown approximately to scale in (b). The J_4 and J_6 distortions of Saturn are shown in (c) and (d), exaggerated by about 10 and 100 times, respectively. (Figure courtesy William Hubbard, Univ. Ariz.)

from observations of spacecraft or stellar occultations. Distortion of level surfaces cannot be described simply by ellipses. Instead, the distortion is more complex and must be described by a power series of shapes, as illustrated in Fig. 2. The most obvious distortion of a spherical planet (Fig. 2a) is illustrated in Fig. 2b. More subtle distortions are described by harmonic coefficients of ever increasing degree, as illustrated in Figs. 2c and 2d.

A nonrotating, fluid planet would have no J_{2n} terms in its gravitational potential. Thus, the gravitational harmonics provide information on how the shape of a planet responds to rotating-frame forces arising from its own spin. Since the gravitational harmonics depend on the distribution in mass of a particular planet, they cannot be easily compared between planets. Instead a dimensionless linear response coefficient, Λ_2, is used to compare the response of each jovian planet to rotation. To lowest order in the square of the angular planetary rotation rate, ω^2, $\Lambda_2 \approx J_2/q$, where $q = \omega^2 a^3/GM$. Table 1 lists the Λ_2 calculated for each planet. The jovian planets rotate rapidly enough that the nonlinear response of the planet to rotation is also important and must be considered by computer models.

Because the gravitational harmonics provide information about the planet's response to rotation, interpretation of the harmonics requires accurate knowledge of the rotation rate of the planet. Before the space age, observations of atmo-

spheric features as they rotated around the planet provided rotation periods. This method, however, is subject to errors introduced by winds and weather patterns in the planet's atmosphere. Instead, rotation rates are now found from the rotation rate of the magnetic field of each planet, generally as measured by the *Voyager* spacecraft (radio emissions arising from charged particles in Jupiter's magnetosphere can be detected by radio telescopes on Earth). This approach assumes that convective motions deep in the electrically conducting interior of the planet generate the magnetic field and that the field's rotation consequently follows the rotation of the bulk of the interior. Measuring Saturn's magnetic field rotation rate is particularly difficult because the field is nearly symmetric about the rotation axis of the planet. Indeed, in 2006, data from the *Cassini* spacecraft led to a revision in the previously accepted rotation period by 1%, and the new value, shown in Table 1, may still not reflect the true rotation of the deep interior.

2.2 Atmosphere

The observable atmospheres of the jovian planets provide further constraints on planetary interiors. First, the atmospheric temperature at 1 bar pressure, or T_1, constrains the temperature of the deep interior. The interior temperature distribution of the jovian planets is believed to follow a specified pressure–temperature path known as an adiabat. For an adiabat, knowledge of the temperature and pressure at a single point uniquely specifies the temperature as a function of pressure at all other points along the adiabat. Thus, T_1 gives information about the temperature structure throughout the convective interior of the planet. Both the amount of sunlight that the atmosphere absorbs and the amount of heat carried by convection, up from the interior of the planet to the atmosphere, control T_1. For each planet, save Uranus, T_1 is higher than expected if the atmosphere were simply in equilibrium with sunlight. In fact, these atmospheres are heated from below as energy is transported upward from the slowly cooling planetary interiors. The measured heat flow ranges from 0.3 W m^{-2} at Neptune, to 2.0 W m^{-2} at Saturn, to 5.4 W m^{-2} at Jupiter. Uranus has no detectable internal heat flow.

Second, the composition of the observable atmosphere also holds clues to the internal composition. This is because of the supercritical nature of the jovian atmospheres. The principal component of the jovian atmosphere, hydrogen, does not undergo a vapor–liquid phase change above 33 K. Because the planets are everywhere warmer than this temperature, the observed atmosphere is directly connected to the deep interior. Knowledge of the composition of the top of the atmospheres therefore provides some insight to the composition at depth. [*See* ATMOSPHERES OF THE GIANT PLANETS.]

The *Galileo* spacecraft entry probe returned direct measurements of the composition of Jupiter's atmosphere. The composition of the remaining planetary atmospheres is inferred from spectroscopy. In planetary science, compositions are often stated relative to "solar" abundances. Solar abundances are the relative quantities of elements present in the solar nebula at the time of planetary formation. The solar abundances of hydrogen and helium are about 70% and 28% by mass, respectively. Oxygen, carbon, nitrogen, and the other elements make up the remainder. These elements are collectively called the **heavy elements** to distinguish them from hydrogen and helium. Measurements of the rate at which the atmospheric pressure decreases with height in these atmospheres require that hydrogen and helium must be the dominant components of the atmospheres of all four jovian planets. Spectroscopy supports this conclusion and gives the relative abundance of hydrogen and helium. The helium mass fraction of each atmosphere, Y, is listed in Table 1. The heavier elements are generally enriched in the jovian atmospheres over their solar abundances, which must be explained by any formation scenario for these planets.

2.3 Magnetic Field

All four jovian planets possess a magnetic field. Jupiter's is large and complex; Saturn's is less complex and smaller. The magnetic fields of both Uranus and Neptune are very complex: They deviate substantially from a dipole, and their field axes are tilted strongly with respect to their rotation axes. The only known mechanism for producing global planetary magnetic fields, the hydromagnetic dynamo process, requires nonuniform motion of a large electrically conductive region. Convection in the highly conductive interior of the jovian planets is presumed responsible for formation of their fields. The level of complexity of each field plausibly relates to the depth of the electrically conducting region. Magnetic fields formed by relatively small, deep sources may be simpler and smaller than fields formed by large, shallow dynamos. [*See* PLANETARY MAGNETOSPHERES.]

3. Equations of State

3.1 Overview

Beyond observations of the planets themselves, a second major ingredient in interior models is an **equation of state**, or EOS. An EOS is a group of equations—derived from laboratory observations and theory—that relate the pressure (P) of a mixture of materials to its temperature (T), composition (x), and density (ρ). Any attempt to model the interior structure of a giant planet must rely on an EOS. The construction of accurate equations of state is a primary activity in planetary interior modeling.

For an ideal gas, the well-known EOS is $P = nkT$. Here k is Boltzman's constant, and n is the number density of the

gas. The composition of an ideal gas does not affect the pressure; only the number of molecules and atoms in a given volume, n, enters the equation. Under the conditions of high temperature and pressure found in the interiors of the giant planets, atoms and molecules interact strongly with one another, thus violating the conditions under which the ideal gas EOS holds. Additionally, the typical pressures reached in the interiors of the giant planets (tens to hundreds of megabars) are also sufficient to modify the electronic structure of individual atoms and molecules. This further adds to the challenge of understanding the EOS. In short, the properties of planetary materials at high pressures will differ substantially from those encountered in their low-pressure, and more familiar, forms. In practice, the behavior of planetary materials must be understood from both experiments and theory.

For pressures less than about 1–3 Mbar, depending on the material, shock wave experiments provide guidance in the construction of equations of state. In these experiments, a high-velocity projectile is fired into a container holding a sample of the material under study. The thermodynamically irreversible nature of shock compression causes both high temperatures and high pressures in the sample. Alternatively, this shock compression can be achieved with powerful lasers. High-speed measuring devices record the temperatures, pressures, and densities achieved during the brief experiments. A photograph of a shock tube at Lawrence Livermore National Laboratory, used extensively for planetary work, is shown in Fig. 3.

FIGURE 3 The 60-foot-long, two-stage light-gas gun at Lawrence Livermore National Laboratory. This apparatus is used to obtain equation of state, shock temperature, and electrical conductivity data for planetary liquids (H_2, He, H_2O, NH_3, and various mixtures). An experiment begins when a projectile is fired from the gun on the left side of the photo and ends with the impact of a second projectile, accelerated by gas compressed by the first, on the target sample at the extreme right. (Photo courtesy William Nellis, LLNL.)

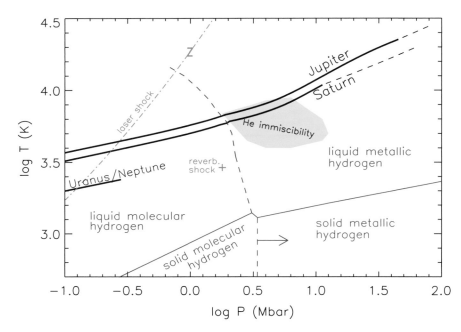

FIGURE 4 Phase diagram for hydrogen, the main constituent of Jupiter and Saturn. The approximate domains of liquid metallic hydrogen and molecular hydrogen are shown along with approximate interior temperature profiles for the Jovian planets. The shaded area indicates the approximate region in the interior of Saturn and, possibly, Jupiter where helium and metallic hydrogen cannot coexist in equilibrium. The locations of phase boundaries are highly uncertain except for the liquid to solid transition of H_2. A "+" marks the highest pressure at which the conductivity of hydrogen has been measured. "Laser shock" shows the pressure–temperature curve where single shock experiments reach. The "Z" marks the highest pressures attained in experiments where single shocks were created by accelerated metal plates. Most of the interior of Jupiter and Saturn exists at temperatures and pressures greater than can currently be probed in laboratory experiments.

The temperatures and pressures reached in these experiments are the closest that terrestrial laboratories can come to reliably duplicating the conditions in the interiors of the jovian planets. For Jupiter, the experiments model conditions about 90% of the way out from the planet's center. The experiments can equal pressures found at about 70% of Saturn's radius and 50% of Uranus and Neptune's. There is currently a controversy regarding the compressibility of hydrogen at the molecular-to-metallic transition near pressures of 1 Mbar. The measured density of shock-compressed liquid deuterium (a heavy isotope of hydrogen) differs by 50% between data sets using shocks produced by intense lasers and data sets obtained using shocks produced by projectiles. We will see later that this discrepancy is our greatest uncertainty in understanding the interior of Jupiter, and it is also important for Saturn.

Diamond anvils are used in another type of experiment to squeeze microscopically small samples of planetary materials to very high pressure. These experiments are most easily conducted at room temperature, making them less applicable to the interiors of the jovian planets.

3.2 Hydrogen

For pressures less than about 1 Mbar, the behavior of molecular hydrogen, H_2, is understood fairly well from theory and the shock experiments. At higher pressures such as those encountered deeper in the interiors of Jupiter and Saturn, the hydrogen molecules are squeezed so closely together that they begin to lose their individual identities. Under these conditions, the hydrogen undergoes a phase transition to a metallic, pressure-ionized state commonly called metallic hydrogen. In giant planets, this metallic hydrogen is fluid, not solid. A shock wave experiment suggests that this transition occurs near 1.4 Mbar at 3000 K; however, more work is needed to fully understand this phase transition. Some theoretical calculations show that the transition is continuous and may not be complete until a pressure of 10 Mbar, while others predict an abrupt, discontinuous (first-order) transition from the molecular to the metallic phase.

In Jupiter and Saturn, liquid metallic hydrogen consists of a dense mixture of ionized protons and electrons at temperatures over about 10,000 K. The EOS of liquid metallic hydrogen is understood well theoretically for pressures above about 10 Mbar, but the EOS is not well constrained from 1 to 10 Mbar, the transition region. A hydrogen phase diagram and temperature–pressure profiles for each giant planet are shown in Fig. 4. Because the detailed behavior of hydrogen near the phase transition itself is not known, various simplifying assumptions must be made when considering these regions of giant planets. The EOS in this region is typically based on a mixture of theory and interpolation.

3.3 Helium

Helium has not been as well studied as hydrogen, but shock wave data do provide information to several hundred kilobars. Above that pressure, theory must guide models of the behavior of this element. Though the equations of state of hydrogen and helium individually are reasonably understood, the behavior of mixtures of these two constituents is less well constrained. This is a serious theoretical void because the hydrogen–helium mixture composes most of the mass of Jupiter and Saturn and is an important component at Uranus and Neptune.

Current calculations of the behavior of hydrogen and helium mixtures show that helium is not soluble in hydrogen at all mass fractions and temperatures. At the temperatures predicted in the interior of Saturn, hydrogen and helium do not mix. According to this model, droplets of helium-rich material are constantly forming in the molecular to metallic transition region of the planet. Because they are more dense than their surroundings, the drops fall to deeper, warmer levels of the envelope, where temperatures may be high enough to again allow mixing. Thus, at certain depths in Saturn's interior, it is always raining helium. This remarkable conclusion is discussed in Section 5.3 in the context of the Saturn interior models.

3.4 Ices

The term **ices** is applied to mixtures of volatile elements in the form of water (H_2O), methane (CH_4), and ammonia (NH_3) in solar proportions, not necessarily present as intact molecules. Ices are a primary constituent of Uranus and Neptune but are less abundant in Jupiter and Saturn. As the planetary interior temperatures are over several thousand Kelvin, they are present as liquids. Shock wave data on a mixture of water, isopropanol, and ammonia (dubbed "synthetic Uranus") have helped establish the equation of state of this material at pressures less than about 2 Mbar and temperatures less than about 4000 K. These experiments helped confirm that ices are a primary constituent of Uranus and Neptune. The shock wave data on this mixture show that, at pressures exceeding ∼200 kbar, the planetary ice constituents ionize to form an electrically conductive fluid. At pressures ≥1 Mbar, the ice constituents dissociate, and the EOS becomes quite "stiff," meaning the density is not particularly sensitive to the pressure.

3.5 Rock

The remaining planetary constituents are lumped into the category **rock**. Rock is presumed to consist of a solar mixture of silicon, magnesium, and iron, with uncertain additions of oxygen and the remaining elements. Although the rock equation of state is not well known, it is also expected to be quite "stiff." The lack of a detailed rock EOS is not a serious limitation for planetary interior models because the rock component is not a major fraction of the mass.

3.6 Mixtures

Because all the planetary components—including gas, ice, and rock—are likely mixed throughout the interiors, equations of state of such mixtures are required for interior modeling. Hydrogen–helium mixtures, considered earlier, may not exist at all temperatures, pressures, and concentrations. The solubility of other mixtures, for example, rock or oxygen in metallic hydrogen, is less well known. From the limited data, it appears that the planetary constituents other than hydrogen and helium do mix well under the temperature and pressure conditions typically found in planetary interiors. This is because delocalization of electrons at high pressure diminishes the well-defined intermolecular bonds present at lower pressures. Thus, the separation of planetary materials into distinct layers of "pure" rock or ice is highly unlikely. If correct, such considerations also have important cosmogonic implications. For example, the rock cores of the planets likely did not "settle" from an initially well-mixed planet, but instead the gaseous components likely collapsed onto a preexisting rocky nucleus that formed in the protosolar nebula.

Since the EOS of all possible mixtures has not been studied, either experimentally or theoretically, approximations must be employed. One approximation, the additive volume law, weights the volumes of individual components in a mixture by their mass fraction. An implication of such approximations is that the computed densities of mixtures of rock, ice, and gas can be similar to that of pure ice. Thus, it is not currently possible to differentiate between models of Uranus and Neptune with mantles of pure ice and models with mantles of a mixture of rock, ice, and gas.

4. Interior Modeling

In addition to an equation of state for the material in the interior of a planet, two more components are required to produce an interior model. The temperature and composition in the interior as a function of pressure, $T(P)$ and $x(P)$, must also be known. (These quantities are described as functions of pressure because the pressure increases monotonically toward the center of the planet.) The first of these ingredients, $T(P)$, is not difficult to find. If the jovian planets are fully convective in their interiors, transporting internal heat to the surface by means of convection, the relation between temperature and pressure in their interiors is known as an adiabat. An adiabat has the property that knowledge of a single temperature and pressure at any point allows specification of T as a function of P at any other point (assuming the material's EOS is known). The temperature and pressure in the convecting region of each Jovian atmosphere have been measured so a unique $T(P)$ relation for each planet can be found.

More difficult to specify is the variation in composition through each planet, $x(P)$. The composition of each planet's atmosphere is known, but there is no guarantee that this composition is constant throughout the planet. Earth's core, for example, has a very different composition from the crust.

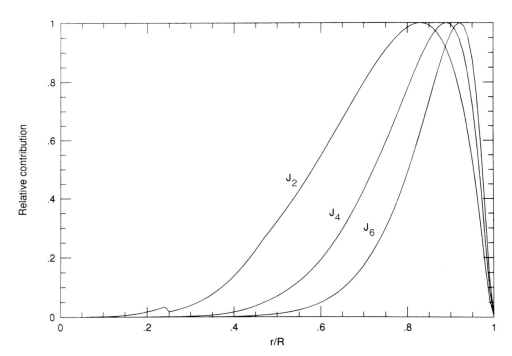

FIGURE 5 Gravitational harmonics are computed from integrals over density and powers of radius of a rotating planet. The curves illustrate the integrands for the harmonics J_2, J_4, and J_6 of a Saturn interior model. Higher-degree terms are proportional to the interior structure in regions progressively closer to the surface. All curves have been normalized to unity at their maximum value. The bump in the J_2 curve near 0.2 is due to the presence of the core.

For the jovian planets, an $x(P)$ relation is typically guessed, an interior model computed, and the results compared to the observational constraints. With multiple iterations, a variation in composition with pressure that is compatible with the observations is eventually found.

The combination of these three ingredients, an equation of state $P = P(T, x, \rho)$, a temperature–pressure relation, $T = T(P)$, and a composition–pressure relation, $x = x(P)$, completely specifies pressure as only a function of density, $P = P(\rho)$. Because the jovian planets are believed to be fluid to their centers, the pressure and density are also related by the equation of hydrostatic equilibrium (with a first-order correction for a rotating planet):

$$\frac{\partial P}{\partial r} = -\rho(r)g(r) + \frac{2}{3}r\omega^2\rho(r)$$

where g is the gravitational acceleration at radius r and ω is the angular rotation rate. This relation simply says that, at equilibrium, the pressure gradient force at each point inside the planet must support the weight of the material at that location. Combining the equation of hydrostatic equilibrium with the $P(\rho)$ relation finally allows determinations of the variation of density with radius in a given planetary model, $\rho = \rho(r)$.

The computed model must then satisfy all the observational constraints discussed in Section 2. Total mass and radius of the model are easily tested. The response of the model planet to rotation and the resulting gravitational har-

monics must be calculated and compared with observations. Figure 5, which shows the relative contribution versus the depth from the center of the planet, illustrates the regions of a Saturn model that contribute to the calculation of the gravitational harmonics J_2, J_4, and J_6. Higher degree modes provide information about layers of the planet progressively closer to the surface.

The construction of computer models that meet all the observational constraints and use realistic equations of state requires several iterations, but the calculation does not strain modern computers. The current state-of-the-art is to calculate dozens of interior models, while varying the many parameters within theoretically or experimentally determined boundaries. An example is the uncertainties in the equations of state of hydrogen and helium that reflect the differences between experimental data and theory. The size and composition of the heavy element core, as well as the heavy element enrichment in the envelope, are also varied with different equations of state for ices and rocks. Only a subset of all the models considered will fit all available planetary constraints, and these models are taken as successful descriptions of the planets. However, by necessity, each modeler begins with an ad hoc set of assumptions that limit the range of models that can be calculated. This inherent limitation of models should always be borne in mind when considering their results, although recent modeling efforts do examine a wider range of possible models, using fewer a priori assumptions about the interiors. The consensus for the structure of jovian planet interior models is presented in the next section.

5. Planetary Interior Models

5.1 General Overview

Even early "cosmographers" recognized that the giant planets of the solar system were distinct from the inner terrestrial planets. The terrestrial planets have mean densities of 4000–5000 kg m^{-3}, intermediate between the density of rocks and iron, whereas the giant planets have mean densities closer to that of water (1000 kg m^{-3}), between 700 and 1700 kg m^{-3}. From this single piece of information, it is clear that the bulk composition of the giant planets must be substantially different from that of the terrestrial planets.

It has been known since the 1940s that if the interiors of Jupiter and Saturn are "cold," the primary component of these planets must be hydrogen. In this context, "cold" means that the densities throughout the interior must not deviate significantly from the values they would assume at the same pressures if the temperature was 0 K. The approximation is relevant because the behavior of substances at 0 K and high pressure can be calculated analytically. Hydrogen is then a likely dominant constituent because, at the high pressures prevalent in the interiors of Jupiter and Saturn, it would be a metallic fluid with a density of about 1000 kg m^{-3}, not the more familiar molecular gas. Because the density of "cold" metallic hydrogen is close to the bulk densities of Jupiter and Saturn, it was recognized as a plausible major constituent of these planets.

Mass–radius calculations provide a more compelling demonstration of the dominance of hydrogen in the interiors of Jupiter and Saturn. For a given composition, there is a unique relation between the radius of a spherical body in hydrostatic equilibrium and its mass. These relations can be calculated analytically for all elements at high pressure and zero temperature. Although the interiors of jovian planets are not at zero temperature, they are cool when measured on an atomic temperature scale. This is adequate for a qualitative calculation, but zero-temperature equations of state are insufficiently accurate for the calculation of detailed interior models.

Mass-radius curves for several likely planetary constituents are shown in Fig. 6. For low masses, the interior pressures are small compared to intermolecular forces and the volume of an object is just proportional to its mass, thus $R \propto M^{1/3}$. This is a realm with which we are familiar in daily life. At much larger masses, the greater interior pressures ionize the material, liberating many electrons. In this regime, $R \propto M^{-1/3}$; when mass is added to an object, it shrinks. For intermediate masses where the curves meet, there is a region where the radius is not highly sensitive to the mass. At sufficiently high masses, the hydrogen in the core of the object will undergo fusion, the temperature will rise, and the zero-temperature relations shown in Fig. 6 are no longer applicable. However, for planets and white

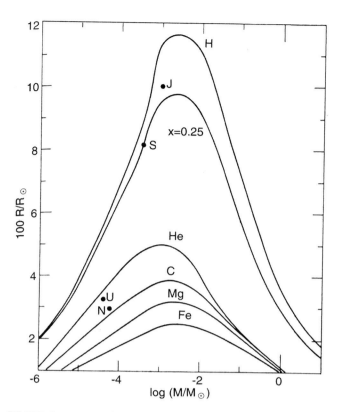

FIGURE 6   Mass–radius curves for objects of various compositions at zero temperature. Curve labeled $x = 0.25$ is for an approximately solar mixture of hydrogen and helium. Points J, S, U, and N represent Jupiter, Saturn, Uranus, and Neptune, respectively. Radius is in units of hundredths of a solar radius and mass is in units of solar masses ($1R_\odot = 6.96 \times 10^5$ km and $1 M_\odot = 1.99 \times 10^{33}$ g). Jupiter and Saturn are clearly composed predominantly of hydrogen and helium; Uranus and Neptune must have a large complement of heavier elements.

dwarf stars, Fig. 6 is applicable. An important consequence of these considerations is that for any given composition, there is a maximum radius that a planet can have. For solar composition, the maximum radius is about 80,000 km for a planet with about four times Jupiter's mass.

The total mass and radius of each Jovian planet are plotted on Fig. 6 as well. This figure immediately proves that Jupiter must be composed primarily of hydrogen and helium. The maximum radii of planets composed of heavier, cosmically abundant elements are all much smaller. For example, only if Jupiter were very hot and very thermally expanded could carbon be a dominant constituent. But Jupiter's observed heat flux rules out a very hot (>10^7 K) internal state. Thus, Jupiter must primarily consist of a mixture of hydrogen and helium. Saturn's position on the graph implies a greater abundance of elements heavier than hydrogen, but it is still a primarily hydrogen bulk composition. Uranus and Neptune lie well below the mass–radius curve for hydrogen, thus revealing an appreciable component of heavier elements in their interiors. In Section 5.5, we will discuss giant planets around other stars, where the only planetary properties we can determine are radii and masses. For these planets, we can get a good estimate of the percentage of their mass that is made of hydrogen and helium, compared to the heavy elements.

Though the mass–radius relations clearly reveal the bulk composition of Jupiter and Saturn, they do not reveal information about the distribution of material inside the planet. It is here that the shape and gravitational harmonics enter the calculation. The response coefficient Λ_2 measures the response of the planet to its own rotation. For a uniform, hydrogen-rich material, $\Lambda_2 = 0.17$. Values smaller than 0.17 indicate a reduced gravitational response to rotation compared with that of the uniform composition hydrogen-rich planet. Such a reduced response results when more of the mass of the planet is concentrated in a dense core. Thus, smaller values of Λ_2 imply greater degrees of central condensation.

Λ_2 varies (see Table 1) from 0.16 for Jupiter to 0.11 for Saturn. The mass–radius relations show that the jovian planets are not pure hydrogen, and their Λ_2 values suggest that they are more centrally condensed than a solar–composition hydrogen–helium object. Hence the heavier constituents are not uniformly distributed in the radius but are concentrated toward the center of each planet. Jupiter exhibits the least central condensation; Saturn and Uranus are most centrally condensed. Thus, we begin to construct an elementary interior model.

Finally, the gravitational harmonics, J_2, J_4, and J_6, probe the detailed variation of the various planetary constituents. To simplify the interpretation of these harmonics, early interior models tended to employ three distinct compositional zones: an inner rocky core, an icy core surrounding the rock one, and a hydrogen/helium envelope. More modern models allow the composition of various zones to vary gradually between layers and allow the outer envelopes to be enriched over solar abundance. The primary unknowns to be found from interior modeling are the size of the rocky/icy core and the abundance of helium and heavy elements in the envelope.

5.2 Jupiter

Jupiter contains more mass than that of all the other planets combined. Because Jupiter's gravitational harmonics are also best known, it serves as a test bed for theoretical understanding of jovian interiors. The observed physical characteristics of Jupiter are listed in Table 1. From *Galileo* Entry Probe data, abundance of methane in Jupiter's atmosphere is about 3.5 times the solar abundance and the abundance of ammonia is about three times solar. Water does not show such enrichment, but it has been argued that

the *Galileo* Entry Probe fell into an anomalously dry region of Jupiter's atmosphere.

The general structure of Jupiter's interior was briefly described in Section 1. Modern interior models attempt to determine specifically the degree of enrichment of heavy elements in the hydrogen/helium envelope of the planet. The atmospheric enrichment of methane and ammonia provides some indication that heavy element enrichment in the deeper interior may be expected. Jupiter's Λ_2 implies that Jupiter is not homogeneous but is slightly centrally condensed. Indeed, detailed modeling has shown that Jupiter's current core is less than 10 Earth masses and that there may not be a core at all. The size and composition of jovian planet cores and the amount of heavy element enrichment in the envelopes have bearing on the scenarios by which they are supposed to have formed.

The variations of density with a radius for two typical Jupiter models are shown in Fig. 7. It should be emphasized that these are two Jupiter models that are consistent with all available constraints. Other, equally valid interior models exist. Figure 8 shows the mass of heavy elements in the cores and hydrogen–helium envelopes for a large number of Jupiter and Saturn models. Any model within the solid red line is a valid interior model for Jupiter, given the current uncertainties in the EOS of hydrogen. Models within the hashed line area are tentatively preferred, given the most recent experimental EOS data. The majority of Jupiter's heavy elements are found within the hydrogen–helium envelope, not within the core. The models also account for uncertainty related to the unknown composition of the core, which is likely some mixture of ice and rock.

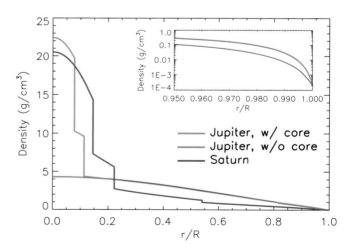

FIGURE 7   Density as a function of normalized radius for Jupiter and Saturn models. A helium deficit in molecular hydrogen regions and corresponding helium enrichment in metallic regions is responsible for the small density change near 0.55 r/R is Saturn. For Jupiter, a model with and without a core is shown. For the models with a core, the core is assumed to be ices overlying rock.

A clear trend of Jupiter modeling over the past 30 years is that as we have gained better knowledge of the EOS of hydrogen, the calculated mass of the core has shrunk.

Surrounding the core is an envelope of hydrogen and helium. The temperature and pressure at the bottom of the hydrogen–helium envelope is near 20,000 K and 40 Mbar for typical models. The gravitational harmonics require the envelope to be denser at each pressure level than a model that has only a solar mixture of elements. Thus, the envelope must be enriched in heavy elements compared to a purely solar composition. The total mass of heavy elements is constrained between 10 and 40 Earth masses. If Jupiter had only a solar abundance of heavy elements, this value would be 6 Earth masses. This means that, averaged throughout the planet, Jupiter is enriched in heavy elements over solar abundances by a factor of 1.5 to 6.

Jupiter's atmospheric abundance of helium, $Y = 0.238 \pm 0.007$, is less than the solar abundance of about 0.28. This depletion is likely an indication that the process of helium differentiation, described more fully in Section 6, may have recently begun on Jupiter. The interior models do not provide a sufficiently clear view into the interior structure to determine if this is the case. The inferred interior structure is, however, compatible with limited helium differentiation.

Hydrogen and helium compose about 90% of Jupiter's mass. Most of the hydrogen exists in the form of metallic hydrogen. Jupiter is the largest reservoir of this material in the solar system. Convection in the metallic hydrogen interior is likely responsible for the generation of Jupiter's magnetic field. The transition from molecular to metallic hydrogen takes place about 10,000 km beneath the cloud tops, compared to about 30,000 km at Saturn. The exceptionally large volume of metallic hydrogen is likely responsible for the great strength of Jupiter's magnetic field. The relative proximity of the electrically conductive region to the surface may explain why Jupiter's magnetic field is more complex than Saturn's.

5.3 Saturn

The observational constraints for Saturn are listed in Table 1. Although Saturn has less than one-third of Jupiter's mass, it has almost the same radius. This is a consequence of the relative insensitivity of radius to mass for hydrogen planets in Jupiter and Saturn's mass range (see Fig. 6). Saturn's atmosphere, like Jupiter's, is enriched in methane and ammonia. The atmosphere's carbon enrichment (in the form of methane) was recently determined to be 7 times the solar abundance. There is also evidence that Saturn's atmosphere has less helium than Jupiter's but the uncertainties are large because there has never been a Saturn entry probe. Since there is no known process by which Saturn could have accreted less helium than Jupiter, another process must be at work. As noted in Section 2.1, Saturn's true rotation rate is uncertain; the following discussion is based on interior mod-

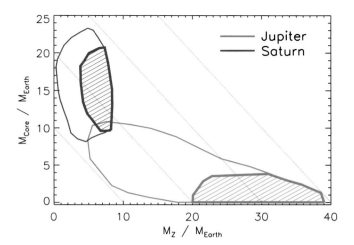

FIGURE 8   The inferred distribution of heavy elements in the interiors of Jupiter and Saturn. The masses of these planets' cores (M_{core}) are shown as a function of the masses of heavy elements in their hydrogen–helium envelopes, M_Z. The hashed regions show current preferred models based on new experimental shock data, which shows that hydrogen is less compressible than previously thought. However, all models within the larger boundary are viable at this time. In general, Saturn has more heavy elements in the central core and less in the envelope than does Jupiter.

els computed by assuming the previously accepted rotation period of 10 hours, 39 minutes, and 22.4 seconds.

Saturn's interior is grossly similar to Jupiter's. The biggest difference is that it is clear that Saturn has a core of 10 to 20 Earth masses. A sample Saturn model is shown in Fig. 7. Temperatures inside Saturn are also cooler. In the model shown in Fig. 7, the temperature and pressure at the base of the metallic hydrogen envelope are 9000 K and 10 Mbar. There is strong evidence that Saturn's envelope, like Jupiter's, is enriched in heavy elements over solar abundance. The mass of the core and the heavy elements in the hydrogen–helium envelope, are also shown in Fig. 8. Again, all models within the solid curve are plausible, but the hashed regions shows models that are currently preferred. The total mass fraction of heavy elements in Saturn is about $2\frac{1}{2}$ times greater than in Jupiter. On the whole, Saturn is enhanced in heavy elements by a factor of 6–14, relative to the Sun. This may be an indication that more condensed icy material was available to be incorporated into Saturn at its location in the solar nebula. Nevertheless, as at Jupiter, hydrogen and helium are the dominant component of Saturn's mass (~75%).

Saturn's somewhat low atmospheric helium abundance implies that the process of helium differentiation (see Section 6) has begun inside the planet. This process results in removal of helium from the outer molecular hydrogen envelope of the planet and enhancement of helium in the deep interior. Thus, the helium fraction should increase with depth in Saturn's interior. The inferred density structure is consistent with this widely accepted explanation for Saturn's low atmospheric helium abundance. If helium is presumed to be uniformly depleted from the outer molecular envelope of the planet, it can be self-consistently accounted for in the deeper interior. The unmixed helium may have actually been removed from molecular *and* metallic regions of the planet and settled down on top of the core. The models lack the sensitivity to confirm that this is definitely the case, however.

The inferred interior structure of Saturn is most consistent with the giant planet formation scenario known as nucleated collapse. In this scenario, a nucleus of rock and ice first forms in the solar nebula. When the nucleus has grown to about 10 $M_\oplus$, the gas of the nebula collapses down upon the core, thus forming a massive hydrogen–helium envelope surrounding a rock/ice core. Planetesimals that accrete later in time cannot pass through the thick atmosphere surrounding the core. Instead, they break up and dissolve into the hydrogen–helium envelope. This scenario accounts for both the core of the planet and the enrichment of heavy elements in the envelope. It is possible that Jupiter formed via a different mechanism, such as the direct gravitational collapse of nebular gas. It is perhaps more likely that both Jupiter and Saturn had larger cores that were partially dredged up by convective plumes over the past 4.5 billion years. This mechanism could plausibly be more efficient in the hotter interior of Jupiter, where convection is more vigorous.

5.4 Uranus and Neptune

Before the *Voyager* encounters, Uranus and Neptune were assumed to have similar interior structures. This assumption was well justified given their similar radii, masses, atmospheric compositions, and location in the outer solar system. Uranus and Neptune were modeled as having three distinct layers: an inner rocky core, a large icy mantle, and a methane-rich hydrogen–helium atmosphere. Little more could be said with precision because their atmospheric oblateness and interior rotation rates were not accurately known.

Upon its arrival at Uranus in 1986 and Neptune in 1989, *Voyager 2* provided the measurements needed to constrain interior models and provide individual identities for each planet. *Voyager* observed the structure of the magnetic field of both planets and measured their rotation rates. In both cases, the fields were off-center, tilted dipoles of similar strengths. *Voyager* also measured the higher order components of the gravitational fields of both planets. The abundance of carbon in both atmospheres is about 30 times the solar value. Although Uranus and Neptune have similar radii and masses, the differences are such that the mean density of Neptune is 24% higher than the mean density of Uranus.

Voyager data revealed that, though similar, the interior structures of the two planets are not identical. As with Jupiter and Saturn, Λ_2 provides information on the distribution of mass inside each planet. If Uranus and Neptune had a similar distribution of mass in their interiors, their Λ_2 parameters would be similar. As Table 1 shows, for Uranus $\Lambda_2 = 0.119$, whereas for Neptune $\Lambda_2 = 0.136$. Neptune's larger value of Λ_2 implies that it is less centrally condensed than Uranus. Models show that this difference can be understood in terms of equal relative amounts of ice, rock, and gas that are simply distributed differently within the two planets. The two planets also follow virtually the same pressure–density law, another indication that they have very similar composition and structure.

Models (Fig. 9) of Uranus and Neptune's interior begin with a hydrogen-rich atmosphere that extends from the observable cloud tops to about 85% of Neptune's radius and 80% of Uranus'. The composition in this region does not vary significantly from the hydrogen-rich atmospheric composition. Near 0.3 Mbar and 3000 K ($0.85R_{\text{Neptune}}$ and $0.80R_{\text{Uranus}}$), the density rises rapidly to over 1000 kg m^{-3}. The density then increases steadily into the deep interior of both planets, where the pressure reaches 6 Mbar at 7000 K. The variation of density with pressure in this region is very similar to that found in the laboratory shock wave experiments on the artificial "icy" mixture known as synthetic Uranus. The composition of this region is thus undoubtedly predominantly icy. However, since the density of rock/ice/gas mixtures can mimic the density of pure ice, the exact composition cannot be known with precision. Any hydrogen present in the deep interior would be in the metallic phase.

Interestingly, Uranus and Neptune models that do not have rock cores can be constructed. Other models with cores as large as $1M_\oplus$ are also consistent with the available data.

The total mass of hydrogen and helium in Uranus and Neptune is about $2M_\oplus$, compared to about $300M_\oplus$ at Jupiter. Given the relatively small amounts of gas compared to ices in Uranus and Neptune, these planets are aptly termed ice giants, whereas Jupiter and Saturn are indeed gas giants.

Shock compression measurements show that the fluids of the hot, ice-rich region of Uranus and Neptune are expected to be substantially ionized and dissociated. The large electrical conductivities of such fluids, coupled with the modest convective velocities predicted for the interiors of Uranus and Neptune, can generate and sustain the observed magnetic fields of the planets. One possible explanation for the complexity of their magnetic fields is that the electrically conductive region of these planets is comparatively close (within about 4000 km) to the cloud tops, a consequence of the ionization behavior of water, ammonia, and methane. This is consistent with the trend in field complexity seen at Jupiter and Saturn.

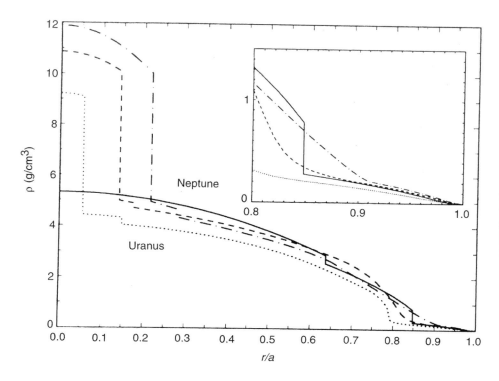

FIGURE 9 Density as a function of normalized radius for three Neptune and one Uranus interior models. The solid, dashed, and dot-dashed curves represent the range of possible Neptune models. Note the wide variety of acceptable core sizes, ranging from a model with no core to a model with a core extending to 20% of Neptune's radius. The dotted curve represents a single Uranus model. Because of Neptune's greater mass, it is everywhere denser than Uranus at the same relative radius. The inset shows the region of transition from a hydrogen-rich atmosphere to the icy mantle in more detail.

Uranus and Neptune likely represent failed gas giant planets. The time to accrete solid objects onto the growing ice and rock planetary cores was much longer in the outer solar nebula than at the orbital distances of Jupiter and Saturn. Thus, Uranus and Neptune took longer to grow. By the time the nebular gas was swept away, these planets had not yet grown massive enough to capture substantial amounts of hydrogen and helium gas from the nebula. Perhaps if the nebular gas had persisted for a longer time, Uranus and Neptune would have grown large enough to complete the capture of a hydrogen–helium envelope. In that case, these planets might now more closely resemble the current Jupiter and Saturn.

5.5 Extrasolar Giant Planets

With current technology, we can learn very little of the physical state of the over 200 planets (as of summer 2006) that have been detected around other stars. The minimum masses of planets are obtained by observing the motion of the parent star induced by the gravitational tug of the planet. But just knowing the minimum mass (since orbital inclinations are unknown) does not tell us much about the structure of a planet. However, there are 10 planets in orbit around other stars for which we have derived accurate masses *and* radii. The radii can be measured if the extrasolar planetary system has a favorable alignment, and the planet passes in front of its parent star (a transit), blocking a small fraction of the star's light. The planet's orbital inclination is then constrained to be essentially edge on, so the mass is then also known. As was shown in Fig. 6, with a deter-

mination of only the mass and radius of a planet, we can get to a reasonable understanding of its interior composition.

New theoretical procedures will have to be developed to understand the structure of giant planets that are 100–200 times closer to their stars than Jupiter is to the Sun and hence receive intense stellar irradiation. This irradiation slows the contraction of a giant planet with time. Our understanding is progressing, but the results are already very surprising. Fig. 10 shows the radius and mass (with

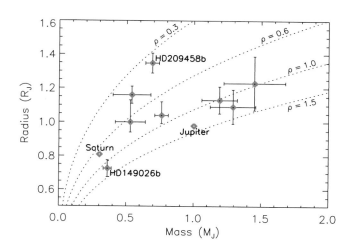

FIGURE 10 Planetary radius as a function of mass for the 10 transiting extra-solar giant planets, Jupiter, and Saturn. HD 209458b and HD 149026b, the transiting planets with the largest and smallest radii, respectively, to date, as well as several others of note, are labeled. Curves of constant bulk density (mass/volume) are shown.

observational error bars) of the transiting planets, as compared to Jupiter and Saturn. The density of these planets can be read from the dotted curves. There are already at least 2 peculiar planets in this sample. One planet, HD 209458b, has a radius 20% larger than expected, and one, HD 149026b, has a radius 20% smaller than expected for a mostly hydrogen–helium composition. Small radii can reasonably be attributed to a large fraction of the planet mass (likely around two-thirds) being made up of rocks and ices. Indeed, HD 149026b is the only extra-solar planet we know for certain has a core. The planets that are more massive than Jupiter, which one would assume would be denser than Jupiter, due to larger self-compression, are modestly less dense than Jupiter, showing that stellar irradiation indeed does slow the contraction of these transiting planets.

However, the planet HD 209458b, the first transiting planet discovered, is quite puzzling. Since it has by far the largest radius of the planets discovered to date, it *may* uniquely have some additional internal energy source that keeps it inflated. Most explanations invoke some sort of tidal dissipation related the planet's close orbit around the parent star. Others involve the penetration of stellar energy to deep regions of the atmosphere by dynamical processes. This is still an open question that is a very active area of research.

6. Jovian Planet Evolution

The amount of energy radiated by each of the jovian planets, except possibly for Uranus, is greater than the amount of energy that they receive from the Sun (see Table 1). This internal heat source is too large to be explained by decay of radioactive elements in the rock cores of the planets. Temperatures, even in the deep interior, are far below the 1,000,000 K required for thermonuclear fusion. The source of the excess energy is gravitational potential energy that was converted to heat during the planets' formation and stored in their interiors. [*See* THE ORIGIN OF THE SOLAR SYSTEM.]

The potential energy of gas and solids in the solar nebula was converted to thermal energy when they were accreted onto the forming planet. Over time, the planets radiated energy into space and cooled, slowly losing their primordial energy content. Thus, all four jovian planets were initially warmer than they are now. During the early evolutionary stages, the planets contracted as they cooled, thereby releasing even more gravitational potential energy. Today, the planets all cool at essentially constant radius because the internal pressures depend only slightly on temperature. The coupled contraction and cooling is known as Kelvin–Helmholtz cooling.

Evolutionary models test whether Kelvin–Helmholtz cooling can account for the current observed heat flows of the jovian planets. In these calculations, a series of sequentially cooler planetary interior models is created, with the last model representing the present-day planet. The time elapsed between each static model is calculated, and thus the evolutionary age of the planet found.

Models predict that Jupiter should have cooled from an initially hot state (accompanied by an atmospheric temperature greater than about 600 K) to its current temperature in about 4.5 billion years. This is about the age of the solar system, so the Kelvin–Helmholtz model is judged a success for Jupiter. For Saturn, however, the model is less successful. The models suggest that Saturn, with its current heat flow, should be about 2 billion years old. Because there is no reason to believe that Saturn formed 2.5 billion years later than Jupiter, another heat source must be adding to Saturn's Kelvin–Helmholtz luminosity. This leads to the hypothesis that differentiation of helium in the interior provides additional thermal energy to the planet.

The helium depletion hypothesis holds that as Saturn has cooled from an initially warmer state with the solar abundance of helium throughout, its interior reached the point (near 2 Mbar and 8000 K) at which hydrogen and helium no longer mix in all proportions. Like oil and water in salad dressing, the hydrogen and helium are separating into different phases.

As the helium-rich drops form in Saturn's envelope and fall to deeper, warmer layers of its interior, the helium eventually again mixes with hydrogen. Over time, this rainfall is depleting the supply of helium in the outer envelope and visible atmosphere and enriching the helium content deeper in the interior, close to the core. The overall planetary inventory of helium remains constant. This model is compatible with the observed depletion at Saturn. Jupiter, with a warmer interior and with smaller helium depletion, has apparently only recently begun this process.

This process of helium differentiation liberates gravitational potential energy as the drops fall. The helium droplets in Saturn may be raining down very far into the planet, possibly all the way down the core. No other process can simultaneously explain Saturn's anomalously high heat flow and the observed atmospheric depletion of helium. Observations from the *Cassini* spacecraft will help allow for a better determination of the helium abundance (and the abundances of many other compounds) in Saturn's atmosphere.

The problem for Uranus and Neptune is somewhat different. The Kelvin–Helmholtz hypothesis predicts ages of the correct order of magnitude for Uranus and Neptune, but the ages are too large. In other words, the model predicts that these planets should have higher heat flows at the current time than they are observed to have. The problem is most severe for Uranus, which has no detectable heat flow. There are several possible resolutions to this contradiction.

One possibility is that gradients in the composition of Uranus with radius have served to impede convection in the deep interior. Composition gradients, for example, a gradual increase in the rock abundance with depth, can severely limit heat flow from the planet. In such a case, only the outermost layers could transport energy by convection to the atmosphere and cool effectively to space, thus producing a lower than expected heat flow. More of Neptune's interior than that of Uranus might be convective, thus explaining its higher current heat flow. Of course if this hypothesis were correct, then the existing interior models of these planets would have to be revised because an initial assumption that the planets are fully convective would have been violated. Inhibition of convection in the deep interior by this mechanism has been proposed as one explanation for the strong nondipole component of both planets' magnetic fields.

Currently it is thought that the highly irradiated extrasolar giant planets evolve in much the same way as Jupiter, except that their interiors cool, and the planets contract, more slowly. The incident stellar flux leads to a radiative zone with a shallow temperature gradient to pressures of up to ~1 kbar, 1000 times deeper than in Jupiter. This limits how quickly the interior heat flux can escape from the planet. Finding more transiting planets, with a variety of radii, masses, and orbital separations will allow for a better understanding of how stellar irradiation effects the cooling of giant planets.

7. Future Directions

Models of jovian planetary interiors have constrained the mass of each planet's core and the approximate composition of their envelopes. These results have provided important constraints on the processes by which these planets form. In turn, formation models place limits on the mass, composition, and evolution of the solar nebula. Further progress, however, requires even tighter limits on the interior structure of these planets. Sufficiently detailed interior models may even provide constraints on the equation of state of hydrogen. Because Jupiter is the largest reservoir of metallic hydrogen in the solar system, it may potentially resolve issues such as the exact pressure of the transition between molecular and metallic hydrogen.

One might expect that future, more accurate measurements of each planet's gravitational harmonics would help to address questions such as these. The higher order moments, however, are most sensitive to the density distribution in the outer 10 or 20% of the planetary radius. Thus, little additional information about the deep interior is likely to be forthcoming from such observational improvements. The higher order harmonics do, however, provide some information about the state of rotation of the outer layers and may help address questions regarding the degree of differential rotation in the jovian planets. For example, it is unknown if Jupiter rotates completely as a solid body, or if different cylindrical regions of its interior rotate at different rates. NASA has recently selected a New Frontiers mission called *Juno* that will travel to Jupiter to answer this

and other questions. The spacecraft will be placed into a low polar orbit such that the spacecraft will readily be able to measure additional higher order harmonics up to J_{12}, which will allow for a determination of the planet's interior rotation. In addition, the spacecraft will observe microwave emission from below the "weather layer" of the planet's atmosphere (100 bars) to determine the deep abundance of water and ammonia. Also, the planet's magnetic field will be mapped in unprecedented detail. Together, these new measurements should shed additional light on the structure of the planet.

Further improvements in delineating the equations of state of jovian planetary components will help to clarify their interior structures. More complete knowledge of the behavior of planetary constituents and their mixtures at high pressure will enable more accurate interior models to be constructed. Nevertheless, dramatic changes in understanding are unlikely to result from such improvements. Only significantly new and different sources of information offer the potential of providing fundamentally new insights into the interior structure of these planets.

Jovian seismology is one promising new avenue of research into these planetary interiors. Much of our knowledge of the interior structure of the Earth arises from study of seismic waves that propagate through the interior of the planet. The speed and trajectory of these waves carry information about the composition and structure of the Earth's interior. During the collisions of the fragments of comet Shoemaker–Levy 9 with Jupiter, several experiments attempted to detect seismic waves launched by the impacts. If these waves had been detected, they would have provided a direct probe into the interior structure of Jupiter.

Another avenue for jovian seismology is to detect resonant acoustic modes trapped inside Jupiter. The frequency of a given jovian oscillation mode depends on the interior structure of the planet within the region in which the mode propagates. Thus, measurement of the frequencies of a variety of modes would provide information on the overall interior structure of the planet. The study of such modes on the Sun, a science known as helioseismology, has revolutionized our knowledge of the solar interior. In the past 20 years, a number of groups have attempted to detect the jovian oscillations with various techniques. However, in all cases the observations and data analysis are difficult, and interpretation of the results has been limited by the restricted number of observing nights on large telescopes. Future observational advances may allow unambiguous detection of jovian oscillations.

As they would at Jupiter, oscillations of Saturn would perturb the external gravitational field of the planet. Though there is yet no way to detect such perturbations at Jupiter, this may be possible at Saturn. Saturn's rings are excellent detectors of faint gravitational perturbations, and thus the possibility arises of using Saturn's rings as a seismometer. There is some evidence that certain wave features in

Saturn's innermost C-ring may be produced by oscillation modes of the planet. Further spacecraft observations are required to confirm this hypothesis, however, and work is currently underway analyzing new data from the *Cassini* spacecraft.

Definitive detection of oscillations of any jovian planet would first serve to accurately determine the core size and rotation profile of the planet. Because such determinations would remove two sources of uncertainty surrounding the interior structure, more information could then be gleaned from the traditional interior model constraints. Seismology might also help to constrain more accurately the location of the transition from molecular to metallic hydrogen in Jupiter's interior. If so, seismology may ultimately provide the tightest constraints on the hydrogen equation of state and interior structure of jovian planets.

Together with a refined understanding of the interiors of our solar system's giant planets, additional understanding of giant planet interiors will come from extra-solar planets. Determinations of the radii of transiting extra-solar planets will allow us to build up a statistical sample to learn how the radius of planets change as a function of mass, age, the amount of heavy elements available in the system (which can be estimated from spectra of the parent star), and the amount of irradiation the planet receives from its parent star. The interiors of giant planets will likely yield many additional surprises as more extra-solar planets are found. [*See* EXTRA-SOLAR PLANETS.]

Bibliography

Guillot, T. (2005). *Annu. Rev. Earth Planet. Sci.* **33**, 493–530.

Guillot, T., Stevenson, D. J., Hubbard, W. B., and Saumon, D. (2004). In "Jupiter: The Planet, Satellites and Magnetosphere" (F. Bagenal, T. E. Dowling, and W. B. McKinnon, eds.). Cambridge Planetary Science, Vol. 1, pp. 35–57. Cambridge Univ. Press, Cambridge, England.

Hubbard, W. B., Burrows, A., and Lunine, J. I. (2002). *Annu. Rev. Astron. Astrophys.* **40**, 103–136.

Podolak, M., Hubbard, W. B., and Stevenson, D. J. (1991). In "Uranus" (J. Bergstralh, E. Minor, and M. S. Matthews, eds.), pp. 29–61. University of Arizona Press, Tucson.

Stevenson, D. J. (1982). *Annu. Rev. Earth Planet. Sci.* **10**, 257–295.

Zapolsky, H. S., and Salpeter, E. E. (1969) *Astrophy. J.* **158**, 809–813.

Io: The Volcanic Moon

Rosaly M.C. Lopes

Jet Propulsion Laboratory
California Institute of Technology
Pasadena, California

CHAPTER 22

Io, the innermost of Jupiter's four Galilean satellites, is the only body outside the Earth so far known to have large-scale active volcanism. Io's heat flow is much higher than the Earth's, and at least one of its many active volcanoes erupts lavas that are hotter than any erupted on the Earth today.

1. Introduction

Io (Fig. 1) was discovered by the Italian scientist and astronomer Galileo Galilei on January 8, 1610, and was named after one of the ancient Roman god Jupiter's illicit lovers. The discovery of active volcanism was made by the *Voyager 1* spacecraft, which flew close to Io in 1979. Images showed volcanic plumes up to 300 km in height and a vividly colored surface dominated by large **caldera**-like and flowlike features. The study of Io's remarkable volcanism has continued since then using observations by telescopes on Earth, by the *Hubble Space Telescope*, and, from 1996–2002, by the *Galileo* spacecraft.

Io's unusual spectroscopic characteristics, due to its volcanic activity and widespread covering of sulfur dioxide, were recognized in the 1970s. In 1979, just prior to the two *Voyager* flybys, Io's 4:2:1 orbital resonance with Europa and Ganymede was predicted to induce severe tidal heating and subsequent active volcanism on Io. The two *Voyager* spacecraft confirmed the prediction that Io is volcanically active.

Io's size (Table 1) is similar to that of the Earth's Moon but its density is higher, indicating that there is more iron in Io's interior than in the Moon's. Io's mantle composition is thought to be predominantly **silicates**. However sulfur compounds are abundant on the surface. After the discovery of volcanism in 1979, a major question was the composition of the erupting material: silicates or sulfur? Ground-based observations and, later, *Galileo*'s results showed that Io's eruption temperatures were too high to be sulfur, but sulfur flows may also exist on the surface. Sulfur dioxide is ubiquitous on Io's surface, and sulfur and sulfur dioxide are known to be present in Io's volcanic plumes.

Io's **heat flow** (Table 1) is very large compared with that of the Earth and other planets. Io's heat flow is about 200 times what could be expected from heating due to the decay of radioactive elements, illustrating how crucial tidal heating is to driving Io's active volcanism. The effect of Io's volcanic eruptions extends well beyond the surface, and there is considerable interaction of Io with the jovian magnetic field. Io has both a patchy, very low density atmosphere and an ionosphere. Sulfur dioxide is the main constituent of the atmosphere, and it is thought to be supplied largely by volcanic plumes, with a lesser amount coming from evaporation of frost deposits on the surface. An important discovery made during the *Galileo* mission was Io's aurora, caused

FIGURE 1 Io imaged by *Galileo*'s Solid State Imaging System on September 7, 1996, at a range of about 487,000 km. The image is centered on the side of Io that always faces away from Jupiter. The black and bright red materials correspond to the most recent volcanic deposits. The near-infrared filter makes Jupiter's atmosphere (in the background) look blue. The active volcano Prometheus is seen as a dark sinuous feature near the right-center of the disk.

TABLE 1	Io's Basic Orbital and Physical Properties

Mean radius: 1821.6 ± 0.5 km
Bulk density: 3528 ± 3 kg m^{-3}
Orbital period: 1.769 days
Orbital eccentricity: 0.0041
Orbital inclination: 0.037
Orbital distance a: 421,800 km
Rotational period: synchronous with orbit
Maximum moment of inertia: 0.3769 ± 0.0004
Potential Love number k^2: 1.292 ± 0.003
Mass: $(8.9320 \pm 0.0013) \times 10^{22}$ kg
Surface gravity: 1.80 m s^{-2}
Global average heat flow: >2.5 W m^{-2}
Radius of core: 656 km (if pure iron)
              947 km (iron and iron sulfide mixture)
Surface equatorial magnetic field strength: <50 nT
Geometric albedo: 0.62
Local topographic relief: up to $\sim$17 km
Active volcanic centers: at least 166
Typical surface temperature (away from hot spots): 85 K
   (night) to 140 K (day)
Atmospheric pressure: $<10^{-9}$ bar, higher at locations
   of plumes

Source: Lopes and Williams (2005).

by collisions between Io's atmospheric gases and energetic charged particles trapped in Jupiter's magnetic field.

Io has an ionosphere and a thin atmosphere. Materials escaping from Io form a cloud of neutrals along Io's orbital path. Escaping materials also populate the Io torus, a doughnut-shaped region along Io's path, made up of ionized particles of sulfur and sulfur dioxide held by Jupiter's powerful magnetic field.

Io is therefore a wonderful natural laboratory for the study of geological and geophysical processes, and its location within Jupiter's magnetic field makes it a rich source for studies of the fields and particles environment in space.

2. Io Exploration

Since its discovery in 1610, Io has been important to our understanding of the solar system, along with the other Galilean satellites, Europa, Ganymede, and Callisto. After a few observations, Galileo Galilei concluded that the four objects were not stars as he originally thought, but satellites in orbit around Jupiter. Galileo's studies of the motion of the newly discovered satellites had a profound effect on human history becoming, along with Galileo's discovery of the phases of Venus, key evidence in favor of the Copernican theory of the universe. Another major step for science came in 1675, when Danish astronomer Olaus Romer noted that the times of the eclipses and occultations of the four moons by Jupiter showed a phase shift with a periodicity of about 6.5 months. He concluded that, when Jupiter is at opposition (when Jupiter and Earth are closest, on the same side as the Sun), light from the jovian system must travel a distance of approximately 4 astronomical units (AU) to reach Earth. However, when Jupiter is at conjunction (when Earth and Jupiter are farthest apart, on opposite sides of the Sun), light traveled about 6 AU on its journey to Earth. Romer concluded that this phase shift in the arrival time of jovicentric events meant that light has a finite velocity, and he used the motions of the Galilean satellites to determine the speed of light.

In 1805, Laplace demonstrated that the Galilean satellites have an orbital configuration (known as the Laplace resonance), which suggested a special dynamical relationship among Io, Europa, and Ganymede. For each time that Ganymede orbits Jupiter, Europa orbits almost exactly twice, and Io orbits four times. Later studies would reveal that this resonance plays a key role in the existence of active volcanism on Io.

Even before the first close-up images of Io were returned by *Voyager* in 1979, there were indications that Io was a remarkably different world from our Moon and other moons in the solar system. Telescopic observations showed that Io's brightness varied according to its position in its orbit, suggesting that the moon always keeps one face toward Jupiter (now referred to as the subjovian hemisphere). During the mid-20th century, photometric and color data showed that Io is the reddest object in the solar system and has a marked color variation with orbital phase angle. These observations also showed Io to be very different from the other Galilean satellites (and most other satellites in the outer solar system) because of the absence of water bands in its spectra.

The peculiar nature of Io's surface became more evident in 1964, when astronomers A. P. Binder and D. P. Cruikshank reported an anomalous brightening of Io's surface as it emerged from eclipse. This first report of "post-eclipse brightening" and the suggestion of a possible atmosphere spurred more telescopic observations but, even though the presence of an atmosphere was confirmed, post-eclipse brightening has remained controversial and has not been confirmed to this day.

The first evidence of an electromagnetic link between Io and the jovian magnetosphere was put forward in 1964 by E. K. Bigg, who found that bursts of decametric radio emission by Jupiter were apparently controlled by Io's orbital position. Models of electrodynamic interaction between Jupiter and Io addressed the coupling mechanism between Io and Jupiter's inner magnetosphere.

The first spacecraft to flyby the Jupiter system was *Pioneer 10* in 1973. These observations revealed that Io has an ionosphere and thin atmosphere. *Pioneer* measurements also showed a cloud of neutrals along Io's orbital path. Ground-based measurements in the mid-1970s revealed ionized sulfur emission in the inner jovian magnetosphere, but on the opposite side of Jupiter from the position of Io at the time. Subsequent studies revealed this to be a plasma torus. The Io torus is a doughnut-shaped trail along Io's orbital path, made up almost exclusively of various charged states of sulfur and oxygen, thought to be derived from the break-up of volcanic sulfurous compounds (SO_2 and S_2). The ionized particles are held within the torus by Jupiter's magnetic field, in a similar way to the mechanism that holds charged particles in the Van Allen radiation belts around the Earth.

The first clues to Io's bulk composition came from measurements of Io's mean radius using a stellar occultation and from mass derived from the *Pioneer* flyby in 1973. The bulk composition of 3.54 g cm^{-3} indicated silicates were dominant on Io, but the surface's high albedo and cold temperatures indicated frosts. Telescopic observations using improved spectral reflectance techniques were used to attempt to determine Io's surface composition, and polysulfides were suggested as a possible coloring agent for the surface. The idea of sulfur on Io was strongly supported by laboratory experiments by W. Wamsteker, which showed that sulfur and its compounds matched the strong UV absorption and reflectance spectrum of Io, suggesting that these compounds might be abundant on Io's surface. However, the discovery of a strong absorption band near 4 μm could not be explained by sulfur. It was later found to be due to sulfur dioxide (SO_2), which is now known to be the dominant compound covering Io's surface. Other key discoveries during the 1970s were those of the Io sodium cloud in 1973 by R. Brown and in 1975 of a potassium cloud by L. Trafton.

The first indications of volcanic activity were given by infrared photometry and radiometry that showed higher brightness temperatures at 10 μm than at 20 μm, but the thinking at the time was that Io was a cold and dead world, and these observations remained puzzling. However, shortly before *Voyager 1* arrived at the Jupiter system in March 1979, several scientists published works that, in retrospect, are suggestive of active volcanism. In 1978, R. Nelson and B. Hapke reported a spectral edge at 0.33 μm and proposed that sulfur was the major contributor to this spectral feature. They suggested that the presence of allotropes of sulfur explain this and several other spectral features, and that these allotropes could be produced by melting yellow sulfur and subsequently quenching it, possibly "in the vicinity of a volcanic fumarole or hot spring." Astronomers F. Witteborn and colleagues reported a telescopic observation of an intense temporary brightening of Io in the infrared wavelengths from 2 to 5 μm. They explained it, although with some skepticism, as thermal emission caused by part of Io's surface being at a temperature of about 600 K, much hotter than the average expected daytime temperature of about 130 K. A few days before the *Voyager 1* flyby of Io, a seminal theoretical paper by Stan Peale and colleagues was published. They had studied the tidal stresses generated within Io as a result of the gravitational "tugs" from Jupiter and Europa. Their calculations showed that the possible heat generated by tidal stresses was in the order of 10^{13} W, much greater than heat that could be released from normal radioactive decay. Their prediction—that Io might have "widespread and recurrent volcanism"—was spectacularly confirmed by *Voyager 1*.

Active volcanoes were not immediately obvious in the first images returned by *Voyager 1*. The most striking aspect of Io shown in the first images was its colorful surface, with yellows, oranges, reds, and blacks. Scientists on the imaging team nicknamed Io the "pizza moon" and suggested the colors were likely due to large quantities of sulfur on the surface. Another surprising aspect was the absence of impact craters. The obvious conclusion was that Io's surface was very young and the craters must have been obliterated—but how? The answer came soon after, when a navigation engineer at the Jet Propulsion Laboratory, Linda Morabito, noticed a peculiar umbrella-shaped feature emanating from Io's limb in one of the images that was taken to aid navigation of the spacecraft (Fig. 2). The pattern turned out to be an eruption plume rising about 260 km above the surface. A second plume was found on the same image, and more plumes were seen upon close examination of various other images. Additional evidence for active volcanism came from another of *Voyager*'s instruments, the infrared interferometer spectrometer (IRIS), which detected enhanced thermal

FIGURE 2 Io's active volcanoes were discovered from this image, taken by *Voyager 1* on March 8, 1979, looking back 4.5 million km. The Pele plume is seen on the lower right, rising nearly 300 km above the surface. The bright spot near the terminator (shadow between day and night) is the top of the Loki plume, illuminated by the Sun.

FIGURE 3 *Voyager 1* image showing the Loki plume on the limb and the heart-shaped Pele plume deposit in the lower part of the image. When *Voyager 2* arrived 18 weeks later, the "heart" had become an oval, as material from the plume had filled out the area.

emission from parts of Io's surface—some areas had temperatures of about 400 K, much higher than the rest of the surface, which has noontime equatorial temperatures of about 107–124 K. When one of the hot areas was found to coincide with one of the plumes, there was no doubt that active volcanism was taking place.

Eighteen weeks after *Voyager 1*'s dramatic discovery, the companion spacecraft, *Voyager 2*, flew close to Io. Intense activity was still taking place, but significant changes had occurred between the two flybys, including the cessation of the largest plume, Pele (Fig. 3), and the altered shape of the deposits associated with this plume. An area of about 10,000 km^2 had been filled in, presumably by fresh material falling down from the plume. It became evident that dramatic changes of Io's surface could occur over short timescales.

Initial analysis of the *Voyager* observations showed 9 plumes and 9 hot spots, though not all plumes coincided with hot spots and vice versa. "Hot spot" is a term used by Io researchers to define a region of enhanced thermal emission, a sign of active volcanism. The *Voyager* IRIS experiment did not observe the whole surface, so it was suspected that other hot spots existed. The surface showed many features with morphologies similar to volcanic landforms on other planets, such as calderas (volcanic craters) and flows.

After the two *Voyager* spacecraft left the Jupiter system on their way to Saturn and beyond, the study of Io's

volcanism was continued from Earth by astronomers using infrared detectors mounted on telescopes. These observations showed that brightenings and fadings of hot spots occur, indicating variations in the level of volcanic activity. Observations by W. Sinton, R. Howell, J. Spencer, J. Rathbun, and other astronomers have shown that Io's most powerful hot spot, Loki, has brightenings that switch on in 1 month or less and last several months before fading. Telescopic observations were also used to analyze the reflected light from Io's surface to determine surface composition, confirming that it was dominated by sulfur dioxide (SO_2). Io was also observed by the *International Ultraviolet Explorer* satellite and by the *Hubble Space Telescope.*

The first spacecraft to orbit Jupiter was *Galileo*, which was able to image Io and monitor its volcanic activity from 1996 through early 2002. Galileo was designed to orbit the planet Jupiter for 2 years (1996–1997) and collect data on the planet's atmosphere, its moons, rings, and magnetic field. However, the failure of *Galileo*'s high-gain antenna to deploy (a problem discovered while the spacecraft was on its way to Jupiter) drastically reduced the quantity of images and data that could be returned to Earth. However, the mission still accomplished its objectives, thanks to the successful reconfiguration of the spacecraft's software to utilize a lower gain antenna and perform data compression on board.

These measures, along with changes made in the Deep Space Network (DSN), maximized the amount of data that could be returned. *Galileo* observations were so successful and spectacular that two mission extensions to gather additional data were approved. The *Galileo Europa Mission* (*GEM*) lasted from 1998–1999 and the *Galileo Millennium Mission* (*GMM*) from 2000–2002. These extensions were particularly important for Io because all the high-resolution remote sensing observations obtained by Galileo were collected during these mission extensions. The close Io flybys during which high spatial resolution remote sensing observations were collected happened in October and November 1999, February 2000, and August and October 2001. The main remote sensing instruments observing Io were the Solid State Imaging System (SSI), the Near-Infrared Mapping Spectrometer (NIMS), and the Photopolarimeter Radiometer (PPR).

3. Io's Surface

Io's phenomenal volcanic activity makes it the most geologically active object in the solar system. Remote sensing observations from *Galileo* revealed its surface in unprecedented detail and substantially changed our understanding of Io's geology and geophysics. Galileo's remote sensing instruments (visible and infrared) were used to study the surface features and volcanic activity, while the tracking of the spacecraft itself provided new constraints on the interior. Gravity measurements from tracking indicated that Io has a large iron/iron sulfide core and a silicate mantle. Galileo's close flybys of Io failed to reveal an intrinsic magnetic field, suggesting that little core convection is taking place.

The surface of Io contains three primary types of features (Fig. 4): (1) broad, flat, layered plains, which are partially covered with visible, diffuse **pyroclastic materials**; (2) volcanic structures including **paterae** (caldera-like depressions), **flucti** (lava flow fields), and **tholi** (shield volcanoes and other positive-relief structures); and (3) mountains of volcano-tectonic origin. The complementary imaging coverage of *Galileo* and *Voyager* has allowed these features to be mapped in a global scale, thus giving us a window into not only local but also global processes.

Between the paterae, mountains, and other major geologic features, Io's surface appears smooth except for scarps that cut across the plains. Some scarps are linear and occur in parallel groups, which suggest a tectonic origin. Other scarps, however, are irregular and appear to be erosional, sometimes forming a series of mesas or large plateaus. The presence of these features on Io is somewhat puzzling because of the lack of a significant atmosphere or flow of liquid water. Sulfur and sulfur dioxide, possibly escaping explosively from a subterranean "aquifer," have been suggested as the main eroding agent on Io's surface.

Volcanic features dominate Io's surface, and the volcanoes cover a wide range of sizes and present varying charac-

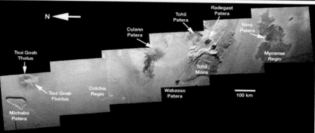

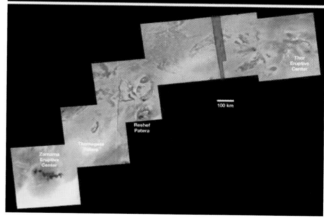

FIGURE 4   Mosaics of images acquired by *Galileo*'s camera of three regions that accentuate the different types of geologic materials and terrains on Io. (Top) The Chaac-Camaxtli region shows paterae in various sizes, shapes, and colors, indicating varying volcanic and tectonic influences on their formation (from Williams et al., 2002, *J. Geophys. Res.* **107**, 5068). (Middle) The Culann-Tohil region, which contains paterae, fluctii (lava flow fields), a mountain (Tohil Mons), and a volcanic construct (Tsui Goab Tholus). (From Williams et al., 2004, *Icarus* **169**, 80–97). (Bottom) The Zamama-Thor region, dominated by two eruptive centers, Zamama and Thor. Zamama has a long lava flow field and Thor was the site of the tallest eruptive plume seen on Io. (From Williams et al., 2005, *Icarus* **177**, 69–88). (Figure courtesy of David Williams, Alfred McEwen, and Moses Milazzo.)

teristics such as power output, persistency of activity, and association with plumes. Interestingly, most of Io's volcanoes manifest themselves as caldera-like depressions, referred to as paterae. Unlike terrestrial volcanoes, those on Io rarely build large topographic structures such as tall shields (like Mauna Loa) or stratovolcanoes (like Mount St. Helens). There are only a few tholi scattered across Io.

Io's surface shows a few remarkably large flucti. The lava flow field from the Amirani volcano is ~300 km long, the largest active flow field known in the solar system. Io's large lava flows are possibly analogs of the continental flood basalt lavas on Earth, such as the Columbia River Basalts in the United States. These ancient terrestrial flows were never directly observed, but they are suspected of producing major climatic effects.

A major question about Ionian volcanism after *Voyager* was the nature of volcanism—whether sulfur or silicates were predominant. Although temperature measurements from Galileo clearly showed that many hot spots have temperatures far too high for sulfur, the possibility that some sulfur flows occur on the surface cannot be ruled out. At the time of the *Voyager* flybys, Carl Sagan argued that the colorful flows around Io's Ra Patera volcano were sulfur (Fig. 5). Unfortunately, the flows could not be studied by *Galileo* as the area had been covered over by new eruptions before *Galileo*'s first observations in 1996. However, other locations may have sulfur flows. Most Ionian flows appear dark, but a few location show pale yellow or white flows that may well be molten sulfur. D. Williams and colleagues proposed that flows radiating from Emakong Patera may be sulfur and that low-temperature liquid sulfur (~450 K) could explain many of the morphological features seen around Emakong Patera, such as a meandering

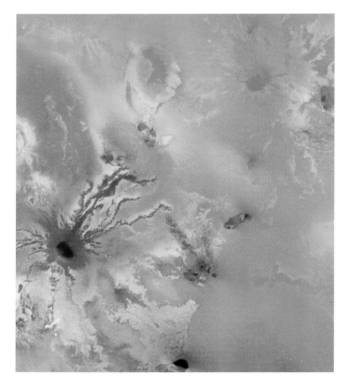

FIGURE 5 This image of Ra Patera volcano was taken by *Voyager 1* at a range of 128,500 km (77,100 miles). The width of the picture is about 1000 km. Ra Patera is the dark spot with the irregular radiating pattern of flows, which were interpreted as being sulfur.

channel 105 km in length that appears to feed a gray-white flow some 270 km in length. Infrared measurements using *Galileo* NIMS indicated temperatures less than 400 K inside Emakong caldera, and much cooler (below the instrument's detection capabilities) over the flows. However, *Galileo*'s instruments could not distinguish between sulfur flows or cooled silicates coated by bright sulfurous materials after erupting. One possibility, suggested by R. Greeley and colleagues, based on studies of a sulfur flow at Mauna Loa in 1984, is that rising silicate magma may melt sulfur-rich country rock as it nears the surface, producing "secondary" sulfur flows (as opposed to "primary" flows that originate from molten magmas at depth). Sulfur dioxide is ubiquitous on Io, and the colorful surface is thought to be the result of **sulfur allotrope** deposits, making the possibility of secondary sulfur flows likely. *Galileo* data of the volcanoes Balder and Tohil Paterae suggest that sulfur dioxide could be mobilized as "flows" in very cold regions. However, the presence of sulfur and SO_2, flows on Io have not been confirmed and whether these flows exist on Io remains an open question.

The most common type of volcanic feature on Io is the patera. Although the origin of paterae is still somewhat uncertain, they are thought to be similar to terrestrial volcanic calderas, formed by collapse over shallow magma chambers following partial removal of magma. Some paterae show angular shapes that suggest some structural control, indicating that they may be structural depressions that were later used by magma to travel to the surface. At least 400 Ionian paterae have been mapped. Their average diameter is ~40 km, but Loki, the largest patera known in the solar system, is over 200 km in diameter. In contrast, the largest caldera on Earth, Yellowstone, is ~80 km by 50 km in size. The larger sizes of the Ionian features probably reflect the much larger sizes of magma chambers.

Mountains are major structural landforms on Io and tower over the surrounding plains. Ionian mountains are defined as steep-sided landforms rising more than ~1 km over the plains. At least 115 mountains have now been identified and mapped. Io's mountains rise, on average, about 6 km high, with the highest rising 17 km above the surrounding plains. *Galileo* images revealed that many mountains are partly or completely surrounded by debris aprons, plateaus, and layered plains. Mountains appear to be unstable and are thought to be relatively short-lived features. They are not active volcanoes, but their origin is still uncertain. Various models have been proposed to explain the origin of the mountains. Their asymmetrical shapes suggest the uplift and rotation of crustal blocks, implying that compressional uplift is probably the dominant mechanism.

Neither the volcanic features nor the mountains appear to follow a distinct global pattern such as seen on the Earth, suggesting that, on Io, surface expressions of internal dynamics are subtle. However, the distribution of mountains and paterae is not totally random; both features are concentrated toward lower latitudes and follow a bimodal distribution with longitude. According to P. Schenk and colleagues, the greatest frequency of mountains occurs in two large antipodal regions near the equator at about 65° and 265°. In contrast, J. Radebaugh and colleagues studied the distribution of Ionian patera, and although they found the paterae to follow a similar bimodal distribution, the highest concentrations are 90° out of phase with that of the mountains. When only the hot spots known to be currently or recently active are studied, their distribution appears random, though no active (or inactive) volcanic centers have been detected at latitudes greater than 78°. The distribution pattern for volcanic centers is consistent with the pattern of heat flow from tidal heating in Io's asthenosphere predicted from simulations. The anticorrelation in the distribution of mountains and volcanic centers is further evidence that the two are not related, but the reasons for the anticorrelation are still unknown.

Because of the dynamic nature of Io's volcanism, its surface appearance can change in dramatic ways over time. Detectable changes occurred in the years between the *Voyager* and *Galileo* observations (1979–1995); however, many surface changes at the timescale of months have also been detected. One example is the change in the Pele plume deposit between the two *Voyager* flybys, which were spaced about 4 months apart. Surface changes are mostly due to new volcanic eruptions, particularly sulfur and sulfur dioxide from volcanic plumes and pyroclastic (ash and **tephra**) deposits. Other changes include new lava flows, increases in the area of flows, and changes in surface color. Volcanic materials have been observed to fade or disappear due to burial, alteration, radiation exposure, or erosion. Most surface changes have been detected at visible wavelengths; however, within individual volcanic centers, changes in temperature and sulfur dioxide coverage have been detected at infrared wavelengths. Most surface changes are localized and take place inside dark volcanic paterae that cover only 1.4% of Io's surface, or are ephemeral volcanic plume deposits that fade or change color on timescales of a few months to years. One surprise from the first *Galileo* observations was that Io's surface appearance remained largely the same since the last *Voyager* flyby. Based on the changes observed between the two *Voyager* flybys (4 months apart), major changes were expected in the years between *Voyager* and *Galileo*. Instead, more than 90% of Io's surface remained unchanged between *Voyager* (1979) and the end of the prime *Galileo* mission (1999).

Localized changes from major eruptions, however, can be dramatic. Two of these were particularly useful in the study of surface changes from *Galileo*. The eruption of the Pillan volcanic center in 1997 left a conspicuous "black eye" on Io's surface (Fig. 6), covering an area of about 200,000 km² and reaching distances up to 260 km from the source (Fig. 6). Later observations from *Galileo*'s SSI showed a spectral absorption at 0.9 μm in these and other

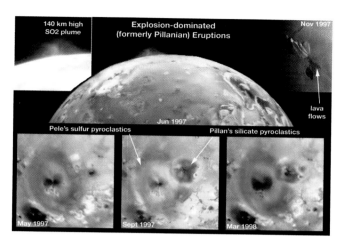

FIGURE 6 Explosion-dominated or Pillanian eruptions on Io occur in relatively brief (few months or less), intense outbursts that produce very high (possibly ultramafic) temperatures, plumes (top left), and rapidly emplaced lava flows (top right). Plumes can reach great heights (several hundred kilometers), and their deposits have produced black (Pillan), red (Pele, Tvashtar), and white (Thor) rings. These compositions are thought to be associated with silicate, sulfur, and SO₂ pyroclastic materials, respectively. (Figure courtesy of David Williams.)

dark materials on Io, suggesting silicate composition (most likely magnesium-rich orthopyroxene). The dark deposit at Pillan slowly faded between 1997 and 1999 as it was covered by red sulfurous deposits from nearby Pele.

Surface colors are the most easily observed manifestations of surface change. Galileo results brought new insights into the intriguing question of what causes the vivid colors of Io's surface. The global distribution of the different color deposits gives some clues to their origin and Galileo's repeated flybys allowed observations at different illumination angles, which affect how colors appear in images. Io's surface has four primary color units: most of the surface is yellow (about 40%), white-gray (about 27%), or red-orange (about 30%), while black deposits are localized around volcanic centers. Red and orange materials are interpreted as deposits of short-chain sulfur molecules (S_3, S_4). These are concentrated at latitudes higher than 30° north and south and, where they are thought to result from the breakdown of sulfur (cyclo-S_8) by charged particle irradiation. These red deposits at high latitudes appear to last longer than those at equatorial regions. At lower latitudes, patches of red materials are associated with hot spots and plumes and are thought to be formed by condensation from sulfur-rich plumes. These red plume deposits are ephemeral, lasting perhaps a few years if the deposit is not replenished.

The yellow materials that cover a lot of the surface are interpreted to be sulfur (cyclo-S_8), with or without a covering of sulfur dioxide (SO_2) frosts deposited by plumes, or alternatively polysulfur oxide and sulfur dioxide without large quantities of elemental sulfur. White-gray materials are interpreted to be composed of coarse- to moderate-grained sulfur dioxide that condensed from plumes and later recrystallized. Black areas (<2% of surface) mostly correlate with active hot spots and occur as patera floors, lava flow fields, or as dark diffuse materials near or surrounding active vents. These materials are most consistent with magnesium-rich orthopyroxene, indicative of silicate lava flows or lava lakes (within paterae) or diffuse silicate pyroclastic deposits near paterae. Perhaps the most intriguing materials on Io's surface are the small greenish yellow deposits seen in a few isolated patches in or near active vents, which are thought to be composed of either sulfur compounds contaminated by iron or silicates such as olivine or pyroxene with or without sulfur-bearing contaminants.

Detection of other substances on Io's surface has been difficult because sulfur dioxide condensed from volcanic plumes blankets most of the surface and hinders detection of other species. *Galileo* NIMS detected a broad absorption at about 1 μm, which had been seen from telescopic observations. However, it is still not known what substance this spectral absorption is due to, though NIMS observations showed that it is anticorrelated with recently emplaced lavas. NIMS also detected local patches of almost pure SO_2, in one case, in Balder Patera, topographically confined, raising the possibility that it was emplaced as a fluid.

4. Io's Volcanic Eruptions

Shortly after the *Voyager* mission, the major controversy about Io's volcanic activity concerned the nature of the volcanism: sulfur or silicates? Io's surface colors were interpreted as sulfur deposits and this, among other factors, made the sulfur volcanism hypothesis attractive. One way to distinguish between sulfur and silicate volcanism is to measure the temperature of the molten material because sulfur has a lower melting temperature than silicate lavas. Sulfur volcanism would not produce temperatures exceeding ~700 K (427°C), whereas basaltic lavas on Earth range from 1300 to 1450 K (1027–1177°C). The temperatures of the hot spots measured by the *Voyager* IRIS instrument were relatively low (below ~650 K) and could be consistent with either molten sulfur or silicates. However, *Voyager* instruments lacked the sensitivity and wavelength coverage needed to detect small areas at higher temperatures; hence, *Voyager* was "seeing" only the cooler areas, perhaps cooling silicate lava flows. Between the *Voyager* observations in 1979 and the *Galileo* observations that started in 1996, several of Io's hot spots were detected by ground-based telescopes. Temperature measurements using infrared detectors mounted on telescopes showed higher temperatures than had been measured by IRIS—such as 900 K reported by T. Johnson and colleagues in 1988, and 1225 and 1500 K reported by G. Veeder and colleagues in 1991. These measurements are consistent with silicate magmas but not with sulfur volcanism.

Galileo included much more sensitive instruments than *Voyager*, such as the SSI system sensitive from ~400 to 1000 nm wavelengths and the NIMS sensitive from 700 to 5200 nm, but both had limitations. SSI was able to detect only spots hotter than ~700 K and only when Io was in eclipse (in Jupiter's shadow) to eliminate reflected and scattered light. NIMS had the ideal spectral coverage for detecting both the temperatures and spectral reflectances expected from silicate lavas, but it had limited spatial resolution (120 km or more) except during the close Io flybys. However, *Galileo*'s instruments soon showed the hot spot temperatures to be indeed consistent with silicate rather than sulfur volcanism. The greatest surprise was the detection of very high temperature volcanism on Io when *Galileo*'s NIMS and SSI instruments observed a vigorous eruption at Pillan in 1997 (Fig. 6). The results, reported by A. McEwen and colleagues in 1998, provided evidence of temperatures exceeding 1500 K at several hot spots and, in the case of the Pillan eruption, temperatures of about 1800 K. The Pillan eruption temperatures are higher than any seen on lavas erupting on Earth now and in recent times. It is possible that very high temperature lavas (>1500 K) are typical for Io, although more rigorous measurements are required. The question remains open whether the Pillan eruption, because it was so unusually vigorous, allowed the detection of large areas at very high temperatures, or whether the eruption was unusual in its composition. A third possibility that cannot be ignored is that errors were underestimated in the Pillan temperature calculations.

Nevertheless, it is clear that several Ionian eruptions detected by *Galileo* had minimum eruption temperatures hotter than current terrestrial basaltic eruptions; what types of lavas were erupted on Io? The most popular explanation is that the lavas are ultramafic (komatiite-like) in composition. Komatiites and komatiitic basalts are ultramafic volcanic rocks on Earth that are rich in magnesium and dominated by olivine or pyroxene. Color data on Io's dark volcanic materials obtained from *Galileo* indicate the presence of orthopyroxene. Komatiitic lavas are perhaps the closest analogs to the lavas erupted at Pillan. These lavas have very rarely been erupted on Earth since the Proterozoic, about 1.8 billion years ago. Therefore, studying Io's current volcanism may lead to a better understanding of the emplacement of lavas on the ancient Earth.

Another hypothesis to explain Io's hottest eruption is superheating. Magma can be superheated by rapid ascent from a deep, high-pressure source. Melting temperatures of dry silicate rocks increase with pressure; therefore, the erupted lava can be significantly hotter than its melting temperature at surface pressure. Rapid ascent of basaltic magmas resulting in ~100° Celsius of superheating should be possible. However, no record of such an eruption is known on Earth.

It is important to note that at present there are no direct measurements of the composition of Io's lavas. The most critical question about Io's volcanism—the composition of the erupting magma and crust—remains open.

Not all Ionian eruptions are vigorous like Pillan. On Earth, volcanic eruptions are often classified depending on their character—effusive, explosive, very explosive—and these eruptions are often named after volcanoes or locations where they have been studied (e.g., Hawaiian, Icelandic, Strombolian, Pelean). High-resolution observations and temporal data on Io's hot spots showed that some broad generalizations can also be made for Ionian eruptions. The majority of Ionian eruptions can be placed in three classes initially designated "Promethean," "Pillanian," and "Lokian," though a single hot spot can exhibit more than one eruption style over time. Explosion-dominated (Pillanian) eruptions (Fig. 6) have an intense, short-lived phase that may correspond to the outbursts detected from Earth. These eruptions originate from either paterae or fissures and produce extensive dark lava flow fields and dark pyroclastic deposits through short-lived, high effusion rate, vigorous activity. These events may or may not include eruption episodes with large (>200 km high) explosive plumes, which can produce large plume deposits such as that around Pillan itself.

Less intense but more persistent flow-dominated eruptions (Promethean) are named after the Prometheus hot spot (Fig. 7), which has a persistent plume about 100 km high, active during both *Voyager* encounters in 1979 and throughout the *Galileo* mission. Surprisingly, distant images obtained by Galileo in 1996 showed that the Prometheus plume site had moved about 80 km west since 1979, but its size and appearance had not changed. A new lava flow linked the old and new plume sites. Images and infrared observations obtained in 1999 showed that the main vent

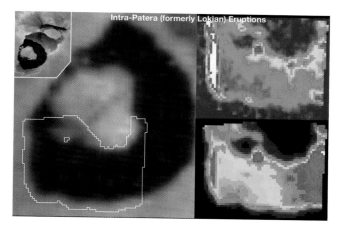

FIGURE 8 Intra-patera or Lokian eruptions are confined to paterae (caldera-like depressions) and are thought to represent the resurfacing of paterae floors by lava flows or overturning lava lakes. The style of eruption is exemplified by Loki (inset upper left, *Voyager* image). In this montage, *Galileo* NIMS images track the temperature changes across the floor of Loki. The top right image is at 2 μm, showing the hottest areas, while the bottom image, at 4 μm, shows the distribution of cooler areas. Note that the center of Loki (which appears white in the SSI image) is cold at infrared wavelengths.

of this volcano was near the *Voyager* plume site and that the plume, not the volcano, had moved west. The plume's movement was modeled in terms of the interaction between the advancing hot lava and the underlying sulfur dioxide snowfield by Susan Kieffer and colleagues. The movement of lava flows on Earth over marshy ground can give rise to small, short-lived explosive activity, but nothing on the scale of the Prometheus plume has ever been observed. This type of eruption may be common on Io and, once the flow stops moving, the plume eventually shuts off, as has been observed at the Amirani volcano. The lava flows associated with these eruptions can be quite extensive and are thought to be emplaced through repeated small breakouts of lava, similar to the slowly emplaced flow fields at Kilauea in Hawaii.

Intrapatera (Lokian) eruptions (Fig. 8) are confined within the caldera-like paterae. These eruptions are thought to be lava lakes, some of which are possibly overturning. Observations from *Galileo* flybys showed that lava lakes are abundant on Io, and they may be a significant mechanism for heat loss from the interior. Io's most powerful hot spot, Loki, is thought to be a giant lava lake that perhaps undergoes periodic overturning, leading to brightenings that have been observed from Earth for decades. Many other hot spots on Io appear to be persistent lava lakes.

5. Heat Flow and Interior

Observations of Io have also provided knowledge about the satellite's interior, where the tidal heat is being dissipated, driving the volcanic eruptions. Observations from

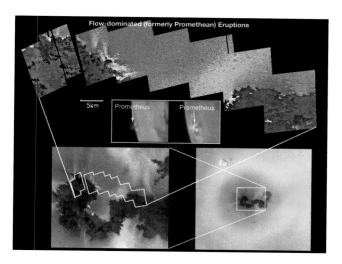

FIGURE 7 Flow-dominated or Promethean eruptions are relatively long-lived (months to years) and are associated with long-lived plumes and flow fields. In this montage, *Galileo* SSI views of the Prometheus plume (center) are surrounded by increasingly higher resolution views of the Prometheus flow field. Note also the bright plume deposit forming an annulus in the lower right image.

Earth and from spacecraft have shown that the heat that comes out of Io and is radiated into space, called the heat flow, is very large compared with that of the Earth and other planets. The heat flow is measured at infrared wavelengths and the portion due to reflected sunlight is calculated and subtracted, taking into account Io's surface albedo. The difference is the heat flow due to volcanism and originates in Io's interior.

The first estimate of Io's heat flow was done by D. Matson and colleagues in 1981, who reported a value of 2 ± 1 W m^{-2}. In 1991, G. Veeder and colleagues reported measurements of Io's heat flow compiled from 10 years of ground-based photometric observations in the range 5–20 µm and estimated a minimum average heat flow of 2.5 W m^{-2}. The latest estimates of Io's heat flow are from J. Spencer and colleagues in 2002, using *Galileo* PPR measurements. They reported 2.2 ± 0.9 W m^{-2}, which is in close agreement with the first estimate by Matson. This range of values is very high even compared to geothermal and volcanic areas on Earth such as Yellowstone. Io's heat flow is about 200 times what could be expected from heating due to the decay of radioactive elements and illustrates how crucial tidal heating is to drive Io's active volcanism. The heat flow is not uniform over the surface; in fact, it is dominated by the Loki hot spot.

An unresolved problem for Io is that there is a discrepancy between observed values of heat flow and theoretical estimates expected from steady-state tidal heating models over the course of Io's history. The current estimates of heat flow from observations are about twice the predicted value. If the theoretical estimates are correct, then Io's heat flow must have varied over time due to its orbital evolution. Other studies also suggest that Io's current heat flow and tidal heating rate are higher than the long-term equilibrium value. One suggestion is that Io is spiraling slowly inward, losing more energy from internal dissipation than it gains from Jupiter's tidal torque. The resolution of the apparent discrepancy between the observed and theoretical heat flow will have important implications for understanding not only the evolution of Io but also that of Europa and Ganymede.

What is happening deep within Io? Our studies of Io's interior, including the lithosphere and mantle, are still in their early stages, but the *Galileo* spacecraft made some significant contributions. The properties of Io's interior determine how tidal forces deform the body. Therefore, one can use measurements of the deformation (variations in shape) of Io to get information on internal structure. Data obtained from images and radio tracking of the spacecraft as it came close to Io during several flybys provided this information. Variations in the spacecraft's motion revealed distortions in Io's gravitational field, which provided evidence that Io is differentiated.

Io is about the same size as the Earth's Moon, but it has a higher density, indicating that there is more iron on Io than on the Moon. On the basis of density alone, it can be inferred that Io has a large metallic core. The size of this core can be inferred from the density and spacecraft measurements of Io's shape, assuming Io is in hydrostatic equilibrium. Work by M. Segatz and colleagues using *Voyager* observations revealed the basic structure of the interior, Galileo measurements have been used to refine this knowledge. Io is thought to be a 2-layer body, consisting of a large metallic core of iron and sulfur and a silicate mantle. *Galileo's* magnetometer instrument failed to reveal a magnetic field which can be interpreted as evidence that that Io's core is either completely solid or completely liquid. Because Io's mantle is hot, it seems likely that Io has no magnetic field because it has a completely liquid core that is kept from cooling and convecting by the surrounding hot mantle. Other key measurements made by *Galileo*, including the discovery of widespread, high-temperature, silicate volcanism and tall mountains have contributed to a model of Io's interior. If the mountains are formed as thrust blocks, then Io's lithosphere must be at least as thick as the tallest mountains ($\sim$15 km).

The discovery of very high temperature volcanism on Io has strong implications for the interior. The idea that Io's crust is ultramafic (magnesium-rich) seems inconsistent with the well-understood process of magmatic **differentiation**. Heat flow on Io is sufficiently high that Io was expected to have undergone partial melting and differentiation hundreds of times, producing a low-density crust, depleted in heavy elements like magnesium (as mantle rocks begin to melt, the first component to melt has a lower density and segregates and rises toward the surface, while the heavier components sink). One possibility, proposed by L. Kezthelyi and colleagues in 2004, is that Io has a completely molten core and a crystal-rich ("mushy") magma ocean. Widespread ultramafic volcanism would be a natural consequence of this model because the upper mantle would consist of orthopyroxene-rich magma with about the same density as the overlying crust. As lavas are deposited, the crustal layers sink and are eventually mixed back into the magma ocean, so a low-density crust cannot form. However, this model may not allow for sufficient tidal heat generation to occur, and thus may be inconsistent with the heat flow observed. Another possibility, suggested by W. Moore and colleagues in 2005, is that local processes such as tidal forcing through cracks may account for the very high temperatures ($>$1400 K) observed at some hot spots on Io.

6. Atmosphere, Torus, and the Jupiter Environment

Io orbits Jupiter at a distance of about 421,800 km, which is deep within the jovian magnetopause. Io has both a patchy but relatively large atmosphere and an ionosphere, and there is considerable interaction of Io with the jovian magnetic field. Io's atmospheric density is low (about 10^{-9} bar), equivalent to good laboratory vacuums on Earth, but the density is greater at the locations of active volcanic plumes. The main constituent of the atmosphere is SO_2, which is supplied largely by volcanic plumes, with a lesser amount coming from evaporation of the SO_2 frost deposits on the surface. Io's low gravity allows some of the atmosphere to escape, but it is continuously replenished by volcanic outgassing.

Since the time of the *Voyager* flybys, Io has been known to produce volcanic plumes hundreds of kilometers high, which serve as an efficient delivery mechanism for gas and dust particles into the magnetosphere and the space surrounding Io, although only a relatively minor amount of atmospheric gas is lost to space. The dynamics of Io's plumes are very complex, particularly because models of plume emplacement have to take into account the very low atmospheric pressure on Io.

One of the last surprises from Galileo observations was the detection of four large plumes at high northern latitudes. Prior to 2000, there was no detection of high latitude plumes by Galileo, though deposits on the surface indicated that plume activity had occurred in the past. The largest plume known on Io (500 km high) was detected from images obtained in August 2001, shortly after the *Galileo* spacecraft had flown through it. Observations by the plasma science experiment indicated the presence of SO_2 molecules in the plume. This in situ measurement is consistent with others that show the presence of SO_2 in plumes and SO_2 frost in plume deposits. Sulfur (S_3 and S_4), in addition to SO_2, was detected in the Pele plume from measurements made from the *Hubble Space Telescope* by J. Spencer and colleagues.

The temperatures of the frost deposits on Io's surface are sufficiently low that cold-trapping of SO_2 by condensation is a very important process. Some material does escape Io, forming a corona and neutral clouds, and the Io torus further away. The corona refers to the region within Io's gravitational pull, where bound and escaping atoms and molecules populate a low-density shell. The neutral clouds of sodium, oxygen, and sulfur extend from the corona to distances of many times the radius of Jupiter.

An important discovery made during the *Galileo* mission was Io's aurora. The aurora (Fig. 9) was detected through color eclipse imaging with the camera while Io was in Jupiter's shadow. The vivid colors detected (red, green, and blue) are caused by collisions between Io's atmospheric gases and energetic charged particles trapped in Jupiter's magnetic field. The green and red emissions are probably produced by mechanisms similar to those in Earth's polar regions that produce terrestrial aurorae. The green (actually yellow) glow comes from emission from sodium ions, whereas the red glow is associated with oxygen ions. The bright blue glows mark the sites of dense plumes of volcanic vapor and may represent the locations where Io is electrically connected to Jupiter via a flux tube.

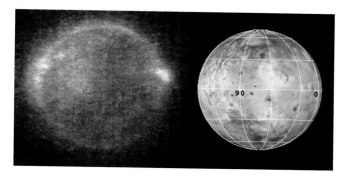

FIGURE 9   Io aurora. Bright blue glows represent sulfur dioxide excited by electrical currents flowing between Io and Jupiter in the flux tube. The blue glow on the right is over the Acala hot spot, which is thought to be the site of a "stealth plume" (composed of mostly gas and therefore hard to detect from images). Red and green (actually yellow) glows represent atomic oxygen and sodium, respectively.

Observations from the *Pioneer* spacecraft were the first to reveal a cloud of neutrals along Io's orbital path. The most easily observed of these neutral clouds around Io is the sodium cloud. The cloud is populated by sodium atoms escaping Io at about 2.6 km s^{-1}. It appears as a diffuse yellowish emission produced by scattered light from volcanic plumes and Io's lit crescent. This emission comes from neutral sodium atoms within Io's extensive material halo that scatter sunlight at the yellow wavelength of about 589 nm. Although neutral sodium atoms are most easily detectable by spectroscopy, it has been determined through extensive Earth-based telescopic studies that sodium is a minor component of the neutral material escaping from Io. The primary neutral elements in the cloud are oxygen and sulfur, which are thought to have dissociated from sulfur dioxide gas (SO_2) expelled from many of Io's active volcanoes at a rate of $\sim$1 ton s^{-1}. So far, sodium has not been detected on Io's surface or plumes, but its existence is inferred because of its detection in the cloud.

The Io torus is a doughnut-shaped trail about 143,000 km wide along Io's orbital path. The torus is made up almost exclusively of various charged states of sulfur and oxygen, thought to be derived from the break-up of volcanic SO_2 and S_2. The ionized particles are held within the torus by Jupiter's magnetic field, in much the same way that charged particles are held in the Van Allen radiation belts around the Earth. Measurements made by the *Galileo* spacecraft during its close flybys showed that the plasma in the torus is slowed by Io's ionosphere, redirected around Io and then reaccelerated in Io's wake. Other *Galileo* measurements showed that Io strongly perturbs Jupiter's magnetic field. These perturbations vary with time, suggesting that Io's variable volcanic activity influences the density of the plasma torus and the strength of its interactions with the jovian magnetic field. [*See* PLANETARY MAGNETOSPHERES.]

7. Outstanding Questions and Future Exploration

On September 21, 2003, the *Galileo* mission came to the end after 14 years in space, when the spacecraft disintegrated in the dense atmosphere of Jupiter. The demise of the spacecraft was planned, since its onboard propellant was nearly depleted and it was considered prudent to avoid any chance of impact with Jupiter's moon Europa in the future, which could have happened if the spacecraft had been left in orbit around Jupiter. Although *Galileo*'s mission significantly advanced our knowledge of Io, the failure of the high-gain antenna to open (and subsequent low data rates) prevented all but a very small part of Io's surface to be imaged at high resolution. Future exploration by spacecraft is needed to reveal Io's surface in detail at a variety of different wavelengths and to answer many outstanding questions. The geometry of *Galileo*'s orbit around Jupiter resulted in lack of coverage at high resolution of Io's Jupiter-facing side, which should be a priority for future missions to observe. However, even parts of the surface previously imaged are likely to change because of the dynamic nature of Io, so new missions will always reveal new features. At the time of writing, there are no missions to Io planned, but there are plans in place to observe Io during the Jupiter flyby of the *New Horizons* spacecraft in early 2007, while the spacecraft is on its way to Pluto and the Kuiper Belt. Although many of the outstanding questions may only be answered by a dedicated mission to Io, significant advances are possible from missions of opportunity, such as *New Horizons*, and from ground-based and space telescope–based programs. Io has been successfully observed by the *Hubble Space Telescope* and can potentially be observed from the Spitzer Infrared Telescope Facility as well as from future orbiting telescopes. Ground-based observations using Adaptive Optics, such as those by Imke de Pater and Frack Marchis, have been a major step forward in the study of Io because the spatial resolution of these observations can now rival some of those obtained from *Galileo*.

One of the most significant questions raised by *Galileo* concerns the nature of Io's high-temperature volcanism. If ultramafic compositions are involved, as is the current favored hypothesis to explain the very high temperatures, it is difficult to explain how the magma composition would have stayed ultramafic throughout Io's history because differentiation would have been expected, leading to evolved types of magmas such as those we find in present-day Earth. It is possible that the current style of volcanic activity is a geologically recent phenomenon (i.e., Io has only recently attained its resonant orbit with resulting tidal heating) or that the response of Io's lithosphere-mantle to tidal heating has prevented extreme differentiation. Perhaps the magmas are not ultramafic, but are basaltic, possibly superheated during ascent. Compositional measurements of Io's fresh magmas would be invaluable for future missions to obtain. Another intriguing question considers what the **volatiles** in the magma are dissolved in. The presence of explosive volcanism on Io is evident from the plumes and dark deposits that are thought to be ash and magma fragments. On Earth, the most common volatile is water, on Io sulfur and sulfur dioxide have been detected in the plumes, but are there other compounds?

Other aspects of Io's geology are also intriguing. How are the mountains formed? Nearly half of Io's mountains are located adjacent to volcanic centers (paterae), but they do not appear to be part of the volcanic system. What can that relationship tell us about Io's crust? Questions also abound about Io's atmosphere and the interaction between Io and the jovian magnetosphere, particularly the recognition of the flux tube that allows the transfer of charged particles between the two bodies. What are the sources of the atoms, neutrals, and ions that are released into the plasma torus and magnetosphere, and what physical processes allow them to escape? These are just some of the key questions that have developed about Io after *Galileo* and that require further analysis of existing data sets, and probably further data obtained by new missions.

Bibliography

Bagenal, F., McKinnon, W., and Dowling, T., eds. (2004). "Jupiter: Planet, Satellites and Magnetosphere." Cambridge Univ. Press. Cambridge, UK

Geissler, P. E. (2003). Volcanic activity on Io during the Galileo era. *Annu. Rev. Earth Planet. Sci.* **31**, 175–211.

Johnson, T. V. (1999). Io. In "The New Solar System," 4th Ed. (J. K. Beatty, C. C. Petersen, and A. Chaikin, eds.), pp. 241–252.

Kargel, J. S., et al. (2003). Extreme volcanism on Io: Latest insights at the end of the *Galileo* era. *EOS* **84**, no. 33.

Kivelson, M.G., Khurana, K. K., Russell, C. T., and Walker, R. J. (2001). Magnetic signature of a polar pass over Io. *EOS* **82**.

Lopes, R., and Gregg, T. K., eds. (2004). "Volcanic Worlds: Exploring the Solar System Volcanoes." Praxis Publishing Company (Springer-Verlag).

Lopes, R., and Spencer, J. R., eds. (In press). "Io after Galileo." Praxis Publishing Company (Springer-Verlag).

Lopes-Gautier, R. (1999). Volcanism on Io. In "Encyclopedia of Volcanoes" (H. Sigurdsson et. al., eds.), pp. 709–726. Academic Press. San Diego

Lopes, R., and Williams, D. (2005). Io after *Galileo*. *Rept. Progr. Phys.* **68**, 303–340.

McEwen, A. S., Lopes-Gautier, R., Keszthelyi, L., and Kieffer, S.W. (2000). Extreme volcanism on Jupiter's moon Io. In "Environmental Effects on Volcanic Eruptions: From Deep Oceans to Deep Space" (J. Zimbelman and T. Gregg, eds.), pp.179–204. Plenum. New York

Morrison, D., ed. (1982). "Satellites of Jupiter." Univ. Arizona Press, Tucson.

Nelson, R. M. (1997): Io. In "Encyclopedia of Planetary Sciences" (J. H. Shirley and R. W. Fairbridge, eds.), pp. 345–351. Chapman and Hall.

Spencer, J. R., and Schneider, N. M. (1996). Io on the eve of the *Galileo* mission. *Annu. Rev. Earth Planet. Sci.* **24**, 125–190.

Europa

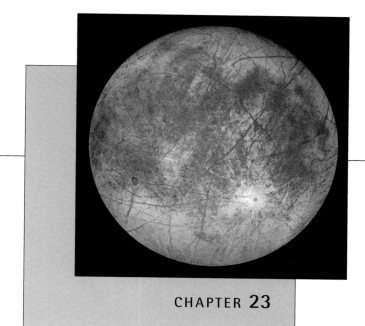

Louise M. Prockter

Johns Hopkins University
Applied Physics Laboratory
Laurel, Maryland

Robert T. Pappalardo

Jet Propulsion Laboratory
California Institute of Technology
Pasadena, California

CHAPTER **23**

1. Introduction and Exploration History

Europa and her sibling satellites were famously discovered by Galileo in 1610, and less famously by Simon Marius at essentially the same time, but it took almost 4 centuries before any detailed views of their surfaces were seen and the grandeur of the **Galilean satellites** was revealed. In the 1960s, ground-based telescopic observations determined that Europa's surface composition is dominated by water ice, as are most other solid bodies in the far reaches of the solar system.

The *Pioneer 10* and *11* spacecraft flew by Jupiter in the 1970s, but the first spacecraft to image the surfaces of Jupiter's moons in detail were the *Voyager* twins. *Voyager 1*'s closest approach to Jupiter occurred in March 1979, and *Voyager 2*'s, in July of the same year. Both *Voyager*s passed farther from Europa than from any of the other Galilean satellites, with the best imaging resolution limited to 2 km per **pixel**. These images revealed a surface brighter than that of the Earth's moon, crossed with numerous bands, ridges, cracks, and a surprising lack of large impact craters or high-standing topography (Fig. 1). Despite the distance from which the images were acquired, they were of sufficiently high resolution that researchers noted some of the dark bands had opposite sides that matched each other extremely well, like pieces of a jigsaw puzzle. These cracks had separated, and **ductile** dark icy material appeared to have flowed into the opened gaps. This suggested that the surface could have once been mobile. The relative youth of Europa's surface is demonstrated by a lack of large impact craters—*Voyager* images showed only a handful—which are expected to build up over time as a planetary surface is constantly bombarded by meteorites over billions of years, until the surface is covered in craters (such as on Mercury). A lack of craters implies that something has erased them—such as volcanic (or in Europa's case, **cryovolcanic**) flows or **viscous relaxation** of the icy crust. Researchers studying the *Voyager* data also noted that the patterns of some of the longest linear features on the surface did not fit with predicted simple models of global stresses that might arise from tidal interactions with Jupiter. However, if the shell was rotated back a few tens of degrees, the patterns fit exceptionally well to a model of **nonsynchronous rotation**, by which the icy surface had slowly migrated with respect to the satellite's tidal axes. This mechanism probably requires a ductile layer between the surface ice and the deeper interior. Combined with the observations of dark bands, there were tantalizing hints that perhaps Europa had a warm interior at some time in the past, and perhaps still today. Increasingly sophisticated theoretical models of **tidal heating** of Europa, discussed later in more detail, suggested that a global subsurface ocean might exist within Europa today.

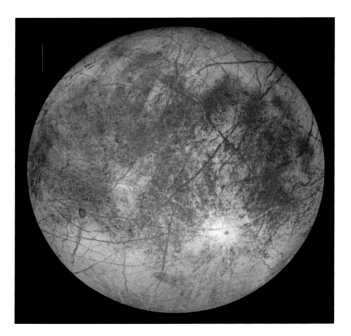

FIGURE 1   Global view of Europa's trailing hemisphere, acquired early in the *Galileo* spacecraft's tour of the jovian system. The colors in this image have been enhanced to show detail. This image shows the extent of the reddish-brown material that mottles Europa's surface, along with numerous linear features, many of them over 1000 km in length. Two large impact craters are also visible: Pwyll, surrounded by bright rays, is thought to be the youngest crater on Europa, and Callanish is the large circular feature toward the west. (NASA/JPL.)

Such intriguing findings meant there was much anticipation for the *Galileo* mission, which launched from the Space Shuttle *Atlantis* in 1989 and entered orbit around Jupiter in 1995. The primary mission included dropping a probe into Jupiter's atmosphere, as well as observations of all the satellites and Jupiter's atmosphere and local environment. Despite severe data rate limitations during the mission (because the spacecraft's main antenna did not open), the data from *Galileo* were so intriguing that the mission was extended in 1997 for a further 3 years, in order to make 8 further close flybys of Europa, and also to study its volcanically active neighbor, Io. Data from the extended *Galileo* Europa Mission afforded many more high-resolution images of Europa, as well as magnetic data that strongly imply the presence of a briny layer beneath the surface today.

The physical and orbital properties of Europa are summarized in Table 1.

2. Formational and Compositional Models

During the formation of our solar system, the growing gas giant planet Jupiter pulled material from the solar nebula. It is now understood that, in contrast to early models of satellite **accretion**, the solids of the Jovian **subnebula** were probably grabbed from the solar nebula in nearly primordial form, and that the subnebula may have been gas-poor. Thus, the material incorporated into the Galilean satellites was probably similar in composition to the asteroids of the outer asteroid belt, containing ice, **anhydrous silicates**, **carbonaceous** material, and nickel–iron metal alloy.

The Galilean satellites formed by aggregation of these solids, with the proportion of ice varying with distance from the warm protoplanet Jupiter. Io formed relatively close to Jupiter, so it was not able to accrete and retain significant amounts of water ice. As the next moon outward, Europa formed as a mostly rocky satellite (density = 3.0), able to accrete sufficient volatiles to form a ~100 km thick outer layer of H_2O. In the colder reaches of the jovian subnebula, Ganymede and Callisto formed with near-equal amounts

TABLE 1	Properties of Europa
Discovered	**1610**
Discoverers	Galileo Galilei, Simon Marius
Mean distance from Jupiter	671,100 km
Radius	1560.8 ± 0.5 km
Mass	$(4.8017 \pm 0.000014) \times 10^{22}$ kg
Density	3.014 ± 0.005 g/cm³
Orbital period	85 hours (3.551 Earth days)
Rotational period	85 hours (3.551 Earth days)
Orbital eccentricity	0.0094
Orbital inclination	0.469°
Visual geometric albedo	0.68
Escape velocity	2.026 km/s
Spacecraft visitors	*Voyager 1* (March 1979)
	Voyager 2 (July 1979)
	Galileo (July 1994 to 2000)
Predicted average surface Temperature	~50K (poles) to ~110K (equator)

of rock and ice. If the jovian subnebula were cold enough, some lower temperature condensates such as CO_2 could have been incorporated as Europa and the other Galilean satellites formed.

Europa's early heat of accretion combined with heat from radioactive decay would have warmed the satellite's interior and formed a primordial ocean, which was likely reduced and sulfidic. Thermal and geochemical evolution would have caused some oxidation of the ocean through time, forming sulfates. Refined models of Europa's accretion and chemical evolution are bringing improved understanding of the satellite's initial conditions.

3. Stress Mechanisms and Global Tectonic Patterns

3.1 Europa's Internal Structure

Although numerous models exist for the thickness and thermal state of Europa's shell, there is a lack of information regarding key input parameters such as the thermal and mechanical properties of its ice. These models can put bounds on the shell thickness and make predictions about how it has varied over time. Such predictions can be compared to geological and geophysical measurements. Measurements of Europa's gravity field from the *Galileo* spacecraft constrain its **moment of inertia**. Further assumptions of likely composition and density of its internal layers suggest that Europa has a ~100 km thick layer of H_2O overlying a rocky **mantle**, which surrounds an iron-rich **core** (Fig. 2).

The most definitive evidence that there is an ocean within Europa at the present time comes from the *Galileo* spacecraft's magnetometer, combined with theoretical studies. Because Jupiter's powerful magnetic field is tilted by 10° relative to the planet's equatorial plane in which the satellites rotate, the satellites experience Jupiter's magnetic field as time-varying. For Europa, each 5.5 hours the satellite finds itself alternately above then below the magnetic equator of Jupiter. Surprisingly, the Galileo magnetometer measured a magnetic field in the vicinity of Europa, which alternately flips to oppose the external jovian magnetic field. This implies that Europa is behaving as a conductor, generating an induced magnetic field in response to the jovian field. Modeling of the *Galileo* observations suggests that there is a conductive layer—probably a briny ocean—possibly many tens of kilometers deep, within the outer portion of Europa.

The thickness of the ice shell overlying the ocean is significant for models of Europa's thermal evolution, geological processes, and astrobiology. Future missions will want to sample material from the ocean to understand Europa's potential for life, but the means to accomplish this task are dependent upon the ease by which material from the ocean can be accessed, and the ways in which this material may have been processed. Because the thickness of Europa's

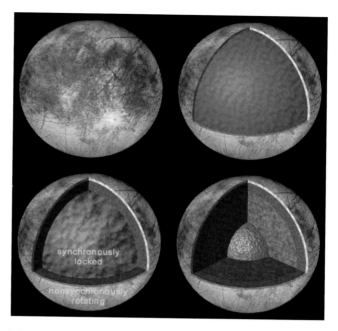

FIGURE 2 Interior structure of Europa. Rocky mantle (brown) and iron-rich core (gray) are synchronously locked in position with respect to Jupiter, but the ice shell (white) may rotate nonsynchronously—slightly faster than the interior—if decoupled by the water layer (blue). Layer thicknesses are not to scale. (NASA/JPL/Brown University.)

ice shell cannot be determined from gravity and magnetic data, we must search for clues in the geophysical history and geological record preserved in the icy surface.

3.2 Tidal Evolution

The principal energy source that heats Europa's interior and drives its tectonics today comes from its orbital interaction with Jupiter. An **orbital resonance** occurs when two satellites have orbital periods that are related by integer relationships, allowing them to exert a gravitational influence over each other and affecting the eccentricity of their orbits. The three Galilean satellites are involved in the **Laplace resonance**, in which the orbital periods of Ganymede:Europa:Io are in a near 1:2:4 ratio, but more important, the mutual **conjunctions** of the Io–Europa pair and of the Europa–Ganymede pair **precess** around Jupiter at precisely the same rate. Like a child on a swing pushed at the optimal moment, the recurring mutual conjunctions force and maintain eccentricities in their orbits (Fig. 3). Although it is not known exactly when or how the moons came to form the precise clock that is the Laplace resonance, one model suggests this resonance was progressively achieved after Io moved outward into a near 2:1 resonance with Europa, and then the Io–Europa pair moved outward until Ganymede was captured into its own near 2:1 resonance with Europa. Europa's forced eccentricity is key to its youthful and complex surface, as will be described in the following sections.

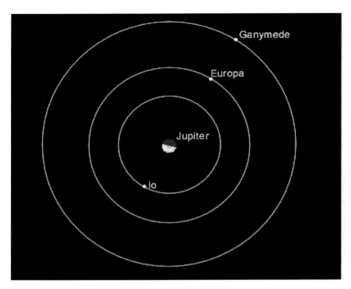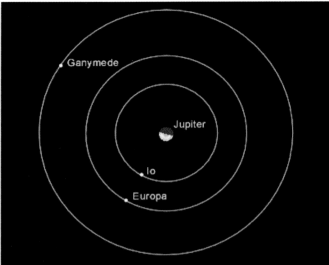

FIGURE 3   Mutual conjunctions resulting from the Laplace resonance of Io, Europa, and Ganymede. Io completes approximately 4 orbits to every 2 of Europa's, while Ganymede orbits approximately once during the same time period. (Left) Mutual conjunction of the Europa–Ganymede pair. (Right) Io and Europa experience a mutual conjunction one Io day later, while Europa has moved along half an orbit, and Ganymede has progressed through one-quarter of its orbit. The resonance forces Europa's eccentricity to be nonzero, causing tidal heating and geological consequences.

3.3 Tidal Heating

Europa's Laplace resonance with siblings Io and Ganymede causes it to have a slightly **eccentric** orbit ($e = 0.0094$). This eccentricity causes Europa to move closer to and farther from Jupiter (at **perijove** and **apojove**, respectively) as it moves along its 85 hour orbit, causing the satellite to undergo increasing and decreasing gravitational pull from Jupiter. At the same time, Europa undergoes **libration** as it orbits Jupiter, its tidal bulge necessarily rocking from side to side as the moon's orbital velocity changes but the rotation rate stays constant. Europa deforms by ~1–30 m over each orbital period (Fig. 3), and the dissipation of **strain** energy resulting from this deformation causes the interior to warm. Dissipation of tidal energy can happen in several ways, such as by friction along **faults**, turbulence at liquid–solid boundaries, and **viscoelastic heating** at the scale of individual ice grains. It is likely that a great degree of tidal energy is currently dissipated at the base of Europa's icy shell, just above the interface between the ice and the underlying liquid ocean, where the ice is warmest and most deformable on the time scale of the satellite's orbit. This regular input of energy is believed to be sufficient to keep Europa's ocean liquid.

3.4 Diurnal Stressing

In addition to heating, the tides induced by Europa's eccentric orbit are believed to be responsible for the majority of its tectonic processes, including formation of the cracks on its surface. As Europa orbits, its radial and librational deformation results in **diurnal stresses** (so named because Europa's day is equal to its orbital period). These stresses are relatively small (= 0.1 MPa), but they are apparently sufficient to crack Europa's ice shell, producing regions of extensional and compressional stresses that migrate across the surface, changing in direction and magnitude as Europa moves through its eccentric orbit. The magnitude of distortion and thus stress due to tidal flexing depends on a satellite's interior structure. If Europa's shell were completely frozen, there would be very little distortion overall, with a tidal amplitude of about 1 m, whereas if there is a liquid water ocean beneath the ice shell, the surface is predicted to distort by up to 30 m during an orbit. (In comparison, the Earth's rocky moon, which has a relatively cold interior, deforms by ~10 cm over each orbit due to tides raised by the Earth.)

Diurnal stresses have been invoked to explain some of Europa's unusual surface features, such as cracks and ridges (Fig. 4), and likely contribute to the relatively youthful surface age. Most dramatically, diurnal stresses can explain the unusual **cycloidal** shapes of some ridges and bands on Europa, due to the changing direction and magnitude of stresses, as further discussed later. Moreover, diurnal stresses tend to rotate anticlockwise in the northern hemisphere, and clockwise in the southern. These stress rotations are likely responsible for the observed preponderance of left-lateral **strike-slip** faults observed in Europa's northern hemisphere and right-lateral strike-slip faults in the southern.

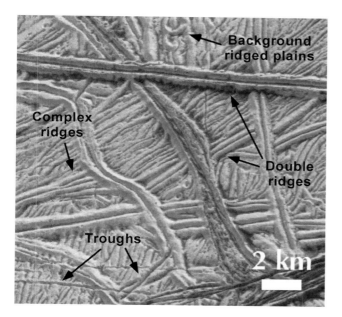

FIGURE 4 *Galileo* image of a typical portion of Europa's ridged plains, showing ridges and troughs criss-crossing and overprinting each other. Double ridges are most common, but complex ridges with more than two crests also occur, along with simple troughs. The background terrain is so heavily overprinted that it is no longer possible to distinguish individual features. (NASA/JPL.)

3.5 Nonsynchronous Rotation

If a satellite is in a perfectly circular orbit around its primary, it keeps the same face toward its parent planet, and this tidal bulge remains fixed relative to the planet. Moreover, most of the solar system's large satellites, including the Earth's moon, rotate synchronously, keeping one hemisphere always facing their parent planet. Because of Europa's eccentricity, however, a net torque tends to cause Europa to rotate slightly faster than synchronously. Europa's massive rocky interior is expected to maintain a permanent mass asymmetry to counter this effect, so beneath the icy shell, Europa is probably synchronously locked (as is the Earth's moon). However, since Europa's icy shell is decoupled from the rocky interior—likely by liquid water—the ice shell can rotate independently, and slightly faster than the interior (Fig. 2). The rate of this "nonsynchronous rotation" is not known, and it might not be constant through time, but a lower limit for one complete rotation of the shell is thought to be in the region of 10,000 years, based on comparisons of *Voyager* and *Galileo* images. Regardless of the actual rate of rotation, the nonsynchronous stresses are expected to be large—many times larger than the diurnal stresses and potentially sufficient to open deep (kilometer-scale) cracks in the ice shell. The orientations of Europa's major lineaments do not correspond to the current patterns expected from tidal stresses alone, but if the shell is backrotated by moving it "back in time" by ~30° westward in longitude,

there is an overall good fit of lineaments to the predicted stresses. This implies that Europa's observable global-scale lineaments may have formed over about 60° of nonsynchronous rotation of the ice shell. Mapping of crosscutting relationships among lineaments in some areas of Europa suggest that the ice shell has completed at least one full rotation, and may have also undergone a small amount of **polar wander** (i.e., tilt relative to the spin axis).

4. Landforms on Europa

Europa exhibits two primary types of terrain: the bright ridged plains criss-crossed by bright and dark linear features and mottled terrain, which shows evidence for **endogenic** disruption and modification of the surface. Each type of terrain and the **morphologies** of their constituent **landforms** are discussed in detail next.

4.1 Ridges, Troughs, and Bands

Europa's linear features are ubiquitous, covering most of the satellite's surface. These landforms exist at a variety of sizes and scales, and exhibit a number of different morphologies, some of which have not been observed on any other solar system body. Many of these linear features have overprinted and offset one another, sometimes by several kilometers, making it difficult to piece together the history of the surface. Some ridges have shallow topographic depressions and/or fine-scale fractures alongside them, which are suggestive of loading of the lithosphere either by the weight of the ridge material from above or from withdrawal of material from below. Understanding how ridged plains form and evolve is important to the question of where liquid water exists within Europa, how it is involved in the formation of surface landforms, and possible niches for life.

4.1.1 INDIVIDUAL TROUGHS

The simplest of Europa's landforms, troughs (commonly called "cracks") may be several hundred kilometers long and less than a few hundred meters wide (Fig. 4). They can have subtle rims or none, and are generally V-shaped, suggesting an origin as tension fractures. Some have undergone mass wasting along their sides, and some troughs have elevated flanks and appear to be transitional forms between simple troughs and double ridges. As discussed previously, Europa's troughs probably originate by tensile cracking due to diurnal and nonsynchronous stresses.

4.1.2 DOUBLE AND COMPLEX RIDGES

The most ubiquitous landform on Europa, ridges are most commonly found in a "double ridge" form, with two parallel ridges separated by an axial V-shaped trough (Fig. 4). Double ridges can be from ~0.5 to ~2 km wide, and some span thousands of kilometers in length. Slopes of these ridges tend to be near the **angle of repose** (~30° BA, and at some, preexisting topography can be traced up the flank, suggesting that they may have formed by upwarping of the surface. Most ridges tend to be relatively linear, or only gently curved, and mass wasting is prevalent along ridge flanks. The cycloidal ridges discussed later in this section have notably arcuate shapes, but are otherwise morphologically identical to other double ridges on Europa. Another form of ridge is the "complex" ridge, which may have from three or more subparallel ridge crests (Fig. 4). Some complex ridges appear to be sets of several double ridges, running parallel to, or in some cases intertwined with, each other, while others seem to be composed of bundles of ridge crests separated by intervening troughs.

Ridge formation on Europa is not yet fully understood, and several models have been suggested. Europa's cracks are probably modified by other processes to form ridges with distinct and uniform crests. In one model, ridges form through the buildup of cryovolcanic material erupted from fissures, as is the case with many terrestrial eruptions such as those that form Hawaiian volcanoes. A major drawback with this model is that it is hard to explain the remarkable uniformity of ridge crests, and the distinct V-shaped trough along their axes. An alternative model suggests that double ridges form in response to cracking and subsequent rise of warm or compositionally buoyant ice. Possibly aided by tidal heating, the buoyant ice intrudes and lifts the surface to form ridges. Although this model does explain some observations, such as why some ridge flanks have preexisting terrain running up them, it does not explain how multiple ridges might form within complex ridges.

Another model proposes that the ridges form in a manner similar to pressure ridges in arctic sea ice. In this model, cracks created by diurnal tidal stresses allow water to seep up from the ocean below, filling the crack and partially freezing into a slurry. It is envisioned that diurnally varying tidal stresses would then push the crack margins back together, and this partially frozen ice is easily smashed up, forming a jumbled pile of ice that squeezes out of the crack. Although the process that forms pressure ridges is understood well on the Earth's sea ice, where the ice is thin and ocean currents cause movement of the ice, it is unknown whether Europa's ice is sufficiently thin and mobile as to pull apart atop the liquid layer. Even if Europa's ice is thin, it is not clear that this model can explain the morphology of Europa's ridges, including distinct V-shaped troughs along double ridges, their uniform parallel ridge crests, and the apparently upwarped features along some ridge flanks.

Alternative models have suggested that the cracks instead penetrate upward from the ocean into the ice shell, and that liquid injected into cracks from beneath then upwarps the surface. This model could explain the general morphology of ridges, but it has difficulty explaining the uniformity of the crests and the morphologies of complex ridges.

The model that seems to best fit the observations of ridge morphology is one in which a fracture forms along the surface and then undergoes **shear stresses** and strike-slip motion as a result of Europa's diurnal tides. This strike-slip motion along the crack produces frictional heating as the walls of the crack rub past each other, warming the subsurface ice (Fig. 5). This shear heating may trigger warm ice

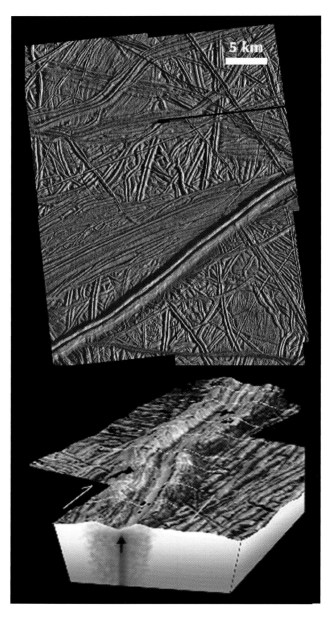

FIGURE 5 Double ridges are ubiquitous on Europa, but their origin is not well understood. In the shear heating model, strike-slip, or shear, motion along a fracture results from diurnal tidal stresses. Friction between the fracture walls warms the ice, softening or partially melting it. Warm ice close to the fracture rises buoyantly, upwarping the ridge crests. Downward drainage of melt may aid formation of the axial depression. (Image: NASA/JPL. Diagram: Topography courtesy B. Giese, DLR.)

to well up beneath the crack, forming ridges, and perhaps inducing partial melting beneath the ridge axis at the same time. This model predicts that a ridge a few hundred meters high could be built by upwelling warm ice in only a decade or so, and because both sides of the crack are subject to the heating, the ridges would be expected to be of uniform width and height, as is observed. If shear heating were sufficient to induce partial melting below the ridge, it would tend to drain downward, perhaps forming the V-shaped axial trough above. This model would soften the ice along the ridge, perhaps enabling contractional deformation to occur in response to compressional stress. A model in which contraction occurs across ridges may be viable based on reconstruction of preexisting features and kinematic arguments. If ridges do hide contraction along initially extensional structures, as in the shear heating model, then ridges could help to balance the abundant extension on Europa that is represented by its pull-apart bands, as discussed later. We may ultimately find that ridge formation is a combination of several models, but they currently remain an enigma.

4.1.3 CYCLOIDAL RIDGES

While most double ridges are linear in overall planform, cycloidal ridges are shaped like a chain of distinct arcs (Fig. 6). Cycloidal ridges and some other structures on Europa's surface are likely explained by the action of diurnal stresses. If a fracture propagates slowly enough—at about walking speed—the rotation of **tensile stresses** over a Europan day occurs on a timescale such that the propagating fracture can be affected by these changing stresses, tracing out an arc instead of a straight path. As Europa moves in its orbit, the tensile diurnal stresses will drop below the critical value needed for fracture propagation, until the next orbit, when tensile diurnal stresses again increase above the critical value for fracture propagation, generating the next cycloid arc. This model requires that the diurnal stresses needed to crack the ice and create cycloidal fractures be relatively small, just a few tens of kilopascals. Ridges would evolve from cycloidal fractures in a manner similar to the formation of other ridges, and some pull-apart bands with scalloped margins may have pulled apart along cycloidal ridges. Tides imparted by Jupiter's gravitational pull would be insufficient to crack the surface into cycloidal patterns if Europa had no ocean, so the presence of the cycloid features is strong argument for the existence of an underlying ocean at the time the fractures formed.

4.1.4 TRIPLE BANDS

One specific type of lineament consists of a bright central ridge, flanked by patchy, diffuse, low-albedo margins, hence the term "triple band" (Fig. 7). These are most commonly larger ridges. It has been suggested that the dark flanks were created by the eruption of icy cryovolcanic material (similar to some explosive volcanoes on Earth), which either seeped out along the ridge flanks or rained dark **pyroclastic** material onto the surface alongside the ridge. Another possibility is that **intrusions** of ice that is warmer than its surroundings might result in local **sublimation** of icy surface materials leaving a layer of more **refractory** dark deposits. The unusual brightness of the central ridge relative to the flanks is as yet unexplained: It may be coated by frost or depleted in dark materials.

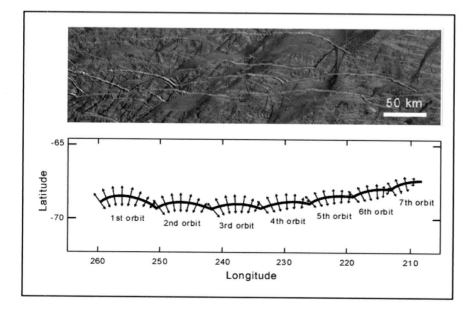

FIGURE 6 (Top) Cycloidal ridges on Europa. (Bottom) Model for cycloidal ridge formation, in which each arc forms during one orbital cycle. A crack initiates when stresses reach a critical value and then propagates slowly enough that the changing diurnal stress field affects its orientation, causing it to curve. When the stress drops below a critical level, the crack ceases propagation until the stresses are once again sufficiently large to reinitiate cracking. When the crack reinitiates, Europa has moved along in its orbit, and the stresses are now in a different orientation, leading to a sharp cusp as the next arc begins to propagate. (Top: NASA/JPL. Bottom: After Hoppa et al., 1999.)

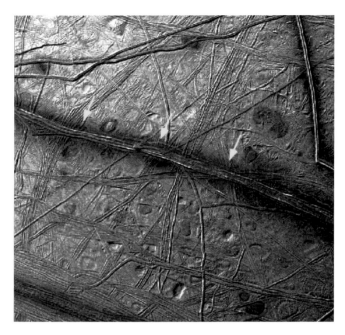

FIGURE 7 An example of a triple band, consisting of a central ridge about 5 km wide that is flanked on each side by diffuse, dark material (arrows). This material can be patchy and discontinuous, and may be related to cryovolcanic eruptions during formation of the band, although the exact mechanism is poorly understood. (NASA/JPL.)

4.1.5 PULL-APART BANDS

Polygonal dark and gray bands on Europa's surface have margins that can be closed together almost perfectly, reconstructing structures that were apparently laterally displaced when the bands formed along fractures. Many bands are bounded along their margins by an individual ridge, suggesting that a double ridge was split along its axis during band formation (Fig. 8). These structures have been termed "pull-apart" bands, and are a clear indication of movement of a brittle surface layer atop a more viscous, yet mobile, subsurface. Where the bands pulled apart, dark, probably low-**viscosity** subsurface material moved up to fill the gap. Limited topographic data across bands suggest that many stand somewhat higher than the surrounding terrain, consistent with formation by upwelling buoyant ice, rather than liquid water. Bands have been shown to have brightened over time, possibly because of frost deposition or radiation damage, leading to a wide range of brightnesses ranging from relatively dark, through gray, to as bright as the brightest background plains on the surface.

Almost all bands exhibit bilateral symmetry, with V-shaped central troughs and hummocky textures, and some have zones of ridges and troughs parallel to the central axis, which may include faults (Fig. 9). Morphological comparisons between bands and terrestrial midocean ridges suggest that band formation may have been analogous to

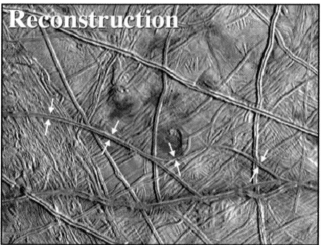

FIGURE 8 Points along a large gray band (arrows, top) can be reconstructed if the band is removed (arrows, bottom), with the preexisting terrain matching up perfectly along the margins. The band appears to have exploited two existing double ridges during its formation. Reconstructions like this show that a completely new surface has been created by band formation, suggesting that they represent a considerable amount of extension of Europa's surface. (After Prockter et al., 2002.)

seafloor spreading centers on the Earth, where plates are pulled apart and new volcanic material erupts along the spreading axis. Features on both planets exhibit central troughs and subparallel ridges where volcanic (in Europa's case, cryovolcanic) material has apparently erupted intermittently through the spreading process. Newly formed terrain at terrestrial midocean ridges undergoes normal faulting as it cools and moves away from the ridge crest. On Europa, normal faults parallel to the central axis may have similarly formed as new band material cooled sufficiently for faulting to take place (Fig. 9).

The major difference between terrestrial plate tectonics and Europan band formation is the lack of subduction zones on Europa. Thus, because band formation has clearly resulted in a large amount of extension (many tens of percent in some areas), there must be some mechanism for balancing this extension elsewhere on Europa's surface. Some fraction of Europa's extension is related to net global expansion, as would be the case if the ice shell were thickening with time.

4.1.6 FOLDS

Analysis of high-resolution images of Europa has identified regional-scale contractual folds in a handful of regions on Europa. The most apparent are identified in the band Astypalaea Linea, where they appear as subtle hills and valleys; several have warped the band at a wavelength of ~25 km, with fine-scale fractures along the crests of the hills and small compressional ridges within the valleys (Fig. 10). It is unlikely that such folds can represent the primary mechanism by which the icy satellite's considerable surface

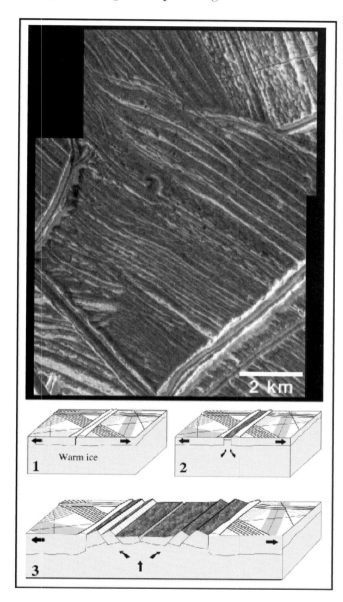

FIGURE 9   Model for band formation suggests they are analogous to midocean ridges on Earth. (Top) Distinct morphological zones are mirrored on either side of the central axis of this band. Closest to a central trough, the terrain is hummocky and relatively fine-textured. Further away from the axis, the terrain breaks into normal faults. These types of terrains are also found at spreading centers on midocean ridges, leading to suggestions that Europa's bands form in a similar way (cracking followed by extension), allowing new, warmer ice to well up to fill the gap. As this material cools and moves away from the central axis, it thickens enough that it can form normal faults. This process is analogous to the way new seafloor forms on Earth. (After Prockter et al., 2002.)

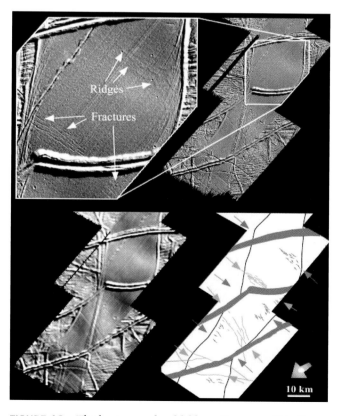

FIGURE 10   The best example of folds on Europa, within the gray band Astapalea Linea. (Top) Close examination of the band reveals fine-scale ridges and fractures. (Bottom left) If a low-pass filter is applied to the image, the 25 km wavelength folds can be distinguished. (Bottom right) Map showing sets of ridges (blue arrows) within the fold valleys and fractures (green arrows) that mark the fold crests. (Prockter and Pappalardo, 2000.)

extension has been accommodated. Other features (such as ridges), along with net global expansion (from freezing of its ice shell), may play important roles, but the mystery of how Europa's surface extension is balanced is yet to be solved.

4.2 Lenticulae and Chaos

Much of Europa's surface is covered with dark terrain with a mottled appearance, termed "mottled terrain" from *Voyager* images. High-resolution *Galileo* images show that in these areas the surface has been endogenically disrupted at small and large scales.

4.2.1 LENTICULAE

Many areas of Europa's surface are disrupted by subcircular to elliptical pits, spots, and domes, and microchaos regions (collectively termed "lenticulae"), which are ~10–15 km in diameter, with a variety of morphologies (Fig. 11). Domes can be convex with upwarped but unbroken margins where they meet the plains. Pits are topographically low areas where the surface has downwarped while preserving the preexisting terrain. Many of these features are associated with dark plains material that **embays** surrounding valleys in the ridged terrain, so it was probably relatively fluid when emplaced. Spots were apparently flooded with dark plains material. Lenticulae known as "microchaos" typically consist of a fine-scale hummocky material, including embedded small plates of preexisting material, commonly with some associated dark plains material. These microchaos regions resemble the larger chaos terrains described later.

Although a range of dome, pit, and spot sizes exists, there is a strong preferred diameter of ~10 km. This consistency in size and the range in their morphologies suggests that

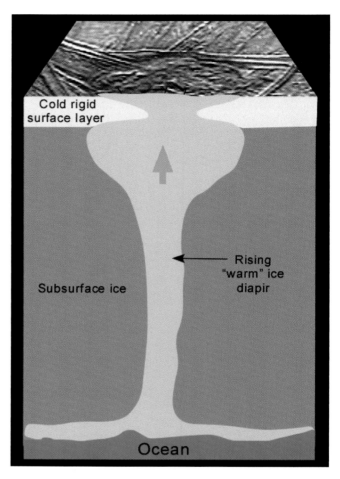

FIGURE 12   Model for the formation of lenticulae through diapiric upwelling of buoyant warm ice.

they are genetically related; the size and range are consistent with an origin from convective upwelling of buoyant ice **diapirs** within Europa's icy shell (Fig. 12). Convection is predicted within a tidally heated ice shell greater than about 20 km thick overlying a liquid water ocean. The ice may be either thermally buoyant (commonly referred to by the counterintuitive term "warm" ice) or compositionally buoyant, where the rising diapiric ice is "clean" relative to its surroundings. Compositional buoyancy of diapirs is possible if they are cleaned out of low-melting-temperature substances (e.g., salts, see Section 5), allowing the clean ice to be more buoyant than the surrounding salty ice. In this model, domes form by buoyant diapirs that would reach and break through the surface, and pits may form when a diapir does not quite make it to the surface but softens and/or melts out impurities from the ice above it, allowing the surface to sag downward.

The range of morphologies and levels of degradation that are observed in microchaos regions supports the suggestion that upwelling diapirs may partially melt pockets of briny ice

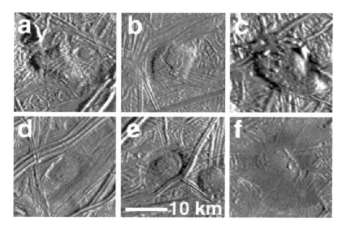

FIGURE 11   Lenticulae are found in a range of morphologies, including domes (a, b), microchaos (c), pits (d), and combinations of these morphologies (e), which may or may not have dark plains material associated with them (f). (After Pappalardo et al., 1998.)

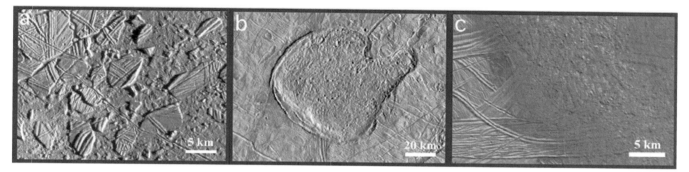

FIGURE 13 Examples of chaos on Europa. (a) Conamara Chaos exhibits distinct plates of preexisting terrain (see also Fig. 14). (b) Murias Chaos, a region of fine-textured chaos that has apparently overflowed its margins on one side, depressing the surrounding plains. (c) The edge of Thrace Macula, showing fine-textured material and a hint of preexisting terrain within, suggesting that the preexisting plains material has disaggregated in place. Dark plains material from Thrace has embayed the surrounding ridged plains, suggesting that it was relatively fluid when it was emplaced. (After Prockter et al., 2004.)

as they rise to the surface, causing disaggregation into matrix material and local flooding by dark, low-viscosity melt.

4.2.2 CHAOS

Chaos regions are areas in which kilometer-scale blocks of existing ridged plains material have translated and rotated with respect to one another within a mixed-albedo matrix of hummocky material (Fig. 13). The matrix material can be low-lying or high-standing relative to the surrounding plains. In one area, Conamara Chaos, at least 60% of the

preexisting terrain has been replaced with or converted into matrix material, and the matrix in part stands above the surroundings. Some of the broken plates have been rotated and/or moved by as much as several kilometers. These plates can be reassembled like a jigsaw puzzle, reconstructing portions of preexisting ridges and troughs (Fig. 14), although much of the original surface has been destroyed.

Chaos regions have been interpreted as places where Europa's heat flow has been enhanced, and where local melt-through of ocean water to the surface may have occurred. In such a model, the blocks are analogous to

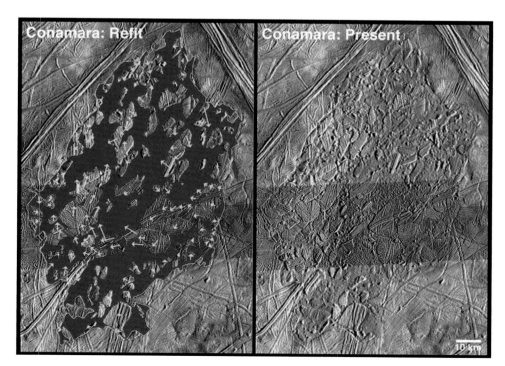

FIGURE 14 Broken plates of material within the hummocky matrix of Conamara Chaos (left) can be reconstructed into their original positions like a jigsaw puzzle, by matching up older lineaments. This exercise shows that the plates may have moved by several kilometers (arrows show approximate amount of displacement), indicating that the matrix material was originally mobile. Most of the original terrain is missing, however, and may have been subsumed or disaggregated during formation of the chaos. (After Spaun et al., 1998.)

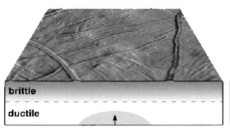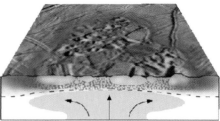

FIGURE 15   Model showing how rising diapir may impinge upon brine-rich ice (reddish material) lowering the melting temperature and thermally disaggregating the surface (right). (See Collins et al., 2000.)

icebergs floating buoyantly on top of the watery matrix. This model requires that Europa has a very thin shell, less than ~6 km; otherwise, the warm base of the ice shell would flow to maintain its thickness faster than the ice shell could melt from below. Moreover, this model requires that the ocean is only weakly **stratified** in temperature and salinity because if stratification were strong, then heat could not be transferred from the ocean floor to the base of the ice shell. In addition, a large, concentrated mantle heat source would need to be stable for hundreds of years. If Europa has a tidal energy budget that scales to Io's (i.e., an icy shell overlying the Io-like tidally heated mantle), then it could potentially have sufficient heat sources for surface melt-through, but the actual level of mantle activity is unknown.

A proposed alternative model for chaos formation is analogous to that for lenticulae, where ice diapirs have risen buoyantly through the ice crust, breaking or otherwise interacting with the surface (Fig. 15). This mechanism would explain why some chaos areas stand several hundred meters above the surrounding plains, something that is hard to explain if they formed atop liquid water, but feasible if buoyant diapirs rose to their level of neutral buoyancy. It has been suggested that partial melting of a salty ice shell could allow surface material to flow. If matrix material is a mixture of disaggregated ice and low-melting-temperature brines, then partial melting could explain the apparent fluidity of materials associated with many chaos regions and the mobility of blocks. Given the morphological similarities between lenticulae and chaos, it seems entirely plausible that they have similar origins through diapiric upwelling. It is possible that chaos terrains form from a number of separate lenticulae that link together by fractures, forming distinct plates that can separate and mobilize. The diapiric model for chaos formation is not the whole story, however; it does have difficulty explaining partial melting of the matrix because initially warm ice diapirs would be expected to cool significantly as they approach the surface, before they would be able to rotate and translate surface crustal blocks. It is possible that tidal heating would concentrate in the warm ice of a rising diapir, countering its cooling. If so, chaos would represent yet another manifestation of the tidal effects imposed by Jupiter and the Laplace resonance.

4.2.3 IMPACT STRUCTURES

Although formed in ice, rather than silicate rock, Europa's craters have the same range of morphological features as craters on other bodies, including bowl shapes, central peaks, bright ray systems, and **secondary crater** fields. Europa's craters are shallower than those formed on silicate bodies, however, probably because of viscous relaxation of the ice in which they form. Another difference is the size at which the transition from simple, bowl-shaped craters to more complex craters with central peaks occurs. On the Moon, a rocky body with similar gravity, this transition occurs at ~15–20 km in diameter, while on Europa it occurs at only ~5–6 km, presumably because the ice crust is relatively weak compared to rock. Simple or bowl-shaped craters are too small to undergo rim collapse or other significant modifications during formation.

The 24 km diameter crater Pwyll (Fig. 16) is thought to be the youngest large impact crater on Europa because it exhibits a bright ray system that extends for over ~1000 km and can be seen in global views of the satellite. These rays

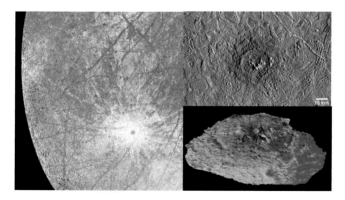

FIGURE 16   Several views of the crater Pwyll. (Left) Global color-enhanced view showing bright ejecta and rays from material thrown over 1000 km by the impact. Their white color indicates exposure of fresh, icy material. (Top right) Higher resolution image of Pwyll, showing distinct rim and central peak, along with ejecta around the impact. (Bottom right) Topographic model of Pwyll, created from stereo imaging. This shows the fresh but shallow topography of the crater. (NASA/JPL/DLR.)

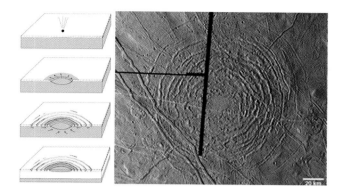

FIGURE 17 (Right) The Tyre impact structure has numerous rings around a smoother central region. Outside the rings are many small craters, which are secondary craters caused by ejecta from the impact. This false-color image highlights the reddish material associated with the Tyre rings. (Left) Model for formation of multiring basins on icy satellites. This model may be applicable to structures like Tyre, which are thought to have formed in a relatively thin, brittle layer (white) over a fluid layer (gray). (Image: NASA/JPL.)

overlie everything in their path, and their brightness suggests they are so young that they have not been darkened or significantly eroded by charged particle irradiation or micrometeorite bombardment. Pwyll's distinctive topography and central peak imply that it formed in relatively solid ice, rather than a thin layer of ice overlying liquid water or slush.

One of Europa's largest impact structures is Tyre (Fig. 17), with a diameter of ~44 km. Tyre, along with one other known feature named Callanish, are multiringed structures, somewhat analogous to impact basins on the terrestrial worlds. Tyre's rim crest is difficult to identify and it exhibits a complex interior with a smooth, bright central patch interpreted to be impact melt, or frozen remnants of fluid material that may have been emplaced from below during the impact event. Tyre's most striking characteristic is its concentric troughs and fractures, tectonic features resulting from the impact process. These structures are thought to originate when an impact occurs into relatively fluid material, allowing for rapid collapse and infill of the **transient crater**, and dragging the cold and brittle overlying crust inward to break along concentric faults (Fig. 17). *Galileo* Near-Infrared Mapping Spectrometer (NIMS) observations of Tyre show that dark material associated with the troughs is similar to the reddish material seen elsewhere on the surface.

Europa's simple and complex craters have morphologies consistent with impact into a solid (though warm and weak) ice target. In contrast, the larger (~40 km) impacts inferred to have formed Tyre and Callanish, with their distinctive rough topography and concentric ring systems, imply penetration of the transient crater to a fluid layer at a depth of

~20 km. These observations are consistent with Europa's solid ice shell being ~20 km thick, overlying a fluid layer that is probably Europa's liquid water ocean.

5. Surface Composition and Thermal State

It has long been known from Earth-based telescopic observations that Europa's surface is predominantly covered with water ice, as amply confirmed by *Galileo*'s NIMS instrument. However, the composition is distinctly different in Europa's darker regions, which are associated with many landforms such as ridges and chaos (Fig. 18). This material is thought to contain impurities such as **hydrated** salts, along with a reddish component.

Spectra from the NIMS instrument show highly distorted water bands in the dark regions, indicative of one or more hydrated minerals (Fig. 19). These deposits have been interpreted to indicate the presence of hydrated salt minerals, sulfates, and possibly carbonates. Some thermal evolution models of Europa predict large quantities of magnesium sulfate hydrate within the ocean, and mixtures of this and sodium hydrate are predicted on Europa. Another candidate for the surface material is sulfuric acid (H_2SO_4) hydrate (more commonly known to us as battery acid). Although all of these candidate compounds are colorless, irradiation of the surface by charged particles may be the reason why Europa's dark areas appear reddish. Because of its proximity to Jupiter, Europa's surface is constantly

FIGURE 18 Composite image of a false-color NIMS infrared image overlain on a monochrome camera image. Blue areas represent relatively clean, icy surfaces, while redder areas have high concentrations of dark, non-ice materials, which may be from a subsurface ocean. The infrared image is about 400 km across. (NASA/JPL.)

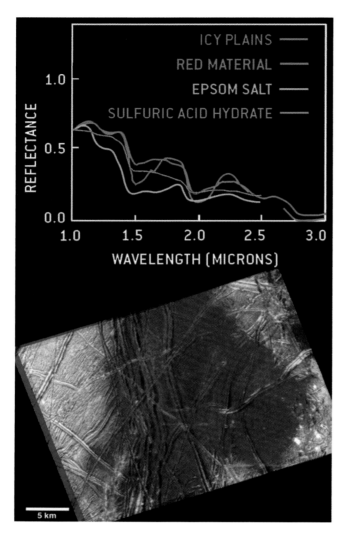

FIGURE 19   Infrared spectrum (top after Pappalardo, 1999, Scientific American) of Castalia Macula (bottom), one of the reddest, darkest spots on Europa. The spectrum of the icy plains material is distinct from the red material of the spot, which is more similar to something like epsom salt or hydrated sulfuric acid. (Image: NASA/JPL.)

irradiated by ions and electrons. This radiation is sufficient to rip apart molecules of water ice and other compounds, allowing them to recombine in a process known as **radiolysis**. This could allow sulfur ions (at least some of which likely originate on Io) and sulfur-containing compounds such as sulfuric acid to synthesize long molecular chains that are ochre in color. These sulfur chains may be responsible for the reddish color of material that has been emplaced on the surface relatively recently. Sulfuric acid itself could result from the breakdown and recombination of ice and sulfur dioxide frost, which has also been detected on Europa.

Generally the stratigraphically youngest features on Europa are the darkest, implying that the darkening and red-

dening process is rapid relative to the age of observable surface features. (Also, the fact that Europa's older features are relatively bright implies that some other process brightens features over time, as discussed below.) Whatever their specific origin, the close association of these hydrated minerals with areas of presumed surface disruption suggests they are related to endogenic processes and may have originated in the subsurface ocean.

Strong absorptions in the infrared region of the spectrum by Europa's H_2O-bearing minerals easily mask the signatures of minor constituents; however, hydrogen peroxide (H_2O_2) is observed and is probably a radiolysis product of water ice. An ultraviolet absorber identified on the trailing side of Europa has been attributed to sulfur from Io, delivered to Europa's surface via the jovian magnetosphere.

The *Galileo* spacecraft carried a Photopolarimeter Radiometer (PPR) instrument that showed that temperatures at low latitudes are in the range 86–132 K, with higher temperatures where the surface is dark, and colder temperatures where it is bright. This inverse correlation between brightness and temperature holds on a global scale, but significant local temperature variations are inferred below the spatial resolution of the PPR instrument. These may be due to local-scale variations in surface physical properties, and a distinct anomaly around the crater Pwyll may imply a relatively warm ejecta blanket. Other thermal variations such as lower than expected temperatures on the equator at dusk are harder to explain. These may be due to variations in grain sizes and structures of ice, but endogenic heat fluxes indicative of interior activity cannot be ruled out.

6. Surface Physical Processes

Processes affecting Europa's surface materials are dominated by thermal processing and radiation bombardment, with meteorite bombardment playing a lesser role.

Jupiter's magnetosphere sweeps up and traps particles including electrons, protons, and heavy ions such as S and O. Because of its close proximity to Jupiter, these particles result in a high-energy (<10 MeV) radiation flux at the surface of Europa. The heavy ions in particular are responsible for **sputtering**, where molecules are physically blasted from the surface, creating an **exosphere** of sputtered products, including sodium and low-energy electrons. There is much still to be learned about the effects of irradiation of ices and the stability of hydrated salt minerals at Europa's surface temperatures.

Europa's water ice exhibits a variety of grain sizes and is particularly abundant and fine-grained (<100 μm diameter) between ±60° latitude on the leading side, but it is less abundant with coarser grains (>400 μm) on the trailing hemisphere. The polar regions have a mixture of particles with a range of grain sizes. Bright regions of Europa's surface are topped with a 1 μm layer of **amorphous** ice, which is probably the result of radiolytic disruption of the regular crystalline structure. This disruption can be counteracted by thermal annealing and recrystallization, but these processes are impeded by Europa's cold surface temperatures.

Sputtering and **thermal desorption** act to remove water from water ice and hydrated minerals. These water molecules may either escape to space, but more typically, recondense elsewhere on the satellite as frost. This deposition will vary depending on temperature and surface albedo, so the frost will be more likely to be deposited at high latitudes and on bright surfaces than at the warmer equatorial latitudes on darker materials. Frost deposition may be responsible for the brightening and whitening of Europa's dark reddish surface features over time. In addition, radiolysis itself may cause chemical changes that brighten the surface.

Mass wasting, which is movement of material downslope under the influence of gravity, is less significant on Europa than on the other Galilean satellites because the surface is so young. Mass wasted material is commonly dark and is likely the non-ice debris that remains after the surrounding ice has been removed by sputtering and sublimation. This lag material may be salts and impactor contaminants.

Sublimation (Fig. 20) has played a significant role in shaping and muting the topography of Callisto and, to a lesser extent, Ganymede, and has occurred only in darker warmer regions on Europa, potentially including where warm material has been in close proximity to the surface. Sublimation lags have been suggested to result as water molecules are driven off by the intrusion of warm water or ice along ridges and chaos regions. This process has been suggested as the origin of low albedo spots along triple bands and of dark material along the flanks of ridges. Some craters have dark material in their floors, which is consistent with a thermal lag produced during the impact cratering process, when very hot material from the impact would have rapidly sublimated any water ice off the surface, and by the downslope movement of dark lag material onto the crater floor.

FIGURE 20   Sublimation of water molecules by sunlight results in a dark lag deposit, which can move down slopes to collect at their bases. This process appears to have occurred at the cliff in the center of this high-resolution *Galileo* image. The water molecules may be "cold-trapped" on brighter, icier surfaces, forming frost deposits. (Image: NASA/JPL.)

7. Surface Age and Evolution

7.1 Surface Age

Europa's surface age can be coarsely estimated from the number of large impact craters on its surface, if accurate estimates of the impactor flux can be made. Modeling of the dynamics of small solar system bodies suggests that the impactor population at Jupiter's orbit is dominated by comets, specifically, **Jupiter-family comets**. From the paucity of large (>10 km diameter) craters on Europa, this model implies a surface age of ~60 Ma (million years), with uncertainties of about a factor of 3.

Another way to estimate Europa's age is to use estimates of ice sputtering, which occurs when high-energy particles swept along with Jupiter's magnetic field impact Europa's surface, causing ice particles to be dislodged, most of which then escape to space. This process has a number of uncertainties, but measurements from *Galileo*'s Energetic Particle Detector (EPD) have lead to estimates that a couple of centimeters to over half a meter of ice may be removed every million years. High-resolution imaging has shown numerous examples of topography on vertical scales of tens of meters; this observation is consistent with an age similar to that predicted by the comet impactor model.

The dearth of impact craters on Europa makes it an excellent place to study the ratio of primary to secondary craters, something that is very difficult to accomplish on heavily cratered bodies like the Moon and Mercury. Although high-resolution imaging of Europa's surface is limited, studies suggest that most of the small (<1 km diameter) craters on Europa are secondaries.

Comparisons of images from *Voyager* and *Galileo*, acquired 20 years apart, show no definitive evidence for current activity on Europa's surface, although such comparisons are hampered by a lack of high-resolution global image data taken at similar lighting geometries. Similarly, searches for plumes such as those observed on Io and Enceladus have proved unsuccessful. Nevertheless, if the surface is only ~60 Ma old, it seems likely that Europa may still be active today.

7.2 Surface History and Geological Evolution

Mapping of Europa's landforms and their interactions with each other yields a time history, or **stratigraphy**, of surface evolution and shows whether the **resurfacing** style has changed over time. Several areas across Europa's surface have been mapped at a variety of scales, and there does appear to have been a change in geological activity and style through the decipherable time-history of the surface. The oldest type of terrain is the "ridged plains," a mélange of ridges and linear structures that are uniformly bright overall, and in which it is difficult to pick out distinct feature types. Bands are intermediate in the stratigraphic column, while chaos and lenticulae are among the youngest surface features, commonly disrupting bands and ridged plains. Troughs and double ridges have formed throughout Europa's surface history and crosscut bands and some lenticulae and chaos. There appears to have been more activity in the earlier surface record, with a waning in the number and width of features in the later stratigraphy.

Stratigraphic mapping therefore suggests that Europa's geological style has generally changed over time, from ridged plains formation, to band formation, to chaos and lenticulae formation, with the activity level simultaneously waning. The mechanism for this change is uncertain, but one plausible model that fits the observations is one in which Europa's ocean is slowly cooling, such that the ice above it is thickening as the ocean freezes out. After the ice shell reaches a critical thickness, **solid-state convection** may be initiated, allowing ice diapirs to be convected toward the surface. A thickening ice shell could be related to a waning intensity of geological activity since the surface is expected to be more mobile if the ice shell is thinner.

Because Europa's surface is probably relatively young, such a fundamental change in style might seem unlikely over the last ~1% of the satellite's history, and we must speculate on its activity over the rest of its ~4.5 billion year existence. Four possible scenarios have been proposed (Fig. 21): (1) Europa resurfaces itself in a steady-state and relatively constant, but patchy style; (2) Europa is at a unique time in its history, having undergone a recent major resurfacing event; (3) global resurfacing is episodic or sporadic; or (4) the satellite's surface is actually much older than our

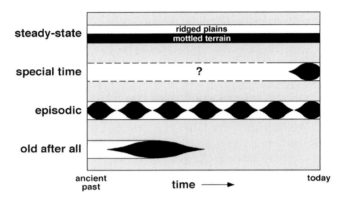

FIGURE 21 Possible schematic evolutionary models for Europa's surface. White represents epochs dominated by ridged plains formation, and black represents mottled terrain formation. Current analyses suggest that Europa is either at a special time in its history or, more likely, that it undergoes episodic resurfacing. (After Pappalardo et al., 1999.)

cratering models suggest. From the standpoint of the dynamical evolution of the Galilean satellite system, there is good reason to believe that Europa's surface evolution could be cyclical (i.e., scenario above). As participants in the Laplace resonance, the orbital characteristics of Io, Europa, and Ganymede are inherently linked to each other, and also to their interior thermal characteristics. Io experiences the greatest amount of tidal heating and largely drives the predicted cycling. The eccentric orbit of Io can cause a great amount of tidal heating, which tends to drive its orbit toward circularity, and in turn decreases its tidal heating. The decreased tidal heating causes Io to cool, but it also allows its eccentricity to increase again, thereby increasing the tidal heating and Io's temperature, thus completing the cycle. This cyclical evolution of Io's tidal heating and orbital characteristics pulls Europa (and Ganymede) along for the ride through the Laplace resonance. In this way, Europa can experience cyclical variations in its orbital characteristics and tidal heating on time scales of perhaps 100 Ma, and therefore may resurface itself on approximately these timescales.

The coupled thermal and orbital evolution of the Galilean satellites can cause significant variations in the thickness of Europa's ice shell and level of geological activity through time. In this scenario, Io and Europa are currently in a diminishing phase of activity. The observed surface characteristics of Europa may represent the latest, waning stage of a long cyclical thermal and geological history.

8. Astrobiological Potential

Based on our terrestrial view, the primary ingredients for life are water, organic compounds, and chemical energy.

Europa may have all three: water of the ocean, organic compounds that have been delivered to the satellite, and chemical energy from radiolysis and possibly chemosynthesis. The evidence for liquid water within Europa is strong, as discussed earlier, and Europa's sub-ice ocean may have a greater volume than that of all Earth's surface water. Cometary and asteroidal impactors have rained onto the surfaces of the Galilean satellites throughout solar system history. Just as Ganymede and Callisto have been darkened by impactor material, similar material must have been delivered to Europa, where its young and bright surface implies that much of this material is now incorporated into the ice shell and ocean. Moreover, the original accretion of Europa may have delivered carbon in the form of CO and CO_2.

Metabolic reactions within living cells depend upon chemical reactions between **oxidants** and **reductants**. For animals, this depends on taking in oxygen, which is combined with sugars to produce CO_2 and water. For plants, CO_2 is combined with water to form sugars and oxygen. In extreme environments on Earth, and possibly within Europa, more exotic materials such as hydrogen sulfide (H_2S), formaldehyde (HCOH), methane (CH_4), or even sulfuric acid (H_2SO_4) can be key to metabolism. The key is that chemical disequilibrium must exist, which organisms then exploit to create the energy needed for life.

Whether Europa has sufficient chemical energy to support life is the most significant unknown in understanding Europa's potential for life. Irradiation of surface ice can form molecules of oxygen and hydrogen, with most of the hydrogen floating away but much of the oxygen and other oxidants remaining behind, like a condensed out atmosphere frozen into the uppermost centimeters of ice. If these oxidants can be delivered to the ice shell and ocean, they may be able to power the chemical reactions necessary for life. Some of these oxidants will be churned into the upper meter of ice by small impacts. Geological processes such as chaos formation may be able to deliver near-surface materials to the ocean, but the means of surface-ocean communication remain poorly understood. Some oxygen and hydrogen is also produced within the ice shell and ocean by radioactive decay of potassium, but this alone could not provide much energy for life.

If Europa's rocky mantle is tidally heated, then hydrothermal systems could exist on Europa's ocean floor. On Earth, hot chemical-laden water pours into the oceans, delivering organic materials and reductants into the water. If hydrothermal systems exist at the bottom of Europa's ocean, and if oxidants are delivered from the ice shell above, then the necessary chemical disequilibrium that could be used by life exists.

Another important consideration is whether Europa's interior environment is stable enough through time, such that if life ever developed it would still exist today. Europa's ocean may have persisted for aeons thanks to internal radioactive heating and the warming resulting from Jupiter's gravitational tug. However, the internal heating induced by the Laplace resonance is not necessarily ancient, and (as discussed earlier) the intensity of tidal heating may have varied (perhaps cyclically) through time. It is an open question whether chemical energy sources for life exist within Europa and have been sufficiently stable to support life through time. Even if life does not exist within Europa today, it may have existed in the past.

9. Future Exploration

The unique requirements for studying Europa—primarily the harsh radiation environment around Jupiter, and the fuel needed to get a spacecraft into orbit around the satellite—make any mission there both technically and financially challenging. Nevertheless, the possibilities for life on this icy moon are sufficiently intriguing that such a mission has a high priority within the scientific community. Key scientific questions remain to be answered, including whether there is indeed a liquid water ocean, the characteristics and composition of this ocean, the means of surface-ocean exchange, and whether Europa can support life. A spacecraft in orbit around Europa could make continuous gravimetric and topographic measurements of the tidal bulge and magnetic measurements of the conductive layer below the ice shell. Ice-sounding radar would be able to sense shallow water deposits including partial melt and may be able to probe to the bottom of the ice if it is relatively cold and thin. The only way to acquire an unambiguous measurement of the thickness of Europa's ice shell (short of actually drilling through it) is to make seismic measurements, by landing a seismometer on the surface.

The composition of Europa's surface is not well known, so high spectral and spatial compositional measurements are also needed to understand Europa's evolution and surface processes. Experiments designed to determine Europa's potential for life are best made with a lander on the surface, either a stationary scientific laboratory or a rover. Spacecraft data has yielded tantalizing insights into Europa's history and evolution, but there is still much we do not know. Further exploration is the only way we will learn Europa's deepest secrets.

Bibliography

General

Greeley, R., Chyba, C. F., Head III, J. W., McCord, T. B., McKinnon, W. B., Pappalardo, R. T., and Figueredo, P. (2004). Geology of Europa. In "Jupiter: The Planet, Satellites, and Magnetosphere" (F. Bagenal, ed.), pp. 329–363. Cambridge Univ. Press, Cambridge, United Kingdom.

Moore, J. M., Asphaug, E., Belton, M. J. S., Bierhaus, B., Breneman, H. H., Brooks, S. M., Chapman, C. R., Chuang, F. C., Collins, G. C., Giese, B., Greeley, R., Head, J. W., Kadel, S., Klaasen, K. P., Klemaszewski, J. E., Magee, K. P., Moreau, J., Morrison, D., Neukum, G., Pappalardo, R. T., Phillips, C. B., Schenk, P. M., Senske, D. A., Sullivan, R. J., Turtle, E. P., and Williams, K. K. (2001). Impact features on Europa: Results from the *Galileo* Europa Mission. *Icarus* **151**, 93–111.

Schenk, P. M., Chapman, C. R., Zahnle, K., and Moore, J. M. (2004). Ages and interiors: The cratering record of the Galilean satellites. In "Jupiter: The Planet, Satellites, and Magnetosphere" (F. Bagenal, ed.), pp. 427–457. Cambridge Univ. Press, Cambridge, United Kingdom.

Specific Topics

Bierhaus, E., Chapman, C., Merline, W., Brooks, S., and Asphaug, E. (2001). Pwyll secondaries and other small craters on Europa. *Icarus* **153**, 264–276.

Carlson, R.W., Johnson, R.E., Anderson, M.S. (1999). Sulfuric acid on Europa and the radiolytic sulfur cycle. *Science* **286**, 97–99.

Chyba, C. F., and Phillips, C. B. (2001). Possible ecosystems and the search for life on Europa. *Proc. Nat. Acad. Sci. USA* **98**, 801–804.

Collins, G. C., Head, J. W., Pappalardo, R. T., and Spaun, N. A. (2000). Evaluation of models for the formation of chaotic terrain on Europa. *J. Geophys. Res.* **105**, 1709–1716.

Greenberg, R., Geissler, P. E., Hoppa, G., Tufts, B.R., Durda, D. D., Pappalardo, R., Head, J. W., Greeley, R., Sullivan, R., and Carr, M. H. (1998). Tectonic processes on Europa: Tidal stresses, mechanical response, and visible features. *Icarus* **135**, 64–78.

Head, J. W., and Pappalardo, R. T. (1999). Brine mobilization during lithospheric heating on Europa: Implications for formation of chaos terrain. *J. Geophys. Res.* **104**, 27, 143–27, 156.

Pappalardo, R. T., Head, J. W., Greeley, R., Sullivan, R. J., Pilcher, C., Schubert, G., Moore, W., Carr, M. H., Moore, J. M., Belton, M. J. S., and Goldsby, D.L. (1998). Geological evidence for solid-state convection in Europa's ice shell. *Nature* **391**, 365–368.

Prockter, L. M., Head, J. W., Pappalardo, R. T., Sullivan, R. J., Clifton, A. E., Giese, B., Wagner, R., and Neukum, G. (2002). Morphology of europan bands at high resolution: A mid-ocean ridge-type rift mechanism. *J. Geophys. Res.* **107**, 10.1029/2000JE001458.

Schenk, P. M., and McKinnon, W. B. (1989). Fault offsets and lateral crustal on Europa—Evidence for a mobile ice shell. *Icarus* **79**, 75–100.

Schenk, P., and Pappalardo, R. (2004). Topographic variations in chaos on Europa: Implications for diapiric formation. *Geophys. Res. Lett.* **31**, 10.1029/2004GL019978.

Weiss, J. W. (2004). Planetary parameters. In "Jupiter: The Planet, Satellites & Magnetosphere" (F. Bagenal et al., eds.), pp. 699–706. Cambridge Univ. Press Cambridge, United Kingdom.

Zahnle, K., Schenk, P., Levison, H., and Dones, L. (2003). Cratering rates in the outer solar system. *Icarus* **163**, 263–289.

Ganymede and Callisto

Geoffrey Collins

Wheaton College
Norton, Massachusetts

Torrence Johnson

Jet Propulsion Laboratory
California Institute of Technology
Pasadena, California

CHAPTER 24

G anymede and Callisto (Fig. 1) are the largest and outermost of Jupiter's four Galilean satellites. Similar in size to Mercury, and with surfaces dominated by dirty water ice, they are prime examples of planet-sized icy bodies. Though Ganymede and Callisto are neighbors and share many bulk characteristics such as size and density, they have followed divergent evolutionary paths. Their interior structure and surface geology provide insight into which processes are common and which are unique in the development of a large icy world.

1. Exploration

1.1 Discovery

Ganymede and Callisto were discovered by Galileo Galilei in 1610, when he first trained his telescope on Jupiter and shortly thereafter published his results in the *Siderius Nuncius*. Along with Io and Europa, they became the first natural satellites, other than the Moon, known to science. Galileo immediately recognized the significance of the "new stars" traveling with Jupiter and changing their positions every night. The orbits of what are now known as the Galilean satellites were rapidly calculated and found to be essentially circular and in the same plane as Jupiter's equator. Because Galileo made these observations centuries ago, his records

of satellite eclipses provide a long timeline to compare with modern measurements, and they are still used to constrain calculations of the dynamical evolution of Jupiter's satellite system under the influences of tidal dissipation and the satellites' mutual gravitational interactions.

1.2 Astronomical Observations

The Galilean satellites are large enough to exhibit distinct discs (on the order of ~1 arc second in angular diameter) when viewed through even moderate power telescopes, and it was thus known from simple geometry that they must be bodies comparable in size to the Moon. Precise measurements of their sizes proved difficult with conventional astronomical techniques, with published estimates from different observers disagreeing significantly. Even these relatively uncertain size estimates were sufficient, when combined with the satellites' brightness, to indicate that their surfaces are highly reflective compared with that of the Moon.

In the two decades leading up to the first spacecraft exploration of the Jupiter system, astronomical techniques advanced rapidly, particularly in the area of sensors in the visible and near-infrared spectral range (~0.3–2.5 μm). The pioneering planetary astronomer Gerard Kuiper used early infrared detectors to show that Ganymede's reflectance at 2 μm was much lower than in the visible range and suggested

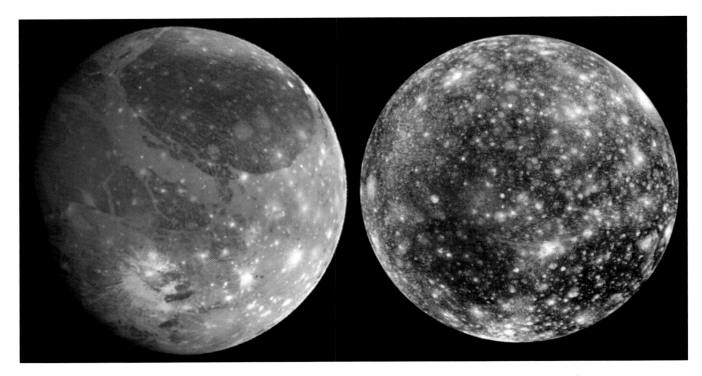

FIGURE 1 Global views of Ganymede (left) and Callisto (right), obtained by the camera on the *Galileo* spacecraft.

that water ice might be responsible. Vassily Moroz, a planetary scientist working at the Crimea Observatory in the Soviet Union, made even more detailed infrared color measurements and concluded that water ice was the best explanation for Ganymede's spectrum.

At the spectral resolution and signal-to-noise ratio of these pioneering measurements, however, a conclusive identification of the surface composition could not be made because several other candidate materials, including ices of carbon dioxide and ammonia, were known to have absorptions in the same part of the infrared spectrum. The issue was settled conclusively for Ganymede and Europa by a team led by Carl Pilcher, then at MIT, who published the first high-resolution infrared reflection spectra for these satellites in 1972 and compared them in detail with laboratory spectra of ices at low temperature. All the significant absorption features in the 1- to 2.5-μm region matched spectra of water ice and ruled out any major contribution from other ices. Callisto's spectrum also displays water ice and hydrated silicate features, although the water signature is subdued compared with Ganymede's strong water ice **spectral absorptions,** due to the larger amount of dark material mixed with the ice on Callisto's surface.

Figure 2 shows a compilation of the best telescopic spectra of Ganymede and Callisto compared with Io and Europa. The dominant features in all the spectra except Io's are the deep absorptions at wavelengths longer than 1 μm due to the presence of hydrated materials and water ice. Laboratory studies of water ice reflectance and theoretical simu-

lations of spectra from mixtures of material have demonstrated that the observed spectra can be explained by water ice/frost, mixed with varying amounts of a spectrally neutral darker component with a reddish color in the visible portion of the spectrum (i.e., one having absorption at ultraviolet and blue wavelengths). The nature of the non-water-ice component in the satellites' surfaces is still under investigation, but spectra of different regions on both satellites taken by the Near-Infrared Mapping Spectrometer (NIMS) instrument on the *Galileo* mission are providing clues to the identification of this material (see Section 3).

1.2.1 MASSES AND DENSITIES

The mass of a distant planetary object is normally impossible to determine from remote astronomical observations alone, unless it happens to have a companion whose orbit can be determined, as is the case for the giant planets with their satellite systems, the Pluto/Charon system, and more recently numerous asteroids and several trans-Neptunian objects. The Galilean satellites represent a more difficult case. They do not themselves have satellites, but the mutual gravitational attraction among these large satellites produces significant and measurable changes in their orbits about Jupiter. The mathematician Pierre Laplace studied these interactions in the late 19th century, and subsequent developments in this new branch of dynamical astronomy permitted reasonable estimates of the satellites' masses to be made in the early 20th century. When combined with the

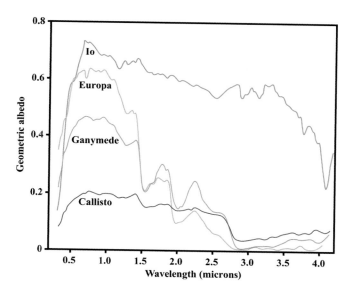

FIGURE 2 Compilation of the best telescopic spectra of Ganymede (green) and Callisto (blue) compared with Io (red) and Europa (yellow). (Modified from R. Clark and T. McCord, *Icarus*, v. 41, pp. 323–339, 1980).

still uncertain size estimates, the best estimates of masses prior to 1970 suggested that the inner satellites, Io and Europa, had rock-like densities, similar to the Moon's, and that Ganymede and Callisto appeared to be less dense, suggesting the possible presence of large amounts of ice in their constituent materials.

In 1972, observations of a stellar occultation by Ganymede from two stations on the Earth provided the first high-precision measurement of its diameter. This was closely followed by the first spacecraft exploration of Jupiter by the *Pioneer 10* and *11* missions in 1973 and 1974, which greatly improved the mass estimates of the satellites from tracking the gravitational perturbations in the spacecraft trajectories caused by the satellites. This led to the first accurate determination of Ganymede's density of about 1900 kg/m^3, adding more evidence to the hypothesis that its bulk composition is a mixture of rock and ice. The *Voyager 1* and *2* Jupiter encounters in 1979 provided even more data on the satellite's masses and accurate determination of their shapes and volumes. These data showed Callisto is very similar to Ganymede in its bulk properties, with a density of about 1800 kg/m^3. Interior structure models, taking into account the high-pressure behavior of water ice, show that the average bulk composition for both satellites is a mixture of 50–60% (by mass) anhydrous silicate "rock" with water ice.

1.3 Spacecraft Exploration

Seven spacecraft have visited the Jupiter system to date: *Pioneer 10* and *11*, *Voyager 1* and *2*, *Galileo*, *Ulysses*, and *Cassini*. *Ulysses*, a joint European Space Agency/NASA

mission to study the Sun's environment at high latitudes, made measurements of Jupiter's magnetic fields, radiation belts, and dust environment but did not study the satellites directly. *Cassini*, on its way to its rendezvous with Saturn, flew by Jupiter in 2000 and returned spectacular observations of its atmosphere and **magnetosphere,** but its trajectory was too far from the Galilean satellite system to provide high-resolution views of the satellites.

The first Jupiter missions, *Pioneer 10* and *11*, were designed to provide the first reconnaissance of the system and to establish the intensity of the radiation belts. The *Pioneer* program's major contribution to knowledge of Ganymede and Callisto, as mentioned earlier, was improving the mass estimates of the satellites, leading to the first precision bulk density measurements.

In 1979, *Voyager 1* and *2*, with powerful remote sensing payloads and close targeted flybys of each Galilean satellite, provided the first in-depth reconnaissance of the satellites and set the stage for the geological and geophysical exploration of these worlds. *Voyager's* cameras showed that Ganymede and Callisto, alike in many large-scale properties, have divergent geological histories (Fig. 1). Callisto's surface is heavily cratered at all scales, from large impact scars over a 1000 km in diameter down to craters a few kilometers in diameter, the smallest scale resolvable on Callisto by the *Voyager* cameras. This battered, uniform surface stands in stark contrast to Ganymede's varied landscape. Ganymede's surface can be divided into two distinct types of terrain, based on a sharp albedo contrast. The darker areas (named "dark terrain") are heavily cratered and exhibit **palimpsests,** much like the surface of Callisto. The brighter parts of Ganymede's surface (named "bright terrain") form wide lanes through the dark terrain and are less heavily cratered, implying a younger surface. *Voyager* images showed the bright terrain to have some areas that appeared to be smooth, while other areas exhibit sets of parallel ridges and troughs.

One of the major objectives of the *Galileo* mission was to perform detailed observations of the big satellites. The mission design allowed multiple close flybys at ranges 100–1000 times closer than the *Voyager* encounters, enabling high-resolution studies of their surfaces and detailed measurements of their gravity fields and interactions with Jupiter's magnetospheric environment. High-resolution images of the different terrains first identified by *Voyager* have illuminated their origins, described in detail in subsequent sections of this chapter. The close flybys also enabled more detailed spectroscopic observations, which identified some of the non-water-ice components on the satellite surfaces, including carbon dioxide embedded in the surface and evidence for carbon compounds.

Repeated close flybys enabled *Galileo* to make precision gravity and magnetic measurements, resulting in several major discoveries. First, Ganymede has a strongly layered internal structure, with heavier rock and metal concentrated in the center, whereas Callisto has a more homogenous structure. Second, Ganymede was found to have a relatively strong internal magnetic field, creating its own "mini-magnetosphere" embedded within Jupiter's vast magnetosphere. Finally, the interactions of Ganymede and Callisto with Jupiter's rotating, tilted magnetic field show that both satellites exhibit an induced magnetic field interpreted as evidence for an electrically conducting liquid water ocean beneath their icy crusts.

2. Interiors

2.1 Interior Structures

The ice/rock bulk composition inferred for Ganymede and Callisto from their densities led to the natural suggestion that even modest heating from accretion and the decay of radioactive elements in the rock fraction would melt ice in the interior and lead to differentiated interiors—that is, a layered structure with the denser rock and metal constituents concentrated closer to the center of the satellite with the ice in the outer layers. Most analyses following the *Voyager* mission operated on the assumption that Ganymede and Callisto had similar differentiated interior structures, but the data to test this assumption would not come until the *Galileo* mission.

Determining the interior structure of a planetary object is intrinsically difficult, particularly from remote observations alone. Most of the information about the interior of our Earth, for instance, comes from over a century of study of seismic data, where waves created by earthquakes travel deep through the Earth and provide clues to the density and composition throughout the interior [see EARTH AS A PLANET: SURFACE AND INTERIOR]. So far the only other world for which we have seismic data is the Moon, acquired with seismometers left by the *Apollo* astronauts [see THE MOON].

An extremely important quantity that can be used to assess the distribution of mass inside an object is its moment of inertia, a dimensionless number; a sphere with uniform density throughout has a moment of inertia of 0.4, with lower values indicating increasing degrees of mass concentration near the center. The moments of inertia for Ganymede and Callisto were measured indirectly by the radio experiment on *Galileo*, which measured the perturbations of the spacecraft's trajectory as it flew by the satellites at low altitude. Although perfect spheres with different moments of inertia have identical external gravity fields, the key to this experiment is that the distribution of mass in the interior of a satellite does affect the way its shape is perturbed from a perfect sphere by rotation and tides. The rotation rates of Ganymede and Callisto, although slow by terrestrial standards (a little over a week for Ganymede, and over two weeks for Callisto), are still

sufficient to cause a slight equatorial bulge and polar flattening, whereas Jupiter's strong gravity raises tidal bulges on the sub- and anti-Jupiter hemispheres. The combination of these two effects leads to distinctly nonspherical components to the external gravity field (in mathematical terms, the description of the satellites' gravity in a spherical harmonic expansion contains significant J_2 and C_{22} terms). The magnitude of these nonspherical terms is dependent on the degree of internal mass **differentiation,** and they are related directly to the moment of inertia as long as the object responds to spin and tidal distortion as a fluid would (i.e., hydrostatically).

The surprising results of the Galileo tracking experiment showed that Ganymede and Callisto have distinctly different interiors. The derived moments of inertia for both satellites were lower than they would be for bodies of uniform density, as expected. Ganymede's measured value of 0.31 is so low that it implies essentially complete separation of its water ice from the heavier rock and metal. However, Callisto has a significantly larger moment of inertia, 0.35. This is small enough to imply some differentiation, but too large to be compatible with full separation of light and heavy components. Callisto probably has some significant portion of its interior composed of a rock–ice mixture.

The measured moments of inertia can be combined with the values for the mean density, the size, and the properties of ice and rock under pressure to construct models of the satellites' interiors that match all the known quantities. Figure 3 shows the best current estimates of their internal structures. Ganymede is shown with a three-layer structure: a metallic core, a rock mantle, and a deep water ice upper layer; Callisto is shown with a two layer structure: a large rock–ice core, with the fraction of dense material increasing toward the center, and an upper ice-rich layer.

2.2 Internal Oceans

A major question regarding these icy worlds is whether they possess subsurface oceans of liquid water. This intriguing possibility was first raised in the early 1970s by planetary geochemist John Lewis, who pointed out that radioactive heating of the satellites' interiors might result in their internal temperatures reaching the water ice melting point at some depth below their surfaces. With the satellites' densities known, a relatively simple calculation of internal temperature from the heating produced by the decay of radioactive nuclides in the rock fraction (primarily U, Th, and K) shows that indeed the ice melting point should be reached about 75 to 100 km below the surface. More detailed calculations are complicated by several additional factors.

The behavior of water as a function of temperature and pressure is complex. At the surface of the Earth, only normal low-density ice, known as Ice-I, exists. It floats in liquid water and melts at 273 K (0°C, 32°F). Increased pressure decreases the melting temperature, but under terrestrial

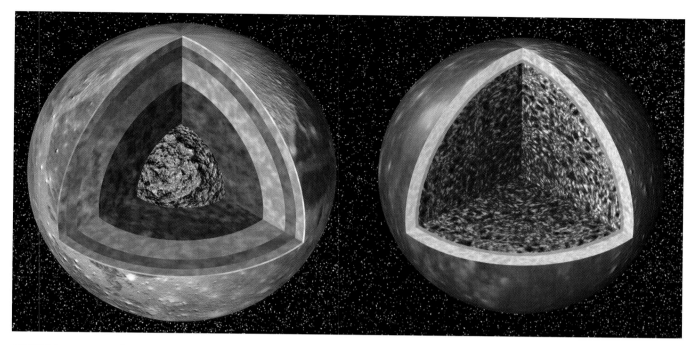

FIGURE 3   Cutaway diagrams showing current models for the interior structures of Ganymede and Callisto based on Galileo gravity data. Ganymede (left) is highly differentiated, with a molten iron core surrounded by a rocky mantle, in turn surrounded by a thick outer layer of ice. An interior ocean of liquid water may exist sandwiched between the surface layer of Ice-I and the higher pressure phases of ice below. Callisto (right) has an interior composed of a mixture of rock and ice, slowly increasing in density toward the center. The outermost layer is relatively clean water ice, with a liquid water ocean at its base. (Zareh Gorjian and Eric De Jong, NASA/Jet Propulsion Laboratory.)

conditions the solid form remains low-density Ice-I. Laboratory studies show that, at the high pressures reached deep in the interiors of icy satellites the size of Ganymede and Callisto, ice transforms to various high-density forms over a wide range of temperatures (Fig. 4). These phases of ice, as they are known, are denser than liquid water and would sink in a liquid ocean. Calculations of the temperature and pressure as a function of depth within the satellites show that if temperatures reach the required melting point, the resulting subsurface oceans would be strange indeed—a liquid layer sandwiched between low-density Ice-I on the top and high-density Ice-III on the bottom (or a mixture of high-density ice and rock in the case of Callisto).

The other major complication is whether the interior will ever actually warm up to the ice melting point. The simple calculations that reach the melting temperature are based on the heat produced by radioactive decay, escaping the interior by thermal conduction through the ice crust. However, as the temperature of ice approaches the melting point within the satellite, another heat transfer process comes into play, convection. Ice near its melting point is not stiff and brittle, but can flow and deform under pressure, particularly over long periods of time. In geophysical terms, it becomes a low-viscosity solid. Low-viscosity ice under some

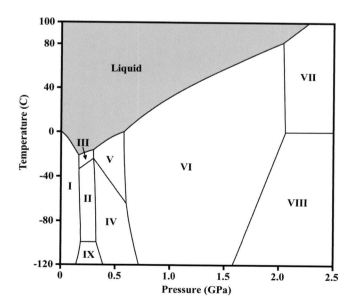

FIGURE 4   Phase diagram of water ice. At low pressures near the surface, Ice-I is less dense than liquid water. At higher pressures, ice converts to denser phases, with higher melting points.

conditions can begin to convect, with warmer, lower-density ice rising toward the surface, exchanging with cooler higher-density ice sinking into the interior. This glacially slow solid ice circulation is similar to what occurs in the Earth's rock mantle. The important point is that it is much more efficient at transporting heat than conduction alone. In simple terms, as the ice heats up from the radioactive energy from below, it will begin to convect, taking heat to the surface, but never allowing the temperature within the ice to rise above the melting point. Under these conditions, even if an ocean formed early in the satellite's history, the convection process in the ice crust could rapidly freeze it solid.

A final complication is the issue of ammonia, NH_3. In many formation models, ammonia is a possible minor constituent of the icy satellites. If present, it has a major effect on the melting point of a water–ammonia mixture, depressing the temperature at which a liquid can exist to about 173 K, a hundred degrees below the point at which pure water melts. Although ammonia has not been detected on the surfaces of the satellites, even small amounts can affect the results of theoretical thermal and convection calculations, and most discussions of the satellites' interiors include both ammonia and nonammonia cases.

Whether convective cooling "wins" over heating determines whether a liquid ocean at the present time can exist. The calculations for interior models including convection are quite complex and depend on some properties of ice that are poorly known. Current models for Ganymede and Callisto show that liquid layers are possible under some conditions, but these models cannot definitively demonstrate their existence. Measurements pointing strongly to the presence of liquid oceans in both satellites came from an unexpected source—magnetic field measurements made by the Galileo mission, which is the subject of the next section.

2.3 Magnetic Fields

The *Galileo* magnetometer experiment had two major objectives for studying Ganymede and Callisto: (1) to determine whether they possess intrinsic magnetic fields of their own and (2) to study the interactions of the satellites with Jupiter's huge and powerful magnetosphere. These two objectives are closely coupled because the satellites orbit deep within the region of space controlled by Jupiter's magnetic field and its associated trapped radiation and plasma (a tenuous ionized gas made up of electrons, protons and positively charged ions). Measurements of magnetic fields in the vicinity of the large satellites thus must take into account the large background field from Jupiter, which is continually changing due to Jupiter's fast rotation sweeping the field past the satellites, and magnetic perturbations from large-scale electrical currents flowing within the magnetosphere. Once these effects are measured and understood, the experimenters can search for the smaller perturbations in the local magnetic field produced by any intrinsic field and from local currents set up by the interactions of the satellites and their tenuous atmospheres with the magnetosphere.

2.3.1 INTRINSIC FIELDS

On the very first *Galileo* close encounter with Ganymede, the space physics instruments detected strong evidence for both an intrinsic field and complex interactions with Jupiter's environment. As the spacecraft flew by the satellite, the magnetometer recorded a marked change in both the magnitude and direction of the magnetic field. At the same time, the plasma wave spectrometer (which receives natural radio "noise" produced by the interactions of charged particles and magnetic fields) showed sharp changes in the nature of the radio signals it received, coinciding closely in time with the observed magnetic deflections. To the investigators on these experiments, these observations were familiar, a "fingerprint" indicating the spacecraft had passed though a planetary magnetosphere. Due to the complexities discussed earlier, it took observations on subsequent flybys to confirm the discovery, but it soon became clear that Ganymede possesses a relatively strong intrinsic field, oriented in the opposite sense to Jupiter's field, which produces a "mini-magnetosphere" embedded within Jupiter's magnetosphere.

An intrinsic field at Ganymede was not totally unexpected. UCLA space physicist Margaret Kivelson, the head of the Galileo magnetometer team, suggested prior to the *Galileo* mission that the big satellites might be able to generate their own internal fields. Nevertheless, the discovery of an intrinsic field at Ganymede raises a number of issues for our understanding of planetary magnetic field generation [*see* PLANETARY MAGNETOSPHERES].

How planetary fields, including the Earth's, are generated and maintained is an active area of research. It is believed that some form of what is called a "geodynamo" is responsible for producing a magnetic field within a planetary core. The exact requirements for generating a field by this dynamo process in a given planet are the subject of debate. Ganymede's internal field is consistent with its high degree of differentiation and favors a three-layer model with a metallic iron/iron-sulfide core. However, merely having a metallic iron core is not sufficient to produce a planetary magnetic field. Although the Earth's field and other planetary fields are frequently described in textbooks as "bar magnet" fields, this only describes the field's mathematical description (having a dipolar—N and S—configuration with field lines connecting the poles). The bar magnet analogy is misleading in terms of the source of the field, since it has long been known that iron will lose its magnetization at the temperatures typical of planetary cores (temperatures above the Curie point, at which a magnetic material loses its magnetism).

Current theories of planetary dynamos suggest that the basic requirement for generating a field is continual convective motion of an electrically conducting fluid. Theoretical models of Ganymede's thermal evolution suggest that it could have a fluid, electrically conducting, iron, or iron-sulfide core at the present time. However, the same models show that, although there could have been convective motion in the fluid core early in Ganymede's history, at present the core should be stable against convection, and thus it will not produce motion of the sort required by dynamo models. So the source of Ganymede's field is still not clear. Possibilities that have been discussed in the literature include some event, such as tidal heating, stirring up the core in recent geological history and producing a magnetic field today. Another possibility is that the timescale of heating Ganymede is longer than earlier models suggest, and that the required conditions for convection and planetary dynamo formation have only recently been reached in the core.

2.3.2 INDUCTION FIELDS AND OCEANS

Callisto shows no evidence for an intrinsic dipole field like the one observed at Ganymede. When *Galileo* flew close to Callisto, the magnetometer recorded perturbations to the background field, but comparisons of data from several encounters showed there was no pattern consistent with single dipole field. However, when the investigators correlated the data with Callisto's position with respect to Jupiter's field, they found another intriguing pattern. Since Jupiter's dipole field is tilted about 10° to the rotation axis, the background field seen by a satellite orbiting in the equatorial plane exhibits a periodic rocking motion. The observed magnetic perturbations correlated with times when this tilt was at different angles.

The key to understanding this type of perturbation lies in the basic theory of electromagnetism: Moving magnetic fields can produce electrical currents and electrical currents can produce magnetic fields (electromagnets and electric motors are among the practical applications of this principle). A classic laboratory physics experiment demonstrates that an electrically conducting sphere (such as a copper ball), when placed in an oscillating magnetic field, will produce a magnetic field (an induced field) countering the imposed field by setting up electrical currents in the surface of the sphere. The magnetometer investigators found that the Callisto perturbations closely matched those expected for an induction field in response to the changing Jupiter field. In other words, Callisto was acting as if it had an electrically conducting layer at or under its surface.

What is the conducting layer on Callisto? The electrical conductivity required to produce the observed perturbations is much larger than the known conductivities of ice or rock, the major surface constituents. Going back to the theoretical possibility of a subsurface ocean, the investigators found a possible explanation for Callisto's behavior. The electrical conductivity of salty ocean water is in the right range to produce the required induction field. Although an indirect argument, these magnetic results are the best evidence to date that the hypothesized ocean exists under

Callisto's icy crust. After this discovery, investigators looked closely at the Ganymede magnetic data and found that there are small deviations from the best-fit intrinsic dipole model, which indicate the presence of an induced field from a conducting ocean layer on Ganymede as well. This same type of induced magnetic field evidence was used to infer a liquid water ocean under the ice on Europa, but ironically the signature of a conducting layer on Callisto is stronger than on Europa because the background field is smaller at Callisto.

2.4 Formation and Evolution

The Galilean satellites have been viewed as a sort of "miniature solar system" since the time of Galileo. Their coplanar, nearly circular orbits strongly suggest that they formed as part of the same process that formed Jupiter. The water-rich, low-density composition of the outer satellites, Ganymede and Callisto, compared with the rock-rich, high-density inner satellites, Io and Europa, suggest that there was a gradient in the conditions within in the circumplanetary gas and dust nebula from which the satellites formed, much like the gradient in the solar system as a whole that produced rocky inner planets and volatile-rich outer planets.

The mixed rock–ice composition of the big icy satellites is very similar to that expected from condensation from a nebula with solar composition. Early models for the formation of the satellites envisioned a gasdust circumplanetary nebula, which was heated by the growing Jupiter at the center. In this scenario, Ganymede and Callisto formed in the cooler outer portion of the system, under conditions similar to the surrounding solar nebula, which permitted the condensation of water ice. Io and Europa, on the other hand, formed further inside the nebula, under warmer conditions with little to no condensation of water.

This relatively simple jovian subnebular theory explains the major characteristics of the system in the context of the formation of Jupiter itself. However, there are problems with the details of the model when the evolution of the forming satellites is considered. One problem is that, as the satellites form, they are subjected to drag from remaining gas and dust in the subnebula. This drag can quickly cause a proto-satellite to spiral inward and be swallowed up by the growing proto-Jupiter. Current calculations show that this is a serious problem with early forms of the subnebular models because the timescale for the accretion of the satellites is much shorter than the times for dissipation of the nebula and decay of the satellite orbits. Another issue is the differences between Ganymede and Callisto. In the simple subnebula accretion models, they should have similar histories. However, *Voyager* and *Galileo* observations show major differences in their interior structures and geologic histories. The most difficult point to reconcile is that Callisto's incompletely differentiated interior implies a longer accretion time to prevent accretional heating from triggering melting and differentiation.

The latest formation models attempt to address these issues in a number of ways. Current models for Jupiter's formation suggest that the jovian subnebula interacts strongly with the surrounding solar nebula as the growing giant planet opens a "gap" in the solar nebula and material is continually fed from the solar nebula surroundings to the outer parts of the subnebula. This class of models can account for longer satellite accretion times and allow them to form without being dragged into the proto-Jupiter. Another type of formation model proposes that the inner satellites formed in a hot dense subnebula but avoided destruction by opening gaps themselves in the subnebula, slowing their orbital decay. In this type of model, Callisto forms more slowly in a thinner outer nebula environment, accounting for some of its differences.

A final factor that may have affected the apparently different histories of Ganymede and Callisto is the existence of what is known as the Laplace resonance condition. This is a dynamical relationship between the orbital periods of the inner three satellites, first studied by the French mathematician Laplace in the 19th century. Io, Europa, and Ganymede currently exhibit a simple numerical relationship (1:2:4) in their orbital periods, causing them to perturb each other's orbits continually, resulting in significantly noncircular orbits. It is the existence of these noncircular orbits that causes tidal heating in each of these satellites, resulting most notably in the violent volcanic activity on Io, which has the largest dose of tidal heating due to its proximity to Jupiter [see IO: THE VOLCANIC MOON]. Callisto does not participate in this celestial dance and apparently has never experienced tidal heating.

Despite the Laplace resonance condition, Ganymede does not currently experience significant tidal heating because of its distance from Jupiter and the relatively small degree of noncircularity of its orbit. However, calculations of the dynamical evolution of the satellite system suggest that Ganymede's orbit may have been more eccentric at times in the past, possibly resulting in a pulse of tidal heating, which could have triggered differentiation and/or stirred up the core and started magnetic field generation. Even though the question of why Ganymede and Callisto have experienced such different interior and geological evolution has not been conclusively solved, it seems likely that the key to the solution lies in some combination of differences in formation and accretion conditions and their subsequent orbital evolution.

3. Surface Materials

3.1 Composition of Surfaces

As noted in the discussion of astronomical discoveries, water ice was identified as a primary surface constituent on the surface of Ganymede and Callisto (and Europa as well) in the 1970s by obtaining infrared spectra of these bodies.

Seen with the eye, the surfaces are darker and redder than pure water ice, so there must be some other material mixed with the ice, but the composition of this material has been difficult to determine. Based on analogy to meteorite and asteroid spectra as well as cosmochemical arguments, most researchers have assumed that the nonwater component of the surface is similar to the material found in primitive, carbon-rich meteorites—a mixture of hydrated silicates (clays) and dark, complex organic compounds (dubbed tholins by the astronomer Carl Sagan, who studied the production of organic material in laboratory simulations of planetary environments). Laboratory studies of ice and mineral mixtures show that even small amounts of dark material will disproportionately lower the reflectance (albedo) of the mixture and damp out the spectral signature of water ice, producing reflectances consistent with the observed spectra of the satellites. Unfortunately, the more subtle spectral signatures of the dark minerals are themselves obscured in the mixed spectra by the much stronger water features, making identification of the dark constituents difficult.

The near-infrared mapping spectrometer on Galileo provided new insights into the composition of the nonwater constituents. This instrument not only covered the spectral range accessible to Earth-based telescopes but also returned spectra in the 3- to 5-μm spectral region. This part of the infrared spectrum is inaccessible from the surface of the Earth due to strong absorptions in the Earth's atmosphere by water vapor and carbon dioxide. It is also a key part of the spectrum for studying non-water-ice components mixed into the satellite surfaces, since water ice is essentially black at these wavelengths and whatever signal is seen arises primarily from the non-water-ice component of the surface mixture.

NIMS spectra of Ganymede and Callisto indeed proved their value in the 3- to 5-μm range, exhibiting a number of detectable absorption features (see Fig. 5). The strongest feature is a relatively sharp absorption of infrared light centered at about the 4.25-μm wavelength, with weaker, but still easily detectable, absorptions at 3.88, 4.05, and 4.57 μm. There is also a weak absorption seen centered near 3.4 μm. These absorptions are seen in the spectra from both satellites but are most easily seen in the Callisto spectra, where there is more of the dark material exposed on the surface.

The 4.25-μm feature has been identified as being caused by the presence of CO_2 on the surface. The location of the center of the absorption indicates that the CO_2 is not in the form of either a solid ice or liquid, but rather occurs in microdeposits, bonded to some other material in the soil. The 4.57-μm absorption is believed to be due to a carbon–nitrogen compound based on its frequency, which corresponds to that expected for C≡N (a triple bond of carbon and nitrogen). The weaker features near 3.4 μm are also believed to be due to carbon bonds with hydrogen (C–H hydrocarbons). These features have also been identified in space spectra of interstellar ice grains obtained by the European Space Agency's Infrared Space Observatory mission.

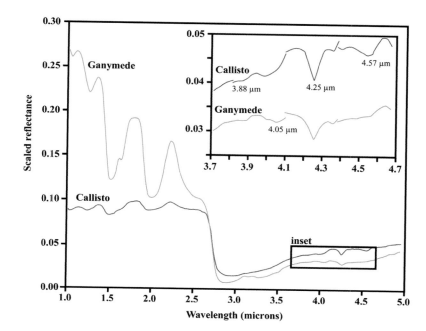

FIGURE 5 *Galileo* NIMS spectra of Ganymede and Callisto, showing absorption of infrared light by various surface materials. See text for details. (Modified from T. McCord et al., J. Geophys. Res., v. 103, pp. 8603–8626, 1998.)

Recent observations of the satellites of Saturn by a similar instrument on the *Cassini* spacecraft show nearly identical CO_2 absorption and similar $C\equiv N$ and C–H features [*see* PLANETARY SATELLITES]. The other two absorption features (3.88 and 4.05 μm) seen on Ganymede and Callisto appear to be unique to the Jupiter system and are thought to be due to S–H bonds and sulfur dioxide, respectively.

3.2 Surface and Atmosphere Interactions with Local Environment

The environment around Jupiter is awash in radiation from charged particles trapped in Jupiter's intense magnetic field. Since they are trapped in the field, which rotates rapidly with Jupiter's spin, the particles sweep past the satellites in their comparatively slow orbits. Thus, the side of a satellite facing away from the direction of orbital motion (the trailing hemisphere) is exposed to a much higher dose of radiation than the leading hemisphere. When the charged particles strike the surfaces of the satellites, they can send surface molecules flying (a process called sputtering), and they can break molecular bonds in the surface material, causing new chemical reactions to occur and creating new compounds.

Laboratory studies suggest that the CO_2, $C\equiv N$, and C–H features seen in the spectra of Ganymede, Callisto, and other icy satellites in the solar system may have a common origin due to charged particle irradiation of minerals containing potassium cyanide and possibly other cyanogens (carbon–nitrogen bearing compounds). Irradiation of these compounds by energetic particles in the presence of water ice is believed to be an important source of the CO_2 found embedded in the mineral/ice matrix on the surfaces of Ganymede and Callisto. On Callisto, the distribution of CO_2 mapped by NIMS shows a marked concentration on the trailing hemisphere of the satellite, as would be expected from radiation-induced CO_2 production.

Sulfur is an important ion in the Jupiter system, continually supplied to space by the escape of sulfur and sulfur-dioxide gases from volcanic Io. The sulfur becomes ionized and joins the low-energy plasma streaming through Jupiter's magnetosphere, which then washes up on the other satellite surfaces. Implantation of sulfur into the surfaces of the icy Galilean satellites has been suggested in the past as the likely reason for the low reflectance of the satellites in the ultraviolet part of the spectrum. The sulfur-induced infrared absorption features are also plausible results of bombarding the icy surfaces with sulfur-rich plasma. However, the distribution of SO_2 on the surfaces of Ganymede and Callisto do not show a strong leading–trailing hemisphere asymmetry as one would expect, indicating that perhaps there is also sulfur in the ice bedrock.

Both Ganymede and Callisto exhibit other effects from their continual bombardment by charged particles in Jupiter's magnetosphere. Both molecular oxygen, O_2, and ozone, O_3, have been detected in telescopic spectra of their surfaces. These oxygen compounds appear to exist as microscopic bubbles trapped in the matrix of the icy surface material and have also been attributed to irradiation by charged particles.

In addition to the frozen and trapped gases in their surfaces, Ganymede and Callisto have very tenuous atmospheres. On Ganymede, Hubble Space Telescope spectra have identified molecular oxygen, and the *Galileo* ultraviolet instrument detected a thin veil of hydrogen in the surrounding space. These gases are apparently produced by a combination of **sublimation** and sputtering from the icy surface. Callisto's atmosphere is similarly of very low density, and the only detectable gas so far has been CO_2

identified in NIMS spectra. A combination of thermal segregation and sputtering has also been invoked to explain the distribution of water frost on Ganymede, where the equatorial zone most impacted by magnetospheric plasma has less visible surface frost, and the high-latitude regions appear to be coated with frost that may have migrated there from the equatorial areas and been retained there due to lower temperatures. Ganymede's frosty polar caps (the northern cap is visible in Fig. 1) closely follow the region where Ganymede's magnetosphere becomes connected with the external Jupiter magnetosphere, indicating that charged particles play an important role in creating this feature.

Another interaction with the external environment of the satellites is micrometeoroid bombardment. All the satellites are exposed to the flux of tiny grains from interplanetary space striking their surfaces. In the cases of Ganymede and Callisto, the interplanetary particle fluxes are enhanced due to Jupiter's gravity. *Galileo* carried a sensitive detector that measured the surrounding dust environment as the spacecraft orbited Jupiter. It also made measurements on the close passes by the satellites to sample the population or dust particles near the satellites. The dust investigators found that both satellites have a population of small (micrometer-sized) particles loosely bound by gravity in the space surrounding the satellites. These measurements are consistent with icy dust grains that have been blasted off the satellite surfaces by the impact of interplanetary micrometeorites.

The ices on the surfaces of Ganymede and Callisto are weakly warmed by the Sun and are exposed to near-vacuum conditions. Even at the cold temperatures in the Jupiter system, ice will slowly sublime and escape as a gas. Water ice could sublimate at a rate of meters per million years, but on Ganymede and Callisto it is soon choked off by a blanket of non-ice dust, since the surface ice is not pure and the dust does not sublimate. Sublimation will occur millions of times faster for SO_2 ice, and CO_2 ice will sublimate thousands of times faster than that, so incorporation of these compounds into the ice bedrock will drive the sublimation erosion process much faster.

3.3 Regolith

Bright ice crystals and dark non-ice dust both exist on the surfaces of Ganymede and Callisto, but they are largely segregated from each other. If one were to pick up a sample of the loose surface material (the regolith), it would probably be composed of mostly ice or mostly dust, and not a mixture of the two. High-resolution images show very high albedo contrasts over small spatial scales, with relatively pure icy material outcropping in patches surrounded by blankets of dark non-ice material. This effect is most pronounced on Callisto (Fig. 6). It appears that the ice bedrock is composed of a mixture of ice and non-ice dust. When a fresh outcrop of bedrock is exposed at the surface, the ice will begin to

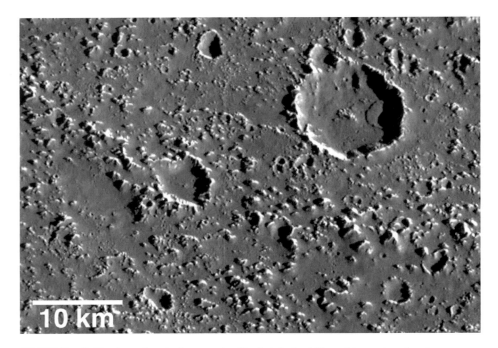

FIGURE 6 Callisto's surface is characterized by bright icy hills and impact crater rims surrounded by blankets of dark dust. This surface is thought to result from sublimation of an ice/dust mixture, leaving a lag deposit of loose dust in the low areas and depositing bright frost on steep slopes and hilltops. Note the raised tongue of a landslide deposit consisting of loose dark material emanating from the shadowed wall in the prominent crater in the northeastern section of the image. This area is located within the Asgard impact basin.

sublimate into the near-vacuum atmosphere. Enhanced solar heating of dark material drives faster sublimation of ice from that material, further darkening the material, while any reflective surface will serve as a cold trap, building up a layer of frost and further brightening the material. These positive feedbacks lead to the effective segregation of ice and non-ice materials in the surface regolith.

This process also operates on Ganymede, but it has not modified the surface to the same extent seen on Callisto. Perhaps there is more dark material mixed in with the surface ice on Callisto, or perhaps the ice on Callisto includes a higher proportion of volatile SO_2 and CO_2 ices that enhance the rate of sublimation. Patches of frost are often seen on steep slopes facing away from the sun. Bright terrain on Ganymede has a higher thermal inertia than the dark terrain, indicating that much of the bright ice exposed in this terrain must be more solid or compacted than the loose dust that covers dark terrain.

On both Ganymede and Callisto, the dark material is found filling the topographic lows, while bright material covers the slopes and hilltops. Part of the reason for this is due to the fact that reflection and emission of light from surrounding terrain tends to make topographic lows slightly warmer, but much of this effect is due to the loose, dusty nature of the dark material. Buildup of loose dark material on steep slopes due to sublimation leads to avalanches of the material into topographic lows. The large crater in the

northeastern corner of Fig. 6 has a thick tongue of material flowing over the floor, this is a deposit of regolith that has slid from the steep eastern wall of the crater, and the shape of the deposit indicates that it slid downhill as a dry avalanche of loose debris. Images of dark terrain on Ganymede also show chutes on steep slopes where material has slid downhill, with dark material piled along the bottoms of the slopes. Bright terrain on Ganymede shows the same effect, with dark material filling in the valleys of the grooved terrain, between bright steep icy slopes. The dark dust often appears to form a thick, smooth blanket on Callisto and Ganymede dark terrain, but there are many small craters that penetrate through the dark material, indicating that the layer of loose dark dust may only be meters deep before a solid layer of ice/dust mixture is reached.

4. Impact Craters

4.1 Crater Structures

Ganymede and Callisto exhibit a wide variety of impact features, including some types unique to these large icy satellites. The smallest craters imaged on the two moons have a classic bowl-shaped morphology as is the case for small craters on any planet. At a diameter of 2–3 km, central peaks begin to appear (Fig. 7a), again following the normal morphological progression for most planets. However, as crater

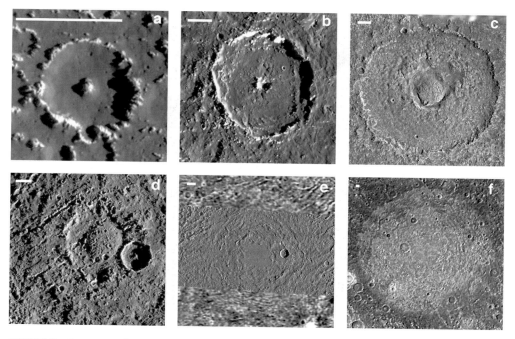

FIGURE 7 Diversity of impact crater morphologies on Ganymede and Callisto. All scale bars are 10 km long, and illumination is from the right. (a) central-peak crater on Callisto; (b) central-pit crater on Callisto; (c) central-dome crater Melkart on Ganymede; (d) anomalous dome crater Har on Callisto; (e) penepalimpsest Buto Facula on Ganymede; (f) palimpsest Memphis Facula on Ganymede.

diameter increases beyond 35 km, instead of the transition to larger central peaks or peak rings seen on the inner planets, large craters on Ganymede and Callisto exhibit central pits (Fig. 7b). Young craters undergo another transition at about 60-km diameter, where the central pits begin to exhibit round domes of material in their centers (Fig. 7c). These central domes have fractured surfaces reminiscent of lava domes, and they may be formed by rapid extrusion of warm, viscous ice into the center of the crater just after its formation. Most large craters on Ganymede and Callisto are very shallow, especially the older craters, indicating that warm subsurface ice has flowed in toward the crater depressions and bowed their floors back up to the topographic level of their surroundings (a process known as viscous relaxation). Some central-dome craters have been so flattened that they do not exhibit any obvious rim structure; the central dome and surrounding pit wall are the only obvious structures remaining (Fig. 7d). A few large craters, known as penepalimpsests, exhibit only subdued topographic rings, with a smooth patch in the middle (Fig. 7e). Where these occur on Callisto and the dark terrain of Ganymede, they show up as a distinct circular patch of bright material against the dark background. Still other bright circular patches, which are almost completely flat (except sometimes an outward facing scarp can be seen around the outside and a depressed smooth patch found in the middle), are found within Ganymede's dark terrain (and a few exist on Callisto) (Fig. 7f). These features are called palimpsests, a word for an ancient piece of parchment where the writing has been erased. In a similar way these large ancient craters have almost been erased by the process of viscous relaxation.

On the Moon, the largest craters form multiring basins [see THE MOON]. On Ganymede, one large basin called Gilgamesh shares similar characteristics with the lunar basins: a smooth central region surrounded by large irregular massifs, which is in turn surrounded by a few large concentric mountain ranges. However, most large impact basins on Ganymede and Callisto exhibit a distinctly different morphology, with a large palimpsest in the middle surrounded by many evenly spaced concentric rings. The best example of such an impact basin is Valhalla on Callisto, which is about 1000 km across and exhibits about 20 concentric rings around its central bright palimpsest (Fig. 8). Most basin rings on Callisto are troughs that appear to have formed by extension of the surface material. These multiring structures are thought to form as subsurface material rapidly flows in from the sides to fill the center of the impact basin, pulling a thin brittle veneer of surface material inward.

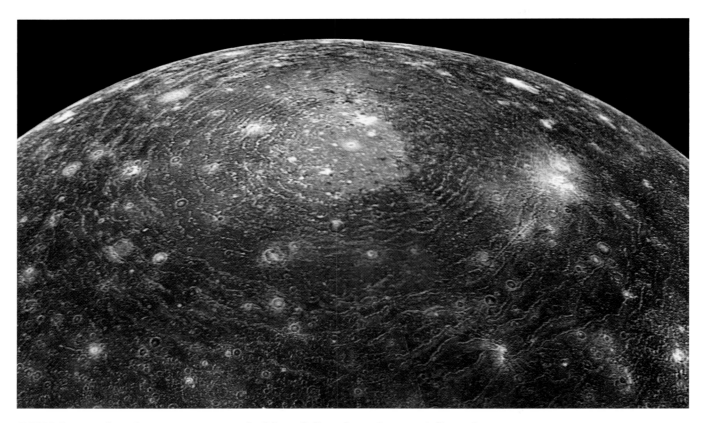

FIGURE 8   View from the *Voyager 1* spacecraft of the Valhalla multiring basin on Callisto. The extensive system of concentric troughs surrounding the impact site is over 3000 km across.

The viscous relaxation and modification of craters can inform us about the nature of the subsurface ice during and after crater formation. Older craters are distinctly shallower, showing the action of viscous relaxation through time. However, there is not a continuum of viscously relaxed craters as one might expect if this process was ongoing at a constant rate. Instead, it appears that early craters relaxed quickly, while more recent craters are being preserved in a stiff material. This implies that heat flow was higher in the past, allowing warm ice to flow just below the surface early in solar system history, while more recently the subsurface ice has become colder and stiffer. There is overlap in size between central-dome craters, penepalimpsests, and palimpsests, implying that impacts of similar energy formed all these morphologies at different times. Palimpsests are found only in the most ancient terrains, whereas central-dome craters appear to be relatively young. Again, it appears that palimpsests formed early when the subsurface was warm and flowed easily, penepalimpsests record a time when the ice was cooling, and dome craters have formed more recently in a thicker layer of cold stiff ice. On Ganymede, the formation of the bright terrain appears to mark an important transition in crater morphology, with no palimpsests or Valhalla-type multiring basins being formed after the formation of bright terrain. Thus, it appears that heat flows were higher on Ganymede until the period of bright terrain formation, and Ganymede's subsurface became colder and stiffer after that period.

4.2 Distribution of Craters and Surface Ages

Variations in the areal density of impact craters are observed on Ganymede and Callisto, giving us information about the population of impactors and the relative ages of different surfaces. In general, the highest crater densities are found on the dark terrain of Ganymede and the plains of Callisto. Bright terrain on Ganymede has a much lower density of craters than the dark terrain, supporting the view that it formed substantially later. The only areas on Callisto with lower crater densities are the interiors of impact craters and large multiring basins, where the surface age has been reset by the impact.

Translating the areal density of impact craters into absolute ages of different surfaces on Ganymede and Callisto is a tricky proposition. On the Moon, this can be accomplished by correlating areas of varying crater density on the lunar surface with physical samples of those surface materials that have been returned to Earth and that can be precisely dated in the laboratory using radioisotope techniques. Since we have no surface samples from the Galilean satellites, we cannot directly date them. In addition, we cannot be sure that the same population of debris that impacted the Moon also impacted the Galilean satellites, so it is dangerous to simply translate crater densities between these two different parts of the solar system. In general, it is agreed that

the surface of Callisto and the dark terrain on Ganymede represent primordial surfaces, formed shortly after the formation of the planets, 4.5 billion years ago. Bright terrain on Ganymede could have formed shortly after that, or it could have formed only a billion years ago. The current best guess from crater statistics is that bright terrain most likely formed at some time during the middle half of solar system history, but obtaining an exact age is likely to remain elusive for a long time.

Since Ganymede and Callisto are tidally locked and always have the same side facing Jupiter, it is expected that they should gather more of the debris coming from outside the Jupiter system on the sides facing forward in their orbital motion (the bug on the windshield effect), and thus there should be more craters on their leading hemispheres than on their trailing hemispheres. Callisto does exhibit such an asymmetry in crater density, but the asymmetry on Ganymede is much weaker. One hypothesis to explain this is that Ganymede's outer ice shell has rotated with respect to Jupiter in the past and has become locked to Jupiter more recently, whereas Callisto's surface has always been locked with respect to Jupiter. Another piece of evidence to support this view comes from the study of split comets. In 1994, we witnessed the impact of comet Shoemaker–Levy 9 into Jupiter—this comet had been disrupted into a string of fragments by a close encounter with Jupiter before the impact. If such a string of comet fragments hit one of the satellites on its way out of the Jupiter system, it would form a line of closely spaced impact craters called a catena, and these are in fact observed on the surfaces on Ganymede and Callisto. On Callisto, all the catenae are on the Jupiter-facing hemisphere, as one would expect from the impact of a comet on its way out of the system after a close brush with Jupiter. On Ganymede, one third of the catenae are found on the other hemisphere, which would be impossible unless Ganymede's ice shell had rotated in the past.

5. Tectonism and Volcanism

The surface record of tectonic and volcanic activity is the most obvious difference between Ganymede and Callisto. Most of Ganymede's surface has been reworked by some combination of these processes, whereas Callisto's surface may be untouched. Next, we separately consider the roles of tectonism and volcanism in the extensively resurfaced bright terrain of Ganymede, the marginally resurfaced dark terrain of Ganymede, and the relatively pristine surface of Callisto.

5.1 Bright Terrain

Bright terrain covers two thirds of Ganymede's surface and is composed of a dense network of intersecting and overlapping areas of parallel ridges and troughs, termed grooved

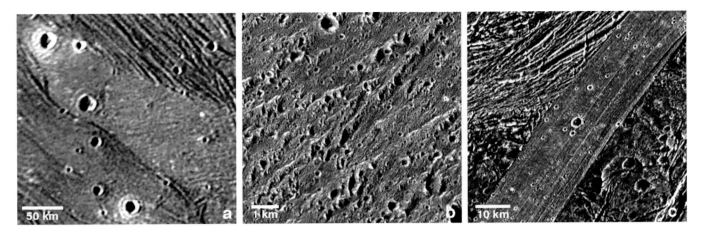

FIGURE 9   Views of smooth bright terrain on Ganymede. (a) *Voyager 1* image of smooth bright
terrain in Harpagia Sulcus; (b) Galileo high-resolution image of smooth terrain from the center of
(a), showing ridges and hills not visible in regional-scale views; (c) Galileo image of Arbela Sulcus,
a narrow lane of smooth terrain cutting through the dark terrain of Nicholson Regio.

terrain, and other areas with more subdued topography,
termed smooth terrain.

Smooth terrain may occur either as patches bounded by
grooved terrain on all sides or as lanes of smooth material tens of kilometers wide cutting across bright and dark
terrain. In either case, the terrain appears to be smooth
in kilometer-resolution regional images (Fig. 9a), leading
to the hypothesis that it formed by low-viscosity cryovolcanic flows flooding the underlying terrain. At higher resolution, it becomes apparent that the smooth terrain is not
so smooth after all. In some areas, it appears to be a flat
plain crossed by ridges or sets of aligned hills (Fig. 9b).
In other areas, especially where the smooth terrain occurs
as narrow lanes, it appears to be a flat or gently undulating surface crossed by parallel dark lineations, which may
be narrow valleys formed by tensile fracturing of the ice
(Fig. 9c). The presence of parallel sets of ridges and valleys in smooth terrain suggests that tectonism plays an important role in shaping this terrain, in addition to possible
cryovolcanism.

Though cryovolcanism is an attractive explanation for
the smooth, flat areas found within smooth terrain, its role
has not been conclusively demonstrated. No obvious volcanic constructs or flows have been observed, though it is
unclear if we know what an ice volcano is really supposed
to look like. A few features that may possibly be volcanic
calderas have been observed (Fig. 10), but most areas of
smooth terrain exhibit no such features. While the smooth
regions shown in Fig. 10 are topographic lows, as one would
expect if they were troughs filled by low viscosity volcanic
flows, the smooth regions shown in Fig. 9 have been found
to lie locally higher than parts of their immediate surroundings. Another possible interpretation for the linear bands of
smooth terrain is that they formed through separation and

spreading of the crust in a manner analogous to Europa's
gray bands. In either case, the formation of smooth terrain
appears to involve the extrusion of liquid water or warm ice
from Ganymede's subsurface.

Tectonism plays a more obvious role in the formation
of grooved terrain. In kilometer-resolution images, grooved
terrain is characterized by parallel valleys and ridges spaced
about 5–10 km apart (Fig. 11a). At higher resolution, each
ridge and valley is itself composed of many smaller ridges
and valleys (Fig 11b). Each of these smaller ridges is thought
to be a fault block, a piece of the icy crust that has been
separated from its surroundings by faults and then moved
and tilted as it slid along those faults. The shapes and intersections of the faults are suggestive of a style of faulting
known as tilt-block normal faulting, in which many parallel faults slice the upper portion of the crust into roughly
rectangular blocks that then tilt over and slide against each
other as the crust extends, much like books sliding over
on a bookshelf when a bookend is removed. This style of
faulting creates parallel ridges with a sawtooth topographic
profile, matching the triangular ridges on Ganymede with
their sharp crests, frosty upper slopes, and dark V-shaped
valleys between them.

In a few places on Ganymede, large impact craters have
been cut by these networks of faults (Fig. 12). Since almost
all craters are formed in a roughly circular shape, these
cut craters offer an opportunity to measure directly how
the crust has deformed as their shape becomes progressively distorted by motion along the faults. Measurements
of these craters confirm that the development of grooved
terrain on Ganymede is dominated by extensional tectonics. At the extreme end of the spectrum, some small parts
of the crust appear to have been pulled apart to more than
twice their original width, but in most cases the extension

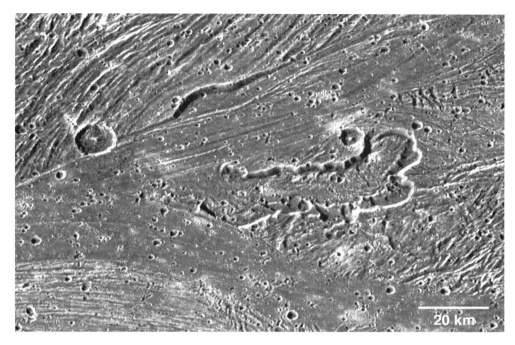

FIGURE 10   This irregular pit is one of several found along the edges of smooth terrain in Sippar Sulcus on Ganymede. Faint curved ridges within the pit suggest flow folding in the surface of a cryovolcanic flow emanating from the closed end of the pit and flowing out the open end, into the surrounding smooth terrain.

appears to be more moderate. Circumstantial evidence exists for contractional deformation in a few areas, but it is not widespread.

These observations force us to ask how the crust of Ganymede could have undergone a large amount of extension with very little evidence for contraction to balance it out. There are a few possible solutions to this conundrum. One solution is that Ganymede actually expanded during the formation of grooved terrain. Differentiation of Ganymede's interior or melting of high pressure ices can serve to increase the volume of the satellite, leading to an increase in surface area and thus stretching of the crust.

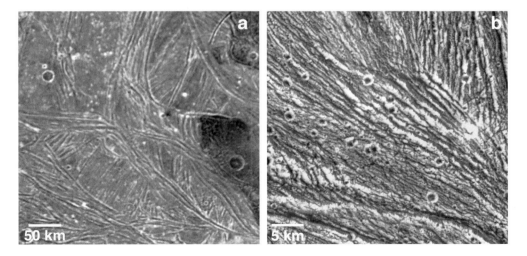

FIGURE 11   Views of bright grooved terrain on Ganymede. (a) *Voyager 2* imaged this region of grooved terrain in Uruk Sulcus; (b) *Galileo* imaged these grooves in the central part of (a), shown at ten times the scale.

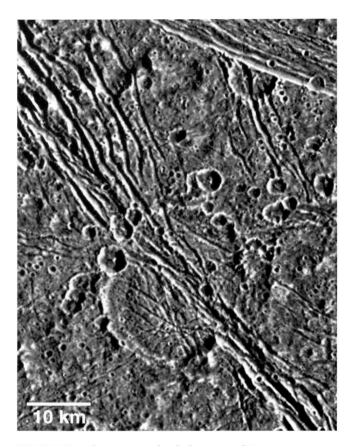

10 km

FIGURE 12   This crater in the dark terrain of Marius Regio on Ganymede has been cut by faults extending from a nearby region of grooved terrain. Measurement of the postdeformation shape of the crater demonstrates that the faults have extended the terrain and also horizontally translated the eastern part northward relative to the western part.

Alternatively, we may have missed seeing the contractional features on Ganymede either because the crust shortened mostly in a ductile fashion, leaving few obvious surface features, or because we simply don't recognize contractional features formed in ice.

In places where narrow lanes of grooved terrain cut across dark terrain, it is clear that bright grooved terrain can form simply by extension of the dark terrain, without the cryovolcanism that may have taken place in smooth terrain. Faulting can serve to erase the impact craters on the dark terrain by slicing them up and making their rims unrecognizable. Brightening of the terrain can occur by breaking through the dark regolith layer and exposing bright subsurface ice along the fault scarps.

5.2 Dark Terrain

Aside from the swaths of bright grooved terrain that cut across the dark terrain and small peripheral fractures ad-

jacent to the grooved terrain, dark terrain on Ganymede is primarily dominated by systems of arcuate to linear features known as furrows. Furrows are usually composed of two bright ridges spaced 10–20 km apart, with a dark trough in between them. Most furrows are arranged in concentric sets of arcs (Fig. 13), indicating that they are probably ancient multiring basins that originally resembled Valhalla on Callisto but are now sliced up into fragments by the formation of bright terrain. This interpretation is aided by some small furrow systems that appear to have an impact basin in the center. Some sparse systems of linear furrows appear to radiate out from a point rather than being concentric arcs. The origin of these radial systems is unclear.

There was speculation based on *Voyager* images that some areas of dark terrain had a splotchy appearance due to patches of dark cryovolcanic material oozing onto the surface. At higher resolution, however, these splotchy areas were revealed to be dark plains interrupted by networks of bright fractures.

5.3 Callisto

The story of tectonism and volcanism on Callisto is reminiscent of Ganymede's dark terrain. All of the obvious tectonic features are arranged in concentric rings and surround large impact basins. The rings are high scarps or deep troughs with sharp boundaries (Fig. 14). The scarps and troughs are formed by faults that have extended the crust by a small amount during the formation of the impact basin (see Section 4). Multiring basins on Callisto are examples of the ancient impact basins that formed the furrow systems on Ganymede before they were broken apart. There is also a strange system of troughs near Callisto's North Pole that seems to radiate out from a point. Unfortunately, the center of this system on Callisto was never imaged at high resolution, so the origin of these features remains mysterious, as does the origin of similar radial furrow systems on Ganymede. Early speculation that smooth dark patches on Callisto might be cryovolcanic in origin has been largely dispelled by evidence that Callisto has a loose regolith that smoothes over the underlying terrain like a thick dark blanket.

6. Unanswered Questions and Future Exploration

Several interesting unanswered questions remain about Ganymede and Callisto. Since the general properties of these satellites appear to be so similar, understanding the processes and events that have driven their interior evolution and geological records to different states is an important problem in comparative planetology. In the interiors of

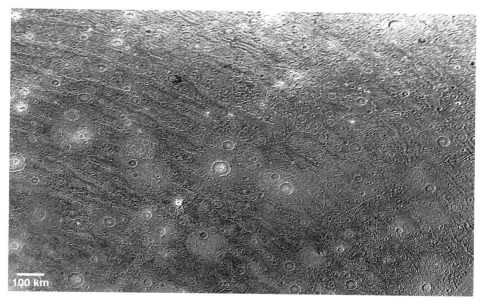

FIGURE 13 Furrows in Galileo Regio on Ganymede arc gently from northwest to southeast. These are thought to be the Ganymede equivalent of the concentric rings found around impact basins such as Valhalla on Callisto (see Fig. 8).

the satellites, making Ganymede hot enough to generate a magnetic dynamo and keeping Callisto cold enough to not differentiate are both challenging problems for our understanding of planetary geophysics. The oceans of liquid water that exist within these bodies, sandwiched between different phases of ice, are exotic phenomena in themselves. On the surfaces of Ganymede and Callisto, we still don't have a clear idea of the composition of some of the materials that are mixed in with the water ice; neither do we know which of those materials come from the interiors of the satellites, which ones come from their external environments, and which ones are the products of chemical reactions and radiation processing at the surface.

The unfortunate failure of the main antenna on the *Galileo* spacecraft left us without a complete global reconnaissance of these bodies at a level of detail sufficient to resolve features at the scale of a kilometer or less. The small target areas that Galileo imaged at high-resolution revolutionized our understanding of these bodies, but much of their surfaces will remain relatively unknown for the near future. No new missions are currently being planned to explore Ganymede and Callisto, though plans for a

FIGURE 14 Oblique view over the surface of Callisto, looking over the edge of one of the concentric ring scarps of the Valhalla impact basin.

possible Europa-orbiting spacecraft probably would involve the use of Ganymede and Callisto for gravitational assistance into Europa orbit. Such a mission would provide several serendipitous opportunities to gather more information about these mysterious twin moons of Jupiter.

Bibliography

Bagenal, F., Dowling, T., and McKinnon, W., eds. (2004). "Jupiter: The Planet, Satellites, and Magnetosphere." Cambridge University Press, Cambridge.

Pappalardo, R.T., Collins, G.C., Head, J.W., Helfenstein, P., McCord, T., Moore, J.M., Prockter, L.M., Schenk, P.M., and Spencer, J. (2004). Geology of Ganymede. In "Jupiter: The Planet, Satellites & Magnetosphere" (F. Bagenal et al., eds.), pp. 363–396, Cambridge University Press, Cambridge.

Moore, J.M., Chapman, C.R., Bierhaus, E.B., Greeley, R., Chuang, F.C., Klemaszewski, J., Clark, R.N., Dalton, J.B., Hibbitts, C.A., Schenk, P.M., Spencer, J.R., Wagner, R., Callisto. (2004) In "Jupiter: The Planet, Satellites, and Magnetosphere." (F. Bagenal, T.E. Dowling, and W.B. McKinnon, eds.), Cambridge University Press, Cambridge, pp. 397–426.

Pappalardo, R.T. (1990) Ganymede and Callisto, "The New Solar System." 4th ed. (J.K. Beatty, C.C. Petersen, and A. Chaikin, eds.), Sky Publishing, Cambridge.

Johnson, T.V. (2004). A look at the Galilean satellites after the Galileo Mission, Physics Today, 57(4), 77–83.

Titan

Athena Coustenis

LESIA, Observatoire de Meudon
Paris, France

CHAPTER 25

1. Introduction

1.1 Titan's Discovery, First Observations, and Models

Titan, Saturn's biggest satellite (second in size among the satellites in our solar system), has attracted the eye of astronomers preferentially ever since its discovery by Dutch astronomer Christiaan Huygens on March 25, 1655. Titan orbits around Saturn at a distance of 1,222,000 km (759,478 mi) in a synchronous rotation, taking 15.9 days to complete. As Titan follows Saturn on its trek around the Sun, one Titanian year equals about 30 Earth years. The sunlight that reaches such distances is only 1/100th of that received by the Earth. Titan is therefore a cold and dark place, but a fascinating one.

It has been known for a long time that Titan possesses a substantial atmosphere: Catalan astronomer Jose Comas i Solà claimed in 1908 to have observed **limb-darkening** on Titan. Due to its thick atmosphere, Titan subtends 0.8 **arcsec** in the sky, and it was thought to be the largest of the satellites in the solar system. This explains the name it was given (following a proposition by Herschel, who suggested names of gods associated with Saturn for naming its satellites), until the advent of the *Voyager* missions that showed Ganymede to be a few kilometers larger. Today, we know that this massive atmosphere is the one most similar to the Earth's among the other objects of our solar system

as N_2 is its major constituent and it is host to a complex organic chemistry.

In 1925, Sir James Jeans showed that Titan could have kept an atmosphere, in spite of its small size and weak gravity, because some of the constituents which could have been present in the proto-solar nebula (ammonia, argon, neon, molecular nitrogen and methane) would not escape. It was realized later that although ammonia (NH_3) is in solid phase at the current Titan temperatures and could not in principle contribute to its present atmosphere, it could have evaporated in the early atmosphere and been converted into N_2 at the end of the accretion period when the environment was warmer.

On the other hand, methane (CH_4), the second most abundant constituent on Titan, is gaseous at present Titan's atmospheric temperature range and, unlike molecular nitrogen, exhibits strong absorption bands in the infrared. These bands were first detected in 1944 by Gerard Kuiper of Chicago University. Ethane (C_2H_6), monodeuterated methane (CH_3D), ethylene (C_2H_4) and acetylene (C_2H_2) were also discovered later.

Prior to spacecraft observations, two models were popular: a "thin methane" atmosphere model, which favored methane as the main component (about 90%) and predicted surface conditions of $T = 86$ K for 20 mbar as well as a temperature inversion in the higher atmospheric levels, illustrated by the presence of emission features of hydrocarbon gases in the infrared spectrum of Titan; and a "thick nitrogen" atmosphere model, which was based on the assumption that ammonia dissociation should produce molecular nitrogen (transparent in the visible and infrared spectrum) in large quantities and held that the surface temperature and pressure could be quite high (200 K for 20 bars). Independent of these two models, an explanation of the high observed ground temperatures was advanced: a pronounced greenhouse effect, resulting essentially from H_2–H_2 pressure-induced **opacity** at **wavelengths** higher than 15 μm. This opacity blocks the thermal emission reflected by the surface, thus creating a heat-up of the lower part of the atmosphere, as found on Earth.

1.2 Titan's Exploration

Titan has since then been extensively studied from the ground and from space. In the latter case, Titan was "blessed" by several space mission encounters in the course of the planetary exploration in our solar system. [*See also* PLANETARY EXPLORATION MISSIONS.]

Although the *Pioneer 11* spacecraft was the first to take a close look at the giant planets Jupiter and Saturn, it flew by Titan at a considerable distance of 363,000 km on September 2, 1979. The *Voyager* missions that followed were also dedicated to an extended study of the outer solar system. The *Voyager 1* (V1) spacecraft (launched in 1977) arrived in the Saturnian system and made its closest approach of Titan on November 12, 1980, at a distance of only 6969 km (4394 miles) to the satellite's center. *Voyager 2* flew by Titan 9 months later but at a distance a hundred times greater (663,385 km) so that the *Voyager 1* encounter was the closest a man-made machine ever came to Titan until 2004.

Titan's visible appearance at the time was unexciting—an orange ball, completely covered by thick haze, which allowed no visibility of the surface (Fig. 1a). The most obvious feature seen by *Voyager* was a difference in the brightness of the two hemispheres. This difference is of the order of 25% at blue wavelengths and falls to a few percent in the ultraviolet and at red wavelengths. This so-called north–south asymmetry (NSA) is probably related to circulation in the atmosphere pushing haze from one hemisphere to the other. The altitude of unity vertical optical depth is of the order of 100 km. Also noticeable was a dark ring above the north (winter) pole. This feature, termed the polar hood, extending from 70° to 90° north latitude, was most prominent at blue and violet wavelengths, and it has since then been suggested that it may be associated with lack of illumination in the polar regions during the winter (since the subsolar latitude goes up to 26.4°) and/or subsidence in global circulation.

Besides the images, the *Voyager* instrument also allowed for the determination of the chemical composition and temperature structure. The latter and other basic parameters for Titan (Table 1) were provided by the **radio-occultation**

FIGURE 1   Titan observed in 1980 with the cameras of *Voyager 1* in the visible and in 2004 with the *Cassini* ISS camera (Team Leader: C. C. Porco) at 0.94 μm. In the first case, the bland appearance of the satellite belies a complex world. The only features apparent in the images taken by *Voyager* were the detached haze layers, the dark polar hood and generally a difference in brightness between the two hemispheres. In contrast one of the most recent images by *Cassini* shows Titan's surface features. (Image Credit: NASA/JPL.)

Voyager experiment obtained by the Radio Science Subsystem (RSS). Titan's surface radius was found to be 2575 ± 2 km, with a surface temperature of 94 ± 2 K and a pressure of about 1.44 bar.

After *Voyager*, scientists had to wait for about 25 years before getting another close look at Titan. *Cassini–Huygens* is a very ambitious mission, planned in the 1980s already. It is an extremely successful collaboration between ESA and NASA (with contribution from 17 countries), composed of an orbiter and a probe (*Huygens*). Although the mission's objectives span the entire Saturnian system, Titan is a privileged target for *Cassini* (as for *Voyager* before it), and the mission is designed to address our principal questions about this satellite during its 6-year duration from 2004 onwards. The spacecraft is equipped with 18 science instruments (12 on the orbiter and 6 carried by the probe), gathering both remote sensing and in situ data. It communicates through one high-gain and two low-gain antennas. Power is

TABLE 1	Titan's Orbital and Body Parameters, and Atmospheric Properties
Surface radius	2575 km
Mass	1.35×10^{23} kg (= 0.022 × Earth)
Mean density	1880 kg^{-3}
Distance from Saturn	1.23×10^9 m (= 20 Saturn radii)
Distance from Sun	9.546 AU
Orbital period	15.95 days
around Sun	29.5 years
Obliquity	26.7°
Surface temperature	93.6 K
Surface pressure	1.467 bar

provided through three radioisotope thermoelectric generators (or RTGs).

The 5650 kg (6 ton) *Cassini–Huygens* spacecraft was launched successfully on October 15, 1997, from the Kennedy Space Center at Cape Canaveral at 4:43 A.M. EDT. Because of its massive weight, *Cassini* could not be sent directly to Saturn but used the "gravity assist" technique to gain the energy required by looping twice around the Sun. This allowed it to also perform flybys by Venus (April 26, 1998, and June 24, 1999), Earth (August 18, 1999), and Jupiter (December 30, 2000). *Cassini–Huygens* reached Saturn in July 2004 and performed a flawless Saturn Orbit Insertion (SOI), becoming trapped forever in orbit like one of Saturn's moons.

The *Cassini* instruments have since then returned a great amount of data concerning the Saturnian system. During its 4 year nominal mission, the *Cassini* orbiter will make about 40 flybys of Titan, some as close as 1000 km (*Voyager 1* flew by at 4400 km) from the surface. *Cassini* will perform direct measurements with the visible, infrared, and radar instruments designed to perform in situ (on-site) studies of elements of Saturn, its atmosphere, moons, rings, and magnetosphere. One set of instruments studies the temperatures in various locations, the plasma levels, the neutral and charged particles, the surface composition, the atmospheres and rings, the solar wind, and even the dust grains in the Saturn system, while another performs spectral mapping for high-quality images of the ringed planet, its moons, and its rings.

Additionally, the mission saw the deployment of the European-built *Huygens* probe. After release from the *Cassini* orbiter, on December 25, 2004, this 300 kg probe plunged into Titan's atmosphere on January 14, 2005, at 11:04 UTC and descended through it by means of several parachute brakes (Fig. 2), which slowed the probe from supersonic speeds of 6 km/s during entry and down to 5 m/s at impact. The five batteries onboard the probe lasted much longer than expected, allowing *Huygens* to collect descent data for 2 hours and 27 minutes and surface data for 1 hour and 12 minutes. During its descent, *Huygens'* camera returned more than 750 images, while the probe's other instruments sampled Titan's atmosphere to help determine its composition and structure. The telemetry data from *Huygens* was stored onboard *Cassini's* Solid State Recorders (SSRs) at a rate of 8 kbits/s, while the spacecraft was at an altitude of 60,000 km from Titan. Although some data from Huygens was lost during its transmission to *Cassini* through a stream called Channel A, in the end all of the measurements were recovered because Titan's weak signal was captured by Earth-based radio telescopes!

As well as measuring the atmosphere and surface properties, the probe took samples of the haze and gases. These in situ measurements complement the remote-sensing data recorded from the orbiter.

The *Cassini–Huygens* mission has already provided a wealth of data. The analysis is in the first stages, and the *Cassini* orbiter promises to unveil yet more of Titan's secrets in the years to come. What follows is an attempt to provide the reader with a precise account of current information on Titan's environment from all available means of investigation.

2. The Atmosphere of Titan

The most interesting feature of Titan, as has been argued previously, is its amazing atmosphere, a close analog to the Earth's primitive gas envelope according to some theories, but it is located almost ten times further away from the Sun.

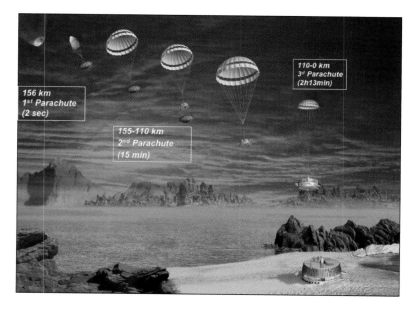

FIGURE 2 *Huygens* descent profile in Titan's atmosphere. The three parachutes that helped brake the descent and reduce the speed to about 5 m/s on the surface are shown. The total descent lasted 2 h 28 mn. The probe spent 1 h 12 mn on the surface. The signal from Huygens received on Earth via radiotelescopes was a total of 5 h 42 mn including 3 h 14 mn at the surface.

2.1 Thermal Structure

The first definitive measurement of the atmospheric temperature structure was made by *Voyager*. The V1/RRS radio-occultation experiment provided density and temperature profiles in Titan's atmosphere from refractivity measurements. Titan's temperature profile was measured in situ on January, 14, 2005 by the *Cassini–Huygens* Atmospheric Structure Instrument (HASI) at the probe's landing site (15°S, 192°W) from 1400 km in altitude down to the surface, where 93.65 ± 0.25 K were measured for a surface pressure of 1467 ± 1 mbar. As *Voyager* did before, HASI found Titan's atmosphere to exhibit the features that characterize the Earth's thermal structure: the atmospheric layers include an exosphere, a mesosphere, a stratosphere and a troposphere, with two major temperature inversions at 40 and 250 km, corresponding to the tropopause and stratopause, associated with temperatures of 70.43 K (min) and 186 K (max), respectively (Fig. 3). At the same time, the Composite Infrared Radiometer Spectrometer (CIRS) on the orbiter took spectra that confirmed the presence of a stratopause around 310 km of altitude for a maximum temperature of 186 K. Another inversion region, less contrasted than the previous ones and corresponding to the mesopause can be found at 490 km (for 152 K).

The HASI data furthermore yield more precise and new information on the upper part of the Titan atmosphere, the thermosphere, where several temperature fluctuations are observed due to dynamical (gravity and tidal) phenomena. Indeed, gravity waves signatures of 10–20 K in amplitude were recorded above 500 km around an average temperature of 170 K. HASI moreover found a lower ionospheric layer between 140 and 40 km, with electrical conductivity peaking near 60 km. A tentative detection of lightning is being investigated.

Besides the *Huygens* measurements, few constraints are available for the temperature structure in Titan's higher atmosphere. The *V1*/UVS experiment recorded a temperature of 186 ± 20 K at 1265 km during a solar occultation for a methane mixing ratio of 8 ± 3% toward 1125 km, placing the homopause level at around 925 ± 70 km. A value of 183 ± 11 K near 450 km was derived from the July 3, 1989, stellar occultation of Titan. The occultation of star 28 Sgr by Titan was observed from places as widely dispersed as Israel, the Vatican, and Paris. This rare event provided information in the 250–500 km altitude range. A mean **scale height** of 48 km at 450 km altitude (∼3 mbar level) was inferred. This allowed the mean temperature to be constrained at that level to between 149 and 178 K.

From *V1* infrared disk-resolved measurements, temperature latitudinal variations were already demonstrated to exist in Titan's stratosphere. At that time, a maximal temperature decrease of 17 K at the 0.4-mbar level (225 km in altitude) was observed between 5°S (the warmest region in the *Voyager* data) and 70°N, whereas the temperature dropped by only 3 K from 5°S to 53°S. The coldest temperatures, found at high northern latitudes, were associated with enhanced gas concentration and haze opacity (as this may be caused by more efficient cooling) or/and **dynamical inertia**. CIRS mapped stratospheric temperatures over much of Titan in the latter half of 2004, when it was early southern summer on Titan (solstice was in October 2002). The warmest temperatures are near the equator. Temperatures are moderately colder at high southern latitudes, by 4–5 K near 1 mbar, but they are coldest at high latitudes in the north, where it is winter.

Titan was also found to have a quite extended ionosphere, due to the lack of a strong intrinsic global magnetic field. Charged particles in the rarified upper atmosphere are then exposed to bombardment by the solar wind and by

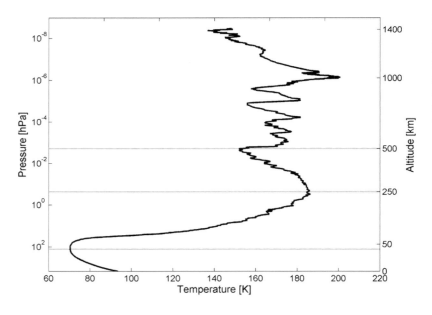

FIGURE 3 Titan's temperature profile as derived from Huygens/HASI measurements. The several large inversion layers in the upper atmosphere correspond to gravity waves. The inversion at around 40 km marks the tropopause, whereas the one at around 250 km is due to the stratopause. (Adapted from Fulchignoni et al., 2005; Nature 438, 8 Dec. 2005, 785–791).

particles precipitated from Saturn's magnetosphere (creating an ionospheric layer between 700 and 2700 through which *Cassini* flew during some of its lower flybys of Titan), as well as by cosmic rays from outer space (producing a second layer between 40 and 140 km). *Cassini* found that more than 10% of the ionosphere is made up of ionized hydrocarbon molecules chemically similar to compounds such as ethylene, propyne, and diacetylene and that this population is lost to space at important rates.

2.2 Chemical Composition

Indeed, the nature of Titan's atmosphere finally emerged as a combination of the two pre-*Voyager* models. Molecular nitrogen (N$_2$, detected by the UV spectrometer) is by far the major component of the atmosphere (average of ~95%). The presence of methane (the next most abundant molecule with abundances ranging from 0.5 to 3.4% in the stratosphere and from 4 to 8 % at the surface), traces of hydrogen, and a host of organic gases were inferred from emission bands observed in the infrared interferometer spectrometer (IRIS) spectra, which cover the 200–1500 cm^{-1} spectral region with a spectral resolution of 4.3 cm^{-1}, and later confirmed in the Infrared Space Observatory (ISO) and CIRS observations, that afforded higher spectral resolution. In 1997 ISO Short Wavelength Spectrometer (SWS) spectra provided a good determination of the chemical abundance on Titan and also the first detection of water vapor in Titan's atmosphere from 2 emission lines around 40 μm, for an associated mole fraction derived at 400 km of altitude of about 10^{-8}. ISO also found the first hint of the presence of benzene (C$_6$H$_6$) at 674 cm^{-1} for a mole fraction on the order of a few 10^{-10}. Since then, the benzene detection has been confirmed by *Cassini*/CIRS. The water vapor abundance, although seemingly small, implies a water influx on Titan significantly superior to what might be expected based on local and interplanetary sources alone (rather in favor of Saturn). By including the laboratory spectra of these gases in radiative transfer calculations, the abundances of all of these constituents can be estimated (Table 2). The spatial

TABLE 2	**Chemical Composition of Titan's Atmosphere Today from *Cassini-Huygens* Results Unless Otherwise Indicated**
Constituent	**Mole Fraction (atm. altitude level)**
Major	
Molecular nitrogen, N$_2$	0.98
Methane, CH$_4$	4.9 × 10^{-2} (surface)
	1.4–1.6 × 10^{-2} (stratosphere)
Monodeuterated methane, CH$_3$D	6 × 10^{-6} (in CH$_3$D, in stratosphere.)
Argon, ^{36}Ar	2.8 × 10^{-7}
^{40}Ar	4.3 × 10^{-5}
Minor	
Hydrogen, H$_2$	~0.0011
Ethane, C$_2$H$_6$	1.5 × 10^{-5} (around 130 km)
Propane, C$_3$H$_8$	5 × 10^{-7} (around 125 km)
Acetylene, C$_2$H$_2$	4 × 10^{-6} (around 140 km)
Ethylene, C$_2$H$_4$	1.5 × 10^{-7} (around 130 km)
Methylacetylene, CH$_3$C$_2$H	6.5 × 10^{-9} (around 110 km)[a]
Diacetylene, C$_4$H$_2$	1.3 × 10^{-9} (around 110 km)[a]
Cyanogen, C$_2$N$_2$	5.5 × 10^{-9} (around 120 km)[a]
Hydrogen cyanide, HCN	1.0 × 10^{-7} (around 120 km)[a]
	5 × 10^{-7} (around 200 km)[b]
	5 × 10^{-6} (around 500 km)[b]
Cyanoacetylene, HC$_3$N	1 × 10^{-9} (around 120 km)[a]
	1 × 10^{-7} (around 500 km)[b]
Acetonitrile, CH$_3$CN	1 × 10^{-8} (around 200 km)[c]
	1 × 10^{-7} (around 500 km)
Water, H$_2$O	8 × 10^{-9} (at 400 km)[d]
Carbon monoxide, CO	4 × 10^{-5} (uniform profile)[e]
Carbon dioxide, CO$_2$	1.5 × 10^{-8} (around 120 km)

[a] Increasing in the North.
[b] From ground-based heterodyne microwave observations.
[c] Only observed from the ground.
[d] From ISO observations.
[e] From Cassini and ground-based data.

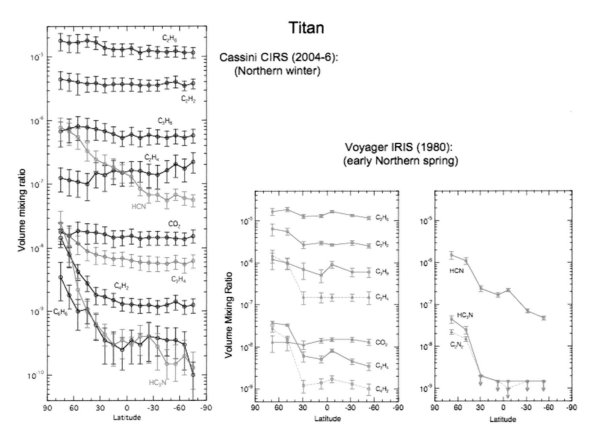

FIGURE 4 Titan's chemical composition and variations in almost one Titan year (30 Earth years) from *Voyager 1* IRIS observations in 1980 to *Cassini* CIRS (2004) measurements in 2004. Note that the enhancement at the North Pole is not as pronounced now as it was during the *Voyager* encounter, because winter is only beginning at the present in Titan's North Pole.

distribution (latitudinal and vertical) of these constituents was also retrieved (Fig. 4). The vertical distributions generally increase with altitude, confirming the prediction of photochemical models that these species form in the upper atmosphere and then diffuse downward in the stratosphere. Below the condensation level of each gas, the distributions are assumed to decrease following the respective vapor saturation law.

Ground-based high-resolution heterodyne millimeter observations of Titan offered the opportunity to determine vertical profiles and partial mapping in some cases of HCN, CO, HC_3N, and CH_3CN, which showed that the nitrile abundances increase with altitude. Subsidence causes the abundance of these species to decrease in the lower atmosphere.

Curiously, the bulk composition of Titan was more difficult to determine than the abundances of the trace constituents. *Cassini–Huygens* finally allowed firm determinations for the major components: *Huygens* Gas Chromatograph Mass Spectrometer (GCMS) found a methane mole fraction of 1.41×10^{-2} in the stratosphere, increasing below the tropopause and reaching $4.9\,5\,10^{-2}$ near the sur-

face, in good agreement with the stratospheric CH_4 value inferred by CIRS on the *Cassini* orbiter ($1.6 \pm 0.5 \times 10^{-2}$) and the surface estimate given by the *Huygens* Descent Images Spectral Radiometer (DISR) spectra (also 5%). The GCMS also saw a rapid increase of the methane signal after landing, which suggests that liquid methane exists on the surface, together with other trace organic species, including cyanogen, benzene, ethane, and carbon dioxide. The only noble gas detected to date is argon, found in the form of primordial ^{36}Ar (2.8×10^{-7}) and its radiogenic isotope ^{40}Ar (4.32×10^{-5}) by GCMS. The low abundance of primordial noble gases on Titan implies that nitrogen was originally captured as NH_3 rather than N_2. Subsequent photolysis may have created the N_2 atmosphere we see today.

Isotopic ratios were determined from *Cassini* and *Huygens* instruments: $^{12}C/^{13}C$ (82.3 ± 1), $^{14}N/^{15}N$ (measured in situ in N_2, 183 ± 5, which is 1.5 times less than on Earth) and D/H (measured in situ in H_2, $2.3 \pm 0.5 \times 10^{-4}$, from the GCMS, and in CH_4, 1.2×10^{-4}, from remote sensing of infrared spectra recorded aboard the Cassini orbiter with CIRS). It is believed that nitrogen was initially brought in Titan in the form of NH_3 and converted into N_2 by photolysis of the early atmosphere. The measured $^{14}N/^{15}N$ implies that a substantial part of N_2, from 2 to 10 times the mass of the early atmosphere, escaped over 4.5 billions of years. The D/H ratio is very important for Titan cosmogonical models. A lower value for D/H in methane ($\sim 1.2 \times 10^{-4}$) was found from the analysis of *Cassini* observations of the ν_6 monodeuterated methane (CH_3D) band at 8.6 μm, confirming a value found from *Voyager* data analyses. Both D/H values tend to suggest a deuterium enrichment in Titan's atmosphere with respect to the proto-solar value as well as in that of the giant planets (D/H $\sim$ 2–3.4 $\times 10^{-5}$). The interpretation of this enrichment is related to the evolution of CH_4 in the atmosphere. The key point is that CH_4 is continuously photodissociated so that, in the absence of a substantial reservoir, it would entirely vanish from the atmosphere in 10–50 Myr. Imaging, infrared, and visible observations from the orbiter rule out the presence of a global ocean containing a large amount of CH_4 on the surface of Titan. It is thus likely that methane outgasses from time to time from the interior of the satellite. Two scenarios for the origin of the internal CH_4 have been proposed. One scenario advocates that CH_4 was chemically produced from H_2O and CO_2 trapped in the planetesimals that formed Titan and which easily condensed in the solar nebula. However, this does not explain the detection of ^{36}Ar in the atmosphere in an amount higher than that which could possibly have been trapped in the silicated core. A more plausible scenario argues that CH_4 and ^{36}Ar were present as ices or **clathrates** in the cool solar nebula and were incorporated in Titan planetesimals. This is consistent with the assumption that CH_4 was enriched in deuterium by ion-molecules reactions in the presolar cloud, the resulting D/H in CH_4 then being at least partly preserved in icy grains falling onto the solar nebula and—since no deuterium fractionation can occur in the interior—reflecting the value observed in the atmosphere of the satellite today.

2.3 Dynamical Processes

2.3.1 ZONAL CIRCULATION

At the time of the *Voyager* encounter, Titan's northern hemisphere was coming out of winter. During the *Cassini* observations in 2006, Titan's northern hemisphere was halfway into winter.

The general faintly banded appearance of Titan's haze suggests rapid zonal motions (i.e., winds parallel to the equator). This impression is reinforced by the infrared temperature maps, which show very small contrasts in the longitudinal direction and rather large ones (of around 20 K) between the equator and the winter pole. The mean zonal winds inferred from this temperature field are weakest at high southern latitudes and increase toward the north, with maximum values at and mid-northern latitudes (20–40N) of about 160 m s^{-1}. On Titan, pressure gradients are in **cyclostrophic balance** with centrifugal forces.

Stellar occultations are another indirect means to obtain the zonal winds. The atmospheric oblateness due to the zonal winds can be constrained from the analysis of the central flash, the increase of the signal at the center of the shadow (when the star is behind Titan) due to the focusing of the atmospheric rays at the limb. On July 3, 1989, Titan occulted the bright K-type star 28 Sgr, and fast zonal winds were derived close to 180 ms^{-1} at high southern latitudes and close to 100 ms^{-1} at low latitudes. Other occultations occurred on December, 20, 2001, and November, 14, 2003. They seem to suggest a seasonal variation with respect to 1989. In 2001, a strong 220 ms^{-1} jet was located at 60°N, with lower winds extending between 20°S and 60°S, and a much slower motion at midlatitudes. The CIRS data suggest that the strongest northern winds have migrated closer to the equator with respect to previous measurements, while the southern winds have weakened.

Space and occultation wind measurements could not provide the wind direction, a crucial factor for the *Huygens* probe mission, so different teams of ground-based observers tried to measure the zonal winds directly using alternative methods. The first measurement of prograde winds (in the sense of the rotation of the surface) was performed using infrared heterodyne spectroscopy of Doppler-shifted ethane emission lines. The measured winds were on the order of 210 ± 150 ms^{-1} between 7 and 0.1 mbar, a result that has since been refined. Other Doppler studies probing somewhat different levels also found prograde winds, using millimeter-wavelength interferometry of nitrile lines or high-resolution spectroscopy of Fraunhofer solar absorption lines in the visible. The recent advances in **adaptive optics** also allowed for the first detections of tropospheric clouds from the ground, mainly at circumpolar southern latitudes, but so far Titan winds remain poorly constrained due to the sparse data set of cloud positions. Better spatially resolved *Cassini/International Space Station* (*ISS*) observations only indicate slow eastward motions, which, extrapolated to the equator under the assumption of solid-body rotation, yield 19±15 ms^{-1} at around 25 km altitude. Finally, in 2005, the *Huygens* probe provided ground-truth measurements of the wind magnitude and direction in the lower stratosphere and troposphere. The Doppler wind experiment shows a marked decrease of winds with decreasing altitude, from 100 ms^{-1} at 140 km down to about nil at 80 km, then an increase up to 40 ms^{-1} at 60 km before decreasing again to null zonal velocity at the surface.

2.3.2 LATITUDINAL AND TEMPORAL VARIATIONS IN THE ATMOSPHERE OF TITAN AS EVIDENCE OF MERIDIONAL CIRCULATION

Periodic change of Titan's disk-integrated brightness has been monitored from Earth-based observations since the 1970s. Spatially resolved observations, starting with *Voyager*, have provided an interpretation of the periodic changes of the disk-integrated brightness as the combined action of the high inclination of the rotation axis and the seasonally varying north–south asymmetry. The NSA that *Voyager 1* observed in 1980, with a darker northern hemisphere in visible light, has since been observed to reverse, as Titan's season shifted from northern spring to present-day northern winter. When the *Hubble Space Telescope (HST)* first observed Titan in 1994, a little over a quarter of a Titan year after the *Voyager* encounters, the northern hemisphere was found to be brighter than the southern hemisphere. The turnover was later also found to occur gradually, starting at higher altitudes in the atmosphere.

Modeling with a two-dimensional general circulation model provided a qualitative description of the seasonal variations of the haze, where both the gradual inversion of the asymmetry and the detached haze layer can be explained by a seasonally varying **Hadley circulation**. The meridional wind in the upper branch of the Hadley cell is stronger close to the production zone (at 450 km) than below, and particles there are more rapidly transported toward the pole, where they sink. The asymmetry thus reverses first at higher altitudes. But this is not the only effect. As the season changes, shortly after equinox, the circulation reverses and an ascending motion sets in where the particles were previously descending. At the time of the transition, the polar haze, which was previously descending, is then redistributed about a scale height below the production zone, becoming physically separated from the freshly created particles aloft.

Meridional variations were also established for the gases in Titan's stratosphere, and these are also tightly coupled with the circulation. The molecular abundances found by *Cassini* at this era indicate an enhancement for some species in the stratosphere at high latitudes, albeit not as dramatic as at the time of the *Voyager* encounter (Fig. 4). The difference in magnitude between the *Voyager 1* and the *Cassini* eras may be due to the difference in seasons, and it will be exciting to await the arrival of northern spring equinox toward the end of the *Cassini* mission and to measure the meridional variations then to see if we return to the IRIS inferences.

In the meantime, such latitudinal contrasts observed in the chemical trace species may be explained by invoking photochemical and dynamical reasons. The UV radiation from the Sun acts on methane and nitrogen to form radicals that combine into nitriles and the higher hydrocarbons. This production occurs in the mesosphere at high altitudes (above 300 km or 0.1 mbar). Eddy mixing transports these molecules into the lower stratosphere and troposphere where most of them condense. Photodissociation by UV radiation occurs on timescales ranging from days to thousands of years. The combination of these processes leads to a vertical variation in the mixing ratio, which usually increases with height towards the production zone. Three-dimensional computation of **actinic fluxes** suggests that

this mechanism alone cannot explain the latitudinal contrasts and that circulation must intervene. Simulations coupling photochemistry and atmospheric dynamics provide a consistent view: Competition between rapid sinking of air from the upper stratosphere in the winter polar vortex and latitudinal mixing controls the vertical distribution profiles of most species. The magnitude of the polar enrichment is controlled by downwelling over the winter pole, which brings enriched air from the production zone to the stratosphere, and by the level of condensation. Short-lived species are more sensitive to the downwelling due to steeper vertical composition gradients and exhibit higher contrasts.

In the stratosphere, the calculated radiative relaxation time is longer than the Titan season, so the temperature contrasts should be symmetric about the equator. That they are not indicates that the Hadley circulation must be connected with the lower atmosphere, where the time constant is much longer. This is consistent with the small thermal contrasts of 2–3 K in the troposphere, which suggest an efficient heat redistribution. Since Titan's slow rotation and small radius rule out nonaxisymmetric processes, such as **baroclinic** eddies, as a preferred mechanism for heat transport, considerable meridional motions must be inferred. Latitudinal contrasts would be much larger if heat were not being transported poleward by Hadley advection.

Another phenomenon was first reported in 2001 from adaptive optics data taken in 1998. A diurnal change was found, manifested in an east–west asymmetry, with a brighter morning limb observed on Titan on several occasions. This dawn haze enhancement could be due to an accumulation of condensates during the Titan night (8 Earth days, though the superrotation of Titan's atmosphere would lead to shorter nights for stratospheric clouds).

2.3.3 A THREE–DIMENSIONAL VIEW AND WAVES

Meridional contrasts are apparent in Titan's atmospheric distributions of composition, haze, and temperature, and their seasonal variability is proof for a strong coupling with an underlying meridional circulation that has never been directly detected.

The superrotation observed in the stratosphere, a dynamical state in which the averaged angular momentum is much greater than that corresponding to corotation with the surface, is difficult to explain and has defied our understanding in the much better documented Venus case, the paradigm of a slowly rotating body with an atmosphere in rapid rotation. In recent studies, such a process has been identified under the form of planetary waves, forced by instabilities in the equatorward flank of the high-latitude jet. Two factors play a key role in facilitating the acceleration process. On the one hand, high altitude absorption processes decouple upper atmosphere dynamics from dissipation occurring at the surface layer, while on the other hand the slow rotation allows the Hadley cell to reach high latitudes by reducing centrifugal forces in the poleward branch. A strong seasonal cycle due to Titan's obliquity of 26.7° was also established: During most of the Titan year, the meridional motion is dominated by a large Hadley cell extending from the winter to the summer pole, with the symmetric two-cell configuration typical of equinoxes occurring only in a limited transition period. In the model, the jet is located close to 60° in the winter hemisphere, while the summer zonal circulation is close to solid body rotation.

The radiative time constant is long in the troposphere, but the surface has a smaller thermal inertia, so the surface temperature does respond to seasonal forcing, albeit by only a few Kelvin. This surface temperature variation is sufficient to reverse the circulation pattern of the Hadley circulation after the equinox when the Sun moves to the opposite hemisphere. Also the development of convective methane clouds is partly ascribed to seasonal surface heating. The reversal of the Hadley circulation may play an important role in the methane "hydrological" cycle because the vertical and horizontal transport of methane would vary seasonally.

Direct evidence for wave processes in Titan's atmosphere remains scarce, despite their importance in the maintenance of superrotation. Because baroclinic processes are excluded, waves essentially **barotropic** in nature should be expected as the principal carrier of momentum from high to low latitudes. Modeling predicts **wavenumber**-2 waves with an amplitude of the zonal component about 10% of the mean wind speed, and in principle they can be inferred from horizontal maps of temperature and trace species exhibiting strong latitudinal contrasts. The first *Cassini*/CIRS temperature maps at 1.8 mbar do show spatial inhomogeneity, but long time series and better spatial coverage are needed to constrain spatial and temporal variations.

Another relevant nonaxisymmetric phenomenon in Titan's troposphere is the gravitational tide exerted by Saturn. The eccentric orbit of Titan around Saturn gives rise to a tidal force, resulting in periodical oscillation in the atmospheric pressure and wind with a period of a Titan day (16 days), among which the most notable effect is the periodical reversal of the north–south component of the wind. In the lower atmosphere, the effect of this tide is modest, with a maximum temperature amplitude about 0.3 K and winds of $2 \, \text{m s}^{-1}$.

Temperature inversions have been detected in both the *Huygens* HASI measurements and in stellar occultation data. Inversion layers were present close to 510 km altitude in HASI and 2003 occultation data, and at 425 and 455 km in 1989 occultation light curves. Vertical wavelengths were on the order of 100 km.

2.4 Haze and Clouds on Titan

It was recognized quite early that another important aspect of Titan's atmosphere was the presence of aerosols.

Pre-*Cassini* models treated the dissociation of methane molecules by solar actinic radiation, followed by chemical combination to heavier hydrocarbons that condense into particles. The cloud physics models with sedimentation and coagulation predicted a strong increase in haze density with decreasing altitude.

2.4.1 TITAN'S HAZE

The analysis of high-phase *Voyager* images indicated aerosol radii between 0.2 and 0.5 μm. These "smog" particles form a layer that enshrouds the entire globe of Titan and stretches from the surface to an altitude of about 200 km. A detached haze layer at 340–360 km altitude with large, compact, irregular dark particles was also found. The small haze particles required by *Voyager* measurements (radii less than or equal to 0.1 μm) produce a strong increase in optical depth with decreasing wavelength shortward of 1 μm. To fit the observations in the methane bands, it was necessary to remove the haze permitted by the cloud physics calculations at altitudes below about 70–90 km (called cut-off altitude) by invoking condensation of organic gases produced at high altitudes as they diffused down to colder levels. The condensation of many organic gases produced by photochemistry at high altitudes on Titan seemed consistent with this view. The next step in the development of Titan haze models included the use of fractal aggregate particles composed of several tens of small (0.06 μm in radius) monomers to produce strong linear polarization. Monomers composed of 45 aggregates with an effective radius of about 0.35 μm matched the *Voyager* observations.

Starting from the upper atmosphere, the *Cassini* ISS camera showed a faint thin haze layer that encircles the denser stratospheric haze (Fig. 1b) and could be the equivalent of the "detached haze layer" observed by *Voyager* 25 years ago, except for the difference in altitudes: The thin current haze layer is indeed located 150–200 km higher than the one seen by *Voyager*. Current models are still unable to render the complexity of seasonal phenomena or circulation patterns on Titan, which could be responsible for such an upward shift.

Cassini images also show a multilayer structure in the north polar hood region and, in some cases, at lower latitudes. These features could be due to gravity waves that have been detected on Titan at lower altitudes. Some of these layers may be related to the two global inversion layers observed in stellar occultations of Titan above 400 km in altitude.

The nature of the haze aerosols measured by *Huygens*/DISR during the descent through Titan's lower atmosphere came as a surprise to scientists recalling the results from *Pioneer* and *Voyager*, as well as predictions by cloud physics models with sedimentation and coagulation. The new observations estimate the monomer radius to be 0.05 μm, in good agreement with previous values. However, contrary to previous assumptions, the DISR data seem to show that the size of the aggregate particles is several times as large as previously supposed.

In addition, measurements by the DISR violet photometer extend the optical measurements of the haze to wavelengths as short as the band from 350 to 480 nm, also helping to constrain the size of the haze particles. The number density of the haze particles does not increase with depth nearly as dramatically as predicted by the older cloud physics models. In fact, the number density increases by only a factor of a few over the altitude range from 150 km to the surface. This implies that vertical mixing is much less than had been assumed in the older models where the particles were distributed approximately as the gas is with altitude. In any event, the clear space at low altitudes, which was suggested earlier, was not observed.

The methane mole fraction of 1.4–1.6% measured in the stratosphere by the CIRS and the GCMS is consistent with the DISR spectral measurements. At very low altitudes (20 m), DISR and the GCMS measured 5 ± 1% for the methane mole fraction.

2.5 Clouds

Cassini–Huygens has provided new information on the role of methane and the methane cycle in Titan's atmosphere. The relative humidity of methane (about 50%) at the surface found by DISR and the evaporation witnessed by the GCMS show that fluid flows have existed and will probably again exist on the surface, implying precipitation of methane through the atmosphere.

Although some discussion took place as to whether Titan's lower atmosphere could support convection and as to whether methane was supersaturated, there is clear evidence today that clouds exist in Titan's troposphere, although in general they tend to appear higher than expected and are mostly restricted to high southern latitudes.

Methane clouds in Titan's troposphere were first suspected from variability in the methane spectrum observed from the ground. Direct imaging of clouds on Titan has been achieved from Earth-based observatories since the turning of the century. Most of the currently detected clouds are located in Titan's southern hemisphere, as expected given the season on Titan (summer in the south), which means that solar heating is concentrated there as are rising motions. Other than the large, bright South Pole system observed for the past 5 years or so, discrete clouds detected at midlatitudes are infrequent, small and short-lived (*Cassini* Visual and Infrared Mapping Spectrometer (VIMS) observations tend to indicate that they rise quickly to the upper troposphere and dissipate through rain within an hour). Keck and Gemini data indicate that they tend to cluster near 350°W and 40°S. They may be related to some surface–atmosphere exchange (such as geysering or **cryovolcanism**) because they don't seem to be easily explained by a shift in global circulation. A dozen or so large-scale zonal streaks have also been

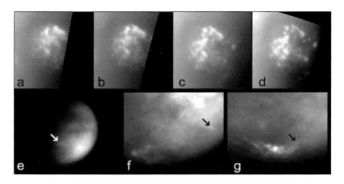

FIGURE 5 Titan's meteorology observed with *Cassini*/ISS. (a–d) A sequence of four methane continuum (IRP0-IR3, 928 nm) images showing the temporal evolution over the period 05:05–09:38 of the Titan south polar cloud field on 2 July 2004. (e–g) Three examples of discrete midlatitude clouds (arrows) for which motions have been tracked in CB3 images. Image e: 38°S, 81°W (29 May 2004); this image was also viewed through an infrared polarizing filter. Image f: 43° S, 67° W (23 October 2004). Image g: 65°S, 110°W (25 October 2004). (From Porco et al., 2005, *Nature* **434**, 159–168. Image Credit: NASA/JPL.)

observed by *Cassini* preferentially at low southern latitudes and mostly between 50 and 200°W.

The large south polar system has been visible consistently essentially in the near-infrared (at 2.12 μm for instance) since 1999, while no previous indication of it was ever reported. It was extremely bright in 2001–2002, and recent *Cassini* images have shown that it is disappearing (indeed it was visible only during the few first Titan flybys and not afterwards, see Fig. 5). Its shape is irregular and changing with time, recently resembling more a cluster of smaller-scale clouds than a large compact field. Should it prove that this system's life was indeed on the order of 5–6 years (fairly close to a Titan season), stringent constraints can be retrieved on seasonal and circulation patterns on Titan. The cloud made a reappearance in 2006.

Note that DISR reported no definite detection of clouds during its descent through Titan's atmosphere. However, the data are compatible with a thin haze layer at an altitude of 21 km, which could be due to methane condensation.

3. The Surface of Titan

To the eyes of the public and many scientists, the most important features revealed by the *Cassini–Huygens* mission were those found on Titan's surface, finally observed in close-up by the orbiter since 2004 and even in situ conditions by the Huygens probe instruments on January 14, 2005. The spaceship has offered detailed views of Titan's surface in the visible and the near-infrared with its camera, the mapping spectrometer, and radar. Descending through the atmosphere, the *Huygens* probe returned fantastic images of a first-seen domain, the farthest location a human-made vessel has ever landed upon. Although we still haven't exactly determined the nature of all the surface constituents, the combination of the information retrieved by all the observing teams will eventually force Titan to uncover its mysterious soil. Undoubtedly the signs of dried lakes, volcanoes, and channels on Titan's surface were unexpected. They offer an even more amazing view of a land much fantasized on.

3.1 Pre-*Cassini* Glimpses of an Exotic Ground

To the *Voyager* camuas, the surface of Titan was obscured by the dense haze in the atmosphere. Glimpses of what lay below were revealed afterwards by ground-based radar and infrared images from *HST* and ground-based observatories.

Theory argued that unless methane supersaturation conditions prevailed on Titan, the organics present in the atmosphere should condense at some level in the lower stratosphere and precipitate out, ending up on Titan's surface and coat the ground in large proportions. Based on the surface conditions believed to prevail on Titan, liquid methane—and its principal by-product, ethane—is expected to exist and could even form an ocean, and in the troposphere, methane clouds (formed by saturation of methane gas) might cause rains. The degree of saturation in the lower atmosphere, however, was unknown, so the methane abundance was difficult to determine.

On the other hand, much of the outer part of the solid body of the satellite must, to be consistent with the observed mean density, consist of a thick layer of ice. The ethane ocean model, developed in 1983, was aesthetically appealing and compatible with all the *Voyager*-era data. It has since then long been abandoned in view of the spectroscopic and imaging evidence for a heterogeneous surface and the radar echoes indicating the presence of solid material.

Indeed, a shallow, global ocean was shown to be inconsistent with the constraints imposed by Titan's orbital characteristics. The tidal action on an ocean less than 100 m deep would have dissipated Titan's **eccentricity** of 0.03 (where 0 is circular and 1 is parabolic) long ago. Furthermore, the first remote-sensing technique to be used for sounding Titan's surface, radar, indicated that the surface may be nonuniform but mostly solid with at most small lakes. Indeed, the radar echos obtained in 1990 using the National Radio Astronomy Observatory's Very Large Array in New Mexico combined as a receiver of the signal transmitted to Titan by the NASA Goldstone radio telescope in California were among the first evidence against the global ocean model of the surface. Radar measurements from Arecibo Observatory in Puerto Rico in 2003, however, revealed a specular component at 75% (12 of 16) of the regions observed (globally distributed in longitude at about 26°S), which was interpreted as indicative of the existence of dark, liquid hydrocarbon on Titan's surface. The idea of a widespread surface liquid was challenged in more recent observations from the ground, which failed to find any such signatures and proposed instead that very flat solid surfaces could be causing the radar evidence. The nature and extent of the exchange of condensable species between the atmosphere and the surface and the equilibrium which exists between the two is a key science topic.

More compelling evidence against a global hydrocarbon ocean on Titan came from spectroscopic data in the near-infrared (0.8–5 μm). This part of Titan's spectrum, like that of the giant planets, is dominated by the methane absorption bands. At short (blue) wavelengths, light is strongly absorbed by the reddish haze particles. At red wavelengths, light is scattered by the haze, although the column optical depth is still high. In the near-infrared, the haze becomes increasingly more transparent (since the haze particles are smaller than the wavelength), although absorption by methane in a number of bands is very strong. Where the methane absorption is weak, clear regions or "**windows**," situated near 4.8, 2.9, 2.0, 1.6, 1.28, 1.07, 0.94 and 0.83 μm, permit the sounding of the deep atmosphere and perhaps of the surface (Fig. 6). In between these windows, contrary to the giant planets, solar flux is not totally absorbed but scattered back through the atmosphere by stratospheric aerosols, especially at short wavelengths. The near-infrared spectrum is thus potentially extremely rich in information on the atmosphere and surface of Titan.

Titan's near-infrared spectrum was used to investigate Titan's surface in terms of detailed radiative transfer models of the near-infrared spectrum. This study indicated a surface **albedo** inconsistent with a global ocean and a surface reflectivity that showed a change in Titan's albedo precisely correlated with Titan's rotation.

The observations all agreed: The **geometric albedo** of Titan, measured over one orbit (16 days), shows significant

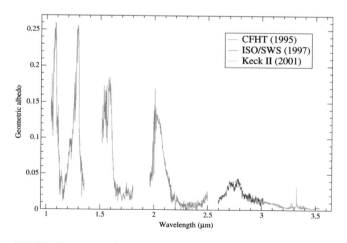

FIGURE 6 Titan's albedo observed from ground-based observatories such as the Very Large Telescope in Chile and the Keck Telescope in Hawaii, as well as with the satellite ISO (in the 2.75 micron window, where the terrestrial turbulence doesn't allow us to observe Titan from the ground). The spectrum exhibits several strong methane absorption bands, but also "windows" where the methane absorption is weak enough to allow for the lower atmosphere and surface to be probed.

variations indicative of a brighter leading hemisphere and a darker trailing one. The leading side corresponds to Titan's Greatest Eastern Elongation (GEE) at about 90° Longitude of the Central Meridian (LCM—as opposed to geographical longitude, which is about 210°), when Titan rotates synchronously with Saturn; the trailing side is near 270° LCM or Greatest Western Elongation (GWE). The longitude at which this "bright" behavior is found was also subsequently identified in Titan images as a bright large area near the equator (see hereafter). At conjunctions (i.e., on the hemispheres facing Saturn and its opposite), the albedo was similar, of intermediate values between the maximum appearing near 120° LCM and the minimum near 230° LCM. As a consequence, Titan's surface had then to be heterogeneous and rather "dry" with the hydrocarbon ocean stored in the porous, uppermost few kilometers of **methane clathrate** or water ice, "bed rock."

The Titan surface spectrum seemed to indicate the presence of two lower-albedo regions near 1.6 and 2 μm (with respect to the continuum near 1 μm). These are also found in Hyperion and Callisto data where they are due to the water ice bands. The existence of a second (or more) surface component(s) was advocated by the orbital variations. It could be spectrally neutral or not and mixed with water ice zonally or intimately. Complex organics (tholins) show a neutral and fairly bright spectrum in the near-infrared, in agreement with high absolute albedos, but should be distributed uniformly with longitude. Hydrocarbon lakes or ices, silicate components, and other dark material are possible. Another possibility would be that the orbital variations may be due to longitudinal differences in the ice morphology (fresh or old, big or small particles, etc.).

Another technique, high-resolution imaging with the possibility to resolve Titan's disk, offered further constraints on the Titan surface problem. Starting in 1994, two sets of data taken independently and with different methods were conclusively analyzed and presented to the public. The images showed clearly extensive quasi-permanent features, which were furthermore too bright to be hydrocarbon liquid. The heterogeneity of Titan's surface, indicated in the near-infrared and with radar lightcurves, was graphically revealed by observations of Titan's surface using the *Hubble Space Telescope* and adaptive optics technique.

On Titan images obtained with the *Hubble Space Telescope,* features were made discernible on Titan's surface. Maps were produced of the surface in the 940 nm and 1070 nm windows, showing in more detail the bright leading and dark trailing sides, with notably a large (2500 × 4000 km) bright region, at 114°E and 10°S (nowadays known as Xanadu, this region has also a peculiar spectral behavior in that it appears bright at all investigated wavelengths (0.9, 1.1, 1.3, 1.6 and 2.0 μm), which may be indicative of an ice-covered mountain or something equivalent), as well as at a number of less bright regions. Subsequent *HST* data have confirmed the initial findings with more extensive mapping

at 1.6 and 2.0 μm and allowed identification of spectrally distinct surface units, which may indicate regions of different composition.

At the same time, images taken using the adaptive optics system at the 3.6-m European Southern Observatory (ESO) Telescope at Chile, showed the same bright region at the equator and near 120° orbital longitude but also revealed a north–south hemispheric asymmetry apparent on Titan's darker side. Adaptive optics is now a generally adopted method, and such systems exist in almost all the large Earth-based telescopes. Prior to the *Cassini* encounter, the adaptive optics system at the Canadian French Hawaiian Telescope on top of Mauna Kea and its twin at the Very Large Telescope (VLT) in Chile, as well as the Keck telescope, were applied to Titan and returned some of the most interesting and ground-breaking images of the satellite (Fig. 7). The contrast on the adaptive optics images can achieve 50% under good observing conditions.

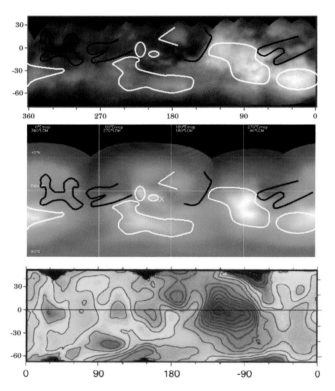

FIGURE 7   Three maps of Titan's surface taken with the *Cassini*/ISS at 0.94 μm (upper panel); the adaptive optics system NAOS at the VLT at 1.28 μm (middle panel) and the *HST* NICMOS at 1.6 μm (lower panel). The surface features are coherent from one data set to the other. The bright areas dominate Titan's leading hemisphere, while the darker ones prevail on the other side. Xanadu Regio is observed near 110° LCM. The *Huygens* landing site is marked with an "X" near 192° LCM and 10°S. (Porco et al., 2005, *Nature* **434**, 159–168; Coustenis et al., 2005; Icarus **177**, 89–105; Meier et al., 2000, Icarus **145**, 462–473).

It was also then essentially demonstrated that Titan's surface was much more complex than initially thought and that the "dark" hemisphere was—fortunately (because it was soon found out that the *Huygens* probe was not going to land where initially scheduled, close to the bright region, but rather on the trailing side)—not all that dark, showing some fine structure with bright areas.

3.2 The View from the Orbiter

The ISS and VIMS cameras confirmed these results and showed that the borders of these regions were linear but not smooth and that dramatic changes in surface albedo could be noted in the maps produced by these measurements (Fig. 7). It is notable how well the distribution of bright and dark areas agrees among these three maps. The best resolution achieved by ISS was of a few kilometers on Titan's surface. The large bright area around the equator first observed by the *HST* and the adaptive optics in 1994 was resolved and finely observed by *Cassini* instruments. It is centered at 10°S and 100°W and officially named Xanadu Regio. The midlatitude regions around the equator on Titan were found to be rather uniformly bright, while the southern pole is relatively dark. What exactly is causing the albedo variations is still uncertain. A plausible candidate for the darker regions could be accumulations of hydrocarbons (in liquid or solid form), precipitating down from the atmosphere.

These variations are more readily attributed to the presence on the surface of constituents with different albedos rather than topography, although contribution from the latter is also expected. The reason is that the *Cassini* camera observing at 0.94 μm cannot see shadows and also Titan's icy bulk does not plead for high topographic structures on the surface (mountains should not exceed 3 km or so).

For the brighter regions, the task of interpreting the data is more difficult. It has been hypothesized that they could be associated with some topography and more exposed ice content, and this tends to be in agreement with findings by the *Huygens*/DISR instrument whose stereoscopic imaging revealed that the brighter terrain was also more elevated than the darker, smoother, and lower ice regions. The exact ice constituent that can satisfy the constraints imposed by all the observations is not easy to determine, hydrocarbon ice has been invoked on the basis of Xanadu appearing bright at all the near-infrared wavelengths observed to date.

A bright circular structure (about 30 km in diameter) found in the VIMS hyperspectral images is interpreted as a cryovolcanic dome in an area dominated by extension. The VIMS team hypothesized that the dry channels observed on Titan are related to upwelling "hot ice" and contaminated by hydrocarbons that vaporize as they get close to the surface (to account for the methane gas in the atmosphere), which are similar to those mechanisms operating for silicate volcanism on Earth (using tidal heating as an

energy source) and which may lead to flows of non-H_2O ices on Titan's surface. Following such eruptions, methane rain could produce the dendritic dark structures seen by Cassini–Huygens. If these structures are indeed channels, they could have dried out due to the short timescale for methane dissociation in the atmosphere. Studying volcanism on Titan (if Cassini definitely yields evidence for it) is important to understand not only the thermal history of Titan (which must surely have evolved differently because it differs in its incorporation of volatiles from the Galilean satellites) but also how volatiles—in particular, methane— were delivered to the surface.

Titan's present environment is very placid—tidal currents are weak; rainfall, if it occurs, is soft; and the diurnal temperature contrasts are small (and therefore winds are gentle). The solubility of ice in hydrocarbons is smaller than that of most rocks in water. Thus, except where the surface is more susceptible to erosion, due to organic deposits or perhaps water–ammonia ice, Titan's topography should not be significantly modified by erosion.

The *Cassini* instruments have found no obvious evidence for a heavy craterization on the bright or the dark areas of Titan so far. A few features interpreted as impact craters have been announced to date: *Cassini*'s RADAR and VIMS saw a 440-km diameter impact crater on Titan during two separate flybys in early 2005. The coloring of the feature indicates that its terrain is rough, with different material for the crater floor and the ejecta and tilted toward the radar during the observations. The multiringed impact basin was named Circus Maximus by the science team. A smaller crater of about 40 km was also observed, exhibiting a parabola-shaped ejecta blanket. In spite of the detection of a third crater-like feature, such formations, identified by the RADAR, VIMS, or the ISS are rare. This may mean that the surface of Titan is young (less than a billion years) or highly eroded/modified.

Other features observed by the *Cassini* orbiter include areas covered with analogs to terrestrial dunes in a set of linear dark features visible across a large part of the RADAR swath to the west of the large crater. These formations are aligned west to east covering hundreds of kilometers and rising to about 100 m. They are expected to have formed by a process similar to that on Earth, but the nature of this "sand" is quite different, consisting of fine grains of ice or organic material, rather than of silicates. The winds responsible for these structures (about 0.5 m/s on the surface) should primarily be attributable to the influence of Saturn, through tidal forces 400 times greater than on Earth and could easily move the Titanian "sand" in this world of low gravity.

Additionally, the RADAR onboard *Cassini* has discovered lakes sprinkled over the high northern altitudes of Titan (Fig. 8). In the images recorded, a variety of dark patches is observed, some of which extended outward (or inward) by means of channels, seemingly carved by liquid.

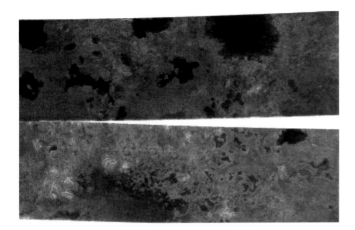

FIGURE 8 *Cassini* RADAR images (P.I. Elachi) of Titan's surface in synthetic aperture mode taken on July 21, 2006, and showing the highly contrasted terrain with a variety of geological features like the dark areas which are most probably hydrocarbon lakes. The top radar image is centered at 80°N, 92°W and measures about 420 km by 150 km. The lower one is centered at 78°N, 18°W and measures about 475 km by 150 km. The most resolved features in these images are about 500 m across. (Image Credit: NASA/JPL/Space Science Institute.)

The missing reservoir of liquid methane or ethane, which scientists have speculated on for a long time, may indeed—at least partly—be found in such areas.

3.3 In Situ Data: Landing on Titan

On January 14, 2005, the *Huygens* probe manufactured by ESA landed at 10.3°S and 192.3°W on Titan, providing the "ground truth" for the orbital measurements in terms of composition, structure, and geomorphology. The probe flew over an icy surface and then floated down and drifted eastward for about 160 km. Several of the instruments on board contributed to our knowledge of Titan's surface conditions.

The HASI instrument measured the surface temperature and the pressure at the landing site to be 93.65 ± 0.25 K and 1467 ± 1 bar, respectively. The fact that the surface is solid but unconsolidated was verified by all the data. The first part of the probe to touch the surface was the Surface Science Package (SSP) penetrometer whose data are now interpreted as indicative of the probe first hitting one of the icy pebbles littering the landing area before sinking into the softer, darker ground material. The SSP detected the ground from 88 m in altitude by acoustic sounding, revealing a relatively smooth, but not flat surface for which our best current hypothesis is gravel, wet sand, wet clay, or lightly packed snow. With a landing speed of about 5 m/s the front of the probe followed and penetrated the surface, then slid slightly before settling to allow the DISR camera to take several pictures of a Mars-like landscape, complete with a dark riverbed and brighter pebbles.

No evidence for liquid was found at the *Huygens* landing site, but the surface is expected to be very humid because methane evaporation (a 40% increase of the abundance) was measured by the GCMS after landing. Thus, either the methane liquid reservoir may not be so far below the surface, but located instead in niches close to the exposed ground, or perhaps *Huygens* landed on Titan at a "dry" season when the rivers and lakes that may exist near the equator were empty but that could be flowing with hydrocarbons at a different era. Also, the presence of hydrocarbon lakes close to the North Pole, may also imply that there are seasonal phenomena that distribute the liquid on the ground. Nevertheless, *Huygens* landed on an organic-rich surface, with trace organic species such as cyanogens and ethane detected on the ground.

In spite of some misadventures (loss of the sun sensor measurements, of about half the images from Channel B and the probe's erratic motion), the DISR imager and spectrometer gathered a precious set of data both in spectroscopy and imaging. Starting from the first surface image at 49 km, down to the unprecedented-quality snapshots of the *Huygens* landing site, and through the lamp-on data recorded below 700 m in altitude, this instrument played a decisive part in untangling the enigma of Titan's surface morphology and lower atmospheric content. Panoramic mosaics constructed from a set of images taken at different altitudes show brighter regions separated by lanes or lineaments of darker material, interpreted as channels, which come in short stubby features or more complex ones with many branches (Fig. 9). This latter dendritic network can

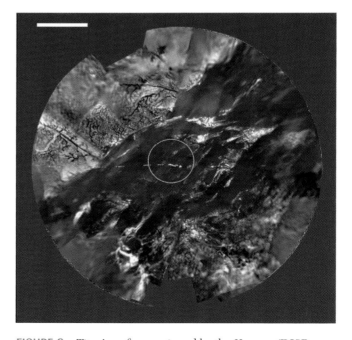

FIGURE 9 Titan's surface as viewed by the *Huygens*/DISR cameras from a distance of 8 km in altitude. (Tomasko et al., 2005; *Nature*, **438**, 465–778, 8 Dec. 2005. Image Credit: ESA/JPL University of Arizona.)

be caused by rainfall creating drainage channels, implying a liquid source somewhere or at some times on Titan's surface. The former stubby channels are wider and rectilinear. They often start or end in dark circular areas suggesting dried lakes or pits. No obvious crater features were observed.

Stereoscopic analysis was performed on the DISR images indicating that the bright area cut with the dendritic systems is 50–200 m higher than the large darker plane to the south. If the latter feature is a dried lakebed, it seems too large by Earth standards to have been created by the creeks and channels seen on the images and could be due to larger rivers or a catastrophic event in the past. The dark channels visible in Fig. 9 could be due to liquid methane irrigating the bright elevated terrains before being carried through the channels to the region offshore in southeasterly flows. This migration toward the lower regions probably leads to water ice being exposed along the upstream faces of the ridges. The slopes are generally on the order of 30°. Some of the bright linear streaks seen on the images could be due to icy flows from the interior of Titan emerging through fissures.

The images taken after the probe had landed on Titan's surface show a dark riverbed strewn with brighter round rocks. These "stones," which are 15 cm in diameter at most, could possibly be hydrocarbon-coated water ice pebbles (Fig. 10).

The spectra acquired during the descent gave information on the atmospheric properties (Table 1) and on the surface properties. Indeed, it was shown from spectral reflectance data of the region seen from the probe that the differences in albedo were related to differences in topography, which in turn can be connected to the spectral behavior of the ground constituents. Thus, the higher brighter regions were also found to be redder than the lowland lakebeds. The regions near the mouths of the rivers are also redder than the lake regions. The spectra taken by DISR are compatible with the presence of water ice on Titan's surface, something that had already been suggested from ground-based observations. The most intriguing feature found in the spectra was, however, the featureless quasi-linear unidentified blue slope observed between 830 and 1420 nm. No combination of any ice and organic material from laboratory measurements has been adequate in reproducing this characteristic. The jury is still out on the constituent(s) that create(s) this signature.

Although many questions still remain about the sequence of flooding and the formation of all the complex structures observed by DISR, these data tend to clear the picture we have of Titan today and at the same time enhance the impression that by studying Saturn's satellite we're looking at an environment resembling the Earth more closely than any other place in our solar system.

No "little orange men" were photographed on Titan. The public is very interested about a possible past, present, or future life on Titan. One of the elements in the negative

FIGURE 10   Titan's surface after the landing of the *Huygens* probe. The icy pebbles are at most 15 cm in diameter, and the darker riverbed is thought to be methane-wet sand (Tomasko et al., 2005; *Nature* 8 Dec. 2005. Image Credit: ESA/JPL University of Arizona.)

response (at least so far as the present or past life is concerned) was found by the GCMS in the $^{13}C/^{14}C$ isotopic ratio (around 82), which showed that no active biota exist on Titan and that the methane on Titan is not produced by life (a biological origin would have required the isotopic ratio to be in the 92–96 range).

The reality pictured by the *Cassini–Huygens* instruments went beyond anything that has been speculated about Titan's surface. The diversity of the terrain includes impact craters, dark plains with some brighter flows, mysterious linear black features possibly related to winds, sand dunes, snow dunes and a host of possible actors (solids, winds, liquids, ices, volcanism, etc.). Titan has proven to be a much more complex world than originally thought and much tougher to unveil.

4. Looking Ahead

Much like Earth, a greenhouse effect exists on Titan; it is produced essentially by methane, with contributions by nitrogen and hydrogen, which have important consequences on the surface temperature. Methane is normally photolyzed in Titan's atmosphere, and unless it can be replenished by a large reservoir on or beneath the surface, it is bound to disappear in a few million years. In such a case, the surface temperature would drop below the condensation point for nitrogen, and Titan's atmosphere would collapse. Should the absorptivity of the surface increase subsequently (e.g., due to the accumulation of organics), the surface temperature might once again rise and cause the reevaporation of methane and nitrogen, thus rebuilding the atmosphere. Such cycles have been hypothesized to occur on Titan.

On the other hand, should the methane supply become abundant, a small perturbation in the solar flux received on Titan (such as is expected when the Sun becomes a red giant and then a dwarf) would produce a dramatic warming of the climate, raising the temperature on the surface and the pressure to values as high as 180 K (twice what we have today) for several bars. It is not inconceivable to imagine that some day in the distant future, conditions on Titan one day may very closely resemble those found on our own planet today.

In the meantime, the *Cassini* mission has demonstrated the complexity of this world and our need to further investigate it in order to better comprehend our solar system. Beyond the extended *Cassini* mission (2010), discussions on future missions to Titan are already underway.

Although future ideas for Titan missions are not mature (after all, *Cassini* is still on the spot), a prominent concept is the use of "aerobots," or intelligent balloons, to explore a variety of Titan locations seems to be favored. Titan may have many more surprises in store for us.

Acknowledgments

The author wishes to thank D. Gautier, M. Hirtzig, F. Ferri, T. Krimigis, D. Luz, and T. Tokano for inputs and discussions.

Bibliography

Coustenis, A., and Taylor, F. (2007). "Titan: An Earth-like Moon," 2nd Ed. World Scientific Publishers, Singapore.

Several articles in *Science* and *Nature* issues of 2004 and 2005 describe in detail the *Cassini–Huygens* mission first findings. In particular: *Cassini* arrives at Saturn; *Science*, 307: February 25, 2005, and Imaging of Titan from the Cassini spacecraft. *Nature*, 434: March 10, 2005, and *Nature*, 438: December 8. 2005. Porco et al., *Nature*, 10 March 2005, **434**; *Nature*, 8 December 2005, Imaging of Titan from the *Cassini* spacecraft, **438**.

Major Web sites: http://saturn.jpl.nasa.gov/home/index.cfm http:// www.esa.int/SPECIALS/Cassini–Huygens/index.html

Triton

William B. McKinnon
Washington University
St. Louis, Missouri

Randolph L. Kirk
U.S. Geological Survey
Flagstaff, Arizona

CHAPTER 26

1. Introduction

Triton is the major moon of the planet Neptune. It is also one of the most remarkable bodies in the solar system (Fig. 1). Its orbit is unusual, circular and close to Neptune, but highly inclined to the planet's equator (by 157°). Furthermore, Triton's sense of motion is retrograde, meaning it moves in the opposite direction to Neptune's spin (Fig. 2). Triton's history therefore must have been quite different from those of "regular" satellites, such as the moons of Jupiter, which orbit in a prograde sense in their primary's equatorial plane. The modern consensus is that Triton originally formed in solar orbit and was subsequently captured by Neptune's gravity.

Like nearly all solar system satellites, tides have slowed Triton's spin period to be coincident with its orbital period and shifted its spin axis to be perpendicular to its orbital plane. Consequently, one hemisphere of Triton permanently faces Neptune. The combination of Neptune's axial tilt (29.6°) and Triton's inclined orbit gives Triton a complicated and extreme seasonal cycle. In the distant geological past, tides associated with Triton's capture may have strongly heated and transformed its interior.

Although discovered soon after Neptune, little was learned about Triton until the modern telescopic era, and even so, most of the information we have was acquired during the *Voyager 2* encounter with Neptune in 1989.

Triton is a relatively large moon (1352 km in radius), larger than all of the middle-sized satellites of Saturn and Uranus (200 to 800 km in radius), but not quite as large as the biggest icy satellites—the Galilean satellites and Titan (1570 to 2630 km in radius). It is a relatively dense world (close to 2 g cm^{-3}), rock-rich, but with a substantial proportion of water and other ices. Ices comprise its reddish visible surface (Fig. 1), and the freshness of the ices cause Triton to be one of the most reflective bodies in the solar system (its total, or Bond albedo, is $\approx$0.85). This, combined with the satellite's distance from the Sun (30 AU), make Triton's surface a very cold place ($\approx$38° K). Yet, despite these frigid surface conditions, *Voyager 2* discovered a thin atmosphere of nitrogen surrounding the satellite (14 μ bar surface pressure, where 1 bar is the approximate surface pressure of Earth's atmosphere). Triton's atmosphere is dense enough to support clouds and hazes and to transport particles across Triton's surface. It is also changing; since 1989 it has been warming and increasing in total mass and pressure.

As with all solid planets and satellites, Triton's history is written into the geological record of its surface. Triton, however, is a geologically young body. Most of the approximately 40% of the satellite's surface that was imaged by *Voyager* at sufficient resolution tells us that it is sparsely cratered. No heavily cratered terrains survive from early solar system times, an absence that may reflect an epoch of severe tidal heating. The geologic terrains that do survive are unique in

FIGURE 1 Digital photomosaic of Triton, centered on the Neptune-facing hemisphere at 15°N, 15°E. The latitude of the subsolar point at the time of the *Voyager* encounter was –45°, so the north polar region was in darkness. Triton's surface is covered with deposits of solid nitrogen with small admixtures of radiation reddened and darkened methane; the bluish tinge is characteristic of fresh frosts. Because Triton's spin is tidally locked to Neptune, the eastern hemisphere is also the leading hemisphere in its orbit. (Courtesy of the NASA Planetary Data System Photojournal.)

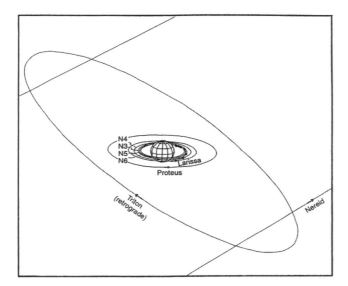

FIGURE 2 Orbits of Neptune's family of satellites, except distant irregulars. Shown is a perspective view along a line of sight inclined 18° to Neptune's equatorial plane. The innermost satellites are all relatively small and were not discovered until *Voyager 2* passed through the Neptune system. They orbit in Neptune's equatorial plane, while much more massive Triton circles outside them in an inclined, retrograde orbit. All the satellites have virtually circular orbits except for Nereid. The apparent crossing of Nereid's and Triton's orbits is an artifact of the projection. [From J.S. Kargel (1997). *In* "Encyclopedia of Planetary Sciences" (J.H. Shirley and R.W. Fairbridge, eds.). Chapman & Hall, London.]

the Solar System. At least three major terrain types can be distinguished: smooth, walled, and terraced plains; an enigmatic "cantaloupe" terrain; and a hemispheric-scale polar deposit or cap. The polar cap is thought to be predominantly solid nitrogen. Other ices that have been identified on Triton are, in approximate order of abundance, H_2O, CO_2, CO, CH_4, and C_2H_6. The cap is a site of present-day geological activity, in particular the eruption of plumes or geysers of gas and fine particles.

In the following sections, Triton will be described in greater detail with emphasis on its geology, the interaction of its icy surface and atmosphere (including the plumes), and its probable origin and violent early evolution.

2. Discovery and Orbit

Acting on the mathematical prediction of Urbain Le Verrier, the planet Neptune was first identified at the Berlin Observatory on September 23, 1846. It was announced in England on October 1. On that day, Sir John Herschel, son of the discoverer of Uranus, wrote to William Lassell, asking him to look for any satellites of the new planet "with all possible expedition," using his own 24-inch reflector. Lassell was a brewer by profession, but he made his own telescopes and was a keen visual observer. Herschel was no doubt seeking to ease some of the sting of Neptune's being found by continental astronomers, given that he was aware of the independent prediction of Neptune's position by John Couch Adams and the unsuccessful search for the planet from English soil. Lassell wasted no time, making his first observations on October 2, and on October 10, 1846, he discovered Triton.

By 1930 it was established that Triton was a most unusual moon. Orbiting at 14.3 Neptune radii, or R_N (using the modern value of 24,760 km for Neptune's equatorial radius), the orbit was circular inasmuch as this could be measured, but distinctly retrograde compared with Neptune's prograde spin or sense of orbital motion. It was also *alone*. No new satellites would be found for over 100 years. The early contrast with the regular satellite systems of Jupiter, Saturn, and Uranus could hardly have been greater. [*See* NEPTUNE.]

The year 1930 also marked the discovery of the dwarf planet Pluto. It was soon determined that Pluto actually crosses inside the orbit of Neptune for about 20 years of its 248 year orbital revolution. Although Pluto's orbit is also substantially inclined so that it does not actually intersect Neptune's, British astronomer R.A. Lyttleton argued that differential precession of the orbits could cause them to intersect, either in the future or in the past. In 1936 he published a paper that theoretically explored the possibility that such an orbital configuration once did exist, and that Pluto was in reality an escaped satellite of Neptune. Although intriguing, planetary scientists now reject this early theory. The modern view of Triton's origin, and that of Pluto and other bodies in the deep outer solar system, is discussed later in this chapter. [*See* PLUTO.]

3. Pre-*Voyager* Astronomy

3.1 Radius, Mass, and Spectra

Through the telescope, Triton is a faint, 14th magnitude object, never more than 17 seconds of arc from Neptune. Consequently, physical studies of the satellite from the ground have historically been very difficult. Showing no visible disk, only crude limits could be put on its size, or mass, for many years. But mid-20th century estimates implied Triton was one of the largest moons in the solar system and massive, possibly *the* most massive moon in the solar system. Triton was clearly a moon of mystery.

The first real breakthrough occurred in 1978, when infrared detector technology had improved to the point that a methane (CH_4) band was detected in Triton's infrared spectrum. Soon more bands were found. The relative depths of the new bands, plus their variability as Triton orbited Neptune, indicated that much (if not all) of the methane detected was in solid form, that is, an ice on the surface of Triton. Ices on the surface implied that Triton might be a relatively bright, smaller world, rather than a darker, larger body of the same visual magnitude. [*See* TITAN.]

Methane ice on the surface also offered a potential explanation for Triton's reddish visual color. Experiments had shown that when solid methane is irradiated by solar ultraviolet rays, or bombarded by charged particles, it turns pink or red as hydrogen is driven off and the remaining carbon and hydrogen form various carbonaceous compounds. Continued radiation or charged particle bombardment ultimately turns methane into a blackish carbon-rich residue, however, so Triton's persistent redness also implied the satellite's methane ice is refreshed on a relatively short time scale.

3.2 Seas of Liquid Nitrogen?

An even more amazing discovery was made in the early 1980s. A single infrared spectral feature was found at ≈2.15 μm (see Fig. 3), a feature that could not be attributed to any of the usual spectral suspects (CH_4, H_2O, silicates, etc.). Nitrogen (N_2) does have an absorption at this wavelength, and because *Voyager 1* had recently determined the dominant atmospheric gas on Titan to be N_2 (not CH_4), finding nitrogen on Triton was not far fetched. The amount of nitrogen gas required to account for the absorption was quite large, however, as nitrogen, a homonuclear diatomic molecule, is a very poor absorber of infrared light. The astronomers concluded that in order to get the neces-

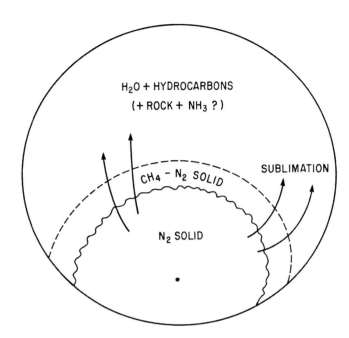

FIGURE 3   A pre-*Voyager* prediction for the state of Triton's surface. A subliming N_2 ice cap is centered on the illuminated south pole (dot). [From J.I. Lunine and D.J. Stevenson (1985). *Nature* **317**, 238–240.]

sary pathlength for the absorption, the nitrogen had to be in condensed form, either solid or liquid. Liquid nitrogen was the favored interpretation, and a fantastic vista emerged—a satellite covered with a global or near-global sea of liquid nitrogen, along with methane-ice-coated islands or even floating methane "icebergs"!

The "problem" with liquid nitrogen is that it freezes at zero pressure at about 63 K. For Triton to have a global ocean at that temperature requires (1) Triton absorb most of the sunlight striking it (have a low albedo) and (2) Triton's surface radiate infrared heat very inefficiently (have a low emissivity). For this and other reasons, planetary chemists offered a competing concept for Triton's surface, one in which both the nitrogen and methane were solid and distributed nonuniformly (Fig. 3). Because of nitrogen's great volatility, it was argued that crystals of up to cm size could grow on Triton's surface over a season and so provide the pathlength for the 2.15-μm absorption. Methane, as in all the spectroscopic models, would be only a minor component; it dominates Triton's near-infrared spectrum by virtue of the relative strength of its absorptions.

The difference between the two models for Triton's surface had important implications for the atmosphere and for Triton's seasons. Triton's seasonal cycle is complicated. Because of the precession of its orbit, its seasons vary in intensity and length, and in the decades before the *Voyager* encounter Triton was moving towards the peak of maximal southern summer (Fig. 4). Correspondingly, Triton's northern hemisphere was (and is) enduring prolonged darkness.

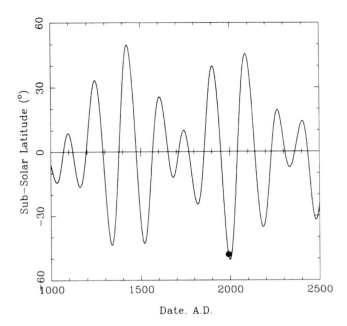

FIGURE 4 Seasonal excursion of the subsolar latitude on Triton. Dot shows the subsolar latitude at the time of the *Voyager* encounter. [From R.L. Kirk *et al.* (1995). *In* "Neptune and Triton" (D.P. Cruikshank, ed.). University of Arizona Press, Tucson.]

The possibility of long-term cold traps at both poles, with strong seasonal atmospheric flows from pole to pole, was recognized. The illustration in Figure 4 was in fact based in part on an analogy with Mars, with N_2 replacing CO_2 as the dominant, and condensable, atmospheric constituent, and CH_4 replacing H_2O as the secondary, less volatile component (an analogy that is strengthening, as will be discussed later). Specifically, a large cap of solid nitrogen was predicted for the south pole, sublimating slowly in the feeble summer sun.

3.3 Similarities with Pluto

As Triton was coming into clearer astronomical focus in the 1980s, parallel developments were occurring for other outer solar system bodies, especially Pluto. Methane ice had been discovered on Pluto prior to Triton, and overall, Pluto's visible and near-infrared spectrum bore a strong resemblance to that of Triton, although Pluto's methane absorptions were deeper. Their common bond was reinforced by their similar visual magnitudes (Pluto and its moon together are only ~0.3 magnitudes fainter than Triton when referenced to a common distance and solar phase angle).

Pluto's fundamental properties (mass and radius) were by the time of the *Voyager 2* encounter with Neptune (and Triton) relatively well constrained. Pluto's relatively large satellite, Charon, had been discovered in 1978, which allowed determination of the mass of the Pluto–Charon system by means of Kepler's Third Law. Careful monitoring of

the Pluto-Charon system's lightcurve, plus observations of the occultation of a star by Pluto in 1988, established that Pluto's radius lay between 1150 and 1200 km. Pluto turned out to be a smallish, bright, more-or-less ice-covered world, and a relatively dense one as ice-rock bodies go, close to 2 gm cm^{-3}. [*See* PLUTO.]

Pluto's density corresponds to a rock/ice ratio of about 70/30, and is, curiously, close to what is predicted for a body accreted in the deep cold reaches of the outer solar system. According to current thinking, the solar nebula at that distance from the Sun, when the Sun and planets were forming, was relatively cold and unprocessed. The outer nebula thus retained many of the chemical signatures of the interstellar gas and dust (molecular cloud) that was the ultimate source of the nebula. Specifically, carbon would be in the form of organic matter and carbon monoxide (CO) gas. CO is very volatile, and the solar nebula was unlikely to have ever been cold enough for it to condense in bulk (though small amounts could be adsorbed on or trapped in water ice). The key point is that volatile CO ties up oxygen that would otherwise be available to form water ice. Therefore, bodies formed in the outer solar system, but not near a giant planet, should have relatively *high* rock/water-ice ratios. In contrast, in the high-pressure environment near a giant planet, CO combines with H_2 to make H_2O and CH_4, which can both condense. The resulting satellites are predicted to be much icier, with rock/ice ratios of 50/50 or less. [*See* THE ORIGIN OF THE SOLAR SYSTEM.]

That Pluto was so rock-rich was one line of reasoning that pointed to Pluto being an original solar-orbiting body and not an escaped satellite of Neptune. Dynamical evidence against Pluto being an escaped satellite also accumulated. By the 1980s it was being argued that Triton and Pluto should be considered as two independent solar system bodies, with independent histories. The link between the two, in terms of brightness (and presumably size) and composition, was that they formed in the same region—the outer solar nebula near or beyond Neptune. Essentially, they are surviving examples of large outer solar system protoplanets. Pluto became locked in a dynamical resonance with Neptune, which preserved its peculiar orbital geometry, while Triton was later captured by Neptune's gravity. [*See* PLUTO.]

If the analogy with Pluto is correct, then Triton should also be rock-rich. If Triton had a relatively bright, icy surface like Pluto, Triton's visible magnitude implied it would probably be somewhat larger, but its density would be similar to that of Pluto–Charon. Of course, the surface state and thus the size of Triton could not be pinned down before the *Voyager* encounter, but the consequences for Triton of being captured (as has been alluded to) were potentially spectacular. These include intense tidal heating and wholesale melting of the satellite. These ideas were appreciated by the planetary community on the eve of the *Voyager 2* encounter. So with the observational and theoretical backdrop just described, and with the promise of resolution of fundamental questions and the revelation of novelty, anticipation was high.

4. *Voyager 2* Encounter

Future history will no doubt record the *Voyager* project as one of humankind's great journeys of discovery. Originally conceived as a "grand tour" of all the giant planets and Pluto, the *Mariner*-class spacecraft that were eventually launched in 1977 (and renamed *Voyager*) were only designed to encounter Jupiter and Saturn. If they worked, though, a highly capable complement of remote sensing instruments for the planets and satellites and *in situ* detectors for the magnetospheres and plasmaspheres would be carried into the outer solar system for the first time. Two spacecraft allowed for different encounter strategies, better satellite coverage, and modification of the second flyby to reflect discoveries made by the first.

At Saturn *Voyager 1* was targeted to pass close to Titan, a trajectory that sent it out of the ecliptic plane afterward. The trajectory of *Voyager 2* was carefully chosen to preserve the grand tour option, whereby each successive encounter would boost the spacecraft to a higher velocity and in just the right direction to reach the next giant planet, which were fortuitously arranged in the 1980s. That *Voyager 2* would reach Uranus, and then Neptune, was the decided wish of the entire planetary science community.

There was no guarantee *Voyager 2* would survive the complete 12-year trip from the Earth to Neptune, many years past its design life. Problems did develop. One radio receiver went out and its backup was failing, and the articulated scan platform, upon which the remote sensing instruments were mounted, could no longer move as easily as before. Nevertheless, in August 1988, after successful encounters at Jupiter, Saturn, and Uranus, and Neptune, *Voyager 2* sent back images of Neptune and Triton that were, for the first time, sharper than the best images taken by groundbased telescopes.

Each new *Voyager* encounter increased scientific and public awareness of the richness of the Solar System. The *Voyager 2* flyby of Neptune and Triton in late August 1989, was going to be the last, and proved to be perhaps the most exciting of all. But there was one last hurdle. In order to get to Triton, *Voyager 2* would have to pass very close to Neptune's north pole in order for Neptune's gravity to bend its trajectory southward (Fig. 5). This would be dangerously close (only 5000 km from the cloudtops) and in an unknown and potentially dangerous environment. To everyone's relief, *Voyager 2* made it past Neptune without incident just after midnight on August 25 (PDT), counting the more than four hours it took for *Voyager's* radio signals to reach Earth. Five hours later it passed within 40,000 km of Triton, sending back a sequence of beautiful, mind-boggling images. These images form much of the basis for understanding, to

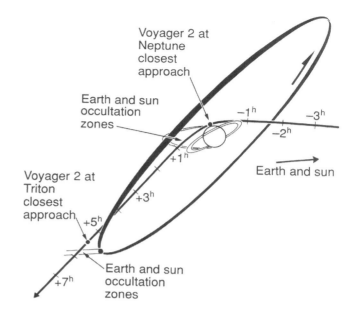

FIGURE 5   The trajectory of *Voyager 2* through the Neptune system. [From C.R. Chapman and D.P. Cruikshank (1995). *In* "Neptune and Triton" (D.P. Cruikshank, ed.). University of Arizona Press, Tucson.]

the extent we do, Triton's geology and surface-atmosphere interactions.

5. General Characteristics

Voyager 2 determined Triton to be even smaller, brighter, and hence colder than anticipated (Table 1). Its average geometric albedo of ≈0.7 is extreme even for an icy satellite. Triton's global appearance was revealed during the approach sequence (Fig. 6). The view, mainly of the southern hemisphere, showed extensive bright polar materials, a bright equatorial fringe with streamers extending to the northeast, and darker low northern latitudes. Radio tracking of *Voyager* yielded a very precise mass for Triton, which when combined with the size, gave a very precise density of ≈2.065 gm cm^{-3}. This density is essentially identical to that of the Pluto–Charon system.

With size and mass known, internal structural models can be created based on a set of plausible chemical components; for bodies formed in the outer solar system these would be rock, metal, ices, and carbonaceous matter. Such models provide context and to some extent guide interpretations of geological history. A calculation for Triton is illustrated in Figure 7. Given that little direct information exists on the internal makeup of Triton, the model shown simply matches Triton's density and assumes the interior is hydrostatic (follows the fluid pressure–depth relation) and differentiated (the major chemical components are separated according to density). These last two assumptions are empirically consistent with Triton's surface appearance, which

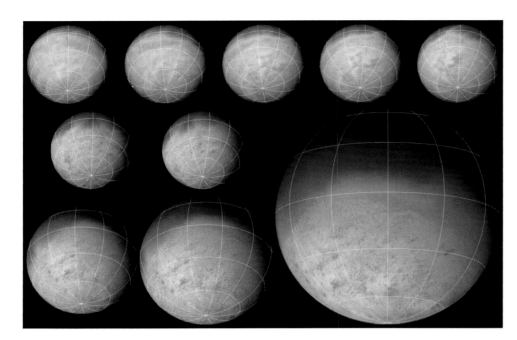

FIGURE 6 Triton approach sequence, overlaid with a latitude-longitude grid. Details on Triton's surface unfold dramatically as the resolution changes from about 60 km/pixel at a distance of 5 million km for the image in the upper left to about 5 km/pixel at a distance of 0.5 million km for the image in the lower right. Mainly looking at the southern hemisphere, Triton rotates retrograde (counterclockwise) over an observational period of 4.3 days. (Courtesy of Alfred McEwen, University of Arizona.)

indicates a prolonged history of melting and separation of icy phases. In the model, ice, structurally represented by the most abundant solar system ice (H_2O) forms a deep mantle around a rock + metal core. A metallic (Fe, Ni, and probably S) inner core is also shown. The proportions of rock and metal in the core are fixed to solar composition (carbonaceous chondrite) values, because relatively involatile rock and metal should have been completely condensed in the outer solar system. Melting and separation of metal from rock are justified by theoretical arguments for intense tidal heating in Triton's past, and by the example of Ganymede, where the *Galileo* orbiter's discovery of a dipole magnetic field demands that such an inner metallic core exists.

Whether Triton is also a magnetized body depends on when its tidal heating ended, but *Voyager 2* passed too far away to tell. Triton is, however, a sufficiently rock-rich body that solid-state convection in its icy mantle should be occurring today, powered by the heat released by the decay of U, Th, and ^{40}K in its rocky core. Its icy mantle should also be warm enough to mobilize lower-melting-point ices such

TABLE 1	Properties of Triton
Radius, R	1352.5 km
Mass, M	2.140×10^{22} kg
Surface gravity, g	0.78 m sec^{-2}
Mean density, ρ	2065 kg m^{-3}
Percent rock + metal by mass	65–70%
Distance from Neptune	354.8×10^3 km = 14.33 R$_N$
Distance from Sun	30.058 AU
Orbit Period	5.877 days
Orbit Period around Sun	164.8 yr
Eccentricity	0.0000(16)
Inclination (present)	156.8°
Geometric albedo (average)	0.70
Bond Albedo (average)	0.85
Surface Temperature	38 K (1989), 39 K (2003)
Surface Composition	$N_2, H_2O, CO_2, CO, CH_4,$ C_2H_6 ices
Surface Atmospheric Pressure	14 µbar (1989), 19 µbar (2003)
Atmospheric Composition	N_2, minor CH_4
Tropopause Height	8 km

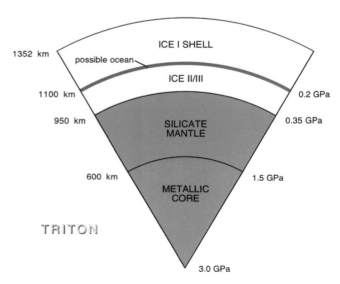

FIGURE 7 Internal structure model for present-day Triton.

as ammonia and methanol, which are among the minor ices a body formed in solar orbit might have accreted. And if Triton formed in solar orbit, it should have also accreted a large carbonaceous component, upwards of 10% by mass if comets such as Halley are a guide. But with or without these additional components, the heat flow from Triton today is sufficient to maintain an internal water layer or "ocean." Similar oceans have been discovered within the large jovian moons Europa, Ganymede, and Callisto by *Galileo*, so there is no fundamental reason why Triton would not possess one as well.

Voyager 2 confirmed the presence of nitrogen ice on Triton's surface. Specifically, a thin nitrogen atmosphere was detected with a surface pressure and temperature consistent with N_2 gas in vapor pressure equilibrium with N_2 ice (see Section 7). All of Triton's surface appears to be icy; even the darker northern hemisphere shown in Figures 1 and 6 has a geometric albedo of ~0.55. Nitrogen is obviously very volatile, and theoretical models show nitrogen ice grains on Triton's surface can rapidly (over many decades) anneal and densify into a transparent glaze or sheet. It is thought that such a nitrogen glaze covers much of Triton; the bright equatorial fringe may be an unannealed frost deposit of other ices.

Triton's surface appearance also appears to be variable on short time scales. Between 1977 and the *Voyager 2* flyby, Triton become remarkably less red, particularly at shorter wavelengths (Figure 8). Presumably, deposition of fresh nitrogen ice and frost have obscured more reddish surface ice in this interval. On any other moon, this would be a major event. On Triton, with its (presumably) active geology, extreme seasons, and sublimation, transport, and condensa-

tion of highly volatile ices in response to both, it seems an almost forgone conclusion that the satellite's global color, if not its overall brightness (and thus its surface temperature and atmospheric pressure), are not constant. Changes over time in Triton's methane spectral absorptions and ultraviolet albedo have also been noted over the years.

As mentioned earlier, the overall redness of the ice (Fig. 1) is thought to be due to UV and charged particle processing of CH_4 (along with N_2), which can yield darker, redder chromophores—heavier hydrocarbons, nitriles, and other polymers. CH_4 exists as an atmospheric gas as well as a surface ice. *Voyager's* ultraviolet spectrometer solar occultation experiment determined the CH_4 mole fraction at the base of the atmosphere to be ~2 to 6×10^{-4}, near or at saturation for 38 K. Dark streaks and patches on the polar cap and elsewhere may be methane-rich; if they are depleted of N_2 ice, they should be warmer than the global mean surface temperature, which is buffered by the latent heat of nitrogen condensation/sublimation.

The nature and chemistry of Triton's surface ices have been determined by advanced ground-based spectroscopy (Figure 9). In 1991 astronomers detected the spectral absorptions of CO and CO_2 ice, along with CH_4 and N_2 ice, on Triton. Later work confirmed the presence of water ice, and most recently, ethane ice has been detected. The shapes of the absorption bands are so well determined that the abundances, grain sizes, and degree of mixing of various components can be modeled. It turns out that CH_4 and CO ice are dissolved in solid solution with the far more abundant N_2 ice, which covers about 55% of Triton's surface. The CH_4 abundance relative to N_2 is about 0.1% and CO abundance of half that. CO is an important tracer of outer solar nebula or cometary chemistry (as discussed in Section 3.3), but

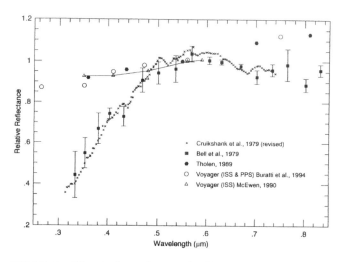

FIGURE 8   Historical visual spectral reflectance of Triton. The differences between the data for 1977 and 1989 are evidence for changes on the surface of Triton. *Voyager* ISS and PPS refer to the imaging camera and the photopolarimeter, respectively. [From R.H. Brown *et al.* (1995) *In* "Neptune and Triton" (D.P. Cruikshank, ed.). University of Arizona Press, Tucson.]

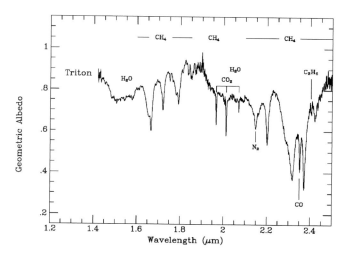

FIGURE 9   Modern, high-resolution, near-infrared telescopic reflectance spectra of Triton. Absorptions due to individual species are indicated. The spectral resolution ($\lambda/\Delta\lambda$) is a remarkable 800. [From D.P. Cruikshank (2005). *Space Sci. Rev.* **116**, 421–439.]

is not expected to survive in giant planet satellite-forming nebulae. The detection of CO thus directly supports a capture origin for Triton. Some discrete CH_4 patches probably also exist, and the ethane ice is one of the "heavier hydrocarbons" predicted to form from methane. CO_2 and H_2O are distributed as discrete units covering the complementary 45% of the rest of the surface. Within these units, CO_2 ice particles represent about 10–20% of the material present. Water ice and CO_2 ice thus represent the composition of Triton's involatile "bedrock."

The geology revealed by the *Voyager* encounter is as remarkable as it was unprecedented. The surface is almost wholly **endogenic** in nature. Intrusive and extrusive volcanism (calderas, flows, diapirs, etc.) dominates the landscape outside the polar terrain, with tectonic structures (mainly ridges) being decidedly subsidiary. Impact cratering is an even more minor process. Triton's surface is geologically young and has apparently been active up until recent times. Triton's topography can be rugged, but does not exceed a kilometer or so in vertical scale (and usually no more than a few 100 m), due to the inherent mechanical weakness of most of the ices that comprise its surface. Polar ices appear to bury much of this topography, and so may in this sense constitute a true polar cap. It is usually assumed that this cap is mostly nitrogen, similar to the surface ice. Details of Triton's geology are pursued in the following section.

Triton's atmosphere is unique as well. It is too thin and cold for radiative processes to play a dominant role. Heat is transported by conduction throughout most of its vertical extent, which is by definition a **thermosphere**, up to an exobase of ~950 km, where the mean free path of N_2 molecules equals the pressure/density scale height. The thermospheric temperature is a nearly constant 102 ± 3 K above ~300 km altitude, and is set by a balance between absorption of solar and magnetospheric energy in a well-developed ionosphere between ~250 and 450 km altitude and both radiation to space by a trace of CO and photochemically produced HCN below ~100 km and downward conduction to the cold, 38 K surface. The lowermost atmosphere is characterized by an interhemispheric, seasonal condensation flow. Turbulence near the ground forces the temperature profile to follow a convective, nitrogen-saturated lapse rate of ~−0.1 K km^{-1} up to an altitude of ~8 km (as determined by observations of clouds, hazes, and plume heights; Section 7), forming a **troposphere** or "weather layer." Unlike in the atmosphere of the Earth and other planets, there is no intervening radiatively controlled **stratosphere** between Triton's troposphere and thermosphere.

6. Geology

Triton's surface, at least the 40% seen by *Voyager* at resolutions useful for geological analysis, can be roughly separated into three distinct regions or terrains: smooth, walled, and terraced plains; cantaloupe terrain; and bright polar materials. Each terrain is characterized by unique landforms and geological structures. Substantial variations within each terrain do occur, and the boundaries between each are in many locations gradational, but in general the classification of Triton's surface at any point is unambiguous. Certain geological structures are common to nearly all terrains, specifically, the tectonic ridges and fissures, and impact craters, naturally, can form anywhere.

Although Triton's surface is composed almost entirely of ices, many of the individual geological structures can be readily interpreted as variations of structures terrestrial planet geologists would find familiar, such as volcanic vents, lava flows, and fissures. The volcanic features in particular have inspired a designation "**cryovolcanic**" in order to distinguish them from those formed by traditional silicate magmatic processes. The physics and physical chemistry are fundamentally the same, however, whether one deals with silicate or icy volcanism. There are in addition geological structures and features on Triton that are unusual and *not* readily interpretable in terms of terrestrial analogues. Some defy explanation altogether.

6.1 Undulating, High Plains

Plains units are found on Triton's eastern or leading hemisphere (referring to the sense of orbital motion, to the right in Figure 1) and to the north of the polar terrain boundary. Figure 10 shows a regional close-up of various plains near the terminator in the center of Figure 1. To the bottom and right of the image are flat-to-undulating smooth plains centered around circular depressions or linear arrangements of rimless pits. These plains are relatively high-standing and bury preexisting topography, with edges that may be well defined or diffuse. There is little doubt that these high plains are the result of icy volcanism and that the various pits and circular structures are the vents from which this material emanated. In general, eruptions along deep-seated fissures or rifts often manifest as a series of vents, and the irregular, ~85-km wide circular depression toward the lower left resembles a terrestrial volcanic caldera complex. This feature, Leviathan Patera (all the features on Triton have been given names drawn from the world's aquatic mythologies), sits at the vertex of two linear eruption trends. Towards the terminator (northeast), one of these trends is anchored by another caldera-like depression of similar scale.

Volcanic activity on the Earth often occurs in cycles, whereby magma formed by partial melting in the mantle rises due to buoyancy, accumulates at intermediate, crustal levels to form a magma chamber, and subsequently erupts; things are then quiescent until the magma is replenished and the cycle begins anew. The loss of magma volume often leads to collapse of the vent region over the magma chamber, forming a caldera. Cycles of eruption and collapse can

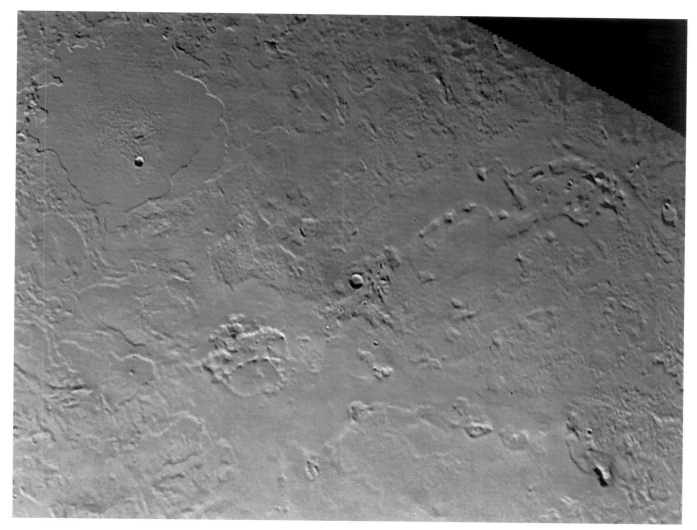

FIGURE 10 Young volcanic region on Triton. Towards the bottom and right, smooth undulating flows apparently emanate from complex caldera-like depressions and linear alignments of volcanic pits and vents, burying preexisting topography. At the upper left, terraced plains surround an exceptionally level plain, Ruach Planitia. This region, 675 km across, is very sparsely cratered. (Courtesy of NASA/Paul Schenk, Lunar and Planetary Institute.)

create some complex forms, but calderas are generally composed of quasicircular elements. The two paterae (from the Latin for saucer) in Figure 10 are clearly of the caldera type in which renewed volcanism has occurred, because both are partially buried by younger icy lavas.

The compositions of the icy lavas are, strictly speaking, unknown. *Voyager 2* carried no remote sensing instruments designed to determine compositions. The icy plain-forming lavas shown in Fig. 10 were clearly viscous enough to form thick enough deposits to bury preexisting topography of a few hundred meters elevation.

The favored composition for viscous lavas on icy satellites has long been ammonia-water. As outlined by pioneering planetary chemist J.S. Lewis, ammonia (NH_3) is the chemically stable form of nitrogen in a low-temperature gas of solar composition, and when condensed forms various hydrates with water ice, all of which have low melting points. Triton would not have accreted much ammonia if it formed in solar orbit, because N_2 would have been the dominant original form of nitrogen in the outer solar nebula for the same reasons CO and organic material were favored over CH_4 (see Section 3.3), but it still would have acquired some NH_3 based on cometary compositions (up to a percent or two compared with water). A water-rich NH_3–H_2O mixture (0 to 33 mole% NH_3) would be composed of frozen H_2O and ammonia dihydrate ($NH_3 \cdot 2H_2O$), which yields a lowest-melting-point (or eutectic) melt at ~177 K at pressures typical of Triton's mantle. This melt (or cryolava) is ammonia-rich (about 32%) and has a viscosity similar to some types of basaltic magma.

Comets also contain a host of other exotic, presumably interstellar, ices, some of which may have been important in Triton's geological history. For example, methanol (CH_3OH) pushes the minimum melting temperature of ammonia-water ice down to ~152 K, and the resulting lava is even more viscous, equivalent to certain types of silicic lavas on Earth. The range of viscosities available to liquids in the H_2O–NH_3–CH_3OH system is compatible with the appearance of the undulating smooth plains seen in Fig. 10. [*See* PHYSICS AND THE CHEMISTRY OF COMETS.]

The abundances of original ices may also have been altered, and new ices created altogether, during Triton's tidal heating epoch (see Section 8). For example, *copious* NH_3 and CO_2 may have been chemically produced within Triton provided there was a sufficient supply of nitrogen and carbon. Despite these exciting possibilities, neither NH_3 nor any complex, exotic ices have yet been discovered by ground-based spectroscopy (Section 5).

6.2 Walled and Terraced Plains

Shown in the northwest corner of Figure 10 is an ~175-km wide, remarkably flat plain, Ruach Planitia, that is bounded on all sides by a rougher plains unit that rises in one or more topographic steps (scarps) from the plain floor. It is one of four so-called walled plains identified on Triton; these are generally quasicircular in outline, with typical relief across the bounding steps or scarps of ~200 m. Ruach Planitia and the other walled plains are the flattest places seen on Triton, which implies infill by a very fluid lava or other liquid. Clusters of irregular, coalesced pits towards the centers of these plains have been likened to eruptive vents or drainage pits.

The planitia themselves have been likened to calderas, but they are generally much larger than the nearby paterae and do not resemble them structurally. Specifically, there is no evidence for collapse at the periphery of any of the walled plains. Rather, the outline of the inward-facing scarps is indented and crenulate, with islands of the bounding plains occurring in the interior. If anything, the outlines of walled plains resemble eroded shorelines. How erosion occurred and under what environmental conditions on Triton is unclear. If the fluid that filled the planitia was responsible for the erosion, it does not explain the similar outline of the plains that overlap the eastern edge of Ruach Planitia (Fig. 10), which gives this area a terraced appearance and indicates that the rougher plains were laid down in layers. A distinct possibility is that the layers are composed at least in part of a more friable or volatile material, and that over time (or with higher heat flows) the layers disintegrated and the scarps formed by retreat. Similar processes of mass wasting, removal, and scarp retreat are believed responsible for the etched plains of the martian south polar highlands and similar terrains on Io.

6.3 Smooth Plains and Zoned Maculae

Other plains units can be seen in Figure 11, as well as the transition to the bright polar materials. At the top left is a hummocky terrain, composed of a maze of depressions and bulbous mounds. Stratigraphically, it is older than the volcanic plains to the north that overlap it, and appears older (more degraded) as well. The hummocky terrain gives way to a much smoother plains unit to the south. At the available resolution it is unclear whether this smoothness is due to volcanic flooding, volcanic or condensation mantling, or some other form of degradation. These hummocky and smoother units are the most heavily cratered regions on Triton, but by solar system standards are not heavily cratered at all.

Among Triton's most perplexing geological features are the large zoned maculae (spots) close to the eastern limb are shown in Figure 11. Each such macula consists of a smooth, relatively dark patch or patches surrounded by a brighter annulus or aureole. The width of any given annulus tends to be relatively constant (20 to 30 km for the three major

FIGURE 11 Southeastern limb of Triton, showing (from top) hummocky terrain, smooth terrain, and bright polar terrain. A prominent bulbous ridge zigzags across the top, and distinct bright-ringed dark features of uncertain origin, termed maculae (spots), are seen at the right, and more faintly, along the limb and in the bright terrain. The largest crater on Triton, the 27-km diameter, central-peaked Mozamba, is to the left of the largest prominent macula, Zin. (Courtesy of NASA/Paul Schenk, Lunar and Planetary Institute.)

maculae shown in Fig. 11). The maculae betray almost no topographic expression, and so must vary in height across their extents by no more than a few tens of meters. The darkness and redness (Fig. 1) of the central patches implies the presence of carbonaceous material, which probably means some methane ice is present. The brightness of the annuli is similar to that of the bright terrain, so they may consist of similar ices (predominantly N_2).

The extreme eastern limb shown in Figure 11 is composed of a mosaic of maculae, and much of the bright terrain in the rest of the image contains similar, though generally less distinct, features (see also Fig. 1). Perhaps the maculae are outliers of the southern polar cap, which should have been retreating at the season observed (late southern spring). Furthermore, another walled plain can be seen along the middle left edge of the frame. Its eastern rim is incomplete, and breaks down into a region of small mesas. If this planitia were filled with bright ice, it would passably resemble, in plan and in albedo, the bright terrains to the south, especially those near the boundary with the smoother plains. The resemblance would be further improved if the planitia are bowed upwards, for which there is independent topographic evidence (see Fig. 10). Perhaps the maculae are planitia underneath, and the mysterious erosive process that cut back the planitia scarps has operated more extensively on Triton.

6.4 Cantaloupe Terrain, Ridges, and Fissures

The entire western half of Triton's non-polar surface shown in Figure 1 is termed cantaloupe terrain, as it appears covered by large dimples and criss-crossed by prominent quasi-linear ridges. Much of the terrain displays a well-ordered structural pattern: at high resolution the dimples become a network of interfering, closely spaced, elliptical and kidney-shaped depressions, termed cavi (Fig. 12). Unlike impact craters, the cavi are of roughly uniform size, ~25-to-35 km in diameter, and do not overlap or crosscut. They are clearly internal in origin, but the leading explanation is not volcanism, but *diapirism*.

Diapirism is triggered by a gravitational instability involving a less dense material rising through overlying denser material. The required buoyancy may be thermal or compositional. Probably the best known terrestrial examples of **diapirs** are salt domes, in which a layer of salt rises as a series of individual blobs, or diapirs, through overlying denser sedimentary strata. In one region of extreme dryness, the Great Kavir in central Iran, the salt diapirs breach the surface, rotating and pushing the overlying strata to the side. The shapes, close spacing, and interference relations of the diapirs of the great Kavir in fact bear a significant resemblance to the cavi.

The implications of a diapiric origin for cantaloupe terrain are that Triton possesses distinct crustal layering, and based on the spacing of the cavi, that the overlying denser

FIGURE 12   Cantaloupe terrain at the bottom and polar terrain at the top, in this high-resolution *Voyager* image taken from a distance of 40,000 kilometers. Each cantaloupe "dimple" is about 25–35 kilometers across. A tectonic ridge and fissure set runs through the cantaloupe terrain, probably formed by the extension of Triton's icy crust. Towards the south (upper right), smooth materials, and beyond them, brighter ice, appear to mostly bury cantaloupe and fissure topography. (Courtesy of NASA/JPL.)

layer or layers is ~20 km thick. This crustal layer could simply be a weaker ice (possibly ammonia rich) that responded to heating from below, or it may be an ice denser than the ammonia-water ices presumably below (such as CO_2 ice).

Triton's surface is crosscut by a system of ridges and fissures, which are best expressed in the cantaloupe terrain (Fig. 1). The ridges occur in a variety of forms: pairs of low, parallel ridges bounding a central trough, ~6–8 km across crest-to-crest and a few hundred meters high; similar but wider ridge-bounded troughs with one or more medial ridges (one, Slidr Sulcus, can be seen in Fig. 12); and single, broad, bulbous ridges (e.g., Fig. 11). The fissures, which are less numerous, appear to be simple, long, narrow valleys only 2–3 km wide. All of these fundamentally tectonic features appear to result from extension and/or strike-slip faulting of Triton's surface. The medial ridges may be due to dike-like intrusions of icy material, and the bulbous appearance of some may be due to overflow of such injected ice, which could also be a source for smooth plains deposits.

Ridges on Triton bear more than a passing resemblance to those on Europa, and a similar mechanical origin has been proposed.

6.5 Bright Polar Terrains

Most of Triton seen by *Voyager* is actually bright terrain of one type or another, but the imagery is generally not of sufficient quality for geological analysis. Interpretations are further confused by the numerous dark streaks, plumes, and clouds. Nevertheless, the bright terrains represent substantial, not superficial deposits. The view shown in Figure 12 looks across the edge of the cantaloupe terrain, into a band of subdued or mantled cantaloupe-like topography, and then into brighter materials beyond. Cantaloupe-like topographic elements and sections of a linear ridge appear engulfed by bright ice, probably up to a few hundred meters in thickness. The important questions are whether the bright ice thickness increases into the interior of the bright materials in the distance, and does it become sufficiently deep to qualify as a true polar cap.

Low-resolution imagery shows that quasicircular elements can be made out at many locations well within the bright materials. Ridges also cross into the bright terrains, and one bright lineament is seen close to the south pole. The implication is that much of the polar topography is incompletely buried. On the other hand, there are extensive bright, featureless regions as well (up to several 100 km across), which indicate either complete burial at these locations or obscuration by clouds. Overall thickness of the bright polar ice is therefore probably less than 1 km, but even if not organized as a uniform ice cap or sheet, a thick deposit of a volatile ice such as N_2 could be warm and deformable enough at its base to flow laterally. Although not literally a polar cap, much of the bright polar terrains may behave as if glaciated.

6.6 Geological History

It is notable that the volcanic province shown in Figure 10 is one of two similar ones, with the second occurring to the southeast and together stretching across 1000 km of Triton's surface. The alignments of volcanic vents in both provinces suggest extension and rifting of Triton's relatively strong icy outer shell, or lithosphere. The volcanic plains shown in Figure 10 are also very sparsely cratered (the largest crater visible is 16 km across), much less cratered than, say, the lunar maria. Estimates of the rate at which comets bombard Triton suggest that these provinces are no more than 300 million years old, and possibly much less. A broad region of Triton's sublithospheric mantle was thus hot and partially molten very late in solar system history, and probably remains so. Such internal warmth is also consistent with a deep subsurface ocean (Fig. 7).

The high volcanic plains postdate most of the other terrains on Triton. They stratigraphically overlie the terraced plains to the west and the hummocky plains to the east. The terraced plains grade into and appear to superpose the cantaloupe terrain. The relative age of the cantaloupe terrain cannot be determined by traditional crater counting methods, because the rugged topography there prevents reliable crater identification in *Voyager* images. Stratigraphically, however, cantaloupe terrain appears to be the oldest unit on Triton. The linear ridges obviously postdate the cantaloupe terrain, yet some ridges fade into the terraced plains to the east and another is discontinuous as it crosses the hummocky and smooth plains near the equator to the east (Figs. 1 and 11); no ridges cut the high volcanic plains.

The eastern hummocky and smooth plains comprise the most heavily cratered region on Triton, and when due account is taken of the concentration of cometary impacts on Triton's leading hemisphere, appears to be somewhat older than the high volcanic plains to the north and northwest. The cantaloupe terrain, then, must be even older. The hummocky terrain may be a degraded version of cantaloupe terrain. Indeed, cantaloupe terrain has been suggested to underlie much of Triton's surface. (For example, cantaloupe-like topography extends well south into the bright region of the trailing hemisphere.)

The youngest surfaces on Triton, naturally, involve the mobile materials of the bright terrains. These probably include the zoned maculae of the eastern hemisphere. The geological substrate upon which the bright materials reside may of course be older. The walled plains themselves are locally the youngest stratigraphic units. Ruach Planitia and a larger planitia immediately to the west are less cratered than the high volcanic plains, albeit with a large statistical uncertainty. The filling of these walled plains may thus represent the most recent volcanic activity on the hemisphere of Triton seen by *Voyager*.

7. Atmosphere and Surface

7.1 Atmosphere

Triton is one of only seven solid bodies in the Solar System with an appreciable atmosphere, and one of only four in which the major component of the atmosphere also condenses onto the surface. Triton's atmosphere is composed primarily of nitrogen. The complicated oscillation of the subsolar latitude with time drives an exchange of N_2 and trace species between the atmosphere and surface frost deposits in the two hemispheres that is equally complicated and as yet not fully understood. Internal heating (which is comparatively important because of Triton's extreme distance from the Sun and large proportion of rocky materials containing radioactive elements) and even glacier-like creep of solid nitrogen caps may also play important roles in the interaction of atmosphere and surface. [See IO; MARS: Atmosphere: History and Surface Interaction.]

As described in Section 3, spectroscopic evidence prior to the *Voyager 2* encounter indicated that nitrogen existed on Triton in condensed form. *Voyager* showed Triton to be much smaller, brighter, and colder than had been guessed. Surface temperatures could be inferred from the visible reflectivity as well as measured directly by the Infrared Interferometer Spectrometer (IRIS). Occultations (passage of the spacecraft or a star behind Triton) observed by the Ultraviolet Spectrometer (UVS) and Radio Science Subsystem (RSS) probed different parts of the atmosphere, revealing its temperature and density, from which pressure and composition could be deduced. These investigations revealed a consistent picture of a surface and lowermost atmosphere at about 38 K. The pressure at the surface was only 14 microbars, indicating that the gas was in equilibrium with solid nitrogen at the same temperature. The thermal structure of the lower atmosphere is not well constrained, but the temperature probably reaches a minimum at about 8 km height, above which it increases to about 100 K in the upper atmosphere because of heat deposited from space and conducted downwards. In meteorological parlance, Triton's thermosphere directly overlays its troposphere.

The *Voyager* images and occultation data revealed a variety of condensates in the lower atmosphere. Most of the atmosphere contains a diffuse haze that can be seen against the background of space at Triton's limbs, and which probably consists of hydrocarbons and nitriles produced by the action of sunlight on trace gases such as methane. Discrete clouds were also seen at the limbs and against the unlit part of the satellite beyond the terminator, where they formed east-west trending "crescent streaks" roughly 10 km wide, a few hundred kilometers long, and 1 to 3 km above the surface. At the limbs, clouds could be distinguished from haze by being optically thicker and localized both in height (10 km or less) and in horizontal extent (patchy, and mainly concentrated at mid to high southern latitudes, where they cover a third of the limb). The sharper upper boundary to the clouds suggests that they consist of condensed nitrogen rather than involatile solids like the haze.

The crescent streaks provide clues to atmospheric motion by their east-west orientation and the apparent eastward motion of the largest, highest cloud seen. Further clues come from markings on the surface. Over 100 dark "streaks" were seen in the southern hemisphere, mainly between latitudes of 15° and 45°S. The streaks range from 4 to over 100 km in length, and many are fan-shaped. The vast majority extend to the northeast from their narrow end (presumably the origin point); a smaller number are directed westward. These streaks are extremely similar to "wind tails" that are common on Mars and are seen on the Earth and Venus as well. On these other bodies, wind tails are created by deposition (or sometimes erosion) of loose material by localized eddies downwind of topographic fea-

tures. It was initially difficult to understand how wind tails could form on Triton, however, because the atmosphere is so thin that even the slightest tendency for dust grains to stick to one another would prevent their being lifted by the wind.

The interpretation of the surface streaks as wind-created was nevertheless strengthened by the discovery, shortly after closest encounter, that some of the streak-like features were actually atmospheric phenomena. Stereoscopic viewing of images obtained from varying angles as *Voyager 2* passed by Triton (Fig. 13) revealed that, although the majority of the streaks were on the surface (or at least too low to measure their altitude, less than 1 km), at least two had an altitude of roughly 8 km. These features were subsequently named Mahilani Plume (48°S 2°E, with a very narrow, straight cloud 90–150 km long) and Hili Plume (57°S 28°E, actually a cluster of several plumes with broadly tapering clouds up to 100 km long). Thus, it is clear that winds on Triton *do* transport suspended material, but the question is *how* the material becomes suspended.

The plumes were entirely unexpected, and explaining their vigorous activity became a major focus of research as described below. What is clearest is that they complete a coherent picture of winds on Triton at the time of the encounter. Unlike most surface streaks, both plume clouds extend westward from their apparent sources (the plumes proper—narrow, possibly unresolved vertical columns linking the horizontal plume clouds with the surface). Images of Mahilani appear to show kilometer-sized "clumps" within the cloud moving westward at 10–20 m sec^{-1}, and elongation of the cloud from 90 to 150 km at a similar speed. Thus, putting all the descriptions above together (crescent streak clouds, dark surface streaks, and plume tails), the wind is northeast nearest the surface, eastward at intermediate altitudes, and westward at 8 km, the top of the troposphere.

This is precisely the circulation pattern predicted at the time of encounter, the height of summer in the southern hemisphere. Heating by sunlight is presently causing solid nitrogen in the south to sublimate (evaporate); meanwhile in the colder north, the atmosphere is precipitating. Because of the rotation of Triton once every 5.877 days, however, the wind does not blow directly from south to north to make up the difference. Instead, gas is transported northward only in a thin skin of atmosphere near the surface (the Ekman layer) in which the flow is northeastward. The atmosphere above the 1-km-thick Ekman layer circulates from west to east. The westward flow at the altitude of the plumes can be explained if Triton's atmosphere is slightly warmer over the equator than at the south pole (perhaps because the equator is darker), in which case the temperature gradient will drive a thermal wind that causes the eastward flow to weaken and eventually change to westward flow with increasing altitude.

Basic properties of the plumes can be inferred from the images. The plume clouds do not settle out visibly (no more

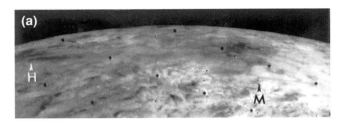

FIGURE 13 (a) *Voyager 2* image of the southern polar region of Triton in which geyser-like eruptions were discovered. Here plumes are viewed obliquely with Hili (H) and Mahilani (M) plumes marked. (b) Highly magnified images of Mahilani plume on Triton, taken from increasingly oblique angles and at increasing resolution (top to bottom). The images have been projected onto a spherical surface with a viewing geometry similar to that at the top. The increasing parallax from top to bottom makes the plume "stem" appear to grow taller. (Courtesy of NASA/Alfred McEwen, University of Arizona.)

than the ∼1-km resolution of the best images) over their length, so the suspended particles must be smaller than about 5 mm. From this particle size and the width and contrast of the clouds—about 5% darker when seen against Triton—one can further infer the amount of solids: about 10 kg sec^{-1} must be discharged if the material is dark or twice as much if it is bright. (Bright material in a cloud would appear relatively dark against Triton's very bright surface, though not as dark as intrinsically dark material. However, bright particles deposited from such a cloud would not show up as a dark streak on the surface.) The cloud moves horizontally at the wind speed, 10–20 m sec^{-1}, but the vertical velocity in the plume must be significantly faster because the plumes are not blown visibly askew by the wind. The columns may be just barely resolved in the best images. Thus the plumes may be 2 km across or perhaps smaller. The source area must have similar (or smaller) dimensions. Little or no structure is visible in the columns, though a "sheath" of descending material around the plume has been described by some authors. The active lifetime of the plumes can be estimated at a few Earth years: shorter, and *Voyager* would have been unlikely to see any plumes active; longer, and active plumes should have been more numerous compared with surface streaks.

7.2 Plume Models

Numerous attempts have been made to model the plumes in order to answer the questions of where the particulates, the gas suspending them, and the energy to drive the gas flow originate. Most models have taken their cue from the presence of the active plumes (and surface streaks) at mid to high southern latitudes at a season when the sun was almost directly overhead (Fig. 14), and assumed that the plumes are somehow solar powered. It is also possible, however, that Triton's internal heat drives the plumes and that their location is determined not by the sun but by a local enhancement of this heat source (i.e., by cryovolcanic activity) or by the thickness of the nitrogen "cap," the equivalent area of the northern hemisphere being hidden in darkness during the encounter.

It is conceivable that the plumes are purely an atmospheric phenomenon. One early suggestion was that the plumes are dust devils, localized regions of spinning and ascending hot atmosphere formed above patches on the surface that are bare of N_2 frost and that can therefore be heated by the sun to higher temperatures than their frosty surroundings. Tritonian dust devils would, however, have difficulty picking up dust from the surface and becoming visible, simply because their winds are not strong enough. If the hot areas on the ground were not only nitrogen-free but contained methane frost, though, they would give off clouds of methane gas. Being lighter than nitrogen, this methane would ascend, and might partially recondense in the atmosphere, making the rising plume visible. Falling

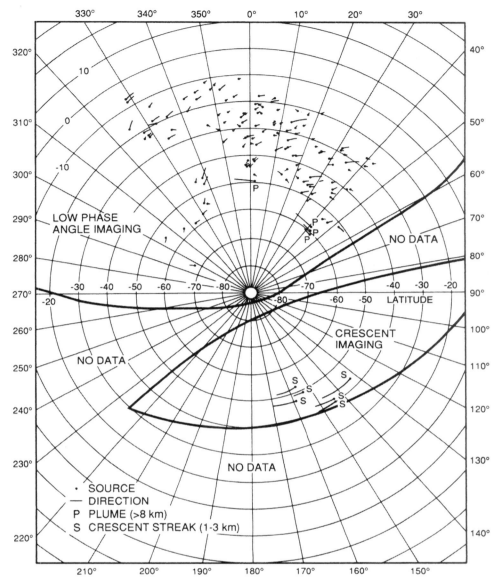

FIGURE 14 The geographic distribution and orientation of wind streaks, crescent streaks, and plumes on Triton as seen by *Voyager*. The latitude and longitude of each feature source is plotted as a dot; tails indicate streak or plume length and direction. [From C.J. Hansen *et al.* (1990) *Science* **250**, 421–424.]

back onto the ground, the methane frost would over time darken from exposure to radiation, explaining the surface wind streaks. Although this model ingeniously solves the problem of how such a gently rising plume picks up or generates enough solids to become visible, there is the fundamental objection that such a plume would be blown sideways (as dust devils on the Earth and Mars are). A final variation on these types of plume model suggests that it is nitrogen rather than methane that is ascending and condensing. Of the same composition as the rest of the atmosphere, the plume in this model would be buoyant only because it is warm. Condensation during its ascent could release enough heat to accelerate the plume substantially, but the nitrogen must somehow start off fast enough to pick up dust and to

avoid being blown sideways by winds near the base of the plume.

How could a plume of nitrogen gas get started? One possibility is that they are geysers. Like geysers on earth, which consist of water and water vapor, those on Triton would be eruptions of volatile material that has been heated underground. Whereas the water in terrestrial geysers starts as a liquid and partially boils, however, Tritonian geysers would start as hot gas that would partially condense as it expanded to the ambient pressure. This expansion could drive a gas flow powerful enough to pick up dust and form the observed plumes. Solar-powered nitrogen geysers have been studied in some detail. The pieces of the model are as follows:

7.2.1 PLUMES AS JETS

The energy needed to drive the plumes is determined by how much gas is involved and how fast it has to be erupted. The worst case assumption is that the nitrogen does not condense as it rises. Instead of becoming buoyant and accelerating, it is denser than its surroundings because of any inert dust entrained in it and the small amount of N_2 (several percent by mass) that crystallizes immediately upon eruption. The plume is therefore slowed both by gravity and by interaction with the atmosphere around it. How high it will rise depends on both the size of the eruption and its speed, and can be calculated based on laboratory simulations. As an example, a jet with a diameter of 20 m, a velocity of 230 m sec^{-1}, and 5% solids by mass will reach the observed altitude of 8 km on Triton. The plumes might be this small, but they could be as big as 1–2 km in diameter, in which case they could be somewhat slower. As discussed, plumes erupting more slowly could also reach 8 km if condensation continues after eruption, but in either case the plume will stop at about 8 km because of the increasing atmospheric temperature (buoyancy) above this altitude.

7.2.2 ERUPTION VELOCITY AND TEMPERATURE

Both the initial velocity of the gas and the amount of solid nitrogen that will condense can be calculated from the initial and final temperatures and the thermodynamic properties of nitrogen. The example given above (5% solids, 230 m sec^{-1}) is attained for nitrogen expanding freely (no change in entropy) and cooling from 42 K to 38 K. Thus, the subsurface gas must be heated about 4 K to power the geyser to the right altitude. We also learn from this calculation that the 10-to 20-kg sec^{-1} of solids estimated to be feeding the plumes is accompanied by as much as 400 kg sec^{-1} of gas. Given the latent heat of sublimation of nitrogen, about 100 megawatts of power is needed to convert solid to gas at this rate.

7.2.3 TEMPERATURE OF A SOLID-STATE GREENHOUSE

The "greenhouse effect" usually describes heating of the Earth's atmosphere (or that of another planet) when sunlight at visible wavelengths penetrates the atmosphere before being absorbed, but longer wavelength thermal radiation is absorbed by the atmosphere and cannot escape to space as easily. A similar effect can take place in a transparent solid, for example, nitrogen ice on Triton. The amount of solar energy is not great at Triton's distance from the Sun, but nitrogen is an excellent thermal insulator and the deeper the sunlight is absorbed the warmer the subsurface will get. A 6-m thick layer of clear nitrogen ice over a dark subsurface layer would actually melt at the base, while even a 4-m layer would blow itself apart because the hot ice would produce gas at a pressure higher than the weight of the solid above. (This cannot be how plumes originate, however, because the production of gas would cease very quickly as chunks of the ruptured layer cooled.) Heating by 4 K can be achieved with a greenhouse layer only 1–2 m thick.

7.2.4 SUBSURFACE ENERGY TRANSPORT

What happens after sunlight is absorbed below Triton's surface and before hot gas is erupted? As just estimated, 100 megawatts are needed to heat the gas in a typical plume. This is the amount of power deposited by sunlight on a region of Triton about 10 km in diameter, much bigger than the 1- to 2-km size of the plume sources. We can therefore conclude that gas (or energy to produce gas by sublimation) is stored over time and then released quickly, or is transported horizontally from the larger area to the geyser, or both. Somewhat counterintuitively, gas is not mainly "stored" in voids in the nitrogen ice, but is produced on demand from hot ice, while heat transport is mainly carried by flowing gas rather than ordinary thermal conduction. Nitrogen ice can give off more than 100,000 times its own volume of gas as it cools just 4 K. If there are voids in the solid nitrogen, this gas will flow to colder areas and recondense, warming them by releasing its latent heat. Depending on the size of such void spaces, the gas flow can transport energy hundreds of times more efficiently than conduction. Not only could flow between meter-sized blocks of solid readily supply a geyser, but when a path to the surface was first opened eruption would be vigorous at first and decline over a period of about a year, roughly the estimated lifetime of the plumes. Energy transport by production of gas, its flow through pores, and recondensation at colder points is known on Earth: "heat pipes" containing a condensable gas (with a wick to return the liquid to the hot end) conduct heat better than metal and are used for baking potatoes from the inside out and for controlling the temperature of spacecraft, including *Voyager*! How a suitably fractured layer of nitrogen ice, overlain by a clear, gas-tight greenhouse layer, might form on Triton is discussed in the next section.

The idea of solar-powered geysers thus seems extremely promising, though much work remains to take the separate pieces that have been modeled so far and make sure that they fit together. Internally powered geysers (more similar to their terrestrial counterparts) have not been studied nearly as thoroughly, but other possibilities exist. As discussed below, the nitrogen "polar caps" on Triton may be so thick near their center (over a kilometer) that they begin to melt at the base. Liquid N_2 finding its way to the surface could erupt as a boiling geyser, with more than enough energy to power the plumes. Gases other than nitrogen could also be erupted from deeper in Triton's water-ice mantle, driven by internal heating. Most recently, a similar solar-powered geyser model has been proposed for the formation of dark spots, "spiders," and fans at high southern latitudes on Mars. In this case the polar cap material and geyser gas is not N_2, but CO_2.

7.3 Polar Cap and Climate

We turn now from the plumes to a consideration of how Triton's surface frosts and atmosphere change over time. Here, too, the *Voyager* images yielded a surprise: at the height of southern hemisphere summer (Fig. 4), most of the southern hemisphere was covered with a bright deposit (a polar cap), but the visible portion of the northern, winter hemisphere was darker. Models of the redistribution of N_2 frost with the seasons can be constructed with varying degrees of complexity, but a fundamental expectation is that the summer hemisphere should have less of a polar cap than the winter one!

The basic physics of seasonal frost-distribution models is as follows.

1. The whole atmosphere and all frosted areas are at very nearly the same temperature. If a frosted area were colder, more nitrogen would condense there and release of latent heat would raise the temperature. Conversely, a warm frost area would be cooled by sublimation. Winds would quickly even out the atmospheric pressure and temperature.

2. At this fixed temperature, sublimation occurs where frosts are exposed to the sun and condensation where the average input of solar energy is less. Sublimation/condensation rates can be calculated from the amount of sunlight absorbed at each point on Triton.

3. Bare (unfrosted) areas can be warmer than the atmosphere and frosts (if they are dark and/or well exposed to the sun) but they cannot be colder, or frost would immediately condense on them.

Using the albedo of the surface as measured by *Voyager*, models indicate that frost in most of the southern hemisphere is currently subliming, thinning the surface deposits. Nitrogen is presumably being deposited in the northern hemisphere and in a few of the brightest areas of the south where little sunlight is absorbed. Stellar occultations since *Voyager* have shown that Triton's surface pressure (and thus atmospheric mass) has measurably increased, to around 19 μbar! By inference the surface temperature of the nitrogen ice, which controls the atmospheric pressure, has also increased by 1–2 K. But what about the long run? By assuming that frost has some given albedo and that the surface underneath has some other albedo, one can model the redistribution of nitrogen over long periods. A layer of nitrogen frost about a meter thick is moved back and forth as the sun shines on one hemisphere and the other, and the pressure and temperature of the atmosphere change as well. Notably, such models predict that all nitrogen deposited in the southern hemisphere the last time it was winter there would have resublimated before *Voyager* arrived. Corre-

spondingly, the northern hemisphere should be extensively frosted.

How can these predictions be reconciled with observation? The frost might deposit mainly in shadows and on north-facing slopes where *Voyager* could not see it, or it could be glassy and transparent, hence invisible. There is some evidence for the last possibility, from laboratory observations of condensing nitrogen, calculations of the rate at which loose frost grains would merge or anneal into a dense, transparent layer, and even from observations of the light-scattering properties of Triton's equator. Such suggestions would each explain the dark, apparently frost-free northern hemisphere, but the bright "cap" in the south must be explained as well. Perhaps it is a much thicker deposit of nitrogen that never completely sublimes away (this is certainly the impression one gets geologically). Although nitrogen frost may be very transparent when first annealed, changing temperatures will make the residual cap expand and contract, fracturing it and making it appear bright. Thus, we are led to the idea of a clear, uncracked (i.e., gas-tight) seasonal frost layer over a thick, fractured permanent cap: precisely the kind of layering hypothesized above to explain the plumes as solar-powered geysers.

What controls the size of the residual cap, and why is one not seen in the north? A good candidate is solid-state creep, or flow, of the thick nitrogen deposit, similar to the flow of glaciers and spreading of polar caps on the Earth and Mars. Models based on terrestrial polar caps, combined with estimates of the rate at which solid nitrogen would flow, suggest that the permanent cap is about a kilometer thick at the center. Cap spreading also prevents the eventual disappearance of the seasonal frosts predicted by the models discussed above. Because the pole always receives less sunlight than the edges of the seasonal frost deposits, more frost will be deposited at the pole than at the edges, maintaining the cap. There may be a northern as well as a southern permanent cap. If this northern cap extends less than 45° from the pole, it would lie in the dark portion of Triton unseen by *Voyager*. The southern permanent cap might be larger because of hemispheric differences in the heat released from Triton's interior, or it might also extend only 45°, in which case the bright deposits extending almost to the equator have still to be explained. Some of this bright material may be nitrogen "snow" that condenses in the atmosphere into grains that are too big to anneal on a seasonal time scale into a transparent layer. It should be apparent from this discussion that, as with the plumes, we seem to have many pieces of the puzzle of the polar caps (and perhaps a few spurious pieces of unrelated puzzles), but they have yet to be assembled into a final picture of Triton's surface-atmosphere interaction.

Additional clues to the behavior of volatiles on Triton are presently being gathered from Earth-based spectroscopic measurements, and by the occultation of stars by Triton. As noted above, Triton's atmosphere is changing, becoming thicker and slightly warmer. Strong winds aloft are also indicated. Surface-atmosphere interactions on the polar caps of Mars are also being studied in great detail, and the lessons learned can be applied to Triton. And of course, continued monitoring of Pluto provides a valuable second case against which to test theoretical models.

8. Origin and Evolution

Triton and Pluto turn out to be remarkably similar in size, density, and in surface and atmospheric compositions as well. There is little doubt that they share a common heritage. Moreover, they are not isolated in the outer solar system. An entirely new reservoir of minor planets has been found orbiting near and beyond Neptune—the Kuiper Belt. The first Kuiper Belt object was found in 1992, and as of this writing over 1200 have been discovered. A number are as large as Pluto or Triton. The largest, Eris, has a density similar to that of Triton, and methane and nitrogen ice on its surface. [*See* Kuiper Belt Objects: Physical Studies.]

The link between Triton, Pluto, and the Kuiper Belt is strengthened by what is known of the orbital dynamics of this region. For example, a number of Kuiper Belt objects share the same dynamical resonance with Neptune that Pluto occupies (this orbital resonance prevents encounters between Neptune and Pluto, and is one of the strong arguments against the Pluto-as-escaped-satellite hypothesis). In this sense, Pluto and its companion "Plutinos" are more like the Trojan or Hilda groups of asteroids (which are locked in orbital resonances with Jupiter), only that Pluto-Charon is the clearly dominant member of its group.

Dynamical calculations show that Pluto and its companions were probably swept into this orbital resonance as Neptune's orbit expanded early in solar system history. During this time the flux close to Neptune of bodies orbiting near and beyond Neptune would have been quite high, and even today Neptune continues to deplete the inner Kuiper Belt population, the short period comets being one result. It is perhaps not surprising then that Neptune should have had a catastrophic encounter with at least one escapee from the Kuiper Belt: Triton. [*See* Cometary Dynamics.]

Satellite capture does not occur easily. Generally, objects passing near a planet leave with the same speed that they came in with. Even complicated trajectories called temporary gravitational captures (enjoyed by Comet Shoemaker-Levy 9) are just that, temporary. To be permanently captured, a cosmic body must lose energy (velocity) by running into or through something. In Triton's case, it could have collided with another stray body just passing by Neptune, but the probability of this having happened is quite low. Because Triton orbits close to Neptune, in the region usually occupied by regular satellites, it is much more likely that it ran into a regular satellite or its precursor protosatellite disk.

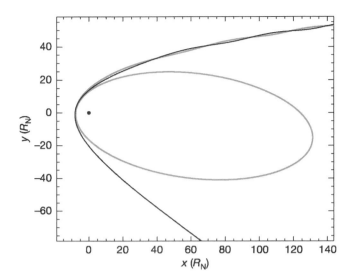

FIGURE 15 A possible capture mechanism for Triton. In this example "exchange capture" calculation, an equal mass Triton binary approaches from the upper left, and is disrupted by tides from Neptune. One member of the binary is captured into an elliptical orbit with a semimajor axis of ≈ 70 R_N, while the other escapes (R_N is Neptune's radius). [From C.B. Agnor and D.P. Hamilton (2006) *Nature* **441**, 192–194.]

A recent, alternative model proposes that Triton was once part of a binary, and when it passed too close to Neptune, strong tides from the planet split the binary in two. One member of the binary escaped back into solar orbit, while the other stayed behind in Neptune orbit (Fig. 15). In this case, the captured member of the binary loses orbital energy to the escaping member. While this may at first glance seem far-fetched, we now know that a good fraction of Kuiper belt objects are binaries, and tidal stripping close to a much more massive planet such as Neptune simply requires a close passage. Although permanent capture of one of the original binary members is not assured, the probability is much greater than, say, being captured by colliding with an original Neptune satellite.

The inclination of Triton's postcapture orbit depends on the initial encounter geometry, and is essentially random. Triton could have ended up either prograde or retrograde. After capture, Triton's orbital evolution would be strongly influenced by tides. Every time Triton reapproached Neptune, Neptune's gravity would raise a tidal bulge on Triton. The periodic rise and fall of the bulge would dissipate energy as heat, which would be extracted from the energy of Triton's orbit. Because the tidal couple between Triton's bulge and Neptune would be (on average) radial, no change in Triton's orbital angular momentum would occur. Based on these constraints, and ignoring for the moment any encounters with original satellites, Triton's orbital configuration after capture would evolve as depicted in Figure 16. Triton may have begun with a semimajor axis of 1000 R_N

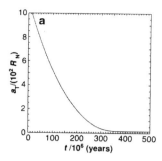
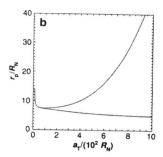

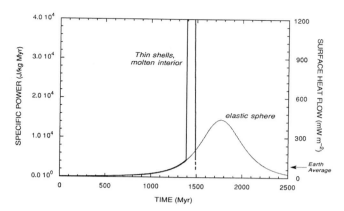

FIGURE 16 (a) Example evolution of Triton's semimajor axis, a_T, as a function of time, t, due to tidal dissipation within Triton. (b) Evolution of Triton's minimum and maximum periapse distance, r_p (the closest point to Neptune in its orbit), as a function of semimajor axis due to the combined influence of semiannual solar perturbations and tidal dissipation. The periapse distance oscillates between the two curves shown. [Adapted from P. Goldreich *et al.* (1989). *Science* **245**, 500–504.]

FIGURE 17 Power dissipated per unit mass and surface heat flow for Triton as its post-capture orbit shrinks and circularizes. Two models are shown. One assumes Triton remains a uniform, undifferentiated sphere, while the second allows for melting. In both cases the time scales are longer than in the calculations in Figure 16a, due to updated parameters for Triton, but the periapse variations as a function of semimajor axis in Figure 16b are unchanged. The thin shells model is more realistic than the elastic sphere, but even here the meltdown of Triton has been artificially suppressed; in reality a thermal runaway probably occurs much earlier. [From W.B. McKinnon *et al.* (1995). *In* "Neptune and Triton" (D.P. Cruikshank, ed.). University of Arizona Press, Tucson.]

or greater, or it may have begun closer in, such as in the example shown in Figure 15. The important point is that early on Triton's **periapse** (the closest point to Neptune in its orbit) would lie as low as half its present semimajor axis. Triton's tidal evolution probably took 100 million years or longer, so there would have been sufficient time for Triton's orbit to evolve through and interact with any preexisting satellites.

This point is emphasized in Figure 16b, which includes the periodic effects of solar tides on Triton's evolving orbit. If Triton's initial capture orbit was very large and eccentric, its periapse would have fluctuated, and may have periodically been as low as 5 R_N! Triton would have had ample opportunity for collisions with Neptune's original satellites (if they were like Uranus' today), possibly accreting them in the process. It may also have scattered original satellites into distant orbits, caused them to crash into Neptune, or perhaps even ejected them from Neptune altogether. There is now nothing left of Neptune's original system (if it indeed existed) other than the inner satellites and Nereid. The inner satellites all lie within 5 R_N, however, which is perfectly consistent with this capture scenario. Nereid may also be a survivor of this orbital mayhem. Little is known about this distant moon, save its size (~340 km in diameter), reflectivity (~20%), and presence of surface water ice, but these facts make Nereid more akin to a regular satellite than a dark captured asteroid or comet.

The end state of Triton's orbital evolution is an extremely circular orbit. As such, the orbital energy potentially dissipated by tides within Triton represents an absolutely enormous reservoir, about 10^4 kJ kg^{-1}. It is sufficient to completely melt all the ice, rock, and metal within Triton ten times over. The magnitude of Triton's temperature change, however, depends on the heating rate, and somewhat on the size of the initial capture orbit. Two such models are

illustrated in Figure 17. Tidal heating after capture in either model is at first modest, as the satellite spends most of its time far from Neptune. As its semimajor axis shrinks and its orbital period decreases, the average heating rate begins to rise. The epoch of greatest heating occurs when the relative change in semimajor axis is the greatest (because orbital energy is inversely proportional to semimajor axis), roughly when the semimajor axis drops below 100 R_N. Because the orbit can only evolve as fast as the tides can convert orbital energy to heat, the response of Triton to tidal flexing is crucial. If Triton responds as a dissipative elastic sphere, then the semimajor axis drops continuously (Fig. 16a) and the tidal heating rises and then falls smoothly as the orbit becomes more circular (Fig. 17, elastic sphere model). The calculations in Figures 16a and 17 are actually for two different elastic sphere models, but are shown here to illustrate a range of possible time scales.

A dissipative elastic sphere is clearly an idealized and oversimplified model for Triton. Triton is in reality a complex rock, metal, organic matter, water-ice, and volatile ice body. The volatile ices especially should be melted and mobilized within Triton early in its history (e.g., ammonia), with or without tidal heating. A partially molten body is a particularly dissipative body, so when capture occurs and tidal heating begins, heat concentrates in the partially liquid regions. This causes more melting, which makes the body more dissipative, which results in greater tidal heating.

Thus, within a few hundred million years after capture, Triton in all probability went through an episode of runaway melting. This is schematically illustrated in Figure 17, where in the model labeled thin shells Triton melts spontaneously when enough energy has been accumulated to do so (in reality the runaway occurs much earlier). Thereafter Triton is a nearly totally molten, but still dissipative body. Its tidal heating curve rises and falls sharply over the course of ~100 million years.

During this epoch of extreme tidal heating Triton's heat flow is an amazing ~ 2–4 W m^{-2}, equal or greater than that measured today from Io. Its surface temperature is governed by this flux, and corresponds to a blackbody temperature of 80–90 K. During and after this epoch there would likely have been large chemical exchanges between the global oceanic mantle with its dissolved volatiles and the hot rock core below. Much of Triton's volatiles may have been driven into a massive atmosphere. Atmospheric components plausibly include CO, CH_4, CO_2, and NH_3, or even H_2 (from photolysis of methane or ammonia or as a minor component in Triton's original ice). Conservative assumptions yield an atmospheric greenhouse with surface temperatures well above 100 K; more extreme possibilities allow for surface temperatures greater than 200 K.

A most intriguing aspect of raising a massive greenhouse atmosphere by tidal heating is that it may persist well after the tidal heating input has tapered off and Triton's interior has begun to freeze. It may only collapse after enough of it has been lost to space due to solar-UV-heating-driven hydrodynamic escape, which could have taken in excess of 1 billion years. While the atmosphere existed it would have kept Triton's surface warmer, and enhanced the geological mobility of the satellite's surface layers. Unfortunately, there are as yet no definitive indicators of the atmosphere's former presence (e.g., ancient aeolian or fluvial features, peculiar crater shapes, etc.). If a thick atmosphere existed, Triton's continued geological activity has obscured the evidence.

Regardless, once tidal heating ended, Triton's interior should have begun to freeze. It would probably have taken a few 100 million years to do so, but even today such freezing would not be complete. Triton's ice mantle is probably warm enough, due to radiogenic heating from the core, that any ammonia- and methanol-rich fluids are stable (perhaps in an internal ocean), and Triton's inner core of alloyed iron, nickel, and sulfur should likewise be warm enough (more than ≈1250 K) to allow for a eutectic liquid mixture of those elements.

The possible persistence of cryomagmas in Triton's mantle due solely to radiogenic heating has raised the question as to whether any of the geological observations in Section 6 actually *demand* that Triton was massively tidally heated. Certainly, solar-powered plume models do not require Triton to be internally active at all. Triton's surface, on the other hand, is so peculiar (in the sense of being unique or special). Furthermore, the extent and intensity of the geological activity recorded there is only seen on satellites that are undergoing active and substantial tidal heating (Io, Europa, and Enceladus). While no ironclad argument can be made, Triton's geology and chemistry in all likelihood indicate that it did indeed experience massive tidal heating.

The proof of Triton's history and provenance requires further exploration of this extraordinary body. For example, determination of the compositions of Triton's icy lavas, and terrains in general, would be key constraints. Detailed exploration of the Neptune system by spacecraft is also a technically feasible proposition, given recent and projected technological advances. Instruments and electronics are being increasingly miniaturized, thereby requiring smaller launch vehicles. Missions to Triton can also take advantage of innovative flight strategies, such as using aerobraking in the Neptune atmosphere to go into initial Neptune orbit. Thereafter a complement of advanced instruments can be trained on Triton during repeated encounters, filling out our picture of this amazing satellite.

As for Triton's ultimate future, as a retrograde satellite its orbit is actually decaying due to tides it raises on Neptune. In the 1960s it was estimated that Triton would closely approach Neptune and be torn apart by tides in a geologically short time. Present estimates imply less peril: Triton's orbit will probably shrink by no more than 15% over the next 5 billion years, giving Triton plenty of time for further geological and atmospheric adventures.

Bibliography

Beatty, J.K., *et al.* (1999). "The New Solar System, 4th Ed." Sky Publishing, Cambridge, MA.

Cruikshank, D.P. (ed.) (1995). "Neptune and Triton." University of Arizona Press, Tucson.

Greeley, R., and Batson, R. (1997). "NASA Atlas of the Solar System." Cambridge University Press, Cambridge, UK.

Littmann, M. (2004). "Planets Beyond: Discovering the Outer Solar System." Dover Publications, Mineola, NY.

Morrison, D., and Owen, T. (2003). "The Planetary System" 3d ed. Addison Wesley, San Francisco, CA. NASA Planetary Photojournal. http://photojournal.jpl.nasa.gov/index.html.

Rothery, D.A. (1999). "Satellites of the Outer Planets: Worlds in Their Own Right: 2d ed. Oxford Univ. Press, NY.

Smith, B.A., and the *Voyager* Imaging Team (1989). *Voyager 2* at Neptune: Imaging science results. *Science* **246**, 1422–1449.

Planetary Rings

Carolyn C. Porco

Space Science Institute
Boulder, Colorado

Douglas P. Hamilton

University of Maryland
College Park, Maryland

CHAPTER **27**

1. Introduction

Planetary rings are those strikingly flat and circular appendages embracing all the giant planets in the outer Solar System: Jupiter, Saturn, Uranus, and Neptune. Like their cousins, the spiral galaxies, they are formed of many bodies, independently orbiting in a central gravitational field. Rings also share many characteristics with, and offer invaluable insights into, flattened systems of gas and colliding debris that ultimately form solar systems. Ring systems are accessible laboratories capable of providing clues about processes important in these circumstellar disks, structures otherwise removed from us by nearly insurmountable distances in space and time. Like circumstellar disks, rings have evolved to a state of equilibrium where their random motions perpendicular to the plane are very small compared to their orbital motions. In Saturn's main rings (Fig. 1), for example, orbital speeds are tens of km/sec while various lines of evidence indicate random motions as small as a few millimeters per second. The ratio of vertical to horizontal dimensions of the rings is consequently extreme: one part in a million or less, like a huge sheet of paper spread across a football field.

Rings, in general, find themselves in the **Roche zone** of their mother planet, that region within which the tidal effects of the planet's gravity field prevent ring particles, varying in size from micron-sized powder to objects as big as

houses, from coalescing under their own gravity into larger bodies. Rings are arranged around planets in strikingly different ways despite the similar underlying physical processes that govern them. Gravitational tugs from satellites account for some of the structure of densely-packed massive rings [*see* SOLAR SYSTEM DYNAMICS: REGULAR AND CHAOTIC MOTION], while nongravitational effects, including solar radiation pressure and electromagnetic forces, dominate the dynamics of the fainter and more diffuse dusty rings. Spacecraft flybys of all of the giant planets and, more recently, orbiters at Jupiter and Saturn, have revolutionized our understanding of planetary rings. New rings have been discovered and many old puzzles have been resolved. Other problems, however, stubbornly persist and, as always, new questions have been raised. Despite significant advances over the past decade, it is still the case that most ring structure remains unexplained.

2. Sources of Information

2.1 Planetary Spacecraft

While rings have been observed from the surface of the Earth ever since Galileo Galilei discovered two curious blobs near Saturn in 1610, the study of planetary rings did not emerge as the rich field of scientific investigation it is today until the *Voyager* spacecraft made their historic tours

FIGURE 1   Saturn and it main ring system in near natural color as seen from *Voyager*. From bottom, the satellites Rhea, Dione, and Tethys are visible against the darkness of space, with Mimas just above them on Saturn's bright limb. Shadowing abounds in this image: black dots cast by Mimas and Tethys are visible on Saturn's disk, the planet blocks light from getting to the rings at lower right, and the foreground rings paint a dark band on the planet's cloudtops. From the outside are the bright A and B rings separated by the Cassini Division. The narrow Encke Gap in the outer A ring is also visible, as is the dark C ring near set the planet.

of the outer Solar System in the 1980s. Not even the two *Pioneer* spacecraft, the first human artifacts to pass through the realms of Jupiter and Saturn in the mid to late 1970s, hinted at the enormous array of phenomena to be found within these systems.

Voyager 1 arrived first at Jupiter in March 1979, followed by *Voyager 2* four months later. After its encounter with Saturn in November 1980, *Voyager 1* was placed on a trajectory that took it out of the Solar System; *Voyager 2* encountered Saturn in August 1981 and then sailed on to reach Uranus in January 1986, and Neptune, its last planetary target, in August 1989. Each spacecraft was equipped with a suite of instruments collectively capable of covering a wide range of wavelength and resolution. Tens of thousands of images

of planetary ring systems in the outer Solar System were acquired by the *Voyager* cameras at geometries and resolutions impossible to obtain from the ground. Also, occultations of bright stars by the rings were observed from the spacecraft, and occultations by the rings of the spacecraft telemetry radio signals were observed from the Earth; both produced maps of the radial architecture of the rings at spatial scales of ~100 m. In addition to these remote-sensing observations, local (or *in situ*) measurements were made of charged particles, plasma waves, and, indirectly, impacts of micron-sized meteoroids as each spacecraft flew through the ring regions of each planet. These data sets contributed in varying degrees to the picture that ultimately emerged of the unique character and environment of the ring systems surrounding the giant planets.

The *Galileo* spacecraft, launched in 1989, became the first artificial satellite of Jupiter in December 1995 and remained in orbit until September 2003 when, fuel running out and instruments ailing, it was directed to crash into the giant planet. Images of the Jovian ring system are few but have improved resolution and image quality significantly over those obtained by *Voyager*. Galileo resolved one of three separate ring components imaged by *Voyager*—the Gossamer Ring—into two distinct structures and clarified the intimate relationship between these components and the nearby orbiting satellites (Table 1, Fig. 2).

The Cassini spacecraft, orbiting Saturn since July 1, 2004, is the best ring-imaging machine built by humans to date. Cassini carries a host of remote imaging and *in situ* instruments that are currently making detailed observation of Saturn, its moons, rings, and magnetosphere. One author of this chapter (CCP) is also the leader of the visual imaging instrument that returned many of the figures displayed in this chapter. Other imaging instruments cover infrared and ultraviolet wavelengths, the radio science package will perform new occultation experiments, and numerous *in situ* experiments are studying local properties of dust, plasma, and magnetic fields.

2.2 Earth-Based Observations

In the past two decades, Earth-based telescopic facilities and instrumentation have become increasingly sophisticated and sensitive; key advances include 10-m class telescopes, active adaptive optics that instantaneously correct for variations in the Earth's atmosphere, and ever-larger arrays of digital CCDs sensitive to visual and infrared light. The Hubble Space Telescope (HST), placed in orbit around the Earth in 1990, nicely complements ground-based instruments by providing unparalleled sharp views and ultraviolet capabilities. Clever observers have taken advantage of these advances, as well as unique geometric opportunities to push beyond spacecraft discoveries, despite the severe distance handicap. These advances have proven invaluable for furthering the study of planetary rings.

TABLE 1   Locations of Major Ring Components. The inner limits of Jupiter's Amalthea and Thebe rings are poorly constrained. The Saturnian F ring has multiple narrow strands that are part of a continuous spiral ring in addition to the bright core listed above. The uranian η ring has a diffuse component that extends ~55 km beyond the ring. R1 and R2 were discovered by HST in 2003. New Saturnian rings discovered by Cassini that lie exterior to the main rings are given below; those within the main rings have been omitted.

Planet	Ring Component	Radial Location or (width) in km	Optical Depth
Jupiter (Radius: 71,492 km)	Halo	89,400–123,000	10^{-6}
	Main	123,000–128,940	10^{-6}
	Amalthea Ring	140,000?–81,000	10^{-7}
	Thebe Ring	140,000?–221,900	10^{-7}
	Thebe Extension	221,900–280,000	10^{-8}
Saturn (Radius: 60,330 km)	D	67,000–74,500	10^{-5}
	C	74,500–92,000	0.05–0.35
	B	92,000–17,580	0.4–>3
	Cassini division	117,580–122,200	0–0.1
	A	122,200–136,780	0.4–1.0
	R/2004 S1 (Atlas)	137,630	$\sim 10^{-4}$
	R/2004 S2	138,900	$\sim 10^{-5}$
	F	140,200 (~1)	0.1–1
	R/2006 S1 (Janus/Epimetheus)	151,500	?
	G	166,000–173,000	10^{-6}
	R/2006 S2 (Pallene)	212,000	?
	E	181,000–483,000	10^{-6}
Uranus (Radius: 26,200 km)	1986 U2R	~38,000	?
	6	41,837 (1.5)	0.3
	5	42,234 (~2)	0.5
	4	42,571 (~2)	0.3
	α	44,718 (4–10)	0.4
	β	45,661 (5–11)	0.2
	η	47,176 (1.6)	<0.4
	γ	47,627 (1–4)	>0.3
	δ	48,300 (3–7)	0.5
	λ	50,024 (~2)	0.1
	ε	51,149 (20–96)	0.5–2.3
	R2	66,100–69,900	10^{-8}
	R1	86,000–103,000	10^{-8}
Neptune (Radius: 25,225 km)	Galle	41,000–43,000	10^{-4}
	Le Verrier	53,200 (<100)	0.01
	Lassell	53,200–59,100	10^{-4}
	Arago	57,200 (~10)	?
	Adams	62,933 (15–100)	0.01–0.1

2.2.1 STELLAR OCCULTATIONS

A stellar occultation occurs when, as viewed from Earth, a bright star passes behind a planetary ring system. These events occur rarely, typically last for hours, and can yield data on the location of ring features that rival spacecraft resolutions. The Uranian ring system was discovered in 1977 by stellar occultation, and the first hint of the Neptunian ring arcs also came during such an event. The value of these observations is dramatically illustrated by the 1989 occultation of a particularly bright star by Saturn's ring system that revealed numerous ring features to a precision of 2 km, produced an important refinement of Saturn's pole

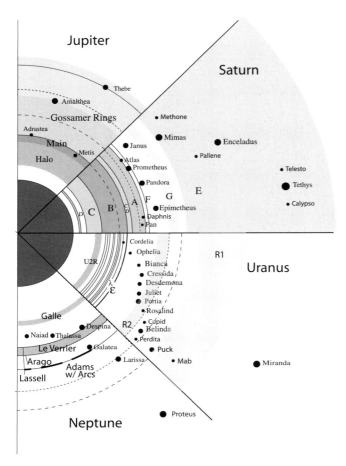

FIGURE 2 A graphic schematic of the ring-moon systems of the giant planets scaled to a common planetary radius (compare with Table 1). The planet is the solid central circle, ring regions are shaded, and nearby satellites are plotted at the correct relative distances. Dotted lines indicate the Roche radius for a satellite density of 0.9 g/cm³, and dashed lines show the position of synchronous orbit where an object's orbital period matches the planetary rotation period. The Roche radius is outside the synchronous distance for Jupiter and Saturn but inside it for Uranus and Neptune due to the more rapid spins of the larger planets. (Figures courtesy of Judith K. Burns)

position, and allowed the two million-year precession period of the pole to be measured for the first time. The prevalence of collisions amongst particles in Saturn's main rings causes the rings to be extremely thin and exactly perpendicular to Saturn's pole, enabling this interesting observation; this is perhaps the longest-period astronomical motion measured to date.

2.2.2 RING PLANE CROSSINGS

Ring plane crossings (RPXs), those times when the plane containing a planet's rings sweeps over the Earth or the Sun as the planet moves along its orbital path, are unique observational opportunities. Near these special times, the Sun

and the Earth can be on opposite sides of the ring plane, above or below it by just a few degrees or tenths of a degree. The near edge-on aspect of planetary rings in this geometry and our view of the unilluminated side drastically reduces the glare of sunlight scattered off or through the rings and allows nearby faint objects to be much more easily seen. Five small satellites of Saturn were discovered during past RPXs: Janus (in 1966) and Epimetheus, Telesto, Calypso, and Helene (in 1980). [*See* OUTER PLANET ICY SATELLITES] Saturn's outer dusty E ring (Fig. 2) was also discovered during the 1966 RPX and its strange bluish color revealed in the 1980 RPX. The most recent crossing, which occurred from 1995–1996, showed the F-ring (Fig. 2) to be slightly tilted, revealed a number of clumps in the F-ring that appear and disappear, constrained the thickness of the main rings to be less than 1.5 km (the apparent thickness of the outer F-ring), recovered several tiny satellites not seen since the *Voyager* flybys, and further refined Saturn's pole position and its precession rate. In addition, light filtered through the optically thin regions of the rings has allowed these diffuse structures to be studied in a unique way.

Ring plane crossings occur twice per orbit, roughly every 6, 15, 43, 82 years for Jupiter, Saturn, Uranus, and Neptune, respectively. Upcoming RPXs for these planets occur in 2009, 2009, 2007, and 2046, making the next few years an exciting time for ring scientists. One author of this chapter (DPH) has been involved in RPX observations of Jupiter, Saturn, Uranus, and even Mars (which is predicted to have an extremely faint ring derived from material lofted from its two small moons).

2.3 Numerical Studies

Continuous advances in the speed and design of desktop computers have made numerical studies of ring systems an essential tool for investigating dynamically important factors that are not easily treated by analytical methods. Numerical methods have been used to simulate a myriad of ring processes, including the collisional and gravitational interactions among orbiting ring particles, the effects of micrometeoroid impacts onto the rings, the behavior of small charged ring particles under the influence of rotating magnetic fields, and the evolution of debris resulting from a catastrophic disruption of a satellite orbiting close to or within a planet's Roche zone. Key algorithm advances over the past decades include energy-preserving "symplectic" codes, which can efficiently integrate the exact forces arising in a collection of interacting bodies, and significantly faster "tree" codes, optimized for large collections of interacting bodies, which employ clever approximations to the exact equations of motion. Numerical models are an important tool for scientists seeking to understand the physical processes active in known ring features. These simulations, when targeted well, can also make testable predictions, in some cases steering observers toward refining their observational strategies.

3. Overview of Ring Structure

Rings are characterized by an enormous variety of structural detail, only some of which has been attributed successfully to known physical processes, either internal or external to the rings (Section 4). Looking across all four ring systems, however, we do find trends and commonalities. In particular, we now recognize three main types of planetary rings in the Solar System. First are broad massive rings, replete with fine-scale structure. Some of this structure is produced by embedded moonlets and some by interactions with both nearby and distant satellites. Saturn's extensive main ring system provides the only example of this type (Fig. 1). The second ring type consists of sets of sharply defined narrow rings, interspersed with small moons. These narrow structures are found primarily at Uranus and Neptune, although Saturn has interesting examples: its F-ring and numerous ringlets in the fainter C and D rings (Fig. 2). Finally, all of the giant planets have broad relatively featureless sheets of dusty debris that are usually found in close association with small source satellites. Jupiter's ring system provides the best understood archetype, but numerous additional examples are found around each of the other giant planets.

3.1 Jupiter

The particles comprising the diffuse tenuous rings of Jupiter almost certainly have their origin in the release of dust from each of the four moonlets—Adrastea, Metis, Amalthea, and Thebe—embedded in the rings. These small rocky objects are continually pummeled by bits of space debris that are accelerated to high relative speeds by Jupiter's intense gravity. When struck by this flotsam, puffs of dust are ejected from the moonlet surfaces. The main ring of Jupiter has a small normal optical depth, $\tau_N \sim 10^{-6}$, in tiny (<10 μm) particles; the optical depth may be even smaller for large (>1 mm) particles (Fig. 3). The main ring has a relatively

FIGURE 4 A *Galileo* image showing Jupiter's main ring (lower panel) and main ring plus interior halo (top panel). Note the patchiness of the main ring, hinting at further complexity.

sharp outer edge suspiciously coincident with the orbit of Adrastea; just interior to this, the satellite Metis creates a depression in ring brightness. The fact that the main ring extends only inward from the small source satellites strongly suggests that ring particles drift inward. A ~20,000-km vertically thick toroidal ring, or halo, lies interior to the main ring (Fig. 4). Its normal optical depth is comparable to the main ring, a fact that is consistent with inward drift. It took the arrival of the *Galileo* spacecraft to show that the diffuse material exterior to the main ring was, in fact, split into two components, each associated with a small moon (Fig. 5). As with the main ring, these gossamer rings extend primarily inward from their source moons Thebe and Amalthea and, moreover, have vertical thicknesses that exactly correspond to the vertical motions of the inclined moons (Fig. 6). An extremely faint outer extension to the Thebe ring is composed of particles on significantly eccentric orbits. *Cassini's*

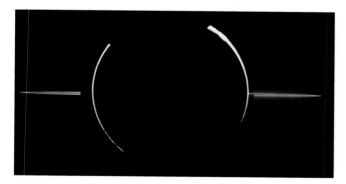

FIGURE 3 A *Voyager* mosaic of images taken from Jupiter's shadow looking back toward the Sun. Sunlight traces out the edge of the planet's atmosphere and the distribution of micron-sized dust in its main ring. The gap between one ring arm and the planet on the right is due to Jupiter's shadow; the gap in both arms on the left is an artifact from the stitching together of multiple images.

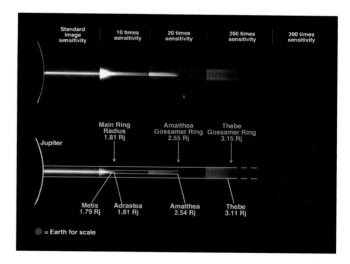

FIGURE 5 A mosaic of *Galileo* images enhanced to bring out faint jovian ring features. The main ring shows up clearly in standard images, while the jovian halo and Amalthea ring become apparent only in enhanced images. The outermost Thebe ring appears only in images with the greatest sensitivity.

FIGURE 6 A schematic of Jupiter, its innermost four moonlets, and its ring components (shown in different colors) as determined by *Voyager*, *Galileo*, and ground-based observations. Note that the thickness of the inner Halo component is due to an electromagnetic effect operating on dusty grains, while the vertical extension of the Gossamer rings have a more prosaic cause: the tilted orbits of the source satellites themselves.

flyby of Jupiter revealed that, similarly, the vertical motions of metris, and possibly Adrastea. The jovian ring particles have reddish colors, suggestive of a silicate or carbonaceous composition, just like the embedded moonlets.

3.2 Uranus and Neptune

Ground and space-based observations reveal ten, narrow, sharp-edged continuous rings encircling Uranus (Fig. 7). Interspersed amongst these features are broad dusty swaths of material best seen when Voyager was looking back at Uranus from a vantage point further from the Sun (Fig. 8). In addition, HST has recently detected two distant and extremely faint dust sheets similar to those around Jupiter (R1 and R2 in Fig. 2). Most of the narrow rings are eccentric and some are tilted relative to Uranus' equator plane by a few hundredths of a degree. Since a ring of colliding debris left to itself would spread in radius, rings with sharp edges require some confining mechanism. In the case of the outermost ring, ε, gravitational perturbations from two small neighboring satellites on opposite sides of the ring play a key role (Fig. 7). If the mass of the satellites dominate that of the ring, then radial spreading is significantly slowed because the spreading is now applied to the total mass of the system: ring plus satellites. The situation is analogous to a pair of runners standing back to back. The runners can separate rapidly if unopposed, but if each is forced to push an automobile ahead of him or her, they separate much more slowly. It is suspected that the other Uranian rings may also have so-called shepherding satellites, but because these objects have not been spotted yet, they must be smaller than *Voyager* and now Hubble Space Telescope limits of ~ 10 km.

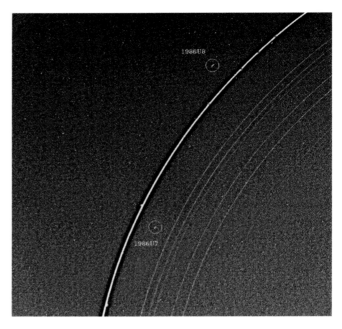

FIGURE 7 The outermost ε ring of Uranus, shepherded by the small satellites Cordelia (1986U7) and Ophelia (1986U8). The ε ring is noticeably brighter and wider than the other uranian rings. Heading inward, the first triplet of rings are δ, γ, and η; the next pair are β and α, and the final triplet (barely visible) are the 4, 5, and 6 rings. The satellites are smeared azimuthally by their orbital motion during the exposure. [*See* PLANET SATELLITES]

FIGURE 8 A comparison of *Voyager 2* images of the uranian rings taken looking away from the Sun (upper panel) and toward the Sun (lower panel). The latter geometry highlights rings composed of small dust grains. Short line segments in the lower panel are star trails; these attest to the long exposure time needed to highlight the faint dusty features. Note that not all rings features line up perfectly, implying eccentric orbits, particularly in the case of the ε ring (far left). Note that the narrow λ ring and many broad dusty features are visible only in the lower panel. The bright feature visible at the extreme right of the lower plot is the 1986 U2R ring.

HST observations of Uranus in 2003 discovered two new moons of this size; when the instrument is trained on Uranus during the upcoming Uranian ring plane crossing, moonlets as small as six km should be revealed. It will be interesting to see if these observations find that some of the missing shepherds are, in fact, loyally tending their flocks.

Returning to the ε ring, it is thought that some combination of internal self-gravity and interparticle collisions is probably responsible for maintaining the ring's eccentric shape (Fig. 8) and tilted aspect. These effects must be strong enough to enforce uniform precession, since the ring is observed to change its orientation in space as if it were a rigid body. Several other of the less massive and less optically thick uranian rings (e.g., δ and γ) are also tilted and eccentric, and the λ ring has an unexplained five-lobed azimuthal pattern. These mysteries are all waiting to be solved.

Two broad diffuse rings and two narrow denser ones encircle Neptune (Fig. 9). The outermost one, the Adams ring, contains the set of discrete, clustered, narrower- and denser-than-average arc segments for which Neptune has become famous (Fig. 10). The Adams ring is at least partially confined, both radially and azimuthally, by a single satellite Galatea. Other small satellites orbit in and amongst the Neptunian rings (Fig. 2) in a configuration that is somewhat reminiscent of the Jovian system.

Extensive sheets of icy powder, like fine snow, particularly conspicuous when backlit by the Sun, fill in the ring systems of Uranus (Fig. 8) and, possibly, Neptune (Fig. 9). These structures, though poorly understood, are probably similar to the more-extensively observed dusty rings of Jupiter and Saturn. There are significant differences though, as the optical depths vary by nearly a factor of a million from the extremely tenuous uranian R1 and R2 rings, through the not-so-faint jovian and saturnian dust sheets, to the more robust structures, like Galle and Lassell, located near Uranus and Neptune (Table 1). Hopefully these enigmatic structures will become better understood over the next several years.

3.3 Saturn

Finally, the rings of Saturn (Figs. 1 and 11), containing as much mass as the 200-km radius Saturnian satellite, Mimas, are home to almost all the ring phenomena described earlier and more: empty gaps in the rings whose widths vary with longitude (Fig. 12), narrow uranian-like rings (Fig. 13), ghostly time-variable radial markings called spokes (Figs. 14 and 15), spiral corrugations and density enhancements that tightly wind around the planet while slowly diminishing in amplitude (Fig. 16), and more. The ring system has now fallen under the sharp scrating of the *Cassini* spacecraft, and significant advances in the survey of its ring phemonology have been made as a result. Saturn's rings are the only ones whose composition is known with certainty: they are made predominantly of water ice, whereas rocky material seems most likely at Jupiter, and mixtures of ammonia

FIGURE 9 A long exposure of Neptune (on the right) and its ring system. The salt and pepper splotches are due to cosmic ray hits. Midway between the bright Le Verrier and the outermost Adams rings is the much fainter Arago ringlet and the broad Lassell ring extending inward to Le Verrier. The innermost ring, Galle, is also visible.

and methane ices coated with carbon are plausible constituents of the much darker rings of Uranus and Neptune. The main saturnian rings consist of the classical components seen from Earth: A, B, and C (Figs. 1 and 11). The narrow F ring (Fig. 13) immediately outside the main rings was discovered by *Pioneer* and has been the subject of intense investigation and speculation; the innermost D ring and the tenuous G ring were not clearly identified as rings until *Voyager* arrived in the system in 1980. Hidden from ground-based telescopes by its intrinsically low optical depth and the bright glare from nearby Saturn, the D ring has recently been revealed by *Cassini* to be extremely complex and dynamic (Fig. 17). Structures seen by *Voyager* 1980 are absent

FIGURE 10   The brightest two neptunian rings, Le Verrier (inner curve) and Adams (outer curve) are revealed in this *Voyager* image. Neptune is overexposed to lower left, indicating the difficulties faced in searching for faint features near planets. A short-exposure crescent-shaped Neptune has been overlaid to indicate the planet's true size and phase. Three of the famous ring arcs are visible in the outer Adams ring, while the Le Verrier ring has no such features.

FIGURE 11   A beautiful natural-color view of Saturn's rings from *Cassini*. From upper left are the dark C ring, with intricate substructure, the bright sandy-colored B ring, the dark Cassini division, and the grayish A ring. The narrow Encke gap and the narrow faint F ring are clearly visible, about equidistant from the A ring's outer edge. Saturn's rings are made primarily of water ice. Since pure water ice is white, the different colors in the rings probably reflect varying amounts of contamination by exogenic materials such as rock or carbon compounds.

in *Cassini* images today and vice versa. Strange periodicities near the C-ring boundary may hint at the cause of the dramatic drop in optical depth that occurs there. The very broad outer E ring, whose particle number density peaks at the orbit of Enceladus, appears to be produced from particles liberated from the satellite's interior by volcanic processes (Fig. 18). Its nature, and that of the G ring, has been delineated with increasing accuracy by Earth-based observations made during the ring-plane crossing events in 1995. *Cassini's* onboard dust detector finds that the E ring extends out nearly to the orbit of Titan, over 500,000 km beyond the outer visible boundary listed in Table 1.

4. Ring Processes

The fact that certain architectural details are common to all ring systems speaks of common physical processes operating within them. To date, only a subset of planetary ring features can be confidently explained. Here we break down the physical processes believed to be responsible for the creation of ring features into two categories: internal and external. Internal processes are present, to some extent, in all rings, while external processes arise when we consider the particular environments in which rings systems are located.

4.1 Dense Rings: Internal Processes

Dense rings with closely packed constituent particles are shaped strongly by collisions and self-gravity; in the denser parts of Saturn's rings, individual particles experience collisions hourly, upwards of 10 times per circuit of Saturn. Faint dusty rings, by contrast, are relatively unaffected by these processes; for example in Jupiter's outer gossamer rings, a dust grain might orbit the planet 10 million times (for 10,000 years) before experiencing a collision and the effects of self-gravity are similarly reduced. This subsection covers the physics that plays a role in the densest rings of Saturn, Uranus, and Neptune.

Two physical concepts underlie the internal workings of dense ring systems: the presence of a forced systematic change in orbital speeds across the rings (the so-called Kepler shear), and the dissipation of orbital energy that arises from the presence and the inelastic nature of

FIGURE 12 This *Cassini* image is a close up of the lit face of Saturn's A ring showing exquisite details in the Encke gap. Several faint narrow ringlets are visible; the brightest central one is coincident with the orbit of the tiny moon Pan. The wavy inner edge of the gap and the spiral structures wrapping inward are also caused by Pan. The waves on the inner gap edge lead Pan, while similar waves on the outer gap edge (not seen) trail it.

collisions among ring particles. Collisions between particles, which occur regularly due to differential orbital speeds, force random motions amongst the particles. These random motions can also be diminished in collisions, as energy is lost to the chipping, cracking, compaction, and sound propagation through the particles. A balance is struck, with the details determined by the number of collisions forced by Kepler shear and the inexorable loss of energy during these and subsequent inelastic collisions. Collisional processes can also alter ring particle sizes and shapes, resulting in the erosion and smoothing of surfaces in some cases and the accretion or sticking of particles in others.

Significant progress has been made in the theory of dense rings by treating the rings as fluids; this prescription is called kinetic theory. Kinetic theory shows that collisional equilibrium is achieved after several orbital periods and yields a monotonically decreasing relation between the particle random velocities, $\sim v$, and the overall optical depth, τ. That is, in steady state each ring region is characterized by a particular optical depth (or surface mass density Σ) and has a typical value for the random velocities of its constituent particles. At low τ, the random velocities and ring thickness tend to be larger, while at high τ the reverse is true. The details of the equilibrium depend on the kinematic viscosity, v of the ring particles, a quantity that measures the tendency for a fluid to resist shear flow. Like the coefficients of friction for sliding and rolling bodies (e.g., sleds and cars), v must usually be empirically determined.

In a disk system of colliding particles following Kepler orbits, the faster particles are on the inside, and so collisions naturally transfer angular momentum outward across the disk. Kinetic theory shows that the rate of flow is related to the product Σv, which is crudely the number of collisions times the effect of a single collision. Thus narrow rings must spread in time, unless another process prevents them from doing so. This conclusion can also be reached by realizing that a narrow ring has more orbital energy than a broad ring with the same mass and angular momentum. Since collisions always deplete orbital energy and do not affect the total angular momentum, all rings are inexorably driven to spread toward the lower energy state.

FIGURE 13   This *Cassini* image reveals details of Saturn's mysterious F ring that lies just outside the A ring (lower right). The bright core of the F ring stands out crisply, embedded in parallel bands of fainter material. A wispy, ribbon-like feature points accusingly at the inner shepherding moonlet, Prometheus, whose eccentric orbit brings it near enough to the ring to strip some material away.

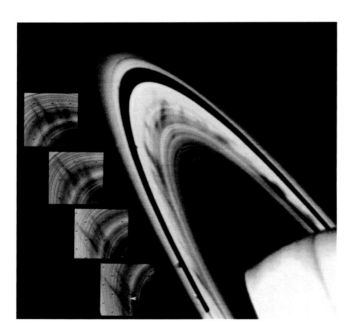

FIGURE 14   A *Voyager* image of dark spokes seen against Saturn's sunlit B ring. Small dust particles appear dark under this lighting condition, hinting at the still poorly understood physical processes behind spoke creation. The inset panels show the change of a given feature with time.

Distinctly different ring regions can exist in near-equilibrium (but the entire ring will still spread) if they have similar values of $\Sigma \nu$. Basic kinetic theory, for example, predicts that two ring regions with different values of the optical depth might have the same $\Sigma \nu$, allowing distinctly different contiguous ring regions to potentially coexist in equilibrium. This mechanism was regarded as a possibility for explaining the large degree and variety of ring structure in Saturn's B ring (Fig. 11) until laboratory measurements on ice particles indicated that the particles were stickier than expected. This implied that the rings were less extended vertically and the particle number densities larger than originally believed, so much so that the precepts of simple kinetic theory were violated.

When it was recognized that very dense rings with highly inelastic collisions violate the principles of kinetic theory on which much of ring theory was based, it became necessary to introduce a new effect into the theory: the transport of angular momentum (via sound waves) across a tightly packed system of orbiting particles. The result of adding this effect, which becomes important in high τ regions, was to change the dependence of $\Sigma \nu$ on τ to a monotonically increasing function. On the basis of this conclusion, it seemed impossible for ring regions of differing optical depth, and therefore

FIGURE 15   A three-panel *Cassini* image of bright spokes seen against the dark side of Saturn's B ring. Small dust particles appear bright under this lighting condition. The motion of the spokes can be seen clearly by comparing the three panels.

differing natural angular momentum flow (proportional to $\Sigma\nu$), to exist stably side by side.

Other possibilities, however, have been suggested to explain the fine-scale structure within a dense ring like the saturnian B ring. These suggestions include adjacent narrow ring regions alternating in behavior between a liquid and a solid and the possibility that density waves may be driven to the point of instability in very dense ring regions. A sea of embedded bodies too small to open gaps and too faint to be noticed by spacecraft could control much of the structure; Pan (responsible for the Encke Gap) and Daphnis (the newly-discovered Keeler-Gap moonlet), may be just the tip of this particular iceberg (Fig. 19). Additional evidence for tiny embedded moonlets comes from particles

FIGURE 16   This is a narrow-angle *Cassini* image of the dark side of Saturn's A ring. Amazing detail of the Prometheus 12:11 density wave in the lower left part of the image and the Mimas 5:3 bending wave to the upper right are apparent. These features wind around Saturn literally dozens of times before fading into invisibility.

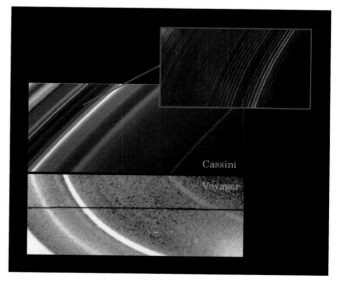

FIGURE 17   A comparison of *Voyager* and *Cassini* images of Saturn's inner D ring (see Table 1). Some differences in the two images are apparent; the brightest *Voyager* ring appears to have shifted inward in the new *Cassini* data. The regular pattern shown in the inset appears just inward of the C ring (bright upper and lower left corners). It has been suggested that this feature is due to collision of a meteoroid into a C- or D-ring parent body just 20 years ago.

FIGURE 18   This dramatic *Cassini* image of Saturn's icy satellite Enceladus shows tiny particles ejected violently from near the satellite's south pole. These icy grains are destined to join Saturn's diffuse outer E ring.

organized into theoretically-predicted "propeller" shapes, which are beginning to be found in *Cassini* images.

Saturn's outer A ring also exhibits a strange so-called quadrupole asymmetry that manifests itself as alternating 90-degree swaths of brighter and darker regions. This asymmetry has been seen optically and with Earth-based radar, and is best explained by narrow wakes in the ring, oriented obliquely at a given angle. These wakes are thought to be caused by the gravitational clumping of ring particles into temporary agglomerations, as might be expected of material near the edge of the Roche zone (see Fig. 2).

Our understanding of very dense rings is far from complete, and we must regard the bulk of the exquisite structure in Saturn's main ring system as mostly unexplained. New data, mysteries, and ideas have started to emerge from the *Cassini* mission though, so in time new insights and explanations will follow.

4.2 External Causes of Ring Structure

All planetary rings interact with their local environment via long-range forces, and they are also subject to incident

FIGURE 19   A *Cassini* image of the Saturn's outer A ring. The Encke gap slashes a diagonal through the center of the frame and the narrower Keeler gap is also visible at upper right. Both features arise from the action of embedded moons; Pan (centered in the images) opens the Encke gap while Daphnis (not visible) is the cause of the Keeler gap. Many of the bright lines running across the image are resonant features forced by external satellites.

mass fluxes from interplanetary debris. External gravitational forces from other satellites and the nonspherical shape of the planet itself can imprint wavelike signatures in dense planetary rings. Faint dusty rings are also subject to solar radiation pressure, electromagnetic interactions, and different kinds of drag forces. Finally, an external flux of interplanetary debris strikes satellites embedded in rings as well as larger ring particles, cratering their surfaces and ejecting large amounts of additional ring material. This incident debris can also color, chip, erode, and catastrophically fragment ring particles.

4.2.1 EXTERNAL GRAVITATIONAL FORCES

All rings in the Solar System circle planets that are somewhat flattened due to their rapid spin rates. An extra gravitational perturbation arises from this planetary oblateness and slightly adjusts a ring particle's oscillation frequencies in the radial, vertical, and azimuthal directions. The main outcome is orbital precession, which causes tilted and/or elliptical orbits to slowly shift their spatial orientations. More dramatic effects occur for time-variable gravitational forces such as those arising from orbiting satellites and, potentially, a spinning lumpy planet. The perturbations are concentrated at discrete orbital locations known as resonances, where a frequency of external forcing matches a natural orbital frequency of the system. Some forcing frequencies match a natural radial frequency and affect the ring's surface density in a systematic way; others match a natural vertical frequency and lead to warped corrugations in the ring. In both cases, resonances enable the external perturber to exchange energy and angular momentum with particular locations in the ring.

Operating over sufficiently long time scales, satellites can create a staggering variety of features in planetary rings. The degree to which external perturbations on ring particle orbits will create visible disturbances in a broad featureless disk system depends on the ring's natural ability to keep up with the rate of change in angular momentum imposed on it by the external perturbation. If the angular momentum is removed or deposited by external means at a rate that is less than the ring's ability to transport it away from the excitation region (proportional to $\Sigma \nu$), then the ring response will take the form of a wave. If the rate of removal or deposition is greater, however, then the rings will respond by opening a gap, i.e., the particles themselves must physically move, carrying angular momentum with them, to accommodate the external driving force.

The satellite Mimas is responsible for the strongest resonances within Saturn's rings; it causes the Cassini Division, the 4700-km gap between the A and B rings (Fig. 1). Two smaller but closer moons, Janus and Epimetheus, cause the sharp outer edge of the A ring. Detailed inspection of these ring edges by *Voyager* and *Cassini* reveal two- and seven-lobed patterns of radial oscillations, signatures of the

specific resonances responsible, but *Cassini* has found significant and complex deviations from these simple patterns. These two dense rings contain many additional examples of features caused by external perturbations of satellites. For example, the 320-km-wide Encke gap in the outer A ring (Fig. 12) is believed to be maintained against collisional diffusion by the gravitational perturbations of the 20-km-diameter satellite, Pan, orbiting within it; radial oscillations of characteristic azimuthal wavelength ~0.7° seen along the edges of this gap are also attributable to this small satellite. Density and bending waves are seen throughout the rings—these are radial and vertical disturbances that wrap around the planet multiple times on tightly wound spirals (Fig. 16). Such waves are created by gravitational resonances too weak to open gaps; features due to Mimas, Janus, Epimetheus, Pandora and Prometheus have been known since the *Voyager* flybys. *Cassini* has identified numerous additional examples, including ones due to tiny Atlas and Pan (Fig. 19). With few exceptions, the best understood features in Saturn's main rings are due to gravitational resonances.

There has been some success at linking narrow rings to nearby shepherding satellites. At Uranus, it is clear that the particles within the ε ring are shepherded in their movement around the planet by the gravitational perturbations of two small satellites on either side of it, Cordelia and Ophelia. At Saturn, the F ring (Fig. 13) is flanked by two small satellites, although the larger and more massive of the two is closer to the ring, in contrast to expectations. And the action of a single satellite, Galatea, may confine Nepune's Adams ring and its intriguing arcs (Fig. 10). A resonance with Galatea forces a coherent 30-km amplitude radial distortion to travel through the arcs at the orbital speed of the satellite. This particular resonance also seems capable of confining the arcs both in radius and azimuth—one satellite doing double duty—although it alone may not be sufficient to explain the observed configuration of arcs. Small, kilometer-sized bodies embedded within the ring or arcs might assist Galatea in arc confinement as well as slow the rapid retreat of the arcs from the satellite. Unfortunately, satellites of this size are well below the detection limit in Voyager images. If smaller satellites are discovered in close proximity to this or other narrow features, then dense narrow rings may be, fundamentally, not very different from Saturn's dense broad rings. If, however, the uranian and neptunian rings maintain their narrowness in some other way, then their internal dynamics, like their appearances, may be quite distinct from their broad saturnian cousins.

4.2.2 RADIATION AND ELECTROMAGNETIC FORCES

Small dust grains accumulate electric charges in planetary magnetospheres by running into trapped electrons and ions and by interacting with solar photons. These grains can be affected by electromagnetic forces that arise from their motion relative to the spinning magnetic field of the host planet. Additionally, the absorption, reemission, and scattering of solar photons by dust grains impart small momentum kicks to orbiting material that can, over long enough times, cause significant orbital changes. These are the two dominant nongravitational forces active in ring systems. Additionally, much weaker drag forces arising from the physical interaction of dust grains with photons, orbiting ions and atoms, and other smaller dust grains cause orbits to slowly spiral into the planet or, in some cases, to slowly drift away from it. All of these nongravitational forces, acting in concert with gravitational ones, cause long-period eccentricity and, to a lesser extent, inclination oscillations in faint dusty rings where collisions are rare. These effects are seen most clearly in Saturn's E ring, whose icy particles are thought to be ejected from newly discovered volcanic vents on the satellite Enceladus (Fig. 18). Despite this single source, the perturbation forces spread ring material hundreds of thousands of kilometers inward and outward to form a broad, relatively flat, and nearly featureless structure known as the E ring, the largest ring in the Solar System (Fig. 2).

Jupiter's magnetic field is ten times stronger than that of any other planet, and so it is no surprise that its dusty ring components are all strongly affected by electromagnetic processes. Because the magnetic field is also asymmetric (unlike Saturn's), electromagnetic resonances analogous to satellite gravitational resonances discussed above are active at particular locations in Jupiter's ring. For example, as discussed previously, ring particles are created by impacts into the four small satellites that populate the inner jovian system, and these grains subsequently evolve inward. A pair of electromagnetic resonances await the evolving grains, acting as sentinels guarding the approach to the King of the Planets. The first, at the inner edge of the main ring, imparts inclinations to the ring particles and creates the vertically extended jovian halo (Fig. 4). The second imposes still higher inclinations at the inner edge of the visible halo.

Other dusty rings at Uranus and Neptune may behave similarly; the upcoming 2007 uranian ring plane crossing will provide an excellent opportunity to search for faint vertically extended structures.

4.2.3 EXTERNAL MASS FLUXES

Yet another possibility for externally influencing ring structure arises from the redistribution of mass and angular momentum caused by meteoroid bombardment of the rings. Saturn's rings present a large surface area—twice that of the planet itself—to the hail storm of interplanetary debris raining down on them. The total mass falling onto the rings over billions of years may be greater than the mass of the rings themselves; this process is therefore likely to be a major contributor to ring erosion and modification.

Numerical simulations of the process indicate that sand-blasted ring particles should drift inward by up to several centimeters per year. This rate depends sensitively on the amount of material impacting the rings, a quantity that is presently poorly constrained. Potentially, though, the entire C ring of Saturn could decay into the planet in $\sim 10^8$ years. Because the ejecta from each impact is distributed preferentially in one direction, meteoroid bombardment provides a mechanism for altering radial structure. This is especially true when the initial radial distribution of mass is grossly non-uniform, such as near an abrupt and large change in optical depth. The shapes of the inner edges of the A and B rings and features near them can be explained roughly by this process and may take as little as $\sim 10^7$ to 10^8 years to evolve to their currently observed configurations. These results hint that other structural features in ring systems may also be explainable by this process.

The impacts of micrometeoroids onto Saturn's rings have also been proposed as the first step in the production of spokes, those ghostly patchy features in the B ring that come and go while revolving around Saturn (Figs. 14 and 15). Spokes are almost certainly powder-sized ice debris that have been lifted off bigger ring particles; the elevation mechanism is believed to involve electromagnetic forces acting on charged dust grains. Details of spoke formation and evolution depend on Saturn's orbital period, a fact that strongly indicates the importance of electromagnetic interactions between the dust and the planet's magnetic field.

5. Ring Origins

Three distinct scenarios have been suggested for the origin of rings: (1) rings may be the inner unaccreted remnants of the circumplanetary nebulae that ultimately formed the satellite systems surrounding each planet; (2) they may be the remnant debris from satellites that have tidally evolved inward toward the Roche zone, were completely disrupted by cometary or meteoroid impacts, and then spread quickly into a ring system, replete with small embedded satellites; or (3) they may be the result of the disruption of an icy planetesimal in heliocentric orbit that strayed too close to the planet, was torn apart by planetary tides, and subsequently evolved into a ring/satellite system. We discuss the pros and cons of each of these possibilities in turn.

Saturn's main rings, far more massive than all other ring systems put together, would appear to have the best chance of being primordial. Several lines of circumstantial evidence, however, indicate that this may not be so. First, the presence of the large moonlets Pan and Daphnis in the Encke and Keeler gaps shows that a certain amount of accretion would have to have occurred in a primordial disk. Why would the larger of these moons be closer to the planet where tidal forces limiting accretion are stronger? Additional evidence against primordial rings rests on the calculation of the rate of separation expected in the orbits of satellites and ring particles locked in gravitational resonance, e.g., the predicted recession of the small ring shepherds from the A ring due to their resonant interactions with ring particles. Simple inverse extrapolation of these rates brings the nearest of these satellites to the edge of the rings roughly 10^7 years ago. Estimates for the lifetime of all rings against erosion and darkening by micrometeoroid impacts yield similar time scales. On the basis of these arguments, ancient, and certainly unchanging, ring systems seem unlikely. Certain aspects of these theoretical models, however, are extremely uncertain and additional, as yet unidentified, processes may also be active. Thus, arguments both for or against ancient, but ever-changing, rings are still inconclusive.

The second possibility is somewhat more appealing at first glance. The large number of satellites presently orbiting each of the giant planets, and the ever-increasing discoveries of icy planetesimals found in the Kuiper Belt (a suspected source of planet-crossing bodies), indicate sufficient fodder for ring creation. The interpretation of the crater populations on the surfaces of outer Solar System satellites suggests that satellite disruption must have been a common event in the past. [See KUIPER BELT.] Jupiter's ring cleanly fits the second scenario, as the ring components are far less massive than the embedded satellites and, as far as we know, all structures are consistent with debris launched from these four objects. The individual particles in dusty rings, in general, have ages of well under a million years, as a variety of processes remove dust grains on these or appreciably faster timescales. Thus they must be replenished from known or unseen sources. At Uranus and Neptune there is also sufficient mass, even today, in ring-region satellites to create the present ring systems. But the possibility of creating Saturn's massive ring system in the recent past from satellite disruption is rather low, as Mimas-sized bodies near the Roche zone are nonexistent now and probably were rare in the past.

Finally, the fate of Comet Shoemaker-Levy 9, captured by Jupiter and torn into a long train of fragments, led to renewed interest in the idea of a ruptured planetesimal origin for rings. This is the weakest of the three scenarios, as it is expected that most of the debris from such an event would escape the planet or evolve to collide with it or its larger satellites before mutual collisions amongst the debris itself could damp the system down to a flat circular ring. Furthermore, the frequency with which large icy planetesimals pass near planetary cloudtops is too low to make tidal disruption a plausible scenario. Thus youthful rings appear more likely at Jupiter, Uranus, and Neptune, while the origin of Saturn's massive ring system remains an unsolved mystery.

6. Prospects for the Future

Further improvements in ground-based observing facilities and instrumentation can be expected in the future, but the most spectacular advances in the study of rings will certainly come when the vast quantity of data returning from the *Cassini* spacecraft is fully digested. New saturnian satellites well below the *Voyager* detection limit ($r \sim 6$ km), both internal (Daphnis) and external (Methone, Pallene) to the rings have already been detected (Fig. 2). High-resolution maps of the rings' composition and radial structure, and detailed studies of time-variable features are currently being undertaken. The figures in this chapter highlight some of the exciting first discoveries.

The ring systems of today offer invaluable insights into the processes operating in primordial times in the flattened circumsolar disk that ultimately formed the solar system. Yet almost all the results on the internal workings of Saturn's rings that will come from *Cassini*—the collisional frequency and elasticity of ring particles, the kinematic viscosity, and self-gravity—will be made on the basis of inference, as direct imaging of ring particles and their interactions will be impossible from the trajectory that *Cassini* will follow through the Saturn system. [See THE ORIGIN OF THE SOLAR SYSTEM.]

For this reason, it is likely that in the not-too-distant future we will dispatch, to follow in the wake of *Cassini*, small spacecraft capable of hovering over the rings of Saturn or orbiting within one of the large ring gaps. Views of the rings from these unique vantage points will capture individual ring particles—large and small—in the act of colliding, chipping, breaking, and coalescing. Observations like these will give planetary scientists an unprecedented opportunity to view details of these key processes that were probably also active in the solar nebula disk from which our solar system formed.

To follow up on our initial exploration of the outer solar system, orbiter missions to Uranus and Neptune are sorely needed. These missions, currently in the early planning stages, will raise our knowledge of distant ring systems up to the level of those of Jupiter and Saturn and allow meaningful comparisons to be made. Why does Saturn alone have a massive resplendent ring system? What new rings await discovery at Uranus and Neptune? Closely monitoring the timeless ballet danced by planetary rings and their satellite companions will ultimately reveal the underlying music to which they move. Perhaps one day in the far future, a cometary impact may rip a small satellite of Uranus or Neptune asunder, wreathing one or the other of the blue planets in a beautiful broad ring system to rival Saturn's.

In the next few decades, entirely new ring systems are likely to be detected around extrasolar giant planets; these will almost certainly show new forms and provide new hints about the dynamical forces that shape these elegant structures. Future generations of planetary ring enthusiasts will have much to look forward to and can expect many further surprises.

Bibliography

The Planetary Rings Node, administered by NASA's Planetary Data System, has a wealth of information, images, and movies at http://pds-rings.seti.org/.

The Cassini Imaging Central Laboratory for Operations (CICLOPS) hosts the website of the Cassini Imaging Team, http://ciclops.org, the source of all high resolution images returned by Cassini on Saturn's rings.

Burns, J.A. (1999). "Planetary Rings." In "The New Solar System," (J.K. Beatty and A. Chaikin, Eds.), 4th Ed., Sky Publishing Corporation and Cambridge University Press, Cambridge, MA, pp. 221–240.

Burns, J.A., Hamilton, D.P., and Showalter, M.R. (2001). "Dusty Rings and Circumplanetary Dust." In "Interplanetary Dust," (E.Grün, B.A.S. Gustafson, S.F. Dermott and H. Fechtig, Eds.), Springer Verlag, Berlin, pp. 641–725.

Burns, J.A., Hamilton, D.P., and Showalter, M.R. (2002). "Bejeweled Worlds" *Scientific American*, February issue, pp. 66–73.

Burns, J.A., Simonelli, D.P., Showalter, M.R., Hamilton, D.P., Esposito, L.W., Porco, C.C., and Throop, H. 2003. "Jupiter's Ring-Moon System." In "Jupiter, the Planet, Satellites, and Magnetosphere," (F. Bagenal, T. Dowling, and W.B. McKinnon, Eds.), Cambridge Planetary Science Series, pp. 241–262.

Esposito, L.W. (2006). "Planetary Rings," Cambridge University Press, Cambridge, U.K.

French, R.G., Nicholson, P.D., Porco, C.C., and Marouf, E.A. (1991). "Dynamic and Structure of the Uranian Rings." In "Uranus" (J.T. Bergstrahl, E.D. Miner, and M.S. Matthews, Eds.), University of Arizona Press, Tucson, pp. 327–409.

Porco, C.C., Nicholson, P.D., Cuzzi, J.N., Lissauer, J.J., and Esposito, L.W. (1995). "Neptune and Triton". In "Neptune and Triton" (D. Cruikshank, M.S. Matthews, and A.M. Schumann Eds.). Univ. of Arizona Press, Tucson, pp. 703–804.

Porco, C.C., et al. (2003). Cassini Imaging of Jupiter's Atmosphere, Satellites, and Rings. *Science* 299, 1541–1547.

Porco, C.C., et al. (2005). Cassini Imaging Science: Initial Results on Saturn's Rings and Small Satellites. *Science* 307, 1226–1236.

Planetary Magnetospheres

Margaret Galland Kivelson

University of California
Los Angeles, California

Fran Bagenal

University of Colorado, Boulder
Boulder, Colorado

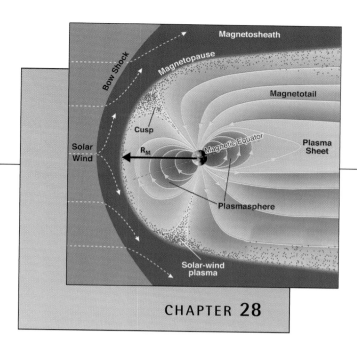

CHAPTER **28**

1. What is a Magnetosphere?

The term **magnetosphere** was coined by T. Gold in 1959 to describe the region above the **ionosphere** in which the magnetic field of the Earth controls the motions of charged particles. The magnetic field traps low-energy plasma and forms the Van Allen belts, torus-shaped regions in which high-energy ions and electrons (tens of keV and higher) drift around the Earth. The control of charged particles by the planetary magnetic field extends many Earth radii into space but finally terminates near 10 Earth radii in the direction toward the Sun. At this distance, the magnetosphere is confined by a low-density, magnetized plasma called the **solar wind** that flows radially outward from the Sun at supersonic speeds. Qualitatively, a planetary magnetosphere is the volume of space from which the solar wind is excluded by a planet's magnetic field. (A schematic illustration of the terrestrial magnetosphere is given in Fig. 1, which shows how the solar wind is diverted around the magnetopause, a surface that surrounds the volume containing the Earth, its distorted magnetic field, and the plasma trapped within that field.) This qualitative definition is far from precise. Most of the time, solar wind plasma is not totally excluded from the region that we call the magnetosphere. Some solar wind plasma finds its way in and indeed many important dynamical phenomena give clear evidence of intermittent direct links between the solar wind and the plasmas governed by a

planet's magnetic field. Moreover, unmagnetized planets in the flowing solar wind carve out cavities whose properties are sufficiently similar to those of true magnetospheres to allow us to include them in this discussion. Moons embedded in the flowing plasma of a planetary magnetosphere create interaction regions resembling those that surround unmagnetized planets. If a moon is sufficiently strongly magnetized, it may carve out a true magnetosphere completely contained within the magnetosphere of the planet.

Magnetospheric phenomena are of both theoretical and phenomenological interest. Theory has benefited from the data collected in the vast plasma laboratory of space in which different planetary environments provide the analogue of different laboratory conditions. Furthermore, magnetospheric plasma interactions are important to diverse elements of planetary science. For example, plasma trapped in a planetary magnetic field can interact strongly with the planet's atmosphere, heating the upper layers, generating neutral winds, ionizing the neutral gases and affecting the ionospheric flow. Energetic ions and electrons that precipitate into the atmosphere can modify atmospheric chemistry. Interaction with plasma particles can contribute to the isotopic fractionation of a planetary atmosphere over the lifetime of a planet. Impacts of energetic charged particles on the surfaces of planets and moons can modify surface properties, changing their albedos and spectral properties. The motions of charged dust grains in a planet's environment

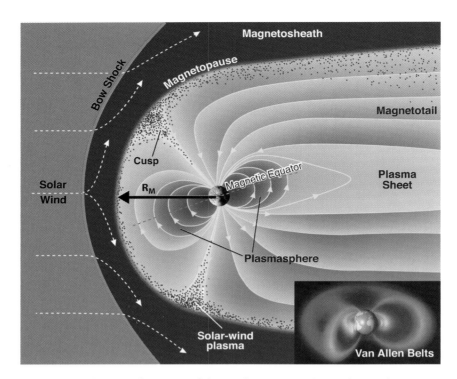

FIGURE 1 Schematic illustration of the Earth's magnetosphere. The Earth's magnetic field lines are shown as modified by the interaction with the solar wind. The solar wind, whose flow speed exceeds the speeds at which perturbations of the field and the plasma flow directions can propagate in the plasma, is incident from the left. The pressure exerted by the Earth's magnetic field excludes the solar wind. The boundary of the magnetospheric cavity is called the magnetopause, its nose distance being R_M. Sunward (upstream) of the magnetopause, a standing bow shock slows the incident flow, and the perturbed solar wind plasma between the bow shock and the magnetopause is called the magnetosheath. Antisunward (downstream) of the Earth, the magnetic field lines stretch out to form the magnetotail. In the northern portion of the magnetotail, field lines point generally sunward, while in the southern portion, the orientation reverses. These regions are referred to as the northern and southern lobes, and they are separated by a sheet of electrical current flowing generally dawn to dusk across the near-equatorial magnetotail in the plasmasheet. Low-energy plasma diffusing up from the ionosphere is found close to Earth in a region called the plasmasphere whose boundary is the plasmapause. The dots show the entry of magnetosheath plasma that originated in the solar wind into the magnetosphere, particularly in the polar cusp regions. Inset is a diagram showing the 3-dimensional structure of the Van Allen belts of energetic particles that are trapped in the magnetic field and drift around the Earth. [the New Solar System, (eds. Kelly Beatty et al.), CUP/Sky Publishing] Credit: Steve Bartlett; Inset: Don Davis.

are subject to both electrodynamic and gravitational forces; recent studies of dusty plasmas show that the former may be critical in determining the role and behavior of dust in the solar nebula as well as in the present-day solar system.

In Section 2, the different types of magnetospheres and related interaction regions are introduced. Section 3 presents the properties of observed planetary magnetic fields and discusses the mechanisms that produce such fields. Section 4 reviews the properties of plasmas contained within magnetospheres, describing their distribution, their sources, and some of the currents that they carry. Section 5 covers magnetospheric dynamics, both steady and "stormy."

Section 6 addresses the interactions of moons with planetary plasmas. Section 7 concludes the chapter with remarks on plans for future space exploration.

2. Types of Magnetospheres

2.1 The Heliosphere

The solar system is dominated by the Sun, which forms its own magnetosphere referred to as the **heliosphere**. [*See* THE SUN.] The size and structure of the heliosphere are governed by the motion of the Sun relative to the local interstellar medium, the density of the interstellar plasma, and the pressure exerted on its surroundings by the outflowing solar wind that originates in the solar corona. [See THE SOLAR WIND.] The corona is a highly ionized gas, so hot that it can escape the Sun's immense gravitational field and flow outward at supersonic speeds. Through much of the heliosphere, the solar wind speed is not only supersonic but also much greater than the **Alfvén speed** ($v_A = B/(\mu_0\rho)^{1/2}$), the speed at which rotational perturbations of the magnetic field propagate along the magnetic field in a magnetized plasma. (Here B is the magnetic field magnitude, μ_0 is the magnetic permeability of vacuum, and ρ is the mass density of the plasma.)

The solar wind is threaded by magnetic field lines that map back to the Sun. A useful and picturesque description of the field contained within a plasma relies on the idea that if the conductivity of a plasma is sufficiently large, the magnetic field is frozen into the plasma and field lines can be traced from their source by following the motion of the plasma to which it is frozen. Because the roots of the field lines remain linked to the rotating sun (the sun rotates about its axis with a period of approximately 25 days), the field lines twist in the form of an Archimedean spiral as illustrated in Fig. 2. The outflow of the solar wind flow along the direction of the Sun's motion relative to the interstellar plasma is terminated by the forces exerted by the interstellar plasma. Elsewhere the flow is diverted within the boundary of the heliosphere. Thus, the Sun and the solar wind are (largely) confined within the heliospheric cavity; the heliosphere is the biggest of the solar system magnetospheres.

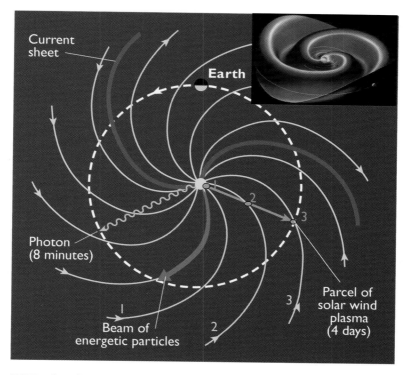

FIGURE 2   The magnetic field of the Sun is carried by the solar wind away from the Sun and is wound into a spiral. The heliospheric current sheet (colored magenta in the inset 3-dimensional diagram) separates magnetic fields of opposite polarities and is warped into a "ballerina skirt" by combined effects of the Sun's spin and the tilt of the magnetic field. The main diagram (2-dimensional projection) shows a cut through the heliosphere in the ecliptic plane. In the ecliptic plane, the radial flow of the solar wind and the rotation of the Sun combine to wind the solar magnetic field (yellow lines) into a spiral. A parcel of solar wind plasma (traveling radially at an average speed of 400 km/s) takes about 4 days to travel from the Sun to Earth's orbit at 1 AU. The dots and magnetic field lines labeled 1, 2, and 3 represent snapshots during this journey. Energetic particles emitted from the Sun travel much faster (beamed along the magnetic field) reaching the Earth in minutes to hours. Traveling at the speed of light, solar photons reach the Earth in 8 minutes. Credit: J. A. Van Allen and F. Bagenal, 1999, Planetary magnetospheres and the interplanetary medium, in "The New Solar System," 4th Ed. (Beatty, Petersen, and Chaikin, eds.), Sky Publishing and Cambridge Univ. Press.

Our knowledge of the heliosphere beyond the orbits of the giant planets was for decades principally theoretical, but data acquired by *Voyager 1* and *2* since their last planetary encounters in 1989 have provided important evidence of the structure of the outer heliosphere. The solar wind density continues to decrease as the inverse square of the distance from the Sun; as the plasma becomes sufficiently tenuous, the pressure of the interstellar plasma impedes its further expansion. The solar wind slows down abruptly across a shock (referred to as the termination shock) before reaching the **heliopause,** the boundary that separates the solar wind from the interstellar plasma. (The different plasma regimes are schematically illustrated in Fig. 3.)

Voyager 1 encountered the termination shock on December 16, 2004, at a distance of 94 AU (AU is an astronomical unit, equal to the mean radius of Earth's orbit or about 1.5×10^8 km) from the Sun and entered the heliosheath, the boundary layer between the termination shock and the heliopause. The encounter with the termination shock had long been anticipated as an opportunity to identify the processes that accelerate a distinct class of cosmic rays, referred to as anomalous cosmic rays (ACRs). ACRs are extremely energetic, singly charged ions (energies of the order of 10 MeV/nucleon) produced by ionization of interstellar neutrals. The mechanism that accelerates them to high energy is not established. Some models propose that these particles are ionized and accelerated near the termination shock, but the *Voyager* data show no sign of a change in the energy spectrum or the intensity of the flux across the termination shock; thus, the acceleration mechanism remains a mystery.

Various sorts of electromagnetic waves and plasma waves have been interpreted as coming from the termination shock or the heliopause. Bursts of radio emissions that do not weaken with distance from known sources within the solar systems were observed intermittently by *Voyager* between 1983 and 2004. They are thought to be emissions generated when an interplanetary shock propagating outward from the Sun reaches the heliopause. Plasma waves driven by electron beams generated at the termination

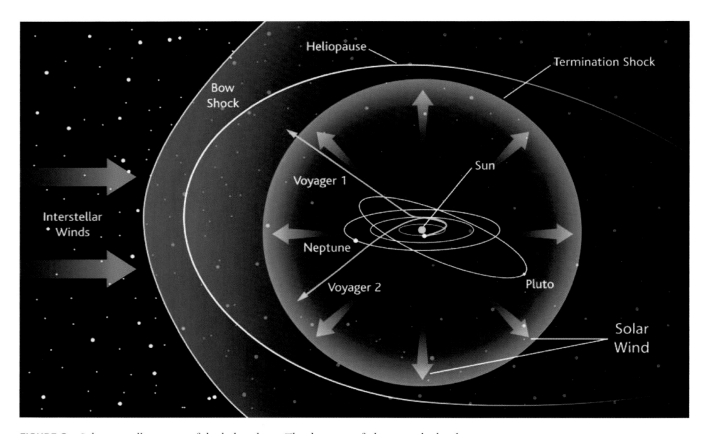

FIGURE 3 Schematic illustration of the heliosphere. The direction of plasma in the local interstellar medium relative to the Sun is indicated, and the boundary between solar wind plasma and interstellar plasma is identified as the heliopause. A broad internal shock, referred to as the termination shock, is shown within the heliopause. Such a shock, needed to slow the outflow of the supersonic solar wind inside of the heliopause, is a new feature in this type of magnetosphere. Beyond the heliopause, the interstellar flow is diverted around the heliosphere and a shock that slows and diverts flow probably exists. Credit: L. A. Fisk, 2005, Journey into the unknown beyond, *Science* **2016** (September 23), 309, www.sciencemag.org.

TABLE 1		Properties of the Solar Wind and Scales of Planetary Magnetospheres								
	Mercury	**Venus**	**Earth**	**Mars**	**Jupiter**	**Saturn**	**Uranus**	**Neptune**	**Pluto**
Distance, a_{planet} (AU)^a	0.31–0.47	0.723	1^b	1.524	5.2	9.5	19	30	30–50
Solar wind density (amu cm^{-3})^b	35–80	16	8	3.5	0.3	0.1	0.02	0.008	0.008–0.003
Radius, R_P (km)	2,439	6,051	6,373	3,390	71,398	60,330	25,559	24,764	1,170 ($\pm$33)
Surface magnetic field, B_0 (Gauss = 10^{-4} T)	3×10^{-3}	$<2 \times 10^{-5}$	0.31	$<10^{-4}$	4.28	0.22	0.23	0.14	?
R_{MP} (R_{Planet})	1.4–1.6 R_M	—	10 R_E	—	42 R_J	19 R_S	25 R_U	24 R_N	
Observed size of magnetosphere (km)	1.4 R_M 3.6×10^3	—	8–12 R_E 7×10^4	—	50–100 R_J 7×10^6	16–22 R_S 1×10^6	18 R_U 5×10^5	23–26 R_N 6×10^5	?

^{a} 1 AU = 1.5×10^8 km.

^{b} The density of the solar wind fluctuates by about a factor of 5 about typical values of $\rho_{\text{sw}} \sim [(8 \text{ amu cm}^{-3})/a^2_{\text{planet}}]$.

^{c} Magnetopause nose distance, R_{MP} is calculated using $R_{\text{MP}} = (B_0^2/2\mu_0\rho u^2)^{1/6}$ for typical solar wind conditions of ρ_{sw} given above and $u \sim 400$ km s^{-1}. For outer planet magnetospheres, this is usually an underestimate of the actual distance.

shock and propagating inward along the spiral field lines of the solar wind have also been identified. As *Voyager* continues its journey out of the solar system, it should encounter the heliopause and enter the shocked interstellar plasma beyond. One can predict that new surprises await discovery.

2.2 Magnetospheres of the Unmagnetized Planets

Earth has a planetary magnetic field that has long been used as a guide by such travelers as scouts and sea voyagers. However, not all of the planets are magnetized. Table 1 summarizes some key properties of some of the planets including their surface magnetic field strengths. The planetary magnetic field of Mars is extremely small, and the planetary magnetic field of Venus is nonexistent. [*See* MARS and VENUS: SURFACE AND INTERIOR.] The nature of the interaction between an unmagnetized planet and the supersonic solar wind is determined principally by the electrical conductivity of the body. If conducting paths exist across the planet's interior or ionosphere, then electric currents flow through the body and into the solar wind where they create forces that slow and divert the incident flow. The diverted solar wind flows around a region that is similar to a planetary magnetosphere. Mars and Venus have ionospheres that provide the required conducting paths The barrier that separates planetary plasma from solar wind plasma is referred to as an **ionopause.** The analogous boundary of the magnetosphere of a magnetized planet is called a magnetopause. Earth's Moon, with no ionosphere and a very low conductivity surface, does not deflect the bulk of the solar wind incident on it. Instead, the solar wind runs directly into the surface, where it is absorbed. [*See* THE MOON.] The absorption leaves the region immediately downstream of the Moon in the flowing plasma (the wake) devoid of plasma, but the void fills in as solar wind plasma flows toward the

center of the wake. The different types of interaction are illustrated in Fig. 4.

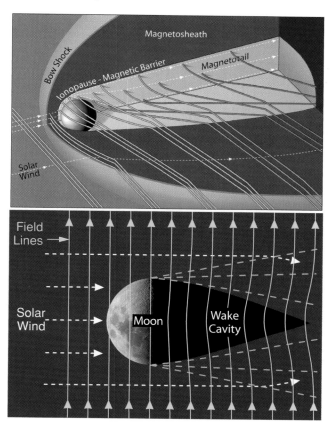

FIGURE 4   Schematic illustrations of the interaction regions surrounding, top, a planet like Mars or Venus, which is sufficiently conducting that currents close through the planet or its ionosphere (solar magnetic field lines are shown in yellow to red and are draped behind the planet) and, bottom, a body like the Moon, which has no ionosphere and low surface and interior conductivity. Credit: Steve Bartlett.

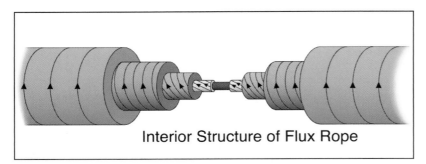

Interior Structure of Flux Rope

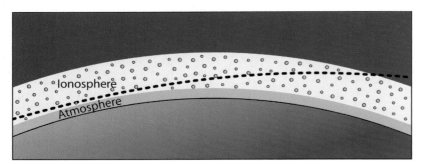

Distribution of Flux Ropes

FIGURE 5   Schematic illustration of a flux rope, a magnetic structure that has been identified in the ionosphere of Venus (shown as black dots within the ionosphere) and extensively investigated (a low-altitude pass of the *Pioneer Venus Orbiter* is indicated by the dashed curve). The rope (see above) has an axis aligned with the direction of the central field. Radially away from the center, the field wraps around the axis, its helicity increasing with radial distance from the axis of the rope. Structures of this sort are also found in the solar corona and in the magnetotails of magnetized planets. Credit: Steve Bartlett.

The magnetic structure surrounding Mars and Venus has features much like those found in a true magnetosphere surrounding a strongly magnetized planet. This is because the interaction causes the magnetic field of the solar wind to drape around the planet. The draped field stretches out downstream (away from the Sun), forming a magnetotail. The symmetry of the magnetic configuration within such a tail is governed by the orientation of the magnetic field in the incident solar wind, and that orientation changes with time. For example, if the interplanetary magnetic field (IMF) is oriented northward, the east–west direction lies in the symmetry plane of the tail and the northern lobe field (see Fig. 1 for the definition of lobe) points away from the Sun, while the southern lobe field points toward the Sun. A southward-oriented IMF would reverse these polarities, and other orientations would produce rotations of the symmetry axis.

Much attention has been paid to magnetic structures that form in and around the ionospheres of unmagnetized planets. Magnetic flux tubes of solar wind origin pile up at high altitudes at the day side ionopause where, depending on the solar wind dynamic pressure, they may either remain for extended times, thus producing a magnetic barrier that diverts the incident solar wind, or penetrate to low altitudes in localized bundles. Such localized bundles of magnetic flux are often highly twisted structures stretched out along the direction of the magnetic field. Such structures, referred to as flux ropes, are illustrated in Fig. 5.

Although Mars has only a small global scale magnetic field and interacts with the solar wind principally through currents that link to the ionosphere, there are portions of the surface over which local magnetic fields block the ac-

cess of the solar wind to low altitudes. It has been suggested that "mini-magnetospheres" extending up to 1000 km form above the regions of intense crustal magnetization in the southern hemisphere; these mini-magnetospheres protect portions of the atmosphere from direct interaction with the solar wind. As a result, the crustal magnetization may have modified the evolution of the atmosphere and may still contribute to the energetics of the upper atmosphere.

2.3 Interactions of the Solar Wind with Asteroids, Comets and Pluto

Asteroids are small bodies (<1000 km radius and more often only tens of kilometers) whose signatures in the solar wind were first observed by the *Galileo* spacecraft in the early 1990s. [*See* MAIN-BELT ASTEROIDS.] Asteroid-related disturbances are closely confined to the regions near to and downstream of the magnetic field lines that pass through the body, and thus the interaction region is fan-shaped as illustrated in Fig. 6 rather than bullet-shaped like Earth's magnetosphere. Unlike Earth's magnetosphere, there is no shock standing ahead of the disturbance in the solar wind. The signature found by *Galileo* in the vicinity of the asteroid Gaspra suggested that the asteroid is magnetized at a level similar to the magnetization of meteorites. Because the measurement locations were remote from the body, its field was not measured directly, and it is possible that the putative magnetic signature was a fortuitous rotation of the interplanetary magnetic field. Data from other asteroids do not establish unambiguously the strength of their magnetic

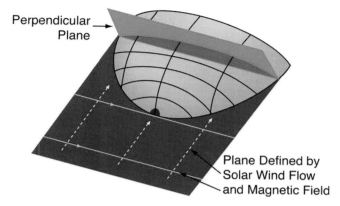

Perpendicular
Plane

Plane Defined by
Solar Wind Flow
and Magnetic Field

FIGURE 6   Schematic of the shape of the interaction between an asteroid and the flowing solar wind. The disturbance spreads out along the direction of the magnetic field downstream of the asteroid. The disturbed region is thus fan-shaped, with greatest spread in the plane defined by the solar wind velocity and the solar wind magnetic field. The curves bounding the intersection of that plane with the surface and with a perpendicular plane are shown. Credit: Steve Bartlett.

fields. A negligibly small magnetic field was measured by the *NEAR–Shoemaker* mission close to and on the surface of asteroid Eros, possibly because it is formed of magnetized rocks of random orientation. Although there will be no magnetometer on the *DAWN* spacecraft that will make measurements at Ceres and Vesta, other missions under discussion would add to our knowledge of asteroid magnetic properties. We may some day have better determinations of asteroidal magnetic fields and be able to establish how they interact with the solar wind.

Comets are also small bodies. The spectacular appearance of an active comet, which can produce a glow over a large visual field extending millions of kilometers in space on its approach to the Sun, is somewhat misleading because comet nuclei are no more than tens of kilometers in diameter. It is the gas and dust released from these small bodies by solar heating that we see spread out across the sky. Some of the gas released by the comet remains electrically neutral, with its motion governed by purely mechanical laws, but some of the neutral matter becomes ionized either by photoionization or by exchanging charge with ions of the solar wind. The newly ionized cometary material is organized in interesting ways that have been revealed by spacecraft measurements in the near neighborhood of comets Halley, Giacobini–Zinner, and Borrelly. Figure 7 shows schematically the types of regions that have been identified. Of particular interest is that the different gaseous regions fill volumes of space many orders of magnitude larger than the actual solid comet. The solar wind approaching the comet first encounters the expanding neutral gases blown off the comet. As the neutrals are ionized by solar photons, they extract momentum from the solar wind, and the flow slows a bit. Passing through a shock that further decelerates the

flow, the solar wind encounters ever-increasing densities of newly ionized gas of cometary origin, referred to as pickup ions. Energy is extracted from the solar wind as the pickup ions are swept up, and the flow slows further. Still closer to the comet, in a region referred to as the cometopause, a transition in composition occurs as the pickup ions of cometary origin begin to dominate the plasma composition. Close to the comet, at the **contact surface,** ions flowing away from the comet carry enough momentum to stop the flow of the incident solar wind. Significant asymmetry of the plasma distribution in the vicinity of a comet may arise if strong collimated jets of gas are emitted by the cometary nucleus. Such jets have been observed at Halley's comet and at comet Borrelly.

Pluto is also a small body even though it has been classified as a planet (until 2006). Pluto's interaction with the solar wind has not yet been observed, but it is worth speculating about what that interaction will be like in order to test our understanding of comparative planetology. [*See* Pluto.] The solar wind becomes tenuous and easily perturbed at large distances from the Sun (near 30 AU), and either escaping gases or a weak internal magnetic field could produce an interaction region many times Pluto's size. At some phases of its 248-year orbital period, Pluto moves close enough to the Sun for its surface ice to sublimate, producing an atmosphere and possibly an ionosphere. Models of Pluto's atmosphere suggest that the gases would then escape and flow away from the planet. If the escape flux is high, the solar wind interaction would then appear more like a comet than like Venus or Mars. Simulations show a very asymmetric shock surrounding the interaction region for a small but possible neutral escape rate. Pluto's moon, Charon, may serve as a plasma source within the magnetosphere, and this could have interesting consequences of the type addressed in Section 6 in relation to the moons of Jupiter and Saturn. As is the case for small asteroids and comets, ions picked up in the solar wind at Pluto have **gyroradii** and ion inertial lengths that are large compared with the size of the obstacle, a situation that adds asymmetry and additional complexity to the interaction. For most of its orbital period, Pluto is so far from the Sun that its interaction with the solar wind is more likely to resemble that of the Moon, with absorption occurring at the sunward surface and a void developing in its wake. It seems unlikely that a small icy body will have an internal magnetic field large enough to produce a magnetospheric interaction region, but one must recognize that actual observations of the magnetic fields of small bodies have repeatedly challenged our ideas about magnetic field generation.

2.4 Magnetospheres of Magnetized Planets

In a true magnetosphere, the scale size is set by the distance, R_{MP}, along the planet–Sun line at which the sum of the pressure of the planetary magnetic field and the pressure exerted by plasma confined within that field balance the dynamic pressure of the solar wind. (The dynamic pressure

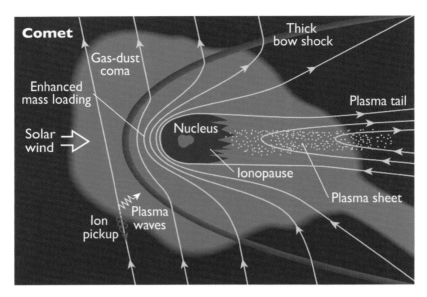

FIGURE 7 Schematic illustration of the magnetic field and plasma properties in the neighborhood of a comet. The length scale is logarithmic. The nucleus is surrounded by a region of dense plasma into which the solar wind does not penetrate. This region is bounded by a contact surface. Above that lies an ionopause or cometopause bounding a region in which ions of cometary origin dominate. Above this, there is a transition region in which the solar wind has been modified by the addition of cometary ions. As ions are added, they must be accelerated to become part of the flow. The momentum to accelerate the picked-up ions is extracted from the solar wind; consequently, in the transition region, the density is higher and the flow speed is lower than in the unperturbed solar wind. The newly picked up ions often generate plasma waves. The region filled with cometary material is very large, and it is this region that imposes the large-scale size on the visually observable signature of a comet. Spacecraft observations suggest that there is no shock bounding the cometary interaction region because the effects of ion pickup serve to slow the flow below the critical sound and Alfvén speeds without the need for a shock transition. Similar to Venus-like planets, the solar wind magnetic field folds around the ionopause, producing a magnetic tail that organizes the ionized plasma in the direction radially away from the Sun and produces a distinct comet tail with a visual signature. The orientation of the magnetic field in the tail is governed by the solar wind field incident on the comet, and it changes as the solar wind field changes direction. Dramatic changes in the structure of the magnetic tail are observed when the solar wind field reverses direction. Credit: J. A. Van Allen and F. Bagenal, 1999, Planetary magnetospheres and the interplanetary medium, in "The New Solar System," 4th Ed. (Beatty, Petersen, and Chaikin, eds.), Sky Publishing and Cambridge Univ. Press.

is ρu^2 where ρ is the mass density and u is its flow velocity in the rest frame of the planet. The thermal and magnetic pressures of the solar wind are small compared with its dynamic pressure.) Assuming that the planetary magnetic field is dominated by its dipole moment and that the plasma pressure within the magnetosphere is small, one can estimate R_{MP} as $R_{MP} \approx R_P (B_0^2/2\mu_0\rho u^2)^{1/6}$. Here B_0 is the surface equatorial field of the planet and R_P is its radius. Table 1 gives the size of the magnetosphere, R_{MP}, for the different planets and shows the vast range of scale

sizes both in terms of the planetary radii and of absolute distance.

Within a magnetosphere, the magnetic field differs greatly from what it would be if the planet were placed in a vacuum. The field is distorted, as illustrated in Fig. 1, by currents carried on the magnetopause and in the plasma trapped within the magnetosphere. Properties of the trapped plasma and its sources are discussed in Section 4. An important source of magnetospheric plasma is the solar wind. Figure 1 makes it clear that, along most of the boundary, solar wind plasma would have to move across magnetic field lines to enter the magnetosphere. The **Lorentz force** of the magnetic field opposes such motion. However, shocked solar wind plasma of the magnetosheath easily penetrates the boundary by moving along the field in the polar cusp. Other processes that enable solar wind plasma to penetrate the boundary are discussed in Section 5.

3. Planetary Magnetic Fields

Because the characteristic time scale for **thermal diffusion** is greater than the age of the solar system, the planets tend to have retained their heat of formation. At the same time, the characteristic time scale for diffusive decay of a magnetic field in a planetary interior is much less than the age of the planets. Consequently, primordial fields and permanent magnetism on a planetary scale are small and the only means of providing a substantial planetary magnetic field is an internal dynamo. For a planet to have a magnetic dynamo, it must have a large region that is fluid, electrically conducting and undergoing convective motion. The deep interiors of the planets and many larger satellites are expected to contain electrically conducting fluids: terrestrial planets and the larger satellites have differentiated cores of liquid iron alloys; at the high pressures in the interiors of the giant planets Jupiter and Saturn, hydrogen behaves like a liquid metal; for Uranus and Neptune, a water–ammonia–methane mixture forms a deep conducting "ocean." [See INTERIORS OF THE GIANT PLANETS.] The fact that some planets and satellites do not have dynamos tells us that their interiors are stably stratified and do not convect or that the interiors have solidified. Models of the thermal evolution of terrestrial planets show that as the object cools, the liquid core ceases to convect, and

further heat is lost by conduction alone. In some cases, such as the Earth, convection continues because the nearly pure iron solidifies out of the alloy in the outer core, producing an inner solid core and creating compositional gradients that drive convection in the liquid outer core. The more gradual cooling of the giant planets also allows convective motions to persist.

Of the eight planets, six are known to generate magnetic fields in their interiors. Exploration of Venus has provided an upper limit to the degree of magnetization comparable to the crustal magnetization of the Earth suggesting that its core is stably stratified and that it does not have an active dynamo. The question of whether Mars does or does not have a weak internal magnetic field was disputed for many years because spacecraft magnetometers had measured the field only far above the planet's surface. The first low-altitude magnetic field measurements were made by *Mars Global Surveyor* in 1997. It is now known that the surface magnetic field of Mars is very small ($|\mathbf{B}| < 10\,\mathrm{nT}$ or 1/3000 of Earth's equatorial surface field) over most of the northern hemisphere but that in the southern hemisphere there are extensive regions of intense crustal magnetization as already noted. Pluto has yet to be explored. Models of Pluto's interior suggest it is probably differentiated, but its small size makes one doubt that its core is convecting and any magnetization is likely to be remanent. Earth's moon has a negligibly small planet-scale magnetic field, though localized regions of the surface are highly magnetized. Jupiter's large moons are discussed in Section 6.

The characteristics of the six known planetary fields are listed in Table 2. Assuming that each planet's magnetic field has the simplest structure, a dipole, we can characterize the magnetic properties by noting the equatorial field strength (B_0) and the tilt of the axis with respect to the planet's spin axis. For all the magnetized planets other than Mercury, the surface fields are on the order of a Gauss = 10^{-4} T, meaning

TABLE 2	Planetary Magnetic Fields					
	Mercury	**Earth**	**Jupiter**	**Saturn**	**Uranus**	**Neptune**
Magnetic moment, (M_{Earth})	4×10^{-4}	1^a	20,000	600	50	25
Surface magnetic field						
At dipole equator (Gauss)	0.0033	0.31	4.28	0.22	0.23	0.14
Maximum/minimum[b]	2	2.8	4.5	4.6	12	9
Dipole tilt and sense[c]	$+14°$	$+10.8°$	$-9.6°$	$-0.0°$	$-59°$	$-47°$
Obliquity[d]	$0°$	$23.5°$	$3.1°$	$26.7°$	$97.9°$	$29.6°$
Solar wind angle[e]	$90°$	$67–114°$	$87–93°$	$64–117°$	$8–172°$	$60–120°$

[a] $M_{\mathrm{Earth}} = 7.906 \times 10^{25}$ Gauss cm^3 = 7.906×10^{15} Tesla m^3.
[b] Ratio of maximum surface field to minimum (equal to 2 for a centered dipole field).
[c] Angle between the magnetic and rotation axes.
[d] The inclination of the equator to the orbit.
[e] Range of angle between the radial direction from the Sun and the planet's rotation axis over an orbital period.

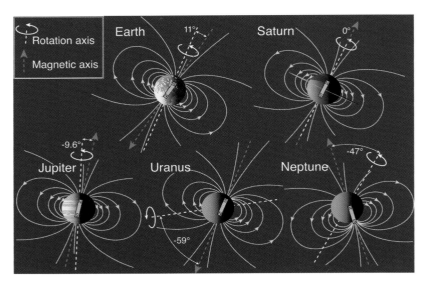

FIGURE 8   Orientation of the planets' spin axes and their magnetic fields (magnetic field lines shown in yellow) with respect to the ecliptic plane (horizontal). The larger the angle between these two axes, the greater the magnetospheric variability over the planet's rotation period. The variation in the angle between the direction of the solar wind (close to radial from the Sun) and a planet's spin axis over an orbital period is an indication of the degree of seasonal variability. Credit: Steve Bartlett.

that their dipole moments are of order $4\pi\mu_o{}^{-1}R_p{}^3\,10^{-4}$ T, where R_p is the planetary radius (i.e., the dipole moments scale with planetary size). The degree to which the dipole model is an oversimplification of more complex structure is indicated by the ratio of maximum to minimum values of the surface field. This ratio has a value of 2 for a dipole. The larger values, particularly for Uranus and Neptune, are indications of strong nondipolar contributions to the planets' magnetic fields. Similarly, the fact that the magnetic axes of these two planets are strongly tilted (see Fig. 8) also suggests that the dynamos in the icy giant planets may be significantly different than those of the planets with aligned, dipolar planetary magnetic fields.

The size of a planet's magnetosphere (R_{MP}) depends not only on the planet's radius and magnetic field but also on the ambient solar wind density, which decreases as the inverse square of the distance from the Sun. (The solar wind speed is approximately constant with distance from the Sun.) Thus, it is not only planets with strong magnetic fields that have large magnetospheres but also the planets Uranus and Neptune whose weak magnetic fields create moderately large magnetospheres in the tenuous solar wind far from the Sun. Table 1 shows that the measured sizes of planetary magnetospheres generally agree quite well with the theoretical R_{MP} values. Jupiter, where the plasma pressure inside the magnetosphere is sufficient to further "inflate" the magnetosphere, is the only notable exception. The combination of a strong internal field and relatively low solar wind density at 5 AU makes the magnetosphere of Jupiter a huge object—about 1000 times the volume of the Sun, with a tail that extends at least 6 A.U. in the antisunward direction, beyond the orbit of Saturn. If the jovian magnetosphere were visible from Earth, its angular size would be much larger than the size of the Sun, even though it is at least 4 times farther away. The magnetospheres of the other giant plan-

ets are smaller (although large compared with the Earth's magnetosphere), having similar scales of about 20 times the planetary radius, comparable to the size of the Sun. Mercury's magnetosphere is extremely small because the planet's magnetic field is weak and the solar wind close to the Sun is very dense. Figure 9 compares the sizes of several planetary magnetospheres.

Although the size of a planetary magnetosphere depends on the strength of a planet's magnetic field, the configuration and internal dynamics depend on the field orientation (illustrated in Fig. 8). At a fixed phase of planetary rotation, such as when the dipole tilts toward the Sun, the orientation of a planet's magnetic field is described by two angles (tabulated in Table 2): the tilt of the magnetic field with respect to the planet's spin axis and the angle between the planet's spin axis and the solar wind direction, which is generally within a few degrees of being radially outward from the Sun. Because the direction of the spin axis with respect to the solar wind direction varies only over a planetary year (many Earth years for the outer planets), and the planet's magnetic field is assumed to vary only on geological time scales, these two angles are constant for the purposes of describing the magnetospheric configuration at a particular epoch. Earth, Jupiter and Saturn have small dipole tilts and small obliquities. This means that changes of the orientation of the magnetic field with respect to the solar wind over a planetary rotation period and seasonal effects, though detectable, are small. Thus, Mercury, Earth, Jupiter, and Saturn have reasonably symmetric, quasi-stationary magnetospheres, with the first three exhibiting a small wobble at the planetary rotation period owing to their ~10° dipole tilts. In contrast, the large dipole tilt angles of Uranus and Neptune imply that the orientation of their magnetic fields with respect to the interplanetary flow direction varies greatly over a planetary rotation period, resulting in highly asymmetric

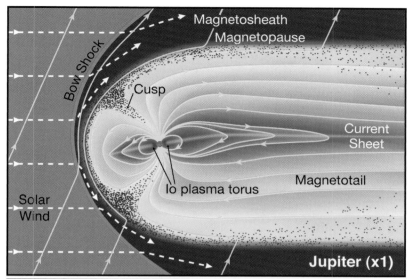

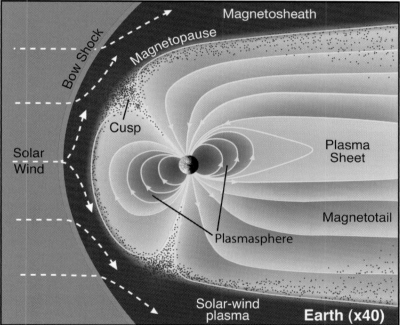

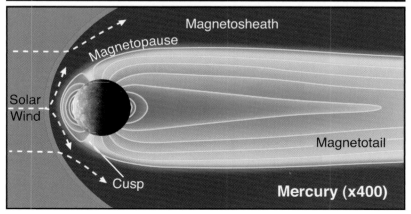

FIGURE 9 Schematic comparison of the magnetospheres of Jupiter, Earth, and Mercury. Relative to the Jupiter schematic, the one for Earth is blown up by a factor of 40, and the one for Mercury is blown up by a factor of 400. The planetary radii are given in Table 1. Credit: Steve Bartlett.

magnetospheres that vary at the period of planetary rotation. Furthermore, Uranus' large obliquity means that the magnetospheric configuration will undergo strong seasonal changes over its 84-year orbit.

4. Magnetospheric Plasmas

4.1 Sources of Magnetospheric Plasmas

Magnetospheres contain considerable amounts of plasma, electrically charged particles in equal proportions of positive charge on ions and negative charge on electrons, from various sources. The main source of plasma in the solar system is the Sun. The solar corona, the upper atmosphere of the Sun (which has been heated to temperatures of 1–2 million Kelvin), streams away from the Sun at a more or less steady rate of 10^9 kg s^{-1} in equal numbers (8×10^{35} s^{-1}) of electrons and ions. The boundary between the solar wind and a planet's magnetosphere, the magnetopause, is not entirely plasma-tight. Wherever the interplanetary magnetic field has a component antiparallel to the planetary magnetic field near the magnetopause boundary, magnetic **reconnection** (discussed in Section 5) is likely to occur, and solar wind plasma can enter the magnetosphere across the magnetopause. Solar wind material is identified in the magnetosphere by its energy and characteristic composition of protons (H$^+$) with ~4% alpha particles (He^{2+}) and trace heavy ions, many of which are highly ionized.

A secondary source of plasma is the ionosphere. Although ionospheric plasma is generally cold and gravitationally bound to the planet, a small fraction can acquire sufficient energy to escape up magnetic field lines and into the magnetosphere. In some cases, field-aligned potential drops accelerate ionospheric ions and increase the escape rate. Ionospheric plasma has a composition that reflects the composition of the planet's atmosphere (e.g., abundant O$^+$ for the Earth and H$^+$ for the outer planets).

The interaction of magnetospheric plasma with any natural satellites or ring particles that are embedded in the magnetosphere must also be considered; sources of this type can generate significant quantities of plasma. The outermost layers of a satellite's atmosphere can be ionized by interacting with the magnetospheric plasma. Energetic particle sputtering of the satellite surface or atmosphere produces ions of lower energy than the incident energy through a direct interaction but also can create an extensive cloud of neutral atoms that are subsequently ionized, possibly far from the satellite. The distributed sources of water-product ions (totaling ~2 kg s^{-1}) in the magnetosphere of Saturn suggest that energetic particle sputtering of the rings and icy satellites is an important process. Although the sputtering process, which removes at most a few microns of surface ice per thousand years, is probably insignificant in geological terms, sputtering has important consequences for the optical properties of the satellite or ring surfaces.

Table 3 summarizes the basic characteristics of plasmas measured in the magnetospheres of the planets that have detectable magnetic fields. The composition of the ionic species indicates the primary sources of magnetospheric plasma: satellites in the cases of Jupiter, Saturn, and Neptune; the planet's ionosphere in the case of Uranus. In the magnetospheres where plasma motions are driven by the solar wind, solar wind plasma enters the magnetosphere, becoming the primary source of plasma in the case of Mercury's small magnetosphere and a secondary plasma source at Uranus and Neptune. At Earth, both the ionosphere and the solar wind are important sources. Earth's moon remains well beyond the region in which sputtering or other plasma effects are important. In the magnetospheres where plasma flows are dominated by the planet's rotation (Jupiter, Saturn, and within a few R_E of Earth's surface), the plasma is

TABLE 3		Plasma Characteristics of Planetary Magnetospheres					
		Mercury	**Earth**	**Jupiter**	**Saturn**	**Uranus**	**Neptune**
Maximum density (cm^{-3})		~1	1–4000	>3000	~100	3	2
Composition		H$^+$	O$^+$, H$^+$	O^{n+}, S^{n+}	O$^+$, H$_2$O$^+$H$^+$	H$^+$	N$^+$, H$^+$
Dominant source		Solar wind	Ionosphere[a]	Io	Rings, Enceladus, Tethys, Dione	Atmosphere	Triton
Strength (ions/s)		?	2×10^{26}	>10^{28}	>10^{26}	10^{25}	10^{25}
(kg/s)			5	700	2	0.02	0.2
Lifetime		Minutes	Days[a] Hours[b]	10–100 days	30 days– years	1–30 days	~1 day
Plasma motion		Solar wind driven	Rotation[a] Solar wind[b]	Rotation	Rotation	Solar wind + rotation	Rotation (+ solar wind?)

[a] Inside plasmasphere.
[b] Outside plasmasphere.

confined by the planet's strong magnetic field for many days so that densities can become relatively high.

4.2 Energy

Plasmas of different origins can have very different characteristic temperatures. Ionospheric plasma has a temperature on the order of ~10,000 K or ~1 eV, much higher than temperature of the neutral atmosphere from which it formed (<1000 K) but much lower than the ~1 keV temperature characteristic of plasmas of solar wind origin, which are heated as they cross the bow shock and subsequently thermalized. Plasmas from satellite sources extract their energy from the planet's rotation through a complicated process. When the neutrals are ionized, they experience a Lorentz force as a result of their motion relative to the surrounding plasma; this force accelerates both ions and electrons, which then begin to gyrate about the magnetic field at a speed equal to the magnitude of the neutral's initial velocity relative to the flowing plasma. At the same time, the new ion is accelerated so that its bulk motion (the motion of the instantaneous center of its circular orbit) moves at the speed of the incident plasma, close to corotation with the planet near the large moons of Jupiter and Saturn. Because the electric field pushes them in opposite directions, the new ion and its electron separate after ionization. Hence a radial current develops as the ions are "picked up" by the magnetic field and the associated Lorentz force at the equator acts to accelerate the newly ionized particles to the local flow speed. The radial current in the near equatorial region is linked by field-aligned currents to the planet's ionosphere where the Lorentz force is in the direction opposite to the planet's rotation (i.e., in a direction that slows (insignificantly) the ionospheric rotation speed). Thus, the planet's angular momentum is tapped electrodynamically by the newly ionized plasma.

In the hot, tenuous plasmas of planetary magnetospheres, collisions between particles are very rare. By contrast, in the cold, dense plasmas of a planet's ionosphere, collisions allow ionospheric plasmas to conduct currents and cause ionization, charge exchange, and recombination. Cold, dense, collision-dominated plasmas are expected to be in thermal equilibrium, but such equilibrium was not originally expected for the hot, tenuous collisionless plasmas of the magnetosphere. Surprisingly, even hot, tenuous plasmas in space are generally found not far from equilibrium (i.e., their particle distribution functions are observed to be approximately **Maxwellian**, though the ion and electron populations often have different temperatures). This fact is remarkable because the source mechanisms tend to produce particles whose initial energies fall in a very narrow range. Although time scales for equilibration by means of **Coulomb collisions** are usually much longer than transport time scales, a distribution close to equilibrium is achieved by interaction with waves in the plasma. Space

plasmas support many different types of plasma waves, and these waves grow when free energy is present in the form of non-Maxwellian energy distributions, unstable spatial distributions, or anisotropic velocity–space distributions of newly created ions. Interactions between plasma waves and particle populations not only bring the bulk of the plasma toward thermal equilibrium but also accelerate or scatter suprathermal particles.

Plasma detectors mounted on spacecraft can provide detailed information about the particles' velocity distribution, from which bulk parameters such as density, temperature, and flow velocity are derived, but plasma properties are determined only in the vicinity of the spacecraft. Data from planetary magnetospheres other than Earth's are limited in duration and spatial coverage so there are considerable gaps in our knowledge of the changing properties of the many different plasmas in the solar system. Some of the most interesting space plasmas, however, can be remotely monitored by observing emissions of electromagnetic radiation. Dense plasmas, such as Jupiter's plasma torus, comet tails, Venus's ionosphere, and the solar corona, can radiate collisionally excited line emissions at optical or UV wavelengths. Radiative processes, particularly at UV wavelengths, can be significant sinks of plasma energy. Figure 10 shows an image of optical emission from the plasma that forms a ring deep within Jupiter's magnetosphere near the orbit of its moon, Io (see Section 6). Observations of these emissions give compelling evidence of the temporal and spatial variability of the Io plasma torus. Similarly, when magnetospheric particles bombard the planets' polar atmospheres, various auroral emissions are generated from radio to x-ray wavelengths and these emissions can also be used for remote monitoring of the system. [See ATMOSPHERES OF THE GIANT PLANETS.] Thus, our knowledge of space plasmas is based on combining the remote sensing of plasma phenomena with available spacecraft measurements that provide "ground truth" details of the particles' velocity distribution and of the local electric and magnetic fields that interact with the plasma.

4.3 Energetic Particles

Significant populations of particles at keV–MeV energies, well above the energy of the thermal population, are found in all magnetospheres. The energetic particles are largely trapped in long-lived radiation belts (summarized in Table 4) by the strong planetary magnetic field. Where do these energetic particles come from? Since the interplanetary medium contains energetic particles of solar and galactic origins an obvious possibility is that these energetic particles are "captured" from the external medium. In most cases, the observed high fluxes are hard to explain without identifying additional internal sources. Compositional evidence supports the view that some fraction of the thermal plasma is accelerated to high energies, either by tapping the rotational

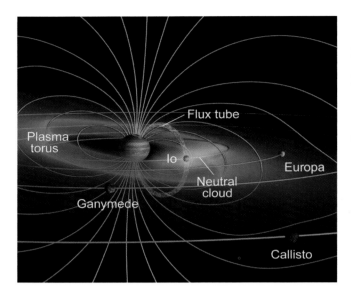

FIGURE 10 The ionization of an extended atmosphere of neutral atoms (yellow) around Jupiter's moon Io is a strong source of plasma, which extends around Jupiter in a plasma torus. Electrical currents generated in the interaction of Io with the surrounding plasma couple the moon to Jupiter's atmosphere where they stimulate auroral emissions. The main ring of auroral emissions is associated with currents generated as the plasma from the Io torus spreads out into the vast, rotating magnetosphere of Jupiter. Credit: John Spencer.

energy of the planet, in the cases of Jupiter and Saturn, or by acceleration in the distorted and dynamic magnetic field in the magnetotails of Earth, Uranus, and Neptune. In a nonuniform magnetic field and particularly in a rapidly rotating magnetosphere, the ions and electrons drift at different speeds around the planet, producing an azimuthal electric current. If the energy density of the energetic particle populations is comparable to the magnetic field energy density, the azimuthal current produces magnetic perturbations that significantly modify the planetary magnetic field. Table 4 shows that this occurs at Jupiter and Saturn, where the high particle pressures inflate and stretch out the magnetic field and generate a strong azimuthal current in the magnetodisc. Even though Uranus and Neptune have significant

radiation belts, the energy density of particles remains small compared with the magnetic field and the azimuthal current is very weak. In Earth's magnetosphere, the azimuthal current, referred to as the **ring current,** is extremely variable, as discussed in Section 5. Relating the magnetic field produced by the azimuthal current to the kinetic energy of the trapped particle population (scaled to the dipole magnetic energy external to the planet), we find that even though the total energy content of magnetospheres varies by many orders of magnitude and the sources are very different, the net particle energy builds up to only 1/1000 of the magnetic field energy in each magnetosphere. Earth, Jupiter, and Saturn all have energetic particle populations close to this limit. The energy in the radiation belts of Uranus and

TABLE 4	Energetic Particle Characteristics in Planetary Magnetospheres				
	Earth	**Jupiter**	**Saturn**	**Uranus**	**Neptune**
Phase space density[a]	20,000	200,000	60,000	800	800
Plasma beta[b]	<1	>1	>1	~0.1	~0.2
Ring current, ΔB (nT)[c]	10–200	200	10	<1	<0.1
Auroral power (W)	10^{10}	10^{14}	10^{11}	10^{11}	$<10^{8}$

[a] The phase space density of energetic particles (in this case 100 MeV/Gauss ions) is measured in units of $(cm^2 s\ sr\ MeV)^{-1}$ and is listed near its maximum value.
[b] The ratio of the thermal energy density to magnetic energy density of a plasma, $\beta = nkT/(B^2\mu_0)$. These values are typical for the body of the magnetosphere. Higher values are often found in the tail plasma sheet and, in the case of the Earth, at times of enhanced ring current.
[c] The magnetic field produced at the surface of the planet due to the ring current of energetic particles in the planet's magnetosphere.

Neptune is much below this limit, perhaps because it is harder to trap particles in nondipolar magnetic fields.

Where do these energetic particles go? Most appear to diffuse inward toward the planet. Loss processes for energetic particles in the inner magnetospheres are ring and satellite absorption, charge exchange with neutral clouds, and scattering by waves so that the particles stream into the upper atmospheres of the planets where they can excite auroral emission and deposit large amounts of energy, at times exceeding the local energy input from the sun.

The presence of high fluxes of energetic ions and electrons of the radiation belts must be taken into account in designing and operating spacecraft. At Earth, relativistic electron fluxes build to extremely high levels during magnetically active times referred to as storm times. High fluxes of relativistic electrons affect sensitive electronic systems and have caused anomalies in the operation of spacecraft. The problem arises intermittently at Earth but is always present at Jupiter. Proposed missions to Jupiter's moon Europa must be designed with attention to the fact that the energetic particle radiation near Europa's orbit is punishingly intense.

5. Dynamics

Magnetospheres are ever-changing systems. Changes in the solar wind, in plasma source rates, and in energetic cosmic ray fluxes can couple energy, momentum, and additional particle mass into the magnetosphere and thus drive magnetospheric dynamics. Sometimes the magnetospheric response is direct and immediate. For example, an increase of the solar wind dynamic pressure compresses the magnetosphere. Both the energy and the pressure of field and particles then increase even if no particles have entered the system. Sometimes the change in both field and plasma properties is gradual, similar to a spring being slowly stretched. Sometimes, as for a spring stretched beyond its breaking point, the magnetosphere responds in a very nonlinear manner, with both field and plasma experiencing large-scale, abrupt changes. These changes can be identified readily in records of magnetometers (a magnetometer is an instrument that measures the magnitude and direction of the magnetic field), in scattering of radio waves by the ionosphere or emissions of such waves from the ionosphere, and in the magnetic field configuration, plasma conditions and flows, and energetic particle fluxes measured by a spacecraft moving through the magnetosphere itself.

Auroral activity is the most dramatic signature of magnetospheric dynamics and it is observed on distant planets as well as on Earth. Records from ancient days include accounts of the terrestrial aurora (the lights flickering in the night sky that inspired fear and awe), but the oldest scien-

tific records of magnetospheric dynamics are the measurements of fluctuating magnetic fields at the surface of the Earth. Consequently, the term **geomagnetic activity** is used to refer to magnetospheric dynamics of all sorts. Fluctuating magnetic signatures with time scales from seconds to days are typical. For example, periodic fluctuations at frequencies between ~1 mHz and ~1 Hz are called magnetic pulsations. In addition, impulsive decreases in the horizontal north–south component of the surface magnetic field (referred to as the H-component) with time scales of tens of minutes occur intermittently at latitudes between 65° and 75° often several times a day. The field returns to its previous value typically in a few hours. These events are referred to as **substorms.** A signature of a substorm at a ~70° latitude magnetic observatory is shown in Fig. 11. The H-component decreases by hundreds to 1000 nT (the Earth's surface field is 31,000 nT near the equator). Weaker signatures can be identified at lower and higher latitudes. Associated with the magnetic signatures and the current systems that produce them are other manifestations of magnetospheric activity including particle precipitation and auroral

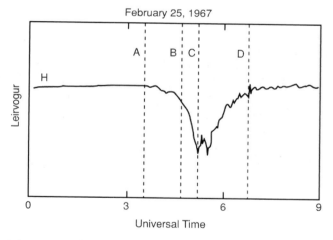

February 25, 1967

FIGURE 11 The variation of the H component of the surface magnetic field of the Earth at an auroral zone station at 70° magnetic latitude plotted versus universal time in hours during a 9-hour interval that includes a substorm. Perturbations in H typically range from 50 to 200 nT during geomagnetic storms. Vertical lines mark: A, The beginning of the growth phase during which the magnetosphere extracts energy from the solar wind, and the electrical currents across the magnetotail grow stronger. B, The start of the substorm expansion phase during which currents from the magnetosphere are diverted into the auroral zone ionosphere and act to release part of the energy stored during the growth phase. Simultaneously, plasma is ejected down the tail to return to the solar wind. C, The end of the substorm onset phase and the beginning of the recovery phase during which the magnetosphere returns to a stable configuration. D, The end of the recovery phase.

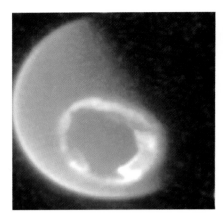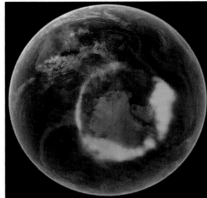

FIGURE 12   (Left) The image shows Earth's aurora observed with the Far Ultraviolet Imaging System on the *IMAGE* spacecraft during a major geomagnetic storm that occurred on July 15, 2000. The picture was obtained when the *IMAGE* spacecraft was at a distance of 7.9 Earth radii, and was looking down onto the northern polar region. The Sun is to the left. The auroral emissions are from molecular nitrogen that is excited by precipitating electrons. Photo credit: S. Mende and H. Frey, University of California, Berkeley. (Right) An ultraviolet image of aurora overlaid on a NASA visible image of the Earth. The aurora occurred during a strong geomagnetic storm on September 11, 2005. Photo credit: NASA. http://earthobservatory.nasa.gov/Newsroom/NewImages/images.php3?img_id=17165.

activation in the polar region and changes within the magnetosphere previously noted.

The auroral activity associated with a substorm can be monitored from above by imagers on spacecraft. The dramatic intensification of the brightness of the aurora as well as its changing spatial extent can thereby be accurately determined. Figure 12 shows an image of the aurora taken by the Far Ultraviolet Imaging System on the IMAGE spacecraft on July 15, 2000. Note that the intense brightness is localized in a high latitude band surrounding the polar regions. This region of auroral activity is referred to as the auroral oval. Only during very intense substorms does the auroral region move far enough equatorward to be visible over most of the United States.

The intensity of substorms and other geomagnetic activity is governed to some extent by the speed of the solar wind but of critical importance is the orientation of the magnetic field embedded in the solar wind incident on a magnetized planet. The fundamental role of the magnetic field in the solar wind may seem puzzling. It is the orientation of the interplanetary magnetic field that is critical, and at Earth it is normally tilted southward when substorm activity is observed. The issue is subtle. Magnetized plasma flowing through space is frozen to the magnetic field. The high conductivity of the plasma prevents the magnetic field from diffusing through the plasma, and, in turn, the plasma particles are bound to the magnetic field by a "$\mathbf{v} \times \mathbf{B}$" Lorentz force that causes the particles to spiral around a

field line. How, then, can a plasma ion or electron move from a solar wind magnetic field line to a magnetospheric field line?

The coupling arises through a process called reconnection, which occurs when plasmas bound on flux tubes with oppositely directed fields approach each other sufficiently closely. The weak net field at the interface may be too small to keep the plasma bound on its original flux tube and the field connectivity can change. Newly linked field lines will be bent at the reconnection location. The curvature force at the bend accelerates plasma away from the reconnection site. At the day side magnetopause, for example, solar wind magnetic flux tubes and magnetospheric flux tubes can reconnect in a way that extracts energy from the solar wind and allows solar wind plasma to penetrate the magnetopause. A diagram first drawn in a French café by J. W. Dungey in 1961 (and reproduced frequently thereafter) provides the framework for understanding the role of magnetic reconnection in magnetospheric dynamics (Fig. 13). Shown in the diagram on the top are southward-oriented solar wind field lines approaching the day side magnetopause. Just at the nose of the magnetosphere, the northern ends of the solar wind field lines break their connection with the southern ends, linking instead with magnetospheric fields. Accelerated flows develop near the reconnection site. The reconnected field lines are dragged tailward by their ends within the solar wind, thus forming the tail lobes. When the magnetic field of the solar wind points strongly northward

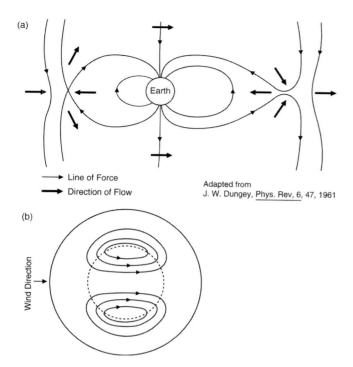

→ Line of Force
⟶ Direction of Flow

Adapted from
J. W. Dungey, Phys. Rev. 6, 47, 1961

Wind Direction

FIGURE 13 Adapted from the schematic view of reconnection sketched by J. W. Dungey in 1961. (a) A noon–midnight cut through the magnetosphere showing from left to right, in addition to two dipole-like field lines (rooted at two ends in the Earth): a solar wind field line with plasma flowing earthward; a newly reconnected pair of field lines, one of solar wind origin and one dipole-like field line, with plasma flowing toward the reconnection point from two sides near the midplane and accelerated both north and south away from the reconnection point; two reconnected field lines with one end in the solar wind and one end in the Earth flowing over the polar caps; two field lines about to reconnect in the magnetotail carried by plasma flow toward the midplane of the diagram; and a newly reconnected field line moving further away from the Earth in the solar wind. (b) A view down on the northern polar cap showing flow lines moving from day to night near the center, above the auroral zone, and returning to the day side at latitudes below the auroral zone.

at Jupiter or Saturn, reconnection is also thought to occur at the low latitude day side magnetopause, but the full process has not yet been documented by observations, although there is some evidence that auroral displays intensify at Jupiter as at Earth when magnetopause reconnection is occurring. At Earth, if the reconnection is persistent, disturbances intensify. Energetic particle fluxes increase and move to low latitudes and the ring current (see Section 4.3) intensifies.

If day side reconnection occurs at Earth, the solar wind transports magnetic flux from the day side to the night side. The path of the foot of the flux tube crosses the center of the polar cap, starting at the polar edge of the day side auroral zone and moving to the polar edge of the night side auroral zone as shown schematically in Fig. 13a. Ultimately that flux must return, and the process is also shown, both in the magnetotail where reconnection is shown closing a flux tube that had earlier been opened on the day side and in the polar cap (Fig. 13b) where the path of the foot of the flux tube appears at latitudes below the auroral zone, carrying the flux back to the day side. In the early stage of a substorm (between A and B in Fig. 11), the rate at which magnetic flux is transported to the night side is greater than the rate at which it is returned to the day side. This builds up stress in the tail, reducing the size of the region within the tail where the magnetic configuration is dipole-like and compressing the plasma in the plasma sheet (see Fig. 1). Only after reconnection starts on the night side (at B in Fig.11) does flux begin to return to the day side. Complex magnetic structures form in the tail as plasma jets both earthward and tailward from the reconnection site. In some cases, the magnetic field appears to enclose a bubble of tailward-moving plasma called a **plasmoid**. At other times, the magnetic field appears to twist around the earthward- or tailward-moving plasma in a flux rope (see Fig. 5). Even on the day side magnetopause, twisted field configurations seem to develop as a consequence of reconnection, and, because these structures are carrying flux tailward, they are called **flux transfer events.**

The diversity of the processes associated with geomagnetic activity, their complexity and the limited data on which studies of the immense volume of the magnetosphere must be based have constrained our ability to understand details of substorm dynamics. However, both new research tools and anticipated practical applications of improved understanding have accelerated progress toward the objective of being able to predict the behavior of the magnetosphere during a substorm. The new tools available in this century include a fleet of spacecraft in orbit around and near the Earth (*ACE*, *Wind*, *Polar*, *Geotail*, *Cluster*, *Double Star*, and several associated spacecraft) that make coordinated measurements of the solar wind and of different regions within the magnetosphere, better instruments that make high time resolution measurements of particles and fields, spacecraft imagers covering a broad spectral range, ground radar systems, and networks of magnetometers. The anticipated applications relate to the concept of forecasting **space weather** much as we forecast weather on the ground. An ability to anticipate an imminent storm and take precautions to protect spacecraft in orbit, astronauts on space stations, and electrical systems on the surface (which can experience power surges during big storms) has been adopted as an important goal by the space science community, and improvements in our understanding of the dynamics of the magnetosphere will ultimately translate into a successful forecasting capability.

Dynamical changes long studied at Earth are also expected in the magnetospheres of the other planets. In passes through Mercury's magnetosphere, the *Mariner* spacecraft observed substorms that lasted for minutes. These will be investigated by the *Messenger* spacecraft in the next decade. Substorms or related processes should also occur at the outer planets, but the time scale for global changes in a system is expected to increase as its size increases. For a magnetosphere as large as Jupiter's, the equivalent of a substorm is not likely to occur more often than every few days or longer, as contrasted with several each day for Earth. Until December 1995 when *Galileo* began to orbit Jupiter, no spacecraft had remained within a planetary magnetosphere long enough to monitor its dynamical changes. Data from Galileo's 8-year orbital reconnaissance of Jupiter's equatorial magnetosphere demonstrate unambiguously that this magnetosphere like that of Earth experiences intermittent injections of energetic particles and, in the magnetotail, unstable flows correlated with magnetic perturbations of the sort that characterize terrestrial substorms. Yet the source of the disturbances is not clear. The large energy density associated with the rotating plasma suggests that centrifugally driven instabilities must themselves contribute to producing these dynamic events. Plasma loaded into the magnetosphere near Io may ultimately be flung out down the magnetotail, and this process may be intermittent, possibly governed both by the strength of internal plasma sources and by the magnitude of the solar wind dynamic pressure that determines the location of the magnetopause. Various models have been developed to describe the pattern of plasma flow in the magnetotail as heavily loaded magnetic flux tubes dump plasma on the night side, but it remains ambiguous what aspects of the jovian dynamics are internally driven and what aspects are controlled by the solar wind.

Whether or not the solar wind plays a role in the dynamics of the jovian magnetosphere, it is clear that a considerable amount of solar wind plasma enters Jupiter's magnetosphere. One way to evaluate the relative importance of the solar wind and Io as plasma sources is to estimate the rate at which plasma enters the magnetosphere when day side reconnection is active and compare that estimate with the few $\times 10^{28}$ ions/s whose source is Io. If the solar wind near Jupiter flows at 400 km/s with a density of 0.5 particles/cm^3, it carries $\sim 10^{31}$ particles/s onto the circular cross section of a magnetosphere with $>50\ R_J$ radius. If reconnection is approximately as efficient as it is at Earth, where a 10% efficiency is often suggested, and if a significant fraction of the solar wind ions on reconnected flux tubes enter the magnetosphere, the solar wind source could be important, and, as at Earth, the solar wind may contribute to the variability of Jupiter's magnetosphere. *Galileo* data are still being analyzed in the expectation that answers to the question of how magnetospheric dynamics are controlled are contained in the archives of the mission.

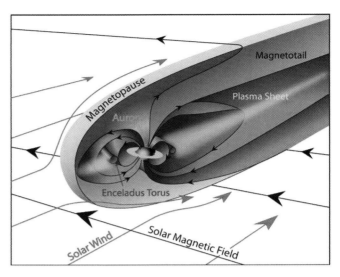

FIGURE 14 Saturn's magnetosphere shown in three dimensions. Water vapor from Enceladus' plumes is dissociated and ionized to form a torus of plasma that diffuses out into an equatorial plasma sheet. Credit: Steve Bartlett

Cassini arrived at Saturn in mid 2004. Earlier passes through the magnetosphere (*Voyager 1* and *2* and *Pioneer 10*) were too rapid to provide insight into the dynamics of Saturn's magnetosphere or even to identify clearly the dominant sources of plasma (Fig. 14). Periodic features had been found in the magnetometer data, and the intensity of radio emissions in the kilometric wavelength band varied at roughly the planetary rotation period, but the source of the periodic variations was unclear because Saturn's magnetic moment is closely aligned with its spin axis, and there is no evident longitudinal asymmetry. The *Cassini* data confirm the strong periodic variation of field and particle properties. There is evidence that the period changes slowly, which makes it likely that the source of periodicity is not linked to the deep interior of the planet, but there is not yet consensus on the source of the periodicity.

The energetic particle detector on *Cassini* is capable of "taking pictures" of particle fluxes over large regions of the magnetosphere. The technique relies on the fact that if an energetic ion exchanges charge with a slow-moving neutral, a fast-moving neutral particle results. The energetic particle detector then acts like a telescope, collecting energetic neutrals instead of light and measuring their intensity as a function of the look direction. The images show that periodic intensifications and substorm-like acceleration are present at Saturn.

It is still uncertain just how particle transport operates at Saturn and how the effects of rotation compare in importance with convective processes imposed by interaction with the solar wind. It seems quite possible that with a major source of plasma localized close to the planet (see discussion of the plume of Enceladus in Section 6), Saturn's magnetospheric dynamics will turn out to resemble those of Jupiter more closely than those of Earth. Observations scheduled in the coming years will surely bear on this speculation.

6. Interactions with Moons

Embedded deeply within the magnetosphere of Jupiter, the four Galilean moons (Io, Europa, Ganymede, and Callisto whose properties are summarized in Table 5) are immersed in magnetospheric plasma that corotates with Jupiter (i.e., flows once around Jupiter in each planetary spin period). At Saturn, Titan, shrouded by a dense atmosphere, is also embedded within the flowing plasma of a planetary magnetosphere. [See TITAN.] In the vicinity of these moons, interaction regions with characteristics of induced or true magnetospheres develop. The scale of each interaction region is linked to the size of the moon and to its electromagnetic properties. Ganymede, Callisto, and Titan are similar in size to Mercury; Io and Europa are closer in size to Earth's Moon. Io is itself the principal source of the plasma in which it is embedded. Approximately 1 ton(s) of ions is introduced into Jupiter's magnetosphere by the source at Io, thus creating the Io plasma torus alluded to in Section 4. The other moons, particularly Europa and Titan, are weaker plasma sources.

The magnetospheric plasma sweeps by the moons in the direction of their orbital motion because the Keplerian orbital speeds are slow compared with the speed of local plasma flow. Plasma interaction regions develop around the moons, with details depending on the properties of the moon. Only Ganymede, which has a significant internal magnetic moment, produces a true magnetosphere.

The interaction regions at the moons differ in form from the model planetary magnetosphere illustrated in Fig. 1. An important difference is that no bow shock forms upstream of the moon. This difference can be understood by recognizing that the speed of plasma flow relative to the moons is smaller than either the sound speed or the Alfvén speed, so that instead of experiencing a sudden decrease of flow speed across a shock surface, the plasma flow can be gradually deflected by distributed pressure perturbations upstream of a moon. The ratio of the thermal pressure to the magnetic pressure is typically small in the surrounding plasma, and this minimizes the changes of field geometry associated with the interaction. Except for Ganymede, the magnitude of the magnetic field changes only very near the moon. Near each of the unmagnetized moons the magnetic field rotates because the plasma tied to the external field slows near the body but continues to flow at its unperturbed speed both above and below. The effect is that expected if the field lines are "plucked" by the moon. The regions containing rotated field lines are referred to as Alfvén wings. Within the Alfvén wings, the field connects to the moon and its surrounding ionosphere. Plasma on these flux tubes is greatly affected by the presence of the moon. Energetic particles may be depleted as a result of direct absorption, but low-energy plasma densities may increase locally because the moon's atmosphere serves as a plasma source. In many cases, strong plasma waves, a signature of anisotropic or non-Maxwellian particle distributions, are observed near the moons.

In the immediate vicinity of Io, both the magnetic field and the plasma properties are substantially different from those in the surrounding torus because Io is a prodigious source of new ions. The currents associated with the ionization process greatly affect the plasma properties in Io's immediate vicinity. When large perturbations were first observed near Io it seemed possible that they were signatures of an internal magnetic field, but multiple passes established that the signatures near Io can be interpreted purely in terms of currents flowing in the plasma.

Near Titan, the presence of an extremely dense atmosphere and ionosphere also results in a particularly strong

| TABLE 5 | Properties of Major Moons of Jupiter and Saturn |

Moon	Orbit Distance (R_P)	Rotation Period (Earth days)[a]	Radius (km)	Radius of Core (moon radii)[b]	Mean Density (kg/m^3)	Surface B at Dipole Equator (nT)	Approx. Average B_{ext} (nT)[c]
Io	5.9	1.77	1821	0.25–0.5	3550	≤200	−1900
Europa	9.4	3.55	1570		2940	0 or small	−420
Ganymede	15	7.15	2631	0.25–0.5	1936	750	−90
Callisto	26	16.7	2400		1850	0 or small	−30
Titan	20	15.9	2575		1900	0 or small	−5.1

[a] Jupiter's rotation period is 9 hours 55 minutes, so corotating plasma moves faster than any of the moons.

[b] Core densities can be assumed in the range from 5150 to 8000 kg/m^3. This corresponds to maximum and minimum core radii, respectively.

[c] The magnetic field of Jupiter at the orbits of the moons oscillates in both magnitude and direction at Jupiter's rotation period of 9 hours 55 minutes. The average field over a planetary rotation period is southward oriented (i.e., antiparallel to Jupiter's axis of rotation). Neither the orbits nor the spin axes of the moons are significantly inclined to Jupiter's equatorial plane, so we use averages around the moon's orbit from the model of Khurana and Kivelson (1997).

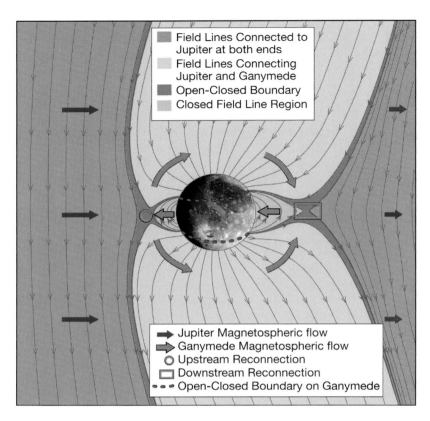

FIGURE 15   A schematic view of Ganymede's magnetosphere embedded in Jupiter's magnetospheric field in a plane that is normal to the direction of corotation flow. The thick purple line that bounds the region in which field lines link to Ganymede is the equivalent of the magnetopause and the polar cusp in a planetary magnetosphere. Credit: Steve Bartlett.

interaction whose effects on the field and the flow were observed initially by *Voyager 1*; the region will be explored thoroughly by the *Cassini* orbiter. Saturn's magnetospheric field drapes around the moon's ionosphere much as the solar wind field drapes to produce the magnetosphere of Venus, a body that like Titan has an exceptionally dense atmosphere.

Saturn's tiny moon, Enceladus, orbiting deep within the magnetosphere at 4 R_S, has proved to be a significant source of magnetospheric heavy ions. Alerted by anomalous draping of the magnetic field to the possibility that high-density ionized matter was present above the south pole of the moon, the trajectory of *Cassini* was modified to enable imaging instruments to survey the region. A plume of vapor, largely water, was observed to rise far above the surface. This geyser is a major source of Saturn's magnetospheric plasma and thus plays a role much like that of Io at Jupiter.

One of the great surprises of the *Galileo* mission was the discovery that Ganymede's internal magnetic field not only exists but is strong enough to stand off the flowing plasma of Jupiter's magnetosphere and to carve out a bubble-like magnetospheric cavity around the moon. A schematic of the cross section of the magnetosphere in the plane of the background field and the upstream flow is illustrated in Fig. 15. Near Ganymede, both the magnetic field and the plasma properties depart dramatically from their values in the surroundings. A true magnetosphere forms with a distinct magnetopause separating the flowing jovian plasma

from the relatively stagnant plasma tied to the moon. Within the magnetosphere, there are two types of field lines. Those from low latitudes have both ends linked to Ganymede and are called closed field lines. Little plasma from sources external to the magnetosphere is present on those field lines. The field lines in the polar regions are linked at one end to Jupiter. The latter are the equivalent of field lines linked to the solar wind in Earth's magnetosphere and are referred to as open field lines. On the open field lines, the external plasma and energetic charged particles have direct access to the interior of the magnetosphere. The particle distributions measured in the polar regions are extremely anisotropic because the moon absorbs a large fraction of the flux directed toward its surface. Where the energetic particles hit the surface, they change the reflectance of the ice, so the regions of open field lines can be identified in images of Ganymede's surface and compared with the regions inferred from magnetic field models. The two approaches are in good agreement. As expected, the angular distribution of the reflected particles has also been found to be modified by Ganymede's internal dipole field.

Ganymede's dipole moment is roughly antiparallel to Jupiter's, implying that the field direction reverses across the near equatorial magnetopause. This means that magnetic reconnection is favored. Should future missions allow a systematic study of this system, it will be of interest to learn whether with steady upstream conditions reconnection occurs as a steady process or whether it occurs with some periodic or aperiodic modulation.

7. Conclusions

We have described interactions between flowing plasmas and diverse bodies of the solar system. The interaction regions all manifest some of the properties of magnetospheres. Among magnetospheres of magnetized planets, one can distinguish (a) the large, symmetric, and rotation-dominated magnetospheres of Jupiter and Saturn; (b) the small magnetosphere of Mercury where the only source of plasma is the solar wind that drives rapid circulation of material through the magnetosphere [*see* MERCURY]; and (c) the moderate-sized and highly asymmetric magnetospheres of Uranus and Neptune, whose constantly changing configuration does not allow substantial densities of plasma to build up. The Earth's magnetosphere is an interesting hybrid of the first two types, with a dense corotating plasmasphere close to the planet and tenuous plasma, circulated by the solar wind driven convection, in the outer region. All of these magnetospheres set up bow shocks in the solar wind. The nature of the interaction of the solar wind with nonmagnetized objects depends on the presence of an atmosphere that becomes electrically conducting when ionized. Venus and Mars have tightly bound atmospheres so that the region of interaction with the solar wind is close to the planet on the sunward-facing side, with the interplanetary magnetic field draped back behind the planet to form a magnetotail. Bow shocks form in front of both these magnetospheres. The regions on the surface of Mars where strong magnetization is present produce mini-magnetospheres whose properties are being explored. Comets cause the solar wind field to drape much as at Venus and Mars; they produce clouds extended over millions of kilometers. The interaction of the solar wind with the cometary neutrals weakens or eliminates a bow shock. Small bodies like asteroids disturb the solar wind without setting up shocks. Within the magnetospheres of Saturn and Jupiter, the large moons interact with the subsonic magnetospheric flow, producing unique signatures of interaction with fields that resist draping. No shocks have been observed in these cases.

The complex role of plasmas trapped in the magnetosphere of a planetary body must be understood as we attempt to improve our knowledge of the planet's internal structure, and this means that the study of magnetospheres links closely to the study of intrinsic properties of planetary systems. Although our understanding of the dynamo process is still rather limited, the presence of a planetary magnetic field has become a useful indicator of properties of a planet's interior. As dynamo theory advances, extensive data on the magnetic field may provide a powerful tool from which to learn about the interiors of planets and large satellites. For example, physical and chemical models of interiors need to explain why Ganymede has a magnetic field while its neighbor of similar size, Callisto, does not and why Uranus and Neptune's magnetic fields are highly nondipolar and tilted while Jupiter's and Saturn's fields are nearly dipolar and aligned.

Continued exploration of the plasma and fields in the vicinity of planets and moons is needed to reveal features of the interactions that we do not yet understand. We do not know how effective reconnection is in the presence of the strong planetary fields in which the large moons of Jupiter are embedded. We have not learned all we need to know about moons as sources of new ions in the flow. We need many more passes to define the magnetic fields and plasma distributions of some of the planets and all of the moons because single passes do not provide constraints sufficient to determine more than the lowest order properties of the internal fields. Temporal variability of magnetospheres over a wide range of times scales makes them inherently difficult to measure, especially with a single spacecraft. Spurred by the desire to understand how the solar wind controls geomagnetic activity, space scientists combine data from multiple spacecraft and from ground-based instruments to make simultaneous measurements of different aspects of the Earth's magnetosphere or turn to multiple spacecraft missions like *Cluster* and *Themis* and the much anticipated Magnetospheric Multiscale Mission. As it orbited Jupiter, the *Galileo* spacecraft mapped out different parts of the jovian magnetosphere, monitoring changes and measuring the interactions of magnetospheric plasma with the Galilean satellites. *Cassini* in orbit around Saturn will provide even more complete coverage of the properties of another magnetosphere and its interaction with Titan. The properties of the magnetic and plasma environment of Mars are still being clarified by spacecraft measurements. *Messenger* is en route to Mercury where it will go into orbit with instruments that will characterize the mysterious magnetic field of this planet. And finally, Pluto beckons as the prototype of an important new group of solar system bodies; the dwarf planets. It is sure to interact with the solar wind in an interesting way. As new technologies lead to small, lightweight instruments, we look forward to missions of the new millennium that will determine if Pluto or Charon have magnetic fields and help us understand the complexities of magnetospheres large and small throughout the solar system.

Bibliography

Bagenal, F. (1992). Giant planet magnetospheres. *Ann. Rev. Earth Planet. Sci.* **20,** 289.

Bagenal, F., Dowling, T., and McKinnon, W., eds. (2004). "Jupiter: The Planet, Satellites and Magnetosphere." Cambridge Univ. Press, Cambridge, U.K.

Cheng, A. F., and Johnson, R. E. (1989). Effects of magnetosphere interactions on origin and evolution of atmospheres. In

"Origin and Evolution of Planetary and Satellite Atmospheres" (S. K. Atreya, J. B. Pollack, and M. S., Matthews, eds.). Univ. Arizona Press, Tucson.

Kivelson, M.G. (2006) Planetary magnetospheres. In "Handbook of Solar-Terrestrial Environment" (Y. Kamide and A. C.-L. Chian, eds.) In Press. Springer Verlag, New York.

Kivelson, M.G., and Russell, C. T., eds. (1995). "Introduction to Space Physics." Cambridge Univ. Press, Cambridge, U.K.

Luhmann, J. G. (1986). The solar wind interaction with Venus. *Space Sci. Rev.* **44**, 241.

Luhmann, J. G., Russell, C. T., Brace, L. H., and Vaisberg, O. L. (1992). The intrinsic magnetic field and solar wind interaction of Mars. In "Mars" (Kieffer et al., eds.). Univ. Arizona Press, Tucson.

Russell, C. T., Baker, D. N., and Slavin, J. A. (1988). The magnetosphere of Mercury, In "Mercury" (Vilas, Chapman, and Matthews, eds.). Univ. Arizona Press, Tucson.

Van Allen, J.A., and Bagenal, F. (1999). Planetary magnetospheres and the interplanetary medium. In "The New Solar System," 4th Ed. (Beatty, Petersen and Chaikin, eds.). Sky Publishing and Cambridge Univ. Press, Cambridge, U.K.

Pluto

S. Alan Stern
Southwest Research Institute
Boulder, Colorado

CHAPTER 29

P luto is the ninth planet and the prototype of the dwarf planets so common in the Kuiper Belt and beyond. It is in an elliptical, 248 year orbit that ranges from 29.5 to 49.5 **Astronomical Units** (AU) from the Sun. Its largest satellite, Charon, is close enough to Pluto in size that the pair are widely considered to be a double planet. Pluto's two other known satellites, Nix and Hydra, which orbit beyond Charon but in Charon's orbital plane, are both relatively small. Almost nothing is known about Nix and Hydra save their orbits, approximate sizes, and their neutral, Charon-like colors. Both Pluto and Charon are rich in ices, but their surface compositions, **albedos**, and colors are very different. Unlike Charon, Pluto is known to possess distinct surface markings, polar caps, and an atmosphere. Major questions under study about the Pluto system include the fate of Pluto's atmosphere, the degrees of internal activity Pluto and Charon exhibit, and the origin of the system.

1. Historical Background

1.1 Overview

Pluto was discovered in February 1930, at Lowell Observatory in Flagstaff, Arizona. This discovery was made by Clyde Tombaugh (1906–1997), an observatory staff assistant working on a search for a long-suspected perturber of the orbits of Uranus and Neptune. That search, which was first begun in 1905 by the observatory's founder, Percival Lowell, never located the large object originally being searched for because the positional discrepancies of Uranus and Neptune which prompted that search were fictitious. Still, the search for Lowell's "Planet X" resulted in the discovery of the tiny planet Pluto, which itself heralded the discovery of the Kuiper Belt some 70+ years later.

Within a year of Pluto's discovery, its orbit was well determined. That orbit is both eccentric and highly inclined to the plane of the **ecliptic**, compared to the orbits of the other planets (see Table 1). However, no important discoveries about Pluto's physical properties were made until the early 1950s. This lack of information was largely due to the difficulty of observing Pluto with the scientific instruments available in the 1930s and 1940s. Between 1953 and 1976, however, technological advances in photoelectric photometry made possible several important findings. Among these were the discovery of Pluto's ~6.387-day rotation period, the discovery of Pluto's reddish surface color, and the discovery of Pluto's high axial tilt, or **obliquity**.

Between 1976 and 1989, the pace of discoveries increased more dramatically. In rapid succession, there was the discovery of methane (CH_4) on Pluto's surface; the detection of Pluto's largest satellite Charon; the prediction,

TABLE 1	Pluto's Heliocentric Orbit
Orbital Element	**Value**
Semimajor Axis, a	39.44 AU
Orbital period, P	247.688 year
Eccentricity, e	0.254
Inclination, i	17.14°
Longitudinal ascending node, ω	110.29°
Longitudinal perihelion, ω	223.94°
Perihelion epoch, T	05.1 September 1989 UT

Note: Osculating elements on JD 2449000.5, referred to the mean ecliptic and equinox of J2000.0.

detection, and then study of a set of once-every-124-year mutual eclipse events between Pluto and Charon; and the occultation by Pluto of a bright star, confirming the presence of an atmosphere. In addition, the 1989 *Voyager 2* encounter with the Neptune system gave us detailed insights into the object believed to be Pluto's closest analog in the solar system, Triton, thereby showing how complex and scientifically interesting Pluto would be under close scrutiny by spacecraft. In the 1990s, it was discovered that Pluto's surface consists of a complex mixture of low-temperature volatile ices, that this surface displays large-scale bright and dark units, and that Pluto's atmosphere consists primarily of nitrogen gas, with trace amounts of carbon monoxide and only a trace of methane. Additionally, Pluto's small moons Nix and Hydra were discovered, and Pluto's context in the solar system became understood only after the discoveries of many smaller objects in the region of the solar system beyond Neptune called the Kuiper Belt.

1.2 The Discovery of Pluto's Three Satellites

Charon (pronounced correctly as "Kharon," but more colloquially pronounced as "Sharon") was discovered by J. W. Christy and R. S. Harrington on a series of photographic plates made in 1978 at the U.S. Naval Observatory's Flagstaff Station in Arizona. Interestingly, these images were taken less than 4 miles from Lowell Observatory, where Pluto had been discovered 48 years before. Charon was apparent on the 1978 Naval Observatory images as a bump or elongation in Pluto's apparent shape. This elongation of Pluto had occasionally been seen on photographic plates made in the 1960s, but it had not been recognized to be a satellite. This was because the elongation of Pluto's image by Charon was attributed to turbulence in the Earth's atmosphere causing a distortion of Pluto's point-like image (the two are <1 **arc second** (arcsec) apart, and blended together by atmospheric seeing). What Christy and Harrington recognized in 1978 was that although Pluto was

Regarding the 2006 IAU Planetary Definition and Pluto

In August 2006, a motion passed the International Astronomical Union's (IAU's) General Assembly in Prague which defined "dwarf planets" as those bodies in heliocentric orbit that are large enough to be rounded by self gravity and thus reach a state of approximate hydrostatic equilibrium. The IAU further required that "planets" fit a context-dependent criterion, in that a planet must have cleared its orbital neighborhood. Since Pluto and all dwarf planets fail this test, the IAU currently does not consider Pluto a planet. However, based on the inclusion of the dynamical clearing clause and its restriction to planets being objects that orbit the Sun and not other stars, the IAU definition of planethood has been criticized as narrowly constructed, technically flawed, poorly worded, biased against size with increasing heliocentric distance, and at odds with other classification schemes in astronomy that rely only on the intrisic properties of the object in question. These and other criticisms have come from planetary scientists, astronomers, teachers, and lay people, many of whom have elected to neglect the IAU definition. It is not known if the IAU definition will be widely adopted, or how long the IAU definition will survive before it is modified. For those reasons, here we continue to refer to Pluto as a planet in this chapter. An international scientific congress to further assess the definition of planets is planned for 2007.

distorted, none of the stars in the photographs were! This led them to look for a periodicity in the elongations. The recognition that the bump was in fact a close-in satellite was made when it was determined that this bump regularly cycled around Pluto in a 6.39-day period, which matched Pluto's rotation period, implying that the elongation was due to an object that circled Pluto.

In the first few months after Charon's discovery, Christy and Harrington determined that Charon's orbit is synchronous with Pluto's rotation and also in Pluto's equatorial plane, and therefore highly inclined to the plane of the ecliptic. During that same year, 1978, Leif Andersson recognized that Pluto's orbital motion would cause Charon's orbital plane to sweep through the line of sight to the Earth for a period of several years every half Pluto orbit, or 124 terrestrial years. Mutual eclipses (also called mutual events) would then begin occurring every 3.2 days (half Charon's orbit period). These eclipses were predicted to progress over a period of 5 to 6 years, from shallow, partial events to central events lasting up to 5 hours, then to recede again to shallow grazing events. It was widely recognized that such a series of mutual eclipses and occultations would be scientifically valuable events. Fortuitously, these mutual events began occurring in 1984 and ended in 1990. These events

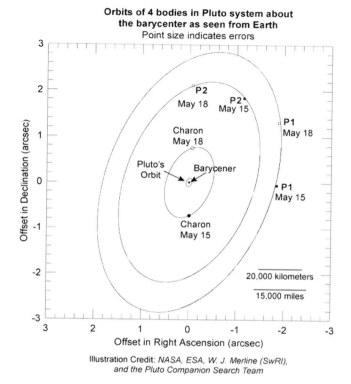

Orbits of 4 bodies in Pluto system about the barycenter as seen from Earth
Point size indicates errors

Illustration Credit: NASA, ESA, W. J. Merline (SwRI), and the Pluto Companion Search Team

FIGURE 1   The Pluto system. This figure shows the orbits of Charon, Nix, and Hydra around Pluto as they appeared from Earth ca. 2005, the year Nix and Hydra were discovered. (Adapted from Weaver et al., 2006, *Nature* **439**, 943.)

(described in Section 3) yielded a wealth of data on both Pluto and Charon. Searches for other satellites of Pluto were made in the early and mid 1990s by ground-based observatories and *Hubble Space Telescope* (*HST*) images obtained for other Pluto studies, but no moons were detected.

In May 2005, however, a much more sensitive, dedicated satellite search by an *HST* observing team led by H. A. Weaver and S. A. Stern yielded the detections of two small satellites. These bodies, which were subsequently named Nix and Hydra, orbit in circular orbits in Pluto's equatorial plane, as Charon does. Nix and Hydra orbit Pluto somewhat further out than Charon, with Nix being near 48,700 km from Pluto's center and Hydra near 64,800 km. These orbits are close to or in resonance with Charon, with the Charon:Nix:Hydra periods being very close to or at 1:4:6. Figure 1 depicts the satellite orbits of Nix and Hydra in relation to Charon. Based on their observed magnitudes, we can make reasonable assumptions about their albedos yield size estimates of approximately 40–160 km diameters for both. Initial color measurements made with *HST* indicate both satellites are neutrally reflecting, much like Charon. No compositional or lightcurve results are available on either satellite as of late 2006.

The *HST* images that revealed Nix and Hydra have also been used to search for other satellites; none were found.

From these data, it is possible to say that Pluto does not have any other satellites close to Nix and Hydra's brightness anywhere beyond Charon's orbit; inside Charon's orbit, such bodies could remain undetected, but they are not expected for theoretical reasons relating to Charon's outward migration during tidal despinning.

2. Pluto's Orbit and Spin

2.1 Pluto's Heliocentric Orbit

Relative to the eight previously discovered planets, Pluto's orbit is unusually eccentric (eccentricity $e \approx 0.25$), highly inclined (inclination $i \approx 17°$), and large (semimajor axis $a \approx 39.4$ AU). Pluto's orbit period is 248 years, during which the planet ranges from inside Neptune's orbit (Pluto's perihelion is near 29.7 AU) to nearly 49.5 AU. The Pluto–Charon **barycenter** passed its once-every-248-year perihelion at 05.1 ± 0.1 September 1989 UT; this will not occur again until A.D. 2236.

Current orbit integrations using osculating elements are able to predict Pluto's position to 0.5 arcsec accuracy over timescales of a decade. The fact that Pluto's perihelion is closer to the Sun than Neptune's orbit is quite unusual: No other known planet in the solar system crosses the orbit of another. The large change in Pluto's heliocentric distance as it moves around the Sun causes the surface **insolation** on Pluto and Charon to vary by factors of 3, which has important implications for Pluto's atmosphere (see Section 6). Pluto's perihelion lies slightly inside Neptune's orbit.

In the mid-1960s, it was discovered through computer simulations that Pluto's orbit librates in a 2:3 resonance with Neptune, which prevents mutual close approaches between the objects. This discovery has been verified by a series of increasingly longer and more accurate simulations of the outer solar system now exceeding 4×10^9 years. It is likely that Pluto was caught in this resonance and had its orbital eccentricity and inclination amplified to current values as Neptune migrated outward during the clearing of the outer solar system by the giant planets.

Pluto and Neptune can never closely approach one another, owing to this resonance, and the fact that the argument of Pluto's perihelion (i.e., the angle between the perihelion position and the position of its ascending node) librates (i.e., oscillates) about 90° with an amplitude of approximately 23°. This ensures that Pluto is never near perihelion when it is in conjunction with Neptune. Thus, Pluto is "protected" because Neptune passes Pluto's longitude only near Pluto's aphelion, never allowing Neptune and Pluto to come closer than ≈ 17 AU. Indeed, Pluto approaches Uranus more closely than Neptune, with a minimum separation of ≈ 11 AU, but still too far to significantly perturb its orbit.

In the late 1980s, it was discovered that Pluto's orbit exhibits a high degree of sensitivity to initial conditions. This is called "orbital chaos" by modern dynamicists. This discovery of a formal kind of chaos in Pluto's orbit does not imply Pluto undergoes frequent, dramatic changes. However, it does mean that Pluto's position is unpredictable on very long timescales. The timescale for this dynamical unpredictability has been established to be 2×10^7 years by Jack Wisdom, Gerald Sussman, and their co-workers.

2.2 Pluto's Lightcurve, Rotation Period, and Pole Direction

As previously indicated, since the mid-1950s Pluto's photometric brightness has been known to vary regularly with a period of about 6.387 days; more precisely, this period is 6.387223 days. Despite Pluto's faintness as seen from Earth, its period was easily determined using photoelectric techniques because the planet displays a large lightcurve amplitude, 0.35 magnitudes at visible wavelengths, which is equivalent to 38%.

Since at least 1955, it has also been known that Pluto's lightcurve is exhibiting an increase in its amplitude with time. Although the 6.387223 day period is identical to Charon's orbit period, Charon's photometric contribution is too small to account for the lightcurve's amplitude. This in turn implies that the lightcurve's structure is caused by surface features on Pluto. Figure 2 shows the shape of the combined Pluto–Charon lightcurve and its evolution over the past few decades.

The first study of Pluto's polar obliquity (or tilt relative to its orbit plane) was reported in 1973. By assuming that the variation of the lightcurve amplitude from the 1950s to the early 1970s was caused by a change in the aspect angle from which we see Pluto's spin vector from Earth, it was then determined that Pluto has a high obliquity (i.e., $90 \pm 40°$). In 1983, additional observations allowed the obliquity to be refined to $118.5 \pm 4°$. Even more recently, the results of the Pluto–Charon mutual events (or eclipses, see following discussion) have given a very accurate value of $122 \pm 1°$; Pluto's corresponding pole position lies near declination $-9°$, right ascension $312°$ (equator and equinox of 1950).

It is important to note, however, that torques on the Pluto–Charon pair cause Pluto's obliquity to oscillate between $\sim105°$ and $\sim130°$ with an $\sim3.7 \times 10^6$ year period. Thus, although Pluto presently reaches perihelion with its pole vector nearly normal to the Sun and roughly coincident with the orbit velocity vector, this configuration is only coincidental. The pole position executes a $360°$ circulation with a 3.7×10^6 year precession period.

2.3 Charon's Orbit and the System Mass

The discovery that Charon orbits Pluto with a period equal to Pluto's rotation period immediately implied the pair has

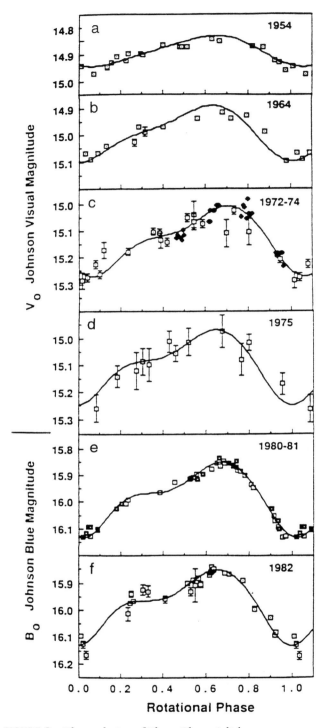

FIGURE 2   The evolution of Pluto–Charon's lightcurve over several decades. (Adapted from R. L. Marcialis, 1988, *Astronom. J.* **95**, 941.)

reached spin-orbit synchronicity. This is an unprecedented situation among the planets in the solar system.

Table 2 gives a solution to Charon's orbital elements obtained from various data. This fit relies on a semimajor axis

TABLE 2   Charon's Orbit[a]

Orbital Element	Value
Semimajor axis, a	19636 ± 8 km
Orbital period, P	6.387223 ± 0.00002 days
Eccentricity, e	0.0076 ± 0.003
Inclination, i	$96.2 \pm 0.3°$
Longitudinal perihelion, ω	$222.99 \pm 0.5°$
Mean anomaly, M	$34.84 \pm 0.35°$

[a]These elements are referred to the epoch 17 January 13.0 UT 1993. See D. J. Tholen and M. W. Buie, 1997, in "Pluto & Charon" (S. A. Stern and D. J. Tholen, eds.), Univ. Arizona Press, Tucson.

determination of $a = 19{,}636 \pm 8$ km derived from ground-based and *Hubble Space Telescope* data; it is statistically indistinguishable from ground-based results obtained in the mid-1980s of $a = 19{,}640 \pm 320$ and of $a = 19{,}558 \pm 153$ km.

Based on Charon's known orbital period and the 19,636 km semimajor axis, the system's (i.e., combined Pluto + Charon) mass is $1.47 \pm 0.002 \times 10^{25}$ g; this is very small, just $2.4 \times 10^{-3}\ M_{\text{Earth}}$.

Data from the mutual events showed that unless Charon's orbit has a very special orientation relative to Earth, Charon's orbital eccentricity is very low. Recently, *HST* observations have shown that Charon's orbital eccentricity is nonzero, with a best-estimated value of 0.0076. The fact that the orbit is not precisely circular indicates some disequilibrium forces have disturbed it from the zero value expected from tidal evolution. It is most likely that the disturbance causing this is generated by occasional close encounters between the Pluto–Charon system and 100-km class Kuiper Belt objects (see Section 8), but it may also be related to perturbations by Nix and Hydra.

3. The Mutual Events

3.1 Background

After Charon's discovery, the realization that mutual eclipses between Pluto and Charon would soon occur opened up the possibility of studying the Pluto–Charon system with the powerful data analysis techniques developed for eclipses between binary stars. Initial predictions by Leif Andersson indicated that the events could begin as early as 1979. As Charon's orbit pole position was refined, however, the predicted onset date moved to 1983–1986 (this was fortuitous because knowledge of the pole could have changed to indicate that the events had already just ended in the mid-1970s!). After a multiyear effort by several groups to detect the onset of these events, the first definitive eclipse

detections were made on 17 February 1985 by Richard Binzel at McDonald Observatory and were confirmed during an event 3.2 days later on 20 February 1985 by David Tholen at Mauna Kea. These first, shallow events ($\sim$0.01–0.02% in depth) revealed Pluto and Charon grazing across one another as seen from Earth.

The very existence of these eclipses proved the hypothesis (by 1985 widely accepted) that Charon was in fact a satellite, rather than some incredible topographic high on Pluto. The mutual eclipses persisted until October 1990, and dozens of events were observed. Important results from the 1985–1990 mutual events included reconstructed surface "maps" of Pluto and Charon; individual albedos, colors, and spectra for each object; and improvements in Charon's orbit. First, however, was the opportunity to use event timing to accurately determine the radii of Pluto and Charon.

3.2 Radii and Average Density of Pluto and Charon

Prior to the mutual events, the radii of Pluto and Charon were highly uncertain. Because Pluto and Charon remained unresolved in terrestrial telescopes (their apparent diameters are each <0.1 arcsec), direct measurements of their diameters were not available. A well-observed, near-miss occultation of Pluto in 1965 had constrained Pluto's radius to be <3400 km, but no better observations were available until the mutual events. However, circumstantial evidence that Pluto was smaller than 3400 km was inferred from the combination of Pluto's V astronomical $\approx$14 magnitude and the 1976 discovery of CH_4 frost (see Section 4.3), which exhibits an intrinsically high albedo. The small system mass determined after the discovery of Charon in 1978 strengthened this inference, but Pluto's radius was still uncertain within the bounds 900–2200 km.

The first concrete data to remedy the situation came when a fortuitous **stellar occultation** by Charon was observed on 07 April 1980. The 50 second length of the star's disappearance, observed by a 1 m telescope at Sutherland, South Africa, gave a value for Charon's radius of 605 ± 20 km. Improved results from subsequent stellar occultations yield 603.5 ± 3 km.

As noted earlier, accurate radius measurements for Pluto resulted from both stellar occultations and from fits of mutual event lightcurves, yielding solutions between 1150 and 1200 km. The range of uncertainty in Pluto's radius, which is significantly larger than in Charon's, results primarily from uncertainties in Pluto's atmospheric depth.

The two striking implications of the small radii and comparable masses of Pluto and Charon are (1) that Pluto is a very small planet—even smaller than the seven largest planetary satellites (the Moon, Io, Europa, Ganymede, Callisto, Titan, and Triton), and (2) that Pluto and Charon form the only known example of a binary planet (with the system barycenter outside of Pluto). Based on the radii and the total mass of the pair, it is possible to derive an average density of 2.03 ± 0.06 g cm^{-3}, where the error bar is dominated by the uncertainty in the radius of Pluto. Any density of 1.8 g cm^{-3} or higher implies that the system is compositionally dominated by rocky material, probably hydrated chondrites, as opposed to ices. This result and its implications will be discussed in more detail in Section 5.

4. Pluto's Surface Properties and Appearance

Pluto's surface properties have been studied since the 1950s. Photometric, spectroscopic, and polarimetric techniques have been applied, and the explorable wavelength regime has expanded from the ground-based window to the reflected IR and the **space ultraviolet**. Thermal-IR and millimeter-wave measurements have also been made.

4.1 Albedo and Color

Two of the most basic photometric parameters one desires to know for any solid body are its albedo and color. Accurate knowledge of Pluto's albedo was obtained only after the onset of the mutual events because until then Pluto's radius was unknown, and there was no definitive way of removing Charon's contribution.

The very first report of eclipse detections revealed a factor of 2 difference in depth between partial eclipses of Charon and Pluto, indicating Pluto's geometric albedo is substantially higher than Charon's. Once the eclipse season was complete, a more complete data set became available for analysis. Comprehensive models for the analysis of mutual event lightcurve data simultaneously solve for the individual radii of Pluto and Charon, the individual albedos, and Charon's orbital elements. The modeling of these parameters is complicated by solar phase angle effects, the presence of shadows during eclipse events, and instrumental and timing uncertainties. To derive the albedo lightcurve for Pluto alone, Pluto's albedo at the longitude of the total superior eclipses (in which Charon was completely hidden) must first be determined; albedos at other rotational epochs are then derived from this anchor point, assuming Charon's **rotational lightcurve** contributes only a small constant to the combined Pluto + Charon lightcurve. The assumption that a constant Charon contribution can be removed is not unreasonable, because (1) its geometric cross section is small (one fourth) of Pluto's, and (2) its eclipsed hemisphere has a geometric albedo only about 50–60% of Pluto's. However, *HST* observations have shown that Charon does vary somewhat in brightness ($\approx$8%) as it rotates on its axis. Analysis of a large set of mutual event data in the way just described has found that Pluto's maximum, disk-integrated, B-bandpass ($\sim$4360 Å) geometric albedo is 0.61. Rotational variations cause this albedo to range from values as low as 0.44 to values as high as 0.61 as Pluto rotates.

Information on Pluto's color comes from both photometry and the mutual events. As described in Section 1, Pluto's visible-bandpass color slope has been known to be red since the 1950s. Analysis of premutual event photometry yields B-V and U-B color differences of 0.84 and 0.31, respectively, for Pluto + Charon. There is only weak evidence that this value has changed since the 1950s when photoelectric measurements were first made. Eclipse data have revealed that the B-V color of Pluto itself is very close to 0.85 astronomical magnitudes. By comparison, this color is much less red than the **refractory** surfaces of Mars (B-V = 1.36) and Io (B-V = 1.17), and slightly redder than its closest analog in the solar system, Triton (B-V = 0.72).

4.2 Solar Phase Curve

The photometric behavior of a planet or satellite as it changes in brightness on approach to opposition can be used to derive surface scattering properties, and therefore its microphysical properties. Knowledge of the complete solar phase curve is also required to transform geometric albedos into bolometric Bond albedos. *HST* observations in the 1990s gave linear phase coefficients for Pluto and Charon of 0.029 ± 0.001 magnitudes/deg and 0.866 ± 0.008 magnitudes/deg, respectively.

Pluto's maximum solar phase angle ($\phi_{\max}$) as seen from Earth is just $\approx 1.9°$. Therefore, no measurements of the large-angle scattering behavior have been possible. Without measurements at large phase angles, no definitive determination of Pluto's phase integral q or Bond albedo A can be made. However, some improvement in estimates of q and A could become possible if the *Cassini* spacecraft is able to obtain Pluto phase curve observations from Saturn orbit, where $\phi_{\max} \approx 18°$. However, what is really needed are flyby spacecraft measurements of Pluto at high phase angles. For the present, the best available phase integral to use for Pluto is probably Triton's (Pluto and Triton also have similar linear phase coefficients). Triton's q has been measured by *Voyager*, giving $q = 1.2$ (at green wavelengths) to 1.5 (at violet wavelengths). If Pluto is similar, then its surface may have Bond albedos ranging from 0.3 to 0.7.

4.3 Surface Composition

Progress in understanding Pluto's surface composition required the development of sensitive detectors capable of making moderate spectral resolution measurements in the infrared, where most surface ices show diagnostic spectral absorptions. Although this technology began to be widely exploited as early as the 1950s in planetary science, Pluto's faintness (e.g., 700 times fainter than the jovian Galilean satellites) delayed compositional discoveries about it until the mid-1970s.

The first identification of a surface constituent on Pluto was the discovery by Dale Cruikshank, Carl Pilcher, and David Morrison in 1976 of CH_4 ice absorptions between 1 and 2 μm (a wavelength of 1 μm = 10,000 Å). Cruikshank et al. made this discovery using infrared photometers equipped with customized, compositionally diagnostic filters. In their report, Cruikshank et al. also presented evidence against the presence of strong H_2O and NH_3 absorptions in Pluto's spectrum. Confirmation of the methane detection came in 1978 and 1979 when both additional CH_4 absorption bands and true IR spectra of Pluto became available.

In mid-1992, another breakthrough occurred when Toby Owen, Dale Cruikshank, and other colleagues made observations using a new, state-of-the-art IR spectrometer at the UK Infrared Telescope (UKIRT) on Mauna Kea. These data revealed the presence of both N_2 and CO ices on Pluto. These molecules are much harder to detect than methane because they produce much weaker spectral features. Their presence on Pluto indicates the surface is chemically more heterogeneous, and more interesting than had previously been thought. Because N_2 and CO are orders of magnitude more **volatile** (i.e., have higher vapor pressures) than CH_4, their presence also implies they play a highly important role in Pluto's annual atmospheric cycle. Abundance inversions of Pluto reflectance spectra make clear that N_2 dominates the composition of much of Pluto's surface, with CO and CH_2 being trace constituents.

In 2006, ethane (C_2H_6) was detected on Pluto's surface. This and other hydrocarbons and nitriles had long been predicted to reside on Pluto as a result of photochemical and radiological processing of Pluto's surface ices and atmosphere. Future surface reflectance studies are expected to yield additional surface constituent detections.

Rotationally resolved spectra of Pluto's CH_4 absorption bands have been reported by a number of groups. Their studies showed that Pluto's methane is present at all rotational epochs, but the band depths are correlated with the lightcurve so that the minimum absorption occurs at minimum light. Mutual event spectroscopy has now demonstrated that Charon is not the cause of this variation, since Charon's surface is devoid of detectable CH_4 absorptions (see Section 6). This important discovery suggests that Pluto's dark regions could contain reaction products resulting from the photochemical or radiological conversion of methane and N_2 to complex nitriles and higher hydrocarbons.

We thus have the following basic picture of Pluto's surface composition: CH_4 appears rotationally ubiquitous, but with its surface coverage more widespread in regions of high albedo. In many areas, the methane is dissolved in a matrix of other ices, but in some locations the CH_4 is seen as pure ice. CO and N_2 have also been detected. In the bright areas of the planet where these ices are thought to mainly be located, N_2 dominates the surface abundance, and the CO is more abundant than the previously known (but more spectroscopically detectable) CH_4. Ethane, a byproduct of CH_4 chemistry, was detected in 2006. Pluto's strong lightcurve and red color demonstrate that other widespread, probably involatile surface constituents exist. This may either be due to rocky material, or hydrocarbons resulting from radiation processing of the CH_4 due to long-term exposure to ultraviolet sunlight, or both. Whether the volatile frost we are seeing is a surface veneer or the major component of Pluto's crust has not yet been established.

4.4 Surface Temperature

Results from the *Infrared Astronomical Satellite* (*IRAS*) indicated that Pluto's perihelion-epoch surface temperature was in the range of 55 to 60 K, close to that expected in radiative equilibrium with solar insolation. However, it has subsequently become appreciated that the situation on Pluto's surface is more complicated.

One line of evidence for this conclusion comes from millimeter-wave measurements of Pluto's Rayleigh–Jeans blackbody spectrum. Such measurements, reported first by Wilhelm Altenhoff and collaborators, and then later by Alan Stern, Michel Festou, and David Weintraub, and independently confirmed by David Jewitt, indicate that a significant fraction of Pluto's surface is significantly colder than 60 K, most likely in the range 35–42 K. A second line of evidence came in 1994 from high-resolution spectroscopy of the temperature-sensitive 2.15 μm N_2 ice absorption band, which Kimberley Tryka and her co-workers found indicates a surface temperature of about 40 K for the widespread nitrogen ices on Pluto. As described in Stern et al. reported in 1993, although the surface pressure of N_2 is not well known, it must be less than $\approx$60 μbar. This is consistent with an N_2 ice temperature of $\approx$40 K, assuming vapor pressure equilibrium between the N_2 ice and the atmosphere. This, combined with the *IRAS* measurements, led to the conclusion that Pluto's surface must exhibit both warm and cold regions. This was subsequently confirmed by rotationally resolved studies of Pluto's thermal emission spectrum by the *Infrared Space Observatory* (*ISO*) and the *Spitzer Infrared Space Telescope Facility* (*SIRTF*). These space telescopes also revealed that Pluto's coldest regions are correlated with bright surface units, and that the warmer regions are correlated with darker surface units with lower abundances of sublimating ices.

It is now well established that Pluto's surface temperature varies from place to place on the surface, with $\approx$40 K regions where N_2 ice is sublimating and $\approx$55–60 K regions where N_2 ice is not present in great quantities. The strong temperature contrasts across Pluto's surface imply strong wind speeds and significant lateral transport of material across the surface.

4.5 Surface Appearance and Markings

Because Pluto is less than 0.1 arcsec across as seen from Earth, its disk could not be resolved until the advent of the *Hubble Space Telescope*. However, evidence for surface markings has been available since the mid-1950s, when lightcurve modulation was first detected. Because Pluto is large enough to be essentially spherical (and indeed, mutual event and stellar occultation data show it actually is), the distinct variation in this lightcurve must be related to large-scale albedo features.

From the lightcurve in Fig. 2, it can be seen that Pluto's surface must contain at least three major longitudinal provinces. Information on the latitudinal distribution of albedo can be gained by observing the evolution of this lightcurve as Pluto moves around its orbit while the pole position remains inertially fixed, assuming, of course, that the surface albedo distribution is time invariant.

The most complete mapping products obtained from photometric data inversions (variously using rotational lightcurves and mutual event lightcurves) have been obtained by two teams. The first team, led by Marc Buie of Lowell Observatory has used both mutual event lightcurves and rotational lightcurve data compiled from 1954 to 1986 to compute a complete map of Pluto. The second group, consisting of Eliot Young and Richard Binzel, of MIT and SwRI, numerically fit a spherical harmonic series to each element of a finite element grid using the Charon transit mutual event lightcurve data as the model input. Because Young and Binzel used only mutual event data, their map is limited to the hemisphere of Pluto that Charon eclipses. Because the two groups used different data sets and different numerical techniques, their results are complementary and serve to check one another on the Charon-facing hemisphere they share in common.

These two maps are shown in Fig. 3. There are differences between the two maps, but it must be remembered that each map has intrinsic noise. The common features of these maps are (1) a very bright south polar cap, (2) a dark band over mid-southern latitudes, (3) a bright band over mid-northern latitudes, (4) a dark band at high northern latitudes, and (5) a northern polar region that is as bright

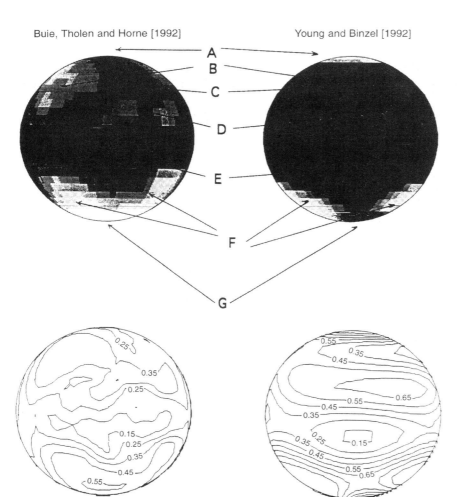

FIGURE 3 Two maps of Pluto's Charon-facing hemisphere. The map on the left was derived by M. Buie, K. Horne, and D. Tholen using both mutual event and lightcurve data. The map on the right was derived by E. Young and R. Binzel from their mutual event data. Although the fine details of these maps differ, their gross similarities are striking. See R. P. Binzel et al., 1997, in "Pluto & Charon (S.A. Stern and D.J. Tholen, eds.), Univ. Arizona Press, Tucson, for additional details.

as the southern cap. Later results, including some color information, were subsequently obtained by Eliot Young and colleagues.

In 1990, the *Hubble Space Telescope* imaged Pluto, but owing to its then-severe optical aberrations, these images, obtained by R. Albrecht and a team of collaborators cleanly separated Pluto and Charon, but it did not reveal significant details about the surface of Pluto. After *HST* was repaired by an astronaut crew in late-1993, its optics were good enough to resolve crude details on Pluto's surface. And in mid-1994, it obtained the first actual images of Pluto that revealed significant details about Pluto's surface. These images were made by Alan Stern, Marc Buie, and Laurence Trafton using the Faint Object Camera (FOC) of the *Hubble Space Telescope*. The 20-image *HST* data set is longitudinally complete and rotationally resolved and obtained at both blue and ultraviolet wavelengths. The various images that *HST* obtained were combined to make blue and UV maps of the planet, such as the one shown in Fig. 4. The *HST* images and derived maps reveal that Pluto has (1) a highly variegated surface, (2) extensive, bright, asymmetric polar regions, (3) large midlatitude and equatorial spots, and (4) possible linear features hundreds of kilometers in extent. The dynamic range of albedo features across the planet detected at the FOC's resolution in both the 410 and 278 nm bandpasses exceeds 5:1. New *HST* images were obtained in 2002 by a team led by Marc Buie using *HST*'s Advanced Camera for Surveys (ACS), but the results from these observations had not been published as of late 2006.

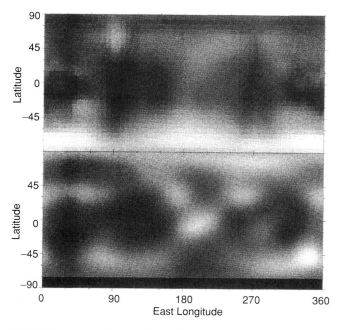

FIGURE 4 A map derived from direct imaging of Pluto using *HST* images made in 1994. (Adapted from Stern et al., 1997, *Astronom. J.* **113**, 827.)

5. Pluto's Interior and Bulk Composition

5.1 Density

To determine the separate densities of Pluto and Charon, one must either obtain precise astrometric measurements that detect the barycentric wobble between Pluto and Charon or use orbit solutions for Pluto's small satellites. Since 1992, both *HST* and ground-based measurements have been gathered to address the mass ratio, and therefore the relative masses and densities of Pluto and Charon. These are very difficult measurements. The best available density determination for Pluto is due to Buie and co-workers, who analyzed the orbits of Nix and Hydra in a 2006 publication that gave 2.03 ± 0.06 g cm^{-3} for a reference radius of 1153 km.

5.2 Bulk Composition and Internal Structure

The 1980s discovery that the Pluto–Charon system's average density is near 2 g cm^{-3} was a major surprise resulting from the mutual events. Many scientific papers had previously predicted values closer to the density of water ice (~ 1 g cm^{-3}), or even lower. Thus, contrary to earlier thinking, the Pluto–Charon pair is known to be mass-dominated by rocky material. Based on this information, a three-component model for Pluto's bulk composition and internal structure can be derived. In such a model, Pluto's bulk density is assumed to consist of three of the most common condensates in the outer solar system: water ice ($\rho = 1.00$ g cm^{-3}), "rock" ($2.8 < \rho < 3.5$ g cm^{-3}, depending upon its degree of hydration), and methane ice ($\rho = 0.53$ g cm^{-3}).

From three-component models, it is believed that Pluto's rock fraction is in the range of 60–80%, with preferred values close to 70%. By comparison, the large (e.g., $R > 500$ km) icy satellites of Jupiter, Saturn, and Uranus have typical rock fractions in the range 50–60% by mass. Only Io, Europa, and Triton rival Pluto in terms of their computed rock content. Pluto's high rock (i.e., nonvolatile) mass fraction is in contrast to the $\approx$50:50% rock:ice ratio predicted for objects formed from solar nebula material according to many nebular chemistry models and our present-day understanding of the nebular C/O ratio. This high rock fraction indicates that the nebular material from which Pluto formed was CO-rich rather than CH$_4$-rich. As such, roughly half of the available nebular oxygen should have gone into CO, rather than H$_2$O formation, which in turn would lead to a high rock:ice ratio.

There are two possible ways out of the apparent nebular chemistry dilemma imposed by Pluto's high rock fraction. One is that Pluto's minimum estimated radius of 1150 km may be too small; a value near 1200 km, as suggested by some stellar occultation models, would solve the problem. Alternatively, William McKinnon and the late Damon Simonelli independently suggested that a giant impact may

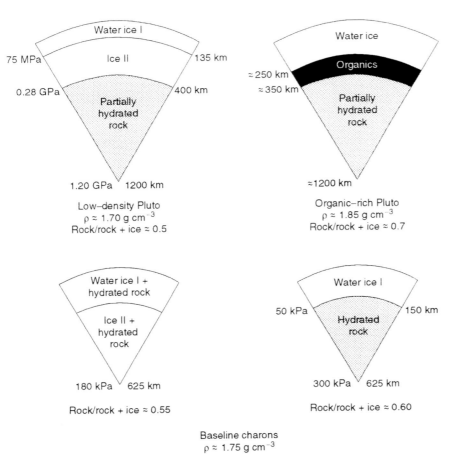

FIGURE 5   Typical interior structural models for Pluto and Charon. Adapted from W. B. McKinnon et al., 1997, in "Pluto & Charon" (S. A. Stern and D. J. Tholen, eds.), Univ. Arizona Press, Tucson.

have induced volatile loss from an already differentiated Pluto, which may have raised Pluto's rock fraction somewhat (perhaps 20%) to reach its present value. As we discuss in Section 8, such an impact is thought to be responsible for the formation of Pluto's satellite system.

The gross internal thermal structure of Pluto depends on several factors, virtually all of which are uncertain. These include material viscosities in the interior, the internal convection state, the actual rock fraction and radioisotope content, and the internal density distribution (i.e., most fundamentally, the differentiation state). It would appear likely that Pluto's deep interior reaches temperatures of at least 100–200 K, but not much higher. Whether or not Pluto is warm enough to exhibit convection in its ice mantle depends on both the internal thermal structure and the radial location of water ice in its interior.

Based on the results just given and laboratory equations of state, Pluto's central pressure can be estimated to lie between 0.6 and 0.9 GPa (gigapascals) if the planet is undifferentiated, or 1.1–1.4 GPa if differentiation has occurred. As such, the high-pressure water ice phase Ice VI is expected in the deep interior if the planet has not differentiated. If differentiation has occurred, as is likely, then a higher pressure

form of water ice called Ice II may be present, but only near the base of the convection layer. If Pluto did differentiate, then its gross internal structure may be represented by a model like that shown in Fig. 5.

6. Pluto's Atmosphere

6.1 Atmospheric Composition

The existence of an atmosphere on Pluto was strongly suspected after the discovery of methane on its surface in 1976, largely because at the predicted surface temperatures (~40–60 K), sufficient methane vapor pressure should obtain to constitute a significant atmosphere. This circumstantial argument was supported by the high reflectivity of Pluto's surface, which suggested some kind of resurfacing—most plausibly due to volatile laundering through an orbitally cyclic atmosphere. Still, however, there was no definitive evidence for an atmosphere until the late 1980s.

The formal proof of Pluto's atmosphere came from the occultation of a 12th magnitude star by Pluto in 1988, by providing the first direct observational evidence for an atmosphere. The best measurements of the occultation were obtained by Robert Millis and James Elliot, and their various MIT, Lowell Observatory, and Australian collaborators. These teams used both NASA's mobile, *Kuiper Airborne Observatory* (which contained a 36-inch diameter telescope) and ground-based telescopes to observe the occultation event. They discovered that light from the star was diminished far more gradually than it would be from an airless body. The apparent extinction of starlight observed during the occultation was caused by atmospheric refraction (i.e., the degree of bending of the starlight by the atmosphere), which varies with height. The rate at which the refractivity of the atmosphere varies with altitude depends on the ratio of atmospheric temperature (T) to atmospheric mean molecular weight (m). The 1989 Pluto occultation data implied $T/m = 3.7 \pm 0.7$ K/g at and above an altitude of 1215 km. If the atmosphere were composed entirely of methane ($m = 16$ g/mole), the implied atmospheric temperature would be 60 K, whereas an N_2 or CO atmosphere ($m = 28$ g/mole) would be at a temperature near 106 K.

From the stellar occultation data alone it was impossible to separately determine the mean atmospheric molecular weight and temperature of Pluto's atmosphere. However, theoretical calculations of the atmospheric temperature made by Roger Yelle and Jonathan Lunine of the University of Arizona indicated a value of 106 K in the upper atmosphere, under a variety of assumed compositions. This is relatively high compared with the surface temperature (~40–55 K) because the efficiency at which the atmosphere radiates and cools is very small. An upper atmospheric temperature near 106 K implies that the atmospheric mean molecular weight is close to 28 g/mole. This is consistent with an atmosphere dominated by N_2 or CO gas, with trace amounts of other species.

The detection of N_2 ice absorption features on Pluto's surface (see Section 4), coupled with the discovery by *Voyager 2* that Triton's atmosphere also consists predominantly of N_2 and only a trace of CO, suggests that Pluto's atmosphere is likely to be N_2-dominated. Nevertheless, if the high-temperature (106 K) atmospheric model is correct, then at least a few percent methane is thought to be required because methane (which is efficient at atmospheric heating) is thought to be responsible for the elevated atmospheric temperatures.

A nitrogen-dominated atmosphere with only a minor amount of methane was significantly strengthened in 1994 when Leslie Young and colleagues at MIT detected CH_4 gas in Pluto's atmosphere for the first time. This discovery, which was made possible by sensitive, high-resolution IR spectroscopy of the 2.3 μm CH_4 band system, indicated a total methane mixing ratio of <1%, and perhaps as little as 0.1% in the atmosphere. Subsequent high-resolution observations by Young and colleagues revealed that the CO abundance in Pluto's atmosphere must also be very very low.

6.2 Atmospheric Structure

The 1988 occultation data exhibited interesting behavior at altitudes below 1215 km, as is shown in Fig. 6. The starlight, which was decreasing gradually at higher altitudes, dropped suddenly to a value close to zero below this level; this is called lightcurve steepening. The drop is still not as sudden as would be expected from the setting of a star behind the limb of an airless planet, however. Two possible explanations have been proposed for this change.

In one model, the steepening was caused by the presence of aerosol hazes in the lower atmosphere. (Condensation clouds can be ruled out as an explanation for the aerosol layer because of the temperature structure of the atmosphere.) Because reproducible albedo features have been seen on Pluto's surface, any such aerosol layer must be transparent when viewed from above, but relatively opaque when viewed horizontally. The aerosols must also extend around most of the planet since the steepening of the occultation

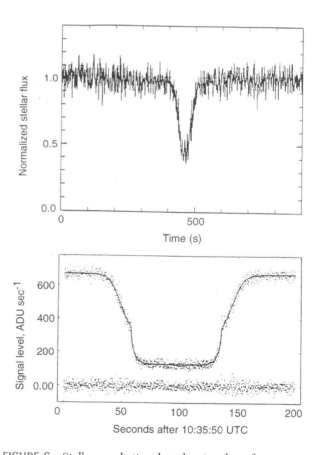

FIGURE 6   Stellar occultation data showing the refractive signature of Pluto's atmosphere and the steepening of the lightcurve around the half-light level that is discussed in the article. The upper panel is a ground-based data product; the lower panel was obtained from the *Kuiper Airborne Observatory*. (Adapted from Elliot et al., 1989, *Icarus*, **77**, 148.)

lightcurve was seen in both immersion and emersion. It has been suggested that the aerosols could be "photochemical smog" similar to the aerosols discovered on Titan and Triton (and a distant cousin to the air pollution in the industrial basins on Earth, such as Los Angeles).

In a second model, the sudden drop in the brightness of starlight below 1215 km was caused by a change in the vertical thermal structure of the atmosphere near the half-light level. Such a gradient is not unexpected from theoretical modeling (see earlier discussion) because atmospheric temperatures are expected to be higher than surface temperatures. Changes in atmospheric temperature cause a variation of refractivity with height in the atmosphere that could be manifested as the accelerated diminution of starlight seen in the occultation.

The haze layer and temperature gradient explanations imply differences in the way that the color of starlight changes during an occultation. Future occultations may help decide between the two explanations if simultaneous observations can be made at two or more well-separated wavelengths. If the temperature gradient explanation is correct, Pluto's surface radius is likely near 1206 ± 11 km. If the haze layer explanation is correct, Pluto's surface radius is more difficult to determine, but it is probably closer to 1180 km.

In either case, the occultation implies a radius that is a few percent larger than the mutual event solution (1151 km); in the case of the haze model, the radius cannot be much less than 1180 km or else the haze would be so thick as to completely obscure the surface. Clearly, there is a discrepancy between the radii determined from the occultation and those derived from the mutual events, which future research will have to resolve.

The subsequent well-observed occultation of a star by Pluto occurred on 20 July 2002. Both large fixed telescopes and small portable instruments observed the event. Fortuitously, yet another event occurred on 21 August 2002, which was successfully observed from large telescopes on Mauna Kea. From these events, it was determined that the "kink" or "knee" seen in the 1988 data is largely absent from the 2002 data, implying that large changes in Pluto's atmospheric thermal structure, or its haze profile, or both, occurred during the intervening interval.

Further analysis of the data reveals that the pressure in Pluto's atmosphere more than doubled between 1988 and 2002. This is likely due to pressure fluctuations associated with seasonal change, and may even be related to instabilities in the atmosphere prior to complete atmospheric collapse. Further observations will be required to sort this out.

Yet another occultation was observed on 12 June 2006. This event showed that the lightcurve kink near the half-light level remained less distinct than in 1988 and that the turbulence level in Pluto's lower atmosphere had increased.

Fortunately, Pluto is now moving through the dense star fields of Sagittarius, and several more occultation events are expected to be observed between 2007 and 2012.

6.3 Atmospheric Escape

A particularly interesting feature of Pluto's atmosphere is the very rapid rate at which it escapes to space. Because of Pluto's low mass and consequently weak gravitational binding energy, combined with the 100 K gas temperature in Pluto's upper atmosphere, sufficiently energetic molecules at the top of the atmosphere are able to escape the gravitational pull entirely. This can result in a condition called **hydrodynamic escape**, in which the high-altitude atmosphere achieves an internal thermal energy greater than the planetary gravitational potential energy acting on the atmosphere.

The time-averaged rate of escape from Pluto's atmosphere is likely to be of order 1–5×10^{27} molecules/second. This corresponds to a total loss of up to several kilometers of material from the surface over the age of the solar system.

Escape rate estimates also indicate that the present escape rate may be so high that Pluto's tenuous atmosphere may be lost to the escape process (thus requiring replenishment from sublimating surface ices) on timescales possibly as short as a few hundred years. Relatively speaking, the atmosphere of Pluto is escaping at a rate far greater than any other planetary atmosphere in the solar system!

Another interesting feature of Pluto's atmosphere is its strong orbital variability. This is driven by the fact that the strength of solar heating varies by a factor of almost 4 around Pluto's orbit, which in turn causes the vapor pressures of N_2, CO, and CH_4 to vary by factors of hundreds to thousands. Therefore, unlike any other planet, Pluto's atmosphere is thought to be essentially seasonal, with the perihelion pressure being many many times the aphelion pressure. Indeed, some models predict that between 2010 and 2020, just 2 or 3 decades after perihelion, Pluto's atmosphere will largely condense onto the surface, a condition called atmospheric collapse.

7. Charon

As previously described, Pluto's largest satellite, Charon, was discovered in 1978. Charon's radius of ≈ 604 km is about half of Pluto's, implying its mass is most likely between 10 and 14% of Pluto's. By comparison, typical satellite:planet mass ratios are 1000:1 or greater, and even the mass ratio of the Moon to the Earth is only 81:1.

The Pluto–Charon mutual event observations resulted in several key discoveries. These included the fact (1) that Charon's average visible surface albedo is 30–35%, much lower than Pluto's, and (2) that Charon's visible surface color

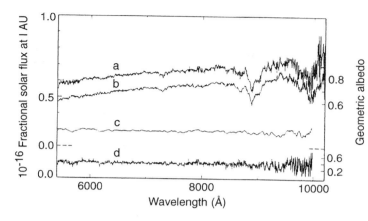

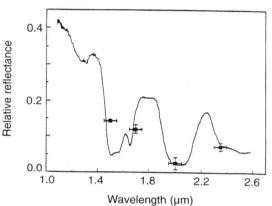

FIGURE 7   Pluto and Charon spectra. Top panel shows spectra of (a) Pluto + Charon made prior to eclipse; (b) Pluto-only after second contact with Charon hidden; (c) Charon-only smoothed to 80 Å resolution resulting from the subtraction of (a) − (b); and (d) the raw Charon-only spectrum resulting from the subtraction of (a) − (b). Notice that the strong methane absorption bands present in Pluto's spectrum are not detected in the Charon-only spectrum. (Adapted from Fink and DiSanti, 1988.) Bottom panel shows Marcialis et al.'s (1987) detection of water ice in Charon's reflectance spectrum (data points) against a laboratory spectrum of water ice at 55 K. (Adapted from Marcialis et al., 1987, *Science* **237**, 1349.)

is quite neutral, unlike Pluto's clearly reddish tint. Another major set of advances that resulted from the eclipse events was the first set of constraints on Charon's basic surface composition. These came from the subtraction of spectra made just prior to eclipse events from those made when Charon was completely hidden behind Pluto. The resulting "net" spectrum thus contains the Charon-only signal. As shown in Fig. 7, this technique has been applied both in the visible (0.55–1.0 μm) and infrared (1–2.5 μm) bandpasses. The visible light data show that Charon's surface does not display the prominent CH_4 absorption bands that Pluto does, indicating that Charon's surface has little or (more likely) no substantial methane on it.

Additionally, there is no evidence for strong absorptions due to a number of other possible surface frosts, including CO, CO_2, H_2S, N_2, or NH_4HS, on Charon. The IR spectra of Charon do show that Charon does, however, display clear evidence of water ice absorptions, which Pluto does not. It is tempting to speculate (as some authors have) that Charon may have lost its volatiles through the escape of a primordial atmosphere or by heating resulting from its formation in a giant impact.

Since the launch of *Hubble Space Telescope*, it has been possible to routinely separate Charon's light from Pluto's,

and to learn Charon's phase coefficient, UV albedo, and rotational lightcurve. Most notably among these, Marc Buie and Dave Tholen have determined that Charon displays a small but significant lightcurve variation near 8% as it rotates on its axis.

Because the major identified surface constituent of Charon is water ice, which is not volatile at the expected 50–60 K surface radiative equilibrium temperature at perihelion, one does not expect Charon to have an atmosphere. The fact that CH_4 is not present on the surface supports this expectation. However, absence of evidence is not the same as evidence of absence. One published interpretation of the 1980 Charon stellar occultation claims there is some evidence for a weak atmospheric refraction signal. To definitively resolve the issue of Charon's atmosphere, either a better-observed stellar occultation event or a spacecraft flyby is required.

In 2001, groups led by Mike Brown of Cal Tech and Will Grundy of Lowell Observatory used infrared ground-based telescopes to find spectroscopic evidence for both crystalline water ice and ammonia (NH_3) or ammonium hydrates on Charon's surface. If these identifications are correct, they imply the possibility of recent geologic activity on Charon.

TABLE 3	Pluto and Charon Comparison	
Parameter	Pluto	Charon
Rotation period	6.387223 days	6.387223 days
Radius	1150–1220 km	602–606 km
Density	≈ 2.1 gm cm^{-3}	≈ 1.3 gm cm^{-3}
Perihelion, V_0	13.6 magnitude	15.5 magnitude
Mean B geometric albedo	0.55	0.38
Rotational lightcurve	38%	8%
B-V color	0.85 mag	0.70 mag
V-I color	0.84 mag	0.70 mag
Known surface ices	CH_4, N_2, CO	H_2O, NH_3?
Atmosphere	Confirmed	None detected

Table 3 compares some basic facts about Pluto and Charon.

8. The Origin of Pluto's Satellite System

In the past few years, much progress has been made in understanding Pluto's likely origin and its context in the outer solar system. This work began with theoretical consideration in the late 1980s and early 1990s, and was advanced considerably by the discovery of numerous 100 to 1600 km diameter objects in the Kuiper Belt, where Pluto also resides.

Any scenario for the origin of Pluto must of course provide a self-consistent explanation for the major attributes of the Pluto–Charon system. These include (1) the existence of the exceptionally low, ~8.5:1 planet:satellite mass ratio of Pluto:Charon; (2) the synchronicity of Pluto's rotation period with Charon's orbit period; (3) Pluto's inclined, elliptical, Neptune-resonant orbit; (4) the high axial obliquity of Pluto's spin axis and Charon's apparent alignment to it; (5) Pluto's small mass (~10^{-4} of Uranus's and Neptune's); (6) Pluto's high rock content—the highest among all the outer planets and their major satellites; and (7) the dichotomous surface compositions of Pluto and Charon. This formidable list of constraints on origin scenarios is very clearly dominated by Charon's presence, the unique dynamical state of the binary, and the low mass of Pluto/Charon compared to other planets.

8.1 The Origin of Pluto's Satellite System

Several scenarios have been examined for the origin of the Pluto–Charon system. These include coaccretion in the solar nebula, mutual capture via an impact between proto-Pluto and proto-Charon, and rotational fission. Gravitational capture of Pluto by Charon without physical contact is not dynamically viable. The formation of Pluto and Charon together in a subnebular collapse is not considered realistic because of their small size; standard planetary formation theory suggests bodies in the Pluto and Charon size class formed via solid-body accretion of planetesimals. Similarly, the rotational fission hypothesis is unlikely to be correct because the Pluto–Charon system has too much angular momentum per unit mass to have once been a single body.

The more likely explanation for the origin of the Pluto system is an inelastic collision between two bodies, Pluto and proto-Charon, which were on intersecting heliocentric orbits. A similar scenario has been proposed for the origin of the Earth–Moon binary, based in part on its relatively high mass ratio (81:1) and high specific angular momentum. In the collision theory, Pluto and the Charon-impactor formed independently by the accumulation of small planetesimals and then suffered a chance collision that dissipated enough energy to permit binary formation.

An important qualitative difference between the Pluto–Charon and Earth–Moon giant impacts is that the relative collision velocities, and hence the impact energies of the Pluto–Charon event, were much smaller. This enormously reduced the thermal consequences of the collision. Thus, whereas the Earth may have been left molten by the Mars-sized impactor necessary to have created the Moon, the proto-Charon impactor would probably have raised Pluto's global mean temperature by no more than 50 to 75 K. This would have been insufficient to melt either body, but may have been sufficient to induce the internal differentiation of either. It would have also produced a substantial short-lived, hot, volatile atmosphere with intrinsically high escape rates. Such an escaping atmosphere could have interacted with the Charon-forming orbital debris, and also perhaps affected Pluto's present-day volatile content.

Until recently, only scaling calculations showing the plausibility of the giant impact hypothesis has been performed, and it was accepted largely because it is the only scenario that remains at all viable given the various constraints—most particularly the high specific angular momentum of the binary. In 2005, however, Robin Canup of the Southwest Research Institute published the first detailed giant impact simulations demonstrating the viability of Pluto–Charon formation owing to the collision of Pluto with another large body. Canup's work further demonstrated that the most promising candidate impacts involved an oblique collision by an impactor with 30–100% of Pluto's mass, approaching at a relative speed up to 1 km/s.

The discovery of Nix and Hydra in 2005 yielded additional support for the giant impact hypothesis. This support comes in two forms: the fact that all three satellites orbit in a single orbital plane and the near or perfect orbital period resonance of the three. The orbital coplanarity would be unlikely for other satellite formation mechanisms like capture, but naturally result from the giant impact scenario. The orbital resonance line of evidence naturally suggests the three bodies were together caught up in the outward tidal migration of Charon following its formation closer in.

8.2 The Origin of Pluto Itself

The presence of volatile ices, including methane, nitrogen, and carbon monoxide on Pluto, and water and other ices on Charon, argues strongly for their formation in the outer solar system. The average density and consequent high rock content of these two bodies also argues for formation from the outer solar nebula, rather than from planetary subnebula material. As described earlier, it is thought that the two objects (or more precisely Pluto and a Charon-progenitor) formed independently and subsequently collided, thus forming the binary either through direct, inelastic capture or through the accretion of Charon from debris put in orbit around Pluto by the impact.

The first widely discussed theory for Pluto's origin was R.A. Lyttleton's 1936 suggestion, which was based on the fact that Pluto's orbit is Neptune-crossing. In Lyttleton's well-remembered scenario, Pluto was formerly a satellite of Neptune, ejected via a close encounter between itself and the satellite Triton. According to Lyttleton, this encounter also reversed the orbit of Triton. Variants on the "origin-as-a-former-satellite-of-Neptune" hypothesis were later proposed. However, all these scenarios were dealt a serious blow by the discovery of Charon, which severely complicates the Pluto-ejection problem by requiring either (1) Charon to also be ejected from the Neptune system in such a way that it enters orbit around Pluto, or (2) Charon to be formed far beyond Neptune where Pluto currently orbits and then captured into orbit around Pluto (presumably by a collision).

Other strong objections to scenarios like Lyttleton's also exist. First among these is the fact that any object ejected from orbit around Neptune would be Neptune-crossing and therefore subject to either accretion or rapid dynamical demise. It is implausible that such an object would be transferred to the observed 2:3 Neptune:Pluto resonance, because stable 2:3 libration orbits are dynamically disconnected in orbital phase space from orbits intersecting Neptune. Further, because Pluto is less massive than Triton by about a factor of 2, it is impossible for Pluto to reverse Triton's orbit to a retrograde one, as is observed. Further still, Pluto's rock content is so high that it is unlikely that Pluto formed in a planetary subnebula. Of course, none of these facts were known until decades after Lyttleton made his original (and then quite logical) suggestion that Pluto might be a former satellite of Neptune.

As described in Section 2, it is likely that Pluto was caught in the 2:3 resonance and had its orbital eccentricity and inclination amplified to current values as Neptune migrated outward during the clearing of the outer solar system by the giant planets.

The heliocentric formation/giant collision scenario described earlier for the origin of the satellites can account for most of the major attributes of the system, including the elliptical, Neptune-crossing orbit, the high axial obliquities, and the ≈8.5:1 mass ratio. Further, the present tidal equilibrium state would naturally be reached by Pluto and Charon in 10^8–10^9 years—a small fraction of the age of the solar system.

Still, such a scenario begs two questions. First, why is Pluto so small? And, second, how could Pluto and the Charon-progenitor, alone in over 10^3 AU^3 of space, "find" each other in order to execute a mutual collision? That is, the giant impact hypothesis still fails to explain (1) the existence of Pluto and Charon themselves; (2) the very small masses of Pluto and Charon compared to the gas giants in general, and Neptune and Uranus in particular; (3) the fact that the collision producing the impact was highly unlikely; and (4) the system's position in the Neptune resonance.

In 1991, Alan Stern of the Southwest Research Institute suggested that the solution to (1)–(3) lies in the possibility that Pluto and Charon were members of a large population (300–3000) of small ($\sim 10^{25}$ g) **ice dwarf** planets present during the accretion of Uranus and Neptune in the 20–30 AU zone. Such a population would make likely the Pluto–Charon collision, as well as three otherwise highly unlikely occurrences in the 20–30 AU region: the capture of Triton into retrograde orbit and the tilting of Uranus and Neptune. Similar conclusions based on different considerations were reached by William McKinnon of Washington University in the late 1980s. According to this work, the vast majority of the ice dwarfs were either scattered (with the comets) to the Oort cloud or ejected from the solar system altogether by perturbations from Neptune and Uranus. Only Pluto–Charon and Triton remain in the 20–30 AU zone today, specifically because they are trapped in unique dynamical niches that protect them against loss to such strong perturbations.

If this is correct, it implies that Pluto, Charon, and Triton are important "relics" of a very large population of small planets, dubbed ice dwarfs first by Stern, which by number (but not mass) dominate the planetary population of the solar system. As such, these three bodies would no longer appear as isolated anomalies in the outer solar system and would be genetic relations from an ancient, ice dwarf ensemble, and therefore worthy of intense study as a new and valuable class of planetary body unto themselves.

8.3 The Context of Pluto in the Outer Solar System

When the existence of the ice dwarf population was first suggested, the solar system beyond Neptune appeared to only be inhabited by Pluto and the numerous comets scattered out of the planetary region during the accretion of the giant planets.

Since late 1992, however, our concept of the outer solar system has evolved considerably, owing to a rapid set of discoveries of faint (i.e., 22nd–25th astronomical magnitude) largish bodies orbiting between 30 and 50 AU in what is known as the Edgeworth–Kuiper Belt. The first such objects were detected by David Jewitt and Jane Luu using the University of Hawaii's 2.2 m telescope on Mauna Kea.

As of this writing at the end of 2006, over 1000 small worlds with diameters of 100 to 2000 km have been discovered in the Kuiper Belt, including some objects that have clearly been scattered out of the giant planets' region. Many of these are apparently in the 2:3 mean-motion resonance with Neptune that Pluto also occupies. The largest discovered Kuiper Belt body is almost Pluto's size. Some have satellites. Beyond the Kuiper Belt, in the so-called scattered belt, lie other large bodies, including 2003 UB313 (EoS) which is slightly larger than Pluto.

Because the Kuiper Belt census obtained to date has covered only a tiny fraction of the ecliptic sky, it is estimated that many times the discovered population exists. Current models of the population of the region between 30 and 50 AU from the Sun now indicate that some 100,000 or more objects with diameters larger than approximately 100 km and perhaps several billion comets 1–20 km in diameter reside there. The total mass of bodies currently in the 30–50 AU zone may amount to as much as 0.01 $M_\oplus$, exceeding the mass of the Asteroid Belt by more than an order of magnitude.

Interestingly, various collisional evolution models that have been developed by Don Davis (PSI), Paolo Farinella (deceased), Alan Stern (SwRI), and Scott Kenyon (Harvard) have provided strong evidence that the 100 km diameter and larger bodies detected in the Edgeworth–Kuiper Belt could not have grown there in the age of the solar system, unless the mass of the primordial Kuiper Belt region was many times higher—in the range of 10 to perhaps 50 Earth masses.

Both the discovery of the rapidly expanding cohort of objects found in the 30–50 AU zone, and the circumstantial evidence that this region of the solar system was much more heavily populated when the solar system was young, finally provide a context for Pluto (and the putative Charon-progenitor as well). We now see that Pluto did not form in isolation and does not exist so today. Instead, Pluto is simply one of a large number of significant miniplanets that grew in the region beyond Neptune when the solar system was young. Pluto's presence there today is in large measure due to its location in the stable 2:3 resonance with Neptune. The question now has moved from why a small planet like Pluto formed in isolation, to why a large population of objects hundreds and thousands of kilometers in diameter formed in the 30–50 AU zone without progressing to the formation of a larger planet there. Perhaps the answer lies in the influence of "nearby" Neptune.

Bibliography

Binzel, R. P. (1990). Pluto. *Sci. Am.*, **252** (6), 50–58.

Stern, S. A. (1992). The Pluto–Charon system. *Ann. Rev. Astron. Astrophys.* **30**, 185–233.

Stern, S. A., and Mitton, J. (2005). "Pluto and Charon: Ice-Dwarfs on the Ragged Edge of the Solar System," 2nd Ed. John Wiley & Sons, New York.

Stern, S. A., and Tholen, D. J., eds. (1997). "Pluto and Charon." Univ. Arizona Press, Tucson.

Tombaugh, C. W., and Moore, P. (1980). "Out of the Darkness: The Planet Pluto." Stackpole Books, Harrisburg, Pennsylvania.

Whyte, A. J. (1980). "The Planet Pluto." Pergamon Press Ltd., Toronto, Canada.

Physics and Chemistry of Comets

John C. Brandt

Department of Physics and Astronomy
University of New Mexico
Albuquerque, New Mexico

CHAPTER 30

The spectacular sight of a bright comet with a tail stretching across the sky (Fig. 1) prompts questions about the nature of the object and the physical processes at work. The current era is one of major comet research, with several space missions to comets producing pioneering results. The images and data that are becoming available often prompt new questions and challenge old ideas. Two decades have passed since the first space missions to comets, and comet science has reached a level of maturity that was unimaginable not long ago.

1. Space Missions to Comets

Many lines of evidence indicate that the source of all cometary phenomena is a rather small central body called the **nucleus**. Typical dimensions are in the range 1–10 km. Viewing an object, say, 3 km across from a distance of 0.2 AU (or 3×10^7 km) means that the object subtends an angle of 1/50th of an arc second. Typical resolution from ground-based observatories is about 1.0 arc second and, for large telescopes, is due to the effects of the Earth's atmosphere. Mountaintop observatories in good locations can do better, and the *Hubble Space Telescope (HST)* has a resolution of about 0.1 arc seconds. From Earth, except in extraordinary circumstances, the nucleus cannot be resolved, and no

detail on the surface can be seen. The solution is to send spacecraft with imaging systems close to the cometary nuclei. In situ measurements of gas, dust, plasma, magnetic fields, and energetic particles can be obtained while the spacecraft is near the comet. The imaging and in situ data provide a major source of information on comets.

Table 1 summarizes completed missions to comets. Of course, analysis often continues for years. In this section, only the missions with imaging are discussed. The missions to comet Halley in 1986 were collectively called the Halley Armada, and three of them had imaging. Two *VEGA* spacecraft were sent by the Soviet Union and passed within 8890 km (*VEGA 1*) and 8030 km (*VEGA 2*) of the comet. The images from the *VEGA*s are valuable, but they were somewhat noisy and were taken from larger distances than those taken by *Giotto*.

The European Space Agency (ESA) sent the *Giotto* spacecraft to pass the nucleus of Halley's comet within 596 km. The spacecraft carried the Halley Multicolor Camera (HMC), which obtained images of the nucleus until approximately the time of closest approach when it was damaged by the impacts of dust particles. Figure 2A is an overall view of the nucleus composed of 68 individual images. The nucleus was not spherical but was a potato-shaped object with a long axis of approximately 15 km and short axes of approximately 7 km. The nucleus showed features

FIGURE 1   Comet Hale–Bopp on 8 April 1997, showing the whitish dust tail and the blue plasma tail. (Courtesy of H. Mikuz, Crni Vrh Observatory, Slovenia.)

that appeared to be valleys, hills, and craters. The average albedo or reflectivity of the surface was only 0.04; the surface was very dark. The **jets** containing the dust and gas emission from the nucleus came from approximately 10% of the entire surface and were active when their location was in sunlight. The direction of emission was generally sunward.

The National Aeronautics and Space Administration (NASA) sent the *Deep Space 1* spacecraft to within about 2171 km of the nucleus of comet Borrelly on 22 September 2001 (Fig. 2B shows a close-up view of the nucleus). The long axis of the nucleus is approximately 8 km and the short axes are about 3.2 km. The surface showed features and was also very dark. The albedo varied between 0.01 and 0.03 over the surface. The jets with dust and gas emission, which were clearly seen, occupied 10% or less of the surface area.

NASA's *Stardust* spacecraft passed within approximately 236 km of comet Wild 2 on 2 January 2004. Excellent images were obtained, and an example, taken just after closest approach, is shown in Fig. 2C. The nucleus is roughly a rounded body with a diameter of 4 km. Jets of dust and gas emission were seen, and the albedo determined was 0.03 ± 0.015. Features with steep slopes have been identified, providing clues to the history of the surface. The main goal of the *Stardust* mission, which is to return to Earth dust samples collected in the comet's **coma**, has been achieved with the return of samples that parachuted to the Utah desert on 15 January 2006 (see the discussion in Section 4).

The *Deep Impact* impactor spacecraft collided with comet Tempel 1 on 4 July 2005. The impactor spacecraft separated from the flyby spacecraft 24 hours before impact, and the flyby spacecraft passed the nucleus at a distance of 500 km. An image of the nucleus taken from the impactor is shown in Fig. 2D. The average diameter of the nucleus is close to 6.0 km, the longest dimension is 7.6 km, and

TABLE 1	Missions to Comets		
Spacecraft	**Comet**	**Encounter Date**	**Imaging**
International Cometary Explorer (ICE)	Giacobini–Zinner	11 September 1985	No
VEGA 1	Halley	6 March 1986	Yes
Suisei	Halley	8 March 1986	No
VEGA-2	Halley	9 March 1986	Yes
Sakigake	Halley	11 March 1986	No
Giotto	Halley	14 March 1986	Yes
ICE	Halley	25 March 1986	No
Giotto Extended Mission (GEM)	Grigg–Skjellerup	10 July 1992	No
Deep Space 1	Borrelly	22 September 2001	Yes
Stardust	Wild 2	2 January 2004	Yes
Deep Impact	Tempel 1	4 July 2005	Yes

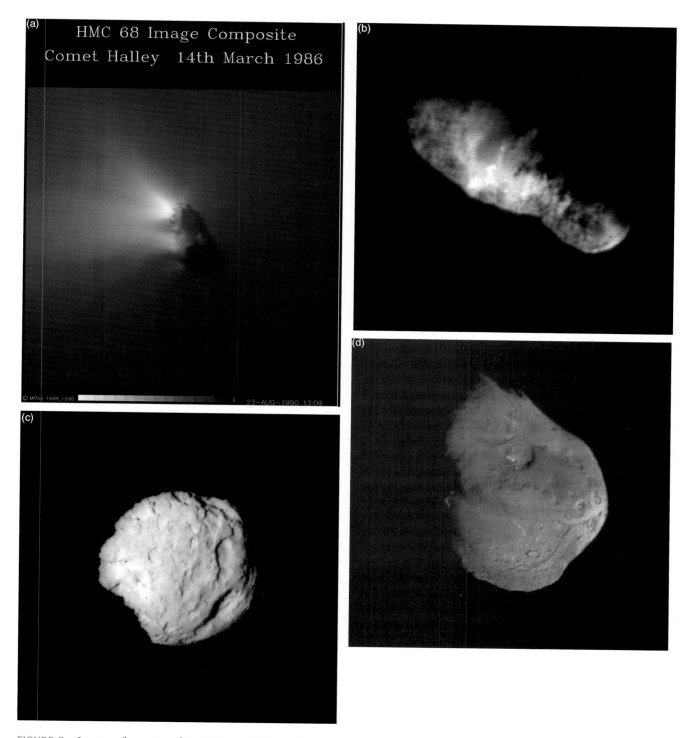

FIGURE 2   Images of comet nuclei. A. Comet Halley nucleus composite. (Courtesy of H. U. Keller, Max-Planck-Institut für Aeronomie, Katlenburg-Lindau, Germany © MPAE.) B. Comet Borrelly. (Courtesy of NASA/JPL.) C. Comet Wild 2. (Courtesy of NASA and the *Stardust* Mission Team.) D. Comet Tempel 1. (Courtesy of NASA/JPL-Caltech/UMD.)

the shortest dimension is 4.9 km. The surface shows both smooth and rough terrain, scarps (a line of cliffs usually produced by faulting), and impact craters. The surface is generally homogeneous in color and albedo, which varied from 0.02 to 0.06, and the temperature of the surface indicates an equilibrium with sunlight. Observations on approach detected numerous short outbursts that can be associated with specific regions on the surface.

FIGURE 3 Spectacular image of comet Tempel 1 taken from *Deep Impact's* flyby spacecraft 67 seconds after the impactor spacecraft's impact. The linear spokes of light radiate away from the impact site. Light from the collision site saturated the camera's detector. Compare with Figure 2D. (Courtesy of NASA/JPL-Caltech/UMD.)

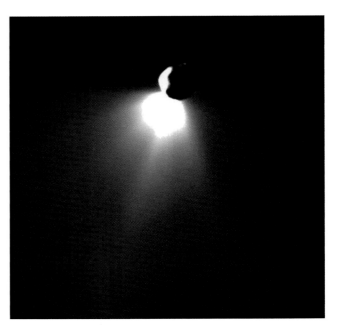

FIGURE 4 Image of comet Tempel 1 taken from *Deep Impact's* flyby spacecraft 50 minutes after impact showing the plume of ejected material. The comet's nucleus is mostly in shadow with the sunlit portion visible on the right-hand side. (Courtesy of NASA/JPL-Caltech/UMD.)

The impactor spacecraft delivered 19 GJ of kinetic energy to comet Tempel 1. The spectacular impact is shown in Fig. 3, and a view of the ejecta plume containing $\sim 10^6$ kg of material is shown in Fig. 4. In addition to observations from *Deep Impact*, the event was extensively observed by ground-based and space-based observatories. By 9 July the comet had returned to its pre-impact state; the impact crater has not been seen. The ejecta consisted of fine particles (1–100 μm) and individual species, including water, **water ice**, carbon, carbon dioxide, hydrocarbons, and crystalline silicates. The spectra of the ejecta are a good match to the spectra of material ejected from comet Hale–Bopp and to the dusty disk spectrum of a young stellar object.

All these images confirm the basic view of the nucleus as a single, sublimating (direct-phase transition from the solid to the gas state) body as proposed by F. L. Whipple. As the solid body approaches the Sun, energy supplied by solar radiation raises the temperature of the near-surface layers, sublimation of ices (mostly water ice) takes place, and the emission of gas and entrained dust produces the large features seen in the sky.

Before leaving space missions, it is important to note that ESA's *Rosetta* mission to comet Churyumov–Gerasimenko was launched on 2 March 2004 to begin its 10-year journey to the comet. The plan is for the main spacecraft to spend approximately 2 years in the vicinity of the comet and to place a lander on the surface.

2. A Brief History of Comet Studies

The realization that the nucleus of a comet was a single, sublimating body prior to the confirmation by direct imaging was the result of several lines of reasoning. In the 17th century, it was known that the part of a comet's orbit near the Sun could often be accurately represented by a parabola with the Sun at the focus. This idea was used by Isaac Newton to determine a parabolic orbit for the comet of 1680. Edmond Halley refined the calculation and showed that an ellipse of high eccentricity very accurately represented the comet's orbit. Comet orbits generally are ellipses with high eccentricities. Halley continued to determine the orbits of comets and found that the orbits of comets observed in 1531, 1607, and 1682 were quite similar and had periods of approximately 75 to 76 years. This was the basis of his famous prediction that the comet that now bears his name would return in 1758.

A complication in the detailed orbit calculations was that Jupiter and Saturn would perturb the orbit through their gravitational attraction. Halley's comet passed perihelion in early 1759. The successful prediction of the return of Halley's comet began the development of celestial mechanics and the positional astronomy of comets that flourished in the 18th and 19th centuries. But the orbit of comet Encke presented another problem. The comet had a very short period of 3.3 years. Many orbits were observed, and it would

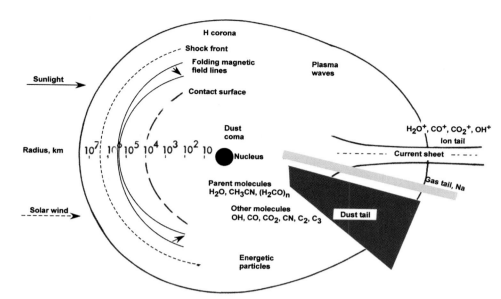

FIGURE 5   Summary schematic on a logarithmic scale of cometary features and phenomena. (Reprinted with permission from John C. Brandt and Robert D. Chapman, "Introduction to Comets," 2nd Ed., Cambridge Univ. Press, Cambridge, United Kingdom. Copyright © Cambridge University Press, 2004.)

typically arrive at perihelion about 0.1 day early. The only explanation for this behavior was some sort of nongravitational force, and the only version that has stood the test of time is a "rocket effect" produced by the ejection of material in a preferential direction. Such an effect was suggested by F. W. Bessel based on his observations of a sunward plume of material in Halley's comet in 1835. But how would such a plume of material be produced?

Another problem was the persistence of comets after many passes through the inner solar system. Comets are rich in water ice (discussed later), and small icy clumps or a surface layer of ice on dust grains would not persist.

The Whipple model solves these problems by postulating that the nucleus is a single, rotating, icy body. Ices are poor conductors of heat, and only a relatively thin layer is lost during a perihelion passage. The rocket effect is produced by the reaction force on the nucleus due to the sublimating ices. Historically, the mass loss due to sublimation of ices was assumed to come preferentially from the afternoon side. Just as on Earth, the warmer temperatures would occur in the afternoon, and the sublimation rate is higher. This type of mass loss would accelerate or retard the comet in its orbit. This basic type of nongravitational force model was used for decades and was successful in producing accurate ephemeris predictions. Nevertheless, the basic model is not realistic when complications are considered, such as the mass loss occurring in jets and precession of the rotation axis. Physically sound models require detailed models of the outgassing surface features and the nucleus rotation. The sublimation of the ices produces the gas molecules that form the gas coma and subsequently the **plasma tail**. When the ices sublimate, the embedded dust particles are released to form the dust coma and the **dust tail**. The dust particles that are not carried away or that fall back onto the nucleus form an insulating crust on the surface.

The bright coma and tails of comets are the features that distinguish them from other solar system objects. Their study was greatly facilitated during the 20th century by the development of photography. Images and spectra of comets could be accurately recorded and analyzed. The gas and dust comas could extend to approximately 10^5–10^6 km. The nucleus and the coma surrounding it form the comet's head. Dust tails could achieve lengths of roughly 10^7 km, and plasma tails often could achieve lengths of tenths of AU (or several times 1.5×10^7 km). In exceptional cases, plasma tails can exceed 1 AU (or 1.5×10^8 km) in length. Figure 5 shows a summary of comet features. Subsequent sections present the physical processes that produce features with these large dimensions, all originating from the small icy bodies shown in Section 1.

Traditionally, comet orbits were classified as short period or long period with the dividing line at periods (P) of 200 years. Currently, three groups of comets classified by their orbits are considered. The Jupiter family contains comets with periods $P \leq 20$ years. These orbits are direct (in the same sense as the Earth's revolution around the Sun) and generally have low inclinations with respect to the plane of the ecliptic. Halley-type comets have periods $20 < P \leq 200$ years. The long-period comets have $P > 200$ years, and their orbital inclinations to the plane of the ecliptic are approximately isotropic.

3. Physics of the Nucleus

The basic physical process—the one that ultimately produces the cometary features (e.g., the tails)—is sublimation of ices. Sublimation is the phase transition that goes directly from the solid to the gaseous state without passing through the liquid state. The evidence for the ice composition of the nucleus—80–90% H_2O (water) ice; roughly 10% CO (carbon monoxide) ice; and small amounts of other ices—is presented in Section 6. The ice in a cometary interior is almost surely amorphous ice. This comes about because ices formed by condensation on a surface at low temperatures do not have energy available to change into the crystalline forms that minimize energy.

When the water ice or snow sublimates, a water vapor is produced, and embedded dust particles are released. The energy sources for the sublimation are solar radiation, ice phase transitions, and radioactive decay. Solar radiation deposits energy on the surface or in the near-surface layers. This energy affects the deeper layers by producing a heat wave that moves inward. The transition from amorphous ice to crystalline ice releases energy. Amorphous ice undergoes a transition to cubic ice at approximately 137 K, and cubic ice undergoes a transition to hexagonal ice at approximately 160 K. Model calculations usually treat both transitions as a single energy release event. Radioactive decay is primarily from short-lived isotopes, such as ^{26}Al. This source is most important in the deep interior and when the nucleus is far from the Sun and diffusion of volatiles could result.

Insight into the production of cometary features and the energy balance for the surface region is illustrated in Fig. 6. Considered here is the simple case of energy input from solar radiation only, with no heat wave into the interior. When a comet is far from the Sun, the energy balance is achieved by the solar radiant energy being reradiated by blackbody (infrared) radiation. The temperature of the surface layers is not high enough to produce significant sublimation. At intermediate distances, the surface temperature is high enough for sublimation, and the solar radiant energy input is balanced both by blackbody reradiation and by sublimation. At closer distances to the Sun, the surface temperature increases further, and essentially the entire solar radiation input is balanced by sublimation. Of course, blackbody reradiation takes place, but it is small in terms of the energy balance. For water ice, sublimation becomes important around 3 AU, and the energy balance (primarily through sublimation) occurs near 1 AU. This copious production of material drives cometary activity and produces cometary features as described later.

Naturally, there are complications to this simple picture. When the surface layer ices are sublimated, not all of the dust is liberated, and a porous dust mantle is formed. The mantle insulates the ices beneath the surface. This idea has been confirmed observationally. Infrared observations of the surface layers indicate temperatures reasonably close to values expected for a nonsublimating, low-albedo object bathed by sunlight. These temperatures are much higher than the temperatures for sublimating water ice. The ice sublimation probably takes place a few centimeters below the surface. Also, there is no reason to believe that sublimation takes place uniformly over the surface. Regions of enhanced sublimation are expected, a view consistent with

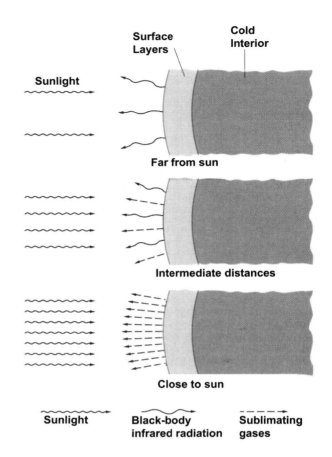

FIGURE 6   Energy-balance regimes for different distances from the Sun. No heat flow into the interior is considered. Only the principal components of the energy balance are shown. Some sublimation occurs far from the Sun and some blackbody reradiation occurs close to the Sun. (Reprinted with permission from John C. Brandt and Robert D. Chapman, "Introduction to Comets," 2nd Ed., Cambridge Univ. Press, Cambridge, United Kingdom. Copyright © Cambridge University Press, 2004.)

the images of comet nuclei that show dust and gas emission predominantly in jets. These jets can produce some of the surface features on the nucleus, and, along with impact craters, they can produce an irregular shape for the nucleus.

Figure 7 shows how the surface layers of a comet can become stratified and illustrates the potential complexity of accurate modeling. These layers include many intermediate stages—from the pristine composition of the deep interior to the ejected gas and dust—and these must be modeled accurately. The details of the gas flow through the porous dust layers are important. In recent years, the trend has been to think of the nucleus as a fairly porous body. The porosity is defined as the fraction of the volume occupied by the pores, and values of roughly 0.5 are often discussed. At present, such values can apply to some, but probably not all, comet nuclei.

The rotation of comet nuclei provides an example of how complex some situations can become. Given the extensive

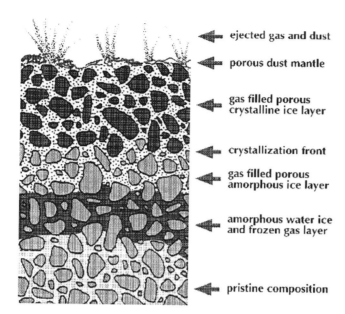

FIGURE 7 Schematic showing the layered structure of a cometary nucleus from the pristine composition up to the porous dust mantle. The vertical scale is arbitrary. (Courtesy of D. Prialnik, Department of Geophysics and Planetary Science, Tel Aviv University, Israel: from Prialnik, 1997–1999, Modeling gas and dust release from comet Hale–Bopp, *Earth, Moon, and Planets*, 77: 223–230, Figure 1. Copyright © 1999, with kind permission of Springer Science and Business Media.)

FIGURE 8 The complex rotation of comet Halley's nucleus through one full sequence. The images read left to right starting at top left. The time between images is 0.25 days and the sequence repeats after *approximately* 7.25 days. The five active areas (jets) are marked as low-albedo features. (Courtesy of M. J. S. Belton, Belton Space Initiatives.)

ground-based observations of comet Halley and the close-up images taken by *VEGA 1*, *VEGA 2*, and *Giotto*, the determination of the rotation was expected to be straightforward. An initial complication was the reports of different periods of brightness variation. Sorting things out was a major effort. In short, the rotation was complex, and a model with five jets was needed to reproduce the observations. Figure 8 shows views of the rotating nucleus through an entire period. The solution was consistent with a constant internal density.

The rotation state determined for comet Halley is interesting because it is not in the lowest rotational energy state for a given angular momentum. This would be rotation only around the short axis. The excited rotational state is probably not primordial because estimates of the relaxation time due to frictional dissipation in the comet's interior are in the range 10^6–10^8 years. It is probably due to jet activity or splitting of the nucleus.

The splitting of comet nuclei has been observed many times. A recent example is the case of comet LINEAR in early August 2000 (see Fig. 9). Large pieces and fragments of the nucleus are visible in the images. Most of the fragments have an estimated size of less than 500 m. This is an example of "spontaneous" splitting (i.e., there is no apparent correlation with orbital parameters or time in the orbit relative to perihelion). This type of splitting occurs for roughly 10% of dynamically new comets on the first perihelion passage. Splitting can also occur when the nucleus

passes close enough to the Sun or a planet and is tidally disrupted. Comet Shoemaker–Levy 9 passed close to Jupiter in July 1992. The disruption produced about 20 fragments (see Fig. 10). These crashed into Jupiter over several days in July 1994. The tide-induced splittings have been used to estimate the tensile strength of the nuclei, and very low values were found. The units of tensile strength are force per unit area (N m^{-2}) or the pascal (Pa). The inferred values from splittings are in the range 10^2–10^4 Pa. For comparison, rocks have values $\sim 4 \times 10^6$ Pa, and the value for steel is $\sim 4 \times 10^8$ Pa.

The splittings are consistent with the view of the cometary interior as being porous, having a weak structure, and perhaps consisting of agglomerated building blocks called cometesimals. Available evidence indicates that the interior consists of volatile ices (mostly H_2O ices, probably amorphous) and dust. The interior does not appear to be differentiated, the compositions are surprisingly uniform, and the ratio of ice to dust does not vary with depth.

Cometary outbursts may be related to splittings. In a major outburst, the brightness of a comet increases by a factor typically of 6–100, and the outburst lasts for weeks. The observational evidence indicates that the increase in

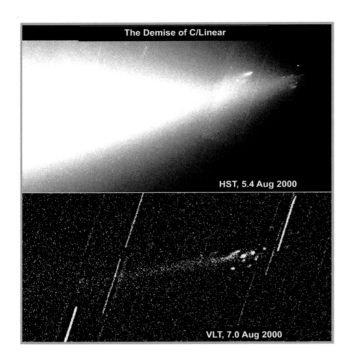

FIGURE 9 Splitting of comet LINEAR. (Top) The *Hubble Space Telescope* (*HST*) image on 5.4 August 2000 showing the dust tail (extending to the left) and several large remnants near the tip of the tail; 22 July 2000 is the estimated date of disintegration of the nucleus. (Bottom) The *Very Large Telescope* (*VLT*) image on 7.0 August 2000 showing fragments. Image processing was used to suppress light from the diffuse tail. The streaks are star trails. (Courtesy of H. Weaver, Johns Hopkins University; C. Delahodde, O. Hainaut, R. Hook, European Southern Observatory; Z. Levay, Space Telescope Science Institute; and the *HST/VLT* observing team; NASA/ESA, ESO.)

brightness is due to an increase in the number of dust particles that scatter sunlight. Comet Halley displayed an extraordinary outburst on 12 February 1991 when it was 14.3 AU from the Sun.

Splitting exposes fresh ice surfaces and hence produces enhanced loss of material. An impact from an interplanetary boulder would have much the same effect. A plausible mechanism not involving splitting or impacts uses the crystallization of amorphous ices as the energy source. On this picture, a heat wave propagates inward, triggering the energy release from the amorphous ice and producing pockets of gas that break through to the surface to produce the outburst. This mechanism is plausible for the outburst in comet Halley (mentioned earlier) and in comets that have repeated outbursts, such as comet Schwassmann–Wachmann 1.

A summary schematic of a comet nucleus is shown in Fig. 11. Because our knowledge of interiors is insecure, the figure presents processes at work rather than a specific interior. Some hints about nucleus structure have already come from the *Stardust* images of comet Wild 2. The nucleus shows a highly structured surface that can be described as pockmarked. Some of the features, possibly impact craters, have steep slopes, and the surface must have some cohesive strength. How did these sharp features persist if layers were peeled off by sublimation during every perihelion passage? For comet Wild 2, the answer lies in its orbital history. Comet Wild 2 was captured into its current (Jupiter-family) orbit by a close encounter with Jupiter only 30 years ago. With an orbital period of 6.4 years, this comet has probably made only a handful of passes through the inner solar system. By comparison, comet Halley has probably made

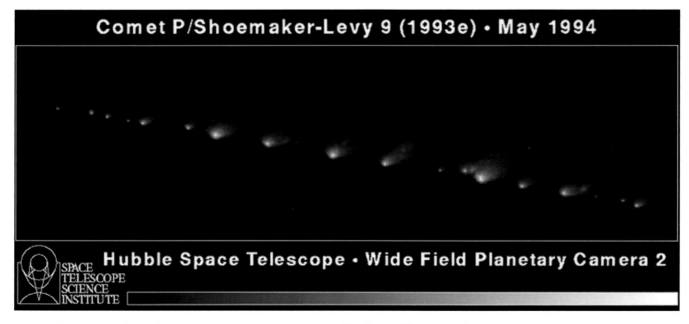

FIGURE 10 Comet Shoemaker–Levy 9 on 17 May 1994 as imaged by the *Hubble Space Telescope*. The fragments extended for over 1.1 million km. (Courtesy of H. A. Weaver and T. E. Smith, Space Telescope Science Institute/NASA.)

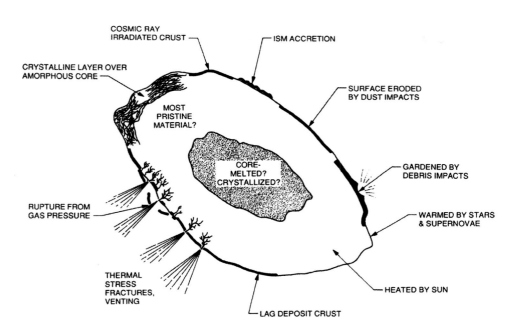

FIGURE 11   Schematic of a
cometary nucleus illustrating the
physical processes at work.
(Courtesy of P. R. Weissman,
NASA-JPL.)

hundreds or thousands of inner solar system passes and thus
has a surface smoothed by many sublimation episodes. The
surface of comet Wild 2 appears young in terms of sublimation exposure. The steep slopes, which imply some cohesive
strength, mean that the surface does not resemble a pile of
material held together by gravity. The results from *Deep
Impact* have raised new questions and begun the process of
understanding the nucleus. The surface geology of comet
Tempel 1 shows clearly distinct layers that seem to be discrete blocks like geologic strata. The surfaces of the three
Jupiter-family comets (Borrelly, Wild 2, and Tempel 1) are
quite different, and this fact challenges the notion of a typical comet. Analysis of the ejecta and its evolution yields
the following results. The fine particles seen in the ejecta
must be from a surface layer at least tens of meters deep.
The tensile strength was estimated at 65 Pa or less. This is
comparable to the strength of talcum powder. The density
of the nucleus is about $0.6\,g\,cm^{-3}$ meaning that the interior
must be porous with some 50–70% of the volume consisting
of empty space.

4. Coma and Hydrogen Cloud

The gas and dust liberated by the sublimation of the ice
is the origin of comet features with large dimensions. The
coma is the essentially spherical cloud around the nucleus of
neutral molecules and dust particles. It is visible in images of
comets with low gas-production rates (Fig. 12) or in short-
exposure images of comets with high production rates. The
dusty gas expands at speeds of $\approx 1\,km\,sec^{-1}$, and the flow
is transonic in that the flow begins subsonic and becomes
supersonic. This is similar to the flow of the solar wind. [*See*

THE SOLAR WIND.] Because the gas is dragging the dust
along, the gas flows faster than the dust. Images and in situ
measurements show that the material emission from the
nucleus is structured into jets in the near-nucleus region.
Well away from the nucleus, this structure is not usually
important. The size of comas can range up to 10^5–10^6 km.

Most of our observations and measurements of composition in comets refer to the coma region. For some species,
the variation with radial distance from the nucleus can be
modeled by including creation and destruction mechanisms
for parent and daughter molecules. For a molecular gas
expanding radially at constant speed, the density falls off
as r^{-2} (r is the distance from the nucleus), and the surface
brightness (proportional to an integral along a line of sight

FIGURE 12   Comet Giacobini–Zinner on 26 October 1959
showing the coma and a plasma tail extending some 450,000 km.
(Photograph by E. Roemer, University of Arizona: official U. S.
Navy photograph.)

through the coma) falls off as ρ^{-1} (ρ is the projected distance from the nucleus). The slope on a log B (brightness) versus log ρ plot would be -1. Shallower slopes indicate a creation process, and steeper slopes indicate a destruction process. This behavior is observed in molecules such as C_2. These results and results from more detailed modeling lead to an important conclusion.

The molecules measured and observed in the coma are not necessarily the molecules coming directly from the nucleus, but they are part of a chain of creation and destruction of species, presumably from complex molecules in the nucleus to progressively simpler molecules with increasing distance from the nucleus. Thus, the molecules observed are simply the ones that are caught at some specific distance from the nucleus or with the method of observation.

Calculations that include the various changes in composition with the goal of understanding the composition of the original material from the nucleus are very complex and must include gas-phase reactions and photolytic (involving photons) reactions as well as possible interactions between the gas and dust. While progress has been made, final reso-

lution of this problem may require measurements obtained at a cometary surface. As discussed in Section 6, the knowledge of the bulk composition of comets seems secure and is consistent with condensation from a cloud that initially had solar abundances.

Table 2 (in Section 6) lists chemical species observed spectroscopically and measured by mass spectrometry in comets. This table shows the variety of species in comets and the similarity to interstellar material. This relationship is discussed in Section 6.

The **hydrogen cloud** around comets is much larger than the coma but was not observed until the 1970s. Its existence was predicted in 1968 by L. Biermann. Observations above the Earth's atmosphere were needed because the hydrogen cloud is best seen in Lyman-α (121.6 nm), the resonance line of hydrogen. Figure 13 shows the hydrogen cloud of comet Hale–Bopp along with a visible light image. The huge size of the cloud is shown by the yellow disk at the right. This disk is the angular size of the Sun at the comet's distance. The hydrogen cloud has the largest size; however, smaller clouds of oxygen and carbon are also seen.

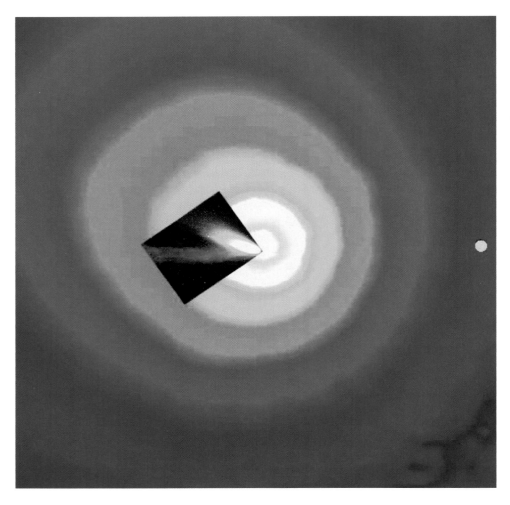

FIGURE 13   Hydrogen Lyman-a image taken on 1 April 1997 showing the hydrogen cloud of comet Hale–Bopp (contours in shades of blue) along with a visible image showing the plasma and dust tails. The image is approximately 40° on a side. The small yellow disk shows the angular size of the Sun and the solar direction. (Courtesy of M. Combi, University of Michigan; visual photograph by Dennis di Cicco and *Sky and Telescope*.)

Modeling the outflow of hydrogen (the lifetime of the H atoms is determined primarily by the proton flux in the solar wind) to produce the observed cloud size shows that the required outflow speed is 8 km sec^{-1}. This is much larger than the outflow speed in the coma, ≈ 1 km sec^{-1}. An additional energy source is needed. If H_2O were photodissociated, a speed of 19 km sec^{-1} would result, and this value is too high. The likely scenario is that OH is produced by photodissociation and then is further dissociated into H outside the thermalization region. These H atoms and the thermalized H atoms from H_2O photodissociation combine to give the deduced outflow speed of 8 km sec^{-1}.

The outflow rate of hydrogen, Q_H, provides a good surrogate for the total gas production rate from a comet. For large comets, this rate can approach 10^{31} atoms sec^{-1}, and the general range is 10^{27}–10^{30} atoms sec^{-1}. The heliocentric variation is roughly $r_h^{-1.3}$ (r_h = the heliocentric distance). This expression follows the practice of basing variations on the value at 1 AU (where comets are most easily observed) and using a power law to give the heliocentric variation.

Early dust measurements were made in the coma of comet Halley by dust detectors on the *VEGA* spacecraft and on *Giotto*. Three basic types of dust composition were found. The CHON particles have only the light elements **C**arbon, **H**ydrogen, **O**xygen, and **N**itrogen. The silicate particles are rich in Silicon, Magnesium, and Iron. The third type is essentially a mixture of the CHON and silicate types. The differential size distribution can be represented by a power law in size, r^a, with $a \sim -3.5$, for grain sizes greater than 20 μm. This implies that most of the dust mass is emitted in large grains. There was also evidence for large numbers of small dust grains down to sizes of 0.01 μm. The results are compatible with the sizes needed in models of the dust tail.

Interest in the dust particles from the coma has increased with the return to Earth of the dust collected by *Stardust*. Some of the coma dust particles may be similar to the fluffy particles collected in the Earth's upper atmosphere, the interplanetary dust particles (IDP) or "Brownlee particles." But, having particles collected in the coma and available for analysis in the laboratory opens a whole new era. The sample return portion of the *Stardust* mission to comet Wild 2 was accomplished by catching the particles in an ultra low-density glass-like material called aerogel. The collection exceeded expectations with thousands of particles embedded in the aerogel. The mineral structure has been preserved for many of the grains. Some first results indicate the presence of high-temperature minerals such as olivine, one of the most common minerals in the universe. It certainly did not form inside the comet's cold body. It probably formed near the Sun or from hot regions around other stars. In any event, the discovery that cometary material contains substances formed in hot and cold environments adds a new constraint to formation scenarios.

5. Tails

The dust and gas in the coma are the raw materials for the comet's tails. The prominent dust and gas (plasma) tails are the traditional identifying characteristic of comets. Dust tails are flat, curved structures and, compared to plasma tails, are relatively featureless. They can reach lengths $\sim 10^7$ km.

Dust particles, once they are decoupled from the coma gas, are in independent orbits around the Sun. But the solar gravitational attraction is not the full value because the dust particles generally stream away from the Sun. An extra force, solar radiation pressure, is acting on the particles. Because both solar gravity and radiation pressure vary as r_h^{-2}, the orbit is determined by initial conditions and an effective gravity. The parameter μ is the ratio of the net force on the tail particle to the gravitational force. Or, the parameter $(1 - \mu)$ gives the normalized nongravitational force.

For a constant emission rate of dust particles with a single size or a small range of sizes, the *syndyne* (or same force) from the Bessel–Bredichin theory is a good description. The tails are tangent to the radius vector (the prolonged Sun–comet line) at the head, and the curvature of the tail increases with decreasing $(1 - \mu)$. An important concept is the fact that the shape of a particle's orbit is not the observed shape of the tail. The observed tail shape is the locations of dust particles at a specific time of previously emitted particles.

Another case from the Bessel–Bredichin theory is the *synchrone* (or same time). It is produced by particles with many sizes [or values of $(1 - \mu)$] being emitted at the same time. These features are rectilinear, and the angle with the radius vector increases with time. This type of feature is occasionally observed as synchronic bands.

In practice, comets emit dust particles with a range of sizes and at a rate that varies with time. Several computational approaches that accurately model observed dust tails with reasonable assumptions are available. The size distribution generally peaks at a diameter around 1 μm.

Besides the synchronic bands (mentioned earlier), fine structure in the form of *striae* occasionally appear in dust tails. They are a system of parallel, narrow bands found at large distances from the head. So far, striae appear at heliocentric distances greater than 1 AU and always after perihelion. Figure 14 shows a spectacular example in comet Hale–Bopp. Currently, there is no satisfactory explanation. Organization by the solar wind's magnetic field acting on electrically charged dust particles or dust particle fragmentation has been proposed.

Two other dust features are sometimes observed. Antitails or sunward spikes are produced by large dust particles in the plane of the comet's orbit. These particles do not experience the relatively large force that sends the smaller dust particles into the dust tail. They remain near the comet and, when seen in projection, appear to point in the

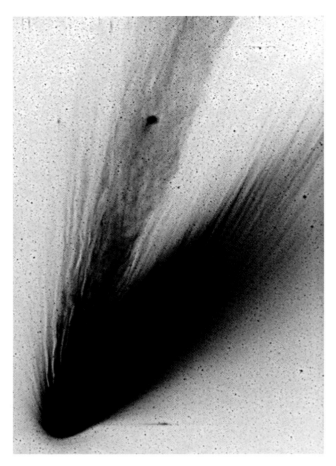

FIGURE 14 Comet Hale–Bopp on 17 March 1997 showing well-defined striae in the dust tail at right. The plasma tail is at left. (Courtesy of Kurt Birkle, Max-Planck-Institute für Astronomie, Heidelberg, Germany.)

sunward direction. If the Earth is close to the plane of the comet's orbit, a sunward spike is observed. If the Earth is away from the orbital plane but reasonably close, a sunward fan is observed.

The most famous sunward spike of the 20th century was observed in comet Arend–Roland during April 1957 (Fig. 15). Comets Kohoutek (December 1973/January 1974) and Halley (February 1986) also showed sunward spikes. Some of these are produced by large ejection speeds in the sunward direction, but most only appear to be sunward in projection.

The neck-line structure is a long, narrow dust feature observed when the comet is past perihelion and the Earth is close to the comet's orbital plane. Dust particles emitted from the comet at low speeds are, in fact, in orbit around the Sun. These orbits return to the orbital plane to produce a dust concentration. The neck-line structure has been observed in comets Bennett, Halley, and Hale–Bopp (Fig. 16). The neck-line structure in comet Halley was stable and was a major feature for over a month in May and June 1986.

FIGURE 15 Comet Arend–Roland on 25 April 1957 showing the sunward spike. (Photo © UC Regents/Lick Observatory.)

Sodium gas tails were observed in earlier comets, and comet Hale–Bopp displayed a dramatic example. Figure 17 shows the long, narrow sodium tail. There is also a wide sodium tail superimposed on the dust tail. The source for the narrow tail is probably sodium-bearing molecules in the inner coma that are dissociated. The source for the wide tail is probably the dust tail itself.

Sodium tails may well be a common feature of comets. Comet Hale–Bopp's nucleus was very large, with a diameter 60 ± 20 km. Estimates for the total gas production rate

FIGURE 16 Comet Hale–Bopp on 6 June 1997 showing the neck-line structure, the narrow feature extending to the left from the head. (Image taken by G. Pizarro, European Southern Observatory.)

FIGURE 17   Images of comet Hale–Bopp in April 1997. The left-hand image records the fluorescence emission from sodium atoms and clearly shows the thin, straight sodium tail. Compare to the right-hand image, which shows the traditional plasma and dust tails. (Courtesy of Gabriele Cremonese, INAF-Astronomical Observatory Padova, and the Isaac Newton Team.)

near perihelion are as high as 10^{31} molecules s^{-1}. Visibility of the sodium tail was enhanced by the sodium atom's high oscillator strength (one of the highest in nature), but the exceptional brightness of comet Hale–Bopp greatly increased the likelihood of observing the sodium tails.

The plasma tails of comets are long and generally straight and show a great deal of fine structure that constantly changes. They are typically 10^5–10^6 km wide, and the lengths recorded optically are routinely several tenths of AU (or several times 1.5×10^7 km). The structure of the plasma tail may extend much farther. Measurements of magnetic fields and ions made on board the *Ulysses* spacecraft have detected the signature of comet Hyakutake's plasma tail 550 million km (or 3.7 AU) from the head.

These tails are composed of electron-molecular ion plasmas. As the neutral molecules in the coma flow outward, they are ionized. Photoionization is the traditional process and easiest to include in models. Impact ionization by solar wind and cometary electrons and ionization by charge exchange also need to be considered. The result is to produce

the molecular ions H_2O^+, OH^+, CO^+, CO_2^+, CH^+, and N_2^+. Images of plasma tails, particularly those taken with photographic emulsion, usually show the plasma tail a bright blue because of strong bands of CO^+ (e.g., see Fig. 1).

These molecular ions cannot continue their simple outward flow because they encounter the solar wind magnetic field. The Larmor radius gives the radius for an ion spiraling around the magnetic field lines, and a typical value is ∼100 km. Thus, the solar wind and the cometary ions are joined together. The magnetic field lines are said to be loaded with the addition of the pickup ions and their motion slows down. This effect is strong near the comet and weak well away from the comet. The effect causes the field lines loaded with ions to wrap around the comet like a folding umbrella. This behavior is observed. These bundles of field lines loaded with molecular ions form the plasma tail. The central, dense part of the plasma tail contains a current sheet separating the field lines of opposite magnetic polarity. Because the tail is formed by an interaction with the solar wind flow, the tail points approximately antisunward but makes an angle of a few degrees with the prolonged radius vector opposite to the comet's orbital motion. The flow direction is given by the aberration angle produced by the solar wind speed and the comet's motion perpendicular to the radius vector. This aberration effect was used by L. Biermann to discover the solar wind in 1951 and to estimate its speed. H. Alfvén introduced the magnetic field into the interaction and gave the basic view of plasma tails presented here. Spacecraft measurements have verified this view. Note that plasma tails usually should be considered as attached to the head of the comet. This contrasts with dust tails where the tail emanates from the head region but the dust particles are on independent orbits. Additional complications from the interaction with the solar wind are a bow shock and plasma waves, which are present over very large volumes of space.

The interaction between the solar wind and a comet is clearly shown in Fig. 18, which is a plot of results from the ion analyzer on the *Deep Space 1* mission. The undisturbed

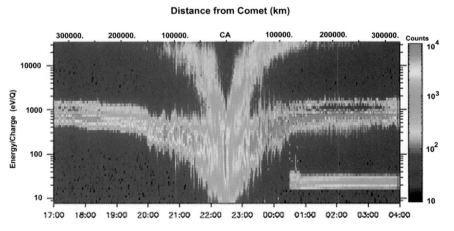

FIGURE 18   Plasma results from comet Borrelly measured by the ion analyzer on the *Deep Space 1* mission. The times refer to 22–23 September 2001. The bar at lower right was produced by xenon ions from the spacecraft thruster. See text for discussion. (Courtesy of Los Alamos National Laboratory.)

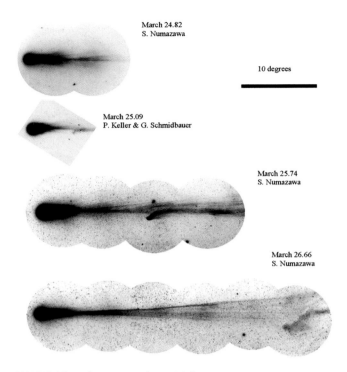

March 24.82
S. Numazawa

10 degrees

March 25.09
P. Keller & G. Schmidbauer

March 25.74
S. Numazawa

March 26.66
S. Numazawa

FIGURE 19 The spectacular 1996 disconnection event in comet Hyakutake. The 24.82 March, 25.74 March, and 26.66 March images appeared on the July 1996 cover of *Sky and Telescope* and are courtesy of *Sky and Telescope* and S. Numazawa, Japan. The 25.09 March image is courtesy of P. Keller and G. Schmidbauer, Ulysses Comet Watch. (Image sequence courtesy of the Ulysses Comet Watch.)

solar wind flow is shown at approximately 700 eV/Q, and it steadily decreases toward closest approach (CA) as the solar wind flow is loaded by the addition of cometary ions. The situation reverses as the spacecraft passes through the comet. The higher energy ions are the pick-up molecules from the comet.

The exception to the picture of plasma tails usually being attached to the comet's head is when disconnection events (DEs) occur. Here, the entire plasma tail disconnects from the head and drifts away. The comet forms a new plasma tail. Many DEs have been observed over the last century, and Fig. 19 shows a spectacular example in comet Hyakutake. DEs occur when a comet crosses the heliospheric current sheet (HCS). The HCS is an important feature in the solar wind. It separates "hemispheres" of opposite magnetic polarity and is, in essence, the magnetic equator of the heliosphere. When a comet crosses the HCS, the field lines being captured by the comet (as described earlier) are of opposite polarity. Thus, field lines of opposite polarity are pressed together in the comet causing the field lines to be severed by the process of magnetic reconnection. The old plasma is no longer attached to the head and moves away. Meanwhile, the comet develops a new plasma tail. The sequence is a regular process and repeats at each HCS crossing.

The HCS separates the heliosphere into regions of opposite magnetic polarity and defines the latitudinal structure of the solar wind. Well away from solar maximum, the solar wind is organized into a dense, gusty, slow equatorial region and a less dense, steady, fast polar region. These solar wind properties are clearly reflected in plasma tails. In the polar region, plasma tails have a smooth appearance, show aberration angles corresponding to a fast solar wind, and do not exhibit DEs. In the equatorial region, plasma tails have a disturbed appearance, show aberration angles corresponding to a slow solar wind, and exhibit DEs.

Although cometary X rays properly belong in the coma discussion, they are included here because they are produced by a solar wind interaction. X rays in the energy range 0.09–2.0 keV were unexpectedly discovered in comet Hyakutake; see Fig. 20 for a false-color X-ray image of comet LINEAR. When databases were searched, several more comets were seen as X-ray sources. X-ray emission is an expected phenomenon of all comets.

The principal mechanism is charge exchange between heavy minor species in the solar wind and neutral molecules in the coma. The heavy species in the solar wind are multiply ionized. For example, six-time ionized oxygen can

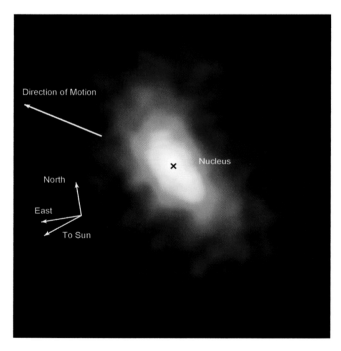

Direction of Motion

Nucleus

North

East

To Sun

FIGURE 20 False-color rendering of an X-ray image of comet LINEAR obtained on 14 July 2000 by the Chandra X-Ray Observatory. (Courtesy of C. M. Lisse, University of Maryland, College Park; D. J. Christian, Queens University, Belfast, United Kingdom: K. Dennerl, Max-Planck-Institut für Extraterrestrische Physik, Garching, Germany; and S. J. Wolk, Chandra X-Ray Center, Harvard-Smithsonian Center for Astrophysics.)

charge exchange with a neutral molecule to produce an ionized molecule and a five-time ionized oxygen in an excited state. X-ray lines are produced when the excited ions spontaneously decay. Spectroscopic X-ray observations have confirmed this mechanism. Some contribution to the total flux may come from electron–neutral thermal bremsstrahlung.

6. Comet Chemistry

The overall chemical composition of comets seems to be rather uniform. Exceptions to this general statement are discussed later. Ultraviolet spectra of comets (see Fig. 21) are dominated by the hydrogen (H) Lyman-α line at 121.6 nm and by the hydroxyl (OH) bands at 309.0 nm. This is certainly compatible with the conclusion that the nucleus is composed of roughly 80–90% water ice, 10% carbon monoxide (CO), and many minor constituents.

Table 2 lists species in comets that have been observed spectroscopically or measured in situ by mass spectrometers on spacecraft. The list is not exhaustive.

Providing a detailed explanation of the abundances of these species is a formidable task and is subject to many processes in the coma. But, as argued by W. F. Huebner, the situation is comprehensible if we assume a condensation process in the primordial solar nebula at a temperature of 30 K and solar abundances except for H and N. The abundance of hydrogen is determined by the capability to chemically bind to other species. Much is lost from the solar system. Some nitrogen is also lost; for example, when N_2 is formed, the nitrogen is in a form that is not chemically active. A gas mixture consisting of C, O, Mg, Si, S, and Fe in solar abundances with reduced amounts of H and N can condense into molecules at 30 K. The silicates Fe_2SiO_4 and Mg_2SiO_4 are formed from Fe, Mg, Si, and O. Then, the remainder of O goes into H_2O and into HCO and CO-compounds. Finally, the remainder of the C, N, and S goes into HCNS-compounds.

The result of this fairly straightforward condensation sequence is a material that, when formed into a substantial solid body, resembles comets. By mass, the relative abundance of H_2O:silicates:carbonaceous molecules plus hydrocarbons is approximately 1:1:1. Also, by mass, the abundances of ices:dust is about 1:1.

The temperature of 30 K used in the previous discussion is not only the appropriate temperature for the condensation sequences, but it is also consistent with direct determinations of the interior temperatures of cometary nuclei using the *ortho*- to *para*-hydrogen ratio (OPR). Hydrogen in water (and some other compounds) can have the spin of their nuclei in the same direction (*ortho*-water) or in the opposite direction (*para*-water). The OPR depends on the temperature of the water molecules at the time of formation, and the OPR can only be changed by chemical

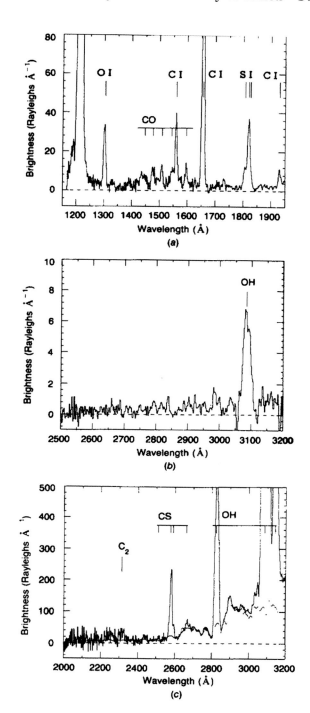

FIGURE 21 *International Ultraviolet Explorer* (*IUE*) spectra of comet Halley. (a) Spectrum on 9 March 1986: the very strong line close to 1200 Å is the Lyman-α line of neutral hydrogen. (b) Spectrum on 12 September 1985. (c) Spectrum on 11 March 1986. (Courtesy of P. D. Feldman, Johns Hopkins University.)

reactions. Thus, the ice can be sublimated in a comet's subsurface layers and flow through the crust into the coma while retaining its original OPR.

Infrared measurements of the OPR for comets Halley, Hale–Bopp, and LINEAR are all consistent with an interior

TABLE 2	Measured and Observed Species in Comets

Atoms + Molecules	Ions
H, C, O, S, Na, Fe, Ni, CO, CS, NH, OH, C_2, $^{12}C^{13}C$, CH, CN, ^{13}CN, S_2, SO, H_2, CO_2, HDO, CHO, HCN, DCN, $H^{13}CN$, OCS, SO_2, C_3, NH_2, H_2O, H_2S, HCO, H_2CS, C_2H_2, HNCO, H_2CO, CH_4, HC_3N, CH_3OH, CH_3CN, NH_2CHO, C_2H_6	C^+, N^+, O^+, Na^+, CO^+, CH^+, CN^+, OH^+, NH^+, H_2O^+, HCO^+, CO_2^+, C_3^+, CH_2^+, H_2S^+, NH_2^+, HCN^+, DCN^+, CH_3^+, H_3O^+, H_3S^+, NH_3^+, C_3H^+, CH_4^+, H_3CO^+, CH_5^+, $C_3H_3^+$

temperature near 30 K. These results are important in discussing formation scenarios. The existence of S_2 in comets may require a formation temperature as low as 15 K. While there is some uncertainty in the exact temperature, cold temperatures are required.

A monumental study using narrow-band photometry with major results for the chemical compositions of comets was led by astronomer M. F. A'Hearn. Standardized techniques were used to characterize 85 comets with filters that covered emission bands from CN, C_2, C_3, OH, and NH as well as selected continuum regions. As with the ultraviolet results described previously, the compositions are surprisingly uniform. Barring some unusual event, a comet's production of gases and dust from orbit to orbit (and position in the orbit) is essentially the same. This implies a basically homogeneous interior. When the sample of comets was divided into old and new comets based on their orbital properties, no compositional differences were found.

Still, there were significant exceptions to the similarity in compositions. A class of comets shows depletions in the carbon chain molecules C_2 and C_3 relative to CN. Comet Giacobini–Zinner is the prototype for this class. Almost all the members of this class are Jupiter-family comets, but not all Jupiter-family comets are members of the class. Are the compositional differences due to formation in different regions in the solar nebula or to some kind of physical processing for comets with a different orbital history? Recent observations show differences in the following way. Comets Halley, Hyakutake, and Hale–Bopp were extensively observed, and their compositions are similar to those in the cores of dense interstellar clouds. The observed composition of comet LINEAR shows depletions in CO, CH_4, C_2H_6, and CH_3OH. These are highly volatile species, and a plausible scenario could place the comet's formation in the warmer Jupiter–Saturn region of the solar nebula. Most comets such as Halley, Hyakutake, and Hale–Bopp are believed to have formed in the cooler Uranus–Neptune region. An inconsistency arises with the measurement of the interior temperature of comet LINEAR (using a variant of the OPR discussed earlier) where a result close to 30 K was found. Thus, comet LINEAR may have formed under essentially the same conditions as the other comets.

Even though the gross compositions of comets are similar, chemical diversity is an established fact. Because comets are surely formed over a range of heliocentric distances and because they have a variety of orbital histories, diversity could arise from formation conditions and from postformation processing. The relative importance of the two is to be determined.

7. Formation and Ultimate Fate of Comets

The icy bodies of the solar system formed as part of the process that produced the Sun, the terrestrial planets, and the giant planets. The icy bodies include some of the asteroids (including the Centaurs, which are bodies with eccentric orbits generally between Saturn and Neptune), comets, and Kuiper Belt Objects (KBOs). [See KUIPER BELT OBJECTS: PHYSICAL STUDIES.]

The solar system is thought to have formed from the collapse of an interstellar gas cloud. The collapse process produced a newly formed star with a circumstellar disk of gas and dust, the solar nebula. [See THE ORIGIN OF THE SOLAR SYSTEM.] As discussed in Section 6, cometary material can condense at temperatures of roughly 30 K. Models of the early solar nebula have temperatures of roughly 30 K in the Uranus–Neptune region, and it is reasonable to conclude that comets formed near there, meaning that the material condensed and agglomerated into comet-sized (most with radii in the range 1–10 km) bodies. Note, however, that the uncertainty in the temperatures for models of the presolar nebula is approximately a factor of 2.

But the story does not end there because most comets are not in the Uranus–Neptune region today. Dynamical processes dispersed the icy bodies. Gravitational perturbations by the giant planets sent some of the comets to large distances from the Sun and some into the inner solar system. The latter comets faded long ago. Many of the comets sent to large distances escaped from the solar system, but the ones that are barely bound form a roughly spherical cloud with dimensions of 10^4–10^5 AU. This is the cloud of comets, the Oort cloud, postulated by J. Oort many years ago. It is the source of the long-period comets ($P > 200$

years). They are perturbed and sent into the inner solar system by passing stars, passing giant molecular clouds, and the tidal gravitational field of the Milky Way galaxy. [*See* COMETARY DYNAMICS.]

Further study indicates that the Oort cloud probably has two components: the spherical outer cloud discussed previously and a more flattened inner cloud. The inner cloud is probably the source of the Halley-type comets ($20 < P \leqq 200$ years). Comets from this region can reach the inner solar system and be captured into stable orbits. The boundary between the inner and outer Oort cloud is at approximately 20,000 AU.

The Jupiter-family comets ($P \leqq 20$ years) cannot come from the Oort cloud. Their origin requires a close-in, flattened source. This is the Kuiper Belt, now believed to be the source of Jupiter-family comets. Studies of scattering processes within the Kuiper Belt show that objects that can be captured into stable orbits with the orbital characteristics of Jupiter-family comets are produced. Most observed KBOs are much larger than observed Jupiter-family comets, but this is almost surely due to observational selection. It is reasonable to assume that the size distribution of objects in the Kuiper Belt includes comets. Note that most KBOs are currently found with semimajor axes between 35 and 50 AU. They were not always there but were moved outward along with the outward migration of Uranus and Neptune early in the history of the solar system. The sharp outer boundary for the region of the KBOs was thought to originally be at about 30 AU; it is now at 50 AU. Some KBOs (the scattered population) are found well beyond 50 AU. Two KBOs with semimajor axes of 230 AU are known. The trans-Neptunian object Sedna has a semimajor axis of 526 AU. If it is a KBO, it could indicate additional objects at large distances. [*See* KUIPER BELT: DYNAMICS.]

Figure 22 is a summary schematic that attempts to tie together the ideas for the Kuiper Belt, inner Oort cloud, and outer Oort cloud as the source regions for the Jupiter-family, Halley-type, and long-period comets, respectively. The flaring of the line near 10^4 AU indicates that structure interior to this point is believed to be flattened, while the structure exterior to this point is essentially spherical. There is no evidence to suggest that the boundaries between regions are sharp.

The dynamical processes that involve comets eject many of them from the solar system. Some estimates suggest that the number lost can be as high as 30–100 for every comet in the Oort cloud. There are many stars similar to the Sun in the solar neighborhood and throughout the galaxy, and if the formation of comets is an integral part of star and planetary system formation, there should be many interstellar comets. Some of these should pass through the solar system. They would reveal themselves by having clearly hyperbolic orbits. A quantitative calculation yields the result that six or more comets should have traveled through the solar system at distances within the orbit of Mars during the past 150 years. None has been observed so far.

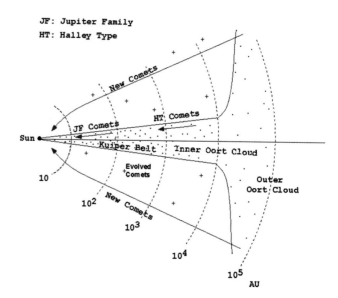

FIGURE 22 Schematic of the Kuiper Belt and inner and outer Oort cloud as source regions for comets. See text for discussion. (After Fernández; reprinted with permission from John C. Brandt and Robert D. Chapman, 2004, "Introduction to Comets," 2nd Ed., Cambridge Univ. Press, Cambridge, United Kingdom. Copyright © Cambridge University Press.)

Active comets have a limited life because the volatile materials sublimated away are not replenished. Eventually, the volatiles are gone and the body is inactive. Such objects would be classified as asteroids, and some "asteroids" are clearly dead comets because examples of the transition from comet to asteroid have been documented. [*See* NEAR-EARTH OBJECTS; MAIN-BELT ASTEROIDS.]

Remnants of comets in the solar system include the dust particles on bound Keplerian orbits that, along with an asteroidal contribution, constitute the cloud that produces the zodiacal light from scattered sunlight. The remnants also include the meteoroid streams that produce meteor showers. These streams have long been known to be closely associated with the orbits of comets. Perturbations distribute the rocky or dusty pieces of the comet along its orbit. When the Earth encounters the stream, the pieces enter our upper atmosphere and are observed as meteor showers.

Infrared observations of comets show many long trails of dust, and several were associated with known comets. Figure 23 shows the long dust trail of comet Tempel 2. The false-color image from the *Infrared Astronomical Satellite* (*IRAS*) was constructed from 12, 60, and 100 μm scans. The dust trail is the thin blue line stretching from the comet's head at upper left to lower right. The particle sizes are estimated to be in the range 1 mm–1 cm. These dust trails appear to be meteoroid streams in the making. [*See* SOLAR SYSTEM DUST.]

Comets can also be destroyed by collisions with the Sun, moon, planets, and satellites. The collision of the train of fragments from comet Shoemaker–Levy 9 (see Fig. 10) with

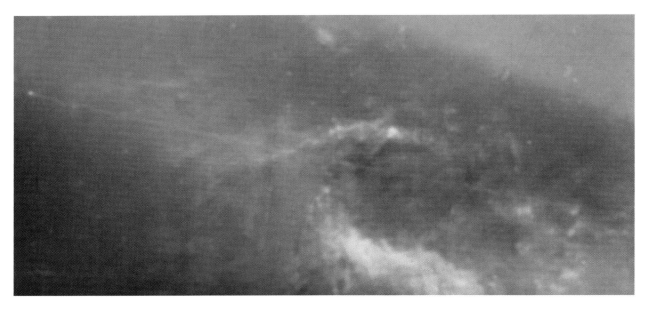

FIGURE 23   *Infrared Astronomical Satellite* false-color image constructed from infrared scans showing the long dust trail of comet Tempel 2. The trail appears as the thin blue line stretching from the comet's head at upper left to lower right. (Courtesy of Mark Sykes, Planetary Science Institute.)

Jupiter in July of 1994 is a spectacular example. Collisions of comets with Earth have been invoked as a source of terrestrial water and possibly a source of complex organic molecules that could be important for the origin of life. At present, there is no consensus on these ideas.

8. Summary

Comets are a diverse population of icy, sublimating bodies that display large-scale phenomena. The central body, the nucleus, has typical dimensions of 1–10 km. The bulk composition is mostly H_2O ice and dust, and the details of the minor constituents may hold clues to the origin of comets and the formation of the solar system. The physical processes involved—sublimation of ices in the interior, the flow of gases away from the nucleus, the dissociation and ionization of molecules, and the interaction with the solar wind—continue to provide challenges for scientists. Comets are important to our understanding of other solar system phenomena such as meteors and the zodiacal light. Many problems in comet physics can be solved only by sending spacecraft to the immediate vicinity for close-up imaging and in situ measurements. The past few years have seen several space missions to comets and an extraordinary increase in our knowledge of comets and their diversity. The interiors of comets are not well understood, but results from the *Deep Impact* mission provide an important first step. *Deep Impact* showed that comet Tempel 1's nucleus is porous and that at least the outer layers are gravitationally bound and have very low tensile strength.

Ultimately, samples of cometary material must be returned to Earth for analysis in the laboratory. This has begun with the return of dust particle samples from the *Stardust* mission in 2006. Although the *Rosetta* mission to comet Churyumov–Gerasimenko is expected to greatly expand our knowledge of comets, with the main spacecraft spending an extended time period near the comet and the lander spacecraft landing on and anchoring itself to the nucleus, the return of icy materials to Earth for analysis is far in the future.

Bibliography

Brandt, J. C. (1999). Comets. In "The New Solar System," (J. K. Beatty, C. C. Petersen, and A. Chaikin, eds.). Sky Publishing, Cambridge, Massachusetts.

Brandt, J. C., and Chapman, R. D. (2004). "Introduction to Comets," 2nd Ed. Cambridge Univ. Press, Cambridge, United Kingdom.

"*Deep Impact* at comet Tempel 1." *Science* **310** (14 October 2005), Special Section, 257–283.

Fernández, J. A. (2005). "Comets: Nature, Dynamics, Origin, and their Cosmogonical Relevance." Springer, Dordrecht.

Festou, M., Keller, H. U., and Weaver, H. A., eds. (2005). "Comets II." Univ. Arizona Press, Tucson.

Huebner, W. F., ed. (1990). "Physics and Chemistry of Comets." Springer-Verlag, Berlin.

"*Stardust* at Comet Wild 2." *Science* **304** (18 June 2004), Special Section, 1760–1780.

Yeomans, D. K. (1991). "Comets: A Chronological History of Observations, Science, Myth, and Folklore." Wiley, New York.

Comet Populations and Cometary Dynamics

Harold F. Levison
Luke Dones

Southwest Research Institute
Boulder, Colorado

CHAPTER **31**

The Solar System formed from a collapsing cloud of dust and gas. Most of this material fell into the Sun. However, since the primordial cloud had a little bit of angular momentum or spin, a flattened disk also formed around the Sun. This disk contained a small amount of mass, as compared to the Sun, but most of the cloud's original angular momentum. This disk, known as the **protoplanetary nebula**, contained the material from which the planets, satellites, asteroids, and comets formed.

The first step in the planet formation process was that the dust, which contained ice in the cooler, distant regions of the nebula, settled into a thin central layer within the nebula. Although the next step has not been fully explained (*see* THE ORIGIN OF THE SOLAR SYSTEM), as the dust packed itself into an ever-decreasing volume of space, larger bodies started to form. First came the objects called **planetesimals** (meaning small planets), which probably ranged in size from roughly a kilometer across to tens of kilometers across. As these objects orbited the Sun, they would occasionally collide with one another and stick together. Thus, larger objects would slowly grow. This process continued until the planets or the cores of the gas giant planets formed. (*See* INTERIORS OF THE GIANT PLANETS.)

Fortunately for us, planet formation was a messy process and was not 100% efficient. There are a large number of remnants floating around the Solar System. Today we call these small bodies comets and asteroids. These pieces of refuse of planet formation are interesting because they can tell us a lot about how the planets formed. For example, because comets and asteroids are the least chemically processed objects in the Solar System (there is a lot of chemistry that happens on planets), studying their composition tells us about the composition of the protoplanetary nebula.

From our perspective, however, comets and asteroids are most interesting because their orbits can tell us the story of how the planets came together. Just as blood spatters on the wall of a murder scene can tell as much, or more, about the event than the body itself, the orbits of asteroids and comets play a pivotal role in unraveling the planetary system's sordid past.

In this chapter we present the story of where comets originated, where they have spent most of their lives, and how they occasionally evolve through the planetary system and move close enough to the Sun to become the spectacular objects we sometimes see in the night sky.

However, to tell this story, we must work backwards because the majority of observational information we have about these objects comes from the short phase when they are close to the Sun. The rest of the story is gleaned by combining this information with computer-generated dynamical models of the Solar System. Thus, in Section 1 we start with a discussion of the behavior of the orbits of comets. In Section 2 we present a classification scheme for comets.

This step is necessary because, as we will show, there are really two stories here. Comets can follow either one of them, but we must discuss each of them separately. In Section 3, we describe the cometary reservoirs that are believed to exist in the Solar System today. In addition, we discuss our current understanding of how these reservoirs came to be. We conclude in Section 4.

1. Basic Orbital Dynamics of Comets

For the most part, comets follow the basic laws of orbital mechanics first set down by Johannes Kepler and Isaac Newton. These are the same laws that govern the orbits of the planets. In this section, we present a brief overview of the orbits of small bodies in the Solar System. (For a more detailed discussion, *see* SOLAR SYSTEM DYNAMICS: REGULAR AND CHAOTIC MOTION.)

In the Solar System there are eight major planets, many smaller dwarf planets, and vast numbers of smaller bodies, each acting to perturb gravitationally the orbits of the others. The major planets in the Solar System follow nearly circular orbits. They also all lie in nearly the same plane, and so it has been long assumed that the planets formed in a disk. The planets never get close to each other. So, the first-order gravitational effect of the planets on one another is that each applies a torque on the other's orbit, as if the planets were replaced by rings of material smoothly distributed along their orbits. These torques cause both the longitude of perihelion, $\bar{\omega}$, and longitude of the ascending node, Ω, to precess. In particular, $\dot{\bar{\omega}} > 0$ and $\dot{\Omega} < 0$. The periods associated with these frequencies range from 47,000 to 2,000,000 years in the outer planetary system. Because the masses of the planets are much smaller than the Sun's mass, this is much longer than the orbital periods of the major planets, which are all less than 170 years.

There are four main differences between the orbits of the comets that we see and those of the planets. First, unlike planets, visible comets usually are on eccentric orbits, and so they tend to cross the orbits of the planets. So, they can suffer close encounters with the planets. While these encounters sometimes lead to direct collisions, like the impact of the comet D/Shoemaker-Levy 9 on Jupiter in 1994, more frequently the planet acts as a gravitational slingshot, scattering the comet from one orbit to another. The solid curve in Figure 1 shows the temporal evolution of comet 95P/Chiron's semimajor axis according to a numerical integration of the comet's orbit (black curve). This comet currently has $a = 14$ AU, which means it is between Saturn and Uranus, $e = 0.4$, and $i = 7°$. All the changes seen in the figure are due to gravitational encounters with the giant planets. Individual distant encounters lead to small changes, while close encounters lead to large changes. According to this integration, the comet will be ejected from the Solar System by a close encounter with Jupiter in 675,000 years.

This calculation illustrates that the orbits of objects on planet-crossing orbits, and thus the comets that we see, are generally unstable. This means that, on timescales very

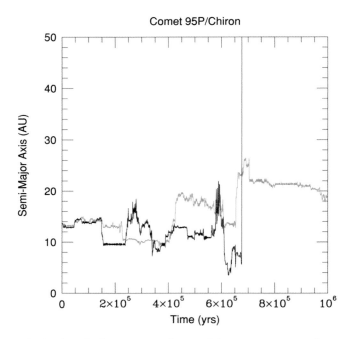

FIGURE 1 The long-term evolution of the semimajor axis of comet 95P/Chiron (black curve) and a *clone* of this comet (red curve). These trajectories were determined by numerically integrating the equations of motion of these comets, the Sun, and the four giant planets. The clone was an object with almost the exact same initial conditions as 95P/Chiron, but the position was offset by 1 cm. The fact that the two trajectories diverge shows that the orbit is chaotic.

short compared to the age of the Solar System, most of these objects will be ejected from the Solar System by a gravitational encounter with a planet, or hit the Sun or a planet. (Some comets appear to disintegrate spontaneously, for reasons that are not well understood.) So, the comets that we see could not have formed on the orbits that we see them on, because if they had, they would no longer be there. They must have formed, or at least been stored, for long periods of time in a reservoir or reservoirs where their orbits are long-lived and they remain cold enough so that their volatiles are, for the most part, preserved. These reservoirs are mainly hidden from us because they are far from the Sun. We discuss cometary reservoirs in more detail in Section 3.

Figure 1 also shows that cometary orbits are formally **chaotic**. If the Solar System consisted of only the Sun and one planet, interacting through Newton's law of gravity, the planet's orbit would remain a Keplerian ellipse for all time. The distance between the planet and the Sun would vary periodically, akin to a pendulum. This is an example of **regular** motion. For regular motion, if there were two planetary systems that were exactly the same, except that the position of the planet was slightly offset in one versus the other, this offset would increase linearly with time. However, if three or more bodies are present in the system, **chaos** is possible, meaning that any offset between two nearly identical systems would increase exponentially. In certain cases, such as if the orbit of a comet or asteroid crosses that of a planet, chaos leads to gross unpredictability. That is, in these cases it is impossible to foretell, even qualitatively, the orbit of a comet or asteroid very far into their future or past.

For example, in Figure 1, the black curve shows the predicted evolution of 95P/Chiron's semimajor axis, using its nominal orbit. The red curve shows the evolution of an object ("the clone") that initially had exactly the same velocity as 95P/Chiron, and an initial position that differed by 1 cm! In less than a million years, a tiny fraction of the age of the Solar System, the orbits are totally different. One clone has been ejected from the Solar System, while the other continues to orbit within the planetary region. This **sensitivity to initial conditions** means that we can never predict where any object in the Solar System will be over long periods of time. By "long periods" we mean at most tens of millions of years for the planets, but for many comets less than a few hundred years. On timescales longer than this, we can only make statistical statements about the ultimate fate of small bodies on chaotic orbits.

The chaotic nature of cometary orbits has important implications for our study of cometary reservoirs. Once we determine the current orbit of a comet, it would be ideal if we could calculate how the orbit has changed with time and trace it backward to its source region. Thus, by studying the physical characteristics of these comets, we could determine what the cometary reservoirs are like. Unfortunately, the unpredictability of chaotic orbits affects orbital integrations that go backward in time as well as those that go forward in time. Thus, it is impossible to follow a particular comet backward to its source region. To illustrate this point, consider the analogy of an initially evacuated room with rough walls and a large open window into which molecules are injected through a narrow hose. Once the system has reached a steady state (i.e., the number of molecules entering through the hose is equal to the number leaving through the window), suppose that the position and velocity of all the particles in the room were recorded, but with less than perfect accuracy. If an attempt were made to integrate the system backwards, the small errors in our initial positions and velocities would be amplified every time a molecule bounced off a wall. Eventually, the particles would have "forgotten" their initial state, and thus, in our backwards simulation of the gas, more particles would leave through the window than through the hose, simply because the window is bigger. In our case, injection through the hose corresponds to a comet's leaving its reservoir, and leaving through the window corresponds to the many more avenues of escape available to a comet.

So, it is not possible to directly determine which comet comes from which reservoir. Therefore, the only way to use visible comets to study reservoirs is to dynamically model the behavior of comets after they leave the reservoir, and follow these hypothetical comets through the Solar System, keeping track of where they go and what kind of comets they become. By comparing the resulting orbital element distribution of the hypothetical comets to real comet types, we can determine, at least statistically, which type of comets come from which reservoir.

A second major difference between cometary and planetary orbits is that many comets are active. That is, since they are mainly made of dust (or rock) and water ice, and water ice only sublimates within ~4 AU of the Sun, comets that get close to the Sun spew out large amounts of gas and dust. This activity is what makes comets so noticeable and beautiful in the night sky. However, outgassing also acts like a rocket engine that can push the comet around and change its orbit. The most obvious effect of these so-called **nongravitational forces** is to change the orbital period of the comet. For example, nongravitational forces increase the orbital period (ΔP) of comet 1P/Halley by roughly 4 days every orbit.

The magnitude, direction, and variation with time of nongravitational forces are functions of the details of an individual comet's activity. Most of the outflow is in the sunward direction; however, the thermal inertia of the spinning nucleus delays the maximum outgassing toward the afternoon hemisphere. Thus, there is a nonradial component of the force. This delay is a function of the angle between the equator of the cometary nucleus and its orbital plane and will vary with time due to seasonal effects. Also, localized jetting can also produce a nonradial force on the comet and will also change the spin state and orientation of the nucleus.

As a result, there is a huge variation of nongravitational forces from comet to comet. For example, for many comets there is no measurable nongravitational force because they are large and/or relatively inactive. Some active comets, like Halley, have nongravitational forces that behave similarly from orbit to orbit. For yet other comets, the magnitude of these forces has been observed to change over long periods of time. A good example of this type of behavior is comet 2P/Encke, which had $\Delta P = -0.13$ days in the early nineteenth century, but now has ΔP of -0.008 days.

In general it is possible to describe the nongravitational accelerations $\vec{a}_{ng}$ that a comet experiences by:

$$\vec{a}_{ng} = g(r) \left[A_1 \hat{r} + A_2 \hat{t} + A_3 \hat{n} \right],$$

where the A's are constants fit to each comet's behavior, r is the instantaneous heliocentric distance, and $\hat{r}$, $\hat{n}$, and $\hat{t}$ are unit vectors in the radial direction, the direction normal to the orbit of the comet, and the transverse direction, respectively. The value $g(r)$ is related to the gas production rate as a function of heliocentric distance and is usually given as:

$$g(r) = 0.111262 \left(\frac{r}{r_0} \right)^{-2.15} \left[1 + \left(\frac{r}{r_0} \right)^{5.093} \right]^{-4.6142},$$

where the parameter $r_0 = 2.808$ AU is the heliocentric distance at which most of the solar radiation goes into sublimating water ice.

A third difference between a planetary orbit and a cometary orbit arises because visible comets tend to be on eccentric (sometimes very eccentric) orbits and on orbits that are inclined with respect to the ecliptic (sometimes even **retrograde** orbits with inclinations greater than 90°). The rates at which the **apse** and **node** of a comet ($\bar{\omega}$ and Ω) precess depend upon the comet's eccentricity and inclination. Thus, although cometary orbits precess, like the orbits of the planets, their behavior can be very different from the subtle behavior of the planets. Of particular interest, if the inclination of a comet is large, it can find itself in a situation in which, on average, $\dot{\bar{\omega}} = \dot{\Omega}$, i.e., $\bar{\omega}$ and Ω are said to be in resonance with one another. Since these two frequencies are linked to changes in eccentricity and inclination, this resonance allows eccentricity and inclination to become coupled, and allows each to undergo huge changes at the expense of the other. And, since a comet's semimajor axes is preserved in this resonance, changes in inclination also lead to changes in perihelion distance.

An example of this so-called **Kozai resonance** can be seen in the behavior of comet 96P/Machholz 1 (Fig. 2). 96P/Machholz 1 currently has an eccentricity of 0.96 and an inclination of 60°. Its perihelion distance, q, is currently 0.12 AU, well within the orbit of the planet Mercury. Figure 2 shows the evolution of the orbit of 96P/Machholz 1 over the next few thousand years. The Kozai resonance is responsible for the slow, systematic oscillations in both inclination and eccentricity (or q, which equals $a \times (1 - e)$). These oscillations are quite large; the inclination varies between roughly 10° and 80°, while the perihelion distance gets as large as 1 AU. According to these calculations, the Kozai resonance will drive this comet into the Sun ($e = 1$) in less than 12,000 years! Similarly, the Kozai resonance was important in driving comet D/Shoemaker-Levy 9 to collide with Jupiter. However, in that case, the comet had been captured into orbit around Jupiter, and the oscillations in i and e were with respect to the planet, not the Sun.

The final gravitational effect that we want to discuss in this section is the effect that the galactic environment has on cometary orbits. Up to this point, our discussion has assumed that the Solar System was isolated from the rest of the Universe. This, of course, is not the case. The Sun, along with its planets, asteroids, and comets, is in orbit within the Milky Way Galaxy, which contains hundreds of billions of stars. Each of these stars is gravitationally interacting with the members of the Solar System. Luckily for the planets, the strength of the Galactic perturbations varies as a^2, so the effects of the Galaxy are not very important for objects that orbit close to the Sun. However, if a comet has a semimajor axis larger than a few thousand AU, as some do (see Section 2), the Galactic perturbations can have a major effect on its orbit.

For example, Figure 3 shows a computer simulation of the evolution through time of the orbit of a hypothetical comet with an initial semimajor axis of 20,000 AU, roughly 10% of the distance to the nearest star. (For scale remember

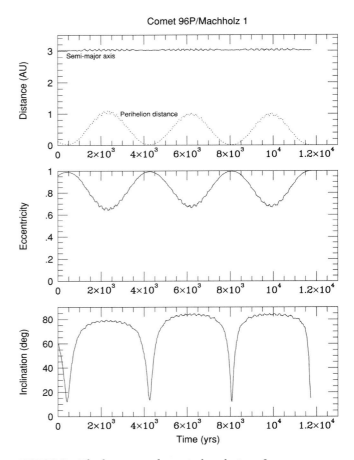

FIGURE 2 The long-term dynamical evolution of comet 96P/Machholz 1, which is currently in a Kozai resonance. Three panels are shown. The top presents the evolution of the comet's semimajor axis (solid curve) and perihelion distance (dotted curve). The middle and bottom panels show the eccentricity and inclination, respectively. Because of the Kozai resonance, the eccentricity and inclination oscillate with the same frequency, but are out of phase (i.e., eccentricity is large when inclination is small and vice versa). According to this calculation, this comet will hit the Sun in less than 12,000 years.

that Neptune is at 30 AU.) For the sake of discussion, it is useful to divide the evolution into two superimposed parts: (1) a slow secular change in perihelion distance (i.e., eccentricity) and inclination, and (2) a large number of small, but distinct jumps leading to a **random walk** in the orbit.

The secular changes are due to the smooth background gravitational potential of the Galaxy as a whole. If we define a rectangular coordinate system ($\tilde{x}$, $\tilde{y}$, $\tilde{z}$), centered on the Sun, such that $\tilde{x}$ points away from the galactic center, $\tilde{y}$ points in the direction of the galactic rotation, and $\tilde{z}$ points toward the south, it can be shown that the acceleration of a comet with respect to the Sun is

$$a_{gal} = \Omega_0^2 \left[(1 - 2\delta)\tilde{x}\hat{\tilde{x}} - \tilde{y}\hat{\tilde{y}} - \left(\frac{4\pi G_{\rho 0}}{\Omega_0^2} - 2\delta \right)\tilde{z}\hat{\tilde{z}} \right],$$

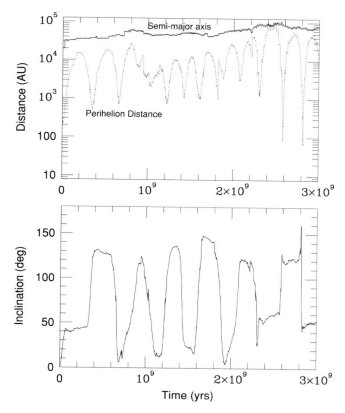

FIGURE 3   The long-term dynamical evolution of a fictitious object initially at 20,000 AU from the Sun under the gravitational perturbations of the Galaxy. Two panels are shown. The top presents the evolution of the comet's semimajor axis (solid curve) and perihelion distance (dotted curve; recall that $e = 1 - q/a$). The bottom panel shows the inclination.

where $\Omega_0 = 27.2 \pm 0.9$ km/s/kpc is the Sun's angular speed about the Galactic center, $\delta \equiv -\frac{A+B}{A-B}$ and $A = 14.5 \pm 1.5$ km/s/kpc and $B = -12 \pm 3$ km/s/kpc are Oort's constants of Galactic rotation, $\rho_0 = 0.1 M_\odot$ pc^{-3} is the density of the galactic disk in the solar neighborhood, and G is the gravitational constant. The value of δ is usually assumed to be zero.

Due to the nature of the above acceleration, it acts as a torque on the comet. As a result, the smooth part of the Galactic perturbations can change a comet's eccentricity and inclination, but not its semimajor axis. In addition, the eccentricity and inclination oscillate in a predictable way. In this example, in Figure 3 the oscillation period is approximately 300 million years. However, this period scales as $a^{-3/2}$, and thus the oscillations are faster for large semimajor axes. The small jumps are due to the effects of individual stars passing close to the Sun. Since these stars can come in from any direction, the kick that the comet feels can affect all the orbital elements, including the semimajor axis. The apparent random walk of the comet's semimajor axis seen in the figure is due to this effect.

2. Taxonomy of Cometary Orbits

The first step toward understanding a population is to construct a classification scheme that allows one to place like objects with like objects. This helps us begin to construct order from the chaos. However, before we talk about comet classification, we need to make the distinction between what we see and what is really out there. As we describe in much more detail below, most of the comets that we see are on orbits that cross the orbits of the planets. For example, the most famous comet, 1P/Halley (the "1P" stands for the first known *periodic* comet, see below), has $q = 0.6\,AU$ and an **aphelion distance** (farthest distance from the Sun) of 35 AU. Thus, it crosses the orbits of all the planets except Mercury. But planet-crossing comets represent only a very small fraction of the comets in the Solar System, because we can only easily see those comets that get close to the Sun.

Comets are very small compared to the planets. As a result, we cannot see comets very far away. For example, 1P/Halley, a relatively large comet, is a roughly (American) football-shaped object roughly 16 km long and 8 km wide. The farther away an object is, the fainter it is. The brightness (b) of a light-bulb decreases as the square of the distance d from the observer ($b \propto 1/d^2$). However, this is not true for objects in the Solar System that shine by reflected sunlight. To first approximation, the brightness of a solid sphere seen from the Earth is proportional to $1/(d_\odot^2\, d_\oplus^2)$, where $d_\odot$ and $d_\oplus$ are the distance between the object and the Sun and Earth, respectively. As objects get farther from the Sun, they get less light from the Sun and so reflect less (that is the $1/d_\odot^2$ term). Also, the further they get from us, the fainter they appear (that is the $1/d_\oplus^2$ term). In the outer Solar System, $d_\oplus$ and $d_\odot$ are nearly equal and thus $b \sim 1/d^4$.

It is even worse for a comet since it is not simply a solid sphere. As described above, as a comet approaches the Sun, its ice begins to sublimate. The resulting gas entrains dust from the comet's surface, forming a halo known as the **coma**. Because the dust is made of small objects with a lot of surface area, it can reflect a lot of sunlight. So, this cometary activity makes the comet much brighter. Observational studies show that as a comet approaches the Sun, its brightness typically increases as $1/(d_\odot^4\, d_\oplus^2)$! The result of all this activity is that it can make an object that would normally be very difficult to see, even through a telescope, into a body visible with the naked eye. Thus, we know of only a very small fraction of comets in the Solar System and this sample is **biased** because it represents only those objects that get close to the Sun. However, before we can try to understand the population as a whole, we need to first try to understand the part that we see.

The practice of developing a classification scheme or taxonomy is widespread in astronomy, where it has been applied to everything from Solar System dust particles to clusters of galaxies. Classification schemes allow us to put the objects of study into a structure in which we can look for correlations between various physical parameters and begin to develop evolutionary models. In this way, classification schemes have played a crucial role in advancing our understanding of the universe. However, we must be careful not to confuse these schemes with reality. In many cases, we are forcing a classification scheme on a continuum of objects. Then we argue over where to draw the boundaries. The fact that we astronomers find cubbyholing objects convenient does not imply that the universe will necessarily cooperate. With this caveat in mind, in the remainder of this section we present a scheme for the classification of cometary orbits.

Historically, comets have been divided into two groups: long-period comets (with periods greater than 200 years) and short-period comets (with $P < 200$ years). This division was developed to help observers determine whether a newly discovered comet had been seen before. Since orbit determinations have been reliable for only about 200 years, it may be possible to link any comet with a period less than this length of time with previous apparitions. Conversely, it is very unlikely to be possible to do so for a comet with a period greater than 200 years, because even if it had been seen before, its orbit determination would not have been accurate enough to prove the linkage. Thus this division has no physical justification and is now of historical interest only. Unfortunately, there does not yet exist a physically meaningful classification scheme for comets that is universally accepted. Nonetheless, such schemes exist. Here we present a scheme developed by one of the authors roughly 10 years ago. A flowchart of this scheme is shown in Figure 4.

The first step is to divide the population of comets into two groups. Astronomers have found that the most physically reasonable way of doing this is to employ the so-called **Tisserand parameter**, which is defined as

$$T \equiv a_J/a + 2\sqrt{(1-e^2)a/a_J}\ \cos\ i,$$

where a_J is Jupiter's semimajor axis. This parameter is an approximation to the **Jacobi constant**, which is an **integral of the motion** in the **circular restricted three-body problem**. The circular restricted three-body problem, in turn, is a well-understood dynamical problem consisting of two massive objects (mainly the Sun and Jupiter in this context) in circular orbits about one another, with a third, very small, body in orbit about the massive pair. If, to zeroth order, a comet's orbit is approximately a perturbed Kepler orbit about the Sun, then, to first order, it is better approximated as the small object in the circular restricted three-body problem with the Sun and Jupiter as the massive bodies. This means that as comets gravitationally scatter off Jupiter or evolve due to processes like the Kozai resonance, T is approximately conserved. The Tisserand parameter is also a measure of the relative velocity between a comet and Jupiter during close encounters, $v_{rel} \sim v_J\sqrt{3-T}$, where v_J is Jupiter's orbital speed around the Sun. Objects with $T > 3$ cannot cross Jupiter's orbit in the circular restricted case, being confined to orbits either totally interior or totally exterior to Jupiter's orbit.

Figure 5 shows a plot of inclination versus semimajor axis for known comets. Astronomers put the first division in our classification scheme at $T = 2$. Objects with $T > 2$ are shown as open circles in the figure, while those with $T < 2$ are the filled circles. The bodies with $T > 2$ are confined to low inclinations. Thus, we call these objects **ecliptic comets**. We call the $T < 2$ objects **nearly-isotropic**

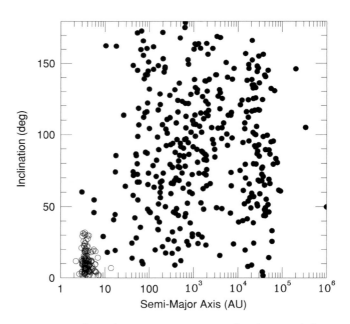

FIGURE 5 The inclination–semimajor axis distribution of all comets in the 2003 version of Marsden and Williams' *Catalogue of Cometary Orbits*. Comets with $T > 2$ are marked by the open circles, while comets with $T < 2$ are indicated by the filled circles.

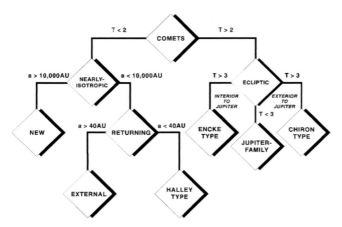

FIGURE 4 A flow chart showing the cometary classification scheme used in this chapter.

comets to reflect their broad inclination distribution. We now discuss each of these in turn.

2.1 Nearly-Isotropic Comets

Nearly-isotropic comets (hereafter NICs) are divided into two groups: dynamically "new" comets and "returning" comets. This division is one that has its roots in the dynamics of these objects and is based on the distribution of their semimajor axes, a. Figure 6 shows a histogram of $1/a$, which is proportional to orbital binding energy $E = -\frac{GM_\odot}{2a}$. These values of semimajor axes were determined by numerically integrating the observed trajectory of each comet backwards in time to a point before it entered the planetary system. Taken at face value, a comet with $1/a < 0$ is unbound from the Sun, i.e., it follows a hyperbolic orbit. However, all of the negative values of $1/a$ are due to errors in orbit determination either due to poor astrometry or uncertainties in the estimates of the nongravitational forces. Thus, we have yet to discover a comet from interstellar space. The fraction of comets that suffer from this problem is small and we will ignore them for the remainder of this chapter.

The most striking feature of this plot is the peak at about $1/a \sim 0.00005$ AU^{-1}, i.e., $a \sim 20,000$ AU. In 1950, this feature led Jan Oort to conclude that the Solar System is surrounded by a spherically symmetric cloud of comets, which we now call the Oort cloud. The peak in the $1/a$ distribution of NICs is fairly narrow. And yet, the typical kick that a comet receives when it passes through the planetary system is approximately ±0.0005 AU^{-1}, i.e., a factor of 10 larger than the energy of a comet initially in the peak (Fig. 6). Thus it is unlikely that a comet that is in the peak when it first passes through the Solar System will remain there

during successive passes. We conclude from this argument that comets in the peak are dynamically "new" in the sense that this is the first time that they have passed through the planetary system.

Comets not in the peak ($a \lesssim 10,000$ AU) are most likely objects that have been through the planetary system before. Comets with $a \ll 20,000$ AU that are penetrating the planetary system for the first time cannot make it into the inner Solar System where we see them as active comets without first encountering a planet (see Section 3.1 for a more complete discussion). Therefore, we should expect to see few comets directly from the Oort cloud with semimajor axes smaller than this value. We can conclude that a NIC not in the peak is a comet that was initially in it but has evolved to smaller a during previous passes through the planetary system. These comets are called "returning" comets. The boundary between new and returning comets is usually placed at $a = 10,000$ AU.

Returning comets are, in turn, divided into two groups based on their dynamics. Long-term numerical integrations of the orbits of returning comets show that a significant fraction of those with semimajor axes less than about 40 AU are temporarily trapped in what are called **mean motion resonances** with one of the giant planets during a significant fraction of the time they spend in this region of the Solar System. Such a resonance is said to occur if the ratio of the orbital period of the comet to that of the planet is near the ratio of two small integers. For example, on average Pluto orbits the Sun twice every time Neptune orbits three times. So, Pluto is said to be in the 2:3 mean motion resonance with Neptune. Comet 109P/Swift-Tuttle, with a semimajor axis of 26 AU, is currently trapped in a 1:11 mean motion resonance with Jupiter. Mean motion resonances can have a large effect on the orbital evolution of comets because they can change eccentricities and inclinations, as well as protecting the comet from close encounters with the planet it is resonating with. This is true even if the comet is only temporarily trapped. In our classification scheme, comets that have a small enough semimajor axis to be able to be trapped in a mean motion resonance with a giant planet are designated as **Halley-type** comets, named for its most famous member comet 1P/Halley. Returning comets that have semimajor axes larger than this are known as **external** comets. Although it is not really clear exactly where the boundary between these two type of comets should be, we place the boundary at $a = 40$ AU.

2.2 Ecliptic Comets

Recall that ecliptic comets are those comets with $T > 2$. These comets are further divided into three groups. Comets with $2 < T < 3$ are generally on Jupiter-crossing orbits and are dynamically dominated by that planet. Thus, we call these **Jupiter-family** comets. This class contains most of the known ecliptic comets. As described above, comets with

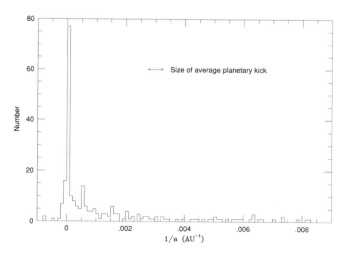

FIGURE 6 The distribution of inverse semimajor axis a, which measures the strength with which comets are gravitationally bound to the Solar System, for the known nearly-isotropic comets.

$T > 3$ cannot cross the orbit of Jupiter and thus should not be considered members of the Jupiter family. A comet that has $T > 3$ and whose orbit is interior to that of Jupiter is designated a *Encke-type*. This class is named after its best-known member, 2P/Encke. 2P/Encke is a bright, active comet that is decoupled from Jupiter. Its aphelion distance is only 4.2 AU.

A comet that has $T > 3$ and has a semimajor axis larger than that of Jupiter is known as a **Chiron-type**, again named after its best-known member, 95P/Chiron. As we discussed in Section 1.2, Chiron has a semimajor axis of 14 AU and a perihelion distance of 8 AU, putting it well beyond the grasp of Jupiter. Indeed, 95P/Chiron is currently dynamically controlled by Saturn. Although 95P/Chiron has a weak coma and is designated as a comet by the International Astronomical Union (IAU), it is also considered to be part of a population of asteroids known as **Centaurs**, which are found on orbits beyond Jupiter and that cross the orbits of the giant planets. The IAU distinguishes between a comet and an asteroid based on whether an object is active or not. This distinction is therefore not dependent on an object's dynamical history or where it came from. Thus, Chiron is simply a member of the Centaurs, of which there are currently a few dozen known members. For the remainder of this chapter, we will not distinguish between the **Chiron-type** comets and the Centaur asteroids, and will call both Centaurs.

2.3 Orbital Distribution of Comets

Figure 7 shows the location of the comet classes described above as a function of their Tisserand parameter and semimajor axis. Also shown is the location of all comets in the 2003 version of Marsden and Williams' *Catalogue of Cometary Orbits*. The major classes of ecliptic and nearly isotropic comets are defined by T and are independent of a. The ranges of these two classes are thus shown with arrows only. The extent of the subclasses is shown by different shadings. Also shown is the location of all the comets with $1/a > 0$ in the catalog. The white curve shows the relationship of T versus a for a comet with $q = 2.5$ AU and $i = 0$. Comets above and to the left of this line have $q > 2.5$ AU and thus are difficult to detect. By far, most comets in the plot are new or returning NICs. The second largest group consists of the Jupiter-family comets.

We end this section with a short discussion of the robustness of this classification scheme. Long-term orbital integrations show that comets rarely change their primary class (*ecliptic* versus *nearly isotropic*), but do frequently change their subclass (i.e., *new* versus *returning* or *Jupiter-family* versus *Chiron-type*). This result suggests that ecliptic comets and nearly isotropic comets come from different source reservoirs. In particular, as we will now describe, the NICs come from the Oort cloud, while the ecliptic comets are thought to originate in a structure that we call the **scattered disk**.

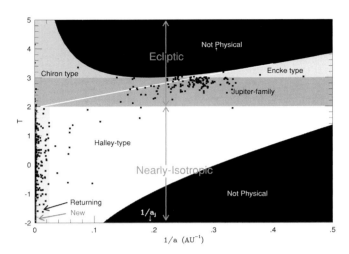

FIGURE 7 The location of the classes in our adopted comet taxonomy as a function of the Tisserand parameter (T) and semimajor axis (a). The major classes of ecliptic and nearly isotropic comets are defined by their values of T. The ranges of these two classes are thus shown with arrows only. The extent of each subclass is shown by different shadings. Also shown is the location of all the comets with $1/a > 0$ in the 2003 version of Marsden and Williams' *Catalogue of Cometary Orbits*. The white curve shows the relationship of T versus a for a comet with $q = 2.5$ AU and $i = 0$. Comets above and to the left of this line have $q > 2.5$ AU and thus are difficult to detect.

3. Comet Reservoirs

As we discussed above, the active comets that we see are on unstable, short-lived orbits because they cross the orbits of the planets. For example, the median dynamical lifetime of a Jupiter-family comet (defined as the span of time measured from when a comet first evolves onto Jupiter-family comet-type orbit until it is ejected from the Solar System, usually by Jupiter) is only about 300,000 years. So, these comets must have been stored in one or more reservoirs, presumably outside the planetary region, for billions of years before being injected into the inner Solar System where they can be observed. These reservoirs are far from the Sun (and they would have to be in order to store an ice ball for 4 billion years), and thus much of what we know about them has been learned by studying the visible comets and linking them to their reservoirs through a theoretical investigation of the orbital evolution of comets. As we currently understand things, there are two main cometary reservoirs: the Oort cloud and the scattered disk. We discuss each of these separately.

3.1 The Oort Cloud

Nearly isotropic comets originate in the Oort cloud, which is a nearly spherical distribution of comets (at least in the outer regions of the cloud), centered on the Sun. The position of its outer edge is defined by the Solar System's tidal truncation radius at about 100,000–200,000 AU from the Sun. At these distances, the gravitational effect of stars and other material in the Galaxy can strip a comet away from the Solar System. This edge can be seen in the distribution of NICs shown in Figure 6. For reasons described below, we have no direct information about the location of the Oort cloud's inner edge, but models of Oort cloud formation (see Section 3.3) predict that it should be between 2,000 and 5,000 AU.

The orbits of comets stored in the Oort cloud evolve due to the forces from the Galaxy. As shown in Figure 3, the primary role of the Galaxy is to change the angular momentum of the comet's orbit, causing large changes in the inclination and, more importantly, the perihelion distance of the comet. Occasionally, a comet will evolve so that its perihelion distance falls to within a few AU of the Sun, thus making it visible as a new nearly isotropic comet. As we discussed above, the new comets that we see have semimajor axes larger than 20,000 AU, as illustrated by the spike in Figure 6. This led Jan Oort to suggest that the inner edge of the Oort cloud was at this location. However, this turns out not to be the case. In order for us to see a new comet from the Oort cloud, it has to get close to the Sun, which generally means that its perihelion distance, q, must be less than 2 or 3 AU.[1] However, during the perihelion passage before the one on which we see a comet for the first time, its perihelion distance must have been outside the realm of the gas giants ($q \gtrsim 15$ AU), because if the comet had q near either Jupiter or Saturn when it was near perihelion, it would have received a kick from the planets that would have knocked it out of the spike. Thus, new comets can only come from the region in the Oort cloud in which the Galactic tides are strong enough that the change in perihelion in one orbit (Δq) is greater than $\sim$10 AU. It can be shown that the timescale on which a comet's perihelion changes is

$$\tau_q = 6.6 \times 10^{14} \text{ yr } a^{-2} \, \Delta q \, / \, \sqrt{q},$$

in the current galactic environment where a, Δq, and q are measured in AU. Thus, only those objects for which τ_q is larger than the orbital period can become a visible new comet. For $\Delta q = 10$ AU and $q = 15$ AU, this occurs when $a \gtrsim 20,000$ AU.

The above result does not imply that Oort comets far inside of 20,000 AU do not contribute to the population of nearly isotropic comets. In fact, they do. It is simply that these objects do not become active comets until their orbits have been significantly modified by the giant planets.

[1] Comets are sometimes discovered at larger perihelion distances because the comet is unusually active due to the sublimation of ices, such as carbon monoxide, that are more volatile than water ice. The current record holder, the new comet C/2003 A2 Gleason, had $q = 11$ AU.

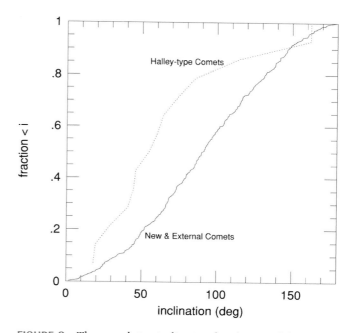

FIGURE 8   The cumulative inclination distribution of the nearly-isotropic comets in Marsden and Williams' catalog. We divide the population into two groups: Halley-types ($a < 40$ AU) and a combination of new and external comets.

Some become returning comets. Indeed, from modeling the inclination distribution of the Halley-type comets, we think that some objects from the inner regions of the Oort cloud eventually become NICs.

Figure 8 shows the cumulative inclination distribution for a combination of new and external comets (solid curve) and Halley-type comets (dotted curve). The solid curve is what would be expected from an isotropic Oort cloud. The curve follows a roughly sin (i) distribution, which has a median inclination of 90° and thus has equal numbers of prograde and retrograde orbits. It is these data that astronomers use to argue that the outer Oort cloud is basically spherical.

The inclination distribution of the Halley-type comets is quite different from that of the rest of the NICs. Almost 80% of Halley-type comets are on prograde orbits ($i < 90°$); the median inclination is only 55°. Numerical simulations of the evolution of comets from the Oort cloud to Halley-type orbits show that the inclination distribution of the comets is approximately conserved during the capture process. This means that the source region for these comets should have the same inclinations, on average, as the dotted curve in Figure 8. The only way to reconcile this with the roughly spherical shape of the outer Oort cloud is if the inner regions of the Oort cloud are flattened into a disk-like structure. Indeed, simulations suggest that the inner Oort cloud must have a median inclination of between 10 and 50° for it to match the observed inclination distribution of Halley-type comets. Figure 9 shows an artist's conception of what the Oort cloud may look like in cross-section.

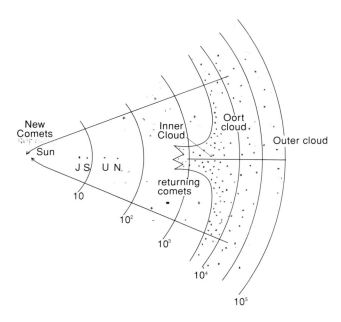

FIGURE 9 An artist's conception of the structure of the Oort cloud. In particular, the locations of the inner and outer edges of the Oort cloud, and where the cloud is flattened, are shown with respect to the location of the giant planets. Note that the radial distance from the Sun is spaced logarithmically. The location of the returning comets and the source for the new comets are also illustrated.

3.2 The Scattered Disk

To start the discussion of the scattered disk, we turn our attention back to Figure 5, which shows the semimajor axis–inclination distribution of the known comets. There is a clear concentration of comets on low-inclination orbits near $a \sim 4$ AU. Indeed, 27% of all the comets in the catalog lie within this concentration. As we described above, we call these objects ecliptic comets, and most are Jupiter-family comets.

Until the 1980s, the origin of these objects was a mystery. Even at that time it was recognized that the inclination distribution of comets does not change significantly as they evolve from long-period orbits inward. This is a problem for a model in which these comets originate in the Oort cloud, as most astronomers believed, because the median inclination of the Jupiter family is only $11°$. So, dynamicists argued that Jupiter-family comets could not come from the Oort cloud, but must have originated in a flattened structure. Indeed, it was suggested that these objects originated in a disk of comets that extends outward from the orbit of Neptune. Spurred on by this argument, observers discovered the first trans-Neptunian object in 1992. Although this object is about a million times more massive then the typical ecliptic comet (it needs to be much larger than a typical comet, or we would not have seen it that far away), it was soon recognized that it was part of a population of objects both large and small—mainly small.

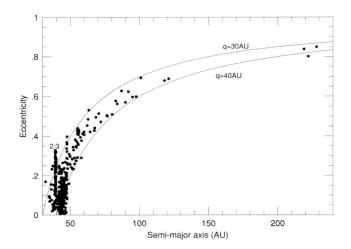

FIGURE 10 The eccentricity–semimajor axis distribution for the known trans-Neptunian objects with good orbits as of November 2005. We truncated the plot at 250 AU in order to resolve the inner regions better. Two curves of constant perihelion distance (q) are shown. In addition, the location of Neptune's 2:3 mean motion resonance is marked.

Since 1992, the trans-Neptunian region has been the focus of intense research, and over a thousand objects are now known to reside there. The diversity (both physical and dynamical) of its objects make it one of the most puzzling and fascinating places in the Solar System. As such, a complete discussion is beyond the scope of this chapter and, indeed, chapters on the Kuiper Belt are dedicated to this topic [*See* KUIPER BELT: DYNAMICS; KUIPER BELT OBJECTS: PHYSICAL STUDIES]. For our purposes, it suffices to say that the trans-Neptunian region is inhabited by at least two populations of objects that roughly lie in the same region of physical space, but have very different dynamical properties. These are illustrated in Figure 10, which shows the semimajor axis and eccentricity of all known trans-Neptunian objects with good orbits as of November 2005.

The first population of interest consists of those objects which are on orbits that are stable for the age of the Solar System. These objects mostly have perihelion distances (q) larger than 40 AU, or are in mean motion resonances with Neptune. Of particular note are the bodies in Neptune's 2:3 mean motion resonance, which are marked in the figure. Pluto is a member of this group. Even though some objects in the resonances are on orbits that cross the orbit of Neptune, they are stable because the resonance protects them from close encounters with that planet. All in all, we call this population the **Kuiper Belt**.[2]

The second population is mainly made up of objects with small enough perihelion distances that Neptune can push

---

[2] There are two meanings of the phrase "Kuiper Belt" in the literature. There is the one employed above. In addition, some researchers use the phrase to describe the entire trans-Neptunian region. In this case the term "classical Kuiper Belt" is used to distinguish the stable regions. We prefer the former definition.

them around as they go through perihelion. Because of this characteristic, we call this population the **scattered disk**. These are mainly nonresonant objects with $q < 40$ AU. [*See* KUIPER BELT: DYNAMICS for a more detailed definition.] Although most of the trans-Neptunian objects thus far discovered are members of the Kuiper Belt as defined here, it turns out that this is due to observational bias, and the Kuiper Belt and scattered disk contain roughly the same amount of material. In particular, the scattered disk contains about a billion objects that are comet-sized (roughly kilometer-sized) or larger.

Since the scattered disk is a dynamically active region, objects are slowly leaking out of it with time. Indeed, models of the evolution of scattered disk objects show that the scattered disk contained about 100 times more objects when it was formed roughly 4 billion years ago than it does today (see below). Objects can leave the scattered disk in two ways. First, they can slowly evolve outward in semimajor axis until they get far enough from the Sun that Galactic tides become important. These objects then become part of the Oort cloud. However, most of the objects evolve inward onto Neptune-crossing orbits. Close encounters with Neptune can then knock an object out of the scattered disk. Roughly one comet in three that becomes Neptune-crossing, in turn, evolves through the outer planetary system to become a Jupiter-family comet for a small fraction of its lifetime.

Figure 11 shows what we believe to be the evolution of a typical scattered disk object as it follows its trek from the scattered disk to the Jupiter family and out again. The figure shows this evolution in the perihelion distance (q) – aphelion distance (Q) plane. The positions are joined by blue lines until the object first became "visible" (which we take to be $q < 2.5$ AU) and are linked in red thereafter. Initially, the object spent considerable time in the scattered disk, i.e., with perihelion near the orbit of Neptune (30 AU) and aphelion well beyond the planetary system. However, once an object evolves inward, it tends to be under the dynamical control of just one planet. That planet will scatter it inward and outward in a random walk, typically handing it off to the planet directly interior or exterior to it. Because of the roughly geometric spacing of the giant planets, comets tend to have eccentricities of about 25% between "handoffs" and spend a considerable amount of time with perihelion or aphelion near the semimajor axis of Saturn, Uranus, or Neptune.

However, once comets have been scattered into the inner Solar System by Jupiter, they can have much larger eccentricities as they evolve back outward. The postvisibility phase of the object in Figure 11 is reasonably typical of Jupiter-family comets, with much larger eccentricities than the previsibility comets and perihelion distances near Jupiter or Saturn. This object was eventually ejected from the Solar System by a close encounter with Saturn.

Numerical models, like the one used to create Figure 11, show that most of the ecliptic comets and Centaurs most

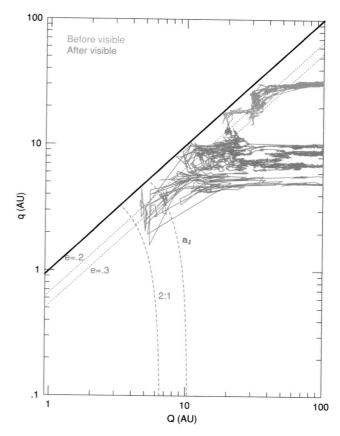

FIGURE 11 The orbital evolution of a representative object originating in the scattered disk. In particular, the locations of the object's orbit in the $q - Q$ (perihelion-aphelion) plane are joined by blue lines until the object became "visible" ($q < 2.5$ AU) and are linked in red thereafter. The sampling interval was every 10,000 years in the previsibility phase and every 1000 years thereafter. Also shown in the figure are three lines of constant eccentricity at $e = 0$, 0.2, and 0.3. In addition, we plot two dashed curves of constant semimajor axis, one at Jupiter's orbit and one at its 2:1 mean motion resonance. Note that it is impossible for an object to have $q > Q$, so objects cannot move into the region above and to the right of the solid diagonal line.

likely originated in the scattered disk. Figure 12 shows the distribution of the ecliptic comets derived from these simulations. The figure is a contour plot of the relative number of comets per square AU in perihelion-aphelion ($q - Q$) space. Also shown are the locations of 95P/Chiron and 2P/Encke (big dots marked "C" and "E", respectively), and the known Jupiter-family comets (small gray dots).

There are two well defined regions in Figure 12. Beyond approximately $Q = 7$ AU, there is a ridge of high density extending diagonally from the upper right to the center of the plot, near $e \approx 0.25$. The peak density in this ridge drops by almost a factor of 100 as it moves inward, having a minimum where the semimajor axes of the comets are the same as Jupiter's (shown by a dotted curve and marked with a_J). This region of the plot is inhabited mainly by the Centaurs. Inside of $Q \approx 7$ AU, the character of the distribution is quite different. Here there is a ridge of high density extending vertically in the figure at $Q \sim 5–6$ AU that extends over a wide range of perihelion distances. Objects in this region are the Jupiter-family comets. This characteristic of a very narrow distribution in Q is seen in the real Jupiter-family comets and is a result of the narrow range in T which, in turn, comes from the low to moderate inclinations and eccentricities of bodies in the scattered disk.

Figure 12 shows the relationship between the Centaurs and the Jupiter-family comets and illustrates the distribution of objects throughout the outer Solar System. The simulations predict that the inclinations of this population should be small everywhere, which is consistent with observations.

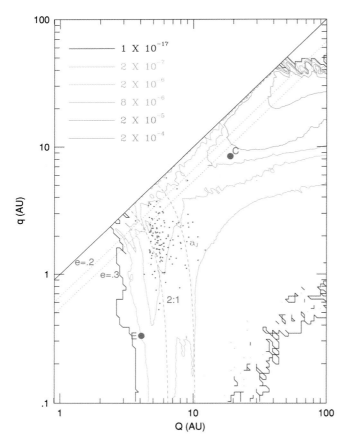

FIGURE 12 A contour plot of the relative distribution of ecliptic comets in the solar system as a function of aphelion (Q) and perihelion (q). The units are the fraction of comets per square AU in $q - Q$ space. Also shown in the figure are three lines of constant eccentricity at $e = 0$ (solid), 0.2, and 0.3 (both dotted). In addition, we plot two dashed curves of constant semimajor axis, one at Jupiter's orbit and one at its 2:1 mean motion resonance. They gray dots labeled "E" and "C" show the locations of comets 2P/Encke and 95P/Chiron. The small gray dots show the orbits of the Jupiter-family comets.

3.3 Formation of the Oort Cloud and Scattered Disk

Let us take stock of where we have come thus far. Active comets can be divided into two groups based on the value of the Tisserand parameter, T. The nearly isotropic comets have $T < 2$ and originate in the Oort cloud. The ecliptic comets have $T > 2$ and originate in the scattered disk. The Oort cloud is a population of comets that lie very far from the Sun, with semimajor axes extending from tens of thousands of AU down to thousands of AU. It also is roughly spherical in shape. The scattered disk, on the other hand, lies mainly interior to ~1000 AU and is flattened. It may be surprising, therefore, that modern theories suggest that both of these structures formed as a result of the same process and therefore the objects in them formed in the same region of the Solar System.

First, we must address why we think that these structures did not form where they are. The answer has to do with the comets' eccentricities and inclinations. Although comets are much smaller than planets, they probably formed in a similar way. The Solar System formed from a huge cloud of gas and dust that initially collapsed to a protostar surrounded by a disk. The comets, asteroids, and planets formed in this disk. However, initially the disk only contained very small solid objects, similar in size to particles of smoke, and much smaller than comets. Although it is not clear how these objects grew to become comet-sized, all the processes thus far suggested require that the relative velocity between the dust particles was small. This, in turn, requires the dust particles to be on nearly circular, coplanar orbits. So, the eccentric and inclined orbits of bodies in the cometary reservoirs must have arisen because they were dynamically processed from the orbits in which they were formed to the orbits in which they are found today.

Astronomers generally agree that comets originally formed in the region of the Solar System now inhabited by the giant planets. Although comets formed in nearly circular orbits, their orbits were perturbed by the giant planets as the planets grew and/or the planets' orbits evolved. Figure 13 shows the behavior of a typical comet as it evolves into the Oort cloud. At first, the comet is handed off from planet to planet, remaining in a nearly circular orbit (Region 1 in the figure). However, eventually Neptune scatters the body outward. It then goes through a period of time when its semimajor axis is changing due to encounters with Neptune (Region 2). During this time its perihelion distance is near the orbit of Neptune, but its semimajor axis can become quite large. (If this reminds you of the scattered disk, it should.) When the object gets into the region beyond 10,000 AU, galactic perturbations lift its perihelion out of the planetary system, and it is then stored in the Oort cloud for billions of years (Region 3).

Figure 14 shows the result of a numerical model of the formation of the Oort cloud and scattered disk. The simulation followed the orbital evolution of a large number of

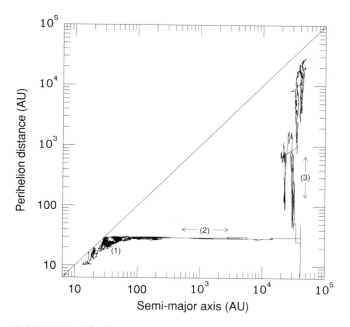

FIGURE 13 The dynamical evolution of an object as it evolves into the Oort cloud. The object was initially in a nearly circular orbit between the giant planets. Its evolution follows three distinct phases. During Phase 1 the object remains in a relatively low eccentricity orbit between the giant planets. Neptune eventually scatters it outward, after which the object undergoes a random walk in semimajor axis (Phase 2). When it reaches a large enough semimajor axis, galactic perturbations lift its perihelion distance to large values (Phase 3).

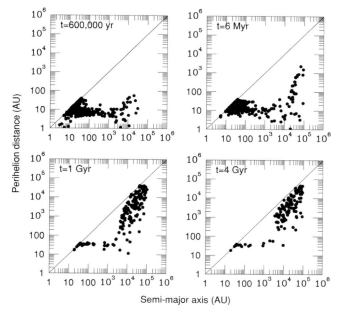

FIGURE 14 Four snapshots of comets in a simulation of the formation of the scattered disk and the Oort cloud.

comets initially placed on nearly-circular, low-inclination orbits between the giant planets, under the gravitational influence of the Sun, the four giant planets, and the Galaxy. The major steps of Oort cloud formation can be seen in this figure. Initially the giant planets start scattering objects to large semimajor axes. By 600,000 years, a massive scattered disk has formed, but only a few objects have evolved far enough outward that Galactic perturbations are important.

At $t = 6$ million years the Oort cloud is beginning to form. The Galactic perturbations have started to raise the perihelion distances of the most distant comets, but a complete cycle in q has yet to occur (see Fig. 3). Note that the scattered disk is still massive. By 1 billion years, the Oort cloud beyond 10,000 AU is inhabited by objects on moderate-eccentricity orbits (i.e., where $a \sim q$). Note also that a scattered disk still exists. There is also a transition region between ~2,000 AU and ~ 5,000 AU, where objects are beginning to have their perihelia lifted by the Galaxy, but have not yet undergone a complete cycle in perihelion distance. By 4 billion years, the Oort cloud is fully formed and extends from 3000 AU to 100,000 AU. The scattered disk can easily be seen extending from Neptune's orbit outward. If our current understanding of comet reservoirs is correct, these are the two source reservoirs of all the known visible comets.

The above calculations assume that the Sun has always occupied its current Galactic environment, i.e., it is isolated and not a member of a star cluster. However, almost all stars form in dense clusters. The gravitational effects of such a star cluster on a growing Oort cloud is similar to that of the Galaxy except that the torques are much stronger. This would lead to an Oort cloud that is much more compact if the Sun had been in such an environment at the time that the cloud was forming. However, models of the dynamical evolution of star clusters show that the average star spends less than 5 million years in such an environment and the giant planets might take that long to form. Additionally, even if the planets formed very quickly, Figure 14 shows that the Oort cloud is only partially formed after a few million years. In particular, only those objects that originated in the Jupiter-Saturn region have evolved much in semimajor axis. Therefore, the Oort cloud probably formed in two stages. Before ~5 Myr a dense *first generation* Oort cloud formed from Jupiter-Saturn planetesimals at roughly $a \sim 1,000$ AU due to the effects of the star cluster. After the Sun left the cluster, a normal Oort cloud formed at $a \sim 10,000$ AU from objects that originated beyond Saturn. Figure 15 shows an example of such an Oort cloud as determined from numerical experiments. There is some observational evidence that the Solar System contains a first generation Oort cloud. In 2004, the object known as Sedna was discovered. Sedna has $a = 468\,AU$ and $q = 76\,AU$, placing it well beyond the planetary region. Numerical experiments have shown that the most likely way to get objects with perihelion distances as large as Sedna is through external torques (as in Fig. 15).

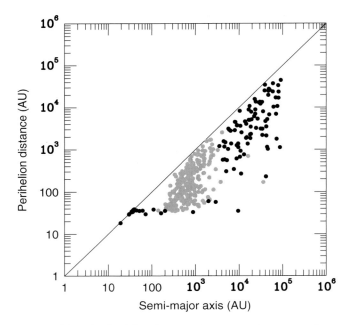

FIGURE 15 The final distribution of comets in the scattered disk and the Oort cloud according to a numerical experiment where the Sun spent 3 Myr in a star cluster. The grey and black dots refer to objects that formed interior to or exterior to 14 AU.

And, since the current Galactic environment is too weak to place Sedna on its current orbit, Sedna's orbit probably formed when the Sun was in its birth star cluster. If true, Sedna's orbit represents the first observational constraint we have concerning the nature of this star cluster. If such a structure really exists, it does not contribute to the population of observed comets because it is in a part of the Solar System which is currently stable: objects in this region do not get close to the planets and the Galactic tides are too weak.

4. Conclusions

Comets are only active when they get close to the Sun. However, they must come from more distant regions of the Solar System where it is cold enough for them to survive the age of the Solar System without sublimating away. Dynamical simulations of cometary orbits argue that there are two main source regions in the Solar System. One, known as the Oort cloud, is a roughly spherical structure located at heliocentric distances of thousands to tens of thousands of AU. The nearly isotropic comets come from this reservoir. The scattered disk is the other important cometary reservoir. It is a disk-shaped structure that extends outward from the orbit of Neptune. The ecliptic comets come from the scattered disk.

However, there are substantial reasons to believe that these two cometary reservoirs are not primordial structures and that their constituent members formed elsewhere and were dynamically transported to their current locations. Indeed, current models suggest that objects in both the Oort cloud and scattered disk formed in the region between the giant planets and were delivered to their current locations by the action of the giant planets as these planets formed and evolved. Comets, therefore, represent the leftovers of planet formation and contain vital clues to the origin of the Solar System.

Bibliography

Brandt, J. C., R. D. Chapman 2004. *Introduction to Comets*, 2nd ed. Cambridge University Press, 450 pp.

British Astronomical Association 2006. BAA comet section. Updated January 6, 2006. http://www.ast.cam.ac.uk/~jds/.

Dones, L., Weissman, P. R., Levison, H. F., Duncan, M. J. 2004. Oort cloud formation and dynamics. In *Comets II*, Festou, M. C., H. U. Keller, H. A. Weaver, eds., pp. 153–174.

Duncan, M., Levison, H., Dones, L. 2004. Dynamical evolution of ecliptic comets. In *Comets II*, Festou, M. C., H. U. Keller, H. A. Weaver, eds., pp. 193–204.

Fern'ández, J. A. 2005. *Comets–Nature, Dynamics, Origin, and their Cosmogonical Relevance*. Springer, 383. pp.

Fernández, Y. 2006. List of Jupiter-family and Halley-family comets. Updated January 5, 2006. http://www.physics.ucf.edu/~yfernandez/cometlist.html.

Festou, M. C., H. U. Keller, H. A. Weaver, eds. 2004. *Comets II*. Univ. Arizona Press, 745. pp.

Jet Propulsion Laboratory 2005. JPL solar system dynamics. Updated October 4, 2005. http://ssd.jpl.nasa.gov/.

Kinoshita, K. 2006. Comet orbit home page. Updated January 4,2006. http://www9.ocn.ne.jp/~comet/.

Kresák, L. 1982. Comet discoveries, statistics, and observational selection. In *Comets*, ed. L. L. Wilkening, Univ. Arizona Press, Tucson, pp. 56–82.

Kronk, G. W. 1999. *Cometography: A Catalog of Comets*. Volume 1: Ancient to 1799. Cambridge Univ. Press, New York, 563 pp.

Kronk, G. W. 2003. *Cometography: A Catalog of Comets*. Volume 2: 1800–1899. Cambridge Univ. Press, New York, 852 pp.

Kronk, G. W. 2006. Cometography. http://cometography.com/.

Marsden, B. G., Z. Sekanina, D. K. Yeomans 1973. Comets and nongravitational forces. V. *Astron. J.* **78**, 211–225.

Marsden, B. G., G. V. Williams 2003. *Catalogue of Cometary Orbits*, 15th ed. Cambridge, Mass., Smithsonian Astrophysical Observatory. 169 pp.

Minor Planet Center 2006. IAU: Minor Planet Center. Updated January 7, 2006. http://www.cfa.harvard.edu/iau/mpc.html.

Oort, J. H. 1950. The structure of the cloud of comets surrounding the solar system and a hypothesis concerning its origin. *Bull Astron. Inst. Neth.* **11**, 91–110.

Kuiper Belt: Dynamics

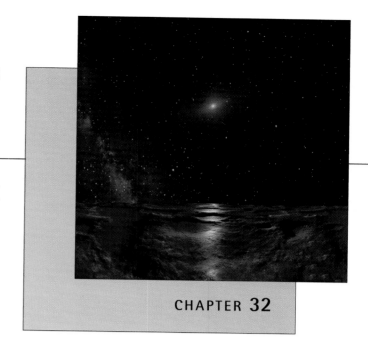

Alessandro Morbidelli

Observatoire de la Côte d'Azur
Nice, France

Harold F. Levison

Southwest Research Institute
Boulder, Colorado

CHAPTER **32**

The name Kuiper Belt is generically referred to a population of small bodies, the orbits of which have a **semimajor axis**—and hence orbital period—larger than those of Neptune. It can be viewed as a second Asteroid Belt, but located at the outskirts of the solar system. The Kuiper Belt objects—having formed at large distances from the Sun—are rich in water ice and other volatile chemical compounds and have physical properties similar to those of comets. Indeed, the existence of the Kuiper Belt was first deduced from observations of the Jupiter-family comets, a population with short orbital periods and small to moderate orbital inclinations, of which the Kuiper Belt is the source.

In 14 years since the discovery of the first object, about 1200 Kuiper Belt objects have been detected. Of these, ~700 objects have been observed for more than 2 years, a necessary condition to compute their orbital parameters with significant precision. The results of this detailed observational exploration of the Kuiper Belt structure have provided several surprises. Indeed, it was expected that the Kuiper Belt preserved the pristine conditions of the protoplanetary disk. But it is now evident that this picture is not correct: The disk has been affected by a number of processes that have altered its original structure.

The Kuiper Belt may thus provide us with a large number of clues to understand what happened in the outer solar system during the primordial ages. Potentially, the Kuiper Belt might teach us more about the formation of the giant planets than the planets themselves. And, as in a domino game, a better knowledge of giant planets formation would inevitably boost our understanding of the subsequent formation of the solar system as a whole. Consequently, Kuiper Belt research is now considered a top priority of modern planetary science.

1. Historical Perspective

Since its discovery in 1930, Pluto has traditionally been viewed as the last vestige of the planetary system—a lonely outpost at the edge of the solar system, orbiting beyond Neptune with a 248 year period. Pluto receives very little light from the Sun (being almost 40 times farther from the Sun on average than the Earth) and thus it is very cold. The view was that it was a distant, isolated, and unfriendly place, with nothing of substance beyond it.

Pluto itself has always appeared to be an oddity among the planets. Traditionally, the planets are divided into two main groups. The first group, the terrestrial planets, formed in the inner regions of the solar system where the material from which the planets were made was too warm for water and other volatile gases to be condensed as ices. These planets, which include the Earth, are small and rocky. Farther out from the Sun, the cores of the planets grew from a combination of rock and condensed ices and captured significant amounts of nebula gas. These are the jovian planets, the giants of the solar system; they most likely do not have solid surfaces. But, then there is Pluto, unique, small (its radius is only ~1180 km, only two thirds that of the Earth's Moon) and made of a mix of rock and frozen ices.

The planets formed in a disk of material that originally surrounded the Sun. As the Sun formed from the collapse of its parent molecular cloud, it faced a problem. The cloud had a slight spin and as it collapsed, the spin rate had to increase in order to conserve **angular momentum**. The cloud could not form a single star with the amount of angular momentum it possessed, so it shed a disk of material that contained very little mass (as compared with the mass of the Sun), but most of the angular momentum of the system. As such, the planets formed in a narrow disk structure; the plane of that disk is known as the invariable plane. But, then there is Pluto, unique, having an orbital inclination of 15.6° with respect to the invariable plane.

The orbits of the planets are approximately ellipses with the Sun at one focus. As the planets formed in the original circumsolar disk, they tended to evolve onto orbits that were well separated from one another. This was required so that their mutual gravitational attraction would not disrupt the whole system. (Or to put is another way, if our system had not formed that way, we would not be here to talk about it!) But, then there is Pluto, unique, having an orbit that crosses the orbit of its nearest neighbor, Neptune.

So, the historical view was that Pluto was an oddity in the solar system. Unique for its physical makeup and size as well as its dynamical niche. But, this view changed in September 1992 with the announcement of the discovery of the first of a population of small (compared to planetary bodies) objects orbiting beyond the orbit of Neptune, in the same region as Pluto. Since that time, over 1000 objects with radii between a few tens and ~1000 km have been discovered. One object, 136199 Eris (previously known under the provisional designation 2003 UB_{313}), even turned out to be 10% larger than Pluto. Moreover, a modeling of the detection efficiency of the performed surveys suggests that there are approximately 1,000,000 objects larger than a few tens of kilometers occupying this region of space, approximately between 30 and 50 AU from the Sun. There are almost certainly many more smaller ones. As discussed in more detail in the following sections, these objects likely have a similar physical makeup to that of Pluto, and many have similar orbital characteristics. Thus, in the last decade, Pluto has been transformed from an oddity, to the founding member of what is perhaps the most populous class of objects in the planetary system.

The discovery of the Kuiper Belt, as it has come to be known, represents a revolution in our thinking about the solar system. First predicted on theoretical grounds and later confirmed by observations, the Kuiper Belt is the first totally new class of bodies to be discovered in the solar system since the first asteroid was found on New Year's day, 1801. Its discovery is on a par with the discovery of the solar wind and the planetary magnetospheres in the 1950s and 1960s, and it has radically changed our view of the outer solar system.

Speculation on the existence of a trans-Neptunian disk of icy objects dates back over 90 years. In the early 1900, Campbell, Aitken, and Leuschner considered the possibility of trans-Neptunian planets and speculated on the orbital distribution of small bodies in the outer planetary system. In the 1940s and early 1950s, a more comprehensive approach to the problem was made independently by Kenneth Edgeworth and Gerard Kuiper. They noticed that if one were to grind up the giant planets and spread out their masses to form a disk, then this disk would have a very smooth distribution, with a density that slowly decreases as the distance from the Sun increases. That holds until Neptune, at which point there is an apparent edge beyond which there was thought to be nothing except tiny Pluto. Edgeworth and Kuiper suggested that perhaps this edge was not real. Perhaps the disk of planetesimals (i.e., small bodies, potentially precursors of planet formation) that formed the planets extended past Neptune, but the density was too low or the formation times too long to form large planets. If so, they argued, these planetesimals should still be there in nearly circular orbits beyond Neptune. Unfortunately, Edgeworth's contribution was overlooked until recently, and thus this disk has come to be known as the Kuiper Belt.

The idea of a trans-Neptunian disk received little attention for many years. The objects in the hypothetical disk were too faint to be seen with the telescopes of the time, so there was no way to prove or disprove their existence. Comet dynamicists showed that the lack of detectable perturbations on the orbit of Halley's comet limited the mass of such a disk to no more than 1.3 Earth masses $(M_\oplus)$ if it was at 50 AU from the Sun.

However, the idea was resurrected in 1980 when Julio Fernandez proposed that a cometary disk beyond Neptune could be a possible source reservoir for the short-period comets (those with orbital periods <200 years). Subsequent dynamical simulations showed that a comet belt beyond Neptune is the most plausible source for the low inclination subgroup of the short-period comets, named the Jupiter-family comets. This work led observers to search for Kuiper Belt objects. With the discovery of the first object, 1992 QB_1 by D. Jewitt and J. Luu, the Kuiper Belt ceased to be a speculation and became a concrete entity of the solar system.

2. Basic Orbital Dynamics

Much of the story of the Kuiper Belt to date involves the distribution of the orbits of its members. In this section, we present a brief overview of the important aspects of the orbits of small bodies in the solar system. [For a more detailed discussion, *see* SOLAR SYSTEM DYNAMICS: REGULAR AND CHAOTIC MOTION.]

The most basic problem of orbital dynamics is the two-body problem: a planet, say, orbiting a star. The orbit's trajectory is an ellipse with the Sun at one of the foci. Energy, angular momentum, and the orientation of the ellipse are conserved quantities. The semimajor axis, a, of the ellipse is a function of the orbital energy. The **eccentricity**, e, of the ellipse is a function of the energy and the angular momentum. For a particular semimajor axis, the angular momentum is a maximum for a circular orbit, $e = 0$. These two-body orbits are known as Keplerian orbits.

A Keplerian orbit is characterized by its semimajor axis and eccentricity, as well as by three angles that describe the orientation of the orbital ellipse in space. The first, known as the inclination, i, is the angle between the angular momentum vector of the orbit and some reference direction for the system. In our solar system, the reference direction is usually taken as the angular momentum vector of the Earth's orbit (which defines the ecliptic plane, the reference plane), but it is sometimes taken to be the angular momentum vector of all the planetary orbits combined (which defines the invariable plane).

The point where the orbit passes through the reference plane in an "upward" direction is called the ascending node. Thus, the second orientation angle of the orbit is the angle between the ascending node and some reference direction in the reference plane, as seen from the Sun. In our solar system, the reference direction is usually taken to be the direction toward the vernal equinox. This angle is known as the longitude of the ascending node, Ω.

The third and final orientation angle is the angle between the ascending node and the point where the orbit is closest to the Sun (known as **perihelion**), as seen from the Sun. It is called the *argument of perihelion*, ω. Another useful angle, known as the *longitude of perihelion*, $\tilde{\omega}$, is defined to be $\omega + \Omega$.

The first-order gravitational effect of the planets on one another is that each applies a torque on the other's orbit, as if the planets were replaced by rings of material smoothly distributed along their orbits. This torque causes both the longitude of perihelion, $\tilde{\omega}$, and the longitude of the ascending node, Ω, to rotate slowly, a motion called *precession*. For a given planet, the precession of $\tilde{\omega}$, is typically dominated by one frequency. The same is true for Ω, although the dominant frequency is different. The periods associated with these frequencies range from 4.6×10^4 to 2×10^6 years in the outer planetary system. This is much longer than the orbital periods of the planets (164 years for Neptune).

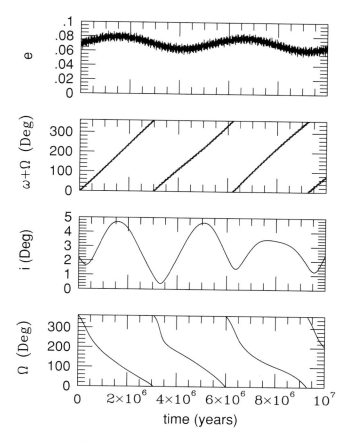

FIGURE 1   The temporal evolution of the orbit of the first Kuiper Belt object found, 1992 QB$_1$. As described in the text, the eccentricity, e, and inclination, i, oscillate, while the longitude of the ascending node, Ω, and the longitude of perihelion $\tilde{\omega} = \omega + \Omega$ circulate.

The orbit of a small object in the solar system, when it is not being strongly perturbed by a close encounter with a planet or is not located near a resonance (see later), is usually characterized by slow oscillations of e and i and a circulation (i.e., continuous change) in $\tilde{\omega}$ and Ω. The variation in the eccentricity is coupled with the $\tilde{\omega}$ precession and the variation in the inclination is coupled with the $\tilde{\omega}$ precession. Figure 1 shows this behavior for the first discovered Kuiper Belt object, 1992 QB$_1$.

The behavior of objects that are in a resonance can be very dramatic. There are two types of resonances that are known to be important in the Kuiper Belt. The most basic is known as a mean-motion resonance. A mean-motion resonance is a commensurability between the orbital period of two objects. That is, the ratio of the orbital periods of the two bodies in question is a ratio of two (usually small) integers. Perhaps the most well-known and important example of a mean-motion resonance in the solar system is the one between Pluto and Neptune.

As noted earlier, one of the unique aspects of Pluto's orbit is that when Pluto is at perihelion, it is closer to the Sun than Neptune. Normally, this configuration would, sooner or later, lead to close encounters between the two planets that would eventually scatter Pluto away. However, close encounters do not occur because Pluto is locked in a mean-motion resonance where it goes around the Sun twice every time Neptune goes around three times. So, every time Pluto crosses the trajectory of Neptune, the giant planet is always in one of three specific locations on its orbit, all very far away from the crossing point. This resonance is known as the 2:3 mean-motion resonance.

The other type of resonance that is important in the Kuiper Belt is called a secular resonance. There are actually two types of secular resonances. The first, which was discussed earlier, is a resonance between the precession rates of the longitudes of perihelion. As discussed, this can lead to changes in eccentricity. These resonances are identified by the Greek letter v with a numbered subscript that indicates the resonant planet (1 for Mercury through 9 for Pluto). In the Kuiper Belt, the perihelion secular resonance with Neptune, or v_8, is most important. The other type of secular resonance occurs when the small body's nodal precession

rate is the same as for a planet. This type of resonance can cause significant changes in the inclination of the orbit of the small body. These resonances are identified by v_{1x}, where x is the number of the resonant planet. For example, the nodal resonance with Neptune is the v_{18}.

The dynamical structure of the Kuiper Belt has been sculpted by a combination of mean-motion and secular resonances and by the evolution of these resonances during the formation of Uranus and Neptune. We come back to this issue in Sections 3 and 7.

3. Orbital and Dynamical Structure of the Trans–Neptunian Population

Figure 2 shows the distribution of the objects with semimajor axis larger than 30 astronomical units (AU) whose orbits have been determined from observations spanning over at least 3 years.

A glance at the figure reveals that the orbits of the trans-Neptunian objects can be very diverse. The majority

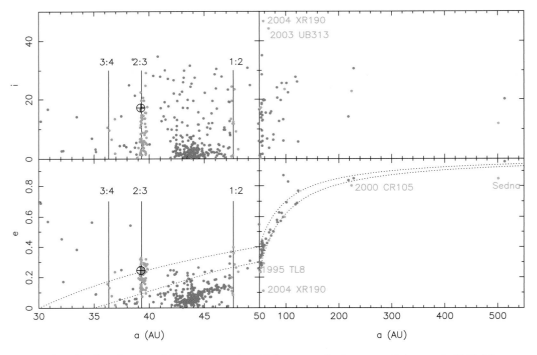

FIGURE 2 The distribution of the objects with well-determined orbits, as to February 1, 2005. The upper and lower panels show respectively the inclination and the eccentricity vs. semimajor axis. Two different semimajor axis scales are used to illustrate the Kuiper Belt (left panels) and the scattered disk (right panels) distributions. Red dots correspond to the scattered disk, magenta dots to the extended scattered disk, blue dots to the classical Kuiper Belt, and green dots to the resonant populations. The big, crossed circle denotes the orbit of Pluto. The vertical lines labeled 3:4, 2:3, and 1:2 mark the location of the corresponding mean-motion resonances with Neptune. The two dotted curves on the lower panels correspond to perihelion distances $q = 30$ AU and $q = 35$ AU on the left, and $q = 30$ AU and $q = 38$ AU on the right. These curves approximately bound the scattered disk orbital distribution.

of the discovered objects are clustered in the 36–48 AU range, but several others form a "tail" structure extending beyond 50 AU. Their semimajor axes range up to several 100 AU. Their perihelion distances are generally between 30 and 38 AU (dotted curves in the bottom right panel of Fig. 2), so that on average the orbital eccentricities increase with semimajor axes. These objects are dynamically unstable because they suffer sufficiently close encounters with Neptune. At each encounter, they receive an impulse-like acceleration, which changes the semimajor axis of their orbits. The perihelion distance remains roughly constant during an encounter, so that the eccentricity changes together with the semimajor axis. Thus, under the scattering gravitational action of Neptune, these objects move in a sort of random walk in the region confined by the dotted curves. For this reason, this population of objects is now called the **scattered disk**. The name "disk" is justified, because the orbital inclinations, although large, are significantly smaller than $90°$, giving this population a disk-like structure.

Up to now, we have used "Kuiper Belt" to denote generically the population of objects with $a > 30$ AU. However, the existence of the scattered disk suggests that we should reserve the name "Kuiper Belt" for the population of objects that do not suffer encounters with Neptune and therefore have orbits that either do not significantly change with time, or do so very slowly. Adopting this definition, the objects of the Kuiper Belt are plotted with blue and green dots in Fig. 2, while the scattered disk objects are plotted in red. As one sees, scattered disk objects can also have $a < 50$ AU, provided that they have a small perihelion distance and are not in one of the most prominent mean-motion resonances with Neptune (indicated by the vertical lines labeled 3:4, 2:3 and 1:2 in Fig. 2). In Fig. 2, the scattered disk seems to be outnumbered by the Kuiper Belt bodies. However, the scattered disk objects are more difficult to discover, given that most of them have very elongated orbits and spend most of the time very far from the Sun. Accounting for this difficulty, astronomers have estimated that the scattered disk and the Kuiper Belt should constitute roughly equal populations.

All solar system bodies should have accreted on quasi-circular orbits. This is a necessary condition for small planetesimals being able to stick together and form larger objects. Indeed, if the eccentricities are large, the relative encounter velocities are such that, upon collisions, planetesimals do not grow, but fragment into smaller pieces. This consideration suggests that the scattered disk objects formed much closer to Neptune, on quasi-circular orbits, and have been transported outward by the scattering action of that planet. The fact that the scattering action is still continuing implies that the origin of the scattered disk does not necessarily require that the primordial solar system was different from the current one.

However, recent observations have revealed that, in addition to the Kuiper Belt and the scattered disk, there is a third category of objects, represented with magenta dots in Fig. 2. Their orbital distribution mimics that of the scattered disk objects, but their perihelion distance is somewhat larger, so that they avoid the scattering action of Neptune. Their orbits do not significantly change over the age of the solar system. Among the objects with these orbital properties are 1995 TL$_8$ ($a \sim 52$ AU, $q \sim 40$ AU), 2000 CR$_{105}$ ($a \sim 225$ AU, $q \sim 44$ AU), 90377 Sedna ($a \sim 500$ AU, $q \sim 76$ AU), and the recently discovered 136199 Eris ($a \sim 67.5$ AU, $q = 38$ AU, the largest Trans-Neptunian Object (TNO) known so far) and 2004 XR$_{190}$ ($a \sim 57.4$ AU, $q \sim 51$ AU—exceptional for its inclination of about $45°$). For the previously listed reasons, these bodies also should have formed closer to Neptune on much more circular orbits, and presumably they have been transported outward through close encounters with the planet. However, given that they do not undergo close encounters now, their existence suggests that the solar system was different in the past (either the planetary orbits were different or the environment was different—rogue planets, passing stars, etc.), so that the scattered disk extended further out in perihelion distance during the primordial times. We will come back to this in Section 7.

If we look at Fig. 2 in more detail (left panels), the Kuiper Belt can also be subdivided in a natural way in subpopulations. Several objects (green dots) are located in mean-motion resonances with Neptune. As explained in Section 2, the mean-motion resonances provide a protection mechanism, so that resonant objects can avoid close encounters with Neptune even if their perihelion distance is smaller than 30 AU, as it is in the case of Pluto. For this reason, resonant Kuiper Belt objects can be on much more elliptic orbits than the nonresonant ones, the eccentricities of the former ranging up to 0.35. The objects in the 2:3 mean-motion resonance with Neptune are usually called the Plutinos (because they share the same resonance as Pluto), while those in the 1:2 resonance are sometimes called twotinos. In Fig. 2, the resonant population seems to constitute a substantial fraction of the Kuiper Belt population. However, resonant objects are easier to discover because at perihelion they come closer to the Sun than the nonresonant ones. When accounting for this fact, astronomers estimate that, all together, the objects in mean-motion resonances constitute about 10% of the total Kuiper Belt population.

The nonresonant Kuiper Belt objects (blue dots) are usually referred to as classical. This adjective is attributed because their orbital distribution is the most similar to what the astronomers were expecting, before the discoveries of trans-Neptunian objects began: that of a disk of objects on stable, low-eccentricity, nonresonant orbits. However, even the classical population has unexpected properties. Their eccentricities are moderate—a necessary condition to avoid encounters with Neptune, given that they are not protected by any resonant mechanism. Nevertheless, the eccentricities are definitely larger than those of the protoplanetary disk in which the objects had to form. Some mechanism must have excited the eccentricities, making them grow from almost zero to the current values.

The same is true, and even more striking, for the inclinations (top panel of Fig. 2). The inclinations are related to the relative encounter velocities among the objects, so that the Kuiper Belt bodies had to grow in a razor-thin disk. Despite this, the current inclinations range up to 30–40°. Figure 2 gives the impression that large inclination bodies are a modest fraction among the classical objects. However, one should take into account that the discovery surveys have been concentrated near the ecliptic plane, so that large inclination bodies have a lower probability of being discovered than low inclination ones. Accounting for this selection effect, astronomers have computed that the real inclination distribution of the classical objects is bimodal (Fig. 3). There is a cluster of objects with inclination smaller than 4° and a second group of objects with a very distended inclination distribution. The former constitute what is now usually called the cold population and the latter the hot population. The adjectives "hot" and "cold" do not refer to physical temperature (it is always very cold out there) but to the encounter velocities inside each population, in an analogy with gas kinetic theory. The cold and the hot populations should contain roughly the same number of objects.

The last striking property of the Kuiper Belt is its outer edge. Figure 2 shows that the belt ends at the location of the 1:2 mean motion resonance with Neptune. For several years, the astronomers suspected that this edge is only apparent, due to the fact that more distant objects are more difficult to discover. However, with an increasing statistical sample, it turned out that this is not true. It has been shown that more distant objects should have been discovered by now, unless either (1) the Kuiper Belt population steeply decays in number beyond 48–50 AU or (2) the maximal size of the objects beyond this limit is much smaller than that in the observed Kuiper Belt. For various reasons, astronomers tend to favor hypothesis (1): the existence of a physical outer edge of the Kuiper Belt.

An important issue is to understand which of the orbital properties discussed earlier is due to the dynamical processes that are still occurring in the Kuiper Belt or not. For instance, do the eccentricities and the inclinations slowly grow due to some dynamical phenomenon? Are the low eccentricity objects beyond 48 AU unstable? If these are the cases, then the existence of large eccentricities and inclinations, as well as the outer edge of the Kuiper Belt could be simply explained. In the opposite case, these properties—like the existence of the **extended scattered disk**—reveal that the solar system was different in the past.

Dynamical astronomers have studied in great detail the dynamics beyond Neptune, using numerical simulations and semianalytic models. Figures 4 and 5 show maps of the dynamical lifetime of trans-Neptunian bodies on a wide range of initial semimajor axes, eccentricities, and inclinations. These maps have been computed numerically, by simulating the evolution of thousands of massless particles under the gravitational perturbations of the giant planets. The latter have been assumed to be initially on their current orbits. Each particle was followed until it suffered a close encounter with Neptune. Objects encountering Neptune, would then evolve in the scattered disk for a time of order $\sim 10^8$ years, until they are transported by planetary encounters into the inner planets region, or are ejected to the Oort cloud or to interstellar space. This issue is described in more detail in Section 6.

In Fig. 4, the colored strips indicate the length of time required for a particle to encounter Neptune as a function of its initial semimajor axis and eccentricity. The initial inclination of the particles was set equal to 1°. Strips that are colored yellow represent objects that survive for the length of the simulation, 4×10^9 years, the approximate age of the solar system. As can be seen in the figure, the Kuiper Belt can be expected to have a complex structure, although the general trends are readily explained. Objects with perihelion distances less than $\sim$35 AU (shown as a red curve) are unstable, unless they are near, and presumably librating about, a mean-motion resonance with Neptune (Section 2). Indeed, the results in Fig. 4 show that many of the Neptunian mean-motion resonances (shown in blue) are stable for the age of the solar system. Objects with semimajor axes between 40 and 42 AU are unstable. This is presumably due to the presence of three overlapping secular resonances that occur in this region of the solar system: two with Neptune and one with Uranus.

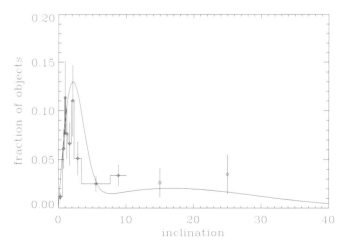

FIGURE 3   The inclination distribution (in deg) of the classical Kuiper Belt after observational biases have been subtracted, according to the work of M. Brown. The points with error bars show the model-independent estimate constructed from a limited subset of confirmed classical belt bodies, while the smooth line shows a best fit bimodal population model. In this model $\sim$60% of the objects have $i > 4°$.

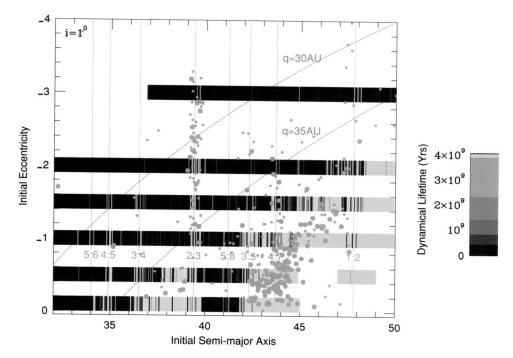

FIGURE 4 The dynamical lifetime for small particles in the Kuiper Belt derived from 4 billion year integrations by M. Duncan, H. Levison, and M. Budd. Each particle is represented by a narrow vertical strip of color, the center of which is located at the particle's initial eccentricity and semimajor axis (initial orbital inclination for all objects was $1°$). The color of each strip represents the dynamical lifetime of the particle. Strips colored yellow represent objects that survive for the length of the integration, 4×10^9 years. Dark regions are particularly unstable on these timescales. For reference, the locations of the important Neptune mean-motion resonances are shown in blue and two curves of constant perihelion distance, q, are shown in red. The orbital distribution of the real objects is also plotted. Big dots correspond to objects with $i < 4°$, and small dots to objects with larger inclination. Remember that the dynamical lifetime map has been computed assuming $i = 1°$.

Indeed, secular resonances appear to play a critical role in ejecting particles from this region of the Kuiper Belt. This can be better seen in Fig. 5, which is an equivalent map, but plotted relative to the initial semimajor axis and inclination for particles with initial eccentricity of 0.01. Also shown are the locations of the Neptune longitude of perihelion secular resonance (in red) and the Neptune longitude of the ascending node secular resonance (in yellow). It is important to note that much of the clearing of the Kuiper Belt occurs where these two resonances overlap. This includes the low inclination region between 40 and 42 AU, which is indeed depleted of bodies (compare with Fig. 2). The Neptune mean-motion resonances are also shown (in green).

It is interesting to compare the numerical results to the current best orbital elements of the known Kuiper Belt objects. This comparison is also made in Fig. 4, where the observed objects with good orbital determination are overplotted with green dots. Big dots refer to bodies with $I < 4°$, consistent with the low inclination at which the sta-

bility map has been computed. Small dots refer to objects with larger inclination and are plotted only for completeness. The conclusion is that most observed objects (with the exception of scattered disk bodies) are associated with stable zones. Their orbits do not significantly change over the age of the solar system. Thus, their current excited eccentricities and inclination cannot be obtained from primordial circular and coplanar orbits in the framework of the current planetary system orbital configuration. Likewise, the region beyond the 1:2 mean-motion resonance with Neptune is totally stable. Thus, the absence of bodies beyond 48 AU cannot be explained by current dynamical instabilities.

Therefore, it is evident that the orbital structure of the Kuiper Belt has been sculpted by mechanisms that are no longer at work, but presumably were active when the solar system formed. The main goal of dynamical astronomers interested in the Kuiper Belt is to uncover these mechanisms and from them deduce, as far as possible, how the solar system formed and early evolved.

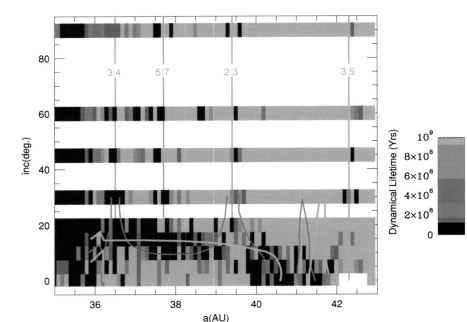

FIGURE 5 The dynamical lifetime for test particles with initial eccentricity of 0.01 derived from 1 billion year integrations by M. Duncan, H. Levison, and M. Budd. This plot is similar to that of Fig. 4 except that coordinates are semimajor axis and inclination (instead of semimajor axis and eccentricity) and a different color table was used for the solid bars. In addition, the red and yellow curves show the locations of Neptune longitude of perihelion secular resonances (v_8) and the Neptune longitude of the ascending node secular resonances (v_{18}), respectively. The green lines show the location of the important Neptune mean motion resonances.

4. Correlations Between Physical and Orbital Properties

The existence of two distinct classical Kuiper Belt populations, called the hot ($i > 4°$) and cold ($i < 4°$) classical populations, could be caused in one of two general manners. Either a subset of an initially dynamically cold population was excited, leading to the creation of the hot classical population, or the populations are truly distinct and formed separately.

One manner in which we can attempt to determine which of these scenarios is more likely is to examine the physical properties of the two classical populations. If the objects in the hot and cold populations are physically different, it is less likely that they were initially part of the same population.

The first suggestion of a physical difference between the hot and the cold classical objects came from the observation that the intrinsically brightest classical belt objects (those with lowest **absolute magnitudes**) are preferentially found with high inclination. Figure 6 shows the distribution of the classical objects in an inclination vs. absolute magnitude diagram. As one sees, for an absolute magnitude $H > 5.5$, there is a given proportion between the number of objects discovered in the cold and the hot populations respectively. This ratio is completely different for $H < 5.5$, where cold population objects are almost absent. All the biggest classical objects, such as, for instance, 50000 Quaoar, 20000 Varuna, 19521 Chaos, 28978 Ixion, 2005 FY$_9$, and 2003 EL$_{61}$ have inclinations larger than 5°. Their median inclination is 12°. It has been argued that this is a result

of an observational bias because the brightest objects have been discovered in wide field surveys not confined around the ecliptic, which are thus more likely to find large inclination objects than the deep ecliptic surveys that detected the fainter bodies. However, a recent survey for bright objects, which covered ~70% of the ecliptic, found many hot classical objects but few cold classical objects, confirming that the effect illustrated in Fig. 6 is real.

The second possible physical difference between hot and cold classical Kuiper Belt objects is their colors. With the name "color" astronomers generically refer to the slope of the spectrum of the light reflected by a trans-Neptunian object at visible wavelengths, relative to that of the light emitted by the Sun. "Red" objects reflect more at long than at short wavelengths, while "gray" objects have a more or less uniform reflectance. Colors relate in a poorly understood manner to objects' surface composition. It has been shown and repeatedly confirmed that, for the classical belt, the inclination, and possibly the perihelion distance, is correlated with color. In essence, the low inclination classical objects tend to be redder than higher inclination objects. More interestingly, colors naturally divide into distinct red and gray populations at precisely the location of the divide between the inclinations of the hot and cold classical objects. These populations differ at a 99.9% confidence level. Interestingly, the cold classical population also differs in color from the Plutinos and the scattered objects at the 99.8 and 99.9% confidence level, respectively, while the hot classical population appears identical in color to these other populations.

The possibility remains, however, that the colors of the objects, rather than being markers of different populations,

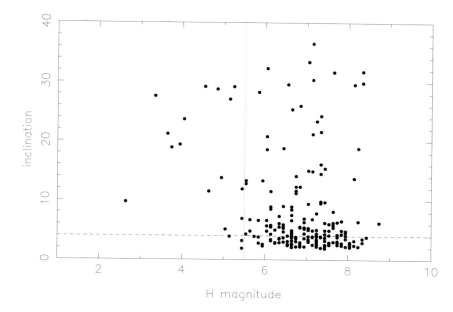

FIGURE 6 The inclination of the classical Kuiper Belt objects as a function of their absolute magnitude. The horizontal dashed line at $i = 4°$ separates the cold from the hot population. The vertical dotted line is plotted at $H = 5.5$. The distribution on the left side of the dotted line is clearly different from that on the right-hand side. The largest classical objects are all in the hot population.

are actually *caused* by the different inclinations. For example, it has been suggested that the higher average impact velocities of the high inclination objects could cause large-scale resurfacing by fresh water ice and carbonaceous materials, which could be gray in color. However, a similar color-inclination trend should be observed also among the plutinos and the scattered disk objects, which is not the case. A careful analysis shows that there is no clear correlation between average impact velocity and color.

In summary, the significant color and size differences between the hot and cold classical objects imply that these two populations are physically different in addition to being dynamically distinct.

5. Size Distribution of the Trans–Neptunian Population and Total Mass

As briefly described in Section 1, the disk out of which the planetary system accreted was created as a result of the Sun shedding angular momentum as it formed. As the Sun condensed from a molecular cloud, it left behind a disk of material (mostly gas with a little bit of dust) that contained a small fraction of the total mass but most of the angular momentum of the system. It is believed that the initial solid objects in the protoplanetary disk were pebble-sized, of the order of centimeters in size. These objects formed larger objects through a process of accretion to form asteroids and comets, which in turn accreted to form planets (or the cores of the giant planets which then accreted gas directly from the solar nebula). Understanding this process is one of the main goals of astronomy today.

There are few clues in our planetary system about this process. We know that the planets formed, and we know how big they are. Unfortunately, the planets have been so altered by internal and external processes that they preserve almost no record of their formation process. Luckily, we also have the Asteroid Belt, the Kuiper Belt, and the scattered disk. These structures contain the best clues to the planet formation process because they are regions where the process started, but for some reason, did not run to completion (i.e., a large planet). Thus, the size distribution of objects in these regions may show us how the processes progressed with time and (hopefully) what stopped them. The Kuiper Belt and the scattered disk are perhaps the best places to learn about the accretion process.

Because the size of the object is not a quantity that can be easily measured (one needs to make hypotheses on the intrinsic reflectivity of the objects, or albedo), and the absolute magnitude is readily obtained from the observations, astronomers generally prefer the absolute magnitude distribution, instead of the size distribution. The magnitude distribution is usually given in the form

$$Log\, N(< H) \propto H^{a}, a > 0$$

where N is the cumulative number of objects brighter than absolute magnitude H. The slope of this distribution, a, contains important clues about the physical strengths, masses, and orbits of the objects involved in the accretion process.

For example, there are two extremes to the accretion process. If two large, strong objects collide at low velocities, then the amount of kinetic energy in the collision is

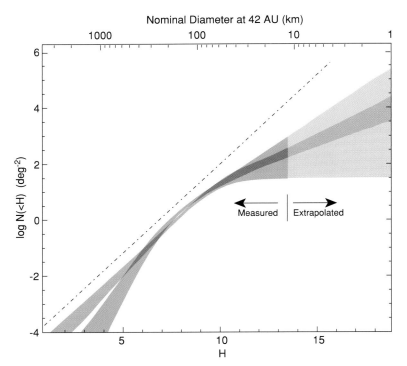

FIGURE 7 The cumulative magnitude distribution of the cold population (red) and of the hot population and scattered disk (green) according to a recent analysis by G. Bernstein and collaborators. A turnover of the magnitude distribution is detected around $H \sim 10$. The slope of the magnitude distribution is very uncertain beyond this limit.

small compared to the amount of energy holding the objects together. In this case, the objects merge to form a larger object. If two small, weak objects collide at high velocities, then the energy in the collision overpowers the gravitational and material binding energies. In this case, the objects break apart, forming a large number of much smaller objects. In realistic models of the Kuiper Belt with a range of sizes and velocities, we expect small objects to fragment and large objects to grow. This produces a size distribution with $a \sim 0.4$–0.5 at small sizes and a much steeper slope at large sizes where accretion is important.

Statistics of discoveries of Kuiper Belt objects (Fig. 7) suggest that the absolute magnitude distribution is indeed very steep for $H < 9$ (approximately equivalent to a diameter $D > 100$ km), with $a \sim 0.6$–0.7, and then turns over toward a significantly shallower slope. Interestingly, the hot and the cold classical population seems to have two different values of a in the steep part. More precisely, the hot population and the scattered disk have a shallower magnitude distribution than the cold population (Fig. 7). This is consistent with the fact that the largest bodies are all in the hot population, and yet the hot and cold populations and the scattered disk contain roughly the same number of bodies bigger than 100 km.

The value of a in the shallow part of the magnitude distribution beyond $H \sim 10$ is very uncertain. Only few surveys with the most powerful telescopes could probe this region, but they have discovered very few objects. The results are therefore affected by small number statistics. It is possible that $a < 0.5$ in some magnitude range. In fact, in the

Asteroid Belt, the magnitude distribution is wavy, and the canonical values of 0.4–0.5 of a is only a mean value. It is possible that the magnitude distribution in the Kuiper Belt is wavy as well, and that the range $10 < H < 14$ corresponds to the very shallow part of one of these waves.

It is possible to integrate under the magnitude distribution shown in Fig. 7 in order to estimate the total mass in the Kuiper Belt between 30 and 50 AU. Such an integration with limits between $R = 1$ km and 1200 km (the approximate radius of Pluto) and assuming a density of 1 g cm^{-3}, shows that the total mass is a few hundredths of an Earth mass. Given the uncertainties, it is possible that the mass is of order of 0.1 $M_{\oplus}$, but not significantly larger.

As with many scientific endeavors, the discovery of new information tends to raise more questions than it answers. Such is the case with the preceding mass estimate. Edgeworth's and Kuiper's original arguments for the existence of the Kuiper Belt were based on the idea that it seemed unlikely that the disk of planetesimals that formed the planets would have abruptly ended at the current location of the outermost known planet. An extrapolation into the Kuiper Belt (between 30 and 50 AU) of the current surface density of nonvolatile material in the outer planets region predicts that there should originally have been about 30 $M_{\oplus}$ of material there. However, as stated previously, our best estimate is over 200 times less than that figure!

Edgeworth's and Kuiper's argument is not the only indication that the mass of the primordial Kuiper Belt had to be significantly larger in the past. Models of collisional accretion show that it is not possible for objects with radii greater than about 30 km to form in the current Kuiper Belt, at least by pairwise accretion, over the age of the solar system. The current surface density of solid material is too low to accrete bodies larger than this size. However, the models show that objects the size of 1992 QB$_1$ could have grown in a more massive Kuiper Belt (provided that, as already said in Section 3, the mean orbital eccentricities of the accreting objects were much smaller than the current ones). A Kuiper Belt of at least several Earth masses is required in order for 100 km sized objects to have formed.

The same applies even more strongly to the accretion of Pluto and Charon. For those two bodies to have grown to their current sizes in the trans-Neptunian region, there must have originally been a far more massive Kuiper Belt. A massive and dynamically cold primordial Kuiper Belt is also required by the models that attempt to explain the formation of the observed numerous binary Kuiper Belt objects.

Therefore, the general formation picture of an initial massive Kuiper Belt appears secure, and understanding the ultimate fate of the 99% (or 99.9%) of the initial Kuiper Belt mass that appears to be no longer in the Kuiper Belt is a crucial step in reconstructing the history of the outer solar system.

6. Ecliptic Comets

As described in Section 1, the current renaissance in Kuiper Belt research was prompted by the suggestion that the Jupiter-family comets originated there. We now know that there are mainly two populations of small bodies beyond Neptune: the Kuiper Belt and the scattered disk. Which one is the dominant source of these comets?

To answer this question, we need to examine a few considerations on the origin of the scattered disk. We have seen in Section 3 that the bodies in the scattered disk have intrinsically unstable orbits. The close encounters with Neptune move them in semimajor axis, until they either evolve into the region with $a < 30$ AU or reach the Oort cloud at the frontier of the solar system. In both of these cases, the bodies are removed from the scattered disk. Despite this possibility of dynamical removal, we still observe scattered disk bodies today. How can this be?

There are a priori two possibilities. The first one is that the scattered disk population is sustained in a sort of steady state by the bodies escaping from the Kuiper Belt. This means that on a timescale comparable to that for the dynamical removal of scattered disk bodies, new bodies enter the scattered disk from the Kuiper Belt. For example, a similar situation occurs for the population of near-Earth asteroids (NEAs). NEA dynamical lifetimes are only of a few million years because they intersect the orbits of the terrestrial planets. Nevertheless, the population remains roughly constant because new asteroids enter the NEA population from the Main Asteroid Belt at the same rate at which old NEAs are eliminated.

The second possibility is that the scattered disk that we see today is only what remains of a much more numerous population that has been decaying in number since planetary formation. Numerical simulations show that roughly 1% of the scattered disk bodies can survive in the scattered disk for the age of the solar system. Thus, the primordial scattered disk population should have been about 100 times more numerous.

Which of these possibilities is true? In the first case, we would expect that the Kuiper Belt is much more populated than the scattered disk. For instance, the Asteroid Belt contains about 1000 times more objects than the NEA population, at comparable sizes. However, observations indicate that the scattered disk and the Kuiper Belt contain roughly the same number of objects. Thus, the second possibility has to be true. Scattered disk objects most likely formed in the vicinity of the current positions of Uranus and Neptune. When these planets grew massive, they scattered them away from their neighborhoods. In this way, a massive scattered disk of about 10 $M_\oplus$ formed. What we see today is just the last vestige of that primordial population, which is still decaying in number.

The fact that the scattered disk is not sustained in steady state by the Kuiper Belt, but it is still decaying, implies that the scattered disk provides more objects to the giant planet region ($a < 30$ AU) than it receives from the Kuiper Belt. Thus, the outflow from the scattered disk is more important than the outflow from the Kuiper Belt. This implies that the scattered disk, not the Kuiper Belt, is the dominant source of Jupiter family comets.

A significant amount of research has gone into understanding the dynamical behavior of objects that penetrate into the $a < 30$ AU region from the scattered disk. These studies show that the encounters with the planets spread them throughout the planetary system. These objects are usually called ecliptic comets, even if at large distances from the Sun they typically do not show any cometary activity. The distribution of these objects as predicted by numerical integrations is shown in Fig. 8.

The ecliptic comets that get close to the Sun become active. When their semi-major axis is smaller than that of Jupiter, they are called Jupiter-family comets. It is somewhat surprising that about a third of the objects leaving the scattered disk in the simulations spend at least some of their time as Jupiter-family comets. The Jupiter-family comets that we see today are, in majority, small, $R \lesssim 10$ km. However, if our understanding of the size-distribution of these objects is correct (see Section 5), we should expect to see a 100 km sized Jupiter-family comet about 0.4% of the time. What a show that would be!

Those ecliptic comets between Jupiter and Neptune are called the Centaurs (only the largest of which are observable). The simulations predict that there are $\sim 10^6$ ecliptic

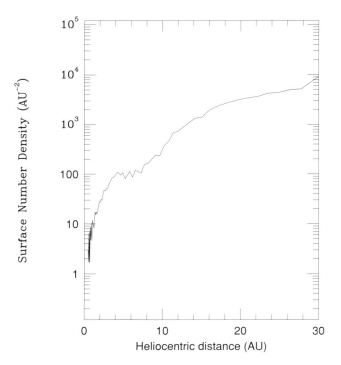

FIGURE 8   The surface number density (on the plane of the ecliptic) of ecliptic comets as determined from numerical integrations by M. Duncan and H. Levison. There are approximately 10^6 comets larger than about 1 km in radius in this population.

comets larger than about 1 km in radius currently in orbits between the giant planets.

7. The Primordial Sculpting of the Trans–Neptunian Population

In the previous sections, we have seen that many properties of the Kuiper Belt cannot be explained in the framework of the current solar system:

1. The existence of the resonant populations
2. The excitation of the eccentricities in the classical belt
3. The coexistence of a cold and a hot population with different physical properties
4. The presence of an apparent outer edge at the location of the 1:2 mean-motion resonance with Neptune
5. The mass deficit of the Kuiper Belt
6. The existence of the extended scattered disk population

These puzzling aspects of the trans-Neptunian population reveal that it has been sculpted when the solar system was different, due to mechanisms that are no longer at work. Like detectives on the scene of a crime, trying to reconstruct

what happened from the available clues, the astronomers try to reconstruct how the solar system formed and evolved from the traces left in the structure of the Kuiper Belt.

Planet migration has been the first aspect of the primordial evolution of which the astronomers found a signature in the Kuiper Belt. Once the planets formed and the gas disappeared, the planetesimals that failed to be incorporated in the planets' cores had to be removed from the planets' vicinity by the gravitational scattering action of the planets themselves. If a planet scatters a planetesimal outward, the latter gains energy. Because of energy conservation, the planet has to lose energy, moving slightly inward. The opposite happens if the planet scatters the planetesimal toward the inner solar system. A planet is much more massive than a planetesimal, thus the displacement of the planet is infinitesimal. However, if the number of planetesimals is large, and their total mass is comparable to that of the planet, the final effect on the planet is not negligible. This is a general process. We now come to what should have happened in our solar system.

Numerical simulations show that only a small fraction of the planetesimals originally in the vicinity of Neptune was scattered outward: About 1% ended up in the scattered disk, and 5%, in the Oort cloud. The remaining 94% of the planetesimals eventually were scattered inward toward Jupiter. The latter, given its large mass, ejected from the solar system almost everything that came to cross its orbit. Thus, the net effect was that Neptune took energy away from the planetesimals and moved outward, while Jupiter gave energy to them and moved inward. Numerical simulations show that Saturn and Uranus also moved outward. Following Neptune's migration, the mean-motion resonances with Neptune also migrated outward, sweeping the primordial Kuiper Belt until they reached their present position. During this process, some of the Kuiper Belt objects swept by a mean-motion resonance could be captured into resonance. Once captured, these bodies had to follow the resonance in its migration, while their eccentricity had to steadily grow. Thus, the planetesimals that were captured first, ended up on very eccentric resonant orbits, while those captured last could preserve a small eccentricity inside the resonance. Numerical simulations show that this process produces an important population of resonant bodies inside all the main mean-motion resonances with Neptune. To reproduce the observed range of eccentricities of resonant bodies, Neptune had to migrate more than 7 AU, thus starting not further than 23 AU (see Fig. 9). The existence of resonant bodies in the Kuiper Belt thus provides a strong indication that planet migration really happened.

However, as Fig. 9 also shows, several important properties of the Kuiper Belt cannot be explained by this simple model invoking resonance sweeping through a dynamically cold, radially extended disk. The eccentricity of the classical belt is only moderately excited, and the inclination remains very cold. The planetesimals are only relocated, from the

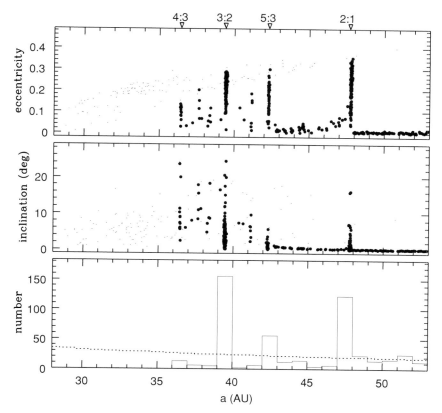

FIGURE 9 Final distribution of the Kuiper Belt bodies according to a simulation of the sweeping resonances scenario by R. Malhotra. The simulation is done by numerical integrating, over a 200 million year timespan, the evolution of 800 test particles on initial quasi-circular and coplanar orbits. The planets are forced to migrate (Jupiter, −0.2 AU; Saturn, 0.8 AU; Uranus, 3 AU; Neptune, 7 AU) and reach their current orbits on an exponential timescale of 4 million years. Large solid dots represent "surviving" particles (i.e., those that have not suffered any planetary close encounters during the integration time); small dots represent the "removed" particles at the time of their close encounter with a planet. In the lowest panel, the solid line is the histogram of semimajor axis of the surviving particles; the dotted line is the initial distribution. Most of the initial mass of the Kuiper Belt is simply relocated from the classical belt to the resonant populations. The mass lost is only a small fraction of the total mass.

classical belt to the resonances, and only a minority of them are lost, which cannot explain the mass depletion of the belt. Finally, the region beyond the 1:2 mean-motion resonance is unaffected by planet migration, and therefore the existence of an outer edge requires a different explanation.

Four plausible models have been proposed so far to explain the formation of an outer edge: (1) the outer part of the disk was destroyed by the passage of a star; (2) it was photoevaporated by the radiation emitted by massive stars originally in the neighborhood of the Sun; (3) planetesimals beyond some threshold distance could not grow because of the enhanced turbulence in the outer disk which prevented the accumulation of solid material; and (4) distant dust particles and/or planetesimals migrated to smaller heliocentric distance during their growth, as a consequence of gas drag, thus forming sizeable objects only within some threshold distance from the Sun. The first two scenarios require that the Sun formed in a dense stellar environment, consistent with recent observations showing that stars tend to form in clusters which typically disperse in about 100 million years. The entire protoplanetary disk—both the gas and the planetesimal components—would be truncated by these mechanisms. However, the protoplanetary disks that we see around other stars, even in dense stellar associations, are typically much larger than 50 AU. Thus, the history of our proto–solar system disk was not typical. The third and the fourth scenarios, conversely, form

a truncated planetesimal disk out of an extended gaseous disk. Therefore, they are more consistent with observations, which are sensitive only to the gas and dust components, and do not detect the location of planetesimals.

Whatever mechanism formed the edge, it is intriguing that the latter is now at the location of a resonance with Neptune, despite the fact that Neptune did not play any role in the edge formation. Is this a coincidence? Probably not. It may suggest that originally the outer edge of the planetesimal disk was well inside 48 AU, and that the migration of Neptune pushed somehow a small fraction of the disk planetesimals beyond the disk's original boundary. These pushed-out planetesimals are now identified with the current members of the Kuiper Belt. The fact that the Kuiper Belt is mass deficient all over its radial extent (36–48 AU), in addition suggests that the original edge was inside 36 AU. In fact, if the original edge had been somewhere in the 36–48 AU range, we would see a discontinuity in the current radial mass distribution of the Kuiper Belt, which is not the case. An edge of the planetesimal disk close to 30 AU also helps to explain why Neptune stopped there and did not continue its outer migration beyond this limit.

Several mechanisms have been identified to push beyond the original disk edge a small fraction (of order 0.1%) of the disk's planetesimals, and to implant them on stable Kuiper Belt orbits. They are described next. More mechanisms might be identified in the future.

As Neptune moved through the disk on a quasi-circular orbit, it scattered the planetesimals with which it had close encounters. Through multiple encounters, some planetesimals were transported outward on eccentric, inclined orbits. A small fraction of these objects still exist today and constitute the scattered disk. Occasionally, some scattered disk objects entered a resonance with Neptune. Resonances can modify the eccentricity of the orbits. If decreased, the perihelion distance is lifted away from the planet; the sequence of encounters stops, and the body becomes "decoupled" from Neptune like a Kuiper Belt object. If Neptune had not been migrating, the eccentricity would have eventually increased back to Neptune-crossing values—the dynamics being reversible—and the sequence of encounters would have restarted again. Neptune's migration broke the reversibility so that some of the decoupled bodies managed to escape from the resonances and remained permanently trapped in the Kuiper Belt. These bodies preserved the large inclinations acquired during the Neptune-encountering phase, and they can now be identified with the "hot" component of the Kuiper Belt population.

At the same time, while Neptune was migrating through the disk, its 1:2 and 2:3 resonances swept through the disk, capturing a fraction of the disk planetesimals as explained earlier. When the 1:2 resonance passed beyond the edge of the disk, it kept carrying its load of objects. Because the migration of Neptune was presumably not a perfectly smooth process, the resonance was gradually dropping objects during its outward motion. Therefore, like a farmer seeding as he advances through a field, the resonance disseminated its previously trapped bodies all along its way up to its final position at about 48 AU. This explains the current location of the outer edge of the Kuiper Belt. Because the 1:2 resonance does not significantly enhance the orbital inclinations, the bodies transported by the resonance preserved their initially small inclination and can now be identified with the cold component of the Kuiper Belt.

This scenario, reproduced in numerical simulations, explains qualitatively the orbital properties of the trans-Neptunian population, but it has difficulties explaining why the hot and the cold classical populations have different physical properties. Indeed, the members of these two populations should have formed more or less in the same region of the disk, although they followed two different dynamical evolutions toward the Kuiper Belt.

An alternative possibility is that the hot population formed as explained earlier, but the cold population formed in situ, where it is now observed. Thus, the formation places being well separated, the corresponding physical properties could be different. However, this model has difficulties explaining how the cold population lost most of its primordial mass. It has been proposed that the objects grind down to dust in a collisional cascade process, but the latter has not been shown to be really effective, and seems inconsis-

tent with a number of constraints, such as the existence of binary objects with large separations, or the total number of comet-sized bodies in the scattered disk. Moreover, if the cold population is local, one is faced again with the problem of explaining why the outer edge of the population is exactly at the location of the 1:2 resonance with Neptune.

A further possibility may be offered by a recent model, on the evolution of the outer solar system, that has been developed in order to explain the origin of the so-called Late Heavy Bombardment (LHB) of the terrestrial planets. The latter is a cataclysmic period characterized by huge impact rates on all planets that occurred between 4.0 billion and 3.8 billion years ago, namely about 600 million years after planet formation. In this model, the giant planets are assumed to be initially on quasi-circular and coplanar orbits, with orbital separations significantly smaller than the current ones. In particular, Saturn is assumed to be closer to Jupiter than their mutual 1:2 resonance (they are now close to the 2:5 resonance). The planetesimal disk is assumed to exist only from about 1.5 AU beyond the location of the outermost planet, up to $\sim$35 AU, with a total mass of $\sim$35 $M_{\oplus}$. With this setting, the planetesimals at the inner edge of the disk acquire Neptune-scattered orbits on a timescale of a few million years. Consequently, the migration of the giant planets proceeds at very slow rate, governed by the slow escape rate of planetesimals from the disk. This slow migration continues for hundreds of millions of years, until Jupiter and Saturn cross their mutual 1:2 resonance. This resonance crossing excites their eccentricities, which destabilizes the planetary system as a whole. The planetary orbits become chaotic and start to approach each other. Both Uranus and Neptune are scattered outward, onto large eccentricity orbits ($e \sim 0.3-0.4$) that penetrate deeply into the disk. This destabilizes the full planetesimal disk and triggers the LHB. The interactions with the planetesimals damp the planetary eccentricities, stabilizing the planetary system once again, and forcing a residual short radial migration of the planets, which eventually reach final orbits when most of the disk has been eliminated. Simulations show that this model is consistent with the current orbital architecture of the giant planets of the solar system.

In this model, objects can be implanted into the current Kuiper Belt during the large eccentricity phase of Neptune. In fact, the full Kuiper Belt is unstable at that time, so that it can be visited by objects that leave the original planetesimal disk when the latter is destabilized. When Neptune's eccentricity is damped, the Kuiper belt becomes stable so that the objects which, by chance, are in the Kuiper Belt region at that time, become trapped forever. Because the large eccentricity phase of Neptune is short, the inclinations of these objects remain predominantly small, consistent with the cold population of the current Kuiper Belt. The objects with the largest inclinations, conversely, are captured later, during the final bit of Neptune's migration, as explained before. As Fig. 10 shows, this model reproduces the structure

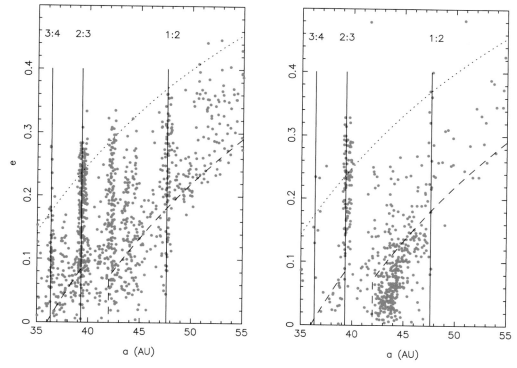

FIGURE 10   The distribution of semimajor axes and eccentricities in the Kuiper Belt. Left panel: Result of a simulation based on the recent model on the origin of the LHB. Right panel: The observed distribution. The model reproduces fairly well the outer edge of the Kuiper Belt at the 1:2 resonance with Neptune, the characteristic shape of the (a, e) distribution of the classical belt, the scattered and the extended scattered disks, and the resonant populations. The vertical solid lines mark the main resonance with Neptune. The dotted curve denotes perihelion distance equal to 30 AU, and the dashed curve delimits the region above which only high inclination objects or resonant objects can be stable over the age of the solar system. The overabundance of objects above this curve in the simulation is therefore an artificial consequence of the fact that the final orbits of the giant planets are not exactly the same as the real ones.

of the Kuiper Belt remarkably well. It is also consistent with its low mass because in the simulations the probability of capture in the Kuiper Belt is roughly of 1/1000 (which predicts a final mass of 0.03 $M_{\oplus}$). Moreover, the Kuiper Belt objects with final low inclinations and those with final large inclinations are found to come predominantly from different portions of the original planetesimal disk (respectively outside and inside 29 AU), which can explain, at least at a qualitative level, the correlations with physical properties.

We finally come to the issue of the origin of the extended scattered disk. Simulations show that, in the same process described earlier, bodies are also delivered to orbits with moderate semimajor axis and perihelion distance, like that of the extended scattered disk object 1995 TL$_8$ and its companions. The origin of Sedna is probably different. The key issue is that bodies with comparably large perihelion distances (~80 AU) but smaller semimajor axis (a < 500 AU) have never been discovered despite the more favorable observational conditions. Therefore, they probably do not exist. If this is true, and the population of extended scattered

disk bodies with large perihelion distance starts only beyond several hundreds of AUs, then an "external" perturbation is required. The best candidate is a stellar passage at about 1000 AU from the Sun, lifting the perihelion distance of the distant members of the primordial, massive scattered disk. Such a stellar encounter is very unlikely in the framework of the current galactic environment of the Sun, but it would have been probable if the Sun formed in a moderately dense cluster, as mentioned earlier.

8. Concluding Remarks

At the time of the first edition of this encyclopedia, 60 objects had been discovered in the Kuiper Belt. Now, we know 20 times more objects, and our view of the Kuiper Belt has become much more precise. It is now clear that the trans-Neptunian population has been sculpted in the primordial phases of the history of the solar system, by processes that are no longer at work.

It has been argued that the explanation of the most important observed properties of the trans-Neptunian population require a "cocktail" with three ingredients: (1) a truncated planetesimal disk; (2) a dense galactic environment, favoring stellar passages at about 1000 AU from the Sun; and (3) the outward migration of Neptune with, presumably, a phase of large eccentricity of the planetary orbits. Some problems still remain open, and the details of some mechanisms have still to be understood, but the basic composition of the cocktail appears quite secure. This is a big step forward with respect to our understanding of solar system formation, before the discovery of the Kuiper Belt.

What is next? The upcoming generation of telescopic surveys will probably increase by another order of magnitude the number of discovered trans-Neptunian objects with good orbits within a decade. Thus, in the third edition of the encyclopedia, we will probably have a different story to tell. It is unlikely that our view will totally change with the new discoveries (or at least we hope so!), but certainly there will be surprises. We are anxious to know more precisely the absolute magnitude distributions of the various subpopulations of the Kuiper Belt, their color properties, the real nature of the outer edge and its exact location, the orbital distribution of the extended scattered disk. This information will allow us to refine the scenarios outlined earlier, possibly to reject some and design new ones, in an attempt to read with less uncertainty the history of our solar system that is written out there.

Bibliography

Bernstein, G. M., Trilling, D. E., Allen, R. L., Brown, M. E., Holman, M., and Malhotra, R. (2004). The size distribution of trans-Neptunian bodies. *Astron. J.* **128**, 1364–1390.

Brown M. (2001). The inclination distribution of the Kuiper Belt. *Astron. J.* **121**, 2804–2814.

Gomes R. S. (2003). The origin of the Kuiper Belt high inclination population. *Icarus* **161**, 404–418.

Gomes, R., Levison, H. F., Tsiganis, K., and Morbidelli, A. (2005). Origin of the cataclysmic Late Heavy Bombardment period of the terrestrial planets. *Nature* **435**, 466–469.

Levison H. F., and Stern S. A. (2001). On the size dependence of the inclination distribution of the Main Kuiper Belt. *Astron. J.* **121**, 1730–1735.

Levison, H. F., and Morbidelli, A. (2003). The formation of the Kuiper Belt by the outward transport of bodies during Neptune's migration. *Nature* **426**, 419–421.

Malhotra R. (1995). The origin of Pluto's orbit: Implications for the solar system beyond Neptune. *Astron. J.* **110**, 420–432.

Morbidelli, A., Brown, M.E., and Levison, H. F. (2003). The Kuiper Belt and its primordial sculpting. *Earth Moon and Planets* **92**, 1–27.

Morbidelli, A., and Levison, H. F. (2004). Scenarios for the origin of the orbits of the trans-Neptunian objects 2000 CR_{105} and 2003 VB_{12} (Sedna). *Astron. J.* **128**, 2564–2576.

Trujillo C. A., and Brown M. E. (2001). The radial distribution of the Kuiper Belt. *Astrophys. J.* **554**, 95–98.

Trujillo C. A., and Brown M. E. (2002). A correlation between inclination and color in the classical Kuiper Belt. *Astroph. J.* **566**, 125–128.

Kuiper Belt Objects: Physical Studies

Stephen C. Tegler

Northern Arizona University
Flagstaff, Arizona

CHAPTER 33

Our Solar System began as a slowly spinning cloud of gas and dust about 4.5 billion years ago. As gravity caused the cloud to shrink in size, conservation of angular momentum required it to spin faster and evolve into a thin disk of gas, ice, and dust surrounding the young Sun. In the outer region of the disk, cold material accreted to first form boulder-sized objects, then mountain-sized objects, and then comet nucleus-sized (1–10 km) objects. Eventually, a small number of objects reached the size of planetary cores. Two cores eventually grew in size to become Uranus and Neptune. As Uranus and Neptune grew in size, their gravitational influence stopped the numerous remaining smaller objects from forming an additional large planet.

The first hint of a debris disk of icy material in the outer Solar System came in 1930 with the discovery of Pluto by Clyde Tombaugh of Lowell Observatory in Flagstaff, Arizona. It soon became clear that Pluto was much smaller than any other planet in the Solar System. Pluto's small size did not follow the pattern of planetary properties—four small, rocky terrestrial planets (Mercury, Venus, Earth, and Mars) close to the Sun followed by four giant, hydrogen-rich Jupiter-like planets (Jupiter, Saturn, Uranus, and Neptune) farther from the Sun. Why wasn't Pluto a giant like the other Jupiter-like planets? In 1978, J. W. Christy and R. S. Harrington added to the inventory of small bodies beyond Neptune by discovering Pluto's satellite, Charon (pronounced either "Kharon" or "Sharon"), on images taken at the U.S. Naval Observatory's Flagstaff station. Figure 1 illustrates the small sizes of Pluto and Charon by comparing them to the dimensions of the United States.

Perhaps the most important clue to solving the mystery of Pluto's small size came in 1988, when Martin Duncan, Thomas Quinn, and Scott Tremaine presented an extensive series of numerical simulations of the evolution of comet orbits due to the gravitational perturbations of the giant planets. Their simulations provided a dynamical proof that a belt in the outer Solar System is a far more likely source of Jupiter-family comets than the Oort cloud. The calculations set David Jewitt and Jane Luu of the University of Hawaii looking for the belt. In 1992, they discovered an object much smaller and fainter than Pluto and Charon orbiting beyond Neptune. At the present time, ~1000 objects ranging in size from a large comet nucleus to Pluto are known. It is now clear that Pluto, Charon, and the numerous smaller objects are what remain of the ancient disk of icy debris that did not accrete into a giant, Jupiter-like planet beyond the orbit of Neptune (Fig. 2). The discovery of an object slightly larger than Pluto by Michael Brown of the California Institute of Technology in 2003 triggered the International Astronomical Union (IAU) to downgrade Pluto from its status as a

FIGURE 1 The diameter of Pluto (2302 km) and its moon Charon (1186 km) in comparison to the size of the United States. The diameters of Pluto and Charon are each smaller than the diameter of Earth's Moon, 3476 km. (Courtesy of Dan Boone and NAU Bilby Research Center)

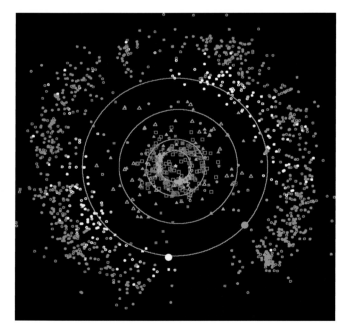

FIGURE 2 Positions of known bodies in the outer Solar System on September 24, 2006. The orbits of the outer planets are shown in light blue. The location of Neptune is marked by a large blue circle on the outermost orbit. Pluto is marked by a large white circle. Kuiper Belt objects (KBOs) are marked as red (classical KBOs), white (Plutinos), and magenta (SDOs) circles. Centaur objects are marked as orange triangles. Comets are marked as blue squares. (Courtesy of Minor Planet Center)

planet in 2006. The icy debris disk beyond Neptune is commonly called the **Kuiper Belt** in honor of Dutch-American astronomer Gerard P. Kuiper, who postulated its existence in 1951.

There are several dozen icy bodies that make up a class of objects closely related to Kuiper Belt objects (KBOs). These **Centaur objects** are recent escapees from the Kuiper Belt. They are on elliptical orbits about the Sun that cross the near-circular orbits of Saturn, Uranus, and Neptune. Within a few tens of millions of years after a Centaur object escapes from the Kuiper Belt, the giant planets scatter it out of the Solar System, into the Sun or a planet, or cause it to migrate into the region of the terrestrial planets where it becomes a Jupiter-family comet. Centaurs are quite important because they come closer to the Sun and Earth than KBOs. By virtue of their "close" approach, many of them become bright enough for certain physical studies that are not possible on fainter KBOs. However, it is important to remember that Centaur objects experience a warmer environment than KBOs, and the warmth may alter their physical and chemical properties away from their initial properties at the time of their formation in the Kuiper Belt.

By studying KBOs and Centaurs, we are studying the preserved building blocks of a planet, and we can therefore shed some light on the process of planet building in our Solar System as well as extrasolar planetary systems. After more than a decade of study, fundamental physical properties of KBOs and Centaurs—diameter, **albedo**, period of rotation, shape, mass, and surface composition—are being measured with accuracy.

1. Discovering Kuiper Belt and Centaur Objects

Centaur objects and KBOs orbit the Sun every 30 to 330 years. In addition to their own intrinsic motion about the Sun, the Earth's motion about the Sun imparts an apparent (parallactic) motion on these objects as well. These two motions distinguish KBOs and Centaurs from the multitude of background stars and galaxies and thereby make it possible to discover them.

Figure 3 illustrates the motion of the Centaur 1994 TA against the "fixed" pattern of much more distant stars and galaxies. Each panel of Figure 3 is a 300-sec exposure of 1994 TA taken with the Keck II 10-m telescope and a charge coupled device (CCD) camera in October 1998. The image on the right was taken about an hour after the image on the left.

2. Naming Objects

After a Centaur object or KBO is discovered, it needs a name. The IAU is responsible for naming celestial objects.

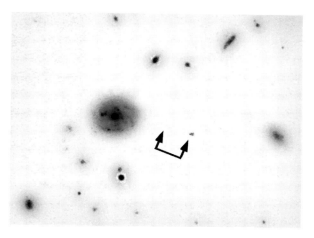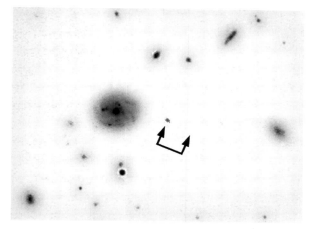

FIGURE 3   The motion of a Solar System object relative to the background stars and galaxies. Two 300-sec images of the Centaur 1994 TA taken with the Keck 10-m telescope and a CCD camera in October 1998. The image on the right was taken about an hour after the image on the left. Because of the Earth's and 1994 TA's revolution about the Sun, 1994 TA moves relative to the fixed pattern of background stars and galaxies. Such motion is how we discover KBOs and Centaur objects; however, the motion complicates physical studies of a known KBO or Centaur when the image of the KBO or Centaur comes close to an image of a background star or galaxy.

The IAU names KBOs and Centaurs the same way it names asteroids. Upon discovery, an object is given a preliminary designation consisting of a four-digit number indicating the year of discovery, a letter to indicate the half month of discovery, another letter to indicate the order of discovery within the half month, and another number to indicate the number of times the second letter was repeated within the half month period. For example, the provisional name of the KBO 2002 LM_{60} tells us the object was discovered between June 1 and 15 of 2002. After the orbit of a KBO about the Sun becomes well enough known that it isn't likely to be lost, the KBO is given a number. It can take observations over several years to establish a good orbit for a KBO. The number of 2002 LM_{60} is 50,000. No other Solar System object has the number 50,000. After an object receives a number, it receives a name. For example, 2002 LM_{60} is known as Quaoar. In this case, the same KBO has three names. After an object has a number and name, it's rarely called by its provisional name. If a KBO or Centaur object has a number or name, we know that its orbit about the Sun is well established and there is very little chance of losing it.

3. Databases of Known Objects

The IAU maintains an Internet listing of known KBOs and Centaurs as well as elements that describe their orbits about the Sun. In addition, the IAU, Lowell Observatory, and NASA's Jet Propulsion Laboratory in Pasadena, California, provide Internet tools that enable observers to figure out where to point telescopes to see a specific KBO or Centaur object on a specific night. Links to these tools are given in Table 1.

4. Dynamical Classes

It is possible to divide KBOs into dynamical classes. This section provides a brief discussion of the classes and likely

TABLE 1	KBO and Centaur Internet Tools
Institution	**Web Address**
Lowell Observatory	http://asteroid.lowell.edu/cgi-bin/koehn/asteph
NASA/JPL	http://ssd.jpl.nasa.gov/?horizons
IAU	http://cfa-www.harvard.edu/iau/MPEph/MPEph.html

interconnections between the classes. A thorough discussion of KBO dynamics can be found in the chapter by Harold Levison in this volume. [*See* COMET POPULATIONS AND COMETARY DYNAMICS.]

Classical KBOs are on orbits with perihelion distances, q, larger than 40 AU, semimajor axes, a, between 42 and 45 AU, eccentricities, e, less than 0.1, and inclination angles, i, less than 10°. It appears that classical KBOs did not experience strong perturbations by Neptune and hence they probably formed at or near their present location.

Resonant KBOs are a subset of classical KBOs that became trapped in mean motion resonances during the primordial migration of the planets. The process of resonance trapping tends to increase the **eccentricity** and inclination of trapped objects. Plutinos are objects trapped in the 2:3 mean motion resonance of Neptune at a = 39.6 AU, just like Pluto.

Scattered disk objects (SDOs) are thought to have originated in the primordial inner belt, a < 40 AU, and the primordial Uranus-Neptune region. They were subsequently scattered by Neptune onto orbits with large inclination angles, i > 15°, large eccentricities, e > 0.3, and large semimajor axes, a > 45 AU.

Centaur objects are on outer planet crossing orbits with q > 5.2 AU and a < 30.1 AU. Relatively recent gravitational interactions between SDOs and Neptune, and to a lesser extent between classical KBOs and Neptune, result in Centaur objects. Because Centaur objects cross the orbits of the outer planets, they are dynamically unstable and have mean lifetimes of $\sim 10^6$ years. As mentioned above, some Centaurs evolve into Jupiter-family comets, others are ejected from the Solar System, and yet others impact the giant planets. In addition, some Jupiter-family comets evolve back into Centaurs.

5. Brightness

5.1 Apparent Magnitude

The first physical property measured for a KBO is typically its brightness. A KBO is brightest in visible light (4000–8000 Å) by virtue of the sunlight it reflects toward the Earth. It is possible to isolate the brightness of a KBO in a particular bandpass by placing a colored glass filter in front of a CCD camera at the focal plane of a telescope. For example, a blue, green, or red filter in front of a CCD camera makes it possible to measure the brightness of blue, green, or red light from a KBO, i.e., its B (λ_{center} = 4500 Å), V (λ_{center} = 5500 Å), or R (λ_{center} = 6500 Å) magnitudes. Table 2 lists V magnitudes of the brightest KBOs. At the other extreme of brightness, Gary Bernstein used the Hubble Space Telescope (HST) to discover and measure the brightness of the faintest known KBO, V $\sim$ 28. The Centaur in Figure 3, 1994 TA, has V = 24.31 ± 0.05. For comparison, the Sun has V = −26.74 and the faintest star visible in the sky with the unaided eye has V $\sim$ 6.

5.2 Luminosity Function

There are many more faint KBOs than bright KBOs. Figure 4 comes from KBO discoveries made by a number of surveys, and shows the number of KBOs per unit magnitude per square degree on the sky near the **ecliptic** plane as a function of brightness (R-band magnitude), a luminosity function. Surveys find $\sim$100 KBOs with 27 < R < 28, $\sim$2 KBOs with 23 < R < 24, and only $\sim$0.001 KBOs with 19 < R < 20, all per square degree of sky. For reference, the full Moon occupies $\sim$ one-quarter of a square degree of sky and the Sun has R = −27.10.

5.3 Absolute Magnitude

The apparent magnitude of a KBO or Centaur depends on its heliocentric distance, r, and geocentric distance, Δ, in AU. For example, a KBO receding from the Sun and Earth will become fainter and its apparent magnitude will become larger in value. The absolute magnitude, H, of a KBO is a way to compare the *intrinsic* brightness of one KBO with another KBO and it does not depend on distance. The absolute magnitude of the same KBO receding from the Sun and Earth will not change. The absolute magnitude of a KBO is the brightness it would have if it were located at a distance of 1 AU from the Sun and 1 AU from the Earth, and had a Sun-KBO-Earth (phase) angle, α, of 0°. The relation

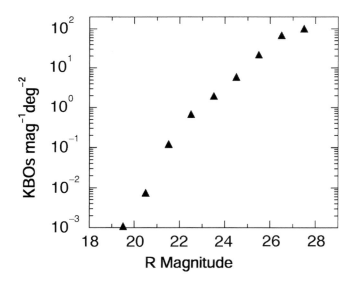

FIGURE 4 Number of KBOs per unit magnitude interval per square degree of sky vs. R-band magnitude. There are many more faint KBOs than bright KBOs. This is typical of small body populations in the Solar System that have been collisionally processed. (Courtesy of Gary Bernstein)

between absolute magnitude, H_v, and apparent magnitude, V, is given by

$$H_v = V - 5\log(r\Delta) + 2.5\log\left[(1-G)\Phi_1(\alpha) + G\Phi_2(\alpha)\right],$$

where the last term of the equation is an empirical phase function that describes how H_v of an object varies with phase angle. $G = 0.15$ and Φ_1 and Φ_2 given by

$$\Phi_i(\alpha) = \exp\left[-A_i\left(\tan\frac{1}{2}\alpha\right)^{B_i}\right]$$

where $i = 1$ and 2, $A_1 = 3.33$, $B_1 = 0.63$, $A_2 = 1.87$, and $B_2 = 1.22$ seem most appropriate for KBOs. H_v values for discovered KBOs and Centaurs range from about -1 to 15. Table 2 lists KBOs with the brightest H_v values.

6. Diameter

Size is among the most fundamental physical properties of an astronomical object, yet we are only beginning to get accurate diameter measurements for KBOs and Centaurs. The most direct way to measure the diameter of a KBO, D (in km), is to measure its angular diameter, θ (in arc sec), and geocentric distance, Δ (in AU). Geometry gives

$$D = 727\Delta\theta.$$

Unfortunately, KBOs and Centaurs have sufficiently small values for D and sufficiently large values for Δ that the resulting values for θ are too small for measurement even by the HST. Michael Brown pushed the HST to its limits and measured $\theta = 0.0343 \pm 0.0014$ arc sec for the KBO Eris, which was at a geocentric distance of 96.4 AU at the time of their observations. They found a diameter of 2400 ± 100 km for Eris, making it slightly larger than Pluto, $D = 2302$ km.

For KBOs and Centaurs with θ too small for measurement, it is possible to estimate their diameters from their brightness,

$$p\Phi D^2 = 9 \times 10^{16} r^2 \Delta^2 10^{0.4(m-V)},$$

where, as before, r is the heliocentric distance in AU and Δ is the geocentric distance in AU, m is the V-band brightness of the Sun (-26.74), V is the brightness of the KBO, p is the albedo of the object, and

$$\Phi = \left[(1-G)\Phi_1(\alpha) + G\Phi_2(\alpha)\right].$$

Since Jupiter-family comets come from Centaurs and the Kuiper Belt, most KBO diameter estimates assume an albedo similar to albedo measurements for a handful of Jupiter-family comets, i.e., $p = 0.04$. Diameter estimates from V magnitudes for about 100 objects range between $D = 25$ km for 2003 BH$_{91}$ to $D = 2400$ km for Eris. KBO and Centaur object diameters on the scale of Figure 1 range from the tiniest specks to Pluto.

The assumption of a comet-like albedo, although reasonable, is dangerous because Jupiter-family comets come much closer to the Sun than KBOs and Centaur objects. The frequent close proximity of short-period comets to the Sun results in the sublimation of H_2O ice and produces surfaces largely covered by a dark, refractory-rich, lag deposit. The surfaces of Jupiter-family comets may have chemical and physical properties quite different from the surfaces of Centaurs and KBOs. Charon has a relatively large albedo of 0.37. If we assume that a KBO has $p = 0.04$, but it is actually has $p = 0.4$, we will estimate a diameter that is more than three times too large. Measurements of albedos are essential for accurate measurements of KBO and Centaur object diameters.

7. Albedo

By measuring the brightness of sunlight *reflected* from a KBO at visible wavelengths and the brightness of heat *emitted* by the same KBO at thermal infrared wavelengths, it is possible to disentangle albedo from diameter, and thereby measure separate values for both quantities. The Spitzer Space Telescope, an infrared telescope in orbit about the Sun, is enabling John Stansberry of the University of Arizona, Dale Cruikshank of NASA's Ames Research Center, William Grundy of Lowell Observatory, and John Spencer of Southwest Research Institute to observe much fainter levels of heat from KBOs and Centaurs than is possible with telescopes on the Earth. As a result of their work, we have accurate diameters and albedos for more than a dozen KBOs (Table 2).

8. Brightness Variation

KBOs and Centaurs may have weak internal constitutions (i.e., rubble pile type interiors) due to fracturing by past impacts between objects. In other words, it is possible that KBOs and Centaurs are nearly strengthless bodies, held together primarily by their own self-gravity. If so, then some objects may deform from spheres into triaxial ellipsoids with axes $a > b > c$ as a result of their rotation.

The rotation of an ellipsoid can result in periodic variation of its projected area on the sky and hence a periodic variation of the sunlight it reflects and its brightness (Fig. 5). Monitoring such a brightness variation can result in a wealth of physical data about the object (e.g., its period of rotation, shape, and perhaps even its density and porosity).

TABLE 2	KBO Magnitudes, Albedos, and Diameters[1]					
Name	Number	Prov Des	V^2	H_v^3	p_v^4	D^5
Triton			13.5	−1.2	75	2707
Eris	136199	2003 UB_{313}	18.7	−1.1	>70	<2600
Pluto	134340		14.0	−0.7	61	2290
	136472	2005 FY_9	17.0	0.1	70–90	1250–1650
Charon			15.9		37	1242
	136108	2003 EL_{61}	17.5	0.4	55–75	1000–1600
Sedna	90377	2003 VB_{12}	21.1	1.20	>8.5	<1800
Orcus	90482	2004 DW	19.3	2.3	27	1000
Quaoar	50000	2002 LM_{60}	19.2	2.7	12	1300
	55637	2002 UX_{25}	19.9	3.6	10	900
	55565	2002 AW_{197}	20.2	3.6	12	734
	90568	2004 GV_9	19.8	3.7	15	700
Varuna	20000	2000WR_{106}	20.1	3.9	14	586
Ixion	28978	2001 KX_{76}	19.9	4.0	19	480
Huya	38628	2000 EB_{173}	19.5	5.1	6.6	500
	47171	1999 TC_{36}	19.6	5.4	7.9	405
	15874	1996 TL_{66}	20.9	5.5	>1.8	<958
	15789	1993 SC	22.4	7.3	3.5	398
	15875	1996 TP_{66}	21.1	7.4	1.1	406
	29981	1999 TD_{10}	21.1	9.1	5.3	88

[1] Courtesy John Stansberry.
[2] V-band magnitude.
[3] Absolute magnitude in V-band.
[4] Visual Albedo in units of percentage from Spitzer Space Telescope and ISO observations.
[5] Diameter in km from Spitzer Space Telescope and ISO observations

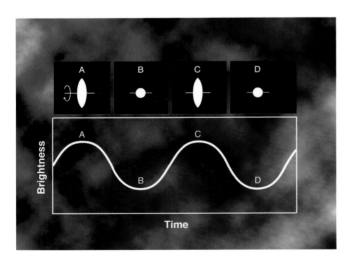

FIGURE 5   (a) The rotation of a non-spherical KBO or Centaur object results in a periodic variation of the object's projected area on the plane of the sky and hence a periodic variation in its brightness. (b) Brightness vs. time (lightcurve) for the rotation of a non-spherical object. During one rotation of the object, it goes through two maxima (points A and C) and two minima (points B and D) in brightness. (Courtesy of Ron Redsteer and NAU Bilby Research Center)

8.1 Period of Rotation

If we can determine the form of the periodic brightness variation (lightcurve) for a KBO or Centaur, they can determine its period of rotation. Figure 6 shows a plot of V magnitude vs. time in hours for the Centaur Pholus. At the time of observation in 2003, Pholus was ~18 AU from the Sun, nearly the same distance as Uranus. We see that the two maxima are of nearly equal brightness, but one minimum (at ~5 hr) is ~0.03 magnitude (3%) fainter than the other minimum (at 0 hr). The pattern of two maxima and two minima repeats every 9.980 ± 0.002 hr, Pholus' period of rotation on its axis.

Determining the period of rotation for a KBO or Centaur takes a significant amount of telescope time. In the case of Pholus, each of the 99 points in Figure 6 represents a brightness measurement from a 300 sec CCD image. Because of their faintness, measurements of KBO and Centaur **lightcurves** require telescopes with moderately large apertures, typically with diameters ≥2 m. A large amount of time on moderate-size telescopes is difficult to obtain, so periods of rotations are available for only a handful of objects (Table 3). Groups led by William Romanishin of the

TABLE 3		Rotation Periods and Lightcurve Amplitudes			
Name	**Number**	**Prov Des**	**Class**[1]	**Period**[2]	**Δm**[3]
	136108	2003 EL$_{61}$	kbo	3.9154	0.28
	15820	1994 TB	kbo	6.0, 7.0	0.30
Varuna	20000	2000 WR$_{106}$	kbo	6.34	0.42
	26308	1998 SM$_{165}$	kbo	7.1	0.45
	32929	1995 QY$_9$	kbo	7.3	0.60
	19255	1994 VK$_8$	kbo	7.8, 8.6, 9.4, 10.4	0.42
	19308	1996 TO$_{66}$	kbo	7.9	0.25
	47932	2000 GN$_{171}$	kbo	8.329	0.61
	33128	1998 BU$_{48}$	kbo	9.8, 12.6	0.68
Pholus	5145	1992 AD	cen	9.980	0.60
	40314	1999 KR$_{16}$	kbo	11.858, 11.680	0.18
		2001 QG$_{298}$	kbo	13.7744	1.14

[1] Dynamical class. Kuiper Belt object (kbo) or Centaur object (cen).
[2] Period of rotation in hours. Multiple entries indicate possible periods.
[3] Peak to trough amplitude in magnitudes.

University of Oklahoma and Scott Sheppard of the University of Hawaii are responsible for many lightcurve measurements. They find periods of rotation between 4 and 14 hours (Table 3).

8.2 Amplitude

In Figure 6, Pholus has a maximum brightness of $V_{max} = 20.09$ and a minimum brightness of $V_{min} = 20.69$, i.e., each time through its repeating pattern it has a maximum brightness variation or lightcurve amplitude of $\Delta m = V_{min} - V_{max} = 0.60$ magnitude. Since

$$\Delta m = 2.5 \log \frac{F_{max}}{F_{min}},$$

the ratio of maximum to minimum brightness is $F_{max}/F_{min} = 1.74$. From Table 3, we see that KBOs and Centaurs exhibit $0.1 \le \Delta m \le 1.1$ magnitude.

8.3 Shape

If a KBO or Centaur lightcurve is due to the rotation of a triaxial ellipsoid about its shortest axis, c, we can in principle determine its shape (i.e., the ratio of its axes a/b and c/b). How? As a KBO or Centaur orbits the Sun, we observe it at different aspect angles. Aspect angle is the angle between lines originating at the center of the body and toward the Earth and the north rotational pole of the body. Figure 7 illustrates how a change in aspect angle results in a change in lightcurve amplitude. At point A, we are looking

at the object equator-on (aspect angle of 90°), and we see a lightcurve with an amplitude as large as it gets for the object. A quarter of a revolution about the Sun later, at point B, we are looking down the rotation axis of the body (aspect angle of 0°), and we won't see any brightness variation. If we

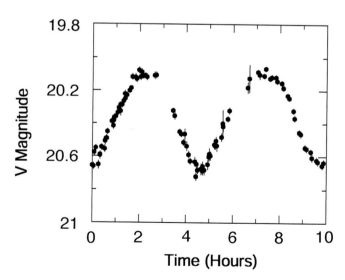

FIGURE 6 Lightcurve for the Centaur Pholus. The brightness pattern of two nearly equal brightness maxima and two brightness minima (at ~0 hr and ~5 hr) that differ by 0.03 magnitude (3%) repeats every 9.980 hr. The maximum peak-to-trough brightness variation of Pholus is 0.60 magnitude. From the lightcurve, we know that Pholus rotates once about its axis every 9.980 hr.

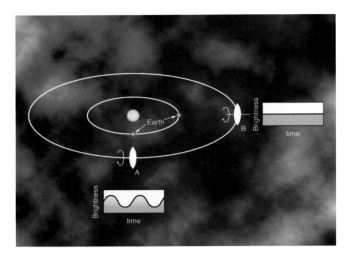

FIGURE 7 Changes in aspect angle result in changes in lightcurve amplitude. When the object is at point A, the angle between two lines originating at the center of the body and toward the Earth and the north rotational pole of the body, the aspect angle, is 90°. We see the object with an equator-on aspect and the lightcurve of the object has its maximum amplitude. When the body moves to point B, the aspect angle is 0° and we are looking down on the rotational pole. At point B, the object does not exhibit any brightness variation; its lightcurve amplitude is zero. By monitoring changes in the lightcurve amplitude of a body as it orbits the Sun, it is possible to calculate the shape of the body. (Courtesy of Ron Redsteer and NAU Bilby Research Center)

observe the amplitude change of a body's lightcurve over a significant portion of the object's revolution about the Sun, it's possible to use a computer to search through all possible combinations of shape and orientation of the rotation axis to find a shape that best simulates the observed amplitude changes.

Figure 8 shows three lightcurves of Pholus from 1992, 2000, and 2003. The x-axis is labeled with the rotational phase of Pholus. The rotational phase interval of 0 to 1 is equal to a time interval of 9.980 hr, the time it takes Pholus to complete one rotation about its axis. The amplitude of the lightcurves grew from 0.15 to 0.39 to 0.60 magnitude. A computer search of orientations of the rotation axis and shapes for Pholus yields four possible orientations for the rotational axis, all with the same shape of $a/b = 1.9$ and $c/b = 0.9$. Pholus appears to have a significantly elongated shape.

The amplitude measurements of Pholus span little more than 10% of its 92-year period of revolution about the Sun. Confirmation of the shape for Pholus will require additional amplitude measurements two or three decades into the future.

KBO shape measurements require amplitude measurements over more than a century. Yet, we can still say something about the shapes without waiting a century. For

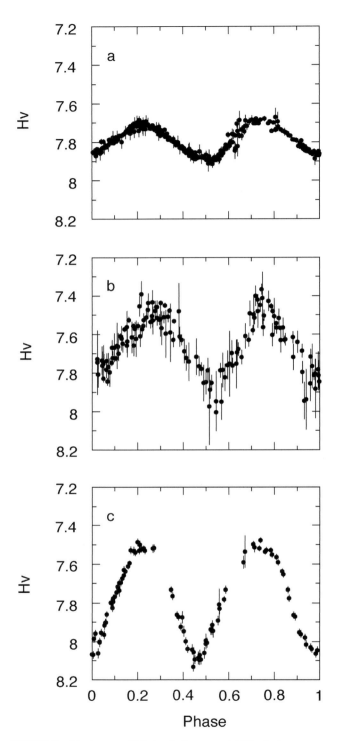

FIGURE 8 Evolution of Pholus lightcurve. (a) The lightcurve observed by Marc Buie and Bobby Bus in 1992 has an amplitude of 0.15 magnitude. (b) The lightcurve observed by Tony Farnham in 2000 has an amplitude of 0.39 magnitude. (c) The lightcurve observed by Bill Romanishin and Guy Consolmagno in 2003 has an amplitude of 0.60 magnitude. The period in 1992, 2000, and 2003 remained constant at 9.980 hr; however, the increasing amplitude indicates that we were seeing Pholus more equator-on with each passing year between 1992 and 2003.

example, the KBO Varuna has $\Delta m = 0.42 \pm 0.02$ magnitude. If we assume we are seeing Varuna with an "equator-on" aspect (i.e., an aspect angle of $90°$), which corresponds to the largest possible lightcurve amplitude, we can relate the amplitude of the lightcurve to an axial ratio,

$$\Delta m = 2.5 \log \frac{a}{b}.$$

Such an assumption gives $a/b = 1.5$ for Varuna. Since we don't know if they are viewing Varuna "equator-on," Δm and a/b are lower limits. At the present time, all KBO axial ratios are lower limits.

8.4 Density

Besides periods of rotation and shapes, lightcurves allow us to estimate densities for KBOs and Centaurs. If nonspherical shapes are the result of rotational deformation, then we can use the formalism developed by Chandrasekhar that relates the period of rotation and shape to the density of a strengthless ellipsoid. Application of Chandrasekhar's formalism to Pholus, the object with the best shape estimation ($a/b = 1.9$ and $c/b = 0.9$) and period of rotation ($P_{rot} = 9.980$ hr), gives an average density of $\rho_{avg} = 0.5$ g cm^{-3}. It is interesting to note that the similar-sized Saturnian satellites Janus, Epimetheus, Prometheus, and Pandora have average densities of 0.61, 0.64, 0.42, and 0.52 g cm^{-3}. In the case of KBO Varuna, its assumed shape ($a/b = 1.5$ and $c/b = 0.7$) and period of rotation ($P_{rot} = 6.3442$ hr) yield $\rho_{avg} = 1.0$ g cm^{-3}. Average densities ≤ 1 g cm^{-3} suggest that KBOs and Centaurs likely have ice-rich and porous interiors.

8.5 Porosity

Porosity is the fraction of void space in a KBO or Centaur. If a KBO is some mixture of ice, refractory material (dust), and empty space, the average density of a KBO is given by

$$\rho_{avg} = f_i \rho_i + f_r \rho_r,$$

where f_i and f_r are the fractional volumes occupied by icy and refractory material and ρ_i and ρ_r are the densities of icy and refractory material. In addition, the sum of the parts must equal the whole, so

$$f_i + f_r + f_v = 1,$$

where f_v is the fraction of void space or the porosity. The fraction of total mass locked up in refractories, ψ, is given by

$$\psi = \frac{\rho_r f_r}{\rho_r f_r + \rho_i f_i}.$$

By combining the above three equations, it is possible to obtain an algebraic expression for the porosity of a KBO or Centaur,

$$f_v = 1 - \frac{\rho_{avg}}{\rho_i}\left[1 + \psi\left(\frac{\rho_i}{\rho_r} - 1\right)\right].$$

Assuming reasonable values of $\rho_{avg} = 1$ g cm^{-3}, $\rho_i = 1$ g cm^{-3}, $\rho_r = 2$ g cm^{-3}, and $\psi = 0.5$ for Varuna, Jewitt and Sheppard estimate that Varuna has a porosity ~ 0.25. For comparison, beach sand has $f_v \sim 0.4$ and basaltic lunar regolith has $0.4 < f_v < 0.7$.

9. Composition

9.1 Surface Color

An early expectation was that all KBOs should exhibit a similar red surface color. Why? Initially, KBOs were thought to form over a small range of heliocentric distances where the temperature in the young solar nebula was the same. The similar temperature suggested that KBOs formed out of the solar nebula with the same mixture of molecular ices and the same ratio of dust to icy material. In addition, their similar formation distance from the Sun suggested that KBOs should experience a similar evolution. Specifically, the irradiation of surface CH_4 ice by solar ultraviolet light and solar wind particles should have converted some surface CH_4 ice into red, complex, organic molecules. By their nature, the complex organic molecules were expected to absorb more incident blue sunlight than red sunlight. Therefore, the light reflected from the surfaces of KBOs was expected to consist of a larger ratio of red to blue light than the incident sunlight. It was a surprise to find KBOs exhibit a range of surface colors rather than just red colors. At one extreme, some KBOs reflect sunlight equally at all wavelengths (i.e., exhibit neutral or gray surface colors). On the Johnson-Kron-Cousins photometric system such KBOs have B-R = 1.0. At the other extreme, some KBO have extraordinary red colors, i.e., B-R = 2.0.

Because it's a painstaking process to measure the color of a KBO or Centaur, taking as much as three hours of telescope time to obtain an accurate color for a single object, the first color surveys consisted of only 10 to 20 objects. These small samples lumped KBOs and Centaurs together and resulted in a controversy. Some groups found their samples to exhibit a uniform distribution of colors from gray (B-R = 1.0) to extraordinarily red (B-R = 2.0). These groups suggested that KBOs and Centaurs experienced a steady reddening of their surfaces by solar radiation, and occasional impacts by smaller objects punctured the red surfaces and excavated gray, interior material. Such a radiation-reddening and impact-graying mechanism would explain the uniform distribution of colors. Another group found that their sample of KBOs and Centaurs divided into two distinct color groups—gray objects with $1.0 < $ B-R $ < 1.4$ and red objects with $1.5 < $ B-R $ < 2.0$. They found almost no objects with $1.4 < $ B-R $ < 1.5$. They did not have a physical explanation for their surprising result. Everyone agreed KBOs and Centaurs did not exhibit only red surface colors.

As the groups pressed hard at telescopes to measure more surface colors and test their initial findings, sample sizes grew from 10 to more than 100 objects. Once the sample sizes became large enough, it became apparent that different dynamical classes of KBOs had different color signatures.

9.1.1 CENTAUR OBJECTS

Figure 9a shows a histogram of the number of objects vs. B-R color for a sample of 22 Centaur objects. Fourteen objects have B-R $ < 1.3$ and eight objects have B-R $ > 1.7$. Notice there are no objects with $1.3 < $ B-R $ < 1.7$. Is it possible that Centaurs actually exhibit a uniform distribution of B-R colors and either insufficient sampling or chance is responsible for the apparent split into two B-R color groups? Application of statistical tests like the "dip test" tell us that the probability of making observations in Figure 9a for an actual uniform distribution of B-R colors is about 1 in 100. The split into two B-R color groups appears to be real. Unlike the earlier controversy, two groups, one led by Nuno Peixinho and the other by this author, find the same highly unusual split. What makes the split so unusual is that there doesn't seem to be any other physical property that correlates with the color of a Centaur object. For example, if Centaurs that came closest to the Sun were all gray, we might suspect that the warmth of the Sun was chemically or physically altering the surfaces and graying them. But there is no statistically significant correlation between color and **perihelion distance** or any other orbital element.

9.1.2 CLASSICAL KBOS

Figure 9b shows a histogram of the number of objects vs. B-R color for a sample of 21 classical KBOs. All 21 classical objects have B-R $ > 1.5$, i.e., there are no gray objects at all in the sample. Classical KBOs exhibit the color signature originally expected for all KBOs.

9.1.3 SCATTERED DISK OBJECTS

Figure 9c shows a histogram of the number of objects vs. B-R color for a sample of 20 SDOs. Seventeen of the 20 objects exhibit B-R $ < 1.5$. There appears to be a deficit of red objects among this group.

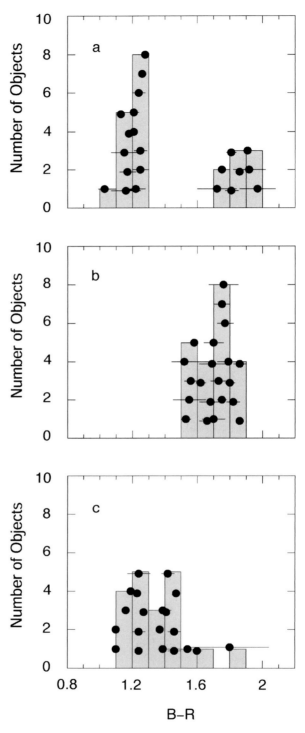

FIGURE 9 Correlations between colors and orbital properties of KBOs and Centaurs. (a) A sample of 22 Centaurs neatly divide into two color groups; 14 objects exhibit B-R $ < 1.3$ and eight objects exhibit B-R $ > 1.7$. Surprisingly, there are no Centaurs with $1.3 < $ B-R $ < 1.7$. (b) All 21 objects of a sample of classical KBOs with $q > 40$ AU, $e < 0.1$ and $e < 10°$ are all red (B-R $ > 1.5$). (c) A sample of 20 SDOs are mostly gray (B-R $ < 1.5$). The mechanisms responsible for these correlations between color and orbital properties are not well understood yet.

9.1.4 REASONS FOR COLOR PATTERNS

What could cause these color signatures? One possibility is the radiation-reddening and impact-graying mechanism discussed earlier in Section 1. However, such a mechanism should result in a uniform distribution of B-R colors for Centaurs and not two clusters of B-R colors. In addition, gray impact craters and their ejecta blankets would be randomly distributed on the surface so that one hemisphere might have more than another, resulting in measurable color changes as the object rotates. However, repeated and random measurements of individual rotating KBOs and Centaurs give the same B-R color. Also, extensive observations of Pholus suggest that it has a highly homogeneous surface color. Figures 10a and 10b show the R-band brightness and B-band brightness of Pholus as a function of a single rotation phase taking 9.980 hr. Figure 10c is the difference of 10a from 10b, yielding the B-R color across the entire surface of Pholus as it makes one rotation about its axis. The solid horizontal line is the average of the points. The dashed lines are plus or minus one standard deviation, $\sigma = 0.04$. Any variation in the B-R surface color of Pholus must be smaller than 0.04 magnitude (4%). Again, there is no evidence of gray impact craters on a radiation-reddened surface.

Another possibility is that the colors of KBOs are the remaining signature of a temperature-induced, primordial composition gradient. The small, rocky terrestrial planets (Mercury, Venus, Earth, and Mars) close to the Sun and the giant, hydrogen-rich gas giant planets (Jupiter, Saturn, Uranus, and Neptune) farther away from the Sun are the result of such a gradient. In the inner Solar System, temperatures were so high that only metal and rock forming elements could condense from the nebular gas to form small, rocky, and metal-rich solids. At and beyond the orbit of Jupiter, the hydrogen-dominated nebular gas was cold enough for the H_2O to condense out. We may be seeing a similar effect on the colors of KBOs and Centaurs. We now suspect KBOs did not all form at about the same distance from the Sun. Perhaps the red classical KBOs formed farther out in the nebula where it was cold enough to hang on to their CH_4 ice reddening agent. Perhaps the gray KBOs formed closer to the Sun and were not able to hang on to their CH_4 ice reddening agent.

Additional work is necessary to figure out whether the radiation-reddening and collisional-graying mechanism, the temperature-gradient mechanism, or some other mechanism is responsible for the colors of KBOs and Centaurs.

9.2 Spectroscopy

There are only a handful of KBOs and Centaurs that are known to exhibit ice absorption bands in their spectra. H_2O-ice bands are seen in the spectra of Charon, 19308 (1996 TO_{66}), Varuna, Quaoar, Orcus, Pholus, and Chariklo. CH_4-ice bands are seen in the spectra of Pluto, Neptune'

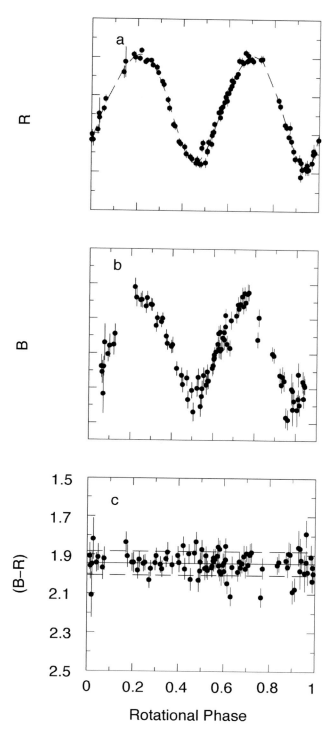

FIGURE 10 Homogeneous B-R surface color of Pholus. (a) R-band magnitude vs. rotation phase. The x-axis spans a time interval of 9.980 hr. (b) B-band magnitude vs. rotational phase. (c) Difference between above two panels yield B-R color vs. rotational phase. The solid line is the average of the 94 points. The dashed lines are plus or minus one standard deviation, σ, of 0.04 magnitude. Any variation in the surface color of Pholus as it completes one rotation on its axis must be less than 0.04 magnitude (4%). Pholus exhibits a homogeneous surface color.

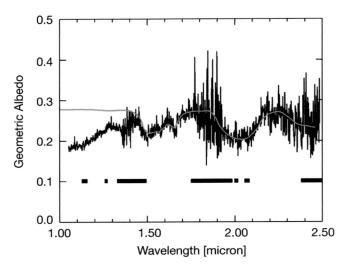

FIGURE 11 Near-infrared reflection spectrum of Quaoar (black) compared to a spectrum of H_2O ice (red). The broad absorption bands near 1.5 μm and 2.0 μm reveal the presence of H_2O ice on the surface of Quaoar. The narrow absorption band near 1.65 μm indicates the presence of crystalline H_2O ice and is not present in amorphous ice. (Courtesy of David Jewitt and Jane Luu)

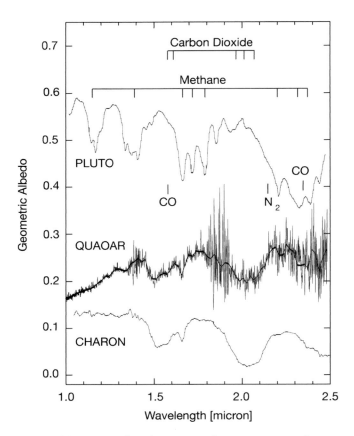

FIGURE 12 Near-infrared spectrum of Quaoar compared to near-infrared spectra of Pluto and Charon. The spectra of Quaoar and Charon are similar in that they exhibit three strong H_2O-ice absorption bands at 1.5 μm, 1.65 μm, and 2.0 μm, but no CH_4-ice bands. The spectrum of Pluto exhibits strong CH_4 ice bands. (Courtesy of David Jewitt and Jane Luu)

satellite Trition, which may be a captured KBO, Eris, and 136472 (2005 FY_9).

Perhaps one of the most intriguing spectroscopic results comes from David Jewitt and Jane Luu's observations of Quaoar. Specifically, they find not only the H_2O ice bands at 1.5 and 2.0 μm, but they also find another H_2O band at 1.65 μm (Figure 11). The later band suggests the surprising result that the H_2O-ice has a crystalline rather than an amorphous structure. The H_2O molecules of crystalline ice have a periodic structure whereas the H_2O molecules of amorphous ice do not. Crystalline H_2O on Quaoar is a surprise because Quaoar's maximum surface temperature is only ~50°K. At such a low temperature, it is difficult for the H_2O molecules to arrange themselves into a coordinated structure of a crystal lattice; somewhere around 100°K, amorphous ice arranges itself into an ordered crystalline lattice. In other words, the 1.65-μm band suggests that the H_2O-ice on Quaoar was somehow heated to temperatures above 100°K.

An intriguing possibility for the source of the "warm" H_2O on Quaoar is NH_3-H_2O volcanism. Long ago, long-lived radioactive elements heated the interior of Quaoar, and that heat may still be propagating through its interior. The heat may have been sufficient to create a melt of H_2O and NH_3. The lower density melt may have percolated upward, perhaps forming fluid-filled cracks all the way or nearly all the way to the surface in the surrounding, higher density icy-rock mixture. Eventually, the cooling "lava" containing crystalline H_2O ice and crystalline am-

monium hydrate might become exposed by occasional impacts on Quaoar's surface. What makes this mechanism even more intriguing is that Jewitt and Luu claim there is evidence for an ammonia hydrate band in their spectra of Quaoar. Ammonia-water volcanism as the source of the crystalline H_2O ice is highly speculative. Some other mechanism, not requiring a warm interior and volcanoes, may explain the presence of the crystalline H_2O ice on Quaoar.

Figure 12 illustrates that Quaoar has a spectrum similar to Charon, but quite different from Pluto. Quaoar and Charon exhibit the 1.5- and 2.0-μm H_2O-ice bands as well as the 1.65-μm crystalline band, but none of the strong CH_4 ice bands seen on Pluto. Note that Quaoar has the 1.65-μm band despite having a larger semimajor axis, a = 43.6 AU, than Pluto, a = 39.8 AU.

Another intriguing spectroscopic result comes from Javier Licandro's observations of 136472 (2005 FY_9). He finds that CH_4-ice bands in the spectra of 136472 (2005 FY_9) are much deeper than the CH_4-ice bands in the spectra

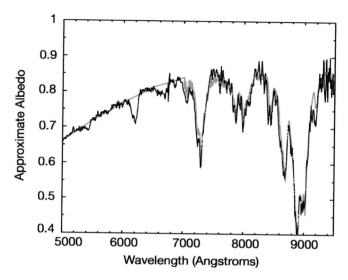

FIGURE 13 Optical spectrum of 136472 (2005 FY$_9$) (black line) and a Hapke model of pure CH$_4$-ice (red line). The CH$_4$ absorption bands of 136472 (2005 FY$_9$) are blue shifted by 3.25 ± 2.25 Å relative to the pure CH$_4$ model indicating the presence of another molecular ice, possibly N$_2$, CO, or Ar.

of Pluto, implying that the abundance of CH$_4$ on the surface of 136472 (2005 FY$_9$) could be higher than on the surface of Pluto. This author finds the CH$_4$-ice bands in his spectrum and Javier Licandro's spectrum of 136472 (2005 FY$_9$) are blueshifted by 3.25 Å relative to the positions of pure CH$_4$-ice bands (Figure 13). Such a shift suggests the presence of another ice component on the surface of 136472 (2005 FY$_9$), possibly N$_2$-ice, CO-ice, or Ar. In addition, Licandro finds CH$_4$-ice bands blueshifted in a spectrum of Eris.

It is odd that some KBOs exhibit strong CH$_4$ bands and others exhibit strong H$_2$0 bands. Pluto and Charon are part of the same system, yet they exhibit very different spectra. Perhaps the difference is due to Pluto's size, it may have experienced some form of methane ice volcanism. In the end, we may find only the largest KBOs exhibit CH$_4$-ice bands. Eris, Pluto, and possibly 2005 FY$_9$ are the three largest KBOs and they all exhibit CH$_4$-ice bands.

10. KBO Binaries

In 2001, Christian Veillet announced the discovery of two components to the KBO 1998 WW$_{31}$. Over the next few years, Keith Noll used the superior imaging resolution of HST to observe 122 KBOs for additional binaries. His survey was sensitive to binaries with separations ≥0.15 arc sec and a magnitude difference between components ≤1 magnitude. Noll discovered six more binaries. Currently, 22 KBO binaries are known (Table 4).

10.1 System Mass

Two KBOs of a binary pair revolve about their common center of mass. However, it is far more convenient to observe the position of the fainter of the two components as it makes a complete revolution about the brighter component on the plane of the sky, i.e., to observe the apparent relative orbit. Figure 14 illustrates the apparent relative orbit of 1998 WW$_{31}$. The true orbit of the KBO binary system will not happen to lie exactly in the plane of the sky. Hence, the apparent relative orbit is merely a projection of the true relative orbit onto the plane of the sky. Techniques exist to determine the inclination of the true orbit relative to the plane of the sky. Once the period of revolution, P, and the semimajor axis, a, of the true relative orbit are known, it is possible to use Kepler's Third Law to calculate the combined mass of the binary system,

$$m_1 + m_2 = \frac{4\pi^2 a^3}{GP^2}.$$

From the HST observations in Figure 14, Veillet and Noll found that 1998 WW$_{31}$ has a true relative orbit with a semimajor axis of 22,300 km, an eccentricity of 0.8, and a period of revolution of the fainter component about the brighter component of 574 days. The 1998 WW$_{31}$ system has a combined mass of 2.7×10^{18} kg, much smaller than the Pluto-Charon system combined mass of 1.46×10^{22} kg. Table 4 lists the true relative orbital properties and combined masses for the better studied binary systems.

10.2 Mutual Events

Between 1985 and 1990, Pluto's orbital motion about the Sun caused the Pluto-Charon orbital plane to sweep through the line of sight to the Earth. As a result, mutual eclipses (also known as mutual events) occurred every 3.2 days (half of Charon's orbital period). Because of the mutual events, observers were able to accurately measure diameters of 2302 ± 12 km and 1186 ± 26 km for Pluto and Charon, and with the total mass of the binary, they were able to derive an average density for the system of 1.95 ± 0.10 g cm^{-3}.

A key objective of current binary KBO work is to discover as many binaries as possible and to determine their orbits sufficiently well to predict when the onset of mutual events will occur. By observing KBO mutual events, we will obtain radii and density measurements that only a spacecraft encounter could improve upon. At present, no KBO binary orbit (other than Pluto and Charon) is known well enough to predict the onset of a mutual event with confidence.

10.3 Origin of KBO Binaries

Two of the most unusual features of KBO binaries, compared to main belt asteroid and near-Earth asteroid binaries,

TABLE 4   Binary KBOs[1]

Name	Number	Prov Des	a[2]	e[3]	Period[4]	Mass[5]
Resonant						
Pluto/Charon	134340		19,636	0.0076	6.38722	14,710
	47171	1999 TC$_{36}$	7,640		50.4	13.9
	26308	1998 SM$_{165}$	11,310		130	6.78
Classical						
		2005 EO$_{304}$				
		2003 UN$_{284}$				
		2003 QY$_{90}$				
		2001 QW$_{322}$				
		2000 CQ$_{114}$				
		2000 CF$_{105}$				
		1999 OJ$_4$				
		1998 WW$_{31}$	22,300	0.82	574	2.7
	134860	2000 OJ$_{67}$				
	88611	2001 QT$_{297}$	27,300	0.240	825	2.3
	80806	2000 CM$_{105}$				
	79360	1997 CS$_{29}$				
	66652	1999 RZ$_{253}$	4,660	0.46	46.263	3.7
	58534	1997 CQ$_{29}$	8,010	0.45	312	0.42
Scattered						
		2001 QC$_{298}$	3,690		19.2	10.8
Eris	136199	2003 UB$_{313}$				
	136108	2003 EL$_{61}$	49,500	0.05	49.12	4,200
	82075	2000 YW$_{134}$				
	48639	1995 TL$_8$				

[1] Courtesy Keith Noll.
[2] Semimajor axis in km.
[3] Eccentricity.
[4] Period in days.
[5] Mass in units of 10^{18} kg.

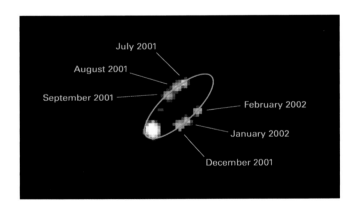

FIGURE 14   Binary KBO. The apparent orbit of the fainter component of 1998 WW$_{31}$ relative to the brighter component on the plane of the sky. (Courtesy of Christian Veillet, Keith Noll, and NASA)

are the wide separation and similar diameter of each pair of components. These unusual features make it unlikely that collisions between two KBOs created each binary system, as in the case of the Earth and the Moon. Similarly, it isn't likely that one KBO gravitationally captured another KBO to form a binary system. A mechanism put forth by Stuart Weidenschilling suggests that it is possible to create a loosely bound KBO binary by collision and capture in the presence of a third body. His mechanism requires many more KBOs than are seen today; perhaps such a mechanism operated long ago in a more densely populated Kuiper Belt (see the next section). Peter Goldreich put forth a mechanism wherein capture takes place during a close encounter as a result of the dynamical friction with the many surrounding small bodies. Each of these mechanisms produces its signature on the population of binaries we see today. For example, Weidenschilling's mechanism favors the production of wide binary pairs, and Goldreich's mechanism favors the production of closer pairs. Only the discovery of many more binaries will allow us to determine whether either of these mechanisms or some other mechanism is responsible for the formation of KBO binaries.

11. Mass of the Kuiper Belt

What's the mass of the entire Kuiper Belt? Gary Bernstein combined his HST survey for the faintest KBOs with ground-based telescope surveys for brighter KBOs, and assumed KBOs have an albedo of 0.04 and a density of 1 g cm^{-3}, to estimate a Kuiper Belt mass of ~3 percent of the Earth's mass, or about 14 times the mass of Pluto. A major source of uncertainty in his mass estimate is the uncertainty in the albedos and densities of KBOs.

It appears that the Kuiper Belt did not always have a mass of ~3 percent of the Earth's mass. Specifically, the present number of KBOs per AU^3 is too small to grow KBOs larger than ~100 km in diameter by accretion in less time than the age of the Solar System. Since 1000-km sized KBOs exist, it is likely that the Kuiper Belt initially had many more KBOs per AU^3 than today. Calculations by Alan Stern suggest that the initial Kuiper Belt probably had a mass ten times the mass of the Earth, and as Neptune grew to a fraction of its present size, it stirred KBOs from their initial circular orbits to more eccentric orbits, resulting in frequent disruptive, rather than accretive collisions especially between KBOs smaller than 40 to 60 km in diameter. These collisions probably eroded the Kuiper belt mass down to its current value.

12. New Horizons

Because astronomers can discover and then measure the physical properties of many KBOs, their work is important because it gives us a global view of the Kuiper Belt and context for in situ spacecraft measurements. In January of 2006, NASA's New Horizons spacecraft departed Earth on a journey that will culminate in the first flyby of the Pluto-Charon system in 2015, and hopefully the first flyby of a KBO sometime before 2020. The $500 million spacecraft weighs only 416 kg (917 lb) and has four instrument packages: (1) a CCD camera, (2) an ultraviolet, optical, and near-infrared imaging-spectrometer, (3) a charged particle detector, and (4) a radio telescope.

These instrument packages will provide in-depth observations impossible with telescopes on and near the Earth. For example, if New Horizons comes within a few thousand kilometers to a KBO, it could image the surface of the KBO with a resolution of 25 m $pixel^{-1}$. For comparison, HST can only image a KBO at 42 AU with a resolution of about 1200 km $pixel^{-1}$.

What kind of surface might the spacecraft image? If New Horizons visits a small KBO, perhaps it will image a surface with numerous craters, suggestive of an ancient surface bombarded by other small bodies (KBOs and comets) over the age of the Solar System? On the other hand, if New Horizons visits a large KBO, perhaps it will see few craters on the surface, suggestive of some process erasing older craters. Perhaps the images of a large KBO will show long linear features in an icy crust, and some roughly round basins that appear flooded by liquids from the interior, much like the Voyager spacecraft images of Triton. Perhaps the spacecraft will catch a geyser erupting, and shooting a plume of gas and ice above the surface.

There are some problems concerning a New Horizon's flyby of a KBO. The spacecraft trajectory is fixed since first it will fly by Pluto. In addition, the spacecraft has a limited fuel supply for adjusting its trajectory after the Pluto encounter. At present, none of the almost 1000 currently-known KBOs are close to the spacecraft's trajectory. A flyby of a KBO by New Horizons depends on discovering a candidate close to the spacecraft's trajectory. Perhaps New Horizons will have enough fuel to visit one of the smaller (50 km diameter) and more common KBOs. The chances for the spacecraft visiting one of the larger (1000 km diameter) and rarer KBOs appear slim at the moment.

13. Future Work

It is likely that future work on the physical properties of KBOs and Centaurs will be driven by future state-of-the-art observatories. The 6-m James Webb Space Telescope (JWST) near the L2 point will be able to obtain images and spectra of very large numbers of KBOs and Centaurs from 0.6 to 27 μm. It should be possible to measure diameters, albedos, surface colors, and optical and infrared spectra for many more objects than possible today. A large increase in the number of objects with physical property measurements will make it possible to look for statistically significant correlations between many more physical properties than possible with today's telescopes, and thereby better constrain the important formation and evolution mechanisms in the outer Solar System.

Large ground-based telescopes of the future will likely play a big role in the field too. For example, the Giant Magellan Telescope (GMT), a configuration of six off-axis 8.4-m mirror segments around a central on-axis segment that is equivalent to a filled aperture 21.4 meters in diameter, and the Thirty Meter Telescope (TMT), a configuration of more than 700 hexagonal-shaped mirror segments that is equivalent to a filled aperture 30 meters in diameter, will make it possible to obtain higher signal precision optical and infrared spectra than possible with current 10 meter telescopes. Better spectra and models will make it possible to map surface concentration of ices (e.g., the CH_4/N_2

concentration) as a function of depth and as a function of rotational phase over the surfaces of numerous objects. Such measurements will provide a wealth of data for constraining cosmochemistry models of the outer Solar System. Finally, the Atacama Large Millimeter Array (ALMA), a configuration of about sixty-four 12-meter antennas located at an elevation of 16,400 feet in Chile, may reveal extra-solar Kuiper Belts for comparison with our Kuiper Belt. ALMA may provide density and temperature profiles as well as chemical measurements through the detection of spectral lines in the belts. ALMA may initiate a new field of study, comparative Edgeworth-Kuiper Belt object ology, i.e., comparative EKO-logy.

Bibliography

Davies, J. (2001). *Beyond Pluto: Exploring the Outer Limits of the Solar System*. Cambridge University Press.

Jewitt, D. (2005). Kuiper Belt. http://www.ifa.hawaii.edu/faculty/jewitt/kb.html

Solar System Dust

Eberhard Grün

Max-Planck-Institut für Kernphysik
Heidelberg, Germany
Hawaii Institute of Geophysics and Planetology
Honolulu, Hawaii

CHAPTER 34

Solar system dust is finely divided particulate matter that exists between the planets. Sources of this dust are larger meteoroids, comets, asteroids, the planets, and their satellites and rings; there is interstellar dust sweeping through the solar system. These cosmic dust particles are also often called micrometeoroids and range in size from assemblages of a few molecules to tenth- millimeter-sized grains, above which size they are called meteoroids. Because of their small sizes, forces additional to solar and planetary gravity affect their trajectories. Radiation pressure and the interactions with ubiquitous magnetic fields disperse dust particles in space away from their sources. In this way, micrometeoroids become messengers of their parent bodies in distant regions of the solar system. Because of their small sizes, a tablespoon of finely dispersed micrometer-sized dust grains scatter about 10 million times more light than a single **meteoroid** of the same mass. Therefore, a tiny amount of dust becomes recognizable, while the parent body from which it derived may remain undetected.

1. Introduction

One of the earliest known phenomena caused by solar system dust is the zodiacal light. **Zodiacal light** is a prominent light phenomenon that is visible to the human eye in the morning and evening sky in nonpolluted areas (Fig. 1). Already in 1683, Giovanni Domenico Cassini presented the correct explanation of this phenomenon: It is sunlight scattered by dust particles orbiting the Sun. The relation to other "dusty" interplanetary phenomena, like comets, was soon suspected. Comets shed large amounts of dust, visible as dust tails, during their passage through the inner solar system. The genetic relation between meteors and comets was already known in the 19th century. Meteoroids became the link between interplanetary dust and the larger objects: meteorites, asteroids, and comets.

Cosmic dust can have different appearances in different regions of the solar system. It consists not only of **refractory** rocky or metallic material as in stony and iron meteorites, but also of **carbonaceous material**; dust in the outer solar system can even be ice particles.

Individual dust particles in interplanetary space have much shorter lifetimes than the age of the solar system. Several dynamic effects disperse the material in space and in size (generally going from bigger to smaller particles). Therefore, interplanetary dust must have contemporary sources, namely, bigger objects like meteoroids, comets, and asteroids in interplanetary space but also planetary satellites and rings. In addition there are dust particles immersed in the local interstellar cloud through which the solar system currently passes that penetrate the planetary system.

Dust is often a synonym for dirt, which is annoying and difficult to quantify. This is also true for interplanetary dust. Astronomers who want to observe extra–solar system objects have to fight separating the foreground scattered light from the zodiacal light. Theoreticians who want to model

FIGURE 1   A wedge of interplanetary dust. The dusk twilight sky (pink) toward the northwest shows zodiacal light (blue), framed by the Pleiades (upper left), Comet Hale–Bopp (upper right), and Mercury in Aries (left of center above horizon). (Courtesy M. Fulle.)

interplanetary dust have the difficulty of representing these particles by simplified models, for example, a spherical particle of uniform composition and optical properties of a pure material. True interplanetary dust particles can be very different from these simple models (Fig. 2).

Another practical aspect of dust is its danger to technical systems. A serious concern of the first spaceflights was the hazard from meteoroid impacts. Among the first instruments flown in space were simple dust detectors, many of which were unreliable devices that responded not only to impacts but also to mechanical, thermal, or electrical interference. A dust belt around Earth was initially suggested, which was dismissed only years later when instruments had developed enough to suppress this noise by several orders of magnitude. Modern dust detectors are able to reliably measure dust impact rates from a single impact per month up to a thousand impacts per second.

In the early days of spaceflight, measures were taken to protect spacecraft against the heavy bombardment by meteoroids. The bumper shield concept found its ultimate verification in the European Space Agency's *Giotto* mission to comet Halley. This spacecraft was designed to survive impacts of particles of up to 1 g mass at an impact speed of 70 km/s. These grains carry energies comparable to cannon balls that are 1000 times more massive. Heavy metal armor was not possible because spacecraft are notoriously

lightweight. The *Giotto* bumper shield combined a 1-mm-thick aluminum sheet positioned 23 cm in front of a 7-cm-thick lightweight composite rear shield. A dust particle that struck the thin front sheet was completely vaporized. The vapor cloud then expanded into the empty space between the two sheets and struck the rear shield, where its energy was absorbed by being distributed over a large area. In this way, the 2.7-m^2 front surface of the spacecraft was effectively protected by armor that weighed only 50 kg.

Only recently has the dust hazard become important again, because of man-made **space debris** in Earth orbit. Each piece of equipment carried into space becomes, after disruption by an explosion due to malfunctioning batteries or fuel systems or by an impact, the source of small projectiles, which endanger other satellites. Some estimates indicate that, in 50 years, the continuous increase in man-made space activity will lead to a runaway effect that will make the near-Earth space environment unhabitable to humans and equipment.

However, we are not concerned with this aspect of interplanetary dust; rather, the topic of this chapter is interplanetary dust as an exciting object of astrophysical research. Through its wide distribution over the solar system, cosmic dust can tell stories about its parents (comets, asteroids, even interstellar matter) that otherwise are not easily accessible. This view, however, requires that dust particles be traced back to their origins. To do this, we must understand their dynamics. Dust particles not only follow the gravitational pull of the Sun and the planets but also feel the interplanetary magnetic field and the electromagnetic radiation that fills the solar system. In addition, they interact with the solar wind and with other dust particles that they encounter in space, generally at high speeds. These collisions lead to erosion or to disruption of both particles, thus generating many smaller particles. The dynamics of interplanetary dust cannot be described solely in terms of position and velocity; their size or mass must also be considered.

2. Observations

Different methods are available to study cosmic dust (Fig. 3). They are distinguished by the size or mass range of particles that can be studied. The earliest methods were ground-based zodiacal light and **meteor** observations. Fifty years ago, radar observations of meteor trails became available. With the onset of spaceflight, in situ detection by space instrumentation provided new information on small interplanetary dust particles. Among the first reliable instruments were simple penetration detectors; modern impact ionization detectors allow not only the detection but also the chemical analysis of micrometeoroids. Deep space probes have identified micrometeoroids in interplanetary space from 0.3 to 18 AU from the Sun. Natural (e.g., lunar samples) and artificial surfaces exposed to **micrometeoroid**

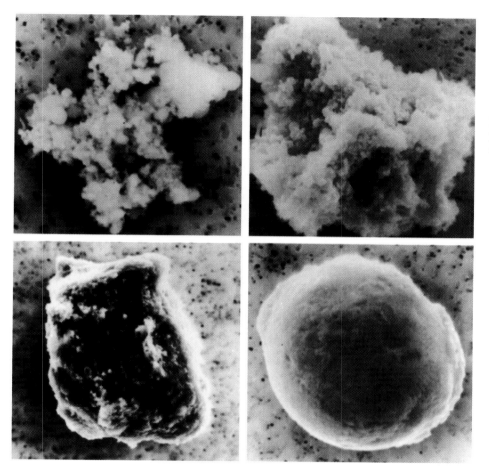

FIGURE 2   Interplanetary dust particles collected in the stratosphere by NASA's cosmic dust program. Three grains are of chondritic composition and of various degrees of compactness, and there is one Fe–S–Ni sphere (lower right). The widths of the photographs are 15 µm (first and third photos, clockwise from upper left) and 30 µm (second and fourth photos). (Courtesy of NASA.)

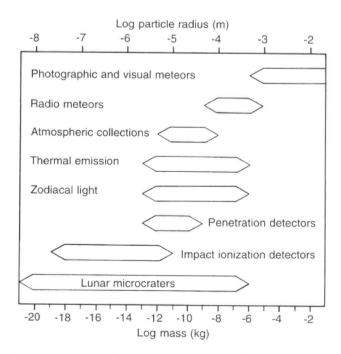

FIGURE 3   Comparison of meteoroid sizes and masses covered by different observational methods.

impacts have been returned from space and analyzed. High-flying aircraft have collected from the stratosphere dust that was identified as extraterrestrial material and that was analyzed by the most advanced microanalytic tools. Modern space-based infrared observatories now allow the observation of the thermal emission from interplanetary dust in the outer solar system.

2.1 Meteors

Looking up at the clear night sky, one can record about 10 faint meteors (or shooting stars in colloquial language) per hour. Once in a while, a brighter streak or trail of light or "fireball" will appear. Around the year 1800, the extraterrestrial nature of meteors was established when triangulation was used to deduce their height and speed. This technique is still used in modern meteor research by employing specifically equipped cameras and telescopes. About 50 years ago, radar techniques were also developed to observe faint meteor trains even during daylight.

Visible meteors result when centimeter-sized meteoroids enter the Earth's atmosphere at a speed greater than 10 km/s. At this speed, the energy of motion, which is

TABLE 1   Major Meteor Showers, Date of Shower Maximum, Radiant in Celestial Coordinates, Geocentric Speed (km/s), Maximum Hourly Rate, Parent Objects[a]

Name	Date	Radiant[b]		Speed	Rate	Parent Object[c]
		RA	DEC			
Quadrantids	Jan. 3	230	+49	42	140	
April Lyrids	Apr. 22	271	+34	48	10	Comet1861 I Thatcher
Eta Aquarids	May 3	336	−2	66	30	P/Halley
June Lyrids	June 16	278	+35	31	10	
S. Delta Aquarids	July 29	333	−17	41	30	
Alpha Capricornids	July 30	307	−10	23	30	P/Honda-Mrkos-Pajdusakova
S. Iota Aquarids	Aug. 5	333	−15	34	15	
N. Delta Aquarids	Aug. 12	339	−5	42	20	
Perseids	Aug. 12	46	+57	59	400 (1993)	P/Swift-Tuttle
N. Iota Aquarids	Aug. 20	327	−6	31	15	
Aurigids	Sept. 1	84	+42	66	30	Comet1911 II Kiess
Giacobinids	Oct. 9	262	+54	20	10	P/Giacobini-Zinner
Orionids	Oct. 21	95	+16	66	30	P/Halley
Taurids	Nov. 3	51	+14	27	10	P/Encke
Taurids	Nov. 13	58	+22	29	10	P/Encke
Leonids	Nov. 17	152	+22	71	3,000 (1966)	P/Tempel-Tuttle
Geminids	Dec. 14	112	+33	34	70	Phaeton
Ursids	Dec. 22	217	+76	33	20	P/Tuttle

[a] After A. F. Cook (1973), In "Evolutionary and Physical Properties of Meteoroids" (C. L. Hemenway, P. M. Millman, and A. F. Cook, eds.), NASA SP-319, 183–191.

[b] RA, right ascension, and DEC, declination, in degrees.

[c] If known, short-period comets are indicated by P/.

converted to heat, is sufficient to totally vaporize the meteoroid. During the deceleration of the meteoroid in the atmosphere at about 100 km altitude, the meteoroid will heat up and atoms from its outer surface will be ablated until it is completely evaporated. A luminous train several kilometers in length follows the meteoroid. It is this ionized and luminous atmospheric gas and material from the meteoroid that is visible and that scatters radar signals. From triangulation of the meteor train by ground stations (several cameras or a radar station), the preatmospheric meteoroid orbit is obtained with high accuracy.

During the atmospheric entry of objects larger than several tens of kilograms or about 10 cm in diameter, a surface layer of several centimeters in thickness will burn away, and the object will be decelerated. That which reaches Earth's surface is called a **meteorite**. Meteorites of 1 kg to several tons are sufficiently decelerated and fall on Earth with the interior little altered by atmospheric entry. These meteorites are the source of our earliest knowledge about extraterrestrial material. [See METEORITES; NEAR-EARTH OBJECTS; PLANETARY IMPACTS.]

Much of the ablated material from a meteor will condense again into small droplets, which will cool down and form cosmic spherules that subsequently rain down to Earth. These cosmic spherules can be found and identified in abundance in deep-sea sediments and on the large ice masses of Greenland, the Arctic, and Antarctica. An average of 40 tons of extraterrestrial material per day in the form of fine dust falls onto the surface of Earth.

At certain times, meteor showers can be observed at a rate that is a hundred (and more) times higher than the average sporadic meteor rate (Table 1). Figure 4 shows several meteors in a photograph of the night sky taken on 17 November 1966. The visible rate was about one meteor per second. Because all of these meteoroids travel on parallel trajectories, to an observer they seem to arrive from a common point in the sky (the radiant), which in this case lies in the constellation Leo. Therefore, this meteor shower is called the Leonid shower.

The explanation for the yearly occurrence of meteor showers is that all meteoroids in one stream closely follow a common elliptic orbit around the Sun but are spread out all along the orbit. Each year when the Earth crosses this orbit on the same day, some meteoroids of the stream hit the atmosphere and cause the shower.

Many meteor streams have orbits similar to those of known comets (cf. Table 1). It is a generally accepted view that meteor streams are derived from comets. Millimeter-to centimeter-sized particles that are emitted from comets at low speeds (m/s) are not visible in the normal comet tail

FIGURE 4 An unusually strong meteor shower (Leonid) was observed on 17 November 1966. The meteor trails seem to radiate from the constellation Leo.

but form so-called comet trails along a short segment of the comet's orbit. Their different speeds will slowly spread the particles out over the full orbit. Infrared observations by the *Infrared Astronomical Satellite (IRAS)* identified many such trails connected to short-period comets. Gravitational interactions with planets and collisions with other cosmic dust particles will scatter meteoroids out of the stream, and they will become part of the sporadic background cloud of meteoroids. The fact that some meteor showers display strong variations of their intensities indicates that they are young streams that are still concentrated in a small segment of the parent's orbit. The parent comet of the Leonids, the periodic Comet Tempel–Tuttle, has the same periodicity of 33.3 years. The parent object of one of the strongest yearly meteor showers, the Geminids, is 3200 Phaeton, which had been previously classified as an asteroid because it shows no cometary activity. However, its association with a meteor stream indicates that it is an inactive, dead comet that at some time in the past emitted large quantities of meteoroids. [*See* Physics and Chemistry of Comets; Cometary Dynamics; Near-Earth Objects.]

Fewer than one out of ten thousand radar meteors has been identified to be caused by interstellar meteoroids that pass through the solar system on a **hyperbolic orbit**. Their heliocentric speed is significantly higher than the solar system escape speed, confirming that they are of truly interstellar origin. The radius of these interstellar meteoroids is about 20 μm. These particles have been found to arrive generally from southern ecliptic latitudes with enhanced fluxes from discrete sources.

2.2 Interplanetary Dust Particles

There is another "window" through which extraterrestrial material reaches the surface in a more or less undisturbed state. Small **interplanetary dust particles (IDPs)** of a few to 50 μm in diameter are decelerated in the tenuous atmosphere above 100 km. At this height, the deceleration is so gentle that the grains will not reach the temperature of substantial evaporation ($T \sim 800°C$), especially, since these small particles have a high surface area-to-mass ratio that enables them to effectively radiate away excessive heat. These dust particles subsequently sediment through the atmosphere and become accessible to collection and scientific examination. The abbreviation IDP (or "Brownlee particle" after Don Brownlee, who first reliably identified their extraterrestrial nature) is often used for such extraterrestrial particles that are collected in Earth's atmosphere.

Early attempts to collect IDPs by rockets above about 60 km were not successful because of the very low influx of micrometeoroids into the atmosphere and the short residence times of IDPs at these altitudes. More successful were airplane collections in the stratosphere at or above 20 km. At this height, the concentration of 10-μm-diameter particles is about 10^6 times higher than in space and terrestrial contamination of this sized particles is still low. Only micrometer- and submicrometer-sized terrestrial particles (e.g., from volcanic eruptions) can reach these altitudes in significant amounts. Another type of interference is caused by man-made contamination: About 90% of all collected particles in the 3- to 8-μm size range are aluminum oxide spheres, which are products of solid rocket fuel exhaust. Because of this overwhelming contamination problem for small particles, the lower size limit of IDPs collected by airplanes is a few micrometers in diameter.

Since 1981, IDP collection by airplanes has been routinely performed by NASA using high-flying aircraft, which can cruise at 20 km for many hours. On its wings it carries dust collectors that sweep huge amounts of air. Dust particles stick to the collector surface that is coated with silicone oil. After several hours of exposure, the collector is retracted into a sealed storage container and returned to the laboratory. There, all particles are removed from the collector plate, the silicone oil is washed off, and the particles are preliminarily examined and catalogued. Individual IDPs can be ordered for further scientific investigation. A wide variety of microanalytic tools is used to examine and analyze IDPs. Scanning electron microscopes (SEM) can image atomic lattice layer structures. Focused ion beams in combination with a SEM are used for sample preparation and secondary ion mass spectrometers (SIMS) can measure the distribution of individual elements and isotopes at submicrometer resolution, deriving important information on the mineralogy of the samples.

According to their elemental composition, IDPs come in three major types: chondritic, 60% (cf. Table 2);

TABLE 2	Average Elemental Composition (All Major and Selected Minor and Trace Elements) of Several Chondritic IDPs Is Compared with C1 Chondrite Composition[a]			
Element	C1	IDP	Variation	T_c
Mg	1,071,000	0.9	0.6–1.1	1067
Si	1,000,000	1.2	0.8–1.7	1311
Fe	900,000	1	1	1336
S	515,000	0.8	0.6–1.1	648
Al	84,900	1.4	0.8–2.3	1650
Ca	61,100	0.4	0.3–0.6	1518
Ni	49,300	1.3	1.0–1.7	1354
Cr	13,500	1.1	0.9–1.4	1277
Mn	9,550	1.1	0.8–1.6	1190
Cl	5,240	3.6	2.8–4.6	863
K	3,770	2.2	2.0–2.5	1000
Ti	2,400	1.5	1.3–1.7	1549
Co	2,250	1.9	1.2–2.9	1351
Zn	1,260	1.4	1.1–1.8	660
Cu	522	2.8	1.9–4.2	1037
Ge	119	2.3	1.6–3.4	825
Se	62	2.2	1.6–3.0	684
Ga	38	2.9	2.1–3.9	918
Br	12	34	23–50	690

[a] The IDP abundances are normalized to iron (Fe) and to C1. C1 abundance is normalized to Si = 1,000,000 condensation temperatures T_c (°C). From E. K. Jessberger et al. (1992), *Earth Planet. Sci. Lett.* **112**, 91.

iron–sulfur–nickel, 30%; and mafic silicates (iron–magnesium–rich silicates, i.e., olivine and pyroxene), 10%. Most chondritic IDPs are porous aggregates, but some smooth chondritic particles are found as well. Chondritic aggregates may contain varying amounts of carbonaceous material of unspecified composition. Table 2 shows a significant enrichment in volatile (low condensation temperature) elements when compared to C1 chondrites. This observation is being used to support the argument that these particles consist of some very primitive solar system material that had never seen temperatures above about 500°C, as is the case for some cometary material. This and compositional similarity with comets argue for a genetic relation between comets and IDPs.

A remarkable feature of IDPs is their large variability in isotopic composition. Extreme isotopic anomalies have been found in some IDPs. Under typical solar system conditions, only fractions of a percent of isotopic variations can occur. These huge isotopic variations indicate that some grains are not homogenized with other solar system material but have preserved much of their presolar character. Submicrometer-sized grains known as GEMS (glass with embedded metal and sulfides) are major constituents of the chondritic porous class of IDPs. Several GEMS with nonsolar oxygen isotopic compositions were identified, confirming that at least some are indeed presolar grains. These amorphous interstellar silicates are considered one of the fundamental building blocks of the solar system.

2.3 Zodiacal Light

The wedge-shaped appearance of the zodiacal light (see Fig. 1) demonstrates its concentration in the **ecliptic plane**. For an observer on Earth, the zodiacal light extends in the ecliptic all the way around to the antisolar direction, however, at strongly reduced intensities. In the direction opposite to the Sun, this light forms a hazy area of a few degrees in dimension known as the gegenschein, or counterglow. If seen from outside the solar system, the zodiacal dust cloud would have a flattened, lenticular shape that extends along the ecliptic plane about seven times farther from the Sun than perpendicular to the ecliptic plane.

The brightness of zodiacal light is the result of light scattered by a huge number of particles in the direction of observation. The observed zodiacal brightness is a mean value, averaged over all sizes, compositions, and structures of particles along the line of sight. Zodiacal light brightness can be traced clearly into the solar corona. However, most of this dust is foreground dust close to the observer because of a favorable scattering function. Nevertheless, the vicinity of the Sun is of considerable interest for zodiacal light measurements because it is expected that close to the Sun the temperature of the dust rises, and the dust particle starts to sublimate, first the more volatile components and closer to the Sun even the refractory ones. Inside about four solar radii distance, dust should completely sublimate. Some observers have found a sharp edge of a dust-free zone at four solar radii; others have not seen a sharp edge. Perhaps the inner edge of the zodiacal cloud may change in time.

The large-scale distribution of the zodiacal dust cloud is obtained from zodiacal light measurements onboard interplanetary spacecraft spanning a distance ranging from 0.3 to approx. 3 AU from the Sun. Even though the intensity decreases over this distance by a factor 150, the spatial density of dust needs only to decrease by a factor 15. The radial dependence of the number density is slightly steeper than an inverse distance dependence. From zodiacal light measurements, a slight inclination of about 3° of the symmetry plane of zodiacal light with respect to the ecliptic plane has been determined.

At visible wavelengths, the spectrum of the zodiacal light closely follows the spectrum of the Sun. A slight reddening (i.e., the ratio of red and blue intensities is larger for zodiacal light than for the Sun) indicates that the majority of particles are larger than the mean visible wavelength of 0.54 μm. In fact, most of the zodiacal light is scattered by 10- to 100-μm-sized particles. Therefore, the dust seen by zodiacal light is only a subset of the interplanetary dust cloud. Submicrometer- and micrometer-sized particles, as well as millimeter and bigger particles, are not well represented by the zodiacal light at optical wavelengths.

Above about 1 μm in wavelength, the intensities in the solar spectrum rapidly decrease. The zodiacal light spectrum follows this decrease until about 5 μm, above which the thermal emission of the dust particles prevails. Because of the low albedo (fraction of incident sunlight reflected back and scattered in all directions is smaller than 10%) of interplanetary dust particles, most visible radiation (>90%) is absorbed and emitted at infrared wavelengths. The maximum of the thermal infrared emission from the zodiacal dust cloud lies between 10 and 20 μm. From the thermal emission observed by the *IRAS* and Cosmic Background Explorer (COBE) satellites, an average dust temperature at 1 AU distance from the Sun between 0°C and 20°C has been derived. Some spatial structure has been observed at thermal infrared wavelengths. Asteroid bands mark several asteroids families as significant sources of solar system dust just as comet trails identify dust emitted from individual comets.

Optical and infrared observations of other extraterrestrial dusty phenomena have also provided important insights into the zodiacal complex. Cometary and asteroidal dust is considered to be an important source of the zodiacal cloud. The study of circumplanetary dust and rings has stimulated much research in the dynamics of dust clouds. Interstellar dust is believed to be the ultimate source of all refractory material in the solar system. Circumstellar dust clouds like the one around β-Pictoris are "zodiacal clouds" of their own right. The study of which may eventually give information on extra solar planetary systems. [*See* INFRARED VIEWS OF THE SOLAR SYSTEM FROM SPACE; PLANETARY RINGS; EXTRA-SOLAR PLANETS.]

FIGURE 5 Microcraters on the glassy surface of a lunar sample. Bright spallation zones surround circular central pits.

2.4 Lunar Microcraters and the Near–Earth Dust Environment

The size distribution of interplanetary dust particles is represented by the lunar microcrater record. Microcraters on lunar rocks have been found ranging from 0.02 μm to millimeters in diameter (Fig. 5). Laboratory simulations of high-velocity impacts on lunar-like materials have been performed to calibrate crater sizes with projectile sizes and impact speeds. Submicrometer- to centimeter-sized projectiles have been used with speeds above several kilometers per second. The typical impact speed of interplanetary meteoroids on the Moon is about 20 km/s. For the low-mass particles, electrostatic dust accelerators that reach projectile speeds of up to 100 km/s were used. The high-mass projectiles were accelerated with light-gas guns, which reached speeds up to about 10 km/s. For the intermediate mass range, plasma drag accelerators reached impact speeds of 20 km/s. The crater diameter to projectile diameter ratio varies from 2 for the smallest microcrater to about 10 for centimeter-sized projectiles.

The difficulty in deriving the impact rate from a crater count on the Moon is that the degree to which rocks shield other rocks and thus the exposure time of any surface is generally unknown. Therefore, the crater size or meteoroid distribution has to be normalized with the help of an impact rate or meteoroid flux measurement obtained by other means. In situ detectors or recent analyses of impact plates that were exposed on NASA's *Long Duration Exposure Facility* to the meteoroid flux for several years provided this flux calibration (Fig. 6). Flux of the smallest particles dominates, and the mass flux of meteoroids peaks at 10^{-5} g. The total mass density of interplanetary dust at 1 AU is 10^{-16} g/m^3 and the total mass of the zodiacal cloud inside Earth's orbit is between 10^{16} and 10^{17} kg, which corresponds

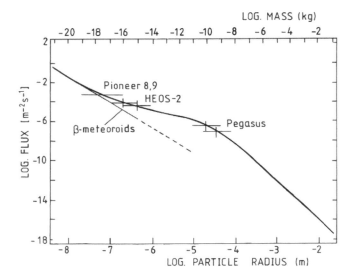

FIGURE 6 Cumulative flux of interplanetary meteoroids on a spinning flat plate at 1 AU distance from the Sun. The solid line has been derived from lunar microcrater statistics, and it is compared with satellite and spaceprobe measurements.

to the mass of a single object (comet or asteroid) of about 20 km in diameter.

In low Earth orbit the meteoroid flux is about a factor of two higher than in deep space because of the Earth's gravitational concentration. However, micrometer-sized natural meteoroids are outnumbered (by a factor of three) by man-made space debris. Craters produced by space debris particles are identified by chemical analyses of residues in the craters. Residues have been found from space materials and signs of human activities in space, such as paint flakes, plastics, aluminum, titanium, and human excretion.

2.5 In Situ Dust Measurements

Complementary to ground-based and astronomical dust observations are in situ observations by dust impact detectors on board interplanetary spacecraft. In situ measurements have been performed in interplanetary space between 0.3 and 18 AU heliocentric distance (Table 3).

Two types of impact detectors were mainly used for interplanetary dust measurements: penetration detectors and impact ionization detectors. Penetration detectors record the mechanical destruction from a dust particle's impact, for example, a 25- or 50-μm-thick steel film has a detection threshold of 10^{-9} or 10^{-8} g (approx. 10 or 20 μm radius) at a typical impact speed of 20 km/s. At lower impact speeds the minimum detectable particle mass is bigger and vice versa. A more sensitive penetration detector is the PVDF (PolyVinylidine Fluoride) film. PVDF is a polarized material (i.e., all dipolar molecules in the material are aligned so that they are pointing in the same direction). When a dust particle impacts the film, it excavates some polarized material. This depolarization generates an electric signal, which is then detected. The pulse height of the signal is a function of the mass and speed of the dust particle. A typical measurement range is from 10^{-13} to 10^{-9} g (1–10 μm radius).

The most sensitive dust detectors are impact ionization detectors. Figure 7 shows a photo of the dust detector flown on the *Cassini* spacecraft. The detector has an aperture of 0.1 m² and is based on the impact ionization effect: A dust particle that enters the detector and hits the hemispherical target in the back at speeds above 1 km/s will produce an impact crater and part or all of the projectile's material will vaporize. Because of the high temperature at the impact site some electrons are stripped off atoms and molecules and generate a vapor that is partially ionized. These ions and electrons are separated in an electric field within the detector and collected by electrodes. Coincident electric

TABLE 3	In Situ Dust Detectors in Interplanetary Space: Distance of Operation, Mass Sensitivity, and Sensitive Area.			
Spacecraft	**Year of Launch**	**Distances (AU)**	**Mass Threshold (g)**	**Area (m²)**
Pioneer 8	1967	0.97–1.09	2×10^{-13}	0.0094
Pioneer 9	1968	0.75–0.99	2×10^{-13}	0.0074
HEOS 2	1972	1	2×10^{-16}	0.01
Pioneer 10	1972	1–18	2×10^{-9}	0.26
Pioneer 11	1973	1–10	10^{-8}	0.26
Helios 1/2	1974/76	0.3–1	10^{-14}	0.012
Galileo	1989	0.7–5.3	10^{-15}	0.1
Hiten	1990	1	10^{-15}	0.01
Ulysses	1990	1–5.4	10^{-15}	0.1
Cassini	1997	0.7–10	2×10^{-16}	0.1
Nozomi	1998	1–1.5	10^{-15}	0.01

FIGURE 7   The *Cassini* cosmic dust analyzer consists of two types of dust detectors—the high rate detector (HRD) and the dust analyzer (DA). The cylindrical DA (upper center) has a diameter of 43 cm. The bottom of the sensor contains the hemispherical impact target; in the center are charge-collecting electrodes and the multiplier for measurement of the mass impact spectrum. Two entrance grids sense the electric charge of incoming dust grains. The detector records impacts of submicrometer- and micrometer-sized dust particles above an impact speed of 1 km/s. HRD consists of two circular film detectors that record impacts of micrometer-sized dust particles at a rate of 10,000 per second. The detectors are carried by the electronics box that is mounted on top of a turntable bolted to the spacecraft.

pulses on these electrodes signal the impact of a high-velocity dust particle. The strength and the wave form of the signal are measures of the mass and speed of the impacting particle. The small central part of the *Cassini* detector is a time-of-flight mass spectrometer: A high electric field between the target and a grid 3 mm in front of the target accelerates the ions at high speed. During the flight between the grid and the ion collector, ions of different masses separate and arrive at different times at the multiplier. The lightest ions arrive first and the heavier ones appear later. In this way, a mass spectrum that represents the elementary composition of the dust grain is measured. Entrance grids in front of the target pick up any electric charge of dust particles. Measurements of the electric charge on interplanetary dust particles have been accomplished for the first time by the *Cassini* detector. Dust detectors incorporating a mass spectrometer have been flown on the *Helios* spacecraft, the *Giotto* and *VEGA* missions to Comet Halley, the *Stardust* mission to Comet Wild 2, and the *Cassini* mission to Saturn. Electrostatic dust accelerators are used to calibrate these detectors with micrometer- and submicrometer-sized projectiles at impact speeds of up to about 100 km/s.

2.5.1 INTERPLANETARY DUST

The radial profile of the dust flux in the inner solar system between 1 and 0.3 AU from the Sun has been determined by the *Helios 1* and *2* space probes. Three dynamically different interplanetary dust populations have been identified in the inner solar system. First, particles that orbit the Sun in low-eccentricity orbits had already been detected by the *Pioneer 8/9* and *HEOS 2* dust experiments. They relate to particles originating in the asteroid belt and spiraling under the **Poynting–Robertson** effect toward the Sun. Second, there are particles on highly eccentric orbits that have, in addition, large semimajor axes and that derive from short-period comets. Third, the *Pioneer 8/9* dust experiments detected a significant flux of small particles, which were called β-meteoroids, from approximately the solar direction. Existence of these particles was recently confirmed by measurements with the Japanese *Hiten* satellite.

Recently, the *Galileo* and *Ulysses* spacecraft carried dust detectors through interplanetary space between the orbits of Venus and Jupiter and above the ecliptic plane. Swing-bys of Venus and Earth (two times) were necessary to give the heavy *Galileo* spacecraft (mass of 2700 kg) the necessary boost to bring it to Jupiter within 6 years of flight time, where it became the first man-made satellite of this giant planet. The *Ulysses* spacecraft, being much lighter (mass of 375 kg), made the trajectory to Jupiter within 1.5 years. In a swing-by of Jupiter, the *Ulysses* spacecraft was brought into an orbit almost perpendicular to the ecliptic plane that carried it under the South Pole, through the ecliptic plane, and over the North Pole of the Sun.

Interplanetary dust measurements were obtained by the *Galileo* spacecraft in the ecliptic plane between Venus's orbit and the Asteroid Belt. The dust impact rate was generally higher closer to the Sun than it was farther away. After all planetary flybys, the spacecraft moved away from the Sun. At these times, the impact rate was more than an order of magnitude higher than before the flyby when the spacecraft moved toward the Sun. This observation is explained by the fact that interplanetary dust inside the asteroid belt orbits the Sun on low-inclination (<30°) and in low-eccentricity **bound orbits.** Thus, the detector that looks away from the Sun all the time, detects more dust impacts when the

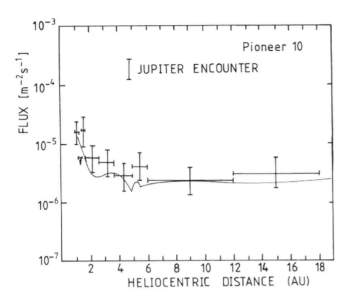

FIGURE 8   Flux of meteoroids with masses $>8 \times 10^{-10}$ kg (about 10 μm in size) in the outer solar system measured by the *Pioneer 10* penetration detector. At 18 AU from the Sun, the instrument quit operation. The measurements are in agreement with a model of constant spatial dust density in the outer planetary system. [From D. H. Humes (1980), *J. Geophys. Res.* **85**, 5841–5852.]

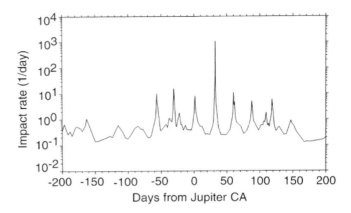

FIGURE 9   Dust impact rate observed by the *Ulysses* dust detector during the 400 days around the closest approach to Jupiter (CA, 8 February 1992). At the beginning and end of the period shown, Ulysses was 240 million km (1.6 AU) from Jupiter, while at CA the distance was only 450,000 km. Except for the flux peak at CA, when bigger particles were detected, the peaks at other times consisted of submicrometer-sized dust particles.

spacecraft moves in the same direction (outward) than in the opposite case when the spacecraft moves inward. The spatial dust density follows roughly an inverse radial distance dependence. Close passages of the asteroids Gaspra and Ida did not exhibit increased dust impact rates.

In the outer solar system, the dust detectors on board *Pioneers 10/11*, *Galileo*, *Ulysses*, and recently *Cassini* measured the flux of interplanetary dust particles. The flux of micrometer-sized particles decreased from 1 AU going outward. No sign of a flux enhancement in the Asteroid Belt was detected. Outside Jupiter's orbit, *Pioneer 10* recorded a flat flux profile (Fig. 8), which indicates a constant spatial density of micrometer-sized dust in the outer solar system. This observation has been interpreted to be due to the combined input of dust from the Kuiper Belt and comets like Halley and Schwassmann-Wachmann 1.

In interplanetary space, the highest dust fluxes have been observed near comets. So far, four comets were visited by spacecraft that carried dust detectors: Comets P/Giacobini-Zinner, P/Halley, P/Grigg-Skjellerup, and Wild 2. Specially optimized dust analyzers have been used to study Comet Halley's dust. Chemical analyses showed that, in addition to the expected dust particles consisting of silicates, a large fraction of cometary dust consists of carbonaceous materials. Extreme isotopic anomalies have been found to exist in some of these particles. Similar compositions are expected for interplanetary dust. [See PHYSICS AND CHEMISTRY OF COMETS; METEORITES.)

2.5.2 PLANETARY DUST STREAMS

Inside a distance of about 3 AU from Jupiter, both *Ulysses* and *Galileo* spacecraft detected unexpected swarms of submicrometer-sized dust particles arriving from the direction of Jupiter. Figure 9 shows the strongly time-variable dust flux observed by *Ulysses* during its flyby of Jupiter. About one month after its closest approach to Jupiter, *Ulysses* encountered the most intense dust burst at about 40 million km from Jupiter. For about 10 hours, the impact rate of submicrometer-sized particles increased by a factor 1000 above the background rate. The similarity of the impact signals and the sensor-pointing directions indicated that the particles in the burst were moving in collimated streams at speeds of several 100 km/s. Even stronger and longer lasting dust streams were observed in 1995 by the *Galileo* dust detector during its approach to Jupiter. Dust measurements inside the jovian magnetosphere showed a modulation of the small particle impact rate with a period of 10 hours, which is the rotation period of Jupiter and its magnetic field. Positively charged dust particles in the 10-nm size range coupled to the magnetic field and are thrown out of Jupiter's magnetosphere in the form of a warped dust sheet. Sources of these dust particles are the volcanoes on Jupiter's moon Io and to a smaller extent Jupiter's ring. During *Cassini*'s flyby of Jupiter, this phenomenon was also observed, and mass spectra of the particles were obtained. Both sodium chloride and sulfurous components were identified in the mass spectra, which is consistent with spectral measurements of Io's volcano-induced environment.

At Saturn, *Cassini* observed dust streams emanating from this system as well. In this case, Saturn's dense A ring and the extended E ring have been identified as sources. The ejection mechanism is very similar to that acting at Jupiter. Freshly generated nanometer-sized dust grains get charged and—if the charge is positive—thrown out by Saturn's magnetic field. In some parts of the magnetosphere, dust particles become negatively charged; these particles remain bound to the magnetic field and stay in the vicinity of Saturn. The stream particles primarily consist of silicate materials that imply that the particles are the contamination of icy ring material rather than the ice particles themselves. [*See* Planetary Rings.]

2.5.3 INTERSTELLAR DUST IN THE HELIOSPHERE

The solar system is currently passing through a region of low-density, weakly ionized interstellar material in our galaxy, which shows a larger abundance of heavy refractory elements in the gas phase such as iron, magnesium, and silicon than in cold dense interstellar clouds. Interstellar dust is part of the interstellar medium, although it has not been directly observed by astronomical means in the tenuous local interstellar cloud. Interstellar dust is formed as stardust in the cool atmospheres of giant stars and in nova and supernova explosions.

More than a decade ago, interstellar dust was positively identified inside the planetary system. At the distance of Jupiter, the dust detector on board the *Ulysses* spacecraft detected impacts predominantly from a direction that was opposite to the expected impact direction of interplanetary dust grains. The impact velocities exceeded the local solar system escape velocity, even if radiation pressure effects were considered. The motion of interstellar grains through the solar system is parallel to the flow of neutral interstellar hydrogen and helium gas, both traveling at a speed of 26 km/s with respect to the Sun. The interstellar dust flow persisted at higher latitudes above the ecliptic plane, even over the poles of the Sun, whereas interplanetary dust is strongly concentrated toward the ecliptic plane (Fig. 10).

Since that time, *Ulysses* monitored the stream of interstellar dust grains through the solar system at higher latitudes. It was found that the flux of small interstellar grains showed some variation with the period of the solar cycle, which indicates a coupling of the flux to the solar wind magnetic field. Interstellar dust has initially been identified outside 3 AU out to Jupiter's distance. However, refined analyses showed that both *Cassini* and *Galileo* recorded several hundred interstellar grains in the region between 0.7 and 3 AU from the Sun. Even in the *Helios* dust data interstellar grains were identified down to 0.3 AU distance from the Sun.

The radii of clearly identified interstellar grains range from 0.1 μm to above 1 μm with a maximum at about 0.3 μm. Even bigger interstellar particles have been reliably

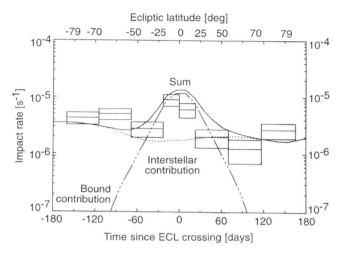

FIGURE 10 *Ulysses* dust impact rate observed around the time of its ecliptic plane crossing (ECL). ECL occurred on 12 March 1995 at a distance of 1.3 AU from the Sun. The boxes indicate the mean impact rates and their uncertainties. The top scales give the spacecraft latitude. Model calculations of the impact rate during *Ulysses'* south to north traverse through the ecliptic plane are shown by the lines. Contributions from interplanetary dust on bound orbits and interstellar dust on hyperbolic trajectories and the sum of both are displayed. From these measurements, it is concluded that interstellar dust is not depleted to a distance of 1.3 AU from the Sun.

identified by their hyperbolic speeds in radar meteor observations. The flow direction of these bigger particles varies over a much wider angular range than that of small (submicrometer-sized) grains observed by spacecraft. The deficiency of small grain masses (<0.3 μm) compared to astronomically observed interstellar dust indicates a depletion of small interstellar grains in the heliosphere.

There are significant differences in the particle sizes that were recorded at different heliocentric distances. Measurements of the interstellar particle mass distribution revealed a lack of small grains inside 3 AU heliocentric distance. Measurements by *Cassini* and *Galileo* in the distance range between 0.7 and 3 AU showed that interstellar particles were bigger than 0.5 μm with increasing masses closer to the Sun. The flux of these bigger particles did not exhibit temporal variations due to the solar wind magnetic field like the flux of smaller particles observed by *Ulysses*. The trend of increasing masses of particles continues as demonstrated by *Helios* measurements, which recorded particles of about 1 μm radius down to 0.3 AU. These facts support the idea that the interstellar dust stream is filtered by both radiation pressure and electromagnetic forces. It is concluded that interstellar particles with optical properties of grains consisting of astronomical silicates or organic refractory materials are consistent with the observed radiation pressure effect.

3. Dynamics and Evolution

3.1 Gravity and Keplerian Orbits

In the planetary system, solar gravitation determines the orbits of all bodies larger than dust particles for which other forces become important. But even for dust, gravity is an important factor. Near planets, planetary gravitation takes over. However, the basic orbital characteristics remain the same. Two types of orbits are possible: bound and unbound orbits around the central body. Circular and elliptical orbits are bound to the Sun; the planets exert only small disturbances to these orbits. Planets, asteroids, and comets move on such orbits. Objects on unbound orbits will eventually leave the solar system. Typically, interstellar dust particles move on unbound, hyperbolic orbits through the solar system. Similarly, interplanetary particles are unbound to any planetary system and traverse it on hyperbolic orbits with respect to the planet. [See SOLAR SYSTEM DYNAMICS.]

A Keplerian orbit is a conic section that is characterized by its semimajor axis a, eccentricity e, and inclination i. The Sun (or a planet) is in one focus. The **perihelion** distance (closest to the Sun) is given by $q = a(1 - e)$. Circular orbits have eccentricity $e = 0$, elliptical orbits have $0 < e < 1$, and hyperbolic orbits have $e > 1$ and a is taken negative. The **aphelion** distances (furthest from the Sun) are finite only for circular and elliptical orbits. The inclination is the angle between the orbit plane and the ecliptic (i.e., the orbit plane of Earth).

Dust particles in interplanetary space move on very different orbits, and several classes of orbits with similar characteristics have been identified. One class of meteoroids moves on orbits that are similar to those of asteroids, which peak in the Asteroid Belt. Another class of orbits that represents the majority of zodiacal light particles has a strong concentration toward the Sun. Both orbit populations have low to intermediate eccentricities $(0 < e < 0.6)$ and low inclinations $(i < 40°)$. These asteroidal and zodiacal core populations satisfactorily describe meteors, the lunar crater size distribution, and a major portion of zodiacal light observations. Also, spacecraft measurements inside 2 AU are well represented by the core population. [See MAIN-BELT ASTEROIDS.]

3.2 Radiation Pressure and the Poynting–Robertson Effect

Electromagnetic radiation from the Sun (most intensity is in the visible wavelength range at $\lambda_{max} = 0.5$ μm) being absorbed, scattered, or diffracted by any particulate exerts pressure on this particle. Because solar radiation is directed outward from the Sun, radiation pressure is also directed away from the Sun. Thus, gravitational attraction is reduced by the radiation pressure force. Both radiation pressure and gravitational forces have an inverse square

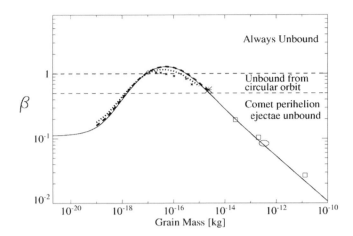

FIGURE 11 Ratio β of the radiation pressure force over solar gravity as a function of particle radius. Values are given for particles made of astronomical silicates (from Gustafson et al., 2001) in various shapes: sphere (solid curve), long cylinders (dashes), and flat plates (dots).

dependence on the distance from the Sun. Radiation pressure depends on the cross section of the particle and gravity on the mass; therefore, for the same particle, the ratio β of radiation pressure, F_R, over gravitational force, F_G, is constant everywhere in the solar system and depends only on particle properties: $\beta = F_R/F_G \sim Q_{pr}/s\rho$, where Q_{pr} is the efficiency factor for radiation pressure, s is the particle radius, and ρ is its density.

Figure 11 shows the dependence of β on the particle size for different shapes. For big particles $(s \gg \lambda_{max})$, radiation pressure force is proportional to the geometric cross section giving rise to the $1/s$-dependence of β. At particle sizes comparable to the wavelength of sunlight $s \approx \lambda_{max}$, β-values peak and decline for smaller particles as their interaction with light decreases.

A consequence of the radiation pressure force is that particles with $\beta > 1$ are not attracted by the Sun but rather are repelled by it. If such particles are generated in interplanetary space either by a collision or by release from a comet, they are expelled from the solar system on hyperbolic orbits. But even particles with β values smaller than 1 will leave the solar system on hyperbolic orbits if their speed at formation is high enough so that the reduced solar attraction can no longer keep the particle on a bound orbit. If a particle that is released from a parent body moving on a circular orbit has $\beta > 1/2$, then it will leave on a hyperbolic orbit. These particles are termed **beta-meteoroids**.

Because of the finite speed of light $(c \approx 300,000$ km/s$)$ radiation pressure does not act perfectly radial but has an aberration in the direction of motion of the particle around the Sun. Thus, a small component (approximately proportional to v/c, where v is the speed of the particle) of the radiation pressure force always acts against the orbital motion reducing its orbital energy. This effect is called Poynting–Robertson effect. As a consequence of this drag force, the particle is decelerated. This deceleration is largest at its perihelion distance where both the light pressure and the velocity peak. Consequently, the eccentricity (aphelion distance) is reduced, and the orbit is circularized. Subsequently, the particle spirals toward the Sun, where it finally sublimates.

The lifetime τ_{PR} of a particle on a circular orbit that spirals slowly to the Sun is given by $\tau_{PR} = 7 \times 10^5 \rho s r^2 / Q_{pr}$, where τ_{PR} is in years, r is given in AU, and all other quantities are in SI units. Even a centimeter-sized ($s = 0.01$ m), stony ($\rho = 3000$ kg/m^3, $Q_{pr} \approx 1$) particle requires only 21 million years to spiral to the Sun if it is not destroyed by an earlier collision. This example shows that all interplanetary dust had to be recently generated; no dust particles remain from the times of the formation of the solar system. The dust we find today had to be stored in bigger objects (asteroids and comets), which have sufficient lifetimes.

The effect of solar wind impingement on particulates is similar to radiation pressure and Poynting–Robertson effect. Although direct particle pressure can be neglected with respect to radiation pressure, solar wind drag is about 30% of Poynting–Robertson drag.

Particle orbits that evolve under Poynting–Robertson drag will eventually cross the orbits of the inner planets and, thereby, will be affected by planetary gravitation. During the orbit evolution of particles, resonances with planetary orbits may occur even if the orbit periods of the particle and the planet are not the same but form a simple integer ratio. This effect is largest for big particles, the orbits of which evolve slower and which spend more time near the resonance position. Density enhancements of interplanetary dust have been found (i.e., the Earth resonant ring was identified in *IRAS* data and later confirmed by data from the Cosmic Background Explorer satellite *COBE*).

Dust near other stars will also evolve under Poynting–Robertson effect and form a dust disk around this star. Such a disk has been found around many stars (e.g., β-Pictoris). There is an ongoing search in this disk for resonance enhancements that would indicate planets around this star [*See* EXTRA-SOLAR PLANETS.]

3.3 Collisions

Mutual high-speed ($v > 1$ km/s) collisions among dust particles lead to grain destruction and generation of fragments. By these effects, dust grains are modified or destroyed, and many new fragment particles are generated in interplanetary space. From impact studies in stony material, we know that, at a typical collision speed of 10 km/s, an impact crater is formed on the surface of the target particle if it is more than 50,000 times more massive than the projectile. This mass ratio is strongly speed- and material-dependent. A typical impact crater in brittle stony material (Fig. 5) consists of a central hemispherical pit surrounded by a shallow

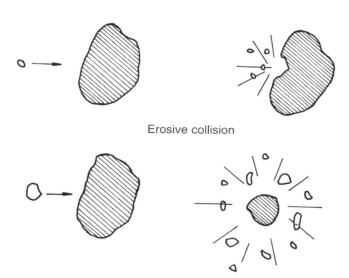

FIGURE 12 Schematics of meteoroid collisions in space. If the projectile is very small compared to the target particle, only a crater is formed in the bigger one. If the projectile exceeds a certain size limit, the bigger particle is also shattered into many fragments. The transition from one type to another is abrupt.

spallation zone. The largest ejecta particle (from the spallation zone) can be many times bigger than the projectile; however, it is emitted at a very low speed on the order of meters per second. The total mass ejected from an impact crater at an impact speed of 10 km/s is about 500 times the projectile mass.

However, if the target particle is smaller than the stated limit, the target will be catastrophically destroyed. The material of both colliding particles will be transformed into a huge number of fragment particles (Fig. 12). Thus, catastrophic collisions are a very effective process for generating small particles in interplanetary space. It has been found that interplanetary particles bigger than about 0.1 mm in diameter will be destroyed by a catastrophic collision rather than transported to the Sun by Poynting–Robertson drag.

3.4 Charging of Dust and Interaction with the Interplanetary Magnetic Field

Any meteoroid in interplanetary space will be electrically charged, and several competing charging processes determine the actual charge of a meteoroid (Fig. 13). Irradiation by solar ultraviolet (UV) light frees photoelectrons, which leave the grain. Electrons and ions are collected from the ambient solar wind plasma. Energetic ions and electrons then cause the emission of secondary electrons. Whether electrons or ions can reach or leave the grain depends on their energy and on the polarity and electrical potential of the grain. Because of the predominance of the photoelectric effect in interplanetary space, meteoroids are mostly charged positively at a potential of a few volts. Only at times

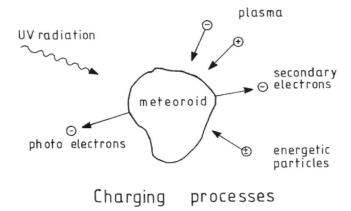

FIGURE 13   Charging processes of meteoroids in interplanetary space. UV radiation releases photoelectrons, electrons, and ions from the solar wind plasma, and they are collected; the impact of energetic particle radiation releases secondary electrons.

of very high solar wind densities does the electron flux to the particle dominate and the particle gets charged negatively. The final charging state is reached when all currents to and from the meteoroid cancel. The timescale for charging is seconds to hours depending on the size of the particle; small particles charge slower. Electric charges on dust particles in interplanetary space have been measured by the *Cassini* Cosmic Dust Analyzer. These measurements indicate a dust potential of +5 V. In the dense plasma of the inner Saturnian magnetosphere, dust particles at −2 V potential have been found.

The outward-streaming (away from the Sun) solar wind carries a magnetic field away from the Sun. Due to the rotation of the Sun (at a period of 25.7 days), magnetic field lines are drawn in a spiral, like water from a lawn sprinkler. The polarity of the magnetic field can be positive or negative depending on the polarity at the base of the field line in the solar corona, which varies spatially and temporally. For an observer or a meteoroid in interplanetary space, the magnetic field sweeps outward at the speed of the solar wind (400 to 600 km/s). [*See* THE SUN.] In the magnetic reference frame, the meteoroid moves inward at about the same speed because its orbital speed is comparatively small. The **Lorentz force** on a charged dust particle near the ecliptic plane is mostly either upward or downward depending on the polarity of the magnetic field. Near the ecliptic plane, the polarity of the magnetic field changes at periods (days to weeks) that are much faster than the orbital period of an interplanetary dust particle, and the net effect of the Lorentz force on micrometer-sized particles is small. Only secular effects on the bigger zodiacal particles are expected to occur, which could have an effect on the symmetry plane of the zodiacal cloud close to the Sun. For nanometer-sized particles, like the ones that have been found in the dust streams, the Lorentz force dominates all other forces, and

as a result the particles gyrate about the magnetic field lines and are eventually convected with the solar wind out of the solar system.

The overall polarity of the solar magnetic field changes with the solar cycle of 11 years. For one solar cycle, positive magnetic polarity prevails away from the ecliptic in the northern hemisphere and negative polarity in the southern hemisphere. Submicrometer-sized interstellar particles that enter the solar system are deflected either toward the ecliptic plane or away from it depending on the overall polarity of the magnetic field. Interstellar particles entering the heliosphere from one direction at a speed of 26 km/s need about 20 years (two solar cycles) to get close to the Sun. Therefore, trajectories of small interstellar grains (0.1 μm in radius) are strongly diverted: In some regions of space, their density is strongly increased; in others, they are depleted. At the time of the initial *Ulysses* and *Galileo* measurements (1992 to 1996), the overall solar magnetic field had changed to the unfavorable configuration; therefore, only big (micrometer-sized) interstellar particles reached the positions of *Ulysses* and *Galileo*. By 2003, the magnetic field had changed to the focusing configuration and the interstellar dust flux had recovered. [*See* THE SOLAR WIND.]

3.5 Evolution of Dust in Interplanetary Space

Forces acting on interplanetary particles are compared in Table 4. The force from solar gravity depends on the mass of the particle; therefore, it depends on the size as $F_G \sim s^3$. Radiation pressure depends on the cross section of the particle, hence $F_R \sim s^2$. The electric charge on a dust grain depends on the size directly, as does the Lorentz force $F_L \sim s$. Therefore, these latter forces become more dominating at smaller dust sizes. At a size comparable to the wavelength of visible light ($s \sim 0.5$ μm), radiation pressure is dominating gravity, and below that size the Lorentz force dominates the particles' dynamics. Though gravity is attractive to the Sun, radiation pressure is repulsive. The net effect of solar wind interactions on small particles is that they are convected out of the solar system.

Besides energy-conserving forces, there are also dissipative forces: the Poynting–Robertson effect and the ion drag from the solar wind. They cause a loss of orbital energy and force particles to slowly spiral to the Sun, where they eventually evaporate. These atoms and molecules become ionized and are flushed out of the solar system by the solar wind.

Figure 14 shows the flow of meteoritic matter through the solar system as a function of the meteoroid size. There is a constant input of mass from comets and asteroids. From the intensity enhancement of zodiacal light toward the Sun, it was deduced that, inside 1 AU, significant amounts of mass have to be injected by short-period comets into the zodiacal cloud. While comets shed their debris over a large

TABLE 4	Comparison of Various Forces Acting on Dust Particles of Size *s* Under Typical Interplanetary Conditions at 1 AU Distance from the Sun[a]				
s (μm)	F_G (N)	F_R (N)	F_L (N)	F_{PR} (N)	F_{ID} (N)
0.01	9×10^{-23}	1.4×10^{-21}	$\mathbf{1.5 \times 10^{-20}}$	1.4×10^{-25}	4×10^{-26}
0.1	9×10^{-20}	$\mathbf{1.4 \times 10^{-19}}$	$\mathbf{1.5 \times 10^{-19}}$	$\mathbf{1.4 \times 10^{-23}}$	4×10^{-24}
1	$\mathbf{9 \times 10^{-17}}$	1.4×10^{-17}	1.5×10^{-18}	1.4×10^{-21}	4×10^{-22}
10	$\mathbf{9 \times 10^{-14}}$	1.4×10^{-15}	1.5×10^{-17}	1.4×10^{-19}	4×10^{-20}
100	$\mathbf{9 \times 10^{-11}}$	1.4×10^{-13}	1.5×10^{-16}	1.4×10^{-17}	4×10^{-18}

[a] *Notes:* Dominating forces are in bold. Subscripts refer to gravity, radiation pressure, Lorentz force, Poynting–Robertson drag, and ion drag.

range of heliocentric distances but preferentially close to the Sun, asteroid debris is mostly generated in the Asteroid Belt, between 2 and 4 AU from the Sun. Collisions dominate the fate of big particles and are a constant source of smaller fragments. Meteoroids in the range of 1 to 100 μm are dragged by the Poynting–Robertson drag to the Sun. Smaller fragments are driven out of the solar system by radiation pressure and Lorentz force.

Estimates of the mass loss from the zodiacal cloud inside 1 AU give the following numbers. About 10 tons per second are lost by collisions from the big (meteor-sized) particle population. A similar amount (on the average) has to be replenished by cometary and asteroidal debris. Nine tons per second of the collisional fragments are lost as small particles to interstellar space, and the remainder of 1 ton per second is carried by the Poynting–Robertson effect toward the Sun, evaporates, and eventually becomes part of the solar wind. Interstellar dust transiting the solar system becomes increasingly important farther away from the Sun. At 3 AU from the Sun, the interstellar dust flux seems to already dominate the flux of submicrometer- and micrometer-sized interplanetary meteoroids.

4. Future Studies

New techniques will generate new insights. These techniques will include innovative observational methods, new space missions to unexplored territory, and new experimental and theoretical methods to study the processes affecting solar system dust. Questions to address are: the composition (elemental, molecular, and isotopic) and spatial distribution of interplanetary dust; the quantitative understanding of effects or processes affecting dust in interplanetary space; and the quantitative determination of the contributions from different sources (asteroids, comets, planetary environs, and interstellar dust).

Analyses of brightness measurements at infrared wavelengths up to 200 μm by the *COBE* satellite result in refined models of the distribution of dust mostly outside 1 AU. Spectrally resolved observations of asteroids, comets, and zodiacal dust by the infrared space observatories (*ISO* and *Spitzer*) show the genetic relation between these larger bodies and interplanetary dust. Improved observations of the inner zodiacal light and the edge of the dust-free zone around the Sun will provide some clues to the composition of zodiacal dust. Optical and infrared observations of

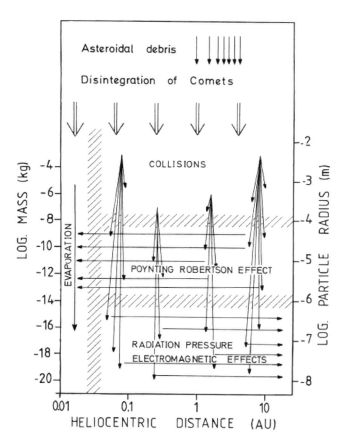

FIGURE 14   Mass flow of meteoric matter through the solar system. Most of the interplanetary dust is produced by collisions of large meteoroids, which represent a reservoir continually being replenished by disintegration of comets or asteroids. Most of it is blown out of the solar system as submicrometer-sized grains. The remainder is lost by evaporation after being driven close to the Sun by the Poynting–Robertson effect. In addition to the flow of interplanetary matter shown, there is a flow of interstellar grains through the planetary system.

extrasolar systems will bring new insights to zodiacal clouds around other stars.

Interplanetary space missions presently under way that carry dust detectors are the *Ulysses* and *Cassini* missions. *Ulysses* has probed the space above the poles of the Sun and outside 1 AU and continues its study of the interplanetary dust cloud at times of high solar activity. *Galileo* and *Cassini* had become the first man-made satellites of Jupiter and Saturn, respectively, and are studying their dust environments. The detailed study of cometary and interstellar dust is the goal of NASA's *Stardust* mission, which returned samples of dust from Comet Wild 2 in early 2006. The Japanese *Hyabusa* mission collected dust from Asteroid Itokawa and is on its return to Earth. The European Space Agency's *Rosetta* mission will follow Comet Churyumov Gerasimenko through its perihelion and investigate its release of dust to interplanetary space.

Dust particles, like photons, are born at remote sites in space and time, and carry from there information that may not be accessible to direct investigation. From knowledge of the dust particles' birthplace and the particles' bulk properties, we can learn about the remote environment out of which the particles were formed. This approach is called dust astronomy and is carried out by means of dust telescopes on dust observatories in space. Targets for dust telescopes are dust from the local interstellar medium, cometary, asteroidal dust, and space debris. Dust particles' trajectories are determined by the measurements of the electric charge signals that are induced when the charged grains fly through charge-sensitive grid systems. Modern in situ dust detectors are capable of providing mass, speed, and physical and chemical information of dust grains in space. A dust telescope can, therefore, be considered as a combination of detectors for dust particle trajectories along with detectors for physical and chemical analysis of dust particles. Both dust trajectory sensors and large-area dust analyzers have been developed recently and await their use in space.

In near-Earth space, ambitious new techniques will be applied to collect meteoritic material that is not accessible by other methods. High-speed meteoroid catchers, which permit the determination of the trajectory as well as the recovery of material for analysis in ground laboratories, are under development. A cosmic dust collector is currently being flown on the International Space Station (ISS).

Laboratory studies are instrumental in improving our understanding of planetary and interplanetary processes in which dust plays a major role. The study of dust–plasma interactions is a new and expanding field that is attracting considerable attention. New phenomena are expected to occur when plasma is loaded with large amounts of dust. Processes of this type are suspected to play a significant role in cometary environments, in planetary rings, and in protoplanetary disks.

Bibliography

Green, S. F., Williams, I. P., McDonnell, J. A. M., and McBride, N., eds. (2002). "Dust in the Solar System and Other Planetary Systems." Pergamon, Amsterdam.

Grün, E., Gustafson, B. A. S., Dermott, S., and Fechtig, H., eds. (2001). "Interplanetary Dust." Springer, Heidelberg.

Gustafson, B. A. S., Greenberg, J. M., Kolokolova, L., Xu, Y., Stognienko, R. (2001) Interactions with electromagnetic radiation: Theory and laboratory simulations. In "Interplanetary Dust" (E. Grün, B. A. S. Gustafson, S. F. Dermott, H. Fechtig, eds.), Springer, Heidelberg.

Gustafson, B. A. S., and Hanner, M. S., eds. (1996). "Physics, Chemistry, and Dynamics of Interplanetary Dust," Conference Series Vol. 104. Astronomical Society of the Pacific, San Francisco.

Leinert, C., and Grün, E. (1990) In "Physics of the Inner Heliosphere" (R. Schwenn and E. Marsch, eds.), pp. 207–275. Springer-Verlag, Berlin.

Levasseur-Regourd, A. C., and Hasegawa, H., eds. (1991). "Origin and Evolution of Interplanetary Dust." Kluwer, Dordrecht.

X-Rays in the Solar System

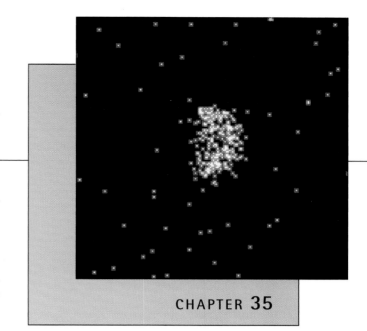

Anil Bhardwaj

Space Physics Laboratory
Vikram Sarabhai Space Centre
Trivandrum, India

Carey M. Lisse

Applied Physics Laboratory
Johns Hopkins University
Laurel, Maryland

CHAPTER **35**

1. Introduction

The usually defined range of X-ray photons spans $\sim$0.1–100 keV. Photons in the lower (<5 keV) end of this energy range are termed soft X-rays. In space, X-ray emission is generally associated with high temperature phenomena, such as hot plasmas of 1 million to 100 million K and above in stellar coronae, accretion disks, and supernova shocks. However, in the solar system, X-rays have been observed from bodies that are much colder, $T < 1000$ K. This makes the field of planetary X-rays a very interesting discipline, where X-rays are produced from a wide variety of objects under a broad range of conditions.

The first planetary X-rays detected were terrestrial X-rays, discovered in the 1950s. The first attempt to detect X-rays from the moon in 1962 failed, but it discovered the first extrasolar source, Scorpius X-1, which resulted in the birth of the field of X-ray astronomy. In the early 1970s, the *Apollo 15* and *16* missions studied fluorescently scattered X-rays from the Moon. Launch of the first X-ray satellite UHURU in 1970 marked the beginning of satellite-based X-ray astronomy. The subsequently launched X-ray observatory *Einstein* discovered, after a long search, X-rays from Jupiter in 1979. Before 1990, the three objects known to emit X-rays were Earth, Moon, and Jupiter. In 1996, *Rontgensatellit* (*ROSAT*) made an important contribution

to the field of planetary X-rays by discovering X-ray emissions from comets. This discovery revolutionized the field of solar system X-rays and highlighted the importance of solar wind charge exchange (SWCX) mechanism in the production of X-rays in the solar system, which will be discussed in this chapter in various sections.

Today the field of solar system X-rays is very dynamic and in the forefront of new research. During the last few years, our knowledge about the X-ray emission from bodies within the solar system has significantly improved. The advent of higher resolution X-ray spectroscopy with the *Chandra* and *XMM-Newton* X-ray observatories (and now the next generation *SWIFT* and *Suzaku* observatories that are coming on-line in 2005–2006) has been of great benefit in advancing the field of planetary X-ray astronomy. Several new solar system objects are now known to shine in the X-ray (Fig. 1). At Jupiter, Saturn, Venus, Mars, and Earth, nonauroral disk X-ray emissions have been observed. The first soft X-ray observation of Earth's aurora by *Chandra* shows that it is highly variable, and the Jovian aurora is a fascinating puzzle that is just beginning to yield its secrets. The nonauroral X-ray emissions from Jupiter, Saturn, and Earth, and those from disks of Mars, Venus, and the Moon are mainly produced by scattering of solar X-rays. The X-ray emission from comets, the heliosphere, the geocorona, and the Martian halo are all largely driven by charge exchange between highly charged

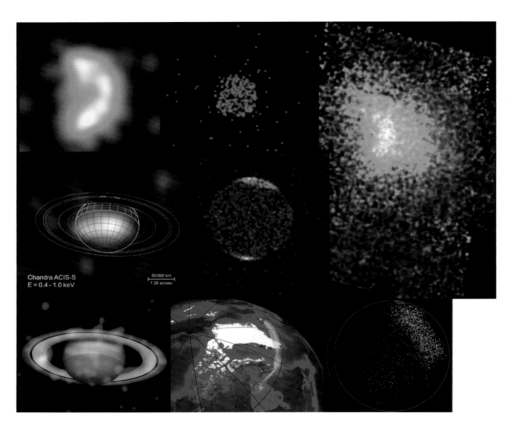

FIGURE 1 *Chandra* montage of solar system X-ray sources. Clockwise, from upper left: *Chandra* images of Venus, Mars, comet C/Ikeya–Zhang 2001, Jupiter, and Saturn. Bottom panel, left, Saturn rings, middle, Earth, and right, Moon.

highly charged minor (heavy) ions in the solar wind and gaseous neutral species in the bodies' atmosphere.

This chapter surveys the current understanding of X-ray emission from the solar system bodies. We start our survey locally, at the Earth, move to the Moon and the nearby terrestrial planets, and then venture out to the giant planets and their moons. Next, we move to the small bodies, comets and asteroids, found between the planets, and finally we study the emission from the heliosphere surrounding the whole solar system. An overview is provided on the main source mechanisms of X-ray production at each object. For further detail, readers are referred to the bibliography provided at the end of the chapter and references therein.

2. Earth

2.1 Auroral Emissions

Precipitation of energetic charged particles from the magnetosphere into Earth's auroral upper atmosphere leads to ionization, excitation, dissociation, and heating of the neutral atmospheric gas. Deceleration of precipitating particles during their interaction with atom and molecules in the atmosphere results in the production of continuous spectrum of X-ray photons, called bremsstrahlung (*bremsstrahlung* is a German word for braking radiation). The main X-ray production mechanism in the Earth's auroral zones, for energies above $\sim$3 keV, is electron bremsstrahlung; therefore, the X-ray spectrum of the aurora has been found to be very useful in studying the characteristics of energetic electron precipitation. In addition, particles precipitating into the Earth's upper atmosphere give rise to discrete atomic emission lines in the X-ray range. The characteristic inner-shell line emissions for the main species of the Earth's atmosphere are all in the low-energy range (Nitrogen Kα at 0.393 keV, Oxygen Kα at 0.524 keV, Argon Kα at 2.958 keV, and Kβ at 3.191 keV). Very few X-ray observations have been made at energies where these lines emit.

While charged particles spiral around and travel along the magnetic field lines of the Earth, the majority of the X-ray photons in Earth's aurora are directed normal to the field, with a preferential direction toward the Earth at higher energies. Downward propagating X-rays cause additional ionization and excitation in the atmosphere below the altitude where the precipitating particles have their peak energy deposition. The fraction of the X-ray emission that is moving away from the ground can be studied using satellite-based imagers (e.g, AXIS on *UARS* and PIXIE on *POLAR* spacecraft).

Auroral X-ray bremsstrahlung has been observed from balloons and rockets since the 1960s and from spacecraft since the 1970s. Because of absorption of the low-energy

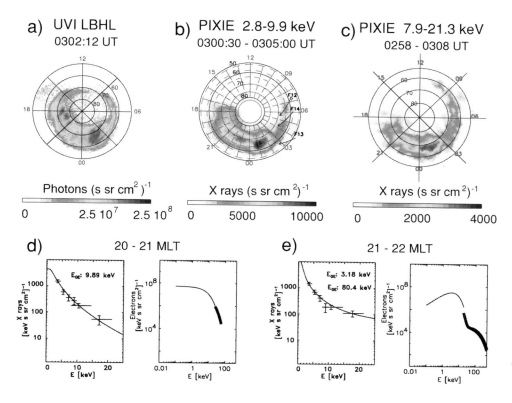

FIGURE 2 Earth's aurora. Polar satellite observation on July 31, 1987. (a) UVI and (b, c) PIXIE images in two different energy bands. (d) Left: The measured X-ray energy spectrum where an estimated X-ray spectrum produced by a single exponential electron spectrum with e-folding energy 9.89 keV is shown to be the best fit to the measurements. Right: The electron spectrum derived from UVI and PIXIE, where thin line is UVI contribution, thick line is PIXIE contribution. Both plots are averages within a box within 20–21 magnetic local time and 64°–70° magnetic latitude. (e) Same as (d) but within 21–22 MLT, where X rays produced by a double exponential electron spectrum is shown to be the best fit to the X-ray measurements. (From Østgaard et al., *JGR*, 106, 26081, 2001.)

X-rays propagating from the production altitude (∼100 km) down to balloon altitudes (35–40 km), such measurements were limited to >20 keV X-rays. Nevertheless, these early omnidirectional measurements of X-rays revealed detailed information of temporal structures from slowly varying bay events to fast pulsations and microburst.

The PIXIE instrument aboard *POLAR* is the first X-ray detector that provides true two-dimensional global X-ray image at energies >3 keV. In Fig. 2 two images taken by PIXIE in two different energy bands. The auroral X-ray zone can be clearly seen. Data from the PIXIE camera have shown that the X-ray bremsstrahlung intensity statistically peaks at midnight, is significant in the morning sector, and has a minimum in the early dusk sector. During solar substorms X-ray imaging shows that the energetic electron precipitation brightens up in the midnight sector and has a prolonged and delayed maximum in the morning sector due to the scattering of magnetic-drifting electrons and shows an evolution significantly different than viewing in the UV emissions.

During the onset/expansion phase of a typical substorm the electron energy deposition power is about 60–90 GW, which produces 10–30 MW of bremsstrahlung X-rays. By combining the results of PIXIE with the UV imager aboard *POLAR*, it has been possible to derive the energy distribution of precipitating electrons in the 0.1–100 keV range with a time resolution of about 5 min (see Fig. 2). Because these energy spectra cover the entire energy range important for

the electrodynamics of the ionosphere, important parameters like Hall and Pedersen conductivity and Joule heating can be determined on a global scale with larger certainties than parameterized models can do. Electron energy deposition estimated from global X-ray imaging also give valuable information on how the constituents of the upper atmosphere, like NO_x, is modified by energetic electron precipitation.

Limb scans of the nighttime Earth at low- to mid-latitude by the X-ray astronomy satellite *HEAO-1* in 1977, in the energy range 0.15–3 keV, showed clear evidence of the $K\alpha$ lines for nitrogen and oxygen sitting on top of the bremsstrahlung spectrum. Recently, the High-Resolution Camera (HRC-I) aboard the *Chandra* X-ray Observatory imaged the northern auroral regions of the Earth in the 0.1- to 10-keV X-ray range at 10 epochs (each ∼20 min duration) between December 2003 and April 2004. These first soft X-ray observations of Earth's aurora (see Fig. 3) showed that it is highly variable (intense arcs, multiple arcs, diffuse patches, at times absent). Also, one of the observations showed an isolated blob of emission near the expected cusp location. Modeling of the observed soft X-ray emissions suggests that it is a combination of bremsstrahlung and characteristic K-shell line emissions of nitrogen and oxygen in the atmosphere produced by electrons. In the soft X-ray energy range of 0.1–2 keV, these line emissions are ∼5 times more intense than the X-ray bremsstrahlung.

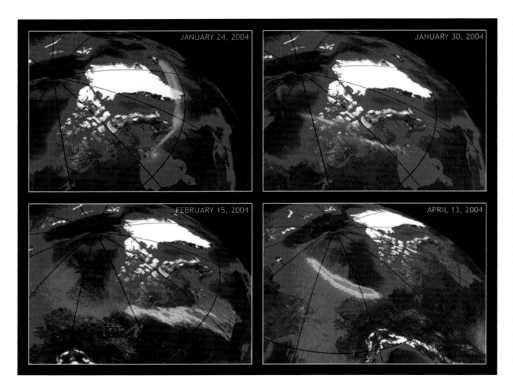

FIGURE 3 Earth's aurora. Four X-rays images (shown on the same brightness scale) of the north polar regions of Earth obtained by *Chandra* HRC-I on different days (marked at the top of each image), showing large variability in soft (0.1–10.0 keV) X-ray emissions from Earth's aurora. The bright arcs in these *Chandra* images show low-energy X-rays generated during auroral activity. The images—seen here superimposed on a simulated image of Earth—are from an approximately 20-minute scan during which *Chandra* was pointed at a fixed point in the sky while the Earth's motion carried the auroral region through the field of view. Distance from the North Pole to the black circle is 3,340 km. (From Bhardwaj et al., 2006, *J. Atmos. Sol-Terr. Phys.*, and http://chandra.harvard.edu/press/05_releases/press_122805.html.)

2.2 Nonauroral Emissions

The nonauroral X-ray background above 2 keV from the Earth is almost completely negligible except for brief periods during major solar flares. However, at energies below 2 keV, soft X-rays from the sunlit Earth's atmosphere have been observed even during quiet (nonflaring) Sun conditions. The two primary mechanisms for the production of X-rays from the sunlit atmosphere are: (1) Thomson (coherent) scattering of solar X-rays from the electrons in the atomic and molecular constituents of the atmosphere, and (2) the absorption of incident solar X-rays followed by the resonance fluorescence emission of characteristic K lines of nitrogen, oxygen, and argon. During flares, solar X-rays light up the sunlit side of the Earth by Thomson and fluorescent scattering; the X-ray brightness can be comparable to that of a moderate aurora.

Around 1994, the Compton Gamma Ray Observatory (CGRO) satellite detected a new type of X-ray source from the Earth. These are very short-lived (1 ms) X-ray and γ-ray bursts ($\sim$25 keV to 1 MeV) from the atmosphere above thunderstorms, whose occurrence is also supported by the more recent Reuvan Ramaty High Energy Solar Spectroscopic Imager (RHESSI) observations. It has been suggested that these emissions are bremsstrahlung from upward-propagating, relativistic (MeV) electrons generated in a runaway electron discharge process above thunderclouds by the transient electric field following a positive cloud-to-ground lightning event.

3. The Moon

X-Ray emissions from the Earth's nearest planetary body, the Moon, have been studied in two ways: close up from lunar orbiters (e.g., *Apollo 15* and *16*, *Clementine*, and *SMART-1*), and more distantly from Earth-orbiting X-ray astronomy telescopes (e.g., *ROSAT* and *Chandra*). Lunar X-rays result mainly from fluorescence of sunlight by the surface, in addition to a low level of scattered solar radiation and a very low level of bremsstrahlung from solar wind electrons impacting the surface. Thus, X-ray fluorescence studies provide an excellent way to determine the elemental composition of the lunar surface by remote sensing, since at X-ray wavelengths the optical properties of the surface are dominated by its elemental abundances. Elemental abundance maps produced by the X-ray spectrometers on the *Apollo 15* and *16* orbiters were limited to the equatorial regions but succeeded in finding geochemically interesting variations in the relative abundances of Al, Mg, and Si. Although the energy resolution of the *Apollo* proportional counters was low, important results were obtained, such as the enhancement of Al/Si in the lunar highlands relative to the mare. Recently, the D-CIXS instrument on *SMART-1* has obtained abundances of Al, Si, Fe, and even Ca at 50-km resolution from a 300-km altitude orbit about the Moon. Upcoming missions planned for launch in 2007–2008 by Japan (*SELENE*), India (*Chandrayaan-1*), and China (*Chang'e*) will each carry X-ray spectrometers to obtain further improved maps of the Moon's elemental composition, at ~20-to 50-km resolution from ~100- to 200-km altitude polar orbits.

Early observations from Earth orbit were made using the *ROSAT*. A marginal detection by the Advanced Satellite for Cosmology and Astrophysics (ASCA) is also reported. Figure 4a shows the *ROSAT* images of the Moon, the right image is data from a lunar occultation of the bright X-ray source GX5-1. The power of the reflected and fluoresced X-rays observed by *ROSAT* in the 0.1- to 2-keV range coming from the sunlit surface was determined to be only 73 kW. The faint but distinct lunar night side emissions (100 times less bright than the day side emissions) were until recently a matter of controversy. Earlier suggestions had the night side X-rays produced by bremsstrahlung of solar wind electrons of several hundred eV impacting the night side of the Moon on its evening (leading) hemisphere. However, this was before the GX5-1 data were acquired, which clearly show lunar night side X-rays from the early morning (trailing) hemisphere as well. A new, much better and accepted explanation is that the heavy ions in the solar wind charge exchange with geocoronal and interstellar H atoms that lie between the Earth and Moon resulting in foreground X-ray emissions between *ROSAT* and the Moon's dark side. This was confirmed by *Chandra* ACIS observations in 2001 (see Fig. 4c).

The July 2001 *Chandra* observations also provide the first remote measurements that clearly resolve discrete K-shell fluorescence lines of O, Mg, Al, and Si on the sunlit side of the Moon (see Fig. 4b). The observed O–K line photons correspond to a flux of 3.8×10^{-5} photons/s/cm²/arcmin² (3.2×10^{-14} erg/s/cm²/arcmin²). The Mg–K, Al–K, and Si–K lines each had roughly 10% as many counts and 3% as much flux as O–K line, but statistics were inadequate to draw any conclusions regarding differences in element abundance ratios between highlands and maria. More recent *Chandra* observations of the Moon used the photon counting, high spatial resolution HRC-I imager to look for albedo variations due to elemental composition differences between highlands and maria. The observed albedo contrast was noticeable, but very slight, making remote elemental mapping difficult.

4. Venus

The first X-ray observation of Venus was obtained by *Chandra* in January 2001. It was expected that Venus would be an X-ray source due to two processes: (1) charge exchange interactions between highly charged ions in the solar wind and the Venusian atmosphere and (2) scattering of solar X-rays in the Venusian atmosphere. The predicted X-ray luminosities were ~0.1–1.5 MW for the first process, and ~35 MW for the second one, with an uncertainty factor of about two. The *Chandra* observation of 2001 consisted of two parts: grating spectroscopy with LETG/ACIS-S and

direct imaging with ACIS-I. This combination yielded data of high spatial, spectral, and temporal resolution. Venus was clearly detected as a half-lit crescent, exhibiting considerable brightening on the sunward limb (Fig. 5); the LETG/ACIS-S data showed that the spectrum was dominated by O–Kα and C–Kα emission, and both instruments indicated temporal variability of the X-ray flux. An average luminosity of 55 MW was found, which agreed well with the theoretical predictions for scattered solar X-rays. In addition to the C–Kα and O–Kα emission at 0.28 and 0.53 keV, respectively, the LETG/ACIS-S spectrum also showed evidence for N–Kα emission at 0.40 keV. An additional emission line was indicated at 0.29 keV, which might be the signature of the C 1s $\rightarrow \pi^*$ transition in CO_2. The observational results are consistent with fluorescent scattering of solar X-rays by the majority species in the Venusian atmosphere, and no evidence of the 30 times weaker charge exchange interactions was found. Simulations showed that fluorescent scattering of solar X-rays is most efficient in the Venusian upper atmosphere at heights of ~120 km, where an optical depth of one is reached for incident X-rays with energy 0.2–0.9 keV.

The appearance of Venus is different in optical light and X-rays. The reason for this is that the optical light is reflected from clouds at a height of 50–70 km, while scattering of X-rays takes place at higher regions extending into the tenuous, optically thin parts of the thermosphere and exosphere. As a result, the Venusian sun-lit hemisphere appears surrounded by an almost transparent luminous shell in X-rays, and Venus looks brightest at the limb because more luminous material is there. Because X-ray brightening depends sensitively on the density and chemical composition of the Venusian atmosphere, its precise measurement will provide direct information about the atmospheric structure in the thermosphere and exosphere. This opens up the possibility of using X-ray observations for monitoring the properties of these regions that are difficult to investigate by other means, as well as their response to solar activity. In 2007, *Chandra* will reobserve Venus during its best window for 2 years, while the *MESSENGER* spacecraft, flying by on its way to Mercury, and the *Venus Express* spacecraft in Venusian orbit probe the temperature, density, pressure, and composition of the Venusian atmosphere.

5. Mars

The first X-rays from Mars were detected on 4 July 2001 with the ACIS-I detector onboard *Chandra*. In the *Chandra* observations, Mars showed up as an almost fully illuminated disk (Fig. 6). An indication of limb brightening on the sunward side, accompanied by some fading on the opposite side, was observed. The observed morphology and X-ray luminosity of ~4 MW, about 10 times less than at Venus,

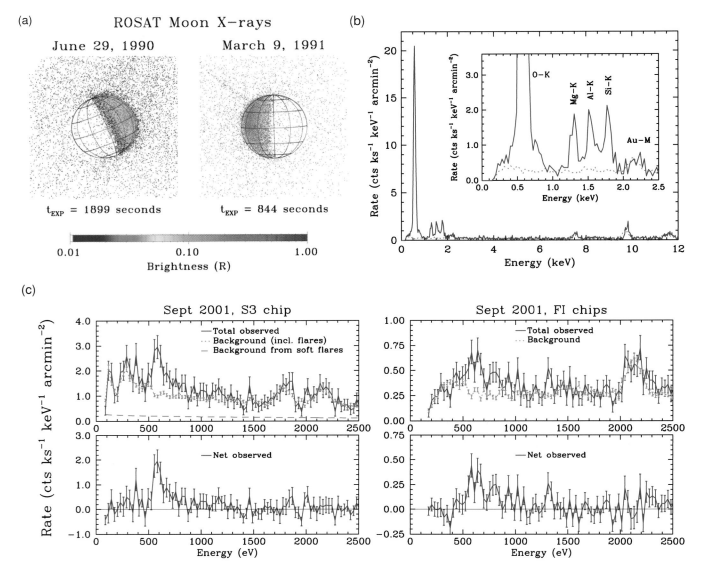

FIGURE 4   The Moon. (a) *ROSAT* soft X-ray (0.1–2 keV) images of the Moon at first (left side) and last (right side) quarter. The day side lunar emissions are thought to be primarily reflected and fluoresced sunlight, while the faint night side emissions are foreground due to charge exchange of solar wind heavy ions with H atoms in Earth's exosphere. The brightness scale in R assumes an average effective area of 100 cm² for the *ROSAT PSPC* over the lunar spectrum. [From Bhardwaj et al., 2002, ESA-SP-514, 215–226.] (b) *Chandra* spectrum of the bright side of the Moon. The green dotted curve is the detector background. K-shell fluorescence lines from O, Mg, Al, and Si are shifted up by 50 eV from their true values because of residual optical leak effects. Features at 2.2, 7.5, and 9.7 keV are intrinsic to the detector. [From Wargelin et al., 2004, *Astrophys. J.*, **607**, 596–610.] (c) Observed and background-subtracted spectra from the September 2001 *Chandra* observation of the dark side of the Moon, with 29-eV binning. Left panel is from the higher-QE but lower-resolution ACIS S3 CCD; right panel shows the higher resolution ACIS front-illuminated (FI) CCDs. Oxygen emission from charge exchange is clearly seen in both spectra, and energy resolution in the FI chips is sufficient that O Lyman α is largely resolved from O Kα. High-n H-like O Lyman lines are also apparent in the FI spectrum, along with what is likely Mg Kα around 1340 eV. (From Wargelin et al., 2004, *Astrophys. J.*, **607**, 596–610.)

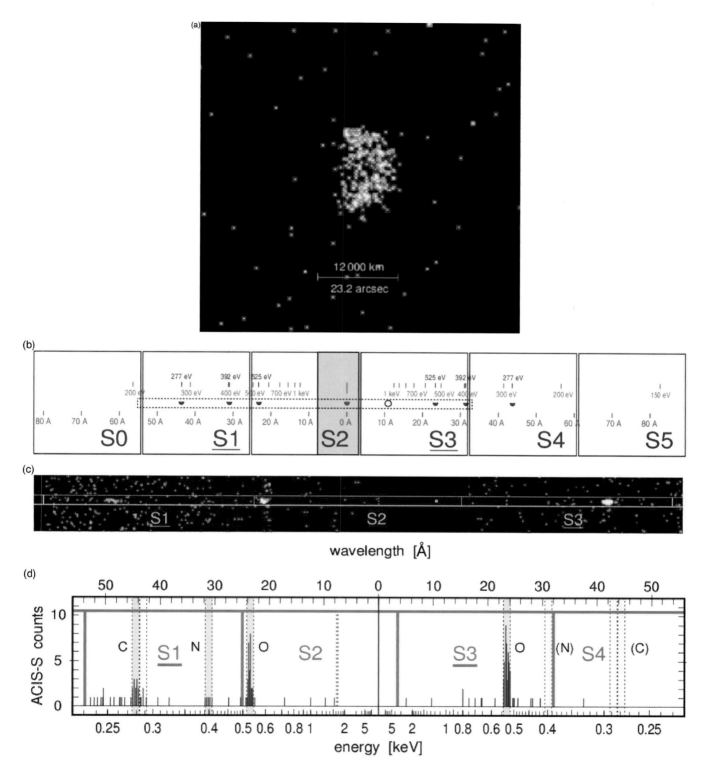

FIGURE 5 Venus. (a) First X-ray image of Venus, obtained with *Chandra* ACIS-I on 13 January 2001. The X-rays result mainly from fluorescent scattering of solar X-rays on C and O in the upper Venus atmosphere, at heights of 120–140 km. In contrast to the Moon, the X-ray image of Venus shows evidence for brightening on the sunward limb. This is caused by the scattering that takes place on an atmosphere and not on a solid surface. [From Dennerl et al., 2002, *Astron. Astrophys.*, **386**, 319]. (b) Expected LETG spectrum of Venus on the ACIS-S array. Energy and wavelength scales are given along the dispersion direction. Images of Venus are drawn at the position of the C, N, and O fluorescence lines, with the correct size and orientation. The dashed rectangle indicates the section of the observed spectrum shown below. (c) Observed spectrum of Venus, smoothed with a Gaussian function with $\sigma = 20''$. The two bright crescents symmetric to the center are images in the line of the O-Kα fluorescent emission, while the elongated enhancement at left is at the position of the C-Kα fluorescent emission line. The Sun is at bottom. (d) Spectral scan along the region outlined above. Scales are given in keV and Å. The observed C, N, and O fluorescent emission lines are enclosed by dashed lines; the width of these intervals matches the size of the Venus crescent (22.8''). (From Dennerl et al., 2002, *Astron. Astrophys.*, **386**, 319.)

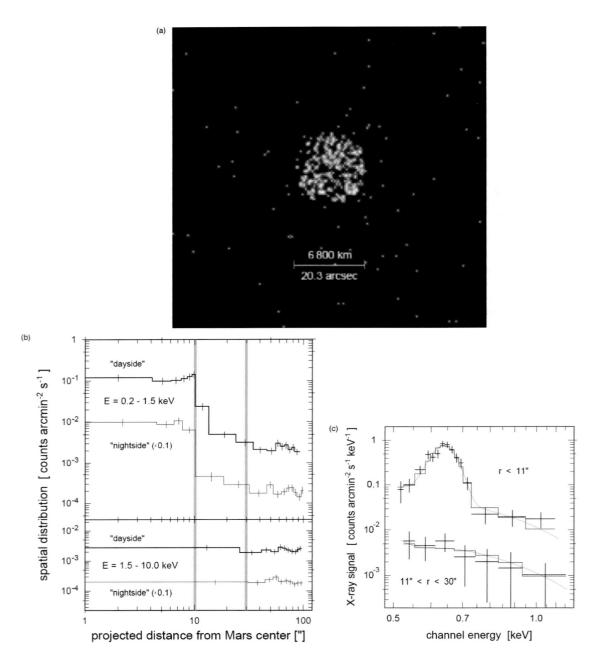

FIGURE 6 Mars. (a) First X-ray image of Mars, obtained with *Chandra* ACIS-I. The X-rays result mainly from fluorescent scattering of solar X-rays on C and O in the upper Mars atmosphere, at heights of 110–130 km, similar to Venus. The X-ray glow of the Martian exosphere is too faint to be directly visible in this image. (From Dennerl, 2002, *Astronomy and Astrophysics*, **394**, 1119–1128.) (b) Spatial distribution of the photons around Mars in the soft ($E = 0.2$–1.5 keV) and hard ($E = 1.5$–10.0 keV) energy range, in terms of surface brightness along radial rings around Mars, separately for the day side (offset along projected solar direction >0) and the night side (offset <0); note, however, that the phase angle was only 18.2°. For better clarity the night side histograms were shifted by one decade downward. The bin size was adaptively determined so that each bin contains at least 28 counts. The thick vertical lines enclose the region between one and three Mars radii. (c) X-Ray spectra of Mars (top) and its X-ray halo (bottom). Crosses with 1-σ error bars show the observed spectra; the model spectra, convolved with the detector response, are indicated by gray curves (unbinned) and by histograms (binned as the observed spectra). The spectrum of Mars itself is characterized by a single narrow emission line (this is most likely the O-Kα fluorescence line at 0.53 keV (the apparent displacement of the line energy is due to optical loading). At higher energies, the presence of an additional spectral component is indicated. The spectral shape of this component can be well modeled by the same 0.2 keV thermal bremsstrahlung emission which describes the spectrum of the X-ray halo. (From Dennerl, 2002, *Astron. Astrophys.*, **394**, 1119–1128.)

was consistent with fluorescent scattering of solar X-rays in the upper Mars atmosphere. The X-ray spectrum was dominated by a single narrow emission line caused by O $K\alpha$ fluorescence.

Simulations suggest that scattering of solar X-rays is most efficient between 110 km (along the subsolar direction) and 136 km (along the terminator) above the Martian surface. This behavior is similar to that seen on the Venus. No evidence for temporal variability or dust-related emission was found, which is in agreement with fluorescent scattering of solar X-rays as the dominant process responsible for the Martian X-ray. A gradual decrease in the X-ray surface brightness between 1 and ~3 Mars radii is observed (see Fig. 6). Within the limited statistical quality of the low flux observations, the spectrum of this region (halo) resembled that of comets: suggesting that they are caused by charge exchange interactions between highly charged heavy ions in the solar wind and neutrals in the Martian exosphere (corona). For the X-ray halo observed within 3 Mars radii, excluding Mars itself, the *Chandra* observation yielded a flux of about 1×10^{-14} erg cm^{-2} s^{-1} in the energy range 0.5–1.2 keV, corresponding to a luminosity of 0.5 ± 0.2 MW for isotropic emission, which agrees well with that expected theoretically for solar wind charge exchange mechanism.

The first *XMM-Newton* observation of Mars in November 2003 confirmed the presence of the Martian X-ray halo and made a detailed analysis of its spectral, spatial, and temporal properties. High-resolution spectroscopy of the halo with *XMM-Newton RGS* revealed the presence of numerous (~12) emission lines at the positions expected for deexcitation of highly ionized C, N, O, and Ne atoms, the dominant atomic species in the Martian atmosphere. The He-like O multiplet was resolved and found to be dominated by the spin-forbidden magnetic dipole transition 2 $^3S_1 \rightarrow 1\ ^1S_0$, confirming that charge exchange process is at the origin of the emission. This was the first definite detection of charge exchange induced X-ray emission from the exosphere of another planet.

The *XMM-Newton* observation confirmed that the fluorescent scattering of solar X-rays from the Martian disk is clearly concentrated on the planet, and is directly correlated with the solar X-ray flux levels. On the other hand, the Martian X-ray halo was found to extend out to ~8 Mars radii, with pronounced morphological differences between individual ions and ionization states. While the emission from ionized oxygen (Fig. 7c) appears to be concentrated in two distinct blobs a few thousand kilometers above the Martian poles, with larger heights for O^{7+} than for O^{6+}, the emissions from ionized carbon (Fig. 7f) exhibit a more band-like structure without a pronounced intensity dip at the position of Mars. The halo emission exhibited pronounced variability, but, as expected for solar wind interactions, the variability of the halo did not show any correlation with the solar X-ray flux.

6. Jupiter

6.1 Auroral Emission

Like the Earth, Jupiter emits X-rays both from its aurora and its sunlit disk. Jupiter's ultraviolet auroral emissions were first observed by the *International Ultraviolet Explorer* (*IUE*) and soon confirmed by the *Voyager 1* Ultraviolet Spectrometer as it flew through the Jupiter system in 1979 (see Bhardwaj and Gladstone, 2000 for review). The first detection of the X-ray emission from Jupiter was also made in 1979; the satellite-based Einstein observatory detected X-rays in the 0.2–3.0 keV energy range from both poles of Jupiter, due to the aurora. Analogous to the processes on Earth, it was expected that Jupiter's X-rays might originate as bremsstrahlung by precipitating electrons. However, the power requirement for producing the observed emission with this mechanism (10^{15}–10^{16} W) is more than two orders of magnitude larger than the input auroral power available as derived from *Voyager* and *IUE* observations of the ultraviolet aurora. (The strong Jovian magnetic field excludes the bulk of the solar wind from penetrating close to Jupiter, and the solar wind at Jupiter at 5.2 AU is 27 times less dense than at the Earth at 1 AU.) Precipitating energetic sulfur and oxygen ions from the inner magnetosphere, with energies in the 0.3–4.0 MeV/nucleon range, was suggested as the source mechanism responsible for the production of X-rays on Jupiter. The heavy ions are thought to start as neutral SO and SO$_2$ emitted by the volcanoes on Io into the jovian magnetosphere, where they are ionized by solar UV radiation, and then swept up into the huge dynamo created by Jupiter's rotating magnetic field. The ions eventually become channeled onto magnetic field lines terminating at Jupiter's poles, where they emit X-rays by first charge stripping to a highly ionized state, followed by charge exchange and excitation through collisions with H$_2$.

ROSAT's observations of Jupiter X-ray emissions supported this suggestion. The spatial resolution of these early observations was not adequate to distinguish whether the emissions were linked to source regions near the Io torus of Jupiter's magnetosphere (inner magnetosphere) or at larger radial distances from the planet. The advent of *Chandra* and *XMM-Newton* X-ray observatories revolutionized our thinking about Jupiter's X-ray aurora. High-spatial resolution (<1 arcsec) observations of Jupiter with the *Chandra* in December 2000 (see Fig. 8) revealed that most of Jupiter's northern auroral X-rays come from a "hot spot" located significantly poleward of the UV auroral zones (20–30 R_J), and not at latitudes connected to the inner magnetosphere. The hot spot is fixed in magnetic latitude (60–70°) and longitude (160–180° system III longitude) and occurs in a region where anomalous infrared and ultraviolet emissions (the so-called flares) have also been observed. On the other hand, auroral X-rays from the south (70–80° S latitude) spread almost halfway across the planet (~300–360° and 0–120°

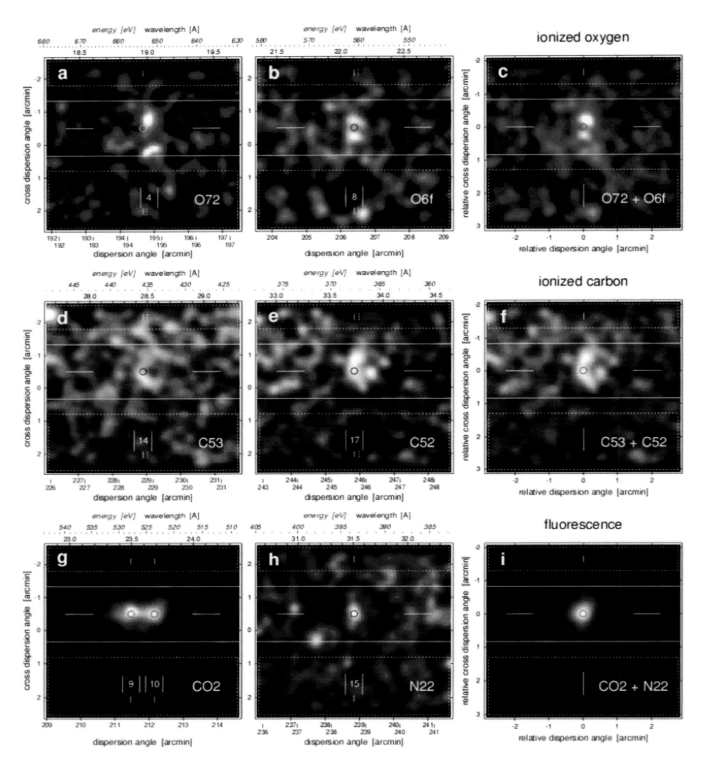

FIGURE 7 Mars. *XMM-Newton's* RGS images of Mars and its halo in the individual emission lines of ionized oxygen (top row), ionized carbon (middle row), and fluorescence of CO_2 and N_2 molecules (bottom row). The images were corrected for exposure variations, were binned into 2"×2" pixels and smoothed with a Gaussian function with $\sigma = 8$"×8". All are displayed at the same angular scale; the dynamic scale, however, was individually adjusted. (From Dennerl et al., 2006, *Astron. Astrophys.*, **451**, 709–722)

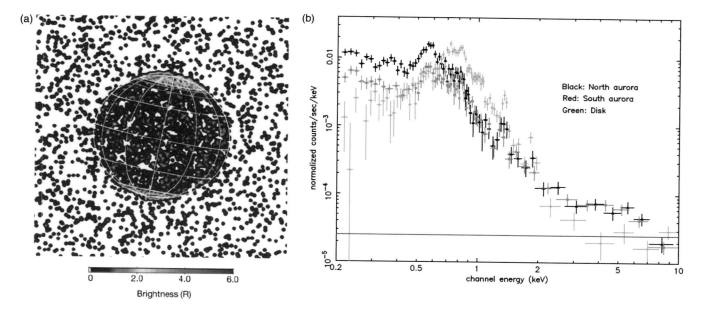

FIGURE 8 Jupiter. (a) Detailed X-ray morphology first obtained with Chandra HRC-I on 18 Dec. 2000, showing bright X-ray emission from the polar 'auroral' regions, indicating the high-latitude position of the emissions, and a uniform distribution from the low-latitude 'disk' regions. [from Gladstone et al., Nature, 415, 1000, 2002]. (b) Combined XMM-Newton EPIC spectra from the Nov. 2003 observation of Jupiter. Data points for the North and South aurorae are in black and red respectively. In green is the spectrum of the low-latitude disk emission. Differences in spectral shape between auroral and disk spectra are clear. The presence of a high energy component in the spectra of the aurorae is very evident, with a substantial excess relative to the disk emission extending to 7 keV. The horizontal blue line shows the estimated level of the EPIC particle background. [from Branduardi-Raymont et al., ESA SP-604, Vol. 1, pp. 15–20, 2006].

longitude). The location of the auroral X-rays connects along magnetic field lines to regions in the jovian magnetosphere well in excess of 30 jovian radii from the planet, a region where there are insufficient S and O ions to account for the X-ray emission. Acceleration of energetic ions was invoked to increase the phase space distribution, but now the question was whether the acceleration involved outer magnetospheric heavy ions or solar wind heavy ions.

Surprisingly, *Chandra* observations also showed that X-rays for jovian aurora pulsate with a periodicity that is quite systematic (approximately 45-min period) at times (in December 2000) and irregular (20–70 min range) at other times (in February 2003). The 45-min periodicity is highly reminiscent of a class of Jupiter high-latitude radio emissions known as quasi-periodic radio bursts, which had been observed by *Ulysses* in conjunction with energetic electron acceleration in Jupiter's outer magnetosphere. During the 2003 *Chandra* observation of Jupiter, the *Ulysses* radio data did not show any strong 45-min quasi-periodic oscillations, although variability on time scales similar to that in X-rays was present. *Chandra* also found that X-rays from the north and south auroral regions are neither in phase nor in antiphase, but that the peaks in the south are shifted from those in the north by about 120° (i.e., one-third of a planetary rotation).

A clear temporal association of the X-ray emission intensity with a jovian UV flare has been observed during a simultaneous *Hubble Space Telescope* and *Chandra* observation in February 2003. However, the spatial correlation was not as expected. The X-rays did increase in time in a manner consistent with the ultraviolet flare, but rather than peak at the ultraviolet flare location they were peaked in a morphologically associated region, the "kink," which most likely magnetically maps to the dusk flank of Jupiter's magnetosphere.

The *Chandra* and *XMM-Newton* spectral observations have now established that soft (∼0.1–2 keV) X-rays from jovian aurora are line emissions, which are consistent with high-charge states of precipitating heavy (C, O, S) ions, and not a continuum as might be expected from electron bremsstrahlung (see Fig. 8). *XMM-Newton* has provided spectral information on the X-rays from Jupiter, which is somewhat better than *Chandra*. The RGS on *XMM-Newton* clearly resolves the strongest lines in the spectra, while the EPIC camera has provided images of the planets in the strong OVII and OVIII lines present in the jovian auroral emissions. The spectral interpretation of *Chandra* and *XMM-Newton* observations is consistent with a source due to energetic ion precipitation that undergoes acceleration to attain energies of >1 MeV/nucleon before impacting the jovian upper atmosphere. This is also supported by modeling studies.

Recently, *XMM-Newton* and *Chandra* data have suggested that there is a higher (>2 keV) energy component present in the spectrum of Jupiter's aurora; they found it to be variable on timescales of days. The observed spectrum and flux, at times, appears consistent with that predicted from bremsstrahlung of energetic electrons precipitating from the magnetosphere, but at energies greater than 2 keV (at lower energies bremsstrahlung still fall short by an order of magnitude). The variability suggests a link to changes in the energy distribution of the precipitating magnetospheric electrons and may be related to the solar activity at the time of observation.

6.2 Nonauroral (Disk) Emission

The existence of low-latitude "disk" X-ray emission from Jupiter was first recognized in *ROSAT* observations made in 1994. These X-rays were initially thought to be the result of precipitation of energetic S and O ions from Jupiter's inner radiation belts into the planet's atmosphere. Later, as for the inner planets, it was suggested that elastic scattering of solar X-rays by atmospheric neutrals (H_2) and fluorescent scattering of carbon K-shell X-rays from CH_4 molecules located below the jovian homopause was the source of the disk X-rays.

A general decrease in the overall X-ray brightness of Jupiter observed by *ROSAT* over the years 1994–1996 was found to be coincident with a similar decay in solar activity index (solar 10.7 cm flux). A similar trend is seen in the data obtained by *Chandra* in 2000 and 2003; Jupiter disk was about 50% dimmer in 2003 compared to that in 2000, which is consistent with variation in the solar activity index. First direct evidence for temporal correlation between jovian disk X-rays and solar X-rays is provided by *XMM-Newton* observations of Jupiter in November 2003, which demonstrated that day-to-day variation in disk X-rays of Jupiter are synchronized with variation in the solar X-ray flux, including a solar flare that has a matching feature in the jovian disk X-ray light curve. *Chandra* observations of December 2000 and February 2003 also support this association between light curves of solar and planetary X-rays. However, there is an indication of higher X-ray counts from regions of low surface magnetic field in the *Chandra* data, suggesting the presence of some particle precipitation.

The higher spatial resolution observation by Chandra has shown that nonauroral disk X-rays is relatively more spatially uniform than the auroral X-rays (Fig. 8). Unlike the $\sim$40 ± 20-min quasi-periodic oscillations seen in auroral X-ray emission, the disk emission does not show any systematic pulsations. There is a clear difference between the X-ray spectra from the disk and auroral region on Jupiter; the disk spectrum peaks at higher energies (0.7– 0.8 keV) than the aurorae (0.5–0.6 keV) and lacks the high–energy component (above $\sim$ 3 keV) present in the latter (see Fig. 8).

7. Galilean Satellites

The jovian *Chandra* observations on 25–26 November 1999 and 18 December 2000 discovered X-ray emission from the Galilean satellites (Fig. 9). These satellites are very faint when observed from Earth orbit (by *Chandra*), and the detections of Io and Europa, although statistically very significant, were based on $\sim$10 photons each! The energies of the detected X-ray events ranged between 300 and 1890 eV and appeared to show a clustering between 500 and 700 eV, suggestive of oxygen K-shell fluorescent emission. The estimated power of the X-ray emission was 2 MW for Io and 3 MW for Europa. There were also indications of X-ray emission from Ganymede. X-Ray emission from Callisto seems likely at levels not too far below the CXO sensitivity limit because the magnetospheric heavy ion fluxes are an order of magnitude lower than at Ganymede and Europa, respectively.

The most plausible emission mechanism is inner (K shell) ionization of the surface (and incoming magnetospheric) atoms followed by prompt X-ray emission. Oxygen should be the dominant emitting atom either in an SiO_x (silicate) or SO_x (sulfur oxides) surface (Io) or on an icy one (the outer Galilean satellites). It is also the most common heavy ion in the jovian magnetosphere. The extremely tenuous atmospheres of the satellites are transparent to X-ray photons with these energies, as well as to much of the energy range of the incoming ions. However, oxygen absorption in the soft X-ray is strong enough that the X-rays must originate within the top 10 micrometers of the surface in order to escape. Simple estimates suggest that excitation by incoming ions dominates over electrons and that the X-ray flux produced is within a factor of 3 of the measured flux. The detection of X-ray emission from the Galilean satellites thus provides a direct measure of the interactions of the magnetosphere of Jupiter with the satellite surfaces. An intriguing possibility is placement of an imaging X-ray spectrometer on board a mission to the Jupiter system. If such an instrument was in orbit around a Galilean satellite (e.g., Europa or Ganymede), even though it would be immersed in a fierce radiation environment, it would be able to map the elemental abundances of the surface for elements from C through Fe.

8. Io Plasma Torus

The Io Plasma Torus (IPT) is known to emit at extreme ultraviolet (EUV) energies and below, but it was a surprise when *Chandra* discovered that it was also a soft X-ray source. The 1999 jovian *Chandra* observations

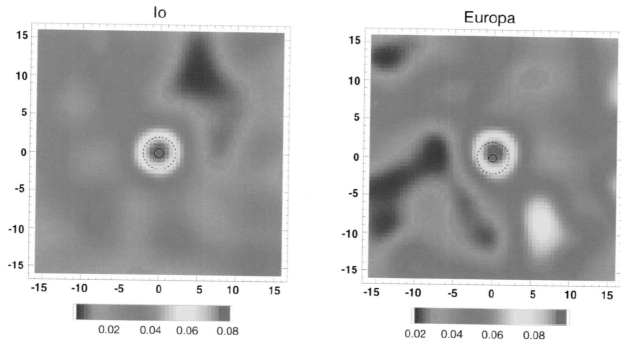

FIGURE 9 Galilean Moons. Chandra X-ray images of Io and Europa (0.25 keV < E < 2.0 keV) from November, 1999 observations. The images have been smoothed by a two-dimensional gaussian with $\sigma = 2.46$ arcsec (5 detector pixels). The axes are labeled in arcsec (1 arcsec $\sim= 2995$ km) and the scale bar is in units of smoothed counts per image pixel (0.492 by 0.492 arcsec). The solid circle shows the size of the satellite (the radii of Io and Europa are 1821 km and 1560 km, respectively), and the dotted circle the size of the detect cell. [from Elsner et al., *Astrophysical Journal*, 572, 1077–1082, 2002].

detected a faint diffuse source of soft X-rays from the region of the IPT. The 2000 *Chandra* image, obtained with the HRC-I camera (Fig. 10), exhibited a dawn-to-dusk asymmetry similar to that seen in the EUV. Figure 10 shows the background-subtracted *Chandra*/ACIS-S IPT spectrum for 25–26 November 1999. This spectrum shows evidence for line emission centered on 574 eV (very near a strong O VII line), together with a very steep continuum spectrum at the softest X-ray energies. Although formed from the same source, the spectrum is different than from the jovian aurora because the energies, charge states, and velocities of the ions in the torus are much lower—the bulk ions have not yet been highly accelerated. There could be contributions from other charge states because current plasma torus models consist mostly of ions with low charge states, consistent with photoionization and ion-neutral charge exchange in a low-density plasma and neutral gas environment. The 250–1000 eV energy flux at the telescope aperture was 2.4×10^{-14} erg cm^{-2} s^{-1}, corresponding to a luminosity of 0.12 GW. Although bremsstrahlung from nonthermal electrons might account for a significant fraction of the continuum X-rays, the physical origin of the observed IPT X-ray emission is not yet fully understood. The 2003 jovian *Chandra* observations also detected X-ray emission from the IPT, although at a fainter level than in 1999 or 2000. The morphology exhibited the familiar dawn-to-dusk asymmetry.

9. Saturn

The production of X-rays at Saturn was expected because, like the Earth and Jupiter, Saturn was known to possess a magnetosphere and energetic electrons and ions particles within it; however, early attempts to detect X-ray emission from Saturn with *Einstein* in December 1979 and with *ROSAT* in April 1992 were negative and marginal, respectively. Saturnian X-rays were unambiguously observed by *XMM-Newton* in October 2002 and by the *Chandra* X-ray Observatory in April 2003. In January 2004, Saturn was again observed by the *Chandra* ACIS-S in two exposures, one on 20 January and other on 26–27 January, with each observation lasting for about one full Saturn rotation. The X-ray power emitted from Saturn's disk is roughly one-fourth of that from Jupiter's disk, which is consistent with Saturn being twice as far as Jupiter from Sun and Earth.

The January 2004 Chandra observation showed (Fig. 11) that X-rays from Saturn are highly variable—a factor of 2 to 4 variability in brightness over 1 week. These observations also revealed X-rays from Saturn's south polar cap on January 20 (see Fig. 11, left panel), which are not evident in the January 26 observation (see Fig. 11, right panel) and in earlier Chandra observations. X-rays from the south polar cap region were present only in the 0.7–1.4 keV energy band, in contrast with Jupiter's X-ray aurora for which the

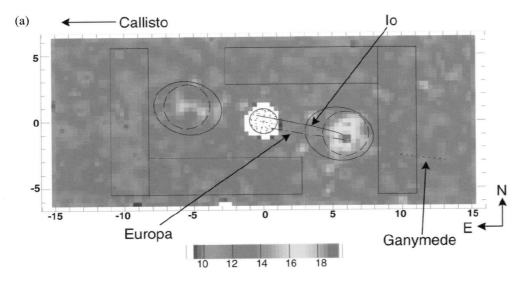

(a) ← Callisto | Io | Europa | Ganymede | N | E

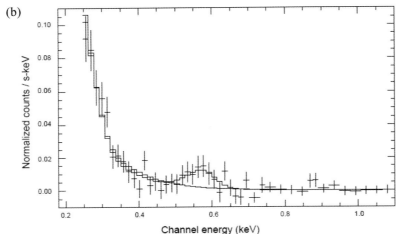

(b)

FIGURE 10   Plasma Torus. (a) Chandra/HRC-I image of the IPT (2000 December 18). The image has been smoothed by a two-dimensional Gaussian with $\sigma = 7.38''$ (56 HRC-I pixels). The axes are labeled in units of Jupiter's radius, R_J, and the scale bar is in units of smoothed counts per image pixel. The paths traces by Io, Europa, and Ganymede are marked on the image. Callisto is off the image to the dawn side, although the satellite did fall within the full microchannel plate field of view. The regions bounded by rectangles were used to determine background. The regions bounded by dashed circles or solid ellipses were defined as source regions. (b) Chandra/ACIS-S spectrum for the Io Plasma Torus from November 1999. The solid line presents a model fit for the sum of a power-law spectrum and a Gaussian line, while the dashed line represents just a pure power law spectrum. The line is consistent with K-shell flurorescent emission from oxygen ions. [From Elsner et al., *Astrophysical Journal*, 572, 1077–1082, 2002].]

emission is mostly in the bands 0.3–0.4 keV and 0.6–0.7 keV. Because of this, it is likely that the X-ray emission from the south polar cap is unlikely to be auroral in nature, and more likely that they are an extension of the disk X-ray emission of Saturn. Any emission from the north polar cap region was blocked by Saturn's rings.

As is the case for Jupiter's disk, X-ray emission from Saturn seems likely to be due to the scattering of the incident solar X-ray flux. An X-ray flare has been detected from the nonauroral disk of Saturn during the *Chandra* observation on 20 January 2004. Taking light travel time into account, this X-ray flare from Saturn coincided with an M6-class flare emanating from a sunspot that was clearly visible from both Saturn and Earth. Moreover, the lightcurve for the X-rays from Saturn was very similar to that of the solar X-ray flux. This was the first direct evidence suggesting that Saturn's disk X-ray emission is principally controlled by processes happening on the Sun. Further, a good correla-

tion has been observed between Saturn X-rays and F10.7 solar activity index: suggesting a solar connection.

The spectrum of X-rays from Saturn's disk is very similar to that from Jupiter's disk. Saturn's disk spectrum measured on 20 January 2004 is quite similar to that measured on 14–15 April 2003 in the 0.3–0.6 keV range. However, at energies 0.6–1.2 keV, the former is stronger by a factor of 2 to 4. This is probably due to the nature of the M6-class solar X-ray flare on 20 January, with a corresponding hardening of the solar X-ray flux driving Saturn's X-ray emission.

10. Rings of Saturn

The rings of Saturn, known to be made of mostly water (H_2O) ice, are one of the most fascinating objects in our solar system. Recently, the discovery of X-rays from the rings

(a)

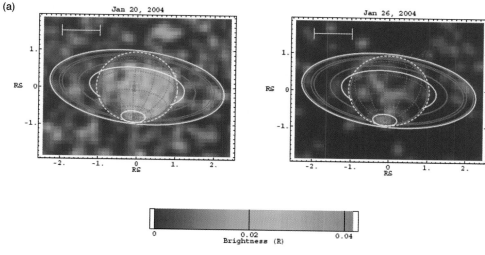

(b)

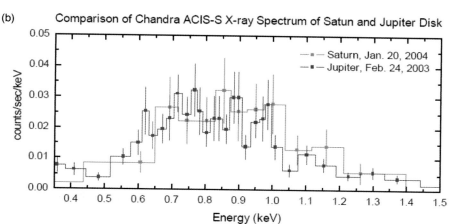

FIGURE 11 Saturn. (a) Chandra ACIS X-ray 0.24–2.0 keV images of Saturn on January 20 and 26, 2004. Each continuous observation lasted for about one full Saturn rotation. The horizontal and vertical axes are in units of Saturn's equatorial radius. The white scale bar in the upper left of each panel represents 10?. The two images, taken a week apart and shown on the same color scale, indicate substantial variability in Saturn's X-ray emission. [from Bhardwaj et al. Astrophysical Journal Letters, 624, L121–L124 2005]. (b) Disk X-ray spectrum of Saturn (red curve) and Jupiter (blue curve). Values for Saturn spectrum are plotted after multiplying by a factor of 5. [from Bhardwaj, Advances in Geosciences, vol.3, 215–230, 2006].

of Saturn was made from the *Chandra* ACIS-S observations of the Saturnian system conducted in January 2004 and April 2003. X-rays from the rings are dominated by emission in a narrow (~130 eV wide) energy band of 0.49–0.62 keV (Fig. 12). This band is centered on the oxygen Kα fluorescence line at 0.53 keV, suggesting that fluorescent scattering of solar X-rays from oxygen atoms in the surface of H_2O icy ring material is the likely source mechanism for ring X-rays. The X-ray power emitted by the rings in the 0.49–0.62 keV band on 20 January 2004 is 84 MW, which is about one-third of that emitted from the Saturn disk in the 0.24- to 2.0-keV band. The projected rings have about half the surface area of the Saturn disk, consistent with this ratio. During 14–15 April 2003, the X-ray power emitted by the rings in the 0.49- to 0.62-keV band is about 70 MW.

Figure 12 shows the X-ray image of the Saturnian system in January 2004 in the 0.49- to 0.62-keV band, the energy range where X-rays from the rings are unambiguously detected. The observations of January 2004 also suggested that, similar to Saturn's X-ray emission, the ring X-rays are highly variable—a factor of 2–3 variability in brightness over

1 week. There is an apparent asymmetry in X-ray emission from the east (morning) and west (evening) ansae of the rings (see Fig. 12a). However, when the *Chandra* ACIS-S data set of January 2004 and April 2003 is combined, the evidence for asymmetry is not that strong.

11. Comets

The discovery of high-energy X-ray emission in 1996 from C/1996 B2 (Hyakutake) created a new class of X-ray–emitting objects. Observations since 1996 have shown that the very soft ($E < 1$ keV) emission is due to an interaction between the solar wind and the comet's atmosphere, and that X-ray emission is a fundamental property of comets. Theoretical and observational work has demonstrated that charge exchange collision of highly charged heavy solar wind ions with cometary neutral species is the best explanation for the emission. The X-rays are extremely easy to detect because the neutral atmosphere of a comet is large and extended and gravitationally unbound, intercepting a

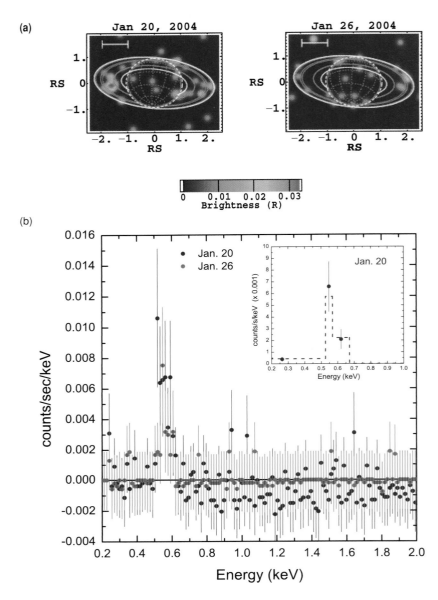

FIGURE 12 Saturn's Rings. (a) Chandra ACIS X-ray images of the Saturnian system in the 0.49–0.62 keV band on 2004 January 20 and 26–27. The X-ray emission from the rings is clearly present in these restricted energy band images; the emission from the planet is relatively weak in this band (see Fig. 11(a) for an X-ray image of the Saturnian system in the 0.24–2.0 keV band). (b) Background-subtracted Chandra ACIS-S3–observed X-ray energy spectrum for Saturn's rings in the 0.2–2.0 keV range on 2004 January 20 and 26–27. The cluster of X-ray photons in the ∼0.49–0.62 keV band suggests the presence of the oxygen Kα line emission at 0.53 keV in the X-ray emission from the rings. The inset shows a Gaussian fit (peak energy = 0.55 keV, σ = 140 eV), shown by the dashed line, to the ACIS-observed rings' spectrum on January 20. Each spectral point (filled circle with error bar) represents ≥10 measured events. The spectral fitting suggests that X-ray emissions from the rings are predominantly oxygen Kα photons. [from Bhardwaj et al., *Astrophys. J. Lett.*, 627, L73-L76, 2005].

large amount of solar wind ions as they stream away from the Sun. The observed characteristics of the emission can be organized into the following four categories: (1) spatial morphology, (2) total X-ray luminosity, (3) temporal variation, and (4) energy spectrum. Any physical mechanism that purports to explain cometary X-ray emission must account for all of these characteristics.

X-Ray and EUV images of C/1996 B2 (Hyakutake) made by the *ROSAT* and *EUVE* satellites look very similar (Fig. 13). Except for images of C/1990 N1 and C/Hale-Bopp 1995 O1, all EUV and X-ray images of comets have exhibited similar spatial morphologies. The emission is largely confined to the sunward side of the cometary coma; almost no emission is found in the extended tails of dust or

plasma. The peak X-ray brightness gradually decreases with increasing cometocentric distance r with a dependence of about r^{-1}. The brightness merges with the soft X-ray background emission at distances that exceed 10^4 km for weakly active comets, and can exceed 10^6 km for the most luminous comets. The region of peak emission is crescent-shaped with a brightness peak displaced towards the Sun from the nucleus. The distance of this peak from the nucleus appears to increase with increasing values of Q (total gas production rate); for Hyakutake, it was located at r_{peak} ∼2 × 10^4 km.

The observed X-ray luminosity, L_x, of C/1996 B2 (Hyakutake) was 4 × 10^{15} ergs s^{-1} for an aperture radius at the comet of 1.2 × 10^5 km. (Note that the photometric luminosity depends on the energy bandpass and the observational aperture at the comet. The quoted value assumes a *ROSAT* photon emission rate of $P_X \sim 10^{25}$ s^{-1} (0.1–0.6 keV), in comparison to the *EUVE* estimate of $P_{EUV} \sim 7.5 \times 10^{25}$ s^{-1} (0.07–0.18 keV and 120,000-km aperture). A positive correlation between optical and X-ray luminosities was demonstrated using observations of several comets having similar gas (Q_{H_2O})-to-dust ($Af\rho$) emission rate ratios. L_x correlates more strongly with the gas production rate Q_{gas} than it does with $L_{opt} \sim Q_{dust} \sim Af\rho$. Particularly dusty comets, like Hale–Bopp, appear to have less X-ray emission than would be expected from their overall optical luminosity L_{opt}. The peak X-ray surface brightness decreases with increasing heliocentric distance r, independent of Q, although the total luminosity appears roughly independent of r. The maximum soft X-ray luminosity observed for a comet to date is $\sim 2 \times 10^{16}$ erg s^{-1} for C/Levy at 0.2–0.5 keV.

Photometric lightcurves of the X-ray and EUV emission typically show a long-term baseline level with superimposed impulsive spikes of a few hours' duration, and maximum amplitude 3 to 4 times that of the baseline emission level. Figure 13 demonstrates the strong correlation found between the time histories of the solar wind proton flux (a proxy for the solar wind minor ion flux), the solar wind magnetic field intensity, and a comet's X-ray emission, for the case of comet 2P/Encke 19997. Comparison of the *ROSAT* and *EUVE* luminosity of C/1996 B2 (Hyakutake) with time histories of the solar wind proton flux, oxygen ion flux, and solar X-ray flux, showed a strongest correlation between the cometary emission and the solar wind oxygen ion flux, a good correlation between the comet's emission and the solar wind proton flux, but no correlation between the cometary emission and the solar X-ray flux.

Until 2001, all published cometary X-ray spectra had very low spectral energy resolution ($\Delta E/E \sim 1$ at 300–600 eV), and the best spectra were those obtained by *ROSAT* for C/1990 K1 (Levy) and by *Beppo*SAX for comet C/ 1995 O1 (Hale–Bopp). These observations were capable of showing that the spectrum was very soft (characteristic thermal bremsstrahlung temperature $kT \sim 0.23 \pm 0.04$ keV) with intensity increasing toward lower energy in the 0.01- to 0.60-keV energy range and established upper limits to the contribution of the flux from K-shell resonance fluorescence of carbon at 0.28 keV and oxygen at 0.53 keV. However, even in these "best" spectra, continuum emission (such as that produced by the thermal bremsstrahlung mechanism) could not be distinguished from a multiline spectrum, such as would result from the SWCX mechanism. Nondetections of comets C/Hyakutake, C/Tabur, C/Hale–Bopp, and 55P/Tempel–Tuttle using the XTE PCA (2–30 keV) and ASCA SIS (0.6–4 keV) imaging spectrometers were consistent with an extremely soft spectrum.

Higher resolution spectra of cometary X-ray emission have now appeared in the literature. The *Chandra* X-ray Observatory (CXO) measured soft X-ray spectra from comet C/1999 S4 (LINEAR) over an energy range of 0.2–0.8 keV, and with a full width half maximum energy resolution of $\Delta E = 0.11$ keV (Fig. 13). The spectrum is dominated by line emission from C^{+4}, C^{+5}, O^{+6}, and O^{+7} excited ions, not by continuum. A spectrum of comet C/1999 T1 (McNaught–Hartley) showed similar line emission features, with a somewhat higher ratio of OVII to OVIII emission, and emission due to Ne^{+9}. A new spectrum of comet 2P/Encke shows a very different ratio of line emission in the C^{+4}, C^{+5}, O^{+6}, and O^{+7} lines, due to the collisionally thin nature of the low activity coma, and the unusual postshock charge state of the solar wind at the time of observation. Line emission is also found in *XMM-Newton* spectra of comet C/1999 T1 (McNaught–Hartley) and, more recently, in CXO spectra of C/2001 WM1 (Lincoln Near-East Asteroid Research, Linear) and C/2002 Ikeya–Zhang. An *XMM-Newton* spectrum of C/2001 WM1 (LINEAR) shows characteristic SWCX X-ray signatures in unprecedented detail.

From other work, there are suggestions of charge exchange line emission from other species than C^{+4}/C^{+5}, O^{+6}/O^{+7}, and Ne^{+9}. A reanalysis of archival *EUVE* Deep Survey spectrometer spectra suggests EUV line emission features from comet C/1996 B2 (Hyakutake) due to O^+, O^{+5}, O^{+4}, C^{+4}, O^{+6}, C^{+5}, He$^+$, and Ne^{+7}. It has been suggested that emission lines are attributable to Mg and Si in C/McNaught–Hartley, and He^{+2} in C/Hale–Bopp, although these remain unconfirmed and controversial due to the sensitivity of the results on the details of the instrumental background subtraction. Hints of possible emission due to N^{+6} at 425 eV contributing to a reduced 380/450 eV ratio were found in *Chandra* observations of 2P/Encke in 2003.

Numerical simulations of the solar wind interaction with Hyakutake including SWCX have been used to generate X-ray images. A global magneto-hydrodynamic (MHD) model and a hydrodynamic model were used to predict solar wind speeds and densities in addition to the X-ray emission around a comet. The simulated X-ray images are similar to the observed images. Recent work has shown that by determining the location of the emission maximum in the collisionally thick case, the neutral gas production rate can be determined in 5 comets observed by *ROSAT* and *XMM-Newton*. On comet WM1, the position of the cometary bow shock has been determined using the location of rapid changes in the first and second derivatives of the flux with distance from the nucleus.

It is not clear that the emission pattern always follows the plasma structures. New work suggests that the crescent-shaped, sunward offset morphology is found only for comets with coma dense enough to be in the collisionally thick regime—for low activity comets, the emission will be maximal wherever the coma has its maximum density, typically at the nucleus. This may explain the unusual emission

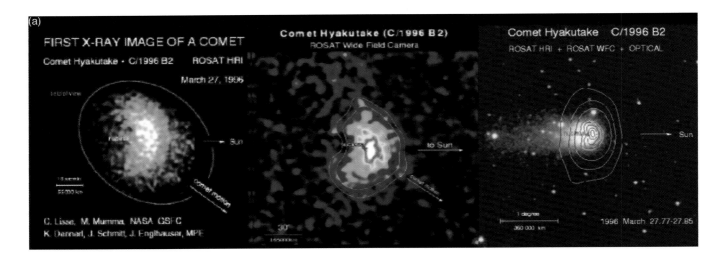

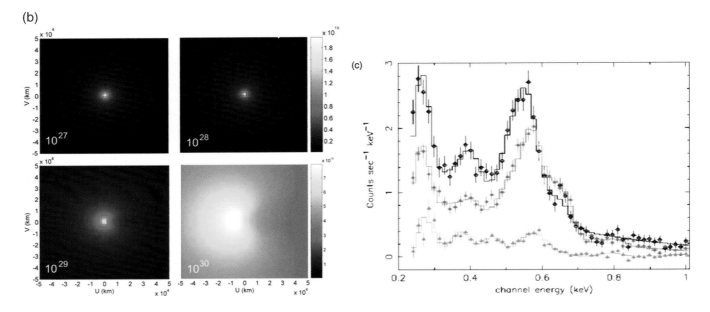

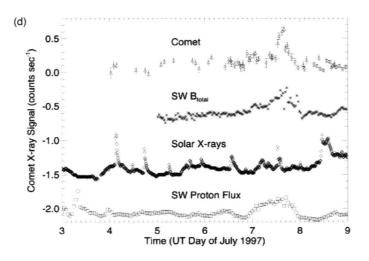

morphologies seen in comets like d'Arrest 1997 and 2P/Encke 2003.

Up until now, the temporal variation of the solar wind dominated the observed behavior on all but the longest timescales of weeks to months. A "new" form of temporal variation has recently been demonstrated in the *Chandra* observations of comet 2P/Encke 2003, wherein the observed X-ray emission is modulated at the 11.1-hour period of the nucleus rotation. Rotational modulation of the signal should be possible only in collisionally thin (to SWCX) comae with weak cometary activity, where a change in the coma neutral gas density can directly affect power density of cometary X-ray.

Driven by the solar wind, cometary X-rays provide an observable link between the solar corona, where the solar wind originates, and the solar wind where the comet resides. Once we have understood the SWCX mechanism's behavior in cometary comae in sufficient detail, we will be able to use comets as probes to measure the solar wind throughout the heliosphere. This will be especially useful in monitoring the solar wind in places hard to reach with spacecraft—such as over the solar poles, at large distances above and below the ecliptic plane, and at heliocentric distances greater than a few AU. For example, about one-third of the observed soft X-ray emission is found in the 530- to 700-eV oxygen O^{+7} and O^{+6} lines; observing photons of this energy will allow studies of the oxygen ion charge ratio of the solar wind, which is predicted to vary significantly between the slow and fast solar winds at low and high solar latitudes, respectively.

12. Asteroids

X-Rays from asteroids have been studied by experiments on two *in situ* missions, the X-ray/gamma-ray spectrometer (XGRS) on the Near Earth Asteroid Rendezvous (NEAR)–Shoemaker mission to asteroid 433 Eros, and the X-ray spectrometer (XRS) on the *Hayabusa* mission to asteroid 25143 Itokawa. The only attempt to detect X-rays from an asteroid was a 10-ks distant, remote observation by *Chandra* on 11 December 2001 of 1998 WT24, but it was unsuccessful. The results of the in situ observations show X-ray emission due to fluorescence and scattering of incident solar X-rays, similar to the emission seen from the surface of the airless Moon. In fact, the best measurements were obtained during a strong solar flare, when the incident solar X-rays were highly amplified. As for the Moon, X-ray spectroscopy of resonantly scattered solar X-rays can be used to map the elemental composition of the surface.

NEAR–Shoemaker entered Eros orbit on 14 February 2000 and completed a 1-year long mission around it. Eros at 33 × 13 × 13 km in size is the second largest near-Earth asteroid, and its "day" is 5.27 hours long. Eros exhibits a heavily cratered surface with one side dominated by a huge, scallop-rimmed gouge; a conspicuous sharp, raised rimmed crater occupies the other side. The XRS part of the XGRS detected X-rays in the 1- to 10-keV energy range to determine the major elemental composition of Eros' surface. The XRS observed the asteroid in low orbit (<50 km) during 2 May–12 August 2000 and again during 12 December 2000–2 February 2001. These observations suggest that

---

FIGURE 13   The Rich Behavior of X-Ray Emission Seen From Comets. (a) Cometary X-ray Emission Morphology. Images of C/Hyakutake 1996B2 on 26 - 28 March 1996 UT: ROSAT HRI 0.1 - 2.0 keV X-ray, ROSAT WFC .09 - 0.2 keV extreme ultraviolet, and visible light, showing a coma and tail, with the X-ray emission contours superimposed. The Sun is towards the right, the plus signs mark the position of the nucleus, and the orbital motion of the comet is towards the lower right in each image. [From Lisse et al., Science 292, 1343 – 1348, 1996]. (b) Morphology as a function of comet gas production rate (given in terms of molecules sec^{-1} in the lower right of each panel). Note the decreasing concentration of model source function and the increasing importance of diffuse halo emission in the extended coma as the gas production rate increases. [from Lisse et al., *Astrophys. J.*, **635**, 1329-1347, 2005]. (c) Chandra ACIS medium resolution CCD X-ray spectra of the X-ray emission from three comets. All curves show ACIS-S3 measurements of the 0.2 – 1.5 keV pulse height spectrum, as measured in direct detection mode. with ±1 error bars and the best-fit emission line + thermal bremsstrahlung model convolved with the ACIS-S instrument response as a histogram. The positions of several possible atomic lines are noted. Pronounced emission due to O^{7+} and O^{6+} is evident at 560 and 660 eV, and for C^{5+}, C^{4+}, and N^{5+} emission lines at 200 - 500 eV. Best-fit model lines at 284, 380, 466, 552, 590, 648, 796, and 985 are close to those predicted for charge exchange between solar wind C^{+5}, C^{+6}, C^{+6}/N^{+6}, O^{+7}, O^{+7}, O^{+8}, O^{+8}, and Ne^{+9} ions and neutral gases in the comet's coma. (Black) ACIS spectra of C/LINEAR 1999 S4 (circles), from Lisse et al. (2001). (Red) Comet McNaught-Hartley spectra (squares), after Krasnopolsky et al. 2003. (Green) 2P/Encke spectrum taken on 24 Nov 2003, multiplied by a factor of 2. The C/1999 S4 (LINEAR) and C/McNaught-Hartley 2001 observations had an average count rate on the order 20 times as large, even though Encke was closer to Chandra and the Earth when the observations were being made. Note the 560 complex to 400 eV complex ratio of 2 to 3 in the two bright, highly active comets, and the ratio of approximately 1 for the faint, low activity comet Encke. [from Lisse et al., op. cit 2005]. (d) Temporal trends of the cometary X-ray emission. Lightcurve, solar wind magnetic field strength, solar wind proton flux, and solar X-ray emission for 2P/Encke 1997 on 4-9 July 1997 UT. All error bars are ± 1. D - HRI light curve, 4-8 July 1997. ◇ - EUVE scanner Lexan B light curve 6 - 8 July 1997 UT, taken contemporaneously with the HRI observations, and scaled by a factor of 1.2. Also plotted are the WIND total magnetic field B$_{total}$ (°), the SOHO CELIAS/SEM 1.0 - 500 Å solar X-ray flux (◇), and the SOHO CELIAS solar wind proton flux (boxes). There is a strong correlation between the solar wind magnetic field/density and the comet's emission. There is no direct correlation between outbursts of solar X-rays and the comet's outbursts. [from Lisse et al., *Icarus* **141**, 316-330, 1997].

elemental ratios for Mg/Si, Al/Si, Ca/Si, and Fe/Si on Eros are most consistent with a primitive chondrite and give no evidence of global differentiation. The S/Si ratio is considerably lower than that for a chondrite and is most likely due to surface volatilization ("space weathering"). The overall conclusion is that Eros is broadly "primitive" in its chemical composition and has not experienced global differentiation into a core, mantle, and crust, and that surface effects cause the observed departures from chondritic S/Si and Fe/Si.

Hayabusa reached the asteroid 25143 Itokawa on 12 September 2005. The first touchdown occurred on 19 November 2005. The observations made during the touchdown, a period of relatively enhanced solar X-ray flux, returned an average elemental mass ratio of Mg/Si = 0.78 ± 0.07 and Al/Si = 0.07 ± 0.03. These early results suggest that, like Eros, asteroid Itokawa's composition can be described as an ordinary chondrite, although occurrence of some differentiation cannot be ruled out.

The composition and structure of the rocks and minerals in asteroids provides critical clues to their origin and evolution and are a fundamental line of inquiry in understanding the asteroids, of which more than 20,000 have been detected and catalogued. It is interesting to note that for both Eros and Itokawa the compositions derived by remote X-ray observations using spacecraft in close proximity to the asteroid seem consistent with those found using Earth-based optical and infrared spectroscopy.

13. Heliosphere

The solar wind flow starts out slowly in the corona but becomes supersonic at a distance of few solar radii. The gas cools as it expands, falling from $\sim 10^6$ K down to about 10^5 K at 1 AU. The average properties of the solar wind at 1 AU are proton number density ~ 7 cm^{-3}, speed ~ 450 km s^{-1}, temperature $\sim 10^5$ K, magnetic field strength ~ 5 nT, and Mach number ~ 8. However, the composition and charge state distribution far from the Sun are "frozen-in" at coronal values due to the low collision frequency outside the corona. The solar wind contains structure, such as slow (400 km s^{-1}) and fast (700 km s^{-1}) streams, which can be mapped back to the Sun. The solar wind "terminates" in a shock called the heliopause, where the ram pressure of the streaming

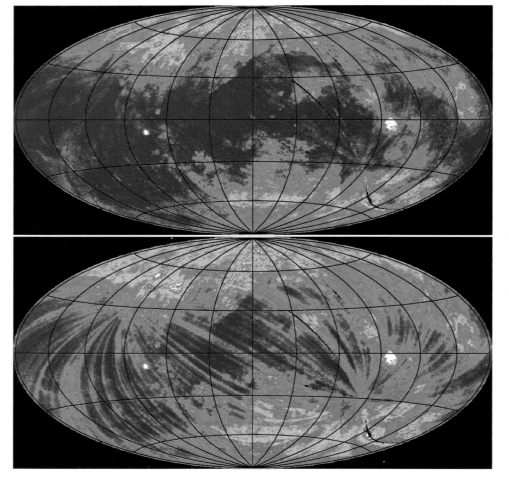

FIGURE 14 Heliosphere. Upper panel) ROSAT All-Sky Survey map of the cosmic X-ray background at 1/4 keV. The data are displayed using an Aitoff projection in Galactic coordinates centered on the Galactic center with longitude increasing to the left and latitude increasing upwards. Low intensity is indicated by purple and blue while red indicates higher intensity. Lower panel) Same as above except the contaminating longterm enhancements (SWCX emission) were not removed. The striping is due to the survey geometry where great circles on the sky including the ecliptic poles were scanned precessing at $\sim 1°$ per day. [from Snowden et al., *Astrophys. J.*, **485**, 125–135 1997].

TABLE 1 Summary of the Characteristics of Soft X-Ray Emission from Solar System Bodies

Object	Emitting Region	Power Emitted[a]	Special Characteristics	Possible Production Mechanism
Earth	Auroral atmosphere	10–30 MW	Correlated with magnetic storm and substorm activity	Bremsstrahlung from precipitating electrons + characteristic line emission from atmospheric neutrals due to electron impact
Earth	Nonauroral atmosphere	40 MW	Correlated with solar X-ray flux	Scattering of solar X-rays by atmosphere
Jupiter	Auroral atmosphere	0.4–1 GW	Pulsating (~20–60 min) X-ray hot spot in north polar region	Energetic ion precipitation from magnetosphere and/or solar wind + electron bremsstrahlung
Jupiter	Nonauroral atmosphere	0.5–2 GW	Relatively uniform over disk	Resonant scattering of solar X-rays + possible ion precipitation from radiation belts
Moon	Dayside surface	0.07 MW	Correlated with solar X-rays	Scattering and fluorescence due to solar X-rays by the surface elements on dayside.
	Night side (geocoronal)		Night side emissions are ~1% of the day side	SWCX with geocorona
Comets	Sunward-side coma	0.2–1 GW	Intensity peaks in sunward direction, ~10^5–10^6 km ahead of cometary nucleus, and is correlated with solar wind parameters	SWCX with cometary neutrals
Venus	Sunlit atmosphere	50 MW	Emissions from ~120 to 140 km above the surface	Fluorescent scattering of solar X-rays by C and O atoms in the atmosphere
Mars	Sunlit atmosphere	1–4 MW	Emissions from upper atmosphere at heights of 110–130 km	Fluorescent scattering of solar X-rays by C and O atoms in the upper atmosphere
	Exosphere	1–10 MW	Emissions extend out to ~8 Mars radii	SWCX with Martian corona
Io	Surface	2 MW	Emissions from upper few micrometers of the surface	Energetic jovian magnetospheric ions impact on the surface
Europa	Surface	3 MW	Emissions from upper few micrometers of the surface	Energetic jovian magnetospheric ions impact on the surface
Io plasma torus	Plasma torus	0.1 GW	Dawn-dusk asymmetry observed	Electron bremsstrahlung + ?
Saturn	Sunlit disk	0.1–0.4 GW	Varies with solar X-rays	Scattering of solar X-rays + Electron bremsstrahlung ?
Rings of Saturn	Surface	80 MW	Emissions confined to a narrow energy band around at 0.53 keV.	Fluorescent scattering of solar X-rays by atomic oxygen in H_2O ice + ?
Asteroid	Sunlit surface		Emissions vary with solar X-ray flux	Fluorescent scattering of solar X-rays by elements on the surface
Heliosphere	Entire heliosphere	10^{16} W	Emissions vary with solar wind	SWCX with heliospheric neutrals

[a] The values quoted are values at the time of observation. X-rays from all bodies are expected to vary with time. For comparison, the total X-ray luminosity from the Sun is 10^{20} W. SWCX ≡ solar wind charge exchange = charge exchange of heavy, highly ionized solar wind ions with neutrals.

solar system. It has been demonstrated that roughly half of the observed 0.25-keV X-ray diffuse background can be attributed to this process (see Fig. 14). SoHO observations of neutral hydrogen Lyman alpha emission show a clear asymmetry in the ISM flow direction, with a clear deficit of neutral hydrogen in the downstream direction of the incoming neutral ISM gas, most likely created by SWCX ionization of the ISM. The analogous process applied to other stars has been suggested as a means of detecting stellar winds. Also a strong correlation between the solar wind flux density and the *ROSAT* "long-term enhancements," systematic variations in the soft X-ray background of the *ROSAT* X-ray detectors has been shown. Photometric imaging observations of the lunar night side by *Chandra* made in September 2001 do not show any lunar night side emission above a SWCX background. The soft X-ray emission detected from the dark side of the Moon, using *ROSAT*, would appear to be attributable not to electrons spiraling from the sunward to the dark hemisphere, as proposed earlier, to SWCX in the column of heliosphere between the Earth and the Moon.

Just as charge exchange–driven X-rays are emitted throughout the heliosphere, similar emission must occur within the astrospheres of other stars with highly ionized stellar winds that are located within interstellar gas clouds that are at least partially neutral. Although very weak, in principle, this emission offers the opportunity to measure mass-loss rates and directly image the winds and astrospheres of other main sequence late-type stars. Imaging would provide information on the geometry of the stellar wind, such as whether outflows are primarily polar, azimuthal, or isotropic, and whether or not other stars have analogs of the slow (more ionized) and fast (less ionized) solar wind streams.

14. Summary

Table 1 summarizes our current knowledge of the X-ray emissions from the planetary bodies that have been observed to produce soft X-rays. Several other solar system bodies, including Titan, Uranus, Neptune, and inner-icy satellites of Saturn, are also expected to be X-ray sources, but they are yet to be detected. During its flyby in 2008–2009, NASA's *MESSENGER* spacecraft will measure X-rays from Mercury by the onboard XRS experiment. Such measurements will continue after insertion of *MESSENGER* in the Herminian orbit in 2011. The *MESSENGER* XRS will provide information on elemental composition in the Mercury surface by observing the Kα line of elements present that are induced by solar X-rays as well as by high-energy electron precipitation.

Acknowledgments

A large part of this chapter is based on the review article by Bhardwaj et al. (2006), which is a collective effort of several authors, and we deeply acknowledge all the authors of that paper. We also thank the entire solar system X-ray community whose work have led to this review.

Bibliography

Bhardwaj, A. (2006). X-Ray emission from Jupiter, Saturn, and Earth: A short review. In "Advances in Geosciences," vol. 3, pp. 215–230. World Scientific, Singapore.

Bhardwaj, A., Elsner, R. F., Gladstone, G. R., Cravens, T. E., Lisse, C. M., Dennerl, K., Branduardi-Raymont, G., Bradford, J., Wargelin, J., Hunter Waite, Jr., J., Robertson, I., Ostgaard, N., Beiersdorfer, P., Snowden, S. L., and Kharchenko, V. (2006). X-Rays from solar system bodies. *Planetary and Space Science*, in press.

Dennerl, K., Lisse, C. M., Bhardwaj, A., Burwitz, V., Englhauser, J., Gunnels, H., Holmström, M., Jansen, F., Kharchenko, V., and Rodríguez-Pascual, P. M. (2006). First observation of Mars with *XMM-Newton*. High resolution X-ray spectroscopy with RGS. *Astronomy and Astrophysics* **451** (2), 709–722.

Dennerl, K., Burwitz, V., Englhauser, J., Lisse, C., and Wolk, S. (2002). Discovery of X-rays from Venus with *Chandra*. *Astronomy and Astrophysics* **386**, 319–330; A&A homepage.

Krasnopolsky, V. A., Greenwood, J. B., and Stancil, P. C. (2004). X-Ray and extreme ultraviolet emission from comets. *Space Science Review* **113**, 271–374.

Lisse, C. M., Cravens, T. E., and Dennerl, K. (2004). X-Ray and extreme ultraviolet emission from comets. In "Comet II" (M. C. Festou, H. U. Keller, and H. A. Weaver, eds.), pp. 631–643. Univ. Arizona Press, Tucson.

The Solar System at Ultraviolet Wavelengths

Amanda R. Hendrix

Jet Propulsion Laboratory
California Institute of Technology
Pasadena California

Robert M. Nelson

Jet Propulsion Laboratory
California Institute of Technology
Pasadena California

Deborah L. Domingue

Applied Physics Laboratory
Johns Hopkins University
Laurel, Maryland

CHAPTER 36

Ultraviolet imaging and spectroscopy are powerful tools for probing planetary atmospheres and surfaces. In this chapter, we review the significant contributions to our understanding of the solar system that have been made by ultraviolet observing methods. We cover results from the near-ultraviolet (NUV, 2000–3500 Å), far-ultraviolet (FUV, 1000–2000 Å), and extreme-ultraviolet (EUV, 500–1000 Å) wavelength ranges. These are shorter than visible and near-infrared wavelengths and involve photons of increasingly higher energy. Ultraviolet observations therefore provide unique insight into planetary processes involving more energetic processes that cannot be studied using photons of longer wavelength. Many of the solar system observations in the ultraviolet have been performed by Earth-orbiting telescopes, such as the *International Ultraviolet Explorer* (*IUE*) and the *Hubble Space Telescope* (*HST*). We also review results from ultraviolet instruments on *Galileo*, *Cassini*, *Voyager*, and other spacecraft. Each planet in the solar system except Mercury has been observed in the ultraviolet by Earth-orbiting telescopes. Many of the larger planetary satellites, selected asteroids, and comets have also been observed. This data set has provided important information regarding the atmospheres and surfaces of solar system objects and the processes shaping their compositions.

1. A Brief History of Ultraviolet Astronomy

The ultraviolet spectral region is important to the entire community of astronomers, from those who study nearby objects such as Earth's Moon to those who study objects at the edge of the observable universe. From the perspective of a planetary astronomer, the spectral information is important for determining the composition of, and understanding the physical processes that are occurring on, the surfaces and atmospheres of solar system objects.

Prior to the dawn of the space age, spectrophotometry of solar system objects at wavelengths shorter than ~3000 Å had long been desired in order to complement observations made by ground-based telescopes at longer wavelengths. However, the presence in Earth's atmosphere of ozone, a strong absorber of ultraviolet light between 2000 and 3000 Å, and molecular oxygen (O_2), which is the dominant ultraviolet absorber below 2000 Å, prevented astronomers of the 1950s and earlier from observing the universe in this important spectral region.

The ultraviolet wavelengths of astronomical sources became observable midway through the last century when instruments could be deployed above Earth's atmosphere. A rocket or spacecraft provides a platform from which astronomical observations can be made where the light being collected has not been subjected to absorption from Earth's atmospheric gases. Thus, the space revolution dramatically enhanced the ability of astronomers to access the full spectrum of electromagnetic radiation emitted by celestial objects.

In the 1950s, a series of rocket-flown instruments began slowly to reveal the secrets of the ultraviolet universe. The first photometers and spectrometers were flown on unstabilized Aerobee rockets. They remained above the ozone layer for several tens of minutes while they scanned the sky at ultraviolet wavelengths. By the early 1960s, spectrometers on three-axis-stabilized platforms launched by rockets on suborbital trajectories were able to undertake observations with sufficient resolution such that individual spectral lines could be resolved in the target bodies.

Shortly thereafter, the military spacecraft designated 1964-83C carried an ultraviolet spectrometer into Earth orbit. This was followed closely by NASA's launch of the first Orbiting Astronomical Observatory (OAO) satellite in 1966. These space platforms permitted long-duration observations compared to what was possible from a rocket launch on a suborbital trajectory. By 1972, the third spacecraft of the OAO series was launched. It was designated the *Copernicus* spacecraft and was an outstanding success.

In Europe, a parallel pattern of development for exploring the ultraviolet sky was under way using sounding rockets followed by orbiting spacecraft. In 1972, the European Space Research Organization launched an Earth-orbiting spacecraft (TD-1A) dedicated to ultraviolet stellar astronomy. Such developments set the stage in the 1970s for a joint U.S.–European collaboration, the *International Ultraviolet Explorer* satellite.

The *IUE* spacecraft was launched in 1978 into a geosynchronous orbit over the Atlantic Ocean. From there it could be controlled from ground stations in Greenbelt Maryland in the United States or in Villefranca, Spain, by engineers from NASA or ESA. It functioned continuously from launch until it was terminated in 1996 and its capabilities taken over by instruments on the *Hubble Space Telescope*. *IUE* spectra were recorded in two wavelength ranges of 1150–1950 Å and 1900–3200 Å, at either high or low spectral resolution. *IUE* had no imaging capability, though spatial discrimination was possible within the largest (10 × 20 arcsec oval) spectrograph entrance aperture.

Additional Earth-orbiting satellites with ultraviolet observing capabilities were launched in the early 1990s. These include NASA's *Extreme Ultraviolet Explorer* satellite (*EUVE*) and the joint U.S.–European *Hubble Space Telescope*. *HST* is in a low-Earth orbit, allowing upgrades to the facilities by astronauts. However, the low orbit reduced the observational duty cycle to 50–60% of that of *IUE* in high orbit. UV spectroscopy with *HST* has been performed with the Goddard High-Resolution Spectrograph (GHRS), the Faint-Object Spectrograph (FOS), the Space Telescope Imaging Spectrograph (STIS), and the Advanced Camera

for Surveys (ACS). In 1990 and 1995, the *Hopkins Ultraviolet Telescope* (*HUT*) was flown aboard the U.S. space shuttle as part of the Astro Observatory. The *Far-Ultraviolet Spectroscopic Explorer* (*FUSE*) was launched in 1999 and has spectroscopic capabilities in the 900–1200 Å wavelength range.

Many interplanetary spacecraft missions have included ultraviolet instruments in their payloads. *Pioneers 10* and *11*, which were launched in 1970 and 1973, respectively, included ultraviolet photometers among their scientific instruments. [*See* PLANETARY EXPLORATION MISSIONS.] These two spacecraft were the first to safely pass through the Asteroid Belt and fly by Jupiter and Saturn. *Mariner 6* and 7, Mars flyby missions launched in 1969, and *Mariner 9*, the first spacecraft to orbit Mars, launched in 1971, all carried ultraviolet spectrometers. *Mariner 10*, which flew by Mercury three times in 1974 and 1975, carried two extreme ultraviolet spectrometers (an **airglow** spectrometer and an occultation spectrometer) to measure the planet's exospheric composition. *Mariner 10* also made measurements of Earth's Moon after launch. *Pioneer Venus*, which was launched in 1978, was the first U.S. mission dedicated to the exploration of the planet Venus. It included an ultraviolet spectrometer among its instrument package. Soviet spacecraft missions *Vega 1* and *Vega 2*, launched in 1985, dropped two descent probes into Venus' atmosphere, which included a French–Russian ultraviolet spectroscopy experiment. The *Voyager* project sent two spacecraft that included ultraviolet spectrometers within their instrument payloads, both launched in 1977, to the outer solar system. *Voyager 2* was the first spacecraft to fly by all four of the jovian planets (Jupiter, Saturn, Uranus, and Neptune). In 1989, the *Galileo* spacecraft was launched. This spacecraft was the first dedicated mission to the Jupiter system, and it included within its scientific instrument payload two ultraviolet spectrometers, the EUV (extreme ultraviolet spectrometer, which operated between 500 and 1400 Å) and the UVS (the ultraviolet spectrometer that covered the 1150–4300 Å wavelength range). En route to Jupiter, the *Galileo* spacecraft collected ultraviolet spectra as it flew by Venus, the Moon, and the asteroids Gaspra and Ida. The *Cassini* mission, launched in 1997, includes the Ultraviolet Imaging Spectrograph (UVIS) and arrived at Saturn for a 4 year tour in June 2004. The *Nozomi* spacecraft, launched in 1998, carried two UV instruments; measurements were made of the Moon en route to Mars. (*Nozomi* unfortunately failed to enter Mars orbit.) *Mars Express*, in orbit since December 2003, has an ultraviolet instrument called SPICAM as part of its payload. In 2004, the *MErcury Surface, Space ENvironment, GEochemistry, and Ranging* (*MESSENGER*) mission launched on its way to Mercury carrying an ultraviolet-visible (1150 to 6000 Å) spectrometer. Also in 2004, Rosetta with its ALICE ultraviolet instrument (covering 700–2050 Å) was launched en route to comet 67P/Churyumov–Gerasimenko, set to enter orbit in 2014. The *New Horizons* mission to Pluto, launched in January 2006, carries the next-generation version ALICE instrument, covering 520–1870 Å. *Venus Express* has been in orbit at Venus since April 2006 and employs the SPICAV/SOIR ultraviolet and infrared instrument for atmospheric studies.

2. Nature of Solar System Astronomical Observations

Most astronomers observe objects that have their own intrinsic energy source, such as stars and galaxies. However, the majority of the observations undertaken by planetary astronomers are of targets that do not emit their own radiation but are observable principally because they reflect the sunlight that falls on them or emit energy as a result of various physical processes. The measured spectrum of a body can thus reveal significant information on the composition of, and processes occurring within, planetary surfaces and atmospheres. The measured spectrum includes absorption features that can determine or constrain the composition of a surface or atmosphere, or emission features that suggest excitation processes in a gas or thermal emission from solids. The measured spectrum often displays solar features (either emission features or spectral continuum). To study the spectrum of the body itself, the solar spectrum is divided out, resulting in what is known as the spectral reflectance. The variation of the reflectance or **geometric albedo** (the reflectance at zero phase angle) as a function of wavelength is used to measure the strength of absorption features, from which the abundance of spectrally active species can be estimated.

Measuring reflected light at ultraviolet wavelengths can pose some interesting problems for instrument designers. First, instrument spectral sensitivity becomes weaker with decreasing wavelength and so does the Sun's energy output. The energy output of the Sun changes by a factor of 10^3 between the EUV and NUV spectral ranges, which until recently exceeded the dynamic range of UV detectors.

Second, in order to obtain the spectral reflectance of a body, a solar spectrum must be measured, which is no easy task in the ultraviolet range. Furthermore, below 1800 Å, the spectrum of the Sun is variable. Therefore, a simultaneous spectrum of the Sun (or the reflection spectrum from an object whose spectrum is well understood) must be gathered at the same time that any ultraviolet observations are undertaken.

Lastly, particularly when performing measurements from an Earth-orbiting observatory such as *IUE* or *HST*, solar system objects change positions against the background of stars during the course of an individual observation. In most cases, special tracking rates must be calculated prior to each observing run in order to know the change of the position of the target with time. Inaccurately calculated tracking rates can cause the observed target to drift from the instrument's field of view, thus adding noise and uncertainty to a measurement.

3. Observations of Planetary Atmospheres

With the exception of the innermost planet Mercury (which possesses a surface-bounded exosphere, similar to that observed on the Moon), all the planets in the solar system (and a few planetary satellites) are surrounded by detectable atmospheres. All the planets with atmospheres absorb ultraviolet light, and as a result ultraviolet observations provide information on the composition of, and processes that are occurring in, the object's atmosphere.

In general, the atmospheres of the terrestrial planets (Mercury, Venus, Earth, and Mars) are considered secondary atmospheres because they evolved after the primordial atmospheres were lost. However, the atmospheres of the four jovian planets (Jupiter, Saturn, Uranus, and Neptune), because of their strong gravitational attraction and comparatively low temperatures, retained the primordial elements, particularly hydrogen and helium. From ground-based observations, methane (CH_4) and ammonia (NH_3) were identified in the atmospheres of the giant planets and therefore atmospheric processes were suspected of producing a host of daughter products that can be detected at ultraviolet wavelengths. [*See* ATMOSPHERES OF THE GIANT PLANETS.]

Sunlight entering a planetary atmosphere can experience or initiate a wide variety of processes that contribute to the total energy emitted by the object and observed by an astronomical facility. The objects described previously all possess atmospheres that contribute significantly to their spectral behavior. Astronomical observations of such bodies search for and measure the depths of absorption bands in the spectrum or emission bands due to atmospheric interactions with both solar photons and energetic particles that originate from the solar wind or the planet's magnetosphere. These bands are unique to specific gases; thus, it is possible to identify or eliminate particular gases as candidate materials in the atmospheres of these objects. The interpretation of an ultraviolet spectrum can be an arduous task, given that the bands and lines observed in the spectrum may arise from a combination of processes. These include:

1. Single and multiple scattering of photons by aerosols such as haze and dust (**Mie scattering**) and gas (**Rayleigh scattering/Raman scattering**) in the planetary atmosphere.

2. Absorption of the incident ultraviolet solar light by atmospheric species.

3. Stimulation of an atmospheric gas by incident sunlight and emission by **fluorescence**, chemiluminescence, or resonant scattering.

4. Photoionization and photodissociation reactions that produce a reaction product in an excited state.

5. Excitation of gas by precipitation of magnetospheric particles.

Each of these processes is associated with a well-understood physical mechanism, the details of which are beyond the scope of this chapter. The reader is referred to the Bibliography and other chapters in this volume.

Two significant methods of studying atmospheres at UV wavelengths are stellar occultations (observing a star as a body passes in front of the star and measuring the stellar flux as it is diminished) and reflection/airglow measurements (measurements of the backscatter of the solar continuum either by Rayleigh–Raman atomic/molecular scattering or by Mie scattering from atmospheric aerosols). The atmospheric species and density can be constrained by studying the occulted stellar spectrum as it passes through the atmosphere. Limited-wavelength facilities can identify some but not all of the constituents present and processes ongoing in a planetary atmosphere. The ultraviolet data from Earth-orbiting satellites have been used in combination with ground-based observations at other wavelengths and with observations by other spacecraft (including flyby missions) to develop an understanding of the atmospheres of planetary objects. The following discussion summarizes the results of those bodies in the solar system that possess atmospheres.

3.1 Mercury and the Moon

Both Mercury and the Moon have very tenuous atmospheres that are often referred to as surface-bounded exospheres. The atoms in these atmospheres do not collide with each other; rather, they bounce from place to place on the surface. The *Mariner 10* ultraviolet airglow experiment detected hydrogen, helium, and oxygen atoms as constituents in Mercury's exosphere. No molecules were detected. The pressure of Mercury's atmosphere was determined to be about 10^{-12} bar (compared to the 1 bar atmospheric pressure at sea level on Earth). Ground-based telescopic observations in the visible have identified resonant scattering emission features attributed to sodium, potassium, and calcium as well. Observations of Mercury's exospheric sodium demonstrate that it is spatially and temporally variable, and the variability is not solely related to interactions with the solar environment. Sources for the known exospheric species include impact vaporization, ion sputtering, thermal and photon stimulated desorption, crustal outgassing, and neutralization of solar wind ions. The relative importance of these production mechanisms has been debated, but they predict the existence of several species (such as Ar, Si, Al, Mg, Fe, S, and OH) that have yet to be detected. The *MESSENGER* spacecraft carries an ultraviolet spectrometer as part of the Mercury Atmosphere and Surface Composition Spectrometer (MASCS) instrument package. This spectrometer operates from 1150 to 6000 Å and its goal is to map the constituents of the atmosphere and provide information to relate them to specific source and production mechanisms.

The known lunar atmospheric species present in detectable abundances are Ar, He, Na, and K. So far, only upper limits on other species have been set using UV wavelengths. In the UV, the lunar atmosphere was initially studied at FUV wavelengths by the *Apollo 17* UVS, which provided upper limits on the number density of H, H_2, O, C, N and CO. More recently, *HST* FOS NUV observations of the region away from the surface of the Moon resulted in upper limits on OH, Al, Si, and Mg abundances.

3.2 Venus

For more than half a century, the very dense Venus atmosphere has been known to be composed principally of carbon dioxide (CO_2) based on the existence of strong spectral absorption features in the near-infrared spectrum. Several layers of clouds many kilometers thick composed of sulfuric acid completely cloak the surface. Although these clouds obscure the surface at visual and ultraviolet wavelengths, the *Magellan* spacecraft used radar to construct maps of the volcanic surface. Atmospheric measurements in the UV have been performed by sounding rockets and spacecraft, including *IUE*, the *Pioneer Venus* orbiter, and *HUT*. An image from *Pioneer Venus* is shown in Fig. 1. (This 3650 Å image is just outside the strict definition of the NUV range.) Ultraviolet images of Venus' atmosphere show distinctive cloud patterns; in particular, a horizontal "Y"-shaped cloud feature (discovered by *Mariner 10* Venus scientists in 1974) is visible near the equator. This feature may suggest atmospheric waves, analogous to high and low pressure cells on Earth. Bright clouds toward Venus' poles appear to follow latitude lines. The polar regions are bright, possibly showing a haze of small particles overlying the main clouds. The dark regions show the location of enhanced sulfur dioxide near the cloud tops.

Within a few years of launch, *IUE* identified several important trace constituents, including nitric oxide (NO), and confirmed the presence of several others, such as sulfur dioxide (SO_2). The *Vega 1* and *2* probes measured local ultraviolet absorptions due primarily to SO_2 and aerosols. Ultraviolet reflectance spectra obtained during two sounding rocket observations in 1988 and 1991 found that SO_2 is the primary spectral absorber between 1900 and 2300 Å and that sulfur monoxide (SO) is also present in Venus' atmosphere. The EUV instrument aboard the *Galileo* spacecraft

FIGURE 1 *Pioneer Venus* Orbiter Cloud Photopolarimeter (OCPP) image of Venus at 3650 Å.

observed Venus in the extreme ultraviolet wavelength range (550–1250 Å) during its flyby. It detected emissions due to helium, ionized oxygen, atomic hydrogen, and an atomic hydrogen–atomic oxygen blend. In 1994, an extreme ultraviolet spectrograph (EUVS) was launched aboard a sounding rocket to observe the Venusian atmosphere from 825 to 110 Å. The EUVS identified several species, including N I, N II, N_2, H I, O I, and O II. The results of the EUVS measurements are consistent with earlier observations by *IUE*, *Pioneer Venus*, *Venera 11* and *12*, and the *Galileo* EUV spectrometer. The *EUVE* provided the first full EUV (70–760 Å) spectrum of Venus in 1998 and made brightness measurements on the He I (584 Å) and O II (539 Å) lines. The FUV spectrum of Venus is dominated by the CO Fourth-positive band system, as well as by neutral oxygen and carbon features, and has been measured by *HUT* (820–1840 Å) in 1995 and by *Cassini* UVIS.

IUE spectra of the Venus day side and night side obtained while Venus was near elongation displayed SO_2 absorptions at 2080–2180 Å, which when combined with the **column densities** reported by the *Pioneer Venus* orbiter and with ground-based observations, are a measure of the SO_2 **mixing ratio** with altitude and its variation at the top of the cloud deck. This provides information on its variation in spatial distribution and permits models to be constructed of the planet's atmospheric dynamics. Observations

of the Venus night side with *Pioneer Venus* orbiter and *IUE* detected the Venus nightglow, which is caused by the emission bands of nitric oxide (NO). Because of the short lifetime of NO on the night side, this finding implies the rapid day side–night side transport of material in the Venus atmosphere. Observations of the Venus day side have led to the discovery that the dayglow emission is carbon monoxide fluorescence, probably due to fluorescent scattering of solar Lyman-alpha radiation.

3.3 Mars

The atmosphere of Mars, like that of Venus, is dominated by carbon dioxide, and also consists of small amounts of N_2, H_2O, and their photochemical products. Mars' atmosphere is much less dense than Venus's atmosphere and is relatively transparent at most wavelengths. Therefore, ultraviolet to infrared observations of Mars reveal information about both its atmosphere and its surface. The observations of Mars by the UVS instruments on *Mariner 6* and *7* were the first to reveal the ultraviolet dayglow of that planet; later observations by *Mariner 9* confirmed and extended these results. NUV spectra (Fig. 2a) revealed the presence of CO (a-X) Cameron bands, CO^+ (B-X), CO_2^+ (A-X), and CO_2^+ (B-X) features. FUV spectra displayed oxygen features at 1304 and 1356 Å, neutral carbon emission features at 1561 and 1657

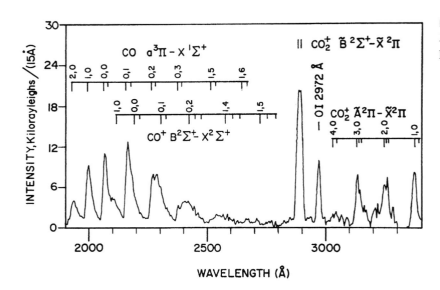

FIGURE 2a   Mars dayglow as measured by the *Mariner 9* UVS. (Figure reproduced with permission from Elsevier.)

Å, and the CO (A-X) at fourth-positive bands. All these UV features of the martian airglow are products of processes involving Mars' CO_2 atmosphere. The CO_2^+ band systems are the result of a combination of photoionization excitation of CO_2 and fluorescent scattering of CO_2^+, and the CO (a-X) and (A-X) bands are due to photon or electron dissociative excitation of CO_2. The presence of escaping hydrogen (Ly-α at 1216 Å), suggested atomic hydrogen within the atmosphere, and also suggested the accumulation of oxygen and loss of water.

More recent (1995) *HUT* observations (820–1840 Å) confirmed these early *Mariner* results (Fig. 2b), and *FUSE* measurements detected molecular hydrogen (H_2) emission features at 1070 and 1167 Å for the first time. *EUVE* provided the first measurements of helium (584 Å) within the martian atmosphere. These helium observations have been used to set constraints on outgassing processes. Recent UV observations from the *Mars Express* spacecraft have made the first detection of the martian nightglow, revealing nitric oxide emission features similar to those seen on Venus, auroral emissions associated with crustal magnetic field features, and high altitude CO_2 ice clouds.

Mariner 6 and 7 UVS observations revealed the presence of the Hartley band of ozone (O_3), centered near 2550 Å. The feature was detected at the south polar cap, through ratios of south polar spectra to low latitude spectra (Fig. 2c). Further observations with the UVS on the *Mariner 9* orbiter revealed that the north and south polar ozone amount

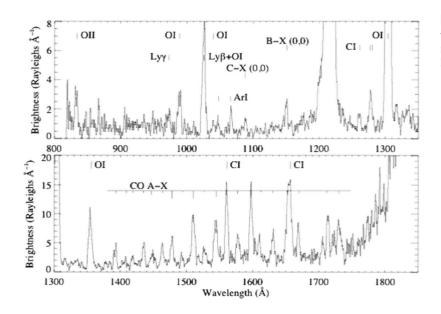

FIGURE 2b   Mars spectrum at EUV–FUV wavelengths as measured by *HUT*. (Figure reproduced with permission from AAS Publications.)

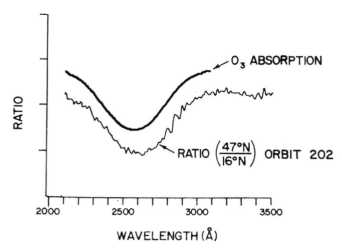

FIGURE 2c Ozone at high latitudes on Mars as measured by *Mariner 9* UVS. A high-latitude spectrum is shown ratioed to a low-latitude spectrum, compared with a laboratory spectrum of the Hartley band of O_3. (Figure reproduced with permission from Elsevier.)

varied with season. Ozone densities were highest in winter and lowest in summer, anticorrelated with atmospheric water vapor content. The correlation between higher amounts of ozone with a cold, clean, dry atmosphere led to the conclusion that ozone is formed through the combination of atomic and molecular oxygen, both of which are more readily present when less water is available. Subsequent *HST* observations have studied the seasonal variation of atmospheric ozone at low latitudes and have linked low latitude ozone abundance variations across the martian perihelion–aphelion cycle with the large annual water vapor variation due to the eccentricity of Mars' orbit. *HST* observations have also been used to study atmospheric aerosol (dust and cloud) opacities. These UV observations demonstrate the critical function that is performed by the small amounts of H_2O in the martian atmosphere, which control the buildup of CO and O_2, and sustain the stability of CO_2.

3.4 Jupiter

Jupiter, the target of numerous Earth-based observations as well as spacecraft flybys, is composed of 90% hydrogen and 10% helium, with small amounts of ammonia and methane. The uppermost layers of the atmosphere are observable in the ultraviolet wavelength range and display products of photochemical processes.

The first FUV spectra of Jupiter were measured in sounding rocket experiments in the late 1960s and early 1970s. These early measurements displayed H_2 Lyman and Werner band emissions and hinted at the presence of absorption features due to C_2H_2, C_2H_4 and NH_3. Stellar occultation observations by the *Voyager* UVS were significant in providing measurements of upper atmospheric

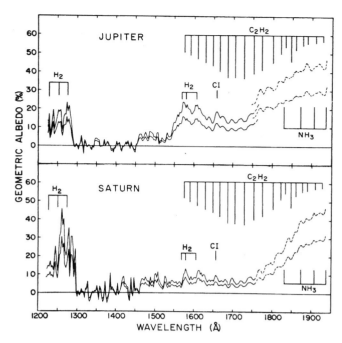

FIGURE 3a FUV geometric albedos of Jupiter and Saturn as measured by *IUE*. The albedos are derived from composite spectra of the planets between 1978 and 1980. The upper and lower curves for each planet correspond to assumptions of uniform and cosine-limb-darkened disks, respectively. The dashed lines represent data that are uncertain in magnitude owing to the subtraction of scattered light and are regarded as upper limits. (Figure reproduced with permission from AAS Publications.)

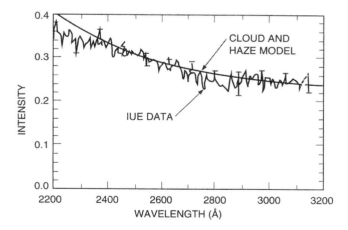

FIGURE 3b The spectral geometric albedo of Jupiter as measured by *IUE*. The smooth solid line is the best fit from a model that assumes a layer of haze particles with single-scattering albedo of 0.42 that overlie a cloud deck with geometric albedo of 0.25. (Figure reproduced with permission from Elsevier.)

temperatures. Early *IUE* observations confirmed that C_2H_2 absorption bands are the dominant features in the 1650–1850 Å wavelength region. Figure 3a shows the geometric albedo of Jupiter in the FUV wavelength range, displaying C_2H_2 and NH_3 features, derived using a composite spectrum of Jupiter from the 1978–1980 time frame.

The spectral geometric albedo of Jupiter as measured by *IUE* at NUV wavelengths is shown in Fig. 3b. Most of this spectral behavior is attributable to hazes that are high above the cloud deck. The best-fit model (solid line in the figure) to the data occurs for a jovian cloud deck with a geometric albedo of 0.25 and for a haze composed of particles with a single scattering albedo of 0.42. Though such a result may not be able to provide an unambiguous identification of the materials that compose the haze, it can constrain the eligible candidate materials that are suggested by other observations.

IUE observations also permitted an ammonia–hydrogen mixing ratio to be calculated, and it was found to be 5×10^{-7}. The fact that *IUE* was able to observe the absorption features of these species indicates that they are above the jovian tropopause, where the clouds create an opaque barrier to light emitted from the material underneath and hence make spectral identification of the underlying material impossible. In July 1994, the comet Shoemaker–Levy 9 (SL-9) collided with Jupiter. It was not until the impact of the fragmented comet that studies of this underlying material became possible. The *EUVE* satellite observed Jupiter before, during, and after this event. *EUVE* found that 2 to 4 hours after the impact of several of the larger fragments, the amount of neutral helium temporarily increased by a factor of ~10. This transient increase is attributed to the interaction of sunlight with the widespread high-altitude remnants of the plumes from the larger impacts. *HST* also observed this event with the GHRS and the Faint-Object Camera (FOC). The ultraviolet spectra obtained by *HST* of Jupiter after the collision of SL-9 identified approximately 10 species of molecules and atoms in the perturbed atmosphere, many of which had never been detected before in Jupiter's atmosphere. Among these were S_2, CS_2, CS, H_2S, and S^+, which are believed to be derived from a sulfur-bearing parent molecule native to Jupiter. The observations also detected stratospheric ammonia (NH_3). Neutral and ionized metals, including Mg II, Mg I, Si I, Fe I, and Fe II, were also observed in emission and are believed to be from the SL-9 comet fragments. The surprising observation was the absence of absorptions due to oxygen-containing molecules.

A major focus of study at UV wavelengths is the polar regions of Jupiter and their impressive exhibit of auroral activity. Jupiter's auroral displays are the most energetic in the solar system. FUV measurements were first made using the *Voyager* UVS, and subsequent observations were performed by *IUE*. The far-ultraviolet emissions are dominated by the hydrogen Lyman-alpha and the H_2 Lyman

and Werner system bands. Synoptic observations of these ultraviolet emissions using *IUE* have shown that they vary with Jupiter's magnetic (not planetary) longitude, and hence these emissions are magnetospheric phenomena. *IUE* observations have been used to construct a spatial map of the Lyman-alpha emission and the data indicate that the emitting material is upwelling at about 50 m/s relative to the surrounding material. More intensive ultraviolet observations with *FUSE* and *HST* instruments GHRS, FOC, STIS, and ACS have measured the temporal variability within the aurora and temperature variations within the auroral ovals seen at both poles. These variations are reflections of possible distortions in the magnetic field of Jupiter. *HST* measured the first detection of reversed Lyman-alpha emissions, which are linked to variable atomic hydrogen. Estimates of vertical column densities ($1–5 \times 10^{16}$ cm^{-2}) of atomic hydrogen above the auroral source have been made. *HST* has also detected ultraviolet emission from a superthermal hydrogen population. The *Galileo* spacecraft EUV and UVS spectrometers also observed Jupiter's aurora. These observations have placed constraints on the vertical distribution of methane (CH_4) in Jupiter's atmosphere. Slant methane column abundances are estimated to be 2×10^{16} cm^{-2} in the north and 5×10^{16} cm^{-2} in the south based on the *Galileo* observations. *Cassini* UVIS measurements showed that Jupiter's aurora responded strongly to the compression events produced when large solar coronal mass ejections reached Jupiter's magnetosphere. Figure 4 displays Jupiter's UV aurorae as imaged by *HST* in 1998. Evident in Fig. 4 is the auroral "footprint" of Io, where the field line intersecting Io connects to the planet, revealing the magnetospheric relationship between the planet and the moon. Magnetic footprints of the other moons also exist but are less obvious in this image.

Bright H Ly-α emissions have been observed from Jupiter's equatorial region, and the source of this "equatorial bulge" has been debated. The source is likely a combination of charged particle excitation and solar resonance scattering and fluorescence. The emission has been shown to be consistent with resonant scattering of solar Ly-α with a large planetary line width, requiring a fractional (~1%) suprathermal population of fast H atoms in the uppermost atmosphere. The fast atoms are likely due to dissociative excitation of molecular hydrogen (H_2).

3.5 Saturn

Like Jupiter, Saturn's atmosphere is dominated by hydrogen and helium, with traces of water, ammonia, and methane. The far-ultraviolet spectrum of Saturn was first measured in sounding rocket experiments in 1978. Absorption features in the ultraviolet spectrum of Saturn that have been associated with acetylene (C_2H_2) in the upper atmosphere were discovered using early *IUE* measurements. Figure 3a displays the FUV geometric albedo of Saturn derived using

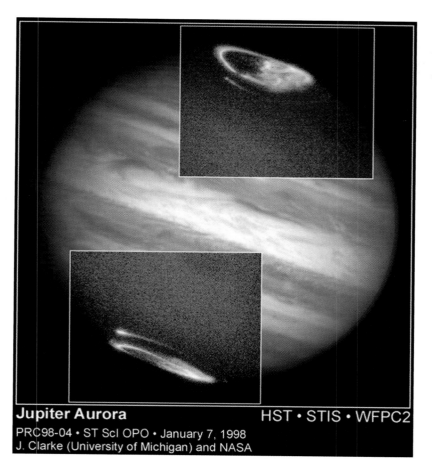

Jupiter Aurora

HST • STIS • WFPC2

PRC98-04 • ST Scl OPO • January 7, 1998
J. Clarke (University of Michigan) and NASA

FIGURE 4 *HST* image of Jupiter aurora. The magnetic "footprint" of Io, marking the location where magnetic field lines joining the moon and Jupiter connect with the planet, is also seen as a bright spot with a tail outside the main auroral oval. (Image credit: J. Clarke, NASA.)

a composite of *IUE* Saturn spectra from the 1978–1980 time period. The mixing ratio of the acetylene is about 1×10^{-7}. Although acetylene is a well-known strong absorber of ultraviolet radiation, it alone cannot explain the low UV spectral geometric albedo of Saturn that has been reported by *IUE*. Other ultraviolet-absorbing materials must be present. Comparisons of laboratory spectra of C_2H_2, PH_3, AsH_3, and GeH_4 with the *IUE* observations show that the best-fit model for Saturn's atmospheric ultraviolet spectrum includes absorptions by C_2H_2, H_2O, CH_4, C_2H_6, PH_3, and GeH_4. The distribution of PH_3 and GeH_4 decreases with increasing altitude in these models, suggesting that ultraviolet photolysis is an important process occurring at higher altitudes.

Pole-to-pole mapping studies of the hydrogen Lyman-alpha emission across Saturn's disk led to the discovery of pronounced spatial asymmetries in the emission. Other observations of hydrogen do not find a variation in intensity with rotational period as with Jupiter. There is no rotational bulge in the Lyman-alpha emission as seen on Jupiter. This is probably due to the fact that Saturn's magnetic pole is coincident with the rotational pole, whereas in Jupiter's case the poles are offset.

Like Jupiter, Saturn displays auroral activity. On both planets this auroral activity also creates aerosols that are detectable in the ultraviolet as dark-absorbing regions. *HST* FOC ultraviolet observations discovered a dark oval encircling the north magnetic pole that is spatially coincident with the aurora detected by the *Voyager* UVS. *Voyager 2* ultraviolet PPS measurements also demonstrate a geographical correlation between the auroral zones of Jupiter and Saturn with UV-dark polar regions. Additional ultraviolet observations with the *HST* FOC of Saturn's northern ultraviolet aurora and polar haze support the hypothesis that the polar haze particles are composed of hydrocarbon aerosols produced during H_2^+ auroral activity. More recent *HST* UV imaging of Saturn's aurorae (Fig. 5) shows that they behave differently from Jupiter's aurorae, varying in brightness and shifting in latitude.

3.6 Uranus

Uranus presents a unique observational circumstance to the inner solar system observer because of the fact that its pole is inclined 89° to the ecliptic and that at the present position in its 84 year orbit about the Sun it presents its pole

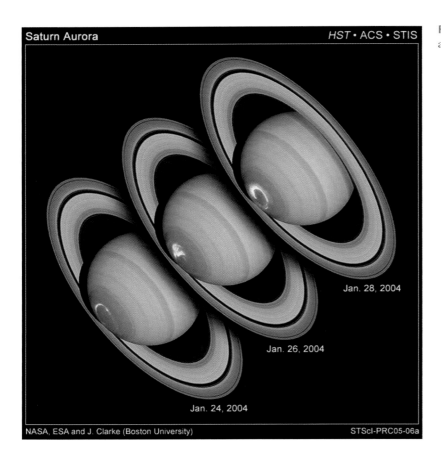

FIGURE 5   *HST* images of Saturn's varying aurorae. (Image credit: J. Clarke, NASA.)

to Earth. This unusual inclination, combined with its great distance from Earth, makes it impossible to use an Earth-based instrument to undertake pole-to-pole comparisons as was done with Jupiter and Saturn. Uranus has a geometric albedo at NUV wavelengths of about 0.5, more than twice that of Jupiter and Saturn. This suggests that additional absorbers are present in the jovian and Saturanian atmospheres that are not present in the atmosphere of Uranus. Both Uranus and Neptune possess hot thermospheres and stratospheres that are substantially clear of hydrocarbons and other heavy constituents, making the UV albedos higher than for Jupiter and Saturn. A sharp increase in measured reflectance intensity at wavelengths longward of 1500 Å is indicative of acetylene (C_2H_2) present in the atmosphere of Uranus.

Voyager 2 spacecraft observations of Uranus found a very small internal heat source compared to the large internal heat sources found at Jupiter and Saturn. This suggests that there is very little atmospheric mixing driven by heating and buoyancy in the Uranian atmosphere. Thus, ultraviolet observations are able to sense a deeper region of the atmosphere.

The ultraviolet emissions from Uranus' atmosphere have been measured by *IUE* and the *Voyager* UVS. To increase the signal-to-noise ratio, *IUE* observers used principally low-resolution observations and binned broad-wavelength regions together to search for broadband absorbers at ultraviolet wavelengths. Analysis of the *IUE* observations detected acetylene absorptions, which were also detected on Jupiter and Saturn. Based on these observations, the mixing ratio of the acetylene is estimated to be 3×10^{-8}. Analysis of the *Voyager* UVS observations of H_2 band ultraviolet airglow emissions shows aurora at both magnetic poles, which are offset from the rotational poles by $\sim 60°$. The auroral emissions on Uranus are very localized in magnetic longitude and do not form complete auroral ovals as are seen on Jupiter and Saturn.

3.7 Neptune

Neptune is so distant that only broadband ultraviolet measurements are possible from Earth orbit. The geometric albedo of Neptune measured by *IUE* is 0.5, which, like that of Uranus, is twice that of Jupiter and Saturn. Below 1500 Å, Neptune's albedo is reduced by the higher hydrocarbon abundance carried into its stratosphere by its more vigorous vertical transport. Most of the important data for Neptune at ultraviolet wavelengths have come from the UVS onboard the *Voyager 2* spacecraft and from *HST*. CH_4 and C_2H_6 abundances inferred from the *Voyager* UVS solar occultation experiment are between 0.0006 and 0.005 mole fraction for CH_4 in the lower stratosphere (with a mixing ratio of 5–100×10^{-5}) and a density of C_2H_6 estimated to reach $3 \times 10^9 \, cm^{-3}$.

In 1994, *HST* imaged Neptune in six broadband filters, one of which was in the ultraviolet. The goal of these observations was to study the cloud structure on Neptune and compare the measurements with the observations made by *Voyager 2*. The *HST* images showed that the Great Dark Spot seen by *Voyager* no longer existed, but a new large dark feature of comparable size had appeared in the northern latitudes. *HST* measurements also detected weak carbon monoxide lines at 1992 and 2063 Å, suggesting a mixing ratio of $\sim 3 \pm 2 \times 10^{-6}$ in the upper troposphere.

Voyager 2 UVS measurements tentatively identified weak auroral emissions at Neptune's South Pole, interpreted as H_2 emissions. The North Pole was not observed by *Voyager*, so it is unknown whether that hemisphere displays aurora.

3.8 Pluto

Pluto and its large satellite Charon are at a great distance from Earth and are quite small compared to the four gas-giant planets that populate the outer solar system. UV spectroscopy is a potentially rich source of information about these icy bodies due to the atmospheric chemistry that is likely occurring. The presence of methane in Pluto's atmosphere suggests that photochemical products should include hydrocarbons and nitriles, detectable at UV wavelengths. *IUE* obtained a few spectra of these objects and observed that the ultraviolet albedos vary with rotation. The amplitude of the rotational variation as measured at ultraviolet wavelengths by *IUE* is greater than the rotational variation measured at longer wavelengths by Earth-based observers, consistent with the presence of an absorbing material that is spectrally active in the 3200–4800 Å wavelength range; the geometric albedo of Pluto in the NUV is spectrally flat. The composition of the absorbing material is unknown. [*See* PLUTO.]

Observations made with the FOS on *HST* were used to determine upper limits on Pluto's predicted atmospheric species C_4H_2, C_6H_2, HC_3N, and C_4N_2 of 1.6×10^{16}, 1.8×10^{16}, 2.7×10^{16}, and $4 \times 10^{16} \, cm^{-2}$, respectively. The ultraviolet spectrum of Pluto's satellite, Charon, was also measured by the *HST* FOS and was found to have a spectrally flat geometric albedo in the NUV; Charon's spectrum does not exhibit any absorption or emission features that provide compositional clues.

The Pluto–Charon system is the target of the upcoming *New Horizons* flyby mission, which will include an FUV imaging spectrograph (which operates from 520 to 1870 Å) to probe the atmospheres and surfaces of these distant worlds.

3.9 Galilean Satellites

Jupiter's moon Io has one of the most unique atmospheres in the solar system: The primary sources of the atmosphere are volcanic emissions and sublimation of SO_2 frost on the surface. The result is a tenuous, patchy atmosphere made up of SO_2, SO, S_2, S, and O; trace species include Na, K, Cl, NaCl, and H. Gaseous SO_2 can be deposited onto the surface at night or during eclipse by Jupiter. Material is also lost to the torus as the ionized material sweeps by Io. Gaseous SO_2 was discovered at Io by the IRIS instrument on *Voyager*; since that discovery, much study of Io's atmospheric processes has been made at UV wavelengths. The *IUE*, *HST*, and the *Galileo* UVS have made measurements of Io at near-UV wavelengths (2000–3500 Å). Far-ultraviolet observations from *HST* and *Cassini* UVIS identified emissions from neutral oxygen and sulfur and have been used in mapping the distribution of the SO_2 atmosphere.

Associated with Io is a plasma torus, or a donut-shaped ion cloud centered at Io's orbital radius. This torus has been studied by *Pioneer*, *Voyager*, *IUE*, *HST*, *EUVE*, *HUT*, *FUSE*, *Galileo*, and *Cassini*. Oxygen, sulfur, and sodium ions are the major constituents of the torus, and protons are present at $\sim 10\%$ abundance; chlorine ions have also been detected. The torus is not uniform, and the density of ions shows various asymmetries dependent on Io's position and dawn–dusk timings, in addition to temporal variations.

Intriguing auroral features are a consequence of Io's SO_2 atmosphere, resulting from electron impact excitation of atomic oxygen and sulfur, and electron dissociation and excitation of SO_2, and have been observed at visible and FUV wavelengths. The Io flux tubes (IFT) and the Io plasma torus are the two primary sources of electrons in the Io environment. Due to the $10°$ tilt of Jupiter's magnetic field, Io is alternately above and below the magnetic equator (depending upon Io's System III jovian magnetic longitude, λ_{III}), the primary region of the torus electrons. Furthermore, the tangent points between field-aligned electrons and Io's atmosphere change as Jupiter rotates. The interaction of the torus electrons and the IFT electrons with Io's atmosphere has been detected at FUV wavelengths. Equatorial spots (Fig. 6) have been observed to wobble up and down, reflecting the changing location of the IFT tangent points in time. The equatorial spot on the antijovian hemisphere has been measured to be brighter than the spot on the subjovian hemisphere, likely due to the motion of electrons through Io's atmosphere by the Hall effect, with hotter electrons on the antijovian side. [*See* IO: THE VOLCANIC MOON.]

Observations with the *HST* GHRS have detected atomic oxygen emissions at 1304 and 1356 Å from Jupiter's satellite Europa, which have been interpreted as evidence for a tenuous O_2 atmosphere about this satellite. The source of this oxygen atmosphere is likely sputtering of the icy surface by corotating magnetospheric particles. These emission

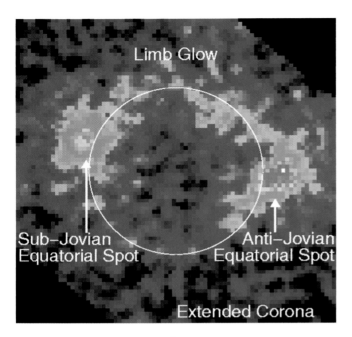

FIGURE 6 Io imaged at 1356 Å by *HST* STIS. Equatorial "spots" are the result of interaction between electrons flowing along the Io flux tubes with Io's SO_2 atmosphere. (Figure courtesy of K. Retherford.)

features were also measured by the *Cassini* UVIS instrument during the Jupiter system flyby in 1999–2000 (Fig. 7). Similar oxygen emission features were detected by *HST* at Ganymede, though it was found that Ganymede's emissions are restricted primarily to the polar regions. The emis-

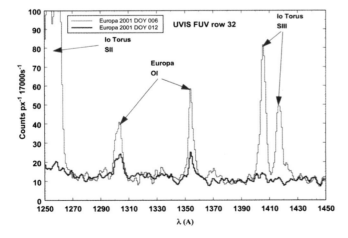

FIGURE 7 The FUV spectrum of Europa as measured by *Cassini* UVIS on January 6, 2001, and January 12, 2001. Neutral oxygen emission features appear at 1304 and 1356 Å in both observations. Io torus emission features from ionized sulfur also appear. The presence of the O I features is due to electron dissociation and excitation of a tenuous O_2 atmosphere at Europa. (Figure reproduced with permission from Elsevier.)

sions are auroral features produced by dissociative excitation of O_2 by electrons traveling along the field lines of Ganymede's own magnetosphere. *HST* imaged Ganymede's auroral emissions at 1356 Å and found them to be temporally and longitudinally variable (Fig. 8). *Galileo's* UVS detected hydrogen escaping from Ganymede, possibly due to sputtering of Ganymede's surface by charged particles. Callisto, in contrast to Ganymede and Europa, does not exhibit oxygen emission features. An analysis of *HST* measurements found that the oxygen and CO emission features, expected after the discovery of CO_2 gas, are so faint that Callisto's interaction with the magnetosphere is like that of a unipolar inductor, and that another species such as O_2 is likely abundant, enhancing the ionosphere and its conductivity. Callisto's ionosphere is apparently of sufficient conductivity to reduce the flow of plasma into its atmosphere and inhibits oxygen emission features in contrast to Ganymede and Europa.

3.10 Titan and Triton

Saturn's satellite Titan and Neptune's satellite Triton are among the largest satellites in the solar system. In addition, they are far from the Sun; therefore, the reduced solar energy allows the atmospheric gases to remain cold enough that they cannot easily escape by thermal processes. [*See* TITAN; TRITON.]

Titan is a solar system curiosity due to its very thick (~1.5 bar) nitrogen atmosphere, which prevents UV observations of the surface. Ground-based and *Voyager* spacecraft observations have identified methane (CH_4) as a significant constituent of Titan's atmosphere. Analyses of *IUE* and *HST* observations of Titan at NUV wavelengths have placed constraints on the properties of Titan's high-altitude haze and the abundances of simple organic compounds such as acetylene (C_2H_2). At FUV wavelengths, observations by *Cassini* UVIS demonstrate the presence of molecular and atomic nitrogen based on emission features due to electron dissociation and excitation (Fig. 9).

Triton's surface contains N_2, CO, and CH_4 frosts, which are highly volatile and in a continual state of exchange between the atmosphere and surface. *IUE* observations of distant Triton likely tested the limits of *IUE's* sensitivity. The photopolarimeter on *Voyager 2* measured an albedo of 0.59 on all sides of Triton. *HST* FOS observations in 1993 detected broad apparent absorption features centered near 2750 Å and between 2000 and 2100 Å. The FOS analysis also led to mixing ratio upper limits for atmospheric constituents of OH, NO, and CO of 3×10^{-6}, 8×10^{-5}, and 1.5×10^{-2}, respectively. *HST* STIS observations from August to September 1999 showed that Triton's albedo in the 2500–3200 Å range was 15–30% brighter, and also spectrally redder, than measured by the *HST* FOS in 1993, suggesting that Triton's NUV albedo undergoes changes on timescales shorter than the seasonal cycle. Such changes

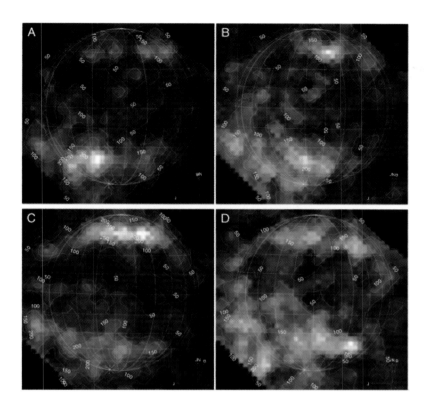

FIGURE 8   *HST* STIS images of Ganymede's UV aurora. The images represent neutral oxygen at 1356 Å and reveal brightening in the polar regions that is variable over the four *HST* orbits during which the imaging took place. (Figure reproduced with permission from AAS Publications.)

may be due to bright frost deposition or to the emplacement of a relatively dark UV material. At this point, the source of the dramatic, short-timescale changes in UV brightness on Triton remains unknown.

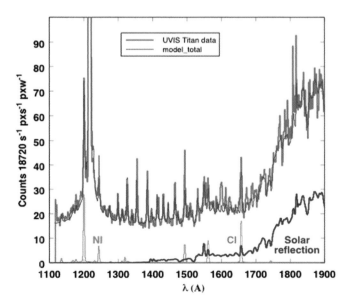

FIGURE 9   Titan's FUV spectrum as measured by the *Cassini* UVIS. The spectrum includes an overall continuum due to reflected solar light, in addition to emission features due to electron dissociation and excitation of molecular and atomic nitrogen. (Figure courtesy of D. Shemansky.)

4. Observations of Solid Surfaces

Many solid-state materials that make up the surfaces of solar system objects exhibit spectral absorption features, and thus it is possible to identify or constrain the abundance of solid components on the surfaces of these objects. This is accomplished by comparing the spectral geometric albedo of the object with the reflection spectrum of the solid-state materials as measured in the laboratory. The following discussion focuses on ultraviolet observations of solid surfaces throughout the solar system. Light is reflected from particulate surfaces by volume scattering and surface scattering. At longer visible and near-IR wavelengths, volume scattering dominates in most materials. At shorter NUV and FUV wavelengths, surface scattering dominates. Absorptions in the NUV are generally due to charge-transfer and result in rather broad absorption features, compared with absorption features in the near-IR, usually due to weaker electronic transitions. Many non-ice materials are absorbing in the NUV, including organics, sulfur compounds, and many refractory materials. There is usually an absorption edge between 3000 and 3700 Å. In contrast, ices such as H_2O and CO_2 are not very absorbing at visible wavelengths and at NUV wavelengths. (Water and CO_2 ice exhibit strong absorption features near 1650 Å.) Atmophereless bodies in the solar system experience weathering, whereby their surfaces are affected by bombardment of micrometeorites as well as charged particles, either from the solar wind in the case of the Moon and asteroids or from the magnetosphere

TABLE 1	Ultraviolet Geometric Albedos of the Galilean Satellites		
	2600–2700 Å	**2800–3000 Å**	**3000–3200 Å**
Io (leading)	0.015 ± 0.001	0.017 ± 0.001	0.042 ± 0.001
Io (trailing)	0.028 ± 0.002	0.030 ± 0.005	0.038 ± 0.003
Europa (leading)	0.213 ± 0.004	0.347 ± 0.010	0.407 ± 0.020
Europa (trailing)	0.118 ± 0.002	0.164 ± 0.004	0.222 ± 0.006
Ganymede (leading)	0.15 ± 0.007	0.190 ± 0.009	0.200 ± 0.001
Ganymede (trailing)	0.050 ± 0.003	0.060 ± 0.004	0.080 ± 0.008
Callisto (leading)	0.040 ± 0.008	0.049 ± 0.001	0.066 ± 0.002
Callisto (trailing)	0.056 ± 0.002	0.064 ± 0.002	0.105 ± 0.008

of the parent planet in the case of the icy satellites of the outer solar system. Particularly in the case of icy surfaces, radiation-aged surfaces tend to be darker at UV wavelengths. Thus, UV observations of icy surfaces can be used to indicate ice "freshness," or amount of contamination.

4.1 Galilean Satellites

The first in-depth ultraviolet studies of the Galilean satellites (Io, Europa, Ganymede, and Callisto) were accomplished with the use of the *IUE* satellite. Subsequent disk-integrated observations with *HST* supported the initial findings of *IUE*, in addition to adding to our knowledge of the composition of the surfaces of these satellites. Galileo studies contributed to disk-resolved studies of these bodies.

The very high spatial resolution provided by *Voyager* and *Galileo* visible images shows that the Galilean satellites, particularly Io, are variegated in color on continental scales. Compositional information may be derived from high spectral resolution studies from Earth or near Earth orbit because the satellite's synchronous rotation permits any given full-disk observation of a satellite to be associated with a uniquely defined hemisphere of that particular object. The extension of the available spectral range to shorter wavelengths with ultraviolet telescopes enhances this data set by permitting the identification of more absorption features, thereby providing further constraints on the compositional models that have been developed. [*See* PLANETARY SATELLITES.]

The Galilean satellites are phase-locked, so that one hemisphere (the subjovian, central longitude of 0° W) faces Jupiter at all times. The leading hemisphere is the side that faces the direction of motion of the satellite in its orbit and is centered on 90°W longitude, while the trailing hemisphere has a central longitude of 270°W. The corotating charged particles of Jupiter's magnetosphere have orbital speeds greater than those of the moons, so that the plasma sweeps by the moons, impacting primarily the trailing sides. An in-depth study of several hundred *IUE* spectra of the Galilean satellites revealed significant hemispheric UV spectral asymmetries that are indicative of composi-

tional variations. Ratios of spectra from the leading and trailing hemispheres are a useful tool for studying hemispheric compositional variations.

Table 1 lists the geometric albedos of the Galilean satellites in three NUV wavelength bands, for the leading and trailing hemispheres. This table displays the significant hemispheric differences in brightness exhibited by these moons. Io's leading hemisphere is brighter than the trailing hemisphere only in the longest NUV wavelength band. Shortward of 3000 Å, Io's trailing side has a higher albedo than its leading side, just the opposite of what is seen when Io is observed at visible wavelengths. (*IUE*'s precursor, *OAO-2*, measured Io's albedo at 2590 Å to be just 3%, in marked contrast to its 70% albedo at visible wavelength. This result was so unusual that it remained in doubt until *IUE* confirmed it by measuring Io's spectrum in this spectral range.) Io's reversal in brightness associated with orbital phase is more pronounced than for any other object in the solar system, and proved to be important in efforts to determine the surface composition variation in longitude across Io's surface. It can be directly inferred from the Io data that there is a longitudinally asymmetric distribution of a spectrally active surface component on Io's surface. The material was determined to be sulfur dioxide (SO_2) frost, which is strongly absorbing shortward of ~3200 Å and very reflective longward of that wavelength, as a result of *IUE* observations. Sulfur dioxide frost is in greatest abundance on the leading hemisphere of Io, and it is in least abundance on the trailing hemisphere. Figure 10 shows a Galileo UVS spectrum that displays Io's dramatic increase in albedo between 2000 and 3300 Å.

Europa and Ganymede exhibit a variation in brightness at NUV wavelengths with orbital phase that is in the same sense as the variation reported at the visible wavelengths; at all NUV wavelengths, these objects are brighter on their leading sides than on their trailing sides. A gradual decrease in albedo toward shorter wavelengths occurs on both hemispheres of both objects. The ratio of *IUE* spectra of Europa's trailing hemisphere to its leading hemisphere led to the discovery of an absorption feature present primarily on the trailing hemisphere centered near 2800 Å. The absorption

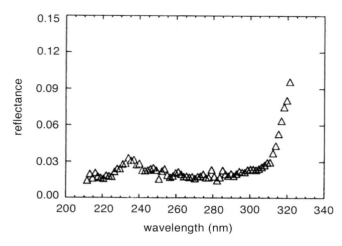

FIGURE 10 The NUV reflectance spectrum of Io, displaying the dramatic increase in brightness with wavelength starting at ~320 nm, due to the presence of SO_2 frost on the surface. This spectrum was measured by the *Galileo* UVS. The original discovery of SO_2 frost on the surface of Io in NUV spectra was made using *IUE* spectra. (Figure reproduced with permission from AAS Publications.)

feature, displayed in Fig. 11a, was attributed to an S—O bond and was suggested to be due to implantation of sulfur ions into the ice lattice on the trailing hemisphere. *HST* measurements confirmed the absorption feature, and it was suggested that the feature was similar to laboratory spectra of SO_2 frost on water ice. Subsequent disk-resolved *Galileo* UVS measurements showed that the 2800 Å absorption feature is strongest in regions associated with visibly dark terrain. These locations have also been found to have relatively high concentrations of non-ice material, interpreted to be hydrated sulfuric acid or hydrated salt minerals. An additional *Galileo* discovery was the presence of hydrogen peroxide (H_2O_2) on Europa, primarily in regions of lower non-ice concentrations, such as on the leading hemisphere (Fig. 11b). A study of Europa spectra from the early *IUE* era (1978–1984) compared to the late *IUE* era (1995–1996) suggested a temporal variation in Europa's leading and antijovian hemisphere spectra that may be linked with variations in H_2O_2 abundances as a result of temporal variability in the space environment.

IUE spectra of Ganymede's trailing hemisphere ratioed to the leading hemisphere revealed the presence of a possible absorption feature centered close to 2600 Å, though the signal was approaching the *IUE* detection limits. It was suggested that ozone (O_3) in the ice could explain the apparent absorption feature. Subsequent *HST* measurements confirmed the presence of the O_3 absorption feature in the ice lattice on the trailing hemisphere. Disk-resolved observations of Ganymede from *Galileo* showed that the O_3 feature was strongest in the polar regions, and at large solar zenith angles, suggesting a connection with the magnetic field lines, or with photolysis or ice temperatures. Figure 11c displays the O_3 absorption feature as measured

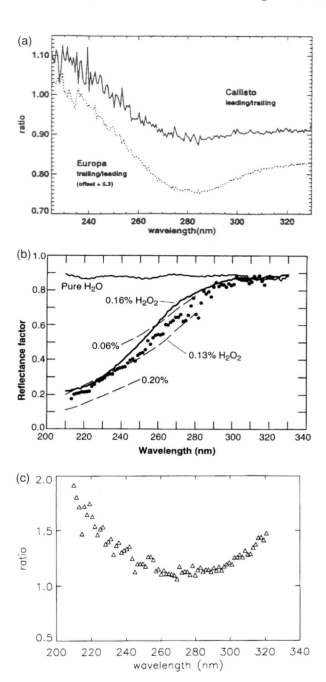

FIGURE 11 Spectra of significant absorption features on the icy Galilean satellites. (a) SO_2 absorption features on Europa and Callisto obtained by ratioing the spectra of the trailing to leading hemisphere (Europa) and the spectra of the leading to trailing hemisphere (Callisto). (Figure reproduced with permission from AAS Publications.) (b) Hydrogen peroxide (H_2O_2) absorption feature as measured on Europa by the *Galileo* UVS. Also shown are mixture models for varying amounts of H_2O_2 in a water ice mixture, and the spectrum of pure H_2O ice. (Figure reproduced with permission from AAAS/Science.) (c) Ozone (O_3) absorption feature as measured on Ganymede by *Galileo* UVS. Shown is the ratio of a spectrum from the north polar region to a region on the leading hemisphere. The broad absorption feature mimics O_3 in water ice. (Figure reproduced with permission from AAS Publications.)

by the *Galileo* UVS. Galileo measurements also found that the UV absorption feature associated with H_2O_2 is present also on Ganymede and is anticorrelated with the O_3 concentrations.

Ground-based observations of Callisto have found that its albedo varies with **orbital phase angle** in the opposite sense to that of Europa and Ganymede (i.e., its trailing side has a higher albedo than its leading side). This is also true at NUV wavelengths. The albedo of Callisto decreases shortward of 5500 Å and continues to decrease throughout the NUV. Its albedo at all wavelengths is lower than the albedo of Europa and Ganymede. Analysis of many of the *IUE* spectra, in addition to *HST* spectra, shows a broad, weak absorption at 2800 Å similar to that seen on Europa (Fig. 11a). These observations suggest the presence of SO_2 in a few leading hemisphere regions. The source of this SO_2 is not well understood and may be linked with implantation of neutral sulfur flowing outward from Io. The H_2O_2 absorption feature seen by *Galileo* UVS at Europa and Ganymede is also seen at Callisto, though the absorption feature is weaker and its distribution is less obvious.

4.2 Saturnian Satellites

All of Saturn's large and medium-sized satellites, like the Galilean satellites and Earth's Moon, are in synchronous rotation (with the exception of Phoebe and Hyperion). At visual wavelengths, Tethys, Dione, and Rhea all have leading side albedos that are 10–20% higher than those of their trailing sides, which suggests that there are longitudinal differences in chemical/mineralogical abundance and/or composition in the optically active regoliths of these objects. The hemispheric albedo asymmetry of Iapetus at visual wavelengths is extremely large (the trailing side is brighter by a factor of 5). Infrared observations of the large satellites of Saturn have identified water ice as the principal absorbing species of the optically active surface of Mimas, Enceladus, Tethys, Dione, and Rhea and the trailing (bright) hemisphere of Iapetus. The leading (dark) hemisphere of Iapetus does not show spectral features consistent with water ice

and has an infrared spectrum that is nearly featureless. The albedo of water ice alone is too high for the surfaces of the satellites to be covered only by this material. Other materials must be present in varying amounts to explain the albedos of all the Saturnian satellites. In the case of the dark hemisphere of Iapetus, the darkening material is most probably the dominant specie on the surface. Observations at improved spectral resolution and extended spectral range are required to identify these absorbers on the surfaces of the Saturnian satellites. The suite of instruments on the *Cassini* spacecraft will likely make progress in determining the other species that characterize the surfaces of these moons.

A limited number of UV observations of the Saturnian satellites have been undertaken from Earth orbit. Mimas and Enceladus are too close to Saturn to obtain useful spectra from Earth: There is too much scattered light from Saturn. The next satellites, in order of distance from Saturn, are Tethys, Dione, Rhea, and Iapetus; these moons were first successfully measured at UV wavelengths by *IUE* and later observed by *HST*. The ultraviolet geometric albedos for the Saturnian satellites were calculated for the three UV wavelength bandpasses using a combination of *IUE* and *HST* data. These are shown in Table 2.

The UV albedo of Tethys (~60%) is the highest of the Saturnian satellites and is comparable to the high visual albedo reported by *Voyager* and ground-based visual observations. The leading side of Dione is ~30% brighter than its trailing side, similar to the brightness variation reported from ground-based visual wavelength observations. The leading side of Rhea is ~60% brighter than its trailing side at the NUV wavelengths. This is more than the ~20% observed from the ground at visual wavelengths and may signal the presence of a UV absorption feature focused on the trailing hemisphere.

The UV albedo of Iapetus is consistent with the albedos reported at longer wavelengths from ground-based observations and the *Voyager* spacecraft. In the UV, as in the visual, the leading side of Iapetus is extremely absorbing, and the trailing side is comparable to the trailing side albedos

TABLE 2	Ultraviolet Geometric Albedos of the Saturnian Satellites		
	2400–2700 Å	**2800–3000 Å**	**3000–3200 Å**
Tethys (leading)	0.52 ± 0.02	0.57 ± 0.02	0.61 ± 0.02
Tethys (trailing)	0.39 ± 0.01	0.46 ± 0.01	0.50 ± 0.02
Dione (leading)	0.58 ± 0.05	0.57 ± 0.03	0.60 ± 0.03
Dione (trailing)	0.24 ± 0.05	0.32 ± 0.04	0.33 ± 0.04
Rhea (leading)	0.43 ± 0.05	0.46 ± 0.03	0.46 ± 0.03
Rhea (trailing)	0.25 ± 0.09	0.28 ± 0.04	0.30 ± 0.04
Iapetus (leading)	0.040 ± 0.02	0.040 ± 0.01	0.041 ± 0.01
Iapetus (trailing)	0.298 ± 0.03	0.308 ± 0.02	0.32 ± 0.02

of other Saturnian satellites. The leading side albedo is ~6 times lower than the trailing side albedo at the IUE wavelengths, somewhat greater than the 5 times darker reported at visual wavelengths. The spectral absorber that darkens the leading hemisphere of Iapetus is more absorbing toward shorter wavelengths. Efforts to identify this absorber should focus on a similar decrease in reflectance in the laboratory spectrum of any candidate absorber.

The broadband UV albedos reported by *IUE* and *HST* observations of the Saturnian satellites confirm the suggested differences in chemical/mineralogical composition on Dione, Rhea, and Iapetus that the longer wavelength observations imply. In the case of Dione, these observations indicate that there are no strong UV absorptions in the unidentified materials on the satellite's surface. In the case of Rhea and Iapetus, the UV absorption becomes greater toward longer wavelengths. This may be due to a gradual decrease in reflectance or may be the effect of an absorption band.

HST observations of Dione and Rhea (Figs. 12a and 12b) have detected an absorption similar to the 2600 Å absorption feature detected by *HST* on Ganymede and has been attributed to the presence of ozone on both satellites. Like Ganymede, Dione and Rhea orbit within the magnetosphere of their planet. Ozone on these satellites is likely a product of radiolysis. Iapetus, orbiting outside the magnetosphere, does not exhibit the 2600 Å absorption feature. However, because the hemispheric albedo dichotomy on Iapetus is greater at NUV wavelengths compared to the visual wavelengths, the possibility of an absorption feature somewhere between 2400 and 5600 Å can be inferred.

More recently, with the arrival of the *Cassini* spacecraft at the Saturn system, FUV measurements of the icy satellites have been made with the UVIS. The FUV spectra of the icy satellites are dominated by the strong water absorption feature at ~1650 Å. At wavelengths shortward of ~1650 Å, the icy satellites are extremely dark due to the presence of water ice, and in fact are darker than the surrounding interplanetary hydrogen (IPH). *Cassini* UVIS images of the icy satellites therefore reveal both the day and night sides of the icy satellites. Figure 13a displays the FUV reflectance spectrum of Phoebe, one of the outermost satellites of Saturn; Phoebe's spectrum is compared with the UVIS-measured reflectance spectrum of the Moon. Figure 13b shows the FUV image of Phoebe; the visible wavelength image is shown in Fig. 13c for comparison.

4.3 Enceladus

Enceladus, not easily observed from Earth orbit due to its proximity to Saturn and the rings, has recently been the target of key UV measurements from the *Cassini* spacecraft in orbit around Saturn. Enceladus has long intrigued scientists because it is the brightest object in the solar system; *Voyager* images revealed vast regions that were evidently crater-free, suggesting recent resurfacing by geologic

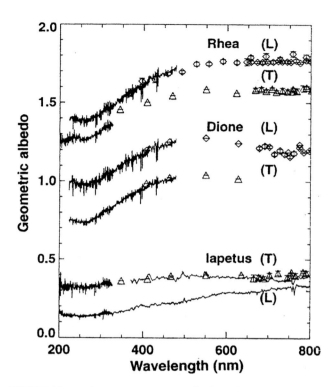

FIGURE 12a The NUV geometric albedos of the leading (L) and trailing (T) hemispheres of Rhea, Dione, and Iapetus as measured by *HST* FOS (thick solid lines). Also shown are longer wavelength albedos (thin lines and discrete points) from other sources. The Rhea albedos are offset by 1.0 and the Dione albedos are offset by 0.5. The spectrum of Iapetus' leading hemisphere has been scaled by a factor of 2.5. (Figure reproduced with permission from Nature Publishing Group.)

activity. Furthermore, its orbit at the densest part of the broad, tenuous E-ring has suggested that Enceladus could somehow be the source of the E-ring ice particles.

Ultraviolet measurements from *HST* detected the hydroxyl radical, OH, in emission (3085 Å) in the Saturn

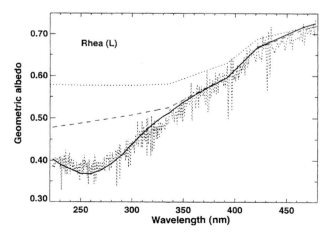

FIGURE 12b Rhea's leading hemisphere albedo (from Fig. 12a) with a model including O_3. (Figure reproduced with permission from Nature Publishing Group.)

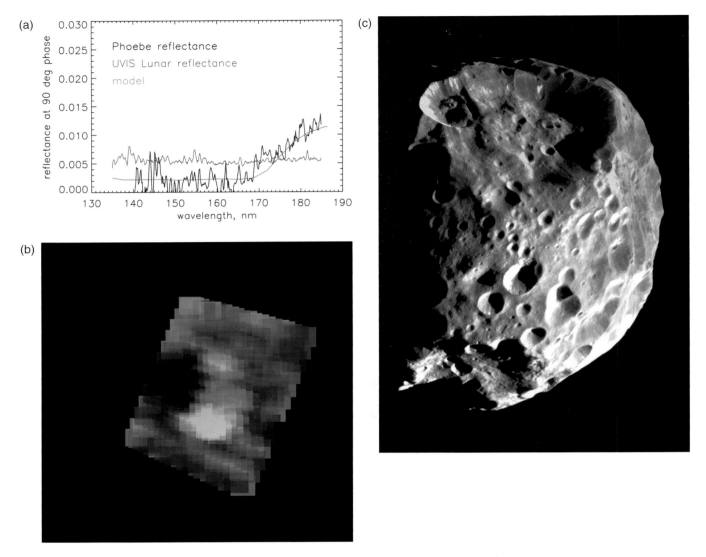

FIGURE 13   Saturn's moon Phoebe as measured by the *Cassini* UVIS. (a) Phoebe's FUV spectral reflectance measured at 90° **solar phase angle**. Phoebe's spectrum is overplotted with a model (red) that includes H_2O frost and dark, carbonaceous material. Also shown is the lunar albedo (blue) as measured by UVIS at the same phase angle. Note that water ice on Phoebe makes it even darker than the Moon at short FUV wavelengths, and brighter at long FUV wavelengths. (b) FUV wavelength image of Phoebe. Red represents 1216 Å where interplanetary hydrogen is bright throughout the solar system. Phoebe's water ice makes it darker than the background IPH on both the illuminated and the dark hemispheres so that the entire disk is visible. Blue colors represent longer FUV wavelengths and show that the brightness varies across Phoebe's surface due to solar incidence, topography, and compositional variations. (c) Visible wavelength image of Phoebe from *Cassini*. (Figures reproduced with permission from AAAS/Science; ISS image courtesy of NASA.)

system, primarily near the orbit of Tethys. Similarly, *Cassini* UVIS measured neutral oxygen (at 1304 Å), in varying amounts, with the greatest abundances near the orbit of Enceladus. The presence of OH and O suggested that H_2O is produced by erosion of the inner icy satellites of Saturn by micrometeoroid bombardment and is then broken down by photodissociation to produce the neutral species. However, the amounts of H_2O necessary to produce the observed

OH and O abundances were not consistent with sputtering rates; an additional source of H_2O was needed—and remained a mystery until *Cassini* observations of Enceladus in 2005.

The *Cassini* spacecraft, through unique multi-instrument observations, discovered active water plumes on Enceladus. A stellar occultation by *Cassini* UVIS measured the presence of water vapor above the limb of the

South Pole. Similar UV occultations of other regions of the moon had found no evidence of any gases, indicating that the vapor was locally confined to the south polar region. Surface temperatures, measured by the far-IR spectrometer, were found to be anomalously high; measurements by the magnetometer, mass spectrometer, dust detector, and near-IR spectrometer on *Cassini* confirmed the presence of gaseous species and ice grains being expelled from Enceladus' south polar hot spot. It is surprising that such a small, icy body is currently geologically active! The cause of the south polar hot spot and associated plumes is under investigation, and Enceladus remains a primary observational target of the *Cassini* mission.

4.4 Uranian Satellites

The five major satellites of Uranus—Miranda, Ariel, Umbriel, Titania, and Oberon—are a suite of icy satellites that are situated at about the limit at which *IUE* was able to confidently return spectral information. They are so faint that it is not even possible to divide the *IUE* wavelength range into several bands, as was done with the jovian and Saturnian satellites. All the spectral information is integrated into one wavelength range, and a geometric albedo can be determined.

The Uranian satellites are in an orbital plane that is parallel with the Uranian equator, and the pole of Uranus' orbit is tilted such that, at the present time, it is pointed toward Earth. Therefore, only the poles of one hemisphere of the satellites of Uranus are observable with *IUE*, and hence it is not possible to construct orbital phase curves and leading/trailing side ratio spectra.

IUE was able to observe Oberon, Uranus's brightest satellite. The *IUE* result proved to be an important and independent confirmation of results from the *Voyager 2* photopolarimeter experiment. The ultraviolet geometric albedo of Oberon was found to be 0.19–0.025, an excellent confirmation of the earlier *Voyager 2* PPS result of 0.17.

Spectra from 2200 to 4800 Å were obtained with the *HST* FOC for the Uranian satellites Ariel, Titania, and Oberon. The inner Uranian satellites Miranda and Puck were also observed from 2500 to 8000 Å with the *HST* FOC. The geometric albedos for Ariel, Titania, and Oberon display a broad, weak absorption at 2800 Å, similar to the feature seen on Europa and Callisto. Although this absorption feature on the Galilean satellites has been attributed to SO_2, it has been attributed to OH on the Uranian satellites. Both SO_2 and OH produce an absorption feature near 2800 Å, however, the molecule OH (a by-product of the photolysis and radiolysis of water) is unstable at the surface temperatures of the Galilean satellites but is stable at the colder surface temperatures of the Uranian satellites. No detection of the 2600 Å ozone feature seen on Ganymede, Dione, and Rhea has been detected in any of the Uranian satellite spectra.

4.5 A Comparison of Icy Satellite Systems

The ultraviolet observations of planetary satellites can be integrated with the results of observations at longer wavelengths to provide a comparative assessment of the families of large planetary satellites in the solar system. The *IUE*- and *HST*-determined photometric properties of the larger planetary satellites of Jupiter, Saturn, and Uranus are shown in Fig. 14 as a plot of ultraviolet-to-infrared color ratio versus ultraviolet geometric albedo. The geometric albedos of the Saturnian satellites indicate that in this system there is a wide variation in UV geometric albedo. The Galilean and Uranian satellites have photometric properties that are common within each group. This is consistent with the hypothesis that the surface modification processes that have occurred are similar within the Galilean and Uranian satellite systems, but the two systems have surface modification processes that are distinct from each other. The diverse nature of the photometric properties of the Saturnian satellites

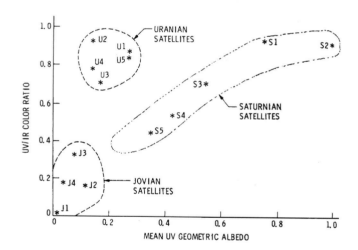

FIGURE 14 Comparisons of geometric albedos of the Galilean, Saturnian, and Uranian satellites. For the Galilean satellites and for Dione, Rhea, and Tethys, the UV albedos are from *IUE* and *HST*. For Mimas and Enceladus, the UV albedos are from *Voyager* images and are of longer wavelength (3500 Å) than those from *IUE* and *HST*. The infrared albedos of the Galilean and Saturnian satellites are from ground-based spectrophotometry. All the Uranian satellite data are from the *Voyager* Photopolarimeter. With the exception of the slight difference in wavelength from Mimas and Enceladus noted previously, all the wavelength ranges are similar. Jovian and Uranian satellites each have distinct color ratios that distinguish the two groups of satellites from each other. The Saturnian satellites have little albedo similarity among themselves. This is consistent with the hypothesis that the surfaces of all the icy Galilean satellites are being modified by a common process. Likewise, the Uranian satellite surfaces may also have a common process of surface modification that is different from the jovian system. The Saturnian satellite system may not have a common process of surface modification, or the system may have been disturbed and the albedos of the satellites altered.

suggests that no common surface modification process is altering the surfaces of the satellites or, alternatively, that the entire satellite system was recently modified in some way and the slow process of space weathering has not had time to restore the surfaces to a common photometric state. A possibility is that coating by E-ring grains, an ongoing process, effectively keeps the optical surface of the Saturnian satellites young and unweathered.

In general, the NUV spectra of icy satellites are dominated by weathering products. Radiolysis and photolysis are extremely important processes at the surfaces of these satellites, and products of these processes are apparent at NUV wavelengths. This is evidenced by the presence of SO_2, O_3, and H_2O_2 in the surfaces of the icy Galilean satellites (and O_3 in the surfaces of some of the icy Saturnian satellites). The icy Galilean satellites are all relatively dark at UV wavelengths. However, water ice, the primary constituent of these surfaces, is bright in the NUV. Therefore, another material must be responsible for the ultraviolet absorption of the icy Galilean satellites. The most likely darkening agents are elemental sulfur and sulfur-bearing compounds originating from the very young and active surface of Io, which are transported as ions outward from Io's orbit by jovian magnetospheric processes. These energetic ions and neutrals interact with the icy surfaces of Europa, Ganymede, and Callisto and cause the ices to become darkened at UV wavelengths. This process competes with other processes of surface modification such as infall of interplanetary debris.

In contrast to Jupiter's UV-dark icy satellites, Saturn's icy satellites, particularly those orbiting closer to Saturn, are relatively bright. This may be related to the presence of the large and tenuous E-ring. Mimas, Enceladus, Tethys, Dione, and possibly Rhea all orbit Saturn within this broad ring of tiny icy grains. The relative velocities between the E-ring particles and the icy satellites may explain the overall brightness, as well as the large-scale longitudinal albedo patterns on the icy satellites. Mimas and Enceladus are both slightly darker on the leading hemispheres than on the trailing hemispheres, possibly because the E-ring particles sweep by the trailing hemispheres, brightening them. The leading hemispheres of Tethys and Dione are brighter than the trailing hemispheres because their leading hemispheres sweep by the E-ring particles. Furthermore, Saturn's magnetosphere is different from Jupiter's and appears to be dominated by neutrals rather than by electrons and other charged particles. This difference may have an effect on the ice chemistry that occurs within the surfaces of the icy satellites because much of the ice chemistry occurring on the icy Galilean satellites is the result of charged particle bombardment. The Saturnian satellites are less spectrally red in the NUV than the Galilean satellites, and H_2O_2 does not appear to be present, in contrast to the Jovian satellites. These differences could be a result of the different charged particle environments in the Saturn and Jupiter magnetospheres.

4.6 Pluto and Charon

The first spatially resolvable images of Pluto and its satellite Charon were obtained by the *HST* FOC at visible (4100 Å) and ultraviolet (2780 Å) band passes. The image resolution is sufficient to show the presence of large, longitudinally asymmetric polar cap regions in addition to a variety of albedo markings. The combination of UV and visible images were used to look for regions of clean ice and nonclean (contaminated) ice—either radiation darkened or sites where atmospheric chemistry products were deposited. No positive identification of solids on the surfaces of Pluto and Charon has been made at UV wavelengths. The cleanest ice (bright in UV and VIS bandpasses) was found in a location at the equator, though overall the equator was found to be heavier in contaminated ice than mid latitudes. The north polar region was found to have the cleanest ice.

4.7 Asteroids and Comets

At ultraviolet wavelengths, asteroids have been studied by *IUE* and *HST* from Earth-based orbit. The *Galileo* spacecraft obtained NUV spectra of Ida and Gaspra during its travels through the Asteroid Belt. Finally, the *Mariner* 9 UVS obtained spectra of Phobos and Deimos, the martian moons that are likely captured asteroids.

The *IUE* satellite obtained ultraviolet observations of ~45 main-belt asteroids in the wavelength range between 2300 and 3250 Å. The geometric albedos for these objects are consistently low, and three major asteroid taxonomic classes seen in the visible persist into the ultraviolet. Analysis of the *IUE* asteroid data shows that the asteroids observed have ultraviolet albedos that range from 0.02 for C-class asteroids to 0.08 for M-class asteroids; albedos of S-class asteroids are intermediate. Analysis of a set of *IUE* Vesta observations covering more than one rotation of Vesta indicates that this unusual asteroid displays UV albedo variations across the surface such that the UV lightcurve of Vesta is opposite that of the visible lightcurve. Such a spectral reversal is consistent with a hemispheric dichotomy in composition and/or a variation in geologic age (due to space weathering) across the surface. A study of *IUE* measurements of S-class asteroids has found that the UV spectral slope may be an important indicator of space weathering and ultimately exposure age; the strong decrease in albedo that is typical in silicates at NUV wavelengths appears to lessen with exposure. Measurements at visible-infrared wavelengths suggested the presence of hydrated minerals on Ceres; a search for OH (emissions at ~3085 Å) in *IUE* spectra, however, found none. *HST* FOC, and more recently ACS, images of Ceres at UV wavelengths were the first well-resolved images of this largest asteroid. Albedo variations are detectable across the surface. The brightness of the surface in the FUV may rule out the presence of a large amount of water ice; an absorption band with a central wavelength of 2800 Å may be present but its cause is unidentified. Analysis of *Mariner 9* UVS spectra of Phobos and Deimos show these bodies to be spectrally similar to carbonaceous chondrites. However, analysis of *HST* FOS data of these moons at UV-visible wavelengths, compared with FOS spectra of a C-type asteroid and a D-type asteroids, showed the martian moons to be more similar to the D-type asteroid than to the C-type asteroid.

Both *HST* and *IUE* observed the Centaur asteroid 2060 Chiron, a possible former resident of the Kuiper Belt. Neither instrument detected emission from gaseous species at ultraviolet wavelengths, in contrast to CN emissions that have been reported at visible wavelengths. The UV albedo of Chiron is similar to that of some of the Saturnian and Uranian satellites. In particular, Chiron's UV/IR color and ultraviolet albedo are very similar to those of Dione. [*See* KUIPER BELT OBJECTS: PHYSICAL STUDIES.]

Observations of comets at ultraviolet wavelengths are extremely useful for measuring fluorescence of solar photons by important atomic and molecular species, and thereby studying relative abundances of the vaporizing species and probing the photochemical and physical processes acting in the densest regions of the coma. UV observations of comets were first accomplished by sounding rockets and the *OAO* satellite prior to the launch of *IUE*. These observations established the emission of hydroxyl ion at near the limit of ground-based observations, 3085 Å. This is consistent with a cometary composition dominated by water ice; the hydroxyl ion is a product of exposure to solar radiation.

IUE observed more than 50 comets (~400 individual spectra). *IUE*'s photometric constancy provided the ability to compare observations of comets that appeared several years apart. Those observed range from short-period comets with aphelion near Jupiter to long-period comets that may be first-time visitors to our solar system.

All the comets observed by *IUE* have displayed the 3085 Å hydroxyl line, which is consistent with water ice being a major part of comet composition. Although all comets appear to have similar principal compositional components (water), each has different amounts of trace components, including carbon dioxide, ammonia, and methane, detected by *IUE* and *HST*. Gas production rates have been derived for species such as H_2O, CS_2, and NH_3. Several comets that were observed over a long period of time exhibited differences in their dust-to-gas ratios from one observation to the next, consistent with a variation as a function of heliocentric distance.

The first detection of diatomic elemental sulfur in a comet was seen in comet IRAS–Araki–Alcock. The lifetime of the diatomic sulfur in the cometary atmosphere is quite short (~500 seconds). This makes sulfur a useful tracer of the dynamics of the tenuous cometary atmosphere, which appears during the short time that the comet is near the Sun. Analysis of the S I triplet emission band near 1814 Å in cometary comae spectra taken with *IUE* and the *HST* FOS

shows that cometary sulfur, which is present and stored in a variety of volatile species, is depleted in abundance compared to solar abundances. The detection of CS at ultraviolet wavelengths in comae is attributed to the presence of CS_2 in the comet. Sulfur detected in the comae in excess of the sulfur attributable to CS_2 is assumed to originate from H_2S and nuclear atomic sulfur in the comet. Using this assumption, models have been used to measure total sulfur versus water abundances, which range between ~0.001 and ~0.01. [*See* PHYSICS AND CHEMISTRY OF COMETS.]

Ultraviolet observations using *IUE*, *HST*, and *FUSE* have also detected ultraviolet CO Cameron band emissions from comets, which is useful for measuring the CO_2 production rate. This rate derived from *IUE* observations of comet 1P/Halley agrees with the rate measured in situ by the spacecraft *Giotto*. These *HST* and *IUE* observations suggest that the level of activity of a comet may be linked to its CO abundance; however, this is based on a small sample of the comet population. *FUSE* measurements of C/2001 A2 (LINEAR) displayed H_2 emission lines of the Lyman system at 1071.6, 1118.6, and 1166.8 Å, in addition to CO features that suggested both a hot and a cold component of CO, the hot component likely being due to excitation of CO_2, the cold component being attributed to fluorescent scattering of CO or to electron impact excitation of CO.

4.8 The Moon and Mercury

The first UV observations of the Moon were made at FUV wavelengths using the instrument aboard the *Apollo 17* orbiter. It was noted in these measurements that the lunar maria regions, darker than the highlands at visible wavelengths, are brighter than the highlands in the FUV. This was the first indication of the so-called spectral reversal, which was also detected at EUV wavelengths using measurements by the *EUVE*. This phenomenon is attributed to the concept that FUV measurements probe just the outer layers of the grains (surface scattering, as opposed to volume scattering measured at longer wavelengths), and that space weathering processes may cause the lunar grains to be covered with a fine coating. Lunar samples measured in the laboratory support this idea: Lunar soils (presumably more weathered) show the spectral reversal, while ground-up lunar rocks (presumably less weathered) do not. *Galileo* UVS measurements in the NUV showed that the maria are darker than the highlands and that the spectral reversal must occur at a wavelength shorter than ~2200 Å. The *HUT* measurement of the lunar surface (a region near Flammarion-C, a border area between mare and highlands) at FUV wavelengths indicated an albedo of ~4% with a slight increase in brightness toward shorter wavelengths. Because of the different spectral behavior at UV versus visible wavelengths, ratio images of UV to visible color images and visible reflectance spectra are used to map spectral trends related to opaque mineral abundance and the combined effects of

FeO content and soil maturity. From the *Apollo* samples, it is known that the dominant opaque mineral is ilmenite, which is high in Ti content. Thus, UV/visible ratio images have been used to map Ti content variation in the lunar mare basalts.

The *Mariner 10* spacecraft carried a color imager that included a near UV filter (3550 Å). Mercury image ratios (UV/visible) have been used to map spectral trends associated with geologic features, using similar methods as used on lunar images. A lower UV/visible ratio suggests more FeO, or more mature soil. Spectrally neutral opaque minerals (such as ilmenite) tend to lead to a higher UV/visible ratio. Mercurian regions believed to be volcanic in origin have been found to have FeO amounts slightly less than average, consistent with ancient lava flows.

4.9 Planetary Rings

The rings of Saturn were successfully observed by *IUE* in a series of observations between 1982 and 1985. The spectrum of the rings in the 1600–3100 Å range is dominated by the water ice absorption edge at ~1650 Å. More recently, *Cassini* UVIS has made higher resolution observations of Saturn's rings; an image is shown in Fig. 15. This image shows a combination of the UV reflectance and transmission of the ring system. The red-colored region at the left is

FIGURE 15 Saturn's rings as imaged by *Cassini* UVIS. This false color two-dimensional representation of Saturn's Cassini Division and A ring was generated from UVIS data obtained during a radial scan of the rings immediately after Saturn Orbit Insertion as *Cassini* flew over the rings. To generate the image, azimuthal symmetry was assumed. Although there are azimuthal variations in the structure of the rings, they are smaller than the 100 km resolution of this image. Red represents Lyman-alpha emission from interplanetary hydrogen (1216 Å) and shines through gaps and optically thin parts of the ring. Green and blue represent reflection of solar ultraviolet light longward of the water ice absorption edge near 1650 Å. (Figure courtesy J. Colwell.)

the Cassini Division with a mean opacity of about 0.1, and the thin bright band near the outer edge of the rings is the 300 km wide Encke gap. Brighter blue-green regions indicate cleaner water ice (less absorption by non-ice species). The A ring material is cleaner than the *Cassini* Division and the abundance of water ice is seen to increase near the outer edge of the A ring.

5. Conclusions

The importance of ultraviolet solar system science has been exhibited through discoveries and continuing studies spanning the topics of atmospheric and auroral science, surface composition and space weathering. Ultraviolet observations of solar system surfaces and atmospheres have been made possible by the *IUE* and *HST* orbiting telescopes, along with *FUSE*, *HUT*, and *EUVE*, and have been substantially complemented by interplanetary missions such as *Voyager*, *Mariner*, *Galileo*, and *Cassini*. The *IUE* spacecraft provided the astronomical community with the first stable long-term (spanning nearly two decades) observing platform in space, from which astronomers have been able to study regions of the spectrum that are inaccessible from telescopes on Earth's surface. This foundation, with the support of ultraviolet spectrometers incorporated into the payloads of deep space missions, filled an observational void that had existed since the dawn of astronomy. These observations have led to important new discoveries and have provided tests of physical models that have been developed based on ground-based observations. *IUE*'s observing capability was surpassed by *HST*, which has provided the astronomical community with the opportunity to look at fainter and more distant solar system objects and has led to new discoveries in the ultraviolet spectrum. The future of Earth-based orbiting UV telescopes is unclear, but such UV instruments, with ever-improving spectroscopic and imaging capabilities, are vital to understanding solar system objects and complementing longer-wavelength observations.

Bibliography

Barth, C. A. (1985). The photochemistry of the atmosphere of Mars. In "The Photochemistry of Atmospheres." Academic Press, San Diego.

Chamberlain, J. W., and Hunten, D. M. (1987). "Theory of Planetary Atmospheres: An Introduction to Their Physics and Chemistry." Academic Press, New York.

Nelson, R. M., Lane, A. L., Matson, D. L., Fanale, F. P., Nash, D. B., and Johnson, T. V. (1980). Io: Longitudinal distribution of SO_2 frost. *Science* **210**, 784–786.

Nelson, R. M., and Lane, A. L. (1987). In "Exploring the Universe with the IUE Satellite" (Y. Kondo, ed.). D. Reidel, Dordrecht, The Netherlands.

Infrared Views of the Solar System from Space

Mark V. Sykes

Planetary Science Institute
Tucson, Arizona

CHAPTER **37**

Since 1983, a series of telescopes operating in the thermal infrared have been launched into Earth orbit and now heliocentric orbit. The images and other data returned have resulted in the discovery of new phenomena in the solar system and a new perspective on the processes within it. These observations have focused on comets, asteroids, and interplanetary dust because the major planets and Earth's Moon were too bright to be observed.

1. Introduction

At night we see objects in the solar system by the sunlight they reflect. The Moon, planets, comets, and (with the help of telescopes) asteroids and distant Kuiper Belt objects are visible to the extent that they efficiently reflect that light, coupled with their apparent size. Small particles and dust are basically invisible with the exception of the zodiacal light seen near before sunrise and after sunset at certain times of the year and the interplanetary particles that give off light as they burn up as meteors in the Earth's upper atmosphere. At thermal wavelengths, the sky is dramatically different (Fig. 1). The otherwise invisible dust now dominates the view and "familiar" phenomena like comets have a very different appearance that has changed our understanding of their nature.

At thermal wavelengths, we are looking at the objects themselves as sources of light, instead of reflected light from another source like the Sun. All objects in the universe radiate heat. The energy distribution of this radiation with wavelength is a function of the temperature of the source. The Sun, at a temperature of more than 5000 K, radiates primarily at visual wavelengths and appears yellow. Colder sources radiate at longer wavelengths. Thus, the heating element of an electric stove appears orange-red.

Objects in space (e.g., asteroids, comets, and planets) also radiate, but they do so at wavelengths much longer than can be detected by the human eye. This region of the spectrum (generally beyond 5 μm to the submillimeter) is referred to as the thermal infrared. Analysis of the thermal radiation from an object can tell us much about its composition and other physical properties including thermal inertia and grain size distribution and characteristics.

Observing this radiation from ground-based telescopes is complicated by thermal emission from the telescope itself and the atmosphere, both of which are much brighter than the astrophysical sources being observed. This has been compared to observing a star in the daytime with the telescope on fire.

Techniques that allowed objects within tiny patches of sky to be observed by ground-based and aircraft-borne

FIGURE 1   At wavelengths visible to the human eye, the night sky (**above**) is dominated by black space and point-like stars (Courtesy of A. Mellinger). From space, in the thermal infrared the same area of sky (**below**) is dominated by clouds of interstellar dust and extended solar system structures. Both images span 30×20 degrees near the First Point of Ares. The false-color thermal image was constructed from scans made by the Infrared Astronomical Satellite. Interstellar dust, known as 'cirrus' is cold (indicated by red). Warm (blue) interplanetary dust reveals rings of dust around the solar system arising from asteroid collisions (one of which is seen as the broad band extending diagonally across the top of the image), and long contrail-like structures consisting of cometary debris (one is seen below the band).

telescopes were developed. Somewhat larger strips of the sky were observed by small aperture, rocket-borne telescopes with tantalizing results. Only by getting above the atmosphere with a cooled telescope would it be possible to study the sky on a large scale at these wavelengths. This was achieved on January 26, 1983, with the launch of the *Infrared Astronomical Satellite* (*IRAS*). It was the first in a series of space-based infrared telescopes, the latest of which include the *Spitzer Space Telescope* and Japanese *Akari* satellite (Fig. 2, Table 1).

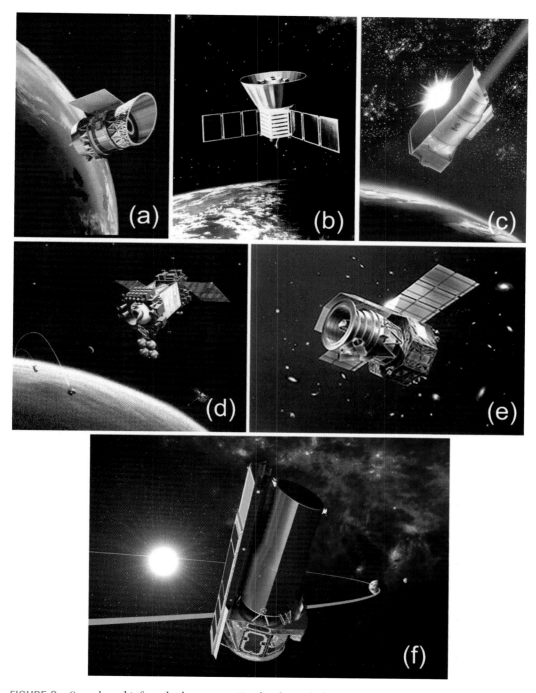

FIGURE 2   Spacebased infrared telescopes in Earth orbit include (a) the Infrared Astronomical Satellite (IRAS) in 1983, (b) the Cosmic Background Explorer, 1989–1990, (c) the European Infrared Space Observatory (ISO), 1995–1998, (d) the US Air Force Midcourse Space Experiment (MSX), 1996–1997, and (e) the Japanese Akari spacecraft, launched in 2006. The Spitzer Space Telescope (f) was launched into a heliocentric orbit in 2003 and is expected to operate for five years.

TABLE 1	Space-Based Telescopes Operating in the Thermal Infrared			
Spacecraft	**Launch Date**	**End of Mission**[a]	**Aperture (cm)**	**Wavelength Coverage (μm)**
IRAS	January 1983	November 1983	57	12–100
COBE (DIRBE)	November 1989	September 1990	19	1.25–240
ISO	November 1995	May 1998	60	2.5–240
MSX	April 1996	September 1997	33	8.3–21.3
Spitzer	August 2003	+5 years	85	3.6–106
Akari	April 2006	+18 months	68.5	1.7–180

[a] Loss of cryogen and thermal infrared sensitivity.

These telescopes and their detectors were cryogenically cooled to minimize the noise introduced by the telescope and detectors themselves. In general, their operating temperatures need to be well below that of the sources they wish to observe. For solar system objects, this means well below 20 K to study the Kuiper Belt and beyond (Fig. 3). This is accomplished by carrying a reservoir of liquid helium (having a temperature between a fraction of a degree and several degrees Kelvin), which has a finite lifetime before it is expended. At that point, the telescope and detectors warm up and loose their sensitivity.

Since the launch of *IRAS*, infrared detectors have become increasingly sensitive, thus able to study fainter and fainter sources. At the same time, the different spacecraft have operated in different modes in order to focus on different science questions. *IRAS* was primarily a survey instrument, mapping out the complete celestial sphere almost three times. Since the sky had not been mapped at thermal wavelengths, this was a mission of discovery. The *Cosmic Background Experiment* (*COBE*) was also a survey instrument, with the primary goal of understanding the distribution of the cosmic background radiation from the Big Bang. One of its instruments, the Diffuse Infrared Background Experiment (DIRBE) operated at thermal and near-infrared wavelengths at lower spatial resolution than *IRAS*. The *Infrared Space Observatory* (*ISO*) and *Midcourse Space Experiment* (*MSX*) were primarily pointing instruments, designed to measure specific targets or map out small regions of sky in detail. *Spitzer* is also primarily a pointing and mapping instrument. *Akari* is planned to spend a portion of its mission generating the first thermal map of the sky since *IRAS* and *COBE* and to use the remainder of its time conducting pointed observations. Even though all these missions have been designed to address primarily astrophysical questions, they have been a great boon to our understanding of solar system phenomena.

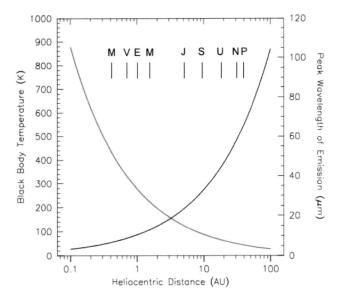

FIGURE 3 Bodies decrease in temperature with increasing distance from the Sun as shown by the red curve (which assumes that a body absorbs all incident sunlight). The corresponding wavelength at which the thermal emission spectrum peaks is shown by the black curve. At a given heliocentric distance, bodies that reflect increasing amounts of sunlight have lower equilibrium temperatures and will emit an increasing fraction of their thermal energy at longer wavelengths. For reference, distances of the planets are denoted by their first letters.

2. A New View of the Zodiacal Dust Cloud and its Sources

When we think of the solar system, the image that often comes to mind is the textbook picture of planets orbiting the Sun on concentric orbits, asteroids between Mars and Jupiter and the occasional comet flying by. However, in the inner solar system, we are immersed in a cloud of dust that we see sometimes on the horizon as the zodiacal light (Fig. 4) and sometimes in the direction opposite the Sun as the gegenschein. [*See* SOLAR SYSTEM DUST.]

The zodiacal light is caused by the scattering of sunlight off of small particles near the Earth's orbit viewed

FIGURE 4   The zodiacal light from Mauna Kea, Hawaii. It is seen most prominently after sunset in the spring and before dawn in autumn at northern latitudes. (Courtesy M. Ishiguro, ISAS.)

when the geometry is optimal. Comets were long thought to be the origin of the zodiacal cloud. However, estimates of dust production by short-period comets fell far short of that needed to maintain the cloud in steady state against losses from particles spiraling into the Sun. This mechanism, where the absorbtion and reemission of solar radiation continually decreases particle velocity, is called Poynting–Robertson drag.

A cometary cloud would have to be replenished by the occasional capture of "new," highly active comets into short-period orbits. Comet Encke was suggested as one such possible source in the past. Asteroid collisions have also been considered to be a source of interplanetary dust, and a significant fraction of interplanetary dust particles (IDPs) collected by high-altitude aircraft are thought to be consistent with such an origin, but there were few observational constraints on estimates of their relative contribution to the cloud as a whole.

At thermal wavelengths, interplanetary dust is seen around the sky, peaking about the ecliptic plane (Fig. 5). It appears brighter as we look closer to the Sun (where it is warmer and more dense, hence giving off more thermal radiation). Within this broad band of dust, there are structures related to dust-producing processes not seen before the advent of space-based infrared telescopes. The most prominent of these structures are the dust bands—parallel rings of dust straddling the plane of the ecliptic (Fig. 5). These bands arise from collisions in the Asteroid Belt. When asteroids collide, the resultant fragments are ejected with velocities that are small compared to the orbital velocity of

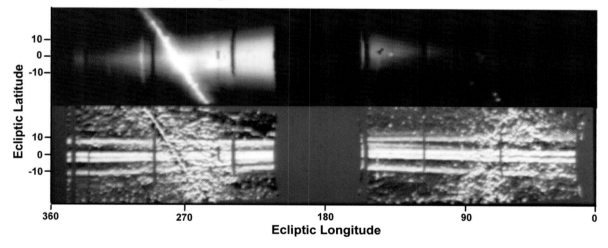

FIGURE 5   The zodiacal cloud (top) is seen extending from 0° to 360° in ecliptic latitude from right to left, constructed from scans of the ecliptic plane by IRAS. Ecliptic latitudes between 30° and −30° are shown. The diagonal structure crossing the ecliptic plane near 90° and 270° longitude is the galactic plane. Where the cloud is bright and wide (in latitude), the sky is being scanned at lower solar elongations, picking up the brighter thermal emissions of the warmer dust that lies closer to the Sun. As the satellite scans further away from the Sun at higher solar elongations, it is looking through less dust near the Earth and seeing a greater fraction of colder fainter dust. When filtered to remove its broad component (bottom), the zodiacal cloud reveals dust bands, located out in the asteroid belt and surrounding the inner solar system. Parallax results in their separation being smaller at lower solar elongations, where they are seen at a greater distance. Other solar system structures include dust trails.

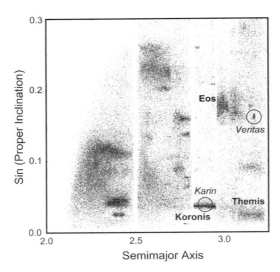

FIGURE 6   When a large enough asteroid is disrupted, its fragments are identified as other asteroids having similar orbital elements. The distribution of proper elements of asteroids in the main asteroid belt (below) reveals many of these groupings referred to as families. The principal asteroid families, first identified by Kiyotsugu Hirayama in 1914, are Themis, Koronis, and Eos. The Karin (within Koronis) and Veritas families arose from the disruption of smaller asteroids within the past 10 million years and were identified as the sources of the two most prominent pairs of dust bands by D. Nesvorny.

the original asteroid. Consequently, the orbits of the fragments are close to each other, forming a "family" of smaller asteroids (Fig. 6). All asteroid orbits precess like tops because of the gravitational influence of Jupiter. Small differences in the semimajor axes of the debris orbits cause them to precess at slightly different rates, so that over time, while their semimajor axes, orbital inclinations, and eccentricities remain roughly the same, their nodes become randomized.

They are still identifiable as families, but the volume of space they fill is a torus.

These fragments continue to experience collisions and generate smaller and smaller pieces that fill the torus, whose cross section is shown in Fig. 7, which peaks in number density in its corners. A torus of asteroid dust, observed from Earth's orbit, would have the appearance of parallel bands of dust, straddling the ecliptic (the bands closer to the Sun overlapping those further from the Sun along our line of sight).

Dust production from collisions is continuous down to sizes at which they are finally removed from the production region by radiation forces. When the fragments are around 1 μm in size they are immediately ejected from the solar system along hyperbolic orbits. These are known as β-meteoroids. Otherwise the solar radiation field and solar wind act as a friction to the particle's orbital motion (Poynting–Robertson drag), and it will slowly spiral past the orbit of the Earth into the Sun. It is thought that the dust ultimately vaporizes and is incorporated into the Sun or recondenses into small particles that are then lost to the solar system as β-meteoroids.

Poynting–Robertson drag stretches out the small particle component of the torus (Fig. 8), which retains is number density peak near its greatest distance from the ecliptic plane at a given heliocentric distance. When viewed in the thermal infrared from Earth's orbit, it still results in the appearance of distinct parallel bands straddling the ecliptic over all longitudes.

Initially, the dust bands were thought to be associated with the principal Hirayama asteroid families because of the proximity of their apparent latitudes with the orbital inclinations of those groups. These families are thought to have arisen from the catastrophic disruption of asteroids 100–250 km in diameter over a billion years previously. If the Asteroid Belt as a whole was grinding down

FIGURE 7   When the nodes of an orbit are randomized, they fill a torus. (Left) Tori associated with the principal Hirayama asteroid families would appear as parallel rings when viewed from Earth's orbit. (Right) Viewed in cross-section, particle number densities are maximum near the outer surface and are highest near the corners.

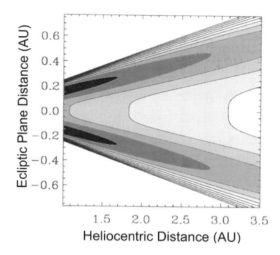

FIGURE 8 As hypothetical interplanetary dust particles originating in the Eos family torus migrate in towards the Sun, they contribute to the overall zodiacal cloud. The density contours of their contribution is shown, with darker regions corresponding to increasing particle number densities. The x axis is heliocentric distance in AU, and the y axis is roughly the distance above the ecliptic plane. As the particles evolve to smaller heliocentric distances, the number density increases, and the extrema near the upper and lower edges of the cloud component is maintained. Viewed from the Earth, as we scan from the pole to the ecliptic, the column density of particles increases as we approach an angle near their average orbital inclination resulting in the appearance of a pair of parallel dust bands. (Coutesy of W. Reach).

and generating dust, then it would follow that the most dust would be generated in the regions of greatest asteroid concentration—the largest asteroid families. Assuming this dust to be the main source of the zodiacal cloud, the cloud itself would be something expected to change slowly over much of the age of the solar system. An alternative hypothesis proposed that the dust bands arose from more recent collisions of smaller asteroids and that the zodiacal cloud was highly variable over time. David Nesvorny and colleagues identified the sources of the two most prominent pairs of dust bands as the Karin and Veritas families and determined that the collisions forming these families occurred within the past 10 million years. This demonstrates that the zodiacal cloud, once assumed to be in relative steady state, may vary substantially over time as dust production in a given family slowly declines as more and more of its mass is ground up and removed by radiation forces and a new random collision creates a family of debris that generates more dust. It is interesting that a faint inner pair of bands is still associated with the very ancient Themis family, the largest asteroid family, which was formed by the catastrophic disruption of a 240 km diameter asteroid billions of years ago and may still be producing dust today.

3. A Ring of Dust Around the Earth's Orbit

As the sky was being mapped for the first time in the thermal infrared by *IRAS*, something odd was noticed: The sky always seemed to be a few percent brighter in the direction opposite the Earth's motion about the Sun than in the direction of the Earth's motion. Since the satellite orbited above the terminator of the Earth and was facing different parts of the sky as the Earth orbited the Sun, if it was a difference in the actual sky brightness, eventually the satellite would see that difference flip when viewed from the other side of Earth's orbit. That did not happen. The "trailing" sky was always brighter. It made no sense that the Earth could be tracked by a large orbiting cloud—such a cloud would not be stable and disperse. Unable to come up with a satisfactory explanation, it was thought to be a strange calibration problem.

In 1993, a graduate student, Sumita Jayaraman, calculated that particles evolving from the Asteroid Belt past the Earth under Poynting–Robertson drag would have that orbital decay interrupted as a consequence of resonance interactions with the Earth (where the ratio of particle and Earth orbital periods is a ratio of integers). This dust would pile up for a while before continuing its sunward spiral, forming a ring around the Earth's orbit. The resonant ring has a clump (about 10% enhancement in density over the background zodiacal cloud) always trailing the Earth by about 0.2 AU in its orbit (Fig. 9). This resonant structure represents a volume through which particles are circulating around the Sun, to be distinguished from a cloud of self-attracting particles. This explained the *IRAS* mystery. The existence of the resonance ring structure was confirmed by *COBE* observations.

The *Spitzer Space Telescope* is in a heliocentric orbit slowly trailing further and further behind the Earth. At the end of its nominal 5 year mission, *Spitzer* will end up 0.6 AU behind the Earth and will have completely traveled through the trailing cloud, allowing for its 3-dimensional structure to be probed in detail. These observations will place tight constraints on the production and evolution of particles from the Asteroid Belt in the size range sensitive to resonance with Earth motion, and provide insights into how such structures in dust disks about other stars may provide details about planets imbedded in those disks. [*See* EXTRA-SOLAR PLANETS.]

4. Comets and Their Nature

Comets are members of the solar system that have been known since ancient times. At visible wavelengths they are characterized by distinctive tails of micrometer-sized dust particles ejected from the nucleus and pushed away under radiation pressure, a coma of gas, ice, and dust surrounding the nucleus, and sometimes an ion tail of gas molecules

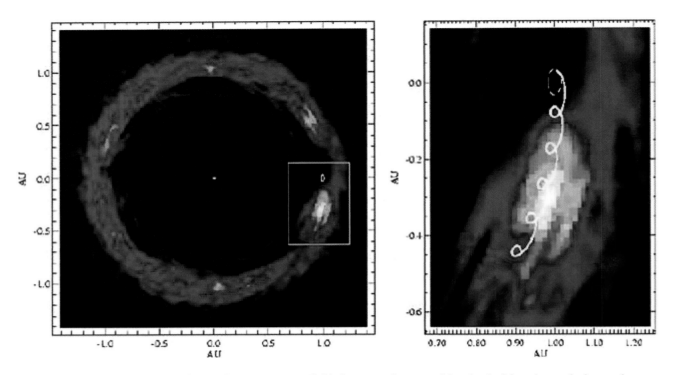

FIGURE 9   A simulated image of the Earth's resonant ring (left) showing a close-up of the cloud of dust that trails the Earth in its orbit through the year. The ellipse shows the orbit of the Earth in a rotating reference frame. The resolution of the image is 0.01 AU in the X and Y directions. The cloud is modeled using 12 μm spherical particles of astrophysical silicate. The proximity of the clump of dust behind the Earth near its orbit explains why scans of the sky behind the Earth were always brighter than scans in front of the Earth at the same solar elongation angle. Over the course of its mission, Spitzer will travel through the dust cloud trailing Earth (right). The 'loops' of Spitzer's orbital path and the oval of Earth's motion arises from the small eccentricity of the orbits. (Figure courtesy of S. Jayaraman.)

carried away by the solar wind (its typically blue color arising from electron recombination events) (Fig. 10). In the 1950s, Fred Whipple developed the standard model of comet nuclei as bodies largely of ice with a mixture of some dust—a "dirty snowball," which explained the nongravitational components of their motion. He also linked their activity to the maintenance of the zodiacal cloud, which required constant replenishment as its constituent particles spiraled into the Sun under Poynting–Robertson drag. [*See* PHYSICS AND CHEMISTRY OF COMETS.]

Comets are known to eject large particles from their association with many meteor streams. These particles spread over a comet's orbit and are scattered within its plane. If the comet orbit happens to extend inside the Earth's, these particles will be seen as meteors as the Earth passes through their orbital plane. Because they are striking the Earth's atmosphere from the same direction, meteor streams seem to come from a particular location in the sky. This is called the radiant. Analysis of meteors as they burn up in the Earth's atmosphere indicates that particles associated with known comets have low (<1 g/cm^3) to modest (<2.5 g/cm^3) mass densities.

FIGURE 10   Comet West on March 9, 1976, less than half an AU from the Sun after perihelion, exhibiting classic dust (white) and ion (blue) tails. (Image by J. Laborde.)

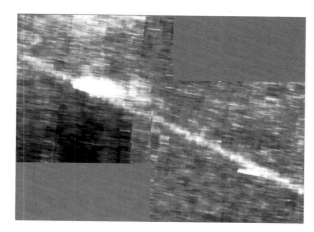

FIGURE 11   The most prominent dust trail in 1983 was associated with the short period comet Tempel 2. The dust coma and tail appear as the fish to the dust trail's stream. Trails are characteristically narrow (as a consequence of the small relative velocities of the constituent dust relative to the nucleus of the parent comet) and trace out a portion of the comet's orbit. The particles ahead of the comet (to the left) are preferentially larger than those following the comet.

Viewed from space in the thermal infrared, many short-period comets were found to have very extensive, narrow trails consisting of millimeter- to centimeter-sized particles extending degrees to tens of degrees across the sky (Fig. 11). The narrowness of the trails is due to the low velocities with which their constituent particles are ejected. They retain a record of comet emission history over a period of years to centuries. For comets having perihelia interior to the Earth's orbit, trails represent the birth of a meteor stream. The number density of particles within them are such that were the Earth to pass through one, there would be a "meteor storm" equal to or exceeding the famous Leonid storms of 1833 and 1966. First discovered by *IRAS*, it was inferred that **dust trails** were common to short-period comets. Continuing surveys by *Spitzer* suggest this is the case (Fig. 12).

Space-based infrared observations revealed that comets possessed far more dust than had been thought. Classical "gassy" comets such as P/Encke were found to possess both a significant large particle dust coma and trail (Fig. 13). Encke's trail was found to extend over 80° of its orbit. It was determined that the ejection of large particles into trails was the principal mechanism by which comets lose mass. These particles quickly devolitilize after leaving the comet nucleus; this means that most of the comet's mass loss is in refractory particles.

The discovery of cometary dust trails is changing the picture of comet nuclei from being primarily icy bodies to objects more akin to "frozen mudballs" because of their much higher than expected fraction of refractory dust. The fraction of dust-to-gas in comet nuclei provides important information about where the comets formed and how they evolve, once captured into short-period orbits.

FIGURE 12   Spitzer has detected the first new dust trails in the infrared since IRAS. Shown are P/Johnson (top) and P/Shoemaker-Levy 3 (bottom). Spitzer is confirming the commonality of such large particle emissions across the short-period comet population. (Figure courtesy of W. Reach.)

Dust-to-gas mass ratios corresponding to the canonical dirty snowball model range between 0.1 and 1. If we were to compress comet nucleus material so that refractories have a density of 3 g/cm^3 and volatiles had a density of 1 g/cm^3, this would give us a nucleus in which 3–33% of the volume consisted of refractory material.

FIGURE 13   Comet Encke (left) is considered a classical 'gassy' comet based on visible wavelength observations showing only a gas coma and no dust tail. (Image courtesy of J. Scotti.) An ISO map (right) of P/Encke and its trail at 11.5 μm, evidencing anisotropic emission and requiring the spin axis of the nucleus to lie nearly in the orbital plane. The inferred dust-to-gas mass ratio of 10–30 is even higher than that inferred from IRAS observations. (Figure courtesy of W. Reach.)

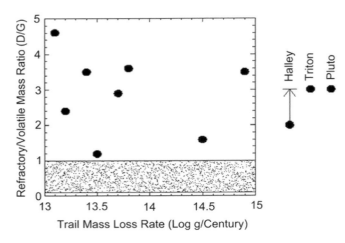

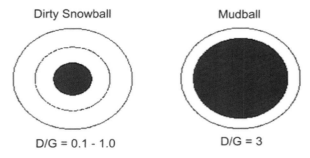

FIGURE 15   Dust to gas mass ratios are shown for comets having detected dust trails. For comparison, values are shown for Halley, Triton, and Pluto. The shaded area spans the "canonical" ratios between 0.1 and 1.

FIGURE 14   The canonical "dirty-snowball" model of comets, inferred from groundbased observations at visible wavelengths, is compared to the "frozen-mudball" model inferred from spacebased observations in the thermal infrared. All refractories are collected at the center in both cases.

This picture is based largely on ground-based observations of dust at visual wavelengths, sensitive to particles within a decade or so of 1 μm in size. These observations underestimate the mass fluence of dust from comets.

Most of cometary mass loss appears to be in much larger (and dark) particles, which are difficult to detect at visual wavelengths. This conclusion was also reached after the European *Giotto* spacecraft was struck by a small number of large particles as it flew by comets Halley and Grigg–Skellerup.

Analysis of the *IRAS* observations of eight trails indicates that short-period comets lose their mass primarily in refractory particles in the mm to cm diameter size range. An average dust-to-gas mass ratio of 3 was calculated (Fig. 14). This was the upper limit inferred for Halley by *Giotto* (with a nominal value of 2). Assuming the same densities for refractories and volatiles as previously, this corresponds to a comet nucleus that is 75% refractive by mass and 50% by volume (Fig. 15). Mixing equal volumes of dirt and water in a backyard experiment demonstrates the apt description of such a mixture as a mudball.

These dust-to-gas ratios also provide insight into the formation location of short-period comets. Dynamical considerations have lead investigators to focus on the proto-Uranus and proto-Neptune regions as that location. Significant amounts of ice have long suggested the outer solar system as the source of short-period comets. Consideration has also been given to their formation beyond the solar system, for instance in molecular clouds. Models of comets forming in such interstellar locations yield comets dominated by their volatile components, contradicting inferences drawn from space-based thermal infrared observations. On the other hand, it is very interesting that both

Pluto and Triton have effective dust-to-gas mass ratios that are identical to the average comet values determined from *IRAS* and *Giotto* (Fig. 15). This is not unexpected if Pluto and Triton accumulated from proto-comets in the vicinity of Neptune's orbit.

The existence of dust trails indicates that short-period comets are losing mass more rapidly than previously thought. Hence, their lifetime against sublimation may be shorter. A greater fraction of refractory material, however, would allow for the rapid formation of a nonvolatile mantle that is difficult to blow off, progressively choking off cometary activity. Such a mantle was apparent in the *Giotto* images of the Halley nucleus, which was near perihelion at the time. When activity is choked off, the comet would look like an asteroid until such time as sufficient pressure built up from subsurface ices to break through the crust in a burst of resumed activity. The discovery in August 1992 that asteroid 4015 was actually comet P/Wilson–Harrington (last seen in outburst in 1949) provided the first hard evidence of such "dormant" comets in the inner solar system.

In addition to trails associated with known short-period comets, IRAS also detected trails having no known source (Fig. 16). Unfortunately, since these were discovered in the data long after the mission had ended, it was not possible to follow up the *IRAS* observations with observations from the ground in order to determine their orbits. So these objects are now lost. However, assuming a cometary origin, the numbers of these "**orphan trails**" suggest that there may be twice as many short-period comets as previously recorded, with the majority of them being less active and hence more difficult to detect by traditional means. The serendipitous detection of orphans requires a space-based thermal infrared survey of the sky having sufficient spatial resolution. Such a survey is planned for the ongoing Japanese *Akari* mission and the future NASA *Wide-field Infrared Survey Explorer* (*WISE*).

Space-based thermal infrared telescopes have also provided direct compositional information about comets through spectroscopic observations raising questions about

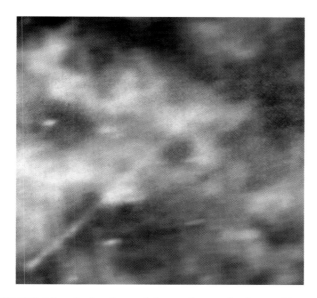

FIGURE 16 The brightest of the "orphan trails" detected by IRAS, seen against a background of interstellar clouds. Blue elongated sources are stars distorted by the rectangular shape of the detectors. Orphan trails are probably associated with comets never before detected.

the conditions in the early solar system when they formed. ISO made the first detections of crystalline silicates and CO_2 in its observations of the long-period comet Hale–Bopp and Jupiter-family comet P/Hartley 2. Crystalline silicates are formed at high temperatures not associated with the outer solar system. *Spitzer* measured the spectra of pristine material excavated by the Deep Impact event on P/Tempel 1 finding materials never before seen in comets such as carbonates and clay (which form in the presence of liquid water) as well as metal sulfides, polycyclic aromatic hydrocarbons, and crystalline silicates. Liquid water is also not expected to be present in the outer solar system. These observations suggest perhaps substantial radial mixing of materials forming in hot and cold environments before the comets begin to accrete.

5. Asteroid Physical Properties

Over a million asteroids with diameters greater than 1 km reside primarily in a belt between Mars and Jupiter. [*See* MAIN-BELT ASTEROIDS.] The asteroid population extends interior to the orbit of the Earth and beyond Jupiter. These objects are the scattered and disrupted remains of an early population of protoplanets whose continued growth was interrupted early on by mutual gravitational stirring, the growth of a massive Jupiter, or both. The size distribution of asteroids provides insight into their origins and evolution through collisions and dynamical processes. Asteroid sizes are also important to determine the hazard of near-Earth

objects. Asteroids are almost all unresolved to telescopes, and their brightnesses are insufficient for determining their sizes unless their albedos are known. By combining visible and thermal infrared observations of an asteroid, the diameter and albedo of an asteroid can be simultaneously determined. The difficulty of making radiometric (thermal) observations of asteroids from ground-based telescopes made for slow growth in the number for which physical properties could be determined. Space-based surveys in the thermal infrared greatly increased the numbers of asteroids for which albedos and diameters were available, providing new information about the composition of the Asteroid Belt.

Because most space-based infrared telescopes have operated in a pointed mode, targeting specific objects or locations for observations, and few have engaged in surveys, most asteroid albedos and diameters were derived from the *IRAS* survey in 1983. This survey resulted in 8210 observations of 2004 asteroids. A more limited survey by *MSX* in 1996 resulted in observations of 168 asteroids. By and large, such surveys rely upon the detection of asteroids in known orbits, since they are unable to provide sufficient astrometry to determine the orbits of newly discovered asteroids.

During the *IRAS* mission, there were tens of thousands of asteroids having known orbits. More than two decades later that number has increased by an order of magnitude. Coupled with greatly increased detector sensitivity, an all-sky survey is no longer necessary to produce an even larger catalog of asteroid diameters and albedos. Over the 5 year nominal mission of *Spitzer*, it is estimated that about 25,000 serendipitous asteroid measurements will be made.

Known asteroid diameters inferred from *IRAS* are shown in Fig. 17 as a function of heliocentric distance. The absence of small asteroids with increasing heliocentric distance is a consequence of the limits of detector sensitivity. The absence of larger asteroids beyond 4 AU, however, is real. This indicates that either the inner and outer belt/Jupiter

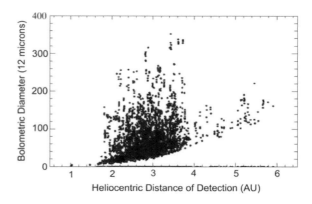

FIGURE 17 Asteroid diameters versus heliocentric distance of detection. The lower limit of detected asteroid sizes reflects the sensitivity limits of the IRAS detectors. The outer asteroid belt is shown to have few large asteroids compared to the inner belt.

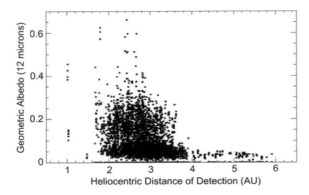

FIGURE 18   Geometric albedo versus heliocentric distance of known asteroids detected by IRAS. The high albedo asteroids are located almost exclusively in the inner portion of the main asteroid belt.

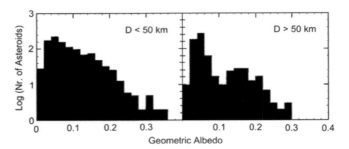

FIGURE 19   Large and small asteroids evidence different albedo distributions.

Trojan populations had a very different collisional history, or that their origins are different.

Albedo provides insight into composition. Meteorite studies show that very dark surfaces arise from largely carbonaceous materials, while high albedo surfaces are associated with silicic compositions lacking such carbonaceous material. IRAS confirmed that most C-type asteroids (thought to be carbonaceous) are indeed dark compared to the "stony" S-type asteroids, and that there is a trend toward darker asteroid surfaces with increasing heliocentric distance (Fig. 18). This is consistent with the view that there is not only a residual primordial composition gradient through the Asteroid Belt, but that inner belt asteroids (predominantly S-type) were significantly processed by heating in the early solar system, while the outermost asteroids have experienced little heating and have retained a more "primitive" mineralogy.

New observations often result in as many new questions as new answers, and the IRAS asteroid observations are no exception. Prior to IRAS, ground-based thermal observations had been preferentially made of the largest asteroids. It was noticed that there was a bimodal distribution in the inferred albedos, which was consistent with the main-belt asteroid population being dominated by dark C-type and bright S-type asteroids. IRAS added large numbers of observations of smaller asteroids and it was found that they had an albedo distribution quite different from the larger asteroids. Small asteroids have a unimodal distribution spanning the total range of albedos inferred for the large asteroids (Fig. 19). Since the small asteroids are fragments of larger asteroids, this might imply that surface minerologies are not representative of interior minerologies or that significant "space weathering" may affect the surface spectra of the larger bodies.

There has been some question as to whether all asteroid families mark the site of a past catastrophic disruption or whether in some cases asteroids might be clumped together

due to dynamical forces such as gravitational perturbations on their orbits by Jupiter. Albedo distributions can also provide clues to the origin of some asteroid families. Assuming the parent to have been compositionally homogeneous, the fragments should exhibit similar spectral properties. On the other hand, members of purely dynamical clusters would not be expected to have similar compositions. IRAS scanned enough of the members of the largest families to show that family members had albedos more similar to each other than to the background asteroids nearby, giving support, in those cases, to the asteroid breakup hypothesis (Fig. 20).

Thermal spectra can provide information including thermal inertia and composition. Spitzer observations of Jupiter Trojan asteroids have yielded results suggesting fine-grained silicates in a relatively low-density, perhaps "fairy-castle" matrix. Does this indicate a possible cometary origin? Evidence for complex carbon compounds on primitive asteroids has also been detected. A Spitzer survey of M-class asteroids, spectrally similar at visible wavelengths, indicates major differences in thermal inertia among them, which could indicate significant differences in surface densities (due to relative age and collisional processing) or conductivity (some M-types are metallic, some may be stony). Thermal infrared observations of asteroids from space-based facilities are revealing an increasingly diverse population of objects.

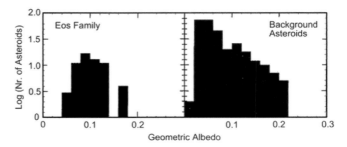

FIGURE 20   A comparison of the albedo histograms of the Eos asteroid family and non-family members near the same location in the asteroid belt supports the hypothesis that the family members derived from a single parent body.

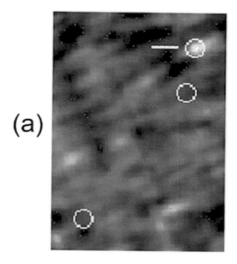

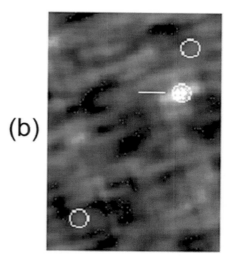

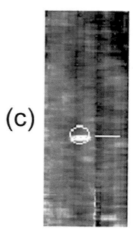

FIGURE 21   Pluto–Charon were detected moving across the infrared sky by IRAS. These images were constructed from 60 micron scans for (a) July 13, (b) July 23–24, and (c) August 16 in 1983. The predicted positions of Pluto–Charon at each of these times are indicated by circles. The August 16 position is the lower left circle in (a) and (b).

6. Pluto and Beyond

Thermal radiation from Pluto and its moon Charon was first detected by *IRAS* (Fig. 21). The thermal flux of the system was consistent with that of a rapidly rotating graybody having an equatorial temperature of ~60 K. This information in combination with ground-based spectroscopic measurements and albedo maps derived from the mutual eclipses between Pluto and its moon between 1984 and 1990 has provided important insights into the nature and dynamics of the surface of Pluto. [*See* PLUTO.]

When methane was first detected in visible wavelengths on the surface of these objects, it was thought that Pluto must be completely covered by the frozen ice, and would be isothermal because of the transport of heat as highly insolated locations would be cooled by sublimation and less insolated locations would be warmed by the condensation of atmospheric methane. Charon was thought to be a less likely location of such a coating of methane frost because of its lower gravity, from which methane would be expected to escape over time.

The detection of an extended atmosphere from a stellar occultation in 1988 and the subsequent detection in the near-infrared of nitrogen ice on Pluto's surface required that the volatile surface ices be dominated by nitrogen with a small fraction of the more spectroscopically active methane, and that these surface ices must be very cold, ~35 K.

The spectroscopic and *IRAS*-derived temperatures appear to contradict each other. Nitrogen ice at the warmer radiometric value would produce an enormous atmosphere that would have been evident in the occultation observations. The surface albedo maps, however, show that Pluto's surface is segregated into bright and dark regions with bright ices generally at higher latitudes. A high-albedo surface is bright at visible wavelengths (reflected light) and faint at thermal wavelengths, while a dark surface is faint at visible wavelengths and bright in the thermal. On Pluto, visible wavelength spectroscopy samples primarily the bright icy polar regions of the planet while space-based thermal observations are dominated by the dark equatorial region.

IRAS tells us that the volatile nitrogen ice—from which the atmosphere derives—is segregated on the surface of Pluto, away from the warmer regions, giving rise to the thermal emission detected by the satellite. These warmer regions are probably a mixture of water ice and carbonaceous residue resulting from the radiation processing of methane ice over the age of the solar system. The dark regions are not contributing significantly to the atmospheric gases, which is consistent with a water/organic composition that would have negligible vapor pressure at the temperature inferred.

Spitzer spectroscopic studies of Pluto are expected to provide rotationally resolved information on the complex organics comprising Pluto's dark regions, as well as constituents of its bright ice regions. These observations will complement the spatially resolved studies, in reflected light, of the *New Horizons* mission as it passes by Pluto and Charon in 2016.

Spitzer has been trained on trans-Neptunian objects beyond Pluto, finding that their albedos vary considerably. Many have been determined to have high albedos from the fact that they have not been detected by the telescope, requiring brighter, hence cooler and thermally fainter surfaces. Nominally, an undifferentiated primitive surface exposed to galactic cosmic radiation over the age of the solar system would be expected to be covered with dark complex organics. Bright surfaces suggest substantial evolution that allows volatile ices such as nitrogen to migrate to the surface—perhaps evidence of collisional activity or heating resulting in differentiation. Even nondetections can provide essential insights into the nature of things.

Beyond the Kuiper Belt lies the question of what additional parts of the solar system wait to be discovered. The sky is filled with cirrus-like structures (Fig. 1) and begs the question of whether any of them are local to the solar system.

IRAS surveyed 96% of the sky twice and 75% of the sky a third time. Images from these surveys were used to conduct the first-ever parallactic survey of the sky. Images of the same location taken weeks to several months apart were compared to search for reflex motion of extended sources. There were a couple of exciting possible detections, but in the end, nothing was identified in a volume extending 100 AU in some directions and 1000 AU in others—within the sensitivity of *IRAS*. With the greater sensitivity of current radiometric detectors and the larger apertures of space-based telescopes, the question asked with *IRAS* can be asked again

and again, perhaps one day with a positive result or a definitive answer.

7. An Exciting Future

At the time this article is being written, *Spitzer* is only half way through its nominal mission, and the initial results are just being published and digested. Detailed scans are being made of the zodiacal plane to test models of dust band origins, search for evidence of additional bands, and make sample searches for trail structures. Potential noncometary dust trails, first identified in *IRAS* observations, have been recovered, and efforts are underway to determine their specific origins. It is possible that they are the debris from asteroid collisions only tens of thousands of years past. *Spitzer* continues its surveys of Kuiper Belt objects, to find the extent of nonuniformity of this population. It is conducting surveys of comets, finding interesting variations in their large dust production. It is conducting surveys of the outer satellites of Jupiter, Saturn, Uranus, and Neptune. *Akari*, successfully launched, is just beginning its mission to conduct pointed observations and a new survey of the sky at thermal wavelengths. An entirely new suite of comets will be approaching perihelion (for maximum brightness) than at the time of *IRAS*, and the potential for identifying orphan trails is renewed. There is the potential of a longer baseline all-sky survey by *WISE*. We face the prospect of a wealth of new data and new perspective on the solar system, building on what we have learned over the past quarter century and what we are continuing to learn from infrared views of the solar system.

The Solar System at Radio Wavelengths

Imke de Pater

University of California, Berkeley
Berkeley, California

William S. Kurth

University of Iowa
Iowa City, Iowa

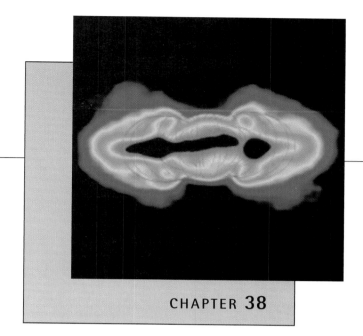

CHAPTER **38**

1. Introduction

Ground-based radio astronomical observations of planetary objects provide information that is complementary to that obtained at other (visual, infrared, ultraviolet) wavelengths. We distinguish between thermal and nonthermal emissions. Thermal radio emission originates from a body's surface (or more appropriately subsurface) and/or atmosphere, and **nonthermal radio emissions** are produced by charged particles in a planet's magnetosphere. The thermal emission can be used to deduce the structure and composition of a planet's atmosphere and surface layers; the nonthermal radiation provides information about its magnetic field and charged particle distributions therein. Ground-based radio astronomy is essentially limited to frequencies above about 10 MHz because of the shielding effects of Earth's ionosphere at lower frequencies. Space-based measurements extend the frequency range of solar system radio astronomy as low as a few kHz. In this chapter, we discuss radio emissions from a few kHz up to $\gtrsim 500$ GHz. Since we cover over 9 orders of magnitude in frequency, we can include only brief summaries of a select number of topics.

Instrumentation

A radio telescope consists of an antenna and a receiver. The antenna can be a simple monopole, dipole, or parabolic dish (Fig. 1). The sensitivity of the antenna depends upon many factors, but the most important are the effective aperture and system temperature. The effective aperture depends upon the size of the dish and the aperture efficiency. The sensitivity of the telescope increases when the effective aperture increases and/or the system temperature decreases.

The response of an antenna as a function of direction is given by its antenna pattern, which consists of a "main" lobe and a number of smaller "side" lobes, as depicted in Fig. 2a. The resolution of the telescope depends upon the angular size of the main lobe. It is common to express the main lobe width as the angle between the directions for which the power is half that at lobe maximum; this is referred to as the half power beam width. This angle depends upon the size of the dish and the observing wavelength: For a uniform illumination, the beam width is approximately λ/D radians, with D the dish diameter in the same units as the wavelength λ. Space-borne radio telescopes at low frequencies usually are composed of one or more long cylindrical elements since dish antennas are prohibitive in terms of mass.

The resolution of a radio telescope can be improved by connecting the outputs of two antennas which are separated by a distance S, at the input of a radio receiver. The VLA (Very Large Array) in Socorro, New Mexico, consists of a Y-shaped track, with 9 antennas along each of the arms (Fig. 1b). This telescope thus provides 351 individual interferometer pairs, each of which has its own instantaneous resolution along its projected (on the sky) baseline S'. Such

Encyclopedia of the Solar System 2e ©2007 by Academic Press. All rights of reproduction in any form reserved.

FIGURE 1   (a) The Cambridge Low-Frequency Synthesis Telescope (CLFST) is an east–west aperture synthesis radio telescope, which currently operates at 151 MHz. It consists of 60 yagi antennas on a 4.6 km baseline. The telescope is located at the Mullard Radio Astronomy Observatory. In addition to the astronomical 6C and 7C 151 MHz catalogues, it also participated in a worldwide campaign to observe Jupiter (Fig. 18). (Courtesy Mullard Radio Astronomy Observatory.) (b) Aerial photograph of the Very Large Array of radio telescopes in New Mexico. (Image Courtesy NRAO/AUI Very Large Array is a facility of the National Radio Astronomy Observatory, operated by the Associated Universities, Inc. (AUI), under contract with the National Science Foundation.) (c) The former Berkeley–Illinois–Maryland Association (BIMA) array at the Hat Creek Radio Observatory in its most compact configuration. The telescopes each have a diameter of 6 m, and operate at millimeter wavelengths. (Courtesy Seth Shostak/SETI Institute.) The BIMA array is now being merged with the Owens Valley Radio Observatory (OVRO) millimeter array into the Combined Array for Research in Millimeter-wave Astronomy (CARMA) at Cedar Flat in eastern California.

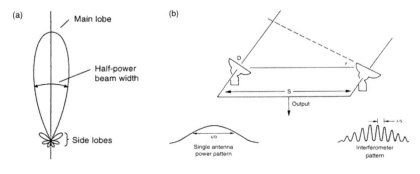

FIGURE 2   (a) A generic antenna pattern consists of a "main" lobe and a number of smaller "side" lobes, as depicted in the figure. The half power beam width is the full width at half power (FWHP). (After J. D. Kraus, 1986, "Radio Astronomy," Cygnus Quasar Books, Powell, Ohio.) (b) Top: Geometry of a two-element interferometer. Bottom: Antenna response for a single element of the interferometer (left) and response of the interferometer (right) to an unresolved radio source. (S. Gulkis and I. de Pater, 2002, Radio astronomy, planetary, "Encyclopedia of Physical Science and Technology," vol. 13, 3rd Ed., Academic Press, pp 687–712.)

an array of antennas is needed to construct an image that shows both the large and small scale structure of a radio source. At short spacings, the entire object can be "seen," but details on the planet are washed out due to the low resolution of such baselines. At longer baselines, details on the planet can be distinguished, but the large-scale structure of the object gets resolved out, and hence would be invisible on the image unless short spacing data are included as well. Hence, arrays of antennas are crucial to image an object.

2. Thermal Emission from Planetary Bodies

2.1 Thermal or Blackbody Radiation

Any object with a temperature above absolute zero emits a continuous spectrum of electromagnetic radiation at all frequencies, which is its thermal or "blackbody" radiation. A blackbody radiator is defined as an object that absorbs all radiation that falls on it at all frequencies and all angles of incidence; none of the radiation is reflected. **Blackbody radiation** can be described by Planck's radiation law, which, at radio wavelengths, can usually be approximated by the Rayleigh–Jeans law:

$$B_\nu(T) = \frac{2\nu^2}{c^2} kT \tag{1}$$

where $B_\nu T$ is the brightness (W/m^2/Hz/sr), ν the frequency (Hz), T the temperature (K), k Boltzmann's constant (1.38×10^{-23} J/deg(K)) and c the velocity of light (3×10^8 m/s). With a radio telescope, one measures the **flux density** emitted by the object. A common unit is the flux unit or Jansky, where 1 Jy $= 10^{-26}$ W/m^2/Hz. This flux density can be related to the temperature of the object:

$$S = \frac{abT}{4.9 \times 10^6 \lambda^2} \quad \text{Jy} \tag{2}$$

with λ the observing wavelength (in m), $2a$ and $2b$ are the equatorial and polar diameters (in arc seconds), and T the temperature (in K). Usually, planets do not behave like a blackbody, and the temperature T in Eq. (2) is called the brightness temperature, defined as the temperature of an equivalent blackbody of the same brightness.

2.2 Radio Emission from a Planet's (Sub)surface

Radio observations can be used to extract information about the (sub)surface layers of planetary bodies. The temperature structure of the (sub)surface layers of airless bodies depends upon a balance between solar insolation, heat transport within the crust, and reradiation outward. The fraction of the solar flux absorbed by the surface depends upon the object's albedo, A, while the energy radiated by the surface (at a given temperature) depends upon its emissiv-

ity, e (which is 1 for a blackbody, $e = 1 - A$). During the day, a planet's surface heats up and reaches its peak temperature at noon or early afternoon (the exact time depends upon the body's thermal inertia—see later); at night the object cools off. Its lowest temperature is reached just before sunrise. Because it takes time for the heat to be carried downward, there will be a phase lag in the diurnal heating pattern of the subsurface layers with respect to that at the surface, and the amplitude of the variation will be suppressed. At night, heat is carried upward and radiated away from the surface. Hence, while during the day the surface is hotter than the subsurface layers, at night the opposite is true.

The amplitude and phase of the diurnal temperature variations and the temperature gradient with depth in the crust are largely determined by the thermal inertia and the thermal skin depth of the material. The thermal inertia, γ, measures the ability of the surface layers to store energy, and depends on the thermal conductivity K, the density ρ, heat capacity C: $\gamma = \sqrt{K\rho C}$.

The amplitude of diurnal temperature variations is largest at the surface, and decreases exponentially into the subsurface, with an e-folding scale length equal to the thermal skin depth:

$$L_t = \sqrt{\left(\frac{KP}{\pi\rho C}\right)}. \tag{3}$$

where P is the rotational period.

For the terrestrial planets, using thermal properties of lunar soils and the proper rotation rates, the skin depths are of order a few centimeters (Earth and Mars) to a few tens of centimeters (Moon, Mercury, and Venus, because of their slow rotation). The $1/e$ depth to which a radio wave at wavelength λ probes into the subsurface is given by

$$L_r = \lambda/(2\pi\sqrt{\varepsilon_r} \tan \Delta) \tag{4}$$

where ε_r is the real part of the dielectric constant, and $\tan \Delta$ is the "loss tangent" (or absorptivity) of the material—the ratio of the imaginary to the real part of the dielectric constant. Radio waves typically probe ~10 wavelengths into the crust. By observing at different wavelengths, one can thus determine the diurnal heating pattern of the Sun in the subsurface layers. Such observations can be used to constrain thermal and electrical properties of the crustal layers. The thermal properties relate to the physical state of the crust (e.g., rock versus dust), while the electrical properties are related to the mineralogy of the surface layers (e.g., metallicity).

2.3 Radio Emission from a Planet's Atmosphere

Radio spectra of a planet's atmosphere can be interpreted by comparing observed spectra with synthetic spectra, which are obtained by integrating the equation of radiative transfer through a model atmosphere:

$$B_\nu(T_D) = 2 \int_0^1 \int_0^\infty B_\nu(T) e^{(-\tau/\mu)} d(\tau/\mu) d\mu \qquad (5)$$

where $B_\nu(T_D)$ can be compared to the observed disk-averaged **brightness temperature**. The brightness $B_\nu(T)$ is given by the Planck function, and the optical depth τ_ν (z) is the integral of the total absorption coefficient over the altitude range z at frequency ν. The parameter μ is the cosine of the angle between the line of sight and the local vertical. By integrating over μ, one obtains the disk-averaged brightness temperature, to be compared to the observed brightness temperature.

Before the integration in Eq. (5) can be carried out, the atmospheric structure, as composition and temperature-pressure profile, needs to be defined. Over our region of interest, the temperature structure can often be approximatel by an adiabatic lapse rate. The temperature, pressure, and composition of an atmosphere are related to one another through an equation of state, such as the ideal gas law. Cloud formation and chemical alteration of constituents due to, for example, photolysis (breakup of molecules by ultraviolet sunlight), all need to be considered when making a model atmosphere. The shape of absorption/emission lines depend on the temperature and pressure of the environment, and may vary from relatively narrow lines (e.g., Mars, Venus, Titan, Io) to broad quasi-continuum spectra (e.g., giant planets).

2.4 Terrestrial Planets and the Moon

2.4.1 THE MOON

Lunar radio astronomy dates back to the mid-1940s, well before the first *Apollo* landing on the Moon. Since the mid-1970s, after a decade of "neglect," there was renewed interest in lunar radio astronomy since radio receivers had improved substantially and laboratory measurements of *Apollo* samples provided a ground-truth for several sites on the Moon. By using lunar core samples, one could determine a density profile of the soil with depth near the landing sites, as well as the complex dielectric constant of a variety of rocks and powders. Both are essential parameters in modeling radio observations of the Moon.

A microwave image of the full Moon reveals that the maria are ~5 K warmer than the highlands. This may result from a difference in albedo (the maria are darker than the highlands), radio emissivity and/or the microwave opacity. Lunar samples suggest that the microwave opacity in the highlands is somewhat (factor of ~2) lower than in the maria, so that deeper cooler layers are probed in the lunar highlands compared to the maria during full Moon (as observed); at new Moon the temperature contrast should

be reversed (no observations have yet been reported), since the temperature increases with depth at night.

2.4.2 MERCURY

Radio images of Mercury show a brightness variation across the disk, which displays the history of solar insolation. At short wavelengths, where shallow layers are probed, the day side temperature is usually highest. However, when deeper layers are probed, the diurnal heating pattern is less obvious, and one can distinguish two relatively "hot" regions, one at longitude 0° and one at 180°. This hot–cold pattern results from Mercury's 3/2 spin-orbit resonance: Mercury rotates three times around its axis for every two revolutions around the Sun. This, combined with the planet's large orbital eccentricity, leads to factor-of-2.5 variation in the average diurnal insolation as a function of longitude. Mercury's peak (noon) surface temperature varies between 700 K for longitudes facing the Sun at perihelion (longitudes 0° and 180°) to 570 K 90° away. While the surface temperature responds almost instantaneously to changes in illumination, the subsurface layers do not, and this variation in solar insolation remains imprinted at depths well below the surface.

Figure 3a shows a radio image at 3.6 cm, probing a depth of ~70 cm. The two hot regions discussed previously are

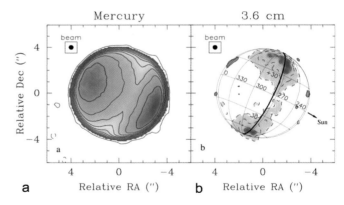

FIGURE 3 (a) The 3.6 cm thermal emission from Mercury observed with the VLA. Contours are at 42 K intervals (10% of maximum), except for the lowest contour, which is at 8 K (dashed contours are negative). The beamsize is 0.4″ or 1/10 of a Mercurian radius. Note the two so-called hot regions, discussed in the text. (b) A residual map of Mercury, which shows the residuals after subtracting a model image from the observed map. We further indicated the geometry of Mercury during the observation, as the direction to the Sun, and the morning terminator (heavy line). The hot regions have been modeled extremely well, since they do not show up in this residual map. However, we see instead large negative (blue) temperatures near the poles and along the morning side of the terminator. These are likely caused by shadows on the surface resulting from local topography, such as craters. Contour intervals are in steps of 10 K, which is roughly 3 times the rms noise in the image. (D. L. Mitchell and I. de Pater, 1994, Microwave imaging of Mercury's thermal emission: Observations and models, *Icarus* **110**, 2–32.)

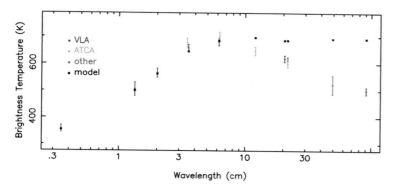

FIGURE 4   A radio spectrum of Venus. At short wavelengths, one probes approximately down to Venus' cloud layers. The brightness temperature increases when deeper warmer layers are probed and decreases again at wavelengths long enough to probe down into Venus' surface. (B. J. Butler and R. J. Sault, 2003, Long wavelength observations of the surface of Venus, *IAUSS*, 1E, 17B.)

clearly visible. These are modeled well, as exemplified by the map in Fig. 3b, which shows the difference between the observed map and a thermal model. The viewing geometry is superimposed on the latter image. The negative temperatures near the poles and along the terminator are indicative of areas colder than predicted in the model. This is likely caused by surface topography, which causes permanent shadowing at high latitudes and transient effects in the equatorial regions, where crater floors and hillsides are alternately in shadow and sunlight as the day progresses. Some crater floors near the poles are permanently shadowed, and radar observations have revealed evidence for the existence of water ice in such crater floors.

Radio spectra and images, together with *Mariner 10* infrared (IR) data show that Mercury's surface properties are quite similar to those of the Moon, except for the microwave opacity, which is ∼2–3 smaller than that of most lunar samples. This suggests a low ilmenite ($FeTiO_3$) content, the mineral that is largely responsible for the dark appearance of the lunar maria. The absence of iron (Fe) and titanium (Ti) bearing minerals from Mercury's surface suggests this planet to be largely devoid of basalt, which, if true, contains clues as to its volcanic past.

2.4.3 VENUS AND MARS

Venus and Mars have atmospheres which consist of over 95% carbon dioxide gas (CO_2). Other than having a similar composition, the atmospheres are very different. The surface pressure on Venus is approximately 90 times larger than that on Earth, while that on Mars is ∼140 times smaller. The shear amount of CO_2 gas on Venus provides so much opacity that Venus' surface can only be probed at wavelengths longward of ∼6 cm, whereas Mars's atmosphere is essentially transparent at most radio wavelengths. On both planets, CO_2 gas is photodissociated (molecules are broken up) by sunlight at high altitudes into carbon monoxide (CO) and oxygen (O). CO gas has strong rotational transitions at millimeter wavelengths, which can be utilized to determine the atmospheric temperature profile and the CO abundance on Venus and Mars in the altitude regions probed.

Radio astronomical observations of Venus go back to the mid-1950s, when measurements at a wavelength of 3 cm indicated a brightness temperature of over 560 K, well above that expected (∼300 K) from a terrestrial analog. In the early 1960s, Carl Sagan postulated this high temperature to result from a strong greenhouse effect in Venus' atmosphere. A full **radio spectrum** (Fig. 4) reveals that the planet's surface is probed at a wavelength of ∼6 cm, with a surface temperature close to ∼700 K. At longer wavelengths, one probes below the surface, where the observed brightness temperatures are well below predicted values—by up to 200 K, an effect that is not (yet) understood.

The CO 1–0 (3 mm) and 2–1 (1 mm) rotational transitions have been observed routinely. Since CO is formed in the upper part of the atmosphere, the line is seen in absorption against the warm continuum background on both Venus and Mars (Fig. 5). On Venus, one probes the so-called mesosphere in these transitions, a region between the massive lower atmosphere (altitudes ≲70 km), in which the radiative

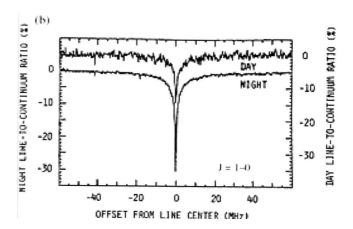

FIGURE 5   Spectra of Venus in the $J = 1-0$ line. The upper curve is for the day side hemisphere (when Venus is near superior conjunction), the lower curve is for the night side hemisphere (when Venus is near inferior conjunction). (F. P. Schloerb, 1985, Millimeter-wave spectroscopy of solar system objects: Present and future, "Proceedings of the ESO–IRAM–Onsala Workshop on (Sub)millimeter Astronomy," P. A. Shaver and K. Kjar, eds., Aspenas, Sweden, 17–20 June 1985, pp. 603–616.)

time constant is much greater than a solar day, and the upper atmosphere (altitudes $\gtrsim 120$ km), which has a low heat capacity. In contrast to the lower atmosphere, a strong day-to-night gradient in temperature exists above the mesosphere, which leads to strong winds from the day to the night side. This is very different from the retrograde zonal winds observed in the visible cloud layers. These mesospheric winds likely carry CO, formed on the day side upon photodissociation of CO_2, to the night side of the planet. Therefore, contrary to expectations, the spectra in Fig. 5 show the night side line to be much deeper and narrower than the day side line, suggestive of a large concentration of CO at high altitudes on the night side of the planet.

On Mars, the CO mixing ratio, $CO/CO_2 \sim 10^{-3}$, is much less than expected from theories on photolysis of CO_2 and subsequent recombination of CO and O. This recombination proceeds faster in the presence of chemistry involving hydroxyl radicals (OH), derived from water vapor. Regular photolysis of water in the martian atmosphere may be too slow, however. New ideas being pursued include the creation of OH by electric fields in martian dust storms. The reaction $CO + OH \rightarrow CO_2 + H$ frees up H, which eventually may lead to the formation of hydrogen peroxide (H_2O_2), a strong oxidizer. Dust storms prevail in Mars' lower atmosphere, and the formation of OH in such storms occurs close to the surface where the water abundance is highest. Hence, for this mechanism to efficiently remove CO from the upper atmosphere, the mixing of CO throughout the atmosphere should be an efficient process.

2.5 Giant Planets

2.5.1 RADIO SPECTRA

At millimeter to centimeter wavelengths, one typically probes altitudes in the giant planet atmospheres from within to well below (pressure of tens of bars) the cloud layers. Representative microwave spectra are shown in Figs. 6a (Jupiter) and 6b (Uranus). They generally show an increase in brightness temperature with increasing wavelength beyond 1.3 cm, due to the combined effect of a decrease in opacity at longer wavelengths, and an increase in temperature at increasing depth in the planet. The main source of opacity is ammonia (NH_3) gas, which has a broad absorption band at 1.3 cm. At longer wavelengths (typically >10 cm) absorption by water vapor and droplets becomes important, while at short millimeter wavelengths the contribution of collision induced absorption by molecular hydrogen becomes noticeable. On Uranus and Neptune, there is additional absorption by hydrogen sulfide (H_2S) and (perhaps) phosphine (PH_3) gas.

The composition of all four giant planets is dominated by H_2 and He gases, while the condensable gases CH_4, NH_3, H_2S, and H_2O constitute only a small fraction of the total. These gases, however, determine much of the "weather"

on these planets. Although only cloud tops are seen "visually," thermochemical equilibrium calculations reveal the presence of a number of cloud layers deeper in the atmosphere, as depicted in Figs. 6c and 6d: an aqueous ammonia solution cloud, water ice, a cloud of ammonium hydrosulfide particles ($NH_3 + H_2S \rightarrow NH_4SH$ around 250 K), ammonia and/or hydrogen sulfide ice, and methane ice. The "visible" cloud layers on Jupiter and Saturn are composed of ammonia ice, while Uranus and Neptune are cold enough to allow condensation of methane gas.

To first approximation, the spectra of both Jupiter and Saturn resemble those expected for a solar composition atmosphere, while the spectra of Uranus and Neptune indicate a depletion of ammonia gas compared to the solar value by ~two orders of magnitude. As shown in Fig. 6d, this depletion has been explained via formation of an extensive NH_4SH cloud, which is discussed in more detail later.

The thermal emission from all four giant planets has been imaged with the VLA. To construct high signal-to-noise images, the observations are integrated over several hours, so that the maps are smeared in longitude and only reveal brightness variations in latitude. The observed variations have typically been attributed to spatial variations in opacity (NH_3, H_2S gases), as caused by a combination of atmospheric dynamics and condensation at higher altitudes. Below we briefly discuss findings for each planet individually.

2.5.2 JUPITER

In situ observations by the *Galileo* probe revealed that the NH_3 and H_2S abundances in Jupiter's deep atmosphere ($P \gtrsim 8$ bar) are 3–4 times solar, while radio spectra (Fig. 6) show a subsolar abundance of NH_3 gas at pressures $P < 2$ bar. The apparent decrease in the NH_3 abundance at higher altitudes may be caused by dynamical processes, but the jury is still out on this.

Radio images of Jupiter clearly show bright zonal bands across the disk (Fig. 7a), which coincide with the brown belts seen at visible wavelengths. These bands have a higher brightness temperature, likely due to a lower opacity in the belts relative to the zonal regions, so deeper warmer layers are probed in the belts. This phenomenon is suggestive of gas rising up in the zones; when the temperature drops below ~140 K, ammonia gas condenses out. In the belt regions, the air, now depleted in ammonia gas (i.e., dry air), descends. This general picture agrees with that suggested from analyses of visible and infrared data. Note, though, that the radio data probe the gas from which the clouds condense, while visible and infrared data are sensitive primarily to the cloud particles. Thus, the base level of the clouds is determined through radio observations, whereas the cloud tops are probed at optical and infrared wavelengths.

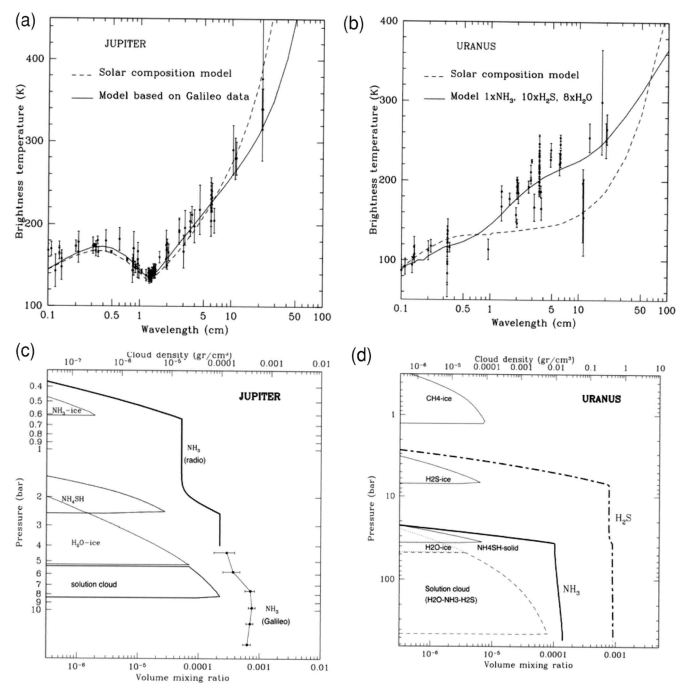

FIGURE 6   (a) Microwave spectrum of Jupiter, with superposed models for a solar composition atmosphere (dashed line), and one in which the altitude profiles for the condensable gases were based upon the *Galileo* probe data (solid line). (b) Microwave spectrum of Uranus, with superposed models for a solar composition atmosphere (dashed line), and one in which H_2S and H_2O gases are enhanced by a factor ~10 above solar values. In these models, ammonia gas is significantly depleted at higher altitudes in Uranus atmosphere through formation of NH_4SH, so that deeper warmer levels are probed. (c) Cloud structure in Jupiter's atmosphere as calculated assuming thermochemical equilibrium. The altitude profile of ammonia gas, based on *Galileo* and ground-based radio data, is superposed. (d) Cloud structure in Uranus's atmosphere as calculated assuming thermochemical equilibrium and CH_4, H_2O and H_2S abundances 30 times solar. The altitude profiles for H_2S and NH_3 gas are indicated. (I. de Pater and J. J. Lissauer, forthcoming, "Planetary Sciences," Rev. Ed., Cambridge Univ. Press.)

(a)   (b)

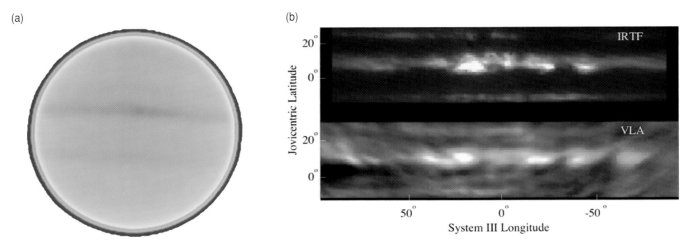

FIGURE 7   (a) A radio photo of Jupiter at a wavelength of 2 cm, integrated over 6–7 hours, so any longitudinal structure is smeared out. The data were obtained with the VLA on 25 January 1996. The angular resolution is 1.4," which was 0.044 Jupiter radii at the time of the observations. (I. de Pater et al., 2001, Comparison of Galileo probe data with ground-based radio measurements, *Icarus* **149**, 66–78.) (b) A comparison of the North Equatorial Belt of Jupiter at an infrared wavelength of 5 μm (IRTF, NASA's Infrared Telescope Facility) and radio wavelength of 2 cm (VLA). The latter image is constructed from the same data as displayed in Fig. 7a, but using novel new data reduction techniques. (R. J. Sault et al., 2004, Longitude-resolved imaging of Jupiter at λ = 2 cm, *Icarus* **168**, 336–343.)

In recent years, an algorithm has been developed to construct longitude-resolved images of Jupiter, and these maps reveal, for the first time, hot spots at radio wavelengths that are strikingly similar to those seen in the infrared. An example of Jupiter's North Equatorial Belt is shown in Fig. 7b. At radio wavelengths, the hot spots indicate a relative absence of NH₃ gas, whereas they suggest a lack of cloud particles in the infrared. Models show that ammonia must be de-

pleted down to pressure levels of ~5 bar in the hot spots, the approximate altitude of the water cloud.

2.5.3 SATURN

Images of Saturn's microwave emission at different viewing geometries are shown in Fig. 8. The planet itself is visible through its thermal emission. The emission from the

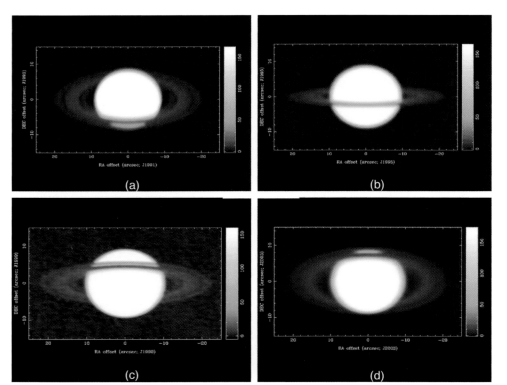

FIGURE 8   Radio photographs of Saturn at 2 and 3.6 cm, at different viewing aspects of the planet. (a) 3.6 cm, 1990; (b) 2 cm, 1994; (c) 2 cm, 1998; (d) 3.6 cm, 2002. (Dunn, D.E., I. de Pater, and L.A. Molnar, 2006. Examining the wake structure in Saturn's rings from microwave observations over varying ring opening angles and wave lengths. (*Icarus*, in press.)

planet's rings is dominated by Saturn's thermal radiation reflected off the ring particles. Only a small fraction of the radiation at centimeter wavelengths is thermal emission from the rings themselves.

Like on Jupiter, radio spectra of the atmospheric emission can be interpreted in terms of its ammonia abundance and local variations therein with altitude and latitude. The ammonia and hydrogen sulfide abundances on Saturn are likely ~3 times more enhanced than on Jupiter. The latitudinal structure on Saturn's disk, presumably caused by latitudinal variations in microwave opacity, changes considerably over time.

The classical A, B and C rings, with the Cassini Division, are clearly visible on Fig. 8. The inner B ring is brightest, with a brightness temperature of ~10 K. At 1–3 mm the temperature rises to ~20–25 K. In front of the planet, the rings block out part of Saturn's radio emission, resulting in an absorption feature. From this feature one can determine the optical depth of the rings, which is approximately 1 in the B ring. The West (right) ring ansa is usually somewhat brighter than the East side, which has been attributed to the presence of gravitational 'wakes', which are 10–100 m sized density enhancements behind large ring particles which, because of Keplerian shear, travel at an angle to the big particle's orbit. Similar asymmetries have been seen in the A ring in front of the planet.

A combination of radio and radar data show that the ring particles have sizes from ~1 cm up to ~5–10 m, where the number of particles, N, at a given size, R, varies approximately as $N \sim R^{-3}$. Such a particle size distribution would be expected from a collisionally evolved population of particles.

2.5.4 URANUS AND NEPTUNE

Radio spectra of Uranus and Neptune (Fig. 6b) suggest an overall depletion of ammonia gas in their upper atmospheres, by roughly 2 orders of magnitude compared to the solar nitrogen value. This apparent depletion is likely caused by a nearly complete removal of NH_3 gas in the upper atmosphere through the formation of NH_4SH. This is possible if H_2S is considerably (factor of >5) enhanced above solar S. Radio models predict enhancements by a factor of ~10 on Uranus and ~30 on Neptune. Good fits to Uranus' spectrum are obtained if NH_3 is close to the solar N abundance in Uranus' deep atmosphere. However, ammonia gas must be depleted in Neptune's atmosphere to match radio spectra. Nitrogen on Neptune may therefore be present in the form of both N_2 and NH_3, rather than only in the form of ammonia gas. An alternative idea that is advocated by some researchers is based on a large uptake of ammonia in the icy giant's ionic oceans, deep in their interiors.

Uranus is unique among the planets in having its rotation axis closely aligned with the plane in which the planet orbits the Sun. With its orbital period of 84 years, the seasons on Uranus last 21 years. During the *Voyager* encounter, in 1986, Uranus' south pole was facing the Sun (and us). Since that time, this pole is slowly moving out of sight, while the north pole is coming into view. Uranus brightness temperature has been monitored since 1966. A pronounced increase in brightness temperature was noticed when the south pole came into view, followed by a decrease when the pole moved away again (Fig. 6b). These measurements suggest that Uranus' south pole is considerably warmer than the equatorial region, a theory later confirmed by radio images from the VLA. Figure 9 on the following page shows one such image taken in the summer of 2003, along with an image at near-infrared wavelengths (1.6 μm) taken with the adaptive optics system on the Keck telescope. The VLA image shows that the south pole is brightest. It also shows enhanced brightness in the far-north (to the right on the image). At near-infrared wavelengths, Uranus is visible in reflected sunlight. The bright regions are clouds at high (upper troposphere) altitudes. The bright band around the south pole is at the lower edge of the VLA-bright south polar region. Air in this band may rise up, with condensables forming clouds, and descend over the pole. At radio wavelengths, this dry air allows us to probe deeper warmer layers in Uranus' atmosphere.

On Neptune we also see the poles (at least the visible south pole) to be the hottest region on the planet, indicative of a similarity in atmospheric dynamics between the two ice giants.

2.6 Major Satellites and Small Bodies

2.6.1 GALILEAN SATELLITES

Radio spectra of the Galilean satellites are diverse. The brightness temperature at infrared wavelengths can be related directly to the satellite's albedo, and hence Callisto, with its relatively low albedo ($A = 0.13$) is warmer than Io and Europa. The brightness temperature at radio wavelengths is determined by the physical temperature and radio emissivity, e, of the subsurface, $e = 1 - a$, with a the radar geometric albedo. The observed brightness temperatures for Ganymede and, in particular, Europa are well below the physical temperature of the subsurface layers. This measurement is consistent with the high radar albedo for these objects: $a = 0.33$ for Ganymede and 0.65 for Europa. These high albedos and consequently low emissivities and radio brightness temperatures are likely caused by coherent backscattering in fractured ice.

Since the detection of an ionosphere around Io by the *Pioneer 10* spacecraft in 1973, this satellite is known to posses a tenuous atmosphere. The first detection of a global atmosphere was obtained in 1990, where a rotational line of sulfur dioxide (SO_2) gas was measured at 222 GHz. Io is the only object with an atmosphere dominated by SO_2 gas, the origin of which can ultimately be attributed to the satellite's volcanism.

(a)

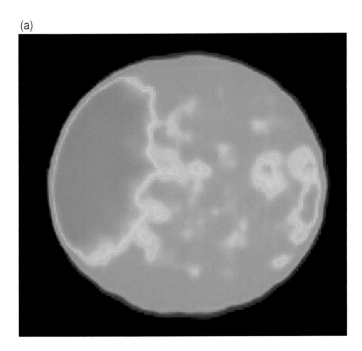

(b)

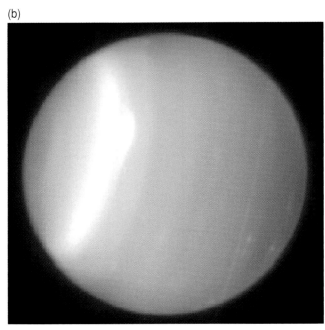

FIGURE 9 (a) VLA image of Uranus at 2 cm wavelength taken in the summer of 2003. Note the hot (red) poles. (M. D. Hofstadter and B. J. Butler, 2003, Seasonal change in the deep atmosphere of Uranus. *Icarus* **165**, 168–180.) (b) Infrared (1.6 μm) image of Uranus taken with the Keck adaptive optics system in October 2003. The polar collar around the south pole (left in figure) lines up with the edge of the hot pole seen at radio wavelengths. Several cloud features are visible in the infrared image, and the thin line near the right is Uranus' ring system. Hammel, H. B., I. de Pater, S. Gibband, G. W. Lockwood, and K. Rages, 2005. Uranus in 2003: Zonal winds, banded structures, and discrete features. *Icarus*, **175**, 534–545.

Part of the gas is of direct volcanic origin, and part is driven by subliming SO_2 frost, which itself is a product of volcanic eruptions. Several SO_2, as well as SO, lines have now been observed, which have been used to derive Io's atmospheric structure. The surface pressure is of the order of a few, perhaps up to 40 nbar, covering 5–20% of the surface, and the atmosphere may be relatively hot, 500–600 K at 40 km altitude on the trailing, and 250–300 K on the leading hemisphere.

2.6.2 TITAN

Of all solar system bodies, Titan's atmosphere is most similar to that of Earth, being dominated by nitrogen gas and with a surface pressure 1.5 times that on Earth. Methane gas, with an abundance of a few percent, has a profound effect on the atmosphere. [*See* TITAN.] Photolysis and subsequent chemical reactions lead to the formation of hydrocarbons and nitriles. Because CO and the nitriles HCN, HC_3N (cyanoacetylene), and CH_3CN (acetonitrile) have several transitions at (sub)millimeter wavelengths (Fig. 10), radio observations can be used to constrain the vertical distributions of these species. As expected from photochemical models, their abundances increase with altitude and are highest in the stratosphere.

Disk-resolved spectra, such as obtained with the Submillimeter Array (SMA) and the IRAM Plateau de Bure Interferometer, also contain information on the zonal wind profile. Although 12 μm spectroscopic measurements had already suggested the winds to be prograde at ~100–300 km altitude, the radio data confirmed the direction of the winds and reported more precise values for the wind speeds in the upper stratosphere (160 ± 60 m/s at ~200–400 km altitude), and lower mesosphere (60 ± 20 m/s at ~350–550 km). At lower altitudes, the winds were determined via the Doppler Wind Experiment on the *Huygens* probe, when it went down through Titan's atmosphere. The radio signal from the probe (communication to the *Cassini* orbiter) was recorded by the very long baseline interferometry (VLBI) network. Winds in Titan's atmosphere affected the horizontal velocity of the probe during its descent, which was measured by ground-based radio telescopes through a shift in the probe's transmitted frequency (Doppler shift). These measurements revealed weak prograde winds near the surface, rising to ~100 m/s at 100–150 km altitude, with a substantial drop (down to a few m/s at most) near 60–80 km altitude.

The isotopic carbon and nitrogen ratios were first determined from ground-based radio data, and subsequently confirmed/improved by instruments on board the *Cassini* spacecraft and *Huygens* probe. The $^{12}C/^{13}C$ isotope ratio

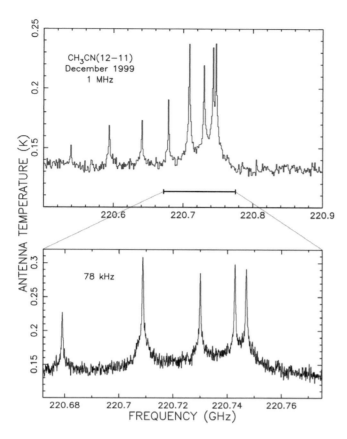

FIGURE 10 A spectrum of Titan's 12–11 transition of CH_3CN, taken in December 1999 with the IRAM 30-m telescope in Spain. The upper panel shows the spectrum at 1 MHz resolution; the lower panel shows it at 78 kHz. (A. T. Marten et al., 2002, New millimeter heterodyne observations of Titan: Vertical distributions of nitriles HCN, HC_3N, CH_3CN, and the isotopic ratio $^{15}N/^{14}N$ in its atmosphere, *Icarus* **158**, 532–544.)

on Titan is very similar to that on Earth (89), while $^{14}N/^{15}N$ was measured to be several times less than the terrestrial value of 272. This has been explained by a large loss of Titan's primitive atmosphere over time, which would lead to an isotopic fractionation in nitrogen. In contrast, the similar-to-Earth value in $^{12}C/^{13}C$ hints at a continuous or periodic replenishment of methane gas into Titan's atmosphere, such as could happen, for example, through cryovolcanism or a methane cycle akin to the hydrology cycle on Earth.

Radiometry maps of Titan obtained with the *Cassini* spacecraft can be used with radar and infrared measurements to better constrain the surface composition and compactness. Observations show variations in brightness temperature up to $\sim$10 K, which are more or less anticorrelated with infrared brightness (i.e., the infrared/optically bright areas have a low radio brightness temperature). The *Cassini* radar team suggests the optically bright, radio-cold areas perhaps are composed of fractured or porous ice (as on Europa and Ganymede), and the optically dark, radio-warm regions are composed of an organic sludge, or perhaps more solid water ice (higher dielectric constant).

2.6.3 ASTEROIDS AND TRANS-NEPTUNIAN OBJECTS

In analogy with the terrestrial planets, a comparison of multiwavelength radio data of small airless bodies with thermophysical models provides information on the (sub)surface properties of the material, as composition and compactness. Radio spectra of several main-belt asteroids suggest that these bodies are typically overlain by a layer of fluffy (highly porous) dust a few centimeters thick, as on the Moon and Mercury.

It has been challenging to observe trans-Neptunian (TNO) or Kuiper Belt objects (KBO) at radio wavelengths, including Pluto, due to their small angular extent and low surface temperature. Much progress has been made in the past decades, however. For an object in radiative equilibrium, with an albedo of $\sim$0.6, the surface temperature should be approximately 50 K, consistent with the 53–59 K temperatures for Pluto as measured by IRAS at 60 and 100 μm. Since at radio wavelengths one probes approximately 10 wavelengths deep into the surface, a brightness temperature of $\sim$40 K is expected.

Observations of Pluto with the 30-m IRAM telescope at millimeter wavelengths revealed brightness temperatures closer to 30 K, indicative of a low radio emissivity ($e \approx$ 0.6–0.7), similar to that seen on Ganymede. Such a low emissivity can be reconciled with a surface composed of icy grains, and hence relatively high porosity.

Radio measurements of KBOs have been used to determine the size and albedo of several of the largest objects (Quaoar, Ixion, Varuna, 2002 AW197), in concert with optical measurements and the so-called Standard Thermal Model (STM) to interpret the data. Although one has to be aware of the assumptions made in the STM, which can lead to over- or underestimates of the size and albedo, such measurements are usually our only means to get a reasonable size estimate for these objects.

2.7 Comets

Radio observations of comets provide information that complements studies at other wavelengths. Continuum measurements are sensitive to the thermal emission from a cometary nucleus and of large dust grains in its coma, while spectroscopic observations provide information on the "parent" molecules in a comet's coma. Upper limits to the radio continuum emission of a few comet nuclei suggest that the temperature gradient in the nucleus may be very steep, or, alternatively, that the emission is substantially suppressed by subsurface scattering.

Quasi-continuum spectra reveal a spectral slope that is steeper than that of blackbody thermal emission, yet smaller than that expected from Rayleigh scattering from small

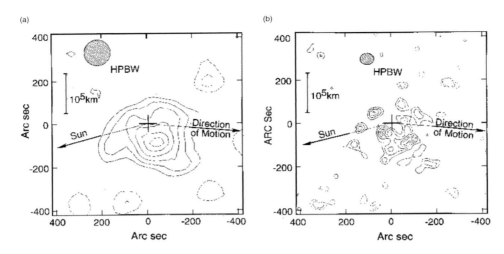

FIGURE 11 Contour plots of comet Halley, November 13–16, 1985. The image is taken at the peak flux density of the line (0.0 km/s in the reference frame of the comet). The left side shows a low-resolution image (3′), and the right side shows a high-resolution image (1′), after the data for both dates were combined. Contour levels for the low-resolution image are 4.9, 7.8, 10.8, 13.7, 16.7, and 18.6 mJy/beam. For the high-resolution image, they are 4.4, 4.4, 6.0, 7.7, 9.3, and 10.4 mJy/beam. Dashed contours indicate negative values. The beam size, a linear scale, the direction of motion, and the direction to the Sun are indicated in the figures. The cross indicates the position of the nucleus at the time of the observations. (I. de Pater et al., 1986, The brightness distribution of OH around comet Halley, *Astrophys. J. Lett.* **304**, L33–L36.)

particles. These data thus hint at the presence of grains with sizes in the (sub)millimeter range.

The most significant advances in cometary radio research have been obtained from spectroscopic studies. The cometary nucleus consists primarily of water ice, which sublimates off the surface when the comet approaches the Sun. After about a day, H_2O dissociates into OH and H. Since the early 1970s, the 18 cm OH line has been observed and monitored in many comets. Such observations are important because they provide indirect information on the production rate, and time variability therein, of water, a molecule that remains difficult to observe on a routine basis.

The OH line is sometimes seen in emission, and at other times in absorption against the galactic background. The OH emission is maser emission (i.e., stimulated emission from molecules in which the population of the various energy levels is inverted, so that the higher energy level is overpopulated compared to the lower energy level). This population inversion is caused by absorption of solar photons at ultraviolet (UV) wavelengths. However, this excitation process depends on the comet's velocity with respect to the Sun (heliocentric velocity), the so-called Swings effect. If the heliocentric velocity is such that solar Fraunhofer (absorption) lines are Doppler shifted into the OH excitation frequency, the molecule is not excited. In that case, OH will absorb 18 cm photons from the galactic background and be seen in absorption against the galactic background. If the line is excited, background radiation at the same wavelength (18 cm) will trigger its deexcitation, and the line is

seen in emission (maser or stimulated emission). With radio interferometers, the OH emission can be imaged. Such images have, for example, revealed the so-called quenching region directly, a region around the nucleus where collisions between particles thermalize the energy levels of OH molecules, so they no longer produce maser emission (Fig. 11).

One of the strengths of radio astronomy is the detection of "parent" molecules in a cometary coma, molecules that evolve directly from its icy surface. Such observations are crucial for our understanding of a comet's composition, and, indirectly, on the conditions in the early solar nebula from which our planetary system formed because cometary nuclei have presumably not been altered by excessive heating or high pressures. A growing number of molecular species have been detected at radio wavelengths. Figure 12 shows the time evolution of observed production rates for a large number of gases, subliming from comet Hale–Bopp (C/1995 O2). Only the most volatile materials sublime at heliocentric distances $r \gtrsim 5$ AU, while OH (from H_2O) becomes dominant at $r < 3$ AU.

Whereas most molecules sublime directly off the cometary nucleus, some gases, such as carbon monoxide and formaldehyde, are also released from dust grains in a comet's coma. With the advent of new powerful (sub)millimeter arrays, much improved images of the spatial distribution of parent molecules in a cometary coma can be obtained. Figure 13 shows a composite of the formaldehyde emission from comet Hale–Bopp, as observed with the

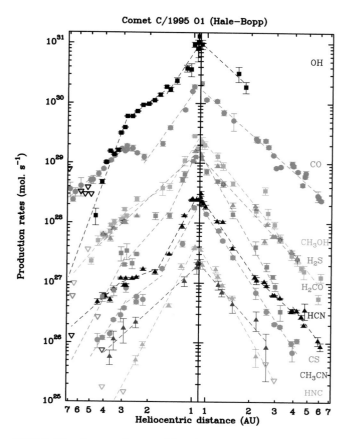

Comet C/1995 O1 (Hale–Bopp)

FIGURE 12 Time evolution of the observed production rates of comet C/Hale–Bopp as a function of heliocentric distance, with superposed fitted power laws (dashed lines). (N. Biver et al., 1999, Post-perihelion observations of the distant gaseous activity of comet C/1995 O1 (Hale–Bopp) with the Swedish–ESO Submillimeter Telescope (SEST), *Asteroids, Comets and Meteors.*)

ARO 12 m telescope and the Berkeley–Illinois–Maryland Association (BIMA) array. Transitions at several frequencies are shown, as well as a contour map from BIMA at 72.8 GHz (in bold) superposed on the ARO 225.7 GHz image. These observations show that formaldehyde indeed originates both from the nucleus and in the coma, where the coma-source appears dominated by a single fragment in this case.

ESA's *Rosetta* spacecraft, currently on its way to comet 67P/Churyumov–Gerasimenko, carries a microwave instrument, MIRO, with receivers centered at 190 and 562 GHz. Upon rendezvous at a heliocentric distance of 3.5 AU, Rosetta will move with the comet down to perihelion near 1.3 AU. MIRO is one of the instruments that will observe the comet during this time. It has broadband channels on both receivers to measure near-surface temperatures and temperature gradients in the comet's nucleus. Particularly exciting is the high spectral resolution spectrometer connected to the 562 GHz receiver, which will measure several major volatile species (H_2O, CO, CH_3OH, and NH_3) at

extreme high spatial (down to 5 m at the comet's surface) and spectral resolution. These measurements will provide unprecedented information on the outgassing of the comet as a function of heliocentric distance.

3. Nonthermal Radiation

Nonthermal planetary radio emissions are usually produced by electrons spiraling around magnetic field lines. Until the era of spacecraft missions, we had only received **nonthermal radio emissions** from the planet Jupiter, and these were usually limited to frequencies ≳ 10 MHz, since radiation at lower frequencies is blocked by Earth's ionosphere. Strong radio bursts at frequencies below 40 MHz were attributed to emission via the cyclotron maser instability in which auroral electrons with energies of a few to several keV power the emission, while radiation at frequencies ≳ 100 MHz was interpreted as synchrotron radiation, emitted by high energy (MeV range) electrons trapped in Jupiter's radiation belts, a region in Jupiter's magnetic field analogous to the Earth's **Van Allen belts**. Like Earth, the magnetic fields of the four giant planets resemble to first approximation that of a dipole magnetic field. Despite several searches, no positive detections of nonthermal radio emissions from any of the other three giant planets were made until the *Voyager* spacecraft approached these objects. Now we know that all four giant planets as well as Earth are strong radio sources at low frequencies (kilometric wavelengths). Jupiter's moon Ganymede is also a source of nonthermal radio emissions. The strongest planetary radio emissions usually originate near the auroral regions and are intimately related to auroral processes.

A graph of the average normalized spectra of the auroral radio emissions from the four giant planets and Earth is displayed in Fig. 14. All data are adjusted to a distance of 1 AU. Jupiter is the strongest low-frequency radio source, followed by Saturn, Earth, Uranus, and Neptune. In Sections 3.3–3.7, we discuss the emissions from each planet.

3.1 Low-Frequency Emissions

3.1.1 CYCLOTRON MASER EMISSIONS

Radio emission at frequencies of a few kHz to 40 MHz (for Jupiter) is usually attributed to electron cyclotron maser radiation, emitted by keV (nonrelativistic) electrons in the auroral regions of a planet's magnetic field. The radiation is emitted at the frequency that electrons spiral around the local magnetic field lines (the cyclotron or Larmor frequency):

$$\nu_L = \frac{qB}{2\pi m_e c} \qquad (6)$$

Comet Hale–Bopp: H₂CO Spectra and Maps

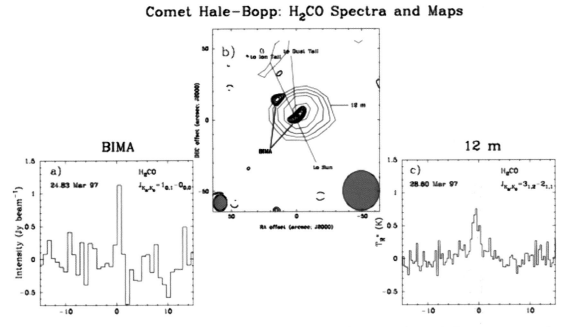

FIGURE 13 Images and spectra of H_2CO in comet C/1995 O1 (Hale–Bopp) taken with the BIMA array and ARO 12 m telescope on different days and in different transitions, as indicated in each panel. Panel b shows a contour map from BIMA at 72.8 GHz (in bold; from spectrum in panel a), superposed on the ARO 225.7 GHz image (from spectrum in panel c). The synthesized beam for BIMA is shown in the lower left, and that of ARO appears in the lower right. (S. N. Milam et al., Formaldehyde in comets C/1995 O1 (Hale–Bopp), C/2002 T7 (LINEAR), and C/2001 Q4 (NEAT): Investigating the cometary origin of H_2CO, *Astrophys. J.* **649**, 1169–1177.)

with q the elemental charge, B the magnetic field strength, m_e the electron mass, and c the speed of light. Propagation of the radiation depends on the interaction of the radiation with the local plasma, or charged particle population. The oscillation of these particles, as caused by the electromagnetic properties of the plasma, leads to a complex interaction between the propagating radiation (the electromagnetic waves) and the local plasma. For example, the radiation can escape its region of origin only if the local cyclotron frequency is larger than the electron plasma frequency:

$$\nu_e = \left(\frac{4\pi N_e q^2}{m_e} \right)^{1/2} \qquad (7)$$

with N_e the electron density. Hence, the plasma frequency is the frequency at which electrons oscillate about their equilibrium positions in the absence of a magnetic field. This similarly sets the limit for propagation through Earth's ionosphere at ~ 10 MHz. If the local cyclotron frequency is less than the electron plasma frequency, the waves are locally trapped and amplified, until it reaches a region from where it can escape. The cyclotron maser instability also

requires a large ratio of ν_L/ν_e. The auroral regions in planetary magnetospheres are characterized by such conditions. The mode of propagation (or polarization) of auroral radio emissions is in the so-called extraordinary (X) sense, and the polarization (direction of the electric vector of the radiation) depends upon the direction of the magnetic field. The emission is right-handed circularly polarized (RH) if the field at the source is directed toward the observer and left-handed circularly polarized (LH) if the field points away from the observer.[1]

Cyclotron radiation is emitted in a dipole pattern, where the lobes are bent in the forward direction. The resulting emission is like a hollow cone pattern, as displayed in Fig. 15. The radiation intensity is zero along the axis of the

[1] Circular polarization is in the RH sense when the electric vector of the radiation in a plane perpendicular to the magnetic field direction rotates in the same sense as a RH screw advancing in the direction of propagation. Thus, rotation is counterclockwise when propagation is toward and viewed by the observer. RH polarization is defined as positive; LH, as negative. In some cases, the radio emissions propagate in the ordinary (O) magneto-ionic mode. In this mode the polarization is reversed. The theory of the cyclotron maser instability does admit the possibility of emission in the ordinary mode. However, it is less common.

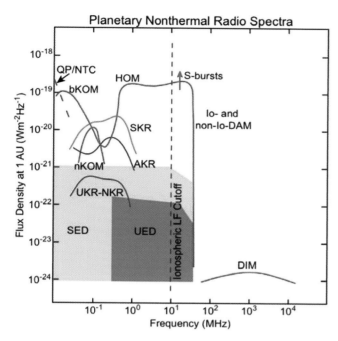

FIGURE 14   A comparison of the peak flux density spectrum of the kilometric continuum radio emissions of the four giant planets and Earth. All emissions were scaled such that the planets appear to be at a distance of 1 AU. Jovian emissions shown include quasi-periodic bursts (QP), nonthermal continuum (NTC), broadband and narrowband kilometric radiation (bKOM, nKOM), hectometric radiation (HOM), decametric radiation (DAM), and decimetric radiation (DIM). Saturn's kilometric radiation is designated SKR, and its electrostatic discharge emissions are labeled SED. Terrestrial auroral kilometric radiation is designated AKR. UKR and NKR refer to kilometric radiation from Uranus and Neptune, respectively. Uranus' electrostatic discharges are labeled UED. (Adapted from P. Zarka and W. S. Kurth, 2005, Radio wave emission from the outer planets before *Cassini, Space Sci. Rev.* **116**, 371–397.)

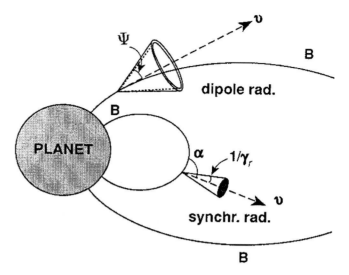

FIGURE 15   Radiation patterns in a magnetic field. Indicated are the hollow cone pattern caused by cyclotron (dipole) radiation from nonrelativistic electrons in the auroral zone. The electrons move outward along the planet's magnetic field lines. The hollow cone opening half-angle is given by Ψ. At low magnetic latitudes, in the Van Allen belts, the filled radiation cone of a relativistic electron is indicated. The angle between the particle's direction of motion and the magnetic field, commonly referred to as the particle's pitch angle, α, is indicated on the sketch. The emission is radiated into a narrow cone with a half width of $1/\gamma$. (I. de Pater and J. J. Lissauer, 2001, "Planetary Sciences," Cambridge Univ. Press.)

cone, in the direction of the particle's parallel motion, and reaches a maximum at an angle Ψ. Theoretical calculations show that Ψ is very close to $90°$. Observed opening angles, however, can be much smaller, down to $\sim 50°$, which has been attributed to refraction of the electromagnetic waves as they depart from the source region.

The cyclotron maser instability derives energy from a few keV electrons, which have distribution functions with a positive slope in the direction perpendicular to the magnetic field. Recent observations in the source of Earth's auroral kilometric radiation reveal "horseshoe"-shaped electron distributions that provide a highly efficient (of order 1%) source of free energy for the generation of the radio waves. This distribution is thought to be the result of parallel electric fields in the auroral acceleration region, the loss of small pitch-angle electrons to the planetary atmosphere,

and trapping of reflected electrons. Radio emissions generated in planetary magnetospheres by this mechanism often display a bewildering array of structure on a frequency-time spectrogram including narrowband tones that rise or fall in frequency, sharp cutoffs, and more continuum-like emissions. While it is generally accepted that emissions that rise or fall in frequency are related to tiny sources moving down or up the magnetic field line (hence, to regions with higher or lower cyclotron frequencies), there is no generally accepted theoretical explanation for the fine structure.

3.1.2 OTHER TYPES OF LOW–FREQUENCY RADIO EMISSIONS

While the radio emissions generated by the cyclotron maser instability are, by far, the most intense in any planetary magnetosphere, other types of radio emissions do occur that are of interest. Perhaps the most ubiquitous of these is the so-called nonthermal continuum radiation that arises from the conversion of wave energy in electrostatic waves near the source plasma frequency to radio waves, usually propagating in the ordinary mode. There are arguments for both linear and nonlinear conversion mechanisms. The term "continuum" was originally assigned to this class of emissions because they can be generated at very low frequencies and can be trapped in low-density cavities in the outer portions of the magnetosphere when the surrounding solar wind density is higher. The mixture of multiple sources at different frequencies and multiple reflections off the moving walls of the magnetosphere tend to homogenize the spectrum. However, at higher frequencies, these emissions are often created as narrowband emissions from narrowband electrostatic bands at the upper hybrid resonance frequency on density gradients in the inner magnetosphere and can propagate directly away from the source, yielding a complex narrowband spectrum. These emissions were first discovered at Earth and have been found at all of the magnetized planets. Furthermore, emissions of this nature are also produced by Ganymede's magnetosphere.

Another type of planetary radio emission is closely related to a common solar emission mechanism, the conversion of Langmuir waves to radio emissions at either the plasma frequency or its harmonic. The Langmuir waves are common features of the solar wind upstream of a planetary bow shock, which arises from the interaction of the supersonic flow of solar wind plasma past the planets. This mechanism is a nonlinear mechanism involving three waves: the Langmuir wave, the radio wave, and either a low-frequency wave in the case of emission near the plasma frequency or a second Langmuir wave in the case of harmonic emission. The resulting emissions are weak, narrowband emissions.

3.1.3 ATMOSPHERIC LIGHTNING

Radio emissions from planets are sometimes associated with atmospheric lightning. The lightning discharge, in addition to producing the visible flash, also produces broad, impulsive radio emissions. If the spectrum of this impulse extends above the ionospheric plasma frequency and if absorption in the atmosphere is not too great, a remote observer can detect the high-frequency end of the spectrum. The "interference" detected with an AM radio on Earth during a thunderstorm is the same phenomenon.

3.2 Synchrotron Radiation

Synchrotron radiation is emitted by relativistic electrons gyrating around magnetic field lines. In essence, this emission consists of photons emitted by the acceleration of electrons as they execute their helical trajectories about magnetic field lines. The emission is strongly beamed in the forward direction (see Fig. 15) within a cone $1/\gamma$:

$$\frac{1}{\gamma} = \sqrt{1 - \frac{v^2}{c^2}} \qquad (8)$$

with v the particle's velocity and c the speed of light. The relativistic beaming factor $\gamma = 2E$, with E the energy in

MeV. The radiation is emitted over a wide range of frequencies, but shows a maximum at 0.29 v_c, with the critical frequency, v_c, in MHz:

$$v_c = 16.08 E^2 B \qquad (9)$$

where the energy E is in MeV and the field strength B is in Gauss. The emission is polarized, where the direction of the electric vector depends on the direction of the local magnetic field. Jupiter is the only planet for which this type of emission has been observed. It has been mapped by ground-based radio telescopes and by *Cassini* to provide some of the most comprehensive, though indirect, information about Jupiter's intense radiation belts.

3.3 Earth

The terrestrial version of the cyclotron maser emission, commonly referred to as auroral kilometric radiation (AKR), has been studied both at close range and larger distances by many Earth-orbiting satellites. The radiation is very intense; the total power is 10^7 W, sometimes up to 10^9 W. The intensity is highly correlated with geomagnetic substorms, thus it is indirectly modulated by the solar wind. It originates in the night side auroral regions and in the day side polar cusps at low altitudes and high frequencies and spreads to higher altitudes and lower frequencies. Typical frequencies are between 100 and 600 kHz. Since AKR is generated by auroral electrons, it can be used as a proxy for auroral activity. And, since numerous in situ studies of the terrestrial auroral electron populations and the resulting radio emissions have been carried out, we can apply our understanding of this emission process to similar emissions at other planets where in situ studies have not yet been carried out.

Earth is also the source of nonthermal continuum radiation. Below the solar wind plasma frequency this radiation is trapped within the magnetosphere. The spectrum is relatively smooth down to the local plasma frequency, typically in the range of a few kHz, where the emission cuts off at the ordinary mode cutoff. A few observations of this emission also show an extraordinary mode cutoff. Hence, the emission is either generated in both polarizations, or some of the initially dominant ordinary mode is converted into the extraordinary mode via reflections or other interactions with the magnetospheric medium. Above the solar wind plasma frequency, typically at a few tens of kHz, the "continuum" radiation spectrum exhibits a plethora of narrowband emissions; some of these extend well into the range of a few hundred kHz.

While not as important as the auroral radio emissions from an energetics point of view, the low-frequency limit of the continuum radiation at the plasma frequency provides an accurate measure of the plasma density, an often difficult

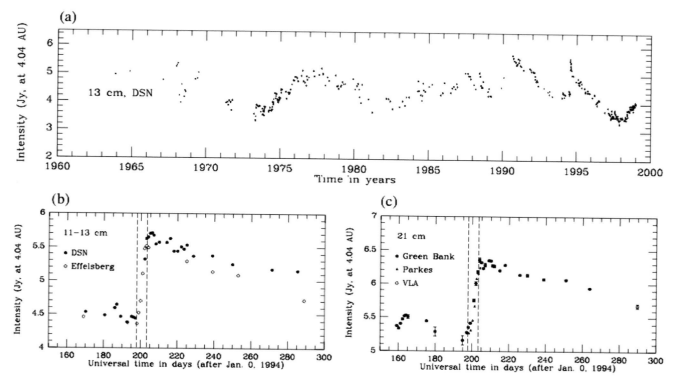

FIGURE 16 Time variability in Jupiter's radio emission. Panel (a) shows the radio intensity at a wavelength of 13 cm between the years 1963 and 1998. (Courtesy M. J. Klein.) Panels (b) and (c) show Jupiter's radio intensity at 11–13 and 21 cm, respectively, during 1994 up to the summer of 1995. The impact of comet D/Shoemaker–Levy 9 with Jupiter occurred in July of 1994 (indicated by the vertical dashed lines). (I. de Pater and J. J. Lissauer, 2001, "Planetary Sciences," Cambridge Univ. Press.)

measurement for a plasma instrument because of spacecraft charging effects.

3.4 Jupiter's Synchrotron Radiation

Jupiter is the only planet from which we receive synchrotron radiation. The variation in total intensity and polarization characteristics during one jovian rotation (the so-called beaming curves) indicate that Jupiter's magnetic field is approximately dipolar in shape, offset from the planet by roughly one tenth of a planetary radius toward a longitude of 140°, and inclined by ~10° with respect to the rotation axis. Most electrons are confined to the magnetic equatorial plane. The magnetic north pole is in the northern hemisphere, tipped toward a longitude of 200°. The total flux density of the planet varies significantly over time (Fig. 16). These variations seem to be correlated with solar wind parameters, in particular the solar wind ram pressure, suggesting that the solar wind may influence the supply and/or loss of electrons into Jupiter's inner magnetosphere. In addition to variations in the total flux density, the radio spectrum changes as well (Fig. 18).

An image of Jupiter's synchrotron radiation obtained with the VLA in 1994 is shown in Fig. 17a. This image was obtained at a wavelength of 20 cm and has a spatial resolution of ~ 6″ or 0.3 R_J. Since Jupiter's synchrotron radiation is optically thin, one can use tomography to extract the 3-dimensional distribution of the radio emissivity from data obtained over a full jovian rotation. The example in Fig. 17b shows that most of the synchrotron radiation is concentrated near the magnetic equator, which, due to the higher order moments in Jupiter's field, is warped like a potato chip. The secondary emission regions, apparent at high latitudes in Fig. 17a, show up as rings of emission north and south of the main ring. These emissions are produced by electrons at their mirror points and reveal the presence of a rather large number of electrons bouncing up and down field lines that thread the magnetic equator at ~2.5 jovian radii. This emission may be "directed" by the moon Amalthea. A fraction of the electrons near Amalthea's orbit undergoes a change in their direction of motion, caused perhaps by interactions with low-frequency plasma waves near Amalthea (such plasma noise was detected by the *Galileo* spacecraft when it crossed Amalthea's orbit), and through interactions with dust in Jupiter's rings, while regular synchrotron radiation losses also lead to small changes in an electron's direction of motion.

Figure 18 shows radio spectra of Jupiter's synchrotron radiation from 74 MHz up to ≳20 GHz. These spectra show that the electrons in Jupiter's radiation belts do not follow

(a)

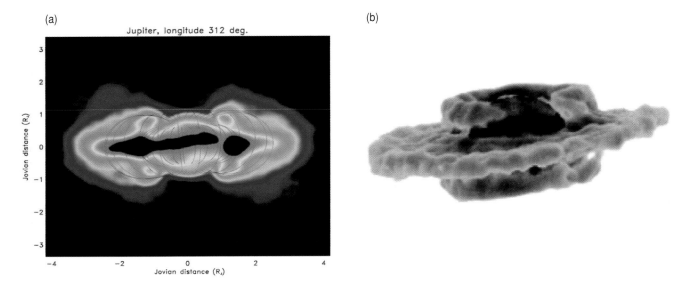

(b)

FIGURE 17   (a) Radio photograph of Jupiter's decimetric emission at a wavelength of 20 cm, and a central meridian longitude of $\lambda_{cml} \sim 312°$. Magnetic field lines at equatorial distances of 1.5 and 2.5 Jupiter radii are superposed. Field lines are shown every 15°, between $\lambda_{cml} - 90°$ and $\lambda_{cml} + 90°$. The image was taken with the VLA in June 1994. The resolution is 0.3 Jupiter radii, roughly the size of the high latitude emission regions. [I. de Pater et al., 1997, Synchrotron evidence for Amalthea's influence on Jupiter's electron radiation belt, *J. Geoph. Res.*, **102** (A10), 22,043–22,064; Copyright 1997 American Geophysical Union. Reproduced/modified by permission of American Geophysical Union.] (b) Three-dimensional reconstruction of Jupiter's nonthermal radio emissivity, from VLA data taken in June 1994, as seen from Earth at $\lambda_{cml} = 140°$ ($D_E = -3°$). The planet is added as a black sphere in this visualization. (I. de Pater and R. J. Sault, 1998, An intercomparison of 3-D reconstruction techniques using data and models of Jupiter's synchrotron radiation. *J. Geophys. Res. Planets* **103** (E9), 19,973–19,984; Copyright 1998 American Geophysical Union. Reproduced/modified by permission of American Geophysical Union.)

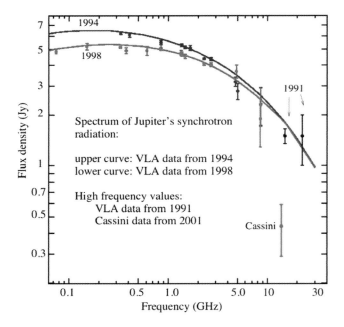

FIGURE 18   Jupiter's radio spectrum as measured in September 1998 and June 1994. Superposed are various model calculations. (Adapted from I. de Pater et al., 2003, Jupiter's radio spectrum from 74 MHz up to 8 GHz. *Icarus* **163**, 434–448, and I. de Pater and D. E. Dunn, 2003, VLA Observations of Jupiter's synchrotron radiation at 15 and 22 GHz, *Icarus* **163**, 449–455.

a simple $N(E) \propto E^{-a}$ power law. Well outside the synchrotron radiation region, beyond Io's orbit at 6 jovian radii, the electron energy spectrum appears to follow a double power law, $N(E) \propto E^{-0.5}(1 + E/100)^{-3}$, consistent with in situ measurements by the *Pioneer* spacecraft. Processes as radial diffusion, pitch angle scattering, synchrotron radiation losses, and absorption by moons and rings change the electron spectrum. The radio spectra superposed on the data were derived from such models.

Early in the 20th century (~1930), Jupiter captured a comet, now known as comet D/Shoemaker–Levy 9. During a close encounter with the planet, this comet was ripped apart by Jupiter's strong tidal force into over 20 pieces. These comet fragments, all in orbit about Jupiter, were discovered by the Shoemaker–Levy comet hunting team in May 1993. About a year later, from July 16 to 22 (1994), all comet fragments hit Jupiter. These events were widely observed, at wavelengths across the entire electromagnetic spectrum. At infrared wavelength these impacts were incredibly bright, while at optical wavelengths the impact sites were visible as dark spots with even the smallest telescopes. This collision also triggered large temporary changes in Jupiter's synchrotron radiation. The total flux density increased by ~20% (Fig. 16), the radio spectrum hardened, and the spatial brightness distribution changed considerably (Fig. 19a, b). These changes were brought about by a

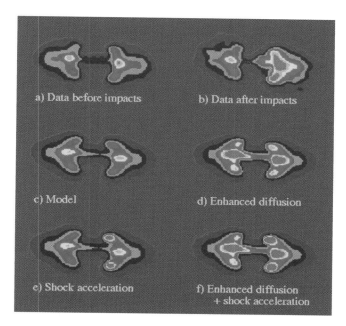

FIGURE 19 Real and synthetic false color images at a wavelength of 20 cm (1.5 GHz) of Jupiter following the impacts of comet D/Shoemaker–Levy 9 with the planet. (a and b) Observations of the synchroptron radiation before (June 1994) and after several impacts (19 July 1994), respectively. (c) Theoretical emission based on a model of the ambient relativistic electron distribution within a multipole magnetic field configuration. (d) Theoretical synchrotron radiation after an enhancement in the radial diffusion coefficient by a factor of a few million. (e) Enhancement in the theoretical synchrotron radiation, as produced from just shock acceleration. (f) Theoretical synchrotron radiation using the shock model and radial diffusion combined. (S. H. Brecht et al., 2001, Modification of the jovian radiation belts by Shoemaker–Levy 9: An Explanation of the data, *Icarus* **151**, 25–38.)

complex interaction of the radiating particles with shocks and electromagnetic waves induced in the magnetosphere by the series of cometary impacts. Results from models simulating the effects are shown in Figs. 19c–f.

3.5 Jupiter at Low Frequencies

Jupiter has the most complex low-frequency radio spectrum of all the planets. Examples of most of these are shown in Fig. 20 and are discussed in this section.

3.5.1 DECAMETRIC AND HECTOMETRIC RADIO EMISSIONS

From the ground, Jupiter's decametric (DAM) emission, confined to frequencies below 40 MHz, has routinely been observed since its discovery in the early 1950s, occasionally down to frequencies of 4 MHz. The upper-frequency cutoff is determined by the local magnetic field strength in the

auroral regions: 40 MHz for RH emissions translates into ~14 Gauss in the north polar region, and 20 MHz for LH into ~7 Gauss in the south.

The dynamic spectra in the frequency–time domain are extremely complex, but well ordered. On time scales of minutes, the emission displays a series of arcs, like open or closed parentheses (Fig. 20). Within one storm, the arcs are all oriented the same way. The emissions have been interpreted as coherent cyclotron emissions. The satellite Io appears to modulate some of the emissions: Both the intensity and the probability of the occurrence of bursts increase when Io is at certain locations in its orbit with respect to Jupiter and the observer. The non-Io emission originates near Jupiter's aurora, and is produced by electrons that travel along magnetic field lines from the middle-to-outer magnetosphere toward Jupiter's ionosphere. Particles that enter the atmosphere are "lost." These may locally excite atoms and molecules through collisions, which upon de-excitation are visible as aurora at UV and IR wavelengths. Other electrons are reflected back along the field lines, and produce DAM, where their motion along the field line is reflected in the form of arcs in the radio emission (i.e., a drift with frequency). The Io-dependent emissions are produced at or near the footprints of the magnetic flux tube passing through Io (similar, but much weaker, emissions originate along the flux tubes passing through Ganymede, and perhaps Callisto).

Hectometric (HOM) emissions are, in many ways, indistinguishable from DAM except that they are found at lower frequencies, from a few hundred kHz to a few MHz, with a local maximum near 1 MHz. The source region of HOM must be further from Jupiter than the DAM source. Otherwise, like DAM, HOM is predominantly emitted in the extraordinary mode and is likely generated by the cyclotron maser instability.

Because the dipole moment of Jupiter is tilted by some 10° from the rotational axis, most jovian radio emissions exhibit a strong rotational modulation. Given that Jupiter is a gas giant, this modulation is thought to be the best indicator of the rotation of the deep interior of the planet. The rotation period of the interior is important, for example, because this provides a rotating coordinate system against which the atmospheric winds can be measured. Because these radio observations have been recorded over many decades of time, analysis of these data lead to an extremely accurate determination of Jupiter's rotation period, $9^h 55^m 29^s.6854$.

3.5.2 KILOMETRIC RADIO EMISSIONS

Between a few kHz up to 1 MHz various spacecraft detected both broadband (bKOM) and narrowband (nKOM) kilometric radiation from Jupiter (Fig. 20). The lower frequency cutoff for bKOM, ~20 kHz (sometimes down to ~5 kHz) is likely set by propagation of the radiation through the Io

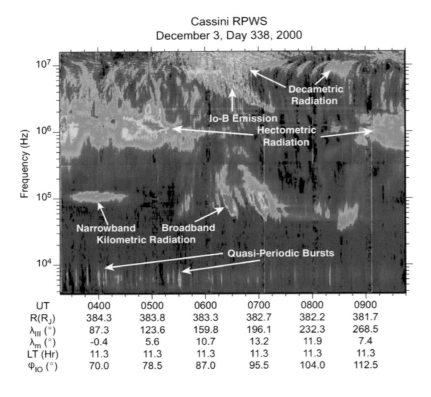

Cassini RPWS
December 3, Day 338, 2000

UT	0400	0500	0600	0700	0800	0900
R(R$_J$)	384.3	383.8	383.3	382.7	382.2	381.7
λ_{III} (°)	87.3	123.6	159.8	196.1	232.3	268.5
λ_m (°)	-0.4	5.6	10.7	13.2	11.9	7.4
LT (Hr)	11.3	11.3	11.3	11.3	11.3	11.3
φ_{IO} (°)	70.0	78.5	87.0	95.5	104.0	112.5

FIGURE 20 A representative dynamic spectrum of several of Jupiter's low-frequency radio emissions. The color bar is used to relate the color to the intensity of the emission. The emission is plotted as a function of frequency (along the y-axis) and time (along the x-axis). (A. Lecacheux, 2001, Radio Observations During the Cassini Flyby of Jupiter, in *Planetary Radio Emissions* V, edited by H. O. Rucker, M. L. Kaiser, and Y. Leblanc, Austrian Academy of Sciences Press, Vienna, pp. 1–13.)

plasma torus. The source of these emissions is at high magnetic latitudes and appears fixed in local time. The forward lobe near the north magnetic pole is of opposite polarization than a "back lobe" of the same source. The nKOM emissions last longer (up to a few hours) than bKOM, are confined to a smaller frequency range, 50–180 kHz, and show a smooth rise and fall in intensity. The recurrence period for nKOM events suggests the source lags behind Jupiter's rotation by 3–5%, which was the first indication that this emission, in contrast to any other low-frequency emissions, is produced by distinct sources near the outer edge of the plasma torus. *Galileo* and *Ulysses* studies have shown that these emissions occur as a part of an apparently global magnetospheric dynamic event. There is a sudden onset of these emissions, they are visible for a few to several planetary rotations, and finally, they fade away.

3.5.3 VERY LOW FREQUENCY EMISSIONS

The *Voyager* spacecraft detected continuum radiation in Jupiter's magnetosphere at frequencies below 20 kHz, both in its escaping and trapped form. As discussed in Section 3.1, radiation can be trapped inside the magnetic cavity if it cannot propagate through the high plasma density magnetosheath. This trapped emission has been observed from a few hundred Hz up to ~5 kHz. Occasionally, it has been detected up to 25 kHz, suggesting a compression of the mag-

netosphere caused by an increased solar wind ram pressure. Outside the magnetosphere the lower frequency cutoff of the freely propagating radiation corresponds to the plasma frequency in the magnetosheath and appears to be well correlated with the solar wind ram pressure. This escaping component is characterized by a complex narrowband spectrum, attributed to a linear or nonlinear conversion of electrostatic waves near the plasma frequency into freely propagating electromagnetic emissions. The linear mechanism favors ordinary mode radiation, but the trapped emission appears to be a mix of both ordinary and extraordinary radiation, perhaps from the multiple reflections off high density regions in the magnetosphere and at the magnetopause.

The quasi-periodic (QP), or jovian type III emissions (in analogy to solar type III bursts, because of their similar dispersive spectral shape) often occur at intervals of 15 and 40 min as observed by *Ulysses*, but neither *Galileo* nor *Cassini* found particularly dominant periodicities at these or other intervals (see Fig. 20). The emission likely originates near the poles. Simultaneous measurements by the Galileo *and* Cassini spacecraft, both in the solar wind but at different locations, observed similar QP characteristics, suggestive of a strobe light pattern rather than a search light rotating with the planet. Within the magnetosphere, the QP bursts can then appear as enhancements of the continuum emission. At the magnetosheath, the lower frequency components of the bursts are dispersed by the higher density plasma, which produces the characteristic type III spectral shape. The 40-minute QP bursts were correlated with energetic ($\sim$1 MeV) electrons observed by *Ulysses*. *Chandra* detected similar periods in X-rays from the auroral region, although not directly correlated with QP bursts themselves. Such observations suggest that the QP bursts are related to an important particle acceleration process, but the details of the relationship and the details of the process remain elusive.

3.5.4 GANYMEDE

Jupiter's satellite Ganymede has its own magnetosphere embedded within Jupiter's magnetic field. It presents a rich plasma wave spectrum, similar to that expected from a planetary magnetosphere. It also is the source of nonthermal narrowband radio emissions at 15–50 kHz, very similar to the escaping continuum emissions from Jupiter. The more intense cyclotron maser emission, seen from the auroral regions of all giant planets and Earth, is absent, however. This is almost certainly because the electron plasma frequency is greater than the cyclotron frequency; hence, the cyclotron maser instability does not operate.

3.6 Saturn

Saturn's nonthermal radio spectrum consists of several components, as displayed in Fig. 21, and discussed in the following section.

3.6.1 SATURN KILOMETRIC RADIO EMISSIONS

Saturn's kilometric radiation (SKR) is characterized by a broad band of emission, 100% circularly polarized, covering the frequency range from 20 kHz up to several hundred kHz. When displayed in the frequency–time domain, it is sometimes organized in arc-like structures, reminiscent of Jupiter's DAM arcs (see Fig. 21a). *Cassini* has revealed some fine structure characteristic of cyclotron maser emissions (Fig. 21b). As on Earth, the SKR source appears to be fixed at high latitudes primarily in the local morning to noon sector, but it also appears at other local times. The SKR intensity is strongly correlated with the solar wind ram pressure, perhaps suggesting a continuous transfer of the solar wind into Saturn's low-altitude polar cusps. In fact, a detailed comparison between high-resolution *Hubble Space Telescope* (*HST*) images of Saturn's aurora with SKR suggests a strong correlation between the intensity of UV auroral spots and SKR.

Even though the emission is highly variable over time, a clear periodicity at $10^h\ 39^m\ 24^s \pm 7^s$ was derived from the *Voyager* data, which was adopted as the planet's rotation period. Because the emission is tied to Saturn's magnetic field, which is axisymmetric, the cause of the modulation remains

a mystery, although it may be indirect evidence of higher order moments in Saturn's magnetic field. Even more mysterious, however, is that the SKR modulation period measured by *Ulysses* and *Cassini* varies by 1% or more (several minutes) on timescales of a few years or less. Clearly, this change in period cannot represent a change in the planet's rotation itself, but there is no commonly accepted explanation.

3.6.2 VERY LOW FREQUENCY EMISSIONS

While the spacecraft was within Saturn's magnetosphere, it detected low-level continuum radiation (trapped radiation) at frequencies below 2–3 kHz (VLF, very Low Frequency). At higher frequencies, the emission can escape and appears to be concentrated in narrow frequency bands. It is believed that both the "trapped" and narrowband radio emissions are generated by the same mechanism, that is, mode conversion from electrostatic waves near the upper hybrid resonance frequency. However, the source location has not been determined. In particular, one source that has been suggested is related to Saturn's icy moons.

During the passage of the *Cassini* spacecraft through the inner region of the Saturnian system on July 1, 2004, the Radio and Plasma Wave Science (RPWS) instrument detected many narrowband emissions in a plasma density minimum over the A and B rings. These have been shown to be propagating in the z-mode, at least partially. It is not clear how these narrowband emissions are related, if at all, to those measured well beyond the planet.

3.6.3 SATURN ELECTROSTATIC DISCHARGES

Saturn electrostatic discharges (SEDs) are strong, impulsive events, which last for a few tens of milliseconds from a few hundred kHz to the upper frequency limit of the *Voyager* planetary radio astronomy experiment (40.2 MHz), and are also detected by the *Cassini* spacecraft. Structure in individual bursts can be seen down to the *Voyager* time resolution limit of 140 μs, which suggests a source size less than 40 km. During the *Voyager* era, episodes of SED emissions occurred approximately every $10^h\ 10^m$, distinctly different from the periodicity in SKR. In contrast to SKR, the SED source is fixed relative to the planet-observer line. The emissions are likely electrostatic discharge events as a counterpart of lightning flashes in Saturn's atmosphere. Some SED episodes have been linked directly to cloud systems observed in Saturn's atmosphere by the *Cassini* spacecraft. *Cassini*, however, has found SEDs to be much less common, generally speaking, than *Voyager*. Cassini can go months without seeing the discharges. Perhaps it may be a seasonal effect or related to the extent of ring shadowing on the atmosphere (or ionosphere, if propagation is an issue). *Cassini* should continue to observe through similar seasonal and ring

(a)

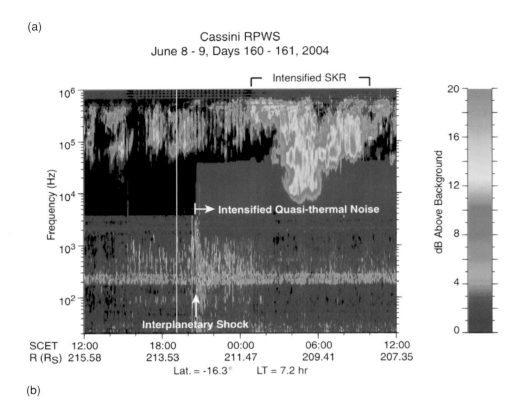

Cassini RPWS
June 8 - 9, Days 160 - 161, 2004

FIGURE 21 (a) Dynamic spectra of Saturn's SKR emission. This illustrates a dramatic intensification of the SKR in response to an interplanetary shock that passed *Cassini* at about 20:30 on June 8, 2004. (b) A high temporal and spectral resolution record of SKR obtained by *Cassini*. This spectrogram illustrates the complex structure and variations in the SKR spectrum, which is also typical of cyclotron maser emissions at Jupiter and Earth. (After Kurth, et al., 2005, High Spectral and Temporal Resolution Observations of Saturn Kilometric Radiation. *Geophys. Res. Lett.* **32**, L20S07, doi:1029/2005 GL022648.)

(b)

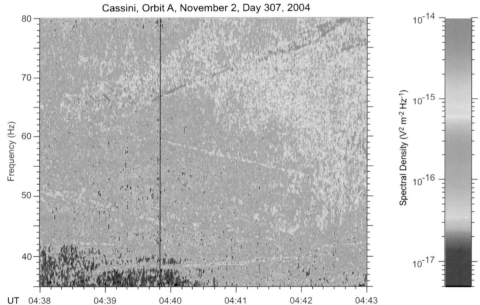

Cassini, Orbit A, November 2, Day 307, 2004

shadowing conditions to the *Voyager* era late in *Cassini*'s orbital tour, so such speculation can be tested.

3.7 Uranus and Neptune

Like Saturn's radio emissions, both smooth and bursty components are apparent in the radio emissions from Uranus and Neptune, and these emissions probably originate in the southern auroral regions of the planets. Note, though, that the magnetic fields of these planets are inclined by large angles (47° for Uranus, 59° for Neptune) with respect to their rotational axes, and hence the auroral regions are not near the rotation poles. The periodicity of the emissions leads to the determination of the rotation periods of both planets, 17.24 ± 0.01 hours for Uranus and 16.11 ± 0.02 hours for Neptune. The upper bound to the frequency of the emissions is determined by (and indicative of) the planets' surface magnetic field strength.

From Uranus, we have also received impulsive bursts, similar to the SED events of Saturn, which are referred to as UED or Uranus electrostatic discharge events. They were fewer in number and less intensive than the SEDs. If these emissions are caused by lightning, the lower frequency cutoff suggests peak ionospheric electron densities on the day side of $\sim 6 \times 10^5$ cm^{-3}. In addition to the broadband emissions, both planets also emit trapped continuum and narrowband radiation.

4. Future of Ground-Based Radio Astronomy for Solar System Research

This chapter highlighted the value of radio observations for planetary atmospheres (composition, dynamics), surface composition and structure, comets (parent molecules, source of material, outgassing), and magnetospheres (magnetic field configurations, particle distributions). Momentarily, many exciting projects are not quite doable with existing telescopes. The prospects for the future, however, when new large arrays come on-line, are spectacular. Planetary science may be advanced in significant ways with these arrays.

At millimeter wavelengths, the BIMA and Owens Valley Radio Observatory (OVRO) arrays are combined (and expanded) into the Combined Array for Research in Millimeter-wave Astronomy (CARMA), located at Cedar Flat in eastern California, at ~8000 ft altitude. The Atacama Large Millimeter Array (ALMA) is being built in Chili, jointly by the United States, Canada, Europe, and Chili. The Smithsonian Submillimeter Array (SMA) is already in existence, and has produced interesting scientific results; in this chapter we highlighted some of its results on Titan. At longer wavelengths, the Allan Telescope Array (ATA), operating at ~0.5-~10 GHz, is being built in California by the SETI institute and UC Berkeley, with funding from Paul Allan. Several low-frequency arrays are either under construction (the Low Frequency Array LOFAR in the Netherlands) or being planned, while the ultimate Square Kilometer Array (SKA) is under discussion in many countries.

These new arrays open up a wealth of potential observations for planetary research, in all areas. For example, several millimeter telescopes observed the apparition of comet Hale–Bopp, with fantastic results, as described in Section 2.7. ALMA will enable detection of hundreds of asteroids, "bare" cometary nuclei, emissions from molecular "jets" from comets at high spatial and time resolution,

Io's volcanic plumes, Titan's hydrocarbon chemistry, and "proto-Jupiters" in nearby stellar systems. We expect, besides simple detection experiments, to actually carry out scientific research in these areas, such as to determine the mass and chemical composition of protoplanets. SKA will improve maps at centimeter wavelengths by orders of magnitude; it will enable mapping thermal emissions from giant planets in minutes of time and obtain maps of Jupiter's synchrotron emission at many wavelengths quasi-simultaneously. At lower frequencies, below 40 MHz, arrays such as LOFAR will allow, for the first time, mapping of Jupiter's decametric emissions, and pinpoint its sources with high accuracy.

Bibliography

Berge, G. L., and Gulkis, S. (1976). Earth based radio observations of Jupiter: Millimeter to meter wavelengths. In "Jupiter" (T. Gehrels, ed.), pp. 621–692, Univ. Arizona Press, Tucson.

Butler, B. J., Campbell, D. B., de Pater, I., and Gary, D. E. (2004). Solar system science with SKA. *New Astronomy Reviews*, **48** (11–12), 1511–1535.

Carr, T. D., Desch, M. D., and Alexander, J. K. (1983). Phenomenology of magnetospheric radio emissions. In "Physics of the Jovian Magnetosphere" (A. J. Dessler, ed.), pp. 226–284. Cambridge Univ. Press, Cambridge, United Kingdom.

Crovisier, J., and Schloerb, F. P. (1991). The study of comets at radio wavelengths. In "Comets in the Post-Halley Era" (R. L. Newburn and J. Rahe, eds.); a book as a result from an international meeting on Comets in the Post-Halley Era, Bamberg, April 24–28, 1989, 149–174.

de Pater, I., Schulz, M., and Brecht, S. H. (1997). Synchrotron evidence for Amalthea's influence on Jupiter's electron radiation belt. *J. Geoph. Res.* **102** (A10), 22,043–22,064.

de Pater, I., Butler, B., Green, D. A., Strom, R., Millan, R., Klein, M. J., Bird, M. K., Funke, O., Neidhofer, J., Maddalena, R., Sault, R. J., Kesteven, M., Smits, D. P., and Hunstead, R. (2003). Jupiter's radio spectrum from 74 MHz up to 8 GHz. *Icarus* **163**, 434–448.

Desch, M. D., Kaiser, M. L., Zarka, P., Lecacheux, A., LeBlanc, Y., Aubier, M., and Ortega-Molina, A. (1991). Uranus as a radio source. In "Uranus" (J. T. Bergstrahl, A. D. Miner, and M. S. Matthews, eds.), pp. 894–925. Univ. Arizona Press, Tucson.

Gulkis, S., and de Pater, I. (2002). Radio astronomy, planetary. In "Encyclopedia of Physical Science and Technology," vol., 13, 3rd Ed., pp. 687–712. Academic Press.

Harrington, J., de Pater, I., Brecht, S. H., Deming, D., Meadows, V. S., Zahnle, K., and Nicholson, P. D. (2004). Lessons from Shoemaker–Levy 9 about Jupiter and planetary impacts. In "Jupiter: Planet, Satellites & Magnetosphere" (F. Bagenal, T. E. Dowling, and W. McKinnon, eds.), pp. 158–184. Cambridge Univ. Press, Cambridge, United Kingdom.

Kaiser, M. L., Desch, M. D., Kurth, W. S., Lecacheux, A., Genova, F., Pederson, B. M., and Evans, D. R. (1984). Saturn as a radio source. In "Saturn" (T. Gehrels and M.S. Matthews, eds.), pp. 378–415. Univ. Arizona Press, Tucson.

Kraus, J. D. (1986). "Radio Astronomy." Cygnus Quasar Books, Powell, Ohio.

Kurth, W. S., Hospodarsky, G. B., Gurnett, D. A., Cecconi, B., Louarn, P., Lecacheux, A., Zarka, P., Rucker, H. O., Boudjada, M., and Kaiser, M. L. (2005). High Spectral and Temporal Resolution Observations of Saturn Kilometric Radiation. *Geophys. Res. Lett.*, **32**, L20S07, doi:10.1029/2005GL022648.

Perley, R. A., Schwab, F. R., and Bridle, A. H. (1989). Synthesis imaging in radio astronomy, NRAO Workshop no. 21, Astronomical Society of the Pacific.

Thompson, A. R., Moran, J. M., and Swenson, Jr., G. W. (1986). "Interferometry and Synthesis in Radio Astronomy." John Wiley and Sons, New York.

Zarka, P. (1998). Auroral radio emissions at the outer planets: Observations and theories. *J. Geophys. Res.* **103**, 20, 159–20, 194.

Zarka, P., and Kurth, W. S. (2005). Radio wave emission from the outer planets before *Cassini*. *Space Sci. Rev.* **116**, 371–397.

Zarka, P., Pederson, B. M., Lecacheux, A., Kaiser, M. L., Desch, M. D., Farrell, W. M., and Kurth, W. S. (1995). Radio emissions from Neptune. In "Neptune" (D. Cruikshank, ed.), pp. 341–388. Univ. Arizona Press, Tucson.

New Generation Ground-Based Optical/ Infrared Telescopes

Alan T. Tokunaga

Robert Jedicke

Institute for Astronomy
University of Hawaii
Honolulu, Hawaii

CHAPTER 39

The telescope is a crucial tool for astronomers. This chapter gives an overview of the recent advances in ground-based telescope construction and instrumentation for visible and infrared wavelengths, which have spurred extraordinary advances in our understanding of the solar system. Although space-based observatories such as the Hubble Space Telescope and the Spitzer Space Telescope have also immensely enriched our understanding of the solar system we live in, the results from space observatories are discussed elsewhere in this encyclopedia. Astronomers strive to build ever-larger telescopes in order to collect as much light as possible. While cosmologists need the large collecting area of telescopes to study the distant universe, solar system astronomers need the large collecting area to study both nearby small objects and faint objects at the limits of our solar system, and to exploit the high angular resolution they provide. We discuss future telescope projects that promise to make further discoveries possible in the next few decades and offer the prospect of studying solar systems other than our own. Advances in instrumentation have in equal measure revolutionized the way astronomy is done.

We discuss two major advances in this chapter: the advent of the large-format solid-state detector for visible and infrared wavelengths and the development of adaptive optics. The development of large-format arrays has led to ambitious digital sky surveys. These surveys allow searches for

objects that may collide with Earth and are leading to a fundamental understanding of the early history of our solar system. The development of adaptive optics is reaching maturity and is allowing routine observations to be made at the diffraction-limit at the largest telescopes in the world. Thus the limitation on image sharpness imposed by the atmosphere since the invention of the telescope is now removed with adaptive optics.

1. Introduction

The telescope has played a critical role in planetary science from the moment of its use by Galileo in 1608. The observations that he made of the craters on our Moon and the moons of Jupiter were the first astronomical discoveries made with a telescope. The development of larger refracting and reflecting telescopes led to the seminal discoveries of the rings of Saturn, asteroids, the outer planets Uranus and Neptune, new satellites of Mars and the outer planets, and Pluto by 1930.

Although spacecraft missions have revolutionized our understanding of the solar system (of which there are many examples in this encyclopedia), ground-based telescopes continue to play a very important role in making new discoveries, and this is the focus of this chapter. The discovery of the first Kuiper Belt Object (KBO) was made in 1992 on the University of Hawaii 2.2-m telescope. Tremendous advances have been made in detecting KBOs since then: presently over 900 KBOs have been discovered. Using several of the largest telescopes in the world, it was recently found that the largest KBO known, 2003 UB_{313}, has methane ice on its surface and a moon (Fig. 1). This finding has challenged our definition of what is considered to be a planet in our solar system. Another recent result was the discovery of comets among the main-belt asteroids. The most recent of these, asteroid 118401 was discovered by the 8-m Gemini-North telescope. Two other comets in the main belt were detected previously by other astronomers, and many more such comets are now thought to exist in the asteroid main belt. If this is confirmed then such comets were likely the main source of water delivered to the Earth during its formation. A final example is the Near-Earth Object (NEO) designated 2004 MN_4, which was discovered with the University of Arizona's 2.3-m telescope. For a short time at the end of December 2004, this NEO had the highest probability of any yet found for colliding with Earth (Fig. 3). These discoveries demonstrate the importance of ground-based astronomy, and they will no doubt provide the scientific motivation for future missions.

Solar system astronomers typically use telescopes built for other fields of astronomy. However, during the 1970s, NASA constructed ground-based telescopes to support its planetary missions. NASA funded the construction of the 2.7-m McDonald telescope, the University of Hawaii 2.2-m telescope, and the 3.0-m NASA Infrared Telescope Facility (IRTF) to provide mission support, but currently only the IRTF continues to be funded by NASA for that purpose. NASA also provides funding for searches for NEOs as part of a Congressional directive.

Telescopes are designed to collect and focus starlight onto a detector. While conceptually simple, ground-based observers have to contend with limitations imposed by physics, the atmosphere, and technology. First, the collecting area of a telescope is limited in size. The largest optical telescope in the world presently has an equivalent collecting area of an 11.8-m diameter mirror. Although larger telescopes could be built, there are serious technical and financial difficulties to overcome. Larger telescopes not only allow more light to be collected and put onto the detector, they also allow sharper images to be obtained at the diffraction limit of the telescope. Second, the atmosphere limits observations to specific observing "windows" where the atmosphere is transparent, and the wavelength range 25 μm to 350 μm is largely inaccessible to ground-based observers because of water absorption bands. Third, for infrared observations, the thermal emission of the atmosphere at wavelengths longer than 2.5 μm greatly reduces the sensitivity of observations. To overcome the problems of atmospheric absorption and ther-

mal emission, it is necessary to go to high-mountain sites such as Mauna Kea in Hawaii and Atacama in Chile, or to use balloons, aircraft, or spacecraft. Fourth, atmospheric seeing typically limits the sharpness of images to 0.25–0.5 arcseconds at the best high-altitude sites. To achieve

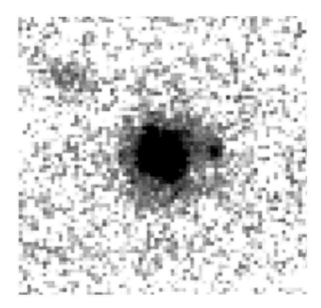

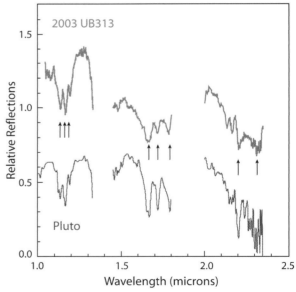

FIGURE 1 (a) Image of KBO UB_{313} obtained with the 10-m Keck II telescope with a laser guide star adaptive optics system. With a diameter estimated to be about 2400 km, it is the largest KBO known and is slightly larger than Pluto. It was recently named Eris. This image shows that UB_{313} has a satellite, as does Pluto. (b) A near-infrared spectrum of UB_{313} and Pluto. The spectrum of Pluto was obtained with the 8-m Gemini North telescope. Both objects have methane ice on their surface (methane ice absorption marked with arrows), thus strengthening the idea that there is a common origin for these objects. (Courtesy of M. Brown and C. Trujillo.)

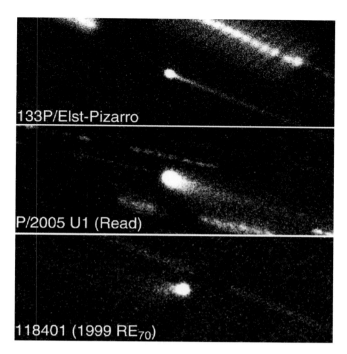

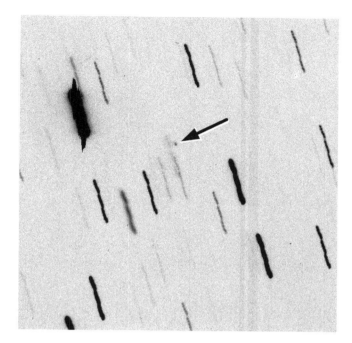

FIGURE 2 Images of known comets in the asteroid main belt taken with the University of Hawaii 2.2-meter telescope. These objects are known as the main-belt comets and are a fundamentally new class of comets. The fuzzy appearance of these comets are due to reflected light from dust particles that are ejected by a volatile material, most likely sublimating water ice. (Courtesy of H. Hsieh and D. Jewitt.)

diffraction-limited imaging, one must employ special techniques that actively reduce it many times per second. One such technique, called adaptive optics, is discussed later in Section 4.

Very large and low-noise visible and infrared detector arrays have been developed in the past decade, and this advance has been as significant as improvement of telescope construction in providing greater observing capability. An important capability of large-format detector arrays has been to allow large sky surveys to be undertaken. The key objectives of these sky surveys are to detect asteroids that may present an impact hazard to Earth and to complete the reconnaissance of KBOs. The major challenges of these survey projects are obtaining large enough detector arrays to provide the field-of-view required, and analyzing and storing the tremendous amounts of data that they generate.

In this chapter, we discuss very large telescopes that have been developed in the past 15 years to maximize collecting area, optimize image quality, and achieve diffraction-limited imaging with techniques to reduce the atmospheric turbulence. We also discuss sky survey telescopes that take advantage of the large-format detectors for the detection of solar system objects.

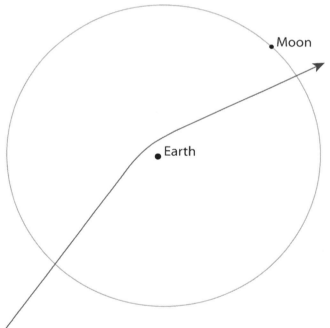

FIGURE 3 (a) Image of the asteroid 99942 Apophis. When it was discovered during its last close approach to the Earth in 2004, it had a significant probability of striking the Earth in the future. Subsequent observations show that it will pass within 5.6 Earth radii of the Earth in 2029 (see panel b). However, the future trajectory of the asteroid cannot be predicted well and the asteroid will have to be carefully monitored with ground-based telescopes. The diameter of the asteroid is about 250 m. Close passages by an asteroid of this size are estimated to occur about once in 1300 years. (Courtesy of R. Tucker, D. Tholen, and F. Bernardi.)

2. Advances in the Construction of Large Telescopes and in Image Quality

The Hale 5.1-m telescope went into operation in 1949. It represented the culmination of continual telescope design improvements since the invention of the reflecting telescope by Newton in 1668. The basic approach was to scale up and improve design approaches that were used previously. Figure 4 shows the increase in telescope aperture with time. After the completion of the Hale telescope, astronomers recognized that building larger telescopes would require completely new approaches. Simple scaling of the classical techniques would lead to primary mirrors that would be too massive and an observatory (including the dome enclosure) that would be too costly to build. Since the 1990s, a number of ground-breaking approaches have been tried, and the barrier imposed by classical telescope design has been broken. Table 1 shows a list of telescopes with apertures greater than 5 meters. Some of the telescopes listed in Table 1 are still under development.

Major technical advances that have led to the development of large telescopes include:

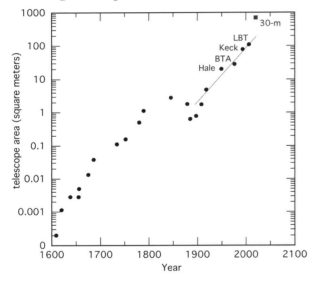

FIGURE 4 Increase in telescope area with time. Only the area of the largest telescopes at each time period is shown, so this indicates the envelope of maximum telescope area as a function of year. The time for the telescope area to double is about 26 years from the invention of the telescope in 1608 to the current year. However the doubling time has decreased from about 1900 to the present. The solid line shows a doubling of telescope aperture about every 19 years. The next jump in aperture size is likely to be in the range of 20–50 meters. For comparison the square symbol shows a 30-m class telescope in the year 2020, and this indicates an even shorter doubling time. The increase in telescope area is due to advances in telescope construction technology and the willingness of society to bear the costs. How much longer can this increase in telescope area continue on the ground? (See Racine 2004, Pub. Astron. Soc. Pacific, vol. 116, p. 77) for data on the growth of telescope aperture with time.)

(1) Advances in computer-controlled hardware allows correction for flexure of the primary mirror. This has permitted thinner mirrors to be used, reducing the mass of the mirror and the total mass of the telescope. For example, the mass of the ESO Very Large Telescope 8.2-m primary mirror is 23 tons with an aspect ratio (mirror diameter to mirror thickness ratio) of 46. This is a very thin mirror compared with the 5.1-m Hale telescope, which has a weight of 14.5 tons and an aspect ratio of 9.

(2) Altitude-azimuth (alt-az) mounts reduce the size of the required telescope enclosure. An 8-m alt-az telescope can fit into the same size enclosure as a 4-m equatorial telescope. An alt-az telescope requires computer-controlled pointing and tracking on two axes (whereas the traditional mount requires tracking on only a single axis). The Hale telescope is the largest equatorial telescope ever built. All larger and more recent telescopes use alt-az mountings. Figure 5 illustrates the basic types of telescope mounts, and Figure 6 shows examples of the equatorial and alt-az mounts.

(3) Advances in mirror casting and computer-controlled mirror polishing allow the production of larger primary mirrors with shorter focal lengths. A shorter focal length allows the telescope structure to be smaller, thus lowering the weight and cost of the telescope. It also greatly reduces the cost of the dome enclosure. The state-of-the-art in short focal length primary mirrors are those with a focal length to diameter ratio (f/no) of 1.14 installed in the Large Binocular Telescope. This can be compared to the Hale telescope primary mirror that has an f/no of 3.3. The smaller telescope structure with reduced mass requires less time to reach thermal equilibrium, and its lower mass makes it easier to move. This is extremely important in achieving the best image quality and to efficiently reposition in the telescope.

(4) Advances in reducing dome seeing led to significant improvement in image quality. Dome seeing is caused by temperature differences within the dome, especially differences between the mirror and the surrounding air. To reduce dome seeing, it is necessary to flush the dome with outside air at night, refrigerate it during the daytime, and cool the primary mirror to about 0.5°C below the ambient air temperature. Dome seeing is so important that large telescope projects use wind tunnel experiments to determine what type of dome design to employ. Careful attention to dome design is critical in eliminating dome seeing and achieving the very best seeing at the observatory site. Figure 6b shows an innovative approach to providing dome flushing by providing slits in the dome.

(5) Advances in telescope construction have led to novel methods of reducing the cost of building extremely large telescopes. For example, the 10-m Keck telescopes have segmented mirrors to make up the primary mirror (Fig. 6c). Although this technique had been used to build radio telescopes, the difficulty of making the segments and

TABLE 1　Telescopes with Apertures Greater than 5 Meters

(1) Aperture (m)	(2) Circular Aperture Equivalent (m)	(3) Telescope Name	(4) Location	(5) Date of Operation	(6) primary f/no	(7) Mirror Type	(8) Mirror Aspect Ratio	(9) Mounting Type	(10) Ref.
2 × 8.4	11.8	Large Binocular Telescope (LBT)	Mt. Graham, Arizona	(2006)	1.14	Honeycomb	9.4	Alt-Az	1
11 × 9.4 Hexagon	10.0	Keck I	Mauna Kea, Hawaii	1993	1.75	Segmented	133	Alt-Az	2
11 × 9.4 Hexagon	10.0	Keck II	Mauna Kea, Hawaii	1996	1.75	Segmented	133	Alt-Az	2
11 × 9.4 Hexagon	10.0	Gran Telescopio Canarias (GTC)	La Palma, Canary Islands	(2007)	1.65	Segmented	125	Alt-Az	3
11 × 10 Hexagon	9.2	Hobby-Eberley Telescope	Mt. Fowlkes, Texas	1997	1.4	Segmented	200	Azimuth only	4
11 × 10 Hexagon	9.2	Southern African Large Telescope (SALT)	Sutherland South Africa	2005	1.4	Segmented	200	Azimuth only	5
	8.2	Subaru	Mauna Kea, Hawaii	1999	1.8	Meniscus	41	Alt-Az	6
	8.2	Very Large Telescope (VLT) UT1 Antu	Cerro Paranal, Chile	1998	1.75	Meniscus	46	Alt-Az	7
	8.2	Very Large Telescope (VLT) UT2 Kueyen	Cerro Paranal, Chile	1999	1.75	Meniscus	46	Alt-Az	7
	8.2	Very Large Telescope (VLT) UT3 Melipal	Cerro Paranal, Chile	2000	1.75	Meniscus	46	Alt-Az	7
	8.2	Very Large Telescope (VLT) UT4 Yepun	Cerro Paranal, Chile	2000	1.75	Meniscus	46	Alt-Az	7
	8.0	Gemini North	Mauna Kea, Hawaii	1998	1.8	Meniscus	40	Alt-Az	8
	8.0	Gemini South	Cerro Pachon, Chile	2000	1.8	Meniscus	40	Alt-Az	8
	6.5	MMT Conversion	Mt. Hopkins, Arizona	1999	1.25	Honeycomb	9	Alt-Az	9
	6.5	Magellan I - Walter Baade	Cerro Manqui, Chile	2000	1.25	Honeycomb	9	Alt-Az	10
	6.5	Magellan II - Landon Clay	Cerro Manqui, Chile	2002	1.25	Honeycomb	9	Alt-Az	10
	6.0	Large Zenith Telescope (LZT)	Vancouver, Canada	2005	1.5	Liquid Hg	n/a	Fixed	11
	6.0	Bol'shoi Teleskop Azimultal'nyi (BTA)	Mt. Pastukhova, Russia	1977	4	Solid	6	Alt-Az	12
	5.1	Hale	Mt. Palomar, California	1949	3.3	Honeycomb	8	Equatorial	13

References

(1) http://lbto.org/ (2) http:// http://www.keckobservatory.org// (3) http://www.gtc.iac.es/ (4) http://www.as.utexas.edu/mcdonald/het/het.html, (5) http://www.salt.ac.za/, (6) http://www.naoj.org/, (7) http://www.eso.org/, (8) http://www.gemini.edu/, (9) http://www.mmto.org/, (10) http://www.ociw.edu/magellan/magellan.html, (11) http://www.astro.ubc.ca/LMT/, (12) http://www.sao.ru/Doc-en/index.html, (13) http://astro.caltech.edu/observatories/palomar/

This table is adapted from J.M. Hill's web site: http://abell.as.arizona.edu/~hill/list/bigtel99.htm.

Column (1). The aperture is the diameter of the primary that can collect light. Unless specified, the number given is the diameter of a circular aperture. The LBT consists of two 8.4-m mirrors that are on a single mount and the light from both mirrors are combined to form a single image. The Keck, HET, and SALT telescopes have primary mirrors that are made from hexagonal segments. The primary mirror has a hexagonal shape and the largest and smallest widths of the hexagon are given.

Column (2). This is the diameter of the equivalent circular aperture equal to the total light collecting area of the telescope. For the HET and SALT telescopes this is the maximum equivalent circular aperture that is accepted by the prime focus optics. The LBT, Keck, and VLT observatories can combine light from the mirrors for use as an interferometer. This mode of observations is not considered in this table for the purpose of determining the equivalent circular aperture.

Column (5). Year that science operations started. Parentheses denote year science operations expected.

Column (6). Primary mirror f/no, which is equal to the focal length of the telescope divided by the mirror diameter.

Column (7). Honeycomb: Primary mirror that is lightened with a honeycomb structure in the back. Segmented: Primary mirror is made out of hexagonal segments. Meniscus: Single thin concave mirror. Liquid Hg: Liquid mercury mirror. Parabolic shape is obtained by spinning the mirror. Solid: Thick mirror with no light-weighting.

Column (8). The aspect ratio is the primary mirror diameter divided by the mirror (or segment) thickness.

Column (9). The azimuth only and fixed telescope mounts conduct observations by tracking object in the focal plane of the telescope. For such telescopes the telescope is fixed but the instrumentation tracks the object.

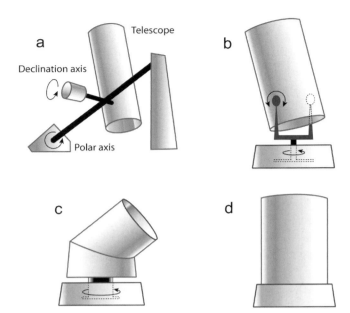

FIGURE 5   Schematic of different telescope mounts: (a) equatorial, (b) alt-az, (c) azimuth-only, (d) fixed. The Hale 5.1-m telescope was the last large telescope to be built with an equatorial mount. The equatorial mount has one axis aligned to the rotation axis of the Earth. (Note: there are many types of equatorial mounts. The Hale telescope uses a type known as the horseshoe equatorial mount.) All fully steerable large telescopes utilize the alt-az mount, such as the Keck, Gemini, VLT, and Subaru telescopes (see Table 1). In the alt-az mount, the azimuth axis points to the zenith with a perpendicular altitude axis. Two large telescopes built specially for spectroscopy use the azimuth-only mount—the Hobby-Eberly and the South African Large Telescope. The telescope moves only in azimuth and is fixed in declination. The only large telescope to date that uses a fixed mount (the telescope points only to the zenith) is the Large Zenith Telescope, and it uses a liquid mercury mirror.

FIGURE 6a   Hale 5.1-m telescope. The last large telescope to be built in the "classical style" with an equatorial mount, a culmination of about 280 years of development of the reflecting telescope. © 2005 Gigapxl Project

the high-precision alignment at visible wavelengths presented formidable obstacles. Fortunately, the problems of fabricating segmented mirrors and aligning them were solved. The hexagonal mirror segments have a thickness of 75 mm, and so the aspect ratio of the 10-m primary is 133 and the total weight of the glass required is 14.4 tons, about the same weight as the 5-m Hale telescope. Another novel approach uses two 8.4-m primary mirrors on a single structure as in the Large Binocular Telescope (Fig. 6d). A third approach involves building a telescope with a fixed vertical elevation. Stars move past the prime focus and are tracked for a limited time. This approach has limitations but is much less expensive to build. Two projects (the Hobby-Eberly Telescope and the South African Large Telescope) have adopted this design to achieve 9-m class telescopes at about 15–20% of the cost of an equivalent alt-az telescope. An even less expensive approach is to simply stare at the zenith with a liquid mercury mirror as demonstrated by the Large Zenith Telescope.

Large telescopes generally employ one of three different types of primary mirror fabrication. These are (1) Segmented mirrors. Each segment is figured appropriately and all segments are aligned so as to act as a single mirror. (2) Thin meniscus mirror using low expansion glass. Such mirrors are made as thin as possible to be light weight and to have a short thermal time constant (thus coming into equilibrium with the atmospheric temperature quickly). (3) Thick honeycomb mirror using borosilicate glass. The advantage of using borosilicate glass instead of low expansion glass is that the former is much cheaper. The disadvantage of borosilicate glass is that the mirror temperature needs to be controlled more carefully. All of these types of primary mirror fabrication approaches have been proven successful. Column (7) in Table 1 shows the type of mirror used.

All large telescopes use active optics to control the shape of the primary mirror. Active optics is the slow adjustment of a mirror to correct aberrations in the image. These

FIGURE 6b 8-m Gemini South telescope. Instruments are mounted on the back of the telescope. These instruments are on the telescope all of the time so that instrument changes can be made very quickly. The dome has vents to allow flushing of the dome by the night air. This allows the telescope and dome to quickly reach equilibrium with the air temperature. (Courtesy of Gemini Observatory/AURA)

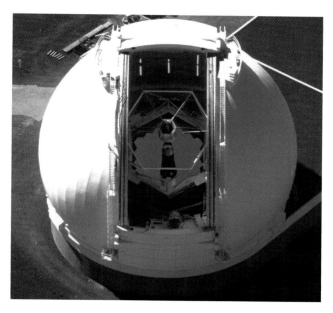

FIGURE 6c 10-m Keck telescope. This image shows one of the two Keck telescopes. The primary mirror consists of 36 hexagonal segments that are aligned to optical precision. The instruments are located on a platform on two sides of the telescope facing the declination bearings. Light from the two telescopes can be combined to provide angular resolution equivalent to an 85 m telescope. (Courtesy R. Wainscoat.)

FIGURE 6d Large Binocular Telescope consisting of two 8.4-m primary mirrors. First light with a single mirror took place in in 2005 and the second mirror was installed in 2006. The light-gathering power of the two primary mirrors combined is equivalent to a 11.8-m telescope. Both mirrors are on a single structure and the light from both mirrors is combined for imaging, spectroscopy, and interferometry. The combined light from the two mirrors will have the angular resolution of a 22.8 m telescope when the LBT is used as an interferometer. (Courtesy of the Large Binocular Telescope Observatory)

adjustments are not fast enough to correct for the atmospheric turbulence but they can correct for flexure in the telescope structure and for temperature changes (which will cause the telescope structure to expand and contract). The process for doing this is illustrated in Figure 7. A star is required for the active optics system to be able to compute the deformations on the primary mirror that are needed to correct the image. Although Figure 7 illustrates the case for a single mirror, a similar approach is employed for correcting the surface figure of a segmented primary mirror, although the details are quite different.

Efforts to escape the harmful effects of the Earth's atmosphere have led to telescopic observations using balloons, aircraft, and rockets. Although we do not discuss space observatories in this article, we note here that a major program undertaken by NASA and the German Aerospace Center (DLR) is to fly a 2.5-meter telescope in the stratosphere using a Boeing 747SP aircraft. At this high altitude it will be possible to observe throughout the 25 μm to 350 μm wavelength range that is inaccessible from the ground. This facility will provide long-term access to a critical wavelength range that otherwise would only be exploited infrequently with spacecraft.

We do not know what ultimately will be the largest ground-based telescope to be built (see Fig. 4). The

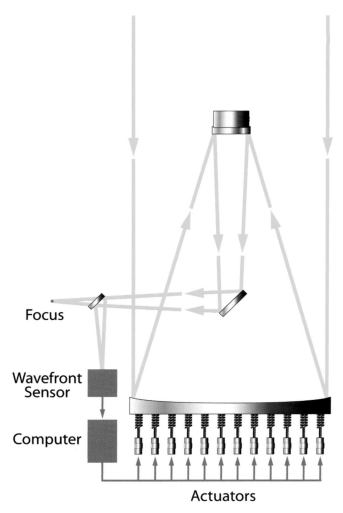

Focus

Wavefront Sensor

Computer

Actuators

FIGURE 7 Schematic of an active optics system. Starlight from the telescope is sent to a beamsplitter that simultaneously sends light to the focus and to a wavefront sensor. The computer analyses the output of the wavefront sensor and sends control signals to the primary and secondary mirrors to correct any aberrations in the image. (Courtesy of C. Barbieri.)

limitations arise from the need to be diffraction limited, the difficulty of building a suitable enclosure, and the cost. To be competitive with space observatories, all large telescopes must work at the diffraction limit using adaptive optics. But the need to be diffraction limited will ultimately cause adaptive optics systems to be too complex on an extremely large telescope. An enclosure is necessary to keep the disturbance by wind to acceptable levels, and the cost to build and operate the telescope will be enormous. At some point, it may be more cost effective to go into space, where gravity and the weather are not factors driving the design. This has been estimated to be at approximately 70-m in diameter. This argument applies to fully steerable telescopes,

not to designs such as the Hobby-Eberly Telescope or the Large Zenith Telescope.

The drive to build ever-larger telescopes is motivated by the need to collect as much light as possible and thereby increase the signal-to-noise (S/N) ratio of observations. One can derive that for a diffraction-limited telescope and a detector that is background-limited, the S/N in a given integration time is proportional to:

$$S/N \approx (A * \eta / \varepsilon)^{0.5} / (FWHM), \tag{1}$$

where A is the area of the telescope, η is the total transmission of the optics and the detector quantum efficiency, ε is the background emission, and FWHM is the full width at half maximum of a stellar image. η takes into account all of the light losses that occurs from the reflection of the mirrors and transmission losses of lenses as light propagates from the telescope to the detector. In order to minimize these losses it is necessary to utilize high reflection coatings on mirrors and lenses as well as to minimize the number of lenses. The detector quantum efficiency is the fraction of light that is absorbed by the detector material. This is near the theoretical maximum of 1.0 at visual wavelengths and about 0.8–0.9 for the 1–15 μm wavelength range. The background emission, ε, arises from the sky emission lines at visual wavelengths and thermal background from the telescope and sky at wavelengths longer than 2 μm. To reduce the thermal emission from the telescope, it is necessary to have the highest reflectivity mirrors available and to reduce or eliminate the thermal emission from the secondary mirror. The latter is often accomplished by forming an image of the secondary within the instrument and then blocking it with a cooled metal plate. Then the infrared detector will only sense the thermal emission from the sky and the object being observed.

After maximizing η and reducing ε as much as possible, one can only increase the telescope area and reduce the FWHM to further increase the S/N. Reducing the image FWHM requires decreasing the dome seeing to the absolute minimum, building on sites that have good atmospheric seeing, and working at the diffraction-limit of the telescope. Astronomical sites in Hawaii, Chile, and La Palma are prime locations for large telescopes due to the good seeing they offer as well as having good weather conditions.

Figure 8 shows the advances in image quality that have been achieved. The development of adaptive optics has led to the ability to work at the diffraction limit in the near-infrared and to achieve improvements in S/N given by equation 1. Adaptive optics is discussed in Section 4. The advances in constructing large telescopes coupled with reducing dome seeing and adaptive optics have provided the means for studying the surfaces of some KBOs and larger planetary satellites (see Fig. 1). Ground-based telescopes provide the discoveries that pose new questions

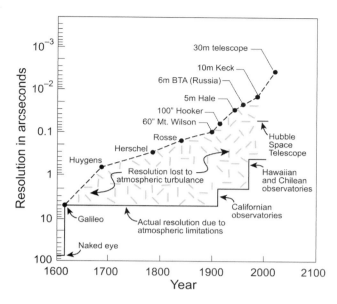

FIGURE 8   Improvement in angular resolution at optical wavelengths. The development of adaptive optics has permitted diffraction-limited observations from ground-based observatories since 1990, largely eliminating the effects of the atmosphere. The dashed line shows the theoretical diffraction-limited resolution for the telescope. The solid line shows the seeing limit imposed by the atmosphere. Improvements were obtained by going to very good seeing sites. The resolution of the Hubble Space Telescope is shown, (From P. Bely, 2003.)

and motivation for future planetary missions. This is likely to continue in the coming decades as the push to build ever-larger telescopes continues.

Several groups in the US are proposing the next leap in technology to a telescope in the 20–30-m class, and the engineering studies have started. One proposal is the Thirty-Meter Telescope, an international consortium consisting of research groups in the US and Canada (http://www.tmt.org/). This project proposes to build a telescope similar in concept to the Keck telescopes that will have over 700 hexagonal segments composing the primary mirror. As the name implies, the collecting area is equivalent to a circular mirror 30 m in diameter. The other project is the Giant Magellan Telescope, which is supported by a group of public and private institutions in the US (http://www.gmto.org/). This telescope concept consists of seven 8.4-m mirrors to create a single telescope with the collecting area equivalent to a 21.4-m circular mirror. The European Southern Observatory is also considering an even larger telescope concept (see http://www.eso.org/projects/owl/). Thus it seems inevitable that a ground-based telescope larger than 10 m will be built.

3. Advances with Detector Arrays

Initial observations with telescopes were conducted solely with the human eye (still much recommended for the non-professional), but the advantages of using photographic plates to record and archive observations of the sky were quickly exploited beginning in the 1850s. Photographic plates were eventually supplemented with electronic devices like the photomultiplier tube, which amplified the signal from stars by about one million. At infrared wavelengths, there were specialized detectors that employed bolometers, photovoltaic devices, and photoconductive devices. However, photographic plates were a necessity for recording high-resolution images of large areas of sky and recording spectra with a wide wavelength range.

Images recorded by photographic plates depend on the chemical reaction that is induced by a photon of light. Although the efficiency of the photographic plate in converting a photon to an image is only a few percent, it allows quantitative measurements to be made on the brightness of stars and the strength of spectral lines. Most importantly, the information is archived on the photographic plate for future use. This was absolutely necessary for the development of astrophysics.

The next technological revolution came with the invention of the charge-coupled device (CCD) in 1973. CCDs are composed of millions of picture elements, or pixels. Each pixel is a single detector and is capable of converting photons to electrons. The accumulated electrons can then be sent to an amplifier to be "read out" and recorded by a computer. CCD technology is employed in digital cameras, and just as digital photography is gradually replacing photography, a similar transformation has taken place in astronomy.

The impact of the CCD on astronomy was immediately apparent after its first use. CCDs have two major advantages over the photographic plate: the capability to directly record photons with an efficiency of 80–90% and to store data electronically. The stored data can then be processed with a computer. Until recently, the main deficiency of the CCD relative to the photographic plate was the relatively small amount of sky that could be covered. However, the recent development of very large CCD mosaics now permits larger areas of sky to be covered by a CCD than by a photographic

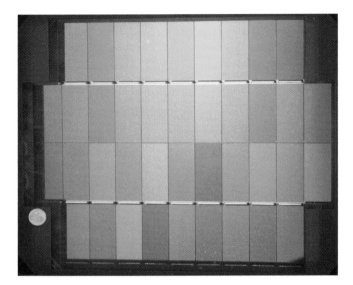

FIGURE 9 Large CCD mosaic installed in MegaCam, a prime focus camera at the Canada-France-Hawaii. This mosaic consists of 40 CCDs, each with 9.5 million pixels. In total the camera has 380 million pixels, the largest mosaic CCD currently in use. This camera is capable of generating 100 billion bytes (100 gigabytes) per night. Larger mosaic cameras are being planned. Each telescope of the Pan-STARRS survey telescope will have a 1.4-Gigapixel camera and the Large Synoptic Survey Telescope will have a single 3.2-Gigapixel camera. (Courtesy of CFHT)

plate. The rapid development of computing power and disk storage has made it practical to use large CCD mosaics. While astronomers have worked hard to develop CCD technology that is optimized for astronomy, they are fortunate that the consumer market has driven the development of the necessary computing power and storage. Figure 9 shows an example of a state-of-the-art large format CCD.

There has been a similar revolution in the development of infrared arrays. The first infrared arrays for astronomy were used in the early 1980s. While initially very modest in size (32 × 32 pixels), infrared arrays now typically contain a million pixels. There are several significant differences between CCDs and infrared arrays. One is that a CCD has a single readout amplifier, while an infrared array has one readout amplifier per pixel. The electrons in a CCD are transferred to a single readout amplifier (hence the origin of the term "charge transfer"). Only a single readout amplifier is needed since the readout electronics and the detector material are made out of the same semiconductor material. In an infrared array, the detector material and the readout amplifier have to be made out of different materials, so each pixel must have a separate amplifier. A second difference is that the infrared arrays must be cooled to much lower temperatures. CCDs can operate effectively at about −30 to −40°C. Infrared arrays must be cooled to liquid nitrogen (−196°C) or liquid helium (−269°C) temperatures.

We show in Figure 10 an example of Saturn imaged at a wavelength of 18 micrometers. At these wavelengths, we are observing the thermal emission (heat) from the planet. Thus temperatures can be measured in the atmosphere of Saturn and for the dust particles in the rings.

The development of large-format CCDs and infrared arrays has enabled astronomers to undertake large-scale digital sky surveys at visible and infrared wavelengths, just as the use of large photographic plates enabled the first deep sky surveys over 50 years ago.

4. Advances in Adaptive Optics

Adaptive optics (AO) is a technique that removes the atmospheric disturbance and allows a telescope to achieve

FIGURE 10 Image of Saturn and its rings obtained in 2004 with the 10-m Keck I telescope at a wavelength of 17.6 micrometers. This is a false color image, where higher signal levels are shown lighter. At these wavelengths we are seeing the heat radiated by the atmosphere and rings of Saturn. The South pole has an elevated temperature (−182 C) compared to its surrounding. This is likely due to the fact that the South pole has been illuminated by the sun for the past 15 years. (Courtesy of G. Orton, JPL).

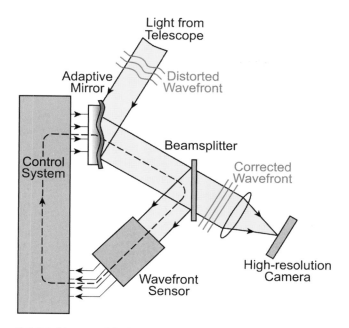

Light from
Telescope

Adaptive
Mirror

Distorted
Wavefront

Control
System

Beamsplitter

Corrected
Wavefront

High-resolution
Camera

Wavefront
Sensor

FIGURE 11   Simplified diagram of an AO system. Light from the telescope is collimated and sent to an adaptive or deformable mirror. If there were no atmospheric turbulence, the wavefront of the light would be perfectly straight and parallel. The light is then reflected to a beamsplitter, where part of the light is reflected to the wavefront sensor. The wavefront sensor measures the distortion of the wavefront and sends a correction signal to the adaptive mirror. The adaptive mirror is capable of changing its shape to remove the deformations in the light wave caused by the atmospheric turbulence. In this way the light with a corrected wavefront reaches the high-resolution camera, where a diffraction-limited image is formed. (Courtesy of C. Max)

diffraction-limited imaging from the ground. This is critical in achieving the maximum S/N given in equation (1). The basic idea of AO is to first measure the amount of atmospheric disturbance, then correct for it before the light reaches the camera. A schematic of how this can be done is shown in Figure 11.

The effect of using AO is dramatic. It is like taking the telescope into space. An impressive example of how AO can improve image quality is shown in Figure 12. AO has been essential for detecting binary asteroids. With it over 60 systems have been found, and the first triple system was recently found as shown in Figure 13.

AO requires a star or another object bright enough to use for rapidly and accurately measuring the incoming wavefront. If the object of interest is not bright enough, then it is necessary to use a nearby bright star. This limits the sky coverage, since not every region of the sky will have a bright enough star nearby. If there is no nearby bright star, then it is necessary to use a laser guide star. A laser is pointed in the same direction as the telescope and is used to excite a thin layer of sodium atoms in the Earth's ionosphere (at an

altitude of 90 km). This provides a point source that acts as an artificial star for the AO system.

Figure 14 shows a laser guide star being used at the Keck Observatory. This laser guide star system was used to detect the satellite of the largest KBO known (see Fig. 1).

With AO we can look forward to the exploration of other solar systems. Figure 15 shows a faint object next to a brighter object that is thought to have a mass 5 times that of Jupiter—a planet. This is one of the first planetary-mass objects to be imaged. Most planets are found by detecting radial velocity variations in the star they are orbiting. About 160 planets have already been detected by the radial velocity method and there is a possibility to detect Earth-mass planets around nearby low-mass stars. We can expect future planetary systems to be discovered, and thus to be able to study the physical characteristics of other solar systems for the first time. The study of extrasolar planets is a key science area for all large telescopes.

5. Sky Survey Telescopes

Although large telescope projects tend to get a lot of attention, recently there has been a corresponding quantum jump in the construction of visible and infrared survey telescopes. This has been made possible by the availability of large-format CCD and infrared arrays. In addition, the discovery of the Kuiper Belt has led to fundamental advances in our understanding of how our solar system formed. There is a great need to continue the survey of the Kuiper Belt because detailed knowledge of the size and orbit distributions of these objects will allow us to test theories of the orbital migration of the outer planets (Jupiter, Saturn, Uranus, Neptune), the origin of the short-period comets, and the cause of the late heavy bombardment of the inner solar system.

There is also an increased awareness that it is important to identify asteroids and comets that could collide with Earth (see Fig. 3). In 1998 the Congress of the United States directed NASA to identify within 10 years at least 90% of NEOs larger than 1 km that may collide with Earth. There are a number of scientific benefits that arise from the NEO surveys, including determining the origin of NEOs, identifying interesting NEOs that could be visited by spacecraft, improving our knowledge of the numbers and sizes of the asteroids in the main asteroid belt, and the discovery of new comets.

The reason that the discovery of all NEOS larger than 1 km is important is because if such an object collides with Earth the consequences will be catastrophic. If it is possible to predict that there will be a collision, it may be possible to divert the asteroid so that it misses Earth. The earlier such a prediction can be made, the more likely it is that the diversion is possible. This is a case in which there is a practical use for astronomy, and it is very fitting.

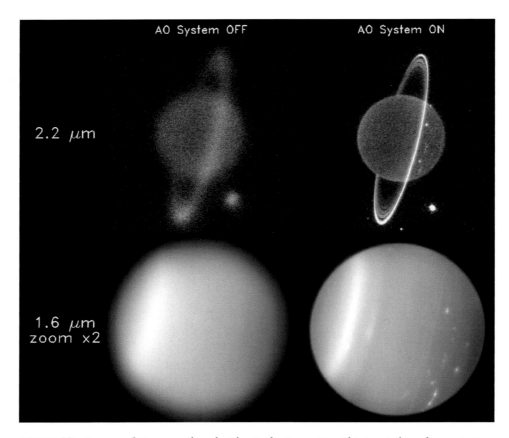

FIGURE 12 Images of Uranus with and without adaptive optics. This is a striking demonstration of the effectiveness of adaptive optics in removing atmospheric turbulence. One can also see that the signal-to-noise is greatly enhanced because light is concentrated into a diffraction-limited image with adaptive optics, thus greatly increasing the ability to detect faint spots and cloud structure. At a wavelength of 1.6 micrometers, we are seeing reflected light from low-altitude clouds while at 2.2 micrometers the high-altitude clouds are revealed. The planet is much darker at 2.2 micrometers due to absorption of methane gas in the atmosphere. This allows a much longer exposure and for the rings to be seen clearly. The point-like cloud features at 2.2 micrometers show that in certain places turbulence is very strong and is pushing material from lower altitudes into the stratosphere. (Courtesy of H. B. Hammel, I. de Pater, and the W. M. Keck Observatory.)

A number of programs are underway in the US and other countries that meet or exceed the requirements set by Congress. Table 2 shows a partial list of sky survey programs that are currently in progress or planned. Current productivity of various programs is shown in Figure 16, which shows all NEOs discovered irrespective of size. While the NASA directive is aimed at identifying NEOs larger than 1 km diameter, many NEOs smaller than 1 km are also discovered due to the sensitivity of the search programs and because small objects that come very close to Earth may be bright enough to be detected. A recent NEO, 2005 WX, approached to within 1.3 million km of the Earth and had an estimated diameter of only 10 m!

The number of known NEOS has been increasing due to the larger number of funded survey programs and advances in detector arrays that have allowed much larger areas of sky to be covered in a single exposure. The number of NEOs discovered as a function of time is shown in Figure 16. Note that while the total number of asteroids discovered is still increasing at a rapid rate, the number of new asteroids larger than 1 km discovered each year is decreasing. This is a result of the fact that the remaining unknown NEAs are intrinsically more difficult to detect. Their size and orbit distribution is different from the known population due to observational selection effects in the population of known objects. It is likely that existing survey programs (see Table 2) will just miss the goal of discovering at least 90% of all near-Earth asteroids larger than 1 km by 2008 as mandated by Congress. However, when the next generation surveys (see Table 2) come online within the next decade they will quickly complete the inventory of NEAs larger than 1 km.

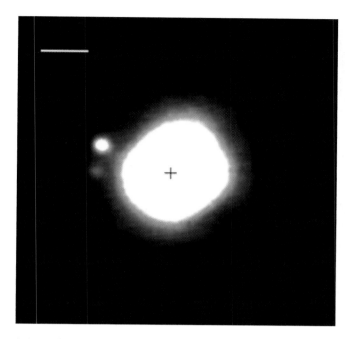

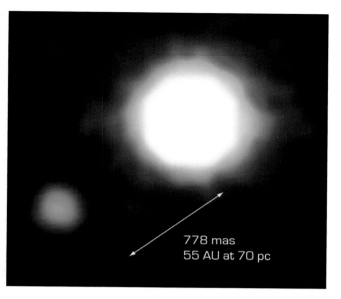

FIGURE 13   Image of the asteroid 87 Sylvia showing its two satellites. This image was taken with the European Southern Observatory 8-m Very Large Telescope at 2.2 micrometers with an adaptive optics system. The cross marks the location of the asteroid and the scale bar shown is 0.25 arcseconds. The diameter of 87 Sylvia is about 280 km, and the diameters of the satellites are about 7 and 14 km. The orbits of the satellites were measured in order to determine a density of about 1.2 grams/cm^3 for 87 Sylvia—only 20% higher than the density of water. Thus 87 Sylvia is likely to have a rubble pile internal structure with 20-60% of its volume being empty. (Courtesy of F. Marchis.)

FIGURE 15   Infrared image of 2M 1207 (a brown dwarf and planet binary system) obtained with one of the 8.2-m VLT telescopes. The brown dwarf (white) is 100 times brighter than the planet (red) and both are emitting heat left over from their formation. Their masses are estimated to be 25 and 5 Jupiter masses. In this image the infrared colors at wavelengths 3.8, 2.2, and 1.6 microns are portrayed as red, green, and blue, respectively. The separation of the objects in the sky is 0.78 arcseconds and this corresponds to a physical separation of 55 AU. (Courtesy Gael Chauvin / ESO).

FIGURE 14   Sodium laser guide star in use at Keck II. The laser operates at a wavelength of 5890 Angstroms (0.589 micrometers), and the laser light is propagated through a smaller telescope attached to the Keck telescope. It excites sodium atoms in a layer in the Earth's atmosphere at an altitude of 90 km. The sodium atoms emit light at the same wavelength as the laser and this is viewed as an artificial star by the telescope. (This is a long exposure photograph. The laser guide star is barely visible with the naked eye from this angle. The lights of the island of Hawaii are below the clouds. (Courtesy of Jean-Charles Cuillandre.)

TABLE 2   **Summary of Sky Survey Telescopes**

Survey	Status	Aperture (m)	f/no	Field-of-view (degree²)	Magnitude limit	Speed (degree² per hour)	Ref.
CSS – Mt. Lemmon	operational	1.5	2.0	1.3	21	20	1
CSS – Catalina Schmidt	operational	0.68	1.9	8	19.5	150	1
CSS – Siding Spring Uppsala Schmidt	operational	0.5	3.5	4.2	19.5	75	1
LINEAR	operational	2 × 1.0	2.2	2.0	19.4	1200	2
LONEOS (Schmidt)	operational	0.44	1.9	8.3	19.3	106	3
LONEOS (USNO)	in development	1.3	2.4	1.3	21.4	15	3
NEAT (Palomar)	operational	1.2	1.5	9.5	22.5	85	4
NEAT (MSSS)	operational	1.2	3.0	2.3	19.7	40.5	4
NEAT (Schmidt)	in development	1.2	2.5	9.4	~20.0	50	4
Spacewatch (Mosaic)	operational	0.93	3.0	2.9	21.5	160	5
Spacewatch (1.8 m)	operational	1.82	2.7	0.32	22.5	8.9	5
Pan-STARRS (Hawaii)	in development	4 × 1.8	4	3.0	24.0	700	6
Discovery Channel Telescope (Lowell)	in development	4.0	2.2	3.1	21.8	110	7
Large Synoptic Survey Telescope	proposed	6.9	1.25	7.0	24.0	2500	8

References: (1) Catalina Sky Survey, http://www.lpl.arizona.edu/css/, (2) Lincoln Near Earth Asteroid Research, http://www.ll.mit.edu/LINEAR/, (3) Lowell Observatory Near-Earth-Object Search, http://asteroid.lowell.edu/asteroid/loneos/loneos1.html, (4) Near-Earth Asteroid Tracking, http://neat.jpl.nasa.gov/, (5) http://spacewatch.lpl.arizona.edu/, (6) Panoramic Survey Telescope & Rapid Response System, http://pan-starrs.ifa.hawaii.edu/public/, (7) http://www.lowell.edu/DCT/, (8) http://www.lsst.org/

1. Field-of-view is the area of sky covered in a single exposure.

2. Magnitude limit is the faintest star recorded at visible wavelengths.

3. Speed is the rate at which observations can be carried out. One can see that of the operational facilities, LINEAR covers the most sky per hour (1200 degree²/hour) but the faintest stars it can observe at this speed is 19.4 mag. The Spacewatch (1.8 m) telescope can observe stars that are 3 magnitudes fainter but at a speed of only 8.9 degree²/hour.

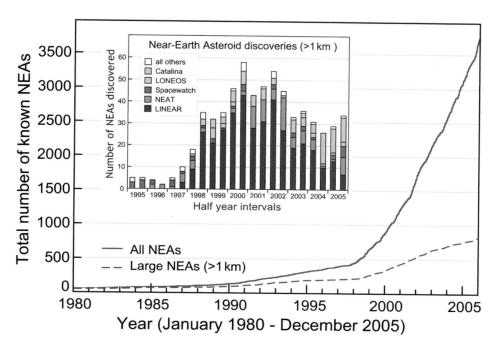

Near-Earth Asteroid discoveries (>1 km)

FIGURE 16 Cumulative discoveries of near-Earth asteroids. The total number of large near-Earth asteroids (larger than 1 km) is increasing at a slower rate since most of the easy-to-detect NEOs have already been discovered. The remaining unknown NEOs are on orbits that are intrinsically more difficult to detect and therefore require a longer time to discover. (From NEO.)(Courtesy of Alan Chamberlin.)

There are three major ground-based sky surveys currently under development or study (see Table 2). The Discovery Channel Telescope is a 4.2-m telescope that is under construction near Flagstaff in Northern Arizona and should be operational by 2009. Another survey telescope that is under development is Pan-STARRS, which consists of four 1.8-m telescopes (with a combined aperture approximately equivalent to a 3.6-m telescope) to perform rapid wide-field surveying of the entire sky on a weekly basis. It is hoped that the full system will be operational by 2010, but a prototype single telescope unit will be operational on Haleakala on Maui by the end of 2007. The proposed Large Synoptic Survey Telescope is currently under engineering and design study and is envisioned to be a monolithic 8.4-m wide-field telescope (with a collecting area equal to a 6.7-m telescope). With its large diameter and fast focal ratio it should be capable of reaching 24th magnitude in single 10-s exposures. Due to their extreme depth and wide-field coverage each of these surveys should reach 99% completion for NEOs larger than 1 km diameter within two years of beginning operation.

6. Concluding Remarks

Space does not allow coverage of all of the relevant subjects related to the vibrant topics of novel telescope construction, optical fabrication techniques, advances in mirror figure control, adaptive optics, and detector improvements at visible and infrared wavelengths. The topics covered in this chapter can only hint at the tremendous advances that have

taken place in recent years and that carry on unabated. Since the invention of the refractive and reflective telescopes by Galileo and Newton, the construction of ground-based telescopes continues to challenge the very best minds in physics and engineering. At the present time there are strong scientific drivers to build larger telescopes in the 20–50 m range. It seems only a matter of time before such extremely large telescopes are built.

Solar system astronomy is driven by the need to have large telescopes in order to study very faint objects in the Kuiper Belt and very faint NEOs that may present a hazard to Earth. It is also necessary to have the highest spatial resolution possible by working at the diffraction limit of large telescopes. This will enable researchers to study the surface and atmospheric features of the outer planets, dwarf planets, and their satellites. Large telescopes also allow the study of exo-planets, and thus bring about a merging of studies of our solar system with those around distant stars.

Another driver of solar system astronomy is to detect and characterize NEOs that may present an impact hazard to the Earth. Numerous sky survey programs are underway to detect at least 90% of all NEOs larger than 1 km, and there is a push at the present time to expand this program to detect at least 90% of all NEOs larger than 140 m. These survey programs will play a significant role in greatly expanding our knowledge of the building blocks of our solar system—the asteroidal and cometary bodies from the inner to the outer reaches of the solar system. These studies are likely to profoundly affect understanding of the formation of our solar system and life itself.

We anticipate continuing growth in telescope and instrument development for at least another generation. It is indeed a period great innovation—a renaissance in telescope building and instrumentation—that we are fortunate to be able to witness and participate in.

Bibliography

Bely, P.Y. (ed.) (2003). *The Design and Construction of Large Optical Telescopes.* Springer-Verlag, New York.

Kitchin, C.R. (2003). *Telescopes and Techniques.* Springer-Verlag, London.

McLean, I. (1997). *Electronic Imaging in Astronomy.* John Wiley & Sons, Chichester.

Tyson, R.K. (2000). *Introduction to Adaptive Optics.* Soc. Of Photo-Optical Instrumentation Eng., Bellingham.

NEO web site: http://neo.jpl.nasa.gov/programs/discovery .html. Racine, R. 2004, "The Historical Growth of Telescope Aperture", Pub. Astron. Soc. Pacific, vol. 116, p. 77.

Zirker, J.B. (2005). *An Acre of Glass: A History and Forecast of the Telescope.* The Johns Hopkins University Press, Baltimore.

Planetary Radar

Steven J. Ostro

Jet Propulsion Laboratory
California Institute of Technology
Pasadena, California

CHAPTER 40

P lanetary radar astronomy is the study of solar system entities (the Moon, asteroids, and comets, as well as the major planets and their satellites and ring systems) by transmitting a radio signal toward the target and then receiving and analyzing the echo. This field of research has primarily involved observations with Earth-based radar telescopes, but it also includes certain experiments with the transmitter and/or the receiver onboard a spacecraft orbiting or passing near a planetary object. However, radar studies of Earth's surface, atmosphere, or ionosphere from spacecraft, aircraft, or the ground are not considered part of planetary radar astronomy. Radar studies of the Sun involve such distinctly individual methodologies and physical considerations that solar radar astronomy is considered a field separate from planetary radar astronomy.

1. Introduction

1.1 Scientific Context

Planetary radar astronomy is a field of science at the intersection of planetology, radio astronomy, and radar engineering. A radar telescope is a radio telescope equipped with a high-power radio transmitter and specialized electronic instrumentation designed to link transmitter, receiver, data acquisition, and telescope-pointing components together in an integrated radar system. The principles underlying

operation of this system are not fundamentally very different from those involved in radars used, for example, in marine and aircraft navigation, measurement of automobile speeds, and satellite surveillance. However, planetary radars must detect echoes from targets at interplanetary distances ($\sim 10^5$–10^9 km) and therefore are the largest and most powerful radar systems in existence.

The advantages of radar observations in astronomy stem from the high degree of control exercised by the observer on the transmitted signal used to illuminate the target. Whereas virtually every other astronomical technique relies on passive measurement of reflected sunlight or naturally emitted radiation, the radar astronomer controls all the properties of the illumination, including its intensity, direction, polarization, and time/frequency structure. The properties of the transmitted waveform are selected to achieve particular scientific objectives. By comparing the properties of the echo to the very well known properties of the transmission, some of the target's properties can be deduced. Hence, the observer is intimately involved in an active astronomical observation and, in a very real sense, performs a controlled laboratory experiment on the planetary target.

Radar delay–Doppler and interferometric techniques can spatially resolve a target whose angular extent is dwarfed by the antenna's beamwidth (that is, its diffraction-limited angular resolution), thereby bestowing a considerable advantage on radar over optical techniques in the study of asteroids, which appear like "point sources" through ground-based optical telescopes. Furthermore, by virtue of the centimeter-to-meter wavelengths employed, radar is sensitive to scales of surface structure many orders of magnitude larger than those probed in visible or infrared regions of the spectrum. Radar is also unique in its ability to "see through" the dense clouds that enshroud Venus and Titan and the glowing gaseous coma that conceals the nucleus of a comet. Because of its unique capabilities, radar astronomy has made notable contributions to planetary exploration for four decades.

1.2 History

Radar technology was developed rapidly to meet military needs during World War II. In 1946, soon after the war's conclusion, groups in the United States and Hungary obtained echoes from the Moon, giving birth to planetary radar astronomy. These early postwar efforts were motivated primarily by interest in electromagnetic propagation through the ionosphere and the possibility of using the Moon as a "relay" for radio communication.

During the next two decades, the need for ballistic missile warning systems prompted enormous improvements in radar technology. This period also saw rapid growth in radio astronomy and the construction of huge radio telescopes. In 1957, the Soviet Union launched *Sputnik* and with it the space age, and in 1958, with the formation by the U.S. Congress of the National Aeronautics and Space Administration (NASA), a great deal of scientific attention turned to the Moon and to planetary exploration in general. During the ensuing years, exhaustive radar investigations of the Moon were conducted at wavelengths from 0.9 to 20 m, and the results generated theories of radar scattering from natural surfaces that still see wide application.

By 1963, improvements in the sensitivity of planetary radars in both the United States and the U.S.S.R. had permitted the initial detections of echoes from the terrestrial planets (Venus, Mercury, and Mars). During this period, radar investigations provided the first accurate determinations of the rotations of Venus and Mercury and the earliest indications for the extreme geologic diversity of Mars. Radar images of Venus have revealed small portions of that planet's surface at increasingly fine resolution since the late 1960s, and in 1979 the Pioneer Venus Spacecraft Radar Experiment gave us our first look at Venus' global distributions of topography, radar reflectivity, and surface slopes. During the 1980s, maps having sparse coverage but resolution down to ~1 km were obtained from the Soviet *Venera 15* and *16* orbiters and from ground-based observations with improved systems. In the early 1990s, the *Magellan* spacecraft radar revealed most of the planet's surface with unprecedented clarity (~100-m resolution), revealing a rich assortment of volcanic, tectonic, and impact features.

The first echoes from a near-Earth asteroid (1566 Icarus) were detected in 1968; it would be nearly another decade before the first radar detection of a main belt asteroid (1 Ceres in 1977), to be followed in 1980 by the first detection of echoes from a comet (Encke). During 1972 and 1973, detection of 13-cm-wavelength radar echoes from Saturn's rings shattered prevailing notions that typical ring particles were 0.1–1.0 mm in size—the fact that decimeter-scale radio waves are backscattered efficiently requires that a large fraction of the particles be larger than a centimeter. Observations by the *Voyager* spacecraft confirmed this fact and further suggested that particle sizes extend to at least 10 m.

In the mid-1970s, echoes from Jupiter's Galilean satellites Europa, Ganymede, and Callisto revealed that the manner in which these icy moons backscatter circularly polarized waves is extraordinarily strange, and totally outside the realm of previous radar experience. We now understand that those echoes were due to high-order multiple scattering from within the top few decameters of the satellites' regoliths, which are orders of magnitude more transparent to radio waves than rocky regoliths.

The late 1980s saw the initial detections of Phobos and Titan, the accurate measurement of Io's radar properties, the discovery of large-particle clouds accompanying comets, dual-polarization mapping of Mars and the icy Galilean satellites, and radar imaging of asteroids that presaged the diversity of these objects' shapes and rotations.

During the 1990s, the novel use of instrumentation and waveforms yielded the first full-disk radar images of the terrestrial planets, revealing the startling presence of radar-bright polar anomalies on Mercury as well as Mars. Similarities between the polarization and albedo signatures of these features and those of the icy Galilean satellites argue persuasively that Mercury's polar anomalies are deposits of water ice in the floors of craters that are perpetually shaded from sunlight by Mercury's low obliquity. On the other hand, conjectures about radar-detectable lunar ice deposits have not been substantiated by radar imaging and topographic mapping. In 1992, the first time-delay-resolved ("ranging") measurements to Ganymede and Callisto were carried out, and delay–Doppler images of the closely approaching asteroid 4179 Toutatis revealed it to be in a very slow, non-principal-axis spin state and provided the first geologically detailed pictures of an Earth-crossing asteroid. The 1990s also saw the first intercontinental radar observations and the beginning of planetary radar experiments in Germany, Japan, and Spain. The Arecibo telescope was upgraded in the mid-1990s, and with the resultant order-of-magnitude improvement in its sensitivity (along with significant improvements in Goldstone hardware and software), a new era of radar contributions to planetary science had begun.

As of July 2006, radar had detected 12 comets and 194 near-Earth asteroids, as well as 112 main-belt asteroids (Table 1). Radar's unique capabilities for trajectory refinement and physical characterization give it a natural role in predicting and preventing collisions with small bodies. During the past few years, radar has discovered the existence

TABLE 1 Radar-Detected Planetary Targets[a]

Moon
Mercury
Venus
Mars

Mars satellite:	Phobos
Jupiter satellites:	Io, Europa, Ganymede, Callisto
Saturn satellites:	Enceladus, Tethys, Dione, Rhea, Titan, Hyperion, Iapetus, Phoebe
Saturn's rings	

Comets:

C/IRAS–Araki–Alcock	(nucleus and coma)
C/1996 B2 Hyakutake	(nucleus and coma)
C/2004 Q2 Machholz	(nucleus and coma)
Catalina (P/2005 JQ5)	(nucleus and coma)
73P/Schwassmann–Wachmann 3 (B,C)	(nucleus and coma)
2P/Encke	(nucleus)
C/1998 K5 LINEAR	(nucleus)
26P/Grigg–Skjellerup	(nucleus)
C/Sugano–Saigusa–Fujikawa	(nucleus)
1P/Halley	(coma)
C/2002 O6 Swan	(coma)
C/2001 LINEAR A2-B	(coma)

Main-belt asteroids:

1 Ceres	50 Virginia	144 Vibilia	393 Lampetia
2 Pallas	53 Kalypso	145 Adeona	405 Thia
3 Juno	54 Alexandra	164 Eva	407 Arachne
4 Vesta	56 Melete	165 Loreley	429 Lotis
5 Astraea	59 Elpis	182 Elsa	434 Hungaria
6 Hebe	60 Echo	192 Nausikaa	444 Gyptis
7 Iris	66 Maja	194 Prokne	455 Bruchsalia
8 Flora	71 Niobe	198 Ampella	463 Lola
9 Metis	78 Diana	211 Isolda	476 Hedwig
12 Victoria	80 Sappho	212 Medea	488 Kreusa
13 Egeria	83 Beatrix	216 Kleopatra	505 Cava
15 Eunomia	84 Klio	220 Stephania	524 Fidelio
16 Psyche	85 Io	224 Oceana	532 Herculina
18 Melpomene	88 Thisbe	225 Henrietta	554 Peraga
19 Fortuna	91 Aegina	230 Athamantis	622 Esther
20 Massalia	97 Klotho	247 Eukrate	654 Zelinda
21 Lutetia	101 Helena	253 Mathilde	690 Wratislavia
22 Kalliope	105 Artemis	266 Aline	694 Ekard
23 Thalia	109 Felicitas	270 Anahita	704 Interamnia
25 Phocaea	111 Ate	288 Glauke	711 Boliviana
27 Euterpe	114 Kassandra	313 Chaldea	785 Zwetana
28 Bellona	127 Johanna	324 Bamberga	796 Sarita
31 Euphrosyne	128 Nemesis	325 Heidelberga	914 Palisana
33 Polyhymnia	129 Antigone	335 Roberta	963 Bezovec
36 Atalante	135 Hertha	336 Lacadiera	1139 Atami
38 Leda	137 Meliboea	354 Eleonora	
41 Daphne	139 Juewa	356 Liguria	
46 Hestia	140 Siwa	363 Padua	
49 Pales	141 Lumen	377 Campania	

Near-Earth asteroids:

433 Eros		
1580 Betulia	1036 Ganymed	1566 Icarus
1620 Geographos	7482 (1994 PC1)	89136 (2001 US16)
1627 Ivar	7753 (1988 XB)	99942 Apophis
1685 Toro	7822 (1991 CS)	100085 (1992 UY4)
1862 Apollo	7889 (1994 LX)	101955 (1999 RQ36)
	8014 (1990 MF)	1990 OS

(Continued)

TABLE 1   Radar-Detected Planetary Targets[a] (*Continued*)

1866 Sisyphus	8201 (1994 AH2)	1991 BN
1915 Quetzalcoatl	9856 (1991 EE)	1994 XD
1917 Cuyo	10115 (1992 SK)	1996 JG
1981 Midas	11066 Sigurd	1998 BY7
2062 Aten	12711 (1991 BB)	1998 KY26
2063 Bacchus	13651 (1997 BR)	1998 ST27
2100 Ra-Shalom	14827 Hypnos	1999 FN19
2101 Adonis	16834 (1997 WU22)	1999 FN53
2201 Oljato	17511 (1992 QN)	1999 LF6
3103 Eger (1982 BB)	22753 (1998 WT)	1999 MN
3199 Nefertiti	22771 (1999 CU3)	1999 NW2
3757 (1982 XB)	23187 (2000 PN9)	1999 RR28
3908 Nyx	25143 Itokawa	1999 TN13
4034 (1986 PA)	26663 (2000 XK47)	1999 TY2
4183 Cuno	29075 (1950 DA)	2000 BD19
4179 Toutatis	30825 1990 TG1	2000 CE59
4197 (1982 TA)	33342 (1998 WT24)	2000 DP107
4486 Mithra	35396 (1997 XF11)	2000 ED14
4544 Xanthus	37655 Illapa	2000 EE104
4660 Nereus	38071 (1999 GU3)	2000 EH26
4769 Castalia	52387 (1993 OM7)	2000 EW70
4953 (1990 MU)	52760 (1998 ML14)	2000 GD2
5189 (1990 UQ)	53319 (1999 JM8)	2000 JS66
5381 Sekhmet	54509 (2000 PH5)	2000 LF3
5604 (1992 FE)	65803 Didymos	2000 QW7
5660 (1974 MA)	65909 (1998 FH12)	2000 RD53
6037 (1988 EG)	66063 (1998 RO1)	2000 UG11
6239 Minos	66391 (1999 KW4)	2000 UK11
6178 (1986 DA)	68950 (2002 QF15)	2000 YA
6489 Golevka	69230 Hermes	2000 YF29
7025 (1993 QA)	85182 (1991 AQ)	2001 AV43
7335 (1989 JA)	85774 (1998 UT18)	2001 BE10
7341 (1991 VK)	85938 (1999 DJ4)	2001 BF10
2001 CP36	2002 KK8	2004 JA27
2001 EB18	2002 NY40	2004 RF84
2001 EC16	2002 SR41	2004 RQ10
2001 FR85	2002 SY50	2004 VB
2001 GQ2	2002 TD60	2004 VG64
2001 JV1	2002 TS69	2004 WG1
2001 KZ66	2002 TZ66	2004 XP14
2001 SE286	2002 VE68	2005 AB
2001 SG276	2003 CY18	2005 CR37
2001 SP263	2003 EP4	2005 ED318
2001 UP	2003 GY	2005 EU2
2001 WM15	2003 HN16	2005 FA
2001 XX4	2003 HM	2005 HB4
2001 YE4	2003 KP2	2005 JE46
2001 YP3	2003 MS2	2005 OE3
2002 AA29	2003 QB30	2005 TD49
2002 AL14	2003 RU11	2005 TF49
2002 AV	2003 SR84	2005 TU50
2002 AY1	2003 SS84	2005 WA1
2002 BG25	2003 TH2	2005 WC1
2002 BM26	2003 TL4	2005 WK56
2002 CE26	2003 UC20	2005 XA
2002 CQ11	2003 YT1	2006 GY2
2002 FC	2004 AD	
2002 FD6	2004 DC	
2002 HK12	2004 FY31	
2002 HW	2004 HX53	

[a] Updated lists of radar-detected asteroids and comets are available at http://echo.jpl.nasa.gov.

of binary systems and extremely rapid rotators among the near-Earth asteroids; detected the nongravitational, thermal-recoil "Yarkovsky" acceleration of a near-Earth asteroid and used the measurement to estimate the asteroid's mass; and discovered the dumbbell shape and metallic composition of a large main-belt asteroid. Four-station radar-interferometry-assisted selection of the landing sites for the *Mars Exploration Rovers*, and a novel two-station "radar speckle displacement" technique has produced ultraprecise measurements of Mercury's spin state that should constrain the nature of the core. As the *Cassini* spacecraft approached the Saturn system, Arecibo echoes revealed surfaces on Titan with the radar signature expected for areas of liquid hydrocarbons. At this writing, *Cassini*'s RADAR instrument is well into its multiyear reconnaissance of Titan and eight other Saturnian satellites, returning the first clear pictures of an utterly strange, geologically young world.

2. Techniques and Instrumentation

2.1 Echo Detectability

How close must a planetary target be for its radar echo to be detectable? For a given transmitted power P_T and **antenna gain** G, the power flux a distance R from the radar will be $P_T G/4\pi R^2$. We define the target's **radar cross section**, σ, as 4π times the backscattered power per unit of solid angle per unit of flux incident at the target. Then, letting λ be the radar wavelength and defining the antenna's effective aperture as $A = G\lambda^2/4\pi$, we have the received power

$$P_R = P_T G A \sigma /(4\pi)^2 R^4 \qquad (1)$$

This power might be much less than the receiver noise power, $P_N = kT_S \ \Delta f$, where k is Boltzmann's constant, T_S is the receiver system temperature, and Δf is the frequency resolution of the data. However, the mean level of P_N constitutes a background that can be determined and removed, so P_R will be detectable as long as it is at least several times larger than the standard deviation of the random fluctuations in P_N. These fluctuations can be shown to have a distribution that, for usual values of Δf and the integration time Δt, is nearly Gaussian with standard deviation $\Delta P_N = P_N /(\Delta f \Delta t)^{1/2}$. The highest signal-to-noise ratio, or SNR $= P_R/\Delta P_N$, will be achieved for a frequency resolution equal to the effective bandwidth of the echo. As discussed in the following, that bandwidth is proportional to $D/\lambda P$, where D is the target's diameter and P is the target's rotation period, so let us assume that $\Delta f \sim D/\lambda P$. By writing $\sigma = \eta\pi D^2/4$, where the **radar albedo** η is a measure of the target's radar reflectivity, we arrive at the following expression for the echo's signal-to-noise ratio:

$$\text{SNR} \sim (\text{System Factor}) (\text{Target Factor})(\Delta t)^{1/2} \qquad (2)$$

where

$$\text{System Factor} \sim P_T A^2/\lambda^{3/2} T_S$$
$$\sim P_T G^2 \lambda^{5/2} / T_S \qquad (3)$$

and

$$\text{Target Factor} \sim \eta D^{3/2} P^{1/2}/R^4 \qquad (4)$$

The inverse-fourth-power dependence of SNR on target distance is a severe limitation in ground-based observations, but it can be overcome by constructing very powerful radar systems.

2.2 Radar Systems

The world has two active planetary radar facilities: the Arecibo Observatory (part of the NSF's National Astronomy and Ionosphere Center) in Puerto Rico and NASA's Goldstone Solar System Radar in California. Radar wavelengths are 13 cm and 70 cm for Arecibo and 3.5 cm and 13 cm for Goldstone. With each instrument, enormously more sensitivity is achievable with the shorter wavelength. The upgraded Arecibo telescope has twice the range and can see three times the volume of Goldstone, whereas Goldstone can see twice as much sky as Arecibo and can track targets at least three times longer. Figure 1 shows the relative

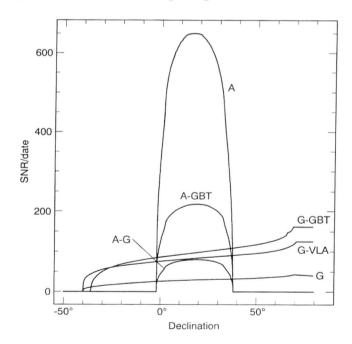

FIGURE 1 Sensitivities of planetary radar systems. Curves plot the single-date, signal-to-noise ratio of echoes from a typical 1-km asteroid at a distance of 0.1 AU for the upgraded Arecibo telescope (A), Goldstone (G), and bistatic configurations using those instruments and the Very Large Array (VLA) or the Greenbank Telescope (GBT).

FIGURE 2 The Arecibo telescope in Puerto Rico. The triangular platform suspended above the 305-m primary reflector supports the azimuth and elevation structures that let the Gregorian feed inside the 26-m radome or the line feed point up to 20° off the zenith. The S-band (2380-MHz, 13-cm) transmitter and front-end receiver are inside the radome. (Courtesy of the NAIC—Arecibo Observatory, a facility of the NSF.)

sensitivities of planetary radar systems as a function of target declination.

The Arecibo telescope (Fig. 2) consists of a 305 m diameter, fixed reflector whose surface is a 51-m-deep section of a 265-m-radius sphere. Movable feeds designed to correct for spherical aberration are suspended from a triangular platform 137 m above the reflector and can be aimed toward various positions on the reflector, enabling the telescope to point within about 20° of the overhead direction (declination 18.3°N). Components of the 1990s upgrade included a megawatt transmitter, a ground screen to reduce noise generated by radiation from the ground, and replacement of most of the old single-frequency line feeds with a Gregorian reflector system (named after the 17th-century mathematician James Gregory) that employs 22-m secondary and 8-m tertiary subreflectors enclosed inside a 26-m radome.

The Goldstone main antenna, DSS-14 (DSS stands for Deep Space Station), is part of NASA's Deep Space Network, which is run by the Jet Propulsion Laboratory (JPL). It is a fully steerable, 70-m, parabolic reflector (Fig. 3). Bistatic experiments using DSS-14 transmissions and reception of echoes at DSS-13, a 34-m antenna 22 km away, have been conducted on several very close targets. Bistatic observations between Arecibo and Goldstone, or using transmission from Arecibo or Goldstone and reception at the 100-m Greenbank Telescope (GBT) in West Virginia, have proven advantageous for the Moon, the inner planets, outer planet satellites, and nearby asteroids and comets.

Figure 4 is a simplified block diagram of a planetary radar system. A waveguide switch, a movable subreflector, or a moveable mirror system is used to place the antenna in a transmitting or receiving configuration. The heart of the transmitter is one or two **klystron** vacuum-tube amplifiers. In these tubes, electrons accelerated by a potential drop of some 60 kV are magnetically focused as they enter the first of five or six cavities. In this first cavity, an oscillating elec-

FIGURE 3 The 70-m Goldstone Solar System Radar main antenna, DSS-14, in California. The 3.5-cm planetary radar transmitter and front-end receivers are inside the lowest cone near the focus of the antenna, which is fully steerable.

tric field at a certain radio frequency (RF, e.g., 2380 MHz for Arecibo or 8560 MHz) modulates the electrons' velocities and hence their density and energy flux. Subsequent resonant cavities enhance this velocity bunching (they

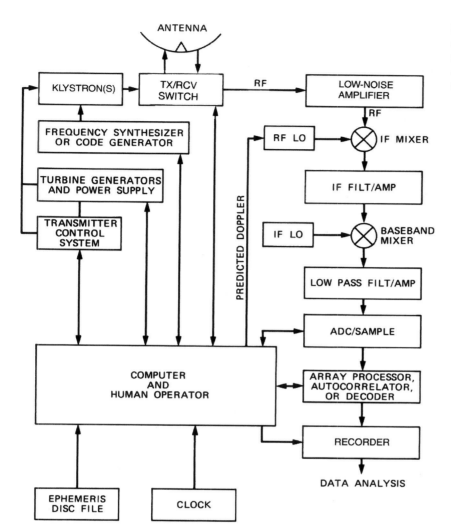

FIGURE 4   Block diagram of a planetary radar system. RF LO and IF LO denote radio frequency and intermediate frequency local oscillators, and ADC denotes analog-to-digital converter.

constitute what is called a cascade amplifier), and about half of the input DC power is converted to RF power and sent out through a waveguide to the antenna feed system and radiated toward the target. The other half of the input power is waste heat and must be transported away from the klystron by cooling water. The impact of the electrons on the collector anode generates dangerous X-rays that must be contained by heavy metal shielding surrounding the tube, a requirement that further boosts the weight, complexity, and hence cost of a high-power transmitter.

In most single-antenna observations, one transmits for a duration near the roundtrip propagation time to the target (i.e., until the echo from the beginning of the transmission is about to arrive) and then receives for a similar duration. In the "front end" of the receiving system, the echo signal is amplified by a cooled, low-noise amplifier and converted from RF down to intermediate frequencies (IF, e.g., 30 MHz), for which transmission line losses are small and passed from the proximity of the antenna feed to a remote control room containing additional stages of signal-processing equipment, computers, and digital recorders.

The signal is filtered, amplified, and converted to frequencies low enough for analog voltage samples to be digitized and recorded. The frequency down-conversion can be done in several stages using analog devices called superheterodyne mixers, but in recent years it has become possible to do this digitally, at increasingly higher frequencies. The nature of the final processing prior to recording of data on a hard disk or magnetic tape depends on the nature of the radar experiment and particularly on the time/frequency structure of the transmitted waveform. Each year, systems for reducing and displaying echoes in "real time" and techniques for processing recorded data are becoming more ambitious as computers get faster.

2.3 Echo Time Delay and Doppler Frequency

The time between transmission of a radar signal and reception of the echo is called the echo's roundtrip **time delay**, τ, and is of order $2R/c$, where c is the speed of light, by definition equal to 299,792,458 m sec^{-1}. Because planetary targets are not points, even an infinitesimally short transmitted pulse would be dispersed in time delay, and the total extent $\Delta \tau_{TARGET}$ of the distribution $\sigma(\tau)$ of echo power (in units of radar cross section) would be D/c for a sphere of diameter D and in general depends on the target's size and shape.

The translational motion of the target with respect to the radar introduces a **Doppler shift** $\Delta \nu$ in the frequency of the transmission. Both the time delay and the Doppler shift of the echo can be predicted in advance from the target's **ephemeris**, which is calculated using the geodetic position of the radar and the orbital elements of Earth and the target. The predicted Doppler shift can be removed electronically by continuously tuning the local oscillator used, for example, for RF-to-IF frequency conversion (see Fig. 4). Sometimes it is convenient to "remove the Doppler on the uplink" by modulating the transmission so that echoes return at a fixed frequency. The predicted Doppler (i.e., the predicted rate of change of the delay) must be accurate enough to avoid smearing out the echo in delay, and this requirement places stringent demands on the quality of the observing ephemeris. Time and frequency measurements are critical because the delay/Doppler distribution of echo power is the source of fine spatial resolution and also can be used to refine the target's orbit. Reliable, precise time/frequency measurements are made possible by high-speed data acquisition systems and stable, accurate clocks and frequency standards.

Because different parts of the rotating target will have different velocities relative to the radar, the echo will be dispersed in Doppler frequency as well as in time delay. The basic strategy of any radar experiment always involves measurement of some characteristic(s) of the function $\sigma(\tau, \nu)$, perhaps as a function of time and perhaps using more than one combination of transmitted and received polarizations. Ideally, one would like to obtain $\sigma(\tau, \nu)$ with very fine resolution, sampling that function within intervals whose dimensions $\Delta \tau \times \Delta \nu$ are minute compared to the echo dispersions $\Delta \tau_{TARGET}$ and $\Delta \nu_{TARGET}$. Figure 5 shows the geometry of delay-resolution cells and Doppler-resolution cells for a spherical target and sketches their relation to $\sigma(\tau)$ and $\sigma(\nu)$.

2.4 Radar Waveforms

In the simplest radar experiment, the transmitted signal is a highly monochromatic, unmodulated, continuous wave (cw) signal. Analysis of the received signal comprises Fourier transformation of a series of time samples and yields an estimate of the echo power spectrum $\sigma(\nu)$, but it contains no information about the distance to the target or $\sigma(\tau)$. To avoid **aliasing,** the sampling rate must be at least as large as the bandwidth of the low-pass filter (see Fig. 4) and usually is comparable to or larger than the echo's intrinsic dispersion $\Delta \nu_{TARGET}$ from Doppler broadening. Fast Fourier transform (FFT) algorithms greatly speed the calculation of discrete spectra from time series and are ubiquitous in

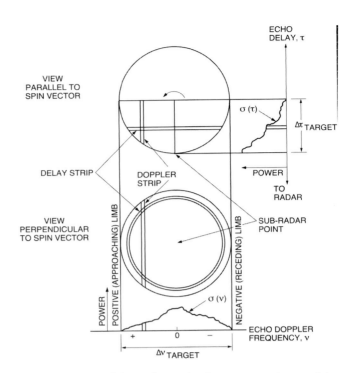

FIGURE 5   Time-delay and Doppler-frequency resolution of the radar echo from a rotating sphere.

radar astronomy. In a single FFT operation, a string of N time samples taken at intervals of Δt seconds is transformed into a string of N spectral elements with frequency resolution $\Delta \nu = 1/(N \Delta t)$.

To obtain delay resolution, one must apply some sort of time modulation to the transmitted waveform. For example, a short-duration pulse of cw signal lasting 1 μs would provide delay resolution of 150 m. However, the echo would have to compete with the noise power in a bandwidth of order 1 MHz (i.e., the reciprocal of 1 μs), so the echo power from many consecutive pulses would probably have to be summed to yield a detection. One would not want these pulses to be too close together, however, or there would be more than one pulse incident on the target at once, and interpretation of echoes would be insufferably ambiguous. Thus, one arranges the pulse repetition period t_{PRP} to exceed the target's intrinsic delay dispersion $\Delta \tau_{TARGET}$, ensuring that the echo will consist of successive, nonoverlapping "replicas" of $\sigma(\tau)$ separated from each other by t_{PRP}. To generate this "pulsed cw" waveform, the transmitter is switched on and off while the frequency synthesizer (see Fig. 4) maintains phase coherence from pulse to pulse. Then Fourier transformation of time samples taken at the same position within each of N successive replicas of $\sigma(\tau)$ yields the power spectrum of echo from a certain delay resolution cell on the target. This spectrum has an unaliased bandwidth of $1/t_{PRP}$ and a frequency resolution of $1/(Nt_{PRP})$. Repeating this process for a different position within each replica of $\sigma(\tau)$ yields the power spectrum for echo from a different delay resolution cell, and in this manner one obtains the delay–Doppler image $\sigma(\tau, \nu)$.

In practice, instead of pulsing the transmitter, one usually codes a cw signal with a sequence of 180° phase reversals and cross-correlates the echo with a representation of the code (e.g., using the decoder in Fig. 4), thereby synthesizing a pulse train with the desired values of Δt and t_{PRP}. With this approach, one optimizes SNR because it is much cheaper to transmit the same average power continuously than to pulse the transmitter. Most modern ground-based radar as-

tronomy observations employ cw or repetitive, phase-coded cw waveforms.

A limitation of coherent-pulsed or repetitive, binary-phase-coded cw waveforms follows from combining the requirement that there never be more than one echo received from the target at any instant (i.e., that $t_{PRP} > \Delta\tau_{TARGET}$) with the antialiasing frequency requirement that the rate $(1/t_{PRP})$ at which echo from a given delay resolution cell is sampled be no less than the target bandwidth $\Delta\nu_{TARGET}$. Therefore, a target must satisfy $\Delta\tau_{TARGET}\Delta\nu_{TARGET} < 1$ or it is "overspread" (Table 2) and cannot be investigated

| TABLE 2 | Characteristics of Selected Planetary Radar Targets[a] |

Target	Minimum Echo Delay[b] (min)	Radar Cross Section (km²)	Radar Albedo, η_{OC}	Circular polarization ratio, μ_C	Maximum Dispersions[c] Delay (ms)	Doppler (Hz)	Product
Moon	0.04	6.6×10^5	0.07	0.1	12	60	0.7
Mercury	9.1	1.1×10^6	0.06	0.1	16	110	2
Venus	4.5	1.3×10^7	0.11	0.1	40	110	4
Mars	6.2	2.9×10^6	0.08	0.3	23	7600	170
Phobos	6.2	22	0.06	0.1	0.1	100	10^{-2}
1 Ceres	26	2.7×10^4	0.05	0.0	3	3100	9
2 Pallas	25	1.7×10^4	0.08	0.0	2	2000	4
12 Victoria	15	2.3×10^3	0.22	0.1	0.5	590	3
16 Psyche	28	1.4×10^4	0.31	0.1	0.8	2200	2
216 Kleopatra	20	7.1×10^3	0.44	0.0	?	750	?
324 Bamberga	13	2.9×10^3	0.06	0.1	0.8	230	0.2
1685 Toro	2.3	1.7	0.1	0.2	0.02	14	10^{-4}
1682 Apollo	0.9	0.2	0.1	0.4	0.01	16	10^{-4}
2100 Ra-Shalom	3.0	1.1	0.2	0.3	0.01	5	10^{-4}
2101 Adonis	1.5	0.02	<0.3	1.0	?	2	?
4179 Toutatis	0.4	1.0	0.18	0.3	0.01	1	10^{-5}
4769 Castalia	0.6	0.2	0.15	0.3	0.01	10	10^{-4}
6178 1986DA	3.4	2.4	0.6	0.1	12	15	0.2
1998 KY26	0.09	2.5×10^{-5}	0.01 to 0.1	0.5	0.0001	15	10^{-6}
25143 Itokawa	0.4	0.01–0.02	0.1	0.2	0.003	1	10^{-6}
Comet IRAS-Araki-Alcock							
nucleus	0.5	2.4	0.04?	0.1	?	4	?
coma	0.5	0.8	?	0.01	?	600	?
Comet Hyakutake	(C/1996 B2)						
nucleus	1.7	0.11	?	0.5	?	12	?
coma	1.7	1.3	?	< 1	?	3000	?
Io	66	2×10^6	0.2	0.5	12	2400	29
Europa	66	8×10^6	1.0	1.5	10	1000	11
Ganymede	66	1×10^7	0.6	1.4	18	850	15
Callisto	66	5×10^6	0.3	1.2	16	330	5
Saturn's rings	134	10^8–10^9	0.7	0.5	1600	6×10^5	10^6

[a] Typical 3.5- to 13-cm values. Question marks denote absence of radar data or of prior information about target dimensions.

[b] For asteroids and comets, this is the minimum echo time delay for radar observations to date.

[c] Doppler dispersion for transmitter frequency of 2380 MHz (13 cm). The product of the dispersions in delay and Doppler is the overspread factor at 2380 MHz.

completely and simultaneously in delay and Doppler without aliasing, at least with the waveforms discussed so far. Various degrees of aliasing may be "acceptable" for overspread factors less than about 10, depending on the precise experimental objectives and the exact properties of the echo.

How can the full delay–Doppler distribution be obtained for overspread targets? Frequency-swept ("chirped") and frequency-stepped waveforms have seen limited use in planetary radar; the latter approach has been used to image Saturn's rings. Another technique uses a nonrepeating, binary-phase-coded cw waveform; in this case the received signal for any given delay cell is decoded by multiplying it by a suitably lagged replica of the entire, very long code. Developed for observations of the highly overspread ionosphere, this "coded-long-pulse" or "random-code" waveform redistributes delay-aliased echo power into an additive white-noise background. The SNR is reduced accordingly, but this penalty is acceptable for strong targets.

3. Radar Measurements and Target Properties

3.1 Albedo and Polarization Ratio

A primary goal of the initial radar investigation of any planetary target is estimation of the target's radar cross section, σ, and its normalized radar cross section or "radar albedo," $\eta = \sigma/A_P$, where A_P is the target's geometric projected area. Since the radar astronomer selects the transmitted and received polarizations, any estimate of σ or η must be identified accordingly. The most common approach is to transmit a circularly polarized wave and then to use separate receiving systems for simultaneous reception of the same sense of circular polarization as transmitted (i.e., the SC sense) and the opposite (OC) sense. The handedness of a circularly polarized wave is reversed on normal reflection from a smooth dielectric interface, so the OC sense dominates echoes from targets that look smooth at the radar wavelength. In this context, a surface with minimum radius of curvature very much larger than λ would "look smooth." SC echo power can arise from single scattering from rough surfaces, multiple scattering from smooth surfaces or subsurface heterogeneities (e.g., particles or voids), or certain subsurface refraction effects. The **circular polarization ratio**, $\mu_C = \sigma_{SC}/\sigma_{OC}$, is thus a useful measure of near-surface structural complexity or "roughness." When linear polarizations are used, it is convenient to define the ratio $\mu_L = \sigma_{OL}/\sigma_{SL}$, which would be close to zero for normal reflection from a smooth dielectric interface. For all radar-detected planetary targets, $\mu_L < 1$ and $\mu_L < \mu_C$. Although the OC radar albedo, η_{OC}, is the most widely used gauge of radar reflectivity, some radar measurements are reported in terms of the total power (OC + SC = OL + SL) radar

albedo η_T, which is four times the geometric albedo used in optical planetary astronomy. A smooth metallic sphere would have $\eta_{OC} = \eta_{SL} = 1$, a geometric albedo of 0.25, and $\mu_C = \mu_L = 0$.

If μ_C is close to zero (see Table 2), its physical interpretation is unique, as the surface must be smooth at all scales within about an order of magnitude of λ and there can be no subsurface structure at those scales within several $1/e$ power absorption lengths, L, of the surface proper. In this special situation, we may interpret the radar albedo as the product $g\rho$, where ρ is the Fresnel power-reflection coefficient at normal incidence and the backscatter gain g depends on target shape, the distribution of surface slopes with respect to that shape, and target orientation. For most applications to date, g is between 1.0 and 1.1, so the radar albedo provides a reasonable first approximation to ρ. Both ρ and L depend on interesting characteristics of the surface material, including bulk density, porosity, particle size distribution, and metal abundance.

If μ_C is as large as $\sim \frac{1}{3}$ (e.g., Mars and typical near-Earth asteroids), then much of the echo arises from some backscattering mechanism other than single, coherent reflections from large, smooth surface elements. Possibilities include multiple scattering from buried rocks or from the interiors of concave surface features such as craters, or reflections from very jagged surfaces with radii of curvature much less than a wavelength. Most planetary targets have values of μ_C of only a few tenths, so their surfaces are dominated by a component that is smooth at centimeter to meter scales.

3.2 Dynamical Properties from Delay–Doppler Measurements

Consider radar observation of a point target a distance R from the radar. As noted earlier, the "roundtrip time delay" between transmission of a pulse toward the target and reception of the echo would be $\tau = 2\,R/c$. It is possible to measure time delays to within 10^{-7} s with standard planetary radar setups. Actual delays encountered range from about 2.5 s for the Moon to about 2.5 hr for Saturn's rings and satellites. For a target distance of about one astronomical unit (AU), the time delay is about 1000 s and can be measured with a fractional timing uncertainty at least as fine as 10^{-9}, that is, with the same fractional precision as the definition of the speed of light.

Because the target is in motion and has a line-of-sight component of velocity toward the radar of V_{LOS}, the target will "see" a frequency that, to first order in V_{LOS}/c, equals $f_{TX}(1 + V_{LOS}/c)$, where f_{TX} is the transmitter frequency. The target reradiates the Doppler-shifted signal, and the radar receives an echo whose frequency is, again to first order, given by $f_{TX}(1 + 2V_{LOS}/c)$. That is, the total Doppler shift in the received echo is $V_{LOS}/(\lambda/2)$, so a 1-Hz Doppler shift corresponds to a velocity of half a wavelength per second (e.g., 6.3 cm sec^{-1} for $\lambda 12.6$ cm). It is not difficult to measure echo frequencies to within 0.01 Hz, so V_{LOS} can be estimated with a precision finer than 1 mm s^{-1}. Actual values of V_{LOS} for planetary radar targets can be as large as several tens of kilometers per second, so radar velocity measurements have fractional errors as low as 10^{-8}. At this level, the second-order (special relativistic) contribution to the Doppler shift becomes measurable; in fact, planetary radar observations provided the initial experimental verification of the second-order term.

By virtue of their high precision, radar measurements of time delay and Doppler frequency are very useful in refining our knowledge of various dynamical quantities. The first delay-resolved radar observations of Venus, during 1961–1962, yielded an estimate of the light-second equivalent of the astronomical unit that was accurate to one part in 10^6, constituting a thousandfold improvement in the best results achieved with optical observations alone. Subsequent radar observations provided additional refinements of about two more orders of magnitude. In addition to determining the scale of the solar system precisely, these observations greatly improved our knowledge of the orbits of Earth, Venus, Mercury, and Mars, and were essential for the success of the first interplanetary missions. Radar observations contribute to maintaining the accuracy of planetary ephemerides for objects in the inner solar system, and have been useful in dynamical studies of Jupiter's Galilean satellites. For newly discovered near-Earth asteroids, whose orbits must be estimated from optical astrometry that spans short arcs, a few radar observations can mean the difference between being able to find the object during its next close approach and losing it entirely.

Precise interplanetary time-delay measurements have allowed increasingly decisive tests of physical theories for light, gravitational fields, and their interactions with matter and each other. For example, radar observations verify general relativity's prediction that for radar waves passing nearby the Sun, echo time delays are increased because of the distortion of space by the Sun's gravity. The extra delay is ∼100 μs if the angular separation of the target from the Sun is several degrees. (The Sun's angular diameter is about half a degree.) Because planets are not point targets, their echoes are dispersed in delay and Doppler, and the refinement of dynamical quantities and the testing of physical theories are tightly coupled to estimation of the mean radii, the topographic relief, and the radar-scattering behavior of the targets. The key to this entire process is resolution of the distributions of echo power in delay and Doppler. In the next section, we will consider inferences about a target's dimensions and spin vector from measurements of the dispersions $\Delta\tau_{TARGET}$ and $\Delta\nu_{TARGET}$ of the echo in delay and Doppler. Then we will examine the physical information

contained in the functional forms of the distributions $\sigma(\tau)$, $\sigma(\nu)$, and $\sigma(\tau,\nu)$.

3.3 Dispersion of Echo Power in Delay and Doppler

Each backscattering element on a target's surface returns echo with a certain time delay and Doppler frequency (see Fig. 5). Since parallax effects and the curvature of the incident wave front are negligible for most ground-based observations (but not necessarily for observations with spacecraft), contours of constant delay are intersections of the surface with planes perpendicular to the line of sight. The point on the surface with the shortest echo time delay is called the subradar point; the longest delays generally correspond to echoes from the planetary limbs. The difference between these extreme delays is called the dispersion, $\Delta\tau_{TARGET}$, in $\sigma(\tau)$ or simply the "delay depth" of the target.

If the target appears to be rotating, the echo will be dispersed in Doppler frequency. For example, if the radar has an equatorial view of a spherical target with diameter D and apparent rotation period P, then the difference between the line-of-sight velocities of points on the equator at the approaching and receding limbs would be $2\pi D/P$. Thus, the dispersion of $\sigma(\nu)$ would be $\Delta\nu_{TARGET} = 4\pi D/\lambda P$. This quantity is called the bandwidth, B, of the echo power spectrum. If the view is not equatorial, the bandwidth is simply $(4\pi D \sin\alpha)/\lambda P$, where α is the "aspect angle" between the instantaneous spin vector and the line of sight. Thus, a radar bandwidth measurement furnishes a joint constraint on the target's size, rotation period, and pole direction.

In principle, **echo bandwidth** measurements obtained for a sufficiently wide variety of directions can yield all three scalar coordinates of the target's intrinsic (i.e., sidereal) spin vector $\mathbf{W}$. This capability follows from the fact that the apparent (synodic) spin vector $\mathbf{W}_{APP}$ is the vector sum of $\mathbf{W}$ and the contribution ($\mathbf{W}_{SKY} = (d\mathbf{e}/dt) \times \mathbf{e}$, where the unit vector $\mathbf{e}$ points from the target to the radar) from the target's plane-of-sky motion. Variations in $\mathbf{e}$, $d\mathbf{e}/dt$, and hence $\mathbf{W}_{SKY}$, all of which are known, lead to measurement of different values of $\mathbf{W}_{APP} = \mathbf{W} + \mathbf{W}_{SKY}$, permitting unique determination of all three scalar components of $\mathbf{W}$.

What if the target is not a sphere but instead is irregular and nonconvex? In this situation, which is most applicable to small asteroids and cometary nuclei, the relationship between the echo power spectrum and the target's shape is shown in Fig. 6. We must interpret D as the sum of the distances r_+ and r_- from the plane ψ_0 containing the line of sight and the spin vector to the surface elements with the greatest positive (approaching) and negative (receding) line-of-sight velocities. In different words, if the planes ψ_+ and ψ_- are defined as being parallel to ψ_0 and tangent to the target's approaching and receding limbs, then ψ_+ and ψ_- are at distances r_+ and r_- from ψ_0. Letting f_0, f_+, and

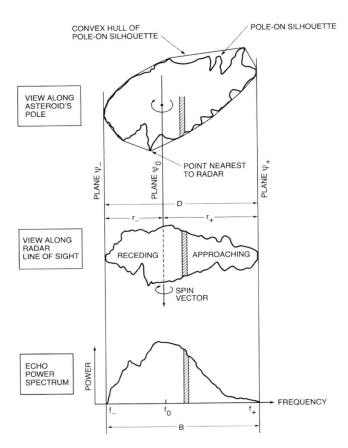

FIGURE 6 Geometric relations between an irregular, nonconvex rotating asteroid and its echo power spectrum. The plane ψ_0 contains the asteroid's spin vector and the asteroid-radar line. The crosshatched strip of power in the spectrum corresponds to echoes from the cross-hatched strip on the asteroid.

f_- be the frequencies of echoes from portions of the target intersecting ψ_0, ψ_+, and ψ_-, we have $B = f_+ - f_-$. Note that f_0 is the Doppler frequency of hypothetical echoes from the target's center of mass and that any constant-Doppler contour lies in a plane parallel to ψ_0.

It is useful to imagine looking along the target's pole at the target's projected shape, that is, its pole-on silhouette S. D is simply the width, or "breadth," of this silhouette (or, equivalently, of the silhouette's convex envelope or "hull," H) measured normal to the line of sight (see Fig. 6). In general, r_+ and r_- are periodic functions of rotation phase ϕ and depend on the shape of H as well as on the projected location of the target's center of mass, about which H rotates. If the radar data thoroughly sample at least $180°$ of rotational phase, then in principle one can determine $f_+(\phi)$ and $f_-(\phi)$ completely, and can recover H as well as the astrometrically useful quantity f_0. For many small, near-Earth asteroids, pronounced variations in $B(\phi)$ reveal highly noncircular pole-on silhouettes (see Fig. 7 and Section 3.12).

3.4 Rotations of Mercury and Venus

The principles described previously were applied in the early 1960s to yield the first accurate determination of the rotations of Venus and Mercury (Fig. 8). Results in both cases were completely unexpected. Venus' rotation is retrograde with a 243-day sidereal period that is close to the value (243.16 days) characterizing a resonance with the relative orbits of Earth and Venus, wherein Venus would appear from Earth to rotate exactly four times between successive inferior conjunctions with the Sun. However, two decades of ground-based observations and ultimately images obtained by the *Magellan* spacecraft have conclusively demonstrated nonresonant rotation: The average period computed from reliable published measurements is $243.0185 + 0.0001$ days.

For Mercury, long imagined on the basis of optical observations to rotate once per 88-day revolution around the Sun, radar bandwidth measurements (see Fig. 8) demonstrated direct rotation with a period (59 days) equal to two-thirds of the orbital period. In 1976, Peale showed that the measurement of the obliquity of the planet, the amplitude of its longitude librations, and the second-degree gravitational harmonics are sufficient to determine the size and state of Mercury's core. A new radar technique called radar speckle displacement (RSD) is attempting to measure Mercury's spin state with unprecedented accuracy. The planet is illuminated with a monochromatic (cw) transmission from Goldstone, and echo is received at Goldstone and the Greenbank Telescope (GBT) in West Virginia. Cross-correlation of the time series from the two sites directly constrains the instantaneous spin rate and orientation, permitting measurement of spin rate variations to 1 part in 10^5, one-tenth the size of the spin rate signature expected from longitude librations.

3.5 Topography on the Moon and Inner Planets

For the Moon, Mercury, Mars, and Venus, topography along the subradar track superimposes a modulation on the echo delay above or below that predicted by ephemerides, which generally are calculated for a sphere with the object's assumed mean radius. Prior to spacecraft exploration of these objects, there were radar-detectable errors in the radii estimates as well as in the target's predicted orbit. These circumstances required that an extended series of measurements of the time delay of the echo's leading edge be folded into a computer program designed to estimate simultaneously parameters describing the target's orbit, mean radius, and topography. These programs also contain parameters from models of wave propagation through the interplanetary medium or the solar corona, as well as parameters used to test general relativity, as noted earlier.

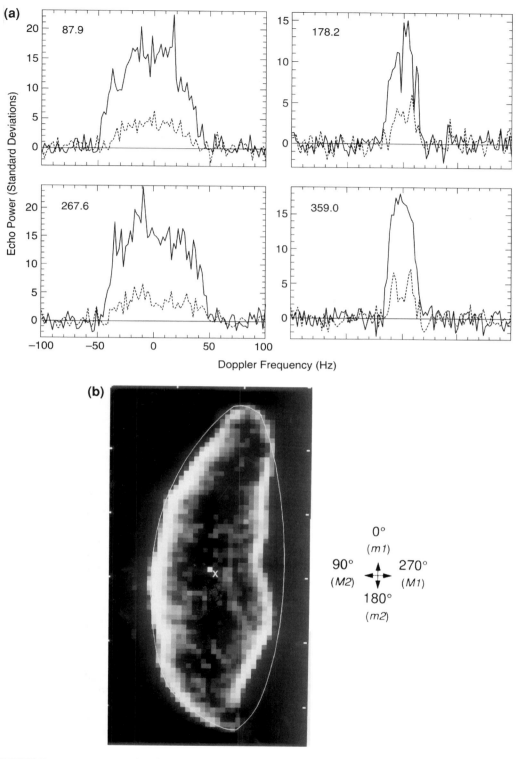

FIGURE 7 Constraints on the shape of near-Earth asteroid 1620 Geographos from Goldstone 3.5-cm radar echoes. (a) Spectra obtained at phases of bandwidth extrema. OC (solid curve) and SC (dotted curve) echo power is plotted versus Doppler frequency. (b) Comparison of an estimate of the hull (H) on the asteroid's pole-on silhouette (S) with an estimate of S itself. The white curve is the cw estimate of H and the X marks the projected position of the asteroid's center of mass (COM) with respect to H. That curve and the X are superposed on an estimate of S from delay–Doppler images; the bright pixel is the projection of the COM determined from analysis of those images. The absolute scales and relative rotational orientations of the two figures are known: Border ticks are 1 km apart. The offset between the X and the bright pixel is a measure of the uncertainty in our knowledge of the COM's delay–Doppler trajectory during the experiment. In the diagram at right, the arrows point to the observer at phases of lightcurve maxima ($M1$, $M2$) and minima ($m1$, $m2$). (From S. J. Ostro et al., 1996, *Icarus* **121**, 46–66.)

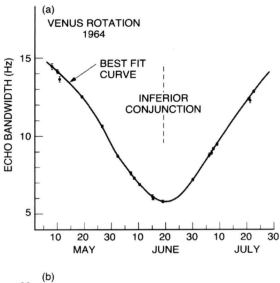

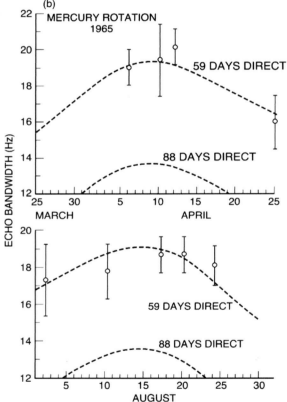

FIGURE 8 Measurements of echo bandwidth (i.e., the dispersion of echo power in Doppler frequency) used to determine the rotations of (a) Venus and (b) Mercury. (From R. B. Dyce, G. H. Pettengill, and I. I. Shapiro, 1967, *Astron. J.* **72**, 351–359.)

Radar has been used to obtain topographic profiles across the Moon and the inner planets. For example, Fig. 9 shows a three-dimensional reconstruction of topography derived from altimetric profiles obtained for Mars in the vicinity of the giant shield volcano Arsia Mons. The altimet-

ric resolution of the profiles is about 150 m (1 μs in delay), but the surface resolution, or footprint, is very coarse (~75 km). The *Magellan* radar altimeter, with a footprint typically 20 km across and vertical resolution on the order of tens of meters, has produced detailed topographic maps of most of Venus, and the *Cassini* radar has revealed an intriguing lack of topographic relief on Titan.

3.6 Angular Scattering Law

The functional forms of the distributions $\sigma(\tau)$ and $\sigma(\nu)$ contain information about the radar-scattering process and the target's surface. Suppose the target is a large, smooth sphere. Then echoes from the subradar region (near the center of the visible disk; see Fig. 5), where the surface elements are nearly perpendicular to the line of sight, would be much stronger than those from the limb regions (near the disk's periphery). This effect is seen visually when one shines a flashlight on a smooth, shiny ball—a bright glint appears where the geometry is right for backscattering. If the ball is roughened, the glint is spread out over a wider area and, in the case of extreme roughness, the scattering would be described as "diffuse" rather than "specular."

For a specular target, $\sigma(\tau)$ would have a steep leading edge followed by a rapid drop. The power spectrum $\sigma(\nu)$ would be sharply peaked at central frequencies, falling off rapidly toward the spectral edges. If, instead, the spectrum were very broad, severe roughness at some scale(s) comparable to or larger than λ would be indicated. In this case, knowledge of the echo's polarization properties would help to ascertain the roughness scale(s) responsible for the absence of the sharply peaked spectral signature of specular scattering.

By inverting the delay or Doppler distribution of echo power, one can estimate the target's average angular **scattering law**, $\sigma_0(\theta) = d\sigma/dA$, where dA is an element of surface area and θ is the "incidence angle" between the line of sight and the normal to dA. For the portion of the echo's "polarized" (i.e., OC or SL) component that is specularly scattered, $\sigma_0(\theta)$ can be related to statistics describing the probability distribution for the slopes of surface elements. Examples of scattering laws applied in planetary radar astronomy are the Hagfors law,

$$\sigma_0(\theta) \sim C(\cos^4 \theta + C \sin^2 \theta)^{-3/2} \quad (5)$$

the Gaussian law,

$$\sigma_0(\theta) \sim [C \exp(-C \tan^2 \theta)]/\cos^4 \theta \quad (6)$$

and the Cosine law,

$$\sigma_0(\theta) \sim (C + 1) \cos^{2C} \theta \quad (7)$$

where $C^{-1/2} = S_0 = <\tan^2 \theta>^{1/2}$ is the adirectional rms slope.

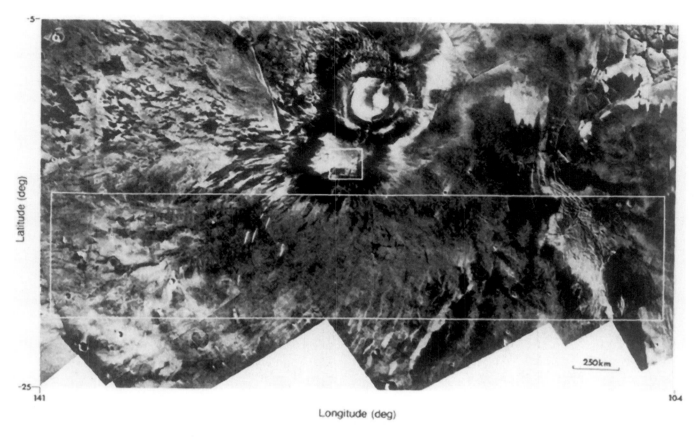

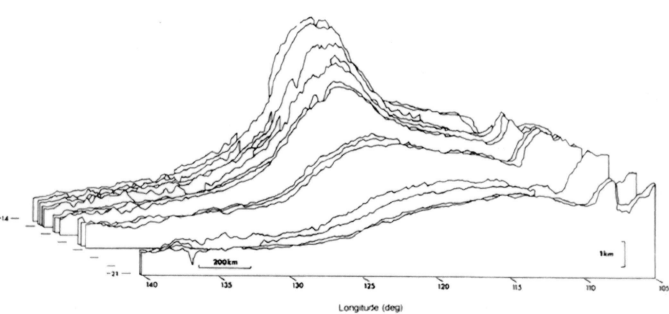

FIGURE 9   Topographic contours for the southern flank (large rectangle) of the Martian shield volcano Arsia Mons, obtained from radar altimetry. (From L. Roth, G. S. Downs, R. S. Saunders, and G. Schubert, 1980, *Icarus* **42**, 287–316.)

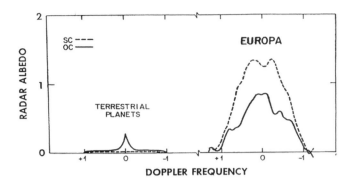

FIGURE 10 Typical 13-cm echo spectra for the terrestrial planets are compared to echo spectra for Jupiter's icy moon Europa. The abscissa has units of half the echo bandwidth.

Echoes from the Moon, Mercury, Venus, and Mars are characterized by sharply peaked OC echo spectra (Fig. 10). Although these objects are collectively referred to as "quasi-specular" radar targets, their echoes also contain a diffusely scattered component and have full-disk circular polarization ratios averaging about 0.07 for the Moon, Mercury, and Venus, but ranging from 0.1 to 0.4 for Mars, as discussed next.

Typical rms slopes obtained at decimeter wavelengths for these four quasi-specular targets are around $7°$ and consequently these objects' surfaces have been described as "gently undulating." As might be expected, values estimated for S_0 increase as the observing wavelength decreases. For instance, estimates of S_0 for the Moon increase from $\sim 4°$ at

20 m to $\sim 8°$ at 10 cm, to $\sim 33°$ at 1 cm. At visual wavelengths, the Moon shows no trace of a central glint, that is, the scattering is entirely diffuse. This phenomenon arises because the lunar surface (Fig. 11) consists of a regolith (an unconsolidated layer of fine-grained particles) with much intricate structure at the scale of visible wavelengths. At decimeter wavelengths, the ratio of diffusely scattered power to quasi-specularly scattered power is about one-third for the Moon, Mercury, and Venus, but two to three times higher for Mars. This ratio can be determined by assuming that all the SC echo is diffuse and then calculating the diffusely scattered fraction (x) of OC echo by fitting to the OC spectrum a model based on a "composite" scattering law, for example, $S_0(\theta) = x\sigma_{DIF}(\theta) + (1-x)\sigma_{QS}(\theta)$. Here $\sigma_{QS}(\theta)$ might be the Hagfors law and usually $\sigma_{DIF}(\theta) = \cos^m \theta$; when this is done, estimated values of m usually fall between 1 (geometric scattering, which describes the visual appearance of the full Moon) and 2 (Lambert scattering).

For the large, spheroidal asteroids 1 Ceres and 2 Pallas (see Section 3.12), the closeness of μ_C to zero indicates quasi-specular scattering, but the OC spectra, rather than being sharply peaked, are fit quite well using a Cosine law with C between 2 and 3, or a Gaussian law with C between 3 and 5, and here we can interpret the diffuse echo as due to the distribution of surface slopes, with $S_0 > 20°$. OC echo spectra obtained from asteroid 4 Vesta and Jupiter's satellite Io have similar shapes, but these objects' substantial polarization ratios ($\mu_C \sim 0.3$ and 0.5, respectively) suggest that small-scale roughness is at least partially responsible for the diffuse echoes. Circular polarization ratios between 0.5 and

FIGURE 11 Structure on the lunar surface near the *Apollo 17* landing site. Most of the surface is smooth and gently undulating at scales much larger than a centimeter. This smooth component of the surface is responsible for the predominantly quasi-specular character of the Moon's radar echo at $\lambda \gg 1$ cm. Wavelength-scale structure produces a diffuse contribution to the echo. Wavelength-sized rocks are much more abundant at $\lambda \sim 4$ cm than at $\lambda \sim 10$ m (the scale of the boulder being inspected by astronaut H. Schmitt), and hence diffuse echo is more substantial at shorter wavelengths.

FIGURE 12 This lava flow near Sunset Crater in Arizona is an example of an extremely rough surface at decimeter scales and is similar to terrestrial flows yielding large circular polarization ratios at decimeter wavelengths.

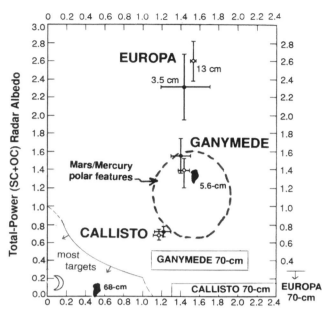

FIGURE 13 Radar properties of Europa, Ganymede, and Callisto compared to those of some other targets. The icy Galilean satellites' total-power radar albedos do not depend on wavelength between 3.5 and 13 cm, but plummet at 70 cm. Solid symbols shaped like Greenland indicate properties of that island's percolation zone at 5.6 and 68 cm. The domain of most of the bright polar features on Mars and Mercury is sketched.

1.0 have been measured for several asteroids (see Table 2) and parts of Mars and Venus, implying extreme decimeter-scale roughness, perhaps analogous to terrestrial lava flows (Fig. 12). Physical interpretations of the diffusely scattered echo employ information about albedo, scattering law, and polarization to constrain the size distributions, spatial densities, and electrical properties of wavelength-scale rocks near the surface, occasionally using the same theory of multiple light scattering applied to radiative transfer problems in other astrophysical contexts.

3.7 Jupiter's Icy Galilean Satellites

Europa, Ganymede, and Callisto have extraordinary 3.5- and 13-cm radar properties. Their reflectivities are enormous compared with those of the Moon and inner planets (see Table 2); Europa is the extreme example (Fig. 10), with an OC radar albedo (1.0) as high as that of a metal sphere. Since the radar and visual albedos and estimates of fractional water frost coverage increase by satellite in the order Callisto–Ganymede–Europa, the presence of water ice has long been understood to be somehow responsible for the unusually high reflectivities even though ice is less radar-reflective than silicates. In spite of the satellites' smooth appearances in *Voyager* and *Galileo* high-resolution images, a diffuse scattering process and hence a high degree of near-surface structure at centimeter to meter scales is indicated by broad spectral shapes and large linear polarization ratios ($\mu_L \sim 0.5$).

The most peculiar aspect of the satellites' echoes is their circular polarization ratios, which exceed unity. That is, in contrast to the situation with other planetary targets, the scattering largely preserves the handedness, or helicity, of the transmission. Mean values of μ_C for Europa,

Ganymede, and Callisto are about 1.5, 1.4, and 1.2, respectively. Wavelength dependence is negligible from 3.5 to 13 cm, but dramatic from 13 to 70 cm (Fig. 13). Significant polarization and/or albedo features are present in the echo spectra and in a few cases correspond to geologic features in *Voyager* and *Galileo* images.

The icy satellites' echoes are due not to external surface reflections but to subsurface "volume" scattering. The high radar transparency of ice compared with that of silicates permits deeper radar sounding, longer photon path lengths, and higher-order scattering from regolith heterogeneities; radar is seeing Europa, Ganymede, and Callisto in a manner that the Moon has never been seen. The satellites' radar behavior apparently involves the coherent backscatter effect, which accompanies any multiple-scattering process; occurs for particles of any size, shape, and refractive index; and was first discovered in laboratory studies of the scattering of electrons and of light. Coherent backscatter yields strong echoes and $\mu_C > 1$ because the incident, circularly polarized wave's direction is randomized before its helicity is randomized and also before its power is absorbed and because photons traveling along identical paths but in opposite directions interfere constructively. The vector-wave theory of coherent backscatter accounts for the unusual radar signatures in terms of high-order, multiple anisotropic scattering from within the upper few decameters of the regoliths, which the radar sees as an extremely low-loss, disordered random medium. Inter- and intrasatellite albedo variations probably are due to variation in ice purity.

As sketched in Fig. 13, there are similarities between the icy Galilean satellites' radar properties and those of the radar-bright polar caps on Mars, features inside perpetually shadowed craters at the poles of Mercury (see Section 3.9), and the percolation zone in the Greenland ice sheet. However, the subsurface configuration in the Greenland zone, where the scattering heterogeneities are "ice pipes" produced by seasonal melting and refreezing, are unlikely to resemble those on the satellites. Therefore, unique models of subsurface structure cannot be deduced from the radar signatures of any of these terrains.

3.8 Radar Mapping of Spherical Targets

The term "radar image" usually refers to a measured distribution of echo power in delay, Doppler, and/or up to two angular coordinates. The term "radar map" usually refers to a display in suitable target-centered coordinates of the residuals with respect to a model that parameterizes the target's size, shape, rotation, average scattering properties, and possibly its motion with respect to the delay–Doppler ephemerides. Knowledge of the dimensions of the Moon and inner planets has long permitted conversion of radar images to maps of these targets. For small asteroids, the primary use of images is to constrain the target's shape (see Section 3.12).

As illustrated in Fig. 5, intersections between constant-delay contours and constant-Doppler contours on a sphere constitute a "two-to-one" transformation from the target's surface to delay–Doppler space. For any point in the northern hemisphere, there is a conjugate point in the southern hemisphere at the same delay and Doppler. Therefore, the source of echo in any delay–Doppler resolution cell can be located only to within a twofold ambiguity. This north–south ambiguity can be avoided completely if the radar beamwidth (~2 arcmin for Arecibo at 13 cm or Goldstone at 3.5 cm) is comparable to or smaller than the target's apparent angular radius, as in the case of observations of the Moon (angular radius ~ 15 arcmin). Similarly, no such ambiguity arises in the case of side-looking radar observations from spacecraft (e.g., *Magellan* or *Cassini*) for which the geometry of delay–Doppler surface contours differs somewhat from that in Fig. 5. For ground-based observations of Venus and Mercury, whose angular radii never exceed a few tens of arcseconds, the separation of conjugate points is achievable by either offsetting the pointing to place a null of the illumination pattern on the undesired hemisphere or interferometrically, using two receiving antennas, as follows.

The echo waveform received at either antenna from one conjugate point will be highly correlated with the echo waveform received at the other antenna from the same conjugate point. However, echo waveforms from the two conjugate points will be largely uncorrelated with each other, no matter where they are received. Thus, echoes from two conjugate points can, in principle, be distinguished by cross-correlating echoes received at the two antennas with themselves and with each other, and performing algebraic manipulations on long time averages of the cross product and the two self products.

The echo waveform from a single conjugate point will experience slightly different delays in reaching the two antennas, so there will be a phase difference between the two received signals, and this phase difference will depend only on the geometrical positions of the antennas and the target. This geometry will change as the Earth rotates, but it will change very slowly and in a predictable manner. The antennas are best positioned so contours of constant phase difference on the target disk are as orthogonal as possible to the constant-Doppler contours, which connect conjugate points. Phase difference hence becomes a measure of north–south position, and echoes from conjugate points can be distinguished on the basis of their phase relation.

The total number of "fringes," or cycles of phase shift, spanned by the disk of a planet with diameter D and a distance R from the radar is approximately $(D/R)(b_{PROJ}/\lambda)$, where b_{PROJ} is the projection of the interferometer baseline normal to the mean line of sight. For example, Arecibo interferometry linked the main antenna to a 30.5-m antenna about 11 km farther north. It placed about seven fringes on Venus, quite adequate for separation of the north–south ambiguity. The Goldstone main antenna has been linked to smaller antennas to perform multielement interferometry, which permit one to solve so precisely for the north–south location of a given conjugate region that one can obtain the region's elevation relative to the mean planetary radius. Goldstone interferometry of the Moon's polar regions has produced both topographic maps and backscatter maps (Fig. 14) using somewhat more advanced radar techniques.

In constructing a radar map, the unambiguous delay–Doppler distribution of echo power is transformed to planetocentric coordinates, and a model is fit to the data, using a maximum-likelihood or weighted-least-squares estimator. The model contains parameters for quasi-specular and diffuse scattering as well as prior information about the target's dimensions and spin vector. For Venus, effects of the dense atmosphere on radar wave propagation must also be modeled. Residuals between the data and the best-fit model constitute a radar reflectivity map of the planet (e.g., Fig. 15). Reflectivity variations can be caused by many different physical phenomena, and their proper interpretation demands due attention to the radar wavelength, echo polarization, viewing geometry, prior knowledge about surface properties, and the nature of the target's mean scattering

(a)

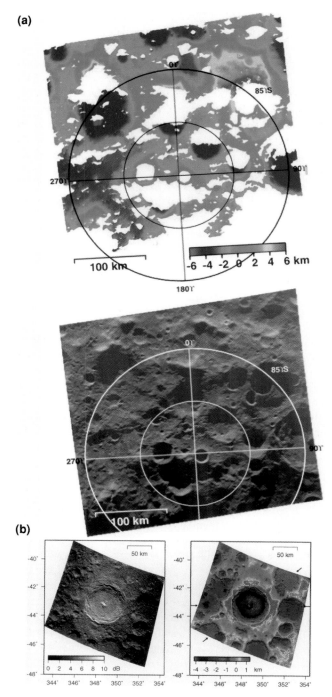

(b)

FIGURE 14   Radar backscatter and digital elevation models of regions on the Moon from Goldstone 3.5-cm, OC, tristatic and quadristatic interferometry. (a) The south polar region. The radar results establish that the interiors of many of the Moon's polar craters are in permanent shadow from solar illumination. (Reprinted with permission from J. M. Margot, D. B. Campbell, R. F. Jurgens, and M. A. Slade, 1999, *Science* **284**, 1658–1660, copyright 1999 American Association for the Advancement of Science.) (b) The crater Tycho. (J. L. Margot, D. B. Campbell, R. F. Jurgens, and M. A. Slade, 1999, *J. Geophys. Res.* **104**, E5, 11875–11882.)

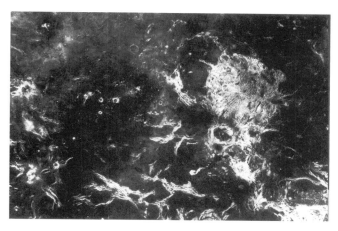

FIGURE 15   Arecibo 13-cm OC delay–Doppler image of Venus. In the middle of the right half of the image is the bright, 1200-km-wide Alpha Regio, a complex of intersecting ridges. Just south of Alpha is the 300-km-diameter circular feature Eve. The three prominent craters in the middle of the left half of the image are seen close-up in Fig. 17d. Courtesy of D. B. Campbell.

behavior. For example, subsurface scattering of an incident circularly polarized signal results in a linearly polarized component in the radar echo due to the differing transmission coefficients at a smooth surface boundary for the horizontally and vertically polarized components of the incident wave. A linearly polarized component in 70-cm echoes from certain topographic features on Venus has been attributed to subsurface echoes from a mantled substrate or from buried rocks.

3.9 Radar Evidence for Ice Deposits at Mercury's Poles

The first full-disk (Goldstone–VLA) radar portraits of Mercury surprisingly showed bright polar features with $\mu_C > 1$, and subsequent delay–Doppler imagery from Arecibo established that the anomalous echoes originate from interiors of craters that are perpetually shaded from sunlight because of Mercury's near-zero obliquity (see Fig. 16). The angle between the orbital planes of Mercury and Earth is 7°, so portions of the permanently shadowed regions are visible to Earth-based radars. At each pole, bright radar features correlate exactly with craters seen in *Mariner 10* images; numerous features also lie in the hemisphere not imaged by that spacecraft.

Similarities between the radar scattering properties of the Mars and Mercury polar anomalies and those of the icy Galilean satellites (see Section 3.7) support the inference that the radar anomalies are deposits of water ice. Temperatures below 120 K in the permanent shadows are expected and are low enough for ice to be stable against sublimation for billions of years. Temperatures several tens of kelvins lower may exist inside high-latitude craters and perhaps

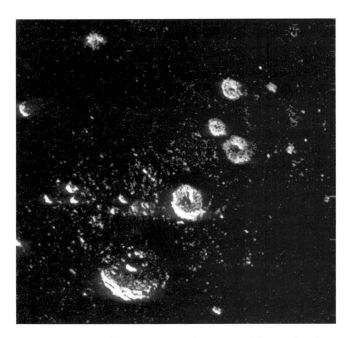

FIGURE 16   Arecibo 13-cm, SC radar image of the north polar region of Mercury. The resolution is 1.5 km, and the image is 395 km wide. The bright features are thought to be ice deposits on permanently shadowed crater floors. (Harmon, J. K., Perillat, P. J., and Slade, M. A., 2001, *Icarus* **149**, 1–15.)

also beneath at least 10 cm of visually bright regolith. Detection of the north polar features at 70-cm wavelength indicates that the deposits may be at least several meters thick. Plausible sources of water on Mercury include comet impacts and outgassing from the interior. It has been noted that most water vapor near the surface is photodissociated, but that some molecules will random-walk to polar cold traps. Ices of other volatiles, including CO_2, NH_3, HCN, and SO_2, might also be present.

There are perpetually shadowed craters at the Moon's poles (Fig. 14), but no convincing radar evidence has been found for ice there. If ice exists on the Moon, it is likely to have low concentrations in the soil.

3.10 Venus Revealed by *Magellan*

The *Magellan* spacecraft entered Venus orbit in August 1990 and during the next two years explored the planet with a single scientific instrument operating as a 13-cm radar imager, altimeter, and thermal radiometer. *Magellan*'s imaging resolution (~100 m) and altimetric resolution (5 to 100 m) improved upon the best previous spacecraft and ground-based measurements by an order of magnitude, and did so with nearly global coverage.

Venus' surface contains a plethora of diverse tectonic and impact features, but its formation and evolution have clearly been dominated by widespread volcanism, whose

legacy includes pervasive volcanic planes, thousands of tiny shield volcanoes, monumental edifices, sinuous lava flow channels, pyroclastic deposits, and pancake-like domes. The superposition of volcanic signatures and elaborate, complex tectonic forms records a history of episodic crustal deformation. The paucity of impact craters smaller than 25 km and the lack of any as small as a few kilometers attests to the protective effect of the dense atmosphere. The multilobed, asymmetrical appearance of many large craters presumably results from atmospheric breakup of projectiles before impact. Atmospheric entrainment and transport of ejecta are evident in very elongated ejecta blankets. Numerous craters are surrounded by radar-dark zones, perhaps the outcome of atmospheric pressure-wave pulverization and elevation of surface material that upon resettling deposited a tenuous and hence unreflective "impact regolith." Figure 17 shows examples of *Magellan* radar images.

3.11 The Radar Heterogeneity of Mars

Ground-based investigations of Mars have achieved more global coverage than those of the other terrestrial targets because the motion in longitude of the subradar point on Mars (whose rotation period is only 24.6 hours) is rapid compared to that on the Moon, Venus, or Mercury, and because the geometry of Mars's orbit and spin vector permits subradar tracks throughout the Martian tropics. The existing body of Mars radar data reveals extraordinary diversity in the degree of small-scale roughness as well as in the rms slope of smooth surface elements. Slopes on Mars have rms values from less than 0.5° to more than 10°. Radar slope estimates, polarization-ratio estimates, and/or multi-station interferometric images have been used in selection of *Viking Lander*, *Mars Pathfinder*, and *Mars Exploration Rover* landing sites (Fig. 18).

Diffuse scattering from Mars is much more substantial than for the other quasi-specular targets, and often accounts for most of the echo power; therefore, the average near-surface abundance of centimeter-to-meter-scale rocks is much greater on Mars than on the Moon, Mercury, or Venus. Features in Mars SC spectra first revealed the existence of regions of extremely small-scale roughness, and the trajectory of these features' Doppler positions versus rotation phase suggested that their primary sources are the Tharsis and Elysium volcanic regions. The best terrestrial analog for this extremely rough terrain might be young lava flows. Goldstone–VLA images of Mars at longitudes that cover the Tharsis volcanic region confirmed that this area is the predominant source of strong SC echoes and that localized features are associated with individual volcanoes. A 2000-km-long band with an extremely low albedo cuts across Tharsis; the radar darkness of this "Stealth" feature probably arises from an under-dense, unconsolidated blanket of pyroclastic deposits ~1 m deep.

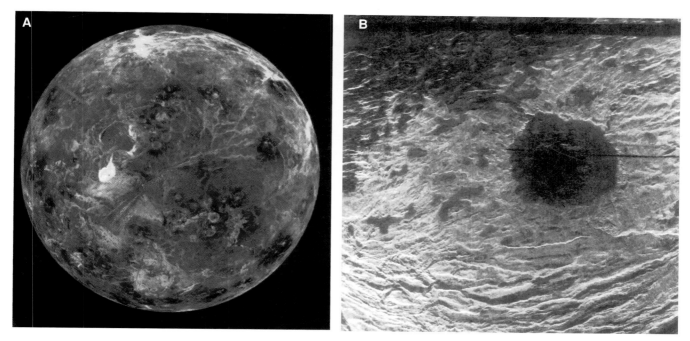

FIGURE 17   Magellan 13-cm, SL radar maps of Venus: (a) Northern-hemisphere projection of mosaics. The North Pole is at the center of the image, with 0° and 90° E. longitudes at the 12 and 9 o'clock positions. Gaps use *Pioneer Venus* data or interpolations. The bright, porkchop-shaped feature is Maxwell Montes, a tectonically produced mountain range first seen in ground-based images. (b) 120-m-resolution map of Cleopatra, a double-ringed impact basin on the eastern slopes of Maxwell Montes. The diameter of the outer ring is about 100 km. (c) Image of a 350-km wide portion of the Atla region of Venus' southern hemisphere showing several types of volcanic features criss-crossed by numerous superimposed, and hence more recent, surface fractures. Various flower-shaped patterns formed from linear fissures or lava flows emanate from circular pits. (d) Mosaic of part of Lavinia showing three large craters, with diameters ranging from 37 to 50 km, that were discovered in Arecibo images (Fig. 16). (e) Pancake-like, ~25-km-diameter, volcanic domes located southeast of Alpha Regio. (NASA/JPL.)

3.12 Asteroids

Radar has been established as the most powerful post-discovery, Earth-based technique for determining the physical properties and orbits of asteroids. As of mid 2006, echoes from 112 main belt asteroids (MBAs) and 194 near-Earth asteroids (NEAs), including 109 Potentially hazardous asteroids (PHAs; see discussion that follows) have provided a wealth of new information about these objects' sizes, shapes, spin vectors, and surface characteristics such as decimeter-scale roughness, topographic relief, regolith porosity, and metal concentration.

3.12.1 DISK-INTEGRATED PROPERTIES

The low polarization ratios and broad spectral shapes of some of the largest MBAs (e.g., 1 Ceres and 2 Pallas) reveal surfaces that are smoother than that of the Moon at decimeter scales but much rougher at some much larger scale. For some asteroids in the 200-km-diameter range (including 7 Iris, 9 Metis, and 654 Zelinda), brightness spikes within narrow ranges of the rotation phase suggest large, flat regions.

There is a 10-fold variation in asteroid radar albedos, implying substantial variations in these objects' surface porosities or metal concentrations, or both. The lowest MBA albedo estimate, 0.04 for Ceres, indicates a lower surface bulk density than that on the Moon. The highest MBA albedo estimates, 0.31 for 16 Psyche and 0.44 for Kleopatra, are consistent with metal concentrations near unity and lunar porosities. These objects might be the collisionally exposed interiors of differentiated asteroids and by far the largest pieces of refined metal in the solar system.

The radar albedo of the 2-km NEA 6178 (1986DA), 0.58, strongly suggests that it is a regolith-free metallic fragment, presumably derived from the interior of a much larger object that melted, differentiated, cooled, and subsequently was disrupted in a catastrophic collision. 1986 DA might be (or have been a part of) the parent body of some iron meteorites. At the other extreme, the range for 1986 JK's radar albedo (0.005 to 0.07) suggests a surface bulk density within a factor of 2 of 0.9 g cm^{-3}. Similarly, the distribution of NEA circular polarization ratios runs from near zero to near unity. The highest values, for 2101 Adonis, 1992QN, 3103 Eger, 3980 1980PA, 2000 EE104, and 2004

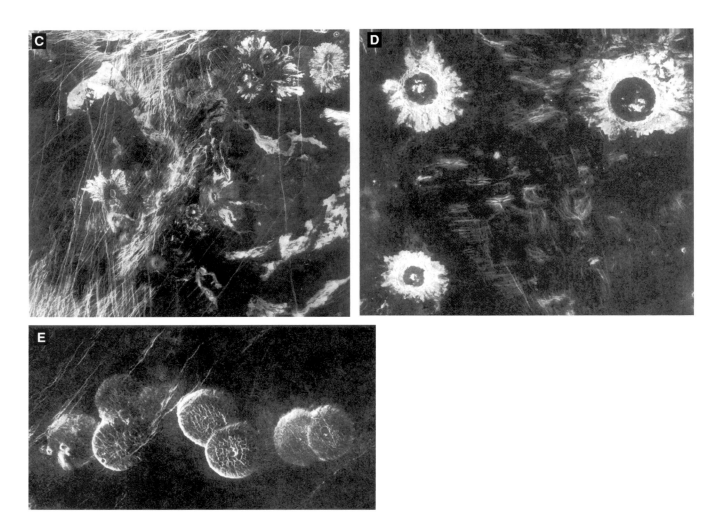

FIGURE 17 (*Continued*)

XP14, indicate extreme near-surface structural complexity, but we cannot distinguish between multiple scattering from subsurface heterogeneities (see Section 3.7) and single scattering from complex structure on the surface.

3.12.2 IMAGING AND SHAPE RECONSTRUCTION

During the past decade, delay–Doppler imaging of asteroids has produced spatial resolution as fine as a decameter. The images generally can be "north–south" ambiguous, that is, they constitute a two-to-one (or even many-to-one) mapping from the surface to the image. However, if the radar is not in the target's equatorial plane, then each surface point has a unique delay–Doppler trajectory as the target rotates. Hence images that provide adequate orientational coverage can be inverted, and in principle one can reconstruct the target's three-dimensional shape as well as its spin state.

The first asteroid radar data set suitable for reconstruction of the target's shape was a 2.5-hour sequence of 64 delay–Doppler images of 4769 Castalia, obtained 2 weeks after its August 1989 discovery. The images, which were taken at a subradar latitude of about 35°, show a bimodal distribution of echo power over the full range of sampled rotation phases, and least-squares estimation of Castalia's three-dimensional shape reveals it to consist of two kilometer-sized lobes in contact. Castalia was the first of several "contact binaries" revealed by radar.

If the radar view is equatorial, unique reconstruction of the asteroid's three-dimensional shape is ruled out, but a sequence of images that thoroughly samples rotation phase can allow unambiguous reconstruction of the asteroid's pole-on silhouette. For example, observations of 1620 Geographos yield several hundred images with ~100-m resolution. The pole-on silhouette's extreme dimensions are in a ratio, $2.76 + 0.21$, that establishes Geographos as the most elongated solar system object imaged so far (see Fig. 7b).

Delay–Doppler imaging of 4179 Toutatis in 1992 and 1996 achieved resolutions as fine as 125 ns (19 m in range)

Mars landing site 3.5 cm radar properties

Viking Lander 1
Chryse Planitia
$\theta_{rms}=4.7°\pm1.8°$ $\mu_c=0.60$

Mars Pathfinder
Ares Valles
$\theta_{rms}=4.7°\pm1.6°$ $\mu_c=0.43$

Spirit
Gusev Crater
$\theta_{rms}=1.6°\pm0.5°$ $\mu_c=0.61$

Opportunity
Meridiani Planum
$\theta_{rms}=1.4°\pm0.4°$ $\mu_c=0.50$

FIGURE 18   *Mars Lander* sites and estimates of 3.5-cm rms slope and circular polarization ratio. The width across the front of the image is about 3 m in the right column and about 2 m in the left column. (NASA/JPL, courtesy of A. F. Haldemann.)

and 8.3 mHz (0.15 mm s^{-1} in radial velocity), placing thousands of pixels on the asteroid. This data set provided physical and dynamical information that was unprecedented for an Earth-crossing object. Extraction of the information in this imaging data set required inversion with a much more comprehensive physical model than in the analysis of Castalia images; free parameters included the asteroid's

shape and inertia matrix, initial conditions for the asteroid's spin and orientation, the radar scattering properties of the surface, and the delay–Doppler trajectory of the center of mass (see Fig. 19).

Toutatis has complex linear features as well as circular crater-like structures down to the several-decameter resolution limit. The features suggest a complex interior

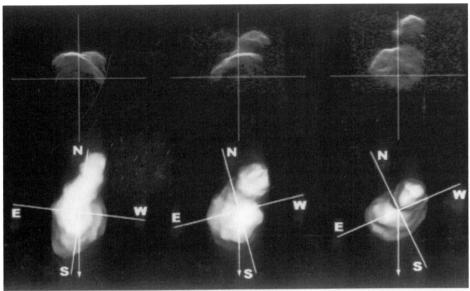

Dec. 1 - 33602   Dec. 2 - 33702   Dec. 3 - 33803

FIGURE 19 High-resolution Goldstone 3.5-cm, OC, delay–Doppler images from each of three observation dates in 1996 and the corresponding plane-of-sky (POS) appearance of the radar model. The crosshairs are 5 km long and centered on Toutatis' center of mass (COM). In the radar images, time delay (range) increases from top to bottom, and Doppler frequency (radial velocity) increases from left to right. The model is rendered with a Lambertian scattering law, with the viewer co-located with the illumination source. The crosshairs are aligned north–south and east–west on the plane of the sky. In each POS frame, the arrow radiating from the COM shows the POS projection of the instantaneous spin vector.

configuration involving monolithic fragments with various sizes and shapes, presumably due to collisions in various energy regimes. Toutatis might be an impact-sculpted, single, coherent body, or it might consist of two separate objects that came together in a gentle collision; the difference in the two lobes' gravitational slopes supports the latter idea (Fig. 20).

Toutatis is rotating in a long-axis non-principal-axis (NPA) spin state (see Fig. 21) characterized by periods of 5.4 days (rotation about the long axis) and 7.4 days (average for long-axis precession about the angular momentum vector). The asteroid's principal moments of inertia are in ratios within 1% of 3.22 and 3.09, and the inertia matrix is indistinguishable from that of a homogeneous body. Such information has yet to be determined for any other small body except the *NEAR-Shoemaker* target 433 Eros and probably is impossible to acquire in a fast spacecraft flyby.

Images of another NPA rotator, 53319 (1999 JM8), reveal an asymmetric, irregularly shaped, 7-km object (Fig. 22). The asteroid's rotation has a dominant 7-day period and is not far from uniform rotation. 1999 JM8 has pronounced topographic relief, prominent facets several kilometers in extent, and numerous crater-like features between ~100 m and 1.5 km in diameter.

Radar images of 6489 Golevka (1991 JX) reveal a half-kilometer object whose shape is extraordinarily angular, with flat sides, sharp edges and corners, and peculiar concavities. Extremely large gravitational slopes in some areas of the radar-derived model indicate the presence of exposed, solid, monolithic rock (Fig. 20). This asteroid, the first sub-kilometer object studied in this much detail, probably is a monolithic collision fragment rather than a rubble pile. Golevka was the target of the first intercontinental radar observations, in June 1995, when Goldstone provided a transmission and echoes were received by the Russian 70-m Evpatoria antenna and also by the Japanese 34-m Kashima antenna. The asteroid's name is made from leading letters of those antennas' names.

Radar has revealed numerous NEAs to have nearly circular pole-on silhouettes [e.g., 1999 RQ36, 7822 (1991 CS), 2100 Ra-Shalom, 1998 ML14 and 1998 FH12]. 1998 ML14 has isolated, several-hundred-meter protrusions on one side, while 1999 RQ36 has no noticeable features anywhere. At the opposite extreme, several NEAs, including 22771 (1999 CU3) and 2003 MS2 have elongated shapes with curious irregularities. Ironically, the dogbone shape of the 235-km-long main-belt object Kleopatra is the most exotic yet discerned by radar (Fig. 20).

Asteroids with visual absolute magnitude $H_V > 21$ (diameters 0.2 km or less) constitute about one-fourth of radar-detected NEAs; the smallest are comparable in size to boulders seen on the surface of Eros. Most of them have rotation periods no longer than an hour and in some cases only a few minutes, but at least two, 2001 EC16 and 2004 XP14, are very slow rotators.

3.12.3 BINARY SYSTEMS

Radar obtained the first undeniable evidence for NEA binary systems and has now imaged 20 of them. Current detection statistics, including evidence from optical

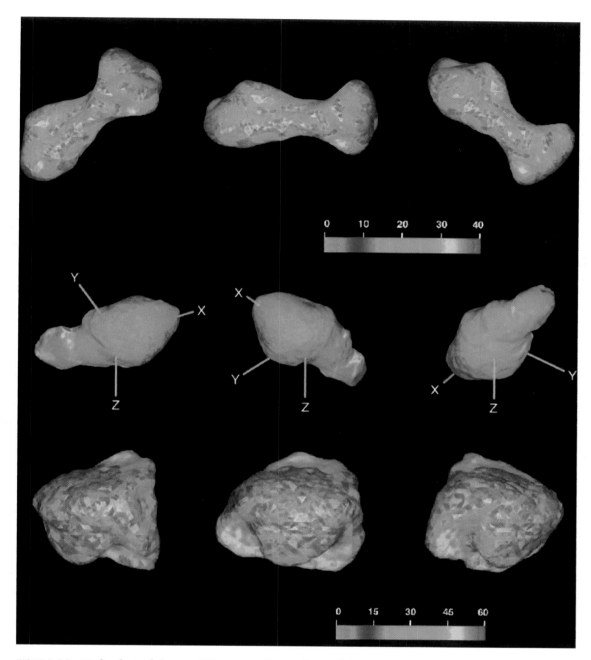

FIGURE 20   Radar-derived shapes of (from top to bottom) main-belt asteroid 216 Kleopatra (maximum model dimension 217 km) and the near-Earth asteroids 4179 Toutatis (4.60 km) and 6489 Golevka (0.685 km), color-coded for gravitational slope (degrees), defined as the acute angle a plumb line would make with the local surface normal. Uniform internal density is assumed.

lightcurves, suggest that between 10 and 20% of PHAs are binary systems.

For 2000 DP107, with an 800-m primary and a 300-m secondary, the orbital period of 1.767 days and orbital semimajor axis of 2620 + 160 m yield a bulk density of 1.7 + 1.1 g cm^{-3} for the primary. For 66391 (1999 KW4), very high-SNR, high-resolution delay–Doppler images characterized

the components and their dynamics in detail (Fig. 23). The resemblance of the primary to a canonical oblate spheroid is striking. For 1998 ST27, the orbital period is several days, the semimajor axis could be as large as 7 km, and the rotation period of the secondary is more than an order of magnitude shorter than its orbital period, the first such case among binary NEAs. 1998 ST27 and the somewhat similar

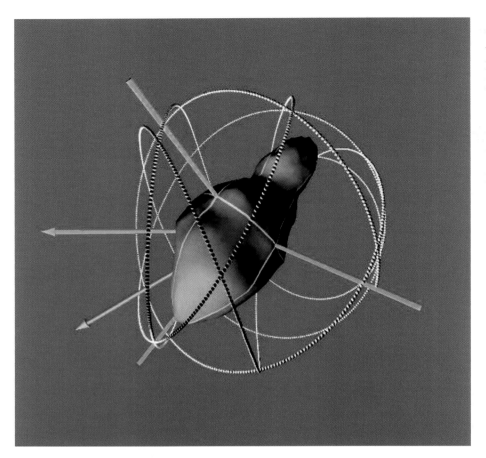

FIGURE 21 Spin state of asteroid 4179 Toutatis derived from radar. The axes with no arrow tips are the asteroid's principal axes of inertia and the vertical arrow is its angular momentum vector. The direction of the spin vector (yellow arrow) relative to the principal axes is a (5.41-day) periodic function. A flashlamp attached to the short axis of inertia and flashed every 15 minutes for 20 days would trace out the intricate path indicated by the small spheres stacked end to end; the path never repeats. Toutatis' spin state differs radically from those of the vast majority of solar system bodies that have been studied, which are in principal-axis spin states. For those objects, the spin vector and angular momentum vector point in the same direction, and the flashlamp's path would be a circle.

binaries 1990 OS and 2003 YT1 may be relatively young systems.

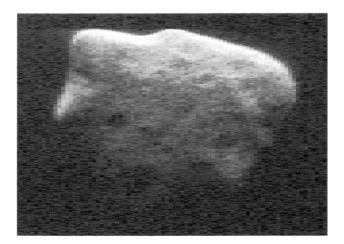

FIGURE 22 Arecibo 13-cm, OC radar image of 53319 (1999 JM8). Radar illumination is from the top. The vertical resolution is 15 m. The horizontal resolution depends on the asteroid's NPA spin state, which is not yet known. [From Benner, L. A. M., et al., 2002, *Meteoritics Planet. Sci.* **37**, 779–792.)

3.12.4 COLLISION PREDICTION AND PREVENTION

The NEA collision hazard has emerged as a primary driving issue in asteroid science. Radar provides very precise astrometric positions and leads to more accurate trajectory predictions than optical data alone. On average, radar has added a third of a millennium to the window of accurate future predictions of PHA close Earth approaches. When radar astrometry is excluded from single-apparition PHA radar + optical orbit solutions, some 40% cannot have their next close approach predicted within the adopted confidence level using only the single apparition of optical data. The net effect of radar for multiapparition cases is to improve the orbit's accuracy. For example, integrations of the radar-refined orbit of 29075 (1950 DA) revealed that in 2880 there could be a potentially hazardous approach that had not been indicated in the half-century arc of preradar optical data. The dominant source of uncertainty in the collision probability involves the Yarkovsky effect, which is the nongravitational "recoil" acceleration of a rotating object due to its anisotropic thermal emission of absorbed sunlight, and which depends on the asteroid's size, shape, mass, rotation, and optical and thermal characteristics.

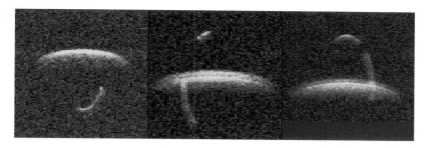

FIGURE 23   Several-hour delay–Doppler time exposures of radar echoes from binary asteroid 66391 (1999 KW4). Distance from Earth increases toward the bottom, and speed from Earth increases toward the left. The motion of the secondary (smaller) component about the primary component is clockwise. Gaps in the trail are due to breaks in the data-taking. The primary appears much wider than the secondary because it is a few times bigger and is rotating much faster. Although the components have the same speeds along the radar line of sight and the same distances from the radar where their echoes overlap, their positions in space are never the same. The components orbit a common center of mass, and each component's average distance from that point is inversely proportional to its mass. The motion of the relatively massive primary is much less obvious than the motion of the secondary, but it can be seen in the double appearance of the primary's top edge in the two time exposures that follow the secondary from in front of the primary to behind it. These Goldstone (8560-MHz, 3.5-cm, OC) images have overall extents of 37.5 μs by 67 Hz (5.6 km by 1.2 m s^{-1}). (NASA/JPL.)

It was suggested that radar-refined orbits with sufficiently long astrometric time bases could permit direct detection of nongravitational acceleration of NEAs due to the Yarkovsky effect. The first such detection was achieved via radar ranging to Golevka. That experiment, which constitutes the first estimation of the mass (and, using the previously derived radar shape model, the density) of a small solitary asteroid using ground-based observations.

3.13 Comets

Because a cometary coma is nearly transparent at radio wavelengths, radar is much more capable of unambiguous detection of a cometary nucleus than are visible-wavelength and infrared methods, and radar observations of several comets (see Table 1) have provided useful constraints on nuclear dimensions. The radar signature of one particular comet (IRAS–Araki–Alcock, which came within 0.03 AU of Earth in May 1983) altered our concepts of the physical nature of these intriguing objects. Echoes obtained at both Arecibo (Fig. 24) and Goldstone have a narrowband component from the nucleus as well as a much weaker broadband component from large particles ejected mostly from the sunlit side of the nucleus. Models of the echoes suggest that the nucleus is very rough on scales larger than a meter, that its maximum overall dimension is within a factor of 2 of 10 km, and that its spin period is 2–3 days. The particles are probably several centimeters in size and account for a significant fraction of the particulate mass loss from the nucleus. Most of them appear to be distributed within ~1000 km of the nucleus, that is, in the volume filled by particles ejected at several meters per second over a few days. The typical particle lifetime may have been this short, or the particle ejection rate may have been highly variable.

In late 1985, radar observations of comet Halley, which was much more active than IRAS–Araki–Alcock, yielded echoes with a substantial broadband component presumed to be from a large-particle swarm, but no narrowband component, a negative result consistent with the hypothesis that the surface of the nucleus has an extremely low bulk density. In 1996, Goldstone obtained 3.5-cm echoes from the nucleus and coma of comet Hyakutake (C/1996 B2). The coma-to-nucleus ratio of radar cross section is about 12 for Hyakutake versus about 0.3 for IAA.

3.14 The Saturn System and First Cassini Results

3.14.1 RINGS

The only radar-detected ring system is quite unlike other planetary targets in terms of both the experimental techniques employed and the physical considerations involved. For example, the relation between ring-plane location and delay–Doppler coordinates for a system of particles traveling in Keplerian orbits is different from the geometry portrayed in Fig. 5. The rings are grossly overspread (see Table 2), requiring the use of frequency-stepped waveforms in delay–Doppler imaging.

Radar determinations of the rings' backscattering properties complement results of the *Voyager* spacecraft radio occultation experiment (which measured the rings' forward scattering efficiency at identical wavelengths) in constraining the size and spatial distributions of ring particles. The rings' circular polarization ratio is ~1.0 at 3.5 cm and ~0.5 at 13 cm, more or less independent of the inclination angle δ between the ring plane and the line of sight. Whereas multiple scattering between particles might cause some of the depolarization, the lack of strong dependence of μ_C on δ suggests that the particles are intrinsically rougher at

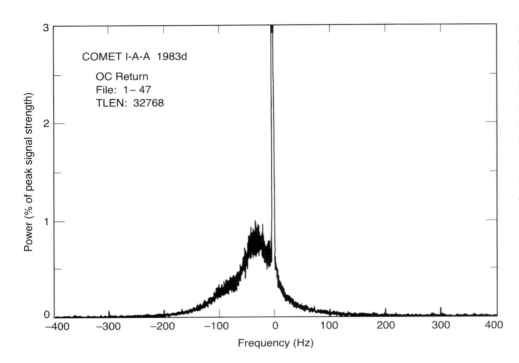

FIGURE 24 Arecibo OC and SC echo spectra obtained for comet IRAS–Araki–Alcock, truncated at 2% of the maximum OC amplitude. The narrowband echo from the nucleus is flanked by broadband echo from large (= 1 cm) particles in a 1000-km-radius cloud surrounding the nucleus. (From Harmon, J. K., Campbell, D. B., Hine, A. A., Shapiro, I. I., and Marsden, B. G., 1989, *Astrophys. J.* **338**, 1071–1093.)

the scale of the smaller wavelength. Delay–Doppler resolution of ring echoes indicates that the portions of the ring system that are brightest optically (the A and B rings) also return most of the radar echoes. The C ring has a very low radar reflectivity, presumably because of either a low particle density in that region or compositions or particle sizes that lead to inefficient scattering.

Recent 13-cm images show a pronounced azimuthal asymmetry in the reflectivity of the A ring. The analogous phenomenon at visual wavelengths is ascribed to gravitational "wakes" generated by individual large ring particles or arising from internal instabilities, which are distorted by Keplerian shear into elongated structures trailing at angles of 70° from the radial direction. The strength of the radar asymmetry may be due to strongly forward-scattering meter-size ice particles and the resultant sensitivity to optical depth variations.

3.14.2 TITAN

Titan's thick, hazy atmosphere poses challenges to visible-wavelength and near-infrared imaging of its surface. *Voyager* and ground-based data indicate a surface temperature and pressure of 94 K and 1.5 bar and show that the atmosphere is mostly N_2 with traces of hydrocarbons and nitriles. Thermodynamic considerations imply a near-surface reservoir of liquid hydrocarbons. Arecibo 13-cm echoes show most of the power to be diffusely scattered, and the longitude dependence of the radar albedo mimics the dependence of the disk-integrated near-IR albedos, indicating that whatever properties of the surface—roughness, composition, etc.—that are responsible for the variation in the

near-IR albedo are also responsible for the variation in the radar cross section. A specular component is present for about 75% of the sub-Earth locations. The most specular echoes (e.g., Fig 25), which are subradar glints that must come from extremely smooth surfaces, have properties consistent with those expected for irregularly shaped 50-km or larger bodies of liquid hydrocarbons.

Titan is the primary target of the *Cassini* mission. *Cassini*'s radar instrument, a 13.8-GHz (2.2-cm)

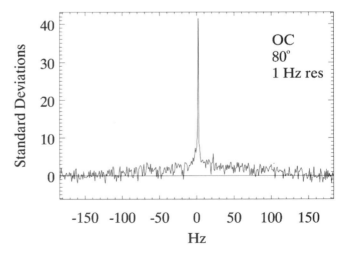

FIGURE 25 Arecibo 13-cm OC radar echo spectrum ot Titan at 1.0-Hz resolution for sub-Earth longitude of 80°. A fit of a composite model with Hagfors and Cosine terms gives an rms slope of 0.2° and a reflection coefficient of 0.023. Titan's echo bandwidth is 325 Hz. (From Campbell, D. B., Black, G. J., Carter, L. M., and Ostro, S. J., 2003, *Science* **302**, 431–434.)

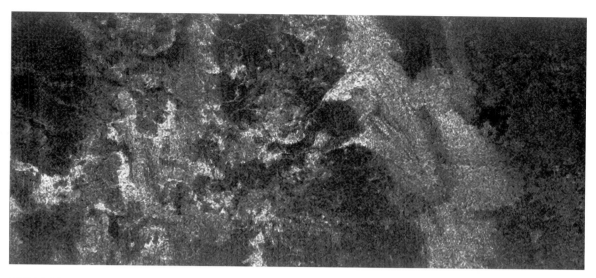

FIGURE 26 *Cassini* RADAR (2.2-cm, SL) images of Titan. The bright, rough region on the left side of the image seems to be topographically high terrain cut by channels and bays. The boundary of the bright (rough) region and the dark (smooth) region appears to be a shoreline. The patterns in the dark area indicate that it may once have been flooded. The image is 175 × 330 km and is centered at 66 S, 356 W. (NASA/JPL.)

synthetic-aperture radar (SAR) imager, altimeter, scatterometer, and radiometer, will operate during around half of the 44 Titan flybys, covering about a fifth of the surface with imaging resolution of 2 km or finer. Scatterometry observations will cover most of the surface, albeit at resolution no finer than tens of kilometers, and a limited number of short altimetry tracks will give regional topographic information.

At this writing, *Cassini* has completed its first 6 Titan SAR flybys (Fig. 26), revealing a surface with low relief and an Earth-like variety of surface features providing evidence for fluvial/pluvial, cryovolcanic, Aeolian, impact, and probably tectonic modification processes. Diverse styles of channels are seen; some suggest that precipitated liquids are collected and transported hundreds of kilometers, others indicate a cryovolcanic origin, and some may be spring-fed. Circular features apparently include cryovolcanic vents as well as a surprisingly small number of impact craters. Regions of dune-like forms that run for hundreds of kilometers establish that particulate matter is available and that there are winds that can transport them. The radar-bright, continent-sized landform Xanadu is revealed in the radar images to be a landmass of Appalachian-sized mountains and valleys cut by channels and marked with craters and dark patches. Such patches, seen in numerous images, are tentatively identified as hydrocarbon lakes or organic sludge.

3.14.3 ICY SATELLITES

Cassini radar and radiometric observations of Saturn's icy satellites yield properties that apparently are dominated by subsurface volume scattering and are similar to those of the icy Galilean satellites. Average radar albedos decrease in

the order Enceladus/Tethys, Rhea, Dione, Hyperion, Iapetus, and Phoebe. This sequence most likely corresponds to increasing contamination of near-surface water ice, which is intrinsically very transparent at radio wavelengths. Plausible candidates for contaminants include ammonia, silicates, metallic oxides, and polar organics. There is correlation of our targets' radar and optical albedos, probably due to variations in the concentration of optically dark contaminants in near-surface water ice and the resulting variable attenuation of the high-order multiple scattering responsible for high radar albedos. Iapetus' 2.2-cm radar albedo is dramatically higher on the optically bright trailing side than the optically dark leading side, whereas 13-cm results show hardly any hemispheric asymmetry and give a mean radar reflectivity several times lower than the reflectivity measured at 2.2 cm. These Iapetus results are understandable if ammonia is much less abundant on both sides within the upper one to several decimeters than at greater depths, and if the leading side's optically dark contaminant is present to depths of at least one to several decimeters. A combination of ion erosion and micrometeoroid gardening may have depleted ammonia from the surfaces of Saturn's icy satellites. Given the hypersensitivity of water ice's absorption length to ammonia concentration, an increase in ammonia with depth could allow efficient 2.2-cm scattering from within the top one to several decimeters while attenuating 13-cm echoes, which would require a 6-fold thicker scattering layer.

4. Prospects for Planetary Radar

There is growing interest in the possibility of a subsurface ocean on Europa and in the feasibility of using an orbiting radar sounder to probe many kilometers below that object's fractured crust using meter- to several-decameter wavelengths. Another possibility is to use radar reflection tomography to construct a three-dimensional image of the interior of an asteroid or comet. Meanwhile, the Mars Advanced Radar for Subsurface and Ionospheric Sounding (MARSIS) on the European Space Agency's *Mars Express* spacecraft has probed the depths of Mars' north and south polar deposits, and the Shallow Subsurface Radar (SHARAD) is about to start searching for subsurface water on Mars from NASA's *Mars Reconnaissance Orbiter*.

Reconnaissance of near-Earth asteroids will occupy ground-based radar astronomy indefinitely. Most of the optically discoverable NEAs traverse the detectability windows of Arecibo and/or Goldstone at least once every few decades, and efforts are under way to increase the NEA discovery rate by more than an order of magnitude. The power of radar observations for orbit refinement and physical characterization motivates radar observations of newly discovered NEAs whenever possible. Eventually the initial radar detection of a new NEA could become an almost daily opportunity.

Bibliography

Black, G. J., Campbell, D. B., and Nicholson, P. D. (2001). Icy Galilean satellites: Modeling radar reflectivities as a coherent backscatter effect. *Icarus* **151**, 167–180.

Butrica, A. J. (1996). "To See the Unseen: A History of Planetary Radar Astronomy," NASA History Series No. SP-4218. NASA, Houston.

Harmon, J. K., Nolan, M. C., Ostro, S. J., and Campbell, D. B. (2004). Radar studies of comet nuclei and grain comae. In "Comets II" (M. Festou, U. Keller, and H. Weaver, eds.), pp. 265–279. Univ. Arizona Press, Tucson.

Magri, C., Ostro, S. J., Rosema, K. D., Thomas, M. L., Mitchell, D. L., Campbell, D. B., Chandler, J. F., Shapiro, I. I., Giorgini, J. D., and Yeomans, D. K. (1999). Mainbelt asteroids: Results of Arecibo and Goldstone radar observations of 37 objects during 1980–1995. *Icarus* **140**, 379–407.

Ostro, S. J. (1993). Planetary radar astronomy. *Rev. Modern Physics* **65**, 1235–1279.

Ostro, S. J., and Giorgini, J. D. (2004). The role of radar in predicting and preventing asteroid and comet collisions with Earth. In "Mitigation of Hazardous Comets and Asteroids" (M. J. S. Belton, D. K. Yeomans, and T. H. Morgan, eds.), pp. 38–65. Cambridge Univ. Press, Cambridge, England.

Ostro, S. J., Hudson, R. S., Benner, L. A. M., Giorgini, J. D., Magri, C., Margot, J.-L., and Nolan, M. C. (2002). Asteroid radar astronomy. In "Asteroids III" (W. Bottke, A. Cellino, P. Paolicchi, and R. P. Binzel, eds.), pp. 151–168. Univ. Arizona Press, Tucson.

Pettengill, G. H., Ford, P. G., Johnson, W. T. K., Raney, R. K., and Soderblom, L. A. (1991). Magellan: Radar performance and data products. *Science* **252**, 260–265.

Shapiro, I. I., Chandler, J. F., Campbell, D. B., Hine, A. A., and Stacy, N. J. S. (1990). The spin vector of Venus. *Astron. J.* **100**, 1363–1368.

Tyler, G. L., Ford, P. G., Campbell, D. B., Elachi, C., Pettengill, G. H., and Simpson, R. A. (1991). Magellan: Electrical and physical properties of Venus' surface. *Science* **252**, 265–270.

Remote Chemical Sensing Using Nuclear Spectroscopy

Thomas H. Prettyman

Los Alamos National Laboratory,
Los Alamos, New Mexico

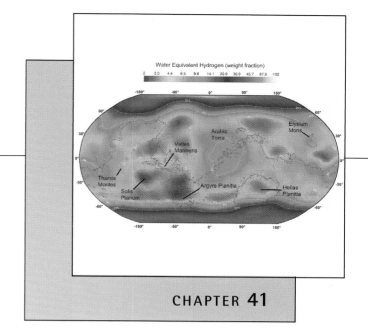

CHAPTER **41**

1. Introduction

Nuclear spectroscopy techniques are used to determine the elemental composition of planetary surfaces and atmospheres. Radiation, including gamma rays and neutrons, is produced steadily by cosmic ray bombardment of the surfaces and atmospheres of planetary bodies and by the decay of radionuclides within the solid surface. The leakage **flux** of gamma rays and neutrons contains information about the abundance of major elements, selected trace elements, and light elements such as H and C. Gamma rays and neutrons can be measured from high altitudes (less than a planetary radius), enabling global mapping of elemental composition by an orbiting spacecraft. Radiation that escapes into space originates from shallow depths (<1 m within the solid surface). Consequently, nuclear spectroscopy is complementary to other surface mapping techniques, such as reflectance spectroscopy, which is used to determine the mineralogy of planetary surfaces.

The main benefit of gamma ray and neutron spectroscopy is the ability to reliably identify elements important to planetary geochemistry and to accurately determine their abundance. This information can be combined with other remote sensing data, including surface thermal inertia and mineralogy, to investigate many aspects of planetary science. This article provides an overview of this burgeoning area of remote sensing. The origin of gamma rays and neutrons, their information content, measurement techniques, and scientific results from the *Lunar Prospector* and *Mars Odyssey* missions are described.

Nuclear reactions and radioactive decay result in the emission of gamma rays with discrete energies, which provide a fingerprint that can uniquely identify specific elements in the surface. Depending on the composition of the surface, the abundance of major rock-forming elements, such as O, Mg, Al, Si, Cl, Ca, Ti, Fe, and radioactive trace elements, such as K, Th, and U can be determined from measurements of the gamma ray spectrum. The geochemical data provided by nuclear spectroscopy can be used to investigate a wide range of topics, including the following:

- Determining bulk composition for comparative studies of planetary geochemistry and the investigation of theories of planetary origins and evolution;
- Constraining planetary structure and differentiation processes by measuring large-scale stratigraphic variations within impact basins that probe the crust and mantle;
- Characterization of regional scale geological units, such as lunar mare and highlands;
- Estimating the global heat balance by measuring the abundance of radioisotopes such as K, Th, and U;

- Measuring the ratio of the volatile element K to the refractory element Th to determine the depletion of volatile elements in the source material from which planets were accreted and to estimate the volatile inventory of the terrestrial planets.

Neutrons are produced by cosmic ray interactions and are sensitive to the presence of light elements within planetary surfaces and atmospheres, including H, C, and N, which are the major constituents of ices as well as elements such as Gd and Sm, which are strong neutron absorbers. In addition, alpha particles are produced by radioactive decay of heavy elements such as U and Th and have been used to identify radon emissions from the lunar surface, possibly associated with tectonic activity.

Close proximity to the planetary body is needed to measure neutrons and gamma rays because their production rate is relatively low. Unlike optical techniques, distances closer than a few hundred kilometers are needed in order to obtain a strong signal. In addition, sensors used for gamma ray and neutron spectroscopy are generally insensitive to incident direction. Consequently, spatial resolution depends on orbital altitude, and higher resolution can be achieved by moving closer to the planet. Regional scale measurements are generally achieved using nuclear spectroscopy, in contrast to the meter to kilometer scale generally achieved by optical remote sensing methods.

Measurements of the solid surface are not possible for planets with thick atmospheres, including the Earth, Venus, and outer planets other than Pluto. Variations in atmospheric composition can be measured and have important implications to understanding seasonal weather patterns. Gamma ray and neutron spectroscopy can be applied to investigate the surfaces of planets with thin atmospheres, such as Mercury, Mars, the Moon, comets, and asteroids. In principle, the satellites of Jupiter and Saturn could be investigated using nuclear spectroscopy; however, the intense radiation environment within the magnetospheres of these planets may be a limiting factor.

X-ray spectroscopy can also be used to determine elemental composition and is complementary to nuclear spectroscopy. Intense bursts of x-rays produced by solar flares cause planetary surfaces to fluoresce. The characteristic x-rays that are emitted can be analyzed to determine the abundance of rock-forming elements such as Fe and Mg. In contrast to nuclear spectroscopy, surface coverage may be limited, especially when solar activity is low; however, high statistical precision for elemental abundances can be achieved during flares. The depth sensitivity of x-ray and nuclear spectroscopy is very different. X-rays are produced much closer to the surface than gamma rays and neutrons. Missions that have used x-ray spectroscopy include *Apollo* and *NEAR* [see Near-Earth Asteroids], and *SMART-1*. The *MESSENGER* mission will use both x-ray and nuclear spectrometers to determine the elemental composition of Mercury, and an x-ray spectrometer will be on the payload of *Chandrayaan-1*, the Indian Space Research Organization's first mission to the Moon.

2. Origin of Gamma Rays and Neutrons

Neutrons and gamma rays are produced by the interaction of energetic particles and cosmic rays with planetary surfaces and atmospheres. While solar energetic particle events can produce copious gamma rays and neutrons, we will focus our attention on **galactic cosmic rays**, which are somewhat higher in energy, penetrate more deeply into the surface, and have a constant flux over relatively long periods of time. **Gamma rays** are also produced steadily by the decay of radioactive elements such as K, Th, and U. A diagram of production and transport processes for neutrons and gamma rays is shown in Fig. 1.

2.1 Galactic Cosmic Rays

Galactic cosmic rays consist primarily of protons with an average flux of about 4 protons per cm^2 per s and with a wide distribution of energies extending to many GeV (Fig. 2; inset). The flux and energy distribution of galactic protons reaching a planetary surface is modulated by the solar cycle [see The Sun]. Sunspot counts are a measure of solar activity (Fig. 2). Higher fluxes of galactic protons are observed during periods of low solar activity. In addition, more low-energy protons penetrate the heliosphere during solar minimum, resulting in a shift in the population towards lower energies. The flux and energy distribution of the cosmic rays are controlling factors in the production rate, energy distribution, and depth of production of **neutrons** and gamma rays. For example, the neutron counting rates at MacMurdo Station in Antarctica are modulated by the solar cycle as shown in Fig. 2.

The GeV-scale energy of galactic protons can be compared to the relatively small binding energy of protons and neutrons in the nucleus (for example, 8.8 MeV/nucleon for ^{56}Fe). High energy interactions with nuclei can be modeled as an intranuclear cascade, in which the energy of the incident particle is transferred to the nucleons, resulting in the emission of secondary particles by spallation, followed by evaporation, and subsequent de-excitation of the residual nuclei. The secondary particles, which include neutrons and protons, undergo additional reactions with nuclei until the initial energy of the cosmic ray is absorbed by the medium. Since most of the gamma ray production is caused by reactions with neutrons, we will focus our attention on how neutrons slow down in matter.

2.2 Fundamentals of Neutron Moderation

Neutrons transfer their energy to the medium through successive interactions with nuclei and are eventually absorbed

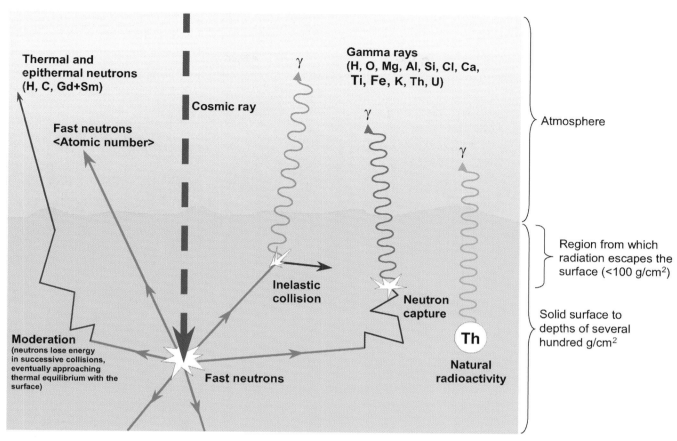

FIGURE 1   Overview of the production of gamma rays and neutrons by cosmic ray interactions and radioactive decay. Fast neutrons produced by high-energy cosmic ray interactions undergo inelastic collisions, resulting in the production of characteristic gamma rays that can be measured from orbit. Neutrons lose energy through successive collisions with nuclei and approach thermal equilibrium with the surface. Thermal and epithermal neutrons provide information about the abundance of light elements, such as H and C, and strong thermal neutron absorbers, such as Gd and Sm. Fast neutrons are sensitive to the average atomic mass of the surface. Gamma rays produced by neutron capture and inelastic scattering can be used to measure the abundance of rock-forming elements, such as O and Fe. Gamma rays are also produced by the decay of long-lived radioisotopes, including K, Th, and U. While cosmic rays can penetrate deep into the surface, the radiation escaping the surface originates from shallow depths, generally less than 100 g/cm^2.

in the surface or atmosphere or escape into space. The process of slowing-down via repeated collisions is known as "moderation." There are three general interaction categories that are important in the context of planetary science: (1) nonelastic reactions, in which the incident neutron is absorbed, forming a compound nucleus, which decays by emitting one or more neutrons followed by the emission of gamma rays; (2) elastic scattering, a process that can be compared to billiard ball collisions for which kinetic energy is conserved; (3) neutron radiative capture, in which the neutron is absorbed and gamma rays are emitted.

The probability that a neutron will interact with a nucleus can be expressed in terms of an effective area of the target nucleus, known as the microscopic cross section, $\{I\sigma/I\}$, which depends on the energy of the neutron (E) and has

units of barns. One barn is 10^{-24} cm^2. Microscopic cross sections for natural Fe are shown, for example, in Fig. 3 for radiative capture, elastic scattering, and inelastic scattering. Inelastic scattering occurs above a threshold determined by the energy required to produce the first excited state of the compound nucleus. The elastic scattering cross section is constant over a wide range of energies. The cross section for radiative capture usually varies as $E^{-1/2}$. Consequently, radiative capture is important at low energies. The sharp peaks that appear at high energy (greater than 100 eV) are resonances associated with the nuclear structure of the Fe isotopes. Neutron inelastic scattering is an important energy loss mechanism at high energies (greater than about 0.5 MeV for most isotopes of interest to planetary science).

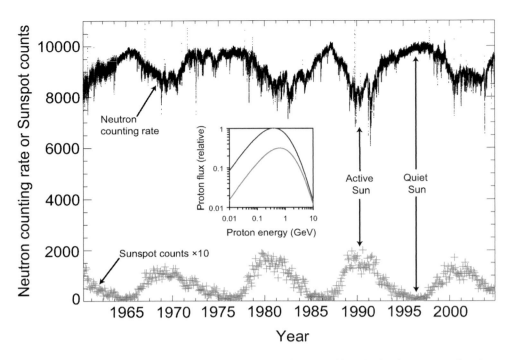

FIGURE 2 The variation of neutron counting rates (with units of hundreds of counts per hour) measured at McMurdo Station in Antarctica as a function of time (neutron monitors of the Bartol Research Institute are supported by NSF grant ATM-0000315). Monthly sunspot counts (multiplied by 10) are shown for comparison (Courtesy SIDC, RWC Belgium, World Data Center for the Sunspot Index, Royal Observatory of Belgium, 1961–2004). During periods of low solar activity (low sunspot counts), low energy galactic cosmic rays penetrate the heliosphere, which results in relatively high neutron production rates. During periods of high solar activity (high sunspot counts), the low energy galactic cosmic rays are cut off, resulting in lower neutron counting rates. The variation in neutron counting rates is about 20% over the solar cycle. Theoretical galactic proton energy spectra within the heliosphere, representative of quiet and active solar years, are shown (*inset*).

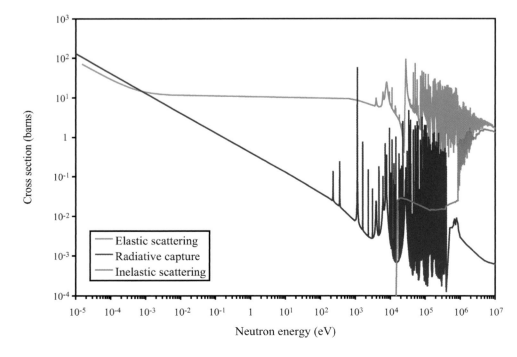

FIGURE 3 Neutron microscopic cross sections for natural Fe (*see text*).

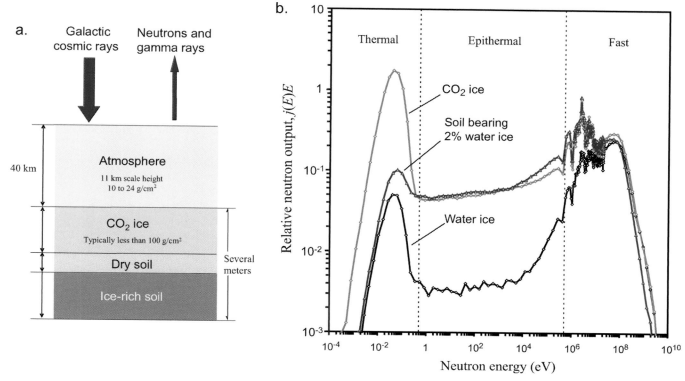

FIGURE 4   (a) Model of the martian surface at high latitudes; (b) The current of neutrons leaking away from Mars for three different solid surface compositions. Neutron energy ranges are indicated. (*See text for details.*)

Under the steady bombardment of cosmic rays, the population of neutrons slowing down in the surface is, on average, constant with time. The steady-state neutron energy-, angle-, and spatial-distributions depend on the composition and stratigraphy of the surface and atmosphere. An important property of the neutron population is the scalar flux (φ), which depends on depth and is given by the product of the speed of the neutrons, v (cm/s), and the number density of neutrons slowing down in the medium (n neutrons per cm^3): $\varphi = nv$, with units of neutrons per cm^2 per s. The rate at which neutrons interact with nuclei is given by the product of the flux of neutrons, the density of the target nuclei (N nuclei per cm^3), and the microscopic cross section: $R = \varphi N \sigma$ (interactions per cm^3 per s).

Cosmic ray showers can be modeled using Monte Carlo methods, in which the random processes of particle production and transport are simulated. The number of times something interesting happens, such as a particle crossing a surface, is tallied. Statistical averages of these interesting events are used to determine different aspects of the particle population such as fluxes and **currents**. Monte Carlo transport simulations generally provide for the following: a description of the cosmic ray source and the target medium (including geometry, composition, and density); detailed

physical models of interaction mechanisms and transport processes (including tabulated data for interaction cross sections); and a system of tallies.

The general purpose code Monte Carlo N-Particle eXtended (MCNPX) developed by Los Alamos National Laboratory provides a detailed model of cosmic ray showers, including the intranuclear cascade and subsequent interactions of particles within the surface and atmosphere. For example, a model of the martian surface used to calculate neutron leakage spectra is shown in Fig. 4a, and includes several layers, representative of the high latitude surface, which is seasonally covered by CO_2 ice due to condensation of atmospheric CO_2 in the polar night, and whose frost-free surface consists of a dry lag deposit covering ice rich soil. The curvature of Mars was included in the MCNPX calculations along with details of the incident galactic proton energy distribution. The goal was to determine the effect of surface parameters on neutron output, including the **column abundance** (g/cm^2) of the layers, their water abundance, and major element composition. The variation of the density of the atmosphere with altitude (the scale height is roughly 11 km) and atmospheric mass were modeled. An accurate treatment of the atmosphere is needed in order to account for variations in neutron production with density by particles such as pions that have very short half-lives. For dense atmospheres, these particles interact more frequently with atmospheric nuclei, resulting in increased neutron production in the atmosphere.

The population of neutrons escaping the surface or atmosphere can be represented as a current, J, which is the ratio of the number of neutrons escaping into space per galactic cosmic ray incident on the planet. The energy distribution of leakage neutrons is given by the current density $j(E)$, which is the number of escaping particles per unit energy per incident cosmic ray, such that $J = \int_0^\infty dE j(E)$. The current density of neutrons leaking away from Mars was calculated by MCNPX for homogeneous solid surfaces consisting of water ice, which is representative of the north polar residual cap; relatively dry soil bearing 2% water ice, which is representative of dry equatorial regions; and CO_2 ice, which is representative of the seasonal polar caps. The relative neutron output, given by the product of the current density and neutron energy is shown in Fig. 4b for each of these materials. Integrating over all energies gives 5, 3, and 1 for the total number of neutrons escaping the surface per incident cosmic ray proton for the CO_2 ice, dry soil, and water ice surfaces, respectively.

The neutron current density spans 14 decades of energy and can be divided into three broad ranges (Fig. 4b), representing different physical processes: (1) Thermal neutrons, which have undergone many collisions, have energies less than about 0.1 eV and are nearly in thermal equilibrium with the surface; (2) epithermal neutrons, which have energies greater than about 0.1 eV and are in the process of slowing down from higher energies; and (3) fast neutrons, including source neutrons and neutrons with energy greater than the threshold for inelastic scattering. Absorption and leakage result in a nonequilibrium energy distribution for the thermal spectrum. Consequently, the most probable neutron energy is slightly higher than would be predicted given the temperature of the surface.

Elastic scattering is the most important loss mechanism for planetary neutron spectroscopy because it provides strong differentiation between H and other more massive nuclei. For elastic scattering, the energy loss per collision varies systematically with atomic mass. The maximum energy that a neutron can lose in a collision is given by fE, where $f = 1 - [(A - 1)/(A + 1)]^2$, E is the energy of the neutron before the collision, and A is the atomic mass of the target nucleus. Thus, a neutron could lose all of its energy in a single collision with hydrogen ($A = 1$), which has roughly the same mass as a neutron. This fact is easily verified by observing head-on collisions in a game of billiards. In contrast, the maximum energy loss in a collision with C, which is the next most massive nucleus of interest in planetary science, is 28%. For Fe, the maximum energy loss per collision is 7%. The average energy loss per collision follows a similar trend. Consequently, for materials that are rich in H, such as water

ice, energy loss by elastic collisions is high and neutrons slow down more quickly than for materials that do not contain H.

For H-rich materials, the population of neutrons that are slowing down is strongly suppressed relative to materials without H. For example, the epithermal current density for the simulated water ice surface in Fig. 4b is considerably lower than either the soil or CO_2 surfaces. The current density of fast neutrons, which have undergone relatively few collisions following their production, are influenced by elastic scattering, but also by variations in neutron production, which depend on the average atomic mass of the medium.

Absorption of neutrons by radiative capture significantly influences the population of thermal neutrons. Elements such as H, Cl, Fe, and Ti have relatively high absorption cross sections and can significantly suppress the thermal neutron flux. C and O have very low absorption cross sections compared to H. Consequently, the thermal neutron output for the water ice in Fig. 4b is suppressed relative to the surfaces containing CO_2 ice and soil.

2.3 Gamma Ray Production and Transport

For galactic cosmic ray interactions, gamma rays are primarily produced by neutron inelastic scattering and radiative capture. De-excitation of residual nuclei produced by these reactions results in the emission of gamma rays with discrete energies. The energies and intensities of the gamma rays provide a characteristic fingerprint that can be used to identify the residual nucleus. Since, in most cases, a residual nucleus can only be produced by a reaction with a specific target isotope, gamma rays provide direct information about the elemental composition of the surface.

For example, neutron inelastic scattering with ^{56}Fe frequently leaves the residual ^{56}Fe nucleus in its first excited state, which transitions promptly to ground state by the emission of an 847 keV gamma ray. The presence of a peak at 847 keV in a planetary gamma ray spectrum indicates that the surface contains Fe. The intensity of the peak is related directly to the abundance of elemental Fe in the surface.

Gamma rays produced by the decay of short-lived neutron activation products and long-lived (primordial) radioisotopes also provide useful information about elemental abundance. Radioactive elements such as K, Th, and U can be detected when present in trace quantities. Most notably, the Th decay chain produces a prominent gamma ray at 2.6 MeV, which can be measured when Th is present in the surface at low levels (>1 ppm).

To illustrate a typical gamma ray leakage spectrum, a Monte Carlo simulation of the lunar gamma ray leakage current induced by galactic cosmic ray protons is shown in Fig. 5. The composition of the surface was assumed to be the mean soil composition from the *Apollo 11* landing site. Contributions from nonelastic reactions and capture are plotted separately. A background component associated primarily

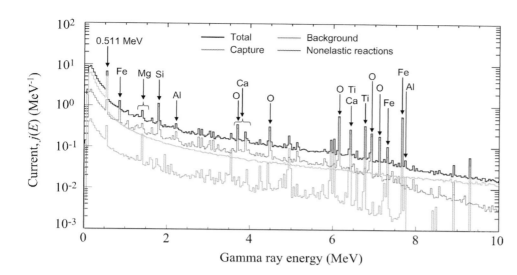

FIGURE 5   The current of gamma rays leaking away from the Moon for a composition representative of the *Apollo 11* landing site.

with the decay of pions is also shown. The peaks correspond to gamma rays that escape into space without interacting with the surface material. The peaks are superimposed on a continuum, which results from the scattering of gamma rays in the surface. The total number of gamma rays escaping the surface per incident cosmic ray proton was 2.7, which is within the range of values for the number of neutrons escaping the martian surface, presented in Section 2.1.

Gamma ray peaks associated with neutron interactions with major elements are labeled with the target element in Fig. 5. The intensity (or area) of each peak is proportional to the product of the abundance of the target element and the number density of neutrons slowing down in the medium. Specifically, the measured intensity (I) of a gamma ray peak with energy E for a selected reaction can be modeled as the product of three terms: $I \propto fyR$, where f accounts for attenuation of gamma rays by intervening surface materials and the variation of detection efficiency with gamma ray energy; y is the number of gamma rays of energy E produced per reaction; and $R = \varphi N \sigma$ is the **reaction rate**, the product of the neutron flux, cross section, and number density of the target element.

Because gamma rays are produced by neutron interactions, the absolute number density or, equivalently, the weight fraction of the target element cannot be determined unless the neutron flux is known. Thus, neutron spectroscopy plays an important role in the analysis of gamma ray data. Relative abundances can be determined without knowledge of the magnitude of the neutron flux. For example, the ratio of Fe to Si abundances can be determined from the ratio of the intensities of the prominent Fe doublet (at 7.65 MeV and 7.63 MeV) the Si gamma ray at 4.93 MeV. Because the magnitude of the attenuation of gamma rays by surface materials depends on gamma ray energy and the distribution of gamma ray production with depth, models

of the depth profile of the neutron flux are needed in order to analyze gamma ray data.

For homogeneous surfaces, accurate results can be obtained for absolute and relative abundances; however, surfaces with strong stratigraphic variations present a difficult challenge for analyzing nuclear spectroscopy data. Compositional layering of major elements on a submeter scale is widespread on Mars as shown, for example, by the *Spirit* and *Opportunity* rovers [see Mars Site Geology and Geochemistry]. In some cases, geophysical assumptions can be made that simplify the analysis and allow quantitative results to be obtained; however, it is often the case that insufficient information is available. In these cases, it is sometimes possible to establish bounds on composition that are useful for geochemical analysis. Development of accurate algorithms for determining elemental abundances, absolute or relative, requires careful synthesis of nuclear physics with constraints from geology, geophysics, and geochemistry.

3. Detection of Gamma Rays and Neutrons

In this section, a simple model of the counting rate observed by orbiting neutron and gamma ray spectrometers is presented along with an overview of radiation detection concepts for planetary science applications.

3.1 Counting Rate Models

The flux of radiation reaching an orbiting spectrometer varies in proportion to the solid angle subtended by the planet at the detector, which depends on orbital altitude. The fractional solid angle of a spherical body is given by

$$\Omega(h) = 1 - \sqrt{1 - R^2/(R+h)^2}, \qquad (1)$$

where h is the orbital altitude and R is the radius. The fractional solid angle varies from 1 at the surface (for $h = 0$) to 0 far away from the planet. For galactic cosmic ray interactions, the flux of gamma rays or neutrons at the orbiting spectrometer is approximately

$$\phi(h) = 1/4 \Phi J \, \Omega(h), \qquad (2)$$

where Φ is the flux of galactic cosmic ray protons far from the planet (about 4 protons/cm^2/s, depending on the solar cycle and location within the heliosphere), and J is the leakage current. Because alpha particles and heavier nuclei of galactic origin contribute to neutron and gamma ray production, Eq. 2 must be multiplied by a factor, approximately 1.4, in order to estimate the total leakage flux.

Eq. 2 can be used, for example, to calculate the flux of neutrons incident on the *Mars Odyssey* neutron spectrometer. The orbital altitude for *Mars Odyssey* is 400 km, and the volumetric mean radius of Mars is 3390 km. The fractional solid angle, given by Eq. 1, is 0.55. The total leakage current for a surface consisting of thick CO_2 ice, representative of the polar seasonal caps during winter, was $J = 5$ (from Section 2.2). Consequently, from Eq. 2, the total flux of neutrons at *Odyssey's* orbit from thick CO_2 deposits is approximately 4 neutrons per cm^2 per s. For a surface that is 100% water, which is representative of the north polar residual cap, J was 1, and the total flux at orbital altitude is expected to be 0.8 neutrons per cm^2 per s.

Radiation detectors, such as the gamma ray and neutron spectrometers on *Mars Odyssey*, count particle interactions and bin them into energy or pulse-height spectra, for example, with units of counts per s per unit energy. For both gamma rays and neutrons, the net counting rate (with units of counts per s) for selected peaks in the spectrum is needed in order to determine elemental abundances.

The flux of particles (gamma rays or neutrons) incident on a spectrometer can be converted to counting rate (C), given the intrinsic efficiency (ε) and projected area (A) of the spectrometer in the direction of the incident particles:

$$C = \varphi(h)\varepsilon A. \qquad (3)$$

The intrinsic efficiency is the probability that an incident particle will interact with the spectrometer to produce an event that is counted. Because particles can pass through the spectrometer without interacting, the intrinsic efficiency is always less than or equal to 1. For example, εA is on the order of 10 cm^2 for the *Mars Odyssey* epithermal neutron detector, which has a maximum projected area of about 100 cm^2. The efficiency-area product (εA) varies with the energy and angle of incidence of the particles. So, the value for εA used in Eq. 3 must be appropriately averaged over neutron energy and direction.

One of the main sources of uncertainty in measured counting rates is statistical fluctuations due to the random nature of the production, transport, and detection of radiation. While a detailed discussion of error-propagation is beyond the scope of this article, the most important result is given here: The statistical uncertainty (precision) in the counting rate is given by $\sigma = \sqrt{C/t}$, where t is the measurement time and C is the mean counting rate. For example, to achieve a precision of (1% $\sigma/C = 0.01$) when $C = 10$ counts per s, which is typical of the epithermal and thermal counting rates measured by the *Mars Odyssey* neutron spectrometer, a counting time of 1000 s is required. Longer counting times are needed when background contributions are subtracted, for example, to determine counting rates for peaks in gamma ray and neutron spectra. Uncertainties in the counting rate due to random fluctuations propagate to the uncertainties in elemental abundance and other parameters determined in the analysis of spectroscopy data. Long counting times are desired to minimize statistical contributions. Alternatively, improved precision can be achieved by increasing the counting rate, which can be accomplished through instrument design, by maximizing, and/or by making measurements at low altitude.

3.2 Gamma Ray and Neutron Detection

Radiation spectrometers measure ionization produced by the interaction of particles within a sensitive volume. Gamma ray interactions produce swift primary electrons that cause ionization as they slow down in the sensitive volume. Neutrons undergo reactions that produce energetic ions and gamma rays. The recoil proton from neutron elastic scattering with hydrogen can produce measurable ionization. The charge liberated by these interactions can be measured using a wide variety of techniques, two of which are illustrated here.

Semiconductor radiation detectors typically consist of a semiconductor dielectric material sandwiched between two electrodes. An electric field is established in the dielectric by applying high voltage across the electrodes. Gamma ray interactions produce free electron-hole pairs which drift in opposite directions in the electric field. As they drift, they induce charge on the electrodes, which is measured using a charge-sensitive preamplifier. The amplitude of the charge pulse, or pulse-height, is proportional to the energy deposited by the gamma ray. Consequently, a histogram of pulse heights, known as a pulse-height spectrum, measured for many interactions provides information about the energy distribution of the incident gamma rays.

For example, a diagram of a high-purity germanium (HPGe) detector is shown in Fig. 6a along with a photograph of an HPGe crystal in Fig. 6b. The closed-end coaxial geometry is designed to minimize trapping of carriers as they drift to the electrodes. To minimize noise due to leakage current, the HPGe must be operated at very low temperatures. The requirement for cooling adds to the mass and complexity of the design for space applications.

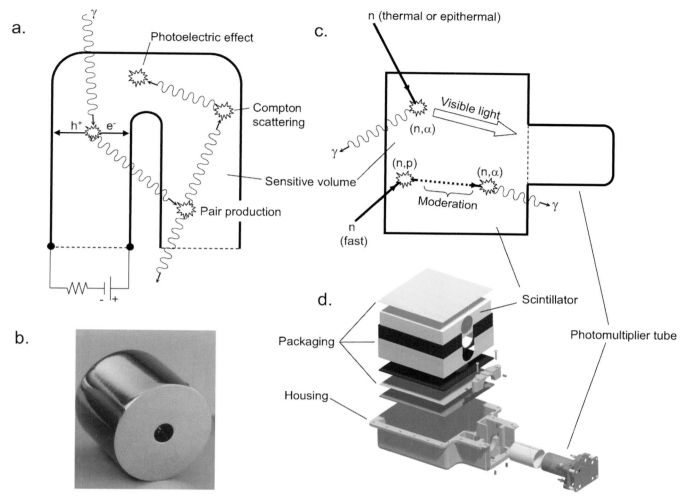

FIGURE 6 (a) Schematic diagram of a coaxial HPGe spectrometer and gamma ray interactions; (b) photograph of a HPGe cystal; (c) diagram of a scintillation-based spectrometer with neutron interactions; and (d) assembly diagram for a boron-loaded plastic scintillator for a flight experiment, including the mechanical structure (including packaging designed to withstand the vibrational environment during launch). (Part b courtesy of AMETEK, Advanced Measurement Technology, Inc., ORTEC Product Line, 801 South Illinois Avenue, Oak Ridge, TN 37830).

A hypothetical gamma ray interaction is superimposed on the diagram in Fig. 6a. Gamma rays undergo three types of interactions: pair production, Compton scattering, and the photoelectric effect. High-energy gamma rays (greater than 1.022 MeV) can undergo pair production, in which the gamma ray disappears and an electron-positron pair is produced. The kinetic energy of the electron and positron is absorbed by the medium. When the positron is annihilated by an electron, two, back-to-back (511 keV) gamma rays are produced, which can undergo additional interactions. In Compton scattering, a portion of the energy of the gamma ray is transferred to an electron. The energy lost by the gamma ray depends on the scattering angle. At low energies, the gamma ray can be absorbed by an electron via the photoelectric effect. All of these interactions vary strongly with the atomic number (Z) and density of the detector material. High Z, high density and a large sensitive volume is desired to maximize the probability that all of the energy of the incident gamma ray is absorbed in the detector.

A pulse height spectrum for a large volume (slightly larger than the crystal flown on *Mars Odyssey*), coaxial HPGe detector is shown in Fig. 7. The gamma rays were produced by moderated neutrons, with an energy distribution similar to the lunar leakage spectrum, incident on an Fe slab. Well-defined peaks corresponding to neutron capture and inelastic scattering with Fe appear in the spectrum. For example, the doublet labeled Fe(1) corresponds to gamma

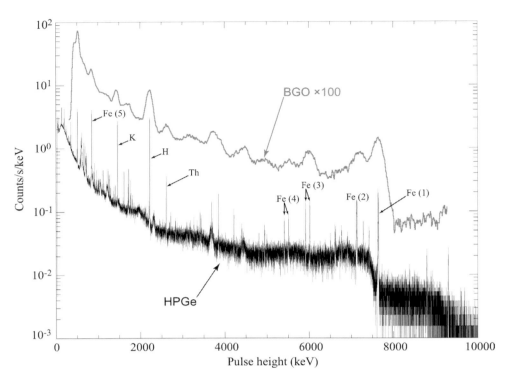

FIGURE 7 Gamma ray spectra acquired by HPGe (black) and BGO (red) spectrometers. To improve visualization, the spectrum for BGO has been multiplied by 100. The source was moderated neutrons, with energy distribution similar to a planetary leakage spectrum, incident on an iron slab. Gamma rays from natural radioactivity in the environment are also visible (from K at 1461 keV and Th at 2615 keV). A gamma ray at 2223 keV from neutron capture with H (from polyethylene in the moderator) is a prominent feature in the HPGe and BGO spectra. Major gamma rays from neutron interactions with Fe that are resolved by the HPGe spectrometer are labeled: (1) 7646- and 7631-keV doublet from neutron capture; (2) their single escape peaks; (3) 6019- and 5921-keV gamma rays from neutron capture; (4) their single escape peaks; and (5) 846.7 keV gamma ray from neutron inelastic scattering. (HPGe spectrum courtesy of S. Garner, J. Shergur, and D. Mercer of Los Alamos National Laboratory).

rays (7646- and 7631-keV) produced by neutron capture with Fe. The peaks labeled Fe(2) are shifted 511 keV lower in energy and correspond to the escape of one of the 511 keV gamma rays produced by pair production in the spectrometer. The continuum that underlies the peaks is caused by external Compton scattering and the escape of gamma rays that scattered in the spectrometer. Gamma rays from neutron capture with H and the radioactive decay of K and Th are also visible.

Scintillators provide an alternative method of detecting ionizing radiation, which can be used for gamma ray and neutron spectroscopy. Scintillators consist of a transparent material that emits detectable light when ionized. The light is measured by a photomultiplier tube or photodiode, which is optically coupled to the scintillator. The amount of light produced and the amplitude of the corresponding charge pulse from the photomultiplier tube and pulse processing circuit is proportional to the energy deposited by the radiation interaction.

A diagram of a boron-loaded, plastic scintillation detector is shown in Fig. 6c along with an assembly diagram of flight sensor (Fig. 6d). Thermal and epithermal neutrons are detected by the $^{10}B(n,\alpha\gamma)^7Li$ reaction. The recoiling reaction products (alpha particle and ^7Li ion) produce ionization equivalent to a 93 keV electron, which makes a well-defined peak in the pulse height spectrum. The area of the peak depends on the flux of incident thermal and epithermal neutrons. Thermal neutrons can be filtered out by wrapping the scintillator in a Cd foil, which strongly absorbs neutrons with energies below about 0.5 eV. Thus, the combination of a bare and Cd-covered scintillator can be used to separately measure contributions from thermal and epithermal neutrons. Above about 500 keV, light is produced by recoiling protons from neutron elastic scattering with hydrogen in the scintillator. Fast neutrons (greater than about 500 keV) can be detected by a prompt pulse from proton recoils followed a short time later by a second pulse, corresponding to neutron capture of the moderated neutron by

[10]B. This characteristic, double-pulse time signature can be used to identify, and separately measure, fast neutron events.

Scintillators are also used routinely for gamma ray spectroscopy. For example, a pulse height spectrum acquired by a bismuth germanate (BGO) scintillator is shown in Fig. 7. The source was exactly the same as measured by the HPGe spectrometer, and the two spectra share similar peak features. Note, however, that the peaks measured by BGO are considerably broader than those measured by HPGe. The width of the peaks is caused by statistical variations in the number of scintillation photons produced in the BGO. Similar dispersion occurs for charge carriers (electrons and holes) in the HPGe crystal; however, the effect is far less pronounced. The pulse height resolution as measured by the full-width-at-half-maximum (FWHM) of the gamma ray peaks is much worse for the BGO than the HPGe. The ability of the HPGe technology to resolve individual peaks is coveted by the planetary spectroscopist; however, the added cost and complexity of HPGe relative to scintillation technology has made scintillators competitive for some missions.

Other technologies that have been flown for gamma ray and neutron detection include ^{3}He ionization chambers (for thermal and epithermal neutron detection on *Lunar Prospector*) and various scintillators, including Tl-doped NaI on NEAR and *Apollo* and Tl-doped CsI on *Phobos*. The *Dawn* mission will fly a new compound semiconductor technology (CdZnTe), which has significantly improved pulse height resolution relative to BGO and, in contrast to HPGe, can be operated at ambient temperatures.

3.3 Spatial Resolution

The spatial resolution that can be achieved by a spectrometer depends on the angular distribution of radiation emitted from the surface, the angular response of the spectrometer, and the altitude of the orbit. The angular response of most spectrometers is roughly isotropic or weakly dependent on incident direction. Consequently, the spectrometer is sensitive to radiation emitted from locations from underneath the spectrometer all the way out to the limb. Due to their increased area, off-nadir regions contribute more to the counting rate than regions directly beneath the spacecraft.

When the spectrometer passes over a point source of radiation on the surface, the counting rate as a function of distance along the orbital path has an approximately Gaussian shape, with the peak occurring when the spacecraft passes over the source. Consequently, the ability of the spectrometer to resolve spatial regions with different compositions depends on the FWHM of the Gaussian, which as a rule of thumb is approximately 1.5 times the orbital altitude. For example, the lowest orbital altitude of *Lunar Prospector* was 30 km for which the spatial resolution was 45 km or 1.5° of arc length. For *Mars Odyssey*, the orbital altitude

was 400 km, and the spatial resolution was approximately 600 km or 10° of arc length.

The broad spatial response of gamma ray and neutron spectrometers must be considered in the analysis and interpretation of data, especially where comparisons to high-resolution data (for example, from optical spectroscopy) are concerned. It may be possible to increase the resolution of a spectrometer by the addition of a collimator, which would add mass to the instrument and also reduce the precision of the measurements. Alternatively, spatial deconvolution and instrument modeling techniques can sometimes be employed to study regions that are smaller in scale than the spatial resolution of the spectrometer.

4. Missions

Since the dawn of space flight, nuclear spectroscopy has been used for a wide variety of applications, from astrophysics to solar astronomy. Orbital planetary science missions with gamma ray and/or neutron spectrometers on the payload are listed in Table 1. While nuclear spectroscopy was used on earlier missions to the Moon, Mars, and the surface of Venus, the first major success was the Apollo Gamma Ray Experiment, which flew on the *Apollo 15* and *16* missions, providing global context for lunar samples. *Phobos II* traveled to Mars and provided a glimpse of the regional composition of the western hemisphere, which includes Tharsis and Valles Marineris. Due to the small size of Eros and high orbital altitudes, the gamma ray spectrometer on *NEAR* provided little useful information about Eros until the *NEAR* landed on the asteroid. Once on the surface, the *NEAR* gamma ray spectrometer acquired data with sufficient precision to determine the abundance of O, Mg, Si, Fe, and K. *NEAR* also had an x-ray spectrometer that provided complementary information about surface elemental composition. The first intended use of neutron spectroscopy for global mapping was on *Mars Observer*, which was lost before reaching Mars.

Lunar Prospector was the first mission to combine gamma ray and neutron spectroscopy to provide accurate, high-precision global composition maps of a planetary body. The missions that followed *Lunar Prospector*, including *2001 Mars Odyssey* and *MESSENGER*, a mission to the planet Mercury, also included neutron and gamma ray spectrometers on the payload. *Dawn*, a mission to the main asteroid belt, and *Selene*, a lunar mission, represent the future of orbital planetary spectroscopy. Both are in preparation for launch in the 2006–2007 timeframe.

5. Science

Lunar Prospector and *Mars Odyssey* acquired high-precision gamma ray and neutron data sets for the Moon and

TABLE 1 Summary of orbital planetary science and exploration missions with gamma ray and/or neutron spectrometers. Missions prior to Apollo, including Luna and Ranger, are not listed.

Mission	Country/ Program	Launch date(s)	Status	Planet or minor body	Orbit	Mapping duration[a]	Gamma ray spectrometer	Neutron spectrometer	Results or Objectives[b]
Apollo 15 and 16	U.S.	26-Jul-1971 16-Apr-1972	Completed	Moon	Equatorial orbit covering 20% of the lunar surface	10.5 days (Apollo 15 and 16 combined)	NaI(Tl) with plastic anti-coincidence shield	None	Maps of major and radioactive elements, including Fe, Th, and Ti.
Phobos II[c]	U.S.S.R.	12-July-1988	Lost during Phobos encounter	Mars and Phobos	Elliptical, equatorial orbit, 900 km periapsis, 80,000 km apoapsis	2 orbits analyzed	CsI(Tl)	None	Abundances for O, Si, Fe, K and Th in two equatorial regions in the western hemisphere
Mars Observer	U.S, NASA Mars Exploration Program	25-Sep-1992	Lost prior to orbital insertion	Mars	400 km altitude circular polar mapping orbit[b]	1 Mars year[b]	HPGe, passively cooled	Boron-loaded plastic scintillators	Global maps major elements and water-equivelent hydrogen (Objectives not achieved)
Near Earth Asteroid Rendezvous (NEAR)	U.S, NASA Discovery Program	17-Feb-1996	Completed mission	Eros	Useful data acquired following successful landing on Eros	7 days on the surface	NaI(Tl) with BGO antico-incidence shield	None	Abundances for O, Mg, Si, Fe, and K
Lunar Prospector	U.S, NASA Discovery Program	6-Jan-1998	Completed mission by planned impact in a south polar crater	Moon	High and low altitude circular polar mapping orbits (100 km and 30 km, respectively)	300 days at high altitude; 220 days at low altitude	BGO with boron-loaded plastic anti-coincidence shield	³He gas proportional counters and boron-loaded plastic scintillator	Discovery of enhanced water-equivalent hydrogen associated with polar cold traps; global maps of major and radioactive elements

Mission	Sponsor	Launch date	Status	Target	Orbit	Duration	Detector	Neutron detector	Objectives
2001 Mars Odyssey	U.S., NASA Mars Exploration Program	7-Apr-2001	Completed primary mission; extended mission underway	Mars	400 km altitude circular polar mapping orbit	Over 2 Mars years completed, Extended mission in progress	HPGe, passively cooled	Boron-loaded plastic scintillators (NS); Stilbene and ^{3}He tubes (HEND)[d]	Distribution of water-equivalent hydrogen and high-latitude stratigraphy; seasonal variations in CO_2 ice and noncondensable gasses; and global maps of major and radioactive elements
MESSENGER	U.S., NASA Discovery Program	3-Aug-2004	Cruise phase	Mercury	Elliptical polar orbit with periapsis at 200 km altitude, 60°N latitude, 15,000 km apoapsis[b]	1 year starting in 2011[b]	HPGe, actively cooled	^{6}Li-loaded glass and boron-loaded plastic scintillators	Maps of elemental composition in the northern hemisphere; search for polar hydrogen deposits
Dawn	U.S., NASA Discovery Program	Summer, 2007	Preparing for launch[e]	Vesta and Ceres	Survey, high, and low altitude circular polar mapping orbits[b]	6 months at each asteroid[b]	CdZnTe and BGO	^{6}Li-loaded glass and boron-loaded plastic scintillators	Global maps of major and radioactive elements and ice constituents (H and C)
Selene	Japan	2007, TBD	Preparing for launch[e]	Moon	100 km circular polar mapping orbit	1 year	HPGe, actively cooled	None	Abundance of major and radioactive elements.

[a] Refers to the time periods during which gamma ray and/or neutron data were acquired.

[b] Objectives are listed for Mars Observer, MESSENGER, Dawn, and Selene.

[c] Neutron and gamma ray spectrometers were flown on Phobos 1, which was launched on 7-July-1988; however, Phobos 1 was lost during the cruise phase of the mission. The Mars 4 and 5 missions (U.S.S.R., 1973) flew identical sodium iodide gamma ray spectrometers. A few gamma ray spectra were acquired by Mars 5 while in an elliptical orbit around Mars (apoapsis 32,560 km, periapsis 1760 km, inclination 35° to the equator).

[d] The high energy neutron detector (HEND) was provided by the Russian Federation.

[e] Future missions that have advanced past the planning stage are listed.

Mars, respectively. Highlights of the science carried out on these missions are presented along with a description of their instrumentation. The Moon and Mars are very different, both in their origin and composition. With the possible exception of polar water ice, the Moon is bone dry and has no atmosphere. The lunar surface has been extensively modified by cratering and basaltic volcanism. Mars has a tenuous atmosphere, extensive water ice deposits, and seasonal CO_2 caps. Volcanic, aqueous, and eolian processes have continued to shape the surface of Mars long past the primordial formation of the crust. The differences between these two bodies will provide the reader with insights into the wealth of information provided by nuclear spectroscopy and the challenges faced in the analysis of the data. For *Lunar Prospector*, emphasis is placed on the combined analysis of neutron and gamma ray data to determine the abundance of major and trace radioactive elements. For *Mars Odyssey*, results from the neutron spectrometer for global water abundance and the seasonal caps are presented.

5.1 Lunar Prospector

Lunar Prospector was a spin-stabilized spacecraft, with the spin axis perpendicular to the plane of the ecliptic. The instruments were deployed on booms to minimize

backgrounds from the spacecraft (Fig. 8a). The payload included a large-volume BGO gamma ray spectrometer (GRS), which was surrounded by a boron-loaded plastic anticoincidence shield (Fig. 8b). The shield served two purposes: (1) to suppress the Compton continuum caused by gamma rays escaping the BGO crystal and to reject energetic particle events; and (2) to measure the spectrum of fast neutrons from the lunar surface using the double-pulse technique described in Section 3.2. Sn- and Cd-covered ³He gas proportional counters were used to detect and separately measure thermal and epithermal neutrons. Gamma ray and neutron spectroscopy data were acquired for long periods of time (Table 1), providing full coverage of the Moon at 100- and 30-km altitude.

The data were analyzed to determine global maps of surface elemental composition. The resulting abundance data were mapped on different spatial scales, depending on the precision of the data and the altitude of the spectrometer. Results of the analysis were submitted to the NASA Planetary Data System and include the following data sets: the abundance of hydrogen from neutron spectroscopy (0.5° **equal angle map**); the average atomic mass from fast neutron spectroscopy (2° **equal area maps**); the abundance of major oxides, including MgO, Al_2O_3, SiO_2, CaO, TiO_2, and FeO, and trace incompatible elements K, Th, and U

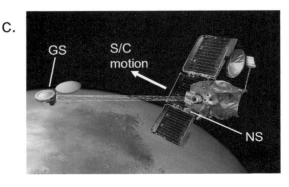

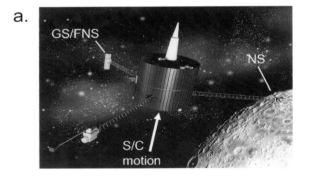

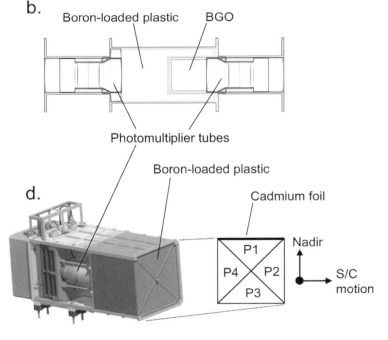

FIGURE 8 (a) Annotated artist's conception of *Lunar Prospector*; (b) Cross sectional view of the gamma ray and fast neutron spectrometer; (c) annotated artist's conception of *2001 Mars Odyssey*; (d) engineering drawing of the Neutron Spectrometer on *Odyssey* cut-away to show the boron loaded plastic scintillators. A schematic diagram of the arrangement of scintillators and their orientation relative to spacecraft motion and nadir is also shown. (Parts a and c courtesy of NASA.)

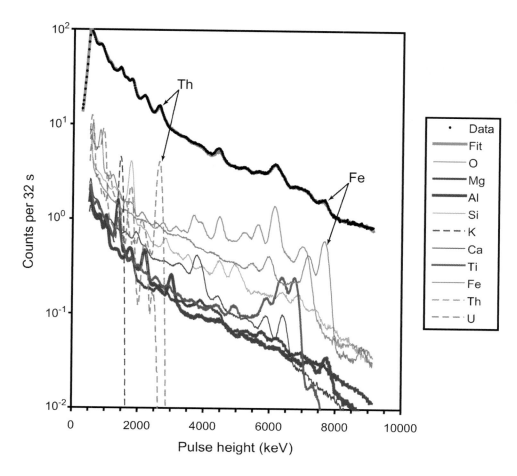

FIGURE 9 *Lunar Prospector* gamma ray spectrum for a 20° equal-area pixel in the western mare is compared to the fitted spectrum and elemental spectral components (*see text for details*).

(2°-, 5°-, and 20°-equal area maps) using a combination of gamma ray and neutron spectroscopy; and the abundance of the rare-earth elements (Gd + Sm) from neutron spectroscopy (0.5° equal angle map).

Perhaps the most significant result of *Lunar Prospector* was the discovery of enhanced hydrogen at the poles in association with craters in permanent shadow, which are thought to be cold traps for water ice. If present, water ice could be an important resource for manned exploration. Consequently, the polar cold traps are a prime target for future missions. Geochemical results from the analysis of neutron and gamma ray spectra fully reveal the dichotomy in the composition of the Moon, with a near side that is enriched in incompatible elements and mafic minerals and a thick far-side crust primarily consisting of plagioclase feldspar.

Global geochemical trends observed by *Lunar Prospector* are not significantly different from trends observed in the sample and meteoritic data; however, there are some notable discrepancies that point to the existence of unique lithologies that are not well represented by the lunar sample data. Interpretation and analysis of the data is ongoing with emphasis on regional studies. For example, the impact that formed the South Pole Aitken (SPA) basin could have excavated into the mantle. Analysis of the composition of

the basin floor may reveal information about the bulk composition of the mantle and lower crust.

The analysis of major and radioactive elements was carried out using a combination of gamma ray and neutron spectroscopy data. A typical gamma ray spectrum is shown in Fig. 9 for a 20° equal-area pixel in the western mare. Two intense, well-resolved peaks, labeled in Fig. 9, were analyzed to determine the abundance of Fe and Th. In addition, a spectral unmixing algorithm similar to those used to analyze spectral reflectance data was developed to simultaneously determine the abundance of all major and radioactive elements from the gamma ray spectrum.

Lunar gamma ray spectra can be modeled as a linear mixture of elemental spectral shapes. The magnitude of the spectral components must be adjusted to account for the nonlinear coupling of gamma ray production to the neutron number density (for neutron capture reactions) and the flux of fast neutrons (for nonelastic reactions). Once the adjustment is made, a linear least squares problem can be solved to determine elemental weight fractions. Fitted elemental spectral shapes are shown, for example, in Fig. 9.

Abundance maps for selected elements determined by *Lunar Prospector* are shown in Fig. 10. To provide context for the elemental abundance maps, a map of topography determined by *Clementine*, superimposed on a shaded

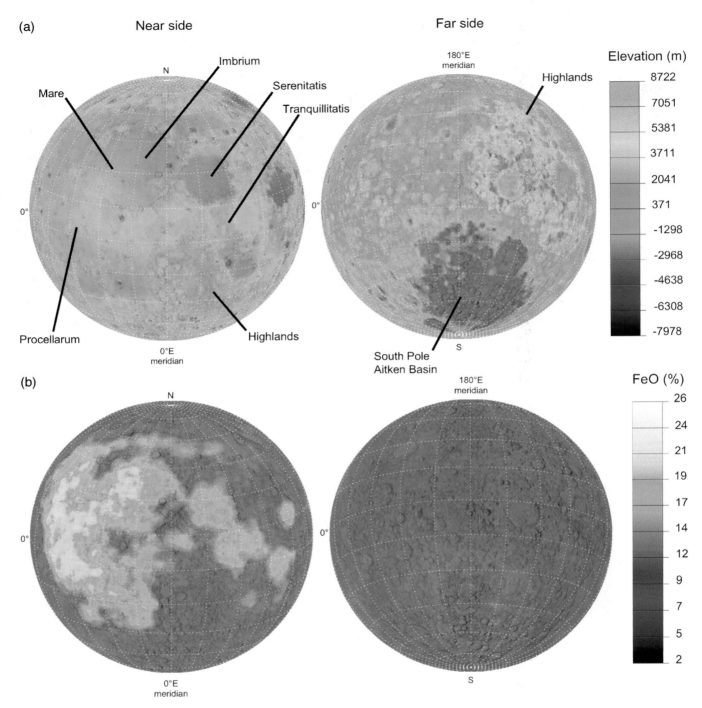

FIGURE 10 Orthographic projections of the lunar near and far sides: (a) Elevation; (b–d) abundance (weight fraction) of selected elements. The map data are superimposed on a shaded relief image. (FeO data courtesy NASA Planetary Data System ; image courtesy United States Geological Survey.)

relief image, is shown in Fig. 10a. The far side includes the feldspathic highlands and the SPA basin. The near side consists of major basins, including Procellarum and Imbrium, which contain mare basalts. The mare basalts are rich in Fe, with the highest concentrations occurring in western Procellarum (Fig. 10b). The low abundance of FeO in the highlands, which are rich in plagioclase feldspar, reflects a significant lunar geochemical trend in which mafic silicate minerals are displaced by plagioclase, which is Fe-poor.

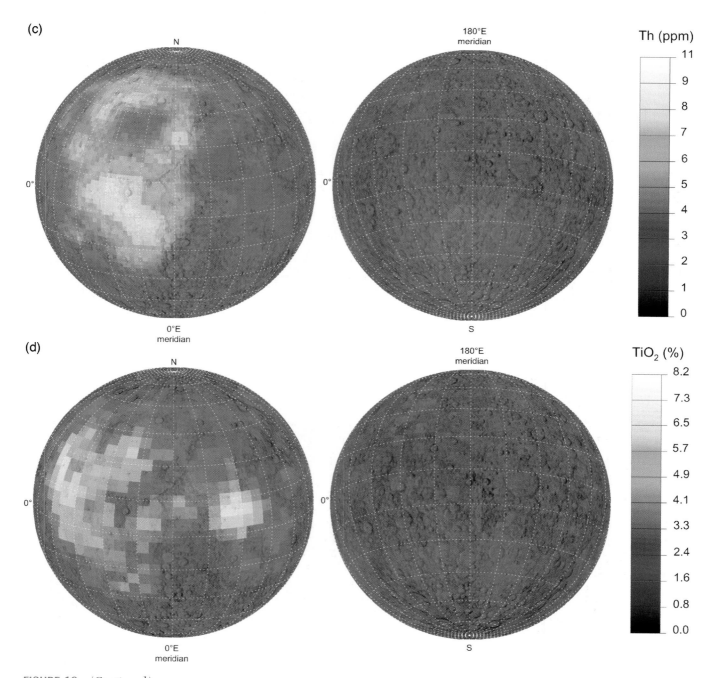

FIGURE 10 *(Continued)*

A large portion of the western near side is enriched in radioactive elements such as Th (Fig. 10c). K, Th, and U are incompatible with major lunar minerals and were likely concentrated in the residual melt during lunar differentiation. Consequently, their distribution on the surface and with depth has important implications to lunar evolution. The association of high Th concentrations with the mare suggests that heating by radioactive elements may have significantly influenced lunar thermal evolution and mare volcanism.

The distribution of TiO_2 is shown in Fig. 10d as a 5° equal area map. The low spatial resolution of the TiO_2 map compared to FeO and Th is a consequence of the relatively low intensity of the Ti gamma rays and their position in the gamma ray spectrum near strong peaks from O and Fe (Fig. 9). The abundance of TiO_2 can be used to classify mare basalts. Strong spatial variations in the abundance of TiO_2, for example, indicate that different source regions and processes were involved in creating the basalts that comprise the mare. The highest concentrations of TiO_2 are found in Tranquillitatis as shown in Fig. 10d; however, high concentrations are also found in western Procellarum. The abundances of Fe and Ti observed in western Procellarum suggest that this region has a unique composition that is not well represented by the lunar samples.

5.2 Mars Odyssey

As of this writing, *2001 Mars Odyssey* is in an extended mission having successfully completed over two Mars years of mapping (each Mars year is 687 days). *Odyssey* is in a circular polar mapping orbit around Mars at an altitude of approximately 400 km (Table 1). The nuclear spectroscopy payload consists of a GRS, a neutron spectrometer (NS), and a Russian-supplied high energy neutron detector (HEND). Gamma ray and neutron spectroscopy data acquired by *Mars Odyssey* provide constraints on geochemistry, the water cycle, climate history, and atmospheric processes, including atmospheric dynamics and atmosphere-surface interactions [see Mars Atmosphere: History and Surface Interaction].

Since the discovery of abundant subsurface **water-equivalent hydrogen** (WEH) at high latitudes, *Odyssey's* gamma ray and neutron spectrometers have continued to provide a wealth of new information about Mars, including the global distribution of near-surface WEH, the elemental composition of the surface, seasonal variations in the composition of the atmosphere at high latitudes, and the column abundance of CO_2 ice in the seasonal caps. This information has contributed to our understanding of the recent history of Mars: The climate is driven strongly by short-term variations in orbital parameters, principally the obliquity, and the surface distribution of surface water-ice is controlled by atmosphere-surface interactions. The discovery of anomalously large amounts of WEH at low latitudes, where water ice is not stable, has stirred considerable debate about the mineral composition of the surface and climate change.

The GRS on *Odyssey* is boom-mounted, passively cooled, HPGe spectrometer, similar in design to the instrument flown on *Mars Observer* (Fig. 8c). The NS is a deck-mounted instrument that consists of a boron-loaded plastic block (roughly 10 cm on a side), which has been diagonally segmented into four prisms and read out by separate photomultiplier tubes (Fig. 8d). The orientation of the spacecraft is constant such that one of the prisms faces nadir (P1), one faces zenith (P3), one faces in the direction of spacecraft motion (P2), and one faces opposite the spacecraft motion (P4). P1 is covered with a Cd foil that prevents thermal neutrons from entering the prism. Consequently, P1 is sensitive to epithermal and fast neutrons originating from the surface and atmosphere.

Neutrons with energy less than the gravitational binding energy of Mars, approximately 0.13 eV, corresponding to an escape speed of about 5000 m/s, travel on parabolic trajectories and return to Mars unless they decay by beta emission. The mean lifetime of a neutron is approximately 900 s. The most probable energy for neutrons in thermal equilibrium with the surface of Mars (for the mean martian temperature of 210 K) is 0.018 eV, which corresponds to a neutron speed of 1860 m/s. Consequently, a significant portion of the thermal neutron population travels on ballistic trajectories and are incident on the spectrometer from above and below. Neutrons that leave the atmosphere with energies less than about 0.014 eV, just below the most probable energy, cannot reach the 400 km orbital altitude of *Odyssey*. Consequently, gravitational binding has a significant effect on the flux and energy distribution of neutrons at *Odyssey's* orbital altitude, and, in contrast to the simplified discussion in Section 3.1, gravitational effects must be accounted for in models of the flux and instrument response.

To separate thermal and epithermal neutrons, the NS makes use of the orbital speed of the spacecraft, which is approximately 3400 m/s, the same speed as a 0.05 eV neutron. Neutrons below the speed of the spacecraft (most of the thermal neutron population) can't catch up to P4. So, P4 is primarily sensitive to epithermal neutrons. In contrast, P2 "rams" into thermal neutrons that arrive at the orbital altitude ahead of the spacecraft. P2 has roughly the same sensitivity as P4 for epithermal neutrons. Consequently, the thermal flux is given by the difference between the counting rates for P2 and P4.

Thermal, epithermal, and fast neutrons are sensitive to surface and atmospheric parameters, including the abundance and stratigraphy of hydrogen in the surface, the presence of strong neutron absorbers such as Cl and Fe in the Martian rocks and soil, the presence of CO_2 ice on the surface, the column abundance of the atmosphere, and the enrichment and depletion of noncondensable gasses, N_2 and Ar, as CO_2 is cycled through the seasonal caps (Table 2). The effect of these parameters on the neutron counting rate can be explored using a simple physical model of the surface and atmosphere as described in Section 2.2 (Fig. 4a). Models of the counting rate are then used to develop algorithms to determine parameters from observations.

For example, the variation of thermal, epithermal, and fast neutron counting rates as a function of water abundance in a homogeneous surface is shown in Fig. 11a. Epithermal and fast neutrons are sensitive to hydrogen (as described in Section 2.2) and their counting rates decrease monotonically with water abundance. Both are insensitive to the abundance of elements in the surface other than hydrogen, as illustrated in Fig. 11a by changing the abundance of Cl, which is a strong thermal neutron absorber. In contrast, thermal neutrons are sensitive to variations in major-element composition and relatively insensitive to hydrogen when the abundance of WEH is less than about 10%. Epithermal neutrons are a good choice for determining the WEH abundance because of their high counting rate and relative insensitivity to other parameters. Measured

		Major	CO_2-free surface	Atmospheric/seasonal
Type	**Energy Range**	**interactions**	**parameters***	**parameters**
Fast	>0.2 MeV	Inelastic scattering, elastic scattering	WEH abundance and stratigraphy, Average atomic mass	Atmospheric mass, CO_2 ice column abundance <100 g/cm²
Epithermal	0.5 eV (Cd cutoff) to 0.2 MeV	Elastic scattering	WEH abundance and stratigraphy	Atmospheric mass, CO_2 ice column abundance up to about 150 g/cm²
Thermal	<0.5 eV (Maxwellian energy distribution)	Elastic scattering, capture (absorption)	WEH abundance, Absorption by Fe, Cl, Ti. Stratigraphy of WEH and absorbers	CO_2 ice column abundance up to about 1000 g/cm², Absorption by N_2 and Ar

TABLE 2 Sensitivity of neutron energy ranges to Mars surface and atmospheric parameters

*The surface in the northern or southern hemisphere during summer following the recession of the seasonal cap.

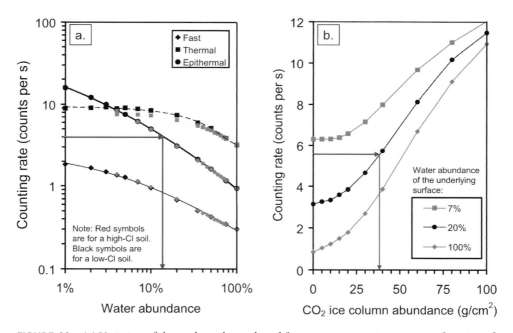

FIGURE 11   (a) Variation of thermal, epithermal, and fast neutron counting rates as a function of water abundance for a soil composition with low Cl abundance (black symbols). The red symbols correspond to a soil with higher Cl abundance, similar to the average composition of soils at the *Pathfinder* landing site. Note that the epithermal and fast neutron counting rates are unaffected by the change in Cl abundance. Because Cl is a strong absorber of thermal neutrons, the thermal neutron counting rate is sensitive to Cl abundance. (b) Variation of epithermal counting rate as a function of CO_2 ice column abundance covering homogeneous surfaces containing 7%, 20%, and 100% water ice (mixed with dry soil). Observed counting rates can be converted directly to water-equivalent hydrogen abundance or CO_2 ice column abundance using the model results in parts a and b as indicated by the arrows. The counting rate during the summer, which is a measure of the water abundance of the underlying surface, must be known in order to select the correct trend for CO_2 ice column abundance.

Water Equivalent Hydrogen (weight fraction)

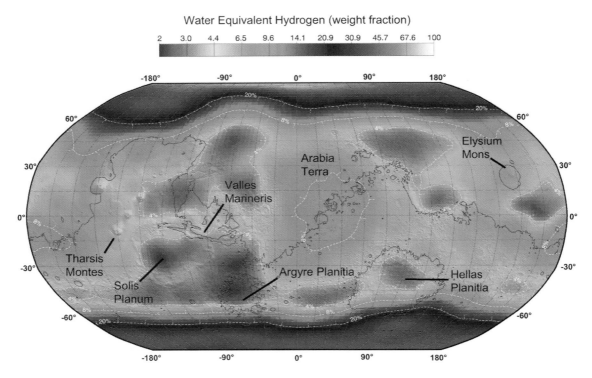

FIGURE 12 Global map of the abundance (weight fraction) of WEH. The map gives a lower bound on the abundance of WEH. Contours for 4%, 8%, and 20% WEH are shown as dashed white lines. The black contour line corresponds to 0 km elevation. The map data are superimposed on a shaded relief image. (Elevation data and shaded relief image courtesy of the NASA Mars Orbiter Laser Altimeter Science Team.)

epithermal counting rates can be converted directly to WEH as indicated by the arrows in Fig. 11a.

A map of WEH determined from measured epithermal counting rates is shown in Fig. 12. In order to avoid contributions from the seasonal CO_2 ice, the northern and southern high latitudes only included counting rates measured during their respective summers. The algorithm for determining WEH included corrections for minor variations in the counting rate due to changes in the atmospheric column abundance with topography. The map gives a lower bound on WEH. Higher WEH abundances could be present if the surface is stratified, for example, with a dry top layer covering a water-rich medium.

The minimum WEH abundance on Mars ranges from 2% in equatorial and midlatitude regions to nearly 100% for the north polar water-ice cap. Low abundances of WEH are found in regions such as northern Argyre Planitia, the midlatitude, southern highlands, Solis Planum, and the eastern flanks of the Tharsis Montes. Correlations between WEH and topography suggest that some aspects of the surface distribution of WEH can be explained by regional and global weather patterns. Moderate WEH abundances (8–10%) can be found in large equatorial regions, for example, in Arabia Terra. Ice stability models predict that water-ice is not stable at equatorial latitudes on Mars under present cli-

mate conditions. Consequently, the moderate abundances of WEH may be in the form of hydrated minerals, possibly as magnesium sulfate hydrate. High abundances of WEH are found at high northern and southern latitudes (poleward of 60°). A detailed analysis of neutron and gamma ray counting rates suggests that the high latitude surface outside of the residual caps consists of soil rich in water-ice covered by a thin layer of dessicated material (soil and rocks). This result is consistent with models that predict that water ice is stable at shallow depths at high latitudes. Similar terrestrial conditions are observed in the Dry Valleys of Antarctica, where ice is stable beneath a dry soil layer that provides thermal and diffusive isolation of the ice from the atmosphere.

Seasonal variations on Mars are driven by its obliquity relative to the orbital plane, which is similar to that of Earth. In the polar night in the northern and southern hemispheres, atmospheric CO_2 condenses to form ice on the surface. Approximately 25% of the martian atmosphere is cycled into and out of the northern and southern seasonal caps. Consequently, the seasonal caps play a major role in atmospheric circulation. The main questions about the seasonal caps that remain unanswered concern the local energy balance, polar atmospheric dynamics, and CO_2 condensation mechanisms. Seasonal parameters constrained by neutron spectroscopy include the column abundance of CO_2

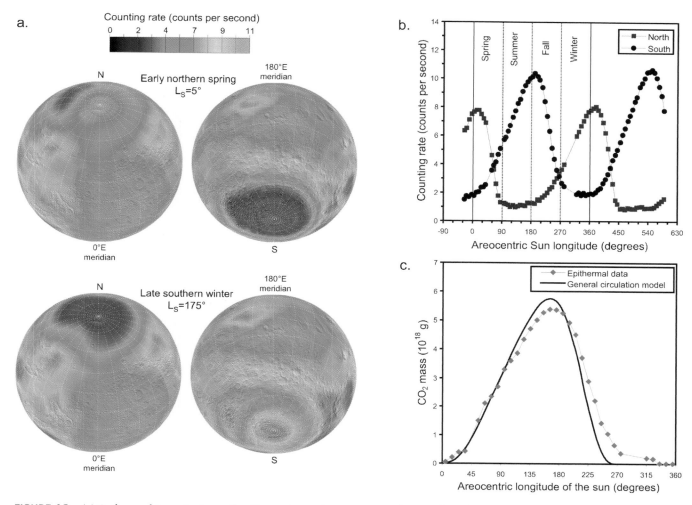

FIGURE 13   (a) Orthographic projections of epithermal counting rates in northern and southern hemispheres of Mars during early northern spring and late southern winter; (b) Epithermal counting rate as a function of time at the north and south pole (poleward of 85°); (c) total mass of CO_2 in the southern seasonal cap poleward of 60°S from a general circulation model is compared to that determined from the epithermal counting data. (General circulation model results courtesy of NASA Ames Research Center and the New Mexico State University Department of Astronomy.)

ice on the surface and the column abundance of noncondensable gasses (N_2 and Ar) in the atmosphere. For example, analyses of gamma ray and neutron spectroscopy data reveal that the southern atmosphere is strongly enriched in N_2 and Ar during cap growth. The observed enrichment may be caused by the formation of a strong polar vortex accompanying the condensation flow, which inhibits **meridional mixing** of the polar atmosphere with lower latitudes.

Based on simulations, the epithermal neutron counting rate generally increases with the column abundance of CO_2 ice on the surface; however, the trend depends on the abundance of water ice in the underlying surface as is shown in Fig. 11b. The sensitivity of epithermal neutrons to CO_2

ice is higher for surfaces that contain more water ice. At high latitudes, the column abundance of CO_2 can be determined from seasonal epithermal counting rates, given the counting rate during summer, when no CO_2-ice is present.

Maps of epithermal counting rates are shown in Fig. 13a. The extent of the seasonal caps can be seen by comparing maps of the northern and southern hemispheres during the two time periods shown in Fig. 13a. For example, during late southern winter, low counting rates are observed in the northern high latitudes, corresponding to the summertime CO_2 frost-free surface, which contains abundant water ice. In early northern spring, elevated epithermal counting rates are observed in the northern hemisphere, corresponding to CO_2 ice on the surface.

During their respective winters, the counting rate at high latitudes increases towards the poles, which indicates that the CO_2 ice column abundance increases with latitude. The observed spatial variation is expected since the polar night lasts longer at higher latitudes and frost has more time to accumulate. The time variation in epithermal counting rates for the north and south poles (poleward of $85°$), shown in Fig. 13b, reveals the cyclic behavior of the seasonal caps during two Mars years. The total inventory of CO_2 in the seasonal caps determined from epithermal counting data is similar to that predicted by general circulation models (GCMs) (for example, see Fig. 13c). The ability to measure the thickness of the CO_2 caps in the polar night is unique to gamma ray and neutron spectroscopy. Local ice column abundances determined by nuclear spectroscopy can be compared to GCM predictions, providing information needed to improve physical models of the seasonal caps and the polar energy balance.

6. Future Prospects

Given the number of orbiters, landers, and rovers targeting Mars and the renewed emphasis on lunar exploration, the Moon and Mars will be the focus of planetary science for years to come. On the Moon, neutron spectrometers may be used on rovers or incorporated into borehole logging tools to search for and characterize water-ice deposits in polar craters. On Mars, gamma ray and neutron spectrometers may be included on rovers, landers, weather stations, and drilling systems for *in situ* determination of composition, for example, to investigate small spatial scale variations in composition and to look for water deep within the Martian surface. In addition, there may be opportunities for low-altitude, high-spatial resolution measurements of selected regions from an airplane or balloon platform. Continued effort is needed to analyze and interpret data already acquired by *Lunar Prospector* and *Mars Odyssey* and to synthesize the information with other data sets to develop a coherent picture of the Moon and Mars. Orbital nuclear spectroscopy will also play an important role on future solar system exploration missions. For example, the *MESSENGER* mission to Mercury and the *Dawn* mission to the main asteroid belt include gamma ray and neutron spectrometers on their payloads.

Bibliography

Boynton, W. V., et al. (2004). The Mars Odyssey gamma-ray spectrometer instrument suite. *Space Science Rev.* **110**, 1–2, 37–83.

Duderstadt, J. J., and L. J. Hamilton (1967). "Nuclear Reactor Analysis" John Wiley & Sons.

Elphic, R. C., et al. (2002). Lunar Prospector neutron spectrometer constraints on TiO2. *J. Geophys. Res.* **107**, E4, 5024, 10.1029/2000JE001460.

Feldman, W. C., et al. (2001). Evidence for water ice near the lunar poles. *J. Geophys. Res.* **106**, 23, 231–23, 251.

Feldman, W. C., et al. (2004). Gamma-ray, neutron, and alpha-particle spectrometers for the Lunar Prospector mission. *J. Geophys. Res. Planets*. **109**, E7, p. E07S06.

Feldman, W. C., et al. (2004). Global distribution of near-surface hydrogen on Mars. *J. Geophys. Res. Planets*. **109**, E9, p E09006.

Knoll, G. F. (1989). "Radiation detection and measurement" 2nd edition, John Wiley & Sons.

Lawrence, D. J., et al. (2000). Thorium abundances on the lunar surface. *J. Geophys. Res.* **105**(E8), 20, 307–20, 331.

Lawrence, D. J., et al. (2002). Iron abundances on the lunar surface as measured by the Lunar Prospector gamma-ray and neutron spectrometers. *J. Geophys. Res.* **107**(E12), 5130, doi:10.1029/2001JE001530.

Pieters, C. M., and P. A. J. Englert, Eds. (1993). "Remote Geochemical Analysis: Elemental and Mineralogical Composition" Cambridge University Press.

Prettyman, T. H., et al. (2004). Composition and structure of the Martian surface at high southern latitudes from neutron spectroscopy. *J. Geophys. Res.* **109**, p. E05001, doi:10.1029/2003JE002139.

Prettyman, T. H., et al. (2006). Elemental composition of the lunar surface: Analysis of gamma ray spectroscopy data from Lunar Prospector. *J. Geophys. Res.* **111**, doi:10.1029/2005JE002656, in press.

Reedy, R. C. (1978). Planetary gamma-ray spectroscopy. Proceedings of the 9th Lunar and Planetary Science Conference, pp. 2961–2984.

Solar System Dynamics: Regular and Chaotic Motion

Jack J. Lissauer
NASA Ames Research Center
Moffett Field, California

Carl D. Murray
Queen Mary, University of London
London, U.K.

CHAPTER 42

1. Introduction: Keplerian Motion

The study of the motion of celestial bodies within our solar system has played a key role in the broader development of classical mechanics. In 1687, Isaac Newton published his *Principia*, in which he presented a unified theory of the motion of bodies in the heavens and on the Earth. Newtonian physics has proven to provide a remarkably good description of a multitude of phenomena on a wide range of length scales. Many of the mathematical tools developed over the centuries to analyze planetary motions in the Newtonian framework have found applications for terrestrial phenomena. The concept of **deterministic** chaos, now known to play a major role in weather patterns on the Earth, was first conceived in connection with planetary motions (by Poincaré, in the late 19th century). Deviations of the orbit of Uranus from that predicted by **Newton's Laws** led to the discovery of the planet Neptune. In contrast, the first major success of Einstein's general theory of relativity was to explain deviations of Mercury's orbit that could not be accounted for by Newtonian physics. But general relativistic corrections to planetary motions are quite small, so this article concentrates on the rich and varied effects of Newtonian gravitation, together with briefer descriptions of non-gravitational forces that affect the motions of some objects in the solar system.

Newton showed that the motion of two spherically symmetric bodies resulting from their mutual gravitational attraction is described by simple conic sections (see Section 2.4). However, the introduction of additional gravitating bodies produces a rich variety of dynamical phenomena, even though the basic interactions between pairs of objects can be straightforwardly described. Even few-body systems governed by apparently simple nonlinear interactions can display remarkably complex behavior, which has come to be known collectively as chaos. On sufficiently long timescales, the apparently regular orbital motion of many bodies in the solar system can exhibit symptoms of this chaotic behavior.

An object in the solar system exhibits chaotic behavior in its orbit or rotation if the motion is sensitively dependent on the starting conditions, such that small changes in its initial state produce different final states. Examples of **chaotic motion** in the solar system include the rotation of the Saturnian satellite Hyperion, the orbital evolution of numerous asteroids and comets, and the orbit of Pluto. Numerical investigations suggest that the motion of the planetary system as a whole is chaotic, although there are no signs of any gross instability in the orbits of the planets. Chaotic motion has probably played an important role in determining the dynamical structure of the solar system.

In this article the basic orbital properties of solar system objects (planets, moons, minor bodies, and dust) and their mutual interactions are described. Several examples are provided of important dynamical processes that occur in the solar system and groundwork is laid for describing some of the phenomena that are discussed in more detail in other articles of this encyclopedia.

1.1 Kepler's Laws of Planetary Motion

By analyzing Tycho Brahe's careful observations of the orbits of the planets, Johannes Kepler deduced the following three laws of planetary motion:

1. All planets move along elliptical paths with the Sun at one focus. The heliocentric distance r (i.e., the planet's distance from the Sun) can be expressed as

$$r = \frac{a(1 - e^2)}{1 + e \cos f},\qquad(1)$$

with a the semimajor axis (average of the minimum and maximum heliocentric distances) and e (the eccentricity of the orbit) $\equiv (1 - b^2/a^2)^{1/2}$, where $2b$ is the minor axis of an ellipse. The true anomaly, f, is the angle between the planet's perihelion (closest heliocentric distance) and its instantaneous position (Fig. 1).

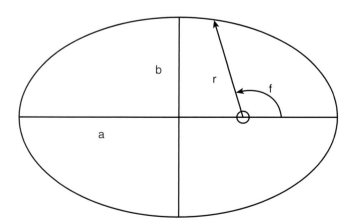

FIGURE 1 Geometry of an elliptical orbit. The Sun is at one focus and the vector **r** denotes the instantaneous heliocentric location of the planet (i.e., r is the planet's distance from the Sun). a is the semimajor axis (average heliocentric distance), and b is the semiminor axis of the ellipse. The true anomaly, f, is the angle between the planet's perihelion (closest heliocentric distance) and its instantaneous position.

2. A line connecting a planet and the Sun sweeps out equal areas ΔA in equal periods of time Δt:

$$\frac{\Delta A}{\Delta t} = \text{constant.}\qquad(2)$$

Note that the value of this constant differs from one planet to the next.

3. The square of a planet's orbital period P about the Sun (in years) is equal to the cube of its semimajor axis a (in AU):

$$P^2 = a^3.\qquad(3)$$

1.2 EllipticaL Motion, Orbital Elements, and the Orbit in Space

The Sun contains more than 99.8% of the mass of the known solar system. The gravitational force exerted by a body is proportional to its mass (Eq. 5), so to an excellent first approximation the motion of the planets and many other bodies can be regarded as being solely due to the influence of a fixed central pointlike mass. For objects like the planets, which are bound to the Sun and hence cannot go arbitrarily far from the central mass, the general solution for the orbit is the ellipse described by Eq. (1). The orbital plane, although fixed in space, can be arbitrarily oriented with respect to whatever reference plane is chosen (such as Earth's orbital plane about the Sun, which is called the **ecliptic**, or the equator of the primary). The inclination, i, of the orbital plane is the angle between the reference plane and the orbital plane and can range from 0 to 180°. Conventionally, bodies orbiting in a direct sense, with orbital angular momentum vectors within 90° of the direction of the Earth's orbital angular momentum (or the rotational angular momentum of the primary), are defined to have inclinations from 0° to 90° and are said to be on prograde orbits. Bodies traveling in the opposite direction are defined to have inclinations from 90° to 180° and are said to be on retrograde orbits. The two planes intersect in a line called the line of nodes and the orbit pierces the reference plane at two locations—one as the body passes upward through the plane (the ascending node) and one as it descends (the descending node). A fixed direction in the reference plane is chosen and the angle to the direction of the orbit's ascending node is called the longitude of the ascending node, Ω. Finally, the angle between the line to the ascending node and the line to the direction of periapse (perihelion for orbits about the Sun, perigee for orbits about Earth) is called the argument of periapse ω. An additional angle, the longitude of periapse $\varpi = \omega + \Omega$ is sometimes used in place of ω. The six orbital elements a, e, i, Ω, ω and f uniquely specify the location of the object in space (Fig. 2). The first

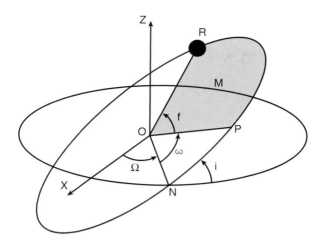

FIGURE 2   Geometry of an orbit in three dimensions. The Sun is at one focus of the ellipse (O) and the planet is instantaneously at location R. The location of the perihelion of the orbit is P. The intersection of the orbital plane ($X - Y$) and the reference plane is along the line ON (where N is the ascending node). The various angles shown are described in the text. The mean anomaly M is an angle proportional to the area OPR swept out by the radius vector **OR** (Kepler's second law).

three quantities (a, e, and i) are often referred to as the principal orbital elements, as they describe the orbit's size, shape, and tilt, respectively.

2. The Two-body Problem

In this section the general solution to the problem of the motion of two otherwise isolated objects in which the only force acting on each body is the mutual gravitational interaction is discussed.

2.1 Newton's Laws of Motion and the Universal Law of Gravitation

Although Kepler's laws were originally found from careful observation of planetary motion, they were subsequently shown to be derivable from Newton's laws of motion together with his universal law of gravity. Consider a body of mass m_1 at instantaneous location $\mathbf{r}_1$ with instantaneous velocity $\mathbf{v}_1 \equiv d\mathbf{r}_1/dt$ and hence momentum $\mathbf{p}_1 \equiv m_1\mathbf{v}_1$. The acceleration $d\mathbf{v}_1/dt$ produced by a net force $\mathbf{F}_1$ is given by Newton's second law of motion:

$$\mathbf{F}_1 = \frac{d(m_1\mathbf{v}_1)}{dt}. \tag{4}$$

Newton's universal law of gravity states that a second body of mass m_2 at position $\mathbf{r}_2$ exerts an attractive force on the

first body given by

$$\mathbf{F}_1 = -\frac{Gm_1m_2}{\mathbf{r}_{12}^3}\mathbf{r}_{12} = -\frac{Gm_1m_2}{\mathbf{r}_{12}^2}\hat{\mathbf{r}}_{12}, \tag{5}$$

where $\mathbf{r}_{12} \equiv \mathbf{r}_1 - \mathbf{r}_2$ is the location of particle 1 with respect to particle 2, $\hat{\mathbf{r}}_{12}$ is the unit vector in the direction of $\mathbf{r}_{12}$, and G is the gravitational constant. Newton's third law states that for every action there is an equal and opposite reaction; thus, the force on each object of a pair is equal in magnitude but opposite in direction. These facts are used to reduce the two-body problem to an equivalent one-body case in the next subsection.

2.2 Reduction to the One–body Case

From the foregoing discussion of Newton's laws, and the two-body problem the force exerted by body 1 *on* body 2 is

$$\frac{d(m_2\mathbf{v}_2)}{dt} = \mathbf{F}_2 = -\mathbf{F}_1 = \frac{Gm_1m_2}{\mathbf{r}_{12}^3}\mathbf{r}_{12} = \frac{Gm_1m_2}{\mathbf{r}_{12}^2}\hat{\mathbf{r}}_{12} \tag{6}$$

Thus, from Eqs. (4) and (6)

$$\frac{d(m_1\mathbf{v}_1 + m_2\mathbf{v}_2)}{dt} = \mathbf{F}_1 + \mathbf{F}_2 = 0. \tag{7}$$

This is of course a statement that the total linear momentum of the system is conserved, which means that the center of mass of the system moves with constant velocity.

Multiplying Eq. (6) by m_1 and Eq. (5) by m_2 and subtracting, the equation for the relative motion of the bodies can be cast in the form

$$\mu_r\frac{d^2\mathbf{r}_{12}}{dt^2} = \mu_r\frac{d^2(\mathbf{r}_1 - \mathbf{r}_2)}{dt^2} = -\frac{G\mu_r M}{\mathbf{r}_{12}^3}\mathbf{r}_{12}, \tag{8}$$

where $\mu_r \equiv m_1m_2/(m_l + m_2)$ is called the reduced mass and $M \equiv m_l + m_2$ is the total mass. Thus, the relative motion is completely equivalent to that of a particle of reduced mass μ_r orbiting a *fixed* central mass M. For known masses, specifying the elements of the relative orbit and the positions and velocities of the center of mass is completely equivalent to specifying the positions and velocities of both bodies. A detailed solution of the equation of motion (8) is discussed in any elementary text on orbital mechanics and in most general classical mechanics books. In the remainder of Section II, a few key results are given.

2.3 Energy, Circular Velocity, and Escape Velocity

The centripetal force necessary to keep an object of mass μ_r in a circular orbit of radius r with speed v_c is $\mu_r v_c^2/r$. Equating this to the gravitational force exerted by the central body of mass M, the circular velocity is

$$v_{\mathrm{c}} = \sqrt{\frac{GM}{r}}. \tag{9}$$

Thus the orbital period (the time to move once around the circle) is

$$P = 2\pi r / v_{\mathrm{c}} = 2\pi \sqrt{\frac{r^3}{GM}}. \tag{10}$$

The total (kinetic plus potential) energy E of the system is a conserved quantity:

$$E = T + V = \frac{1}{2}\mu_r v^2 - \frac{GM\mu_r}{r}, \tag{11}$$

where the first term on the right is the kinetic energy of the system, T, and the second term is the potential energy of the system, V. If $E < 0$, the absolute value of the potential energy of the system is larger than its kinetic energy, and the system is bound. The body will orbit the central mass on an elliptical path. If $E > 0$, the kinetic energy is larger than the absolute value of the potential energy, and the system is unbound. The relative orbit is then described mathematically as a hyperbola. If $E = 0$, the kinetic and potential energies are equal in magnitude, and the relative orbit is a parabola. By setting the total energy equal to zero, the escape velocity at any separation can be calculated:

$$v_{\mathrm{e}} = \sqrt{\frac{2GM}{r}} = \sqrt{2}v_{\mathrm{c}}. \tag{12}$$

For circular orbits it is easy to show [using Eqs. (9) and (11)] that both the kinetic energy and the total energy of the system are equal in magnitude to half the potential energy:

$$T = -\frac{1}{2}V, \tag{13}$$

$$E = -\frac{GM\mu_r}{2r}. \tag{14}$$

For an elliptical orbit, Eq. (14) holds if the radius r is replaced by the semimajor axis a:

$$E = -\frac{GM\mu_r}{2a}. \tag{15}$$

Similarly, for an elliptical orbit, Eq. (10) becomes Newton's generalization of Kepler's third law:

$$P^2 = \frac{4\pi^2 a^3}{G(m_1 + m_2)}. \tag{16}$$

It can be shown that Kepler's second law follows immediately from the conservation of angular momentum, $\mathbf{L}$:

$$\frac{d\mathbf{L}}{dt} = \frac{d(\mu_r \mathbf{r} \times \mathbf{v})}{dt} = 0. \tag{17}$$

2.4 Orbital Elements: Elliptical, Parabolic, and Hyperbolic Orbits

As noted earlier, the relative orbit in the two-body problem is either an ellipse, parabola, or hyperbola depending on whether the energy is negative, zero, or positive, respectively. These curves are known collectively as conic sections and the generalization of Eq. (1) is

$$r = \frac{p}{1 + e \cos f}, \tag{18}$$

where r and f have the same meaning as in Eq. (1), e is the generalized **eccentricity**, and p is a conserved quantity which depends upon the initial conditions. For an ellipse, $p = a(1 - e^2)$, as in Eq. (1)). For a parabola, $e = 1$ and $p = 2q$, where q is the pericentric separation (distance of closest approach). For a hyperbola, $e > 1$ and $p = q(1 + e)$, where q is again the pericentric separation. For all orbits, the three orientation angles i, Ω, and ω are defined as in the elliptical case.

3. Planetary Perturbations and the Orbits of Small Bodies

Gravity is not restricted to interactions between the Sun and the planets or individual planets and their satellites, but rather all bodies feel the gravitational force of one another. Within the solar system, one body typically produces the dominant force on any given body, and the resultant motion can be thought of as a Keplerian orbit about a primary, subject to small perturbations by other bodies. In this section some important examples of the effects of these perturbations on the orbital motion are considered.

Classically, much of the discussion of the evolution of orbits in the solar system used perturbation theory as its foundation. Essentially, the method involves writing the equations of motion as the sum of a part that describes the independent Keplerian motion of the bodies about the Sun plus a part (called the disturbing function) that contains terms due to the pairwise interactions among the planets and minor bodies and the indirect terms associated with the back-reaction of the planets on the Sun. In general, one can then expand the disturbing function in terms of the small parameters of the problem (such as the ratio of the planetary masses to the solar mass, the eccentricities and inclinations, etc.), as well as the other orbital elements of the bodies, including the mean longitudes (i.e., the location of the bodies in their orbits), and attempt to solve the resulting equations for the time-dependence of the orbital elements.

3.1 Perturbed Keplerian Motion and Resonances

Although perturbations on a body's orbit are often small, they cannot always be ignored. They must be included in short-term calculations if high accuracy is required, for example, for predicting stellar occultations or targeting spacecraft. Most long-term perturbations are periodic in nature, their directions oscillating with the relative longitudes of the bodies or with some more complicated function of the bodies' orbital elements.

Small perturbations can produce large effects if the forcing frequency is commensurate or nearly commensurate with the natural frequency of oscillation of the responding elements. Under such circumstances, perturbations add coherently, and the effects of many small tugs can build up over time to create a large-amplitude, long-period response. This is an example of resonance forcing, which occurs in a wide range of physical systems.

An elementary example of resonance forcing is given by the simple one-dimensional harmonic oscillator, for which the equation of motion is

$$m\frac{d^2x}{dt^2} + m\Gamma^2 x = F_0\cos\varphi t. \tag{19}$$

In Eq. (19), m is the mass of the oscillating particle, F_0 is the amplitude of the driving force, Γ is the natural frequency of the oscillator, and φ is the forcing or resonance frequency. The solution to Eq. (19) is

$$x = x_0\cos\varphi t + A\cos\Gamma t + B\sin\Gamma t, \tag{20a}$$

where

$$x_0 \equiv \frac{F_0}{m(\Gamma^2 - \varphi^2)}, \tag{20b}$$

and A and B are constants determined by the initial conditions. Note that if $\varphi \approx \Gamma$, a large-amplitude, long-period response can occur even if F_0 is small. Moreover, if $\varphi = \Gamma$, this solution to Eq. (19) is invalid. In this case the solution is given by

$$x = \frac{F_0}{2m\Gamma}t\sin\Gamma t + A\cos\Gamma t + B\sin\Gamma t. \tag{21}$$

The t in front of the first term at the right-hand side of Eq. (21) leads to **secular** growth. Often this linear growth is moderated by the effects of nonlinear terms that are not included in the simple example provided here. However, some perturbations have a secular component.

Nearly exact orbital commensurabilities exist at many places in the solar system. Io orbits Jupiter twice as frequently as Europa does, which in turn orbits Jupiter twice as frequently as Ganymede does. Conjunctions (at which the bodies have the same longitude) always occur at the same position of Io's orbit (its perijove). How can such commensurabilities exist? After all, the probability of randomly picking a rational from the real number line is 0, and the number of small integer ratios is infinitely smaller still! The answer lies in the fact that orbital resonances may be held in place as stable locks, which result from nonlinear effects not represented in the foregoing simple mathematical example. For example, differential tidal recession (see Section 7.5) brings moons into resonance, and nonlinear interactions among the moons can keep them there.

Other examples of resonance locks include the Hilda asteroids, the Trojan asteroids, Neptune–Pluto, and the pairs of moons about Saturn, Mimas–Tethys and Enceladus–Dione. Resonant perturbation can also force material into highly eccentric orbits that may lead to collisions with other bodies; this is believed to be the dominant mechanism for clearing the Kirkwood gaps in the asteroid belt (see Section 5.1). Spiral density waves can result from resonant perturbations of a self-gravitating particle disk by an orbiting satellite. Density waves are seen at many resonances in Saturn's rings; they explain most of the structure seen in Saturn's A ring. The vertical analog of density waves, bending waves, are caused by resonant perturbations perpendicular to the ring plane due to a satellite in an orbit that is inclined to the ring. Spiral bending waves excited by the moons Mimas and Titan have been seen in Saturn's rings. In the next few subsections these manifestations of resonance effects that do not explicitly involve chaos are discussed. Chaotic motion produced by resonant forcing is discussed later in the chapter.

3.2 Examples of Resonances: Lagrangian Points, and Tadpole and Horseshoe Orbits

Many features of the orbits considered in this section can be understood by examining an idealized system in which two massive (but typically of unequal mass) bodies move on circular orbits about their common center of mass. If a third body is introduced that is much less massive than either of the first two, its motion can be followed by assuming that its gravitational force has no effect on the orbits of the other bodies. By considering the motion in a frame co-rotating with the massive pair (so that the pair remain fixed on a line that can be taken to be the x axis), Lagrange found that there are five points where particles placed at rest would feel no net force in the rotating frame. Three of the so-called **Lagrange points** (L_1, L_2, and L_3) lie along a line joining the two masses m_1 and m_2. The other two Lagrange points

(L_4 and L_5) form equilateral triangles with the two massive bodies.

Particles displaced slightly from the first three Lagrangian points will continue to move away and hence these locations are unstable. The triangular Lagrangian points are potential energy maxima, which are stable for sufficiently large primary to secondary mass ratio due to the Coriolis force. Provided that the most massive body has at least 27 times the mass of the secondary (which is the case for all known examples in the solar system larger than the Pluto–Charon system), the Lagrangian points L_4 and L_5 are stable points. Thus, a particle at L_4 or L_5 that is perturbed slightly will start to "orbit" these points in the rotating coordinate system. Lagrangian points L_4 and L_5 are important in the solar system. For example, the Trojan asteroids in Jupiter's Lagrangian points and both Neptune and Mars confine their own Trojans. There are also small moons in the triangular Lagrangian points of Tethys and Dione, in the Saturnian system. The L_4 and L_5 points in the Earth–Moon system have been suggested as possible locations for space stations.

3.2.1 HORSESHOE AND TADPOLE ORBITS

Consider a moon on a circular orbit about a planet. Figure 3 shows some important dynamical features in the frame corotating with the moon. All five Lagrangian points are indicated in the picture. A particle just interior to the moon's orbit has a higher angular velocity than the moon in the stationary frame, and thus moves with respect to the moon in the direction of corotation. A particle just outside the moon's orbit has a smaller angular velocity, and moves away from the moon in the opposite direction. When the outer particle approaches the moon, the particle is slowed down (loses angular momentum) and, provided the initial difference in semimajor axis is not too large, the particle drops to an orbit lower than that of the moon. The particle then recedes in the forward direction. Similarly, the particle at the lower orbit is accelerated as it catches up with the moon, resulting in an outward motion toward the higher, slower orbit. Orbits like these encircle the L_3, L_4, and L_5 points and are called **horseshoe orbits**. Saturn's small moons Janus and Epimetheus execute just such a dance, changing orbits every 4 years.

Since the Lagrangian points L_4 and L_5 are stable, material can librate about these points individually: such orbits are called **tadpole orbits**. The tadpole **libration** width at L_4 and L_5 is roughly equal to $(m/M)^{1/2}r$, and the horseshoe width is $(m/M)^{1/3}r$, where M is the mass of the planet, m the mass of the satellite, and r the distance between the two objects. For a planet of Saturn's mass, $M = 5.7 \times 10^{29}$ g, and a typical small moon of mass $m = 10^{20}$ g (e.g., an object with a 30-km radius, with density of ~ 1 g/cm^3), at a distance of 2.5 Saturnian radii, the tadpole libration half-width is about 3 km and the horseshoe half-width about 60 km.

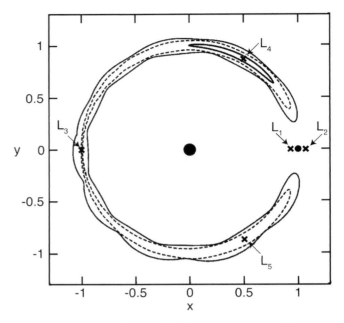

FIGURE 3 Diagram showing the five Lagrangian equilibrium points (denoted by crosses) and three representative orbits near these points for the circular restricted three-body problem. In this example, the secondary's mass is 0.001 times the total mass. The coordinate frame has its origin at the barycenter and corotates with the pair of bodies, thereby keeping the primary (large solid circle) and secondary (small solid circle) fixed on the x axis. Tadpole orbits remain near one or the other of the L_4 and L_5 points. An example is shown near the L_4 point on the diagram. Horseshoe orbits enclose all three of L_3, L_4, and L_5 but do not reach L_1 or L_2. The outermost orbit on the diagram illustrates this behavior. There is a critical curve dividing tadpole and horseshoe orbits that encloses L_4 and L_5 and passes through L_3. A horseshoe orbit near this dividing line is shown as the dashed curve in the diagram.

3.2.2 HILL SPHERE

The approximate limit to a planet's gravitational dominance is given by the extent of its **Hill sphere**,

$$R_{\mathrm{H}} = \left[\frac{m}{3(M + m)} \right]^{1/3} a, \qquad (22)$$

where m is the mass of the planet and M is the Sun's mass. A test body located at the boundary of a planet's Hill sphere is subjected to a gravitational force from the planet comparable to the tidal difference between the force of the Sun on the planet and that on the test body. The Hill sphere essentially stretches out to the L_1 point and is roughly the limit of the Roche lobe (maximum extent of an object held together by gravity alone) of a body with $m \ll M$. Planetocentric orbits that are stable over long periods of time are those well within the boundary of a planet's Hill sphere; all known natural satellites lie in this region. The trajectories

VIII Pasiphae

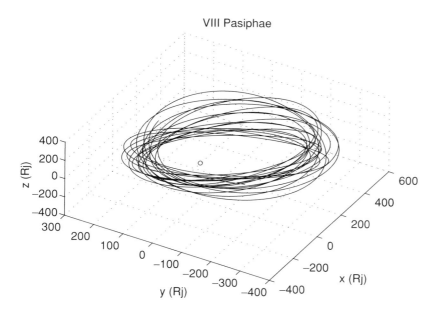

FIGURE 4 The orbit of J VIII Pasiphae, a distant retrograde satellite of Jupiter, is shown as seen in a nonrotating coordinate system with Jupiter at the origin (open circle). The satellite was integrated as a massless test particle in the context of the circular restricted three-body problem for approximately 38 years. The unit of distance is Jupiter's radius, R_J. During the course of this integration, the distance to Jupiter varied from 122 to $548R_J$. Note how the large solar perturbations produce significant deviations from a Keplerian orbit. [Figure reprinted with permission from Jose Alvarellos (1996). "Orbital Stability of Distant Satellites of Jovian Planets," M.Sc. thesis, San Jose State University.]

of the outermost planetary satellites, which lie closest to the boundary of the Hill sphere, show large variations in planetocentric orbital paths (Fig. 4). Stable heliocentric orbits are those that are always well outside the Hill sphere of any planet.

3.3 Examples of Resonances: Ring Particles and Shepherding

In the discussions in Section 2, the gravitational force produced by a spherically symmetric body was described. In this section the effects of deviations from spherical symmetry must be included when computing the force. This is most conveniently done by introducing the gravitational potential $\Phi(\mathbf{r})$, which is defined such that the acceleration $d^2\mathbf{r}/dt^2$ of a particle in the gravitational field is

$$d^2\mathbf{r}/dt^2 = \nabla\Phi. \quad (23)$$

In empty space, the Newtonian gravitational potential $\Phi(\mathbf{r})$ always satisfies Laplace's equation

$$\nabla^2\Phi = 0. \quad (24)$$

Most planets are very nearly axisymmetric, with the major departure from sphericity being due to a rotationally induced equatorial bulge. Thus, the gravitational potential can be expanded in terms of Legendre polynomials instead of the complete spherical harmonic expansion, which would be required for the potential of a body of arbitrary shape:

$$\Phi(r, \phi, \theta) = -\frac{Gm}{r}\left[1 - \sum_{n=2}^{\infty} J_n P_n(\cos\theta)(R/r)^n\right]. \quad (25)$$

This equation uses standard spherical coordinates, so that θ is the angle between the planet's symmetry axis and the vector to the particle. The terms $P_n(\cos\theta)$ are the Legendre polynomials, and J_n are the gravitational moments determined by the planet's mass distribution. If the planet's mass distribution is symmetrical about the planet's equator, the J_n are zero for odd n. For large bodies, J_2 is generally substantially larger than the other gravitational moments.

Consider a particle in Saturn's rings, which revolves around the planet on a circular orbit in the equatorial plane ($\theta = 90°$) at a distance r from the center of the planet. The centripetal force must be provided by the radial component of the planet's gravitational force [see Eq. (9)], so the particle's angular velocity n satisfies

$$rn^2(r) = \left[\frac{\partial\Phi}{\partial r}\right]_{\theta=90°}. \quad (26)$$

If the particle suffers an infinitesimal displacement from its circular orbit, it will oscillate freely in the horizontal and vertical directions about the reference circular orbit with radial (epicyclic) frequency $\kappa(r)$ and vertical frequency $\mu(r)$, respectively, given by

$$\kappa^2(r) = r^{-3}\frac{d}{dr}[(r^2n)^2], \quad (27)$$

$$\mu^2(r) = \left[\frac{\partial^2\Phi}{\partial z^2}\right]_{z=0}. \quad (28)$$

From Eqs. (24)–(28), the following relation is found between the three frequencies for a particle in the equatorial plane:

$$\mu^2 = 2n^2 - \kappa^2. \tag{29}$$

For a perfectly spherically symmetric planet, $\mu = \kappa = n$. Since Saturn and the other ringed planets are oblate, μ is slightly larger and κ is slightly smaller than the orbital frequency n.

Using Eqs. (24)–(29), one can show that the orbital and epicyclic frequencies can be written as

$$n^2 = \frac{GM}{r^3}\left[1 + \frac{3}{2}J_2\left(\frac{R}{r}\right)^2 - \frac{15}{8}J_4\left(\frac{R}{r}\right)^4 \right.$$
$$\left. + \frac{35}{16}J_6\left(\frac{R}{r}\right)^6 + \cdots\right], \tag{30}$$

$$\kappa^2 = \frac{GM}{r^3}\left[1 - \frac{3}{2}J_2\left(\frac{R}{r}\right)^2 + \frac{45}{8}J_4\left(\frac{R}{r}\right)^4 \right.$$
$$\left. - \frac{175}{16}J_6\left(\frac{R}{r}\right)^6 + \cdots\right], \tag{31}$$

$$\mu^2 = \frac{GM}{r^3}\left[1 + \frac{9}{2}J_2\left(\frac{R}{r}\right)^2 - \frac{75}{8}J_4\left(\frac{R}{r}\right)^4 \right.$$
$$\left. + \frac{245}{16}J_6\left(\frac{R}{r}\right)^6 + \cdots\right]. \tag{32}$$

Thus, the oblateness of a planet causes apsides of particle orbits in and near the equatorial plane to precess in the direction of the orbit and lines of nodes of nearly equatorial orbits to regress.

Resonances occur where the radial (or vertical) frequency of the ring particles is equal to the frequency of a component of a satellite's horizontal (or vertical) forcing, as sensed in the rotating frame of the particle. In this case the resonating particle is always near the same phase in its radial (vertical) oscillation when it experiences a particular phase of the satellite's forcing. This situation enables continued coherent "kicks" from the satellite to build up the particle's radial (vertical) motion, and significant forced oscillations may thus result. The location and strengths of resonances with any given moon can be determined by decomposing the gravitational potential of the moon into its Fourier components. The disturbance frequency, $\bar{\omega}$, can be written as the sum of integer multiples of the satellite's angular, vertical, and radial frequencies:

$$\bar{\omega} = jn_s + k\mu_s + \ell\kappa_s, \tag{33}$$

where the azimuthal symmetry number, j, is a nonnegative integer, and k and ℓ are integers, with k being even for horizontal forcing and odd for vertical forcing. The subscript s refers to the satellite. A particle placed at distance $r = r_L$ will undergo horizontal (Lindblad) resonance if r_L satisfies

$$\bar{\omega} - jn(r_L) = \pm\kappa(r_L). \tag{34}$$

It will undergo vertical resonance if its radial position r_v, satisfies

$$\bar{\omega} - jn(r_L) = \pm\mu(r_v). \tag{35}$$

When Eq. (34) is valid for the lower (upper) sign, r_L is referred to as the inner (outer) Lindblad or horizontal resonance. The distance r_v is called an inner (outer) vertical resonance if Eq. (35) is valid for the lower (upper) sign. Since all of Saturn's large satellites orbit the planet well outside the main ring system, the satellite's angular frequency n_s is less than the angular frequency of the particle, and inner resonances are more important than outer ones. When $m \neq 1$, the approximation $\mu \approx n \approx \kappa$ may be used to obtain the ratio

$$\frac{n(r_{L,v})}{n_s} = \frac{j + k + \ell}{j - 1}. \tag{36}$$

The notation $(j + k + \ell)/(j - 1)$ or $(j + k + \ell):(j - 1)$ is commonly used to identify a given resonance.

The strength of the forcing by the satellite depends, to lowest order, on the satellite's eccentricity, e, and inclination, i, as $e^{|\ell|}$ $[\sin i]^{|k|}$. The strongest horizontal resonances have $k = \ell = 0$, and are of the form $j:(j - 1)$. The strongest vertical resonances have $k = 1$, $\ell = 0$, and are of the form $(j + 1):(j - 1)$. The location and strengths of such orbital resonances can be calculated from known satellite masses and orbital parameters and Saturn's gravity field. Most strong resonances in the Saturnian system lie in the outer A ring near the orbits of the moons responsible for them. If $n = \mu = \kappa$, the locations of the horizontal and vertical resonances would consider: $r_L = r_v$. Since, owing to Saturn's oblateness, $\mu > n > \kappa$, the positions r_L and r_v do not coincide: $r_v < r_L$. A detailed discussion of spiral density waves, spiral bending waves, and gaps at resonances produced by moons is presented elsewhere in this encyclopedia. [See PLANETARY RINGS.]

4. Chaotic Motion

4.1 Concepts of Chaos

In the nineteenth century, Henri Poincaré studied the mathematics of the circular restricted **three-body problem**. In this problem, one mass (the secondary) moves in a fixed, circular orbit about a central mass (the primary), while a test massless particle moves under the gravitational effect of both masses but does not perturb their orbits. From this work, Poincaré realized that despite the simplicity of the equations of motion, some solutions to the problem exhibit complicated behavior.

Poincaré's work in celestial mechanics provided the framework for the modern theory of nonlinear dynamics and ultimately led to a deeper understanding of the phenomenon of chaos, whereby dynamical systems described by simple equations can give rise to unpredictable behavior. The whole question of whether or not a given system is stable to sufficiently small perturbations is the basis of the Kolmogorov-Arnol'd-Moser (KAM) theory, which has its origins in the work of Poincaré.

One characteristic of chaotic motion is that small changes in the starting conditions can produce vastly different final outcomes. Since all measurements of positions and velocities of objects in the solar system have finite accuracy, relatively small uncertainties in the initial state of the system can lead to large errors in the final state, for initial conditions that lie in chaotic regions in **phase space**.

This is an example of what has become known as the "butterfly effect," first mentioned in the context of chaotic weather systems. It has been suggested that under the right conditions, a small atmospheric disturbance (such as the flapping of a butterfly's wings) in one part of the world could ultimately lead to a hurricane in another part of the world.

The changes in an orbit that reveal it to be chaotic may occur very rapidly, for example during a close approach to the planet, or may take place over millions or even billions of years. Although there have been a number of significant mathematical advances in the study of nonlinear dynamics since Poincaré's time, the digital computer has proven to be the most important tool in investigating chaotic motion in the solar system. This is particularly true in studies of the gravitational interaction of all the planets, where there are few analytical results.

4.2 The Three-body Problem as a Paradigm

The characteristics of chaotic motion are common to a wide variety of dynamical systems. In the context of the solar system, the general properties are best described by considering the planar circular restricted three-body problem, consisting of a massless test particle and two bodies of masses m_1 and m_2 moving in circular orbits about their common center of mass at constant separation with all bodies moving in the same plane. The test particle is attracted to each mass under the influence of the inverse square law of force given in Eq. (5). In Eq. (16), a is the constant separation of the two masses, and $n = 2\pi/p$ is their constant angular velocity about the center of mass. Using x and y as components of the position vector of the test particle referred to the center of mass of the system (Fig. 5), the equations of motion of the particle in a reference frame

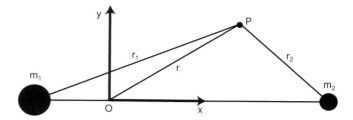

FIGURE 5 The rotating coordinate system used in the circular restricted three-body problem. The masses are at a fixed distance from one another and this is taken to be the unit of length. The position and velocity vectors of the test particle (at point P) are referred to the center of mass of the system at O.

rotating at angular velocity n are

$$\ddot{x} - 2n\dot{y} - n^2 x = -G\left(m_1 \frac{x + \mu_2}{r_1^3} - m_2 \frac{x - \mu_1}{r_2^3}\right), \quad (37)$$

$$\ddot{y} + 2n\dot{x} - n^2 y = -G\left(\frac{m_1}{r_1^3} + \frac{m_2}{r_2^3}\right) y, \quad (38)$$

where $\mu_1 = m_1 a/(m_1 + m_2)$, and $\mu_2 = m_2 a/(m_1 + m_2)$ are constants and

$$r_1^2 = (x + \mu_2)^2 + y^2, \quad (39)$$

$$r_2^2 = (x - \mu_1)^2 + y^2, \quad (40)$$

where r_1 and r_2 are the distances of the test particle from the masses m_1 and m_2, respectively.

These two second-order, coupled, nonlinear differential equations can be solved numerically provided the initial position (x_0, y_0) and velocity $(\dot{x}_0, \dot{y}_0)$ of the particle are known. Therefore the system is deterministic and at any given time the orbital elements of the particle (such as its semimajor axis and eccentricity) can be calculated from its initial position and velocity.

The test particle is constrained by the existence of a constant of the motion called the Jacobi constant, C, given by

$$C = n^2(x^2 + y^2) + 2G\left(\frac{m_1}{r_1} + \frac{m_2}{r_2}\right) - \dot{x}^2 - \dot{y}^2. \quad (41)$$

The values of (x_0, y_0) and $(\dot{x}_0, \dot{y}_0)$ fix the value of C for the system, and this value is preserved for all subsequent motion. At any instant the particle is at some position on the two-dimensional (x, y) plane. However, since the actual orbit is also determined by the components of the velocity $(\dot{x}, \dot{y})$, the particle can also be thought of as being at a particular position in a four-dimensional $(x, y, \dot{x}, \dot{y})$ phase space. Note that the use of four dimensions rather than the customary two is simply a means of representing the position *and* the velocity of the particle at a particular instant in time;

the particle's motion is always restricted to the $x - y$ plane. The existence of the Jacobi constant implies that the particle is not free to wander over the entire 4-D phase space, but rather that its motion is restricted to the 3-D "surface" defined by Eq. (41). This has an important consequence for studying the evolution of orbits in the problem.

The usual method is to solve the equations of motion, convert x, y, $\dot{x}$, and $\dot{y}$ into orbital elements such as semimajor axis, eccentricity, longitude of periapse, and mean longitude, and then plot the variation of these quantities as a function of time. However, another method is to produce a **surface of section**, also called a Poincaré map. This makes use of the fact that the orbit is always subject to Eq. (41), where C is determined by the initial position and velocity. Therefore if any three of the four quantities x, y, $\dot{x}$, and $\dot{y}$ are known, the fourth can always be determined by solving Eq. (41). One common surface of section that can be obtained for the planar circular restricted three-body problem is a plot of values of x and $\dot{x}$ whenever $y = 0$ and $\dot{y}$ is positive. The actual value of $\dot{y}$ can always be determined uniquely from Eq. (41), and so the two-dimensional $(x, \dot{x})$ plot implicitly contains all the information about the particle's location in the four-dimensional phase space. Although surfaces of section make it more difficult to study the evolution of the orbital elements, they have the advantage of revealing the characteristic motion of the particle (regular or chaotic) and a number of orbits can be displayed on the same diagram.

As an illustration of the different types of orbits that can arise, the results of integrating a number of orbits using a mass $m_2/(m_1 + m_2) = 10^{-3}$ and Jacobi constant $C = 3.07$ are described next. In each case, the particle was started with the initial longitude of periapse $\varpi_0 = 0$ and initial mean longitude $\lambda_0 = 0$. This corresponds to $\dot{x} = 0$ and $y = 0$. Since the chosen mass ratio is comparable to that of the Sun-Jupiter system, and Jupiter's eccentricity is small, this

will be used as a good approximation to the motion of fictitious asteroids moving around the Sun under the effect of gravitational perturbations from Jupiter. The asteroid is assumed to be moving in the same plane as Jupiter's orbit.

4.2.1 REGULAR ORBITS

The first asteroid has starting values $x = 0.55$, $y = 0$, $\dot{x} = 0$, with $\dot{y} = 0.9290$ determined from the solution of Eq. (41). Here a set of dimensionless coordinates are used in which $n = 1$, $G = 1$, and $m_1 + m_2 = 1$. In these units, the orbit of m_2 is a circle at distance $a = 1$ with uniform speed $v = 1$. The corresponding initial values of the heliocentric semimajor axis and eccentricity are $a_0 = 0.6944$ and $e_0 = 0.2065$. Since the semimajor axis of Jupiter's orbit is 5.202 AU, this value of a_0 would correspond to an asteroid at 3.612 AU.

Figure 6 shows the evolution of e as a function of time. The plot shows a regular behavior with the eccentricity varying from 0.206 to 0.248 over the course of the integration. In fact, an asteroid at this location would be close to an orbit–orbit resonance with Jupiter, where the ratio of the orbital period of the asteroid, T, to Jupiter's period, T_J, is close to a rational number. From **Kepler's third law** of planetary motion, $T^2 \propto a^3$. In this case, $T/T_J = (a/a_J)^{3/2} = 0.564 \approx 4/7$ and the asteroid orbit is close to a 7:4 resonance with Jupiter. Figure 7 shows the variation of the semimajor axis of the asteroid, a, over the same time interval as shown in Fig. 6. Although the changes in a are correlated with those in e, they are smaller in amplitude and a appears to oscillate about the location of the exact resonance at $a = (4/7)^{2/3} \approx 0.689$. An asteroid in resonance experiences enhanced gravitational perturbations from Jupiter, which can cause regular variations in its orbital elements. The extent of these variations depends on the asteroid's location within the resonance, which is, in turn, is determined by the starting conditions.

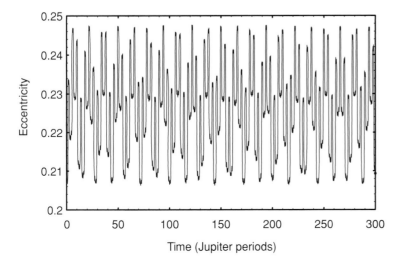

FIGURE 6   The eccentricity as a function of time for an object moving in a regular orbit near the 7:4 resonance with Jupiter. The plot was obtained by solving the circular restricted three-body problem numerically using initial values of 0.6944 and 0.2065 for the semimajor axis and eccentricity, respectively. The corresponding position and velocity in the rotating frame were $x_0 = 0.55$, $y_0 = 0$, $\dot{x} = 0$, and $\dot{y} = 0.9290$.

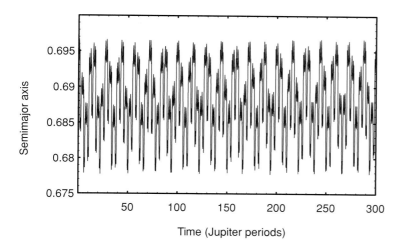

FIGURE 7 The semimajor axis as a function of time for an object using the same starting conditions as in Fig. 6. The units of the semimajor axis are such that Jupiter's semimajor axis (5.202 AU) is taken to be unity.

The equations of motion can be integrated with the same starting conditions to generate a surface of section by plotting the values of x and $\dot{x}$ whenever $y = 0$ with $\dot{y} > 0$ (Fig. 8). The pattern of three distorted curves or "islands" that emerges is a characteristic of resonant motion when displayed in such plots. If a resonance is of the form $(p + q):p$, where p and q are integers, then q is said to be the order of the resonance. The number of islands seen in a surface of section plot of a given resonant trajectory is equal to q. In this case, $p = 4$, $q = 3$ and three islands are visible.

The center of each island would correspond to a starting condition that placed the asteroid at exact resonance where the variation in e and a would be minimal. Such points are said to be fixed points of the Poincaré map. If the starting location was moved farther away from the center, the subsequent variations in e and a would get larger, until eventually some starting values would lead to trajectories that were not in resonant motion.

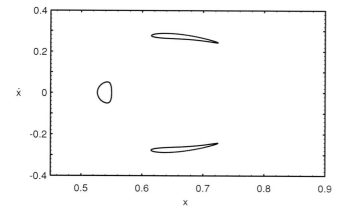

FIGURE 8 A surface of section plot for the same (regular) orbit shown in Figs. 6 and 7. The 2000 points were generated by plotting the values of x and $\dot{x}$ whenever $y = 0$ with positive $\dot{y}$. The three "islands" in the plot are due to the third-order 7:4 resonance.

4.2.2 CHAOTIC ORBITS

Figures 9 and 10 show the plots of e and a as a function of time for an asteroid orbit with starting values $x_0 = 0.56$, $y_0 = 0$, $\dot{x}_0 = 0$, and $\dot{y}$ determined from Eq. (41) with $C = 3.07$. The corresponding orbital elements are $a_0 = 0.6984$ and $e_0 = 0.1967$. These values are only slightly different from those used earlier, indeed the initial behavior of the plots is quite similar to that seen in Figs. 6 and 7. However, subsequent variations in e and a are strikingly different. The eccentricity varies from 0.188 to 0.328 in an irregular manner, and the value of a is not always close to the value associated with exact resonance. This is an example of a chaotic trajectory where the variations in the orbital elements have no obvious periodic or quasi-periodic structure. The anticorrelation of a and e can be explained in terms of the Jacobi constant.

The identification of this orbit as chaotic becomes apparent from a study of its surface of section (Fig. 11). Clearly, this orbit covers a much larger region of phase space than the previous example. Furthermore, the orbit does not lie on a smooth curve, but is beginning to fill an area of the phase space. The points also help to define a number of empty regions, three of which are clearly associated with the 7:4 resonance seen in the regular trajectory. There is also a tendency for the points to "stick" near the edges of the islands; this gives the impression of regular motion for short periods of time.

Chaotic orbits have the additional characteristic that they are sensitively dependent on initial conditions. This is illustrated in Fig. 12, where the variation in e as a function of time is shown for two trajectories; the first corresponds to Fig. 9 (where $x_0 = 0.56$) and the second has $x_0 = 0.56001$. The initial value of $\dot{y}$ was chosen so that the same value of C was obtained. Although both trajectories show comparable initial variations in e, after 60 Jupiter periods it is clear that the orbits have drifted apart. Such a divergence would not occur for nearby orbits in a regular part of the phase space.

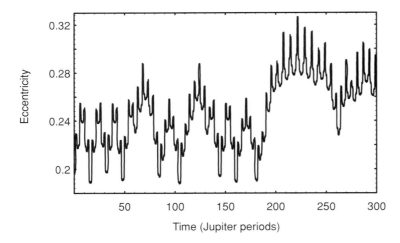

FIGURE 9 The eccentricity as a function of time for an object moving in a chaotic orbit started just outside the 7:4 resonance with Jupiter. The plot was obtained by solving the circular restricted three-body problem numerically using initial values of 0.6984 and 0.1967 for the semimajor axis and eccentricity, respectively. The corresponding position and velocity in the rotating frame were $x_0 = 0.56$, $y_0 = 0$, $\dot{x}_0 = 0$, and $\dot{y} = 0.8998$.

The rate of divergence of nearby trajectories in such numerical experiments can be quantified by monitoring the evolution of two orbits that are started close together. In a dynamical system such as the three-body problem, there are a number of quantities called the **Lyapunov characteristic exponents**. A measurement of the local divergence of nearby trajectories leads to an estimate of the largest of these exponents, and this can be used to determine whether or not the system is chaotic. If two orbits are separated in phase space by a distance d_0 at time t_0, and d is their separation at time t, then the orbit is chaotic if

$$d = d_0 \exp \gamma (t - t_0), \qquad (42)$$

where γ is a positive quantity equal to the maximum Lyapunov characteristic exponent. However, in practice the Lyapunov characteristic exponents can only be derived analytically for a few idealized systems. For practical problems in the solar system, γ can be estimated from the results

of a numerical integration by writing

$$\gamma = \lim_{t \to \infty} \frac{\ln(d/d_0)}{t - t_0} \qquad (43)$$

and monitoring the behavior of γ with time. A plot of γ as a function of time on a log–log scale reveals a striking difference between regular and chaotic trajectories. For regular orbits, $d \approx d_0$ and a log–log plot has a slope of -1. However, if the orbit is chaotic, then γ tends to a constant non-zero value. This method may not always work because γ is defined only in the limit as $t \to \infty$ and sometimes chaotic orbits may give the appearance of being regular orbits for long periods of time by sticking close to the edges of the islands.

If the nearby trajectory drifts too far from the original one, then γ is no longer a measure of the local divergence of the orbits. To overcome this problem, it helps to rescale the separation of the nearby trajectory at fixed intervals. Figure 13 shows $\log \gamma$ as a function of $\log t$ calculated using this

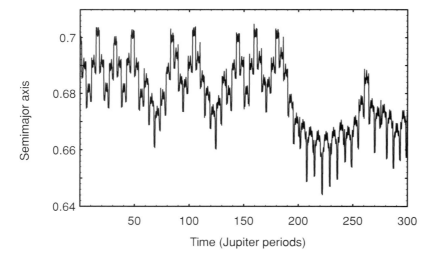

FIGURE 10 The semimajor axis as a function of time for an object using the same starting conditions as in Fig. 9. The units of the semimajor axis are such that Jupiter's semimajor axis (5.202 AU) is taken to be unity.

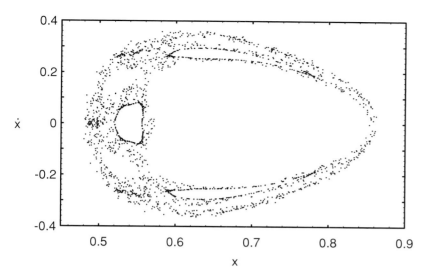

FIGURE 11   A surface of section plot for the same chaotic orbit as shown in Figs. 9 and 10. The 2000 points were generated by plotting the values of x and $\dot{x}$ whenever $y = 0$ with positive $\dot{y}$. The points are distributed over a much wider region of the $(x, \dot{x})$ plane than the points for the regular orbit shown in Fig. 8, and they help to define the edges of the regular regions associated with the 7:4 and other resonances.

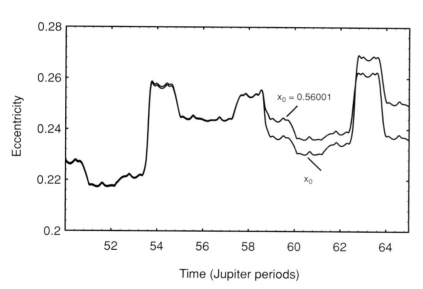

FIGURE 12   The variation in the eccentricity for two chaotic orbits started close to one another. One plot is part of Fig. 9 using the chaotic orbit started with $x_0 = 0.56$, and the other is for an orbit with $x_0 = 0.56001$. Although the divergence of the two orbits is exponential, the effect becomes noticeable only after 60 Jupiter periods.

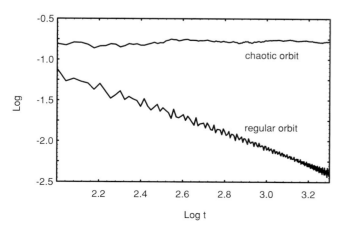

FIGURE 13   The evolution of the quantity γ [defined in Eq. (43)] as a function of time (in Jupiter periods) for a regular ($x_0 = 0.55$) and chaotic ($x_0 = 0.56$) orbit. In this log–log plot, the regular orbit shows a characteristic slope of -1 with no indication of $\log \gamma$ tending toward a finite value. However, in the case of the chaotic orbit, $\log \gamma$ tends to a limiting value close to -0.77.

method for the regular and chaotic orbits described here. This leads to an estimate of $\gamma = 10^{-0.77}$ (Jupiter periods)$^{-1}$ for the maximum Lyapunov characteristic exponent of the chaotic orbit. The corresponding Lyapunov time is given by $1/\gamma$, or in this case ~ 6 Jupiter periods. This indicates that for this starting condition the chaotic nature of the orbit quickly becomes apparent.

It is important to realize that a chaotic orbit is not necessarily unbounded. The maximum Lyapunov characteristic exponent concerns local divergence and provides no information about the global stability of the trajectory. The phrase "wandering on a leash" is an apt description of objects on bounded chaotic orbits—the motion is contained but yet chaotic at the same time. Another consideration is that numerical explorations of chaotic systems have many pitfalls both in how the physical system is modeled and whether or not the model provides an accurate portrayal of the real system.

4.2.3 LOCATION OF REGULAR AND CHAOTIC REGIONS

The extent of the chaotic regions of the phase space of a dynamical system can depend on a number of factors. In the case of the circular restricted three-body problem, the critical quantities are the values of the Jacobi constant and the mass ratio μ_2. In Figs. 14 and 15, ten trajectories are shown for each of two different values of the Jacobi constant. In the first case (Fig. 14), the value is $C = 3.07$ (the same as the value used in Figs. 8 and 11), whereas in Fig. 15 it is $C = 3.13$. It is clear that the extent of the chaos is reduced in Fig. 15. The value of C in the circular restricted problem determines how close the asteroid can get to Jupiter. Larger values of C correspond to orbits with greater minimum distances from Jupiter. For the case $\mu_2 = 0.001$ and $C > 3.04$, it is impossible for their orbits to intersect, although the perturbations can still be significant.

Close inspection of the separatrices in Figs. 14 and 15 reveals that they consist of chaotic regions with regular regions on either side. As the value of the Jacobi constant decreases, the extent of the chaotic separatrices increases until the regular curves separating adjacent resonances are broken down and neighboring chaotic regions begin to merge. This can be thought of as the overlap of adjacent resonances giving rise to chaotic motion. It is this process that permits chaotic orbits to explore regions of the phase space that are inaccessible to the regular orbits. In the context of the Sun–Jupiter–asteroid problem, this observation implies that asteroids in certain orbits are capable of large excursions in their orbital elements.

5. Orbital Evolution of Minor Bodies

5.1 Asteroids

With more than 130,000 accurately determined orbits and one major perturber (the planet Jupiter), the asteroids provide a natural laboratory in which to study the consequences of regular and chaotic motion. Using suitable approximations, asteroid motion can be studied analytically in some special cases. However, it is frequently necessary to resort to numerical integration. [*See* MAIN-BELT ASTEROIDS.]

Investigations have shown that a number of asteroids have orbits that result in close approaches to planets. Of particular interest are asteroids such as 433 Eros, 1033 Ganymed, and 4179 Toutatis, because they are on orbits

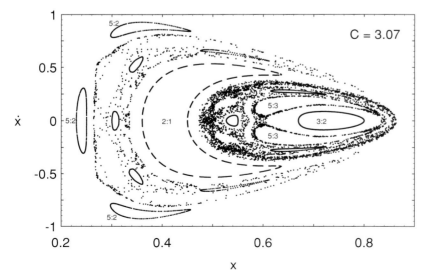

FIGURE 14   Representative surface of section plots for $x_0 = 0.25, 0.29, 0.3, 0.45, 0.475, 0.5, 0.55, 0.56, 0.6$, and 0.8 with $= 0$, $y_0 = 0$, and Jacobi constant $C = 3.07$. Each trajectory was followed for a minimum of 500 crossing points. The plot uses the points shown in Figs. 8 and 11 (although the scales are different), as well as points from other regular and chaotic orbits. The major resonances are identified.

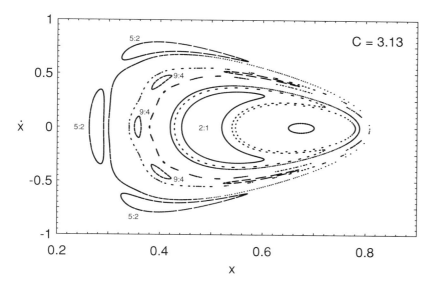

FIGURE 15   Representative surface of section plots for $x_0 = 0.262$, 0.3, 0.34, 0.35, 0.38, 0.42, 0.52, 0.54, 0.7, and 0.78 with $\dot{x}_0 = 0$, $y_0 = 0$, and Jacobi constant $C = 3.13$. Each trajectory was followed for a minimum of 500 crossing points. It is clear from a comparison with Fig. 14 that the phase space is more regular; chaotic orbits still exist for this value of C, but they are more difficult to find. The major resonances are identified.

that bring them close to Earth. One of the most striking examples of the butterfly effect (see Section 4.1) in the context of orbital evolution is the orbit of asteroid 2060 Chiron, which has a perihelion inside Saturn's orbit and an aphelion close to Uranus's orbit. Numerical integrations based on the best available orbital elements show that it is impossible to determine Chiron's past or future orbit with any degree of certainty since it frequently suffers close approaches to Saturn and Uranus. In such circumstances, the outcome is strongly dependent on the initial conditions as well as the accuracy of the numerical method. These are the characteristic signs of a chaotic orbit. By integrating several orbits with initial conditions close to the nominal values, it is possible to carry out a statistical analysis of the orbital evolution. Studies suggest that there is a 1 in 8 chance that Saturn will eject Chiron from the solar system on a hyperbolic orbit, while there is a 7 in 8 chance that it will evolve toward the inner solar system and come under strong perturbations from Jupiter. Telescopic observations of a faint coma surrounding Chiron imply that it is a comet rather than an asteroid; perhaps its future orbit will resemble that of a short-period comet of the Jupiter family.

Numerical studies of the orbital evolution of planet-crossing asteroids under the effects of perturbations from all the planets have shown a remarkable complexity of motion for some objects. For example, the Earth-crossing asteroid 1620 Geographos gets trapped temporarily in a number of resonances with Earth in the course of its chaotic evolution (Fig. 16).

A histogram of the number distribution of asteroid orbits in semimajor axis (Fig. 17) shows that apart from a clustering of asteroids near Jupiter's semimajor axis at 5.2 AU, there is an absence of objects within 0.75 AU of

the orbit of Jupiter. The objects in the orbit of Jupiter are the Trojan asteroids (Section 3.2), which are located $\sim 60°$ ahead of and behind Jupiter.

The cleared region near Jupiter's orbit can be understood in terms of chaotic motion due to the overlap of adjacent resonances. In the context of the Sun–Jupiter–asteroid restricted three-body problem, the perturber (Jupiter) has

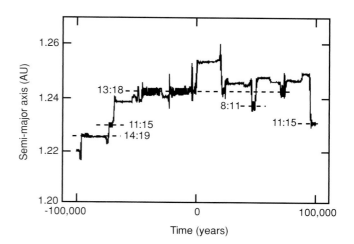

FIGURE 16   A plot of the semimajor axis of the near-Earth asteroid 1620 Geographos over a backward and forward integration of 100,000 years starting in 1986. Under perturbations from the planets, Geographos moves in a chaotic orbit and gets temporarily trapped in a number of high-order, orbit–orbit resonances (indicated in the diagram) with Earth. The data are taken from a numerical study of planet-crossing asteroids undertaken by A. Milani and coworkers. (Courtesy of Academic Press.)

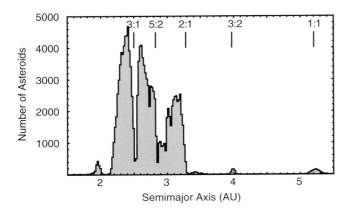

FIGURE 17   A histogram of the distribution of the numbered asteroids with semimajor axis together with the locations of the major jovian resonances. Most objects lie in the main belt between 2.0 and 3.3 AU, where the outer edge is defined by the location of the 2:1 resonance with Jupiter. As well as gaps (the Kirkwood gaps) at the 3:1, 5:2, 2:1, and other resonances in the main belt, there are small concentrations of asteroids at the 3:2 and 1:1 resonances (the Hilda and Trojan groups, respectively).

an infinite sequence of first-order resonances that lie closer together as its semimajor axis is approached. For example, the 2:1, 3:2, 4:3, and 5:4 resonances with Jupiter lie at 3.3, 4.0, 4.3, and 4.5 AU, respectively. Since each $(p + 1):p$ resonance (where p is a positive integer) has a finite width in semimajor axis that is almost independent of p, adjacent resonances will always overlap for some value of p greater than a critical value, p_{crit}. This value is given by

$$p_{crit} \approx 0.51 \left(\frac{m}{m + M} \right)^{-2/7} \qquad (44)$$

where, in this case, m is the mass of Jupiter and M is the mass of the Sun. This equation can be used to predict that resonance overlap and chaotic motion should occur for p values greater than 4; this corresponds to a semimajor axis near 4.5 AU. Therefore chaos may have played a significant role in the depletion of the outer asteroid belt.

The histogram in Fig. 17 also shows a number of regions in the main belt where there are few asteroids. The gaps at 2.5 and 3.3 AU were first detected in 1867 by Daniel Kirkwood using a total sample of fewer than 100 asteroids; these are now known as the Kirkwood gaps. Their locations coincide with prominent Jovian resonances (indicated in Fig. 17), and this led to the hypothesis that they were created by the gravitational effect of Jupiter on asteroids that had orbited at these semimajor axes. The exact removal mechanism was unclear until the 1980s, when several numerical and analytical studies showed that the central regions of these resonances contained large chaotic zones.

The Kirkwood gaps cannot be understood using the model of the circular restricted three-body problem described in Section 4.2. The eccentricity of Jupiter's orbit, although small (0.048), plays a crucial role in producing the large chaotic zones that help to determine the orbital evolution of asteroids. On timescales of several hundreds of thousands of years, the mutual perturbations of the planets act to change their orbital elements and Jupiter's eccentricity can vary from 0.025 to 0.061. An asteroid in the chaotic zone at the 3:1 resonance would undergo large, essentially unpredictable changes in its orbital elements. In particular, the eccentricity of the asteroid could become large enough for it to cross the orbit of Mars. This is illustrated in Fig. 18 for a fictitious asteroid with an initial eccentricity of 0.15 moving in a chaotic region of the phase space at the 3:1 resonance. Although the asteroid can have periods of relatively low eccentricity, there are large deviations and

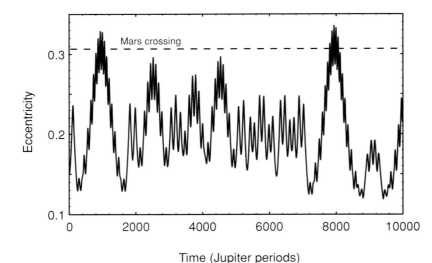

FIGURE 18   The chaotic evolution of the eccentricity of a fictitious asteroid at the 3:1 resonance with Jupiter. The orbit was integrated using an algebraic mapping technique developed by J. Wisdom. The line close to $e = 0.3$ denotes the value of the asteroid's eccentricity, above which it would cross the orbit of Mars. It is believed that the 3:1 Kirkwood gap was created when asteroids in chaotic zones at the 3:1 resonance reached high eccentricities and were removed by direct encounters with Mars, Earth, or Venus.

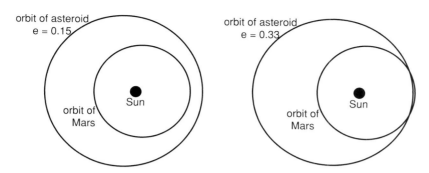

FIGURE 19   The effect of an increase in the orbital eccentricity of an asteroid at the 3:1 Jovian resonance on the closest approach between the asteroid and Mars. For $e = 0.15$, the orbits do not cross. However, for $e = 0.33$, a typical maximum value for asteroids in chaotic orbits, there is a clear intersection of the orbits, and the asteroid could have a close encounter with Mars.

e can reach values in excess of 0.3. Allowing for the fact that the eccentricity of Mars's orbit can reach 0.14, this implies that there will be times when the orbits could intersect (Fig. 19). In this case, the asteroid orbit would be unstable, since it is likely to either impact the surface of Mars or suffer a close approach that would drastically alter its semimajor axis. Although Jupiter provides the perturbations, it is Mars, Earth or Venus that ultimately removes the asteroids from the 3:1 resonance. Figure 20 shows the excellent correspondence between the distribution of asteroids close to the 3:1 resonance and the maximum extent of the chaotic region determined from numerical experiments.

The situation is less clear for other resonances, although there is good evidence for large chaotic zones at the 2:1 and 5:2 resonances. In the outer part of the main belt, large changes in eccentricity will cause the asteroid to cross the orbit of Jupiter before it gets close to Mars. There may also be perturbing effects from other planets. In fact, it is now known that secular resonances have an important role to play in the clearing of the Kirkwood gaps, including the one at the 3:1 resonance. Once again, chaos is involved. Studies of asteroid motion at the 3:2 Jovian resonance indicate that the motion is regular, at least for low values of the eccentricity. This may help to explain why there is a local concentration of asteroids (the Hilda group) at this resonance, whereas others are associated with an absence of material.

Since the dynamical structure of the asteroid belt has been determined by the perturbative effects of nearby planets, it seems likely that the original population was much larger and more widely dispersed. Therefore, the current distribution of asteroids may represent objects that are either recent collision products or that have survived in relatively stable orbits over the age of the solar system.

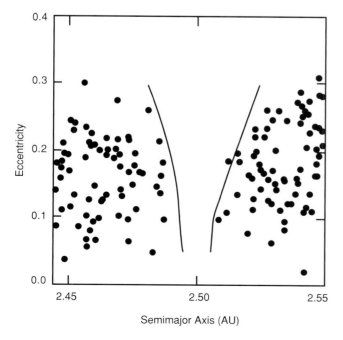

FIGURE 20   The eccentricity and semimajor axes of asteroids in the vicinity of the 3:1 jovian resonance; the Kirkwood gap is centered close to 2.5 AU. The two curves denote the maximum extent of the chaotic zone determined from numerical experiments, and there is excellent agreement between these lines and the edges of the 3:1 gap.

5.2 Meteorites

Most meteorites are thought to be the fragments of material produced from collisions in the asteroid belt, and the reflectance properties of certain meteorites are known to be similar to those of common types of asteroids. Since most collisions take place in the asteroid belt, the fragments have to evolve into Earth-crossing orbits before they can hit Earth and be collected as samples.

An estimate of the time taken for a given meteorite to reach Earth after the collisional event that produced it can be obtained from a measure of its cosmic ray exposure age. Prior to the collisions, the fragment may have been well below the surface of a much larger body, and as such it would have been shielded from all but the most energetic cosmic rays. However, after a collision the exposed fragment would be subjected to cosmic ray bombardment in interplanetary space. A detailed analysis of meteorite samples allows these exposure ages to be measured.

In the case of one common class of meteorites called the ordinary chondrites, the cosmic ray exposure ages are typically less than 20 million years and the samples show little evidence of having been exposed to high pressure, or "shocking." Prior to the application of chaos theory to the origin of the Kirkwood gaps, there was no plausible mechanism that could explain delivery to Earth within the exposure age constraints and without shocking. However, small increments in the velocity of the fragments as a result of the initial collision could easily cause them to enter a chaotic zone near a given resonance. Numerical integrations of such orbits near the 3:1 resonance showed that it was possible for them to achieve eccentricities large enough for them to cross the orbit of Earth. This result complemented previous research that had established that this part of the asteroid belt was a source region for the ordinary chondrites. Another effect that must be considered to obtain agreement between theory and observations is the Yarkovski effect which is discussed below. [See METEORITES.]

5.3 Comets

Typical cometary orbits have large eccentricities and therefore planet-crossing trajectories are commonplace. Many comets are thought to originate in the Oort cloud at several tens of thousands of AU from the Sun; another reservoir of comets, known as the Kuiper belt, exists just beyond the orbit of Neptune. Those that have been detected from Earth are classified as either long period (most of which have made single apparitions and have periods >200 yr) or Halley-type (with orbital periods of 20–200 yr) or Jupiter-family, which have orbital periods < 20 yr. All comets with orbital periods of less than $\sim 10^3$ yr have experienced a close approach to Jupiter or one of the other giant planets. By their very nature, the orbits of comets are chaotic, since the outcome of any planetary encounter will be sensitively dependent on the initial conditions.

Studies of the orbital evolution of the short-period comet P/Lexell highlight the possible effects of close approaches. A numerical integration has shown that prior to 1767 it was a short-period comet with a semimajor axis of 4.4 AU and an eccentricity of 0.35. In 1767 and 1779, it suffered close approaches to Jupiter. The first encounter placed it on a trajectory which brought it into the inner solar system and close (0.0146 AU) to the Earth, leading to its discovery and its only apparition in 1770, whereas the second was at a distance of $\sim$3 Jovian radii. This changed its semimajor axis to 45 AU with an eccentricity of 0.88.

A more recent example is the orbital history of comet Shoemaker-Levy 9 prior to its spectacular collision with Jupiter in 1994. Orbit computations suggest that the comet was first captured by Jupiter at some time during a 9-year interval centered on 1929. Prior to its capture, it is likely that it was orbiting in the outer part of the asteroid belt close to the 3:2 resonance with Jupiter or between Jupiter and Saturn close to the 2:3 resonance with Jupiter. However, the chaotic nature of its orbit means that it is impossible to derive a more accurate history unless prediscovery images of the comet are obtained. [See PHYSICS AND CHEMISTRY OF COMETS; COMETARY DYNAMICS.]

5.4 Small Satellites and Rings

Chaos is also involved in the dynamics of a satellite embedded in a planetary ring system. The processes differ from those discussed in Section 3.1, A because there is a near-continuous supply of ring material and direct scattering by the perturber is now important. In this case, the key quantity is the Hill's sphere of the satellite. Ring particles on near-circular orbits passing close to the satellite exhibit chaotic behavior due to the significant perturbations they receive at close approach. This causes them to collide with surrounding ring material, thereby forming a gap. Studies have shown that for small satellites, the expression for the width of the cleared gap is

$$W \approx 0.44 \left(\frac{m_2}{m_1} \right)^{2/7} a \qquad (45)$$

where m_2 and a are the mass and semimajor axis of the satellite and m_1 is the mass of the planet. Thus, an icy satellite with a radius of 10 km and a density of 1 g cm^{-3} orbiting in Saturn's A ring at a radial distance of 135,000 km would create a gap approximately 140 km wide.

Since such a gap is wider than the satellite that creates it, this provides an indirect method for the detection of small satellites in ring systems. There are two prominent gaps in Saturn's A ring: the $\sim$35-km-wide Keeler gap at 135,800 km and the 320-km-wide Encke gap at 133,600 km. The predicted radii of the icy satellites required to produce these gaps are ~2.5 and ~24 km, respectively. In 1991, an analysis of *Voyager* images by M. Showalter revealed a small satellite, Pan, with a radius of ~10 km orbiting in the Encke gap. In 2005, the moon Daphnis of radius ~3–4 km was discovered in the Keeler gap by the *Cassini* spacecraft. *Voyager 2* images of the dust rings of Uranus show pronounced gaps at certain locations. Although most of the proposed shepherding satellites needed to maintain the narrow rings have yet to be discovered, these gaps may provide indirect evidence of their orbital locations.

6. Long Term Stability of Planetary Orbits

6.1 The *N*–Body Problem

The entire solar system can be approximated by a system of nine planets orbiting the Sun. (Tiny Pluto has been included in most studies of this problem to date, because it was classified as a planet until 2006. But Pluto does not substantially perturb the motions of the eight larger planets.) In a center of mass frame, the vector equation of motion for planet *i* moving under the Newtonian gravitational effect of the Sun and the remaining 8 planets is given by

$$\ddot{\mathbf{r}} = G \sum_{j=0}^{9} m_j \frac{\mathbf{r}_j - \mathbf{r}_i}{r_{ij}^3} (j \neq i), \qquad (46)$$

where $\mathbf{r}_i$ and m_i are the position vector and mass of planet i $(i = 1, 2, \ldots, 9)$, respectively, $\mathbf{r}_{ij} \equiv \mathbf{r}_j - \mathbf{r}_i$, and the subscript 0 refers to the Sun. These are the equations of the N-body problem for the case where $N = 10$, and although they have a surprisingly simple form, they have no general, analytical solution. However, as in the case of the three-body problem, it is possible to tackle this problem mathematically by making some simplifying assumptions.

Provided the eccentricities and inclinations of the N bodies are small and there are no resonant interactions between the planets, it is possible to derive an analytical solution that describes the evolution of all the eccentricities, inclinations, perihelia, and nodes of the planets. This solution, called Laplace–Lagrange secular perturbation theory, gives no positional information about the planets, yet it demonstrates that there are long-period variations in the planetary orbital elements that arise from mutual perturbations. The secular periods involved are typically tens or hundreds of thousands of years, and the evolving system always exhibits a regular behavior. In the case of Earth's orbit, such periods may be correlated with climatic change, and large variations in the eccentricity of Mars are thought to have had important consequences for its climate.

In the early nineteenth century, Pierre Simon de Laplace claimed that he had demonstrated the long-term stability of the solar system using the results of his secular perturbation theory. Although the actual planetary system violates some of the assumed conditions (e.g., Jupiter and Saturn are close to a 5:2 resonance), the Laplace–Lagrange theory can be modified to account for some of these effects. However, such analytical approaches always involve the neglect of potentially important interactions between planets. The problem becomes even more difficult when the possibility of near-resonances between some of the secular periods of the system is considered. However, nowadays it is always possible to carry out numerical investigations of long-term stability.

6.2 Stability of the Solar System

Numerical integrations show that the orbits of the planets are chaotic, although there is no indication of gross instability in their motion provided that the integrations are restricted to durations of 5 billion years (the age of the solar system). The eight planets as well as dwarf planet Pluto remain more or less in their current orbits with small, nearly periodic variations in their eccentricities and inclinations; close approaches never seem to occur. Pluto's orbit is chaotic, partly as a result of its 3:2 resonance with the planet Neptune, although the perturbing effects of other planets are also important. Despite the fact that the timescale for exponential divergence of nearby trajectories (the inverse of the Lyapunov exponent) is about 20 million years, no study has shown evidence for Pluto leaving the resonance.

Chaos has also been observed in the motion of the eight planets, and it appears that the solar system as a whole is chaotic with a timescale for exponential divergence of 4 or 5 million years, although different integrations give different results. However, the effect is most apparent in the orbits of the inner planets. Though there appear to be no dramatic consequences of this chaos, it does mean that the use of the deterministic equations of celestial mechanics to predict the future positions of the planets will always be limited by the accuracy with which their orbits can be measured. For example, some results suggest that if the position of Earth along its orbit is uncertain by 1 cm today, then the exponential propagation of errors that is characteristic of chaotic motion implies that knowledge of Earth's orbital position 200 million years in the future is not possible.

The solar system appears to be "stable" in the sense that all numerical integrations show that the planets remain close to their current orbits for timescales of billions of years. Therefore the planetary system appears to be another example of bounded chaos, where the motion is chaotic but always takes place within certain limits. Although an analytical proof of this numerical result and a detailed understanding of how the chaos has arisen have yet to be achieved, the solar system seems to be chaotic yet stable. When the planetary orbits are integrated forward for timescales for several billion years using the averaged equations of motion, it is found that there is a very small but finite probability that the orbit of Mercury can become unstable and intersect the orbit of Venus. Many challenges remain in understanding how structural stability of planetary systems in the presence of transient and intermittent chaos can be maintained, and this subject remains a rich field for dynamical exploration.

7. Dissipative Forces and the Orbits of Small Bodies

The foregoing sections describe the gravitational interactions between the Sun, planets, and moons. Solar radiation has been ignored, but this is an important force for small particles in the solar system. Three effects can be distinguished: (1) the radiation pressure, which pushes particles primarily outward from the Sun (micron-sized dust); (2) the **Poynting–Robertson drag**, which causes centimeter-sized particles to spiral inward toward the Sun; and (3) the Yarkovski effect, which changes the orbits of meter- to kilometer-sized objects owing to uneven temperature distributions at their surfaces. The latter two effects are relativistic and thus quite weak at solar system velocities, but they can nonetheless be significant as they can lead to secular changes in orbital angular momentum and energy. Each of these effects is discussed in the next three subsections and then the effect of gas drag is examined. In the final subsection the influence of tidal interactions is discussed; this effect (in contrast to the other dissipative effects described in this section) is most important for larger bodies such as moons and planets. [See SOLAR SYSTEM DUST.]

7.1 Radiation Force (Micron-Sized Particles)

The Sun's radiation exerts a force, F_r, on all other bodies of the solar system. The magnitude of this force is

$$F_r = \frac{LA}{4\pi c r^2} Q_{pr}, \qquad (47)$$

where A is the particle's geometric cross section, L is the solar luminosity, c is the speed of light, r is the heliocentric distance, and Q_{pr} is the radiation pressure coefficient, which is equal to unity for a perfectly absorbing particle and is of order unity unless the particle is small compared to the wavelength of the radiation. The parameter β is defined as the ratio between the forces due to the radiation pressure

and the Sun's gravity:

$$\beta \equiv \frac{F_r}{F_g} = 5.7 \times 10^{-5} \frac{Q_{pr}}{\rho R}, \qquad (48)$$

where the radius, R, and the density, ρ, of the particle are in c.g.s. units. Note that β is independent of heliocentric distance and that the solar radiation force is important only for micron- and submicron-sized particles. Using the parameter β, a more general expression for the effective gravitational attraction can be written:

$$F_{geff} = \frac{-(1-\beta)GmM}{r^2}, \qquad (49)$$

that is, the small particles "see" a Sun of mass $(1-\beta)M$. It is clear that small particles with $\beta > 1$ are in sum repelled by the Sun, and thus quickly escape the solar system, unless they are gravitationally bound to one of the planets. Dust which is released from bodies traveling on circular orbits at the Keplerian velocity is ejected from the solar system if $\beta > 0.5$.

The importance of solar radiation pressure can be seen, for example, in comets. Cometary dust is pushed in the antisolar direction by the Sun's radiation pressure. The dust tails are curved because the particles' velocity decreases as they move farther from the Sun, due to conservation of angular momentum. [See COMETARY DYNAMICS; PHYSICS AND CHEMISTRY OF COMETS.]

7.2 Poynting–Robertson Drag (Centimeter-Sized Grains)

A small particle in orbit around the Sun absorbs solar radiation and reradiates the energy isotropically in its own frame. The particle thereby preferentially radiates (and loses momentum) in the forward direction in the inertial frame of the Sun. This leads to a decrease in the particle's energy and angular momentum and causes dust in bound orbits to spiral sunward. This effect is called the Poynting–Robertson drag.

The net force on a rapidly rotating dust grain is given by

$$F_{rad} \approx \frac{LQ_{pr}A}{4\pi c r^2}\left[\left(1 - \frac{2v_r}{c}\right)\hat{r} - \frac{v_\theta}{c}\hat{\theta}\right]. \qquad (50)$$

The first term in Eq. (50) is that due to radiation pressure and the second and third terms (those involving the velocity of the particle) represent the Poynting–Robertson drag.

From this discussion, it is clear that small-sized dust grains in the interplanetary medium are removed: (sub)-micron sized grains are blown out of the solar system, whereas larger particles spiral inward toward the Sun.

Typical decay times (in years) for circular orbits are given by

$$\tau_{\text{P-R}} \approx 400 \frac{r^2}{\beta}, \qquad (51)$$

with the distance r in AU.

Particles that produce the bulk of the zodiacal light (at infrared and visible wavelengths) are between 20 and 200 μm, so their lifetimes at Earth orbit are on the order of 10^5 yr, which is much less than the age of the solar system. Sources for the dust grains are comets as well as the asteroid belt, where numerous collisions occur between countless small asteroids.

7.3 Yarkovski Effect (Meter–Sized Objects)

Consider a rotating body heated by the Sun. Because of thermal inertia, the afternoon hemisphere is typically warmer than the morning hemisphere, by an amount $\Delta T \ll T$. Let us assume that the temperature of the morning hemisphere is $T - \Delta T/2$, and that of the evening hemisphere $T + \Delta T/2$. The radiation reaction upon a surface element dA, normal to its surface, is $dF = 2\sigma T^4 dA/3c$. For a spherical particle of radius R, the Yarkovski force in the orbit plane due to the excess emission on the evening side is

$$F_{\text{Y}} = \frac{8}{3}\pi R^2 \frac{\sigma T^4}{c} \frac{\Delta T}{T} \cos \psi, \qquad (52)$$

where σ is the Stefan–Boltzmann constant and ψ is the particle's **obliquity**, that is, the angle between its rotation axis and orbit pole. The reaction force is positive for an object that rotates in the prograde direction, $0 < \psi < 90°$, and negative for an object with retrograde rotation, $90° < \psi < 180°$. In the latter case, the force enhances the Poynting–Robertson drag.

The Yarkovski force is important for bodies ranging in size from meters to several kilometers. Asymmetric outgassing from comets produces a nongravitational force similar in form to the Yarkovski force. [*See* COMETARY DYNAMICS.]

7.4 Gas Drag

Although interplanetary space generally can be considered an excellent vacuum, there are certain situations in planetary dynamics where interactions with gas can significantly alter the motion of solid particles. Two prominent examples of this process are planetesimal interactions with the gaseous component of the protoplanetary disk during the formation of the solar system and orbital decay of ring particles as a result of drag caused by extended planetary atmospheres.

In the laboratory, gas drag slows solid objects down until their positions remain fixed relative to the gas. In the planetary dynamics case, the situation is more complicated. For example, a body on a circular orbit about a planet loses mechanical energy as a result of drag with a static atmosphere, but this energy loss leads to a decrease in semimajor axis of the orbit, which implies that the body actually speeds up! Other, more intuitive effects of gas drag are the damping of eccentricities and, in the case where there is a preferred plane in which the gas density is the greatest, the damping of inclinations relative to this plane.

Objects whose dimensions are larger than the mean free path of the gas molecules experience Stokes' drag,

$$F_{\text{D}} = -\frac{C_{\text{D}} A \rho v^2}{2}, \qquad (53)$$

where v is the relative velocity of the gas and the body, ρ is the gas density, A is the projected surface area of the body, and C_{D} is a dimensionless drag coefficient, which is of order unity unless the **Reynolds number** is very small. Smaller bodies are subject to Epstein drag,

$$F_{\text{D}} = -A \rho v v' \qquad (54)$$

where v' is the mean thermal velocity of the gas. Note that as the drag force is proportional to surface area and the gravitational force is proportional to volume (for constant particle density), gas drag is usually most important for the dynamics of small bodies.

The gaseous component of the protoplanetary disk in the early solar system is believed to have been partially supported against the gravity of the Sun by a negative pressure gradient in the radial direction. Thus, less centripetal force was required to complete the balance, and consequently the gas orbited less rapidly than the Keplerian velocity. The "effective gravity" felt by the gas is

$$g_{\text{eff}} = -\frac{GM_{\text{S}}}{r^2} - (1/\rho)\frac{dP}{dr}. \qquad (55)$$

To maintain a circular orbit, the effective gravity must be balanced by centripetal acceleration, rn^2. For estimated protoplanetary disk parameters, the gas rotated $\sim$0.5% slower than the Keplerian speed.

Large particles moving at (nearly) the Keplerian speed thus encountered a headwind, which removed part of their angular momentum and caused them to spiral inward toward the Sun. Inward drift was greatest for mid-sized particles, which have large ratios of surface area to mass yet still orbit with nearly Keplerian velocities. The effect diminishes for very small particles, which are so strongly coupled to the gas that the headwind they encounter is very slow. Peak rates of inward drift occur for particles that collide with roughly their own mass of gas in one orbital period. Meter-sized bodies in the inner solar nebula drift inward at a rate of up to 10^6 km/yr! Thus, the material that survives to form the planets must complete the transition from centimeter to kilometer size rather quickly, unless it is confined to a thin dust-dominated subdisk in which the gas is dragged along at essentially the Keplerian velocity.

Drag induced by a planetary atmosphere is even more effective for a given density, as atmospheres are almost entirely pressure supported, so the relative velocity between the gas and particles is high. As atmospheric densities drop rapidly with height, particles decay slowly at first, but as they reach lower altitudes, their decay can become very rapid. Gas drag is the principal cause of orbital decay of artificial satellites in low Earth orbit.

7.5 Tidal Interactions and Planetary Satellites

Tidal forces are important to many aspects of the structure and evolution of planetary bodies:

1. On short timescales, temporal variations in tides (as seen in the frame rotating with the body under consideration) cause stresses that can move fluids with respect to more rigid parts of the planet (e.g., the familiar ocean tides) and even cause seismic disturbances (though the evidence that the Moon causes some earthquakes is weak and disputable, it is clear that the tides raised by Earth are a major cause of moonquakes).

2. On long timescales, tides cause changes in the orbital and spin properties of planets and moons. Tides also determine the equilibrium shape of a body located near any massive body; note that many materials that behave as solids on human timescales are effectively fluids on very long geological timescales (e.g., Earth's mantle).

The gravitational attraction of the Moon and Earth on each other causes tidal bulges that rise in a direction close to the line joining the centers of the two bodies. Particles on the nearside of the body experience gravitational forces from the other body that exceed the centrifugal force of the mutual orbit, whereas particles on the far side experience gravitational forces that are less than the centripetal forces needed for motion in a circle. It is the gradient of the gravitational force across the body that gives rise to the double tidal bulge.

The Moon spins once per orbit, so that the same face of the Moon always points toward Earth and the Moon is always elongated in that direction. Earth, however, rotates much faster than the Earth–Moon orbital period. Thus, different parts of Earth point toward the Moon and are tidally stretched. If the Earth was perfectly fluid, the tidal bulges would respond immediately to the varying force, but the finite response time of Earth's figure causes the tidal bulge to lag behind, at the point on Earth where the Moon was overhead slightly earlier. Since Earth rotates faster than

the Moon orbits, this "tidal lag" on Earth leads the position of the Moon in inertial space. As a result, the tidal bulge of Earth accelerates the Moon in its orbit. This causes the Moon to slowly spiral outward. The Moon slows down Earth's rotation by pulling back on the tidal bulge, so the angular momentum in the system is conserved. This same phenomenon has caused most, if not all, major moons to be in synchronous rotation: the rotation and orbital periods of these bodies are equal. In the case of the Pluto–Charon system, the entire system is locked in a synchronous rotation and revolution of 6.4 days. Satellites in retrograde orbits (e.g., Triton) or satellites whose orbital periods are less than the planet's rotation period (e.g., Phobos) spiral inward toward the planet as a result of tidal forces.

Mercury orbits the Sun in 88 days and rotates around its axis in 59 days, a 3:2 spin–orbit resonance. Hence, at every perihelion one of two locations is pointed at the Sun: the subsolar longitude is either $0°$ or $180°$. This configuration is stable because Mercury has both a large orbital eccentricity and a significant permanent deformation that is aligned with the solar direction at perihelion. Indeed, at $0°$ longitude there is a large impact crater, Caloris Planitia, which may be the cause of the permanent deformation.

3. Under special circumstances, strong tides can have significant effects on the physical structure of bodies. Generally, the strongest tidal forces felt by solar system bodies (other than Sun-grazing or planet-grazing comets) are those caused by planets on their closest satellites. Near a planet, tides are so strong that they rip a fluid (or weakly aggregated solid) body apart. In such a region, large moons are unstable, and even small moons, which could be held together by material strength, are unable to accrete because of tides. The boundary of this region is known as **Roche's limit**. Inside Roche's limit, solid material remains in the form of small bodies and rings are found instead of large moons.

The closer a moon is to a planet, the stronger is the tidal force to which it is subjected. Let us consider Roche's limit for a spherical satellite in synchronous rotation at a distance r from a planet. This is the distance at which a loose particle on an equatorial subplanet point just remains gravitationally bound to the satellite. At the center of the satellite of mass m and radius R_s, a particle would be in equilibrium and so

$$\frac{GM}{r^2} = n^2 r, \tag{56}$$

where $M(\gg m)$ is the mass of the planet. However, at the equator, the particle will experience (i) an excess gravitational or centrifugal force due to the planet, (ii) a centrifugal force due to rotation, and (iii) a gravitational force due to the satellite. If the equatorial particle is *just* in equilibrium, these forces will balance and

$$-\frac{d}{dr}\left(\frac{GM}{r^2}\right) R_s + n^2 r = \frac{Gm}{R_s^2}. \tag{57}$$

In this case, Roche's limit r_{Roche} is given by

$$r_{\text{Roche}} = 3^{1/3} \left(\frac{\rho_{\text{planet}}}{\rho_{\text{s}}} \right)^{1/3} R_{\text{planet}}, \qquad (58)$$

with ρ_{planet} and ρ_{s} are the densities for the planet and satellite, respectively, and R_{planet} is the planetary radius. When a fluid moon is considered and flattening of the object due to the tidal distortion is taken into account, the correct result for a liquid moon (no internal strength) is

$$r_{\text{Roche}} = 2.456 \left(\frac{\rho_{\text{planet}}}{\rho_{\text{s}}} \right)^{1/3} R_{\text{planet}}. \qquad (59)$$

Most bodies have significant internal strength, which allows bodies with sizes $\leq{\sim}100$ km to be stable somewhat inside Roche's limit. Mars's satellite Phobos is well inside Roche's limit; it is subjected to a tidal force equivalent to that in Saturn's B ring.

4. Internal stresses caused by variations in tides on a body in an eccentric orbit or not rotating synchronously with its orbital period can result in significant tidal heating of some bodies, most notably in Jupiter's moon Io. If no other forces were present, this would lead to a decay of Io's orbital eccentricity. By analogy to the Earth–Moon system, the tide raised on Jupiter by Io will cause Io to spiral outward and its orbital eccentricity to decrease. However, there exists a 2:1 mean-motion resonant lock between Io and Europa. Io passes on some of the orbital energy and angular momentum that it receives from Jupiter to Europa, and Io's eccentricity is increased as a result of this transfer. This forced eccentricity maintains a high tidal dissipation rate and large internal heating in Io, which displays itself in the form of active volcanism. [See IO]

7.6 Tidal Evolution and Resonances

Objects in prograde orbits that lie outside the **synchronous orbit** can evolve outward at different rates, so there may have been occasions in the past when pairs of satellites evolved toward an **orbit–orbit resonance**. The outcome of such a resonant encounter depends on the direction from which the resonance is approached. For example, capture into resonance is possible only if the satellites are approaching one another. If the satellites are receding, then capture is not possible, but the resonance passage can lead to an increase in the eccentricity and inclination. In certain circumstances it is possible to study the process using a simple mathematical model. However, this model breaks down near the chaotic separatrices of resonances and in regions of resonance overlap.

It is likely that the major satellites of Jupiter, Saturn, and Uranus have undergone significant tidal evolution and

that the numerous resonances in the Jovian and Saturnian systems are a result of resonant capture. The absence of orbit–orbit resonances among the major moons in the Uranian system is thought to be related to the fact that the oblateness of Uranus is significantly less than that of Jupiter or Saturn. In these circumstances, there can be large chaotic regions associated with resonances and stable capture may be impossible. However, temporary capture into some resonances can produce large changes in eccentricity or inclination. For example, the Uranian satellite Miranda has an anomalously large inclination of $4°$, which is thought to be the result of a chaotic passage through the 3:1 resonance with Umbriel at some time in its orbital history. Under tidal forces, a satellite's eccentricity is reduced on a shorter timescale than its inclination, and Miranda's current inclination agrees with estimates derived from a chaotic evolution. [See PLANETARY SATELLITES.]

8. Chaotic Rotation

8.1 Spin–Orbit Resonance

One of the dissipative effects of the tide raised on a natural satellite by a planet is to cause the satellite to evolve toward a state of synchronous rotation, where the rotational period of the satellite is approximately equal to its orbital period. Such a state is one example of a spin–orbit resonance, where the ratio of the spin period to the orbital period is close to a rational number. The time needed for a near-spherical satellite to achieve this state depends on its mass and orbital distance from the planet. Small, distant satellites take a longer time to evolve into the synchronous state than do large satellites that orbit close to the planet. Observations by spacecraft and ground-based instruments suggest that most regular satellites are in the synchronous spin state, in agreement with theoretical predictions.

The lowest energy state of a satellite in synchronous rotation has the moon's longest axis pointing in the approximate direction of the planet–satellite line. Let θ denote the angle between the long axis and the planet–satellite line in the planar case of a rotating satellite (Fig. 21). The variation of θ with time can be described by equating the time variation of the rotational angular momentum with the restoring torque. The resulting differential equation is

$$\ddot{\theta} + \frac{\omega_0^2}{2r^3} \sin 2(\theta - f) = 0, \qquad (60)$$

where ω_0 is a function of the principal moments of inertia of the satellite, r is the radial distance of the satellite from the planet, and f is the true anomaly (or angular position) of the satellite in its orbit. The radius is an implicit function of time and is related to the true anomaly by the

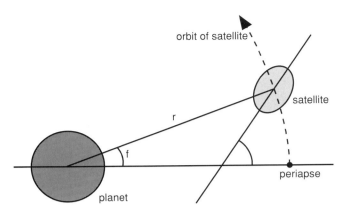

FIGURE 21 The geometry used to define the orientation of a satellite in orbit about a planet. The planet–satellite line makes an angle f (the true anomaly) with a reference line, which is taken to be the periapse direction of the satellite's orbit. The orientation angle, θ, of the satellite is the angle between its long axis and the reference direction.

equation

$$r = \frac{a(1 - e^2)}{1 + e\cos f}, \qquad (61)$$

where a and e are the constant semimajor axis and the eccentricity of the satellite's orbit, respectively, and the orbit is taken to be fixed in space.

Equation (60) defines a deterministic system where the initial values of θ and $\dot{\theta}$ determine the subsequent rotation of the satellite. Since θ and $\dot{\theta}$ define a unique spin position of the satellite, a surface of section plot of $(\theta, \dot{\theta})$ once every orbital period, say at every **periapse** passage, produces a picture of the phase space. Figure 22 shows the resulting surface of section plots for a number of starting conditions using $e = 0.1$ and $\omega_0 = 0.2$. The chosen values of ω_0 and e are larger than those that are typical for natural satellites, but they serve to illustrate the structure of the surface of section; large values of e are unusual since tidal forces also act to damp eccentricity. The surface of section shows large, regular regions surrounding narrow islands associated with the 1:2, 1:1, 3:2, 2:1, and 5:2 spin–orbit resonances at $\dot{\theta} = 0.5, 1, 1.5, 2$, and 2.5, respectively. The largest island is associated with the strong 1:1 resonance and, although other spin states are possible, most regular satellites, including Earth's Moon, are observed to be in this state. Note the presence of diffuse collections of points associated with small chaotic regions at the separatrices of the resonances. These are particularly obvious at the 1:1 spin–orbit state at $\theta = \pi/2$, $\dot{\theta} = 1$. Although this is a completely different dynamical system compared to the circular restricted three-body problem, there are distinct similarities in the types of behavior visible in Fig. 22 and parts of Figs. 14 and 15.

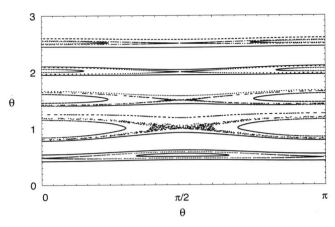

FIGURE 22 Representative surface of section plots of the orientation angle, θ, and its time derivative, $\dot{\theta}$, obtained from the numerical solution of Eq. (59) using $e = 0.1$ and $\omega_0 = 0.2$. The values of θ and $\dot{\theta}$ were obtained at every periapse passage of the satellite. Four starting conditions were integrated for each of the 1:2, 1:1, 3:2, 2:1, and 5:2 spin–orbit resonances in order to illustrate motion inside, at the separatrix, and on either side of each resonance. The thickest "island" is associated with the strong 1:1 spin–orbit state $\theta = 1$, whereas the thinnest is associated with the weak 5:2 resonance at $\theta = 2.5$.

In the case of near-spherical objects, it is possible to investigate the dynamics of spin–orbit coupling using analytical techniques. The sizes of the islands shown in Fig. 22 can be estimated by expanding the second term in Eq. (60) and isolating the terms that will dominate at each resonance. Using such a method, each resonance can be treated in isolation and the gravitational effects of nearby resonances can be neglected. However, if a satellite is distinctly nonspherical, ω_0 can be large and this approximation is no longer valid. In such cases it is necessary to investigate the motion of the satellite using numerical techniques.

8.2 Hyperion

Hyperion is a satellite of Saturn that has an unusual shape (Fig. 23). It has a mean radius of 135 km, an orbital eccentricity of 0.1, a semimajor axis of 24.55 Saturn radii, and a corresponding orbital period of 21.3 days. Such a small object at this distance from Saturn has a large tidal despinning timescale, but the unusual shape implies an estimated value of $\omega_0 = 0.89$.

The surface of section for a *single* trajectory is shown in Fig. 24 using the same scale as Fig. 22. It is clear that there is a large chaotic zone that encompasses most of the spin–orbit resonances. The islands associated with the synchronous and other resonances survive but in a much reduced form. Although this calculation assumes that Hyperion's spin axis remains perpendicular to its orbital plane, studies have shown that the satellite should also be undergoing

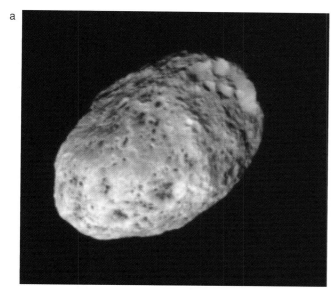

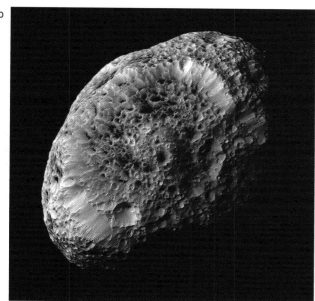

FIGURE 23   Two *Cassini* images of the Saturnian satellite Hyperion show the unusual shape of the satellite, which is one cause of its chaotic rotation. Panel (a) is a true color image, while panel (b) uses false color and has better resolution because it was obtained at closer range. [Courtesy of NASA/JPL/Space Science Institute.]

a tumbling motion, such that its axis of rotation is not fixed in space.

Voyager observations of Hyperion indicated a spin period of 13 days, which suggested that the satellite was not in synchronous rotation. However, the standard techniques that are used to determine the period are not applicable if it varies on a timescale that is short compared with the timespan of the observations. In principle, the rotational period can be deduced from ground-based observations by looking

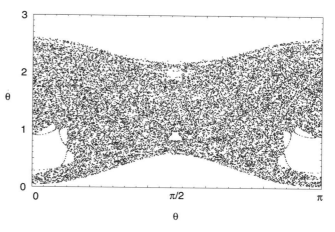

FIGURE 24   A single surface of section plot of the orientation angle, θ, and its time derivative, $\dot{\theta}$, obtained from the numerical solution of Eq. (10) using the values $e = 0.1$ and $\omega_0 = 0.89$, which are appropriate for Hyperion. The points cover a much larger region of the phase space than any of those shown in Fig. 22, and although there are some remaining islands of stability, most of the phase space is chaotic.

for periodicities in plots of the brightness of the object as a function of time (the lightcurve of the object). The results of one such study for Hyperion are shown in Fig. 25. Since there is no recognizable periodicity, the lightcurve is consistent with that of an object undergoing chaotic rotation. Hyperion is the first natural satellite that has been observed to have a chaotic spin state, and results from *Cassini* images confirm this result. Observations and numerical studies of Hyperion's rotation in three dimensions have shown that its

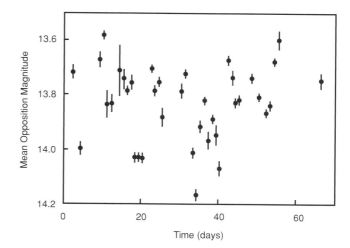

FIGURE 25   Ground-based observations by J. Klavetter of Hyperion's lightcurve obtained over 13 weeks (4.5 orbital periods) in 1987. The fact that there is no obvious curve through the data points is convincing evidence that the rotation of Hyperion is chaotic. (Courtesy of the American Astronomical Society.)

spin axis does not point in a fixed direction. Therefore the satellite also undergoes a tumbling motion in addition to its chaotic rotation.

The dynamics of Hyperion's motion is complicated by the fact that it is in a 4:3 orbit–orbit resonance with the larger Saturnian satellite Titan. Although tides act to decrease the eccentricities of satellite orbits, Hyperion's eccentricity is maintained at 0.104 by means of the resonance. Titan effectively forces Hyperion to have this large value of e and so the apparently regular orbital motion inside the resonance results, in part, in the extent of the chaos in its rotational motion. [See PLANETARY SATELLITES.]

8.3 Other Satellites

Although there is no evidence that other natural satellites are undergoing chaotic rotation at the present time, it is possible that several irregularly shaped regular satellites did experience chaotic rotation at some time in their histories. In particular, since satellites have to cross chaotic separatrices before capture into synchronous rotation can occur, they must have experienced some episode of chaotic rotation. This may also have occurred if the satellite suffered a large impact that affected its rotation. Such episodes could have induced significant internal heating and resurfacing events in some satellites. The Martian moon Phobos and the Uranian moon Miranda have been mentioned as possible candidates for this process. If this happened early in the history of the solar system, then the evidence may well have been obliterated by subsequent cratering events. [See PLANETARY SATELLITES.]

8.4 Chaotic Obliquity

The fact that a planet is not a perfect sphere means that it experiences additional perturbing effects due to the gravitational forces exerted by its satellites and the Sun, and these can cause long-term evolution in its obliquity (the angle between the planet's equator and its orbit plane). Numerical investigations have shown that chaotic changes in obliquity are particularly common in the inner solar system. For example, it is now known that the stabilizing effect of the Moon results in a variation of $\pm 1.3°$ in Earth's obliquity around a mean value of $23.3°$. Without the Moon, Earth's obliquity would undergo large, chaotic variations. In the case of Mars there is no stabilizing factor and the obliquity varies chaotically from $0°$ to $60°$ on a timescale of 50 million years. Therefore an understanding of the long-term changes in a planet's climate can be achieved only by an appreciation of the role of chaos in its dynamical evolution.

9. Epilog

It is clear that nonlinear dynamics has provided us with a deeper understanding of the dynamical processes that have helped to shape the solar system. Chaotic motion is a natural consequence of even the simplest systems of three or more interacting bodies. The realization that chaos has played a fundamental role in the dynamical evolution of the solar system came about because of contemporary and complementary advances in mathematical techniques and digital computers. This coincided with an explosion in our knowledge of the solar system and its major and minor members. Understanding how a random system of planets, satellites, ring and dust particles, asteroids, and comets interacts and evolves under a variety of chaotic processes and timescales, ultimately means that this knowledge can be used to trace the history and predict the fate of other planetary systems.

Bibliography

Burns, J. A. (1987). The motion of interplanetary dust. *In* "The Evolution of the Small Bodies of the Solar System," (Fulchignoni, M. and Kresak, L., eds.) pp. 252–275. Soc. Italiana di Fisica, Bologna, Italy.

Danby, J. M. A. (1992). "Fundamentals of Celestial Mechanics." Willmann–Bell, Richmond, Virginia.

Diacu, F., and Holmes, P. (1996). "Celestial Encounters. The Origins of Chaos and Stability." Princeton Univ. Press, Princeton, NJ.

Duncan, M., and Quinn, T. (1993). The long-term dynamical evolution of the solar system. *Annu. Rev. Astron. Astrophys.* **31**, 265–295.

Dvorak, R., and Henrard, J. (eds.) (1988). "Long Term Evolution of Planetary Systems." Kluwer, Dordrecht, Holland.

Ferraz-Mello, S. (ed.) (1992). "Chaos, Resonance and Collective Dynamical Phenomena in the Solar System." Kluwer, Dordrecht, Holland.

Lichtenberg, A. J., and Lieberman, M. A. (1992). "Regular and Chaotic Dynamics," volume **38** of Applied Mathematical Sciences. Springer-Verlag, New York, second edition.

Lissauer, J. J. (1993). Planet formation. *Annu. Rev. Astron. Astrophys.* **31**, 129–174.

Morbidelli, A. (2002). "Modern Celestial Mechanics." Taylor & Francis, London.

Murray, C. D., and Dermott, S. F. (1999). "Solar System Dynamics." Cambridge University Press, Cambridge.

Peale, S. J. (1976). Orbital resonances in the solar system. *Annu. Rev. Astron. Astrophys.* **14**, 215–246.

Peterson, I. (1993). "Newton's Clock. Chaos in the Solar System." W. H. Freeman, New York.

Roy, A. E., and Steves, B. A. (eds.) (1995). "From Newton to Chaos: Modern Techniques for Understanding and Coping with Chaos in *N*-Body Dynamical Systems." Plenum, New York.

Planetary Impacts

Richard A.F. Grieve
Natural Resources Canada
Ottawa, Canada

Mark J. Cintala*
NASA Johnson Space Center
Houston, Texas

Roald Tagle
Humboldt University
Berlin, Germany

CHAPTER 43

P lanetary impacts have occurred throughout the history of the solar system. Small bodies, such as asteroids and comets, can have their orbits disturbed by gravitational forces, which results in their having a finite probability of colliding with another body or planet. Indeed, the collision of small bodies to form larger bodies was the fundamental process of planetary formation which, in its final stages, involved impacts between planetesimal-sized objects. As the solar system stabilized, the impact rate decreased but was still sufficient as late as ~4.0 billion years ago to produce impact basins with diameters measured in hundreds to thousands of kilometers. As a result, impacts were a major geologic process in early planetary evolution and served to characterize the early upper crusts and surfaces of planetary bodies. Although impacts producing craters 100–200 km in diameter are relatively rare in more recent geologic time, they still occur on timescales of approximately 100 million years. One such event on the Earth marks the boundary between the Cretaceous and Tertiary geologic periods and resulted in the mass extinction of approximately 75% of the species living on Earth 65 million years ago. The 180 km diameter Chicxulub impact crater in the Yucatan, Mexico, is now known to be the site of this global-extinction impact.

1. Impact Craters

1.1 Crater Shape

On bodies that have no atmosphere, such as the Moon, even the smallest pieces of interplanetary material can produce impact craters down to micrometer-sized cavities on individual mineral grains. On larger bodies, atmosphere-induced breakup and deceleration serve to slow smaller impacting objects. On the Earth, for example, impacting bodies with masses below 10^4g can lose up to 90% of their velocity during atmospheric penetration, and the resultant impact pit is only slightly larger than the projectile itself. Atmospheric effects on larger masses, however, are less severe, and the body impacts with relatively undiminished velocity, producing a crater that is considerably larger than the impacting body.

The processes accompanying such events are rooted in the physics of impact, with the differences in response among the various planets largely being due to differences in the properties of the planetary bodies (e.g., surface gravity, atmospheric density, and target composition and strength). The basic shape of virtually all impact craters is

---

* The views expressed by the author are his own and do not represent the views of NASA or any NASA employee.

FIGURE 1   Approximately 1 km diameter, relatively young simple martian crater. Large blocks, ejected late in the cratering process, can be seen on the ejecta near the rim. The ejecta can be differentiated into continuous ejecta and discontinuous ejecta, which appear as separate fingers and braids (*Mars Global Surveyor*).

Simple crater - final form

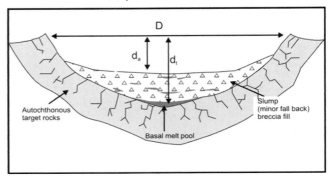

FIGURE 2   Schematic cross section of a simple crater, based on terrestrial observations. D is diameter and d_a and d_t are the depths of the apparent and true crater, respectively. See text for details.

a depression with an upraised rim. With increasing diameter, impact craters become proportionately shallower and develop more complicated rims and floors, including the appearance of central topographic peaks and interior rings.

There are three major subdivisions in shape: simple craters, complex craters, and impact basins Simple impact structures have the form of a bowl-shaped depression with an upraised rim (Fig. 1). An overturned flap of ejected target materials exists on the rim, and the exposed rim, walls, and floor define the apparent crater. Observations at terrestrial impact craters reveal that a lens of brecciated target material, roughly parabolic in cross section, exists beneath the floor of this apparent crater (Fig. 2). This breccia lens is a mixture of different target materials, with fractured blocks set in a finer-grained matrix. These are **allochthonous** materials, having been moved into their present position by the cratering process. Beneath the breccia lens, relatively in-place, or **parautochthonous**, fractured target materials define the walls and floor of what is known as the true crater (Fig. 2). In the case of terrestrial simple craters, the depth to the base of the breccia lens (i.e., the base of the true crater) is roughly twice the depth to the top of the breccia lens (i.e., the floor of the apparent crater).

With increasing diameter, simple craters display signs of wall and rim collapse, as they evolve into complex craters. The diameter at which this transition takes place varies between planetary bodies and is, to a first approximation, an inverse function of planetary gravity. Other variables, such as target strength, and possibly projectile type, and impact angle and velocity, play a role and the transition actually occurs over a small range in diameter. For example, the transition between simple and complex craters occurs in the 15–25 km diameter range on the Moon. The effect of target strength is most readily apparent on Earth, where complex craters can occur at diameters as small as 2 km in sedimentary target rocks, but do not occur until diameters of 4 km, or greater, in stronger, crystalline target rocks.

Complex craters are highly modified structures. A typical complex crater is characterized by a central topographic peak or peaks, a broad, flat floor, and a terraced, inwardly slumped rim area (Fig. 3). Observations at terrestrial complex craters show that the flat floor consists of a sheet of **impact melt** rock and/or **polymict** breccia (Fig. 4). The central region is structurally complex and, in large part, occupied by the central peak, which is the topographic manifestation of a much broader and extensive volume of uplifted rocks that occur beneath the center of complex craters (Fig. 4).

With increasing diameter, a fragmentary ring of interior peaks appears, marking the beginning of the morphologic transition from craters to basins. While a single interior ring is required to define a basin, they can be subdivided further into central-peak basins, with both a peak and ring; peak ring basins (Fig. 5), with a single ring; and multiring basins, with two or more interior rings (Fig. 6). The transition from central-peak basins to peak-ring basins to multiring basins also represents a sequence with increasing diameter. As with the simple to complex crater transition, there is a small amount of overlap in basin shape near transition diameters.

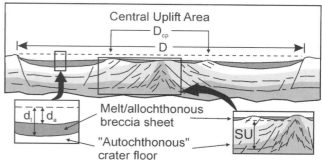

FIGURE 4   Schematic cross section of a complex crater, based on terrestrial observations. Notation is as in Fig. 2, with SU corresponding to the structural uplift and D_{cp}, to the diameter of the central uplift area. Note the preservation of the upper beds (different shades of gray) in the outer portion of the crater floor, indicating excavation was limited to the central area. See text for details.

Thus, at increasing distance from the crater, the final ejecta blanket on the ground includes increasing amounts of local materials. Secondary crater fields, resulting from the impact of larger, coherent blocks and clods of ejecta, surround fresh craters and are particularly evident on bodies with no or thin atmospheres, such as the Moon, Mercury, and Mars. They are often associated with typically bright

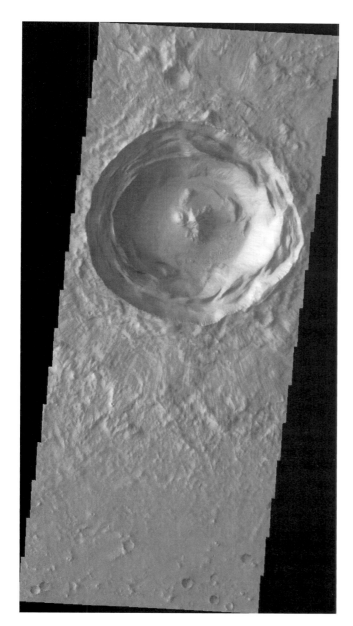

FIGURE 3   Complex central peak crater in the Isidis basin on Mars, with the terraced walls of the crater rim stepping down to a flat floor and a central peak. Also evident are the external rays of continuous (linear) and discontinuous (braided) ejecta on the surrounding terrain. (*Mars Global Surveyor*).

Ejected target material surrounds impact craters and can be subdivided into continuous and discontinuous ejecta facies (Figs. 1 and 3). The continuous deposits are those closest to the crater, being thickest at the rim crest. In the case of simple craters, the net effect of the ejection process is to invert the stratigraphy at the rim. As the distance from the crater rim increases, the ejecta are emplaced at higher velocities and, therefore, land with higher kinetic energies, resulting in the mixing of ejecta with local surface material.

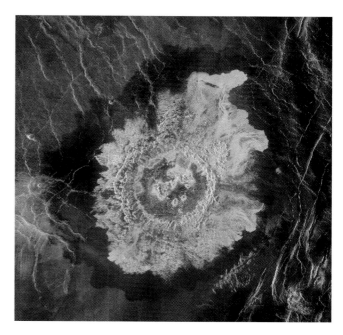

FIGURE 5   The 50 km diameter peak ring basin Barton on Venus, with a discontinuous peak ring. Barton is close to the lower limit of the diameter where peak rings appear in impact craters on Venus and has a discontinuous peak ring (*Magellan*).

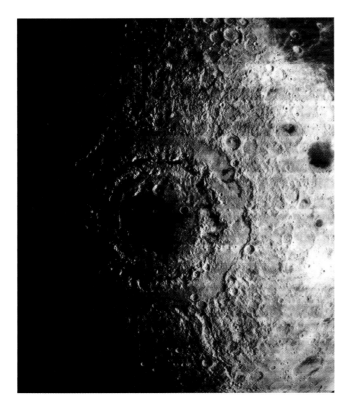

FIGURE 6 With a diameter of ∼900 km in diameter, as defined by the outer ring, the Cordillera mountains, Orientale the youngest and best-preserved multiring basin on the Moon (*Lunar Orbiter*).

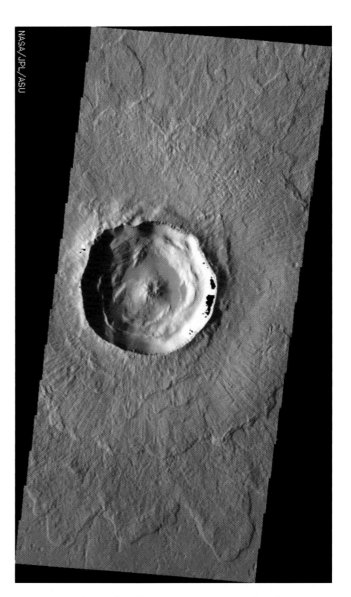

FIGURE 7 This 7.5 km diameter martian central peak crater is close to the transition diameter to complex craters and has a small central peak and simple terraced walls. Ejecta can be discriminated into a fluidized material, which extends farthest and has lobate margins, overlain by a second type of ejecta, which does not extend as far and displays radial linear features (*Mars Global Surveyor*).

or high-albedo "rays" that define an overall radial pattern to the primary crater. Two principal processes have been suggested to explain the rays. The first is a compositional effect, where the ejecta are chemically different from the material on which it is deposited. While this most often results in rays that are brighter than the surrounding material, the reverse can also occur. The second effect is a consequence of "maturity" due to prolonged exposure to "space weathering" agents like radiation and micrometeoroid bombardment on surface materials. [*See* MAIN-BELT ASTEROIDS.] Fresher material excavated by an impact and deposited in the rays is generally brighter than the more mature material of the deposition surface.

Many martian craters display examples of apparently fluidized ejecta (Fig. 7). They have been called "fluidized–ejecta," "rampart," or "pedestal" craters, where their ejecta deposits indicate emplacement as a ground-hugging flow. Most hypotheses on the origin of these features invoke the presence of ground ice (or water), which, upon heating by impact, is incorporated into the ejecta in either liquid or vapor form. This, then, provides lubrication for the mobilized material.

On Venus, impact craters more than 15–20 km in diameter exhibit central peaks and/or peak rings (Fig. 8) and appear, for the most part, to be similar to complex craters and

basins on the other terrestrial planets. Many of the craters smaller than 15 km, however, have rugged, multiple floors or occur as crater clusters. This is attributed to the effects of the dense atmosphere of Venus (surface pressure of ∼90 bar), which effectively crushes and breaks up smaller impacting bodies, so that they result in clusters of relatively shallow craters. Also due to atmospheric effects, there is a deficit in the number of expected craters with diameters up to 35 km, and there are no craters smaller than 3 km in diameter on Venus.

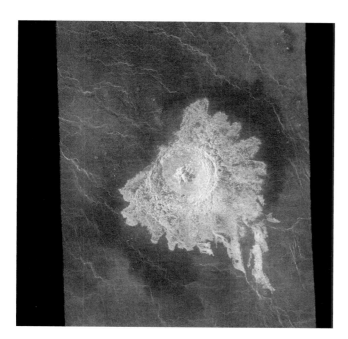

FIGURE 8 Complex venusian central peak crater Aurelia, 32 km in diameter, which exhibits terraced walls, a flat floor, central peaks and long-running lobate flows, particularly in the lower right. Its ejecta pattern is asymmetric, indicating an oblique impact. The crater and the ejecta are also partially surrounded by terrain with a radar dark halo (*Magellan*).

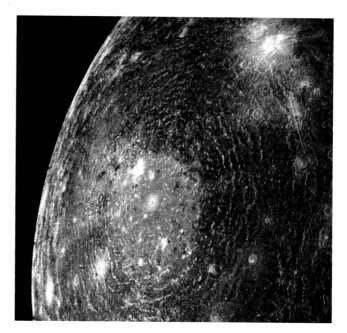

FIGURE 9 The Valhalla multiring basin on Callisto. The overall structure may be as large as 4000 km in diameter, but only the central bright area is believed to be formed directly by impact. The surrounding, multiple scarps were likely formed in response to the subsurface flow of material back toward the initial crater, due to the relatively low internal strength of Callisto (*Voyager*).

In many cases, craters on Venus have ejecta deposits that are visible out to greater distances than expected from simple ballistic emplacement, and the distal deposits are clearly lobate (Fig. 8). These deposits likely owe their origin to entrainment effects of the dense atmosphere and/or the high proportion of impact melt that would be produced on a relatively high-gravity, high–surface temperature planet such as Venus. Another unusual feature on Venus is radar-dark zones surrounding some craters that can extend three to four crater diameters from the crater center (Fig. 8). They are believed to be due to the modification of surface roughness by the atmospheric **shock wave** produced by the impacting body. Small crater clusters have dark halos and dark circular areas where no central crater form has been observed. In these latter cases, the impacting body did not survive atmospheric passage, but the accompanying atmospheric shock wave had sufficient energy to interact with the surface to create a dark, radar-smooth area. [*See* VENUS: SURFACE AND INTERIOR.] The situation is somewhat analogous to the 1908 Tunguska event, when a relatively small body exploded over Siberia at an altitude of ~10 km, and the resultant atmospheric pressure wave leveled some 2000 km² of forest.

Remarkable ring structures occur on the Galilean satellites of Jupiter, Callisto, and Ganymede. The largest is the 4000-km feature Valhalla on Callisto (Fig. 9), which consists of a bright central area up to 800 km in diameter,

surrounded by a darker terrain with bright ridges 20–30 km apart. This zone is about 300 km wide and gives way to an outer zone with **graben** or rift-like features 50–100 km apart. These (very) multiring basins are generally considered to be of impact origin, but with the actual impact crater confined to the central area. The exterior rings are believed to be formed as a result of the original crater puncturing the outer, strong shell, or lithosphere, of these bodies. This permitted the weaker, underlying layer, the asthenosphere, to flow toward the crater, setting up stresses that led to fracturing and the formation of circumscribing scarps and graben.

On Callisto and Ganymede, there is also a unique class of impact craters that no longer have an obvious crater form but appear as bright, or high-albedo, spots on the surfaces of these bodies. These are known as palimpsests and are believed to have begun as complex craters but have had their topography relaxed by the slow, viscous creep of the target's icy crust over time. Palimpsests are old impact features and may have been formed when the icy satellites were young and relatively warm, with a thin crust possibly incapable of retaining significant topography.

Other anomalous crater forms are developed on Ganymede and Callisto. On these icy satellites, most craters larger than 25 km have a central pit or central dome (Fig. 10), rather than a central peak. Pit and dome craters are shallower than other craters of comparable size, and

FIGURE 10   Complex crater Har, 50 km in diameter, on Callisto, with a central dome in place of a central peak. The origin of the central mound is some form of response to the weak icy nature of the target material. A smaller (20 km) and younger central peak complex crater with a central peak, Tindr, occurs on the western rim of Har (*Galileo*).

it has been suggested that the pits are due to the formation of slushy or fluid material by impact melting and the domes are due to uplift of the centers of the craters as a result of layers in the crust with different mechanical properties. The fact that some craters on these icy bodies are anomalous has been ascribed to a velocity effect, as higher impact velocities result in greater melting of the target, or to changes in the mechanical behavior of the crust and its response to impact with time. Interpretations of the origin of the various anomalous crater forms on the icy satellites, however, are generally not well constrained.

1.2 Crater Dimensions

The depth–diameter relations for craters on the terrestrial or silicate planets are given in Table 1. (Relations are in the form $d = aD^b$, where d is apparent depth, D is rim-crest diameter, and units are in kilometers.). Other relations involving parameters such as rim height, rim width, central peak diameter, and central peak height can be found in the literature. Due to the abundant detailed imagery and low rate of crater-modifying process, such as erosion, the best-defined morphometric relations for fresh impact craters are from the Moon.

Simple craters have similar apparent depth–diameter relationships on all the terrestrial planets (Table 1). At first glance, terrestrial craters appear to be shallower than their planetary counterparts. Compared to the other terrestrial planets, erosion is most severe on Earth, and crater rims are rapidly affected by erosion. Few terrestrial craters have well-preserved rims, and it is common to measure terrestrial crater depths with respect to the ground surface, which is known and is assumed to erode more slowly. In the case of other planetary bodies, depths are measured most often by the shadow that the rim casts on the crater floor. That is, the topographic measure is a relative one between the rim crest and the floor. Thus, the measurements of depth for Earth and for other planetary bodies are not exactly the same. For the very few cases in which the rim is well preserved

TABLE 1	Apparent Depth-Diameter Relations for Craters on the Terrestrial Planets		
Planetary Body	**Exponent (b)**	**Coefficient (a)**	**Gravity (cm^{-2})**
	Simple Craters		
Moon	1.010	0.196	162
Mars	1.019	0.204	372
Mercury	0.995	0.199	378
Earth	1.06	0.13	981
	Complex Central Peak Craters		
Moon	0.301	1.044	162
Mars	0.25	0.53	372
Mercury	0.415	0.492	378
Venus	0.30	0.40	891
Earth			
Sedimentary	0.12	0.30	981
Crystalline	0.15	0.43	981

in terrestrial craters, depths from the top of the rim to the crater floor are comparable to those of similar-sized simple craters on the other terrestrial planets.

Unlike simple craters, the depths of complex craters with respect to their diameters do vary between the terrestrial planets (Table 1). While the sense of variation is that increasing planetary gravity shallows final crater depths, this is not a strict relationship. For example, martian complex craters are shallower than equivalent-sized mercurian craters (Table 1), even though the surface gravities of the two planets are very similar. This is probably a function of differences between target materials, with the trapped volatiles and relatively abundant sedimentary deposits making Mars' surface, in general, a weaker target. Mars has also evidence of wind and water processes, which will reduce crater-related topography by erosion and sedimentary infilling. The secondary effect of target strength is also well illustrated by the observation that terrestrial complex craters in sedimentary targets are shallower than those in crystalline targets (Table 1).

Data from the Galileo mission indicates that depth–diameter relationships for craters on the icy satellites Callisto, Europa, and Ganymede have the same general trends as those on the rocky terrestrial planets. Interestingly, the depth–diameter relationship for simple craters is equivalent to that on the terrestrial planets. Although the surface gravities of these icy satellites is only 13–14% of that of the Earth, the transition diameter to complex crater forms occurs at ~3 km, similar to that on the Earth. This may be a reflection of the extreme differences in material properties between icy and rocky worlds. There are also inflections and changes in the slopes of the depth–diameter relationships for the complex craters, with a progressive reduction in absolute depth at diameters larger than the inflection diameter. These anomalous characteristics of the depth–diameter relationship have been attributed to changes in the physical behavior of the crust with depth and the presence of subsurface oceans. [See EUROPA; GANYMEDE AND CALLISTO.]

2. Impact Processes

The extremely brief timescales and extremely high energies, velocities, pressures, and temperatures that accompany impact are not encountered, as a group, in other geologic processes and make studying impact processes inherently difficult. Small-scale impacts can be produced in the laboratory by firing projectiles at high velocity (generally below about 8 km s^{-1}) at various targets. Some insights can also be gained from observations of high-energy, including nuclear explosions. Most recently, "hydrocode" numerical models have been used to simulate impact crater formation. The planetary impact record also provides constraints on the process. The terrestrial record is an important source of

ground-truth data, especially with regard to the subsurface nature and spatial relations at impact craters, and the effects of impact on rocks.

When an interplanetary body impacts a planetary surface, it transfers about half of its kinetic energy to the target. The kinetic energy of such interplanetary bodies is extremely high, with the mean impact velocity on the terrestrial planets for asteroidal bodies ranging from ~12 km s^{-1} for Mars to over ~25 km s^{-1} for Mercury. The impact velocity of comets is even higher. Long-period comets (those with orbital periods greater than 200 years) have an average impact velocity with Earth of ~55 km s^{-1}, whereas short-period comets have a somewhat lower average impact velocity. [See COMETARY DYNAMICS.]

2.1 Crater Formation

On impact, a shock wave propagates back into the impacting body and also into the target. The latter shock wave compresses and heats the target, while accelerating the target material (Fig. 11). The direction of this acceleration is perpendicular to the shock front, which is roughly hemispherical, so material is accelerated downward and outward. Because a state of stress cannot be maintained at a free surface, such as the original ground surface or the edges and rear of the impacting body, a series of secondary release or "rarefaction" waves are generated, which bring the shock-compressed materials back to ambient pressure. As the rarefaction wave interacts with the target material, it alters the direction of the material set in motion by the shock wave, changing some of the outward and downward motions in the relatively near-surface materials to outward and upward, leading to the ejection of material and the growth of a cavity. Directly below the impacting body, however, the two wave fronts are more nearly parallel, and material is still driven downward (Fig. 11).

These motions define the **cratering flow-field** and a cavity grows by a combination of upward ejection and downward displacement of target materials. This "transient cavity" reaches its maximum depth before its maximum radial dimensions, but it is usually depicted in illustrations at its maximum growth in all directions (Fig. 11). At this point, it is parabolic in cross section and, at least for the terrestrial case, has a depth-to-diameter ratio of about 1 to 3. As simple craters throughout the solar system appear to have similar depth–diameter ratios, the 1:3 ratio for the transient cavity can probably be treated as universal.

An asteroidal body of density 3 g cm^{-3} impacting crystalline target rocks at 25 km s^{-1} will generate initial shock velocities in the target faster than 20 km s^{-1}, with corresponding velocities over 10 km s^{-1} for the materials set in motion by the shock wave. The rarefaction wave has an initial velocity similar to that of the shock wave but, because the target materials are compressed by the shock, the rarefaction has a smaller distance to cover to overtake

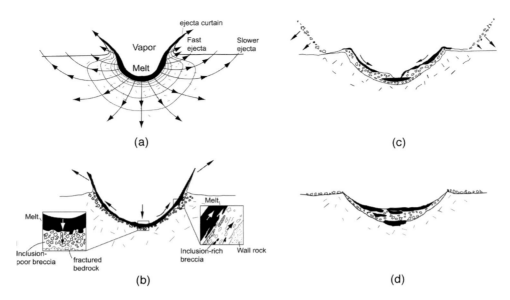

FIGURE 11 Schematic illustration of the formation of a simple crater (Figs. 1 and 2). (a) On impact, the shock wave, indicated by the roughly hemispherical solid lines of shock pressure, propagates into the target rocks. Closer to the point of impact, the combination of the motions imparted by the shock and rarefaction waves has opened up a growing cavity through excavation and displacement of the target rocks. Melted and vaporized material is driven down into this expanding transient cavity. Ultimately, target rocks set in motion by the cratering flow-field will follow the paths outlined by the solid lines with arrows. (b) Close to the end of formation of the transient cavity formed by the cratering flow-field, with melted and shocked target rocks that are moving up the walls on their way to being ejected. (c) The unstable transient cavity walls collapse downward and inward, carrying the lining of melt and shocked target rocks into the cavity and mix them together with the wall rocks to form a breccia deposit. The collapse of the cavity walls also enlarges slightly the diameter of the final crater. (d) Final form of a simple crater with an interior breccia lens. (After Melosh, 1989.)

the moving material and alter its direction of movement. Transient-cavity growth is an extremely rapid event. For example, the formation of a 2.5 km diameter transient cavity will take only about 10 seconds on Earth.

The cratering process is sometimes divided into stages: initial contact and compression, excavation, and modification. In reality, however, it is a continuum with different volumes of the target undergoing different stages of the cratering process at the same time (Fig. 11). As the excavation stage draws to a close, the direction of movement of target material changes from outward to inward, as the unstable transient cavity collapses to a final topographic form more in equilibrium with gravity. This is the modification stage, with collapse ranging from landslides on the cavity walls of the smallest simple craters to complete collapse and modification of the transient cavity, involving the uplift of the center and collapse of the rim area to form central peaks and terraced, structural rims in larger complex craters.

The interior breccia lens of a typical simple crater is the result of this collapse. As the cratering flow comes to an end, the fractured and over-steepened cavity walls become unstable and collapse inward, carrying with them a lining of shocked and melted debris (Fig. 11). The inward-collapsing

walls undergo more fracturing and mixing, eventually coming to rest as the bowl-shaped breccia lens of mixed unshocked and shocked target materials that partially fill simple craters (Fig. 11). The collapse of the walls increases the rim diameter, such that the final crater diameter is about 20% larger than that of the transient cavity. This is offset by the shallowing of the cavity accompanying production of the breccia lens, with the final apparent crater being about half the depth of the original transient cavity (Fig. 11). The collapse process is rapid and probably takes place on timescales comparable to those of transient-cavity formation.

Much of our understanding of complex-crater formation comes from observations at terrestrial craters, where it has been possible to trace the movement of beds to show that central peaks are the result of the uplift of rocks from depth (Fig. 4). Shocked target rocks, analogous to those found in the floors of terrestrial simple craters, constitute the central peak at the centers of complex structures, with the central structure representing the uplifted floor of the original transient cavity. The amount of uplift determined from terrestrial data corresponds to a value of approximately one tenth of the final rim-crest diameter. Further observations at terrestrial complex craters indicate excavation is also limited to the central area and that the transient cavity diameter was about 50–65% of the diameter of the final crater. Radially beyond this, original near-surface units are preserved in the down-dropped annular floor. The rim area is a series of fault terraces, progressively stepping down to the floor (Fig. 3).

Although models for the formation of complex craters are less constrained than those of simple craters, there is a general consensus that, in their initial stages, complex craters were not unlike simple craters. At complex craters, however, the downward displacements in the transient cavity floor observed in simple craters are not locked in and the cavity floor rebounds upward (Fig. 12). As the maximum depth of the transient cavity is reached before the cavity's maximum diameter, it is likely that this rebound and reversal of the flow-field in the center of a complex crater occurs while the diameter of the transient cavity is still growing by excavation (Fig. 12). With the upward movement of material in the transient cavity's floor, the entire rim area of the transient cavity collapses downward and inward (Fig. 12), greatly enlarging the crater's diameter compared to that of the transient cavity. There have been a number of reconstructions of large lunar craters, in which the terraces are restored to their original, pre-impact positions, resulting in estimated transient cavity diameters of about 60% of the final rim-crest diameter. It is clear that

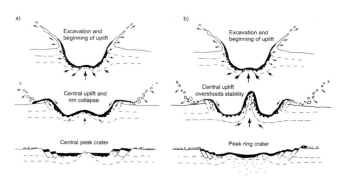

FIGURE 12   Schematic illustration of the formation of complex crater forms: (a) central peak crater (Figs. 3 and 4) and (b) peak ring basin (Fig. 5). Initial excavation and displacement by the cratering flow-field are similar to that of a simple crater (Fig. 11). The downward displacement of the target rocks is permanent, but not locked in, and the floor of the transient cavity is uplifted, even as the transient cavity diameter continues to grow in diameter. As the floor rises, the rim of the transient cavity collapses downward and inward to create a final rim that is a structural set of faulted terraces, considerably enlarging the final rim diameter. Excavation of target material is limited to the central area, and the extensive modification of the transient cavity leads to a final crater with a flat floor and topographically uplifted target material in the center. In the case of the peak ring basin (b), the uplifted material is in excess of what can be accommodated in a central peak and it collapses to form a peak ring. (After Melosh, 1989.)

uplift and collapse, during the modification stage at complex craters, is extremely rapid and that the target materials behave as if they were very weak. A number of mechanisms, including "thermal softening" and "acoustic fluidization," by which strong vibrations cause the rock debris to behave as a fluid, have been suggested as mechanisms to produce the required weakening of the target materials.

There is less of a consensus on the formation of rings within impact basins. The most popular hypothesis for central peak basins is that the rings represent uplifted material in excess of what can be accommodated in a central peak (Fig.12). This may explain the occurrence of both peaks and rings in central peak basins but offers little explanation for the absence of peaks and the occurrence of only rings in peak ring and multiring basins. A number of analogies have been drawn with the formation of "craters" in liquids and semiconsolidated materials such as muds, where the initial uplifted peak of material has no strength and collapses completely, sometimes oscillating up and down several times. At some time in the formation of ringed basins, however, the target rocks must regain their strength, so as to preserve the interior rings. An alternative explanation is that the uplift process proceeds, as in central peak craters, but the uplifted material in the very center is essentially fluid due to impact melting. In large impact events, the depth of impact melting may reach and even exceed the depth of the transient cavity floor. When the transient cavity is uplifted in such events, the central, melted part has no strength and, therefore, cannot form a positive topographic feature, such as a central peak. Only rings from the unmelted portion of the uplifted transient cavity floor can form some distance out from the center (Fig. 5).

2.2 Changes in the Target Rocks

The target rocks are initially highly compressed by the passage of the shock wave, transformed into high-density phases, and then rapidly decompressed by the rarefaction wave. As a result, they do not recover fully to their preshock state but are of slightly lower density, with the nature of their constituent minerals changed. The collective term for these shock-induced changes in minerals and rocks is **shock metamorphism**. Shock metamorphic effects are found naturally in many lunar samples and meteorites and at terrestrial impact craters. They have also been produced in nuclear explosions and in the laboratory, through shock-recovery experiments. No other geologic process is capable of producing the extremely high transient pressures and temperatures required for shock metamorphism, and it is diagnostic of impact.

Metamorphism of rocks normally occurs in planetary bodies as a consequence of thermal and tectonic events originating within the planet. The maximum pressures and temperatures recorded in surface rocks by such metamorphic events in planetary crusts are generally on the order of

1 GPa (10 kb) and 1000°C. During shock metamorphism, materials deform along their "Hugoniot curves," which describe the locus of pressure–volume states achieved by the material while under shock compression. Shock metamorphic effects do not appear until the material has exceeded its "**Hugoniot elastic limit** (HEL)," which is on the order of 5–10 GPa for most geologic materials. This is the pressure–volume point beyond which the shocked material no longer deforms elastically and permanent changes are recorded on recovery from shock compression.

The peak pressures generated on impact control the upper limit of shock metamorphism. These vary with the type of impacting body and target material but are principally a function of impact velocity, reaching into the hundreds to thousands of GPa. For example, the peak pressure generated when a stony asteroidal body impacts crystalline rock at 15 km s^{-1} is over 300 GPa, not much less than the pressure at the center of the Earth (~390 GPa). Shock metamorphism is also characterized by strain rates that are orders of magnitude higher than those produced by internal geologic

processes. For example, the duration of regional metamorphism associated with tectonism on Earth is generally considered to be in the millions of years. In contrast, the peak strains associated with the formation of a crater 20 km in diameter are attained in less than a second.

2.2.1 SOLID EFFECTS

At pressures below the HEL, minerals and rocks respond to shock with brittle deformation, which is manifested as fracturing, shattering, and brecciation. Such features are generally not readily distinguished from those produced by endogenic geologic processes, such as tectonism. There is, however, a unique, brittle, shock-metamorphic effect, which results in the development of unusual, striated, and horse-tailed conical fractures, known as shatter cones (Fig. 13). Shatter cones are best developed at relatively low shock pressures (5–10 GPa) and in fine-grained, structurally homogeneous rocks, such as carbonates, quartzites, and basalts.

(a)

(b)

(c)

(d)

FIGURE 13   Some shock metamorphic effects at terrestrial impact craters. (a) Shatter cones in basalt at the Slate Islands structure, Canada. (b) Photomicrograph of planar deformation features (e.g., in the left grain, thin parallel lines tending upwards to the right) in quartz from the Mistastin structure, Canada. Width of field of view is 0.5 mm, crossed polars. (c) Hand samples of target rocks from the Wanapitei structure, Canada, that are beginning to melt to form mixed mineral glasses and to vesiculate or froth. (d) Outcrop of coherent impact melt rock 80 m high, with columnar cooling joints, at the Mistastin structure, Canada.

Apart from shatter cones, all other diagnostic shock effects are microscopic in character. The most obvious are **planar deformation features** and **diaplectic glasses**. Planar deformation features are intensely deformed, are a few micrometers wide, and are arranged in parallel sets (Fig. 13). They are best known from the common silicate minerals, quartz and feldspar, for which shock-recovery experiments has calibrated the onset shock pressures for particular crystal orientations. They develop initially at ~10 GPa and continue to 20–30 GPa. The increasing effects of shock pressure are mirrored by changes in X-ray characteristics, indicative of the increasing breakdown of the internal crystal structure of individual minerals to smaller and smaller domains.

By shock pressures of ~30–40 GPa, quartz and feldspar are converted to diaplectic (from the Greek, "to strike") glass. These are solid-state glasses, with no evidence of flow, that exhibit the same outline as the original crystal. For this reason, they are sometimes referred to as thetamorphic (from the Greek, "same shape") glasses. The variety produced from plagioclase is known as maskelynite and was originally discovered in the Shergotty meteorite in 1872. The thermodynamics of shock processes are highly irreversible, so the pressure–volume work that is done during shock compression is not fully recovered upon decompression. This residual work is manifested as waste heat and, as a result, shock pressures of 40–50 GPa are sufficient to initiate melting in some minerals (Fig. 13). For example, feldspar grains show incipient melting and flow at shock pressures of ~45 GPa. Melting tends initially to be mineral specific, favoring mineral phases with the highest compressibilities and to be concentrated at grain boundaries, where pressures and temperatures are enhanced by reverberations of the shock wave. As a result, highly localized melts of mixed mineral compositions can arise. The effects of shock reverberations on melting are most obvious when comparing the pressures required to melt particulate materials, such as those that make up the lunar regolith [*see* THE MOON], and solid rock of similar composition. Shock recovery experiments indicate that intergranular melts can occur at pressures as low as 30 GPa in particulate basaltic material, compared to 45 GPa necessary to melt solid basalt.

Most minerals undergo transitions to dense, high-pressure phases during shock compression. Little is known, however, about the mineralogy of the high-pressure phases, as they generally revert to their low-pressure forms during decompression. Nevertheless, metastable high-pressure phases are sometime preserved, as either high-pressure **polymorphs** of preexisting low-pressure phases or high-pressure assemblages due to mineral breakdown. Some known high-pressure phases, such as diamond from carbon or stishovite from quartz (SiO_2), form during shock compression. Others, such as coesite (SiO_2), form by reversion of such minerals during pressure release. Several high-pressure phases that have been noted in shocked me-

teorites, however, are relatively rare at terrestrial craters. This may be due to post-shock thermal effects, which are sufficiently prolonged at a large impact crater to inhibit preservation of metastable phases.

2.2.2 MELTING

The waste heat trapped in shocked rocks is sufficient to result in whole-rock melting above shock pressures of ~60 GPa. Thus, relatively close to the impact point, a volume of the target rocks is melted and can even be vaporized (Figs. 11 and 12). Ultimately, these liquids cool to form impact melt rocks. These occur as glassy bodies in ejecta and breccias, as dikes in the crater floor, as pools and lenses within the breccia lenses of simple craters (Figs. 2 and 11), and as annular sheets surrounding the central structures and lining the floors of complex craters and basins (Figs. 4, 12, and 13). Some terrestrial impact melt rocks were initially misidentified as having a volcanic origin. In general, however, impact melt rocks are compositionally distinct from volcanic rocks. They have compositions determined by a mixture of the compositions of the target rocks, in contrast to volcanic rocks that have compositions determined by internal partial melting of more mafic and refractory progenitors within the planetary body's mantle or crust.

Impact melt rocks can also contain shocked and unshocked fragments of rocks and minerals. During the cratering event, as the melt is driven down into the expanding transient cavity (Figs. 11 and 12), it overtakes and incorporates less-shocked materials such as clasts, ranging in size from small grains to large blocks. Impact melt rocks that cool quickly generally contain large fractions of clasts, while those that cool more slowly show evidence of melting and resorption of the clastic debris, which is possible because impact melts are initially a superheated mixture of liquid melt and vapor. This is another characteristic that sets impact melt rocks apart from volcanic rocks, which are generally erupted at their melting temperature and no higher.

3. Impacts and Planetary Evolution

As the impact flux has varied through geologic time, so has the potential for impact to act as an evolutionary agent. The ancient highland crust of the Moon records almost the complete record of cratering since its formation. Crater counts combined with isotopic ages on returned lunar samples have established an estimate of the cratering rate on the Moon and its variation with time. Terrestrial data have been used to extend knowledge of the cratering rate, at least in the Earth–Moon system, to more recent geologic time. The lunar data are generally interpreted as indicating an exponential decrease in the rate until ~4.0 billion years (Ga) ago, a slower decline for an additional billion years, and a relatively constant rate, within a factor of two, since ~3.0 Ga ago. The actual rate before ~4.0 Ga ago is imprecisely known, as there is the question of whether the ancient lunar highlands reflect all of the craters that were produced (i.e., a production population) or only those that have not been obliterated by subsequent impacts (i.e., an equilibrium population). Thus, it is possible that the oldest lunar surfaces give only a minimum estimate of the ancient cratering rate. Similarly, there is some question as to whether the largest recorded events, represented by the major multiring basins on the Moon, occurred over the relatively short time period of 4.2–3.8 Ga ago (the "called lunar cataclysm") or were spread more evenly with time. [See THE MOON.]

3.1 Impact Origin of Earth's Moon

The impacts of the greatest magnitude dominate the cumulative effects of the much more abundant smaller impacts in terms of affecting planetary evolution. In the case of Earth, this would be the massive impact that likely produced the Moon. Earth is unique among the terrestrial planets in having a large satellite and the origin of the Moon has always presented a problem. The suggestion that the Moon formed from a massive impact with Earth was originally proposed some 30 years ago, but, with the development of complex numerical calculations and more efficient computers, it has been possible more recently to model such an event. Most models involve the oblique impact of a Mars-sized object with the proto-Earth, which produces an Earth-orbiting disk of impact-produced vapor, consisting mostly of mantle material from Earth and the impacting body. This disk, depleted in volatiles and enriched in refractory elements, would cool, condense, and accrete to form the Moon. [See THE MOON.] In the computer simulations, very little material from the iron core of the impacting body goes into the accretionary disk, accounting for the low iron and, ultimately, the small core of the Moon. In addition to the formation of the Moon, the effects of such a massive impact on the earliest Earth itself would have been extremely severe, leading to massive remelting of Earth and loss of any existing atmosphere.

3.2 Early Crustal Evolution

Following planetary formation, the subsequent high rate of bombardment by the remaining "tail" of accretionary debris is recorded on the Moon and the other terrestrial planets and the icy satellites of the outer solar system that have preserved some portion of their earliest crust. Due to the age of its early crust, the relatively large number of space missions, and the availability of samples, the Moon is the source of most interpretations of the effects of such an early, high flux. In the case of the Moon, a minimum of 6000 craters with diameters greater than 20 km are believed to have been formed during this early period. In addition, ~45 impacts produced basins, ranging in diameter from Bailly at 300 km, through the South Pole–Aitken Basin at 2600 km,

to the putative Procellarum Basin at 3500 km, the existence of which is still debated. The results of the *Apollo* missions demonstrate clearly the dominance of impact in the nature of the samples from the lunar highlands. Over 90% of the returned samples from the highlands are impact rock units, with 30–50% of the hand-sized samples being impact melt rocks. The dominance of impact as a process for change is also reflected in the age of the lunar highland samples. The bulk of the near-surface rocks, which are impact products, are in the range of 3.8–4.0 Ga old. Only a few pristine, igneous rocks from the early lunar crust, with ages >3.9 Ga, occur in the *Apollo* collection. Computer simulations indicate that the cumulative thickness of materials ejected from major craters in the lunar highlands is 2–10 km. Beneath this, the crust is believed to be brecciated and fractured by impacts to a depth of 20–25 km.

The large multiring basins define the major topographic features of the Moon. For example, the topography associated with the Orientale Basin (Fig. 6), the youngest multiring basin at ~3.8 Ga and, therefore, the basin with the least topographic relaxation, is over 8 km, somewhat less than Mt. Everest at ~9 km. The impact energies released in the formation of impact basins in the 1000 km size range are on the order of 10^{27}–10^{28} J, one to ten million times the present annual output of internal energy of Earth. The volume of crust melted in a basin-forming event of this size is on the order of a 1×10^6 km^3. Although the majority of crater ejecta is generally confined to within ~2.5 diameters of the source crater, this still represents essentially hemispheric redistribution of materials in the case of an Orientale-sized impact on the Moon.

Following formation, these impact basins localized subsequent endogenic geologic activity in the form of tectonism and volcanism. A consequence of such a large impact is the uplift of originally deep-seated isotherms and the subsequent tectonic evolution of the basin, and its immediate environs is then a function of the gradual loss of this thermal anomaly, which could take as long as a billion years to dissipate completely. Cooling leads to stresses, crustal fracturing, and basin subsidence. In addition to thermal subsidence, the basins may be loaded by later mare volcanism, leading to further subsidence and stress.

All the terrestrial planets experienced the formation of large impact basins early in their histories. Neither Earth nor Venus, however, retains any record of this massive bombardment, so the cumulative effect of such a bombardment on the Earth is unknown. Basin-sized impacts will have also affected any existing atmosphere, hydrosphere, and potential biosphere. For example, the impact on the early Earth of a body in the 500 km size range, similar to the present day asteroids Pallas and Vesta, would be sufficient to evaporate the world's present oceans, if only 25% of the impact energy were used in vaporizing the water. Such an event would have effectively sterilized the surface of Earth. The planet would have been enveloped by an atmosphere of hot rock and water vapor that would radiate heat downward onto the surface, with an effective temperature of a few thousand degrees. It would take thousands of years for the water-saturated atmosphere to rain out and reform the oceans. Models of impact's potential to frustrate early development of life on Earth indicate that life could have survived in a deep marine setting at 4.2–4.0 Ga, but smaller impacts would continue to make the surface inhospitable until ~4.0–3.8 Ga.

3.3 Biosphere Evolution

Evidence from the Earth–Moon system suggests that the cratering rate had essentially stabilized to something approaching a constant value by 3.0 Ga. Although major basin-forming impacts were no longer occurring, there were still occasional impacts resulting in craters in the size range of a few hundred kilometers. The terrestrial record contains remnants of the Sudbury, Canada, and Vredefort, South Africa, structures, which have estimated original crater diameters of ~250 km and ~300 km, respectively, and ages of ~2 Ga. Events of this size are unlikely to have caused significant long-term changes in the solid geosphere, but they likely affected the biosphere of Earth. In addition to these actual Precambrian impact craters, a number of anomalous spherule beds with ages ranging from ~2.0 to 3.5 Ga. have been discovered relatively recently in Australia and South Africa. Geochemical and physical evidence (shocked quartz) indicate an impact origin for some of these beds; at present, however, their source craters are unknown. If, as indicated, one of these spherule beds in Australia is temporally correlated to one in South Africa, its spatial extent would be in excess of 32,000 km².

At present, the only case of a direct physical and chemical link between a large impact event and changes in the biostratigraphic record is at the "Cretaceous–Tertiary boundary," which occurred ~65 million years (Ma) ago. The worldwide physical evidence for impact includes: shock-produced, microscopic planar deformation features in quartz and other minerals; the occurrence of stishovite (a high-pressure polymorph of quartz) and impact diamonds; high-temperature minerals believed to be vapor condensates; and various, generally altered, impact-melt spherules. The chemical evidence consists primarily of a geochemical anomaly, indicative of an admixture of meteoritic material. In undisturbed North American sections, which were laid down in swamps and pools on land, the boundary consists of two units: a lower one, linked to ballistic ejecta, and an upper one, linked to atmospheric dispersal in the impact fireball and subsequent fallout over a period of time. This fireball layer occurs worldwide, but the ejecta horizon is known only in North America.

The Cretaceous–Tertiary boundary marks a mass extinction in the biostratigraphic record of the Earth. Originally, it was thought that dust in the atmosphere from the impact led to global darkening, the cessation of photosynthesis, and cooling. Other potential killing mechanisms have

been suggested. Soot, for example, has also been identified in boundary deposits, and its origin has been ascribed to globally dispersed wildfires. Soot in the atmosphere may have enhanced or even overwhelmed the effects produced by global dust clouds. Recently, increasing emphasis has been placed on understanding the effects of vaporized and melted ejecta on the atmosphere. Models of the thermal radiation produced by the ballistic reentry of ejecta condensed from the vapor and melt plume of the impact indicate the occurrence of a thermal-radiation pulse on Earth's surface. The pattern of survival of land animals 65 Ma ago is in general agreement with the concept that this intense thermal pulse was the first global blow to the biosphere.

Although the record in the Cretaceous–Tertiary boundary deposits is consistent with the occurrence of a major impact, it is clear that many of the details of the potential killing mechanism(s) and the associated mass extinction are not fully known. The "killer crater" has been identified as the ~180 km diameter structure, known as Chicxulub, buried under ~1 km of sediments on the Yucatan peninsula, Mexico. Variations in the concentration and size of shocked quartz grains and the thickness of the boundary deposits, particularly the ejecta layer, point toward a source crater in Central America. Shocked minerals have been found in deposits both interior and exterior to the structure, as have impact melt rocks, with an isotopic age of 65 Ma.

Chicxulub may hold the clue to potential extinction mechanisms. The target rocks include beds of anhydrite ($CaSO_4$), and model calculations for the Chicxulub impact indicate that the SO_2 released would have sent anywhere between 30 billion and 300 billion tons of sulfuric acid into the atmosphere, depending on the exact impact conditions. Studies have shown that the lowering of temperatures following large volcanic eruptions is mainly due to sulfuric-acid aerosols. Models, using both the upper and lower estimates of the mass of sulfuric acid created by the Chicxulub impact, lead to a calculated drop in global temperature of several degrees Celsius. The sulfuric acid would eventually return to Earth as acid rain, which would cause the acidification of the upper ocean and potentially lead to marine extinctions. In addition, impact heating of nitrogen and oxygen in the atmosphere would produce NO_x gases that would affect the ozone layer and, thus, the amount of ultraviolet radiation reaching the Earth's surface. Like the sulfur-bearing aerosols, these gases would react with water in the atmosphere to form nitric acid, which would result in additional acid rains.

The frequency of Chicxulub-size events on Earth is on the order of one every ~100 Ma. Smaller, but still significant, impacts occur on shorter timescales and could affect the terrestrial climate and biosphere to varying degrees. Some model calculations suggest that dust injected into the atmosphere from the formation of impact craters as small as 20 km could produce global light reductions and temperature disruptions. Such impacts occur on Earth with a frequency of approximately two or three every million years but are not likely to have a serious affect upon the biosphere. The most fragile component of the present environment, however, is human civilization, which is highly dependent on an organized and technologically complex infrastructure for its survival. Though we seldom think of civilization in terms of millions of years, there is little doubt that if civilization lasts long enough, it could suffer severely or even be destroyed by an impact event.

Impacts can occur on historical timescales. For example, the Tunguska event in Russia in 1908 was due to the atmospheric explosion of a relatively small body at an altitude of ~10 km. The energy released, based on that required to produce the observed seismic disturbances, has been estimated as being equivalent to the explosion of ~10 megatons of TNT. Although the air blast resulted in the devastation of ~2000 km^2 of Siberian forest, there was no loss of human life. Events such as Tunguska occur on timescales of a thousand of years. Fortunately, 70% of the Earth's surface is ocean and most of the land surface is not densely populated.

4. Planetary Impactors

Apart from inferences from the compositions of asteroids, comets, and meteorites, the specific identification of actual impacting bodies is limited to occasional evidence from samples in or near craters on the Earth and Moon. For the majority of the ~170 impact craters so far identified on the Earth, however, the impactor types are either unknown or the identification is uncertain. The case for the Moon is no better. There are two methods used to determine projectile types: the physical identification of impactor fragments associated with a crater and identification of geochemical traces of an impactor component within impact melt rocks.

4.1 Physical Identification of Impactors

Although there is a widespread belief that the impactor is completely vaporized in large-scale impacts, this is not supported by numerical modeling. For example, at impact angles of ~45° or lower and velocities of 20 km s^{-1}, less than 50% of the impactor's mass vaporizes and the remaining fraction "survives" the impact, as melt or solid, and is deposited within or down range of the crater. Unfortunately, impactor fragments are rarely found associated with terrestrial impact craters. Any exposed remnants of the impactor are strongly affected by weathering processes and are normally destroyed after a few thousand years. As a result, virtually all impactor fragments have been found in the vicinity of very young terrestrial impact craters. Due to the size–frequency relation for impacts, these craters are also relatively small (<1.5 km) and were produced by iron meteorites, as this is the only type of small body that can survive atmospheric passage relatively intact and impact with enough remaining kinetic energy to create a crater.

Nevertheless, under conditions of rapid protection from weathering processes, it may be possible to find other types of impactor remnants associated with larger and older impact structures. This may be the case for a carbonaceous chondrite discovered at the Cretaceous–Tertiary boundary in a sedimentary core from the Pacific Ocean and inferred to be a small fragment of the impactor responsible for the Chicxulub structure. There are two other terrestrial cases where the physical presence of impactor-derived fragments has been inferred in larger impacts: East Clearwater, Canada ($D = 22$ km) and Morokweng, South Africa ($D = 70$ km). In both cases, however, the possible impactor materials have been reprocessed by their residence in impact melt rocks. The melt rocks at these craters have the highest known chemical admixture of impactor material of all terrestrial impact melt rocks (see later). Perhaps surprisingly, although there is no appreciable weathering on the Moon, few impactor fragments have been reported from the *Apollo* collection of lunar samples, although on the basis of geochemistry the lunar regolith is believed to contain a few percent of meteoritic material.

4.2 Chemical Identification of Impactor

The detection of a geochemical component of meteoritic material that has been mixed into impact melt rocks is the more common methodology for the identification of impactor type. Such a component has been detected at a number of terrestrial impact craters, and, in some cases, the impactor type has been identified with some degree of confidence (Table 2). The amount of impactor material in the melt rocks is typically <1%. Exceptions are at Morokweng and East Clearwater, where 7–10% impactor material occurs. The proportion of impactor component that can be incorporated to impact melts depends on the impact conditions, with the highest potential contributions occurring at low velocities and steep impact angles. The geochemical characterization of the incorporated impactor component can be achieved by examining Os isotopes, Cr isotopes, or elemental ratios, mainly the platinum group elements (PGEs), Ni, and Cr.

4.2.1 OS ISOTOPES

Due to the relative enrichment of Re over Os during the differentiation of the Earth's crust from the mantle and the radioactive decay of ^{187}Re to ^{187}Os, the $^{187}Os/^{188}Os$ ratios in terrestrial crustal rocks are higher than in both the Earth's mantle and most extraterrestrial materials. Thus, Os isotope ratios can be used to identify meteoritic components in terrestrial impact melt rock units. Impactor admixtures of less than 0.05% can be detected in the case of an impact into a continental crustal target. It is, however, sometimes not possible to determine whether the noncrustal component is from the Earths' mantle or an extraterrestrial

TABLE 2	Impactor Types at Impact Craters				
Name	**Location**	**Age (Ma)**	**D (km)**	**Impactor Type**	**Evidence**
Henbury	Australia	<0.005	0.16	Iron; type IIIA	M, S
Odessa	United States	<0.05	0.17	Iron; type IA	M
Boxhole	Australia	0.0300 ± 0.0005	0.17	Iron; type IIIA	M
Macha	Russia	<0.007	0.30	Iron	M, S
Aouelloul	Mauritania	3.1 ± 0.3	0.39	Iron	S, Os
Monturaqui	Chile	<1	0.46	Iron; type IA?	M, S
Wolfe Creek	Australia	<0.3	0.88	Ion; type IIIB	M, S
Barringer	United States	0.049 ± 0.003	1.19	Iron; type IA	M, S
New Quebec	Canada	1.4 ± 0.1	3.4	Ordinary chondrite; type L?	S
Brent	Canada	450 ± 30	3.8	Ordinary chondrite; type L or LL	S
Sääksjärvi	Finland	~560	6.0	Stony iron ?	S
Wanapitei	Canada	37.2 ± 1.2	7.5	Ordinary chondrite; type L	S
Bosumtwi	Ghana	1.03 ± 0.02	11	Noncarbonaceous chondrite	S, Os, Cr
Lappajärvi	Finland	77.3 ± 0.4	23	Noncarbonaceous chondrite	S, Cr
Rochechouart	France	214 ± 8	23	Stony iron	S, Cr,
Ries	Germany	15 ± 1	24	No contamination	S
Clearwater East	Canada	290 ± 20	26	Ordinary chondrite; type LL	S
Clearwater West	Canada	290 ± 20	36	No contamination	S
Saint Martin	Canada	220 ± 32	40	No contamination	S
Morokweng	South Africa	145.0 ± 0.8	70	Ordinary chondrite; type LL	M,S, Cr
Popigai	Russia	35 ± 5	100	Ordinary chondrite; type L	S, Cr
Manicouagan	Canada	214 ± 1	100	No contamination	S
Chicxulub§	Mexico	64.98 ± 0.05	170	Carbonaceous chondrite	M ,S, Os, Cr
Serenitatis Basin Spherule beds	Moon	3.9 Ga	740	Ordinary chondrite; type LL	S, Cr
Hamersley Basin	Australia	2.49 Ga	No crater	Enstatite chondrite, type EL ?	S, Cr
Baberton	South Africa	3.1—3.5 Ga	No crater	Carbonaceous chondrite	S, Cr

§ = enrichment in ejecta layer,
S = siderophile elements (PGE, Ni, Au); Cr = chromium isotopes; Os = Os isotopes; M = projectile fragment

source, and the method cannot be used to identify the type of impactor because the variation of the Os isotope ratios between known meteorite types is too small to act as a discriminator.

4.2.2 CR ISOTOPES

Chromium-isotope ratios of extraterrestrial materials differ from those of the Earth and the Moon. It is possible to distinguish between three groups of meteorites on the basis of Cr isotopes: (a) carbonaceous chondrites, (b) enstatite chondrites, and (c) all other types. The relatively high amounts of Cr in terrestrial and lunar rocks, however, restrict the use of this method. The characterization of the impactor type generally needs several percent of contamination, which is not common in terrestrial craters. One exception is measurements on Cretaceous–Tertiary boundary sediments from Stevens Klint, Denmark, and Caravaca, Spain, which have 5–10% extraterrestrial component. These data support the suggestion that the Chicx-

ulub impactor was a carbonaceous chondrite [*See* METEORITES.]

4.2.3 ELEMENTAL RATIOS

Parameters for impactor identification can be derived from ratios of highly siderophile elements (i.e., those associated with Fe), such as the PGEs, Ni, and Au, along with Cr, which is a lithophile element (i.e., associated with Si) element. Relative to most meteorites, these elements are depleted in terrestrial crustal rocks, except where there are concentrations of mafic and ultramafic rocks. It has been argued that the amount of target rock mafic to ultramafic components in the impact melt rocks must be estimated in order to obtain a precise impactor composition. The complete determination of this "indigenous correction" is difficult for most terrestrial craters and essentially impossible for lunar impact craters. It has been have demonstrated, however, that the "indigenous correction" is not required, provided the impactor elemental ratios are calculated by using a mixing

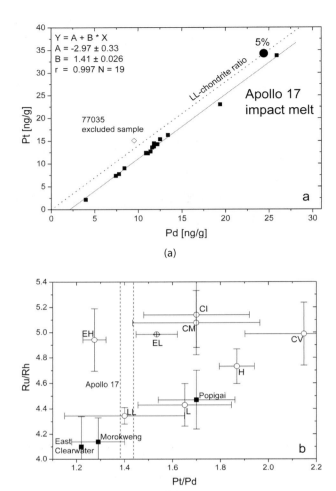

(a)

(b)

FIGURE 14 Identification of impactor type. (a) Pt–Pd ratios and determination of impactor type in lunar impact melt rock from *Apollo 17*. Shown for comparison is the slope of equivalent elemental ratios in LL–chondrite meteorites and where an admixture of 5% of LL–chondrite would plot (black dot). (b) Comparison of elemental ratios relative to few impact craters with different classes of chondrites. Error bars on data points are 1 sigma error bars. The *Apollo 17* impact melt rock appears as vertical dashed lines, as Ru/Rh data are not available.

line, where the elemental ratios of the impactor can be calculated directly from the slope of the mixing line. This is illustrated in Fig. 14 for the melt rocks at Popigai (Russia), Morokweng (South Africa), East Clearwater (Canada), and *Apollo 17* impact melts (Serenitatis) from the Moon. There is essentially no effect of the composition of the target rocks

on the slope of the mixing line, and the resulting projectile elemental ratios can be plotted together with the elemental ratios for the various classes of chondrites to provide a clear discrimination at the level of meteorite class (Fig. 14). It is, however, important to use elemental ratios that allow the best discrimination for a clear identification of the impactor type.

Within the various terrestrial impactor types identified to date, ordinary chondrites are by far the most common (Table 2). The reasons for the relative frequency of ordinary chondrite impactors for the Earth, and likely also the Moon, can be found in the Asteroid Belt. Ordinary chondrites are most likely related to S-class asteroids, which appear to be the most common asteroids in the main belt and among near-Earth asteroids (NEAs), although there is a possible observational bias due to their higher albedo compared to that of carbonaceous chondrites.

Although iron meteorites are responsible for all recent terrestrial craters smaller than 1.5 km in diameter, no unequivocal geochemical signature of iron impactors has yet been identified, at larger impact structures. Some terrestrial craters have no detectable extraterrestrial component in their impact rock units, and it has generally been assumed that the impactors were differentiated achondrites, which are relatively depleted in PGEs and Ni, and, thus, are very difficult to identify in terrestrial impact melt rocks. Differentiated asteroids are relatively rich in Cr and the use of Cr isotopes may be the only method to demonstrate that the impactor was an achondrite.

Although cometary impactors likely play a minor role (1–10% of the total population) in impacts in the Earth–Moon system, their identification is problematic. Their composition is essentially unknown with respect to their very small proportion of refractory elements, such as PGEs. [*See* Physics and Chemistry of Comets.]

Bibliography

French B. M. (1998). "Traces of Catastrophe: A Handbook of Shock-Metamorphic Effects in Terrestrial Meteorite Impact Structures," Lunar and Planetary Institute Contribution 954, Lunar and Planetary Institute, Houston.

Geological Society of America, Special Papers, 293 (1994), 339 (1999), 356 (2002), and 361 (2005).

Melosh, H. J. (1989). "Impact Cratering: A Geologic Process." Oxford Univ. Press, New York.

Spudis, P. D. (1993). "The Geology of Multi-ring Basins: The Moon and Other Planets." Cambridge Univ. Press, Cambridge, United Kingdom.

Planetary Volcanism

CHAPTER **44**

Lionel Wilson
Lancaster University
Lancaster, United Kingdom

V olcanism is one of the major processes whereby a planet transfers heat produced in its interior outward to the surface. Volcanic activity has been directly responsible for forming at least three quarters of the surface rocks of Earth and Venus, all of the surface materials of Jupiter's satellite Io, and extensive parts of the surfaces of Mars, Earth's Moon, and probably Mercury. Investigations of the styles of volcanic activity (e.g., explosive or effusive) on a planet's surface, when viewed in the light of environmental factors such as atmospheric pressure and acceleration due to gravity, provide clues to the composition of the erupted magma and hence, indirectly, to the chemical composition of the interior of the planet and its thermal state and history. Investigations of volcanic features on other planets have been an important spur to the development of an understanding of volcanic processes on Earth.

1. Summary of Planetary Volcanic Features

1.1 Earth

Only in the middle part of the 20th century did it become entirely clear that the ~70% of Earth's surface represented by the crust forming the floors of the oceans consists of geologically very young volcanic rocks. These erupted from long lines of volcanoes, generally located along ridges near the centers of ocean basins, within the last 300 Ma (million

years). Along with this realization came the development of the theory of plate tectonics, which explained the location and distribution of volcanoes over Earth's surface. Volcanoes erupting relatively metal-rich, silica- and volatile-poor magmas (called basalts) tend to concentrate along the midocean ridges, which mark the constructional margins of Earth's rigid crust plates. These magmas represent the products of the partial melting of the mantle at the tops of **convection** cells in which temperature variations cause the solid mantle to deform and flow on very long timescales. Magma compositions are very closely related to the bulk composition of the mantle, which makes up most of Earth's volume outside of the iron-dominated core. The volcanic edifices produced by ocean-floor volcanism consist mainly of relatively fluid (low-**viscosity**) lava flows with lengths from a few kilometers to a few tens of kilometers. Lava flows erupted along the midocean ridges simply add to the topography of the edges of the growing plates as they move slowly (~10 mm/year) away from the ridge crest. [*See* EARTH AS A PLANET: ATMOSPHERE AND OCEANS; EARTH AS A PLANET: SURFACE AND INTERIOR.]

Lavas erupted from vents located some distance away from the ridge crest build up roughly symmetrical edifices that generally have convex-upward shapes and are described, depending on their height-to-width ratio, as shields (having relatively shallow flank slopes) or domes (having relatively steeper flanks). Some of these vent systems are not related to the spreading ridges at all, but instead mark the

FIGURE 1 A Hawaiian-style lava fountain feeding a lava flow and building a cinder cone (Pu'u 'O'o on the flank of Kilauea volcano in Hawai'i). Steaming ground marks the location of the axis of the rift zone along which a dike propagated laterally to feed the vent. (Photograph by P. J. Mouginis-Mark.)

locations of "hot spots" in the underlying mantle, vigorously rising plumes of mantle material from which magmas migrate through the overlying plate. Because the plate moves over the hot spot, a chain of shield volcanoes can be built up in this way, marking the trace of the relative motion. The largest shield volcanoes on Earth form such a line of volcanoes, the Hawaiian Islands, and the two largest of these edifices, Mauna Loa and Mauna Kea, rise ~10 km above the ocean floor and have basal diameters of about 200 km.

Eruptive activity on shield volcanoes tends to be concentrated either at the summit or along linear or arcuate zones radiating away from the summit, called rift zones. The low viscosity of the basaltic magmas released in Hawaiian-style eruptions on these volcanoes (Fig. 1) allows the lava flows produced to travel relatively great distances (a few tens of kilometers), and is what gives shield volcanoes their characteristic wide, low profiles. It is very common for a long-lived reservoir of magma, a magma chamber, to exist at a depth of a few to several kilometers below the summit. This reservoir, which is roughly equant in shape and may be up to 1 to 3 km in diameter, intermittently feeds surface eruptions, either when magma ascends vertically from it in the volcano summit region or when magma flows laterally in a subsurface fracture called a dike, which most commonly follows an established rift zone, to erupt at some distance from the summit. In many cases, magma fails to reach the surface and instead freezes within the fracture it was following, thus forming an **intrusion**. The summit reservoir is fed, probably episodically, from partial melt zones in the mantle beneath. Rare but important events in which a large volume of magma leaves such a reservoir lead to the collapse of the rocks overlying it, and a characteristically steep-sided

crater called a caldera is formed, with a width similar to that of the underlying reservoir.

Volcanoes erupting silica- and volatile-rich magma (andesite or, less commonly, rhyolite) mark the destructive margins of plates, where the plates bend downward to be subducted into the interior and at least partly remelted. These volcanoes tend to form an arcuate pattern (called an island arc when the volcanoes rise from the sea floor), marking the trace on the surface of the zone where the melting is taking place, at depths on the order of 100–150 km. The andesitic magmas thus produced represent the products of the melting of a mixture of subducted ocean floor basalt, sedimentary material that had been washed onto the ocean floor from the continents (which are themselves an older, silica-rich product of the chemical differentiation of Earth), seawater trapped in the sediments, and the primary mantle materials into which the plates are subducted. Thus, andesites are much less representative of the current composition of the mantle. Andesite magmas are rich in volatiles (mainly water, carbon dioxide, and sulfur compounds), and their high silica contents give them high viscosities, making it hard for gas bubbles to escape. As a result, andesitic volcanoes often erupt explosively in Vulcanian-style eruptions, producing localized **pyroclastic** deposits with a range of grain sizes; alternatively, they produce relatively viscous lava flows that travel only short distances (a few kilometers) from the vent. The combination of short flows and localized ash deposits tends to produce steep-sided, roughly conical volcanic edifices.

When large bodies of very silica- and volatile-rich magma (rhyolite) accumulate—in subduction zones or, in some cases, where hot spots exist under continental areas, leading

FIGURE 2 The upper three layers of gray, dark, and bright material are air-fall pyroclastic deposits from the 1875 Plinian eruption of Askja volcano in Iceland. They clearly mantle earlier, dark, more nearly horizontal pyroclastic deposits. (Photograph by L. Wilson.)

to extensive melting of the continental crustal rocks—the potential exists for the occurrence of very large scale explosive eruptions in which finely fragmented magma is blasted at high speed from the vent to form a convecting eruption cloud, called a Plinian cloud, in the atmosphere. These clouds may reach heights up to 50 km, from which pyroclastic fragments fall to create a characteristic deposit spreading downwind from the vent area (Fig. 2). Under certain circumstances, the cloud cannot convect in a stable fashion and collapses to form a fountain-like structure over the vent, which feeds a series of pyroclastic flows—mixtures of incandescent pyroclastic fragments, volcanic gas, and entrained air—that can travel for at least tens of kilometers from the vent at speeds in excess of 100 m/s, eventually coming to rest to form a rock body called ignimbrite. These fall and flow deposits may be so widespread around the vent that no appreciable volcanic edifice is recognizable; however, there may be a caldera, or at least a depression, at the vent site due to the collapse of the surface rocks to replace the large volume of material erupted from depth.

It should be clear from the foregoing descriptions that the distribution of the various types of volcano and characteristic volcanic activity seen on Earth are intimately linked with the processes of plate tectonics. A major finding to emerge from the exploration of the solar system over the last 30 years is that this type of large-scale tectonism is currently confined to the Earth and may never have been active on any of the other bodies. Virtually all of the major volcanic features that we see elsewhere can be related

to the eruption of mantle melts similar to those associated with the midocean ridges and oceanic hot spots on Earth. However, differences between the physical environments (acceleration due to gravity, atmospheric conditions) of the other planets and Earth lead to significant differences in the details of the eruption processes and the deposits and volcanic edifices formed.

1.2 The Moon

During the 1970s, analyses of the samples collected from the Moon by the *Apollo* missions showed that there were two major rock types on the lunar surface. The relatively bright rocks forming the old, heavily cratered highlands of the Moon were recognized as being a primitive crust that formed about 4.5 Ga (billion years) ago by the accumulation of solid minerals at the cooling top of an at least 300 km thick melted layer referred to as a magma ocean. This early crust was extensively modified prior to about 3.9 Ga ago by the impacts of meteoroids and asteroids with a wide range of sizes to form impact craters and basins. Some of the larger craters and basins (the mare basins) were later flooded episodically by extensive lava flows, many more than 100 km long, to form the darker rocks visible on the lunar surface. [*See* THE MOON; PLANETARY IMPACTS.]

Radiometric dating of samples from lava flow units showed that these mare lavas were mostly erupted between 3 and 4 Ga ago, forming extensive, relatively flat deposits inside large basins. Individual flow units, or at least groups of flows, can commonly be distinguished using multispectral remote sensing imagery on the basis of their differing chemical compositions, which give them differing reflectivities in the visible and near-infrared parts of the spectrum. In composition, these lavas are basaltic, and their detailed mineralogy shows that they are the products of partial melting of the lunar mantle at depths between 150 and more than 400 km, the depth of origin increasing with time as the lunar interior cooled. Melting experiments on samples, supported by theoretical calculations based on their mineralogies, show that these lavas were extremely fluid (i.e., had very low viscosities, at least a factor of 3 to 10 less than those of typical basalts on Earth) when they were erupted. This allowed them to travel for great distances, often more than 100 km (Fig. 3) from their vents; it also meant that they had a tendency to flow back into, and cover up, their vents at the ends of the eruptions. Even so, it is clear from the flow directions that the vents were mainly near the edges of the interiors of the basins that the flows occupy. Many vents were probably associated with the arcuate rilles found in similar positions. These are curved grabens, trench-like depressions parallel to the edges of the basins formed as parts of the crust sink between pairs of parallel faults caused by tension. This tension, due to the weight of the lava ponded in the middle of the basin, makes it easier for cracks filled with magma to reach the surface in these places.

FIGURE 3   Lava flows in southwest Mare Imbrium on the Moon. The source vents are off the image to the lower left and the ~300 km long flows extend down a gentle slope toward the center of the mare basin beyond the upper right edge of the frame. (NASA *Apollo* photograph.)

A second class of lunar volcanic features associated with the edges of large basins is the sinuous rilles. These are meandering depressions, commonly hundreds of meters wide, tens of meters deep, and tens of kilometers long, which occur almost entirely within the mare basalts. Some are discontinuous, giving the impression of an underground tube that has been partly revealed by partial collapse of its roof, and these are almost certainly the equivalent of lava tube systems (lava flows whose top surface has completely solidified) on Earth. Other sinuous rilles are continuous open channels all along their length; these generally have origins in source depressions two or three times wider than the rille itself, and become narrower and shallower with increasing downslope distance from the source. At least some of these sinuous rilles appear to have been caused by long-duration lava flows that were very turbulent (i.e., the hot interior was being constantly mixed with the cooler top and bottom of the flow). As a result the flows were able to heat up the pre-existing surface until some of its minerals melted, allowing material to be carried away and an eroded channel to form.

In contrast to the lava flows and lava channels, two types of pyroclastic deposit are recognized on the Moon. There are numerous regions called dark mantles, often roughly circular and up to at least 200 km in diameter, where the fragmental lunar surface regolith is less reflective than usual, and spectroscopic evidence shows that it contains a component of small volcanic particles in addition to the locally derived rock fragments. The centers of these regions are commonly near the edges of mare basins, suggesting that the dark mantle deposits are produced by the same (or similar) source vents as the lava flows. Chemical analyses of the *Apollo* lava samples show that the Moon's mantle is totally devoid of common volatiles like water and carbon dioxide due to its hot origin [*see* **THE MOON**] and suggest that the main gas released from mare lava vents was carbon monoxide, produced in amounts up to a few hundred parts per million by weight as a result of a chemical reaction between free carbon and metal oxides, mainly iron oxide, in the magma as it neared the surface.

Several smaller, dark, fragmental deposits occur on the floor of the old, 90 km diameter impact crater Alphonsus. These patches, called dark haloes, extend for a few kilometers from the rims of subdued craters that are centered on, and elongated along, linear fault-bounded depressions (called linear rilles) on the crater floor. It is inferred that these are the sites of less energetic volcanic explosions.

Localized volcanic constructs such as shield volcanoes and domes are generally rare on the Moon, though more than 200 low, shieldlike features with diameters mainly in the range 3–10 km are found in the Marius region within Oceanus Procellarum, in northeast Mare Tranquillitatis, and in the region between the craters Kepler and Copernicus. Conspicuously absent are edifices with substantial summit calderas. This implies that large, shallow magma reservoirs are very rare, almost certainly a consequence of the difficulty with which very dense magmas rising from the mantle penetrate the low-density lunar crust. However, a few collapse pits with diameters up to 3 km do occur, located near the tops of domes or aligned along linear rilles.

1.3 Mars

About 60% of the surface of Mars consists of an ancient crust containing impact craters and basins. Spectroscopic evidence from orbiting spacecraft suggests that it is composed mainly of volcanic rocks. The other 40% of the planet consists of relatively young, flat, lower lying, plains-forming units that are a mixture of wind-blown sediments, lava flows, and rock debris washed into the lowlands by episodes of water release from beneath the surface. Combining orbital observations with analyses made by the five probes that have so far landed successfully on the surface

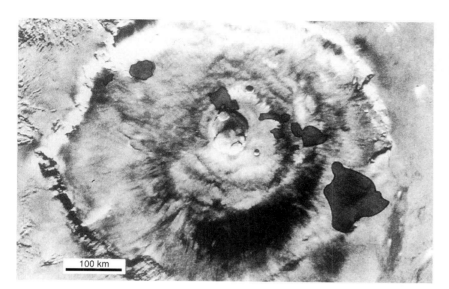

FIGURE 4   The Olympus Mons shield volcano on Mars with the Hawaiian Islands superimposed for scale. (NASA image with overlay by P. J. Mouginis-Mark. Reproduced by permission of the Lunar and Planetary Institute.)

suggests that most of the magmas erupted on Mars are basalts or basaltic andesites. [*See* MARS: SURFACE AND INTERIOR; MARS: LANDING SITE GEOLOGY, MINERALOGY, AND GEOCHEMISTRY.]

The most obvious volcanic features on Mars are four extremely large (~600 km diameter, heights up to >20 km) shield volcanoes (Olympus Mons, Ascraeus Mons, Pavonis Mons, and Arsia Mons) with the same general morphology as basaltic shield volcanoes found on Earth (Fig. 4). There are also about 20 smaller shields on Mars in various stages of preservation. Counts of small impact craters seen in high-resolution (~10 m/pixel) spacecraft images show that the ages of the lava flow units on the volcanoes range from more than 3 Ga to less than ~50 Ma. Complex systems of nested and intersecting calderas are found on the larger shields, implying protracted evolution of the internal plumbing of each volcano, typified by cycles of activity in which a volcano is sporadically active for ~1 Ma and then dormant for ~100 Ma. Individual caldera depressions are up to at least 30 km in diameter, much larger in absolute size than any found on Earth, and imply the presence of very large shallow magma reservoirs during the active parts of the volcanic cycles. The large size of these reservoirs, like that of the volcanoes themselves, is partly a consequence of the low acceleration due to gravity on Mars and partly due to the absence of plate tectonics, which means that a mantle hot spot builds a single large volcano, rather than a chain of small volcanoes as on Earth. The availability of large volumes of melt in the mantle beneath some of the largest shield volcanoes has led to the production of giant swarms of dikes, propagating radially away from the volcanic centers for more than 2000 km in some cases.

Most shields appear to have flanks dominated by lava flows, many more than 100 km long. The flanks of Elysium Mons contain some sinuous channels like the sinuous rilles on the Moon that we think are caused by hot, turbulent, high-speed lavas melting the ground over which they flow. Some of the older and more eroded edifices, like Tyrrhena Patera and Hadriaca Patera, appear to contain high proportions of relatively weak, presumably pyroclastic, rocks. There is a hint, from the relative ages of the volcanoes and the stratigraphic positions of the mechanically weaker layers within them, that pyroclastic eruptions were commoner in the early part of Mars' history. More contentious is the suggestion that some of the plains-forming units, generally interpreted as weathered lava flows, in fact consist of pyroclastic fall or flow deposits.

1.4 Venus

Because of its dense, optically opaque atmosphere, the only detailed synoptic imaging of the Venus surface comes from orbiting satellite-based radar systems. Despite the differences between optical and radar images (radar is sensitive to both the dielectric constant and the roughness of the surface on a scale similar to the radar wavelength), numerous kinds of volcanic features have been unambiguously detected on Venus. Large parts of the planet are covered with plains-forming units consisting of lava flows, having well-defined lobate edges and showing the clear control of topography on their direction of movement (Fig. 5). The lengths (which can be up to several hundred kilometers) and thicknesses (generally significantly less than 30 m, since they are not resolvable in the radar altimetry data) of these flows suggest that they are basaltic in composition. This interpretation is supported by the (admittedly small) amounts of major-element chemical data obtained from six of the Soviet probes that soft-landed on the Venus surface. Some areas show concentrations of particularly long flows called fluctus (Latin for floods). Most of the lava plains, judging

FIGURE 5   A variety of radar-bright lava flows radiate from the summit area down the flanks of a shield volcano on Venus. (NASA *Magellan* image.)

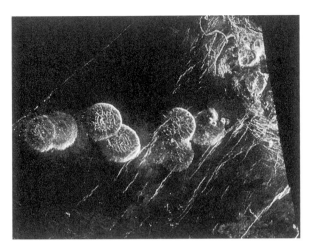

FIGURE 6   A cluster of ~25 km diameter "pancake" domes on Venus. These domes are evidence of the eruption of lava, which is more viscous than that forming the majority of flows on Venus. (NASA *Magellan* image.)

by the numbers of superimposed impact craters, were emplaced within the last ~700 Ma. [*See* VENUS: SURFACE AND INTERIOR].

Many areas within the plains and within other geological units contain groupings (dozens to hundreds) of small volcanic edifices, from less than one to several kilometers in diameter, with profiles that lead to their being classified as shields or domes. These groupings are called shield fields, and at least 500 have been identified. Some of the individual volcanoes have small summit depressions, apparently due to magma withdrawal and collapse, and others are seen to feed lava flows. Quite distinct from these presumably basaltic shields and domes is a class of larger, steep-sided domes (Fig. 6) with diameters of a few tens of kilometers and heights up to ~1 km. The surface morphologies of these domes suggest that most were emplaced in a single episode, and current theoretical modeling shows that their height-to-width ratio is similar to that expected for highly viscous silicic (perhaps rhyolitic) lavas on Earth.

Many much larger volcanic constructs occur on Venus. About 300 of these are classed as intermediate volcanoes and have a variety of morphologies, not all including extensive lava flows. A further 150, with diameters between 100 and about 600 km, are classed as large volcanoes. These are generally broad shield volcanoes that have extensive systems of lava flows and heights above the surrounding plains of up to about 3 km.

Summit calderas are quite common on the volcanoes, ranging in size from a few kilometers to a few tens of kilometers. There are two particularly large volcano-related depressions, called Sacajawea and Colette, located on the upland plateau Lakshmi Planum. With diameters on the order of 200 km and depths of ~2 km, these features appear to represent the downward sagging of the crust over some unusually deep-seated site of magma withdrawal.

Finally, there are a series of large, roughly circular features on Venus, which, though intimately linked with the large-scale tectonic **stresses** acting on the crust (they range from a few hundred to a few thousand kilometers in diameter), also have very strong volcanic associations. These are the coronae, novae, and arachnoids. Though defined in terms of the morphology of circumferential, moatlike depressions, radial fracture systems, and so on, these features commonly contain small volcanic edifices (fields of small shields or domes), small calderas, or lava flows, the latter often apparently fed from elongate vents coincident with the distal parts of radial fractures. In such cases, it seems extremely likely that the main feature is underlain by some kind of magma reservoir that feeds the more distant eruption sites via lateral dike systems.

1.5 Mercury

Much of the surface of Mercury is a heavily cratered ancient terrain like that of the Moon. There are some relatively flat plains-forming units dispersed among the craters, and it is tempting to speculate that these contain lava flows. Half of the surface of the planet was imaged by the flyby probe *Mariner 10*, but the resolution of the images is too poor to allow the lobate edges of any flow units to be identified unambiguously. Earth-based spectroscopic measurements suggest that many of the surface rocks are similar to basalts in composition. In places, patches of materials with these kinds of compositions have shapes consistent with explosive volcanic processes like those that we know occurred early in

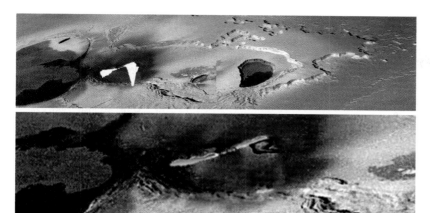

FIGURE 7   The upper part of the figure shows the chain of calderas called Tvashtar Catena on Io, showing a fissure eruption in progress. The high temperature of the lava overloaded the spacecraft imaging system causing "bleeding" of data values down vertical lines of the image. Using later images, the appearance of the eruption as it would have been seen by human eyes was reconstructed as shown in the lower part of the figure. (NASA *Galileo* image.)

the history of the Moon, but this does not in itself guarantee that these materials on Mercury were emplaced volcanically after the era of early intense bombardment that created the craters. [*See* MERCURY.]

1.6 Io

The bulk density of Io is about the same as that of Earth's Moon, suggesting that it has a silicate composition, similar to that of the inner, Earth-like planets. Io and the Moon also have similar sizes and masses, and it might therefore be expected by analogy with the Moon's thermal history that any volcanic activity on Io would have been confined to the first one or two billion years of its life. However, as the innermost satellite of the gas-giant Jupiter, Io is subjected to strong tidal forces. An orbital period resonance driven by the mutual gravitational interactions of Io, Europa, and Ganymede causes the orbit of Io to be slightly elliptical. This, coupled with the fact that it rotates synchronously (i.e., the orbital period is the same as that of the axial rotation), means that the interior of Io is subjected to a periodic tidal flexing. The inelastic part of this deformation generates heat in the interior on a scale that far outweighs any remaining heat source due to the decay of naturally radioactive elements. As a result, Io is currently the most volcanically active body in the solar system. At any one time, there are likely to be up to a dozen erupting vents. Roughly half of these produce lava flows, generally erupted from fissure vents (Fig. 7) associated with calderas located at the centers of very low shield-like features, and half produce umbrella-shaped eruption clouds into which gases and small pyroclasts are ejected at speeds of up to 1000 m/s to reach heights up to 300 km (Fig. 8). [*See* IO: THE VOLCANIC MOON.]

The main gases detected in the eruption clouds are sulfur and sulfur dioxide, and much of the surface is coated with highly colored deposits of sulfur and sulfur compounds that have been degassed from the interior over solar system

history and are now concentrated in the near-surface layers. However, it seems very likely, based on the fluid dynamic and thermodynamic analysis of the eruption clouds, that the underlying cause of the activity is the ascent of very hot basic magmas from the interior of Io. Temperatures up to ~1700–1900 K were initially derived from *Galileo* spacecraft data, suggesting that the magmas might be ultra-basic, similar to the komatiites that erupted on Earth earlier in its history. However, recent reappraisals of the early analyses suggest somewhat lower temperatures, and models of magma ascent on Io show that basalts, made unusually hot by friction effects as they rise through the crust, are more likely candidates. When these magmas, which may themselves have very low volatile contents, reach the surface in places with few volatile deposits, they produce lava flows. However, when they encounter copious deposits of sulfur compounds, they melt and then vaporize the deposits, providing the very high volatile contents needed to drive the violently explosive eruptions. Most of these volatiles condense as they expand

FIGURE 8   An explosive eruption plume on Io. The great height of the plume, more than 100 km, implies that magma is mixing with and evaporating volatile materials (sulfur or sulfur dioxide) on the surface as it erupts. (NASA *Voyager* image.)

and cool, and eventually fall back to the surface, providing the materials to drive future explosive eruptions.

1.7 The Icy Satellites

Many of the satellites of the gas-giant planets have bulk densities indicating that their interiors are mixtures of silicate rocks and the ices of the common volatiles (mainly water, with varying amounts of ammonia and methane). On some of these bodies (e.g., Jupiter's satellites Ganymede and Europa, Uranus' satellite Ariel, Neptune's satellite Triton, and Saturn's large satellite Titan), flowlike features that have many of the morphological attributes of very viscous lava flows are seen. However, there is no spectroscopic evidence for silicate magmas having been erupted onto the surfaces of these bodies, and the flowlike features have forced us to recognize that there is a more general definition of volcanism than that employed so far. [See PLANETARY SATELLITES.]

Volcanism is the generation of partial melts from the internal materials of a body and the transport out onto the surface of some fraction of those melts. In the ice-rich bodies, it is the generation of liquid water from solid ice that mimics the partial melting of rocks, and the ability of this water to erupt at the surface is influenced by the amounts of volatiles like ammonia and methane that it contains. Because the surface temperatures of most of these satellites are very much less than the freezing temperature of water, and because they do not have appreciable atmospheres (except Titan), the fate of any liquid water erupting at the surface is complex. Cooling will produce ice crystals at all boundaries of the flow, and these crystals, being less dense than liquid water, will rise toward the flow surface. Because of the negligible external pressure, evaporation (boiling) will take place within the upper few hundred millimeters of the flow. The vapor produced will freeze as it expands, to settle out as a frost or snow on the surrounding surface. The boiling process extracts heat from the liquid and adds to the rate of ice crystal formation. If enough ice crystals collect at the surface of a flow, they will impede the boiling process, and if a stable ice raft several hundred millimeters thick forms, it will suppress further boiling. Thus, if it is thick enough, a liquid water flow may be able to travel a significant distance from its eruption site. It is even possible that solid ice may form flowlike features on a much longer timescale, in essentially the same way that glaciers are able to flow on Earth.

Thick, glacier-like flow features have been detected in flyby radar images of the surface of Titan taken by the *Cassini* spacecraft in orbit around Saturn. Although they probably consist mainly of water ice, the composition of the other volatile compounds that they may contain is still under debate. One candidate, present as an important addition to the mainly nitrogen atmosphere, is methane. Injection of methane into the atmosphere from cryovolcanic eruptions and its subsequent condensation as "rain" is one possible explanation for the depressions looking strikingly like river valleys imaged on Titan's surface by the *Huygens* lander probe.

If liquid water produced below the surface of an icy satellite contains a large enough amount of volatiles like ammonia or methane, it will erupt explosively at high speed in what, near the vent, is the equivalent of a Plinian eruption. The expanding volatiles will cause the eruption cloud to spread sideways (like the umbrella-shaped plumes on Io) and disperse the water droplets, rapidly freezing to hailstones, over a wide area. If the eruption speed is high enough and the parent body small enough, some of the smaller hailstones may be ejected with escape velocity. Recent data from the *Cassini* spacecraft provide graphic evidence for this process occurring near the South Pole of Saturn's small satellite Enceladus. The orbit of Enceladus is very close to the brightest of Saturn's many rings, the E ring, which appears to be composed of particles of ice. It now seems clear that these are derived directly from Enceladus, having been ejected fast enough to escape from the satellite but not from Saturn itself. [See PLANETARY RINGS.]

1.8 The Differentiated Asteroids

The meteorites that fall to the Earth's surface are fragments ejected from the surfaces of asteroids during mutual collisions. Most of these meteorites are pieces of silicate rock and, even though many have rather simple chemical compositions consistent with their never having been strongly heated, it has long been realized that the mineralogy of some others can only be explained if they are either solidified samples of what was once magma or pieces of what was once a mantle that partially melted and then cooled again after melt was removed from it. Additionally, some meteorites are pieces of a nickel–iron–sulfur alloy that was once molten but subsequently cooled slowly. Taken together, these observations imply that some asteroids went through a process of extensive chemical differentiation by melting to form a crust, mantle, and core. The trace element composition of the meteorites from these differentiated asteroids shows that they were heated by the radioactive decay of a group of short-half-life isotopes that were present at the time the solar system formed, the most important of which was ^{26}Al, which has a half life of $\sim$0.75 Ma. Thus, all the heating, melting, and differentiation must have taken place within an interval of only a few million years. Yet during this brief period, quite small asteroids, only $\sim$100–500 km in diameter, were undergoing patterns where the mantle melts, the melt rises to the surface, and explosive and effusive eruptions occur. Such activity began on Earth, Mars, and Venus many tens of million years later.

Spectroscopic evidence very strongly suggests that the asteroid 4 Vesta is the parent body of one group of surface, crust, and mantle rocks, the Howardite–Eucrite–Diogenite group of meteorites. We have not yet identified any other parent asteroids with as much certainty, but we know from their composition that the Aubrites and the Ureilite meteorites are rocks from the mantles of two different asteroids that had violently explosive eruptions, which ejected what should have become their crustal rocks into space at escape velocity. And the Acapulcoites and Lodranites are rocks from the shallow crust or upper mantle of a body that produced rather small amounts of gas during melting in its mantle so that in these meteorites we see gas bubbles trapped in what was once magma traveling through fractures toward the surface. The importance of these meteorites is that they give us copious samples of the very deep interiors of their parent bodies as well as the surfaces; such samples will not be available for a very long time for Venus and Mars and are rare even for the Earth. [*See* METEORITES; MAIN-BELT ASTEROIDS.]

2. Classification of Eruptive Processes

Volcanic eruption styles on Earth were traditionally classified partly in terms of the observed composition and dispersal of the eruption products. Over the last 20 years, it has been realized that they might be more systematically classified in terms of the physics of the processes involved. This has the advantage that a similar system can be adopted for all planetary bodies, automatically taking account of the ways in which local environmental factors (especially surface gravity and atmospheric pressure) lead to differences in the morphology of the deposits of the same process occurring on different planets.

Eruptive processes are classified as either explosive or effusive. An effusive eruption is one in which lava spreads steadily away from a vent to form one or more lava flows, whereas an explosive eruption is one in which the magma emerging through the vent is torn apart, as a result of the coalescence of expanding gas bubbles, into clots of liquid that are widely dispersed. The clots cool while in flight above the ground and may be partly or completely solid by the time they land to form a layer of pyroclasts. There is some ambiguity concerning this basic distinction between effusive and explosive activity because many lava flows form from the recoalescence, near the vent, of large clots of liquid that have been partly disrupted by gas expansion but that have not been thrown high enough or far enough to cool appreciably. Thus, some eruptions have both an explosive and an effusive component.

There is also ambiguity about the use of the word "explosive" in a volcanic context. Conventionally, an explosion involves the sudden release of a quantity of material that has been confined in some way at a high pressure. Most often the expansion of trapped gas drives the explosion process. In volcanology, the term "explosive" is used not only for this kind of abrupt release of pressurized material but also for any eruption in which magma is torn apart into pyroclasts

that are accelerated by gas expansion, even if the magma is being erupted in a steady stream over a long time period. Eruption styles falling into the first category include Strombolian, Vulcanian, and phreato-magmatic activity, whereas those falling into the second include Hawaiian and Plinian activity. All of these styles are discussed in detail later.

3. Effusive Eruptions and Lava Flows

Whatever the complications associated with prior gas loss, an effusive eruption is regarded as taking place after lava leaves the vicinity of a vent as a continuous flow. The morphology of a lava flow, both while it is moving and after it has come to rest as a solid rock body, is an important source of information about the rheology (the deformation properties) of the lava, which is determined largely by its chemical composition, and about the rate at which the lava is being delivered to the surface through the vent. Because lava flows basically similar to those seen on Earth are so well exposed on Mars, Venus, the Moon and Io, a great deal of effort has been made to understand lava emplacement mechanisms.

In general, lava contains some proportion of solid crystals of various minerals and also gas bubbles. Above a certain temperature called the liquidus temperature, all the crystals will have melted, and the lava will be completely liquid. Under these circumstances, lavas containing less than about 20% by volume of gas bubbles will have almost perfectly Newtonian rheologies, which means that the rate at which the lava deforms, the **strain** rate, is directly proportional to the stress applied to it under all conditions. This constant ratio of the stress to the strain rate is called the Newtonian viscosity of the lava. At temperatures below the liquidus but above the solidus (the temperature at which all the components of the lava are completely solid), the lava in general contains both gas bubbles and crystals and has a non-Newtonian rheology. The ratio of stress to strain rate is now a function of the stress, and is called the apparent viscosity. At high crystal or bubble contents, the lava may develop a nonzero strength, called the yield strength, which must be exceeded by the stress before any flowage of the lava can occur. The simplest kind of non-Newtonian rheology is that in which the increase in stress, after the yield strength is exceeded, is proportional to the increase in strain rate: The ratio of the two is then called the Bingham viscosity, and the lava is described as a Bingham plastic.

The earliest theoretical models of lava flows treated them as Newtonian fluids. Such a fluid released on an inclined plane will spread both downslope and sideways indefinitely (unless surface tension stops it, a negligible factor on the scale of lava flows). Some lavas are channeled by preexisting topography, and so it is understandable that they have not spread sideways. However, others clearly stop spreading sideways even when there are no topographic obstacles, and quickly establish a pattern in which lava moves downhill in a central channel between a pair of stationary banks called levées. Also, lavas do not flow downhill indefinitely after the magma supply from the vent ceases: They commonly stop moving quite soon afterward, often while the front of the flow is on ground with an appreciable slope and almost all the lava is still at least partly liquid. Also, liquid lava present in a channel at the end of an eruption does not drain completely out of the channel: A significant thickness of lava is left in the channel floor. These observations led to the suggestion that no lavas are Newtonian, and attempts were made to model flows as the simplest non-Newtonian fluids, Bingham plastics.

The basis of these models is the idea that the finite thickness of the levées or flow front can be used to determine the yield strength of the lava and that the flow speed in the central channel can be used to give its apparent, and hence Bingham, viscosity. Multiplying the central channel width by its depth and the mean lava flow speed gives the volume flux (the volume per second) being erupted from the vent. Laboratory experiments were used to develop these ideas, and they have been applied by numerous workers to field observation of moving flows on Earth and to images of ancient flows on other planets. For flows on Earth, it is possible to deduce all the parameters just listed; for ancient flow deposits, one can obtain the yield strength unambiguously, but only the product of the viscosity and volume flux can be determined.

There is a possible alternative way to estimate the volume flux if it can be assumed that the flow unit being examined has come to rest because of cooling. An empirical relationship has been established for cooling-limited flows on Earth between the effusion rate from the vent and the length of a flow unit, its thickness, and the width of its active channel. If a flow is treated as cooling-limited when in fact it was not (the alternative being that it was volume-limited, meaning that it came to rest because the magma supply from the vent ceased at the end of the eruption), the effusion rate will inevitably be an underestimate by an unknown amount. Cooling-limited flows can sometimes be recognized because they have breakouts from their sides where lava was forced to form a new flow unit when the original flow front came to rest.

Lava rheologies and effusion rates have been estimated in this way for lava flows on Mars, the Moon, and Venus. It should be born in mind, when assessing these estimates, that a major failing of simple models like the Bingham model is that they assign the same rheological properties to all the material in a flow, whereas it is very likely that lava that has resided in a stationary levée near the vent for a long period will have suffered vastly more cooling than the fresh lava emerging from the vent and will have very different properties. More elaborate models have been evolved since the earliest work, including some that apply to broadly spreading lava lobes that do not have a well-defined levée-channel

structure, but no model yet accounts for all the factors controlling lava flow emplacement. With this caution, the values found suggest that essentially all the lavas studied so far on the other planets have properties similar to those of basaltic to intermediate (andesitic) lavas on Earth. Many of these lavas have lengths up to several hundred kilometers, to be compared with basaltic flow lengths up to a few tens of kilometers on Earth in geologically recent times, and this implies that they were erupted at much higher volume fluxes than is now common on Earth. There is a possibility, however, that some of these flow lengths have been overestimated. If a flow comes to rest so that its surface cools, but the eruption that fed it continues and forms other flow units alongside it, a breakout may eventually occur at the front of the original flow. A new flow unit is fed through the interior of the old flow, and the cooled top of the old flow, which has now become a lava tube, acts as an excellent insulator. As a result, the breakout flow can form a new unit almost as long as the original flow, and a large, complex compound flow field may eventually form in this way. Unless spacecraft images of the area have sufficiently high resolution for the compound nature of the flows to be clear, the total length of the group of flows will be interpreted as the length of a single flow, and the effusion rate will be greatly overestimated.

There are, however, certain volcanic features on the Moon and Mars that may be more unambiguous indicators of high effusion rates: the sinuous rilles. The geometric properties of these meandering channels—widths and depths that decrease away from the source, lengths of tens to a few hundred kilometers—are consistent with the channels being the result of the eruption of a very fluid lava at a very high volume flux for a long time. The turbulent motion of the initial flow, meandering downhill away from the vent, led to efficient heating of the ground on which it flowed, and it can be shown theoretically that both mechanical and thermal erosion of the ground surface are expected to have occurred on a timescale from weeks to months. The flow, which may have been ~10 m deep and moving at ~10 m/s, slowly subsided into the much deeper channel that it was excavating. Beyond a certain distance, the lava would have cooled to the point where it could no longer erode the ground, and it would have continued as an ordinary surface lava flow. The volume eruption rates deduced from the longer sinuous rille channel lengths are very similar to those found for the longest conventional lava flow units; modeling studies show that the turbulence leading to efficient thermal erosion was probably encouraged by a combination of unusually steep slope and unusually low lava viscosity. A few sinuous channels associated with lava plains are visible on Venus, but the lengths of some of the Venus channels are several to ten times as great as those seen on the Moon and Mars. It is not yet clear if the thermal erosion process is capable of explaining these channels by the eruption of low-viscosity basalts, or whether some more exotic volcanic fluid (or some other process) must be assumed.

There are numerous uncertainties in using the foregoing relationships to estimate lava eruption conditions. Thus, there have been many studies of the way heat is transported out of lava flows, taking account of the porosity of the lava generated by gas bubbles, the effects of deep cracks extending inward from the lava surface, and the external environmental conditions—the ability of the planetary atmosphere to remove heat lost by the flow by conduction, convection, and radiation. However, none of these has yet dealt in sufficient detail with turbulent flows, or with the fact that cooling must make the rheological properties of a lava flow a function of distance inward from its outer surface, so that any bulk properties estimated in the ways described earlier can only be approximations to the detailed behavior of the interior of the lava flow. There is clearly some feedback between the way a flow advances and its internal pattern of shear stresses. For example, lava flows on Earth have two basic surface textures. Basaltic flows that have erupted at low effusion rates or while still hot near their vents have smooth, folded surfaces with a texture called pahoehoe (a Hawaiian word), the result of plastic stretching of the outer skin as the lava advances; at higher effusion rates, or at lower temperatures farther from the vent, the surface fractures in a more brittle fashion to produce a very rough texture called 'a'a. A similar but coarser, rough, blocky texture is seen on the surfaces of more andesitic flows. Because there is a possibility of relating effusion rate and composition to the surface roughness of a flow in this way, there is a growing interest in obtaining relatively high resolution radar images of planetary surfaces (and Earth's surface) in which, as in the *Magellan* images of Venus, the returned signal intensity is a function of the small-scale roughness.

4. Explosive Eruptions

4.1 Basic Considerations

Magmas ascending from the mantle on Earth commonly contain volatiles, mainly water and carbon dioxide together with sulfur compounds and halogens. All of these have solubilities in the melt that are both pressure- and temperature-dependent. The temperature of a melt does not change greatly if it ascends rapidly enough toward the surface, but the pressure to which it is subjected changes enormously. As a result, the magma generally becomes saturated in one or more of the volatile compounds before it reaches the surface. Only a small degree of supersaturation is needed before the magma begins to exsolve the appropriate volatile mixture into nucleating gas bubbles. As a magma ascends to shallower levels, existing bubbles grow by decompression, and new ones nucleate. It is found empirically that after the volume fraction of the magma occupied by the bubbles

exceeds some value in the range 65–80%, the foam-like fluid can no longer deform fast enough in response to the shear stresses applied to it and as a result disintegrates into a mixture of released gas and entrained clots and droplets that form the pyroclasts. The eruption is then, by definition, explosive. The pyroclasts have a range of sizes dictated by the viscosity of the magmatic liquid, in turn a function of its composition and temperature, and the rate at which the decompression is taking place, essentially proportional to the rise speed of the magma.

It is not a trivial matter for the volume fraction of gas in a magma to become large enough to cause disruption into pyroclasts. The lowest pressure to which a magma is ever exposed is the planetary surface atmospheric pressure. On Venus, this ranges from about 10 MPa in lowland plains to about 4 MPa at the tops of the highest volcanoes; on Earth, it is about 0.1 MPa at sea level (and 30% less on high volcanoes) but much higher, up to 60 MPa, on the deep ocean floor; on Mars it ranges from about 500 Pa at the mean planetary radius to about 50 Pa at the tops of the highest volcanoes; and it is essentially zero on the Moon and Io. If the magma volatile content is small enough, then even at atmospheric pressure no gas will be exsolved—or at least too little will be exsolved to cause magma fragmentation. Using the solubilities of common volatiles in magmas, calculations show that explosive eruptions can occur on Earth as long as the water content exceeds 0.07 weight percent in basalt. On Mars, the critical level is 0.01 weight percent. On Venus, however, a basalt would have to contain about 2 weight percent water before explosive activity could occur, even at highland sites; this is greater than is common in basalts on Earth and leads to the suggestion that explosive activity may never happen on Venus, at least at lowland sites, or may happen only when some process leads to the local concentration of volatiles within a magma. Examples of this are discussed later. Finally, the negligible atmospheric pressures on the Moon and Io mean that miniscule amounts of magmatic volatiles can in principle cause some kind of explosive activity there.

The preceding discussion assumes that released magmatic volatiles are the only source of explosive activity. However, many Vulcanian and all phreato-magmatic explosive eruptions involve interaction of erupting magma with solid or liquid volatiles already present at the surface (always water or ice on Earth and probably on Mars; mainly sulfur compounds on Io). The total weight fraction of gas in the eruption products in such cases will depend on the detailed nature of the interaction as well as the composition and inherent volatile content of the magma; this is a critical factor in understanding explosive activity on Io.

4.2 Strombolian Activity

Strombolian eruptions, named for the style of activity common on the Italian volcanic island Stromboli, are an excellent example of how the rise speed, gas content, and viscosity of a magma are critical in determining the style of explosive activity that occurs. While the magma as a whole is ascending through a fracture in the planetary crust, bubbles of exsolved gas are rising through the liquid at a finite speed determined by the liquid viscosity and the bubble sizes. If the magma rise speed is negligible, for example, when magma is trapped in a shallow reservoir or a shallow intrusion, and if its viscosity is low, as in the case of a basalt, there may be enough time for gas bubbles to rise completely through the magma and escape into overlying fractures that convey the gas to the surface, where it escapes or is added to the atmosphere if there is one. Subsequent eruption of the residual liquid will be essentially perfectly effusive. If a low-viscosity magma is rising to the surface at a slow enough speed, most of the gas will still escape as bubbles rise to the liquid surface and burst. Because relatively large bubbles (those that nucleated first and have

decompressed most) will rise faster through the liquid than very small bubbles, it is common in some magmas, especially basalts, for large bubbles to overtake and coalesce with small ones. The even larger bubbles produced in this way rise even faster and overtake additional smaller bubbles. In many cases, a runaway situation develops in which a single large bubble completely fills the diameter of the vent system apart from a thin film of magma lining the walls of the fracture. In extreme cases the bubble may have a much greater vertical extent than its width, in which case it is called a slug of gas. As this body of gas emerges at the surface of the slowly rising liquid magma column, it bursts, and a discrete layer of magma forming the upper "skin" of the bubble or slug disintegrates into clots and droplets up to tens of centimeters in size. These are blown outward by the expanding gas (Fig. 9). The pyroclasts produced accumulate around the vent to form a cinder cone that can be up to several tens of meters in size. The time interval between the

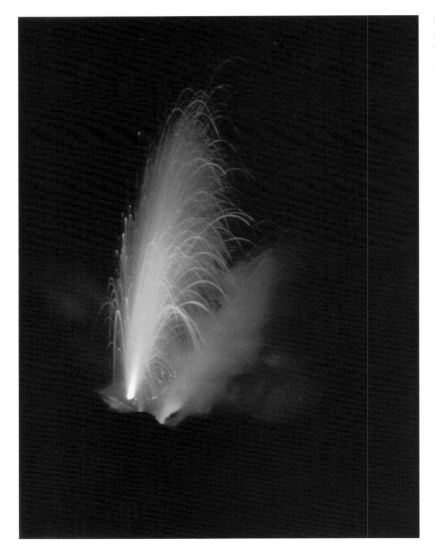

FIGURE 9   Jets of hot gas and entrained incandescent basaltic pyroclasts ejected from a transient Strombolian explosion on the volcano Stromboli in Italy. (Photograph by L. Wilson.)

emergence of successive bubbles or slugs from a vent may range from seconds to at least minutes, making this a distinctly intermittent type of explosive activity. If the largest rising gas bubble does not completely fill the vent, continuous overflow of a lava lake in the vent may take place to form one or more lava flows at the same time that intermittent explosive activity is occurring, resulting in a simultaneously effusive and explosive eruption.

A second method of producing gas slugs has been suggested for some Strombolian eruptions on Earth, in which gas bubbles form during convection in an otherwise stagnant body of magma beneath the surface and drift upward to accumulate into a layer of foam at the top of the magma body. When the vertical extent of the foam layer exceeds a critical value, it begins to collapse. Liquid magma drains from between the bubbles, and these coalesce into a large gas pocket that can now rise through any available fracture to the surface. The argument is that if a fracture was already present, the high effective viscosity of the foam would have inhibited its rise into the fracture, whereas the viscosity of the pure gas is low enough to allow this to occur. If a fracture was not already present, the changing stresses due to the foam collapse may be able to create one.

As long as any volatiles are exsolved from a low-viscosity magma rising sufficiently slowly to the surface, some kind of Strombolian explosive activity, however feeble, should occur at the vent on any planet, even at the high pressures on Venus or on Earth's ocean floors. Strombolian eruptions commonly involve excess pressures in the bursting bubbles of only a few tenths of a megapascal, so that the amount of gas expansion that drives the dispersal of pyroclasts is small. Pyroclast ranges in air on Earth can be several tens to at most a few hundred meters, and ranges would be much smaller in submarine Strombolian events on the ocean floor or on Venus because of the higher ambient pressure. Subaerial Strombolian eruptions on Mars would eject pyroclasts to distances about three times greater than on Earth because of the lower gravity; as a result, the deposits formed would have a tenfold lower relief than on Earth, and so far few examples have been unambiguously identified in spacecraft images.

4.3 Vulcanian Activity

At the other extreme of a slowly rising viscous magma, it is relatively difficult for gas bubbles to escape from the melt. Particularly if the magma stalls as a shallow intrusion, slow diffusion of gas through the liquid and rise of bubbles in the liquid concentrate gas in the upper part of the intrusion, and the gas pressure in this region rises. The pressure rise is greatly enhanced if any volatiles existing near the surface (groundwater on Earth; ground ice on Mars; sulfur or sulfur dioxide on Io) are evaporated. Eventually the rocks overlying the zone of high pressure break under the stress, and the rapid expansion of the trapped gas drives a sudden, discrete

FIGURE 10   A dense cloud of large and small pyroclasts and gas ejected to a height of a few hundred meters in a transient Vulcanian explosion by the volcano Ngauruhoe in New Zealand. (Image courtesy of the University of Colorado in Boulder, Colorado, and the National Oceanic and Atmospheric Administration, National Geophysical Data Center.)

explosion in which fragments of the overlying rock and of the disrupted magma are scattered around the explosion source: This is called Vulcanian activity (Fig. 10), named for the Italian volcanic island Vulcano. Again, as long as any volatiles are released from magma or are present in the near-surface layers of the planet, activity of this kind can occur. Several Vulcanian events on Earth involving fairly viscous magmas have been analyzed in enough detail to provide estimates of typical pressures and gas concentrations. Bombs approaching a meter in size ejected to ranges up to 5 km imply pressures as high as a few megapascals in regions that are tens of meters in size and that have gas mass fractions in the explosion products up to 10%.

On Mars, with the same initial conditions, the lower atmospheric pressure would cause much more gas expansion to accelerate the ejected fragments, and the lower atmospheric density would exert much less drag on them; also the lower gravity would allow them to travel farther for a given initial velocity. The result is that the largest clasts could travel up to 50 km. This means that the roughly circular deposit from a localized, point-source explosion would be spread over an area 100 times greater than on Earth, being on average 100 times thinner. Apart from the possibility that the pattern of small craters produced by the impact of the largest boulders on the surface might be recognized, such a deposit, with almost no vertical relief and having very little influence on the preexisting surface, would almost certainly go unnoticed in even the latest spacecraft images, and indeed no such features have yet been identified. However, if the explosion involves a larger, more complex, and especially elongate vent structure, there would not be such large differences. In the Elysium region of Mars, a large, water-carved channel, Hrad Vallis, has a complex elongate source depression that appears to have been excavated by a Vulcanian explosion when a dike injected a sill into the ice-rich permafrost of the cryosphere—the outer several kilometers of the crust, which is so cold that any H_2O must be present as ice. As heat from the sill magma melted the ice and boiled the resulting water in the cryosphere, violent expansion of the vapor forced intimate mixing of magma and lumps of cryosphere, encouraging ever more vapor production. Soon all the cryosphere above the sill was thrown out in what is called a fuel-coolant explosion (here the fuel is the magma and the coolant is the ice) to produce a deposit extending about 35 km on either side of the 150 km long depression. Residual heat from the magma melted the remaining ice in the shattered cryosphere rocks so that for a while, until it froze again, there was liquid water present to form a characteristic "muddy" appearance in the deposit (Fig. 11).

A Vulcanian explosion on Venus would also be very different from its equivalent on Earth. In this case, however, the high atmospheric pressure would tend to suppress gas expansion and lead to a low initial velocity for the ejecta, and the atmospheric drag would also be high. Pyroclasts that would have reached a range of 5 km on Earth would travel less than 200 m on Venus. On the one hand, this should concentrate the eruption products around the vent and make the deposit more obvious; however, the resolution of the best radar images from *Magellan* is only ∼75 m, and so such a deposit would represent only three or four adjacent pixels, which again would probably not be recognized.

On the Moon, a number of Vulcanian explosion products have been identified. The dark halo craters on the floor of Alphonsus have ejecta deposits with ranges up to 5 km. Since the Moon has a much lower atmospheric pressure than Mars (essentially zero), the preceding analysis suggests at first sight that lunar Vulcanian explosions should eject material to very great ranges. However, the Alphonsus event seems to have involved the intrusion of basaltic magma into the ∼10 m thick layer of fragmental material forming the regolith in this area, and the strength of the resulting mixture of partly welded regolith and chilled basalt was quite low. Thus, only a small amount of pressure buildup occurred before the retaining rock layer fractured. As a

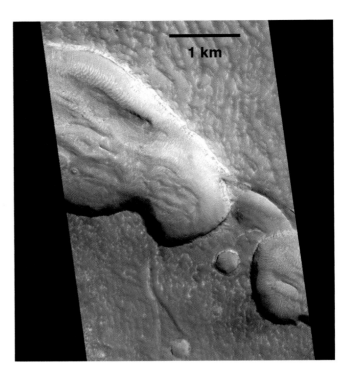

FIGURE 11 Part of the Hrad Vallis depression in the Elysium Planitia area of Mars. The depression is surrounded by a "muddy" deposit and is interpreted to have formed when a volcanic explosion excavated the depression and threw out a mixture of hot rocks and overlying cryosphere—cold rocks containing ice. (NASA *Mars Global Surveyor* image.)

result, the initial speeds of the ejected pyroclasts were low, and their ranges were unusually small.

4.4 Hawaiian Activity

In some cases, especially where low-viscosity basaltic magma travels laterally in dikes at shallow depth, enough gas bubble coalescence and bubble rise occurs for much of the gas to be lost into cracks in the rocks above the dike. Magma then emerges from the vent as a lava flow. However, when basaltic magmas rise mainly vertically at appreciable rates (more than about 1 m/s), some gas bubble coalescence occurs but little gas is lost, and the magma is released at the vent in a nearly continuously explosive manner. A lava fountain, more commonly called a fire fountain, forms over the vent, consisting of pyroclastic clots and droplets of liquid entrained in a magmatic gas stream that fluctuates in its upward velocity on a timescale of a few seconds. The largest clots of liquid, up to tens of centimeters in size, rise some way up the fountain and fall back around the vent to coalesce into a lava pond that overflows to feed a lava flow—the effusive part of the eruption—whereas smaller clasts travel to greater heights in the fountain. Some of the intermediate-sized pyroclasts cool as they fall from the outer parts of the fountain and collect around the lava pond in the vent to build up a roughly conical edifice called an ash

FIGURE 12   A Hawaii eruption from the Pu'u 'O'o vent in Hawaii showing a convecting cloud of gas and small particles in the atmosphere above the 300 m high lava fountain (commonly termed fire fountain) of coarser basaltic pyroclasts. (Photograph by P. J. Mouginis-Mark.)

cone, cinder cone, or scoria cone, the term used depending on the sizes of the pyroclasts involved, ash being smallest. Such pyroclastic cones are commonly asymmetric owing to the influence of the prevailing wind.

Atmospheric gases are entrained into the edge of the fire fountain and heated by contact with the hot pyroclasts and mixing with the hot magmatic gas. In this way, a convecting gas cloud is formed over the upper part of the fountain, and this gas entrains the smallest pyroclasts so that they take part fully in the convective motion. The whole cloud spreads downwind and cools, and eventually the pyroclasts are released again to form a layer on the ground, the smallest particles being deposited at the greatest distances from the vent. This whole process, involving formation of lava flows and pyroclastic deposits at the same time, is called Hawaiian eruptive activity (Fig. 12). This style of activity should certainly have occurred on Mars, but may be suppressed in basaltic magmas on Venus by the high atmospheric pressure, especially in lowland areas, unless, as noted earlier, magma volatile contents are several times higher than is common on Earth.

Figure 13 shows qualitatively how the combination of erupting mass flux and magma gas content in a Hawaiian eruption on Earth determines the nature and size of the possible products: a liquid lava pond at the vent that directly feeds lava flows; a pile of slightly cooled pyroclasts accumulating fast enough to weld together and form a "rootless" lava flow; a cone in which almost all of the pyroclasts are welded together; or a cone formed from pyroclasts that have had time to cool while in flight so that none, or only a few, weld on landing. Attempts have been made to quantify the results in Fig. 13 and extend them to other planetary environments. These results confirm that hot lava ponds around vents on Earth are expected to be no more than a

few tens of meters wide even at very high mass eruption rates. On the Moon, the greater gas expansion due to the lack of an atmosphere causes very thorough disruption of the magma (even at the low gas contents implied by analysis of the *Apollo* samples) and gives the released volcanic gas a high speed. This, together with the lower gravity, allows greater dispersal of pyroclasts of all sizes and provides an explanation of the 100–300 km wide dark mantle deposits

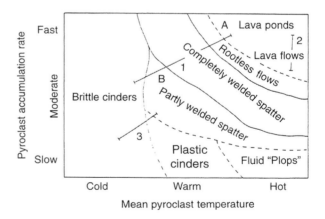

FIGURE 13   Schematic indication of the relative influences of the volatile content and the volume eruption rate of magma on the dispersal and thermal state of pyroclastic material produced in explosive eruptions. (Reprinted from Fig. 5 in the *Journal of Volcanology and Geothermal Research*, Vol. 37, J. W. Head and L. Wilson, Basaltic pyroclastic eruptions: Influence of gas-release patterns and volume fluxes on fountain structure, and the formation of cinder cones, spatter cones, rootless flows, lava ponds and lava flows, pp. 261–271, © 1989, with kind permission of Elsevier Science–NL, Sara Burgerhartstraat 25, 1025 KV Amsterdam, The Netherlands.)

as the products of extreme dispersal of the smallest, 30–100 micrometer-sized particles.

Nevertheless, it appears that hot lava ponds up to ~5 km in diameter could have formed around basaltic vents on the Moon if the eruption rates were high enough—as high as those postulated to explain the long lava flows and sinuous rilles. The motion of the lava in such ponds would have been thoroughly turbulent, thus encouraging thermal erosion of the base of the pond, and this presumably explains why the circular to oval depressions seen surrounding the sources of many sinuous rilles have just these sizes. Similar calculations for the Mars environment show that, as long as eruption rates are high enough, the atmospheric pressure and gravity are low enough on Mars to allow similar hot lava source ponds to have formed there, again in agreement with the observed sizes of depressions of this type that are seen.

Some noticeable differences occur when Hawaiian eruptions take place from very elongate fissure vents. Instead of a roughly circular pyroclastic cone containing a lava pond feeding one main lava flow, a pair of roughly parallel ridges forms, one on either side of the fissure. These are generally called spatter ramparts. Along the parts of the fissure where the eruption rate is highest, pyroclasts may coalesce as they land to form lava flows so that there are gaps in the ramparts from which the flows spread out. One striking example of this has been found so far on Mars (Fig. 14).

FIGURE 14   Mosaic of two images showing fissure vent near Jovis Tholus volcano on Mars. The vent has produced multiple lava flow lobes, probably of basaltic composition. The area shown is 24 km wide. (NASA *Mars Odyssey* image.)

4.5 Plinian Activity

In the case of a basaltic magma that is very rich in volatiles, or (much more commonly on Earth) in the case of a volatile-rich andesitic or rhyolitic magma, fragmentation in a steadily erupting magma is very efficient, and most of the pyroclasts formed are small enough to be thoroughly entrained by the gas stream. Furthermore, the speed of the mixture emerging from the vent, which is proportional to the square root of the amount of gas exsolved from the magma, will be much higher (perhaps up to 500 m/s) than in the case of a basaltic Hawaiian eruption (where speeds are commonly less than 100 m/s). The fire fountain in the vent now entrains so much atmospheric gas that it develops into a very strongly convecting eruption cloud in which the heat content of the pyroclasts is converted into the buoyancy of the entrained gas. The resulting cloud rises to a height that is proportional to the fourth root of the magma eruption rate (and hence the heat supply rate) and that may reach several tens of kilometers on Earth. Only the very coarsest pyroclasts fall out near the vent, and almost all of the erupted material is dispersed over a wide area from the higher parts of the eruption cloud (Fig. 15). This activity is termed Plinian, after Pliny's description of the A.D. 79 eruption of Vesuvius. Not all eruptions of this type produce stable convection clouds. If the vent is too wide or the eruption speed of the magma is too low, insufficient atmospheric gas may be entrained to provide the necessary buoyancy for convection, and a collapsed fountain forms over the vent, feeding large pyroclastic flows or smaller, more episodic pyroclastic surges.

Mars is the obvious place other than Earth to look for explosive eruption products: The low atmospheric pressure encourages explosive eruptions to occur, and the atmospheric density is high enough to allow convecting eruption clouds to form, at least up to ~20 km. However, we think that stable eruption clouds much higher than this cannot form on Mars because the atmosphere becomes too thin to provide the amount of entrained gas that is assumed in current theoretical models. In fact, only one potential fall deposit has yet been identified on Mars with any confidence. This is a region on the flank of the shield volcano Hecates Tholus, where, in contrast to the rest of the volcano, small impact craters appear to be hidden by a blanket of fine material in a region about 50 km wide and at least 70 km long. The sizes of the hidden craters suggest that the deposit is ~100 m thick, giving it a volume of ~65 km³; if we allow for the likely low bulk density of the deposit, this is equivalent to a dense rock volume of 23 km³. The volumes of the four summit depressions on Hecates Tholus range from ~10 to ~30 km³, suggesting that they may be calderas produced by collapse of the summit to compensate for the volume removed from a fairly shallow magma storage reservoir in each of a series of eruptions, the most recent of which produced the deposit described above.

FIGURE 15   The Plinian phase of the explosive eruption of Pinatubo volcano in 1991. A dense cloud of large and small pyroclasts and volcanic gases is ejected at high speed from the vent and entrains and heats the surrounding air. Convections then drives the resulting cloud to a height of tens of kilometers, where it drifts downwind, progressively releasing the entrained pyroclasts. (Photo credit: R. S. Culbreth, U. S. Air Force. Photo courtesy of the National Oceanic and Atmospheric Administration, National Geophysical Data Center.)

Although the high magma gas contents needed suggest that large-scale, steady (Plinian) explosive eruptions are rare on Venus, it is possible to calculate the heights to which their eruption clouds would rise. The high density and temperature of the atmosphere lead to rise heights about a factor of 2 lower than on Earth for the same eruption rate, and very large (at least a few tens of meters) clasts may be transported into near-vent deposits. At distances greater than a few kilometers from the vent, pyroclastic fall deposits will not be very different from those on Earth. A few examples of elongate markings on the Venus surface have been proposed as fall deposits, but no detailed analysis of them has yet been carried out.

The conditions that cause a steady explosive eruption to generate pyroclastic flows instead of feeding a stable, convecting eruption cloud are fairly well understood. If the eruption rate exceeds a critical value (which increases with increasing gas content of the mixture emerging through the vent and decreases with increasing vent diameter), stable convection is not possible whatever the nature of the atmosphere. Because pyroclastic flow formation is linked automatically to high eruption rate and, in general, to high eruption speed, which will encourage a great travel distance, it would not be surprising if large-scale pyroclastic flow deposits distributed radially around a vent were the products of high discharge rate eruptions of gas-rich magmas. Many of the flanking deposits of some martian volcanoes, especially Tyrrhena and Hadriaca Paterae, may have been produced in this way.

Theoretical work has shown that pyroclastic flows on Mars may be able to transport quite large blocks of rock (up to several meters in size, similar to those found on Earth) out of the vent and into nearby deposits. These pyroclast sizes are much greater than those expected in fall deposits on Mars, thus making it potentially possible to distinguish flow and fall deposits in future, high-resolution spacecraft images of martian vents. No equivalent work has yet been carried out for Venus, again mainly because of the expectation that voluminous explosive eruptions may be rare under the high atmospheric pressure conditions.

Short-lived or intermittent explosive eruptions (e.g., Vulcanian explosions, phreato-magmatic explosions, or events in which a gas-rich, high-viscosity lava flow or dome disintegrates into released gas and pyroclasts as a result of excessive gas pressure) can also produce small-scale pyroclastic flows. Because these are shorter lived and have characteristically different grain size distributions, they are called surges. The least well understood aspect of these phenomena is the way in which the magmatic material interacts with the atmosphere. As a result, it is currently almost impossible to predict in detail what the results of this kind of activity on Mars or Venus would look like. Such deposits, by the nature of the way they are generated, would not be very voluminous, however, and so would be spread very thinly, and might not be recognized if they were able to travel far from the vent.

4.6 Phreato–Magmatic Activity

Some types of eruption on Earth are controlled by the vigorous interaction of magma with surface or shallow subsurface water. If an intrusion into water-rich ground causes steam explosions, these are called phreatic events (from the Greek word for a well). If some magma also reaches the surface, the term used is phreato-magmatic, as distinct from normal, purely magmatic eruptions. When the equivalents of Strombolian or Hawaiian explosive events take place from eruption sites located in shallow water, they lead to much greater fragmentation of the magma than usual because of the stresses induced as pyroclasts are chilled by contact with the water. This activity is usually called Surtseyan, named after the eruption that formed the island of Surtsey off the south coast of Iceland. A much more vigorous and long-lived eruption under similar circumstances leads to a pyroclastic fall deposit similar to that of a Plinian event, but again involving greater fragmentation of magma: The result is called phreato-Plinian activity. Since the word "phreatic" does not specifically refer to water as the nonmagmatic volatile involved in these kinds of explosive eruption, it seems safe to apply these terms, as appropriate, to the various kinds of interactions between magma and liquid sulfur or sulfur dioxide forming the plumes currently seen on Io. These eruptions appear to involve about 30% by weight volatiles mixed with the magma; these proportions are close to the optimum for converting the heat of the magma to kinetic energy of the explosion products. Phreatic and phreato-magmatic eruptions should also have occurred on Mars in the distant past if, as many suspect, the atmospheric pressure was high enough to allow liquid water to exist on the surface.

4.7 Dispersal of Pyroclasts into a Vacuum

The conditions in the region above the vent in an explosive eruption on a planet with an appreciable atmosphere (e.g., Venus, Earth, or Mars) are very different from those when the atmospheric pressure is very small (much less than about 1 Pa), as on the Moon or Io. If the mass of atmospheric gas displaced from the region occupied by the eruption products after the magmatic gas has decompressed to the local pressure is much less than the mass of the magmatic gas, there is no possibility of a convecting eruption cloud forming in eruptions that would have been classed as Hawaiian or Plinian on Earth. In the region immediately above the vent, the gas expansion must be quite complex and will involve a series of shock waves. Relatively large pyroclasts will pass through these shocks with only minor deviations in their trajectories, but intermediate-sized particles may follow very complex paths, and few studies have yet been made of these conditions. The magmatic gas eventually expands radially into space, accelerating as it expands and reaching a limiting velocity that depends on its initial temperature. As the density of the gas decreases, its ability to exert a drag force on pyroclasts also decreases. On bodies the size of the Moon, even the smallest particles eventually decoupled from the gas and fell back to the planetary surface, though in gas-rich eruptions on asteroids they were commonly ejected into space.

These are the conditions that led to the formation of the dark mantle deposits on the Moon, with ultimate gas speeds on the order of 500 m/s, leading to ranges up to 150 km for small pyroclasts 30–100 micrometers in size. They are also the conditions that exist now in the eruption plumes on Io, though with an added complication. The driving volatiles in the Io plumes appear to be mainly sulfur and sulfur dioxide, evaporated from the solid or liquid state by intimate mixing with rising basaltic magma in what are effectively phreatomagmatic eruptions. The plume heights imply gas speeds just above the vent of ~1000 m/s, and these speeds are consistent with the plume materials being roughly equal mixtures of basaltic pyroclasts and evaporated surface volatiles. However, as the gas phase expands to very low pressures, both sulfur and sulfur dioxide will begin to condense again, forming small solid particles that rain back onto the surface along with the silicate particles to be potentially recycled again in future eruptions.

A final point concerns pyroclastic eruptions on the smallest atmosphereless bodies, the asteroids. Basaltic partial melts formed within these bodies were erupted at the surface at speeds that depended on the released volatile content. This is estimated to have been as much as 0.2–0.3 weight percent, leading to speeds up to 150 m/s. These speeds are greater than the escape velocities from asteroids with diameters less than about 200 km, and so instead of falling back to the surface, pyroclasts would have been expelled into space, eventually to spiral into the Sun. This process explains the otherwise puzzling fact that we have meteorites representing samples of the residual material left in the mantle of at least two asteroids after partial melting events, but have no meteorites from these asteroids with the expected partial melt composition.

5. Inferences about Planetary Interiors

The presence of the collapse depressions called calderas at or near the summits of many volcanoes on Earth, Mars, Venus, and Io suggests that it is common on all of these bodies for large volumes of magma to accumulate in reservoirs at relatively shallow depths. Theories of magma accumulation suggest that the magma in these reservoirs must have an internal pressure greater than the stress produced in the surrounding rocks by the weight of the overlying crust. This excess pressure may be due to the formation of bubbles by gas exsolution, or to the fact that heat loss from the magma to its cooler surroundings causes the growth of crystals that are less dense than the magmatic liquid and so occupy a larger volume. Most commonly, a pressure increase leads to fracturing of the wall of the reservoir and to the propagation of a magma-filled crack, called a dike, as an intrusion into the surrounding rocks. If the dike reaches the surface, an eruption occurs, and removal of magma from the reservoir allows the wall rocks to relax inward elastically as the pressure decreases. If magma does not reach the surface, the dike propagates underground until either the magma within it chills and comes to rest as its viscosity becomes extremely high, or the pressure within the reservoir falls to the point where there is no longer a great enough stress at the dike tip for rock fracturing to continue.

Under certain circumstances, an unusually large volume of magma may be removed from a shallow reservoir, reducing the internal pressure beyond the point where the reservoir walls behave elastically. Collapse of the overlying rocks may then occur to fill the potential void left by the magma, and a caldera (or, on a smaller scale, a pit crater) will form. The circumstances causing large-volume eruptions on Earth include the rapid eruption to the surface immediately above the reservoir of large volumes of low-density, gas-rich silicic (rhyolitic) magma, and the drainage of magma through extensive lateral dike systems extending along rift zones to distant flank eruption sites on basaltic volcanoes. This latter process appears to have been associated with caldera formation on Kilauea volcano in Hawaii, and it is tempting to speculate that the very large calderas on some of the martian basaltic shield volcanoes (especially Pavonis Mons and Arsia Mons) are directly associated with the large-volume eruptions seen on the distal parts of their rift zones. In contrast, we saw earlier that, at the martian volcano Hecates Tholus, a large explosive summit eruption is implicated in the formation of at least one of its calderas.

The size of a caldera must be related to the volume of the underlying magma reservoir, or more exactly to the volume of magma removed from it in the caldera-forming event. If the reservoir is shallow enough, the diameter of the caldera is probably similar to that of the reservoir. Diameters from 1 to 3 km are common on basaltic volcanoes on Earth and on Venus, with depths up to a few hundred meters implying magma volumes less than about 10 km^3. In contrast, caldera diameters up to at least 30 km occur on several volcanoes on Mars and, coupled with caldera depths up to 3 km, imply volumes ranging up to as much as $10,000 \text{ km}^3$. The stresses implied by the patterns of fractures on the floors and near the edges of some of these martian calderas suggest that the reservoirs beneath them are centered on depths on the order of 10–15 km, about three to four times greater than the known depths to the centers of shallow basaltic reservoirs on Earth. The simplest models of the internal structures of volcanoes suggest that, due to the progressive closing of gas cavities in rocks as the pressure increases, the density of the rocks forming a volcanic edifice should increase, at first quickly and then more slowly, with depth. Rising magma from deep partial melt zones may stall when its density is similar to that of the rocks around it so that it is neither positively nor negatively buoyant, and a reservoir may develop in this way. Because the pressure at a given depth inside a volcano is proportional to the acceleration due to gravity, and because martian gravity is about three times less than that on Earth or Venus, the finding that martian magma reservoirs are centered three to four times deeper than on

Earth is not surprising. However, these simple models do not address the reason for the martian calderas being much more than three times wider than any of those on Earth or many of those on Venus. On Io, we see some caldera-like structures, not necessarily associated with obvious volcanic edifices, that are even wider (but not deeper) than those on Mars, though we have too little information about the internal structure of Io's crust to interpret this observation unambiguously. Much is still not understood about the formation and stability of shallow magma bodies.

Evidence for significant shallow magma storage is conspicuously absent from the Moon. The large volumes observed for the great majority of eruptions in the later part of lunar volcanic history, and the high effusion rates inferred for them, imply that almost all of the eruptions took place directly from large bodies of magma stored at very great depth—at least at the base of the crust and possibly in partial melting zones in the lunar mantle. Not all the dikes propagating up from these depths will have reached the surface, however, and some shallow dike intrusions almost certainly exist. Recent work suggests that many of the linear rilles on the Moon represent the surface deformation resulting from the emplacement of such dikes, having thicknesses of at least 100 m, horizontal and vertical extents of ~100 km, and tops extending to within 1 or 2 km of the surface. Minor volcanic activity associated with some of these features would then be the result of gas loss and small-scale magma redistribution as the main body of the dike cooled.

The emplacement of very large dike systems extending most or all of the way from mantle magma source zones to the surface is not confined to the Moon. It has long been assumed that such structures must have existed to feed the high-volume basaltic lava flow sequences called flood basalts that occur on Earth every few tens of millions of years. These kinds of feature are probably closely related to the systems of giant dikes, tens to hundreds of meters wide and traceable laterally for many hundreds to more than 1000 km, that are found exposed in very ancient rocks on the Earth. The radial patterns of these ancient dike swarms suggest that they are associated with major areas of mantle upwelling and partial melting, with magma migrating vertically above the mantle plume to depths of a few tens of kilometers and then traveling laterally to form the longest dikes. Some of the radial surface fracture patterns associated with the novae and coronae on Venus are almost certainly similar features that have been formed more recently in that planet's geologic history, and on Mars the systems of linear **graben**, some of which show evidence of localized eruptive vents, extending radially from large shield volcanoes, also bear witness to the presence of long-lived mantle plumes generating giant dike swarms. It seems that there may be a great deal of similarity between the processes taking place in the mantles of all the Earth-like planets; it is the near-surface conditions, probably strongly influenced by the current presence of the oceans, that drive the plate tectonic processes distinguishing the Earth from its neighbors.

Bibliography

Cattermole, P. (1994). "Venus: The Geological Story." Johns Hopkins University Press, Baltimore.

Frankel, C. (1996). "Volcanoes of the Solar System." Cambridge Univ. Press, Cambridge, United Kingdom.

Lopes, R. M., and Gregg, T. K. P., eds. (2004). "Volcanic Worlds—Exploring the Solar System's Volcanoes." Springer-Verlag, Berlin, Heidelberg, New York.

Sgurdsson, H. (2000). "Encyclopaedia of Volcanoes." Academic Press, San Diego, California.

Astrobiology

Christopher P. McKay
and
Wanda L. Davis

NASA Ames Research Center
Moffett Field, California

CHAPTER **45**

1. Introduction

Life on Earth is widespread and appears to have been present on the planet since early in its history. Biochemically all life on Earth is similar and seems to share a common origin. Throughout geological history, life has significantly altered the environment of the Earth while at the same time adapting to this environment. It would not be possible to understand the Earth as a planet without the consideration of life. Thus life is a planetary phenomenon that is arguably the most interesting phenomenon observed on planetary surfaces.

Everything we know about life is based on the example of life on Earth. Generalization to other areas or alien forms of life must proceed with this caveat. Although we remain uncertain of the process or the time of its origin, the advent of life on Earth was established within one billion years after the formation of the planet. While life also requires energy and nutrients, liquid water is the single-most defining ecological requirement for life on Earth. Thus a liquid water environment is currently the best indicator of where to search for extraterrestrial life. We do not expect to discover liquid water environments on any of the recently discovered large extrasolar planets because they are too close to their stars. Looking out into the

Solar System, however, we see evidence for liquid water. Europa appears to have a liquid water ocean underneath a global ice surface—the evidence is indirect but persuasive. Enceladus has geysers erupting from its South Polar area presumably powered by subsurface liquid water. There are several lines of evidence that suggest that liquid water existed on Mars in the past. Direct images from orbiting spacecraft show fluvial features on the surface of Mars. Orbital infrared spectrometers have found local regions that show minerals formed in liquid water environments. The Mars Exploration Rovers also have found evidence for past aqueous activity at their landing sites on Mars. Our understanding of life, albeit limited to one example and one planet, would suggest that life is possible on other planets whenever conditions allow for environments like those on Earth—energy, nutrients, and most critically liquid water. This suggests the possibility of early microbial life on Mars and forms the basis for a search for Earth-like planets orbiting other stars. Studies of a second example of life—a second genesis—to which we can compare and contrast terrestrial biochemistry will be the beginning of a more general understanding of life as a process in the universe. This implies a search for not just fossils but a search for the biochemical remains of organisms, dead or alive.

2. What is Life?

Our understanding of life as a phenomenon is currently based only on our study of life on Earth. One of the profound results of biology is the realization that all life forms on Earth share a common physical and genetic makeup. The impression of vast diversity that we experience in nature is a result of manifold variations on a single fundamental biochemistry. The biochemistry of life is based on 20 amino acids and 5 nucleotide bases. Added to this are the few sugars, from which are made the **polysaccharides**, and the simple alcohols and fatty acids that are the building blocks of lipids. This simple collection of primordial biomolecules (Fig. 1) represents the set from which the rest of biochemistry derives.

Except for glycine, the amino acids in Figure 1 can have either left handed (L-) or right handed (D-) symmetry.

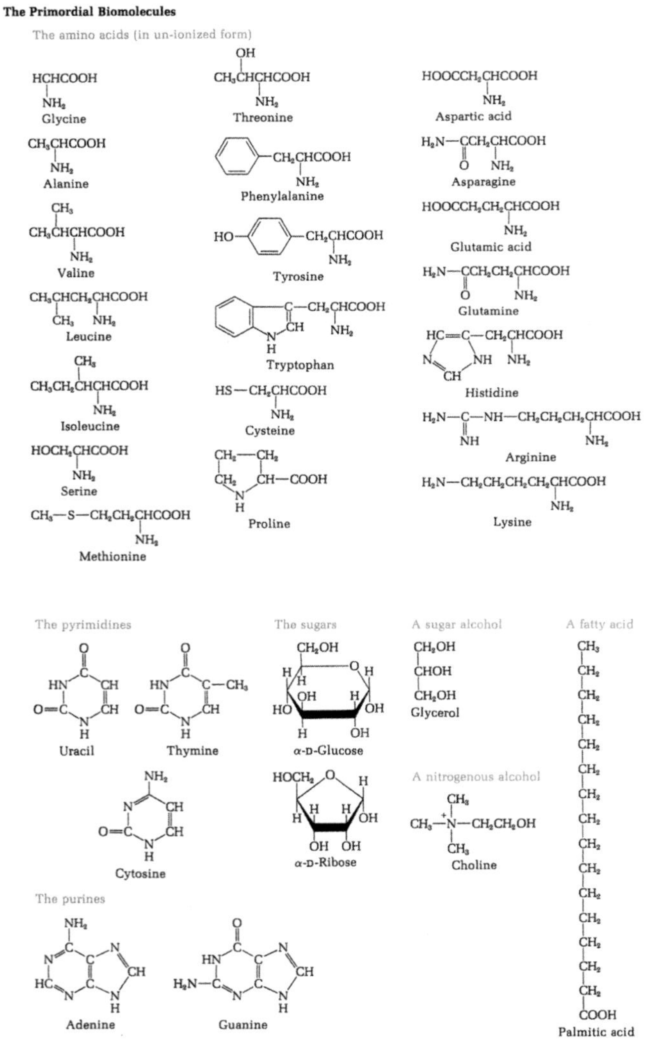

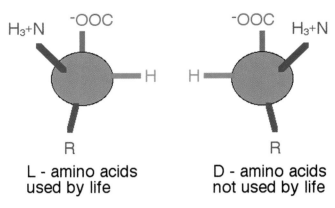

FIGURE 2   The L and D form of the amino acid alanine.

Figure 2 shows the two versions, known as enantiomers (from the Greek *enantios* meaning opposite), for alanine. Life uses only the L-enantiomer to make proteins although there are some bacteria that use certain D-forms in their cell walls, and many others have enzymes that can convert the D-form to the L-form. In addition, L-amino acids other than the 20 listed in Figure 1 are occasionally used in proteins and are sometimes used directly, for example as toxins by fungi and plants. We do not yet understand how and why life acquired a preference for the L-amino acids over the D-amino acids; this is one of the key observations that theories for the origins of life seek to explain.

The genetic material of life—DNA (deoxyribonucleic acid) and RNA (ribonucleic acid) —are both constructed from nucleotide bases that form the alphabet of life's genetic code. In DNA, these are adenine (A), thymine (T), cytosine (C), and guanine (G). In RNA, thymine is replaced by uracil (U). The nucleic acids each provide a four-letter alphabet in which the codes for the construction of proteins are based. This information recording system is found in all living systems.

The biochemical unity of life, in particular the genetic unity, strongly suggests that all living things on Earth descend from a common ancestor. This is the phylogenic unity of life as shown in Figure 3. These genetic trees are obtained by comparing the ribosomal RNA within each organism. Sections within the RNA are remarkably similar within all life forms. These conserved sections show only random point changes and not evolutionary trends. Thus the similarity between the genetic sequences of any two organisms is a measure of their evolutionary distance, or more precisely the time elapsed since they shared a common ancestor. When viewed in this way, life on Earth is divided into three main groups: the eucarya, the bacteria, and the archaea. The eucarya include the multicellular life forms encompassing all plants and animals. The bacteria are the familiar bacteria including intestinal bacteria, common

The Primordial Biomolecules

The amino acids (in un-ionized form)

HCHCOOH | NH₂ | Glycine
CH₃CHCOOH | NH₂ | Alanine
CH₃CHCHCOOH | CH₃ NH₂ | Valine
CH₃CHCH₂CHCOOH | CH₃ NH₂ | Leucine
CH₃CH₂CHCHCOOH | NH₂ | Isoleucine
HOCH₂CHCOOH | NH₂ | Serine
CH₃─S─CH₂CH₂CHCOOH | NH₂ | Methionine

CH₃CHCHCOOH | OH NH₂ | Threonine
─CH₂CHCOOH | NH₂ | Phenylalanine
HO─ ─CH₂CHCOOH | NH₂ | Tyrosine
─C─CH₂CHCOOH | CH NH₂ | Tryptophan
HS─CH₂CHCOOH | NH₂ | Cysteine
CH₂─CH₂ | CH₂ CH─COOH | N H | Proline

HOOCCH₂CHCOOH | NH₂ | Aspartic acid
H₂N─CCH₂CHCOOH | O NH₂ | Asparagine
HOOCCH₂CH₂CHCOOH | NH₂ | Glutamic acid
H₂N─CCH₂CH₂CHCOOH | O NH₂ | Glutamine
HC=C─CH₂CHCOOH | N NH NH₂ | CH | Histidine
H₂N─C─NH─CH₂CH₂CH₂CHCOOH | NH NH₂ | Arginine
H₂N─CH₂CH₂CH₂CH₂CHCOOH | NH₂ | Lysine

The pyrimidines

Uracil | Thymine | Cytosine

The purines

Adenine | Guanine

The sugars

α-D-Glucose | α-D-Ribose

A sugar alcohol

CH₂OH | CHOH | CH₂OH | Glycerol

A nitrogenous alcohol

CH₃ | CH₃─N⁺─CH₂CH₂OH | CH₃ | Choline

A fatty acid

CH₃ ... COOH | Palmitic acid

FIGURE 1   The basic molecules of life.

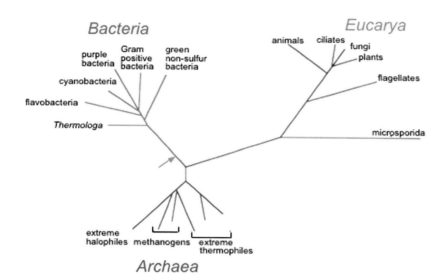

soil bacteria, and the pathogens. The archaea are a different class of microorganisms that are found in unusual and often harsh environments such as hypersaline ponds and H_2-rich anaerobic sediments. All methane-producing microbes are archaea. Archaea are also found in soils and grow on and in humans, producing methane in the gut. Archaea are not known to be human pathogens or to produce substances that are toxic to humans. Why some bacteria but no archaea are pathogenic is not yet understood.

2.1 The Ecology of Life: Liquid Water

In addition to describing the building blocks of life, it is instructive to consider what life does. In this regard it is possible to define a set of ecological or functional requirements for life. There are four fundamental requirements for life on Earth: energy, carbon, liquid water, and a few other elements. These are listed in Table 1 along with the occurrence of these environmental factors in the Solar System.

Energy is required for life from basic thermodynamic considerations. Typically on the Earth this energy is provided by sunlight, which is a thermodynamically efficient (low entropy) energy source. Some limited systems on Earth are capable of deriving their energy from chemical reactions (e.g., methanogenesis, $CO_2 + 4H_2 \rightarrow CH_4 + 2H_2O$) and do not depend on photosynthesis. On Earth these systems are confined to locations where the more typical photosynthetic organisms are not able to grow, and it is not known if an ecosystem that was planetary in scale or survived over billions of years could be based solely on chemical energy. There are no known organisms on Earth that make use of temperature gradients to derive energy. These organisms would be analogous to a Carnot heat engine. Table 2 lists some of the most important metabolic reactions by which living systems generate energy. This list includes autotrophs (which derive energy from nonbiological sources) as well as heterotrophs (which derive energy by the consumption of organic material, usually other life forms).

Elemental material is required for life, and on Earth carbon has the dominant role as the backbone molecule of biochemistry. Life almost certainly requires other elements as well. Life on Earth utilizes a vast array of the elements available on the surface. However, this does not prove that these elements are absolute requirements for life. Other than H_2O and C, the elements N, S, and P are probably the leading candidates for the status of required elements. Table 3 lists the distribution of elements in the cosmos and on the Earth and compares these with the common elements in life.

As indicated in Table 1, sunlight and the elements required for life are common in the Solar System. What appears to be the ecologically limiting factor for life in the Solar System is the stability of liquid water. Liquid water is a necessary requirement for life on Earth. Liquid water is key

TABLE 1	Ecological Requirements for Life
Requirement	**Occurrence in the Solar System**
Energy	Common
Predominately sunlight	Photosynthesis at 100 AU light levels *e.g.*,
Chemical energy	$H_2 + CO_2 \rightarrow CH_4 + H_2O$
Carbon	Common as CO_2 and CH_4
Liquid water	Rare, only on Earth for certain
N, P, S and other elements	Common

TABLE 2	Examples of Metabolic Pathways

Heterotrophy
    1. Fermentation $C_6H_{12}O_6 \rightarrow 2CO_2 + 2C_2H_5OH$
    2. Anaerobic Respiration $C_6H_{12}O_6 + 12NO_3^- \rightarrow 6CO_2 + 6H_2O + 12NO_2^-$
    3. Aerobic Respiration $C_6H_{12}O_6 + 6O_2 \rightarrow 6CO_2 + 6H_2O$
Photoautotrophy
    1. Anoxic photosynthesis $12CO_2 + 12H_2S + h\nu \rightarrow 2C_6H_{12}O_6 + 9S + 3SO_4$
    2. Oxygenic photosynthesis $6CO_2 + 6H_2O + h\nu \rightarrow C_6H_{12}O_6 + 3O_2$
Chemoautotrophy
Anaerobic
    1. Methanogens $CO_2 + 4H_2 \rightarrow CH_4 + 2H_2O$
                            $CO + 3H_2 \rightarrow CH_4 + H_2O$
                            $4CO + 2H_2O \rightarrow CH_4 + 3CO_2$
    2. Acetogens $2CO_2 + 4H_2 \rightarrow CH_3COOH + 2H_2O$
    3. Sulphate Reducers $H_2SO_4 + 4H_2 \rightarrow H_2S + 4H_2O$
    4. Sulfur Reducers $S + H_2 \rightarrow H_2S$
    5. Thionic Denitrifiers $H_2S + 2NO_3^- \rightarrow SO_4^{2-} + H_2O + N_2O$
                            $3S + 4NO_3^- + H_2 \rightarrow 3SO_4^{2-} + 2N_2 + 2H^+$
    6. Iron Reducers $2Fe^{3+} + H_2 \rightarrow 2Fe^{2+} + 2H^+$
Aerobic
    1. Sulfide Oxidizers $2H_2S + 3O_2 \rightarrow 2SO_4S + 2H_2O$
    2. Iron Oxidizers $4FeO + O_2 \rightarrow 2Fe_2O_3$

to biochemistry because it acts as the solvent in which biochemical reactions take place and, furthermore, it interacts with many biochemicals in ways that influence their properties. For example, water forms hydrogen bonds with some parts of a large molecule, the hydrophilic groups, and repels other parts, the hydrophobic groups, thereby forcing these molecules to curl up with their hydrophobic groups in the interior and the hydrophilic groups on the exterior in contact with the water. Certain organisms, notably lichen and some **algae**, are able to use water in the vapor phase if the relative humidity is high enough. Many organisms can continue to metabolize at temperatures well below the freezing point of pure water because their intracellular material contains salts and other solutes that lower the freezing point of the solution. No microorganism currently known is able to obtain water directly from ice. Many organisms, such as the snow algae *Chlamydomonas nivalis* thrive in liquid water associated with ice but in these circumstances the organisms are the beneficiaries of external processes that melt the ice. There is no known occurrence of an organism using metabolic methods to overcome the latent heat of fusion of ice thereby liquefying it.

Because liquid water is universally required for known life and because it appears to be rare in the Solar System, the search for life beyond the Earth begins first with the search for liquid water.

TABLE 3	Elemental Abundances by Mass

1	Cosmic		Earth's Crust		Humans		Bacteria	
1	H	70.7%	O	46.6%	O	64%	O	68%
2	He	27.4	Si	29.7	C	19	C	15
3	O	0.958	Al	8.13	H	9	H	10.2
4	C	0.304	Fe	5.00	N	5	N	4.2
5	Ne	0.174	Ca	3.63	C	1.5	P	0.83
6	Fe	0.126	Na	2.83	P	0.8	K	0.45
7	N	0.110	K	2.59	S	0.6	Na	0.40
8	Si	0.0706	Mg	2.09	K	0.3	S	0.30
9	Mg	0.0656	Ti	0.44	Na	0.15	Ca	0.25
10	S	0.0414	H	0.14	Cl	0.15	Cl	0.12

2.2 Generalized Theories for Life

There have been many attempts at a definition of life, and perhaps such a definition would aid in our investigation for life on other planets and help unravel the origins of life on Earth. However, it is probable that there will never be a simple definition of life and it may not be necessary in a search for life on other worlds. Despite the fundamental unity of biochemistry and the universality of the genetic code, no single definition has proven adequate in describing the single example of life on Earth. Many of the attributes that we would associate with life—for example, self-replication, self-ordering, response to environmental stimuli, can be found in nonliving systems—fire, crystals, and bimetallic thermostats, respectively. Furthermore, various and peculiar life forms such as viruses and giant cell-less slime molds defy even a biological definition of life in terms of the cell or the separation of internal and external environments. In attempting a resolution of this problem, the most useful definition of life is a system that develops Darwinian evolution: reproduction, mutation, and selection (Table 4). This is an answer to the question what does life do?

We are able to answer the questions, what does life need? and what does life do?, even if we do not have a closed form compact definition of life. Thus, the requirements for life listed in Table 1 and the functions of life listed in Table 4 are very general; it is probably unwise to apply more restrictive criteria. For example, for evolution to occur some sort of information storage mechanism is required. However, it is not certain that this information mechanism needs to be a DNA/RNA-based system or even that it be expressed in structures dedicated solely for replication. While on the present Earth, all life uses dedicated DNA and RNA systems for genetic coding, there is evidence that at one time genetic and structural coding were combined into one molecule, RNA. In this so-called RNA world there would have been no distinction between genotype (genetic) and phenotype (structural) molecular replicating systems—both of these processes would have been performed by an RNA-replicating molecule. In present biology, the phenotype is composed of proteins for the most part. This example illustrates the difficulty in determining which aspects of biochemistry are fundamental and which are the result of the peculiarities of life's history on Earth.

In basing our consideration of life on the distribution we observe here on Earth as a general phenomenon, we suffer simultaneously from the problem that there is only one kind of life on this planet while the variety of that life is too complex to allow for precise definitions or characterizations. We can neither extrapolate nor be specific in our theories for life.

Some scientists have suggested that living systems elsewhere in the universe may exhibit vast differences from terrestrial biology and have proposed a variety of alternative life forms. One postulated alternative life form is based on the substitution of ammonia for water. Certainly ammonia is an excellent solvent—in some respects better than water. The range of temperatures over which ammonia is liquid is prevalent in the universe (melting point: $-78°$, normal boiling point $-33°$, liquid at room temperature when mixed with water) and the elements that compose it are abundant in the cosmos. Other scientists have suggested the possibility that silicon may be used as a substitute for carbon in alien life forms. However, silicon does not form polymeric chains either as readily or as long as carbon does and its bonds with oxygen (SiO_2) are much stronger than carbon bonds (CO_2) rendering its oxide essentially inert.

Although speculations of alien life capable of using silicon in place of carbon or ammonia in place of water are intriguing, no specific experiments directed toward alternate biochemistries have been designed. Thus we have no strategies for where or how to search for such alternate life or its fossils. More significantly, these speculations have not contributed to our understanding of life. One can only conclude that our unique understanding of terrestrial life is based on Earth systems, and wide-ranging speculations regarding alternate chemistries are currently too limited to be fruitful. Perhaps some day we will develop general theories for life or, more likely, have many sources of life to compare thereby allowing for complete theories. Basing our theories on Earth-like life should be considered a necessary first approach and not a fundamental limitation.

3. The History of Life on Earth

Several sources of information about the origin of life on Earth include the physical record, the genetic record, the metabolic record, and laboratory simulations. The physical record includes the collection of sedimentary and fossil evidence of life. This record is augmented by theoretical models of the Earth and the Solar System, all of which provide clues to conditions billions of years ago when the origin of life is thought to have occurred. There is also the record stored in the genomes of living systems that comprise the collective gene pool of our planet. Genetic information tells us the path of evolution as shaped by environmental pressures, biological constraints, and random events that connect the earliest genomic organism, the **last universal common ancestor**, and the present tree of life (see Fig. 3). There is also the record of metabolic pathways in the biochemistry of organisms that have evolved in response to changes in the environment

TABLE 4	Properties of Life
Mutation	
Selection	
Reproduction	

while simultaneously causing changes to that environment. All of these records are palimpsests in that they have been overwritten—often repeatedly—over time. Laboratory simulations of prebiotic chemistry—the chemistry assumed present before life—can provide clues to the conditions and chemical solutions leading up to the origin of life. Experiments of DNA/RNA replication sequences can provide clues to the selection process that optimize mutations as well as provide a basic understanding of reproduction. Perhaps one day the process that initiates life will be studied in the laboratory or discovered on another planet.

The major events in the history of life are shown in Figure 4. As the Earth was forming about 4.5 Gyr ago, its surface would have been inhospitable to life. The gravitational energy released by the formation of the planet would have kept surface temperatures too high for liquid water to exist.

Eventually, as the heat flow subsided, rain would have fallen for the first time and life could be sustained in liquid water. However, it is possible that subsequent impacts could have been large enough to sterilize the Earth by melting, excavating, and vaporizing the planetary surface, removing all liquid water. Thus, life may have been frustrated in its early starts. Following a sufficiently large impact, the entire upper crust of the Earth would be ejected into outer space and any remnant left as a magma ocean. Barring these catastrophic events, however, sterilizing the Earth is a difficult task because it is not sufficient to merely heat the surface to high temperatures. At present, microorganisms survive at the bottom of the ocean and even kilometers below the surface of the planet. An Earth-sterilizing impact must not only completely evaporate the oceans but must then heat the surface and subsurface of the Earth such that

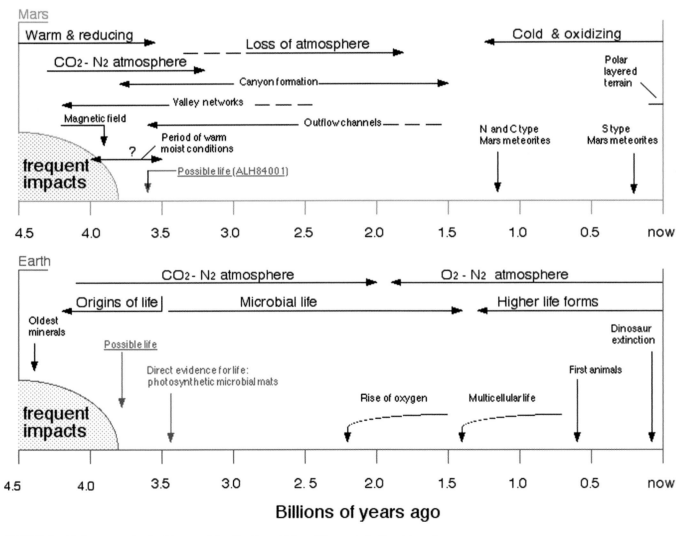

FIGURE 4   Major events in the history of the Earth and Mars. The period of moist surface conditions on Mars may have corresponded to the time during which life originated on Earth. The similarities between the two planets at this time raise the possibility of the origin of life on Mars.

the temperature does not fall anywhere below about 200°C, which is the temperature required for heat sterilization of dry, dormant organisms. This is a difficult requirement because the time it takes heat to diffuse down a given distance scales as the square of the distance. Thus, heat must be applied a million times longer to sterilize to a depth of 1 km compared to a depth of 1 m.

It is not known when the last life-threatening impact occurred on Earth. As shown schematically in Figure 4, the rate of impact, extrapolated from the record on the Moon, rises steeply before 3.8 Gyr ago. It is therefore likely that the Earth was not continuously suitable for life much before 3.8 Gyr ago. There is persuasive evidence, including microbial fossils and **stromatolites**, that microbial life was present on the Earth as early as 3.4 Gyr ago. Stromatolites are large features—often many meters in size—that can be formed by the lithification of laminated microbial mats, (Fig. 5) although physical processes can result in similar forms. Phototactic microorganisms living on the bottom of a shallow lake or ocean shore may be periodically covered with sediment carried in by spring runoff, for example. To retain access to sunlight, the organism must move up through this sediment layer and establish a new microbial zone. After repeated cycles, a layered series of mats are formed by lamination of the sediments containing the organic material. One characteristic of these biogenic mats that distinguishes them from nonbiologically caused layering is that the response is phototactic, not gravitational, so that the layered structure is not usually flat but is more often dome-shaped because covered microorganisms in a lower layer on the periphery of the structure would move more toward the side to reach light. In this way, stromatolites can be distinguished sometimes from similar but nonbiological laminae. Often stromatolites contain microfossils—further testimony to their biological origin.

Microbial life, which is possibly capable of photosynthesis and mobility, appears to have originated early in the history of the Earth, possibly before the end of the late bombardment 3.8 Gyr ago and almost certainly not later than 3.4 Gyr ago. This suggests that the time required for the onset of life was brief. If the Greenland sediments are taken as evidence for life, it suggests that, within the resolution of the geological record, life arose on Earth as soon as a suitable habitat was provided. The microbial mats at 3.4 Gyr ago put an upper limit of 400 million years on the length of time it took for life to arise after clement conditions were present.

In principle, tt is possible to determine which organism on the Earth is the most similar to the last universal common ancestor. To do so, we must determine which organism has changed the least compared to all other organisms. For example, if some **taxon** of organism contains a certain mutation, but many do not, we can trace the mutation to an ancestor common to all organisms in that taxon. Within this related group of organisms, the most primitive

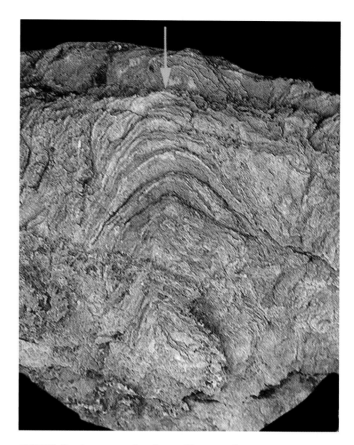

FIGURE 5   A stromatolite formed by cyanobacteria over 1 billion years ago from the Crystal Springs formation, Inyo County, California. Stromatolites are an important form of fossil evidence of life because they form macroscopic structures that could be found on Mars. It is therefore possible that a search for stromatolites near the shores of an ancient Martian lake or bay could be conducted in the near future. Expecting microbial communities to have formed stromatolites on Mars is not entirely misplaced geocentricism. The properties of a microbial mat community that results in stromatolite formation need only be those associated with photosynthetic uptake of CO_2. There are broad ecological properties that we expect to hold on Mars even if the details of the biochemistry and community structure of Martian microbial mats were quite alien compared to their terrestrial counterparts. Within stromatolites, trace microfossils can sometimes be found.

traits can be established based on how widespread they are. Traits that are found in all or most of the major groupings should be primitive, particularly if these traits are found in groups that diverged early. Traits found in only a few recently related groups are probably younger traits. This line of reasoning applied to the entire **phylogenetic** tree would indicate which organism extant today has the most primitive set of traits. This organism would therefore be most similar to the common ancestor. Studies of this type have indicated that the organisms alive today that are most similar, genetically and hence presumably ecologically, to the common ancestor are the thermophilic hydrogen metabolizing bacteria and perhaps the sulfur metabolizing bacteria. The arrow in Figure 3 represents the suggested position of the last common ancestor.

It is important to note here that the last universal common ancestor is not necessarily representative of the first organisms on Earth but was merely the last organism (or group of organisms) from which all life forms today are known to have descended. The common ancestor may have existed within a world of multiple lineages, none of which are in evidence today. If all life on Earth has indeed descended from a sulfur bacterium living in a hot springs environment, this could be the result of at least three possibilities. First it may be the case that hot sulfurous environments are important in the origin of life and the common ancestor may represent this primal cell. Second, the common ancestor may have been a survivor of a catastrophe that destroyed all other life forms. The survival of the common ancestor may have been the result of its ability to live deep within a hydrothermal system. Third, the nature of the common ancestor may be serendipitous with no implications as to origin or evolution of the biosphere.

For over 2 Gyr after the earliest evidence for life, life on the Earth was composed of only microorganisms. There were certainly bacteria and possibly one-celled eukaryotes as well. There seemed to be a major change in the environment of the Earth with the rise of photosynthetically produced oxygen beginning at about 2.5 Gyr ago, reaching significant levels about 1 Gyr ago and culminating about 600 Myr ago. (Figure 4 shows a timeline of Earth's history with these events.) Soon after the development of high levels of oxygen in the atmosphere, multicellular life forms appeared. These rapidly radiated into the major phylum known today (as well as many that have no known living representatives). In time, organisms adapted to land environments in addition to aqueous environments, and plants and animals appeared.

4. The Origin of Life

Numerous and diverse theories for the origin of life are currently under serious consideration within the scientific community. A diagram and classification of current theories for the origin of life on Earth is shown in Figure 6. At the most fundamental level, theories may be characterized within two broad categories: theories that suggest that life originated on Earth (Terrestrial in Fig. 6) and those that suggest that the origin took place elsewhere (Extraterrestrial in Fig. 6). The extraterrestrial or **panspermia** theories suggest that life existed in outer space and was transported by meteorites, asteroids, or comets to a receptive Earth. In this case, the origin of life is not related to environments possible on the early Earth. Along similar lines, life may have been ejected by impacts from another planet in the

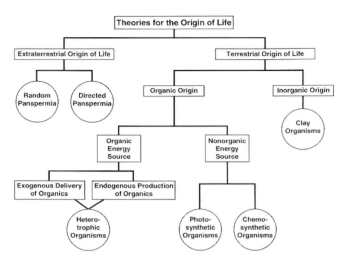

FIGURE 6   Diagrammatic representation and classification of current theories for the origin of life.

Solar System and jettisoned to Earth, or visa versa. Furthermore, it has been suggested in the scientific literature that life may have been purposely directed to Earth (directed panspermia in Fig. 6) by an intelligent species from another planet.

The terrestrial theories are further subdivided into organic origins (carbon-based) and inorganic origins (mineral-based). Mineral-based theories suggest that life's first components were mineral substrates that organized and synthesized clay organisms. These organisms have evolved via natural selection into the organic based life forms visible on Earth today. The majority of theories that do not invoke an extraterrestrial origin require an organic origin for life on Earth. Theories postulating an organic origin suggest that the initial life forms were composed of the same basic building blocks present in biochemistry today, organic material. If life arose in organic form, then there must have been a prebiological source of organics. The **Miller-Urey experiments** and their successors have demonstrated how organic material may have been produced naturally in the primordial environment of Earth (endogenous production in Fig. 6). An alternative to the endogenous production of organics on early Earth is the importation of organic material by celestial impacts and debris—comets, meteorites, interstellar dust particles, and comet dust particles. A comparison of these sources is shown in Table 5. Table 6 lists the organics found in the Murchison meteorite and compares these with the organics produced in a Miller-Urey abiotic synthesis. Organic origins differ mainly in the type of primal energy sources: photosynthetic, chemosynthetic, or heterotrophic. The phototrophs and chemotrophs (collectively called autotrophs) use energy sources that are inorganic (sunlight and chemical energy respectively), whereas heterotrophs acquire their energy by consuming organics (see Table 2).

TABLE 5   Sources of Prebiotic Organics on Early Earth

Source	Energy Dissipation (J yr^{-1})	Organic Production (in a reducing atmosphere) (kg yr^{-1})
Lightning	1×10^{18}	3×10^{9}
Coronal discharge	5×10^{17}	2×10^{8}
Ultraviolet light ($\lambda < 270$ nm)	1×10^{22}	2×10^{11}
Ultraviolet light ($\lambda < 200$ nm)	6×10^{20}	3×10^{9}
Meteor entry shocks	1×10^{17}	1×10^{9}
Meteor post-impact plumes	1×10^{20}	2×10^{10}
Interplanetary Dust	–	6×10^{7}

Hydrothermal vent environments have been suggested for the subsurface origin of chemotrophic life. In the absence of sunlight, these organisms must utilize chemical energy (e.g., $CO_2 + 4H_2 \rightarrow CH_4 + 2H_2O +$ energy). Alternatively, phototrophic life utilizes solar radiation from the surface for prebiotic synthesis. These organisms with the ability to chemosynthesize and photosynthesize can assimilate their own energy from materials in their environment. One feature that the various theories for the origin of life have in common is the requirement for liquid water because the chemistry of even the earliest life requires a liquid water medium. This is true if the primal organism appears fully developed (panspermia), if it engages in organic chemistry, and for the clay inorganic theories.

For many years the standard theory for the origin of life posited a terrestrial organic origin requiring endogenous production of organics leading to the development of heterotrophic organisms, generally known as the primordial "soup" theory. Recently there has been serious consideration for the chemotrophic origin of life, and at the present time the scientific community is split between these two views.

5. Limits to Life

In considering the existence of life beyond the Earth, it is useful to quantitatively determine the limits that life has

TABLE 6   Comparison of the Amino Acids in Murchison Meteorite and in an Electric Discharge Synthesis, Normalized to Glycine

Amino Acid	Murchison Meteorite	Electric Synthesis
Glycine	100	100
Alanine	>50	>50
α-Amino-n-butyric acid	>50	>50
α-Aminoisobutyric acid	10	>50
Valine	10	1
Norvaline	10	10
Isovaline	1	1
Proline	10	0.1
Pipecolic acid	0.1	<1
Aspartic acid	10	10
Glutamic acid	10	1
β-Alanine	1	1
β-Amino-n-butyric acid	0.1	0.1
δ-Aminoisobutyric acid	0.1	0.1
γ-Aminobutyric acid	0.1	1
Sarcosine	1	10
N-Ethyl glycine	1	10
N-Methyl alanine	1	1

TABLE 7	Limits to Life	
Parameter	**Limit**	**Note**
Lower Temperature	$\sim -15^{\circ}C$	Liquid water
Upper Temperature	$113^{\circ}C$	Thermal denaturing of proteins
Low Light	$\sim 10^{-4}S$	Algae under ice and deep sea
pH	1–11	
Salinity	Saturated NaCl	Depends on the salt
Water Activity	0.6	Yeasts and molds
	0.8	Bacteria
Radiation	1–2 Mrad	May be higher for dry or frozen state

been able to reach on this planet with respect to environmental conditions. Life does not exist everywhere on Earth. There are environments on Earth in which life has not been able to effectively colonize even though these environments could be suitable for life. Perhaps the largest life-free zone on Earth is in the polar ice sheets, where there is abundant energy, carbon, and nutrients (from atmospheric deposition) to support life. However, water is available only in the solid form. No organism on Earth has adapted to using metabolic energy to liberate water from ice, even though the energy required per molecule is only $\sim 1\%$ of the energy produced by photosynthesis per molecule. Table 7 lists the limits to life as we currently know them. The lower temperature limit clearly ties to the presence of liquid water, while the higher temperature limit seems to be determined by the stability of proteins, also in liquid water. Life can survive at extremely low light levels corresponding to 100 AU, roughly three times the distance between Pluto and the Sun. Salinity and pH also allow for a wide range. Water activity, effectively a measure of the relative humidity of a solution or vapor, can support life only for values above 0.6 for yeasts, lichens and molds. Bacteria require levels above 0.8. Radiation resistant organisms such as *Deinococcus radiodurans* can easily survive radiation doses of 1–2 Mrad and higher when in a dehydrated or frozen state.

6. Life in the Solar System

Because our knowledge of life is restricted to the unique but varied case found here on Earth, the most practical approach to the search for life on the other planets has been to proceed by way of analogy with life on Earth. The argument for the origin of life on another world is then based on the similarity of other planetary environments with the postulated environments on early Earth. Whatever process led to the establishment of life in one of these environments on Earth could then be logically expected to have led to the origin of life on this comparable world. The more exact the comparison between the early Earth and another planet, the more compelling is the argument by analogy.

This comparative process should be valid for all the theories for the origin of life, ranging from panspermia to the standard theory, listed in Figure 6.

Following this line of reasoning further, we can conclude that if similar environments existed on two worlds and life arose in both of them then these life forms should be comparable in their broad ecological characteristics. If sunlight was the available energy source, CO_2 the available carbon source, and liquid water the solvent, then one could expect phototrophic autotrophs using sunlight to fix carbon dioxide with water as the medium for chemical reactions. Our knowledge of the Solar System suggests that such an environment could have existed on Mars early in its history as well as on Earth early in its history. While life forms independently originating on these two planets would have different biochemical details, they would be recognizably similar in many fundamental attributes. This approach by analogy to Earth life and the early Earth provides a specific search strategy for life elsewhere in the Solar System. The key element of that strategy is the search for liquid water habitats.

Spacecraft have now visited or flown past comets, asteroids, and most of the large worlds in the Solar System except Pluto; however, a spacecraft is en route to Pluto at the time of this writing. Observatory missions have studied all of the major objects in the Solar System as well. We can do a preliminary assessment of the occurrence of liquid water habitats, and indirectly life, in the Solar System.

6.1 Mercury and the Moon

Mercury and the Moon appear to have few prospects for liquid water, now or anytime in the past. These virtually airless worlds have negligible amounts of the volatiles (such as water and carbon dioxide) essential for life. There are no geomorphological features that indicate fluid flow. There is speculation that permanently shaded regions of the polar areas on Mercury can act as traps for water ice. Recent radar data support this hypothesis. However, there is no indication that the pressure and temperature were ever high enough for liquid water to exist at the surface. [*See* MERCURY.]

6.2 Venus

Venus currently has a surface that is clearly inhospitable to life. There is no liquid water on the surface, and the temperature is over 450°C at an atmospheric pressure of 92 times the Earth's. There is water on Venus but only in the form of vapor and clouds in the atmosphere. The most habitable zone on Venus is at the level in the atmosphere where the pressure is about half of the sea level on Earth. At that location, there are clouds composed of about 25% water and 75% sulfuric acid at a temperature of about 25°C; these might be reasonable conditions for life. It is possible therefore to speculate that life can be found, or survive if implanted, in the clouds of Venus. What argues against this possibility is the fact that clouds on Earth that are at similar pressures and temperatures do not harbor life. We do not know of any life forms that thrive in cloud environments. Perhaps the essential elements are there but a stable environment is required. [See VENUS: ATMOSPHERE]

Theoretical considerations suggest that Venus and Earth may have initially had comparable levels of water. In this case Venus may have had a liquid water surface early in its history when it was cooler 4 billion years ago because of the reduced brightness of the fainter early sun. Unfortunately, all record of this early epoch has been erased on Venus and the question of the origin of life during such a liquid water period remains untestable. [See VENUS: SURFACE AND INTERIOR]

6.3 Mars

Of all the extraterrestrial planets and smaller objects in the Solar System, Mars is the one that has held the most fascination in terms of the existence of life. Early telescopic observations revealed Earth-like seasonal patterns on Mars. Large white polar caps that grew in the winter and shrunk in the summer were clearly visible. Regions of the planet's surface near the polar caps appeared to darken beginning at the start of each polar cap's respective spring season and then spread toward the equator. It was natural that these changes, similar to patterns on the Earth, would be attributed to like causes. Hence, the polar caps were thought to be water ice and the wave of darkening was believed to have been caused by the growth of vegetation. The 19th century arguments for the existence of life, and even intelligent life, on Mars culminated in the book *Mars as the Abode of Life* by Percival Lowell in 1908 and in the investigations of the celebrated canals. The Mars revealed by spacecraft exploration is decidedly less alive than Lowell anticipated but its standing as the most interesting object for biology outside Earth still remains.

6.3.1 THE VIKING RESULTS

In 1976 the *Viking* landers successfully reached the Martian surface while the two orbiters circled the planet repeatedly

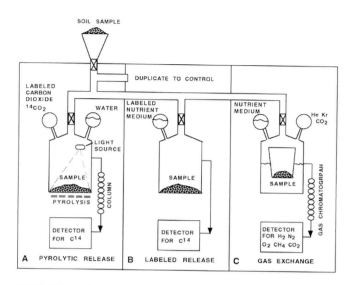

FIGURE 7 Schematic diagram of the *Viking* biology experiments.

photographing and monitoring the surface. The primary objective of the *Viking* mission was the search for microbial life. Previous reconnaissance of Mars by the *Mariner* flyby spacecraft and the photographs returned from the *Mariner 9* orbiter had already indicated that Mars was a cold dry world with a thin atmosphere. There were intriguing features indicative of past fluvial erosion but there was no evidence for current liquid water. It was thought that any life to be found on Mars would be microbial. The *Viking* biology package consisted of three experiments shown schematically in Figure 7.

The Pyrolytic Release (PR) experiment searched for evidence of photosynthesis as a sign of life. The PR was designed to see if Martian microorganisms could incorporate CO_2 under illumination. The experiment could be performed under dry conditions similar to those on the Martian surface or it could be run in a humidified mode. The CO_2 in the chamber was labeled with radioactive carbon which could then be detected in any organic material synthesized during the experiment. The very first run of the pyrolytic release experiment produced a significant response. It was well below the typical response observed when biotic soils from Earth had been tested in the experiment, but it was much larger than the noise level. Subsequent trials did not reproduce this high result, and this initial response was attributed to a startup anomaly, possibly some small prelaunch contamination.

The Gas Exchange (GEx) experiment searched for heterotrophs, which are microorganisms capable of consuming organic material. The GEx was designed to detect any gases that the organisms released as a byproduct of their metabolism, bacterial flatulence. After a sample was placed in the chamber, the soil was first equilibrated with water vapor and then combined with a nutrient solution. At

TABLE 8	A Comparison of GEx O_2 and LR ^{14}C Results[a]			
Sample	GEx O2 (nmol cm^{-3})	Oxidant[b] (KO$_2$ → O$_2$)	LR CO$_2$ (nmol cm^{-3})	Oxidant[b] (H$_2$O$_2$ → O)
Viking 1 (surface)	770	35 ppm/m	~30	1 ppm/m
Viking 2 (surface)	194	10	~30	1
Viking 2 (sub-rock)	70	3	~30	1

[a] After Klein 1979.
[b] Assuming a bulk soil density of 1.5 g cm^{-3}.

prescribed intervals, a sample of the gas above the sample was removed and analyzed by a gas chromatograph.

The GEx results were startling. When the Martian soil was merely exposed to water vapor, it released oxygen gas at levels of 70–700 nanomoles per gram of soil, much larger than could be explained by the release of ambient atmospheric oxygen that had been absorbed onto the soil grains. The GEx results are summarized in Table 8. It was clear that some chemical or biological reaction was responsible for the oxygen release. A biological explanation was deemed unlikely since the reactivity of the soil persisted even after it had been heat sterilized to temperatures of over 160°C. Furthermore, adding the nutrient solution did not change the result that some chemical in the soil was highly reactive with water.

The Labeled Release (LR) experiment also searched for evidence of heterotrophic microorganisms. In the LR experiment, a solution of water containing seven organic compounds was added to the soil. The carbon atoms in each organic compound were radioactive. A radiation detector in the headspace detected the presence of radioactive CO$_2$ released during the experiment. Any carbon metabolism in the soil would be detected as organisms consumed the organics and released radioactive CO$_2$.

When the LR experiment was performed on Mars, there was a steady release of radioactive CO$_2$ (Table 8). When the soil sample was heat sterilized before exposure to the nutrient solution, no radioactive CO$_2$ was detected. The results of the LR experiment were precisely those expected if there were microorganisms in the soil sample. Taken alone, the LR results would have been a strong positive indication for life on Mars.

In addition to the three biology experiments, another instrument, a combination of a gas chromatograph and a mass spectrometer (GCMS), gave information pivotal to the interpretation of the biological results. This instrument received Martian soil samples from the same sampling arm that provided soil to the biology experiments. The sample was then heated to release any organics. The decomposed organics were carried through the gas chromatograph and identified by the mass spectrometer. The only signal was due to cleaning agents used on the spacecraft before launch. No Martian organics were detected. However, the samples were heated to only 500°C, and highly refractory organics would not have been volatilized at this temperature. In addition, it is now known that iron compounds in the soil could have interfered with the release of organics. The limit on the concentration of organics that would remain undetectable by the GCMS was one part per billion. A part per billion of organic material in a soil sample represents over a million individual bacterium, each the size of a typical *Escherichia coli*. This may not seem to rule out a biological explanation for the LR results. However, all life is composed of organic material and it is constantly exuded and processed in the biosphere. On Earth, it is difficult to imagine life without a concomitant matrix of organic material. This apparent absence of organic material is the main argument against a biological interpretation of the positive LR results.

The prevailing explanation for the reactivity of the Martian soil relies on the presence of reactive chemicals in the Martian atmosphere. In particular, hydrogen peroxide (H_2O_2) is assumed to be produced by ultraviolet light in the atmosphere and deposited onto the soil surface. Hydrogen peroxide itself could explain many of the LR results including the loss of reactivity with heating, but it cannot explain the thermally stable results of the GEX. However, peroxide, possibly abetted by ultraviolet radiation could somehow result in the production of the stable reactive chemicals responsible for the release of oxygen upon humidification and the breakdown of organics in the LR experiment. In addition, these reactive chemicals would have broken down any naturally occurring organic material or any material carried in by meteorites on the Martian surface. Table 8 also lists the concentration of oxidant necessary to explain the *Viking* results for typical models of the chemistry of the oxidants.

Amplifying the apparently negative results of the *Viking* biology experiments, the environment of Mars appears to be inhospitable to life. Although the atmosphere contains many of the elements necessary for life—it is composed of 95% CO$_2$ with a few percent N$_2$ and Argon and trace levels of water—the mean surface pressure is less than 1% of sea level pressure on the Earth, and the mean temperature is −60°C. The mean surface pressure is close to the triple point pressure of water, which is the minimum pressure at which a liquid state of water can exist. The low pressures and low temperatures make it unlikely that water will exist as a liquid on Mars. Because of seasonal transport, the available surface water on Mars is trapped as ice in the polar regions. In the locations at low elevation where the pressures and temperatures are sufficient to support liquid water, the surface is desiccated. Even saturated brine solutions cannot exist in equilibrium with the atmosphere near the equator. The absence of liquid water on the surface of Mars is probably the most serious argument against the presence of life anywhere on the surface of the planet. A second significant hazard to life on the Martian surface is the presence of solar ultraviolet light in the wavelengths between 190 and 300 nm. This radiation, which is largely shielded from Earth's surface by atmospheric oxygen and the ozone layer, is highly effective at destroying terrestrial organisms. Wavelengths below 190 nm are absorbed even by the present thin Martian CO_2 atmosphere. Compounding the effects of UV irradiation, and perhaps caused by it, are possible chemical oxidants that are thought to exist in the Martian soil. Such strong oxidants have been suggested as the causative agent for the chemical reactivity observed at the *Viking* sites. [*See* MARS ATMOSPHERE: HISTORY AND SURFACE INTERACTION.]

6.3.2 EARLY MARS

There is considerable evidence that early in its history Mars did have liquid water on its surface. Images from the many orbiters show complex dendritic valley networks that are believed to have been carved by liquid water. These valleys are predominantly found in the heavily cratered, hence ancient, terrains in the southern hemisphere. This would suggest that the period of liquid water on Mars occurred contemporaneously with the end of the last stages of heavy cratering, about 3.8 Gyr ago, the same epoch at which life is thought to have originated on Earth (see Fig. 4). [*See* MARS, SURFACE AND INTERIOR.]

Figure 8 shows part of Nanedi Vallis on Mars. The canyon snakes back and forth, which is characteristic of liquid flow. On the floor of the canyon appears a small channel, which presumably was the flow of the river that carved the canyon. It would have taken considerable flow, although not necessarily continuous flow, for this river to have carved the much larger canyon. This image provides what is perhaps the best evidence from orbit that liquid water flowed on the surface of Mars in stable flow for long periods of time. Figure 9 shows evidence for liquid water form the surface rover missions. The "blueberries" seen at the Meridiani Site are interpreted as concretions formed in liquid water.

The presence of liquid water habitats on early Mars at approximately the time that life is first evident on Earth suggests that life may have originated on Mars during the same time period. Liquid water is the most critical environmental requirement for life on Earth and the general similarity between Earth and Mars leads us to assume that life on

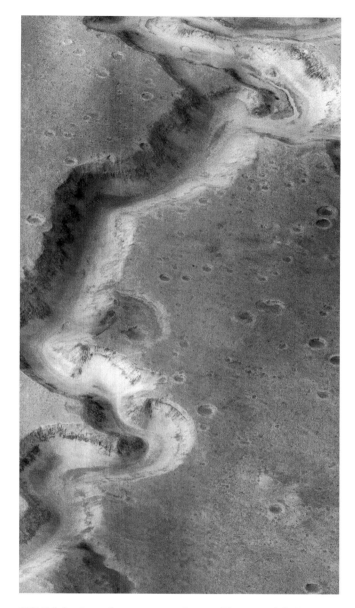

FIGURE 8   Liquid water on another world. Mars *Global Surveyor* image showing Nanedi Vallis in the Xanthe Terra region of Mars. The image covers an area 9.8 km by 18.5 km; the canyon is about 2.5 km wide. This image is the best evidence we have of liquid water anywhere outside the Earth. (Photo from NASA/Malin Space Sciences).

Mars would be similar in this basic environmental requirement. More exotic approaches to life on Mars cannot be ruled out, nor are they supported by any available evidence.

It is interesting to consider how evolution may have progressed on Mars by comparison with the Earth. The history of Earth and Mars are compared in Figure 4, which shows that the period between 4.0 and 3.5 Gyr ago is the time when life is most likely to have evolved on both planets. On Earth, life persists and remains essentially unchanged

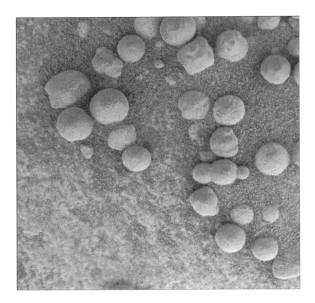

FIGURE 9   Blueberries. The triplet of connected spheres, dubbed blueberries, as seen in this MER *Opportunity* image, is a strong indication that they are concretions formed in the presence of water, not in volcanic eruptions or meteor impacts. Concretions are spherical mineral structures formed by groundwater percolating through porous rocks. On Earth, as concretions grow in close proximity to each other, their outer edges often intersect each other, producing connected spheres. (NASA/JPL)

for several billion years until the cumulative effects of O_2 production induces profound changes on the atmosphere of that planet. On Mars, conditions become unsuitable for life (no liquid water) in a billion years or less. Thus, is it likely that if there were any life on early Mars it remained microbial.

The evidence of liquid water on early Mars, particularly that provided by the valley networks, suggests that the climate on early Mars may have been quite different than at present. It is generally thought that the surface temperature must have been close to freezing, much warmer than the present $-60°C$. These warmer temperatures are thought to have occurred as a result of a greatly enhanced greenhouse due to a thick ($\sim$1-5 atm) CO_2 atmosphere. However, CO_2 condensation may have limited the efficacy of the CO_2 greenhouse but theoretical models indicate that CO_2 clouds or CH_4 could enhance the greenhouse and maintain warmer temperatures.

If Mars did have a thick CO_2 atmosphere, this strengthens the comparison to the Earth, which is thought to have also had a thick CO_2 atmosphere early in its history. The duration of a thick atmosphere on Mars and the concomitant warm, wet surface conditions are unknown but simple climate models suggest that significant liquid water habitats could have existed on Mars for $\sim$0.5 Gyr after the mean surface temperature reached freezing. This model is based on

the presence of deep ice-covered lakes (over 30 m) such as those in the dry valleys of the Antarctic where mean annual temperatures are $-20°C$.

If we divide the possible scenario for the history of water on the surface of Mars into four epochs, the first epoch would have warm surface conditions and liquid water. As Mars gradually loses its thick CO_2 atmosphere, the second and third epochs would be characterized by low temperatures but still relatively high atmospheric pressures. This is because the temperature would drop rapidly as the pressure decreased. During the second epoch, temperatures would rise above freezing during some of the year and liquid water habitats would require a perennial ice-cover. However, by epoch three the temperature would never rise above freezing and the only liquid water would be found in porous rocks with favorable exposures to sunlight. In epoch four the pressure would fall too low for the presence of liquid water.

A point worth emphasizing here is that the biological requirement is for liquid water *per se*. Current difficulties in understanding the composition and pressure of the atmosphere need not lessen the biological importance of the direct evidence for the presence of liquid water. In fact, as we observe in the Antarctic dry valleys, ecosystems can exist when the mean temperatures are well below freezing. Mars need not have ever been above freezing for life to persist.

The particular environment on the early Earth in which life originated is not known. However, this does not pose as serious a problem to the question of the origin of life on Mars as might be expected. The reason is that all of the environments found on the early Earth would be expected to be found on Mars, including hydrothermal sites, hot springs, lakes, oceans (that is planetary scale water reservoirs), volcanoes, tidal pools (solar tides only), marshes, salt flats, and others. Thus, whatever environment or combination of environments needed for life to get started on Earth should have been present on Mars as well, and at the same time.

Since the rationale for life on Mars early in its history is based on analogs with fossil evidence for life on the early Earth it is natural to look to the fossil record on Earth as a guide to how relics of early Martian life might be found. The most persuasive evidence for microbial life on the early Earth comes from stromatolites as discussed before. The resulting structures can be quite large: they are macroscopic fossils generated by microorganisms.

6.3.3 SUBSURFACE LIFE ON MARS

Although there is currently no direct evidence to support speculations about extant life on Mars, there are several interesting possibilities that cannot be ruled out at this time. Protected subsurface niches associated with hydrothermal activity could have continued to support life even after surface conditions became inhospitable. Liquid water could be provided by the heat of geothermal or volcanic activity melting permafrost or other subsurface water sources. Gases from volcanic activity deep in the planet could provide reducing power (as CH_4, H_2, or H_2S) percolating up from below and enabling the development of a microbial community based upon chemolithoautotrophy. An example is a methanogen (or acetogen) that uses H_2 and CO_2 in the production of CH_4. Such ecosystems have been found deep underground on the Earth consuming H_2 produced by the reaction of water with basaltic rock, a plausible reaction for subsurface Mars. However, their existence is neither supported nor excluded by current observations of Mars. Tests for such a subsurface system involve locating active geothermal areas associated with ground ice or detecting trace quantities of reduced atmospheric gases that would leak from such a system. It is interesting to consider the recent reports of CH_4 in the atmosphere of Mars at the tens of ppb level. If these reports are confirmed, it may be that this CH_4 may be related to subsurface biological activity. However nonbiological sources of CH_4 are also possible.

While it certainly seems clear that volcanic activity on Mars has diminished over geological time, intriguing evidence for recent (on the geological time scale) volcanic activity comes from the young crystallization ages (all less than 1 Gyr) of the Shergotty meteorite (and other similar meteorites thought to have come from Mars). Volcanic activity by itself does not provide a suitable habitat for life; liquid water that may be derived from the melting of ground ice is also required. Presumably, the volcanic source in the equatorial region would have depleted any initial reservoir of ground ice and there would be no mechanism for renewal, although there are indications of geologically recent volcano/ground ice interactions at equatorial regions. Closer to the poles, ground ice is stable. It is conceivable that a geothermal heat source could result in cycling of water through the frozen ice-rich surface layers. The heat source would melt and draw in water from any underlying reservoir of groundwater or ice that might exist. [*See* METEORITES.]

Another line of reasoning also supports the possibility of subsurface liquid water. There are outflow channels on Mars that appear to be the result of the catastrophic discharge of subsurface aquifers of enormous sizes. There is evidence based on craters and stratigraphic relations that these have occurred throughout Martian history. If this is the case, then it is possible that intact aquifers remain. This would have profound implications for exobiology (as well as human exploration). Furthermore, it suggests that the debris field and outwash regions associated with the outflow channel may hold direct evidence that life existed within the subsurface aquifer just prior to its catastrophic release.

The collection of available water on Mars in the polar regions naturally suggests that summer warming at the edges of the permanent water ice cap may be a source of meltwater, even if short lived. In the polar regions of Earth, complex microbial ecosystems survive in transient summer meltwater. However, on Mars the temperature and pressures remain too low for liquid water to form. Any energy available is lost from the sublimation of the ice before any liquid is produced. It is unlikely that there are even seasonal habitats at the edge of the polar caps. This situation may be different over longer timescales. Changes in the obliquity axis of Mars can significantly increase the amount of insolation reaching the polar caps in summer. If the obliquity increases to over about 50°, then the increased temperatures, atmospheric pressures, and polar insolation that result may cause summer liquid water meltstreams and ponds at the edge of the polar cap.

The polar regions may harbor remnants of life in another way. Tens of meters beneath the surface, the temperature is well below freezing ($<-70°C$) and does not change from summer to winter. These permafrost zones likely have remained frozen, particularly in the southern hemisphere, since the end of the intense crater formation period. In this case, there may be microorganisms frozen within the permafrost that date back to the time when liquid water was common on Mars, over 3.5 Gyr ago. On Earth permafrost of such age does not exist, but there are sediments in the polar regions that have been frozen for many millions of years. When these sediments are exhumed and samples extracted using sterile techniques, viable bacteria are recovered. The sediments on Mars have been frozen much longer (1000 times) but the temperatures are also much colder; it may be possible that intact microorganisms could be recovered from the Martian permafrost. Natural radiation from U, Th, and K in the soil would be expected to have killed any organisms but their biochemical remains would be available for study. The southern polar region seems like the best site for searching for evidence of ancient microorganisms since the terrain there can be dated to the earliest period of Martian history as determined by the number of observed craters.

6.3.4 METEORITES FROM MARS

Of the thousands of meteorites known, there are over 30 that are thought to have come from Mars. It is certain that these meteorites came from a single source because they all have similar ratios of the oxygen isotopes—values distinct from terrestrial, lunar, or asteroidal ratios. These meteorites can be grouped into four classes. Three of these classes contain all but one of the known Mars meteorites and are known by the name of the type specimen; the S (Shergotty), N (Nakhla), and C (Chassigny) class meteorites. The S, N, and C meteorites are relatively young, having crystallized from lava flows between 200 and 1300 million years ago (see Fig. 4). Gas inclusions in two of the S type meteorites contain gases similar to the present Martian atmosphere as measured by the *Viking* landers, proving that this meteorite, and by inference the others as well, came from Mars. The fourth class of Martian meteorite is represented by the single specimen known as ALH84001. Studies of this meteorite indicate that it formed on Mars about 4.5 Gyr ago in warm, reducing conditions. There are even indications that it contains Martian organic material and appears to have experienced aqueous alteration after formation. This rock formed during the time period when Mars is thought to have had a warm, wet climate capable of supporting life.

It has been suggested that ALH84001 contains evidence for life on Mars based on four observations. (1) Polycyclic aromatic hydrocarbons similar to molecules found in interstellar space are present inside ALH84001. (2) Carbonate globules are found in the meteorite that are enriched in ^{12}C over ^{13}C. The isotopic shift is within the range that on Earth, indicates organic matter derived from biogenic activity. (3) Magnetite and iron-sulfide particles are present that are similar to those produced by microbial activity. (4) Features are seen that could be fossils of microbial life, except that they are much smaller than any bacteria on Earth. As a result of more than a decade of study, most scientists currently prefer a nonbiological explanation for all of these results. Only the magnetite result is generally considered relevant, although not conclusive, evidence related to life.

ALH84001 does not provide convincing evidence of past life on Mars when compared to the multiple lines of evidence for life on Earth 3.4 Gyr ago including fossil evidence. However, the ALH84001 results do provide strong support to the suggestion that conditions suitable for life were present on Mars early in its history. When compared to the SNC meteorites, ALH84001 indicates that Mars experienced a transition from a warm reducing environment with organic material present to a cold oxidizing environment in which organic material was unstable.

6.4 The Giant Planets

The "habitable zone" in the inner Solar System provides the temperature conditions which can support liquid water on a planetary surface, but the outer Solar System is richer in the organic material from which life is made. This comparison is shown in Figure 10, which shows the ratio of carbon to heavy elements (all elements other than H and He) for various objects in the Solar System. Earth is in fact depleted in carbon with respect to the average Solar System value by a factor of about 10^4. It may be interesting then to consider life in the organic rich outer Solar System.

The giant planets Jupiter, Saturn, Uranus, and Neptune, do not have firm surfaces on which water could accumulate and form a reservoir for life. Here the only clement zone would be that region of the clouds in which temperatures were in the range suitable for life. Cloud droplets would provide the only source of liquid water. Such an environment might provide the key elements needed for life as well as an energy source in the form of sunlight. [See ATMOSPHERES OF THE GIANT PLANETS.]

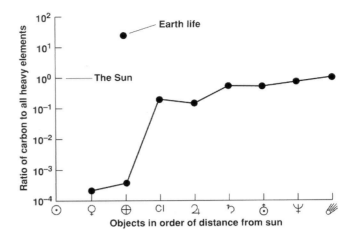

FIGURE 10   Ratio of carbon atoms to total heavy atoms (heavier than He) for various Solar System objects illustrating the depletion of carbon in the inner Solar System. The x-axis is not a true distance scale but the objects are ordered by increasing distance from the sun. Mars is not shown since the size of its carbon reservoir is unknown.

There have been speculations that life, including advanced multicellular creatures, could exist in such an environment. However, such speculations are not supported by considerations of the biological state of clouds on Earth. There are no organisms that have adapted themselves to live exclusively in clouds on Earth even in locations where clouds are virtually always present. This niche remains unfilled on Earth and by analogy is probably unfilled elsewhere in the Solar System.

Following this line of thought leads us to search for environments suitable for life on planetary bodies with surfaces. In the outer Solar System, this focuses us on the moons of the giant planets. Of particular interest are Europa, Titan, and Enceladus.

6.4.1 EUROPA

Europa, one of the moons of Jupiter, appears to be an airless ice-shrouded world. However, theoretical calculations suggest that under the ice surface of Europa there may be a layer of liquid water sustained by tidal heating as Europa orbits Jupiter. The *Galileo* spacecraft imaging showed features in the ice consistent with a subsurface ocean and the magnetometer indicated the presence of a global layer of slightly salty liquid water. The surface of Europa is crisscrossed by streaks that are slightly darker than the rest of the icy surface. If there is an ocean beneath a relatively thin ice layer, these streaks may represent cracks where the water has come to the surface. [See PLANETARY SATELLITES.]

There are many ecosystems on Earth that thrive and grow in water that is continuously covered by ice; these are found in both the Arctic and Antarctic regions. In addition to the polar oceans where sea ice diatoms perform photosynthesis under the ice cover, there are perennially ice-covered lakes in the Antarctic continent in which microbial mats based on photosynthesis are found in the water beneath a 4-m ice cover. The light penetrating these thick ice covers is minimal, about 1% of the incident light. Using these Earth-based systems as a guide, it is possible that sunlight penetrating through the cracks (the observed streaks) in the ice of Europa could support a transient photosynthetic community. Alternatively, if there are hydrothermal sites on the bottom of the Europan ocean, it may be possible that chemosynthetic life could survive there—by analogy to life at hydrothermal vent sites at the bottom of the Earth's oceans. The biochemistry of hydrothermal sites on Earth does depend on O_2 produced at the Earth's surface. On Europa, a chemical scheme like that suggested for subsurface life on Mars would be appropriate ($H_2 + CO_2$).

The main problem with life on Europa is the question of its origin. Lacking a complete theory for the origin of life, and lacking any laboratory synthesis of life, we must base our understanding of the origin of life on other planets on analogy with the Earth. It has been suggested that hydrothermal vents may have been the site for the origin of life on Earth and if this is the case improves the prospects for life in a putative ocean on Europa. However, the early Earth contained many environments other than hydrothermal vents, such as surface hot springs, volcanoes, lake and ocean shores, tidal pools, and salt flats. If any of these environments were the locale for the origin of the first life on Earth, the case for an origin on Europa is weakened considerably.

6.4.2 TITAN

Titan, the largest moon of the planet Saturn, has a substantial atmosphere composed primarily of N_2 and CH_4 with many other organic molecules present. The temperature at the surface is close to $94°K$ and the surface pressure is 1.5 times the pressure of Earth at sea level. The surface does not appear to have expansive oceans as once suggested but numerous small lakes have been discovered in the north polar region. However, the ground beneath the Huygens Probe was wet with liquid CH_4, which was heated by the problem and formed vapor. [See TITAN.]

The spacecraft data from the *Voyager* and *Cassini/Huygens* missions, as well as ground-based studies, indicate that there is an optically thick haze in the upper atmosphere. The haze is composed of organic material, and the atmosphere contains many organic molecules heavier than CH_4. Photochemical models suggest that these organics are produced from CH_4 and N_2 through chemical reactions driven by solar photons and by magnetospheric electrons. The observed organic species and even heavier organic molecules are predicted to result from these chemical transformations. Laboratory simulations of organic reactions in Titan-like gas mixtures produce solid refractory organic substances (tholin) and similar processes are expected to occur in Titan's atmosphere.

Conditions on Titan are much too cold for liquid water to exist, although the pressure is in an acceptable regime. For this reason, it is unlikely that Earth-like life could originate or survive there. The organic material in Titan's atmosphere provides a potential source of energy and the liquid methane on the surface provides a possible liquid medium for life. Life in liquid methane could use active transport and large size to overcome the low solubility of organics in liquid methane and enzymes to catalyze reactions at the low temperatures. If carbon-based life in liquid methane existed on Titan, it could be widespread. With or without life, Titan remains interesting because it is a naturally occurring Miller-Urey experiment in which simple compounds are transformed into more complex organics. A detailed study of this process may yield valuable insight into how such a mechanism might have operated on the early Earth.

There is also some speculation that under unusual conditions Titan may have liquid water on or near the surface. This could have occurred early in its formation when the gravitational energy released by the formation of Titan would have heated it to high temperatures. More recently, impacts could conceivably melt local regions generating warm subsurface temperatures that could last for thousands of years. Whether such brief episodes of liquid water could have led to water-based life remains to be tested.

6.4.3 ENCELADUS

The *Cassini* mission has recently documented geysers erupting from the south polar region of Enceladus. [See PLANETARY SATELLITES.] Associated with this outflow of water, CH_4 is present but no NH_3. The source of the water is considered to be a subsurface liquid water reservoir heated and pressurized by subsurface heat flow. Such a subsurface habitat could support the sort of anaerobic chemoautotrophic life that has been found on Earth. These systems are based on methanogens that consume H_2 produced by geochemical reactions or by radioactive decay. The age or lifetime of any subsurface liquid water on Enceladus is not known, which adds uncertainly to speculations about the origin of life. The theories for the origin of life on Earth, shown in Figure 6, that would apply to Enceladus are panspermia and a chemosynthetic origin of life. The same that would apply to Europa.

If there is subsurface life in the liquid water reservoirs on Enceladus, then the geysers would be carrying these organisms out into space. Here they would quickly become dormant in the cold vacuum of space and would then be killed by solar ultraviolet radiation. But these dead, frozen microbes would still retain the biochemical and genetic molecules of the living forms. Thus a *Stardust*-like mission moving through the plume of Enceladus' geysers might collect lifeforms for return to Earth, which might provide the easiest way to get a sample of a second genesis of life.

6.5 Asteroids

Asteroids seem unlikely locations for life to have originated. Certainly they are too small to support an atmosphere sufficient to allow for the presence of liquid water at the present time. However, asteroids, particularly the so-called carbonaceous type, are thought to contain organic material, thereby playing a role in the delivery of organics to the prebiotic Earth. A more intriguing aspect of some asteroids is the presence of hydrothermally altered materials, which seems to indicate that the asteroids were once part of a larger parent body. Furthermore, conditions on this larger parent body were such that liquid water was present, at least in thin films. Containing both organic material and liquid water, the parent bodies of these asteroids are interesting targets in the search for extraterrestrial lifeforms. However, a thorough assessment of this possibility will require a more detailed study of carbonaceous asteroids in the asteroid belt. Meteorites found on the Earth provide only a glimpse of small fragments of these objects and no signs of extraterrestrial life have been found. But the samples are small and the potential for contamination by Earth life is great.

6.6 Comets

Comets are also known to be rich in organic material. However, unlike asteroids, comets also contain a large fraction of water. In their typical state this water is frozen as ice, which is unsuitable for life processes. As a comet approaches the sun, its surface is warmed considerably, but this leads only to the sublimation of the water ice. Liquid does not form because the pressure at the surface of the comet is much too low.

It has been suggested that soon after their formation the interior of large comets would have been heated by short-lived radioactive elements (^{26}Al) to such an extent that the core would have melted. In this case, there would have been a subsurface liquid water environment similar to that postulated for the present day Europa. Again the question of the origin of life in such an environment rests on the assumption that life can originate in an isolated deep dark underwater setting.

7. How to Search for Life on Mars, Europa, or Enceladus

If we were to find organic material in the subsurface of Mars, or in the ice of Europa, or entrained in the geysers of Enceladus, how could we determine if it was the product of a system of biology or merely abiotic organic material from meteorites or photochemistry? If that life is related to Earth life, it should be easy to detect. We now have very sensitive methods, such as the amplification of DNA and fluorescent antibody markers, for detecting life from Earth. The case of Earth-like life is the easiest but it is also the least interesting. If the life is not Earth-like, then the probes specific to our biology are unlikely to work. We need a general way to determine a biological origin. The question is open and possibly urgent. As we plan missions to Mars and Europa, we may have the opportunity to analyze the remains of alien biology.

One practical approach makes use of the distinction between biochemicals and organic matter that is not dependent on a particular organic molecule but results from considering the pattern of the organics in a sample. Abiotic processes will generate a smooth distribution in molecular types without sharp distinctions between similar molecules, isotopes, or chemical chirality. If we consider a generalized phase space of all possible organic molecules, then for an abiotic production mechanism the relative concentration of different types will be a smooth function. In contrast to abiotic mechanisms, biological production will not involve a wide range of possible types. Instead, biology will select a few types of molecules and build biochemistry up from this restricted set. Thus organic molecules that are chemically very similar may have widely different concentrations in a sample of biological organics. An example of this on Earth is the 20 *amino acids* used in proteins and the selection of life for the left-handed version of these amino acids. To maximize efficiency, life everywhere is likely to evolve this strategy of using a few molecules repeatedly. It may be that other life forms discover the same set of biomolecules that Earth life uses because these are absolutely the most efficient and effective set under any planetary conditions. But it may also be that life elsewhere uses a different set that is optimal given the specific history and conditions of that world. We can search for the repeated use of a set of molecules without knowing in advance what the members of that set will be.

We can apply this approach to the search for biochemistry in the Solar System. Samples of organic material collected from Mars and Europa can be tested for the prevalence of one chirality of amino acid over the other. More generally, a complete analysis of the relative concentration of different types of organic molecules might reveal a pattern that is biological even if that pattern does not involve any of the biomolecules familiar from Earth life. Interestingly, if a sample of organics from Mars or Europa shows a preponderance of D amino acids, this will suggest the presence of extant or extinct life and at the same time show that this life is distinct from Earth life. This same conclusion would apply to any clearly biological pattern that is distinct from the pattern of Earth life. The pattern of biological origin in organic material can potentially persist long after the organisms themselves are dead. Eventually this distinctive pattern will be destroyed as a result of thermal and radiation effects. Below the surface of Mars, both temperature and radiation are low, so this degradation should not be significant. On Europa the intense radiation may destroy the biological signature after several million years at depths to about 1 m below the surface ice.

8. Life about Other Stars

In the Solar System, only our own planet has clear signs of life. Mars, Europa, and Enceladus provide some hopes of finding past or present liquid water but nothing comparable to the richness of water and life on Earth. Our understanding of life as a planetary phenomenon would clearly benefit from finding another Earth-like planet, around another Sun-like star, that harbored life.

One way of formulating the probability of life, and intelligent life, elsewhere in the galaxy is the Drake equation, named after Frank Drake, a pioneer in the search for extraterrestrial intelligence. The equation and the terms used with it are listed in Table 9. The most accurately determined variable in the Drake equation at this time is R_*, the number of stars forming in the galaxy each year. Since we know that there are about 10^{11} stars in our galaxy and that their average lifetime is about 10^{10} years, then $R_* \sim 10$ stars per year. All the other terms are uncertain and can be only estimated by extrapolating from what has occurred on Earth. Estimates by different authors for N, the number of civilizations in the galaxy capable of communicating by radios waves, range from 1 to millions. Perhaps the most uncertain term is L, the length of time that a technologically advanced civilization can survive.

The primary criterion for determining whether a planet can support life is the availability of water in the liquid state. This in turn depends on the surface temperature of the planet which is controlled primarily by the distance to a central star. Life appeared so rapidly on Earth after its formation that it is likely that other planets may only have had to sustain liquid water for a short period of time for life to originate. Planets orbiting a variety of star types could satisfy this criterion at some time in their evolution. The development of advanced life on Earth, and in particular intelligent life, took much longer, almost 4 billion years. Earth maintained habitable conditions for the entire period of time.

Locations about stars in which temperatures are conducive to liquid water for such a long period of time have been called continuously habitable zones (CHZ). Calculations of the CHZ about main sequence stars indicate that the mass of the star must be less than 1.5 times the mass of our sun for the CHZ to persist for more than 2 billion years.

An interesting result of these calculations is that the current habitable zone for the sun has an inner limit at about 0.8 AU and extends out to between 1.3 and 1.6 AU, depending on the way clouds are modeled. Thus, while Venus is not in the habitable zone, Earth and Mars both are. This calculation would suggest that Mars is currently habitable. But we see no indication of life. This is owing to the fact that the distance from the sun is not the only determinant for the presence of liquid water on a planet's surface. The presence of a thick atmosphere and the resultant greenhouse effect is required as well. On Earth the natural greenhouse effect is responsible for warming the Earth by 30°C; without the greenhouse effect the temperature would average −15°C. Mars does not have an appreciable greenhouse effect, and hence its temperature averages −60°C. If Earth were at the same distance from the sun than Mars, it would probably be habitable because of the thermostatic effect of the long-term carbon cycle. This cycle is driven by the burial of carbon in seafloor sediments as organic material and carbonates. The formation of carbonates is due to chemical erosion of the surface rocks. Subduction carries this material to depths where the high temperatures release the sedimentary CO_2 gas, and these gases escape to the surface in volcanoes that lie on the boundary arc of the subduction zones. The thermostatic action of this cycle results because the erosion rate is strongly dependent on temperature. If the temperature were to drop, erosion would slow down. Meanwhile the outgassing of CO_2 would result in a buildup of this greenhouse gas and the temperature would rise. Conversely, higher temperatures would result in higher erosion rates and a lowering of CO_2 again stabilizing the temperature.

Mars became uninhabitable because it lacks plate tectonics and hence has no means of recycling the carbon-containing sediments. As a result, the initial thick

TABLE 9	The Probability of Life, and Intelligent Life, Elsewhere in Galaxy

The Drake Equation $N = R_* \times f_p \times n_e \times f_l \times f_i \times f_c \times L$

N	The number of civilizations in the galaxy.
R_*	The number of stars forming each year in the galaxy.
f_p	The fraction of stars possessing planetary systems.
n_e	The average number of habitable planets in a planetary system.
f_l	The fraction of habitable planets on which life originates.
f_i	The fraction of life forms that develop intelligence.
f_c	The fraction of intelligent life forms that develop advanced technology.
L	The length of time, in years, that a civilization survives.

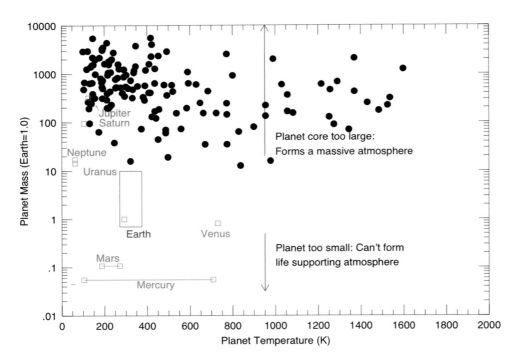

FIGURE 11 Habitable zone (large rectangle) in terms of planetary temperatures and planetary mass. The objects of the Solar System (open squares) are shown as well as the newly discovered extrasolar planets (filled circles).

atmosphere that kept Mars warm has dissipated, presumably into carbonate rocks located on the floor of ancient lake and ocean basins on Mars. Mars lacks plate tectonics because it is too small, 10 times smaller than the Earth, to maintain the active heat flows that drive tectonic activity. The low gravity of Mars and the absence of a magnetic field also contributed to the loss of its atmosphere. Hence, planetary size and its effect on geological activity also play a role in determining the surface temperature and thereby the presence of liquid water and life. Figure 11 shows the habitable zone for a planet in terms of its surface temperature and mass. The planets of the Solar Systems and some of the extrasolar planets discovered as of 2006 are shown.

9. Conclusion

Life is a planetary phenomenon. We see its profound influences on the surface of one planet—the Earth. Its origin, history, present reach, and global scale interactions remain a mystery primarily because we have only one datum. Many questions about life await the discovery of another life form with which to compare. Mars in its early history is probably the best prospective target in the search for extraterrestrial lifeforms, although Europa and Enceladus are also promising candidates because of the likely presence of liquid water beneath a surface ice shell and the possibility of associated hydrothermal vent activity. In any case, it is likely that our true understanding of life is to be found in the exploration of other worlds—both those with and without life forms. We've only just begun to search.

Bibliography

Davis, W. L. and McKay, C. P. Origins of Life (1996). A comparison of theories and application to Mars. *Origins Life Evol. Biosph.* **26**, 61–73.

Goldsmith, D. (1997). The Hunt for Life on Mars, Penguin, New York.

Klein, H. P. (1979). The Viking mission and the search for life on Mars. *Rev. Geophys. and Space Phys.* **17**, 1655–1662.

Knoll, A. H. (2003). Life on a Young Planet: The First Three Billion Years of Evolution on Earth. Princeton University Press, Princeton.

Lederberg, J. (1960). Exobiology: Approaches to life beyond the Earth. *Science* **132**, 393–400.

Lehninger, A. L. (1975). Biochemistry. Worth, New York.

McKay, C. P. (2004). What is life and how do we search for it on other worlds? *PLoS Biol* **2**, 1260–1263.

Miller, S. L. (1992). The prebiotic synthesis of organic compounds as a step toward the origin of life. *In* Major Events in the History of Life (J. W. Schopf, Ed.). pp 1–28. Jones and Bartlett Publishers, Boston.

Shapiro, R. (1986). Origins: A Skeptics Guide to the Creation of Life on Earth, Summit Books.

Tice, M. M. and Lowe, D. R. (2004). Photosynthetic microbial mats in the 3,416-Myr-old ocean. *Nature* **431**, 549–552.

Planetary Exploration Missions

James D. Burke

The Planetary Society
Pasadena, California

CHAPTER **46**

1. Introduction

Immediately upon launching *Sputnik* in 1957, it was clear that technical and political conditions would soon permit humans to realize a dream of centuries—exploring the Moon and planets. With large military rockets plus advanced radio techniques and the dawning skills of robotics, it would be possible eventually to send spacecraft throughout the solar system.

At first, however, the effort mostly failed. Driven by Cold War desires to show superiority in both military and civil endeavor, the Soviet and US governments sponsored hectic attempts to penetrate deep space, using strategic-weapon boosters, cobbled-together upper rocket stages and hastily prepared robotic messengers. In time, as the equipment became more reliable and the management more capable, successes came—but in-flight failures have continued for decades to afflict all deep space programs. Lunar and planetary exploration is barely achievable even with the finest skills.

Here, where our purpose is to trace the development of flight missions, we do not dwell on the failures. The accompanying tables list only those missions that yielded some data in accord with their objectives.

In the early years, the Soviet Union garnered all of the main firsts: the first escape from Earth's gravity, the first man and first woman in orbit, the first lunar impact, the first lunar landing and the first lunar orbit. But the US program came from behind and scored the first data from a planet, Venus, and ultimately the grand prize, the first human exploration of the Moon.

Though Cold War rivalry provided emotional stimulus and government support, both programs were scientific right from the start. The earliest satellites were launched in support of the International Geophysical Year. Every mission carried some instruments to elucidate the character of its target body or region, and this largely continued as more nations and agencies joined the program. As a result, there is now a huge body of data, some of it still unexamined, from flight missions complementing an important archive of ground-based and Earth-orbiting telescopic observations of the Moon, planets, and small bodies in the solar system. In what follows, mission results will be briefly mentioned, with cross references to more extended treatments in other chapters.

Exploration of the Sun's domain by robots and at the Moon by humans has now placed us in a position to build strong hypotheses about the origin and evolution of the solar system and also to begin the study of other such systems as they are discovered. The missions that made this possible are an unprecedented expression, on a grand international scale, of peaceful human values and achievement.

2. Program Evolution

2.1 Launch Services

Sputnik, orbited on 4 October 1957, galvanized a huge response from the United States. Less than 12 years later, two astronauts walked on the Moon. However, in both the USSR and the USA, it was an existing legacy that enabled launch of the first satellites in 1957 and 1958. Strategic weapons programs had had high priority in both nations for many years. *Sputniks*, *Explorers*, and *Discoverers* were launched on early versions of intermediate-range and intercontinental ballistic missile boosters. With modifications and increasingly powerful upper stages added, these boosters have continued to serve in both programs, up to the present day, for sending spacecraft out into the solar system.

Today, while the Russian *Soyuz* and American *Atlas* and *Titan* carry on as direct descendants of the early ICBMs, they are accompanied by *Delta* (an IRBM derivative but later vehicles with the same name are wholly new) the air-launched *Pegasus*, and a whole suite of ex-Soviet vehicles able to launch both smaller and larger robotic spacecraft beyond LEO. The space shuttle was also briefly used as a planetary mission launch vehicle for a period in the late 1980s and early 1990s.

In time, space mission developers in other nations, driven primarily by a desire to have assured, independent access to space but also by a desire for their own organic technology advancement, began to provide their own launch services, at first for low Earth orbit (LEO) missions and later for missions beyond LEO, including geosynchronous (GEO), lunar, interplanetary, and planetary ventures.

In Europe, after some false starts with missile-derived vehicles, the unique Ariane series of rockets, designed exclusively for space, began and has now led to the creation of the powerful *Ariane V*, capable of sending multiton payloads into geosynchronous orbit and beyond.

In Japan, two separate lines of vehicles were developed, one by the Institute of Space and Astronautical Science (ISAS, primarily for science) and one by the National Space Development Agency (NASDA, primarily for applications and technology). ISAS and NASDA are now parts of the Japan Aerospace Exploration Agency (JAXA).

In China, the *Long March* vehicle series began with Soviet-derived technology but soon diverged into a more indigenous form. In India, launch vehicles were developed for both LEO and GEO applications missions.

The first LEO missions with human crews were launched by Soviet and American ICBM-derived rockets. But when it came time to send humans beyond LEO, far larger vehicles were needed. The Moon Race of the 1960s saw the creation of the giant *Saturn V* and *N*-1 (Fig. 1). Both of them have now passed into history.

Following the end of *Apollo* and its Soviet lunar competitor (which never flew successfully), both nations fell back to LEO for human missions and both developed partly reusable launch systems intended to service space stations—the American space shuttle and the Soviet/Russian *Buran*. The shuttle has carried many American human missions into LEO, but *Buran* flew only once, without crew, and was then mothballed. The ancient and reliable, expendable *Soyuz* booster continues to deliver crews, equipment, and supplies to the International Space Station (ISS), a successor to the American Skylab and the Soviet and Russian Salyut and MIR stations.

The search for lower cost launch services, regarded as a key to future space development, has led over decades to the spending of resources equaling billions of dollars in studies and aborted vehicle developments, with as yet no promising result. However, work continues on a variety of approaches including air launch, hybrid air-breathing and rocket propulsion, and alternatively just extreme simplification in booster design.

Even without a radical launch cost reduction, a human breakout into the solar system is conceivable through the use of extraterrestrial resources. With energy and especially materials collected off Earth, in a manner that has come to be called in-situ resource utilization (ISRU), great savings are possible in the mass that must be lifted from Earth. However, this technique has yet to be demonstrated at a large enough scale for its true potential and its real comparative costs to be known.

2.2 Tracking and Data Acquisition

Without some way of delivering robotic mission results to Earth, it does not matter what else works or does not work. In the time before the invention of radio, space science fiction authors assumed that signaling with light beams would be used. In a way they were right: Optical communications using lasers may yet become the method of choice in certain applications. Meanwhile, however, telemetry, tracking and orbit determination, command, and science in deep space are entirely dependent on radio technique.

For the first satellites, tracking stations were improvised based on previous military communications systems. For missions to the Moon and beyond, however, it was necessary to adapt methods used by strategic defense radar developers and radio astronomers. Huge antennas, supersensitive receivers, transmitters with enormous power output, and advanced data recording and processing all were needed.

From the outset, a difference in philosophy guided Soviet and American deep space engineers. In the then secretive USSR, the initial plan was to have spacecraft turn on their transmitters only when over Soviet territory, thus requiring ground stations in only the eastern hemisphere. (In response to that, an American deep space signals intercept site was built in Eritrea.) In the US, on the other hand,

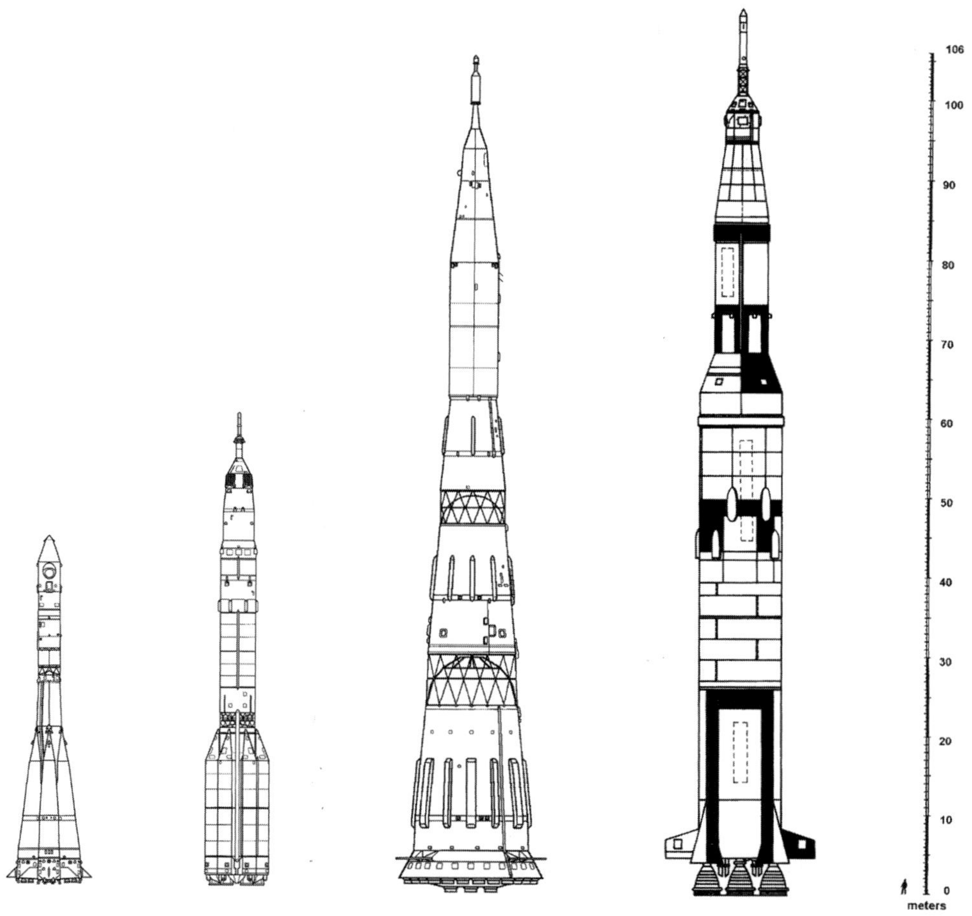

FIGURE 1   Saturn V and N-1: The two largest rockets ever built.

the policy called for continuous contact, meaning that stations would have to be located worldwide, with of course a worldwide ground and space communications system for command, control and data acquisition. That led to the creation of the Deep Space Network (DSN) whose stations today are in California, Spain, and Australia. For *Apollo*, a dedicated network was built, and it was backed up by the DSN plus a few specially equipped radio astronomy sites.

Meanwhile the Soviet system evolved. At first located only in the Crimea, the Soviet network expanded to include sites in the Far East and in the central USSR, plus a fleet of tracking ships offshore. As additional nations joined in exploring the solar system and the cosmos beyond, many more stations were built for both tracking and radio science. Figure 2 shows three examples of the modern deep space stations that now exist in several countries.

2.3 Spacecraft

Sputnik 1 was little more than a ball of batteries plus a beeping transmitter radiating at a frequency that most radio amateurs could tune in. But *Sputnik 2* carried the dog Laika. *Sputnik 3* was, for its time, a large scientific observatory outfitted to investigate the environment just outside Earth's atmosphere. The American *Explorers*, though much smaller, also carried scientific instruments, including the radiation counters that confirmed the existence of trapped charged particles in the Van Allen belts. From that modest beginning, robotic spacecraft in LEO and GEO have evolved into the thousands of diverse science and applications machines that have been sent into orbit. Among these are large, multifunction craft devoted to observing Earth as a planet, such as the European *Envisat* and the American *Terra* and *Aqua*.

Meanwhile, spacecraft designed to explore the solar system beyond Earth underwent a similar evolution. The most important early mission was that of *Luna 3* in 1959, ending centuries of speculation by returning the first images of the Moon's far side. Soon after that, spacecraft design began to elaborate on the features that are essential in interplanetary space: attitude stabilization for pointing cameras and high-gain antennas, capable onboard data handling systems, long-duration power supplies, and long-surviving electronic equipment.

FIGURE 2   Deep Space Network (DSN) stations today are in
(a) California, (b) Spain, and (c) Australia. (Courtesy of NASA).

For human spaceflight, the earliest craft were mainly just capsules capable of sustaining life and returning safely to Earth. But as space stations in LEO and flight beyond LEO became program objectives, more functions became the responsibility of human pilots and other crew members. The *Apollo* and space shuttle designs took full advantage of human capacities, while Soviet missions continued to make more use of teleoperation and onboard automation, as shown by the pilotless flight of *Buran* and the routine automated dockings of *Soyuz* and *Progress* servicing craft with the ISS.

Today, deep space spacecraft design and development is a mature activity as shown by the success of *Soviet Venus* landers, *Apollo* Moon missions, the decades-long *Pioneer* and *Voyager* missions to Jupiter and beyond, the missions to Halley's Comet in 1986, *Galileo* to Jupiter, *Cassini/Huygens* to Saturn, and the fleet of orbiters, landers, and rovers now exploring Mars. But in-flight failure, as in seven Mars

attempts since 1992, is an ever-present threat requiring vigilance and entailing high costs of spacecraft development and operations.

2.4 Operations

In even the earliest lunar and planetary missions, it was necessary to keep track of the spacecraft's trajectory and issue commands for onboard functions both engineering and scientific. Gradually a humans-and-machines art developed, represented today by large rooms full of people and displays backed by buildings full of computers and data systems. Initially centered in main theaters, as missions have become more complex, these facilities have become dispersed, providing work spaces for the many specialized flight management and scientific teams working during a mission. With the Internet and other modern communications available, scientists can now reside at their home institutions and participate in missions in real time.

The latest trend is toward increasing onboard autonomy, which holds the promise of reducing the large staffing needed round the clock to control missions. Some degree of autonomy is needed anyway in deep space, simply because of the round-trip signal times to distant spacecraft, tens of minutes for Mars and Venus, and many hours in the outer solar system.

Operations have become more and more dependent on software whose design and verification now constitute one of the main cost items in each new mission's budget. With the maturing of the operations art have come numerous stories of remarkable rescues when a distant robot (or, as in *Apollo 13*, a human crew) got into trouble, but there are also instances where a mistake on Earth sent a mission to oblivion.

2.5 Reliability and Quality Assurance

A vital part of the deep space exploration art is the creation of systems having but a small chance of disabling failures, plus an ability to work around failures when they do occur. One reason for the high cost of lunar and planetary missions is the need for multiple levels of checking, testing, reviewing, and documentation at every stage from the manufacturing of thousands of tiny components, through assembly into subsystems and systems for both ground and flight, organization of human teams capable of imagining and analyzing failure scenarios and designing around them, and finally launching and controlling a mission during its years or decades of activity.

These costs are aggravated by the nature of deep space exploration as a work of building very complicated things (hardware, software and human-machine complexes) in ones and twos, as distinct from the repetitive manufacture of highly reliable items such as cars or computers whose teething troubles can be eliminated in early prototype testing. In a sense, every lunar or planetary mission is a first effort.

2.6 Management

In the 20th century, as cold and hot warfare became more and more technological, a suite of skills, traditions, and managerial methods grew and created the capability of planning and executing large complicated projects. Many disciplines were involved, ranging from what became known as systems engineering all the way to new ways of organizing academic institutions, industries, and government agencies. The sometimes maligned worldwide military-industrial complex is a product of those developments, and it was the seedbed of the world's deep space programs.

The great lunar contest of the mid-20th century highlighted some stark differences between American and Soviet management methods and organizations. At the outset, both used existing military hardware and existing military ways of working, but over time the programs evolved along different paths. With their head start the Soviets garnered all the early prizes in robotic lunar exploration, but when planning began for human lunar exploration the Soviet system faltered.

Despite a huge and highly capable engineering and industrial base of talented and motivated people, the Soviet human flight lunar enterprise proved unable to solve problems of interagency rivalry and timely decision making, with the result that Apollo won the day. The USSR cut its losses and canceled its program, and *Apollo* soon followed because of pressure on the US federal budget and the lack of the political stimulus of Soviet competition. Decades then passed before lunar robotic exploration resumed, and more decades will pass before humans again bestride the Moon.

3. Sun and Heliosphere

The emphasis in this chapter is on missions to the Moon and planets. However, now that star-planet aggregates are at last being observed as a class of known objects in the cosmos, it is essential for us to include at least a part of the story of missions devoted to our own star as host of a planetary system.

Our tale begins with the International Geophysical Year (IGY). Centuries of ground-based investigations of sunspots and solar and terrestrial magnetism, plus decades of ionospheric and auroral research, had led by the mid-20th century to a drive by scientists for a worldwide campaign of coordinated measurements resembling previous efforts such as international polar years. The new element now was the knowledge that rockets could take instruments beyond Earth's atmosphere and even into orbit. In both the USSR and the US, satellite experiments were planned and announced in support of this goal, and in 1957 and 1958 it was achieved.

Explorer I found an excess of radiation saturating its detector. *Explorer IV* showed that this radiation is due to energetic particles trapped in Earth's magnetic field, the Van Allen belts. Then, in 1962, an instrument aboard *Mariner II*, en route to Venus, confirmed predictions of a fast outward flow of plasma from the Sun—the solar wind, now known to bathe the entire solar system out to the boundary of the heliosphere, where it meets the oncoming, tenuous interstellar medium. *Voyager 1* and *2* are now entering that interaction region, more than 90 AU from the Sun. Over the next 5 to 10 years, they are expected to continue to yield information on phenomena at the outer limits of the Sun's domain.

Meanwhile, over the past five decades, many spacecraft have journeyed into interplanetary space, investigating the particles and fields environment of the solar system or the Sun itself. Some were errant vehicles from planetary misses. The notable *Pioneer* series began in 1959 and continued in the late 1960s. The first international solar mission, *Helios*, a U.S.–German cooperative mission, with interplanetary spacecraft observing the solar wind and radiation, was launched in the mid 1970s (see below).

Now the Sun is continuously observed from space. Currently operating missions include *Ulysses*, *SOHO*, and *ACE*, plus particles-and-fields instruments carried on some new planetary missions. In the aggregate, as described in the chapter on the heliosphere, these investigations have shown a common portrait, with variations, of what happens as the Sun's streaming plasma, coronal mass ejections and electromagnetic radiations interact with the magnetic fields, ionospheres, and atmospheres and surfaces of solar system bodies. These effects are most dramatic when they result in spectacular comet ion tails, but they are also important in causing magnetic storms and driving the evolution of atmospheres due to dissociation of molecules and ionization and sweeping away of atoms.

Study of these interactions as they are imagined to have happened in the ancient past, for example when our star is thought to have gone through a hugely energetic T Tauri phase, enables not only analyses of early planetary history here but also productive reasoning about what may be observed in other star-planet systems as they are found.

Over the history of spaceflight many space-borne investigations, for example surveys of Earth's magnetosphere by missions including *ISEE-2* and *3*, *Interbol*, *Geotail*, *Wind*, *Polar*, and *Cluster* have added to knowledge of the Sun via its interactions with the rest of the solar system. Here we do not dwell on those ventures; instead we focus on missions dedicated to investigating the Sun itself as a star, with improved planetary magnetospheric, ionospheric, and atmospheric knowledge being extra benefits. It is appropriate, however, to observe that the long tradition is vigorously continuing with the worldwide International Heliosphere Year (IHY) due to begin in 2007.

Pioneer 6, 7, 8, 9

These missions, making ingenious use of the technology of their time, employed small spinning spacecraft to obtain a rich harvest of data on the solar wind and other interplanetary phenomena over a period beginning in 1965 and continuing for more than 30 years. (See http://samadhi.jpl.nasa.gov/msl/QuickLooks/pioneer6QL.html)

Helios

Two German spacecraft, launched by NASA *Titan-Centaurs* in 1975 and 1976, explored solar phenomena between Earth's orbit and as close as 0.29 AU from the Sun. An arrangement of mirrors and radiators enabled the spinning spacecraft to survive the consequent extreme heating. (See http://www.linmpi.mpg.de/english/projekte/helios/

Isee-3

Launched in 1978, the *International Sun-Earth Explorer* was a small spacecraft maneuvered into a halo orbit around the L1 libration point, 1.5 million km sunward from Earth, where its x-ray and gamma-ray spectrometers enabled the study of both solar flares and cosmic gamma-ray bursts. In 1982 it was maneuvered onto a trajectory toward Comet Giaccobini-Zinner and renamed the *International Cometary Explorer*. (*See* SMALL BODIES section below.) (See heasarc.gsfc.nasa.gov/docs/heasarc/missions/isee3.html# instrumentation.)

Solar Maximum Mission

Launched in 1980 by Space Shuttle, *SMM* carried a suite of instruments investigating the Sun at the height of the sunspot cycle. Ultraviolet, x-ray, gamma ray, and visible light observations combined to give a picture of the Sun's total radiation and its variations due to flares. The spacecraft failed and was dramatically rescued by a shuttle crew in 1984, whence it continued until atmospheric reentry in 1989. (See umbra.nascom.nasa.gov/smm/.)

Ulysses

Launched in 1990 by the space shuttle with propulsion beyond LEO to send it to Jupiter, ESA's *Ulysses* used the giant planet's gravity to kick its orbit out of the plane of the ecliptic and send the spacecraft back inward, passing over the Sun's poles to survey a region never before explored. Now the craft goes out to the distance of Jupiter's orbit and back to the Sun every five years. Its mission is expected to continue until at least 2007. In addition to its huge yield of information about the Sun, solar magnetism, and the solar wind, Ulysses has observed interstellar dust and interstellar helium atoms in interplanetary space. (See helio.estec.esa.nl/ulysses/.)

Yohkoh

Launched in 1991 from Kagoshima, this mission of ISAS, with contributions from the US and UK, was an x-ray and gamma-ray observatory that gave 10 years of nearly continuous imaging of the solar atmosphere. (See solar.physics.montana.edu/sxt/.)

SOHO

The ESA/NASA Solar and Heliospheric Observatory, launched by an American *Atlas-Centaur* in 1995, orbits about the L1 Lagrangian libration point 1.5 million km

FIGURE 3   The ESA/NASA Solar and Heliospheric Observatory (SOHO) orbits about the L1 Lagrangian libration point 1.5 million km sunward from the Earth, where its 14 instruments continuously observe phenomena relevant to understanding the solar interior, the solar atmosphere, and the solar wind.

sunward from the Earth, where its 14 instruments continuously observe phenomena relevant to understanding the solar interior, the solar atmosphere, and the solar wind (Fig. 3). SOHO's observations are immediately fed to users via the Internet at umbra.nascom.nasa.gov. The mission has already made observations through most of an 11-year solar cycle, and it is expected to continue for several more years. It too survived a massive onboard failure with a dramatic rescue—this time by remote control from Earth. (See: sohowww.estec.esa.nl.)

ACE

The *Advanced Composition Explorer*, a NASA mission with nine instruments and an international team of 20 investigators, was launched by a *Delta II* vehicle in 1997. Like SOHO, it orbits in the L1 region where it continuously surveys the isotopic and elemental composition of particles from the solar corona, the interplanetary medium and interstellar space. In 1998, the ACE data system began providing public, real-time observations that can give warning of solar events that cause geomagnetic storms. (See www.srl.caltech.edu/ACE/,)

TRACE

A small *Explorer* satellite launched in 1998 by the innovative air-launched *Pegasus* rocket system, TRACE provides nearly continuous solar coronal observations with high spatial and temporal resolution, complementing the data from SOHO. (See umbra.nascom.nasa.gov/trace/.)

Genesis

In an audacious venture using gravity assist at Earth and libration orbiting for two years near L1, the *Genesis* mission, launched by a *Delta II* in 2001, in 2004 returned a capsule to Earth bearing actual samples of the solar wind and interplanetary medium embedded in ultraclean collector plates. Due to a failure to signal its parachute to open, the capsule crashed in the Utah desert, but not all was lost: A number of the collector units survived in condition good enough for the recovery of isotopic information and other science data. (See: genesis.lanl.gov.)

RHESSI

Launched by *Pegasus* in 2002, the Reuven Ramaty High Energy Solar Spectroscopic Imager is a small *Explorer* spacecraft dedicated to x-ray and gamma-ray observations for exploring the basic physics of particle acceleration and energy release in solar flares. (See hesperia.gsfc.nasa.gov/hessi/index.html.)

4. Mercury

Mariner 10

Flight to the innermost planet began with *Mariner 10*, launched on 3 November 1973 by an Atlas-Centaur (Fig. 4). It was the first mission to use gravity assist, flying by Venus on 5 February 1974 enroute to Mercury, where it arrived on 29 March. Then using Mercury gravity assist, it flew by again on 21 September 1974 and 16 March 1975, each time passing over the same side of the planet. *Mariner 10*'s images showed a scorched, Moon-like cratered surface, while its infrared and ultraviolet spectrometers recorded mineral composition and its magnetic and plasma instruments surveyed Mercury's surroundings, revealing a weak magnetic field. Precise trajectory analysis

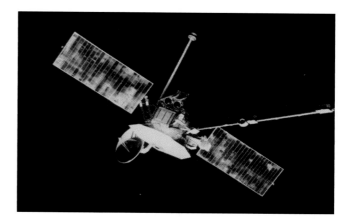

FIGURE 4   Flight to the innermost planet, Mercury, began with *Mariner 10*, launched on 3 November 1973 by an Atlas-Centaur.

confirmed that Mercury has a huge iron core reaching to two thirds of its outer diameter. (See nssdc.gsfc.nasa.gov/nmc/tmp/1973-085A.html.)

Messenger

The *Messenger* spacecraft, launched by a *Delta II* on 2 August 2004, will enter obit about Mercury in 2011 after an Earth gravity assist in 2005, Venus gravity assists in 2006 and 2007, then three Mercury assists in 2008 and 2009. The spacecraft carries a suite of instruments to investigate Mercury's surface and interior composition, its gravity and magnetic fields, its particles and radiation environment and the polar regions where Earth-based radar observations show the possible presence of ices in permanently shadowed craters. (See messenger.jhuapl.edu/.)

5. Venus

Mariner 2

The first mission to return data from another planet, *Mariner 2* in 1962, had amazing escapes from disaster. During ascent its *Atlas* went into uncontrolled rolling and miraculously stopped in an orientation such that the *Agena* upper stage could deliver the spacecraft onto a trajectory toward Venus. En route, the spacecraft survived a series of mortal threats, and shortly after flying by Venus it succumbed to overheating. But during the flyby, as described in the Venus chapters of this encyclopedia, it produced proof of the planet's hellish greenhouse. (See nssdc.gsfc.nasa.gov/nmc/tmp/1962-041A.html.)

Veneras 4 through 16 and Vega

First to enter another atmosphere, *Venera 4* in 1967 carried the emblem of the USSR to Venus. It began the Soviets' most successful interplanetary program. As shown in the table, *Venera* missions of increasing complexity and scientific yield continued to be launched at nearly every celestial mechanics opportunity until 1983, and then in 1985 the two VEGA spacecraft, en route to Halley's comet, delivered balloons into the Venus atmosphere. Scientific results of this decades long exploration are described in the Venus chapter. (See www.russianspaceweb.com/spacecraft_planetary_venus/.)

Mariner 5

Launched two days after *Venera 4* in 1967, *Mariner 5* made flyby observations, including ultraviolet cloud imaging, that revealed the rapid rotation and spiraling equator-to-pole circulation of the Venusian atmosphere. (See nssdc.gsfc.nasa.gov/nmc/tmp/1967-060A.html.)

Mariner 10

During its gravity assist flyby of Venus in 1974 en route to Mercury, *Mariner 10* made observations of the Venusian atmosphere and ionosphere, confirming the equator-to-pole circulation and absence of a magnetosphere. (See nssdc. gsfc.nasa.gov/nmc/tmp/1973-085A.html.)

Pioneer Venus

The two *Pioneer Venus* spacecraft, launched in 1978, had complementary objectives. *Pioneer Venus 1* went into orbit with a radar altimeter to survey the surface though the planet's permanent cloud cover. *Pioneer Venus 2* delivered four probes into the atmosphere to measure its character and composition down to the surface. (See nssdc.nasa.gov/planetary/pioneer_venus.html.)

Vega 1 and 2

Two large Soviet spacecraft *Vega 1* and 2 (Fig. 5) flew by Venus in 1985 en route to close encounters with Halley's comet. Their spherical entry capsules released balloons that were inflated and floated in the Venus atmosphere, returning data for several days. (See www.iki.rssi.ru/IPL/vega.html.)

FIGURE 5 Vega 1, a large Soviet spacecraft, flew by Venus in 1985 en route to Halley's comet.

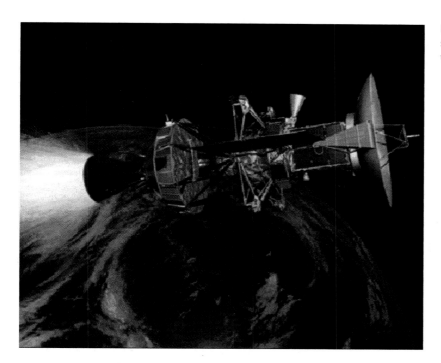

FIGURE 6 *Magellan* was launched in 1989 into a series of orbits enabling it to map the entire planet using synthetic-aperture radar.

Magellan

The ubiquitous clouds of Venus forever hide the planet's surface from outside visual examination. *Venera* landers in 1975–1981 gave close-up surface panoramas and in 1983 radars on the *Venera 15* and *16* orbiters mapped most of the northern hemisphere. Long delayed through years of attempts to gain government approval, *Magellan* (Fig. 6) was finally launched in 1989 into a series of orbits enabling it to map the entire planet using synthetic-aperture radar. Once the radar mission was complete, the spacecraft was moved into a lower orbit to map the Venusian gravity field and to test aerobraking techniques. (See www.jpl.nasa.gov/magellan/.)

Galileo Venus Flyby

En route to Jupiter, the *Galileo* spacecraft performed a gravity assist flyby at Venus in February 1990. Spacecraft observations included infrared imaging of the planet's cloud layers and even surface features, through infrared "windows" in the atmosphere and clouds. (See galileo.jpl.nasa.gov/facts.cfm.)

Venus Express

By modifying the design to cater for the hot environment near Venus, but otherwise using many proven components and operational techniques, ESA was able to mount a low-cost mission to place in Venus orbit a spacecraft based on the successful *Mars Express* to be described later below. Launched by a Russian *Soyuz-Fregat* in 2005, the mission has delivered unique images of Venus's north polar cloud vortex. (See sci.esa.int/venusexpress/.)

6. Earth

Among the thousands of spacecraft launched to date, at least hundreds have made some contributions to the study of our Earth as a planet. Here we make no attempt at a catalog of all those ventures. Instead we highlight a few recent and representative missions that illustrate the state of humans' ongoing endeavor to understand Earth's interior, its oceans and lands, its atmosphere, its evolution and its fate, including that of its biosphere.

Resurs

Soviet and Russian film-return photo-reconnaissance satellites have operated over many years for Earth observation. Civil uses have been publicized since 1979, with increasingly capable camera systems used for both applications and science.

Corresponding US imagery was mostly kept classified until 1995, when much previously secret overhead reconnaissance information was released for public use in historical and scientific studies. (See www.fas.org/spp/guide/russia/earth/resurs-f.)

Galileo Earth Flybys

En route to Jupiter (see Outer Planets section below) the *Galileo* spacecraft made gravity assist passes at Earth in 1990 and 1992. Spectrometric observations were made to simulate a search for evidence of life on an unknown planet, and the data did show an out-of-equilibrium, oxygen-rich atmosphere. (See galileo.jpl.nasa.gov/facts.cfm.)

Terra

Launched in 1999, NASA's *Terra* spacecraft carries five advanced radiometric and spectrometric instruments observing global phenomena of land, oceans and atmosphere. Measuring Earth's radiation budget, its carbon cycle and evolution of its climate and biosphere are main mission goals. (See terra.nasa.gov/.)

Topex/Poseidon and Jason-1

Launched in 1992 and 2001 respectively as parts of a collaboration between NASA and the French national space agency CNES, *Topex/Poseidon* and *Jason-1* use radar altimetry and very precise orbit determination to determine ocean topography, aiding studies of currents, winds and climate effects including El Niño. (See topex-www.jpl.nasa.gov.)

Grace

In a collaboration among NASA, the German space agency DLR and other partners, two small satellites, *Grace*, launched in 2002 use very precise measurements of the distance between them to gain knowledge of the bumps and hollows in Earth's gravity field, leading to information on the exchanges of mass, momentum and energy between oceans and atmosphere. (See www.csr.utexas.edu/grace/.)

Envisat

ESA's 8200-kg Earth observing satellite, *Envisat*, launched in 2002, carries ten large instruments including a synthetic aperture radar, a radar altimeter and a suite of radiometers and spectrometers recording atmospheric, ocean, ice, land and biosphere data, spanning the spectrum from ultraviolet to microwave frequencies. Its polar orbit gives global coverage. (See www.esa.int/export/esaEO/.)

Aqua

NASA's *Aqua* satellite, launched in 2002, carries six radiometric and spectrometric instruments surveying Earth's water cycle, sea and land ice, atmospheric temperature, aerosols and trace gases, and soil moisture, so as to increase understanding of climate and Earth's radiation balance, with both physical and biological influences. (See science.hq.nasa.gov/missions/satellite_aqua/.)

Aura

Launched in 2004, the *Aura* satellite's four instruments complement those of *Terra* and *Aqua* by measuring atmospheric chemistry, including the formation and dissipation of polar ozone holes and the distribution of greenhouse gases. (See science.hq.nasa.gov/missions/satellite_aura/.)

Other Recent Earth Observing Missions

In addition to the major efforts noted here, a host of other orbital remote-sensing missions investigating Earth as a planet with its evolving hydrosphere, cryosphere, atmosphere, biosphere, and magnetosphere have been launched in recent years. Summaries are given at the following Web site with links to pages describing each mission in more detail: science.hq.nasa.gov/missions/earth_sun/.

7. Moon

After centuries of careful naked-eye and telescopic observation from Earth, the Moon has at last become a body to be investigated by robots, visited by human explorers, and perhaps ultimately inhabited by the people of a first outward wave of civilization. At its beginning, scientific lunar exploration was caught up in the great 20th century struggle between the USA and the USSR. With the end of the USSR, the program fell victim to low priority and languished for decades, but now a lively international revival is in progress. Here we list the most important robotic missions of the past, then briefly mention the grand *Apollo* venture and its failed Soviet competitor, and finally remark on the new missions now established in a widening group of countries.

Luna 1, 2, and 3

The *Luna* Soviet missions in 1959 yielded the first escape from Earth's gravity, the first lunar impact, and the first far-side images. (See nssdc.gsfc.nasa.gov/database/MasterCatalog?sc=1959-008A.)

Ranger 7, 8, and 9

After two nonlunar tests and three failed attempts to deliver seismometers to the lunar surface, the NASA *Ranger* missions, launched by *Atlas-Agenas* in 1964 and 1965, yielded thousands of high-resolution television images of the lunar surface showing that all features are mantled by the impact-generated regolith. (See nssdc.gsfc.nasa.gov/planetary/lunar/ranger.html.)

FIGURE 7 *Luna 9*, a Soviet spacecraft, achieved history's first successful lunar touchdowns, delivering image panoramas showing fine surface details.

Zond 3

A Soviet planetary spacecraft, *Zond 3*, launched on a test flight including a lunar flyby, this mission in 1965 returned improved imagery of parts of the Moon's far side. (See nssdc .gsfc.nasa.gov/planetary/lunar/lunarussr.html.)

Luna 9 and 13

After many Soviet lunar failures in 1960–1965, *Luna 9* and *13* in 1966 (Fig. 7) achieved history's first and third successful lunar touchdowns, delivering image panoramas showing fine surface details. (See selena.sai.msu.ru/Home/ Spacecrafts/Luna-9/luna-9e.htm.)

Luna 10, 11, 12, and 14

These Soviet missions, *Luna*, in 1966 and 1968 achieved the first entry into lunar orbit and made some measurements of lunar gravity and geochemistry. (See www.iki.rssi.ru/solar/ eng/luna14.htm.)

Lunar Orbiter 1–5

Designed to image landing sites on the Moon in support of Apollo, the first three of the *Atlas-Agena*-launched *Lunar Orbiter* NASA photographic missions were so successful that the last two were given the expanded task of mapping the entire Moon. (See www.hq.nasa.gov/office/pao/ History/TM-3487/top.htm.)

Surveyor 1, 3, 5, 6, and 7

NASA's *Surveyor 1*, launched by Atlas-Centaur, achieved the first lunar soft landing and returned television mosaics of its surroundings. In addition to imagery, the *Surveyors* in 1966 and 1967 yielded information on the mechanical and chemical properties of the regolith. (See nssdc.gsfc .nasa.gov/planetary/lunar/surveyor.html.)

Zond 5, 6, 7, and 8

The *Zond* Soviet spacecraft, launched from 1968–1970 by large Proton vehicles, flew on circumlunar trajectories, returning to Earth after passing over the Moon's far side. They were test flights for a never-completed human lunar flight program. Payloads consisted of environmental instrumentation and biological specimens including tortoises. The later flights demonstrated an ingenious skip re-entry, dipping briefly into the atmosphere over the Indian Ocean and then traveling on to land in central Asia. (See nssdc .gsfc.nasa.gov/database/MasterCatalog?sc=1970-088A.)

Apollo 8

When in 1961 US President John F. Kennedy called for starting *Apollo*, he had asked his advisors to describe a program in which "we can win" in competition with the USSR. Observation of Soviet lunar launch preparations and test flights led to a decision to send a human crew to the Moon as soon as possible. The risky *Apollo 8* mission in 1968 was the result. It went into lunar orbit with only the Command and Service Modules (CSM) because the lunar landing module (LM) was not yet available. Thus there was no prospect of saving the mission in "LM Lifeboat" mode as had to be done in *Apollo 13* (see below). The *Apollo 8* crew broadcast TV images and a Christmas voice message from lunar orbit, took photos, made visual observations, and returned safely to splashdown in the Pacific Ocean. (See www.lpi.usra.edu/expmoon/Apollo8/Apollo8.html.)

Apollo 10

In the final rehearsal for a lunar landing in 1969 (after *Apollo 9*'s successful Earth-orbiting test of the LM), the *Apollo 10* crew exercised all LM functions in low lunar orbit, rendezvoused with the CSM, and returned to Earth. (See www.lpi.usra.edu/expmoon/Apollo10/Apollo10.html.)

Apollo 11

Apollo 11, the mission that won the greatest peaceful international contest placed, on 20 July 1969, the first human footprints on the Moon. The LM crew gathered rock and soil samples and installed a set of long-lived instruments on

FIGURE 8   Astronaut Pete Conrad examines Surveyor 3's camera and soil-sampler claw in 1969.

the surface. Meanwhile, a photographic survey from the orbiting CSM covered landing sites for future missions. (See-www.hq.nasa.gov/office/pao/History/ap11ann/introduction .htm and nssdc.gsfc.nasa.gov/planetary/lunar.apollo11info .html.)

Apollo 12

An outstanding achievement in 1969 by the *Apollo 12* ground and flight crews is shown in Figure 8. Navigating to a landing within 170 meters of *Surveyor 3*, which had been sitting on the Moon for 31 months, the LM crew walked over to the *Surveyor*, cut off its camera and soil-sampler claw, and returned them to Earth. The mission also brought back a new harvest of rocks, soils, orbital and surface imagery, and other science data. (See www.lpi.usra.edu/expmoon/ Apollo12/Apollo12.html.)

Apollo 13

When the *Apollo 13* spacecraft was en route to the Moon in 1970, an oxygen tank in the service module exploded. The dramatic rescue of the mission during the following week is an epic tale of devotion and ingenuity by the ground and flight crews. Moving out of the crippled CSM into the LM, the crew used the LM descent engine to adjust their trajectory to a circumlunar return to Earth. In the midst of

the emergency, they even managed to obtain some lunar far-side photography. (See nssdc.gsfc.nasa.gov/planetary/ lunar/ap13acc.html.)

Apollo 14

Continuing to expand *Apollo*'s science capabilities, the 1971 Apollo 14 mission's surface exploration included a hand-drawn cart for carrying instruments. (See nssdc.gsfc .nasa.gov/planetary/lunar/apollo14info.html.)

Apollo 15, 16, and 17

During three *Apollo* missions in 1971 and 1972, human lunar scientific exploration showed its real potential. With augmented geological training of astronauts, plus one crew member a professional geologist, plus a rover to carry the LM crew on extended surface traverses, plus a suite of remote sensing instruments on the CSM, these missions yielded a cornucopia of information that is described in the Moon chapter. (See nssdc.gsfc.nasa.gov/planetary/ lunar/apollo17info.html.)

Luna 16, 17, 20, 21, and 24

During the *Apollo* years the USSR had three lunar programs. The first was the robotic science program that began

FIGURE 9   The *Luna 16* spacecraft.

in 1959 and continued with increasing capabilities until 1976. The second was the Proton-launched circumlunar *ZOND* (a name meaning sounder) human-precursor tests. The third was the human lunar landing effort based on the giant N-1 vehicle that failed in four launch attempts.

Lunas 16 through *24* were emissaries of the first program. The Proton-launched *Luna 16, 20,* and *24* (Fig. 9) drilled into the regolith, encapsulated small soil samples and returned them to Earth. *Luna 17* and *21* delivered Lunokhod rovers to the Moon's surface. (See nssdc.gsfc .nasa.gov/database/MasterCatalog?sc=1976-081A.)

Clementine

The mission that revived lunar exploration in 1994 after its decades of stasis, *Clementine*, had an innovative management and technical plan. Proposed as a test of instrument technologies for the American Strategic Defense Initiative, it was sponsored by the Ballistic Missile Defense Organization and NASA, managed by the Naval Research Laboratory, and launched from the Pacific Missile Range on a Titan II-G.

During two months in lunar orbit, it mapped the entire Moon at many wavelengths and hinted at the presence of theoretically predicted excess volatiles, possibly a signature of cold-trapped water ice near the lunar poles. (See www.cmf.nrl.navy.mil/clementine/.)

Lunar Prospector

Launched in 1998 by an Athena solid-fueled vehicle, the NASA *Lunar Prospector* continued the trend toward small, highly capable lunar spacecraft and relatively low mission costs. With neutron, gamma-ray, and alpha-particle spectrometers plus measurements of lunar magnetic and gravity fields, the mission yielded data on the Moon's surface composition and its geochemical and geophysical properties. It

added confidence to the *Clementine* findings of possible polar ices. (See lunar.arc.nasa.gov/project/index.htm.)

Smart-1

ESA's first lunar mission, *Smart-1*, was launched in 2003 with a small, highly advanced spacecraft demonstrating solar-electric propulsion, onboard autonomy, and several new instrument technologies. Spiraling slowly outward from Earth and then inward toward the Moon, the craft was captured by the Moon's gravity late in 2004 and began science operations in lunar orbit in 2005, whence it delivered a fine harvest of imaging and other remote-sensing data until its planned crash into the Moon on 3 September 2006. (See www.esa.int/SPECIALS/SMART-1/.)

New Lunar Missions

Continuing the worldwide revival of interest in the Moon, several robotic lunar orbiting missions are being prepared for launch: Japan's *Lunar-A* for seismic penetrators and *SELENE* for a broad set of remote-sensing objectives; China's *Chang'E-1* for remote sensing and surveying for later landing missions, and India's *Chandrayaan-1* for remote sensing. *SELENE* is an acronym.

In addition, NASA will execute lunar orbital missions both for science and in preparation for a new American space program employing the Moon as a stepping stone toward eventual human exploration of Mars. The first such mission is that of the *Lunar Reconnaissance Orbiter*, a large, multipurpose remote-sensing spacecraft to be launched in 2008 or 2009. Because its launch vehicle has excess payload capacity, the mission will also carry an experiment called LCROSS for observing a planned lunar crash of the vehicle's translunar injection stage. The program is intended to continue with robotic landers and rovers exploring the prospects for use of lunar resources, including the polar ice deposits if they do exist.

8. Mars

With 19th-century telescopic observation showing polar caps and other indications of an atmosphere and changing surface features, Mars became the planet of choice for speculation about other life in the cosmos and about human travel to other worlds.

These pervasive ideas have since driven planetary program priorities with the result that huge resources have been devoted to Martian robotic exploration and to studies of the prospect of human ventures to Mars. But Mars has proved to be a difficult destination: Failure has been an ever present hazard—not only in flight missions but also in the councils where budget decisions are made.

In what follows, we concentrate upon successes, but those must be seen as just the most visible parts of a remarkable, decades-long striving toward a possible breakout of humanity beyond the bounds of Earth.

In addition to the Web page listed for each mission below, a site with a brief story of every publicly acknowledged Mars mission is given at www.planetary.org/learn/missions/marsmissions.html.

Mariner 4

Mars launch opportunities occur about every 26 months. In both the USA and the USSR, the October 1960 window was the favored first chance. The Soviets did launch, with two upper stage vehicle failures. During the 1962 window, the Soviets tried three launches, one of which sent *Mars 1* toward the planet. That spacecraft failed en route. In 1964, NASA launched two *Atlas-Agenas* with one success. *Mariner 4* flew by Mars and returned 22 images of the cratered southern highlands, leading to the impression of a Moon-like Mars, proved false by later missions. (See www.jpl.nasa.gov/missions/past/mariner3-4.html.)

Mariner 6 and 7

Two Mars flyby missions, *Mariner 6 and 7*, launched by *Atlas-Centaurs* in 1969, demonstrated the rapid advance of deep-space data acquisition technology. Their imaging was greatly improved over that of *Mariner 4* in both quality and quantity, and in addition infrared spectrometry gave some first indications of Martian surface compositions. They still covered mainly southern, including polar, ancient landforms, omitting the vast volcanoes and canyons discovered by *Mariner 9*. (See www.jpl.nasa.gov/missions/past/mariner6-7.)

Mars 2 and 3

During the 1971 Mars window, the USSR and USA each launched two missions. The Soviet *Mars 2* and *3* orbiter/landers both arrived successfully into orbit at the planet; *Mars 2* returned some orbital science data but its lander crashed. *Mars 3*, in addition to its orbital operations, delivered its lander with a small tethered mobile platform. But the transmissions from the lander ceased only 20 seconds after touchdown. (See www.earthandspace.org/mars23.htm.)

Mariner 9

The *Atlas-Centaur* carrying *Mariner 8* failed but *Mariner 9* became the most rewarding Mars mission up to its time, waiting out a global dust storm in orbit and then sending imagery of most of the Martian surface until its mission ended in 1972, revealing enormous volcanoes, canyons, apparent river channel networks, sapping collapse features and

clouds, plus imagery of the two small moons, Phobos and Deimos. (See www.jpl.nasa.gov/missions/past/mariner8-9.)

Mars 4, 5, 6, and 7

At the 1973 opportunity, the Soviets made an all out effort to upstage the American *Viking* missions planned for 1975. They launched four large spacecraft, all of which arrived in the vicinity of Mars but each of which ultimately failed for a different reason. *Mars 4* failed to brake into orbit but did return some flyby data; *Mars 5* entered orbit, sent some images and failed after 22 days; *Mars 6* released a lander that failed during descent; *Mars 7*'s lander missed the planet. (See athena.cornell.edu/mars_facts/past_missions_70s.html.)

Viking 1 and 2

In 1975 two large NASA orbiter/landers, *Viking 1 and 2*, were launched by powerful *Titan-Centaurs*. Arriving in June and July 1976, they entered orbit and began surveying for landing sites. The *Viking 1* lander set down in Chryse Planitia on 20 July and the *Viking 2* lander descended to Utopia Planitia on 3 September on the opposite side of Mars. While the orbiters began imaging the whole planet and making spectrometric remote sensing observations, during descent the landers measured atmospheric composition. Then the landers began to operate a suite of instruments for imaging their surroundings and determining meteorological, geological and biological properties. At first, microbial activity was suspected, but eventually most scientists concluded that no life did or could exist in the soil samples. (See www.jpl.nasa.gov/missions/past_missions.cfm.)

Phobos 1 and 2

After a long pause in Martian exploration, in 1988 two large and complex Soviet spacecraft, *Phobos 1* and *2*, were launched by *Proton* vehicles toward the vicinity of Mars. *Phobos 1* was lost en route due to a human error in ground control. *Phobos 2* arrived and began phasing orbits for a rendezvous with the little moon, where it was to make close-up observations and deposit two small landing packages, one of them a hopping rover. Imagery and some other data of Phobos and Mars were obtained, but the spacecraft failed before the landings could occur. (See www.iki.rssi.ru/IPL/phobos.html.)

Mars Pathfinder and Mars Global Surveyor

The 1996 launch window saw the revival of detailed American exploration of Mars. NASA's *Pathfinder* delivered a lander and a small rover, named Sojourner, which explored nearby surroundings in the Ares Vallis outwash plain. The *Global Surveyor* spacecraft entered an eccentric orbit and was delicately aerobraked down into circular mapping orbit over a period of months, the long period being needed due to structural failure of the attachment of one solar panel. The mission has yielded a continuing stream of imaging and other data, revolutionizing scientists' knowledge and modeling of Martian geology and atmospheric processes. (See marsweb.jpl.nasa.gov/missions/.)

Failures of the 1990s

The years 1992, 1998, and 1999 saw three US missions fail during arrival at the planet: *Mars Observer*, *Mars Climate Orbiter*, and *Mars Polar Lander*. An elaborate international Russian mission's launch, *Mars-96*, failed in 1996— a series of events that led in the USA to a management overhaul and in Russia to the end of Mars exploration for the time being. (See marsweb.jpl.nasa.gov/missions/ and www.iki.rssi.ru/mars96/mars96hp.html.)

Mars Odyssey

Mars Odyssey, a NASA orbiter launched in 2001, is instrumented for measurements complementing those of the *Global Surveyor*. With infrared/visible, gamma-ray, and particle spectrometers, it produces thermal imaging enabling evaluation of surface physical properties, subsurface elemental chemistry, and the planet's radiation environment. *Odyssey's* findings have greatly stimulated interpretations of many of Mars's landforms as resulting from the action of subsurface briny water, ice and carbon dioxide. (See marsprogram.jpl.nasa.gov/odyssey/.)

Spirit and Opportunity

In an intense three-year effort, two NASA Mars rover missions, *Spirit* and *Opportunity*, were prepared for the 2003 launch opportunity. Both succeeded, and at the time of writing the two rovers are continuing to make astonishing discoveries in Meridiani Planum and Gusev Crater, on opposite sides of the planet, reinforcing the orbiters' findings of a history dominated by the effects of water. (See marsrovers.jpl.nasa.gov/home/index.html.)

Mars Express

ESA's *Mars Express* orbiter, launched on a Russian *Soyuz-Fregat* vehicle in 2003, delivered the small British *Beagle-2* lander, which failed, and has then gone on at the time of writing to yield excellent imaging, plus spectrometric measurements indicating, among other findings, that there is a correlation between regions of enhanced water vapor and methane concentrations in the atmosphere. Mars Express also carries a ground-penetrating radar for detetecting the signatures of subsurface brines and ices. (See sci.esa.int/science-e/www/area/index.cfm?fareaid=9.)

Mars Reconnaissance Orbiter

Launched in 2005 and delivered in 2006 into aerobraking orbit at Mars, *Mars Reconnaissance Orbiter* is expected to increase by orders of magnitude the quantity and quality of remote-sensing data from Mars, because of its powerful radio system and advanced on-board instruments and system software. Imaging already obtained, while excellent, gives only a small sample of the harvest to come.

Future Mars Missions

The exciting discoveries of the missions listed here and the ongoing debate over the prospect of human missions to Mars have continued to energize an active NASA program. Mars launches are planned for the 2007 and 2009 opportunities. *Phoenix* is a reflight of the failed *Mars Polar Lander*, and *Mars Science Laboratory* is intended to expand on the findings of the Mars rovers, *Spirit* and *Opportunity*. (See marsweb.jpl.nasa.gov/missions/.)

9. Small Bodies

As scientists have come to realize that comets and asteroids contain clues to the ancient history of the solar system— clues largely obliterated by geologic processes in planets and moons—missions to small bodies have increased in importance. Also studies of cratering and meteorite records show that near-Earth asteroids present both a threat and an opportunity. The threat is that of devastating impacts and the opportunity is that of useful resources not found in the Moon. (See www.permanent.com.)

Ice

After completing its solar mission as *ISEE-3* (see Sun and Heliosphere sections) the spacecraft was retargeted and renamed *International Cometary Explorer*. It flew through the tail of Comet Giaccobini-Zinner in 1985, then continued on in heliocentric orbit where it sent low-rate data for the next several years. (See heasarc.gsfc.nasa.gov/docs/heasarc/missions/isee3.html# instrumentation.)

The Halley Armada

As Halley's comet arrived near the Sun in 1986 on its 76-year orbit, it was met by spacecraft from Japan, Europe, and the USSR. Comet enthusiasts lamented the absence of the USA from this once-in-a-lifetime opportunity. Japan's *Suisei* and *Sakigaki* made distant observations of the ultraviolet coma. ESA's *Giotto* passed within 600 km of the nucleus collecting imaging, spectra, and detailed chemical data. The Soviet *VEGA 1* and *2* flew by at intermediate distances after their productive en route encounters with Venus (see

NEAR Spacecraft Configuration

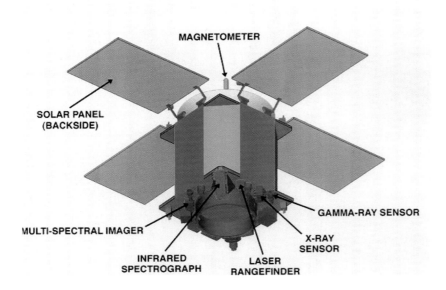

FIGURE 10   The NASA spacecraft *Near-Shoemaker* was launched in 1996.

Venus section). (See www.planetary.org/html/neo/Missions-Research/index.html.)

Galileo En Route Encounters

While en route to Jupiter on its long journey with gravity assists at Venus and Earth, the NASA *Galileo* spacecraft flew by two asteroids, 951 Gaspra in 1991 and 243 Ida in 1993, and obtained close-up imagery, spectra, and other measurements. A highlight of the Ida encounter was the discovery of the tiny moon Dactyl orbiting Ida. (See galileo.jpl.nasa.gov/facts.htm.)

NEAR-Shoemaker

Launched in 1996, the NASA spacecraft *Near-Shoemaker* entered orbit about asteroid 433 Eros in 2000, delivered imagery, spectrometric data, and gravitometric data. After one year in orbit it was commanded to a gentle touchdown, which it survived, even though not designed for landing (Fig. 10). (See near.jhuapl.edu/.)

Deep Space 1

The NASA craft *Deep Space 1* launched in 1998 to demonstrate solar-electric propulsion, autonomous navigation, and other new technologies, encountered asteroid 9969 Braille in 1999, though it only returned a few distant, low-resolution images. Its mission was extended to a close flyby of Comet Borrelly on 22 September 2001 and successfully

imaged the nucleus at visible and infrared wavelengths. (See nmp.jpl.nasa.gov/ds1/.)

Stardust

With the goal of collecting samples of cometary dust and returning them to Earth, NASA's *Stardust* mission, launched in 1999, flew by asteroid 5535 Annefrank in 2002, and encountered Comet Wild 2 in 2004, returning imaging data. The sample return capsule successfully parachuted to Earth on January 4, 2006, with its precious cargo of thousands of cometary (and interstellar) dust particles. (See stardust.jpl.nasa.gov/top.html.)

Hayabusa (Muses-C)

An ISAS mission with assistance from NASA, *Hayabusa* was launched in 2003 and used solar electric propulsion to rendezvous with asteroid 25143 Itokawa in September 2005. It returned multispectral imaging and gravity data and attempted to collect surface samples for return to Earth. The spacecraft is en route to Earth but technical problems have delayed arrival until 2010. (See www.isas.jaxa.jp/e/enterp/missions/hayabusa/index.html.)

Rosetta

Rosetta, an ESA mission launched in March 2004 with an *Ariane V*, is scheduled to arrive at Comet Churyumov-Gerasimenko in 2014 after three Earth gravity assists and one at Mars. It is also targeted to flyby asteroids 2867 Steins in 2008 and 21 Lutetia in 2010. (See rosetta.esa.int/science-e/www/area/index.cfm?fareaid=13.)

Deep Impact

With the goal of determining the physical and chemical makeup of a cometary nucleus, NASA's *Deep Impact* mission, launched in January 2005, successfully delivered a 370-kg projectile to a 10 km/s collision with Comet Tempel 1 on July 4, 2005. Imaging from the impactor and the flyby spacecraft returned the highest resolution pictures of a comet to date and documented the impact event which provided new insights into the nature of cometary nuclei. (See deepimpact.jpl.nasa.gov/mission/facts.html.)

Dawn

Dawn, an ion-propelled spacecraft to be launched in 2007 (after its mission was cancelled and quickly reinstated in 2006) will investigate the surface and interior properties of Vesta and Ceres, the two large asteroids that telescopic observation shows to be quite different from each other. (See dawn.jpl.nasa.gov.)

10. Outer Planets and Moons

In 1610 when Galileo observed four bright specks moving near Jupiter, he set in motion a quest that culminated in the 20th century with history's greatest robotic exploration program, giving never-to-be-repeated first close looks at the giant outer planets and their retinue of moons and rings. (See www.solarsystem.nasa.gov./index.cfm.)

Pioneers 10 and 11

Two NASA missions launched the first two of four human artifacts to escape forever from the Sun's domain. Leaving Earth in 1972, *Pioneer 10* flew by Jupiter in 1973 with imaging and magnetospheric measurements. Its signal continued to be detected at Earth until 2003. After launch in 1973, *Pioneer 11*'s flyby trajectory was adjusted so that, at its encounter in 1974, Jupiter's gravity would fling it onward toward Saturn, where it flew by in 1979. Each spacecraft carried a golden plaque illustrating humans and encoded information on where and when in the cosmos the flight had originated. (See www.solarsystem.nasa.gov/missions/profile.cfm?Sort=Target&Target=Saturn&MCode=Pioneer_11.)

Voyagers 1 and 2

Launched in 1977 by Titan-Centaurs and still operating in 2006, the NASA missions *Voyagers 1* and *2* are a mighty achievement. *Voyager 1* flew by Jupiter in 1979 and Saturn in 1980, whence it is headed toward the heliopause, the boundary between the Sun's realm and that of interstellar space. *Voyager 2* was targeted to a Jupiter flyby and then to Saturn, where Saturn's gravity would send it on to Uranus and Neptune, taking advantage of a planetary alignment that happens at intervals of 173 years. *Voyager 2* passed Uranus in 1986 and Neptune in 1989. The two *Voyagers* returned a vast harvest of imagery, geochemical and geophysical data on the giant planets and their moons and rings, and magnetospheric information. Each one carried a golden phonograph and video record showing characteristics of our planet, its inhabitants, and human civilization. In 2005 *Voyager 1* detected the heliopause, and in 2006 it passed 100 astronomical units from the Sun. (See www.solarsystem.nasa.gov/missions/profile.cfm?Sort=Chron&MCode=Voyager_2&StartYear=2000&EndYear=2009.)

Galileo

NASA's *Galileo* mission was launched by the space shuttle plus the *Inertial Upper Stage* in 1989 after a fraught history of replanning and delays. *Galileo* entered Jovian orbit in 1995, having made one Venus and two Earth gravity assist flybys en route. During the flybys, some science data were collected, including multispectral observations of the Earth and Moon. Galileo performed the first two asteroid flybys and was in position to image the Comet Shoemaker-Levy 9 impacts on Jupiter. At arrival in the Jovian system the spacecraft delivered a probe into the huge planet's atmosphere. Despite the failure of the orbiter's high-gain antenna to deploy, the mission returned a large volume of imaging, spectra and other data on the planet and its moons. In 2003 the craft was commanded to a Jupiter impact, with destruction in Jupiter's atmosphere to keep it from becoming a contamination risk toward any possible biology in the putative subsurface oceans of Europa and Ganymede. (See www.solarsystem.nasa.gov/missions/profile.cfm?Sort=Target&Target=Jupiter&MCode=Galileo.)

Cassini-Huygens

Launched in 1997 by a *Titan-Centaur,* NASA Saturn-orbiter spacecraft *Cassini-Huygens* carried ESA's *Huygens* probe designed to enter the dense atmosphere of the huge moon Titan (Fig. 11). With a 1998 Venus gravity assist and a 2000 Jupiter flyby with some scientific observations en route, the combination entered Saturn orbit in 2004. The probe descended to Titan in 2005, delivered remarkable images and survived impact on the surface for many hours. Both spacecraft returned unique new observations that will cause active scientific analysis and argument for years to come. (See saturn.jpl.nasa.gov.)

FIGURE 11   *Cassini* was launched in 1997.

New Horizons

Launched in 2006 at such a high speed that it will pass Jupiter in less than thirteen months from Earth departure, the *New Horizons* spacecraft is to investigate the surfaces and atmospheres of Pluto and its large moon Charon during a flyby in 2015. After that it is expected to continue functioning for several more years, exploring the mysteries of the Kuiper Belt, that far-out region of the solar system where the first representatives of a likely multitude of small, icy objects have already been discovered.

11. Conclusion

Thus has ended the first, magnificent phase of investigation throughout the Sun's domain. Meanwhile, spaceflight in the inner solar system is reinvigorated as robotic missions to the Moon and Mars take on the purpose, in addition to science, of acting as precursors to renewed human exploration and perhaps ultimately settlement of communities off Earth. And discoveries of giant planets orbiting other stars, more than 150 to date, tell us that exploration of star-and-planet systems has a limitless future.

Extrasolar Planets

Michael Endl

and

William D. Cochran

McDonald Observatory
University of Texas at Austin
Austin, Texas

CHAPTER 47

1. Introduction

Extrasolar planets—planets outside the solar system—were for a long time a mystery for astronomers. Are planets also orbiting other stars than the Sun? Is our solar system unique, or is planet formation a natural by-product of star formation and is our galaxy thus teeming with planets? The answers to these questions eluded astronomers for many centuries. It was only over the past decade that we finally obtained unambiguous evidence for the existence of extrasolar planets. The reason why it took so long to find these objects is the fact that planets are dark objects very close to an extremely bright source, their host star. In visual light, a planet is more than a billion times fainter than a star. But the main problem is not the planet's faintness—today's best telescopes and instruments are sensitive enough—but that the light of the close-by star overwhelms the feeble light coming from the companion. Astronomers had to rely completely on indirect methods to discover and characterize the first extrasolar planets. The most successful method today is the **radial velocity technique**, where tiny variations in the line-of-sight velocity of a star are used to infer the presence of unseen companions. Over the past 10 years, radial velocity surveys have detected more than 150 planetary companions to stars in our galaxy. Most of them are presumably gas giant planets similar to Jupiter and Saturn. The structures of most known extrasolar planetary systems are very different from those in our solar system, with giant planets often very close to the star and a wide range of orbital eccentricities. These observational data resulted in a rethinking of our current understanding and reformulation of our theoretical models how planets form. We might also begin to view our solar system in a different light: Many of the extrasolar planetary systems found so far seem to have undergone far more dynamical evolution than has our own solar system. The next decades of planet search will allow us to determine the frequency of planetary systems similar to ours, and even how abundant possible habitable worlds like our Earth are.

2. Detection Techniques

2.1 Astrometry

Astrometry is the science of positional astronomy, which measures the location of a celestial object and its movement within the plane of the sky. This was one of the first techniques used to search for planets around other stars. As in other indirect methods, astronomers seek to detect the orbit of the central star around the **barycenter** of the star/planet system. The orbit is measured as the change of the position of the star on the plane of the sky, usually compared to a number of more distant background stars, which define an astrometric reference frame.

The amplitude S of an astrometric signal is given by

$$S = \frac{m}{M} \frac{r}{d}$$

where m is the mass of the unseen companion, M the mass of the central star, r is the semimajor axis of the companion's orbit, and d is the distance to the star. For r in AU and d in parsecs, S is given in seconds of arc (1 arcsec is 1/60 of an arcmin, which itself is 1/60 of a degree). For the Sun/Jupiter system with $m/M = 0.001$ and $r = 5.2$ AU the amplitude of the signal would be 0.001 arcsec (1 mas) seen from a distance of 5 pc and 0.5 mas from 10 pc.

The motion of our Sun around the barycenter of our solar system is complicated because of the presence of the other outer planets. Figure 1 shows the astrometric signal due to the Sun's **reflex motion** as seen directly face-on to the ecliptic from a distance of 10 pc.

From the preceding equation, it is obvious that astrometry is more sensitive to companions with large mass ratios (massive planets around less massive stars), at large orbital separations r and around nearby stars (d is small). Because of the r dependence, this technique is a complementary method to the radial velocity technique (which will be discussed next).

As is the case for most of the detection methods, the largest hurdle to overcome in detecting extrasolar planets by

astrometry is the need for very precise measurements and the extreme care required to avoid systematic errors (like instrumental effects) in order to prevent the introduction of spurious signals, which may be misinterpreted as real planets, over a long time baseline.

The astrometric signals for most extrasolar planets are typically less than 1 mas and are beyond the scope of most current state-of-the-art instruments. The European Space Agency's satellite *Hipparcos* was a space mission entirely dedicated to stellar astrometry. Despite the fact that although its precision was not sufficient to detect planetary companions, the *Hipparcos* data placed very useful upper limits on the masses of some companions detected by other methods.

The highest astrometric precision can be achieved by using interferometry. By letting the light, which arrives from the same source at two different locations (two or more telescopes positioned on a well-defined baseline), interfere, one can measure the small difference in the arrival time at these points and thus determine the angle between the source and the baseline very precisely. The Fine Guidance Sensors onboard the *Hubble Space Telescope* (*HST*) can actually be used as an interferometer and they yield currently the best astrometric precision.

2.2 The Radial Velocity Method

Astronomers using the radial velocity technique measure the line of sight component of the space velocity vector of a star (hence the term "radial," i.e., the velocity component along the radius between observer and target). The radial velocity of a star can be determined in absolute values or differentially, if only changes of the velocity are of interest.

In order to measure stellar radial velocities, we rely on the well-known Doppler effect. Depending on whether the star moves toward us or away from us, its light will be blue or red shifted, as compared to a nonmoving source. Such a shift reveals itself as a change in the wavelength position of the absorption lines in the spectrum of the star. Therefore, astronomers use high-resolution spectrometers to perform radial velocity studies. The incoming light of the star is split up into its individual wavelengths, and the spectrum is recorded on a charge-coupled device (CCD) detector. As in astrometry, this method tries to detect the reflex motion of the primary object around the common center of mass with an unseen companion. Only this time this motion reveals itself as a change in the velocity rather than a change in position of the star.

The radial velocity method is traditionally used in stellar astronomy for the discovery and characterization of binary stars. In a binary system, the barycenter of the system is located somewhere between the two stars (the exact location is defined by the mass ratio), and the observed velocity amplitudes are of the order of several kilometer per second. In principle, the same method can be applied to the search

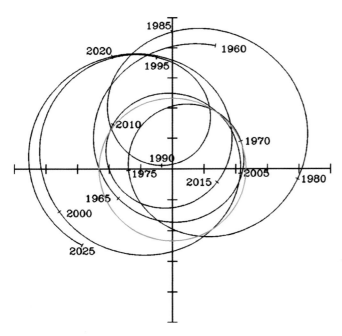

FIGURE 1   The astrometric motion of the center of the Sun (black line) around the barycenter of the solar system due to the gravitational perturbations of the planets, viewed from a point exactly above the ecliptic and from a distance of 10 pc (~ 32.6 light years). One dash mark on the axes is 0.0002 arcsec. The red circle represents the size of the sun.

for extrasolar planets, which induce a much smaller reflex orbit on their host star and produce much smaller velocity amplitudes.

For a system of two gravitationally bound objects m_1 and m_2 in a circular orbit the radial velocity semiamplitude K_1 of m_1 can be calculated by using:

$$K_1 = \frac{(m_2 \sin i)}{(m_1 + m_2)} \sqrt{G \frac{(m_1 + m_2)}{a}}$$

m_1 is the more massive object and m_2 is the less massive secondary companion, i denotes the angle between the orbital plane and the plane of the sky, G is the gravitational constant, and a is the semimajor axis of the orbit. It is immediately clear that for face-on systems ($\sin i = 0$) K_1 is zero.

Using Kepler's famous third law, which relates orbital separation to orbital period, we can recast this:

$$K_1 = \left(\frac{(2\pi G)}{P}\right)^{1/3} \frac{(m_2 \sin i)}{(m_1 + m_2)^{2/3}}$$

We are interested in the case of a planet orbiting the star, where $m_2 \ll m_1$ (and thus $m_1 + m_2 \approx m_1$), which simplifies the equation to

$$K_1 = \left(\frac{(2\pi G)}{P}\right)^{1/3} \frac{(m_2 \sin i)}{m_1^{2/3}}$$

Now we have an expression that relates $m_2 \sin i$ to the observables K_1 (or simply K if only the spectrum of m_1 is detectable) and P. Using units of years for P and m s^{-1} for K, $m_2 \sin i$ is thus given in Jupiter masses by the following expression:

$$m_2 \sin i = K \frac{\left(P m_1^2\right)^{1/3}}{28.4}$$

With a good estimate for m_1 we thus calculate $m_2 \sin i$ for the unseen companion. The $m_2 \sin i$ value represents a lower limit to the true mass of m_2. The $\sin i$ ambiguity is one of the limitations of the radial velocity technique. However, the $m_2 \sin i$ value is probably close to the real value of m_2. Just by assuming a random distribution of orbital planes, we have a 90% statistical probability that m_2 is within a factor of 2.3 of the observed $m_2 \sin i$.

Jupiter induces a K of 12.5 m s^{-1} in the Sun when observed in the plane of its orbit ($\sin i = 1$) and Saturn a K of only 2.8 m s^{-1}. Figure 2 shows the radial velocity of our Sun as it would appear to an astronomer in a different planetary system, who happens to observe the Sun from a point in space that is coplanar to our planetary system. The observed radial velocity signal consists of the superposition of the signals from the individual planets. In Fig. 2, we see the

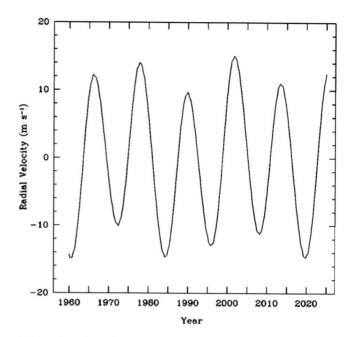

FIGURE 2 The radial velocity of our Sun measured from a point coplanar to the plane of the solar system. The strong signal with a period of 12 years and a semiamplitude of 12.5 m s^{-1} is caused by Jupiter, while the longer periodic and smaller variation is the signal caused by Saturn. The radial velocity variations due to the other planets are negligible.

primary 11.86 year period due to Jupiter, with a modulation due to the orbit of Saturn.

Detection of planets analogous to the two gas giants in our solar system thus calls for measurement uncertainties of a few m s^{-1} or better over many years to decades. More massive planets and also planets at smaller orbital separations produce larger K amplitudes, but the desired velocity precision is still of the order of several m s^{-1}.

Over the past years (even decades) two techniques have been successfully used to attain such a high level of precision: (1) the gas absorption cell technique and (2) the simultaneous Thorium–Argon technique in combination with stabilized spectrometers. In the first method, the star light is passed through a small glass cell that is filled with a suitable gas (in most cases iodine vapor), which superimposes its own dense absorption spectrum onto the stellar spectrum. This reference spectrum not only yields a simultaneous wavelength calibration but can also be used to keep track of the imaging properties of the spectrograph. This allows preventing small changes in the image of the stellar absorption lines, which are caused by fluctuations in the light path from the telescope to the detector, from being misinterpreted as Doppler shifts. In the second technique, the emission spectrum of a Thorium–Argon lamp is imaged parallel to the stellar lines on the CCD frame. Again, this allows a simultaneous wavelength calibration.

This technique only works in combination with stabilized spectrometers. To minimize any instrumental effects, these spectrographs have no movable parts and are placed in pressure- and temperature-stabilized environments. Also, by using optical fibers, the light path from the telescope to the instrument is kept as constant as possible. Both techniques have been demonstrated to reach a radial velocity precision of a few m s^{-1}, and in the best cases even 1 m s^{-1}.

2.3 Transit Photometry

In the special case that the orbital plane of an extrasolar planet is close to perpendicular to the plane of the sky, the planet will appear to move across the disk of the host star. In our own solar system, this phenomenon can be observed from the ground for the two inner planets, Mercury and Venus. Because we cannot spatially resolve the disk of another star, a transiting extrasolar planet can only be observed as a reduction of the light output coming from the star (i.e., by means of precise photometric measurements).

The probability of the visibility of a transit event is a function of both the radius of the star and the planet and its orbital separation a:

$$Transit_{prob} = \frac{\left(R_{star} + R_{planet}\right)}{a}$$

For a random location in our galaxy, the probability is less than 1% to observe transits of the inner terrestrial planets in the solar system and for the outer planets it decreases from 0.1% (Jupiter) to 0.01% (Pluto). But for giant planets orbiting at very small separations ($a \sim 0.04$ AU), the transit probability is around 10%. A transit of such a planet produces a ~1% dip in the so-called lightcurve (i.e., the time series of brightness measurements) of a star. This effect can be detected from the ground with state-of-the-art photometric instruments, which allow a precision of ~0.1%.

Currently numerous ground-based photometric transit surveys are searching for short-periodic giant planets. These surveys usually use small-aperture telescopes with a wide field of view to survey a large amount of stars, typically hundreds or thousands per CCD image. Their results will be discussed in Section 3.

Smaller planets will require higher photometric precisions than ground-based photometry can achieve because of the limitations imposed by our atmosphere. Space-borne observatories on the other hand should be capable of detecting even the miniscule photometric transit of an Earth-like planet orbiting a solar type star at 1 AU.

If photometric data of a transit event can be combined with radial velocity measurements, then the $\sin i$ ambiguity in the planetary mass is removed. Furthermore, the transit depth allows an estimate of the radius of the companion and thus an estimate of the mean density. Comparisons of high-resolution spectroscopic observations during and

outside a planetary transit could possibly reveal spectral signatures of the planetary atmosphere. Clearly, we can gain a tremendous amount of information from planetary transit observations.

2.4 Microlensing

According to Einstein's theory of general relativity, photons are affected by the presence of a gravitational field. Because gravity can be viewed as the changing curvature of the space–time continuum, the path of a photon follows this shape and is "bent" when it passes close to a gravitational potential. In certain geometric cases, this can lead to a focusing of the light from a distant source by a foreground object. This gravitational bending of light has been measured directly during total solar eclipses when the effect of the Sun's gravitational field can be observed as positional changes of stars close to the sun's disk.

In astronomy, this effect is also seen on a much larger scale: Entire galaxies or even clusters of galaxies are acting as massive gravitational lenses for the light of more distant objects in the background. However, as already demonstrated by our Sun, every object with mass can be a gravitational lens: a star, a brown dwarf, or even a planet.

Like the transit method, microlensing is caused by a geometric alignment: when a foreground object (the "lens") moves in front of a more distant background object (the "source"), the light of the source passing close to the lens is bent toward the observer. The observer can see several images of the background object separated by milliseconds of arc, which merge into a full ring called the Einstein ring at the moment the lens is directly in front of the source. Because the gravitational lensing of the source magnifies the image, the total amount of detected light is increased, and the brightness of the distant source is enhanced. The magnification factor depends on the exact geometric situation, and the maximum occurs when the lens is at its smallest projected distance from the source. For microlensing events in our galaxy and for stars acting as both sources as well as lenses, the images cannot be spatially resolved, and only the change in brightness is observed. However, the magnification can be large and theoretically even infinite for point sources. The position where infinite magnification occurs is called a caustic. Because stars are not perfect point sources, the magnification will not be infinite, but it will still be very large.

If the lens is not a single object but a binary, then the caustic is no longer a single point but an extended geometric figure symmetric around the binary axis. Thus, the microlensing technique represents an elegant method to search for planetary companions to stars in our galaxy. Binary lenses reveal themselves by a characteristic shape of their lightcurve (the time series of the brightness measurements during the lensing event). The lightcurve of a binary lens contains sharp peaks of even larger magnification due to the crossing of the caustics. The mass ratio q of the two lenses can be derived from modeling the lightcurve. If the resulting mass ratio is very small (typically $q > 0.001$), the second object in this binary might be a planet.

Like transit surveys, microlensing planet searches have to observe a large number of targets because the probability of observing a single event for a given target is negligible. For microlensing, the situation is even more complex because of the need of a sufficiently large reservoir of lenses moving in front of a high density sample of background sources. Moreover, microlensing events for a particular lens do not repeat. Each microlensing event is a single isolated transient in the lightcurve. There is only one chance to observe it. Hence microlensing surveys monitor regions like the bulge of our galaxy (the central cluster of stars that surround the galactic center) where microlensing events can be observed more frequently.

2.5 Timing Method

The timing method is exceptional with respect to the other techniques because it actually is the method that led to the very first detection of planets outside the solar system. As in the astrometric and radial velocity techniques, the fact that a host star has to orbit the common center of mass with an orbiting planet is utilized to detect the unseen companion. But this time the reflex orbit is observed by the change of the arrival time of signals coming from the star. The change is caused by the difference in the distance the signal has to travel from the source to the observer. If the star is at the location in its orbit where it is the farthest away from Earth then the signal needs the longest time to arrive here, and vice versa for the smallest separation. Because reflex motions due to planets are small compared to the speed of light the changes in arrival time are very small.

The timing method can only be applied to cases where (a) a very short duration signal is emitted by a source with a constant periodicity and (b) the observers are able to measure the arrival time of the signal with very high precision. One astrophysical case where these conditions are met are the so-called pulsars. Pulsars are neutron stars, the end stage in the life of massive stars with 15 and 30 times the mass of the Sun. They are the collapsed core of the star (with about 1.4 times the mass of the Sun) left behind after a supernova explosion. A neutron star is very small with a diameter of only 10–20 km, and hence a very dense object that also rotates very fast. Rotation periods of neutron stars can be as short as milliseconds. Strong magnetic fields produce bipolar jets of radio waves and high-energy radiation like X-rays and gamma rays. Because the magnetic field axis is misaligned with the rotational axis, these stars act like cosmic lighthouses from which we see a pulse every time the jet sweeps over the Earth. Pulsars were first discovered by radio telescopes in 1967, and to the fastest rotators (the

millisecond pulsars) the timing method can be applied to detect orbiting companions.

A second case where the timing method is applicable is stably pulsating white dwarf stars. White dwarfs are the end stage of the life of stars that are not massive enough to form a neutron star (like our Sun). They are also small (about the size of the Earth) and very dense objects. These stars undergo nonradial pulsations for certain temperature ranges that can be detected by precise photometric observations. The periods of these pulsations are of the order of a few minutes. Some of the white dwarfs exhibit the same pulsation modes over decades and are thus suitable targets for the timing method.

2.6 Direct Imaging

Obtaining a direct image of an extrasolar planet is the type of observation the public expects. Besides the obvious advantage of discovering planets with only a few observations, the images might also allow us to characterize the planets in new depth. From the colors and albedos, we might obtain thermal and chemical information. After the direct detection, follow-up observations can be carried out to collect first spectra of the planet.

In many ways, direct imaging of a planet around a nearby star represents the largest challenge in the development of telescope/instrument systems. Surprisingly, it is not the faintness of an irradiated extrasolar planet that is the hurdle to overcome (the *Hubble Space Telescope* would be sensitive enough to detect these faint objects) but rather their proximity to a much brighter source of photons: the planet's own host star.

The distances to even the nearest stars are so large that, due to the perspective, any image of a companion orbiting at separations comparable to our solar system would be located in the side wings of the image of the central object. In the optical, the flux difference between a solar-type star and a giant planetary companion is of the order of a billion. In the infrared, the difference is more of the order of a million (see Fig. 3). But the light coming from the planet is completely overwhelmed by the large amount of scattered light from the star.

There are several techniques to minimize the scattered light from the host star. From the ground, the observations are also affected by atmospheric turbulences, the so-called seeing. Seeing usually prevents telescopes from obtaining images at their theoretical resolving power even at the best observing sites in the world. In the near infrared, atmospheric turbulence can be compensated by an adaptive optics (AO) system. AO systems use wavefront sensors to measure the wavefront errors caused by turbulence in the atmosphere above the telescope and then to adjust the optical path to compensate for these errors using deformable mirrors. This helps to attain images at a spatial resolving power close to the limit set by the diffraction of light. AO

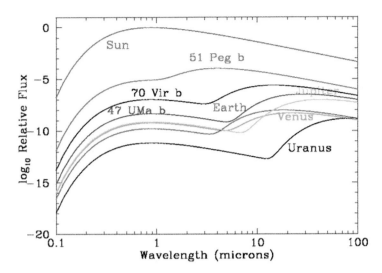

FIGURE 3 The relative flux of planets compared to the Sun's emission as a function of wavelength. Four planets of our solar system and three extrasolar planets (51 Peg b, 70 Vir b, and 47 UMa b) are shown. The difference in flux ranges from 10^{-6} to 10^{-12} in the optical ($< 1 \mu m$) and generally improves toward the infrared ($> 1 \mu m$), where the planet's thermal emission dominates.

imaging improves the situation for high-contrast imaging, but the real goal is to remove the image of the central star entirely from the observations.

The most commonly used instrument to perform this task is the so-called coronagraph. The coronagraph was invented by the French astronomer Bernard Lyot in the 1930s to study the outer parts of the solar atmosphere (the corona) without being totally overwhelmed by the intense glare of the Sun's disk. He managed to remove the light of the Sun's photosphere by introducing an opaque mask (of the same size as the image of the Sun) into the telescope's light path in such a way that it blocked the photons coming from the disk but not from the surrounding environment. This makes a coronagraph the ideal instrument for direct imaging of extrasolar planets. For ground-based searches the highest image quality is achieved by combining a coronagraph with an AO system.

However, no optical system is perfect and even coronagraphic images contain residual scattered light from the central star close to the edge of the opaque mask and other image artifacts produced by diffraction on telescope parts. This makes the detection very close to the central object still very difficult. And even for the nearest stars, the expected angular separations for planetary companions are small compared to the size of typical coronagraphic masks. At a distance of 5.2 pc ($= 17$ light years), an analogous planet to Jupiter would appear 1 arcsec away from the star at maximum projected separation. At 10 pc, the maximum separation is only about 0.5 arcsec. These angles are comparable to the typical dimensions for coronagraphic masks of current state-of-the art instruments (e.g., the *Hubble Space Telescope* instrument Near Infrared Camera and Multi-Object Spectrometer NICMOS has a coronagraphic mask with a diameter of 0.8 arcsec).

The image area around the central obscuration contaminated by scattered light is called the halo, and in very short exposures this halo is resolved into smaller bright and dark spots called "speckles." Speckles are interference phenomena produced by atmospheric seeing and by the superposition of light coming from all parts of a telescope mirror with imperfect smoothness. At the location of a dark speckle, the light of the star is canceled out by destructive interference. The image of a faint companion can be recovered if it is located at the position of a dark speckle, where the light from the star is severely reduced. By taking a great number of short exposures, the companion can be detected by a proper data analysis algorithm simply by the fact that in every image the speckle pattern is different and that a dark speckle never appears at the location of the companion. This method is called dark speckle coronagraphy. In combination with large aperture ground-based telescopes or the next generation space telescope, this method should have the sensitivity to detect extrasolar planets around the nearest stars.

Another technique to achieve high-contrast images is nulling interferometry. In theory, it is possible to combine the wavefronts arriving at two or more telescopes in such a way that a wave maximum coming from one telescope is canceled out by a wave minimum from another telescope. In this way, it produces a null image of the central object while it leaves the light from the circumstellar environment unaffected. First trial runs using ground-based telescopes have already been successfully performed. A nulling interferometer is currently built for the Large Binocular Telescope (LBT), which consists of two 8 m class telescopes mounted side by side on the same support structure. A space-based nulling interferometer operating in the infrared is planned for the second stage of NASA's Terrestrial Planet Finder (TPF) mission and for the European Space Agency's DARWIN mission, with the ultimate goal to image Earth-like planets around nearby stars.

3. Observations of Extrasolar Planets

3.1 The Pulsar Planets

The first discovery of planets outside our solar system was achieved by Alexander Wolszczan and colleagues in the early 1990s. This discovery would set the tone for all subsequent discoveries of extrasolar planets in terms of their strangeness. Wolszczan found the planets orbiting the millisecond pulsar PSR B1257+12 using the Arecibo radio telescope and the timing method described in Section 2.

The first surprise was that the planets orbit a "dead" star, which had undergone a previous supernova explosion. It is unlikely that these planets existed before the star went supernova.

The more plausible scenario is that these planets somehow formed after the explosion. Millisecond pulsars are believed to achieve their high rotation rates due to spin up by in-falling material accreted from a companion star. The planets might have formed during this process in the accretion disk around the pulsar.

The planetary companions to PSR B1257+12 are also remarkable in a different way: they have very small masses. Due to the extreme sensitivity of the timing method in the case of millisecond pulsars (where the arrival time of a pulse can be measured with microsecond precision), even companions with the mass of our Moon or less can be detected.

Table 1 summarizes the properties of the PSR B1257+12 planetary system. Soon after the system was discovered, the mutual gravitational perturbations of the planets on each other were measured, thus confirming that they are indeed planets and not a previously unknown effect, intrinsic to the pulsar.

So far these companions represent the lowest mass objects known to orbit a star other than the Sun. In terms of mass these planets are also the most Earth-like extrasolar planets we know. However they must be barren and dead worlds because of the constant bombardment by high energy radiation coming from the pulsar.

Just recently Wolszczan presented evidence for a fourth object with a mass of only 15% the mass of Pluto orbiting the pulsar at a distance of 2.7 AU. This new object, however, qualifies as an asteroid or comet rather than a planet.

3.2 Planets Around Sun-like Stars: The Success of the Radial Velocity Technique

3.2.1 51 PEGASI: THE FIRST PLANET ORBITING A SOLAR-TYPE STAR

In the fall of 1995, two Swiss astronomers, Michel Mayor and Didier Queloz, stunned the community as well as the public by their announcement of the discovery of the first extrasolar planet around a sun-like star. Their precise radial velocity measurements of the star 51 Pegasi revealed a periodic variation of 4.2 days and an amplitude consistent with an $m \sin i = 0.5$ Jupiter mass companion (Fig. 4). The minimum mass of the object firmly places this companion into the gas giant planet mass range.

However, the extremely short orbital period and small orbital separation of 0.05 AU were surprising in many ways, and alternative explanations for the 51 Peg radial velocity signal were put forward. Stars more evolved than the Sun show similar variability, which is caused by pulsations rather than by Keplerian motion. But 51 Peg passed every test for this type of variability, and soon Mayor and Queloz's claim of having found the very first planet orbiting a "normal" star was generally accepted.

51 Peg b represents the prototype of a new class of planets that soon emerged from the results of the radial velocity surveys, the so-called **hot Jupiters**. Because of their

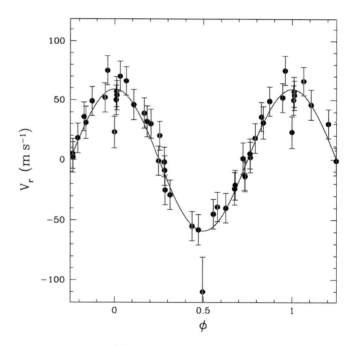

FIGURE 4   The radial velocity measurements (dots with error bars) of the solar-type star 51 Pegasi phased to the orbital period of its planetary companion. The sinusoidal variation is caused by a companion with $m \sin i = 0.5$ Jupiter masses in a circular orbit with $a = 0.05$ AU and an orbital period of 4.2 days. Reproduced with permission from *Nature*.

				TABLE 1	The PSR B1257+12 Planetary System

Planet	$M \sin i$ (Earth mass)	Orbital Period, P (days)	a (AU)
A	0.015	25.34	0.19
B	3.400	66.54	0.36
C	2.8	98.22	0.47

	$m \sin i$ (Jupiter mass)	Orbital Period, P (days)	a (AU)	Eccentricity, e
TABLE 2	**The first 8 Radial Velocity Planets**			
Star				
51 Peg	0.5	4.2	0.05	0
70 Vir	7.4	117	0.48	0.4
47 UMa	2.5	1089	2.09	0.06
ρ 1 Cancri	0.84	14.7	0.12	0.02
τ Boo	4.13	3.3	0.05	0.01
υ And	0.69	4.62	0.06	0.012
16 Cyg B	1.69	799	1.67	0.67
ρ Crb	1.11	39.9	0.23	0.13

close proximity to the host star, these gas giant planets have estimated upper atmosphere temperatures of more than 1000 K.

3.2.2 MORE RADIAL VELOCITY PLANETS

In the 2 years following the discovery of 51 Peg, astronomers from the United States announced the discovery of 7 more extrasolar planets orbiting Sun-like stars. All these detections were based on years of precise radial velocity measurements of these stars using telescopes and spectrographs at Lick, McDonald, and Whipple Observatories. Table 2 lists the first 8 extrasolar planets discovered by the radial velocity technique along with their orbital characteristics.

3.2.3 THE FIRST MULTIPLE PLANETARY SYSTEM

In 1999, the teams of the Lick and Whipple Observatory Doppler surveys announced the discovery of the first extrasolar multiplanetary system around a Sun-like star. The radial velocities of υ Andromedae deviated progressively from the originally derived, single-planet velocity curve, and with the additional years of data it became apparent that a triple Keplerian model is required to describe the complex reflex

motion of this star. In addition to the previously found hot Jupiter, this system contains two more giant planets with $m \sin i = 1.89$ and 3.75 Jupiter masses at separations of 0.8 and 2.53 AU. Also their orbits have significantly nonzero eccentricities (0.28 and 0.27), making this system again quite different from our solar system.

3.2.4 A TRANSITING PLANET

With the discoveries of more and more hot Jupiters, it was just a matter of time until one of them would have a near edge-on orbit so that the planet transits in front of the star. The hot Jupiter companion to HD 209458 was the first transiting extrasolar planet. The planet itself was first discovered by the radial velocity method, but in this case the photometric follow-up observations revealed—for the first time—the characteristically shaped lightcurve of a transiting planet (Fig. 5). With the viewing angle known ($i = 86°$) the $m \sin i$ value transformed into a true mass for the planet of 0.69 Jupiter masses. The depth of the dip in the lightcurve yielded a radius for the companion of 1.4 Jupiter radii. With a known mass and a known radius, a mean density of $0.31\,\mathrm{g\,cm^{-3}}$ was derived. This is an even lower mean density

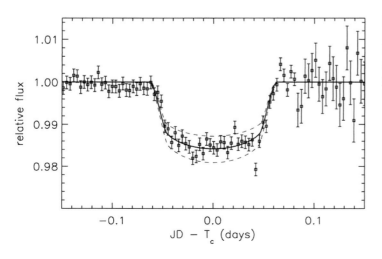

FIGURE 5 The lightcurve of the star HD 209458 showing the reduction in stellar flux by 1.5% due to the transit of its hot Jupiter. From the depth of the transit lightcurve, a radius of 1.4 Jupiter radii was derived, and combined with the planetary mass—determined from radial velocities—a low mean density of 0.31 g/cm^3 was found for this planet.

than Saturn, the planet with the lowest mean density in the solar system. The discovery of the HD 209458 transiting planet represents another milestone in the field of extrasolar planet detection: It demonstrated that the companions discovered by the radial velocity surveys were indeed gas giant planetary companions and not more massive (even stellar) companions seen at a very unfortunate viewing angle.

Using spectroscopic observations obtained with the *Hubble Space Telescope* outside and during the transit, it was even possible to detect the atmosphere of the HD 209548 planet. The *HST* spectra taken during the planetary transit showed a stronger sodium absorption line than the spectra observed without the planet in front of the star. This additional absorption is caused by the sodium in the planet's atmosphere. However, the amount of atmospheric sodium was less than expected from theoretical models, urging the astronomers involved in this study to speculate that a thick cloud cover prevents us from seeing deeper into the planet's atmosphere.

In another *HST* observation of the HD 209458 planet, it was possible to measure hydrogen escaping from the heated upper layers of the planet's atmosphere. The escaping hydrogen gas forms a kind of cometary coma and tail around the planet and is blown away by the radiation and particle wind of the close-by star.

Several additional transiting planets have subsequently been discovered. The most interesting of these are a planet in a 2.2 day orbit around HD 189733 and a planet with a massive rocky core orbiting HD 149026.

3.2.5 GENERAL CHARACTERISTICS OF PLANETS DETECTED BY RADIAL VELOCITY MEASUREMENTS

Over the past decade the radial velocity technique has demonstrated its effectiveness in detecting numerous giant planets and multiplanetary systems around Sun-like stars. At the time of this writing, more than 150 planetary companions were found by the cumulative effort of several Doppler surveys operating in both hemispheres. Several characteristics of these extrasolar planets differ significantly from the giant planets in our solar system.

The gas giants found at very small orbital separation are difficult to explain in terms of their formation. In the classical picture of planet formation, gas giants can only form near (and beyond) the ice line in the protoplanetary nebula. The ice line is the distance from the star where the temperatures in the nebula drop low enough so that ices can condense out and form massive cores (mixed with rocky material and dust grains) onto which nebula gas can accrete on. That is the reason why the common expectation was to find gas giants only at large orbital separations similar to our Jupiter at 5 AU and more. It appears that, in most cases of extrasolar giant planets, moderate to massive orbital migration has occurred, which moved the planets from the place where they have formed to their current location close to the star.

The other remarkable difference is the high eccentricities (*e*) of the orbits of these planets (Fig. 6). Most of the found extrasolar planets have more elongated orbits than the planets in our solar system. Especially for planets at

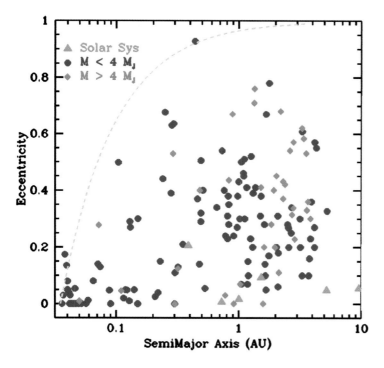

FIGURE 6   The distribution of orbital eccentricities of extrasolar planets (red and blue points) compared to the planets in our solar system (green triangles).

FIGURE 7   Artist conception of the ρ 1 Cancri system. This extrasolar multiplanetary system contains one massive gas giant (pictured in the foreground with a hypothetical moon) at a separation of 5.2 AU, as well as two giant planets close to the star (at 0.12 and 0.24 AU). After the discovery of three gas giant planets in this system, a fourth low-mass planet was found at $a = 0.038$ AU. (Artwork by Lynette Cook.)

larger separations, the eccentricities are distributed quite uniformly and are practically indistinguishable from the eccentricity distribution of stellar binaries. The hot Jupiters have all $e = 0$ (or close to 0) orbits because tidal forces between the star and the planet at these small distances tend to circularize the orbit on much shorter timescales than the typical lifetime of the star. The origin of the nonzero eccentricities is not well understood; possible explanations are a more dynamic formation history than in the case of the solar system, in which mutual dynamical interaction between planet embryos pumped up their eccentricities. Also, planet/disk interactions and gravitational perturbation by stellar companions could be the cause of the higher eccentricities.

Among the known extrasolar planets, 14 multiple systems were detected: 12 systems with 2 planets, one with 3 planets, and one, the ρ 1 Cancri system (Fig. 7), with 4 planets. Some of the planets in these multiple systems show evidence for mean-motion resonances; their orbital periods are equal or close to resonance values (e.g., 2:1 or 5:3).

The mass function of extrasolar planets (Fig. 8) shows a steep rise toward masses of 1 Jupiter mass or less. Thus, although less massive planets are harder (or impossible) to detect by the radial velocity technique, we can expect them to be quite frequent.

About 10% of the stars surveyed by long-term radial velocity programs have detectable giant planets. The majority of these planets orbit stars of the same spectral type (i.e., surface temperature and mass) as the Sun. They usually have orbital separations less than 5 AU; in fact, about half of them reside within the first AU from their host star. But these results also reflect strong observational biases. The radial velocity technique is more sensitive to close-in planets, and it takes a monitoring timescale of over a decade to discover planets beyond 5 AU. Also, stars hotter and more massive than the Sun are not suitable for the radial velocity technique because they tend to have higher rotation rates and much fewer spectral features, which can be used to measure the velocity. And less massive stars than the Sun are fainter, and the effort to collect enough photons to ensure a sufficient data quality increases significantly. Thus, it comes as no surprise that Doppler surveys have traditionally focused on Sun-like stars. As radial velocity programs extend their time baselines and expand their target samples to fainter and lower mass stars, these observational biases will be overcome.

The low-mass star Gliese 876 is the famous "exception from the rule." The star is a so-called M dwarf with a mass of

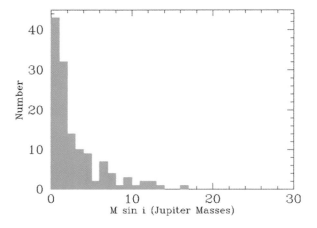

FIGURE 8   The extrasolar planet mass function. A strong general trend toward lower masses is apparent. This trend might indicate that lower mass extrasolar planets are abundant. Planetary companions with $m \sin i$ values larger than 10 Jupiter masses are rarer, despite their better delectability by the radial velocity technique.

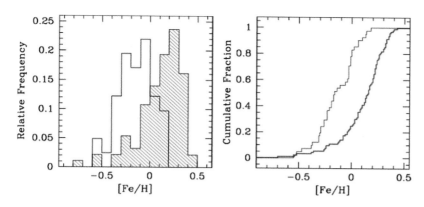

FIGURE 9   The stellar metallicity distribution of planet host stars (in red) compared to a volume limited sample of stars in the solar neighborhood (black). The observed iron abundance ([Fe/H]) is given on a logarithmic scale normalized to the Sun's metal content ([Fe/H] = 0). Stars with [Fe/H] = 0.5 contain a little more than three times the amount of iron, while stars with [Fe/H] = −0.5 have only one third of the Sun's iron abundance. Clearly, stars with detected radial velocity planets tend to be more metal-rich than the average star in the Sun's vicinity.

only 32% of the mass of the Sun. M dwarf stars have masses ranging from roughly 55% to about 0.8% solar masses. Despite the fact that M dwarfs comprise the majority of stars in our galaxy, they only form subsets in the target samples of current radial velocity surveys, due to their faintness. Gliese 876 was found to have a planetary system of two Jupiter-type companions in a 2:1 mean-motion resonance with periods of 30.12 and 61.02 days. This star is so far the only M dwarf known to have gas giant planetary companions. It is also exceptional in a different way: Because of its proximity (15 lightyears), it is the ideal target for astrometry. In 2002, highly precise measurements obtained with the Fine Guidance Sensors onboard the *Hubble Space Telescope* successfully revealed the astrometric signature of the outer planet. By combining the ground-based radial velocity data with the space-based astrometric data, a true mass of 1.9 Jupiter masses was determined for this planet.

There were also a handful of planets detected orbiting giant stars. Giants stars are more evolved than solar-type stars, and their cooler atmospheres have a spectral signature rich in absorption lines. These stars are thus suitable targets for the radial velocity technique. The progenitor stars (i.e., before they evolve into their current giant status) of most giant stars are more massive than the Sun and the successful detection of planetary companions around them is evidence that planet formation can also occur around more massive stars. This is not such a big surprise because several thick dust disks have already been observed around this type of star.

Another interesting correlation emerging from the radial velocity census of extrasolar planets is that their deltectability is a strong function of the metallicity of the host star. Astronomers call every element heavier than helium a metal. **Stellar metallicity** thus means the abundance of all chemical elements in a star besides hydrogen and helium. In general, the element used for the metallicity determination is iron. By measuring the stellar metal content, we can probe the primordial chemical composition of the gas and dust cloud, out of which the star (and presumably its companions) has formed.

It was found that the mean value of the metallicity distribution of planet host stars is offset with respect to the mean metallicity of stars in the solar neighborhood. On average, giant planets are more frequently detected around host stars that are more metal-rich than the solar neighborhood mean (Fig. 9). This can be seen as evidence for the core-accretion model for the formation of gas giants. The efficiency of this model is sensitive to the abundance of heavier elements in the protoplanetary disk (more heavier elements → more cores → more gas giants). Alternatively, this might also be regarded as evidence that orbital migration is a function of the metal content of the planet-forming disk, because close-in planets are easier and faster to detect by radial velocity surveys.

3.2.6 THE HOT NEPTUNES

In 2004, the first radial velocity planets with masses below the gas giant range were discovered. These planets have $m \sin i$ values comparable to the masses of the icy giants of our solar system, Uranus and Neptune. Their very short orbital periods give detectable radial velocity signals despite their low mass. Thus, they have been dubbed hot Neptunes.

So far, we know of three of these type of planets. Two of them reside in already known planetary systems (ρ 1 Cancri and μ Ara) as the innermost companion and the third orbits a low mass M dwarf star (Gliese 436). Table 3 summarizes the general characteristics of the three known hot Neptunes.

The internal composition of these planets remains unknown. They could represent failed gas giants (i.e., planet cores that never acquired, or later lost, their massive gaseous envelopes). Their masses are so low that it becomes unlikely that they consist mostly of H/He. Theoretical model computations indicate that an H/He Neptune so close to its host star would be unstable and would have probably evaporated by now. This supports the notion that the hot Neptunes are

TABLE 3	General Characteristics of the 3 Hot Neptunes			
Star	$m \sin i$ (Earth mass)	Orbital Period P (days)	a (AU)	Notes
ρ 1 Cancri	14.2	2.81	0.04	4 planet system
μ Ara	14.5	9.55	0.09	2 planet system
Gliese 436	21.5	2.64	0.03	M dwarf star

made up mostly of rocky material (possibly surrounded by a small gaseous envelope) and that their discoveries by the radial velocity technique represents another step toward the detection of more Earth-like planets.

3.3 Results of the Other Detection Methods

Although the radial velocity technique is by far the most successful method and pulsar timing was the very first method to detect planetary companions to other stars, they are not the only methods to discover planets. A few extrasolar planets were also detected by transit searches, microlensing surveys, and possibly even by direct imaging.

3.3.1 TRANSIT SEARCHES

The Optical Gravitational Lensing Experiment (OGLE) is a precise photometric survey of millions of stars to search for gravitational lensing events. OGLE is a project from astronomers from Warsaw and Princeton and operates a 1.3 m telescope at Las Campanas Observatory in Chile. The OGLE data can also be used to search for the characteristic flat-bottomed lightcurves of planetary transits. For more than 50,000 stars in the OGLE fields, the quality of the photometry is sufficient to perform this task. In total, more than 100 transit-like events were found in the OGLE data. However, there are a large number of other astrophysical phenomena that can mimic the photometric signal of a planetary transit. Thus, in order to determine whether a transiting object is indeed a planet, it is necessary to obtain spectroscopic follow-up observations to characterize the host star and subsequently to perform radial velocity measurements to derive a mass for the companion. The majority of the OGLE candidates turned out to be false alarms, mostly binary systems with a very hot (or giant) primary star and a very cool secondary star, or binary stars undergoing grazing eclipses.

The first transit candidate which was confirmed as a planet by radial velocity observations was OGLE-TR-56. This object was thus the very first planet discovered by the transit method (HD 209458 b, the first transiting planet, was detected by radial velocities before the photometric transit was observed). The companion to OGLE-TR-56 has a mass of 1.4 Jupiter masses and a radius of 1.2 Jupiter radii. The planet has an extremely short orbital period of only

1.2 days or 29 hours, and the orbital separation is only 0.023 AU. This is even closer to the host star than the hot Jupiters found by radial velocity surveys. Four more transiting planets were revealed by the OGLE data, two of them are "very hot Jupiters" similar to OGLE-TR-56, while the other two have orbital periods typical for hot Jupiters.

In order to survey a sufficiently large number of stars, OGLE monitors areas close to the galactic center. Most of the stars included in these fields are far more distant and thus fainter than the stars included in radial velocity surveys. This makes the spectroscopic follow-up observations more problematic. Large aperture telescopes like the 10 m Keck telescope in Hawaii or the 8 m Very Large Telescope (VLT) in Chile are necessary for this task.

The Trans-Atlantic Exoplanet Survey (TrES) has followed the opposite approach: Instead of studying a few selected fields with many faint stars, it monitors many brighter stars using a whole network of small telescopes distributed over the globe. In 2004, the first planet, TrES-1, was discovered by this program. It completes one orbit in 3 days and has a mass of 0.6 Jupiter masses. Its radius is only slightly larger than Jupiter's: 1.08 Jupiter radii.

Table 4 lists the six extrasolar planets that were found so far by the transit method, and Fig. 10 shows a comparison of their radii and masses with the gas giants of the solar system. The planet orbiting HD 209458 has still the largest radius and lowest mean density of all transiting planets, despite the fact the very hot Jupiters are even closer to their parent star. The larger radius of this gas giant might be the consequence of additional heating, possibly by tidal forces due to a slightly noncircular orbit (although we will see later that this can be ruled out).

3.3.2 A MICROLENSING PLANET

In the summer of 2003, the two microlensing surveys, OGLE and MOA (Microlensing Observations in Astrophysics), independently detected a microlensing event towards the galactic bulge. During the close monitoring of this event, a strong 1 week long deviation from a single lens lightcurve was discovered (see Fig. 11). Careful modeling of the combined photometric data sets showed that the lightcurve of the OGLE 2003-BLG-235/MOA 2003-BLG-53 event is best described by a binary lens model with an extremely small mass ratio of 0.004. In the probable case

TABLE 4	The Extrasolar Planets Discovered by Transit Searches		
Planet	**Mass (Jupiter mass)**	**Radius (Jupiter)**	**Orbital Period, P (days)**
OGLE-TR-56	1.45	1.23	1.21
OGLE-TR-113	1.35	1.08	1.43
OGLE-TR-132	1.19	1.13	1.69
OGLE-TR-10	0.57	1.24	3.1
OGLE-TR-111	0.53	1	4.02
TrES-1	0.61	1.08	3.03

that the primary lens is a normal star, the secondary lens would be a ~1.5 Jupiter mass planet orbiting at ~3 AU. Thus, this event is regarded as the first demonstration of the discovery of an extrasolar planet by microlensing. A better characterization of the star/planet system that caused this lensing event has to await next-generation ground-based or space telescopes, which will be powerful enough to resolve the lens.

3.3.3 INFRARED RADIATION FROM EXTRASOLAR PLANETS

With the launch of NASA's infrared space telescope SPITZER, a new and very interesting spectral window for observations of extrasolar planets became available: the far infrared, where thermal emission dominates the radiation coming from a planet (as shown in Fig. 3). Although SPITZER lacks the spatial resolving power to detect planets, its high sensitivity in the infrared can be used to discern between radiation from a star and its planet.

Two independent teams planned basically the same SPITZER observations: to observe a transiting extrasolar planet during (and out of) secondary eclipse (the time when the planet is directly behind the star and hidden from view). If the amount of infrared radiation measured by SPITZER during the eclipse is less than outside the eclipse, then this difference is the radiation coming from the planet itself. This effect was indeed successfully measured for the two transiting planets HD 209458 b and TrES-1 b. The amount of planetary infrared radiation was used to estimate a "surface" temperature for these two planets: The visible upper atmosphere of HD 209458 b has a temperature of 1130 ± 150 K, and for TrES-1 b the respective value is 1060 ± 50 K. Both values are in good agreement with the expected temperature of a giant planet heated by the intense irradiation of the nearby host star. The exact timing of the secondary eclipse of HD 209458 b also demonstrated that its orbit is indeed circular and that tidal heating cannot be the explanation for its abnormally large radius.

Interestingly (and quite ironically, because the information is obtained by the *lack* of photons), these observations also represent the first unambiguous detections of photons emitted by extrasolar planets.

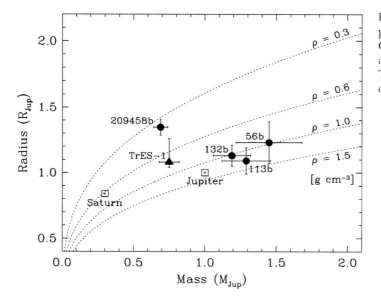

FIGURE 10   The radii and masses of transiting extrasolar planets (HD 209456b, TrES-1, OGLE-TR-132b, OGLE-TR-113b and OGLE-TR-56b) compared to Jupiter and Saturn. The dashed lines are curves of constant density. This shows that HD 209458b has an unusually low mean density.

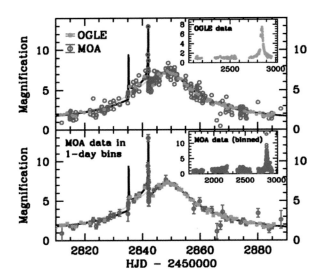

FIGURE 11 The first observed planetary microlensing event: two photometric data sets (MOA survey in blue and OGLE survey in red) were combined to produce this lightcurve. The best fit binary lens model is shown as solid black line. The planetary companion causes the strong double peaked and 1 week long deviation from the broader microlensing lightcurve due to its host star.

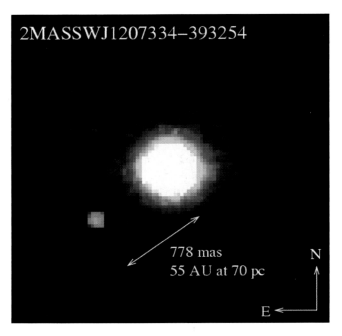

FIGURE 12 An image of an extrasolar planet? The companion (red) to the brown dwarf 2MASSWJ1207334–393254 (blue) might have a mass of only 5 Jupiter masses.

3.3.4 THE FIRST DIRECT IMAGING CANDIDATES

Also imaging searches for extrasolar planets yielded the first candidates for successful detections. In two cases, faint companions were found in deep infrared images taken with the adaptive optics system at the 8 m VLT of the European Southern Observatory in Chile. Fainter background sources, mimicking companions, can be ruled out by the fact that these objects are co-moving with the central star and are thus gravitationally bound to it.

One candidate was found near a young "failed" star, a brown dwarf, which itself is not massive enough to start thermonuclear reactions in its core. The central object (called 2MASSWJ 1207334-393254) has only ~25 times the mass of Jupiter and is located at a distance of approximately 230 lightyears. Its location also gives away its age: The brown dwarf lies within a young star-forming region, the so-called TW Hydrae association. This region contains young stars that are estimated to be only 8 million years old. The fainter (and thus presumably much less massive) companion was detected 0.8 arcsec away from the brown dwarf (Fig. 12). At the distance to the brown dwarf, this transforms to a projected separation of 55 AU. First low-resolution spectra of the companion were also obtained. But is it a planet? This is the tricky part to decide. The observed brightness, colors, and spectral information can be compared to theoretical models for young planets. Such a comparison yields a mass of ~5 Jupiter masses for the fainter companion, a

mass value placing the object firmly within the range of planets. However, some caution remains because the theoretical models for young planets are not calibrated and their uncertainties are difficult to assess. Moreover, the large separation of (at least) 55 AU raises the question how massive the protoplanetary disk around a 25 Jupiter mass object must have been to form such a massive planetary companion so far away from the center.

Another candidate for a directly imaged extrasolar planet is the companion to the star GQ Lupi. This time the central object is really a star similar to the Sun, albeit a lot younger. The age of GQ Lupi is estimated to be between 100,000 and 2 million years. These very young ages for the imaging candidates result from an observational selection effect: Imaging searches specifically target young stars because young planets are much brighter at these early evolutionary stages. The companion appears 0.7 arcsec to the west of GQ Lupi (see Fig. 13), which translates into a projected separation of ~100 AU. A careful comparison of the discovery images with archived images revealed that the companion is indeed a co-moving object (they share the same motion in the plane of the sky, which rules out a background object). Therefore, it can be assumed that it has formed at the same time as the star GQ Lupi has. In this case, the comparison of the available photometric and spectral data with models for young planets yielded a mass between 1 and 3 Jupiter masses for the companion. But again, the remaining uncertainties in the values derived from models are large, and it is difficult

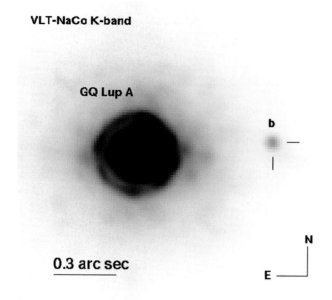

VLT-NaCo K-band

GQ Lup A

b

N

E

0.3 arc sec

FIGURE 13 Another candidate for a first image of an extrasolar planet is the companion (b) to the young star GQ Lupi A.

to distinguish young planets from low-mass brown dwarfs. If GQ Lupi b has indeed formed like a gas giant planet, then it was probably transported from the denser interior of the protoplanetary disk, where sufficient planet-forming material can be found, to its present location of about 100 AU away from the star.

4. Summary and Outlook

The extrasolar planetary systems discovered so far demonstrate that a surprising variety of planetary systems exists in our galaxy. Although we now know that other stars also have planetary companions, and that planet formation is not unique to our star, most of them have characteristics different from the planets in our solar system. The observational results indicate that the majority of extrasolar planetary systems might have had a much more dynamic past than our planetary system. The gas giant planets found at small orbital separations are probably a result of massive orbital migration, while the quasi-random distribution of orbital eccentricities might be caused by more violent and frequent interactions between planets in the early evolutionary stages. The overall mass function of extrasolar planets is steeply rising toward lower masses, and we can extrapolate that less massive planets, which are still undetectable by current techniques, are abundant in the galaxy. We also know today that the metallicity of the star- and planet-forming

nebula has an impact on the structure of planetary systems. Most of the radial velocity planets are detected around stars that are richer in heavier elements than the Sun. The lowest mass planets found around solar-type stars have masses comparable to Neptune and represent the first steps toward finding terrestrial planets.

We truly live in the golden age of discoveries of extrasolar planets. Detection programs using ground-based telescopes will continue to improve their sensitivity and to extend their search spaces, while space telescopes will play a more important and likely an even dominant role in this field in the coming years. There are currently in preparation several space missions that will greatly increase—and even revolutionize—our knowledge of extrasolar planets. To reach the ultimate goal in planet search— to discover Earth-like planet—we will need to use space telescopes.

The Space Interferometry Mission (SIM) has the goal to deliver astrometric measurements with an accuracy of 4 μ arcsec. SIM will be positioned in a so-called Earth-trailing orbit, where it will slowly drift away from our planet to avoid occultations of parts of the sky by the Earth. The impressive astrometric precision of 4 μ arcsec (which is several hundred times better than current techniques) will be attained by optical interferometry. The spacecraft itself consists of a fixed 10 m long boom on which three interferometers are mounted. SIM will be used to survey hundreds of nearby stars for giant planetary companions at large orbital separations. These data will be complementary to the results of the ground-based radial velocity surveys, which are more sensitive to close-in planets. SIM will thus greatly expand the census of Jupiter analogs in the solar neighborhood. It is difficult to predict what the best case astrometric precision of SIM will be, but if it will be close to 1 μ arcsec, then even the discovery of terrestrial planets orbiting the nearest stars will become possible.

KEPLER is a NASA mission specifically dedicated to find Earth-like planets in the habitable zones of other stars. The mission consists of a photometric space telescope of 1 m aperture, which will continuously monitor a specific search field in the sky for planetary transits. Unhampered by the limitations imposed by Earth's atmosphere, its sensitivity should allow us to detect even the miniscule dip in a star's lightcurve caused by the transit of a planet with 1 Earth radius, orbiting at ~1 AU. In order to have a decent chance in finding the transits of extrasolar terrestrial planets, KEPLER will observe more than 100,000 stars simultaneously. After finding one or two transit events (which will be separated by roughly 1 year for a planet at 1 AU) for a given target star, the third and fourth transit will be used to rigorously confirm an orbiting body. KEPLER will be the first mission that will allow us to estimate the frequency of possible habitable Earth-like worlds in our galaxy. The European Space Agency will launch a high-precision

FIGURE 14   Artist conception of the Kepler photometer and spacecraft. A 1 m Schmidt telescope will observe a large section of the sky in the constellation Cygnus continually for over 4 years to detect transits of Earth-sized planets in Earth-like orbits across the disks of solar-type stars. Launch is scheduled for late 2008. (Image courtesy NASA Kepler project.)

photometry spacecraft *COROT* in 2006. While *COROT* will not be able to detect the transits of Earth-size planets, it will be able to detect transits of Neptune-size bodies.

The most complex and ambitious space mission is the Terrestrial Planet Finder. This mission is designed not only to directly image Earth-like planets in the habitable zones of nearby stars, but also to look for the tell-tale signs of an active biosphere on these worlds. NASA has just recently decided to proceed with a two-stage TPF plan: to launch first a large optical telescope equipped with an advanced coronagraph (TPF-C) and then later a more powerful array of infrared space telescopes using interferometry (TPF-I, Fig. 14). Also DARWIN, a mission currently planned by ESA, will be using a space-based infrared interferometry array. These space telescopes will have the capability not only to take direct images of terrestrial planets in the habitable zones of other stars but also to perform first crude spectroscopic follow-up observations of these planets. The spectra of these extrasolar Earths should allow us to determine the chemical composition of their atmosphere. This will allow us to determine if water vapor is abundant, suggesting a planet with oceans, or if green house gases are present, warming the planet by the greenhouse effect. Of particular interest will be whether the spectra also include so-called biosignatures, like absorption by molecular oxygen (or ozone) or methane. These gases are unstable and short-lived and need to be replenished, which, at least in Earth's case, is done primarily by photosynthesis in living organisms. Thus, TPF and DARWIN are planet-finding missions that will even have the potential to answer the eternal question: *Does life exist somewhere else in the cosmos?*

Bibliography

Boss, A. (1998). "Looking for Earths: The Race to Find New Solar Systems." Wiley, New York.

Clark, S. (1998). "Extrasolar Planets: The Search for New Worlds." Wiley, Chichester, United Kingdom.

Croswell, K. (1997). "Planet Quest: The Epic Discovery of Alien Solar Systems." The Free Press, New York.

Mayor, M., and Queloz, D. (1995) A Jupiter-mass companion to a solar-type star. *Nature* **378**, 355.

Appendix

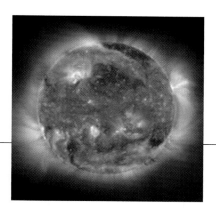

TABLE 1	Planetary Exploration Missions				
Spacecraft	**Source**	**Launch**	**Target**	**Mission**	**Notes**
Pioneer 1	USA	Oct. 11, 1958	Moon	flyby	Reached 70,700 km altitude: missed Moon
Pioneer 2	USA	Nov. 8, 1958	Moon	flyby	Third stage ignition failure
Pioneer 3	USA	Dec. 6, 1958	Moon	flyby	Reached 63,600 km altitude: missed Moon
Luna 1	USSR	Jan. 2, 1959	Moon	flyby	Flew by Moon at 5,998 km distance
Pioneer 4	USA	Mar. 3, 1959	Moon	flyby	Flew by Moon at 60,030 km
Luna 2	USSR	Sept. 12, 1959	Moon	impact	Impacted on Moon
Luna 3	USSR	Oct. 4, 1959	Moon	flyby	Photographed far side of Moon
(Pioneer)	USA	Nov. 26, 1959	Moon	flyby	Payload shroud failed during launch
(Pioneer)	USA	Sept. 25, 1960	Moon	flyby	Second stage malfunction
(unnamed)	USSR	Oct. 10, 1960	Mars	flyby	Failed to achieve Earth orbit
(unnamed)	USSR	Oct. 14, 1960	Mars	flyby	Failed to achieve Earth orbit
(Pioneer)	USA	Dec. 15, 1960	Moon	flyby	First stage exploded
Sputnik 7	USSR	Feb. 4, 1961	Venus	flyby	Failed to depart from low Earth orbit
Venera 1	USSR	Feb. 12, 1961	Venus	flyby	Communications failure in transit
Ranger 1	USA	Aug. 23, 1961	Moon	flyby	Failed to depart from low Earth orbit
Ranger 2	USA	Nov. 18, 1961	Moon	flyby	Failed to depart from low Earth orbit
Ranger 3	USA	Jan. 26, 1962	Moon	impact	Missed Moon by 36,745 km
Orbiting Solar Observatory 1	USA	Mar. 7, 1962	Sun	telescope	Solar observatory in Earth orbit
Ranger 4	USA	Apr. 23, 1962	Moon	impact	Impacted Moon with experiments inoperative
Mariner 1	USA	July 22, 1962	Venus	flyby	Launch failure
Sputnik 19	USSR	Aug. 25, 1962	Venus	flyby	Failed to depart from low Earth orbit
Mariner 2	USA	Aug. 27, 1962	Venus	flyby	Flew by Venus at 34,745 km
Sputnik 20	USSR	Sept. 1, 1962	Venus	flyby	Failed to depart from low Earth orbit
Sputnik 21	USSR	Sept. 12, 1962	Venus	flyby	Failed to depart from low Earth orbit
Ranger 5	USA	Oct. 18, 1962	Moon	impact	Missed Moon by 724 km
Sputnik 22	USSR	Oct. 24, 1962	Mars	flyby	Failed to depart from low Earth orbit
Mars 1	USSR	Nov. 1, 1962	Mars	flyby	Communications failure in transit
Sputnik 24	USSR	Nov. 4, 1962	Mars	flyby	Failed to depart from low Earth orbit
Sputnik 25	USSR	Jan. 4, 1963	Moon	impact	Failed to depart from low Earth orbit
Luna 4	USSR	Apr. 2, 1963	Moon	impact	Missed Moon by 8499 km
Kosmos 21	USSR	Nov. 11, 1963	Venus	test	Failed to depart from low Earth orbit
Ranger 6	USA	Jan. 30, 1964	Moon	impact	Impacted Moon with TV inoperative
Kosmos 27	USSR	Mar. 27, 1964	Venus	flyby	Failed to depart from low Earth orbit
Zond 1	USSR	Apr. 2, 1964	Venus	flyby	Communications failure in transit
Ranger 7	USA	July 28, 1964	Moon	impact	Impacted Moon: returned 4,308 photos
Mariner C	USA	Nov. 5, 1964	Mars	flyby	Shroud failed to separate after launch
Mariner 4	USA	Nov. 28, 1964	Mars	flyby	Flew by Mars: July 15, 1965; returned 21 photos
Zond 2	USSR	Nov. 30, 1964	Mars	flyby	Communications failure in transit
Orbiting Solar Observatory 2	USA	Feb. 3, 1965	Sun	telescope	Solar observatory in Earth orbit
Ranger 8	USA	Feb. 17, 1965	Moon	impact	Impacted moon: returned 7,137 photos
Kosmos 60	USSR	Mar. 12, 1965	Moon	lander	Failed to depart from low Earth orbit
Ranger 9	USA	Mar. 21, 1965	Moon	impact	Impacted moon: returned 5,814 photos
Luna 5	USSR	May 9, 1965	Moon	lander	Landing attempt failed: crashed on Moon
Luna 6	USSR	June 8, 1965	Moon	lander	Missed Moon by 160,935 km
Zond 3	USSR	July 18, 1965	Mars	test	Flew by Moon as test of Mars spacecraft
Orbiting Solar Observatory C	USA	Aug. 23, 1965	Sun	telescope	Solar observatory: failed to orbit
Luna 7	USSR	Oct. 4, 1965	Moon	lander	Retros fired early: crashed on Moon
Venera 2	USSR	Nov. 12, 1965	Venus	probe	Communication failure just prior to Venus arrival
Venera 3	USSR	Nov. 16, 1965	Venus	probe	Communication failure prior to Venus entry
Kosmos 96	USSR	Nov. 23, 1965	Venus	probe	Failed to depart from low Earth orbit
Luna 8	USSR	Dec. 3, 1965	Moon	lander	Retros fired late: crashed on Moon

TABLE 1	Planetary Exploration Missions

Spacecraft	Source	Launch	Target	Mission	Notes
Luna 9	USSR	Jan. 31, 1966	Moon	lander	First lunar soft landing: returned photos
Kosmos 111	USSR	Mar. 1, 1966	Moon	lander	Failed to depart from low Earth orbit
Luna 10	USSR	Mar. 31, 1966	Moon	orbiter	First successful lunar orbiter
Surveyor 1	USA	May 30, 1966	Moon	lander	Lunar soft landing: returned 11,150 photos
Lunar Orbiter 1	USA	Aug. 10, 1966	Moon	orbiter	Lunar photographic mapping
Pioneer 7	USA	Aug. 17, 1966	Solar wind	interplanetary	Monitored solar wind
Luna 11	USSR	Aug. 24, 1966	Moon	orbiter	Lunar orbit science mission
Surveyor 2	USA	Sept. 20, 1966	Moon	lander	Crashed on Moon attempting landing
Luna 12	USSR	Oct. 22, 1966	Moon	orbiter	Lunar photographic mapping
Lunar Orbiter 2	USA	Nov. 6, 1966	Moon	orbiter	Lunar photographic mapping
Luna 13	USSR	Dec. 21, 1966	Moon	lander	Soft lander science mission
Lunar Orbiter 3	USA	Feb. 4, 1967	Moon	orbiter	Lunar photographic mapping
Surveyor 3	USA	Apr. 17, 1967	Moon	lander	Lunar surface science mission
Lunar Orbiter 4	USA	May 4, 1967	Moon	orbiter	Lunar photographic mapping
Kosmos 159	USSR	May 17, 1967	Venus	probe	Possible Venera or Molniya failure in Earth orbit
Venera 4	USSR	June 12, 1967	Venus	probe	Successful atmospheric probe: Oct. 18, 1967
Mariner 5	USA	June 14, 1967	Venus	flyby	Flew by Venus at 3,990 km
Kosmos 167	USSR	June 17, 1967	Venus	probe	Failed to depart from low Earth orbit
Surveyor 4	USA	July 14, 1967	Moon	lander	Communications ceased before landing
Lunar Orbiter 5	USA	Aug. 1, 1967	Moon	orbiter	Lunar photographic mapping
Surveyor 5	USA	Sept. 8, 1967	Moon	lander	Lunar surface science
OSO 4	USA	Oct. 18, 1967	Sun	telescope	Solar observatory in Earth orbit
Surveyor 6	USA	Nov. 7, 1967	Moon	lander	Lunar surface science
Pioneer 8	USA	Dec. 13, 1967	Solar wind	interplanetary	Monitored solar wind
Surveyor 7	USA	Jan. 7, 1968	Moon	lander	Lunar surface science
Zond 4	USSR	Mar. 2, 1968	Moon	test	Unmanned test of Soyuz lunar craft
Luna 14	USSR	Apr. 7, 1968	Moon	orbiter	Mapped lunar gravity field
Zond 5	USSR	Sept. 14, 1968	Moon	test	Circumlunar flyby: spacecraft recovered
Pioneer 9	USA	Nov. 8, 1968	Solar wind	interplanetary	Monitored solar wind
Zond 6	USSR	Nov. 10, 1968	Moon	test	Lunar flyby: precursor of manned flight
Apollo 8	USA	Dec. 21, 1968	Moon	manned	Manned lunar orbiter and return
Venera 5	USSR	Jan. 5, 1969	Venus	probe	Atmospheric entry probe: May 16, 1969
Venera 6	USSR	Jan. 10, 1969	Venus	probe	Atmospheric entry probe: May 17, 1969
OSO 5	USA	Jan. 22, 1969	Sun	telescope	Solar observatory in Earth orbit
Mariner 6	USA	Feb. 24, 1969	Mars	flyby	Flew by Mars: July 31, 1969; returned 75 pictures
Mariner 7	USA	Mar. 27, 1969	Mars	flyby	Flew by Mars: Aug 5, 1969; returned 126 pictures
Apollo 10	USA	May 18, 1969	Moon	manned	Manned lunar orbit test: precursor to landing
Luna 15	USSR	July 13, 1969	Moon	orbiter	Impacted on Moon
Apollo 11	USA	July 16, 1969	Moon	manned	First manned lunar landing and return
Zond 7	USSR	Aug. 8, 1969	Moon	test	Unmanned circumlunar flight and return
OSO 6	USA	Aug. 9, 1969	Sun	telescope	Solar observatory in Earth orbit
Kosmos 300	USSR	Sept. 23, 1969	Moon	test	Possible test of Earth-orbit lunar equipment
Kosmos 305	USSR	Oct. 22, 1969	Moon	test	Aborted lunar landing mission
Apollo 12	USA	Nov. 14, 1969	Moon	manned	Manned lunar landing and return
Apollo 13	USA	Apr. 11, 1970	Moon	manned	Aborted lunar landing: crew returned safely
Venera 7	USSR	Aug. 17, 1970	Venus	lander	First successful lander: Dec. 15, 1970

(Continued)

TABLE 1   Planetary Exploration Missions (*Continued*)

Spacecraft	Source	Launch	Target	Mission	Notes
Kosmos 359	USSR	Aug. 22, 1970	Venus	lander	Failed to depart from low Earth orbit
Luna 16	USSR	Sept. 12, 1970	Moon	sample return	Returned lunar surface samples to Earth
Zond 8	USSR	Oct. 20, 1970	Moon	test	Unmanned circumlunar flight and return
Luna 17	USSR	Nov. 10, 1970	Moon	rover	Lunar rover: Lunokhod
Apollo 14	USA	Jan. 31, 1971	Moon	lander	Manned lunar landing and return
Mariner H	USA	May 8, 1971	Mars	orbiter	Second stage failure at launch
Kosmos 419	USSR	May 10, 1971	Mars	lander	Failed to depart from low Earth orbit
Mars 2	USSR	May 19, 1971	Mars	orbiter	Orbited Mars: Nov. 27, 1971
Mars 2 lander	USSR	May 19, 1971	Mars	lander	Soft landing failed: Nov. 27, 1971
Mars 3	USSR	May 28, 1971	Mars	orbiter	Orbited Mars: Dec. 2, 1971
Mars 3 lander	USSR	May 28, 1971	Mars	lander	First Mars lander: Dec. 2, 1971. Failed after 20 seconds.
Mariner 9	USA	May 30, 1971	Mars	orbiter	First Mars orbiter: Nov. 13, 1971.
Apollo 15	USA	July 26, 1971	Moon	lander	Manned lunar landing and return
P&F satellite	USA	Aug. 4, 1971	Moon	orbiter	Subsatellite deployed by Apollo 15
Luna 18	USSR	Sept. 2, 1971	Moon	return	Crashed on Moon
Luna 19	USSR	Sept. 28, 1971	Moon	orbiter	Photographic mapping mission
OSO 7	USA	Sep. 29, 1971	Sun	telescope	Solar observatory in Earth orbit
Luna 20	USSR	Feb. 14, 1972	Moon	sample return	Lunar surface sample return
Pioneer 10	USA	Mar. 3, 1972	Jupiter	flyby	Jupiter flyby: Dec. 3, 1973
Venera 8	USSR	Mar. 27, 1972	Venus	lander	Landed on Venus: July 22, 1972
Kosmos 482	USSR	Mar. 31, 1972	Venus	lander	Failed to depart from low Earth orbit
Apollo 16	USA	Apr. 16, 1972	Moon	manned	Manned lunar landing and return
P&F satellite	USA	Apr. 19, 1972	Moon	orbiter	Subsatellite deployed by Apollo 16
Apollo 17	USA	Dec. 7, 1972	Moon	manned	Manned lunar landing and return
Luna 21	USSR	Jan. 8, 1973	Moon	rover	Lunar rover: Lunokhod 2
Pioneer 11	USA	Apr. 6, 1973	Jupiter	flyby	Jupiter flyby: Dec. 4, 1974
			Saturn	flyby	Saturn flyby: Sept. 1, 1979
Explorer 49	USA	June 10, 1973	Moon	orbiter	Solar and galactic radio science experiment
Mars 4	USSR	July 21, 1973	Mars	orbiter	Failed to achieve Mars orbit
Mars 5	USSR	July 25, 1973	Mars	orbiter	Orbited Mars: Feb. 12, 1974
Mars 6	USSR	Aug. 5, 1973	Mars	lander	Communications lost just before landing
Mars 7	USSR	Aug. 9, 1973	Mars	lander	Engine failure: missed Mars
Mariner 10	USA	Nov. 3, 1973	Venus	flyby	Flew by Venus: Feb. 5, 1974 en route to Mercury
			Mercury	flyby	Flew by Mercury three times in 1974
Luna 22	USSR	May 29, 1974	Moon	orbiter	Photographic mapper
Luna 23	USSR	Oct. 28, 1974	Moon	sample return	Drill arm damaged: no return attempt
Helios 1	Germany	Dec. 10, 1974	Solar wind	interplanetary	Monitored solar wind and dust
Venera 9	USSR	June 8, 1975	Venus	lander	Landed Oct. 22, 1975
Venera 9 orbiter	USSR	June 8, 1975	Venus	orbiter	Orbited Venus Oct. 22, 1975
Venera 10	USSR	June 14, 1975	Venus	lander	Landed Oct. 25, 1975
Venera 10 orbiter	USSR	June 14, 1975	Venus	orbiter	Orbited Venus Oct. 25, 1975
OSO 8	USA	Jun. 21, 1975	Sun	orbiter	Solar observatory in Earth orbit
Viking 1	USA	Aug. 20, 1975	Mars	orbiter	Orbited Mars: June 19, 1976
Viking 1 lander	USA	Aug. 20, 1975	Mars	lander	Landed on Mars: July 20, 1976; surface science

TABLE 1	Planetary Exploration Missions

Spacecraft	Source	Launch	Target	Mission	Notes
Viking 2	USA	Sept. 9, 1975	Mars	orbiter	Orbited Mars: Aug. 7, 1976
Viking 2 lander	USA	Sept. 9, 1975	Mars	lander	Landed on Mars: Sept. 3, 1976; surface science
Helios 2	Germany	Jan. 15, 1976	Solar wind	interplanetary	Monitored solar wind and dust
Luna 24	USSR	Aug. 9, 1976	Moon	sample return	Lunar surface sample return
Voyager 2	USA	Aug. 20, 1977	Jupiter	flyby	Flew by Jupiter: July 9, 1979
			Saturn	flyby	Flew by Saturn: Aug. 26, 1981
			Uranus	flyby	Flew by Uranus: Jan. 24, 1986
			Neptune	flyby	Flew by Neptune: Aug. 25, 1989
Voyager 1	USA	Sept. 5, 1977	Jupiter	flyby	Flew by Jupiter: Mar. 5, 1979
			Saturn	flyby	Flew by Saturn: Nov. 12, 1980
International Ultraviolet Explorer (IUE)	USA/ESA	Jan. 26, 1978	All	telescope	Ultraviolet observatory in Earth orbit
Pioneer 12	USA	May 20, 1978	Venus	orbiter	Orbited Venus: Dec. 8, 1978
Pioneer 13	USA	Aug. 8, 1978	Venus	probes	Four atmospheric entry probes: Dec. 9, 1978
Venera 11	USSR	Sept. 9, 1978	Venus	lander	Landed: Dec. 25, 1978
Venera 12	USSR	Sept. 14, 1978	Venus	lander	Landed: Dec. 21, 1978
ISEE 3	USA	Aug. 12, 1978	Solar wind	interplanetary	Monitored solar wind; flew through tail of Comet Giacobini-Zinner: Sept. 11, 1985
Solar Max	USA	Feb. 14, 1980	Sun	telescope	Solar observatory in Earth orbit
Venera 13	USSR	Oct. 30, 1981	Venus	lander	Landed: Feb. 27, 1982
Venera 14	USSR	Nov. 4, 1981	Venus	lander	Landed: Mar. 5, 1982
Infrared Astronomical Satellite (IRAS)	USA/UK/ Netherlands	Jan. 25, 1983	All	telescope	Infrared observatory in Earth orbit
Venera 15	USSR	June 2, 1983	Venus	orbiter	Orbited Venus: Oct. 10, 1983; radar mapper
Venera 16	USSR	June 7, 1983	Venus	orbiter	Orbited Venus: Oct. 14, 1983; radar mapper
Vega 1	USSR	Dec. 15, 1984	Venus	lander	Landed: June 11, 1985
	USSR	Dec. 15, 1984	Venus	balloon	Deployed in Venus atmosphere
	USSR	Dec. 15, 1984	Halley	flyby	Flew by Comet Halley at 8,890 km: Mar. 6, 1986
Vega 2	USSR	Dec. 21, 1984	Venus	lander	Landed: June 15, 1985
	USSR	Dec. 21, 1984	Venus	balloon	Deployed in Venus atmosphere
	USSR	Dec. 21, 1984	Halley	flyby	Flew by Comet Halley at 8,030 km: Mar. 9, 1986
Sakigake	Japan	Jan. 8, 1985	Halley	flyby	Distant flyby of Comet Halley: Mar. 11, 1986
Suisei	Japan	Aug. 18, 1985	Halley	flyby	Flew by Comet Halley at 151,000 km: Mar. 8, 1986
Giotto	ESA	July 2, 1985	Halley	flyby	Flew by Comet Halley at 596 km: Mar. 14, 1986
Phobos 1	USSR	July 7, 1988	Phobos	orbiter/lander	Communications lost en route
Phobos 2	USSR	July 12, 1988	Phobos	orbiter/lander	Orbited Mars: Jan. 29, 1989; failed prior to landing
Magellan	USA	May 5, 1989	Venus	orbiter	Orbited Venus: Aug. 10, 1990; radar mapper

(Continued)

TABLE 1   **Planetary Exploration Missions** (*Continued*)

Spacecraft	Source	Launch	Target	Mission	Notes
Galileo Orbiter	USA	Oct. 18, 1989	Jupiter	orbiter	Flew by Venus: Feb. 10, 1990 Flew by Earth: Dec. 8, 1990, Dec. 8, 1992 Flew by asteroid Gaspra: Oct. 29, 1991 Flew by asteroid Ida: Aug. 28, 1993 Orbited Jupiter: Dec. 7, 1995 Impacted Jupiter: Sept. 21, 2003.
Galileo Probe	USA	Oct. 18, 1989	Jupiter	probe	Entered Jupiter's atmosphere: Dec. 7, 1995
Hiten	Japan	Jan. 24, 1990	Moon	flyby	Flew by Moon
Hagormo	Japan	Jan. 24, 1990	Moon	orbiter	Deployed into lunar orbit by Hiten: Mar. 1990
Hubble Space Telescope (HST)	USA/ESA	Apr. 25, 1990	All	telescope	2.4 meter telescope in Earth orbit. Refurbished and new instruments installed by Space Shuttle missions.
Ulysses	ESA/USA	Oct. 6, 1990	Sun	orbiter	Injected into solar polar orbit by Jupiter flyby: Feb. 8, 1992
Yokhoh	Japan	Aug. 30, 1991	Sun	telescope	Solar observatory in Earth orbit
Mars Observer	USA	Sept. 25, 1992	Mars	orbiter	Communications lost en route to Mars
Clementine	USA	Jan. 25, 1994	Moon	orbiter	Orbited Moon on Feb. 19, 1994. Failed after lunar departure.
Infrared Space Observatory (ISO)	ESA	Nov. 17, 1995	All	telescope	Infrared observatory in Earth orbit
Solar Heliospheric Observatory (SOHO)	ESA	Dec. 2, 1995	Sun	telescope	Solar observatory in Earth orbit
Near-Earth Asteroid Rendezvous (NEAR)	USA	Feb. 17, 1996	Eros	orbiter	Flew by asteroid Mathilde: June 27, 1997. Flew by asteroid Eros: Dec. 23, 1998. Orbited asteroid Eros: Feb. 14, 2000. Landed on Eros, Feb. 14, 2001.
Mars Global Surveyor	USA	Nov. 7, 1996	Mars	orbiter	Orbited Mars: Sept. 12, 1997
Mars 96	CIS	Nov. 16, 1996	Mars	orbiter	Failed to depart low Earth orbit
Mars Pathfinder	USA	Dec. 2, 1996	Mars	lander/rover	Landed Mars: July 4, 1997. Deployed rover Sojourner
Cassini	USA	Oct. 15, 1997	Saturn	orbiter/probe	Orbited Saturn: July 1, 2004
Huygens Probe	ESA	Oct. 15, 1997	Titan	probe/lander	Entered Titan atmosphere: Dec. 25, 2004.
Lunar Prospector	USA	Jan. 6, 1998	Moon	orbiter	Orbited Moon: Jan. 11, 1998. Impacted Moon: July 31, 1999.
Nozomi	Japan	July 4, 1998	Mars	orbiter	Failed during cruise to Mars
Deep Space 1	USA	Oct. 24, 1998	Braille	flyby	Flew by asteroid 9969 Braille: July 29, 1999. Flew by comet 19P/Borrelly: Sept. 22, 2001.
Mars Climate Orbiter	USA	Dec. 11, 1998	Mars	orbiter	Lost during orbit insertion maneuver: Sept. 23, 1999
Mars Polar Lander	USA	Jan. 3, 1999	Mars	lander	Crashed on Mars: Dec. 3, 1999
Deep Space 2	USA	Jan. 3, 1999	Mars	penetrators	Crashed on Mars: Dec. 3, 1999
Stardust	USA	Feb. 6, 1999	Comet Wild 2	flyby/sample return	Flew by Comet 81P/Wild 2 Jan. 2, 2004. Returned comet dust samples to Earth: Jan. 2, 2006.
Mars Odyssey	USA	Apr. 7, 2001	Mars	orbiter	Orbited Mars: Oct. 23, 2001.
Genesis	USA	Aug. 8, 2001	Solar wind	sample return	Returned to Earth: Sept. 8, 2004.

TABLE 1	Planetary Exploration Missions				
Spacecraft	**Source**	**Launch**	**Target**	**Mission**	**Notes**
Reuven Ramaty High Energy Solar Spectroscopic Imager (RHESSI)	USA	Feb. 5, 2002	Solar flares	telescope	Solar observatory in Earth orbit
Comet Nucleus Tour (CONTOUR)	USA	July 3, 2002	Comets Encke and SW-1	flyby	Spacecraft failed leaving Earth orbit
Hayabusa	Japan	May 9, 2003	Itokawa	orbiter/sample return	Orbited asteroid 25143 Itokawa: September 2005. En route to Earth with surface sample. Arrival: 2010.
Mars Express	ESA	June 2, 2003	Mars	orbiter	Orbited Mars: Dec. 25, 2003.
Beagle 2	UK	June 2, 2003	Mars	lander	Crashed on Mars: Dec. 25, 2003
Spirit	USA	June 10, 2003	Mars	rover	Landed on Mars: Jan. 3, 2004.
Opportunity	USA	July 7, 2003.	Mars	rover	Landed on Mars: Jan. 24, 2004.
Spitzer Space Telescope	USA	Aug. 25, 2003	All	telescope	Infrared telescope in Earth trailing orbit
SMART 1	ESA	Sept. 27, 2003	Moon	orbiter	Orbited Moon: Nov. 15, 2004. Impacted Moon: Sept. 3, 2006.
Rosetta	ESA	Mar. 2, 2004	Churyumov-Gerasi-menko	orbiter/lander	En route to comet 67P/Churyumov-Gerasimenko. Arrival: 2014. Will fly by asteroids 2867 Steins and 21 Lutetia, plus Earth (3 times) and Mars.
MESSENGER	USA	Aug. 3, 2004	Mercury	orbiter	En route to Mercury. Arrival: Mar. 18, 2011.
Deep Impact	USA	Jan. 12, 2005	Tempel 1	flyby/impactor	Flew by comet 9P/Tempel 1: July 4, 2005. Impactor struck comet nucleus.
Mars Reconnaissance Orbiter (MRO)	USA	Aug. 12, 2005	Mars	orbiter	Orbited Mars: Mar. 10, 2006
Venus Express	ESA	Nov. 9, 2005	Venus	orbiter	Orbited Venus: Apr. 11, 2006.
New Horizons	USA	Jan. 19, 2006	Pluto	flyby	Will arrive: July, 2015
Hinode	Japan	Sept. 22, 2006	Sun	telescope	Solar observatory in Earth orbit
Stereo	USA	Oct. 25, 2006	Sun	telescopes	Dual solar observatories in solar orbit
SELENE	Japan	2007	Moon	orbiter	Will orbit Moon
Dawn	USA	June, 2007	Vesta and Ceres	orbiter	Will orbit asteroids Vesta and Ceres
Phoenix	USA	August, 2007	Mars	lander	Will land on Mars polar cap
Chandrayaan-1	India	March, 2008	Moon	orbiter	Will orbit Moon
Kepler	USA	June, 2008	Extrasolar planets	telescope	Will search for planets around other stars
Mars Science Laboratory	USA	June, 2009	Mars	rovers	Will land on Mars: October, 2010
Juno	USA	June, 2010	Jupiter	orbiter	Will orbit Jupiter
Lunar A	Japan	2010	Moon	orbiter/ penetrators	Will orbit Moon and deploy 2 penetrators to surface

TABLE 2	Selected Astronomical Constants
Astronomical unit, AU	$1.4959787066 \times 10^{11}$ meters
Speed of light, c	2.99792458×10^{8} meters second^{-1}
AU in light time	499.00478353 seconds
Gaussian gravitational constant	0.01720209895 AU$^{3/2}$ day^{-1} solar mass$^{-1/2}$
Gravitational constant, G	6.67259×10^{-11} meters^{3}kg^{-1}sec^{-2}
Mass of the Sun	1.9891×10^{30} kilograms
Mass of the Earth	5.9742×10^{24} kilograms
Solar constant	1368 watts meter^{-2}
Sun-Jupiter mass ratio	1047.3486
Earth-Moon mass ratio	81.30059
Equatorial radius of the Earth	6378.136 kilometers
Obliquity of the ecliptic (J2000)	23°26' 21.412"
Earth sidereal day	23 hours 56 minutes 4.09054 seconds
Sidereal year	365.25636 days
Semimajor axis of the Earth's orbit	1.00000105726665 AU
Parsec, pc	206,264.806 AU
Age of the solar system	4.56×10^{9} years
Age of the galaxy	13×10^{9} years

TABLE 3 Physical and Orbital Properties of the Sun, Planets and Dwarf Planets

Name	Mass kg	Equatorial Radius km	Density g cm^{-3}	Rotation Period	Obliquity degrees	Escape Velocity km sec^{-1}	Semimajor Axis AU	Eccentricity	Inclination degrees	Period years
Sun	1.989×10^{30}	695,500	1.41	25.38–34 d.	7.25*	617.7	—	—	—	—
Mercury	3.302×10^{23}	2,440	5.43	58.646 d.	0.	4.25	0.38710	0.205631	7.0049	0.2408
Venus	4.869×10^{24}	6,052	5.24	243.018 d.	177.33	10.36	0.72333	0.006773	3.3947	0.6152
Earth	5.974×10^{24}	6,378	5.52	23.934 h.	23.45	11.19	1.00000	0.016710	0.0000	1.0000
Mars	6.419×10^{23}	3,397	3.94	24.623 h.	25.19	5.02	1.52366	0.093412	1.8506	1.8808
Ceres[a]	9.47×10^{20}	474	2.1	9.075 h.	—	0.52	2.7665	0.078375	10.5834	
Jupiter	1.899×10^{27}	71,492	1.33	9.925 h.	3.08	59.54	5.20336	0.048393	1.3053	11.862
Saturn	5.685×10^{26}	60,268	0.70	10.656 h.	26.73	35.49	9.53707	0.054151	2.4845	29.457
Uranus	8.682×10^{25}	25,559	1.30	17.24 h.	97.92	21.29	19.1913	0.047168	0.7699	84.018
Neptune	1.028×10^{26}	24,764	1.76	16.11 h.	28.80	23.71	30.0690	0.008586	1.7692	164.78
Pluto[a]	1.314×10^{22}	1,151	2.0	6.387 d.	119.6	1.23	39.4817	0.248808	17.1417	248.4
Eris[a]	1.5×10^{22}	1,200	2.1	—	—	1.29	68.1461	0.432439	43.7408	562.6

*solor obliquity relative to the ecliptic plane

[a]Dwarf planet

Orbital data for January 1, 2000

TABLE 4 Physical and Orbital Properties of the Satellites of the Planets and Dwarf Planets

Name	Semimajor Axis 10³ km	Orbital Eccentricity	Orbital Inclination degrees	Orbital Period days	Mean Radius Km	Mass (10²³ g)	Density g cm⁻³	Year of Discovery	Discovered By
Earth									
Moon	384.40	0.0554	5.16	27.3216	1,737.5	734.9	3.34	—	—
Mars									
Phobos	9.38	0.0151	1.08	0.319	13 × 11 × 9.2	0.000106	1.87	1877	Hall
Deimos	23.46	0.0002	1.79	1.262	7.5 × 6.1 × 5.2	0.000024	1.47	1877	Hall
Jupiter									
J16 Metis	128.0	0.0012	0.02	0.295	20	—	—	1979	*Voyager 1/2*
J15 Adrastea	129.0	0.0018	0.05	0.298	10	—	—	1979	*Voyager 1/2*
J5 Amalthea	181.4	0.0031	0.39	0.498	131 × 73 × 67	0.15	—	1892	Barnard
J14 Thebe	221.9	0.0177	1.07	0.675	50	—	—	1979	*Voyager 1/2*
J1 Io	421.8	0.0041	0.04	1.769	1,821.6	893.2	3.53	1610	Galileo
J2 Europa	671.1	0.0094	0.47	3.551	1,560.8	480.0	3.01	1610	Galileo
J3 Ganymede	1,070.4	0.0011	0.17	7.155	2,631.2	1481.9	1.94	1610	Galileo
J4 Callisto	1,882.7	0.0074	0.19	16.69	2,410.3	1075.9	1.83	1610	Galileo
J13 Leda	11,165	0.1636	27.46	240.92	10	—	—	1974	Kowal
J6 Himalia	11,461	0.1623	27.50	250.56	85	—	—	1904	Perrine
J10 Lysithea	11,717	0.1124	28.30	259.20	18	—	—	1938	Nicholson
J7 Elara	11,741	0.2174	26.63	259.64	43	—	—	1904	Perrine
J12 Ananke	21,276	0.2435	148.89	629.77	10	—	—	1951	Nicholson
J11 Carme	23,404	0.2533	164.91	734.17	23	—	—	1938	Nicholson
J8 Pasiphae	23,624	0.4090	151.43	743.63	30	—	—	1908	Melotte
J9 Sinope	23,939	0.2495	158.11	758.90	19	—	—	1914	Nicholson
J17 Callirrhoe	24,103	0.2828	152.76	758.77	4.3	—	—	1999	Scotti et al.
J18 Themisto	7,284	0.2426	43.26	130.02	4.0	—	—	2000	Sheppard et al.
J19 Megaclite	23,493	0.4197	152.77	752.88	2.7	—	—	2000	Sheppard et al.
J20 Taygete	23,280	0.2525	165.27	732.41	2.5	—	—	2000	Sheppard et al.
J21 Chaldene	23,100	0.2519	165.19	723.70	1.9	—	—	2000	Sheppard et al.
J22 Harpalyke	20,858	0.2268	148.64	623.31	2.2	—	—	2000	Sheppard et al.
J23 Kalyke	23,566	0.2465	165.16	742.03	2.6	—	—	2000	Sheppard et al.
J24 Iocaste	21,061	0.2160	149.43	631.60	2.6	—	—	2000	Sheppard et al.
J25 Erinome	23,196	0.2665	164.93	728.51	1.6	—	—	2000	Sheppard et al.
J26 Isonoe	23,155	0.2471	165.27	726.25	1.9	—	—	2000	Sheppard et al.
J27 Praxidike	20,907	0.2308	148.97	625.38	3.4	—	—	2000	Sheppard et al.
J28 Autonoe	24,046	0.3168	152.42	760.95	2.0	—	—	2001	Sheppard et al.
J29 Thyone	20,939	0.2286	148.51	627.21	2.0	—	—	2001	Sheppard et al.
J30 Hermippe	21,131	0.2096	150.72	633.90	2.0	—	—	2001	Sheppard et al.
J31 Aitne	23,229	0.2643	165.09	730.18	1.5	—	—	2001	Sheppard et al.
J32 Eurydome	22,865	0.2759	150.27	717.33	1.5	—	—	2001	Sheppard et al.

J33 Euanthe	20,797	0.2321	148.91	620.49	1.5	—	—	2001	Sheppard et al.
J34 Euporie	19,304	0.1432	145.77	550.74	1.0	—	—	2001	Sheppard et al.
J35 Orthosie	20,720	0.2808	145.92	622.56	1.0	—	—	2001	Sheppard et al.
J36 Sponde	23,487	0.3121	151.00	748.34	1.0	—	—	2001	Sheppard et al.
J37 Kale	23,217	0.2599	165.00	729.47	1.0	—	—	2001	Sheppard et al.
J38 Pasithee	23,004	0.2675	165.14	719.44	1.0	—	—	2001	Sheppard et al.
J39 Hegemone	23,974	0.3276	155.21	739.60	1.5	—	—	2003	Sheppard
J40 Mneme	21,069	0.2273	151.42	620.04	1.0	—	—	2003	Sheppard and Gladman
J41 Aoede	23,981	0.4322	158.26	761.50	2.0	—	—	2003	Sheppard
J42 Thelxinoe	21,162	0.2206	151.42	628.09	1.0	—	—	2003	Sheppard and Gladman
J43 Arche	23,931	0.2588	165.001	746.39	1.5	—	—	2003	Sheppard
J44 Kallichore	24,034	0.2640	165.50	764.74	1.0	—	—	2003	Sheppard
J45 Helike	21,263	0.1558	154.77	634.77	2.0	—	—	2003	Sheppard
J46 Carpo	16,989	0.4297	51.40	456.10	1.5	—	—	2003	Sheppard
J47 Eukelade	23,661	0.2721	165.48	746.39	2.0	—	—	2003	Sheppard
J48 Cyllene	24,349	0.3189	149.26	751.91	1.0	—	—	2003	Sheppard
S/2003 J2	29,541	0.2255	160.64	979.99	1.0	—	—	2003	Sheppard
S/2003 J3	20,221	0.1970	147.55	583.88	1.0	—	—	2003	Sheppard
S/2003 J4	23,930	0.3618	149.58	755.24	1.0	—	—	2003	Sheppard
S/2003 J5	23,495	0.2478	165.25	738.73	2.0	—	—	2003	Sheppard
S/2003 J9	23,384	0.2632	165.08	733.29	0.5	—	—	2003	Sheppard
S/2003 J10	23,042	0.4295	165.08	716.25	1.0	—	—	2003	Sheppard
S/2003 J12	15,912	0.656	151.91	489.52	0.5	—	—	2003	Sheppard
S/2003 J14	23,614	0.3439	144.51	779.23	1.0	—	—	2003	Sheppard
S/2003 J15	22,627	0.1916	146.51	689.77	1.0	—	—	2003	Sheppard
S/2003 J16	20,963	0.2245	148.534	616.36	1.0	—	—	2003	Sheppard
S/2003 J17	23,001	0.2379	164.92	714.47	1.0	—	—	2003	Sheppard
S/2003 J18	20,514	0.0148	146.06	596.59	1.0	—	—	2003	Sheppard
S/2003 J19	23,533	0.2557	165.16	740.42	1.0	—	—	2003	Sheppard
S/2003 J23	23,563	0.2714	146.31	732.44	1.0	—	—	2003	Sheppard
Saturn									
S18 Pan	133.58	0.0000	0.00	0.575	12.8	—	—	1980	Voyager 1
S15 Atlas	137.67	0.0012	0.00	0.602	19 × 17 × 14	—	—	1980	Voyager 1
S16 Prometheus	139.38	0.0022	0.01	0.613	74 × 50 × 34	—	—	1980	Voyager 1
S17 Pandora	141.72	0.0042	0.05	0.629	55 × 44 × 31	—	—	1980	Voyager 1
S11 Epimetheus	151.41	0.0098	0.35	0.694	69 × 55 × 55	—	—	1979	Pioneer 11
S10 Janus	151.46	0.0068	0.16	0.695	97 × 95 × 77	—	—	1966	Dollfus

(Continued)

TABLE 4 Physical and Orbital Properties of the Satellites of the Planets and Dwarf Planets (*Continued*)

Name	Semimajor Axis 10³ km	Orbital Eccentricity	Orbital Inclination degrees	Orbital Period days	Mean Radius Km	Mass (10²³ g)	Density g cm⁻³	Year of Discovery	Discovered By
S1 Mimas	185.54	0.0196	1.57	0.942	198.8	0.37	1.1	1789	Herschel
S2 Enceladus	238.04	0.0047	0.01	1.370	252.3	0.73	1.1	1789	Herschel
S3 Tethys	294.67	0.0001	1.09	1.888	536.3	6.3	1.0	1684	Cassini
S14 Calypso	294.71	0.0005	1.50	1.888	15 × 8 × 8	—	—	1980	Pascu
S13 Telesto	294.71	0.0002	1.18	1.888	15 × 12 × 8	—	—	1980	Reitsema
S4 Dione	377.42	0.0022	0.03	2.737	562.5	11.0	1.5	1684	Cassini
S12 Helene	377.42	0.0071	0.21	2.737	16.	—	—	1980	Lecacheux
S5 Rhea	527.07	0.0010	0.33	4.518	764.5	23.1	1.2	1672	Cassini
S6 Titan	1,221.87	0.0288	0.28	15.95	2,575.5	1346.	1.88	1655	Huygens
S7 Hyperion	1,500.88	0.0274	0.63	21.28	180 × 140 × 112	15.9	1.0	1848	Bond
S8 Iapetus	3,560.84	0.0283	7.49	79.33	734.5	18.8	1.21	1671	Cassini
S9 Phoebe	12,947.80	0.1635	175.99	550.31	106.6	0.1	—	1898	Pickering
S19 Ymir	23,040.	0.3350	173.12	1315.21	9.	—	—	2000	Gladman
S20 Paaliaq	15,200.	0.3631	45.08	686.93	11.	—	—	2000	Gladman
S21 Tarvos	17,983.	0.5305	33.82	926.23	7.5	—	—	2000	Kavelaars and Gladman
S22 Ijiraq	11,124.	0.3163	46.44	451.43	6	—	—	2000	Kavelaars and Gladman
S23 Suttungr	19,459.	0.1140	175.82	1016.67	3.5	—	—	2000	Gladman and Kavelaars
S24 Kiviuq	11,111.	0.3288	45.70	449.22	8	—	—	2000	Gladman
S25 Mundilfari	18,685.	0.2100	167.32	952.67	3.5	—	—	2000	Gladman and Kavelaars
S26 Albiorix	16,182.	0.4770	34.21	783.46	16.	—	—	2000	Holman and Spahr
S27 Skathi	15,541.	0.2701	152.64	728.21	4.	—	—	2000	Kavelaars and Gladman
S28 Erriapo	17,343.	0.4724	34.69	871.18	5.	—	—	2000	Kavelaars and Gladman
S29 Siarnaq	17,531.	0.2961	46.00	895.55	20.	—	—	2000	Gladman and Kavelaars
S30 Thrymr	20,474.	0.4652	175.97	1094.23	3.5	—	—	2000	Gladman and Kavelaars
S31 Narvi	19,007.	0.4309	145.82	1003.93	3.5	—	—	2003	Sheppard
S32 Methone	194.44	0.0001	0.01	1.01	—	—	—	2004	Cassini
S33 Pallene	212.80	0.0040	0.18	1.15	—	—	—	2004	Cassini
S34 Polydeuces	377. 20	0.0192	0.17	2.74	—	—	—	2004	Cassini
S35 Daphnis	136.50	0.0000	0.00	0.594	—	—	—	2005	Cassini
S/2004 S7	20,999.	0.5299	166.18	1140.28	3.	—	—	2005	Jewitt et al.
S/2004 S8	25,108.	0.2064	170.42	1490.87	3.	—	—	2005	Jewitt et al.
S/2004 S9	20,390.	0.2397	156.38	1086.10	2.5	—	—	2005	Jewitt et al.
S/2004 S10	20,753.	0.2520	166.69	1116.47	3.	—	—	2005	Jewitt et al.
S/2004 S11	17,119.	0.4691	35.01	834.84	3.	—	—	2005	Jewitt et al.
S/2004 S12	19,878.	0.3261	165.28	1046.16	2.5	—	—	2005	Jewitt et al.
S/2004 S13	18,403.	0.2586	168.79	933.45	3.	—	—	2005	Jewitt et al.
S/2004 S14	19,856.	0.3715	165.83	1038.67	3.	—	—	2005	Jewitt et al.

S/2004 S15	19,338.	0.1428	158.56	1005.93	3.	—	—	2005	Jewitt et al.
S/2004 S16	22,453.	0.1364	164.94	1260.28	2.	—	—	2005	Jewitt et al.
S/2004 S17	19,447.	0.1793	168.24	1014.70	2.	—	—	2005	Jewitt et al.
S/2004 S18	20,129.	0.5214	145.21	1083.57	3.5	—	—	2005	Jewitt et al.
S/2006 S1	—	—	—	—	—	—	—	2006	Sheppard et al.
S/2006 S2	—	—	—	—	—	—	—	2006	Sheppard et al.
S/2006 S3	—	—	—	—	—	—	—	2006	Sheppard et al.
S/2006 S4	—	—	—	—	—	—	—	2006	Sheppard et al.
S/2006 S5	—	—	—	—	—	—	—	2006	Sheppard et al.
S/2006 S6	—	—	—	—	—	—	—	2006	Sheppard et al.
S/2006 S7	—	—	—	—	—	—	—	2006	Sheppard et al.
S/2006 S8	—	—	—	—	—	—	—	2006	Sheppard et al.
Uranus									
U6 Cordelia	49.8	0.0003	0.085	0.335	13	—	—	1986	Voyager 2
U7 Ophelia	53.8	0.0099	0.104	0.376	15	—	—	1986	Voyager 2
U8 Bianca	59.2	0.0009	0.193	0.435	21	—	—	1986	Voyager 2
U9 Cressida	61.8	0.0004	0.006	0.464	31	—	—	1986	Voyager 2
U10 Desdemona	62.7	0.0001	0.113	0.474	27	—	—	1986	Voyager 2
U11 Juliet	64.4	0.0007	0.065	0.493	42	—	—	1986	Voyager 2
U12 Portia	66.1	0.0001	0.059	0.513	54	—	—	1986	Voyager 2
U13 Rosalind	69.9	0.0001	0.279	0.558	27	—	—	1986	Voyager 2
U14 Belinda	75.3	0.0001	0.031	0.624	33	—	—	1986	Voyager 2
U15 Puck	86.0	0.0001	0.319	0.762	81	—	—	1985	Voyager 2
U5 Miranda	129.9	0.0013	4.338	1.413	236	0.61	1.1	1948	Kuiper
U1 Ariel	190.9	0.0012	0.041	2.520	579	13.5	1.7	1851	Lassell
U2 Umbriel	266.0	0.0039	0.128	4.144	585	11.7	1.4	1851	Lassell
U3 Titania	436.3	0.0011	0.079	8.706	789	35.3	1.7	1787	Herschel
U4 Oberon	585.3	0.0014	0.068	13.46	761	30.1	1.6	1787	Herschel
U22 Francisco	4,276.	0.1459	145.22	266.56	6	—	—	2001	Holman et al.
U16 Caliban	7,231.	0.1587	140.88	579.73	49	—	—	1997	Gladman et al.
U20 Stephano	8,004.	0.2292	144.11	677.4	10	—	—	1999	Gladman et al.
U21 Trinculo	8,504.	0.2200	167.05	749.2	5	—	—	2001	Holman et al.
U17 Sycorax	12,179.	0.5224	159.40	1288.3	95	—	—	1997	Nicholson et al.
U23 Margaret	14,345.	0.6608	56.63	1687.0	5	—	—	2003	Holman et al.
U18 Prospero	16,256.	0.4448	151.97	1978.3	15	—	—	1999	Holman et al.
U19 Setebos	17,418.	0.5914	158.20	2225.2	15	—	—	1999	Kavelaars et al.
U24 Ferdinand	20,901.	0.3862	169.84	2887.2	6	—	—	2001	Holman et al.
U25 Perdita	76.42	0.0033	0.07	—	13	—	—	1999	Karkoschka / Voyager 2

(Continued)

TABLE 4   Physical and Orbital Properties of the Satellites of the Planets and Dwarf Planets

Name	Semimajor Axis 10³ km	Orbital Eccentricity	Orbital Inclination degrees	Orbital Period days	Mean Radius Km	Mass (10²³ g)	Density g cm⁻³	Year of Discovery	Discovered By
U26 Mab	97.74	0.0025	0.13	—	6	—	—	2003	Showalter and Lissauer
U27Cupid	74.39	0.0	—	—	8	—	—	2003	Showalter and Lissauer
Neptune									
N3 Naiad	48.23	0.0004	4.746	0.294	33	—	—	1989	*Voyager 2*
N4 Thalassa	50.08	0.0002	0.209	0.311	41	—	—	1989	*Voyager 2*
N5 Despina	52.53	0.0002	0.064	0.335	75	—	—	1989	*Voyager 2*
N6 Galatea	61.95	0.0000	0.062	0.429	88	—	—	1989	*Voyager 2*
N7 Larissa	73.55	0.0014	0.205	0.555	97	—	—	1989	*Voyager 2*
N8 Proteus	117.65	0.0005	0.026	1.122	210	—	—	1989	*Voyager 2*
N1 Triton	354.8	0.0000	156.834	5.877	1,353.4	214.0	2.061	1846	Lassell
N2 Nereid	5,513.4	0.7512	7.232	360.14	170	—	—	1949	Kuiper
S/2002 N1	15,728.	0.5711	134.10	1879.71	31	—	—	2002	Holman et al.
S/2002 N2	22,422.	0.2931	48.51	2914.07	22	—	—	2002	Holman et al.
S/2002 N3	23,571.	0.4237	34.74	3167.85	21	—	—	2002	Holman et al.
S/2003 N1	46,695.	0.4499	137.39	9115.91	30	—	—	2003	Jewitt et al.
S/2002 N4	48,387.	0.4945	132.58	9373.99	20	—	—	2002	Holman et al.
Pluto									
P1 Charon	19.40	0.0076	96.16	6.3872	593	16.	1.8	1978	Christy
P2 Nix	48.68	0.0	96.	25.4	40–65	—	—	2005	Weaver et al.
P3 Hydra	64.78	0.0	96.	39.0	30–55	—	—	2005	Weaver et al.
Eris									
Dysnomia	33.	—	—	~14.	200	—	—	2005	Brown et al.

Definition of a Planet

The following Resolutions 5A and 6A, were passed by the International Astronomical Union at its General Assembly in Prague, Czech Republic, on August 24, 2006:

IAU Resolution: Definition of a Planet in the Solar System

Contemporary observations are changing our understanding of planetary systems, and it is important that our nomenclature for objects reflect our current understanding. This applies, in particular, to the designation "planets." The word "planet" originally described "wanderers" that were known only as moving lights in the sky. Recent discoveries lead us to create a new definition, which we can make using currently available scientific information.

RESOLUTION 5A

The IAU therefore resolves that "planets" and other bodies in our Solar System, except satellites, be defined into three distinct categories in the following way:

(1) A "planet"[1] is a celestial body that (a) is in orbit around the Sun, (b) has sufficient mass for its self-gravity to overcome rigid body forces so that it assumes a hydrostatic equilibrium (nearly round) shape, and (c) has cleared the neighbourhood around its orbit.

(2) A "dwarf planet" is a celestial body that (a) is in orbit around the Sun, (b) has sufficient mass for its self-gravity to overcome rigid body forces so that it assumes a hydrostatic equilibrium (nearly round) shape,[2] (c) has not cleared the neighbourhood around its orbit, and (d) is not a satellite.

(3) All other objects[3] except satellites orbiting the Sun shall be referred to collectively as "Small Solar-System Bodies."

[1]The eight "planets" are: Mercury, Venus, Earth, Mars, Jupiter, Saturn, Uranus, and Neptune.

[2]An IAU process will be established to assign borderline objects into either dwarf planet and other categories.

[3]These currently include most of the Solar System asteroids, most Trans-Neptunian Objects (TNOs), comets, and other small bodies.

IAU Resolution: Pluto

RESOLUTION 6A

The IAU further resolves:

Pluto is a "dwarf planet" by the above definition and is recognized as the prototype of a new category of trans-Neptunian objects.

Glossary

Ablation   Removal of material. Meteors ablate during their passage through the atmosphere.

Absolute magnitude (H)   A measure of the brightness of an object. It is defined as the brightness if the object were at 1 AU each from the Sun and the Earth, and viewed at 0 degrees phase angle. For a given albedo, smaller absolute magnitudes correspond to larger objects. A difference of 5 absolute magnitudes corresponds to a factor of 10 in radius for objects with the same albedo.

Accretion   The process of building larger bodies from smaller ones through low velocity collisions where the particles stick to one another, or where the gravity of a relatively large body draws the smaller bodies to it.

Achondrite   Differentiated igneous stony meteorite, apparently solidified from a magma.

Actinic flux   The solar flux used in calculating photodissociation rates, corresponding to the mean intensity at a given point in the atmosphere.

Active optics   The controlled deformation or displacement of optics to compensate for slowly varying effects such as flexure or temperature changes. Typical timescale for updates is longer than 1 second.

Adaptive optics   An observational technique where the phase perturbations induced by the Earth's atmospheric turbulence, responsible for the blur in the images obtained, is corrected in real-time on the incident wavefront reaching the telescope. These perturbations are measured by a wavefront sensor. Opposite phase corrections are then applied using a thin deformable mirror in the pupil plane. The timescale for updates is typically about 1/1000 of a second.

Adiabat   A process occurring without exchange of heat with the surroundings. In an atmosphere, an adiabatic temperature gradient (about -10 K/ km for Venus) is commonly found in regions of rapid vertical motion.

Adiabatic compression   Compression of a gas without exchange of heat. Expansion or compression of rising or sinking air masses in planetary atmospheres is commonly assumed to be driven by adiabatic processes.

Adiabatic temperature lapse rate (or temperature gradient)   For an atmosphere that is marginally unstable to convection, and where there is no heat transfer between the rising and sinking parcels of air with the environment, the temperature profile with altitude follows a so-called adiabat. The dry adiabatic lapse rate on Earth is roughly -7 K/km.

Adsorption   The formation of a thin layer of gas, liquid, or solid on the surface of a solid or, more rarely, a liquid. There are two types. A single layer of molecules, atoms, or ions can be attached to a surface by chemical bonds. Alternatively, molecules can be held onto a surface by weaker physical forces.

Aerosol   In atmospheric physics, aerosol is a generic name for any particle (cloud, dust, haze) suspended in the air, although in the Earth science community the term is usually restricted to apply to haze rather than cloud particles.

Age   Time elapsed since some event at a discrete time, t_o.

Agglutinate   Common particle in the lunar soil, usually about 60 μm in size, consisting of rock, mineral, and glass fragments bonded together by glass (that also contains submicron metal droplets) produced by meteorite impact.

Airglow   The emission of light by an atmosphere. Airglow may result from resonant scattering, fluorescence, impact by charged particles, or radiative decay of atoms, ions, or molecules left in an excited state by some chemical reaction.

Albedo (p)   A ratio of scattered to incident electromagnetic radiation power, most commonly light. It is a unitless measure of a surface or body's reflectivity. The geometric albedo of an astronomical body is the ratio of its total brightness at zero phase angle to that of an idealized fully reflecting, diffusively scattering (Lambertian) disk with the same cross section. The visual geometric albedo refers to this quantity when taking into account only electromagnetic radiation in the visual range. The bond albedo is the fraction of total power in the electromagnetic incident radiation that is scattered back out into space, taking into account all wavelengths. The bond albedo (A) is related to the geometric albedo (p) by the expression $A = pq$, where q is the phase integral.

Encyclopedia of the Solar System 2e ©2007 by Academic Press. All rights of reproduction in any form reserved.

Alfvén speed   The speed of propagation of disturbances in a magnetized plasma that bend a magnetic field without changing its magnitude.

Alfvén waves   A wave propagating in a magnetized plasma in which the magnetic field oscillates transverse to the propagation direction. The propagation speed is given by the Alfvén speed $v_A = B/\sqrt{4\pi\rho}$, where B is the magnetic field strength and ρ is the mass density.

Algae   Any of a large group of mostly aquatic organisms that contain chlorophyll and other pigments and can carry on photosynthesis, but that lack true roots, stems, or leaves; they range from microscopic unicellular organisms to very large multicellular structures.

Aliasing   Overlapping of radar echos at different frequencies or at different time delays.

Alkali plagioclase   An aluminum silicate mineral rich in sodium (see *Feldspar*).

Allochthonous   Describes a rock unit that has been moved into its present location.

Alpha particle   Helium nucleus having mass four times and charge twice that of a proton.

Amino acid   Any organic compound containing an amino acid ($-NH_2$) and a carboxyl ($-COOH$) group; specifically, one of the so-called building blocks of life, a group of 20 such compounds from which proteins are synthesized during ribosomal translation of messenger RNA.

Amorphous   Having no crystalline form.

Angle of repose   The maximum slope at which loose material does not fall downhill.

Angular momentum   Property of orbiting or rotating objects, usually expressed as mvr, where m is the mass, v is the velocity, and r is the distance from the center of rotation. The Earth and Moon have orbital angular momentum on account of their revolution around the Sun and spin angular momentum because of axial rotation. Angular momentum is conserved unless forces act to change it.

Anhydrous silicates   Silicates lacking in water content.

Anomalistic month   The time between successive passages of the Moon through perigee.

Anorthosite   An igneous rock formed almost exclusively of plagioclase. It forms the outer layer of the Moon.

Antenna gain   Ratio of an antenna's sensitivity in the direction toward which it is pointed to its average sensitivity in all directions.

Antipodes   The opposite points on the surface of a sphere, given by a line through the center of the sphere.

Aphelion   The point in the elliptical orbit of a planet, comet, or asteroid farthest from the Sun.

Aphelion distance (Q)   The farthest distance from the Sun of an object in an elliptical orbit, given by $Q = a(1 + e)$, where a is the object's semimajor axis and e is its eccentricity.

Apoapse   Point on an orbit farthest from the center of gravity, called **aphelion** for orbits about the Sun and **apogee** for orbits about the Earth.

Apogee   The point in the orbit of the Moon or an artificial satellite, furthest away from the Earth.

Apojove   The point in an orbit around Jupiter, farthest from the planet.

Apse   Informal synonym for *Longitude of perihelion*.

Arc seconds, arcsec, second of arc   An angle equal to 1/3600 of a degree, or 1/60 of an arc minute. The Sun subtends an angle of ~1919 arcsec on average when viewed from the Earth.

Arc minute, arcmin, minute of arc   An angle equal to 1/60th of a degree, or 60 arc seconds.

Areocentric Sun longitude (L_s)   An angular measure of the Martian year. $L_S = 0°$ corresponds to the vernal equinox, marking the beginning of Spring in the northern hemisphere. $L_S = 90°, 180°$, and $270°$ correspond to the summer solstice, autumnal equinox, and winter solstice, respectively.

Argument of perihelion (ω)   In an orbit around the Sun, the angle between the ascending node and the perihelion point, measured in the body's orbital plane and along its direction of motion.

Asteroid   A rocky, carbonaceous, or metallic body, smaller than a planet and orbiting the Sun. Most asteroids are in semistable orbits between Mars and Jupiter, but others are thrown onto orbits crossing those of the major planets. Also called a **minor planet**.

Asthenosphere   A low viscosity zone that lies between the lithosphere and the mantle.

Astrology   A belief system in which the future of individuals is predicted based upon the date and location of their birth and the positions of the moon and planets relative to the Sun or Earth at specific times.

Astronomical unit (AU)   Commonly thought of as the mean distance of the Earth from the Sun. It is more formally the distance at which a massless particle in a circular orbit would have an orbital period of 1 Gaussian year, equal to $365.256898326\ldots$ days. It is equal to 149.59787066×10^6 km, or about 92.955807×10^6 miles.

Aurora   Atmospheric emissions excited by the precipitation of energetic magnetospheric and solar particles, most frequently at high latitudes.

Autochthonous   Describes a rock unit that has been formed in place.

Autotrophy   Literally, self-feeding; the capacity of an organism to obtain its essential nutrients by synthesizing nonorganic materials from the environment, rather than by consuming organic materials; photosynthetic green plants and chemosynthetic bacteria are examples of autotrophic organisms.

(B-R) color   A color scale for astronomical objects. Light from astronomical objects consists of all the colors of the rainbow: red, orange, yellow, green, blue, indigo, and violet (roygbiv). The

(B-R) color measures the proportion of red to blue light from an object. A red astronomical object has (B-R) ∼ 2, the Sun has (B-R) = 1.03, and a blue astronomical object has (B-R) ∼ 0.

Bar Unit of pressure, equal to 10^6 dyn/cm^2 or 10^5 Pa; the standard sea level pressure of the Earth's atmosphere is 1.013 bar. Typical planetary interior pressures are measured in megabars (Mbar) or 10^6 bar.

Baroclinic instability A 3-dimensional process common in the midlatitude troposphere wherein cold polar air pushes underneath hot low-latitude air, which transports heat toward the poles and produces complex circulation patterns that generate much of the winter rainfall in the midlatitudes.

Baroclinic, barotropic Barotropic is a region of uniform temperature distribution; a lack of fronts. Everyday being similar (hot and humid with no cold fronts to cool things off) would be a barotropic type of atmosphere, such as we find at tropical latitudes. In a baroclinic region, distinct air mass regions exist. Fronts separate warmer from colder air. In a synoptic scale baroclinic environment, you will find the polar jet, troughs of low pressure (midlatitude cyclones), and frontal boundaries. There are clear density gradients in a baroclinic environment caused by the fronts. Mid-latitude cyclones are found in a baroclinic environment.

Barycenter The center of mass of a system of two or more gravitationally bound (orbiting) bodies.

Basalt An igneous rock primarily composed of plagioclase and pyroxene. On Earth, oceanic crust is primarily basaltic in composition.

Beta meteoroid Small meteoroid for which the solar radiation pressure force is comparable to solar gravitational attraction and hence leaves the solar system on an unbound orbit.

Blackbody radiation Continuous spectrum of electromagnetic radiation emitted by an object that absorbs all radiation incident on it.

Bolide A meteoric fireball.

Bond albedo Ratio of the total radiation reflected in all directions from a solar system object to the total incidence flux.

Bound orbit Circular or elliptical orbit about a central body (e.g., Sun, planet).

Breccia Rock composed of fragments derived from previous generations of rocks.

Bremsstrahlung Electromagnetic radiation that is emitted when an energetic electron is deflected by an ion. It is also called free-free emission because both the electron and ion are free in an ionized plasma. The term is borrowed from German and means "braking radiation" because the deflected electron loses energy (by the emitted photon) and is slowed down.

Brightness temperature The temperature a body would have if it were a blackbody producing the same brightness as the observed object at the same wavelength. It can also be defined as the radiant intensity scaled to units of temperature by $\lambda^2/2k$ where λ is wavelength, and k is Boltzmann's constant.

CAI: calcium-aluminum inclusion Minor component of primitive meteorites composed of refractory minerals.

Caldera Large volcanic crater, usually greater than 1 km in diameter. A caldera is many times the size of any associated vent(s). Calderas are formed either by collapse (most often) or explosion.

Carbonaceous (C-type) material Carbon-silicate material rich in simple organic compounds, such as that found in carbonaceous meteorites, which are believed to be among the most primitive (unaltered since their formation in the solar nebula) objects found in the solar system. They contain complex carbon compounds (hydrocarbons, amino acids), made mostly from the elements C, H, O, and N. C-type material is low albedo, spectrally flat and exists on the surfaces of several outer planet moons.

Catalytic cycle Series of chemical reactions facilitated by a substance that remains unchanged.

Cd-cutoff: The neutron capture cross section for natural cadmium (Cd) is very high for thermal neutrons, but drops sharply for energies greater than about 0.5 eV, which is sometimes referred to as the "Cd cutoff" energy. Consequently, Cd is an excellent filter that absorbs thermal neutrons, but allows epithermal and fast neutrons to pass through.

Centaur A small body in a heliocentric orbit whose average distance from the Sun lies between the orbits of Jupiter and Neptune, and that has a Tisserand parameter with respect to Jupiter greater than 3. Typically, the orbits of Centaurs also cross one or more of the orbits of the other giant planets. Centaurs are part of the population of ecliptic comets They are most likely derived from the Kuiper belt and the Scattered disk. Eventually, some Centaurs may evolve into the terrestrial planets zone and become short-period comets.

Chaotic motion A dynamical situation in which the error in the prediction of the long-term motion of a body grows exponentially with time. This exponential growth leads to an inability to predict the location of the body. Chaotic motion can be confined to fairly narrow regions of space so that the orbit of the body will not change much over time. This is the case for the planets in our solar system. However, for most comets and some asteroids that we observe, chaotic motion leads to sudden and drastic changes in the orbit of the body.

Chaotic terrain Areas of the martian surface where the ground has collapsed to form a surface of jostled blocks standing 1–2 km below the surrounding terrain.

Charge–coupled device (CCD) A solid-state device used to record light electronically. A typical CCD has thousands to millions of tiny detectors arranged in a grid pattern. Each detector element is called a pixel. These devices record images electronically. CCDs have completely replaced photographic plates in astronomy due to their very high efficiency in capturing light.

Chasmata Term used in planetary geology to refer to long, relatively narrow, steep-sided troughs.

Chemoautotrophy The capacity of an autotrophic (self-feeding) organism to derive the energy required for its growth from certain chemical reactions (e.g., methanogenesis) rather than from photosynthesis; some bacterial forms are chemoautotrophic organisms.

Chiron-type comet A Centaur that displays cometary activity.

Chlorofluorocarbons (CFCs) Various compounds made with the halogens chlorine and fluorine. Their stability made them favored refrigerants until it was discovered that this also makes them efficient atmospheric ozone destroyers.

Chondrite: Undifferentiated stony or carbonaceous meteorite, usually containing chondrules or their fragments.

Chondrule Approximately spherical, millimeter-sized droplet formed by partial or complete melting and quenching prior to incorporation into undifferentiated meteorites.

Chromophore Any coloring material.

Chromosphere Lower atmosphere of the Sun, above the photosphere and beneath the transition region, with a vertical height extent of about 2000 km and a temperature range of 6000–20,000 K.

Circular polarization ratio Ratio of radar echo power received in the same sense of circular polarization as transmitted (the SC sense) to that received in the opposite (OC) sense.

Circular restricted three-body problem A special case of the problem of calculating the gravity-controlled motion of three bodies. In the circular restricted three-body problem, the two massive bodies follow circular orbits about each other, and the mass of the third body is negligible. The motion of comets and asteroids can often be approximated with the circular restricted three-body problem with the Sun and Jupiter, or sometimes the Sun and Neptune, as the two massive bodies.

Clathrate, clathrate compound, or cage compound A chemical substance consisting of a crystalline lattice of one type of molecule trapping and containing a second type of molecule. A clathrate therefore is a material that is a weak composite, with molecules of suitable size captured in spaces left by the other crystalline molecule. Water ice often forms clathrates with more volatile molecules.

Column abundance The product of density (g/cm^3) and geometric thickness (cm). Measures the mass per unit area of an atmospheric or surface layer.

Column density The number of molecules above a column of unit area in an atmosphere.

Coma The freely outflowing atmosphere of gas and dust around the nucleus of a comet. The nucleus and coma of a comet together are often called the head.

Comet A body containing a significant fraction of ices, smaller than a planet or dwarf planet and orbiting the Sun, usually in a highly eccentric orbit. Most comets are stored beyond the planetary system in two large reservoirs: the Kuiper belt beyond the orbit of Neptune and the Oort cloud at near-interstellar distances. Comets become "active" when their ices sublimate and carry gas and dust into the coma.

Comet dust trail A contrail-like structure extending behind a comet close to its orbit, and sometimes a short distance ahead of the comet, consisting of large particles ($\beta < \sim 10^{-3}$) emitted at low velocities from the nucleus. Trails are distinguished from comet tails, which consist of much smaller particles, more sensitive to solar radiation pressure.

Cometary mantle or crust A layer of refractory material covering some or all of a comet nucleus' surface. When thick enough, a cometary mantle will choke off outgassing over that area.

Conjunction Occurs when two or more planetary bodies appear in the same area of the sky.

Conservation of angular momentum Fundamental physical law requiring that the quantity of angular momentum, p, be conserved (constant) for objects in orbit around a primary body: $p = mvr$, were m is mass, v is velocity, and r is the distance from the primary body, and for rotating objects. Angular momentum is not conserved in the presence of modifying torques.

Contact surface In the vicinity of a comet, the surface that separates outflowing cometary plasma from the slowed solar wind that is approaching the comet.

Convection Transport of energy by mass motion. In turbulent regions of planetary atmospheres and interiors, rising parcels of hot air (or rock) and sinking parcels of cool air (or rock) transport energy outward from the interior.

Core The central part of a differentiated planet, satellite or asteroid. Terrestrial planets have nickel-iron cores. Jovian planets have rocky-iron cores.

Coriolis acceleration Component of the acceleration on a rotating planet that acts perpendicular to the motion and balances the horizontal pressure gradient in an atmosphere or ocean. It causes circulation around high- and low-pressure centers. It is strongest in the polar regions and weakest in the tropics.

Corona Upper atmosphere of the Sun, extending above the transition region out into the heliosphere, with a dominant temperature of 1 million to 2 million K in the lowest 100,000 km. The corona is visible during solar eclipses, and extends outward many solar radii.

Coronae Circular to oval feature surrounded by concentric ridges and fractures.

Coronal mass ejection (CME) Magnetic instabilities in the solar corona that lead to eruption of filaments, prominences, and magnetic flux ropes, which propagate as ejected mass out into the heliosphere, often accompanied by flare phenomenon. Known as an interplanetary coronal mass ejection (ICME) when observed in the solar wind far from the Sun.

Coulomb interaction or collision The interaction of charged particles at large distances through the Coulomb force. The interaction is sometimes called a Coulomb collision because it produces a change of particle momentum similar to that in a conventional collision.

Cratering flow-field   Movement of target materials in an impact event in response to the passage of the shock and rarefaction, or decompression, waves.

Critical point   Temperature and pressure for a given material above which there is no distinction between the liquid and gas phases.

Crust   The chemically distinct, less dense, outer shell of a planet or satellite formed by melting of the interior.

Cryovolcanism   Volcanism where the volcanic materials are melted ices, such as water, ammonia, and methane, as distinguished from the common high-melting point volcanic materials of the terrestrial planets, such as basalt and rhyolite. The melted ices freeze on the surface, forming "lava flows" composed of ice.

Cryovolcano   An icy volcano. See cryovolcanism.

Cumulate   Plutonic igneous rock composed of crystals accumulated by floating or sinking in the silicate melt, or magma.

Current   Strictly, the product of the velocity and number density of particles. The dot product of the current and the unit normal vector of a surface yields the net number of particles crossing the surface per unit area. In the chapter "Remote Chemical Sensing Using Nuclear Spectroscopy," current is taken to be the net number of particles crossing out of a planetary surface per cosmic ray, which is dimensionless.

Cycloidal   Motion of a cycloid, which is the curve defined by a fixed point on a wheel as it rolls.

Cyclostrophic balance   On Earth, the surface rotation typically surpasses the zonal winds, and the pressure gradient force generated by the unequal solar fluxes at low and high latitudes is balanced by the Coriolis force, in what is called a geostrophic balance. On Titan (and Venus), the opposite is true, and pressure gradients are balanced by strong centrifugal forces arising from the rapid rotation of the atmosphere. This balance, typical of cyclones, is called cyclostrophic.

Degenerate matter   Matter at very high pressures where the normal atomic structure is destroyed.

Deterministic system   Dynamical system in which the individual bodies move according to fixed laws described mathematically in the form of equations of motion. A deterministic system can still give rise to chaotic, unpredictable motion because of the finite precision with which any physical measurement or numerical computation can be made.

Deuterium/deuterated   Heavy form of the hydrogen atom, consisting of one proton and one neutron. A deuterated molecule, such as CH_3D, deuterated methane, has one or more deuterium atoms in place of hydrogen.

Diapir   A body of rock or ice that has moved upward due to buoyancy, attaining an inverted teardrop or pear shape, and piercing and displacing the overlying layers.

Diaplectic glass   Glass phase produced from minerals by the destruction of internal structural order, without melting, by the passage of a shock wave.

Differentiation   Melting and fractionation of a planet, moon or asteroid into multiple layers or zones of different chemical composition; e.g., core, mantle and crust. High-density materials sink and low-density materials float.

Diffraction   A physical process in which light from different parts of a mirror or lens interfere with each other. As a result of diffraction, the image of a star is not absolutely sharp; instead the full width at half maximum (FWHM) of the image is given by $\theta = 0.252*\lambda/D$ arcseconds, where the wavelength is given by λ in micrometers and D is the telescope diameter in meters. In the absence of atmospheric seeing, the image of a star would be diffraction-limited if the optics were polished sufficiently well.

Dipole magnetic field   The shape of the field lines around a short bar magnet. This field can be visualized by sprinkling iron filings on a piece of paper on top of a bar magnet.

Direct or prograde motion   Orbital or axial motion of a body in the solar system that is counterclockwise as seen from north of the ecliptic.

Diurnal stresses   Twice-daily forces exerted on a planet, which are derived from gravitational forces between the planet and a satellite.

Doppler shift   Difference between the frequencies of the radar echo and the transmission, caused by the relative velocity of the target with respect to the radar. Also, the shift in frequency of spectral lines due to the motion of a light source toward or away from an observer.

Drainage basin   Geomorphic entity that contains a drainage network. Typically a bowl-shaped catchment in humid areas, drainage basins in arid regions can be quite flat. Drainage patterns typically reflect the topography of the drainage basins that contain them.

D-type material   Primordial, low-albedo material thought to be rich in organic compounds. It is redder than C-type material.

Ductile   Pliable or elastic.

Dust tail   The broad, relatively featureless tail of a comet consisting of micron-sized dust particles being driven away from the nucleus and coma by solar radiation pressure. The dust particles are on independent orbits around the Sun under reduced gravity. The dust tail appears whitish or yellowish from sunlight scattered by the dust particles.

Dwarf planet   A new term created by the IAU in 2006 to describe bodies orbiting the Sun that are round (in hydrostatic equilibrium) but are not massive enough to have cleared their zones. Dwarf planets include Ceres, Pluto, and Eris (2003 UB_{313}). Other main-belt asteroids and Kuiper belt objects are potential candidates for this classification as more is learned about them.

Dynamical inertia   The increase in the radiative time constant due to mixing of more massive, deeper layers of an atmosphere. The thin atmosphere at high altitudes would be expected to respond rapidly to changes in sunlight, but if there is substantial vertical circulation, the changes will occur more slowly as the mixing increases the effective mass of the layer under consideration.

Eccentricity   A measure of the departure of an orbit from circular. For an elliptical orbit, the eccentricity, e, is equal to $(1 - b^2/a^2)^{1/2}$, where a and b are the semimajor and semiminor axes of the ellipse, respectively. Circular orbits have $e = 0$; elliptical orbits have $0 < e < 1$; radial and parabolic orbits have $e = 1$; and hyperbolic orbits have $e > 1$.

Echo bandwidth   Dispersion in Doppler frequency of a radar echo, i.e., the width of the echo power spectrum.

Ecliptic, ecliptic plane   The plane of the Earth's orbit around the Sun. The planets, most asteroids, short-period comets, Kuiper belt objects and Scattered-disk objects follow orbits with small or moderate inclinations (or tilts) relative to the ecliptic.

Ecliptic comet   A comet with a Tisserand parameter with respect to Jupiter greater than 2. They generally have a small or moderate inclination to the ecliptic. The designation of such objects as comets does not necessarily imply visible cometary activity because comets typically do not become active until they pass well within Jupiter's orbit. The term "comet" assumes that the body contains a substantial fraction of water ice. Ecliptic comets include Centaurs and Jupiter-family comets.

Ekman layer   Idealized model of the planetary boundary layer in which the mean flow (as in an atmosphere or ocean) is modified near the ground by friction (either laminar or turbulent). The resulting variation of flow speed and direction with height is described by the Ekman spiral.

Elongation   The angular distance between the Sun and a planet or other solar system body as viewed from the Earth.

Embays   To form a protective barrier.

Emissivity   The ratio of radiant energy flux from a material to that from a blackbody at the same temperature. A blackbody is an ideal material that absorbs all radiant energy incident upon it and emits radiant energy at the maximum possible rate per unit area at each wavelength for a given temperature. A blackbody has an emissivity of 1 across the entire spectrum. Real materials have an emissivity between 0 and 1 for a given wavelength.

Encke-type comet   An ecliptic comet whose entire orbit is interior to the orbit of Jupiter and which has a Tisserand parameter with respect to Jupiter greater than 3.

Endogenic   Forming from within

ENSO   El Niño, "the child," and the Southern Oscillation. El Niño is the episodic appearance of warm water off the coast of South America, often at Christmas time, that devastates Peruvian fishing (usually one fifth of the world's catch), causes drought conditions in Australia, and weakens the monsoon in India. The Southern Oscillation is the historical name for the global (not just southern) atmosphere-ocean oscillation for which El Niño years are the extreme.

Entropy   Broadly, the degree of disorder, or randomness in a system; in thermodynamics, a measure of the amount of heat energy in a closed system that is not available to do work. In a condition of low entropy (high efficiency), the system will convert to energy a large portion of the heat transferred to it from an external source (no actual system can utilize 100% of the heat it receives).

Enzymes   Proteins that catalyze, or accelerate, chemical reactions.

Equal angle map   For mapping, the surface of a planet is subdivided into spatial elements called pixels on which quantities such as counting rates and elemental abundances are specified. In an equal area map, all of the pixels span the same angle in latitude and longitude. Consequently, a parallel near a pole is divided into the same number of pixels as a parallel at the equator and the area of the pixels varies with latitude.

Equal area map   The longitude-range for pixels at different latitudes is adjusted so that all of the pixels have approximately the same area. Consequently, a parallel near the pole is divided into fewer pixels than at the equator. The span of pixels in latitude and longitude at the equator is used to specify the map.

Equation of state   Equation relating the pressure of a given material to its temperature and density, typically derived from experimental and theoretical considerations.

Equilibrium vapor pressure   The ambient pressure of the gas phase over a condensed phase when the gas and condensed phase are in thermodynamic equilibrium (i.e., when the rate of condensation from gas to ice equals the rate of sublimation from ice to gas). In effect, vapor pressure is a measure of the amount of gas an ice or liquid layer at a specified temperature will evolve in a closed container (or planetary atmosphere). Vapor pressures are extremely sensitive functions of temperature and are also related to the composition and structure of a given ice or liquid.

Escape velocity   Minimum velocity required to escape from the surface of a body to infinity.

Europium anomaly and Eu*   Because the Moon is highly reduced, europium is divalent on the Moon and hence is mostly separated from the other smaller trivalent rare earth elements because it is concentrated in plagioclase feldspar. The degree of enrichment or depletion is given by Eu/Eu*, where Eu is the measured abundance and Eu* is the abundance expected if Eu had the same relative concentration as the neighboring rare earth elements, samarium and gadolinium.

Exosphere   The outermost part of an atmosphere, characterized by very low densities and very long mean-free paths, and usually isothermal.

Extended scattered disk   Collection of objects with orbits with semimajor axis >50 AU, large eccentricity, and perihelion distance large enough to avoid destabilizing encounters with Neptune. The apparent similarity with the orbits of objects in the scattered disk suggests that the latter extended further in perihelion distance in the past, due to a different orbital architecture of the planets or of the environment of the solar system. The most prominent members of the extended scattered disk population are 2000 CR$_{105}$ and 90377 Sedna.

External comet A returning comet with a semimajor axis greater than ∼34.2 AU. External comets have Tisserand parameters with respect to Jupiter less than 2. Also known as a **long-period comet**.

Extrasolar planet A planetary companion to a star other than the Sun.

Feldspar A common group of aluminum silicate minerals.

Filaments Near-horizontal magnetic field lines on the Sun suspended above magnetic inversion lines that are filled with cool and dense chromospheric mass, seen on the solar disk.

Flares A magnetic instability in the solar corona that impulsively releases large energies that go into heating of coronal and chromospheric plasma, as well as into acceleration of high-energy particles. A flare is usually accompanied by impulsive emission in gamma rays, hard x-rays, soft x-rays, EUV, and radio emission.

Fluctus (pl., flucti) Term meaning (on Io) a volcanic flow field.

Fluorescence Photons emitted immediately after electron decay. The electron had been elevated to a higher energy state by external stimulation of their parent atoms, ions, and molecules. In planetary atmospheres, the external stimulation is usually sunlight or electrons.

Flux density Power per unit area and per unit frequency interval received from an object. The units of flux density are Janskies: 1 Jy = 10^{-26} W m^{-2} Hz^{-1}.

Flux transfer event A localized spatial region in which magnetic reconnection links the solar wind magnetic field to a planetary magnetic field producing a configuration that transports flux from the day side to the night side of the planet.

Flux The flux of particles (denoted φ with units of cm^{-2}s^{-1}) given by the product of the speed of the neutrons, v (cm/s), and the number density (particles per cm^3).

Fractionation Separation of elements or isotopes based on their masses or chemistry.

Frequency–time spectrogram A graph of the emission intensity as a function of frequency and time. Usually the intensity is shown on a gray scale ranging from black to white or one of several color schemes, with frequency plotted along the vertical y-axis and time along the horizontal x-axis.

Galactic cosmic rays Energetic particles, including photons, electrons, protons, and heavy ions, that originate outside the heliosphere.

Galilean satellites The four major satellites of Jupiter: Io, Europa, Ganymede and Callistio, discovered by Galileo in 1610.

Gamma ray A high energy quantum of electromagnetic radiation (photon) emitted by nuclear transitions. Gamma rays originate from nuclear processes, such as radioactive decay and the de-excitation of residual nuclei produced by nuclear reactions.

Gas chromatography A chemical technique for separating gas mixtures, in which the gas is passed through a long column containing a fixed absorbent phase that separates the gas into its component parts.

Gas drag Drag force experienced by a solid object when it moves through a surrounding gas.

Gaussian year The orbital period of a massless particle in a circular orbit with a semimajor axis of 1 AU, equal to 365.256898326 . . . days. Formally, the Gaussian year is defined as $2p/k$, where k is the Gaussian gravitational constant, 0.01720209895.

Geochemistry The study of the chemical components of the lithosphere of the Earth and other planets, chemical processes and reactions that produce and modify rocks and soils, and the cycles of matter and energy that transport chemical components in space and time.

Geodesy The measurement and representation of Earth's topography, its gravitational field and geodynamic phenomena (e.g., polar motion, tides, and crustal motion) in 3-dimensional, time-varying space.

Geomagnetic activity Disturbances in the magnetized plasma of a magnetosphere associated with fluctuations of the surface field, auroral activity, reconfiguration and changing flows within the magnetosphere, strong ionospheric currents, and particle precipitation into the ionosphere.

Geomagnetic storm The response of the Earth to the arrival of an interplanetary medium disturbance, usually associated with a CME.

Geomagnetism The Earth's magnetic field, which is approximately a magnetic dipole, with the magnetic poles offset from the corresponding geographic poles by approximately 11.3°, and extending several tens of thousands of kilometers into space.

Geometric albedo Ratio of the brightness at a phase angle of zero degrees (full illumination) compared with a diffuse, perfectly reflecting disk of the same size and under the same illumination conditions.

Geomorphology Science of landscape analysis. Geomorphic investigations deal with the processes and timescales of landscape formation and degradation.

Geospace The Earth's magnetosphere and upper atmosphere, including the ionosphere.

Graben A long, usually linear fault trough (valley) produced by subsidence between two inward dipping boundary faults. It is the result of extensional stresses in a body's upper crust.

Granite Light-colored intrusive rock containing more than 50% silica. On Earth, continents are largely granite and other high silica rocks.

Gravitational focusing The tendency of an object's trajectory to curve toward a massive body due to gravitational attraction.

Gravitational instability   Spontaneous collapse of a portion of a protoplanetary disk due to mutual gravitational attraction. This can refer to either the solid or gaseous component of the disk.

Greenhouse effect   Heating of a planetary surface above the temperature that it would have been in the absence of an atmosphere. The atmosphere transmits solar radiation in the visible, but impedes the escape of thermal infrared energy (usually due to absorbing clouds), thus creating the increased temperature.

Gyro radius   Radius of the orbit of a charged particle gyrating in a magnetic field.

Gyrofrequency   The frequency of the circular motion of a charged particle perpendicular to a magnetic field.

Habitable zone   The region of space around a star in which a geologically active, rocky planet can maintain liquid water on its surface.

Hadley circulation   A major component of atmospheric circulation driven directly by latitude-averaged heat sources and sinks. Warm air rises in regions near the equator, flows poleward at higher altitudes, and loses heat in the colder, higher latitude regions. The cooler, denser air then descends and has a flow component near the surface back toward the low-latitude heat source, which completes a circulation cell. The near-surface and high-altitude branches of the flow have eastward ("trade wind") and westward components, respectively, arising from Coriolis forces. When the heat source is located on the equator, the Hadley circulation tends to be symmetric about the equator, but the Hadley circulation is asymmetric about the equator if the heat source is located off the equator, as occurs during solstice seasons on Earth and Mars.

Halley-type comet   A returning comet with a semimajor axis less than ~34.2 AU. Halley-type comets have Tisserand parameters with respect to Jupiter less than 2.

Heat flow   Heat emitted (or received) at the surface of a body that is ultimately radiated to (or absorbed from) space.

Heavy elements   In astrophysics, all elements other than hydrogen and helium.

Heliocentric   A Sun-centered coordinate system.

Heliopause   Interface between the heliosphere and the interstellar plasma; the outer boundary of the heliosphere.

Heliosphere   The cavity carved in the interstellar plasma by the solar wind, containing the solar system and plasma and magnetic field of solar origin;

Heliospheric current sheet   The surface in interplanetary space separating solar wind flows of opposite magnetic polarity; the interplanetary extension of the solar magnetic equator.

Heliospheric magnetic field   Remnant of the solar magnetic field dragged into interplanetary space by the solar wind.

Heterotrophy   Literally, other-feeding; the condition of an organism that is not able to obtain nutrients by synthesizing nonorganic materials from the environment, and that therefore must consume other life forms to obtain the organic products necessary for life; e.g., animals, fungi, most bacteria.

Hill sphere   Region around a secondary in which the secondary's gravity is more influential for the motion of a particle about the secondary than is the tidal influence of the primary.

Hilly and lineated terrain   The broken-up surface of Mercury at the antipode of the Caloris impact basin.

Homopause   Level in an atmosphere, above the stratosphere, at which gases cease being uniformly mixed and separate by diffusion, with the lighter elements diffusing upward.

Horseshoe orbits   Librating orbits encircling the L_3, L_4, and L_5 Lagrangian points in the circular restricted three-body problem. These orbits appear to be shaped like horseshoes in the frame rotating with the mean motion of the system.

Hot Jupiter   An extrasolar gas giant planet at a very small orbital separation of 0.03–0.05 AU from its host star and with an orbital period of a few days. The proximity of the discovered hot Jupiters to their host stars is probably a result of inward orbital migration.

Hot poles   The alternating perihelion subsolar points on Mercury at the 0° and 180° meridians.

Hot spots   Regions of enhanced thermal emission on Io, a marker of volcanic activity. The term does not imply a particular eruption mechanism.

Hugoniot elastic limit   Stress at which a rock or mineral's response to shock changes from elastic to plastic. Stresses above the Hugoniot elastic limit cause the rock or mineral to deform plastically.

Hydrated   A mineral in which water molecules or hydroxyl radicals are attached to the crystalline structure.

Hydrodynamic escape   A limiting case of atmospheric escape that occurs when the escape rate is so rapid that the atmosphere at high altitudes reaches an outward velocity comparable to the speed of sound. This occurs if the thermal energy of the gas molecules becomes comparable to the gravitational binding energy. Hydrodynamic escape allows the upper atmosphere of a planet to escape wholesale, as opposed to the usually slower processes of Jeans-type thermal leakage or solar wind ion pickup.

Hydrogen cloud   The huge cloud of atomic hydrogen surrounding most comets. The hydrogen cloud is produced by the dissociation of water and the hydroxyl molecule (OH).

Hydrostatic equation   Relationship that says pressure is equal to the weight of gas or liquid above the level of interest.

Hyperbolic orbit   Unbound orbit in which the object escapes the gravitational attraction of the central body: examples are orbits of beta meteoroids and interstellar grains.

Hypsometry   Geodetic observations of terrain elevations with respect to sea level.

Ice dwarf   The term given to the planetesimals believed to have been created in large numbers during the formation of the giant planets and later scattered to the Oort cloud or ejected from the solar system by close encounters with the forming giant planets. Pluto and Triton are thought to be among the largest remnants of this population.

Ice   Mixture of water, ammonia, methane, and other volatile compounds in the interiors of jovian planets, not literally in the form of condensed "ice."

IDP   Interplanetary dust particle, collected by aircraft in the stratosphere.

Impact melt   Melt of target rocks resulting from the waste heat generated in an impact event. When solidified, it can be either glassy or crystalline and contain clasts of rock and mineral debris from unmelted portions of the target.

Inclination   The angle between the plane of the orbit of a planet, comet, or asteroid and the ecliptic plane, or between a satellite's orbit plane and the equatorial plane of its primary. Inclination takes on values between $0°$ and $180°$.

Insolation   The flux of sunlight at all wavelengths falling on a body. For the Earth this amounts to a flux of 1.368×10^6 ergs cm^{-2} s^{-1}.

Integral of the motion   Any function of the position and velocity coordinates of an object that remains constant with time along all orbits. In the circular restricted three-body problem, the Jacobi constant is an integral of the motion. The Jacobi constant can be approximated by the Tisserand parameter.

Intercrater plains   The oldest plains on Mercury that occur in the highlands and formed during the period of late heavy bombardment.

Intrusion   Geological structure of igneous material that forces its way into an existing formation.

Invariable plane   The plane passing through the center of mass of the solar system, which is perpendicular to its total angular momentum vector. The invariable plane is inclined $0.5°$ to the orbital plane of Jupiter and $1.6°$ to the ecliptic.

Ionopause   The surface separating ionospheric plasma and the solar wind in the vicinity of an unmagnetized planet.

Ionosphere   Outer portion of an atmosphere where charged particles are abundant.

Isolation mass   The mass of a planetary embryo if it sweeps up all the accessible solid material in its vicinity.

Jacobi constant   An integral of the motion in the circular restricted three-body problem. It is proportional to the total orbital energy of the small body in a reference frame rotating with the two massive bodies.

Jeans escape   The process by which fast (energetic or hot) molecules of an atmosphere escape into space. The energy distribution of a gas at a given temperature has a hot tail—a few atoms moving faster than the rest. If, at an altitude where collisions between molecules are rare, the molecules in the hot tail move faster than the local escape velocity, they can escape to space. This process is fastest for hot atmospheres of light gases (hydrogen, helium) on bodies with low gravity.

Jets   The observed, collimated emission of gas and dust that occurs in restricted areas on the surface of a cometary nucleus. Jets are usually active on the sunlit side of the nucleus.

Joule heating   Heating that occurs when a current flows through a resistive medium. In the high atmospheres of the giant planets, it may be an important process in heating the atmosphere to high temperature as currents of charged particles driven by magnetospheric electric fields collide with the neutral atmosphere atoms, which provide resistance.

Jovian planet   A planet like Jupiter, which is composed mostly of hydrogen, with helium and other gases, but possibly with a silicate/iron core. Also called a gaseous or a giant planet. The jovian planets are Jupiter, Saturn, Uranus, and Neptune.

Jupiter-family comet   An ecliptic comet with a Tisserand parameter between 2 and 3. It is typically on a low to moderate inclination orbit, with a semimajor axis less than that of Jupiter's orbit. Most Jupiter-family comets are in orbits that cross or closely approach Jupiter's orbit.

K or kelvin   Unit of absolute temperature. The freezing and boiling points of water are 273.16 K and 373.16 K, respectively.

Keplerian orbit   The path that a body would follow if it were subject only to the gravitational attraction of its primary, e.g. a planet orbiting the Sun, a satellite orbiting a planet.

Keplerian velocity   The speed with which a solid body moves on a circular orbit about a larger body.

Kepler's laws   Three rules that describe the unperturbed motion of planets about the Sun (and of moons about planets): (1) Planets move on elliptical paths with the Sun at one focus. (2) An imaginary line from the Sun to a planet sweeps out area at a constant rate. (3) The square of a planet's orbital period varies as the cube of the semimajor axis of its orbit.

Kirkwood gaps   Zones in the asteroid belt that have been depleted of objects due to mean-motion orbital resonances with Jupiter.

Klystron   Vacuum-tube amplifier used in planetary radar transmitters.

Kozai resonance   A resonance where an object's nodal precession rate is equal in magnitude and direction to its periapse precession rate. Objects within a Kozai resonance undergo oscillations in eccentricity and inclination that are out of phase (i.e., when one increases, the other decreases). Kozai resonances affect the motion of Pluto and some comets and asteroids in the solar system.

Kuiper belt   Generally used to refer to the population of trans-Neptunian bodies, i.e., those with semimajor axes >30 AU. In a more detailed classification, which partitions the trans-Neptunian population into the Kuiper belt, the scattered disk and the extended scattered disk, the name "Kuiper belt" is associated with a collection of bodies on essentially stable, low inclination, low eccentricity orbits. Almost all Kuiper belt objects discovered so far have semimajor axes <50 AU, which argues for the Kuiper belt having an outer edge at approximately that location.

Lagrangian points   The five locations in the circular restricted three-body problem at which the net gravitational and centrifugal forces in the frame rotating with the massive bodies is zero. The first three Lagrangian points, L_1, L_2, and L_3, lie on the line connecting the massive bodies; all three colinear Lagrangian points are unstable. The L_4 and L_5 Lagrangian points each make equilateral triangles with the two massive bodies; orbits about the triangular Lagrangian points are stable to small perturbations provided the ratio between the masses of the two bodies is ≥ 27.

Landforms   Natural physical features of a planet's surface.

Langmuir probe   Instrument used to measure electron and ion densities. The external sensor is usually a stiff wire and the current is measured as different voltages are applied.

Laplace resonance   Occurs when three or more orbiting bodies have a simple integer ratio between their orbital periods.

Last universal common ancestor   The hypothetical latest living organism from which all currently living organisms descend.

Latent heat   Heat that is released or absorbed during a phase change, i.e., vapor or liquid or ice to ice or liquid or vapor. Latent heat contributes to heating and cooling the atmosphere in regions where ice and liquid clouds form and dissipate. It also contributes to the heat capacity of a parcel of gas/cloud and therefore influences the adiabatic temperature gradient.

Libration   A small oscillation around an equilibrium configuration, such as the angular change in the face that a synchronously rotating satellite presents toward the focus of its orbit.

Lightcurve   A graph of an object's brightness versus time. Since asteroids and cometary nuclei are usually not perfect spheres, the observed projected area of the object varies as the object rotates. The time difference between the peaks of the lightcurve provide a measure of the object's rotation rate and the shape of the lightcurve can be statistically modeled to derive the object's shape.

Limb-darkening   The darkening of the observed edges a planetary disk or a star. This may be due to the scattering properties of the surface (if, for example, it is a strongly backscattering surface, like an icy one) or more usually to the presence of an optically thick atmosphere. It is often characterized by an exponent k, the Minnaert exponent, for a scattering law of the form $I = I_0 \mu^k \mu_0^{k-1}$, where μ and μ_0 are the cosines of the angle between the normal at a given point and the observer and sun respectively and I_0 is the brightness of the center of the disk. $k = 0.5$ corresponds to a flat disk (rather like the moon), while $k = 1$ is a Lambertian disk with strong limb-darkening. $k < 0.5$ corresponds to limb-brightening, typical of a scattering but optically thin region above an absorbing (dark) region in the atmosphere.

Lithophile   Material made of elements that are commonly found in rocks, such as Si, O, Al, Ca, and Fe; derived from the Greek, meaning "rock-loving."

Lithosphere   The rigid outer shell of a planetary body, generally including a chemically distinct crust and part of the upper mantle; the lithosphere is rheologically defined and so its thickness depends strongly on temperature.

Lobate scarp   A long sinuous cliff (see *Thrust fault*).

Longitude of perihelion ($\bar{\omega}$)   The sum of the longitude of the ascending node and the argument of perihelion.

Longitude of the ascending node (Ω)   The nodes of an orbit are the points where the orbit crosses some reference plane, usually the ecliptic. The ascending node is where the orbit crosses the reference plane from south to north. The longitude of the ascending node is the angle between the location of the ascending node and some standard direction in the reference frame, usually the direction of the vernal equinox.

Long-period comet   A comet with an orbital period of more than 200 years. Some long-period comets have orbital periods of millions of years.

Lorentz force   Force exerted by a magnetic field on a moving charged particle. This force is always perpendicular to the motion of the particle.

Lyapunov exponent   Measure of the rate of divergence of two nearby trajectories in a system. A positive Lyapunov exponent is associated with chaotic motion, and its inverse gives an estimate of the timescale for exponential separation of nearby orbits.

Macroscopic cross section   The product of the number density (number per cm^3) of the target nuclei and the microscopic cross section. The macroscopic cross section has units of cm^{-1} and gives the probability per unit path length that a particle will undergo an interaction, for example, in a planetary surface.

Mafic   Dense, Fe- and Mg-rich silicate minerals, such as those that dominate the mantle; usually refers to basalts and other refractory igneous rock types.

Magnetic reconnection   A magnetic instability that can be triggered in the solar corona, where the topology and connectivity of magnetic field lines change; believed to be the primary cause of flares and coronal mass ejections.

Magnetic storm   A prolonged interval of intense geomagnetic activity often lasting for days.

Magnetopause   The outer boundary of a magnetosphere between the solar wind region and a planets' magnetic field region, where a strong thin current generally flows.

Magnetosheath   The region between a planetary bow shock and magnetopause in which the shocked solar wind plasma flows around the magnetosphere.

Magnetosphere   The region of space around a planet or satellite dominated by its intrinsic magnetic field and associated charged particles.

Magnitude   A logarithmic unit of brightness. Large magnitude values correspond to faint objects. The Sun, the faintest star visible with the unaided eye, and the faintest Kuiper belt object seen with the *Hubble Space Telescope* have magnitude values of -26.74, $+6$, and $+28$, respectively. For every change by five magnitudes, the brightness changes by a factor of 100. One magnitude equals a factor of $100^{1/5}$ or ~ 2.5119 in brightness. All magnitudes are scaled to the flux of Alpha Lyrae, also named Vega, which is designated as magnitude 0.

Main sequence   When stars are plotted on a graph of their luminosity versus their surface temperature (or color), most stars fall along a line extending from high-luminosity, high–surface temperature stars, to low- luminosity, low–surface temperature stars. This plot is known as the Hertzsprung–Russell diagram, and the line is known as the main sequence. Stars spend the majority of their lifetimes on the main sequence, during which they produce energy by hydrogen fusion within their cores.

Mantle   Portion of the interior of a body between the core and the lithosphere that is hot enough to allow material to flow by solid-state creep, allowing fluid-like behavior to occur on time scales of 10^8–10^9 years.

Mantle convection   Movement of material within the Earth's mantle occurs because the density of constituent rock is related to temperature. Buoyant hot material thus tends to rise and cooler material tends to sink over time. In our experience, convection is most commonly associated with liquids or gases, however, deep within the Earth, plastic and even solid rock under pressure can convect, and thus transport heat away from the core, through the mantle, and ultimately toward the surface.

Mare (pl., maria)   Latin word for "sea," used first by Galileo to refer to the dark patches on the lunar surface, now known to be basaltic lava flows.

Mascons   Regions of the Moon of excess mass concentrations per unit area, identified by positive gravity anomalies and associated with basalt-filled multiring basins.

Mass wasting   The downslope movement of rock, regolith, and soil under the influence of gravity.

Maxwellian distribution   The distribution of particle velocities for a gas in thermal equilibrium.

Mean-motion resonance   An orbital resonance in which the orbital periods of the bodies involved are in a simple integer ratio. For example, Pluto is in a 2:3 mean-motion resonance with Neptune; it completes two orbits around the Sun for every three of Neptune.

Meridional circulation   Motions of the atmosphere in the plane defined by the vertical and latitudinal coordinates. Atmospheric motions in the vertical and north–south directions participate in the meridional circulation.

Meridional mixing   Mixing of the atmosphere along meridians (lines of constant longitude), for example, between polar regions and mid-latitudes.

Metallic hydrogen   High-pressure (≥ 1.4 Mbar) metallic form of hydrogen found in the interiors of Jupiter and Saturn.

Meteor   ight phenomenon that results from the entry of a meteoroid from space into Earth's atmosphere.

Meteorite   Meteoroid that has reached the surface of Earth or another planet without being completely vaporized.

Meteoroid   A small fragment of an asteroid or comet that is in interplanetary space. When a meteoroid enters a planetary atmosphere and begins to glow from friction with the atmosphere, it is called a **meteor**. A fragment that survives atmospheric entry and can be recovered on the ground is called a **meteorite**.

Micrometeoroid   Meteoroid smaller than about 0.1 mm in size.

Micron, micrometer, or μm   One millionth of a meter.

Microscopic cross section   An effective area that gives the probability that a particle (for example, a neutron or gamma ray) will undergo a reaction with a target nucleus (or atom). The microscopic cross section has units of barns per nucleus. One barn is 10^{-24} cm^2.

Mie scattering   The scattering of sunlight by atmospheric particles such as aerosols.

Miller–Urey experiments   Laboratory experiments in which mixtures of gases representing the composition of planetary atmospheres or the Earth's early atmosphere were placed in sealed vessels and exposed to various forms of energy such as UV and energetic particles. In general the experiments tended to produce more complex combinations of molecules from the initial gases.

Mineral   Naturally-occurring substance of specified chemical composition and physical properties having a characteristic atomic structure and/or crystalline form.

Minor planet   Another term for an asteroid.

Mixing ratio   Fractional mass of a particular component of an intimate mixture.

Molecular cloud   Cold, dense, region of the interstellar medium containing molecular hydrogen: H_2, often the site of star formation.

Moment of inertia   Quantity that is the measure of the density distribution within a planet, specifically the tendency for an increase of density with depth. It has a value of 0.400 for a sphere of uniform density.

Moon   A body in orbit around another larger body, known as the primary, such as a planet, dwarf planet, or asteroid. Also called a **satellite**.

Morphology   Study of the shape of landforms on a planetary surface.

Near-Earth object (NEO)   Any object, such as an asteroid or comet, orbiting the Sun with a perihelion distance less than 1.3 AU.

Nearly isotropic comet (NIC)   A population of comets with orbits that are randomly inclined to the ecliptic plane. By definition, the Tisserand parameter of NICs is less than 2. Also known as a **long-period comet**.

Neutron   A neutral particle with mass similar to that of the proton. Neutrons and protons are the primary constituents of the atomic nucleus. Neutrons liberated from the nucleus by nuclear reactions decay by beta emission with a mean lifetime of 900 seconds.

New comet   A nearly isotropic or long-period comet that is entering the planetary region for the first time since it was placed in the Oort cloud. Dynamically new comets are usually taken to be nearly isotropic comets with original semimajor axes greater than 10,000 AU.

Newton's laws   Three laws of motion and one of gravity that describe aspects of the physical world: (1) A body remains at rest or in uniform motion unless it is acted upon by an external force. (2) The acceleration of a body is directly proportional to the force acting upon it and inversely proportional to its mass. (3) For every action, there exists an equal and opposite reaction. (Gravity) The gravitational attraction between any two spherically symmetric objects is proportional to the product of their masses and inversely proportional to the square of the distance between their centers.

Node   One of the two points where a body's orbit crosses a reference plane, such as the ecliptic.

Nongravitational force   A force not due to gravity that acts on comets and asteroids and that can significantly alter their orbits. The most important nongravitational force for comets is the reaction force due to the outgassing of volatile materials from the day side of the nucleus.

Nonsynchronous rotation   The state of a satellite whose rotation period is not equal to its orbital period.

Nonthermal escape   Atmospheric escape of gases in processes that do not depend on the temperature of the bulk upper atmosphere. Nonthermal escape can occur when a neutral species is photoionized by solar extreme ultraviolet radiation and recombines with an electron to form a fast neutral atom. This is called photochemical escape. Alternatively, a fast ion can impart its charge to a neutral atom through collision or charge exchange, and become a fast neutral atom with escape velocity. Today, photochemical escape is important for the loss of carbon, oxygen, and nitrogen from Mars.

Nonthermal radio emission   Radio emission produced by processes other than those which produce thermal emission. In particular, in planetary science we are concerned with cyclotron and synchrotron radiation. Cyclotron radiation is emitted by (nonrelativistic) electrons, often in the auroral (near-polar) regions of a planet's magnetic field at the frequency of gyration around the magnetic field lines (cyclotron frequency). The emission is like a hollow cone pattern. Synchrotron radiation is produced by relativistic (i.e., particle velocity approaches the speed of light) electrons. This radiation is strongly beamed in the direction in which the particle is moving. There are other types of nonthermal radio emissions that involve coupling from various plasma wave modes to radio waves.

Nucleosynthesis   The creation of stable and unstable isotopes in stars.

Nucleus   The central cometary body that is the source of all the other cometary features. The nucleus is composed of volatile materials, primarily water ice, and dust particles composed of both silicates and organics. Sublimation produces molecular gases and releases the dust particles. The nucleus has a typical radius of 1–10 km.

Oblateness   A measure of the amount to which the shape of a planet or other body differs from a perfect sphere.

Obliquity   The angle between a planet's equator and its orbital plane. Earth's current obliquity of 23.5° is sufficient to cause seasons.

Observational bias   The effect that some astronomical objects are easier to discover and observe than others, generally because they are brighter. The observed sample of comets suffers from severe bias because, in general, only those comets that pass well within the orbit of Jupiter become bright enough to be observable.

Occultation   Obscuration of a body brought about by the passage of another body in front of it. Occultations of stars by planets or by planetary rings allow the observer to probe the atmospheric structure of the planet or the structure of the rings. Occultation of stars by asteroids allow the observer to determine the size of the asteroid.

Oersted   Unit of magnetic intensity in the centimeter–gram–second system, equivalent to the gauss.

Oligarchic growth   Self-regulated stage of planet formation that follows runaway growth.

Oort cloud   A large reservoir of several times 10^{12} cometary nuclei surrounding the planetary system and extending from a few thousand AU to about 100,000 AU from the Sun. The outer Oort cloud, beyond ~10,000 AU, is roughly spherical; the inner Oort cloud is flattened toward the ecliptic plane. The existence of the Oort cloud is inferred from the semimajor axis distribution of long-period comets.

Opacity   The ability of an atmosphere to absorb (or sometimes scatter) radiation. Also called optical depth. A beam of monochromatic radiation passing through an atmosphere with an optical depth of one will have its intensity reduced by a factor of e (= 2.718), while an optical depth of 4 absorbs or scatters 98% of the radiation. Opacity is a function of wavelength as well as of the pressure, temperature, and composition of the region of the atmosphere under consideration.

Opposition effect   Nonlinear surge in brightness as a celestial object approaches being viewed at zero phase angle.

Opposition   The position of a superior planet, a comet or an asteroid when it is opposite the Sun in the sky, i.e., when its elongation approaches 180°.

Optical depth   Measure of the integrated extinction of light along a path through a medium, such as an atmosphere or the disk of particles forming a ring. Normal optical depth refers to the extinction along a path perpendicular to the ring plane.

Orbit   The path of a planet, asteroid, or comet around the Sun, or of a satellite around its primary. Most bodies are in closed elliptical orbits. Some comets and asteroids are thrown on to hyperbolic orbits, which are not closed and which will escape the solar system.

Orbital elements   The six parameters that uniquely specify an object's orbit and it location within the orbit. Two parameters, semimajor axis and eccentricity, enumerate the size and shape of the orbit. Three angles, inclination, longitude of the ascending node, and argument of perihelion, describe the orbit's orientation in space. Finally, the mean anomaly specifies the position of the object along the orbit.

Orbital phase angle   Angular position of a satellite in orbit about its primary object, measured counterclockwise when viewed from the north.

Orbit–orbit resonance   Condition in which two objects have orbital periods in the ratio of small integers. Orbit–orbit resonances are commonly found between Jupiter and minor planets in the asteroid belt, between Neptune and bodies in the Kuiper belt, and in the satellite systems of Jupiter and Saturn. Also known as a **mean-motion resonance**.

Orogenic, orogeny   Process of mountain building, with uplift generally occurring as a result of tectonic plate collisions.

Orphan trail   A dust trail that does not appear to be connected to any cometary source. This might arise as a consequence of planetary perturbations causing a shift in a comet's orbit and disconnecting it from a more distant portion of its dust trail.

Outflow channels   Large channels that start full size and have few if any tributaries. They may be up to several tens of kilometers across and thousands of kilometers long and are believed to have been formed by large floods.

Oxidants   Chemical compound that readily transfers oxygen atoms.

Palimpsest   Flattened, circular bright patches on Ganymede and Callisto that are believed to be the remnants of ancient large impact structures.

Panspermia   A theory by which life spreads through the solar system and the galaxy by spores carried on dust grains or small particles.

Parallax   The apparent change in the position of a nearby star on the celestial sphere when measured from opposite sides of the Earth's orbit, usually given in seconds of arc.

Parautochthonous   Describes a rock unit that has been moved only slightly into its present location.

Parsec   The distance at which a star would have a parallax of 1 arcsec, equal to 206,264.8 AU, or 3.261631 lightyears, abbreviated: pc. One thousand parsecs are equal to a kiloparsec, abbreviated: kpc.

Patera (pl., paterae)   A collective term for a variety of unusual, saucer-shaped, shallow volcanic constructs that often have a central crater or caldera.

Periapse   Point on an orbit closest to the central body, called **perihelion** for orbits about the Sun and **perigee** for orbits about the Earth.

Perigee   The closest point to the Earth of the elliptical orbit of the Moon or an artificial satellite.

Perihelion distance (q)   The closest distance to the Sun an object reaches in its orbit, given by $q = a(1 - e)$, where a is the object's semimajor axis and e is its eccentricity.

Perihelion   Point in a heliocentric orbit when it is closest to the Sun.

Perijove   The point in an orbit around Jupiter when the object is closest to the planet.

Period of late heavy bombardment   The period of intense bombardment of the inner solar system after planetary formation. It may have been a catastrophic bombardment that lasted only about 100 million years and peaked at ~3.9 billion years ago, or a longer bombardment that ended 3.8 billion years ago.

Periodic comet   Traditionally, a comet with an orbital period of less than 200 years; also known as a short-period comet.

Permafrost zone   Near-surface zone within which temperatures are always below 0°C. It may or may not contain ground ice.

Petrology   The study of the nature and history of mineralogic phases and chemical compositions of rocks, and conclusions regarding their origins. One aim of mineralogy and petrology is to decipher the history of igneous and metamorphic rocks.

Phase angle   Angle between the Sun, a given object, and the observer with the object at the vertex.

Phase function   The curve describing the change in brightness of a body as a function of the phase angle, the angle between the observer, the body, and the Sun. Usually expressed in astronomical magnitudes per degree.

Phase integral   Integrated value of the function that describes the directional scattering properties of a surface.

Phase space   Multidimensional space in which the coordinates are, for example, the positions and the velocities.

Photoautotrophy   The capacity of an autotrophic (self-feeding) organism to derive the energy required for its growth from sunlight by means of photosynthesis; green plants are photoautotrophic.

Photolysis   Process that occurs when a molecule absorbs light of sufficiently high energy (usually ultraviolet light) and breaks apart.

Photosphere   A thin, 300-km thick layer above the solar surface from where most of the optical emission (white light) is irradiated, with a temperature of ~6000 K.

Phototactic/phototaxis   The movement of an organism in response to light, either toward or away from the source; e.g., certain microorganisms are phototactic and will migrate in the direction of sunlight.

Phylogenetic   Refers to organisms that are related to each other through evolution.

Physiographic   Referring to the physical appearance of the landscape.

Pixel   A "picture element." One element in a CCD or infrared detector array.

Planar deformation features   Planar, micrometer-sized bands of intense deformation or glass that occur in minerals due to the passage of a shock wave.

Planet   According to the new IAU definition passed in 2006, a planet must have three qualities: (1) it must be round, indicating its interior is in hydrostatic equilibrium; (2) it must orbit the Sun; and (3) it must have gravitationally cleared its zone of other debris. According to the definition, our solar system has eight planets: Mercury, Venus, Earth, Mars, Jupiter, Saturn, Uranus and Neptune. Although most astronomers have accepted the new definition, some are campaigning to have it changed. See The Solar System and Its Place in the Galaxy and Pluto and Charon for more discussion.

Planetary embryo   Large solid body formed by runaway and oligarchic growth.

Planetesimal   A small solid body formed in the early solar system by accretion of dust and ice (if present) near the central plane of the solar nebula. The terrestrial planets, asteroids, comets, and cores of the giant planets are generally thought to have formed through the accretion and aggregation of planetesimals.

Plasma   Ionized medium in which electrons have been stripped from neutral matter to make a gas of charged ions and electrons.

Plasma beta   Ratio of gas pressure to magnetic field pressure within a plasma.

Plasma tail   The narrow, highly structured tail consisting of molecular ions (and electrons) confined to magnetic field lines wrapped around the head of the comet. The plasma tail is normally attached to the head region. The exception is when disconnection events occur. The orientation of the tail, approximately anti-sunward, is produced by the solar wind interaction. The plasma tail appears blue because of resonance scattering of sunlight from ionized carbon monoxide molecules.

Plasmoid   A region within a magnetosphere in which plasma is confined by a magnetic structure that is not directly linked to the planet.

Plate tectonics   The system of rigid plates, tens of kilometers thick, that move over the surface of the Earth, causing mountain belts to form at convergent zones.

Plume   A hot blob of material that rises through the mantle, typically causing uplift of the surface and volcanism.

Polar wandering   Changes in the direction of the magnetic pole relative to its orientation in space.

Polymict   A rock unit consisting of fragments of various pre-existing rock units.

Polymorph   Crystal form of a mineral that has a different crystal structure from that of the original mineral.

Polysaccharide   Any of a group of carbohydrates consisting of long chains of simple sugars; e.g., starch, glycogen.

Poynting–Robertson effect, P-R drag   Drag on interplanetary particles caused by their interaction with solar radiation, which causes the particles to lose orbital angular momentum and to spiral in towards the Sun.

Precess, Precession   The slow, smooth increase or decrease of an angle. For example, the axes of the planets' perihelion directions change, or precess, taking tens of thousands to millions of years to complete an entire cycle. Pole precession is the slow, periodic, and conical motion of the rotation axis of a spinning body.

Precession of the equinoxes   The slow rotation of the equinoxes with respect to the stars. It has a period of about 26,000 years for the Earth.

Primary body   Celestial body (usually the Sun or a planet) around which a planet or a moon, respectively, or secondary body, orbits.

Primitive meteorite   See *Chondrite*.

Prominences   Cool and dense mass structures suspended above the chromosphere, observed above the solar limb, which are called filaments when seen on the solar disk.

Protoplanetary nebula A   disk of gas and dust that surrounds a newborn star, from which the planets, asteroids, and comets are thought to form.

Protostar   A star in the process of formation, which is luminous due to the release of gravitational potential energy from the infall of nebula material.

P-wave velocity   Seismic body wave velocity associated with particle motion (alternating compression and expansion) in the direction of wave propagation.

Pyroclastic materials   Fragmented materials ejected during an explosive volcanic eruption, including ash, pumice, and rock fragments.

Radar albedo   Ratio of a target's radar cross section in a specified polarization to its projected area, hence a measure of the target's radar reflectivity.

Radar cross section   Most common measure of a target's scattering efficiency, equal to the projected area of that perfect metal sphere that would give the same echo power as the target if observed at the target's location.

Radial velocity technique   Observational method used to detect stellar reflex motions by measuring the line-of-sight component of a star's space velocity vector. If the radial velocity can be measured with a precision of a few m s^{-1}, then the reflex motion due to planetary companions become detectable. Today, this is the most successful method for finding extrasolar planets.

Radiation belts   Toroidal zones containing charged particles that are magnetically trapped in a planetary dipole field. The Van Allen belts around the Earth include ions and electrons with energies from hundreds of keV to tens of MeV.

Radio spectrum   A graph of the brightness temperature as a function of wavelength or frequency.

Radiolysis   The dissociation of molecules by high-energy radiation.

Radiometric modeling Measures the thermal emission of an asteroid to provide an estimate of the asteroid's surface temperature and albedo. A dark asteroid, for example, would absorb more of the visible sunlight because it has a low albedo, but it would radiate that additional energy at thermal wavelengths, showing a warmer surface temperature. Combined data on thermal "temperature," and visible reflectance can provide the albedo of an object and an estimate of its size.

Radio occultation The passing of a radio beam through a planet's atmosphere. Attenuation and refraction (bending) of the beam, generally by phase delay, can be used to measure the density of electrons in the planet's ionosphere and the density of the gas in its atmosphere. The abrupt cutoff of the signal can also be used to make a precise measurement of the planet's radius.

Raman scattering Inelastic scattering of sunlight by gas molecules in an atmosphere, such that the scattered photon is shifted in frequency.

Random walk A series of movements in which the direction and size of each move is randomly determined.

Rayleigh scattering The scattering of sunlight by gas molecules in an atmosphere.

Reaction notation The notation for nuclear reactions given by $T(i, p)R$, where T is the target nucleus, i is the incident particle, p indicates the particles(s) produced by the reaction, and R is the product nucleus. For example, the notation for neutron inelastic scattering with ^{56}Fe is ^{56}Fe(n,n'γ), where n' is the scattered neutron and γ is the associated gamma ray. Similarly, neutron capture with ^{56}Fe is denoted ^{56}Fe(n,γ)^{57}Fe, where the product isotope has been appended.

Reaction rate The rate (R) at which gamma rays or neutrons interact with nuclei is given by the product of the flux (ϕ), the number density of the target nuclei (N nuclei per cm^3) and the microscopic cross section (ϕ) for the selected reaction: $R = \varphi N \sigma$ (interactions per unit time).

Reconnection A process in which the magnetic configuration changes as if two field lines were broken and reconnected in a new configuration. This can occur when two plasmas containing oppositely directed magnetic fields flow toward each other.

Reductants Compounds or catalysts that result in the loss of an electron.

Reflectance spectroscopy The study of the physical and mineralogical properties of materials over the wavelength range of reflected electromagnetic radiation. Light interacts with the atoms and crystal structure of materials producing a diagnostic set of absorptions and reflectances.

Refraction Bending of a light ray as it traverses a boundary, for example, between air and glass or between space and an atmosphere.

Refractories Materials not deformed or damaged by high temperatures. Classic refractories are high-melting oxides, like silica and alumina, but also carbides, nitrides, sulfides, and pure carbon. In our terminology, refractories are materials that are not modified by space conditions (temperature and vacuum) in the inner solar system. The opposite are volatile materials, e.g., ices that rapidly sublimate close to the Sun.

Refractory inclusion See **CAI**.

Regolith The outermost unconsolidated fragmental layer on some airless planets, satellites and asteroids that results from the breakup of rocks by repeated impacts of meteoroids.

Regular motion A trajectory that does not display chaos.

Regular satellite A satellite with low orbital eccentricity and inclination.

Resonance A situation in which two orbiting bodies have orbital frequencies (related to the time they take to complete their orbits or for their orbits to precess) that are in a simple integer ratio. Objects in resonance exert a regular gravitational influence on each other. Depending upon the particular resonance involved, resonance can either stabilize an orbit, as in the case of Pluto, or destabilize an orbit as near the Kirkwood gaps in the asteroid belt. Strong satellite resonances open gaps at particular locations in broad planetary rings; weaker ones drive radial and vertical wave trains.

Retrograde, retrograde motion Orbital or rotational motion in the solar system that is clockwise as seen from north of the ecliptic. Nearly isotropic comets with inclinations greater than 90° have retrograde orbits. Triton is in a retrograde orbit around Neptune. Venus is in retrograde rotation.

Returning comet A nearly isotropic or long-period comet that is returning to the planetary region for at least the second time. Returning comets are usually taken to be nearly isotropic comets with semimajor axes less than 10,000 AU.

Reynolds number Dimensionless number that governs the conditions for the occurrence of turbulence in fluids.

Ring current A current carried by energetic particles that flows at radial distances beyond a few planetary radii in the near-equatorial regions of a planetary magnetosphere.

Roche limit, Roche zone The distance from a planet or the Sun, within which another body will be disrupted because tidal forces from the planet exceed the self-gravity of the smaller body, unless the material strength of the body is strong enough to hold it together. For nonrotating bodies of equal density and zero strength, the Roche limit is about 2.2 planetary radii.

Rock Mixture of iron, silicon, magnesium, and other refractory elements found in the interiors of jovian planets.

Rotational lightcurve A graph depicting the variation in brightness of an object verus time as it rotates on its axis. This variation can be caused by nonsphericity (i.e., shape effects) or albedo markings; for objects as large as Pluto and Charon, albedo markings usually dominate. See also **lightcurve**.

Runaway growth Stage of planetary growth in which the largest planetesimals grow rapidly while most others remain small.

Satellite A body in orbit around a planet, dwarf planet, or an asteroid. Also called a **moon**.

Scale height The vertical distance over which atmospheric pressure or density falls by $1/e = 0.368$; equal to kT/mg, where

k is Boltzmann's constant, T is temperature, m is the mean mass of the gas, and g is the acceleration of gravity.

Scattered disk   A collection of $\sim 10^9$ icy planetesimals in high-eccentricity, low-to-moderate inclination orbits beyond Neptune. Scattered disk objects typically have semimajor axes of order 50-100 AU. The scattered disk is probably the primary source of the Jupiter-family comets. Scattered disk objects may have escaped the Kuiper belt billions of years ago, and/or may be scattered Uranus-Neptune planetesimals.

Scattering law   Function giving the dependence of a surface element's radar cross section on viewing angle.

Scintillator A   transparent material that coverts the kinetic energy of charged particles, such as electrons produced by gamma ray interactions or alpha particles and recoil protons produced by neutron reactions, into flashes of light detectable by a photomultiplier tube or photodiode. A wide variety of organic and inorganic materials scintillate and can be used for radiation detection and spectroscopy.

Secondary crater   A crater produced by the impact of blocks of ejecta from a primary impact by a comet or asteroid.

Secular   Continuing or changing over a long period of time.

Secular perturbations   Long-term changes to the orbit of a body caused by the distant gravitational perturbations of the planets and other bodies.

Secular resonance   Near-commensurability among the frequencies associated with the precessions of the line of nodes and/or apsides.

Seeing   Blurringof the image of an astronomical object caused by turbulence in the Earth's atmosphere Atmospheric seeing at the very best observatory sites, such as Mauna Kea, is about 0.5 arcsec and can be as good as 0.25 arcsec. Seeing can be improved using **adaptive optics**.

Semiconductor   Semiconductors, such as germanium, silicon, and CdZnTe, can be used to detect gamma rays. Swift electrons produced by Compton and photoelectric interactions ionize the semiconductor, producing electron-hole pairs. The electrons and holes drift under the influence of an applied electric field to electrical contacts. As they drift, the electrons and holes induce charge on contacts, which can be measured by a charge-sensitive preamplifier. The amplitude of the charge pulse is proportional to the energy deposited by the gamma ray, which enables semiconductors to be used for spectroscopy.

Semimajor axis (a)   Commonly thought of as the mean distance of the orbit of a body from its primary. More formally, it is one half of the longer of the two axes of an ellipse describing the orbit, that passes through both foci of the ellipse.

Semiminor axis (b)   One half of the minor axis (short diameter) of an elliptical orbit.

Sensitivity to initial conditions   A situation in which a tiny change in an object's initial state (position and/or velocity) will make a big change in its final trajectory. Sensitivity to initial conditions is a necessary condition for chaos.

Separatrix   Boundary of a resonance, separating resonant or librating motion inside the resonance from nonresonant or circulating motion outside.

Shield volcano   Broad volcano with a large summit pit formed by collapse and gently sloping flanks, built mainly from overlapping, fluid, basaltic lava flows.

Shock   A discontinuous, nonlinear change in pressure commonly associated with supersonic motion in a gas, plasma, or solid.

Shock metamorphism   Permanent physical, chemical, and mineralogical changes in rocks, resulting from the passage of a shock wave.

Shock wave   Compressional wave, resulting from an impact or explosion, which travels at supersonic velocities.

Short-lived isotopes   Radioactive isotopes with half-lives much shorter than the age of the solar system.

Short-period comet   A comet with an orbital period <200 years. Short-period comets include **Jupiter-family** and **Halle-type** comets.

Sidereal   Relative to the fixed stars. A sidereal period is the orbital period of a planet around the Sun relative to the stars. A sidereal year is the orbital period of the Earth around the Sun relative to the stars.

Siderophile element   Element that tends to join with iron and is predominantly found in a planet's core.

Silicate   A compound containing silicon and oxygen.

Smooth plains   The youngest plains on Mercury with a relatively low impact crater abundance.

SNC meteorites   Group of meteorites (Shergotty–Nakhla–Chassigny) believed to be derived from Mars because of their young ages, basaltic composition, and inclusion of gases with the same composition as the Martian atmosphere.

Solar activity cycle   Cycle of ~ 11 year duration characterized by waxing and waning of various forms of solar activity such as sunspots, flares, and coronal mass ejections.

Solar corona   The hot, tenuous outer atmosphere of the Sun from which the solar wind originates.

Solar energetic particles   Ions (usually protons) and electrons generated in solar flares and in the corona and solar wind by shock waves, with energies above hundreds of keV for electrons and above an MeV per nucleon for ions.

Solar flare   A disturbance in the solar atmosphere characterized by a sudden, localized enhancement in electromagnetic emission from visible to x-ray wavelengths.

Solar nebula   The cloud of dust and gas out of which the Sun and planetary system formed.

Solar phase angle   See **phase angle**.

Solar wind   A magnetized, highly ionized plasma that flows radially out from the solar corona at supersonic and super-Alfvénic speed.

Solidus   Line or surface in a phase diagram below which the system is completely solid.

Space debris   Man-made particulates littered in space.

Space ultraviolet   That part of the ultraviolet electromagnetic spectrum that can only be observed from space because the Earth's atmosphere is opaque at those wavelengths; commonly thought of as the region below wavelengths of 3000 Å.

Space weather   The variable level of geomagnetic activity controlled by the conditions in the solar wind.

Space weathering   A process acting on the surface of planetary and asteroidal bodies that changes their surface optical properties over time.

Spectral absorption   A particular wavelength of light that is selectively absorbed by a particular material. Patterns of absorptions can serve as "fingerprints" to remotely identify surface or atmospheric materials.

Spin–orbit resonance   Simple numerical relationship between the spin period of a planet or satellite and its orbital period. Most natural satellites in the solar system are in the 1:1 spin–orbit resonance, also called the synchronous spin state.

Sputtering   An atmospheric loss process that occurs when ions that have been picked up by the magnetic field embedded in the solar wind impact a planetary atmosphere and undergo charge exchange. Charge exchange neutralizes the ions, which can impart their large energies to surrounding particles by collision. Upward-directed energetic particles can then escape. This process may have been important on Mars after it lost its magnetic field and its upper atmosphere was no longer shielded from the solar wind. Sputtering can also occur when energetic particles from the solar wind or a planetary magnetosphere strike the surface of an airless planet or satellite and cause atoms of the surface materials to escape.

Stellar metallicity   The amount of chemical elements, heavier than hydrogen and helium, contained in a star. Observations indicate that stellar metallicity is a critical factor in the efficiency of the formation and/or orbital migration of extrasolar planets.

Stellar occultation   When a planet or asteroid passes in front of a star and the star is briefly hidden from view. Such events can be used to probe the size and also the atmospheric structure of the planet (or asteroid) doing the occulting, or the structure of rings around a planet.

Stellar reflex motion   The movement of a star along its orbit around the barycenter of a star/companion system. If a star has no companions (stars or planets), the barycenter coincides with the star's own center of mass, and no reflex motion exists.

Strain   Forces acting in opposite directions, pulling materials apart.

Stratigraphy   Study of rock layers.

Stratosphere   Region in an atmosphere overlying the troposphere that is strongly stabilized against convection by heating because of the absorption of ultraviolet radiation from the Sun. *Stratum* is Latin for "layer."

Stratosphere, mesosphere   Region (also together called middle atmosphere) whose temperature is controlled by radiative balance. On Earth, it extends from about 10 to 95 km, and on Venus, from 65 to 95 km.

Stream structure   Pattern of alternating flows of low- and high-speed solar wind.

Stromatolite   A geological feature formed by the conversion of loose, unconsolidated sediment into a coherent layer, as a result of the growth, movement, or activity of microorganisms; e.g., blue-green algae. Microfossils associated with stromatolite formation are an important form of evidence for early life on Earth, and thus a search for stromatolites could undertaken on other planets in sites where liquid water might have accumulated.

Subaerial   Referring to landscapes, such as islands or continents, that are exposed to the air.

Sublimation   The phase change of a solid directly to gas, as in the conversion of ice directly into vapor.

Substorm   The elementary disturbance of the magnetosphere that produces geomagnetic activity.

Sulfur allotropes   Sulfur cooled rapidly from different temperatures, resulting in different colors.

Superior geocentric conjunction   The point in a planetary satellite's orbit where it is directly opposite Earth, such that the satellite lies on a straight line connecting Earth, the planet, and the satellite.

Surface of section   Means of studying the regular or chaotic nature of an orbit by plotting a sequence of points in two dimensions that can represent all or part of the coordinates of the orbit in phase space.

Synchronous orbit   An orbit whose period is equal to the rotation period of the primary.

Synchronous rotation   Dynamical state caused by tidal interactions in which a satellite presents the same face toward the primary, because the satellite's rotation period is equal to its orbital period.

Synodic period   For an inferior planet, the time between successive conjunctions. For a superior planet, the interval between successive oppositions. For the Sun, the time taken for one revolution of the Sun as seen from Earth.

Synodic rotation period   Apparent rotation period of a target that is moving relative to the observer (who may also be moving), to be distinguished from the sidereal rotation period measured with respect to the fixed stars.

Tadpole orbits   Orbits that librate about the stable L_4 or L_5 triangular Lagrangian points in the restricted three-body problem. These orbits appear to be shaped like tadpoles in the frame rotating with the mean-motion of the massive bodies.

Taxon   Grouping of organisms or bodies with similar characteristics.

Tectonic framework   The global or large-scale pattern of fractures and folds formed by crustal deformation.

Tephra   Generic term for all volcanic fragments that are explosively ejected from a volcano.

Termination shock   A discontinuity in the outer heliosphere where the solar wind slows from supersonic to subsonic as it interacts with the interstellar plasma.

Terminator   The boundary between the illuminated and nonilluminated parts of a planet, satellite, asteroid, or cometary nuclei.

Terrane   A particular type of terrain. Generally used to denote the kind of terrain dominated or formed by a particular geomorphic process regime, such as a volcanic terrane or an aeolian terrane.

Terrestrial planet   A planet like the Earth with an iron core and a silicate mantle and crust. The terrestrial planets are Mercury, Venus, Earth, and Mars.

Tesserae   Intensely deformed terrain cut by at least two directions of ridges and/or grooves.

Thermal desorption   Process of heating to drive off volatile gases.

Thermal diffusion   Heat transport resulting from a temperature gradient in a solid body.

Thermal emission   Electromagnetic radiation emitted by a body, typically at infrared wavelengths, due to its temperature.

Thermal inertia   A material property that is a measure of the time it takes for the material to respond to temperature changes. Thermal inertia is mathematically defined in terms of the physical properties of the material as $(k\rho\, C)^{1/2}$, where k is the thermal conductivity, ρ is the bulk density, and C is the specific heat capacity.

Thermal radio emission   Continuous radio emission from an object that results from the object's temperature. Blackbody radiation is a form of thermal radio emission.

Thermal wind   A wind shear developed in one direction due to a temperature gradient in an orthogonal direction.

Thermosphere, exosphere   Outer parts of an atmosphere, heated by ionizing radiation and cooled by conduction. The exosphere is essentially isothermal and is also characterized by very long mean free paths.

Tholeiitic   Referring to basaltic rocks generally found on the ocean floor, erupted from oceanic ridge zones or from shield volcanoes. Such rocks are considered in the mafic family.

Tholus (pl., tholi)   Dome or shield. Small tholi are scattered across Io.

Three-body problem   Problem of the motion of three bodies moving under their mutual gravitational attraction. In the restricted three-body problem, the third body is considered to have negligible mass such that it does not affect the motion of the other two bodies.

Thrust fault   A fault where the block on one side of the fault plane has been thrust up and over the opposite block by horizontal compressive forces.

Tidal heating   Energy deposited in a satellite due to the dissipation of energy from tidal deformation.

Time delay   Time between transmission of a radar signal and reception of the echo.

Tisserand parameter   A nearly-conserved quantity in the circular restricted three-body problem. The Tisserand parameter for comets with respect to Jupiter is used to recognize returning comets even if their orbits were changed by a close approach to Jupiter, and to classify their orbits.

Transient crater   The crater excavated in a hyper-velocity impact, prior to the collapse of the surrounding crater walls.

Transition region   The vertical zone in the Sun where the temperature climbs from 20,000 K above the cool chromosphere to 1 million K in the hot corona.

Triaxial ellipsoid   A 3-dimensional surface defined by three axes that are elliptical in cross section and used to describe the shape of a body.

Troposphere   Lowest level of an atmosphere dominated by vertical mixing and often containing clouds, where temperature falls off with height at close to the neutrally stable (adiabatic) lapse rate. Earth's troposphere contains 80% of the mass of its atmosphere and most of the water vapor, and consequently most of the weather. Terminated at the top by the **tropopause**. On Earth, the troposphere extends to 14 km (equatorial) and 9 km (polar); on Venus, to 65 km. *Tropos* is Greek for "turning."

Turbulent concentration   The concentration of large numbers of similarly sized particles in stagnant regions in a turbulent gas.

Type I migration   Gradual inward spiraling of a planet as it loses angular momentum via gravitational interactions with nebular gas. This affects planetary embryos and rocky planets.

Type II migration   Change in the size of a planet's orbit when the planet is massive enough to clear a gap in the disk. Migration is typically inward.

Van Allen belts   Region in the Earth's magnetic field, inside of ~4 Earth's radii, filled with energetic particles. Other magnetized planets have similar radiation belts

Vernal equinox   The direction of the Sun as viewed from the Earth as it crosses the celestial equator moving northward. On Earth, the vernal equinox denotes the beginning of spring in the northern hemisphere.

Viscoelastic heating   Heating (such as tidal heating) produced by nonrecoverable (permanent) deformation in the viscous portion of a viscoelastic body, i.e., a body that behaves with both viscous and elastic components of deformation

Viscosity   Property of a fluid that resists flow; fluid dynamic stiffness or, in a sense, internal friction. For lava flows that typically have remarkably little excess energy above their solidus, viscosity can be the determining factor for the magnitude and morphology of lava flow fields associated with volcanoes and is often exponentially dependent on the core temperature of the flow.

Viscous relaxation   Process whereby topographic features become subdued over time due to the flow of the surrounding geologic material.

Volatile   Any substance that outgasses or produces a significant vapor pressure at a given temperature. Ice is a volatile on Earth ($T = 270$–300 K), but involatile in the outer solar system ($T < 100$ K). By contrast, the ices of CH_4, CO, and N_2 are volatile throughout the planetary region wherever $T > 30$ K. Also, chemical compounds or elements contained in magmas that are generally released as gases to the atmosphere during a volcanic eruption.

Vorticity   A measure of the circulation of a region of the atmosphere. Spots having cyclonic vorticity rotate counterclockwise in the northern hemisphere and clockwise in the southern hemisphere. Terrestrial hurricanes have cyclonic vorticity. The large stable spots on the giant planets are anticyclones.

Warm poles   The alternating aphelion subsolar points on the surface of Mercury at the $90°$ and $270°$ meridians.

Water ice   The primary volatile constituent of comets. Water ice comes in three forms: amorphous, cubic, and hexagonal. Amorphous ice, believed to be the form in the deep interior of cometary nuclei, is characteristic of ices formed at very low temperatures. It has no crystalline structure. At higher temperatures, typically 100–150 K, energy is available to convert the ice to the lower energy cubic form; this transition releases energy. A similar transition from cubic ice to hexagonal ice also releases energy at $\sim$180 K. Cubic ice and hexagonal ice are collectively known as crystalline ice. The water ice nearest to the surface of the cometary nucleus is thought to be hexagonal.

Water-equivalent hydrogen (WEH)   Gamma ray and neutron spectrometers are sensitive only to the abundance of hydrogen, which is sometimes expressed as the equivalent weight fraction of water. If all of the hydrogen is in the form of H_2O, then the relationship between the weight fraction of hydrogen (w_H) and the weight fraction of water (w_{water}) is $w_{water} = 9w_H$.

Wavenumber, wavelength   Wavenumber is the inverse of wavelength, having units of inverse length. In spectroscopy, the wavenumber ν of electromagnetic radiation is defined as $\nu = 1/\lambda$, where λ is the wavelength in vacuum. This quantity is commonly specified in cm^{-1}, called a reciprocal centimeter, or inverse centimeter.

Western boundary current   Strong ocean current that runs along the western edge of an ocean basin as a result of the much slower eastward group velocity of Rossby waves (planetary waves) relative to the westward group velocity. The Gulf Stream is a well-known example.

Window   A spectral region in a planetary atmosphere that is relatively transparent between two regions that have higher opacity. A window region can be important for remote sensing of a planetary surface and for limiting the extent of a greenhouse effect.

Yarkovsky effect   A nongravitational force that arises from the asymmetric thermal reradiation of incident sunlight on the surface of a rotating body, that can lead to significant orbital evolution of kilometer-sized and smaller objects.

Zodiacal cloud   The cloud of interplanetary dust in the solar system, lying close to the ecliptic plane. The dust in the zodiacal cloud comes from both comets and asteroids.

Zodiacal light   Diffuse glow seen on the Earth in the west after twilight and in the east before dawn, that appears wedge-shaped and lies along the ecliptic. It is widest near the horizon and is caused by the reflection of sunlight from the myriads of interplanetary dust particles concentrated in the ecliptic plane.

Index

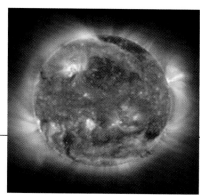